CHILTON®

ASIAN
DIAGNOSTIC SERVICE
2006 EDITION
VOLUME III
Lexus
Scion
Subaru
Suzuki
Toyota

THOMSON
DELMAR LEARNING

Australia • Canada • Mexico • Singapore • Spain • United Kingdom • United States

CHILTON®
ASIAN
DIAGNOSTIC SERVICE
2006 Edition
Volume III
Lexus, Scion, Subaru, Suzuki, Toyota

Vice President,
Technology Professional Business Unit:
Gregory L. Clayton

Publisher,
Technology Professional Business Unit:
David Koontz

Director of Marketing:
Beth A. Lutz

Production Director:
Patty Stephan

Editorial Assistant:
Rebecca Rokitowski

Production Manager:
Andrew Crouth

Marketing Manager:
Brian McGrath

Marketing Coordinator:
Jennifer Stall

Publishing Coordinator:
Paula Baillie

Sr. Content Project Manager:
Elizabeth C. Hough

Managing Editor:
Terry Blomquist

Editors:
Terry Blomquist
Tim Crain
Nick D'Andrea

Graphical Designer:
Melinda Possinger

ISBN: **1-4180-2915-7**

NOTICE TO THE READER

TABLE OF CONTENTS

SECTIONS

USING THIS INFORMATION

Organization

To find where a particular model section or procedure is located, look in the Table of Contents. Main topics are listed with the page number on which they may be found. Following the main topics is a listing of all of the subjects within the section and their page numbers.

Manufacturer and Model Coverage

This product covers 1996-2006 Asian models that are produced in sufficient quantities to warrant coverage, and which have technical content available from the vehicle manufacturers before our publication date. Although this information is as complete as possible at the time of publication, some manufacturers may make changes which cannot be included here. While striving for total accuracy, the publisher cannot assume responsibility for any errors, changes, or omissions that may occur in the compilation of this data.

Part Numbers & Special Tools

Part numbers and special tools are recommended by the publisher and vehicle manufacturer to perform specific jobs. Before substituting any part or tool for the one recommended, you must be completely satisfied that neither your personal safety, nor the performance of the vehicle will be endangered.

ACKNOWLEDGEMENT

The publisher would like to express appreciation to the following vehicle manufacturers for their assistance in producing this publication. No further reproduction or distribution of the material in this manual is allowed without the expressed written permission of the vehicle manufacturers and the publisher. Fuji Heavy Industries Ltd., including Subaru Motors Ltd., Suzuki Motor Corporation, Toyota Motor Sales USA, including Lexus, Scion, and Toyota Divisions.

PRECAUTIONS

Before servicing any vehicle, please be sure to read all of the following precautions, which deal with personal safety, prevention of component damage, and important points to take into consideration when servicing a motor vehicle:

- Always wear safety glasses or goggles when drilling, cutting, grinding or prying.
- Steel-toed work shoes should be worn when working with heavy parts. Pockets should not be used for carrying tools. A slip or fall can drive a screwdriver into your body.
- Work surfaces, including tools and the floor should be kept clean of grease, oil or other slippery material.
- When working around moving parts, don't wear loose clothing. Long hair should be tied back under a hat or cap, or in a hair net.
- Always use tools only for the purpose for which they were designed. Never pry with a screwdriver.
- Keep a fire extinguisher and first aid kit handy.
- Always properly support the vehicle with approved stands or lift.
- Always have adequate ventilation when working with chemicals or hazardous material.

- Carbon monoxide is colorless, odorless and dangerous. If it is necessary to operate the engine with vehicle in a closed area such as a garage, always use an exhaust collector to vent the exhaust gases outside the closed area.
- When draining coolant, keep in mind that small children and some pets are attracted by ethylene glycol antifreeze, and are quite likely to drink any left in an open container, or in puddles on the ground. This will prove fatal in sufficient quantity. Always drain the coolant into a sealable container.
- To avoid personal injury, do not remove the coolant pressure relief cap while the engine is operating or hot. The cooling system is under pressure; steam and hot liquid can come out forcefully when the cap is loosened slightly. Failure to follow these instructions may result in personal injury. The coolant must be recovered in a suitable, clean container for reuse. If the coolant is contaminated it must be recycled or disposed of correctly.
- When carrying out maintenance on the starting system be aware that heavy gauge leads are connected directly to the battery. Make sure the protective caps are in place when

maintenance is completed. Failure to follow these instructions may result in personal injury.
- Do not remove any part of the engine emission control system. Operating the engine without the engine emission control system will reduce fuel economy and engine ventilation. This will weaken engine performance and shorten engine life. It is also a violation of Federal law.
- Due to environmental concerns, when the air conditioning system is drained, the refrigerant must be collected using refrigerant recovery/recycling equipment. Federal law requires that refrigerant be recovered into appropriate recovery equipment and the process be conducted by qualified technicians who have been certified by an approved organization, such as MACS, ASI, etc. Use of a recovery machine dedicated to the appropriate refrigerant is necessary to reduce the possibility of oil and refrigerant incompatibility concerns. Refer to the instructions provided by the equipment manufacturer when removing refrigerant from or charging the air conditioning system.

• Always disconnect the battery ground when working on or around the electrical system.

• Batteries contain sulfuric acid. Avoid contact with skin, eyes, or clothing. Also, shield your eyes when working near batteries to protect against possible splashing of the acid solution. In case of acid contact with skin or eyes, flush immediately with water for a minimum of 15 minutes and get prompt medical attention. If acid is swallowed, call a physician immediately. Failure to follow these instructions may result in personal injury.

• Batteries normally produce explosive gases. Therefore, do not allow flames, sparks or lighted substances to come near the battery. When charging or working near a battery, always shield your face and protect your eyes. Always provide ventilation. Failure to follow these instructions may result in personal injury.

• When lifting a battery, excessive pressure on the end walls could cause acid to spew through the vent caps, resulting in personal injury, damage to the vehicle or battery. Lift with a battery carrier or with your hands on opposite corners. Failure to follow these instructions may result in personal injury.

• Observe all applicable safety precautions when working around fuel. Whenever servicing the fuel system, always work in a well-ventilated area. Do not allow fuel spray or vapors to come in contact with a spark, open flame, or excessive heat (a hot drop light, for example). Keep a dry chemical fire extinguisher near the work area. Always keep fuel in a container specifically designed for fuel storage; also, always properly seal fuel containers to avoid the possibility of fire or explosion. Do not smoke or carry lighted tobacco or open flame of any type when working on or near any fuel related components.

• Fuel injection systems often remain pressurized, even after the engine has been turned OFF. The fuel system pressure must be relieved before disconnecting any fuel lines. Failure to do so may result in fire and/or personal injury.

• The evaporative emissions system contains fuel vapor and condensed fuel vapor. Although not present in large quantities, it still presents the danger of explosion or fire. Disconnect the battery ground cable from the battery to minimize the possibility of an electrical spark occurring, possibly causing a fire or explosion if fuel vapor or liquid fuel is present in the area. Failure to follow these instructions can result in personal injury.

• The EPA warns that prolonged contact with used engine oil may cause a number of skin disorders, including cancer! You should make every effort to minimize your exposure to used engine oil. Protective gloves should be worn when changing oil. Wash your hands and any other exposed skin areas as soon as possible after exposure to used engine oil. Soap and water, or waterless hand cleaner should be used.

• Some vehicles are equipped with an air bag system, often referred to as a Supplemental Restraint System (SRS) or Supplemental Inflatable Restraint (SIR) system. The system must be disabled before performing service on or around system components, steering column, instrument panel components, wiring and sensors. Failure to follow safety and disabling procedures could result in accidental air bag deployment, possible personal injury and unnecessary system repairs.

• Always wear safety goggles when working with, or around, the air bag system. When carrying a non-deployed air bag, be sure the bag and trim cover are pointed away from your body. When placing a non-deployed air bag on a work surface, always face the bag and trim cover upward, away from the surface. This will reduce the motion of the module if it is accidentally deployed.

• Electronic modules are sensitive to electrical charges. The ABS module can be damaged if exposed to these charges.

• Brake pads and shoes may contain asbestos, which has been determined to be a cancer-causing agent. Never clean brake surfaces with compressed air. Avoid inhaling brake dust. Clean all brake surfaces with a commercially available brake cleaning fluid.

• When replacing brake pads, shoes, discs or drums, replace them as complete axle sets.

• When servicing drum brakes, disassemble and assemble one side at a time, leaving the remaining side intact for reference.

• Brake fluid often contains polyglycol ethers and polyglycols. Avoid contact with the eyes and wash your hands thoroughly after handling brake fluid. If you do get brake fluid in your eyes, flush your eyes with clean, running water for 15 minutes. If eye irritation persists, or if you have taken brake fluid internally, immediately seek medical assistance.

• Clean, high quality brake fluid from a sealed container is essential to the safe and proper operation of the brake system. You should always buy the correct type of brake fluid for your vehicle. If the brake fluid becomes contaminated, completely flush the system with new fluid. Never reuse any brake fluid. Any brake fluid that is removed from the system should be discarded. Also, do not allow any brake fluid to come in contact with a painted or plastic surface; it will damage the paint.

• Never operate the engine without the proper amount and type of engine oil; doing so will result in severe engine damage.

• Timing belt maintenance is extremely important! Many models utilize an interference-type, non-freewheeling engine. If the timing belt breaks, the valves in the cylinder head may strike the pistons, causing potentially serious (also time-consuming and expensive) engine damage.

• Disconnecting the negative battery cable on some vehicles may interfere with the functions of the on-board computer system(s) and may require the computer to undergo a relearning process once the negative battery cable is reconnected.

• Steering and suspension fasteners are critical parts because they affect performance of vital components and systems and their failure can result in major service expense. They must be replaced with the same grade or part number or an equivalent part if replacement is necessary. Do not use a replacement part of lesser quality or substitute design. Torque values must be used as specified during reassembly to ensure proper retention of these parts.

INTRODUCTION TO OBD SYSTEMS

1

INTRODUCTION TO OBD

Contents

Notes & Cautions

Before servicing any vehicle, please be sure to read all of the following precautions, which deal with personal safety, prevention of component damage, and important points to take into consideration when servicing a motor vehicle:

- Observe all applicable safety precautions when working around fuel. Whenever servicing the fuel system, always work in a well-ventilated area. Do NOT allow fuel spray or vapors to come in contact with a spark, open flame, or excessive heat (a hot drop light, for example). Keep a dry chemical fire extinguisher near the work area. Always keep fuel in a container specifically designed for fuel storage; also, always properly seal fuel containers to avoid the possibility of fire or explosion. Refer to the additional fuel system precautions that follow.
- Fuel injection systems often remain pressurized, even after the engine has been turned OFF. The fuel system pressure must be relieved before disconnecting any fuel lines. Failure to do so may result in fire and/or personal injury.
- Brake fluid often contains Polyglycol Ethers and Polyglycols. Avoid contact with the eyes and wash your hands thoroughly after handling brake fluid. If you do get brake fluid in your eyes, flush your eyes with clean, running water for 15 minutes. If eye irritation persists, or if you have taken brake fluid internally, IMMEDIATELY seek medical assistance.
- The EPA warns that prolonged contact with used engine oil may cause a number of skin disorders, including cancer. You should make every effort to minimize your exposure to used engine oil. Protective gloves should be worn when changing oil. Wash your hands and any other exposed skin areas as soon as possible after exposure to used engine oil. Soap and water, or waterless hand cleaner should be used.
- The air bag system must be disabled (negative battery cable disconnected and/or air bag system main fuse removed) for at least 30 seconds before performing service on or around system components, steering column, instrument panel components, wiring and sensors. Failure to follow safety and disabling procedures could result in accidental air bag deployment, possible personal injury and unnecessary system repairs.
- Always wear safety goggles when working with, or around, the air bag system. When carrying a non-deployed air bag, be sure the bag and trim cover are pointed away from your body. When placing a non-deployed air bag on a work surface, always face the bag and trim cover upward, away from the surface. This will reduce the motion of the module if it is accidentally deployed. Refer to the additional air bag system precautions later in this section.
- Disconnecting the negative battery cable on some vehicles may interfere with the functions of the on-board computer system(s) and may require the computer to undergo a relearning process once the negative battery cable is reconnected.
- It is critically important to observe all instructions regarding ground disconnects, ignition switch positions, etc., in each diagnostic routine provided. Ignoring these instructions can result in false readings, damage to electronic components or circuits, or personal injury.

Preliminary Diagnostics

HISTORY OF OBD SYSTEMS

Starting in 1978, several vehicle manufacturers introduced a new type of control for several vehicle systems and computer control of engine management systems. These computer-controlled systems included programs to test for problems in the engine mechanical area, electrical fault identification and tests to help diagnose the computer

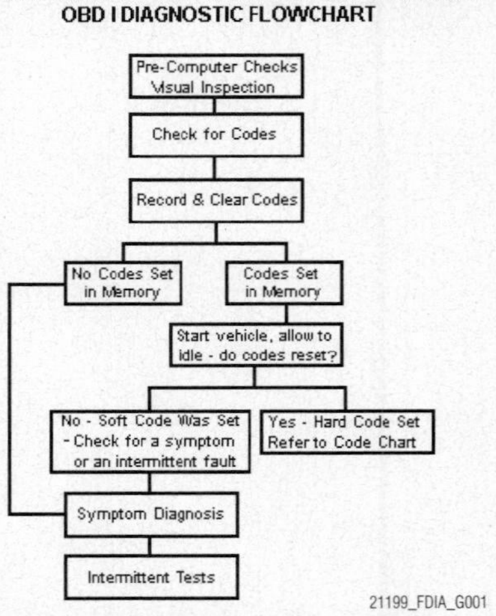

Fig. 1 OBD I diagnostic flow chart

OBD I SYSTEM DIAGNOSTICS

One of the most important things to understand about the automotive repair industry is the fact that you have to continually learn new systems and new diagnostic routines (the test procedures designed to isolate a problem on a vehicle system). For OBD I and II systems, a diagnostic routine can be defined as a procedure (a series of steps) that you follow to find the cause of a problem, make a repair and then verify the problem is fixed.

CHANGES IN DIAGNOSTIC ROUTINES

In some cases, a new Engine Control system may be similar to an earlier system, but it can have more indepth control of vehicle emissions, input and output devices and it may include a diagnostic "monitor" embedded in the engine controller designed to run a thorough set of emission control system tests.

OBD I Diagnostic Flowchart

See Figure 1.

The OBD I Diagnostic Flowchart on this page can be used to find the cause of problems related to Engine Control system trouble codes or driveability symptoms detected on OBD I systems. It includes a step-by-step procedure to use to repair these systems. Compare this flowchart with the one used on OBD II systems.

The steps in this flow chart should be followed as described (from top to bottom).

- Do the Pre-Computer Checks.

- Check for any trouble codes stored in memory.

- Read the trouble codes - If trouble codes are set, record them and then clear the codes.

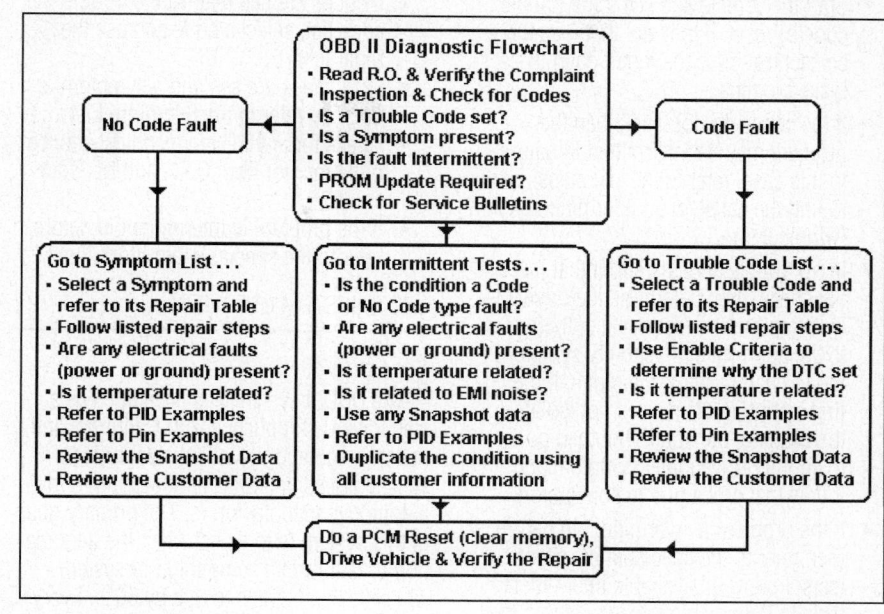

21199_FDIA_G002

Fig. 2 OBD II diagnostic flow chart

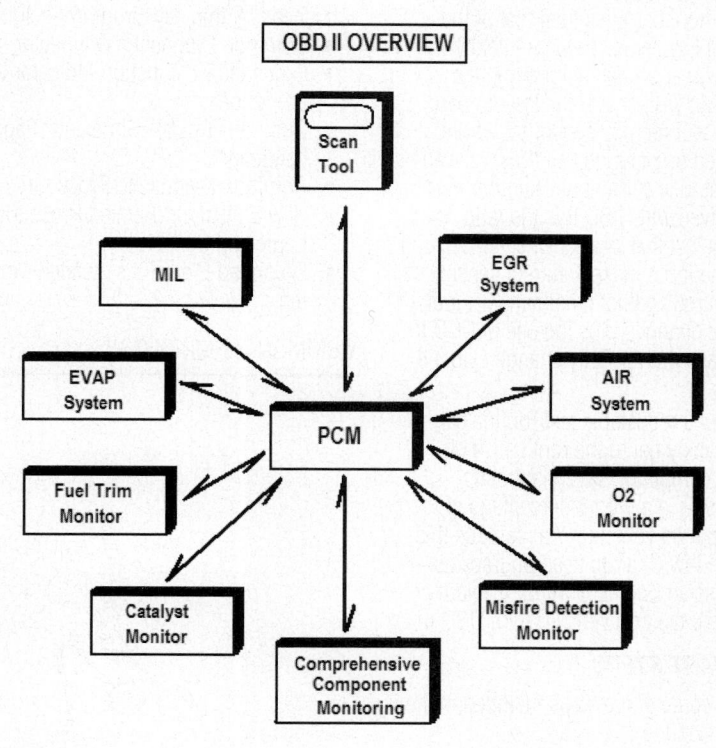

21199_FDIA_G003

Fig. 3 PCM inputs and outputs

- Start the vehicle and see if the trouble code(s) reset. If they do, then use the correct trouble code repair chart to make the repair.
- If the codes do not reset, than the problem may be intermittent in nature. In this case, refer to the test steps used to find the cause of an intermittent fault (wiggle test).
- In no trouble codes are found at the initial check, then determine if a driveability symptom is present. If so, then refer to the approriate driveability symptom repair chart to make the repair. If the first symptom chart does not isolate the cause of the condition, then go on to another driveability symptom and follow that procedure to conclusion.
- If the problem is intermittent in nature, then refer to the special intermittent tests. Follow all available intermittent tests to determine the cause of this type of fault (usually an electrical connection problem).

OBD II System Diagnostics

See Figure 2.

The diagnostic approach used in OBD II systems is more complex than that of the one for OBD I systems. This complexity will effect how you approach diagnosing the vehicle. On an OBD II system, the onboard diagnostics will identify sensor faults (i.e., open, shorted or grounded circuits) as well as those that lose calibration. Another new test that arrived with OBD II is the rationality test (a test that checks whether the value for one input makes rational sense when compared against other sensor input values). The changes plus the use of OBD II Monitors have dramatically changed OBD II diagnostics.

The use of a repeatable test routine can help you quickly get to the root cause of a customer complaint, save diagnostic time and result in a higher percentage of properly repaired vehicles. You can use this Diagnostic Flow Chart to keep on track as you diagnose an Engine Control problem or a base engine fault on vehicles with OBD II.

FLOW CHART STEPS

Here are some of the steps included in the Diagnostic Routine:

- Review the repair order and verify the customer complaint as described
- Perform a Visual Inspection of underhood or engine related items
- If the engine will not start, refer to No Start Tests

- If codes are set, refer to the trouble code list, select a code and use the repair chart
- If no codes are set, and a symptom is present, refer to the Symptom List
- Check for any related technical service bulletins (for both Code and No Code Faults)
- If the problem is intermittent in nature, refer to the special Intermittent Tests

OBD II SYSTEM OVERVIEW

See Figure 3.

The OBD II system was developed as a step toward compliance with California and Federal regulations that set standards for vehicle emission control monitoring for all automotive manufacturers. The primary goal of this system is to detect when the degradation or failure of a component or system will cause emissions to rise by 50%. Every manufacturer must meet OBD II standards by the 1996 model year. Some manufacturers began programs that were OBD II mandated as early as 1992, but most manufacturers began an OBD II phase-in period starting in 1994.

The changes to On-Board Diagnostics influenced by this new program include:

- Common Diagnostic Connector
- Expanded Malfunction Indicator Light Operation
- Common Trouble Code and Diagnostic Language
- Common Diagnostic Procedures
- New Emissions-Related Procedures, Logic and Sensors
- Expanded Emissions-Related Monitoring

COMMON TERMINOLOGY

OBD II introduces common terms, connectors, diagnostic language and new emissions-related monitoring procedures.

The most important benefit of OBD II is that all vehicles will have a common data output system with a common connector. This allows equipment Scan Tool manufacturers to read data from every vehicle and pull codes with common names and similar descriptions of fault conditions. In the future, emissions testing will require the use of an OBD II certifiable Scan Tool.

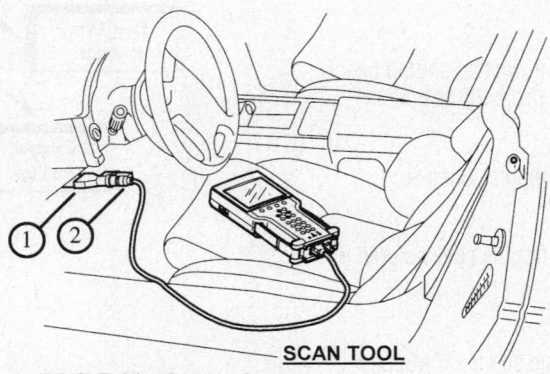

1. DLC Cable Connection
2. SAE 16/19 Pin Adapter

Fig. 4 Typical scan tool hook up

21199_FDIA_G004

Diagnostic Tools & Circuit Testing

HAND TOOLS & METER OPERATION

To effectively use this or any diagnostic information, you should have a solid understanding of how to operate required tools and test equipment.

SCAN TOOLS

See Figure 4.

Vehicle manufacturers designed their computers to have an accessible data line where a diagnostic tester could retrieve data on sensors and the status of operation for components.

These testers became known in the automotive repair industry as "Scan Tools" because they scanned the data on the computers and provided information for the technician.

The Scan Tool is your basic tool link into the on-board electronic control system of the vehicle. Scan Tools are equipped with, or have separate software cards, for each OEM needed to be diagnosed. In this case, always secure a scan tool that has the latest OEM-specific diagnostic software included. Spend some time in the scan tool user's manual to ensure you know how to properly operate the tool and how to select the necessary programs required for full and proper diagnostics.

MALFUNCTION INDICATOR LAMP

Emission regulations require that a Malfunction Indicator Lamp (MIL) be illuminated when an emissions related fault is detected and that a Diagnostic Trouble Code be stored in the vehicle controller (PCM) memory.

When the MIL is illuminated, it is an indication of a problem within one of the electronic components or circuits. When the scan tool is attached to the Data Link Connector (DLC) in the vehicle, it can access the DTCs. In some situations, without the use of a scan tool, the MIL can be activated to flash a series of long and short flashes, which correspond to the numbering of the DTC.

OBD II guidelines define when an emissions-related fault will cause the MIL to activate and set a Diagnostic Trouble Code (DTC). There are some DTCs that will not cause the MIL to illuminate. OBD II guidelines determine how quickly the onboard diagnostics must be able to identify a fault, set the trouble code in memory and activate the MIL (lamp).

ELECTRONIC CONTROLS

You should have a basic knowledge of electronic controls when performing test procedures to keep from making an incorrect

IDENTIFYING THE PROBLEM 2

Table of Contents

Problem Identification

INTRODUCTION

System Control Modules

See Figures 1 and 2.

Before attempting diagnosis of the Electronic Engine Control system, familiarize yourself with the basics of how the system is designed to operate. It consists of a central processing unit: Powertrain Control Module (PCM), Engine Control Module (ECM), Transmission Control Module (TCM) and/or the Body Control Module (BCM). These units are the "heart" of the electronic control systems on the vehicle. In some cases, these units are integral with one another, and on some applications, they are separate. As you get deeper into actual diagnostic testing, you will find out which units are used on the vehicle you are testing.

The PCM is a digital computer that contains a microprocessor. The PCM receives input signals from various sensors and switches that are referred to as PCM inputs. Based on these inputs, the PCM adjusts various engine and vehicle operations through devices that are referred to as PCM outputs. Examples of the input and output devices are shown in the graphic.

Powertrain Subsystems

A key to the diagnosis of the PCM and its subsystems is to determine which subsystems are on a vehicle. Examples of typical subsystems are:

- Cranking & Charging System
- Emission Control Systems
- Engine Cooling System
- Engine Air/Fuel Controls
- Exhaust System
- Ignition System
- Speed Control System
- Transaxle Controls

WHERE TO BEGIN

See Figure 3.

Diagnosis of engine performance or drivability problems on a vehicle with an onboard computer requires that you have a logical plan on how to approach the problem. The "Six Step Test Procedure" is designed to provide a uniform approach to repair any problems that occur in one or more of the vehicle subsystems.

The diagnostic flow built into this test procedure has been field-tested for several years at dealerships - it is the starting point when a repair is required!

It should be noted that a commonly overlooked part of the "Problem Resolution" step is to check for any related Technical Service Bulletins.

Six-Step Test Procedure

The steps outlined as follows were defined to help you determine how to perform a proper diagnosis. Refer to the flow chart that outlines the Six Step Test Procedure as needed. The recommended steps include:

Verify The Complaint & Check For TSBs

To verify the customer complaint, the technician should understand the normal operation of the system. Conduct a thorough visual and operational inspection, review the service history, detect unusual sounds or odors, and gather diagnostic trouble code (DTC) information resources to achieve an effective repair.

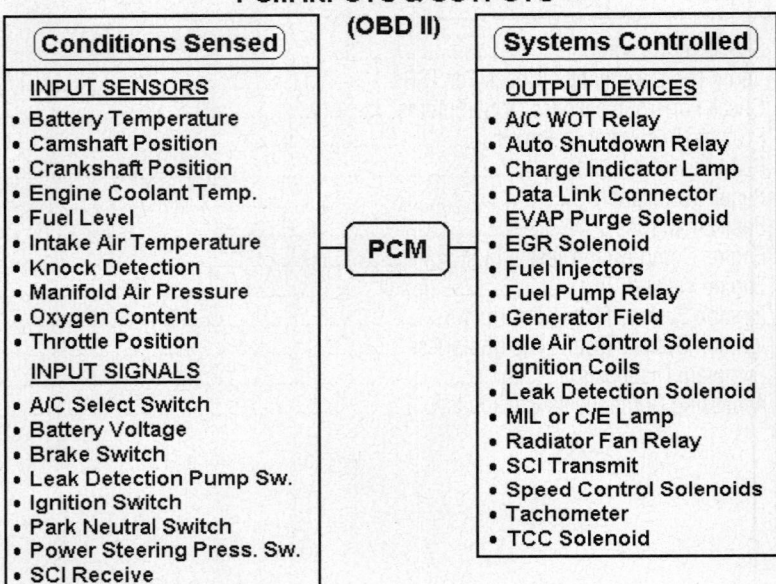

PCM INPUTS & OUTPUTS (OBD II)

Conditions Sensed

INPUT SENSORS
- Battery Temperature
- Camshaft Position
- Crankshaft Position
- Engine Coolant Temp.
- Fuel Level
- Intake Air Temperature
- Knock Detection
- Manifold Air Pressure
- Oxygen Content
- Throttle Position

INPUT SIGNALS
- A/C Select Switch
- Battery Voltage
- Brake Switch
- Leak Detection Pump Sw.
- Ignition Switch
- Park Neutral Switch
- Power Steering Press. Sw.
- SCI Receive
- Speed Control Switches

PCM

Systems Controlled

OUTPUT DEVICES
- A/C WOT Relay
- Auto Shutdown Relay
- Charge Indicator Lamp
- Data Link Connector
- EVAP Purge Solenoid
- EGR Solenoid
- Fuel Injectors
- Fuel Pump Relay
- Generator Field
- Idle Air Control Solenoid
- Ignition Coils
- Leak Detection Solenoid
- MIL or C/E Lamp
- Radiator Fan Relay
- SCI Transmit
- Speed Control Solenoids
- Tachometer
- TCC Solenoid

Fig. 1 An example of OBD II input and output devices

21199_FDIA_G005

PCM LOCATION EXAMPLE

SPEED CONTROL SERVO   WINDSHIELD WASHER FLUID BOTTLE

POWER DISTRIBUTION CENTER   PCM   BATTERY

Fig. 2 Typical PCM location

21199_FDIA_G006

This check should include videos, newsletters, and any other information in the form of TSBs or Dealer Service Bulletins. Analyze the complaint and then use the recommended Six Step Test Procedure. Utilize the wiring diagrams and theory of operation articles. Combine your own knowledge with efficient use of the available service information.

Verify the cause of any related symptoms that may or may not be supported by one or more trouble codes. There are various checks that can be performed to Engine Controls that will help verify the cause of a related symptom. This step helps to lead you in an organized diagnostic approach.

Check For Trouble Codes Or Symptoms

Determine if the problem is a Code or a No Code Fault. Then refer to the appropriate published service diagnostic information to make the repair.

Problem Resolution & Repair

Once the problem component or circuit has been properly identified and verified using published diagnostic procedures, make any needed repairs or replacement to restore the vehicle to proper working order. If the condition has set a DTC, follow the designated repair chart to make an effective repair. If there is not a DTC set, but you can determine specific symptoms that are evident during the failure, select the symptom from the symptom tables and follow the diagnostic paths or suggestions to complete the repair or refer to the applicable component or system in service information.

If the vehicle does not set a DTC and has only intermittent operating failures or concerns, to resolve an intermittent fault, perform the following steps:

• Observe trouble codes, DTC modes and freeze frame data.

• Evaluate the symptoms and conditions described by the customer.

• Use a check sheet to identify the circuit or electrical system component.

• Many Aftermarket Scan Tools and Lab Scopes have data capturing features.

PCM Reset

It is a good idea, prior to tracing any faults, to clear the DTCs, attempt to replicate the condition and see if the same DTC resets. Also, once any repairs are made, it will be necessary to clear the DTC(s) - PCM Reset - to ensure the repair has totally resolved the problem. For procedures on PCM Reset, see DIAGNOSTIC TROUBLE CODES.

Repair Verification

Once a repair is completed, the next step is to verify the vehicle operates properly and that the original symptom was corrected. Verification Tests, related to specific DTC diagnostic steps, can be used to verify a repair.

Base Engine Tests

To determine that an engine is mechanically sound, certain tests need to be performed to verify that the correct A/F mixture enters the engine, is compressed, ignited, burnt, and then discharged out of the exhaust system. These tests can be used to help determine the mechanical condition of the engine.

To diagnose an engine-related complaint, compare the results of the Compression, Cylinder Balance, Engine Cylinder Leakage (not included) and Engine Vacuum Tests.

Engine Compression Test

The Engine Compression Test is used to determine if each cylinder is contributing its equal share of power. The compression readings of all the cylinders are recorded and then compared to each other and to the manufacturer's specification (if available).

Cylinders that have low compression readings have lost their ability to seal. It this type of problem exists, the location of the compression leak must be identified. The leak can be in any of these areas: piston, head gasket, spark plugs, and exhaust or intake valves.

The results of this test can be used to determine the overall condition of the engine and to identify any problem cylinders as well as the most likely cause of the problem.

> ** **CAUTION**
>
> **Prior to starting this procedure, set the parking brake, place the gear selector in P/N and block the drive wheels for safety. The battery must be fully charged.**

COMPRESSION TEST PROCEDURE

1. Allow the engine to run until it is fully warmed up.

2. Remove the spark plugs and disable the Ignition system and the Fuel system for safety. Disconnecting the CKP sensor harness connector will disable both fuel and ignition (except on NGC vehicles).

3. Carefully block the throttle to the wide-open position.

4. Insert the compression gauge into the cylinder and tighten it firmly by hand.

5. Use a remote starter switch or ignition key and crank the engine for 3-5 complete engine cycles. If the test is interrupted for any reason, release the gauge pressure and retest. Repeat this test procedure on all cylinders and record the readings.

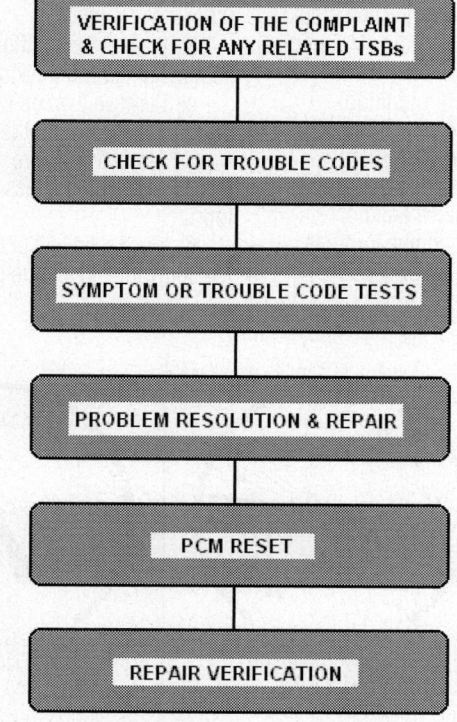

SIX STEP TROUBLESHOOTING PROCEDURE

VERIFICATION OF THE COMPLAINT & CHECK FOR ANY RELATED TSBs

CHECK FOR TROUBLE CODES

SYMPTOM OR TROUBLE CODE TESTS

PROBLEM RESOLUTION & REPAIR

PCM RESET

REPAIR VERIFICATION

Fig. 3 Six-step diagnostic procedure

21199_FDIA_G007

The lowest cylinder compression reading should not be less than 70% of the highest cylinder compression reading and no cylinder should read less than 100 psi.

EVALUATING THE TEST RESULTS

To determine why an individual cylinder has a low compression reading, insert a small amount of engine oil (3 squirts) into the suspect cylinder. Reinstall the compression gauge and retest the cylinder and record the reading. Review the explanations that follow.

Reading is higher - If the reading is higher at this point, oil inserted into the cylinder helped to seal the piston rings against the cylinder walls. Look for worn piston rings.

Reading did not change - If the reading didn't change, the most likely cause of the low cylinder compression reading is the head gasket or valves.

Low readings on companion cylinders - If low compression readings were recorded from cylinders located next to each other, the most likely cause is a blown head gasket.

Readings are higher than normal - If the compression readings are higher than normal, excessive carbon may have collected on the pistons and in the exhaust areas. One way to remove the carbon is with an approved brand of "Top Engine Cleaner."

➡ **Always clean spark plug threads and seat with a spark plug thread chaser and seat cleaning tool prior to reinstallation. Use anti-seize compound on aluminum heads.**

Engine Vacuum Tests

An engine vacuum test can be used to determine if each cylinder is contributing an equal share of power. Engine vacuum, defined as any pressure lower than atmospheric pressure, is produced in each cylinder during the intake stroke. If each cylinder produces an equal amount of vacuum, the measured vacuum in the intake manifold will be even during engine cranking, at idle speed, and at off-idle speeds.

Engine vacuum is measured with a vacuum gauge calibrated to show the difference between engine vacuum (the lack of pressure in the intake manifold) and atmospheric pressure. Vacuum gauge measurements are usually shown in inches of Mercury (in. Hg).

➡ **In the tests described in this article, connect the vacuum gauge to an intake manifold vacuum source at a point below the throttle plate on the throttle body.**

ENGINE CRANKING VACUUM TEST PROCEDURE

The Engine Cranking Vacuum Test can be used to verify that low engine vacuum is not the cause of a No Start, Hard Start, Starts and Dies or Rough Idle condition (symptom).

The vacuum gauge needle fluctuations that occur during engine cranking are indications of individual cylinder problems. If a cylinder produces less than normal engine vacuum, the needle will respond by fluctuating between a steady high reading (from normal cylinders) and a lower reading (from the faulty cylinder). If more than one cylinder has a low vacuum reading, the needle will fluctuate very rapidly.

1. Prior to starting this test, set the parking brake, place the gearshift in P/N and block the drive wheels for safety. Then block the PCV valve and disable the idle air control device.

2. Disable the fuel and/or ignition system to prevent the vehicle from starting during the test (while it is cranking).

3. Close the throttle plate and connect a vacuum gauge to an intake manifold vacuum source. Crank the engine for three seconds (do this step at least twice).

The test results will vary due to engine design characteristics, the type of PCV valve and the position of the AIS or IAC motor and throttle plate. However, the engine vacuum should be steady between 1.0–4.0 in. Hg during normal cranking.

ENGINE RUNNING VACUUM TEST PROCEDURE

See Figure 4.

1. Allow the engine to run until fully warmed up. Connect a vacuum gauge to a clean intake manifold source. Connect a tachometer or Scan Tool to read engine speed.

2. Start the engine and let the idle speed stabilize. Raise the engine speed rapidly to just over 2000 rpm. Repeat the test (3) times. Compare the idle and cruise readings.

EVALUATING THE TEST RESULTS

If the engine wear is even, the gauge should read over 16 in. Hg and be steady. Test results can vary due to engine design and the altitude above or below sea level.

Ignition System Tests–Distributor

This next section provides an overview of ignition tests with examples of Engine Analyzer patterns for a Distributor Ignition System.

PRELIMINARY INSPECTION

1. Perform these checks prior to connecting the Engine Analyzer:

2. Check the battery condition (verify that it can sustain a cranking voltage of 9.6v).

3. Inspect the ignition coil for signs of damage or carbon tracking at the coil tower.

4. Remove the coil wire and check for signs of corrosion on the wire or tower.

5. Test the coil wire resistance with a DVOM (it should be less than 7 k/ohm per foot).

6. Connect a low output spark tester to the coil wire and engine ground. Verify that

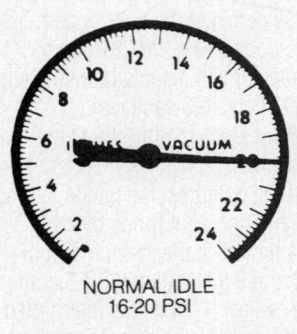

NORMAL IDLE
16-20 PSI

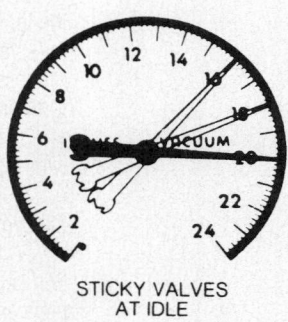

STICKY VALVES
AT IDLE

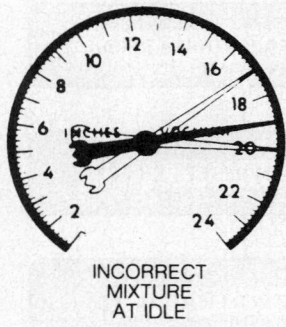

INCORRECT
MIXTURE
AT IDLE

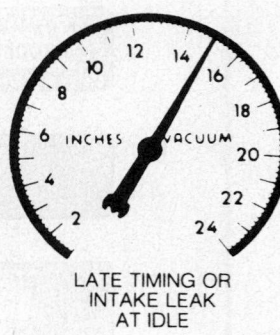

LATE TIMING OR
INTAKE LEAK
AT IDLE

21199_FDIA_G006

Fig. 4 Engine running vacuum test

the ignition coil can sustain adequate spark output while cranking for 3-6 seconds.

7. Connect the Engine Analyzer to the Ignition System, and choose Parade display. Run the engine at 2000 RPM, and note the display patterns, looking for any abnormalities.

Ignition System Tests–Distributorless

Perform the following checks prior to connecting the Engine Analyzer:

1. Check the battery condition (verify that it can sustain a cranking voltage of 9.6v).

2. Inspect the ignition coils for signs of damage or carbon tracking at the coil towers.

3. Remove the secondary ignition wires and check for signs of corrosion.

4. Test the plug wire resistance with a DVOM (specification varies from 15-30 k/ohm).

5. Connect a low output spark tester to a plug wire and to engine ground. Verify that the ignition coil can sustain adequate spark output for 3-6 seconds.

SECONDARY IGNITION SYSTEM SCOPE PATTERNS (V6 ENGINE)

See Figure 5.

1. Connect the Engine Analyzer to the ignition system.

2. Turn the scope selector to view the "Parade Display" of the ignition secondary.

3. Start the engine in Park or Neutral and slowly increase the engine speed from idle to 2000 rpm.

4. Compare actual display to the examples in the illustration.

Symptom Diagnosis

To determine whether vehicle problems are identified by a set Diagnostic Trouble Code, you will first have to connect a proper scan tool to the Data Link Connector and retrieve any set codes. See DIAGNOSTIC TROUBLE CODES for information on retrieving and reading codes.

If no codes are set, the problem must be diagnosed using only vehicle operating symptoms. A complete set of "No Code" symptoms is found in the SYMPTOM DIAGNOSIS (NO CODES).

DO NOT attempt to diagnose driveability symptoms without having a logical plan to use to determine which engine control system is the cause of the symptom - this plan should include a way to determine which systems do NOT have a problem! Remember, there are 2 kinds of NO CODE conditions:

• Symptom diagnosis, in which a continuous problem exists, but no DTC is set as a result. Therefore, only the operating symptoms of the vehicle can be used to pinpoint the root cause of the problem.

• Intermittent problem diagnosis, in which the problem does not occur all the time and does not set any DTCs.

• Both of these NO CODE conditions are covered in the SYMPTOM DIAGNOSIS.

Accessing Components & Circuits

See Figures 6 and 7.

Every vehicle and every diagnostic situation is different. It is a good idea to first determine the best diagnostic path to follow using flow charts, wiring diagrams, TSBs, etc. Part of choosing steps is to determine how time-consuming and effective each step will be. It may be easy to access a component or circuit in one vehicle, but difficult in

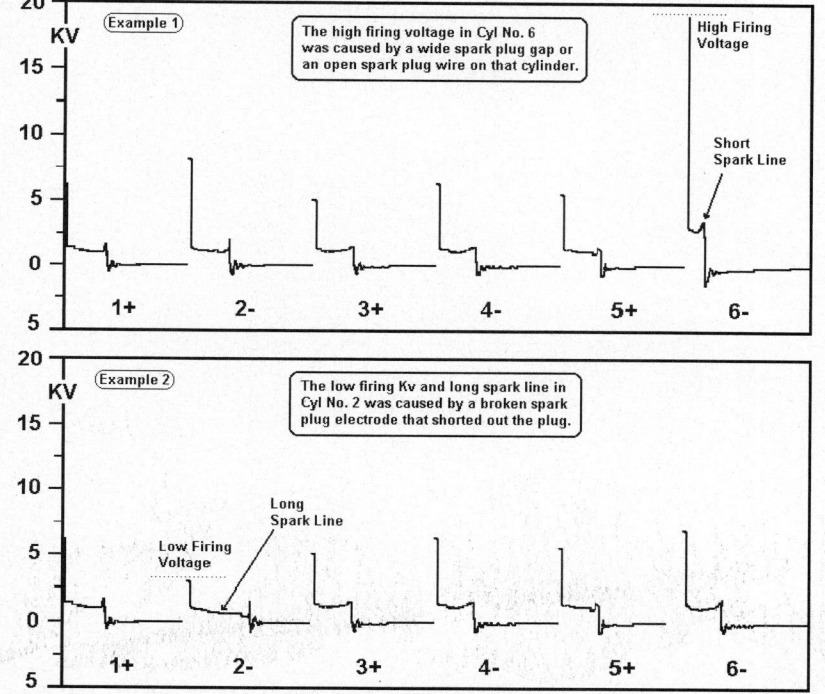

Fig. 5 Secondary ignition system (V6 engine)   21199_FDIA_G009

21199_FDIA_G010

Fig. 6 Circuits located at the back of the PCM connector

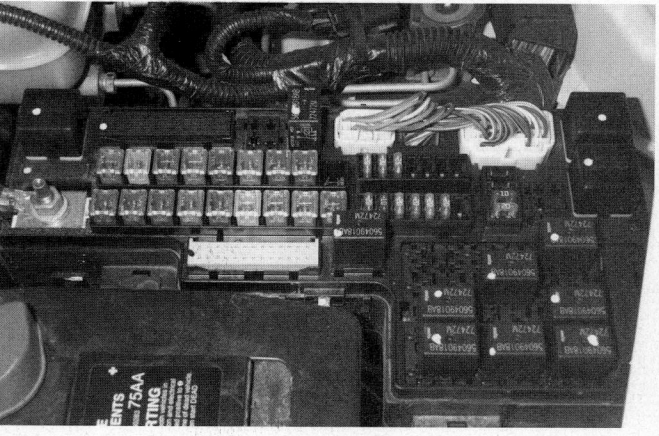

21199_FDIA_G010

Fig. 7 Typical underhood fuse block

another. Many circuits are integrated into a large harness and are difficult to test. Many components are inaccessible without disassembly of unrelated systems.

In the graphic, you will note that the protective covers have been removed from the PCM connectors, and any circuit can be easily identified and back probed. In other cases, PCM access is difficult, and it may be easier to access circuits at the component side of the harness.

Another important point to remember is that any circuit or component controlled by a relay or fused circuit can be monitored from the appropriate fuse box.

There is generally more than one of each type of relay or fuse. Therefore, swapping a suspect relay from another system may be more efficient than testing the relay itself. Relays and fuses may also be removed and replaced with fused jumper wires for testing circuits. Jumper wires can also provide a loop for inductive amperage tests.

Choosing the easiest way has its limitations, however. Remember that an appropriate signal on a PCM controlled circuit at an actuator means that the signal at the PCM is also good. However, a sensor signal at the sensor does not necessarily mean that the PCM is receiving the same signal. Think about the direction flow through a circuit, and not just what signal is appropriate, to save time without making costly assumptions.

INTRODUCTION TO OBD DIAGNOSTIC SYSTEMS

Table of Contents

OBD Systems

The California Air Resources Board (CARB) began regulating On-Board Diagnostic (OBD) systems for vehicles sold in California beginning with the 1988 model year. The initial requirements, known as OBD I, required the identification of the likely area of a fault with regard to the fuel metering system, EGR system, emission-related components and the PCM. Implementation of this new vehicle emission control monitoring regulation was done in several phases.

OBD I SYSTEMS

A Malfunction Indicator Lamp (MIL) labeled Check Engine Lamp or Service Engine Soon was required to illuminate and alert the driver of a fault, and the need to service the emission controls. A Diagnostic Trouble Code (DTC) was required to assist in identifying the system or component associated with the fault. If the fault that caused the MIL goes away, the MIL will go out and the code associated with the fault will disappear after a predetermined number of ignition cycles.

Following extensive research, CARB determined that by the time an Emission System component failed and caused the MIL to illuminate, that the vehicle could have emitted excess emissions over a long period of time. CARB also concluded that semi-annual or annual tailpipe tests were not catching enough of the vehicles with Emission Control systems operating at less than normal efficiency.

To take advantage of improvements in vehicle manufacturer adaptive and failsafe strategies, CARB developed new requirements designed to monitor the performance of Emission Control components, as well as to detect circuit and component hard faults. The new diagnostics were designed to operate under normal driving conditions, and the results of its tests would be viewable without any special equipment.

OBD II SYSTEMS

Beginning in the 1994 model year, both CARB and the EPA mandated Enhanced OBD systems, commonly known as OBD II. The objectives of OBD II were to improve air quality by reducing high in-use emissions caused by emission-related faults, reduce the time between the occurrence of a fault and its detection and repair, and assist in the diagnosis and repair of an emissions-related fault.

Differences Between OBD I & OBD II

As with OBD I, if an emission related problem is detected on a vehicle with OBD II, the MIL is activated and a code is set. However, that is the only real similarity between these systems. OBD II procedures that define emissions component and system tests, code clearing and drive cycles are more comprehensive than tests in the OBD I system.

Powertrain Control Module

The PCM in the OBD II system monitors almost all Emission Control systems that affect tailpipe or evaporative emissions. In most cases, the fault must be detected before tailpipe emissions exceed 1.5 times applicable 50K or 100K-mile FTP standards. If a component exceeds emission levels or fails to operate within the design specifications, the MIL is illuminated and a code is stored within two OBD II drive cycles.

The OBD II test runs continuously or once per trip (it depends on the driving mode requirement). Tests are run once per drive cycle during specific drive patterns called trips. Codes are stored in the PCM memory when a fault is first detected. In most cases, the MIL is turned on after two trips with a fault present. If the MIL is "on", it will go off after three consecutive trips if the same fault does not reappear. If the same fault is not detected after 40 engine warmup periods, the code will be erased (Fuel and Misfire faults require 80 warmup cycles).

OBD II Standardization

OBD II diagnostics require the use of a standardized Diagnostic Link Connector (DLC), standard communication protocol and messages, and standardized trouble codes and terminology. Examples of this standardization are Freeze Frame Data and I/M Readiness Monitors.

Changes in MIL Operation

An important change for OBD II involves when to activate the MIL. The MIL must be activated by at least the second trip if vehicle emissions could exceed 1.5 times the FTP standard. If any single component or system failure would allow the emissions to exceed this level, the MIL is activated and a related code is stored in the PCM.

1994 OBD II Phase-In Systems

Starting in 1994 some manufacturers began to "phase-in" the OBD II system on certain vehicles. The OBD II "phase-in" system on these vehicles included the use of a Misfire Monitor that operated with a "lower threshold" Misfire Detection system

designed to monitor misfires without setting any codes. In addition, the EVAP Monitor was not operational on these vehicles.

1996 & Later OBD II Systems

By the 1996 model year, all California passenger cars and trucks up to 14,000 lb. GVWR, and all Federal passenger cars and trucks up to 8,600 lb. GWVR were required to comply with the CARB-OBD II or EPA OBD requirements. The requirements applied to diesel and gasoline vehicles, and were phased in on alternative-fuel vehicles.

Diagnostic Test Modes

The "test mode" messages available on a Scan Tool are listed below:
- Mode $01: Used to display Powertrain Data (PID data)
- Mode $02: Used to display any stored Freeze Frame data
- Mode $03: Used to request any trouble codes stored in memory
- Mode $04: Used to request that any trouble codes be cleared
- Mode $05: Used to monitor the Oxygen sensor test results
- Mode $06: Used to monitor Non-Continuous Monitor test results
- Mode $07: Used to monitor the Continuous Monitor test results
- Mode $08: Used to request control of a special test (EVAP Leak)
- Mode $09: Used to request vehicle information (INFO MENU)

Onboard Diagnostics

See Figure 1.

The Diagnostic Repair Chart should be used as follows:
- Trouble Code Diagnosis - Refer to the Code List or electronic media for a repair chart for a particular trouble code.
- Driveability Symptoms - Refer to the Driveability Symptom List in manuals or in electronic media.
- Intermittent Faults - Refer to the Intermittent Test Procedures.
- OBD II Drive Cycles - Refer to the Comprehensive Component Monitor or a Main Monitor drive cycle article.

OBD SYSTEM TERMINOLOGY

It is very important that service technicians understand terminology related to OBD II test procedures. Several of the essential OBD II terms and definitions are explained in the following text.

Two-Trip Detection

Frequently, an emission system or component must fail a Monitor test more

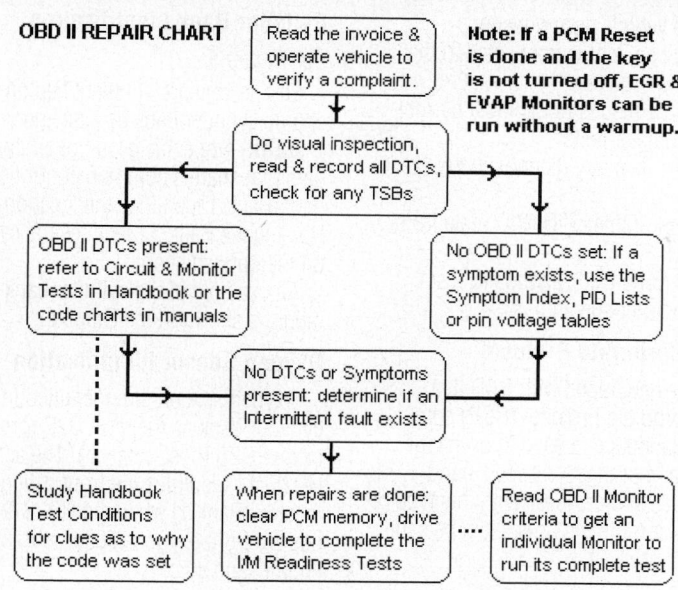

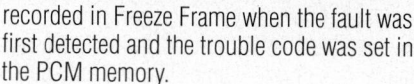

Fig. 1 OBD II repair chart

than once before the MIL is activated. In these cases, the first time an OBD II Monitor detects a fault during any drive cycle it sets a pending code in the PCM memory.

A pending code, which is read by selecting DDL from the Scan Tool menu, appears when Memory or Continuous codes are read. In order for a pending code to cause the MIL to activate, the original fault must be repeated under similar conditions.

This is a critical issue to understand as a pending code could remain in the PCM for a long time before the conditions that caused the code to set reappear. This type of OBD II trouble code logic is frequently referred to as the "Two-Trip Detection Logic".

➡ **Codes related to a Misfire fault and Fuel Trim can cause the PCM to activate the MIL after one trip because these codes are related to critical emission systems that could cause emissions to exceed the federally mandated limits.**

Similar Conditions

If a pending code is set because of a Misfire or Fuel System Monitor fault, the vehicle must meet similar conditions for a second trip before the code matures the PCM activates the MIL and stores the code in memory. Refer to prior Note for exceptions to this rule. The meaning of similar conditions is important when attempting to repair a fault detected by a Misfire or Fuel System Monitor.

To achieve similar conditions, the vehicle must reach the following engine running conditions simultaneously:

- Engine speed must be within 375 RPM of the speed when the trouble code set.
- Engine load must be within 10% of the engine load when the trouble code set.
- Engine warmup state must match a previous cold or warm state.

Summary—Similar conditions are defined as conditions that match the conditions

recorded in Freeze Frame when the fault was first detected and the trouble code was set in the PCM memory.

OBD II Warmup Cycle

See Figure 2.

The meaning of the expression warmup cycle is important. Once the fault that caused an OBD II trouble code to set is gone and the MIL is turned off, the PCM will not erase that code until after 40 warmup cycles. This is the purpose of the warmup cycle: To help clear stored codes.

However, trouble codes related to a Fuel system or Misfire fault require that 80 warmup cycles occur without the fault reappearing before codes related to these monitors will be erased from the PCM memory.

➡ **A warmup cycle is defined as vehicle operation (after an engine off and cool-down period) when the engine temperature rises to at least 40°F and reaches at least 160°F.**

Malfunction Indicator Lamp

If the PCM detects an emission related component or system fault for two consecutive drive cycles on OBD II systems, the MIL is turned on and a trouble code is stored. The MIL is turned off if three consecutive drive cycles occur without the same fault being detected.

Most trouble codes related to a MIL are erased from memory after 40 warmup periods if the same fault is not repeated. The MIL can be turned off after a repair by using the Scan Tool PCM Reset function.

Freeze Frame Data

See Figure 3.

The term Freeze Frame is used to describe the engine conditions that are recorded in PCM memory at the time a Monitor detects an emissions related fault. These conditions include fuel control state, spark timing, engine speed and load.

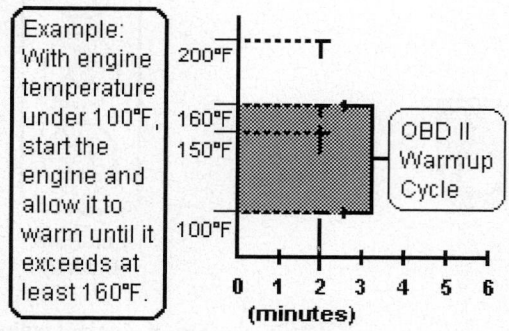

Fig. 2 OBD II warmup cycle

SCAN TOOL DISPLAY	
Freeze Frame Data	
Fuel Sys Status	OL
Load Value	14%
ECT Deg F	+175°F
SHRTFT Adapt	+1.5%
MAP "Hg	18.1"
Engine RPM	750
DTC Priority	01

21199_FDIA_G014

Fig. 3 Scan tool freeze frame

Freeze Frame data is recorded when a system fails the first time for two-trip type faults. The Freeze Frame Data will only be overwritten by a different fault with a "higher emission priority."

Diagnostic Trouble Codes

The OBD II system uses a Diagnostic Trouble Code (DTC) identification system established by the Society of Automotive Engineers (SAE) and the EPA. The first letter of a DTC is used to identify the type of computer system that has failed as shown below:

- The letter 'P' indicates a Powertrain related device
- The letter 'C' indicates a Chassis related device
- The letter 'B' indicates a Body related device
- The letter 'U' indicates a Data Link or Network device code.

The first DTC number indicates a generic (P0xxx) or manufacturer (P1xxx) type code. A list of trouble codes is included.

The number in the hundreds position indicates the specific vehicle system or subgroup that failed (i.e., P0300 for a Misfire code, P0400 for an emission system code, etc.).

Data Link Connector

See Figure 4.

Vehicles equipped with OBD II use a standardized Data Link Connector (DLC). It is typically located between the left end of the instrument panel and 12 inches past vehicle centerline. The connector is mounted out of sight from vehicle passengers, but should be easy to see from outside by a technician in a kneeling position (door open). However, not all of the connectors are located in this exact area.

The DLC is rectangular in design and capable of accommodating up to 16 terminals. It has keying features to allow easy connection to the Scan Tool. Both the DLC and Scan Tool have latching features used to ensure that the Scan Tool will remain con-

nected to the vehicle during testing.

Once the Scan Tool is connected to the DLC, it can be used to:

- Display the results of the most current I/M Readiness Tests
- Read and clear any diagnostic trouble codes
- Read the Parameter ID (PID) data from the PCM
- Perform Enhanced Diagnostic Tests (manufacturer specific)

Standard Corporate Protocol

On vehicles equipped with OBD II, a Standard Corporate Protocol (SCP) communication language is used to exchange bi-directional messages between stand-alone modules and devices. With this type of system, two or more messages can be sent over one circuit.

OBD II Monitor Software

The Diagnostic Executive contains software designed to allow the PCM to organize and prioritize the Main Monitor tests and procedures, and to record and display test results and diagnostic trouble codes.

The functions controlled by this software include:

- To control the diagnostic system so the vehicle continues to operate in a normal manner during testing.
- To ensure the OBD II Monitors run during the first two sample periods of the Federal Test Procedure.
- To ensure that all OBD II Monitors and their related tests are sequenced so that required inputs (enable criteria) for a particular Monitor are present prior to running that particular Monitor.
- To sequence the running of the Monitors to eliminate the possibility of different Monitor tests interfering with each other or upsetting normal vehicle operation.
- To provide a Scan Tool interface by coordinating the operation of special tests or data requests.

Cylinder Bank Identification

See Figure 5.

Engine sensors are identified on each engine cylinder bank as explained next.

Bank—A specific group of engine cylinders that share a common control sensor (e.g., Bank 1 identifies the location of Cyl. No. 1 while Bank 2 identifies the cylinders on the opposite bank).

An example of the cylinder bank configuration is shown in the Graphic.

Oxygen Sensor Identification

Oxygen sensors are identified in each cylinder bank as the front O2S (pre-catalyst) or rear O2S (post-catalyst). The acronym HO2S-11 identifies the front oxygen sensor located (Bank 1) while the HO2S-21 identifies the front oxygen sensor in Bank 2 of the engine, and so on.

OBD II Monitor Test Results

Generally, when an OBD II Monitor runs and fails a particular test during a trip, a pending code is set. If the same Monitor detects a fault for two consecutive trips, the MIL is activated and a code is set in PCM memory. The results of a particular Monitor test indicate that an emission system or component failed: NOT the circuit that failed!

To determine where the fault is located; follow the correct code repair chart, symptom diagnosis or intermittent test. The code and symptom repair charts are the most efficient way to repair an OBD II system.

➡ **Two important pieces of information that can help speed up a diagnosis are code conditions (including all enable criteria), and the parameter information (PID) stored in the Freeze Frame at the time a trouble code is set and stored in memory.**

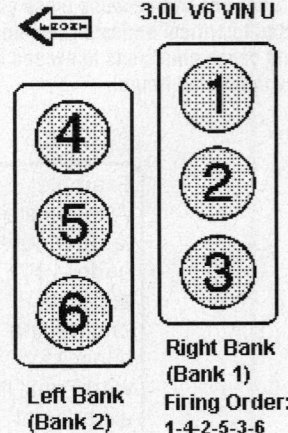

3.0L V6 VIN U

Left Bank (Bank 2)

Right Bank (Bank 1) Firing Order: 1-4-2-5-3-6

21199_FDIA_G016

Fig. 5 Typical cylinder bank identification (V6 engine)

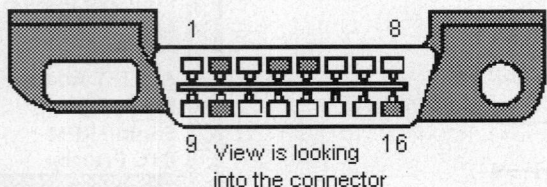

DATA LINK CONNECTOR

1   8

9   16
View is looking into the connector

Courtesy of Ford Motor Co.

21199_FDIA_G015

Fig. 4 Typical data link connector

Adaptive Fuel Control Strategy

The PCM incorporates an Adaptive Fuel Control Strategy that includes an adaptive fuel control table stored to compensate for normal changes in fuel system devices due to age or engine wear.

During closed loop operation, the Fuel System Monitor has two methods of attempting to maintain an ideal A/F ratio of 14:7 to 1 (they are referred to as short term fuel trim and long term fuel trim).

➡ **If a fuel injector, fuel pressure regulator or oxygen sensor is replaced the, memory in the PCM should be cleared by a PCM Reset step so that the PCM will not use a previously learned strategy.**

Short Term Fuel Trim

Short term fuel trim (SHRTFT) is an engine operating parameter that indicates the amount of short term fuel adjustment made by the PCM to compensate for operating conditions that vary from the ideal A/F ratio condition. A SHRTFT number that is negative (-15%) means that the HO2S is indicating a richer than normal condition to the PCM, and that the PCM is attempting to lean the A/F mixture. If the A/F ratio conditions are near ideal, the SHRTFT number will be close to 0%.

Long Term Fuel Trim

Long term fuel trim (LONGFT) is an engine parameter that indicates the amount of long term fuel adjustment made by the PCM to correct for operating conditions that vary from ideal A/F ratios. A LONGFT number that is positive (+15%) means that the HO2S is indicating a leaner than normal condition, and that it is attempting to add more fuel to the A/F mixture. If A/F ratio conditions are near ideal, the LONGFT number will be close to 0%. The PCM adjusts the LONGFT in a range from -35 to +35%. The values are in percentage on a Scan Tool.

Enable Criteria

The term enable criteria describe the conditions necessary for any of the OBD II Monitors to run their diagnostic tests. Each Monitor has specific conditions that must be met before it will run its test.

Enable criteria information can be different for each vehicle and engine type. Examples of trouble code conditions for DTC P0460 and P1168 are shown below:

Code information includes any of the following examples:

- Air Conditioning Status
- BARO, ECT, IAT, TFT, TP and Vehicle Speed sensors
- Camshaft (CMP) and Crankshaft (CKP) sensors
- Canister Purge (duty cycle) and Ignition Control Module Signals
- Short (SHRTFT) and Long Term (LONGFT) Fuel Trim Values
- Transmission Shift Solenoid On/Off Status

Drive Cycle

The term drive cycle has been used to describe a drive pattern used to verify that a trouble code, driveability symptom or intermittent fault had been fixed. With OBD II systems, this term is used to describe a vehicle drive pattern that would allow all the OBD II Monitors to initiate and run their diagnostic tests. For OBD II purposes, a minimum drive cycle includes an engine startup with continued vehicle operation that exceeds the amount of time required to enter closed loop fuel control.

OBD II Trip

The term OBD II Trip describes a method of driving the vehicle so that one or more of the following OBD II Monitors complete their tests:

- Comprehensive Component Monitor (completes anytime in a trip)
- Fuel System Monitor (completes anytime during a trip)
- EGR System Monitor (completes after accomplishing a specific idle and acceleration period)
- Oxygen Sensor Monitor (completes after accomplishing a specific steady state cruise speed for a certain amount of time)

OBD II Drive Cycle

The ambient or inlet air temperature must be from 40-100°F to initiate the OBD II drive cycle. Allow the engine to warm to 130°F prior to starting the test.

Connect the Scan Tool prior to beginning the drive cycle. Some tools are designed to emit a three-pulse beep when all of the OBD II Monitors complete their tests.

➡ **The IAT PID must be from 50-100°F to start the drive cycle. If it is less than 50°F at any time during the highway part of the drive cycle, the EVAP Monitor may not complete. The engine should reach 130°F before starting before attempting to verify an EVAP system fault. Disengage the PTO before proceeding (PTO PID will show OFF) if applicable. For the EVAP Running Loss system, verify FLI PID is at 15-85%. Some Monitors require very specific idle and acceleration steps.**

Drive Cycle Procedure

The primary intention of the OBD II drive cycle is to clear a specific DTC. The drive cycle can also be used to assist in identifying any OBD II concerns present through total Monitor testing. Perform all of the Vehicle Preparation steps.

Connect a Scan Tool and have an assistant watch the Scan Tool I/M Readiness Status to determine when the Catalyst, EGR, EVAP, Fuel System, O2 Sensor, Secondary AIR and Misfire Monitors complete.

OBD II SYSTEM MONITORS

Comprehensive Component Monitor

OBD II regulations require that all emission related circuits and components controlled by the PCM that could affect emissions are monitored for circuit continuity and out-of-range faults. The Comprehensive Component Monitor (CCM) consists of four different monitoring strategies: two for inputs and two for output signals. The CCM is a two trip Monitor for emission faults on most vehicles.

Input Strategies

One input strategy is used to check devices with analog inputs for opens, shorts, or out-of-range values. The CCM accomplishes this task by monitoring A/D converter input voltages. The analog inputs monitored include the ECT, IAT, MAF, TP and Transmission Range Sensors signals.

DTC	Trouble Code Title & Conditions
	EVAP System Small Leak Conditions: Cold startup, engine running at off-idle conditions, then the PCM detected a small leak (a leak of more than 0.040") in the EVAP system.
	FRP Sensor in Range but Low Conditions: Engine running, then the PCM detected that the FRP sensor signal was out-of-range low. Scan Tool Tip: Monitor the FRP PID for a value below 80 psi (551 kPa).

A second input strategy is used to check devices with digital and frequency inputs by performing rationality checks. The PCM uses other sensor readings and calculations to determine if a sensor or switch reading is correct under existing conditions. Some tests run continuously, some only after actuation.

Output Strategies

An Output State Monitor in the PCM checks outputs for opens or shorts by observing the control voltage level of the related device. The control voltage is low with it on, and high with the device off.

IAC Motor Test

The PCM monitors the IAC system in order to "learn" the closed loop correlation it needs to reposition the IAC solenoid (a rationality check).

Catalyst Efficiency Monitor

The Catalyst Monitor is a PCM diagnostic run once per drive cycle that uses the downstream heated Oxygen Sensor (HO2S-12) to determine if a catalyst falls below a minimum level of effectiveness in its ability to control exhaust emissions. The PCM uses a program to determine the catalyst efficiency based on the oxygen storage capacity of the catalytic converter.

Catalyst Monitor Operation

See Figure 6.

The Catalyst Monitor is a diagnostic that tests the oxygen storage capacity of the catalyst. The PCM determines the capacity by comparing the switching frequency of the rear oxygen sensor to the switching frequency of the front oxygen sensor. If the catalyst is okay, the switching frequency of the rear oxygen sensor will be much slower than the frequency of the front oxygen sensor.

However, as the catalyst efficiency deteriorates its ability to store oxygen declines. This deterioration causes the rear oxygen sensor to switch more rapidly. If the PCM detects the switching frequency of the rear oxygen sensor is approaching the frequency of the front oxygen sensor, the test fails and a pending code is set. If the PCM detects a fault on consecutive trips (from two to six consecutive trips) the MIL is activated, and a trouble code is stored in the PCM memory.

The Catalyst Monitor runs after startup once a specified time has elapsed and the vehicle is in closed loop. The amount of time is subject to each PCM calibration. Certain inputs (enable criteria) from various engine sensors (i.e., CKP, ECT, IAT, TPS and VSS) are required before the Catalyst Monitor can run.

Once the Catalyst Monitor is activated, closed loop fuel control is temporarily transferred from the front oxygen sensor to the rear oxygen sensor. During the test, the Monitor analyzes the switching frequency of both sensors to determine if a catalyst has degraded.

Catalyst Efficiency Monitor

CATALYST TEST–STEADY STATE CATALYST EFFICIENCY TEST

The PCM transfers the input for closed loop fuel control from the front HO2S-11 to the rear HO2S-21 during this test. The PCM measures the output frequency of the rear HO2S. This "test frequency" indicates the current oxygen storage capacity of the converter. The slower the frequency of the test result, the higher the efficiency of the converter.

CATALYST TEST–CALIBRATED FREQUENCY TEST

In Part 2 of the test a second frequency is calculated based on engine speed and load. This frequency serves as a high limit threshold for the test frequency. If the PCM detects the test frequency is less than the calibrated frequency the catalyst passes the test. If the frequency is too high, the converter or system has failed (a pending code is set).

The sequence of counting the front and rear O2S switches continues until the drive cycle completes. The ratio of total HO2S-21 switches to the total of the HO2S-11 switches is calculated. If the switch ratio is over the stored threshold, the catalyst has failed and a code is set.

CATALYTIC MONITOR REPAIR VERIFICATION TRIP

See Figure 7.

Start the engine, and drive in stop and go traffic for over 20 minutes. (Ambient air temperature must be over 50ºF to run this test). Drive at speeds from 25-40 mph (6 times) and then at cruise for five minutes.

POSSIBLE CAUSES OF A CATALYST EFFICIENCY FAULT

- Base Engine faults (engine mechanical)
- Exhaust leaks or contaminated fuel

EGR System Monitor

The EGR System Monitor is a PCM diagnostic run once per trip that monitors EGR system component functionality and components for faults that could cause vehicle tailpipe levels to exceed 1.5 times the FTP Standard. A series of sequenced tests is used to test the system.

HO2S-12 WAVEFORM EXAMPLES

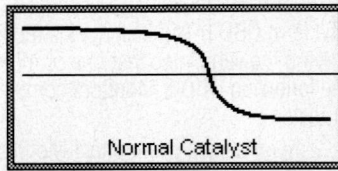

Normal Catalyst

High Storage Capacity - Okay

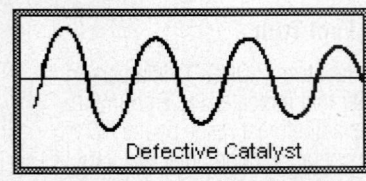

Defective Catalyst

Low Storage Capacity - Not Okay

21199_FDIA_G019

Fig. 6 Typical rear oxygen sensor waveform

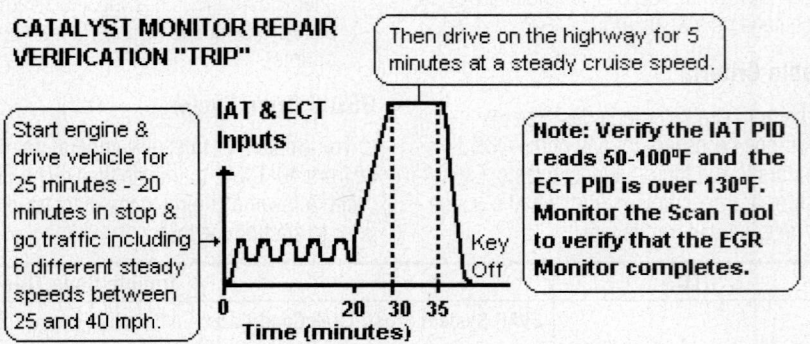

21199_FDIA_G020

Fig. 7 Typical catalyst monitor trip

Possible Causes of an EGR System Failure

See Figure 8.

- Leaks or disconnects in upstream or downstream vacuum hoses
  - Damaged DPFE or EGR EVP sensor
  - Plugged or restricted DPFE or EGR VP sensor or orifice assembly

Evap System Monitor

The EVAP System Monitor is a PCM diagnostic run once per trip that monitors the EVAP system in order to detect a loss of system integrity or leaks in the system (anywhere from 0.020" to 0.040" in diameter).

Possible Causes of an EVAP System Failure

- Cracks, leaks or disconnected hoses in the fuel vapor lines, components or plastic connectors or lines
- Backed-out or loose connectors to the Canister Purge solenoid
- Fuel filler cap (gas cap) loose or missing
- PCM has failed

On-Board Refueling Vapor Recovery System

An On-Board Refueling Vapor Recovery (ORVR) system is used on late model vehicles to recover fuel vapors during vehicle refueling.

SYSTEM OPERATION

The operation of the ORVR system during refueling is described next:

- The fuel filler pipe forms a seal to stop vapors from escaping the fuel tank while liquid is entering the tank (liquid in the 1" diameter tube blocks fuel vapor from rushing back up the fuel filler pipe).
- The fuel vapor control valve controls the flow of vapors out of the tank (it closes when the liquid level reaches a height associated with the fuel tank usable capacity). The fuel vapor control valve:
  a. Limits the total amount of fuel dispensed into the fuel tank.
  b. Prevents liquid gasoline from exiting the fuel tank when submerged (and also when tipped well beyond a horizontal plane as part of the vehicle rollover protection in an accident).
  c. Minimizes vapor flow resistance in a refueling condition.
- Fuel vapor tubing connects the fuel vapor control valve to the EVAP canister. This routes the fuel tank vapors (that are displaced by the incoming fuel) to the canister.

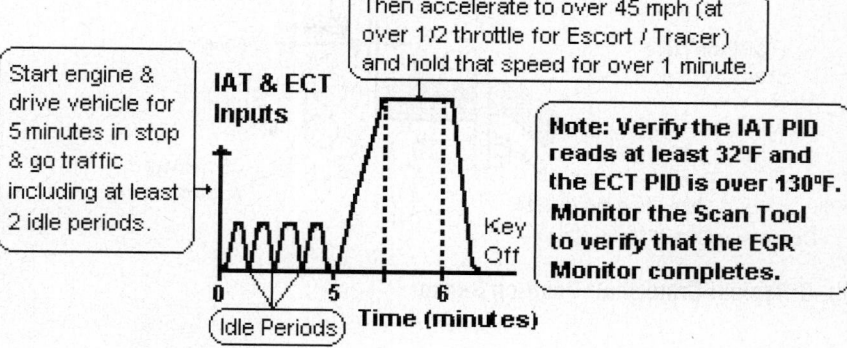

Fig. 8 Typical EGR monitor

- A check valve in the bottom of the pipe prevents any liquid from rushing back up the fuel filler pipe during liquid flow variations associated with the filler nozzle shut-off.
- Between refueling events, the charcoal canister is purged with fresh air so that it may be used again to store vapors accumulated during engine soak periods or subsequent refueling events. The vapors drawn from the canister are consumed in the engine.

Evap Monitor Test Conditions

The PCM allows canister purge to occur when the engine is warm, at wide open or part throttle (as long as the engine is not overheated). The engine can be in open or closed loop fuel control during purging.

Fuel System Monitor

The Fuel System Monitor is a PCM diagnostic that monitors the Adaptive Fuel Control system. The PCM uses adaptive fuel tables that are updated constantly and stored in long term memory (KAM) to compensate for wear and aging in the fuel system components.

FUEL SYSTEM MONITOR OPERATION

Once the PCM determines all the enable criteria has been are met (ECT, IAT and MAF PIDs in range and closed loop enabled), the PCM uses its adaptive strategy to "learn" changes needed to correct a Fuel system that is biased either rich or lean. The PCM accomplishes this task by monitoring Short Term and Long Term fuel trim in closed loop mode.

LONG AND SHORT TERM FUEL TRIM

Short Term fuel trim is a PCM parameter identification (PID) used to indicate Short Term fuel adjustments. This parameter is expressed as a percentage and its range

of authority is from -10% to +10%. Once the engine enters closed loop, if the PCM receives a HO2S signal that indicates the A/F mixture is richer than desired, it moves the SHRTFT command to a more negative range to correct for the rich condition.

If the PCM detects the SHRTFT is adjusting for a rich condition for too long a time, the PCM will "learn" this fact, and move LONGFT into a negative range to compensate so that SHRTFT can return to a value close to 0%. Once a change occurs to LONGFT or SHRTFT, the PCM adds a correction factor to the injector pulsewidth calculation to adjust for variations. If the change is too large, the PCM will detect a fault.

➡ **If a fuel injector, fuel pressure regulator, etc. is replaced, clear the KAM and then drive the vehicle through the Fuel System Monitor drive pattern to reset the fuel control table in the PCM.**

Misfire Detection Monitor

The Misfire Monitor is a PCM diagnostic that continuously monitors for engine misfires under all engine positive load and speed conditions (accelerating, cruising and idling). The Misfire Monitor detects misfires caused by fuel, ignition or mechanical misfire conditions. If a misfire is detected, engine conditions present at the time of the fault are written to the Freeze Frame Data. These conditions overwrite existing data.

Misfire Monitor Operation

See Figure 9.

The Misfire Monitor is designed to measure the amount of power that each cylinder contributes to the engine. The amount of contribution is calculated based upon measurements determined by crankshaft acceleration (TDC of compression stroke to

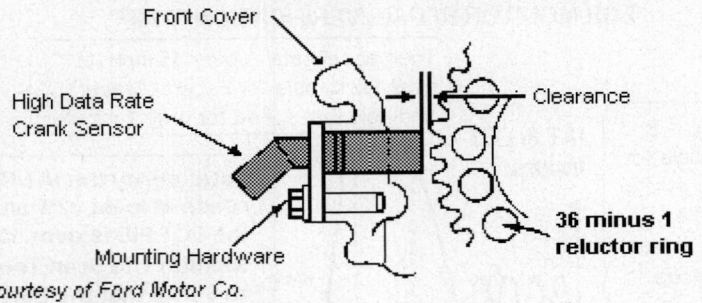

CRANKSHAFT POSITION SENSOR EXAMPLE

Front Cover

High Data Rate
Crank Sensor

Clearance

36 minus 1
reluctor ring

Mounting Hardware

Courtesy of Ford Motor Co.

21199_FDIA_G031

Fig. 9 Typical Crankshaft Position Sensor

BDC of the power stroke) for each cylinder. This calculation requires accurate measurement of the crankshaft angle. Crankshaft angle measurement is determined using a low data rate system on 4-Cyl engines. The high data rate system is used to determine crankshaft angle on all other engines.

Catalyst Damaging Misfire (One-Trip Detection)

If the PCM detects a Catalyst Damaging Misfire, the MIL will flash once per second within 200 engine revolutions from the point where misfire is detected. The MIL will stop flashing and remain on if the engine stops misfiring in a manner that could damage the catalyst.

High Emissions Misfire (Two-Trip Detection)

A High Emissions Misfire is set if a misfire condition is present that could cause the tailpipe emissions to exceed the FTP emissions standard by 1.5 times. If this fault is detected for two consecutive trips under similar engine speed, load and temperature conditions, the MIL is activated. It is also activated if a misfire is detected under similar conditions for two non-consecutive trips that are not 80 trips apart.

State Emissions Failure Misfire (Two-Trip Detection)

A State Emissions Failure Misfire is set if the misfire is sufficient to cause the vehicle to fail a State Inspection or Maintenance (I/M) Test. This fault is determined by identifying misfire percentages that would cause a "durability demonstration vehicle" to fail an Inspection Maintenance (I/M) Test. If the Misfire Monitor detects the fault for two consecutive trips with the engine at similar engine speed, load and temperature conditions, the MIL is activated and a code is set. The MIL is also activated if this type of misfire is detected under similar conditions

for two non-consecutive trips of not more than 80 trips apart.

➡ **Some vehicles set Misfire codes because of an early version of OBD II hardware and software. If a misfire code is set and the cause of the fault is not found, clear the code and retest. Search the TSB list for possible answers or contact the dealer.**

Misfire Detection

See Figure 10.

The Misfire Monitor uses the CKP sensor signals to detect an engine misfire. The amount of contribution is calculated based upon measurements determined by crankshaft acceleration from each cylinder's power stroke.

The PCM performs various calculations to detect individual cylinder acceleration rates. If acceleration for a cylinder deviates beyond the average variation of acceleration for all cylinders, a misfire is detected.

Faults detected by the Misfire Monitor:
- Engine mechanical faults, restricted intake or exhaust system
- Dirty or faulty fuel injectors, loose or damaged injector connectors
- The vehicle has been run low on fuel or run until it ran out of fuel

MISFIRE MONITOR REPAIR VERIFICATION "TRIP"

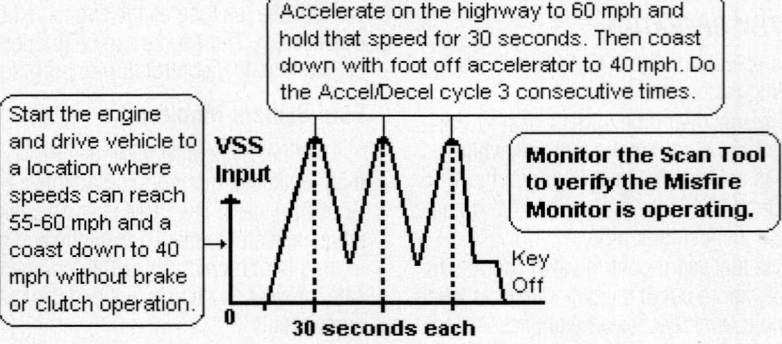

Accelerate on the highway to 60 mph and hold that speed for 30 seconds. Then coast down with foot off accelerator to 40 mph. Do the Accel/Decel cycle 3 consecutive times.

Start the engine and drive vehicle to a location where speeds can reach 55-60 mph and a coast down to 40 mph without brake or clutch operation.

VSS
Input

Monitor the Scan Tool to verify the Misfire Monitor is operating.

Key
Off

0   30 seconds each

21199_FDIA_G032

Fig. 10 Misfire Detection Monitor

Oxygen Sensor Monitor

The Oxygen Sensor Monitor is a PCM diagnostic designed to monitor the front and rear oxygen sensor for faults or deterioration that could cause tailpipe emissions to exceed 1.5 times the FTP standard. The front oxygen sensor voltage and response time are also monitored.

HO2S Monitor Operation

Fuel System and Misfire Monitors must be run and complete before the PCM will start the HO2S Monitor. Additionally, parts of the HO2S Sensor Monitor are enabled during the KOER Self-Test. The HO2S Monitor is run during each drive cycle after the CKP, ECT, IAT and MAF sensor signals are within a predetermined range.

Fixed Frequency Closed Loop Test

See Figure 11.

The HO2S Monitor constantly monitors the sensor voltage and frequency. The PCM detects a high voltage condition by comparing the HO2S signal to a preset level.

FIXED FREQUENCY TEST

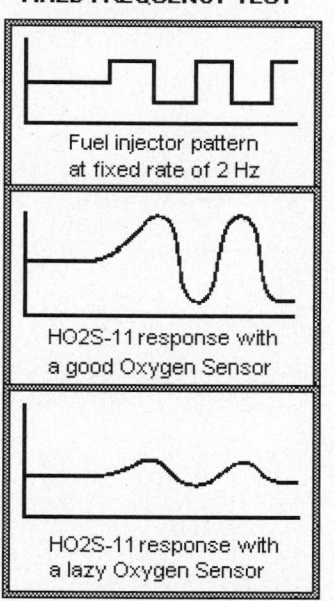

Fuel injector pattern at fixed rate of 2 Hz

HO2S-11 response with a good Oxygen Sensor

HO2S-11 response with a lazy Oxygen Sensor

21199_FDIA_G033

Fig. 11 Fixed Frequency Test

A Fixed Frequency Closed Loop Test is used to check the HO2S voltage and frequency. A sample of the HO2S signal is checked to determine if the sensor is capable of switching properly or has a slow response time (referred to as a lazy sensor).

Oxygen Sensor Heater Monitor

The Oxygen Sensor Heater Monitor is a PCM diagnostic designed to monitor the Oxygen Sensor Heater and its related circuits for faults.

OXYGEN SENSOR HEATER MONITOR OPERATION

The Oxygen Sensor Heater Monitor performs its task by detecting whether the proper amount of O2 sensor voltage change occurred as the HO2S Heater is turned from "on" to "off" with the engine in closed loop. The time it takes for the HO2S-11 and HO2S-12 signal to switch (the response time) is constantly monitored by the Oxygen Sensor Monitor. Once the Oxygen Sensor Heater Monitor is enabled, if the switch time for the HO2S-11 or HO2S-12 signal is too long, the PCM fails the test, the MIL is activated and a trouble code is set.

➡ **Response time is defined as the amount of time it takes for a HO2S signal to switch from Rich to Lean, and then Lean to Rich.**

FRONT AND REAR OXYGEN SENSOR HEATER OPERATION

Both upstream and downstream Oxygen sensors are used on the OBD II system. These sensors are designed with additional protection around the ceramic core to protect them from condensation that could crack them if the heater is turned on with condensation present.

The HO2S heaters are not turned on until the ECT sensor signal indicates that the engine is warm. The delay period can last for as long as 5 minutes from startup. The delay allows any condensation in the Exhaust system to evaporate.

Faults detected by the HO2S or HO2S Heater Monitor:

- A fault in the HO2S, the HO2S heater or its related circuits
- A fault in the HO2S connectors (look for moisture tracking)
- A defective Power Control Module

Air Injection System Monitor

The Air Injection System Monitor is an OBD diagnostic controlled by the PCM that monitors the Air Injection (AIR) system. The Oxygen Sensor Monitor must run and complete before the PCM will run this test. The PCM enables this test during AIR system operation after certain engine conditions are met and these enable criteria are met:

- Crankshaft Position sensor signal must be present
- ECT and IAT sensor input signals must be within limits

AIR MONITOR–ELECTRIC PUMP DESIGN

The AIR Monitor consists of these Solid State Monitor tests:

- A check of the Solid State relay for electrical faults.
- A check of the secondary side of the relay for electrical faults.
- A test to determine if the AIR system can inject additional air.

AIR MONITOR–MECHANICAL PUMP DESIGN

The AIR Monitor for the mechanical (belt-driven air pump) design uses two Output State Monitor configurations to perform two different circuit tests. One test is used to check for faults in the Secondary Air Bypass (AIRB) solenoid circuit. The normal function of the AIRB solenoid and valve assembly is to dump air into the atmosphere.

A second test is used to check for electrical faults in the Secondary Air Divert (AIRD) solenoid. The normal function of the AIRD solenoid and valve assembly is to direct the air either upstream or downstream.

FUNCTIONAL CHECK

See Figure 12.

An AIR system functional check is done at startup with the AIR pump on or during a hot idle period if the startup part of the test was not performed. A flow test is included that uses the HO2S signal to indicate the presence of extra air injected into the exhaust stream.

Diagnostic Trouble Codes

In the Diagnostic Trouble Code charts for the specific manufacturers you will see the following terms in the left column of the chart:

1. 1T–This means the code was activated when the PCM recognized the problem the first time it occurred.

2. 2T–This means the code was activated when the PCM recognized the problem and set the code after it occurred two times.

3. CCM–This means that the code and system affected is an emission related device and has a Comprehensive Component Monitor (CCM) tracking it.

4. MIL: Yes–This means that the Malfunction Indicator Light will be displayed.

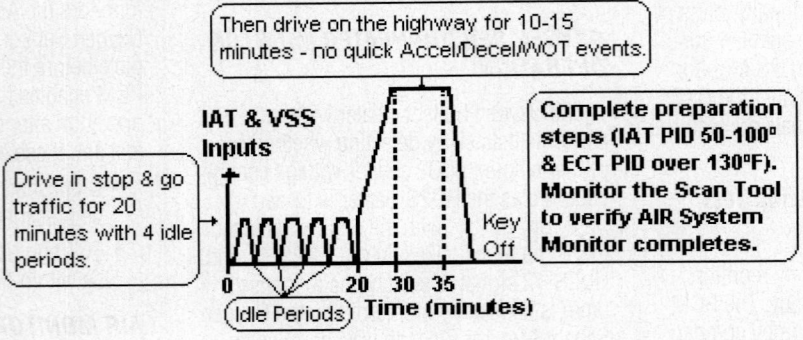

SECONDARY AIR MONITOR REPAIR VERIFICATION "TRIP"

Then drive on the highway for 10-15 minutes - no quick Accel/Decel/WOT events.

IAT & VSS Inputs

Drive in stop & go traffic for 20 minutes with 4 idle periods.

Complete preparation steps (IAT PID 50-100° & ECT PID over 130°F). Monitor the Scan Tool to verify AIR System Monitor completes.

Key Off

Idle Periods   Time (minutes)

0   20 30 35

21199_FDIA_G036

Fig. 12 Secondary AIR monitor

SYMPTOM DIAGNOSIS (NO CODES)

4

Table of Contents

What To Do When There Are No DTCs

Do not attempt to diagnose a Drivability Symptoms without having a logical plan to use to determine which Engine Control system is the cause of the symptom - this plan should include a way to determine which systems do not have a problem! Drivability symptom diagnosis is a part of an organized approach to problem solving and repair.

DRIVABILITY SYMPTOM INDEX TABLE

To use this list, locate the symptom that matches a particular problem and refer to the areas to test. The items listed under each symptom may not apply to all models, engines or vehicle systems. The repair steps indicate what vehicle component or system to test.

➡ The Drivability Symptoms in this list are intended to be generic. While they apply to most vehicles, some vehicles may not have all of the components listed. Refer to other Chilton repair information and electronic media for specific tests.

Symptom Test Table

Symptom Description	Suggested Areas to Test
Test 1 - No Start, Hard Start Condition • No Crank • Hard Start, Long Crank, Erratic Crank • Stall After Start • No Start, Normal Crank • No Start, MIL is off (if the VREF shorts to ground)	- Check battery, battery circuits to starter - Check for a damaged flywheel, engine compression, base timing and minimum air rate - Check for a failed fuel pump relay - Check for distributor rotor "punch-through" - Check for a faulty ignition control module (ICM) - Check for a VREF circuit shorted to ground - Check SKIM (security system) with a Scan Tool
Test 2 - Rough Idle or Stalls Condition • Low or slow idle speed • Fast idle speed • Hunting or rolling idle speed • Slow return to idle speed • Stalls or almost stalls	- Check for engine vacuum leaks - Check the condition of the PCV valve and lines - Check for excessive carbon buildup - Check for a restricted exhaust - Check base idle speed, check for low fuel pressure - Check the throttle linkage for sticking or binding
Test 3 - Runs Rough Condition • At idle speed • During acceleration • At cruise speed • During deceleration	- Check for engine vacuum leaks at intake manifold - Check condition of ignition secondary components - Check base timing and idle speed settings - Check for low or high fuel pressure - Check for dirty, leaking or shorted fuel injectors - Check for excessive carbon buildup on valves
Test 4 - Cuts-out, Misses Condition • At idle speed • During acceleration • At cruise speed • During deceleration	- Check for engine vacuum leaks at intake manifold - Check condition of ignition secondary components - Check that spark timing advance is available - Check for low or high fuel pressure - Check for dirty, leaking or shorted fuel injectors - Check for excessive carbon buildup on valves
Test 5 - Bucks, Jerks Condition • During acceleration • At cruise speed • During deceleration	- Check for engine vacuum leaks at intake manifold - Check condition of ignition secondary components - Check that spark timing advance is available - Check for low or high fuel pressure - Check for dirty, leaking or shorted fuel injectors - Check operation of the TCC solenoid, brake switch

Symptom Diagnosis Test 1 — No Start, Hard Start Condition

➡ **If there is no spark output or fuel pressure available, check for a failed fuel pump relay, no power to the PCM, or loss of the ignition reference signal to the PCM.**

PRELIMINARY CHECKS

Prior to starting this symptom test routine, inspect these underhood items:

• Check battery charge and condition, starter current draw.
• Verify the starter relay operation and that the engine cranks (turns over).
• Verify the check engine light (MIL) operation - if it does not activate, check the PCM power and ground circuits, and check for 5v supply at the MAP or TP sensor.
• Check Air Intake system for restrictions (inspect air inlet tubes, air filter for dirt, etc.).
• Check the status of the Smart Key Immobilizer System (SKIM) with the Scan Tool.

Test 1 Chart

Step	Action	Yes	No
1	**Step Description: No Start Condition Only** » Check battery cables, state of charge. » If the engine does not rotate, inspect for a locked engine (hydrostatic lockup condition). » Does the engine crank normally?	Go to Step 2.	Repair the fault in the battery, starter, or Base Engine. Retest for the symptom when all repairs are done.
2	**Step Description: Check the Fuel System** » Verify that the pump operates at key on. » Check the fuel pump relay operation. If the relay does not operate, check for blown fuse. » Inspect pump for a leak-down condition » Test fuel pressure, volume and quality. » Test the operation of the fuel regulator. » Are there any faults in the Fuel system?	Make needed repairs.	Go to Step 3.
3	**Step Description: Check the Ignition System** » Inspect ignition secondary components for damage (look for rotor "punch-through"). » Inspect the coils for signs of spark leakage at coil towers or primary connections. » Check the spark output with a spark tester. » Test Ignition system with an engine analyzer. » Are there any faults in the Ignition system?	Make repairs to the Ignition system. Then retest the symptom.	Go to Step 4.
4	**Step Description: Check the Exhaust System** » Check Exhaust system for leaks or damage. » Check the Exhaust system for a restriction using the Vacuum or Pressure Gauge Test (e.g., exhaust backpressure reading should not exceed 1.5 psi at cruise speeds). » Are there any faults in the Exhaust system?	Make repairs to the Exhaust system. Then retest the symptom.	Go to Step 5.
5	**Step Description: Check the MAP Sensor** » Disconnect the MAP sensor and attempt to start the engine. » Does the engine start and run normally?	Replace the MAP sensor. Retest for the symptom when repairs are completed.	Go to Step 6.
6	**Step Description: Check for a Hot Engine** » Check for signs of an engine overheating condition related to a Hard Start Symptom. » Does the engine appear to be overheated?	Make the repairs to correct the hot engine and then retest for the symptom when done.	Go to Step 7.
7	**Step Description: Check ECT Sensor PID** » Connect a Scan Tool and turn the key to on. » Read the ECT sensor (compare to chart). » Has the ECT sensor shifted out of range?	Replace the ECT sensor. Then retest for the symptom when all repairs are completed.	Go to Step 8.
8	**Step Description: Check the PCV System** » Inspect the PCV system components for broken parts or loose connections. » Test the operation of the PCV valve. » Are there any faults in the PCV system?	Repair the PCV system. Refer to the PCV system tests. Retest the symptom when all repairs are done.	Go to Step 9.
9	**Step Description: Check the EVAP System** » nspect for damaged or disconnected EVAP system components. » Inspect for a fuel saturated charcoal canister. » Are there any faults in the EVAP system?	Refer to the EVAP system tests. Retest for the symptom when all repairs are completed.	Go to Step 10.
10	**Step Description: Test the Base Engine** » Check the engine compression. » Test valve timing and timing chain condition. » Check for a worn camshaft or valve train. » Check for any large intake manifold leaks. » Are there any faults in the Base Engine?	Repair the Base Engine. Refer to the Base Engine Tests. Retest symptom when done.	Return to Step 2 to repeat the test steps in this series to locate and repair the "No Start, Hard Start" condition.

Symptom Diagnosis Test 2 — Rough, Low or High Idle Speed Condition

➥ **If the vehicle has a rough idle and the base timing, idle speed and the IAC (or AIS) motor operates properly, check the engine for excessive carbon buildup.**

PRELIMINARY CHECKS

Prior to starting this symptom test routine, inspect these underhood items:
- All related vacuum lines for proper routing and integrity.
- All related electrical connectors and wiring harnesses for faults (Wiggle Test).
- Check the throttle linkage for a sticking or binding condition.
- Air Intake system for restrictions (air inlet tubes, dirty air filter, etc.).
- Search for any technical service bulletins related to this symptom.
- Turn the key to off. Unplug the MAP sensor connection and restart the engine to recheck for the idle concern. If the condition is gone, replace the MAP sensor.

Test 2 Chart

Step	Action	Yes	No
1	**Step Description: Verify the rough idle or stall** » Does the engine have a warm engine rough idle, low idle or high idle condition in P or N?	Go to Step 2.	Fault is intermittent. Return to the Symptom List and select another fault.
2	**Step Description: Verify idle speed & timing** » Verify the base timing is within specifications » Verify that the base idle speed is set properly » Are the timing and idle speed set properly?	Go to Step 3.	Set the base idle speed and timing to the specifications and then retest for the symptom.
3	**Step Description: Check AIS / IAC Operation** » Check the AIS or IAC motor operation » Inspect the AIS/IAC housing in throttle body for restricted passages. Clean as needed. » Set the parking brake, block the drive wheels and turn the A/C off. Install the Scan Tool. » IAC Motor Tester - Turn the key off and then connect the IAC tester to the IAC valve. » Start the engine and use the IAC tester to extend and retract the IAC valve. » ATM Test - Start the engine. Use the tool to change the speed from min-idle to 1500 rpm. » Did the idle speed change as commanded?	Install an Aftermarket Noid light and check the operation of the PCM and AIS or IAC motor circuits. Check the motor for signs of open or shorted circuits. Replace the IAC motor or PCM as needed or make repairs to the IAC motor wiring. If all are okay, go to Step 4.	If the AIS/IAC motor passages are clean and engine speed did not change as described when the AIS/IAC motor was extended and retracted, replace the AIS/IAC motor. Then retest for the condition.
4	**Step Description: Check/compare PID values** » Connect Scan Tool & turn off all accessories. » Start the engine and allow it to fully warmup. » Monitor all related PIDs on the Scan Tool. » Verify the P/N switch input in gear and Park. » Check the O2S operation with a Lab Scope. » Are all PIDs within normal range?	Go to Step 5. Note: An IAC motor count of over 80 indicates the pintle is extended and an IAC count of (0) indicates the pintle is retracted.	One or more of the PIDs are out of range when compared to "known good" values. Make repairs to the system that is out of range, then retest for the symptom.

5	**Step Description: Check the Ignition System** » Inspect the coils for signs of spark leakage at coil towers or primary connections. » Check the spark output with a spark tester. » Test Ignition system with an engine analyzer. » Were any faults found in the Ignition system?	Make repairs as needed	Go to Step 6.
6	**Step Description: Check the Fuel System** » Inspect the Fuel delivery system for leaks. » Test the fuel pressure, quality and volume. » Test the operation of the pressure regulator. » Were any faults found in the Fuel system?	Make repairs as needed	Go to Step 7.
7	**Step Description: Check the Exhaust System** » Check Exhaust system for leaks or damage. » Check the Exhaust system for a restriction using the Vacuum or Pressure Gauge Test (e.g., exhaust backpressure reading should not exceed 1.5 psi at cruise speeds). » Were any faults found in Exhaust System?	Make repairs to the Exhaust system. Then retest the symptom.	Go to Step 8.
8	**Step Description: Check the PCV System** » Inspect the PCV system components for broken parts or loose connections. » Test the operation of the PCV valve. » Were any faults found in the PCV system?	Make repairs to the PCV system. Refer to the PCV system tests. Then retest for the condition.	Go to Step 9.
9	**Step Description: Check the EVAP System** » Inspect for damaged or disconnected EVAP system components or a saturated canister. » Were any faults found in the EVAP system?	Make repairs to EVAP system. Retest for the condition.	Go to Step 10.
10	**Step Description: Check the Base Engine** » Test the engine compression. » Test valve timing and timing chain condition. » Check for a worn camshaft or valve train. » Check for any large intake manifold leaks. » Were any faults found in the Base Engine?	Make repairs as needed to the Base Engine. Refer to the Base Engine tests. Then retest for the condition when repairs are completed.	Go to Step 2 and repeat the tests from the beginning to locate and repair the cause of the "Rough, Low or High Idle Speed" condition.

Symptom Diagnosis Test 3 — Runs Rough Condition

PRELIMINARY CHECKS

Prior to starting this symptom test routine, inspect these underhood items:

- All related vacuum lines for proper routing and integrity
- Air Intake system for restrictions (air inlet tubes, dirty air filter, etc.)
- Search for any technical service bulletins related to this symptom.

Test 3 Chart

Step	Action	Yes	No
1	**Step Description: Verify engine runs rough** » Start the engine and allow it to idle in P or N. » Does the engine run rough when warm in Park or Neutral position?	Check for any stored codes. If codes are set, repair codes and retest. If no codes are set, go to Step 3.	Go to Step 2.
2	**Step Description: Condition does not exist!** » Inspect various underhood items that could cause an intermittent Runs Rough condition (i.e., dirt in the throttle body, vacuum leaks, IAC motor connections, etc.). » Were any problems located in this step?	Correct the problems. Do a PCM reset and engine "idle relearn" procedure. Then verify the "runs rough" condition is repaired.	The problem is not present at this time. It may be an intermittent problem.
3	**Step Description: Check/compare PID values** » Connect a Scan Tool to the test connector. » Turn off all accessories. » Start the engine and allow it to fully warmup. » Monitor all related PIDs on the Scan Tool. » Were all PIDs within their normal range?	Go to Step 4. Note: The IAC motor should read from 5-50 counts. Check the LONGFT reading for a large shift into the negative range (due to a rich condition).	One or more of the PIDs are out of range when compared to "known good" values. Make repairs to the system that is out of range, then retest for the symptom.
4	**Step Description: Check the Ignition System** » Inspect the coils for signs of spark leakage at coil towers or primary connections. » Check the spark output with a spark tester. » Test Ignition system with an engine analyzer. » Were any faults found in the Ignition system?	Make repairs as needed	Go to Step 5.
5	**Step Description: Check the Fuel System** » Inspect the Fuel delivery system for leaks. » Test the fuel pressure, quality and volume. » Test the operation of the pressure regulator. » Were any faults found in the Fuel system?	Make repairs as needed	Go to Step 6.
6	**Step Description: Check the Exhaust System** » Check Exhaust system for leaks or damage. » Check the Exhaust system for a restriction using the Vacuum or Pressure Gauge Test (e.g., exhaust backpressure reading should not exceed 1.5 psi at cruise speeds). » Were any faults found in Exhaust System?	Make repairs to the Exhaust system. Then retest the symptom.	Go to Step 7.
7	**Step Description: Check the PCV System** » Inspect the PCV system components for broken parts or loose connections. » Test the operation of the PCV valve. » Were any faults found in the PCV system?	Make repairs to the PCV system. Refer to the PCV system tests. Then retest for the condition.	Go to Step 9.
8	**Step Description: Check the EVAP System** » Inspect for damaged or disconnected EVAP system components or a saturated canister. » Were any faults found in the EVAP system?	Make repairs to EVAP system. Retest for the condition.	Go to Step 10.
9	**Step Description: Check Engine Condition** » Test the engine compression. » Test valve timing and timing chain condition. » Check for a worn camshaft or valve train. » Check for any large intake manifold leaks. » Were any faults found in the Base Engine?	Make repairs as needed to the Base Engine. Refer to the Base Engine tests. Then retest for the condition when repairs are completed.	Return to Step 2 and repeat the tests from the beginning to locate and repair the cause of the "Runs Rough" condition.

Symptom Diagnosis Test 4 — Cuts-out or Misses Condition

PRELIMINARY CHECKS

Prior to starting this symptom test routine, inspect these underhood items:
- All related vacuum lines for proper routing and integrity
- Search for any technical service bulletins related to this symptom.

Test 4 Chart

Step	Action	Yes	No
1	**Step Description: Verify Cuts-out condition** » Start the engine and attempt to verify the Cuts-out or misses condition. » Does the engine have a cuts-out condition?	Check for any stored codes. If codes are set, repair codes and retest. If no codes are set, go to Step 3.	Go to Step 2.
2	**Step Description: Condition does not exist!** » Inspect various underhood items that could cause an intermittent Cuts-out condition (i.e., EVAP, Fuel or Ignition system components). » Were any problems located in this step?	Correct the problems. Do a PCM reset and "Fuel Trim Relearn" procedure. Then verify condition is repaired.	The problem is not present at this time. It may be an intermittent problem.
3	**Step Description: Check/compare PID values** » Connect a Scan Tool to the test connector. » Turn off all accessories. » Start the engine and allow it to fully warmup. » Monitor all related PIDs on the Scan Tool (i.e., ECT IAC Counts and LONGFT at idle). » Were all PIDs within their normal range?	Go to Step 4. Note: The IAC motor should be from 5-50 counts. Watch fuel trim (%) for a large shift into the negative (-) range (due to a rich condition).	One or more of the PIDs are out of range when compared to "known good" values. Make repairs to the system that is out of range, then retest for the symptom.
4	**Step Description: Check the Ignition System** » Inspect the coils for signs of spark leakage at coil towers or primary connections. » Check the spark output with a spark tester. » Test Ignition system with an engine analyzer. » Were any faults found in the Ignition system?	Make repairs as needed	Go to Step 5.
5	**Step Description: Check the Fuel System** » Inspect the Fuel delivery system for leaks. » Test the fuel pressure, quality and volume. » Test the operation of the pressure regulator. » Were any faults found in the Fuel system?	Make repairs as needed	Go to Step 6.
6	**Step Description: Check the Exhaust System** » Check Exhaust system for leaks or damage. » Check the Exhaust system for a restriction using the Vacuum or Pressure Gauge Test (e.g., exhaust backpressure reading should not exceed 1.5 psi at cruise speeds). » Were any faults found in Exhaust System?	Make repairs to the Exhaust system. Then retest the symptom.	Go to Step 7.
7	**Step Description: Check the PCV System** » Inspect the PCV system components for broken parts or loose connections. » Test the operation of the PCV valve. » Were any faults found in the PCV system?	Make repairs to the PCV system. Then retest for the condition.	Go to Step 8.
8	**Step Description: Check the EVAP System** » Inspect for damaged or disconnected EVAP system components » Check for a saturated EVAP canister. » Were any faults found in the EVAP system?	Make repairs to EVAP system. Retest for the condition.	Go to Step 9.
9	**Step Description: Check the AIR system** » Inspect AIR system for broken parts, leaking valves or disconnected hoses. » Test the operation of Secondary AIR system. » Were any faults found in the AIR system?	Make repairs as needed. Refer to the Secondary AIR system tests. Retest for the condition.	Go to Step 10.
10	**Step Description: Check Engine Condition** » Test the engine compression. » Test valve timing and timing chain condition. » Check for a worn camshaft or valve train. » Check for any large intake manifold leaks. » Were any faults found in the Base Engine?	Make repairs as needed to the Base Engine. Refer to the Base Engine tests. Then retest for the condition when repairs are completed.	Go to Step 2 and repeat the tests from the beginning to locate and repair the cause of the "Cuts Out or Misses" condition.

Symptom Diagnosis Test 5 — Surge Condition

PRELIMINARY CHECKS

1. Discuss how the operation of the torque converter clutch (TCC) or air conditioning compressor can affect the "feel" of the vehicle during normal operation. Refer to the information in the Owner's Manual to explain how these devices normally operate.
2. Search for any technical service bulletins related to this symptom.

Test 5 Chart

Step	Action	Yes	No
1	**Step Description: Verify the surge condition** » Drive the vehicle and attempt to verify that the vehicle surges at cruise speeds. » Does the engine have a surge condition?	Check for any stored codes. If codes are set, repair codes and retest. If no codes are set, go to Step 3.	Go to Step 2.
2	**Step Description: Condition does not exist!** » Inspect various underhood items that could cause an intermittent surge condition (check for leaks in the MAP sensor vacuum lines). » Were any problems located in this step?	Correct the problems. Do a PCM reset and "Fuel Trim Relearn" procedure. Then verify condition is repaired.	The problem is not present at this time. It may be an intermittent problem.
3	**Step Description: Check/compare PID values** » Connect a Scan Tool to the test connector. » IStart the engine and allow it to fully warmup. » Monitor all related PIDs on Scan Tool (HO2S switching, LONGFT, and the TCC operation) » Compare VSS PID reading to speedometer. » Were all PIDs within their normal range?	Go to Step 4. Note: Verify that the front HO2S responds quickly to throttle changes. Check for silicon contamination on the front HO2S (this can cause a rich A/F signal).	One or more of the PIDs are out of range when compared to "known good" values. Make repairs to the system that is out of range, then retest for the symptom.
4	**Step Description: Check the Ignition System** » Inspect the coils for signs of spark leakage at coil towers or primary connections. » Check the spark output with a spark tester. » Test Ignition system with an engine analyzer. » Were any faults found in the Ignition system?	Make repairs as needed	Go to Step 5.
5	**Step Description: Check the Fuel System** » Inspect the Fuel delivery system for leaks. » Test the fuel pressure, quality and volume. » Test the operation of the pressure regulator. » Were any faults found in the Fuel system?	Make repairs as needed	Go to Step 6.
6	**Step Description: Check the Exhaust System** » Check Exhaust system for leaks or damage. » Check the Exhaust system for a restriction using the Vacuum or Pressure Gauge Test (e.g., exhaust backpressure reading should not exceed 1.5 psi at cruise speeds). » Were any faults found in Exhaust System?	Make repairs to the Exhaust system. Then retest the symptom.	Return to Step 2 and repeat the tests from the beginning to locate and repair the cause of the "Surge" condition.

INTERMITTENT TESTS

Many trouble code repair charts end with a result that reads "Fault Not Present at this Time." What this expression means is that the conditions that were present when a code set or drivability symptom occurred are no longer there or were not met. In effect, the problem was present at least once, but is not present at this time. However, it is likely to return in the future, so it should be diagnosed and repaired if at all possible.

One way to find an intermittent problem is to gather the information that was present when the problem occurred. In the case of a Code Fault, this can be done in two ways: by capturing the data in Snapshot or Movie mode or by driver observations.

The PCM has to detect the fault for a specific period of time before a trouble code will set. While intermittent problems may appear to be occasional in nature, they usually occur under specific conditions. Therefore, you should identify and duplicate these conditions. Since intermittent faults are difficult to duplicate, a logical routine (checklist) must be followed when attempting to find the faulty component, system or circuit. The tests on the next page can be used to help find the cause of an intermittent fault.

Some intermittent faults occur due to a loose connection, wiring problem or warped circuit board. An intermittent fault can also be caused by poor test techniques that cause damage to the male or female ends of a connector.

Test for Loose Connectors

To test for a loose or damaged connection, take the male end of a connector from another wiring harness and carefully push it into the "suspect" female terminal to verify that the opening is tight. There should be some resistance felt as the male connector is inserted in the terminal connection.

The Wiggle Test

See Figures 1 and 2.

A wiggle test can be used to locate the cause of some intermittent faults. The sensor, switch or the PCM wiring can be back-probed, as shown, while the test is done.

During testing, move or wiggle the suspect device, connector or wiring while watching for a change.

If the DVOM has a Min/Max record mode, use this mode during the test.

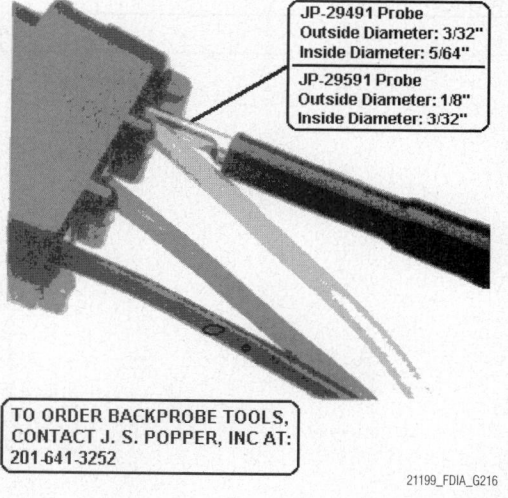

JP-29491 Probe
Outside Diameter: 3/32"
Inside Diameter: 5/64"

JP-29591 Probe
Outside Diameter: 1/8"
Inside Diameter: 3/32"

TO ORDER BACKPROBE TOOLS, CONTACT J. S. POPPER, INC AT: 201-641-3252

21199_FDIA_G216

Fig. 1 Backprobing a connector

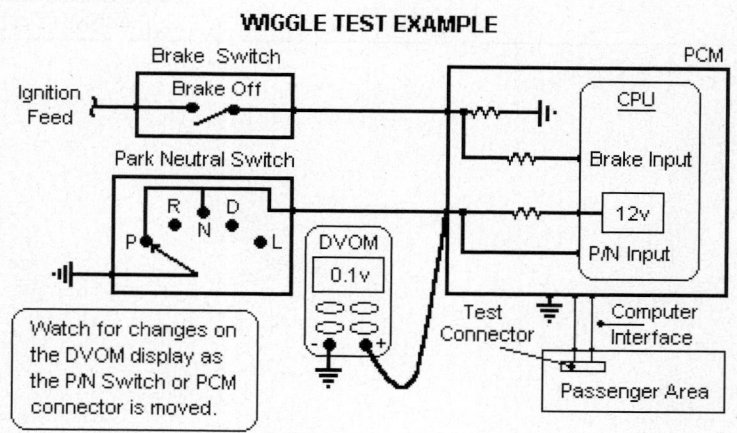

WIGGLE TEST EXAMPLE

Watch for changes on the DVOM display as the P/N Switch or PCM connector is moved.

21199_FDIA_G216

Fig. 2 Wiggle Test Example

Diagnosis And Testing - Vehicle Does Not Fill

CONDITION	POSSIBLE CAUSES	CORRECTION
Pre-Mature Nozzle Shut-Off	Defective fuel tank assembly components.	Fill tube improperly installed (sump)
		Fill tube hose pinched.
		Check valve stuck shut.
		Control valve stuck shut.
	Defective vapor/vent components.	Vent line from control valve to canister pinched.
		Vent line from canister to vent filter pinched.
		Canister vent valve failure (requires double failure, plugged to NVLD and atmosphere).
		Leak detection pump failed closed.
		Leak detection pump filter plugged.
	On-Board diagnostics evaporative system leak test just conducted.	Canister vent valve vent plugged to atmosphere.
		Engine still running when attempting to fill (System designed not to fill).
	Defective fill nozzle.	Try another nozzle.
Fuel Spits Out Of Filler Tube.	During fill.	See Pre-Mature Shut-Off.
	At conclusion of fill.	Defective fuel handling component. (Check valve stuck open).
		Defective vapor/vent handling component.
		Defective fill nozzle.

LEXUS
DIAGNOSTIC TROUBLE CODES

5

TABLE OF CONTENTS

OBD II VEHICLE APPLICATIONS

LEXUS

ES300
1996–2003
3.0L V6 MFI 1MZ-FE ....VIN F

ES330
2004–06
3.3L V6 MFI 3MZ-FE ....VIN A

GS300
1996–2006
3.0L I6 MFI 2JZ-GE ....VIN D

GS400
1998–2000
4.0L V8 MFI 1UZ-FE................................VIN H

GS430
2001–2006
4.3L V8 MFI 3UZ-FE................................VIN N, L

GX470
2003–2006
4.7L V8 MFI 2UZ-FE................................VIN T

IS250
2006
2.5L V6 MFI 4GR-FSE ....VIN K

IS300
2001–2005
3.0L I6 MFI 2JZ-GE ....VIN D

IS350
2006
3.5L V6 MFI 2GR-FSE ....VIN E

LS400
1996–2000
4.0L V8 MFI 1UZ-FE................................VIN H

LS430
2001–2006
4.3L V8 MFI 3UZ-FE................................VIN N

LX450
1996-1997
4.5L I6 MFI 1FZ-FE ....VIN J

LX470
1998–2006
4.7L V8 MFI 2UZ-FE................................VIN T

RX300
1999–2003
3.0L V6 MFI 1MZ-FE................................VIN F

RX330
2004–2006
3.3L V6 MFI 3MZ-FE................................VIN A

SC300
1996–2000
3.0L I6 MFI 2JZ-GE ....VIN D

SC400
1996–2000
4.0L V8 MFI 1UZ-FE................................VIN H

SC430
2002–2006
4.3L V8 MFI 3UZ-FE................................VIN N

DIAGNOSTIC TROUBLE CODES

OBD II Trouble Code List (P0xxx Codes)

DTC	Trouble Code Title, Conditions & Possible Causes:
DTC: P2A00 **2T CCM, MIL: Yes** **1996, 1997, 1998, 1999, 2000, 2001, 2002, 2003, 2004, 2005, 2006** **Models:** ES300, RX300, ES330, RX330, GS400, GS430, GX470, LS400, LS430, SC400, SC430 **Engines:** 3.0L VIN F, 3.3L VIN A, 4.0L VIN H, 4.3L VIN N, L, 4.7L VIN T **Transmissions:** All	**Air Fuel Sensor (Bank 1 Sensor 1) Signal Slow Response Conditions:** Vehicle driven at cruise speed at over 1400 rpm in closed loop at 60 mph, and the PCM detected an unexpected voltage condition on the Bank 1 Air Fuel Sensor 1 (AFS1) circuit. **Possible Causes:** • A/F sensor connector is damaged or loose • A/F sensor circuit is open or shorted, or the sensor has failed • A/F sensor heater is damaged or has failed • A/F sensor heater relay circuit is open or the relay has failed • Fuel delivery component has failed (fuel pressure regulator, one or more fuel injectors is leaking or severely restricted) • Induction system problems (air leaks or restricted air filter) • PCM has failed
DTC: P2A03 **2T CCM, MIL: Yes** **1996, 1997, 1998, 1999, 2000, 2001, 2002, 2003, 2004, 2005, 2006** **Models:** ES300, RX300, ES330, RX330, **Engines:** 3.0L VIN F, 3.3L VIN A **Transmissions:** All	**Air Fuel Sensor (Bank 2 Sensor 1) Signal Slow Response Conditions:** Vehicle driven at cruise speed at over 1400 rpm in closed loop at 60 mph, and the PCM detected an unexpected voltage condition on the Bank 2 Air Fuel Sensor 1 (AFS1) circuit. **Possible Causes:** • A/F sensor connector is damaged or loose • A/F sensor circuit is open or shorted, or the sensor has failed • A/F sensor heater is damaged or has failed • A/F sensor heater relay circuit is open or the relay has failed • Fuel delivery component has failed (fuel pressure regulator, one or more fuel injectors is leaking or severely restricted) • Induction system problems (air leaks or restricted air filter) • PCM has failed
DTC: P0010 **2T CCM, MIL: Yes** **1996, 1997, 1998, 1999, 2000, 2001, 2002, 2003, 2004, 2005, 2006** **Models:** ES300, RX300, ES330, RX330, **Engines:** 3.0L VIN F, 3.3L VIN A **Transmissions:** All	**VVT Oil Control Circuit Malfunction (Bank 1) Conditions:** Key on or engine running; and the PCM detected an unexpected voltage condition on the VVT Oil Control Valve Bank 1 circuit. The VVT system controls the intake camshaft in order to provide optimal valve timing during all conditions based signals from the ECT, IAT and TP sensor. The VVT regulates the intake camshaft angle using oil pressure through the Oil Control Valve. This results in the relative position between the camshaft and crankshaft to become optimal. The result is higher torque, better fuel economy and low emissions. **Possible Causes:** • OCV assembly connector is damaged or loose • OCV assembly control circuit is open or shorted to ground • OCV assembly is damaged or has failed • PCM has failed
DTC: P0011 **2T CCM, MIL: Yes** **1996, 1997, 1998, 1999, 2000, 2001, 2002, 2003, 2004, 2005, 2006** **Models:** ES300, RX300, ES330, RX330, **Engines:** 3.0L VIN F, 3.3L VIN A **Transmissions:** All	**Camshaft Position 'A' Over-Advanced Or System Performance (Bank 1) Conditions:** Engine started, ECT sensor more than 158°F, vehicle driven at an engine speed of 400-4000 rpm, and the PCM detected the valve timing did not change from the "current" valve timing, or the valve timing remain fixed during testing. The VVT system controls the intake camshaft in order to provide optimal valve timing during all conditions based signals from the ECT, IAT and TP sensor. The VVT regulates the intake camshaft angle using oil pressure through the Oil Control Valve. This results in the relative position between the camshaft and crankshaft to become optimal. The result is better engine torque, fuel economy and lower emissions. **Possible Causes:** • Engine valve timing malfunction • Camshaft timing oil control valve unit is damaged or has failed • PCM or VVT ECM has failed
DTC: P0012 **2T CCM, MIL: Yes** **1996, 1997, 1998, 1999, 2000, 2001, 2002, 2003, 2004, 2005, 2006** **Models:** ES300, RX300, ES330, RX330, **Engines:** 3.0L VIN F, 3.3L VIN A **Transmissions:** All	**Camshaft Position 'A' Over-Retarded (Bank 1) Conditions:** Engine started, ECT sensor more than 158°F, vehicle driven at an engine speed of 400-4000 rpm, and the PCM detected the valve timing did not change from the "current" valve timing, or that the valve timing remain fixed during the test period. The VVT system controls the intake camshaft in order to provide optimal valve timing during all conditions. The VVT regulates the intake camshaft angle using oil pressure through the OCV. This causes the relative position between the camshaft and crankshaft to become optimal. The result is improved engine torque, better fuel economy and low emissions. **Possible Causes:** • Engine valve timing malfunction • Camshaft timing oil control valve unit is damaged or has failed • PCM or VVT ECM has failed

DTC	Trouble Code Title, Conditions & Possible Causes
DTC: P0016 **2T CCM, MIL: Yes** **1996, 1997, 1998, 1999, 2000, 2001, 2002, 2003, 2004, 2005, 2006** **Models:** ES300, RX300, ES330, RX330, **Engines:** 3.0L VIN F, 3.3L VIN A **Transmissions:** All	**Camshaft Position-Crankshaft Position (Bank 1 Sensor A) Conditions:** Engine started, engine running, and the PCM detected a deviation between the crankshaft position sensor signal and the VVT Sensor 1 signal during the test period. The crankshaft position (NE) sensor consists of a magnet, iron core and pickup coil. The NE sensor signal plate, installed on the crankshaft-timing pulley, has 34 teeth. This sensor generates 34 signals for each engine revolution. The PCM detects the crankshaft angle and engine speed based on the NE signal. It detects the correct cylinder based on signals from the VVT 1 sensor along signals from the crankshaft position sensor. **Possible Causes:** • Engine valve timing problem • Engine timing belt mechanical problem (skipped teeth or belt) • PCM has failed
DTC: P0018 **2T CCM, MIL: Yes** **1996, 1997, 1998, 1999, 2000, 2001, 2002, 2003, 2004, 2005, 2006** **Models:** ES300, RX300, ES330, RX330 **Engines:** 3.0L VIN F, 3.3L VIN A **Transmissions:** All	**Camshaft Position-Crankshaft Position (Bank 2 Sensor A) Conditions:** Engine started, engine running, and the PCM detected a deviation between the crankshaft position sensor signal and the VVT Sensor 2 signal during the test period. The crankshaft position (NE) sensor consists of a magnet, iron core and pickup coil. The NE sensor signal plate, installed on the crankshaft-timing pulley, has 34 teeth. This sensor generates 34 signals for each engine revolution. The PCM detects the crankshaft angle and engine speed based on the NE signal. It detects the correct cylinder based on signals from the VVT 2 sensor along signals from the crankshaft position sensor. **Possible Causes:** • Engine valve timing problem • Engine timing belt mechanical problem (skipped teeth or belt) • PCM has failed
DTC: P0020 **2T CCM, MIL: Yes** **1996, 1997, 1998, 1999, 2000, 2001, 2002, 2003, 2004, 2005, 2006** **Models:** ES300, RX300, ES330, RX330 **Engines:** 3.0L VIN F, 3.3L VIN A **Transmissions:** All	**Camshaft Position Sensor Actuator 'A' Circuit (Bank 2) Conditions:** Key on or engine running; and the PCM detected an unexpected voltage condition on the Camshaft Position Sensor 'A' Bank 2 circuit. The VVT system controls the intake camshaft in order to provide optimal valve timing during all conditions based signals from the ECT, IAT and TP sensor. The VVT regulates the intake camshaft angle using oil pressure through the OCV. The result is that relative position between the camshaft and crankshaft becomes optimal. The engine has better torque, fuel economy and lower emissions. **Possible Causes:** • OCV assembly connector is damaged or loose • OCV assembly control circuit is open or shorted to ground • OCV assembly is damaged or has failed • PCM has failed
DTC: P0021 **2T CCM, MIL: Yes** **1996, 1997, 1998, 1999, 2000, 2001, 2002, 2003, 2004, 2005, 2006** **Models:** ES300, RX300, ES330, RX330 **Engines:** 3.0L VIN F, 3.3L VIN A **Transmissions:** All	**Camshaft Position 'A' Timing Over-Advanced Or System Performance (Bank 2) Conditions:** Engine started, ECT sensor more than 158°F, vehicle driven at an engine speed of 400-4000 rpm, and the PCM detected the valve timing did not change from the "current" valve timing, or the valve timing remain fixed during testing. The VVT system controls the intake camshaft in order to provide optimal valve timing during all conditions. The VVT regulates the intake camshaft angle using oil pressure through the Oil Control Valve. The result is that relative position between the camshaft and crankshaft becomes optimal. The engine has better torque, fuel economy and lower emissions. **Possible Causes:** • Engine valve timing malfunction • Camshaft timing oil control valve unit is damaged or has failed • PCM or VVT ECM has failed
DTC: P0022 **2T CCM, MIL: Yes** **1996, 1997, 1998, 1999, 2000, 2001, 2002, 2003, 2004, 2005, 2006** **Models:** ES300, RX300, ES330, RX330 **Engines:** 3.0L VIN F, 3.3L VIN A **Transmissions:** All	**Camshaft Position 'A' Timing Over-Retarded (Bank 2) Conditions:** Engine started, ECT sensor over 158°F, engine speed of 400-4000 rpm, and the PCM detected the valve timing did not change from the "current" valve timing, or the valve timing remain fixed. The VVT system controls the intake camshaft in order to provide optimal valve timing during all conditions. The VVT regulates the intake camshaft angle using oil pressure through the OCV. The result is that relative position between the camshaft and crankshaft becomes optimal. The engine has better torque, fuel economy and lower emissions. **Possible Causes:** • Engine valve timing malfunction • Camshaft timing oil control valve unit is damaged or has failed • PCM or VVT ECM has failed
DTC: P0031 **1T CCM, MIL: Yes** **1996, 1997, 1998, 1999, 2000, 2001, 2002, 2003, 2004, 2005, 2006** **Models:** LX450, LX470, GS400, GS430, GX470, LS400, LS430, SC400, SC430 **Engines:** 4.0L VIN H, 4.3L VIN N, L, 4.5L VIN J, 4.7L VIN T **Transmissions:** All	**Heated Oxygen Sensor (Bank 1 Sensor 1) Heater Circuit Low Code Conditions:** Engine started, and the PCM detected the HO2S-11 heater control circuit indicated less than 0.20 amps during the CCM test period. **Possible Causes:** • HO2S-11 heater control circuit is open • HO2S-11 heater assembly is damaged or has failed • PCM has failed

DTC	Trouble Code Title, Conditions & Possible Causes
DTC: P0032 **1T CCM, MIL: Yes** **1996, 1997, 1998, 1999, 2000, 2001, 2002, 2003, 2004, 2005, 2006** **Models:** LX450, LX470, GS400, GS430, GX470, LS400, LS430, SC400, SC430 **Engines: 4.0L VIN H, 4.3L VIN N, L, 4.5L VIN J, 4.7L VIN T** **Transmissions: All**	**Heated Oxygen Sensor (Bank 1 Sensor 1) Heater Circuit High Input Conditions:** Engine started, and the PCM detected the HO2S-11 heater control circuit indicated more than 2.0 amps during the CCM test period. **Possible Causes:** • HO2S-11 heater control circuit is shorted to ground • HO2S-11 heater assembly is damaged or has failed • PCM has failed
DTC: P0036 **1T CCM, MIL: Yes** **1996, 1997, 1998, 1999, 2000, 2001, 2002, 2003, 2004, 2005, 2006** **Models:** ES300, RX300, ES330, RX330 **Engines:** 3.0L VIN F, 3.3L VIN A **Transmissions:** All	**Heated Oxygen Sensor (Bank 1 Sensor 2) Heater Circuit Malfunction Conditions:** Engine started, and the PCM detected the HO2S-12 heater control circuit indicated more than 2.35 amps or less than 0.20 amps. **Possible Causes:** • HO2S-12 heater control circuit is open or shorted to ground • HO2S-12 heater assembly is damaged or has failed • HO2S-12 heater power circuit open (test power from EFI relay) • PCM has failed
DTC: P0037 **2T CCM, MIL: Yes** **1996, 1997, 1998, 1999, 2000, 2001, 2002, 2003, 2004, 2005, 2006** **Models:** LX450, LX470, GS400, GS430, GX470, LS400, LS430, SC400, SC430 **Engines:** 4.0L VIN H, 4.3L VIN N, L, 4.5L VIN J, 4.7L VIN T **Transmissions:** All	**Heated Oxygen Sensor (Bank 1 Sensor 2) Heater Circuit Low Input Conditions:** Engine started, and the PCM detected the HO2S-12 heater control circuit indicated less than 0.20 amps during the CCM test period. **Possible Causes:** • HO2S-12 heater control circuit is open • HO2S-12 heater assembly is damaged or has failed • PCM has failed
DTC: P0038 **2T CCM, MIL: Yes** **1996, 1997, 1998, 1999, 2000, 2001, 2002, 2003, 2004, 2005, 2006** **Models:** LX450, LX470, GS400, GS430, GX470, LS400, LS430, SC400, SC430 **Engines: 4.0L VIN H, 4.3L VIN N, L, 4.5L VIN J, 4.7L VIN T** **Transmissions: All**	**Heated Oxygen Sensor (Bank 1 Sensor 2) Heater Circuit High Input Conditions:** Engine started, and the PCM detected the HO2S-12 heater control circuit indicated more than 2.0 amps during the CCM test period. **Possible Causes:** • HO2S-12 heater control circuit is shorted to ground • HO2S-12 heater assembly is damaged or has failed • PCM has failed
DTC: P0051 **2T CCM, MIL: Yes** **1996, 1997, 1998, 1999, 2000, 2001, 2002, 2003, 2004, 2005, 2006** **Models:** LX450, LX470, GS400, GS430, GX470, LS400, LS430, SC400, SC430 **Engines:** 4.0L VIN H, 4.3L VIN N, L, 4.5L VIN J, 4.7L VIN T **Transmissions:** All	**Heated Oxygen Sensor (Bank 2 Sensor 1) Heater Circuit Low Input Conditions:** Engine started, and the PCM detected the HO2S-21 heater control circuit indicated less than 0.20 amps during the CCM test period. **Possible Causes:** • HO2S-21 heater control circuit is open • HO2S-21 heater assembly is damaged or has failed • PCM has failed
DTC: P0052 **2T CCM, MIL: Yes** **1996, 1997, 1998, 1999, 2000, 2001, 2002, 2003, 2004, 2005, 2006** **Models:** LX450, LX470, GS400, GS430, GX470, LS400, LS430, SC400, SC430 **Engines:** 4.0L VIN H, 4.3L VIN N, L, 4.5L VIN J, 4.7L VIN T **Transmissions:** All	**Heated Oxygen Sensor (Bank 2 Sensor 1) Heater Circuit High Input Conditions:** Engine started, and the PCM detected the HO2S-21 heater control circuit indicated more than 2.0 amps during the CCM test period. **Possible Causes:** • HO2S-21 heater control circuit is shorted to ground • HO2S-21 heater assembly is damaged or has failed • PCM has failed

DTC	Trouble Code Title, Conditions & Possible Causes
DTC: P0057 **2T CCM, MIL: Yes** **1996, 1997, 1998, 1999, 2000, 2001, 2002, 2003, 2004, 2005, 2006** **Models:** LX450, LX470, GS400, GS430, GX470, LS400, LS430, SC400, SC430 **Engines:** 4.0L VIN H, 4.3L VIN N, L, 4.5L VIN J, 4.7L VIN T **Transmissions:** All	**Heated Oxygen Sensor (Bank 2 Sensor 2) Heater Circuit Low Input Conditions:** Engine started, and the PCM detected the HO2S-22 heater control circuit indicated less than 0.20 amps during the CCM test period. **Possible Causes:** • HO2S-22 heater control circuit is open • HO2S-22 heater assembly is damaged or has failed • PCM has failed
DTC: P0058 **2T CCM, MIL: Yes** **1996, 1997, 1998, 1999, 2000, 2001, 2002, 2003, 2004, 2005, 2006** **Models:** LX450, LX470, GS400, GS430, GX470, LS400, LS430, SC400, SC430 **Engines:** 4.0L VIN H, 4.3L VIN N, L, 4.5L VIN J, 4.7L VIN T **Transmissions:** All	**Heated Oxygen Sensor (Bank 2 Sensor 2) Heater Circuit High Input Conditions:** Engine started, and the PCM detected the HO2S-22 heater control circuit indicated more than 2.0 amps during the CCM test period. **Possible Causes:** • HO2S-22 heater control circuit is shorted to ground • HO2S-22 heater assembly is damaged or has failed • PCM has failed
DTC: P0100 **1T CCM, MIL: Yes** **1995** **Models:** All **Engines:** All **Transmissions:** All	**Mass Airflow Sensor Circuit Malfunction Conditions:** Engine running at under 4000 rpm, and the PCM detected an unexpected voltage condition on the MAF sensor circuit. The MAF sensor on this engine includes a hot wire assembly with an air temperature sensor, platinum hot wire and control unit mounted in a plastic housing. This airflow meter works on the principle that the hot wire and temperature sensor located in the intake air bypass of the housing detect any changes in the (incoming) air temperature. **Possible Causes:** • MAF sensor signal circuit is open, shorted to ground or power • MAF sensor ground circuit is open between sensor and ground • MAF sensor power circuit is open (check the power to the relay) • MAF sensor has failed, or the PCM has failed
DTC: P0100 **1T CCM, MIL: Yes** **1996, 1997, 1998, 1999, 2000, 2001, 2002, 2003, 2004, 2005, 2006** **Models:** All **Engines:** All **Transmissions:** All	**Mass Airflow Sensor Circuit Malfunction Conditions:** Engine started, engine running at under 4000 rpm, and the PCM detected an unexpected low or high voltage condition on the Mass Airflow (MAF) sensor circuit for over 3 seconds during the CCM test. The MAF sensor on this engine includes a hot wire assembly with an air temperature sensor, platinum hot wire and control unit mounted in a plastic housing. This airflow meter works on the principle that the hot wire and temperature sensor located in the intake air bypass of the housing detect any changes in the (incoming) air temperature. **Possible Causes:** • MAF sensor signal circuit is open, shorted to ground or power • MAF sensor ground circuit is open between sensor and ground • MAF sensor power circuit is open (check the power to the relay) • MAF sensor has failed, or the PCM has failed
DTC: P0101 **2T CCM, MIL: Yes** **1995** **Models:** All **Engines:** All **Transmissions:** All	**Mass Airflow Sensor Signal Range/Performance Conditions:** DTC P0100 not set; engine speed under 900 rpm, throttle valve closed, ECT sensor more than 158°F, and the PCM detected the MAF sensor was more than 2.20v; or with the engine speed over 1500 rpm, the throttle valve closed and the TP sensor above 0.63v, the MAF sensor was below 1.06v. This airflow meter works on the principle where a temperature sensor and hot wire in the intake air bypass area detect any changes in the incoming air. **Possible Causes:** • MAF sensor signal circuit is open, shorted to ground or power • MAF sensor is contaminated, damaged or it has failed • PCM has failed
DTC: P0101 **2T CCM, MIL: Yes** **1996, 1997, 1998, 1999, 2000, 2001, 2002, 2003, 2004, 2005, 2006** **Models:** All **Engines:** All **Transmissions:** All	**Mass Airflow Sensor Signal Range/Performance Conditions:** DTC P0100 not set, engine speed under 900 rpm, throttle valve closed, ECT sensor over 158°F, and the PCM detected the MAF was above 2.20v; or with the engine speed over 1500 rpm, the throttle valve closed and the TP sensor over 0.63v, the MAF sensor was less than 1.06v. This airflow meter works on the principle that the hot wire and temperature sensor in the intake air bypass of the housing detect any changes in the (incoming) air temperature **Possible Causes:** • MAF sensor signal circuit is open, shorted to ground or power • MAF sensor is contaminated, damaged or it has failed • PCM has failed

DTC	Trouble Code Title, Conditions & Possible Causes
DTC: P0102 **2T CCM, MIL: Yes** **1996, 1997, 1998, 1999, 2000, 2001, 2002, 2003, 2004, 2005, 2006** **Models:** LX450, LX470, GS400, GS430, GX470, LS400, LS430, SC400, SC430 **Engines:** 4.0L VIN H, 4.3L VIN N, L, 4.5L VIN J, 4.7L VIN T **Transmissions:** All	**Mass Airflow Sensor Circuit Low Input Conditions:** DTC P0100 not set, engine started, and the PCM detected an unexpected low voltage condition on the MAF sensor circuit during the CCM test period. This airflow meter works on the principle that the hot wire and temperature sensor located in the intake air bypass of the housing detect any changes in the (incoming) air temperature **Possible Causes:** • MAF sensor signal circuit is open or shorted to ground • MAF sensor is contaminated, damaged or it has failed • PCM has failed
DTC: P0103 **2T CCM, MIL: Yes** **1996, 1997, 1998, 1999, 2000, 2001, 2002, 2003, 2004, 2005, 2006** **Models:** LX450, LX470, GS400, GS430, GX470, LS400, LS430, SC400, SC430 **Engines:** 4.0L VIN H, 4.3L VIN N, L, 4.5L VIN J, 4.7L VIN T **Transmissions:** All	**Mass Airflow Sensor Circuit High Input Conditions:** DTC P0100 not set, engine started, and the PCM detected an unexpected high voltage condition on the MAF sensor circuit during the CCM test period. This airflow meter works on the principle that the hot wire and temperature sensor located in the intake air bypass of the housing detect any changes in the (incoming) air temperature **Possible Causes:** • MAF sensor signal circuit is shorted to power • MAF sensor is contaminated, damaged or it has failed • PCM has failed
DTC: P0110 **1T CCM, MIL: Yes** **1995** **Models: All** **Engines: All** **Transmissions:** All	**Intake Air Temperature Sensor Circuit Malfunction Conditions:** Key on or engine running; and the PCM detected an unexpected "low" or "high" voltage on the IAT sensor circuit (Scan Tool reads less than -40°F or low voltage or more than 284°F for high voltage). **Note: The IAT sensor is built into the MAF sensor on some engines.** **Possible Causes:** • IAT sensor signal circuit is open between the sensor and PCM • IAT sensor signal circuit is shorted to ground or to VREF • IAT sensor ground circuit is open between sensor and ground • IAT sensor is contaminated, damaged or has failed • PCM has failed
DTC: P0110 **1T CCM, MIL: Yes** **1996, 1997, 1998, 1999, 2000, 2001, 2002, 2003, 2004, 2005, 2006** **Models: All** **Engines: All** **Transmissions:** All	**Intake Air Temperature Sensor Circuit Malfunction Conditions:** Engine started, and the PCM detected an unexpected low or high voltage on the IAT sensor circuit (Scan Tool reads -40°F or 284°F). This sensor is located inside the MAF sensor. **Possible Causes:** • IAT sensor signal circuit is open, shorted to ground or VREF • IAT sensor ground circuit is open between sensor and ground • IAT sensor is contaminated, damaged or has failed • PCM has failed
DTC: P0112 **1T CCM, MIL: Yes** **1996, 1997, 1998, 1999, 2000, 2001, 2002, 2003, 2004, 2005, 2006** **Models:** LX450, LX470, GS400, GS430, GX470, LS400, LS430, SC400, SC430 **Engines:** 4.0L VIN H, 4.3L VIN N, L, 4.5L VIN J, 4.7L VIN T **Transmissions:** All	**Intake Air Temperature Sensor Circuit Low Input Conditions:** Key on or engine running; and the PCM detected an unexpected low voltage condition on the IAT sensor circuit for over 500 ms. **Possible Causes:** • IAT sensor connector is damaged (it may be shorted internally) • IAT sensor ground circuit is shorted to ground • IAT sensor is damaged or has failed • PCM has failed
DTC: P0113 **1T CCM, MIL: Yes** **1996, 1997, 1998, 1999, 2000, 2001, 2002, 2003, 2004, 2005, 2006** **Models:** LX450, LX470, GS400, GS430, GX470, LS400, LS430, SC400, SC430 **Engines:** 4.0L VIN H, 4.3L VIN N, L, 4.5L VIN J, 4.7L VIN T **Transmissions:** All	**Intake Air Temperature Sensor Circuit High Input Conditions:** Key on or engine running; and the PCM detected an unexpected high voltage condition on the IAT sensor circuit for over 500 ms. **Possible Causes:** • IAT sensor connector is damaged (it may be open internally) • IAT sensor ground circuit is open • IAT sensor is damaged or has failed • PCM has failed

DTC	Trouble Code Title, Conditions & Possible Causes
DTC: P0115 **2T CCM, MIL: Yes** **1995** **Models: All** **Engines: All** **Transmissions:** All	**Engine Coolant Temperature Sensor Circuit Malfunction Conditions:** Key on or engine running; and the PCM detected an unexpected "low" or "high" voltage on the ECT sensor circuit (Scan Tool reads less than -40°F or low voltage or more than 284°F for high voltage). **Possible Causes:** • ECT sensor signal circuit is open, shorted to ground or VREF • ECT sensor ground circuit is open between sensor and ground • ECT sensor is contaminated, damaged or has failed • PCM has failed
DTC: P0115 **2T CCM, MIL: Yes** **1996, 1997, 1998, 1999, 2000,** **2001, 2002, 2003, 2004, 2005,** **2006** **Models: All** **Engines: All** **Transmissions:** All	**Engine Coolant Temperature Sensor Circuit Malfunction Conditions:** Key on or engine running; and the PCM detected an unexpected "low" or "high" voltage on the ECT sensor circuit (Scan Tool reads less than -40°F or low voltage or more than 284°F for high voltage). **Possible Causes:** • IAT sensor signal circuit is open, shorted to ground or VREF • ECT sensor ground circuit is open between sensor and ground • ECT sensor is contaminated, damaged or has failed • PCM has failed
DTC: P0116 **2T CCM, MIL: Yes** **1996, 1997, 1998, 1999, 2000,** **2001, 2002** **Models: All** **Engines: All** **Transmissions:** All	**Engine Coolant Temperature Sensor Range/Performance Conditions:** Engine runtime over 20 minutes; and the PCM detected the ECT sensor indicated less than 95°F during the test. **Note: Check the condition and operation of the Cooling system.** **Possible Causes:** • Check for a low coolant level or an incorrect coolant mixture • ECT sensor signal circuit or ground circuit has high resistance • ECT sensor is contaminated, damaged or out-of-calibration • PCM has failed
DTC: P0116 **2T CCM, MIL: Yes** **1995** **Models: All** **Engines: All** **Transmissions:** All	**Engine Coolant Temperature Sensor Range/Performance Conditions:** Engine started, engine runtime over 20 minutes, and the PCM detected the ECT sensor indicated less than 95°F during the test. **Note: Check the condition and operation of the Cooling system.** **Possible Causes:** • Check for a low coolant level or an incorrect coolant mixture • ECT sensor signal circuit has high resistance • ECT sensor ground circuit has high resistance • ECT sensor is contaminated, damaged or out-of-calibration • PCM has failed
DTC: P0116 **2T CCM, MIL: Yes** **2003, 2004, 2005, 2006** **Models: All** **Engines: All** **Transmissions:** All	**Engine Coolant Temperature Sensor Range/Performance Conditions:** Engine started, ECT sensor from 95°F to 140°F, IAT sensor more than 19.9°F, vehicle driven with several changes in the VSS signals, and the PCM detected the ECT sensor signal did not increase more than 37.4°F after engine was started and the test period completed. **Possible Causes:** • Check for problems in the cooling system (i.e., coolant, the fan) • ECT sensor signal circuit or ground circuit has high resistance • ECT sensor is contaminated, damaged or has failed • PCM has failed
DTC: P0117 **2T CCM, MIL: Yes** **1996, 1997, 1998, 1999, 2000,** **2001, 2002, 2003, 2004, 2005,** **2006** **Models: LX450, LX470, GS400,** **GS430, GX470, LS400, LS430,** **SC400, SC430** **Engines:** 4.0L VIN H, 4.3L VIN N, L, 4.5L VIN J, 4.7L VIN T **Transmissions:** All	**Engine Coolant Temperature Sensor Circuit Low Input Conditions:** Key on or engine running; and the PCM detected an unexpected low voltage condition on the ECT sensor (Scan Tool reads below -40°F). **Possible Causes:** • ECT sensor signal circuit is shorted to ground • ECT sensor is damaged or has failed • PCM has failed
DTC: P0117 **2T CCM, MIL: Yes** **1996, 1997, 1998, 1999, 2000,** **2001, 2002, 2003, 2004, 2005,** **2006** **Models: LX450, LX470, GS400,** **GS430, GX470, LS400, LS430,** **SC400, SC430** **Engines:** 4.0L VIN H, 4.3L VIN N, L, 4.5L VIN J, 4.7L VIN T **Transmissions:** All	**Engine Coolant Temperature Sensor Circuit High Input Conditions:** Key on or engine running; and the PCM detected an unexpected high voltage condition on the ECT sensor (Scan Tool reads over 284°F). **Possible Causes:** • ECT sensor signal circuit is open • ECT sensor ground circuit is open • ECT sensor is damaged or has failed • PCM has failed

DTC	Trouble Code Title, Conditions & Possible Causes
DTC: P0120 **2T CCM, MIL: Yes** **1995** **Models:** All **Engines:** All **Transmissions:** All	**TP Sensor or Switch 'A' Circuit Malfunction Conditions:** Key on or engine running; and the PCM detected the TP sensor indicated less than 0.1v with the closed throttle position switch off (in open position), or the TP sensor input indicated 4.9v at any time. **Possible Causes:** • TP sensor signal circuit open or shorted to ground • TP sensor ground circuit is open • TP sensor power circuit is open (check VREF circuit at PCM) • TP sensor is damaged or has failed • PCM has failed
DTC: P0120 **2T CCM, MIL: Yes** **1996, 1997, 1998, 1999, 2000,** **2001, 2002** **Models:** All **Engines:** All **Transmissions:** All	**TP Sensor or Switch 'A' Circuit Malfunction Conditions:** Key on or engine running; and the PCM detected the TP sensor indicated less than 0.1v with the closed throttle position switch off (open position), or the TP sensor was 4.9v at any time. **Possible Causes:** • TP sensor signal circuit is open or shorted to ground • TP sensor ground circuit is open, or the power circuit is open • TP sensor is damaged or has failed
DTC: P0120 **2T CCM, MIL: Yes** **1996, 1997, 1998, 1999, 2000,** **2001, 2002, 2003** **Models:** ES300, RX300 **Engines:** 3.0L VIN F **Transmissions:** All	**TP Sensor or Switch 'A' Circuit Malfunction Conditions:** Engine started, and the PCM detected the TP sensor signal (VTA) was under 0.10v or over 4.90v. The TP sensor, mounted on the throttle body, detects the Throttle Valve opening angle (0.30v-0.70v closed). The PCM uses this signal for A/F ratio correction and power increase changes during all modes of engine operation. **Possible Causes:** • VTA sensor signal circuit is open or shorted to ground • VTA sensor ground circuit is open • VTA power circuit (VREF) is open • TP sensor is damaged or has failed • PCM has failed
DTC: P0120 **2T CCM, MIL: Yes** **1996, 1997, 1998, 1999, 2000,** **2001, 2002, 2003, 2004, 2005,** **2006** **Models:** ES300, RX300, ES330, RX330, LX450, LX470, GS400, GS430, GX470, LS400, LS430, SC400, SC430 **Engines:** 3.0L VIN F, 3.3L VIN A, 4.0L VIN H, 4.3L VIN N, L, 4.5L VIN J, 4.7L VIN T **Transmissions:** All	**Throttle Pedal Position Sensor/Switch 'A' Circuit Malfunction Conditions** Key on or engine running; and the PCM detected the TP sensor 'A' Signal indicated less than 0.10v with the throttle position closed (the switch is open), or the TP sensor signal indicated more than 4.9v at any time. The Electric TP Sensor is mounted on the throttle body. It has two sensors (the electrical throttle system does not use a cable). **Possible Causes:** • TP sensor signal circuit open is or shorted to ground • TP sensor ground circuit is open • TP sensor power circuit is open (test VREF circuit at the PCM) • TP sensor is damaged or has failed • PCM has failed
DTC: P0121 **2T CCM, MIL: Yes** **1995** **Models:** All **Engines:** All **Transmissions:** All	**TP Sensor or TP Switch 'A' Signal Range/Performance Conditions:** Engine started, VSS input exceeds 19 mph at least once, and the PCM detected that the TP sensor input was out of the applicable range with the VSS reading between 0 and 30 mph. **Possible Causes:** • TP sensor signal circuit open or shorted to ground (intermittent) • TP sensor is loose at it mounting or the throttle is binding • TP sensor is damaged or has failed (perform a sweep test) • PCM has failed
DTC: P0121 **2T CCM, MIL: Yes** **1996, 1997, 1998, 1999, 2000,** **2001, 2002, 2003, 2004, 2005,** **2006** **Models:** ES300, RX300, ES330, RX330 **Engines:** 3.0L VIN F, 3.3L VIN A **Transmissions:** All	**TP Sensor or TP Switch 'A' Signal Range/Performance Conditions:** Vehicle speed more than 19 mph at least once; and the PCM detected the TP sensor input was out of the applicable range with the VSS input reading between 30 mph and 0 mph. **Possible Causes:** • TP sensor signal circuit open or shorted to ground (intermittent) • TP sensor is loose at it mounting or the throttle is binding • TP sensor is damaged or has failed (perform a sweep test) • PCM has failed

DTC	Trouble Code Title, Conditions & Possible Causes
DTC: P0121 **2T CCM, MIL: Yes** **1996, 1997, 1998, 1999, 2000,** **2001, 2002, 2003, 2004, 2005,** **2006** **Models:** LX450, LX470, GS400, GS430, GX470, LS400, LS430, SC400, SC430 **Engines:** 4.0L VIN H, 4.3L VIN N, L, 4.5L VIN J, 4.7L VIN T **Transmissions:** All	**Throttle Pedal Position Sensor Switch 'A' Signal Range/Performance Conditions:** Engine started; and the PCM detected the difference between the TP sensor VTA1 and VTA2 signal was out-of-range. The TP sensor, mounted on the throttle body, detects the Throttle Valve opening angle (about 0.70v with the throttle closed). The PCM uses the VTA signal for air/fuel ratio and power increase correction. The Electric TP Sensor is mounted on the throttle body. It has two sensors (the electrical throttle system does not use a cable). **Possible Causes:** • TP sensor connector is damaged or it is open • TP sensor is damaged or has failed • PCM has failed
DTC: P0122 **2T CCM, MIL: Yes** **1996, 1997, 1998, 1999, 2000,** **2001, 2002, 2003, 2004, 2005,** **2006** **Models:** LX450, LX470, GS400, GS430, GX470, LS400, LS430, SC400, SC430 **Engines:** 4.0L VIN H, 4.3L VIN N, L, 4.5L VIN J, 4.7L VIN T **Transmissions:** All	**Throttle Pedal Position Sensor/Switch 'A' Circuit Low Input Conditions:** Engine started, and the PCM detected an unexpected low voltage (below 0.20v) on the VTA1 signal circuit. The TP sensor detects the Throttle Valve opening angle (0.70v with the throttle closed). The Electric TP Assembly has 2 sensors (this system does not use a throttle cable). **Possible Causes:** • Throttle control motor and sensor unit is damaged or failed • TP sensor VC (VREF) circuit is open, or the VTA1 signal circuit is shorted to ground • PCM has failed
DTC: P0123 **2T CCM, MIL: Yes** **1996, 1997, 1998, 1999, 2000,** **2001, 2002, 2003, 2004, 2005,** **2006** **Models:** LX450, LX470, GS400, GS430, GX470, LS400, LS430, SC400, SC430 **Engines:** 4.0L VIN H, 4.3L VIN N, L, 4.5L VIN J, 4.7L VIN T **Transmissions:** All	**Throttle Pedal Position Sensor/Switch 'A' Circuit High Input Conditions:** Engine started, and the PCM detected an unexpected high voltage (over 4.80v) on the VTA1 signal circuit. The TP sensor detects the Throttle Valve opening angle (0.70v with the throttle closed). The Electric TP Assembly has 2 sensors (this system does not use a throttle cable). **Possible Causes:** • Throttle control motor and sensor unit is damaged or failed • TP sensor VC (VREF) circuit is shorted to the VTA1 circuit • VTA1 ground circuit is open or the VTA1 signal circuit is open • PCM has failed
DTC: P0125 **1T CCM, MIL: Yes** **1995** **Models:** All **Engines:** All **Transmissions:** All	**Insufficient Coolant Temperature For Closed Loop: Conditions:** DTC P0115 and P0116 not set, engine started, engine running, ECT sensor more than 140°F, vehicle driven to a speed of 25-62 mph at an engine speed over 1500 rpm, throttle valve not fully closed, and the PCM detected the HO2S signal (internal value) did not change. **Possible Causes:** • Check the operation of the thermostat (it may be stuck open) • ECT sensor signal circuit has high resistance • ECT sensor is out-of-calibration, skewed, or it has failed • Inspect for low coolant level or an incorrect coolant mixture
DTC: P0125 **1T CCM, MIL: Yes** **1996, 1997, 1998, 1999, 2000,** **2001, 2002** **Models:** All **Engines:** All **Transmissions:** All	**Insufficient Coolant Temperature For Closed Loop: Conditions:** **California Models** DTC P0115 and P0116 not set, engine started, ECT sensor more than 140°F, vehicle driven to a speed of 25-62 mph at an engine speed over 1500 rpm, throttle valve not fully closed, and the PCM detected the A/FS or HO2S signal (internal value) did not change. **Federal Models** Engine started, ECT sensor more than 140°F, vehicle driven to a speed of 25-62 mph at an engine speed over 1400 rpm, throttle valve not fully closed, and the PCM detected the HO2S-11 input did not exceed 450 mv at least during a test period of 1.5 minutes. **Possible Causes:** • Check the operation of the thermostat (it may be stuck open) • ECT sensor signal circuit has high resistance • ECT sensor has failed • Inspect for low coolant level or an incorrect coolant mixture

DTC	Trouble Code Title, Conditions & Possible Causes
DTC: P0125 **1T CCM, MIL: Yes** **2003, 2004, 2005, 2006** **Models: All** **Engines: All** **Transmissions:** All	**Insufficient Coolant Temperature For Closed Loop Conditions:** **California Models** ECT sensor less than 19.4°F, engine started, engine runtime over 20 minutes, and the PCM detected the ECT sensor indicated 68°F or less; or with the ECT sensor from 19.4°F to 50.0°F at startup, engine runtime over 5 minutes, the PCM detected the ECT sensor indicated 68°F or less; or with the ECT sensor more than 19.4°F at startup, and the engine runtime over 5 minutes, the PCM detected the ECT sensor signal was 68°F or less; or with the ECT sensor over 50°F at startup, engine runtime over 2 minutes, the PCM detected the ECT sensor signal did not reach 86°F during the CCM test period. **Possible Causes:** • Check the operation of the thermostat (it may be stuck open) • ECT sensor signal circuit has high resistance • ECT sensor has failed • Inspect for low coolant level for an incorrect coolant mixture
DTC: P0128 **2T CCM, MIL: Yes** **2000, 2001, 2002, 2003, 2004,** **2005, 2006** **Models: All** **Engines: All** **Transmissions:** All	**Thermostat System Malfunction Conditions:** Engine started, ECT sensor signal below 140°F at startup, and the PCM detected the ECT sensor did not reach 167°F after the warmup period expired (engine runtime 5-10 minutes). **Possible Causes:** • Check the operation of the thermostat (it may be stuck open) • ECT sensor is out-of-calibration or skewed • Inspect for low coolant level or for an incorrect coolant mixture
DTC: P0130 **2T CCM, MIL: Yes** **1995** **Models: All** **Engines: All** **Transmissions:** All	**HO2S-11 (Bank 1 Sensor 1) Circuit Malfunction Conditions:** Engine started, engine warmup completed, engine idling in closed loop, and the PCM detected the HO2S-11 signal was fixed at 400 mv or higher, or that it was fixed at less than 550 mv during the test. **Possible Causes:** • HO2S signal circuit is open between the sensor and the PCM • HO2S signal circuit is shorted to sensor or chassis ground • HO2S signal circuit is shorted to VREF or system power (B+) • HO2S is damaged, contaminated or it has failed
DTC: P0130 **2T CCM, MIL: Yes** **1996, 1997, 1998, 1999, 2000,** **2001, 2002** **Models: All** **Engines: All** **Transmissions:** All	**HO2S-11 (Bank 1 Sensor 1) Circuit Malfunction Conditions:** Engine warmup completed, engine idling in closed loop, and the PCM detected the HO2S-11 signal was fixed at 400 mv or higher, or that it was fixed at less than 550 mv during the test. **Possible Causes:** • HO2S signal circuit is open between the sensor and the PCM • HO2S signal circuit is shorted to sensor or chassis ground • HO2S signal circuit is shorted to VREF or system power (B+) • HO2S is damaged, contaminated or it has failed • PCM has failed
DTC: P0130 **2T CCM, MIL: Yes** **1996, 1997, 1998, 1999, 2000,** **2001, 2002, 2003, 2004, 2005,** **2006** **Models:** LX450, LX470, GS400, GS430, GX470, LS400, LS430, SC400, SC430 **Engines:** 4.0L VIN H, 4.3L VIN N, L, 4.5L VIN J, 4.7L VIN T **Transmissions:** All	**HO2S-11 (Bank 1 Sensor 1) Circuit Malfunction Conditions:** Engine started, engine warmup completed, engine idling in closed loop, and the PCM detected the HO2S-11 signal was fixed at 400 mv or higher, or that it was fixed at less than 550 mv during the test. **Possible Causes:** • HO2S signal circuit is open between the sensor and the PCM • HO2S signal circuit is shorted to sensor ground or to VREF • HO2S is damaged, contaminated or it has failed
DTC: P0133 **2T O2S, MIL: Yes** **1995** **Models: All** **Engines: All** **Transmissions:** All	**HO2S-11 (Bank 1 Sensor 1) Slow Response Conditions:** Engine started, engine idling in closed loop, and the PCM detected the HO2S-11 response time from rich-to-lean or from lean-to-rich was 1.1 second or longer during the CCM test. **Note: This fault must be detected at least 3 times at idle speed.** **Possible Causes:** • HO2S signal circuit is open or shorted to ground • HO2S element is contaminated, or HO2S heater has failed • Intake air leaks, exhaust manifold leaks or PCV system leaks • MAF sensor out of calibration (it may be dirty or contaminated)

DTC	Trouble Code Title, Conditions & Possible Causes
DTC: P0133 2T O2S, MIL: Yes 1996, 1997, 1998, 1999, 2000, 2001, 2002 **Models:** All **Engines:** All **Transmissions:** All	**HO2S-11 (Bank 1 Sensor 1) Slow Response Conditions:** Engine started, engine idling in closed loop, and the PCM detected the HO2S-11 response time from rich-to-lean or from lean-to-rich was 1.1 second or longer during the CCM test. **Note: This fault must be detected at least 3 times at idle speed.** **Possible Causes:** • HO2S signal circuit is open or shorted to ground • HO2S element is contaminated, or HO2S heater has failed • Intake air leaks, exhaust manifold leaks or PCV system leaks • MAF sensor out of calibration (it may be dirty or contaminated)
DTC: P0133 2T O2S, MIL: Yes 1996, 1997, 1998, 1999, 2000, 2001, 2002, 2003, 2004, 2005, 2006 **Models:** LX450, LX470, GS400, GS430, GX470, LS400, LS430, SC400, SC430 **Engines:** 4.0L VIN H, 4.3L VIN N, L, 4.5L VIN J, 4.7L VIN T **Transmissions:** All	**HO2S-11 (Bank 1 Sensor 1) Slow Response Conditions:** Engine started, engine idling in closed loop, and the PCM detected the HO2S-11 response time from rich-to-lean or from lean-to-rich was 1.1 second or longer during the CCM test. **Note: This fault must be detected at least 3 times at idle speed.** **Possible Causes:** • HO2S signal circuit is open or shorted to ground • HO2S element is contaminated, or HO2S heater has failed • Intake air leaks, exhaust manifold leaks or PCV system leaks • MAF sensor out of calibration (it may be dirty or contaminated)
DTC: P0134 2T O2S, MIL: Yes 1996,1997, 1998, 1999, 2000, 2001, 2002, 2003, 2004, 2005, 2006 **Models:** ES300, RX300, ES330, RX330 **Engines:** 3.0L VIN F, 3.3L VIN A **Transmissions:** All	**Air Fuel Sensor 1 (Bank 1 Sensor 1) Circuit No Activity Detected Conditions:** Engine started, engine runtime over 140 seconds, vehicle driven at a steady speed of 25-81 mph at over 1500 rpm with the throttle valve open, and the PCM detected the A/FS-11 signal did not indicate rich (more than 450 mv) after 65 seconds under these conditions. **Possible Causes:** • Air leaks present in the PCV valve, hoses or hose connections • A/FS1 connector is damaged or loose • A/FS1 signal circuit is open or shorted to ground • A/FS1 is damaged, contaminated or it has failed • Fuel Control component faults (sticking injector, low pressure) • Gas leaks in the exhaust system in front of the Oxygen sensor • PCM has failed
DTC: P0134 2T O2S, MIL: Yes 1996, 1997, 1998, 1999, 2000, 2001, 2002, 2003, 2004, 2005, 2006 **Models:** LX450, LX470, GS400, GS430, GX470, LS400, LS430, SC400, SC430 **Engines:** 4.0L VIN H, 4.3L VIN N, L, 4.5L VIN J, 4.7L VIN T **Transmissions:** All	**HO2S-11 (Bank 1 Sensor 1) Circuit No Activity Detected Conditions:** Engine started, engine runtime over 140 seconds, vehicle driven at a steady speed of 25-81 mph at over 1500 rpm with the throttle valve open, and the PCM detected the HO2S-11 signal did not indicate rich (more than 450 mv) after 65 seconds under these conditions. **Possible Causes:** • Air leaks present in the PCV valve, hoses or hose connections • Fuel Control component faults (sticking injector, low pressure) • Gas leaks in the exhaust system in front of the Oxygen sensor • HO2S connector is damaged or loose • HO2S signal circuit is open or shorted to ground • HO2S is damaged, contaminated or it has failed • PCM has failed
DTC: P0135 2T CCM, MIL: Yes 1995 **Models:** All **Engines:** All **Transmissions:** All	**HO2S-11 (Bank 1 Sensor 1) Heater Circuit Malfunction Conditions:** Engine started, engine running, and the PCM detected the HO2S-11 heater current exceeded 2 amps, or that it was 0.25 amps or less. **Possible Causes:** • HO2S heater control circuit is open or shorted to ground • HO2S heater control circuit is shorted to power • HO2S heater power circuit is open (check power from the relay) • HO2S heater is damaged or has failed • PCM has failed
DTC: P0135 2T CCM, MIL: Yes 1996, 1997, 1998, 1999, 2000, 2001, 2002 **Models:** All **Engines:** All **Transmissions:** All	**HO2S-11 (Bank 1 Sensor 1) Heater Circuit Malfunction Conditions:** Engine started, engine running, and the PCM detected the HO2S-11 heater current exceeded 2 amps, or that it was 0.25 amps or less. **Possible Causes:** • HO2S heater control circuit is open or shorted to ground • HO2S heater control circuit is shorted to power • HO2S heater power circuit is open (check power from the relay) • HO2S heater is damaged or has failed • PCM has failed

DTC	Trouble Code Title, Conditions & Possible Causes
DTC: P0135 **2T CCM, MIL: Yes** **1996,1997, 1998, 1999, 2000,** **2001, 2002, 2003, 2004, 2005,** **2006** **Models:** ES300, RX300, ES330, RX330 **Engines:** 3.0L VIN F, 3.3L VIN A **Transmissions:** All	**Air Fuel Sensor 1 (Bank 1 Sensor 1) Heater Circuit Malfunction Conditions:** Engine started, engine running, and the PCM detected the A/FS-11 heater current indicated over 8 amps, or it was less than 0.25 amps. **Possible Causes:** • A/FS1 heater control circuit is open or shorted to ground • A/FS1 heater control circuit is shorted to power • A/FS1 heater power circuit is open (check power from relay) • A/FS1 heater is damaged or has failed • PCM has failed
DTC: P0136 **2T CCM, MIL: Yes** **1995** **Models:** All **Engines:** All **Transmissions:** All	**HO2S-12 (Bank 1 Sensor 2) Circuit Malfunction Conditions:** Engine started, vehicle driven to a speed of over 25 mph while in closed loop, and the PCM detected the HO2S-12 signal remained fixed at more than 400 mv, or that it was fixed at less than 600 mv. **Possible Causes:** • HO2S signal circuit is open or shorted to ground • HO2S signal circuit shorted to VREF or system power (B+) • HO2S is damaged, contaminated or it has failed • PCM has failed
DTC: P0136 **2T CCM, MIL: Yes** **1996, 1997, 1998, 1999, 2000,** **2001, 2002** **Models:** All **Engines:** All **Transmissions:** All	**HO2S-12 (Bank 1 Sensor 2) Circuit Malfunction Code Conditions:** Vehicle driven to a speed of over 25 mph while in closed loop, and the PCM detected the HO2S-12 signal remained fixed at more than 400 mv, or it was fixed at less than 600 mv. **Possible Causes:** • HO2S signal circuit is open or shorted to ground • HO2S signal circuit shorted to VREF or system power (B+) • HO2S is damaged, contaminated or it has failed
DTC: P0136 **2T CCM, MIL: Yes** **1996, 1997, 1998, 1999, 2000,** **2001, 2002, 2003, 2004, 2005,** **2006** **Models:** LX450, LX470, GS400, GS430, GX470, LS400, LS430, SC400, SC430 **Engines:** 4.0L VIN H, 4.3L VIN N, L, 4.5L VIN J, 4.7L VIN T **Transmissions:** All	**HO2S-12 (Bank 1 Sensor 2) Circuit Malfunction Conditions:** Vehicle driven to a speed of over 25 mph while in closed loop, and the PCM detected the HO2S-12 signal was fixed at more than 400 mv, or that it was fixed at less than 600 mv. **Possible Causes:** • HO2S signal circuit is open or shorted to ground • HO2S signal circuit shorted to VREF or system power (B+) • HO2S is damaged, contaminated or it has failed
DTC: P0136 **2T CCM, MIL: Yes** **1996,1997, 1998, 1999, 2000,** **2001, 2002, 2003, 2004, 2005,** **2006** **Models:** ES300, RX300, ES330, RX330, GS400, GS430, GX470, LS400, LS430, SC400, SC430 **Engines:** 3.0L VIN F, 3.3L VIN A, 4.0L VIN H, 4.3L VIN N, L, 4.7L VIN T **Transmissions:** All	**HO2S-12 (Bank 1 Sensor 2) Circuit Malfunction Conditions:** Engine started, vehicle driven to a speed of over 31 mph while in closed loop, and the PCM detected the HO2S-12 signal indicated more than 400 mv, or it indicated less than 500 mv during the test. **Possible Causes:** • HO2S signal circuit is open or shorted to ground • HO2S signal circuit shorted to VREF or system power (B+) • HO2S is damaged, contaminated or it has failed
DTC: P0141 **2T CCM, MIL: Yes** **1995** **Models:** All **Engines:** All **Transmissions:** All	**HO2S-12 (Bank 1 Sensor 2) Heater Circuit Malfunction Conditions:** Engine started, engine running, and the PCM detected the HO2S-12 heater current exceeded 2 amps, or that it was 0.25 amps or less. **Possible Causes:** • HO2S heater control circuit is open or shorted to ground • HO2S heater control circuit is shorted to power • HO2S heater power circuit is open (check power from the relay) • HO2S heater is damaged or has failed • PCM has failed
DTC: P0141 **2T CCM, MIL: Yes** **1996, 1997, 1998, 1999, 2000,** **2001, 2002** **Models:** All **Engines:** All **Transmissions:** All	**HO2S-12 (Bank 1 Sensor 2) Heater Circuit Malfunction Conditions:** Engine started, engine running, and the PCM detected the HO2S-12 heater current exceeded 2 amps, or that it was 0.25 amps or less. **Possible Causes:** • HO2S heater control circuit is open, shorted to ground or power • HO2S heater power circuit is open (check power from the relay) • HO2S heater is damaged or has failed • PCM has failed

DTC	Trouble Code Title, Conditions & Possible Causes
DTC: P0150 **2T CCM, MIL: Yes** **1995** **Models:** All **Engines:** All **Transmissions:** All	**HO2S-21 (Bank 2 Sensor 1) Circuit Malfunction Conditions:** Engine warmup completed, engine idling in closed loop, and the PCM detected the HO2S-21 signal was fixed at 400 mv or higher, or that it was fixed from 400-550 mv during the test. **Possible Causes:** • HO2S signal circuit is open between the sensor and the PCM • HO2S signal circuit is shorted to sensor or chassis ground • HO2S is damaged, contaminated or it has failed
DTC: P0150 **2T CCM, MIL: Yes** **1996, 1997, 1998, 1999, 2000,** **2001, 2002, 2003, 2004, 2005,** **2006** **Models:** LX450, LX470, GS400, GS430, GX470, LS400, LS430, SC400, SC430 **Engines:** 4.0L VIN H, 4.3L VIN N, L, 4.5L VIN J, 4.7L VIN T **Transmissions:** All	**HO2S-21 (Bank 2 Sensor 1) Circuit Malfunction Conditions:** Engine warmup completed, engine idling in closed loop, and the PCM detected the HO2S-21 signal was fixed at 400 mv or higher, or that it was fixed at less than 550 mv during the test. **Possible Causes:** • HO2S signal circuit is open between the sensor and the PCM • HO2S signal circuit is shorted to sensor ground or to VREF • HO2S is damaged, contaminated or it has failed
DTC: P0150 **2T CCM, MIL: Yes** **1996, 1997, 1998, 1999, 2000,** **2001, 2002, 2003, 2004, 2005,** **2006** **Models:** ES330, RX330, LX450, LX470, GS300, IS250, IS300, IS350, LS 300, SC300, GS400, GS430, GX470, LS400, LS430, SC400, SC430 **Engines:** 2.5L VIN K, 3.0L VIN D, 3.0L VIN F, 3.3L VIN A, 3.5L VIN E, 4.0L VIN H, 4.3L VIN N, L, 4.5L VIN J, 4.7L VIN T **Transmissions:** All	**HO2S-21 (Bank 2 Sensor 1) Circuit Malfunction Conditions:** Engine started, engine warmup completed, engine idling in closed loop, and the PCM detected the HO2S-21 signal was fixed at 400 mv or higher, or that it was fixed from 400-550 mv during the test. **Possible Causes:** • HO2S signal circuit is open between the sensor and the PCM • HO2S signal circuit is shorted to sensor or chassis ground • HO2S is damaged, contaminated or it has failed • PCM has failed
DTC: P0153 **2T O2S, MIL: Yes** **1995** **Models:** All **Engines:** All **Transmissions:** All	**HO2S-21 (B2 S1) Slow Response Conditions:** Engine started, engine idling in closed loop, and the PCM detected the HO2S-21 response time from rich-to-lean or from lean-to-rich was 1.1 second or longer during the CCM test. **Note: This fault must be detected at least 3 times at idle speed.** **Possible Causes:** • HO2S signal circuit is open or shorted to ground • HO2S element is contaminated, or HO2S heater has failed • Intake air leaks, exhaust manifold leaks or PCV system leaks • MAF sensor out of calibration (it may be dirty or contaminated)
DTC: P0153 **2T O2S, MIL: Yes** **1996, 1997, 1998, 1999, 2000,** **2001, 2002, 2003, 2004, 2005,** **2006** **Models:** ES330, RX330, LX450, LX470, GS300, IS250, IS300, IS350, LS 300, SC300, GS400, GS430, GX470, LS400, LS430, SC400, SC430 **Engines:** 2.5L VIN K, 3.0L VIN D, 3.0L VIN F, 3.3L VIN A, 3.5L VIN E, 4.0L VIN H, 4.3L VIN N, L, 4.5L VIN J, 4.7L VIN T **Transmissions:** All	**HO2S-21 (B2 S1) Slow Response Conditions:** Engine started, engine idling in closed loop, and the PCM detected the HO2S-21 response time from rich-to-lean or from lean-to-rich was 1.1 second or longer during the CCM test. **Note: This fault must be detected at least 3 times at idle speed.** **Possible Causes:** • HO2S signal circuit is open or shorted to ground • HO2S element is contaminated, or HO2S heater has failed • Intake air leaks, exhaust manifold leaks or PCV system leaks • MAF sensor out of calibration (it may be dirty or contaminated)

DTC	Trouble Code Title, Conditions & Possible Causes
DTC: P0154 **2T O2S, MIL: Yes** **1996,1997, 1998, 1999, 2000, 2001, 2002, 2003, 2004, 2005, 2006** **Models:** ES300, RX300, ES330, RX330, **Engines:** 3.0L VIN F, 3.3L VIN A **Transmissions:** All	**Air Fuel Sensor 2 (Bank 2 Sensor 1) Circuit No Activity Detected Conditions:** Engine started, engine runtime over 140 seconds, vehicle driven at a steady speed of 25-81 mph at over 1500 rpm with the throttle valve open, and the PCM detected the Bank 2 A/FS-21 signal did not indicate rich (more than 450 mv) after 65 seconds during the test. **Possible Causes:** • Air leaks present in the PCV valve, hoses or hose connections • A/FS1 connector is damaged or loose • A/FS1 signal circuit is open or shorted to ground • A/FS1 is damaged, contaminated or it has failed • Fuel Control component faults (sticking injector, low pressure) • Gas leaks in the exhaust system in front of the Oxygen sensor
DTC: P0154 **2T O2S, MIL: Yes** **1996, 1997, 1998, 1999, 2000, 2001, 2002, 2003, 2004, 2005, 2006** **Models:** LX450, LX470, GS400, GS430, GX470, LS400, LS430, SC400, SC430 **Engines:** 4.0L VIN H, 4.3L VIN N, L, 4.5L VIN J, 4.7L VIN T **Transmissions:** All	**HO2S-21 (Bank 2 Sensor 1) Circuit No Activity Detected Conditions:** Engine runtime over 140 seconds, VSS from 25-81 mph at over 1500 rpm with throttle valve open, and the PCM detected the HO2S-21 signal did not go over 450 mv after 65 seconds. **Possible Causes:** • Air leaks present in the PCV valve, hoses or hose connections • Fuel Control component faults (sticking injector, low pressure) • Gas leaks in the exhaust system in front of the Oxygen sensor • HO2S-21 signal circuit is open or shorted to ground • HO2S-21 is damaged, contaminated or it has failed • PCM has failed
DTC: P0155 **2T CCM, MIL: Yes** **1995** **Models:** All **Engines:** All **Transmissions:** All	**HO2S-21 (B2 S1) Heater Circuit Malfunction Conditions:** Engine started, engine running, and the PCM detected the HO2S-21 heater current exceeded 2 amps, or that it was 0.25 amps or less. **Possible Causes:** • HO2S heater control circuit is open, shorted to ground or power • HO2S heater power circuit is open (check power from the relay) • HO2S heater is damaged or has failed • PCM has failed
DTC: P0155 **2T CCM, MIL: Yes** **1996, 1997, 1998, 1999, 2000, 2001, 2002, 2003, 2004, 2005, 2006** **Models:** ES330, RX330, LX450, LX470, GS300, IS250, IS300, IS350, LS 300, SC300, GS400, GS430, GX470, LS400, LS430, SC400, SC430 **Engines:** 2.5L VIN K, 3.0L VIN D, 3.0L VIN F, 3.3L VIN A, 3.5L VIN E, 4.0L VIN H, 4.3L VIN N, L, 4.5L VIN J, 4.7L VIN T **Transmissions:** All	**HO2S-21 (B2 S1) Heater Circuit Malfunction Conditions:** Engine started, engine running, and the PCM detected the HO2S-21 heater current exceeded 2 amps, or that it was 0.25 amps or less. **Possible Causes:** • HO2S heater control circuit is open or shorted to ground • HO2S heater control circuit is shorted to power • HO2S heater power circuit is open (check power from the relay) • HO2S heater is damaged or has failed • PCM has failed
DTC: P0155 **2T CCM, MIL: Yes** **1996,1997, 1998, 1999, 2000, 2001, 2002, 2003, 2004, 2005, 2006** **Models:** ES300, RX300, ES330, RX330, **Engines:** 3.0L VIN F, 3.3L VIN A **Transmissions:** All	**Air Fuel Sensor 2 (Bank 2 Sensor 1) Heater Circuit Malfunction Conditions:** Engine started, engine running, and the PCM detected the A/FS-21 heater current indicated over 8 amps, or it was less than 0.25 amps. **Possible Causes:** • A/FS1 heater control circuit is open, shorted to ground or power • A/FS1 heater power circuit is open (check power from relay) • A/FS1 heater is damaged or has failed • PCM has failed

DTC	Trouble Code Title, Conditions & Possible Causes
DTC: P0156 **2T CCM, MIL: Yes** **1996, 1997, 1998, 1999, 2000,** **2001, 2002, 2003, 2004, 2005,** **2006** **Models:** LX450, LX470, GS400, GS430, GX470, LS400, LS430, SC400, SC430 **Engines:** 3.0L VIN F, 3.3L VIN A, 4.0L VIN H, 4.3L VIN N, L, 4.5L VIN J, 4.7L VIN T **Transmissions:** All	**HO2S-22 (Bank 2 Sensor 2) Circuit Malfunction Conditions:** Engine started, engine warmup completed, engine idling in closed loop, and the PCM detected the HO2S-22 signal was fixed at 400 mv or higher, or that it was fixed from 400-550 mv during the test. **Possible Causes:** • HO2S signal circuit is open between the sensor and the PCM • HO2S signal circuit is shorted to sensor or chassis ground • HO2S is damaged, contaminated or it has failed
DTC: P0161 **2T CCM, MIL: Yes** **1996, 1997, 1998, 1999, 2000,** **2001, 2002, 2003, 2004, 2005,** **2006** **Models:** LX450, LX470, GS400, GS430, GX470, LS400, LS430, SC400, SC430 **Engines:** 4.0L VIN H, 4.3L VIN N, L, 4.5L VIN J, 4.7L VIN T **Transmissions:** All	**HO2S-22 (B2 S2) Heater Circuit Malfunction Conditions:** Engine started, engine running, and the PCM detected the HO2S-22 heater current exceeded 2 amps, or that it was 0.25 amps or less. **Possible Causes:** • HO2S heater control circuit is open, shorted to ground or power • HO2S heater power circuit is open (check power from the relay) • HO2S heater is damaged or has failed • PCM has failed
DTC: P0161 **2T CCM, MIL: Yes** **1996, 1997, 1998, 1999, 2000,** **2001, 2002, 2003, 2004, 2005,** **2006** **Models:** LX450, LX470, GS400, GS430, GX470, LS400, LS430, SC400, SC430 **Engines:** 4.0L VIN H, 4.3L VIN N, L, 4.5L VIN J, 4.7L VIN T **Transmissions:** All	**HO2S-22 (B2 S2) Heater Circuit Malfunction Conditions:** Engine started, engine running, and the PCM detected the HO2S-22 heater current exceeded 2 amps, or it was less than 0.25 amps. **Possible Causes:** • HO2S heater control circuit is open, shorted to ground or power • HO2S heater power circuit is open (check power from the relay) • HO2S heater is damaged or has failed • PCM has failed
DTC: P0170 **2T FUEL, MIL: Yes** **1995** **Models:** All **Engines:** All **Transmissions:** All	**Fuel System Too Rich or Too Lean (Bank 1) Conditions:** DTC P0100, P0101, P0110, P0115, P0120, P0121, P0130, P0133, P0136, P0135, P0136, P0141, P0153, P0155, P0201-206, P0300, P0301-P0306, P0401, P0402 and P0441 not set, engine running in closed loop at a stable engine speed, and the PCM detected the lean or rich fuel trim correction value was more than or less than a calibrated limit in memory. **Possible Causes:** • Air leaks present in the exhaust manifold or exhaust pipes • Air being drawn in from leaks in engine gaskets or other seals • Base engine "mechanical" fault affecting one or more cylinders • Fuel control sensor is out of calibration (i.e., ECT, IAT or MAF) • Fuel delivery system supplying too much or too little fuel at idle or cruise (e.g., faulty fuel pump or dirty, restricted fuel filter) • One or more fuel injectors is dirty, leaking or stuck open/closed • HO2S element is contaminated, damaged or it has failed
DTC: P0171 **2T FUEL, MIL: Yes** **1996, 1997, 1998, 1999, 2000,** **2001, 2002** **Models:** All **Engines:** All **Transmissions:** All	**Fuel System Too Lean (Bank 1) Conditions:** DTC P0100, P0101, P0105, P0110, P0115, P0120, P0121, P0130, P0133, P0136, P0135, P0136, P0141, P0151, P0156, P0161, P0300, P0301-P0306, P0440, P0500 and P0505 not set, ECT sensor more than 158°F, vehicle driven at a constant speed of less than 62 mph with the engine speed over 1500 rpm, and the PCM detected the lean fuel trim correction value was over the limit. **Possible Causes:** • A/FS or HO2S is contaminated, deteriorated or it has failed • Air leaks after the MAF sensor, or in the EGR or PCV system • Base engine "mechanical" fault affecting one or more cylinders • Exhaust leaks located in front of the A/FS or HO2S location • Fuel system supplying too little fuel during cruise or idle (faulty fuel pump or fuel filter) • Fuel injector (one or more) dirty or pressure regulator has failed • Vehicle driven low on fuel or until it ran out of fuel

DTC	Trouble Code Title, Conditions & Possible Causes
DTC: P0171 **2T FUEL, MIL: Yes** **1996, 1997, 1998, 1999, 2000, 2001, 2002, 2003, 2004, 2005, 2006** **Models:** ES300, RX300, ES330, RX330, **Engines:** 3.0L VIN F, 3.3L VIN A **Transmissions:** All	**Fuel System Too Lean (Bank 1) Conditions:** DTC P0100, P0101, P0105, P0110, P0115, P0120, P0121, P0130, P0133, P0136, P0135, P0136, P0141, P0151, P0156, P0161, P0300, P0301-P0306, P0440, P0500 and P0505 not set, vehicle speed less than 62 mph with the engine speed over 1500 rpm, ECT sensor more than 158°F, and the PCM detected the lean fuel trim correction value was over the limit. **Possible Causes:** • A/FS or HO2S is contaminated, deteriorated or it has failed • Air leaks after the MAF sensor, or in the EGR or PCV system • Base engine "mechanical" fault affecting one or more cylinders • Exhaust leaks located in front of the A/FS or HO2S location • Fuel system supplying too little fuel during cruise or idle (faulty fuel pump or fuel filter) • Fuel injector (one or more) dirty or pressure regulator has failed • Vehicle driven low on fuel or until it ran out of fuel
DTC: P0171 **2T FUEL, MIL: Yes** **1996, 1997, 1998, 1999, 2000, 2001, 2002, 2003, 2004, 2005, 2006** **Models:** LX450, LX470, GS400, GS430, GX470, LS400, LS430, SC400, SC430 **Engines:** 4.0L VIN H, 4.3L VIN N, L, 4.5L VIN J, 4.7L VIN T **Transmissions:** All	**Fuel System Too Lean (Bank 1) Conditions:** DTC P0100, P0101, P0105, P0110, P0115, P0120, P0121, P0130, P0133, P0136, P0135, P0136, P0141, P0151, P0156, P0161, P0300, P0301-P0306, P0440, P0500 and P0505 not set, ECT sensor more than 158°F, vehicle driven at a constant speed of less than 62 mph with the engine speed over 1500 rpm, and the PCM detected the lean fuel trim correction value was over the limit. **Possible Causes:** • A/FS or HO2S is contaminated, deteriorated or it has failed • Air leaks after the MAF sensor, or in the EGR or PCV system • Base engine "mechanical" fault affecting one or more cylinders • Exhaust leaks located in front of the A/FS or HO2S location • Fuel control sensor is out of calibration (i.e., ECT, IAT or MAP) • Fuel delivery system supplying too little fuel during cruise or idle periods (e.g., faulty fuel pump or dirty, restricted fuel filter) • Fuel injector (one or more) dirty or pressure regulator has failed • Vehicle driven low on fuel or until it ran out of fuel
DTC: P0172 **2T FUEL, MIL: Yes** **1996, 1997, 1998, 1999, 2000, 2001, 2002** **Models:** All **Engines:** All **Transmissions:** All	**Fuel System Too Rich (Bank 1) Conditions:** DTC P0100, P0101, P0105, P0110, P0115, P0120, P0121, P0130, P0133, P0136, P0135, P0136, P0141, P0151, P0156, P0161, P0300, P0301-P0306, P0440, P0500 and P0505 not set, ECT sensor more than 158°F, vehicle driven at a constant speed of less than 62 mph with the engine speed over 1500 rpm, and the PCM detected the rich fuel trim correction value was over the limit. **Possible Causes:** • A/FS or HO2S is contaminated, deteriorated or it has failed • Base engine "mechanical" fault affecting one or more cylinders • EVAP system component has failed or canister fuel saturated • Exhaust leaks located in front of the A/FS or HO2S location • Fuel control sensor is out of calibration (i.e., ECT, IAT or MAF) • Fuel delivery system supplying too much fuel during cruise or idle periods (e.g., faulty fuel pump, or faulty pressure regulator) • Fuel injector(s) is leaking or stuck partially open (one or more)
DTC: P0172 **2T FUEL, MIL: Yes** **1996, 1997, 1998, 1999, 2000, 2001, 2002, 2003, 2004, 2005, 2006** **Models:** LX450, LX470, GS400, GS430, GX470, LS400, LS430, SC400, SC430 **Engines:** 4.0L VIN H, 4.3L VIN N, L, 4.5L VIN J, 4.7L VIN T **Transmissions:** All	**Fuel System Too Rich (Bank 1) Conditions:** DTC P0100, P0101, P0105, P0110, P0115, P0120, P0121, P0130, P0133, P0136, P0135, P0136, P0141, P0151, P0156, P0161, P0300, P0301-P0306, P0440, P0500 and P0505 not set, ECT sensor more than 158°F, vehicle driven at a constant speed of less than 62 mph with the engine speed over 1500 rpm, and the PCM detected the rich fuel trim correction value was over the limit. **Possible Causes:** • A/FS or HO2S is contaminated, deteriorated or it has failed • Base engine "mechanical" fault affecting one or more cylinders • EVAP system component has failed or canister fuel saturated • Exhaust leaks located in front of the A/FS or HO2S location • Fuel control sensor is out of calibration (i.e., ECT, IAT or MAF) • Fuel delivery system supplying too much fuel during cruise or idle periods (e.g., faulty fuel pump, or faulty pressure regulator) • Fuel injector(s) is leaking or stuck partially open (one or more)

DTC	Trouble Code Title, Conditions & Possible Causes
DTC: P0172 **2T FUEL, MIL: Yes** **1996, 1997, 1998, 1999, 2000, 2001, 2002, 2003, 2004, 2005, 2006** **Models:** ES300, RX300, ES330, RX330 **Engines:** 3.0L VIN F, 3.3L VIN A **Transmissions:** All	**Fuel System Too Rich (Bank 1) Conditions:** DTC P0100, P0101, P0105, P0110, P0115, P0120, P0121, P0130, P0133, P0136, P0135, P0136, P0141, P0151, P0156, P0161, P0300, P0301-P0306, P0440, P0500 and P0505 not set, ECT sensor more than 158°F, vehicle driven at a constant speed of less than 62 mph with the engine speed over 1500 rpm, and the PCM detected the rich fuel trim correction value was over the limit. **Possible Causes:** • A/FS or HO2S is contaminated, deteriorated or it has failed • Base engine "mechanical" fault affecting one or more cylinders • EVAP system component has failed or canister fuel saturated • Exhaust leaks located in front of the A/FS or HO2S location • Fuel control sensor is out of calibration (i.e., ECT, IAT or MAF) • Fuel delivery system supplying too much fuel during cruise or idle periods (e.g., faulty fuel pump, or faulty pressure regulator) • Fuel injector(s) is leaking or stuck partially open (one or more)
DTC: P0174 **2T FUEL, MIL: Yes** **1996, 1997, 1998, 1999, 2000, 2001, 2002, 2003, 2004, 2005, 2006** **Models:** ES330, RX330, GS400, GS430, GX470, LS400, LS430, SC400, SC430 **Engines:** 3.3L VIN A, 4.0L VIN H, 4.3L VIN N, L, 4.7L VIN T **Transmissions:** All	**Fuel System Too Lean (Bank 2) Conditions:** DTC P0100, P0101, P0105, P0110, P0115, P0120, P0121, P0136, P0141, P0151, P0156, P0161, P0300, P0301-P0306, P0440, P0500, P0505, P1130, P1133, P1135, 1150, 53 and P1155 not set, ECT sensor more than 158°F, vehicle driven at a constant speed of less than 62 mph with the engine speed over 1500 rpm, and the PCM detected the lean fuel trim correction value was over the limit. **Possible Causes:** • A/FS or HO2S is contaminated, deteriorated or it has failed • Air leaks after the MAF sensor, or in the EGR or PCV system • Base engine "mechanical" fault affecting one or more cylinders • Exhaust leaks located in front of the A/FS or HO2S location • Fuel control sensor is out of calibration (i.e., ECT, IAT or MAF) • Fuel delivery system supplying too little fuel during cruise or idle periods (e.g., faulty fuel pump or dirty, restricted fuel filter) • Fuel injector (one or more) dirty or pressure regulator has failed • MAF sensor is contaminated, out-of-calibration or damaged • Vehicle driven low on fuel or until it ran out of fuel
DTC: P0174 **2T FUEL, MIL: Yes** **1996, 1997, 1998, 1999, 2000, 2001, 2002, 2003** **Models:** ES300, RX300 **Engines:** 3.0L VIN F **Transmissions:** All	**Fuel System Too Lean (Bank 2) Conditions:** DTC P0100, P0101, P0105, P0110, P0115, P0120, P0121, P0136, P0141, P0151, P0156, P0161, P0300, P0301-P0306, P0440, P0500, P0505, P1130, P1133, P1135, 1150, 53 and P1155 not set, ECT sensor more than 158°F, vehicle driven at a constant speed of less than 62 mph with the engine speed over 1500 rpm, and the PCM detected the lean fuel trim correction value was over the limit. **Possible Causes:** • A/FS or HO2S is contaminated, deteriorated or it has failed • Air leaks after the MAF sensor, or in the EGR or PCV system • Base engine "mechanical" fault affecting one or more cylinders • Exhaust leaks located in front of the A/FS or HO2S location • Fuel control sensor is out of calibration (i.e., ECT, IAT or MAF) • Fuel delivery system supplying too little fuel during cruise or idle periods (e.g., faulty fuel pump or dirty, restricted fuel filter) • Fuel injector (one or more) dirty or pressure regulator has failed • MAF sensor is contaminated, out-of-calibration or damaged • Vehicle driven low on fuel or until it ran out of fuel
DTC: P0175 **2T FUEL, MIL: Yes** **1996, 1997, 1998, 1999, 2000, 2001, 2002, 2003, 2004, 2005, 2006** **Models:** ES330, RX330 GS400, GS430, GX470, LS400, LS430, SC400, SC430 **Engines:** 3.3L VIN A, 4.0L VIN H, 4.3L VIN N, L, 4.7L VIN T **Transmissions:** All	**Fuel System Too Rich (Bank 2) Conditions:** DTC P0100, P0101, P0105, P0110, P0115, P0120, P0121, P0136, P0141, P0151, P0156, P0161, P0300, P0301-P0306, P0440, P0500, P0505, P1130, P1133, P1135, 1150, 53 and P1155 not set, ECT sensor more than 158°F, vehicle driven at a constant speed of less than 62 mph with the engine speed over 1500 rpm, and the PCM detected the rich fuel trim correction value was over the limit. **Possible Causes:** • A/FS or HO2S is contaminated, deteriorated or it has failed • Base engine "mechanical" fault affecting one or more cylinders • EVAP system component has failed or canister fuel saturated • Exhaust leaks located in front of the A/FS or HO2S location • Fuel control sensor is out of calibration (i.e., ECT, IAT or MAF) • Fuel delivery system supplying too much fuel during cruise or idle periods (e.g., faulty fuel pump, or faulty pressure regulator) • Fuel injector(s) is leaking or stuck partially open (one or more)

DTC	Trouble Code Title, Conditions & Possible Causes
DTC: P0175 **2T FUEL, MIL: Yes** **1996,1997, 1998, 1999, 2000,** **2001, 2002, 2003** **Models:** ES300, RX300, **Engines:** 3.0L VIN F **Transmissions:** All	**Fuel System Too Rich (Bank 2) Conditions:** DTC P0100, P0101, P0105, P0110, P0115, P0120, P0121, P0136, P0141, P0151, P0156, P0161, P0300, P0301-P0306, P0440, P0500, P0505, P1130, P1133, P1135, 1150, 53 and P1155 not set, ECT sensor more than 158°F, vehicle driven at a constant speed of less than 62 mph with the engine speed over 1500 rpm, and the PCM detected the rich fuel trim correction value was over the limit. **Possible Causes:** • A/FS or HO2S is contaminated, deteriorated or it has failed • Base engine "mechanical" fault affecting one or more cylinders • EVAP system component has failed or canister fuel saturated • Exhaust leaks located in front of the A/FS or HO2S location • Fuel control sensor is out of calibration (i.e., ECT, IAT or MAF) • Fuel delivery system supplying too much fuel during cruise or idle periods (e.g., faulty fuel pump, or faulty pressure regulator) • Fuel injector(s) is leaking or stuck partially open (one or more)
DTC: P0220 **2T CCM, MIL: Yes** **1996, 1997, 1998, 1999, 2000,** **2001, 2002, 2003, 2004, 2005,** **2006** **Models:** LX450, LX470, GS400, GS430, GX470, LS400, LS430, SC400, SC430 **Engines:** 4.0L VIN H, 4.3L VIN N, L, 4.5L VIN J, 4.7L VIN T **Transmissions:** All	**Throttle Pedal Position Sensor/Switch 'B' Circuit Malfunction Conditions:** Key on or engine running; and the PCM detected the TP sensor 'B' Signal indicated less than 0.50v with the throttle position closed (the switch is open), or the TP sensor signal indicated more than 4.9v at any time. The Electric TP Sensor is mounted on the throttle body. It has two sensors (the electrical throttle system does not use a cable). **Possible Causes:** • TP sensor signal circuit open or shorted to ground • TP sensor ground circuit is open • TP sensor power circuit is open (check VREF circuit at PCM) • TP sensor is damaged or has failed • PCM has failed
DTC: P0220 **2T CCM, MIL: Yes** **1996, 1997, 1998, 1999, 2000,** **2001, 2002, 2003, 2004, 2005,** **2006** **Models:** LX450, LX470, GS400, GS430, GX470, LS400, LS430, SC400, SC430 **Engines:** 4.0L VIN H, 4.3L VIN N, L, 4.5L VIN J, 4.7L VIN T **Transmissions:** All	**Throttle Pedal Position Sensor Switch B Circuit Low Input Conditions:** Key on or engine running; and the PCM detected an unexpected low voltage condition (less than 0.50v) on the VTA2 circuit. The Electric TP Sensor is mounted on the throttle body. It has two sensors (the electrical throttle system does not use a cable). **Possible Causes:** • Electric TP sensor connector is damaged or shorted • Electric TP sensor circuit is shorted to ground • Electric TP sensor is damaged or has failed • PCM has failed
DTC: P0223 **2T CCM, MIL: Yes** **1996, 1997, 1998, 1999, 2000,** **2001, 2002, 2003, 2004, 2005,** **2006** **Models:** LX450, LX470, GS400, GS430, GX470, LS400, LS430, SC400, SC430 **Engines:** 4.0L VIN H, 4.3L VIN N, L, 4.5L VIN J, 4.7L VIN T **Transmissions:** All	**Throttle Pedal Position Sensor Switch B Circuit High Input Conditions:** Key on or engine running; and the PCM detected an unexpected high voltage condition (more than 4.97v) on the VTA2 circuit. The Electric TP Sensor is mounted on the throttle body. It has two sensors (the electrical throttle system does not use a cable). **Possible Causes:** • Electric TP sensor connector is damaged or open • Electric TP sensor circuit is open or shorted to VREF • Electric TP sensor is damaged or has failed • PCM has failed
DTC: P0230 **2T CCM, MIL: Yes** **1996, 1997, 1998, 1999, 2000,** **2001, 2002, 2003, 2004, 2005,** **2006** **Models:** LX450, LX470, GS400, GS430, GX470, LS400, LS430, SC400, SC430 **Engines:** 4.0L VIN H, 4.3L VIN N, L, 4.5L VIN J, 4.7L VIN T **Transmissions:** All	**Fuel Pump Primary Circuit Malfunction Conditions:** Engine started; and the PCM detected an unexpected voltage on the Fuel Pump Primary control circuit (from ST terminal to Starter Relay coil and to the STA terminal of the PCM). **Possible Causes:** • Circuit opening relay is damaged or has failed • Fuel pump relay control circuit is open or shorted to ground • Fuel pump relay is damaged or has failed • Fuel pump is damaged or has failed • PCM has failed

DTC	Trouble Code Title, Conditions & Possible Causes
DTC: P0300 **2T MISFIRE, MIL: Yes** **1996, 1997, 1998, 1999, 2000, 2001, 2002, 2003, 2004, 2005, 2006** **Models:** ES300, RX300, ES330, RX330, LX450, LX470, GS400, GS430, GX470, LS400, LS430, SC400, SC430 **Engines:** 3.0L VIN F, 3.3L VIN A, 4.0L VIN H, 4.3L VIN N, L, 4.5L VIN J, 4.7L VIN T **Transmissions:** All	**Multiple Cylinder Misfire Detected Conditions:** **Trouble Code Conditions** DTC P0100, P0101, P0102, P0103, P0105, P0110, P0112, P0113, P0115, P0117, P0118, P0120, P0121, P0122, P0123, P0125, P0335, P0340, P0500, P0505 and P0510 not set, engine started, vehicle driven to a speed of over 3 mph for 1 minute, and the PCM detected a misfire rate of 1-2% (High Emissions 2T), or a misfire rate of 6-30% (Catalyst Damaging 1T) in two or more cylinders. **Note: If the misfire is severe, the MIL will flash on/off on the 1st trip! Look at the misfire ratio for all of the cylinders on the Scan Tool. The cylinder with the highest misfire ratio should be checked first!** **Possible Causes:** • Air leak in the intake manifold, or in the EGR or PCV system • Base engine mechanical fault that affects two or more cylinders • EGR valve is stuck open or the PCV system has a vacuum leak • Fuel delivery component fault that affects two or more cylinders (e.g., contaminated, dirty or sticking fuel injectors) • Ignition system fault (coil or plug) that affects several cylinders • Mass airflow meter is contaminated, or its signal is out of range
DTC: P0301 **2T MISFIRE, MIL: Yes** **1996, 1997, 1998, 1999, 2000, 2001, 2002, 2003, 2004, 2005, 2006** **Models:** ES300, RX300, ES330, RX330, LX450, LX470, GS400, GS430, GX470, LS400, LS430, SC400, SC430 **Engines:** 3.0L VIN F, 3.3L VIN A, 4.0L VIN H, 4.3L VIN N, L, 4.5L VIN J, 4.7L VIN T **Transmissions:** All	**Cylinder 1 Misfire Detected Conditions:** **Trouble Code Conditions** DTC P0100, P0101, P0102, P0103, P0105, P0110, P0112, P0113, P0115, P0117, P0118, P0120, P0121, P0122, P0123, P0125, P0335, P0340, P0500, P0505 and P0510 not set, engine started, vehicle driven to a speed of over 3 mph for 1 minute, and the PCM detected a misfire rate of 1-2% (High Emissions 2T), or a misfire rate of 6-30% (Catalyst Damaging 1T) in Cylinder 1. **Note: If the misfire is severe, the MIL will flash on/off on the 1st trip!** **Possible Causes:** • Base engine mechanical fault that affects only Cylinder 1 • EGR valve is stuck open or the PCV system has a vacuum leak • Fuel component fault that affects only Cylinder 1 (a contaminated or sticking injector) • Ignition system problem (coil or plug) that affects Cylinder 1
DTC: P0302 **2T MISFIRE, MIL: Yes** **1996, 1997, 1998, 1999, 2000, 2001, 2002, 2003, 2004, 2005, 2006** **Models:** ES300, RX300, ES330, RX330, LX450, LX470, GS400, GS430, GX470, LS400, LS430, SC400, SC430 **Engines:** 3.0L VIN F, 3.3L VIN A, 4.0L VIN H, 4.3L VIN N, L, 4.5L VIN J, 4.7L VIN T **Transmissions:** All	**Cylinder 2 Misfire Detected Conditions:** **Trouble Code Conditions** DTC P0100, P0101, P0102, P0103, P0105, P0110, P0112, P0113, P0115, P0117, P0118, P0120, P0121, P0122, P0123, P0125, P0335, P0340, P0500, P0505 and P0510 not set, engine started, vehicle driven to a speed of over 3 mph for 1 minute, and the PCM detected a misfire rate of 1-2% (High Emissions 2T), or a misfire rate of 6-30% (Catalyst Damaging 1T) in Cylinder 2. **Note: If the misfire is severe, the MIL will flash on/off on the 1st trip!** **Possible Causes:** • Base engine mechanical fault that affects only Cylinder 2 • EGR valve is stuck open or the PCV system has a vacuum leak • Fuel component fault that affects only Cylinder 2 (a contaminated or sticking injector) • Ignition system problem (coil or plug) that affects Cylinder 2
DTC: P0303 **2T MISFIRE, MIL: Yes** **1996, 1997, 1998, 1999, 2000, 2001, 2002, 2003, 2004, 2005, 2006** **Models:** ES300, RX300, ES330, RX330, LX450, LX470, GS400, GS430, GX470, LS400, LS430, SC400, SC430 **Engines:** 3.0L VIN F, 3.3L VIN A, 4.0L VIN H, 4.3L VIN N, L, 4.5L VIN J, 4.7L VIN T **Transmissions:** All	**Cylinder 3 Misfire Detected Conditions:** **Trouble Code Conditions** DTC P0100, P0101, P0102, P0103, P0105, P0110, P0112, P0113, P0115, P0117, P0118, P0120, P0121, P0122, P0123, P0125, P0335, P0340, P0500, P0505 and P0510 not set, engine started, vehicle driven to a speed of over 3 mph for 1 minute, and the PCM detected a misfire rate of 1-2% (High Emissions 2T), or a misfire rate of 6-30% (Catalyst Damaging 1T) in Cylinder 3. **Note: If the misfire is severe, the MIL will flash on/off on the 1st trip!** **Possible Causes:** • Base engine mechanical fault that affects only Cylinder 3 • EGR valve is stuck open or the PCV system has a vacuum leak • Fuel component fault that affects only Cylinder 3 (a contaminated or sticking injector) • Ignition system problem (coil or plug) that affects Cylinder 3

DTC	Trouble Code Title, Conditions & Possible Causes
DTC: P0304 **2T MISFIRE, MIL: Yes** **1996,1997, 1998, 1999, 2000, 2001, 2002, 2003, 2004, 2005, 2006** **Models:** ES300, RX300, ES330, RX330, LX450, LX470, GS400, GS430, GX470, LS400, LS430, SC400, SC430 **Engines:** 3.0L VIN F, 3.3L VIN A, 4.0L VIN H, 4.3L VIN N, L, 4.5L VIN J, 4.7L VIN T **Transmissions:** All	**Cylinder 4 Misfire Detected Conditions:** **Trouble Code Conditions** DTC P0100, P0101, P0102, P0103, P0105, P0110, P0112, P0113, P0115, P0117, P0118, P0120, P0121, P0122, P0123, P0125, P0335, P0340, P0500, P0505 and P0510 not set, engine started, vehicle driven to a speed of over 3 mph for 1 minute, and the PCM detected a misfire rate of 1-2% (High Emissions 2T), or a misfire rate of 6-30% (Catalyst Damaging 1T) in Cylinder 4. **Note: If the misfire is severe, the MIL will flash on/off on the 1st trip!** **Possible Causes:** • Base engine mechanical fault that affects only Cylinder 4 • EGR valve is stuck open or the PCV system has a vacuum leak • Fuel component fault that affects only Cylinder 4 (a contaminated or sticking injector) • Ignition system problem (coil or plug) that affects Cylinder 4
DTC: P0305 **2T MISFIRE, MIL: Yes** **1996,1997, 1998, 1999, 2000, 2001, 2002, 2003, 2004, 2005, 2006** **Models:** ES300, RX300, ES330, RX330, LX450, LX470, GS400, GS430, GX470, LS400, LS430, SC400, SC430 **Engines:** 3.0L VIN F, 3.3L VIN A, 4.0L VIN H, 4.3L VIN N, L, 4.5L VIN J, 4.7L VIN T **Transmissions:** All	**Cylinder 5 Misfire Detected Conditions:** **Trouble Code Conditions** DTC P0100, P0101, P0102, P0103, P0105, P0110, P0112, P0113, P0115, P0117, P0118, P0120, P0121, P0122, P0123, P0125, P0335, P0340, P0500, P0505 and P0510 not set, engine started, vehicle driven to a speed of over 3 mph for 1 minute, and the PCM detected a misfire rate of 1-2% (High Emissions 2T), or a misfire rate of 6-30% (Catalyst Damaging 1T) in Cylinder 5. **Note: If the misfire is severe, the MIL will flash on/off on the 1st trip!** **Possible Causes:** • Base engine mechanical fault that affects only Cylinder 5 • EGR valve is stuck open or the PCV system has a vacuum leak • Fuel delivery component fault that affects only Cylinder 5 (e.g., a contaminated, dirty or sticking fuel injector) • Ignition system problem (coil or plug) that affects Cylinder 5
DTC: P0306 **2T MISFIRE, MIL: Yes** **1996,1997, 1998, 1999, 2000, 2001, 2002, 2003, 2004, 2005, 2006** **Models:** ES300, RX300, ES330, RX330, LX450, LX470, GS400, GS430, GX470, LS400, LS430, SC400, SC430 **Engines:** 3.0L VIN F, 3.3L VIN A, 4.0L VIN H, 4.3L VIN N, L, 4.5L VIN J, 4.7L VIN T **Transmissions:** All	**Cylinder 6 Misfire Detected Conditions:** **Trouble Code Conditions** DTC P0100, P0101, P0102, P0103, P0105, P0110, P0112, P0113, P0115, P0117, P0118, P0120, P0121, P0122, P0123, P0125, P0335, P0340, P0500, P0505 and P0510 not set, engine started, vehicle driven to a speed of over 3 mph for 1 minute, and the PCM detected a misfire rate of 1-2% (High Emissions 2T), or a misfire rate of 6-30% (Catalyst Damaging 1T) in Cylinder 6. **Note: If the misfire is severe, the MIL will flash on/off on the 1st trip!** **Possible Causes:** • Base engine mechanical fault that affects only Cylinder 6 • EGR valve is stuck open or the PCV system has a vacuum leak • Fuel delivery component fault that affects only Cylinder 6 (e.g., a contaminated, dirty or sticking fuel injector) • Ignition system problem (coil or plug) that affects Cylinder 6 • Mass airflow meter is contaminated, or its signal is out of range
DTC: P0307 **2T MISFIRE, MIL: Yes** **1996, 1997, 1998, 1999, 2000, 2001, 2002, 2003, 2004, 2005, 2006** **Models:** LX450, LX470, GS400, GS430, GX470, LS400, LS430, SC400, SC430 **Engines:** 4.0L VIN H, 4.3L VIN N, L, 4.5L VIN J, 4.7L VIN T **Transmissions:** All	**Cylinder 7 Misfire Detected Conditions:** **Trouble Code Conditions** DTC P0100, P0101, P0102, P0103, P0105, P0110, P0112, P0113, P0115, P0117, P0118, P0120, P0121, P0122, P0123, P0125, P0335, P0340, P0500, P0505 and P0510 not set, engine started, vehicle driven to a speed of over 3 mph for 1 minute, and the PCM detected a misfire rate of 1-2% (High Emissions 2T), or a misfire rate of 6-30% (Catalyst Damaging 1T) in Cylinder 7. **Note: If the misfire is severe, the MIL will flash on/off on the 1st trip!** **Possible Causes:** • Base engine mechanical fault that affects only Cylinder 7 • EGR valve is stuck open or the PCV system has a vacuum leak • Fuel delivery component fault that affects only Cylinder 7 (e.g., a contaminated, dirty or sticking fuel injector) • Ignition system problem (coil or plug) that affects Cylinder 7 • Mass airflow meter is contaminated, or its signal is out of range
DTC: P0308 **2T MISFIRE, MIL: Yes** **1996, 1997, 1998, 1999, 2000, 2001, 2002, 2003, 2004, 2005, 2006** **Models:** LX450, LX470, GS400, GS430, GX470, LS400, LS430, SC400, SC430 **Engines:** 4.0L VIN H, 4.3L VIN N, L, 4.5L VIN J, 4.7L VIN T **Transmissions:** All	**Cylinder 8 Misfire Detected Conditions:** **Trouble Code Conditions** DTC P0100, P0101, P0102, P0103, P0105, P0110, P0112, P0113, P0115, P0117, P0118, P0120, P0121, P0122, P0123, P0125, P0335, P0340, P0500, P0505 and P0510 not set, engine started, vehicle driven to a speed of over 3 mph for 1 minute, and the PCM detected a misfire rate of 1-2% (High Emissions 2T), or a misfire rate of 6-30% (Catalyst Damaging 1T) in Cylinder 8. **Note: If the misfire is severe, the MIL will flash on/off on the 1st trip!** **Possible Causes:** • Base engine mechanical fault that affects only Cylinder 8 • EGR valve is stuck open or the PCV system has a vacuum leak • Fuel delivery component fault that affects only Cylinder 8 (e.g., a contaminated, dirty or sticking fuel injector) • Ignition system problem (coil or plug) that affects Cylinder 8 • Mass airflow meter is contaminated, or its signal is out of range

DTC	Trouble Code Title, Conditions & Possible Causes
DTC: P0325 **1T CCM, MIL: Yes** **1995** **Models:** All **Engines:** All **Transmissions:** All	**Knock Sensor 1 Circuit Malfunction Conditions:** Engine started, vehicle driven with the engine speed over 2000 rpm, and the PCM detected an unexpected voltage condition on the Knock Sensor 1 (KS1) circuit during the CCM test. **Possible Causes:** • Verify that the Knock Sensor (KS) is tightened to specification • Knock sensor signal circuit is open or shorted to ground • Knock sensor signal circuit is shorted to VREF or system power • Knock sensor is damaged or has failed • PCM has failed
DTC: P0325 **1T CCM, MIL: Yes** **1996, 1997, 1998, 1999, 2000,** **2001, 2002** **Models:** All **Engines:** All **Transmissions:** All	**Knock Sensor 1 Circuit Malfunction Conditions:** Engine started, vehicle driven with the engine speed over 2000 rpm, and the PCM detected an unexpected voltage condition on the Knock Sensor 1 (KS1) circuit during the CCM test. **Possible Causes:** • Verify that the Knock Sensor (KS) is tightened to specification • Knock sensor signal circuit is open or shorted to ground • Knock sensor signal circuit is shorted to VREF or system power • Knock sensor is damaged or has failed • PCM has failed
DTC: P0325 **1T CCM, MIL: Yes** **2003, 2004, 2005, 2006** **Models:** All **Engines:** All **Transmissions:** All	**Knock Sensor 1 Circuit Malfunction Conditions:** Engine started, vehicle driven with the engine speed over 2000 rpm, and the PCM did not detect a signal on the Knock Sensor 1 circuit. **Possible Causes:** • Knock sensor signal circuit is open or shorted to ground • Knock sensor signal circuit is shorted to VREF or system power • Knock sensor is damaged, not tightened properly or has failed • PCM has failed
DTC: P0326 **1T CCM, MIL: Yes** **1995** **Models:** All **Engines:** All **Transmissions:** All	**Knock Sensor 2 Circuit Malfunction Conditions:** Engine started, engine running at 1200 rpm or higher, and the PCM detected an unexpected voltage condition on the Knock Sensor 2 (KS2 circuit during the CCM test. **Possible Causes:** • Verify that the Knock Sensor (KS) is tightened to specification • Knock sensor signal circuit is open or shorted to ground • Knock sensor signal circuit is shorted to VREF or system power • Knock sensor is damaged or has failed • PCM has failed
DTC: P0327 **1T CCM, MIL: Yes** **1996, 1997, 1998, 1999, 2000,** **2001, 2002, 2003, 2004, 2005,** **2006** **Models:** GS400, GS430, GX470, LS400, LS430, SC400, SC430 **Engines:** 4.0L VIN H, 4.3L VIN N, L, 4.7L VIN T **Transmissions:** All	**Knock Sensor 1 Circuit Low Input (Bank 1) Conditions:** Engine started, engine speed from 1500-5500 rpm, and the PCM detected an unexpected low voltage on the Knock Sensor 1 circuit. **Possible Causes:** • Verify that the Knock Sensor (KS) is tightened to specification • Knock sensor signal circuit is shorted to ground • Knock sensor is damaged or has failed • PCM has failed
DTC: P0328 **1T CCM, MIL: Yes** **1996, 1997, 1998, 1999, 2000,** **2001, 2002, 2003, 2004, 2005,** **2006** **Models:** GS400, GS430, GX470, LS400, LS430, SC400, SC430 **Engines:** 4.0L VIN H, 4.3L VIN N, L, 4.7L VIN T **Transmissions:** All	**Knock Sensor 1 Circuit High Input (Bank 1) Conditions:** Engine started, engine speed from 1500-5500 rpm, and the PCM detected an unexpected high voltage on the Knock Sensor 1 circuit. **Possible Causes:** • Verify that the Knock Sensor (KS) is tightened to specification • Knock sensor signal circuit is open or shorted to power • Knock sensor is damaged or has failed • PCM has failed

DTC	Trouble Code Title, Conditions & Possible Causes
DTC: P0330 **1T CCM, MIL: Yes** **1995** **Models:** All **Engines:** All **Transmissions:** All	**Knock Sensor 2 Circuit Malfunction Conditions:** Engine started, vehicle driven with the engine speed over 1200 rpm, and the PCM detected an unexpected voltage condition on the Knock Sensor 1 (KS1) circuit during the CCM test. **Possible Causes:** • Verify that the Knock Sensor (KS) is tightened to specification • Knock sensor signal circuit is open or shorted to ground • Knock sensor signal circuit is shorted to VREF or system power • Knock sensor is damaged or has failed • PCM has failed
DTC: P0330 **2T CCM, MIL: Yes** **1996, 1997, 1998, 1999, 2000,** **2001, 2002** **Models:** All **Engines:** All **Transmissions:** All	**Knock Sensor 2 Circuit Malfunction Conditions:** Engine started, vehicle driven with the engine speed over 2000 rpm, and the PCM detected an unexpected voltage condition on the Knock Sensor 1 (KS1) circuit during the CCM test. **Possible Causes:** • Verify that the Knock Sensor (KS) is tightened to specification • Knock sensor signal circuit is open, shorted to ground or VREF • Knock sensor is damaged or has failed • PCM has failed
DTC: P0330 **1T CCM, MIL: Yes** **1996,1997, 1998, 1999, 2000,** **2001, 2002, 2003, 2004, 2005,** **2006** **Models:** ES300, RX300, ES330, RX330, LX450, LX470, GS400, GS430, GX470, LS400, LS430, SC400, SC430 **Engines:** 3.0L VIN F, 3.3L VIN A, 4.0L VIN H, 4.3L VIN N, L, 4.5L VIN J, 4.7L VIN T **Transmissions:** All	**Knock Sensor 2 Circuit Malfunction Conditions:** Engine started, vehicle driven with the engine speed over 2000 rpm, and the PCM did not detect a signal on the Knock Sensor 2 circuit. **Possible Causes:** • Knock sensor signal circuit is open, shorted to ground or VREF • Knock sensor is damaged, not tightened properly or has failed • PCM has failed
DTC: P0332 **1T CCM, MIL: Yes** **1996, 1997, 1998, 1999, 2000,** **2001, 2002, 2003, 2004, 2005,** **2006** **Models:** GS400, GS430, GX470, LS400, LS430, SC400, SC430 **Engines:** 4.0L VIN H, 4.3L VIN N, L, 4.7L VIN T **Transmissions:** All	**Knock Sensor 2 Circuit Low Input (Bank 2) Conditions:** Engine started, engine speed from 1500-5500 rpm, and the PCM detected an unexpected low voltage on the Knock Sensor 2 circuit. **Possible Causes:** • Verify that the Knock Sensor (KS) is tightened to specification • Knock sensor signal circuit is shorted to ground • Knock sensor had failed, or the PCM has failed
DTC: P0333 **1T CCM, MIL: Yes** **1996, 1997, 1998, 1999, 2000,** **2001, 2002, 2003, 2004, 2005,** **2006** **Models:** GS400, GS430, GX470, LS400, LS430, SC400, SC430 **Engines:** 4.0L VIN H, 4.3L VIN N, L, 4.7L VIN T **Transmissions:** All	**Knock Sensor 2 Circuit High Input (Bank 2) Conditions:** Engine speed from 1500-5500 rpm, and the PCM detected an unexpected high voltage on the Knock Sensor 2 circuit. **Possible Causes:** • Verify that the Knock Sensor (KS) is tightened to specification • Knock sensor signal circuit is open or shorted to power • Knock sensor has failed, or the PCM has failed
DTC: P0335 **2T CCM, MIL: Yes** **1995** **Models:** All **Engines:** All **Transmissions:** All	**Crankshaft Position Sensor 'A' Circuit Malfunction Conditions:** Engine cranking; and the PCM did not detect any CKP Sensor 'A' signals, or with the engine speed over 600 rpm, it did not receive any CKP sensor signals, or the CKP signal was lost. **Possible Causes:** • CKP Sensor 'A' signal circuit is open or shorted to ground • CKP Sensor 'A' signal ground circuit is open • CKP Sensor 'A' signal is shorted to VREF or system power • CKP Sensor 'A' is damaged or has failed • PCM has failed

DTC	Trouble Code Title, Conditions & Possible Causes
DTC: P0335 **1T CCM, MIL: Yes** **1996, 1997, 1998, 1999, 2000, 2001, 2002, 2003, 2004, 2005, 2006** **Models:** All **Engines:** All **Transmissions:** All	**Crankshaft Position Sensor 'A' Circuit Malfunction Conditions:** Engine cranking; and the PCM did not detect any CKP Sensor 'A' signals, or with the engine speed over 600 rpm, it did not receive any CKP sensor signals, or the CKP signal was lost. **Possible Causes:** • CKP Sensor 'A' signal circuit is open, shorted to ground or shorted to system power • CKP Sensor 'A' signal ground circuit is open • CKP Sensor 'A' is damaged or has failed
DTC: P0339 **2T CCM, MIL: Yes** **1996, 1997, 1998, 1999, 2000, 2001, 2002, 2003, 2004, 2005, 2006** **Models:** ES330, RX330, LX450, LX470, GS400, GS430, GX470, LS400, LS430, SC400, SC430 **Engines:** 3.3L VIN A, 4.0L VIN H, 4.3L VIN N, L, 4.5L VIN J, 4.7L VIN T **Transmissions:** All	**Crankshaft Position Sensor 'A' Circuit Intermittent Conditions:** Engine started, STA signal indicating "off", engine runtime over 3 seconds since STA switched from "on" to "off", engine speed over 1000 rpm, and the PCM did not detect any CKP Sensor 'A' signals for 500 ms. The crankshaft position (NE) sensor consists of a magnet, iron core and pickup coil. The NE sensor signal plate, which has 34 teeth, installed on the crankshaft-timing pulley. This sensor generates 34 signals for each engine revolution. The PCM detects the crankshaft angle and engine speed based on the NE signal. **Possible Causes:** • CKP sensor signal circuit is open, shorted to ground or power • CKP Sensor signal ground circuit is open • Crankshaft timing pulley is damaged or out of alignment • CKP Sensor has failed, or the PCM has failed
DTC: P0340 **2T CCM, MIL: Yes** **1995** **Models:** All **Engines:** All **Transmissions:** All	**Camshaft Position Sensor Circuit Malfunction Conditions:** Engine cranking; and the PCM did not detect any CMP sensor signals, or with the engine speed over 600 rpm, it did not detect any CMP signals or the CMP signal was interrupted. **Possible Causes:** • CMP sensor signal circuit is open or shorted to ground • CMP sensor signal ground circuit is open • CMP sensor signal is shorted to VREF or system power • CMP sensor is damaged or has failed • PCM has failed
DTC: P0340 **2T CCM, MIL: Yes** **1996, 1997, 1998, 1999, 2000, 2001, 2002** **Models:** All **Engines:** All **Transmissions:** All	**Camshaft Position Sensor Circuit Malfunction Conditions:** Engine cranking; and the PCM did not detect any CMP sensor signals, or with the engine speed over 600 rpm, it did not detect any CMP signals, or the CMP signal was interrupted. **Possible Causes:** • CMP sensor signal circuit is open or shorted to ground • CMP sensor signal ground circuit is open • CMP sensor signal is shorted to VREF or system power • CMP sensor is damaged or has failed • PCM has failed • TSB EG010-02 (4/02) contains information related to this code
DTC: P0340 **2T CCM, MIL: Yes** **2003, 2004, 2005, 2006** **Models:** All **Engines:** All **Transmissions:** All	**Camshaft Position Sensor Circuit Malfunction Conditions:** Engine cranking; and the PCM did not detect any CMP sensor signals, or with the engine speed over 600 rpm, it did not detect any CMP signals, or the CMP signal was interrupted. **Possible Causes:** • CMP sensor signal circuit is open, shorted to ground or power • CMP sensor signal ground circuit is open • CMP sensor has failed, or the PCM has failed
DTC: P0341 **2T CCM, MIL: Yes** **2003, 2004, 2005, 2006** **Models:** All **Engines:** All **Transmissions:** All	**Camshaft Position Sensor 'A' Signal Range/Performance Conditions:** Engine cranking; and the PCM detected twelve (12) or more Camshaft Position Sensor 'A' (Bank 1) signals during the test. **Possible Causes:** • CMP sensor signal circuit is open, shorted to ground or power • CMP sensor pulley is damaged, or timing belt has jumped teeth • CMP sensor is damaged or has failed • PCM has failed
DTC: P0345 **2T CCM, MIL: Yes** **1996, 1997, 1998, 1999, 2000, 2001, 2002, 2003, 2004, 2005, 2006** **Models:** ES300, RX300, ES330, RX330, **Engines:** 3.0L VIN F, 3.3L VIN A **Transmissions:** All	**Camshaft Position Sensor 'A' Signal Range/Performance Conditions:** Engine cranking; and the PCM detected twelve (12) or more CMP Sensor 'A' (Bank 2) signals during the test. The Left Hand VVT Camshaft Position sensor consists of a magnet, and a circuit board in which a Magnetic Resistive (MR) device is mounted. The VVT signal plate includes three (3) protrusions on its outer surface. **Possible Causes:** • VVT sensor signal circuit is open, shorted to ground or power • VVT sensor pulley is damaged, or timing belt has jumped teeth • VVT sensor has failed, or the PCM has failed

DTC	Trouble Code Title, Conditions & Possible Causes
DTC: P0346 **2T CCM, MIL: Yes** **1996, 1997, 1998, 1999, 2000, 2001, 2002, 2003, 2004, 2005, 2006** **Models:** ES300, RX300, ES330, RX330 **Engines:** 3.0L VIN F, 3.3L VIN A **Transmissions:** All	**Camshaft Position Sensor 'A' Signal Range/Performance Conditions:** Engine cranking; and the PCM detected twelve (12) or more CMP Sensor 'A' (Bank 2) signals during the test. The Left Hand VVT Camshaft Position sensor consists of a magnet, and a circuit board in which a Magnetic Resistive (MR) device is mounted. The VVT signal plate includes three (3) protrusions on its outer surface. **Possible Causes:** • VVT sensor signal circuit is open, shorted to ground or power • VVT sensor pulley is damaged, or timing belt has jumped teeth • VVT sensor had failed, or the PCM has failed
DTC: P0351 **1T CCM, MIL: Yes** **1996, 1997, 1998, 1999, 2000, 2001, 2002, 2003, 2004, 2005, 2006** **Models:** ES300, RX300, ES330, RX330, LX450, LX470, GS400, GS430, GX470, LS400, LS430, SC400, SC430 **Engines:** 3.0L VIN F, 3.3L VIN A, 4.0L VIN H, 4.3L VIN N, L, 4.5L VIN J, 4.7L VIN T **Transmissions:** All	**Ignition Coil No. 1 Primary/Secondary Circuit Malfunction Conditions:** Engine started, and the PCM did not detect a change in the IGF signal on the Ignition Coil No. 1 IGF circuit. This engine uses a Direct Ignition (DI) system where one coil is used to fire one cylinder. The coil high-energy secondary wire is connected to one spark plug. If P0351 to P0356 are all set, check for an open/shorted IGF circuit. **Possible Causes:** • IGT1 circuit is open or shorted to ground • Ignition Coil No. 1 is damaged or it has failed • Problem present in the Ignition System • PCM has failed
DTC: P0352 **1T CCM, MIL: Yes** **1996, 1997, 1998, 1999, 2000, 2001, 2002, 2003, 2004, 2005, 2006** **Models:** ES300, RX300, ES330, RX330, LX450, LX470, GS400, GS430, GX470, LS400, LS430, SC400, SC430 **Engines:** 3.0L VIN F, 3.3L VIN A, 4.0L VIN H, 4.3L VIN N, L, 4.5L VIN J, 4.7L VIN T **Transmissions:** All	**Ignition Coil No. 2 Primary/Secondary Circuit Malfunction Conditions:** Engine started, and the PCM did not detect a change in the IGF signal on the Ignition Coil No. 2 IGF circuit. This engine uses a Direct Ignition (DI) system where one coil is used to fire one cylinder. The coil high-energy secondary wire is connected to one spark plug. If P0351 to P0356 are all set, check for an open/shorted IGF circuit. **Possible Causes:** • IGT2 circuit is open or shorted to ground • Ignition Coil No. 2 is damaged or it has failed • Problem present in the Ignition System • PCM has failed
DTC: P0353 **1T CCM, MIL: Yes** **1996, 1997, 1998, 1999, 2000, 2001, 2002, 2003, 2004, 2005, 2006** **Models:** ES300, RX300, ES330, RX330, LX450, LX470, GS400, GS430, GX470, LS400, LS430, SC400, SC430 **Engines:** 3.0L VIN F, 3.3L VIN A, 4.0L VIN H, 4.3L VIN N, L, 4.5L VIN J, 4.7L VIN T **Transmissions:** All	**Ignition Coil No. 3 Primary/Secondary Circuit Malfunction Conditions:** Engine started, and the PCM did not detect a change in the IGF signal on the Ignition Coil No. 3 IGF circuit. This engine uses a Direct Ignition (DI) system where one coil is used to fire one cylinder. The coil high-energy secondary wire is connected to one spark plug. If P0351 to P0356 are all set, check for an open/shorted IGF circuit. **Possible Causes:** • IGT3 circuit is open or shorted to ground • Ignition Coil No. 3 is damaged or it has failed • Problem present in the Ignition System • PCM has failed
DTC: P0354 **1T CCM, MIL: Yes** **1996, 1997, 1998, 1999, 2000, 2001, 2002, 2003, 2004, 2005, 2006** **Models:** ES300, RX300, ES330, RX330, LX450, LX470, GS400, GS430, GX470, LS400, LS430, SC400, SC430 **Engines:** 3.0L VIN F, 3.3L VIN A, 4.0L VIN H, 4.3L VIN N, L, 4.5L VIN J, 4.7L VIN T **Transmissions:** All	**Ignition Coil No. 4 Primary/Secondary Circuit Malfunction Conditions:** Engine started, and the PCM did not detect a change in the IGF signal on the Ignition Coil No. 4 IGF circuit. This engine uses a Direct Ignition (DI) system where one coil is used to fire one cylinder. The coil high-energy secondary wire is connected to one spark plug. If P0351 to P0356 are all set, check for an open/shorted IGF circuit. **Possible Causes:** • IGT4 circuit is open or shorted to ground • Ignition Coil No. 4 is damaged or it has failed • Problem present in the Ignition System • PCM has failed

DTC	Trouble Code Title, Conditions & Possible Causes
DTC: P0355 **1T CCM, MIL: Yes** **1996, 1997, 1998, 1999, 2000,** **2001, 2002, 2003, 2004, 2005,** **2006** **Models:** ES300, RX300, ES330, RX330, LX450, LX470, GS400, GS430, GX470, LS400, LS430, SC400, SC430 **Engines:** 3.0L VIN F, 3.3L VIN A, 4.0L VIN H, 4.3L VIN N, L, 4.5L VIN J, 4.7L VIN T **Transmissions:** All	**Ignition Coil No. 5 Primary/Secondary Circuit Malfunction Conditions:** Engine started, and the PCM did not detect a change in the IGF signal on the Ignition Coil No. 5 IGF circuit. This engine uses a Direct Ignition (DI) system where one coil is used to fire one cylinder. The coil high-energy secondary wire is connected to one spark plug. If P0351 to P0356 are all set, check for an open/shorted IGF circuit. **Possible Causes:** • IGT5 circuit is open or shorted to ground • Ignition Coil No. 5 is damaged or it has failed • Problem present in the Ignition System • PCM has failed
DTC: P0356 **1T CCM, MIL: Yes** **1996, 1997, 1998, 1999, 2000,** **2001, 2002, 2003, 2004, 2005,** **2006** **Models:** ES300, RX300, ES330, RX330, LX450, LX470, GS400, GS430, GX470, LS400, LS430, SC400, SC430 **Engines:** 3.0L VIN F, 3.3L VIN A, 4.0L VIN H, 4.3L VIN N, L, 4.5L VIN J, 4.7L VIN T **Transmissions:** All	**Ignition Coil No. 6 Primary/Secondary Circuit Malfunction Conditions:** Engine started, and the PCM did not detect a change in the IGF signal on the Ignition Coil No. 6 IGF circuit. This engine uses a Direct Ignition (DI) system where one coil is used to fire one cylinder. The coil high-energy secondary wire is connected to one spark plug. If P0351 to P0356 are all set, check for an open/shorted IGF circuit. **Possible Causes:** • IGT6 circuit is open or shorted to ground • Ignition Coil No. 6 is damaged or it has failed • Problem present in the Ignition System • PCM has failed
DTC: P0357 **1T CCM, MIL: Yes** **1996, 1997, 1998, 1999, 2000,** **2001, 2002, 2003, 2004, 2005,** **2006** **Models:** LX450, LX470, GS400, GS430, GX470, LS400, LS430, SC400, SC430 **Engines:** 4.0L VIN H, 4.3L VIN N, L, 4.5L VIN J, 4.7L VIN T **Transmissions:** All	**Ignition Coil No. 7 Primary/Secondary Circuit Malfunction Conditions:** Engine started, and the PCM did not detect a change in the IGF signal on the Ignition Coil No. 7 IGF circuit. This engine uses a Direct Ignition (DI) system where one coil is used to fire one cylinder. The coil high-energy secondary wire is connected to one spark plug. If P0351 to P0358 are all set, check for an open/shorted IGF circuit. **Possible Causes:** • IGT7 circuit is open or shorted to ground • Ignition Coil No. 7 is damaged or it has failed • Problem present in the Ignition System • PCM has failed
DTC: P0358 **1T CCM, MIL: Yes** **1996, 1997, 1998, 1999, 2000,** **2001, 2002, 2003, 2004, 2005,** **2006** **Models:** LX450, LX470, GS400, GS430, GX470, LS400, LS430, SC400, SC430 **Engines:** 4.0L VIN H, 4.3L VIN N, L, 4.5L VIN J, 4.7L VIN T **Transmissions:** All	**Ignition Coil No. 8 Primary/Secondary Circuit Malfunction Conditions:** Engine started, and the PCM did not detect a change in the IGF signal on the Ignition Coil No. 8 IGF circuit. This engine uses a Direct Ignition (DI) system where one coil is used to fire one cylinder. The coil high-energy secondary wire is connected to one spark plug. If P0351 to P0358 are all set, check for an open/shorted IGF circuit. **Possible Causes:** • IGT8 circuit is open or shorted to ground • Ignition Coil No. 8 is damaged or it has failed • Problem present in the Ignition System • PCM has failed
DTC: P0385 **2T CCM, MIL: Yes** **1996, 1997, 1998, 1999, 2000,** **2001, 2002, 2003, 2004, 2005,** **2006** **Models:** GS300, IS250, IS300, IS350, LS 300, SC300 **Engines:** 2.5L VIN K, 3.0L VIN D, 3.5L VIN E **Transmissions:** All	**Crankshaft Position Sensor 'B' Circuit Malfunction Conditions:** Engine started, engine speed more than 600 rpm, it did not detect any Crankshaft Position (CKP) Sensor 'B' signals during the test. **Possible Causes:** • CKP Sensor 'B' signal circuit is open, shorted to ground or shorted to system power • CKP Sensor 'B' signal ground circuit is open • CKP Sensor 'B' has failed, or the PCM has failed

DTC	Trouble Code Title, Conditions & Possible Causes
DTC: P0401 **2T EGR, MIL: Yes** **1995** **Models:** All **Engines:** All **Transmissions:** All	**EGR System Insufficient Flow Detected Conditions:** Engine started, vehicle driven at a speed of over 50 mph with the engine running in closed loop at a steady throttle for 3-5 minutes, and the PCM detected the EGR gas temperature signal was not 106-140°F higher than the ambient air temperature signal in the test. **Possible Causes:** • EGR gas temperature sensor circuit open or shorted to VREF • EGR gas temperature sensor is damaged or has failed • EGR valve is stuck in closed position • EGR valve vacuum hose is disconnected or leaking • PCM has failed
DTC: P0401 **2T EGR, MIL: Yes** **1996, 1997, 1998, 1999, 2000, 2001, 2002** **Models:** All **Engines:** All **Transmissions:** All	**EGR System Insufficient Flow Detected Conditions:** Engine started, vehicle driven at a speed of over 50 mph with the engine running in closed loop at a steady throttle for 3-5 minutes, and the PCM detected the EGR sensor signal was not 106-140°F higher than the ambient air temperature in the EGR Monitor test. **Possible Causes:** • EGR gas temperature sensor circuit open or shorted to VREF • EGR gas temperature sensor is damaged or has failed • EGR valve is stuck in closed position • EGR valve vacuum hose is disconnected or leaking
DTC: P0402 **2T EGR, MIL: Yes** **1995** **Models:** All **Engines:** All **Transmissions:** All	**Excessive EGR Flow Detected Conditions:** Engine started cold (ECT sensor signal less than 86°F), engine running without load at less than 4000 rpm, and the PCM detected the EGR sensor indicated a high value during the EGR Cutoff Test, or it detected the EGR valve was open during all driving conditions. **Possible Causes:** • EGR gas temperature sensor circuit open or shorted to ground • EGR gas temperature sensor is damaged or has failed • EGR Vacuum Switching Valve (VSV) is damaged or has failed • EGR valve stuck partially or EGR valve stuck fully open • EGR VSV control circuit is open or shorted to system power • PCM has failed
DTC: P0402 **2T EGR, MIL: Yes** **1996, 1997, 1998, 1999, 2000, 2001, 2002** **Models:** All **Engines:** All **Transmissions:** All	**Excessive EGR Flow Detected Conditions:** Engine started cold (ECT sensor signal less than 86°F), engine running without load at less than 4000 rpm, and the PCM detected the EGR sensor indicated a high value during the EGR Cutoff test; or it detected the EGR valve was open during all driving conditions. **Possible Causes:** • EGR gas temperature sensor circuit open or shorted to ground • EGR gas temperature sensor is damaged or has failed • EGR Vacuum Switching Valve (VSV) is damaged or has failed • EGR valve stuck partially or EGR valve stuck fully open • EGR VSV control circuit is open or shorted to system power • PCM has failed
DTC: P0420 **2T CAT, MIL: Yes** **1995** **Models:** All **Engines:** All **Transmissions:** All	**Catalyst Efficiency Below Normal (Bank 1) Conditions:** DTC P0100, P0101, P0105, P0110, P0115, P0120, P0121, P0335, P0340 and P0500 not set, vehicle driven to a speed of 45-60 mph at 2500-3000 rpm in closed loop for 3-5 minutes, and the PCM detected the voltage amplitudes of the rear and front HO2S were too similar. **Possible Causes:** • Air leaks at the exhaust manifold or in the exhaust pipes • Catalytic converter is damaged, contaminated or has failed • Front HO2S or rear HO2S is contaminated with fuel or moisture • Front HO2S or the rear HO2S is loose in its mounting hole • Front HO2S is older (aged) than the rear HO2S (HO2S is lazy)
DTC: P0420 **2T CAT, MIL: Yes** **1996, 1997, 1998, 1999, 2000, 2001, 2002** **Models:** All **Engines:** All **Transmissions:** All	**Catalyst Efficiency Below Normal (Bank 1) Conditions:** DTC P0100, P0101, P0105, P0110, P0115, P0120, P0121, P0335, P0340 and P0500 not set, engine started, vehicle driven to a speed of 45-60 mph at 2500-3000 rpm in closed loop for 3-5 minutes, and the PCM detected the voltage amplitudes of the rear HO2S-12 and front HO2S-11 were similar during the Catalyst Monitor test. **Possible Causes:** • Air leaks at the exhaust manifold or in the exhaust pipes • Catalytic converter is damaged, contaminated or has failed • Front HO2S or rear HO2S is contaminated with fuel or moisture • Front HO2S or the rear HO2S is loose in its mounting hole • Front HO2S is older (aged) than the rear HO2S (HO2S is lazy)

DTC	Trouble Code Title, Conditions & Possible Causes
DTC: P0420 **2T CAT, MIL: Yes** **2003, 2004, 2005, 2006** **Models:** All **Engines:** All **Transmissions:** All	**Catalyst Efficiency Below Normal (Bank 1) Conditions:** DTC P0100, P0101, P0102, P0103, P0110, P0112, P0113, P0115, P0116, P0117, P0118, P0120, P0121, P0122, P0123, P0335, P0340 and P0500 not set, engine started, vehicle driven to a speed of 45-60 mph at 2500-3000 rpm in closed loop for 3-5 minutes, and the PCM detected too much variation in the voltage amplitudes of the HO2S-12 signal (Bank 1). **Possible Causes:** • Catalytic converter is damaged, contaminated or has failed • Front A/FS or rear HO2S is contaminated with fuel or moisture • Front A/FS or the rear HO2S is loose in its mounting hole • Front A/FS is older (aged) than the rear HO2S (HO2S is lazy) • Gas leaks at the exhaust manifold or in the exhaust pipes
DTC: P0430 **2T CAT, MIL: Yes** **1996, 1997, 1998, 1999, 2000, 2001, 2002, 2003, 2004, 2005, 2006** **Models:** ES300, RX300, ES330, RX330, LX450, LX470, GS400, GS430, GX470, LS400, LS430, SC400, SC430 **Engines:** 3.0L VIN F, 3.3L VIN A, 4.0L VIN H, 4.3L VIN N, L, 4.5L VIN J, 4.7L VIN T **Transmissions:** All	**Catalyst Efficiency Below Normal (Bank 2) Conditions:** DTC P0100, P0101, P0102, P0103, P0110, P0112, P0113, P0115, P0116, P0117, P0118, P0120, P0121, P0122, P0123, P0335, P0340 and P0500 not set, engine started, vehicle driven to a speed of 45-60 mph at 2500-3000 rpm in closed loop for 3-5 minutes, and the PCM detected too much variation in the voltage amplitudes of the Rear HO2S-12 for Bank 2 during the Catalyst Monitor test. **Possible Causes:** • Catalytic converter is damaged, contaminated or has failed • Front A/FS or rear HO2S is contaminated with fuel or moisture • Front A/FS or the rear HO2S is loose in its mounting hole • Front A/FS is older (aged) than the rear HO2S (HO2S is lazy) • Gas leaks at the exhaust manifold or in the exhaust pipes
DTC: P0440 **2T EVAP, MIL: Yes** **1996, 1997, 1998, 1999, 2000, 2001, 2002** **Models:** All **Engines:** All **Transmissions:** All	**EVAP Control System Large Leak (0.080") Detected Conditions:** Engine started cold (ECT sensor signal less 86°F), engine runtime over 20 minutes in closed loop, vehicle driven to a speed of 55-60 mph, and the PCM detected the fuel tank pressure indicated the same value as the atmospheric pressure in the EVAP Monitor test. **Possible Causes:** • Charcoal canister is clogged, loaded with fuel or with moisture • Fuel filler cap missing, loose (not tightened) or the wrong part • Fuel tank over-fill check valve cracked or damaged • Fuel tank seal leaking, fuel tank cracked or damaged/leaking • Fuel vapor hoses/tubes blocked or restricted, or fuel vapor control valve tube or fuel vapor vent valve assembly blocked • Vacuum hose or tubing cracked, damaged or disconnected • Vapor Pressure sensor incorrectly installed • PCM has failed
DTC: P0441 **2T EVAP, MIL: Yes** **1995** **Models:** All **Engines:** All **Transmissions:** All	**EVAP System Incorrect Purge Flow Detected Conditions:** Cold engine startup requirement met (ECT sensor signal less 86°F), engine running at cruise speed for over 3 minutes, VSS input from 55-60 mph, and the PCM detected the canister pressure did not decrease during purge or it remained too low during purge cutoff conditions. **Possible Causes:** • Charcoal canister is clogged, loaded with fuel or with moisture • Fuel tank over-fill check valve cracked or damaged • Fuel tank seal leaking, fuel tank cracked or damaged/leaking • Fuel vapor hoses/tubes blocked or restricted, or fuel vapor control valve tube or fuel vapor vent valve assembly blocked • Vacuum hose or tubing cracked, damaged or disconnected • VSV for the canister purge solenoid is open or shorted to ground, or the purge solenoid is damaged or has failed (closed) • VSV for vapor pressure sensor circuit is open, shorted to ground, or sensor has failed • PCM has failed

DTC	Trouble Code Title, Conditions & Possible Causes
DTC: P0441 **2T EVAP, MIL: Yes** **1996, 1997, 1998, 1999, 2000, 2001, 2002, 2003, 2004, 2005, 2006** **Models:** All **Engines:** All **Transmissions:** All	**EVAP System Incorrect Purge Flow Detected Conditions:** ECT sensor less than 86°F at startup, vehicle driven at 55-60 mph for 2-3 minutes, and the PCM detected the EVAP canister pressure did not decrease during purge conditions, or it remained too low during purge cutoff conditions. The vapor pressure sensor, VSV for the canister closed valve (CCV) and the VSV for the vapor-switching valve are used to detect EVAP system faults. The PCM closes the CCV and opens the VSV for vapor switching valve to cause an increase in vacuum in the EVAP system. Once the vacuum reaches a certain point, the PCM closes the VSV to test system operation. **Possible Causes:** • Charcoal canister is clogged, loaded with fuel or with moisture • Fuel tank over-fill check valve cracked or damaged • Fuel tank seal leaking, fuel tank cracked or damaged/leaking • Fuel vapor hoses/tubes blocked or restricted, or fuel vapor control valve tube or fuel vapor vent valve assembly blocked • Vacuum hose or tubing cracked, damaged or disconnected • Vapor pressure sensor is damaged or has failed • VSV circuit for the canister purge, VSV for the CCV or the VSV for the pressure switching valve is open or shorted to ground • VSV for the vapor pressure sensor circuit is open or shorted to ground, or the vapor pressure sensor is damaged or has failed
DTC: P0442 **2T EVAP, MIL: Yes** **1998, 1999, 2000, 2001, 2002, 2003, 2004, 2005, 2006** **Models:** All **Engines:** All **Transmissions:** All	**EVAP System Small Leak (0.040") Detected Conditions:** Engine started; IAT sensor signal from 39-86°F, fuel tank level from 25-75% for 10 seconds, and the PCM detected the EVAP system was unable to hold a specified vacuum level for a set period of time. After the system is purged, the PCM shuts off the VSV for the purge valve to seal the vacuum in the system, and then monitors the increase in pressure in the system. The pressure should increases slowly. If it increases at too fast a rate, then this code is set. **Possible Causes:** • Canister Purge valve is damaged, leaking or has failed • Charcoal canister is loaded with fuel or moisture • Fuel filler cap loose, cross-threaded, incorrect part or damaged • Fuel tank is cracked (leaking), or a leak exists in the 'O' ring • Fuel tank pressure sensor is damaged or has failed • Fuel vapor line(s), fuel pipes or hoses damaged or leaking • PCM has failed
DTC: P0442 **2T EVAP, MIL: Yes** **1996, 1997, 1998, 1999, 2000, 2001, 2002, 2003, 2004, 2005, 2006** **Models:** All **Engines:** All **Transmissions:** All	**EVAP Vent Control Solenoid Circuit Malfunction Conditions:** Engine started, engine running at cruise speed under light load conditions, VSV for vapor pressure switching valve "on", and the PCM detected a lack of vacuum continuity between the vapor pressure sensor, the charcoal canister and the fuel tank; or with the VSV for the vapor pressure switching valve "off", it detected the pressure in the fuel tank remained at atmospheric pressure; or with the VSV for the CCV "on", it detected the pressure in the charcoal canister and the fuel tank remained near atmospheric pressure. The PCM closes the VSV for the vapor-switching valve, and this action blocks any air from entering the fuel tank side of the system. The pressure rise on the fuel tank under these conditions is minimal. If there was no change in pressure, the PCM determines the VSV for the vapor-switching valve did not close, and it will set this code. **Possible Causes:** • Charcoal canister is clogged, loaded with fuel or with moisture • Fuel tank over-fill check valve cracked or damaged • Fuel tank seal leaking, fuel tank cracked, damaged or leaking • Fuel vapor hoses/tubes blocked or restricted, or fuel vapor control valve tube or fuel vapor vent valve assembly blocked • Vacuum hose or tubing cracked, damaged or disconnected • Vapor pressure sensor is damaged or has failed • VSV circuit for the canister purge, VSV for the CCV or the VSV for the pressure switching valve is open or shorted to ground • VSV for the vapor pressure sensor circuit is open or shorted to ground, or the vapor pressure sensor is damaged or has failed
DTC: P0450 **2T CCM, MIL: Yes** **1996, 1997, 1998, 1999, 2000, 2001, 2002, 2003, 2004, 2005, 2006** **Models:** All **Engines:** All **Transmissions:** All	**EVAP Vapor Pressure Sensor Circuit Malfunction Conditions:** Engine started, engine runtime less than 10 seconds since startup, and the PCM detected the Vapor Pressure sensor value was less than -3.5 kPa (-1.0 in. Hg), or the Vapor Pressure sensor was more than or equal to 1.5 kPa (0.4 in. Hg) during testing. The PCM uses the Vapor Pressure Sensor, VSV for the Canister Closed valve and VSV for the Pressure Switching valve to find faults in this system. **Possible Causes:** • Vapor pressure sensor signal circuit open or shorted to ground • Vapor pressure sensor ground circuit is open • Vapor pressure sensor power circuit is open • Vapor pressure sensor is damaged or has failed • PCM has failed

DTC	Trouble Code Title, Conditions & Possible Causes
DTC: P0451 **2T CCM, MIL: Yes** **1996, 1997, 1998, 1999, 2000, 2001, 2002, 2003, 2004, 2005, 2006** Models: All Engines: All Transmissions: All	**EVAP Vapor Pressure Sensor Range/Performance Conditions:** Engine started, engine at idle speed with the VSS indicating 0 mph, VSV for Vapor Switching valve "off", and the PCM detected too much change in the pressure sensor value, or the pressure sensor value equaled the opening value of the charcoal canister. The PCM uses the Vapor Pressure Sensor, VSV for the Canister Closed valve and VSV for Pressure Switching valve to find faults in the system. **Possible Causes:** • Vapor pressure sensor vacuum hoses loose or damaged • Vapor pressure sensor is damaged or has failed • PCM has failed
DTC: P0452 **2T CCM, MIL: Yes** **1996, 1997, 1998, 1999, 2000, 2001, 2002, 2003, 2004, 2005, 2006** Models: LX450, LX470, GS400, GS430, GX470, LS400, LS430, SC400, SC430 Engines: 4.0L VIN H, 4.3L VIN N, L, 4.5L VIN J, 4.7L VIN T Transmissions: All	**EVAP Vapor Pressure Sensor Circuit Low Input Conditions:** Engine started; VSV for vapor pressure switching valve "off"; and the PCM detected an unexpected low voltage condition on the vapor pressure sensor circuit during the CCM test. **Possible Causes:** • Vapor pressure sensor connector is damaged or open • Vapor pressure sensor circuit is open • Vapor pressure sensor is damaged or has failed • PCM has failed
DTC: P0453 **2T CCM, MIL: Yes** **1996, 1997, 1998, 1999, 2000, 2001, 2002, 2003, 2004, 2005, 2006** Models: LX450, LX470, GS400, GS430, GX470, LS400, LS430, SC400, SC430 Engines: 4.0L VIN H, 4.3L VIN N, L, 4.5L VIN J, 4.7L VIN T Transmissions: All	**EVAP Vapor Pressure Sensor Circuit High Input Conditions:** Engine started; VSV for vapor pressure switching valve "off"; and the PCM detected an unexpected high voltage condition on the vapor pressure sensor circuit during the CCM test. **Possible Causes:** • Vapor pressure sensor connector is damaged or shorted • Vapor pressure sensor circuit is shorted to VREF • Vapor pressure sensor is damaged or has failed • PCM has failed
DTC: P0456 **2T EVAP, MIL: Yes** **1996, 1997, 1998, 1999, 2000, 2001, 2002, 2003, 2004, 2005, 2006** Models: ES330, RX330, LX450, LX470, GS400, GS430, GX470, LS400, LS430, SC400, SC430 Engines: 3.3L VIN A, 4.0L VIN H, 4.3L VIN N, L, 4.5L VIN J, 4.7L VIN T Transmissions: All	**EVAP System Very Small Leak (0.020") Detected Conditions:** Engine started; IAT sensor signal from 39-86°F, fuel tank level from 25-75% for 10 seconds, and the PCM detected the EVAP system was unable to hold a specified vacuum level for a set period of time. After the system is purged, the PCM shuts off the VSV for the purge valve to seal the vacuum in the system, and then monitors the increase in pressure in the system. The pressure should increase slowly. If it increases at too fast a rate, this code is set. **Possible Causes:** • Canister Purge valve is damaged, leaking or has failed • Charcoal canister is loaded with fuel or moisture • Fuel filler cap loose, cross-threaded, incorrect part or damaged • Fuel tank is cracked (leaking), or a leak exists in the 'O' ring • Fuel tank pressure sensor is damaged or has failed • Fuel tank overfill check valve is cracked or is damaged • Fuel vapor line(s), fuel pipes or hoses damaged or leaking • PCM has failed
DTC: P0500 **2T EVAP, MIL: Yes** **1995** Models: All Engines: All Transmissions: All	**Vehicle Speed Sensor Circuit Malfunction Conditions:** Engine started, ECT sensor more than 158°F, vehicle driven with the engine speed from 1500-5500 rpm, and the PCM did not receive any VSS signals from the Combination Meter during the CCM test. **Possible Causes:** • VSS signal circuit is open between the meter and the PCM • VSS signal circuit shorted to ground between meter and PCM • VSS signal circuit shorted to VREF or system power • Combination Meter is damaged or has failed • PCM has failed
DTC: P0500 **1T CCM, MIL: Yes** **1996** Models: All Engines: All Transmissions: A/T	**Vehicle Speed Sensor Circuit Malfunction Conditions:** Engine started, ECT sensor more than 158°F, P/N switch indicating "off", vehicle driven at an engine speed of 1500-3500 rpm, and the PCM did not receive any VSS signals from the Combination Meter. **Possible Causes:** • VSS signal circuit is open between the meter and the PCM • VSS signal circuit shorted to ground between meter and PCM • VSS signal circuit shorted to VREF or system power • Combination Meter is damaged or has failed • PCM has failed

DTC	Trouble Code Title, Conditions & Possible Causes
DTC: P0500 **1T CCM, MIL: Yes** **1997, 1998, 1999, 2000, 2001, 2002** **Models:** All **Engines:** All **Transmissions:** A/T	**Vehicle Speed Sensor Circuit Malfunction Conditions:** Engine started, engine speed more than 2350 rpm, P/N switch indicating "off" for 1 second, throttle angle equal to or less than 13 degrees (°), and the PCM did not detect any VSS signals during the test period (Scan Tool PID ECT = Electronic Controlled Transaxle). **Possible Causes:** • VSS signal circuit is open, shorted to ground or to power (B+) • Combination Meter is damaged or has failed • PCM has failed • TSB EG004-02 (2/02) contains information related to this code
DTC: P0500 **1T CCM, MIL: Yes** **1996, 1997, 1998, 1999, 2000, 2001, 2002** **Models:** All **Engines:** All **Transmissions:** M/T	**Vehicle Speed Sensor Circuit Malfunction Conditions:** Engine started, ECT sensor signal over 158°F, vehicle driven with the engine speed from 1500-5500 rpm, and the PCM did not receive any VSS signals from the Combination Meter. **Possible Causes:** • VSS signal circuit is open between the meter and the PCM • VSS signal circuit shorted to ground or to VREF or power (B+) • Combination Meter is damaged or has failed • PCM has failed
DTC: P0500 **2T CCM, MIL: Yes** **2003, 2004, 2005, 2006** **Models:** All **Engines:** All **Transmissions:** A/T	**Vehicle Speed Sensor Circuit Malfunction Conditions:** Engine started, vehicle driven with the engine speed from 1500-5500 rpm and back to idle several times, and the PCM did not receive any VSS signals during the test. The VSS (No.1) assembly outputs a 4-pulse signal for every revolution of the rotor shaft, which is generated by the transmission output shaft via the driven gear. **Possible Causes:** • VSS signal circuit is open between the meter and the PCM • VSS signal circuit shorted to ground between meter and PCM • VSS No. 1 is damaged or has failed • Combination Meter is damaged or has failed • PCM has failed
DTC: P0500 **2T CCM, MIL: Yes** **2003, 2004, 2005, 2006** **Models:** All **Engines:** All **Transmissions:** M/T	**Vehicle Speed Sensor Circuit Malfunction Conditions:** No TP sensor codes set, engine runtime 2 seconds with the ECT sensor more than 132°F and IAT sensor more than 50°F, P/N switch indicating 'P' or 'N', TP angle less than 13° with the engine speed less than 2350 rpm; or TP angle less than 21° with the engine speed less than 2680 rpm; or TP angle less than 30° with the engine speed less than 2835 rpm; or TP angle less than 30° with the engine speed less than 3250 rpm; and the PCM detected the engine speed was equal or more than the VSS signal speed for 500 ms during testing. **Possible Causes:** • ABS speed sensor connector is damaged or open • ABS speed sensor circuit is open or shorted to ground • ABS speed sensor is damaged or has failed • Combination Meter is damaged or has failed • PCM or the ABS controller has failed
DTC: P0503 **2T CCM, MIL: Yes** **1996, 1997, 1998, 1999, 2000, 2001, 2002, 2003, 2004, 2005, 2006** **Models:** LX450, LX470, GS400, GS430, GX470, LS400, LS430, SC400, SC430 **Engines:** 4.0L VIN H, 4.3L VIN N, L, 4.5L VIN J, 4.7L VIN T **Transmissions:** All	**Vehicle Speed Sensor 'A' Signal Erratic Or Intermittent Conditions:** Engine started, vehicle driven through several transmission shifts and braking events; and the PCM detected an interruption in the VSS signal from the Combination Meter, or the signal was too high. The speed sensor for the ABS detects the wheel speeds and sends the signals to the ABS ECU. The ECU converts these signals into a 4-pulse signal and outputs this signal to the Combination Meter. **Possible Causes:** • ABS ECU is damaged or has failed • Combination Meter is damaged or has failed • VSS signal circuit is open between the meter and the PCM • VSS signal circuit shorted to ground between meter and PCM
DTC: P0504 **2T CCM, MIL: Yes** **1996, 1997, 1998, 1999, 2000, 2001, 2002, 2003, 2004, 2005, 2006** **Models:** LX450, LX470, GS400, GS430, GX470, LS400, LS430, SC400, SC430 **Engines:** 4.0L VIN H, 4.3L VIN N, L, 4.5L VIN J, 4.7L VIN T **Transmissions:** All	**Brake Switch 'A' To 'B' Correlation Malfunction Conditions:** Key on or engine running; brake pedal released, and the PCM detected the STP signal indicated "off" while the ST1 signal also indicated "off". The stoplight switch signal is used to prevent the engine from stalling when the brakes are applied suddenly. The stoplight switch uses a duplex system (STP and ST1 signals). **Possible Causes:** • Stoplight switch signal circuit is shorted to power • Stoplight switch assembly is damaged or shorted • PCM has failed

DTC	Trouble Code Title, Conditions & Possible Causes
DTC: P0505 **2T CCM, MIL: Yes** **1995** **Models:** All **Engines:** All **Transmissions:** All	**Idle Control System Malfunction Conditions:** Engine started, engine running at idle speed in closed loop, and the PCM detected the Actual Idle Speed was more than 100-200 rpm above or below the Target Idle Speed. **Note: RSO is the acronym for Rotary Solenoid (IAC) Open Coil and RSC is the acronym for the Rotary Solenoid (IAC) Close Coil.** **Possible Causes:** • RSC or RSO control circuit is open, shorted to ground or power • RSO or RSO power circuit is open (test power from EFI relay) • IAC valve is contaminated, damaged or has failed
DTC: P0505 **2T CCM, MIL: Yes** **1996, 1997, 1998, 1999, 2000, 2001, 2002** **Models:** All **Engines:** All **Transmissions:** All	**Idle Control System Malfunction Conditions:** Engine started, engine running at idle speed in closed loop, and the PCM detected the Actual Idle Speed was more than 100-200 rpm above or below the Target Idle Speed. **Note: RSO is the acronym for Rotary Solenoid (IAC) Open Coil and RSC is the acronym for the Rotary Solenoid (IAC) Close Coil.** **Possible Causes:** • RSC/RSO connector is damaged, open or shorted • RSC or RSO control circuit is open, shorted to ground or power • RSO or RSO power circuit is open (test power from EFI relay) • IAC valve is contaminated
DTC: P0505 **2T CCM, MIL: Yes** **2003, 2004, 2005, 2006** **Models:** All **Engines:** All **Transmissions:** All	**Idle Control System Malfunction Conditions:** Engine started, engine running at idle speed n closed loop, and the PCM detected the Actual Idle Speed was more than 100-200 rpm above or below the Target Idle Speed. A Rotary solenoid type of ISC valve is located in front of the air intake chamber and intake air bypassing the throttle valve is directed to the Intake Air Control (IAC) valve via a passage. The PCM controls the idle speed by regulating the amount of intake air volume that bypasses the throttle valve. **Possible Causes:** • Air Induction system leaks (check for intake manifold leaks) • Air leaks in the PCV system (at the valve or its related hoses) • Throttle body assembly is damaged or has failed • PCM has failed
DTC: P0510 **2T CCM, MIL: Yes** **1995** **Models:** All **Engines:** All **Transmissions:** All	**Closed Throttle Position Switch Circuit Malfunction Conditions:** Engine started, and the PCM detected throttle switch input did not change from Off to On during a normal driving period. Refer to Circuit Test or code repair chart to test code. **Possible Causes:** • Closed throttle position switch signal circuit is open or grounded • Closed throttle position switch signal circuit is shorted to power • Closed throttle position switch or TP sensor damaged or failed • PCM has failed
DTC: P0510 **2T CCM, MIL: Yes** **1996, 1997** **Models:** All **Engines:** All **Transmissions:** All	**Closed Throttle Position Switch Circuit Malfunction Conditions:** Engine started, vehicle driven with VSS signals received, and the PCM did not detect any change in the Closed Throttle Position switch status (from off to on, or from on to off). **Possible Causes:** • Closed throttle position switch signal circuit is open or grounded • Closed throttle position switch signal circuit is shorted to power • Closed throttle position switch or TP sensor damaged or failed • PCM has failed
DTC: P0511 **2T CCM, MIL: Yes** **1996, 1997, 1998, 1999, 2000, 2001, 2002, 2003, 2004, 2005, 2006** **Models:** ES300, RX300, ES330, RX330, **Engines:** 3.0L VIN F, 3.3L VIN A **Transmissions:** All	**Idle Control Valve Circuit Malfunction Conditions:** Engine running at idle speed n closed loop, and the PCM detected an unexpected voltage condition on the IAC valve circuit. A Rotary solenoid valve, in front of the air intake chamber, allows intake air to bypass the throttle valve and be directed to the Intake Air Control (IAC) valve through a passage. This configuration allows the PCM to control the engine idle speed by regulating the amount of intake air volume that bypasses the throttle valve. **Possible Causes:** • IAC valve connector is damaged or loose • IAC valve control circuit is open, shorted to ground or power • IAC valve is damaged or has failed • PCM has failed

DTC	Trouble Code Title, Conditions & Possible Causes
DTC: P0513 **2T PCM, MIL: Yes** **1996,1997, 1998, 1999, 2000, 2001, 2002, 2003, 2004, 2005, 2006** **Models:** ES300, RX300, ES330, RX330, **Engines:** 3.0L VIN F, 3.3L VIN A **Transmissions:** All	**Unmatched Key Code Detected Conditions:** Key inserted, and the PCM detected that a key with an unregistered key code had been inserted into the ignition lock. **Possible Causes:** • An unmatched ignition key was inserted
DTC: P0560 **2T CCM, MIL: Yes** **2003, 2004, 2005, 2006** **Models:** All **Engines:** All **Transmissions:** All	**System Voltage (Backup Power Circuit) Malfunction Conditions:** Key on or engine running; and the PCM detected an unexpected low voltage condition on the Backup Power Circuit during the test. **Possible Causes:** • Battery backup circuit is open between battery and the PCM • PCM has failed
DTC: P0571 **2T CCM, MIL: Yes** **1996, 1997, 1998, 1999, 2000, 2001, 2002, 2003, 2004, 2005, 2006** **Models:** GS400, GS430, GX470, LS400, LS430, SC400, SC430 **Engines:** 4.0L VIN H, 4.3L VIN N, L, 4.7L VIN T **Transmissions:** All	**Brake Switch 'A' Circuit Malfunction Conditions:** Engine started, vehicle driven to cruise speed and then back to idle speed several times, and the PCM did not detect any change in the Brake Switch 'A' circuit status. The signal from this switch is used to determine when the brakes have been applied, and to determine the Fuel Cutoff engine speed during some types of braking operations. **Possible Causes:** • Stoplight switch signal circuit is shorted to power • Stoplight switch assembly is damaged or shorted • PCM has failed
DTC: P0604 **1T PCM, MIL: Yes** **1996, 1997, 1998, 1999, 2000, 2001, 2002, 2003, 2004, 2005, 2006** **Models:** LX450, LX470, GS400, GS430, GX470, LS400, LS430, SC400, SC430 **Engines:** 4.0L VIN H, 4.3L VIN N, L, 4.5L VIN J, 4.7L VIN T **Transmissions:** All	**PCM Internal Control Module Random Access Memory Processing Error Conditions:** Key on, and the PCM detected a processing error in the Internal Control Module Random Access Memory (RAM) function. **Possible Causes:** • Clear the codes and retest for this code. If the same code resets, substitute a known good control module and retest. If the trouble code is gone, the original PCM has failed. • TSB TC002-03 (6/03) contains information related to this code
DTC: P0606 **1T PCM, MIL: Yes** **1996, 1997, 1998, 1999, 2000, 2001, 2002, 2003, 2004, 2005, 2006** **Models:** LX450, LX470, GS400, GS430, GX470, LS400, LS430, SC400, SC430 **Engines:** 4.0L VIN H, 4.3L VIN N, L, 4.5L VIN J, 4.7L VIN T **Transmissions:** All	**ECM/PCM Processing Error Conditions:** Key on, and the PCM detected a processing error occurred. **Possible Causes:** • Clear the codes and retest for this code. If the same code resets, substitute a known good control module and retest. If the trouble code is gone, the original PCM has failed. • TSB TC002-03 (6/03) contains information related to this code
DTC: P0607 **1T PCM, MIL: Yes** **1996, 1997, 1998, 1999, 2000, 2001, 2002, 2003, 2004, 2005, 2006** **Models:** LX450, LX470, GS400, GS430, GX470, LS400, LS430, SC400, SC430 **Engines:** 4.0L VIN H, 4.3L VIN N, L, 4.5L VIN J, 4.7L VIN T **Transmissions:** All	**Control Module Performance Conditions:** Key on, and the PCM detected a performance problem occurred. **Possible Causes:** • Clear the codes and retest for this code. If the same code resets, substitute a known good control module and retest. If the trouble code is gone, the original PCM has failed. • TSB TC002-03 (6/03) contains information related to this code

DTC	Trouble Code Title, Conditions & Possible Causes
DTC: P0617 **1T CCM, MIL: Yes** **1996, 1997, 1998, 1999, 2000, 2001, 2002, 2003, 2004, 2005, 2006** **Models:** LX450, LX470, GS400, GS430, GX470, LS400, LS430, SC400, SC430 **Engines:** 4.0L VIN H, 4.3L VIN N, L, 4.5L VIN J, 4.7L VIN T **Transmissions:** All	**Starter Relay Circuit High Input Conditions:** Engine started, engine speed over 1000 rpm, system voltage over 10.5v, and the PCM detected the Starter Motor signal indicated high. **Possible Causes:** • Park/Neutral switch assembly is damaged or it has failed • Ignition switch is damaged or has failed • PCM has failed
DTC: P0657 **1T CCM, MIL: Yes** **1996, 1997, 1998, 1999, 2000, 2001, 2002, 2003, 2004, 2005, 2006** **Models:** LX450, LX470, GS400, GS430, GX470, LS400, LS430, SC400, SC430 **Engines:** 4.0L VIN H, 4.3L VIN N, L, 4.5L VIN J, 4.7L VIN T **Transmissions:** All	**PCM Actuator Supply Voltage Circuit Malfunction Conditions:** Key on or engine running; and the PCM detected an unexpected voltage condition on the Actuator Supply Voltage circuit. **Possible Causes:** • Actuator supply voltage circuit is open • Clear the codes and retest for this code. If the same code resets, substitute a known good control module and retest. If the trouble code is gone, the original PCM has failed. • TSB TC002-03 (6/03) contains information related to this code
DTC: P0705 **2T CCM, MIL: Yes** **1996,1997, 1998, 1999, 2000, 2001, 2002, 2003, 2004, 2005, 2006** **Models:** ES300, RX300, ES330, RX330, GS400, GS430, GX470, LS400, LS430, SC400, SC430 **Engines:** 3.0L VIN F, 3.3L VIN A, 4.0L VIN H, 4.3L VIN N, L **Transmissions:** A/T	**A/T Range Sensor Circuit (PRNDL) Malfunction Conditions:** Key on or engine running; and the PCM detected simultaneous "on" signals (N, 2, L or R) from the Transmission Range sensor circuit. The P/N switch indicates "on" whenever the shift lever is in the 'N' or 'P' position. When it is "on", the NSW circuit to the PCM is grounded to chassis ground through the starter motor relay, and reads 0.00v. When the shift lever is in 'R', 'D' or 'L' position, the switch is "off" and the NSW circuit reads 12.0v. When the shift lever is moved from the 'N' to the 'D' position, the PCM uses this signal to air/fuel ratio correction and idle speed control (estimated control) functions. **Possible Causes:** • Park/Neutral switch assembly is shorted • Park/Neutral switch assembly is damaged or has failed • PCM has failed.
DTC: P0710 **1T CCM, MIL: Yes** **1996, 1997, 1998, 1999, 2000, 2001, 2002** **Models:** All **Engines:** All **Transmissions:** A/T	**Transmission Fluid Temperature Sensor Circuit Malfunction Conditions:** Engine started, and the PCM detected the TFT sensor indicated less than -40°F, or after the engine runtime exceeded 15 minutes, it indicated more than 300°F during the CCM test. **Possible Causes:** • TFT sensor signal circuit is open, shorted to ground or shorted to system power • TFT sensor is damaged or has failed • PCM has failed
DTC: P0711 **1T CCM, MIL: Yes** **1996, 1997, 1998, 1999, 2000, 2001, 2002** **Models:** All **Engines:** All **Transmissions:** A/T	**Transmission Fluid Temperature Sensor Performance Conditions:** Engine started, engine runtime over 15 seconds, and the PCM detected the ambient air temperature and transmission fluid temperature varied by more than 40°F, or after an engine runtime of 20 minutes and 6.2 miles were traveled, it indicated less than 50°F. **Possible Causes:** • TFT sensor signal circuit or ground circuit has high resistance • TFT sensor has failed, or the PCM has failed
DTC: P0715 **1T CCM, MIL: Yes** **1996, 1997, 1998, 1999, 2000, 2001, 2002, 2003, 2004, 2005, 2006** **Models:** LX450, LX470, GS400, GS430, GX470, LS400, LS430, SC400, SC430 **Engines:** 4.0L VIN H, 4.3L VIN N, L, 4.5L VIN J, 4.7L VIN T **Transmissions:** A/T	**A/T Input Shaft/Direct Clutch Speed Sensor Circuit Malfunction Conditions:** Engine started, P/N switch indicating "off", vehicle driven in 1st, 2nd or 3rd gear or in Overdrive, the SS1, SS2 and SL (shift valves) and VSS all operating normally, and the PCM detected the ISS signal indicated 100 rpm or less during the CCM Rationality test. **Possible Causes:** • O/D direct clutch speed sensor circuit is open • O/D direct clutch speed sensor circuit is shorted to ground • O/D direct clutch speed sensor is damaged or has failed • PCM has failed

DTC	Trouble Code Title, Conditions & Possible Causes
DTC: P0717 **1T CCM, MIL: Yes** **1996, 1997, 1998, 1999, 2000, 2001, 2002, 2003, 2004, 2005, 2006** **Models:** ES300, RX300, ES330, RX330, **Engines:** 3.0L VIN F, 3.3L VIN A **Transmissions:** A/T	**A/T Input Shaft Speed Sensor 'A' Circuit No Signal Conditions:** No SS1, SS2, SL shift solenoid or VSS codes set, engine started, P/N switch indicating "off", gear position indicating 1st, 2nd, 3rd gear, or in O/D, no gear change occurring, and the PCM detected the Input Shaft Speed sensor was below 300 rpm, or over 1000 rpm for 4 seconds. **Possible Causes:** • Direct clutch connector is damaged (it may be open or shorted) • Direct clutch sensor signal circuit is open or shorted • Direct clutch speed sensor is damaged or has failed • PCM has failed
DTC: P0724 **1T CCM, MIL: Yes** **1996, 1997, 1998, 1999, 2000, 2001, 2002, 2003, 2004, 2005, 2006** **Models:** ES300, RX300, ES330, RX330, **Engines:** 3.0L VIN F, 3.3L VIN A **Transmissions:** A/T	**A/T Torque Converter Clutch Shift Solenoid Performance Conditions:** Engine started, vehicle driven to over 50 mph and then back to idle speed several times, and the PCM detected that TCC lockup did not occur, or that the TCC remained in lockup position in the "off" range. The PCM uses signals from the CKP, MAF and TP sensors to monitor engagement of the Lockup Clutch solenoid to find a fault. **Possible Causes:** • Lockup clutch solenoid is damaged or has failed • Shift solenoid SL is damaged or has failed (mechanical fault) • Shift solenoid SL is stuck in "on" or "off" position • Valve body is blocked or stuck
DTC: P0743 **2T CCM, MIL: Yes** **1996, 1997, 1998, 1999, 2000, 2001, 2002, 2003, 2004, 2005, 2006** **Models:** ES300, RX300, ES330, RX330, **Engines:** 3.0L VIN F, 3.3L VIN A **Transmissions:** A/T	**A/T Torque Converter Clutch Shift Solenoid Circuit Malfunction Conditions:** Vehicle driven to over 50 mph and then back to idle speed several times; and the PCM detected an unexpected voltage on the Shift Solenoid (SL) valve control circuit. The Shift Solenoid valve is turned On/Off" through commands from the PCM to control the hydraulic pressure acting on the lockup relay valve (and it controls the operation of the lockup clutch). **Possible Causes:** • Shift solenoid SL connector is damaged or loose • Shift solenoid SL control circuit is open or shorted to ground • Shift solenoid SL is damaged or has failed (electrical fault) • PCM has failed
DTC: P0750 **2T CCM, MIL: Yes** **1995** **Models:** All **Engines:** All **Transmissions:** All	**A/T Shift Solenoid 1 or 'A' Malfunction (Mechanical) Conditions:** Engine started, vehicle driven under normal driving conditions, and the PCM detected the Actual gear ratio and the Required gear ratio did not match during the CCM Rationality test. **Possible Causes:** • A/T component problems (i.e., in clutch, brake or gears) • SS1 or SSA is damaged, stuck "open" or stuck "closed" • Transmission valve body is clogged, dirty or stuck
DTC: P0750 **2T CCM, MIL: Yes** **1996, 1997, 1998, 1999, 2000, 2001, 2002** **Models:** All **Engines:** All **Transmissions:** A/T	**A/T Shift Solenoid 1 or 'A' Malfunction (Mechanical) Conditions:** Engine started, vehicle driven under normal driving conditions, and the PCM detected the Actual gear ratio and the Required gear ratio did not match during the CCM Rationality test. **Possible Causes:** • A/T component problems (i.e., in clutch, brake or gears) • SS1 or SSA is damaged, stuck "open" or stuck "closed" • Transmission valve body is clogged, dirty or stuck
DTC: P0751 **8T CCM, MIL: Yes** **1996, 1997, 1998, 1999, 2000, 2001, 2002, 2003, 2004, 2005, 2006** **Models:** ES300, RX300, ES330, RX330, **Engines:** 3.0L VIN F, 3.3L VIN A **Transmissions:** A/T	**A/T Shift Solenoid 'A' Signal Range/Performance Conditions:** Engine started, vehicle driven to a speed over 50 mph, and the PCM detected the Actual gear position did not match the Desired gear position during the CCM test period. The PCM uses inputs from the VSS and Direct Clutch speed sensor to determine the actual gear position (i.e., 1st, 2nd, 3rd or O/D gear). **Possible Causes:** • SSA control circuit is open or shorted to ground • SSA control circuit is shorted to system power (B+) • SSA is damaged or has failed (an electrical fault) • PCM has failed
DTC: P0753 **8T CCM, MIL: Yes** **1995** **Models:** All **Engines:** All **Transmissions:** All	**A/T Shift Solenoid 1 or 'A' Circuit Malfunction Conditions:** Engine started, engine running during normal driving conditions; and the PCM detected the Shift Solenoid 1 or 'A' (SS1/SSA) control circuit voltage was "high" with the solenoid "on", or the SS1 or SSA control circuit voltage was "low" with the SS1/SSA commanded "off". **Possible Causes:** • SS1 or SSA control circuit is open or shorted to ground • SS1or SSA control circuit is shorted to system power (B+) • SS1 or SSA is damaged or has failed (an electrical fault) • PCM has failed

DTC	Trouble Code Title, Conditions & Possible Causes
DTC: P0753 **8T CCM, MIL: Yes** **1996, 1997, 1998, 1999, 2000, 2001, 2002** **Models:** All **Engines:** All **Transmissions:** A/T	**A/T Shift Solenoid 1 or 'A' Circuit Malfunction Conditions:** Engine started, engine running during normal driving conditions; and the PCM detected the Shift Solenoid 1 or 'A' (SS1/SSA) control circuit voltage was "high" with the solenoid "on", or the SS1 or SSA control circuit voltage was "low" with the SS1/SSA commanded "off". **Possible Causes:** • SS1 or SSA control circuit is open, shorted to ground or shorted to system power (B+) • SS1 or SSA has failed, or the PCM has failed
DTC: P0753 **8T CCM, MIL: Yes** **2003, 2004, 2005, 2006** **Models:** All **Engines:** All **Transmissions:** A/T	**A/T Shift Solenoid 'A' Circuit Malfunction Conditions:** Engine started, engine running during normal driving conditions; and the PCM detected an unexpected voltage condition on the Shift Solenoid 'A' control circuit during the CCM test. **Possible Causes:** • SSA control circuit is open, shorted to ground or to power (B+) • SSA is damaged or has failed (an electrical fault) • PCM has failed
DTC: P0755 **2T CCM, MIL: Yes** **1995** **Models:** All **Engines:** All **Transmissions:** All	**A/T Shift Solenoid 'B' Malfunction (Mechanical) Conditions:** Engine started, vehicle driven under normal driving conditions, and the PCM detected the Actual gear ratio and the Required gear ratio did not match during the CCM Rationality test. **Possible Causes:** • A/T component problems (i.e., in clutch, brake or gears) • SSB is damaged, stuck "open" or stuck "closed" • Transmission valve body is clogged, dirty or stuck
DTC: P0755 **2T CCM, MIL: Yes** **1996, 1997, 1998, 1999, 2000, 2001, 2002** **Models:** All **Engines:** All **Transmissions:** A/T	**A/T Shift Solenoid 'B' Malfunction (Mechanical) Conditions:** Engine started, vehicle driven under normal driving conditions, and the PCM detected the Actual gear ratio and the Required gear ratio did not match during the CCM Rationality test. **Possible Causes:** • A/T component problems (i.e., in clutch, brake or gears) • SSB is damaged, stuck "open" or stuck "closed" • Transmission valve body is clogged, dirty or stuck
DTC: P0756 **8T CCM, MIL: Yes** **1996, 1997, 1998, 1999, 2000, 2001, 2002, 2003, 2004, 2005, 2006** **Models:** ES300, RX300, ES330, RX330, **Engines:** 3.0L VIN F, 3.3L VIN A **Transmissions:** A/T	**A/T Shift Solenoid 'B' Signal Range/Performance Conditions:** Vehicle driven to a speed over 50 mph, and the PCM detected the Actual gear position did not match the Desired gear position. The PCM uses inputs from the VSS and Direct Clutch speed sensor to determine the actual gear position (i.e., 1st, 2nd, 3rd or O/D gear). **Possible Causes:** • SSB control circuit is open, shorted to ground or to power (B+) • SSB is damaged or has failed (an electrical fault) • PCM has failed
DTC: P0758 **8T CCM, MIL: Yes** **1995** **Models:** All **Engines:** All **Transmissions:** All	**A/T Shift Solenoid 2 or 'B' Circuit Malfunction Conditions:** Engine started, engine running during normal driving conditions, and the PCM detected the Shift Solenoid 2 or 'B' (SS2/SSB) control circuit voltage was "high" with the solenoid "on", or the SS2 or SSB control circuit voltage was "low" with the SS2/SSB commanded "off". **Possible Causes:** • SS2 or SSB control circuit is open or shorted to ground • SS2or SSB control circuit is shorted to system power (B+) • SS2 or SSB is damaged or has failed (an electrical fault) • PCM has failed
DTC: P0758 **1T CCM, MIL: Yes** **1996, 1997, 1998, 1999, 2000, 2001, 2002** **Models:** All **Engines:** All **Transmissions:** A/T	**A/T Shift Solenoid 2 or 'B' Circuit Malfunction Conditions:** Engine started, engine running during normal driving conditions, and the PCM detected the Shift Solenoid 2 or 'B' (SS2/SSB) control circuit voltage was "high" with the solenoid "on", or the SS2 or SSB control circuit voltage was "low" with the SS2/SSB commanded "off". **Note: This problem must occur eight (8) times during one trip before this code is triggered.** **Possible Causes:** • SS2 or SSB control circuit is open or shorted to ground • SS2or SSB control circuit is shorted to system power (B+) • SS2 or SSB is damaged or has failed (an electrical fault) • PCM has failed

DTC	Trouble Code Title, Conditions & Possible Causes
DTC: P0758 **8T CCM, MIL: Yes** **1996,1997, 1998, 1999, 2000, 2001, 2002, 2003, 2004, 2005, 2006** **Models:** ES300, RX300, ES330, RX330 **Engines:** 3.0L VIN F, 3.3L VIN A **Transmissions:** A/T	**A/T Shift Solenoid 'B' Circuit Malfunction Conditions:** Engine started, engine running during normal driving conditions, and the PCM detected an unexpected voltage condition on the Shift Solenoid 'B' control circuit during the CCM test. **Possible Causes:** • SSB control circuit is open, shorted to ground or to power (B+) • SSB is damaged or has failed (an electrical fault) • PCM has failed
DTC: P0770 **2T CCM, MIL: Yes** **1995** **Models: All** **Engines: All** **Transmissions:** A/T	**A/T Shift Solenoid 'E' (SL) Malfunction (Mechanical) Conditions:** Engine started, vehicle driven at a speed of over 50 mph for 1-3 minutes, and the PCM detected the transmission lockup function did not occur in the "lockup" range, or the "lockup" function was "on" during periods when it should have been "off" during the CCM test. **Note: The A/T clutch or A/T brake clutch will slip with this code set.** **Possible Causes:** • A/T component problems (i.e., in clutch, brake or gears) • SL (shift lockup) is damaged, stuck "open" or stuck "closed" • Transmission valve body is clogged, dirty or stuck
DTC: P0770 **1T CCM, MIL: Yes** **1996, 1997, 1998, 1999, 2000, 2001, 2002** **Models: All** **Engines: All** **Transmissions:** A/T	**A/T Shift Solenoid 'E' (SL) Malfunction (Mechanical) Conditions:** Engine started, vehicle driven at a speed of over 50 mph for 1-3 minutes, and the PCM detected the transmission lockup function did not occur in the "lockup" range, or the "lockup" function was "on" during periods when it should have been "off" during the CCM test. **Note: The A/T clutch or A/T brake clutch will slip with this code set.** **Possible Causes:** • A/T component problems (i.e., in clutch, brake or gears) • SL (shift lockup) is damaged, stuck "open" or stuck "closed" • Transmission valve body is clogged, dirty or stuck
DTC: P0773 **1T CCM, MIL: Yes** **1996, 1997, 1998, 1999, 2000, 2001, 2002, 2003, 2004, 2005, 2006** **Models: All** **Engines: All** **Transmissions:** A/T	**A/T Shift Solenoid 'E' (SL) Circuit Malfunction Conditions:** Engine started, engine running during normal driving conditions, and the PCM detected the Shift Lockup (SL) control circuit voltage was "high" with the solenoid "on", or the SL control circuit voltage was "low" with the SL commanded "off" during the CCM Rationality test. **Possible Causes:** • SL control circuit is open or shorted to ground • SL control circuit is shorted to system power (B+) • SL is damaged or has failed (an electrical fault) • PCM has failed
DTC: P0850 **1T CCM, MIL: Yes** **1996, 1997, 1998, 1999, 2000, 2001, 2002, 2003, 2004, 2005, 2006** **Models:** ES300, RX300, ES330, RX330, GS400, GS430, GX470, LS400, LS430, SC400, SC430 **Engines:** 3.0L VIN F, 3.3L VIN A, 4.0L VIN H, 4.3L VIN N, L, 4.7L VIN **Transmissions:** A/T	**A/T Park/Neutral Switch Circuit Malfunction Conditions:** Engine started, vehicle at 1500-2500 rpm at a speed over 43 mph for at least 30 seconds, and the PCM detected a continuous "on" "N" signal from the P/N switch. The P/N switch indicates "on" whenever the shift lever is in the 'N' or 'P' position. When it is "on", the NSW circuit to the PCM is grounded to chassis ground through the starter motor relay, and reads 0.00v. When the shift lever is in 'R', 'D' or 'L' position, the switch is "off" and the NSW circuit reads 12.0v. When the shift lever is moved from the 'N' to the 'D' position, the PCM uses this signal to air/fuel ratio correction and idle speed control functions. **Possible Causes:** • Park/Neutral switch assembly is shorted • Park/Neutral switch assembly is damaged or has failed • PCM has failed.

OBD II Trouble Code List (P1xxx Codes)

DTC	Trouble Code Title, Conditions & Possible Causes
DTC: P1100 **1T CCM, MIL: Yes** **1995, 1996, 1997, 1998, 1999, 2000, 2001, 2002, 2003, 2004, 2005, 2006** **Models:** GS300, LS 300, SC300 **Engines:** 3.0L VIN D **Transmissions:** All	**Barometric Pressure Sensor Circuit Malfunction Conditions:** Key on or engine running; and the PCM detected an unexpected voltage condition on the Barometric Pressure (BARO) sensor circuit during the CCM test. **Possible Causes:** • BARO sensor signal circuit is open or shorted to ground • BARO sensor signal circuit shorted to VREF or system power • BARO sensor power circuit is open between sensor and PCM • BARO sensor ground circuit is open between sensor and PCM • BARO sensor is damaged or has failed • PCM has failed
DTC: P1120 **1T CCM, MIL: Yes** **1996, 1997, 1998, 1999, 2000, 2001, 2002, 2003, 2004, 2005, 2006** **Models:** GS300, IS250, IS300, IS350, LS 300, SC300, LX450, LX470, GS400, GS430, GX470, LS400, LS430, SC400, SC430 **Engines:** 2.5L VIN K, 3.0L VIN D, 3.5L VIN E, 4.0L VIN H, 4.3L VIN N, L, 4.5L VIN J, 4.7L VIN T **Transmissions:** All	**Accelerator Pedal Position Sensor Circuit Malfunction Conditions:** Engine started; and the PCM detected the APP sensor VPA reading was equal to or less than 0.2v with a VPA2 reading of equal to or less than 0.5v; or the VPA reading was equal to or more than 4.7v; or the VPA reading indicated from 0.2-1.8v with the VPA2 reading equal to or more than 4.97v; or the VPA reading minus the VPA2 reading was less than 0.02v; or the VPA 2 reading minus the VPA reading was less than 0.02v for 5 seconds. **Possible Causes:** • APP sensor circuit is open or shorted to ground • APP sensor circuit is shorted to VREF • APP sensor power circuit is open between sensor and the PCM • APP sensor ground circuit is open between sensor and PCM • APP sensor is damaged or has failed • PCM has failed
DTC: P1121 **1T CCM, MIL: Yes** **1996, 1997, 1998, 1999, 2000, 2001, 2002, 2003, 2004, 2005, 2006** **Models:** GS300, IS250, IS300, IS350, LS 300, SC300, LX450, LX470, GS400, GS430, GX470, LS400, LS430, SC400, SC430 **Engines:** 2.5L VIN K, 3.0L VIN D, 3.5L VIN E, 4.0L VIN H, 4.3L VIN N, L, 4.5L VIN J, 4.7L VIN T **Transmissions:** All	**ETCS Accelerator Pedal Position Sensor Performance Conditions:** Engine started, engine running, and the PCM detected the difference between the Electronic Throttle Control System (ETCS) APP sensor VPA and VPA2 readings was less than 0.7v or 1.7v for more than 2 seconds during the CCM test. **Possible Causes:** • APP sensor is damaged or has failed • Throttle assembly or throttle linkage is binding or has failed • PCM has failed
DTC: P1125 **1T CCM, MIL: Yes** **1996, 1997, 1998, 1999, 2000, 2001, 2002, 2003, 2004, 2005, 2006** **Models:** GS300, IS250, IS300, IS350, LS 300, SC300, LX450, LX470, GS400, GS430, GX470, LS400, LS430, SC400, SC430 **Engines:** 2.5L VIN K, 3.0L VIN D, 3.5L VIN E, 4.0L VIN H, 4.3L VIN N, L, 4.5L VIN J, 4.7L VIN T **Transmissions:** All	**ETCS Throttle Control Motor Circuit Malfunction Conditions:** Engine started, engine running, and the PCM detected the motor duty cycle was equal to or more than 80% with a current level less than 0.5A; or the throttle motor current level was more than 16A, or the motor current level was equal to or more than 7A for 600 ms. **Possible Causes:** • ETCS throttle motor control (+) circuit is open, shorted to ground or to power (B+) • ETCS throttle motor control (-) circuit is open, shorted to ground or to power (B+) • ETCS throttle motor is damaged or has failed • PCM has failed

DTC	Trouble Code Title, Conditions & Possible Causes
DTC: P1126 **1T CCM, MIL: Yes** **1996, 1997, 1998, 1999, 2000,** **2001, 2002, 2003, 2004, 2005,** **2006** **Models:** GS300, IS250, IS300, IS350, LS 300, SC300, LX450, LX470, GS400, GS430, GX470, LS400, LS430, SC400, SC430 **Engines:** 2.5L VIN K, 3.0L VIN D, 3.5L VIN E, 4.0L VIN H, 4.3L VIN N, L, 4.5L VIN J, 4.7L VIN T **Transmissions:** All	**Magnetic Clutch Circuit Malfunction Conditions:** Engine started; and the PCM detected the magnetic clutch current was equal to or greater than 1.4 amps, or it was in a range of 0.8 amps to 1.4 amps for 1.5 seconds during the test. **Possible Causes:** • ETCS magnetic clutch (+) circuit is open, shorted to ground or to power (B+) • ETCS magnetic clutch (-) circuit is open, shorted to ground or to power (B+) • ETCS magnetic clutch is damaged or has failed • PCM has failed
DTC: P1127 **1T CCM, MIL: Yes** **1996, 1997, 1998, 1999, 2000,** **2001, 2002, 2003, 2004, 2005,** **2006** **Models:** GS300, IS250, IS300, IS350, LS 300, SC300, LX450, LX470, GS400, GS430, GX470, LS400, LS430, SC400, SC430 **Engines:** 2.5L VIN K, 3.0L VIN D, 3.5L VIN E, 4.0L VIN H, 4.3L VIN N, L, 4.5L VIN J, 4.7L VIN T **Transmissions:** All	**ETCS Actuator Power Source Circuit Malfunction Conditions:** Key on and the PCM detected a fault in the Electric Throttle Control system power source. The PCM shuts off power to the throttle motor and magnetic clutch (throttle valve is fully closed by a return spring, so the accelerator pedal can be opened with the throttle valve). **Possible Causes:** • Battery connections are dirty or loose • ETCS power source circuit is open (check the ETCS fuse) • PCM has failed
DTC: P1128 **1T CCM, MIL: Yes** **1996, 1997, 1998, 1999, 2000,** **2001, 2002, 2003, 2004, 2005,** **2006** **Models:** GS300, IS250, IS300, IS350, LS 300, SC300, LX450, LX470, GS400, GS430, GX470, LS400, LS430, SC400, SC430 **Engines:** 2.5L VIN K, 3.0L VIN D, 3.5L VIN E, 4.0L VIN H, 4.3L VIN N, L, 4.5L VIN J, 4.7L VIN T **Transmissions:** All	**ETCS Throttle Control Motor Lock Malfunction Conditions:** Engine started, engine running, and the PCM detected the throttle control motor position was "locked" during normal operation of the throttle control motor during the CCM test. **Possible Causes:** • ETCS throttle motor is damaged or has failed • Throttle body assembly is damaged or stuck closed • PCM has failed
DTC: P1129 **1T CCM, MIL: Yes** **1996, 1997, 1998, 1999, 2000,** **2001, 2002, 2003, 2004, 2005,** **2006** **Models:** GS300, IS250, IS300, IS350, LS 300, SC300, LX450, LX470, GS400, GS430, GX470, LS400, LS430, SC400, SC430 **Engines:** 2.5L VIN K, 3.0L VIN D, 3.5L VIN E, 4.0L VIN H, 4.3L VIN N, L, 4.5L VIN J, 4.7L VIN T **Transmissions:** All	**Electric Throttle Control System (ETCS) Malfunction Conditions:** Engine started, engine running, and the PCM detected the Actual throttle opening angle varied too much from the Target throttle opening angle during the CCM Rationality test. **Possible Causes:** • ETCS (system) is damaged or has failed • PCM has failed
DTC: P1130 **2T O2S, MIL: Yes** **1998, 1999, 2000, 2001, 2002** **Models:** All **Engines:** All **Transmissions:** All	**A/F Sensor-11 (Bank 1 Sensor 1) Circuit Malfunction Conditions:** DTC P1135 not set, engine started, engine warmup completed, and the PCM detected the A/F sensor signal was more than 3.8v or that it was less than 2.8v; or the A/F sensor signal was fixed at 3.30v; or the PCM detected an unexpected "high" or "low" voltage condition in the A/F sensor signal circuit during the CCM test. **Possible Causes:** • A/FS signal circuit or ground circuit is open (intermittent fault) • A/FS is contaminated, damaged or has failed • PCM has failed

DTC	Trouble Code Title, Conditions & Possible Causes
DTC: P1133 **1T O2S, MIL: Yes** **1998, 1999, 2000, 2001, 2002** **Models:** All **Engines:** All **Transmissions:** All	**A/F Sensor-11 (Bank 1 Sensor 1) Slow Response Conditions:** DTC P1135 not set, vehicle speed over 38 mph at over 1600 rpm for 1-3 minutes, and the PCM detected the A/F sensor response times deteriorated below an acceptable value. **Possible Causes:** • A/F sensor signal circuit or ground circuit is open • A/F sensor power circuit is open • A/F sensor is contaminated, damaged or has failed • Intake air leaks, exhaust manifold leaks or PCV system leaks • MAF sensor out of calibration (it may be dirty or contaminated) • PCM has failed
DTC: P1135 **2T CCM, MIL: Yes** **1998, 1999, 2000, 2001, 2002** **Models:** All **Engines:** All **Transmissions:** All	**A/F Sensor-11 (Bank 1 Sensor 1) Heater Circuit Malfunction Conditions:** Engine started, engine running, and the PCM detected the A/F Sensor-11 (A/FS-11) heater current was less than 0.25A, or that it was more than 8.0A at anytime during the CCM test. **Possible Causes:** • A/F sensor heater circuit or ground circuit is open • A/F sensor heater power circuit is open • A/F sensor heater is damaged or has failed • PCM has failed
DTC: P1150 **2T CCM, MIL: Yes** **1998, 1999, 2000, 2001, 2002** **Models:** All **Engines:** All **Transmissions:** All	**A/F Sensor-21 (Bank 2 Sensor 1) Circuit Malfunction Conditions:** DTC P1155 not set, engine running in closed loop, and the PCM detected the A/F sensor signal was more than 3.8v or less than 2.8v; or the A/F sensor signal was fixed at 3.30v; or it detected an unexpected "high" or "low" voltage condition on the A/F Sensor signal circuit. **Possible Causes:** • A/FS signal circuit or ground circuit is open (intermittent fault) • A/FS power circuit is open (an intermittent fault) • A/FS is contaminated, damaged or has failed • PCM has failed
DTC: P1153 **2T O2S, MIL: Yes** **1998, 1999, 2000, 2001, 2002** **Models:** All **Engines:** All **Transmissions:** All	**A/F Sensor-21 (Bank 2 Sensor 1) Slow Response Conditions:** DTC P1155 not set; vehicle speed over 38 mph at over 1600 rpm for 1-3 minutes, and the PCM detected the A/FS-21 response time had deteriorated below an acceptable value. **Possible Causes:** • A/F sensor signal circuit or ground circuit is open • A/F sensor power circuit is open • A/F sensor is contaminated, damaged or has failed • Intake air leaks, exhaust manifold leaks or PCV system leaks • MAF sensor out of calibration (it may be dirty or contaminated)
DTC: P1155 **2T CCM, MIL: Yes** **1998, 1999, 2000, 2001, 2002** **Models:** All **Engines:** All **Transmissions:** All	**A/F Sensor-21 (Bank 2 Sensor 1) Heater Circuit Malfunction Conditions:** Engine started, engine running, and the PCM detected the A/F Sensor-21 (A/FS-21) heater current was less than 0.25A, or that it was more than 8.0A at anytime during the CCM test. **Possible Causes:** • A/F sensor heater circuit or ground circuit is open • A/F sensor heater power circuit is open • A/F sensor heater is damaged or has failed • PCM has failed
DTC: P1200 **2T CCM, MIL: Yes** **1996, 1997, 1998, 1999, 2000,** **2001, 2002, 2003, 2004, 2005,** **2006** **Models:** GS300, IS250, IS300, IS350, LS 300, SC300 **Engines:** 2.5L VIN K, 3.0L VIN D, 3.5L VIN E **Transmissions:** All	**Fuel Pump Relay/ECU Circuit Malfunction Conditions:** Engine speed less than 1000 rpm; and the PCM detected an unexpected voltage condition on the fuel pump, fuel pump relay ECU input circuit, or on the fuel pump Diagnostic circuit. **Possible Causes:** • Fuel Pump ECU control circuit is open or shorted to ground • Fuel Pump ECU control circuit is shorted to system power (B+) • Fuel Pump ECU power circuit is open (check power from relay) • Fuel Pump ECU is damaged or has failed • Fuel Pump is damaged or has failed • PCM has failed
DTC: P1300 **1T CCM, MIL: Yes** **1996, 1997, 1998, 1999, 2000,** **2001, 2002, 2003, 2004, 2005,** **2006** **Models:** All **Engines:** All **Transmissions:** All	**Igniter Circuit Malfunction Conditions:** Engine started, engine running, and the PCM did not detect any IGF signals after detecting four IGT1 signals during the CCM test. **Possible Causes:** • IGT or IGF signal circuit is open between the igniter and PCM • IGT or IGF signal circuit is shorted to ground • Igniter is damaged or has failed • PCM has failed

DTC	Trouble Code Title, Conditions & Possible Causes
DTC: P1305 **1T CCM, MIL: Yes** **1996, 1997, 1998, 1999, 2000, 2001, 2002, 2003, 2004, 2005, 2006** **Models:** All **Engines:** All **Transmissions:** All	**Igniter No. 2 Circuit Malfunction Conditions:** Engine started; and the PCM did not detect any IGF2 signals after detecting two IGT2 signals. **Note: This vehicle uses a Coil-On-Plug design Ignition System.** **Possible Causes:** • IGT or IGF signal circuit is open or shorted to ground • Igniter is damaged or has failed • PCM has failed
DTC: P1310 **1T CCM, MIL: Yes** **1996, 1997, 1998, 1999, 2000, 2001, 2002, 2003, 2004, 2005, 2006** **Models:** All **Engines:** All **Transmissions:** All	**Igniter No. 3 Circuit Malfunction Conditions:** Engine started; and the PCM did not detect any IGF3 signals after detecting two IGT3 signals during the test. **Note: This vehicle uses a Coil-On-Plug design Ignition System.** **Possible Causes:** • IGT or IGF signal circuit is open or shorted to ground • Igniter is damaged or has failed •
DTC: P1315 **1T CCM, MIL: Yes** **1996, 1997, 1998, 1999, 2000, 2001, 2002, 2003, 2004, 2005, 2006** **Models:** All **Engines:** All **Transmissions:** All	**Igniter No. 4 Circuit Malfunction Conditions:** Engine started; and the PCM did not detect any IGF4 signals after detecting 2 IGT4 signals during the CCM test. **Note: This vehicle uses a Coil-On-Plug design Ignition System.** **Possible Causes:** • IGT or IGF signal circuit is open or shorted to ground • Igniter is damaged or has failed • PCM has failed
DTC: P1320 **1T CCM, MIL: Yes** **1996, 1997, 1998, 1999, 2000, 2001, 2002, 2003, 2004, 2005, 2006** **Models:** All **Engines:** All **Transmissions:** All	**Igniter No. 5 Circuit Malfunction Conditions:** Engine started; and the PCM did not detect any IGF5 signals after detecting IGT5 signals during the CCM test. **Note: This vehicle uses a Coil-On-Plug design Ignition System.** **Possible Causes:** • IGT or IGF signal circuit is open between the igniter and PCM • IGT or IGF signal circuit is shorted to ground • Igniter is damaged or has failed • PCM has failed
DTC: P1325 **1T CCM, MIL: Yes** **1996, 1997, 1998, 1999, 2000, 2001, 2002, 2003, 2004, 2005, 2006** **Models:** All **Engines:** All **Transmissions:** All	**Igniter No. 6 Circuit Malfunction Conditions:** Engine started; and the PCM did not detect any IGF6 signals after detecting IGT6 signals during the CCM test. **Note: This vehicle uses a Coil-On-Plug design Ignition System.** **Possible Causes:** • IGT or IGF signal circuit is open between the igniter and PCM • IGT or IGF signal circuit is shorted to ground • Igniter is damaged or has failed • PCM has failed
DTC: P1330 **1T CCM, MIL: Yes** **1996, 1997, 1998, 1999, 2000, 2001, 2002, 2003, 2004, 2005, 2006** **Models:** LX450, LX470, GS400, GS430, GX470, LS400, LS430, SC400, SC430 **Engines:** 4.0L VIN H, 4.3L VIN N, L, 4.5L VIN J, 4.7L VIN T **Transmissions:** All	**Igniter No. 7 Circuit Malfunction Conditions:** Engine started; and the PCM did not detect any IGF7 signals after detecting IGT7 signals during the CCM test. **Note: This vehicle uses a Coil-On-Plug design Ignition System.** **Possible Causes:** • IGT or IGF signal circuit is open between the igniter and PCM • IGT or IGF signal circuit is shorted to ground • Igniter is damaged or has failed • PCM has failed
DTC: P1335 **1T CCM, MIL: Yes** **1995** **Models:** All **Engines:** All **Transmissions:** All	**Crankshaft Position Sensor Circuit Malfunction Conditions:** Engine speed over 1000 rpm, and the PCM did not detect any CKP (NE) sensor signals. **Possible Causes:** • CKP sensor (+) circuit is open or shorted to ground • CKP sensor (-) circuit is open or shorted to ground • CKP sensor (+), (-) circuit is shorted to VREF or system power • CKP sensor is damaged or has failed • PCM has failed

DTC	Trouble Code Title, Conditions & Possible Causes
DTC: P1335 **1T CCM, MIL: Yes** **1996, 1997, 1998, 1999, 2000, 2001, 2002, 2003, 2004, 2005, 2006** **Models:** All **Engines:** All **Transmissions:** All	**Crankshaft Position Sensor Circuit Malfunction Conditions:** Engine speed over 1000 rpm, and the PCM did not detect any CKP (NE) sensor signals for 500 ms. This code may be set by an intermittent loss of the CKP sensor signal. **Possible Causes:** • CKP sensor (+) circuit is open or shorted to ground • CKP sensor (-) circuit is open or shorted to ground • CKP sensor (+), (-) circuit is shorted to VREF or system power • CKP sensor is damaged or has failed • PCM has failed
DTC: P1340 **1T CCM, MIL: Yes** **1996, 1997, 1998, 1999, 2000, 2001, 2002, 2003, 2004, 2005, 2006** **Models:** LX450, LX470, GS400, GS430, GX470, LS400, LS430, SC400, SC430 **Engines:** 4.0L VIN H, 4.3L VIN N, L, 4.5L VIN J, 4.7L VIN T **Transmissions:** All	**Igniter No. 8 Circuit Malfunction Conditions:** Engine started; and the PCM did not detect any IGF8 signals after detecting IGT8 signals during the CCM test. **Note: This vehicle uses a Coil-On-Plug design Ignition System.** **Possible Causes:** • IGT or IGF signal circuit is open between the igniter and PCM • IGT or IGF signal circuit is shorted to ground • Igniter is damaged or has failed • PCM has failed
DTC: P1400 **1T CCM, MIL: Yes** **1995, 1996, 1997, 1998, 1999, 2000, 2001, 2002, 2003, 2004, 2005, 2006** **Models:** GS300, LS 300, SC300 **Engines:** 3.0L VIN D **Transmissions:** All	**Sub-Throttle Position Sensor Circuit Malfunction Conditions:** Engine started; and the PCM detected the Sub-Throttle Position sensor (VTA2) signal was less than 0.25v with the Closed Throttle Position switch "off", or more than 4.90v at any time. **Possible Causes:** • VTA2 signal circuit is open or shorted to ground • Sub-Throttle position sensor VREF or ground circuit is open • Sub-Throttle position sensor is damaged or has failed • PCM has failed
DTC: P1401 **2T CCM, MIL: Yes** **1995, 1996, 1997, 1998, 1999, 2000, 2001, 2002, 2003, 2004, 2005, 2006** **Models:** GS300, LS 300, SC300 **Engines:** 3.0L VIN D **Transmissions:** All	**Sub-Throttle Position Sensor Range/Performance Conditions:** Vehicle speed over 3 mph, and the PCM detected the difference between the Throttle Position sensor angle and Sub-Throttle Position sensor angle was over 35 degrees. **Possible Causes:** • Sub-Throttle position sensor is damaged or has failed • Throttle linkage or throttle body is binding or sticking • PCM has failed
DTC: P1405 **2T CCM, MIL: Yes** **1995, 1996, 1997, 1998, 1999, 2000, 2001, 2002, 2003, 2004, 2005, 2006** **Models:** GS300, LS 300, SC300 **Engines:** 3.0L VIN D **Transmissions:** All	**Turbo Pressure Sensor Circuit Malfunction Conditions:** Engine started, engine running, and the PCM detected an unexpected voltage condition on the Turbo Boost Pressure sensor circuit during the CCM test. **Possible Causes:** • Boost Pressure sensor signal circuit open or shorted to ground • Boost Pressure sensor power or ground circuit is open • Boost Pressure sensor is damaged or has failed • PCM has failed
DTC: P1406 **2T CCM, MIL: Yes** **1995, 1996, 1997, 1998, 1999, 2000, 2001, 2002, 2003, 2004, 2005, 2006** **Models:** GS300, LS 300, SC300 **Engines:** 3.0L VIN D **Transmissions:** All	**Turbo Pressure Sensor Circuit Malfunction Conditions:** Engine started, engine running, MAF sensor more than 1.3 g/sec, and the PCM detected the Turbo Pressure sensor signal was less than 1.2v; or with the MAF sensor less than 0.45 g/sec, the Turbo Pressure sensor indicated more than 4.2v during the CCM test. **Possible Causes:** • Boost Pressure sensor signal circuit open or shorted to ground • Boost Pressure sensor power or ground circuit is open • Boost Pressure sensor is damaged or has failed • PCM has failed
DTC: P1410 **2T CCM, MIL: Yes** **1996, 1997, 1998, 1999, 2000, 2001, 2002, 2003, 2004, 2005, 2006** 2003Models: ES300, RX300, ES330, RX330 **Engines:** 3.0L VIN F, 3.3L VIN A **Transmissions:** All	**EGR Valve Position Sensor Circuit Malfunction Conditions:** Key on or engine running; and the PCM detected an unexpected voltage condition on the EGR EVP sensor circuit during the test. **Possible Causes:** • EVP sensor signal circuit is open between sensor and the PCM • EVP sensor signal circuit is shorted to ground • EVP sensor power (VREF) circuit is open • EVP sensor ground circuit is open • EVP sensor is damaged or has failed • PCM has failed

DTC	Trouble Code Title, Conditions & Possible Causes
DTC: P1411 2T CCM, MIL: Yes 1996, 1997, 1998, 1999, 2000, 2001, 2002, 2003, 2004, 2005, 2006 **Models:** ES300, RX300, ES330, RX330 **Engines:** 3.0L VIN F, 3.3L VIN A **Transmissions:** All	**EGR Valve Position Sensor Range/Performance Conditions:** Engine started; ECT sensor signal below 131°F, and the PCM detected the EGR sensor signal was under 0.35v or it indicated a value equal to or greater than 1.65v for 7 seconds. **Possible Causes:** • EVP sensor signal circuit is open between sensor and the PCM • EVP sensor signal circuit is shorted to ground • EVP sensor power (VREF) circuit is open • EVP sensor ground circuit is open • EVP sensor is damaged or has failed • PCM has failed
DTC: P1500 1T CCM, MIL: Yes 1995 **Models:** All **Engines:** All **Transmissions:** All	**Starter Signal Circuit Malfunction Conditions:** Engine cranking; and the PCM detected that it did not receive a signal from the starter signal circuit during the test period. **Possible Causes:** • Engine starter signal circuit is open or shorted to ground • Engine starter signal circuit is shorted to system power (B+) • Starter has failed • PCM has failed
DTC: P1500 1T CCM, MIL: Yes 1996, 1997, 1998, 1999, 2000, 2001, 2002 **Models:** All **Engines:** All **Transmissions:** All	**Starter Signal Circuit Malfunction Conditions:** Engine cranking and the PCM detected it did not receive a signal on the starter signal circuit. **Possible Causes:** • Engine starter signal circuit is open or shorted to ground • Engine starter signal circuit is shorted to system power (B+) • Starter has failed • PCM has failed
DTC: P1511 2T CCM, MIL: Yes 1995, 1996, 1997, 1998, 1999, 2000, 2001, 2002, 2003, 2004, 2005, 2006 **Models:** GS300, LS 300, SC300 **Engines:** 3.0L VIN D **Transmissions:** All	**Turbo Boost Pressure Low Malfunction Conditions:** Engine speed more than 2600 rpm in closed loop, then during a WOT event, the PCM detected +740 mmHg or more of intake pipe pressure during the CCM Rationality test. **Possible Causes:** • Actuator for the Waste Gate valve, Intake Air Control valve, Exhaust Gas Control valve is damaged, binding or has failed • Intake Air system is clogged or leaking • VSV control circuit for the Waste Gate valve, Intake Air Control valve, Exhaust Gas Control valve is open • PCM has failed
DTC: P1512 2T CCM, MIL: Yes 1995, 1996, 1997, 1998, 1999, 2000, 2001, 2002, 2003, 2004, 2005, 2006 **Models:** GS300, LS 300, SC300 **Engines:** 3.0L VIN D **Transmissions:** All	**Turbo Boost Pressure High Malfunction Conditions:** Engine speed less than 3400 rpm in closed loop, then during a WOT event, the PCM detected +150 mmHg or less of intake pipe pressure during the CCM Rationality test. **Possible Causes:** • Waste Gate valve, Intake Air Control valve, Exhaust Gas Control valve has failed • Intake Air system is clogged or leaking • VSV control circuit for the Waste Gate valve, Intake Air Control valve, Exhaust Gas Control valve is shorted to ground • PCM has failed
DTC: P1520 1T CCM, MIL: Yes 1998, 1999, 2000, 2001, 2002, 2003, 2004, 2005, 2006 **Models:** All **Engines:** All **Transmissions:** All	**Stop Light Switch Circuit Malfunction Conditions:** Vehicle driven to a speed of over 19 mph several times; and the PCM detected the Stop Light signal status did not change at least once under these operating conditions. **Possible Causes:** • Stop light switch is shorted to ground • Stop light switch is damaged or has failed • PCM has failed
DTC: P1525 1T CCM, MIL: Yes 1996, 1997, 1998, 1999 **Models:** LX450, LX470 **Engines:** 4.5L VIN J, 4.7L VIN T **Transmissions:** All	**Cruise Control Switch Circuit Malfunction Conditions:** Engine started, vehicle driven to a speed of over 35 mph, and the PCM detected an unexpected voltage condition on the Cruise Control switch circuit during the CCM test. **Possible Causes:** • Cruise control switch signal circuit is shorted to ground • Cruise control switch is damaged or has failed • PCM has failed
DTC: P1566 1T CCM, MIL: Yes 1996, 1997, 1998, 1999 **Models:** LX450, LX470 **Engines:** 4.5L VIN J, 4.7L VIN T **Transmissions:** All	**Cruise Control Main Switch Circuit Malfunction Conditions:** Engine started, vehicle driven to a speed of over 35 mph, and the PCM detected an unexpected voltage condition on the Cruise Control Main switch circuit during the CCM test. **Possible Causes:** • Cruise control Main switch signal circuit is open • Cruise control Main switch is damaged or has failed • PCM has failed

DTC	Trouble Code Title, Conditions & Possible Causes
DTC: P1600 **1T CCM, MIL: Yes** **1995** **Models:** All **Engines:** All **Transmissions:** All	**PCM Battery Backup Circuit Malfunction Conditions:** Key on, and the PCM detected an unexpected voltage in the Battery Backup circuit (KAM circuit) during the CCM test. **Note: The PCM will not store any other codes with this code set.** **Possible Causes:** • Battery backup circuit is open (check EFI fuse and fuse link) • Battery terminals are corroded or loose • PCM has failed
DTC: P1600 **1T CCM, MIL: Yes** **1996, 1997, 1998, 1999, 2000,** **2001, 2002, 2003, 2004, 2005,** **2006** **Models:** All **Engines:** All **Transmissions:** All	**PCM Battery Backup Circuit Malfunction Conditions:** Key on, and the PCM detected an unexpected voltage in the Battery Backup circuit (KAM circuit) during the CCM test. **Note: The PCM will not store any other codes with this code set.** **Possible Causes:** • Battery backup circuit is open (check EFI fuse and fuse link) • Battery terminals are corroded or loose • PCM has failed
DTC: P1605 **1T CCM, MIL: Yes** **1995, 1996** **Models:** All **Engines:** All **Transmissions:** All	**Knock Control CPU Malfunction Conditions:** Engine started, engine running, and the PCM detected a problem in the Knock Control portion of the controller during the test period. **Possible Causes:** • Clear the codes and determine if this code resets. If the same trouble code (P1605 resets), the PCM has failed. • TSB TC002-03 (6/03) contains information related to this code
DTC: P1630 **1T CCM, MIL: Yes** **1996, 1997, 1998, 1999, 2000,** **2001, 2002, 2003, 2004, 2005,** **2006** **Models:** GS300, IS250, IS300, IS350, LS 300, SC300 **Engines:** 2.5L VIN K, 3.0L VIN D, 3.5L VIN E **Transmissions:** All	**Traction Control System Malfunction Conditions:** Engine runtime over 5 seconds, and the PCM detected an unexpected voltage condition on the Traction Control system circuit, or it received a signal that a TRAC problem existed. **Possible Causes:** • ETC+ or ETC signal circuit is open, shorted to ground or power • EFI+ or EFI- signal circuit is open, shorted to ground or power • Throttle Control ECU or the PCM has failed
DTC: P1633 **1T CCM, MIL: Yes** **1996, 1997, 1998, 1999, 2000,** **2001, 2002, 2003, 2004, 2005,** **2006** **Models:** GS300, IS250, IS300, IS350, LS 300, SC300 **Engines:** 2.5L VIN K, 3.0L VIN D, 3.5L VIN E **Transmissions:** All	**Engine Throttle Control System Circuit Malfunction Conditions:** Engine started, engine running, and the PCM detected a problem in the ETCS portion of the circuit located in the engine control module. **Possible Causes:** • Clear the codes and determine if this code resets. If the same trouble code (P1633 resets), the PCM has failed. • TSB TC002-03 (6/03) contains information related to this code
DTC: P1700 **2T CCM, MIL: Yes** **1996, 1997, 1998, 1999, 2000,** **2001, 2002, 2003, 2004, 2005,** **2006** **Models:** ES300, RX300, ES330, RX330, LX450, LX470 **Engines:** 3.0L VIN F, 3.3L VIN A, 4.5L VIN J, 4.7L VIN T **Transmissions:** A/T	**Vehicle Speed Sensor '2' Circuit Malfunction Conditions:** Vehicle driven to a speed of over 6 mph, P/N switch indicating "off" (not in Park or Neutral position) for over 4 seconds, TR switch indicating a position other than Neutral, and the PCM did not detect any VSS2 signals after detecting at least 4 VSS1 signals. **Note: This problem must occur at least 500 times (continuously) in order for this trouble code to set.** **Possible Causes:** • VSS signal (+) or (-) circuit is open, shorted to ground or power • VSS is damaged or has failed • PCM has failed
DTC: P1705 **1T CCM, MIL: Yes** **1995** **Models:** All **Engines:** All **Transmissions:** All	**A/T Direct Clutch Speed Sensor Circuit Malfunction Conditions:** Vehicle running, P/N switch at "off" and the PCM detected a Direct Clutch Speed sensor output of 300 rpm or less. **Possible Causes:** • Direct clutch speed sensor signal (+) or (-) circuit is open • Direct clutch speed sensor signal (+) or (-) shorted to ground • Direct clutch speed sensor (+) or (-) signal is shorted to power • Direct clutch is damaged or has failed • PCM has failed

DTC	Trouble Code Title, Conditions & Possible Causes
DTC: P1765 **1T CCM, MIL: Yes** **1995** **Models: All** **Engines: All** **Transmissions:** All	**A/T Linear Shift Solenoid Circuit Malfunction Conditions:** Engine speed over 500 rpm, P/N switch indicating off, and the PCM detected that the Linear Shift Solenoid valve current flow was less than 0.2 amps during the CCM test. **Possible Causes:** • SLT shift solenoid control circuit is open or shorted to ground • Linear shift solenoid control circuit is shorted to power • Linear shift solenoid is damaged or has failed • PCM has failed
DTC: P1780 **1T CCM, MIL: Yes** **1995** **Models: All** **Engines: All** **Transmissions:** All	**Park/Neutral Position Switch Circuit Malfunction Conditions:** Engine started, engine runtime over 30 seconds, and the PCM detected two or more P/N inputs (Drive, Neutral, 2nd, Low or Reverse) at the same time; or with the engine speed from 1500-2500 rpm with the VSS indicating over 50 mph, it detected the P/N switch indicated "on" for 30 seconds during the CCM test. **Possible Causes:** • P/N switch is shorted to ground • P/N switch is out-of-adjustment, damaged or has failed • PCM has failed
DTC: P1780 **1T CCM, MIL: Yes** **1996, 1997, 1998, 1999, 2000, 2001, 2002** **Models: All** **Engines: All** **Transmissions:** All	**Park/Neutral Position Switch Circuit Malfunction Conditions:** Engine started, engine runtime over 30 seconds, and the PCM detected two or more P/N inputs (Drive, Neutral, 2nd, Low or Reverse) at the same time; or with the engine speed from 1500-2500 rpm with the VSS indicating over 50 mph, it detected the P/N switch indicated "on" for 30 seconds during the CCM test. **Possible Causes:** • P/N switch is shorted to ground or wiring harness is shorted • P/N switch is out-of-adjustment, damaged or has failed • PCM has failed

OBD II Trouble Code List (P2xxx Codes)

DTC	Trouble Code Title, Conditions & Possible Causes:
DTC: P2102 **1T CCM, MIL: Yes** **1996, 1997, 1998, 1999, 2000, 2001, 2002, 2003, 2004, 2005, 2006** **Models:** LX450, LX470, GS400, GS430, GX470, LS400, LS430, SC400, SC430 **Engines:** 4.0L VIN H, 4.3L VIN N, L, 4.5L VIN J, 4.7L VIN T **Transmissions:** All	**Throttle Actuator Control Motor Circuit Low Input Conditions:** Engine started, throttle control motor output duty cycle at 80% or higher, and the PCM detected the throttle motor current was less than 0.5 amps for 2 seconds. The PCM controls the motor position in order to open and close the throttle valve. The opening angle of the throttle valve is sensed by the TP sensor mounted on the throttle body. The PCM uses the TP sensor signal to control the Throttle Valve opening angle (throttle motor) to respond to driving conditions. **Possible Causes:** • Throttle motor connector is damaged or open • Throttle control motor circuit is open • Throttle control motor is damaged or has failed • PCM has failed
DTC: P2103 **1T CCM, MIL: Yes** **1996, 1997, 1998, 1999, 2000, 2001, 2002, 2003, 2004, 2005, 2006** **Models:** LX450, LX470, GS400, GS430, GX470, LS400, LS430, SC400, SC430 **Engines:** 4.0L VIN H, 4.3L VIN N, L, 4.5L VIN J, 4.7L VIN T **Transmissions:** All	**Throttle Actuator Control Motor Circuit High Input Conditions:** Engine started, throttle control motor output duty cycle at 80% or higher, and the PCM detected the throttle motor current was more than 10.0 amps for 600 ms. The PCM controls the motor position in order to open and close the throttle valve. The opening angle of the throttle valve is sensed by the TP sensor mounted on the throttle body. The PCM uses the TP sensor signal to control the Throttle Valve opening angle (throttle motor) to respond to driving conditions. **Possible Causes:** • Throttle motor connector is damaged or shorted • Throttle control motor circuit is shorted • Throttle control motor is damaged or has failed • PCM has failed
DTC: P2111 **1T CCM, MIL: Yes** **1996, 1997, 1998, 1999, 2000, 2001, 2002, 2003, 2004, 2005, 2006** **Models:** LX450, LX470, GS400, GS430, GX470, LS400, LS430, SC400, SC430 **Engines:** 4.0L VIN H, 4.3L VIN N, L, 4.5L VIN J, 4.7L VIN T **Transmissions:** All	**Throttle Actuator Control System Stuck Open Conditions:** Key on or engine started, and the PCM detected the throttle control motor position is stuck open. The PCM controls the motor position in order to open and close the throttle valve. The opening angle of the throttle valve is sensed by the TP sensor mounted on the throttle body. The PCM uses the TP sensor signal to control the Throttle Valve opening angle (throttle motor) to respond to driving conditions. **Possible Causes:** • Throttle control motor circuit is open • Throttle control motor is damaged or has failed • Throttle body or throttle valve is damaged or has failed
DTC: P2112 **1T CCM, MIL: Yes** **1996, 1997, 1998, 1999, 2000, 2001, 2002, 2003, 2004, 2005, 2006** **Models:** LX450, LX470, GS400, GS430, GX470, LS400, LS430, SC400, SC430 **Engines:** 4.0L VIN H, 4.3L VIN N, L, 4.5L VIN J, 4.7L VIN T **Transmissions:** All	**Throttle Actuator Control System Stuck Closed Conditions:** Key on or engine started, and the PCM detected the throttle control motor position is stuck closed. The PCM controls the motor position in order to open and close the throttle valve. The opening angle of the throttle valve is sensed by the TP sensor mounted on the throttle body. The PCM uses the TP sensor signal to control the Throttle Valve opening angle (throttle motor) to respond to driving conditions. **Possible Causes:** • Throttle control motor circuit is shorted • Throttle control motor is damaged or has failed • Throttle body or throttle valve is damaged or has failed
DTC: P2118 **1T CCM, MIL: Yes** **1996, 1997, 1998, 1999, 2000, 2001, 2002, 2003, 2004, 2005, 2006** **Models:** LX450, LX470, GS400, GS430, GX470, LS400, LS430, SC400, SC430 **Engines:** 4.0L VIN H, 4.3L VIN N, L, 4.5L VIN J, 4.7L VIN T **Transmissions:** All	**Throttle Actuator Control Motor Current Performance Conditions:** Key on or engine started, and the PCM detected an unexpected low voltage condition (open circuit) on the ETCS power source circuit. Battery positive voltage is applied to the +BM circuit of the PCM under both Key on and Key off conditions. **Possible Causes:** • ETCS power source circuit is open • PCM has failed

DTC	Trouble Code Title, Conditions & Possible Causes
DTC: P2119 **1T CCM, MIL: Yes** **1996, 1997, 1998, 1999, 2000, 2001, 2002, 2003, 2004, 2005, 2006** **Models:** LX450, LX470, GS400, GS430, GX470, LS400, LS430, SC400, SC430 **Engines:** 4.0L VIN H, 4.3L VIN N, L, 4.5L VIN J, 4.7L VIN T **Transmissions:** All	**Throttle Actuator Control Throttle Body Performance Conditions:** Engine started, and the PCM detected Actual throttle opening angle continued to vary greatly from the Target opening angel. The idle speed on this vehicle is controlled by the Electronic Throttle Control system (ETCS). This system includes a throttle control motor to operate the throttle valve, a throttle position sensor to detect the accelerator pedal position, and the PCM to control the ETCS and one-valve design of throttle body. The PCM controls this motor in order to control the throttle valve opening to achieve its target speed. **Possible Causes:** • ETCS throttle control system • PCM has failed
DTC: P2120 **1T CCM, MIL: Yes** **1996, 1997, 1998, 1999, 2000, 2001, 2002, 2003, 2004, 2005, 2006** **Models:** LX450, LX470, GS400, GS430, GX470, LS400, LS430, SC400, SC430 **Engines:** 4.0L VIN H, 4.3L VIN N, L, 4.5L VIN J, 4.7L VIN T **Transmissions:** All	**Throttle Pedal Position Sensor/Switch 'D' Circuit Malfunction Conditions:** Engine started, and the PCM detected VPA1 signal indicated less than 0.20v while the VPA2 signal indicated over 0.97 degrees, or the VPA1 signal indicated more than 4.80v for 500 ms. This system (ETCS) does not use a throttle cable. The Accelerator Pedal Position (APP) sensor is mounted on the accelerator pedal bracket. It includes two sensors to detect the accelerator position, and to detect any faults in the APP sensor or its related circuits. **Possible Causes:** • APP sensor signal circuit is open or shorted to ground • APP sensor is damaged or has failed • PCM has failed
DTC: P2121 **1T CCM, MIL: Yes** **1996, 1997, 1998, 1999, 2000, 2001, 2002, 2003, 2004, 2005, 2006** **Models:** LX450, LX470, GS400, GS430, GX470, LS400, LS430, SC400, SC430 **Engines:** 4.0L VIN H, 4.3L VIN N, L, 4.5L VIN J, 4.7L VIN T **Transmissions:** All	**Accelerator Pedal Position Sensor Signal Performance Conditions:** Engine started, IDL signal "off", and the PCM detected the difference between the VPA and VPA2 signal was out-of-range for 2 seconds. This system (ETCS) does not use a throttle cable. The Accelerator Pedal Position (APP) sensor is mounted on the accelerator pedal bracket. It includes two sensors to detect the accelerator position, and to detect any faults in the APP sensor or its related circuits. **Possible Causes:** • APP sensor is damaged or has failed • PCM has failed
DTC: P2122 **1T CCM, MIL: Yes** **1996, 1997, 1998, 1999, 2000, 2001, 2002, 2003, 2004, 2005, 2006** **Models:** LX450, LX470, GS400, GS430, GX470, LS400, LS430, SC400, SC430 **Engines:** 4.0L VIN H, 4.3L VIN N, L, 4.5L VIN J, 4.7L VIN T **Transmissions:** All	**Throttle Pedal Position Sensor/Switch 'D' Circuit Low Input Conditions:** Engine started, and the PCM detected VPA1 signal was less than 0.20v while the VPA2 signal indicated over 0.97 degrees for 500 ms. This system (ETCS) does not use a throttle cable. The Accelerator Pedal Position (APP) sensor is mounted on the accelerator pedal bracket. It includes two sensors to detect the accelerator position, and to detect any faults in the APP sensor or its related circuits. **Possible Causes:** • APP sensor signal circuit is shorted to ground • APP sensor is damaged or has failed • PCM has failed
DTC: P2123 **1T CCM, MIL: Yes** **1996, 1997, 1998, 1999, 2000, 2001, 2002, 2003, 2004, 2005, 2006** **Models:** LX450, LX470, GS400, GS430, GX470, LS400, LS430, SC400, SC430 **Engines:** 4.0L VIN H, 4.3L VIN N, L, 4.5L VIN J, 4.7L VIN T **Transmissions:** All	**Throttle Pedal Position Sensor/Switch 'D' Circuit High Input Conditions:** Engine started, and the PCM detected VPA1 signal indicated over 4.80v for 2 seconds. This system (ETCS) does not use a throttle cable. The Accelerator Pedal Position (APP) sensor is mounted on the accelerator pedal bracket. It includes two sensors to detect the accelerator position, and to detect any faults in the APP sensor. **Possible Causes:** • APP sensor signal circuit is open • APP sensor is damaged or has failed • PCM has failed

DTC	Trouble Code Title, Conditions & Possible Causes

DTC: P2125
1T CCM, MIL: Yes
1996, 1997, 1998, 1999, 2000, 2001, 2002, 2003, 2004, 2005, 2006
Models: LX450, LX470, GS400, GS430, GX470, LS400, LS430, SC400, SC430
Engines: 4.0L VIN H, 4.3L VIN N, L, 4.5L VIN J, 4.7L VIN T
Transmissions: All
	Throttle Pedal Position Sensor/Switch 'E' Circuit Malfunction Conditions:
	Engine started, and the PCM detected VPA2 signal indicated less than 0.50v while the VPA1 signal indicated over 0.97 degrees, or the VPA1 signal was more than 4.80v or less than 0.20v for 500 ms. This system (ETCS) does not use a throttle cable. The Accelerator Pedal Position (APP) sensor is mounted on the accelerator pedal bracket. It includes two sensors to detect the accelerator position, and to detect any faults in the APP sensor or its circuits.
	Possible Causes:
	  • APP sensor signal circuit is open or shorted to ground
	  • APP sensor is damaged or has failed
	  • PCM has failed

DTC: P2127
1T CCM, MIL: Yes
1996, 1997, 1998, 1999, 2000, 2001, 2002, 2003, 2004, 2005, 2006
Models: LX450, LX470, GS400, GS430, GX470, LS400, LS430, SC400, SC430
Engines: 4.0L VIN H, 4.3L VIN N, L, 4.5L VIN J, 4.7L VIN T
Transmissions: All
	Throttle Pedal Position Sensor/Switch 'E' Circuit Low Input Conditions:
	Engine started, and the PCM detected VPA2 signal was less than 0.20v while the VPA1 signal indicated over 0.97 degrees for 500 ms. The ETCS does not use a throttle cable. The Accelerator Pedal Position sensor is mounted on the accelerator pedal bracket. It includes two sensors to detect the accelerator position or any faults in the APP sensor or its circuits.
	Possible Causes:
	  • APP sensor signal circuit is shorted to ground
	  • APP sensor has failed, or the PCM has failed

DTC: P2128
1T CCM, MIL: Yes
1996, 1997, 1998, 1999, 2000, 2001, 2002, 2003, 2004, 2005, 2006
Models: LX450, LX470, GS400, GS430, GX470, LS400, LS430, SC400, SC430
Engines: 4.0L VIN H, 4.3L VIN N, L, 4.5L VIN J, 4.7L VIN T
Transmissions: All
	Throttle Pedal Position Sensor/Switch 'E' Circuit High Input Conditions:
	Engine started, and the PCM detected VPA1 signal was over 4.80v or under 0.20v for 2 seconds. This system (ETCS) does not use a throttle cable. The Accelerator Pedal Position (APP) sensor is mounted on the accelerator pedal bracket. It includes two sensors to detect the accelerator position, and any faults in the APP sensor.
	Possible Causes:
	  • APP sensor signal circuit is open
	  • APP sensor is damaged or has failed
	  • PCM has failed

DTC: P2135
1T CCM, MIL: Yes
1996, 1997, 1998, 1999, 2000, 2001, 2002, 2003, 2004, 2005, 2006
Models: LX450, LX470, GS400, GS430, GX470, LS400, LS430, SC400, SC430
Engines: 4.0L VIN H, 4.3L VIN N, L, 4.5L VIN J, 4.7L VIN T
Transmissions: All
	Throttle Pedal Position Sensor/Switch 'A'/'B' Voltage Correlation Conditions:
	Engine started, and the PCM detected the value of the VPA1 signal less the VPA2 was less than 0.02v, or the VPA1 signal was less than 0.20v with the VPA2 signal less than 0.50v for 400-500 ms. This system (ETCS) does not use a throttle cable. The Accelerator Pedal Position (APP) sensor is mounted on the accelerator pedal bracket. It includes two sensors to detect the accelerator position, and to detect any faults present in the APP sensor.
	Possible Causes:
	  • APP sensor signal circuit is open
	  • APP sensor is damaged or has failed
	  • PCM has failed

DTC: P2138
1T CCM, MIL: Yes
1996, 1997, 1998, 1999, 2000, 2001, 2002, 2003, 2004, 2005, 2006
Models: LX450, LX470, GS400, GS430, GX470, LS400, LS430, SC400, SC430
Engines: 4.0L VIN H, 4.3L VIN N, L, 4.5L VIN J, 4.7L VIN T
Transmissions: All
	Throttle Pedal Position Sensor/Switch 'D'/'E' Voltage Correlation Conditions:
	Engine started, and the PCM detected the value of the VPA1 signal less the VPA2 was less than 0.02v, or the VPA1 signal was less than 0.20v with the VPA2 signal less than 0.50v for 2 seconds. This system (ETCS) does not use a throttle cable. The Accelerator Pedal Position (APP) sensor is mounted on the accelerator pedal bracket. It includes two sensors to detect the accelerator position, and to detect any faults present in the APP sensor.
	Possible Causes:
	  • APP sensor signal circuit is open
	  • APP sensor is damaged or has failed
	  • PCM has failed

DTC	Trouble Code Title, Conditions & Possible Causes
DTC: P2195 1T CCM, MIL: Yes 1996, 1997, 1998, 1999, 2000, 2001, 2002, 2003, 2004, 2005, 2006 Models: ES300, RX300, ES330, RX330 Engines: 3.0L VIN F, 3.3L VIN A, 3.3L VIN A Transmissions: All	**Air Fuel Sensor 1 (Bank 1 Sensor 1) Signal Stuck "Lean" Conditions:** Vehicle speed from 25-87 mph at over 1500 rpm with the throttle valve open, and the PCM detected the Air Fuel sensor signal indicated more than 3.80v for 10 seconds. **Possible Causes:** • A/FS1 signal circuit is open or shorted to ground • A/FS1 is damaged, contaminated or it has failed • Air induction system is severely restricted • Fuel Control component problems (e.g., low fuel pressure, or one or more severely restricted fuel injectors) • PCM has failed
DTC: P2196 1T CCM, MIL: Yes 1996, 1997, 1998, 1999, 2000, 2001, 2002, 2003, 2004, 2005, 2006 Models: ES300, RX300, ES330, RX330, Engines: 3.0L VIN F, 3.3L VIN A Transmissions: All	**Air Fuel Sensor 1 (Bank 1 Sensor 1) Signal Stuck "Rich" Conditions:** Vehicle speed from 25-87 mph at over 1500 rpm with the throttle valve open, and the PCM detected the Air Fuel sensor signal indicated more than 2.80v for 10 seconds. **Possible Causes:** • A/FS1 signal circuit is open or shorted to ground • A/FS1 is damaged, contaminated or it has failed • Air induction system is leaking (check for PCV system leaks) • Fuel component problem (high fuel pressure, leaking regulator or a leaking injector) • PCM has failed
DTC: P2196 1T CCM, MIL: Yes 1996, 1997, 1998, 1999, 2000, 2001, 2002, 2003, 2004, 2005, 2006 Models: ES300, RX300, ES330, RX330, Engines: 3.0L VIN F, 3.3L VIN A Transmissions: All	**Air Fuel Sensor 1 (Bank 2 Sensor 1) Signal Stuck "Lean" Conditions:** Vehicle speed from 25-87 mph at over 1500 rpm with the throttle valve open, and the PCM detected the Air Fuel sensor signal indicated more than 3.80v for 10 seconds. **Possible Causes:** • A/FS1 signal circuit is open or shorted to ground • A/FS1 is damaged, contaminated or it has failed • Air induction system is severely restricted • Fuel component problem (e.g., low fuel pressure, or a severely restricted fuel injector) • PCM has failed
DTC: P2198 1T CCM, MIL: Yes 1996, 1997, 1998, 1999, 2000, 2001, 2002, 2003, 2004, 2005, 2006 Models: ES300, RX300, ES330, RX330, Engines: 3.0L VIN F, 3.3L VIN A Transmissions: All	**Air Fuel Sensor 1 (Bank 2 Sensor 1) Signal Stuck "Rich" Conditions:** Vehicle driven at a steady speed of 25-87 mph at over 1500 rpm with the throttle valve open, and the PCM detected the Air Fuel sensor signal indicated more than 2.80v for 10 seconds. **Possible Causes:** • A/FS1 signal circuit is open or shorted to ground • A/FS1 is damaged, contaminated or it has failed • Air induction system is leaking (check for PCV system leaks) • Fuel Control component problems (e.g., high fuel pressure, a leaking pressure regulator, one or more leaking fuel injectors) • PCM has failed
DTC: P2237 1T CCM, MIL: Yes 1996, 1997, 1998, 1999, 2000, 2001, 2002, 2003, 2004, 2005, 2006 Models: ES300, RX300, ES330, RX330, Engines: 3.0L VIN F, 3.3L VIN A Transmissions: All	**Air Fuel Sensor 1 (Bank 1 Sensor 1) Pumping Current Signal Open Conditions:** Vehicle speed from 25-87 mph at over 1500 rpm with the throttle valve open, and the PCM detected the A/F Sensor AF+ signal was less than 0.50v or more than 4.80v for 5 seconds. **Possible Causes:** • A/FS1 signal circuit is open or shorted to ground • A/FS1 is damaged, contaminated or it has failed • A/FS1 heater assembly is damaged or its circuit has failed • A/FS1 heater relay is damaged or has failed • PCM has failed
DTC: P2240 1T CCM, MIL: Yes 1996, 1997, 1998, 1999, 2000, 2001, 2002, 2003, 2004, 2005, 2006 Models: ES300, RX300, ES330, RX330, Engines: 3.0L VIN F, 3.3L VIN A Transmissions: All	**Air Fuel Sensor 1 (Bank 2 Sensor 1) Pumping Current Signal Open Conditions:** Engine started, vehicle driven at a steady speed of 25-87 mph at over 1500 rpm with the throttle valve open, and the PCM detected the Air Fuel Sensor AF+ signal indicated less than 0.50v or indicated more than 4.80v for 5 seconds. **Possible Causes:** • A/FS1 signal circuit is open or shorted to ground • A/FS1 is damaged, contaminated or it has failed • A/FS1 heater assembly is damaged or its circuit has failed • A/FS1 heater relay is damaged or has failed • PCM has failed

DTC	Trouble Code Title, Conditions & Possible Causes
DTC: P2725 **1T CCM, MIL: Yes** **1996, 1997, 1998, 1999, 2000, 2001, 2002, 2003, 2004, 2005, 2006** **Models:** ES300, RX300, ES330, RX330, **Engines:** 3.0L VIN F, 3.3L VIN A **Transmissions:** All	**A/T Pressure Control Solenoid 'E' Circuit Malfunction Conditions:** Engine started, engine warmup period completed, gearshift selector in 'P' or 'N', engine speed 500 rpm or more, and the PCM detected that current flowed to the Shift Solenoid (SLN) control circuit for over 1 second. The Shift Solenoid SLN controls the hydraulic pressure acting on the accumulator control valve when gears are shifted in order to provide smooth gear shifting. **Possible Causes:** • Shift solenoid (SLN) connector is damaged or loose • Shift solenoid (SLN) control circuit is open or shorted to ground • Shift solenoid (SLN) is damaged or has failed (electrical fault) • PCM has failed

LEXUS
COMPONENT TESTING

6

TABLE OF CONTENTS

Component Locations

ES300

3.0L V6 MFI 1MZ-FE VIN F

1996

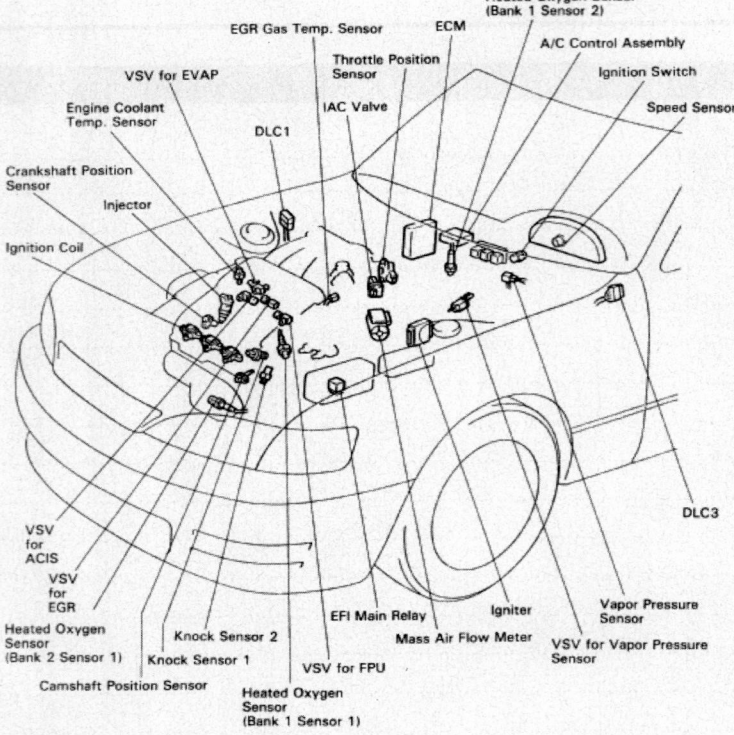

EGR Gas Temp. Sensor
ECM
Heated Oxygen Sensor (Bank 1 Sensor 2)
Throttle Position Sensor
A/C Control Assembly
VSV for EVAP
Ignition Switch
IAC Valve
Speed Sensor
Engine Coolant Temp. Sensor
DLC1
Crankshaft Position Sensor
Injector
Ignition Coil
DLC3
VSV for ACIS
VSV for EGR
EFI Main Relay
Igniter
Vapor Pressure Sensor
Heated Oxygen Sensor (Bank 2 Sensor 1)
Knock Sensor 2
Mass Air Flow Meter
VSV for Vapor Pressure Sensor
Knock Sensor 1
Camshaft Position Sensor
VSV for FPU
Heated Oxygen Sensor (Bank 1 Sensor 1)

29157_LEXU_G0001

1997–98

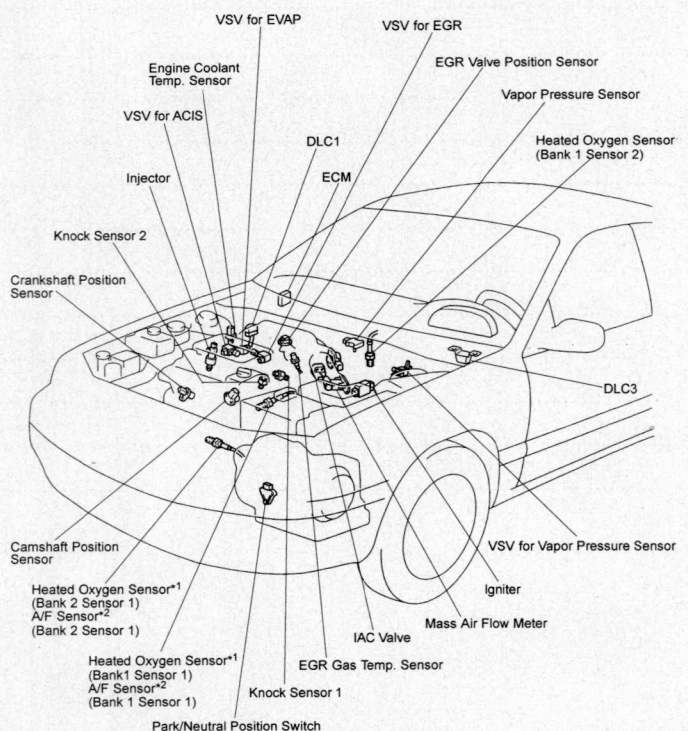

VSV for EVAP
VSV for EGR
Engine Coolant Temp. Sensor
EGR Valve Position Sensor
VSV for ACIS
Vapor Pressure Sensor
DLC1
Heated Oxygen Sensor (Bank 1 Sensor 2)
Injector
ECM
Knock Sensor 2
Crankshaft Position Sensor
DLC3
Camshaft Position Sensor
VSV for Vapor Pressure Sensor
Heated Oxygen Sensor*1 (Bank 2 Sensor 1)
A/F Sensor*2 (Bank 2 Sensor 1)
Igniter
Heated Oxygen Sensor*1 (Bank1 Sensor 1)
A/F Sensor*2 (Bank 1 Sensor 1)
IAC Valve
Mass Air Flow Meter
EGR Gas Temp. Sensor
Knock Sensor 1
Park/Neutral Position Switch

*1 : Except California Specification vehicles
*2 : Only for California Specification vehicles

29157_LEXU_G0002

1999–2001

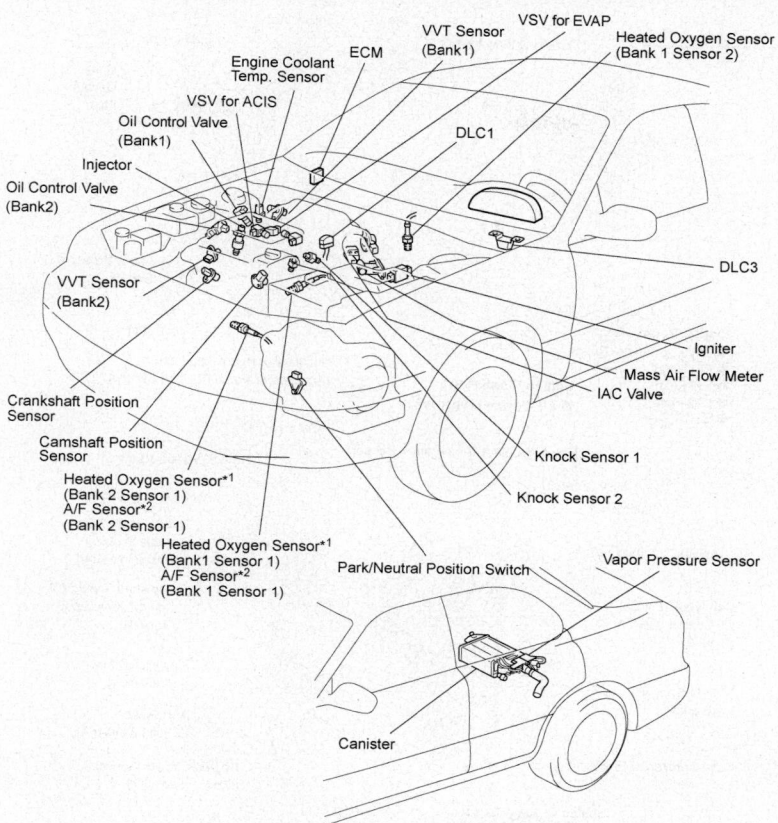

VVT Sensor (Bank1)

VSV for EVAP

Heated Oxygen Sensor (Bank 1 Sensor 2)

Engine Coolant Temp. Sensor

ECM

VSV for ACIS

DLC1

Oil Control Valve (Bank1)

Injector

Oil Control Valve (Bank2)

DLC3

VVT Sensor (Bank2)

Igniter

Mass Air Flow Meter

IAC Valve

Crankshaft Position Sensor

Camshaft Position Sensor

Knock Sensor 1

Knock Sensor 2

Heated Oxygen Sensor*1 (Bank 2 Sensor 1)
A/F Sensor*2 (Bank 2 Sensor 1)

Heated Oxygen Sensor*1 (Bank1 Sensor 1)
A/F Sensor*2 (Bank 1 Sensor 1)

Park/Neutral Position Switch

Vapor Pressure Sensor

Canister

29157_LEXU_G0004

2002-03

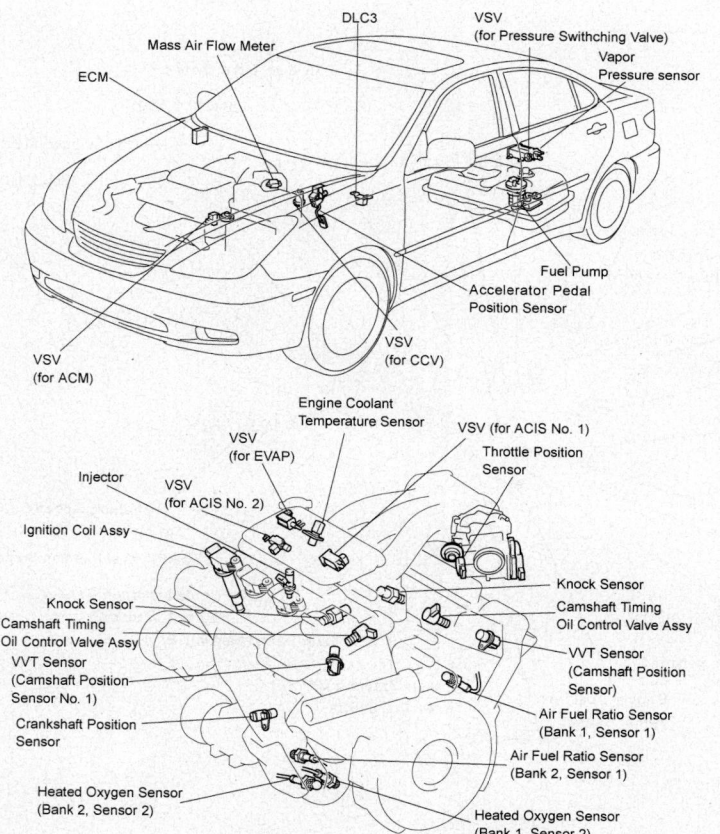

Mass Air Flow Meter

DLC3

VSV (for Pressure Swithching Valve)

Vapor Pressure sensor

ECM

Fuel Pump

Accelerator Pedal Position Sensor

VSV (for CCV)

VSV (for ACM)

Engine Coolant Temperature Sensor

VSV (for EVAP)

VSV (for ACIS No. 1)

Throttle Position Sensor

Injector

VSV (for ACIS No. 2)

Ignition Coil Assy

Knock Sensor

Knock Sensor

Camshaft Timing Oil Control Valve Assy

Camshaft Timing Oil Control Valve Assy

VVT Sensor (Camshaft Position Sensor No. 1)

VVT Sensor (Camshaft Position Sensor)

Crankshaft Position Sensor

Air Fuel Ratio Sensor (Bank 1, Sensor 1)

Air Fuel Ratio Sensor (Bank 2, Sensor 1)

Heated Oxygen Sensor (Bank 2, Sensor 2)

Heated Oxygen Sensor (Bank 1, Sensor 2)

29157_LEXU_G0003

ES330

3.3L V6 MFI 3MZ-FE VIN A

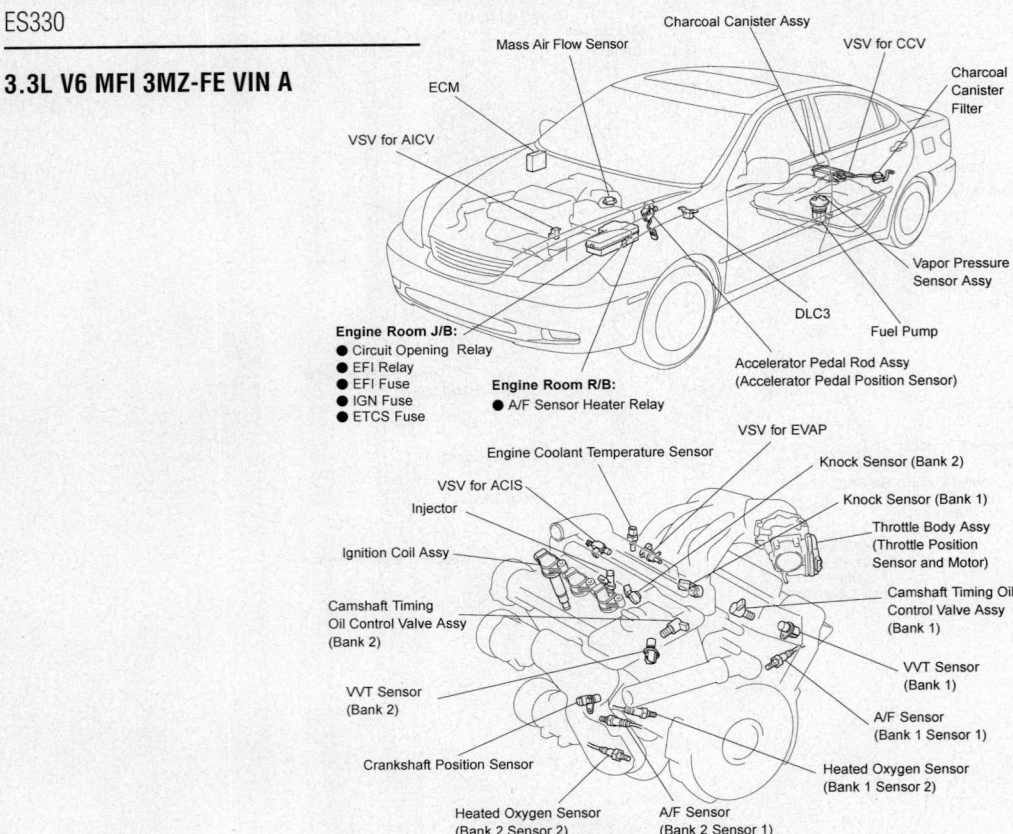

Mass Air Flow Sensor
Charcoal Canister Assy
VSV for CCV
Charcoal Canister Filter
ECM
VSV for AICV
Vapor Pressure Sensor Assy
DLC3
Fuel Pump
Accelerator Pedal Rod Assy (Accelerator Pedal Position Sensor)

Engine Room J/B:
● Circuit Opening Relay
● EFI Relay
● EFI Fuse
● IGN Fuse
● ETCS Fuse

Engine Room R/B:
● A/F Sensor Heater Relay

VSV for EVAP
Engine Coolant Temperature Sensor
Knock Sensor (Bank 2)
VSV for ACIS
Knock Sensor (Bank 1)
Injector
Throttle Body Assy (Throttle Position Sensor and Motor)
Ignition Coil Assy
Camshaft Timing Oil Control Valve Assy (Bank 2)
Camshaft Timing Oil Control Valve Assy (Bank 1)
VVT Sensor (Bank 1)
VVT Sensor (Bank 2)
A/F Sensor (Bank 1 Sensor 1)
Crankshaft Position Sensor
Heated Oxygen Sensor (Bank 1 Sensor 2)
Heated Oxygen Sensor (Bank 2 Sensor 2)
A/F Sensor (Bank 2 Sensor 1)

29157_LEXU_G0005

GS300

3.0L I6 MFI 2JZ-GE VIN D
1996–97

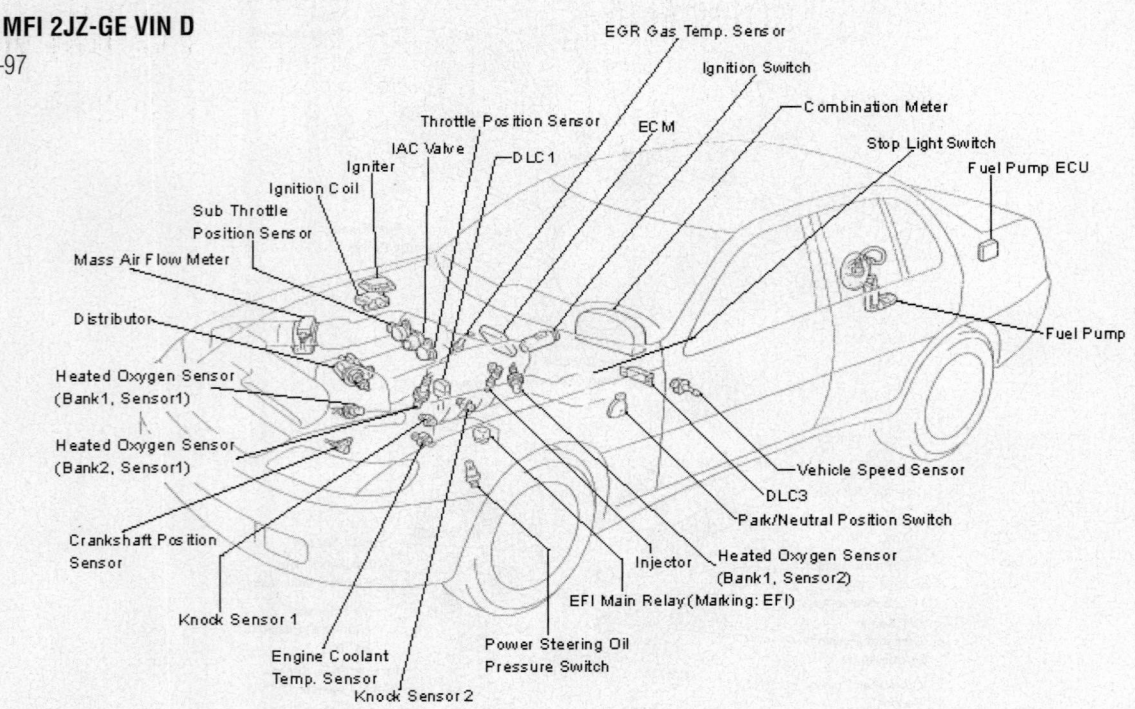

EGR Gas Temp. Sensor
Ignition Switch
Throttle Position Sensor
ECM
Combination Meter
IAC Valve
Stop Light Switch
Igniter
DLC 1
Fuel Pump ECU
Ignition Coil
Sub Throttle Position Sensor
Mass Air Flow Meter
Distributor
Fuel Pump
Heated Oxygen Sensor (Bank1, Sensor1)
Heated Oxygen Sensor (Bank2, Sensor1)
Vehicle Speed Sensor
DLC3
Park/Neutral Position Switch
Crankshaft Position Sensor
Injector
Heated Oxygen Sensor (Bank1, Sensor2)
Knock Sensor 1
EFI Main Relay (Marking: EFI)
Engine Coolant Temp. Sensor
Power Steering Oil Pressure Switch
Knock Sensor 2

29157_LEXU_G0006

1998–2001

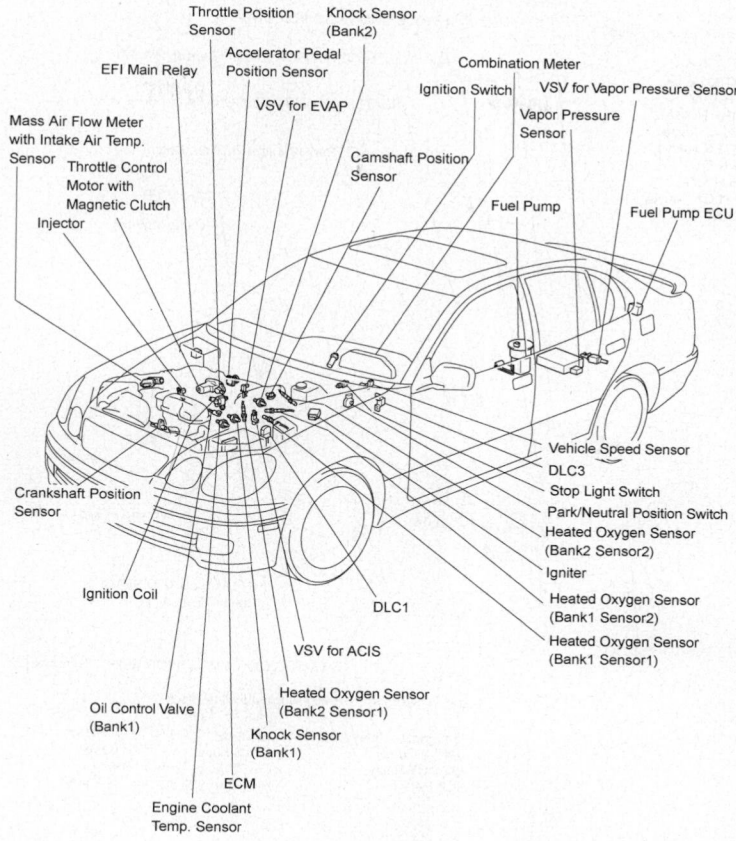

Throttle Position Sensor

Knock Sensor (Bank2)

EFI Main Relay

Accelerator Pedal Position Sensor

VSV for EVAP

Combination Meter

Ignition Switch

VSV for Vapor Pressure Sensor

Mass Air Flow Meter with Intake Air Temp. Sensor

Throttle Control Motor with Magnetic Clutch

Camshaft Position Sensor

Vapor Pressure Sensor

Injector

Fuel Pump

Fuel Pump ECU

Crankshaft Position Sensor

Vehicle Speed Sensor

DLC3

Stop Light Switch

Park/Neutral Position Switch

Heated Oxygen Sensor (Bank2 Sensor2)

Igniter

Ignition Coil

DLC1

Heated Oxygen Sensor (Bank1 Sensor2)

Heated Oxygen Sensor (Bank1 Sensor1)

VSV for ACIS

Oil Control Valve (Bank1)

Heated Oxygen Sensor (Bank2 Sensor1)

Knock Sensor (Bank1)

ECM

Engine Coolant Temp. Sensor

29157_LEXU_G0007

2002–05

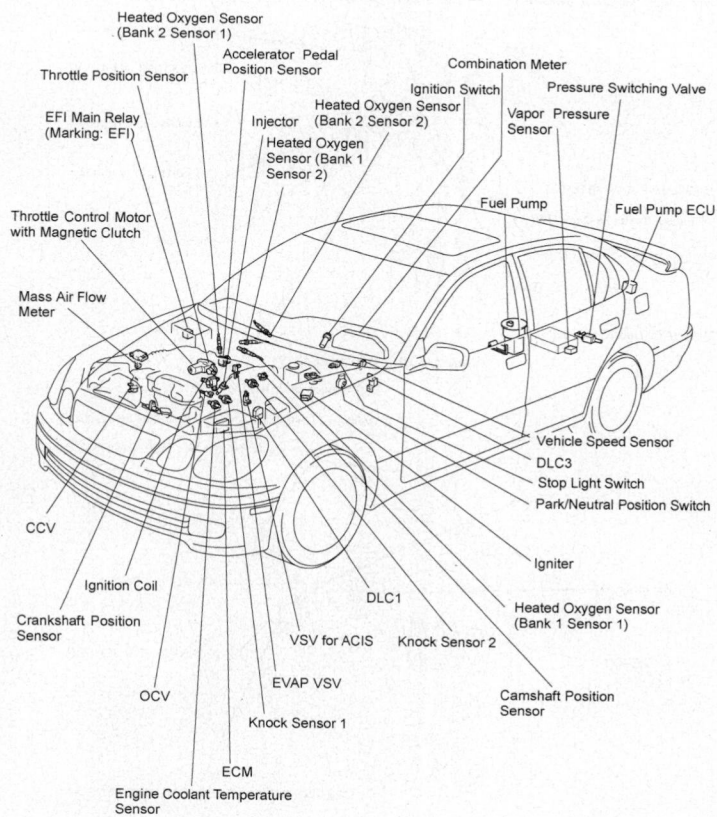

Heated Oxygen Sensor (Bank 2 Sensor 1)

Throttle Position Sensor

Accelerator Pedal Position Sensor

Combination Meter

EFI Main Relay (Marking: EFI)

Injector

Ignition Switch

Pressure Switching Valve

Heated Oxygen Sensor (Bank 2 Sensor 2)

Vapor Pressure Sensor

Heated Oxygen Sensor (Bank 1 Sensor 2)

Throttle Control Motor with Magnetic Clutch

Fuel Pump

Fuel Pump ECU

Mass Air Flow Meter

Vehicle Speed Sensor

DLC3

Stop Light Switch

Park/Neutral Position Switch

CCV

Igniter

Ignition Coil

DLC1

Heated Oxygen Sensor (Bank 1 Sensor 1)

Crankshaft Position Sensor

VSV for ACIS

Knock Sensor 2

OCV

EVAP VSV

Camshaft Position Sensor

Knock Sensor 1

ECM

Engine Coolant Temperature Sensor

29157_LEXU_G0008

2006

Engine Room No. 1
Relay Block, Junction Block

- STARTER Relay
- P/I-B H-fuse
- LH J/B-B H-fuse
- RH J/B-B H-fuse
- STARTER H-fuse
- MAIN H-fuse
- E/G-B H-fuse
- GLW PLG1 H-fuse
- IG2 MAIN Fuse

VACUUM SWITCHING VALVE FOR EVAP

AIR FUEL RATIO SENSOR (Bank 1 Sensor 1)

HEATED OXYGEN SENSOR (Bank 1 Sensor 2)

Park / Neutral Position Switch

CANISTER

- Pump Module

Air Filter

FUEL PUMP

AIR FUEL RATIO SENSOR
(Bank 2 Sensor 1)

HEATED OXYGEN SENSOR (Bank 2 Sensor 2)

FUEL PUMP RESISTOR

ECM

Engine Room No. 2 Relay Block, Junction Block

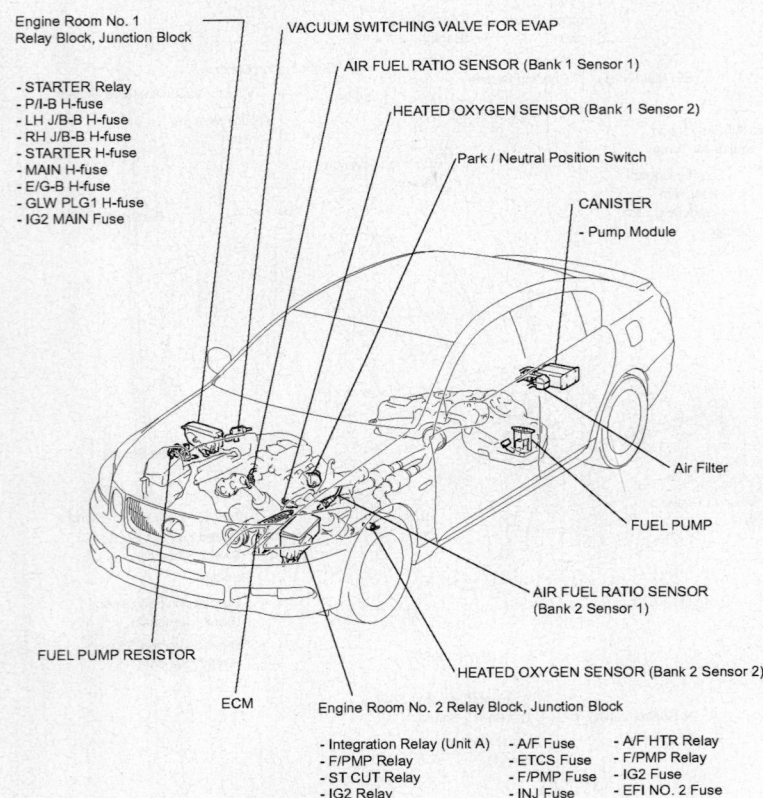

- Integration Relay (Unit A)	- A/F Fuse	- A/F HTR Relay
- F/PMP Relay	- ETCS Fuse	- F/PMP Relay
- ST CUT Relay	- F/PMP Fuse	- IG2 Fuse
- IG2 Relay	- INJ Fuse	- EFI NO. 2 Fuse

29157_LEXU_G0009

CAMSHAFT TIMING OIL CONTROL VALVE ASSEMBLY

VVT Sensor for Exhaust Camshaft

THROTTLE BODY

COLD START INJECTOR

Air Intake Control Valve

VVT Sensor for Exhaust
Camshaft

Ignition Coil

MASS AIR FLOW METER

VVT Sensor for Intake Camshaft

Injector Driver (EDU)

FUEL INJECTOR

CAMSHAFT TIMING OIL CONTROL VALVE ASSEMBLY

CRANKSHAFT POSITION SENSOR

VVT Sensor for Intake Camshaft

CAMSHAFT TIMING OIL CONTROL VALVE ASSEMBLY

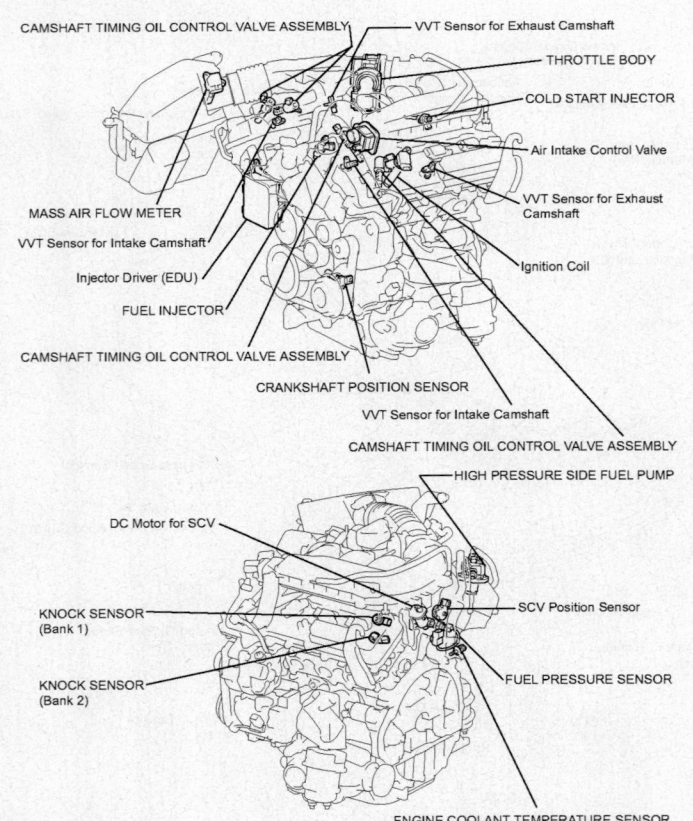

HIGH PRESSURE SIDE FUEL PUMP

DC Motor for SCV

KNOCK SENSOR
(Bank 1)

SCV Position Sensor

KNOCK SENSOR
(Bank 2)

FUEL PRESSURE SENSOR

ENGINE COOLANT TEMPERATURE SENSOR

29157_LEXU_G0010

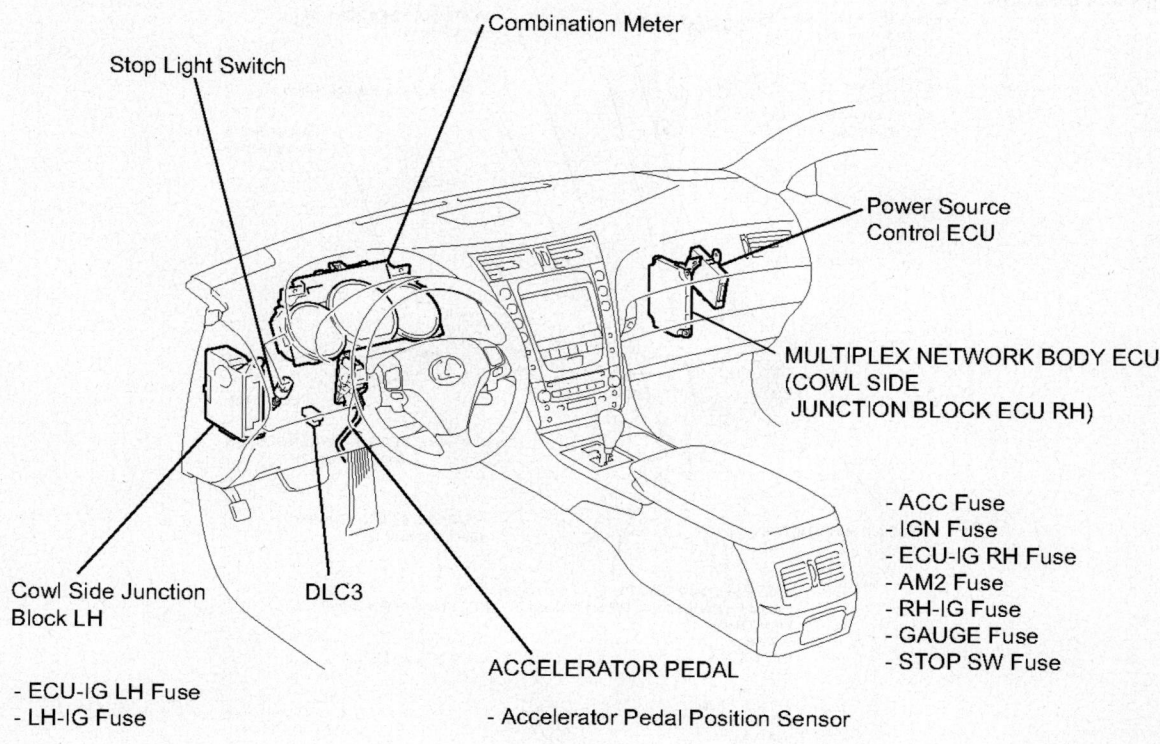

Stop Light Switch

Combination Meter

Power Source
Control ECU

MULTIPLEX NETWORK BODY ECU
(COWL SIDE
 JUNCTION BLOCK ECU RH)

- ACC Fuse
- IGN Fuse
- ECU-IG RH Fuse
- AM2 Fuse
- RH-IG Fuse
- GAUGE Fuse
- STOP SW Fuse

Cowl Side Junction
Block LH

DLC3

ACCELERATOR PEDAL

- Accelerator Pedal Position Sensor

- ECU-IG LH Fuse
- LH-IG Fuse

29157_LEXU_G0011

GS400

4.0L V8 MFI 1UZ-FE VIN H

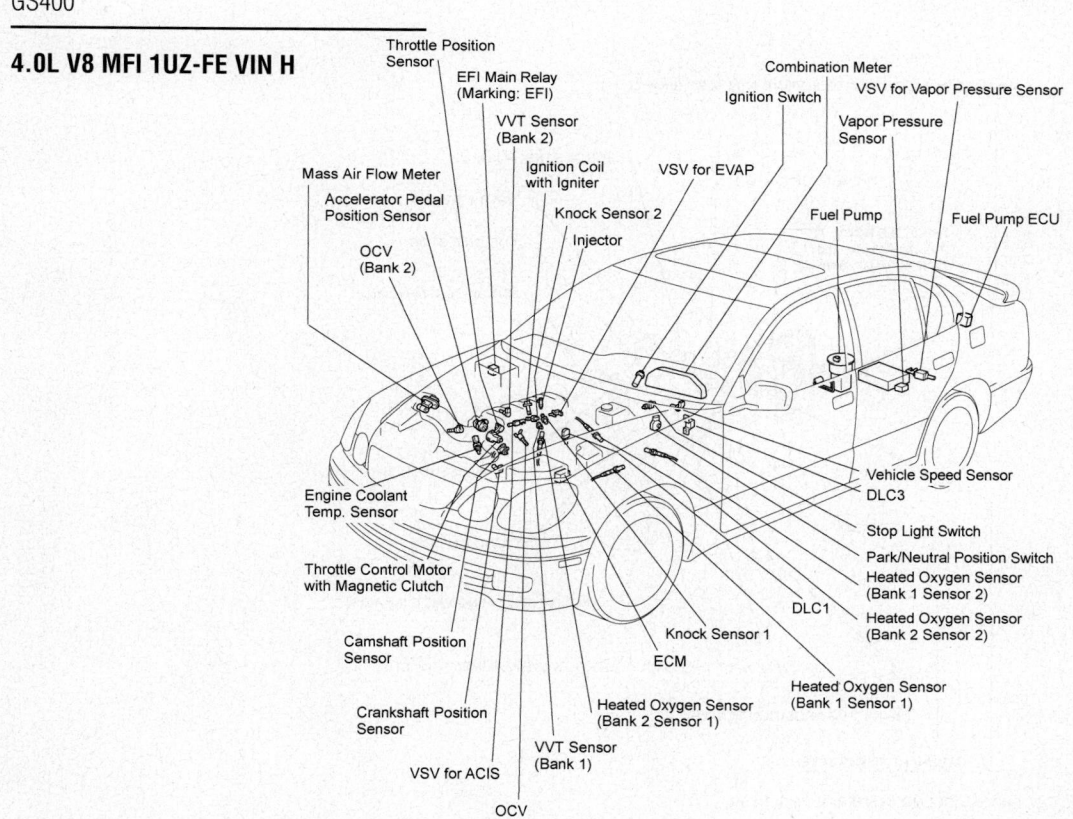

Throttle Position
Sensor

EFI Main Relay
(Marking: EFI)

VVT Sensor
(Bank 2)

Mass Air Flow Meter

Accelerator Pedal
Position Sensor

OCV
(Bank 2)

Ignition Coil
with Igniter

Knock Sensor 2

Injector

VSV for EVAP

Combination Meter

Ignition Switch

VSV for Vapor Pressure Sensor

Vapor Pressure
Sensor

Fuel Pump

Fuel Pump ECU

Engine Coolant
Temp. Sensor

Throttle Control Motor
with Magnetic Clutch

Camshaft Position
Sensor

Crankshaft Position
Sensor

VSV for ACIS

OCV
(Bank 1)

VVT Sensor
(Bank 1)

Heated Oxygen Sensor
(Bank 2 Sensor 1)

ECM

Knock Sensor 1

DLC1

Vehicle Speed Sensor
DLC3
Stop Light Switch
Park/Neutral Position Switch
Heated Oxygen Sensor
(Bank 1 Sensor 2)
Heated Oxygen Sensor
(Bank 2 Sensor 2)

Heated Oxygen Sensor
(Bank 1 Sensor 1)

29157_LEXU_G0012

GS430

4.3L V8 MFI 3UZ-FE VIN N, L

Engine Room No. 1 Relay Block, Junction Block
- IG2 MAIN H-Relay
- STARTER H-Relay
- ALT H-Fuse
- E/G-B H-Fuse
- P/I-B H-Fuse
- RH J/B-AM H-Fuse
- STARTER Fuse

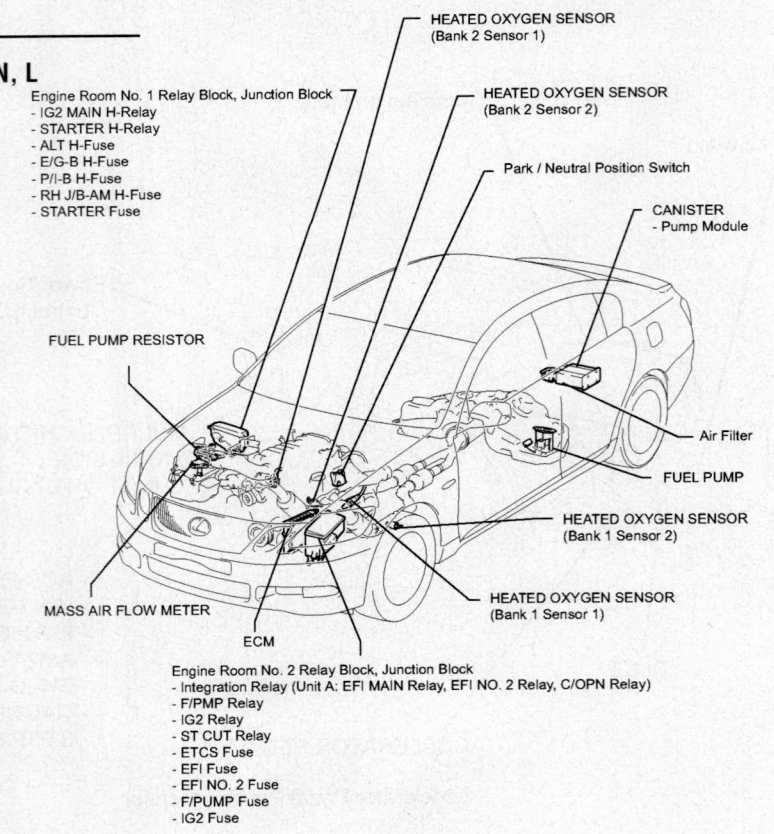

HEATED OXYGEN SENSOR
(Bank 2 Sensor 1)

HEATED OXYGEN SENSOR
(Bank 2 Sensor 2)

Park / Neutral Position Switch

CANISTER
- Pump Module

FUEL PUMP RESISTOR

Air Filter

FUEL PUMP

HEATED OXYGEN SENSOR
(Bank 1 Sensor 2)

HEATED OXYGEN SENSOR
(Bank 1 Sensor 1)

MASS AIR FLOW METER

ECM

Engine Room No. 2 Relay Block, Junction Block
- Integration Relay (Unit A: EFI MAIN Relay, EFI NO. 2 Relay, C/OPN Relay)
- F/PMP Relay
- IG2 Relay
- ST CUT Relay
- ETCS Fuse
- EFI Fuse
- EFI NO. 2 Fuse
- F/PUMP Fuse
- IG2 Fuse

29157_LEXU_G0013

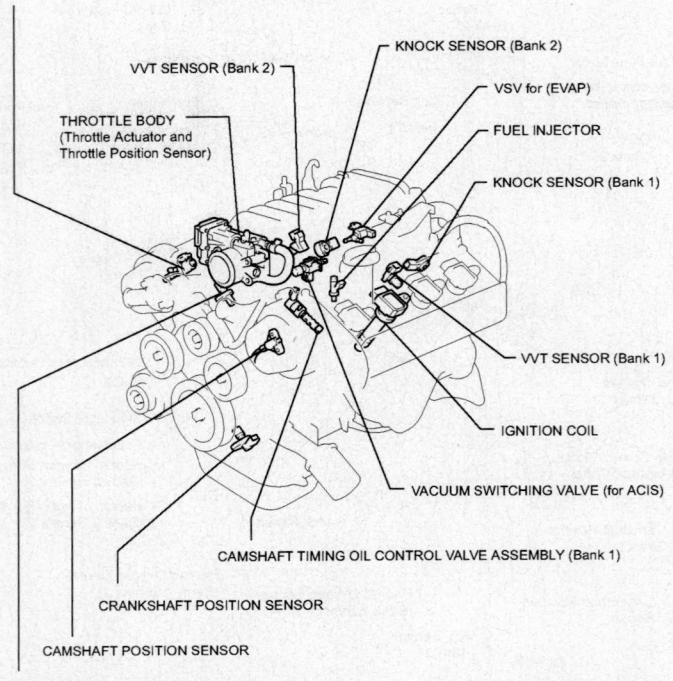

CAMSHAFT TIMING OIL CONTROL VALVE ASSEMBLY (Bank 2)

KNOCK SENSOR (Bank 2)

VVT SENSOR (Bank 2)

VSV for (EVAP)

THROTTLE BODY
(Throttle Actuator and
Throttle Position Sensor)

FUEL INJECTOR

KNOCK SENSOR (Bank 1)

VVT SENSOR (Bank 1)

IGNITION COIL

VACUUM SWITCHING VALVE (for ACIS)

CAMSHAFT TIMING OIL CONTROL VALVE ASSEMBLY (Bank 1)

CRANKSHAFT POSITION SENSOR

CAMSHAFT POSITION SENSOR

ENGINE COOLANT TEMPERATURE SENSOR

29157_LEXU_G0014

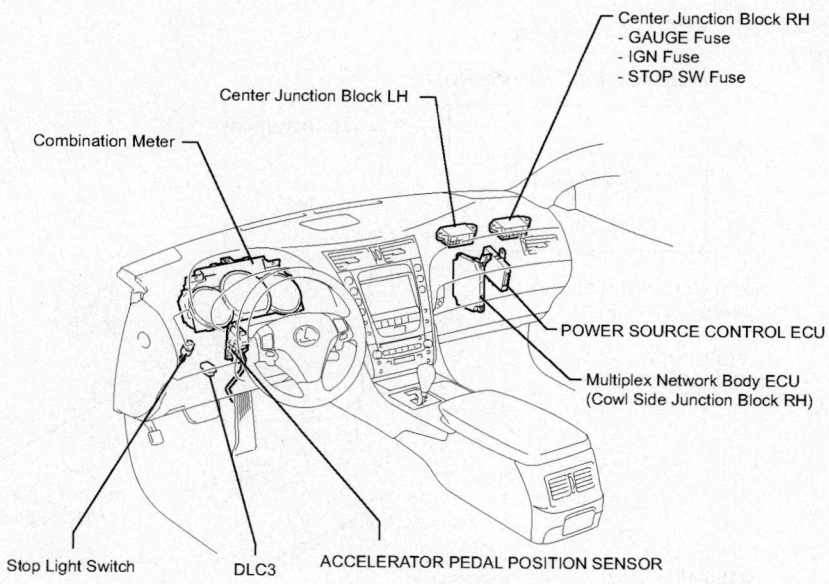

Center Junction Block RH
- GAUGE Fuse
- IGN Fuse
- STOP SW Fuse

Center Junction Block LH

Combination Meter

POWER SOURCE CONTROL ECU

Multiplex Network Body ECU
(Cowl Side Junction Block RH)

Stop Light Switch

DLC3

ACCELERATOR PEDAL POSITION SENSOR

29157_LEXU_G0015

GX470

4.7L V8 MFI 2UZ-FE VIN T

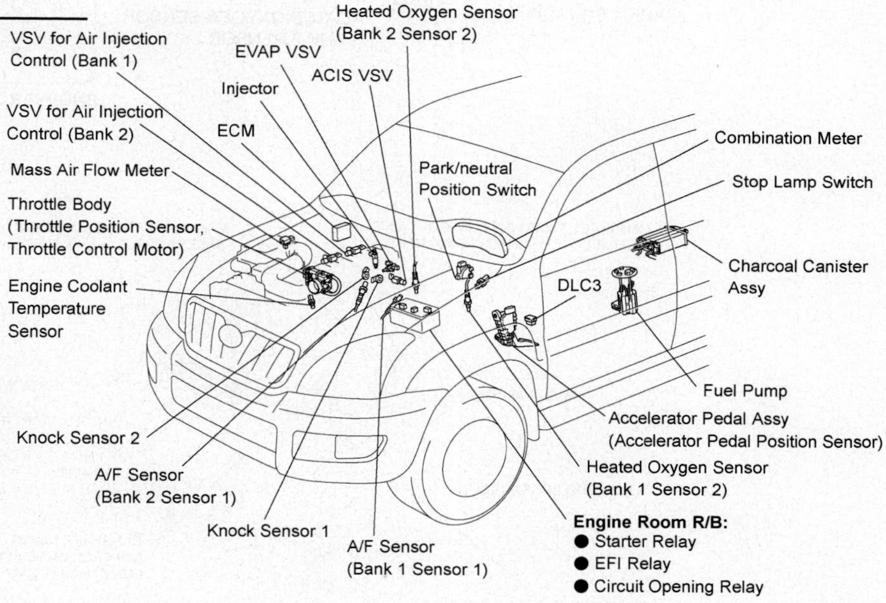

Heated Oxygen Sensor
(Bank 2 Sensor 2)

VSV for Air Injection
Control (Bank 1)

EVAP VSV

ACIS VSV

Injector

VSV for Air Injection
Control (Bank 2)

ECM

Mass Air Flow Meter

Park/neutral
Position Switch

Combination Meter

Stop Lamp Switch

Throttle Body
(Throttle Position Sensor,
Throttle Control Motor)

Charcoal Canister
Assy

Engine Coolant
Temperature
Sensor

DLC3

Fuel Pump

Accelerator Pedal Assy
(Accelerator Pedal Position Sensor)

Knock Sensor 2

A/F Sensor
(Bank 2 Sensor 1)

Heated Oxygen Sensor
(Bank 1 Sensor 2)

Knock Sensor 1

A/F Sensor
(Bank 1 Sensor 1)

Engine Room R/B:
● Starter Relay
● EFI Relay
● Circuit Opening Relay

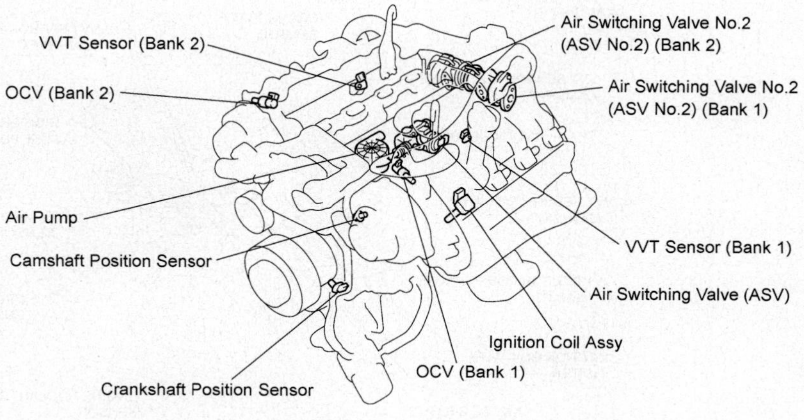

VVT Sensor (Bank 2)

Air Switching Valve No.2
(ASV No.2) (Bank 2)

OCV (Bank 2)

Air Switching Valve No.2
(ASV No.2) (Bank 1)

Air Pump

VVT Sensor (Bank 1)

Camshaft Position Sensor

Air Switching Valve (ASV)

Ignition Coil Assy

Crankshaft Position Sensor

OCV (Bank 1)

IS250

2.5L V6 MFI 4GR-FSE VIN K

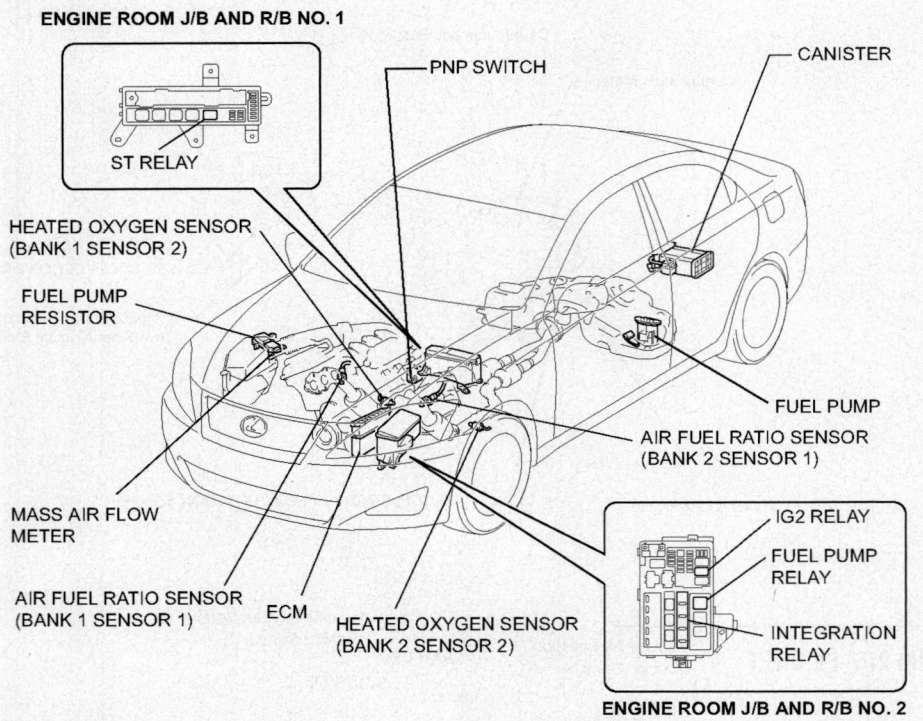

ENGINE ROOM J/B AND R/B NO. 1

ST RELAY

PNP SWITCH

CANISTER

HEATED OXYGEN SENSOR
(BANK 1 SENSOR 2)

FUEL PUMP
RESISTOR

FUEL PUMP

AIR FUEL RATIO SENSOR
(BANK 2 SENSOR 1)

MASS AIR FLOW
METER

IG2 RELAY

FUEL PUMP
RELAY

INTEGRATION
RELAY

AIR FUEL RATIO SENSOR
(BANK 1 SENSOR 1)

ECM

HEATED OXYGEN SENSOR
(BANK 2 SENSOR 2)

ENGINE ROOM J/B AND R/B NO. 2

29157_LEXU_G0017

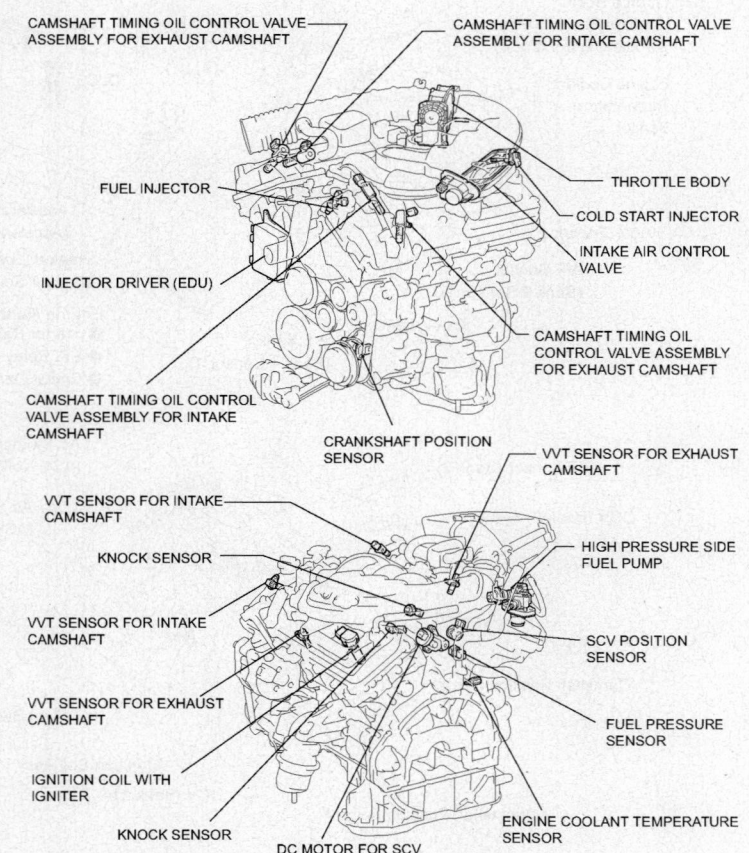

CAMSHAFT TIMING OIL CONTROL VALVE
ASSEMBLY FOR EXHAUST CAMSHAFT

CAMSHAFT TIMING OIL CONTROL VALVE
ASSEMBLY FOR INTAKE CAMSHAFT

FUEL INJECTOR

THROTTLE BODY

COLD START INJECTOR

INTAKE AIR CONTROL
VALVE

INJECTOR DRIVER (EDU)

CAMSHAFT TIMING OIL
CONTROL VALVE ASSEMBLY
FOR EXHAUST CAMSHAFT

CAMSHAFT TIMING OIL CONTROL
VALVE ASSEMBLY FOR INTAKE
CAMSHAFT

CRANKSHAFT POSITION
SENSOR

VVT SENSOR FOR EXHAUST
CAMSHAFT

VVT SENSOR FOR INTAKE
CAMSHAFT

HIGH PRESSURE SIDE
FUEL PUMP

KNOCK SENSOR

VVT SENSOR FOR INTAKE
CAMSHAFT

SCV POSITION
SENSOR

VVT SENSOR FOR EXHAUST
CAMSHAFT

FUEL PRESSURE
SENSOR

IGNITION COIL WITH
IGNITER

KNOCK SENSOR

DC MOTOR FOR SCV

ENGINE COOLANT TEMPERATURE
SENSOR

29157_LEXU_G0018

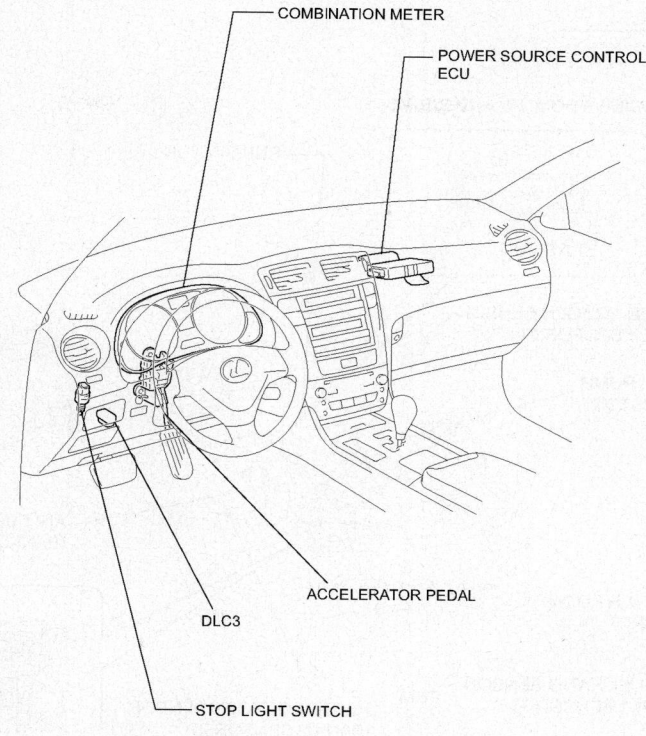

COMBINATION METER

POWER SOURCE CONTROL
ECU

ACCELERATOR PEDAL

DLC3

STOP LIGHT SWITCH

29157_LEXU_G0019

IS300

3.0L I6 MFI 2JZ-GE VIN D

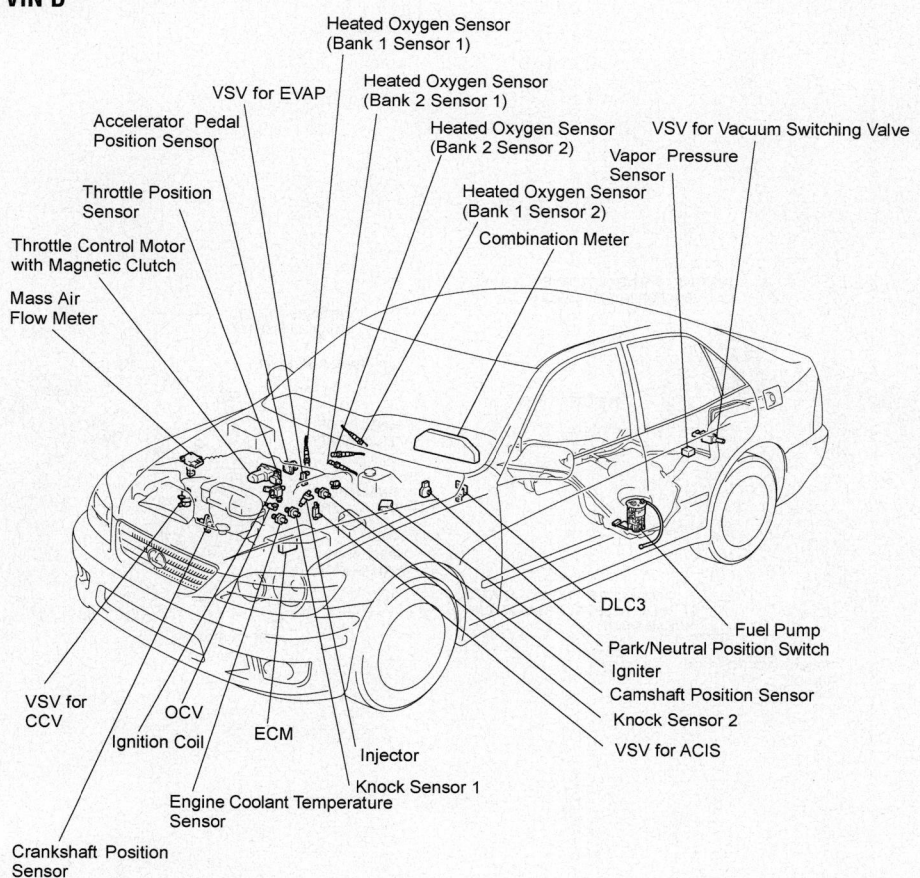

Heated Oxygen Sensor
(Bank 1 Sensor 1)

VSV for EVAP

Heated Oxygen Sensor
(Bank 2 Sensor 1)

Accelerator Pedal
Position Sensor

Heated Oxygen Sensor
(Bank 2 Sensor 2)

VSV for Vacuum Switching Valve

Vapor Pressure
Sensor

Throttle Position
Sensor

Heated Oxygen Sensor
(Bank 1 Sensor 2)

Throttle Control Motor
with Magnetic Clutch

Combination Meter

Mass Air
Flow Meter

DLC3

Fuel Pump

Park/Neutral Position Switch

Igniter

Camshaft Position Sensor

VSV for
CCV

OCV

Knock Sensor 2

Ignition Coil

ECM

VSV for ACIS

Injector

Knock Sensor 1

Engine Coolant Temperature
Sensor

Crankshaft Position
Sensor

29157_LEXU_G0020

IS350

3.5L V6 MFI 2GR-FSE VIN E

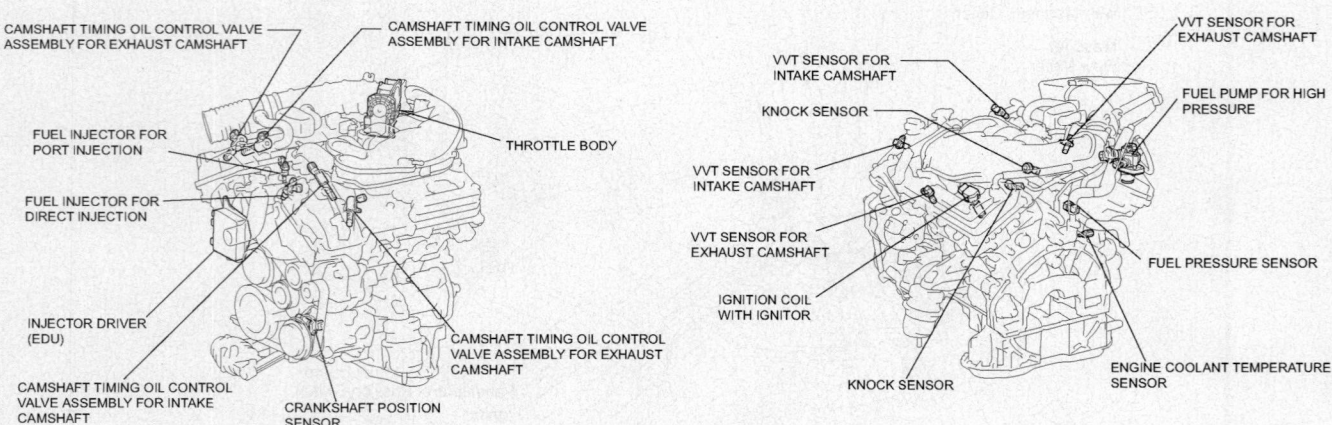

ENGINE ROOM J/B AND R/B NO. 1

ST RELAY

PNP SWITCH

CANISTER

HEATED OXYGEN SENSOR
(BANK 1 SENSOR 2)

FUEL PUMP
RESISTOR

FUEL PUMP

AIR FUEL RATIO SENSOR
(BANK 2 SENSOR 1)

MASS AIR FLOW
METER

IG2 RELAY

FUEL PUMP
RELAY

INTEGRATION
RELAY

AIR FUEL RATIO SENSOR
(BANK 1 SENSOR 1)

ECM

HEATED OXYGEN SENSOR
(BANK 2 SENSOR 2)

ENGINE ROOM J/B AND R/B NO. 2

29157_LEXU_G0021

CAMSHAFT TIMING OIL CONTROL VALVE
ASSEMBLY FOR EXHAUST CAMSHAFT

CAMSHAFT TIMING OIL CONTROL VALVE
ASSEMBLY FOR INTAKE CAMSHAFT

FUEL INJECTOR FOR
PORT INJECTION

THROTTLE BODY

FUEL INJECTOR FOR
DIRECT INJECTION

INJECTOR DRIVER
(EDU)

CAMSHAFT TIMING OIL CONTROL
VALVE ASSEMBLY FOR EXHAUST
CAMSHAFT

CAMSHAFT TIMING OIL CONTROL
VALVE ASSEMBLY FOR INTAKE
CAMSHAFT

CRANKSHAFT POSITION
SENSOR

VVT SENSOR FOR
EXHAUST CAMSHAFT

VVT SENSOR FOR
INTAKE CAMSHAFT

KNOCK SENSOR

FUEL PUMP FOR HIGH
PRESSURE

VVT SENSOR FOR
INTAKE CAMSHAFT

VVT SENSOR FOR
EXHAUST CAMSHAFT

IGNITION COIL
WITH IGNITOR

FUEL PRESSURE SENSOR

KNOCK SENSOR

ENGINE COOLANT TEMPERATURE
SENSOR

29157_LEXU_G0022

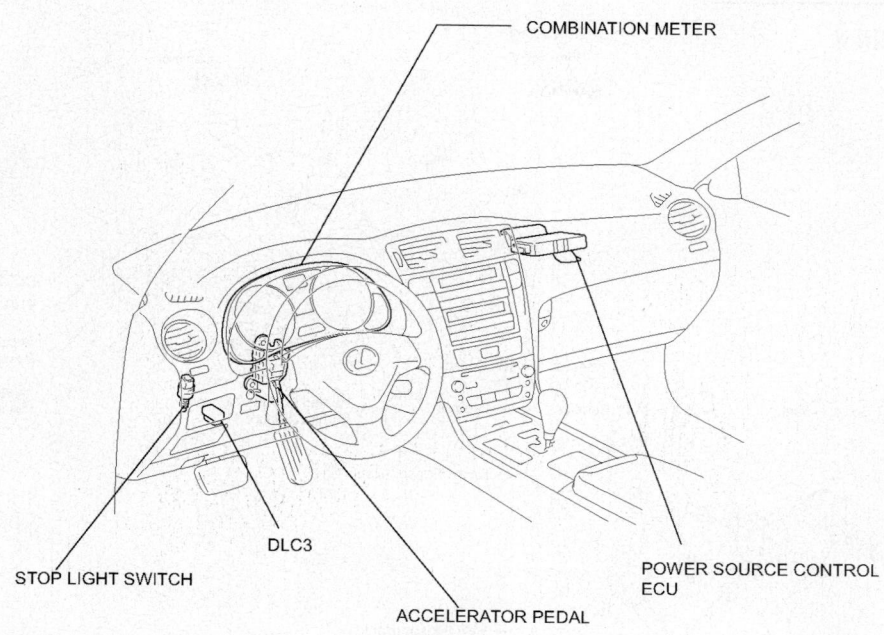

COMBINATION METER

STOP LIGHT SWITCH

DLC3

ACCELERATOR PEDAL

POWER SOURCE CONTROL ECU

29157_LEXU_G0023

LS400

4.0L V8 MFI 1UZ-FE VIN H

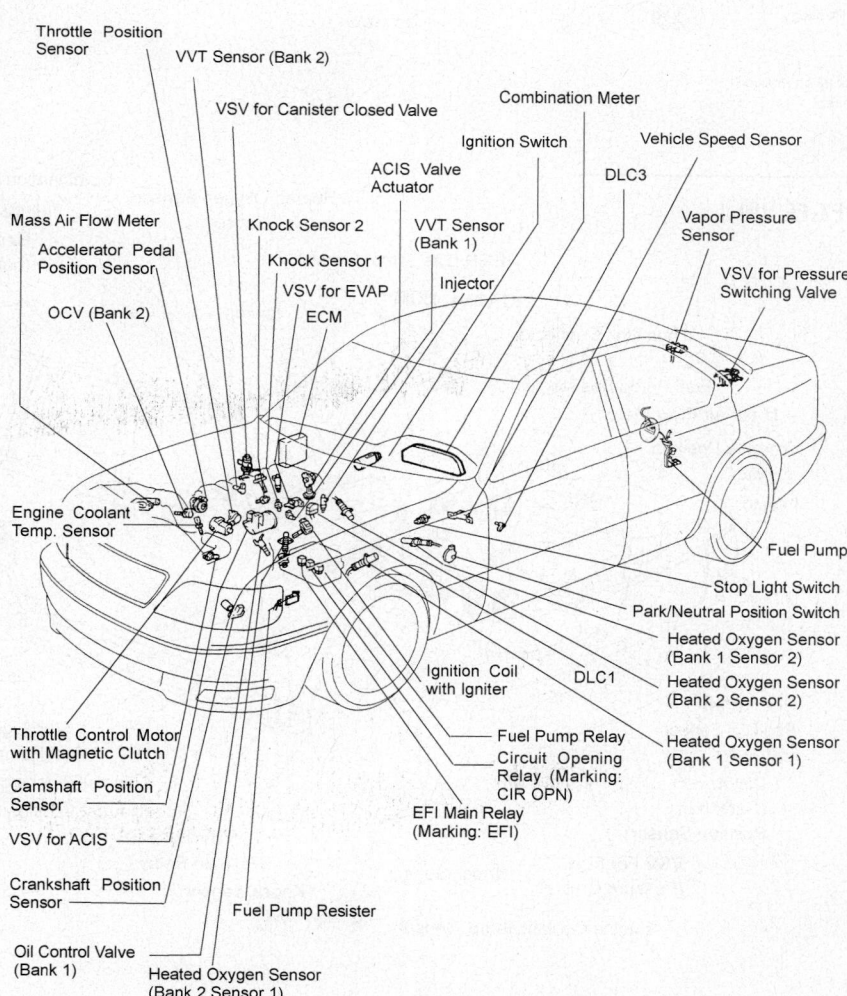

Throttle Position Sensor

VVT Sensor (Bank 2)

VSV for Canister Closed Valve

Combination Meter

Ignition Switch

ACIS Valve Actuator

DLC3

Vehicle Speed Sensor

Mass Air Flow Meter

Knock Sensor 2

VVT Sensor (Bank 1)

Vapor Pressure Sensor

Accelerator Pedal Position Sensor

Knock Sensor 1

VSV for Pressure Switching Valve

OCV (Bank 2)

VSV for EVAP

Injector

ECM

Engine Coolant Temp. Sensor

Fuel Pump

Stop Light Switch

Park/Neutral Position Switch

Heated Oxygen Sensor (Bank 1 Sensor 2)

Ignition Coil with Igniter

DLC1

Heated Oxygen Sensor (Bank 2 Sensor 2)

Throttle Control Motor with Magnetic Clutch

Fuel Pump Relay

Heated Oxygen Sensor (Bank 1 Sensor 1)

Camshaft Position Sensor

Circuit Opening Relay (Marking: CIR OPN)

VSV for ACIS

EFI Main Relay (Marking: EFI)

Crankshaft Position Sensor

Fuel Pump Resister

Oil Control Valve (Bank 1)

Heated Oxygen Sensor (Bank 2 Sensor 1)

29157_LEXU_G0024

LS430

4.3L V8 MFI 3UZ-FE VIN N

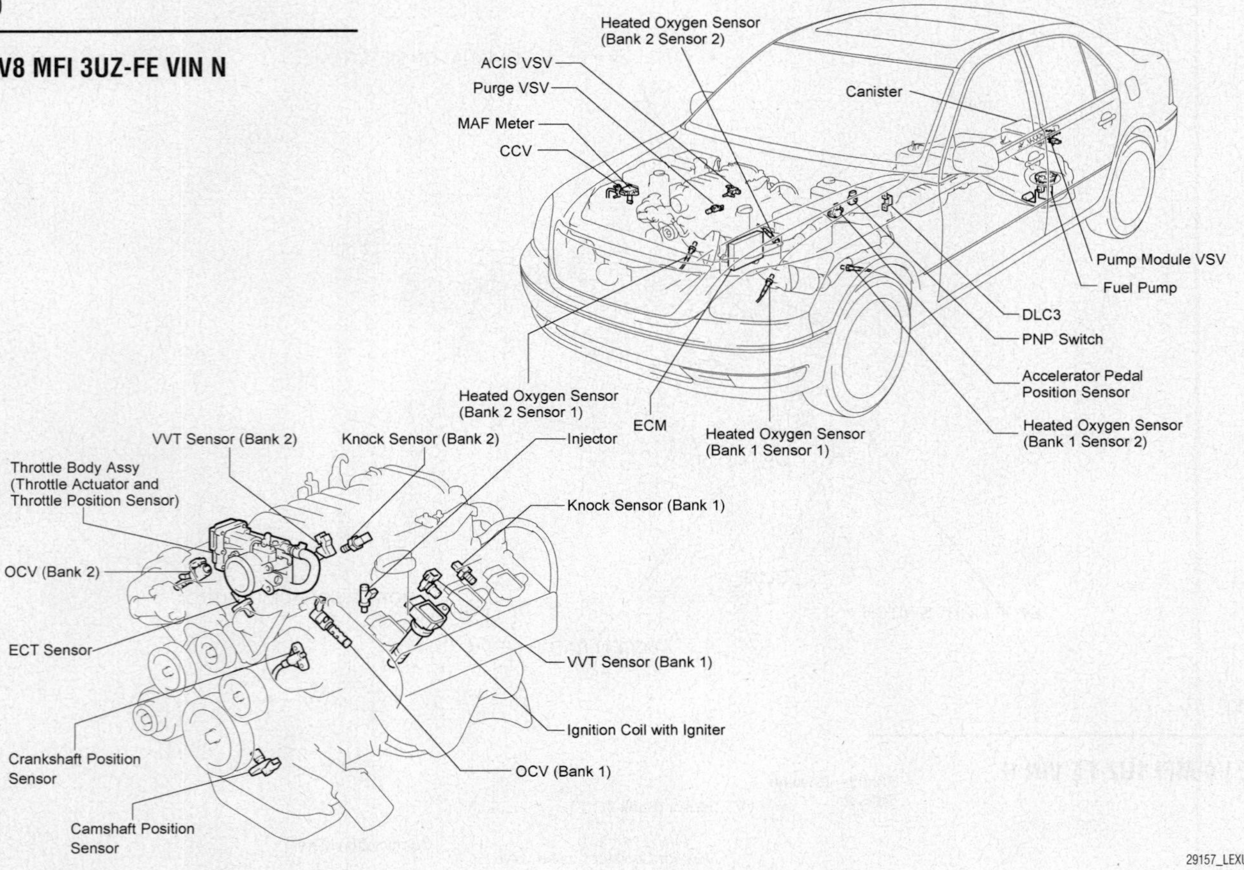

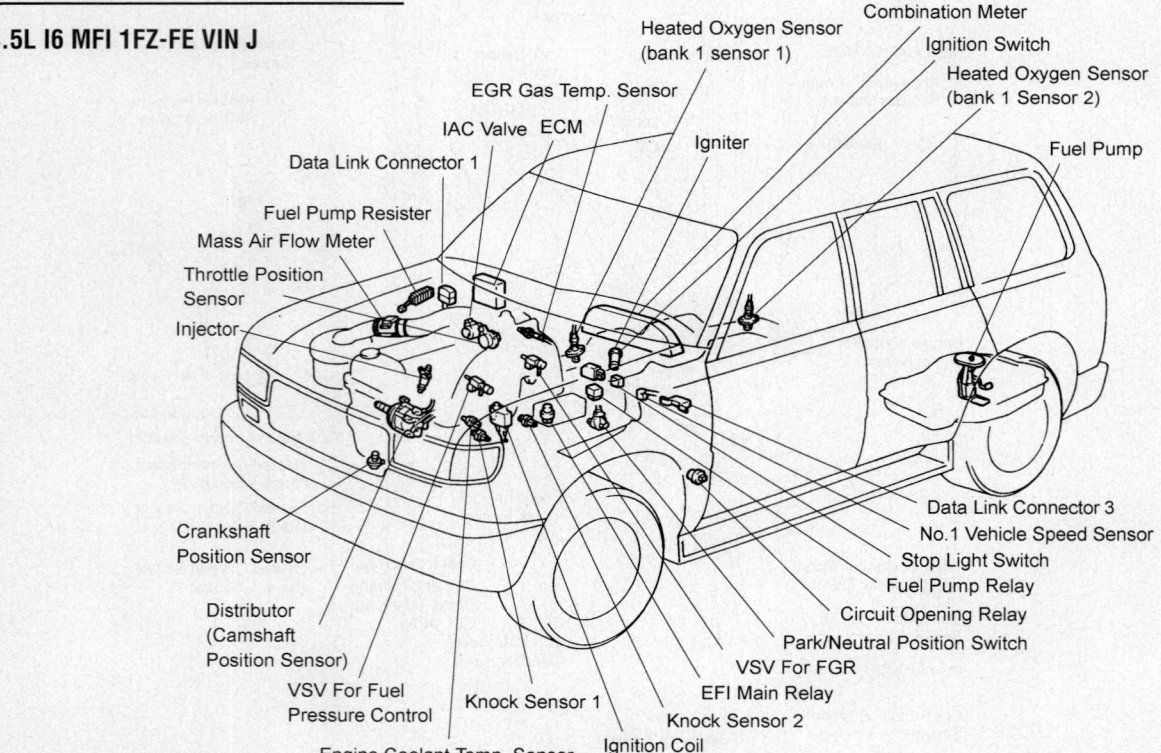

Heated Oxygen Sensor (Bank 2 Sensor 2)
ACIS VSV
Purge VSV
MAF Meter
CCV
Canister
Pump Module VSV
Fuel Pump
DLC3
PNP Switch
Accelerator Pedal Position Sensor
Heated Oxygen Sensor (Bank 1 Sensor 2)
Heated Oxygen Sensor (Bank 2 Sensor 1)
ECM
Heated Oxygen Sensor (Bank 1 Sensor 1)

VVT Sensor (Bank 2)
Knock Sensor (Bank 2)
Injector
Throttle Body Assy (Throttle Actuator and Throttle Position Sensor)
Knock Sensor (Bank 1)
OCV (Bank 2)
ECT Sensor
VVT Sensor (Bank 1)
Crankshaft Position Sensor
Ignition Coil with Igniter
Camshaft Position Sensor
OCV (Bank 1)

29157_LEXU_G0025

LX450

4.5L I6 MFI 1FZ-FE VIN J

Heated Oxygen Sensor (bank 1 sensor 1)
Combination Meter
Ignition Switch
Heated Oxygen Sensor (bank 1 Sensor 2)
EGR Gas Temp. Sensor
IAC Valve ECM
Igniter
Fuel Pump
Data Link Connector 1
Fuel Pump Resister
Mass Air Flow Meter
Throttle Position Sensor
Injector
Crankshaft Position Sensor
Distributor (Camshaft Position Sensor)
VSV For Fuel Pressure Control
Knock Sensor 1
Engine Coolant Temp. Sensor
Ignition Coil
Knock Sensor 2
EFI Main Relay
VSV For FGR
Park/Neutral Position Switch
Circuit Opening Relay
Fuel Pump Relay
Stop Light Switch
No.1 Vehicle Speed Sensor
Data Link Connector 3

29157_LEXU_G0026

LX470

4.7L V8 MFI 2UZ-FE VIN T

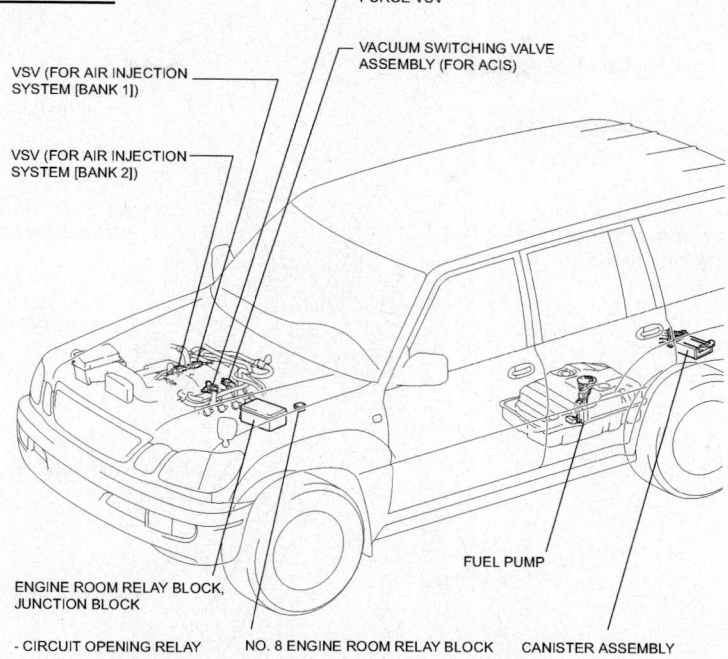

PURGE VSV

VACUUM SWITCHING VALVE
ASSEMBLY (FOR ACIS)

VSV (FOR AIR INJECTION
SYSTEM [BANK 1])

VSV (FOR AIR INJECTION
SYSTEM [BANK 2])

FUEL PUMP

ENGINE ROOM RELAY BLOCK,
JUNCTION BLOCK

- CIRCUIT OPENING RELAY
- EFI RELAY
- FUEL PUMP RELAY
- STARTER RELAY
- EFI OR ECD NO. 1 FUSE
- ETCS FUSE

NO. 8 ENGINE ROOM RELAY BLOCK

- A/F RELAY

CANISTER ASSEMBLY

- CANISTER PUMP MODULE

29157_LEXU_G0027

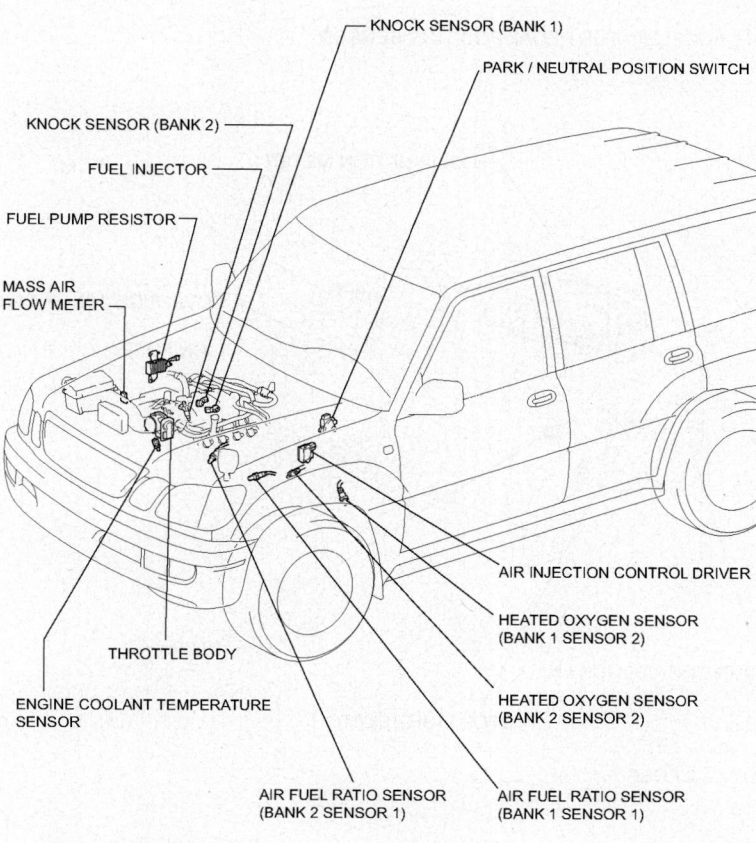

KNOCK SENSOR (BANK 1)

PARK / NEUTRAL POSITION SWITCH

KNOCK SENSOR (BANK 2)

FUEL INJECTOR

FUEL PUMP RESISTOR

MASS AIR
FLOW METER

AIR INJECTION CONTROL DRIVER

HEATED OXYGEN SENSOR
(BANK 1 SENSOR 2)

THROTTLE BODY

HEATED OXYGEN SENSOR
(BANK 2 SENSOR 2)

ENGINE COOLANT TEMPERATURE
SENSOR

AIR FUEL RATIO SENSOR
(BANK 2 SENSOR 1)

AIR FUEL RATIO SENSOR
(BANK 1 SENSOR 1)

29157_LEXU_G0028

PRESSURE SENSOR

VVT SENSOR (BANK 2)

NO. 2 AIR SWITCHING VALVE (BANK 2)

NO. 2 AIR SWITCHING VALVE (BANK 1)

CAMSHAFT TIMING OIL CONTROL VALVE ASSEMBLY (BANK 2)

IGNITION COIL

VVT SENSOR (BANK 1)

AIR SWITCHING VALVE

AIR PUMP

CAMSHAFT POSITION SENSOR

CRANKSHAFT POSITION SENSOR

CAMSHAFT TIMING OIL CONTROL VALVE ASSEMBLY (BANK 1)

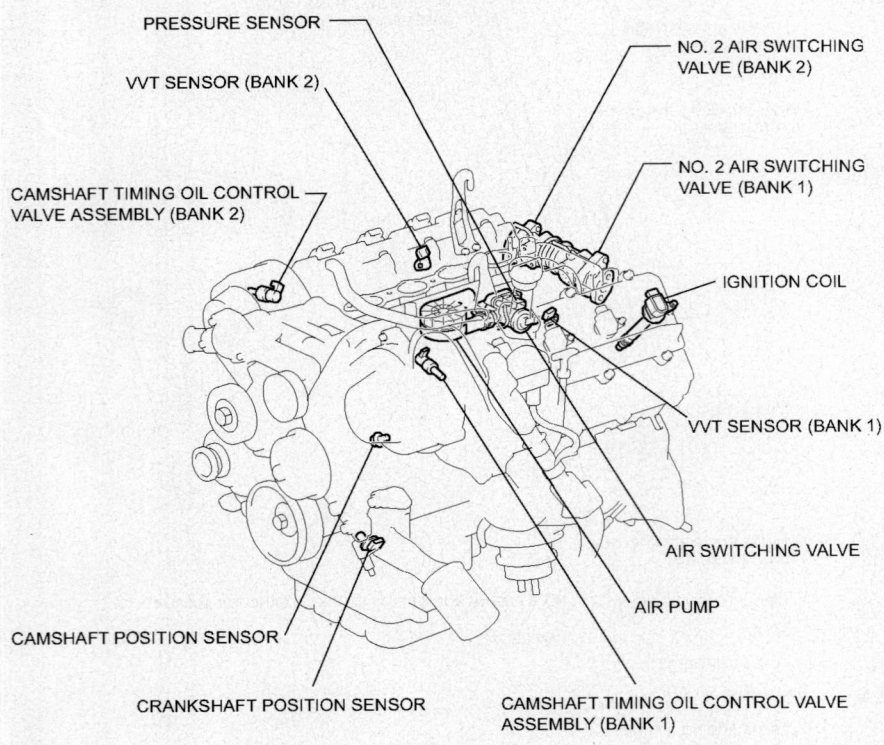

29157_LEXU_G0029

ACCELERATOR PEDAL POSITION SENSOR

COMBINATION METER

ECM

COWL SIDE JUNCTION BLOCK RH

- IGN FUSE

COWL SIDE JUNCTION BLOCK LH

- STOP FUSE

- EFI OR ECD NO. 2 FUSE

STOP LIGHT SWITCH

DLC3

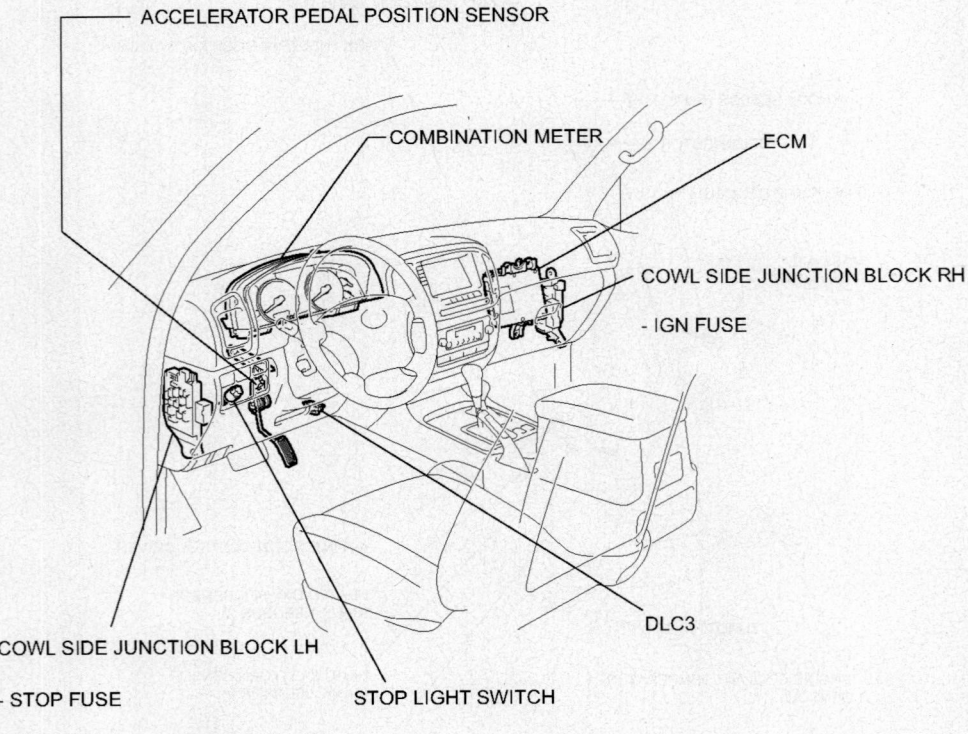

29157_LEXU_G0030

RX300

3.0L V6 MFI 1MZ-FE VIN F

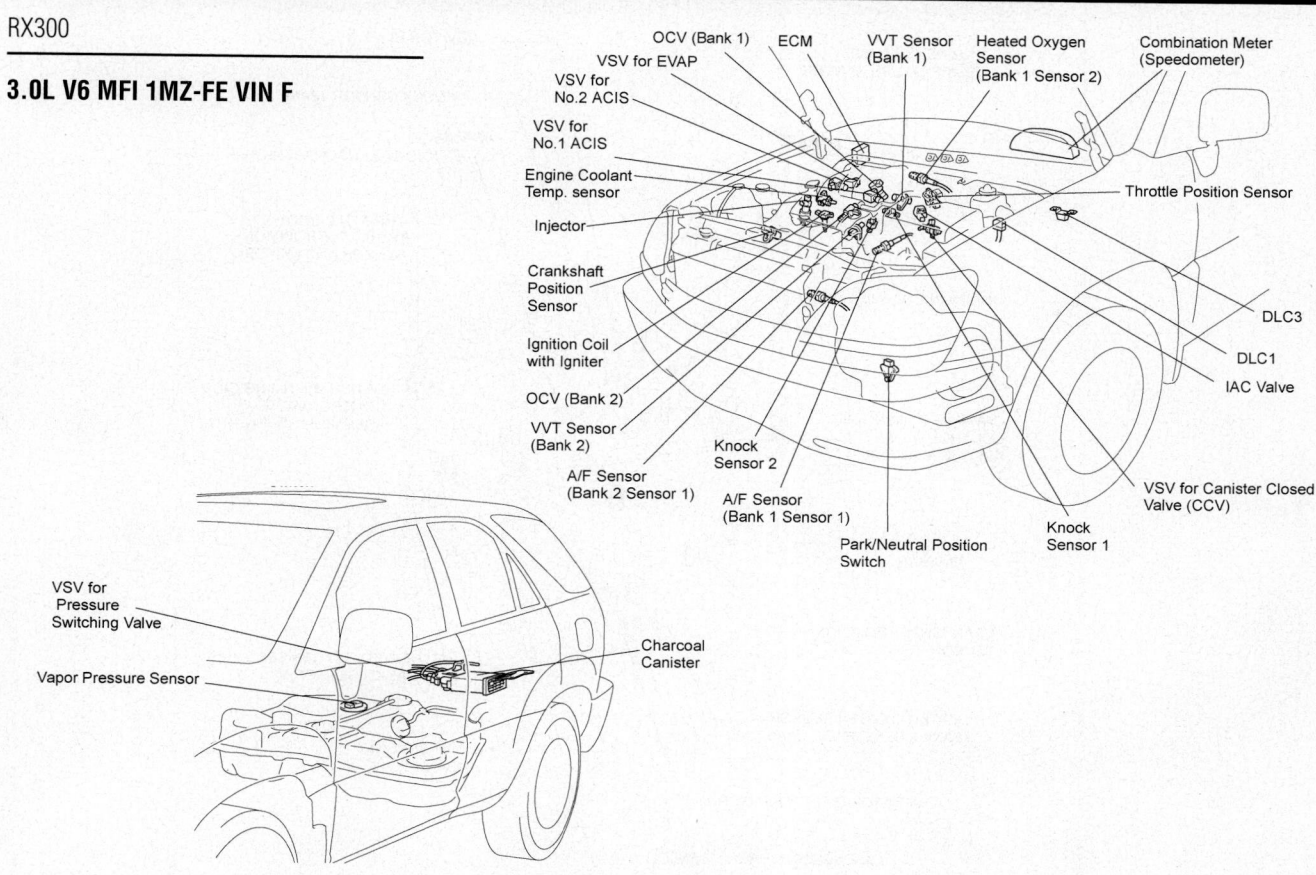

OCV (Bank 1)
ECM
VVT Sensor (Bank 1)
Heated Oxygen Sensor (Bank 1 Sensor 2)
Combination Meter (Speedometer)
VSV for EVAP
VSV for No.2 ACIS
VSV for No.1 ACIS
Engine Coolant Temp. sensor
Injector
Crankshaft Position Sensor
Ignition Coil with Igniter
OCV (Bank 2)
VVT Sensor (Bank 2)
A/F Sensor (Bank 2 Sensor 1)
A/F Sensor (Bank 1 Sensor 1)
Knock Sensor 2
Park/Neutral Position Switch
Knock Sensor 1
Throttle Position Sensor
DLC3
DLC1
IAC Valve
VSV for Canister Closed Valve (CCV)

VSV for Pressure Switching Valve
Vapor Pressure Sensor
Charcoal Canister

29157_LEXU_G0031

RX330

3.3L V6 MFI 3MZ-FE VIN A

CHARCOAL CANISTER ASSEMBLY
VSV FOR CCV
R/B NO. 1: CIRCUIT OPENING RELAY
CHARCOAL CANISTER FILTER
COMBINATION METER ASSEMBLY
MASS AIR FLOW METER
VAPOR PRESSURE SENSOR ASSEMBLY
ECM
FUEL PUMP
VSV FOR ACM
INSTRUMENT PANEL J/B:
- IGN FUSE
DLC3
VSV FOR AICV
ACCELERATOR PEDAL ROD
(ACCELERATOR PEDAL POSITION SENSOR)

ENGINE ROOM R/B AND FUSIBLE LINK BLOCK:

- EFI RELAY

- AIR FUEL RATIO SENSOR HEATER RELAY

- EFI NO. 1 FUSE

- EFI NO. 2 FUSE

- ETCS FUSE

29157_LEXU_G0032

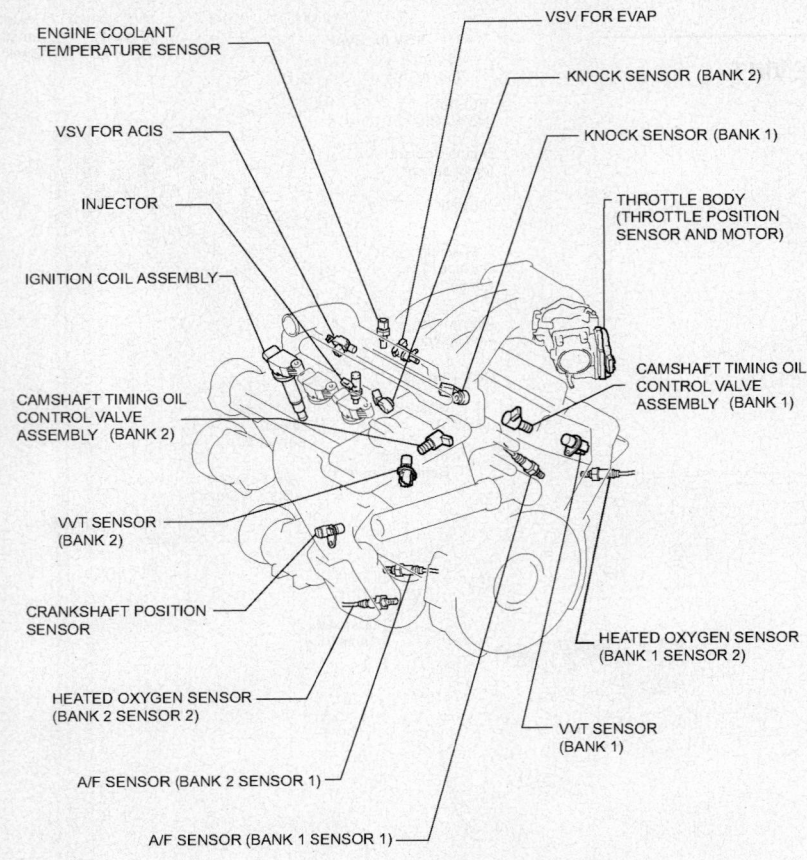

ENGINE COOLANT
TEMPERATURE SENSOR

VSV FOR ACIS

INJECTOR

IGNITION COIL ASSEMBLY

CAMSHAFT TIMING OIL
CONTROL VALVE
ASSEMBLY (BANK 2)

VVT SENSOR
(BANK 2)

CRANKSHAFT POSITION
SENSOR

HEATED OXYGEN SENSOR
(BANK 2 SENSOR 2)

A/F SENSOR (BANK 2 SENSOR 1)

A/F SENSOR (BANK 1 SENSOR 1)

VSV FOR EVAP

KNOCK SENSOR (BANK 2)

KNOCK SENSOR (BANK 1)

THROTTLE BODY
(THROTTLE POSITION
SENSOR AND MOTOR)

CAMSHAFT TIMING OIL
CONTROL VALVE
ASSEMBLY (BANK 1)

HEATED OXYGEN SENSOR
(BANK 1 SENSOR 2)

VVT SENSOR
(BANK 1)

29157_LEXU_G0033

SC300

3.0L I6 MFI 2JZ-GE VIN D

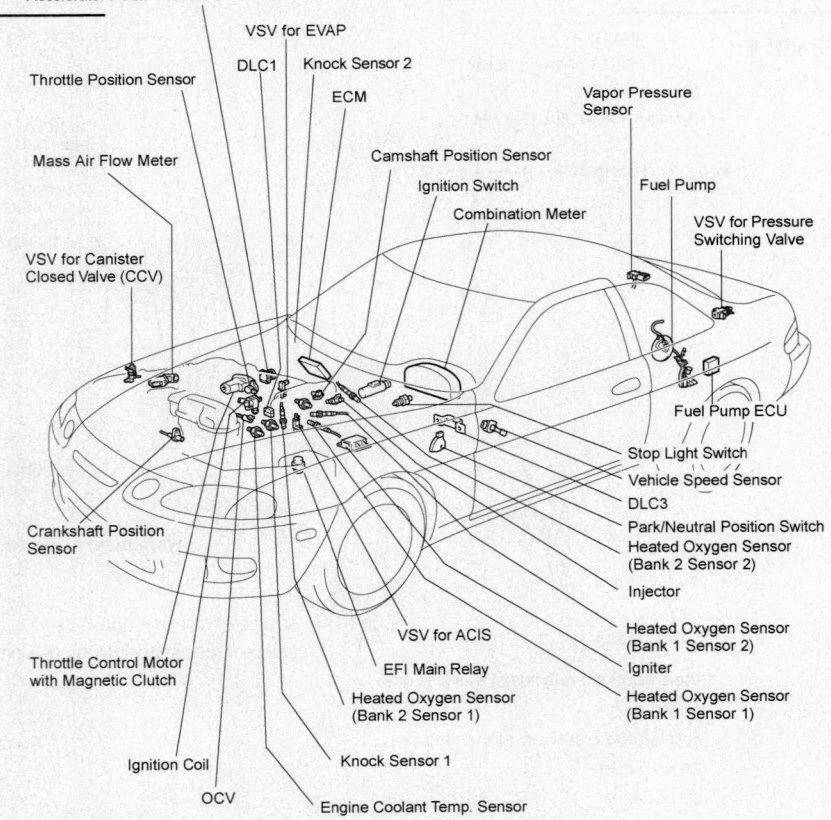

Accelerator Pedal Position Sensor

VSV for EVAP

DLC1

Throttle Position Sensor

ECM

Knock Sensor 2

Mass Air Flow Meter

Camshaft Position Sensor

Ignition Switch

Combination Meter

Vapor Pressure
Sensor

Fuel Pump

VSV for Pressure
Switching Valve

VSV for Canister
Closed Valve (CCV)

Crankshaft Position
Sensor

Throttle Control Motor
with Magnetic Clutch

Ignition Coil

OCV

VSV for ACIS

EFI Main Relay

Heated Oxygen Sensor
(Bank 2 Sensor 1)

Knock Sensor 1

Engine Coolant Temp. Sensor

Fuel Pump ECU

Stop Light Switch

Vehicle Speed Sensor

DLC3

Park/Neutral Position Switch

Heated Oxygen Sensor
(Bank 2 Sensor 2)

Injector

Heated Oxygen Sensor
(Bank 1 Sensor 2)

Igniter

Heated Oxygen Sensor
(Bank 1 Sensor 1)

29157_LEXU_G0034

SC400

4.0L V8 MFI 1UZ-FE VIN H

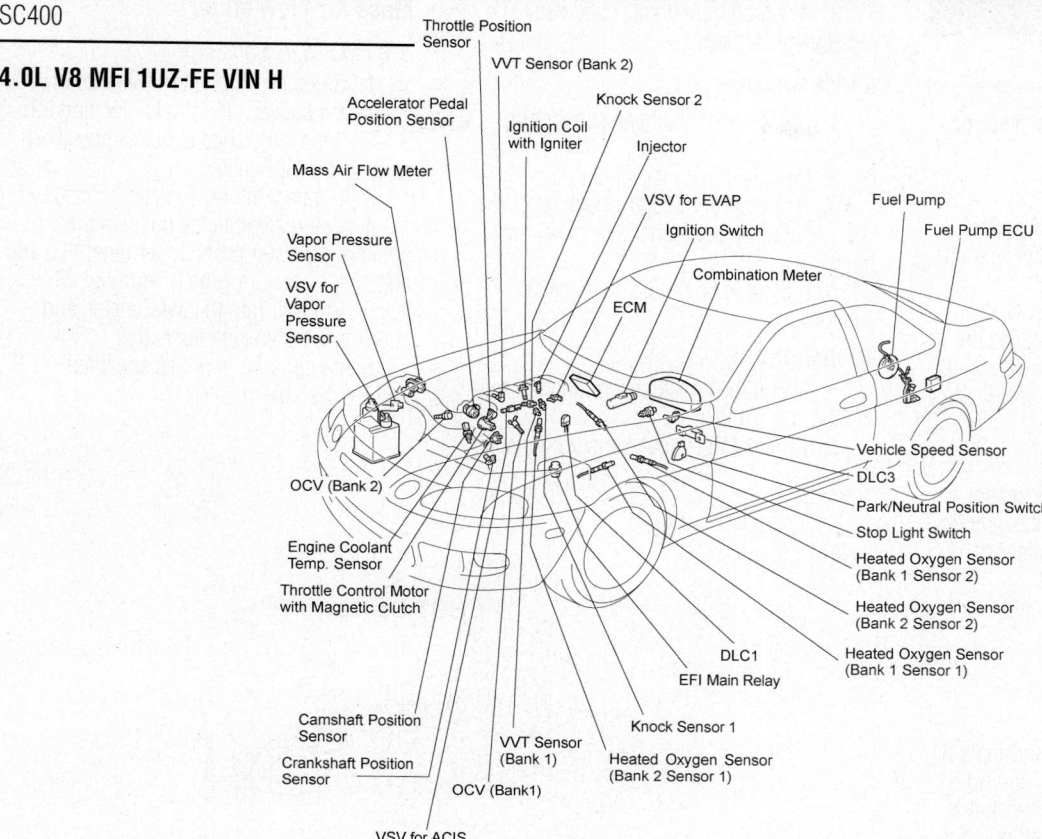

Throttle Position Sensor
VVT Sensor (Bank 2)
Knock Sensor 2
Accelerator Pedal Position Sensor
Ignition Coil with Igniter
Injector
Mass Air Flow Meter
VSV for EVAP
Ignition Switch
Fuel Pump
Fuel Pump ECU
Vapor Pressure Sensor
Combination Meter
VSV for Vapor Pressure Sensor
ECM
OCV (Bank 2)
Vehicle Speed Sensor
DLC3
Park/Neutral Position Switch
Engine Coolant Temp. Sensor
Stop Light Switch
Throttle Control Motor with Magnetic Clutch
Heated Oxygen Sensor (Bank 1 Sensor 2)
Heated Oxygen Sensor (Bank 2 Sensor 2)
Heated Oxygen Sensor (Bank 1 Sensor 1)
DLC1
EFI Main Relay
Camshaft Position Sensor
Knock Sensor 1
Crankshaft Position Sensor
VVT Sensor (Bank 1)
Heated Oxygen Sensor (Bank 2 Sensor 1)
OCV (Bank1)
VSV for ACIS

29157_LEXU_G0035

SC430

4.3L V8 MFI 3UZ-FE VIN N

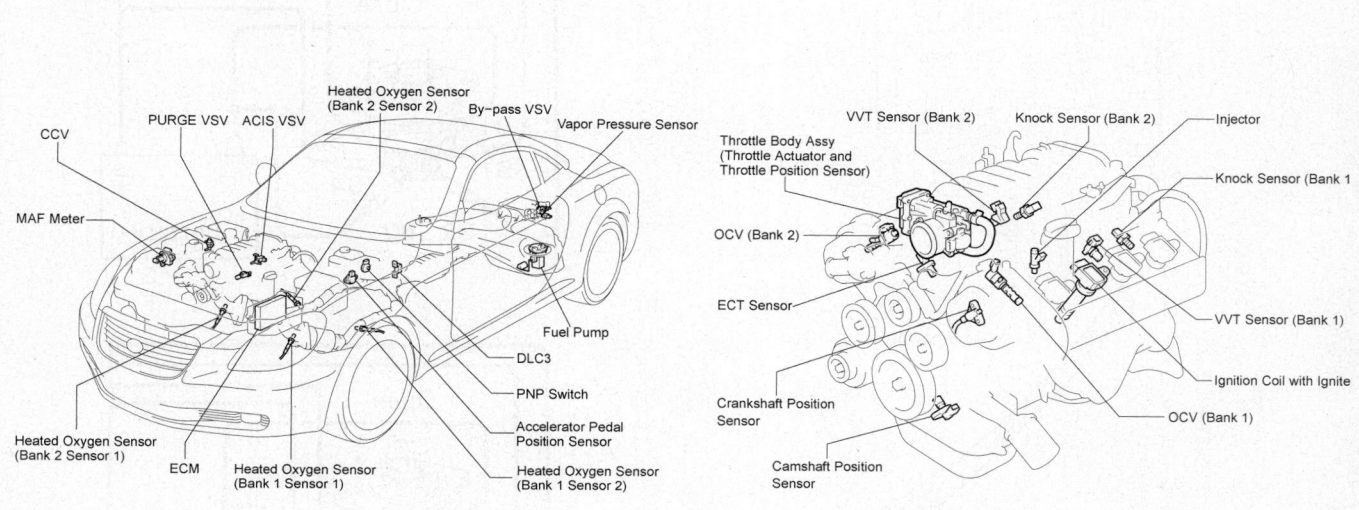

CCV
PURGE VSV
ACIS VSV
Heated Oxygen Sensor (Bank 2 Sensor 2)
By-pass VSV
Vapor Pressure Sensor
VVT Sensor (Bank 2)
Knock Sensor (Bank 2)
Injector
Throttle Body Assy (Throttle Actuator and Throttle Position Sensor)
Knock Sensor (Bank 1
MAF Meter
OCV (Bank 2)
ECT Sensor
VVT Sensor (Bank 1)
Fuel Pump
Ignition Coil with Ignite
DLC3
PNP Switch
Crankshaft Position Sensor
OCV (Bank 1)
Heated Oxygen Sensor (Bank 2 Sensor 1)
ECM
Accelerator Pedal Position Sensor
Heated Oxygen Sensor (Bank 1 Sensor 1)
Heated Oxygen Sensor (Bank 1 Sensor 2)
Camshaft Position Sensor

29157_LEXU_G0036

Component Testing

1FZ-FE VIN J

Engine Coolant Temperature Sensor

See Figures 1 and 2.

1. Drain engine coolant.
2. Disconnect the sensor connector.
3. Using a 19 mm deep socket wrench, remove the sensor and gasket.
4. Using an ohmmeter, measure the resistance between terminals. Refer to the illustration.
5. If the resistance is not as specified, replace the sensor.

To install:

6. Install a new gasket to the sensor.
7. Using a 19 mm deep socket wrench, install the sensor. Tighten the sensor to 17.5 ft. lbs. (24.5 Nm).
8. Connect the sensor connector.
9. Refill with engine coolant.

Idle Air Control Valve

See Figure 3.

1. Apply battery voltage to terminals B1 and B2, and while repeatedly grounding S1-S2-S3-S4-S1 in sequence, and check that the valve moves toward the closed position.
2. Apply battery voltage to terminals B1 and B2, and while repeatedly grounding S4-S3-S2-S1-S4 in sequence, and check that the valve moves toward the open position.

3. If operation is not as specified, replace the IAC valve.

Knock Sensor

1. Disconnect the knock sensor connector.
2. Remove the knock sensor.
3. Using an ohmmeter, check that there is no continuity between the terminal and body.
4. If there is continuity, replace the sensor.

To install:

5. Install the knock sensor and tighten to 33 ft. lbs. (44 Nm).
6. Connect the knock sensor connector.

Mass Air Flow Meter

See Figures 4, 5 and 6.

1. Using an ohmmeter, measure the resistance between terminals THA and E2.
2. If the resistance is not as specified, replace the MAF meter.
3. Connect the MAF meter connector.
4. Using a voltmeter, connect the positive (+) tester probe to terminal VG, and negative (-) tester probe to terminal E3.
5. Blow air into the MAF meter, and check that the voltage fluctuates.
6. If operation is not as specified, replace the MAF meter.

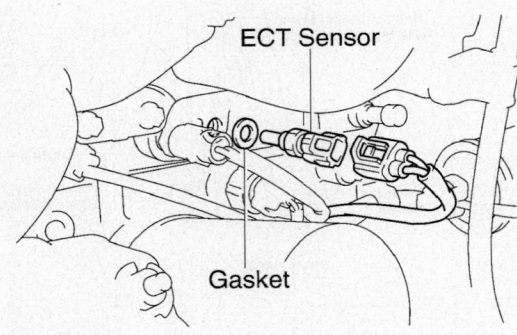

29157_LEXU_G0039

Fig. 1 Remove the ECT sensor

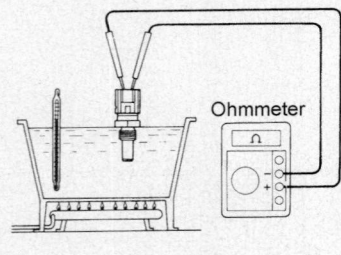

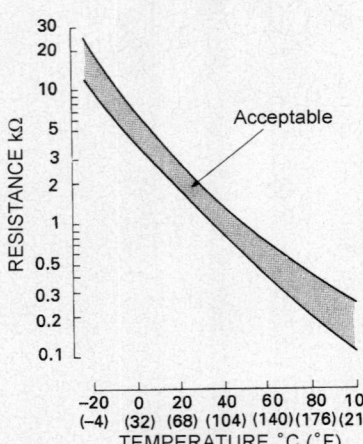

29157_LEXU_G0040

Fig. 2 ECT temperature/resistance graph

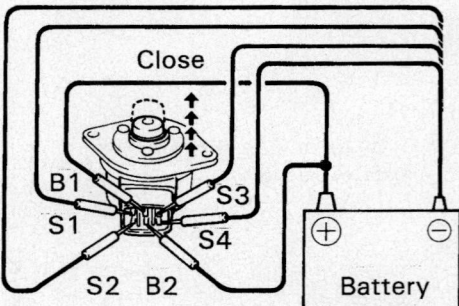

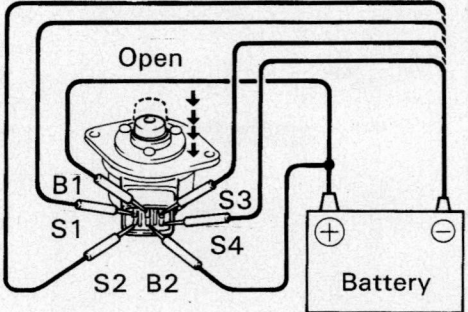

29157_LEXU_G0041

Fig. 3 Testing the IAC valve

Heated Oxygen Sensor

See Figure 7.

1. Disconnect the heated oxygen sensor connector.

2. Using an ohmmeter, measure the resistance between terminals +B and HT.

3. Resistance should be 11 - 16 ohms at 20°C (68°F) If resistance is not as specified, replace the heated oxygen sensor.

4. Reconnect the heated oxygen sensor connector.

Throttle Position Sensor

See Figures 8, 9, 10 and 11.

1. Apply vacuum to the throttle opener.

2. Insert a 0.50 mm (0.020 in.) or 0.75 mm (0.030 in.) feeler gauge between the throttle stop screw and stop lever.

3. Using an ohmmeter, measure the resistance between each terminal.

4. If necessary, adjust throttle position sensor as follows:

   a. Loosen the 2 set screws of the sensor.

   b. Apply vacuum to the throttle opener.

   c. Insert a 0.62 mm (0.024 in.) feeler gauge between the throttle stop screw and stop lever.

   d. Connect the test probe of an ohmmeter to the terminals IDL and E2 of the sensor.

   e. Gradually turn the sensor clockwise until the ohmmeter deflects, and secure it with the 2 set screws.

   f. Recheck the continuity between terminals IDL and E2.

1MZ-FE VIN F

Engine Coolant Temperature Sensor

See Figure 12.

1. Drain engine coolant.

2. Disconnect the ECT sensor connector.

3. Using a 19 mm deep socket wrench, remove the ECT sensor and gasket.

4. Using an ohmmeter, measure the resistance between the terminals.

5. If the resistance is not as specified, replace the ECT sensor.

To install:

6. Install a new gasket to the ECT sensor.

7. Using a 19 mm deep socket, install the ECT sensor. Tighten to 14 ft. lbs. (20 Nm).

8. Connect the ECT sensor connector.

9. Refill with engine coolant.

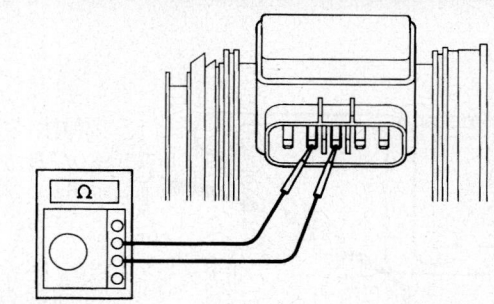

Fig. 4 Terminals THA and E2

29157_LEXU_G0042

Between terminals	Resistance	Temperature
THA - E2	10 - 20 kΩ	-20 °C (-4°F)
THA - E2	4 - 7 kΩ	0°C (32°F)
THA - E2	2 - 3 kΩ	20°C (68°F)
THA - E2	0.9 - 1.3 kΩ	40°C (104°F)
THA - E2	0.4 - 0.7 kΩ	60°C (140°F)
THA - E2	0.2 - 0.4 kΩ	80°C (176°F)

Fig. 5 Temperature/resistance table

29157_LEXU_G0043

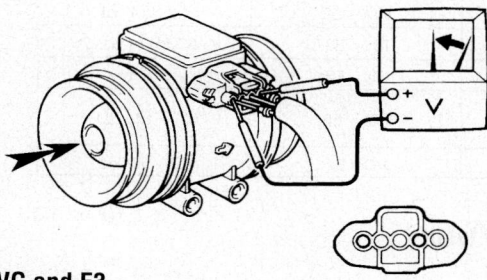

Fig. 6 Terminals VG and E3

29157_LEXU_G0044

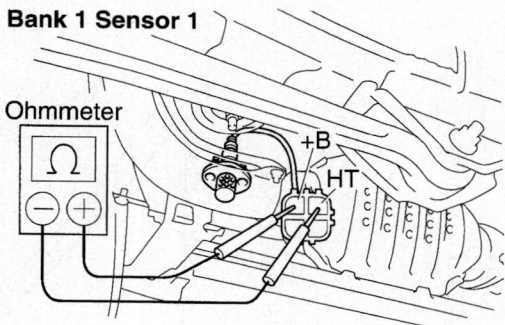
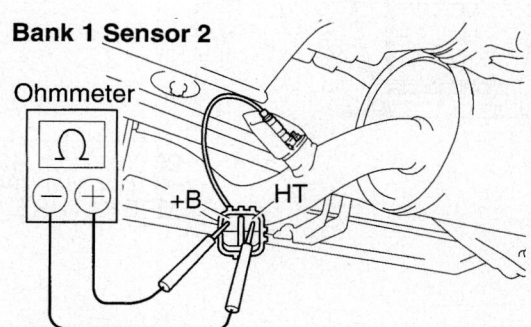

Bank 1 Sensor 1

Ohmmeter

Bank 1 Sensor 2

Ohmmeter

Fig. 7 Testing the Heated Oxygen Sensors

29157_LEXU_G0045

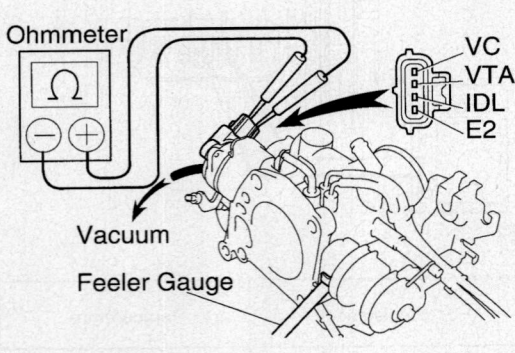

Fig. 8 Throttle Position Sensor testing

29157_LEXU_G0046

Clearance between lever and stop screw	Between terminals	Resistance
0 mm (0 in.)	VTA - E2	0.2 - 5.7 kΩ
0.50 mm (0.020 in.)	IDL - E2	2.3 kΩ or less
0.75 mm (0.030 in.)	IDL - E2	Infinity
Throttle valve fully open	VTA - E2	2.0 - 10.2 kΩ
-	VC - E2	2.5 - 5.9 kΩ

29157_LEXU_G0047

Fig. 9 Resistance table

Fig. 10 Throttle Position Sensor adjustment

29157_LEXU_G0048

Clearance between lever and stop screw	Continuity (IDL - E2)
0.50 mm (0.020 in.)	Continuity
0.75 mm (0.030 in.)	No continuity

29157_LEXU_G0049

Fig. 11 Recheck the TPS adjustment

Idle Air Control Valve

See Figure 13.

1. Connect the positive (+) lead from the battery to terminal +B and negative (-) lead to terminal RSC, and check that the valve is closed.

2. Connect the positive (+) lead from the battery to terminal +B and negative (-) lead to terminal RSO, and check that the valve is open.

3. If operation is not as specified, replace the IAC valve.

Oxygen Sensor

See Figures 14 and 15.

1. Disconnect the oxygen sensor connector.

2. Using an ohmmeter, measure the resistance between the terminals +B and HT. Resistance specification: 11 - 16 ohms at 20°C (68°F).

3. If the resistance is not as specified, replace the sensor.

4. Reconnect the oxygen sensor connector.

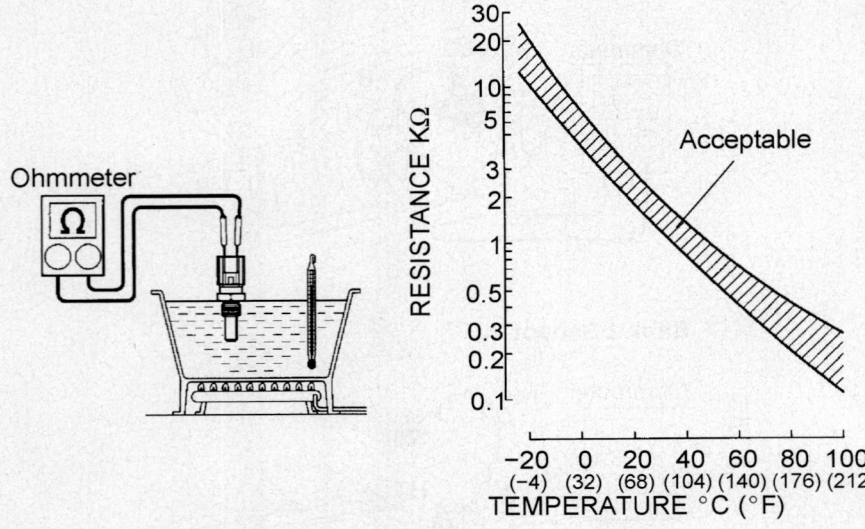

29157_LEXU_G0050

Fig. 12 ECT testing

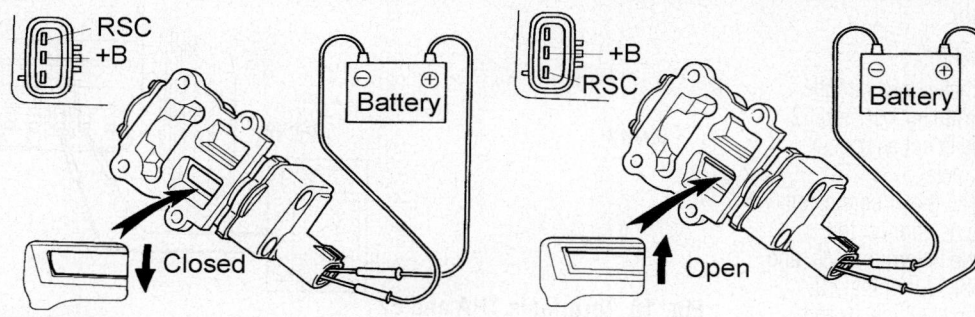

Fig. 13 IAC valve testing

29157_LEXU_G0051

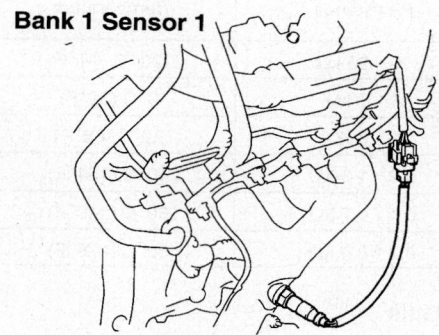

Bank 1 Sensor 1

Bank 2 Sensor 1

Fig. 14 Oxygen Sensor Connector locations

29157_LEXU_G0052

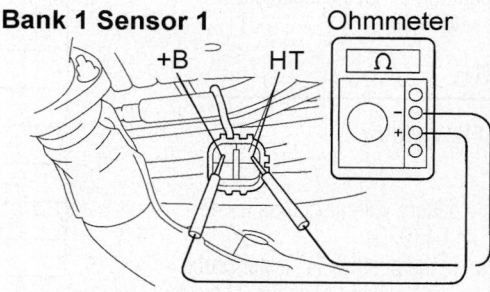

Bank 1 Sensor 1

Bank 2 Sensor 1

Fig. 15 Oxygen Sensor testing

29157_LEXU_G0053

Throttle Position Sensor

See Figures 16 and 17.

1. Apply vacuum to the throttle opener.

2. Disconnect the sensor connector.

3. Insert a feeler gauge between the throttle stop screw and stop lever.

4. Using an ohmmeter, measure the resistance between each terminal.

5. If the resistance is not as specified, replace the sensor.

6. Reconnect the sensor connector.

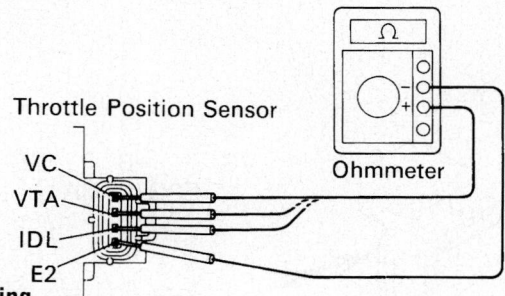

Fig. 16 TPS testing

29157_LEXU_G0054

Clearance between lever and stop screw	Between terminals	Resistance
0 mm (0 in.)	VTA - E2	0.28 - 6.4 kΩ
0.35 mm (0.014 in.)	IDL - E2	0.5 kΩ or less
0.70 mm (0.028 in.)	IDL - E2	Infinity
Throttle valve fully open	VTA - E2	2.0 - 11.6 kΩ
-	VC - E2	2.7 - 7.7 kΩ

Fig. 17 TPS resistance table

29157_LEXU_G0055

Mass Air Flow Meter

See Figures 18, 19 and 20.

1. Using an ohmmeter, measure the resistance between terminals THA and E2.

2. If the resistance is not as specified, replace the MAF meter.

3. Connect the MAF meter connector.

4. Using a voltmeter, connect the positive (+) tester probe to terminal VG, and negative (-) tester probe to terminal E3.

5. Blow air into the MAF meter, and check that the voltage fluctuates.

6. If operation is not as specified, replace the MAF meter.

1UZ-FE VIN H

Mass Air Flow Meter

See Figure 21.

1. Apply battery voltage across terminals 3 (+B) and 4 (E2G).

2. Connect the positive (+) tester probe to terminal 5 (VG), and the negative (-) tester probe to terminal 4 (E2G).

3. Blow air into the MAF meter, and check if the voltage fluctuates.

4. Using an ohmmeter, measure the resistance between terminals 2 (THA) and 1 (E2).

Knock Sensor

1. Using an ohmmeter, measure the resistance between terminals. Resistance should be 120–280 KOhms.

2. If the resistance is not as specified, replace the sensor.

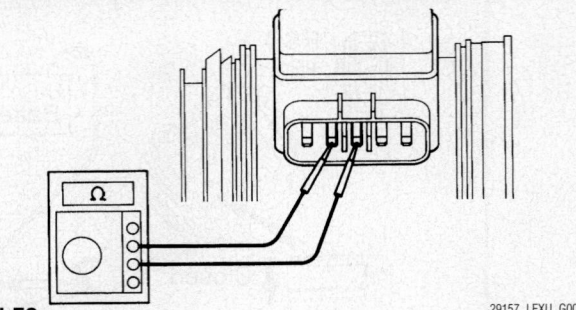

Fig. 18 Terminals THA and E2

29157_LEXU_G0042

Between terminals	Resistance	Temperature
THA - E2	10 - 20 kΩ	-20 °C (-4 °F)
THA - E2	4 - 7 kΩ	0 °C (32 °F)
THA - E2	2 - 3 kΩ	20 °C (68 °F)
THA - E2	0.9 - 1.3 kΩ	40 °C (104 °F)
THA - E2	0.4 - 0.7 kΩ	60 °C (140 °F)
THA - E2	0.2 - 0.4 kΩ	80 °C (176 °F)

29157_LEXU_G0043

Fig. 19 Temperature/resistance table

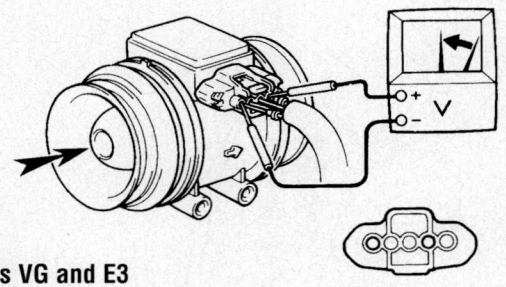

Fig. 20 Terminals VG and E3

29157_LEXU_G0044

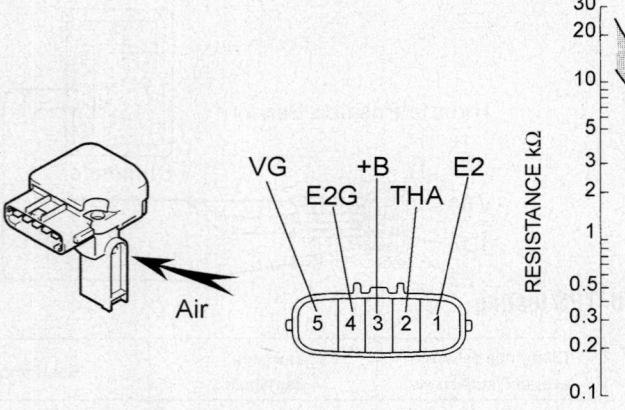

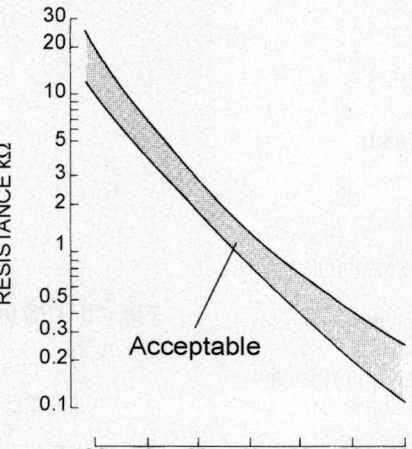

29157_LEXU_G0067

Fig. 21 MAF meter testing

Coolant Temperature Sensor

See Figure 22.

1. Measure the resistance between the terminals.

2. If the result is not as specified, replace the ECT.

Heated Oxygen Sensor

1. Disconnect the oxygen sensor connector.

2. Using an ohmmeter, measure the resistance between the terminals +B and HT. Resistance: 11 - 16 ohms at 20°C (68°F)

3. If the resistance is not as specified, replace the sensor. Tighten to 14 ft. lbs. (20 Nm).

4. Reconnect the oxygen sensor connector.

2GR-FSE VIN E

Crankshaft Position Sensor

1. Using an ohmmeter, measure the resistance between the terminals.

2. Resistance: 1,630–2,740 ohms cold and 2,065–3,225 ohms hot.

Coolant Temperature Sensor

See Figure 23.

1. Measure the resistance between the terminals.

2. If the result is not as specified, replace the ECT.

Knock Sensor

1. Using an ohmmeter, measure the resistance between terminals. Resistance should be 120–280 KOhms.

2. If the resistance is not as specified, replace the sensor.

Mass Air Flow Meter

See Figure 24.

1. Apply battery voltage across terminals 1 (+B) and 2 (E2G).

2. Using a voltmeter, connect the positive (+) tester probe to terminal VG, and the negative (-) tester probe to terminal E2G.

3. Blow air into the MAF meter, and check that the voltage fluctuates.

4. Measure the resistance between terminals 4 (THA) and 5 (E2).

5. If the result is not as specified, replace the MAF meter.

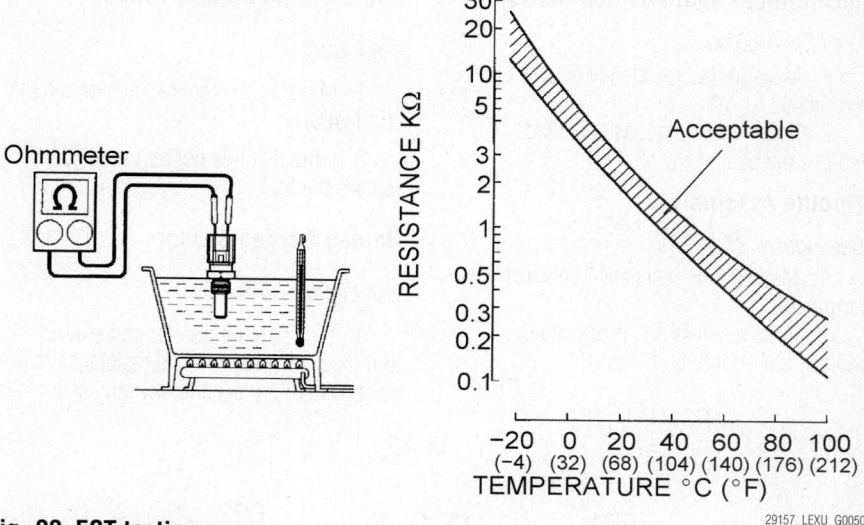

Fig. 22 ECT testing

29157_LEXU_G0050

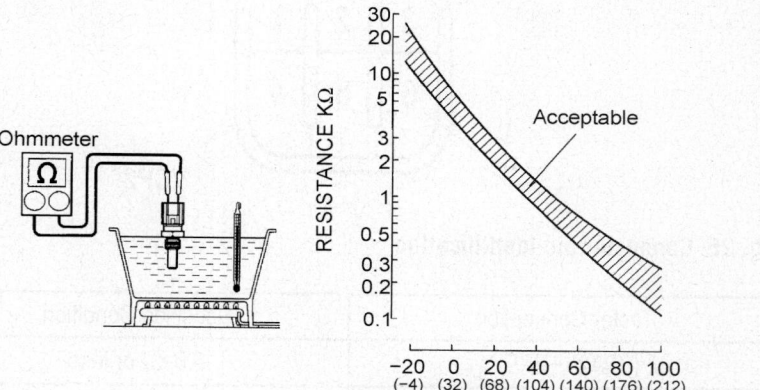

Fig. 23 ECT testing

29157_LEXU_G0050

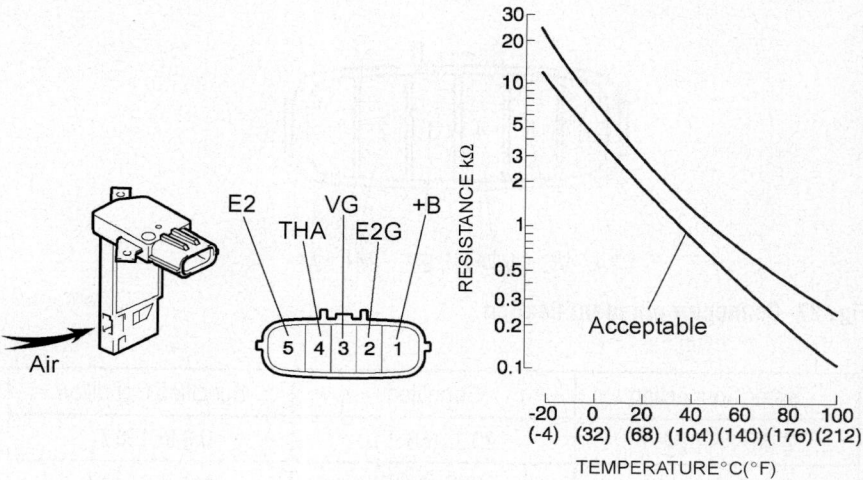

29157_LEXU_G0060

Fig. 24 Connector pin identification and Resistance table

Accelerator Pedal Position Sensor

See Figures 25 and 26.

1. Measure the resistance between the terminals.

2. If the result is not as specified, replace the pedal assy.

Throttle Assembly

See Figures 27 and 28.

1. Measure the resistance between the terminals.

2. If the result is not as specified, replace the throttle body assy.

Coolant Temperature Sensor

See Figure 29.

1. Measure the resistance between the terminals.

2. If the result is not as specified, replace the ECT.

Heated Oxygen Sensor

See Figures 30 and 31.

1. Measure the resistance between terminals +B and HT. If the resistance is not as specified, replace the oxygen sensor.

2JZ-GE VIN D

Coolant Temperature Sensor

See Figure 32.

1. Measure the resistance between the terminals.

2. If the result is not as specified, replace the ECT.

Knock Sensor

See Figure 33.

1. Disconnect the knock sensor connector.

2. Remove the sensor.

3. Using an ohmmeter, check that there is no continuity between

the terminal and body.

4. If there is continuity, replace the sensor.

To install:

5. Install the sensor and tighten to 33 ft. lbs. (44 Nm).

6. Connect the sensor connector.

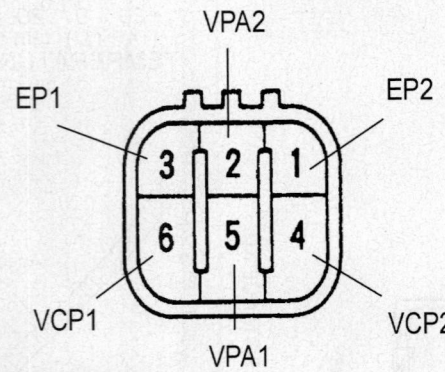

Fig. 25 Connector pin identification

29157_LEXU_G0056

Tester Connection	Specified Condition
2 (VPA2) – 3 (EP1)	5.0 kΩ or less
5 (VPA1) – 1 (EP2)	5.0 kΩ or less
6 (VCP1) – 3 (EP1)	2.25 to 4.75 kΩ
4 (VCP2) – 1 (EP2)	2.25 to 4.75 kΩ

29157_LEXU_G0057

Fig. 26 Resistance table

Fig. 27 Connector pin identification

29157_LEXU_G0058

Tester Connection	Condition	Specified Condition
2 (M+) – 1 (M–)	20°C (68°F)	0.3 to 100 Ω
5 (VC) – 3 (E2)	20°C (68°F)	1.2 to 3.2 kΩ

29157_LEXU_G0059

Fig. 28 Resistance table

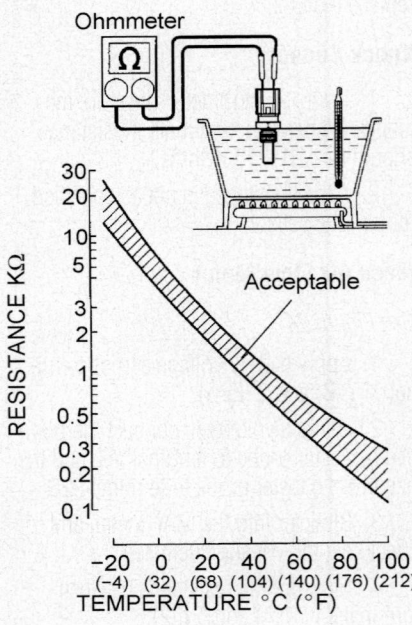

Fig. 29 ECT testing

29157_LEXU_G0050

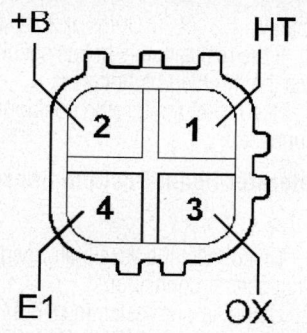

Fig. 30 Connector terminal identification

20°C (68°F)	11 - 16 Ω
800°C (1,472°F)	23 - 32 Ω

Fig. 31 Temperature/resistance table

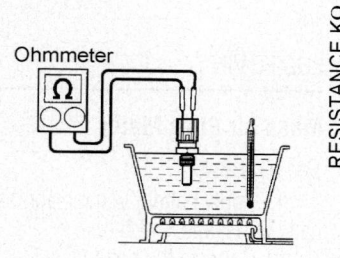

Fig. 32 ECT testing

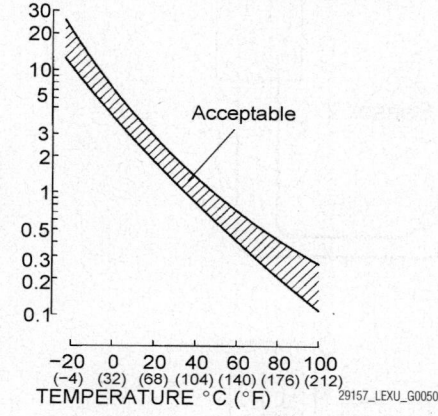

Fig. 33 Testing the knock sensor

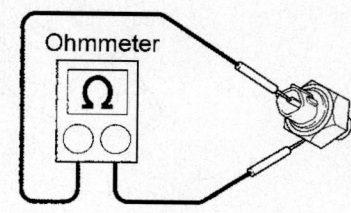

Mass Air Flow Meter

See Figures 34 and 35.

1. Disconnect the negative battery cable.

2. Disconnect the MAF meter connector.

3. Remove the 2 bolts, MAF meter and gasket.

4. Using an ohmmeter, measure the resistance between terminals THA and E2.

5. If the resistance is not as specified, replace the MAF meter.

6. Connect the MAF meter connector.

7. Connect the negative (-) terminal cable to the battery.

8. Turn the ignition switch ON.

9. Using a voltmeter, connect the positive (+) tester probe to terminal VG, and negative (-) tester probe to terminal E2G.

10. Blow air into the MAF meter, and check that the voltage fluctuates. If operation is not as specified, replace the MAF meter.

11. Turn the ignition switch OFF.

12. Disconnect the negative (-) terminal cable from the battery.

13. Disconnect the MAF meter connector.

14. Install the gasket to the MAF meter.

15. Install the MAF meter with the 2 bolts.

16. Connect the negative battery cable.

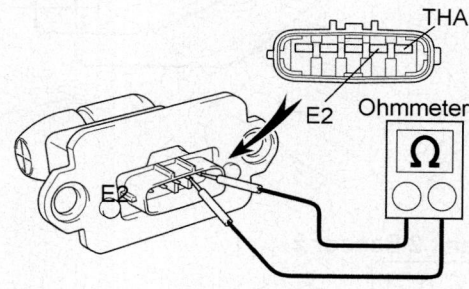

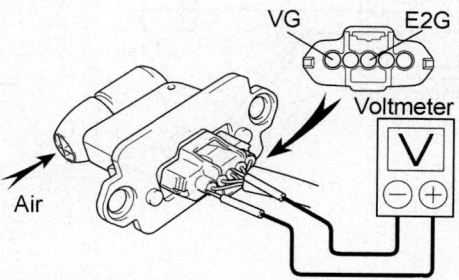

Fig. 34 Testing the MAF meter

Terminals	Resistance	Temperature
THA-E2	13.6 - 18.4 kΩ	-20 °C (-4°F)
THA-E2	2.21 - 2.69 kΩ	20°C (68°F)
THA-E2	0.493 - 0.667 kΩ	60°C (140°F)

Fig. 25 Resistance table

Heated Oxygen Sensor

See Figures 36 and 37.

1. Measure the resistance between terminals +B and HT. If the resistance is not as specified, replace the oxygen sensor.

Throttle Position Sensor

See Figure 38.

1. Disconnect the throttle position sensor connector.
2. Using an ohmmeter, measure the resistance between terminal VC and E2.

Resistance: 1.2 - 3.2 KOhms at 20°C (68°F).

3. If the resistance is not as specified, replace the throttle position sensor.
4. Reconnect the throttle position sensor connector.

Accelerator Pedal Position Sensor

See Figure 39.

1. Disconnect the accelerator pedal position sensor connector.
2. Using an ohmmeter, measure the resistance between terminal VC and E2.
3. Reconnect the accelerator pedal position sensor connector. Resistance: 1.2 - 3.2 KOhms at 20°C (68°F).
4. If the resistance is not as specified, replace the accelerator pedal position sensor.

2UZ-FE VIN T

Mass Air Flow Meter

See Figure 40.

1. Apply battery voltage across terminals 3 (+B) and 4 (E2G).
2. Connect the positive (+) tester probe to terminal 5 (VG), and the negative (-) tester probe to terminal 4 (E2G).
3. Blow air into the MAF meter, and check if the voltage fluctuates.

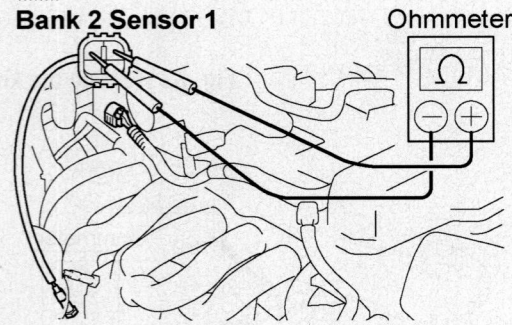

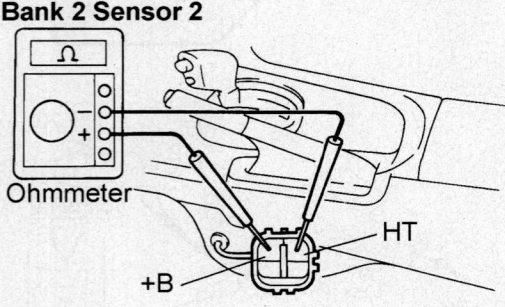

Fig. 36 Testing the Oxygen Sensors   29157_LEXU_G0064

20°C (68°F)	11 - 16 Ω
800°C (1,472°F)	23 - 32 Ω

Fig. 37 Temperature/resistance table   29157_LEXU_G0038

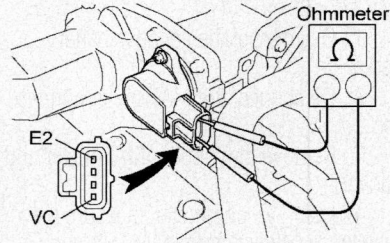

Fig. 38 Testing the TPS   29157_LEXU_G0065

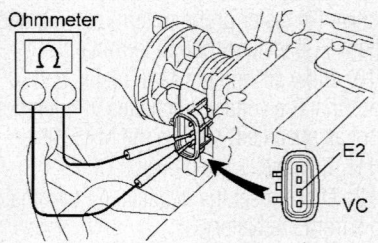

29157_LEXU_G0066

Fig. 39 Testing the Accelerator Pedal Position Sensor

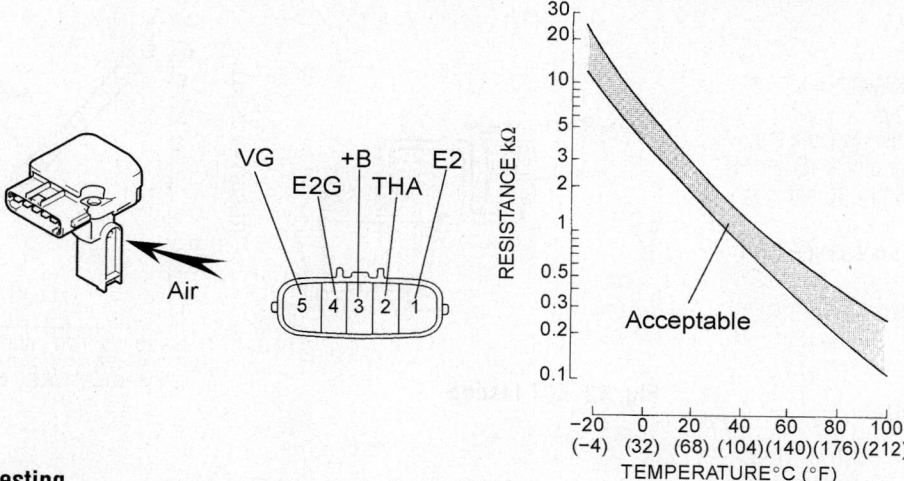

Fig.40 MAF meter testing

4. Using an ohmmeter, measure the resistance between terminals 2 (THA) and 1 (E2).

Knock Sensor

1. Using an ohmmeter, measure the resistance between terminals. Resistance should be 120–280 KOhms.

2. If the resistance is not as specified, replace the sensor.

Coolant Temperature Sensor

See Figure 41.

1. Measure the resistance between the terminals.

2. If the result is not as specified, replace the ECT.

Heated Oxygen Sensor

1. Disconnect the oxygen sensor connector.

2. Using an ohmmeter, measure the resistance between the terminals +B and HT. Resistance: 11 - 16 ohms at 20°C (68°F)

3. If the resistance is not as specified, replace the sensor. Tighten to 14 ft. lbs. (20 Nm).

4. Reconnect the oxygen sensor connector.

3MZ-FE VIN A

Engine Coolant Temperature Sensor

See Figure 42.

1. Drain engine coolant.

2. Disconnect the ECT sensor connector.

3. Using a 19 mm deep socket wrench, remove the ECT sensor and gasket.

4. Using an ohmmeter, measure the resistance between the terminals.

5. If the resistance is not as specified, replace the ECT sensor.

To install:

6. Install a new gasket to the ECT sensor.

7. Using a 19 mm deep socket, install the ECT sensor. Tighten to 14 ft. lbs. (20 Nm).

8. Connect the ECT sensor connector.

9. Refill with engine coolant.

Idle Air Control Valve

See Figure 43.

1. Connect the positive (+) lead from the battery to terminal +B and negative (-) lead to terminal RSC, and check that the valve is closed.

2. Connect the positive (+) lead from the battery to terminal +B and negative (-) lead to terminal RSO, and check that the valve is open.

3. If operation is not as specified, replace the IAC valve.

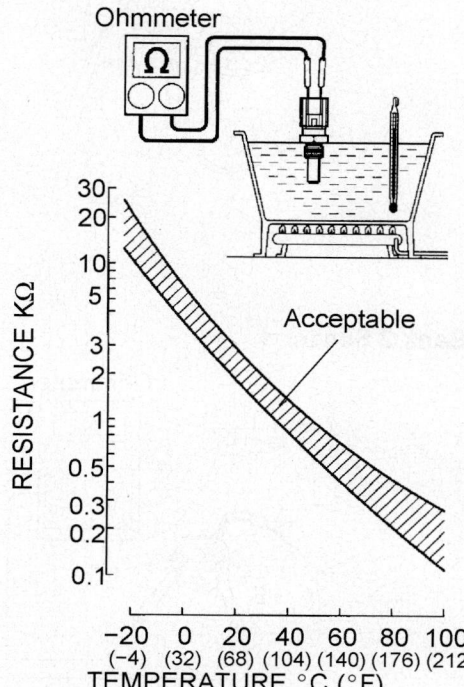

Fig. 41 ECT testing

Oxygen Sensor

See Figures 44 and 45.

1. Disconnect the oxygen sensor connector.

2. Using an ohmmeter, measure the resistance between the terminals +B and HT. Resistance specification: 11 - 16 ohms at 20°C (68°F).

3. If the resistance is not as specified, replace the sensor.

4. Reconnect the oxygen sensor connector.

Throttle Position Sensor

See Figures 46 and 47.

1. Apply vacuum to the throttle opener.

2. Disconnect the sensor connector.

3. Insert a feeler gauge between the throttle stop screw and stop lever.

4. Using an ohmmeter, measure the resistance between each terminal.

5. If the resistance is not as specified, replace the sensor.

6. Reconnect the sensor connector.

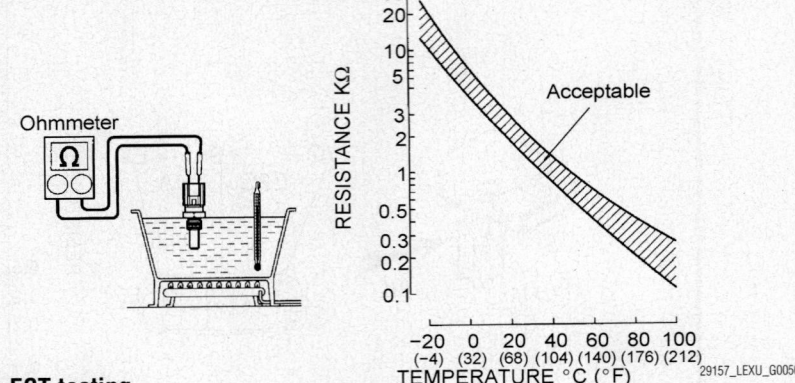

Fig. 42 ECT testing

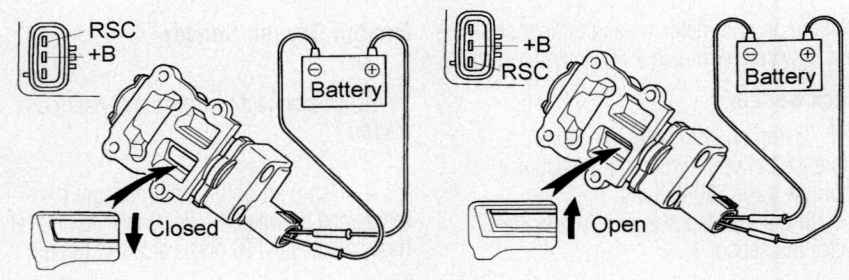
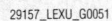

Fig. 43 IAC valve testing

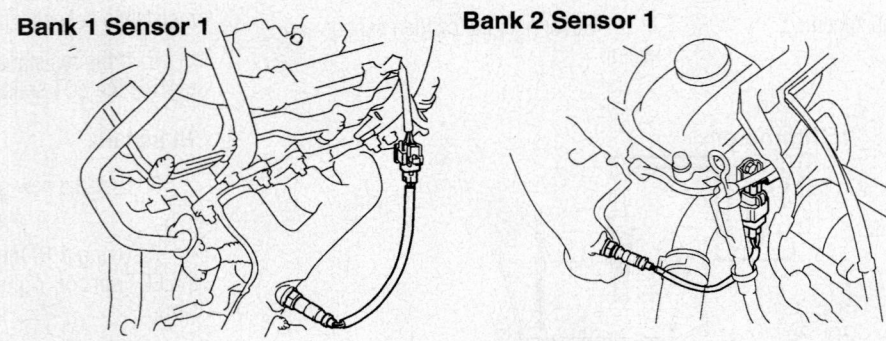

Fig. 44 Oxygen Sensor Connector locations

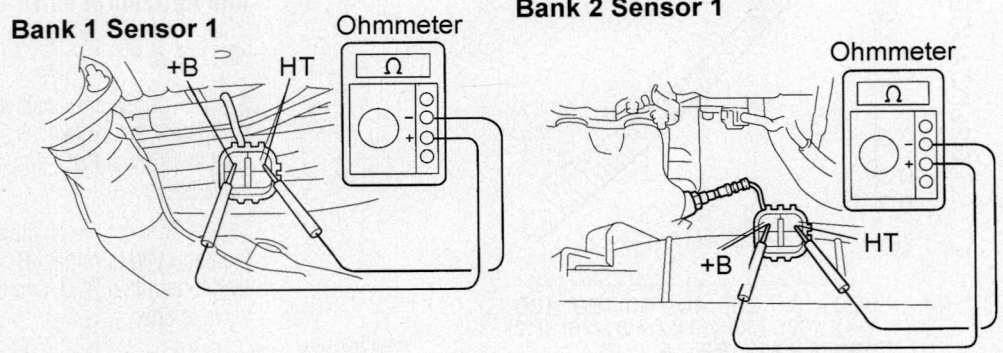

Fig. 45 Oxygen Sensor testing

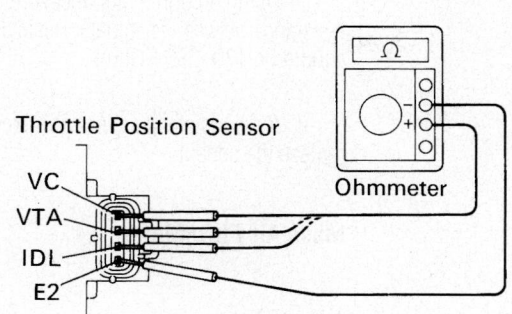

Fig. 46 TPS testing

29157_LEXU_G0054

Clearance between lever and stop screw	Between terminals	Resistance
0 mm (0 in.)	VTA - E2	0.28 - 6.4 kΩ
0.35 mm (0.014 in.)	IDL - E2	0.5 kΩ or less
0.70 mm (0.028 in.)	IDL - E2	Infinity
Throttle valve fully open	VTA - E2	2.0 - 11.6 kΩ
-	VC - E2	2.7 - 7.7 kΩ

Fig. 47 TPS resistance table

29157_LEXU_G0055

Mass Air Flow Meter

See Figures 48, 49 and 50.

1. Using an ohmmeter, measure the resistance between terminals THA and E2.
2. If the resistance is not as specified, replace the MAF meter.
3. Connect the MAF meter connector.
4. Using a voltmeter, connect the positive (+) tester probe to terminal VG, and negative (-) tester probe to terminal E3.
5. Blow air into the MAF meter, and check that the voltage fluctuates.
6. If operation is not as specified, replace the MAF meter.

3UZ-FE VIN N, L

Mass Air Flow Meter

See Figure 51.

1. Apply battery voltage across terminals 3 (+B) and 4 (E2G).
2. Connect the positive (+) tester probe to terminal 5 (VG), and the negative (-) tester probe to terminal 4 (E2G).
3. Blow air into the MAF meter, and check if the voltage fluctuates.
4. Using an ohmmeter, measure the resistance between terminals 2 (THA) and 1 (E2).

Knock Sensor

1. Using an ohmmeter, measure the resistance between terminals. Resistance should be 120–280 KOhms.
2. If the resistance is not as specified, replace the sensor.

Coolant Temperature Sensor

See Figure 52.

1. Measure the resistance between the terminals.
2. If the result is not as specified, replace the ECT.

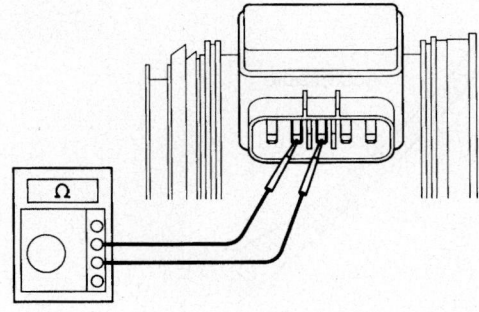

Fig. 48 Terminals THA and E2

29157_LEXU_G0042

Between terminals	Resistance	Temperature
THA - E2	10 - 20 kΩ	-20 °C (-4°F)
THA - E2	4 - 7 kΩ	0°C (32°F)
THA - E2	2 - 3 kΩ	20°C (68°F)
THA - E2	0.9 - 1.3 kΩ	40°C (104°F)
THA - E2	0.4 - 0.7 kΩ	60°C (140°F)
THA - E2	0.2 - 0.4 kΩ	80°C (176°F)

Fig. 49 Temperature/resistance table

29157_LEXU_G0043

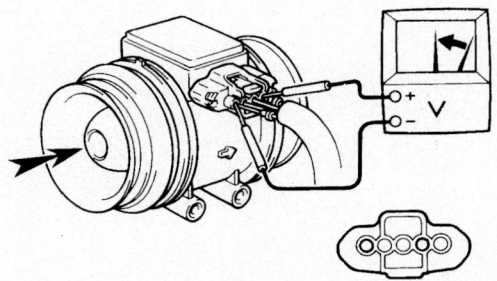

Fig. 50 Terminals VG and E3

29157_LEXU_G0044

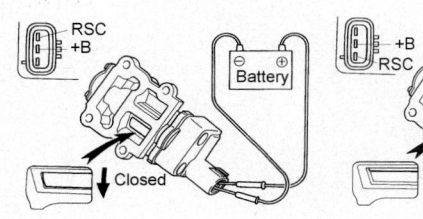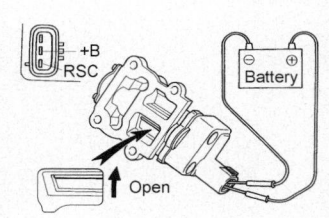

Fig. 51 MAF meter testing

29157_LEXU_G0051

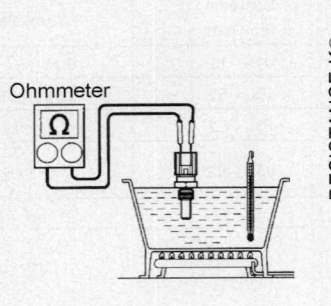

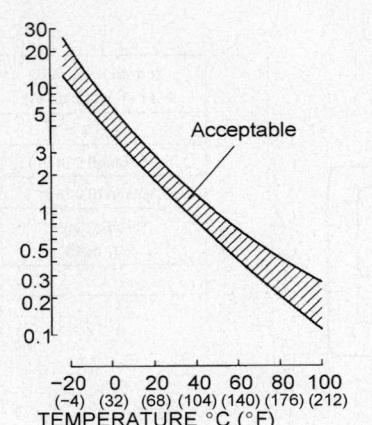

29157_LEXU_G0050

Fig. 52 ECT testing

Heated Oxygen Sensor

1. Disconnect the oxygen sensor connector.

2. Using an ohmmeter, measure the resistance between the terminals +B and HT. Resistance: 11 - 16 ohms at 20°C (68°F).

3. If the resistance is not as specified, replace the sensor. Tighten to 14 ft. lbs. (20 Nm).

4. Reconnect the oxygen sensor connector.

4GR-FSE VIN K

Crankshaft Position Sensor

1. Using an ohmmeter, measure the resistance between the terminals.
2. Resistance: 1,630–2,740 ohms cold and 2,065–3,225 ohms hot.

Coolant Temperature Sensor

See Figure 53.
1. Measure the resistance between the terminals.
2. If the result is not as specified, replace the ECT.

Knock Sensor

1. Using an ohmmeter, measure the resistance between terminals. Resistance should be 120–280 KOhms.

2. If the resistance is not as specified, replace the sensor.

Mass Air Flow Meter

See Figure 54.

1. Apply battery voltage across terminals 1 (+B) and 2 (E2G).

2. Using a voltmeter, connect the positive (+) tester probe to terminal VG, and the negative (-) tester probe to terminal E2G.

3. Blow air into the MAF meter, and check that the voltage fluctuates.

4. Measure the resistance between terminals 4 (THA) and 5 (E2).

5. If the result is not as specified, replace the MAF meter.

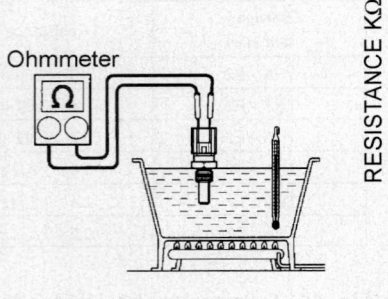

29157_LEXU_G0050

Fig. 53 ECT testing

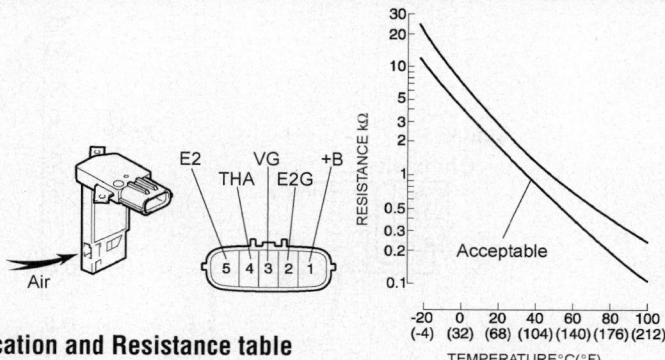

Fig. 54 Connector pin identification and Resistance table

29157_LEXU_G0060

Accelerator Pedal Position Sensor

See Figures 55 and 56.

1. Measure the resistance between the terminals.

2. If the result is not as specified, replace the pedal assy.

Throttle Assembly

See Figures 57 and 58

1. Measure the resistance between the terminals.

2. If the result is not as specified, replace the throttle body assy.

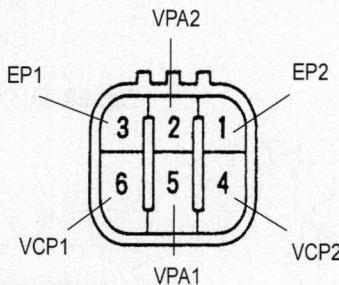

Fig. 55 Connector pin identification

29157_LEXU_G0056

Tester Connection	Specified Condition
2 (VPA2) – 3 (EP1)	5.0 kΩ or less
5 (VPA1) – 1 (EP2)	5.0 kΩ or less
6 (VCP1) – 3 (EP1)	2.25 to 4.75 kΩ
4 (VCP2) – 1 (EP2)	2.25 to 4.75 kΩ

Fig. 56 Resistance table

29157_LEXU_G0057

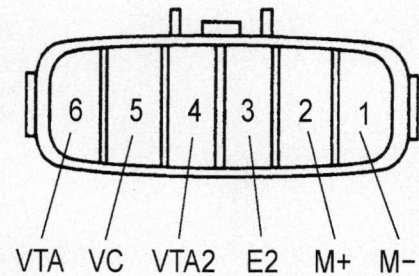

29157_LEXU_G0058

Fig. 57 Connector pin identification

Tester Connection	Condition	Specified Condition
2 (M+) – 1 (M–)	20°C (68°F)	0.3 to 100 Ω
5 (VC) – 3 (E2)	20°C (68°F)	1.2 to 3.2 kΩ

29157_LEXU_G0059

Fig. 58 Resistance table

Coolant Temperature Sensor

See Figure 59.

1. Measure the resistance between the terminals.

2. If the result is not as specified, replace the ECT.

Heated Oxygen Sensor

See Figures 60 and 61.

1. Measure the resistance between terminals +B and HT. If the resistance is not as specified, replace the oxygen sensor.

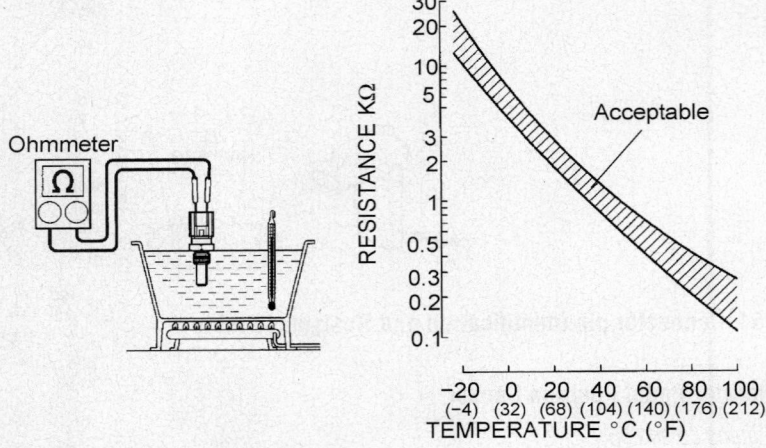

Fig. 59 ECT testing

29157_LEXU_G0050

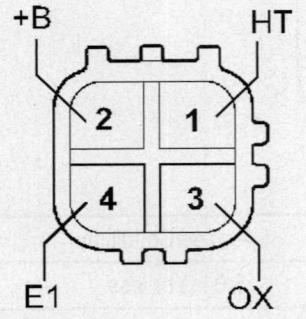

Fig. 60 Connector terminal identification

29157_LEXU_G0061

20°C (68°F)	11 - 16 Ω
800°C (1,472°F)	23 - 32 Ω

Fig. 61 Temperature/resistance table

29157_LEXU_G0038

LEXUS
PIN CHARTS

7

PIN CHARTS

Introduction

A Pin Voltage Table is a term used to describe a table that identifies PCM pins, wire colors of the PCM circuits, circuit descriptions and "known good" values for devices that connect to the PCM. These tables include the following information:

- Signals from various sensors (ECT, IAT, MAP, TPS, etc.)
- Signals from various switches (PNP, PSP, WOT, etc.)
- Signals from oxygen sensors (O2S, HO2S)
- Signals from output devices (IAC, INJ, TCC, etc.)
- Power & ground signals

Pin Voltage Tables

Information contained within the Pin Voltage Tables can be used to:

- Test circuits for open, short to power or short to ground faults
- Check the operation of a component before or after a repair

- Check the operation of a component or system by viewing signals on PCM input/output circuits with a DVOM or Lab Scope

Using a Breakout Box

There are several Breakout Box (BOB) designs available for use to test the PCM and its input and output circuits. However, all of them require removal of the wire harness to the PCM so that the BOB can be installed between the PCM and wire harness connector. Several breakout boxes require the use of overlays in order to allow the tool to be used on more than one year or engine type. Always verify that the correct adapter and overlays are used to prevent connection to the wrong circuits and a misdiagnosis.

Power and Ground Circuit Checks

Measurements made at the BOB are accomplished via test leads and probes from the DVOM or a Lab Scope. If any of the terminals on the PCM or BOB are damaged

or loose, test measurements made at the Breakout Box will be inaccurate. To verify the PCM battery power and ground circuits are normal (correct) at the BOB, test the condition of the circuit between the battery negative (-) post and these circuits prior to starting a test sequence.

Diagnosis with Pin Voltage Tables

See Figure 1.

Once an actual PCM pin voltage reading is recorded, it can be compared to an example from a vehicle with "known good" values. In the example shown the Value at Hot Idle for the EVP sensor signal (0.4v) is the "known good" value.

Wire Color Changes

Every effort has been made to obtain and list the correct circuit wire colors for all vehicles. However, running changes from the vehicle manufacturer can cause the wrong colors to be listed.

PCM Pin #	W/Color	Circuit Description (60-Pin)	Value at Hot Idle
27	BN/LG	EVP Sensor Signal	0.4v

Fig. 1 Example

ES 300 PIN CHARTS

1996 3.0L V6 VIN F (A/T-ECT) 16 Pin Connector

PCM Pin #	Wire Color	Circuit Description (16 Pin)	Value at Hot Idle
2 (Cal)	BK/RD	EVAP Purge Solenoid (VSV)	12v or 0v
3	GN/RD	MIL (lamp) Control	MIL Off: 12v, On: 1v
5	GY/BK	Check Connector	12-14v
6	RD/YL	Intake Air Solenoid	12v or 0v
7	RD/BK	MAF Sensor Ground	<0.050v
8	WT/BL	EVAP Vapor Pressure (VSV)	12v or 0v
10	BL/RD	HO2S-21 (B2 S1) Heater	1v (Heater On)
11	BL/BK	HO2S-11 (B1 S1) Heater	1v (Heater On)
12	BK/WT	EGR Solenoid Control (VSV)	12v or 0v
13	PK	EVAP Vapor Pressure Sensor	2.9-3.1v (with hose off)
14	GN/RD	EGR Gas Temp. Sensor	3.5-4.0v
16	BR	Sensor Ground	<0.050v

1996 3.0L V6 VIN F (A/T-ECT) 22 Pin Connector

PCM Pin #	Wire Color	Circuit Description (22 Pin)	Value at Hot Idle
1	BL/RD	Sensor VREF (VC)	4.9-5.1v
4	WT/BL	OD Clutch Speed Sensor (-)	AC pulse signals
5	WT	CKP Sensor Signal (NE+)	AC pulse signals
6	OR/BL	CKP Sensor Signal (NE-)	<0.050v
7	BK/YL	TP Sensor Signal	0.3-0.8v
8	RD	MAF Sensor Signal	1.1-1.5v
9	YL/BL	OD Clutch Speed Sensor (+)	AC pulse signals
13	WT	HO2S-11 (B1 S1) Signal	0.1-1.1v
14	WT	Knock Sensor 1 Signal	<0.075v AC
15	WT	Knock Sensor 2 Signal	<0.075v AC
16	WT/BL	CMP Sensor Signal (G+)	AC pulse signals
18	GN	Circuit Opening Relay (FC)	0-3v, off-idle: 12v
19	RD/BL	HO2S-21 (B2 S1) Signal	0.1-1.1v
20	GN/BK	ECT Sensor Signal	At 180°F: 0.51v
21	BL/BK	IAT Sensor Signal	At 100°F: 2.60v
22	BR	Sensor Ground	<0.050v

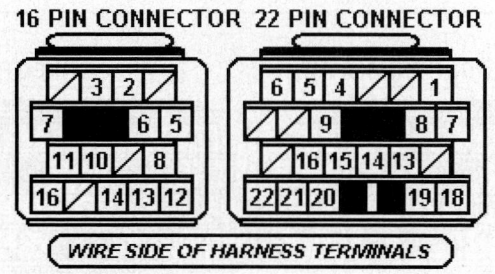

16 PIN CONNECTOR 22 PIN CONNECTOR

WIRE SIDE OF HARNESS TERMINALS

05533_ADIA_G002

Pin Connector Graphic

1996 3.0L V6 VIN F (A/T-ECT) 28 Pin Connector

PCM Pin #	Wire Color	Circuit Description (28 Pin)	Value at Hot Idle
5	PK	A/C Amplifier Signal (ACT)	Clutch On: 12v, Off: 1.5v
9	RD	Cooling Fan Relay	Relay Off: 12v, On: 1v
12	PK/YL	Vehicle Speed Sensor	At 55 mph: 48 Hz
13	BK	Tachometer Signal (TACO)	Pulse Signals
14	BK/YL	Battery Direct	12-14v
20	BK/YL	A/C Amplifier Signal (AC1)	Clutch On: 1.5v, Off: 12v
21	BK/RD	Defogger Idle Up Signal	Switch On: 12v, Off: 0v
23	BK/OR	EFI Main Relay Power	12-14v
24	GN/WT	Stop Light Switch Signal	Brake Off: 0v, On: 12v
25	PK/BK	HO2S-12 (B1 S2) Heater	1v (Heater On)
26	BK	HO2S-12 (B1 S2) Signal	0.1-1.1v
28	WT	Data Link Connector	12v

1996 3.0L V6 VIN F (A/T-ECT) 34 Pin Connector

PCM Pin #	Wire Color	Circuit Description (34 Pin)	Value at Hot Idle
3	YL/GN	A/T-ECT Solenoid (SL-)	In Lockup: 12-14v
5	GN	Injector 6 Control	2.0-3.3 ms
6	RD	Injector 5 Control	2.0-3.3 ms
7	BL	Injector 4 Control	2.0-3.3 ms
8	GY	Injector 3 Control	2.0-3.3 ms
9	YL	Injector 2 Control	2.0-3.3 ms
10	WT	Injector 1 Control	2.0-3.3 ms
11	PK	A/T-ECT Solenoid (S1)	In 3rd or OD: 1v
12	WT/RD	Igniter Signal (IGF)	Digital Signal: 0-5-0v
13	BK/RD	Starter Switch Signal	9-11v (cranking)
14	BK/WT	Neutral Start Switch	In P/N: 9-11v (cranking:
15	GY/BK	Igniter Transistor 3 Control	7% duty cycle
16	YL/RD	Igniter Transistor 2 Control	7% duty cycle
17	PK/BK	A/T-ECT Solenoid (S2)	1st or OD: 1v
22	YL/BK	IAC Signal (RSC)	Pulse Signals
23	GN/BK	IAC Signal (RSO)	Pulse Signals
24	BL/BK	Igniter Transistor 1 Control	7% duty cycle
25	BK/YL	Fuel Pressure Up Solenoid	Hot Restart: 12-14v
27	BL/YL	A/T-ECT Solenoid (SL+)	In Lockup: 12-14v
28, 33-34	WT/BK	Power Ground	<0.1v
32	BL/WT	Closed Throttle Switch	1v, off-idle: 12v

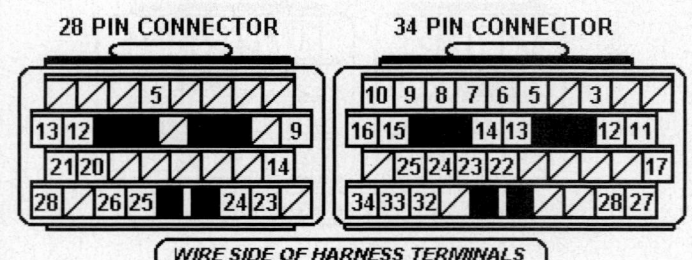

28 PIN CONNECTOR   **34 PIN CONNECTOR**

WIRE SIDE OF HARNESS TERMINALS

05533_ADIA_G003

Pin Connector Graphic

1997 3.0L V6 MFI VIN F (A/T-ECT) 16 Pin Connector

PCM Pin #	Wire Color	Circuit Description (16 Pin)	Value at Hot Idle
2	BK/RD	EVAP Purge Solenoid (VSV)	12v or 0v
3	GN/RD	MIL (lamp) Control	MIL Off: 12v, On: 1v
5	BL	Data Link Connector	12-14v
6	BL	Intake Air Solenoid	12v or 0v
7	RD/BK	MAF Sensor Ground	<0.050v
8	WT/BL	EVAP Vapor Pressure (VSV)	12v or 0v
9	BK/YL	Cooling Fan Relay	Relay Off: 12v, On: 1v
10	BL/RD	HO2S-21 (B2 S1) Heater	1v (Heater on)
11	BL/BK	HO2S-11 (B1 S1) Heater	1v (Heater on)
12	BK/WT	EGR Solenoid Control (VSV)	12v or 0v
13	PK	EVAP Vapor Pressure Sensor	2.9-3.1v (with hose off)
14	GN/RD	EGR Gas Temp. Sensor	3.5-4.0v
15	BK/OR	EGR Valve Position Sensor	0.4-1.6v
16	BR	Sensor Ground	<0.050v

1997 3.0L V6 MFI VIN F (A/T-ECT) 22 Pin Connector

PCM Pin #	Wire Color	Circuit Description (22 Pin)	Value at Hot Idle
1	BL/RD	Sensor VREF (VC)	4.9-5.1v
2	YL/GN	A/T-ECT Solenoid (SLN-)	During shifting: 12v
4	BL	OD Clutch Speed Sensor (-)	AC pulse signals
5	BK/RD	CKP Sensor Signal (NE+)	AC pulse signals
6	BL	CKP Sensor Signal (NE-)	<0.050v
7	BK/YL	TP Sensor Signal	0.3-0.8v
8	RD	MAF Sensor Signal	1.1-1.5v
9	YL	OD Clutch Speed Sensor (+)	AC pulse signals
13	WT	HO2S-11 (B1 S1) Signal	0.1-1.1v
14	WT	Knock Sensor 1 Signal	<0.075v AC
15	WT	Knock Sensor 2 Signal	<0.075v AC
17	BK/WT	CMP Sensor Signal (G+)	AC pulse signals
18	GN	Circuit Opening Relay (FC)	0-3v, off-idle: 12v
19	BL/BK	HO2S-21 (B2 S1) Signal	0.1-1.1v
20	GN/BK	ECT Sensor Signal	At 180°F: 0.51v
21	BL/BK	IAT Sensor Signal	At 100°F: 2.60v
22	BR	Sensor Ground	<0.050v

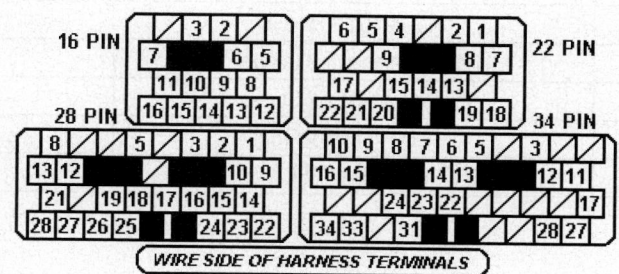

WIRE SIDE OF HARNESS TERMINALS

Pin Connector Graphic

05533_ADIA_G004

1997 3.0L V6 MFI VIN F (A/T-ECT) 28 Pin Connector

PCM Pin #	Wire Color	Circuit Description (28 Pin)	Value at Hot Idle
1	YL/BL	A/T Select Switch Low	In Low: 12v, Others: 0v
2	BK/RD	Mirror Heater Switch Signal	Switch On: 12v, Off: 0v
3	GN	Tail Light Switch Signal	Switch On: 12v, Off: 0v
5	PK	A/C Amplifier Signal (ACT)	Clutch On: 12v, Off: 1.5v
8	WT	Data Link Connector	12v
9	RD	Cooling Fan Relay	Relay Off: 12v, On: 1v
10	GN/YL	A/T Select Switch 2nd	In 2nd: 12v, Others: 0v
12	PK	Vehicle Speed Sensor	At 55 mph: 48 Hz
13	BK	Tachometer Signal (TACO)	Pulse Signals
14	BK/YL	Battery Direct	12-14v
15	RD/BK	A/T Select Switch Reverse	In 'R': 12v, Others: 0v
16	BK/YL	A/C Amplifier Signal (AC1)	Clutch On: 1.5v, Off: 12v
17	PK/BK	HO2S-12 (B1 S2) Heater	1v (Heater On)
18	BK	HO2S-12 (B1 S2) Signal	0.1-1.1v
19	BR/WT	ABS/Traction ECU (NEO)	Pulse Signals
21	BK/RD	Defogger Idle Up Signal	Switch On: 12v, Off: 0v
22	BL/RD	A/T Pattern Selector Switch	Norm: 0v, PWR: 12v
23	LG	EFI Main Relay Power	12-14v
24	GN/WT	Stop Light Switch Signal	Brake Off: 0v, On: 12v
25	PK	ABS/Traction ECU (EFI-)	Pulse Signals
26	LG	ABS/Traction ECU (EFI+)	Pulse Signals
27	GN/RD	ABS/Traction ECU (TRC-)	Pulse Signals
28	GN	ABS/Traction ECU (TRC+)	Pulse Signals

1997 3.0L V6 MFI VIN F (A/T-ECT) 34 Pin Connector

PCM Pin #	Wire Color	Circuit Description (34 Pin)	Value at Hot Idle
3	GN	A/T-ECT Solenoid (SLN-)	During shifting: 12v
5	GN	Injector 6 Control	2.0-3.3 ms
6	RD	Injector 5 Control	2.0-3.3 ms
7	BL	Injector 4 Control	2.0-3.3 ms
8	BK	Injector 3 Control	2.0-3.3 ms
9	YL	Injector 2 Control	2.0-3.3 ms
10	WT	Injector 1 Control	2.0-3.3 ms
11	PK	A/T-ECT Solenoid (S1)	In 3rd or OD: 1v
12	WT/RD	Igniter Signal (IGF)	Digital Signal: 0-5-0v
13	BK/RD	Starter Switch Signal	9-11v (cranking)
14	BK/WT	Neutral Start Switch	In P/N: 9-11v (cranking)
15	GN/RD	Igniter Transistor 3 Control	7% duty cycle
16	YL/RD	Igniter Transistor 2 Control	7% duty cycle
17	GN/YL	A/T-ECT Solenoid (S2)	1st or OD: 1v
22	YL/BK	IAC Signal (RSC)	Pulse Signals

1997 3.0L V6 MFI VIN F (A/T-ECT) 34 Pin Connector, *continued*

23	GN/BK	IAC Signal (RSO)	Pulse Signals
24	BL/BK	Igniter Transistor 1 Control	7% duty cycle
27	BL/YL	A/T-ECT Solenoid (SL)	In Lockup: 12-14v
28	WT/BK	Power Ground	<0.1v
31	PK/BK	PSP Switch Signal	Straight: 12v, Turning: 0v
33	WT/BK	Power Ground	<0.1v
34	WT/BK	Power Ground	<0.1v

1998 3.0L V6 W/O EIS-TC VIN F Federal 16 Pin Connector

PCM Pin #	Wire Color	Circuit Description (16 Pin)	Value at Hot Idle
2	BK/RD	EVAP Purge Solenoid (VSV)	12v or 0v
3	GN/RD	MIL (lamp) Control	MIL Off: 12v, On: 1v
5	BL	Data Link Connector	12v
6	RD/YL	Intake Air Control Solenoid	12v or 0v
7	RD/BK	MAF Sensor Ground	<0.050v
8	WT/BL	EVAP Vapor Pressure (VSV)	12v or 0v
9	BK/YL	Cooling Fan 1 Relay	Relay Off: 12v, On: 1v
10	BL/RD	HO2S-21 (B2 S1) Heater	1v (Heater on)
11	BL/BK	HO2S-11 (B1 S1) Heater	1v (Heater on)
12	BK/WT	EGR Solenoid Control (VSV)	12v or 0v
13	PK	EVAP Vapor Pressure Sensor	2.9-3.1v (with hose off)
14	GN/RD	EGR Gas Temp. Sensor	3.5-4.0v
15	RD/WT	EGR Valve Position Sensor	0.4-1.6v
16	BR	Sensor Ground	<0.050v

1998 3.0L V6 W/O EIS-TC VIN F Federal 22 Pin Connector

PCM Pin #	Wire Color	Circuit Description (22 Pin)	Value at Hot Idle
1	BL/RD	Sensor VREF (VC)	4.9-5.1v
2	YL/GN	A/T-ECT Solenoid (SL-)	In Lockup: 12-14v
4	BL	OD Clutch Speed Signal (-)	AC pulse signals
5	BK/RD	CKP Sensor Signal (NE+)	AC pulse signals
6	BL	CKP Sensor Signal (NE-)	<0.050v
7	BK/YL	TP Sensor Signal	0.3-0.8v
8	RD	MAF Sensor Signal	1.1-1.5v
9	YL	OD Clutch Speed Signal (+)	AC pulse signals
13	WT	HO2S-11 (B1 S1) Signal	0.1-1.1v
14	WT	Knock Sensor 1 Signal	<0.075v AC
15	WT	Knock Sensor 2 Signal	<0.075v AC
17	BK/WT	CMP Sensor Signal (G+)	AC pulse signals
18	GN	Circuit Opening Relay (FC)	0-3v, off-idle: 12v
19	RD/BL	HO2S-21 (B2 S1) Signal	0.1-1.1v
20	GN/BK	ECT Sensor Signal	At 180°F: 0.51v

1998 3.0L V6 W/O EIS-TC VIN F Federal 22 Pin Connector, *continued*

21	BL/BK	IAT Sensor Signal	At 100°F: 2.60v
22	BR	Sensor Ground	<0.050v

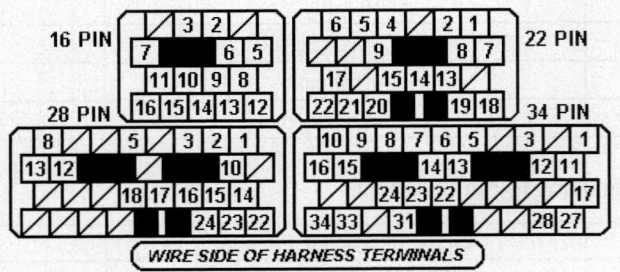

05533_ADIA_G005

Pin Connector Graphic

1998 3.0L V6 W/O EIS-TC VIN F Federal 28 Pin Connector

PCM Pin #	Wire Color	Circuit Description (28 Pin)	Value at Hot Idle
1	YL	A/T Select Switch Low	In Low: 12v, Others: 0v
2	BK/RD	Mirror Heater Switch Signal	Switch On: 12v, Off: 0v
3	GN	Tail Light Switch Signal	Switch On: 12v, Off: 0v
5	PK	A/C Amplifier Signal (ACT)	Clutch On: 12v, Off: 1.5v
8	WT	Data Link Connector	Activated: Pulses
10	GN/YL	A/T Select Switch 2nd	In 2nd: 12v, Others: 0v
12	PK	Vehicle Speed Sensor	At 55 mph: 48 Hz
13	BK	Tachometer Signal (TACO)	Pulses
14	BK/YL	Battery Direct	12-14v
15	RD/BK	A/T Select Switch Reverse	In 'R': 12v, Others: 0v
16	BK/YL	A/C Amplifier Signal (AC1)	Clutch On: 1.5v, Off: 12v
17	PK/BK	HO2S-12 (B1 S2) Heater	1v (Heater on)
18	BK	HO2S-12 (B1 S2) Signal	0.1-1.1v
22	BL/RD	A/T Pattern Selector Switch	Norm: 0v, PWR: 12v
23	LG	EFI Main Relay Power	12-14v
24	GN/WT	Stop Light Switch Signal	Brake Off: 0v, On: 12v

1998 3.0L V6 W/O EIS-TC VIN F Federal 34 Pin Connector

PCM Pin #	Wire Color	Circuit Description (34 Pin)	Value at Hot Idle
1	BR	Sensor Ground	<0.050v
3	GN	A/T-ECT Solenoid (SL-)	In Lockup: 12-14v
5	GN	Injector 6 Control	2.0-3.3 ms
6	RD	Injector 5 Control	2.0-3.3 ms
7	BL	Injector 4 Control	2.0-3.3 ms
8	BK	Injector 3 Control	2.0-3.3 ms
9	YL	Injector 2 Control	2.0-3.3 ms
10	WT	Injector 1 Control	2.0-3.3 ms
11	PK	A/T-ECT Solenoid (S1)	In 3rd or OD: 1v
12	WT/RD	Igniter Signal (IGF)	Digital Signal: 0-5-0v
13	BK/RD	Starter Switch Signal	9-11v (cranking)

1998 3.0L V6 W/O EIS-TC VIN F Federal 34 Pin Connector, *continued*

14	BK/WT	Neutral Start Switch	In P/N: 9-11v (cranking)
15	GN/RD	Igniter Transistor 3 Control	7% duty cycle
16	YL/RD	Igniter Transistor 2 Control	7% duty cycle
17	GN/YL	A/T-ECT Solenoid (S2)	1st or OD: 1v
22	YL/BK	IAC Signal (RSC)	Pulse Signals
23	GN/BK	IAC Signal (RSO)	Pulse Signals
24	BL/BK	Igniter Transistor 1 Control	7% duty cycle
27	BL/YL	A/T-ECT Solenoid (SL+)	In Lockup: 12-14v
28	WT/BK	Power Ground	<0.1v
31	PK/BK	PSP Switch Signal	Straight: 12v, Turning: 0v
33	WT/BK	Power Ground	<0.1v
34	WT/BK	Power Ground	<0.1v

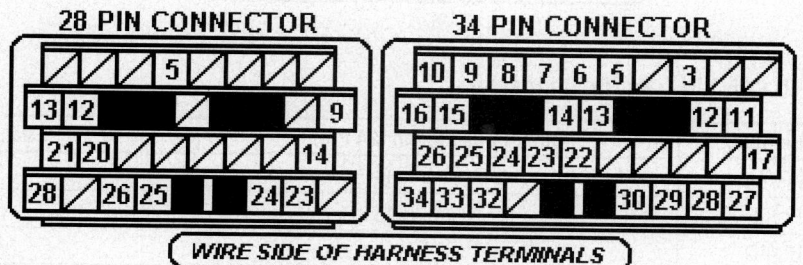

05533_ADIA_G001

Pin Connector Graphic

1998-99 3.0L V6 W/EIS-TC CAL VIN F 17 Pin Connector

PCM Pin #	Wire Color	Circuit Description (17 Pin)	Value at Hot Idle
4	BK/RD	Transponder Amplifier Code	Inserting key: pulses
5	YL	Transponder Amplifier Signal	Inserting key: pulses
10	GN	Transponder Amplifier Signal	Inserting key: pulses
11	RD/YL	Unlock Warning Switch	No key: 4-5v
13	BR	Sensor Shield Ground	<0.050v
16	PK	Security Indicator Light	Digital Signal

1998-99 3.0L V6 W/EIS-TC CAL VIN F 22 Pin Connector

PCM Pin #	Wire Color	Circuit Description (22 Pin)	Value at Hot Idle
1	BK/YL	Battery Direct	12-14v
2	BK/OR	Ignition Switch Power	12-14v
3	GN	Circuit Opening Relay (FC)	0-3v, off-idle: 12v
6	GN/RD	MIL (lamp) Control	MIL Off: 12v, On: 1v
7	BK/RD	Starter Switch Signal	9-11v (cranking)
8	BK/OR	EFI Main Relay Power	12-14v
9	WT/BL	EVAP Vapor Pressure (VSV)	12v or 0v
11	WT	SIL Signal (Scan Tool)	Digital Signal
13	GN	ABS/Traction ECU (TRC+)	Pulses
14	LG	ABS/Traction ECU (EFI+)	Pulses

1998-99 3.0L V6 W/EIS-TC CAL VIN F 22 Pin Connector, _continued_

15	GN/WT	Stop Light Switch Signal	Brake Off: 0v, On: 12v
16	LG	EFI Main Relay Power	12-14v
17	PK	EVAP Vapor Pressure Sensor	2.9-3.1v (with hose off)
18	BK/RD	Mirror Heater Switch Signal	Switch On: 12v, Off: 0v
19	GN	Tail Light Switch Signal	Switch On: 12v, Off: 0v
20	GN/RD	ABS/Traction ECU (TRC-)	Pulses
21	PK	ABS/Traction ECU (EFI-)	Pulses

05533_ADIA_G006

Pin Connector Graphic

1998-99 3.0L V6 W/EIS-TC CAL VIN F 24 Pin Connector

PCM Pin #	Wire Color	Circuit Description (24 Pin)	Value at Hot Idle
1	WT/BK	Power Ground	<0.1v
2	BL/RD	Sensor VREF	4.9-5.1v
3	BL/RD	AFS-11 (B1 S1) Heater	1v (Heater On)
4	BK/RD	AFS-21 (B2 S1) Heater	1v (Heater On)
5	WT	Injector 1 Control	2.0-3.3 ms
6	YL	Injector 2 Control	2.0-3.3 ms
7	BK/RD	EVAP Purge Solenoid (VSV)	12v or 0v
8	WT/BK	Sensor Ground	<0.050v
9	PK/BK	PSP Switch Signal	Straight: 12v, Turning: 0v
10	RD	MAF Sensor Signal	1.1-1.5v
11	BK/WT	AFS-11 (B1 S1) Signal (+)	3.0-3.6v
12	BR	AFS-21 (B2 S1) Signal (+)	3.0-3.6v
13	GN/RD	EGR Gas Temp. Sensor	3.5-4.0v
14	GN/BK	ECT Sensor Signal	At 180°F: 0.51v
16	BK/RD	CKP Sensor Signal (NE+)	AC pulse signals
17	BR	Shield Ground	<0.050v
18	BR	Sensor Ground	<0.050v
19	RD/BK	MAF Sensor Ground	<0.050v
20	BL	AFS-11 (B1 S1) Signal (AF-)	Fixed at 3v
21	BK/RD	AFS-21 (B2 S2) Signal (-)	Fixed at 3v
22	BL/BK	IAT Sensor Signal	At 100°F: 2.60v
23	BK/YL	TP Sensor Signal	0.3-0.8v
24	BL	CKP/CMP Sensor Ground (-)	<0.050v

1998-99 3.0L V6 W/EIS-TC CAL VIN F 28 Pin Connector

PCM Pin #	Wire Color	Circuit Description (28 Pin)	Value at Hot Idle
8	BK	HO2S-12 (B1 S2) Signal	0.1-1.1v
9	PK/BK	HO2S-12 (B1 S2) Heater	1v (Heater on)
13	PK	A/C Amplifier Signal (ACT)	Clutch On: 12v, Off: 1.5v
14	BL/RD	A/C Amplifier Signal (THWO)	Pulse Signals
16	BR/WT	ABS/Traction ECU (NEO)	Pulse Signals
20	BK/WT	Neutral Start Switch	In P/N: 9-11v (cranking)
22	PK	Vehicle Speed Sensor	At 55 mph: 48 Hz
25	BK/YL	A/C Amplifier Signal (AC1)	Clutch On: 1.5v, Off: 12v
27	BK	Tachometer Signal (TACO)	Pulse Signal

Standard Colors and Abbreviations

Abbreviation	Color	Abbreviation	Color	Abbreviation	Color
BK	Black	GY	Gray	RD	Red
BL	Blue	GN	Green	TN	Tan
BR	Brown	LG	Light Green	VT	Violet
DB	Dark Blue	OR	Orange	WT	White
DG	DK Green	PK	Pink	YL	Yellow

1998-99 3.0L V6 W/EIS-TC Calif. VIN F 31 Pin Connector

PCM Pin #	Wire Color	Circuit Description (31 Pin)	Value at Hot Idle
1	BK	Injector 3 Control	2.0-3.3 ms
2	BL	Injector 4 Control	2.0-3.3 ms
3	RD	Injector 5 Control	2.0-3.3 ms
4	GN	Injector 6 Control	2.0-3.3 ms
6	BL	Data Link Connector (TE)	12-14v
10	BK/WT	CMP Sensor Signal (G+)	AC pulse signals
11	BL/BK	Igniter Transistor 1 Control	7% duty cycle
12	YL/RD	Igniter Transistor 2 Control	7% duty cycle
13	GN/RD	Igniter Transistor 3 Control	7% duty cycle
15	YL/BK	IAC Signal (RSC)	Pulse Signals
16	GN/BK	IAC Signal (RSO)	Pulse Signals
17	RD/YL	ACIS Solenoid Control (VSV)	12v or 0v
18	BK/WT	EGR Solenoid Control (VSV)	12v or 0v
21	WT/BK	Power Ground	<0.1v
22	RD/WT	EGR Valve Position Sensor	0.4-1.6v
23 ('98)	BR	Sensor Ground	<0.050v
25	WT/RD	Igniter Signal (IGF)	Digital Signal: 0-5-0v
27	WT	Knock Sensor 1 Signal	<0.075v AC
28	WT	Knock Sensor 2 Signal	<0.075v AC
29	BK/YL	Cooling Fan 1 Relay	Relay Off: 12v, On: 1v
30	WT/BK	Power Ground	<0.1v
31	WT/BK	Power Ground	<0.1v

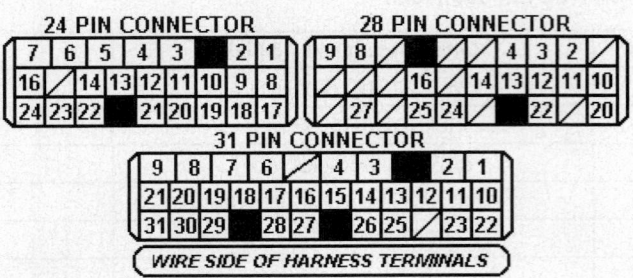

Pin Connector Graphic

1998-99 3.0L V6 W/EIS VIN F Federal 17 Pin Connector

PCM Pin #	Wire Color	Circuit Description (17 Pin)	Value at Hot Idle
4	BK/RD	Transponder Amplifier Code	Inserting key: pulses
5	YL	Transponder Amplifier Signal	Inserting key: pulses
10	GN	Transponder Amplifier Signal	Inserting key: pulses
11	RD/YL	Unlock Warning Switch	No Key: 4-5v
13	BR	Sensor Shield Ground	<0.050v
16	PK	Security Indicator Light	Pulses

1998-99 3.0L V6 W/EIS VIN F Federal 22 Pin Connector

PCM Pin #	Wire Color	Circuit Description (22 Pin)	Value at Hot Idle
1	BK/YL	Battery Direct	12-14v
2	BK/OR	Ignition Switch Power	12-14v
3	GN	Circuit Opening Relay (FC)	0-3v, off-idle: 12v
6	GN/RD	MIL (lamp) Control	MIL Off: 12v, On: 1v
7	BK/RD	Starter Switch Signal	9-11v (cranking)
8	BK/OR	EFI Main Relay Power	12-14v
9	WT/BL	EVAP Vapor Pressure (VSV)	12v or 0v
11	WT	SIL Signal (Scan Tool)	Digital Signal
13	GN	ABS/Traction ECU (TRC+)	Pulses
14	LG	ABS/Traction ECU (EFI+)	Pulses
15	GN/WT	Stop Light Switch Signal	Brake Off: 0v, On: 12v
16	LG	EFI Main Relay Power	12-14v
17	PK	EVAP Vapor Pressure Sensor	2.9-3.1v (with hose off)
18	BK/RD	Mirror Heater Switch Signal	Heater On: 12-14v
19	GN	Tail Light Switch Signal	Switch On: 12v, Off: 0v
20	GN/RD	ABS/Traction ECU (TRC-)	Pulses
21	PK	ABS/Traction ECU (EFI-)	Pulses

Pin Connector Graphic

1998-99 3.0L V6 W/EIS VIN F Federal 24 Pin Connector

PCM Pin #	Wire Color	Circuit Description (24 Pin)	Value at Hot Idle
2	BL/RD	Sensor VREF	4.9-5.1v
3	BL/BK	HO2S-11 (B1 S1) Heater	1v (Heater on)
4	BL/RD	HO2S-21 (B2 S1) Heater	1v (Heater on)
5	WT	Injector 1 Control	2.0-3.3 ms
6	YL	Injector 2 Control	2.0-3.3 ms
7	BK/RD	EVAP Purge Solenoid (VSV)	12v or 0v
9	PK/BK	PSP Switch Signal	Straight: 12v, Turning: 0v
10	RD	MAF Sensor Signal	1.1-1.5v
11	WT	HO2S-11 (B1 S1) Signal	0.1-1.1v
12	RD/BL	HO2S-21 (B2 S1) Signal	0.1-1.1v
13	GN/RD	EGR Gas Temp. Sensor	3.5-4.0v
14	GN/BK	ECT Sensor Signal	At 180°F: 0.51v
16	BK/RD	CKP Sensor Signal (NE+)	AC pulse signals
17	BR	Shield Ground	<0.050v
18	BR	Sensor Ground	<0.050v
19	RD/BK	MAF Sensor Ground	<0.050v
22	BL/BK	IAT Sensor Signal	At 100°F: 2.60v
23	BK/YL	TP Sensor Signal	0.3-0.8v
24	BL	CKP/CMP Sensor Ground (-)	<0.050v

1998-99 3.0L V6 W/EIS VIN F Federal 28 Pin Connector

PCM Pin #	Wire Color	Circuit Description (28 Pin)	Value at Hot Idle
8	BK	HO2S-12 (B1 S2) Signal	0.1-1.1v
9	PK/BK	HO2S-12 (B1 S2) Heater	1v (Heater on)
13	PK	A/C Amplifier Signal (ACT)	Clutch On: 12v, Off: 1.5v
14	BL/RD	A/C Amplifier Signal (THWO)	Pulse Signals
16	BR/WT	ABS/Traction ECU (NEO)	Pulse Signals
20	BK/WT	Neutral Start Switch	In P/N: 9-11v (cranking)
22	PK	Vehicle Speed Sensor	At 55 mph: 48 Hz
25	BK/YL	A/C Amplifier Signal (AC1)	Clutch On: 1.5v, Off: 12v
27	BK	Tachometer Signal (TACO)	Pulse Signals

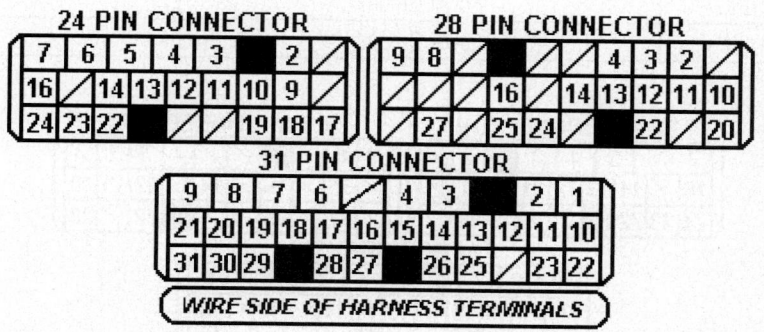

Pin Connector Graphic

05533_ADIA_G008

Standard Colors and Abbreviations

Abbreviation	Color	Abbreviation	Color	Abbreviation	Color
BK	Black	GY	Gray	RD	Red
BL	Blue	GN	Green	TN	Tan
BR	Brown	LG	Light Green	VT	Violet
DB	Dark Blue	OR	Orange	WT	White
DG	DK Green	PK	Pink	YL	Yellow

1998-99 3.0L V6 W/EIS VIN F Federal 31 Pin Connector

PCM Pin #	Wire Color	Circuit Description (31 Pin)	Value at Hot Idle
1	BK	Injector 3 Control	2.0-3.3 ms
2	BL	Injector 4 Control	2.0-3.3 ms
3	RD	Injector 5 Control	2.0-3.3 ms
4	GN	Injector 6 Control	2.0-3.3 ms
5	---	Not Used	---
6	BL	Data Link Connector (TE)	12v
7-9	---	Not Used	---
10	BK/WT	CMP Sensor Signal (G+)	AC pulse signals
11	BL/BK	Igniter Transistor 1 Control	7% duty cycle
12	YL/RD	Igniter Transistor 2 Control	7% duty cycle
13	GN/RD	Igniter Transistor 3 Control	7% duty cycle
14	---	Not Used	---
15	YL/BK	IAC Signal (RSC)	Pulse Signals
16	GN/BK	IAC Signal (RSO)	Pulse Signals
17	RD/YL	ACIS Solenoid Control (VSV)	12v or 0v
18	BK/WT	EGR Solenoid Control (VSV)	12v or 0v
19-20	---	Not Used	---
21	WT/BK	Power Ground	<0.1v
22	RD/WT	EGR Valve Position Sensor	0.4-1.6v
23 ('98)	BR	Sensor Ground	<0.050v
25	WT/RD	Igniter Signal (IGF)	Digital Signal: 0-5-0v
26	---	Not Used	---
27	WT	Knock Sensor 1 Signal	<0.075v AC
28	WT	Knock Sensor 2 Signal	<0.075v AC
29	BK/YL	Cooling Fan 1 Relay Control	Relay Off: 12v, On: 1v
30	WT/BK	Power Ground	<0.1v
31	WT/BK	Power Ground	<0.1v

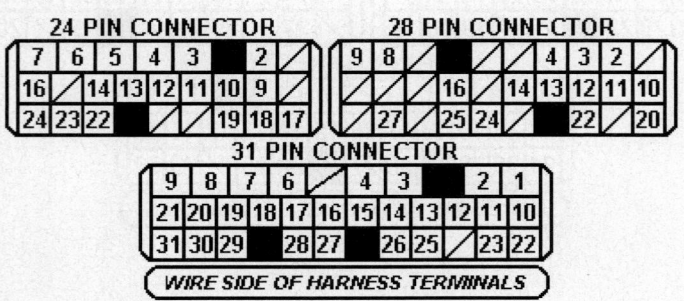

Pin Connector Graphic

05533_ADIA_G008

2000-01 3.0L V6 DOHC VIN F (A/T-ECT) E4 31P Connector

PCM Pin #	Wire Color	Circuit Description (31 Pin)	Value at Hot Idle
1	YL	Injector 3 Control	1.6-2-9 ms
2	WT	Injector 4 Control	1.6-2-9 ms
3	GN	Injector 5 Control	1.6-2-9 ms
4	GN	Injector 6 Control	1.6-2.9 ms
5	RD	VVT Solenoid Control (OC1-)	12v or 0v
6	RD/BK	VVT Solenoid Control (OC1+)	12v or 0v
7	PK	A/T-ECT Solenoid (S1)	In 3rd or OD: 1v
8-9	---	Not Used	---
10	BK/WT	CMP Sensor Signal (RH+)	AC pulse signals
11	GY	Igniter Transistor 1 Control	6°, at 55 mph: 8° dwell
12	BK/RD	Igniter Transistor 2 Control	6°, at 55 mph: 8° dwell
13	LG/BK	Igniter Transistor 3 Control	6°, at 55 mph: 8° dwell
14	BL/YL	Igniter Transistor 4 Control	6°, at 55 mph: 8° dwell
15	BL	Igniter Transistor 5 Control	6°, at 55 mph: 8° dwell
16	LG	Igniter Transistor 6 Control	6°, at 55 mph: 8° dwell
17	RD/YL	ACIS 1 Control (VSV)	12v or 0v
18	RD/WT	VVT Solenoid Control (OC2-)	12v or 0v
19	YL	A/T-ECT Solenoid (SLN-)	Pulse Signals
20	PK/BL	A/T-ECT Solenoid (SLN+)	Pulse Signals
21	WT/BK	Power Ground (E01)	<0.1v
22	BK/WT	CMP Sensor Signal (LH+)	AC pulse signals
23	GN	O/D Clutch Speed Sensor (-)	AC pulse signals
24	PK/BL	O/D Clutch Speed Sensor (+)	AC pulse signals
25	BK/YL	Igniter Signal (IGF)	Digital Signal: 0-5-0v
26	GN/BK	IAC Signal (RSO)	Pulse Signals
27	WT	Knock Sensor 1 Signal	0.075v AC
28	WT	Knock Sensor 2 Signal	0.075v AC
29	RD/BL	VVT Solenoid Control (OC2+)	12v or 0v
30	WT/BK	Power Ground (E03)	<0.1v
31	WT/BK	Power Ground (E04)	<0.1v

E4 31-PIN CONNECTOR

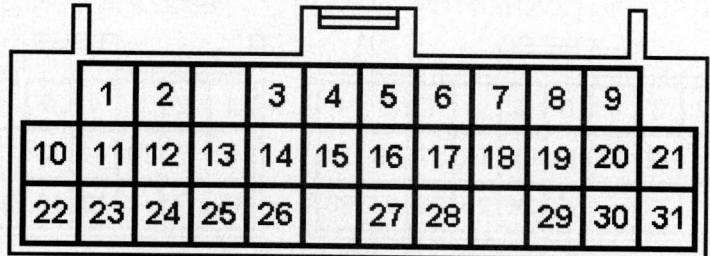

WIRE SIDE OF HARNESS TERMINALS

05533_ADIA_G009

Pin Connector Graphic

2000-01 3.0L V6 DOHC VIN F (A/T-ECT) E5 24P Connector

PCM Pin #	Wire Color	Circuit Description (24 Pin)	Value at Hot Idle
1	WT/BK	Power Ground (E04)	<0.1v
2	BL/RD	Sensor VREF (VC)	4.9-5.1v
3	BK/RD	AFS-11 (B1 S1) Heater	1v (Heater On)
4	BK/RD	AFS-21 (B2 S1) Heater	1v (Heater On)
5	BL	Injector 1 Control	1.6-2-9 ms
6	RD	Injector 2 Control	1.6-2-9 ms
7	BK/RD	EVAP Purge Solenoid (VSV)	12v or 0v
8	WT/BK	Power Ground (E05)	<0.1v
9	BK/RD	PSP Switch Signal	Straight: 12v, Turned: 0v
10	RD	MAF Sensor Signal	1-1.1v
11	BK/RD	AFS-11 (B1 S1) Signal (+)	3.0-3.6v
12	BK/WT	AFS-21 (B2 S1) Signal (+)	3.0-3.6v
13	---	Not Used	---
14	GN/YL	ECT Sensor Signal	At 180°F: 0.51v
15	WT/RD	VSV ACIS 2 Control	12v or 0v
16	BK/WT	CKP Sensor Signal (NE+)	310-330 Hz
17	BR	Shield Ground (E1)	<0.050v
18	WT	Sensor Ground	<0.050v
19	RD/BK	MAF Sensor Ground (E2G)	<0.050v
20	BR	AFS-11 (B1 S1) Signal (AF-)	3.0-3.6v
21	LB	AFS-21 (B2 S1) Signal (-)	3.0-3.6v
22	BL/BK	IAT Sensor Signal	0.5-3.4v
23	LG	TP Sensor Signal	0.53-1.27v
24	BL	CKP/CMP Sensor Ground (-)	<0.050v

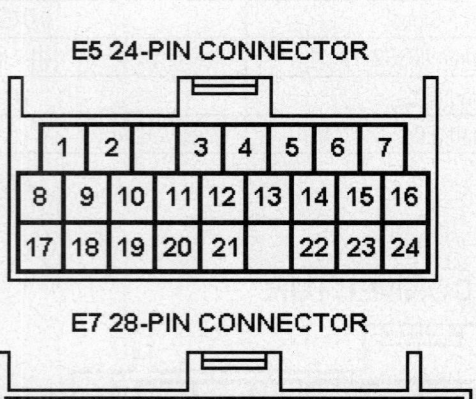

E5 24-PIN CONNECTOR

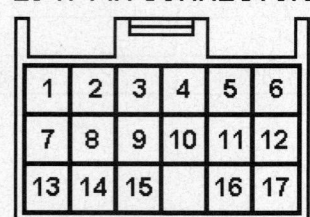

E6 17-PIN CONNECTOR

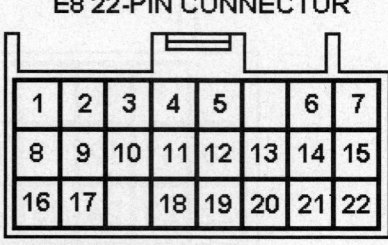

E7 28-PIN CONNECTOR

E8 22-PIN CONNECTOR

WIRE SIDE OF HARNESS TERMINALS

05533_ADIA_G010

Pin Connector Graphic

2000-01 3.0L V6 DOHC VIN F (A/T-ECT) E6 17P Connector

PCM Pin #	Wire Color	Circuit Description (17 Pin)	Value at Hot Idle
1	GN/YL	A/T-ECT Solenoid (S2)	1st or OD: 1v
2-7	---	Not Used	---
8	RD/BK	A/T Select Switch Reverse	In 'R': 12v, Others: 0v
9-11	---	Not Used	---
12	GN/OR	Overdrive Main Switch	Switch Off: 12v, On: 1v
13	YL	A/T Select Switch Low	In Low: 12v, Others: 0v
14	GN/YL	A/T Select Switch 2nd	In 2nd: 12v, Others: 0v
15	GN/WT	A/T-ECT Solenoid (SL+)	In Lockup: 12-14v

2000-01 3.0L V6 DOHC VIN F (A/T-ECT) E7 28P Connector

PCM Pin #	Wire Color	Circuit Description (28 Pin)	Value at Hot Idle
1-2	---	Not Used	---
3	PK/BK	EVAP Vapor Pressure (VSV)	12v or 0v
4	GN	Tail Light Switch Signal	Switch On: 12v, Off: 0v
5	BL	Data Link Connector (TE1)	12-14v
6	LG/RD	A/C Magnetic Clutch (ACMG)	Clutch Off: 0v, On: 12v
7	YL/RD	A/C Amplifier Signal (AC1)	Clutch On: 1.5v, Off: 12v
8	BK	HO2S-12 (B1 S2) Signal	0.1-1.1v
9	LG	HO2S-12 (B1 S2) Heater	1v (Heater On)
10-12	---	Not Used	---
13	BK	Mirror Heater Switch Signal	Switch Off: 1.5v, On: 12v
14	YL/GN	A/C Amplifier Signal (THWO)	Pulse Signals
15	---	Not Used	---
16	YL/BK	ABS/Traction NEO Signal	Pulse Signals
17	---	Not Used	---
18	BL/YL	TXCT Ignition Signal	Inserting key: pulses
19	---	Not Used	---
20	BK/WT	Neutral Start Switch (NSW)	In P/N: 9-11v (cranking)
21	---	Not Used	---
22	VT/WT	Speedometer Indicator	At 55 mph: 48 Hz
23	RD/YL	Unlock Warning Switch	Key In: 1.5v, Out: 4.5v
24	GY	Cruise Control Signal (OD1)	At Cruise in OD: 12v
25	YL	Cruise Control Signal (IDLO)	1.5v, off-idle: 12v
26	GN	Cooling Fan 1 Relay Control	Relay Off: 12v, On: 1v
27	BK	Tachometer Signal (TACO)	Pulse Signals
28	PK/GN	Ignition Switch Code	Inserting key: pulses

2000-01 3.0L V6 DOHC VIN F (A/T-ECT) E8 22P Connector

PCM Pin #	Wire Color	Circuit Description (22 Pin)	Value at Hot Idle
1	BK/RD	Direct Battery	12-14v
2	BK/OR	Ignition Switch Power	12-14v
3	GN/BK	Circuit Opening Relay (FC)	0-0.3v, off-idle: 12-14v
4	WT	SIL Signal (Scan Tool)	Digital Signal
5	---	Not Used	---
6	BK/YL	MIL (lamp) Control	MIL Off: 12v, On: 1v
7	BK/RD	Starter Signal (STA)	Cranking: 9-11v
8	BL/OR	EFI Main Relay Power	12-14v
9	GN/OR	Overdrive Lamp Control	At Cruise in OD: 1v
10	GN	EVAP Canister Closed Valve	12v or 0v
11-12	---	Not Used	---
13	BR	ABS/Traction: TRC (+) Signal	DC pulse signals
14	PK	ABS/Traction: ENG (+) Signal	DC pulse signals
15	GN/WT	Brake Switch Signal	Brake Off: 0v, On: 12v
16	BK/WT	EFI Main Relay Power	12-14v
17	PK	EVAP Vapor Pressure Sensor	2.9-3.1v (with hose off)
18-19	---	Not Used	---
20	GN	ABS/Traction: TRC (-) Signal	DC pulse signals
21	GY	ABS/Traction: ENG (-) Signal	DC pulse signals
22	VT/WT	Security Indicator Light	Inserting key: pulses

2002-03 3.0L V6 DOHC VIN F (A/T-ECT) E4 31P Connector

PCM Pin #	Wire Color	Circuit Description (31 Pin)	Value at Hot Idle
1	YL	Injector 3 Control	1.6-2-9 ms
2	WT	Injector 4 Control	1.6-2-9 ms
3	BL/RD	Injector 5 Control	1.6-2-9 ms
4	GN	Injector 6 Control	1.6-2.9 ms
5	GR	VVT Solenoid Control (OC1-)	12v or 0v
6	RD/BK	VVT Solenoid Control (OC1+)	12v or 0v
7	PK	A/T-ECT Solenoid (S1)	In 3rd or OD: 1v
8-9	---	Not Used	---
10	BK/WT	CMP Sensor Signal (RH+)	AC pulse signals
11	GY/RD	Igniter Transistor 1 Control	6°, at 55 mph: 8° dwell
12	YL/GY	Igniter Transistor 2 Control	6°, at 55 mph: 8° dwell
13	GY/YL	Igniter Transistor 3 Control	6°, at 55 mph: 8° dwell
14	BL/YL	Igniter Transistor 4 Control	6°, at 55 mph: 8° dwell
15	BL	Igniter Transistor 5 Control	6°, at 55 mph: 8° dwell
16	BK	Igniter Transistor 6 Control	6°, at 55 mph: 8° dwell
17	RD/YL	ACIS 1 Control (VSV)	12v or 0v
18	RD/WT	VVT Solenoid Control (OC2-)	12v or 0v
19	YL	A/T-ECT Solenoid (SLN-)	Pulse Signals
20	PK/BL	A/T-ECT Solenoid (SLN+)	Pulse Signals

02-03 3.0L V6 DOHC VIN F (A/T-ECT) E4 31P Connector, *continued*

21	WT/BK	Power Ground (E01)	<0.1v
22	BK/WT	CMP Sensor Signal (LH+)	AC pulse signals
23	GN	O/D Clutch Speed Sensor (-)	AC pulse signals
24	PK/BL	O/D Clutch Speed Sensor (+)	AC pulse signals
25	BK/YL	Igniter Signal (IGF)	Digital Signal: 0-5-0v
26	GN/BK	IAC Signal (RSO)	Pulse Signals
27	WT	Knock Sensor 1 Signal	0.075v AC
28	WT	Knock Sensor 2 Signal	0.075v AC
29	RD/BL	VVT Solenoid Control (OC2+)	12v or 0v
30	WT/BK	Power Ground (E03)	<0.1v
31	WT/BK	Power Ground (E02)	<0.1v

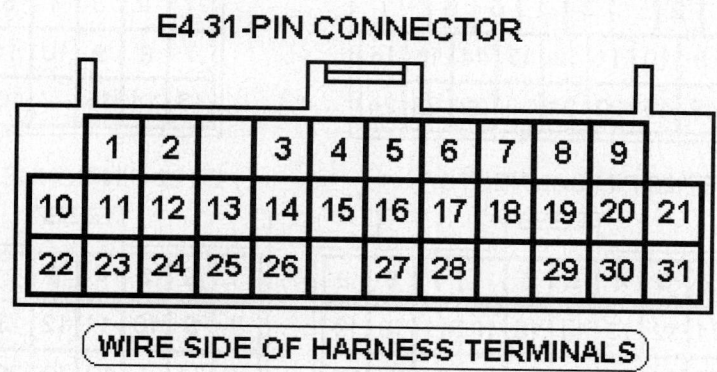

E4 31-PIN CONNECTOR

WIRE SIDE OF HARNESS TERMINALS

05533_ADIA_G009

Pin Connector Graphic

2002-03 3.0L V6 DOHC VIN F (A/T-ECT) E5 24P Connector

PCM Pin #	Wire Color	Circuit Description (24 Pin)	Value at Hot Idle
1	WT/BK	Power Ground (E04)	<0.1v
2	BL/RD	Sensor VREF (VC)	4.9-5.1v
3	GN	AFS-11 (B1 S1) Heater	1v (Heater On)
4	BK/RD	AFS-21 (B2 S1) Heater	1v (Heater On)
5	BL	Injector 1 Control	1.6-2-9 ms
6	RD	Injector 2 Control	1.6-2-9 ms
7	BK	EVAP Purge Solenoid (VSV)	12v or 0v
8	WT/BK	Power Ground (E05)	<0.1v
9	GN	PSP Switch Signal	Straight: 12v, Turned: 0v
10	RD	MAF Sensor Signal	1-1.1v
11	GN	AFS-11 (B1 S1) Signal (+)	3.0-3.6v
12	BK/WT	AFS-21 (B2 S1) Signal (+)	3.0-3.6v
13	---	Not Used	---
14	GN/YL	ECT Sensor Signal	At 180°F: 0.51v
15	WT/RD	VSV ACIS 2 Control	12v or 0v
16	BK/WT	CKP Sensor Signal (NE+)	310-330 Hz

2002-03 3.0L V6 DOHC VIN F (A/T-ECT) E5 24P Connector, *continued*

17	BR	Shield Ground (E1)	<0.050v
18	WT	Sensor Ground	<0.050v
19	RD/BK	MAF Sensor Ground (E2)	<0.050v
20	RD	AFS-11 (B1 S1) Signal (AF-)	3.0-3.6v
21	BL	AFS-21 (B2 S1) Signal (-)	3.0-3.6v
22	BL/BK	IAT Sensor Signal	0.5-3.4v
23	LG	TP Sensor Signal	0.53-1.27v
24	BL	CKP/CMP Sensor Ground (-)	<0.050v

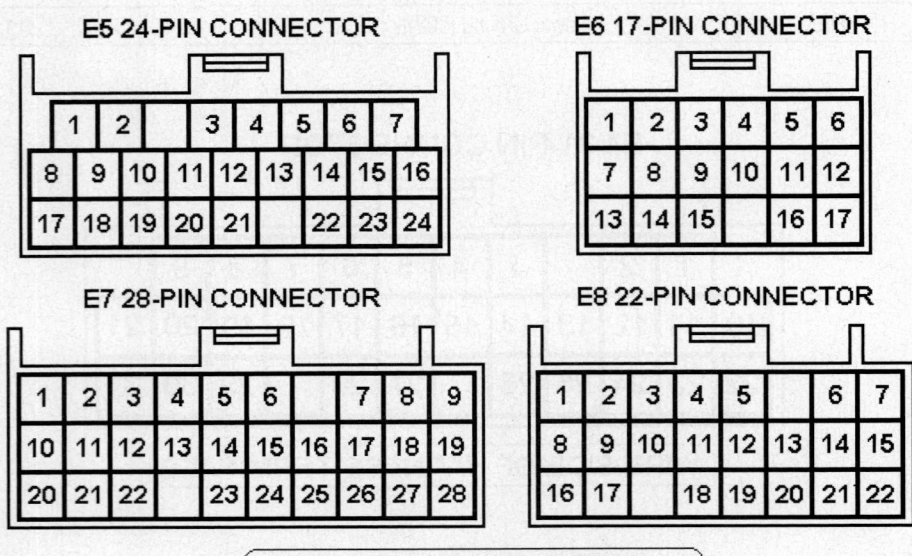

05533_ADIA_G010

Pin Connector Graphic

2002-03 3.0L V6 DOHC VIN F (A/T-ECT) E6 17P Connector

PCM Pin #	Wire Color	Circuit Description (17 Pin)	Value at Hot Idle
1	GN/YL	A/T-ECT Solenoid (S2)	1st or OD: 1v
2-7	---	Not Used	---
8	RD/BK	A/T Select Switch Reverse	In 'R': 12v, Others: 0v
9-10	---	Not Used	---
11	WT/BL	Active Control Engine Mount (VSV)	12v or 0v
12	GN/OR	Overdrive Main Switch	Switch Off: 12v, On: 1v
13	YL	A/T Select Switch Low	In Low: 12v, Others: 0v
14	GN/YL	A/T Select Switch 2nd	In 2nd: 12v, Others: 0v
15	GN/WT	A/T-ECT Solenoid (SL+)	In Lockup: 12-14v

2002-03 3.0L V6 DOHC VIN F (A/T-ECT) E7 28P Connector

PCM Pin #	Wire Color	Circuit Description (28 Pin)	Value at Hot Idle
1	BR	Power Ground (EOM)	<0.1v
2	---	Not Used	---
3	PK/BK	EVAP Vapor Pressure (VSV)	12v or 0v

2002-03 3.0L V6 DOHC VIN F (A/T-ECT) E7 28P Connector, *continued*

4	GN	Tail Light Switch Signal	Switch On: 12v, Off: 0v
5	LG/RD	Data Link Connector (TE1)	12-14v
6	LG/BK	A/C Magnetic Clutch (ACMG)	Clutch Off: 0v, On: 12v
7	YL/RD	A/C Amplifier Signal (AC1)	Clutch On: 1.5v, Off: 12v
8	BK	HO2S-12 (B1 S2) Signal	0.1-1.1v
9	LG	HO2S-12 (B1 S2) Heater	1v (Heater On)
10-12	---	Not Used	---
13	BK	Mirror Heater Switch Signal	Switch Off: 1.5v, On: 12v
14	YL/GN	A/C Amplifier Signal (THWO)	Pulse Signals
15	---	Not Used	---
16	YL/BK	ABS/Traction NEO Signal	Pulse Signals
17	---	Not Used	---
18	BL/YL	TXCT Ignition Signal	Inserting key: pulses
19	---	Not Used	---
20	BK/WT	Neutral Start Switch (NSW)	In P/N: 9-11v (cranking)
21	---	Not Used	---
22	VT/WT	Speedometer Indicator	At 55 mph: 48 Hz
23	RD/YL	Unlock Warning Switch	Key In: 1.5v, Out: 4.5v
24	GY	Cruise Control Signal (OD1)	At Cruise in OD: 12v
25	YL	Cruise Control Signal (IDLO)	1.5v, off-idle: 12v
26	GN	Cooling Fan 1 Relay Control	Relay Off: 12v, On: 1v
27	BK	Tachometer Signal (TACO)	Pulse Signals
28	PK/GN	Ignition Switch Code	Inserting key: pulses

2002-03 3.0L V6 DOHC VIN F (A/T-ECT) E8 22P Connector

PCM Pin #	Wire Color	Circuit Description (22 Pin)	Value at Hot Idle
1	BK/RD	Direct Battery	12-14v
2	BK/OR	Ignition Switch Power	12-14v
3	GN/BK	Circuit Opening Relay (FC)	0-0.3v, off-idle: 12-14v
4	WT	SIL Signal (Scan Tool)	Digital Signal
6	BK/YL	MIL (lamp) Control	MIL Off: 12v, On: 1v
7	BK/RD	Starter Signal (STA)	Cranking: 9-11v
8	BL/OR	EFI Main Relay Power	12-14v
9	GN/OR	Overdrive Lamp Control	At Cruise in OD: 1v
10	GN	EVAP Canister Closed Valve	12v or 0v
13	BR	ABS/Traction: TRC (+) Signal	DC pulse signals
14	VT	ABS/Traction: ENG (+) Signal	DC pulse signals
15	GN/WT	Brake Switch Signal	Brake Off: 0v, On: 12v
16	BK/WT	EFI Main Relay Power	12-14v
17	PK	EVAP Vapor Pressure Sensor	2.9-3.1v (with hose off)
20	GN/WT	ABS/Traction: TRC (-) Signal	DC pulse signals
21	GY	ABS/Traction: ENG (-) Signal	DC pulse signals
22	VT/WT	Security Indicator Light	Inserting key: pulses

ES 330 PIN CHARTS

2004-06 3.3L V6 DOHC VIN A (A/T-ECT) E4 31P Connector

PCM Pin #	Wire Color	Circuit Description (31 Pin)	Value at Hot Idle
1	YL	Injector 3 Control	1.6-2-9 ms
2	WT	Injector 4 Control	1.6-2-9 ms
3	BL/RD	Injector 5 Control	1.6-2-9 ms
4	GN	Injector 6 Control	1.6-2.9 ms
5	GR	VVT Solenoid Control (OC1-)	12v or 0v
6	RD/BK	VVT Solenoid Control (OC1+)	12v or 0v
7	PK	A/T-ECT Solenoid (S1)	In 3rd or OD: 1v
8-9	---	Not Used	---
10	BK/WT	CMP Sensor Signal (RH+)	AC pulse signals
11	GY/RD	Igniter Transistor 1 Control	6°, at 55 mph: 8° dwell
12	YL/GY	Igniter Transistor 2 Control	6°, at 55 mph: 8° dwell
13	GY/YL	Igniter Transistor 3 Control	6°, at 55 mph: 8° dwell
14	BL/YL	Igniter Transistor 4 Control	6°, at 55 mph: 8° dwell
15	BL	Igniter Transistor 5 Control	6°, at 55 mph: 8° dwell
16	BK	Igniter Transistor 6 Control	6°, at 55 mph: 8° dwell
17	RD/YL	ACIS 1 Control (VSV)	12v or 0v
18	RD/WT	VVT Solenoid Control (OC2-)	12v or 0v
19	YL	A/T-ECT Solenoid (SLN-)	Pulse Signals
20	PK/BL	A/T-ECT Solenoid (SLN+)	Pulse Signals
21	WT/BK	Power Ground (E01)	<0.1v
22	BK/WT	CMP Sensor Signal (LH+)	AC pulse signals
23	GN	O/D Clutch Speed Sensor (-)	AC pulse signals
24	PK/BL	O/D Clutch Speed Sensor (+)	AC pulse signals
25	BK/YL	Igniter Signal (IGF)	Digital Signal: 0-5-0v
26	GN/BK	IAC Signal (RSO)	Pulse Signals
27	WT	Knock Sensor 1 Signal	0.075v AC
28	WT	Knock Sensor 2 Signal	0.075v AC
29	RD/BL	VVT Solenoid Control (OC2+)	12v or 0v
30	WT/BK	Power Ground (E03)	<0.1v
31	WT/BK	Power Ground (E02)	<0.1v

E4 31-PIN CONNECTOR

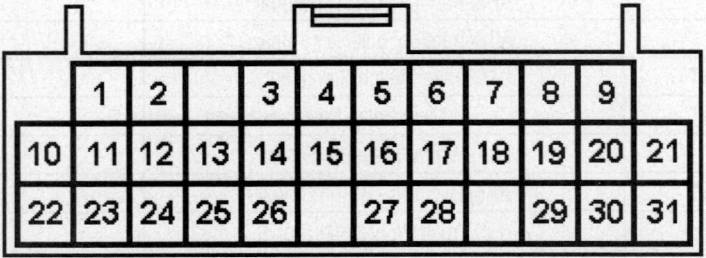

WIRE SIDE OF HARNESS TERMINALS

05533_ADIA_G009

Pin Connector Graphic

2004-06 3.3L V6 DOHC VIN A (A/T-ECT) E5 24P Connector

PCM Pin #	Wire Color	Circuit Description (24 Pin)	Value at Hot Idle
1	WT/BK	Power Ground (E04)	<0.1v
2	BL/RD	Sensor VREF (VC)	4.9-5.1v
3	GN	AFS-11 (B1 S1) Heater	1v (Heater On)
4	BK/RD	AFS-21 (B2 S1) Heater	1v (Heater On)
5	BL	Injector 1 Control	1.6-2-9 ms
6	RD	Injector 2 Control	1.6-2-9 ms
7	BK	EVAP Purge Solenoid (VSV)	12v or 0v
8	WT/BK	Power Ground (E05)	<0.1v
9	GN	PSP Switch Signal	Straight: 12v, Turned: 0v
10	RD	MAF Sensor Signal	1-1.1v
11	GN	AFS-11 (B1 S1) Signal (+)	3.0-3.6v
12	BK/WT	AFS-21 (B2 S1) Signal (+)	3.0-3.6v
13	---	Not Used	---
14	GN/YL	ECT Sensor Signal	At 180°F: 0.51v
15	WT/RD	VSV ACIS 2 Control	12v or 0v
16	BK/WT	CKP Sensor Signal (NE+)	310-330 Hz
17	BR	Shield Ground (E1)	<0.050v
18	WT	Sensor Ground	<0.050v
19	RD/BK	MAF Sensor Ground (E2)	<0.050v
20	RD	AFS-11 (B1 S1) Signal (AF-)	3.0-3.6v
21	BL	AFS-21 (B2 S1) Signal (-)	3.0-3.6v
22	BL/BK	IAT Sensor Signal	0.5-3.4v
23	LG	TP Sensor Signal	0.53-1.27v
24	BL	CKP/CMP Sensor Ground (-)	<0.050v

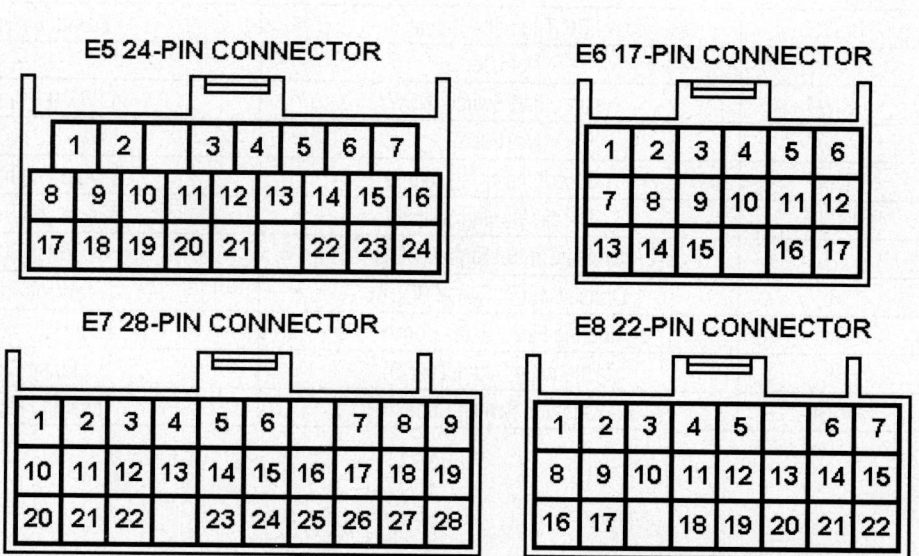

Pin Connector Graphic

2004-06 3.3L V6 DOHC VIN A (A/T-ECT) E6 17P Connector

PCM Pin #	Wire Color	Circuit Description (17 Pin)	Value at Hot Idle
1	GN/YL	A/T-ECT Solenoid (S2)	1st or OD: 1v
2-7	---	Not Used	---
8	RD/BK	A/T Select Switch Reverse	In 'R': 12v, Others: 0v
9-10	---	Not Used	---
11	WT/BL	Active Control Engine Mount (VSV)	12v or 0v
12	GN/OR	Overdrive Main Switch	Switch Off: 12v, On: 1v
13	YL	A/T Select Switch Low	In Low: 12v, Others: 0v
14	GN/YL	A/T Select Switch 2nd	In 2nd: 12v, Others: 0v
15	GN/WT	A/T-ECT Solenoid (SL+)	In Lockup: 12-14v

2004-06 3.3L V6 DOHC VIN A (A/T-ECT) E7 28P Connector

PCM Pin #	Wire Color	Circuit Description (28 Pin)	Value at Hot Idle
1	BR	Power Ground (EOM)	<0.1v
2	---	Not Used	---
3	PK/BK	EVAP Vapor Pressure (VSV)	12v or 0v
4	GN	Tail Light Switch Signal	Switch On: 12v, Off: 0v
5	LG/RD	Data Link Connector (TE1)	12-14v
6	LG/BK	A/C Magnetic Clutch (ACMG)	Clutch Off: 0v, On: 12v
7	YL/RD	A/C Amplifier Signal (AC1)	Clutch On: 1.5v, Off: 12v
8	BK	HO2S-12 (B1 S2) Signal	0.1-1.1v
9	LG	HO2S-12 (B1 S2) Heater	1v (Heater On)
10-12	---	Not Used	---
13	BK	Mirror Heater Switch Signal	Switch Off: 1.5v, On: 12v
14	YL/GN	A/C Amplifier Signal (THWO)	Pulse Signals
15	---	Not Used	---
16	YL/BK	ABS/Traction NEO Signal	Pulse Signals
17	---	Not Used	---
18	BL/YL	TXCT Ignition Signal	Inserting key: pulses
19	---	Not Used	---
20	BK/WT	Neutral Start Switch (NSW)	In P/N: 9-11v (cranking)
21	---	Not Used	---
22	VT/WT	Speedometer Indicator	At 55 mph: 48 Hz
23	RD/YL	Unlock Warning Switch	Key In: 1.5v, Out: 4.5v
24	GY	Cruise Control Signal (OD1)	At Cruise in OD: 12v
25	YL	Cruise Control Signal (IDLO)	1.5v, off-idle: 12v
26	GN	Cooling Fan 1 Relay Control	Relay Off: 12v, On: 1v
27	BK	Tachometer Signal (TACO)	Pulse Signals
28	PK/GN	Ignition Switch Code	Inserting key: pulses

2004-06 3.3L V6 DOHC VIN A (A/T-ECT) E8 22P Connector

PCM Pin #	Wire Color	Circuit Description (22 Pin)	Value at Hot Idle
1	BK/RD	Direct Battery	12-14v
2	BK/OR	Ignition Switch Power	12-14v
3	GN/BK	Circuit Opening Relay (FC)	0-0.3v, off-idle: 12-14v
4	WT	SIL Signal (Scan Tool)	Digital Signal
6	BK/YL	MIL (lamp) Control	MIL Off: 12v, On: 1v
7	BK/RD	Starter Signal (STA)	Cranking: 9-11v
8	BL/OR	EFI Main Relay Power	12-14v
9	GN/OR	Overdrive Lamp Control	At Cruise in OD: 1v
10	GN	EVAP Canister Closed Valve	12v or 0v
13	BR	ABS/Traction: TRC (+) Signal	DC pulse signals
14	VT	ABS/Traction: ENG (+) Signal	DC pulse signals
15	GN/WT	Brake Switch Signal	Brake Off: 0v, On: 12v
16	BK/WT	EFI Main Relay Power	12-14v
17	PK	EVAP Vapor Pressure Sensor	2.9-3.1v (with hose off)
20	GN/WT	ABS/Traction: TRC (-) Signal	DC pulse signals
21	GY	ABS/Traction: ENG (-) Signal	DC pulse signals
22	VT/WT	Security Indicator Light	Inserting key: pulses

GS 300 PIN CHARTS

1996 3.0L I6 MFI VIN D (All) 80 Pin Connector

PCM Pin #	Wire Color	Circuit Description (80 Pin)	Value at Hot Idle
41	BL/RD	Sensor VREF (VC)	4.9-5.1v
43	YL	TP Sensor Signal	0.3-0.8v
44	BL/YL	ECT Sensor Signal	At 180°F: 0.51v
45	PK/BL	IAT Sensor Signal	At 100°F: 2.60v
46	BR/YL	EGR Gas Temp. Sensor	3.5-4.0v
47	RD/BL	Rear HO2S Signal	0.1-1.1v
48	WT	Front HO2S Signal	0.1-1.1v
49	WT	Knock Sensor 2 Signal	<0.075v AC
50	WT	Knock Sensor 1 Signal	<0.075v AC
57	RD/WT	Igniter Signal (IGT)	Digital Signal: 0-5-0v
58	RD/YL	Igniter Signal (IGF)	Digital Signal: 0-5-0v
64	RD	Closed Throttle Switch	1v, at off-idle: 12v
65	BR/BK	Sensor Ground	<0.050v
66	YL/RD	Mass Airflow Sensor	1.0-1.8v
69	BR	Shield Ground	<0.050v
72	BK/YL	Rear HO2S Heater	1v (Heater On)
73	BK/BL	Front HO2S Heater	1v (Heater On)
74	PK	EVAP Purge Solenoid (VSV)	12v or 0v
75	PK	EGR Solenoid Control (VSV)	1v (Heater On)
76	BK/WT	A/T Neutral Start Switch	In P/N: 9-11v (cranking)

1996 3.0L I6 MFI VIN D (All) 80 Pin Connector, *continued*

77	BK	Starter Switch Signal	9-11v (cranking)
78	BR	Power Ground	<0.1v
79	BR	Power Ground	<0.1v
80	BR	Power Ground	<0.1v

Standard Colors and Abbreviations

Abbreviation	Color	Abbreviation	Color	Abbreviation	Color
BK	Black	GY	Gray	RD	Red
BL	Blue	GN	Green	TN	Tan
BR	Brown	LG	Light Green	VT	Violet
DB	Dark Blue	OR	Orange	WT	White
DG	DK Green	PK	Pink	YL	Yellow

1997 3.0L I6 MFI VIN D (All) 40 Pin Connector

PCM Pin #	Wire Color	Circuit Description (40 Pin)	Value at Hot Idle
1	BK/OR	Ignition Switch Power	12-14v
2	PK	Speed Sensor 1 Signal	Moving: 0-5-0v
3	YL	Kickdown Switch	Open: 12v, Closed: 0v
4	GN/WT	Stop Light Switch Signal	Brake Off: 0v, On: 12v
6	BL/BK	MIL (lamp) Control	MIL Off: 12v, On: 1v
8	GN	Data Link Connector (SDL)	0v
9	GN/RD	A/T Select Switch 2nd	In 2nd: 12v, Others: 0v
10	GN/BK	A/T Select Switch Low	In Low: 12v, Others: 0v
12	GN/BK	Cruise Control ECU	At Cruise in OD: 12v
15	RD/YL	Idle Up Diode Signal	Switch On: 12v, Off: 0v
18	GN/YL	A/T Pattern Selector Switch	Norm: 0v, PWR: 12v
20	YL/GN	Data Link Connector	12-14v
21	GN	Fuel Pump (Control) ECU	Pulse Signals
22	PK/WT	Fuel Pump (DI) ECU	Pulse Signals
23	WT/GN	A/C Magnetic Clutch (ACMG)	Clutch Off: 0v, On: 12v
24	BK/YL	EFI Main Relay Power	12-14v
25	WT/BL	Manual Indicator Light	Light Off: 12v, On: 1v
28	PK	Overdrive Main Switch	Switch Off: 12v
30	RD/BL	Sub HO2S Signal	0.1-1.1v
31	BK/RD	EFI Main Relay Power	12-14v
32	BR /WT	PSP Switch Signal	Straight: 12v, Turning: 0v
33	BK/WT	Battery Direct	12-14v
34	BL/RD	A/C Amplifier Signal (AC1)	Clutch On: 1.5v, Off: 12v
36	BR/WT	Sub HO2S Heater	1v (Heater On)

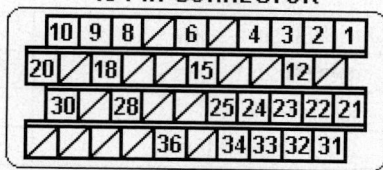

40 PIN CONNECTOR

WIRE SIDE OF HARNESS TERMINALS

05533_ADIA_G095

Pin Connector Graphic

1997 3.0L I6 MFI VIN D (All) 80 Pin Connector

PCM Pin #	Wire Color	Circuit Description (80 Pin)	Value at Hot Idle
3	GN	Vehicle Speed 2 Sensor (-)	Moving: 0-5-0v
4	BK	CKP Sensor Signal (NE2+)	AC pulse signals
5	WT	CKP Sensor Signal (NE2-)	<0.050v
7	GN	Distributor Signal (G-)	<0.050v
8	BK/RD	A/T-ECT Solenoid (SL)	In Lockup: 12-14v
9	RD/BL	A/T-ECT Solenoid (S2)	1st or OD: 1v
10	WT/RD	A/T-ECT Solenoid (S1)	In 3rd or OD: 1v
15	RD/BK	Injector 6 Control	2.0-3.3 ms
16	RD	Injector 5 Control	2.0-3.3 ms
17	RD/WT	Injector 4 Control	2.0-3.3 ms
18	BL	Injector 3 Control	2.0-3.3 ms
19	BL/RD	Injector 2 Control	2.0-3.3 ms
20	RD/BL	Injector 1 Control	2.0-3.3 ms
23	RD	Vehicle Speed 2 Sensor (+)	Moving: 0-5-0v
24	BL/BK	A/T Oil Temperature Sensor	0.5-4.5v
25	WT	Distributor Signal (G2+)	AC pulse signals
26	RD	Distributor Signal (G1+)	AC pulse signals
27	BK	Distributor Signal (NE+)	AC pulse signals
28	BR	MAF Sensor Ground	<0.050v
32	RD/BK	ISC Motor (ISC4)	Pulse Signals
33	BL/BK	ISC Motor (ISC3)	Pulse Signals
34	GN/WT	ISC Motor (ISC2)	Pulse Signals
35	YL/BK	ISC Motor (ISC1)	Pulse Signals
39	GN/YL	Intake Air Control Solenoid	12v or 0v

1997 3.0L I6 MFI VIN D (All) 80 Pin Connector

PCM Pin #	Wire Color	Circuit Description (80 Pin)	Value at Hot Idle
41	BL/RD	Sensor VREF (VC)	4.9-5.1v
43	YL	TP Sensor Signal	0.3-0.8v
44	BL/YL	ECT Sensor Signal	At 180°F: 0.51v
45	GN/BK	IAT Sensor Signal	At 100°F: 2.60v
46	BR/YL	EGR Gas Temp. Sensor	3.5-4.0v
47	RD/BL	HO2S-21 (B2 S1) Signal	0.1-1.1v

1997 3.0L I6 MFI VIN D (All) 80 Pin Connector, *continued*

48	WT	HO2S-11 (B1 S1) Signal	0.1-1.1v
49	WT	Knock Sensor 2 Signal	<0.075v AC
50	WT	Knock Sensor 1 Signal	<0.075v AC
57	RD/WT	Igniter Signal (IGT)	Digital Signal: 0-5-0v
58	RD/YL	Igniter Signal (IGF)	Digital Signal: 0-5-0v
64	RD/BK	Closed Throttle Switch	1v, at off-idle: 12v
65	WT/BK	Sensor Ground	<0.050v
66	YL/RD	Mass Airflow Sensor	1.0-1.8v
69	BR	Shield Ground	<0.050v
71	GN	HO2S-11 (B1 S1) Heater	1v (Heater On)
72	BK/YL	HO2S-21 (B2 S1) Heater	1v (Heater On)
73	WT/BL	Fuel Pressure Up Solenoid	1v (at hot restart)
74	PK	EVAP Purge Solenoid (VSV)	12v or 0v
75	PK	EGR Solenoid Control (VSV)	1v (Heater On)
76	BK/WT	A/T Neutral Start Switch	In P/N: 9-11v (cranking)
77	BK	Starter Switch Signal	9-11v (cranking)
78	BR	Power Ground	<0.1v
79	BR	Power Ground	<0.1v
80	BR	Power Ground	<0.1v

1998 3.0L I6 MFI VIN D (A/T-ECT) 31 Pin Connector

80 PIN CONNECTOR

WIRE SIDE OF HARNESS TERMINALS

05533_ADIA_G096

Pin Connector Graphic

PCM Pin #	Wire Color	Circuit Description (31 Pin)	Value at Hot Idle
1	BL	Injector 3 Control	2.0-3.3 ms
2	RD/WT	Injector 4 Control	2.0-3.3 ms
3	RD	Injector 5 Control	2.0-3.3 ms
4	RD/BK	Injector 6 Control	2.0-3.3 ms
5	GN/YL	Intake Air Control Solenoid	12v or 0v
7	BK	Throttle Control Motor (M-)	Pulse Signals
8	WT	Throttle Control Motor (M+)	Pulse Signals
9	BR	Sensor Ground	<0.050v
10	BL	CMP Sensor Signal (G+)	AC pulse signals
11	RD/WT	Igniter Transistor 1 Control	Digital Signal: 0-5-0v
12	LG	Igniter Transistor 2 Control	Digital Signal: 0-5-0v
13	GN/RD	Igniter Transistor 3 Control	Digital Signal: 0-5-0v
16	GN	PSP Switch Signal	Straight: 12v, Turning: 0v

17	YL/BK	Cam Timing Oil signal (OCV-)	Pulse Signals
18	WT/RD	Cam Timing Oil signal (OCV+)	Pulse Signals
19	YL	Throttle Control Motor (CL-)	Pulse Signals
20	BL	Throttle Control Motor (CL+)	Pulse Signals
21	WT/BR	Power Ground	<0.1v
22	WT	CKP/CMP Sensor Ground (-)	<0.050v
23	BK	CKP Sensor Signal (NE+)	AC pulse signals
24	BK/WT	Neutral Start Switch	In P/N: 9-11v (cranking)
25	RD/YL	Igniter Signal (IGF)	Digital Signal: 0-5-0v
27	WT	Knock Sensor 2 Signal	<0.075v AC
28	WT	Knock Sensor 1 Signal	<0.075v AC
30	BR	Shield Ground	<0.050v
31	WT/BR	Power Ground	<0.1v

1998 3.0L I6 MFI VIN D (A/T-ECT) 22 Pin Connector

PCM Pin #	Wire Color	Circuit Description (22 Pin)	Value at Hot Idle
1	BK/WT	Battery Direct	12-14v
4	GN	Fuel Pump (Control) ECU	Pulse Signals
5	PK/WT	Fuel Pump (DI) ECU	Pulse Signals
6	BL/BK	MIL (lamp) Control	MIL Off: 12v, On: 1v
7	BL/RD	EFI Main Relay Power BM+	12-14v
8	BK/RD	EFI Main Relay Power B2+	12-14v
9	BK/OR	Ignition Switch Power	12-14v
10	BK/YL	EFI Main Relay Power	12-14v
11	BK	SIL Signal (Scan Tool)	Digital Signal
16	BK/RD	EFI Main Relay Power	12-14v
22	WT/BK	ECM/Data Link Ground	<0.050v

1998 3.0L I6 MFI VIN D (A/T-ECT) 28 Pin Connector

PCM Pin #	Wire Color	Circuit Description (28 Pin)	Value at Hot Idle
1	BL/RD	A/C Amplifier Signal (AC1)	Clutch On: 1.5v, Off: 12v
2	BK	Starter Switch Signal	9-11v (cranking)
4	GN/WT	Stop Light Switch Signal	Brake Off: 0v, On: 12v
5	RD	Data Link Connector	12-14v
8	WT	HO2S-12 (B1 S2) Signal	0.1-1.1v
13	WT/GN	A/C Magnetic Clutch (ACMG)	Clutch Off: 0v, On: 12v
16	RD/BK	A/T Select Switch Reverse	In 'R': 12v, Others: 0v
17	GN/RD	A/T Select Switch Drive	In 'D': 12v, Others: 0v
18	BR/YL	EVAP Vapor Pressure Sensor	2.5-3.1v (with hose off)
23	BL	Cruise Control ECU	At Cruise in OD: 12v
24	RD/YL	Cruise Control ECU	At Cruise in OD: 12v
25	WT	Vehicle Speed Sensor	At 55 mph: 48 Hz
26	RD/YL	Idle Up Diode Signal	Switch On: 12v, Off: 0v
28	PK	Overdrive Main Switch	Switch Off: 12v, On: 0v

1998 3.0L I6 MFI VIN D (A/T-ECT) 17 Pin Connector

PCM Pin #	Wire Color	Circuit Description (17 Pin)	Value at Hot Idle
1	WT/RD	A/T-ECT Solenoid (S1)	In 3rd or OD: 1v
2	RD/BL	A/T-ECT Solenoid (S2)	1st or OD: 1v
4	RD	OD Clutch Sensor Signal (+)	Pulse Signals
5	RD/YL	Speed Sensor 2 Signal (+)	Pulse Signals
7	GN/WT	A/T-ECT Solenoid (SLU+)	Pulse Signals
8	YL/GN	A/T-ECT Solenoid (SLN+)	Pulse Signals
9	GN/RD	A/T-ECT Solenoid (SLT+)	Pulse Signals
10	RD	OD Clutch Sensor Signal (-)	Pulse Signals
11	BL/YL	Speed Sensor Signal 2 (-)	Pulse Signals
13	BL/RD	A/T-ECT Solenoid (SLU-)	Pulse Signals
14	PK	A/T-ECT Solenoid (SLN-)	Pulse Signals
15	RD/BK	A/T-ECT Solenoid (SLT-)	Pulse Signals
16	OR	C/C Indicator Light	Light Off: 12v, On: 1v
17	BL/BK	A/T Oil Temperature Sensor	0.5-4.5v

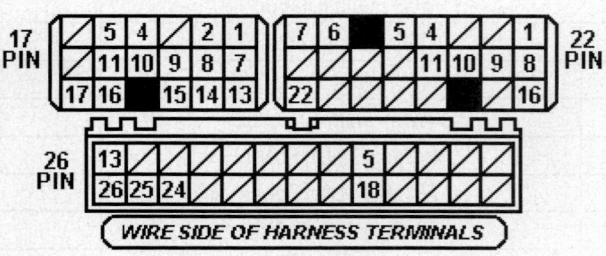

Pin Connector Graphic

1998 3.0L I6 MFI VIN D (A/T-ECT) 26 Pin Connector

PCM Pin #	Wire Color	Circuit Description (26 Pin)	Value at Hot Idle
5	GN/YL	A/T Pattern Selector Switch	Norm: 0v, PWR: 12v
13	RD/BK	EVAP Vapor Pressure (VSV)	12v or 0v
18	WT/BL	Manual Indicator Light	Light Off: 12-14v
24	WT	HO2S-22 (B2 S2) Signal	0.1-1.1v
25	GN/YL	HO2S-22 (B2 S2) Heater	1v (Heater On)
26	BL/WT	HO2S-12 (B1 S2) Heater	1v (Heater On)

1998 3.0L I6 MFI VIN D (A/T-ECT) 24 Pin Connector

PCM Pin #	Wire Color	Circuit Description (24 Pin)	Value at Hot Idle
1	WT/BK	Power Ground	<0.1v
2	BL/RD	Sensor VREF	4.9-5.1v
3	BK/YL	HO2S-21 (B2 S1) Heater	1v (Heater On)
4	GN	HO2S-11 (B1 S1) Heater	1v (Heater On)
5	RD/BL	Injector 1 Control	2.0-3.3 ms
6	BL/RD	Injector 2 Control	2.0-3.3 ms
7	YL	EVAP Purge Solenoid (VSV)	12v or 0v

1998 3.0L I6 MFI VIN D (A/T-ECT) 24 Pin Connector, *continued*

10	YL/RD	MAF Sensor Signal	1.1-1.5v
11	RD/BL	HO2S-21 (B2 S1) Signal	0.1-1.1v
12	WT	HO2S-11 (B1 S1) Signal	0.1-1.1v
13	BL/RD	Mirror Heater Switch Signal	Heater On: 12-14v
14	BL	ECT Sensor Signal	At 180°F: 0.51v
15	GN	Accel Position Sensor (VPA)	0.25-0.9v
16	WT	Accel Position Sensor (VPA2)	1.8-2.7v
17	BR	Sensor Ground	<0.050v
18	BR	Sensor Ground	<0.050v
19	BR	MAF Sensor Ground	<0.050v
20	BL/YL	A/T Select Switch 2nd	In 2nd: 12v, Others: 0v
21	GN/BK	A/T Select Switch Low	In Low: 12v, Others: 0v
22	GN/WT	IAT Sensor Signal	At 100°F: 2.60v
23	YL	TP Sensor (VTA1)	0.4-1.0v
24	RD/BK	TP Sensor (VTA2)	2.0-2.9v

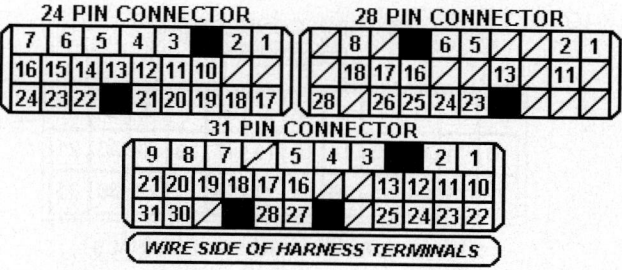

05533_ADIA_G098

Pin Connector Graphic

GS 400 PIN CHARTS

1996-2000 4.0L 1UZ-FE V8 VIN T E5 31P Connector

PCM Pin #	Wire Color	Circuit Description (31 Pin)	Value at Hot Idle
1	BL	Injector 3 Control	2.0-3.3 ms
2	RD	Injector 4 Control	2.0-3.3 ms
3	GN	Injector 5 Control	2.0-3.3 ms
4	RD/BL	Injector 6 Control	2.0-3.3 ms
5	WT	Injector 7 Control	2.0-3.3 ms
6	BK/WT	Injector 8 Control	2.0-3.3 ms
7	WT	Throttle Control Motor (M-)	Pulse Signals
8	RD	Throttle Control Motor (M+)	Pulse Signals
9	WT/BK	Power Ground (ME01)	<0.1v
10	RD	CMP Sensor Signal (G+)	AC pulse signals
11	BK	Igniter Transistor 1 Control	7% duty cycle
12	RD	Igniter Transistor 2 Control	7% duty cycle
13	BL	Igniter Transistor 3 Control	7% duty cycle
14	GN	Igniter Transistor 4 Control	7% duty cycle
15	YL	Igniter Transistor 5 Control	7% duty cycle

1996-2000 4.0L 1UZ-FE V8 VIN T E5 31P Connector, *continued*

16	BK/YL	Igniter Transistor 6 Control	7% duty cycle
17	WT	Knock Sensor 2 Signal (right)	<0.075v AC
18	BK	Knock Sensor 1 Signal (left)	<0.075v AC
19-20	---	Not Used	---
21	WT/BK	Power Ground (E01)	<0.1v
22	GN	CMP/CKP Sensor Signal (-)	<0.050v
23	BL	CKP Sensor Signal (NE+)	AC pulse signals
24	BL	Throttle Control Motor (CL-)	Pulse Signals
25	BK/BL	Igniter Transistor 7 Control	7% duty cycle
26	BL/BK	Igniter Transistor 8 Control	7% duty cycle
27	BK/WT	Igniter Signal (IGF1)	Digital Signal: 0-5-0v
28	BK/RD	Igniter Signal (IGF2)	Digital Signal: 0-5-0v
29	GN	Throttle Control Motor (CL+)	Pulse Signals
30	BR	Shield Ground (GE01)	<0.050v
31	WT/BK	Power Ground (E02)	<0.1v

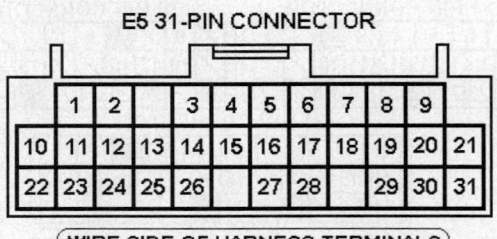

E5 31-PIN CONNECTOR

WIRE SIDE OF HARNESS TERMINALS

05533_ADIA_G120

Pin Connector Graphic

1996-2000 4.0L 1UZ-FE V8 VIN T E6 24P Connector

PCM Pin #	Wire Color	Circuit Description (24 Pin)	Value at Hot Idle
1	WT/BK	Power Ground (E03)	<0.1v
2	BL/RD	Sensor VREF (VC)	4.9-5.1v
3	YL	HO2S-21 (B2 S1) Heater	1v (Heater on)
4	RD	HO2S-11 (B1 S1) Heater	1v (Heater on)
5	YL	Injector 1 Control	2.0-3.3 ms
6	BK	Injector 2 Control	2.0-3.3 ms
7	BL/BK	EVAP Purge Solenoid (VSV)	12v or 0v
8	RD	ABS ECU Signal	Digital Signals
9	RD/BK	Accelerator Position Sensor 2	1.8-2.7v
10	BL/YL	Mass Airflow Sensor (VG)	1.1-1.5v
11	WT	HO2S-21 (B2 S1) Signal	0.1-1.1v
12	BK	HO2S-11 (B1 S1) Signal	0.1-1.1v
13	RD/YL	TP Sensor Signal (VTA)	0.4-1.0v
14	GN/BK	ECT Sensor Signal (THW)	At 180°F: 0.51v
15-16	---	Not Used	---
17	BR	Power Ground (E1)	<0.1v
18	BR/WT	Sensor Ground (E2)	<0.050v

1996-2000 4.0L 1UZ-FE V8 VIN T E6 24P Connector, *continued*

19	GN/WT	MAF Sensor Ground (EVG)	<0.050v
20	YL/BK	TP Sensor Signal (VTA2)	2.0-2.9v
21	RD	Accelerator Position Sensor 1	0.3-0.90v
22	YL/BK	IAT Sensor Signal (THA)	At 100°F: 2.60v
23 (01'-02')	RD/BK	HO2S-22 (B2 S2) Heater	1v (Heater on)
24 (01'-02')	BL	HO2S-12 (B1 S2) Heater	1v (Heater on)

1996-2000 4.0L V8 1UZ-FE VIN T E7 17P Connector

PCM Pin #	Wire Color	Circuit Description (17 Pin)	Value at Hot Idle
1	RD	A/T ECT Solenoid (S1)	In 3rd or OD: 1v
2	WT	A/T ECT Solenoid (S2)	In 1st or OD: 1v
3	GN	A/T ECT Solenoid (SL)	In Lockup: 12-14v
4	BL/RD	Direct Clutch Speed Input (+)	AC pulse signals
5	RD	A/T Vehicle Speed Sensor (+)	AC pulse signals
6-8	---	Not Used	---
9	GN/WT	A/T-ECT Solenoid (SLT+)	Pulse Signals
10	BL/WT	Direct Clutch Speed Input (-)	AC pulse signals
11	GN	A/T Vehicle Speed Sensor (-)	Pulse Signals
12-14	---	Not Used	---
15	GNBK	A/T-ECT Solenoid (SLT-)	Pulse Signals
16	---	Not Used	---
17	GN/YL	A/T Oil Temperature Sensor	At 68°F: 4-5v

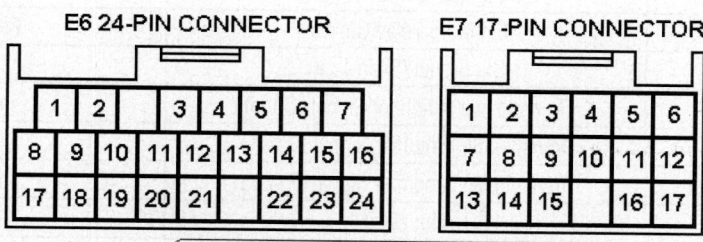

E6 24-PIN CONNECTOR   E7 17-PIN CONNECTOR

WIRE SIDE OF HARNESS TERMINALS

05533_ADIA_G121

Pin Connector Graphic

1996-2000 4.0L 1UZ-FE V8 VIN T E8 28P Connector

PCM Pin #	Wire Color	Circuit Description (28 Pin)	Value at Hot Idle
5	PK/RD	Data Link Connector (TE1)	12-14v
6	GN/WT	Stop Light Switch Signal	Brake Off: 0v, On: 12v
7	RD/BK	HO2S-22 (B2 S2) Heater	1v (Heater on)
8	BL	HO2S-12 (B1 S2) Heater	1v (Heater on)
9	---	Not Used	---
10	BL	EVAP Vapor Pressure Valve (VSV)	12v or 0v
11	---	Not Used	---
12	GN/WT	Defogger & Tail Light Switch	Switch Off: 0v, On: 12v
13	BL/BK	A/C Amplifier Signal (ACT)	Clutch On: 12v, Off: 1.5v
14	YL/BK	A/C Amplifier Signal (THWO)	A/C Off: 12v, On: 1v

1996-2000 4.0L 1UZ-FE V8 VIN T E8 28P Connector, *continued*

15	VT	Vehicle Speed Sensor	At 55 mph: 48 Hz
16	BK	Tachometer Signal (TACO)	Pulse Signals
17	BK/RD	Starter Switch Signal	9-11v (cranking)
18	BK	HO2S-12 (B1 S2) Signal	0.1-1.1v
19	---	Not Used	---
20	BK/WT	Neutral Start Switch (NSW)	In P/N: 0-3.0v
21	WT/BK	Power Ground (EOM)	<0.1v
22	BL/BK	EVAP Vapor Pressure Sensor (PTNK)	2.5-3.1v (with cap off)
23-24	---	Not Used	---
25	WT/GN	A/C Amplifier Signal (AC1)	Clutch On: 1.5v, Off: 12v
26	---	Not Used	---
27	WT	HO2S-22 (B2 S2) Signal	0.1-1.1v
28	---	Not Used	---

1996-2000 4.0L 1UZ-FE V8 VIN T E9 22P Connector

PCM Pin #	Wire Color	Circuit Description (22 Pin)	Value at Hot Idle
1	RD/YL	Direct Battery	12-14v
2-3	---	Not Used	---
4	GN/RD	Fuel Pump Signal (DI)	12-14v
5	GN/WT	Fuel Control Switch Signal	Closed: 0v, Open: 12v
6	WT	Malfunction Indicator Lamp Control	MIL Off: 12v, On: 1v
7	YL/BK	BM (+) Power	12-14v
8	BK/YL	EFI Main Relay B1+	12-14v
9	BK/RD	Ignition Switch Power (IGSW)	12-14v
10	BK/WT	EFI Main Relay Control	Relay Off: 0v, On: 12v
11	PK/WT	SIL Signal (Scan Tool)	12v
12	BL/BK	Transponder Amplifier Signal (Code)	Inserting key: pulses
13	VT/GN	Transponder Amplifier Signal (RXCK)	Inserting key: pulses
14	RD/YL	Transponder Amplifier Signal (RXCK)	Inserting key: pulses
15	---	Not Used	---
16	BK/YL	EFI Main Relay B+	12-14v
18-19	---	Not Used	---
20	RD/BK	Unlock Warning Switch	Key In: 1.5v, Out: 4-5v
21	GN/RD	Center ECU LED Signal	LED Off: 12v, On: 1v
22	---	Not Used	---

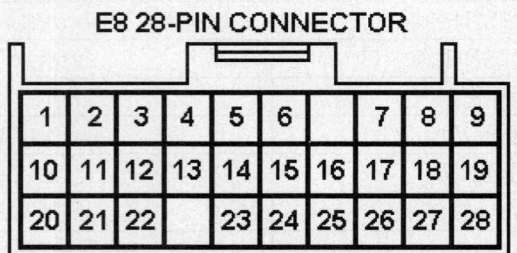

E8 28-PIN CONNECTOR

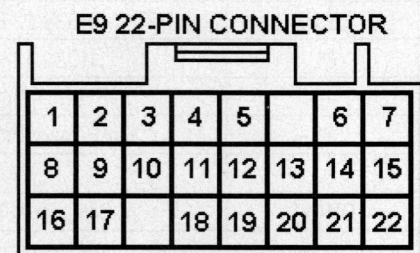

E9 22-PIN CONNECTOR

05533_ADIA_G122

Pin Connector Graphic

GS 430 PIN CHARTS

2001-06 4.7L 3UZ-FE V8 VIN N E5 31P Connector

PCM Pin #	Wire Color	Circuit Description (31 Pin)	Value at Hot Idle
1	BL	Injector 3 Control	2.0-3.3 ms
2	RD	Injector 4 Control	2.0-3.3 ms
3	GN	Injector 5 Control	2.0-3.3 ms
4	RD/BL	Injector 6 Control	2.0-3.3 ms
5	WT	Injector 7 Control	2.0-3.3 ms
6	BK/WT	Injector 8 Control	2.0-3.3 ms
7	WT	Throttle Control Motor (M-)	Pulse Signals
8	RD	Throttle Control Motor (M+)	Pulse Signals
9	WT/BK	Power Ground (ME01)	<0.1v
10	RD	CMP Sensor Signal (G+)	AC pulse signals
11	BK	Igniter Transistor 1 Control	7% duty cycle
12	RD	Igniter Transistor 2 Control	7% duty cycle
13	BL	Igniter Transistor 3 Control	7% duty cycle
14	GN	Igniter Transistor 4 Control	7% duty cycle
15	YL	Igniter Transistor 5 Control	7% duty cycle
16	BK/YL	Igniter Transistor 6 Control	7% duty cycle
17	WT	Knock Sensor 2 Signal (right)	<0.075v AC
18	BK	Knock Sensor 1 Signal (left)	<0.075v AC
19-20	---	Not Used	---
21	WT/BK	Power Ground (E01)	<0.1v
22	GN	CMP/CKP Sensor Signal (-)	<0.050v
23	BL	CKP Sensor Signal (NE+)	AC pulse signals
24	BL	Throttle Control Motor (CL-)	Pulse Signals
25	BK/BL	Igniter Transistor 7 Control	7% duty cycle
26	BL/BK	Igniter Transistor 8 Control	7% duty cycle
27	BK/WT	Igniter Signal (IGF1)	Digital Signal: 0-5-0v
28	BK/RD	Igniter Signal (IGF2)	Digital Signal: 0-5-0v
29	GN	Throttle Control Motor (CL+)	Pulse Signals
30	BR	Shield Ground (GE01)	<0.050v
31	WT/BK	Power Ground (E02)	<0.1v

E5 31-PIN CONNECTOR

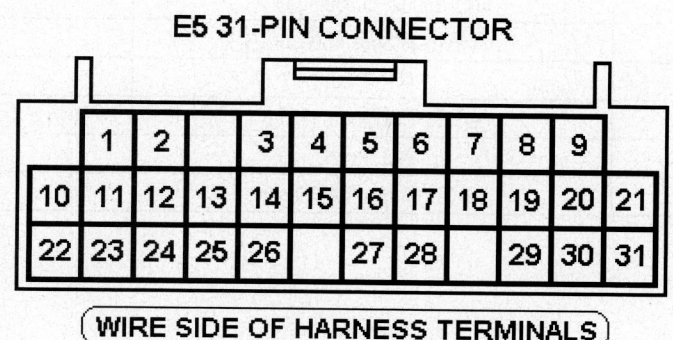

WIRE SIDE OF HARNESS TERMINALS

05533_ADIA_G120

Pin Connector Graphic

2001-06 4.7L 3UZ-FE V8 VIN N E6 24P Connector

PCM Pin #	Wire Color	Circuit Description (24 Pin)	Value at Hot Idle
1	WT/BK	Power Ground (E03)	<0.1v
2	BL/RD	Sensor VREF (VC)	4.9-5.1v
3	YL	HO2S-21 (B2 S1) Heater	1v (Heater on)
4	RD	HO2S-11 (B1 S1) Heater	1v (Heater on)
5	YL	Injector 1 Control	2.0-3.3 ms
6	BK	Injector 2 Control	2.0-3.3 ms
7	BL/BK	EVAP Purge Solenoid (VSV)	12v or 0v
8	RD	ABS ECU Signal	Digital Signals
9	RD/BK	Accelerator Position Sensor 2	1.8-2.7v
10	BL/YL	Mass Airflow Sensor (VG)	1.1-1.5v
11	WT	HO2S-21 (B2 S1) Signal	0.1-1.1v
12	BK	HO2S-11 (B1 S1) Signal	0.1-1.1v
13	RD/YL	TP Sensor Signal (VTA)	0.4-1.0v
14	GN/BK	ECT Sensor Signal (THW)	At 180°F: 0.51v
15-16	---	Not Used	---
17	BR	Power Ground (E1)	<0.1v
18	BR/WT	Sensor Ground (E2)	<0.050v
19	GN/WT	MAF Sensor Ground (EVG)	<0.050v
20	YL/BK	TP Sensor Signal (VTA2)	2.0-2.9v
21	RD	Accelerator Position Sensor 1	0.3-0.90v
22	YL/BK	IAT Sensor Signal (THA)	At 100°F: 2.60v
23 (01'-02')	RD/BK	HO2S-22 (B2 S2) Heater	1v (Heater on)
24 (01'-02')	BL	HO2S-12 (B1 S2) Heater	1v (Heater on)

2001-06 4.7L 3UZ-FE V8 VIN N E7 17P Connector

PCM Pin #	Wire Color	Circuit Description (17 Pin)	Value at Hot Idle
1	RD	A/T ECT Solenoid (S1)	In 3rd or OD: 1v
2	WT	A/T ECT Solenoid (S2)	In 1st or OD: 1v
3	GN	A/T ECT Solenoid (SL)	In Lockup: 12-14v
4	BL/RD	Direct Clutch Speed Input (+)	AC pulse signals
5	RD	A/T Vehicle Speed Sensor (+)	AC pulse signals
6-8	---	Not Used	---
9	GN/WT	A/T-ECT Solenoid (SLT+)	Pulse Signals
10	BL/WT	Direct Clutch Speed Input (-)	AC pulse signals
11	GN	A/T Vehicle Speed Sensor (-)	Pulse Signals
12-14	---	Not Used	---
15	GNBK	A/T-ECT Solenoid (SLT-)	Pulse Signals
16	---	Not Used	---
17	GN/YL	A/T Oil Temperature Sensor	At 68°F: 4-5v

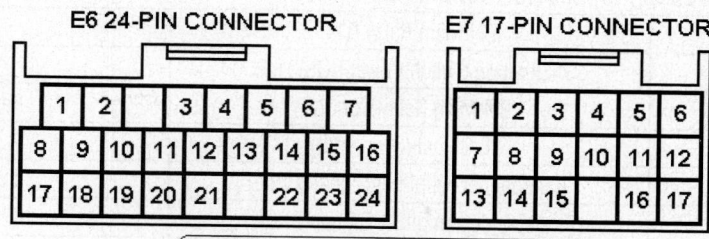

E6 24-PIN CONNECTOR   E7 17-PIN CONNECTOR

WIRE SIDE OF HARNESS TERMINALS

05533_ADIA_G121

Pin Connector Graphic

2001-06 4.7L 3UZ-FE V8 VIN N E8 28P Connector

PCM Pin #	Wire Color	Circuit Description (28 Pin)	Value at Hot Idle
5	PK/RD	Data Link Connector (TE1)	12-14v
6	GN/WT	Stop Light Switch Signal	Brake Off: 0v, On: 12v
7	RD/BK	HO2S-22 (B2 S2) Heater	1v (Heater on)
8	BL	HO2S-12 (B1 S2) Heater	1v (Heater on)
9	---	Not Used	---
10	BL	EVAP Vapor Pressure Valve (VSV)	12v or 0v
11	---	Not Used	---
12	GN/WT	Defogger & Tail Light Switch	Switch Off: 0v, On: 12v
13	BL/BK	A/C Amplifier Signal (ACT)	Clutch On: 12v, Off: 1.5v
14	YL/BK	A/C Amplifier Signal (THWO)	A/C Off: 12v, On: 1v
15	VT	Vehicle Speed Sensor	At 55 mph: 48 Hz
16	BK	Tachometer Signal (TACO)	Pulse Signals
17	BK/RD	Starter Switch Signal	9-11v (cranking)
18	BK	HO2S-12 (B1 S2) Signal	0.1-1.1v
19	---	Not Used	---
20	BK/WT	Neutral Start Switch (NSW)	In P/N: 0-3.0v
21	WT/BK	Power Ground (EOM)	<0.1v
22	BL/BK	EVAP Vapor Pressure Sensor (PTNK)	2.5-3.1v (with cap off)
23-24	---	Not Used	---
25	WT/GN	A/C Amplifier Signal (AC1)	Clutch On: 1.5v, Off: 12v
26	---	Not Used	---
27	WT	HO2S-22 (B2 S2) Signal	0.1-1.1v
28	---	Not Used	---

2001-06 4.7L 3UZ-FE V8 VIN N E9 22P Connector

PCM Pin #	Wire Color	Circuit Description (22 Pin)	Value at Hot Idle
1	RD/YL	Direct Battery	12-14v
2-3	---	Not Used	---
4	GN/RD	Fuel Pump Signal (DI)	12-14v
5	GN/WT	Fuel Control Switch Signal	Closed: 0v, Open: 12v
6	WT	Malfunction Indicator Lamp Control	MIL Off: 12v, On: 1v
7	YL/BK	BM (+) Power	12-14v

2001-06 4.7L 3UZ-FE V8 VIN N E9 22P Connector, *continued*

8	BK/YL	EFI Main Relay B1+	12-14v
9	BK/RD	Ignition Switch Power (IGSW)	12-14v
10	BK/WT	EFI Main Relay Control	Relay Off: 0v, On: 12v
11	PK/WT	SIL Signal (Scan Tool)	12v
12	BL/BK	Transponder Amplifier Signal (Code)	Inserting key: pulses
13	VT/GN	Transponder Amplifier Signal (RXCK)	Inserting key: pulses
14	RD/YL	Transponder Amplifier Signal (RXCK)	Inserting key: pulses
15	---	Not Used	---
16	BK/YL	EFI Main Relay B+	12-14v
18-19	---	Not Used	---
20	RD/BK	Unlock Warning Switch	Key In: 1.5v, Out: 4-5v
21	GN/RD	Center ECU LED Signal	LED Off: 12v, On: 1v
22	---	Not Used	---

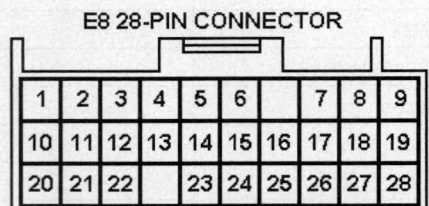

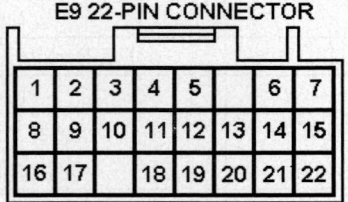

05533_ADIA_G122

Pin Connector Graphic

GX 470 PIN CHARTS

2003-2006 4.7L 2UZ-FE V8 VIN T E5 34 Pin Connector

PCM Pin #	Wire Color	Circuit Description (34 Pin)	Value at Hot Idle
1	YL	Injector 1 Control	2.0-3.3 ms
2	BK	Injector 2 Control	2.0-3.3 ms
3	BL	Injector 3 Control	2.0-3.3 ms
4	RD	Injector 4 Control	2.0-3.3 ms
5	GN	Injector 5 Control	2.0-3.3 ms
6	WT/BK	Power Ground (E02)	<0.1v
7	WT/BK	Power Ground (E01)	<0.1v
8	RD	Igniter Transistor 2 Control	7% duty cycle
9	BK	Igniter Transistor 1 Control	7% duty cycle
10	BL/BK	Igniter Transistor 8 Control	7% duty cycle
11	GN	Igniter Transistor 4 Control	7% duty cycle
12	YL	Igniter Transistor 5 Control	7% duty cycle
13	BK/BL	Igniter Transistor 7 Control	7% duty cycle
15	RD/GN	A/C Relay Control (ACCR)	Relay Off: 12v, On: 1v
16	BK/WT	Neutral Start Switch (NSW)	In P/N: 0-3.0v
17	BL/RD	Starter Switch Signal (STA)	In P/N: 0-3.0v
18	BL/RD	Sensor VREF (VC)	4.9-5.1v
19	GN/BK	ECT Sensor Signal (THW)	At 180°F: 0.51v
20	YL/BK	IAT Sensor Signal (THA)	At 100°F: 2.60v
21	RD/YL	TP Sensor Signal (VTA1)	0.4-1.0v

2003-2006 4.7L 2UZ-FE V8 VIN T E5 34 Pin Connector, *continued*

22, 32	---	Not Used	---
23	BK/RD	Igniter Signal (IGF2)	Digital Signal: 0-5-0v
24	BK/WT	Igniter Signal (IGF1)	Digital Signal: 0-5-0v
25	BL	Igniter Transistor 3 Control	7% duty cycle
26	BK/YL	Igniter Transistor 6 Control	7% duty cycle
27	BL/RD	EVAP Canister Closed Valve (CCV)	12v or 0v
28	BR/WT	Sensor Ground (E2)	<0.050v
29	GN/WT	MAF Sensor Ground (E2G)	<0.050v
30	BL/YL	Mass Airflow Sensor (VG)	1.1-1.5v
31	YL/BK	TP Sensor Signal (VTA2)	2.0-2.9v
33	GN/WT	Fuel Pump Relay Control (FPR)	Relay Off: 12v, On: 1v
34	BL/BK	EVAP Purge Solenoid (VSV)	12v or 0v

2003-2006 4.7L 2UZ-FE V8 VIN T E6 35 Pin Connector

PCM Pin #	Wire Color	Circuit Description (35 Pin)	Value at Hot Idle
1	BK	Knock Sensor 1 Signal (KNK1 - left)	<0.075v AC
2	WT	Knock Sensor 2 Signal (KNK2 - right)	<0.075v AC
3	RD/BL	Injector 6 Control	2.0-3.3 ms
4	RD	HO2S-11 (B1 S1) Heater (HT1A)	1v (Heater on)
5	BL	HO2S-12 (B1 S2) Heater (HT1B)	1v (Heater on)
7-8, 20	---	Not Used	---
9	BK/WT	Start Signal (STAR)	In P/N: 0-3.0v
10	WT	ECT Solenoid Control (S2)	12v or 0v
11	RD	ECT Solenoid Control (S1)	12v or 0v
12	GN/BK	ECT Solenoid Control (SLT-)	12v or 0v
13	GN/WT	ECT Solenoid Control (SLT+)	12v or 0v
16	PK/BK	A/T Solenoid Control (SL2-)	Pulse Signals
17	PK/BK	A/T Solenoid Control (SL2+)	Pulse Signals
18	RD/WT	A/T Solenoid Control (SL1-)	Pulse Signals
19	RD/BL	A/T Solenoid Control (SL1+)	Pulse Signals
21	WT	HO2S-22 (B2 S2) Signal (OX2B)	0.1-1.1v
22	WT	HO2S-21 (B2 S1) Signal (OX2A)	0.1-1.1v
23	BK	HO2S-11 (B1 S1) Signal (OX1A)	0.1-1.1v
24	BL	A/T Oil Temperature Sensor 2 (THO2)	At 68°F: 4-5v
25	RD/BK	HO2S-22 (B2 S2) Heater (HT2B)	1v (Heater on)
26	RD	ECT Vehicle Speed Sensor (SP2+)	Pulse Signals
27	BL	ECT Turbine Speed Sensor (NT-)	Pulse Signals
28	---	Not Used	---
29	BK	HO2S-12 (B1 S2) Signal (OX1B)	0.1-1.1v
30-31	---	Not Used	---
32	GN/YL	A/T Oil Temperature Sensor 1 (THO1)	At 68°F: 4-5v
33	YL	HO2S-21 (B2 S1) Heater (HT2A)	1v (Heater on)
34	GN	ECT Vehicle Speed Sensor (SP2-)	Pulse Signals
35	WT	ECT Turbine Speed Sensor (NT+)	Pulse Signals

2003-2006 4.7L 2UZ-FE V8 VIN T E7 32 Pin Connector

PCM Pin #	Wire Color	Circuit Description (32 Pin)	Value at Hot Idle
1	BR	Power Ground (E1)	<0.1v
2	WT	Throttle Control Motor (M-)	Pulse Signals
3	RD	Throttle Control Motor (M+)	Pulse Signals
4	WT/BK	Power Ground (ME01)	<0.1v
5	BK/WT	Injector 8 Control	2.0-3.3 ms
6	WT	Injector 7 Control	2.0-3.3 ms
7	WT/BK	Power Ground (E03)	<0.1v
11	YL/GN	Neutral Detection Switch (L4)	Switch Open: 0v, Closed: 12v
12	BK/WT	Start Switch Signal (STSW)	Cranking: 9-11v
13-14, 18-20	---	Not Used	---
15	BK	A/T Solenoid Control (SLU-)	Pulse Signals
16	PK/GN	A/T Solenoid Control (SLU+)	Pulse Signals
17	BR	Shield Ground (GE01)	<0.050v
21	BK/OR	Generator Control (RL)	12v
22, 26	---	Not Used	---
23	BL	A/C Lock Sensor (LCK)	12v or 0v
24	GN	CKP Sensor Signal (NE-)	<0.050v
25	BL	CKP Sensor Signal (NE+)	AC pulse signals
27	RD	CMP Sensor Signal (G2+)	AC pulse signals
28-31	---	Not Used	---
32	GN	CMP Sensor Signal (G2-)	AC pulse signals

2003-2006 4.7L V8 2UZ-FE VIN T E8 35 Pin Connector

PCM Pin #	Wire Color	Circuit Description (35 Pin)	Value at Hot Idle
1	WT/BK	Power Ground (HP)	<0.1v
2, 7-8	---	Not Used	---
3	GN	A/T Select Switch 2nd Signal (2L)	In 2nd: 12v, Others: 0v
4	BK/BL	Low Detection Switch (L4)	Switch Open: 0v, Closed: 12v
5	BL/WT	ECT Pattern Switch 2nd Position (SNW1)	2nd Position: 12v
6	YL/BK	ETCS Power (+BM)	12-14v
9	GN/BK	Shift Lock ECU Control (D)	12v or 0v
10	GN/YL	Shift Lock ECU Control (D)	12v or 0v
11	RD/BK	A/T Select Switch Reverse Signal	In 'R': 1v, Others: 12v
12, 15-16	---	Not Used	---
13	YL/RD	A/C Magnetic Clutch Relay (ACMG)	Relay Off: 0v, On: 12v
14	YL/GN	A/C Amplifier Signal (THWO)	A/C Off: 12v, On: 1v
17	VT	Vehicle Speed Sensor (SPD)	At 55 mph: 48 Hz
18	PK/BK	Body Control ECU Signal (MPX1)	Digital Signals
19	GN/WT	Stop Light Switch Signal	Brake Off: 0v, On: 12v
20-22, 25	---	Not Used	---
23	GN/RD	Shift Lock ECU Control (4)	12v or 0v
26	YL	Transponder Amplifier Signal (IMD)	Inserting key: pulses
27	WT	Transponder Amplifier Signal (IMI)	Inserting key: pulses
28	BL/WT	ECT Pattern Switch Power Signal	Power Position: 12v

2003-2006 4.7L V8 2UZ-FE VIN T E8 35 Pin Connector, *continued*

29	BK	Body Control ECU Signal (MPX2)	Digital Signals
30	---	Not Used	---
31	BL/BK	A/C Amplifier Signal (ACT)	Relay Off: 12v, On: 1v
32	PK/BK	A/C Amplifier Signal (THE)	A/C Off: 12v, On: 1v
33	GN/BK	A/C Switch Signal (ACLD)	A/C On: 12v, Off: 0v
34-35	---	Not Used	---

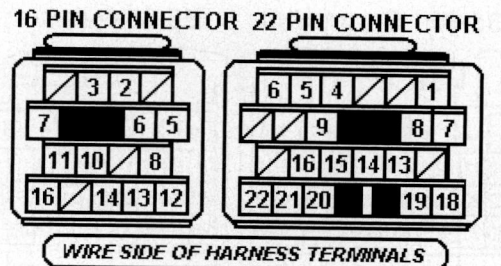

05533_ADIA_G002

Pin Connector Graphic

2003-2006 4.7L 2UZ-FE V8 VIN T E9 31 Pin Connector

PCM Pin #	Wire Color	Circuit Description (31 Pin)	Value at Hot Idle
1	BK/YL	EFI Main Relay Power (EFI Fuse)	12-14v
2	BK/YL	EFI Main Relay Power (EFI Fuse)	12-14v
3	BK/RD	Direct Battery	12-14v
4	BL	EVAP Pressure Switching Valve (VSV)	12v or 0v
5	BK	Tachometer Signal (TACO)	Pulse Signals
6	GN/WT	A/T Select Switch Park Signal (P)	In 'P': 12v, Others: 0v
7	GN/RD	A/T Select Switch Neutral Signal (N)	In 'N': 12v, Others: 0v
8	BK/WT	EFI Main Relay Control	Relay Off: 0v, On: 12v
9	BK/RD	Ignition Switch Power (IGSW)	12-14v
10	BK/WT	Circuit Opening Relay (FC)	0-3v, off-idle: 12v
11	WT	Malfunction Indicator Lamp Control	MIL Off: 12v, On: 1v
12	GN/WT	Body Control ECU	Digital Signals
13	YL/RD	Horn Relay Control	Relay Off: 12v, On: 1v
14	BK	Center Airbag Assembly (F/PS)	Digital Signals
15	WT/BK	Power Ground (EOM)	<0.1v
16	---	Not Used	---
17	WT	Transponder Amplifier Signal (NEO)	Inserting key: pulses
18	VT/WT	SIL Signal (Scan Tool)	Transmitting: pulses
19	WT/RD	Data Link Connector (WFSE)	N/A
20	PK/BK	Data Link Connector (TC)	12v
21	BL/BK	EVAP Vapor Pressure Sensor (PTNK)	2.5-3.1v (with cap off)
22	RD	Accelerator Position Sensor 1 (VPA)	0.3-0.90v
23	RD/BK	Accelerator Position Sensor 2 (VPA2)	1.8-2.7v
24	RD	Traction Control Engine Signal (ENG+)	Pulse Signals
25	YL	Traction Control Signal (TRC+)	Pulse Signals
26	BL/RD	Accelerator Pedal Position Sensor 1 VREF	4.9-5.1v

2003-2006 4.7L 2UZ-FE V8 VIN T E9 31 Pin Connector, *continued*

27	WT	Accelerator Pedal Position Sensor 2 VREF	4.9-5.1v
28	BR/WT	Accelerator Pedal Position Sensor Ground	<0.050v
29	WT/RD	Accelerator Position Sensor 2 (EPA2)	1.8-2.7v
30	GN	Traction Control Engine Signal (ENG-)	Pulse Signals
31	BL	Traction Control Signal (TRC-)	Pulse Signals

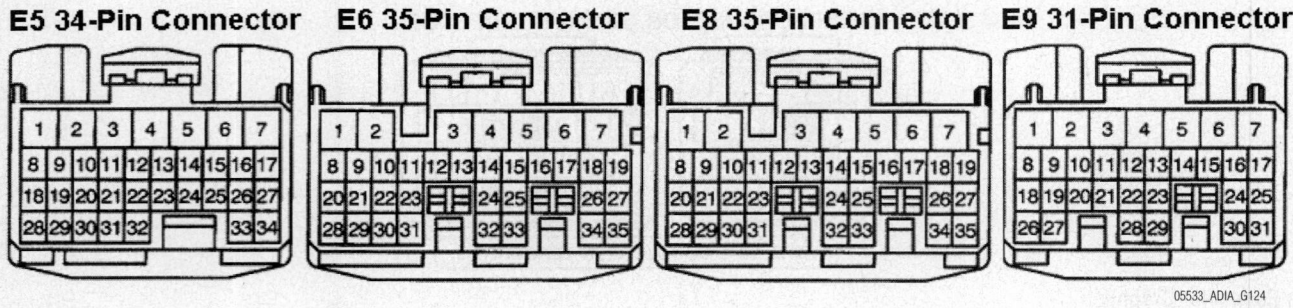

E5 34-Pin Connector   **E6 35-Pin Connector**   **E8 35-Pin Connector**   **E9 31-Pin Connector**

05533_ADIA_G124

Pin Connector Graphic

IS 300 PIN CHARTS

2006 2.5L V6 VIN K E4 34 Pin Connector

PCM Pin #	Wire Color	Circuit Description (34 Pin)	Value at Hot Idle
1	RD/BL	Injector 1 Control	2.0-3.3 ms
2	BK	Injector 2 Control	2.0-3.3 ms
3	LG	Injector 3 Control	2.0-3.3 ms
4	GN	Injector 4 Control	2.0-3.3 ms
5	YL	Injector 5 Control	2.0-3.3 ms
6, 7	WT/BK	Power Ground (E02), (E01)	<0.1v
8	YL/RD	Igniter Transistor 1 Control	7% duty cycle
9	PK/BL	Igniter Transistor 2 Control	7% duty cycle
10	LG	Igniter Transistor 3 Control	7$ duty cycle
11	LG/BK	Igniter Transistor 4 Control	7% duty cycle
12	GY	Igniter Transistor 5 Control	7% duty cycle
13	BL	Igniter Transistor 6 Control	7% duty cycle
14	WT/GN	A/C Relay Control (ACCR)	Relay Off: 12v, On: 1v
15	WT/BL	ACIS Control (VSV)	12v or 0v
16	PK	Neutral Start Switch (NSW)	In P/N: 0-3.0v
17	BK/YL	Starter Switch Signal (STA)	In P/N: 0-3.0v
18	BL/RD	Sensor VREF (VC)	4.9-5.1v
19	BK/BL	ECT Sensor Signal (THW)	At 180ºF: 0.51v
20	RD/BK	IAT Sensor Signal (THA)	At 100ºF: 2.60v
21	GN/BK	TP Sensor Signal (VTA1)	0.4-1.0v
22-23, 25-26	---	Not Used	---
24	WT/RD	Igniter Signal (IGF1)	Digital Signal: 0-5-0v
27	RD/GN	EVAP Canister Closed Valve (CCV)	12v or 0v

2006 2.5L V6 VIN K E4 34 Pin Connector, *continued*

28	BR	Sensor Ground (E2)	<0.050v
29	RD/WT	MAF Sensor Ground (E2G)	<0.050v
30	RD/YL	Mass Airflow Sensor (VG)	1.1-1.5v
31	GN/WT	TP Sensor Signal (VTA2)	2.0-2.9v
33	YL/BK	Fuel Pump Relay Control (FPR)	Relay Off: 12v, On: 1v
34	GN/YL	EVAP Purge Solenoid (VSV)	12v or 0v

2006 2.5L V6 VIN K E5 35 Pin Connector

PCM Pin #	Wire Color	Circuit Description (35 Pin)	Value at Hot Idle
1	BK	Knock Sensor 1 Signal (KNK1)	<0.075v AC
2	GN	Knock Sensor 2 Signal (KNK2)	<0.075v AC
3	BL	Injector 6 Control	2.0-3.3 ms
4	BK/WT	AFS-11 (B2 S1) Heater (HAFL)	1v (Heater on)
5	RD/BL	AFS-11 (B1 S1) Heater (HAFR)	1v (Heater on)
6, 7	WT/BK	Power Ground (E05), (E04)	<0.050v
8	GN	4WD Switch Signal (4WD)	In 4WD: 12v
9	PK	Start Signal (STAR)	In P/N: 0-3.0v
10	BL/WT	ECT Solenoid Control (S2)	12v or 0v
11	BL/RD	ECT Solenoid Control (S1)	12v or 0v
12	BL/BK	ECT Solenoid Control (SLT-)	12v or 0v
13	GN/YL	ECT Solenoid Control (SLT+)	12v or 0v
15	GN	ECT Solenoid Control (SL)	12v or 0v
16	PK/BL	A/T Solenoid Control (SL2-)	Pulse Signals
17	PK/BK	A/T Solenoid Control (SL2+)	Pulse Signals
18	RD/WT	A/T Solenoid Control (SL1-)	Pulse Signals
19	RD/BL	A/T Solenoid Control (SL1+)	Pulse Signals
20	RD	Knock Sensor 2 Ground	<0.050v
21	WT	HO2S-12 (B1 S2) Signal (OX1B)	0.1-1.1v
22	PK	AFS-11 (B1 S1) Signal (AFR+)	3.0-3.6v
23	YL	AFS-21 (B2 S1) Signal (AFL+)	3.0-3.6v
24	BL	A/T Oil Temperature Sensor 2 (THO2)	At 68°F: 4-5v
25	GN	HO2S-12 (B1 S2) Heater (HT1B)	1v (Heater on)
26	GN	ECT Vehicle Speed Sensor (SP2+)	Pulse Signals
27	WT/RD	Turbine Speed Sensor (NCO+)	Pulse Signals
28	WT	Knock Sensor 1 Ground	<0.050v
29	BK	HO2S-22 (B2 S2) Signal (OX2B)	0.1-1.1v
30	BL	AFS-11 (B1 S1) Signal (AFR-)	Fixed at 3.3v
31	BR	AFS-21 (B2 S1) Signal (AFL-)	Fixed at 3.3v
32	GN/YL	A/T Oil Temperature Sensor 1 (THO1)	At 68°F: 4-5v
33	BL	HO2S-22 (B2 S2) Heater (HT2B)	1v (Heater on)
34	RD	ECT Vehicle Speed Sensor (SP2-)	Pulse Signals
35	YL/RD	Turbine Speed Sensor (NCO-)	Pulse Signals

2006 2.5L V6 VIN K E6 32 Pin Connector

PCM Pin #	Wire Color	Circuit Description (32 Pin)	Value at Hot Idle
1	BR	Power Ground (E1)	<0.1v
2	BL	Throttle Control Motor (M-)	Pulse Signals
3	PK	Throttle Control Motor (M+)	Pulse Signals
4	WT/BK	Power Ground (ME01)	<0.1v
5-6	---	Not Used	---
7	WT/BK	Power Ground (E03)	<0.1v
8-9	---	Not Used	---
10	GN/WT	PSP Switch Signal (PSW)	Straight: 12v, Turning: 0v
11	YL/GN	Neutral Detection Switch (L4)	Switch Open: 0v, Closed: 12v
12	BL/YL	Start Switch Signal (STSW)	Cranking: 9-11v
13	BL/RD	Camshaft Timing Control Valve LH (OC2-)	Pulse Signals
14	BL/WT	Camshaft Timing Control Valve LH (OC2+)	Pulse Signals
15	BL/BK	Camshaft Timing Control Valve RH (OC1-)	Pulse Signals
16	GN/YL	Camshaft Timing Control Valve RH (OC1+)	Pulse Signals
17	BR	Shield Ground (GE01)	<0.050v
18-20	---	Not Used	---
21	BK/OR	Generator Control (RL)	12v
22	---	Not Used	---
23	RD/YL	A/C Lock Sensor (LCK)	12v or 0v
24	WT	CKP Sensor Signal (NE-)	<0.050v
25	BK	CKP Sensor Signal (NE+)	AC pulse signals
26	YL	Variable Valve Timing Sensor LH (VV2+)	AC pulse signals
27	RD	Variable Valve Timing Sensor RH (VV1+)	AC pulse signals
28-31	---	Not Used	---
32	WT	CMP Sensor Signal (G2-)	AC pulse signals

2006 2.5L V6 VIN K E7 35 Pin Connector

PCM Pin #	Wire Color	Circuit Description (35 Pin)	Value at Hot Idle
1	WT/BK	Power Ground (HP)	<0.1v
2	BK/YL	A/C Magnetic Clutch Relay (ACMG)	Relay Off: 0v, On: 12v
3	GN	Transmission Control Switch Signal (2L)	In 2nd: 12v, Others: 0v
4	WT/BL	Low Detection Switch (L4)	Switch Open: 0v, Closed: 12v
5	BL/WT	ECT Pattern Switch 2nd Position (SNW1)	2nd Position: 12v
6	GN	ETCS Power (+BM)	12-14v
7	---	Not Used	---
8	BL/YL	Shift Lock ECU Control (L)	12v or 0v
9	PK/BL	4WD Low Indicator Control	Indicator Off: 12v, On: 0v
10	GN/YL	Park Neutral Position Switch (D)	In 'D': 12v, Others: 0v
11	RD/YL	A/T Select Switch Reverse (R)	In 'R': 12v, Others: 0v
12	---	Not Used	---
14	BL/BK	A/C Amplifier Signal (THWO)	A/C Off: 12v, On: 1v
15-16	---	Not Used	---
17	VT/RD	Vehicle Speed Sensor (SPD)	At 55 mph: 48 Hz
18	PK/BK	Body Control ECU Signal (MPX1)	Digital Signals

2006 2.5L V6 VIN K E7 35 Pin Connector, *continued*

19	GN/YL	Stop Light Switch Signal	Brake Off: 0v, On: 12v
20	BL	Shift Lock ECU Control (3)	12v or 0v
21-22	---	Not Used	---
23	GN/RD	Shift Lock ECU Control (4)	12v or 0v
25	PK	A/T Oil Temperature Indicator	Indicator Off: 12v, On: 1v
26	BL/RD	Transponder Amplifier Signal (IMD)	Inserting key: pulses
27	WT/RD	Transponder Amplifier Signal (IMI)	Inserting key: pulses
28	BL/WT	ECT Pattern Switch Power Signal	Power Position: 12v
29	BK	Body Control ECU Signal (MPX2)	Digital Signals
30	---	Not Used	---
31	GR/BK	A/C Amplifier Signal (A/CS)	Relay Off: 12v, On: 1v
32	GY/GN	A/C Amplifier Signal (THE)	A/C Off: 12v, On: 1v
33	BK/RD	A/C Switch Signal (ACLD)	A/C On: 12v, Off: 0v
34-35	---	Not Used	---

2006 2.5L V6 VIN K E8 31 Pin Connector

PCM Pin #	Wire Color	Circuit Description (31 Pin)	Value at Hot Idle
1	BK	EFI Main Relay Power (+B)	12-14v
2	BK	EFI Main Relay Power (+B2)	12-14v
3	BL	Direct Battery	12-14v
4	PK/BL	EVAP Pressure Switching Valve (VSV)	12v or 0v
5	BK/WT	Tachometer Signal (TACO)	Pulse Signals
6	GN/WT	A/T Select Switch Park Signal (P)	In 'P': 12v, Others: 0v
7	GN/RD	A/T Select Switch Neutral Signal (N)	In 'N': 12v, Others: 0v
8	WT/GN	EFI Main Relay Control	Relay Off: 0v, On: 12v
9	BK/OR	Ignition Switch Power (IGSW)	12-14v
10	GR/BK	Circuit Opening Relay (FC)	0-3v, off-idle: 12v
11	RD/BK	Malfunction Indicator Lamp Control	MIL Off: 12v, On: 1v
12	YL/GN	Defogger Switch Signal (ELS)	Defogger Off: 0v, On: 12v
13	GN	Taillight Switch Signal (ELS2)	Taillights Off: 12v, On: 1v
14	BL	Center Airbag Assembly (F/PS)	Digital Signals
15	WT/BK	Power Ground (EOM)	<0.1v
16	---	Not Used	---
17	PK	Transponder Amplifier Signal (NEO)	Inserting key: pulses
18	RD/YL	SIL Signal (Scan Tool)	Transmitting: pulses
19	RD/WT	Data Link Connector (WFSE)	12v
20	PK/BL	Data Link Connector (TC)	12v
21	OR	EVAP Vapor Pressure Sensor (PTNK)	2.5-3.1v (with cap off)
22	WT/RD	Accelerator Position Sensor 1 (VPA)	0.3-0.90v
23	RD/BK	Accelerator Position Sensor 2 (VPA2)	1.8-2.7v
24	RD	Traction Control Engine Signal (ENG+)	Pulse Signals
25	BK	Traction Control Signal (TRC+)	Pulse Signals
26	BK/YL	Accelerator Pedal Position Sensor 1 VREF	4.9-5.1v
27	WT/BL	Accelerator Pedal Position Sensor 2 VREF	4.9-5.1v
28	LG/BK	Accelerator Pedal Position Sensor (EPA)	<0.050v

2006 2.5L V6 VIN K E8 31 Pin Connector, *continued*

29	VT/WT	Accelerator Position Sensor 2 (EPA2)	1.8-2.7v
30	WT	Traction Control Engine Signal (ENG-)	Pulse Signals
31	YL	Traction Control Signal (TRC-)	Pulse Signals

IS 250 PIN CHARTS

2006 2.5L V6 VIN K E4 34 Pin Connector

PCM Pin #	Wire Color	Circuit Description (34 Pin)	Value at Hot Idle
1	RD/BL	Injector 1 Control	2.0-3.3 ms
2	BK	Injector 2 Control	2.0-3.3 ms
3	LG	Injector 3 Control	2.0-3.3 ms
4	GN	Injector 4 Control	2.0-3.3 ms
5	YL	Injector 5 Control	2.0-3.3 ms
6, 7	WT/BK	Power Ground (E02), (E01)	<0.1v
8	YL/RD	Igniter Transistor 1 Control	7% duty cycle
9	PK/BL	Igniter Transistor 2 Control	7% duty cycle
10	LG	Igniter Transistor 3 Control	7$ duty cycle
11	LG/BK	Igniter Transistor 4 Control	7% duty cycle
12	GY	Igniter Transistor 5 Control	7% duty cycle
13	BL	Igniter Transistor 6 Control	7% duty cycle
14	WT/GN	A/C Relay Control (ACCR)	Relay Off: 12v, On: 1v
15	WT/BL	ACIS Control (VSV)	12v or 0v
16	PK	Neutral Start Switch (NSW)	In P/N: 0-3.0v
17	BK/YL	Starter Switch Signal (STA)	In P/N: 0-3.0v
18	BL/RD	Sensor VREF (VC)	4.9-5.1v
19	BK/BL	ECT Sensor Signal (THW)	At 180°F: 0.51v
20	RD/BK	IAT Sensor Signal (THA)	At 100°F: 2.60v
21	GN/BK	TP Sensor Signal (VTA1)	0.4-1.0v
22-23, 25-26	---	Not Used	---
24	WT/RD	Igniter Signal (IGF1)	Digital Signal: 0-5-0v
27	RD/GN	EVAP Canister Closed Valve (CCV)	12v or 0v
28	BR	Sensor Ground (E2)	<0.050v
29	RD/WT	MAF Sensor Ground (E2G)	<0.050v
30	RD/YL	Mass Airflow Sensor (VG)	1.1-1.5v
31	GN/WT	TP Sensor Signal (VTA2)	2.0-2.9v
33	YL/BK	Fuel Pump Relay Control (FPR)	Relay Off: 12v, On: 1v
34	GN/YL	EVAP Purge Solenoid (VSV)	12v or 0v

2006 2.5L V6 VIN K E5 35 Pin Connector

PCM Pin #	Wire Color	Circuit Description (35 Pin)	Value at Hot Idle
1	BK	Knock Sensor 1 Signal (KNK1)	<0.075v AC
2	GN	Knock Sensor 2 Signal (KNK2)	<0.075v AC
3	BL	Injector 6 Control	2.0-3.3 ms
4	BK/WT	AFS-11 (B2 S1) Heater (HAFL)	1v (Heater on)
5	RD/BL	AFS-11 (B1 S1) Heater (HAFR)	1v (Heater on)

2006 2.5L V6 VIN K E5 35 Pin Connector, *continued*

6, 7	WT/BK	Power Ground (E05), (E04)	<0.050v
8	GN	4WD Switch Signal (4WD)	In 4WD: 12v
9	PK	Start Signal (STAR)	In P/N: 0-3.0v
10	BL/WT	ECT Solenoid Control (S2)	12v or 0v
11	BL/RD	ECT Solenoid Control (S1)	12v or 0v
12	BL/BK	ECT Solenoid Control (SLT-)	12v or 0v
13	GN/YL	ECT Solenoid Control (SLT+)	12v or 0v
15	GN	ECT Solenoid Control (SL)	12v or 0v
16	PK/BL	A/T Solenoid Control (SL2-)	Pulse Signals
17	PK/BK	A/T Solenoid Control (SL2+)	Pulse Signals
18	RD/WT	A/T Solenoid Control (SL1-)	Pulse Signals
19	RD/BL	A/T Solenoid Control (SL1+)	Pulse Signals
20	RD	Knock Sensor 2 Ground	<0.050v
21	WT	HO2S-12 (B1 S2) Signal (OX1B)	0.1-1.1v
22	PK	AFS-11 (B1 S1) Signal (AFR+)	3.0-3.6v
23	YL	AFS-21 (B2 S1) Signal (AFL+)	3.0-3.6v
24	BL	A/T Oil Temperature Sensor 2 (THO2)	At 68°F: 4-5v
25	GN	HO2S-12 (B1 S2) Heater (HT1B)	1v (Heater on)
26	GN	ECT Vehicle Speed Sensor (SP2+)	Pulse Signals
27	WT/RD	Turbine Speed Sensor (NCO+)	Pulse Signals
28	WT	Knock Sensor 1 Ground	<0.050v
29	BK	HO2S-22 (B2 S2) Signal (OX2B)	0.1-1.1v
30	BL	AFS-11 (B1 S1) Signal (AFR-)	Fixed at 3.3v
31	BR	AFS-21 (B2 S1) Signal (AFL-)	Fixed at 3.3v
32	GN/YL	A/T Oil Temperature Sensor 1 (THO1)	At 68°F: 4-5v
33	BL	HO2S-22 (B2 S2) Heater (HT2B)	1v (Heater on)
34	RD	ECT Vehicle Speed Sensor (SP2-)	Pulse Signals
35	YL/RD	Turbine Speed Sensor (NCO-)	Pulse Signals

2006 2.5L V6 VIN K E6 32 Pin Connector

PCM Pin #	Wire Color	Circuit Description (32 Pin)	Value at Hot Idle
1	BR	Power Ground (E1)	<0.1v
2	BL	Throttle Control Motor (M-)	Pulse Signals
3	PK	Throttle Control Motor (M+)	Pulse Signals
4	WT/BK	Power Ground (ME01)	<0.1v
5-6	---	Not Used	---
7	WT/BK	Power Ground (E03)	<0.1v
8-9	---	Not Used	---
10	GN/WT	PSP Switch Signal (PSW)	Straight: 12v, Turning: 0v
11	YL/GN	Neutral Detection Switch (L4)	Switch Open: 0v, Closed: 12v
12	BL/YL	Start Switch Signal (STSW)	Cranking: 9-11v
13	BL/RD	Camshaft Timing Control Valve LH (OC2-)	Pulse Signals
14	BL/WT	Camshaft Timing Control Valve LH (OC2+)	Pulse Signals
15	BL/BK	Camshaft Timing Control Valve RH (OC1-)	Pulse Signals
16	GN/YL	Camshaft Timing Control Valve RH (OC1+)	Pulse Signals

2006 2.5L V6 VIN K E6 32 Pin Connector, *continued*

17	BR	Shield Ground (GE01)	<0.050v
18-20	---	Not Used	---
21	BK/OR	Generator Control (RL)	12v
22	---	Not Used	---
23	RD/YL	A/C Lock Sensor (LCK)	12v or 0v
24	WT	CKP Sensor Signal (NE-)	<0.050v
25	BK	CKP Sensor Signal (NE+)	AC pulse signals
26	YL	Variable Valve Timing Sensor LH (VV2+)	AC pulse signals
27	RD	Variable Valve Timing Sensor RH (VV1+)	AC pulse signals
28-31	---	Not Used	---
32	WT	CMP Sensor Signal (G2-)	AC pulse signals

2006 2.5L V6 VIN K E7 35 Pin Connector

PCM Pin #	Wire Color	Circuit Description (35 Pin)	Value at Hot Idle
1	WT/BK	Power Ground (HP)	<0.1v
2	BK/YL	A/C Magnetic Clutch Relay (ACMG)	Relay Off: 0v, On: 12v
3	GN	Transmission Control Switch Signal (2L)	In 2nd: 12v, Others: 0v
4	WT/BL	Low Detection Switch (L4)	Switch Open: 0v, Closed: 12v
5	BL/WT	ECT Pattern Switch 2nd Position (SNW1)	2nd Position: 12v
6	GN	ETCS Power (+BM)	12-14v
7	---	Not Used	---
8	BL/YL	Shift Lock ECU Control (L)	12v or 0v
9	PK/BL	4WD Low Indicator Control	Indicator Off: 12v, On: 0v
10	GN/YL	Park Neutral Position Switch (D)	In 'D': 12v, Others: 0v
11	RD/YL	A/T Select Switch Reverse (R)	In 'R': 12v, Others: 0v
12	---	Not Used	---
14	BL/BK	A/C Amplifier Signal (THWO)	A/C Off: 12v, On: 1v
15-16	---	Not Used	---
17	VT/RD	Vehicle Speed Sensor (SPD)	At 55 mph: 48 Hz
18	PK/BK	Body Control ECU Signal (MPX1)	Digital Signals
19	GN/YL	Stop Light Switch Signal	Brake Off: 0v, On: 12v
20	BL	Shift Lock ECU Control (3)	12v or 0v
21-22	---	Not Used	---
23	GN/RD	Shift Lock ECU Control (4)	12v or 0v
25	PK	A/T Oil Temperature Indicator	Indicator Off: 12v, On: 1v
26	BL/RD	Transponder Amplifier Signal (IMD)	Inserting key: pulses
27	WT/RD	Transponder Amplifier Signal (IMI)	Inserting key: pulses
28	BL/WT	ECT Pattern Switch Power Signal	Power Position: 12v
29	BK	Body Control ECU Signal (MPX2)	Digital Signals
30	---	Not Used	---
31	GR/BK	A/C Amplifier Signal (A/CS)	Relay Off: 12v, On: 1v
32	GY/GN	A/C Amplifier Signal (THE)	A/C Off: 12v, On: 1v
33	BK/RD	A/C Switch Signal (ACLD)	A/C On: 12v, Off: 0v
34-35	---	Not Used	---

2006 2.5L V6 VIN K E8 31 Pin Connector

PCM Pin #	Wire Color	Circuit Description (31 Pin)	Value at Hot Idle
1	BK	EFI Main Relay Power (+B)	12-14v
2	BK	EFI Main Relay Power (+B2)	12-14v
3	BL	Direct Battery	12-14v
4	PK/BL	EVAP Pressure Switching Valve (VSV)	12v or 0v
5	BK/WT	Tachometer Signal (TACO)	Pulse Signals
6	GN/WT	A/T Select Switch Park Signal (P)	In 'P': 12v, Others: 0v
7	GN/RD	A/T Select Switch Neutral Signal (N)	In 'N': 12v, Others: 0v
8	WT/GN	EFI Main Relay Control	Relay Off: 0v, On: 12v
9	BK/OR	Ignition Switch Power (IGSW)	12-14v
10	GR/BK	Circuit Opening Relay (FC)	0-3v, off-idle: 12v
11	RD/BK	Malfunction Indicator Lamp Control	MIL Off: 12v, On: 1v
12	YL/GN	Defogger Switch Signal (ELS)	Defogger Off: 0v, On: 12v
13	GN	Taillight Switch Signal (ELS2)	Taillights Off: 12v, On: 1v
14	BL	Center Airbag Assembly (F/PS)	Digital Signals
15	WT/BK	Power Ground (EOM)	<0.1v
16	---	Not Used	---
17	PK	Transponder Amplifier Signal (NEO)	Inserting key: pulses
18	RD/YL	SIL Signal (Scan Tool)	Transmitting: pulses
19	RD/WT	Data Link Connector (WFSE)	12v
20	PK/BL	Data Link Connector (TC)	12v
21	OR	EVAP Vapor Pressure Sensor (PTNK)	2.5-3.1v (with cap off)
22	WT/RD	Accelerator Position Sensor 1 (VPA)	0.3-0.90v
23	RD/BK	Accelerator Position Sensor 2 (VPA2)	1.8-2.7v
24	RD	Traction Control Engine Signal (ENG+)	Pulse Signals
25	BK	Traction Control Signal (TRC+)	Pulse Signals
26	BK/YL	Accelerator Pedal Position Sensor 1 VREF	4.9-5.1v
27	WT/BL	Accelerator Pedal Position Sensor 2 VREF	4.9-5.1v
28	LG/BK	Accelerator Pedal Position Sensor (EPA)	<0.050v
29	VT/WT	Accelerator Position Sensor 2 (EPA2)	1.8-2.7v
30	WT	Traction Control Engine Signal (ENG-)	Pulse Signals
31	YL	Traction Control Signal (TRC-)	Pulse Signals

IS 300 PIN CHARTS

2001-2005 3.0L I6 MFI VIN D (A/T-ECT) 31 Pin Connector

PCM Pin #	Wire Color	Circuit Description (31 Pin)	Value at Hot Idle
1	BL	Injector 3 Control	2.0-3.3 ms
2	RD/WT	Injector 4 Control	2.0-3.3 ms
3	RD	Injector 5 Control	2.0-3.3 ms
4	RD/BK	Injector 6 Control	2.0-3.3 ms
5	GN/YL	Intake Air Control Solenoid	12v or 0v
7	BK	Throttle Control Motor (M-)	Pulse Signals
8	WT	Throttle Control Motor (M+)	Pulse Signals
9	BR	Sensor Ground	<0.050v

2001-2005 3.0L I6 MFI VIN D (A/T-ECT) 31 Pin Connector, *continued*

10	BL	CMP Sensor Signal (G+)	AC pulse signals
11	RD/WT	Igniter Transistor 1 Control	Digital Signal: 0-5-0v
12	LG	Igniter Transistor 2 Control	Digital Signal: 0-5-0v
13	GN/RD	Igniter Transistor 3 Control	Digital Signal: 0-5-0v
16	GN	PSP Switch Signal	Straight: 12v, Turning: 0v
17	YL/BK	Cam Timing Oil signal (OCV-)	Pulse Signals
18	WT/RD	Cam Timing Oil signal (OCV+)	Pulse Signals
19	YL	Throttle Control Motor (CL-)	Pulse Signals
20	BL	Throttle Control Motor (CL+)	Pulse Signals
21	WT/BR	Power Ground	<0.1v
22	WT	CKP/CMP Sensor Ground (-)	<0.050v
23	BK	CKP Sensor Signal (NE+)	AC pulse signals
24	BK/WT	Neutral Start Switch	In P/N: 9-11v (cranking)
25	RD/YL	Igniter Signal (IGF)	Digital Signal: 0-5-0v
27	WT	Knock Sensor 2 Signal	<0.075v AC
28	WT	Knock Sensor 1 Signal	<0.075v AC
30	BR	Shield Ground	<0.050v
31	WT/BR	Power Ground	<0.1v

2001-2005 3.0L I6 MFI VIN D (A/T-ECT) 22 Pin Connector

PCM Pin #	Wire Color	Circuit Description (22 Pin)	Value at Hot Idle
1	BK/WT	Battery Direct	12-14v
4	GN	Fuel Pump (Control) ECU	Pulse Signals
5	PK/WT	Fuel Pump (DI) ECU	Pulse Signals
6	BL/BK	MIL (lamp) Control	MIL Off: 12v, On: 1v
7	BL/RD	EFI Main Relay Power BM+	12-14v
8	BK/RD	EFI Main Relay Power B2+	12-14v
9	BK/OR	Ignition Switch Power	12-14v
10	BK/YL	EFI Main Relay Power	12-14v
11	BK	SIL Signal (Scan Tool)	Digital Signal
16	BK/RD	EFI Main Relay Power	12-14v
22	WT/BK	ECM/Data Link Ground	<0.050v

2001-2005 3.0L I6 MFI VIN D (A/T-ECT) 28 Pin Connector

PCM Pin #	Wire Color	Circuit Description (28 Pin)	Value at Hot Idle
1	BL/RD	A/C Amplifier Signal (AC1)	Clutch On: 1.5v, Off: 12v
2	BK	Starter Switch Signal	9-11v (cranking)
4	GN/WT	Stop Light Switch Signal	Brake Off: 0v, On: 12v
5	RD	Data Link Connector	12-14v
8	WT	HO2S-12 (B1 S2) Signal	0.1-1.1v
13	WT/GN	A/C Magnetic Clutch (ACMG)	Clutch Off: 0v, On: 12v
16	RD/BK	A/T Select Switch Reverse	In 'R': 12v, Others: 0v
17	GN/RD	A/T Select Switch Drive	In 'D': 12v, Others: 0v
18	BR/YL	EVAP Vapor Pressure Sensor	2.5-3.1v (with hose off)

2001-2005 3.0L I6 MFI VIN D (A/T-ECT) 28 Pin Connector, *continued*

23	BL	Cruise Control ECU	At Cruise in OD: 12v
24	RD/YL	Cruise Control ECU	At Cruise in OD: 12v
25	WT	Vehicle Speed Sensor	At 55 mph: 48 Hz
26	RD/YL	Idle Up Diode Signal	Switch On: 12v, Off: 0v
28	PK	Overdrive Main Switch	Switch Off: 12v, On: 0v

2001-2005 3.0L I6 MFI VIN D (A/T-ECT) 17 Pin Connector

PCM Pin #	Wire Color	Circuit Description (17 Pin)	Value at Hot Idle
1	WT/RD	A/T-ECT Solenoid (S1)	In 3rd or OD: 1v
2	RD/BL	A/T-ECT Solenoid (S2)	1st or OD: 1v
4	RD	OD Clutch Sensor Signal (+)	Pulse Signals
5	RD/YL	Speed Sensor 2 Signal (+)	Pulse Signals
7	GN/WT	A/T-ECT Solenoid (SLU+)	Pulse Signals
8	YL/GN	A/T-ECT Solenoid (SLN+)	Pulse Signals
9	GN/RD	A/T-ECT Solenoid (SLT+)	Pulse Signals
10	RD	OD Clutch Sensor Signal (-)	Pulse Signals
11	BL/YL	Speed Sensor Signal 2 (-)	Pulse Signals
13	BL/RD	A/T-ECT Solenoid (SLU-)	Pulse Signals
14	PK	A/T-ECT Solenoid (SLN-)	Pulse Signals
15	RD/BK	A/T-ECT Solenoid (SLT-)	Pulse Signals
16	OR	C/C Indicator Light	Light Off: 12v, On: 1v
17	BL/BK	A/T Oil Temperature Sensor	0.5-4.5v

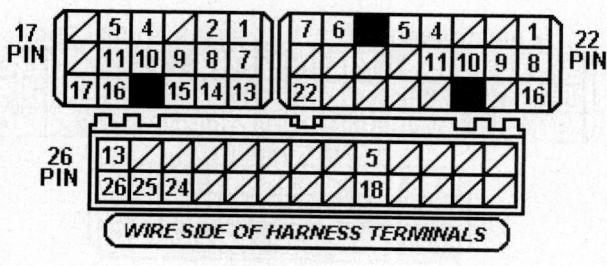

WIRE SIDE OF HARNESS TERMINALS

05533_ADIA_G097

Pin Connector Graphic

2001-2005 3.0L I6 MFI VIN D (A/T-ECT) 26 Pin Connector

PCM Pin #	Wire Color	Circuit Description (26 Pin)	Value at Hot Idle
5	GN/YL	A/T Pattern Selector Switch	Norm: 0v, PWR: 12v
13	RD/BK	EVAP Vapor Pressure (VSV)	12v or 0v
18	WT/BL	Manual Indicator Light	Light Off: 12-14v
24	WT	HO2S-22 (B2 S2) Signal	0.1-1.1v
25	GN/YL	HO2S-22 (B2 S2) Heater	1v (Heater On)
26	BL/WT	HO2S-12 (B1 S2) Heater	1v (Heater On)

2001-2005 3.0L I6 MFI VIN D (A/T-ECT) 24 Pin Connector

PCM Pin #	Wire Color	Circuit Description (24 Pin)	Value at Hot Idle
1	WT/BK	Power Ground	<0.1v
2	BL/RD	Sensor VREF	4.9-5.1v
3	BK/YL	HO2S-21 (B2 S1) Heater	1v (Heater On)
4	GN	HO2S-11 (B1 S1) Heater	1v (Heater On)
5	RD/BL	Injector 1 Control	2.0-3.3 ms
6	BL/RD	Injector 2 Control	2.0-3.3 ms
7	YL	EVAP Purge Solenoid (VSV)	12v or 0v
10	YL/RD	MAF Sensor Signal	1.1-1.5v
11	RD/BL	HO2S-21 (B2 S1) Signal	0.1-1.1v
12	WT	HO2S-11 (B1 S1) Signal	0.1-1.1v
13	BL/RD	Mirror Heater Switch Signal	Heater On: 12-14v
14	BL	ECT Sensor Signal	At 180°F: 0.51v
15	GN	Accel Position Sensor (VPA)	0.25-0.9v
16	WT	Accel Position Sensor (VPA2)	1.8-2.7v
17	BR	Sensor Ground	<0.050v
18	BR	Sensor Ground	<0.050v
19	BR	MAF Sensor Ground	<0.050v
20	BL/YL	A/T Select Switch 2nd	In 2nd: 12v, Others: 0v
21	GN/BK	A/T Select Switch Low	In Low: 12v, Others: 0v
22	GN/WT	IAT Sensor Signal	At 100°F: 2.60v
23	YL	TP Sensor (VTA1)	0.4-1.0v
24	RD/BK	TP Sensor (VTA2)	2.0-2.9v

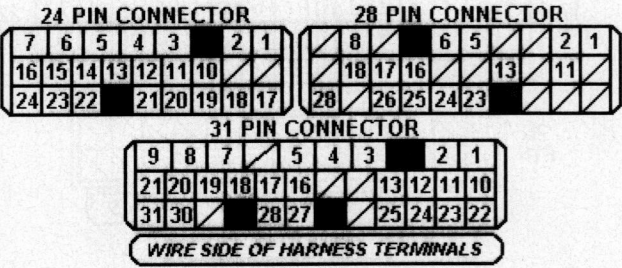

05533_ADIA_G098

Pin Connector Graphic

IS 350 PIN CHARTS

2006 3.5L V6 VIN E E4 34 Pin Connector

PCM Pin #	Wire Color	Circuit Description (34 Pin)	Value at Hot Idle
1	RD/BL	Injector 1 Control	2.0-3.3 ms
2	BK	Injector 2 Control	2.0-3.3 ms
3	LG	Injector 3 Control	2.0-3.3 ms
4	GN	Injector 4 Control	2.0-3.3 ms
5	YL	Injector 5 Control	2.0-3.3 ms
6, 7	WT/BK	Power Ground (E02), (E01)	<0.1v
8	YL/RD	Igniter Transistor 1 Control	7% duty cycle
9	PK/BL	Igniter Transistor 2 Control	7% duty cycle

10	LG	Igniter Transistor 3 Control	7$ duty cycle
11	LG/BK	Igniter Transistor 4 Control	7% duty cycle
12	GY	Igniter Transistor 5 Control	7% duty cycle
13	BL	Igniter Transistor 6 Control	7% duty cycle
14	WT/GN	A/C Relay Control (ACCR)	Relay Off: 12v, On: 1v
15	WT/BL	ACIS Control (VSV)	12v or 0v
16	PK	Neutral Start Switch (NSW)	In P/N: 0-3.0v
17	BK/YL	Starter Switch Signal (STA)	In P/N: 0-3.0v
18	BL/RD	Sensor VREF (VC)	4.9-5.1v
19	BK/BL	ECT Sensor Signal (THW)	At 180°F: 0.51v
20	RD/BK	IAT Sensor Signal (THA)	At 100°F: 2.60v
21	GN/BK	TP Sensor Signal (VTA1)	0.4-1.0v
22-23, 25-26	---	Not Used	---
24	WT/RD	Igniter Signal (IGF1)	Digital Signal: 0-5-0v
27	RD/GN	EVAP Canister Closed Valve (CCV)	12v or 0v
28	BR	Sensor Ground (E2)	<0.050v
29	RD/WT	MAF Sensor Ground (E2G)	<0.050v
30	RD/YL	Mass Airflow Sensor (VG)	1.1-1.5v
31	GN/WT	TP Sensor Signal (VTA2)	2.0-2.9v
33	YL/BK	Fuel Pump Relay Control (FPR)	Relay Off: 12v, On: 1v
34	GN/YL	EVAP Purge Solenoid (VSV)	12v or 0v

2006 3.5L V6 VIN E E5 35 Pin Connector

PCM Pin #	Wire Color	Circuit Description (35 Pin)	Value at Hot Idle
1	BK	Knock Sensor 1 Signal (KNK1)	<0.075v AC
2	GN	Knock Sensor 2 Signal (KNK2)	<0.075v AC
3	BL	Injector 6 Control	2.0-3.3 ms
4	BK/WT	AFS-11 (B2 S1) Heater (HAFL)	1v (Heater on)
5	RD/BL	AFS-11 (B1 S1) Heater (HAFR)	1v (Heater on)
6, 7	WT/BK	Power Ground (E05), (E04)	<0.050v
8	GN	4WD Switch Signal (4WD)	In 4WD: 12v
9	PK	Start Signal (STAR)	In P/N: 0-3.0v
10	BL/WT	ECT Solenoid Control (S2)	12v or 0v
11	BL/RD	ECT Solenoid Control (S1)	12v or 0v
12	BL/BK	ECT Solenoid Control (SLT-)	12v or 0v
13	GN/YL	ECT Solenoid Control (SLT+)	12v or 0v
15	GN	ECT Solenoid Control (SL)	12v or 0v
16	PK/BL	A/T Solenoid Control (SL2-)	Pulse Signals
17	PK/BK	A/T Solenoid Control (SL2+)	Pulse Signals
18	RD/WT	A/T Solenoid Control (SL1-)	Pulse Signals
19	RD/BL	A/T Solenoid Control (SL1+)	Pulse Signals
20	RD	Knock Sensor 2 Ground	<0.050v
21	WT	HO2S-12 (B1 S2) Signal (OX1B)	0.1-1.1v
22	PK	AFS-11 (B1 S1) Signal (AFR+)	3.0-3.6v
23	YL	AFS-21 (B2 S1) Signal (AFL+)	3.0-3.6v
24	BL	A/T Oil Temperature Sensor 2 (THO2)	At 68°F: 4-5v

2006 3.5L V6 VIN E E5 35 Pin Connector, *continued*

25	GN	HO2S-12 (B1 S2) Heater (HT1B)	1v (Heater on)
26	GN	ECT Vehicle Speed Sensor (SP2+)	Pulse Signals
27	WT/RD	Turbine Speed Sensor (NCO+)	Pulse Signals
28	WT	Knock Sensor 1 Ground	<0.050v
29	BK	HO2S-22 (B2 S2) Signal (OX2B)	0.1-1.1v
30	BL	AFS-11 (B1 S1) Signal (AFR-)	Fixed at 3.3v
31	BR	AFS-21 (B2 S1) Signal (AFL-)	Fixed at 3.3v
32	GN/YL	A/T Oil Temperature Sensor 1 (THO1)	At 68°F: 4-5v
33	BL	HO2S-22 (B2 S2) Heater (HT2B)	1v (Heater on)
34	RD	ECT Vehicle Speed Sensor (SP2-)	Pulse Signals
35	YL/RD	Turbine Speed Sensor (NCO-)	Pulse Signals

2006 3.5L V6 VIN E E6 32 Pin Connector

PCM Pin #	Wire Color	Circuit Description (32 Pin)	Value at Hot Idle
1	BR	Power Ground (E1)	<0.1v
2	BL	Throttle Control Motor (M-)	Pulse Signals
3	PK	Throttle Control Motor (M+)	Pulse Signals
4	WT/BK	Power Ground (ME01)	<0.1v
5-6	---	Not Used	---
7	WT/BK	Power Ground (E03)	<0.1v
8-9	---	Not Used	---
10	GN/WT	PSP Switch Signal (PSW)	Straight: 12v, Turning: 0v
11	YL/GN	Neutral Detection Switch (L4)	Switch Open: 0v, Closed: 12v
12	BL/YL	Start Switch Signal (STSW)	Cranking: 9-11v
13	BL/RD	Camshaft Timing Control Valve LH (OC2-)	Pulse Signals
14	BL/WT	Camshaft Timing Control Valve LH (OC2+)	Pulse Signals
15	BL/BK	Camshaft Timing Control Valve RH (OC1-)	Pulse Signals
16	GN/YL	Camshaft Timing Control Valve RH (OC1+)	Pulse Signals
17	BR	Shield Ground (GE01)	<0.050v
18-20	---	Not Used	---
21	BK/OR	Generator Control (RL)	12v
22	---	Not Used	---
23	RD/YL	A/C Lock Sensor (LCK)	12v or 0v
24	WT	CKP Sensor Signal (NE-)	<0.050v
25	BK	CKP Sensor Signal (NE+)	AC pulse signals
26	YL	Variable Valve Timing Sensor LH (VV2+)	AC pulse signals
27	RD	Variable Valve Timing Sensor RH (VV1+)	AC pulse signals
28-31	---	Not Used	---
32	WT	CMP Sensor Signal (G2-)	AC pulse signals

2006 3.5L V6 VIN E E7 35 Pin Connector

PCM Pin #	Wire Color	Circuit Description (35 Pin)	Value at Hot Idle
1	WT/BK	Power Ground (HP)	<0.1v
2	BK/YL	A/C Magnetic Clutch Relay (ACMG)	Relay Off: 0v, On: 12v
3	GN	Transmission Control Switch Signal (2L)	In 2nd: 12v, Others: 0v
4	WT/BL	Low Detection Switch (L4)	Switch Open: 0v, Closed: 12v
5	BL/WT	ECT Pattern Switch 2nd Position (SNW1)	2nd Position: 12v
6	GN	ETCS Power (+BM)	12-14v
7	---	Not Used	---
8	BL/YL	Shift Lock ECU Control (L)	12v or 0v
9	PK/BL	4WD Low Indicator Control	Indicator Off: 12v, On: 0v
10	GN/YL	Park Neutral Position Switch (D)	In 'D': 12v, Others: 0v
11	RD/YL	A/T Select Switch Reverse (R)	In 'R': 12v, Others: 0v
12	---	Not Used	---
14	BL/BK	A/C Amplifier Signal (THWO)	A/C Off: 12v, On: 1v
15-16	---	Not Used	---
17	VT/RD	Vehicle Speed Sensor (SPD)	At 55 mph: 48 Hz
18	PK/BK	Body Control ECU Signal (MPX1)	Digital Signals
19	GN/YL	Stop Light Switch Signal	Brake Off: 0v, On: 12v
20	BL	Shift Lock ECU Control (3)	12v or 0v
21-22	---	Not Used	---
23	GN/RD	Shift Lock ECU Control (4)	12v or 0v
25	PK	A/T Oil Temperature Indicator	Indicator Off: 12v, On: 1v
26	BL/RD	Transponder Amplifier Signal (IMD)	Inserting key: pulses
27	WT/RD	Transponder Amplifier Signal (IMI)	Inserting key: pulses
28	BL/WT	ECT Pattern Switch Power Signal	Power Position: 12v
29	BK	Body Control ECU Signal (MPX2)	Digital Signals
30	---	Not Used	---
31	GR/BK	A/C Amplifier Signal (A/CS)	Relay Off: 12v, On: 1v
32	GY/GN	A/C Amplifier Signal (THE)	A/C Off: 12v, On: 1v
33	BK/RD	A/C Switch Signal (ACLD)	A/C On: 12v, Off: 0v
34-35	---	Not Used	---

2006 3.5L V6 VIN E E8 31 Pin Connector

PCM Pin #	Wire Color	Circuit Description (31 Pin)	Value at Hot Idle
1	BK	EFI Main Relay Power (+B)	12-14v
2	BK	EFI Main Relay Power (+B2)	12-14v
3	BL	Direct Battery	12-14v
4	PK/BL	EVAP Pressure Switching Valve (VSV)	12v or 0v
5	BK/WT	Tachometer Signal (TACO)	Pulse Signals
6	GN/WT	A/T Select Switch Park Signal (P)	In 'P': 12v, Others: 0v
7	GN/RD	A/T Select Switch Neutral Signal (N)	In 'N': 12v, Others: 0v
8	WT/GN	EFI Main Relay Control	Relay Off: 0v, On: 12v
9	BK/OR	Ignition Switch Power (IGSW)	12-14v

2006 3.5L V6 VIN E E8 31 Pin Connector, *continued*

10	GR/BK	Circuit Opening Relay (FC)	0-3v, off-idle: 12v
11	RD/BK	Malfunction Indicator Lamp Control	MIL Off: 12v, On: 1v
12	YL/GN	Defogger Switch Signal (ELS)	Defogger Off: 0v, On: 12v
13	GN	Taillight Switch Signal (ELS2)	Taillights Off: 12v, On: 1v
14	BL	Center Airbag Assembly (F/PS)	Digital Signals
15	WT/BK	Power Ground (EOM)	<0.1v
16	---	Not Used	---
17	PK	Transponder Amplifier Signal (NEO)	Inserting key: pulses
18	RD/YL	SIL Signal (Scan Tool)	Transmitting: pulses
19	RD/WT	Data Link Connector (WFSE)	12v
20	PK/BL	Data Link Connector (TC)	12v
21	OR	EVAP Vapor Pressure Sensor (PTNK)	2.5-3.1v (with cap off)
22	WT/RD	Accelerator Position Sensor 1 (VPA)	0.3-0.90v
23	RD/BK	Accelerator Position Sensor 2 (VPA2)	1.8-2.7v
24	RD	Traction Control Engine Signal (ENG+)	Pulse Signals
25	BK	Traction Control Signal (TRC+)	Pulse Signals
26	BK/YL	Accelerator Pedal Position Sensor 1 VREF	4.9-5.1v
27	WT/BL	Accelerator Pedal Position Sensor 2 VREF	4.9-5.1v
28	LG/BK	Accelerator Pedal Position Sensor (EPA)	<0.050v
29	VT/WT	Accelerator Position Sensor 2 (EPA2)	1.8-2.7v
30	WT	Traction Control Engine Signal (ENG-)	Pulse Signals
31	YL	Traction Control Signal (TRC-)	Pulse Signals

LS 400 PIN CHARTS

1996-2000 4.0L 1UZ-FE V8 VIN T E5 31P Connector

PCM Pin #	Wire Color	Circuit Description (24 Pin)	Value at Hot Idle
1	WT/BK	Power Ground (E03)	<0.1v
2	BL/RD	Sensor VREF (VC)	4.9-5.1v
3	YL	HO2S-21 (B2 S1) Heater	1v (Heater on)
4	RD	HO2S-11 (B1 S1) Heater	1v (Heater on)
5	YL	Injector 1 Control	2.0-3.3 ms
6	BK	Injector 2 Control	2.0-3.3 ms
7	BL/BK	EVAP Purge Solenoid (VSV)	12v or 0v
8	RD	ABS ECU Signal	Digital Signals
9	RD/BK	Accelerator Position Sensor 2	1.8-2.7v
10	BL/YL	Mass Airflow Sensor (VG)	1.1-1.5v
11	WT	HO2S-21 (B2 S1) Signal	0.1-1.1v
12	BK	HO2S-11 (B1 S1) Signal	0.1-1.1v
13	RD/YL	TP Sensor Signal (VTA)	0.4-1.0v
14	GN/BK	ECT Sensor Signal (THW)	At 180°F: 0.51v
15-16	---	Not Used	---
17	BR	Power Ground (E1)	<0.1v
18	BR/WT	Sensor Ground (E2)	<0.050v
19	GN/WT	MAF Sensor Ground (EVG)	<0.050v

1996-2000 4.0L 1UZ-FE V8 VIN T E5 31P Connector, *continued*

20	YL/BK	TP Sensor Signal (VTA2)	2.0-2.9v
21	RD	Accelerator Position Sensor 1	0.3-0.90v
22	YL/BK	IAT Sensor Signal (THA)	At 100°F: 2.60v
23 (01'-02')	RD/BK	HO2S-22 (B2 S2) Heater	1v (Heater on)
24 (01'-02')	BL	HO2S-12 (B1 S2) Heater	1v (Heater on)

1996-2000 4.0L V8 1UZ-FE VIN T E7 17P Connector

PCM Pin #	Wire Color	Circuit Description (17 Pin)	Value at Hot Idle
1	RD	A/T ECT Solenoid (S1)	In 3rd or OD: 1v
2	WT	A/T ECT Solenoid (S2)	In 1st or OD: 1v
3	GN	A/T ECT Solenoid (SL)	In Lockup: 12-14v
4	BL/RD	Direct Clutch Speed Input (+)	AC pulse signals
5	RD	A/T Vehicle Speed Sensor (+)	AC pulse signals
6-8	---	Not Used	---
9	GN/WT	A/T-ECT Solenoid (SLT+)	Pulse Signals
10	BL/WT	Direct Clutch Speed Input (-)	AC pulse signals
11	GN	A/T Vehicle Speed Sensor (-)	Pulse Signals
12-14	---	Not Used	---
15	GNBK	A/T-ECT Solenoid (SLT-)	Pulse Signals
16	---	Not Used	---
17	GN/YL	A/T Oil Temperature Sensor	At 68°F: 4-5v

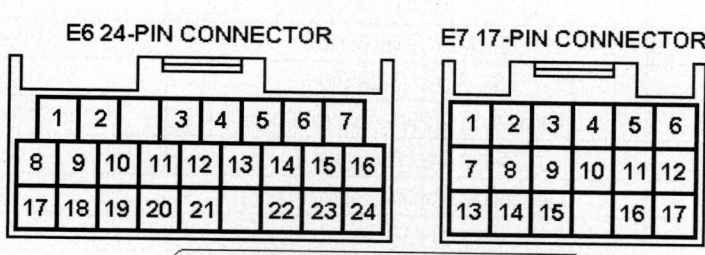

Pin Connector Graphic

1996-2000 4.0L 1UZ-FE V8 VIN T E8 28P Connector

PCM Pin #	Wire Color	Circuit Description (28 Pin)	Value at Hot Idle
5	PK/RD	Data Link Connector (TE1)	12-14v
6	GN/WT	Stop Light Switch Signal	Brake Off: 0v, On: 12v
7	RD/BK	HO2S-22 (B2 S2) Heater	1v (Heater on)
8	BL	HO2S-12 (B1 S2) Heater	1v (Heater on)
9	---	Not Used	---
10	BL	EVAP Vapor Pressure Valve (VSV)	12v or 0v
11	---	Not Used	---
12	GN/WT	Defogger & Tail Light Switch	Switch Off: 0v, On: 12v
13	BL/BK	A/C Amplifier Signal (ACT)	Clutch On: 12v, Off: 1.5v
14	YL/BK	A/C Amplifier Signal (THWO)	A/C Off: 12v, On: 1v
15	VT	Vehicle Speed Sensor	At 55 mph: 48 Hz

1996-2000 4.0L 1UZ-FE V8 VIN T E8 28P Connector, *continued*

16	BK	Tachometer Signal (TACO)	Pulse Signals
17	BK/RD	Starter Switch Signal	9-11v (cranking)
18	BK	HO2S-12 (B1 S2) Signal	0.1-1.1v
19	---	Not Used	---
20	BK/WT	Neutral Start Switch (NSW)	In P/N: 0-3.0v
21	WT/BK	Power Ground (EOM)	<0.1v
22	BL/BK	EVAP Vapor Pressure Sensor (PTNK)	2.5-3.1v (with cap off)
23-24	---	Not Used	---
25	WT/GN	A/C Amplifier Signal (AC1)	Clutch On: 1.5v, Off: 12v
26	---	Not Used	---
27	WT	HO2S-22 (B2 S2) Signal	0.1-1.1v
28	---	Not Used	---

1996-2000 4.0L 1UZ-FE V8 VIN T E9 22P Connector

PCM Pin #	Wire Color	Circuit Description (22 Pin)	Value at Hot Idle
1	RD/YL	Direct Battery	12-14v
2-3	---	Not Used	---
4	GN/RD	Fuel Pump Signal (DI)	12-14v
5	GN/WT	Fuel Control Switch Signal	Closed: 0v, Open: 12v
6	WT	Malfunction Indicator Lamp Control	MIL Off: 12v, On: 1v
7	YL/BK	BM (+) Power	12-14v
8	BK/YL	EFI Main Relay B1+	12-14v
9	BK/RD	Ignition Switch Power (IGSW)	12-14v
10	BK/WT	EFI Main Relay Control	Relay Off: 0v, On: 12v
11	PK/WT	SIL Signal (Scan Tool)	12v
12	BL/BK	Transponder Amplifier Signal (Code)	Inserting key: pulses
13	VT/GN	Transponder Amplifier Signal (RXCK)	Inserting key: pulses
14	RD/YL	Transponder Amplifier Signal (RXCK)	Inserting key: pulses
15	---	Not Used	---
16	BK/YL	EFI Main Relay B+	12-14v
18-19	---	Not Used	---
20	RD/BK	Unlock Warning Switch	Key In: 1.5v, Out: 4-5v
21	GN/RD	Center ECU LED Signal	LED Off: 12v, On: 1v
22	---	Not Used	---

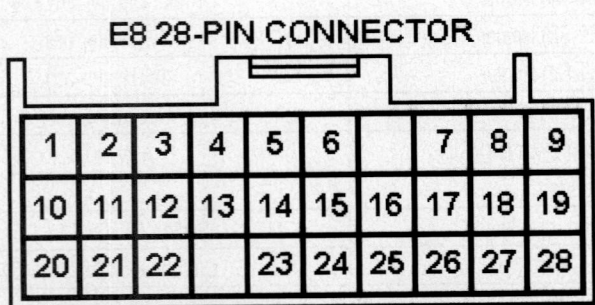

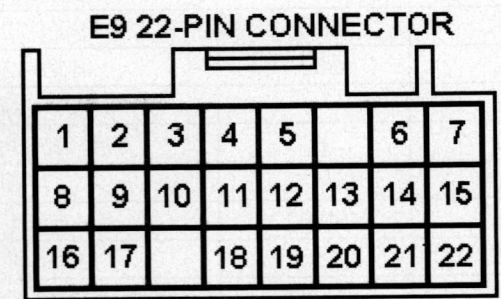

Pin Connector Graphic

05533_ADIA_G122

2001-05 4.3L 3UZ-FE V8 VIN N E5 31P Connector

PCM Pin #	Wire Color	Circuit Description (24 Pin)	Value at Hot Idle
1	WT/BK	Power Ground (E03)	<0.1v
2	BL/RD	Sensor VREF (VC)	4.9-5.1v
3	YL	HO2S-21 (B2 S1) Heater	1v (Heater on)
4	RD	HO2S-11 (B1 S1) Heater	1v (Heater on)
5	YL	Injector 1 Control	2.0-3.3 ms
6	BK	Injector 2 Control	2.0-3.3 ms
7	BL/BK	EVAP Purge Solenoid (VSV)	12v or 0v
8	RD	ABS ECU Signal	Digital Signals
9	RD/BK	Accelerator Position Sensor 2	1.8-2.7v
10	BL/YL	Mass Airflow Sensor (VG)	1.1-1.5v
11	WT	HO2S-21 (B2 S1) Signal	0.1-1.1v
12	BK	HO2S-11 (B1 S1) Signal	0.1-1.1v
13	RD/YL	TP Sensor Signal (VTA)	0.4-1.0v
14	GN/BK	ECT Sensor Signal (THW)	At 180°F: 0.51v
15-16	---	Not Used	---
17	BR	Power Ground (E1)	<0.1v
18	BR/WT	Sensor Ground (E2)	<0.050v
19	GN/WT	MAF Sensor Ground (EVG)	<0.050v
20	YL/BK	TP Sensor Signal (VTA2)	2.0-2.9v
21	RD	Accelerator Position Sensor 1	0.3-0.90v
22	YL/BK	IAT Sensor Signal (THA)	At 100°F: 2.60v
23 (01'-02')	RD/BK	HO2S-22 (B2 S2) Heater	1v (Heater on)
24 (01'-02')	BL	HO2S-12 (B1 S2) Heater	1v (Heater on)

2001-05 4.3L 3UZ-FE V8 VIN N E7 17P Connector

PCM Pin #	Wire Color	Circuit Description (17 Pin)	Value at Hot Idle
1	RD	A/T ECT Solenoid (S1)	In 3rd or OD: 1v
2	WT	A/T ECT Solenoid (S2)	In 1st or OD: 1v
3	GN	A/T ECT Solenoid (SL)	In Lockup: 12-14v
4	BL/RD	Direct Clutch Speed Input (+)	AC pulse signals
5	RD	A/T Vehicle Speed Sensor (+)	AC pulse signals
6-8	---	Not Used	---
9	GN/WT	A/T-ECT Solenoid (SLT+)	Pulse Signals
10	BL/WT	Direct Clutch Speed Input (-)	AC pulse signals
11	GN	A/T Vehicle Speed Sensor (-)	Pulse Signals
12-14	---	Not Used	---
15	GNBK	A/T-ECT Solenoid (SLT-)	Pulse Signals
16	---	Not Used	---
17	GN/YL	A/T Oil Temperature Sensor	At 68°F: 4-5v

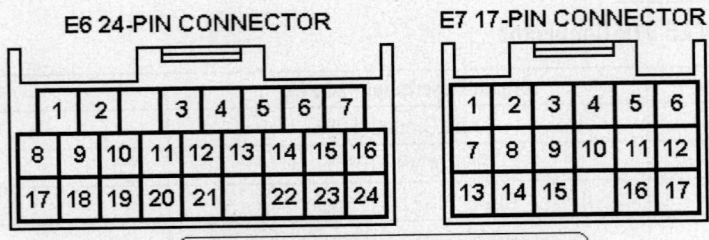

WIRE SIDE OF HARNESS TERMINALS

05533_ADIA_G121

Pin Connector Graphic

2001-05 4.3L 3UZ-FE V8 VIN N E8 28P Connector

PCM Pin #	Wire Color	Circuit Description (28 Pin)	Value at Hot Idle
5	PK/RD	Data Link Connector (TE1)	12-14v
6	GN/WT	Stop Light Switch Signal	Brake Off: 0v, On: 12v
7	RD/BK	HO2S-22 (B2 S2) Heater	1v (Heater on)
8	BL	HO2S-12 (B1 S2) Heater	1v (Heater on)
9	---	Not Used	---
10	BL	EVAP Vapor Pressure Valve (VSV)	12v or 0v
11	---	Not Used	---
12	GN/WT	Defogger & Tail Light Switch	Switch Off: 0v, On: 12v
13	BL/BK	A/C Amplifier Signal (ACT)	Clutch On: 12v, Off: 1.5v
14	YL/BK	A/C Amplifier Signal (THWO)	A/C Off: 12v, On: 1v
15	VT	Vehicle Speed Sensor	At 55 mph: 48 Hz
16	BK	Tachometer Signal (TACO)	Pulse Signals
17	BK/RD	Starter Switch Signal	9-11v (cranking)
18	BK	HO2S-12 (B1 S2) Signal	0.1-1.1v
19	---	Not Used	---
20	BK/WT	Neutral Start Switch (NSW)	In P/N: 0-3.0v
21	WT/BK	Power Ground (EOM)	<0.1v
22	BL/BK	EVAP Vapor Pressure Sensor (PTNK)	2.5-3.1v (with cap off)
23-24	---	Not Used	---
25	WT/GN	A/C Amplifier Signal (AC1)	Clutch On: 1.5v, Off: 12v
26	---	Not Used	---
27	WT	HO2S-22 (B2 S2) Signal	0.1-1.1v
28	---	Not Used	---

2001-05 4.3L 3UZ-FE V8 VIN N E9 22P Connector

PCM Pin #	Wire Color	Circuit Description (22 Pin)	Value at Hot Idle
1	RD/YL	Direct Battery	12-14v
2-3	---	Not Used	---
4	GN/RD	Fuel Pump Signal (DI)	12-14v
5	GN/WT	Fuel Control Switch Signal	Closed: 0v, Open: 12v
6	WT	Malfunction Indicator Lamp Control	MIL Off: 12v, On: 1v
7	YL/BK	BM (+) Power	12-14v
8	BK/YL	EFI Main Relay B1+	12-14v
9	BK/RD	Ignition Switch Power (IGSW)	12-14v

2001-05 4.3L 3UZ-FE V8 VIN N E9 22P Connector, *continued*

10	BK/WT	EFI Main Relay Control	Relay Off: 0v, On: 12v
11	PK/WT	SIL Signal (Scan Tool)	12v
12	BL/BK	Transponder Amplifier Signal (Code)	Inserting key: pulses
13	VT/GN	Transponder Amplifier Signal (RXCK)	Inserting key: pulses
14	RD/YL	Transponder Amplifier Signal (RXCK)	Inserting key: pulses
15	---	Not Used	---
16	BK/YL	EFI Main Relay B+	12-14v
18-19	---	Not Used	---
20	RD/BK	Unlock Warning Switch	Key In: 1.5v, Out: 4-5v
21	GN/RD	Center ECU LED Signal	LED Off: 12v, On: 1v
22	---	Not Used	---

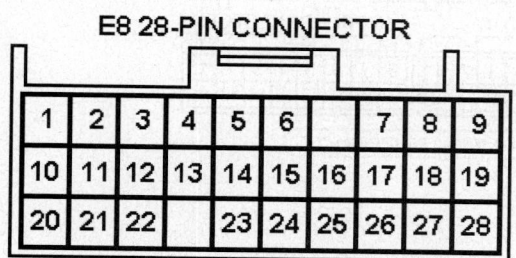

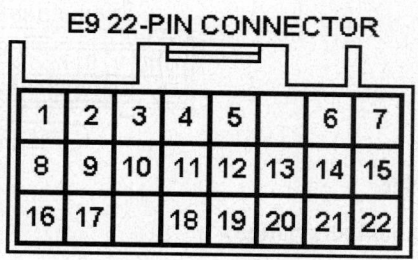

05533_ADIA_G122

Pin Connector Graphic

LX 450 PIN CHARTS

1996-97 4.5L I6 VIN J (A/T-ECT) 12 Pin Connector

PCM Pin #	Wire Color	Circuit Description (12 Pin)	Value at Hot Idle
1	GY/OR	HO2S-11 (B1 S1) Heater	1v (Heater on)
2	PK	A/T Vehicle Speed Sensor (+)	Pulse Signals
4	BL	CKP Sensor Signal (NE+)	AC pulse signals
5	BR	CKP Sensor Signal (NE-)	<0.050v
6	BL	CMP Sensor Signal (G-)	<0.050v
7	GY/OR	HO2S-12 (B1 S2) Heater	1v (Heater on)
8	PK/GN	A/T Vehicle Speed Sensor (-)	Pulse Signals
10	GN	CMP Sensor Signal (G2+)	AC pulse signals
11	RD	CMP Sensor Signal (G1+)	AC pulse signals
12	WT	CMP Sensor Signal (G+)	AC pulse signals

1996-97 4.5L I6 VIN J (A/T-ECT) 16 Pin Connector

PCM Pin #	Wire Color	Circuit Description (16 Pin)	Value at Hot Idle
1	RD/GN	TP Sensor VREF	4.9-5.1v
2	GN	MAF Sensor Signal	1.1-1.5v
3	BL/YL	IAT Sensor Signal (THA)	At 100°F: 2.60v
4	RD/WT	ECT Sensor Signal (THW)	At 180°F: 0.51v
5	PK/BL	HO2S-11 (B1 S1) Signal	0.1-1.1v
6	WT	Knock Sensor 1 Signal	<0.075v AC

1996-97 4.5L I6 VIN J (A/T-ECT) 16 Pin Connector, *continued*

7	GN/OR	Data Link Connector	12-14v
9	BR/BK	Sensor Ground	<0.050v
10	GN/BK	TP Sensor Signal (VTA)	0.3-0.8v
11	GN/WT	Closed Throttle Switch	1v, off-idle: 12v
12	RD/WT	A/T Oil Temperature Sensor	At 68°F: 4-5v
13	WT	HO2S-12 (B1 S2) Signal	0.1-1.1v
14	BK	Knock Sensor 2 Signal	<0.075v AC
16	BR/BK	Sensor Ground	<0.050v

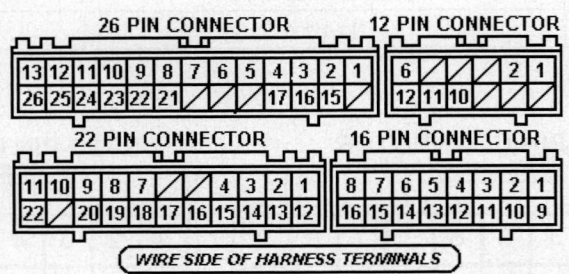

05533_ADIA_117

Pin Connector Graphic

Standard Colors and Abbreviations

Abbreviation	Color	Abbreviation	Color	Abbreviation	Color
BK	Black	GY	Gray	RD	Red
BL	Blue	GN	Green	TN	Tan
BR	Brown	LG	LT Green	VT	Violet
DB	Dark Blue	OR	Orange	WT	White
DG	DK Green	PK	Pink	YL	Yellow

1996-97 4.5L I6 VIN J (A/T-ECT) 22 Pin Connector

PCM Pin #	Wire Color	Circuit Description (22 Pin)	Value at Hot Idle
1	BK/BL	Ignition Switch Power (IGSW)	12-14v
2	RD/YL	Direct Battery	12-14v
3	RD	EFI Main Relay	12-14v
4	YL/RD	Malfunction Indicator Lamp Control	MIL Off: 12v, On: 1v
5	YL/BL	A/T Oil Temperature Lamp	Lamp Off: 12v
6	WT	Data Link Connector (SDL)	0v
7	BK/WT	A/C Magnetic Clutch Relay (ACMG)	Relay Off: 0v, On: 12v
8	BL/WT	Vehicle Speed Sensor	At 55 mph: 48 Hz
9	BK/BL	4WD Detection Transfer (L4)	Switch Closed: 12v
10	BK	Tachometer Signal (TACO)	Pulse Signals
11	BK/RD	Starter Switch Signal	In P/N: 0-3.0v
12	YL	EFI Main Relay B+	12-14v
13	WT/RD	Fuel Pump Relay	Relay Off: 12v, On: 1v
14	GN/WT	Stop Light Switch Signal	Brake Off: 0v, On: 12v
15	RD/BK	A/T Select Switch Reverse	In 'R': 12v, Others: 0v
16	OR	A/T Select Switch 2nd	In 2nd: 12v, Others: 0v

1996-97 4.5L I6 VIN J (A/T-ECT) 22 Pin Connector, *continued*

17	GN/WT	A/T Select Switch Low	In Low: 12v, Others: 0v
18	GN/OR	Cruise Control ECU	At Cruise in OD: 12v
19	PK/BL	Overdrive Main Switch	Switch Off: 12v, On: 1v
20	PK/BK	A/T Pattern Select Switch	Norm: 0v, PWR: 12v
21	YL/BL	4WD Detection Transfer (N)	Open: 12v, Closed: 0v
22	BK/WT	Neutral Start Switch	In P/N: 0-3.0v

1996-97 4.5L I6 VIN J (A/T-ECT) 26 Pin Connector

PCM Pin #	Wire Color	Circuit Description (26 Pin)	Value at Hot Idle
1	YL/RD	Injector 5 Control	2.0-3.3 ms
2	YL	Injector 4 Control	2.0-3.3 ms
3	OR	A/T Pattern Select Switch	Norm: 0v, PWR: 12v
4	BL/YL	ISC Signal (ISC4)	Pulse Signals
5	GN/YL	ISC Signal (ISC3)	Pulse Signals
6	RD/BK	ISC Signal (ISC2)	Pulse Signals
7	RD/GN	ISC Signal (ISC1)	Pulse Signals
8	RD/BL	A/T-ECT Solenoid (SL)	In Lockup: 12-14v
9	RD/YL	A/T-ECT Solenoid (S2)	1st or OD: 1v
10	RD	A/T-ECT Solenoid (S1)	3rd or OD: 1v
11	WT/RD	Injector 2 Control	2.0-3.3 ms
12	WT/BL	Injector 1 Control	2.0-3.3 ms
13	BR	Power Ground	<0.1v
14	RD/WT	Circuit Opening Relay (FC)	0-3v, off-idle: 12v
15	YL/BL	Injector 6 Control	2.0-3.3 ms
16	BR	Power Ground	<0.1v
17	BK/YL	Igniter Signal (IGF)	Digital Signal: 0-5-0v
18	RD/WT	A/T Select Switch 2nd	2nd: 12-14v
19	GN/YL	EGR Gas Temperature Sensor	3.5-4.0v
21	BL/RD	Fuel Pressure Up Solenoid	1v (at hot restart)
22	BL/WT	EGR Solenoid Control (VSV)	12v or 0v
23	BK/GN	Igniter Signal (IGF)	Digital Signal: 0-5-0v
24	BR/BK	Sensor Ground	<0.050v
25	WT/GN	Injector 3 Control	2.0-3.3 ms
26	BR	Power Ground	<0.1v

Standard Colors and Abbreviations

Abbreviation	Color	Abbreviation	Color	Abbreviation	Color
BK	Black	GY	Gray	RD	Red
BL	Blue	GN	Green	TN	Tan
BR	Brown	LG	LT Green	VT	Violet
DB	Dark Blue	OR	Orange	WT	White
DG	DK Green	PK	Pink	YL	Yellow

LX 470 PIN CHARTS

1998 WT/EIS 4.7L V8 VIN T (A/T-ECT) 17 Pin Connector

PCM Pin #	Wire Color	Circuit Description (17 Pin)	Value at Hot Idle
4	BL/RD	Direct Clutch Speed Input (+)	Pulse Signals
5	RD	A/T Vehicle Speed Sensor (+)	Pulse Signals
10	BL/WT	Direct Clutch Speed Input (-)	Pulse Signals
11	GN	A/T Vehicle Speed Sensor (-)	Pulse Signals
17	GN/YL	A/T Oil Temperature Sensor	At 68°F: 4-5v

1998 WT/EIS 4.7L V8 VIN T (A/T-ECT) 28 Pin Connector

PCM Pin #	Wire Color	Circuit Description (28 Pin)	Value at Hot Idle
1	BK	Overdrive Main Switch	Switch Off: 12v, On: 1v
2	RD/BK	A/T Select Switch Reverse	In 'R': 12v, Others: 0v
3	GN	A/T Select Switch 2nd	In 2nd: 12v, Others: 0v
4	GN/BK	A/T Select Switch Low	In Low: 12v, Others: 0v
5	PK/RD	Data Link Connector	12-14v
6	GN/WT	Stop Light Switch Signal	Brake Off: 0v, On: 12v
7	RD/BK	HO2S-22 (B2 S2) Heater	1v (Heater on)
8	BL	HO2S-12 (B1 S2) Heater	1v (Heater on)
9	YL/BK	C/C Indicator Lamp	Lamp On: 0.1v
10	BL	EVAP Vapor Pressure Valve (VSV)	12v or 0v
11	BL/WT	A/T Pattern Select Switch	Norm: 0v, PWR: 12v
12	GN/WT	Idle-Up Signal	Load On: 1v, Off: 12v
13	BL/BK	A/C Amplifier Signal (ACT)	Clutch On: 12v, Off: 1.5v
14	YL/BK	A/C Amplifier Signal (THWO)	Pulse Signals
15	PK	Vehicle Speed Sensor	At 55 mph: 48 Hz
16	BK	Tachometer Signal (TACO)	Pulse Signals
17	BK/RD	Starter Switch Signal	In P/N: 0-3.0v
18	BK	HO2S-12 (B1 S2) Signal	0.1-1.1v
19	GN/YL	A/T Select Switch Drive	In Drive: 12-14v
20	BK/WT	Start Circuit Signal	Cranking: 9-11v
21	WT/BK	Sensor Ground	<0.050v
22	BL/BK	EVAP Vapor Pressure Sensor (PTNK)	2.5-3.1v (with cap off)
25	WT/GN	A/C Amplifier Signal (AC1)	Clutch On: 1.5v, Off: 12v
26	OR	A/T Oil Temperature Lamp	Lamp Off: 12v, On: 1v
27	WT	HO2S-22 (B2 S2) Signal	0.1-1.1v

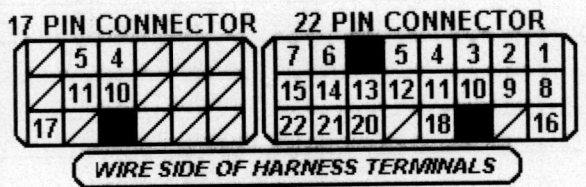

Pin Connector Graphic

1998 WT/EIS 4.7L V8 VIN T (A/T-ECT) 24 Pin Connector

PCM Pin #	Wire Color	Circuit Description (24 Pin)	Value at Hot Idle
1	WT/BK	Power Ground	<0.1v
2	BL/RD	Sensor VREF	4.9-5.1v
3	YL	HO2S-21 (B2 S1) Heater	1v (Heater on)
4	RD	HO2S-11 (B1 S1) Heater	1v (Heater on)
5	YL	Injector 1 Control	2.0-3.3 ms
6	BK	Injector 2 Control	2.0-3.3 ms
7	BL/BK	EVAP Purge Solenoid (VSV)	12v or 0v
8	RD	ABS ECU Signal	Pulse Signals
9	RD/BK	Accelerator Position Sensor 2	1.8-2.7v
10	BL/YL	MAF Sensor Signal (VG)	1.1-1.5v
11	WT	HO2S-21 (B2 S1) Signal	0.1-1.1v
12	BK	HO2S-11 (B1 S1) Signal	0.1-1.1v
13	RD/YL	TP Sensor Signal (VTA1)	0.4-1.0v
14	GN/BK	ECT Sensor Signal (THW)	At 180°F: 0.51v
17	BR	Sensor Ground	<0.050v
18	BR/WT	Sensor Ground	<0.050v
19	GN/WT	MAF Sensor Ground (EVG)	<0.050v
20	YL/BK	TP Sensor Signal (VTA2)	2.0-2.9v
21	RD	Accelerator Position Sensor 1	0.25-0.90v
22	YL/BK	IAT Sensor Signal (THA)	At 100°F: 2.60v

1998 WT/EIS 4.7L V8 VIN T (A/T-ECT) 22 Pin Connector

PCM Pin #	Wire Color	Circuit Description (22 Pin)	Value at Hot Idle
1	RD/YL	Direct Battery	12-14v
2	YL	A/T Pattern Select Switch	2nd: 12-14v
3	BL/RD	A/T Select Switch 2nd	In 2nd: 12v, Others: 0v
4	GN/RD	Fuel Pump Signal (DI)	Pulse Signals
5	GN/WT	Fuel Control Switch Signal	Open: 12v, Closed: 0v
6	WT	Malfunction Indicator Lamp Control	MIL Off: 12v, On: 1v
7	YL/BK	BM (+) Signal	12-14v
8	BK/YL	EFI Main Relay B1+	12-14v
9	BK/RD	Ignition Switch Power (IGSW)	12-14v
10	BK/WT	EFI Main Relay	12-14v
11	PK/WT	SIL Signal (Scan Tool)	12v
12	BL/BK	Transponder Amplifier Signal (Code)	Inserting key: pulses
13	PK/GN	Transponder Amplifier Signal (RXCK)	Inserting key: pulses
14	RD/YL	Transponder Amplifier Signal (RXCK)	Inserting key: pulses
15	YL/GN	A/T Select Switch Park	In P/N: 0-3.0v
16	BK/YL	EFI Main Relay B+	12-14v
18	YL/RD	A/C Amplifier On Signal	Clutch On: 1.5v, Off: 12v
20	RD/BK	Unlock Warning Switch	No Key: 4-5v
21	GN/RD	LED Signal	LED Off: 12v, On: 1v
22	BK/BL	4WD Detection Transfer (L4)	Open: 12v, Closed: 0v

1998 WT/EIS 4.7L V8 VIN T (A/T-ECT) 31 Pin Connector

PCM Pin #	Wire Color	Circuit Description (31 Pin)	Value at Hot Idle
1	BL	Injector 3 Control	2.0-3.3 ms
2	RD	Injector 4 Control	2.0-3.3 ms
3	GN	Injector 5 Control	2.0-3.3 ms
4	RD/BL	Injector 6 Control	2.0-3.3 ms
5	WT	Injector 7 Control	2.0-3.3 ms
6	BK/WT	Injector 8 Control	2.0-3.3 ms
7	WT	Throttle Control Motor (M-)	Pulse Signals
8	RD	Throttle Control Motor (M+)	Pulse Signals
9	WT/BK	Power Ground	<0.1v
10	RD	CMP Sensor Signal (G+)	AC pulse signals
11	BK	Igniter Transistor 1 Control	7% duty cycle
12	RD	Igniter Transistor 2 Control	7% duty cycle
13	BL	Igniter Transistor 3 Control	7% duty cycle
14	GN	Igniter Transistor 4 Control	7% duty cycle
15	YL	Igniter Transistor 5 Control	7% duty cycle
16	BK/YL	Igniter Transistor 6 Control	7% duty cycle
17	WT	Knock Sensor 2 Signal	<0.075v AC
18	BK	Knock Sensor 1 Signal	<0.075v AC
21	WT/BK	Power Ground	<0.1v
22	GN	CMP/CKP Sensor Signal (-)	<0.050v
23	BL	CKP Sensor Signal (NE+)	AC pulse signals
24	BL	Throttle Control Motor (CL-)	Pulse Signals
25	BK/BL	Igniter Transistor 7 Control	7% duty cycle
26	BL/BK	Igniter Transistor 8 Control	7% duty cycle
27	BK/WT	Igniter Signal (IGF1)	Digital Signal: 0-5-0v
28	BK/RD	Igniter Signal (IGF2)	Digital Signal: 0-5-0v
29	GN	Throttle Control Motor (CL+)	Pulse Signals
30	BR	Shield Ground	<0.050v
31	WT/BK	Power Ground	<0.1v

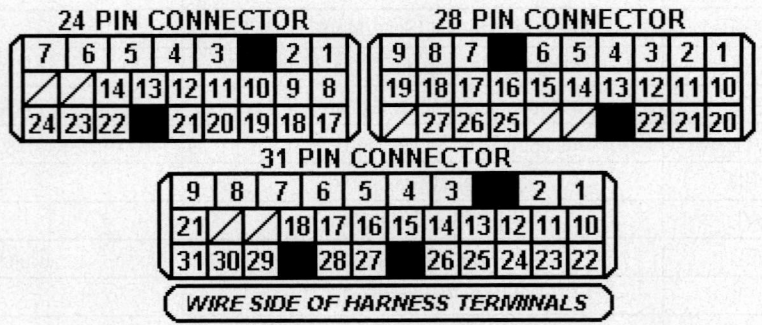

05533_ADIA_G119

Pin Connector Graphic

1999-2002 4.7L 2UZ-FE V8 VIN T E5 31P Connector

PCM Pin #	Wire Color	Circuit Description (31 Pin)	Value at Hot Idle
1	BL	Injector 3 Control	2.0-3.3 ms
2	RD	Injector 4 Control	2.0-3.3 ms
3	GN	Injector 5 Control	2.0-3.3 ms
4	RD/BL	Injector 6 Control	2.0-3.3 ms
5	WT	Injector 7 Control	2.0-3.3 ms
6	BK/WT	Injector 8 Control	2.0-3.3 ms
7	WT	Throttle Control Motor (M-)	Pulse Signals
8	RD	Throttle Control Motor (M+)	Pulse Signals
9	WT/BK	Power Ground (ME01)	<0.1v
10	RD	CMP Sensor Signal (G+)	AC pulse signals
11	BK	Igniter Transistor 1 Control	7% duty cycle
12	RD	Igniter Transistor 2 Control	7% duty cycle
13	BL	Igniter Transistor 3 Control	7% duty cycle
14	GN	Igniter Transistor 4 Control	7% duty cycle
15	YL	Igniter Transistor 5 Control	7% duty cycle
16	BK/YL	Igniter Transistor 6 Control	7% duty cycle
17	WT	Knock Sensor 2 Signal (right)	<0.075v AC
18	BK	Knock Sensor 1 Signal (left)	<0.075v AC
19-20	---	Not Used	---
21	WT/BK	Power Ground (E01)	<0.1v
22	GN	CMP/CKP Sensor Signal (-)	<0.050v
23	BL	CKP Sensor Signal (NE+)	AC pulse signals
24	BL	Throttle Control Motor (CL-)	Pulse Signals
25	BK/BL	Igniter Transistor 7 Control	7% duty cycle
26	BL/BK	Igniter Transistor 8 Control	7% duty cycle
27	BK/WT	Igniter Signal (IGF1)	Digital Signal: 0-5-0v
28	BK/RD	Igniter Signal (IGF2)	Digital Signal: 0-5-0v
29	GN	Throttle Control Motor (CL+)	Pulse Signals
30	BR	Shield Ground (GE01)	<0.050v
31	WT/BK	Power Ground (E02)	<0.1v

E5 31-PIN CONNECTOR

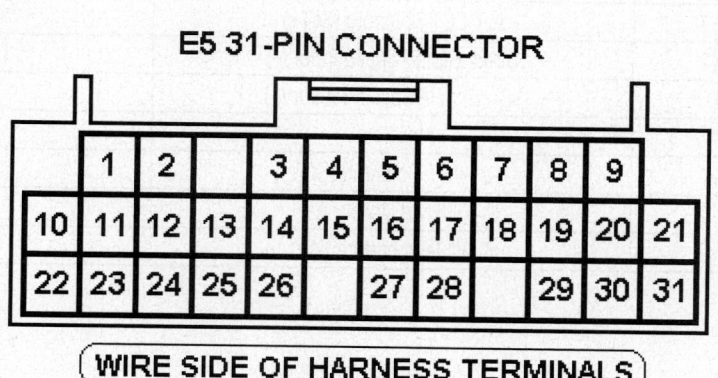

WIRE SIDE OF HARNESS TERMINALS

Pin Connector Graphic

05533_ADIA_G120

1999-2002 4.7L 2UZ-FE V8 VIN T E6 24P Connector

PCM Pin #	Wire Color	Circuit Description (24 Pin)	Value at Hot Idle
1	WT/BK	Power Ground (E03)	<0.1v
2	BL/RD	Sensor VREF (VC)	4.9-5.1v
3	YL	HO2S-21 (B2 S1) Heater	1v (Heater on)
4	RD	HO2S-11 (B1 S1) Heater	1v (Heater on)
5	YL	Injector 1 Control	2.0-3.3 ms
6	BK	Injector 2 Control	2.0-3.3 ms
7	BL/BK	EVAP Purge Solenoid (VSV)	12v or 0v
8	RD	ABS ECU Signal	Digital Signals
9	RD/BK	Accelerator Position Sensor 2	1.8-2.7v
10	BL/YL	Mass Airflow Sensor (VG)	1.1-1.5v
11	WT	HO2S-21 (B2 S1) Signal	0.1-1.1v
12	BK	HO2S-11 (B1 S1) Signal	0.1-1.1v
13	RD/YL	TP Sensor Signal (VTA)	0.4-1.0v
14	GN/BK	ECT Sensor Signal (THW)	At 180°F: 0.51v
15-16	---	Not Used	---
17	BR	Power Ground (E1)	<0.1v
18	BR/WT	Sensor Ground (E2)	<0.050v
19	GN/WT	MAF Sensor Ground (EVG)	<0.050v
20	YL/BK	TP Sensor Signal (VTA2)	2.0-2.9v
21	RD	Accelerator Position Sensor 1	0.3-0.90v
22	YL/BK	IAT Sensor Signal (THA)	At 100°F: 2.60v
23 (01'-02')	RD/BK	HO2S-22 (B2 S2) Heater	1v (Heater on)
24 (01'-02')	BL	HO2S-12 (B1 S2) Heater	1v (Heater on)

1999-2002 4.7L V8 2UZ-FE VIN T E7 17P Connector

PCM Pin #	Wire Color	Circuit Description (17 Pin)	Value at Hot Idle
1	RD	A/T ECT Solenoid (S1)	In 3rd or OD: 1v
2	WT	A/T ECT Solenoid (S2)	In 1st or OD: 1v
3	GN	A/T ECT Solenoid (SL)	In Lockup: 12-14v
4	BL/RD	Direct Clutch Speed Input (+)	AC pulse signals
5	RD	A/T Vehicle Speed Sensor (+)	AC pulse signals
6-8	---	Not Used	---
9	GN/WT	A/T-ECT Solenoid (SLT+)	Pulse Signals
10	BL/WT	Direct Clutch Speed Input (-)	AC pulse signals
11	GN	A/T Vehicle Speed Sensor (-)	Pulse Signals
12-14	---	Not Used	---
15	GNBK	A/T-ECT Solenoid (SLT-)	Pulse Signals
16	---	Not Used	---
17	GN/YL	A/T Oil Temperature Sensor	At 68°F: 4-5v

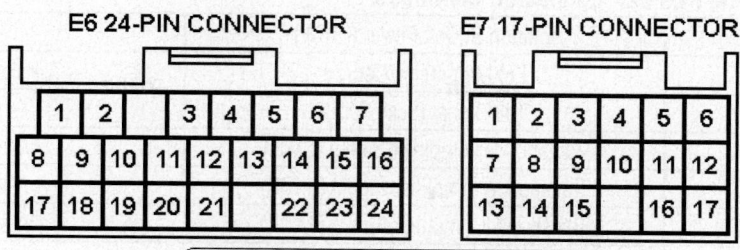

E6 24-PIN CONNECTOR   E7 17-PIN CONNECTOR

WIRE SIDE OF HARNESS TERMINALS

05533_ADIA_G121

Pin Connector Graphic

1999-2002 4.7L 2UZ-FE V8 VIN T E8 28P Connector

PCM Pin #	Wire Color	Circuit Description (28 Pin)	Value at Hot Idle
5	PK/RD	Data Link Connector (TE1)	12-14v
6	GN/WT	Stop Light Switch Signal	Brake Off: 0v, On: 12v
7	RD/BK	HO2S-22 (B2 S2) Heater	1v (Heater on)
8	BL	HO2S-12 (B1 S2) Heater	1v (Heater on)
9	---	Not Used	---
10	BL	EVAP Vapor Pressure Valve (VSV)	12v or 0v
11	---	Not Used	---
12	GN/WT	Defogger & Tail Light Switch	Switch Off: 0v, On: 12v
13	BL/BK	A/C Amplifier Signal (ACT)	Clutch On: 12v, Off: 1.5v
14	YL/BK	A/C Amplifier Signal (THWO)	A/C Off: 12v, On: 1v
15	VT	Vehicle Speed Sensor	At 55 mph: 48 Hz
16	BK	Tachometer Signal (TACO)	Pulse Signals
17	BK/RD	Starter Switch Signal	9-11v (cranking)
18	BK	HO2S-12 (B1 S2) Signal	0.1-1.1v
19	---	Not Used	---
20	BK/WT	Neutral Start Switch (NSW)	In P/N: 0-3.0v
21	WT/BK	Power Ground (EOM)	<0.1v
22	BL/BK	EVAP Vapor Pressure Sensor (PTNK)	2.5-3.1v (with cap off)
23-24	---	Not Used	---
25	WT/GN	A/C Amplifier Signal (AC1)	Clutch On: 1.5v, Off: 12v
26	---	Not Used	---
27	WT	HO2S-22 (B2 S2) Signal	0.1-1.1v
28	---	Not Used	---

1999-2002 4.7L 2UZ-FE V8 VIN T E9 22P Connector

PCM Pin #	Wire Color	Circuit Description (22 Pin)	Value at Hot Idle
1	RD/YL	Direct Battery	12-14v
2-3	---	Not Used	---
4	GN/RD	Fuel Pump Signal (DI)	12-14v
5	GN/WT	Fuel Control Switch Signal	Closed: 0v, Open: 12v
6	WT	Malfunction Indicator Lamp Control	MIL Off: 12v, On: 1v
7	YL/BK	BM (+) Power	12-14v
8	BK/YL	EFI Main Relay B1+	12-14v

1999-2002 4.7L 2UZ-FE V8 VIN T E9 22P Connector, *continued*

9	BK/RD	Ignition Switch Power (IGSW)	12-14v
10	BK/WT	EFI Main Relay Control	Relay Off: 0v, On: 12v
11	PK/WT	SIL Signal (Scan Tool)	12v
12	BL/BK	Transponder Amplifier Signal (Code)	Inserting key: pulses
13	VT/GN	Transponder Amplifier Signal (RXCK)	Inserting key: pulses
14	RD/YL	Transponder Amplifier Signal (RXCK)	Inserting key: pulses
15	---	Not Used	---
16	BK/YL	EFI Main Relay B+	12-14v
18-19	---	Not Used	---
20	RD/BK	Unlock Warning Switch	Key In: 1.5v, Out: 4-5v
21	GN/RD	Center ECU LED Signal	LED Off: 12v, On: 1v
22		Not Used	---

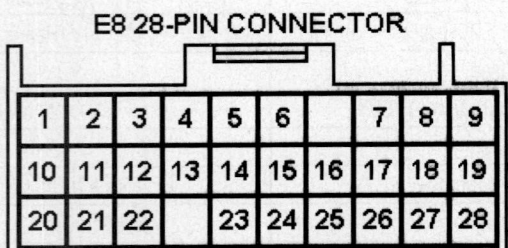

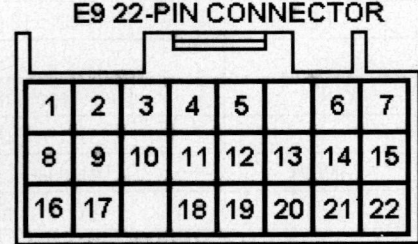

05533_ADIA_G122

Pin Connector Graphic

2003-2006 4.7L 2UZ-FE V8 VIN T E5 34 Pin Connector

PCM Pin #	Wire Color	Circuit Description (34 Pin)	Value at Hot Idle
1	YL	Injector 1 Control	2.0-3.3 ms
2	BK	Injector 2 Control	2.0-3.3 ms
3	BL	Injector 3 Control	2.0-3.3 ms
4	RD	Injector 4 Control	2.0-3.3 ms
5	GN	Injector 5 Control	2.0-3.3 ms
6	WT/BK	Power Ground (E02)	<0.1v
7	WT/BK	Power Ground (E01)	<0.1v
8	RD	Igniter Transistor 2 Control	7% duty cycle
9	BK	Igniter Transistor 1 Control	7% duty cycle
10	BL/BK	Igniter Transistor 8 Control	7% duty cycle
11	GN	Igniter Transistor 4 Control	7% duty cycle
12	YL	Igniter Transistor 5 Control	7% duty cycle
13	BK/BL	Igniter Transistor 7 Control	7% duty cycle
15	RD/GN	A/C Relay Control (ACCR)	Relay Off: 12v, On: 1v
16	BK/WT	Neutral Start Switch (NSW)	In P/N: 0-3.0v
17	BL/RD	Starter Switch Signal (STA)	In P/N: 0-3.0v
18	BL/RD	Sensor VREF (VC)	4.9-5.1v
19	GN/BK	ECT Sensor Signal (THW)	At 180°F: 0.51v
20	YL/BK	IAT Sensor Signal (THA)	At 100°F: 2.60v
21	RD/YL	TP Sensor Signal (VTA1)	0.4-1.0v
22, 32	---	Not Used	---

2003-2006 4.7L 2UZ-FE V8 VIN T E5 34 Pin Connector, *continued*

23	BK/RD	Igniter Signal (IGF2)	Digital Signal: 0-5-0v
24	BK/WT	Igniter Signal (IGF1)	Digital Signal: 0-5-0v
25	BL	Igniter Transistor 3 Control	7% duty cycle
26	BK/YL	Igniter Transistor 6 Control	7% duty cycle
27	BL/RD	EVAP Canister Closed Valve (CCV)	12v or 0v
28	BR/WT	Sensor Ground (E2)	<0.050v
29	GN/WT	MAF Sensor Ground (E2G)	<0.050v
30	BL/YL	Mass Airflow Sensor (VG)	1.1-1.5v
31	YL/BK	TP Sensor Signal (VTA2)	2.0-2.9v
33	GN/WT	Fuel Pump Relay Control (FPR)	Relay Off: 12v, On: 1v
34	BL/BK	EVAP Purge Solenoid (VSV)	12v or 0v

2003-2006 4.7L 2UZ-FE V8 VIN T E6 35 Pin Connector

PCM Pin #	Wire Color	Circuit Description (35 Pin)	Value at Hot Idle
1	BK	Knock Sensor 1 Signal (KNK1 - left)	<0.075v AC
2	WT	Knock Sensor 2 Signal (KNK2 - right)	<0.075v AC
3	RD/BL	Injector 6 Control	2.0-3.3 ms
4	RD	HO2S-11 (B1 S1) Heater (HT1A)	1v (Heater on)
5	BL	HO2S-12 (B1 S2) Heater (HT1B)	1v (Heater on)
7-8, 20	---	Not Used	---
9	BK/WT	Start Signal (STAR)	In P/N: 0-3.0v
10	WT	ECT Solenoid Control (S2)	12v or 0v
11	RD	ECT Solenoid Control (S1)	12v or 0v
12	GN/BK	ECT Solenoid Control (SLT-)	12v or 0v
13	GN/WT	ECT Solenoid Control (SLT+)	12v or 0v
16	PK/BK	A/T Solenoid Control (SL2-)	Pulse Signals
17	PK/BK	A/T Solenoid Control (SL2+)	Pulse Signals
18	RD/WT	A/T Solenoid Control (SL1-)	Pulse Signals
19	RD/BL	A/T Solenoid Control (SL1+)	Pulse Signals
21	WT	HO2S-22 (B2 S2) Signal (OX2B)	0.1-1.1v
22	WT	HO2S-21 (B2 S1) Signal (OX2A)	0.1-1.1v
23	BK	HO2S-11 (B1 S1) Signal (OX1A)	0.1-1.1v
24	BL	A/T Oil Temperature Sensor 2 (THO2)	At 68°F: 4-5v
25	RD/BK	HO2S-22 (B2 S2) Heater (HT2B)	1v (Heater on)
26	RD	ECT Vehicle Speed Sensor (SP2+)	Pulse Signals
27	BL	ECT Turbine Speed Sensor (NT-)	Pulse Signals
28	---	Not Used	---
29	BK	HO2S-12 (B1 S2) Signal (OX1B)	0.1-1.1v
30-31	---	Not Used	---
32	GN/YL	A/T Oil Temperature Sensor 1 (THO1)	At 68°F: 4-5v
33	YL	HO2S-21 (B2 S1) Heater (HT2A)	1v (Heater on)
34	GN	ECT Vehicle Speed Sensor (SP2-)	Pulse Signals
35	WT	ECT Turbine Speed Sensor (NT+)	Pulse Signals

2003-2006 4.7L 2UZ-FE V8 VIN T E7 32 Pin Connector

PCM Pin #	Wire Color	Circuit Description (32 Pin)	Value at Hot Idle
1	BR	Power Ground (E1)	<0.1v
2	WT	Throttle Control Motor (M-)	Pulse Signals
3	RD	Throttle Control Motor (M+)	Pulse Signals
4	WT/BK	Power Ground (ME01)	<0.1v
5	BK/WT	Injector 8 Control	2.0-3.3 ms
6	WT	Injector 7 Control	2.0-3.3 ms
7	WT/BK	Power Ground (E03)	<0.1v
11	YL/GN	Neutral Detection Switch (L4)	Switch Open: 0v, Closed: 12v
12	BK/WT	Start Switch Signal (STSW)	Cranking: 9-11v
13-14, 18-20	---	Not Used	---
15	BK	A/T Solenoid Control (SLU-)	Pulse Signals
16	PK/GN	A/T Solenoid Control (SLU+)	Pulse Signals
17	BR	Shield Ground (GE01)	<0.050v
21	BK/OR	Generator Control (RL)	12v
22, 26	---	Not Used	---
23	BL	A/C Lock Sensor (LCK)	12v or 0v
24	GN	CKP Sensor Signal (NE-)	<0.050v
25	BL	CKP Sensor Signal (NE+)	AC pulse signals
27	RD	CMP Sensor Signal (G2+)	AC pulse signals
28-31	---	Not Used	---
32	GN	CMP Sensor Signal (G2-)	AC pulse signals

2003-2006 4.7L V8 2UZ-FE VIN T E8 35 Pin Connector

PCM Pin #	Wire Color	Circuit Description (35 Pin)	Value at Hot Idle
1	WT/BK	Power Ground (HP)	<0.1v
2, 7-8	---	Not Used	---
3	GN	A/T Select Switch 2nd Signal (2L)	In 2nd: 12v, Others: 0v
4	BK/BL	Low Detection Switch (L4)	Switch Open: 0v, Closed: 12v
5	BL/WT	ECT Pattern Switch 2nd Position (SNW1)	2nd Position: 12v
6	YL/BK	ETCS Power (+BM)	12-14v
9	GN/BK	Shift Lock ECU Control (D)	12v or 0v
10	GN/YL	Shift Lock ECU Control (D)	12v or 0v
11	RD/BK	A/T Select Switch Reverse Signal	In 'R': 1v, Others: 12v
12, 15-16	---	Not Used	---
13	YL/RD	A/C Magnetic Clutch Relay (ACMG)	Relay Off: 0v, On: 12v
14	YL/GN	A/C Amplifier Signal (THWO)	A/C Off: 12v, On: 1v
17	VT	Vehicle Speed Sensor (SPD)	At 55 mph: 48 Hz
18	PK/BK	Body Control ECU Signal (MPX1)	Digital Signals
19	GN/WT	Stop Light Switch Signal	Brake Off: 0v, On: 12v
20-22, 25	---	Not Used	---
23	GN/RD	Shift Lock ECU Control (4)	12v or 0v
26	YL	Transponder Amplifier Signal (IMD)	Inserting key: pulses
27	WT	Transponder Amplifier Signal (IMI)	Inserting key: pulses
28	BL/WT	ECT Pattern Switch Power Signal	Power Position: 12v

29	BK	Body Control ECU Signal (MPX2)	Digital Signals
30	---	Not Used	---
31	BL/BK	A/C Amplifier Signal (ACT)	Relay Off: 12v, On: 1v
32	PK/BK	A/C Amplifier Signal (THE)	A/C Off: 12v, On: 1v
33	GN/BK	A/C Switch Signal (ACLD)	A/C On: 12v, Off: 0v
34-35	---	Not Used	---

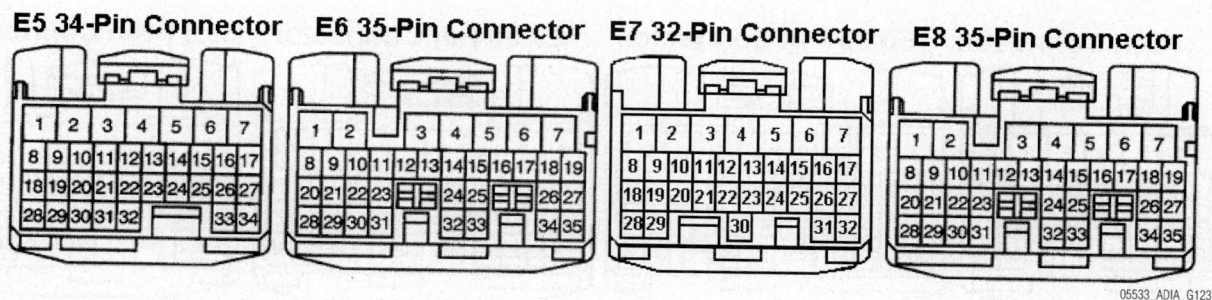

05533_ADIA_G123

Pin Connector Graphic

2003-2006 4.7L 2UZ-FE V8 VIN T E9 31 Pin Connector

PCM Pin #	Wire Color	Circuit Description (31 Pin)	Value at Hot Idle
1	BK/YL	EFI Main Relay Power (EFI Fuse)	12-14v
2	BK/YL	EFI Main Relay Power (EFI Fuse)	12-14v
3	BK/RD	Direct Battery	12-14v
4	BL	EVAP Pressure Switching Valve (VSV)	12v or 0v
5	BK	Tachometer Signal (TACO)	Pulse Signals
6	GN/WT	A/T Select Switch Park Signal (P)	In 'P': 12v, Others: 0v
7	GN/RD	A/T Select Switch Neutral Signal (N)	In 'N': 12v, Others: 0v
8	BK/WT	EFI Main Relay Control	Relay Off: 0v, On: 12v
9	BK/RD	Ignition Switch Power (IGSW)	12-14v
10	BK/WT	Circuit Opening Relay (FC)	0-3v, off-idle: 12v
11	WT	Malfunction Indicator Lamp Control	MIL Off: 12v, On: 1v
12	GN/WT	Body Control ECU	Digital Signals
13	YL/RD	Horn Relay Control	Relay Off: 12v, On: 1v
14	BK	Center Airbag Assembly (F/PS)	Digital Signals
15	WT/BK	Power Ground (EOM)	<0.1v
16	---	Not Used	---
17	WT	Transponder Amplifier Signal (NEO)	Inserting key: pulses
18	VT/WT	SIL Signal (Scan Tool)	Transmitting: pulses
19	WT/RD	Data Link Connector (WFSE)	N/A
20	PK/BK	Data Link Connector (TC)	12v
21	BL/BK	EVAP Vapor Pressure Sensor (PTNK)	2.5-3.1v (with cap off)
22	RD	Accelerator Position Sensor 1 (VPA)	0.3-0.90v
23	RD/BK	Accelerator Position Sensor 2 (VPA2)	1.8-2.7v
24	RD	Traction Control Engine Signal (ENG+)	Pulse Signals
25	YL	Traction Control Signal (TRC+)	Pulse Signals
26	BL/RD	Accelerator Pedal Position Sensor 1 VREF	4.9-5.1v

2003-2006 4.7L 2UZ-FE V8 VIN T E9 31 Pin Connector, *continued*

27	WT	Accelerator Pedal Position Sensor 2 VREF	4.9-5.1v
28	BR/WT	Accelerator Pedal Position Sensor Ground	<0.050v
29	WT/RD	Accelerator Position Sensor 2 (EPA2)	1.8-2.7v
30	GN	Traction Control Engine Signal (ENG-)	Pulse Signals
31	BL	Traction Control Signal (TRC-)	Pulse Signals

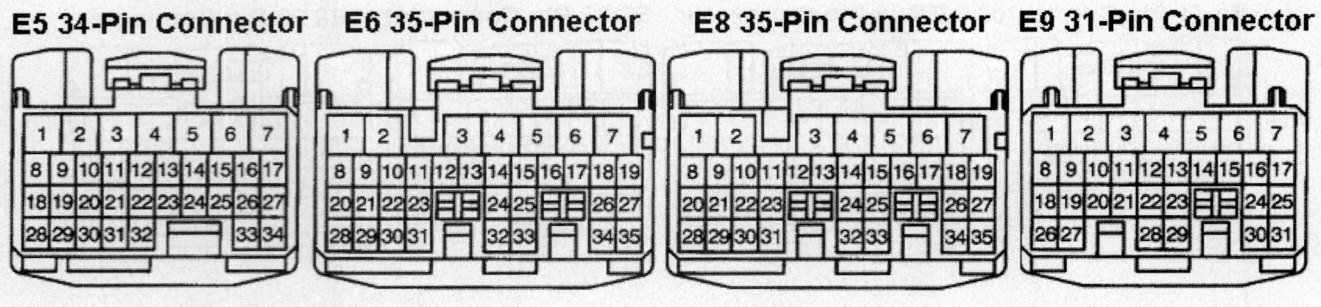

05533_ADIA_G124

Pin Connector Graphic

RX 300 PIN CHARTS

1999-2000 3.0L 1MZ-FE V6 VIN F E8 22 Pin Connector

PCM Pin #	Wire Color	Circuit Description (22 Pin)	Value at Hot Idle
1	RD/BK	Direct Battery	12-14v
2	BK/OR	Ignition Switch Power	12-14v
3	LG/RD	Circuit Opening Relay (FC)	0-3v, at off-idle: 12v
4-5	---	Not Used	---
6	GN/RD	MIL (lamp) Control	MIL Off: 12v, On: 1v
7 ('00)	GY	Starter Switch Signal	Cranking: 0-3v
8	YL/GN	EFI Main Relay Control	Relay Off: 0v, On: 12v
9	WT/RD	EVAP Vapor Pressure (VSV)	12v or 0v
10	---	Not Used	---
11	RD	SIL (Scan Tool Signal)	Digital Signal
12-14	---	Not Used	---
15	GN/WT	Stop Light Switch Signal	Brake Off: 0v, On: 12v
16	BK/RD	EFI Main Relay Power	12-14v
17	BL/RD	EVAP Vapor Pressure Sensor	2.9-3.1v (with hose off)
18	PK	Heated Mirror Switch Signal	Switch Off: 0v, On: 12v
19	GN	Rear Tail Light Switch Signal	Switch Off: 0v, On: 12v
20-22	---	Not Used	---

1999-2000 3.0L 1MZ-FE V6 VIN F E9 28 Pin Connector

PCM Pin #	Wire Color	Circuit Description (28 Pin)	Value at Hot Idle
1	---	Not Used	---
2	RD/BK	A/T Select Switch Reverse	In 'R': 12v, Others: 0v
3	OR	A/T Select Switch 2nd	In 2nd: 12v, Others: 0v
4	BL	Cruise Control ECU (IDLO)	1.5v, off-idle: 12v
5-7	---	Not Used	---
8	WT	HO2S-12 (B1 S2) Signal	0.1.1-1.5v
9	PK/BK	HO2S-12 (B1 S2) Heater	1v (Heater on)
10	---	Not Used	---
11-19	LG/BK	A/C Amplifier Signal (ACT)	Clutch On: 12v, Off: 1.5v
12	YL	A/T Select Switch Low	In Low: 12v, Others: 0v
20	BK/WT	Neutral Start Circuit Signal	In P/N: Cranking: 0-3v
21, 23	---	Not Used	---
22	VT/YL	Vehicle Speed Sensor	At 55 mph: 48 Hz
24	YL/BK	Cruise Control Signal (OD1)	At Cruise in OD: 12v
25	BK/YL	A/C Amplifier Signal (AC1)	Clutch On: 1.5v, Off: 12v
26, 28	---	Not Used	---
27	VT/RD	Tachometer Signal (TACO)	Pulse Signals
28	---	Not Used	---

E8 22-PIN CONNECTOR   E9 28-PIN CONNECTOR

05533_ADIA_G226

Pin Connector Graphic

1999-2000 3.0L 1MZ-FE V6 VIN F E10 17-Pin Connector

PCM Pin #	Wire Color	Circuit Description (17-Pin)	Value at Hot Idle
1-3, 7-10	---	Not Used	---
4	BR/WT	Transponder Amplifier Code	Inserting key: pulses
5	BR/RD	Transponder Amplifier Signal	Inserting key: pulses
6	BR/YL	Transponder Amplifier Signal	Inserting key: pulses
11	BL/BK	Unlock Warning Switch	No Key: 4-5v
13 ('99)	BK/RD	Shield Ground	<0.050v
16	RD/WT	Theft Security Indicator Light	LED Off: 12v, On: 1v

1999-2000 3.0L 1MZ-FE V6 VIN F E11 24-Pin Connector

PCM Pin #	Wire Color	Circuit Description (24-Pin)	Value at Hot Idle
1	WT/BK	Power Ground (E04)	<0.1v
2	YL	Sensor VREF (VC)	4.9-5.1v
3 (Fed)	BL/BK	HO2S-11 (B1 S1) Signal	0.1.1-1.5v
3 (Cal)	BL	AFS-11 (B1 S1) Heater	1v (Heater on)
4 (Fed)	YL/RD	HO2S-21 (B2 S1) Signal	0.1.1-1.5v
4 (Cal)	BL	AFS-21 (B2 S1) Heater	1v (Heater on)
5	BL/RD	Injector 1 Control	1.6-2-9 ms
6	RD	Injector 2 Control	1.6-2-9 ms
7	LG	EVAP Purge Solenoid (VSV)	12v or 0v
8	WT/BK	Power Ground (E05)	<0.1v
9	BK/BL	PSP Switch Signal	Straight: 12v, Turned: 0v
10	PK	MAF Sensor Signal (VG)	1.1-1.5v
11 (Fed)	WT	HO2S-11 (B1 S1) Heater	1v (Heater on)
11 (Cal)	OR	AFS-11 (B1 S1) Signal (AF+)	Fixed at 3.3v
12 (Fed)	BK	HO2S-21 (B2 S1) Heater	1v (Heater on)
12 (Cal)	OR	AFS-21 (B2 S1) Signal (AF+)	Fixed at 3.3v
14	GN/BK	ECT Sensor Signal (THW)	0.5-0.6v
15	GN/YL	A/T: Fluid Temp. Input (THO)	At 68°F: 4-5v
16	BK/RD	CKP Sensor Signal (NE+)	AC pulse signals
17	BR	Shield Ground (E1)	<0.050v
18	BR	Sensor Ground (E2)	<0.050v
19	RD/BK	MAF Sensor Ground (E2G)	<0.050v
20 (Cal)	WT	AFS-11 (B1 S1) Signal (AF-)	Fixed at 3.3v
21 (Cal)	WT	AFS-21 (B2 S1) Signal (AF-)	Fixed at 3.3v
22	BL/YL	IAT Sensor Signal (THA)	0.5-3.4v
23	BL/WT	TP Sensor Signal (VTA1)	0.53-1.27v
24	BL	CMP/CKP Sensor Ground (-)	<0.050v

E10 17-PIN CONNECTOR   **E11 24-PIN CONNECTOR**

(WIRE SIDE OF HARNESS TERMINALS)

05533_ADIA_G227

Pin Connector Graphic

1999-2000 3.0L 1MZ-FE V6 VIN F E12 31-Pin Connector

PCM Pin #	Wire Color	Circuit Description (31-Pin)	Value at Hot Idle
1	YL	Injector 3 Control	1.6-2-9 ms
2	WT	Injector 4 Control	1.6-2-9 ms
3	RD/BL	Injector 5 Control	1.6-2-9 ms
4	GN	Injector 6 Control	1.6-2.9 ms
5	---	Not Used	---
6	BL/WT	DLC 1 Signal (TC)	12v
7	VT	A/T-ECT Shift Solenoid (S1)	In 3rd or OD: 1v
8	BL/BK	A/T-ECT Shift Solenoid (S2)	1st or OD: 1v
9	PK/BL	A/T-ECT Shift Solenoid (SL)	In Lockup: 12-14v
10	BK/WT	CMP Sensor Signal (G22+)	AC pulse signals
11	GY	IGT 1 Control	6°, 55 mph: 8° dwell
12	BR/YL	IGT 2 Control	6°, 55 mph: 8° dwell
13	LG/BK	IGT 3 Control	6°, 55 mph: 8° dwell
14	---	Not Used	---
15	YL/BK	IAC Signal (RSC)	Pulse Signals
16	RD/WT	IAC Signal (RSO)	Pulse Signals
17	RD/YL	ACIS Control (VSV)	12v or 0v
18	GN/RD	Overdrive Lamp Control	At Cruise in OD: 1v
19-20	---	Not Used	---
21	WT/BK	Power Ground (E01)	<0.1v
22	---	Not Used	---
23	G/OR	OD Main Sw. Input (ODMS)	Open: 12v, Closed: 1v
25	WT/RD	Igniter Signal (IGF)	Digital Signal: 0-5-0v
26	---	Not Used	---
27	WT	Knock Sensor 1 Signal (right)	<0.075v AC
28	WT	Knock Sensor 2 Signal (left)	<0.075v AC
29	GN/WT	Cooling Fan 1 Relay	Relay Off: 12v, On: 1v
30	WT/BK	Power Ground (E03)	<0.1v
31	WT/BK	Power Ground (E01)	<0.1v

E12 31-PIN CONNECTOR

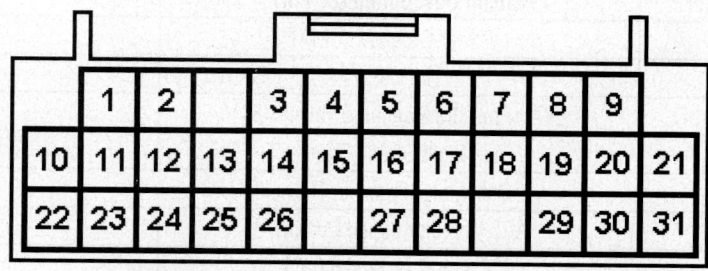

WIRE SIDE OF HARNESS TERMINALS

05533_ADIA_G228

Pin Connector Graphic

Standard Colors and Abbreviations

Abbreviation	Color	Abbreviation	Color	Abbreviation	Color
BK	Black	GY	Gray	RD	Red
BL	Blue	GN	Green	TN	Tan
BR	Brown	LG	LT Green	VT	Violet
DB	Dark Blue	OR	Orange	WT	White
DG	DK Green	PK	Pink	YL	Yellow

2001-03 3.0L 1MZ-FE V6 VIN F (A/T) E8 22 Pin Connector

PCM Pin #	Wire Color	Circuit Description (22 Pin)	Value at Hot Idle
1	RD/BK	Direct Battery	12-14v
2	BK/OR	Ignition Switch Power	12-14v
3	LG/RD	Circuit Opening Relay (FC)	0-3v, at off-idle: 12v
4	RD	SIL (Scan Tool Signal)	12v
5, 11	---	Not Used	---
6	BK/BL	MIL (lamp) Control	MIL Off: 12v, On: 1v
7	GY	Starter Switch Signal	Cranking: 0-3v
8	YL/GN	EFI Main Relay Control	Relay Off: 0v, On: 12v
9	GN/RD	OD Lamp Indicator Control	Lamp Off: 12v, On: 1v
10	BR/RD	EVAP Canister Closed Valve (VSV)	12v or 0v
12	BK	Center Airbag Sensor (F/PS)	Digital Signals
13	OR	TRAC Signal (TRC+)	Pulse Signals
14	GN	TRAC Engine Signal (ENG+)	Pulse Signals
15	GN/WT	Stop Light Switch Signal	Brake Off: 0v, On: 12v
16	BK/RD	EFI Main Relay (B+)	12-14v
17	BL/RD	EVAP Vapor Pressure Sensor (PTNK)	2.9-3.1v (with hose off)
18	PK	Heated Mirror Circuit	Heater On: 12-14v
19	GN	Electric Load Sensor Circuit	Lights On: 12-14v
20	BL/WT	TRAC Signal (TRC-)	Pulse Signals
21	WT	TRAC Engine Signal (ENG-)	Pulse Signals
22	RD/WT	Theft Deterrent Indicator (IMLD)	LED Off: 0v, On:

2001-03 3.0L 1MZ-FE V6 VIN F (A/T) E9 28 Pin Connector

PCM Pin #	Wire Color	Circuit Description (28 Pin)	Value at Hot Idle
1	BK/RD	Power Ground (EOM)	<0.1v
2, 10-12	---	Not Used	---
3	WT/RD	EVAP Pressure Switching Valve (VSV)	12v or 0v
4	GN	Tail Light Switch Signal	Switch Off: 0v, On: 12v
5	PK/BK	Data Link Connector (TC)	12v
6	RD/WT	A/C Amplifier Signal (ACMG)	Relay Off: 12v, On: 1v
7	BL/YL	A/C Amplifier Signal (AC)	Clutch On: 12v, Off: 1.5v
8	WT	HO2S-12 (B1 S2) Signal (OXS)	0.1.1-1.5v
9	BL/WT	HO2S-12 (B1 S2) Heater (HTS)	1v (Heater on)
13	VT	Mirror Heater Switch Signal	Switch Off: 0v, On: 12v
14	GY/BK	A/C Amplifier THWO Signal	A/C On: 12-14v
15-17, 21	---	Not Used	---

2001-03 3.0L 1MZ-FE V6 VIN F (A/T) E9 28 Pin Connector, *continued*

16	YL/BK	Transponder ECU Signal (NEO)	Inserting key: pulses
18	BR/YL	Transponder Amplifier Signal (TXCT)	Inserting key: pulses
19	BR/RD	Transponder Amplifier Signal (RXCK)	Inserting key: pulses
20	BK/WT	Neutral Start Switch Signal (NSW)	Cranking: 9-11v
22	VT/YL	Vehicle Speed Sensor	At 55 mph: 48 Hz
23	BL/BK	Unlock Warning Switch (KSW)	Switch Open:12v, Closed: 0v
24	BK/BL	Cruise Control Signal (OD1)	At Cruise in OD: 12v
25	BL	Cruise Control ECU (IDLO)	1.5v, off-idle: 12v
26	GN/WT	Cooling Fan Relay 1 (CF)	Relay Off: 12v, On: 1v
27	LG/BK	Tachometer Signal (TACO)	Pulse Signals
28	BR/WT	Transponder Amplifier Signal (Code)	Inserting key: pulses

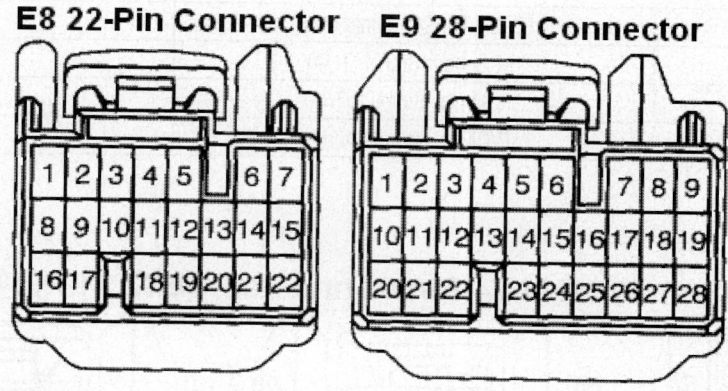

E8 22-Pin Connector   E9 28-Pin Connector

05533_ADIA_G229

Pin Connector Graphic

2001-03 3.0L 1MZ-FE V6 VIN F (A/T) E10 17-Pin Connector

PCM Pin #	Wire Color	Circuit Description (17-Pin)	Value at Hot Idle
1	BL/BK	A/T-ECT Shift Solenoid (S2)	1st or OD: 1v
2-7	---	Not Used	---
8	RD/BK	A/T Select Switch Signal (Reverse)	In 'R': 12v, Others: 0v
9-11	---	Not Used	---
12	GN/OR	Overdrive Main Switch Signal (ODMS)	Open: 12v, Closed: 1v
13	YL	A/T Select Switch Signal (Low)	In Low: 12v, Others: 0v
14	OR	A/T Select Switch Signal (2nd)	In 2nd: 12v, Others: 0v
15	PK/BL	A/T-ECT Shift Solenoid (SL)	In Lockup: 12-14v

2001-03 3.0L 1MZ-FE V6 VIN F (A/T) E11 24-Pin Connector

PCM Pin #	Wire Color	Circuit Description (24-Pin)	Value at Hot Idle
1	WT/BK	Power Ground (E04)	<0.1v
2	YL	Sensor VREF (VC)	4.9-5.1v
3	RD	AFS-11 (B1 S1) Heater (HAFR)	1v (Heater on)
4	BL	AFS-21 (B2 S1) Heater (HAFL)	1v (Heater on)
5	WT	Injector 1 Control	1.6-2-9 ms
6	BK	Injector 2 Control	1.6-2-9 ms
7	LG	EVAP Purge Solenoid (VSV)	12v or 0v

2001-03 3.0L 1MZ-FE V6 VIN F (A/T) E11 24-Pin Connector, *continued*

8	WT/BK	Power Ground (E05)	<0.1v
9	BK/BL	PSP Switch Signal (PS)	Straight: 12v, Turned: 0v
10	PK	MAF Sensor Signal (VG)	1.1-1.5v
11	BR	AFS-11 (B1 S1) Signal (AFR+)	Fixed at 3.3v
12	BL	AFS-21 (B2 S1) Signal (AFL+)	Fixed at 3.3v
13	GN/YL	Transmission Fluid Temperature Sensor	At 68°F: 4-5v
14	GN/BK	ECT Sensor Signal (THW)	0.5-0.6v
15	BK	Intake Air Control 2 (VSV)	12v or 0v
16	BK/WT	CKP Sensor Signal (NE+)	AC pulse signals
17	BR	DLC Ground (E1)	<0.050v
18	WT	Sensor Ground (E2)	<0.050v
19	RD/BK	MAF Sensor Ground (E2G)	<0.050v
20	BK/RD	AFR-11 (B1 S1) Signal (AFR-)	Fixed at 3.3v
21	BK/WT	AFR-21 (B2 S1) Signal (AFL-)	Fixed at 3.3v
22	BL/YL	IAT Sensor Signal (THA)	0.5-3.4v
23	BL/WT	TP Sensor Signal (VTA1)	0.53-1.27v
24	WT	CMP/CKP Sensor Ground (-)	<0.050v

E10 17-Pin Connector E11 24-Pin Connector E12 31-Pin Connector

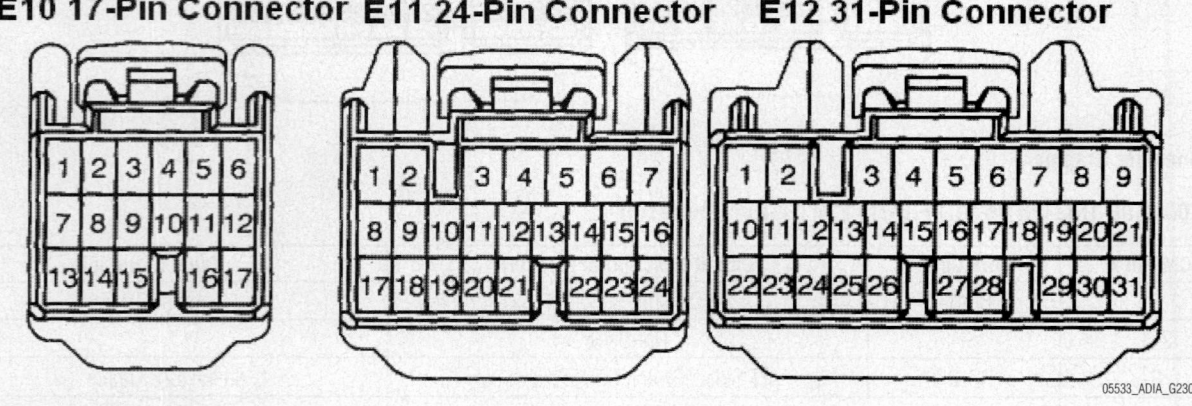

05533_ADIA_G230

Pin Connector Graphic

2001-03 3.0L 1MZ-FE V6 VIN F (A/T) E12 31-Pin Connector

PCM Pin #	Wire Color	Circuit Description (31-Pin)	Value at Hot Idle
1	BK	Injector 3 Control	1.6-2-9 ms
2	BL	Injector 4 Control	1.6-2-9 ms
3	RD	Injector 5 Control	1.6-2-9 ms
4	GN	Injector 6 Control	1.6-2.9 ms
5	BL/BK	VVT Solenoid RH Control (OC1-)	AC Pulse signals
6	RD/BK	VVT Solenoid RH Control (OC1+)	AC Pulse signals
7	WT/GN	A/T-ECT Solenoid S1	In 3rd or OD: 1v
8-9	---	Not Used	---
10	BK	Cam Timing Oil Control Valve RH (VV1+)	Pulse signals (17-18 Hz)
11	BL/RD	IGT 1 Control	6°, 55 mph: 8° dwell

2001-03 3.0L 1MZ-FE V6 VIN F (A/T) E12 31-Pin Connector, *continued*

12	GN/RD	IGT 2 Control	6°, 55 mph: 8° dwell
13	WT/RD	IGT 3 Control	6°, 55 mph: 8° dwell
14	BK/RD	IGT 4 Control	6°, 55 mph: 8° dwell
15	BL/WT	IGT 5 Control	6°, 55 mph: 8° dwell
16	GN/BK	IGT 6 Control	6°, 55 mph: 8° dwell
17	RD/YL	Intake Air Control Valve (ACIS)	12v or 0v
18	BK/WT	VVT Solenoid LH Control (OC2-)	AC Pulse signals
19	GN/WT	A/T-ECT Solenoid (SLN-)	Pulse Signals
20	BK/YL	A/T-ECT Solenoid (SLN+)	Pulse Signals
21	WT/BK	Power Ground (E01)	<0.1v
22	OR	Cam Timing Oil Control Valve LH (VV2+)	Pulse signals (17-18 Hz)
23	YL	Direct Clutch Speed Input (-)	AC pulse signals
24	BL	Direct Clutch Speed Input (+)	AC pulse signals
25	BK	Igniter Signal (IGF)	Digital Signal: 0-5-0v
26	RD/WT	Idle Air Control Valve (RSO)	Pulse Signals
27	WT	Knock Sensor 1 Signal (KNKR)	<0.075v AC
28	WT	Knock Sensor 2 Signal (KNKL)	<0.075v AC
29	RD	VVT Solenoid LH Control (OC2+)	AC Pulse signals
30	WT/BK	Power Ground (E03)	<0.1v
31	WT/BK	Power Ground (E02)	<0.1v

E10 17-Pin Connector   **E11 24-Pin Connector**   **E12 31-Pin Connector**

05533_ADIA_G230

Pin Connector Graphic

RX 330 PIN CHARTS

2004-06 3.3L 3MZ-FE V6 VIN W (A/T) E8 22 Pin Connector

PCM Pin #	Wire Color	Circuit Description (22 Pin)	Value at Hot Idle
1	RD/BK	Direct Battery	12-14v
2	BK/OR	Ignition Switch Power	12-14v
3	LG/RD	Circuit Opening Relay (FC)	0-3v, at off-idle: 12v
4	RD	SIL (Scan Tool Signal)	12v
5, 11	---	Not Used	---
6	BK/BL	MIL (lamp) Control	MIL Off: 12v, On: 1v
7	GY	Starter Switch Signal	Cranking: 0-3v
8	YL/GN	EFI Main Relay Control	Relay Off: 0v, On: 12v

2004-06 3.3L 3MZ-FE V6 VIN W (A/T) E8 22 Pin Connector, *continued*

9	GN/RD	OD Lamp Indicator Control	Lamp Off: 12v, On: 1v
10	BR/RD	EVAP Canister Closed Valve (VSV)	12v or 0v
12	BK	Center Airbag Sensor (F/PS)	Digital Signals
13	OR	TRAC Signal (TRC+)	Pulse Signals
14	GN	TRAC Engine Signal (ENG+)	Pulse Signals
15	GN/WT	Stop Light Switch Signal	Brake Off: 0v, On: 12v
16	BK/RD	EFI Main Relay (B+)	12-14v
17	BL/RD	EVAP Vapor Pressure Sensor (PTNK)	2.9-3.1v (with hose off)
18	PK	Heated Mirror Circuit	Heater On: 12-14v
19	GN	Electric Load Sensor Circuit	Lights On: 12-14v
20	BL/WT	TRAC Signal (TRC-)	Pulse Signals
21	WT	TRAC Engine Signal (ENG-)	Pulse Signals
22	RD/WT	Theft Deterrent Indicator (IMLD)	LED Off: 0v, On:

2004-06 3.3L 3MZ-FE V6 VIN W (A/T) E9 28 Pin Connector

PCM Pin #	Wire Color	Circuit Description (28 Pin)	Value at Hot Idle
1	BK/RD	Power Ground (EOM)	<0.1v
2, 10-12	---	Not Used	---
3	WT/RD	EVAP Pressure Switching Valve (VSV)	12v or 0v
4	GN	Tail Light Switch Signal	Switch Off: 0v, On: 12v
5	PK/BK	Data Link Connector (TC)	12v
6	RD/WT	A/C Amplifier Signal (ACMG)	Relay Off: 12v, On: 1v
7	BL/YL	A/C Amplifier Signal (AC)	Clutch On: 12v, Off: 1.5v
8	WT	HO2S-12 (B1 S2) Signal (OXS)	0.1.1-1.5v
9	BL/WT	HO2S-12 (B1 S2) Heater (HTS)	1v (Heater on)
13	VT	Mirror Heater Switch Signal	Switch Off: 0v, On: 12v
14	GY/BK	A/C Amplifier THWO Signal	A/C On: 12-14v
15-17, 21	---	Not Used	---
16	YL/BK	Transponder ECU Signal (NEO)	Inserting key: pulses
18	BR/YL	Transponder Amplifier Signal (TXCT)	Inserting key: pulses
19	BR/RD	Transponder Amplifier Signal (RXCK)	Inserting key: pulses
20	BK/WT	Neutral Start Switch Signal (NSW)	Cranking: 9-11v
22	VT/YL	Vehicle Speed Sensor	At 55 mph: 48 Hz
23	BL/BK	Unlock Warning Switch (KSW)	Switch Open:12v, Closed: 0v
24	BK/BL	Cruise Control Signal (OD1)	At Cruise in OD: 12v
25	BL	Cruise Control ECU (IDLO)	1.5v, off-idle: 12v
26	GN/WT	Cooling Fan Relay 1 (CF)	Relay Off: 12v, On: 1v
27	LG/BK	Tachometer Signal (TACO)	Pulse Signals
28	BR/WT	Transponder Amplifier Signal (Code)	Inserting key: pulses

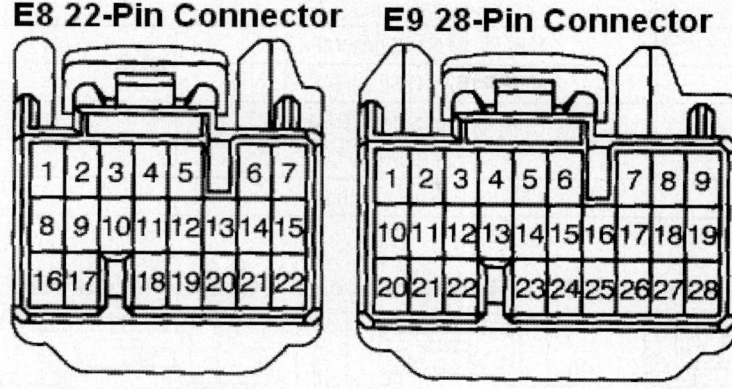

E8 22-Pin Connector **E9 28-Pin Connector**

05533_ADIA_G229

Pin Connector Graphic

2004-06 3.3L 3MZ-FE V6 VIN W (A/T) E10 17-Pin Connector

PCM Pin #	Wire Color	Circuit Description (17-Pin)	Value at Hot Idle
1	BL/BK	A/T-ECT Shift Solenoid (S2)	1st or OD: 1v
2-7	---	Not Used	---
8	RD/BK	A/T Select Switch Signal (Reverse)	In 'R': 12v, Others: 0v
9-11	---	Not Used	---
12	GN/OR	Overdrive Main Switch Signal (ODMS)	Open: 12v, Closed: 1v
13	YL	A/T Select Switch Signal (Low)	In Low: 12v, Others: 0v
14	OR	A/T Select Switch Signal (2nd)	In 2nd: 12v, Others: 0v
15	PK/BL	A/T-ECT Shift Solenoid (SL)	In Lockup: 12-14v

2004-06 3.3L 3MZ-FE V6 VIN W (A/T) E11 24-Pin Connector

PCM Pin #	Wire Color	Circuit Description (24-Pin)	Value at Hot Idle
1	WT/BK	Power Ground (E04)	<0.1v
2	YL	Sensor VREF (VC)	4.9-5.1v
3	RD	AFS-11 (B1 S1) Heater (HAFR)	1v (Heater on)
4	BL	AFS-21 (B2 S1) Heater (HAFL)	1v (Heater on)
5	WT	Injector 1 Control	1.6-2-9 ms
6	BK	Injector 2 Control	1.6-2-9 ms
7	LG	EVAP Purge Solenoid (VSV)	12v or 0v
8	WT/BK	Power Ground (E05)	<0.1v
9	BK/BL	PSP Switch Signal (PS)	Straight: 12v, Turned: 0v
10	PK	MAF Sensor Signal (VG)	1.1-1.5v
11	BR	AFS-11 (B1 S1) Signal (AFR+)	Fixed at 3.3v
12	BL	AFS-21 (B2 S1) Signal (AFL+)	Fixed at 3.3v
13	GN/YL	Transmission Fluid Temperature Sensor	At 68°F: 4-5v
14	GN/BK	ECT Sensor Signal (THW)	0.5-0.6v
15	BK	Intake Air Control 2 (VSV)	12v or 0v
16	BK/WT	CKP Sensor Signal (NE+)	AC pulse signals
17	BR	DLC Ground (E1)	<0.050v
18	WT	Sensor Ground (E2)	<0.050v

2004-06 3.3L 3MZ-FE V6 VIN W (A/T) E11 24-Pin Connector, *continued*

19	RD/BK	MAF Sensor Ground (E2G)	<0.050v
20	BK/RD	AFR-11 (B1 S1) Signal (AFR-)	Fixed at 3.3v
21	BK/WT	AFR-21 (B2 S1) Signal (AFL-)	Fixed at 3.3v
22	BL/YL	IAT Sensor Signal (THA)	0.5-3.4v
23	BL/WT	TP Sensor Signal (VTA1)	0.53-1.27v
24	WT	CMP/CKP Sensor Ground (-)	<0.050v

E10 17-Pin Connector   **E11 24-Pin Connector**   **E12 31-Pin Connector**

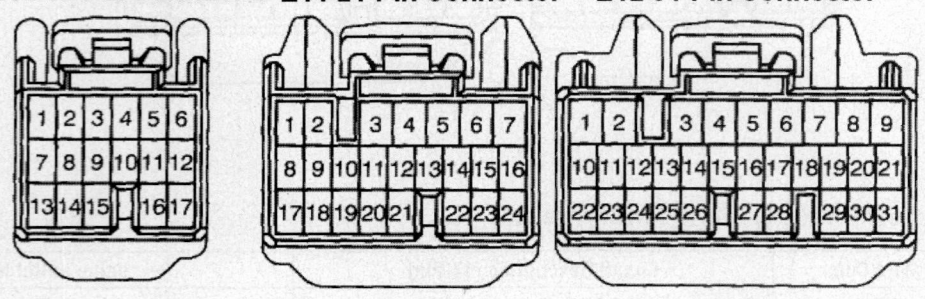

05533_ADIA_G230

Pin Connector Graphic

2004-06 3.3L 3MZ-FE V6 VIN W (A/T) E12 31-Pin Connector

PCM Pin #	Wire Color	Circuit Description (31-Pin)	Value at Hot Idle
1	BK	Injector 3 Control	1.6-2-9 ms
2	BL	Injector 4 Control	1.6-2-9 ms
3	RD	Injector 5 Control	1.6-2-9 ms
4	GN	Injector 6 Control	1.6-2.9 ms
5	BL/BK	VVT Solenoid RH Control (OC1-)	AC Pulse signals
6	RD/BK	VVT Solenoid RH Control (OC1+)	AC Pulse signals
7	WT/GN	A/T-ECT Solenoid S1	In 3rd or OD: 1v
8-9	---	Not Used	---
10	BK	Cam Timing Oil Control Valve RH (VV1+)	Pulse signals (17-18 Hz)
11	BL/RD	IGT 1 Control	6°, 55 mph: 8° dwell
12	GN/RD	IGT 2 Control	6°, 55 mph: 8° dwell
13	WT/RD	IGT 3 Control	6°, 55 mph: 8° dwell
14	BK/RD	IGT 4 Control	6°, 55 mph: 8° dwell
15	BL/WT	IGT 5 Control	6°, 55 mph: 8° dwell
16	GN/BK	IGT 6 Control	6°, 55 mph: 8° dwell
17	RD/YL	Intake Air Control Valve (ACIS)	12v or 0v
18	BK/WT	VVT Solenoid LH Control (OC2-)	AC Pulse signals
19	GN/WT	A/T-ECT Solenoid (SLN-)	Pulse Signals
20	BK/YL	A/T-ECT Solenoid (SLN+)	Pulse Signals
21	WT/BK	Power Ground (E01)	<0.1v
22	OR	Cam Timing Oil Control Valve LH (VV2+)	Pulse signals (17-18 Hz)
23	YL	Direct Clutch Speed Input (-)	AC pulse signals
24	BL	Direct Clutch Speed Input (+)	AC pulse signals

25	BK	Igniter Signal (IGF)	Digital Signal: 0-5-0v
26	RD/WT	Idle Air Control Valve (RSO)	Pulse Signals
27	WT	Knock Sensor 1 Signal (KNKR)	<0.075v AC
28	WT	Knock Sensor 2 Signal (KNKL)	<0.075v AC
29	RD	VVT Solenoid LH Control (OC2+)	AC Pulse signals
30	WT/BK	Power Ground (E03)	<0.1v
31	WT/BK	Power Ground (E02)	<0.1v

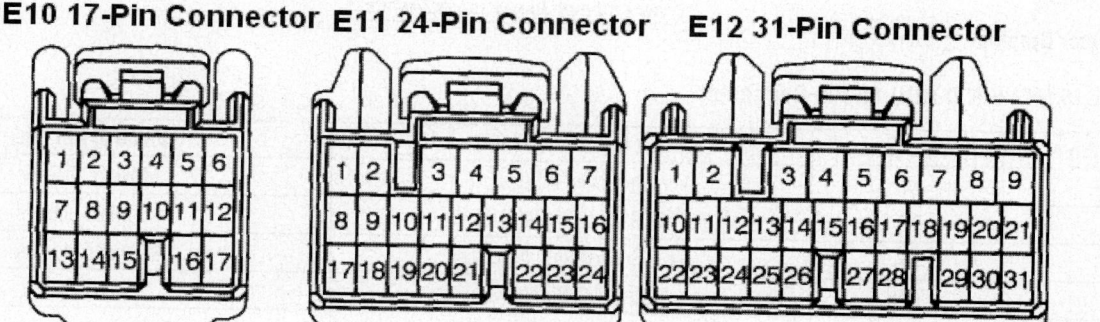

E10 17-Pin Connector E11 24-Pin Connector E12 31-Pin Connector

05533_ADIA_G230

Pin Connector Graphic

SC 300 PIN CHARTS

1996 3.0L I6 MFI VIN D (All) 40 Pin Connector

PCM Pin #	Wire Color	Circuit Description (40 Pin)	Value at Hot Idle
1	BK/OR	Ignition Switch Power	12-14v
2	PK	Speed Sensor 1 Signal	Moving: 0-5-0v
3	YL	Kickdown Switch	Open: 12v, Closed: 0v
4	GN/WT	Stop Light Switch Signal	Brake Off: 0v, On: 12v
6	BL/BK	MIL (lamp) Control	MIL Off: 12v, On: 1v
8	GN	Data Link Connector (SDL)	No Scan Tool: 0v
9	GN/RD	A/T Select Switch 2nd	In 2nd: 12v, Others: 0v
10	GN/BK	A/T Select Switch Low	In Low: 12v, Others: 0v
12	BR/BK	Cruise Control ECU	At Cruise in OD: 12v
15	RD/YL	Idle Up Diode Signal	Switch On: 12v, Off: 0v
18	GN/YL	A/T Pattern Selector Switch	Norm: 0v, PWR: 12v
20	YL/BL	Data Link Connector	12-14v
21	GN	Fuel Pump (Control) ECU	Pulse Signals
22	PK/WT	Fuel Pump (DI) ECU	Pulse Signals
23	WT/GN	A/C Magnetic Clutch (ACMG)	Clutch Off: 0v, On: 12v
24	GY	EFI Main Relay Power	12-14v
25	WT/BL	Manual Indicator Light	Light Off: 12-14v
28	PK/GN	Overdrive Main Switch	Switch Off: 12v, On: 0v
30	RD/BL	Sub HO2S Signal	0.1-1.1v
31	BK/RD	EFI Main Relay Power	12-14v
33	BK/YL	Battery Direct	12-14v

1996 3.0L I6 MFI VIN D (All) 40 Pin Connector, *continued*

34	BL/RD	A/C Amplifier Signal (AC1)	Clutch On: 1.5v, Off: 12v
35	BR/WT	PSP Switch Signal	Straight: 12v, Turning: 0v
36	BR/WT	Sub HO2S Heater	1v (Heater On)

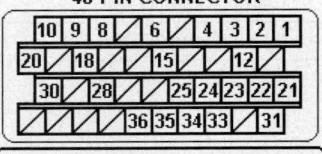

05533_ADIA_G093

Pin Connector Graphic

1996 3.0L I6 MFI VIN D (All) 80 Pin Connector

PCM Pin #	Wire Color	Circuit Description (80 Pin)	Value at Hot Idle
3	GN	Vehicle Speed 2 Sensor (-)	Moving: 0-5-0v
5	WT	CKP Sensor Signal (NE2-)	AC pulse signals
6	BK	CKP Sensor Signal (NE2+)	AC pulse signals
7	GN	Distributor Signal (G-)	<0.050v
8	BK/RD	A/T-ECT Solenoid S3	In Lockup: 12-14v
9	RD/BL	A/T-ECT Solenoid (S2)	1st or OD: 1v
10	WT/RD	A/T-ECT Solenoid (S1)	In 3rd or OD: 1v
15	RD/BK	Injector 6 Control	2.0-3.3 ms
16	RD	Injector 5 Control	2.0-3.3 ms
17	RD/WT	Injector 4 Control	2.0-3.3 ms
18	RD/GN	Injector 3 Control	2.0-3.3 ms
19	RD/YL	Injector 2 Control	2.0-3.3 ms
20	RD/BL	Injector 1 Control	2.0-3.3 ms
23	RD	Vehicle Speed 2 Sensor (+)	Moving: 0-5-0v
24	OR	A/T Oil Temperature Sensor	Varies: 0.5-4.5v
25	WT	Distributor Signal (G2+)	AC pulse signals
26	RD	Distributor Signal (G1+)	AC pulse signals
27	BK	Distributor Signal (NE+)	AC pulse signals
30	BR	MAF Sensor Ground	<0.050v
32	RD/GN	ISC Motor (ISC4)	Pulse Signals
33	GN/OR	ISC Motor (ISC3)	Pulse Signals
34	GN/WT	ISC Motor (ISC2)	Pulse Signals
35	PK/YL	ISC Motor (ISC1)	Pulse Signals
36	WT/BL	Fuel Pressure Up Solenoid	1v (at hot restart)
39	GN/BK	Intake Air Control Solenoid	12v or 0v

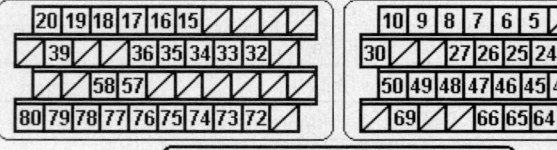

05533_ADIA_G002

Pin Connector Graphic

1996 3.0L I6 MFI VIN D (All) 80 Pin Connector

PCM Pin #	Wire Color	Circuit Description (80 Pin)	Value at Hot Idle
41	BL/RD	Sensor VREF (VC)	4.9-5.1v
43	YL	TP Sensor Signal	0.3-0.8v
44	BL/YL	ECT Sensor Signal	At 180°F: 0.51v
45	PK/BL	IAT Sensor Signal	At 100°F: 2.60v
46	BR/YL	EGR Gas Temp. Sensor	3.5-4.0v
47	RD/BL	Rear HO2S Signal	0.1-1.1v
48	WT	Front HO2S Signal	0.1-1.1v
49	WT	Knock Sensor 2 Signal	<0.075v AC
50	WT	Knock Sensor 1 Signal	<0.075v AC
57	RD/WT	Igniter Signal (IGT)	Digital Signal: 0-5-0v
58	RD/YL	Igniter Signal (IGF)	Digital Signal: 0-5-0v
64	RD	Closed Throttle Switch	1v, at off-idle: 12v
65	BR/BK	Sensor Ground	<0.050v
66	YL/RD	Mass Airflow Sensor	1.0-1.8v
69	BR	Shield Ground	<0.050v
72	BK/YL	Rear HO2S Heater	1v (Heater On)
73	BK/BL	Front HO2S Heater	1v (Heater On)
74	PK	EVAP Purge Solenoid (VSV)	12v or 0v
75	PK	EGR Solenoid Control (VSV)	1v (Heater On)
76	BK/WT	A/T Neutral Start Switch	In P/N: 9-11v (cranking)
77	BK	Starter Switch Signal	9-11v (cranking)
78	BR	Power Ground	<0.1v
79	BR	Power Ground	<0.1v
80	BR	Power Ground	<0.1v

Standard Colors and Abbreviations

Abbreviation	Color	Abbreviation	Color	Abbreviation	Color
BK	Black	GY	Gray	RD	Red
BL	Blue	GN	Green	TN	Tan
BR	Brown	LG	Light Green	VT	Violet
DB	Dark Blue	OR	Orange	WT	White
DG	DK Green	PK	Pink	YL	Yellow

1997 3.0L I6 MFI VIN D (All) 40 Pin Connector

PCM Pin #	Wire Color	Circuit Description (40 Pin)	Value at Hot Idle
1	BK/OR	Ignition Switch Power	12-14v
2	PK	Speed Sensor 1 Signal	Moving: 0-5-0v
3	YL	Kickdown Switch	Open: 12v, Closed: 0v
4	GN/WT	Stop Light Switch Signal	Brake Off: 0v, On: 12v
6	BL/BK	MIL (lamp) Control	MIL Off: 12v, On: 1v
8	GN	Data Link Connector (SDL)	0v
9	GN/RD	A/T Select Switch 2nd	In 2nd: 12v, Others: 0v
10	GN/BK	A/T Select Switch Low	In Low: 12v, Others: 0v
12	GN/BK	Cruise Control ECU	At Cruise in OD: 12v

1997 3.0L I6 MFI VIN D (All) 40 Pin Connector, *continued*

15	RD/YL	Idle Up Diode Signal	Switch On: 12v, Off: 0v
18	GN/YL	A/T Pattern Selector Switch	Norm: 0v, PWR: 12v
20	YL/GN	Data Link Connector	12-14v
21	GN	Fuel Pump (Control) ECU	Pulse Signals
22	PK/WT	Fuel Pump (DI) ECU	Pulse Signals
23	WT/GN	A/C Magnetic Clutch (ACMG)	Clutch Off: 0v, On: 12v
24	BK/YL	EFI Main Relay Power	12-14v
25	WT/BL	Manual Indicator Light	Light Off: 12v, On: 1v
28	PK	Overdrive Main Switch	Switch Off: 12v
30	RD/BL	Sub HO2S Signal	0.1-1.1v
31	BK/RD	EFI Main Relay Power	12-14v
32	BR/WT	PSP Switch Signal	Straight: 12v, Turning: 0v
33	BK/WT	Battery Direct	12-14v
34	BL/RD	A/C Amplifier Signal (AC1)	Clutch On: 1.5v, Off: 12v
36	BR/WT	Sub HO2S Heater	1v (Heater On)

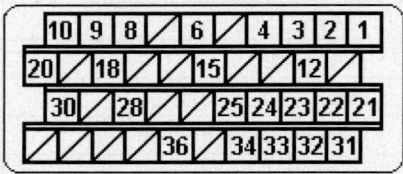

05533_ADIA_G095

Pin Connector Graphic

1997 3.0L I6 MFI VIN D (All) 80 Pin Connector

PCM Pin #	Wire Color	Circuit Description (80 Pin)	Value at Hot Idle
3	GN	Vehicle Speed 2 Sensor (-)	Moving: 0-5-0v
4	BK	CKP Sensor Signal (NE2+)	AC pulse signals
5	WT	CKP Sensor Signal (NE2-)	<0.050v
7	GN	Distributor Signal (G-)	<0.050v
8	BK/RD	A/T-ECT Solenoid (SL)	In Lockup: 12-14v
9	RD/BL	A/T-ECT Solenoid (S2)	1st or OD: 1v
10	WT/RD	A/T-ECT Solenoid (S1)	In 3rd or OD: 1v
15	RD/BK	Injector 6 Control	2.0-3.3 ms
16	RD	Injector 5 Control	2.0-3.3 ms
17	RD/WT	Injector 4 Control	2.0-3.3 ms
18	BL	Injector 3 Control	2.0-3.3 ms
19	BL/RD	Injector 2 Control	2.0-3.3 ms
20	RD/BL	Injector 1 Control	2.0-3.3 ms
23	RD	Vehicle Speed 2 Sensor (+)	Moving: 0-5-0v
24	BL/BK	A/T Oil Temperature Sensor	0.5-4.5v
25	WT	Distributor Signal (G2+)	AC pulse signals
26	RD	Distributor Signal (G1+)	AC pulse signals

1997 3.0L I6 MFI VIN D (All) 80 Pin Connector, *continued*

27	BK	Distributor Signal (NE+)	AC pulse signals
28	BR	MAF Sensor Ground	<0.050v
32	RD/BK	ISC Motor (ISC4)	Pulse Signals
33	BL/BK	ISC Motor (ISC3)	Pulse Signals
34	GN/WT	ISC Motor (ISC2)	Pulse Signals
35	YL/BK	ISC Motor (ISC1)	Pulse Signals
39	GN/YL	Intake Air Control Solenoid	12v or 0v

1997 3.0L I6 MFI VIN D (All) 80 Pin Connector

PCM Pin #	Wire Color	Circuit Description (80 Pin)	Value at Hot Idle
41	BL/RD	Sensor VREF (VC)	4.9-5.1v
43	YL	TP Sensor Signal	0.3-0.8v
44	BL/YL	ECT Sensor Signal	At 180°F: 0.51v
45	GN/BK	IAT Sensor Signal	At 100°F: 2.60v
46	BR/YL	EGR Gas Temp. Sensor	3.5-4.0v
47	RD/BL	HO2S-21 (B2 S1) Signal	0.1-1.1v
48	WT	HO2S-11 (B1 S1) Signal	0.1-1.1v
49	WT	Knock Sensor 2 Signal	<0.075v AC
50	WT	Knock Sensor 1 Signal	<0.075v AC
57	RD/WT	Igniter Signal (IGT)	Digital Signal: 0-5-0v
58	RD/YL	Igniter Signal (IGF)	Digital Signal: 0-5-0v
64	RD/BK	Closed Throttle Switch	1v, at off-idle: 12v
65	WT/BK	Sensor Ground	<0.050v
66	YL/RD	Mass Airflow Sensor	1.0-1.8v
69	BR	Shield Ground	<0.050v
71	GN	HO2S-11 (B1 S1) Heater	1v (Heater On)
72	BK/YL	HO2S-21 (B2 S1) Heater	1v (Heater On)
73	WT/BL	Fuel Pressure Up Solenoid	1v (at hot restart)
74	PK	EVAP Purge Solenoid (VSV)	12v or 0v
75	PK	EGR Solenoid Control (VSV)	1v (Heater On)
76	BK/WT	A/T Neutral Start Switch	In P/N: 9-11v (cranking)
77	BK	Starter Switch Signal	9-11v (cranking)
78	BR	Power Ground	<0.1v
79	BR	Power Ground	<0.1v
80	BR	Power Ground	<0.1v

80 PIN CONNECTOR

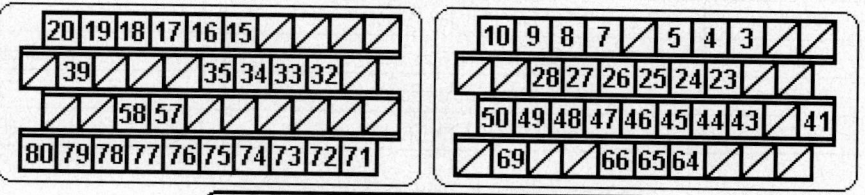

WIRE SIDE OF HARNESS TERMINALS

Pin Connector Graphic

05533_ADIA_G096

1998-2000 3.0L I6 MFI VIN D (A/T-ECT) 31 Pin Connector

PCM Pin #	Wire Color	Circuit Description (31 Pin)	Value at Hot Idle
1	BL	Injector 3 Control	2.0-3.3 ms
2	RD/WT	Injector 4 Control	2.0-3.3 ms
3	RD	Injector 5 Control	2.0-3.3 ms
4	RD/BK	Injector 6 Control	2.0-3.3 ms
5	GN/YL	Intake Air Control Solenoid	12v or 0v
7	BK	Throttle Control Motor (M-)	Pulse Signals
8	WT	Throttle Control Motor (M+)	Pulse Signals
9	BR	Sensor Ground	<0.050v
10	BL	CMP Sensor Signal (G+)	AC pulse signals
11	RD/WT	Igniter Transistor 1 Control	Digital Signal: 0-5-0v
12	LG	Igniter Transistor 2 Control	Digital Signal: 0-5-0v
13	GN/RD	Igniter Transistor 3 Control	Digital Signal: 0-5-0v
16	GN	PSP Switch Signal	Straight: 12v, Turning: 0v
17	YL/BK	Cam Timing Oil signal (OCV-)	Pulse Signals
18	WT/RD	Cam Timing Oil signal (OCV+)	Pulse Signals
19	YL	Throttle Control Motor (CL-)	Pulse Signals
20	BL	Throttle Control Motor (CL+)	Pulse Signals
21	WT/BR	Power Ground	<0.1v
22	WT	CKP/CMP Sensor Ground (-)	<0.050v
23	BK	CKP Sensor Signal (NE+)	AC pulse signals
24	BK/WT	Neutral Start Switch	In P/N: 9-11v (cranking)
25	RD/YL	Igniter Signal (IGF)	Digital Signal: 0-5-0v
27	WT	Knock Sensor 2 Signal	<0.075v AC
28	WT	Knock Sensor 1 Signal	<0.075v AC
30	BR	Shield Ground	<0.050v
31	WT/BR	Power Ground	<0.1v

1998-2000 3.0L I6 MFI VIN D (A/T-ECT) 22 Pin Connector

PCM Pin #	Wire Color	Circuit Description (22 Pin)	Value at Hot Idle
1	BK/WT	Battery Direct	12-14v
4	GN	Fuel Pump (Control) ECU	Pulse Signals
5	PK/WT	Fuel Pump (DI) ECU	Pulse Signals
6	BL/BK	MIL (lamp) Control	MIL Off: 12v, On: 1v
7	BL/RD	EFI Main Relay Power BM+	12-14v
8	BK/RD	EFI Main Relay Power B2+	12-14v
9	BK/OR	Ignition Switch Power	12-14v
10	BK/YL	EFI Main Relay Power	12-14v
11	BK	SIL Signal (Scan Tool)	Digital Signal
16	BK/RD	EFI Main Relay Power	12-14v
22	WT/BK	ECM/Data Link Ground	<0.050v

1998-2000 3.0L I6 MFI VIN D (A/T-ECT) 28 Pin Connector

PCM Pin #	Wire Color	Circuit Description (28 Pin)	Value at Hot Idle
1	BL/RD	A/C Amplifier Signal (AC1)	Clutch On: 1.5v, Off: 12v
2	BK	Starter Switch Signal	9-11v (cranking)
4	GN/WT	Stop Light Switch Signal	Brake Off: 0v, On: 12v
5	RD	Data Link Connector	12-14v
8	WT	HO2S-12 (B1 S2) Signal	0.1-1.1v
13	WT/GN	A/C Magnetic Clutch (ACMG)	Clutch Off: 0v, On: 12v
16	RD/BK	A/T Select Switch Reverse	In 'R': 12v, Others: 0v
17	GN/RD	A/T Select Switch Drive	In 'D': 12v, Others: 0v
18	BR/YL	EVAP Vapor Pressure Sensor	2.5-3.1v (with hose off)
23	BL	Cruise Control ECU	At Cruise in OD: 12v
24	RD/YL	Cruise Control ECU	At Cruise in OD: 12v
25	WT	Vehicle Speed Sensor	At 55 mph: 48 Hz
26	RD/YL	Idle Up Diode Signal	Switch On: 12v, Off: 0v
28	PK	Overdrive Main Switch	Switch Off: 12v, On: 0v

1998-2000 3.0L I6 MFI VIN D (A/T-ECT) 17 Pin Connector

PCM Pin #	Wire Color	Circuit Description (17 Pin)	Value at Hot Idle
1	WT/RD	A/T-ECT Solenoid (S1)	In 3rd or OD: 1v
2	RD/BL	A/T-ECT Solenoid (S2)	1st or OD: 1v
4	RD	OD Clutch Sensor Signal (+)	Pulse Signals
5	RD/YL	Speed Sensor 2 Signal (+)	Pulse Signals
7	GN/WT	A/T-ECT Solenoid (SLU+)	Pulse Signals
8	YL/GN	A/T-ECT Solenoid (SLN+)	Pulse Signals
9	GN/RD	A/T-ECT Solenoid (SLT+)	Pulse Signals
10	RD	OD Clutch Sensor Signal (-)	Pulse Signals
11	BL/YL	Speed Sensor Signal 2 (-)	Pulse Signals
13	BL/RD	A/T-ECT Solenoid (SLU-)	Pulse Signals
14	PK	A/T-ECT Solenoid (SLN-)	Pulse Signals
15	RD/BK	A/T-ECT Solenoid (SLT-)	Pulse Signals
16	OR	C/C Indicator Light	Light Off: 12v, On: 1v
17	BL/BK	A/T Oil Temperature Sensor	0.5-4.5v

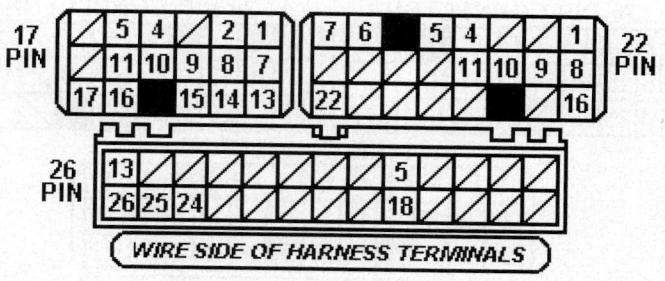

05533_ADIA_G097

Pin Connector Graphic

1998-2000 3.0L I6 MFI VIN D (A/T-ECT) 26 Pin Connector

PCM Pin #	Wire Color	Circuit Description (26 Pin)	Value at Hot Idle
5	GN/YL	A/T Pattern Selector Switch	Norm: 0v, PWR: 12v
13	RD/BK	EVAP Vapor Pressure (VSV)	12v or 0v
18	WT/BL	Manual Indicator Light	Light Off: 12-14v
24	WT	HO2S-22 (B2 S2) Signal	0.1-1.1v
25	GN/YL	HO2S-22 (B2 S2) Heater	1v (Heater On)
26	BL/WT	HO2S-12 (B1 S2) Heater	1v (Heater On)

1998-2000 3.0L I6 MFI VIN D (A/T-ECT) 24 Pin Connector

PCM Pin #	Wire Color	Circuit Description (24 Pin)	Value at Hot Idle
1	WT/BK	Power Ground	<0.1v
2	BL/RD	Sensor VREF	4.9-5.1v
3	BK/YL	HO2S-21 (B2 S1) Heater	1v (Heater On)
4	GN	HO2S-11 (B1 S1) Heater	1v (Heater On)
5	RD/BL	Injector 1 Control	2.0-3.3 ms
6	BL/RD	Injector 2 Control	2.0-3.3 ms
7	YL	EVAP Purge Solenoid (VSV)	12v or 0v
10	YL/RD	MAF Sensor Signal	1.1-1.5v
11	RD/BL	HO2S-21 (B2 S1) Signal	0.1-1.1v
12	WT	HO2S-11 (B1 S1) Signal	0.1-1.1v
13	BL/RD	Mirror Heater Switch Signal	Heater On: 12-14v
14	BL	ECT Sensor Signal	At 180°F: 0.51v
15	GN	Accel Position Sensor (VPA)	0.25-0.9v
16	WT	Accel Position Sensor (VPA2)	1.8-2.7v
17	BR	Sensor Ground	<0.050v
18	BR	Sensor Ground	<0.050v
19	BR	MAF Sensor Ground	<0.050v
20	BL/YL	A/T Select Switch 2nd	In 2nd: 12v, Others: 0v
21	GN/BK	A/T Select Switch Low	In Low: 12v, Others: 0v
22	GN/WT	IAT Sensor Signal	At 100°F: 2.60v
23	YL	TP Sensor (VTA1)	0.4-1.0v
24	RD/BK	TP Sensor (VTA2)	2.0-2.9v

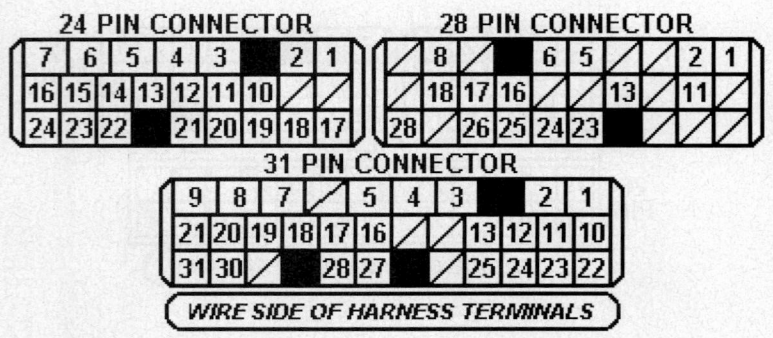

Pin Connector Graphic

05533_ADIA_G098

SC 400 PIN CHARTS

1996-2000 4.0L 1UZ-FE V8 VIN T E5 31P Connector

PCM Pin #	Wire Color	Circuit Description (24 Pin)	Value at Hot Idle
1	WT/BK	Power Ground (E03)	<0.1v
2	BL/RD	Sensor VREF (VC)	4.9-5.1v
3	YL	HO2S-21 (B2 S1) Heater	1v (Heater on)
4	RD	HO2S-11 (B1 S1) Heater	1v (Heater on)
5	YL	Injector 1 Control	2.0-3.3 ms
6	BK	Injector 2 Control	2.0-3.3 ms
7	BL/BK	EVAP Purge Solenoid (VSV)	12v or 0v
8	RD	ABS ECU Signal	Digital Signals
9	RD/BK	Accelerator Position Sensor 2	1.8-2.7v
10	BL/YL	Mass Airflow Sensor (VG)	1.1-1.5v
11	WT	HO2S-21 (B2 S1) Signal	0.1-1.1v
12	BK	HO2S-11 (B1 S1) Signal	0.1-1.1v
13	RD/YL	TP Sensor Signal (VTA)	0.4-1.0v
14	GN/BK	ECT Sensor Signal (THW)	At 180°F: 0.51v
15-16	---	Not Used	---
17	BR	Power Ground (E1)	<0.1v
18	BR/WT	Sensor Ground (E2)	<0.050v
19	GN/WT	MAF Sensor Ground (EVG)	<0.050v
20	YL/BK	TP Sensor Signal (VTA2)	2.0-2.9v
21	RD	Accelerator Position Sensor 1	0.3-0.90v
22	YL/BK	IAT Sensor Signal (THA)	At 100°F: 2.60v
23 (01'-02')	RD/BK	HO2S-22 (B2 S2) Heater	1v (Heater on)
24 (01'-02')	BL	HO2S-12 (B1 S2) Heater	1v (Heater on)

1996-2000 4.0L V8 1UZ-FE VIN T E7 17P Connector

PCM Pin #	Wire Color	Circuit Description (17 Pin)	Value at Hot Idle
1	RD	A/T ECT Solenoid (S1)	In 3rd or OD: 1v
2	WT	A/T ECT Solenoid (S2)	In 1st or OD: 1v
3	GN	A/T ECT Solenoid (SL)	In Lockup: 12-14v
4	BL/RD	Direct Clutch Speed Input (+)	AC pulse signals
5	RD	A/T Vehicle Speed Sensor (+)	AC pulse signals
6-8	---	Not Used	---
9	GN/WT	A/T-ECT Solenoid (SLT+)	Pulse Signals
10	BL/WT	Direct Clutch Speed Input (-)	AC pulse signals
11	GN	A/T Vehicle Speed Sensor (-)	Pulse Signals
12-14	---	Not Used	---
15	GNBK	A/T-ECT Solenoid (SLT-)	Pulse Signals
16	---	Not Used	---
17	GN/YL	A/T Oil Temperature Sensor	At 68°F: 4-5v

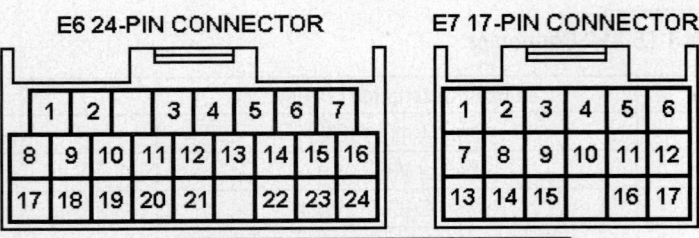

E6 24-PIN CONNECTOR   E7 17-PIN CONNECTOR

WIRE SIDE OF HARNESS TERMINALS

05533_ADIA_G121

Pin Connector Graphic

1996-2000 4.0L 1UZ-FE V8 VIN T E8 28P Connector

PCM Pin #	Wire Color	Circuit Description (28 Pin)	Value at Hot Idle
5	PK/RD	Data Link Connector (TE1)	12-14v
6	GN/WT	Stop Light Switch Signal	Brake Off: 0v, On: 12v
7	RD/BK	HO2S-22 (B2 S2) Heater	1v (Heater on)
8	BL	HO2S-12 (B1 S2) Heater	1v (Heater on)
9	---	Not Used	---
10	BL	EVAP Vapor Pressure Valve (VSV)	12v or 0v
11	---	Not Used	---
12	GN/WT	Defogger & Tail Light Switch	Switch Off: 0v, On: 12v
13	BL/BK	A/C Amplifier Signal (ACT)	Clutch On: 12v, Off: 1.5v
14	YL/BK	A/C Amplifier Signal (THWO)	A/C Off: 12v, On: 1v
15	VT	Vehicle Speed Sensor	At 55 mph: 48 Hz
16	BK	Tachometer Signal (TACO)	Pulse Signals
17	BK/RD	Starter Switch Signal	9-11v (cranking)
18	BK	HO2S-12 (B1 S2) Signal	0.1-1.1v
19	---	Not Used	---
20	BK/WT	Neutral Start Switch (NSW)	In P/N: 0-3.0v
21	WT/BK	Power Ground (EOM)	<0.1v
22	BL/BK	EVAP Vapor Pressure Sensor (PTNK)	2.5-3.1v (with cap off)
23-24	---	Not Used	---
25	WT/GN	A/C Amplifier Signal (AC1)	Clutch On: 1.5v, Off: 12v
26	---	Not Used	---
27	WT	HO2S-22 (B2 S2) Signal	0.1-1.1v
28	---	Not Used	---

1996-2000 4.0L 1UZ-FE V8 VIN T E9 22P Connector

PCM Pin #	Wire Color	Circuit Description (22 Pin)	Value at Hot Idle
1	RD/YL	Direct Battery	12-14v
2-3	---	Not Used	---
4	GN/RD	Fuel Pump Signal (DI)	12-14v
5	GN/WT	Fuel Control Switch Signal	Closed: 0v, Open: 12v
6	WT	Malfunction Indicator Lamp Control	MIL Off: 12v, On: 1v
7	YL/BK	BM (+) Power	12-14v
8	BK/YL	EFI Main Relay B1+	12-14v

1996-2000 4.0L 1UZ-FE V8 VIN T E9 22P Connector, *continued*

9	BK/RD	Ignition Switch Power (IGSW)	12-14v
10	BK/WT	EFI Main Relay Control	Relay Off: 0v, On: 12v
11	PK/WT	SIL Signal (Scan Tool)	12v
12	BL/BK	Transponder Amplifier Signal (Code)	Inserting key: pulses
13	VT/GN	Transponder Amplifier Signal (RXCK)	Inserting key: pulses
14	RD/YL	Transponder Amplifier Signal (RXCK)	Inserting key: pulses
15	---	Not Used	---
16	BK/YL	EFI Main Relay B+	12-14v
18-19	---	Not Used	---
20	RD/BK	Unlock Warning Switch	Key In: 1.5v, Out: 4-5v
21	GN/RD	Center ECU LED Signal	LED Off: 12v, On: 1v
22	---	Not Used	---

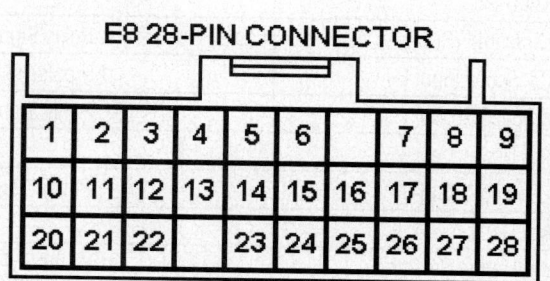

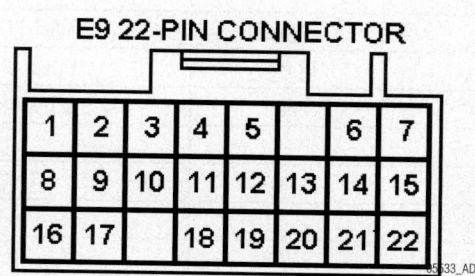

Pin Connector Graphic

SC 430 PIN CHARTS

2001-05 4.3L 3UZ-FE V8 VIN N E5 31P Connector

PCM Pin #	Wire Color	Circuit Description (24 Pin)	Value at Hot Idle
1	WT/BK	Power Ground (E03)	<0.1v
2	BL/RD	Sensor VREF (VC)	4.9-5.1v
3	YL	HO2S-21 (B2 S1) Heater	1v (Heater on)
4	RD	HO2S-11 (B1 S1) Heater	1v (Heater on)
5	YL	Injector 1 Control	2.0-3.3 ms
6	BK	Injector 2 Control	2.0-3.3 ms
7	BL/BK	EVAP Purge Solenoid (VSV)	12v or 0v
8	RD	ABS ECU Signal	Digital Signals
9	RD/BK	Accelerator Position Sensor 2	1.8-2.7v
10	BL/YL	Mass Airflow Sensor (VG)	1.1-1.5v
11	WT	HO2S-21 (B2 S1) Signal	0.1-1.1v
12	BK	HO2S-11 (B1 S1) Signal	0.1-1.1v
13	RD/YL	TP Sensor Signal (VTA)	0.4-1.0v
14	GN/BK	ECT Sensor Signal (THW)	At 180°F: 0.51v
15-16	---	Not Used	---
17	BR	Power Ground (E1)	<0.1v
18	BR/WT	Sensor Ground (E2)	<0.050v
19	GN/WT	MAF Sensor Ground (EVG)	<0.050v

2001-05 4.3L 3UZ-FE V8 VIN N E5 31P Connector, *continued*

20	YL/BK	TP Sensor Signal (VTA2)	2.0-2.9v
21	RD	Accelerator Position Sensor 1	0.3-0.90v
22	YL/BK	IAT Sensor Signal (THA)	At 100°F: 2.60v
23 (01'-02')	RD/BK	HO2S-22 (B2 S2) Heater	1v (Heater on)
24 (01'-02')	BL	HO2S-12 (B1 S2) Heater	1v (Heater on)

2001-05 4.3L 3UZ-FE V8 VIN N E7 17P Connector

PCM Pin #	Wire Color	Circuit Description (17 Pin)	Value at Hot Idle
1	RD	A/T ECT Solenoid (S1)	In 3rd or OD: 1v
2	WT	A/T ECT Solenoid (S2)	In 1st or OD: 1v
3	GN	A/T ECT Solenoid (SL)	In Lockup: 12-14v
4	BL/RD	Direct Clutch Speed Input (+)	AC pulse signals
5	RD	A/T Vehicle Speed Sensor (+)	AC pulse signals
6-8	---	Not Used	---
9	GN/WT	A/T-ECT Solenoid (SLT+)	Pulse Signals
10	BL/WT	Direct Clutch Speed Input (-)	AC pulse signals
11	GN	A/T Vehicle Speed Sensor (-)	Pulse Signals
12-14	---	Not Used	---
15	GNBK	A/T-ECT Solenoid (SLT-)	Pulse Signals
16	---	Not Used	---
17	GN/YL	A/T Oil Temperature Sensor	At 68°F: 4-5v

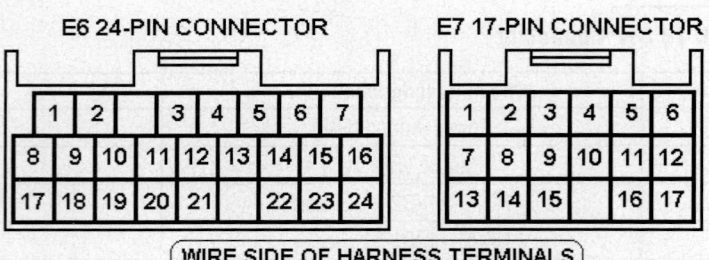

05533_ADIA_G121

Pin Connector Graphic

2001-05 4.3L 3UZ-FE V8 VIN N E8 28P Connector

PCM Pin #	Wire Color	Circuit Description (28 Pin)	Value at Hot Idle
5	PK/RD	Data Link Connector (TE1)	12-14v
6	GN/WT	Stop Light Switch Signal	Brake Off: 0v, On: 12v
7	RD/BK	HO2S-22 (B2 S2) Heater	1v (Heater on)
8	BL	HO2S-12 (B1 S2) Heater	1v (Heater on)
9	---	Not Used	---
10	BL	EVAP Vapor Pressure Valve (VSV)	12v or 0v
11	---	Not Used	---
12	GN/WT	Defogger & Tail Light Switch	Switch Off: 0v, On: 12v
13	BL/BK	A/C Amplifier Signal (ACT)	Clutch On: 12v, Off: 1.5v
14	YL/BK	A/C Amplifier Signal (THWO)	A/C Off: 12v, On: 1v
15	VT	Vehicle Speed Sensor	At 55 mph: 48 Hz
16	BK	Tachometer Signal (TACO)	Pulse Signals
17	BK/RD	Starter Switch Signal	9-11v (cranking)
18	BK	HO2S-12 (B1 S2) Signal	0.1-1.1v
19	---	Not Used	---
20	BK/WT	Neutral Start Switch (NSW)	In P/N: 0-3.0v
21	WT/BK	Power Ground (EOM)	<0.1v
22	BL/BK	EVAP Vapor Pressure Sensor (PTNK)	2.5-3.1v (with cap off)
23-24	---	Not Used	---
25	WT/GN	A/C Amplifier Signal (AC1)	Clutch On: 1.5v, Off: 12v
26	---	Not Used	---
27	WT	HO2S-22 (B2 S2) Signal	0.1-1.1v
28	---	Not Used	---

Pin Connector Graphic

05533_ADIA_G122

2001-05 4.3L 3UZ-FE V8 VIN N E9 22P Connector

PCM Pin #	Wire Color	Circuit Description (22 Pin)	Value at Hot Idle
1	RD/YL	Direct Battery	12-14v
2-3	---	Not Used	---
4	GN/RD	Fuel Pump Signal (DI)	12-14v
5	GN/WT	Fuel Control Switch Signal	Closed: 0v, Open: 12v
6	WT	Malfunction Indicator Lamp Control	MIL Off: 12v, On: 1v
7	YL/BK	BM (+) Power	12-14v
8	BK/YL	EFI Main Relay B1+	12-14v
9	BK/RD	Ignition Switch Power (IGSW)	12-14v
10	BK/WT	EFI Main Relay Control	Relay Off: 0v, On: 12v
11	PK/WT	SIL Signal (Scan Tool)	12v
12	BL/BK	Transponder Amplifier Signal (Code)	Inserting key: pulses
13	VT/GN	Transponder Amplifier Signal (RXCK)	Inserting key: pulses
14	RD/YL	Transponder Amplifier Signal (RXCK)	Inserting key: pulses
15	---	Not Used	---
16	BK/YL	EFI Main Relay B+	12-14v
18-19	---	Not Used	---
20	RD/BK	Unlock Warning Switch	Key In: 1.5v, Out: 4-5v
21	GN/RD	Center ECU LED Signal	LED Off: 12v, On: 1v
22	---	Not Used	---

SCION
DIAGNOSTIC TROUBLE CODES

8

OBD II VEHICLE APPLICATIONS

SCION

tC

2005–2006
  2.4L .... VIN D, E

xA, xB

2004–2006
  1.5L .... VIN T

DIAGNOSTIC TROUBLE CODES

OBD II Trouble Code List (P0xxx Codes)

DTC	Trouble Code Title, Conditions & Possible Causes
DTC: P2A00 2T CCM, MIL: Yes 2005, 2006 **Models:** tC **Engines:** 2.4L VIN D, E **Transmissions:** All	**Air Fuel Sensor (Bank 1 Sensor 1) Signal Slow Response Conditions:** Vehicle driven at cruise speed at over 1400 rpm in closed loop at 60 mph, and the PCM detected an unexpected voltage condition on the Bank 1 Air Fuel Sensor 1 (AFS1) circuit. **Possible Causes:** • A/F sensor connector is damaged or loose • A/F sensor circuit is open or shorted, or the sensor has failed • A/F sensor heater is damaged or has failed • A/F sensor heater relay circuit is open or the relay has failed • Fuel delivery component has failed (fuel pressure regulator, one or more fuel injectors is leaking or severely restricted) • Induction system problems (air leaks or restricted air filter) • PCM has failed
DTC: P0010 2T CCM, MIL: Yes 2004, 2005, 2006 **Models:** xA, xB **Engines:** 1.5L VIN T **Transmissions:** All	**VVT Oil Control Circuit Malfunction (Bank 1) Conditions:** Key on or engine running; and the PCM detected an unexpected voltage condition on the VVT Oil Control Valve Bank 1 circuit. The VVT system controls the intake camshaft in order to provide optimal valve timing during all conditions based signals from the ECT, IAT and TP sensor. The VVT regulates the intake camshaft angle using oil pressure through the Oil Control Valve. This results in the relative position between the camshaft and crankshaft to become optimal. The result is higher torque, better fuel economy and low emissions. **Possible Causes:** • OCV assembly connector is damaged or loose • OCV assembly control circuit is open or shorted to ground • OCV assembly is damaged or has failed • PCM has failed
DTC: P0011 2T CCM, MIL: Yes 2004, 2005, 2006 **Models:** xA, xB **Engines:** 1.5L VIN T **Transmissions:** All	**Camshaft Position 'A' Over-Advanced Or System Performance (Bank 1) Conditions:** Engine started, ECT sensor more than 158°F, vehicle driven at an engine speed of 400-4000 rpm, and the PCM detected the valve timing did not change from the "current" valve timing, or the valve timing remain fixed during testing. The VVT system controls the intake camshaft in order to provide optimal valve timing during all conditions based signals from the ECT, IAT and TP sensor. The VVT regulates the intake camshaft angle using oil pressure through the Oil Control Valve. This results in the relative position between the camshaft and crankshaft to become optimal. The result is better engine torque, fuel economy and lower emissions. **Possible Causes:** • Engine valve timing malfunction • Camshaft timing oil control valve unit is damaged or has failed • PCM or VVT ECM has failed
DTC: P0012 2T CCM, MIL: Yes 2004, 2005, 2006 **Models:** xA, xB **Engines:** 1.5L VIN T **Transmissions:** All	**Camshaft Position 'A' Over-Retarded (Bank 1) Conditions:** Engine started, ECT sensor more than 158°F, vehicle driven at an engine speed of 400-4000 rpm, and the PCM detected the valve timing did not change from the "current" valve timing, or that the valve timing remain fixed during the test period. The VVT system controls the intake camshaft in order to provide optimal valve timing during all conditions. The VVT regulates the intake camshaft angle using oil pressure through the OCV. This causes the relative position between the camshaft and crankshaft to become optimal. The result is improved engine torque, better fuel economy and low emissions. **Possible Causes:** • Engine valve timing malfunction • Camshaft timing oil control valve unit is damaged or has failed • PCM or VVT ECM has failed
DTC: P0016 2T CCM, MIL: Yes 2004, 2005, 2006 **Models:** xA, xB **Engines:** 1.5L VIN T **Transmissions:** All	**Camshaft Position-Crankshaft Position (Bank 1 Sensor A) Conditions:** Engine started, engine running, and the PCM detected a deviation between the crankshaft position sensor signal and the VVT Sensor 1 signal during the test period. The crankshaft position (NE) sensor consists of a magnet, iron core and pickup coil. The NE sensor signal plate, installed on the crankshaft-timing pulley, has 34 teeth. This sensor generates 34 signals for each engine revolution. The PCM detects the crankshaft angle and engine speed based on the NE signal. It detects the correct cylinder based on signals from the VVT 1 sensor along signals from the crankshaft position sensor. **Possible Causes:** • Engine valve timing problem • Engine timing belt mechanical problem (skipped teeth or belt) • PCM has failed
DTC: P0031 1T CCM, MIL: Yes 2004, 2005, 2006 **Models:** tC, xA, xB **Engines:** 1.5L VIN T, 2.4L VIN D, E **Transmissions:** All	**Heated Oxygen Sensor (Bank 1 Sensor 1) Heater Circuit Low Code Conditions:** Engine started, and the PCM detected the HO2S-11 heater control circuit indicated less than 0.20 amps during the CCM test period. **Possible Causes:** • HO2S-11 heater control circuit is open • HO2S-11 heater assembly is damaged or has failed • PCM has failed

DTC	Trouble Code Title, Conditions & Possible Causes
DTC: P0032 **1T CCM, MIL: Yes** **2004, 2005, 2006** **Models:** tC, xA, xB **Engines:** 1.5L VIN T, 2.4L VIN D, E **Transmissions:** All	**Heated Oxygen Sensor (Bank 1 Sensor 1) Heater Circuit High Input Conditions:** Engine started, and the PCM detected the HO2S-11 heater control circuit indicated more than 2.0 amps during the CCM test period. **Possible Causes:** • HO2S-11 heater control circuit is shorted to ground • HO2S-11 heater assembly is damaged or has failed • PCM has failed
DTC: P0037 **2T CCM, MIL: Yes** **2004, 2005, 2006** **Models:** tC, xA, xB **Engines:** 1.5L VIN T, 2.4L VIN D, E **Transmissions:** All	**Heated Oxygen Sensor (Bank 1 Sensor 2) Heater Circuit Low Input Conditions:** Engine started, and the PCM detected the HO2S-12 heater control circuit indicated less than 0.20 amps during the CCM test period. **Possible Causes:** • HO2S-12 heater control circuit is open • HO2S-12 heater assembly is damaged or has failed • PCM has failed
DTC: P0038 **2T CCM, MIL: Yes** **2004, 2005, 2006** **Models:** tC, xA, xB **Engines:** 1.5L VIN T, 2.4L VIN D, E **Transmissions:** All	**Heated Oxygen Sensor (Bank 1 Sensor 2) Heater Circuit High Input Conditions:** Engine started, and the PCM detected the HO2S-12 heater control circuit indicated more than 2.0 amps during the CCM test period. **Possible Causes:** • HO2S-12 heater control circuit is shorted to ground • HO2S-12 heater assembly is damaged or has failed • PCM has failed
DTC: P0100 **1T CCM, MIL: Yes** **2004, 2005, 2006** **Models:** tC, xA, xB **Engines:** 1.5L VIN T, 2.4L VIN D, E **Transmissions:** All	**Mass Airflow Sensor Circuit Malfunction Conditions:** Engine started, engine running at under 4000 rpm, and the PCM detected an unexpected low or high voltage condition on the Mass Airflow (MAF) sensor circuit for over 3 seconds during the CCM test. The MAF sensor on this engine includes a hot wire assembly with an air temperature sensor, platinum hot wire and control unit mounted in a plastic housing. This airflow meter works on the principle that the hot wire and temperature sensor located in the intake air bypass of the housing detect any changes in the (incoming) air temperature. **Possible Causes:** • MAF sensor signal circuit is open, shorted to ground or power • MAF sensor ground circuit is open between sensor and ground • MAF sensor power circuit is open (check the power to the relay) • MAF sensor has failed, or the PCM has failed
DTC: P0101 **2T CCM, MIL: Yes** **2004, 2005, 2006** **Models:** tC, xA, xB **Engines: 1.5L VIN T, 2.4L VIN D, E** **Transmissions:** All	**Mass Airflow Sensor Signal Range/Performance Conditions:** DTC P0100 not set, engine speed under 900 rpm, throttle valve closed, ECT sensor over 158°F, and the PCM detected the MAF was above 2.20v; or with the engine speed over 1500 rpm, the throttle valve closed and the TP sensor over 0.63v, the MAF sensor was less than 1.06v. This airflow meter works on the principle that the hot wire and temperature sensor in the intake air bypass of the housing detect any changes in the (incoming) air temperature **Possible Causes:** • MAF sensor signal circuit is open, shorted to ground or power • MAF sensor is contaminated, damaged or it has failed • PCM has failed
DTC: P0102 **2T CCM, MIL: Yes** **2004, 2005, 2006** **Models:** xA, xB **Engines:** 1.5L VIN T **Transmissions:** All	**Mass Airflow Sensor Circuit Low Input Conditions:** DTC P0100 not set, engine started, and the PCM detected an unexpected low voltage condition on the MAF sensor circuit during the CCM test period. This airflow meter works on the principle that the hot wire and temperature sensor located in the intake air bypass of the housing detect any changes in the (incoming) air temperature **Possible Causes:** • MAF sensor signal circuit is open or shorted to ground • MAF sensor is contaminated, damaged or it has failed • PCM has failed
DTC: P0103 **2T CCM, MIL: Yes** **2004, 2005, 2006** **Models:** xA, xB **Engines:** 1.5L VIN T **Transmissions:** All	**Mass Airflow Sensor Circuit High Input Conditions:** DTC P0100 not set, engine started, and the PCM detected an unexpected high voltage condition on the MAF sensor circuit during the CCM test period. This airflow meter works on the principle that the hot wire and temperature sensor located in the intake air bypass of the housing detect any changes in the (incoming) air temperature **Possible Causes:** • MAF sensor signal circuit is shorted to power • MAF sensor is contaminated, damaged or it has failed • PCM has failed
DTC: P0105 **1T CCM, MIL: Yes** **2005, 2006** **Models:** tC **Engines:** VIN G, 2.4L VIN D, E **Transmissions:** All	**Manifold Air Pressure Sensor Circuit Malfunction Conditions:** Key on or engine running; and the PCM detected the MAP sensor signal was less than 0 kPa or more than 130 kPa in the CCM test. **Possible Causes:** • MAP sensor signal circuit is open, shorted to ground or power • MAP sensor ground circuit is open between sensor and ground • MAP sensor is damaged or has failed • PCM has failed

DTC	Trouble Code Title, Conditions & Possible Causes
DTC: P0106 **1T CCM, MIL: Yes** **2005, 2006** **Models:** tC **Engines:** 2.4L VIN D, E **Transmissions:** All	**Manifold Air Pressure Sensor Signal Range/Performance Conditions:** DTC P0105 not set, engine started, engine speed at 400-1000 rpm, closed throttle switch indicating "on", ECT sensor input over 158°F, and the PCM detected the MAP sensor was above 3.3v (10 sec's), or it less below 1.0v at under 2500 rpm (VTA over 1.82v for 5 sec's). **Possible Causes:** • MAP sensor source vacuum line is leaking or restricted • MAP sensor source vacuum line is plugged at intake manifold • MAP sensor is damaged, out-of-calibration or has failed • PCM has failed
DTC: P0110 **1T CCM, MIL: Yes** **2004, 2005, 2006** **Models:** tC, xA, xB **Engines:** 1.5L VIN T, 2.4L VIN D, E **Transmissions:** All	**Intake Air Temperature Sensor Circuit Malfunction Conditions:** Engine started, and the PCM detected an unexpected low or high voltage on the IAT sensor circuit (Scan Tool reads -40°F or 284°F). This sensor is located inside the MAF sensor. **Possible Causes:** • IAT sensor signal circuit is open, shorted to ground or VREF • IAT sensor ground circuit is open between sensor and ground • IAT sensor is contaminated, damaged or has failed • PCM has failed
DTC: P0112 **1T CCM, MIL: Yes** **2004, 2005, 2006** **Models:** xA, xB **Engines:** 1.5L VIN T **Transmissions:** All	**Intake Air Temperature Sensor Circuit Low Input Conditions:** Key on or engine running; and the PCM detected an unexpected low voltage condition on the IAT sensor circuit for over 500 ms. **Possible Causes:** • IAT sensor connector is damaged (it may be shorted internally) • IAT sensor ground circuit is shorted to ground • IAT sensor is damaged or has failed • PCM has failed
DTC: P0113 **1T CCM, MIL: Yes** **2004, 2005, 2006** **Models:** xA, xB **Engines:** 1.5L VIN T **Transmissions:** All	**Intake Air Temperature Sensor Circuit High Input Conditions:** Key on or engine running; and the PCM detected an unexpected high voltage condition on the IAT sensor circuit for over 500 ms. **Possible Causes:** • IAT sensor connector is damaged (it may be open internally) • IAT sensor ground circuit is open • IAT sensor is damaged or has failed • PCM has failed
DTC: P0115 **2T CCM, MIL: Yes** **2004, 2005, 2006** **Models:** tC, xA, xB **Engines:** 1.5L VIN T, 2.4L VIN D, E **Transmissions:** All	**Engine Coolant Temperature Sensor Circuit Malfunction Conditions:** Key on or engine running; and the PCM detected an unexpected "low" or "high" voltage on the ECT sensor circuit (Scan Tool reads less than -40°F or low voltage or more than 284°F for high voltage). **Possible Causes:** • IAT sensor signal circuit is open, shorted to ground or VREF • ECT sensor ground circuit is open between sensor and ground • ECT sensor is contaminated, damaged or has failed • PCM has failed
DTC: P0116 **2T CCM, MIL: Yes** **2004, 2005, 2006** **Models:** tC, xA, xB **Engines:** 1.5L VIN T, 2.4L VIN D, E **Transmissions:** All	**Engine Coolant Temperature Sensor Range/Performance Conditions:** Engine started, ECT sensor from 95°F to 140°F, IAT sensor more than 19.9°F, vehicle driven with several changes in the VSS signals, and the PCM detected the ECT sensor signal did not increase more than 37.4°F after engine was started and the test period completed. **Possible Causes:** • Check for problems in the cooling system (i.e., coolant, the fan) • ECT sensor signal circuit or ground circuit has high resistance • ECT sensor is contaminated, damaged or has failed • PCM has failed
DTC: P0117 **2T CCM, MIL: Yes** **2004, 2005, 2006** **Models:** xA, xB **Engines:** 1.5L VIN T **Transmissions:** All	**Engine Coolant Temperature Sensor Circuit Low Input Conditions:** Key on or engine running; and the PCM detected an unexpected low voltage condition on the ECT sensor (Scan Tool reads below -40°F). **Possible Causes:** • ECT sensor signal circuit is shorted to ground • ECT sensor is damaged or has failed • PCM has failed

DTC	Trouble Code Title, Conditions & Possible Causes
DTC: P0117 **2T CCM, MIL: Yes** **2004, 2005, 2006** **Models:** xA, xB **Engines:** 1.5L VIN T **Transmissions:** All	**Engine Coolant Temperature Sensor Circuit High Input Conditions:** Key on or engine running; and the PCM detected an unexpected high voltage condition on the ECT sensor (Scan Tool reads over 284°F). **Possible Causes:** • ECT sensor signal circuit is open • ECT sensor ground circuit is open • ECT sensor is damaged or has failed • PCM has failed
DTC: P0120 **2T CCM, MIL: Yes** **2004, 2005, 2006** **Models:** tC, xA, xB **Engines:** 1.5L VIN T, 2.4L VIN D, E **Transmissions:** All	**TP Sensor or Switch 'A' Circuit Malfunction Conditions:** Engine started, and the PCM detected the TP sensor signal (VTA) was under 0.10v or over 4.90v. The TP sensor, mounted on the throttle body, detects the Throttle Valve opening angle (0.30v-0.70v closed). The PCM uses this signal for A/F ratio correction and power increase changes during all modes of engine operation. **Possible Causes:** • VTA sensor signal circuit is open or shorted to ground • VTA sensor ground circuit is open • VTA power circuit (VREF) is open • TP sensor is damaged or has failed • PCM has failed
DTC: P0121 **2T CCM, MIL: Yes** **2004, 2005, 2006** **Models:** tC, xA, xB **Engines:** 1.5L VIN T, 2.4L VIN D, E **Transmissions:** All	**TP Sensor or TP Switch 'A' Signal Range/Performance Conditions:** Vehicle speed more than 19 mph at least once; and the PCM detected the TP sensor input was out of the applicable range with the VSS input reading between 30 mph and 0 mph. **Possible Causes:** • TP sensor signal circuit open or shorted to ground (intermittent) • TP sensor is loose at it mounting or the throttle is binding • TP sensor is damaged or has failed (perform a sweep test) • PCM has failed
DTC: P0125 **1T CCM, MIL: Yes** **2004, 2005, 2006** **Models:** tC, xA, xB **Engines:** 1.5L VIN T, 2.4L VIN D, E **Transmissions:** All	**Insufficient Coolant Temperature For Closed Loop Conditions:** **California Models** ECT sensor less than 19.4°F, engine started, engine runtime over 20 minutes, and the PCM detected the ECT sensor indicated 68°F or less; or with the ECT sensor from 19.4°F to 50.0°F at startup, engine runtime over 5 minutes, the PCM detected the ECT sensor indicated 68°F or less; or with the ECT sensor more than 19.4°F at startup, and the engine runtime over 5 minutes, the PCM detected the ECT sensor signal was 68°F or less; or with the ECT sensor over 50°F at startup, engine runtime over 2 minutes, the PCM detected the ECT sensor signal did not reach 86°F during the CCM test period. **Possible Causes:** • Check the operation of the thermostat (it may be stuck open) • ECT sensor signal circuit has high resistance • ECT sensor has failed • Inspect for low coolant level for an incorrect coolant mixture
DTC: P0128 **2T CCM, MIL: Yes** **2004, 2005, 2006** **Models:** tC, xA, xB **Engines:** 1.5L VIN T, 2.4L VIN D, E **Transmissions:** All	**Thermostat System Malfunction Conditions:** Engine started, ECT sensor signal below 140°F at startup, and the PCM detected the ECT sensor did not reach 167°F after the warm-up period expired (engine runtime 5-10 minutes). **Possible Causes:** • Check the operation of the thermostat (it may be stuck open) • ECT sensor is out-of-calibration or skewed • Inspect for low coolant level or for an incorrect coolant mixture
DTC: P0130 **2T CCM, MIL: Yes** **2004, 2005, 2006** **Models:** tC, xA, xB **Engines:** 1.5L VIN T, 2.4L VIN D, E **Transmissions:** All	**HO2S-11 (Bank 1 Sensor 1) Circuit Malfunction Conditions:** Engine started, engine warm-up completed, engine idling in closed loop, and the PCM detected the HO2S-11 signal was fixed at 400 mv or higher, or that it was fixed at less than 550 mv during the test. **Possible Causes:** • HO2S signal circuit is open between the sensor and the PCM • HO2S signal circuit is shorted to sensor ground or to VREF • HO2S is damaged, contaminated or it has failed
DTC: P0133 **2T O2S, MIL: Yes** **2004, 2005, 2006** **Models:** tC, xA, xB **Engines:** 1.5L VIN T, 2.4L VIN D, E **Transmissions:** All	**HO2S-11 (Bank 1 Sensor 1) Slow Response Conditions:** Engine started, engine idling in closed loop, and the PCM detected the HO2S-11 response time from rich-to-lean or from lean-to-rich was 1.1 second or longer during the CCM test. **Note: This fault must be detected at least 3 times at idle speed.** **Possible Causes:** • HO2S signal circuit is open or shorted to ground • HO2S element is contaminated, or HO2S heater has failed • Intake air leaks, exhaust manifold leaks or PCV system leaks • MAF sensor out of calibration (it may be dirty or contaminated)

DTC	Trouble Code Title, Conditions & Possible Causes
DTC: P0134 **2T O2S, MIL: Yes** **2004, 2005, 2006** **Models:** tC, xA, xB **Engines:** 1.5L VIN T, 2.4L VIN D, E **Transmissions:** All	**Air Fuel Sensor 1 (Bank 1 Sensor 1) Circuit No Activity Detected Conditions:** Engine started, engine runtime over 140 seconds, vehicle driven at a steady speed of 25-81 mph at over 1500 rpm with the throttle valve open, and the PCM detected the A/FS-11 signal did not indicate rich (more than 450 mv) after 65 seconds under these conditions. **Possible Causes:** • Air leaks present in the PCV valve, hoses or hose connections • A/FS1 connector is damaged or loose • A/FS1 signal circuit is open or shorted to ground • A/FS1 is damaged, contaminated or it has failed • Fuel Control component faults (sticking injector, low pressure) • Gas leaks in the exhaust system in front of the Oxygen sensor • PCM has failed
DTC: P0135 **2T CCM, MIL: Yes** **2004, 2005, 2006** **Models:** xA, xB **Engines:** 1.5L VIN T **Transmissions:** All	**HO2S-11 (Bank 1 Sensor 1) Heater Circuit Malfunction Conditions:** Engine started, engine running, and the PCM detected the HO2S-11 heater current exceeded 2 amps, or that it was 0.25 amps or less. **Possible Causes:** • HO2S heater control circuit is open or shorted to ground • HO2S heater control circuit is shorted to power • HO2S heater power circuit is open (check power from the relay) • HO2S heater is damaged or has failed • PCM has failed
DTC: P0135 **2T CCM, MIL: Yes** **2005, 2006** **Models:** tC **Engines:** 2.4L VIN D, E **Transmissions:** All	**Air Fuel Sensor 1 (Bank 1 Sensor 1) Heater Circuit Malfunction Conditions:** Engine started, engine running, and the PCM detected the A/FS-11 heater current indicated over 8 amps, or it was less than 0.25 amps. **Possible Causes:** • A/FS1 heater control circuit is open or shorted to ground • A/FS1 heater control circuit is shorted to power • A/FS1 heater power circuit is open (check power from relay) • A/FS1 heater is damaged or has failed • PCM has failed
DTC: P0136 **2T CCM, MIL: Yes** **2004, 2005, 2006** **Models:** xA, xB **Engines:** 1.5L VIN T **Transmissions:** All	**HO2S-12 (Bank 1 Sensor 2) Circuit Malfunction Conditions:** Vehicle driven to a speed of over 25 mph while in closed loop, and the PCM detected the HO2S-12 signal was fixed at more than 400 mv, or that it was fixed at less than 600 mv. **Possible Causes:** • HO2S signal circuit is open or shorted to ground • HO2S signal circuit shorted to VREF or system power (B+) • HO2S is damaged, contaminated or it has failed
DTC: P0136 **2T CCM, MIL: Yes** **2005, 2006** **Models:** tC **Engines:** 2.4L VIN D, E **Transmissions:** All	**HO2S-12 (Bank 1 Sensor 2) Circuit Malfunction Conditions:** Engine started, vehicle driven to a speed of over 31 mph while in closed loop, and the PCM detected the HO2S-12 signal indicated more than 400 mv, or it indicated less than 500 mv during the test. **Possible Causes:** • HO2S signal circuit is open or shorted to ground • HO2S signal circuit shorted to VREF or system power (B+) • HO2S is damaged, contaminated or it has failed
DTC: P0141 **2T CCM, MIL: Yes** **2004, 2005, 2006** **Models:** tC, xA, xB **Engines:** 1.5L VIN T, 2.4L VIN D, E **Transmissions:** All	**HO2S-12 (Bank 1 Sensor 2) Heater Circuit Malfunction Conditions:** Engine started, engine running, and the PCM detected the HO2S-12 heater current exceeded 2 amps, or that it was 0.25 amps or less. **Possible Causes:** • HO2S heater control circuit is open, shorted to ground or power • HO2S heater power circuit is open (check power from the relay) • HO2S heater is damaged or has failed • PCM has failed
DTC: P0171 **2T FUEL, MIL: Yes** **2005, 2006** **Models:** tC **Engines:** 2.4L VIN D, E **Transmissions:** All	**Fuel System Too Lean (Bank 1) Conditions:** DTC P0100, P0101, P0105, P0110, P0115, P0120, P0121, P0130, P0133, P0136, P0135, P0136, P0141, P0151, P0156, P0161, P0300, P0301-P0306, P0440, P0500 and P0505 not set, vehicle speed less than 62 mph with the engine speed over 1500 rpm, ECT sensor more than 158°F, and the PCM detected the lean fuel trim correction value was over the limit. **Possible Causes:** • A/FS or HO2S is contaminated, deteriorated or it has failed • Air leaks after the MAF sensor, or in the EGR or PCV system • Base engine "mechanical" fault affecting one or more cylinders • Exhaust leaks located in front of the A/FS or HO2S location • Fuel system supplying too little fuel during cruise or idle (faulty fuel pump or fuel filter) • Fuel injector (one or more) dirty or pressure regulator has failed • Vehicle driven low on fuel or until it ran out of fuel

DTC	Trouble Code Title, Conditions & Possible Causes
DTC: P0171 **2T FUEL, MIL: Yes** **2004, 2005, 2006** **Models:** xA, xB **Engines:** 1.5L VIN T **Transmissions:** All	**Fuel System Too Lean (Bank 1) Conditions:** DTC P0100, P0101, P0105, P0110, P0115, P0120, P0121, P0130, P0133, P0136, P0135, P0136, P0141, P0151, P0156, P0161, P0300, P0301-P0306, P0440, P0500 and P0505 not set, ECT sensor more than 158°F, vehicle driven at a constant speed of less than 62 mph with the engine speed over 1500 rpm, and the PCM detected the lean fuel trim correction value was over the limit. **Possible Causes:** • A/FS or HO2S is contaminated, deteriorated or it has failed • Air leaks after the MAF sensor, or in the EGR or PCV system • Base engine "mechanical" fault affecting one or more cylinders • Exhaust leaks located in front of the A/FS or HO2S location • Fuel control sensor is out of calibration (i.e., ECT, IAT or MAP) • Fuel delivery system supplying too little fuel during cruise or idle periods (e.g., faulty fuel pump or dirty, restricted fuel filter) • Fuel injector (one or more) dirty or pressure regulator has failed • Vehicle driven low on fuel or until it ran out of fuel
DTC: P0172 **2T FUEL, MIL: Yes** **2004, 2005, 2006** **Models:** xA, xB **Engines:** 1.5L VIN T **Transmissions:** All	**Fuel System Too Rich (Bank 1) Conditions:** DTC P0100, P0101, P0105, P0110, P0115, P0120, P0121, P0130, P0133, P0136, P0135, P0136, P0141, P0151, P0156, P0161, P0300, P0301-P0306, P0440, P0500 and P0505 not set, ECT sensor more than 158°F, vehicle driven at a constant speed of less than 62 mph with the engine speed over 1500 rpm, and the PCM detected the rich fuel trim correction value was over the limit. **Possible Causes:** • A/FS or HO2S is contaminated, deteriorated or it has failed • Base engine "mechanical" fault affecting one or more cylinders • EVAP system component has failed or canister fuel saturated • Exhaust leaks located in front of the A/FS or HO2S location • Fuel control sensor is out of calibration (i.e., ECT, IAT or MAF) • Fuel delivery system supplying too much fuel during cruise or idle periods (e.g., faulty fuel pump, or faulty pressure regulator) • Fuel injector(s) is leaking or stuck partially open (one or more)
DTC: P0172 **2T FUEL, MIL: Yes** **2005, 2006** **Models:** tC **Engines:** 2.4L VIN D, E **Transmissions:** All	**Fuel System Too Rich (Bank 1) Conditions:** DTC P0100, P0101, P0105, P0110, P0115, P0120, P0121, P0130, P0133, P0136, P0135, P0136, P0141, P0151, P0156, P0161, P0300, P0301-P0306, P0440, P0500 and P0505 not set, ECT sensor more than 158°F, vehicle driven at a constant speed of less than 62 mph with the engine speed over 1500 rpm, and the PCM detected the rich fuel trim correction value was over the limit. **Possible Causes:** • A/FS or HO2S is contaminated, deteriorated or it has failed • Base engine "mechanical" fault affecting one or more cylinders • EVAP system component has failed or canister fuel saturated • Exhaust leaks located in front of the A/FS or HO2S location • Fuel control sensor is out of calibration (i.e., ECT, IAT or MAF) • Fuel delivery system supplying too much fuel during cruise or idle periods (e.g., faulty fuel pump, or faulty pressure regulator) • Fuel injector(s) is leaking or stuck partially open (one or more)
DTC: P0300 **2T MISFIRE, MIL: Yes** **2004, 2005, 2006** **Models:** tC, xA, xB **Engines:** 1.5L VIN T, 2.4L VIN D, E **Transmissions:** All	**Multiple Cylinder Misfire Detected Conditions:** **Trouble Code Conditions** DTC P0100, P0101, P0102, P0103, P0105, P0110, P0112, P0113, P0115, P0117, P0118, P0120, P0121, P0122, P0123, P0125, P0335, P0340, P0500, P0505 and P0510 not set, engine started, vehicle driven to a speed of over 3 mph for 1 minute, and the PCM detected a misfire rate of 1-2% (High Emissions 2T), or a misfire rate of 6-30% (Catalyst Damaging 1T) in two or more cylinders. **Note: If the misfire is severe, the MIL will flash on/off on the 1st trip! Look at the misfire ratio for all of the cylinders on the Scan Tool. The cylinder with the highest misfire ratio should be checked first!** **Possible Causes:** • Air leak in the intake manifold, or in the EGR or PCV system • Base engine mechanical fault that affects two or more cylinders • EGR valve is stuck open or the PCV system has a vacuum leak • Fuel delivery component fault that affects two or more cylinders (e.g., contaminated, dirty or sticking fuel injectors) • Ignition system fault (coil or plug) that affects several cylinders • Mass airflow meter is contaminated, or its signal is out of range

DTC	Trouble Code Title, Conditions & Possible Causes
DTC: P0301 **2T MISFIRE, MIL: Yes** **2004, 2005, 2006** **Models:** tC, xA, xB **Engines:** 1.5L VIN T, 2.4L VIN D, E **Transmissions:** All	**Cylinder 1 Misfire Detected Conditions:** **Trouble Code Conditions** DTC P0100, P0101, P0102, P0103, P0105, P0110, P0112, P0113, P0115, P0117, P0118, P0120, P0121, P0122, P0123, P0125, P0335, P0340, P0500, P0505 and P0510 not set, engine started, vehicle driven to a speed of over 3 mph for 1 minute, and the PCM detected a misfire rate of 1-2% (High Emissions 2T), or a misfire rate of 6-30% (Catalyst Damaging 1T) in Cylinder 1. **Note: If the misfire is severe, the MIL will flash on/off on the 1st trip** **Possible Causes:** • Base engine mechanical fault that affects only Cylinder 1 • EGR valve is stuck open or the PCV system has a vacuum leak • Fuel component fault that affects only Cylinder 1 (a contaminated or sticking injector) • Ignition system problem (coil or plug) that affects Cylinder 1
DTC: P0302 **2T MISFIRE, MIL: Yes** **2004, 2005, 2006** **Models:** tC, xA, xB **Engines:** 1.5L VIN T, 2.4L VIN D, E **Transmissions:** All	**Cylinder 2 Misfire Detected Conditions:** **Trouble Code Conditions** DTC P0100, P0101, P0102, P0103, P0105, P0110, P0112, P0113, P0115, P0117, P0118, P0120, P0121, P0122, P0123, P0125, P0335, P0340, P0500, P0505 and P0510 not set, engine started, vehicle driven to a speed of over 3 mph for 1 minute, and the PCM detected a misfire rate of 1-2% (High Emissions 2T), or a misfire rate of 6-30% (Catalyst Damaging 1T) in Cylinder 2. **Note: If the misfire is severe, the MIL will flash on/off on the 1st trip!** **Possible Causes:** • Base engine mechanical fault that affects only Cylinder 2 • EGR valve is stuck open or the PCV system has a vacuum leak • Fuel component fault that affects only Cylinder 2 (a contaminated or sticking injector) • Ignition system problem (coil or plug) that affects Cylinder 2
DTC: P0303 **2T MISFIRE, MIL: Yes** **2004, 2005, 2006** **Models:** tC, xA, xB **Engines:** 1.5L VIN T, 2.4L VIN D, E **Transmissions:** All	**Cylinder 3 Misfire Detected Conditions:** **Trouble Code Conditions** DTC P0100, P0101, P0102, P0103, P0105, P0110, P0112, P0113, P0115, P0117, P0118, P0120, P0121, P0122, P0123, P0125, P0335, P0340, P0500, P0505 and P0510 not set, engine started, vehicle driven to a speed of over 3 mph for 1 minute, and the PCM detected a misfire rate of 1-2% (High Emissions 2T), or a misfire rate of 6-30% (Catalyst Damaging 1T) in Cylinder 3. **Note: If the misfire is severe, the MIL will flash on/off on the 1st trip!** **Possible Causes:** • Base engine mechanical fault that affects only Cylinder 3 • EGR valve is stuck open or the PCV system has a vacuum leak • Fuel component fault that affects only Cylinder 3 (a contaminated or sticking injector) • Ignition system problem (coil or plug) that affects Cylinder 3
DTC: P0304 **2T MISFIRE, MIL: Yes** **2004, 2005, 2006** **Models:** tC, xA, xB **Engines:** 1.5L VIN T, 2.4L VIN D, E **Transmissions:** All	**Cylinder 4 Misfire Detected Conditions:** **Trouble Code Conditions** DTC P0100, P0101, P0102, P0103, P0105, P0110, P0112, P0113, P0115, P0117, P0118, P0120, P0121, P0122, P0123, P0125, P0335, P0340, P0500, P0505 and P0510 not set, engine started, vehicle driven to a speed of over 3 mph for 1 minute, and the PCM detected a misfire rate of 1-2% (High Emissions 2T), or a misfire rate of 6-30% (Catalyst Damaging 1T) in Cylinder 4. **Note: If the misfire is severe, the MIL will flash on/off on the 1st trip!** **Possible Causes:** • Base engine mechanical fault that affects only Cylinder 4 • EGR valve is stuck open or the PCV system has a vacuum leak • Fuel component fault that affects only Cylinder 4 (a contaminated or sticking injector) • Ignition system problem (coil or plug) that affects Cylinder 4
DTC: P0325 **1T CCM, MIL: Yes** **2004, 2005, 2006** **Models:** tC, xA, xB **Engines:** 1.5L VIN T, 2.4L VIN D, E **Transmissions:** All	**Knock Sensor 1 Circuit Malfunction Conditions:** Engine started, vehicle driven with the engine speed over 1200 rpm, and the PCM detected an unexpected voltage condition on the Knock Sensor 1 (KS1) circuit during the CCM test. **Possible Causes:** • Verify that the Knock Sensor (KS) is tightened to specification • Knock sensor signal circuit is open or shorted to ground • Knock sensor signal circuit is shorted to VREF or system power • Knock sensor is damaged or has failed • PCM has failed
DTC: P0335 **1T CCM, MIL: Yes** **2004, 2005, 2006** **Models:** tC, xA, xB **Engines:** 1.5L VIN T, 2.4L VIN D, E **Transmissions:** All	**Crankshaft Position Sensor 'A' Circuit Malfunction Conditions:** Engine cranking; and the PCM did not detect any CKP Sensor 'A' signals, or with the engine speed over 600 rpm, it did not receive any CKP sensor signals, or the CKP signal was lost. **Possible Causes:** • CKP Sensor 'A' signal circuit is open, shorted to ground or shorted to system power • CKP Sensor 'A' signal ground circuit is open • CKP Sensor 'A' is damaged or has failed

DTC	Trouble Code Title, Conditions & Possible Causes
DTC: P0339 **2T CCM, MIL: Yes** **2005, 2006** **Models:** tC **Engines:** 2.4L VIN D, E **Transmissions:** All	**Crankshaft Position Sensor 'A' Circuit Intermittent Conditions:** Engine started, STA signal indicating "off", engine runtime over 3 seconds since STA switched from "on" to "off", engine speed over 1000 rpm, and the PCM did not detect any CKP Sensor 'A' signals for 500 ms. The crankshaft position (NE) sensor consists of a magnet, iron core and pickup coil. The NE sensor signal plate, which has 34 teeth, installed on the crankshaft-timing pulley. This sensor generates 34 signals for each engine revolution. The PCM detects the crankshaft angle and engine speed based on the NE signal. **Possible Causes:** • CKP sensor signal circuit is open, shorted to ground or power • CKP Sensor signal ground circuit is open • Crankshaft timing pulley is damaged or out of alignment • CKP Sensor has failed, or the PCM has failed
DTC: P0340 **2T CCM, MIL: Yes** **2004, 2005, 2006** **Models:** tC, xA, xB **Engines:** 1.5L VIN T, 2.4L VIN D, E **Transmissions:** All	**Camshaft Position Sensor Circuit Malfunction Conditions:** Engine cranking; and the PCM did not detect any CMP sensor signals, or with the engine speed over 600 rpm, it did not detect any CMP signals, or the CMP signal was interrupted. **Possible Causes:** • CMP sensor signal circuit is open, shorted to ground or power • CMP sensor signal ground circuit is open • CMP sensor has failed, or the PCM has failed
DTC: P0341 **2T CCM, MIL: Yes** **2004, 2005, 2006** **Models:** tC, xA, xB **Engines:** 1.5L VIN T, 2.4L VIN D, E **Transmissions:** All	**Camshaft Position Sensor 'A' Signal Range/Performance Conditions:** Engine cranking; and the PCM detected twelve (12) or more Camshaft Position Sensor 'A' (Bank 1) signals during the test. **Possible Causes:** • CMP sensor signal circuit is open, shorted to ground or power • CMP sensor pulley is damaged, or timing belt has jumped teeth • CMP sensor is damaged or has failed • PCM has failed
DTC: P0351 **1T CCM, MIL: Yes** **2004, 2005, 2006** **Models:** tC, xA, xB **Engines:** 1.5L VIN T, 2.4L VIN D, E **Transmissions:** All	**Ignition Coil No. 1 Primary/Secondary Circuit Malfunction Conditions:** Engine started, and the PCM did not detect a change in the IGF signal on the Ignition Coil No. 1 IGF circuit. This engine uses a Direct Ignition (DI) system where one coil is used to fire one cylinder. The coil high-energy secondary wire is connected to one spark plug. If P0351 to P0356 are all set, check for an open/shorted IGF circuit. **Possible Causes:** • IGT1 circuit is open or shorted to ground • Ignition Coil No. 1 is damaged or it has failed • Problem present in the Ignition System • PCM has failed
DTC: P0352 **1T CCM, MIL: Yes** **2004, 2005, 2006** **Models:** tC, xA, xB **Engines:** 1.5L VIN T, 2.4L VIN D, E **Transmissions:** All	**Ignition Coil No. 2 Primary/Secondary Circuit Malfunction Conditions:** Engine started, and the PCM did not detect a change in the IGF signal on the Ignition Coil No. 2 IGF circuit. This engine uses a Direct Ignition (DI) system where one coil is used to fire one cylinder. The coil high-energy secondary wire is connected to one spark plug. If P0351 to P0356 are all set, check for an open/shorted IGF circuit. **Possible Causes:** • IGT2 circuit is open or shorted to ground • Ignition Coil No. 2 is damaged or it has failed • Problem present in the Ignition System • PCM has failed
DTC: P0353 **1T CCM, MIL: Yes** **2004, 2005, 2006** **Models:** tC, xA, xB **Engines:** 1.5L VIN T, 2.4L VIN D, E **Transmissions:** All	**Ignition Coil No. 3 Primary/Secondary Circuit Malfunction Conditions:** Engine started, and the PCM did not detect a change in the IGF signal on the Ignition Coil No. 3 IGF circuit. This engine uses a Direct Ignition (DI) system where one coil is used to fire one cylinder. The coil high-energy secondary wire is connected to one spark plug. If P0351 to P0356 are all set, check for an open/shorted IGF circuit. **Possible Causes:** • IGT3 circuit is open or shorted to ground • Ignition Coil No. 3 is damaged or it has failed • Problem present in the Ignition System • PCM has failed
DTC: P0354 **1T CCM, MIL: Yes** **2004, 2005, 2006** **Models:** tC, xA, xB **Engines:** 1.5L VIN T, 2.4L VIN D, E **Transmissions:** All	**Ignition Coil No. 4 Primary/Secondary Circuit Malfunction Conditions:** Engine started, and the PCM did not detect a change in the IGF signal on the Ignition Coil No. 4 IGF circuit. This engine uses a Direct Ignition (DI) system where one coil is used to fire one cylinder. The coil high-energy secondary wire is connected to one spark plug. If P0351 to P0356 are all set, check for an open/shorted IGF circuit. **Possible Causes:** • IGT4 circuit is open or shorted to ground • Ignition Coil No. 4 is damaged or it has failed • Problem present in the Ignition System • PCM has failed

DTC	Trouble Code Title, Conditions & Possible Causes
DTC: P0401 **2T EGR, MIL: Yes** **2004, 2005, 2006** **Models:** tC, xA, xB **Engines:** 1.5L VIN T, 2.4L VIN D, E **Transmissions:** All	**EGR System Insufficient Flow Detected Conditions:** Engine started, vehicle driven at a speed of over 50 mph with the engine running in closed loop at a steady throttle for 3-5 minutes, and the PCM detected the EGR sensor signal was not 106-140°F higher than the ambient air temperature in the EGR Monitor test. **Possible Causes:** • EGR gas temperature sensor circuit open or shorted to VREF • EGR gas temperature sensor is damaged or has failed • EGR valve is stuck in closed position • EGR valve vacuum hose is disconnected or leaking
DTC: P0402 **2T EGR, MIL: Yes** **2004, 2005, 2006** **Models:** tC, xA, xB **Engines:** 1.5L VIN T, 2.4L VIN D, E **Transmissions:** All	**Excessive EGR Flow Detected Conditions:** Engine started cold (ECT sensor signal less than 86°F), engine running without load at less than 4000 rpm, and the PCM detected the EGR sensor indicated a high value during the EGR Cutoff test; or it detected the EGR valve was open during all driving conditions. **Possible Causes:** • EGR gas temperature sensor circuit open or shorted to ground • EGR gas temperature sensor is damaged or has failed • EGR Vacuum Switching Valve (VSV) is damaged or has failed • EGR valve stuck partially or EGR valve stuck fully open • EGR VSV control circuit is open or shorted to system power • PCM has failed
DTC: P0420 **2T CAT, MIL: Yes** **2004, 2005, 2006** **Models:** tC, xA, xB **Engines:** 1.5L VIN T, 2.4L VIN D, E **Transmissions:** All	**Catalyst Efficiency Below Normal (Bank 1) Conditions:** DTC P0100, P0101, P0102, P0103, P0110, P0112, P0113, P0115, P0116, P0117, P0118, P0120, P0121, P0122, P0123, P0335, P0340 and P0500 not set, engine started, vehicle driven to a speed of 45-60 mph at 2500-3000 rpm in closed loop for 3-5 minutes, and the PCM detected too much variation in the voltage amplitudes of the HO2S-12 signal (Bank 1). **Possible Causes:** • Catalytic converter is damaged, contaminated or has failed • Front A/FS or rear HO2S is contaminated with fuel or moisture • Front A/FS or the rear HO2S is loose in its mounting hole • Front A/FS is older (aged) than the rear HO2S (HO2S is lazy) • Gas leaks at the exhaust manifold or in the exhaust pipes
DTC: P0440 **2T EVAP, MIL: Yes** **2004, 2005, 2006** **Models:** tC, xA, xB **Engines:** 1.5L VIN T, 2.4L VIN D, E **Transmissions:** All	**EVAP Control System Large Leak (0.080") Detected Conditions:** Engine started cold (ECT sensor signal less 86°F), engine runtime over 20 minutes in closed loop, vehicle driven to a speed of 55-60 mph, and the PCM detected the fuel tank pressure indicated the same value as the atmospheric pressure in the EVAP Monitor test. **Possible Causes:** • Charcoal canister is clogged, loaded with fuel or with moisture • Fuel filler cap missing, loose (not tightened) or the wrong part • Fuel tank over-fill check valve cracked or damaged • Fuel tank seal leaking, fuel tank cracked or damaged/leaking • Fuel vapor hoses/tubes blocked or restricted, or fuel vapor control valve tube or fuel vapor vent valve assembly blocked • Vacuum hose or tubing cracked, damaged or disconnected • Vapor Pressure sensor incorrectly installed • PCM has failed
DTC: P0441 **2T EVAP, MIL: Yes** **2004, 2005, 2006** **Models:** tC, xA, xB **Engines:** 1.5L VIN T, 2.4L VIN D, E **Transmissions:** All	**EVAP System Incorrect Purge Flow Detected Conditions:** ECT sensor less than 86°F at startup, vehicle driven at 55-60 mph for 2-3 minutes, and the PCM detected the EVAP canister pressure did not decrease during purge conditions, or it remained too low during purge cutoff conditions. The vapor pressure sensor, VSV for the canister closed valve (CCV) and the VSV for the vapor-switching valve are used to detect EVAP system faults. The PCM closes the CCV and opens the VSV for vapor switching valve to cause an increase in vacuum in the EVAP system. Once the vacuum reaches a certain point, the PCM closes the VSV to test system operation. **Possible Causes:** • Charcoal canister is clogged, loaded with fuel or with moisture • Fuel tank over-fill check valve cracked or damaged • Fuel tank seal leaking, fuel tank cracked or damaged/leaking • Fuel vapor hoses/tubes blocked or restricted, or fuel vapor control valve tube or fuel vapor vent valve assembly blocked • Vacuum hose or tubing cracked, damaged or disconnected • Vapor pressure sensor is damaged or has failed • VSV circuit for the canister purge, VSV for the CCV or the VSV for the pressure switching valve is open or shorted to ground • VSV for the vapor pressure sensor circuit is open or shorted to ground, or the vapor pressure sensor is damaged or has failed

DTC	Trouble Code Title, Conditions & Possible Causes
DTC: P0442 **2T EVAP, MIL: Yes** **2004, 2005, 2006** **Models:** tC, xA, xB **Engines:** 1.5L VIN T, 2.4L VIN D, E **Transmissions:** All	**EVAP System Small Leak (0.040") Detected Conditions:** Engine started; IAT sensor signal from 39-86°F, fuel tank level from 25-75% for 10 seconds, and the PCM detected the EVAP system was unable to hold a specified vacuum level for a set period of time. After the system is purged, the PCM shuts off the VSV for the purge valve to seal the vacuum in the system, and then monitors the increase in pressure in the system. The pressure should increases slowly. If it increases at too fast a rate, then this code is set. **Possible Causes:** • Canister Purge valve is damaged, leaking or has failed • Charcoal canister is loaded with fuel or moisture • Fuel filler cap loose, cross-threaded, incorrect part or damaged • Fuel tank is cracked (leaking), or a leak exists in the 'O' ring • Fuel tank pressure sensor is damaged or has failed • Fuel vapor line(s), fuel pipes or hoses damaged or leaking • PCM has failed
DTC: P0450 **2T CCM, MIL: Yes** **2004, 2005, 2006** **Models:** tC, xA, xB **Engines:** 1.5L VIN T, 2.4L VIN D, E **Transmissions:** All	**EVAP Vapor Pressure Sensor Circuit Malfunction Conditions:** Engine started, engine runtime less than 10 seconds since startup, and the PCM detected the Vapor Pressure sensor value was less than -3.5 kPa (-1.0 in. Hg), or the Vapor Pressure sensor was more than or equal to 1.5 kPa (0.4 in. Hg) during testing. The PCM uses the Vapor Pressure Sensor, VSV for the Canister Closed valve and VSV for the Pressure Switching valve to find faults in this system. **Possible Causes:** • Vapor pressure sensor signal circuit open or shorted to ground • Vapor pressure sensor ground circuit is open • Vapor pressure sensor power circuit is open • Vapor pressure sensor is damaged or has failed • PCM has failed
DTC: P0451 **2T CCM, MIL: Yes** **2004, 2005, 2006** **Models:** tC, xA, xB **Engines:** 1.5L VIN T, 2.4L VIN D, E **Transmissions:** All	**EVAP Vapor Pressure Sensor Range/Performance Conditions:** Engine started, engine at idle speed with the VSS indicating 0 mph, VSV for Vapor Switching valve "off", and the PCM detected too much change in the pressure sensor value, or the pressure sensor value equaled the opening value of the charcoal canister. The PCM uses the Vapor Pressure Sensor, VSV for the Canister Closed valve and VSV for Pressure Switching valve to find faults in the system. **Possible Causes:** • Vapor pressure sensor vacuum hoses loose or damaged • Vapor pressure sensor is damaged or has failed • PCM has failed
DTC: P0452 **2T CCM, MIL: Yes** **2004, 2005, 2006** **Models:** tC, xA, xB **Engines:** 1.5L VIN T, 2.4L VIN D, E **Transmissions:** All	**EVAP Vapor Pressure Sensor Circuit Low Input Conditions:** Engine started; VSV for vapor pressure switching valve "off"; and the PCM detected an unexpected low voltage condition on the vapor pressure sensor circuit during the CCM test. **Possible Causes:** • Vapor pressure sensor connector is damaged or open • Vapor pressure sensor circuit is open • Vapor pressure sensor is damaged or has failed • PCM has failed
DTC: P0453 **2T CCM, MIL: Yes** **2004, 2005, 2006** **Models:** tC, xA, xB **Engines:** 1.5L VIN T, 2.4L VIN D, E **Transmissions:** All	**EVAP Vapor Pressure Sensor Circuit High Input Conditions:** Engine started; VSV for vapor pressure switching valve "off"; and the PCM detected an unexpected high voltage condition on the vapor pressure sensor circuit during the CCM test. **Possible Causes:** • Vapor pressure sensor connector is damaged or shorted • Vapor pressure sensor circuit is shorted to VREF • Vapor pressure sensor is damaged or has failed • PCM has failed
DTC: P0456 **2T EVAP, MIL: Yes** **2004, 2005, 2006** **Models:** tC, xA, xB **Engines:** 1.5L VIN T, 2.4L VIN D, E **Transmissions:** All	**EVAP System Very Small Leak (0.020") Detected Conditions:** Engine started; IAT sensor signal from 39-86°F, fuel tank level from 25-75% for 10 seconds, and the PCM detected the EVAP system was unable to hold a specified vacuum level for a set period of time. After the system is purged, the PCM shuts off the VSV for the purge valve to seal the vacuum in the system, and then monitors the increase in pressure in the system. The pressure should increase slowly. If it increases at too fast a rate, this code is set. **Possible Causes:** • Canister Purge valve is damaged, leaking or has failed • Charcoal canister is loaded with fuel or moisture • Fuel filler cap loose, cross-threaded, incorrect part or damaged • Fuel tank is cracked (leaking), or a leak exists in the 'O' ring • Fuel tank pressure sensor is damaged or has failed • Fuel tank overfill check valve is cracked or is damaged • Fuel vapor line(s), fuel pipes or hoses damaged or leaking • PCM has failed

DTC	Trouble Code Title, Conditions & Possible Causes
DTC: P0500 **2T CCM, MIL: Yes** **2004, 2005, 2006** **Models:** tC, xA, xB **Engines:** 1.5L VIN T, 2.4L VIN D, E **Transmissions:** All	**Vehicle Speed Sensor Circuit Malfunction Conditions:** Engine started, vehicle driven with the engine speed from 1500-5500 rpm and back to idle several times, and the PCM did not receive any VSS signals during the test. The VSS (No.1) assembly outputs a 4-pulse signal for every revolution of the rotor shaft, which is generated by the transmission output shaft via the driven gear. **Possible Causes:** • VSS signal circuit is open between the meter and the PCM • VSS signal circuit shorted to ground between meter and PCM • VSS No. 1 is damaged or has failed • Combination Meter is damaged or has failed • PCM has failed
DTC: P0505 **2T CCM, MIL: Yes** **2004, 2005, 2006** **Models:** tC, xA, xB **Engines:** 1.5L VIN T, 2.4L VIN D, E **Transmissions:** All	**Idle Control System Malfunction Conditions:** Engine started, engine running at idle speed n closed loop, and the PCM detected the Actual Idle Speed was more than 100-200 rpm above or below the Target Idle Speed. A Rotary solenoid type of ISC valve is located in front of the air intake chamber and intake air bypassing the throttle valve is directed to the Intake Air Control (IAC) valve via a passage. The PCM controls the idle speed by regulating the amount of intake air volume that bypasses the throttle valve. **Possible Causes:** • Air Induction system leaks (check for intake manifold leaks) • Air leaks in the PCV system (at the valve or its related hoses) • Throttle body assembly is damaged or has failed • PCM has failed
DTC: P0511 **2T CCM, MIL: Yes** **2004, 2005, 2006** **Models:** tC, xA, xB **Engines:** 1.5L VIN T, 2.4L VIN D, E **Transmissions:** All	**Idle Control Valve Circuit Malfunction Conditions:** Engine running at idle speed n closed loop, and the PCM detected an unexpected voltage condition on the IAC valve circuit. A Rotary solenoid valve, in front of the air intake chamber, allows intake air to bypass the throttle valve and be directed to the Intake Air Control (IAC) valve through a passage. This configuration allows the PCM to control the engine idle speed by regulating the amount of intake air volume that bypasses the throttle valve. **Possible Causes:** • IAC valve connector is damaged or loose • IAC valve control circuit is open, shorted to ground or power • IAC valve is damaged or has failed • PCM has failed
DTC: P0552 **1T CCM, MIL: Yes** **2004, 2005, 2006** **Models:** xA, xB **Engines:** 1.5L VIN T **Transmissions:** All	**Power Steering Pressure Sensor Circuit Low Input Conditions:** Key on or engine running; and the PCM detected the Power Steering Pressure (PSP) sensor was less than 0.26v during the CCM test. **Possible Causes:** • PSP sensor signal circuit is open or shorted to ground • PSP sensor ground circuit is open • PSP sensor is damaged or has failed • PCM has failed
DTC: P0553 **1T CCM, MIL: Yes** **2004, 2005, 2006** **Models:** xA, xB **Engines:** 1.5L VIN T **Transmissions:** All	**Power Steering Pressure Sensor Circuit High Input Conditions:** Key on or engine running; and the PCM detected the Power Steering Pressure (PSP) sensor was more than 4.90v during the CCM test. **Possible Causes:** • PSP sensor signal circuit is shorted to power • PSP sensor ground circuit is open • PSP sensor is damaged or has failed • PCM has failed
DTC: P0560 **2T CCM, MIL: Yes** **2004, 2005, 2006** **Models:** tC, xA, xB **Engines:** 1.5L VIN T, 2.4L VIN D, E **Transmissions:** All	**System Voltage (Backup Power Circuit) Malfunction Conditions:** Key on or engine running; and the PCM detected an unexpected low voltage condition on the Backup Power Circuit during the test. **Possible Causes:** • Battery backup circuit is open between battery and the PCM • PCM has failed
DTC: P0606 **1T PCM, MIL: Yes** **2004, 2005, 2006** **Models:** tC, xA, xB **Engines:** 1.5L VIN T, 2.4L VIN D, E **Transmissions:** All	**ECM/PCM Processing Error Conditions:** Key on, and the PCM detected a processing error occurred. **Possible Causes:** • Clear the codes and retest for this code. If the same code resets, substitute a known good control module and retest. If the trouble code is gone, the original PCM has failed. • TSB TC002-03 (6/03) contains information related to this code

DTC	Trouble Code Title, Conditions & Possible Causes
DTC: P0607 **1T PCM, MIL: Yes** **2004, 2005, 2006** **Models:** tC, xA, xB **Engines:** 1.5L VIN T, 2.4L VIN D, E **Transmissions:** All	**Control Module Performance Conditions:** Key on, and the PCM detected a performance problem occurred. **Possible Causes:** • Clear the codes and retest for this code. If the same code resets, substitute a known good control module and retest. If the trouble code is gone, the original PCM has failed. • TSB TC002-03 (6/03) contains information related to this code
DTC: P0617 **1T CCM, MIL: Yes** **2004, 2005, 2006** **Models:** tC, xA, xB **Engines:** 1.5L VIN T, 2.4L VIN D, E **Transmissions:** All	**Starter Relay Circuit High Input Conditions:** Engine started, engine speed over 1000 rpm, system voltage over 10.5v, and the PCM detected the Starter Motor signal indicated high. **Possible Causes:** • Park/Neutral switch assembly is damaged or it has failed • Ignition switch is damaged or has failed • PCM has failed
DTC: P0705 **2T CCM, MIL: Yes** **2004, 2005, 2006** **Models:** tC, xA, xB **Engines:** 1.5L VIN T, 2.4L VIN D, E **Transmissions:** A/T	**A/T Range Sensor Circuit (PRNDL) Malfunction Conditions:** Key on or engine running; and the PCM detected simultaneous "on" signals (N, 2, L or R) from the Transmission Range sensor circuit. The P/N switch indicates "on" whenever the shift lever is in the 'N' or 'P' position. When it is "on", the NSW circuit to the PCM is grounded to chassis ground through the starter motor relay, and reads 0.00v. When the shift lever is in 'R', 'D' or 'L' position, the switch is "off" and the NSW circuit reads 12.0v. When the shift lever is moved from the 'N' to the 'D' position, the PCM uses this signal to air/fuel ratio correction and idle speed control (estimated control) functions. **Possible Causes:** • Park/Neutral switch assembly is shorted • Park/Neutral switch assembly is damaged or has failed • PCM has failed.
DTC: P0710 **1T CCM, MIL: Yes** **2005, 2006** **Models:** tC **Engines:** 2.4L VIN D, E **Transmissions:** A/T	**Transmission Fluid Temperature Sensor Circuit Malfunction Conditions:** Engine started, and the PCM detected the TFT sensor indicated less than -40°F, or after the engine runtime exceeded 15 minutes, it indicated more than 300°F during the CCM test. **Possible Causes:** • TFT sensor signal circuit is open, shorted to ground or shorted to system power • TFT sensor is damaged or has failed • PCM has failed
DTC: P0710 **1T CCM, MIL: Yes** **2004, 2005, 2006** **Models:** xA, xB **Engines:** 1.5L VIN T **Transmissions:** A/T	**Transmission Fluid Temperature Sensor Circuit Malfunction Conditions:** Engine started, and the PCM detected the TFT sensor indicated less than -40°F, or with the engine runtime over 15 minutes, it detected the TFT sensor indicated more than 300°F. **Possible Causes:** • TFT sensor signal circuit is open, shorted to ground or shorted to system power • TFT sensor is damaged or has failed • PCM has failed
DTC: P0711 **1T CCM, MIL: Yes** **2005, 2006** **Models:** tC **Engines:** 2.4L VIN D, E **Transmissions:** A/T	**Transmission Fluid Temperature Sensor Performance Conditions:** Engine started, engine runtime over 15 seconds, and the PCM detected the ambient air temperature and transmission fluid temperature varied by more than 40°F, or after an engine runtime of 20 minutes and 6.2 miles were traveled, it indicated less than 50°F. **Possible Causes:** • TFT sensor signal circuit or ground circuit has high resistance • TFT sensor has failed, or the PCM has failed
DTC: P0711 **1T CCM, MIL: Yes** **2004, 2005, 2006** **Models:** xA, xB **Engines:** 1.5L VIN T **Transmissions:** A/T	**Transmission Fluid Temperature Sensor Performance Conditions:** ECT and IAT sensor signals more than 14°F, engine runtime over 12 seconds, then after the vehicle was driven for more than 6.2 miles and the engine runtime exceeded 20 minutes, the PCM detected the TFT sensor remained at a value less than 14°F during the CCM test. **Possible Causes:** • TFT sensor signal circuit has high resistance • TFT sensor is damaged or has failed (it may be contaminated) • PCM has failed
DTC: P0712 **1T CCM, MIL: Yes** **2004, 2005, 2006** **Models:** xA, xB **Engines:** 1.5L VIN T **Transmissions:** A/T	**Transmission Fluid Temperature Sensor Circuit Low Input Conditions:** Engine started, engine running, and the PCM detected the Transmission Fluid Temperature (TFT) sensor indicated a value of more than 284°F for 500 ms in the test. **Possible Causes:** • TFT sensor signal circuit is shorted • TFT sensor is damaged or has failed • PCM has failed

DTC	Trouble Code Title, Conditions & Possible Causes
DTC: P0713 **1T CCM, MIL: Yes** **2004, 2005, 2006** **Models:** xA, xB **Engines:** 1.5L VIN T **Transmissions:** A/T	**Transmission Fluid Temperature Sensor Circuit High Input Conditions:** Engine started, engine running, and the PCM detected the TFT sensor indicated a value of less than -40°F for 500 ms in t he test. **Possible Causes:** • TFT sensor signal circuit is open • TFT sensor is damaged or has failed • PCM has failed
DTC: P0717 **1T CCM, MIL: Yes** **2004, 2005, 2006** **Models:** xA, xB **Engines:** 1.5L VIN T **Transmissions:** A/T	**A/T Turbine Shaft Speed Sensor Circuit No Signal Conditions:** No Shift Solenoid or P/N codes set, engine started, P/N switch indicating "off", gear position indicating 2nd, 3rd gear, or in O/D, no gear change occurring, and the PCM detected the Turbine Shaft Speed (ISS) sensor indicated less than 300 rpm, or more than 1000 rpm for 4 seconds. The PCM detects the rotation speed of the input turbine, and compares the signals from the input turbine speed (NT) sensor to the counter gear speed sensor (NC). The PCM uses this signal to detect the shift time so that it can control the engine torque and hydraulic pressure in response to various driving conditions. **Possible Causes:** • Input shaft speed sensor is damaged or loose • Input shaft speed sensor signal (NT) circuit is open or shorted • Input shaft speed sensor is damaged or has failed • PCM has failed
DTC: P0724 **2T CCM, MIL: Yes** **2005, 2006** **Models:** tC **Engines:** 2.4L VIN D, E **Transmissions:** M/T	**Brake Switch 'B' Circuit High Input Conditions:** Engine started, vehicle driven to cruise speed and then back to idle speed at least 30 times, and the PCM did not detect any change in the Brake Switch 'A' circuit status. The STP 'B' switch signal is used to determine when the brakes have been applied, and to determine the Fuel Cutoff engine speed during periods with the brakes applied. **Possible Causes:** • Stoplight switch signal circuit is shorted to power • Stoplight switch assembly is damaged or shorted • PCM has failed
DTC: P0724 **1T CCM, MIL: Yes** **2004, 2005, 2006** **Models:** xA, xB **Engines:** 1.5L VIN T **Transmissions:** M/T	**Brake Switch 'B' Circuit High Input Conditions:** Engine started, vehicle driven to cruise speed and then back to idle speed at least 30 times, and the PCM did not detect any change in the Brake Switch 'A' circuit status. The STP 'B' switch signal is used to determine when the brakes have been applied, and to determine the Fuel Cutoff engine speed during periods with the brakes applied. **Possible Causes:** • Stoplight switch signal circuit is shorted to power • Stoplight switch assembly is damaged or shorted • PCM has failed
DTC: P0724 **1T CCM, MIL: Yes** **2004, 2005, 2006** **Models:** tC, xA, xB **Engines:** 1.5L VIN T, 2.4L VIN D, E **Transmissions:** A/T	**A/T Torque Converter Clutch Shift Solenoid Performance Conditions:** Engine started, vehicle driven to over 50 mph and then back to idle speed several times, and the PCM detected that TCC lockup did not occur, or that the TCC remained in lockup position in the "off" range. The PCM uses signals from the CKP, MAF and TP sensors to monitor engagement of the Lockup Clutch solenoid to find a fault. **Possible Causes:** • Lockup clutch solenoid is damaged or has failed Shift solenoid SL is damaged or has failed (mechanical fault) • Shift solenoid SL is stuck in "on" or "off" position • Valve body is blocked or stuck
DTC: P0743 **2T CCM, MIL: Yes** **2004, 2005, 2006** **Models:** xA, xB **Engines:** 1.5L VIN T **Transmissions:** A/T	**A/T Torque Converter Clutch Shift Solenoid Circuit Malfunction Conditions:** Vehicle driven to over 50 mph and then back to idle speed several times; and the PCM detected an unexpected voltage on the Shift Solenoid (SL) valve control circuit. The Shift Solenoid valve is turned On/Off" through commands from the PCM to control the hydraulic pressure acting on the lockup relay valve (and it controls the operation of the lockup clutch). **Possible Causes:** • Shift solenoid SL connector is damaged or loose • Shift solenoid SL control circuit is open or shorted to ground • Shift solenoid SL is damaged or has failed (electrical fault) • PCM has failed
DTC: P0750 **2T CCM, MIL: Yes** **2004, 2005, 2006** **Models:** xA, xB **Engines:** 1.5L VIN T **Transmissions:** A/T	**A/T Shift Solenoid 1 or 'A' Malfunction (Mechanical) Conditions:** Engine started, vehicle driven under normal driving conditions, and the PCM detected the Actual gear ratio and the Required gear ratio did not match during the CCM Rationality test. **Possible Causes:** • A/T component problems (i.e., in clutch, brake or gears) • SS1 or SSA is damaged, stuck "open" or stuck "closed" • Transmission valve body is clogged, dirty or stuck

DTC	Trouble Code Title, Conditions & Possible Causes
DTC: P0751 **8T CCM, MIL:** Yes **2004, 2005, 2006** **Models:** xA, xB **Engines:** 1.5L VIN T **Transmissions:** A/T	**A/T Shift Solenoid 'A' Signal Range/Performance Conditions:** Engine started, vehicle driven to a speed over 50 mph, and the PCM detected the Actual gear position did not match the Desired gear position during the CCM test period. The PCM uses inputs from the VSS and Direct Clutch speed sensor to determine the actual gear position (i.e., 1st, 2nd, 3rd or O/D gear). **Possible Causes:** • SSA control circuit is open or shorted to ground • SSA control circuit is shorted to system power (B+) • SSA is damaged or has failed (an electrical fault) • PCM has failed
DTC: P0753 **8T CCM, MIL:** Yes **2005, 2006** **Models:** tC **Engines:** 2.4L VIN D, E **Transmissions:** A/T	**A/T Shift Solenoid 1 or 'A' Circuit Malfunction Conditions:** Engine started, engine running during normal driving conditions; and the PCM detected the Shift Solenoid 1 or 'A' (SS1/SSA) control circuit voltage was "high" with the solenoid "on", or the SS1 or SSA control circuit voltage was "low" with the SS1/SSA commanded "off". **Possible Causes:** • SS1 or SSA control circuit is open, shorted to ground or shorted to system power (B+) • SS1 or SSA has failed, or the PCM has failed
DTC: P0756 **8T CCM, MIL:** Yes **2004, 2005, 2006** **Models:** xA, xB **Engines:** 1.5L VIN T **Transmissions:** A/T	**A/T Shift Solenoid 'B' Signal Range/Performance Conditions:** Vehicle driven to a speed over 50 mph, and the PCM detected the Actual gear position did not match the Desired gear position. The PCM uses inputs from the VSS and Direct Clutch speed sensor to determine the actual gear position (i.e., 1st, 2nd, 3rd or O/D gear). **Possible Causes:** • SSB control circuit is open, shorted to ground or to power (B+) • SSB is damaged or has failed (an electrical fault) • PCM has failed
DTC: P0758 **1T CCM, MIL:** Yes **2004, 2005, 2006** **Models:** xA, xB **Engines:** 1.5L VIN T **Transmissions:** A/T	**A/T Shift Solenoid 2 or 'B' Circuit Malfunction Conditions:** Engine started, engine running during normal driving conditions, and the PCM detected the Shift Solenoid 2 or 'B' (SS2/SSB) control circuit voltage was "high" with the solenoid "on", or the SS2 or SSB control circuit voltage was "low" with the SS2/SSB commanded "off". **Note: This problem must occur eight (8) times during one trip before this code is triggered.** **Possible Causes:** • SS2 or SSB control circuit is open or shorted to ground • SS2 or SSB control circuit is shorted to system power (B+) • SS2 or SSB is damaged or has failed (an electrical fault) • PCM has failed
DTC: P0758 **8T CCM, MIL:** Yes **2005, 2006** **Models:** tC **Engines:** 2.4L VIN D, E **Transmissions:** A/T	**A/T Shift Solenoid 'B' Circuit Malfunction Conditions:** Engine started, engine running during normal driving conditions, and the PCM detected an unexpected voltage condition on the Shift Solenoid 'B' control circuit during the CCM test. **Possible Causes:** • SSB control circuit is open, shorted to ground or to power (B+) • SSB is damaged or has failed (an electrical fault) • PCM has failed
DTC: P0787 **1T CCM, MIL:** Yes **2004, 2005, 2006** **Models:** xA, xB **Engines:** 1.5L VIN T **Transmissions:** A/T	**A/T Shift Timing Solenoid (ST) Circuit Low Input Conditions:** Engine started, engine running during normal driving conditions, and the PCM detected an unexpected low voltage condition on the Shift Timing Solenoid (SL) control circuit at least (4) times during testing. **Possible Causes:** • ST solenoid connector is damaged or shorted • SL solenoid control circuit is shorted to ground • SL solenoid is damaged or has failed • PCM has failed
DTC: P0788 **1T CCM, MIL:** Yes **2004, 2005, 2006** **Models:** xA, xB **Engines:** 1.5L VIN T **Transmissions:** A/T	**A/T Shift Timing Solenoid (ST) Circuit High Input Conditions:** Engine started, engine running during normal driving conditions, and the PCM detected an unexpected high voltage condition on the Shift Timing Solenoid (SL) control circuit at least (4) times during testing. **Possible Causes:** • ST solenoid connector is damaged or loose • SL solenoid control circuit is open or shorted to system power • SL solenoid is damaged or has failed • PCM has failed

DTC	Trouble Code Title, Conditions & Possible Causes
DTC: P0850 **1T CCM, MIL:** Yes **2005, 2006** **Models:** tC **Engines:** 2.4L VIN D, E **Transmissions:** A/T	**A/T Park/Neutral Switch Circuit Malfunction Conditions:** Engine started, vehicle at 1500-2500 rpm at a speed over 43 mph for at least 30 seconds, and the PCM detected a continuous "on" "N" signal from the P/N switch. The P/N switch indicates "on" whenever the shift lever is in the 'N' or 'P' position. When it is "on", the NSW circuit to the PCM is grounded to chassis ground through the starter motor relay, and reads 0.00v. When the shift lever is in 'R', 'D' or 'L' position, the switch is "off" and the NSW circuit reads 12.0v. When the shift lever is moved from the 'N' to the 'D' position, the PCM uses this signal to air/fuel ratio correction and idle speed control functions. **Possible Causes:** • Park/Neutral switch assembly is shorted Park/Neutral switch assembly is damaged or has failed PCM has failed.
DTC: P0973 **1T CCM, MIL:** Yes **2004, 2005, 2006** **Models:** xA, xB **Engines:** 1.5L VIN T **Transmissions:** A/T	**A/T Shift Solenoid 1 Circuit Low Input Conditions:** Engine started, engine running during normal driving conditions, and the PCM detected an unexpected low voltage condition on the Shift Solenoid 1 control circuit at least (4) times. **Possible Causes:** • SS1 (solenoid) connector is damaged or shorted • SS1 (solenoid) control circuit is shorted to ground • SS1 (solenoid) is damaged or has failed • PCM has failed
DTC: P0974 **1T CCM, MIL:** Yes **2004, 2005, 2006** **Models:** xA, xB **Engines:** 1.5L VIN T **Transmissions:** A/T	**A/T Shift Solenoid 1 Circuit High Input Conditions:** Engine started, engine running during normal driving conditions, and the PCM detected an unexpected high voltage condition on the Shift Solenoid 1 control circuit at least (4) times. **Possible Causes:** • SS1 (solenoid) connector is damaged or open • SS1 (solenoid) control circuit is open or shorted to power (B+) • SS1 (solenoid) is damaged or has failed • PCM has failed
DTC: P0976 **1T CCM, MIL:** Yes **2004, 2005, 2006** **Models:** xA, xB **Engines:** 1.5L VIN T **Transmissions:** A/T	**A/T Shift Solenoid 2 Circuit Low Input Conditions:** Engine started, engine running during normal driving conditions, and the PCM detected an unexpected low voltage condition on the Shift Solenoid 2 control circuit at least (4) times during testing. **Possible Causes:** • SS2 (solenoid) connector is damaged or shorted • SS2 (solenoid) control circuit is shorted to ground • SS2 (solenoid) is damaged or has failed • PCM has failed
DTC: P0977 **1T CCM, MIL:** Yes **2004, 2005, 2006** **Models:** xA, xB **Engines:** 1.5L VIN T **Transmissions:** A/T	**A/T Shift Solenoid 2 Circuit High Input Conditions:** Engine started, engine running during normal driving conditions, and the PCM detected an unexpected high voltage condition on the Shift Solenoid 2 control circuit at least (4) times during testing. **Possible Causes:** • SS2 (solenoid) connector is damaged or open • SS2 (solenoid) control circuit is open or shorted to power (B+) • SS2 (solenoid) is damaged or has failed • PCM has failed

OBD II Trouble Code List (P1xxx Codes)

DTC	Trouble Code Title, Conditions & Possible Causes
DTC: P1130 **2T O2S, MIL: Yes** **2004, 2005, 2006** **Models:** tC, xA, xB **Engines:** 1.5L VIN T, 2.4L VIN D, E **Transmissions:** All	**A/F Sensor-11 (Bank 1 Sensor 1) Circuit Malfunction Conditions:** DTC P1135 not set, engine started, engine warm-up completed, and the PCM detected the A/F sensor signal was more than 3.8v or that it was less than 2.8v; or the A/F sensor signal was fixed at 3.30v; or the PCM detected an unexpected "high" or "low" voltage condition in the A/F sensor signal circuit during the CCM test. **Possible Causes:** • A/FS signal circuit or ground circuit is open (intermittent fault) • A/FS is contaminated, damaged or has failed • PCM has failed
DTC: P1133 **1T O2S, MIL: Yes** **2004, 2005, 2006** **Models:** tC, xA, xB **Engines:** 1.5L VIN T, 2.4L VIN D, E **Transmissions:** All	**A/F Sensor-11 (Bank 1 Sensor 1) Slow Response Conditions:** DTC P1135 not set, vehicle speed over 38 mph at over 1600 rpm for 1-3 minutes, and the PCM detected the A/F sensor response times deteriorated below an acceptable value. **Possible Causes:** • A/F sensor signal circuit or ground circuit is open • A/F sensor power circuit is open • A/F sensor is contaminated, damaged or has failed • Intake air leaks, exhaust manifold leaks or PCV system leaks • MAF sensor out of calibration (it may be dirty or contaminated) • PCM has failed
DTC: P1135 **2T CCM, MIL: Yes** **2004, 2005, 2006** **Models:** tC, xA, xB **Engines:** 1.5L VIN T, 2.4L VIN D, E **Transmissions:** All	**A/F Sensor-11 (Bank 1 Sensor 1) Heater Circuit Malfunction Conditions:** Engine started, engine running, and the PCM detected the A/F Sensor-11 (A/FS-11) heater current was less than 0.25A, or that it was more than 8.0A at anytime during the CCM test. **Possible Causes:** • A/F sensor heater circuit or ground circuit is open • A/F sensor heater power circuit is open • A/F sensor heater is damaged or has failed • PCM has failed
DTC: P1300 **1T CCM, MIL: Yes** **2004, 2005, 2006** **Models:** tC, xA, xB **Engines:** 1.5L VIN T, 2.4L VIN D, E **Transmissions:** All	**Igniter No. 1 Circuit Malfunction Conditions:** Engine started, engine running, and the PCM did not detect any IGF1 signals after detecting two IGT1 signals during the CCM test. **Note: This vehicle uses a Coil-On-Plug design Ignition System.** **Possible Causes:** • IGT or IGF signal circuit is open or shorted to ground • Igniter is damaged or has failed • PCM has failed
DTC: P1305 **1T CCM, MIL: Yes** **2004, 2005, 2006** **Models:** tC, xA, xB **Engines:** 1.5L VIN T, 2.4L VIN D, E **Transmissions:** All	**Igniter No. 2 Circuit Malfunction Conditions:** Engine started; and the PCM did not detect any IGF2 signals after detecting two IGT2 signals. **Note: This vehicle uses a Coil-On-Plug design Ignition System.** **Possible Causes:** • IGT or IGF signal circuit is open or shorted to ground • Igniter is damaged or has failed • PCM has failed
DTC: P1310 **1T CCM, MIL: Yes** **2004, 2005, 2006** **Models:** tC, xA, xB **Engines:** 1.5L VIN T, 2.4L VIN D, E **Transmissions:** All	**Igniter No. 3 Circuit Malfunction Conditions:** Engine started; and the PCM did not detect any IGF3 signals after detecting two IGT3 signals during the test. **Note: This vehicle uses a Coil-On-Plug design Ignition System.** **Possible Causes:** • IGT or IGF signal circuit is open or shorted to ground • Igniter is damaged or has failed
DTC: P1315 **1T CCM, MIL: Yes** **2004, 2005, 2006** **Models:** tC, xA, xB **Engines:** 1.5L VIN T, 2.4L VIN D, E **Transmissions:** All	**Igniter No. 4 Circuit Malfunction Conditions:** Engine started; and the PCM did not detect any IGF4 signals after detecting 2 IGT4 signals during the CCM test. **Note: This vehicle uses a Coil-On-Plug design Ignition System.** **Possible Causes:** • IGT or IGF signal circuit is open or shorted to ground • Igniter is damaged or has failed • PCM has failed
DTC: P1335 **1T CCM, MIL: Yes** **2004, 2005, 2006** **Models:** tC, xA, xB **Engines:** 1.5L VIN T, 2.4L VIN D, E **Transmissions:** All	**Crankshaft Position Sensor Circuit Malfunction Conditions:** Engine speed over 1000 rpm, and the PCM did not detect any CKP (NE) sensor signals for 500 ms. This code may be set by an intermittent loss of the CKP sensor signal. **Possible Causes:** • CKP sensor (+) circuit is open or shorted to ground • CKP sensor (-) circuit is open or shorted to ground • CKP sensor (+), (-) circuit is shorted to VREF or system power • CKP sensor is damaged or has failed • PCM has failed

DTC	Trouble Code Title, Conditions & Possible Causes
DTC: P1346 **1T CCM, MIL: Yes** **2004, 2005, 2006** **Models:** tC, xA, xB **Engines:** 1.5L VIN T, 2.4L VIN D, E **Transmissions:** All	**Variable Valve Timing, CMP Sensor Range/Performance Conditions:** Engine started; and the PCM detected a deviation between the CKP sensor and the VVT sensor signals due to a mechanical fault in the timing belt or the VVT sensor during testing. **Possible Causes:** • Worn timing belt (i.e., the belt may be skipping teeth) • PCM has failed
DTC: P1349 **1T CCM, MIL: Yes** **2004, 2005, 2006** **Models:** tC, xA, xB **Engines:** 1.5L VIN T, 2.4L VIN D, E **Transmissions:** All	**Variable Valve Timing System (Bank 1) Conditions:** Engine speed from 400-4000 rpm in closed loop, and the PCM detected the valve timing did not change from its initial position, or the current valve timing was fixed during the CCM test. **Possible Causes:** • Oil control valve is damaged or has failed • Valve timing is not correct • PCM has failed
DTC: P1500 **1T CCM, MIL: Yes** **2004, 2005, 2006** **Models:** tC, xA, xB **Engines:** 1.5L VIN T, 2.4L VIN D, E **Transmissions:** All	**Starter Signal Circuit Malfunction Conditions:** Engine cranking and the PCM detected it did not receive a signal on the starter signal circuit. **Possible Causes:** • Engine starter signal circuit is open or shorted to ground • Engine starter signal circuit is shorted to system power (B+) • Starter has failed • PCM has failed
DTC: P1520 **1T CCM, MIL: Yes** **2004, 2005, 2006** **Models:** tC, xA, xB **Engines:** 1.5L VIN T, 2.4L VIN D, E **Transmissions:** All	**Stop Light Switch Circuit Malfunction Conditions:** Vehicle driven to a speed of over 19 mph several times; and the PCM detected the Stop Light signal status did not change at least once under these operating conditions. **Possible Causes:** • Stop light switch is shorted to ground • Stop light switch is damaged or has failed • PCM has failed
DTC: P1600 **1T CCM, MIL: Yes** **2004, 2005, 2006** **Models:** tC, xA, xB **Engines:** 1.5L VIN T, 2.4L VIN D, E **Transmissions:** All	**PCM Battery Backup Circuit Malfunction Conditions:** Key on, and the PCM detected an unexpected voltage in the Battery Backup circuit (KAM circuit) during the CCM test. **Note: The PCM will not store any other codes with this code set.** **Possible Causes:** • Battery backup circuit is open (check EFI fuse and fuse link) • Battery terminals are corroded or loose • PCM has failed
DTC: P1725 **1T CCM, MIL: Yes** **2004, 2005, 2006** **Models:** xA, xB **Engines:** 1.5L VIN T **Transmissions:** A/T	**A/T Turbine Shaft Speed Sensor Circuit Malfunction Conditions:** No Shift Solenoid or P/N codes set, engine started, P/N switch indicating "off", gear position indicating 2nd, 3rd gear, or in O/D, no gear change occurring, and the PCM detected the Turbine Shaft Speed (ISS) sensor indicated less than 300 rpm for 4 seconds. **Possible Causes:** • Input shaft speed sensor signal (NT) circuit is open or shorted • Input shaft speed sensor is damaged or has failed • PCM has failed
DTC: P1780 **1T CCM, MIL: Yes** **2004, 2005, 2006** **Models:** tC, xA, xB **Engines:** 1.5L VIN T, 2.4L VIN D, E **Transmissions:** All	**Park/Neutral Position Switch Circuit Malfunction Conditions:** Engine started, engine runtime over 30 seconds, and the PCM detected two or more P/N inputs (Drive, Neutral, 2nd, Low or Reverse) at the same time; or with the engine speed from 1500-2500 rpm with the VSS indicating over 50 mph, it detected the P/N switch indicated "on" for 30 seconds during the CCM test. **Possible Causes:** • P/N switch is shorted to ground or wiring harness is shorted • P/N switch is out-of-adjustment, damaged or has failed • PCM has failed
DTC: P1780 **1T CCM, MIL: Yes** **2004, 2005, 2006** **Models:** xA, xB **Engines:** 1.5L VIN T **Transmissions:** All	**A/T Shift Solenoid Valve Circuit Malfunction Conditions:** Engine speed over 500 rpm, P/N switch indicating off, and the PCM detected an unexpected low voltage or unexpected high voltage (12v) on the Shift Solenoid Valve (ST) control circuit. **Possible Causes:** • ST shift solenoid connector is damaged or loose • ST shift solenoid control circuit is open or shorted to ground • ST shift solenoid is damaged or has failed • PCM has failed

OBD II Trouble Code List (P2xxx Codes)

DTC	Trouble Code Title, Conditions & Possible Causes
DTC: P2716 **1T CCM, MIL: Yes** **2004, 2005, 2006** **Models:** xA, xB **Engines:** 1.5L VIN T **Transmissions:** A/T	**A/T Pressure Control Solenoid 'D' Circuit Malfunction Conditions:** Engine speed over 500 rpm, P/N switch indicating off, and the PCM detected an unexpected low voltage (under 0.20v) on the Shift Solenoid Valve (SLT) control circuit, or it detected an unexpected high voltage (over 12.0v) on the SLT solenoid control circuit during the CCM test. **Possible Causes:** • SLT shift solenoid connector is damaged or loose • SLT shift solenoid control circuit is open or shorted to ground • SLT shift solenoid is damaged or has failed • PCM has failed
DTC: P2195 **1T CCM, MIL: Yes** **2004, 2005, 2006** **Models:** tC, xA, xB **Engines:** 1.5L VIN T, 2.4L VIN D, E **Transmissions:** All	**Air Fuel Sensor 1 (Bank 1 Sensor 1) Signal Stuck "Lean" Conditions:** Vehicle speed from 25-87 mph at over 1500 rpm with the throttle valve open, and the PCM detected the Air Fuel sensor signal indicated more than 3.80v for 10 seconds. **Possible Causes:** • A/FS1 signal circuit is open or shorted to ground • A/FS1 is damaged, contaminated or it has failed • Air induction system is severely restricted • Fuel Control component problems (e.g., low fuel pressure, or one or more severely restricted fuel injectors) • PCM has failed
DTC: P2196 **1T CCM, MIL: Yes** **2004, 2005, 2006** **Models:** tC, xA, xB **Engines:** 1.5L VIN T, 2.4L VIN D, E **Transmissions:** All	**Air Fuel Sensor 1 (Bank 1 Sensor 1) Signal Stuck "Rich" Conditions:** Vehicle speed from 25-87 mph at over 1500 rpm with the throttle valve open, and the PCM detected the Air Fuel sensor signal indicated more than 2.80v for 10 seconds. **Possible Causes:** • A/FS1 signal circuit is open or shorted to ground • A/FS1 is damaged, contaminated or it has failed • Air induction system is leaking (check for PCV system leaks) • Fuel component problem (high fuel pressure, leaking regulator or a leaking injector) • PCM has failed

SCION
COMPONENT TESTING

9

TABLE OF CONTENTS

Component Locations

tC

2.4L I4 MFI 2AZ-FE VIN D, E

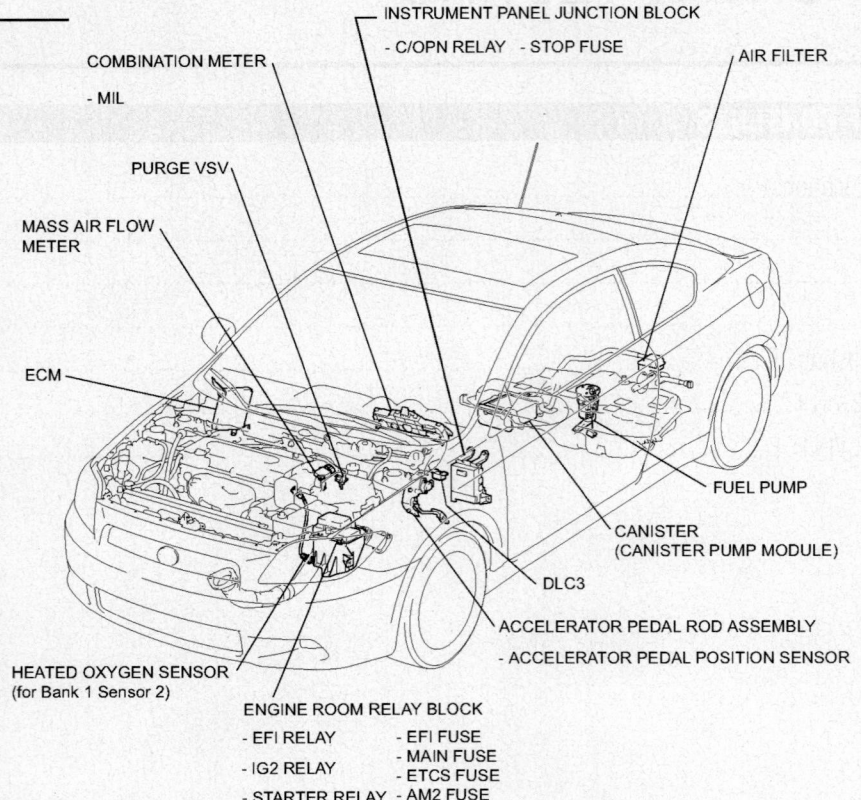

COMBINATION METER
- MIL

INSTRUMENT PANEL JUNCTION BLOCK
- C/OPN RELAY   - STOP FUSE

AIR FILTER

PURGE VSV

MASS AIR FLOW METER

ECM

FUEL PUMP

CANISTER
(CANISTER PUMP MODULE)

DLC3

ACCELERATOR PEDAL ROD ASSEMBLY
- ACCELERATOR PEDAL POSITION SENSOR

HEATED OXYGEN SENSOR
(for Bank 1 Sensor 2)

ENGINE ROOM RELAY BLOCK

- EFI RELAY   - EFI FUSE
                     - MAIN FUSE
- IG2 RELAY   - ETCS FUSE
- STARTER RELAY   - AM2 FUSE

29157_SCIO_G0001

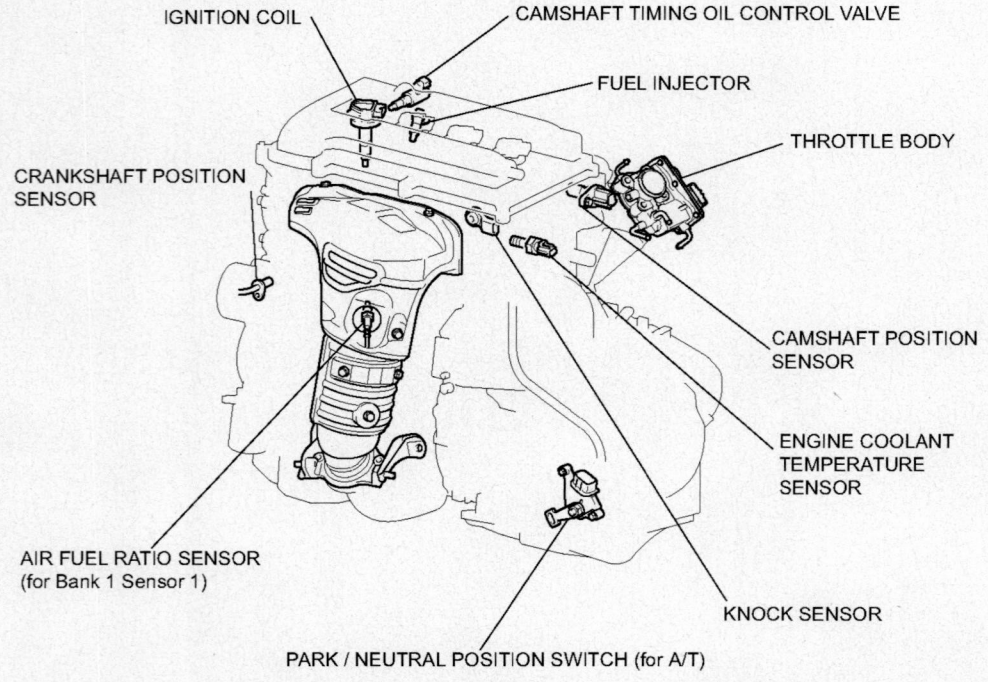

IGNITION COIL

CAMSHAFT TIMING OIL CONTROL VALVE

FUEL INJECTOR

CRANKSHAFT POSITION SENSOR

THROTTLE BODY

CAMSHAFT POSITION SENSOR

ENGINE COOLANT TEMPERATURE SENSOR

AIR FUEL RATIO SENSOR
(for Bank 1 Sensor 1)

KNOCK SENSOR

PARK / NEUTRAL POSITION SWITCH (for A/T)

29157_SCIO_G0002

xA

1.5L I4 MFI 1NZ_FE VIN T

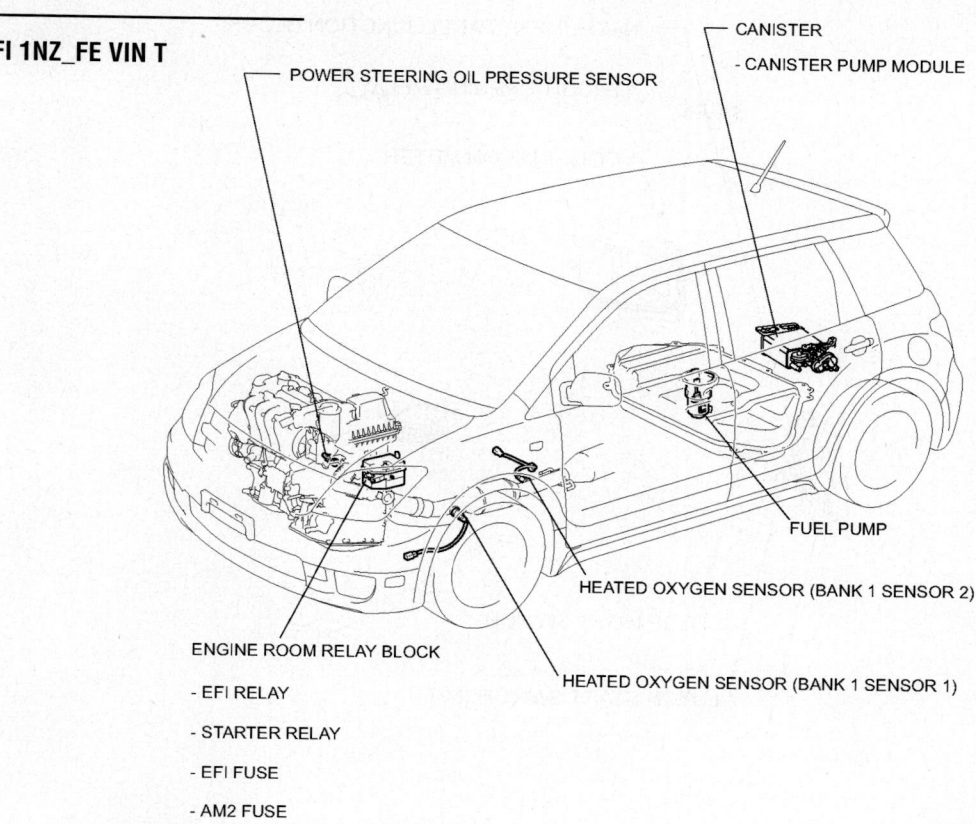

POWER STEERING OIL PRESSURE SENSOR

CANISTER

- CANISTER PUMP MODULE

FUEL PUMP

HEATED OXYGEN SENSOR (BANK 1 SENSOR 2)

HEATED OXYGEN SENSOR (BANK 1 SENSOR 1)

ENGINE ROOM RELAY BLOCK

- EFI RELAY

- STARTER RELAY

- EFI FUSE

- AM2 FUSE

29157_SCIO_G0003

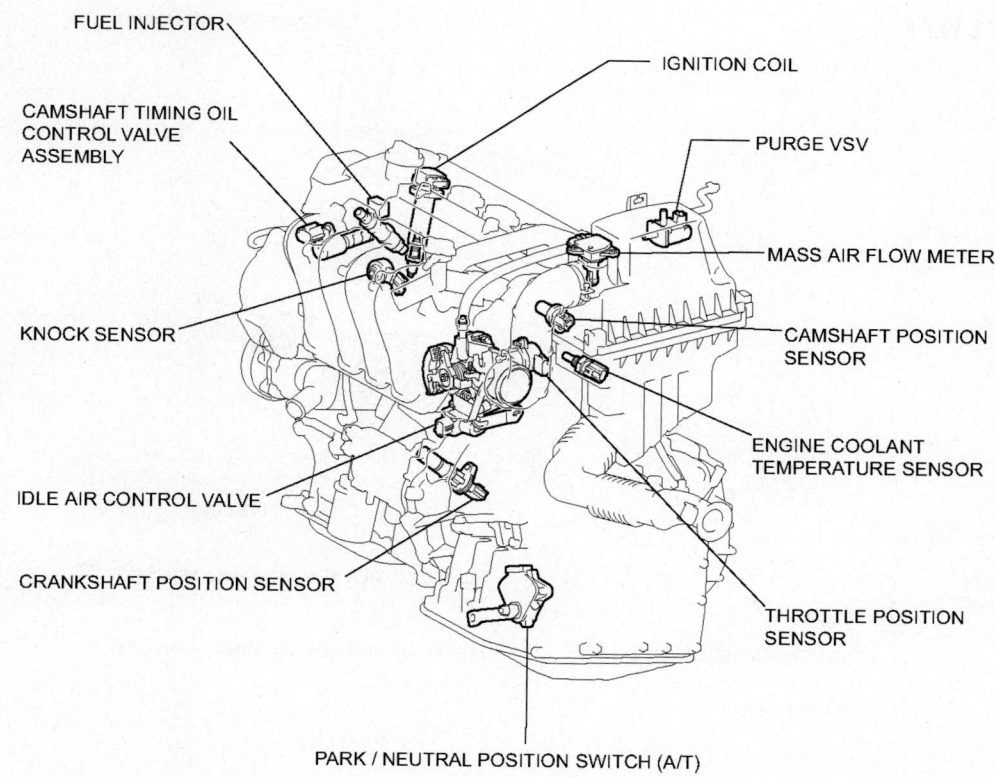

FUEL INJECTOR

IGNITION COIL

CAMSHAFT TIMING OIL
CONTROL VALVE
ASSEMBLY

PURGE VSV

MASS AIR FLOW METER

KNOCK SENSOR

CAMSHAFT POSITION
SENSOR

ENGINE COOLANT
TEMPERATURE SENSOR

IDLE AIR CONTROL VALVE

CRANKSHAFT POSITION SENSOR

THROTTLE POSITION
SENSOR

PARK / NEUTRAL POSITION SWITCH (A/T)

29157_SCIO_G0004

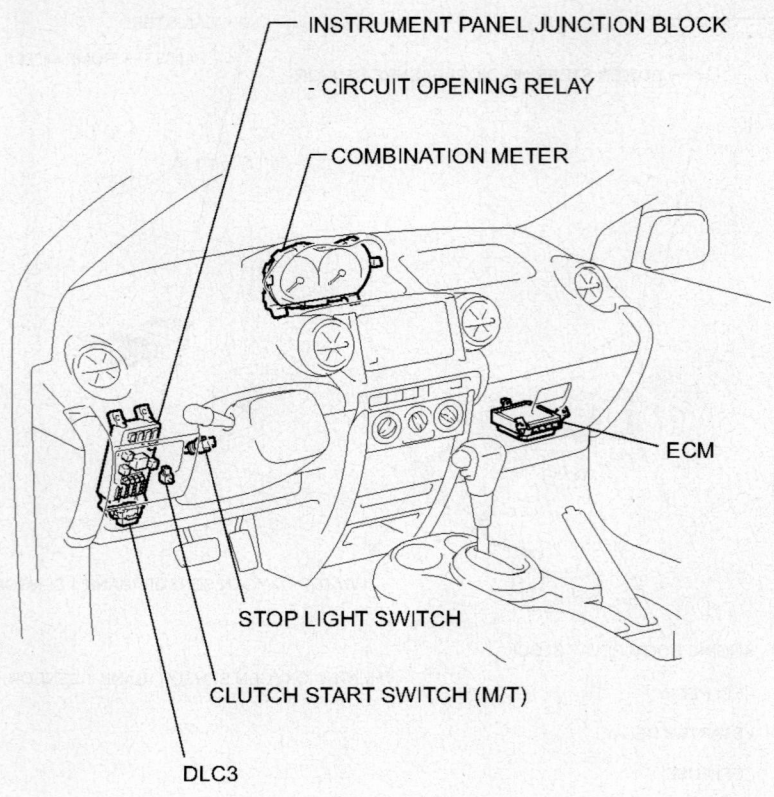

INSTRUMENT PANEL JUNCTION BLOCK

- CIRCUIT OPENING RELAY

COMBINATION METER

ECM

STOP LIGHT SWITCH

CLUTCH START SWITCH (M/T)

DLC3

29157_SCIO_G0005

xB

1.5L I4 MFI 1NZ_FE VIN T

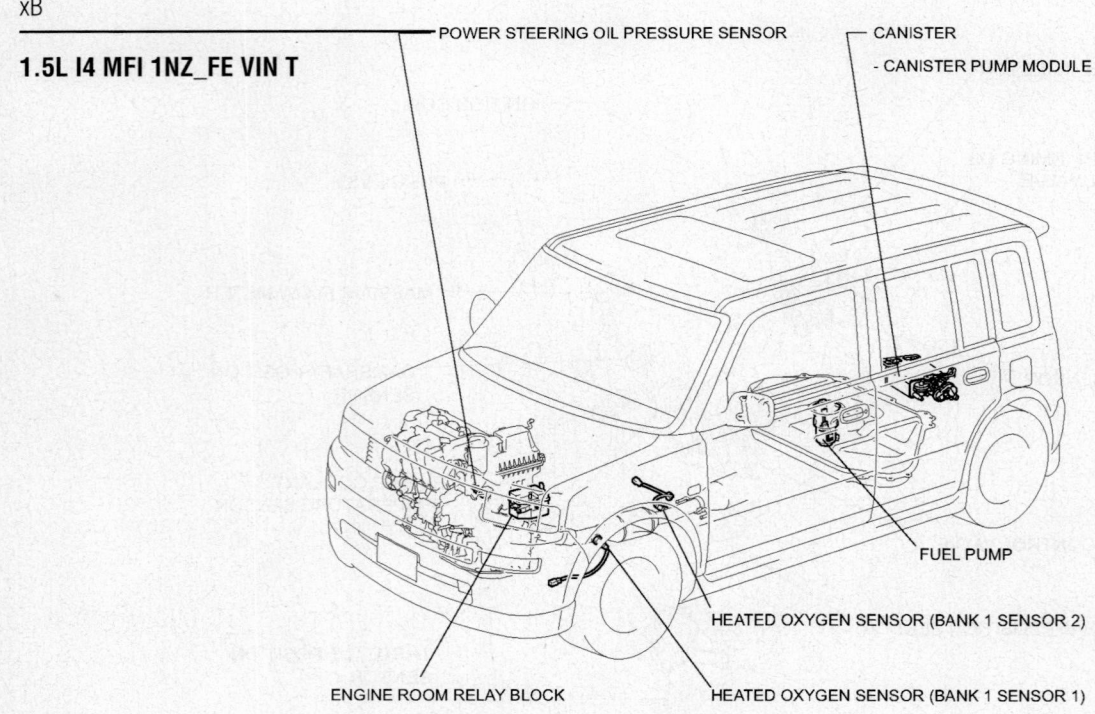

POWER STEERING OIL PRESSURE SENSOR

CANISTER

- CANISTER PUMP MODULE

FUEL PUMP

HEATED OXYGEN SENSOR (BANK 1 SENSOR 2)

ENGINE ROOM RELAY BLOCK

HEATED OXYGEN SENSOR (BANK 1 SENSOR 1)

- STARTER RELAY

- EFI FUSE

- AM2 FUSE

29157_SCIO_G0006

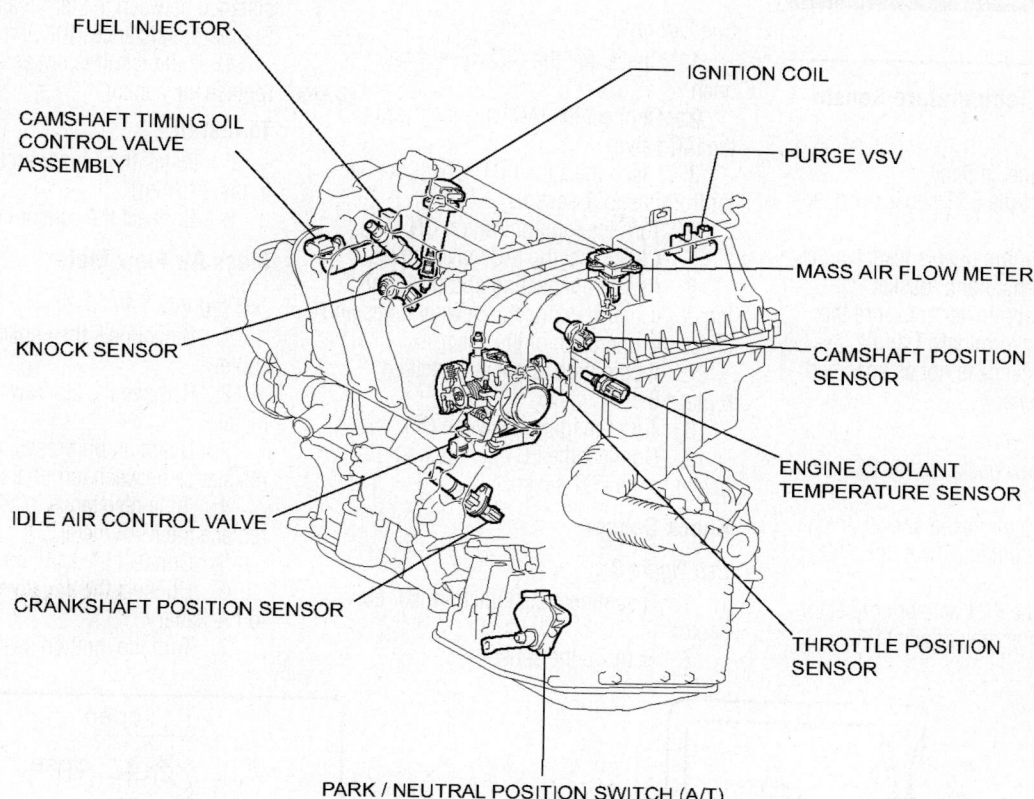

FUEL INJECTOR

IGNITION COIL

CAMSHAFT TIMING OIL
CONTROL VALVE
ASSEMBLY

PURGE VSV

MASS AIR FLOW METER

KNOCK SENSOR

CAMSHAFT POSITION
SENSOR

ENGINE COOLANT
TEMPERATURE SENSOR

IDLE AIR CONTROL VALVE

CRANKSHAFT POSITION SENSOR

THROTTLE POSITION
SENSOR

PARK / NEUTRAL POSITION SWITCH (A/T)

29157_SCIO_G0007

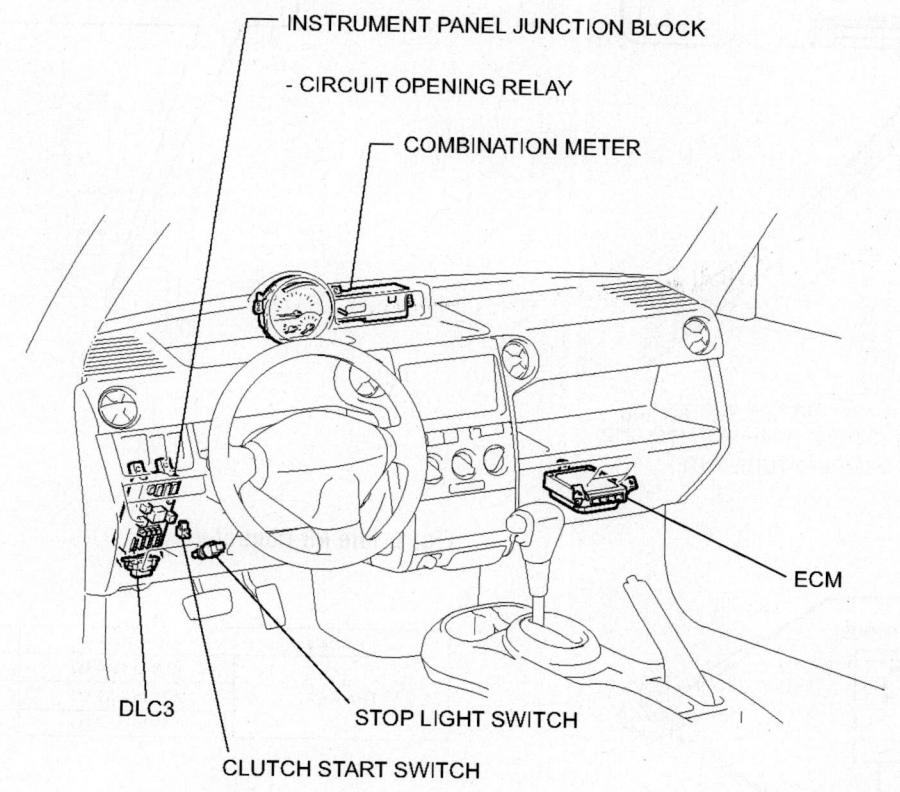

INSTRUMENT PANEL JUNCTION BLOCK

- CIRCUIT OPENING RELAY

COMBINATION METER

ECM

DLC3

STOP LIGHT SWITCH

CLUTCH START SWITCH

29157_SCIO_G0008

Component Testing

1NZ-FE VIN T

Engine Coolant Temperature Sensor

See Figure 1.

1. Drain engine coolant.
2. Disconnect the ECT sensor connector.
3. Using a 19 mm deep socket wrench, remove the ECT sensor and gasket.
4. Using an ohmmeter, measure the resistance between terminals 1 and 2.
5. If the resistance is not as specified, replace the ECT sensor.

To install:

6. Install a new gasket to the ECT sensor.
7. Using a 19 mm deep socket, install the ECT sensor. Tighten to 14 ft. lbs. (20 Nm).
8. Connect the ECT sensor connector.
9. Refill with engine coolant.

Idle Air Control Valve

See Figure 2.

1. Check that the IAC valve is half open.
2. Connect the IAC valve connector to the IAC valve.
3. Disconnect the ECT sensor connector from the ECT sensor.
4. Turn the ignition switch ON.
5. Check that the IAC valve moves.
6. Repeat connecting and disconnecting of IAC valve connector several times and check the operation of the valve.
7. If operation is not as specified, replace the IAC valve.
8. Turn the ignition switch OFF.
9. Connect the ECT sensor connector to the ECT sensor.

Knock Sensor

See Figure 3.

1. Disconnect the knock sensor connector.
2. Remove the sensor.

3. Using an ohmmeter, check resistance between the terminal and body. Standard resistance: 10KOhm or greater.
4. If the result is not as specified, replace the sensor.

To install:

5. Install the sensor and tighten to 29 ft. lbs. (39 Nm).
6. Connect the sensor connector.

Mass Air Flow Meter

See Figures 4 and 5.

1. Disconnect the negative battery cable.
2. Remove the 2 screws and MAF meter.
3. Using an ohmmeter, measure the resistance between terminals THA and E2.
4. If the resistance is not as specified, replace the MAF meter.
5. Connect the MAF meter connector.
6. Connect the negative terminal cable to the battery.
7. Turn the ignition switch ON.

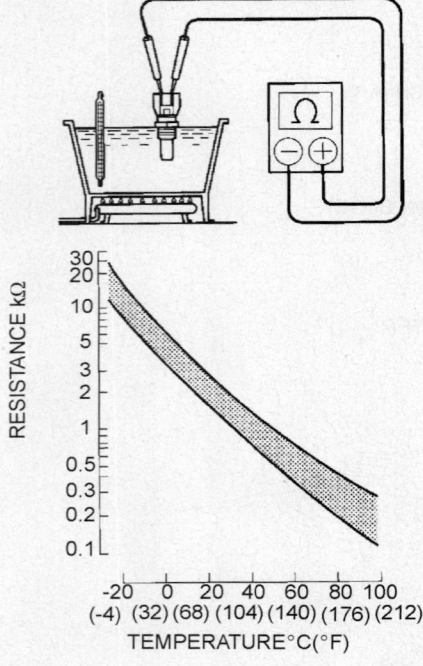

Fig. 1 ECT testing

29157_SCIO_G0011

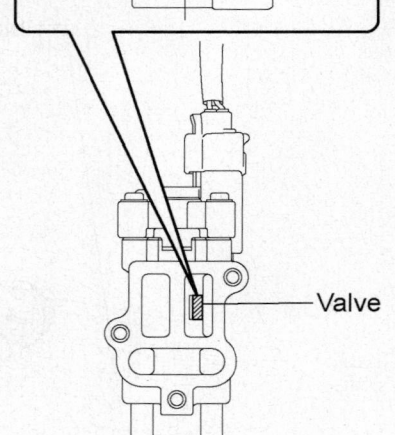

Fig. 2 Idle Air Control Valve

29157_SCIO_G0012

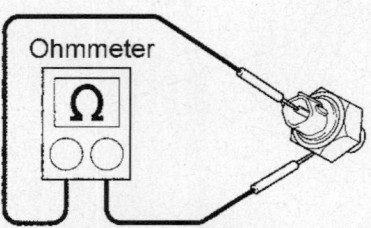

Fig. 3 Testing the knock sensor

29157_SCIO_G0013

Terminals	Resistance	Temperature
THA - E2	13.6 to 18.4 kΩ	-20 °C (-4 °F)
	2.21 to 2.69 kΩ	20 °C (68 °F)
	0.49 to 0.67 kΩ	60 °C (140 °F)

Fig. 4 Temperature/Resistance table

29157_SCIO_G0009

8. Using a voltmeter, connect the positive (+) tester probe to terminal VG, and negative (-) tester probe to terminal E2G.

9. Blow air into the MAF meter, and check that the voltage fluctuates. If the operation is not as specified, replace the MAF meter.

10. Turn the ignition switch OFF.

11. Disconnect the negative terminal cable from the battery.

12. Disconnect the MAF meter connector.

13. Install the MAF meter with the 2 screws.

14. Reconnect MAF meter connector.

Heated Oxygen Sensor

See Figures 6 and 7.

Measure the resistance between terminals +B and HT. If the resistance is not as specified, replace the oxygen sensor.

Throttle Position Sensor

See Figures 8 and 9.

1. Disconnect the TPS connector.

2. Using an ohmmeter, measure the resistance between terminals VC and E2. Standard value: 2.5–6.0 KOhms.

3. Check the resistance between terminals VTA and E2. Resistance should increase as throttle is opened.

4. If the resistance is not as specified, replace the TPS.

2AZ-FE VIN D, E

Mass Air Flow Meter

See Figure 10.

1. Apply battery voltage across terminals 1 (+B) and 2 (E2G).

2. Using a voltmeter, connect the positive (+) tester probe to terminal VG, and negative (-) tester probe to terminal E2G.

3. Blow air into the MAF meter, and check that the voltage fluctuates.

4. Measure the resistance between terminals 4 (THA) and 5 (E2).

5. If the result is not as specified, replace the MAF meter.

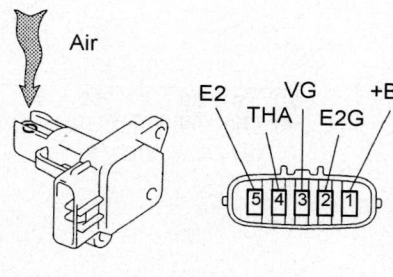

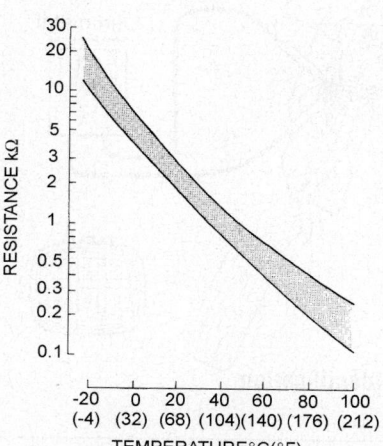

Fig. 5 MAF Terminal Identification and Resistance chart

29157_SCIO_G0010

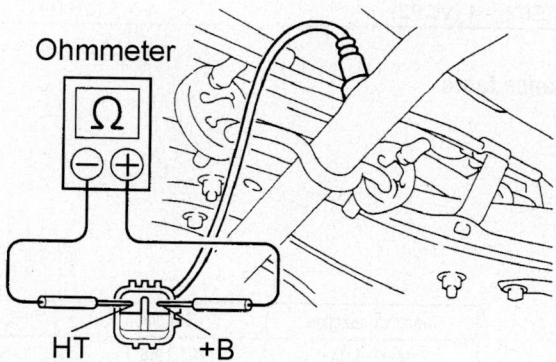

Fig. 6 Oxygen sensor testing

29157_SCIO_G0014

20°C (68°F)	11 - 16 Ω
800°C (1,472°F)	23 - 32 Ω

Fig. 7 Temperature/resistance table

29157_SCIO_G0015

Fig. 8 TPS terminal identification

29157_SCIO_G0016

Throttle Valve Movement	Resistance
Fully open	1.9 to 4.9 kΩ
Fully close	0.2 to 1.0 kΩ

29157_SCIO_G0017

Fig. 9 Resistance between terminals VTA and E2

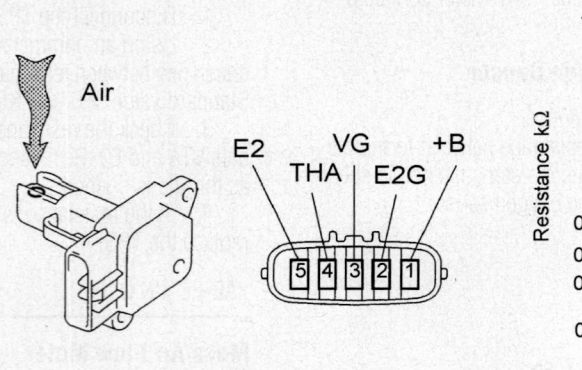

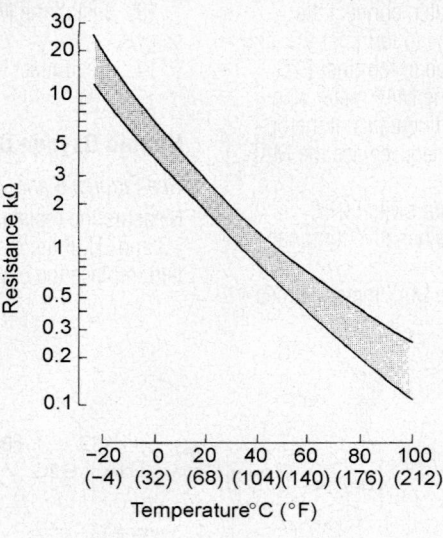

29157_SCIO_G0018

Fig. 10 Connector pin identification and Resistance table

Accelerator Pedal Position Sensor

See Figures 11 and 12.

1. Measure the resistance between the terminals.

2. If the result is not as specified, replace the pedal assy.

Throttle Assembly

See Figures 13 and 14.

1. Measure the resistance between the terminals.

2. If the result is not as specified, replace the throttle body assy.

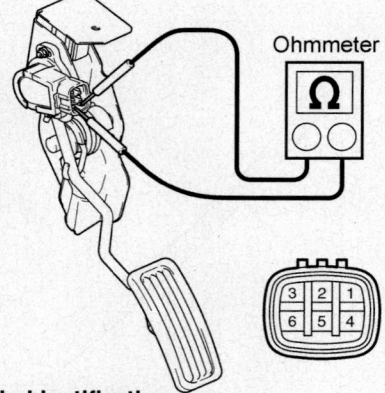

29157_SCIO_G0019

Fig. 11 Connector pin identification

Tester Connection	Specified Condition
3 (EP1) - 6 (VCP1)	1.5 to 6.0 kΩ
1 (EP2) - 4 (VCP2)	1.5 to 6.0 kΩ

29157_SCIO_G0020

Fig. 12 Resistance table

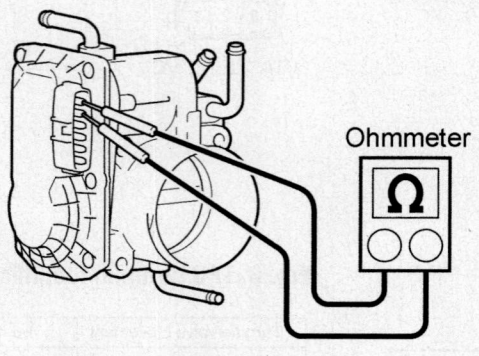

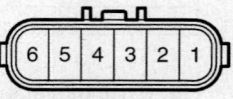

29157_SCIO_G0021

Tester Connection	Condition	Specified Condition
2 (M+) – 1 (M–)	20°C (68°F)	0.3 to 100 Ω
5 (VC) – 3 (E2)	20°C (68°F)	1.2 to 3.2 kΩ

29157_SCIO_G0022

Fig. 13 Connector pin identification   ## Fig. 14 Resistance table

Coolant Temperature Sensor

See Figures 15 and 16.

1. Measure the resistance between the terminals.
2. If the result is not as specified, replace the ECT.

Knock Sensor

1. Using an ohmmeter, measure the resistance between terminals. Resistance should be 120–280 KOhms.
2. If the resistance is not as specified, replace the sensor.

Heated Oxygen Sensor

See Figures 17 and 18.

1. Measure the resistance between terminals +B and HT. If the resistance is not as specified, replace the oxygen sensor.

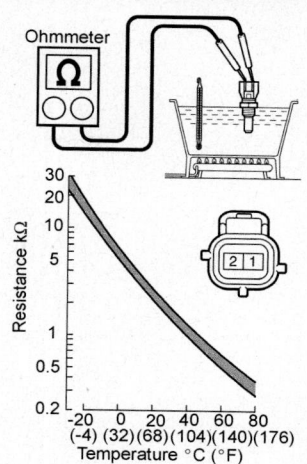

Fig. 15 ECT testing

29157_SCIO_G0023

Tester Connection	Specified Condition
1 (E2) - 2 (THW)	2.32 to 2.59 kΩ at 20°C (68°F)
1 (E2) - 2 (THW)	0.310 to 0.326 kΩ at 80°C (176°F)

29157_SCIO_G0024

Fig. 16 Resistance and Temperature table

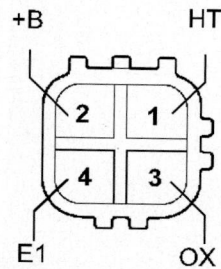

Fig. 17 Connector terminal identification

29157_SCIO_G0025

20°C (68°F)	11 - 16 Ω
800°C (1,472°F)	23 - 32 Ω

29157_SCIO_G0026

Fig. 18 Temperature/resistance table

SCION
PIN CHARTS

TABLE OF CONTENTS

PIN CHARTS

Introduction

A Pin Voltage Table is a term used to describe a table that identifies PCM pins, wire colors of the PCM circuits, circuit descriptions and "known good" values for devices that connect to the PCM. These tables include the following information:

- Signals from various sensors (ECT, IAT, MAP, TPS, etc.)
- Signals from various switches (PNP, PSP, WOT, etc.)
- Signals from oxygen sensors (O2S, HO2S)
- Signals from output devices (IAC, INJ, TCC, etc.)
- Power & ground signals

Pin Voltage Tables

Information contained within the Pin Voltage Tables can be used to:

- Test circuits for open, short to power or short to ground faults
- Check the operation of a component before or after a repair
- Check the operation of a component or system by viewing signals on PCM input/output circuits with a DVOM or Lab Scope

Using a Breakout Box

There are several Breakout Box (BOB) designs available for use to test the PCM and its input and output circuits. However, all of them require removal of the wire harness to the PCM so that the BOB can be installed between the PCM and wire harness connector. Several breakout boxes require the use of overlays in order to allow the tool to be used on more than one year or engine type. Always verify that the correct adapter and overlays are used to prevent connection to the wrong circuits and a misdiagnosis.

Power and Ground Circuit Checks

Measurements made at the BOB are accomplished via test leads and probes from the DVOM or a Lab Scope. If any of the terminals on the PCM or BOB are damaged or loose, test measurements made at the Breakout Box will be inaccurate. To verify the PCM battery power and ground circuits are normal (correct) at the BOB, test the condition of the circuit between the battery negative (-) post and these circuits prior to starting a test sequence.

Diagnosis with Pin Voltage Tables

See Figure 1.

Once an actual PCM pin voltage reading is recorded, it can be compared to an example from a vehicle with "known good" values. In the example shown the Value at Hot Idle for the EVP sensor signal (0.4v) is the "known good" value.

Wire Color Changes

Every effort has been made to obtain and list the correct circuit wire colors for all vehicles. However, running changes from the vehicle manufacturer can cause the wrong colors to be listed.

PCM Pin #	W/Color	Circuit Description (60-Pin)	Value at Hot Idle
27	BN/LG	EVP Sensor Signal	0.4v

Fig. 1 Example

tC PIN CHARTS

2005-06 tC 2.4L I4 VIN D, E E6 31 Pin Connector

PCM Pin #	Wire Color	Circuit Description (31 Pin)	Value at Hot Idle
1	BK/RD	EFI Main Relay Output (B+)	12-14v
2	BL/RD	Battery Direct (+BM)	12-14v
3	BK/YL	Battery Direct (BATT)	12-14v
4	VT	EVAP Vapor Pressure (VSV)	12v or 0v
5	BK/OR	Tachometer Signal (TACO)	DC pulse signals
6-7, 20, 24, 26	---	Not Used	---
8	BK/WT	EFI Main Relay Control	Relay On: 12v, Off: 0v
9	BK/OR	Ignition Switch Power (B+)	12-14v
10	GN/RD	Circuit Opening Relay (FC)	0-3v, off-idle: 12v
11	WT	SIL Signal (Scan Tool)	Digital Signal
12	GN	Tail Light Switch Signal	Switch Off: 1.5v, On: 12v
13	BK/YL	Mirror Heater Switch Signal	Switch Off: 1.5v, On: 12v
14	PK/BK	Data Link Connector (TC)	12-14v
15	BR	Power Ground (EOM)	<0.1v
18	GN/RD	MIL (lamp) Control	MIL Off: 12v, On: 1v
19	RD	Scan Tool (WFSE)	12v
21	PK	EVAP Vapor Pressure Sensor	2.9-3.1v (with hose off)
22	BL/YL	APP Sensor (VPA) Signal	0.8-1.2v
23	WT/RD	APP Sensor (VPA2) Signal	0.8-1.2v
25	RD	APP Sensor VREF	4.5-5.5v
27	RD	APP Sensor VREF	4.5-5.5
28	LG/BK	APP Sensor Ground (EPA)	<0.050v
29	LG	APP Sensor Ground (EPA2)	<0.050v

2005-06 tC 2.4L I4 VIN D, E E7 35 Pin Connector

PCM Pin #	Wire Color	Circuit Description (35 Pin)	Value at Hot Idle
1-3, 5-6, 12-13	---	Not Used	---
4	BL	HO2S-12 (B1 S2) Heater	Heater On: <1v
7	OR	Overdrive Indicator Control	LED Off: 12v, On: 0v
8	YL	A/T Select Switch Low Signal	In Low: 12v, Others: 0v
9	BL/WT	A/T Select Switch 2nd Signal	In 2nd: 12v, Others: 0v
10	WT/BL	A/T Select Switch 'D' Signal	In Drive: 12v, Others: 0v
11	RD/BK	A/T Select Switch 'R' Signal	In 'R': 12v, Others: 0v
14	YL/GN	Hot Engine Lamp Control	Lamp Off: 12v, On: 1v
15	GN/WT	Transponder Amplifier Code	Inserting key: pulses
16	VT	Transponder IMLD Signal	Inserting key: pulses
17	VT/WT	Speedometer Indicator	Moving: 0-5-0v
18, 20-21	---	Not Used	---
19	GN/WT	Stop Light Switch Signal	Brake Off: 0v, On: 12v
22	BK	HO2S-12 (B1 S2) Signal	0.1-1.1v
23-25, 28, 32	---	Not Used	---
26	BL/YL	Theft Deterrent ECU Signal	No key inserted: pulses
27	RD/BL	Transponder Amplifier Code	Inserting key: pulses
29	GN/OR	Overdrive Main Switch	Switch Off: 12v, On: 0v
31	PK/BL	A/C Switch Signal	AC On: 9-14v
33	BK	A/C Switch Signal (Auto AC)	AC On: 9-14v
34	BL	Transponder Amplifier Signal	Key Inserted: <1.5v

2005-06 tC 2.4L I4 VIN D, E E8 32 Pin Connector

PCM Pin #	Wire Color	Circuit Description (24 Pin)	Value at Hot Idle
1	BR	Power Ground (E1)	<0.1v
2	BL/WT	A/C Magnetic Clutch (ACMG)	A/C On: <3.0v
4	WT	Throttle Body Motor (M-)	DC pulse signals
5	BK	Throttle Body Motor (M+)	DC pulse signals
6	WT/BK	Sensor Ground (ME01)	<0.050v
7	WT/BK	Power Ground (E03)	<0.1v
8	WT/BK	Power Ground (E02)	<0.1v
9	GN/WT	Cooling Fan Relay Control	Relay Off: 12v, On: 1v
10	RD/WT	PSP Switch Signal	Straight: 12v, Turning: 0v
11	BK/RD	EVAP Solenoid Control (VSV)	12v or 0v
12	GN	EVAP Canister Closed Valve	12v or 0v
17	BR	Sensor Shield Ground	<0.050v
15	YL	Cam Timing Oil Valve (OCV-)	Pulse signals
16	BK/WT	Cam Timing Oil Valve (OCV+)	Pulse signals
30	YL/BK	Cooling Fan Relay Control	Relay Off: 12v, On: 1v

2005-06 tC 2.4L I4 VIN D, E E9 35 Pin Connector

PCM Pin #	Wire Color	Circuit Description (35 Pin)	Value at Hot Idle
1	WT	Knock Sensor Signal	<0.075v AC
2-3	---	Not Used	---
4	BK/RD	AFS-11 (B1 S1) Heater	1v (Heater On)
5	---	Not Used	---
6	WT/BK	Power Ground (E05)	<0.1v
7	WT/BK	Power Ground (E04)	<0.1v
8	BK/YL	Neutral Start Switch	In P/N: 9-11v (cranking)
9	BK/WT	Starter Switch Signal (STA)	9-11v (cranking)
10	---	Not Used	---
11	YL	A/T-ECT Solenoid (DSL)	In Lockup: 12-14v
12-15	---	Not Used	---
16	BL/RD	A/T-ECT Solenoid (SL2-)	Moving in 3rd or OD: <1v
17	BL/YL	A/T-ECT Solenoid (SL2+)	Moving in 3rd or OD: <1v
18	PK	A/T-ECT Solenoid (SL1-)	Moving in 1st Gear: <1v
19	RD/BK	A/T-ECT Solenoid (SL1+)	Moving in 1st Gear: <1v
18	WT	IAC Signal (RSD)	Pulses
23	OR	AFS-11 (B1 S1) Signal (AF+)	3.0-3.6v
24	RD	MAF Sensor Signal (VG)	1.1-1.5v
26	RD	Counter Gear Speed (NC+)	AC pulse signals
27	BL	Turbine Speed Sensor (NT+)	AC pulse signals
28-30	---	Not Used	---
31	WT	AFS-11 (B1 S1) Signal (AF-)	Fixed at 3v
32	BL/WT	MAF Sensor Ground (E2G)	<0.050v
33	---	Not Used	---
34	GN	Counter Gear Speed (NC-)	AC pulse signals
35	LG	Turbine Speed Sensor (NT-)	AC pulse signals

2005-06 tC 2.4L I4 VIN D, E E10 34 Pin Connector

PCM Pin #	Wire Color	Circuit Description (34 Pin)	Value at Hot Idle
1	BL	Injector 1 Control	2.0-3.3 ms
2	RD	Injector 2 Control	2.0-3.3 ms
3	YL	Injector 3 Control	2.0-3.3 ms
4	WT	Injector 4 Control	2.0-3.3 ms
5	---	Not Used	---
6	WT/BK	Power Ground (E02)	<0.1v
7	WT/BK	Power Ground (E01)	<0.1v
8	RD/WT	Igniter Transistor 1 Control	7% duty cycle
9	PK	Igniter Transistor 2 Control	7% duty cycle
10	LG/BK	Igniter Transistor 3 Control	7% duty cycle
11	BL/YL	Igniter Transistor 2 Control	7% duty cycle
12	---	Not Used	---
13	YL	Sensor VREF (VC)	4.9-5.1v
14-15	---	Not Used	---
16	YL/BK	A/T-ECT Solenoid (SLT-)	During shifting: 12v
17	YL/RD	A/T-ECT Solenoid (SLT+)	During shifting: 12v
18	YL	TP Sensor VREF (VC)	4.9-5.1v
19	GN/YL	ECT Sensor Signal	At 180°F: 0.51v
20	LB	IAT Sensor Signal	At 100°F: 2.60v
21	LG	TP Sensor Signal	0.4-1.0v
22	---	Not Used	---
23	WT/RD	Igniter Signal (IGF)	Digital Signal: 0-5-0v
24-25	---	Not Used	---
26	BK/WT	CMP Sensor Signal (G22+)	AC pulse signals
27	RD	CKP Sensor Signal (NE+)	AC pulse signals
28	BR	Power Ground (E2)	<0.1v
29	---	Not Used	---
30 (TMC)	GN	A/T ATF Sensor Signal	<1.5v at 239°F
30 (TMM)	GN/RD	A/T ATF Sensor Signal	<1.5v at 239°F
31	BK/RD	TP Sensor 2 Signal (VTA2)	2.0-2.9v
32-33	---	Not Used	---
34	GN	CKP/CMP Sensor Ground	<0.050v

xA, xB PIN CHARTS

2004-2006 xA, xB 1.5L I4 DOHC MFI VIN T (All) E4 31 Pin Connector

PCM Pin #	Wire Color	Circuit Description (31 Pin)	Value at Hot Idle
1	BK/RD	EFI Main Relay Power	12-14v
2	BL/BK	Maximum Hot Switch (MHSW)	Switch Open: 0v, Closed: 12v
3	BK/YL	Battery Direct	12-14v
4	---	Not Used	---
5	BK	Tachometer Signal (TACO)	Pulse Signals
6	YL/RD	Cooling Fan 1 Relay (CF)	Relay Off: 12v, On: 1v
7	WT/BL	A/C Single Pressure Switch (FAN)	Switch Open: 12v, Closed: 1v
8-9	---	Not Used	---
10	BK	Circuit Opening Relay (FC)	0-3v, off-idle: 12v
11	YL/RD	MIL (lamp) Control	MIL Off: 12v, On: 1v
12	BL/WT	Heater Sub1 Relay Control (ELS)	Relay Off: 12v, On: 1v
13	---	Not Used	---
14	YL/BK	Center Airbag Assembly	Digital Signals
15-17	---	Not Used	---
18	LG/BK	Scan Tool Signal (SIL)	Digital Signal
19	RD	Scan Tool (WFSE)	12v
20	PK/BL	Data Link Connector (TC)	12-14v
21	BK	EVAP Vapor Pressure Sensor	Fuel Cap Off: 2.9-3.7v
22-31	---	Not Used	---

2004-2006 xA, xB 1.5L I4 DOHC MFI VIN T (All) E5 35 Pin Connector

PCM Pin #	Wire Color	Circuit Description (35 Pin)	Value at Hot Idle
1-2	---	Not Used	---
3	YL/RD	Water Temperature Cool Indicator	Indicator On: 12v, Off: 0v
4	BK/RD	HO2S-12 (B1 S2) Heater	1v (Heater On)
5-6	---	Not Used	---
7	GN/OR	A/T Overdrive "Off" Indicator	Light Off: 12v
8	GN/WT	A/T Select Switch Low	In Low: 12v, Others: 0v
9	GN	A/T Select Switch 2nd	In 2nd: 12v, Others: 0v
10	RD/YL	A/T Select Switch Drive	In 'D': 12v, Others: 0v
11	RD/WT	A/T Select Switch Reverse	In 'R': 12v, Others: 0v
12	---	Not Used	---
13	YL	PSP Switch Signal	Straight: 12v, Turning: 0v
14-16	---	Not Used	---
17	VT/WT	Speed Signal from Combination Meter	At 55 mph: 48 Hz
18	---	Not Used	---
19	GN/WT	Stop Light Switch Signal	Brake Off: 0v, On: 12v
20-27	---	Not Used	---
28	YL/RD	Water Temperature Hot Indicator	Indicator On: 12v, Off: 0v
29	GN/OR	A/T: Overdrive Main Switch	Switch Off: 12v
30	---	Not Used	---
31	BK/WT	A/C Amplifier Signal (AC)	Clutch On: 1.5v, Off: 12v
32	---	Not Used	---
33	BK	A/C Amplifier Signal (ACT)	Clutch On: 12v, Off: 1.5v
34-35	---	Not Used	---

2004-2006 xA, xB 1.5L I4 DOHC MFI VIN T (All) E6 34 Pin Connector

PCM Pin #	Wire Color	Circuit Description (34 Pin)	Value at Hot Idle
1	BK/OR	Injector 1 Control	2.0-3.3 ms
2	BK/YL	Injector 2 Control	2.0-3.3 ms
3	BK/WT	Injector 3 Control	2.0-3.3 ms
4	BK/BL	Injector 4 Control	2.0-3.3 ms
5	BK/RD	IAC Signal (RSD)	Pulse Signals
6	BR	Power Ground (E02)	<0.1v
7	BR	Power Ground (E01)	<0.1v
8	GN/RD	Igniter Transistor 1 Control	Digital Signal: 0-5-0v
9	GN/BK	Igniter Transistor 2 Control	Digital Signal: 0-5-0v
10	GN/OR	Igniter Transistor 3 Control	Digital Signal: 0-5-0v
11	GN/YL	Igniter Transistor 4 Control	Digital Signal: 0-5-0v
12	WT/GN	EVAP Purge Solenoid (VSV)	12v or 0v
16	WT	A/T-ECT Solenoid (SLT-)	Pulse Signals
17	WT/BL	A/T-ECT Solenoid (SLT+)	Pulse Signals
18	RD/WT	Sensor VREF (VC)	4.9-5.1v
19	RD/BL	ECT Sensor Signal (THW)	1-4v (varies with temp.)
20	YL/BK	IAT Sensor Signal	1-4v (varies with temp.)
21	YL/RD	TP Sensor Signal (VTA)	1.1-1.9v
23	YL	Igniter Signal (IGF)	Pulse Signal: 0-5-0v
26	BK	CMP Sensor Signal (G2+)	AC pulse signals
27	OR	CKP Sensor Signal (NE+)	AC pulse signals
28	BR	Sensor Ground (E2)	<0.050v
30	YL/GN	A/T Oil Temperature Sensor (OIL)	At 230°F: 0.95v
34	WT	CKP Sensor Ground (NE-)	<0.050v

2004-2006 xA, xB 1.5L I4 DOHC MFI VIN T (All) E7 35 Pin Connector

PCM Pin #	Wire Color	Circuit Description (35 Pin)	Value at Hot Idle
1	WT	Knock Sensor Signal (KNK1)	<0.075v AC
4	BK/RD	HO2S-11 (B1 S1) Heater	1v (Heater On)
5	BR	Power Ground (E03)	<0.1v
7	BL	Generator Control Signal	12-0-12v
8	BK	A/T Neutral Start Switch (NSW)	Cranking: 9-11v
9	BK/YL	Starter Switch Signal (STA)	9-11v (cranking)
12	BK/WT	A/T-ECT Solenoid (ST)	Pulse Signals
13	GN	A/T-ECT Solenoid (SL)	In Lockup: 12-14v
14	WT/RD	A/T: ECT Solenoid (S2)	1st or OD: 1v
15	WT/GN	A/T-ECT Solenoid (S1)	In 3rd or OD: 1v
21	BK	HO2S-12 (B1 S2) Signal	0.1-1.1v
23	WT	HO2S-11 (B1 S1) Signal	0.1-1.1v
24	PK	MAF Sensor Signal (VG)	1.1-1.5v
27	RD	Turbine Speed Sensor (NT+)	AC pulse signals
28	BR	Power Ground (EC)	<0.1v
29	VT	PSP Switch Signal	Straight: 12v, Turning: 0v
32	VT	MAF Sensor Ground (EVG)	<0.050v
35	BK	Turbine Speed Sensor (NT-)	AC pulse signals

TABLE OF CONTENTS

OBD II VEHICLE APPLICATIONS

SUBARU

Baja
2003-2006
2.5L I4 MFI (SOHC).. VIN 6
2004-2006
2.5L I4 MFI (DOHC)... VIN 6

Forester
1998-2006
2.5L I4 MFI (SOHC).. VIN 6
2004-2006
2.5L I4 MFI (DOHC)... VIN 6

Impreza/Impreza Outback/WRX
1996
1.8L I4 MFI .... VIN 1
1996-1997
1.8L I4 MFI .... VIN 2
2002-2004
2.0L I4 MFI .... VIN 2
1996-2001
2.2L I4 MFI .... VIN 4
1998-2006
2.5L I4 MFI (SOHC).. VIN 6
2004-2006
2.5L I4 MFI (DOHC)... VIN 7

Legacy
1996-1998
2.2L I4 MFI .... VIN 3
1996-1999
2.2L I4 MFI .... VIN 4
1996-2006
2.5L I4 MFI (SOHC).. VIN 6
2004-2006
2.5L I4 MFI (DOHC)... VIN 6

Outback
2000-2006
2.5L I4 MFI (SOHC).. VIN 6
2004-2006
2.5L I4 MFI (DOHC)... VIN 6
2002-2006
3.0L V6 MFI .... VIN 8

SVX
1996-1997
3.3L V6 MFI .... VIN 8

Tribeca
2006
3.0L V6 MFI... VIN 8

DIAGNOSTIC TROUBLE CODES

OBD II Trouble Code List (P0xxx Codes)

DTC	Trouble Code Title, Conditions & Possible Causes:
DTC: P0011 **2T CCM, MIL: Yes** **2004-06** **Engines:** 2.5L SOHC & DOHC, 3.0L **Models:** Baja, Forester, Impreza, WRX, Legacy, Outback, B9 Tribeca **Transmissions:** All	**Intake Camshaft Position Timing (over advanced) Bank 1** Engine stalls. Erroneous idling. **Possible Causes:** • Check any other DTC on display • Using a scan tool, check current data. Check AVCS system. • Oil pipe clog • Clogged oil flow control solenoid valve • Intake camshaft damaged • Timing chain problem
DTC: P0016 **2T CCM, MIL: Yes** **2005-06** **Engines:** 2.5L DOHC, 3.0L **Models:** Forester, Impreza, WRX, B9 Tribeca, Outback **Transmissions:** All	**Crankshaft Position (camshaft position correlation) Bank 1** Engine stalls. Erroneous idling. **Possible Causes:** • Check any other DTC on display • Using a scan tool, check current data. Check AVCS system. • Oil pipe clog • Clogged oil flow control solenoid valve • Intake camshaft damaged • Timing chain problem
DTC: P0018 **2T CCM, MIL: Yes** **2005-06** **Engines:** 2.5L DOHC, 3.0L **Models:** Forester, Impreza, WRX, B9 Tribeca, Outback **Transmissions:** All	**Crankshaft Position (camshaft position correlation) Bank 2** Engine stalls. Erroneous idling. **Possible Causes:** • Check any other DTC on display • Using a scan tool, check current data. Check AVCS system. • Oil pipe clog • Clogged oil flow control solenoid valve • Intake camshaft damaged • Timing chain problem
DTC: P0021 **2T CCM, MIL: Yes** **2004-06** **Engines:** 2.5L DOHC, 3.0L **Models:** Forester, Impreza, WRX, Legacy, Outback, B9 Tribeca **Transmissions:** All	**Intake Camshaft Position Timing (over advanced) Bank 2** Engine stalls. Erroneous idling. **Possible Causes:** • Check any other DTC on display • Using a scan tool, check current data. Check AVCS system. • Oil pipe clog • Clogged oil flow control solenoid valve • Dirty engine oil • Intake camshaft damaged • Timing chain problem
DTC: P0026 **2T CCM, MIL: Yes** **2005-06** **Engines:** 2.5L SOHC, 3.0L **Models:** Forester, Impreza, Legacy, Outback, B9 Tribeca **Transmissions:** All	**Intake Valve Control Solenoid Circuit Range/Performance Bank 1** Erroneous idling. **Possible Causes:** • Check any other DTC on display • Check harness between ECM and variable valve lift diagnosis oil pressure switch connector • Check harness between ECM and variable valve lift diagnosis oil pressure switch connector • Check oil switching solenoid valve
DTC: P0028 **2T CCM, MIL: Yes** **2005-06** **Engines:** 2.5L SOHC, 3.0L **Models:** Forester, Impreza, Legacy, Outback, B9 Tribeca **Transmissions:** All	**Intake Valve Control Solenoid Circuit Range/Performance Bank 2** Erroneous idling. **Possible Causes:** • Check any other DTC on display • Check harness between ECM and variable valve lift diagnosis oil pressure switch connector • Check harness between ECM and variable valve lift diagnosis oil pressure switch connector • Check oil switching solenoid valve

DTC	Trouble Code Title, Conditions & Possible Causes
DTC: P0030 **2T CCM, MIL: Yes** **2002-06** **Engines:** 2.0L, 2.5L SOHC & DOHC, 3.0L **Models:** Baja, Forester, Impreza, WRX, Legacy, Outback, B9 Tribeca **Transmissions:** All	**HO2S Heater Control Circuit Bank 1 Sensor 1** Poor driveability. **Possible Causes:** • Check harness between ECM and front oxygen sensor connector • Check harness between main relay and front oxygen sensor connector • Check front oxygen sensor • Check poor contact
DTC: P0031 **2T CCM, MIL: Yes** **2002-06** **Engines:** 2.0L, 2.5L SOHC & DOHC, 3.0L **Models:** Baja, Forester, Impreza, WRX, Legacy, Outback, B9 Tribeca **Transmissions:** All	**HO2S Heater Control Circuit Low Bank 1 Sensor 1** DTC P0141 and P1132 not set, Engine started, system voltage from 11-16 volts, and the PCM detected an unexpected low voltage condition on the Air Fuel Sensor (A/FS) heater control circuit. **Possible Causes:** • Check power supply to front oxygen sensor • Check ground circuit for ECM • Using a scan tool check current data • Check output signal of ECM • Check front oxygen sensor
DTC: P0032 **2T CCM, MIL: Yes** **2001-06** **Engines:** 2.0L, 2.5L SOHC & DOHC, 3.0L **Models:** Baja, Forester, Impreza, WRX, Legacy, Outback, B9 Tribeca **Transmissions:** All	**HO2S Heater Control Circuit High Bank 1 Sensor 1** Engine started, system voltage from 11-16 volts, and the PCM detected an unexpected high voltage condition on the Air Fuel Sensor (A/FS) heater control circuit during the CCM continuous test. **Possible Causes:** • Check output signal of ECM • Check front oxygen sensor heater current • Check output signal of ECM
DTC: P0037 **2T CCM, MIL: Yes** **2001-06** **Engines:** 2.0L, 2.5L SOHC & DOHC, 3.0L **Models:** Baja, Forester, Impreza, WRX, Legacy, Outback, B9 Tribeca **Transmissions:** All	**HO2S Heater Control Circuit Low Bank 1 Sensor 2** Engine started, system voltage from 11-16 volts, and the PCM detected an unexpected low voltage condition on the HO2S-12 heater control circuit during the CCM continuous test. **Possible Causes:** • Check power supply to rear oxygen sensor • Check ground circuit for ECM • Using a scan tool check current data • Check output signal of ECM • Check rear oxygen sensor
DTC: P0038 **2001-06** **2T CCM, MIL: Yes** **Engines:** 2.0L, 2.5L SOHC & DOHC, 3.0L **Models:** Baja, Forester, Impreza, WRX, Legacy, Outback, B9 Tribeca **Transmissions:** All	**HO2S Heater Control Circuit High Bank 1 Sensor 2** Engine started, system voltage from 11-16 volts, and the PCM detected an unexpected "high" voltage condition on the HO2S-12 heater control circuit during the CCM continuous test. **Possible Causes:** • Check input signal of ECM • Using a scan tool check current data • Check poor contact
DTC: P0050 **2T CCM, MIL: Yes** **2003-06** **Engines:** 3.0L **Models:** B9 Tribeca, Outback **Transmissions:** All	**HO2S Heater Control Circuit Bank 2 Sensor 1** Poor driveability. **Possible Causes:** • Check harness between ECM and front oxygen sensor connector • Check harness between main relay and front oxygen sensor connector • Check front oxygen sensor • Check poor contact
DTC: P0051 **2T CCM, MIL: Yes** **2001-06** **Engines:** 3.0L **Models:** B9 Tribeca, Outback **Transmissions:** All	**HO2S Heater Control Circuit Low Bank 2 Sensor 1** Engine started, system voltage from 11-16 volts, and the PCM detected an unexpected "low" voltage condition on the HO2S-21 heater control circuit during the CCM test. **Possible Causes:** • Check power supply to front oxygen sensor • Check ground circuit for ECM • Using a scan tool check current data • Check output signal of ECM • Check front oxygen sensor

DTC	Trouble Code Title, Conditions & Possible Causes

DTC: P0052
2T CCM, MIL: Yes
2001-06
Engines: 3.0L
Models: B9 Tribeca, Outback
Transmissions: All
	HO2S Heater Control Circuit High Bank 2 Sensor 1
	Engine started, system voltage from 11-16 volts, and the PCM detected an unexpected "high" voltage condition on the HO2S-21 heater control circuit during the CCM test.
	Possible Causes:
	   • Check output signal of ECM
	   • Check front oxygen sensor heater current
	   • Check output signal of ECM

DTC: P0057
2T CCM, MIL: Yes
2005-06
Engines: 3.0L
Models: B9 Tribeca, Outback
Transmissions: All
	HO2S Heater Control Circuit Low Bank 2 Sensor 2
	Poor driveability.
	Possible Causes:
	   • Check power supply to rear oxygen sensor
	   • Check ground circuit for ECM
	   • Using a scan tool check current data
	   • Check output signal of ECM
	   • Check rear oxygen sensor

DTC: P0058
2T CCM, MIL: Yes
2005-06
Engines: 3.0L
Models: B9 Tribeca, Outback
Transmissions: All
	HO2S Heater Control Circuit High Bank 2 Sensor 2
	Poor driveability.
	Possible Causes:
	   • Check input signal of ECM
	   • Using a scan tool check current data
	   • Check poor contact

DTC: P0065
2T CCM, MIL: Yes
2001-06
Engines: 2.5L SOHC
Models: Baja, Forester, Impreza, Legacy, Outback
Transmissions: All
	Air Assisted Injector Control Range/Performance
	Engine started, engine running at idle speed, system voltage from 11-16 volts, and the PCM detected the Air Assist Injector solenoid was not operating correctly.
	Possible Causes:
	   • Check any other DTC on display
	   • Check air assist injector solenoid valve operation
	   • Check air by-pass hoses
	   • Check fuel injector

DTC: P0066
2T CCM, MIL: Yes
2001-06
Engines: 2.5L SOHC
Models: Baja, Forester, Impreza, Legacy, Outback
Transmissions: All
	Air Assisted Injector Control Circuit or Circuit Low
	Engine started, engine running at idle speed, system voltage from 11-16 volts, and the PCM detected an unexpected "low" voltage condition on the Air Assist Injector solenoid circuit during the test.
	Possible Causes:
	   • Check output signal from ECM
	   • Check power supply to air assist injector solenoid valve
	   • Check harness between ECM and air assist injector solenoid valve connector
	   • Check poor contact

DTC: P0067
2T CCM, MIL: Yes
2001-06
Engines: 2.5L SOHC
Models: Baja, Forester, Impreza, Legacy, Outback
Transmissions: All
	Air Assisted Injector Control Circuit High
	Engine started, engine running at idle speed, system voltage from 11-16 volts, and the PCM detected an unexpected "high" voltage condition on the Air Assist Injector solenoid circuit during the test.
	Possible Causes:
	   • Check input signal from ECM
	   • Check output signal from ECM

DTC: P0068
2T CCM, MIL: Yes
2002-06
Engines: 2.0L, 2.5L SOHC & DOHC, 3.0L
Models: Baja, Forester, Impreza, WRX, Legacy, B9 Tribeca, Outback
Transmissions: All
	MAP/MAF- Throttle Position Correlation
	Failure of engine to start.
	Possible Causes:
	   • Check idle switch signal
	   • Check any other DTC on display
	   • Check condition of the sensor
	   • Check condition of the throttle body

DTC: P0076
1T CCM, MIL: Yes
2005-06
Engines: 2.5L SOHC, 3.0L
Models: Forester, Impreza, Legacy, B9 Tribeca, Outback
Transmissions: All
	Intake valve Control Circuit Low Bank 1
	Erroneous idling.
	Possible Causes:
	   • Check harness between ECM and oil switching solenoid valve
	   • Check oil switching solenoid valve

DTC	Trouble Code Title, Conditions & Possible Causes
DTC: P0077 **1T CCM, MIL: Yes** **2005-06** **Engines:** 2.5L SOHC, 3.0L **Models:** Forester, Impreza, Legacy, B9 Tribeca, Outback **Transmissions:** All	**Intake valve Control Circuit High Bank 1** Erroneous idling. **Possible Causes:** • Check harness between ECM and oil switching solenoid valve • Check oil switching solenoid valve
DTC: P0082 **1T CCM, MIL: Yes** **2005-06** **Engines:** 2.5L SOHC, 3.0L **Models:** Forester, Impreza, Legacy, B9 Tribeca, Outback **Transmissions:** All	**Intake valve Control Circuit Low Bank 2** Erroneous idling. **Possible Causes:** • Check harness between ECM and oil switching solenoid valve • Check oil switching solenoid valve
DTC: P0083 **1T CCM, MIL: Yes** **2005-06** **Engines:** 2.5L SOHC, 3.0L **Models:** Forester, Impreza, Legacy, B9 Tribeca, Outback **Transmissions:** All	**Intake valve Control Circuit High Bank 2** Erroneous idling. **Possible Causes:** • Check harness between ECM and oil switching solenoid valve • Check oil switching solenoid valve
DTC: P0100 **1T CCM, MIL: Yes** **1996-97** **Engines:** 1.8L, 2.2L, 3.3L **Models:** Impreza, SVX **Transmissions:** All	**Mass Air Flow Circuit Malfunction** Poor driveability. **Possible Causes:** • Check input signal for ECM • Check power supply to mass air flow sensor • Check harness between ECM and sensor connector
DTC P0101 **2T CCM, MIL: Yes** **1996-06** **Engines:** 1.8L, 2.2L, 2.5L SOHC & DOHC, 3.0L, 3.3L **Models:** Baja, Forester, Impreza, WRX, Legacy, B9 Tribeca, Outback **Transmissions:** All	**Mass or Volume Air Flow Circuit Range/Performance** DTC P0102 and P0103 not set, Engine started, engine running at idle speed, system voltage from 11-16 volts, and the PCM detected the MAF sensor indicated less than 0.3 volt (Scan Tool reads 1.3 gm/s), or it indicated more than 5.0 volts (Scan Tool read over 250 gm/s). **Possible Causes:** • Check any other DTC on display
DTC: P0102 **2T CCM, MIL: Yes** **1997-06** **Engines:** 1.8L, 2.0L, 2.2L, 2.5L SOHC & DOHC, 3.0L, 3.3L **Models:** Baja, Forester, Impreza, WRX, Legacy, B9 Tribeca, Outback **Transmissions:** All	**Mass or Volume Air Flow Circuit Low Input** Engine stalls. Erroneous idling. Poor driving performance. **Possible Causes:** • Using a scan tool, check current data • Check input signal of ECM • Check power supply to sensor • Check harness between ECM and sensor connector • Check poor contact
DTC: P0103 **1T CCM, MIL: Yes** **1997-06** **Engines:** 1.8L, 2.0L, 2.5L SOHC & DOHC, 3.0L, 3.3L **Models:** Baja, Forester, Impreza, WRX, Legacy, B9 Tribeca, Outback **Transmissions:** All	**Mass or Volume Air Flow Circuit High Input** Engine stalls. Erroneous idling. Poor driving performance. **Possible Causes:** • Using a scan tool, check current data • Check harness between ECM and sensor connector
DTC: P0105 **2T CCM, MIL: Yes** **1996-97** **Engines:** 1.8L, 2.2L, 3.3L **Models:** Impreza, SVX **Transmissions:** All	**Pressure Sensor Circuit Malfunction** Poor driveability. **Possible Causes:** • Check input signal for ECM • Check harness between ECM and sensor connector

DTC	Trouble Code Title, Conditions & Possible Causes
DTC: P0106 **2T CCM, MIL: Yes** 1996-04 **Engines:** 1.8L, 2.2L, 2.5L SOHC, 3.0L **Models:** Forester, Impreza, Legacy, Outback **Transmissions:** All	**Pressure Sensor Circuit Range/Performance** DTC P0105, P0106 and P0107 not set, Engine started, vehicle driven to a speed over 3 mph, then back to idle speed, and the PCM detected the Pressure sensor signal was out of the normal range. **Possible Causes:** • Check any other DTC on display • Check air intake system • Check pressure sensor • Check throttle position
DTC: P0107 **1T CCM, MIL: Yes** 1997-06 **Engines:** 1.8L, 2.0L, 2.2L, 2.5L SOHC & DOHC, 3.0L **Models:** Baja, Forester, Impreza, WRX, Legacy, Outback, B9 Tribeca **Transmissions:** All	**Manifold Absolute pressure/Barometric Pressure Circuit Low Input** Key on or engine running; system voltage from 11-16 volts, and the PCM detected the MAP sensor indicated less than 0.20 volt (Scan Tool reads 13.3L kPa) during the CCM test. **Possible Causes:** • Using a scan tool, check current data • Check poor contact • Check output signal of ECM • Check input signal of ECM • Check harness between ECM and sensor connector
DTC: P0108 **1T CCM, MIL: Yes** 1997-06 **Engines:** 1.8L, 2.0L, 2.2L, 2.5L SOHC & DOHC, 3.0L **Models:** Baja, Forester, Impreza, WRX, Legacy, Outback, B9 Tribeca **Transmissions:** All	**Manifold Absolute pressure/Barometric Pressure Circuit Low Input** Key on or engine running; system voltage from 11-16 volts, and the PCM detected the MAP sensor indicated more than 4.50 volts (Scan Tool reads 119.5 kPa) during the CCM test. **Possible Causes:** • Check output signal of ECM • Check input signal of ECM • Check harness between ECM and sensor connector • Check harness between the sensor connector • Check poor contact
DTC: P0111 **2T CCM, MIL: Yes** 1999-06 **Engines:** 2.0L, 2.2L, 2.5L SOHC & DOHC, 3.0L **Models:** Baja, Forester, Impreza, WRX, Legacy, Outback, B9 Tribeca **Transmissions:** All	**Intake Air Temperature Circuit Range/Performance** DTC P0112 and P0113 not set, ECT sensor from 14-86°F, vehicle driven at 20-25 mph at normal operating temperature for at least 10 minutes, and the PCM detected the change in the IAT sensor signal was too small during the CCM Rationality test. **Possible Causes:** • Check any other DTC on display • Check engine coolant temperature
DTC: P0112 **2T CCM, MIL: Yes** 1999-06 **Engines:** 2.0L, 2.2L, 2.5L SOHC & DOHC, 3.0L **Models:** Baja, Forester, Impreza, WRX, Legacy, Outback, B9 Tribeca **Transmissions:** All	**Intake Air Temperature Circuit Low Input** Key on or engine running; and the PCM detected the IAT sensor indicated less than 0.20 volt (Scan Tool reads 257°F) in the CCM test. **Possible Causes:** • Using a scan tool, check current data • Check the harness between the sensor and the ECM connector
DTC: P0113 **1T CCM, MIL: Yes** 1999-06 **Engines:** 2.0L, 2.2L, 2.5L SOHC & DOHC, 3.0L **Models:** Baja, Forester, Impreza, WRX, Legacy, Outback, B9 Tribeca **Transmissions:** All	**Intake Air Temperature Circuit high Input** Key on or engine running; and the PCM detected the IAT sensor indicated more than 4.60 volts (Scan Tool reads -49°F) in the CCM test. **Possible Causes:** • Using a scan tool, check current data • Check the harness between the sensor and the ECM connector
DTC: P0115 **1T CCM, MIL: Yes** 1996-97 **Engines:** 1.8L, 2.2L, 3.3L **Models:** Impreza, SVX **Transmissions:** All	**Engine Coolant Temperature Sensor Malfunction** Key on or engine running; and the PCM detected the ECT sensor signal indicated less than -40°F, or it indicated more than 300°F. **Possible Causes:** • Check harness between sensor and ECM connector

DTC	Trouble Code Title, Conditions & Possible Causes
DTC: P0116 **1T CCM, MIL: Yes** **1997-01** **Engines:** 1.8L, 2.2L, 2.5L SOHC **Models:** Impreza, Forester, Legacy, Outback **Transmissions:** All	**Engine Coolant Temperature Sensor Circuit Low Input** Poor driveability. **Possible Causes:** • Check harness between sensor and ECM connector
DTC: P0117 **1T CCM, MIL: Yes** **1997-06** **Engines:** 1.8L, 2.0L, 2.2L, 2.5L SOHC & DOHC, 3.0L **Models:** Baja, Forester, Impreza, WRX, Legacy, Outback, B9 Tribeca	**Engine Coolant Temperature Sensor Circuit Low Input** Key on or engine running; and the PCM detected the ECT sensor indicated more than 0.10 volt (Scan Tool reads 300°F) in the CCM test. **Possible Causes:** • Using a scan tool, check current data • Check harness between sensor and ECM connector
DTC: P0118 **1T CCM, MIL: Yes** **1998-06** **Engines:** 2.0L, 2.2L, 2.5L SOHC & DOHC, 3.0L **Models:** Baja, Forester, Impreza, WRX, Legacy, Outback, B9 Tribeca **Transmissions:** All	**Engine Coolant Temperature Sensor Circuit High Input** Key on or engine running; and the PCM detected the ECT sensor indicated more than 4.60 volts (Scan Tool reads -40°F) in the CCM test. **Possible Causes:** • Using a scan tool, check current data • Check harness between sensor and ECM connector
DTC: P0120 **1T CCM, MIL: Yes** **1996-97** **Engines:** 1.8L, 2.2L, 3.3L **Models:** Impreza, SVX **Transmissions:** All	**Throttle Position Sensor Circuit Malfunction** Engine started, engine at idle speed, and the PCM detected the TP sensor signal indicated less than 0.1 volt (throttle closed), or it indicated more than 4.9 volts under any type of driving conditions during the test. **Possible Causes:** • Check input signal of ECM • Check harness between ECM and sensor connector • Check harness between sensor and body connector
DTC: P0122 **1T CCM, MIL: Yes** **1997-06** **Engines:** 2.0L, 2.2L, 2.5L SOHC & DOHC, 3.0L **Models:** Baja, Forester, Impreza, WRX, Legacy, Outback, B9 Tribeca **Transmissions:** All	**Throttle/Pedal Position Sensor/Switch "A" Circuit Low Input** Key on or engine running; and the PCM detected the TP sensor indicated less than 0.10 volt (Scan Tool reading of 3.40%) in the test. **Possible Causes:** • Check sensor output • Check poor contact • Check harness between ECM and electronic throttle control (if equipped)
DTC: P0123 **1T CCM, MIL: Yes** **1997-06** **Engines:** 1.8L, 2.0L, 2.2L, 2.5L SOHC & DOHC, 3.0L **Models:** Baja, Forester, Impreza, WRX, Legacy, Outback, B9 Tribeca **Transmissions:** All	**Throttle/Pedal Position Sensor/Switch "A" Circuit Low Input** Key on or engine running; and the PCM detected the TP sensor indicated more than 4.90 volts (Scan Tool reading of 95.6%) in the test. **Possible Causes:** • Check sensor output • Check poor contact • Check harness between ECM and electronic throttle control (if equipped)
DTC: P0125 **2T CCM, MIL: Yes** **1996-06** **Engines:** 1.8L, 2.0L, 2.2L, 2.5L SOHC & DOHC, 3.0L, 3.3L **Models:** Baja, Forester, Impreza, WRX, Legacy, Outback, SVX, B9 Tribeca **Transmissions:** All	Insufficient Coolant Temperature For Closed loop Fuel Control DTC P0115, P0116 and P0117 not set, Engine started, vehicle driven to a speed of 25-40 at over 1500 rpm, and the PCM detected the coolant temperature did not reach closed loop status. **Possible Causes:** • Check any other DTC on display • Check tire size • Check engine coolant • Check thermostat
DTC: P0126 **2T CCM, MIL: Yes** **2005-06** **Engines:** 2.5L SOHC & DOHC, 3.0L **Models:** Baja, Forester, Impreza, WRX, Legacy, Outback, B9 Tribeca **Transmissions:** All	**Insufficient Coolant Temperature For Stable Operation** Hard to start. Erroneous idling. Poor driving performance. **Possible Causes:** • Check any other DTC on display • Check engine coolant temperature sensor

DTC	Trouble Code Title, Conditions & Possible Causes
DTC: P0128 **2T CCM, MIL: Yes** **2001-06** **Engines:** 2.5L SOHC & DOHC, 3.0L **Models:** Baja, Forester, Impreza, WRX, Legacy, Outback, B9 Tribeca **Transmissions:** All	**Coolant Temperature Thermostat (coolant temperature below thermostat regulating temperature)** DTC P0115 and P0116 not set, cold, ECT sensor less than 140°F at startup, IAT sensor more than 50°F at startup (cold engine conditions met), Engine started, engine runtime over 10 minutes, and the PCM detected the ECT sensor did not reach 167°F after a normal warmup period had elapsed. **Possible Causes:** • Check vehicle • Check any other DTC on display • Check engine coolant • Check radiator fan
DTC: P0129 **2T CCM, MIL: Yes** **2002-04** **Engines:** 2.0L, 2.5L SOHC & DOHC, 3.0L **Models:** Baja, Forester, Impreza, WRX, Legacy, Outback **Transmissions:** All	**Barometric Pressure Too Low** Poor driveability. **Possible Causes:** • Check any other DTC on display
DTC: P0130 **2T CCM, MIL: Yes** **1996-06** **Engines:** 1.8L, 2.0L, 2.2L, 2.5L SOHC & DOHC, 3.0L, 3.3L **Models:** Baja, Forester, Impreza, WRX, Legacy, Outback, SVX **Transmissions:** All	**O2 Sensor Circuit Bank 1 Sensor 1** DTC P0131, P0132, P1131, P1132 not set, Engine started, engine running in closed loop at hot idle speed for more than 1 minute, and the PCM detected a problem in the A/F sensor signal was not in a range of 0.85-1.15 volts during the CCM test. **Possible Causes:** • Check any other DTC on display • Using a scan tool, check current data • Check harness between ECM and sensor • Check exhaust system
DTC: P0131 **1T CCM, MIL: Yes** **1999-06** **Engines:** 2.2L, 2.5L SOHC & DOHC, 3.0L **Models:** Baja, Forester, Impreza, WRX, Legacy, Outback, B9 Tribeca **Transmissions:** All	**O2 Sensor Circuit Low Voltage Bank 1 Sensor 1** Engine started, engine running in closed loop at hot idle speed, and the PCM detected an unexpected "low" voltage condition on the A/F sensor signal circuit for over 6 seconds during the CCM test. **Possible Causes:** • Check harness between ECM and sensor
DTC: P0132 **2T CCM, MIL: Yes** **1999-06** **Engines:** 2.2L, 2.5L SOHC & DOHC, 3.0L **Models:** Baja, Forester, Impreza, WRX, Legacy, Outback, B9 Tribeca **Transmissions:** All	**O2 Sensor Circuit Low Voltage Bank 1 Sensor 1** Engine started, engine running in closed loop at hot idle speed, and the PCM detected an unexpected "high" voltage condition on the A/F sensor signal circuit for over 6 seconds during the CCM test. **Possible Causes:** • Check harness between ECM and sensor
DTC: P0133 **2T CCM, MIL: Yes** **1996-06** **Engines:** 1.8L, 2.0L, 2.2L, 2.5L SOHC & DOHC, 3.0L, 3.3L **Models:** Baja, Forester, Impreza, WRX, Legacy, Outback, B9 Tribeca, SVX **Transmissions:** All	**O2 Sensor Circuit Slow Response Bank 1 Sensor 1** Engine started, engine running in closed loop, ECT sensor signal more than 170°F, and the PCM detected the number of A/FS-11 rich-to-lean or lean-to-rich switches was less than a calibrated amount. **Possible Causes:** • Check any other DTC on display • Check exhaust system
DTC: P0134 **1T CCM, MIL: Yes** **2002-06** **Engines:** 2.0L, 2.5L SOHC & DOHC, 3.0L **Models:** Baja, Forester, Impreza, WRX, Legacy, B9 Tribeca, Outback **Transmissions:** All	**O2 Sensor Circuit No Activity Detected Bank 1 Sensor 1** Poor driveability. **Possible Causes:** • Check harness between ECM and sensor connector • Check poor contact

DTC	Trouble Code Title, Conditions & Possible Causes
DTC: P0135 **2T CCM, MIL: Yes** **1996-99** **Engines:** 1.8L, 2.2L, 2.5L SOHC, 3.3L **Models:** Forester, Impreza, Legacy, Outback, SVX **Transmissions:** All	**Front Oxygen Sensor 1 (RH) Heater Circuit Malfunction** Poor driveability. **Possible Causes:** • Check any other DTC on display (P0141 and P0147) • Using a scan tool, check current data • Check output signal from ECM • Check front oxygen sensor 1 (RH)
DTC: P0136 **2T CCM, MIL: Yes** **1996-02** **Engines:** 1.8L, 2.2L, 2.5L SOHC, 3.0L, 3.3L **Models:** Forester, Impreza, Legacy, Outback, SVX **Transmissions:** All	**Rear Oxygen Sensor Circuit Malfunction** DTC P1130 and P1131 not set, Engine started, vehicle driven at a speed of 30-55 mph for 2 minutes, and the PCM detected the HO2S signal was too high or too low, or the HO2S signal did not reach the maximum or minimum voltages, or that it took too long for the HO2S signal to switch from rich to lean during the CCM test period. **Possible Causes:** • Using a scan tool, check current data • Check the harness between the rear sensor and the ECM connector • Check the exhaust system
DTC: P0137 **2T CCM, MIL: Yes** **2002-06** **Engines:** 2.0L, 2.5L SOHC, 3.0L **Models:** Baja, Forester, Impreza, WRX, Legacy, Outback, B9 Tribeca **Transmissions:** All	**O2 Sensor Circuit Low Voltage Bank 1 Sensor 2** Engine started, vehicle driven at a speed of 30-55 mph for 2 minutes, and the PCM detected an unexpected low voltage condition on the HO2S-12 circuit during the CCM test period. **Possible Causes:** • Check any other DTC on display • Using the a scan tool, check current data • Check Harness between ECM and rear oxygen sensor connector • Check exhaust system
DTC: P0138 **2T CCM, MIL: Yes** **2002-06** **Engines:** 2.0L, 2.5L SOHC, 3.0L **Models:** Baja, Forester, Impreza, WRX, Legacy, Outback, B9 Tribeca **Transmissions:** All	**O2 Sensor Circuit High Voltage Bank 1 Sensor 2** Engine started, vehicle driven at a speed of 30-55 mph for 2 minutes, and the PCM detected an unexpected high voltage condition on the HO2S-12 circuit during the CCM test period. **Possible Causes:** • Check any other DTC on display • Using a scan tool, check current data • Check Harness between ECM and rear oxygen sensor connector • Check exhaust system
DTC: P0139 **2T CCM, MIL: Yes** **1996-06** **Engines:** 1.8L, 2.0L, 2.5L SOHC, 3.0L, 3.3L **Models:** Baja, Forester, Impreza, WRX, Legacy, Outback, B9 Tribeca, SVX **Transmissions:** All	**O2 Sensor Circuit High Voltage Bank 1 Sensor 2** DTC P0136 not set, Engine started, vehicle driven to a speed of over 20 mph for 1 minute, and then back to idle, IAT sensor from 14-122°F, and the PCM detected the average HO2S-12 response time was more than 1 second during the Oxygen Sensor Monitor test. **Possible Causes:** • Check any other DTC on display • Check Harness between ECM and rear oxygen sensor connector • Check sensor
DTC: P0140 **2T CCM, MIL: Yes** **2006** **Engines:** 2.5L SOHC & DOHC, 3.0L **Models:** Baja, Forester, Impreza, WRX, Legacy, Outback, B9 Tribeca **Transmissions:** All	**O2 Sensor Circuit No Activity Detected Bank 1 Sensor 2** Poor driveability. **Possible Causes:** • Check any other DTC on display • Using a scan tool, check current data • Check Harness between ECM and rear oxygen sensor connector • Check exhaust system
DTC: P0141 **2T CCM, MIL: Yes** **1996-01** **Engines:** 1.8L, 2.2L, 2.5L SOHC, 3.3L **Models:** Forester, Impreza, Legacy, Outback, SVX **Transmissions:** All	**Rear Oxygen Sensor Heater Circuit Malfunction** Engine started, engine running at idle speed, and the PCM detected the HO2S-12 heater current was below 0.2 amps, or over 2 amps. **Possible Causes:** • Check output signal from ECM • Check power supply to sensor • Check sensor

DTC	Trouble Code Title, Conditions & Possible Causes
DTC: P0142 **2T CCM, MIL: Yes** **1996-97** **Engines:** 3.3L **Models:** SVX **Transmissions:** All	**Front Oxygen Sensor 2 (LH) Circuit Malfunction** Engine started, vehicle speed from 30-55 mph for 2 minutes, and the PCM detected the HO2S signal was excessively high or low, or the HO2S signal did not reach the maximum or minimum voltages or the signal remained between 400-500 mv for too long a time. **Possible Causes:** • Check for other causes affecting exhaust gas • Check the front sensor 2 (LH) • Check Harness between the front oxygen sensor 2 (JH) and the ECM connector
DTC: P0145 **2T CCM, MIL: Yes** **1996-97** **Engines:** 3.3L **Models:** SVX **Transmissions:** All	**Front Oxygen Sensor 2 (LH) Circuit Slow Response** Engine started, engine running in closed loop, ECT sensor signal more than 170ºF, and the PCM detected the HO2S response time to switch between 300-600 mv was too slow, or the rich-to-lean or lean-to-rich switch time was too slow in the Oxygen Sensor Monitor test. Note: This trouble code applies to vehicles produced for California. **Possible Causes:** • Check for DTC P0142 • Check exhaust system
DTC: P0147 **2T CCM, MIL: Yes** **1996-97** **Engines:** 3.3L **Models:** SVX **Transmissions:** All	**Front Oxygen Sensor 2 (LH) Heater Circuit Malfunction** Engine started, engine running at idle speed, and the PCM detected the HO2S-21 heater current indicated less than 0.2 amps, or that it indicated more than 2 amps during the CCM test. **Possible Causes:** • Using a scan tool, check current data • Check the output signal from the ECM • Check the power supply to the front oxygen sensor 2 (LH) • Check the sensor
DTC: P0150 **1T CCM, MIL: Yes** **2003-04** **Engines:** 3.0L **Models:** Outback **Transmissions:** All	**O2 Sensor Circuit Bank 2 Sensor 1** Poor driveability. **Possible Causes:** • Check the harness between the ECM and the front sensor connector • Check the output signal from the ECM
DTC: P0151 **1T CCM, MIL: Yes** **2001-06** **Engines:** 3.0L **Models:** Outback, B9 Tribeca **Transmissions:** All	**O2 Sensor Circuit Low Voltage Bank 2 Sensor 1** Engine started, engine running in closed loop at hot idle speed, and the PCM detected an unexpected "low" voltage condition on the front A/F sensor signal circuit for over 6 seconds during the CCM test. **Possible Causes:** • Check the harness between the ECM and the front sensor connector
DTC: P0152 **1T CCM, MIL: Yes** **2001-06** **Engines:** 3.0L **Models:** Outback, B9 Tribeca **Transmissions:** All	**O2 Sensor Circuit High Voltage Bank 2 Sensor 1** Engine started, engine running in closed loop at hot idle speed, and the PCM detected an unexpected "high" voltage condition on the front A/F sensor signal circuit for over 6 seconds in the CCM test. **Possible Causes:** • Check the harness between the ECM and the front sensor connector
DTC: P0153 **2T CCM, MIL: Yes** **2001-06** **Engines:** 3.0L **Models:** Outback, B9 Tribeca **Transmissions:** All	**O2 Sensor Circuit Slow Response Bank 2 Sensor 1** Engine started, engine running in closed loop, ECT sensor more than 170ºF, and the PCM detected the number of A/FS-21 rich-to-lean or lean-to-rich switches was less than a calibrated amount. **Possible Causes:** • Check any other DTC on display • Check exhaust system
DTC: P0154 **1T CCM, MIL: Yes** **2003-06** **Engines:** 3.0L **Models:** Outback, B9 Tribeca **Transmissions:** All	**O2 Sensor Circuit No Activity Detected Bank 2 Sensor 1** Poor driveability. **Possible Causes:** • Check the harness between the ECM and the front sensor • Check poor contact

DTC	Trouble Code Title, Conditions & Possible Causes
DTC: P0157 **2T CCM, MIL: Yes** **2005-06** **Engines:** 3.0L **Models:** Outback, B9 Tribeca **Transmissions:** All	**O2 Sensor Circuit Low Voltage Bank 2 Sensor 2** Poor driveability. **Possible Causes:** • Check any other DTC on display • Using a scan tool, check data • Check harness between ECM and rear sensor connector • Check harness between rear sensor and ECM connector • Check exhaust system
DTC: P0158 **2T CCM, MIL: Yes** **2005-06** **Engines:** 3.0L **Models:** Outback, B9 Tribeca **Transmissions:** All	**O2 Sensor Circuit High Voltage Bank 2 Sensor 2** Poor driveability. **Possible Causes:** • Check any other DTC on display • Using a scan tool, check data • Check harness between ECM and rear sensor connector • Check harness between rear sensor and ECM connector • Check exhaust system
DTC: P0159 **2T CCM, MIL: Yes** **2005-06** **Engines:** 3.0L **Models:** Outback, B9 Tribeca **Transmissions:** All	**O2 Sensor Circuit Slow Response Bank 2 Sensor 2** Poor driveability. **Possible Causes:** • Check any other DTC on display • Check harness between ECM and rear sensor connector • Check sensor
DTC: P0160 **2T CCM, MIL: Yes** **2006** **Engines:** 3.0L **Models:** Outback, B9 Tribeca **Transmissions:** All	**O2 Sensor Circuit No Activity Detected Bank 2 Sensor 2** Poor driveability. **Possible Causes:** • Check any other DTC on display • Using a scan tool, check data • Check harness between ECM and rear sensor connector • Check harness between rear sensor and ECM connector • Check exhaust system
DTC: P0170 **2T CCM, MIL: Yes** **1996-01** **Engines:** 1.8L, 2.2L, 2.5L SOHC, 3.3L **Models:** Forester, Impreza, Legacy, Outback, SVX **Transmissions:** All	**Fuel Trim Malfunction** DTC P0106, P0107, P0108, P0112, P0113, P0117, P0118, P0120, P0130, P0133, P0135, P0136, P0139, P0141, P0201-204, P0300, P0301-P0304, P0400, P0500, P0506, P0507, P1400 and P1401 not set, engine running in closed loop at a speed of over 3 mph, and the PCM detected the rich or lean Fuel Trim correction amount was more than a threshold stored in memory. **Possible Causes:** • Check exhaust system • Check air intake system • Check fuel pressure • Check engine coolant • Check mass air flow sensor
DTC: P0171 **2T CCM, MIL: Yes** **1999-06** **Engines:** 2.0L, 2.2L, 2.5L SOHC & DOHC, 3.0L **Models:** Baja, Forester, Impreza, WRX, Legacy, Outback, B9 Tribeca **Transmissions:** All	**System To Lean Bank 1** DTC P0106, P0107, P0108, P0112, P0113, P0117, P0118, P0121, P0130, P0133, P0135, P0136, P0139, P0141, P0301-306, P0400, P0500, P0506, P0507 and P1400 not set, engine running in closed loop at more than 30 mph, and the PCM detected the "lean" Fuel Trim correction amount was more than a threshold value in memory. **Possible Causes:** • Check any other DTC on display • Check exhaust system • Check air intake system • Check fuel pressure • Check fuel injector • Check coolant temperature • Check intake manifold pressure sensor • Check intake air temperature sensor

DTC	Trouble Code Title, Conditions & Possible Causes

DTC: P0172
2T CCM, MIL: Yes
1999-06
Engines: 2.0L, 2.2L, 2.5L SOHC & DOHC, 3.0L
Models: Baja, Forester, Impreza, WRX, Legacy, Outback, B9 Tribeca
Transmissions: All
	System To Rich Bank 1
	DTC P0106, P0107, P0108, P0112, P0113, P0117, P0118, P0121, P0130, P0133, P0135, P0136, P0139, P0141, P0301-306, P0400, P0500, P0506, P0507 and P1400 not set, engine running in closed loop at more than 30 mph, and the PCM detected the "rich" Fuel Trim correction amount was more than a threshold value in memory.
	Possible Causes:
	  • Check any other DTC on display
	  • Check exhaust system
	  • Check air intake system
	  • Check fuel pressure
	  • Check fuel injector
	  • Check coolant temperature
	  • Check intake manifold pressure sensor
	  • Check intake air temperature sensor

DTC: P0174
2T CCM, MIL: Yes
2001-06
Engines: 3.0L
Models: Outback, B9 Tribeca
Transmissions: All
	System To Lean Bank 1
	DTC P0106, P0107, P0108, P0112, P0113, P0117, P0118, P0121, P0130, P0133, P0135, P0136, P0139, P0141, P0301-306, P0400, P0500, P0506, P0507 and P1400 not set, engine running in closed loop at more than 30 mph, and the PCM detected the "lean" Fuel Trim correction amount was more than a threshold value in memory.
	Possible Causes:
	  • Check any other DTC on display
	  • Check exhaust system
	  • Check air intake system

DTC: P0175
2T CCM, MIL: Yes
2001-06
Engines: 3.0L
Models: Outback, B9 Tribeca
Transmissions: All
	System To rich Bank 1
	DTC P0106, P0107, P0108, P0112, P0113, P0117, P0118, P0121, P0130, P0133, P0135, P0136, P0139, P0141, P0301-306, P0400, P0500, P0506, P0507 and P1400 not set, engine running in closed loop at more than 30 mph, and the PCM detected the "rich" Fuel Trim correction amount was more than a threshold value in memory.
	Possible Causes:
	  • Check any other DTC on display
	  • Check exhaust system
	  • Check air intake system

DTC: P0180
1T CCM, MIL: Yes
1996
Engines: 1.8L, 2.2L
Models: Impreza
Transmissions: All
	Fuel temperature Sensor "A" Circuit
	Poor driveability.
	Possible Causes:
	  • Using a scan tool, check data
	  • Check harness between the sensor and the ECM

DTC: P0181 **2T CCM, MIL: Yes** **1996-06** **Engines:** 1.8L, 2.0L, 2.2L, 2.5L SOHC & DOHC, 3.0L **Models:** Baja, Forester, Impreza, WRX, Legacy, Outback, B9 Tribeca **Transmissions:** All	**Fuel Temperature Sensor "A" Circuit Range/Performance** DTC P0180 not set, Engine started, vehicle driven to a speed of over 3 mph in closed loop, and the PCM detected the Fuel Temperature Sensor 'A' signal was out-of-range, or the signal was not plausible. **Possible Causes:** • Check any other DTC on display
DTC: P0182 **1T CCM, MIL: Yes** **1997-06** **Engines:** 1.8L, 2.0L, 2.2L, 2.5L SOHC & DOHC, 3.0L **Models:** Baja, Forester, Impreza, WRX, Legacy, Outback, B9 Tribeca **Transmissions:** All	**Fuel Temperature Sensor "A" Circuit Low Input** Key on or engine running; vehicle driven to a speed of over 3 mph, and the PCM detected the Fuel Temperature Sensor 'A' signal was less than 0.10 volt (Scan Tool reads 300°F) during the CCM test. **Possible Causes:** • Using a scan tool, check data

DTC: P0183
1T CCM, MIL: Yes
1997-06
Engines: 1.8L, 2.0L, 2.2L, 2.5L SOHC & DOHC, 3.0L
Models: Baja, Forester, Impreza, WRX, Legacy, Outback, B9 Tribeca
Transmissions: All
	Fuel Temperature Sensor "A" Circuit High Input
	Key on or engine running; vehicle driven to a speed of over 3 mph, and the PCM detected the Fuel Temperature Sensor 'A' signal was more than 4.80 volts (Scan Tool reads -40°F) during the CCM test.
	Possible Causes:
	  • Using a scan tool, check data
	  • Check harness between sensor and ECM connector

DTC	Trouble Code Title, Conditions & Possible Causes
DTC: P0196 **2T CCM, MIL: Yes** **2005-06** **Engines:** 2.5L SOHC, 3.0L **Models:** Forester, Impreza, Legacy, Outback, B9 Tribeca **Transmissions:** All	**Engine Oil Temperature Sensor Circuit Range/Performance** Engine stalls. Erroneous idling. Poor driving performance. **Possible Causes:** • Check any other DTC on display
DTC: P0197 **1T CCM, MIL: Yes** **2005-06** **Engines:** 2.5L SOHC, 3.0L **Models:** Forester, Impreza, Legacy, Outback, B9 Tribeca **Transmissions:** All	**Engine Oil Temperature Sensor Circuit Low** Engine stalls. Erroneous idling. Poor driving performance. **Possible Causes:** • Check harness between sensor and ECM • Check poor contact
DTC: P0198 **1T CCM, MIL: Yes** **2005-06** **Engines:** 2.5L SOHC, 3.0L **Models:** Forester, Impreza, Legacy, Outback, B9 Tribeca **Transmissions:** All	**Engine Oil Temperature Sensor Circuit High** Engine stalls. Erroneous idling. Poor driving performance. **Possible Causes:** • Check harness between sensor and ECM
DTC: P0201 **1T CCM, MIL: Yes** **1996-97** **Engines:** 1.8L, 2.2L, 3.3L **Models:** Impreza, SVX **Transmissions:** All	**Fuel Injector Circuit Malfunction #1** Engine started, engine speed over 500 rpm, and the PCM detected the Injector 1 control circuit was in a "high" state when it should be "low", or it was in a "low" state when it should have be "high". **Possible Causes:** • Check output signal for ECM • Check harness between injector and ECM connector • Check injector • Check power supply line
DTC P0202 **1T CCM, MIL: Yes** **1996-97** **Engines:** 1.8L, 2.2L, 3.3L **Models:** Impreza, SVX **Transmissions:** All	**Fuel Injector Circuit Malfunction #2** Engine started, engine speed over 500 rpm, and the PCM detected the Injector 2 control circuit was in a "high" state when it should be "low", or it was in a "low" state when it should have be "high". **Possible Causes:** • Check output signal for ECM • Check harness between injector and ECM connector • Check injector • Check power supply line
DTC: P0203 **1T CCM, MIL: Yes** **1996-97** **Engines:** 1.8L, 2.2L, **3.3L** **Models:** Impreza, SVX **Transmissions:** All	**Fuel Injector Circuit Malfunction #3** Engine started, engine speed over 500 rpm, and the PCM detected the Injector 3 control circuit was in a "high" state when it should be "low", or it was in a "low" state when it should have be "high". **Possible Causes:** • Check output signal for ECM • Check harness between injector and ECM connector • Check injector • Check power supply line
DTC: P0204 **1T CCM, MIL: Yes** **1996-97** **Engines:** 1.8L, 2.2L, **3.3L** **Models:** Impreza, SVX **Transmissions:** All	**Fuel Injector Circuit Malfunction #4** Engine started, engine speed over 500 rpm, and the PCM detected the Injector 4 control circuit was in a "high" state when it should be "low", or it was in a "low" state when it should have be "high". **Possible Causes:** • Check output signal for ECM • Check harness between injector and ECM connector • Check injector • Check power supply line
DTC: P0205 **1T CCM, MIL: Yes** **1996-97** **Engines: 3.3L** **Models:** SVX **Transmissions:** All	**Fuel Injector Circuit Malfunction #5** Engine started, engine speed over 500 rpm, and the PCM detected the Injector 5 control circuit was in a "high" state when it should be "low", or it was in a "low" state when it should have be "high". **Possible Causes:** • Check output signal for ECM • Check harness between injector and ECM connector • Check injector • Check power supply line

DTC	Trouble Code Title, Conditions & Possible Causes
DTC: P0206 **1T CCM, MIL: Yes** **1996-97** **Engines:** 3.3L **Models:** SVX **Transmissions:** All	**Fuel Injector Circuit Malfunction #6** Engine started, engine speed over 500 rpm, and the PCM detected the Injector 6 control circuit was in a "high" state when it should be "low", or it was in a "low" state when it should have be "high". **Possible Causes:** • Check output signal for ECM • Check harness between injector and ECM connector • Check injector • Check power supply line
DTC: P0222 **1T CCM, MIL: Yes** **2004-06** **Engines:** 2.5L SOHC **& DOHC,** **3.0L** **Models:** Baja, Forester, Impreza, WRX, Legacy, Outback, B9 Tribeca **Transmissions:** All	**Throttle/Pedal Position Sensor/Switch "B" Circuit Low Input** Erroneous idling. Poor driving performance. Engine stalls. **Possible Causes:** • Check sensor output • Check poor contact • Check harness between ECM and electronic throttle control (if equipped) • Check sensor power supply • Check for short circuit in ECM
DTC: P0223 **1T CCM, MIL: Yes** **2004-06** **Engines:** 2.5L SOHC **& DOHC,** **3.0L** **Models:** Baja, Forester, Impreza, WRX, Legacy, Outback, B9 Tribeca **Transmissions:** All	**Throttle/Pedal Position Sensor/Switch "B" Circuit High Input** Erroneous idling. Poor driving performance. Engine stalls. **Possible Causes:** • Check sensor output • Check poor contact • Check harness between ECM and electronic throttle control (if equipped)
DTC: P0230 **2T CCM, MIL: Yes** **2004-06** **Engines:** 2.5L DOHC, 3.0L **Models:** Baja, Forester, Impreza, WRX, Legacy, Outback, B9 Tribeca	**Fuel Pump Primary Circuit** Poor driveability. **Possible Causes:** • Check power supply to fuel pump control unit • Check ground circuit of fuel pump control unit • Check harness between the fuel pump control unit and fuel pump connector • Check poor contact • Check if vehicle has previously ran out of fuel
DTC: P0244 **1T CCM, MIL: Yes** **2002-06** **Engines:** **2.0L,** 2.5L DOHC **Models:** Baja, Forester, Impreza, WRX, Legacy, Outback **Transmissions:** All	**Turbocharger/Supercharger Wastegate Solenoid "A" Range/Performance** Poor driveability. **Possible Causes:** • Check any other DTC on display
DTC: P0245 **1T CCM, MIL: Yes** **2002-06** **Engines:** **2.0L,** 2.5L DOHC **Models:** Baja, Forester, Impreza, WRX, Legacy, Outback **Transmissions:** All	**Turbocharger/Supercharger Wastegate Solenoid "A" Low** Poor driveability. **Possible Causes:** • Check output signal from ECM • Check harness between wastegate control solenoid valve and ECM connector • Check poor contact • Check wastegate control solenoid valve • Check power supply to wastegate control solenoid valve • Check poor contact
DTC: P0246 **1T CCM, MIL: Yes** **2002-06** **Engines:** **2.0L,** 2.5L DOHC **Models:** Baja, Forester, Impreza, WRX, Legacy, Outback **Transmissions:** All	**Turbocharger/Supercharger Wastegate Solenoid "A" High** Poor driveability. **Possible Causes:** • Check output signal from ECM • Check harness between wastegate control solenoid valve and ECM connector • Check poor contact • Check wastegate control solenoid valve

DTC	Trouble Code Title, Conditions & Possible Causes
DTC: P0261 **1T CCM, MIL: Yes** 1997-99 **Engines:** 1.8L, 2.2L, 2.5L SOHC **Models:** Impreza, Forester, Legacy, Outback	**Fuel Injector Circuit Low Input #1** Poor driveability. **Possible Causes:** • Check output signal from ECM • Check harness between injector and ECM connector • Check fuel injector • Check power supply line
DTC: P0262 **1T CCM, MIL: Yes** 1997-99 **Engines:** 1.8L, 2.2L, 2.5L SOHC **Models:** Impreza, Forester, Legacy, Outback **Transmissions:** All	**Fuel Injector Circuit High Input #1** Poor driveability. **Possible Causes:** • Check output signal from ECM • Check harness between injector and ECM connector • Check any other DTC on display • Using a scan tool, check current data • Check air intake system • Check for misfire
DTC: P0264 **1T CCM, MIL: Yes** 1997-99 **Engines:** 1.8L, 2.2L, 2.5L SOHC **Models:** Impreza, Forester, Legacy, Outback **Transmissions:** All	**Fuel Injector Circuit Low Input #2** Poor driveability. **Possible Causes:** • Check output signal from ECM • Check harness between injector and ECM connector • Check fuel injector • Check power supply line
DTC: P0265 **1T CCM, MIL: Yes** 1997-99 **Engines:** 1.8L, 2.2L, 2.5L SOHC **Models:** Impreza, Forester, Legacy, Outback **Transmissions:** All	**Fuel Injector Circuit High Input #2** Poor driveability. **Possible Causes:** • Check output signal from ECM • Check harness between injector and ECM connector • Check any other DTC on display • Using a scan tool, check current data • Check air intake system • Check for misfire
DTC: P0267 **1T CCM, MIL: Yes** 1997-99 **Engines:** 1.8L, 2.2L, 2.5L SOHC **Models:** Impreza, Forester, Legacy, Outback **Transmissions:** All	**Fuel Injector Circuit Low Input #3** Poor driveability. **Possible Causes:** • Check output signal from ECM • Check harness between injector and ECM connector • Check fuel injector • Check power supply line
DTC: P0268 **1T CCM, MIL: Yes** 1997-99 **Engines:** 1.8L, 2.2L, 2.5L SOHC **Models:** Impreza, Forester, Legacy, Outback	**Fuel Injector Circuit High Input #3** Poor driveability. **Possible Causes:** • Check output signal from ECM • Check harness between injector and ECM connector • Check any other DTC on display • Using a scan tool, check current data • Check air intake system • Check for misfire
DTC: P0270 **1T CCM, MIL: Yes** 1997-99 **Engines:** 1.8L, 2.2L, 2.5L SOHC **Models:** Impreza, Forester, Legacy, Outback **Transmissions:** All	**Fuel Injector Circuit Low Input #4** Poor driveability. **Possible Causes:** • Check output signal from ECM • Check harness between injector and ECM connector • Check fuel injector • Check power supply line

DTC	Trouble Code Title, Conditions & Possible Causes
DTC: P0271 **1T CCM, MIL: Yes** **1997-99** **Engines:** 1.8L, 2.2L, 2.5L SOHC **Models:** Impreza, Forester, Legacy, Outback **Transmissions:** All	**Fuel Injector Circuit High Input #4** Poor driveability. **Possible Causes:** • Check output signal from ECM • Check harness between injector and ECM connector • Check any other DTC on display • Using a scan tool, check current data • Check air intake system • Check for misfire
DTC: P0301 **2T CCM, MIL: Yes** **1996-06** **Engines:** 1.8L, 2.0L, 2.2L, 2.5L SOHC & DOHC, 3.0L, 3.3L **Models: Baja,** Forester, Impreza, **WRX,** Legacy, **B9 Tribeca,** Outback, **SVX** **Transmissions:** All	**Cylinder 1 Misfire Detected** DTC P0106, P0107, P0108, P0117, P0118, P0121, P0122, P0123, P0335, P0340 and P0500 not set, Engine started, vehicle driven to a speed of over 3 mph, and the PCM detected a misfire in Cylinder 1 in the 200 (Catalyst) or 1000 rpm (High Emissions) revolution range. Note: If the misfire is severe, the MIL will flash on/off on the 1st trip! **Possible Causes:** • Check any other DTC on display • Check output signal of ECM • Check harness between injector and ECM connector • Check fuel injector • Check power supply line • Check power supply to mass air flow sensor • Check harness between ECM and sensor connector • Check instAllation of camshaft and crankshaft sensors • Check crank plate • Check condition of timing chain • Check fuel level • Check status of MIL light • Check air intake system • Check cause of misfire (engine running) • Check engine condition • Check All cylinders individuAlly, in pairs, in groups and at random
DTC P0302 **2T CCM, MIL: Yes** **1996-06** **Engines:** 1.8L, 2.0L, 2.2L, 2.5L SOHC & DOHC, 3.0L, 3.3L **Models: Baja,** Forester, Impreza, **WRX,** Legacy, **B9 Tribeca,** Outback, **SVX** **Transmissions:** All	**Cylinder 2 Misfire Detected** DTC P0106, P0107, P0108, P0117, P0118, P0121, P0122, P0123, P0335, P0340 and P0500 not set, Engine started, vehicle driven to a speed of over 3 mph, and the PCM detected a misfire in Cylinder 2 in the 200 (Catalyst) or 1000 rpm (High Emissions) revolution range. Note: If the misfire is severe, the MIL will flash on/off on the 1st trip! **Possible Causes:** • Check any other DTC on display • Check output signal of ECM • Check harness between injector and ECM connector • Check fuel injector • Check power supply line • Check power supply to mass air flow sensor • Check harness between ECM and sensor connector • Check instAllation of camshaft and crankshaft sensors • Check crank plate • Check condition of timing chain • Check fuel level • Check status of MIL light • Check air intake system • Check cause of misfire (engine running) • Check engine condition • Check All cylinders individuAlly, in pairs, in groups and at random

DTC	Trouble Code Title, Conditions & Possible Causes
DTC: P0303 **2T CCM, MIL: Yes** **1996-06** **Engines:** 1.8L, 2.0L, 2.2L, 2.5L SOHC & DOHC, 3.0L, 3.3L **Models: Baja,** Forester, Impreza, **WRX,** Legacy, **B9 Tribeca,** Outback, **SVX** **Transmissions:** All	**Cylinder 3 Misfire Detected** DTC P0106, P0107, P0108, P0117, P0118, P0121, P0122, P0123, P0335, P0340 and P0500 not set, Engine started, vehicle driven to a speed of over 3 mph, and the PCM detected a misfire in Cylinder 3 in the 200 (Catalyst) or 1000 rpm (High Emissions) revolution range. Note: If the misfire is severe, the MIL will flash on/off on the 1st trip! **Possible Causes:** • Check any other DTC on display • Check output signal of ECM • Check harness between injector and ECM connector • Check fuel injector • Check power supply line • Check power supply to mass air flow sensor • Check harness between ECM and sensor connector • Check instAllation of camshaft and crankshaft sensors • Check crank plate • Check condition of timing chain • Check fuel level • Check status of MIL light • Check air intake system • Check cause of misfire (engine running) • Check engine condition • Check All cylinders individuAlly, in pairs, in groups and at random
DTC: P0304 **2T CCM, MIL: Yes** **1996-06** **Engines:** 1.8L, 2.0L, 2.2L, 2.5L SOHC & DOHC, 3.0L, 3.3L **Models: Baja,** Forester, Impreza, **WRX,** Legacy, **B9 Tribeca,** Outback, **SVX** **Transmissions:** All	**Cylinder 4 Misfire Detected** DTC P0106, P0107, P0108, P0117, P0118, P0121, P0122, P0123, P0335, P0340 and P0500 not set, Engine started, vehicle driven to a speed of over 3 mph, and the PCM detected a misfire in Cylinder 4 in the 200 (Catalyst) or 1000 rpm (High Emissions) revolution range. Note: If the misfire is severe, the MIL will flash on/off on the 1st trip! **Possible Causes:** • Check any other DTC on display • Check output signal of ECM • Check harness between injector and ECM connector • Check fuel injector • Check power supply line • Check power supply to mass air flow sensor • Check harness between ECM and sensor connector • Check instAllation of camshaft and crankshaft sensors • Check crank plate • Check condition of timing chain • Check fuel level • Check status of MIL light • Check air intake system • Check cause of misfire (engine running) • Check engine condition • Check All cylinders individuAlly, in pairs, in groups and at random
DTC: P0305 **2T CCM, MIL: Yes** **1996-06** **Engines:** 3.0L, 3.3L **Models:** B9 Tribeca, Outback, SVX **Transmissions:** All	**Cylinder 5 Misfire Detected** DTC P0106, P0107, P0108, P0117, P0118, P0121, P0122, P0123, P0335, P0340 and P0500 not set, Engine started, vehicle driven to a speed of over 3 mph, and the PCM detected a misfire in Cylinder 5 in the 200 (Catalyst) or 1000 rpm (High Emissions) revolution range. Note: If the misfire is severe, the MIL will flash on/off on the 1st trip! **Possible Causes:** • Check any other DTC on display • Check output signal of ECM • Check harness between injector and ECM connector • Check fuel injector • Check power supply line • Check power supply to mass air flow sensor • Check harness between ECM and sensor connector • Check instAllation of camshaft and crankshaft sensors • Check crank plate • Check condition of timing chain • Check fuel level • Check status of MIL light • Check air intake system • Check cause of misfire (engine running) • Check engine condition • Check All cylinders individuAlly, in pairs, in groups and at random

DTC	Trouble Code Title, Conditions & Possible Causes
DTC: P0306 **2T CCM, MIL: Yes** **1996-06** **Engines:** 3.0L, 3.3L **Models:** B9 Tribeca, Outback, SVX **Transmissions:** All	**Cylinder 6 Misfire Detected** DTC P0106, P0107, P0108, P0117, P0118, P0121, P0122, P0123, P0335, P0340 and P0500 not set, Engine started, vehicle driven to a speed of over 3 mph, and the PCM detected a misfire in Cylinder 6 in the 200 (Catalyst) or 1000 rpm (High Emissions) revolution range. Note: If the misfire is severe, the MIL will flash on/off on the 1st trip! **Possible Causes:** • Check any other DTC on display • Check output signal of ECM • Check harness between injector and ECM connector • Check fuel injector • Check power supply line • Check power supply to mass air flow sensor • Check harness between ECM and sensor connector • Check instAllation of camshaft and crankshaft sensors • Check crank plate • Check condition of timing chain • Check fuel level • Check status of MIL light • Check air intake system • Check cause of misfire (engine running) • Check engine condition • Check All cylinders individuAlly, in pairs, in groups and at random
DTC: P0325 **1T CCM, MIL: Yes** **1996-01** **Engines:** 1.8L, 2.2L, 2.5L SOHC, 3.3L **Models:** Forester, Impreza, Legacy, Outback, SVX **Transmissions:** All	**Knock Sensor 1 Circuit Malfunction** Engine started, engine speed over 1200 rpm, system voltage from 11-16 volts, and the PCM detected an unexpected "high" voltage condition on the Knock Sensor 1 signal circuit during the CCM test. **Possible Causes:** • Check harness between knock sensor 1 and ECM connector • Check sensor • Check input signal of ECM
DTC: P0327 **1T CCM, MIL: Yes** **2001-06** **Engines:** 2.0L, 2.5L SOHC & DOHC, 3.0L **Models:** Baja, Forester, Impreza, WRX, Legacy, Outback, B9 Tribeca **Transmissions:** All	**Knock Sensor 1 Circuit Low Input Bank 1 or Single Sensor** Engine started, engine speed over 1200 rpm, system voltage from 11-16 volts, and the PCM detected an unexpected "low" voltage condition on the Knock Sensor 1 circuit during the CCM test. **Possible Causes:** • Check harness between sensor and ECM connector • Check sensor • Check sensor instAllation
DTC: P0328 **1T CCM, MIL: Yes** **2001-06** **Engines:** 2.0L, 2.5L SOHC & DOHC, 3.0L **Models:** Baja, Forester, Impreza, WRX, Legacy, Outback, B9 Tribeca **Transmissions:** All	**Knock Sensor 1 Circuit High Input Bank 1 or Single Sensor** Engine started, engine speed over 1200 rpm, system voltage from 11-16 volts, and the PCM detected an unexpected "high" voltage condition on the Knock Sensor 1 circuit during the CCM test. **Possible Causes:** • Check harness between sensor and ECM connector • Check sensor • Check input signal of ECM
DTC: P0330 **1T CCM, MIL: Yes** **1996-97** **Engines:** 3.3L **Models:** SVX **Transmissions:** All	**Knock Sensor 2 Circuit Malfunction** Engine started, engine speed over 1200 rpm, system voltage from 11-16 volts, and the PCM detected an unexpected "high" voltage condition on the Knock Sensor 2 circuit during the CCM test. **Possible Causes:** • Check harness between the sensor and the ECM connector • Check the sensor • Check the input signal for the ECM
DTC: P0332 **1T CCM, MIL: Yes** **2001-06** **Engines:** 3.0L, 3.3L **Models:** Outback, B9 Tribeca **Transmissions:** All	**Knock Sensor 2 Circuit Low Input Bank 2** Engine started, engine speed over 1200 rpm, system voltage from 11-16 volts, and the PCM detected an unexpected "low" voltage condition on the Knock Sensor 2 circuit during the CCM test. **Possible Causes:** • Check harness between the sensor and the ECM connector • Check the sensor • Check sensor instAllation

DTC	Trouble Code Title, Conditions & Possible Causes
DTC: P0333 **1T CCM, MIL: Yes** **2001-06** **Engines:** 3.0L, 3.3L **Models:** Outback, B9 Tribeca **Transmissions:** All	**Knock Sensor 2 Circuit High Input Bank 2** Engine started, engine speed over 1200 rpm, system voltage from 11-16 volts, and the PCM detected an unexpected "high" voltage condition on the Knock Sensor 2 circuit during the CCM test. **Possible Causes:** • Check harness between the sensor and the ECM connector • Check the sensor • Check input signal of ECM
DTC: P0335 **1T CCM, MIL: Yes** **1996-06** **Engines:** 1.8L, 2.0L, 2.2L, 2.5L SOHC & DOHC, 3.0L, 3.3L **Models:** Baja, Forester, Impreza, WRX, Legacy, Outback, B9 Tribeca, SVX **Transmissions:** All	**Crankshaft Position Sensor "A" Circuit** Engine cranking; and then the PCM did not detect any Crankshaft Position (CKP) sensor signals, or the CKP sensor signal was interrupted after then engine was running during the CCM test. Note: The engine will not start without a proper CKP sensor signal. **Possible Causes:** • Check harness between the sensor and the ECM connector • Check condition of the sensor • Check the sensor
DTC: P0336 **2T CCM, MIL: Yes** **1997-06** **Engines:** 1.8L, 2.0L, 2.2L, 2.5L SOHC & DOHC, 3.0L, 3.3L **Models:** Baja, Forester, Impreza, WRX, Legacy, Outback, B9 Tribeca, SVX **Transmissions:** All	**Crankshaft Position Sensor "A" Circuit range/Performance** Engine started, and the PCM did not detect any Crankshaft Position Sensor (CKP) signals, or the CKP sensor signal was interrupted after the engine was running during the CCM test. Note: The engine may not start, or it may stall if it loses the proper CKP sensor signal after is has been running. **Possible Causes:** • Check any other DTC on display • Check condition of the sensor • Check the crankshaft plate • Check timing chain
DTC: P0340 **1T CCM, MIL: Yes** **1996-06** **Engines:** 1.8L, 2.0L, 2.2L, 2.5L SOHC & DOHC, 3.0L, 3.3L **Models:** Baja, Forester, Impreza, WRX, Legacy, Outback, B9 Tribeca, SVX **Transmissions:** All	**Camshaft Position Sensor "A" Circuit Bank 1 or Single Sensor** Engine cranking; and then the PCM did not detect any Camshaft Position (CMP) sensor signals, or the CMP sensor signal was interrupted after the engine was running during the CCM test. Note: The engine may not start without a proper CMP sensor signal. **Possible Causes:** • Check any other DTC on display • Check power supply • Check harness between sensor connector and ECM • Check sensor instAllation and condition • Check the sensor • Check for poor contact
DTC: P0341 **1T CCM, MIL: Yes** **1997-06** **Engines:** 2.0L, 2.2L, 2.5L SOHC & DOHC, 3.0L **Models:** Baja, Forester, Impreza, WRX, Legacy, Outback **Transmissions:** All	**Camshaft Position Sensor "A" Circuit Range/Performance Bank 1 or Single Sensor** Engine started, and the PCM detected the signals from the CMP sensor and CKP sensor were erratic (not plausible). **Possible Causes:** • Check any other DTC on display • Check harness between sensor connector and ECM • Check sensor instAllation and condition • Check the sensor • Check the cam sprocket • Check the instAllation and condition of the timing belt
DTC: P0345 **1T CCM, MIL: Yes** **2004-06** **Engines:** 2.5L DOHC, 3.0L **Models:** Baja, Forester, Impreza, WRX, Legacy, Outback, B9 Tribeca **Transmissions:** All	**Camshaft Position Sensor "A" Circuit Bank 2** **Engine stAlls. Failure of engine to start.** **Possible Causes:** • Check any other DTC on display • Check power supply • Check harness between sensor connector and ECM • Check the sensor • Check poor contact • Check the instAllation and condition of the sensor
DTC: P0400 **2T CCM, MIL: Yes** **1996-06** **Engines:** 1.8L, 2.2L, 2.5L SOHC, 3.0L, 3.3L **Models:** Baja, Forester, Impreza, Legacy, Outback, SVX **Transmissions:** All	**Exhaust Gas Recirculation Flow Malfunction** Engine started, vehicle driven at a steady speed of 52-58 mph for 1-3 minutes in closed loop, and the PCM detected little or no change in Pressure sensor signal after the EGR solenoid was cycled on/off. **Possible Causes:** • Check any other DTC on display • Check EGR control solenoid valve circuit • Check vacuum line • Check EGR system

DTC	Trouble Code Title, Conditions & Possible Causes

DTC P0410
2T CCM, MIL: Yes
1996-06
Engines: 1.8L, 2.5L DOHC
Models: Forester, Impreza
Transmissions: All
	Secondary Air Injection System
	Poor driveability.
	Possible Causes:
	  • Check air pump operation
	  • Check duct between pump and combination valve
	  • Check power supply to pump
	  • Check harness between pump relay and pump connector
	  • Check pump relay
	  • Check pump relay power source
	  • Check harness between ECM and pump relay connector

DTC: P0411 **2T CCM, MIL: Yes** **2006** **Engines:** 2.5L DOHC **Models:** Forester, Impreza **Transmissions:** All	**Secondary Air Injection System Incorrect Flow Detected** Poor driveability. **Possible Causes:** • Check secondary air combination valve
DTC: P0413 **1T CCM, MIL: Yes** **2006** **Engines:** 2.5L DOHC **Models:** Forester, Impreza	**Secondary Air Injection System Switching Valve "A" Circuit Open** Poor driveability. **Possible Causes:** • Check harness between ECM and valve relay #1
DTC: P0414 **1T CCM, MIL: Yes** **2006** **Engines:** 2.5L DOHC **Models:** Forester, Impreza **Transmissions:** All	**Secondary Air Injection System Switching Valve "A" Circuit Shorted** Poor driveability. **Possible Causes:** • Check harness between ECM and valve relay #1
DTC: P0416 **1T CCM, MIL: Yes** **2006** **Engines:** 2.5L DOHC **Models:** Forester, Impreza **Transmissions:** All	**Secondary Air Injection System Switching Valve "B" Circuit Open** Poor driveability. **Possible Causes:** • Check harness between ECM and valve relay #2
DTC: P0417 **1T CCM, MIL: Yes** **2006** **Engines:** 2.5L DOHC **Models:** Forester, Impreza **Transmissions:** All	**Secondary Air Injection System Switching Valve "B" Circuit Shorted** Poor driveability. **Possible Causes:** • Check harness between ECM and valve relay #2
DTC: P0418 **1T CCM, MIL: Yes** **2006** **Engines:** 2.5L DOHC **Models:** Forester, Impreza	**Secondary Air Injection System Switching Valve "A" Circuit Open** Poor driveability. **Possible Causes:** • Check harness between ECM and pump relay

DTC: P0420
2T CCM, MIL: Yes
1996-06
Engines: 1.8L, 2.0L, 2.2L, 2.5L SOHC & DOHC, 3.0L, 3.3L
Models: Baja, Forester, Impreza, WRX, Legacy, Outback, B9 Tribeca, SVX
Transmissions: All
	Catalyst System Efficiency Below Threshold Bank 1
	DTC P0106, P0107, P0108, P0117, P0118, P0121, P0122, P0123, P0131, P0132, P0133, P0134, P0137, P0138, P0140, P0141, P0171, P0172, P0301-P0306, P0335, P0340, P0342, P0401 and P0500 not set, Engine started, vehicle driven at 45-60 mph in closed loop for 3-5 minutes, and the PCM detected the amplitudes of the HO2S-11 and HO2S-12 signals were too similar during the test.
	Possible Causes:
	  • Check any other DTC on display
	  • Check harness between ECM and rear oxygen sensor connector
	  • Check exhaust system

DTC	Trouble Code Title, Conditions & Possible Causes
DTC: P0440 **1T CCM, MIL: Yes** **1996-01** **Engines:** 1.8L, 2.2L, 2.5L SOHC **Models:** Forester, Impreza, Legacy, Outback, SVX **Transmissions:** All	**Evaporative Emission Control System Malfunction** DTC P0106, P0107, P0108, P0112, P0113, P0117, P0118, P0121, P0122, P0123, P0500 ad P1422 not set, ECT and IAT sensors less than 90°F and within 12°F at startup (cold engine conditions), Engine started, system voltage from 11-16 volts, MAP sensor more than 75 kPa, throttle angle from 7-30%, vehicle driven to a speed of 25-55 mph in closed loop for 3 minutes, and the PCM detected it was unable to achieve or maintain a vacuum in the EVAP System during the test. **Possible Causes:** • Check fuel filler cap • Check fuel filler pipe packing • Check purge control valve solenoid • Check evaporative emission control system line • Check canister • Check fuel tank • Check any other mechanical components in the evaporative control system
DTC: P0441 **2T CCM, MIL: Yes** **1996-00** **Engines:** 1.8L, 2.2L, 2.5L SOHC, 3.3L **Models:** Forester, Impreza, Legacy, Outback, SVX **Transmissions:** All	**Evaporative Emission Control System Indirect Purge Flow** DTC P0106, P0107, P0108, P0112, P0113, P0117, P0118, P0121, P0122, P0123 and P0440 not set, ECT and IAT sensors more than 90°F and within 12°F at startup (cold engine conditions), Engine started, engine speed from 600-3500 rpm, system voltage from 11-16 volts, MAP sensor more than 85 kPa, throttle angle more than 14%, Purge solenoid commanded "on", and the PCM detected the EVAP Purge flow was incorrect during the EVAP purge flow test period. **Possible Causes:** • Check any other DTC on display • Check purge control valve solenoid operation • Check purge control valve
DTC: P0442 **2T CCM, MIL: Yes** **2000-06** **Engines:** 2.0L, 2.5L SOHC & DOHC, 3.0L **Models:** Baja, Forester, Impreza, WRX, Legacy, Outback, B9 Tribeca **Transmissions:** All	**Evaporative Emission Control System Leak Detected (smAll leak)** DTC P0106, P0107, P0108, P0117, P0118, P0121, P0122, P0123, P0440, P0441 and P1422 not set, ECT and IAT sensors more than 90°F and within 12°F at startup, system voltage at 11-16 volts, throttle angle over 75%, MAP sensor more than 75 kPa, vehicle driven at a steady cruise speed of less than 60 mph, fuel from 15-85%, and the PCM detected a vacuum decaying condition existed during the test. **Possible Causes:** • Check any other DTC on display • Check fuel filler cap • Check fuel filler pipe packing • Check the drain valve • Check the purge control solenoid valve • Check the pressure control solenoid valve • Check evaporative emission control system line • Check canister • Check fuel tank • Check any other mechanical components in the evaporative control system
DTC: P0443 **2T CCM, MIL: Yes** **1996-01** **Engines:** 1.8L, 2.2L, 2.5L SOHC, 3.3L **Models:** Baja, Forester, Impreza, Legacy, Outback, SVX **Transmissions:** All	**Evaporative Emission Control System Purge Control Valve Circuit Low Input** Key on or engine running; and the PCM detected an unexpected "low" voltage condition on the Purge Control solenoid control circuit. **Possible Causes:** • Check output signal from ECM • Check harness between purge control solenoid valve and ECM connector • Check the purge control solenoid valve • Check power supply to purge control solenoid valve • Check evaporative emission control system line • Check poor contact
DTC: P0444 **1T CCM, MIL: Yes** **2002** **Engines:** 2.5L SOHC **Models:** Impreza **Transmissions:** All	**Evaporative Emission Control System Purge Control Valve Circuit Low Input** Key on or engine running; and the PCM detected an unexpected "low" voltage condition on the Purge Control solenoid control circuit. **Possible Causes:** • Check output signal from ECM • Check harness between purge control solenoid valve and ECM connector • Check the purge control solenoid valve • Check power supply to purge control solenoid valve • Check evaporative emission control system line • Check poor contact

DTC	Trouble Code Title, Conditions & Possible Causes
DTC: P0445 **2T CCM, MIL: Yes** **2001-02** **Engines:** 2.5L SOHC **Models:** Forester, Impreza, Legacy, Outback **Transmissions:** All	**Evaporative Emission Control System Purge Control Valve Circuit High Input** Key on or engine running; and the PCM detected an unexpected high voltage condition on the Purge Control Solenoid control circuit. **Possible Causes:** • Check output signal from ECM • Check harness between purge control solenoid valve and ECM connector • Check poor contact
DTC: P0446 **2T CCM, MIL: Yes** **1996-01** **Engines:** 1.8L, 2.2L, 2.5L SOHC **Models:** Impreza, Legacy, Outback **Transmissions:** All	**Evaporative Emission Control System Purge Control Valve Circuit High Input** Key on or engine running; and the PCM detected an unexpected low voltage condition on the Vent Control solenoid control circuit. **Possible Causes:** • Check poor contact • Check harness between drain valve and ECM connector • Check poor contact • Check power supply to drain valve
DTC: P0447 **2T CCM, MIL: Yes** **2001-06** **Engines:** 2.0L, 2.5L SOHC & DOHC, 3.0L **Models:** Baja, Forester, Impreza, WRX, Legacy, Outback, B9 Tribeca **Transmissions:** All	**Evaporative Emission Control System Purge Control Valve Circuit High Input** Key on or engine running; and the PCM detected an unexpected low voltage condition on the Vent Control solenoid control circuit. **Possible Causes:** • Check output signal from ECM • Check poor contact • Check harness between drain valve and ECM connector • Check drain valve • Check power supply to drain valve
DTC: P0448 **2T CCM, MIL: Yes** **2001-06** **Engines:** 2.0L, 2.5L SOHC & DOHC, 3.0L **Models:** Baja, Forester, Impreza, WRX, Legacy, Outback, B9 Tribeca **Transmissions:** All	**Evaporative Emission Control System Vent Control Circuit Shorted** Key on or engine running; and the PCM detected an unexpected high voltage condition on the Vent Control Solenoid control circuit. **Possible Causes:** • Check input signal from ECM • Check poor contact • Check harness between drain valve and ECM connector • Check drain valve
DTC: P0450 **1T CCM, MIL: Yes** **1996** **Engines:** 1.8L, 2.2L **Models:** Impreza **Transmissions:** All	**Evaporative Emission Control System Pressure Sensor Malfunction** Poor driveability. **Possible Causes:** • Using a scan tool, check current data • Check input signal from ECM • Check harness between ECM and fuel tank pressure sensor connector
DTC: P0451 **1T CCM, MIL: Yes** **1996-06** **Engines:** 1.8L, 2.0L, 2.2L, 2.5L SOHC & DOHC, 3.0L **Models:** Baja, Forester, Impreza, WRX, Legacy, Outback, B9 Tribeca **Transmissions:** All	**Evaporative Emission Control System Pressure Sensor Malfunction** DTC P0450 not set, Engine started, vehicle driven to a speed over 3 mph at more than 1500 rpm, and the PCM detected an out-of-range Pressure sensor signal during the CCM continuous test. **Possible Causes:** • Check any other DTC on display • Check fuel filler cap • Check pressure vacuum line
DTC: P0452 **2T CCM, MIL: Yes** **1997-06** **Engines:** 2.0L, 2.2L, 2.5L SOHC & DOHC, 3.0L **Models:** Baja, Forester, Impreza, WRX, Legacy, Outback, B9 Tribeca **Transmissions:** All	**Evaporative Emission Control System Pressure Sensor Low Input** Key on or engine running; and the PCM detected the EVAP Pressure sensor less than 0.20 volt (Scan Tool reads under -2.8 kPa) during the CCM continuous test. **Possible Causes:** • Using a scan tool, check current data • Check power supply to fuel tank pressure sensor • Check input signal to ECM • Check harness between ECM and coupling connector in rear wiring harness • Check fuel tank cord • Check the purge control solenoid valve • Check poor contact

DTC	Trouble Code Title, Conditions & Possible Causes
DTC P0453 **2T CCM, MIL: Yes** **1997-06** **Engines:** 2.0L, 2.2L, 2.5L SOHC & DOHC, 3.0L **Models:** Baja, Forester, Impreza, WRX, Legacy, Outback, B9 Tribeca **Transmissions:** All	**Evaporative Emission Control System Pressure Sensor Low Input** Key on or engine running; and the PCM detected the EVAP Pressure sensor indicated more than 4.80 volts (Scan Tool reads over +2.8 kPa) during the CCM continuous test. **Possible Causes:** • Using a scan tool, check current data • Check power supply to fuel tank pressure sensor • Check input signal to ECM • Check harness between ECM and coupling connector in rear wiring harness • Check fuel tank cord • Check poor contact • Check harness between ECM and fuel tank pressure sensor connector
DTC: P0456 **2T CCM, MIL: Yes** **2001-06** **Engines:** 2.0L, 2.5L SOHC & DOHC, 3.0L **Models:** Baja, Forester, Impreza, WRX, Legacy, Outback, B9 Tribeca **Transmissions:** All	**Evaporative Emission Control System Pressure Sensor Low Input** DTC P0106, P0107, P0108, P0117, P0118, P0121, P0122, P0123, P0440, P1422 and P1423 not set, ECT and IAT sensors more than 90°F and within 12°F at startup, system voltage at 11-16 volts, throttle angle over 75%, MAP sensor more than 75 kPa, vehicle driven at a steady cruise speed of less than 60 mph, fuel from 15-85%, and the PCM detected a vacuum decaying condition existed during the test. **Possible Causes:** • Check any other DTC on display • Check fuel filler cap • Check fuel filler pipe packing • Check drain valve • Check purge control solenoid valve • Check pressure control solenoid valve • Check harness between ECM and fuel tank pressure sensor connector • Check evaporative emission control system line • Check canister • Check any other mechanical component in the system
DTC: P0457 **2T CCM, MIL: Yes** **2002-06** **Engines:** 2.0L, 2.5L SOHC & DOHC, 3.0L **Models:** Baja, Forester, Impreza, WRX, Legacy, Outback, B9 Tribeca **Transmissions:** All	**Evaporative Emission Control System Leak Detected** Engine started, and the PCM detected an unexpected low voltage condition on the EVAP Purge Controls solenoid control circuit. **Possible Causes:** • Check any other DTC on display • Check fuel filler cap • Check fuel filler pipe packing • Check drain valve • Check purge control solenoid valve • Check pressure control solenoid valve • Check canister • Check fuel tank • Check any other mechanical component in the system
DTC: P0458 **2T CCM, MIL: Yes** **2002-06** **Engines:** 2.0L, 2.5L SOHC & DOHC, 3.0L **Models:** Baja, Forester, Impreza, WRX, Legacy, Outback, B9 Tribeca **Transmissions:** All	**Evaporative Emission System Purge Control Valve Circuit Low** DTC P0106, P0107, P0108, P0117, P0118, P0121, P0122, P0123, P0440, P1422 and P1423 not set, ECT and IAT sensors more than 90°F and within 12°F at startup (cold engine conditions) Engine started, system voltage at 11-16 volts, throttle angle more than 75%, MAP sensor more than 75 kPa, vehicle driven at a steady cruise speed of less than 60 mph, Fuel Level from 15-85%, and the PCM detected a vacuum decaying condition existed during the leak test. **Possible Causes:** • Check output signal of ECM • Check harness between purge control solenoid valve and ECM connector • Check purge control solenoid valve • Check power supply to purge control solenoid valve • Check poor contact
DTC: P0459 **2T CCM, MIL: Yes** **2002-06** **Engines:** 2.0L, 2.5L SOHC & DOHC, 3.0L **Models:** Baja, Forester, Impreza, WRX, Legacy, Outback, B9 Tribeca **Transmissions:** All	**Evaporative Emission System Purge Control Valve Circuit High** **Improper idling.** **Possible Causes:** • Check output signal of ECM • Check harness between purge control solenoid valve and ECM connector • Check purge control solenoid valve • Check poor contact

DTC	Trouble Code Title, Conditions & Possible Causes
DTC: P0461 **2T CCM, MIL: Yes** **1997-06** **Engines:** 2.0L, 2.2L, 2.5L SOHC & DOHC, 3.0L **Models:** Baja, Forester, Impreza, WRX, Legacy, Outback, B9 Tribeca **Transmissions:** All	**Evaporative Emission System Purge Control Valve Circuit High** Engine started, then after the vehicle was driven a predetermined amount of miles, the PCM detected the Fuel Level sensor signal changed less than 0.14 volt, or the PCM detected the Fuel Level sensor signal was too low or tool high during the CCM test. **Possible Causes:** • Check any other DTC on display
DTC: P0462 **2T CCM, MIL: Yes** **1997-06** **Engines:** 2.0L, 2.2L, 2.5L SOHC & DOHC, 3.0L **Models:** Baja, Forester, Impreza, WRX, Legacy, Outback, B9 Tribeca **Transmissions:** All	**Evaporative Emission Control System Leak Detected** Key on or engine running; and the PCM detected the Fuel Level sensor signal indicated less than 0.10 volt at any time during the CCM continuous test. **Possible Causes:** • Check speedometer and tachometer operation • Check input signal of ECM • Check input voltage from ECM • Check harness between ECM and combination meter connector • Check fuel tank cord • Check fuel level sensor • Check fuel sub level sensor
DTC: P0463 **2T CCM, MIL: Yes** **1997-06** **Engines:** 2.0L, 2.2L, 2.5L SOHC & DOHC, 3.0L **Models:** Baja, Forester, Impreza, WRX, Legacy, Outback, B9 Tribeca **Transmissions:** All	**Fuel Level Sensor "A" Circuit High** Key on or engine running; and the PCM detected the Fuel Level sensor signal indicated more than 4.60 volts at any time during the CCM continuous test. **Possible Causes:** • Check speedometer and tachometer operation • Check input signal of ECM • Check input voltage from ECM • Check harness between ECM and fuel tank cord • Check harness between fuel tank cord and ground • Check fuel tank cord • Check fuel level sensor • Check fuel sub level sensor
DTC: P0464 **2T CCM, MIL: Yes** **2001-06** **Engines:** 2.0L, 2.5L SOHC & DOHC, 3.0L **Models:** Baja, Forester, Impreza, WRX, Legacy, Outback, B9 Tribeca **Transmissions:** All	**Fuel level Sensor Circuit Intermittent** DTC P0462 and P0463 not set, Engine started, vehicle driven to a speed of over 3 mph, and the PCM detected an unexpected voltage condition on the Fuel Level Sensor signal circuit in the CCM test. **Possible Causes:** • Check any other DTC on display • Check fuel level sensor • Check fuel level sub sensor
DTC: P0480 **2T CCM, MIL: No** **1997-06** **Engines: 2.2L,** 2.5L SOHC, **3.0L** **Models: Baja,** Forester, Impreza, Legacy, Outback **Transmissions:** All	**Cooling Fan relay 1 Circuit Low Input** Key on or engine running; and the PCM detected an unexpected low or high voltage condition in the Cooling Fan Relay control circuit. **Possible Causes:** • Check output signal from ECM • Check ground short circuit min radiator fan relay 1 control circuit • Check power supply for relay • Check main fan relay • Check for open circuit in main fan relay control circuit • Check poor contact
DTC: P0483 **2T CCM, MIL: No** **1997-06** **Engines: 2.2L,** 2.5L SOHC, **3.0L** **Models: Baja,** Forester, Impreza, Legacy, Outback **Transmissions:** All	**Fan Rationality Check** Engine started, engine runtime over 5 minutes, and the PCM detected a problem with the operation of the Cooling Fan Relay. Note: This code may set if the vehicle idles for too long near a wall. **Possible Causes:** • Check any other DTC on display

DTC	Trouble Code Title, Conditions & Possible Causes
DTC: P0500 **1T CCM, MIL: Yes** **1996-06** **Engines:** 1.8L, 2.2L, 2.5L SOHC **&** **DOHC, 3.0L, 3.3L** **Models:** Baja, Forester, Impreza, WRX, Legacy, Outback, B9 Tribeca, SVX **Transmissions:** All	**Vehicle Speed Sensor** Engine started, vehicle driven to a speed of over 30 mph at medium engine load for 30 seconds, and the PCM did not detect any VSS signals during the CCM Rationality test. **Possible Causes:** • Check DTC of VDC
DTC P0502 **1T CCM, MIL: Yes** **2002-06** **Engines:** 2.0L, 2.5L SOHC & DOHC, 3.0L, **3.3L** **Models:** Baja, Forester, Impreza, WRX, Legacy, Outback **Transmissions:** A/T	**Vehicle Speed Sensor Circuit Low Input** Poor driveability. **Possible Causes:** • Check to be sure vehicle is equipped with automatic transaxle • Check poor contact • Check harness between ECM and TCM connector • Check harness between VSS and ECM connector
DTC: P0503 **1T CCM, MIL: Yes** **2002-06** **Engines:** 2.0L, 2.5L SOHC & DOHC, 3.0L, **3.3L** **Models:** Baja, Forester, Impreza, WRX, Legacy, Outback **Transmissions:** All	**Vehicle Speed Sensor Intermittent/Erratic/High** Poor driveability. **Possible Causes:** • Check speedometer operation in combination meter • Check harness between ECM and combination meter connector
DTC: P0505 **1T CCM, MIL: Yes** **1996-01** **Engines:** 1.8L, 2.2L, 2.5L SOHC, **3.3L** **Models:** Forester, Impreza, Legacy, Outback, SVX **Transmissions:** All	**Idle Control System Malfunction** Engine started, ECT sensor signal more than 140°F, and the PCM detected the Actual idle speed was more than or less than the Target idle speed by too great an amount during the test. **Possible Causes:** • Check air intake system • Check output signal from ECM • Check idle air control solenoid valve • Check harness between ECM and idle air control solenoid valve connector
DTC: P0506 **2T CCM, MIL: Yes** **1996-06** **Engines:** 1.8L, 2.0L, 2.2L, 2.5L SOHC & DOHC, 3.0L, 3.3L **Models:** Baja, Forester, Impreza, WRX, Legacy, Outback, B9 Tribeca, SVX **Transmissions:** All	**Idle Control System RPM Lower Than Expected** Engine started, engine running at idle speed, ECT sensor signal more than 140°F, and the PCM detected the Actual idle speed was 100-200 rpm less than the Target idle speed for over 10 seconds. Note: The engine may stall or start to stall when this code is set. **Possible Causes:** • Check any other DTC on display • Check air cleaner element • Check electronic throttle control
DTC: P0507 **2T CCM, MIL: Yes** **1996-06** **Engines:** 1.8L, 2.0L, 2.2L, 2.5L SOHC & DOHC, 3.0L, 3.3L **Models:** Baja, Forester, Impreza, WRX, Legacy, Outback, B9 Tribeca, SVX **Transmissions:** All	**Idle Control System RPM Higher Than Expected** Engine started, engine running at idle speed, ECT sensor signal more than 140°F, and the PCM detected the Actual idle speed was 100-200 rpm more than the Target idle speed for over 10 seconds. Note: The engine may stall or start to stall when this code is set. **Possible Causes:** • Check any other DTC on display • Check air intake system • Check electronic throttle control
DTC: P0508 **1T CCM, MIL: Yes** **2001-05** **Engines:** **2.0L**, 2.5L DOHC, 3.0L **Models:** Impreza, **WRX**, Outback **Transmissions:** All	**Idle Control System Circuit Low** Key on or engine running; and the PCM detected an unexpected low voltage condition on the Idle Air Control (IAC) solenoid control circuit during the CCM continuous test. **Possible Causes:** • Check output signal from ECM • Check power supply to idle air control solenoid valve • Check harness between ECM and idle air control solenoid valve connector • Check ground circuit of the idle air control solenoid valve • Check poor contact

DTC	Trouble Code Title, Conditions & Possible Causes
DTC: P0509 **1T CCM, MIL: Yes** 2001-05 **Engines: 2.0L,** 2.5L DOHC, 3.0L **Models:** Impreza, **WRX,** Outback	**Idle Control System Circuit High** Key on or engine running; and the PCM detected an unexpected high voltage condition on the Idle Air Control (IAC) solenoid control circuit during the CCM continuous test. **Possible Causes:** • Check throttle cable • Check output signal from ECM
DTC: P0512 **2T CCM, MIL: No** 2003-06 **Engines:** 2.0L, 2.5L SOHC & DOHC, 3.0L **Models:** Baja, Forester, Impreza, WRX, Legacy, Outback, B9 Tribeca **Transmissions:** All	**Starter Request Circuit** Engine cranking; and the PCM detected an unexpected high voltage condition on the Starter Switch circuit. Note: Apply the Parking Brake and turn the key "on". Move the shift selector through all of the shift ranges. If the starter motor turns in any position, the Starter Switch circuit is shorted to system power. **Possible Causes:** • Check starter motor
DTC: P0519 **1T CCM, MIL: Yes** 2002-06 **Engines:** 2.0L, 2.5L SOHC & DOHC, 3.0L **Models:** Baja, Forester, Impreza, WRX, Legacy, Outback, B9 Tribeca **Transmissions:** All	**Idle Control System Malfunction (fail safe)** **Engine keeps running at higher speeds than specified idle speed.** **Possible Causes:** • Check any other DTC on display • Check air intake system • Check electronic throttle control
DTC: P0545 **1T CCM, MIL: Yes** 2002-06 **Engines: 2.0L,** 2.5L DOHC **Models:** Impreza, **WRX,** Legacy, Outback **Transmissions:** All	**Exhaust Gas Temperature Sensor Circuit Low Bank 1 Sensor 1** **hard to start. Improper idle. Poor driving performance.** **Possible Causes:** • Using a scan tool, check current data • Check harness between sensor and ECM connector
DTC: P0558 **1T CCM, MIL: No** 2003 **Engines:** 3.0L **Models:** Outback **Transmissions:** All	**Alternator Circuit Low Input** Poor driveability. **Possible Causes:** • Check harness between alternator and ECM connector
DTC: P0559 **1T CCM, MIL: No** 2003 **Engines:** 3.0L **Models:** Outback	**Alternator Circuit High Input** Poor driveability. **Possible Causes:** • Check harness between alternator and ECM connector
DTC: P0565 **2T CCM, MIL: No** 2002-06 **Engines:** 2.0L, 2.5L SOHC & DOHC, 3.0L **Models:** Baja, Forester, Impreza, WRX, Legacy, Outback **Transmissions:** All	**Cruise Control ON Signal** Poor driveability. **Possible Causes:** • Check harness between TCM and CCM connector • Check input signal for TCM (on AWD vehicles raise wheels off the ground) • Check poor contact
DTC: P0600 1996-06 **Engines:** 1.8L, 2.2L, 2.5L SOHC & DOHC, 3.0L, 3.3L **Models:** Baja, Forester, Impreza, WRX, Legacy, Outback, B9 Tribeca, SVX **Transmissions:** All	**Serial Communication Link** Key on or engine running; and the PCM detected an unexpected low or high voltage condition on the Serial Communications Link circuit. **Possible Causes:** • Check engine start failure (does engine run) • Check illumination of MIL light • Check indication of DTC on display • Using the a scan tool, perform diagnosis

DTC	Trouble Code Title, Conditions & Possible Causes
DTC: P0601 **2T CCM, MIL: Yes** **1996-06** **Engines:** 1.8L, 2.2L, 2.5L SOHC, 3.3L **Models:** Forester, Impreza, Legacy, Outback, SVX **Transmissions:** All	**Internal Control Module Memory Check Sum Error** Key on, and the PCM detected a Memory Check Sum Error during the initial check. Note: The engine will not start when this trouble code is set. **Possible Causes:** • Check any other DTC on display
DTC: P0604 **2T CCM, MIL: Yes** **2003-06** **Engines:** 2.0L, 2.5L SOHC & DOHC, 3.0L **Models:** Baja, Forester, Impreza, WRX, Legacy, Outback, B9 Tribeca **Transmissions:** All	**Internal Control Module Random Access Memory (RAM) Error** Engine stalls. Engine does not start. **Possible Causes:** • Check any other DTC on display
DTC: P0605 **1T CCM, MIL: Yes** **2004-06** **Engines:** 2.5L SOHC & DOHC, 3.0L **Models:** Baja, Forester, Impreza, WRX, Legacy, Outback, B9 Tribeca **Transmissions:** All	**Internal Control Module Read Only Memory (ROM) Error** Erroneous idling. Poor driving performance. **Possible Causes:** • Check input voltage from ECM • Check ground harness
DTC: P0607 **1T CCM, MIL: Yes** **2004-06** **Engines:** 2.5L SOHC & DOHC, 3.0L **Models:** Baja, Forester, Impreza, WRX, Legacy, Outback, B9 Tribeca **Transmissions:** All	**Control Module Performance** Erroneous idling. Poor driving performance. **Possible Causes:** • Check input voltage from ECM • Check ground harness
DTC: P0638 **1T CCM, MIL: Yes** **2004-06** **Engines:** 2.5L SOHC & DOHC, 3.0L **Models:** Baja, Forester, Impreza, WRX, Legacy, Outback, B9 Tribeca **Transmissions:** All	**Throttle Actuator Control range/Performance Bank 1** Erroneous idling. Engine stalls. Poor driving performance. **Possible Causes:** • Check electronic throttle control relay • Check power supply of electronic control relay • Check harness between ECM and electronic throttle control relay • Check sensor output • Check poor contact • Check harness between ECM and electronic throttle control • Check sensor power supply • Check for short in ECM • Check sensor output • Check harness between ECM and electronic throttle control motor • Check electronic throttle control motor harness • Check ground circuit • Check electronic throttle control
DTC: P0691 **2T CCM, MIL: No** **2001-04** **Engines:** 3.0L **Models:** Outback **Transmissions:** All	**Cooling Fan 1 Control Circuit Low** Key on or engine running; and the PCM detected an unexpected low voltage condition on the Cooling Fan Relay control circuit. **Possible Causes:** • Check output signal from ECM • Check for ground short circuit in radiator fan relay 1 control circuit • Check power supply for relay • Check main fan relays • Check for open circuit in main fan relay control circuit • Check poor contact

DTC	Trouble Code Title, Conditions & Possible Causes
DTC: P0691 **2T CCM, MIL: No** **2002-06** **Engines:** 2.0L, 2.5L SOHC & DOHC, 3.0L **Models:** Baja, Forester, Impreza, WRX, Legacy, Outback, B9 Tribeca **Transmissions:** All	**Cooling Fan 1 Control Circuit Low** Radiator fan does not operate properly. Overheating. **Possible Causes:** • Check any other DTC on display
DTC: P0692 **2T CCM, MIL: No** **2001-04** **Engines:** 3.0L **Models:** Outback **Transmissions:** All	**Cooling Fan 1 Control Circuit High** Key on or engine running; and the PCM detected an unexpected high voltage condition on the Cooling Fan Relay control circuit. **Possible Causes:** • Check output signal from ECM • Check for short circuit in radiator fan relay control circuit • Check main fan relay • Check sub fan relay • Check poor contact
DTC: P0692 **2T CCM, MIL: No** **2002-06** **Engines:** 2.0L, 2.5L SOHC & DOHC, 3.0L **Models:** Baja, Forester, Impreza, WRX, Legacy, Outback, B9 Tribeca **Transmissions:** All	**Cooling Fan 1 Control Circuit Low** Radiator fan does not operate properly. Overheating. **Possible Causes:** • Check any other DTC on display
DTC: P0700 **2004-06** **Engines:** 2.5L SOHC. 3.0L **Models:** Baja, Forester, Impreza, Legacy, Outback, B9 Tribeca **Transmissions:** All	**Transaxle Control System (MIL Request)** Vehicle performance issue. **Possible Causes:** • Check engine start failure (does engine run) • Check illumination of MIL light • Check indication of DTC on display • Using the a scan tool, perform diagnosis
DTC: P0703 **2T CCM, MIL: Yes** **1996-06** **Engines:** 1.8L, 2.0L, 2.2L, 2.5L SOHC & DOHC, 3.0L, 3.3L **Models:** Baja, Forester, Impreza, WRX, Legacy, Outback, B9 Tribeca, SVX **Transmissions:** All	**Torque Converter/Brake Switch "B" Circuit** Engine started, vehicle driven to a speed of over 30 mph, then back to idle speed at least two times, then once these conditions were met, the PCM did not detect any change in the Brake Switch status (i.e., the Brake switch did not cycle high and low). **Possible Causes:** • Check operation of brake light • Check harness between TCM and brake light switch connector • Check input signal for TCM • Check poor contact
DTC: P0705 **2T CCM, MIL: No** **1996-05** **Engines:** 1.8L, 2.0L, 2.2L, 2.5L SOHC & DOHC, 3.0L, 3.3L **Models:** Baja, Forester, Impreza, WRX, Legacy, Outback, B9 Tribeca, SVX **Transmissions:** All	**Transmission Range Sensor Circuit (PRNDL input)** Engine started, vehicle driven to a speed over 30 mph, and the PCM detected the gear selector was in Park or Neutral, or it detected multiple Transmission Range (TR) switch signals in Drive position. **Possible Causes:** • Using the a scan tool, read the data
DTC: P0710 **2T CCM, MIL: No** **1996-05** **Engines:** 1.8L, 2.0L, 2.2L, 2.5L SOHC & DOHC, 3.0L, 3.3L **Models:** Baja, Forester, Impreza, WRX, Legacy, Outback, B9 Tribeca, SVX **Transmissions:** All	**Transmission Fluid Temperature Sensor** No shift up to 4th speed after engine warm up. No lock up after engine warm up. Engine started, vehicle drive to a speed over 30 mph for 5-10 minutes, and the PCM detected the TFT sensor indicated less than -40°F, or it detected the TFT sensor indicated more than 300°F during the CCM continuous test. **Possible Causes:** • Using the a scan tool, read the data

DTC	Trouble Code Title, Conditions & Possible Causes
DTC: P0715 **2T CCM, MIL:** No **1997-02** **Engines:** 2.2L, 2.5L SOHC, 3.0L **Models:** Forester, Impreza, Legacy, Outback **Transmissions:** All	**Torque Converter Turbine Speed Sensor Circuit Malfunction** Engine started, vehicle driven to a speed over 3 mph for 1 minute, and the PCM did not detect any signals from the Torque Converter Turbine Speed sensor during the CCM continuous test. **Possible Causes:** • Using the a scan tool, read the data
DTC: P0716 **2T CCM, MIL:** No **2002-05** **Engines:** 2.0L, 2.5L SOHC & DOHC, 3.0L **Models:** Baja, Forester, Impreza, WRX, Legacy, Outback, B9 Tribeca **Transmissions:** All	**Input/Turbine Speed Sensor Circuit Range/Performance** Vehicle performance issue. **Possible Causes:** • Using the a scan tool, read the data
DTC: P0720 **2T CCM, MIL:** No **1996-05** **Engines:** 1.8L, 2.0L, 2.2L, 2.5L SOHC & DOHC, 3.0L, 3.3L **Models:** Baja, Forester, Impreza, WRX, Legacy, Outback, B9 Tribeca, SVX **Transmissions:** All	**Output Speed Sensor Circuit** Engine started, vehicle driven to a speed of over 3 mph for 1 minute, and the PCM did not detect any signals from the Output Speed sensor during the CCM test. **Possible Causes:** • Using a scan tool, read the data
DTC: P0725 **2T CCM, MIL:** No **1996-02** **Engines:** 1.8L, 2.0L, 2.2L, 2.5L SOHC, 3.0L, 3.3L **Models:** Baja, Forester, Impreza, Legacy, Outback, SVX **Transmissions:** All	**Engine Speed Input Circuit Malfunction** Engine started, and the TCM indicated a problem with the Engine Speed signal circuit to the PCM during the CCM continuous test. **Possible Causes:** • Using the a scan tool, read the data
DTC: P0726 **2T CCM, MIL:** No **2002-05** **Engines:** 2.0L, 2.5L SOHC & DOHC, 3.0L **Models:** Baja, Forester, Impreza, WRX, Legacy, Outback **Transmissions:** All	**Engine Speed Input Circuit Range/Performance** Vehicle performance issue. **Possible Causes:** • Using the a scan tool, read the data
DTC: P0731 **2T CCM, MIL:** Yes **1996-06** **Engines:** 1.8L, 2.0L, 2.5L SOHC & DOHC, 3.0L, 3.3L **Models:** Baja, Forester, Impreza, WRX, Legacy, Outback, B9 Tribeca, SVX **Transmissions:** All	**Gear 1 Incorrect Ratio** DTC P0500 not set, Engine started, vehicle driven to a speed of over 3 mph with 1st Gear commanded "on", and the PCM detected an incorrect 1st Gear ratio during the CCM continuous test. **Possible Causes:** • Check any other DTC on display • Check throttle position sensor circuit • Check front VSS circuit • Check torque converter turbine speed sensor circuit • Check poor contact • Check for mechanical problems
DTC: P0732 **2T CCM, MIL:** Yes **1996-06** **Engines:** 1.8L, 2.0L, 2.5L SOHC & DOHC, 3.0L, 3.3L **Models:** Baja, Forester, Impreza, WRX, Legacy, Outback, B9 Tribeca, SVX **Transmissions:** All	**Gear 2 Incorrect Ratio** **Shift point too high or too low. Engine brake not effected in "3" range. Excessive shift shock. Excessive tight corner "braking".** **Possible Causes:** • Check any other DTC on display • Check throttle position sensor circuit • Check front VSS circuit • Check torque converter turbine speed sensor circuit • Check poor contact • Check for mechanical problems

DTC	Trouble Code Title, Conditions & Possible Causes
DTC: P0733 **2T CCM, MIL: Yes** **1996-06** **Engines:** 1.8L, 2.0L, 2.5L SOHC & DOHC, 3.0L, 3.3L **Models:** Baja, Forester, Impreza, WRX, Legacy, Outback, B9 Tribeca, SVX **Transmissions:** All	**Gear 3 Incorrect Ratio** DTC P0500 not set, Engine started, vehicle driven to a speed of over 20 mph with 3rd Gear commanded "on", and the PCM detected an incorrect 3rd Gear ratio during the CCM Rationality test. **Possible Causes:** • Check any other DTC on display • Check throttle position sensor circuit • Check front VSS circuit • Check torque converter turbine speed sensor circuit • Check poor contact • Check for mechanical problems
DTC: P0734 **2T CCM, MIL: Yes** **1996-06** **Engines:** 1.8L, 2.0L, 2.5L SOHC & DOHC, 3.0L, 3.3L **Models:** Baja, Forester, Impreza, WRX, Legacy, Outback, B9 Tribeca, SVX **Transmissions:** All	**Gear 4 Incorrect Ratio** DTC P0500 not set, Engine started, vehicle driven to a speed of over 20 mph with 4th Gear commanded "on", and the PCM detected an incorrect 4th Gear ratio during the CCM Rationality test. **Possible Causes:** • Check any other DTC on display • Check throttle position sensor circuit • Check front VSS circuit • Check torque converter turbine speed sensor circuit • Check poor contact • Check for mechanical problems
DTC: P0740 **2T CCM, MIL: Yes** **1996-04** **Engines:** 2.0L, 2.5L SOHC & DOHC, 3.0L, 3.3L **Models:** Baja, Forester, Impreza, WRX, Legacy, Outback, SVX **Transmissions:** All	**Torque Converter Clutch System Malfunction** Engine started, vehicle driven to a speed over 28 mph, ECT sensor signal more than 140°F, and the PCM detected excessive slippage in the Torque Converter Clutch (TCC) during lockup operation. Inspect SSA for being stuck or for a valve body blockage. **Possible Causes:** • Check any other DTC on display • Check lock up duty solenoid circuit • Check throttle position sensor circuit • Check torque converter turbine speed sensor circuit • Check engine speed input circuit • Check brake light switch circuit • Check ATF temperature sensor circuit • Check poor contact • Check for mechanical problems
DTC: P0741 **2T CCM, MIL: Yes** **2001-06** **Engines:** 2.0L, 2.5L SOHC & DOHC, 3.0L **Models:** Baja, Forester, Impreza, WRX, Legacy, Outback **Transmissions:** All	**Torque Converter Clutch Circuit Performance or Stuck Off** Engine started, vehicle driven to a speed over 30 mph for 1 minute, and the PCM detected a problem while operating the TCC system. **Possible Causes:** • Check any other DTC on display • Check lock up duty solenoid circuit • Check throttle position sensor circuit • Check torque converter turbine speed sensor circuit • Check engine speed input circuit • Check brake light switch circuit and/or inhibitor switch circuit • Check ATF temperature sensor circuit • Check poor contact • Check for mechanical problems
DTC P0743 **2T CCM, MIL: Yes** **1996-05** **Engines:** 1.8L, 2.0L, 2.2L, 2.5L SOHC & DOHC, 3.0L, 3.3L **Models:** Baja, Forester, Impreza, WRX, Legacy, Outback, SVX **Transmissions:** All	**Torque Converter Clutch System (lock up duty solenoid) Electrical** Engine started, vehicle driven to a speed over 30 mph, and the PCM detected an unexpected voltage on the TCC solenoid circuit. **Possible Causes:** • Using the a scan tool, read the data

DTC	Trouble Code Title, Conditions & Possible Causes
DTC: P0748 **2T CCM, MIL: No** **1996-05** **Engines:** 1.8L, 2.0L, 2.2L, 2.5L SOHC & DOHC, 3.0L, 3.3L **Models:** Baja, Forester, Impreza, WRX, Legacy, Outback, SVX **Transmissions:** All	**Pressure Control Solenoid (line pressure duty solenoid) Electrical** Engine started, vehicle driven to a speed over 30 mph, and the PCM detected an unexpected voltage condition on the Pressure Control Solenoid (PCS) control circuit during the CCM continuous test. **Possible Causes:** • Using the a scan tool, read the data
DTC: P0753 **2T CCM, MIL: No** **1996-05** **Engines:** 1.8L, 2.0L, 2.2L, 2.5L SOHC & DOHC, 3.0L **Models:** Baja, Forester, Impreza, WRX, Legacy, Outback **Transmissions:** All	**Shift Solenoid "A" (shift solenoid 1) Electrical** Engine started, vehicle driven with 1st Gear commanded "on", and the PCM detected an unexpected voltage condition on the Shift Solenoid 'A' circuit during the CCM test. **Possible Causes:** • Using a scan tool, read the data
DTC: P0758 **2T CCM, MIL: No** **1996-05** **Engines:** 1.8L, 2.0L, 2.2L, 2.5L SOHC & DOHC, 3.0L, 3.3L **Models:** Baja, Forester, Impreza, WRX, Legacy, Outback, SVX **Transmissions:** All	**Shift Solenoid "B" (shift solenoid 2) Electrical** Engine started, vehicle driven with 2nd Gear commanded "on", and the PCM detected an unexpected voltage condition on the Shift Solenoid 'B' circuit during the CCM test. **Possible Causes:** • Using a scan tool, read the data
DTC: P0760 **2T CCM, MIL: No** 1996-98 **Engines:** 1.8L, 2.2L, 2.5L SOHC, 3.3L **Models:** Forester, Impreza, Legacy, Outback, SVX **Transmissions:** All	**Shift Solenoid "C" (shift solenoid 3) Malfunction** Engine started, vehicle driven with 3rd Gear commanded "on", and the PCM detected a problem in the mechanical operation of the Shift Solenoid 'C' during the CCM test. **Possible Causes:** • Check any other DTC on display • Check inhibiter switch circuit • Check shift solenoid 1 circuit • Check shift solenoid 3 circuit
DTC: P0763 **2T CCM, MIL: No** **1996-98** **Engines:** 1.8L, 2.5L SOHC, 3.3L **Models:** Forester, Impreza, Legacy, Outback, SVX **Transmissions:** All	**Shift Solenoid "C" (shift solenoid 3) Electrical** Engine started, vehicle driven with 3rd Gear commanded "on", and the PCM detected an unexpected voltage condition on the Shift Solenoid 'C' circuit during the CCM test. **Possible Causes:** • Check any other DTC on display
DTC: P0851 **2T CCM, MIL: No** **2002-06** **Engines:** 2.0L, 2.5L SOHC & DOHC, 3.0L **Models:** Baja, Forester, Impreza, WRX, Legacy, Outback **Transmissions:** A/T	**Neutral Switch Input Circuit Low (automatic transaxle)** Erroneous idling. **Possible Causes:** • Check any other DTC on display • Check input signal of ECM • Check harness between ECM and transmission harness connector • Check transmission harness connector • Check inhibitor switch • Check selector cable connection
DTC: P0851 **2T CCM, MIL: No** **2002-06** **Engines:** 2.0L, 2.5L SOHC & DOHC, 3.0L **Models:** Baja, Forester, Impreza, WRX, Legacy, Outback **Transmissions:** M/T	**Neutral Switch Input Circuit Low (manual transaxle)** Engine started, vehicle driven to 30-40 mph, and then back to idle, and the PCM detected an unexpected low voltage condition on the Neutral Position switch circuit during the CCM test. **Possible Causes:** • Check input signal of ECM • Check poor contact • Check neutral safety switch • Check harness between ECM and neutral safety switch connector • Check neutral safety switch ground

DTC	Trouble Code Title, Conditions & Possible Causes
DTC: P0852 **2T CCM, MIL: No** **2002-06** **Engines:** 2.0L, 2.5L SOHC & DOHC, 3.0L **Models:** Baja, Forester, Impreza, WRX, Legacy, Outback **Transmissions:** A/T	**Neutral Switch Input Circuit High (automatic transaxle)** Engine started, vehicle driven to 30-40 mph, and then back to idle, and the PCM detected an unexpected low voltage condition on the Neutral Position switch circuit during the CCM test. **Possible Causes:** • Check any other DTC on display • Check input signal of ECM • Check poor contact • Check harness between ECM and inhibitor switch connector • Check inhibitor switch ground • Check inhibitor switch • Check selector cable connection
DTC: P0852 **2T CCM, MIL: No** **2002-06** **Engines:** 2.0L, 2.5L SOHC & DOHC, 3.0L **Models:** Baja, Forester, Impreza, WRX, Legacy, Outback **Transmissions:** M/T	**Neutral Switch Input Circuit High (manual transaxle)** Engine started, vehicle driven to 30-40 mph, and then back to idle, and the PCM detected an unexpected low voltage condition on the Neutral Position switch circuit during the CCM test. **Possible Causes:** • Check input signal of ECM • Check poor contact • Check harness between ECM and transmission harness connector • Check neutral safety switch ground • Check neutral safety switch
DTC: P0864 **2T CCM, MIL: No** **2002-06** **Engines:** 2.0L, 2.5L SOHC & DOHC, 3.0L **Models:** Baja, Forester, Impreza, WRX, Legacy, Outback **Transmissions:** All	**TCM Communication Circuit Range/Performance** Engine started, vehicle driven to 30-40 mph, and then back to idle, and the PCM detected an unexpected voltage condition on the A/T Diagnosis circuit during the CCM test. **Possible Causes:** • Check driving condition (is AT shift control functioning) • Check accessory
DTC: P0865 **2T CCM, MIL: No** **2002-06** **Engines:** 2.0L, 2.5L SOHC & DOHC, 3.0L **Models:** Baja, Forester, Impreza, WRX, Legacy, Outback **Transmissions:** A/T	**TCM Communication Circuit Low** Engine started, vehicle driven to 30-40 mph, and then back to idle, and the PCM detected an unexpected low voltage condition on the A/T Diagnosis circuit during the CCM test. **Possible Causes:** • Check harness between ECM and TCM Connector • Check output signal for ECM • Check trouble code for automatic transaxle
DTC: P0866 **2T CCM, MIL: No** **2002-06** **Engines:** 2.0L, 2.5L SOHC & DOHC, 3.0L **Models:** Baja, Forester, Impreza, WRX, Legacy, Outback **Transmissions:** A/T	**TCM Communication Circuit High** Engine started, vehicle driven to 30-40 mph, and then back to idle, and the PCM detected an unexpected high voltage condition on the A/T Diagnosis circuit during the CCM test. **Possible Causes:** • Check harness between ECM and TCM Connector • Check poor contact

OBD II Trouble Code List (P1xxx Codes)

DTC	Trouble Code Title, Conditions & Possible Causes:
DTC: P1086 **1T CCM, MIL:** Yes **2002-04** **Engines:** 2.0L, 2.5L DOHC **Models:** Baja, Forester, Impreza, WRX **Transmissions:** All	**Tumble Generated Valve Position Sensor 2 Circuit Low** Engine stalls. Erroneous idling. Poor driving performance. **Possible Causes:** • Using a scan tool, check current data • Check input signal for ECM • Check harness between ECM and tumble generator valve position sensor connector • Check poor contact
DTC: P1087 **1T CCM, MIL:** Yes **2002-04** **Engines:** 2.0L, 2.5L DOHC **Models:** Baja, Forester, Impreza, WRX **Transmissions:** All	**Tumble Generated Valve Position Sensor 2 Circuit High** Engine stalls. Erroneous idling. Poor driving performance. **Possible Causes:** • Using a scan tool, check current data • Check harness between tumble generator valve position sensor and ECM connector • Check harness between throttle position sensor and ECM connector
DTC: P1088 **1T CCM, MIL:** Yes **2002-04** **Engines:** 2.0L, 2.5L DOHC **Models:** Baja, Forester, Impreza, WRX **Transmissions:** All	**Tumble Generated Valve Position Sensor 1 Circuit Low** Engine stalls. Erroneous idling. Poor driving performance. **Possible Causes:** • Using a scan tool, check current data • Check input signal for ECM • Check harness between ECM and tumble generator valve position sensor connector • Check poor contact
DTC: P1089 **1T CCM, MIL:** Yes **2002-04** **Engines:** 2.0L, 2.5L DOHC **Models:** Baja, Forester, Impreza, WRX **Transmissions:** All	**Tumble Generated Valve Position Sensor 1 Circuit High** Engine stalls. Erroneous idling. Poor driving performance. **Possible Causes:** • Using a scan tool, check current data • Check harness between tumble generator valve position sensor and ECM connector
DTC: P1090 **1T CCM, MIL:** Yes **2002-04** **Engines:** 2.0L, 2.5L DOHC **Models:** Baja, Forester, Impreza, WRX **Transmissions:** All	**Tumble Generated Valve System 1 (valve open)** Engine stalls. Erroneous idling. Poor driving performance. **Possible Causes:** • Check any other DTC on display • Check tumble generator valve (RH)
DTC: P1091 **1T CCM, MIL:** Yes **2002-04** **Engines:** 2.0L, 2.5L DOHC **Models:** Baja, Forester, Impreza, WRX **Transmissions:** All	**Tumble Generated Valve System 1 (valve closed)** Engine stalls. Erroneous idling. Poor driving performance. **Possible Causes:** • Check any other DTC on display • Check tumble generator valve (RH)
DTC: P1092 **1T CCM, MIL:** Yes **2002-04** **Engines:** 2.0L, 2.5L DOHC **Models:** Baja, Forester, Impreza, WRX **Transmissions:** All	**Tumble Generated Valve System 2 (valve open)** Engine stalls. Erroneous idling. Poor driving performance. **Possible Causes:** • Check any other DTC on display • Check tumble generator valve (RH)
DTC: P1093 **1T CCM, MIL:** Yes **2002-04** **Engines:** 2.0L, 2.5L DOHC **Models:** Baja, Forester, Impreza, WRX **Transmissions:** All	**Tumble Generated Valve System 2 (valve closed)** Engine stalls. Erroneous idling. Poor driving performance. **Possible Causes:** • Check any other DTC on display • Check tumble generator valve (RH)

DTC	Trouble Code Title, Conditions & Possible Causes
DTC: P1094 **1T CCM, MIL: Yes** **2002-04** **Engines:** 2.0L, 2.5L DOHC **Models:** Baja, Forester, Impreza, WRX **Transmissions:** All	**Tumble Generated Valve Signal 1 Circuit Malfunction** Engine stalls. Erroneous idling. Poor driving performance. **Possible Causes:** • Check harness between ECM and tumble generator valve actuator connector • Check poor contact
DTC: P1095 **1T CCM, MIL: Yes** **2002-04** **Engines:** 2.0L, 2.5L DOHC **Models:** Baja, Forester, Impreza, WRX	**Tumble Generated Valve Signal 1 Circuit Malfunction (short)** Engine stalls. Erroneous idling. Poor driving performance. **Possible Causes:** • Check harness between ECM and tumble generator valve actuator connector
DTC: P1096 **1T CCM, MIL: Yes** **2002-04** **Engines:** 2.0L, 2.5L DOHC **Models:** Baja, Forester, Impreza, WRX **Transmissions:** All	**Tumble Generated Valve Signal 2 Circuit Malfunction (open)** Engine stalls. Erroneous idling. Poor driving performance. **Possible Causes:** • Check harness between ECM and tumble generator valve actuator connector • Check poor contact
DTC: P1097 **1T CCM, MIL: Yes** **2002-04** **Engines:** 2.0L, 2.5L DOHC **Models:** Baja, Forester, Impreza, WRX **Transmissions:** All	**Tumble Generated Valve Signal 2 Circuit Malfunction (short)** Engine stalls. Erroneous idling. Poor driving performance. **Possible Causes:** • Check harness between ECM and tumble generator valve actuator connector
DTC: P1100 **2T CCM, MIL: No** **1996-01** **Engines:** 1.8L, 2.2L, 2.5L SOHC, 3.3L **Models:** Forester, Impreza, Legacy, Outback, SVX **Transmissions:** All	**Starter Switch Circuit Malfunction** Engine cranking; and PCM detected an unexpected "low" voltage condition on the Starter Switch signal during the CCM test. **Possible Causes:** • Check starter motor
DTC: P1101 **2T CCM, MIL: No** **1996-01** **Engines:** 1.8L, 2.2L, 2.5L SOHC, 3.3L **Models:** Forester, Impreza, Legacy, Outback, SVX **Transmissions:** All	**Neutral Position Switch Circuit Malfunction** Engine started, vehicle driven to a speed of over 3 mph, and the PCM detected any change in the voltage status of the Neutral Position switch circuit during the CCM test. **Possible Causes:** • Check input signal for ECM • Check harness between ECM and transmission harness connector • Check inhibitor switch
DTC: P1102 **2T CCM, MIL: No** **1996-99** **Engines:** 1.8L, 2.2L, 2.5L SOHC, 3.3L **Models:** Forester, Impreza, Legacy, Outback, SVX **Transmissions:** All	**Pressure Sources Switching Solenoid Valve Circuit Malfunction** Key on or engine running; and the PCM detected an unexpected voltage condition on the Pressure Sources Solenoid during the test. **Possible Causes:** • Check output signal from ECM • Check harness between ECM and pressure sources switching solenoid valve connector • Check valve • Check power supply to valve
DTC: P1103 **2T CCM, MIL: No** **1996-01** **Engines:** 1.8L, 2.2L, 2.5L SOHC, 3.3L **Models:** Forester, Impreza, Legacy, Outback, SVX **Transmissions:** All	**Engine Torque Control Signal Circuit Malfunction** Key on or engine running; and the TCM did not detect any Engine Torque Control '1' signals on the circuit between the controllers. **Possible Causes:** • Check input signal of ECM • Check harness between ECM and TCM connector

DTC	Trouble Code Title, Conditions & Possible Causes
DTC: P1106 **2T CCM, MIL:** No **1997-01** **Engines:** 2.2L, 2.5L SOHC **Models:** Forester, Impreza, Legacy, Outback **Transmissions:** All	**Engine Torque Control Signal 2 Circuit** Key on or engine running; and the TCM did not detect any Engine Torque Control '2' signals on the circuit between the controllers. **Possible Causes:** • Check input signal of ECM • Check poor contact • Check harness between ECM and TCM connector
DTC: P1107 **2T CCM, MIL:** No **1996** **Engines:** 1.8L, 2.2L **Models:** Impreza **Transmissions:** All	**Air Injection System Diagnosis Solenoid Valve Circuit Malfunction** Vehicle performance issue. **Possible Causes:** • Check transmission type • Check output signal from ECM • Check harness between air injection system solenoid valve and ECM connector • Check valve • Check power supply to the valve
DTC: P1108 **2T CCM, MIL:** No **1996-97** **Engines:** 3.3L **Models:** SVX **Transmissions:** All	**Induction Control Solenoid Valve Circuit Malfunction** Key on or engine running; and the PCM detected an unexpected voltage condition on the Induction Air Control solenoid circuit. **Possible Causes:** • Check output signal from ECM • Check harness between induction control solenoid valve and ECM connector • Check valve • Check power supply to the valve
DTC: P1110 **1T CCM, MIL:** Yes **1999-04** **Engines:** 2.0L, 2.5L SOHC & DOHC, 3.0L **Models:** Baja, Forester, Impreza, WRX, Legacy, Outback **Transmissions:** All	**Atmospheric Pressure Sensor Circuit Low Input** Key on or engine running; and the PCM detected the Atmospheric Pressure sensor signal indicated less than 0.10 volt (Scan Tool reads 0 kPa) for one second during the CCM test. **Possible Causes:** • Using the a scan tool, check current data • Check poor contact • Check input signal for ECM • Check harness between ECM and atmospheric pressure sensor connector
DTC: P1111 **1T CCM, MIL:** Yes **1999-04** **Engines:** 2.0L, 2.5L SOHC & DOHC, 3.0L **Models:** Baja, Forester, Impreza, WRX, Legacy, Outback **Transmissions:** All	**Atmospheric Pressure Sensor Circuit High Input** Poor driveability. **Possible Causes:** • Using the a scan tool, check current data • Check input signal for ECM • Check harness between ECM and pressure sensor connector
DTC: P1112 **2T CCM, MIL:** Yes **1999-02** **Engines:** 2.2L, 2.5L SOHC, 3.0L **Models:** Forester, Impreza, Legacy, Outback **Transmissions:** All	**Atmospheric Pressure Sensor Circuit Range/Performance Problem** Key on or engine running; and the PCM detected the Atmospheric Pressure sensor signal indicated more than 4.90 volts (Scan Tool reads 140 kPa) for 1 second during the CCM test. **Possible Causes:** • Check input signal for ECM • Check harness between ECM and pressure sensor connector • Check any other DTC on display • Check sensor filter • Using the a scan tool, check current data
DTC: P1115 **2T CCM, MIL:** Yes **1997-01** **Engines:** 2.2L, 2.5L SOHC **Models:** Forester, Impreza, Legacy, Outback **Transmissions:** All	**Engine Torque Control Cut Signal Circuit High Input** Key on or engine running; and the TCM detected an unexpected "high" voltage condition on the Engine Torque Control signal circuit between the controllers. **Possible Causes:** • Check output signal from ECM • Check harness between ECM and TCM connector
DTC: P1116 **2T CCM, MIL:** Yes **1997-01** **Engines:** 2.2L, 2.5L SOHC **Models:** Forester, Impreza, Legacy, Outback **Transmissions:** All	**Engine Torque Control Cut Signal Circuit Low Input** Key on or engine running; and the TCM detected an unexpected "low" voltage condition on the Engine Torque Control signal circuit between the controllers. **Possible Causes:** • Check output signal from ECM • Check harness between ECM and TCM connector

DTC	Trouble Code Title, Conditions & Possible Causes
DTC: P1120 **2T CCM, MIL: Yes** **1997-01** **Engines:** 1.8L, 2.2L, 2.5L SOHC **Models:** Forester, Impreza, Legacy, Outback **Transmissions:** All	**Starter Circuit Switch High Input** Engine cranking; and PCM detected an unexpected "high" voltage condition on the Starter Switch circuit during the CCM test. **Possible Causes:** • Check operation of starter motor
DTC: P1121 **2T CCM, MIL: Yes** **1997-01** **Engines:** 1.8L, 2.2L, 2.5L SOHC **Models:** Forester, Impreza, Legacy, Outback **Transmissions:** M/T	**Starter Circuit Switch High Input** Engine started, vehicle driven to a speed of over 3 mph, and the PCM detected an unexpected "low" voltage condition on the A/T Neutral Position switch circuit during the CCM test. **Possible Causes:** • Check transmission type (manual transaxle) • Check input signal of ECM • Check poor contact • Check neutral position sensor • Check harness between ECM and sensor connector • Check any other DTC on display • Check harness between ECM and transmission harness connector • Check inhibitor switch • Check selector cable connection
DTC: P1122 **2T CCM, MIL: No** **1997-99** **Engines:** 1.8L, 2.2L, 2.5L SOHC **Models:** Forester, Impreza, Legacy, Outback **Transmissions:** All	**Pressure Sources Switching Solenoid Valve High Input** Key on or engine running; and the PCM detected an unexpected voltage condition on the Pressure Sources Solenoid during the test. **Possible Causes:** • Check output signal of ECM • Check poor contact • Check harness between ECM and valve connector • Check valve
DTC: P1130 **2T CCM, MIL: Yes** **1999-01** **Engines:** 2.2L, 2.5L SOHC, 3.0L **Models:** Forester, Impreza, Legacy, Outback **Transmissions:** All	**Front Oxygen Sensor Circuit Malfunction** Engine started, engine runtime over 3 minutes, and the PCM detected an unexpected "low" voltage condition on the front Air Fuel Sensor (A/FS) signal circuit for 10 seconds during the CCM test. **Possible Causes:** • Check harness between ECM and sensor connector • Check poor contact
DTC: P1131 **2T CCM, MIL: Yes** **1999-01** **Engines:** 2.2L, 2.5L SOHC, 3.0L **Models:** Forester, Impreza, Legacy, Outback **Transmissions:** All	**Front Oxygen Sensor Circuit Malfunction (short circuit)** Engine started, engine runtime over 3 minutes, and the PCM detected an unexpected "high" voltage condition on the front A/F Sensor (A/FS) signal circuit for 10 seconds during the CCM test. **Possible Causes:** • Check harness between ECM and sensor connector • Check output signal for ECM
DTC: P1132 **2T CCM, MIL: Yes** **1999-01** **Engines:** 2.2L, 2.5L SOHC **Models:** Forester, Impreza, Legacy, Outback **Transmissions:** All	**Front Oxygen Sensor Heater Circuit Low Input** Engine started, engine runtime over 3 minutes, and the PCM detected an unexpected "low" current condition on the front A/F Sensor (A/FS) signal circuit for 10 seconds during the CCM test. If DTC P0141 is also set, the fault is common to both sensors. **Possible Causes:** • Check any other DTC on display • Check power supply circuit of ECM • Check ground circuit of ECM • Check power supply of front sensor • Check sensor
DTC: P1133 **2T CCM, MIL: Yes** **1999-01** **Engines:** 2.2L, 2.5L SOHC **Models:** Forester, Impreza, Legacy, Outback **Transmissions:** All	**Front Oxygen Sensor Heater Circuit High Input** Engine started, engine runtime over 3 minutes, and the PCM detected an unexpected "high" current condition on the front A/F Sensor (A/FS) signal circuit for 10 seconds during the CCM test. **Possible Causes:** • Check output signal from ECM • Check sensor heater current

DTC	Trouble Code Title, Conditions & Possible Causes
DTC: P1134 **2T CCM, MIL: Yes** **1999-06** **Engines:** 2.0L, 2.2L, 2.5L SOHC & DOHC, 3.0L **Models:** Baja, Forester, Impreza, WRX, Legacy, Outback **Transmissions:** All	**Front Oxygen Sensor Micro Computer Problem** Key on or engine running; and the PCM detected an internal problem related to the Micro Computer for the front A/F sensor. **Possible Causes:** • Check any other DTC on display
DTC: P1137 **2T CCM, MIL: Yes** **2001-06** **Engines:** 2.5L SOHC **Models:** Baja, Forester, Impreza, Legacy, Outback **Transmissions:** All	**O2 Sensor Circuit (bank 1 sensor 1)** Poor driveability. **Possible Causes:** • Check any other DTC on display • Using a scan tool, check current data • Check front sensor • Check harness between ECM and sensor • Check exhaust system
DTC: P1139 **2T CCM, MIL: Yes** **1999-02** **Engines:** 2.5L SOHC, 3.0L **Models:** Impreza, Legacy, Outback **Transmissions:** All	**Front Oxygen Sensor 1 Heater Circuit Range/Performance Problem** Key on or engine running; and the PCM detected an internal problem related to the Micro Computer for the front A/F sensor. Note: This diagnostic trouble code is for California models. **Possible Causes:** • Check harness between ECM and sensor connector • Check front sensor • Check poor contact
DTC: P1140 **2T CCM, MIL: Yes** **2001-02** **Engines:** 3.0L **Models:** Outback **Transmissions:** All	**Bank 2 and Sensor 1 Oxygen Sensor (front left) Heater Circuit Range/ Performance Problem** Engine started, engine runtime over 3 minutes, and the PCM detected an unexpected "high" current condition on the front A/F Sensor (A/FS) signal circuit for 10 seconds during the CCM test. **Possible Causes:** • Check harness between ECM and sensor connector • Check front sensor • Check poor contact
DTC: P1141 **2T CCM, MIL: Yes** **1997-99** **Engines:** 1.8L, 2.2L, 2.5L SOHC **Models:** Forester, Impreza, Legacy, Outback **Transmissions:** All	**Mass Air Flow Sensor Circuit Range/Performance Problem (high output)** DTC P0102 and P0103 not set, Engine started, engine running at idle or cruise speed, and the PCM detected an unexpected "high" voltage condition on the MAF sensor circuit during the CCM test. **Possible Causes:** • Check any other DTC on display
DTC: P1142 **2T CCM, MIL: Yes** **1997-99** **Engines:** 1.8L, 2.2L, 2.5L SOHC, 3.0L **Models:** Forester, Impreza, Legacy, Outback **Transmissions:** All	**Throttle Position Sensor Circuit Range/Performance Problem (low input)** DTC P0122 and P0123 not set, Engine started, engine running at idle or cruise speed, and the PCM detected an unexpected "low" voltage condition on the TP sensor circuit during the CCM test. **Possible Causes:** • Check any other DTC on display
DTC: P1143 **2T CCM, MIL: Yes** **1997-99** **Engines:** 1.8L, 2.2L, 2.5L SOHC **Models:** Forester, Impreza, Legacy, Outback **Transmissions:** All	**Pressure Sensor Circuit range/Performance Problem (low input)** DTC P0107 and P0108 not set, Engine started, engine running at idle speed, and the PCM detected the Pressure sensor signal was less than 32 kPa during the CCM test. **Possible Causes:** • Using a scan tool, check current data • Check pressure sensor
DTC: P1144 **2T CCM, MIL: Yes** **1997-99** **Engines:** 1.8L, 2.2L, 2.5L SOHC **Models:** Forester, Impreza, Legacy, Outback **Transmissions:** All	**Pressure Sensor Circuit range/Performance Problem (high input)** DTC P0107 and P0108 not set, Engine started, engine running at idle speed, and the PCM detected the Pressure sensor signal was more than 133 kPa during the CCM test. **Possible Causes:** • Using a scan tool, check current data

DTC	Trouble Code Title, Conditions & Possible Causes
DTC: P1146 **2T CCM, MIL: Yes** **2001-02** **Engines:** 2.5L SOHC, 3.0L **Models:** Forester, Impreza, Legacy, Outback **Transmissions:** All	**Pressure Sensor Circuit range/Performance Problem (high input)** DTC P0107, P0108 and P1112 not set, Engine started, ECT sensor signal more than 160F, and the PCM detected an unexpected "high" voltage condition on the Pressure sensor circuit in the CCM test. **Possible Causes:** • Check any other DTC on display • Check air intake system • Check pressure sensor • Check throttle position
DTC: P1151 **2T CCM, MIL: Yes** **1997-01** **Engines:** 2.2L, 2.5L SOHC **Models:** Forester, Impreza, Legacy, Outback **Transmissions:** All	**Rear Oxygen Sensor Heater Circuit High Input** Poor driveability. **Possible Causes:** • Check input signal for ECM • Check any other DTC on display
DTC: P1152 **2T CCM, MIL: Yes** **2002-06** **Engines:** 2.0L, 2.5L SOHC & DOHC, 3.0L **Models:** Baja, Forester, Impreza, WRX, Legacy, Outback **Transmissions:** All	**Bank 1 and Sensor 2 Oxygen Sensor (front right) Circuit Range/Performance Problem Low Input** Poor driveability. **Possible Causes:** • Check any other DTC on display • Using a scan tool, check current data • Check rear sensor signal • Check exhaust system
DTC: P1153 **2T CCM, MIL: Yes** **2002-06** **Engines:** 2.0L, 2.5L SOHC & DOHC, 3.0L **Models:** Baja, Forester, Impreza, WRX, Legacy, Outback **Transmissions:** All	**Bank 1 and Sensor 1 Oxygen Sensor (front right) Circuit Range/Performance Problem High Input** Poor driveability. **Possible Causes:** • Check any other DTC on display • Using a scan tool, check current data • Check rear sensor signal • Check exhaust system
DTC: P1154 **2T CCM, MIL: Yes** **2002-06** **Engines:** 3.0L **Models:** Outback, B9 Tribeca **Transmissions:** All	**Bank 1 and Sensor 1 Oxygen Sensor (front left) Circuit Range/Performance Problem low Input** Poor driveability. **Possible Causes:** • Check any other DTC on display • Using a scan tool, check current data • Check rear sensor signal • Check exhaust system
DTC: P1155 **2T CCM, MIL: Yes** **2002-06** **Engines:** 3.0L **Models:** Outback, B9 Tribeca **Transmissions:** All	**Bank 2 and Sensor 1 Oxygen Sensor (front left) Circuit Range/Performance Problem High Input** Poor driveability. **Possible Causes:** • Check any other DTC on display • Using a scan tool, check current data • Check rear sensor signal • Check exhaust system
DTC: P1160 **1T CCM, MIL: Yes** **2004-06** **Engines:** 2.5L SOHC, 3.0L **Models:** Baja, Forester, Impreza, Legacy, Outback, B9 Tribeca **Transmissions:** All	**Throttle Actuator Control range/Performance Bank 1** Erroneous idling. Engine stalls. Poor driving performance. **Possible Causes:** • Check electronic throttle control relay • Check power supply of electronic control relay • Check harness between ECM and electronic throttle control relay • Check sensor output • Check poor contact • Check harness between ECM and electronic throttle control • Check sensor power supply • Check for short in ECM • Check sensor output • Check harness between ECM and electronic throttle control motor • Check electronic throttle control motor harness • Check ground circuit • Check electronic throttle control

DTC	Trouble Code Title, Conditions & Possible Causes
DTC: P1207 **1T CCM, MIL: Yes** **2000-01** **Engines:** 2.5L SOHC **Models:** Baja, Forester, Impreza, Legacy, Outback **Transmissions:** All	**Air Assist Injector Solenoid Valve Circuit Low Input** Key on or engine running; and the PCM detected an unexpected "low" voltage condition on the Air Assist Injector Solenoid circuit. **Possible Causes:** • Check output signal from ECM • Check power supply to valve • Check harness between ECM and valve connector • Check poor contact
DTC: P1208 **1T CCM, MIL: Yes** **2000-01** **Engines:** 2.5L SOHC **Models:** Baja, Forester, Impreza, Legacy, Outback **Transmissions:** All	**Air Assist Injector Solenoid Valve Circuit high Input** Key on or engine running; and the PCM detected an unexpected "high" voltage condition on the Air Assist Injector Solenoid circuit. **Possible Causes:** • Check output signal from ECM
DTC: P1301 **1T CCM, MIL: Yes** **2002-06** **Engines:** 2.0L, 2.5L DOHC **Models:** Baja, Forester, Impreza, WRX, Legacy, Outback **Transmissions:** All	**Misfire Detected (high temperature exhaust gas)** Improper idling. Engine stalls. Poor driving performance. **Possible Causes:** • Check any other DTC on display • Using a scan tool, check current data
DTC: P1312 **1T CCM, MIL: Yes** **2002-06** **Engines:** 2.0L, 2.5L DOHC **Models:** Baja, Forester, Impreza, WRX, Legacy, Outback **Transmissions:** All	**Exhaust Gas Temperature Sensor malfunction** Poor driveability. **Possible Causes:** • Check any other DTC on display • Using a scan tool, check current data
DTC: P1325 **1T CCM, MIL: Yes** **1997-01** **Engines:** 2.0L, 2.5L SOHC **Models:** Forester, Impreza, Legacy, Outback **Transmissions:** All	**Knock Sensor Circuit Low Input** Engine started, engine speed more than 200 rpm, and the PCM detected an unexpected "low" voltage condition on the Knock Sensor signal for over 10 seconds during the CCM test. **Possible Causes:** • Check harness between sensor and ECM connector • Check sensor • Check sensor instAllation • Check input signal for ECM
DTC: P1400 **2T CCM, MIL: Yes** **1996-06** **Engines:** 1.8L, 2.2L, 2.5L SOHC & DOHC, 3.0L **Models:** Baja, Forester, Impreza, WRX, Legacy, Outback **Transmissions:** All	**Fuel Tank Pressure Control Solenoid Valve Circuit Low** Key on or engine running; and the PCM detected an unexpected "low" voltage condition on the Fuel Tank Pressure solenoid circuit during the CCM test. **Possible Causes:** • Check output signal of ECM • Check poor contact • Check harness between valve and ECM connector • Check valve • Check power supply to valve
DTC: P1401 **2T CCM, MIL: Yes** **1996** **Engines:** 1.8L, 2.2L **Models:** Impreza **Transmissions:** All	**Fuel Tank Pressure Control System Function Problem** Poor driveability. **Possible Causes:** • Check fuel tank pressure control solenoid valve
DTC: P1402 **2T CCM, MIL: Yes** **1996** **Engines:** 1.8L, 2.2L **Models:** Impreza **Transmissions:** All	**Fuel Level Sensor Circuit Malfunction** Poor driveability. **Possible Causes:** • Check speedometer and tachometer operation • Check ground circuit of combination meter • Check input signal of ECM • Check fuel level sensor • Check ground circuit f fuel level sensor • Check harness between ECM and fuel pump connector • Check harness between combination meter and fuel pump connector • Check combination meter

DTC	Trouble Code Title, Conditions & Possible Causes
DTC: P1410 **1T CCM, MIL: Yes** **2006** **Engines:** 2.5L SOHC **Models:** Forester, Impreza **Transmissions:** All	**Secondary Air Injection System Switching Valve Stuck Open** Poor driveability. **Possible Causes:** • Using the a scan tool, check current data • Check harness between mass air flow and intake air temperature sensor and ECM
DTC: P1418 **1T CCM, MIL: Yes** **2006** **Engines:** 2.5L SOHC **Models:** Forester, Impreza **Transmissions:** All	**Secondary Air Injection System Control "A" Circuit Shorted** Poor driveability. **Possible Causes:** • Using a scan tool, check current data • Check harness between mass air flow and intake air temperature sensor and ECM
DTC: P1420 **2T CCM, MIL: Yes** **1997-06** **Engines:** 1.8L, 2.0L, 2.2L, 2.5L SOHC & DOHC, 3.0L **Models:** Baja, Forester, Impreza, WRX, Legacy, Outback **Transmissions:** All	**Fuel Tank Pressure Control Solenoid Valve Circuit High** Key on or engine running; and the PCM detected an unexpected "high" voltage condition on the Fuel Tank Pressure (FTP) Control solenoid circuit during the CCM test. **Possible Causes:** • Check input signal of ECM • Check poor contact • Check ground circuit fuel level sensor • Check harness between valve and ECM connector • Check harness between combination meter and fuel pump connector • Check valve
DTC: P1421 **2T CCM, MIL: Yes** **1997-99** **Engines:** 1.8L, 2.2L, 2.5L SOHC **Models:** Forester, Impreza, Legacy, Outback **Transmissions:** All	**Exhaust Gas Recirculation Circuit High Input** Poor driving performance at low engine speed. **Possible Causes:** • Check output signal of ECM • Check poor contact • Check harness between EGR solenoid and ECM connector • Check EGR solenoid valve
DTC: P1422 **2T CCM, MIL: Yes** **1997-01** **Engines:** 1.8L, 2.2L, 2.5L SOHC **Models:** Forester, Impreza, Legacy, Outback **Transmissions:** All	**Evaporative Emission Control System Purge Control Valve Circuit High Input** Key on or engine running; and the PCM detected an unexpected "high" voltage condition on the EVAP Purge Control solenoid circuit during the CCM test. **Possible Causes:** • Check output signal of ECM • Check poor contact • Check harness between purge control solenoid and ECM connector • Check purge control solenoid valve
DTC: P1423 **2T CCM, MIL: Yes** **1997-01** **Engines:** 1.8L, 2.2L, 2.5L SOHC **Models:** Forester, Impreza, Legacy, Outback **Transmissions:** All	**Evaporative Emission Control System Vent Control High Input** Key on or engine running; and the PCM detected an unexpected "high" voltage condition on the EVAP Vent Control solenoid circuit during the CCM test. **Possible Causes:** • Check output signal of ECM • Check poor contact • Check harness between vent control solenoid valve and ECM connector • Check vent control solenoid valve
DTC: P1440 **2T CCM, MIL: Yes** **1997-99** **Engines:** 1.8L, 2.2L, 2.5L SOHC **Models:** Forester, Impreza, Legacy, Outback **Transmissions:** All	**Fuel Tank Pressure Control System Function Problem (low input)** Key on or engine running; and the PCM detected an unexpected "low" voltage condition on the Fuel Tank Pressure (FTP) sensor circuit during the CCM test. **Possible Causes:** • Check fuel tank pressure control solenoid valve • Check fuel filler cap • Check fuel filler pipe seal • Check vent control solenoid valve • Check purge control solenoid valve

DTC	Trouble Code Title, Conditions & Possible Causes
DTC: P1441 **2T CCM, MIL: Yes** **1997-99** **Engines:** 1.8L, 2.2L, 2.5L SOHC **Models:** Forester, Impreza, Legacy, Outback **Transmissions:** All	**Fuel Tank Pressure Control System Function Problem (high input)** Key on or engine running; and the PCM detected an unexpected "high" voltage condition on the Fuel Tank Pressure (FTP) sensor circuit during the CCM test. **Possible Causes:** • Check fuel tank pressure control solenoid valve • Check fuel filler cap • Check fuel filler pipe seal • Check vent control solenoid valve • Check purge control solenoid valve • Check canister • Check fuel tank • Check for other mechanical problems
DTC: P1442 **2T CCM, MIL: Yes** **1997-02** **Engines:** 1.8L, 2.2L, 2.5L SOHC **Models:** Forester, Impreza, Legacy, Outback **Transmissions:** All	**Fuel level Sensor Circuit Range/Performance Problem 2** DTC P0461, P0462 and P0463 not set, Engine started, vehicle driven to a speed over 20 mph for 10-20 minutes, and the PCM detected a performance problem with the Fuel Level Sensor signal. **Possible Causes:** • Check any other DTC on display
DTC: P1443 **1T CCM, MIL: Yes** **1997-06** **Engines:** 2.0L, 2.2L, 2.5L SOHC & DOHC, 3.0L **Models:** Baja, Forester, Impreza, WRX, Legacy, Outback, B9 Tribeca **Transmissions:** All	**Vent Control Solenoid Valve Function Problem** DTC P0461, P0462 and P0463 not set, Engine started, vehicle driven at a speed of over 20 mph for 10-20 minutes), and PCM detected the problem in the operation of the EVAP Vent system in the test. **Possible Causes:** • Check any other DTC on display • Check vent line hoses • Check drain valve operation
DTC: P1445 **1T CCM, MIL: Yes** **2000-01** **Engines:** 2.5L SOHC **Models:** Forester, Impreza, Legacy, Outback **Transmissions:** All	**Air Assist Injector Solenoid Valve malfunction** Key on or engine running; and the PCM detected an unexpected voltage condition on the Air Assist Injector solenoid circuit in the test. **Possible Causes:** • Check any other DTC on display • Check air assist injector solenoid valve operation • Check air by passes hoses • Check fuel injector
DTC: P1446 **2T CCM, MIL: Yes** **2002-06** **Engines:** 2.0L, 2.5L SOHC & DOHC, 3.0L **Models:** Baja, Forester, Impreza, WRX, Legacy, Outback **Transmissions:** All	**Fuel Tank Sensor Control Valve Circuit Low** Engine started, and the PCM detected an unexpected low voltage condition on the Atmospheric Pressure valve circuit during the test. **Possible Causes:** • Check output signal from ECM • Check poor contact • Check harness between fuel tank sensor control valve and ECM connector • Check fuel tank sensor control valve • Check power supply to fuel tank sensor control valve
DTC: P1447 **2T CCM, MIL: Yes** **2002-06** **Engines:** 2.0L, 2.5L SOHC & DOHC, 3.0L **Models:** Baja, Forester, Impreza, WRX, Legacy, Outback **Transmissions:** All	**Fuel Tank Sensor Control Valve Circuit High** Engine started, and the PCM detected an unexpected high voltage condition on the Atmospheric Pressure valve circuit during the test. **Possible Causes:** • Check output signal from ECM • Check poor contact • Check harness between fuel tank sensor control valve and ECM connector • Check fuel tank sensor control valve
DTC: P1448 **2T CCM, MIL: Yes** **2002-06** **Engines:** 2.0L, 2.5L SOHC & DOHC, 3.0L **Models:** Baja, Forester, Impreza, WRX, Legacy, Outback **Transmissions:** All	**Fuel Tank Sensor Control Valve Range/Performance** Engine started, vehicle driven at cruise speed and then back to idle, and the PCM detected a problem in the operation of the Atmospheric Pressure valve system during the CCM test period. **Possible Causes:** • Check for other DTC on display • Check fuel filler cap • Check evaporative emission line

DTC	Trouble Code Title, Conditions & Possible Causes

DTC: P1480
2T CCM, MIL: No
2001-02
Engines: 2.5L SOHC, 3.0L
Models: Forester, Impreza, Legacy, Outback
Transmissions: All
	Cooling Fan Relay 1 Circuit High Input
	Key on or engine running; then with the cooling fan commanded "on", the PCM detected an unexpected "high" voltage condition on the Cooling Fan Relay circuit during the CCM test.
	Possible Causes:
	    • Using a scan tool, check current data
	    • Check for short circuit in radiator fan relay control circuit
	    • Check main fan relay
	    • Check sub fan relay
	    • Check poor contact

DTC: P1490
1T CCM, MIL: No
2001-02
Engines: 2.5L SOHC
Models: Forester, Impreza, Legacy, Outback
Transmissions: All
	Thermostat Malfunction
	DTC P0125, P0301, P0302, P0303 and P0304 not set, ECT sensor less than 140°F and IAT sensor more than 50°F at startup, engine runtime over 10 minutes, and the PCM detected the ECT sensor signal did not reach 160°F under these engine operating conditions.
	Possible Causes:
	    • Check vehicle condition
	    • Check other DTC on display
	    • Check engine coolant
	    • Check radiator fan

DTC: P1491
2T CCM, MIL: Yes
2004-06
Engines: 2.5L SOHC & DOHC
Models: Baja, Forester, Impreza, WRX, Legacy, Outback
Transmissions: All
	Positive Crankcase Ventilation (blow by) Function problem
	Erroneous idling.
	Possible Causes:
	    • Check blow by hose
	    • Check harness between PCV diagnosis connector and ECM connector
	    • Inspect PCV diagnosis connector

DTC: P1492
1T CCM, MIL: Yes
2003-06
Engines: 2.5L SOHC
Models: Baja, Forester, Impreza, Legacy, Outback
Transmissions: All
	EGR Solenoid Valve Signal #1 Circuit Malfunction (low input)
	Erroneous idling. Poor driving performance. Engine breathing.
	Possible Causes:
	    • Check power supply to EGR solenoid
	    • Check harness between ECM and EGR solenoid valve connector
	    • Check poor contact

DTC: P1493
1T CCM, MIL: Yes
2003-06
Engines: 2.5L SOHC
Models: Baja, Forester, Impreza, Legacy, Outback
Transmissions: All
	EGR Solenoid Valve Signal #1 Circuit Malfunction (high input)
	Erroneous idling. Poor driving performance. Engine breathing.
	Possible Causes:
	    • Check other DTC on display
	    • Check harness between ECM and EGR solenoid valve connector

DTC: P1494
1T CCM, MIL: Yes
2003-06
Engines: 2.5L SOHC
Models: Baja, Forester, Impreza, Legacy, Outback
Transmissions: All
	EGR Solenoid Valve Signal #2 Circuit Malfunction (low input)
	Erroneous idling. Poor driving performance. Engine breathing.
	Possible Causes:
	    • Check power supply to EGR solenoid
	    • Check harness between ECM and EGR solenoid valve connector
	    • Check poor contact

DTC: P1495
1T CCM, MIL: Yes
2003-06
Engines: 2.5L SOHC
Models: Baja, Forester, Impreza, Legacy, Outback
Transmissions: All
	EGR Solenoid Valve Signal #2 Circuit Malfunction (high input)
	Erroneous idling. Poor driving performance. Engine breathing.
	Possible Causes:
	    • Check other DTC on display
	    • Check harness between ECM and EGR solenoid valve connector

DTC: P1496
1T CCM, MIL: Yes
2003-06
Engines: 2.5L SOHC
Models: Baja, Forester, Impreza, Legacy, Outback
Transmissions: All
	EGR Solenoid Valve Signal #3 Circuit Malfunction (low input)
	Erroneous idling. Poor driving performance. Engine breathing.
	Possible Causes:
	    • Check power supply to EGR solenoid
	    • Check harness between ECM and EGR solenoid valve connector
	    • Check poor contact

DTC	Trouble Code Title, Conditions & Possible Causes
DTC: P1497 **1T CCM, MIL: Yes** **2003-06** **Engines:** 2.5L SOHC **Models:** Baja, Forester, Impreza, Legacy, Outback **Transmissions:** All	**EGR Solenoid Valve Signal #3 Circuit Malfunction (high input)** Erroneous idling. Poor driving performance. Engine breathing. **Possible Causes:** • Check other DTC on display • Check harness between ECM and EGR solenoid valve connector
DTC: P1498 **1T CCM, MIL: Yes** **2003-06** **Engines:** 2.5L SOHC **Models:** Baja, Forester, Impreza, Legacy, Outback **Transmissions:** All	**EGR Solenoid Valve Signal #4 Circuit Malfunction (low input)** Erroneous idling. Poor driving performance. Engine breathing. **Possible Causes:** • Check power supply to EGR solenoid • Check harness between ECM and EGR solenoid valve connector • Check poor contact
DTC: P1499 **1T CCM, MIL: Yes** **2003-06** **Engines:** 2.5L SOHC **Models:** Baja, Forester, Impreza, Legacy, Outback **Transmissions:** All	**EGR Solenoid Valve Signal #4 Circuit Malfunction (high input)** Erroneous idling. Poor driving performance. Engine breathing. **Possible Causes:** • Check other DTC on display • Check harness between ECM and EGR solenoid valve connector
DTC: P1500 **2T CCM, MIL: No** **1996-98** **Engines:** 1.8L, 2.2L, 2.5L SOHC, 3.3L **Models:** Forester, Impreza, Legacy, Outback, SVX **Transmissions:** All	**Radiator Fan Relay 1 Circuit malfunction (fan 1)** Key on or engine running; and the PCM detected an unexpected "high or "low" voltage condition on the Cooling Fan Relay circuit or Sub Fan Relay circuit (for models with Air Conditioning) in the test. **Possible Causes:** • Check any other DTC on display • Check output signal from ECM • Check power supply line for relays • Check main fan relay 1 and sub fan relay • Check for open circuit in radiator fan relay 1 control circuit
DTC: P1501 **2T CCM, MIL: No** **1996-97** **Engines:** 3.3L **Models:** SVX **Transmissions:** All	**Radiator Fan Relay 2 Circuit malfunction (fan 2)** DTC P0505 not set, Engine started, engine running at idle speed, ECT sensor more than 170°F, and the PCM detected the Actual Idle Speed was 100-200 rpm more or less than the Target Idle Speed. Note: When this code is set, the PCM will operate in Fail Safe Mode. **Possible Causes:** • Check any other DTC on display • Check output signal from ECM • Check power supply line for relays • Check main fan relay 2 and sub fan relay • Check for open circuit in radiator fan relay 2 control circuit
DTC: P1502 **2T CCM, MIL: No** **1996-98** **Engines:** 1.8L, 2.2L, 2.5L SOHC, 3.3L **Models:** Forester, Impreza, Legacy, Outback, SVX **Transmissions:** All	**Radiator Fan Function problem** Engine started, ECT sensor signal less than 200°F, and the PCM detected an unexpected "low" voltage condition on the Radiator Fan Relay control circuit during the CCM Rationality test. **Possible Causes:** • Check any other DTC on display
DTC: P1505 **1T CCM, MIL: Yes** **1999-01** **Engines:** 2.2L, 2.5L SOHC **Models:** Impreza, Legacy, Outback **Transmissions:** All	**Idle Control System Circuit High Input** Erroneous idling. Engine stalls. Engine breathing. **Possible Causes:** • Check throttle cable • Check output signal from ECM
DTC: P1507 **1T CCM, MIL: Yes** **1997-02** **Engines:** 2.2L, 2.5L SOHC, 3.0L **Models:** Forester, Impreza, Legacy, Outback **Transmissions:** All	**Idle Control System Malfunction (fail safe)** DTC P0505 not set, Engine started, engine running at idle speed, ECT sensor more than 170°F, and the PCM detected the Actual Idle Speed was 100-200 rpm more or less than the Target Idle Speed. Note: When this code is set, the PCM will operate in Fail Safe Mode. **Possible Causes:** • Check any other DTC on display • Check air intake system

DTC	Trouble Code Title, Conditions & Possible Causes
DTC: P1510 **1T CCM, MIL: Yes** **1997-06** **Engines:** 2.0L, 2.5L SOHC **Models:** Baja, Forester, Impreza, WRX, Legacy, Outback **Transmissions:** All	**ISC Solenoid Valve Signal #1 Circuit Malfunction (low input)** Key on or engine running; and the PCM detected an unexpected "low" condition on the IAC solenoid Signal 1 circuit during the test. **Possible Causes:** • Check power supply to idle air control solenoid valve • Check harness between ECM and idle control solenoid valve connector • Check poor contact
DTC: P1511 **1T CCM, MIL: Yes** **1997-06** **Engines:** 2.0L, 2.5L SOHC **Models:** Baja, Forester, Impreza, WRX, Legacy, Outback **Transmissions:** All	**ISC Solenoid Valve Signal #1 Circuit Malfunction (high input)** Key on or engine running; and the PCM detected an unexpected "low" condition on the IAC solenoid Signal 1 circuit during the test. **Possible Causes:** • Check any other DTC on display • Check ground circuit for ECM • Check harness between ECM and idle control solenoid valve connector
DTC: P1512 **1T CCM, MIL: Yes** **1997-06** **Engines:** 2.0L, 2.5L SOHC **Models:** Baja, Forester, Impreza, WRX, Legacy, Outback **Transmissions:** All	**ISC Solenoid Valve Signal #2 Circuit Malfunction (low input)** Key on or engine running; and the PCM detected an unexpected "low" condition on the IAC solenoid Signal 2 circuit during the test. **Possible Causes:** • Check power supply to idle air control solenoid valve • Check harness between ECM and idle control solenoid valve connector • Check poor contact
DTC: P1513 **2T CCM, MIL: Yes** **1997-06** **Engines:** 2.0L, 2.5L SOHC **Models:** Baja, Forester, Impreza, WRX, Legacy, Outback **Transmissions:** All	**ISC Solenoid Valve Signal #2 Circuit Malfunction (high input)** Key on or engine running; and the PCM detected an unexpected "low" condition on the IAC solenoid Signal 2 circuit during the test. **Possible Causes:** • Check any other DTC on display • Check ground circuit for ECM • Check harness between ECM and idle control solenoid valve connector
DTC: P1514 **2T CCM, MIL: Yes** **1997-06** **Engines:** 2.0L, 2.5L SOHC **Models:** Baja, Forester, Impreza, WRX, Legacy, Outback **Transmissions:** All	**ISC Solenoid Valve Signal #3 Circuit Malfunction (low input)** Key on or engine running; and the PCM detected an unexpected "low" condition on the IAC solenoid Signal 3 circuit during the test. **Possible Causes:** • Check power supply to idle air control solenoid valve • Check harness between ECM and idle control solenoid valve connector • Check poor contact
DTC: P1515 **2T CCM, MIL: Yes** **1997-06** **Engines:** 2.0L, 2.5L SOHC **Models:** Baja, Forester, Impreza, WRX, Legacy, Outback **Transmissions:** All	**ISC Solenoid Valve Signal #3 Circuit Malfunction (high input)** Key on or engine running; and the PCM detected an unexpected "low" condition on the IAC solenoid Signal 3 circuit during the test. **Possible Causes:** • Check any other DTC on display • Check ground circuit for ECM • Check harness between ECM and idle control solenoid valve connector
DTC: P1516 **2T CCM, MIL: Yes** **1997-06** **Engines:** 2.0L, 2.5L SOHC **Models:** Baja, Forester, Impreza, WRX, Legacy, Outback **Transmissions:** All	**ISC Solenoid Valve Signal #4 Circuit Malfunction (low input)** Key on or engine running; and the PCM detected an unexpected "low" condition on the IAC solenoid Signal 4 circuit during the test. **Possible Causes:** • Check power supply to idle air control solenoid valve • Check harness between ECM and idle control solenoid valve connector • Check poor contact
DTC: P1517 **2T CCM, MIL: Yes** **1997-06** **Engines:** 2.0L, 2.5L SOHC **Models:** Baja, Forester, Impreza, WRX, Legacy, Outback **Transmissions:** All	**ISC Solenoid Valve Signal #4 Circuit Malfunction (high input)** Key on or engine running; and the PCM detected an unexpected "low" condition on the IAC solenoid Signal 4 circuit during the test. **Possible Causes:** • Check any other DTC on display • Check ground circuit for ECM • Check harness between ECM and idle control solenoid valve connector

DTC	Trouble Code Title, Conditions & Possible Causes
DTC: P1518 **1T CCM, MIL: No** 2001-06 **Engines:** 2.0L, 2.5L SOHC & DOHC, 3.0L **Models:** Baja, Forester, Impreza, WRX, Legacy, Outback, B9 Tribeca **Transmissions:** All	**Starter Switch Circuit Low Input** **Ignition key in crank position, and the PCM detected an unexpected "low" voltage condition on the Starter Switch circuit during the test. Note: The engine will not start with this condition present.** **Possible Causes:** • Check operation of starter motor
DTC: P1520 **2T CCM, MIL: No** 1997-01 **Engines:** 2.0L, 2.5L SOHC **Models:** Forester, Impreza, Legacy, Outback **Transmissions:** All	**Cooling Fan Relay 1 Circuit High Input** Key on or engine running; and the PCM detected an unexpected "high" voltage condition on the Radiator Fan Relay 1 circuit during the CCM test. **Possible Causes:** • Check output signal of ECM • Check for short circuit in radiator fan relay control circuit • Check main fan relay • Check sub fan relay • Check poor contact
DTC: P1540 **1T CCM, MIL: No** 1997-02 **Engines:** 2.0L, 2.5L SOHC **Models:** Forester, Impreza, Legacy, Outback **Transmissions:** A/T	**Vehicle Speed Sensor Malfunction 2** Engine started, vehicle driven at cruise speed at light to medium engine load for 2 minutes, and the PCM detected it received erratic VSS signals or it detected incorrect Combination Meter signals. **Possible Causes:** • Check transaxle type (automatic) • Check any other DTC on display • Check speedometer operation in combination meter • Check harness between ECM and combination meter connector
DTC: P1544 **1T CCM, MIL: Yes** 2002-06 **Engines:** 2.0L, 2.5L DOHC **Models:** Baja, Forester, Impreza, WRX, Legacy, Outback **Transmissions:** All	**Exhaust Gas Temperature Too High** Erroneous idling. Poor driving performance. **Possible Causes:** • Check any other DTC on display • Check exhaust system
DTC: P1560 **1T CCM, MIL: Yes** 1997-06 **Engines:** 2.0L, 2.0L, 2.5L SOHC & DOHC, 3.0L **Models:** Baja, Forester, Impreza, WRX, Legacy, Outback, B9 Tribeca **Transmissions:** All	**Back-Up Voltage Circuit Malfunction** **Key on, and the PCM detected an unexpected "low" voltage condition on the Battery Backup circuit during the CCM test. Note: The engine will not start with this condition present.** **Possible Causes:** • Check input signal of ECM • Check harness between ECM and main fuse box connector • Check fuse number 13
DTC: P1590 **2T CCM, MIL: No** 2001-02 **Engines:** 2.5L SOHC, 3.0L **Models:** Impreza, Legacy, Outback **Transmissions:** All	**Neutral Position Switch Circuit High Output** **Key on, and the PCM detected an unexpected "high" voltage condition on the A/T Neutral Position Switch circuit during the test.** **Possible Causes:** • Check any other DTC on display • Check input signal of ECM • Check poor contact • Check harness between ECM and inhibitor switch connector • Check switch ground • Check switch • Check selector cable connection
DTC: P1591 **2T CCM, MIL: No** 2001-02 **Engines:** 2.5L SOHC, 3.0L **Models:** Impreza, Legacy, Outback **Transmissions:** All	**Neutral Position Switch Circuit Low Output** **Key on, and the PCM detected an unexpected "low" voltage condition on the A/T Neutral Position Switch circuit during the test. Note: The engine will not start with this condition present.** **Possible Causes:** • Check any other DTC on display • Check input signal of ECM • Check harness between ECM and transmission harness connector • Check switch • Check selector cable connection

DTC	Trouble Code Title, Conditions & Possible Causes
DTC: P1594 2T CCM, MIL: No 2001-02 **Engines:** 2.5L SOHC, 3.0L **Models:** Impreza, Legacy, Outback **Transmissions:** A/T	**Automatic Transaxle Diagnosis Input Signal Circuit Malfunction** **Key on, and the PCM detected an unexpected voltage condition on the A/T Diagnosis Signal circuit (5-volt) during the CCM test.** **Possible Causes:** • Check driving condition (shift control function) • Check vehicle accessories for grounding (car phone, CB etc)
DTC: P1595 2T CCM, MIL: No 2001-02 **Engines:** 2.5L SOHC, 3.0L **Models:** Impreza, Legacy, Outback **Transmissions:** A/T	**Automatic Transaxle Diagnosis Input Signal Circuit Low Input** **Key on, and the PCM detected an unexpected "low" voltage condition on the A/T Diagnosis Signal circuit (5-volt) in the CCM test.** **Possible Causes:** • Check harness between ECM and TCM connector • Check output signal for ECM • Check any other DTC on display
DTC: P1596 2T CCM, MIL: No 2001-02 **Engines:** 2.5L SOHC, 3.0L **Models:** Impreza, Legacy, Outback **Transmissions:** A/T	**Automatic Transaxle Diagnosis Input Signal Circuit High Input** Key on or engine running; and the PCM detected an unexpected "high" voltage condition on the A/T Diagnosis Signal circuit (5-volt) during the CCM test. **Possible Causes:** • Check harness between ECM and TCM connector • Check output signal for ECM • Check poor contact • Check harness between ECM and TCM connector
DTC: P1698 2T CCM, MIL: Yes 2001-06 **Engines:** 2.0L, 2.5L SOHC, 3.0L **Models:** Baja, Forester, Impreza, Legacy, Outback **Transmissions:** All	**Engine Torque Control Cut Signal Circuit Malfunction (low input)** Engine started, and the PCM detected an unexpected "low" voltage condition on the Engine Torque Control Cut Signal circuit (5-volt) during the CCM test. **Possible Causes:** • Check output signal for ECM • Check harness between ECM and TCM connector
DTC: P1699 2T CCM, MIL: Yes 2001-06 **Engines:** 2.0L, 2.5L SOHC, 3.0L **Models:** Baja, Forester, Impreza, Legacy, Outback **Transmissions:** A/T	**Engine Torque Control Cut Signal Circuit Malfunction (high input)** Engine started, and the PCM detected an unexpected "high" voltage condition on the Engine Torque Control Cut Signal circuit (5-volt) during the CCM test. **Possible Causes:** • Check output signal for ECM • Check harness between ECM and TCM connector
DTC: P1700 2T CCM, MIL: No 1996-05 **Engines:** 1.8L, 2.0L, 2.2L, 2.5L SOHC & DOHC, 3.0L, 3.3L **Models:** Baja, Forester, Impreza, WRX, Legacy, Outback, SVX **Transmissions:** A/T	**Throttle Position Sensor Circuit Malfunction For Automatic transmission** Key on or engine running; and the TCM detected an unexpected "low" or "high" voltage condition on the TP Sensor circuit. Note: The TP sensor signal is shared with the PCM on this circuit. **Possible Causes:** • Check any other DTC on display • Check throttle position sensor circuit
DTC: P1701 2T CCM, MIL: No 1996-02 **Engines:** 1.8L, 2.0L, 2.2L, 2.5L SOHC, 3.0L, 3.3L **Models:** Baja, Forester, Impreza, WRX, Legacy, Outback, SVX **Transmissions:** A/T	**Cruise Control Set Signal Circuit Malfunction For Automatic Transmission** Engine started, vehicle driven to a speed over 30 mph, and the PCM detected an unexpected voltage condition on the data line between the Cruise Control Module (CCM) and the PCM. **Possible Causes:** • Check harness between TCM and CCM • Check input signal for TCM
DTC: P1702 2T CCM, MIL: No 1996-01 **Engines:** 1.8L, 2.0L, 2.2L, 2.5L SOHC, 3.3L **Models:** Baja, Forester, Impreza, WRX, Legacy, Outback, SVX	**Automatic Transmission diagnosis Input Signal Circuit Malfunction** Key on or engine running; the PCM detected an unexpected "low" voltage condition on the data line (5-volt) between the Transmission Control Module and the PCM during the CCM test. **Possible Causes:** • Check harness between TCM and CCM

DTC	Trouble Code Title, Conditions & Possible Causes
DTC: P1703 **2T CCM, MIL:** No **1997-02** **Engines:** 2.2L, 2.5L SOHC, 3.3L **Models:** Forester, Impreza, Legacy, Outback **Transmissions:** A/T	**Low Clutch Timing Control Solenoid Valve Circuit Malfunction** Engine started, vehicle driven to a speed of over 3 mph, and the PCM detected an unexpected voltage condition on the A/T Low Clutch Timing Control Solenoid circuit during the CCM test. **Possible Causes:** • Check low clutch timing control solenoid valve circuit
DTC: P1704 **2T CCM, MIL:** No **1997-01** **Engines:** 2.2L, 2.5L SOHC **Models:** Forester, Impreza, Legacy, Outback **Transmissions:** All	**2-4 Brake Timing Control Solenoid Valve Circuit Malfunction** Engine started, vehicle driven to a speed of over 3 mph, and the PCM detected an unexpected voltage condition on the A/T 2-4 Brake Timing Control Solenoid duty cycle circuit during the CCM test. **Possible Causes:** • Check 2-4 brake timing control solenoid valve circuit
DTC: P1705 **2T CCM, MIL:** No **1997-01** **Engines:** 2.2L, 2.5L SOHC **Models:** Forester, Impreza, Legacy, Outback **Transmissions:** All	**2-4 Brake Pressure Control Solenoid Valve (duty solenoid "D") Circuit Malfunction** Engine started, vehicle driven to a speed of over 3 mph, and the PCM detected an unexpected voltage condition on the A/T 2-4 Brake Timing Control Solenoid 'D' duty cycle circuit during the CCM test. **Possible Causes:** • Check 2-4 brake pressure control solenoid valve (duty solenoid "D") circuit
DTC: P1711 **2T CCM, MIL:** No **2001-06** **Engines:** 2.0L, 2.5L SOHC & DOHC, 3.0L **Models:** Baja, Forester, Impreza, WRX, Legacy, Outback **Transmissions:** All	**Engine Torque Control Signal #1 Circuit Malfunction** Engine started, vehicle driven to a speed of over 20 mph, and the PCM detected an unexpected voltage condition on the Engine Torque Control Signal 1 circuit during the CCM continuous test. **Possible Causes:** • Check input signal for ECM • Check poor contact • Check harness between ECM and TCM connector
DTC: P1712 **2T CCM, MIL:** No **2001-06** **Engines:** 2.0L, 2.5L SOHC & DOHC, 3.0L **Models:** Baja, Forester, Impreza, WRX, Legacy, Outback	**Engine Torque Control Signal #2 Circuit Malfunction** Engine started, vehicle driven to a speed of over 20 mph, and the PCM detected an unexpected voltage condition on the Engine Torque Control Signal 2 circuit during the CCM continuous test. **Possible Causes:** • Check input signal for ECM • Check poor contact • Check harness between ECM and TCM connector
DTC: P1722 **2T CCM, MIL:** No **1997-01** **Engines:** 1.8L, 2.2L, 2.5L SOHC **Models:** Forester, Impreza, Legacy, Outback **Transmissions:** A/T	**Automatic Transmission Diagnosis Input Signal Circuit High Input** Key on or engine running; the PCM detected an unexpected high voltage condition on the data line (5-volt) between the Transmission Control Module (TCM) and the PCM. **Possible Causes:** • Check for identification circuit malfunction • Check harness between ECM and TCM connector • Check output signal for ECM • Check poor contact
DTC: P1742 **2T CCM, MIL:** No **1997-01** **Engines:** 1.8L, 2.2L, 2.5L SOHC **Models:** Forester, Impreza, Legacy, Outback **Transmissions:** A/T	**Automatic Transmission Diagnosis Input Signal Circuit Malfunction** Vehicle performance issue. **Possible Causes:** • Check for identification circuit malfunction • Check driving conditions (shift control pattern) • Check output signal for ECM • Check accessories (car phone, CB instAllation, etc)

OBD II Trouble Code List (P2xxx Codes)

DTC	Trouble Code Title, Conditions & Possible Causes:
DTC: P2004 **1T CCM, MIL: Yes** **2005-06** **Engines:** 2.5L DOHC **Models:** Baja, Forester, Impreza, WRX, Legacy, Outback **Transmissions:** All	**Tumble Generated Valve System 1 (valve open)** Vehicle performance issue. **Possible Causes:** • Check any other DTC on display • Check tumble generator valve (RH)
DTC: P2005 **1T CCM, MIL: Yes** **2005-06** **Engines:** 2.5L DOHC **Models:** Baja, Forester, Impreza, WRX, Legacy, Outback **Transmissions:** All	**Tumble Generated Valve System 2 (valve open)** Vehicle performance issue. **Possible Causes:** • Check any other DTC on display • Check tumble generator valve (RH)
DTC: P2006 **1T CCM, MIL: Yes** **2005-06** **Engines:** 2.5L DOHC **Models:** Baja, Forester, Impreza, WRX, Legacy, Outback **Transmissions:** All	**Tumble Generated Valve System 1 (valve closed)** Vehicle performance issue. **Possible Causes:** • Check any other DTC on display • Check tumble generator valve (RH)
DTC: P2007 **1T CCM, MIL: Yes** **2005-06** **Engines:** 2.5L DOHC **Models:** Baja, Forester, Impreza, WRX, Legacy, Outback **Transmissions:** All	**Tumble Generated Valve System 2 (valve closed)** Vehicle performance issue. **Possible Causes:** • Check any other DTC on display • Check tumble generator valve (RH)
DTC: P2008 **1T CCM, MIL: Yes** **2005-06** **Engines:** 2.5L DOHC **Models:** Baja, Forester, Impreza, WRX, Legacy, Outback **Transmissions:** All	**Tumble Generated Valve System 1 Circuit Malfunction (open)** Vehicle performance issue. **Possible Causes:** • Check harness between ECM and tumble generator valve actuator connector • Check poor contact
DTC: P2009 **1T CCM, MIL: Yes** **2005-06** **Engines:** 2.5L DOHC **Models:** Baja, Forester, Impreza, WRX, Legacy, Outback **Transmissions:** All	**Tumble Generated Valve System 1 Circuit Malfunction (closed)** Vehicle performance issue. **Possible Causes:** • Check harness between ECM and tumble generator valve actuator connector
DTC: P2011 **1T CCM, MIL: Yes** **2005-06** **Engines:** 2.5L DOHC **Models:** Baja, Forester, Impreza, WRX, Legacy, Outback **Transmissions:** All	**Tumble Generated Valve System 2 Circuit Malfunction (short)** Vehicle performance issue. **Possible Causes:** • Check harness between ECM and tumble generator valve actuator connector
DTC: P2016 **1T CCM, MIL: Yes** **2005-06** **Engines:** 2.5L DOHC **Models:** Baja, Forester, Impreza, WRX, Legacy, Outback **Transmissions:** All	**Tumble Generated Valve Position Sensor 1 Circuit Low** Engine stalls. Erroneous idling. Poor driving performance. **Possible Causes:** • Using the a scan tool, check current data • Check input signal for ECM • Check harness between ECM and tumble generator valve position sensor connector • Check harness between ECM and throttle position sensor connector • Check poor contact

DTC	Trouble Code Title, Conditions & Possible Causes
DTC: P2017 **1T CCM, MIL: Yes** **2005-06** **Engines:** 2.5L DOHC **Models:** Baja, Forester, Impreza, WRX, Legacy, Outback **Transmissions:** All	**Tumble Generated Valve Position Sensor 1 Circuit High** Engine stalls. Erroneous idling. Poor driving performance. **Possible Causes:** • Using a scan tool, check current data • Check harness between tumble generator valve position sensor and ECM connector
DTC: P2021 **1T CCM, MIL: Yes** **2005-06** **Engines:** 2.5L DOHC **Models:** Baja, Forester, Impreza, WRX, Legacy, Outback **Transmissions:** All	**Tumble Generated Valve Position Sensor 2 Circuit Low** Engine stalls. Erroneous idling. Poor driving performance. **Possible Causes:** • Using a scan tool, check current data • Check input signal for ECM • Check harness between ECM and tumble generator valve position sensor connector • Check poor contact
DTC: P2022 **1T CCM, MIL: Yes** **2005-06** **Engines:** 2.5L DOHC **Models:** Baja, Forester, Impreza, WRX, Legacy, Outback **Transmissions:** All	**Tumble Generated Valve Position Sensor 2 Circuit High** Engine stalls. Erroneous idling. Poor driving performance. **Possible Causes:** • Using a scan tool, check current data • Check harness between tumble generator valve position sensor and ECM connector • Check poor contact
DTC: P2088 **2T CCM, MIL: No** **2004-06** **Engines:** 2.5L DOHC, 3.0L **Models:** Baja, Forester, Impreza, WRX, Legacy, Outback, B9 Tribeca **Transmissions:** All	**OCV Solenoid Valve Signal "A" Circuit Open (bank 1)** Erroneous idling. **Possible Causes:** • Check harness between ECM and oil flow control solenoid valve • Check oil flow control solenoid valve
DTC: P2089 **2T CCM, MIL: No** **2004-06** **Engines:** 2.5L DOHC, 3.0L **Models:** Baja, Forester, Impreza, WRX, Legacy, Outback, B9 Tribeca **Transmissions:** All	**OCV Solenoid Valve Signal "A" Circuit Short (bank 1)** Erroneous idling. **Possible Causes:** • Check harness between ECM and oil flow control solenoid valve • Check oil flow control solenoid valve
DTC: P2092 **2T CCM, MIL: No** **2004-06** **Engines:** 2.5L DOHC, 3.0L **Models:** Baja, Forester, Impreza, WRX, Legacy, Outback, B9 Tribeca **Transmissions:** All	**OCV Solenoid Valve Signal "A" Circuit Open (bank 2)** Erroneous idling. **Possible Causes:** • Check harness between ECM and oil flow control solenoid valve • Check oil flow control solenoid valve
DTC: P2093 **2T CCM, MIL: No** **2004-06** **Engines:** 2.5L DOHC, 3.0L **Models:** Baja, Forester, Impreza, WRX, Legacy, Outback, B9 Tribeca **Transmissions:** All	**OCV Solenoid Valve Signal "A" Circuit Short (bank 2)** Erroneous idling. **Possible Causes:** • Check harness between ECM and oil flow control solenoid valve • Check oil flow control solenoid valve
DTC: P2096 **2T CCM, MIL: Yes** **2004-06** **Engines:** 2.5L SOHC & DOHC, 3.0L **Models:** Baja, Forester, Impreza, WRX, Legacy, Outback, B9 Tribeca **Transmissions:** All	**Post Catalyst Fuel Trim System Too Lean Bank 1** Poor driveability. **Possible Causes:** • Check any other DTC on display • Using the a scan tool, check front oxygen sensor data • Using the Subaru scan tool, check rear oxygen sensor data • Check exhaust system • Check air intake system • Check fuel pressure • Check engine coolant temperature sensor • Check Mass air flow and intake air temperature • Check harness between ECM and front oxygen sensor connector • Using harness between ECM and rear oxygen sensor connector

DTC	Trouble Code Title, Conditions & Possible Causes
DTC: P2097 **2T CCM, MIL: Yes** **2004-06** **Engines:** 2.5L SOHC & DOHC, 3.0L **Models:** Baja, Forester, Impreza, WRX, Legacy, Outback, B9 Tribeca **Transmissions:** All	**Post Catalyst Fuel Trim System Too Rich Bank 1** Poor driveability. **Possible Causes:** • Check any other DTC on display • Using a scan tool, check front oxygen sensor data • Using a scan tool, check rear oxygen sensor data • Check exhaust system • Check air intake system • Check fuel pressure • Check engine coolant temperature sensor • Check Mass air flow and intake air temperature • Check harness between ECM and front oxygen sensor connector • Using harness between ECM and rear oxygen sensor connector
DTC: P2098 **2T CCM, MIL: Yes** **2005-06** **Engines:** 3.0L **Models:** Outback, B9 Tribeca **Transmissions:** All	**Post Catalyst Fuel Trim System Too Lean Bank 2** Poor driveability. **Possible Causes:** • Check any other DTC on display • Using a scan tool, check front oxygen sensor data • Using a scan tool, check rear oxygen sensor data • Check exhaust system • Check air intake system • Check fuel pressure • Check engine coolant temperature sensor • Check Mass air flow and intake air temperature • Check harness between ECM and front oxygen sensor connector • Using harness between ECM and rear oxygen sensor connector
DTC: P2099 **2T CCM, MIL: Yes** **2005-06** **Engines:** 3.0L **Models:** Outback, B9 Tribeca **Transmissions:** All	**Post Catalyst Fuel Trim System Too Rich Bank 2** Poor driveability. **Possible Causes:** • Check any other DTC on display • Using a scan tool, check front oxygen sensor data • Using a scan tool, check rear oxygen sensor data • Check exhaust system • Check air intake system • Check fuel pressure • Check engine coolant temperature sensor • Check Mass air flow and intake air temperature • Check harness between ECM and front oxygen sensor connector • Using harness between ECM and rear oxygen sensor connector
DTC: P2101 **1T CCM, MIL: Yes** **2004-06** **Engines:** 2.5L SOHC & DOHC 3.0L **Models:** Baja, Forester, Impreza, WRX, Legacy, Outback, B9 Tribeca **Transmissions:** All	**Throttle Actuator Control Range/Performance Bank 1** Erroneous idling. Engine stalls. Poor driving performance. **Possible Causes:** • Check electronic throttle control relay • Check power supply of electronic control relay • Check harness between ECM and electronic throttle control relay • Check sensor output • Check poor contact • Check harness between ECM and electronic throttle control • Check sensor power supply • Check for short in ECM • Check sensor output • Check harness between ECM and electronic throttle control motor • Check electronic throttle control motor harness • Check ground circuit • Check electronic throttle control
DTC: P2102 **1T CCM, MIL: Yes** **2004-06** **Engines:** 2.5L SOHC & DOHC, 3.0L **Models:** Baja, Forester, Impreza, WRX, Legacy, Outback, B9 Tribeca **Transmissions:** All	**Throttle Actuator Control Motor Circuit Low** Erroneous idling. Engine stalls. Poor driving performance. **Possible Causes:** • Check electronic throttle control relay • Check power supply of electronic control relay • Check harness between ECM and electronic throttle control relay

DTC	Trouble Code Title, Conditions & Possible Causes
DTC: P2103 **1T CCM, MIL: Yes** **2004-06** **Engines:** 2.5L SOHC & DOHC, 3.0L **Models:** Baja, Forester, Impreza, WRX, Legacy, Outback, B9 Tribeca **Transmissions:** All	**Throttle Actuator Control Motor Circuit High** Poor driveability. **Possible Causes:** • Check electronic throttle control relay • Check for short circuit of the relay and power supply • Check harness between ECM and electronic throttle control relay
DTC: P2109 **1T CCM, MIL: Yes** **2004-06** **Engines:** 2.5L SOHC & DOHC 3.0L **Models:** Baja, Forester, Impreza, WRX, Legacy, Outback, B9 Tribeca **Transmissions:** All	**Throttle/Pedal Position Sensor "A" Minimum Stop Performance** Erroneous idling. Engine stalls. Poor driving performance. **Possible Causes:** • Check electronic throttle control relay • Check power supply of electronic control relay • Check harness between ECM and electronic throttle control relay • Check sensor output • Check poor contact • Check harness between ECM and electronic throttle control • Check sensor power supply • Check for short in ECM • Check sensor output • Check harness between ECM and electronic throttle control motor • Check electronic throttle control motor harness • Check ground circuit • Check electronic throttle control
DTC: P2122 **1T CCM, MIL: Yes** **2004-06** **Engines:** 2.5L SOHC & DOHC 3.0L **Models:** Baja, Forester, Impreza, WRX, Legacy, Outback, B9 Tribeca **Transmissions:** All	**Throttle/Pedal Position Sensor "D" Circuit Low Input** Erroneous idling. Poor driving performance. **Possible Causes:** • Check accelerator pedal position sensor output • Check poor contact • Check harness between ECM and accelerator pedal position sensor • Check power supply of accelerator pedal position sensor • Check sensor
DTC: P2123 **1T CCM, MIL: Yes** **2004-06** **Engines:** 2.5L SOHC & DOHC 3.0L **Models:** Baja, Forester, Impreza, WRX, Legacy, Outback, B9 Tribeca **Transmissions:** All	**Throttle/Pedal Position Sensor/Switch "D" Circuit High Input** Erroneous idling. Poor driving performance. **Possible Causes:** • Check accelerator pedal position sensor output • Check poor contact • Check harness between ECM and accelerator pedal position sensor
DTC: P2127 **1T CCM, MIL: Yes** **2004-06** **Engines:** 2.5L SOHC & DOHC 3.0L **Models:** Baja, Forester, Impreza, WRX, Legacy, Outback, B9 Tribeca **Transmissions:** All	**Throttle/Pedal Position Sensor/Switch "E" Circuit Low Input** Erroneous idling. Poor driving performance. **Possible Causes:** • Check accelerator pedal position sensor output • Check poor contact • Check harness between ECM and accelerator pedal position sensor • Check sensor
DTC: P2128 **1T CCM, MIL: Yes** **2004-06** **Engines:** 2.5L SOHC & DOHC 3.0L **Models:** Baja, Forester, Impreza, WRX, Legacy, Outback, B9 Tribeca **Transmissions:** All	**Throttle/Pedal Position Sensor/Switch "E" Circuit High Input** Erroneous idling. Poor driving performance. **Possible Causes:** • Check accelerator pedal position sensor output • Check poor contact • Check harness between ECM and accelerator pedal position sensor • Check sensor

DTC	Trouble Code Title, Conditions & Possible Causes
DTC: P2135 **1T CCM, MIL: Yes** **2004-06** **Engines:** 2.5L SOHC & DOHC 3.0L **Models:** Baja, Forester, Impreza, WRX, Legacy, Outback, B9 Tribeca **Transmissions:** All	**Throttle/Pedal Position Sensor/Switch "A"/"B" Voltage Rationality** Erroneous idling. Poor driving performance. **Possible Causes:** • Check sensor output • Check poor contact • Check harness between ECM and electronic throttle control • Check sensor power supply • Check for short circuit in ECM • Check sensor output • Check electronic throttle control harness
DTC: P2138 **1T CCM, MIL: Yes** **2002-06** **Engines:** 2.5L SOHC & DOHC 3.0L **Models:** Baja, Forester, Impreza, WRX, Legacy, Outback, B9 Tribeca **Transmissions:** All	**Throttle/Pedal Position Sensor/Switch "D"/"E" Voltage Rationality** Erroneous idling. Poor driving performance. **Possible Causes:** • Check accelerator pedal position sensor output • Check poor contact • Check harness between ECM and sensor • Check sensor • Check sensor output
DTC: P2227 **2T CCM, MIL: Yes** **2002-06** **Engines:** 2.5L SOHC & DOHC 3.0L **Models:** Baja, Forester, Impreza, WRX, Legacy, Outback, B9 Tribeca **Transmissions:** All	**Barometric Pressure Circuit Range/Performance** poor driveability. **Possible Causes:** • Check any other DTC on display
DTC: P2228 **1T CCM, MIL: Yes** **2002-06** **Engines:** 2.5L SOHC & DOHC 3.0L **Models:** Baja, Forester, Impreza, WRX, Legacy, Outback, B9 Tribeca **Transmissions:** All	**Barometric Pressure Circuit Low Input** poor driveability. **Possible Causes:** • Check any other DTC on display
DTC: P2229 **1T CCM, MIL: Yes** **2002-06** **Engines:** 2.5L SOHC & DOHC 3.0L **Models:** Baja, Forester, Impreza, WRX, Legacy, Outback, B9 Tribeca **Transmissions:** All	**Barometric Pressure Circuit High Input** poor driveability. **Possible Causes:** • Check any other DTC on display
DTC: P2431 **2T CCM, MIL: Yes** **2006** **Engines:** 2.5L DOHC **Models:** Forester, Impreza, WRX **Transmissions:** All	**Secondary Air Injection System Air Flow/Pressure Sensor Circuit Range/Performance** poor driveability. **Possible Causes:** • Check any other DTC on display • Using a scan tool, check front oxygen sensor data
DTC: P2432 **1T CCM, MIL: Yes** **2006** **Engines:** 2.5L DOHC **Models:** Forester, Impreza, WRX **Transmissions:** All	**Secondary Air Injection System Air Flow/Pressure Sensor Circuit Low** poor driveability. **Possible Causes:** • Check harness between ECM and valve LH connector
DTC: P2433 **1T CCM, MIL: Yes** **2006** **Engines:** 2.5L DOHC **Models:** Forester, Impreza, WRX **Transmissions:** All	**Secondary Air Injection System Air Flow/Pressure Sensor Circuit High** poor driveability. **Possible Causes:** • Check harness between ECM and valve LH connector

DTC	Trouble Code Title, Conditions & Possible Causes
DTC: P2440 **2T CCM, MIL: Yes** **2006** **Engines:** 2.5L DOHC **Models:** Forester, Impreza, WRX **Transmissions:** All	**Secondary Air Injection System Switching Valve Stock Open (bank 1)** poor driveability. **Possible Causes:** • Check secondary air combination valve operation • Check duct between secondary air pump and secondary air combination valve • Check pipe between secondary air combination valve and cylinder head • Check power supply to valve • Check harness between secondary air combination valve relay and secondary air combination valve connector terminal • Check valve relay • Check harness between ECM and valve relay connector
DTC: P2441 **2T CCM, MIL: Yes** **2006** **Engines:** 2.5L DOHC **Models:** Forester, Impreza, WRX **Transmissions:** All	**Secondary Air Injection System Switching Valve Stock Closed (bank 1)** poor driveability. **Possible Causes:** • Check secondary air combination valve operation • Check duct between secondary air pump and secondary air combination valve • Check pipe between secondary air combination valve and cylinder head • Check power supply to valve • Check harness between secondary air combination valve relay and secondary air combination valve connector terminal • Check valve relay • Check harness between ECM and valve relay connector
DTC: P2442 **2T CCM, MIL: Yes** **2006** **Engines:** 2.5L DOHC **Models:** Forester, Impreza, WRX **Transmissions:** All	**Secondary Air Injection System Switching Valve Stock Open (bank 2)** poor driveability. **Possible Causes:** • Check secondary air combination valve operation • Check duct between secondary air pump and secondary air combination valve • Check pipe between secondary air combination valve and cylinder head • Check power supply to valve • Check harness between secondary air combination valve relay and secondary air combination valve connector terminal • Check valve relay • Check harness between ECM and valve relay connector
DTC: P2443 **2T CCM, MIL: Yes** **2006** **Engines:** 2.5L DOHC **Models:** Forester, Impreza, WRX **Transmissions:** All	**Secondary Air Injection System Switching Valve Stock Closed (bank 2)** poor driveability. **Possible Causes:** • Check secondary air combination valve operation • Check duct between secondary air pump and secondary air combination valve • Check pipe between secondary air combination valve and cylinder head • Check power supply to valve • Check harness between secondary air combination valve relay and secondary air combination valve connector terminal • Check valve relay • Check harness between ECM and valve relay connector
DTC: P2444 **1T CCM, MIL: Yes** **2006** **Engines:** 2.5L DOHC **Models:** Forester, Impreza, WRX **Transmissions:** All	**Secondary Air Injection System Pump Stuck ON** poor driveability. **Possible Causes:** • Check secondary air piping pressure • Check power supply to valve • Check valve relay
DTC: P2503 **1T CCM, MIL: No** **2005-06** **Engines:** 3.0L **Models:** Outback **Transmissions:** All	**Charging System Voltage Low** Vehicle performance issue. **Possible Causes:** • Check harness between alternator and ECM connector
DTC: P2504 **1T CCM, MIL: No** **2005-06** **Engines:** 3.0L **Models:** Outback **Transmissions:** All	**Charging System Voltage High** Vehicle performance issue. **Possible Causes:** • Check harness between alternator and ECM connector

SUBARU
COMPONENT TESTING

12

COMPONENT TESTING

B9 TRIBECA

See Figure 1.

Camshaft Position (CMP) Sensor

OPERATION

The CMP relays the relative camshaft position to the ECM for determining the position and stroke of the No. 1 cylinder. This information is required for proper fuel injection functioning.

REMOVAL & INSTALLATION

1. Disconnect the negative battery cable.
2. Remove the collector cover, as required.
3. Remove the alternator harness from the fuel pipe. Remove the fuel pipe protector.
4. Disconnect the connector from the sensor.
5. Remove the bolt that retains the sensor.
6. Remove the sensor.

To install:

7. Installation is the reverse of the removal procedure.
8. Tighten the sensor to 4.7 ft. lbs.

TESTING

1. Using the Subaru scan tool, check the sensor output waveform.
2. Measurement is taken using connector B134 terminal 11 for the left side and connector B134 terminal 21 for the right side.
3. If abnormality is found, replace the sensor.

Crankshaft Position (CKP) Sensor

LOCATION

The CKP is located in the front of the engine near the crankshaft pulley.

OPERATION

The CKP is a variable reluctance sensor which is used to inform the ECM when the No.1 piston is at top dead center. This information is used for controlling and adjusting ignition and fuel injector timing.

REMOVAL & INSTALLATION

1. Disconnect the negative battery cable.
2. Remove the collector cover, as required.
3. Remove the air intake chamber, if required.
4. Remove the bolt that retains the sensor in place at the cylinder block.
6. Remove the sensor from its mounting.
7. Disconnect the connector from the sensor.

To install:

8. Installation is the reverse of the removal procedure.
9. Tighten the sensor to 4.7 ft. lbs.

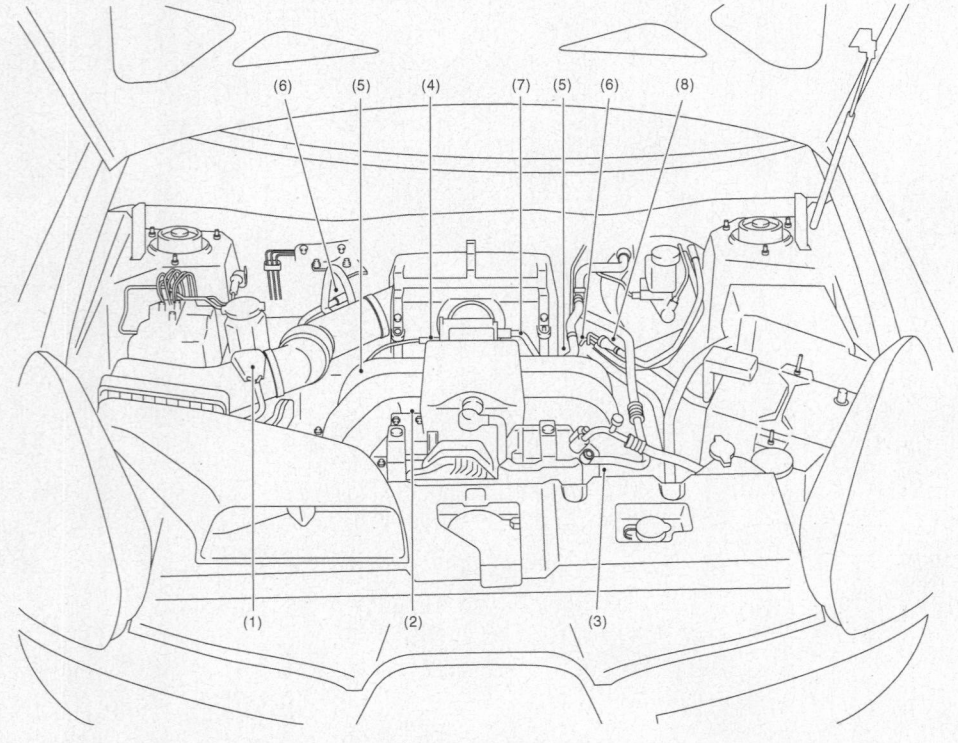

(1)	Mass air flow and intake air temperature sensor	(3)	Engine coolant temperature sensor	(6)	Camshaft position sensor
(2)	Manifold absolute pressure sensor	(4)	Electronic throttle control	(7)	Crankshaft position sensor
		(5)	Knock sensor	(8)	Oil temperature sensor

29157_SUBA_G0001

Fig. 1 Underhood sensor locations—B9 Tribeca

TESTING

1. Remove the sensor from the vehicle.
2. Measure the resistance between the two connector terminals of the sensor, (terminals 1 and 2).
3. If measured value is not within 1–4 kohms, replace the sensor.

Electronic Control Module (ECM)

LOCATION

See Figure 2.

On ECM is located on the passenger's side of the vehicle, underneath the instrument panel glove box area.

OPERATION

The ECM controls the vehicle engine operating system.

REMOVAL & INSTALLATION

1. Disconnect the negative battery cable.
2. Remove the lower inner trim on the passenger's side of the vehicle.
3. Detach the floor mat. Remove the protective cover.
4. Remove the ECM bracket retaining nuts. Remove the clip from the bracket.
5. Disconnect the connectors.

6. Remove the ECM from the vehicle.

To install:

7. Installation is the reverse of the removal procedure.
8. Tighten the retaining screws to 3.7 ft. lbs.

➡ **When replacing the ECM, be careful not to use the wrong part number, as damage to the injection system could occur.**

Engine Coolant Temperature (ECT) Sensor

LOCATION

The ECT is located by the heater outlet fitting or in a cooling passage on the engine, depending upon the particular vehicle.

OPERATION

This component detects the temperature of the engine coolant and relays the information to the electronic control assembly.

REMOVAL & INSTALLATION

1. Disconnect the negative battery cable.
2. Disconnect the connector from the sensor.
3. Remove the sensor from its mounting.

To install:

4. Installation is the reverse of the removal procedure.
5. Tighten the sensor to 13.3 ft. lbs.

TESTING

1. Turn the ignition switch to OFF.
2. Disconnect the connector from the sensor.
3. Turn the ignition switch ON.
4. Using the Subaru scan tool, read the sensor signal data. If the measured value is less than -40 degrees F, replace the sensor.

Heated Oxygen (HO2S) Sensor

LOCATION

See Figure 3.

The front oxygen sensors are located in the front section of the front catalytic converter. The rear oxygen sensors are located in the rear section of the front catalytic converter. There are two front catalytic converters, but only one rear catalytic converter.

OPERATION

The exhaust gas oxygen sensor supplies the electronic control assembly with a signal which indicates either a rich or lean mixture condition, during the engine operation.

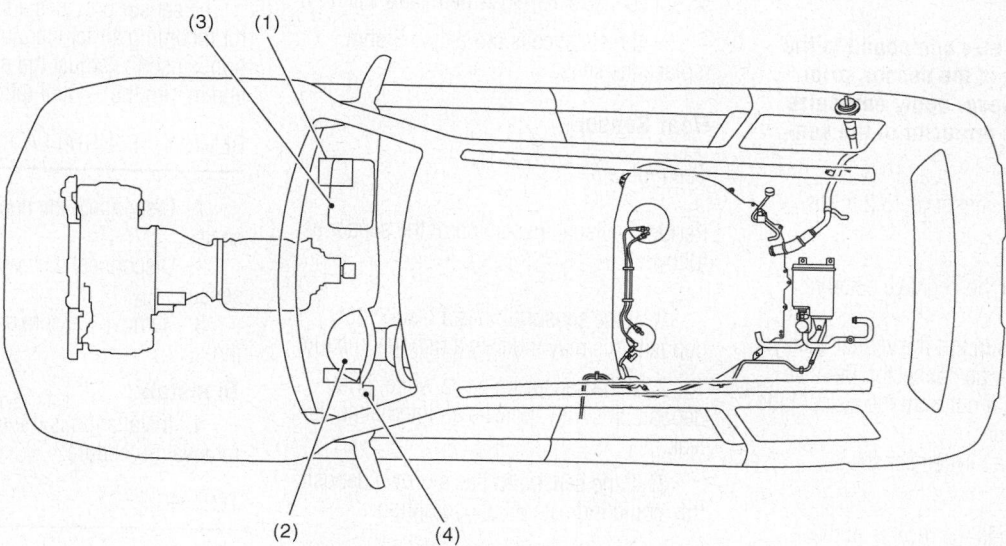

| (1) | Engine control module (ECM) | (3) | Test mode connector |
| (2) | Malfunction indicator light | (4) | Data link connector |

29157_SUBA_G0002

Fig. 2   ECM and related components—B9 Tribeca

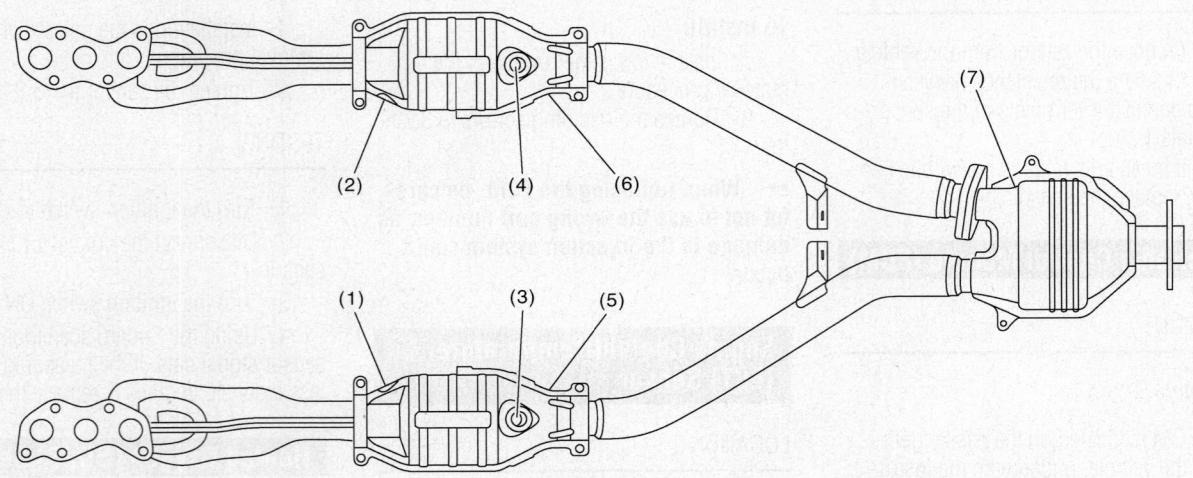

(1) Front oxygen (A/F) sensor LH
(2) Front oxygen (A/F) sensor RH
(3) Rear oxygen sensor LH

(4) Rear oxygen sensor RH
(5) Front catalytic converter LH

(6) Front catalytic converter RH
(7) Rear catalytic converter

29157_SUBA_G0003

Fig. 3 Oxygen sensor location—B9 Tribeca

REMOVAL & INSTALLATION

Front Sensor

1. Disconnect the negative battery cable.
2. Raise and support the vehicle safely.
3. Disconnect the connector.
4. Remove the oxygen sensor.

To install:

5. Installation is the reverse of the removal procedure.

➡ **Apply anti-seize compound to the threaded portion of the sensor, prior to installation. Never apply anti-seize compound to the protector of the sensor.**

6. Tighten the sensor to 15.2 ft. lbs.

Rear Sensor

1. Disconnect the negative battery cable.
2. Raise and support the vehicle safely.
3. Disconnect the connector. Remove the clip by pulling it out from the up per side of the crossmember.
4. Remove the oxygen sensor.

To install:

5. Installation is the reverse of the removal procedure.

➡ **Apply anti-seize compound to the threaded portion of the sensor, prior to installation. Never apply anti-seize compound to the protector of the sensor.**

6. Tighten the sensor to 22.1 ft. lbs. on all engines except the 3.0L engine. On the 3.0L engine, tighten the sensor to 15.2 ft. lbs.

TESTING

Front Sensor

See Figure 4.

1. Measure the resistance between the sensor connector terminals (terminals 1 and 2).

2. If resistance is more than 5 ohm, replace the sensor.

Rear Sensor

See Figure 5.

Perform a visual inspection of the sensor as follows:

1. If the sensor tip has a black/sooty deposit, this may indicate a rich fuel mixture.

2. If the sensor tip has a white, gritty deposit, this may indicate an internal coolant leak.

3. If the sensor tip has a brown deposit, this could indicate oil consumption.

4. Turn the ignition switch OFF.

5. Measure the resistance of the harness between the sensor connector terminals (terminals 1 and 2).

6. If measured value is more than 30ohm, replace the sensor.

Intake Air Temperature (IAT) Sensor

LOCATION

The IAT is mounted in the intake air hose of the air cleaner assembly. On some vehicles this sensor is combined with the mass air flow sensor.

OPERATION

The sensor provides a signal to the ECM for incoming air temperature. This information is used to adjust the injector pulse width and in turn the air/fuel ratio.

REMOVAL & INSTALLATION

1. Disconnect the negative battery cable.
2. Disconnect the connector from the sensor.
3. Remove the sensor from its mounting.

To install:

4. Installation is the reverse of the removal procedure.

TESTING

1. Start the engine and allow it to reach operating temperature.
2. Using the Subaru scan tool, read the sensor signal data.
3. If the measured value is less than -40 degrees F, replace the sensor.

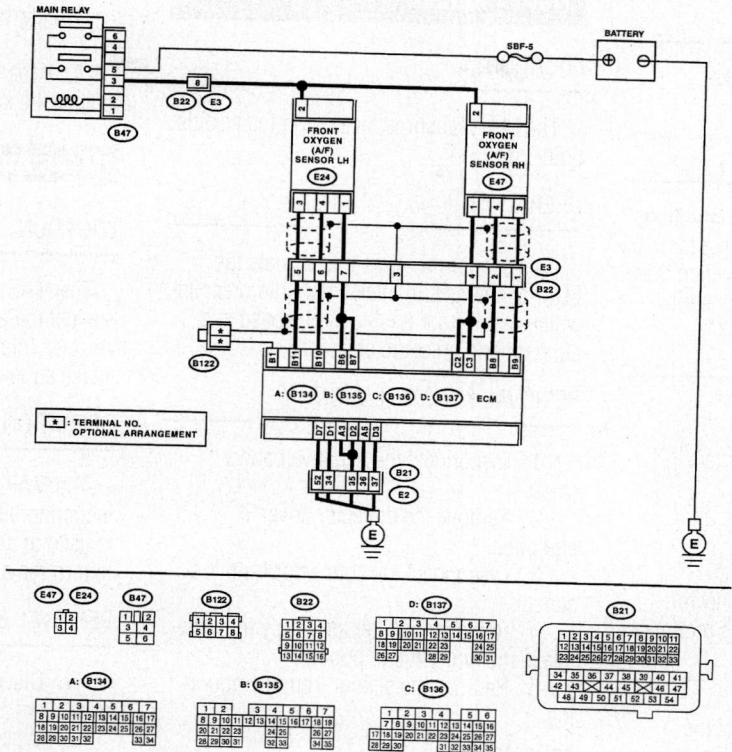

Fig. 4 Front oxygen sensor connector location—B9 Tribeca

29157_SUBA_G0135

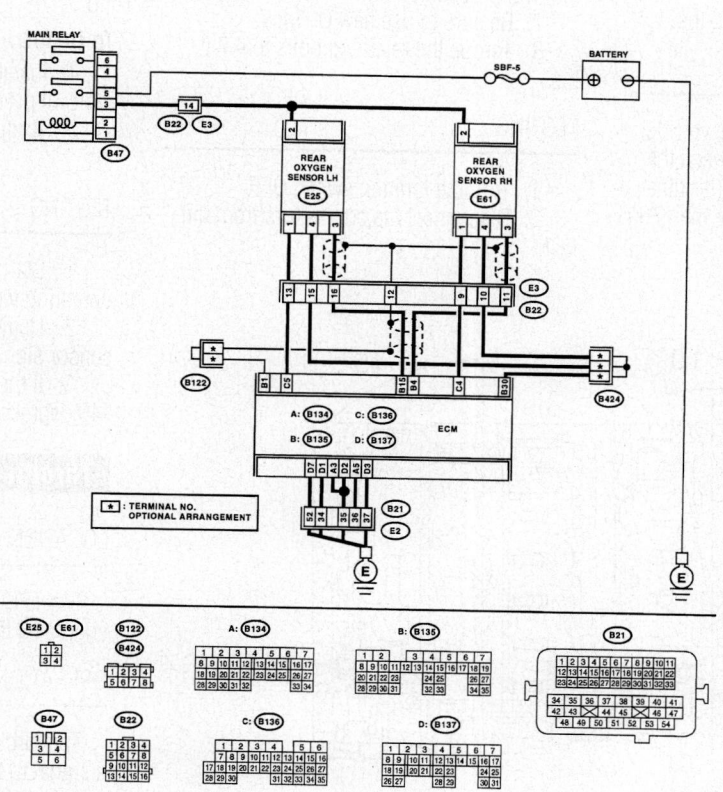

Fig. 5 Rear oxygen sensor connector location—B9 Tribeca

29157_SUBA_G0136

Knock Sensor (KS)

LOCATION

The KS is located under the intake manifold.

OPERATION

The KS is used to detect engine vibrations caused by preignition or detonation and provides information to the ECM, which then retards the timing to eliminate detonation.

REMOVAL & INSTALLATION

See Figure 6.

1. Disconnect the negative battery cable.
2. Remove the collector cover, as required.
3. Remove the intake manifold.
4. Disconnect the sensor connector.
5. Remove the sensor from its mounting.

To install:

6. Installation is the reverse of the removal procedure.
7. Refer to the illustration for proper sensor installation angle.
8. Tighten the sensor to 18.0 ft. lbs.

TESTING

1. Remove the sensor from the vehicle.
2. Measure the resistance between the connector terminals of the sensor, (terminal 1 and 2). If measured value is more than 700 kohms, replace the sensor.

Manifold Absolute Pressure (MAP) Sensor

LOCATION

The MAP sensor is located on the throttle body unit.

OPERATION

This sensor monitors and signals the ECM of changes in intake manifold pressure which result from engine load, speed and atmospheric pressure changes.

REMOVAL & INSTALLATION

1. Disconnect the negative battery cable.
2. Remove the collector cover, if equipped.
3. Disconnect the connector from the sensor.
4. Remove the filter assembly from the intake manifold, if equipped.
5. Remove the sensor from its mounting.

To install:

6. Installation is the reverse of the removal procedure.
7. Be sure to use new O-rings.
8. Torque the retaining bolts to 4.7 ft. lbs.

TESTING

1. Turn the ignition switch OFF.
2. Disconnect the connectors from the ECM.

3. Measure the resistance of the harness between the ECM and the sensor (connector B134 terminal 29 and connector E21 terminal1)
4. If the resistance is less than 1 ohm, replace the sensor.

Mass Air Flow (MAF) Sensor

LOCATION

The MAF is mounted in the intake air hose of the air cleaner assembly. On some vehicles this sensor is combined with the intake air temperature sensor.

OPERATION

The MAF provides a signal to the ECM for incoming air temperature. This information is used to adjust the injector pulse width and in turn the air/fuel ratio.

REMOVAL & INSTALLATION

1. Disconnect the negative battery cable.
2. Disconnect the connector from the sensor.
3. Remove the sensor from its mounting.

To install:

4. Installation is the reverse of the removal procedure.
5. Torque the retaining screw to 0.8 ft. lbs.

TESTING

1. Start the engine and allow it to reach operating temperature.
2. Using the Subaru scan tool, read the sensor signal data.
3. If the measured value is less than -40 degrees F, replace the sensor.

Throttle Position Sensor (TPS)

LOCATION

The TPS is located near the center of the engine by the air cleaner assembly.

OPERATION

The throttle position sensor provides a signal to the ECM that is related to the relative throttle plate position. As the throttle plate moves in relation to driving conditions, a signal is sent to the control unit

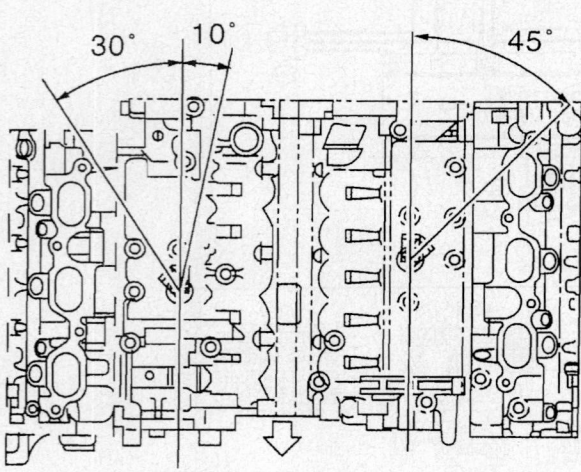

29157_SUBA_G0134

Fig. 6   Knock sensor installation angle—B9 Tribeca

which adjusts the injector pulse width and air/fuel ratio. As the throttle plate is opened further, more air is taken into the combustion chambers, and as a result the relative fuel demand of the engine changes. The throttle position sensor relays this information to the ECM which in alters the fuel amount.

REMOVAL & INSTALLATION

1. Disconnect the negative battery cable.
2. Remove the air intake chamber.
3. Disconnect the sensor connector.
4. Remove the sensor retaining screws. Remove the sensor from its mounting.

To install:

5. Installation is the reverse of the removal procedure.

BAJA

See Figures 7 and 8.

Air Injector Solenoid (AIS) Valve

LOCATION

The sensor is mounted in the intake manifold on the left (passenger's) side of the engine at the top near the ignition coil.

TESTING

1. Turn the ignition switch OFF.
2. Disconnect the connector from the ECM.
3. Measure the resistance between the ECM connectors (connector B134 terminal 18 and connector B134 terminal 19).
4. If the measured value is more than 1 Mohm, repair the electronic throttle control.

Vehicle Speed Sensor (VSS)

LOCATION

Vehicles equipped with automatic transaxle use a front and rear sensors. Both sensors are mounted on the transaxle. Vehicles equipped with manual transaxle use a front sensor. This sensor is mounted on the transaxle.

OPERATION

The sensor provides a signal to the ECM. This signal adjusts engine management parameters that allow the engine to perform at optimal performance.

OPERATION

The vehicle speed sensor sends a signal to the ECM to control. The ECM uses this information to control transmission shift patterns. The vehicle speed sensor is also used to send a speed signal to the speed control servo of the cruise control system.

REMOVAL & INSTALLATION

1. Raise and support the vehicle safely.
2. Place a drip pan below the speed sensor to catch any spilled fluid.
3. Disconnect the connector.
4. Remove the sensor from its mounting.

To install:

5. Installation is the reverse of the removal procedure.
6. Replace any lost fluid.

REMOVAL & INSTALLATION

1. Disconnect the negative battery cable.
2. Disconnect the connector from the air assist injector solenoid valve.
3. Disconnect the air by-pass hoses.
4. Remove the valve from the intake manifold.

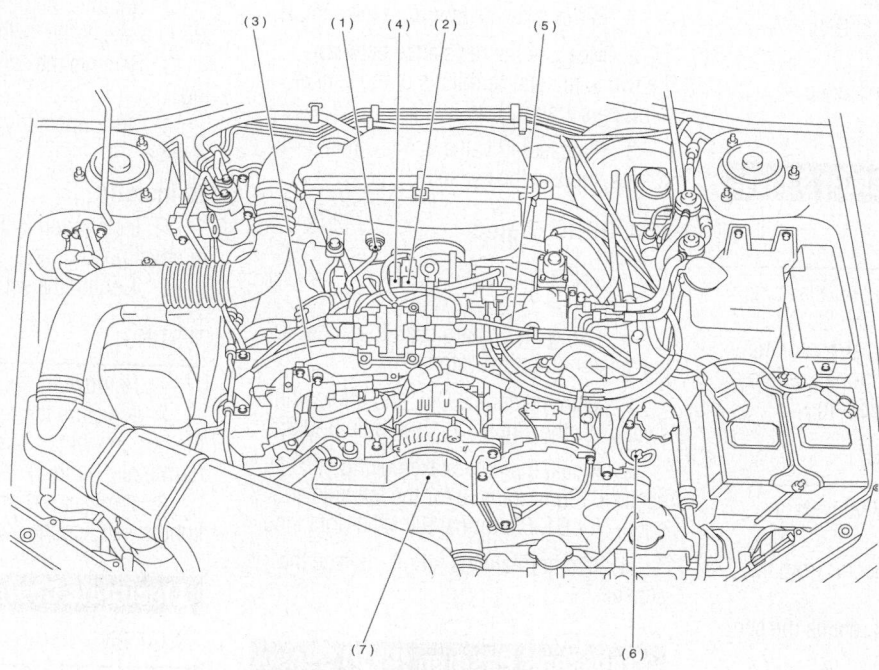

(1) Intake air temperature sensor	(5) Knock sensor
(2) Pressure sensor	(6) Camshaft position sensor
(3) Engine coolant temperature sensor	(7) Crankshaft position sensor
(4) Throttle position sensor	

Fig. 7 Underhood sensor locations—Baja 2.5L SOHC engine

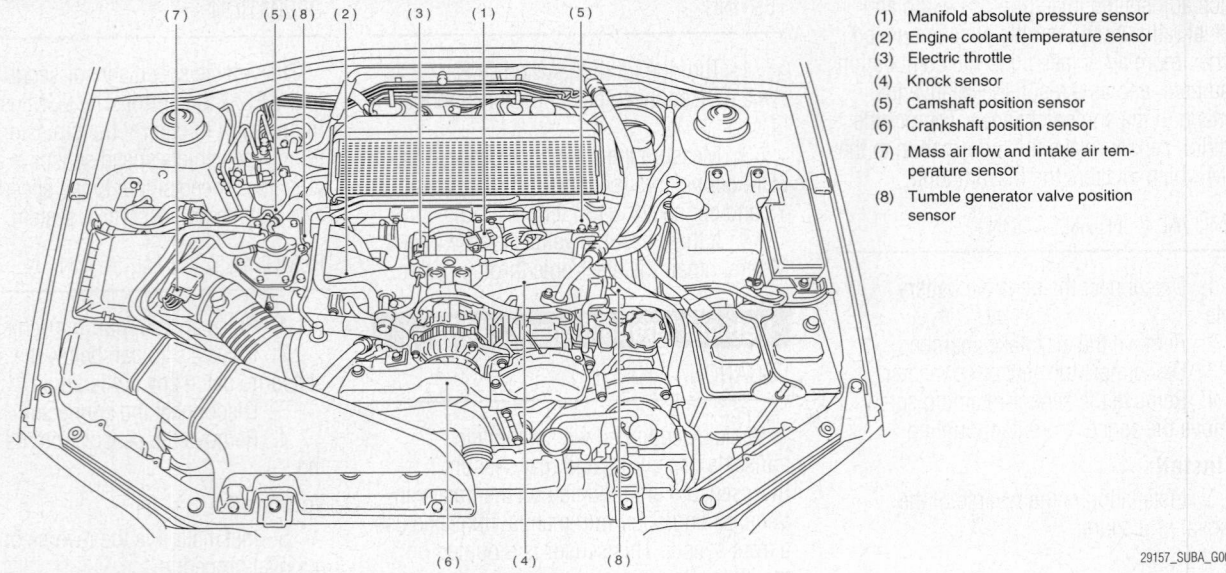

(1) Manifold absolute pressure sensor
(2) Engine coolant temperature sensor
(3) Electric throttle
(4) Knock sensor
(5) Camshaft position sensor
(6) Crankshaft position sensor
(7) Mass air flow and intake air temperature sensor
(8) Tumble generator valve position sensor

29157_SUBA_G0008

Fig. 8 Underhood sensor locations—Baja 2.5L DOHC engine

To install:

5. Installation is the reverse of the removal procedure.

6. Tighten the sensor to 3.7 ft. lbs.

TESTING

1. Turn the ignition switch OFF.

2. Using the Subaru scan tool, connect the test mode connector at the lower portion of the instrument panel (driver's side) to the side of the center console box.

3. Turn the ignition switch ON.

4. Operate the valve.

5. If the valve does not produce an operating sound, replace it.

Camshaft Position (CMP) Sensor

OPERATION

The camshaft position sensor relays the relative camshaft position to the ECM for determining the position and stroke of the No. 1 cylinder. This information is required for proper fuel injection functioning.

REMOVAL & INSTALLATION

1. Disconnect the negative battery cable.

2. Disconnect the connector from the sensor.

3. Remove the bolt that retains the sensor to the sensor support.

4. Remove the bolt that retains the sensor support to the camshaft cap.

5. Remove the sensor and the sensor support as a unit.

6. Separate the sensor from the support.

To install:

7. Installation is the reverse of the removal procedure.

8. Tighten the sensor support to 4.7 ft. lbs.

9. Tighten the sensor to 4.7 ft. lbs.

TESTING

2.5L SOHC Engine

1. Remove the sensor from the vehicle.

2. Measure the resistance between the two connector terminals of the sensor, (terminals 1 and 2).

3. If measured value is not within 1–4 kohms, replace the sensor.

2.5L DOHC Engine

See Figure 9.

1. Using the Subaru scan tool, check the sensor output waveform.

2. With the engine OFF and the ignition switch ON the signal (V) should be 0–0.9.

3. Measurement is taken using connector B135 terminal 8 for the left side and connector B135 terminal 9 for the right side.

4. If abnormality is found, replace the sensor.

Crankshaft Position (CKP) Sensor

LOCATION

The sensor is located in the front of the engine near the crankshaft pulley.

OPERATION

The crankshaft position sensor is a variable reluctance sensor which is used to inform the ECM when the No.1 piston is at top dead center. This information is used for controlling and adjusting ignition and fuel injector timing.

REMOVAL & INSTALLATION

1. Disconnect the negative battery cable.

2. Remove the bolt that retains the sensor in place at the cylinder block.

3. Remove the sensor from its mounting.

4. Disconnect the connector from the sensor.

To install:

5. Installation is the reverse of the removal procedure.

6. Tighten the sensor to 4.7 ft. lbs.

TESTING

1. Remove the sensor from the vehicle.

2. Measure the resistance between the two connector terminals of the sensor, (terminals 1 and 2).

3. If measured value is not within 1–4 kohms, replace the sensor.

Electronic Control Module (ECM)

LOCATION

See Figure 10.

The ECM is located on the passenger's side of the vehicle, underneath the instrument panel glove box area.

OPERATION

The ECM controls the vehicle engine operating system.

REMOVAL & INSTALLATION

1. Disconnect the negative battery cable.

2. Remove the lower inner trim on the passenger's side of the vehicle.

3. Detach the floor mat. Remove the protective cover.

4. Remove the ECM bracket retaining nuts. Remove the clip from the bracket.

5. Disconnect the connectors.

6. Remove the ECM from the vehicle.

To install:

7. Installation is the reverse of the removal procedure.

8. Tighten the retaining screws to 3.7 ft. lbs.

➡ **When replacing the ECM, be careful not to use the wrong part number, as damage to the injection system could occur.**

EGR Valve

LOCATION

On the SOHC engine the EGR valve is mounted on the intake manifold. The DOHC engine does not use a conventional EGR valve.

OPERATION

The EGR valve controls the flow of exhaust gases. The ECM monitors the flow and regulates the valve accordingly.

REMOVAL & INSTALLATION

1. Disconnect the negative battery cable.

2. Disconnect the connector.

3. Remove the valve from the intake manifold.

To install:

4. Installation is the reverse of the removal procedure.

5. Tighten the retaining bolts to 14.0 ft. lbs.

TESTING

See Figure 11.

1. Disconnect the connector from the ECM.

2. Measure the resistance between the ECM connector and chassis ground.

➡ **If DTC code P1492 was set, use connector B134 and terminal 18/chassis ground. If DTC code P1494 was set, use connector B134 and terminal 17/chassis ground. If DTC code P1496 was set, use connector B134 and terminal 16/chassis ground. If DTC code P1498 was set, use connector B134 and terminal 15/chassis ground.**

3. If measured value is more than 1 Mohm, check the ECM connector and the EGR solenoid for poor contact.

4. If contact is good, replace the EGR solenoid valve.

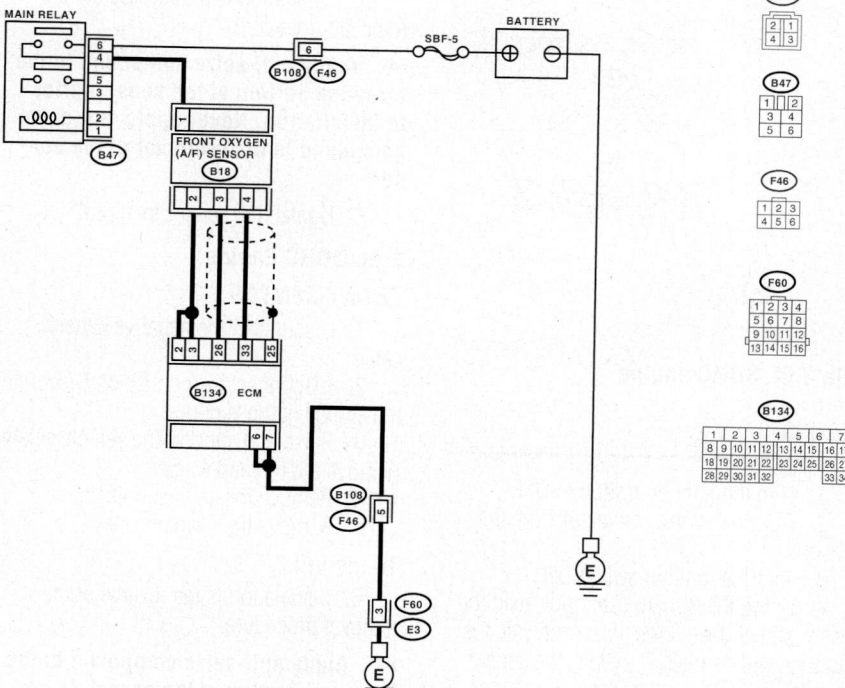

Fig. 9   Camshaft position sensor connector location—Baja 2.5L DOHC engine

29157_SUBA_G0030

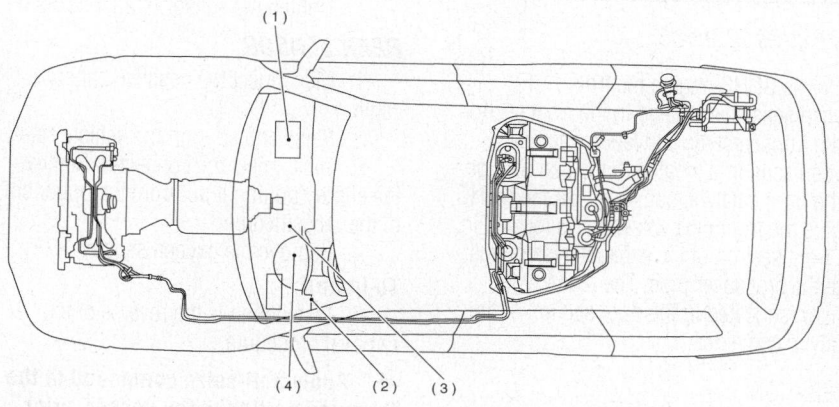

(1) Engine control module (ECM)
(2) Data link connector
(3) Test mode connector
(4) Malfunction indicator light

Fig. 10   ECM and related components—Baja

29157_SUBA_G0004

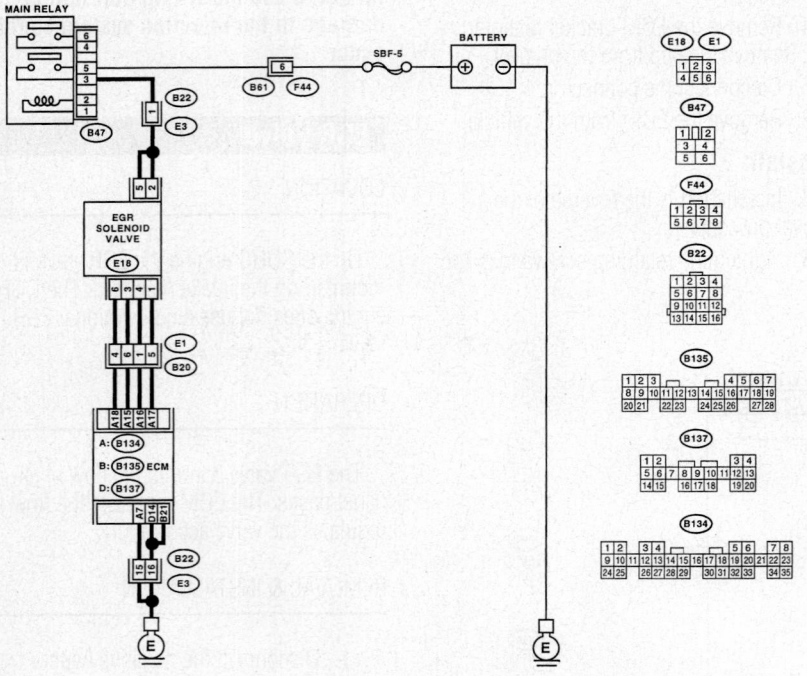

29157_SUBA_G0033

Fig. 11   EGR valve connector location—Baja 2.5L SOHC engine

Engine Coolant Temperature (ECT) Sensor

LOCATION

The sensor is located by the heater outlet fitting or in a cooling passage on the engine, depending upon the particular vehicle.

OPERATION

This component detects the temperature of the engine coolant and relays the information to the electronic control assembly.

REMOVAL & INSTALLATION

1. Disconnect the negative battery cable.
2. Remove the air intake duct and air cleaner case.
3. Disconnect the connector from the sensor.
4. Drain the cooling system, as required.
5. Remove the sensor from its mounting.

To install:

6. Installation is the reverse of the removal procedure.
7. Tighten the sensor to 13.3 ft. lbs.

TESTING

1. Turn the ignition switch to OFF.
2. Disconnect the connector from the sensor.
3. Turn the ignition switch ON.
4. Using the Subaru scan tool, read the sensor signal data. If the measured value is less than -40 degrees F, replace the sensor.

Heated Oxygen (HO2S) Sensor

LOCATION

See Figures 12 and 13.

On the SOHC engine the front (A/F) oxygen sensor is located in the front section of the front catalytic converter. The rear oxygen sensor is located in the rear section of the front catalytic converter. On the DOHC engine the front oxygen sensor is located in the front section of the exhaust system, just past the crossover pipe. The rear oxygen sensor is located in the front section of the catalytic converter.

OPERATION

The exhaust gas oxygen sensor supplies the electronic control assembly with a signal which indicates either a rich or lean mixture condition, during the engine operation.

REMOVAL & INSTALLATION

2.5L SOHC Engine

1. Disconnect the negative battery cable.
2. On front sensor, disconnect the connectors from the engine hanger and the front sensor.
3. Disconnect the connector from the rear sensor.
4. Raise and support the vehicle safely.
5. Remove the oxygen sensor.

To install:

6. Installation is the reverse of the removal procedure.

➡ **Apply anti-seize compound to the threaded portion of the sensor, prior to installation. Never apply anti-seize compound to the protector of the sensor.**

7. Tighten the sensor to 15.2 ft. lbs.

2.5L DOHC Engine

FRONT SENSOR

1. Disconnect the negative battery cable.
2. Disconnect the connector. Disconnect the engine harness clip.
3. Raise and support the vehicle safely. Remove the tire and wheel.
4. Remove the service hole cover.
5. Remove the oxygen sensor.

To install:

6. Installation is the reverse of the removal procedure.

➡ **Apply anti-seize compound to the threaded portion of the sensor, prior to installation. Never apply anti-seize compound to the protector of the sensor.**

7. Tighten the sensor to 22.1 ft. lbs.

REAR SENSOR

1. Disconnect the negative battery cable.
2. Raise and support the vehicle safely.
3. Disconnect the connector. Remove the clip by pulling it out from the upper side of the crossmember.
4. Remove the oxygen sensor.

To install:

5. Installation is the reverse of the removal procedure.

➡ **Apply anti-seize compound to the threaded portion of the sensor, prior to installation. Never apply anti-seize compound to the protector of the sensor.**

6. Tighten the sensor to 22.1 ft. lbs.

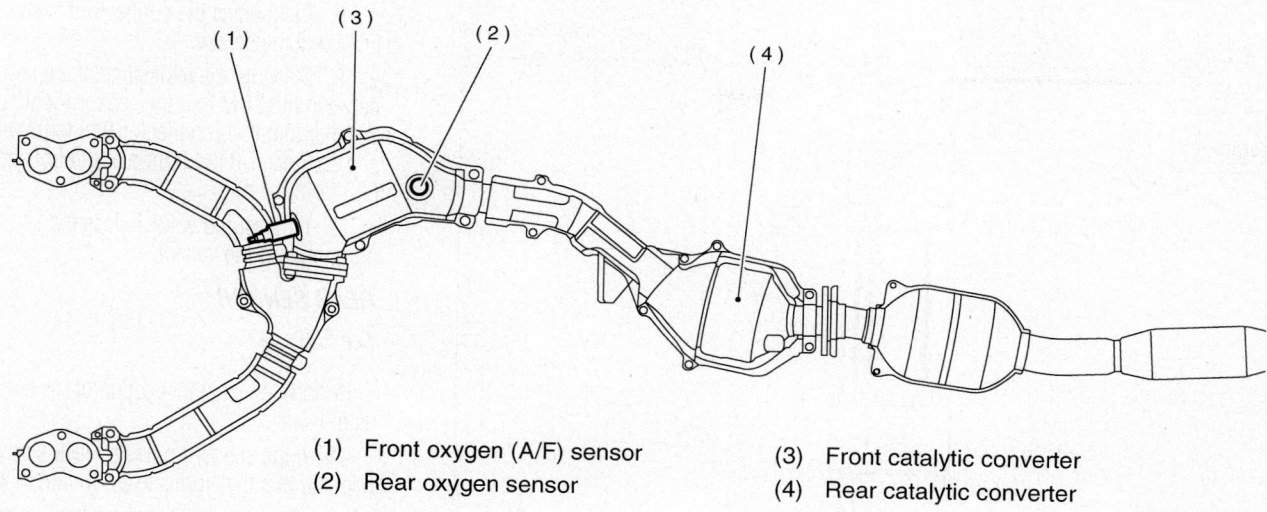

(1) Front oxygen (A/F) sensor
(2) Rear oxygen sensor
(3) Front catalytic converter
(4) Rear catalytic converter

29157_SUBA_G0008

Fig. 12 Oxygen sensor location—Baja 2.5L SOHC engine

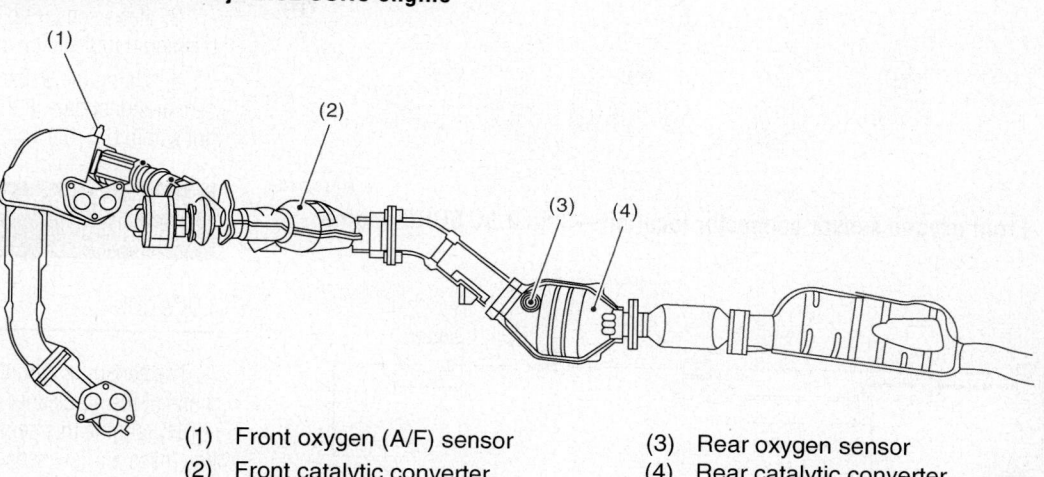

(1) Front oxygen (A/F) sensor
(2) Front catalytic converter
(3) Rear oxygen sensor
(4) Rear catalytic converter

29157_SUBA_G0006

Fig. 13 Oxygen sensor location—Baja 2.5L DOHC engine

TESTING

2.5L SOHC Engine

FRONT SENSOR

See Figure 14.

Perform a visual inspection of the sensor as follows:

1. If the sensor tip has a black/sooty deposit, this may indicate a rich fuel mixture.
2. If the sensor tip has a white, gritty deposit, this may indicate an internal coolant leak.
3. If the sensor tip has a brown deposit, this could indicate oil consumption.
4. Turn the ignition switch to OFF.
5. Disconnect the connectors from the ECM and the sensor.
6. Measure the resistance of the harness between the ECM and the sensor (connector B136 terminal 13/connector E18 terminal 1 and connector B136 terminal 22/connector B18 terminal 2).

7. If measured value is less than 1 ohm, replace the sensor.

REAR SENSOR

See Figure 15.

Perform a visual inspection of the sensor as follows:

1. If the sensor tip has a black/sooty deposit, this may indicate a rich fuel mixture.
2. If the sensor tip has a white, gritty deposit, this may indicate an internal coolant leak.
3. If the sensor tip has a brown deposit, this could indicate oil consumption.
4. Turn the ignition switch to OFF.
5. Disconnect the connectors from the sensor.

6. Turn the ignition switch ON.
7. Measure the voltage between the sensor harness connector and engine ground (connector T6 terminal 4+/engine ground -).
8. If measured value is within 0.2–0.5 volt, replace the sensor.

2.5L DOHC Engine

FRONT SENSOR

See Figure 16.

Perform a visual inspection of the sensor as follows:

1. If the sensor tip has a black/sooty deposit, this may indicate a rich fuel mixture.
2. If the sensor tip has a white, gritty deposit, this may indicate an internal coolant leak.
3. If the sensor tip has a brown deposit, this could indicate oil consumption.

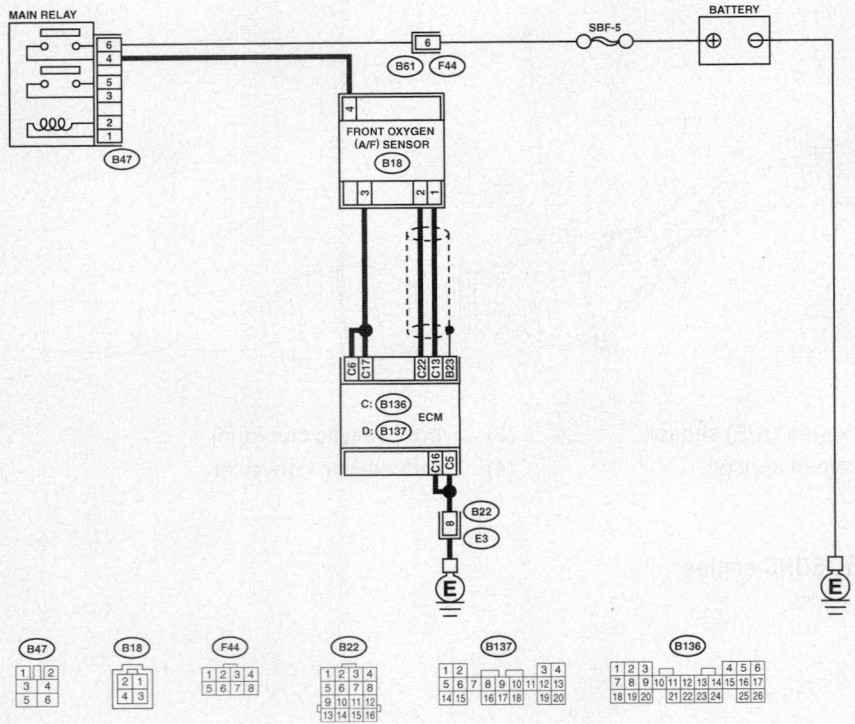

29157_SUBA_G0026

Fig. 14   Front oxygen sensor connector location—Baja 2.5L SOHC engine

29157_SUBA_G0027

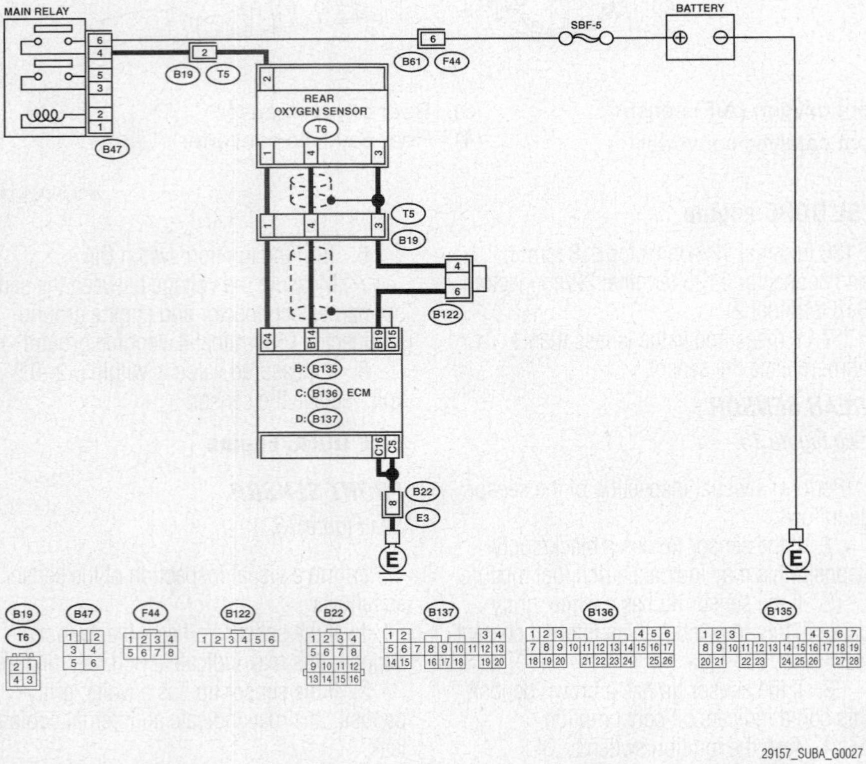

Fig. 15   Rear oxygen sensor connector location—Baja 2.5L SOHC engine

4. Turn the ignition switch to OFF.

5. Disconnect the connectors from the ECM and the sensor.

6. Measure the resistance of the harness between the ECM and the sensor (connector B134 terminal 26/connector B18 terminal 3 and connector B134 terminal 33/connector B18 terminal 4).

7. If measured value is less than 1 ohm, replace the sensor.

REAR SENSOR

See Figure 17.

Perform a visual inspection of the sensor as follows:

1. If the sensor tip has a black/sooty deposit, this may indicate a rich fuel mixture.

2. If the sensor tip has a white, gritty deposit, this may indicate an internal coolant leak.

3. If the sensor tip has a brown deposit, this could indicate oil consumption.

4. Using the Subaru scan tool, read the sensor signal data. If the measured value is not within 0.2–0.4 volt, replace the sensor.

Intake Air Temperature (IAT) Sensor

LOCATION

The sensor is mounted in the intake air hose of the air cleaner assembly. On the DOHC engine this sensor is combined with the mass air flow sensor.

OPERATION

The sensor provides a signal to the ECM for incoming air temperature. This information is used to adjust the injector pulse width and in turn the air/fuel ratio.

REMOVAL & INSTALLATION

1. Disconnect the negative battery cable.

2. Disconnect the connector from the sensor.

3. Remove the sensor from its mounting.

To install:

4. Installation is the reverse of the removal procedure.

5. On the DOHC engine, torque the retaining screw to 0.8 ft. lbs.

TESTING

1. Start the engine and allow it to reach operating temperature.

2. Using the Subaru scan tool, read the sensor signal data. If the measured value is within 167–203 degrees F, replace the sensor.

Knock Sensor (KS)

LOCATION

The sensor is located at the top right (driver's) side of the engine and is positioned on the cylinder block.

OPERATION

The sensor is used to detect engine vibrations caused by preignition or detonation and provides information to the ECM, which then retards the timing to eliminate detonation.

REMOVAL & INSTALLATION

1. Disconnect the negative battery cable.

2. Remove the air cleaner case.

3. On DOHC engine, remove the intercooler.

4. Disconnect the sensor connector.

5. Remove the sensor from its mounting.

To install:

6. Installation is the reverse of the removal procedure.

7. Tighten the sensor to 17.4 ft. lbs.

➡ **The extraction area of the knock sensor wire must be positioned at a 60 degree angle relative to the engine rear.**

TESTING

1. Remove the sensor from the vehicle.

2. Measure the resistance between the connector terminal of the sensor, (terminal 2) and ground.

3. If measured value is less than 400 kohms, replace the sensor.

Manifold Absolute Pressure (MAP) Sensor

LOCATION

The sensor is located on the throttle body unit.

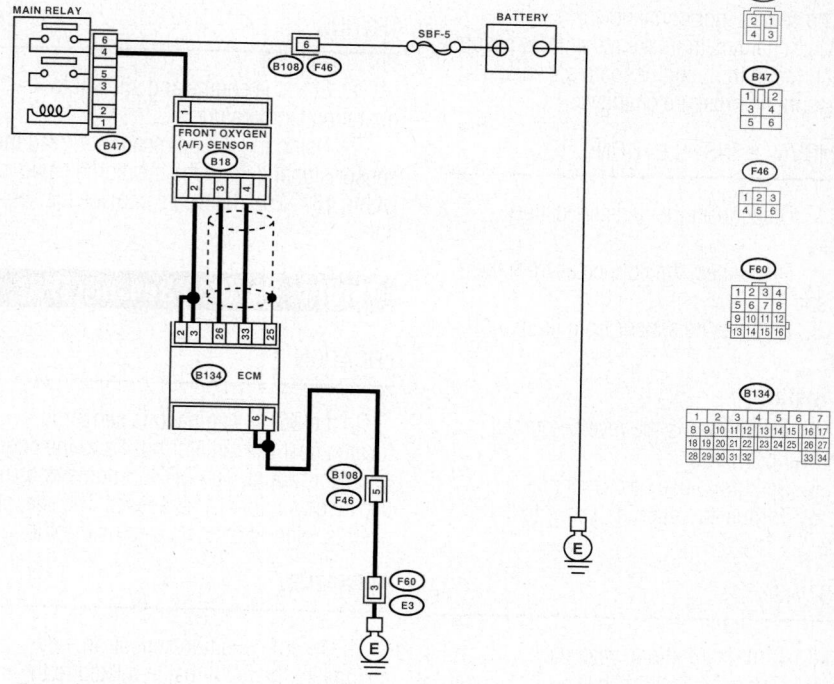

29157_SUBA_G0028

Fig. 16   Front oxygen sensor connector location—Baja 2.5L DOHC engine

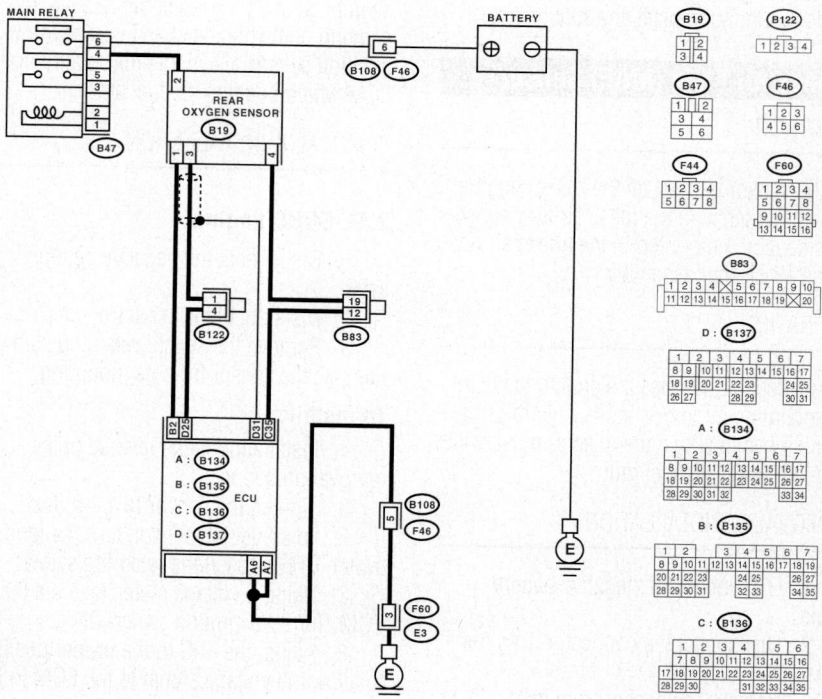

29157_SUBA_G0029

Fig. 17   Rear oxygen sensor connector location—Baja 2.5L DOHC engine

OPERATION

The sensor monitors and signals the ECM of changes in intake manifold pressure which result from engine load, speed and atmospheric pressure changes.

REMOVAL & INSTALLATION

1. Disconnect the negative battery cable.
2. Disconnect the connector from the sensor.
3. Remove the sensor from its mounting.

To install:

4. Installation is the reverse of the removal procedure.
5. Be sure to use new O-rings.
6. Torque the retaining screw to 1.2 ft. lbs.

TESTING

7. Turn the ignition switch ON.
8. Using the Subaru scan tool, operate the LED operation mode.
9. If the idle switch signal comes on and there are no other displayed codes check the condition of the sensor.
10. If the sensor is securely in place on its mounting and the throttle body is secure on its mounting, replace the sensor.

Mass Air Flow (MAF) Sensor

LOCATION

The sensor is used on the DOHC engine. It is combined with the mass air flow sensor. This sensor is mounted in the intake air hose of the air cleaner assembly.

OPERATION

The sensor provides a signal to the ECM for incoming air temperature. This information is used to adjust the injector pulse width and in turn the air/fuel ratio.

REMOVAL & INSTALLATION

1. Disconnect the negative battery cable.
2. Disconnect the connector from the sensor.
3. Remove the sensor from its mounting.

To install:

4. Installation is the reverse of the removal procedure.

---

5. Torque the retaining screw to 0.8 ft. lbs.

TESTING

1. Start the engine and allow it to reach operating temperature.
2. Using the Subaru scan tool, read the sensor signal data. If the measured value is within 167–203 degrees F, replace the sensor.

Throttle Position Sensor (TPS)

LOCATION

On the SOHC engine, this sensor is located near the radiator hose and the engine insulator mount. The DOHC engine does not use a conventional TPS sensor. The sensor used is referred to as an electric throttle.

OPERATION

The throttle position sensor provides a signal to the ECM that is related to the relative throttle plate position. As the throttle plate moves in relation to driving conditions, a signal is sent to the control unit which adjusts the injector pulse width and air/fuel ratio. As the throttle plate is opened further, more air is taken into the combustion chambers, and as a result the relative fuel demand of the engine changes. The throttle position sensor relays this information to the ECM which in alters the fuel amount.

REMOVAL & INSTALLATION

2.5L SOHC Engine

1. Disconnect the negative battery cable.
2. Disconnect the sensor connector.
3. Remove the sensor retaining screws. Remove the sensor from its mounting.

To install:

4. Installation is the reverse of the removal procedure.
5. Tighten the sensor to 1.7 ft. lbs.
6. To adjust the sensor, turn the ignition switch OFF. Loosen the retaining screws.
7. Using a voltage meter, take out the ECM. Turn the ignition switch ON.
8. Adjust the TPS to the proper position to allow the voltage signal to the ECM to be within specification.
9. Specification is 0.45–0.55 volt. Specification is measured at connector B135 and terminal 13.
10. Tighten the retaining screws.

TESTING

2.5L SOHC Engine

See Figure 18.

1. Visually inspect the throttle linkage and throttle for binding and sticking.
2. Measure the resistance of the harness between the TPS connector and engine ground, (connector E13 terminal 3 and ground).
3. If the measured value is more than 1 Mohm, replace the sensor.

2.5L DOHC Engine

See Figure 19.

1. Turn the ignition switch OFF.
2. Disconnect the connector from the ECM.
3. Measure the resistance between the ECM connectors (connector B136 terminal 18/connector B136 and terminal 16).
4. If the measured value is more than 1 Mohm, replace the electric throttle.

Vehicle Speed Sensor (VSS)

LOCATION

Vehicles equipped with automatic transaxle use a front and rear sensors. Both sensors are mounted on the transaxle. Vehicles equipped with manual transaxle use a front sensor. This sensor is mounted on the transaxle.

OPERATION

The vehicle speed sensor sends a signal to the ECM to control. The ECM uses this information to control transmission shift patterns. The vehicle speed sensor is also used to send a speed signal to the speed control servo of the cruise control system.

REMOVAL & INSTALLATION

1. Raise and support the vehicle safely.
2. Place a drip pan below the speed sensor to catch any spilled fluid.
3. Disconnect the connector.
4. Remove the sensor from its mounting.

To install:

5. Installation is the reverse of the removal procedure.
6. Replace any lost fluid.

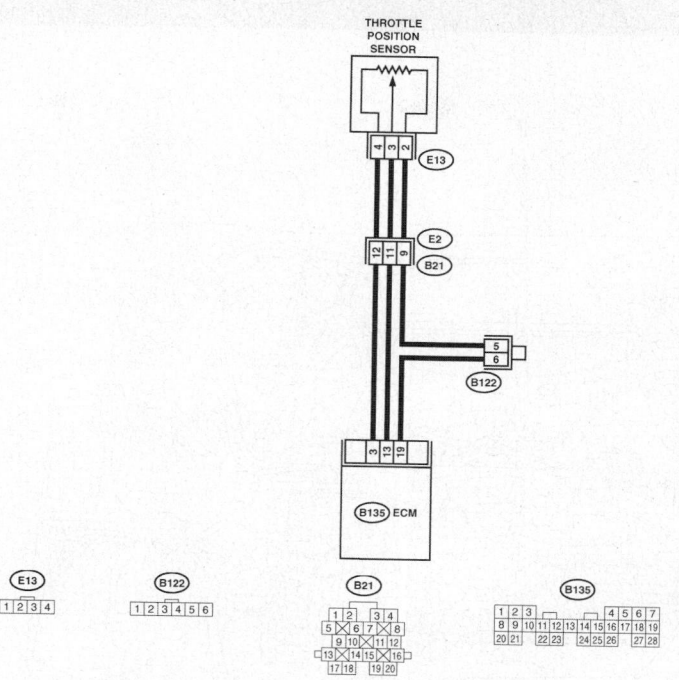

Fig. 18   Throttle position sensor connector location—Baja 2.5L SOHC engine

29157_SUBA_G0031

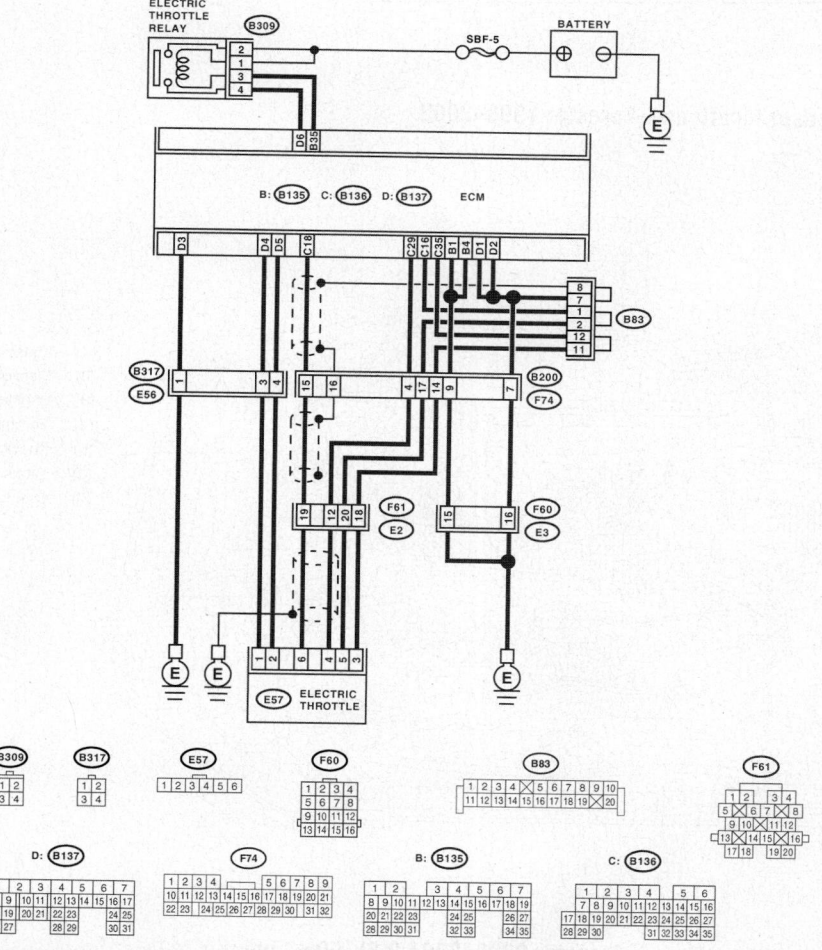

Fig. 19   Electric throttle connector location—Baja 2.5L DOHC engine

29157_SUBA_G0032

FORESTER

See Figures 20 through 25.

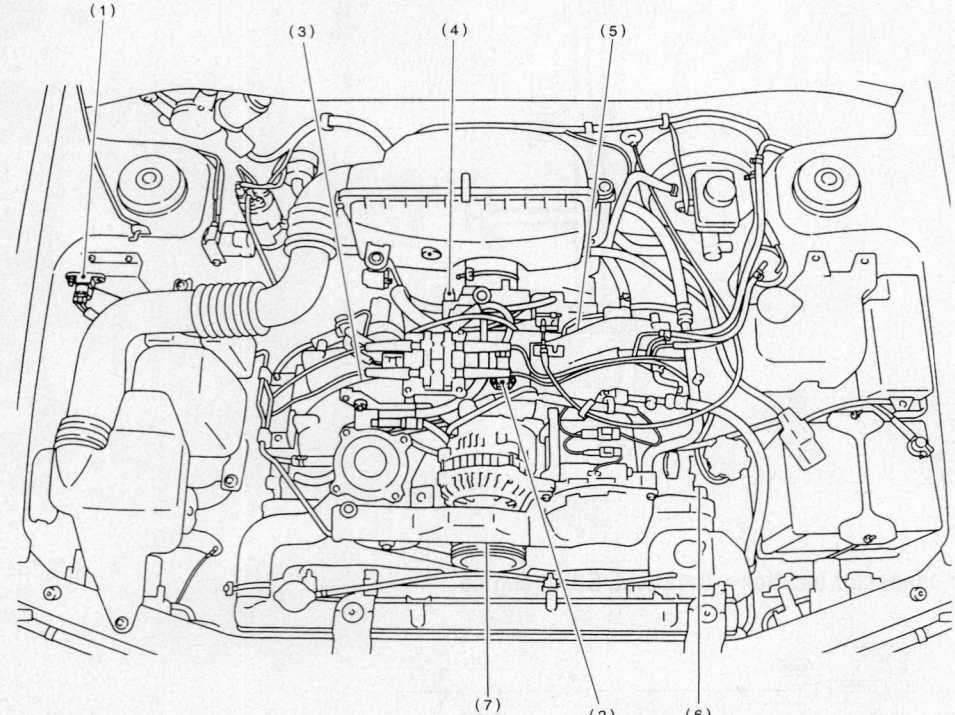

(1) Atmospheric pressure sensor
(2) Intake air temperature and pres-
    sure sensor
(3) Engine coolant temperature sensor

(4) Throttle position sensor
(5) Knock sensor
(6) Camshaft position sensor
(7) Crankshaft position sensor

29157_SUBA_G0013

Fig. 20 Underhood sensor locations—Forester 1998–2002

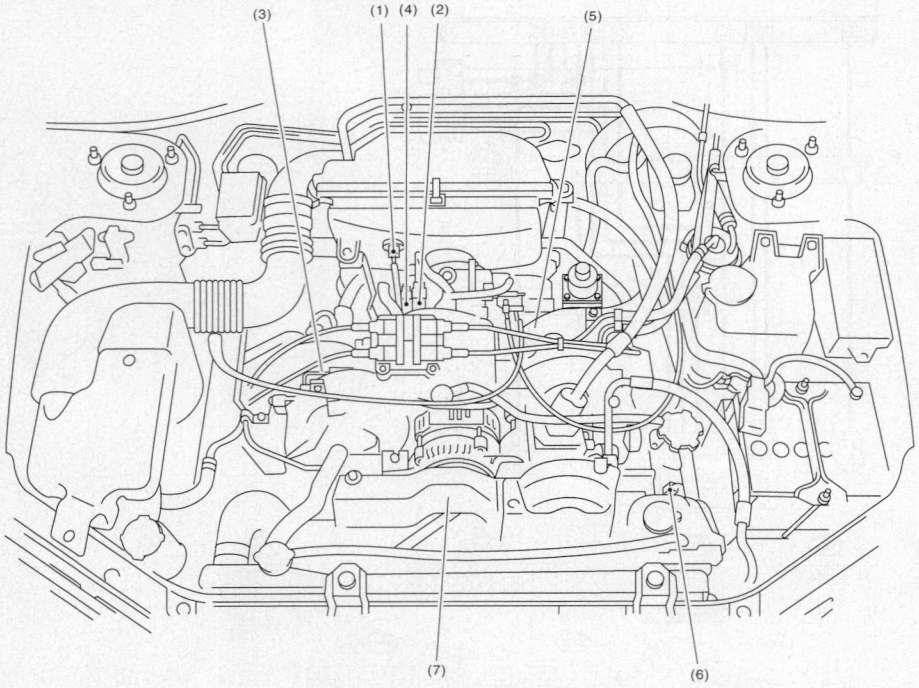

(1) Intake air temperature sensor
(2) Manifold absolute pressure sensor
(3) Engine coolant temperature sensor
(4) Throttle position sensor
(5) Knock sensor
(6) Camshaft position sensor
(7) Crankshaft position sensor

29157_SUBA_G0018

Fig. 21 Underhood sensor locations—Forester 2003–2004 2.5L SOHC engine

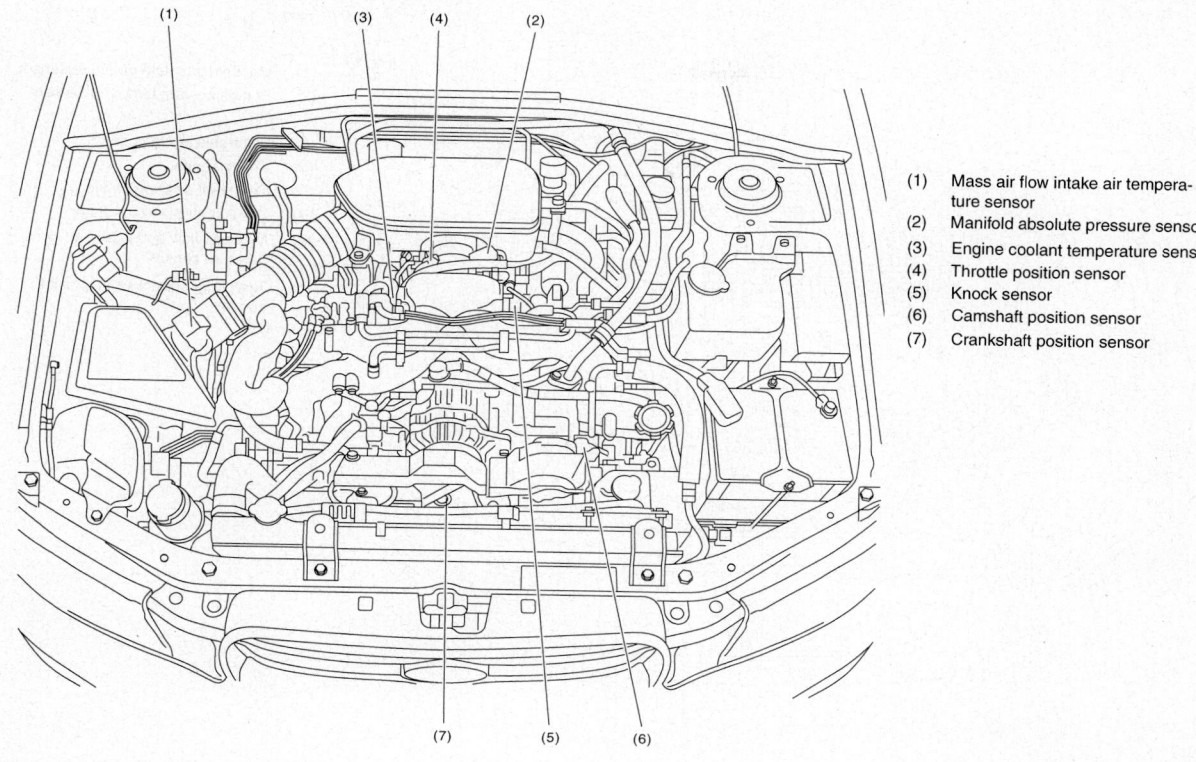

(1) Mass air flow intake air temperature sensor
(2) Manifold absolute pressure sensor
(3) Engine coolant temperature sensor
(4) Throttle position sensor
(5) Knock sensor
(6) Camshaft position sensor
(7) Crankshaft position sensor

29157_SUBA_G0020

Fig. 22   Underhood sensor locations—Forester 2005 2.5L SOHC engine

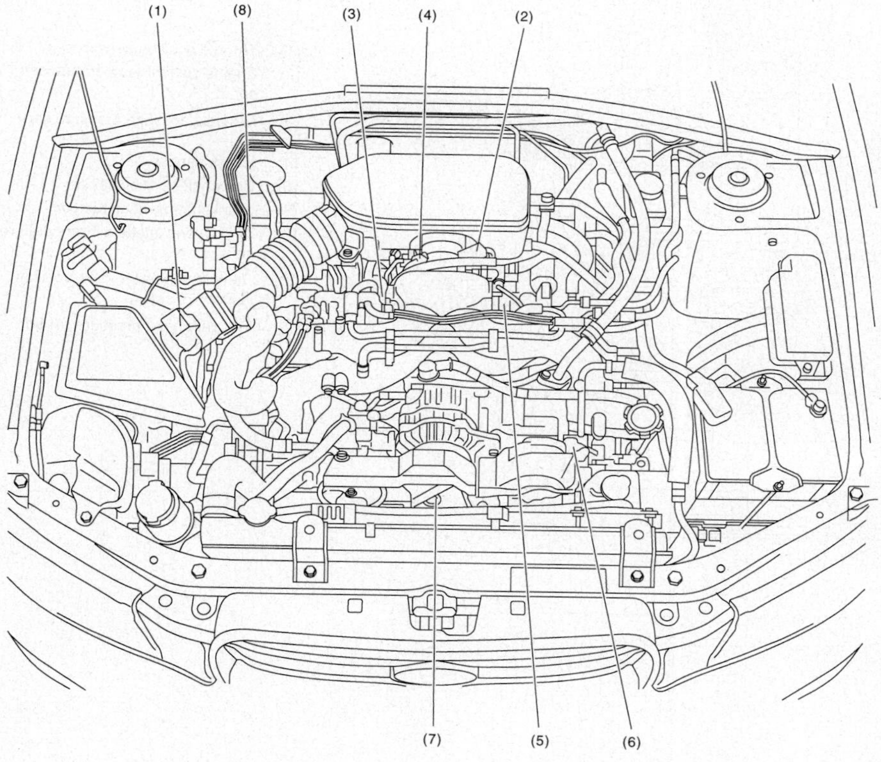

(1) Mass air flow and intake air temperature sensor
(2) Manifold absolute pressure sensor
(3) Engine coolant temperature sensor
(4) Throttle position sensor
(5) Knock sensor
(6) Camshaft position sensor
(7) Crankshaft position sensor
(8) Oil temperature sensor

29157_SUBA_G0016

Fig. 23   Underhood sensor locations—Forester 2006 2.5L SOHC engine

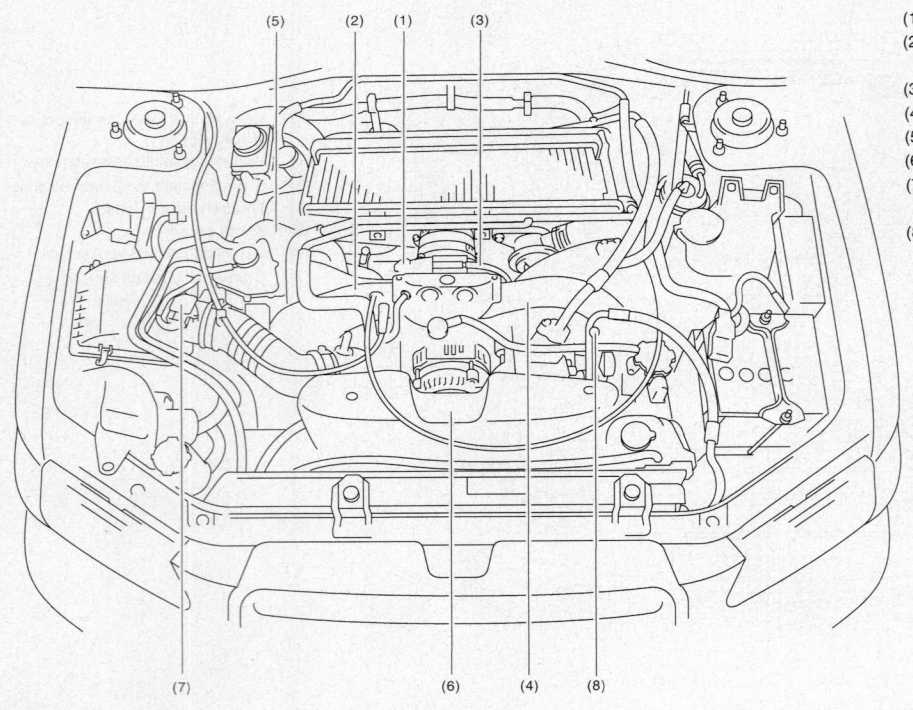

(1) Manifold absolute pressure sensor
(2) Engine coolant temperature sensor
(3) Electronic throttle control
(4) Knock sensor
(5) Camshaft position sensor
(6) Crankshaft position sensor
(7) Mass air flow and intake air temperature sensor
(8) Tumble generator valve position sensor

29157_SUBA_G0022

Fig. 24 Underhood sensor locations—Forester 2004–2005 2.5L DOHC engine

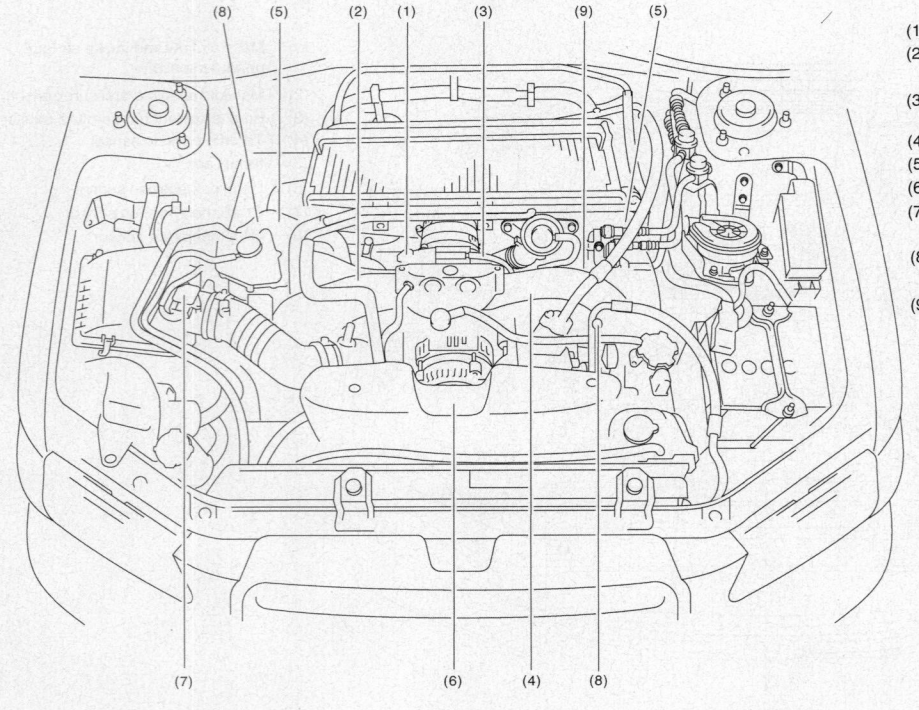

(1) Throttle position sensor
(2) Engine coolant temperature sensor
(3) Manifold absolute pressure sensor
(4) Knock sensor
(5) Camshaft position sensor
(6) Crankshaft position sensor
(7) Air flow and intake air temperature sensor
(8) Tumble generator valve position sensor
(9) Secondary air pressure sensor

29157_SUBA_G0024

Fig. 25 Underhood sensor locations—Forester 2006 2.5L DOHC engine

Camshaft Position (CMP) Sensor

OPERATION

The camshaft position sensor relays the relative camshaft position to the ECM for determining the position and stroke of the No. 1 cylinder. This information is required for proper fuel injection functioning.

REMOVAL & INSTALLATION

1. Disconnect the negative battery cable.
2. Remove the collector cover, as required.
3. Disconnect the connector from the sensor.
4. Remove the bolt that retains the sensor to the sensor support.
5. Remove the bolt that retains the sensor support to the camshaft cap.
6. Remove the sensor and the sensor support as a unit.
7. Separate the sensor from the support.

To install:

8. Installation is the reverse of the removal procedure.
9. Tighten the sensor support to 4.7 ft. lbs.
10. Tighten the sensor to 4.7 ft. lbs.

TESTING

2.5L SOHC Engine

1. Remove the sensor from the vehicle.
2. Measure the resistance between the two connector terminals of the sensor, (terminals 1 and 2).
3. If measured value is not within 1–4 kohms, replace the sensor.

2.5L DOHC Engine

See Figures 26 and 27.

1. Using the Subaru scan tool, check the sensor output waveform.
2. With the engine OFF and the ignition switch ON the signal (V) should be 0–0.9.
3. On 2004–2005 vehicles measurement is taken using connector B135 terminal 8 for the left side and connector B135 terminal 9 for the right side.
4. On 2006 vehicles measurement is taken using connector B134 terminal 21 for the left side and connector B134 terminal 11 for the right side.
5. If abnormality is found, replace the sensor.

Crankshaft Position (CKP) Sensor

LOCATION

The sensor is located in the front of the engine near the crankshaft pulley.

OPERATION

The crankshaft position sensor is a variable reluctance sensor which is used to inform the ECM when the No.1 piston is at top dead center. This information is used for controlling and adjusting ignition and fuel injector timing.

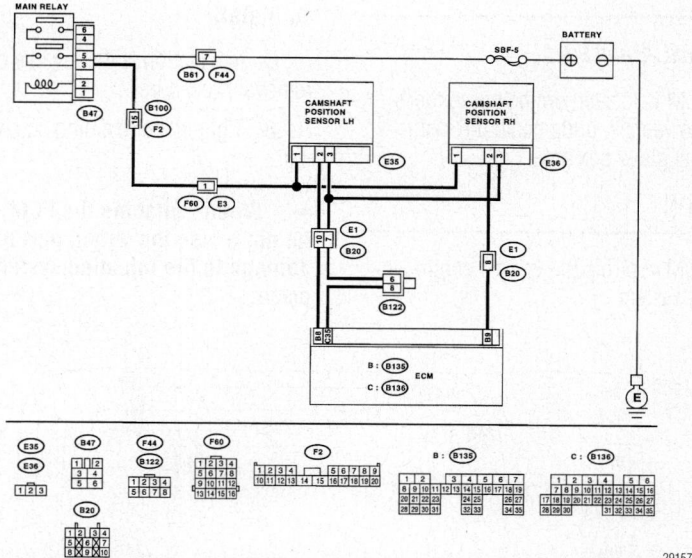

Fig. 26 Camshaft position sensor connector location—Forester 2004–2005 2.5L DOHC engine

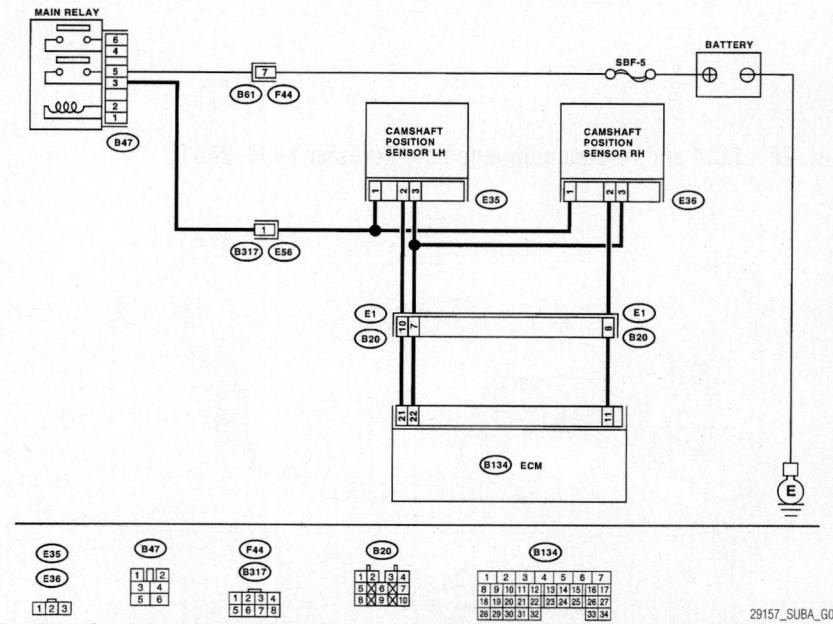

Fig. 27 Camshaft position sensor connector location—Forester 2006 2.5L DOHC engine

REMOVAL & INSTALLATION

1. Disconnect the negative battery cable.
2. Remove the collector cover, as required.
3. Remove the bolt that retains the sensor in place at the cylinder block.
4. Remove the sensor from its mounting.
5. Disconnect the connector from the sensor.

To install:

6. Installation is the reverse of the

removal procedure.

7. Tighten the sensor to 4.7 ft. lbs.

TESTING

1. Remove the sensor from the vehicle.

2. Measure the resistance between the two connector terminals of the sensor, (terminals 1 and 2).

3. If measured value is not within 1–4 kohms, replace the sensor.

Electronic Control Module (ECM)

LOCATION

See Figures 28 and 29.

The ECM is located on the passenger's side of the vehicle, underneath the instrument panel glove box area.

OPERATION

The ECM controls the vehicle engine operating system.

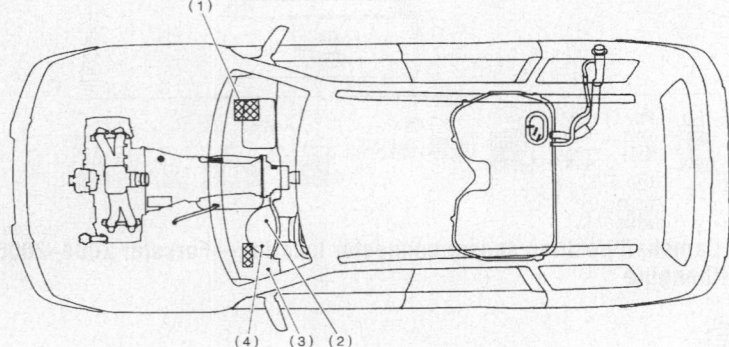

(1) Engine control module (ECM)	(3) Data link connector
(2) CHECK ENGINE malfunction indicator lamp (MIL)	(4) Test mode connector

29157_SUBA_G0012

Fig. 28 ECM and related components—Forester 1998–2001

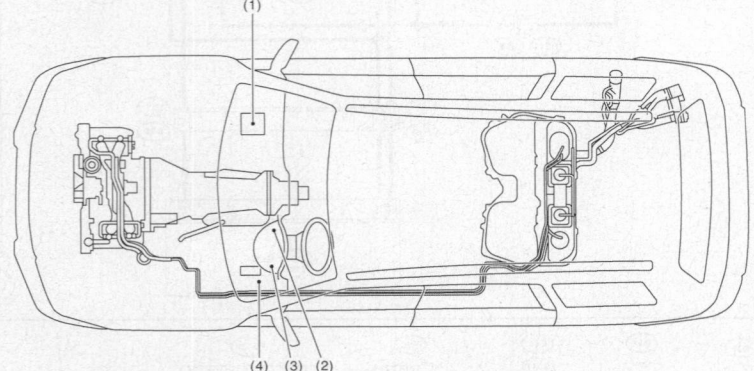

(1) Engine control module (ECM)	(3) Test mode connector
(2) Malfunction indicator light	(4) Data link connector

29157_SUBA_G0015

Fig. 29 ECM and related components—Forester 2002–2006

REMOVAL & INSTALLATION

1. Disconnect the negative battery cable.

2. Remove the lower inner trim on the passenger's side of the vehicle.

3. Detach the floor mat. Remove the protective cover.

4. Remove the ECM bracket retaining nuts. Remove the clip from the bracket.

5. Disconnect the connectors.

6. Remove the ECM from the vehicle.

To install:

7. Installation is the reverse of the removal procedure.

8. Tighten the retaining screws to 3.7 ft. lbs.

➡ **When replacing the ECM, be careful not to use the wrong part number, as damage to the injection system could occur.**

EGR Valve

LOCATION

On the SOHC engine the EGR valve is mounted on the intake manifold. The DOHC engine does not use a conventional EGR valve.

OPERATION

The EGR valve controls the flow of exhaust gases. The ECM monitors the flow and regulates the valve accordingly.

REMOVAL & INSTALLATION

1. Disconnect the negative battery cable.

2. Disconnect the connector.

3. Remove the valve from the intake manifold.

To install:

4. Installation is the reverse of the removal procedure.

5. Tighten the retaining bolts to 14.0 ft. lbs.

TESTING

See Figures 30, 31 and 32.

1. Disconnect the connector from the ECM.

2. Measure the resistance between the ECM connector and chassis ground.

3. On 2004 vehicles, if DTC code P1493 was set, use connector B134 and terminal 18/chassis ground. If DTC code P1495 was set, use connector B134 and terminal 17/chassis ground. If DTC code P1497 was set, use connector B134 and terminal 16/chassis ground. If DTC code P1499 was set, use connector B134 and terminal 15/chassis ground.

4. On 2005 vehicles, if DTC code P1492 was set, use connector B134 and terminal 9/chassis ground. If DTC code P1494 was set, use connector B134 and terminal 10/chassis ground. If DTC code P1496 was set, use connector B134 and terminal 11/chassis ground. If DTC code P1498 was set, use connector B134 and terminal 8/chassis ground.

5. On 2006 vehicles, if DTC code P1492 was set, use connector B134 and terminal 10/chassis ground. If DTC code P1494 was set, use connector B134 and terminal 9/chassis ground. If DTC code P1496 was set, use connector B134 and terminal 8/chassis ground. If DTC code P1498 was set, use connector B134 and terminal 20/chassis ground.

6. If measured value is more than 1 Mohm, check the ECM connector and the EGR solenoid for poor contact.

7. If contact is good, replace the EGR solenoid valve.

Engine Coolant Temperature (ECT) Sensor

LOCATION

The sensor is located by the heater outlet fitting or in a cooling passage on the engine, depending upon the particular vehicle.

OPERATION

This component detects the temperature of the engine coolant and relays the information to the electronic control assembly.

REMOVAL & INSTALLATION

1. Disconnect the negative battery cable.
2. Remove the alternator, as required.
3. Remove the air intake duct and air cleaner case.
4. Disconnect the connector from the sensor.
5. Drain the cooling system, as required.
6. Remove the sensor from its mounting.

To install:

7. Installation is the reverse of the removal procedure.
8. Tighten the sensor to 13.3 ft. lbs.

TESTING

1. Turn the ignition switch to OFF.
2. Disconnect the connector from the sensor.
3. Turn the ignition switch ON.
4. Using the Subaru scan tool, read the sensor signal data. If the measured value is less than -40 degrees F, replace the sensor.

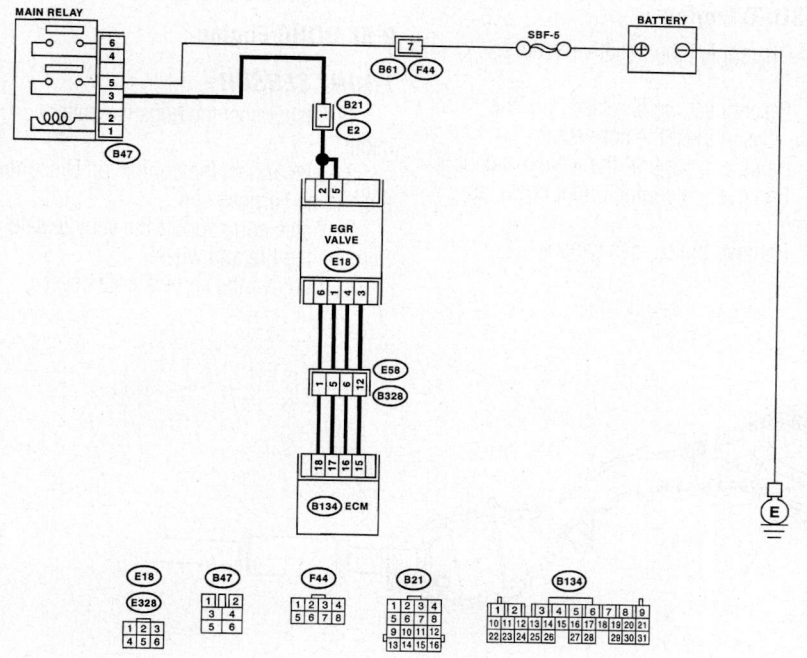

Fig. 30   EGR valve connector location—Forester 2004 2.5L SOHC engine

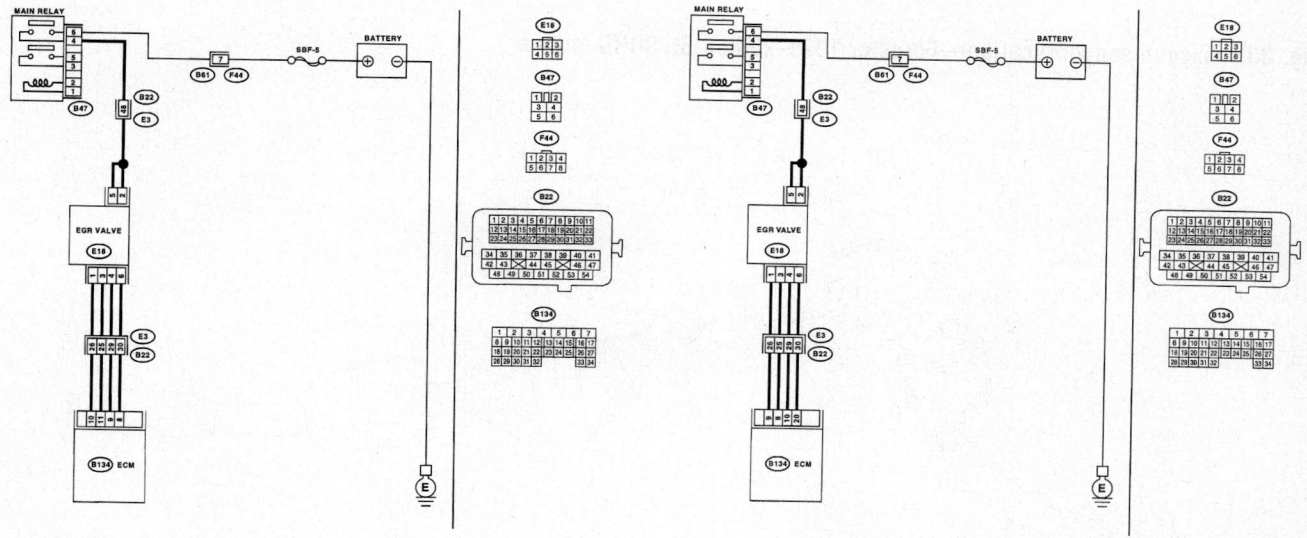

Fig. 31   EGR valve connector location—Forester 2005 2.5L SOHC engine

Fig. 32   EGR valve connector location—Forester 2006 2.5L SOHC engine

Heated Oxygen (HO2S) Sensor

LOCATION

See Figures 33 through 38.

The 1998–2002 vehicles the front (A/F) oxygen sensor is located in the front section of the front catalytic converter. The rear oxygen sensor is located in the rear section of the rear catalytic converter. The 2003–2006 vehicles with SOHC engine the front oxygen sensor is located in the front section of the front catalytic converter. The rear oxygen sensor is located in the rear section of the front catalytic converter. The 2004–2006 DOHC engine the front oxygen sensor is located in the front section of the exhaust system, just past the crossover pipe. The rear oxygen sensor is located in the front section of the rear catalytic converter.

OPERATION

The exhaust gas oxygen sensor supplies the electronic control assembly with a signal which indicates either a rich or lean mixture condition, during the engine operation.

REMOVAL & INSTALLATION

2.5L SOHC Engine

1. Disconnect the negative battery cable.
2. Disconnect the clip fastening the harness. Disconnect the connector.
3. Raise and support the vehicle safely.
4. Remove the engine under cover, as required.
5. Remove the oxygen sensor.

To install:

6. Installation is the reverse of the removal procedure.

➡ **Apply anti-seize compound to the threaded portion of the sensor, prior to installation. Never apply anti-seize compound to the protector of the sensor.**

7. Tighten the sensor to 15.2 ft. lbs.

2.5L DOHC Engine

FRONT SENSOR

1. Disconnect the negative battery cable.
2. Disconnect the connector. Disconnect the engine harness clip.
3. Raise and support the vehicle safely. Remove the tire and wheel.
4. Remove the service hole cover.

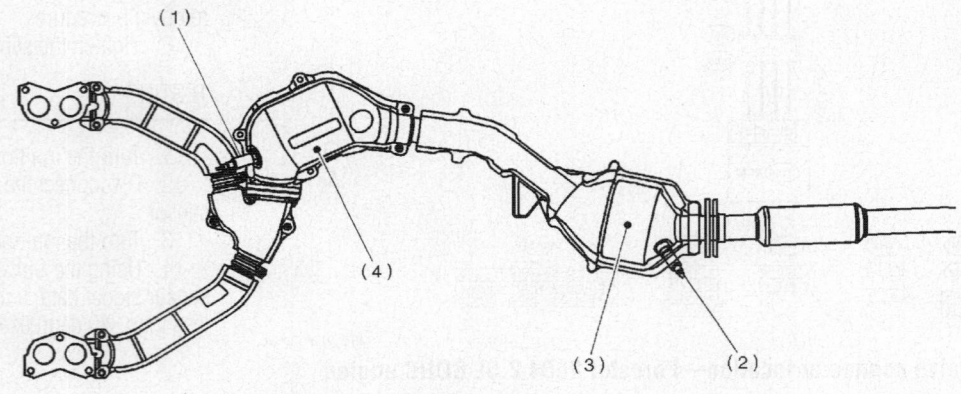

(1)	Front oxygen (A/F) sensor	(3)	Rear catalytic converter
(2)	Rear oxygen sensor	(4)	Front catalytic converter

29157_SUBA_G0014

Fig. 33 Oxygen sensor location—Forester 1998–2002 2.5L SOHC engine

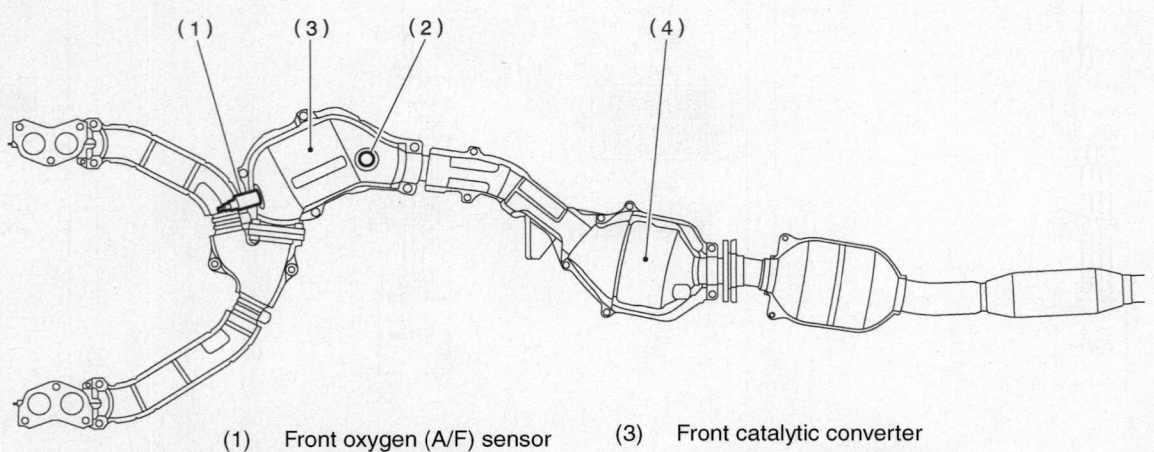

(1)	Front oxygen (A/F) sensor	(3)	Front catalytic converter
(2)	Rear oxygen sensor	(4)	Rear catalytic converter

29157_SUBA_G0019

Fig. 34 Oxygen sensor location—Forester 2003–2004 2.5L SOHC engine

5. Remove the oxygen sensor.

To install:

6. Installation is the reverse of the removal procedure.

➡ **Apply anti-seize compound to the threaded portion of the sensor, prior to installation. Never apply anti-seize compound to the protector of the sensor.**

7. Tighten the sensor to 22.1 ft. lbs.

REAR SENSOR

1. Disconnect the negative battery cable.

2. Raise and support the vehicle safely.

3. Disconnect the connector. Remove the clip by pulling it out from the upper side of the crossmember.

4. Remove the oxygen sensor.

To install:

5. Installation is the reverse of the removal procedure.

➡ **Apply anti-seize compound to the threaded portion of the sensor, prior to installation. Never apply anti-seize compound to the protector of the sensor.**

6. Tighten the sensor to 22.1 ft. lbs.

TESTING

2.5L SOHC Engine

FRONT SENSOR–1998–2001

Perform a visual inspection of the sensor as follows:

1. If the sensor tip has a black/sooty deposit, this may indicate a rich fuel mixture.

2. If the sensor tip has a white, gritty deposit, this may indicate an internal coolant leak.

3. If the sensor tip has a brown deposit, this could indicate oil consumption.

4. Check the exhaust system for defects.

5. Replace the sensor, as required.

FRONT SENSOR–2002

Perform a visual inspection of the sensor as follows:

1. If the sensor tip has a black/sooty deposit, this may indicate a rich fuel mixture.

2. If the sensor tip has a white, gritty deposit, this may indicate an internal coolant leak.

3. If the sensor tip has a brown deposit, this could indicate oil consumption.

4. Turn the ignition switch to OFF.

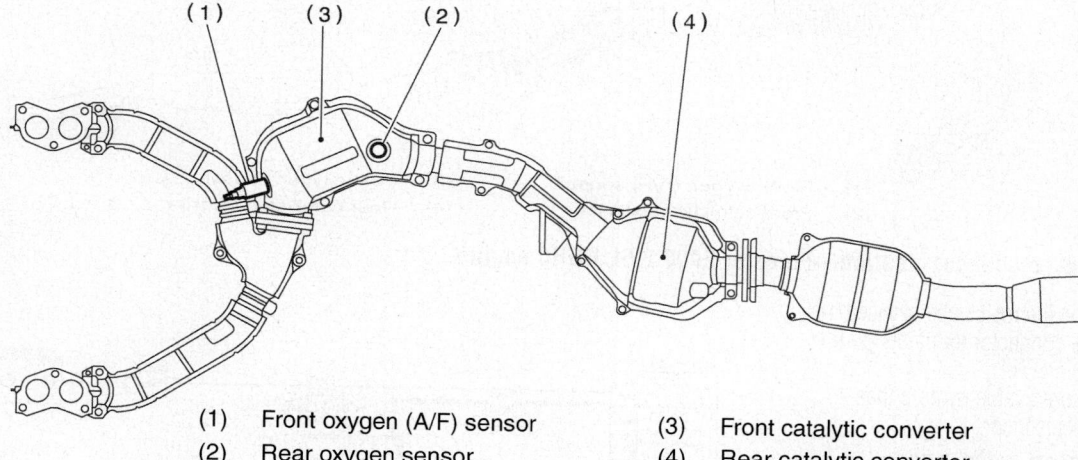

(1)	Front oxygen (A/F) sensor	(3)	Front catalytic converter
(2)	Rear oxygen sensor	(4)	Rear catalytic converter

Fig. 35 Oxygen sensor location—Forester 2005 2.5L SOHC engine

29157_SUBA_G0021

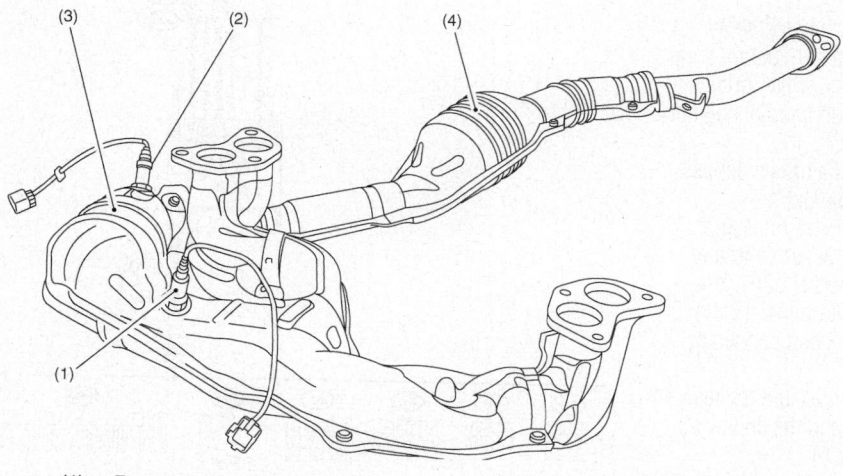

(1)	Front oxygen (A/F) sensor	(3)	Front catalytic converter
(2)	Rear oxygen sensor	(4)	Rear catalytic converter

Fig. 36 Oxygen sensor location—Forester 2006 2.5L SOHC engine

29157_SUBA_G0017

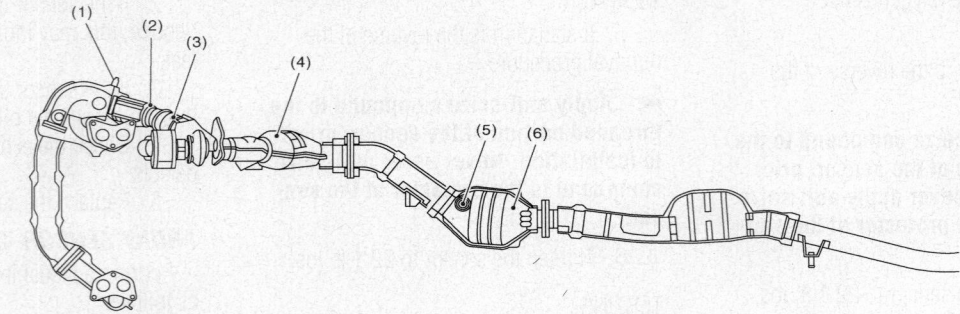

(1)	Front oxygen (A/F) sensor	(3)	Exhaust temperature sensor	(5)	Rear oxygen sensor
(2)	Precatalytic converter	(4)	Front catalytic converter	(6)	Rear catalytic converter

29157_SUBA_G0023

Fig. 37   Oxygen sensor location—Forester 2004–2005 2.5L DOHC engine

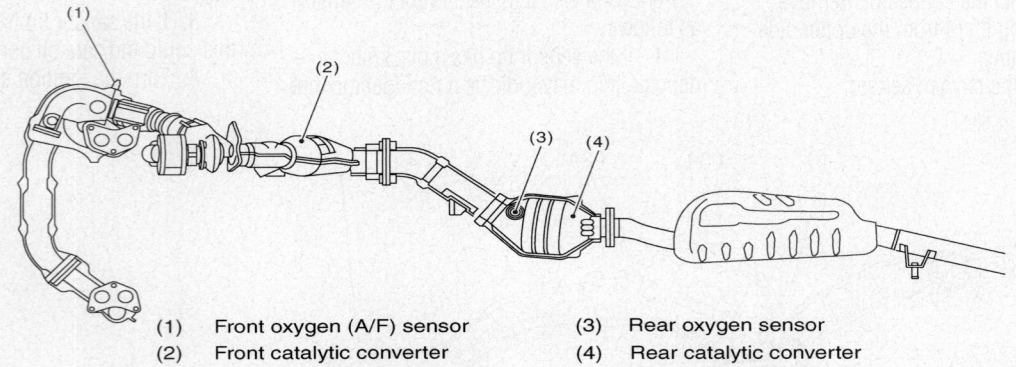

(1)	Front oxygen (A/F) sensor	(3)	Rear oxygen sensor
(2)	Front catalytic converter	(4)	Rear catalytic converter

29157_SUBA_G0025

Fig. 38   Oxygen sensor location—Forester 2006 2.5L DOHC engine

5. Measure the resistance between the sensor and the connector terminals (terminals 2 and 5).

6. If measured value is more than 1 ohm, replace the sensor.

FRONT SENSOR–2003–2006

See Figures 39 through 42.

Perform a visual inspection of the sensor as follows:

1. If the sensor tip has a black/sooty deposit, this may indicate a rich fuel mixture.

2. If the sensor tip has a white, gritty deposit, this may indicate an internal coolant leak.

3. If the sensor tip has a brown deposit, this could indicate oil consumption.

4. On 2003–2004 vehicles, measure the resistance between the sensor and the connector terminals (connector B262/terminal 2 and connector B262/terminal 1) and (connector B262/terminal 3 and connector B262/terminal 4).

5. On 2005 vehicles, measure the resistance between the sensor and the connector terminals (terminals 1 and 6).

6. On 2006 vehicles, measure the resistance between the sensor and the connector terminals (terminals 1 and 4).

7. If measured value is more than 5 ohm, replace the sensor.

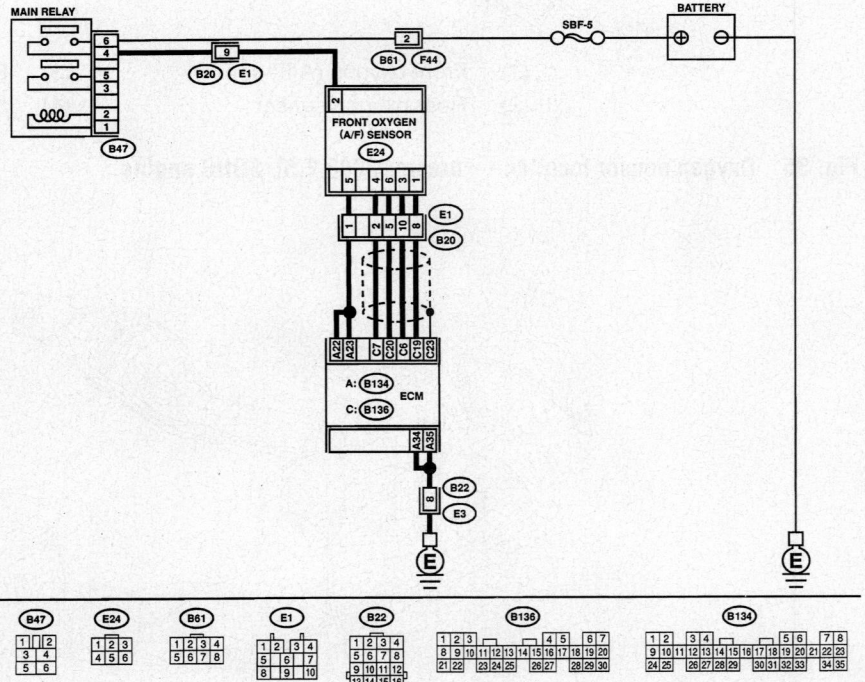

29157_SUBA_G0039

Fig. 39   Front oxygen sensor connector location—Forester 2002 2.5L SOHC engine

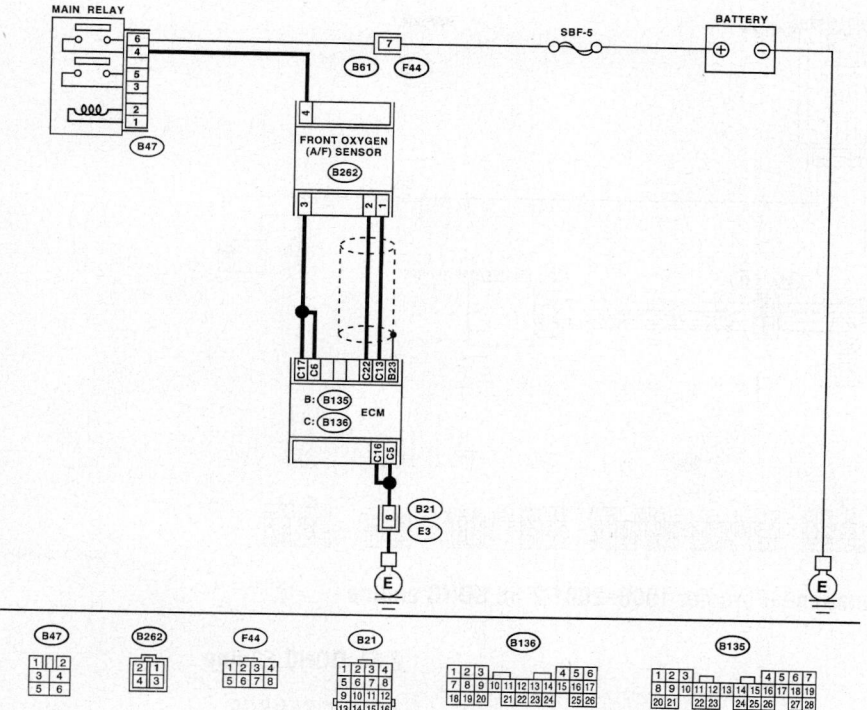

Fig. 40 Front oxygen sensor connector location—Forester 2003–2004 2.5L SOHC engine

29157_SUBA_G0040

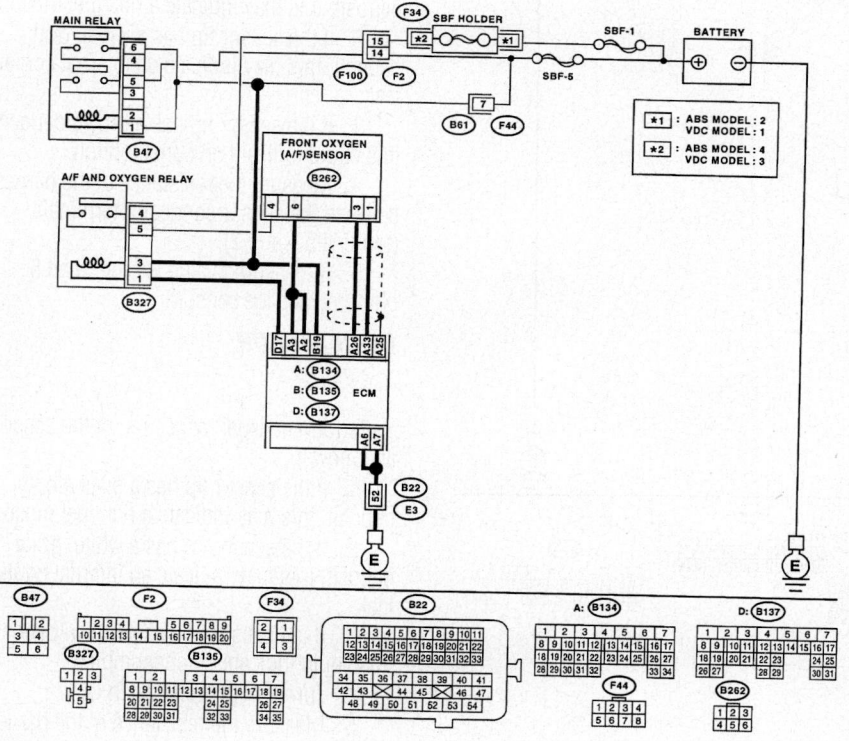

Fig. 41 Front oxygen sensor connector location—Forester 2005 2.5L SOHC engine

29157_SUBA_G0041

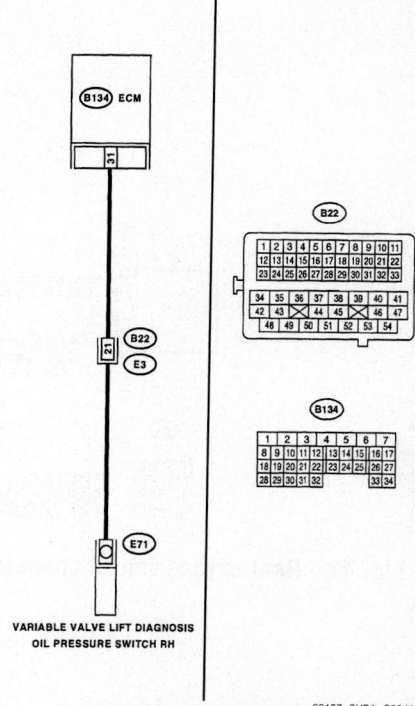

29157_SUBA_G0042

Fig. 42 Front oxygen sensor connector location—Forester 2006 2.5L SOHC engine

REAR SENSOR

See Figures 43, 44 and 45.

Perform a visual inspection of the sensor as follows:

1. If the sensor tip has a black/sooty deposit, this may indicate a rich fuel mixture.

2. If the sensor tip has a white, gritty deposit, this may indicate an internal coolant leak.

3. If the sensor tip has a brown deposit, this could indicate oil consumption.

4. Turn the ignition switch to OFF.

5. Disconnect the connectors from the sensor.

6. Turn the ignition switch ON.

7. On 1998–2001 vehicles, measure the voltage between the sensor harness connector and engine ground (connector E24 terminal 4+/engine ground -). If measured value is more than 0.2 volt, replace the sensor.

8. On 2002–2006 vehicles, measure the voltage between the sensor harness connector terminals (terminals 1 and 2). If measured value is more than 30 ohms, replace the sensor.

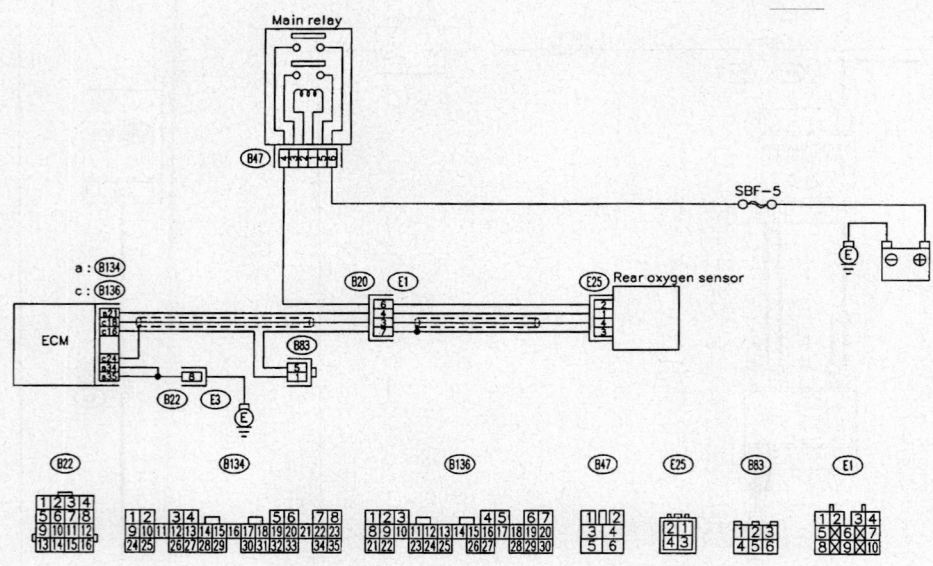

Fig. 43   Rear oxygen sensor connector location—Forester 1998–2001 2.5L SOHC engine

29157_SUBA_G0043

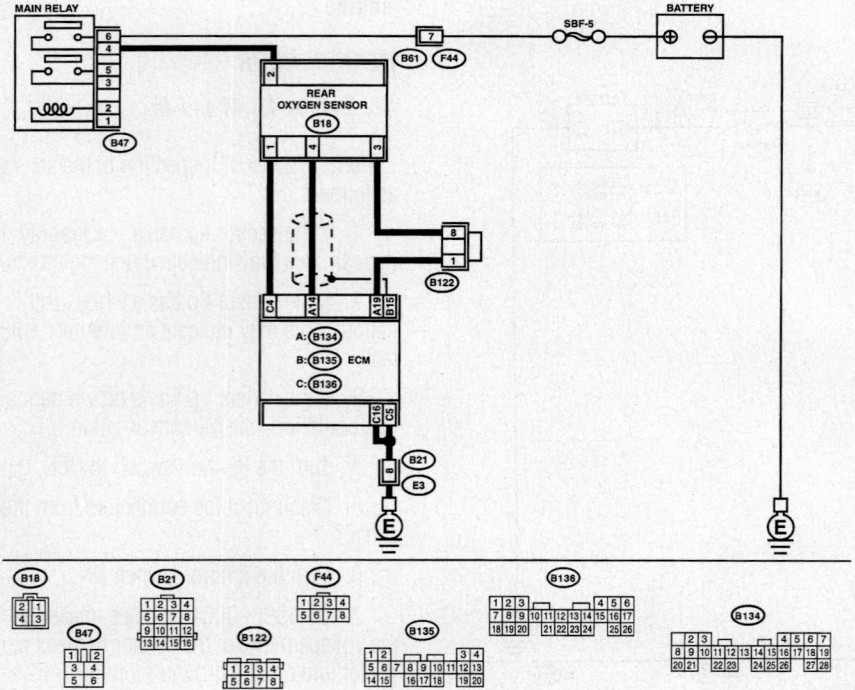

Fig. 44   Rear oxygen sensor connector location—Forester 2002–2004 2.5L SOHC engine

29157_SUBA_G0044

2.5L DOHC Engine

FRONT SENSOR

See Figure 46.

Perform a visual inspection of the sensor as follows:

1. If the sensor tip has a black/sooty deposit, this may indicate a rich fuel mixture.

2. If the sensor tip has a white, gritty deposit, this may indicate an internal coolant leak.

3. If the sensor tip has a brown deposit, this could indicate oil consumption.

4. Measure the resistance of the harness between the sensor connector terminals (terminals 1 and 2).

5. If measured value is more than 5 ohm, replace the sensor.

REAR SENSOR

See Figure 47.

Perform a visual inspection of the sensor as follows:

1. If the sensor tip has a black/sooty deposit, this may indicate a rich fuel mixture.

2. If the sensor tip has a white, gritty deposit, this may indicate an internal coolant leak.

3. If the sensor tip has a brown deposit, this could indicate oil consumption.

4. Turn the ignition switch OFF.

5. Measure the resistance of the harness between the sensor connector terminals (terminals 1 and 2).

6. If measured value is more than 30ohm, replace the sensor.

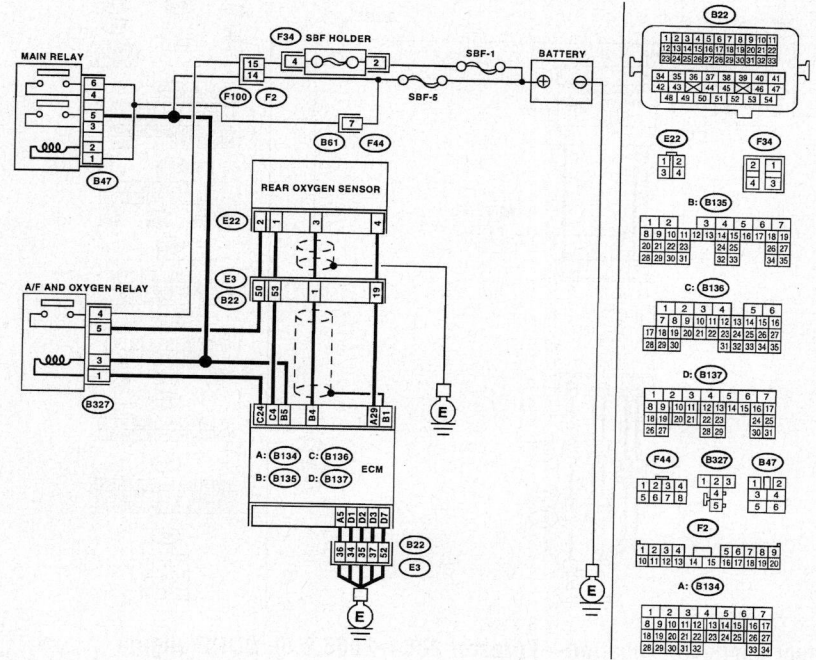

29157_SUBA_G0045

Fig. 45   Rear oxygen sensor connector location—Forester 2005–2006 2.5L SOHC engine

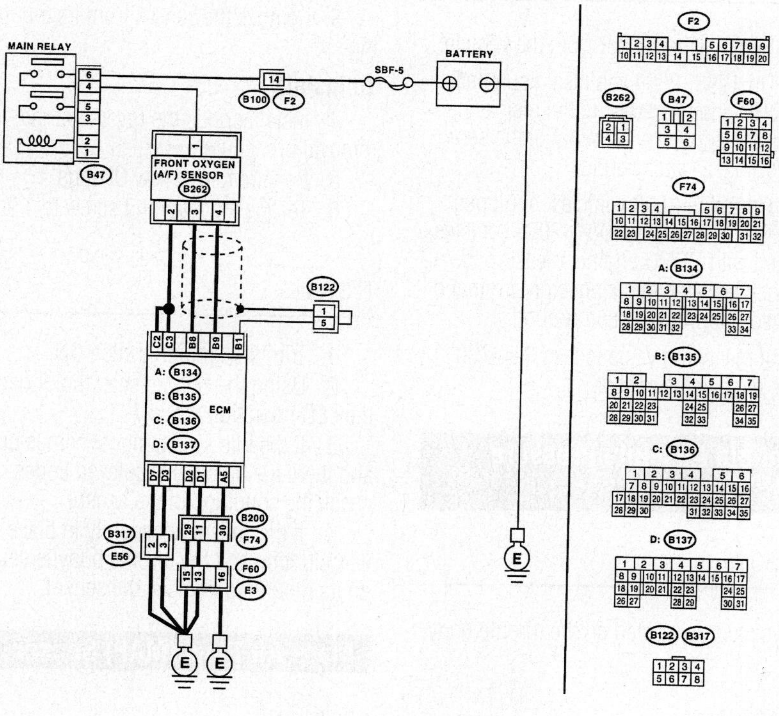

Intake Air Temperature (IAT) Sensor

LOCATION

The sensor is mounted in the intake air hose of the air cleaner assembly. On some vehicles this sensor is combined with the mass air flow sensor.

OPERATION

The sensor provides a signal to the ECM for incoming air temperature. This information is used to adjust the injector pulse width and in turn the air/fuel ratio.

REMOVAL & INSTALLATION

1. Disconnect the negative battery cable.

2. Disconnect the connector from the sensor.

3. Remove the sensor from its mounting.

To install:

4. Installation is the reverse of the removal procedure.

TESTING

1. Start the engine and allow it to reach operating temperature.

29157_SUBA_G0046

Fig. 46   Front oxygen sensor connector location—Forester 2004–2006 2.5L DOHC engine

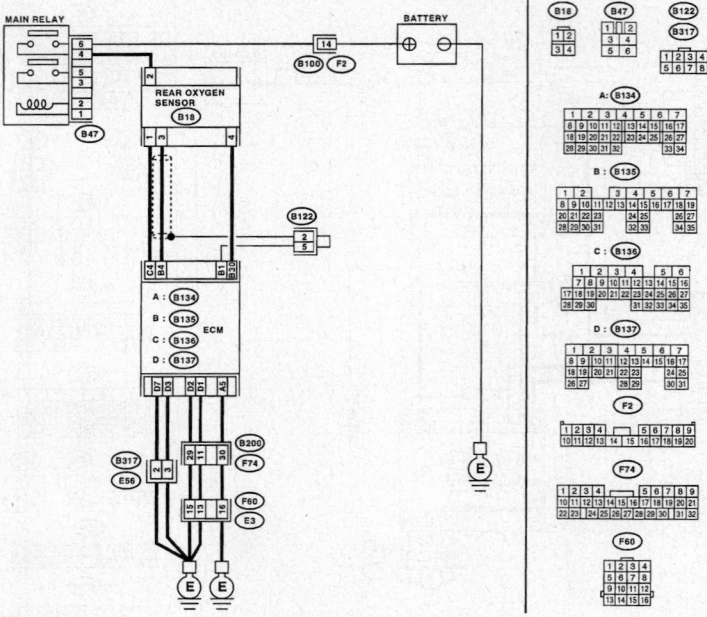

Fig. 47   Rear oxygen sensor connector location—Forester 2004–2006 2.5L DOHC engine

29157_SUBA_G0047

2. Using the Subaru scan tool, read the sensor signal data. If the measured value is less than -40 degrees F, replace the sensor.

Knock Sensor (KS)

LOCATION

The sensor is located at the top right (driver's) side of the engine and is positioned on the cylinder block.

OPERATION

The sensor is used to detect engine vibrations caused by preignition or detonation and provides information to the ECM, which then retards the timing to eliminate detonation.

REMOVAL & INSTALLATION

1. Disconnect the negative battery cable.
2. Remove the air cleaner case.
3. Remove the collector cover, as required.
4. On DOHC engine, remove the intercooler.
5. Disconnect the sensor connector.
6. Remove the sensor from its mounting.

To install:

7. Installation is the reverse of the removal procedure.
8. Tighten the sensor to 17.4 ft. lbs.

➡ **The extraction area of the knock sensor wire must be positioned at a 60 degree angle relative to the engine rear.**

TESTING

1. Remove the sensor from the vehicle.
2. On 1998–2004 vehicles, equipped with SOHC engine, measure the resistance between the connector terminal of the sensor, (terminal 2) and ground.
3. On 2005–2006 vehicles, equipped with SOHC engine and 2004–2006 vehicles, equipped with DOHC engine, measure the resistance between the connector terminal of the sensor, (terminal 1) and ground.
4. If measured value is less than 700 kohms, replace the sensor.

Manifold Absolute Pressure (MAP) Sensor

LOCATION

The sensor is located on the throttle body unit.

OPERATION

The sensor monitors and signals the ECM of changes in intake manifold pressure which result from engine load, speed and atmospheric pressure changes.

REMOVAL & INSTALLATION

1. Disconnect the negative battery cable.
2. Disconnect the connector from the sensor.
3. Remove the sensor from its mounting.

To install:

4. Installation is the reverse of the removal procedure.
5. Be sure to use new O-rings.
6. Torque the retaining screw to 1.2 ft. lbs.

TESTING

1. Turn the ignition switch ON.
2. Using the Subaru scan tool, operate the LED operation mode.
3. If the idle switch signal comes on and there are no other displayed codes check the condition of the sensor.
4. If the sensor is securely in place on its mounting and the throttle body is secure on its mounting, replace the sensor.

Mass Air Flow (MAF) Sensor

LOCATION

The sensor is mounted in the intake air hose of the air cleaner assembly. On some vehicles this sensor is combined with the intake air temperature sensor.

OPERATION

The sensor provides a signal to the ECM for incoming air temperature. This information is used to adjust the injector pulse width and in turn the air/fuel ratio.

REMOVAL & INSTALLATION

1. Disconnect the negative battery cable.
2. Disconnect the connector from the sensor.
3. Remove the sensor from its mounting.

To install:

4. Installation is the reverse of the removal procedure.
5. Torque the retaining screw to 0.8 ft. lbs.

TESTING

1. Start the engine and allow it to reach operating temperature.
2. Using the Subaru scan tool, read the sensor signal data. If the measured value is less than -40 degrees F, replace the sensor.

Throttle Position Sensor (TPS)

LOCATION

On 1998–2004 vehicles, equipped with the 2.5L SOHC engine, the sensor is located near the radiator hose and the engine insulator mount. On 2005–2006 vehicles, equipped with the 2.5L SOHC engine and 2004–2006 vehicles, equipped with the 2.5L DOHC engine, the conventional TPS sensor is not used. The sensor used is referred to as an electric throttle.

OPERATION

The throttle position sensor provides a signal to the ECM that is related to the relative throttle plate position. As the throttle plate moves in relation to driving conditions, a signal is sent to the control unit which adjusts the injector pulse width and air/fuel ratio. As the throttle plate is opened further, more air is taken into the combustion chambers, and as a result the relative fuel demand of the engine changes. The throttle position sensor relays this information to the ECM which in alters the fuel amount.

REMOVAL & INSTALLATION

2.5L SOHC Engine

1. Disconnect the negative battery cable.
2. Disconnect the sensor connector.
3. Remove the sensor retaining screws. Remove the sensor from its mounting.

To install:

4. Installation is the reverse of the removal procedure.
5. Tighten the sensor to 1.7 ft. lbs.
6. To adjust the sensor, turn the ignition switch OFF. Loosen the retaining screws.
7. Using a voltage meter, take out the ECM. Turn the ignition switch ON.
8. Adjust the TPS to the proper position to allow the voltage signal to the ECM to be within specification.
9. Specification is 0.45–0.55 volt. Specification is measured at connector B135 and terminal 13.
10. Tighten the retaining screws.

TESTING

2.5L SOHC Engine

1998–2004

1. Visually inspect the throttle linkage and throttle for binding and sticking.
2. Measure the resistance of the harness between the TPS connector and engine ground, (connector E13 terminal 3 and ground).
3. If the measured value is more than 1 Mohm, replace the sensor.

2.5L DOHC Engine

2004–2005

4. Turn the ignition switch OFF.
5. Disconnect the connector from the ECM.
6. Measure the resistance between the ECM connectors (connector B136 terminal 18/connector B136 and terminal 16).
7. If the measured value is more than 1 Mohm, replace the electric throttle.

2006

8. Measure the voltage between the electronic throttle control connector and ground (connector E57 terminal 6+ and ground).
9. If the measured value is more than 10 volt, replace the sensor.

Vehicle Speed Sensor (VSS)

LOCATION

Vehicles equipped with automatic transaxle use a front and rear sensors. Both sensors are mounted on the transaxle. Vehicles equipped with manual transaxle use a front sensor. This sensor is mounted on the transaxle.

OPERATION

The vehicle speed sensor sends a signal to the ECM to control. The ECM uses this information to control transmission shift patterns. The vehicle speed sensor is also used to send a speed signal to the speed control servo of the cruise control system.

REMOVAL & INSTALLATION

1. Raise and support the vehicle safely.
2. Place a drip pan below the speed sensor to catch any spilled fluid.
3. Disconnect the connector.
4. Remove the sensor from its mounting.

To install:

5. Installation is the reverse of the removal procedure.
6. Replace any lost fluid.

IMPREZA, OUTBACK SPORT, WRX

See Figures 48 through 55.

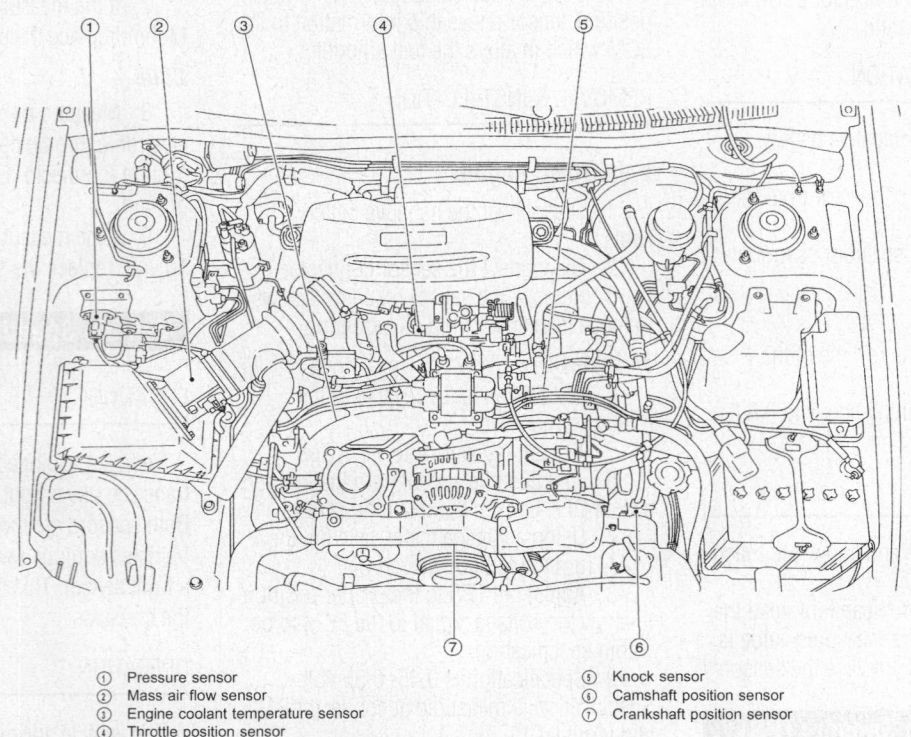

(1) Pressure sensor	(5) Knock sensor
(2) Mass air flow sensor	(6) Camshaft position sensor
(3) Engine coolant temperature sensor	(7) Crankshaft position sensor
(4) Throttle position sensor	

29157_SUBA_G0052

Fig. 48   Underhood sensor locations—Impreza 1996–1997 1.8L engine

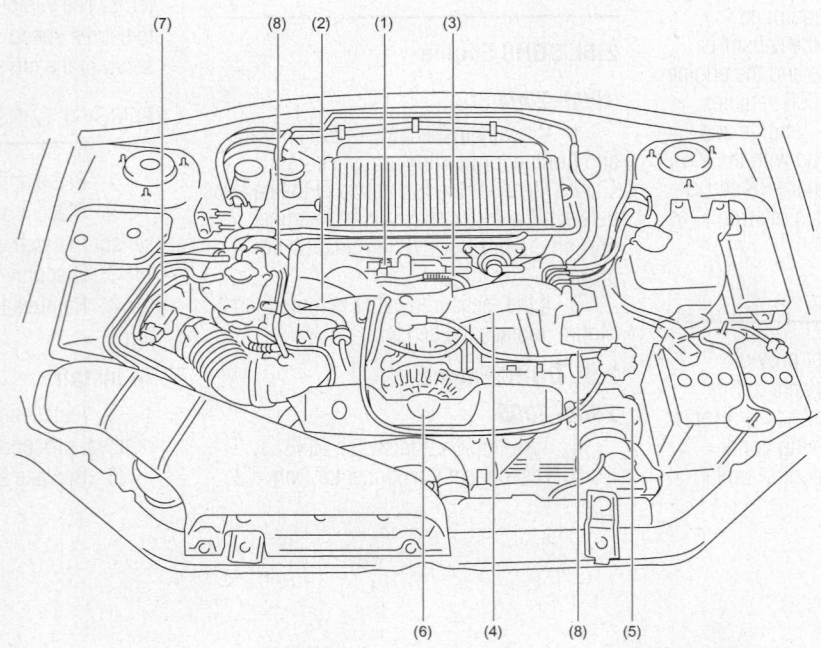

(1) Pressure sensor	(4) Knock sensor	(7) Mass air flow & intake air temperature sensor
(2) Engine coolant temperature sensor	(5) Camshaft position sensor	
(3) Throttle position sensor	(6) Crankshaft position sensor	(8) Tumble generator valve position sensor

29157_SUBA_G0050

Fig. 49   Underhood sensor locations—Impreza 2002–2003 2.0L engine

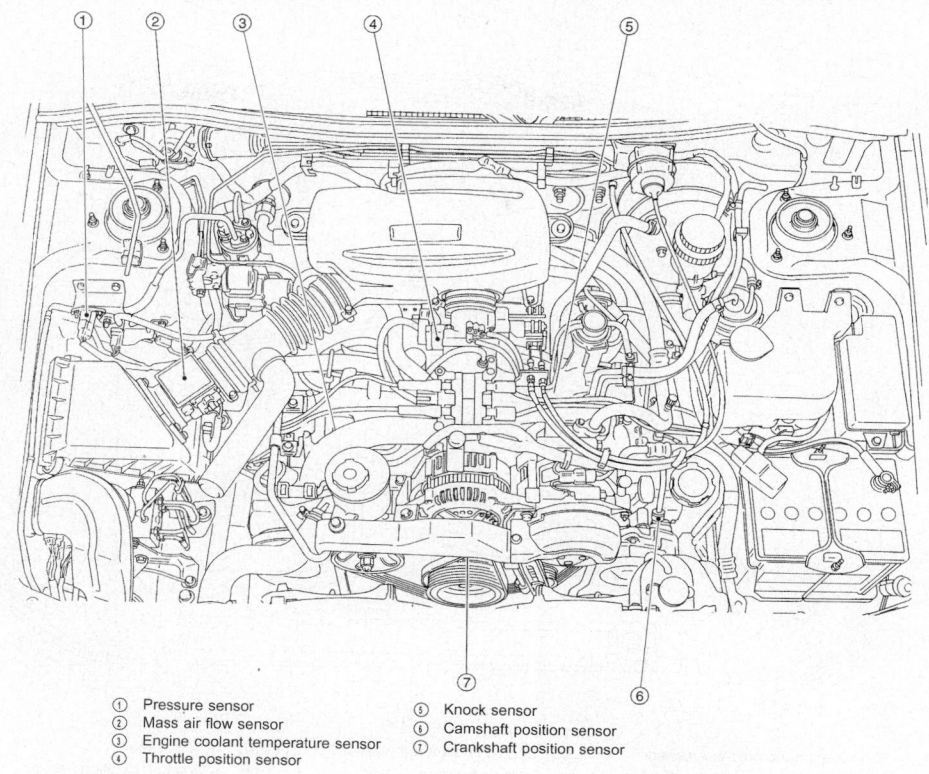

① Pressure sensor
② Mass air flow sensor
③ Engine coolant temperature sensor
④ Throttle position sensor
⑤ Knock sensor
⑥ Camshaft position sensor
⑦ Crankshaft position sensor

Fig. 50 Underhood sensor locations—Impreza 1996–2001 2.2L engine

29157_SUBA_G0055

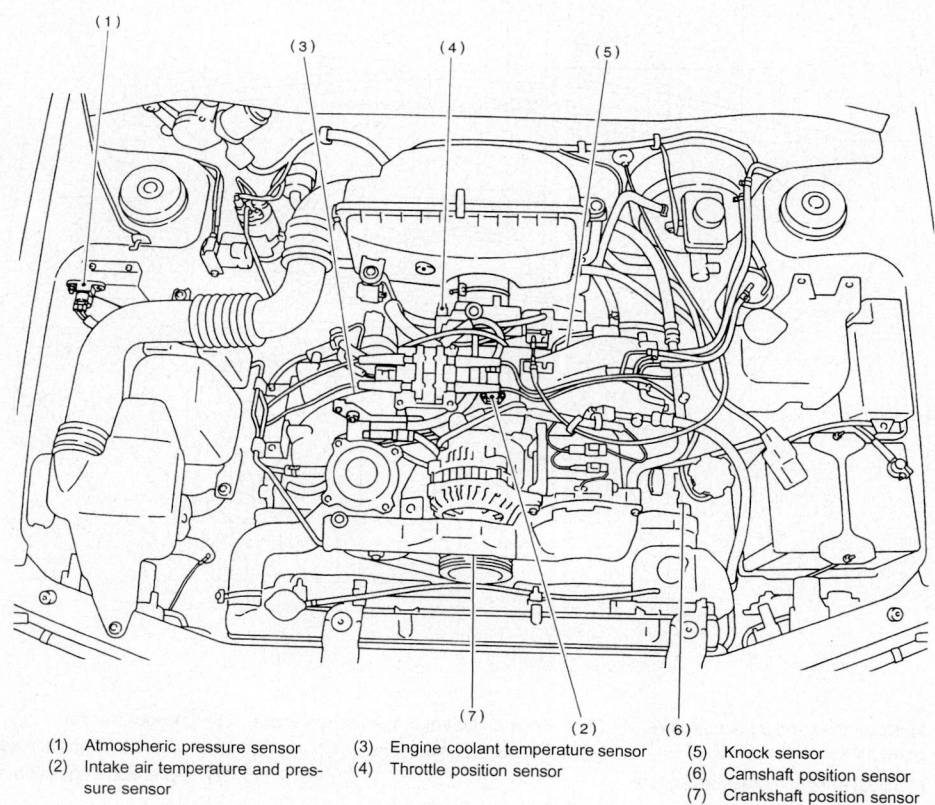

(1) Atmospheric pressure sensor
(2) Intake air temperature and pressure sensor
(3) Engine coolant temperature sensor
(4) Throttle position sensor
(5) Knock sensor
(6) Camshaft position sensor
(7) Crankshaft position sensor

Fig. 51 Underhood sensor locations—Impreza 1998–2001 2.5L SOHC engine

29157_SUBA_G0061

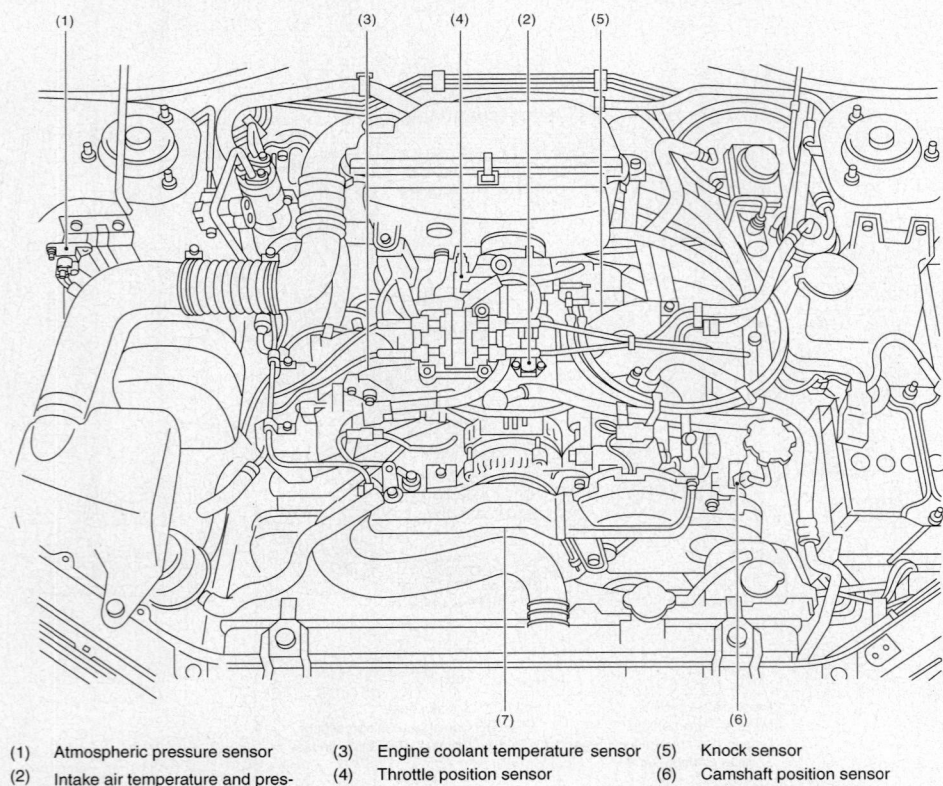

(1)	Atmospheric pressure sensor	(3)	Engine coolant temperature sensor	(5)	Knock sensor
(2)	Intake air temperature and pressure sensor	(4)	Throttle position sensor	(6)	Camshaft position sensor
				(7)	Crankshaft position sensor

29157_SUBA_G0064

Fig. 52 Underhood sensor locations—Impreza 2002 2.5L SOHC engine

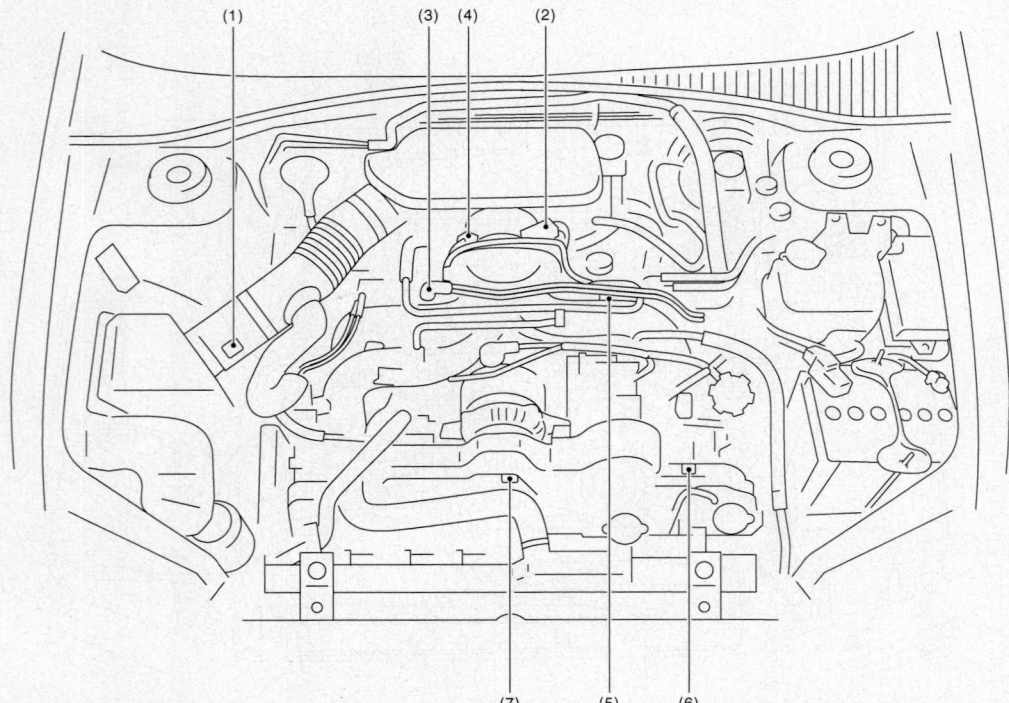

(1)	Mass air flow and intake air temperature sensor	(3)	Engine coolant temperature sensor	(5)	Knock sensor
(2)	Manifold absolute pressure sensor	(4)	Throttle position sensor	(6)	Camshaft position sensor
				(7)	Crankshaft position sensor

29157_SUBA_G0069

Fig. 53 Underhood sensor locations—Impreza 2003–2005 2.5L SOHC engine

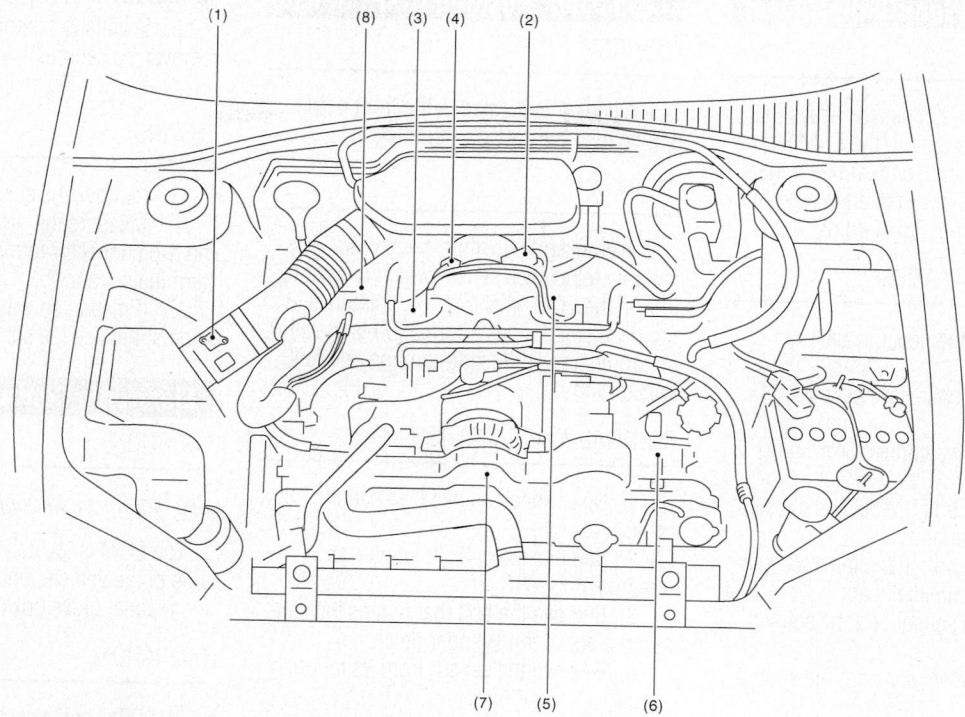

(1)	Mass air flow and intake air temperature sensor	(3)	Engine coolant temperature sensor	(6)	Camshaft position sensor
(2)	Manifold absolute pressure sensor	(4)	Throttle position sensor	(7)	Crankshaft position sensor
		(5)	Knock sensor	(8)	Oil temperature sensor

29157_SUBA_G0066

Fig. 54 Underhood sensor locations—Impreza 2006 2.5L SOHC engine

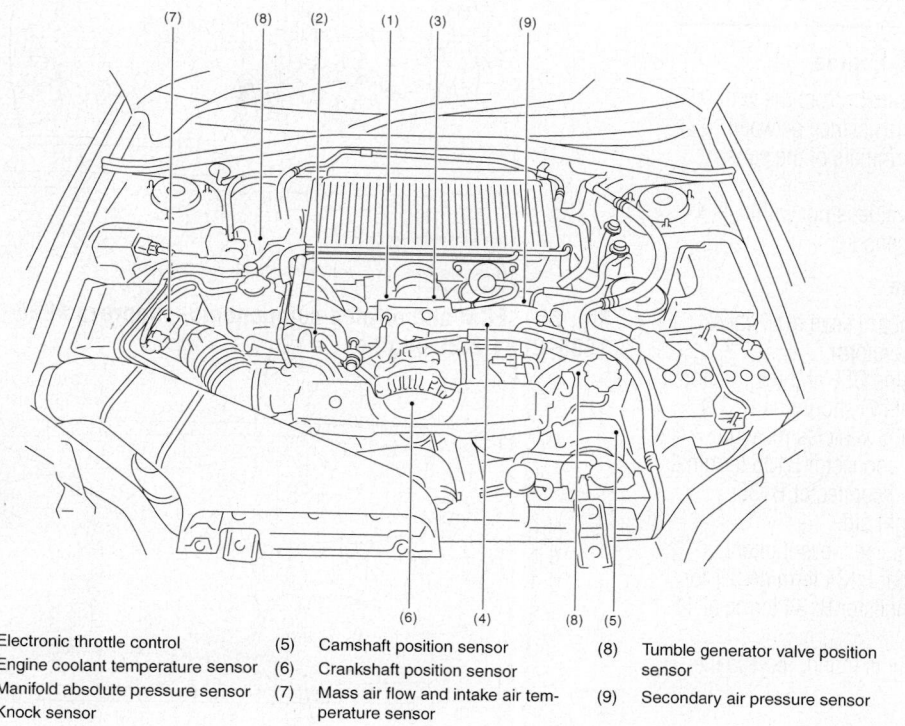

(1)	Electronic throttle control	(5)	Camshaft position sensor	(8)	Tumble generator valve position sensor
(2)	Engine coolant temperature sensor	(6)	Crankshaft position sensor		
(3)	Manifold absolute pressure sensor	(7)	Mass air flow and intake air temperature sensor	(9)	Secondary air pressure sensor
(4)	Knock sensor				

29157_SUBA_G0058

Fig. 55 Underhood sensor locations—Impreza 2004–2006 2.5L DOHC engine

Camshaft Position (CMP) Sensor

OPERATION

The camshaft position sensor relays the relative camshaft position to the ECM for determining the position and stroke of the No. 1 cylinder. This information is required for proper fuel injection functioning.

REMOVAL & INSTALLATION

1. Disconnect the negative battery cable.
2. Remove the collector cover, as required.
3. Disconnect the connector from the sensor.
4. Remove the bolt that retains the sensor to the sensor support.
5. Remove the bolt that retains the sensor support to the camshaft cap.
6. Remove the sensor and the sensor support as a unit.
7. Separate the sensor from the support.

To install:

8. Installation is the reverse of the removal procedure.
9. Tighten the sensor support to 4.7 ft. lbs.
10. Tighten the sensor to 4.7 ft. lbs.

TESTING

Except 2.5L DOHC Engine

1. Remove the sensor from the vehicle.
2. Measure the resistance between the two connector terminals of the sensor, (terminals 1 and 2).
3. If measured value is not within 1–4 kohms, replace the sensor.

2.5L DOHC Engine

1. Using the Subaru scan tool, check the sensor output waveform.
2. With the engine OFF and the ignition switch ON the signal (V) should be 0–0.9.
3. On 2004–2005 vehicles measurement is taken using connector B135 terminal 8 for the left side and connector B135 terminal 9 for the right side.
4. On 2006 vehicles measurement is taken using connector B134 terminal 21 for the left side and connector B134 terminal 11 for the right side.
5. If abnormality is found, replace the sensor.

Crankshaft Position (CKP) Sensor

LOCATION

The sensor is located in the front of the engine near the crankshaft pulley.

OPERATION

The crankshaft position sensor is a variable reluctance sensor which is used to inform the ECM when the No.1 piston is at top dead center. This information is used for controlling and adjusting ignition and fuel injector timing.

REMOVAL & INSTALLATION

1. Disconnect the negative battery cable.
2. Remove the collector cover, as required.
3. Remove the bolt that retains the sensor in place at the cylinder block.
4. Remove the sensor from its mounting.
5. Disconnect the connector from the sensor.

To install:

6. Installation is the reverse of the removal procedure.
7. Tighten the sensor to 4.7 ft. lbs.

TESTING

1. Remove the sensor from the vehicle.
2. Measure the resistance between the two connector terminals of the sensor, (terminals 1 and 2).
3. If measured value is not within 1–4 kohms, replace the sensor.

Electronic Control Module (ECM)

LOCATION

See Figures 56 through 60.

The ECM is located on the passenger's side of the vehicle, underneath the instrument panel glove box area.

OPERATION

The ECM controls the vehicle engine operating system.

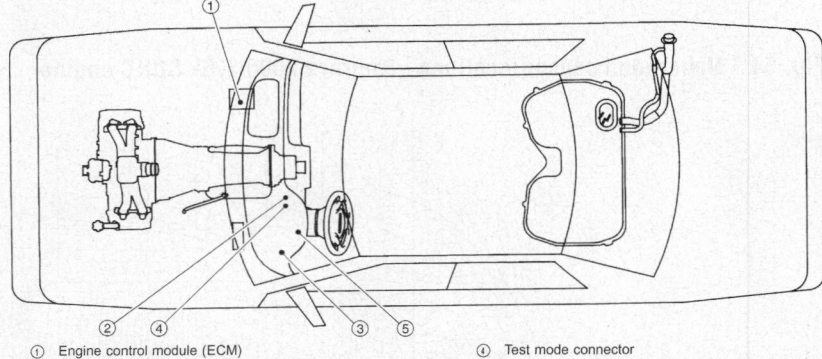

① Engine control module (ECM)
② Data link connector (for Subaru select monitor only)
③ Data link connector (for Subaru select monitor and OBD-II general scan tool)
④ Test mode connector
⑤ CHECK ENGINE malfunction indicator lamp (MIL)

29157_SUBA_G0054

Fig. 56 ECM and related components—Impreza 1996–1997 1.8L engine and 1996–2001 2.2L engine

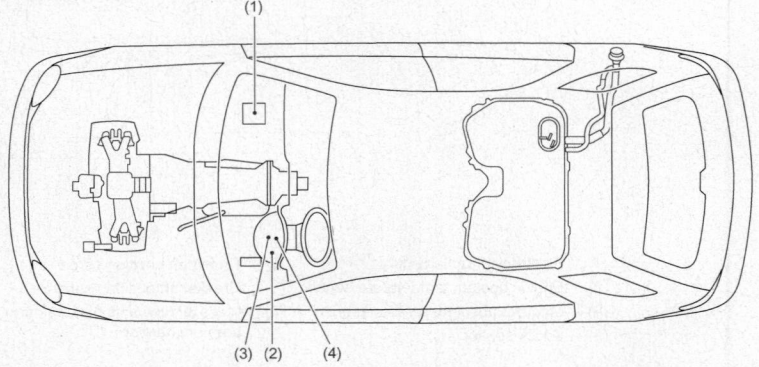

(1) Engine control module (ECM)
(2) CHECK ENGINE malfunction indi-
(3) Data link connector
(4) Test mode connector

29157_SUBA_G0048

Fig. 57 ECM and related components—Impreza 2002–2003 2.0L engine

REMOVAL & INSTALLATION

1. Disconnect the negative battery cable.

2. Remove the lower inner trim on the passenger's side of the vehicle.

3. Detach the floor mat. Remove the protective cover.

4. Remove the ECM bracket retaining nuts. Remove the clip from the bracket.

5. Disconnect the connectors.

6. Remove the ECM from the vehicle.

To install:

7. Installation is the reverse of the removal procedure.

8. Tighten the retaining screws to 3.7 ft. lbs.

➡ **When replacing the ECM, be careful not to use the wrong part number, as damage to the injection system could occur.**

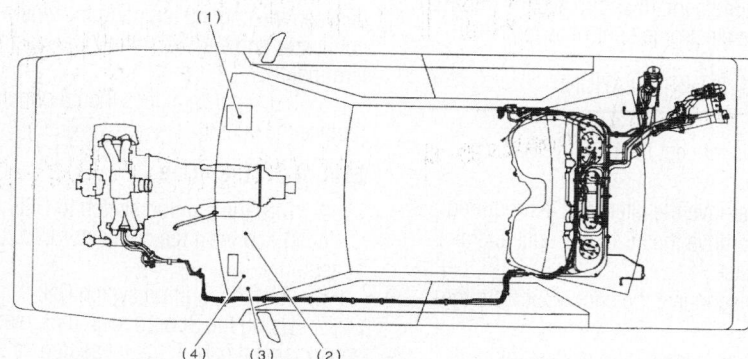

(1) Engine control module (ECM)
(2) Data link connector (for Subaru Select Monitor and OBD-II general scan tool)
(3) Test mode connector
(4) CHECK ENGINE malfunction indicator lamp (MIL)

29157_SUBA_G0060

Fig. 58 ECM and related components—Impreza 1998–2001 2.5L SOHC engine

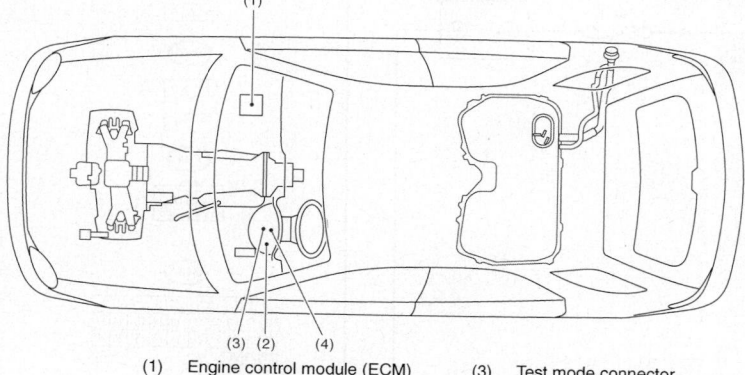

(1) Engine control module (ECM)
(2) CHECK ENGINE malfunction indicator lamp (MIL)
(3) Test mode connector
(4) Data link connector

29157_SUBA_G0063

Fig. 59 ECM and related components—Impreza 2002–2006 2.5L SOHC engine

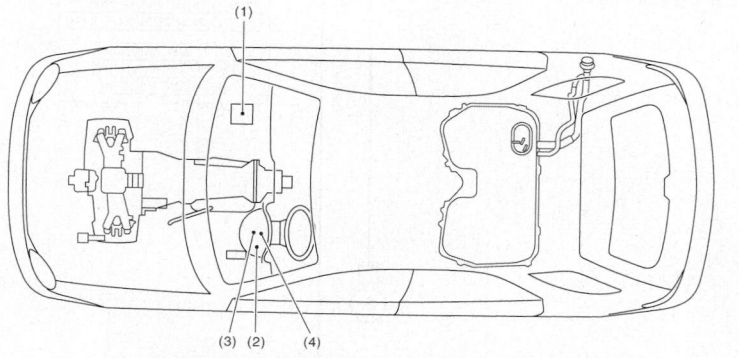

(1) Engine control module (ECM)
(2) Malfunction indicator light
(3) Data link connector
(4) Test mode connector

29157_SUBA_G0057

Fig. 60 ECM and related components—Impreza 2004–2006 2.5L DOHC engine

EGR Valve

LOCATION

The EGR valve is mounted on the intake manifold. The 2.0L, 2.2L and the 2.5L DOHC engines do not use a conventional EGR valve.

OPERATION

The EGR valve controls the flow of exhaust gases. The ECM monitors the flow and regulates the valve accordingly.

REMOVAL & INSTALLATION

1. Disconnect the negative battery cable.

2. Disconnect the connector.

3. On valves equipped with a vacuum hose, disconnect it.

4. Remove the valve from the intake manifold.

5. Discard the gasket.

To install:

6. Installation is the reverse of the removal procedure.

7. Be sure to use a new gasket.

8. Tighten the retaining bolts to 14.0 ft. lbs.

TESTING

1.8L Engine

1. Check and replace vacuum hoses, if they are clogged or brittle.

2. Check the valve for blockage, replace or clean as required.

3. To check the solenoid valve, turn the ignition switch OFF.

4. Connect the Subaru diagnostic tool.

5. Turn the ignition switch to ON.

6. Does the solenoid valve engage? If not, replace it.

2.5L SOHC Engine

See Figure 61.

1. Disconnect the connector from the ECM.

2. Measure the resistance between the ECM connector and chassis ground.

3. On 2004 vehicles, if DTC code P1493 was set, use connector B134 and terminal 18/chassis ground. If DTC code P1495 was set, use connector B134 and terminal 17/chassis ground. If DTC code P1497 was set, use connector B134 and terminal 16/chassis ground. If DTC code P1499 was set, use connector B134 and terminal 15/chassis ground.

4. On 2005 vehicles, if DTC code P1492 was set, use connector B134 and terminal 9/connector E18 and terminal 4. If DTC code P1494 was set, use connector B134 and terminal 10/connector E18 and terminal 6. If DTC code P1496 was set, use connector B134 and terminal 11/connector E18 and terminal 3. If DTC code P1498 was set, use connector B134 and terminal 8/connector E18 and terminal 1.

5. On 2006 vehicles, if DTC code P1492 was set, use connector B134 and terminal 10/chassis ground. If DTC code P1494 was set, use connector B134 and terminal 9/chassis ground. If DTC code P1496 was set, use connector B134 and terminal 8/chassis ground. If DTC code P1498 was set, use connector B134 and terminal 20/chassis ground.

6. If measured value is more than 1 Mohm, check the ECM connector and the EGR solenoid for poor contact.

7. If contact is good, replace the EGR solenoid valve.

Engine Coolant Temperature (ECT) Sensor

LOCATION

The sensor is located by the heater outlet fitting or in a cooling passage on the engine, depending upon the particular vehicle.

OPERATION

This component detects the temperature of the engine coolant and relays the information to the electronic control assembly.

REMOVAL & INSTALLATION

1. Disconnect the negative battery cable.
2. Remove the alternator, as required.
3. Remove the air intake duct and air cleaner case.
4. Disconnect the connector from the sensor.
5. Drain the cooling system, as required.

6. Remove the sensor from its mounting.

To install:

7. Installation is the reverse of the removal procedure.
8. Tighten the sensor to 13.3 ft. lbs.

TESTING

2.2L Engine

1. Turn the ignition switch to OFF.
2. Measure the resistance of the harness between the coolant temperature sensor connector and engine ground (connector E8 terminal 2).
3. If resistance is less than 5 ohms, replace the sensor.

Except 2.2L Engine

4. Turn the ignition switch to OFF.
5. Disconnect the connector from the sensor.
6. Turn the ignition switch ON.
7. Using the Subaru scan tool, read the sensor signal data. If the measured value is less than -40 degrees F, replace the sensor.

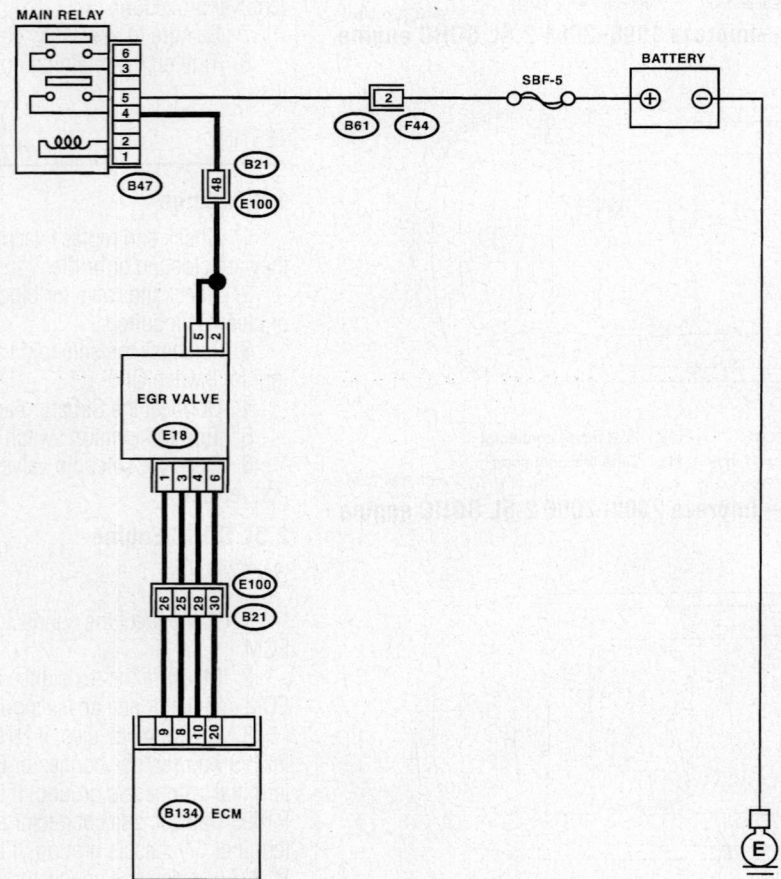

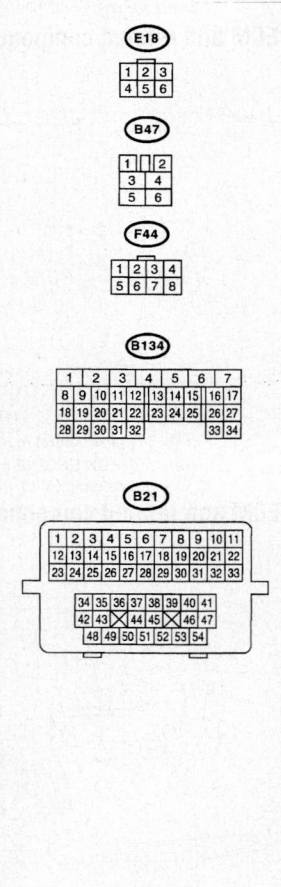

29157_SUBA_G0071

Fig. 61   EGR valve connector location—Impreza 2005–2006 2.5L SOHC engine

Heated Oxygen (HO2S) Sensor

LOCATION

See Figures 62 through 67.

On the 1.8L and 2.2L engine the front (A/F) oxygen sensor is located in the front section of the front catalytic converter. The rear oxygen sensor is located in the rear section of the rear catalytic converter. On the 2.0L engine the front (A/F) oxygen sensor is located in the front section of the exhaust system, just past the crossover pipe. The rear oxygen sensor is located in the front section of the rear catalytic converter. On 1998–2002 vehicles equipped with the 2.5L SOHC engine, the front (A/F) oxygen sensor is located in the front section of the front catalytic converter. The rear oxygen sensor is located in the rear section of the rear catalytic converter. On 2003–2005 vehicles equipped with the 2.5L SOHC engine, the front oxygen sensor is located in the front section of the front catalytic converter. The rear oxygen sensor is located in the rear section of the front catalytic converter. On 2006 vehicles equipped with the 2.5L SOHC

engine, the front oxygen sensor is located at the front section of the front catalytic converter. The rear oxygen sensor is located at the rear section of the front catalytic converter. On 2004–2006 2.5L DOHC engine the front oxygen sensor is located in the front section of the exhaust system, just past the crossover pipe. The rear oxygen sensor is located in the front section of the rear catalytic converter.

OPERATION

The exhaust gas oxygen sensor supplies the electronic control assembly with a signal which indicates either a rich or lean mixture condition, during the engine operation.

REMOVAL & INSTALLATION

1.8L and 2.2L Engines

1. Disconnect the negative battery cable.
2. Remove the necessary components to gain access to the sensor.
3. Disconnect the clip fastening the harness, if equipped.

4. Disconnect the connector.
5. Remove the oxygen sensor.

To install:

6. Installation is the reverse of the removal procedure.

➡ **Apply anti-seize compound to the threaded portion of the sensor, prior to installation. Never apply anti-seize compound to the protector of the sensor.**

7. Tighten the sensor to 15.2 ft. lbs.

2.0L and 2.5L SOHC Engines

1. Disconnect the negative battery cable.
2. Disconnect the clip fastening the harness. Disconnect the connector.
3. Raise and support the vehicle safely.
4. Remove the engine under cover, as required.
5. Remove the oxygen sensor.

To install:

6. Installation is the reverse of the removal procedure.

➡ **Apply anti-seize compound to the threaded portion of the sensor, prior to installation. Never apply anti-seize compound to the protector of the sensor.**

7. Tighten the sensor to 15.2 ft. lbs.

2.5L DOHC Engine

FRONT SENSOR

1. Disconnect the negative battery cable.
2. Disconnect the connector. Disconnect the engine harness clip.
3. Raise and support the vehicle safely. Remove the tire and wheel.
4. Remove the service hole cover.
5. Remove the oxygen sensor.

To install:

6. Installation is the reverse of the removal procedure.

➡ **Apply anti-seize compound to the threaded portion of the sensor, prior to installation. Never apply anti-seize compound to the protector of the sensor.**

7. Tighten the sensor to 22.1 ft. lbs.

REAR SENSOR

1. Disconnect the negative battery cable.
2. Raise and support the vehicle safely.
3. Disconnect the connector. Remove the clip by pulling it out from the upper side of the crossmember.

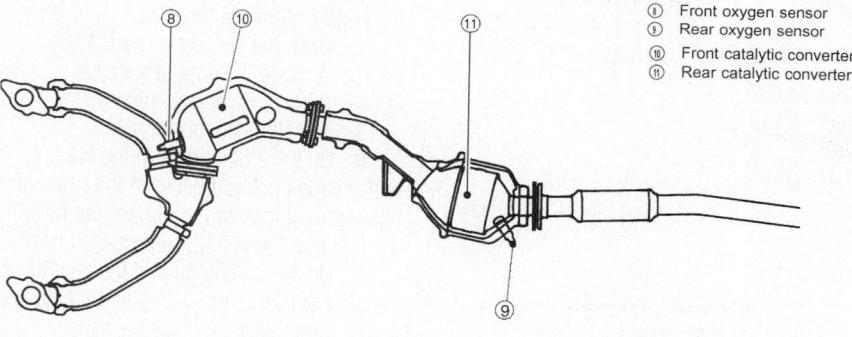

⑧	Front oxygen sensor
⑨	Rear oxygen sensor
⑩	Front catalytic converter
⑪	Rear catalytic converter

29157_SUBA_G0056

Fig. 62 Oxygen sensor location—Impreza 1996–1997 1.8L engine and 1996–2001 2.2L engine

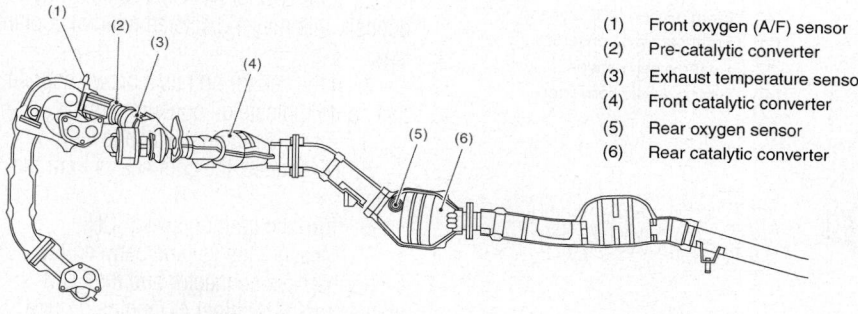

(1)	Front oxygen (A/F) sensor
(2)	Pre-catalytic converter
(3)	Exhaust temperature sensor
(4)	Front catalytic converter
(5)	Rear oxygen sensor
(6)	Rear catalytic converter

29157_SUBA_G0049

Fig. 63 Oxygen sensor location—Impreza 2002–2003 2.0L engine

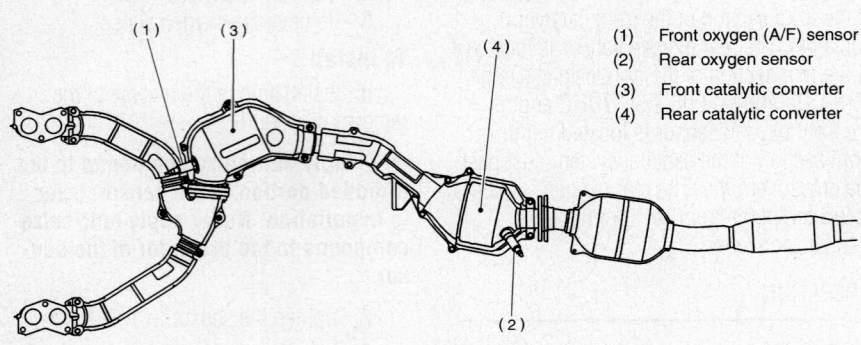

(1) Front oxygen (A/F) sensor
(2) Rear oxygen sensor
(3) Front catalytic converter
(4) Rear catalytic converter

29157_SUBA_G0065

Fig. 64   Oxygen sensor location—Impreza 1998–2002 2.5L SOHC engine

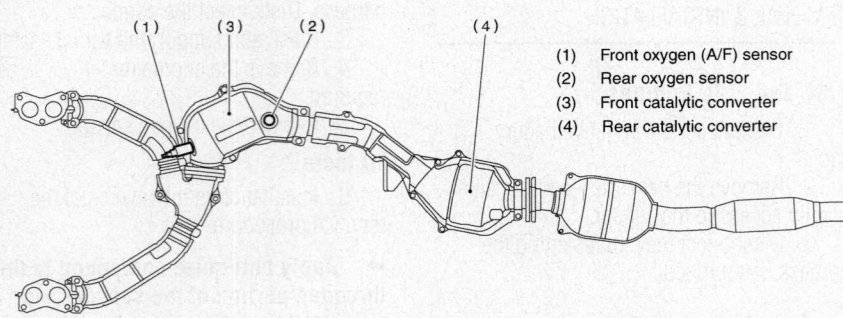

(1) Front oxygen (A/F) sensor
(2) Rear oxygen sensor
(3) Front catalytic converter
(4) Rear catalytic converter

29157_SUBA_G0070

Fig. 65   Oxygen sensor location—Impreza 2003–2005 2.5L SOHC engine

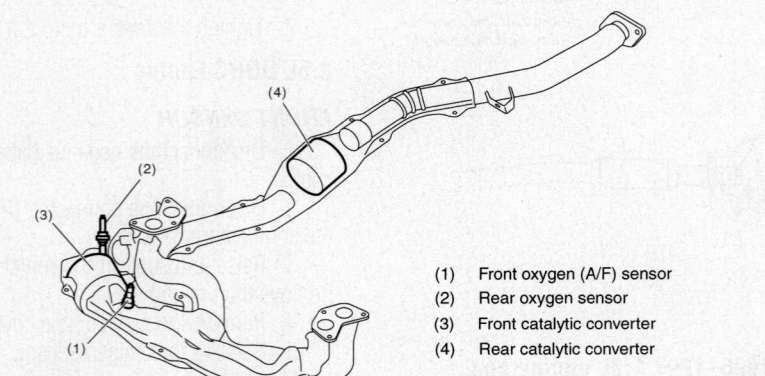

(1) Front oxygen (A/F) sensor
(2) Rear oxygen sensor
(3) Front catalytic converter
(4) Rear catalytic converter

29157_SUBA_G0067

Fig. 66   Oxygen sensor location—Impreza 2006 2.5L SOHC engine

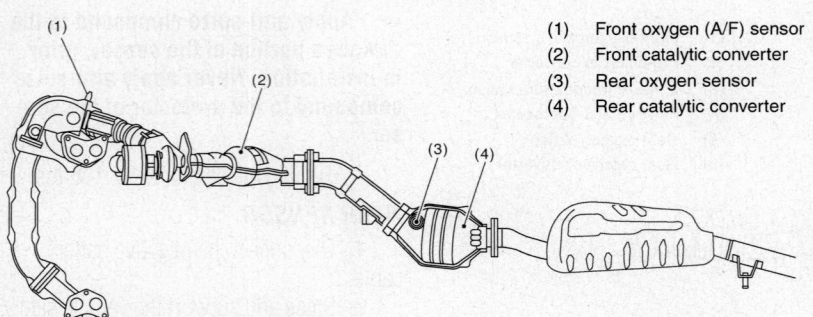

(1) Front oxygen (A/F) sensor
(2) Front catalytic converter
(3) Rear oxygen sensor
(4) Rear catalytic converter

29157_SUBA_G0059

Fig. 67   Oxygen sensor location—Impreza 2004–2006 2.5L DOHC engine

4. Remove the oxygen sensor.

To install:

5. Installation is the reverse of the removal procedure.

➡ **Apply anti-seize compound to the threaded portion of the sensor, prior to installation. Never apply anti-seize compound to the protector of the sensor.**

6. Tighten the sensor to 22.1 ft. lbs.

TESTING

1.8L and 2.2L Engines

FRONT SENSOR

Perform a visual inspection of the sensor as follows:

1. If the sensor tip has a black/sooty deposit, this may indicate a rich fuel mixture.
2. If the sensor tip has a white, gritty deposit, this may indicate an internal coolant leak.
3. If the sensor tip has a brown deposit, this could indicate oil consumption.
4. Check the exhaust system for defects.
5. Turn the ignition switch OFF.
6. Connect the Subaru scan tool. Start the engine. Turn the tool ON.
7. Allow the engine to reach 160 degrees F. Keep the engine speed at 2000–3000 rpm's for one minute.
8. Read the data from the tool.
9. On the 1.8L engine, if the reading is more than 0.1 volt between the value of maximum output and minimum output with the function mode F12, replace the sensor.
10. On the 2.2L engine, if the reading is equal to or more than 0.85 and equal to less than 1.15 in idling, replace the sensor.

REAR SENSOR

Perform a visual inspection of the sensor as follows:

1. If the sensor tip has a black/sooty deposit, this may indicate a rich fuel mixture.
2. If the sensor tip has a white, gritty deposit, this may indicate an internal coolant leak.
3. If the sensor tip has a brown deposit, this could indicate oil consumption.
4. Turn the ignition switch to OFF.
5. Disconnect the connectors from the sensor.
6. Turn the ignition switch ON.
7. Measure the voltage between the sensor harness connector and the ECM (connector T6 terminal 4+/engine ground -). If measured value is more than 0.2 volt, replace the sensor.

2.0L Engine

FRONT SENSOR

Perform a visual inspection of the sensor as follows:

1. If the sensor tip has a black/sooty deposit, this may indicate a rich fuel mixture.

2. If the sensor tip has a white, gritty deposit, this may indicate an internal coolant leak.

3. If the sensor tip has a brown deposit, this could indicate oil consumption.

4. Turn the ignition switch to OFF.

5. Measure the resistance between the sensor and the connector terminals (terminals 2 and 1).

6. If measured value is more than 5 ohm, replace the sensor.

REAR SENSOR

Perform a visual inspection of the sensor as follows:

1. If the sensor tip has a black/sooty deposit, this may indicate a rich fuel mixture.

2. If the sensor tip has a white, gritty deposit, this may indicate an internal coolant leak.

3. If the sensor tip has a brown deposit, this could indicate oil consumption.

4. Turn the ignition switch to OFF.

5. If, measured resistance between the sensor connector terminals is more than 30 ohms, replace the sensor.

2.5L SOHC Engine

FRONT SENSOR–1998–2001

Perform a visual inspection of the sensor as follows:

1. If the sensor tip has a black/sooty deposit, this may indicate a rich fuel mixture.

2. If the sensor tip has a white, gritty deposit, this may indicate an internal coolant leak.

3. If the sensor tip has a brown deposit, this could indicate oil consumption.

4. Check the exhaust system for defects.

5. Measure the resistance between the ECM and chassis ground (connector B136 and terminal 6, connector B136 and terminal 7, connector B136 and terminal 19, connector B136 and terminal 20).

6. If measured value is more than 1 ohm, replace the sensor.

FRONT SENSOR–2002–2004

Perform a visual inspection of the sensor as follows:

1. If the sensor tip has a black/sooty deposit, this may indicate a rich fuel mixture.

2. If the sensor tip has a white, gritty deposit, this may indicate an internal coolant leak.

3. If the sensor tip has a brown deposit, this could indicate oil consumption.

4. Turn the ignition switch to OFF.

5. On 2002 vehicles, measure the resistance between the sensor and the connector terminals (terminals 2 and 5).

6. On 2003–2004 vehicles, measure the resistance between the sensor and the connector terminals (terminals 3 and 4).

7. If measured value is more than 10 ohm, replace the sensor.

FRONT SENSOR–2005–2006

See Figures 68 through 71.

Perform a visual inspection of the sensor as follows:

1. If the sensor tip has a black/sooty deposit, this may indicate a rich fuel mixture.

2. If the sensor tip has a white, gritty deposit, this may indicate an internal coolant leak.

3. If the sensor tip has a brown deposit, this could indicate oil consumption.

4. Measure the resistance between the sensor and the connector terminals (terminals 3 and 4).

5. If measured value is more than 5 ohm, replace the sensor.

REAR SENSOR

See Figures 72 through 75.

Perform a visual inspection of the sensor as follows:

1. If the sensor tip has a black/sooty deposit, this may indicate a rich fuel mixture.

2. If the sensor tip has a white, gritty deposit, this may indicate an internal coolant leak.

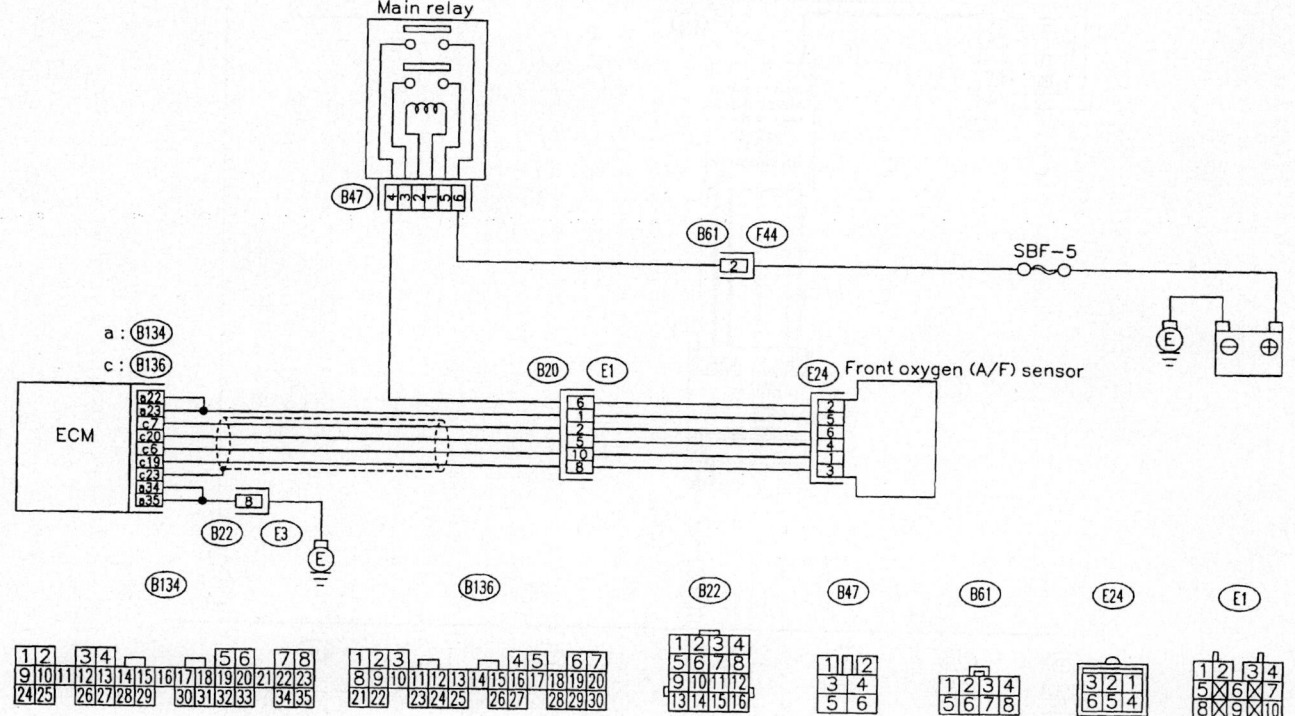

Fig. 68   Front oxygen sensor connector location—Impreza 1998–2001 2.5L SOHC engine

29157_SUBA_G0072

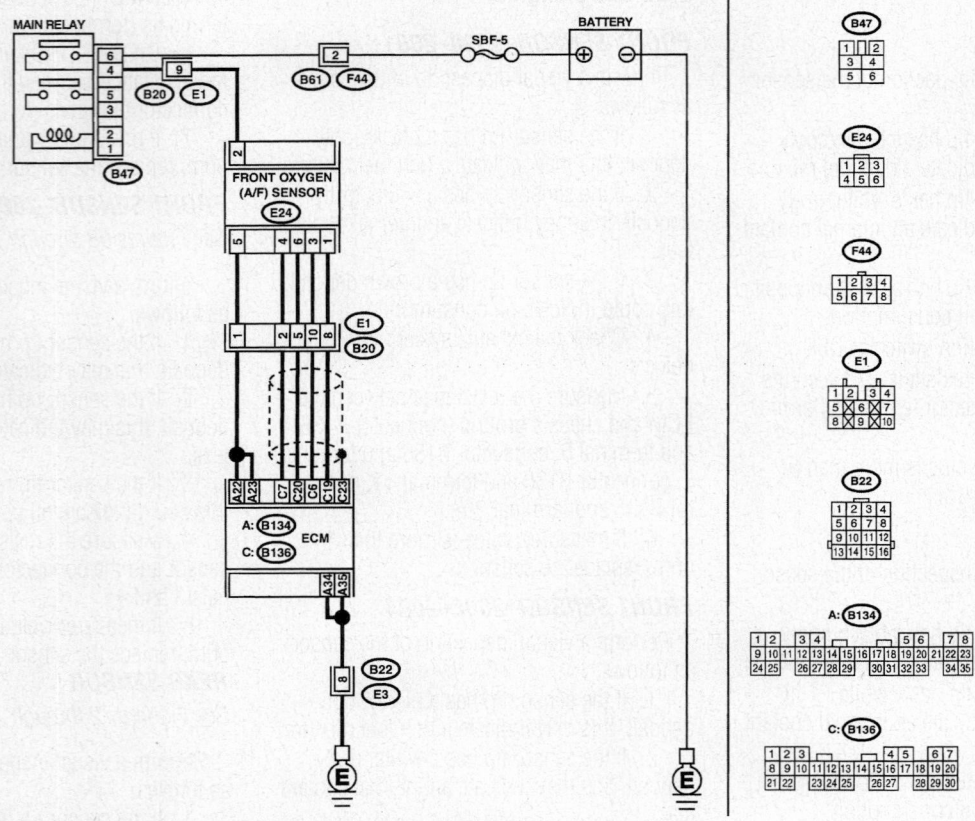

Fig. 69 Front oxygen sensor connector location—Impreza 2002 2.5L SOHC engine

29157_SUBA_G0073

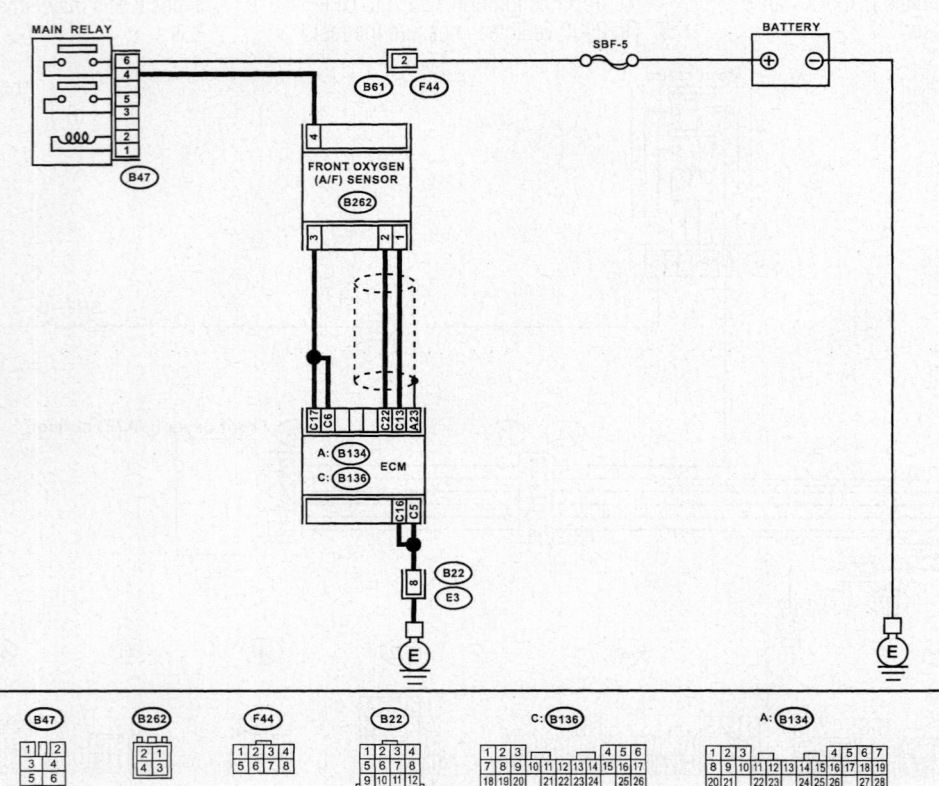

Fig. 70 Front oxygen sensor connector location—Impreza 2003–2004 2.5L SOHC engine

29157_SUBA_G0074

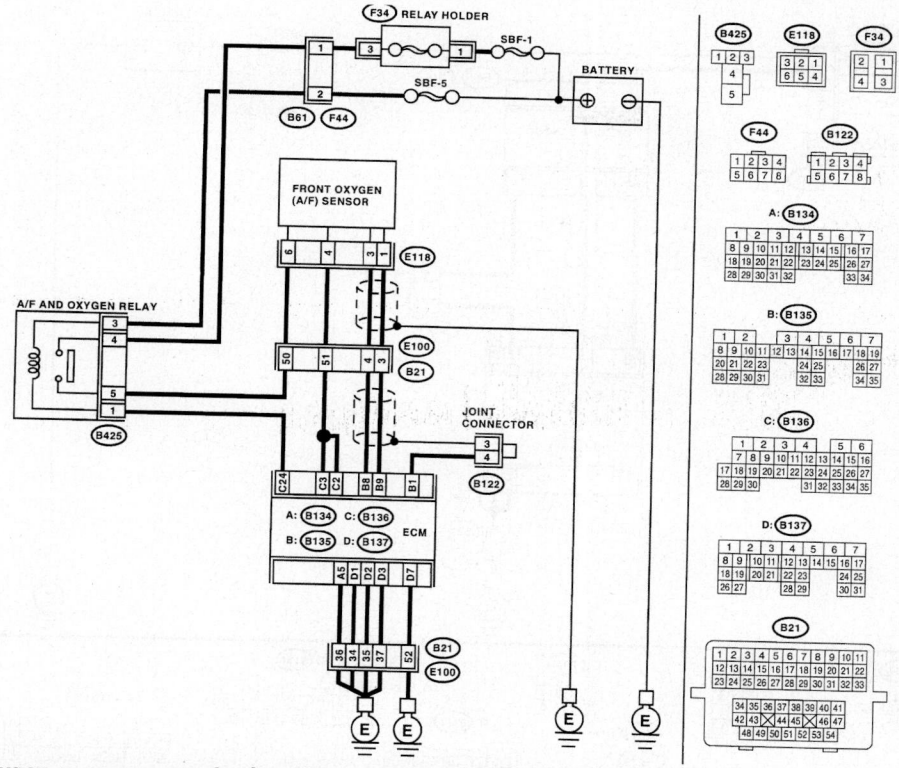

Fig. 71 **Front oxygen sensor connector location—Impreza 2005–2006 2.5L SOHC engine**

29157_SUBA_G0075

3. If the sensor tip has a brown deposit, this could indicate oil consumption.

4. Turn the ignition switch to OFF.

5. Disconnect the connectors from the sensor.

6. On 1998–2001 vehicles, turn the ignition switch ON.

7. On 1998–2001 vehicles, measure the voltage between the sensor harness connector and engine ground (connector E25 terminal 4+/engine ground -). If measured value is more than 0.2 volt, replace the sensor.

8. On 2002–2006 vehicles, measure the voltage between the sensor harness connector terminals (terminals 1 and 2). If measured value is more than 30 ohms, replace the sensor.

2.5L DOHC Engine

FRONT SENSOR

See Figure 76.

Perform a visual inspection of the sensor as follows:

1. If the sensor tip has a black/sooty deposit, this may indicate a rich fuel mixture.

2. If the sensor tip has a white, gritty deposit, this may indicate an internal coolant leak.

3. If the sensor tip has a brown deposit, this could indicate oil consumption.

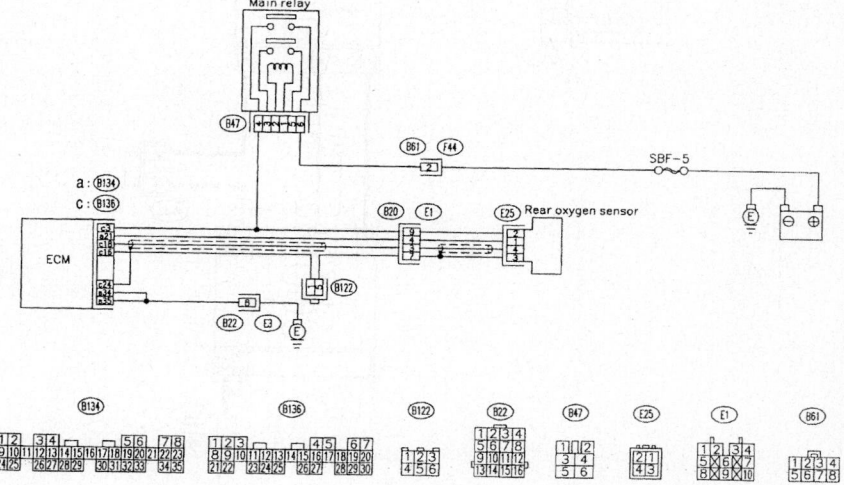

Fig. 72 **Rear oxygen sensor connector location—Impreza 1998–2001 2.5L SOHC engine**

29157_SUBA_G0076

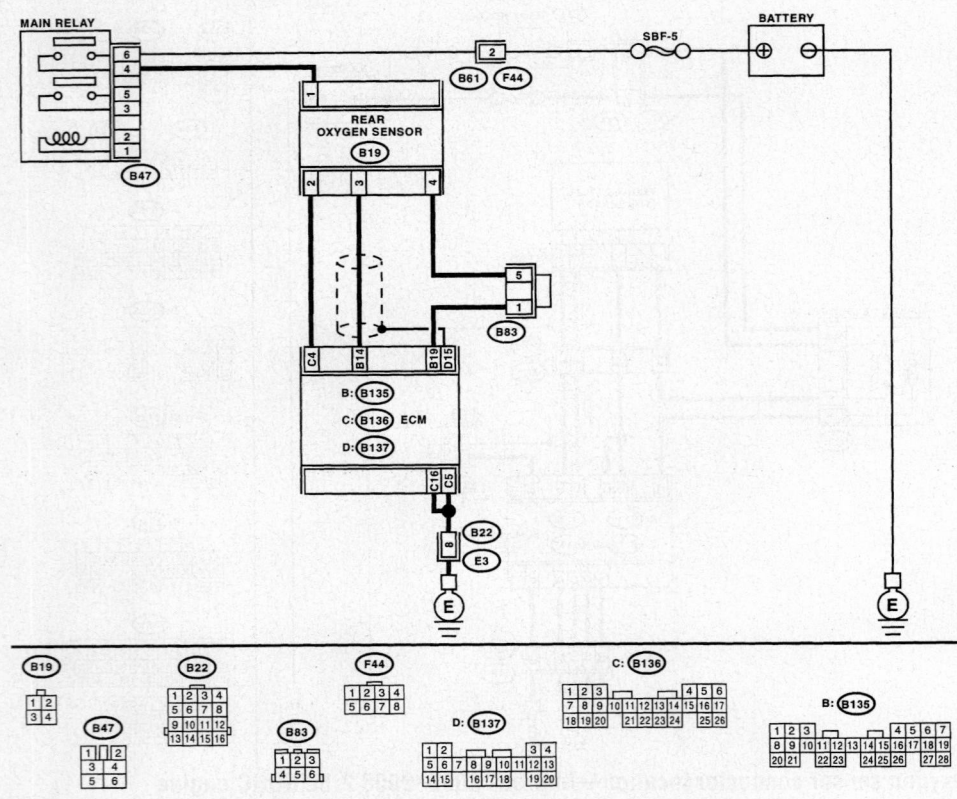

Fig. 73 Rear oxygen sensor connector location—Impreza 2002–2003 2.5L SOHC engine

29157_SUBA_G0077

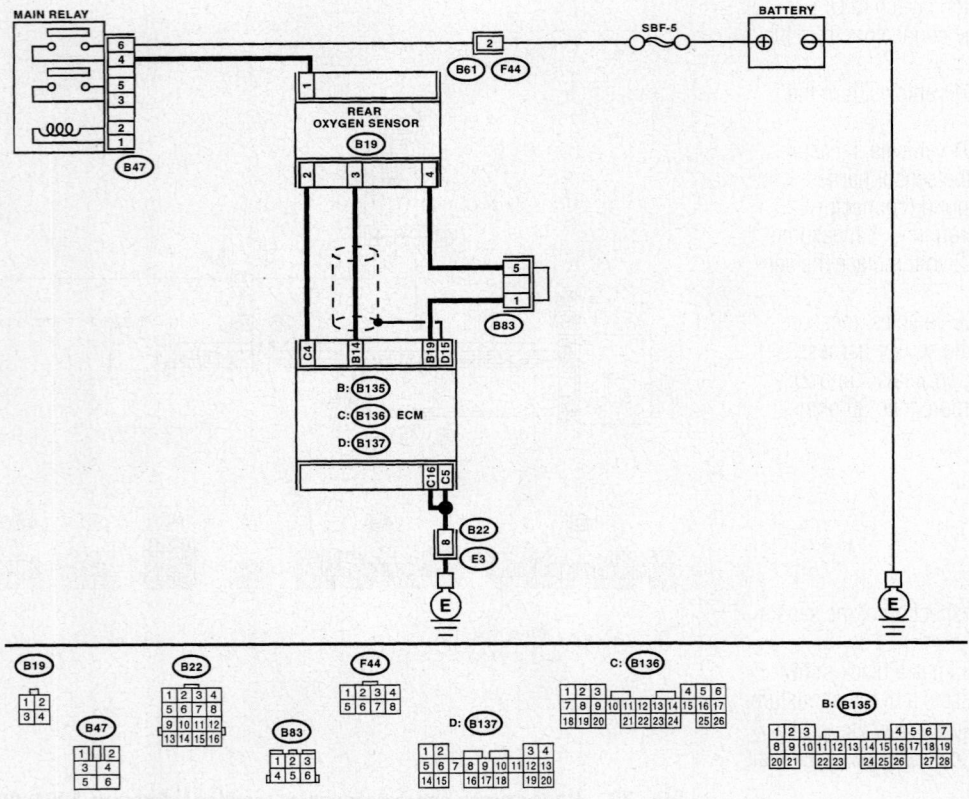

Fig. 74 Rear oxygen sensor connector location—Impreza 2004 2.5L SOHC engine

29157_SUBA_G0078

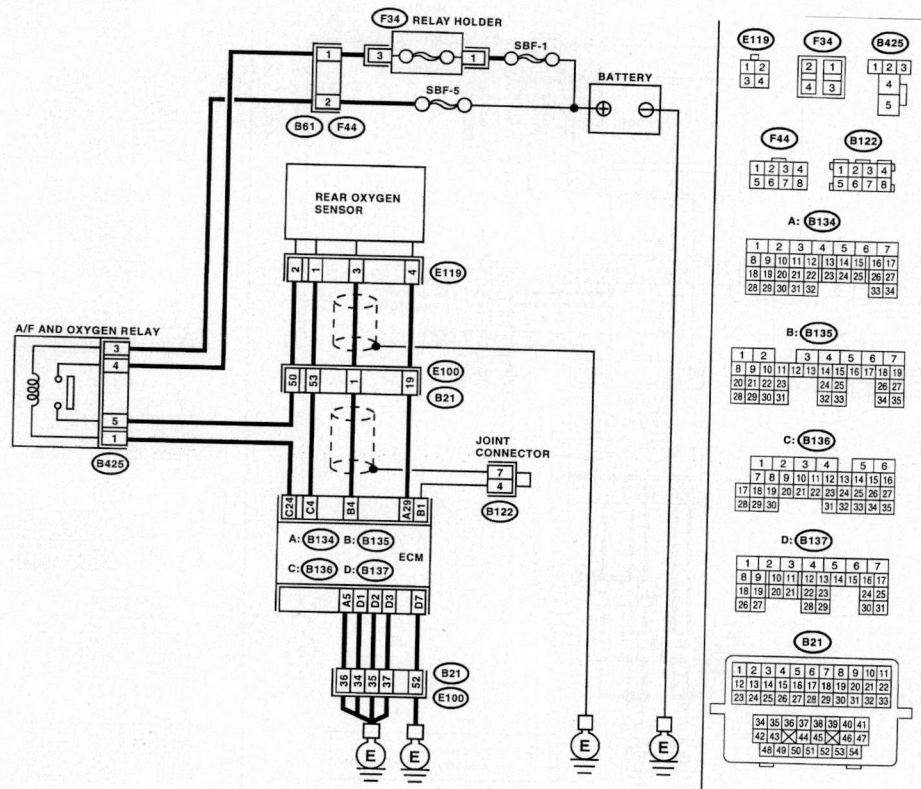

Fig. 75 Rear oxygen sensor connector location—Impreza 2005–2006 2.5L SOHC engine

29157_SUBA_G0079

4. Measure the resistance of the harness between the sensor connector terminals (terminals 1 and 2). 5. If measured value is more than 5 ohm, replace the sensor.

REAR SENSOR

See Figure 77.

Perform a visual inspection of the sensor as follows:

1. If the sensor tip has a black/sooty deposit, this may indicate a rich fuel mixture.

2. If the sensor tip has a white, gritty deposit, this may indicate an internal coolant leak.

3. If the sensor tip has a brown deposit, this could indicate oil consumption.

4. Turn the ignition switch OFF.

5. Measure the resistance of the harness between the sensor connector terminals (terminals 1 and 2). 6. If measured value is more than 30ohm, replace the sensor.

Intake Air Temperature (IAT) Sensor

LOCATION

The sensor is mounted in the intake air hose of the air cleaner assembly. On some

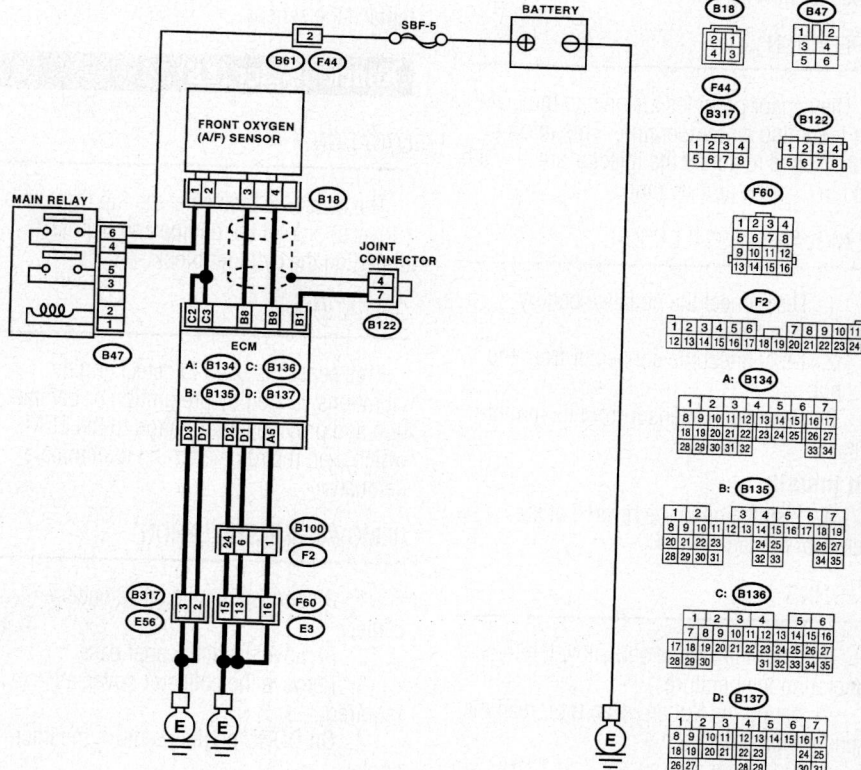

Fig. 76 Front oxygen sensor connector location—Impreza 2.5L DOHC engine

29157_SUBA_G0080

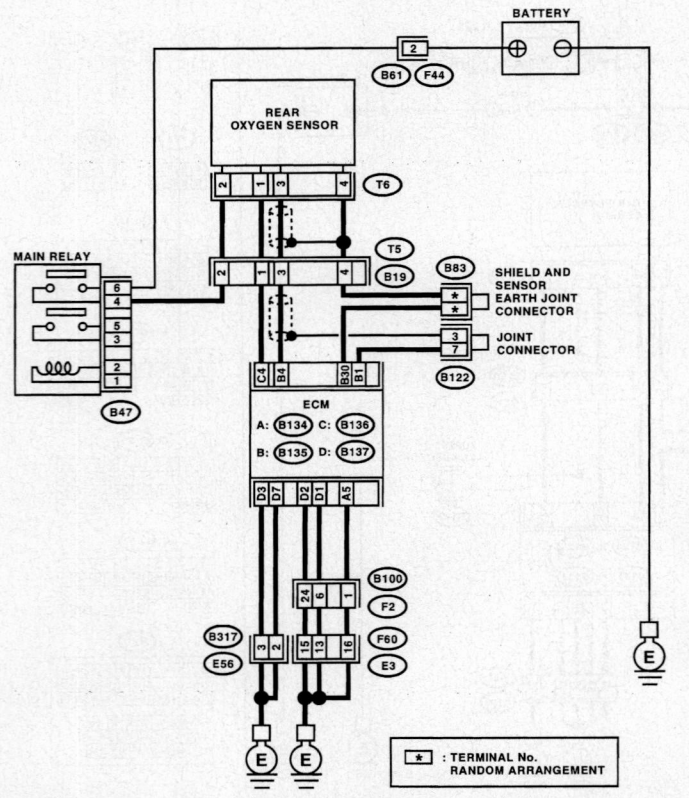
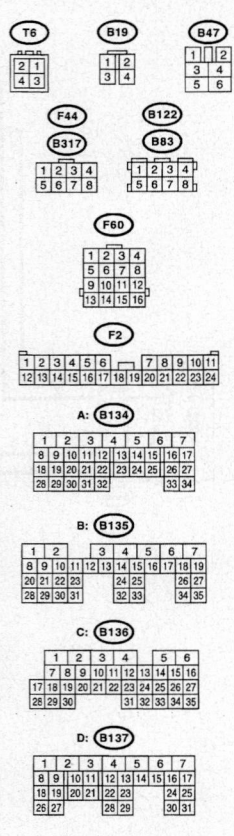

Fig. 77 Rear oxygen sensor connector location—Impreza 2.5L DOHC engine

29157_SUBA_G0081

vehicles this sensor is combined with the mass air flow sensor.

OPERATION

The sensor provides a signal to the ECM for incoming air temperature. This information is used to adjust the injector pulse width and in turn the air/fuel ratio.

REMOVAL & INSTALLATION

1. Disconnect the negative battery cable.
2. Disconnect the connector from the sensor.
3. Remove the sensor from its mounting.

To install:

4. Installation is the reverse of the removal procedure.

TESTING

1. Start the engine and allow it to reach operating temperature.
2. Using the Subaru scan tool, read the sensor signal data.
3. On all engines except 2.5L DOHC engine, if the measured value is less than -40 degrees F, replace the sensor.

4. On the 2.5L DOHC engine, if the measured value is less than -33 degrees F, replace the sensor.

Knock Sensor (KS)

LOCATION

The sensor is located at the top right (driver's) side of the engine and is positioned on the cylinder block.

OPERATION

This sensor is used to detect engine vibrations caused by preignition or detonation and provides information to the ECM, which then retards the timing to eliminate detonation.

REMOVAL & INSTALLATION

1. Disconnect the negative battery cable.
2. Remove the air cleaner case.
3. Remove the collector cover, as required.
4. On DOHC engine, remove the intercooler.
5. Disconnect the sensor connector.

6. Remove the sensor from its mounting.

To install:

7. Installation is the reverse of the removal procedure.
8. Tighten the sensor to 17.4 ft. lbs.

➡ **The extraction area of the knock sensor wire must be positioned at a 60 degree angle relative to the engine rear.**

TESTING

1. Remove the sensor from the vehicle.
2. On 1.8L and 2.2L engines, measure the resistance between the connector terminal of the sensor, (terminal 2) and ground. If measured value is less than 400 kohms, replace the sensor.
3. On 2.0L engine, measure the resistance between the connector terminal of the sensor, (terminal 2) and ground. If measured value is more than 700 kohms, replace the sensor.
4. On 1998–2001 vehicles, equipped with the 2.5L SOHC engine, measure the resistance between the connector terminal of the sensor, (terminal 2) and ground. If measured value is less than 400 kohms, replace the sensor.

5. On 2002–2006 vehicles, equipped with the 2.5L SOHC engine, measure the resistance between the connector terminal of the sensor, (terminal 1) and ground. If measured value is less than 700 kohms, replace the sensor.

6. On 2004–2006 vehicles, equipped with the 2.5L DOHC engine, measure the resistance between the connector terminal of the sensor, (terminal 2) and ground. If measured value is less than 700 kohms, replace the sensor.

Manifold Absolute Pressure (MAP) Sensor

LOCATION

The sensor is located on the throttle body unit.

OPERATION

This sensor monitors and signals the ECM of changes in intake manifold pressure which result from engine load, speed and atmospheric pressure changes.

REMOVAL & INSTALLATION

1. Disconnect the negative battery cable.
2. Disconnect the connector from the sensor.
3. Remove the sensor from its mounting.

To install:

4. Installation is the reverse of the removal procedure.
5. Be sure to use new O-rings.
6. Torque the retaining screw to 1.2 ft. lbs.

TESTING

Except 2.5L DOHC Engine

1. Turn the ignition switch ON.
2. Using the Subaru scan tool, operate the LED operation mode.
3. If the idle switch signal comes on and there are no other displayed codes check the condition of the sensor.
4. If the sensor is securely in place on its mounting and the throttle body is secure on its mounting, replace the sensor.

2.5L DOHC Engine

1. Measure the resistance of the harness between the sensor and engine ground (connector E21 and terminal 1).
2. If the resistance is more than 1 Mohm, replace the sensor.

Mass Air Flow (MAF) Sensor

LOCATION

The sensor is mounted in the intake air hose of the air cleaner assembly. On some vehicles this sensor is combined with the intake air temperature sensor.

OPERATION

The sensor provides a signal to the ECM for incoming air temperature. This information is used to adjust the injector pulse width and in turn the air/fuel ratio.

REMOVAL & INSTALLATION

1. Disconnect the negative battery cable.
2. Disconnect the connector from the sensor.
3. Remove the sensor from its mounting.

To install:

4. Installation is the reverse of the removal procedure.
5. Torque the retaining screw to 0.8 ft. lbs.

TESTING

2.5L SOHC Engine

1. Turn the ignition switch OFF.
2. Disconnect the connector from the ECM.
3. On 2002–2005 vehicles, measure the resistance of the harness between the ECM and the sensor connector (connector B135 terminal 19/connector E21 terminal 1).
4. On 2006 vehicles, measure the resistance of the harness between the ECM and the sensor connector (connector B3 terminal 2/connector B135 terminal 34).
5. If the resistance is less than 1 ohm, replace the sensor.

Except 2.5L SOHC Engine

6. Start the engine and allow it to reach operating temperature.
7. Using the Subaru scan tool, read the sensor signal data.
8. On all engines except 2.5L DOHC engine, if the measured value is less than -40 degrees F, replace the sensor.
9. On the 2.5L DOHC engine, if the measured value is less than -33 degrees F, replace the sensor.

Throttle Position Sensor (TPS)

LOCATION

On the 1996–1997 vehicles, equipped with the 1.8L engine, and 1996–2001 vehicles equipped with the 2.2L engine the sensor is located near the center of the engine by the air cleaner assembly. On the 1996–2001 vehicles, equipped with the 2.2L engine, the sensor is located near the center of the engine by the air cleaner assembly. On the 1998–2004 vehicles, equipped with the 2.5L SOHC engine, the sensor is located near the radiator hose and the engine insulator mount. On 2005–2006 vehicles, equipped with the 2.5L SOHC engine and 2004–2006 vehicles, equipped with the 2.5L DOHC engine, the conventional TPS sensor is not used. The sensor used is referred to as an electric throttle.

OPERATION

The throttle position sensor provides a signal to the ECM that is related to the relative throttle plate position. As the throttle plate moves in relation to driving conditions, a signal is sent to the control unit which adjusts the injector pulse width and air/fuel ratio. As the throttle plate is opened further, more air is taken into the combustion chambers, and as a result the relative fuel demand of the engine changes. The throttle position sensor relays this information to the ECM which in alters the fuel amount.

REMOVAL & INSTALLATION

1.8L and 2.2L Engines

1. Disconnect the negative battery cable.
2. Disconnect the sensor connector.
3. Remove the sensor retaining screws. Remove the sensor from its mounting.

To install:

4. Installation is the reverse of the removal procedure.
5. Tighten the sensor to 1.7 ft. lbs.

2.0L Engine

1. Disconnect the negative battery cable.
2. Remove the intercooler.
3. Disconnect the sensor connector.
4. Remove the sensor retaining screws. Remove the sensor from its mounting.

To install:

5. Installation is the reverse of the removal procedure.
6. Tighten the sensor to 1.2 ft. lbs.

2.5L SOHC Engine

1. Disconnect the negative battery cable.
2. Disconnect the sensor connector.
3. Remove the sensor retaining screws. Remove the sensor from its mounting.

To install:

4. Installation is the reverse of the removal procedure.
5. Tighten the sensor to 1.7 ft. lbs.
6. To adjust the sensor, turn the ignition switch OFF. Loosen the retaining screws.
7. Using a voltage meter, take out the ECM. Turn the ignition switch ON.
8. Adjust the TPS to the proper position to allow the voltage signal to the ECM to be within specification.
9. Specification is 0.45–0.55 volt. Specification is measured at connector B135 and terminal 13.
10. Tighten the retaining screws.

TESTING

2.0L Engine

1. Measure the resistance of the harness between the TPS connector and engine ground, (connector E13 terminal 3 and ground).
2. If the measured value is more than 10 ohm, replace the sensor.

2.2L Engine

1. Start the engine.
2. Using the Subaru scan tool, read the intake manifold absolute pressure signal.

3. If the value is less than 53.3 KPa, replace the sensor.

2.5L SOHC Engine

1998–2004

1. Visually inspect the throttle linkage and throttle for binding and sticking.
2. Measure the resistance of the harness between the TPS connector and engine ground, (connector E13 terminal 3 and ground).
3. If the measured value is more than 1 Mohm, replace the sensor.

2005–2006

4. Measure the voltage between the electronic throttle control connector and ground (connector E57 terminal 6+ and ground).
5. If the measured value is less than 10 volt, replace the sensor.

2.5L DOHC Engine

2004–2005 EXCEPT STI

1. Turn the ignition switch ON.
2. Measure the voltage between the sensor connector and ground (connector E13 and terminal 3).
3. If the voltage is less than 4.7 volts, replace the sensor.

2006 AND STI

4. Measure the voltage between the electronic throttle control connector and ground (connector E57 terminal 6+ and ground).
5. If the measured value is more than 10 volt, replace the sensor.

LEGACY AND OUTBACK

See Figures 78 through 88.

Camshaft Position (CMP) Sensor

OPERATION

The camshaft position sensor relays the relative camshaft position to the ECM for determining the position and stroke of the No. 1 cylinder. This information is required for proper fuel injection functioning.

REMOVAL & INSTALLATION

Except 3.0L Engine

1. Disconnect the negative battery cable.
2. Remove the collector cover, as required.
3. Disconnect the connector from the sensor.
4. Remove the bolt that retains the sensor to the sensor support.

5. Remove the bolt that retains the sensor support to the camshaft cap.
6. Remove the sensor and the sensor support as a unit.
7. Separate the sensor from the support.

To install:

8. Installation is the reverse of the removal procedure.
9. Tighten the sensor support to 4.7 ft. lbs.
10. Tighten the sensor to 4.7 ft. lbs.

3.0L Engine

1. Disconnect the negative battery cable.
2. Remove the collector cover, as required.
3. Remove the alternator harness from the fuel pipe. Remove the fuel pipe protector.
4. Disconnect the connector from the sensor.
5. Remove the bolt that retains the sensor.
6. Remove the sensor.

To install:

7. Installation is the reverse of the removal procedure.
8. Tighten the sensor to 4.7 ft. lbs.

Vehicle Speed Sensor (VSS)

LOCATION

Vehicles equipped with automatic transaxle use a front and rear sensors. Both sensors are mounted on the transaxle. Vehicles equipped with manual transaxle use a front sensor. This sensor is mounted on the transaxle.

OPERATION

The vehicle speed sensor sends a signal to the ECM to control. The ECM uses this information to control transmission shift patterns. The vehicle speed sensor is also used to send a speed signal to the speed control servo of the cruise control system.

REMOVAL & INSTALLATION

1. Raise and support the vehicle safely.
2. Place a drip pan below the speed sensor to catch any spilled fluid.
3. Disconnect the connector.
4. Remove the sensor from its mounting.

To install:

5. Installation is the reverse of the removal procedure.
6. Replace any lost fluid.

TESTING

Except 2.5L DOHC and 3.0L Engines

1. Remove the sensor from the vehicle.
2. Measure the resistance between the two connector terminals of the sensor, (terminals 1 and 2).
3. If measured value is not within 1–4 kohms, replace the sensor.

2.5L DOHC Engine

4. Using the Subaru scan tool, check the sensor output waveform.

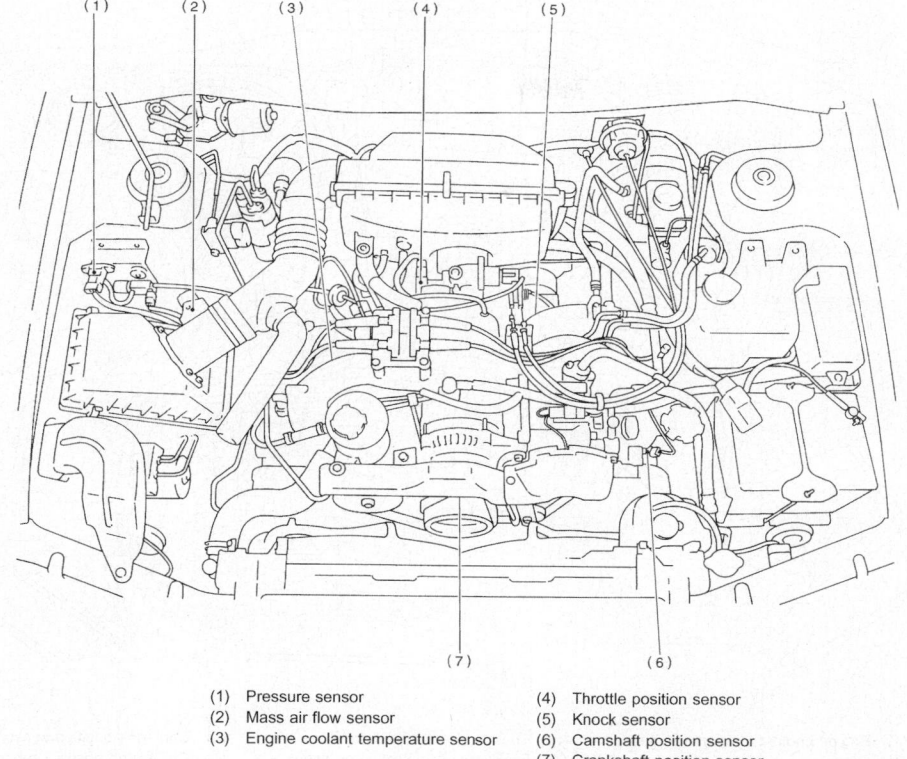

(1) Pressure sensor
(2) Mass air flow sensor
(3) Engine coolant temperature sensor
(4) Throttle position sensor
(5) Knock sensor
(6) Camshaft position sensor
(7) Crankshaft position sensor

29157_SUBA_G0083

Fig. 78 Underhood sensor locations—Legacy 1997–1999 2.2L engine

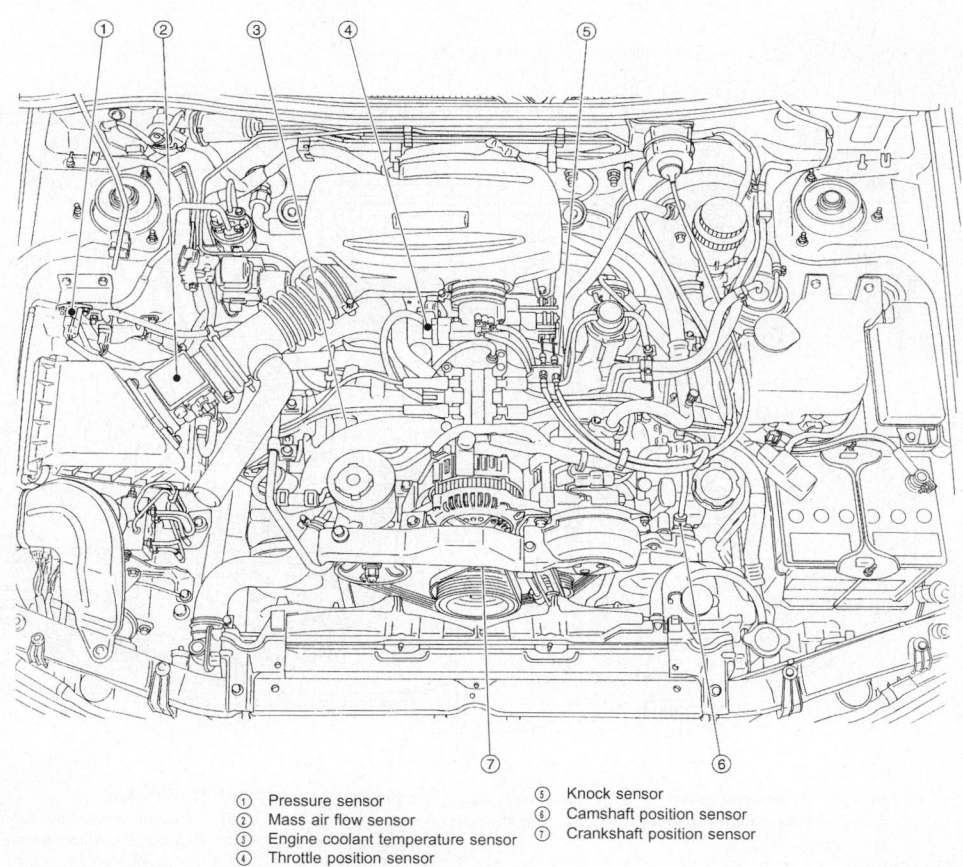

① Pressure sensor
② Mass air flow sensor
③ Engine coolant temperature sensor
④ Throttle position sensor
⑤ Knock sensor
⑥ Camshaft position sensor
⑦ Crankshaft position sensor

29157_SUBA_G0099

Fig. 79 Underhood sensor locations—Legacy 1999–1999 2.5L SOHC engine

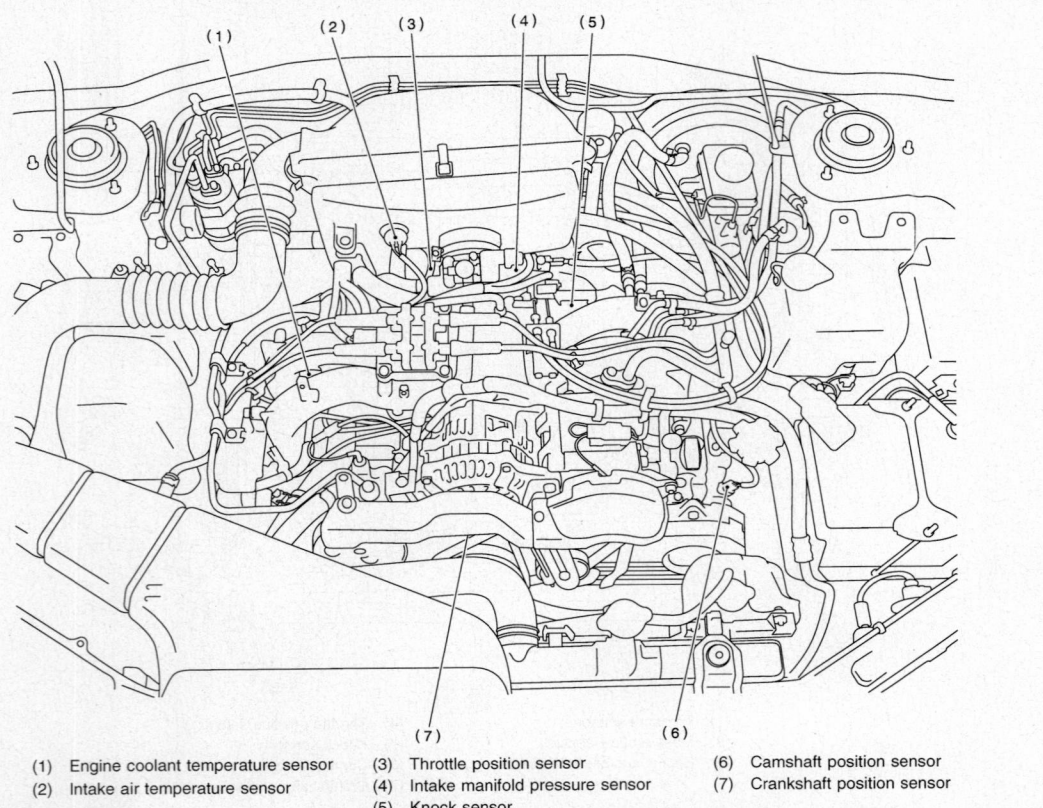

(1) Engine coolant temperature sensor
(2) Intake air temperature sensor
(3) Throttle position sensor
(4) Intake manifold pressure sensor
(5) Knock sensor
(6) Camshaft position sensor
(7) Crankshaft position sensor

29157_SUBA_G0102

Fig. 80 Underhood sensor locations—Legacy and Outback 2000–2001 2.5L SOHC engine with manual transaxle

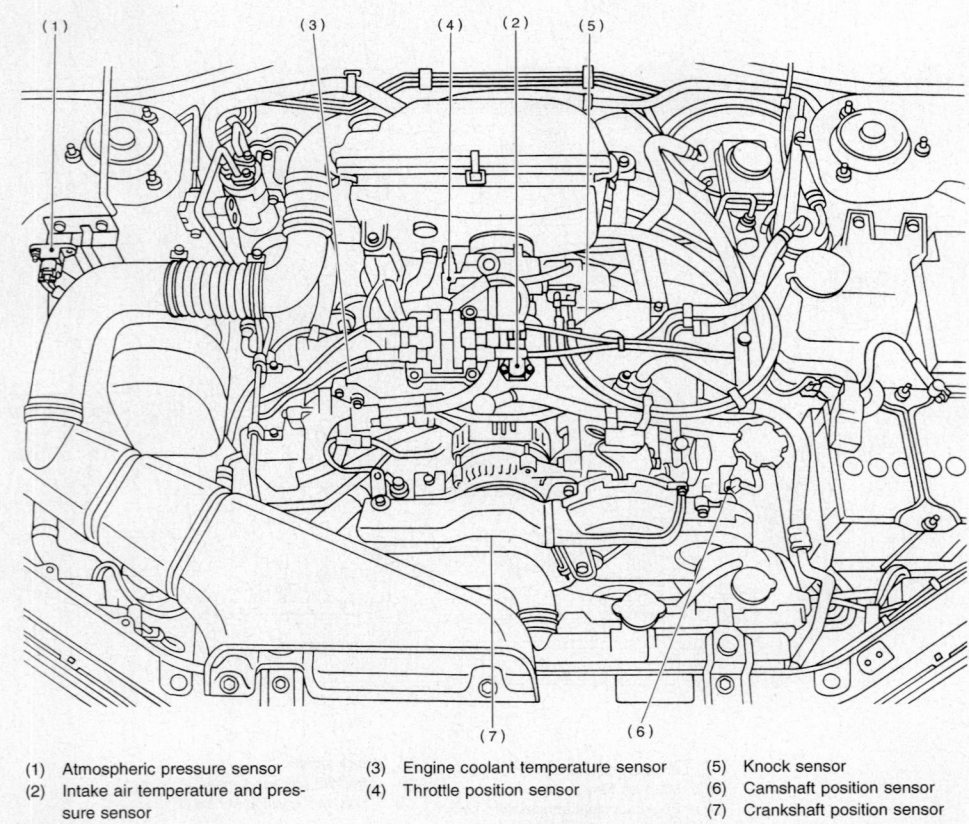

(1) Atmospheric pressure sensor
(2) Intake air temperature and pressure sensor
(3) Engine coolant temperature sensor
(4) Throttle position sensor
(5) Knock sensor
(6) Camshaft position sensor
(7) Crankshaft position sensor

29157_SUBA_G0104

Fig. 81 Underhood sensor locations—Legacy and Outback 2000–2001 2.5L SOHC engine with automatic transaxle

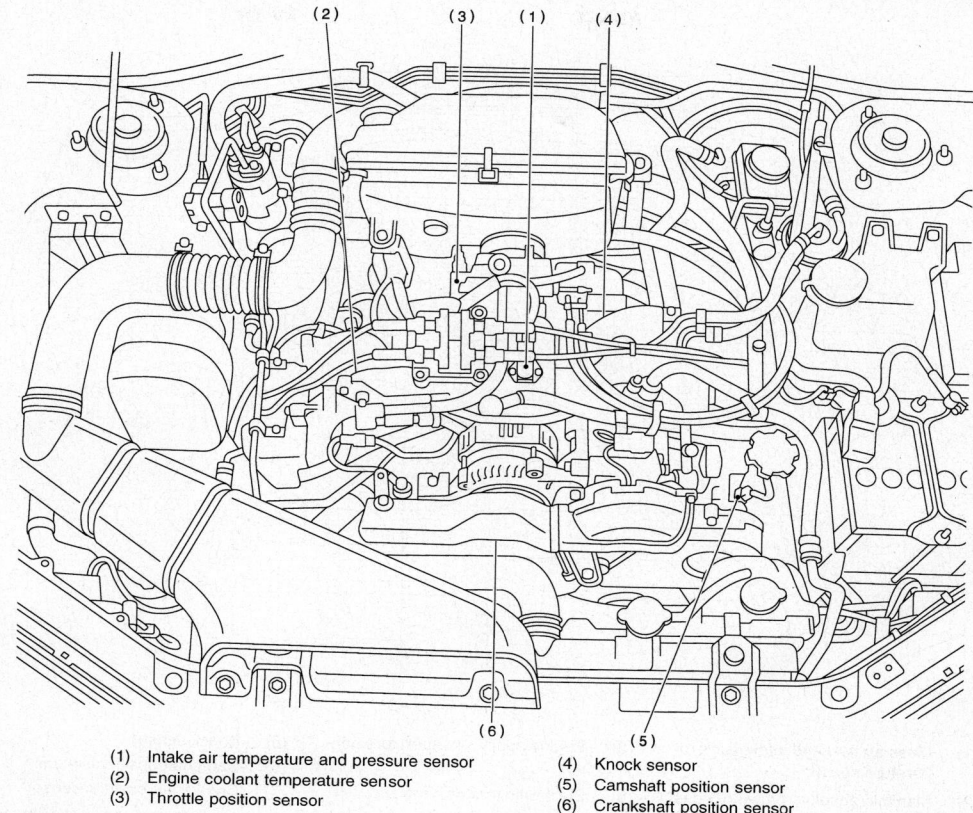

(2)　(3)　(1)　(4)

(6)　(5)

(1) Intake air temperature and pressure sensor
(2) Engine coolant temperature sensor
(3) Throttle position sensor
(4) Knock sensor
(5) Camshaft position sensor
(6) Crankshaft position sensor

Fig. 82　Underhood sensor locations—Legacy and Outback 2002 2.5L SOHC engine

29157_SUBA_G0105

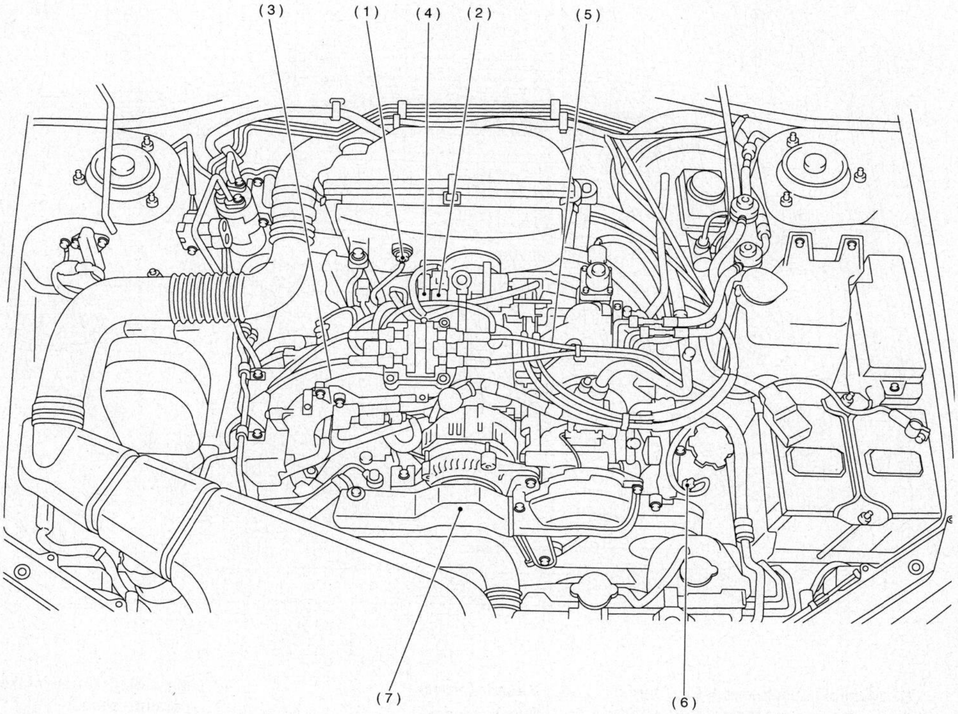

(3)　(1)　(4)　(2)　(5)

(7)　(6)

(1) Intake air temperature sensor
(2) Pressure sensor
(3) Engine coolant temperature sensor
(4) Throttle position sensor
(5) Knock sensor
(6) Camshaft position sensor
(7) Crankshaft position sensor

Fig. 83　Underhood sensor locations—Legacy and Outback 2003–2004 2.5L SOHC engine

29157_SUBA_G0107

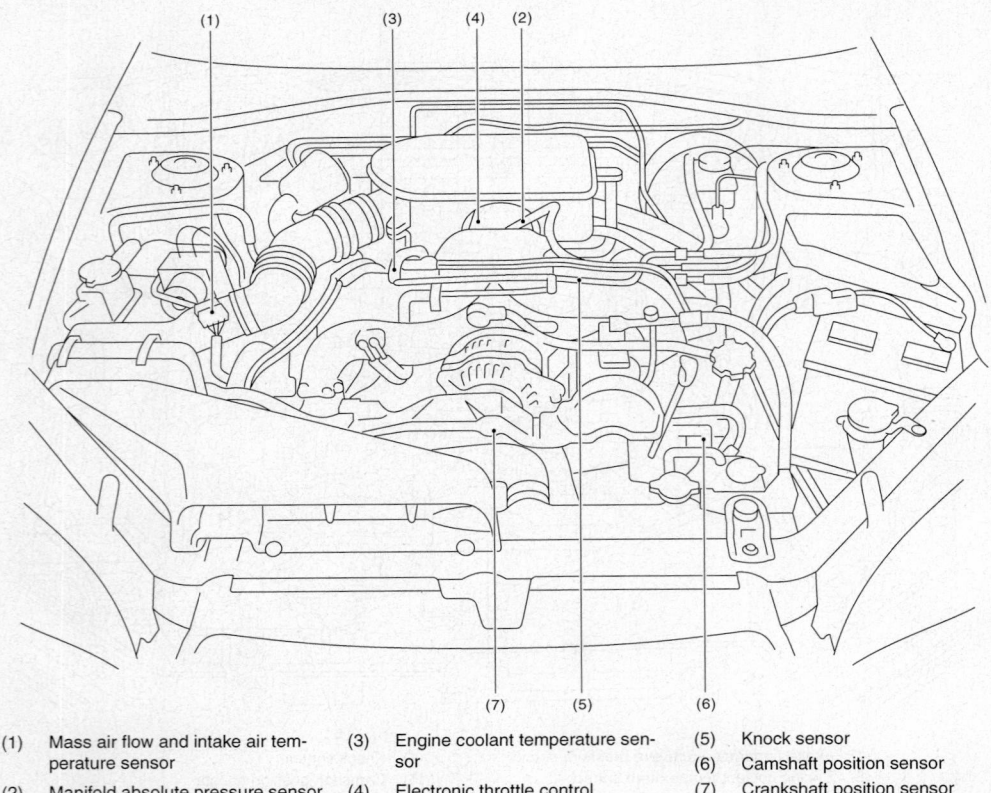

(1) Mass air flow and intake air temperature sensor
(2) Manifold absolute pressure sensor
(3) Engine coolant temperature sensor
(4) Electronic throttle control
(5) Knock sensor
(6) Camshaft position sensor
(7) Crankshaft position sensor

29157_SUBA_G0109

Fig. 84 Underhood sensor locations—Legacy and Outback 2005–2006 2.5L SOHC engine

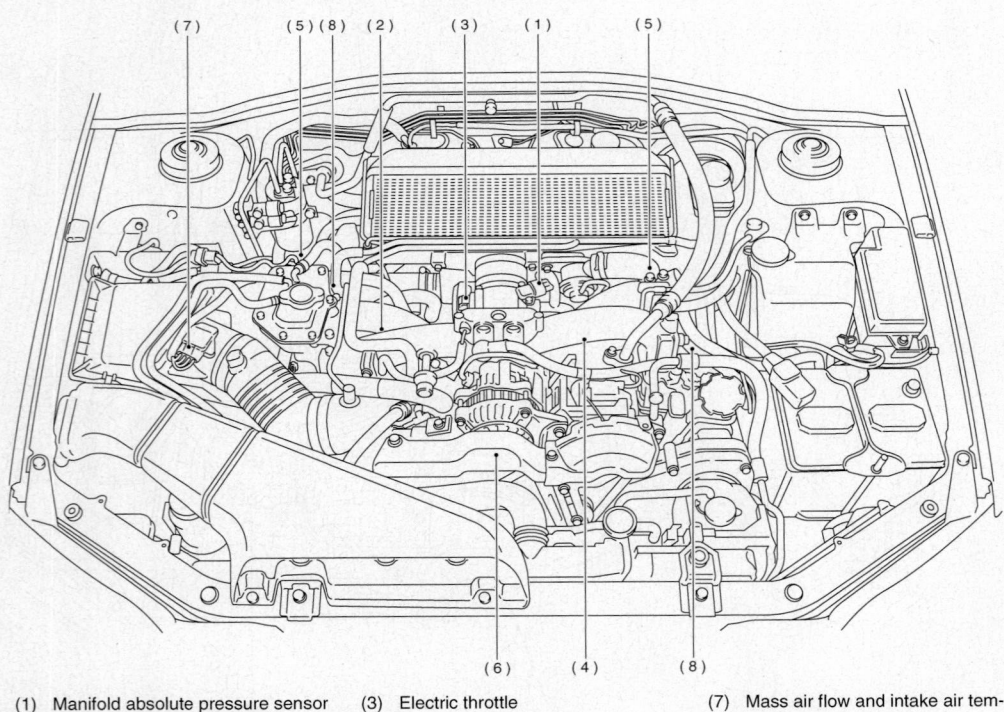

(1) Manifold absolute pressure sensor
(2) Engine coolant temperature sensor
(3) Electric throttle
(4) Knock sensor
(5) Camshaft position sensor
(6) Crankshaft position sensor
(7) Mass air flow and intake air temperature sensor
(8) Tumble generator valve position sensor

29157_SUBA_G0095

Fig. 85 Underhood sensor locations—Legacy and Outback 2004 2.5L DOHC engine

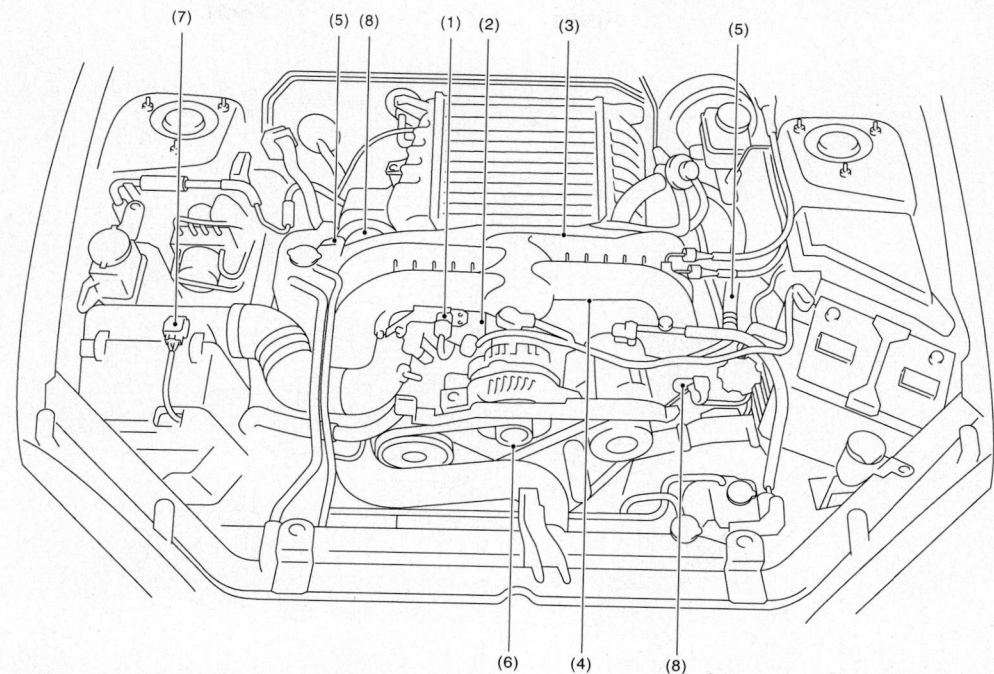

(1) Manifold absolute pressure sensor
(2) Engine coolant temperature sensor
(3) Electronic throttle control
(4) Knock sensor
(5) Intake camshaft position sensor
(6) Crankshaft position sensor
(7) Mass air flow and intake air temperature sensor
(8) Tumble generator valve position sensor

Fig. 86 Underhood sensor locations—Legacy and Outback 2005–2006 2.5L DOHC engine

29157_SUBA_G0092

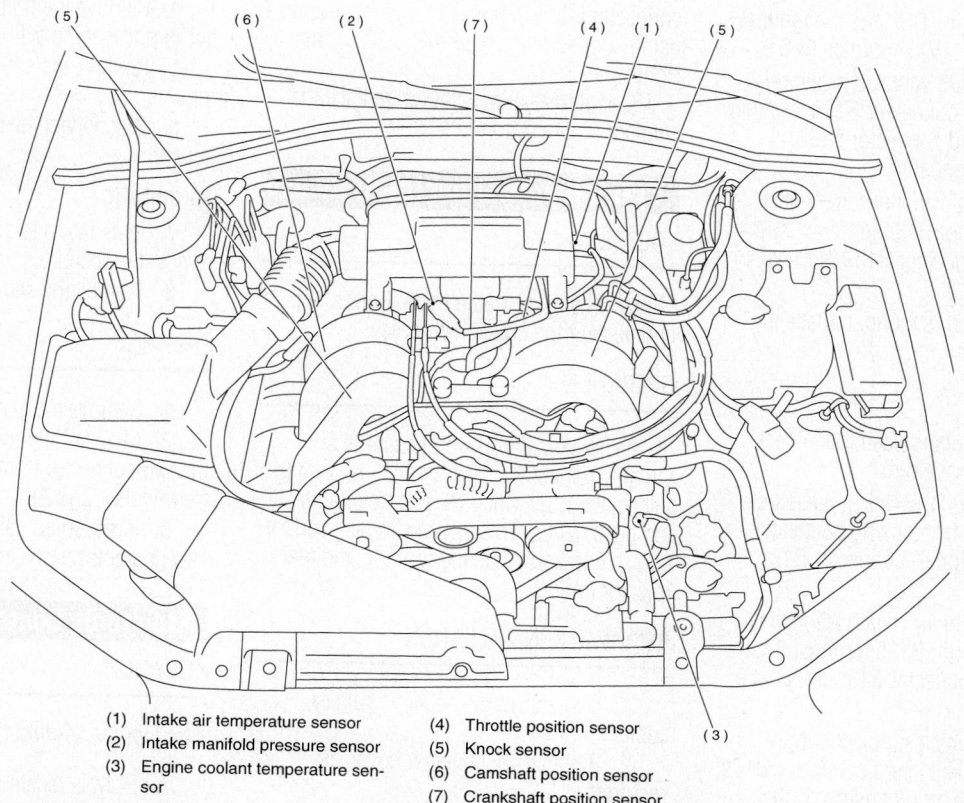

(1) Intake air temperature sensor
(2) Intake manifold pressure sensor
(3) Engine coolant temperature sensor
(4) Throttle position sensor
(5) Knock sensor
(6) Camshaft position sensor
(7) Crankshaft position sensor

Fig. 87 Underhood sensor locations—Legacy and Outback 2002–2004 3.0L engine

29157_SUBA_G0086

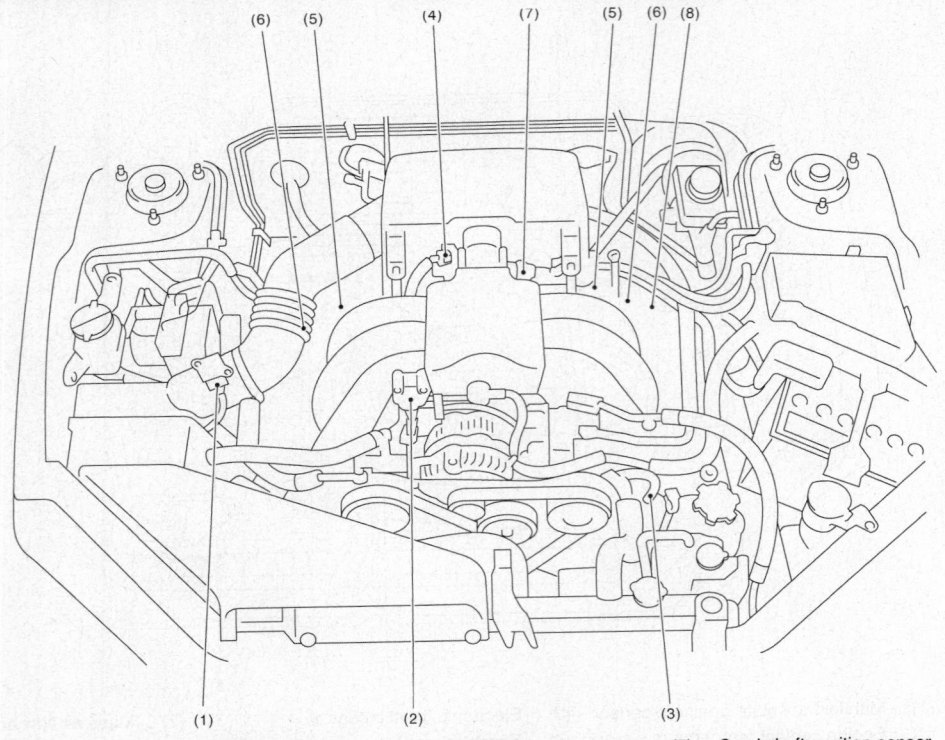

(6) (5) (4) (7) (5) (6) (8)

(1) (2) (3)

(1) Mass air flow and intake air tem-   (4) Electronic throttle control   (7) Crankshaft position sensor
    perature sensor   (5) Knock sensor   (8) Oil temperature sensor
(2) Manifold absolute pressure sensor   (6) Camshaft position sensor
(3) Engine coolant temperature sensor

29157_SUBA_G0089

Fig. 88 Underhood sensor locations—Legacy and Outback 2005–2006 3.0L engine

5. With the engine OFF and the ignition switch ON the signal (V) should be 0–0.9.

6. On 2004–2005 vehicles measurement is taken using connector B135 terminal 8 for the left side and connector B135 terminal 9 for the right side.

7. On 2006 vehicles measurement is taken using connector B134 terminal 21 for the left side and connector B134 terminal 11 for the right side.

8. If abnormality is found, replace the sensor.

3.0L Engine

9. Using the Subaru scan tool, check the sensor output waveform.

10. On 2002–2004 vehicles, measurement for signal+ is taken using connector B135 terminal 1, signal- connector B135 terminal 10.

11. On 2005 vehicles, measurement is taken using connector B135 terminal 8 for the left side and connector B135 terminal 9 for the right side.

12. On 2006 vehicles, power supply measurement is taken using connector B134 terminal 21 and for ground using connector B134 terminal 21, for the left side and power supply measurement is taken using

connector B134 terminal 11 and for ground using connector B134 terminal 22, for the right side.

13. If abnormality is found, replace the sensor.

Crankshaft Position (CKP) Sensor

LOCATION

The sensor is located in the front of the engine near the crankshaft pulley.

OPERATION

The crankshaft position sensor is a variable reluctance sensor which is used to inform the ECM when the No.1 piston is at top dead center. This information is used for controlling and adjusting ignition and fuel injector timing.

REMOVAL & INSTALLATION

1. Disconnect the negative battery cable.

2. Remove the collector cover, as required.

3. As required, remove the air intake chamber.

4. Remove the bolt that retains the sensor in place at the cylinder block.

5. Remove the sensor from its mounting.

6. Disconnect the connector from the sensor.

To install:

7. Installation is the reverse of the removal procedure.

8. Tighten the sensor to 4.7 ft. lbs.

TESTING

1. Remove the sensor from the vehicle.

2. Measure the resistance between the two connector terminals of the sensor, (terminals 1 and 2).

3. If measured value is not within 1–4 kohms, replace the sensor.

Electronic Control Module (ECM)

LOCATION

See Figures 89 through 96.

The ECM is located on the passenger's side of the vehicle, underneath the instrument panel glove box area.

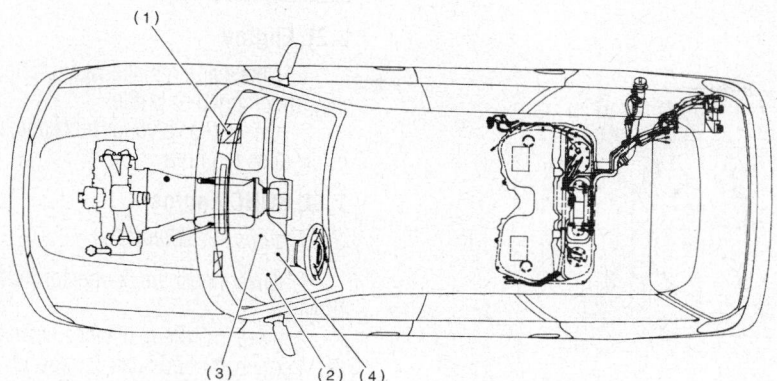

(1) Engine control module (ECM)
(2) Data link connector (for Subaru Select Monitor and OBD-II general scan tool)

(3) Test mode connector
(4) CHECK ENGINE malfunction indicator lamp (MIL)

29157_SUBA_G0082

Fig. 89 ECM and related components—Legacy 1997–1999 2.2L engine

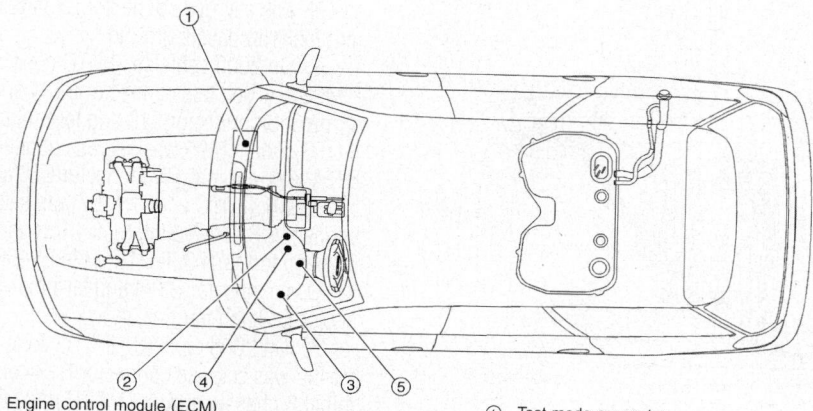

① Engine control module (ECM)
② Data link connector (for Subaru select monitor only)
③ Data link connector (for Subaru select monitor and OBD-II general scan tool)

④ Test mode connector
⑤ CHECK ENGINE malfunction indicator lamp (MIL)

29157_SUBA_G0098

Fig. 90 ECM and related components—Legacy 1997–1999 2.5L SOHC engine

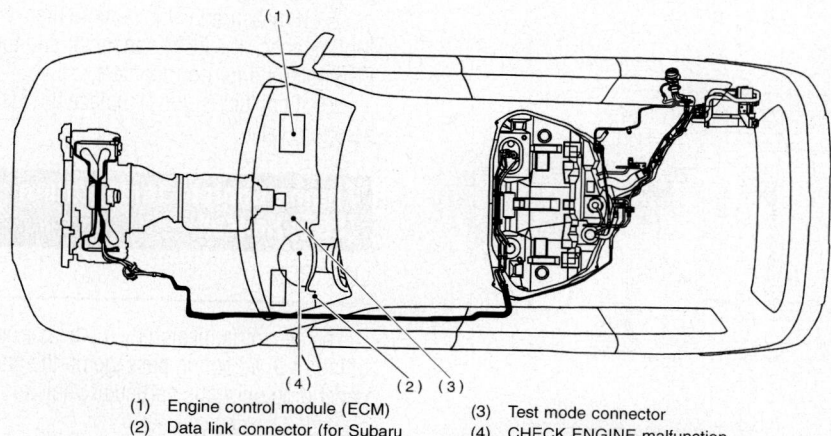

(1) Engine control module (ECM)
(2) Data link connector (for Subaru Select Monitor and OBD-II general scan tool)

(3) Test mode connector
(4) CHECK ENGINE malfunction indicator lamp (MIL)

29157_SUBA_G0101

Fig. 91 ECM and related components—Legacy and Outback 2000–2004 2.5L SOHC engine

OPERATION

The ECM controls the vehicle engine operating system.

REMOVAL & INSTALLATION

1. Disconnect the negative battery cable.
2. Remove the lower inner trim on the passenger's side of the vehicle.
3. Detach the floor mat. Remove the protective cover.
4. Remove the ECM bracket retaining nuts. Remove the clip from the bracket.
5. Disconnect the connectors.
6. Remove the ECM from the vehicle.

To install:

7. Installation is the reverse of the removal procedure.
8. Tighten the retaining screws to 3.7 ft. lbs.

➡ **When replacing the ECM, be careful not to use the wrong part number, as damage to the injection system could occur.**

EGR Valve

LOCATION

The EGR valve is mounted on the intake manifold. The 2.5L DOHC and the 3.0L engines do not use a conventional EGR valve.

OPERATION

The EGR valve controls the flow of exhaust gases. The ECM monitors the flow and regulates the valve accordingly.

REMOVAL & INSTALLATION

1. Disconnect the negative battery cable.
2. Disconnect the connector.
3. On valves equipped with a vacuum hose, disconnect it.
4. Remove the valve from the intake manifold.
5. Discard the gasket.

To install:

6. Installation is the reverse of the removal procedure.
7. Be sure to use a new gasket.
8. Tighten the retaining bolts to 14.0 ft. lbs.

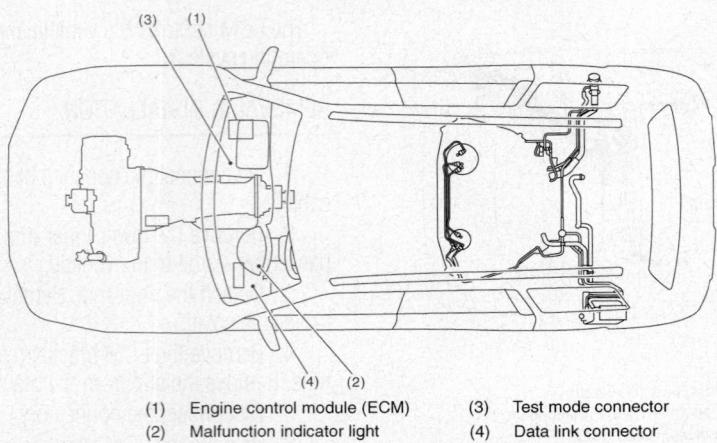

(1)	Engine control module (ECM)	(3)	Test mode connector
(2)	Malfunction indicator light	(4)	Data link connector

29157_SUBA_G0108

Fig. 92 ECM and related components—Legacy and Outback 2005–2006 2.5L SOHC engine

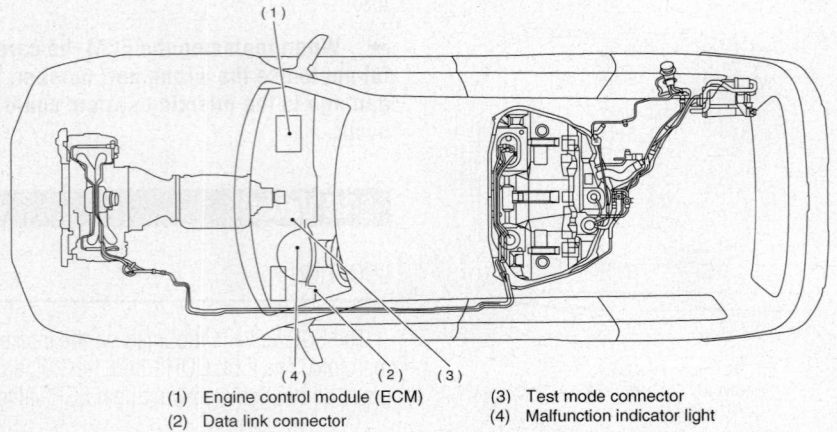

(1)	Engine control module (ECM)	(3)	Test mode connector
(2)	Data link connector	(4)	Malfunction indicator light

29157_SUBA_G0094

Fig. 93 ECM and related components—Legacy and Outback 2004 2.5L DOHC engine

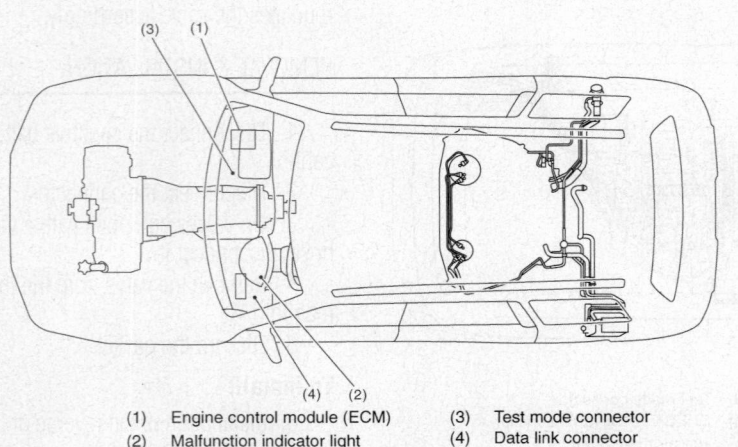

(1)	Engine control module (ECM)	(3)	Test mode connector
(2)	Malfunction indicator light	(4)	Data link connector

29157_SUBA_G0091

Fig. 94 ECM and related components—Legacy and Outback 2005–2006 2.5L DOHC engine

TESTING

2.2L Engine

1. Check and replace vacuum hoses, if they are clogged or brittle.

2. Check the valve for blockage, replace or clean as required.

2.5L SOHC Engine

See Figures 97, 98 and 99.

1. Disconnect the connector from the ECM.

2. Measure the resistance between the ECM connector and chassis ground.

3. On 2004 vehicles, if DTC code P1492 was set, use connector B134 and terminal 18/chassis ground. If DTC code P1494 was set, use connector B134 and terminal 17/chassis ground. If DTC code P1496 was set, use connector B134 and terminal 16/chassis ground. If DTC code P1498 was set, use connector B134 and terminal 15/chassis ground.

4. On 2005 vehicles, if DTC code P1493 was set, use connector B134 and terminal 9/connector E18 and terminal 4. If DTC code P1495 was set, use connector B134 and terminal 10/connector E18 and terminal 6. If DTC code P1497 was set, use connector B134 and terminal 11/connector E18 and terminal 3. If DTC code P1499 was set, use connector B134 and terminal 8/connector E18 and terminal 1.

5. On 2006 vehicles, if DTC code P1492 was set, use connector B134 and terminal 8/chassis ground. If DTC code P1494 was set, use connector B134 and terminal 9/chassis ground. If DTC code P1496 was set, use connector B134 and terminal 10/chassis ground. If DTC code P1498 was set, use connector B134 and terminal 20/chassis ground.

6. If measured value is more than 1 Mohm, check the ECM connector and the EGR solenoid for poor contact.

7. If contact is good, replace the EGR solenoid valve.

Engine Coolant Temperature (ECT) Sensor

LOCATION

The sensor is located by the heater outlet fitting or in a cooling passage on the engine, depending upon the particular vehicle.

OPERATION

This component detects the temperature of the engine coolant and relays the information to the electronic control assembly.

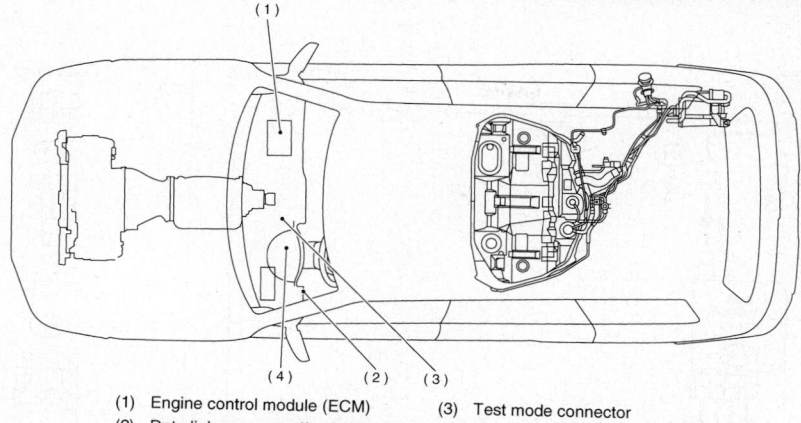

(1)	Engine control module (ECM)	(3)	Test mode connector
(2)	Data link connector (for Subaru Select Monitor and OBD-II general scan tool)	(4)	CHECK ENGINE malfunction indicator lamp (MIL)

Fig. 95 ECM and related components—Legacy and Outback 2002–2004 3.0L engine

29157_SUBA_G0085

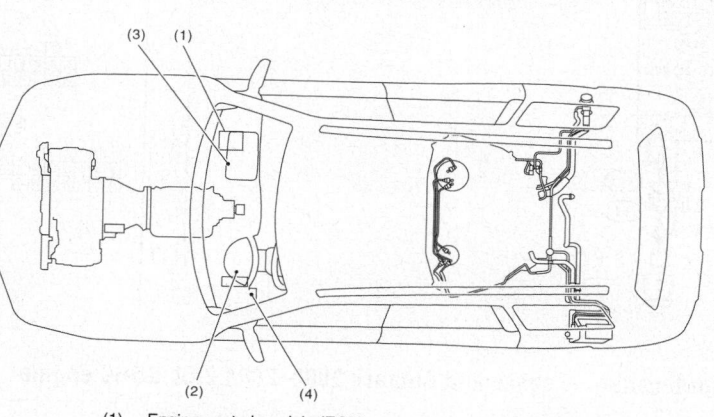

(1)	Engine control module (ECM)	(3)	Test mode connector
(2)	Malfunction indicator light	(4)	Data link connector

Fig. 96 ECM and related components—Legacy and Outback 2005–2006 3.0L engine

29157_SUBA_G0088

REMOVAL & INSTALLATION

1. Disconnect the negative battery cable.
2. Remove the alternator, as required.
3. Remove the air intake duct and air cleaner case, as required.
4. Disconnect the connector from the sensor.
5. Drain the cooling system, as required.
6. Remove the sensor from its mounting.

To install:

7. Installation is the reverse of the removal procedure.
8. Tighten the sensor to 13.3 ft. lbs.

TESTING

2.2L Engine

1. Turn the ignition switch to OFF.
2. Measure the resistance of the harness between the coolant temperature sensor

connector and engine ground (connector E8 terminal 2).

3. If resistance is less than 5 ohms, replace the sensor.

Except 2.2L Engine

7. Turn the ignition switch to OFF.
5. Disconnect the connector from the sensor.
6. Turn the ignition switch ON.
7. Using the Subaru scan tool, read the sensor signal data. If the measured value is less than -40 degrees F, replace the sensor.

Heated Oxygen (HO2S) Sensor

LOCATION

See Figures 100 through 108.

On the 2.2L engine the front (A/F) oxygen sensor is located in the front section of the front catalytic converter. The rear oxygen sensor is located in the rear section of the rear catalytic converter. On 1997–2001 ve-

hicles equipped with the 2.5L SOHC engine, the front (A/F) oxygen sensor is located in the front section of the front catalytic converter. The rear oxygen sensor is located in the rear section of the rear catalytic converter. On 2002–2006 vehicles equipped with the 2.5L SOHC engine, the front oxygen sensor is located in the front section of the front catalytic converter. The rear oxygen sensor is located in the rear section of the front catalytic converter. On 2004–2006 2.5L DOHC engine the front oxygen sensor is located in the front section of the exhaust system, just past the crossover pipe. The rear oxygen sensor is located in the front section of the rear catalytic converter. On 2005–2006 3.0L engine the front oxygen sensors are located in the front section of the front catalytic converter. The rear oxygen sensors are located in the rear section of the front catalytic converter. There are two front catalytic converters, but only one rear catalytic converter.

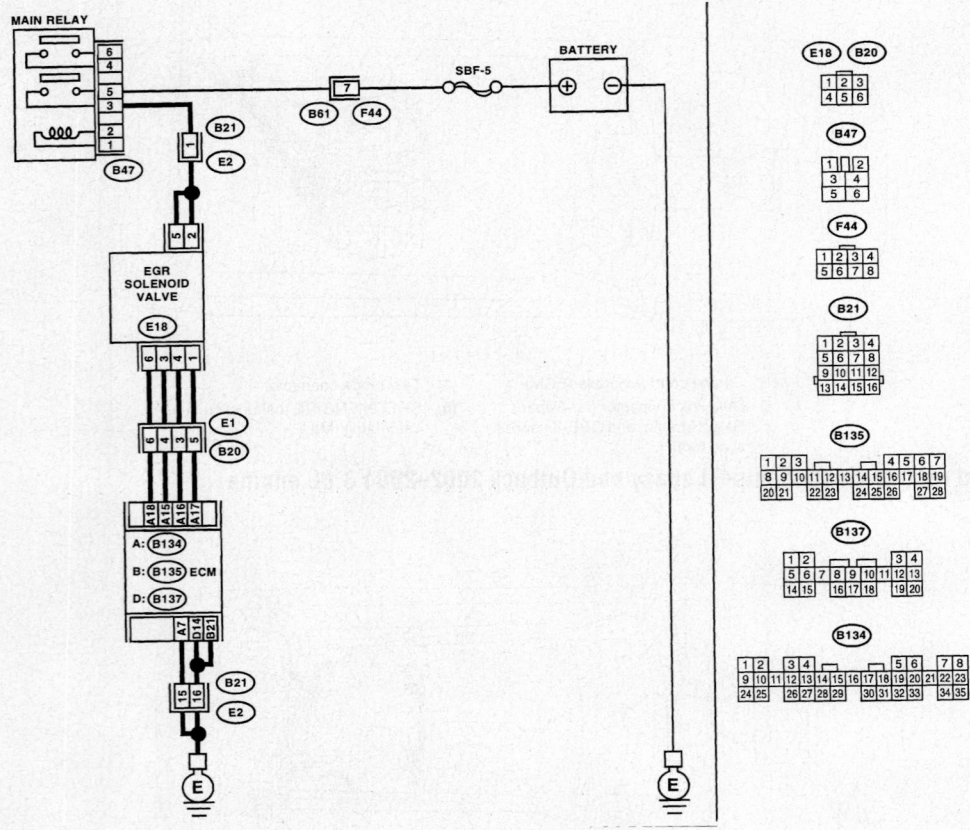

Fig. 97 EGR valve connector location—Legacy and Outback 2003–2004 2.5L SOHC engine

29157_SUBA_G0123

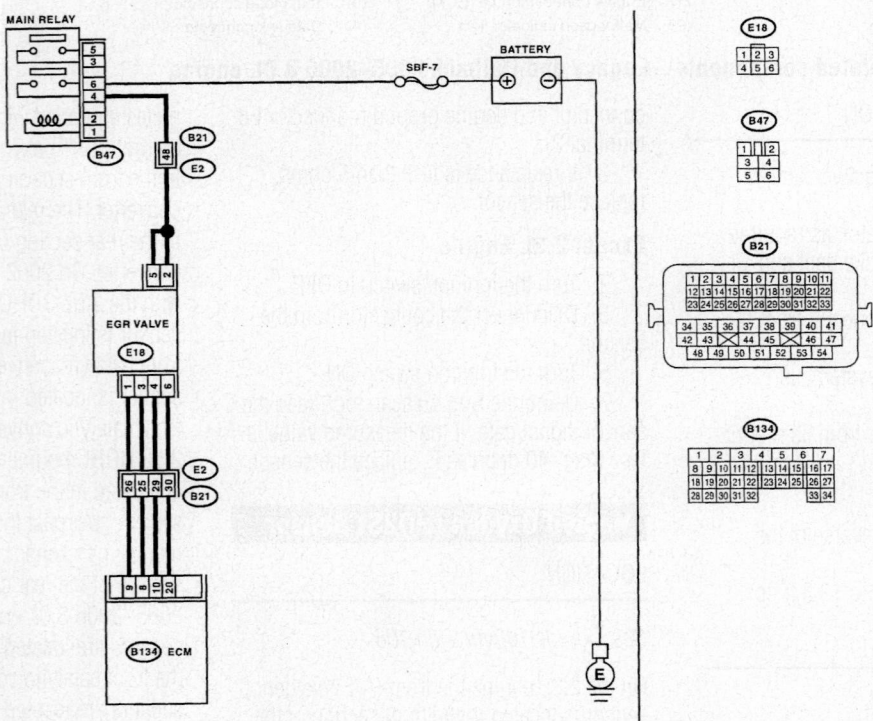

Fig. 98 EGR valve connector location—Legacy and Outback 2005 2.5L SOHC engine

29157_SUBA_G0125

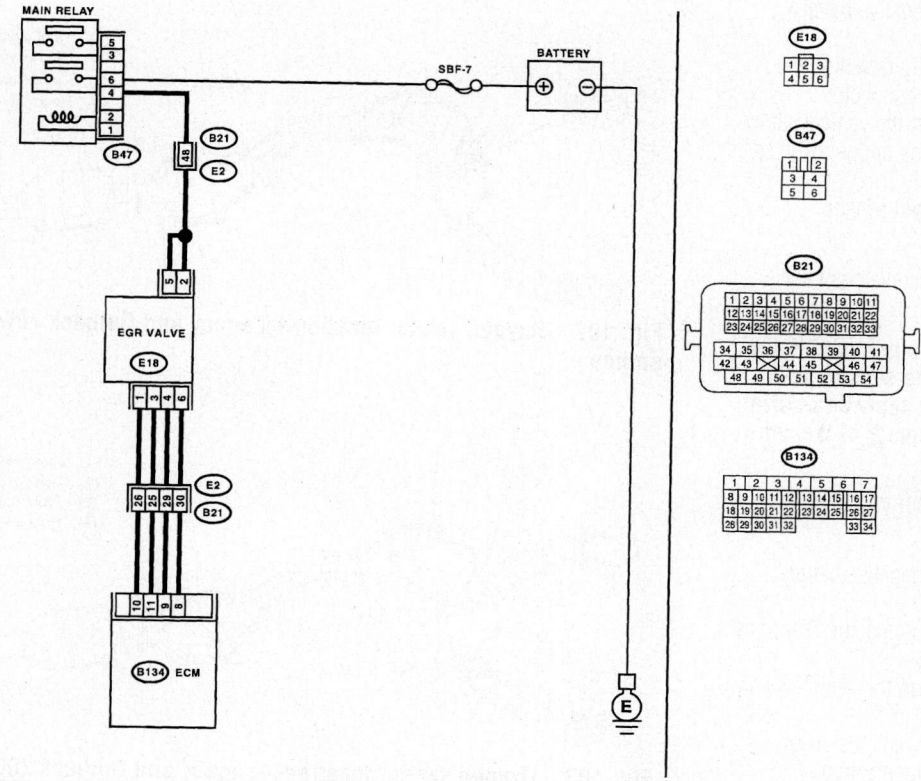

Fig. 99   EGR valve connector location—Legacy and Outback 2006 2.5L SOHC engine

29157_SUBA_G0124

OPERATION

The exhaust gas oxygen sensor supplies the electronic control assembly with a signal which indicates either a rich or lean mixture condition, during the engine operation.

REMOVAL & INSTALLATION

Front Sensor

2.2L ENGINE

1. Disconnect the negative battery cable.

2. Remove the necessary components to gain access to the sensor.

3. Disconnect the clip fastening the harness, if equipped.

4. Disconnect the connector.

5. Remove the oxygen sensor.

To install:

6. Installation is the reverse of the removal procedure.

➡ **Apply anti-seize compound to the threaded portion of the sensor, prior to installation. Never apply anti-seize compound to the protector of the sensor.**

7. Tighten the sensor to 15.2 ft. lbs.

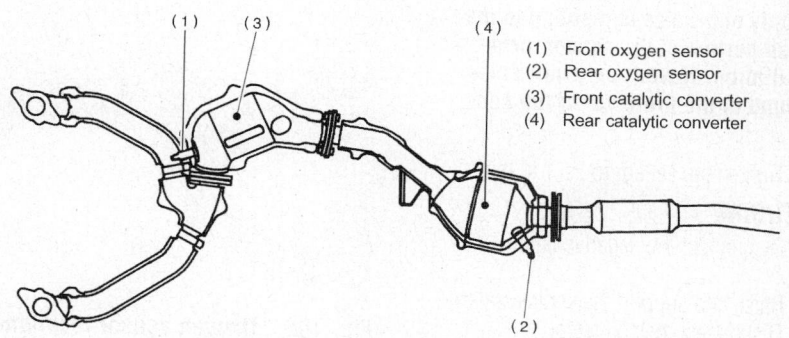

(1) Front oxygen sensor
(2) Rear oxygen sensor
(3) Front catalytic converter
(4) Rear catalytic converter

Fig. 100   Oxygen sensor location—Legacy 1997–1999 2.2L engine

29157_SUBA_G0084

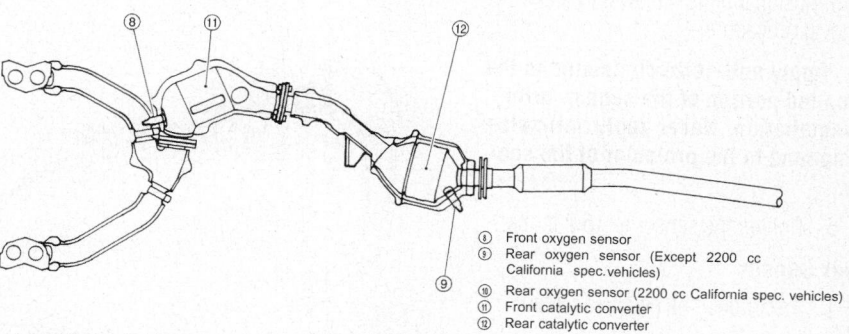

⑧ Front oxygen sensor
⑨ Rear oxygen sensor (Except 2200 cc California spec. vehicles)
⑩ Rear oxygen sensor (2200 cc California spec. vehicles)
⑪ Front catalytic converter
⑫ Rear catalytic converter

Fig. 101   Oxygen sensor location—Legacy 1997–1999 2.5L SOHC engine

29157_SUBA_G0100

2.5L SOHC ENGINE

1. Disconnect the negative battery cable.
2. Disconnect the clip fastening the harness. Disconnect the connector.
3. Raise and support the vehicle safely.
4. Remove the engine under cover, as required.
5. Remove the oxygen sensor.

To install:

6. Installation is the reverse of the removal procedure.

➡ **Apply anti-seize compound to the threaded portion of the sensor, prior to installation. Never apply anti-seize compound to the protector of the sensor.**

7. Tighten the sensor to 15.2 ft. lbs.

2.5L DOHC ENGINE

1. Disconnect the negative battery cable.
2. Disconnect the connector. Disconnect the engine harness clip.
3. Raise and support the vehicle safely. Remove the tire and wheel.
4. Remove the service hole cover.
5. Remove the oxygen sensor.

To install:

6. Installation is the reverse of the removal procedure.

➡ **Apply anti-seize compound to the threaded portion of the sensor, prior to installation. Never apply anti-seize compound to the protector of the sensor.**

7. Tighten the sensor to 22.1 ft. lbs.

3.0L ENGINE

1. Disconnect the negative battery cable.
2. Raise and support the vehicle safely.
3. Disconnect the connector.
4. Remove the oxygen sensor.

To install:

5. Installation is the reverse of the removal procedure.

➡ **Apply anti-seize compound to the threaded portion of the sensor, prior to installation. Never apply anti-seize compound to the protector of the sensor.**

6. Tighten the sensor to 15.2 ft. lbs.

Rear Sensor

1. Disconnect the negative battery cable.
2. Raise and support the vehicle safely.
3. Disconnect the connector. Remove

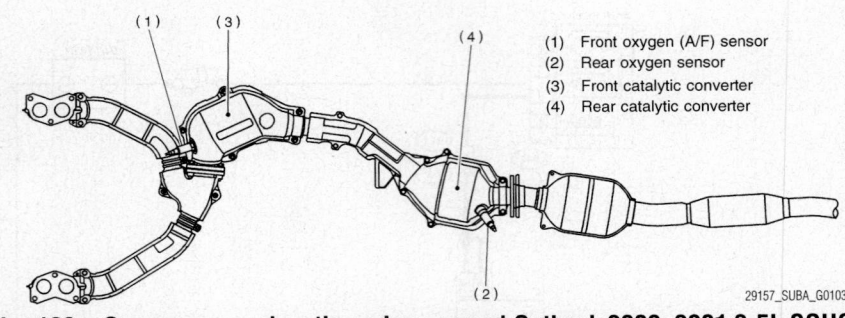

(1) Front oxygen (A/F) sensor
(2) Rear oxygen sensor
(3) Front catalytic converter
(4) Rear catalytic converter

Fig. 102 Oxygen sensor location—Legacy and Outback 2000–2001 2.5L SOHC engine

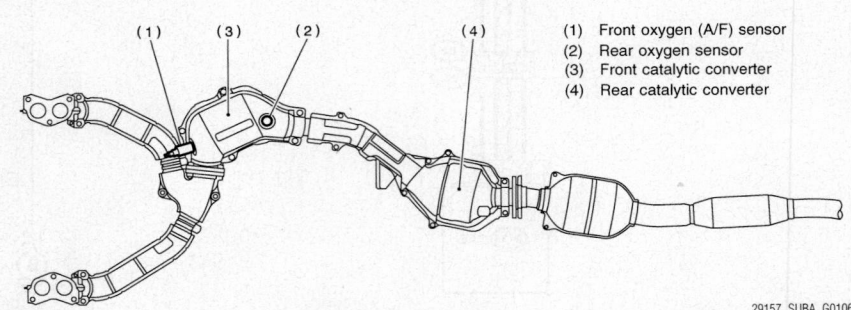

(1) Front oxygen (A/F) sensor
(2) Rear oxygen sensor
(3) Front catalytic converter
(4) Rear catalytic converter

Fig. 103 Oxygen sensor location—Legacy and Outback 2002–2005 2.5L SOHC engine

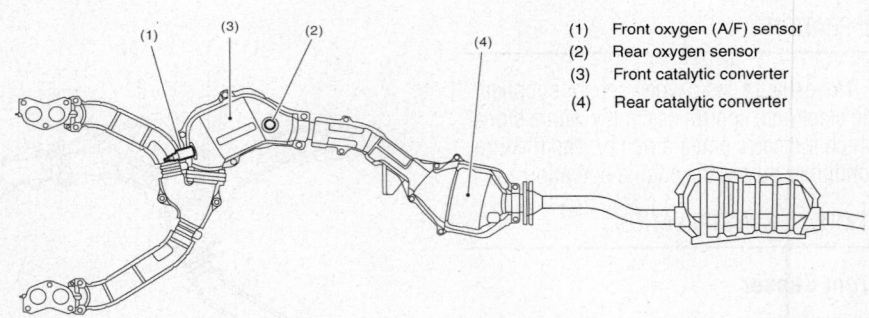

(1) Front oxygen (A/F) sensor
(2) Rear oxygen sensor
(3) Front catalytic converter
(4) Rear catalytic converter

Fig. 104 Oxygen sensor location—Legacy and Outback 2005–2006 2.5L SOHC engine

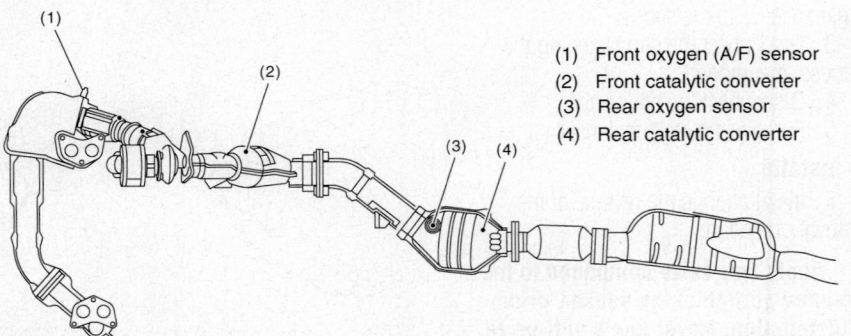

(1) Front oxygen (A/F) sensor
(2) Front catalytic converter
(3) Rear oxygen sensor
(4) Rear catalytic converter

Fig. 105 Oxygen sensor location—Legacy and Outback 2004 2.5L DOHC engine

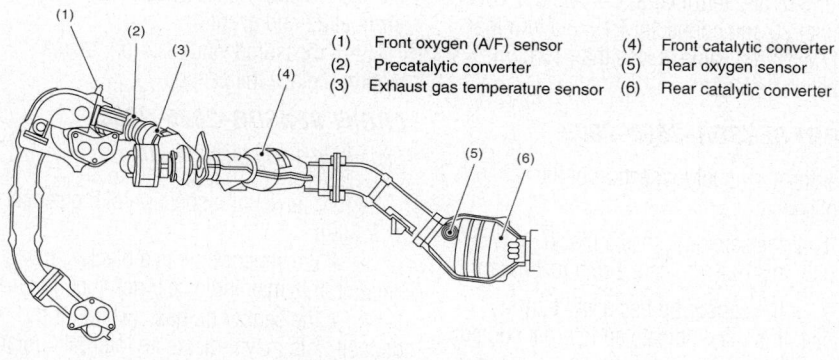

(1) Front oxygen (A/F) sensor   (4) Front catalytic converter
(2) Precatalytic converter   (5) Rear oxygen sensor
(3) Exhaust gas temperature sensor   (6) Rear catalytic converter

29157_SUBA_G0093

Fig. 106 Oxygen sensor location—Legacy and Outback 2004–2006 2.5L DOHC engine

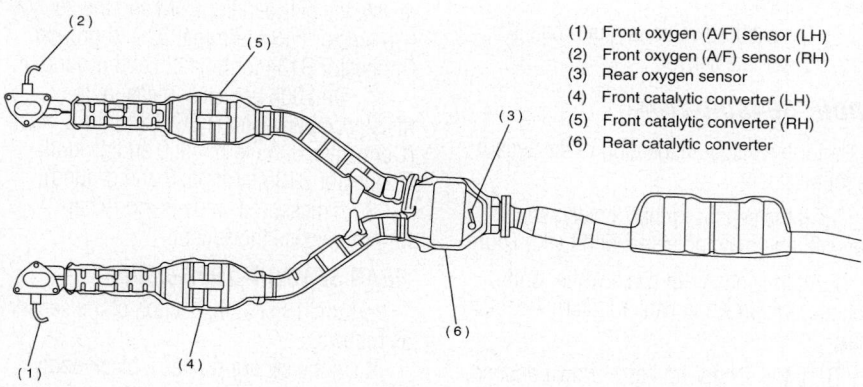

(1) Front oxygen (A/F) sensor (LH)
(2) Front oxygen (A/F) sensor (RH)
(3) Rear oxygen sensor
(4) Front catalytic converter (LH)
(5) Front catalytic converter (RH)
(6) Rear catalytic converter

29157_SUBA_G0087

Fig. 107 Oxygen sensor location—Legacy and Outback 2002–2004 3.0L engine

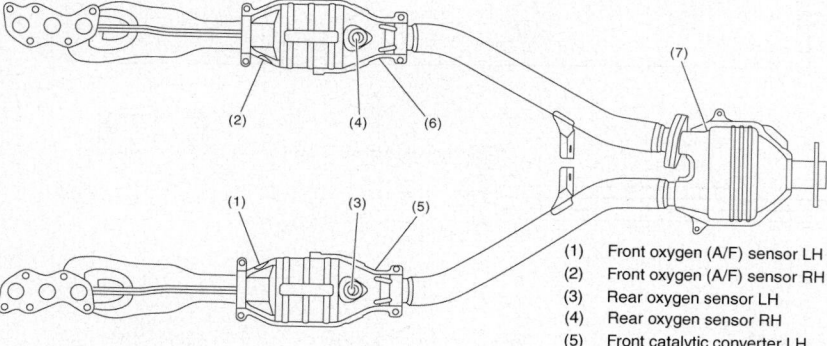

(1) Front oxygen (A/F) sensor LH
(2) Front oxygen (A/F) sensor RH
(3) Rear oxygen sensor LH
(4) Rear oxygen sensor RH
(5) Front catalytic converter LH
(6) Front catalytic converter RH
(7) Rear catalytic converter

29157_SUBA_G0090

Fig. 108 Oxygen sensor location—Legacy and Outback 2005–2006 3.0L engine

the clip by pulling it out from the upper side of the crossmember.

4. Remove the oxygen sensor.

To install:

5. Installation is the reverse of the removal procedure.

➡ **Apply anti-seize compound to the threaded portion of the sensor, prior to installation. Never apply anti-seize compound to the protector of the sensor.**

6. Tighten the sensor to 22.1 ft. lbs. on all engines except the 3.0L engine. On the 3.0L engine, tighten the sensor to 15.2 ft. lbs.

TESTING

2.2L Engine

FRONT SENSOR

Perform a visual inspection of the sensor as follows:

1. If the sensor tip has a black/sooty deposit, this may indicate a rich fuel mixture.

2. If the sensor tip has a white, gritty deposit, this may indicate an internal coolant leak.

3. If the sensor tip has a brown deposit, this could indicate oil consumption.

4. Check the exhaust system for defects.

5. Turn the ignition switch OFF.

6. Connect the Subaru scan tool. Start the engine. Turn the tool ON.

7. Allow the engine to reach 160 degrees F. Keep the engine speed at 2000–3000 rpm's for one minute.

8. Read the data from the tool.

9. If the reading is more than 0.1 volt between the value of max. output and min. output, replace the sensor.

REAR SENSOR

Perform a visual inspection of the sensor as follows:

1. If the sensor tip has a black/sooty deposit, this may indicate a rich fuel mixture.

2. If the sensor tip has a white, gritty deposit, this may indicate an internal coolant leak.

3. If the sensor tip has a brown deposit, this could indicate oil consumption.

4. Turn the ignition switch to OFF.

5. Disconnect the connectors from the sensor.

6. Turn the ignition switch ON.

7. Measure the voltage between the sensor harness connector and engine ground (connector E25 and terminal 4+/engine ground -). If measured value is more than 0.2 volt, replace the sensor.

2.5L SOHC Engine

FRONT SENSOR–1997–2000

Perform a visual inspection of the sensor as follows:

1. If the sensor tip has a black/sooty deposit, this may indicate a rich fuel mixture.

2. If the sensor tip has a white, gritty deposit, this may indicate an internal coolant leak.

3. If the sensor tip has a brown deposit, this could indicate oil consumption.

4. Check the exhaust system for defects.

5. Replace the sensor, as required.

FRONT SENSOR–2001

Perform a visual inspection of the sensor as follows:

1. If the sensor tip has a black/sooty deposit, this may indicate a rich fuel mixture.

2. If the sensor tip has a white, gritty deposit, this may indicate an internal coolant leak.

3. If the sensor tip has a brown deposit, this could indicate oil consumption.

4. Check the exhaust system for defects.

5. If equipped with manual transaxle, replace the sensor, as required.

6. If equipped with automatic transaxle, turn the ignition switch to OFF. Disconnect the connectors from the ECM and the sensor. Measure the resistance of the harness between the ECM and the sensor connector (Connector B136 terminal 6 and connector E18 and terminal 1. Connector B136 terminal 7 and connector E18 and terminal 6. Connector B136 terminal 19 and connector E18 and terminal 3. Connector B136 terminal 20 and connector E18 and terminal 4. If measured value is less than 10 ohm, replace the sensor.

FRONT SENSOR–2002–2003

Perform a visual inspection of the sensor as follows:

1. If the sensor tip has a black/sooty deposit, this may indicate a rich fuel mixture.

2. If the sensor tip has a white, gritty deposit, this may indicate an internal coolant leak.

3. If the sensor tip has a brown deposit, this could indicate oil consumption.

4. Turn the ignition switch to OFF.

5. Measure the resistance between the sensor and the connector terminals (terminals 2 and 5).

6. If measured value is more than 10 ohm, replace the sensor.

FRONT SENSOR–2004

Perform a visual inspection of the sensor as follows:

1. If the sensor tip has a black/sooty deposit, this may indicate a rich fuel mixture.

2. If the sensor tip has a white, gritty deposit, this may indicate an internal coolant leak.

3. If the sensor tip has a brown deposit, this could indicate oil consumption.

4. Measure the resistance between the ECM and chassis ground (Connector B136 terminal 13 and ground. Connector B136 terminal 22 and ground).

5. If measured value is more than 1 Mohm, replace the sensor.

FRONT SENSOR–2005–2006

See Figures 109 through 116.

Perform a visual inspection of the sensor as follows:

1. If the sensor tip has a black/sooty deposit, this may indicate a rich fuel mixture.

2. If the sensor tip has a white, gritty deposit, this may indicate an internal coolant leak.

3. If the sensor tip has a brown deposit, this could indicate oil consumption.

4. Turn the ignition switch OFF.

5. Disconnect the connectors from the ECM and the sensor connector.

6. On 2005 vehicles, measure the resistance between the ECM and the sensor (Connector B134 terminal 33 and ground. Connector B134 terminal 26 and ground).

7. On 2006 vehicles, measure the resistance between the ECM and the sensor (Connector B135 terminal 9 and ground. Connector B135 terminal 8 and ground).

8. If measured value is more than 1 Mohm, replace the sensor.

REAR SENSOR–1997–2000

Perform a visual inspection of the sensor as follows:

1. If the sensor tip has a black/sooty deposit, this may indicate a rich fuel mixture.

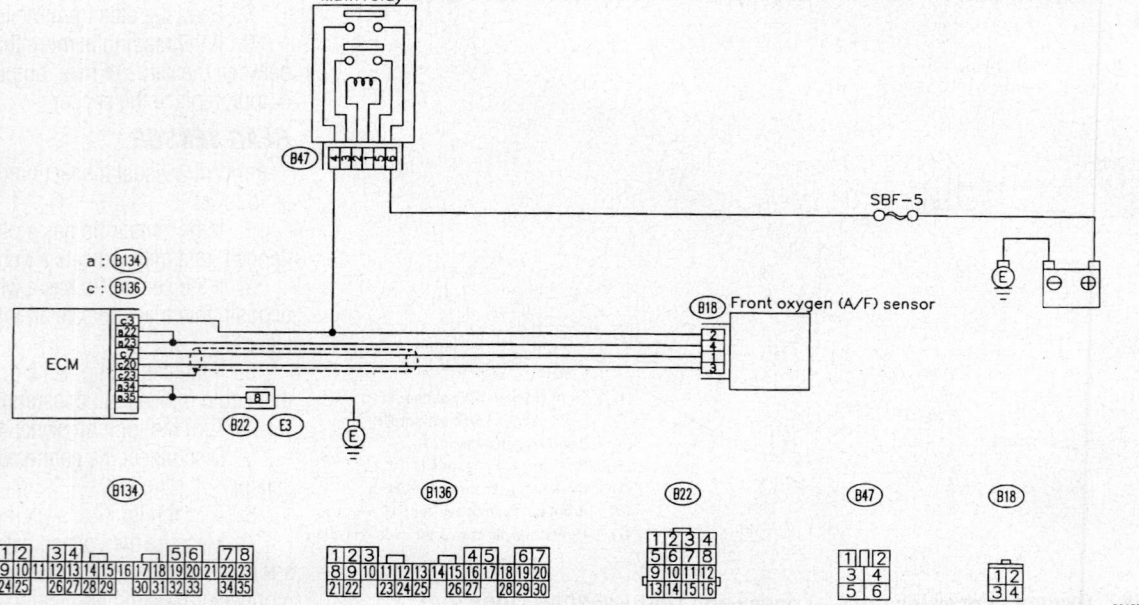

Fig. 109 Front oxygen sensor connector location—Legacy 1997–2000 2.5L SOHC engine with manual transaxle

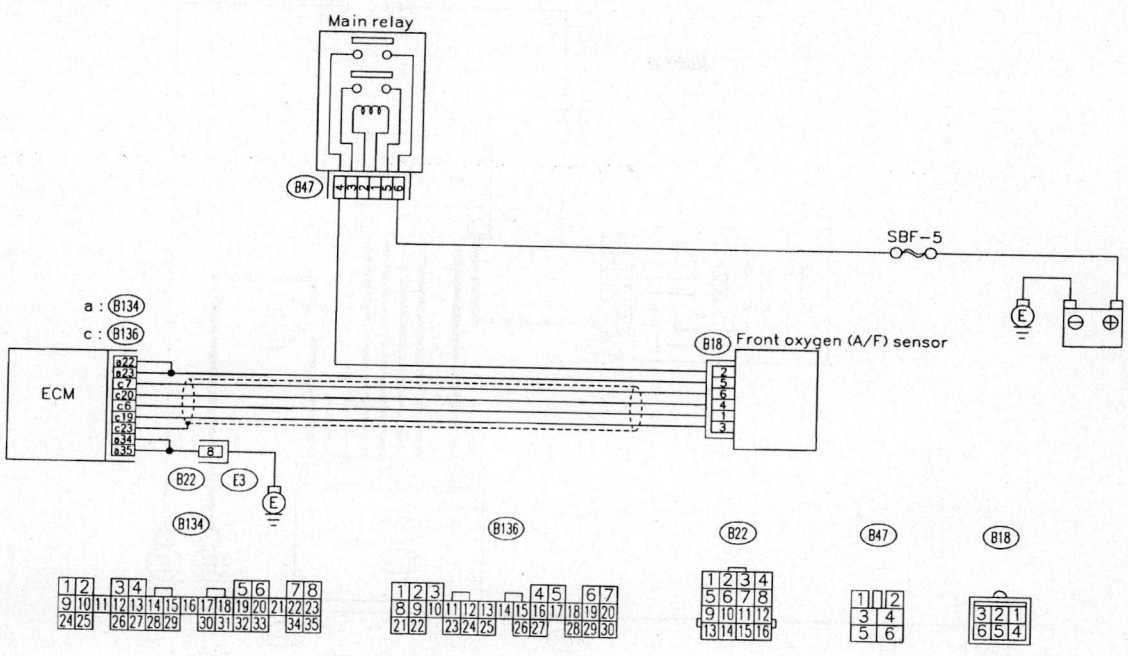

Fig. 110   Front oxygen sensor connector location—Legacy 1997–2000 2.5L SOHC engine with automatic transaxle

29157_SUBA_G0112

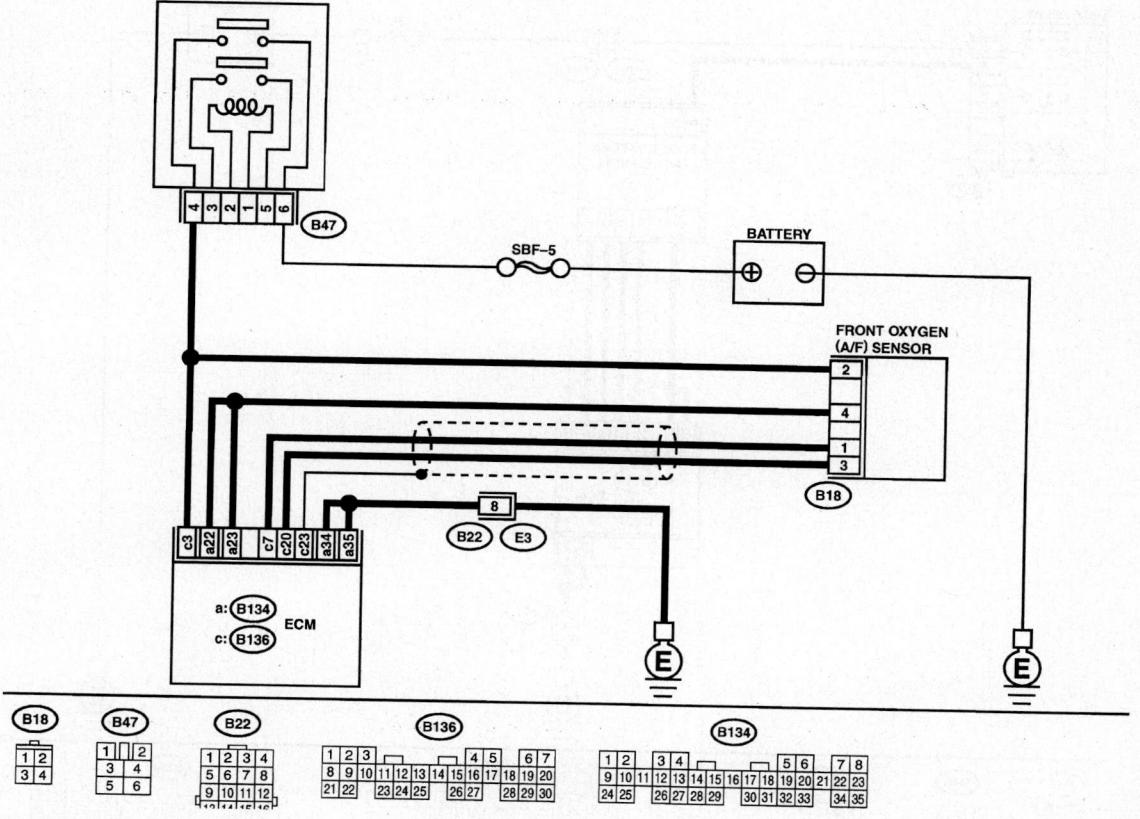

Fig. 111   Front oxygen sensor connector location—Legacy 2001 2.5L SOHC engine with manual transaxle

29157_SUBA_G0113

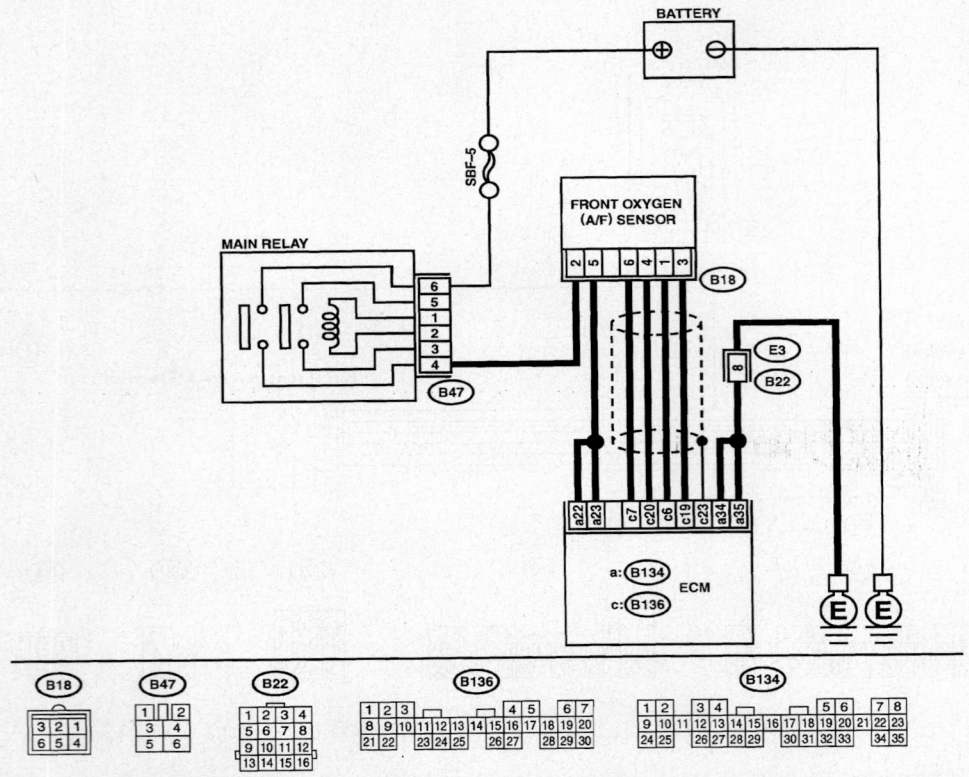

Fig. 112 Front oxygen sensor connector location—Legacy 2001 2.5L SOHC engine with automatic transaxle

29157_SUBA_G0114

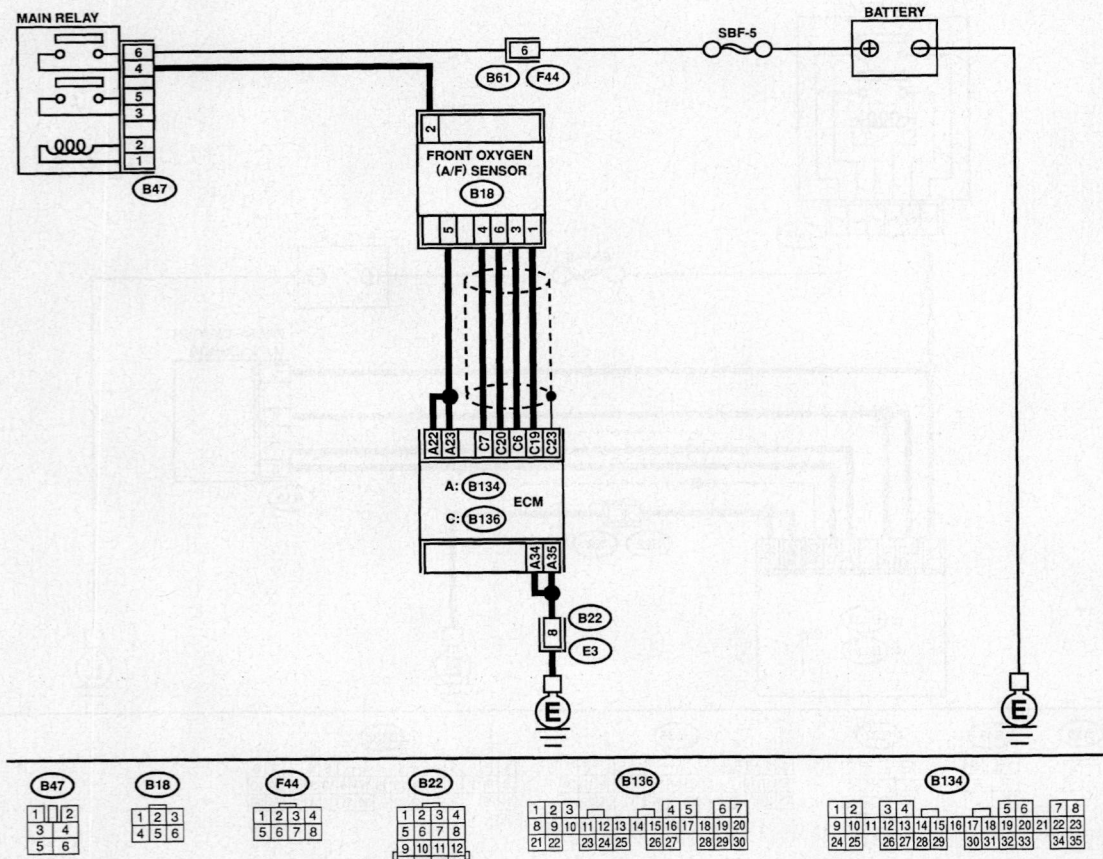

Fig. 113 Front oxygen sensor connector location—Legacy and Outback 2002 2.5L SOHC engine

29157_SUBA_G0115

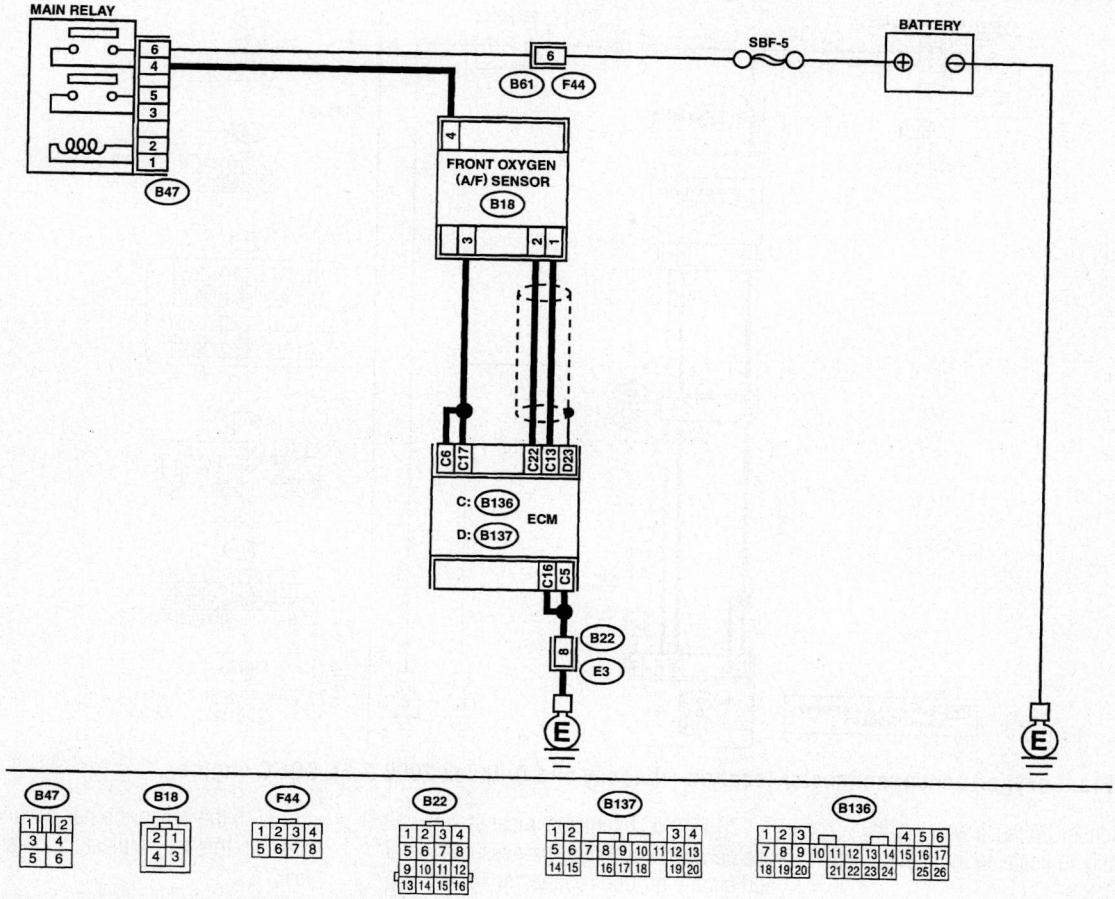

Fig. 114   Front oxygen sensor connector location—Legacy and Outback 2003–2004 2.5L SOHC engine

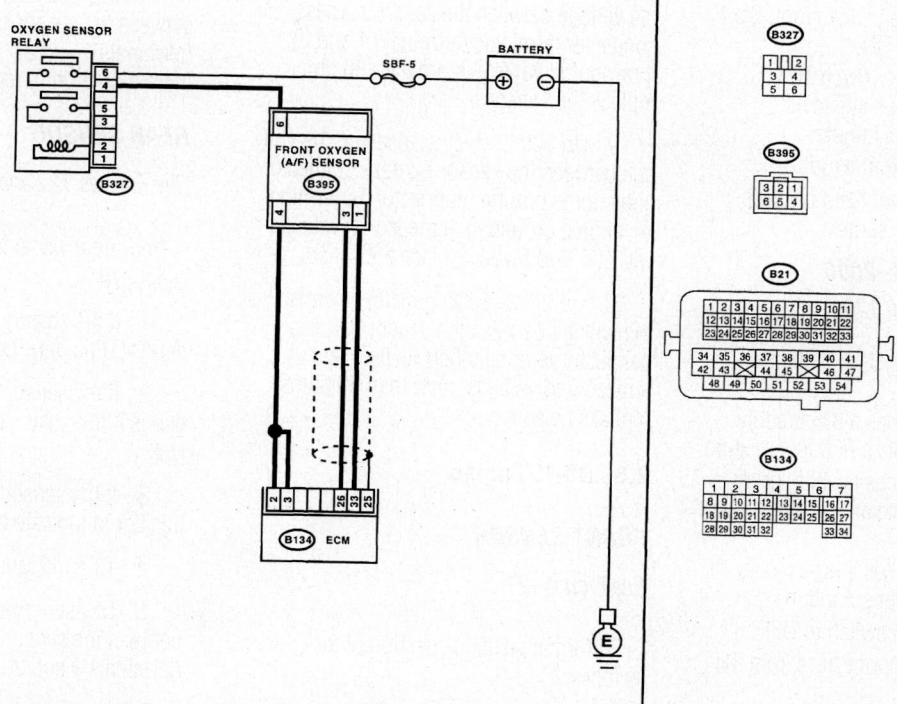

Fig. 115   Front oxygen sensor connector location—Legacy and Outback 2005 2.5L SOHC engine

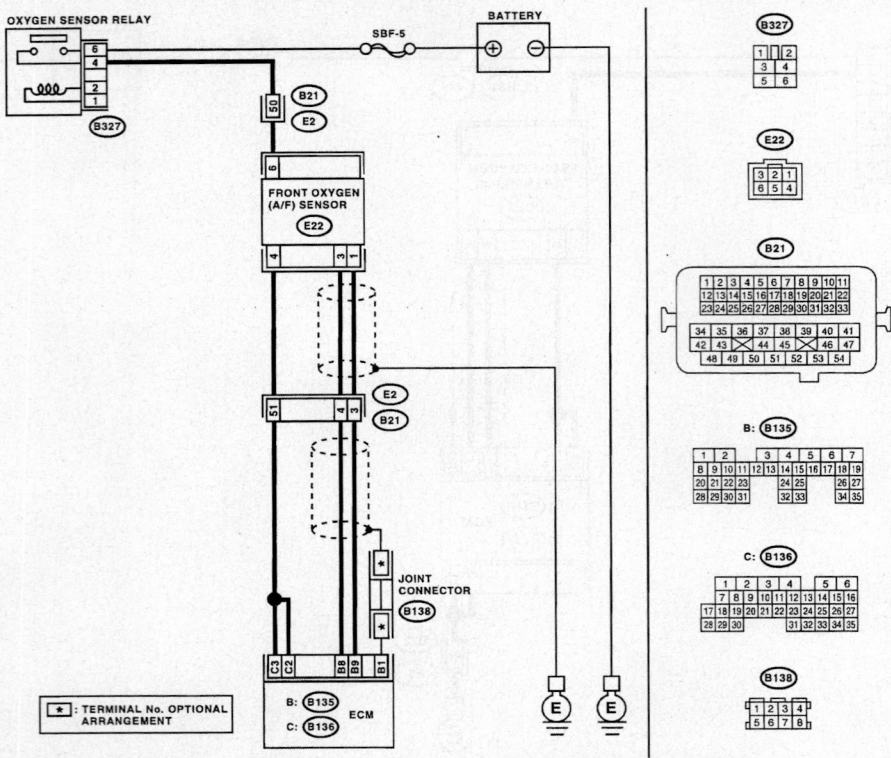

Fig. 116   Front oxygen sensor connector location—Legacy and Outback 2006 2.5L SOHC engine

2. If the sensor tip has a white, gritty deposit, this may indicate an internal coolant leak.

3. If the sensor tip has a brown deposit, this could indicate oil consumption.

4. Turn the engine OFF.

5. Connect the Subaru scan tool. Start the engine. Turn the tool ON.

6. Allow the engine to reach 160 degrees F. Keep the engine speed at 2000–3000 rpm's for one minute.

7. Read the data from the tool.

8. If the reading is not fixed between 0.2–0.4 volt, replace the sensor.

REAR SENSOR–2001–2006

See Figures 117 through 120.

Perform a visual inspection of the sensor as follows:

1. If the sensor tip has a black/sooty deposit, this may indicate a rich fuel mixture.

2. If the sensor tip has a white, gritty deposit, this may indicate an internal coolant leak.

3. If the sensor tip has a brown deposit, this could indicate oil consumption.

4. Turn the ignition switch to OFF.

5. Disconnect the connectors from the sensor.

6. On 2001 vehicles, turn the ignition switch ON.

7. On 2001 vehicles, measure the voltage between the sensor harness connector and engine ground (connector T6 terminal 4+/engine ground -). If measured value is more than 0.2 volt, replace the sensor.

8. On 2002–2003 vehicles, measure the voltage between the sensor harness connector terminals (terminals 1 and 2). If measured value is more than 30 ohms, replace the sensor.

9. On 2004 vehicles, measure the voltage between the sensor harness connector and engine ground (connector T6 terminal 4+/engine ground -). If measured value is within 0.2–0.5 volt, replace the sensor.

10. On 2005–2006 vehicles, measure the voltage between the sensor harness connector terminals (terminals 1 and 2). If measured value is more than 30 ohms, replace the sensor.

2.5L DOHC Engine

FRONT SENSOR

See Figure 121.

Perform a visual inspection of the sensor as follows:

1. If the sensor tip has a black/sooty deposit, this may indicate a rich fuel mixture.

2. If the sensor tip has a white, gritty deposit, this may indicate an internal coolant leak.

3. If the sensor tip has a brown deposit, this could indicate oil consumption.

4. Measure the resistance of the harness between the sensor connector terminals (terminals 1 and 2). 5. If measured value is more than 5 ohm, replace the sensor.

REAR SENSOR

See Figures 122 and 123.

Perform a visual inspection of the sensor as follows:

1. If the sensor tip has a black/sooty deposit, this may indicate a rich fuel mixture.

2. If the sensor tip has a white, gritty deposit, this may indicate an internal coolant leak.

3. If the sensor tip has a brown deposit, this could indicate oil consumption.

4. Turn the ignition switch OFF.

5. Measure the resistance of the harness between the sensor connector terminals (terminals 1 and 2).

6. If measured value is more than 30ohm, replace the sensor.

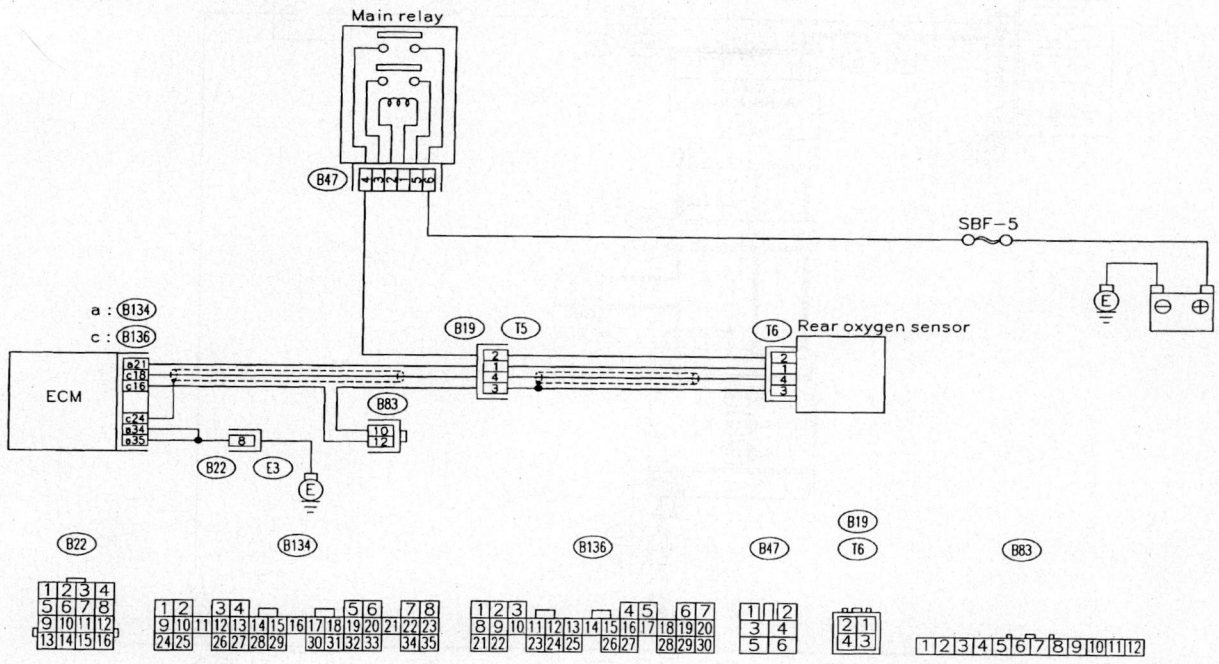

Fig. 117   Rear oxygen sensor connector location—Legacy 1997–2000 2.5L SOHC engine

29157_SUBA_G0119

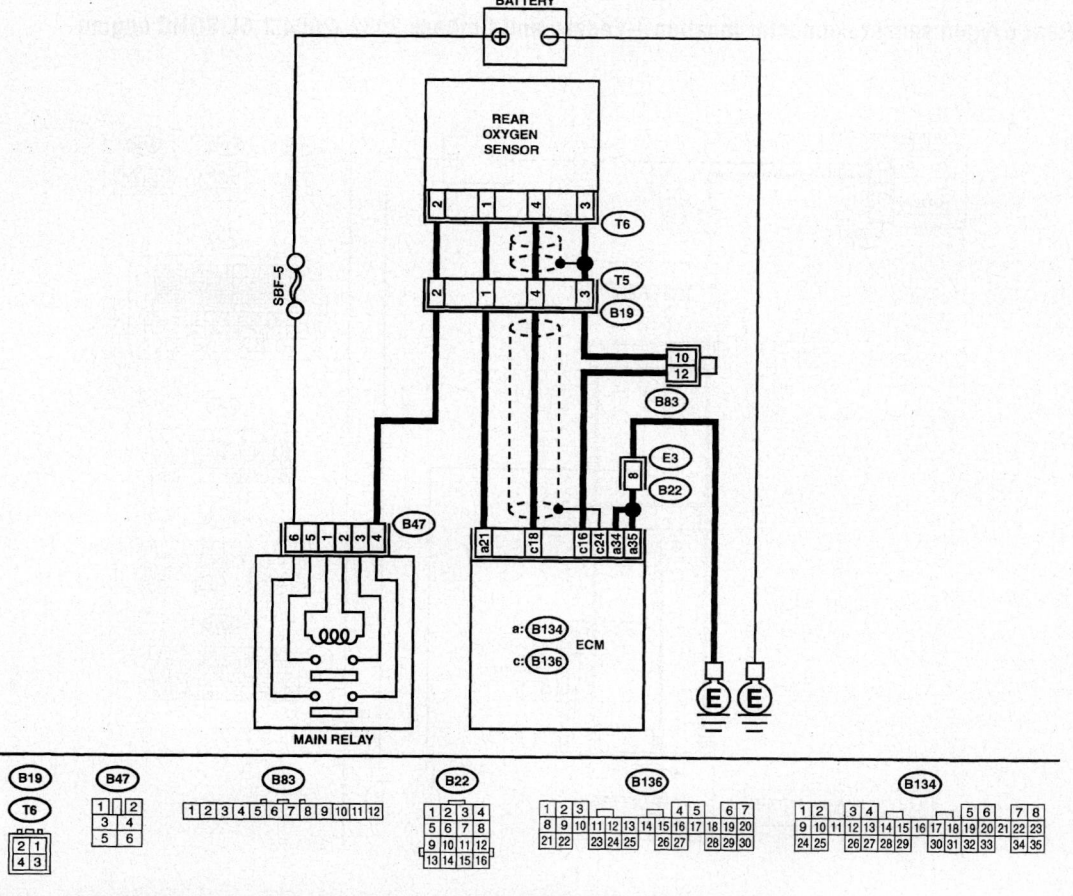

Fig. 118   Rear oxygen sensor connector location—Legacy 2001 2.5L SOHC engine

29157_SUBA_G0120

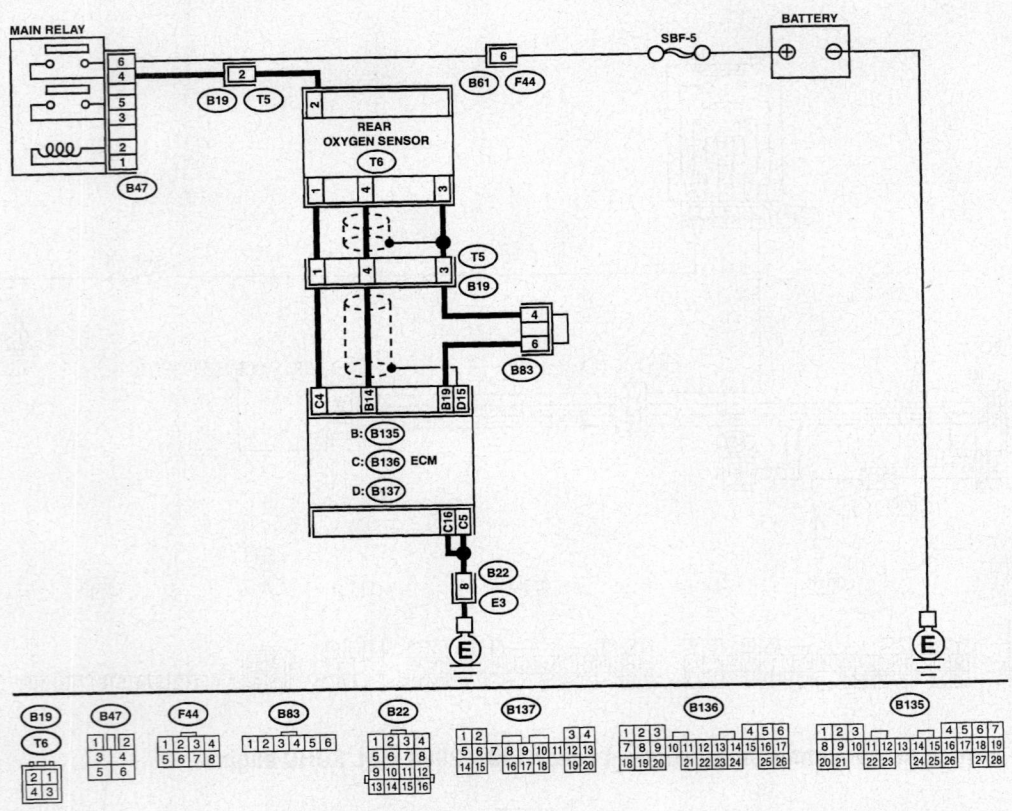

Fig. 119 **Rear oxygen sensor connector location—Legacy and Outback 2002–2004 2.5L SOHC engine**

29157_SUBA_G0121

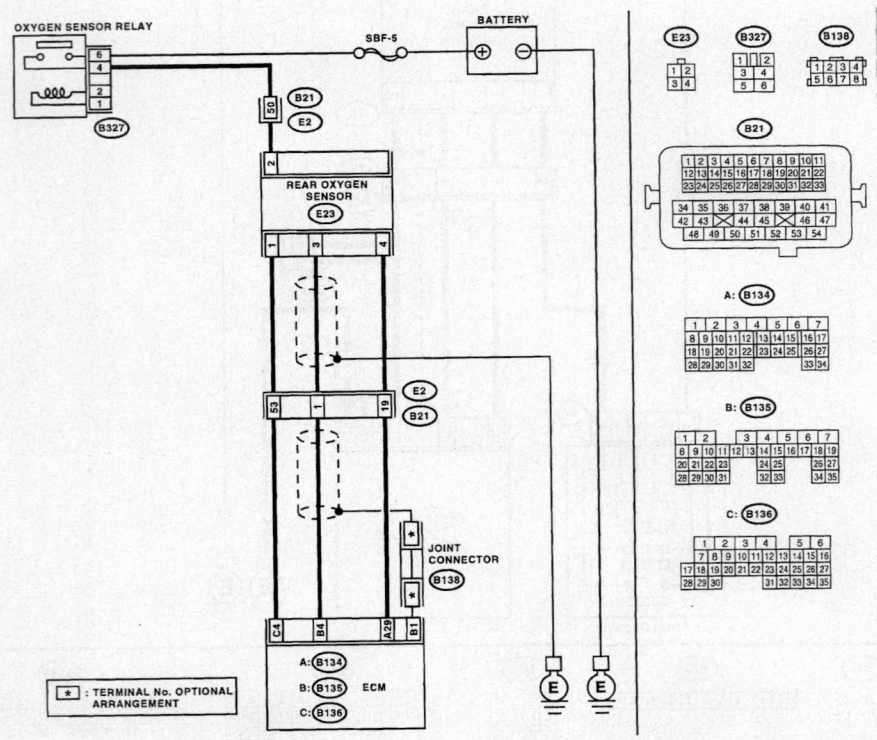

29157_SUBA_G0122

Fig. 120 **Rear oxygen sensor connector location—Legacy and Outback 2005–2006 2.5L SOHC engine**

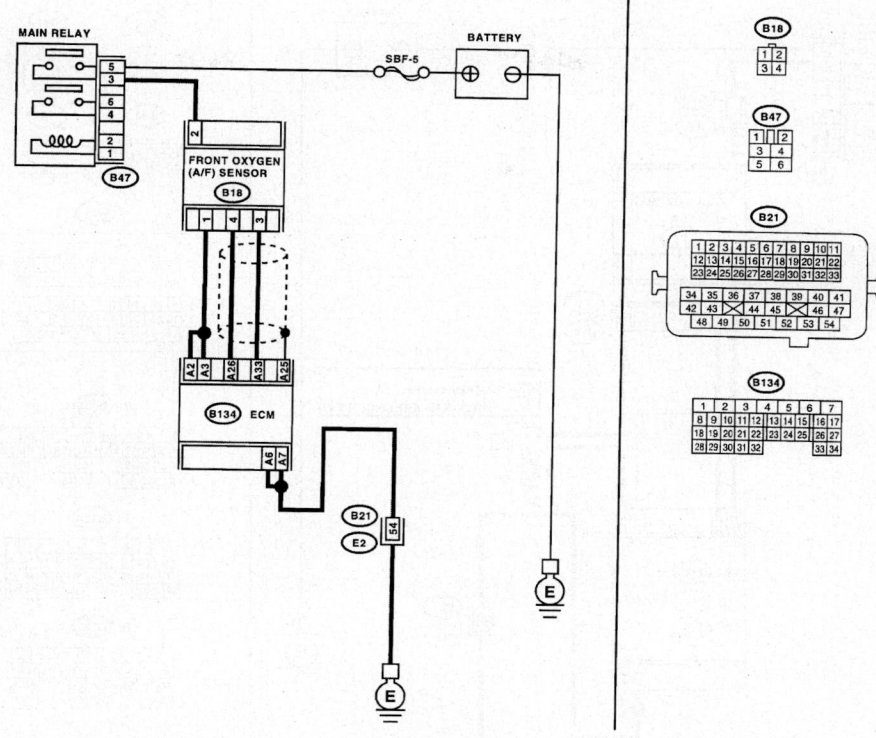

29157_SUBA_G0126

Fig. 121   Front oxygen sensor connector location—Legacy and Outback 2.5L DOHC engine

29157_SUBA_G0127

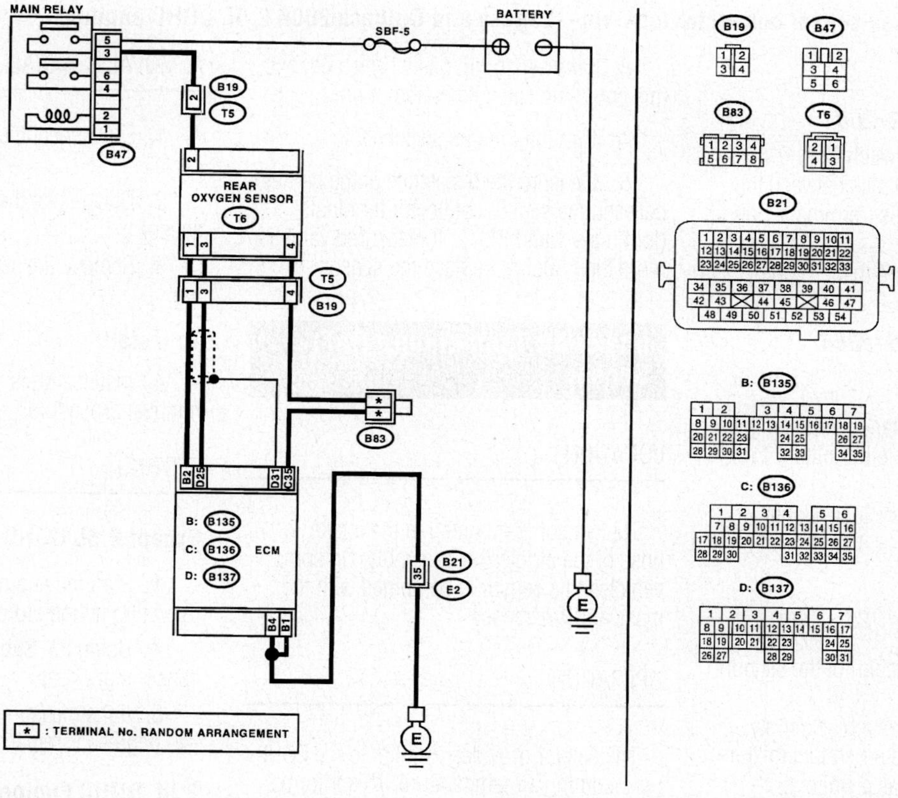

Fig. 122   Rear oxygen sensor connector loca tion—Legacy and Outback 2004–2005 2.5L DOHC engine

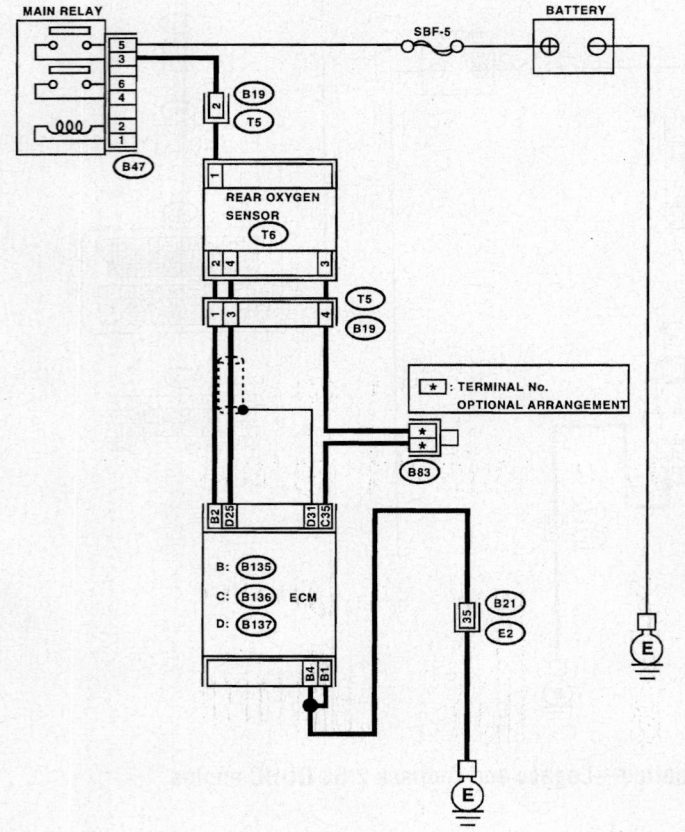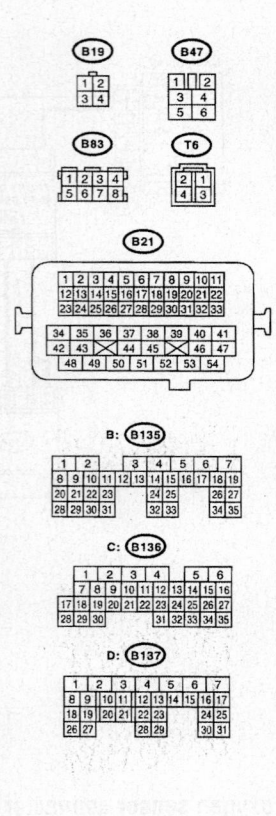

29157_SUBA_G0128

Fig. 123 Rear oxygen sensor connector location—Legacy and Outback 2006 2.5L DOHC engine

3.0L Engine

FRONT SENSOR–2002–2004

1. Turn the ignition switch OFF.

2. Measure the resistance between the sensor connector terminals (terminals 1 and 2).

3. If resistance is more than 10 ohm, replace the sensor.

FRONT SENSOR–2005–2006

See Figures 124 and 125.

4. Measure the resistance between the sensor connector terminals (terminals 1 and 2).

5. If resistance is more than 5 ohm, replace the sensor.

REAR SENSOR

See Figures 126, 127 and 128.

Perform a visual inspection of the sensor as follows:

1. If the sensor tip has a black/sooty deposit, this may indicate a rich fuel mixture.

2. If the sensor tip has a white, gritty deposit, this may indicate an internal coolant leak.

3. If the sensor tip has a brown deposit, this could indicate oil consumption.

4. Turn the ignition switch OFF.

5. Measure the resistance of the harness between the sensor connector terminals (terminals 1 and 2). 6. If measured value is more than 30ohm, replace the sensor.

Intake Air Temperature (IAT) Sensor

LOCATION

The sensor is mounted in the intake air hose of the air cleaner assembly. On some vehicles this sensor is combined with the mass air flow sensor.

OPERATION

The sensor provides a signal to the ECM for incoming air temperature. This information is used to adjust the injector pulse width and in turn the air/fuel ratio.

REMOVAL & INSTALLATION

1. Disconnect the negative battery cable.

2. Disconnect the connector from the sensor.

3. Remove the sensor from its mounting.

To install:

4. Installation is the reverse of the removal procedure.

TESTING

Except 2.5L DOHC Engine

1. Start the engine and allow it to reach operating temperature.

2. Using the Subaru scan tool, read the sensor signal data.

3. If the measured value is less than -40 degrees F, replace the sensor.

2.5L DOHC Engine

1. Turn the ignition switch OFF.

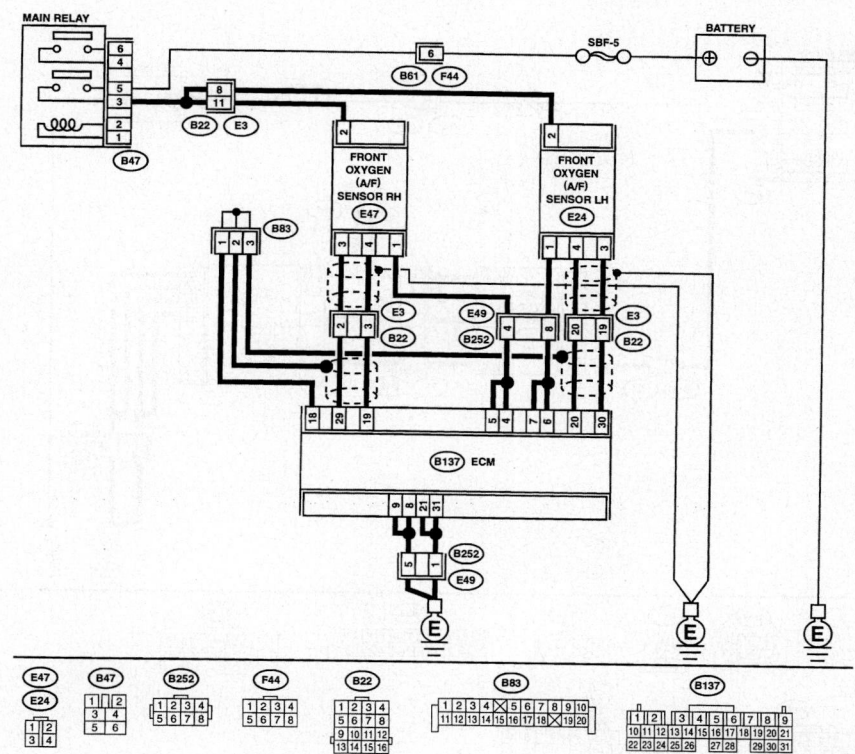

Fig. 124   Front oxygen sensor connector location—Legacy and Outback 2002–2004 3.0L engine

29157_SUBA_G0129

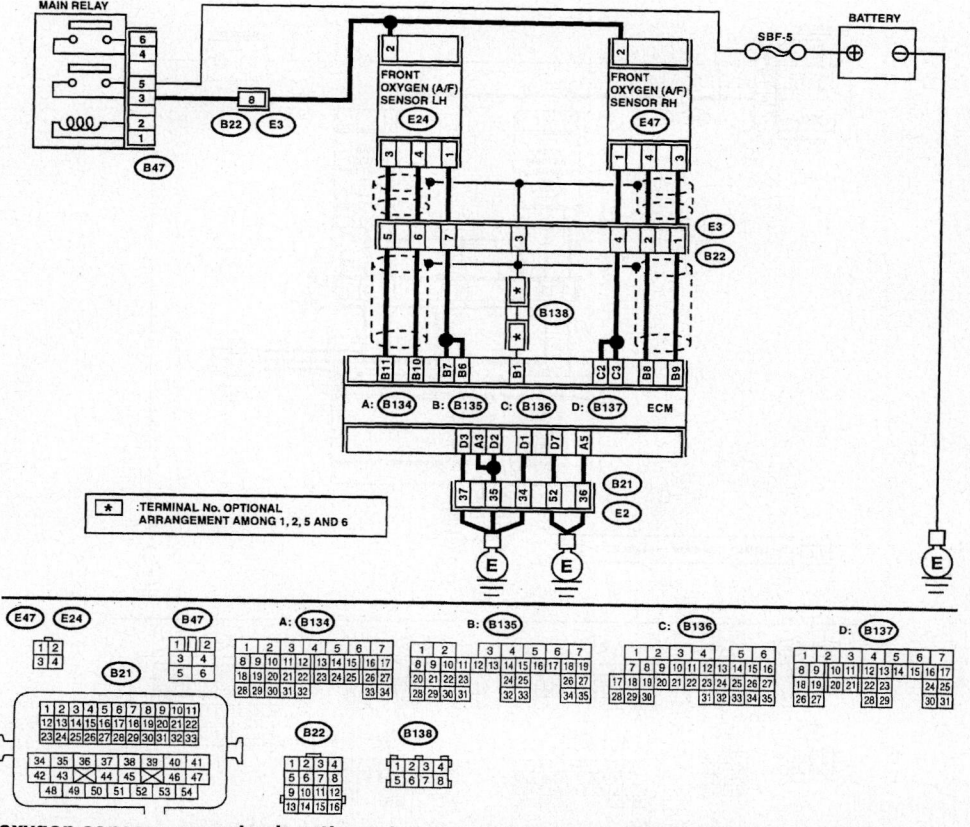

Fig. 125   Front oxygen sensor connector location—Legacy and Outback 2005–2006 3.0L engine

29157_SUBA_G0130

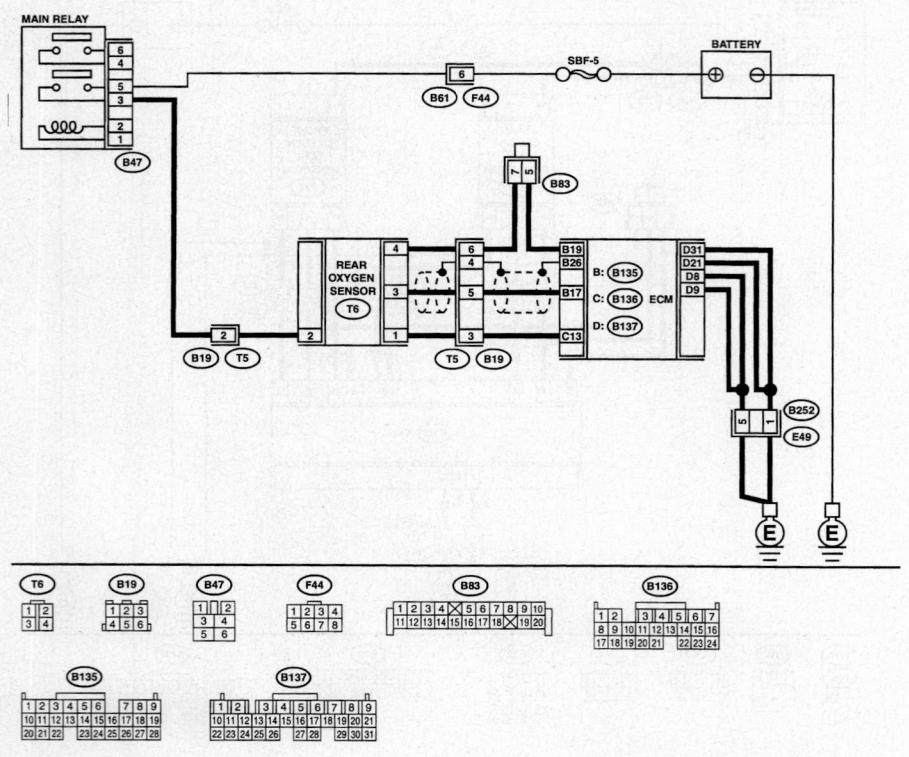

Fig. 126　Rear oxygen sensor connector location—Legacy and Outback 2002–2004 3.0L engine

29157_SUBA_G0131

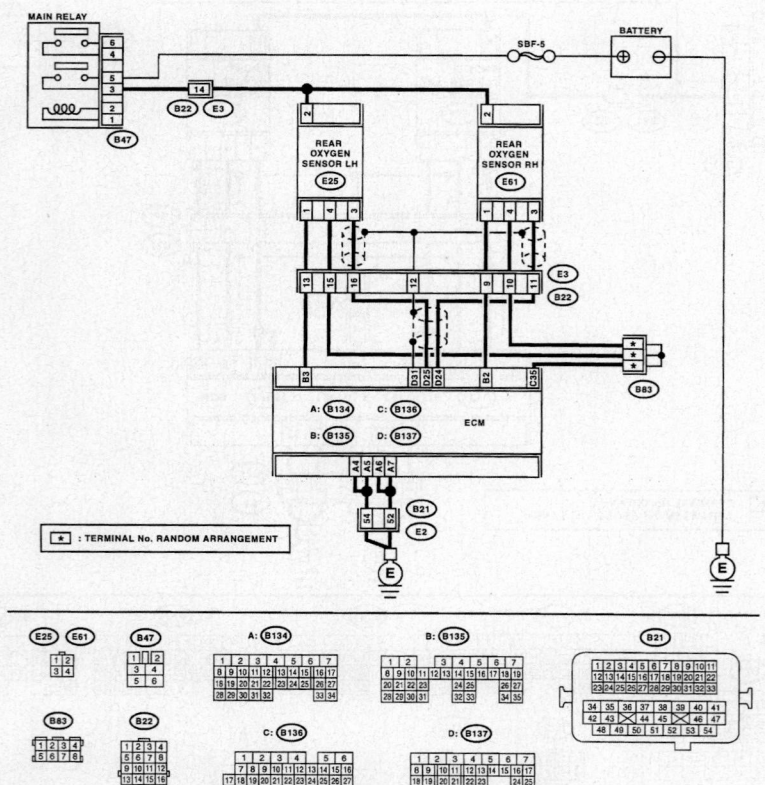

Fig. 127　Rear oxygen sensor connector location—Legacy and Outback 2005 3.0L engine

29157_SUBA_G0132

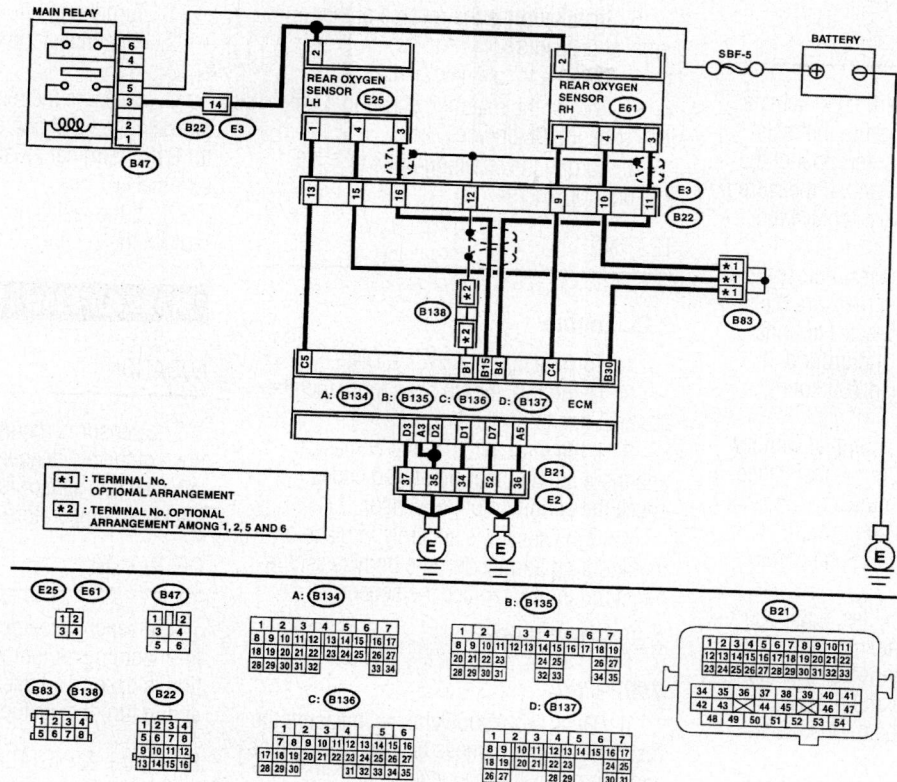

Fig. 128   Rear oxygen sensor connector location—Legacy and Outback 2006 3.0L engine

29157_SUBA_G0133

2. Measure the resistance of the harness between the sensor and engine ground.

3. If measured value is less than 5 ohms, replace the sensor.

Knock Sensor (KS)

LOCATION

The sensor is located at the top right (driver's) side of the engine and is positioned on the cylinder block, on all engines except the 3.0L engine. On the 3.0L engine the sensor is located under the intake manifold.

OPERATION

The sensor is used to detect engine vibrations caused by preignition or detonation and provides information to the ECM, which then retards the timing to eliminate detonation.

REMOVAL & INSTALLATION

Except 3.0L Engine

1. Disconnect the negative battery cable.

2. Remove the air cleaner case.

3. Remove the collector cover, as required.

4. On DOHC engine, remove the intercooler.

5. Disconnect the sensor connector.

6. Remove the sensor from its mounting.

To install:

7. Installation is the reverse of the removal procedure.

8. Tighten the sensor to 17.4 ft. lbs.

➡ **The extraction area of the knock sensor wire must be positioned at a 60 degree angle relative to the engine rear.**

3.0L Engine

See Figure 129.

1. Disconnect the negative battery cable.

2. Remove the collector cover, as required.

3. Remove the intake manifold.

4. Disconnect the sensor connector.

5. Remove the sensor from its mounting.

To install:

6. Installation is the reverse of the removal procedure.

7. Refer to the illustration for proper sensor installation angle.

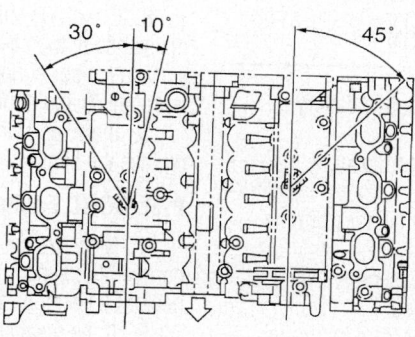

29157_SUBA_G0134

Fig. 129   Knock sensor installation angle—Legacy and Outback 3.0L engine

8. Tighten the sensor to 18.0 ft. lbs.

TESTING

1. Remove the sensor from the vehicle.
2. On 2.2L engine, measure the resistance between the connector terminal of the sensor, (terminal 2) and ground. If measured value is less than 400 kohms, replace the sensor.
3. On 2002–2005 vehicles, equipped with the 2.5L SOHC engine, measure the resistance between the connector terminal of the sensor, (terminal 2) and ground. If measured value is more than 700 kohms, replace the sensor.
4. On 2006 vehicles, equipped with the 2.5L SOHC engine, measure the resistance between the connector terminals of the sensor, (terminal 1 and 2) and ground. If measured value is more than 700 kohms, replace the sensor.
5. On 2004–2006 vehicles, equipped with the 2.5L DOHC engine, measure the resistance between the connector terminal of the sensor, (terminal 2) and ground. If measured value is more than 700 kohms, replace the sensor.
6. On 2004–2006 vehicles, equipped with the 3.0L engine, measure the resistance between the connector terminals of the sensor, (terminal 1 and 2). If measured value is more than 700 kohms, replace the sensor.

Manifold Absolute Pressure (MAP) Sensor

LOCATION

The sensor is located on the throttle body unit.

OPERATION

The sensor monitors and signals the ECM of changes in intake manifold pressure which result from engine load, speed and atmospheric pressure changes.

REMOVAL & INSTALLATION

1. Disconnect the negative battery cable.
2. Remove the collector cover, if equipped.
3. Disconnect the connector from the sensor.
4. Remove the filter assembly from the intake manifold, if equipped.
5. Remove the sensor from its mounting.

To install:

6. Installation is the reverse of the removal procedure.
7. Be sure to use new O-rings.
8. Torque the retaining screw to 1.2 ft. lbs., except 3.0L engine.
9. Torque the retaining bolts to 4.7 ft. lbs. on 3.0L engine.

TESTING

2.2L Engine

1. Turn the ignition switch ON.
2. Using the Subaru scan tool, operate the LED operation mode.
3. If the idle switch signal comes on and there are no other displayed codes check the condition of the sensor.
4. If the sensor is securely in place on its mounting and the throttle body is secure on its mounting, replace the sensor.

2.5L SOHC Engine

1997–2004

1. On 1997–2000 vehicles, measure the resistance of the harness between the sensor and engine ground (connector E21 and terminal 1).
2. On 2001–2002 vehicles, measure the resistance of the harness between the sensor and engine ground (connector E21 and terminal 4).
3. On 2003–2004 vehicles, measure the resistance of the harness between the sensor and engine ground (connector E21 and terminal 2).
4. If the resistance is more than 500 Kohm.

2005–2006

5. Turn the ignition switch ON.
6. Disconnect the connector from the ECM.
7. On 2005 vehicles, measure the resistance of the harness between the ECM and the sensor connector (connector B136 terminal 35 and connector E21 terminal 2).
8. On 2006 vehicles, measure the resistance of the harness between the ECM and the sensor connector (connector B134 terminal 29 and connector E21 terminal 2).
9. If the resistance is less than 1 ohm, replace the sensor.

2.5L DOHC Engine

1. Measure the resistance of the harness between the sensor and engine ground (connector E21 and terminal 1).
2. If the resistance is more than 1 Mohm, replace the sensor.

3.0L Engine

1. Turn the ignition switch OFF.
2. Disconnect the connectors from the ECM.
3. Measure the resistance of the harness between the ECM and the sensor (connector B134 terminal 29 and connector E21 terminal1).
4. If the resistance is less than 1 ohm, replace the sensor.

Mass Air Flow (MAF) Sensor

LOCATION

The sensor is mounted in the intake air hose of the air cleaner assembly. On some vehicles this sensor is combined with the intake air temperature sensor.

OPERATION

The sensor provides a signal to the ECM for incoming air temperature. This information is used to adjust the injector pulse width and in turn the air/fuel ratio.

REMOVAL & INSTALLATION

1. Disconnect the negative battery cable.
2. Disconnect the connector from the sensor.
3. Remove the sensor from its mounting.

To install:

4. Installation is the reverse of the removal procedure.
5. Torque the retaining screw to 0.8 ft. lbs.

TESTING

2.2L Engine

1. Turn the ignition switch OFF. Turn the Subaru scan tool switch OFF.
2. Disconnect the connector from the sensor.
3. Turn the ignition switch ON. Turn the Subaru scan tool switch ON.
4. Read the data from the tool.
5. If the value is more than 5 volts, replace the sensor.

2.5L SOHC Engine

6. Turn the ignition switch OFF.
7. Disconnect the connector from the ECM.
8. On all except 2006 vehicles, measure the resistance of the harness between the

ECM and the sensor connector (connector B135 terminal 19/connector E21 terminal 1).

9. On 2006 vehicles, measure the resistance of the harness between the ECM and the sensor connector (connector B3 terminal 2/connector B135 terminal 34).

10. If the resistance is less than 1 ohm, replace the sensor.

2.5L DOHC Engine

1. Turn the ignition switch OFF.

2. Measure the resistance of the harness between the sensor and engine ground.

3. If measured value is less than 5 ohms, replace the sensor.

3.0L Engine

4. Start the engine and allow it to reach operating temperature.

5. Using the Subaru scan tool, read the sensor signal data.

6. If the measured value is less than -40 degrees F, replace the sensor.

Throttle Position Sensor (TPS)

LOCATION

On the 2.2L engine the sensor is located near the center of the engine by the air cleaner assembly. On the 1997–2004 vehicles, equipped with the 2.5L SOHC engine, the sensor is located near the radiator hose and the engine insulator mount. On 2005–2006 vehicles, equipped with the 2.5L SOHC engine and 2004–2006 vehicles, equipped with the 2.5L DOHC engine, the conventional TPS sensor is not used. The sensor used is referred to as an electric throttle. On the 3.0L engine the sensor is located near the center of the engine by the air cleaner assembly.

OPERATION

The throttle position sensor provides a signal to the ECM that is related to the relative throttle plate position. As the throttle plate moves in relation to driving conditions, a signal is sent to the control unit which adjusts the injector pulse width and air/fuel ratio. As the throttle plate is opened further, more air is taken into the combustion chambers, and as a result the relative fuel demand of the engine changes. The throttle position sensor relays this information to the ECM which in alters the fuel amount.

REMOVAL & INSTALLATION

2.2L Engine

1. Disconnect the negative battery cable.

2. Disconnect the sensor connector.

3. Remove the sensor retaining screws. Remove the sensor from its mounting.

To install:

4. Installation is the reverse of the removal procedure.

55. Tighten the sensor to 1.7 ft. lbs.

2.5L SOHC Engine

1. Disconnect the negative battery cable.

2. Disconnect the sensor connector.

3. Remove the sensor retaining screws. Remove the sensor from its mounting.

To install:

4. Installation is the reverse of the removal procedure.

5. Tighten the sensor to 1.7 ft. lbs.

6. To adjust the sensor, turn the ignition switch OFF. Loosen the retaining screws.

7. Using a voltage meter, take out the ECM. Turn the ignition switch ON.

8. Adjust the TPS to the proper position to allow the voltage signal to the ECM to be within specification.

9. Specification is 0.45–0.55 volt. Specification is measured at connector B135 and terminal 13.

10. Tighten the retaining screws.

3.0L Engine

1. Disconnect the negative battery cable.

2. Remove the air intake chamber.

3. Disconnect the sensor connector.

4. Remove the sensor retaining screws. Remove the sensor from its mounting.

To install:

5. Installation is the reverse of the removal procedure.

TESTING

2.2L Engine

1. Start the engine.

2. Does the Subaru scan tool; indicate DTC P0122 or DTC P0123?

3. If not replace the sensor.

2.5L SOHC Engine

1997–2001

1. Visually inspect the throttle linkage and throttle for binding and sticking.

2. Turn the ignition switch ON.

3. Measure the voltage between the sensor and engine ground, (connector E13 terminal 3 and ground).

4. If the measured value is less than 4.9 volts, replace the sensor.

2002–2004

1. Visually inspect the throttle linkage and throttle for binding and sticking.

2. Measure the resistance of the harness between the TPS connector and engine ground, (connector E13 terminal 3 and ground).

3. If the measured value is more than 1 Mohm, replace the sensor.

2005–2006

4. Turn the ignition switch OFF.

5. Measure the resistance between the electronic throttle control connector and ground (connector E57 terminal 6+ and ground).

6. If the measured value is more than 10 ohm, replace the sensor.

2.5L DOHC Engine

1. Turn the ignition switch OFF.

2. Measure the resistance between the electronic throttle control connector and ground (connector E57 terminal 6+ and ground).

3. If the measured value is more than 10 ohm, replace the sensor.

3.0L SOHC Engine

2002–2003

1. Visually inspect the throttle linkage and throttle for binding and sticking.

2. Turn the ignition switch ON.

3. Measure the voltage between the sensor and engine ground, (connector E13 terminal 3 and ground).

4. If the measured value is less than 4.9 volts, replace the sensor.

2004

5. Visually inspect the throttle linkage and throttle for binding and sticking.

6. Measure the resistance of the harness between the TPS connector and engine ground, (connector E13 terminal 3 and ground).

7. If the measured value is more than 10 ohm, replace the sensor.

2005–2006

8. Turn the ignition switch OFF.

9. Disconnect the connector from the ECM.

10. On 2005 vehicles, measure the resistance between the ECM connectors (connector B136 terminal 18 and connector B136 terminal 16).

11. On 2006 vehicles, measure the resistance between the ECM connectors (connector B134 terminal 18 and connector B134 terminal 19).

12. If the measured value is more than 1 Mohm, repair the electronic throttle control.

Vehicle Speed Sensor (VSS)

LOCATION

Vehicles equipped with automatic transaxle use a front and rear sensors.

SVX

See Figure 130.

Camshaft Position (CMP) Sensor

OPERATION

The camshaft position sensor relays the relative camshaft position to the ECM for determining the position and stroke of the No. 1 cylinder. This information is required for proper fuel injection functioning.

Both sensors are mounted on the transaxle. Vehicles equipped with manual transaxle use a front sensor. This sensor is mounted on the transaxle.

OPERATION

The vehicle speed sensor sends a signal to the ECM to control. The ECM uses this information to control transmission shift patterns. The vehicle speed sensor is also used to send a speed signal to the speed control servo of the cruise control system.

REMOVAL & INSTALLATION

1. Disconnect the negative battery cable.

2. Remove the collector cover, as required.

3. Disconnect the connector from the sensor.

4. Remove the bolt that retains the sensor.

5. Remove the sensor.

REMOVAL & INSTALLATION

1. Raise and support the vehicle safely.

2. Place a drip pan below the speed sensor to catch any spilled fluid.

3. Disconnect the connector.

4. Remove the sensor from its mounting.

To install:

5. Installation is the reverse of the removal procedure.

6. Replace any lost fluid.

To install:

6. Installation is the reverse of the removal procedure.

7. Tighten the sensor to 4.7 ft. lbs.

TESTING

1. Remove the sensor from the vehicle.

2. Measure the resistance between the two connector terminals of the sensor, (terminals 1 and 2).

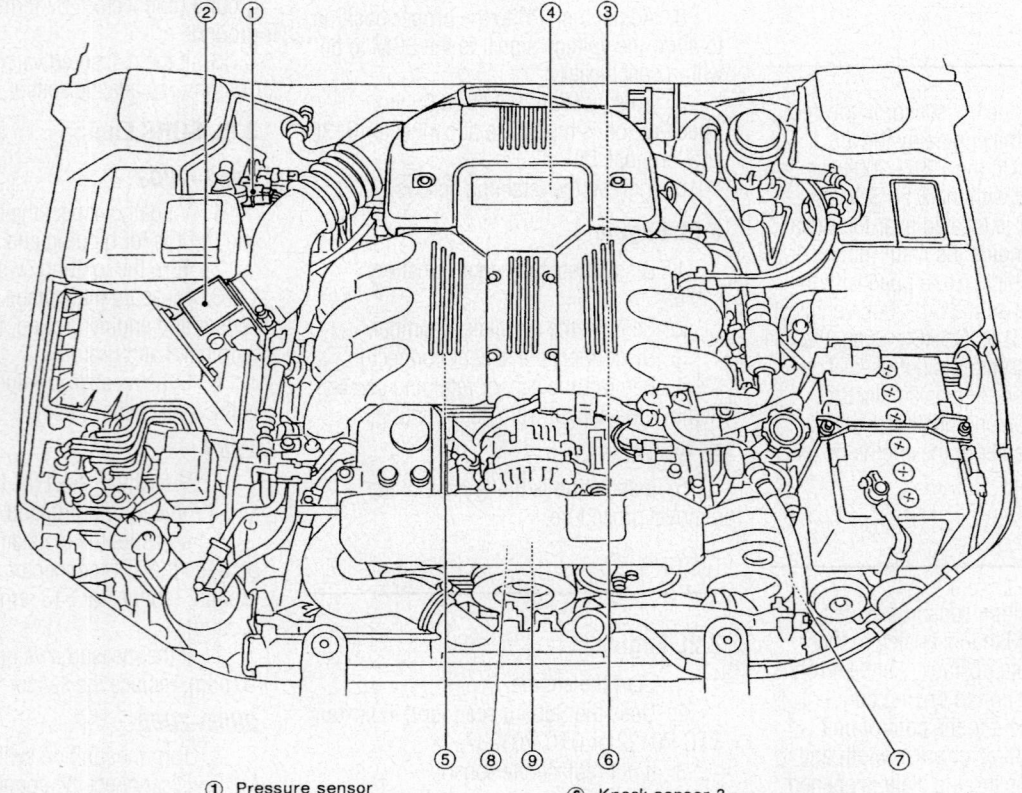

① Pressure sensor
② Mass air flow sensor
③ Engine coolant temperature sensor
④ Throttle position sensor
⑤ Knock sensor 1
⑥ Knock sensor 2
⑦ Camshaft position sensor
⑧ Crankshaft position sensor 1
⑨ Crankshaft position sensor 2

29157_SUBA_G0010

Fig. 130 Underhood sensor locations—SVX

3. If measured value is not within 1–4 kohms, replace the sensor.

Crankshaft Position (CKP) Sensor

LOCATION

The sensor is located in the front of the engine near the crankshaft pulley.

OPERATION

The crankshaft position sensor is a variable reluctance sensor which is used to inform the ECM when the No.1 piston is at top dead center. This information is used for controlling and adjusting ignition and fuel injector timing.

REMOVAL & INSTALLATION

1. Disconnect the negative battery cable.
2. Remove the collector cover, as required.
3. As required, remove the air intake chamber.
4. Remove the bolt that retains the sensor in place at the cylinder block.
5. Remove the sensor from its mounting.
6. Disconnect the connector from the sensor.

To install:

7. Installation is the reverse of the removal procedure.
8. Tighten the sensor to 4.7 ft. lbs.

TESTING

1. Remove the sensor from the vehicle.
2. Measure the resistance between the two connector terminals of the sensor, (terminals 1 and 2).
3. If measured value is not within 1–4 kohms, replace the sensor.

Electronic Control Module (ECM)

LOCATION

See Figure 131.

The ECM is located on the passenger's side of the vehicle, underneath the instrument panel glove box area.

OPERATION

The ECM controls the vehicle engine operating system.

REMOVAL & INSTALLATION

1. Disconnect the negative battery cable.

2. Remove the lower inner trim on the passenger's side of the vehicle.
3. Detach the floor mat. Remove the protective cover.
4. Remove the ECM bracket retaining nuts. Remove the clip from the bracket.
5. Disconnect the connectors.
6. Remove the ECM from the vehicle.

To install:

7. Installation is the reverse of the removal procedure.
8. Tighten the retaining screws to 3.7 ft. lbs.

➡ **When replacing the ECM, be careful not to use the wrong part number, as damage to the injection system could occur.**

EGR Valve

LOCATION

The EGR valve is mounted on the intake manifold.

OPERATION

The EGR valve controls the flow of exhaust gases. The ECM monitors the flow and regulates the valve accordingly.

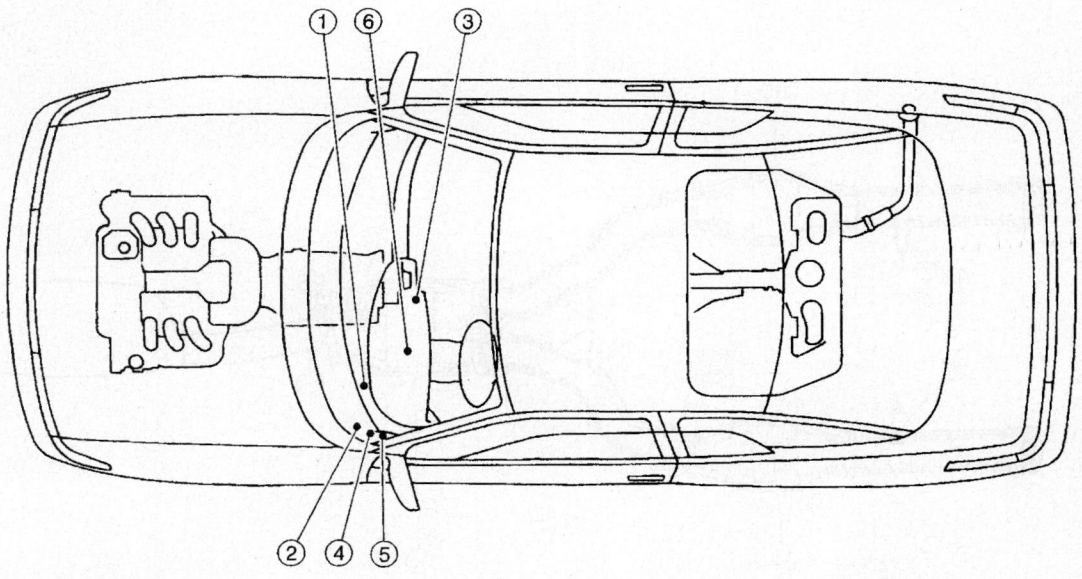

① Engine control module (ECM)
② Data link connector (for Subaru select monitor only)
③ Data link connector (for Subaru select monitor and OBD-II general scan tool)
④ Diagnosis connector (Black)
⑤ Diagnosis terminal
⑥ CHECK ENGINE malfunction indicator lamp (MIL)

29157_SUBA_G0009

Fig. 131 ECM and related components—SVX

REMOVAL & INSTALLATION

1. Disconnect the negative battery cable.
2. Disconnect the connector.
3. On valves equipped with a vacuum hose, disconnect it.
4. Remove the valve from the intake manifold.
5. Discard the gasket.

To install:

6. Installation is the reverse of the removal procedure.
7. Be sure to use a new gasket.
8. Tighten the retaining bolts to 14.0 ft. lbs.

TESTING

1. Check and replace vacuum hoses, if they are clogged or brittle.
2. Check the valve for blockage, replace or clean as required.

Engine Coolant Temperature (ECT) Sensor

LOCATION

The sensor is located by the heater outlet fitting or in a cooling passage on the engine, depending upon the particular vehicle.

OPERATION

This component detects the temperature of the engine coolant and relays the information to the electronic control assembly.

REMOVAL & INSTALLATION

1. Disconnect the negative battery cable.
2. Remove the air intake duct and air cleaner case, as required.
3. Disconnect the connector from the sensor.
4. Drain the cooling system, as required.
5. Remove the sensor from its mounting.

To install:

6. Installation is the reverse of the removal procedure.
7. Tighten the sensor to 13.3 ft. lbs.

TESTING

1. Turn the ignition switch to OFF.
2. Disconnect the connector from the sensor.
3. Turn the ignition switch ON.
4. Using the Subaru scan tool, read the sensor signal data. If the measured value is less than -40 degrees F, replace the sensor.

Heated Oxygen (HO2S) Sensor

LOCATION

See Figure 132.

The front oxygen sensors are located in the front section of the front exhaust pipes. The rear oxygen sensor is located in the rear section of the rear catalytic converter. There are two front catalytic converters, but only one rear catalytic converter.

OPERATION

The exhaust gas oxygen sensor supplies the electronic control assembly with a signal which indicates either a rich or lean mixture condition, during the engine operation.

REMOVAL & INSTALLATION

Front Sensor

1. Disconnect the negative battery cable.
2. Remove the necessary components to gain access to the sensor.
3. Disconnect the clip fastening the harness, if equipped.
4. Disconnect the connector.
5. Remove the oxygen sensor.

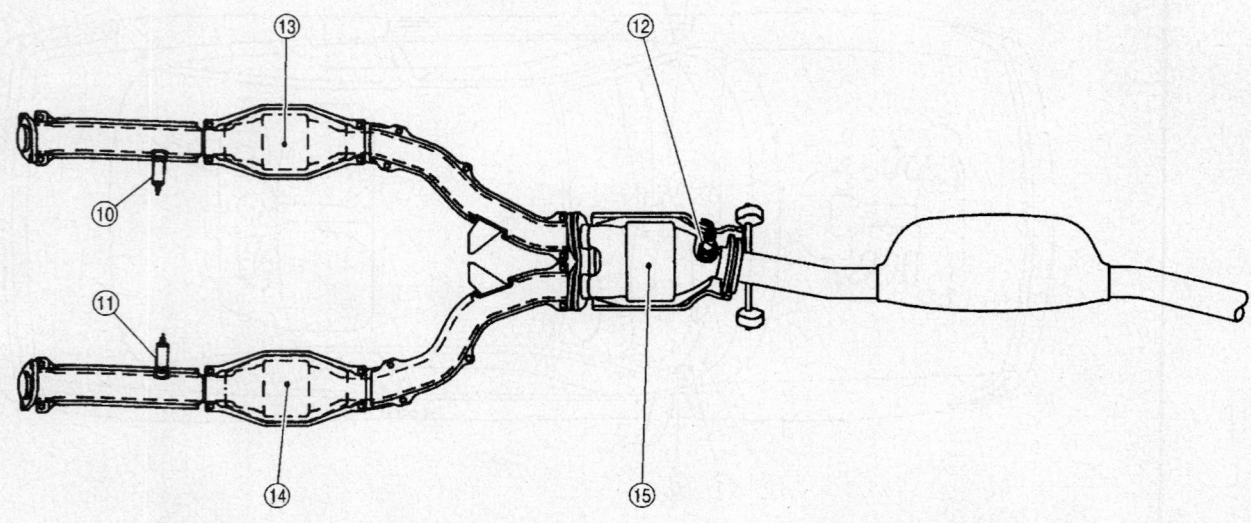

⑩ Front oxygen sensor 1 (RH)
⑪ Front oxygen sensor 2 (LH)
⑫ Rear oxygen sensor
⑬ Front catalytic converter (RH)
⑭ Front catalytic converter (LH)
⑮ Rear catalytic converter

29157_SUBA_G0011

Fig. 132 Oxygen sensor location—SVX

To install:

6. Installation is the reverse of the removal procedure.

➡ **Apply anti-seize compound to the threaded portion of the sensor, prior to installation. Never apply anti-seize compound to the protector of the sensor.**

7. Tighten the sensor to 15.2 ft. lbs.

Rear Sensor

1. Disconnect the negative battery cable.
2. Raise and support the vehicle safely.
3. Disconnect the connector.
4. Remove the oxygen sensor.

To install:

5. Installation is the reverse of the removal procedure.

➡ **Apply anti-seize compound to the threaded portion of the sensor, prior to installation. Never apply anti-seize compound to the protector of the sensor.**

TESTING

Front Sensor

Perform a visual inspection of the sensor as follows:

1. If the sensor tip has a black/sooty deposit, this may indicate a rich fuel mixture.
2. If the sensor tip has a white, gritty deposit, this may indicate an internal coolant leak.
3. If the sensor tip has a brown deposit, this could indicate oil consumption.
4. Check the exhaust system for defects.
5. Turn the ignition switch OFF.
6. Connect the Subaru scan tool. Start the engine. Turn the tool ON.
7. Allow the engine to reach 160 degrees F. Keep the engine speed at 2000–3000 rpm's for one minute.
8. Read the data from the tool.
9. If the reading is more than 0.1 volt between the value of max. output and min. output, replace the sensor.

Rear Sensor

Perform a visual inspection of the sensor as follows:

1. If the sensor tip has a black/sooty deposit, this may indicate a rich fuel mixture.
2. If the sensor tip has a white, gritty deposit, this may indicate an internal coolant leak.

3. If the sensor tip has a brown deposit, this could indicate oil consumption.
4. Turn the ignition switch to OFF.
5. Disconnect the connectors from the sensor.
6. Turn the ignition switch ON.
7. Measure the voltage between the sensor harness connector and engine ground (connector B24 and terminal 4+/engine ground -). If measured value is more than 0.2 volt, replace the sensor.

Intake Air Temperature (IAT) Sensor

LOCATION

The sensor is mounted in the intake air hose of the air cleaner assembly. On some vehicles this sensor is combined with the mass air flow sensor.

OPERATION

The sensor provides a signal to the ECM for incoming air temperature. This information is used to adjust the injector pulse width and in turn the air/fuel ratio.

REMOVAL & INSTALLATION

1. Disconnect the negative battery cable.
2. Disconnect the connector from the sensor.
3. Remove the sensor from its mounting.

To install:

4. Installation is the reverse of the removal procedure.

TESTING

1. Start the engine and allow it to reach operating temperature.
2. Using the Subaru scan tool, read the sensor signal data.
3. If the measured value is less than -40 degrees F, replace the sensor.

Knock Sensor (KS)

LOCATION

The sensor is located at the top right (driver's) side of the engine and is positioned on the cylinder block.

OPERATION

The sensor is used to detect engine vibrations caused by preignition or detonation and provides information to the ECM, which then retards the timing to eliminate detonation.

REMOVAL & INSTALLATION

1. Disconnect the negative battery cable.
2. Remove the air cleaner case.
3. Remove the collector cover, as required.
4. Remove the components in order to gain access to the sensor.
5. Disconnect the sensor connector.
6. Remove the sensor from its mounting.

To install:

7. Installation is the reverse of the removal procedure.
8. Tighten the sensor to 17.4 ft. lbs.

➡ **The extraction area of the knock sensor wire must be positioned at a 60 degree angle relative to the engine rear.**

TESTING

1. Disconnect the connector from the sensor.
2. Measure the resistance between the connector terminals of the sensor, (terminal 1 and 2). If measured value is more than 700 kohms, replace the sensor.

Manifold Absolute Pressure (MAP) Sensor

LOCATION

The sensor is located on the throttle body unit.

OPERATION

The sensor monitors and signals the ECM of changes in intake manifold pressure which result from engine load, speed and atmospheric pressure changes.

REMOVAL & INSTALLATION

1. Disconnect the negative battery cable.
2. Remove the collector cover, if equipped.

3. Disconnect the connector from the sensor.

4. Remove the filter assembly from the intake manifold, if equipped.

5. Remove the sensor from its mounting.

To install:

6. Installation is the reverse of the removal procedure.

7. Be sure to use new O-rings, as required.

TESTING

1. Turn the ignition switch ON.

2. Using the Subaru scan tool, operate the LED operation mode.

3. If the idle switch signal comes on and there are no other displayed codes check the condition of the sensor.

4. If the sensor is securely in place on its mounting and the throttle body is secure on its mounting, replace the sensor.

Mass Air Flow (MAF) Sensor

LOCATION

The sensor is mounted in the intake air hose of the air cleaner assembly. On some vehicles this sensor is combined with the intake air temperature sensor.

OPERATION

The sensor provides a signal to the ECM for incoming air temperature. This information is used to adjust the injector pulse width and in turn the air/fuel ratio.

REMOVAL & INSTALLATION

1. Disconnect the negative battery cable.

2. Disconnect the connector from the sensor.

3. Remove the sensor from its mounting.

To install:

4. Installation is the reverse of the removal procedure.

5. Torque the retaining screw to 0.8 ft. lbs.

TESTING

1. Turn the ignition switch OFF.

2. Disconnect the connector from the ECM.

3. Measure the resistance of the harness between the ECM and the sensor connector (connector B59 terminal 5/connector B1 terminal 4).

4. If the resistance is less than 1 ohm, replace the sensor.

Throttle Position Sensor (TPS)

LOCATION

The sensor is located near the center of the engine by the air cleaner assembly.

OPERATION

The throttle position sensor provides a signal to the ECM that is related to the relative throttle plate position. As the throttle plate moves in relation to driving conditions, a signal is sent to the control unit which adjusts the injector pulse width and air/fuel ratio. As the throttle plate is opened further, more air is taken into the combustion chambers, and as a result the relative fuel demand of the engine changes. The throttle position sensor relays this information to the ECM which in alters the fuel amount.

REMOVAL & INSTALLATION

1. Disconnect the negative battery cable.

2. Remove the collector cover.

3. Disconnect the sensor connector.

4. Remove the sensor retaining screws. Remove the sensor from its mounting.

To install:

5. Installation is the reverse of the removal procedure.

6. Tighten the sensor to 1.7 ft. lbs.

TESTING

1. Turn the ignition switch OFF.

2. Measure the resistance of the harness between the sensor connector and ground (connector E11 terminal 2) and ground.

3. If the measured value is more than 10 ohm, replace the sensor.

Vehicle Speed Sensor (VSS)

LOCATION

Vehicles equipped with automatic transaxle use a front and rear sensors. Both sensors are mounted on the transaxle. Vehicles equipped with manual transaxle use a front sensor. This sensor is mounted on the transaxle.

OPERATION

The vehicle speed sensor sends a signal to the ECM to control. The ECM uses this information to control transmission shift patterns. The vehicle speed sensor is also used to send a speed signal to the speed control servo of the cruise control system.

REMOVAL & INSTALLATION

1. Raise and support the vehicle safely.

2. Place a drip pan below the speed sensor to catch any spilled fluid.

3. Disconnect the connector.

4. Remove the sensor from its mounting.

To install:

5. Installation is the reverse of the removal procedure.

6. Replace any lost fluid.

SUZUKI
DIAGNOSTIC TROUBLE CODES

13

OBD II Vehicle Applications

SUZUKI

Aerio
2002
  2.0L I4 . . . . VIN 4

Esteem
1996-2002
  1.6L I4 . . . . VIN 3
1999-2002
  1.8L I4 . . . . VIN 4

Grand Vitara
1999-2003
  2.5L V6 . . . . VIN 6

Sidekick
1996-1998
  1.6L I4 . . . . VIN 0
1996-1998
  1.8L I4 . . . . VIN 2

Swift
1996-2001
  1.3L I4 . . . . VIN 2

Vitara
1999-2002
  1.6L I4 . . . . VIN 0
1999
  2.0L I4 . . . . VIN 5

X-90
1996-1998
  1.6L I4 . . . . VIN 1

XL-7
2002
  2.7L V6 . . . . VIN 9

Gas Engine OBD II Trouble Code List (P0xxx Codes)

DTC	Trouble Code Title
DTC: P0101 **2T CCM, MIL: Yes** **1999, 2000, 2001, 2002** **Models:** Esteem, Grand Vitara, Sidekick, Vitara, XL-7 **Engines:** 1.6L VIN 0, 1.8L VIN 4, 2.0L VIN 5, 2.5L VIN 6, 2.7L VIN 9 **Transmissions:** All	**Mass Airflow Sensor Range/Performance** Engine started, vehicle driven at a speed of over 35 mph in closed loop, then back to idle speed, altitude less than 8000 ft, ECT sensor from 18-230°F, IAT sensor more than 6.8°F, and the PCM detected the MAF signal was less than 0.82-1.08 lbs/min at 2500 rpm, or that it was less than 0.22-0.44 lbs/min at idle speed during the CCM test. **Possible Causes:** • Intake air system is restricted, or the air filter is clogged • MAF sensor signal or ground circuit has high resistance • MAF sensor is contaminated, damaged or has failed • PCM has failed
DTC: P0102 **1T CCM, MIL: Yes** **1999, 2000, 2001, 2002** **Models:** Esteem, Grand Vitara, Sidekick, Vitara, XL-7 **Engines:** 1.6L VIN 0, 1.8L VIN 4, 2.0L VIN 5, 2.5L VIN 6, 2.7L VIN 9 **Transmissions:** All	**Mass Airflow Sensor Circuit Low Input** Engine started, engine running in closed loop, altitude less than 8000 ft, ECT signal from 18-230°F, IAT signal from 18-122°F, and the PCM detected the MAF signal indicated less than (Scan Tool reads close to '0' lbs/minute) for 1 second during the CCM test. **Possible Causes:** • MAF sensor signal circuit is open or shorted to ground • MAF sensor power circuit is open • MAF sensor is damaged or has failed • PCM has failed
DTC: P0103 **1T CCM, MIL: Yes** **1999, 2000, 2001, 2002** **Models:** Esteem, Grand Vitara, Sidekick, Vitara, XL-7 **Engines:** 1.6L VIN 0, 1.8L VIN 4, 2.0L VIN 5, 2.5L VIN 6, 2.7L VIN 9 **Transmissions:** All	**Mass Airflow Sensor Circuit High Input** Engine started engine running at idle speed in closed loop, and the PCM detected the MAF sensor indicated more than 4.90v (Scan Tool reads 28 lbs/minute) for 1 second during the CCM test. **Possible Causes:** • MAF sensor ground circuit is open • MAF sensor signal is shorted to VREF or system power (B+) • MAF sensor is damaged or has failed • PCM has failed
DTC: P0101 **2T CCM, MIL: Yes** **1996, 1997, 1998** **Models:** Sidekick, X-90 **Engines:** 1.6L VIN 1, 1.6L VIN 4, 1.8L VIN 2 **Transmissions:** All	**Mass Airflow Sensor Range/Performance** Engine started, vehicle driven at a speed of over 35 mph in closed loop, then back to idle speed, altitude less than 9150 ft, IAT sensor from 6.8-158°F, and the PCM determined the intake airflow (from the MAF signal) and the value calculated from the MAP or MDP sensor and the engine speed signals did not agree during the CCM test. **Possible Causes:** • Intake air system is restricted, or the air filter is clogged • MAF sensor signal or ground circuit has high resistance • MAF sensor is contaminated, damaged or has failed • PCM has failed
DTC: P0102 **1T CCM, MIL: Yes** **1996, 1997, 1998** **Models:** Sidekick, X-90 **Engines:** 1.6L VIN 1, 1.6L VIN 4, 1.8L VIN 2 **Transmissions:** All	**Mass Airflow Sensor Circuit Low Input** Engine started, engine running in closed loop for one minute, and the PCM the MAF signal indicated less than 0.30v during the test. **Possible Causes:** • MAF sensor signal circuit is open or shorted to ground • MAF sensor power circuit is open • MAF sensor is damaged or has failed • PCM has failed
DTC: P0103 **1T CCM; MIL: Yes** **1996, 1997, 1998** **Models:** Sidekick, X-90 **Engines:** 1.6L VIN 1, 1.6L VIN 4, 1.8L VIN 2 **Transmissions:** All	**Mass Airflow Sensor Circuit High Input** Engine started, engine running in closed loop for one minute, and the PCM the MAF signal indicated more than 4.90v during the test. **Possible Causes:** • MAF sensor ground circuit is open • MAF sensor signal is shorted to VREF or system power (B+) • MAF sensor is damaged or has failed • PCM has failed
DTC: P0106 **2T CCM; MIL: Yes** **1996, 1997, 1998, 1999** **Models:** Esteem **Engines:** 1.6L VIN 3 **Transmissions:** All	**Manifold Air Pressure Sensor Range/Performance** Engine cranking, altitude less than 8000 ft, ambient temperature over 14°F, IAT signal less than 158°F, ECT sensor from 158-230°F, Closed Throttle switch "on", and the PCM detected the MAP signal differed from the startup MAP signal by less than 1.39 kPa, or with the vehicle driven to over 2000 rpm in closed loop, and then back to idle with the Closed Throttle switch indicating "on", it detected the BARO sensor signal differed from MAP sensor signal by less than 33 kPa during the CCM test. **Possible Causes:** • Air leak present between vacuum passage and the MAP sensor • MAP sensor vacuum line clogged between sensor and manifold • MAP sensor is damaged, out-of-calibration or has failed • PCM has failed

DTC	Trouble Code Title, Conditions & Possible Causes
DTC: P0107 **1T CCM; MIL: Yes** **1996, 1997, 1998, 1999** **Models:** Esteem **Engines:** 1.6L VIN 3 **Transmissions:** A/T, M/T	**MAP Sensor Circuit Low Input (High Vacuum)** Engine started, engine running at idle speed for 1 minute, and the PCM detected the MAP sensor was less than 5 kPa during the test. **Note: If DTC P0122 is set along with this code, the sensor power (VREF) circuit may be open between the sensor and PCM.** **Possible Causes:** • MAP sensor 5-volt power circuit open or shorted to ground • MAP sensor signal circuit is shorted to ground • MAP sensor is damaged or has failed • PCM has failed
DTC: P0108 **1T CCM; MIL: Yes** **1996, 1997, 1998, 1999** **Models:** Esteem **Engines:** 1.6L VIN 3 **Transmissions:** A/T, M/T	**MAP Sensor Circuit High Input (Low Vacuum)** Engine started, engine running at idle speed for 1 minute, and the PCM detected the MAP sensor was more than 130 kPa in the test. **Possible Causes:** • MAP sensor signal circuit is open between sensor and the PCM • MAP sensor ground circuit is open between sensor and ground • MAP sensor signal circuit is shorted to VREF or system power • MAP sensor is damaged or has failed • PCM has failed
DTC: P0106 **2T CCM; MIL: Yes** **1996, 1997, 1998, 1999, 2000, 2001** **Models:** Swift **Engines:** 1.3L VIN 2 **Transmissions:** A/T, M/T	**Manifold Air Pressure Sensor Range/Performance** Engine cranking, altitude less than 8000 ft, ambient temperature over 14°F, IAT signal less than 158°F, ECT sensor from 158-230°F, Closed Throttle switch "on", and the PCM detected the MAP signal differed from the startup MAP signal by less than 1.39 kPa, or with the vehicle driven to over 4000 rpm, then returned to idle speed with Closed Throttle switch "on" and Fuel Cut command "on", it detected the BARO sensor signal differed from MAP sensor signal by less than 33 kPa during the CCM Rationality test. **Possible Causes:** • MAP sensor vacuum line clogged between sensor and manifold • Air leak present between vacuum passage and the MAP sensor • MAP sensor is damaged, out-of-calibration or has failed • PCM has failed
DTC: P0107 **1T CCM; MIL: Yes** **1996, 1997, 1998, 1999, 2000, 2001** **Models:** Swift **Engines:** 1.3L VIN 2 **Transmissions:** A/T, M/T	**MAP Sensor Circuit Low Input (High Vacuum)** Engine started, engine running at idle speed for 1 minute, and the PCM detected the MAP sensor was less than 5 kPa during the test. **Note: If DTC P0122 and P0450 are set with this code, the sensor power (VREF) circuit may be open between the sensor and PCM.** **Possible Causes:** • MAP sensor 5-volt power circuit is open or shorted to ground • MAP Sensor signal circuit is shorted to ground • MAP Sensor is damaged or has failed • PCM has failed
DTC: P0108 **1T CCM; MIL: Yes** **1996, 1997, 1998, 1999, 2000, 2001** **Models:** Swift **Engines:** 1.3L VIN 2 **Transmissions:** A/T, M/T	**MAP Sensor Circuit High Input (Low Vacuum)** Engine started, engine running at idle speed for 1 minute, and the PCM detected the MAP sensor was more than 129.3 kPa in the test. **Note: If DTC P0113, P0118 and P0450 are also set, the sensor ground circuit may be open between the sensor and the PCM.** **Possible Causes:** • MAP sensor signal circuit is open between sensor and the PCM • MAP Sensor ground circuit is open between sensor and PCM • MAP sensor signal circuit is shorted to VREF or system power • MAP Sensor is damaged or has failed • PCM has failed
DTC: P0111 **2T CCM; MIL: Yes** **1998, 1999, 2000, 2001, 2002** **Models:** Esteem, Grand Vitara, Vitara, XL-7, X-90 **Engines:** 1.3L VIN 2, 1.6L VIN 1, 1.6L VIN 3, 1.8L VIN 4, 2.5L VIN 6, 1.6L VIN 0, 2.0L VIN 5, 2.7L VIN 9 **Transmissions:** A/T, M/T	**Intake Air Temperature Sensor Range/Performance** Engine started with the ECT sensor less than 86°F, engine runtime over 5 minutes, and the PCM detected too small an amount of change in the IAT sensor signal under these conditions in the test. **Possible Causes:** • IAT sensor signal circuit has high resistance • IAT sensor ground circuit has high resistance • IAT sensor is contaminated, damaged or has failed • PCM has failed

DTC	Trouble Code Title, Conditions & Possible Causes
DTC: P0112 **1T CCM; MIL: Yes** **1996, 1997, 1998, 1999, 2000, 2001, 2002** **Models:** Esteem, Sidekick, X-90, Grand Vitara, Vitara, XL-7, X-90 **Engines:** 1.6L VIN 1, 1.6L VIN 3, 1.8L VIN 4, 1.6L VIN 0, 1.8L VIN 2, 1.6L VIN 1, 2.5L VIN 6, 2.0L VIN 5, 2.7L VIN 9 **Transmissions:** A/T, M/T	**Intake Air Temperature Sensor Circuit Low Input** Key on or engine running, and the PCM detected an unexpected "low" voltage condition on the IAT sensor circuit (Scan Tool reads 246°F or higher) during the CCM test. **Possible Causes:** • IAT sensor signal circuit is shorted to sensor ground • IAT sensor signal circuit is shorted to chassis ground • IAT sensor is damaged or has failed (it may be shorted) • PCM has failed
DTC: P0113 **1T CCM; MIL: Yes** **1996, 1997, 1998, 1999, 2000, 2001, 2002** **Models:** Esteem, Sidekick, X-90, Grand Vitara, Vitara, XL-7, X-90 **Engines:** 1.6L VIN 1, 1.6L VIN 3, 1.8L VIN 4, 1.6L VIN 0, 1.8L VIN 2, 1.6L VIN 1, 2.5L VIN 6, 2.0L VIN 5, 2.7L VIN 9 **Transmissions:** A/T, M/T	**Intake Air Temperature Sensor Circuit High Input** Key on or engine running, and the PCM detected the IAT sensor was more than 4.90v (Scan Tool reads -40°F or lower) in the test. **Possible Causes:** • IAT sensor signal circuit or ground circuit is open • IAT sensor signal circuit is shorted to VREF or system power • IAT sensor is damaged or has failed • PCM has failed
DTC: P0116 **2T CCM; MIL: Yes** **1999, 2000, 2001, 2002** **Models:** Esteem, Grand Vitara, Vitara, XL-7 **Engines:** 1.6L VIN 3, 1.8L VIN 4, 2.5L VIN 6, 1.6L VIN 0, 2.0L VIN 5, 2.7L VIN 9 **Transmissions:** A/T, M/T	**Engine Coolant Temperature Sensor Range/Performance** Engine started, engine runtime over 5 minutes, and the PCM detected the ECT sensor indicated too small an amount of change when compared to a calibrated amount in memory during the test. **Possible Causes:** • Check the operation of the thermostat (it may be stuck open) • Inspect for low coolant level or an incorrect coolant mixture • ECT sensor signal circuit has high resistance • ECT sensor has failed
DTC: P0117 **1T CCM; MIL: Yes** **1996, 1997, 1998, 1999, 2000, 2001, 2002** **Models:** Esteem, Swift, Sidekick, X-90, Vitara, Grand Vitara, XL-7, X-90 **Engines:** 1.3L VIN 2, 1.6L VIN 1, 1.6L VIN 3, 1.8L VIN 4, 1.6L VIN 0, 1.8L VIN 2, 1.6L VIN 1, 2.5L VIN 6, 2.0L VIN 5, 2.7L VIN 9 **Transmissions:** A/T, M/T	**Engine Coolant Temperature Sensor Circuit Low Input** Key on and engine running, and the PCM detected the ECT signal indicated more than 0.20v (Scan Tool reads 246°F) during the test. **Possible Causes:** • ECT sensor signal circuit is shorted to sensor ground • ECT sensor signal circuit is shorted to chassis ground • ECT sensor is damaged or has failed (it may be shorted) • PCM has failed
DTC: P0118 **1T CCM; MIL: Yes** **1996, 1997, 1998, 1999, 2000, 2001, 2002** **Models:** Esteem, Swift, Sidekick, X-90, Vitara, Grand Vitara, XL-7, X-90 **Engines:** 1.3L VIN 2, 1.6L VIN 1, 1.6L VIN 3, 1.8L VIN 4, 1.6L VIN 0, 1.8L VIN 2, 1.6L VIN 1, 2.5L VIN 6, 2.0L VIN 5, 2.7L VIN 9 **Transmissions:** A/T, M/T	**Engine Coolant Temperature Sensor Circuit High Input** Key on or engine running, and the PCM detected the ECT sensor was more than 4.90v (Scan Tool reads -40°F or lower) in the test. **Possible Causes:** • ECT sensor signal circuit is open between sensor and the PCM • ECT sensor signal circuit is shorted to VREF or system power • ECT sensor ground circuit is open between sensor and PCM • ECT sensor is damaged or has failed • PCM has failed

DTC	Trouble Code Title, Conditions & Possible Causes
DTC: P0121 **2T CCM; MIL: Yes** **1996, 1997, 1998, 1999, 2000,** **2001, 2002** **Models:** Esteem, Swift, Sidekick, X-90, Vitara, Grand Vitara, XL-7, X-90 **Engines:** 1.3L VIN 2, 1.6L VIN 1, 1.6L VIN 3, 1.8L VIN 4, 1.8LVIN 0, 1.8L VIN 2, 1.6L VIN 1, 2.5L VIN 6, 2.0L VIN 5, 2.7L VIN 9 **Transmissions:** A/T, M/T	**Throttle Position Sensor Range/Performance** Engine started, vehicle driven to a speed of 30-40 mph, ambient temperature more than 14°F, IAT sensor less than 122°F, and the PCM detected the difference between the Actual TP sensor signal and the Calculated target signal (from the MAP and TP inputs) was more than a calibrated amount stored in memory during the test. **Possible Causes:** • MAF sensor or TP sensor ground circuit has high resistance • MAF sensor or TP sensor is out-of-calibration • TP sensor is damaged or has failed • PCM has failed
DTC: P0122 **1T CCM; MIL: Yes** **1996, 1997, 1998, 1999, 2000,** **2001, 2002** **Models:** Esteem, Swift, Sidekick, X-90, Vitara, Grand Vitara, XL-7, X-90 **Engines:** 1.3L VIN 2, 1.6L VIN 1, 1.6L VIN 3, 1.8L VIN 4, 1.8LVIN 0, 1.8L VIN 2, 1.6L VIN 1, 2.5L VIN 6, 2.0L VIN 5, 2.7L VIN 9 **Transmissions:** A/T, M/T	**Throttle Position Sensor Circuit Malfunction** Engine started, engine running at idle for 1 minute, and the PCM detected the TP sensor indicated less than 2% during the CCM test. **Possible Causes:** • TP sensor signal circuit is shorted to ground (intermittent fault) • TP sensor power circuit is open between sensor and the PCM • TP sensor is damaged or has failed (perform a "sweep" test) • PCM has failed
DTC: P0123 **1T CCM; MIL: Yes** **1996, 1997, 1998, 1999, 2000,** **2001, 2002** **Models:** Esteem, Swift, Sidekick, X-90, Vitara, Grand Vitara, XL-7, X-90 **Engines:** 1.3L VIN 2, 1.6L VIN 1, 1.6L VIN 3, 1.8L VIN 4, 1.8LVIN 0, 1.8L VIN 2, 1.6L VIN 1, 2.5L VIN 6, 2.0L VIN 5, 2.7L VIN 9 **Transmissions:** A/T, M/T	**Throttle Position Sensor Circuit High Input** Engine started, engine running at idle for 1 minute, and the PCM detected the TP sensor indicated more than 96% during the test. **Possible Causes:** • TP sensor signal circuit is open between sensor and the PCM • TP sensor signal circuit is shorted to VREF (intermittent fault) • TP sensor ground circuit is open between sensor and the PCM • TP sensor is damaged or has failed (perform a "sweep" test) • PCM has failed
DTC: P0103 **2T OBD/ECT; MIL: Yes** **1996, 1997, 1998, 1999, 2000,** **2001, 2002** **Models:** Esteem **Engines:** 1.6L VIN 3, 1.8L VIN 4 **Transmissions:** A/T, M/T	**Insufficient Coolant Temperature For Closed Loop** Engine started, engine runtime over 5 minutes, or until the ECT sensor reaches 104°F, BARO sensor more than 75 kPa, IAT sensor less than 122°F, ambient temperature over 14°F, fuel level from 15-85%, and the PCM detected the ECT sensor signal was too low. **Possible Causes:** • Inspect for low coolant level or an incorrect coolant mixture • Check the operation of the thermostat (it may be stuck open) • ECT sensor signal circuit has high resistance • ECT sensor has failed
DTC: P0125 **2T OBD/ECT; MIL: Yes** **1996, 1997, 1998, 1999, 2000,** **2001** **Models:** Grand Vitara, Vitara, Sidekick, Swift, X-90 **Engines:** 1.3L VIN 2, 1.6L VIN 0, 1.6L VIN 1, 1.8L VIN 2, 2.5L VIN 6, 1.6L VIN 0, 2.0L VIN 5 **Transmissions:** A/T, M/T	**Insufficient Coolant Temperature For Closed Loop** Engine started, vehicle driven at over 35 mph for 15 minutes if the ECT sensor indicated less than 5°F at startup, or vehicle driven for over 5 minutes if the ECT sensor indicated over 5°F at startup, and the PCM detected the ECT sensor did not reach a closed loop value. **Possible Causes:** • Inspect for low coolant level or an incorrect coolant mixture • Check the operation of the thermostat (it may be stuck open) • ECT sensor signal circuit has high resistance • ECT sensor has failed
DTC: P0125 **2T OBD/ECT; MIL: Yes** **1999, 2000, 2001, 2002** **Models:** Vitara, Grand Vitara, XL-7 **Engines:** 1.6L VIN 0, 2.0L VIN 5, 2.5L VIN 6, 2.7L VIN 9 **Transmissions:** A/T, M/T	**Insufficient Coolant Temperature For Closed Loop** Engine started, ECT sensor from 18-230°F, IAT sensor over 18°F, BARO sensor more than 75 kPa, vehicle driven at over 35 mph for 5 minutes under normal stop and go or cruise conditions, then back to idle speed for 5 minutes, and the PCM detected the engine coolant temperature did not reach a value that allows closed loop operation. **Possible Causes:** • Inspect for low coolant level or an incorrect coolant mixture • Check the operation of the thermostat (it may be stuck open) • ECT sensor signal circuit has high resistance • ECT sensor has failed

DTC	Trouble Code Title, Conditions & Possible Causes
DTC: P0131 **2T CCM; MIL: Yes** **1996, 1997, 1998, 1999, 2000, 2001, 2002** **Models:** Esteem, Sidekick, Swift, Vitara, Grand Vitara, XL-7, X-90 **Engines:** 1.3L VIN 2, 1.6L VIN 3, 1.8L VIN 4, 1.6L VIN 0, 1.6L VIN 1, 1.8L VIN 2, 2.5L VIN 6, 2.0L VIN 5, 2.7L VIN 9 **Transmissions:** A/T, M/T	**HO2S-11 (Bank 1 Sensor 1) Circuit Low Input** Engine started, vehicle driven to a speed over 45 mph for 6 minutes at an engine speed from 2500-3000 rpm, followed by a deceleration period of 3 seconds, fuel level from 15-85%, BARO sensor more than 75 kPa, IAT sensor more than 6.8°F, and the PCM detected the maximum HO2S-11 voltage was less than 300 mv during the test. **Possible Causes:** • HO2S signal ground circuit is open • HO2S is contaminated or has failed • Vehicle driven until it ran out of fuel, or while very low on fuel • PCM has failed
DTC: P0132 **2T CCM; MIL: Yes** **1996, 1997, 1998, 1999, 2000, 2001, 2002** **Models:** Esteem, Sidekick, Swift, Vitara, Grand Vitara, XL-7, X-90 **Engines:** 1.3L VIN 2, 1.6L VIN 0, 1.6L VIN 1, 1.6L VIN 3, 1.8L VIN 4, 1.8L VIN 2, 2.5L VIN 6, 2.0L VIN 5, 2.7L VIN 9 **Transmissions:** A/T, M/T	**HO2S-11 (Bank 1 Sensor 1) Circuit High Input** Engine started, vehicle driven to a speed over 45 mph for 6 minutes at an engine speed from 2500-3000 rpm, followed by a deceleration period of 3 seconds, fuel level from 15-85%, BARO sensor more than 75 kPa, IAT sensor more than 6.8°F, and the PCM detected the HO2S-11 signal did not go below 600 mv during the CCM test. **Possible Causes:** • HO2S signal circuit is open or shorted to system power • Fuel supply system is too rich (fuel injector or regulator leaking) • HO2S is contaminated, damaged or has failed • PCM has failed • TSB TS015 (12/95) contains information related to this code
DTC: P0133 **2T OBD/O2S; MIL: Yes** **1996, 1997, 1998, 1999, 2000, 2001, 2002** **Models:** Esteem, Sidekick, Swift, Vitara, Grand Vitara, XL-7, X-90 **Engines:** 1.3L VIN 2, 1.6L VIN 0, 1.6L VIN 1, 1.6L VIN 3, 1.8L VIN 4, 1.8L VIN 2, 2.5L VIN 6, 2.0L VIN 5, 2.7L VIN 9 **Transmissions:** A/T, M/T	**HO2S-11 (Bank 1 Sensor 1) Slow Response** DTC P0131 and P0132 not set, vehicle driven at a speed of 35 mph for 2 minutes in closed loop, then back to idle speed for 2 minutes, ambient air temperature over 14°F, fuel level from 15-85%, BARO sensor more than 75 kPa, IAT sensor more than 14°F, and the PCM detected the HO2S-11 signal response time to change from rich-to-lean or lean-to-rich was more than 1 second during the CCM test. **Possible Causes:** • Air leaks in the intake manifold or exhaust manifold or pipes • IAT sensor or MAF sensor has deteriorated (out of calibration) • HO2S signal circuit is open or shorted to ground (intermittent) • HO2S contaminated with wrong fuel, has deteriorated or failed • PCM has failed • TSB TS015 (12/95) contains information related to this code
DTC: P0134 **2T OBD/O2S; MIL: Yes** **1996, 1997, 1998, 1999, 2000, 2001, 2002** **Models:** Esteem, Sidekick, Swift, Vitara, Grand Vitara, XL-7, X-90 **Engines:** 1.3L VIN 2, 1.6L VIN 0, 1.6L VIN 1, 1.6L VIN 3, 1.8L VIN 4, 1.8L VIN 2, 2.5L VIN 6, 2.0L VIN 5, 2.7L VIN 9 **Transmissions:** A/T, M/T	**HO2S-11 (Bank 1 Sensor 1) No Activity Detected** DTC P0131 and P0132 not set, vehicle driven at a speed of 35 mph for 2 minutes in closed loop, then back to idle speed for 2 minutes, ambient air temperature over 14°F, fuel level from 15-85%, BARO sensor more than 75 kPa, IAT sensor more than 14°F, and the PCM detected the HO2S-11 signal remained "high" or "low" in the test. **Possible Causes:** • Air leaks in the intake manifold or exhaust manifold or pipes • HO2S is contaminated (wrong fuel), has deteriorated or failed • HO2S heater element is damaged or has failed • PCM has failed • TSB TS015 (12/95) contains information related to this code
DTC: P0135 **2T OBD/O2S; MIL: Yes** **1996, 1997, 1998, 1999, 2000, 2001, 2002** **Models:** Esteem, Sidekick, Swift, Vitara, X-90 **Engines:** 1.3L VIN 2, 1.6L VIN 0, 1.6L VIN 1, 1.6L VIN 3, 1.8L VIN 4, 1.8L VIN 2, 2.0L VIN 5 **Transmissions:** A/T, M/T	**HO2S-11 (Bank 1 Sensor 1) Heater Circuit Malfunction** Engine started, engine runtime 1 minute, and the PCM detected the HO2S-11 heater signal was "low" with the heater commanded "off", or that it was "high: with the heater commanded "on" during the test. **Possible Causes:** • HO2S heater control circuit is open or shorted to ground • HO2S heater power circuit is open (check the IG fuse) • HO2S heater control circuit is shorted to power • HO2S heater is damaged or has failed • PCM has failed
DTC: P0135 **2T OBD/O2S; MIL: Yes** **1999, 2000, 2001, 2002** **Models:** Grand Vitara, XL-7 **Engines:** 2.5L VIN 6, 2.7L VIN 9 **Transmissions:** A/T, M/T	**HO2S-11 (Bank 1 Sensor 1) Heater Circuit Malfunction** Engine started, engine runtime over 1 minute, IAT sensor more than 14°F, ECT sensor from 18-230°F, and the PCM detected the HO2S-11 heater current level was more than 5.3A or less than 0.09A. **Possible Causes:** • HO2S heater control circuit is open or shorted to ground • HO2S heater power circuit is open (check the IG fuse) • HO2S heater control circuit is shorted to power • HO2S heater is damaged or has failed • PCM has failed

DTC	Trouble Code Title, Conditions & Possible Causes
DTC: P0136 **2T CCM; MIL: Yes** **1996, 1997, 1998, 1999, 2000, 2001, 2002** **Models:** Esteem, Sidekick, Swift, Vitara, Grand Vitara, XL-7, X-90 **Engines:** 1.3L VIN 2, 1.6L VIN 0, 1.6L VIN 1, 1.6L VIN 3, 1.8L VIN 4, 1.8L VIN 2, 2.5L VIN 6, 2.0L VIN 5, 2.7L VIN 9 **Transmissions:** A/T, M/T	**HO2S-12 (Bank 1 Sensor 2) Circuit Low Input** Engine started, vehicle driven to a speed over 45 mph for 6 minutes at an engine speed from 2500-3000 rpm, followed by a deceleration period of 3 seconds, fuel level from 15-85%, BARO sensor more than 75 kPa, IAT sensor more than 6.8ºF, and the PCM detected the maximum HO2S-12 voltage was less than 600 mv, or the minimum voltage was more than 400 mv, or the maximum HO2S-12 voltage was more than 4.5v during the CCM test. **Possible Causes:** • HO2S signal circuit open (sensor signal reads over 4.5v) • HO2S is shorted to ground • HO2S signal ground circuit is open • HO2S is contaminated or has failed • Vehicle driven until it ran out of fuel, or while very low on fuel • PCM has failed
DTC: P0141 **2T OBD/O2S; MIL: Yes** **1996, 1997, 1998, 1999, 2000, 2001, 2002** **Models:** Esteem, Sidekick, Swift, Vitara, X-90 **Engines:** 1.3L VIN 2, 1.6L VIN 0, 1.6L VIN 1, 1.6L VIN 3, 1.8L VIN 4, 1.8L VIN 2, 2.0L VIN 5 **Transmissions:** A/T, M/T	**HO2S-12 (Bank 1 Sensor 2) Heater Circuit Malfunction** Engine started, engine runtime 1 minute, and the PCM detected the HO2S-12 heater signal was "low" with the heater commanded "off", or that it was "high: with the heater commanded "on" during the test. **Possible Causes:** • HO2S heater control circuit is open or shorted to ground • HO2S heater power circuit is open (check the IG fuse) • HO2S heater control circuit is shorted to power • HO2S heater is damaged or has failed • PCM has failed
DTC: P0141 **2T OBD/O2S; MIL: Yes** **1999, 2000, 2001, 2002** **Models:** Grand Vitara, XL-7 **Engines:** 2.5L VIN 6, 2.7L VIN 9 **Transmissions:** A/T, M/T	**HO2S-12 (Bank 1 Sensor2) Heater Circuit Malfunction** Engine started, engine runtime over 1 minute, IAT sensor more than 14ºF, ECT sensor from 18-230ºF, and the PCM detected the HO2S-12 heater current level was more than 5.3A or less than 0.09A. **Possible Causes:** • HO2S heater control circuit is open or shorted to ground • HO2S heater power circuit is open (check the IG fuse) • HO2S heater control circuit is shorted to power • HO2S heater is damaged or has failed • PCM has failed
DTC: P0151 **2T CCM; MIL: Yes** **1999, 2000, 2001, 2002** **Models:** Grand Vitara, XL-7 **Engines:** 2.5L VIN 6, 2.7L VIN 9 **Transmissions:** A/T, M/T	**HO2S-21 (Bank 2 Sensor 1) Circuit Low Input** Engine started, vehicle driven at a speed of over 35 mph for 2 minutes, then back to idle speed for 1 minute, BARO sensor more than 75 kPa, ECT sensor from 18-230ºF, IAT sensor more than 18ºF, fuel level from 15-85%, and the PCM detected the maximum HO2S-12 voltage was less than 300 mv during the CCM test. **Possible Causes:** • HO2S is shorted to ground • HO2S is contaminated, damaged or it has failed • Vehicle driven until it ran out of fuel, or while very low on fuel • PCM has failed
DTC: P0152 **2T CCM; MIL: Yes** **1999, 2000, 2001, 2002** **Models:** Grand Vitara, XL-7 **Engines:** 2.5L VIN 6, 2.7L VIN 9 **Transmissions:** A/T, M/T	**HO2S-21 (Bank 2 Sensor 1) Circuit High Input** Engine started, vehicle driven at a speed of over 35 mph for 2 minutes, then back to idle speed for 1 minute, BARO sensor more than 75 kPa, ECT sensor from 18-230ºF, IAT sensor more than 18ºF, fuel level from 15-85%, and the PCM detected the maximum HO2S-12 voltage was less than 600 mv during the CCM test. **Possible Causes:** • HO2S signal circuit is open or shorted to system power • Fuel supply system is too rich (fuel injector or regulator leaking) • HO2S is contaminated, damaged or has failed • PCM has failed
DTC: P0153 **2T CCM; MIL: Yes** **1999, 2000, 2001, 2002** **Models:** Grand Vitara, XL-7 **Engines:** 2.5L VIN 6, 2.7L VIN 9 **Transmissions:** A/T, M/T	**HO2S-21 (Bank 2 Sensor 1) Slow Response** Engine started, vehicle driven at a constant speed of over 35 mph for 2 minutes in closed loop, then back to idle speed for 1 minute, ambient air temperature over 14ºF, fuel level from 15-85%, BARO sensor more than 75 kPa, IAT sensor from 18-230ºF, and the PCM detected the response rate of the HO2S-21 signal was too long during the CCM test. **Possible Causes:** • Air leaks in the intake manifold or exhaust manifold or pipes • IAT sensor or MAF sensor has deteriorated (out of calibration) • HO2S signal circuit is open or shorted to ground (intermittent) • HO2S contaminated with wrong fuel, has deteriorated or failed • PCM has failed

DTC	Trouble Code Title, Conditions & Possible Causes
DTC: P0154 **2T CCM; MIL: Yes** **1999, 2000, 2001, 2002** **Models:** Grand Vitara, XL-7 **Engines:** 2.5L VIN 6, 2.7L VIN 9 **Transmissions:** A/T, M/T	**HO2S-21 (Bank 2 Sensor 1) No Activity Detected** Engine started, vehicle driven at a constant speed of over 35 mph for 2 minutes in closed loop, then back to idle speed for 1 minute, ambient air temperature over 14°F, fuel level from 15-85%, BARO sensor more than 75 kPa, IAT sensor more than 18°F, and the PCM detected the HO2S-21 signal did not exceed 0.45v during the test. **Possible Causes:** • Air leaks in the intake manifold or exhaust manifold or pipes • HO2S is contaminated (wrong fuel), has deteriorated or failed • HO2S heater element is damaged or has failed • PCM has failed
DTC: P0155 **2T OBD/O2S; MIL: Yes** **1999, 2000, 2001, 2002** **Models:** Grand Vitara, XL-7 **Engines:** 2.5L VIN 6, 2.7L VIN 9 **Transmissions:** A/T, M/T	**HO2S-21 (Bank 2 Sensor 1) Heater Circuit Malfunction** Engine started, engine runtime over 1 minute, IAT sensor more than 14°F, ECT sensor from 18-230°F, and the PCM detected the HO2S-21 heater current level was more than 5.3A or less than 0.09A. **Possible Causes:** • HO2S heater control circuit is open or shorted to ground • HO2S heater power circuit is open (check the IG fuse) • HO2S heater control circuit is shorted to power • HO2S heater is damaged or has failed • PCM has failed
DTC: P0156 **2T CCM; MIL: Yes** **1999, 2000, 2001, 2002** **Models:** Grand Vitara, XL-7 **Engines:** 2.5L VIN 6, 2.7L VIN 9 **Transmissions:** A/T, M/T	**HO2S-22 (Bank 2 Sensor 2) Range/Performance** Engine started, vehicle driven to a constant speed of over 45 mph for 5 minutes, then back to idle speed for 1 minute, IAT sensor more than 18°F, ECT sensor from 14-230°F, BARO sensor more than 75 kPa, fuel level from 15-85%, and the PCM detected the HO2S-22 signal was more than 4.5v, or the maximum HO2S-22 signal voltage was too low, or that the minimum HO2S-22 signal voltage was more than 300 mv during the CCM test. **Possible Causes:** • HO2S signal circuit open (sensor signal reads over 4.5v) • HO2S is shorted to ground • HO2S signal ground circuit is open • HO2S is contaminated or has failed • Vehicle driven until it ran out of fuel, or while very low on fuel • PCM has failed
DTC: P0161 **2T OBD/O2S; MIL: Yes** **1999, 2000, 2001, 2002** **Models:** Grand Vitara, XL-7 **Engines:** 2.5L VIN 6, 2.7L VIN 9 **Transmissions:** A/T, M/T	**HO2S-22 (Bank 2 Sensor 2) Heater Circuit Malfunction** Engine started, engine runtime over 1 minute, IAT sensor more than 14°F, ECT sensor from 18-230°F, and the PCM detected the HO2S-22 heater current level was more than 11 mA or less than 0.32 mA. **Possible Causes:** • HO2S heater control circuit is open or shorted to ground • HO2S heater power circuit is open (check the IG fuse) • HO2S heater control circuit is shorted to power • HO2S heater is damaged or has failed • PCM has failed
DTC: P0171 **2T; MIL: Yes** **1996, 1997, 1998, 1999, 2000, 2001, 2002** **Models:** Esteem, Sidekick, Swift, Vitara, X-90 **Engines:** 1.3L VIN 2, 1.6L VIN 0, 1.6L VIN 1, 1.6L VIN 3, 1.8L VIN 4, 1.8L VIN 2, 2.0L VIN 5 **Transmissions:** A/T, M/T	**Fuel System Too Lean (Bank 1)** DTC P0101, P0102, P0103, P0106, P0107, P0108, P0112, P0113, P0117, P0118, P0121, P0131, P0132, P0133, P0135, P0136, P0141, P0301-306, P0400, P0400 and P0601 not set, engine started, vehicle driven at a speed of 30-40 mph for 5 minutes at an engine speed from 2500-3000 rpm, BARO sensor more than 75 kPa, IAT sensor from 14-158°F, ambient temperature over 14°F, fuel level from 15-85%, and the PCM detected the Short Term fuel trim was over +20%, or the total of the Short Term and Long Term fuel trims values (added together) exceeded +33% during the test. **Possible Causes:** • Air leaks after the MAF sensor, or in the EGR or PCV system • Base engine "mechanical" fault affecting one or more cylinders • Exhaust leaks before or near where the front HO2S is mounted • Fuel control sensor is out of calibration (i.e., ECT, IAT or MAP) • Fuel delivery system supplying too little fuel during cruise or idle periods (e.g., faulty fuel pump or dirty, restricted fuel filter) • Fuel injector (one or more) dirty or pressure regulator has failed • HO2S is contaminated, deteriorated or it has failed • Vehicle driven low on fuel or until it ran out of fuel

DTC	Trouble Code Title, Conditions & Possible Causes
DTC: P0172 **2T; MIL: Yes** **1996, 1997, 1998, 1999, 2000, 2001, 2002** **Models:** Esteem, Sidekick, Swift, Vitara, X-90 **Engines:** 1.3L VIN 2, 1.6L VIN 0, 1.6L VIN 1, 1.6L VIN 3, 1.8L VIN 4, 1.8L VIN 2, 2.0L VIN 5 **Transmissions:** A/T, M/T	**Fuel System Too Rich (Bank 1)** DTC P0101, P0102, P0103, P0106, P0107, P0108, P0112, P0113, P0117, P0118, P0121, P0131, P0132, P0133, P0135, P0136, P0141, P0301-306, P0400, P0400 and P0601 not set, engine started, vehicle driven at a speed of 30-40 mph for 5 minutes at an engine speed from 2500-3000 rpm, BARO sensor more than 75 kPa, IAT sensor from 14-1588°F, ambient temperature over 14°F, fuel level from 15-85%, and the PCM detected the Short Term fuel trim was over -20%, or the total of the Short Term and Long Term fuel trims values (added together) exceeded -33% during the test. **Possible Causes:** • Base engine "mechanical" fault affecting one or more cylinders • EVAP system component has failed or canister fuel saturated • Fuel control sensor is out of calibration (i.e., ECT, IAT or MAP) • Fuel delivery system supplying too much fuel during cruise or idle periods (e.g., faulty fuel pump, or faulty pressure regulator) • Fuel injector(s) is leaking or stuck partially open (one or more) • HO2S is contaminated, deteriorated or it has failed
DTC: P0171 **2T; MIL: Yes** **1999, 2000, 2001, 2002** **Models:** Grand Vitara, XL-7 **Engines:** 2.5L VIN 6, 2.7L VIN 9 **Transmissions:** A/T, M/T	**Fuel System Too Lean (Bank 1)** DTC P0101, P0102, P0103, P0106, P0107, P0108, P0112, P0113, P0117, P0118, P0121, P0131, P0132, P0133, P0135, P0136, P0141, P0301-306, P0400, P0400 and P0601 not set, engine started, vehicle driven at a speed of 30-40 mph for 5 minutes at an engine speed from 2500-3000 rpm, BARO sensor more than 75 kPa, IAT sensor from 14-158°F, ambient temperature over 14°F, fuel level from 15-85%, and the PCM detected the total fuel trim (Short and Long Term fuel trim total) exceed +31% in the Fuel System test. **Possible Causes:** • Air leaks after the MAF sensor, or in the EGR or PCV system • Base engine "mechanical" fault affecting one or more cylinders • Exhaust leaks before or near where the front HO2S is mounted • Fuel control sensor is out of calibration (i.e., ECT, IAT or MAP) • Fuel delivery system supplying too little fuel during cruise or idle periods (e.g., faulty fuel pump or dirty, restricted fuel filter) • Fuel injector (one or more) dirty or pressure regulator has failed • HO2S is contaminated, deteriorated or it has failed • Vehicle driven low on fuel or until it ran out of fuel
DTC: P0172 **2T; MIL: Yes** **1999, 2000, 2001, 2002** **Models:** Grand Vitara, XL-7 **Engines:** 2.5L VIN 6, 2.7L VIN 9 **Transmissions:** A/T, M/T	**Fuel System Too Rich (Bank 1)** DTC P0101, P0102, P0103, P0106, P0107, P0108, P0112, P0113, P0117, P0118, P0121, P0131, P0132, P0133, P0135, P0136, P0141, P0301-306, P0400, P0400 and P0601 not set, engine started, vehicle driven at a speed of 30-40 mph for 5 minutes at an engine speed from 2500-3000 rpm, BARO sensor more than 75 kPa, IAT sensor from 14-158°F, ambient temperature over 14°F, fuel level from 15-85%, and the PCM detected the total fuel trim exceeded -31 (Short and Long Term fuel trim total) during the test. **Possible Causes:** • Base engine "mechanical" fault affecting one or more cylinders • EVAP system component has failed or canister fuel saturated • Fuel control sensor is out of calibration (i.e., ECT, IAT or MAP) • Fuel delivery system supplying too much fuel during cruise or idle periods (e.g., faulty fuel pump, or faulty pressure regulator) • Fuel injector(s) is leaking or stuck partially open (one or more) • HO2S is contaminated, deteriorated or it has failed
DTC: P0174 **2T; MIL: Yes** **1999, 2000, 2001, 2002** **Models:** Grand Vitara, XL-7 **Engines:** 2.5L VIN 6, 2.7L VIN 9 **Transmissions:** A/T, M/T	**Fuel System Too Lean (Bank 2)** DTC P0101, P0102, P0103, P0106, P0107, P0108, P0112, P0113, P0117, P0118, P0121, P0131, P0132, P0133, P0135, P0136, P0141, P0301-306, P0400, P0400 and P0601 not set, engine started, vehicle driven at a speed of 30-40 mph for 5 minutes at an engine speed from 2500-3000 rpm, BARO sensor more than 75 kPa, IAT sensor from 14-158°F, ambient temperature over 14°F, fuel level from 15-85%, and the PCM detected the total fuel trim (Short and Long Term fuel trim total) **exceed +31% in the Fuel System test.** **Possible Causes:** • Air leaks after the MAF sensor, or in the EGR or PCV system • Base engine "mechanical" fault affecting one or more cylinders • Exhaust leaks before or near where the front HO2S is mounted • Fuel control sensor is out of calibration (i.e., ECT, IAT or MAP) • Fuel delivery system supplying too little fuel during cruise or idle periods (e.g., faulty fuel pump or dirty, restricted fuel filter) • Fuel injector (one or more) dirty or pressure regulator has failed • HO2S is contaminated, deteriorated or it has failed • Vehicle driven low on fuel or until it ran out of fuel

DTC	Trouble Code Title, Conditions & Possible Causes
DTC: P0175 **2T; MIL: Yes** **1999, 2000, 2001, 2002** **Models:** Grand Vitara, XL-7 **Engines:** 2.5L VIN 6, 2.7L VIN 9 **Transmissions:** A/T, M/T	**Fuel System Too Rich (Bank 2)** DTC P0101, P0102, P0103, P0106, P0107, P0108, P0112, P0113, P0117, P0118, P0121, P0131, P0132, P0133, P0135, P0136, P0141, P0301-306, P0400, P0400 and P0601 not set, engine started, vehicle driven at a speed of 30-40 mph for 5 minutes at an engine speed from 2500-3000 rpm, BARO sensor more than 75 kPa, IAT sensor from 14-158°F, ambient temperature over 14°F, fuel level from 15-85%, and the PCM detected the total fuel trim (Short and Long Term fuel trim total) exceed -31% in the Fuel System test. **Possible Causes:** • Base engine "mechanical" fault affecting one or more cylinders • EVAP system component has failed or canister fuel saturated • Fuel control sensor is out of calibration (i.e., ECT, IAT or MAP) • Fuel delivery system supplying too much fuel during cruise or idle periods (e.g., faulty fuel pump, or faulty pressure regulator) • Fuel injector(s) is leaking or stuck partially open (one or more) • HO2S is contaminated, deteriorated or it has failed
DTC: P0300 **2T CCM; MIL: Yes** **1996, 1997, 1998, 1999, 2000, 2001, 2002** **Models:** Esteem **Engines:** 1.6L VIN 3 **Transmissions:** A/T, M/T	**Multiple Cylinder Misfire Detected** DTC P0106, P0107, P0108, P0117, P0118, P0121, P0122, P0123, P0335, P0340 and P0500 not set, BARO sensor over 75 kPa, fuel level from 15-85%, IAT signal less than 158°F, ambient temperature over 14°F, ECT sensor from 14-230°F, engine runtime 1 minute, and the PCM detected a misfire in more than 1 cylinder in the 200 (Catalyst) or 1000-rpm (High Emissions) revolution range. **Note: If the misfire is severe, the MIL will flash on/off on the 1st trip!** **Possible Causes:** • Air leak in the intake manifold, or in the EGR or PCM system • Base engine mechanical fault that affects one or more cylinders • Fuel delivery component fault that affects one or more cylinders (i.e., a contaminated, dirty or sticking fuel injector) • Ignition system problem (coil or plug) in one or more cylinders
DTC: P0300 **2T CCM; MIL: Yes** **1996, 1997, 1998, 1999, 2000, 2001, 2002** **Models:** Esteem **Engines:** 1.8L VIN 4 **Transmissions:** A/T, M/T	**Multiple Cylinder Misfire Detected** DTC P0101, P0102, P0103, P0117, P0118, P0121, P0122, P0123, P0335, P0340 and P0500 not set, BARO sensor over 75 kPa, fuel level from 15-85%, IAT signal from 6.8-158°F, ambient temperature over 14°F, ECT sensor from 14-230°F, engine runtime 1 minute, and the PCM detected a misfire in more than 1 cylinder in the 200 (Catalyst) or 1000-rpm (High Emissions) revolution range. **Note: If the misfire is severe, the MIL will flash on/off on the 1st trip!** **Possible Causes:** • Air leak in the intake manifold, or in the EGR or PCM system • Base engine mechanical fault that affects one or more cylinders • Fuel delivery component fault that affects one or more cylinders (i.e., one or more contaminated, dirty or sticking fuel injectors) • Ignition system problem (coil or plug) in one or more cylinders
DTC: P0300 **2T CCM; MIL: Yes** **1996, 1997, 1998, 1999, 2000, 2001** **Models:** Sidekick, Swift, Vitara, X-90 **Engines:** 1.3L VIN 2, 1.6L VIN 0, 1.6L VIN 1, 1.8L VIN 2, 2.0L VIN 5 **Transmissions:** A/T, M/T	**Multiple Cylinder Misfire Detected** DTC P0101, P0102, P0103, P0117, P0118, P0121, P0122, P0123, P0335, P0340 and P0500 not set, BARO sensor over 75 kPa, fuel level from 15-85%, IAT signal from 6.8-158°F, ambient temperature over 14°F, ECT sensor from 14-230°F, engine runtime 1 minute, and the PCM detected a misfire in more than 1 cylinders in the 200 (Catalyst) or 1000-rpm (High Emissions) revolution range. **Note: If the misfire is severe, the MIL will flash on/off on the 1st trip!** **Possible Causes:** • Air leak in the intake manifold, or in the EGR or PCM system • Base engine mechanical fault that affects one or more cylinders • Fuel delivery component fault that affects one or more cylinders (i.e., one or more contaminated, dirty or sticking fuel injectors) • Ignition system problem (coil or plug) in one or more cylinders
DTC: P0300 **2T CCM; MIL: Yes** **1999, 2000, 2001, 2002** **Models:** Grand Vitara, XL-7 **Engines:** 2.5L VIN 6, 2.7L VIN 9 **Transmissions:** A/T, M/T	**Multiple Cylinder Misfire Detected** DTC P0335 and P0340 not set, altitude less than 9150 feet, fuel level from 15-85%, IAT signal from 6.8-158°F, ambient temperature over 14°F, ECT sensor from 14-230°F, engine runtime over 1 minute, and the PCM detected a misfire in more than 1 cylinders in the 200 (Catalyst) or 1000-rpm (High Emissions) revolution range. **Note: If the misfire is severe, the MIL will flash on/off on the 1st trip!** **Possible Causes:** • Air leak in the intake manifold, or in the EGR or PCM system • Base engine mechanical fault that affects one or more cylinders • Fuel delivery component fault that affects one or more cylinders (i.e., one or more contaminated, dirty or sticking fuel injectors) • Ignition system problem (coil or plug) in one or more cylinders

DTC	Trouble Code Title, Conditions & Possible Causes
DTC: P0301 **2T CCM; MIL: Yes** **1996, 1997, 1998, 1999, 2000** **Models:** Esteem **Engines:** 1.6L VIN 3 **Transmissions:** A/T, M/T	**Cylinder 1 Misfire Detected** DTC P0106, P0107, P0108, P0117, P0118, P0121, P0122, P0123, P0335, P0340 and P0500 not set, BARO sensor over 75 kPa, fuel level from 15-85%, IAT signal from 6.8-158°F, ambient temperature over 14°F, ECT sensor from 14-230°F, engine runtime 1 minute, and the PCM detected a misfire condition in one cylinder in the 200 (Catalyst) or 1000-rpm (High Emissions) revolution range. **Note: If the misfire is severe, the MIL will flash on/off on the 1st trip!** **Possible Causes:** • Air leak in the intake manifold, or in the EGR or PCM system • Base engine mechanical fault that affects only one cylinder • Fuel delivery component fault that affects only one cylinder (i.e., a contaminated, dirty or sticking fuel injector) • Ignition system problem (coil or plug) that affects one cylinder
DTC: P0302 **2T CCM; MIL: Yes** **1996, 1997, 1998, 1999, 2000** **Models:** Esteem **Engines:** 1.6L VIN 3 **Transmissions:** A/T, M/T	**Cylinder 2 Misfire Detected** DTC P0106, P0107, P0108, P0117, P0118, P0121, P0122, P0123, P0335, P0340 and P0500 not set, BARO sensor over 75 kPa, fuel level from 15-85%, IAT signal from 6.8-158°F, ambient temperature over 14°F, ECT sensor from 14-230°F, engine runtime 1 minute, and the PCM detected a misfire condition in one cylinder in the 200 (Catalyst) or 1000-rpm (High Emissions) revolution range. **Note: If the misfire is severe, the MIL will flash on/off on the 1st trip!** **Possible Causes:** • Air leak in the intake manifold, or in the EGR or PCM system • Base engine mechanical fault that affects only one cylinder • Fuel delivery component fault that affects only one cylinder (i.e., a contaminated, dirty or sticking fuel injector) • Ignition system problem (coil or plug) that affects one cylinder
DTC: P0303 **2T CCM; MIL: Yes** **1996, 1997, 1998, 1999, 2000** **Models:** Esteem **Engines:** 1.6L VIN 3 **Transmissions:** A/T, M/T	**Cylinder 3 Misfire Detected** DTC P0106, P0107, P0108, P0117, P0118, P0121, P0122, P0123, P0335, P0340 and P0500 not set, BARO sensor over 75 kPa, fuel level from 15-85%, IAT signal from 6.8-158°F, ambient temperature over 14°F, ECT sensor from 14-230°F, engine runtime 1 minute, and the PCM detected a misfire condition in one cylinder in the 200 (Catalyst) or 1000-rpm (High Emissions) revolution range. **Note: If the misfire is severe, the MIL will flash on/off on the 1st trip!** **Possible Causes:** • Air leak in the intake manifold, or in the EGR or PCM system • Base engine mechanical fault that affects only one cylinder • Fuel delivery component fault that affects only one cylinder (i.e., a contaminated, dirty or sticking fuel injector) • Ignition system problem (coil or plug) that affects one cylinder
DTC: P0304 **2T CCM; MIL: Yes** **1996, 1997, 1998, 1999, 2000** **Models:** Esteem **Engines:** 1.6L VIN 3 **Transmissions:** A/T, M/T	**Cylinder 4 Misfire Detected** DTC P0106, P0107, P0108, P0117, P0118, P0121, P0122, P0123, P0335, P0340 and P0500 not set, BARO sensor over 75 kPa, fuel level from 15-85%, IAT signal from 6.8-158°F, ambient temperature over 14°F, ECT sensor from 14-230°F, engine runtime 1 minute, and the PCM detected a misfire condition in one cylinder in the 200 (Catalyst) or 1000-rpm (High Emissions) revolution range. **Note: If the misfire is severe, the MIL will flash on/off on the 1st trip!** **Possible Causes:** • Air leak in the intake manifold, or in the EGR or PCM system • Base engine mechanical fault that affects only one cylinder • Fuel delivery component fault that affects only one cylinder (i.e., a contaminated, dirty or sticking fuel injector) • Ignition system problem (coil or plug) that affects one cylinder
DTC: P0301 **2T CCM; MIL: Yes** **1999, 2000, 2001, 2002** **Models:** Esteem **Engines:** 1.8L VIN 4 **Transmissions:** A/T, M/T	**Cylinder 1 Misfire Detected** DTC P0101, P0102, P0103, P0117, P0118, P0121, P0122, P0123, P0335, P0340 and P0500 not set, BARO sensor over 75 kPa, fuel level from 15-85%, IAT signal from 6.8-158°F, ambient temperature over 14°F, ECT sensor from 14-230°F, engine runtime 1 minute, and the PCM detected a misfire condition in one cylinder in the 200 (Catalyst) or 1000-rpm (High Emissions) revolution range. **Note: If the misfire is severe, the MIL will flash on/off on the 1st trip!** **Possible Causes:** • Air leak in the intake manifold, or in the EGR or PCM system • Base engine mechanical fault that affects only one cylinder • Fuel delivery component fault that affects only one cylinder (i.e., a contaminated, dirty or sticking fuel injector) • Ignition system problem (coil or plug) that affects one cylinder
DTC: P0302 **2T CCM; MIL: Yes** **1999, 2000, 2001, 2002** **Models:** Esteem **Engines:** 1.8L VIN 4 **Transmissions:** A/T, M/T	**Cylinder 2 Misfire Detected** DTC P0101, P0102, P0103, P0117, P0118, P0121, P0122, P0123, P0335, P0340 and P0500 not set, BARO sensor over 75 kPa, fuel level from 15-85%, IAT signal from 6.8-158°F, ambient temperature over 14°F, ECT sensor from 14-230°F, engine runtime 1 minute, and the PCM detected a misfire condition in one cylinder in the 200 (Catalyst) or 1000-rpm (High Emissions) revolution range. **Note: If the misfire is severe, the MIL will flash on/off on the 1st trip!** **Possible Causes:** • Air leak in the intake manifold, or in the EGR or PCM system • Base engine mechanical fault that affects only one cylinder • Fuel delivery component fault that affects only one cylinder (i.e., a contaminated, dirty or sticking fuel injector) • Ignition system problem (coil or plug) that affects one cylinder

DTC	Trouble Code Title, Conditions & Possible Causes
DTC: P0303 **2T CCM; MIL: Yes** **1999, 2000, 2001, 2002** **Models:** Esteem **Engines:** 1.8L VIN 4 **Transmissions:** A/T, M/T	**Cylinder 3 Misfire Detected** DTC P0101, P0102, P0103, P0117, P0118, P0121, P0122, P0123, P0335, P0340 and P0500 not set, BARO sensor over 75 kPa, fuel level from 15-85%, IAT signal from 6.8-158°F, ambient temperature over 14°F, ECT sensor from 14-230°F, engine runtime 1 minute, and the PCM detected a misfire condition in one cylinder in the 200 (Catalyst) or 1000-rpm (High Emissions) revolution range. **Note:** If the misfire is severe, the MIL will flash on/off on the 1st trip! **Possible Causes:** • Air leak in the intake manifold, or in the EGR or PCM system • Base engine mechanical fault that affects only one cylinder • Fuel delivery component fault that affects only one cylinder (i.e., a contaminated, dirty or sticking fuel injector) • Ignition system problem (coil or plug) that affects one cylinder
DTC: P0304 **2T CCM; MIL: Yes** **1999, 2000, 2001, 2002** **Models:** Esteem **Engines:** 1.8L VIN 4 **Transmissions:** A/T, M/T	**Cylinder 4 Misfire Detected** DTC P0101, P0102, P0103, P0117, P0118, P0121, P0122, P0123, P0335, P0340 and P0500 not set, BARO sensor over 75 kPa, fuel level from 15-85%, IAT signal from 6.8-158°F, ambient temperature over 14°F, ECT sensor from 14-230°F, engine runtime 1 minute, and the PCM detected a misfire condition in one cylinder in the 200 (Catalyst) or 1000-rpm (High Emissions) revolution range. **Note:** If the misfire is severe, the MIL will flash on/off on the 1st trip! **Possible Causes:** • Air leak in the intake manifold, or in the EGR or PCM system • Base engine mechanical fault that affects only one cylinder • Fuel delivery component fault that affects only one cylinder (i.e., a contaminated, dirty or sticking fuel injector) • Ignition system problem (coil or plug) that affects one cylinder
DTC: P0301 **2T CCM; MIL: Yes** **1996, 1997, 1998, 1999, 2000, 2001** **Models:** Sidekick, Swift, Vitara, X-90 **Engines:** 1.3L VIN 2, 1.6L VIN 0, 1.6L VIN 1, 1.8L VIN 2, 2.0L VIN 5 **Transmissions:** A/T, M/T	**Cylinder 1 Misfire Detected** DTC P0106, P0107, P0108, P0117, P0118, P0121, P0122, P0123, P0335, P0340 and P0500 not set, BARO sensor over 75 kPa, fuel level from 15-85%, IAT signal from 6.8-158°F, ambient temperature over 14°F, engine running, and the PCM detected a misfire in one cylinder in the 200 (Catalyst) or 1000-rpm (High Emissions) range. **Note:** If the misfire is severe, the MIL will flash on/off on the 1st trip! **Possible Causes:** • Air leak in the intake manifold, or in the EGR or PCM system • Base engine mechanical fault that affects only one cylinder • Fuel delivery component fault that affects only one cylinder (i.e., a contaminated, dirty or sticking fuel injector) • Ignition system problem (coil or plug) that affects one cylinder
DTC: P0302 **2T CCM; MIL: Yes** **1996, 1997, 1998, 1999, 2000, 2001** **Models:** Sidekick, Swift, Vitara, X-90 **Engines:** 1.3L VIN 2, 1.6L VIN 0, 1.6L VIN 1, 1.8L VIN 2, 2.0L VIN 5 **Transmissions:** A/T, M/T	**Cylinder 2 Misfire Detected** DTC P0106, P0107, P0108, P0117, P0118, P0121, P0122, P0123, P0335, P0340 and P0500 not set, BARO sensor over 75 kPa, fuel level from 15-85%, IAT signal from 6.8-158°F, ambient temperature over 14°F, engine running, and the PCM detected a misfire in one cylinder in the 200 (Catalyst) or 1000-rpm (High Emissions) range. **Note:** If the misfire is severe, the MIL will flash on/off on the 1st trip! **Possible Causes:** • Air leak in the intake manifold, or in the EGR or PCM system • Base engine mechanical fault that affects only one cylinder • Fuel delivery component fault that affects only one cylinder (i.e., a contaminated, dirty or sticking fuel injector) • Ignition system problem (coil or plug) that affects one cylinder
DTC: P0303 **2T CCM; MIL: Yes** **1996, 1997, 1998, 1999, 2000, 2001** **Models:** Sidekick, Swift, Vitara, X-90 **Engines:** 1.3L VIN 2, 1.6L VIN 0, 1.6L VIN 1, 1.8L VIN 2, 2.0L VIN 5 **Transmissions:** A/T, M/T	**Cylinder 3 Misfire Detected** DTC P0106, P0107, P0108, P0117, P0118, P0121, P0122, P0123, P0335, P0340 and P0500 not set, BARO sensor over 75 kPa, fuel level from 15-85%, IAT signal from 6.8-158°F, ambient temperature over 14°F, engine running, and the PCM detected a misfire in one cylinder in the 200 (Catalyst) or 1000-rpm (High Emissions) range. **Note:** If the misfire is severe, the MIL will flash on/off on the 1st trip! **Possible Causes:** • Air leak in the intake manifold, or in the EGR or PCM system • Base engine mechanical fault that affects only one cylinder • Fuel delivery component fault that affects only one cylinder (i.e., a contaminated, dirty or sticking fuel injector) • Ignition system problem (coil or plug) that affects one cylinder
DTC: P0304 **2T CCM; MIL: Yes** **1996, 1997, 1998, 1999, 2000, 2001** **Models:** Sidekick, Swift, Vitara, X-90 **Engines:** 1.3L VIN 2, 1.6L VIN 0, 1.6L VIN 1, 1.8L VIN 2, 2.0L VIN 5 **Transmissions:** A/T, M/T	**Cylinder 4 Misfire Detected** DTC P0106, P0107, P0108, P0117, P0118, P0121, P0122, P0123, P0335, P0340 and P0500 not set, BARO sensor over 75 kPa, fuel level from 15-85%, IAT signal from 6.8-158°F, ambient temperature over 14°F, engine running, and the PCM detected a misfire in one cylinder in the 200 (Catalyst) or 1000-rpm (High Emissions) range. **Note:** If the misfire is severe, the MIL will flash on/off on the 1st trip! **Possible Causes:** • Air leak in the intake manifold, or in the EGR or PCM system • Base engine mechanical fault that affects only one cylinder • Fuel delivery component fault that affects only one cylinder (i.e., a contaminated, dirty or sticking fuel injector) • Ignition system problem (coil or plug) that affects one cylinder

DTC	Trouble Code Title, Conditions & Possible Causes
DTC: P0301 **2T CCM; MIL: Yes** **1999, 2000, 2001, 2002** **Models:** Grand Vitara, XL-7 **Engines:** 2.5L VIN 6, 2.7L VIN 9 **Transmissions:** A/T, M/T	**Cylinder 1 Misfire Detected** DTC P0335 and P0340 not set, BARO sensor more than 75 kPa, fuel level from 15-85%, IAT signal more than 18°F, ECT sensor from 18-230°F, engine running, and the PCM detected a misfire in one cylinder in the 200 (Catalyst) or 1000-rpm (High Emissions) range. **Note: If the misfire is severe, the MIL will flash on/off on the 1st trip!** **Possible Causes:** • Air leak in the intake manifold, or in the EGR or PCM system • Base engine mechanical fault that affects only one cylinder • Fuel delivery component fault that affects only one cylinder (i.e., a contaminated, dirty or sticking fuel injector) • Ignition system problem (coil or plug) that affects one cylinder
DTC: P0302 **2T CCM; MIL: Yes** **1999, 2000, 2001, 2002** **Models:** Grand Vitara, XL-7 **Engines:** 2.5L VIN 6, 2.7L VIN 9 **Transmissions:** A/T, M/T	**Cylinder 2 Misfire Detected** DTC P0335 and P0340 not set, BARO sensor more than 75 kPa, fuel level from 15-85%, IAT signal more than 18°F, ECT sensor from 18-230°F, engine running, and the PCM detected a misfire in one cylinder in the 200 (Catalyst) or 1000-rpm (High Emissions) range. **Note: If the misfire is severe, the MIL will flash on/off on the 1st trip!** **Possible Causes:** • Air leak in the intake manifold, or in the EGR or PCM system • Base engine mechanical fault that affects only one cylinder • Fuel delivery component fault that affects only one cylinder (i.e., a contaminated, dirty or sticking fuel injector) • Ignition system problem (coil or plug) that affects one cylinder
DTC: P0303 **2T CCM; MIL: Yes** **1999, 2000, 2001, 2002** **Models:** Grand Vitara, XL-7 **Engines:** 2.5L VIN 6, 2.7L VIN 9 **Transmissions:** A/T, M/T	**Cylinder 3 Misfire Detected** DTC P0335 and P0340 not set, BARO sensor more than 75 kPa, fuel level from 15-85%, IAT signal more than 18°F, ECT sensor from 18-230°F, engine running, and the PCM detected a misfire in one cylinder in the 200 (Catalyst) or 1000-rpm (High Emissions) range. **Note: If the misfire is severe, the MIL will flash on/off on the 1st trip!** **Possible Causes:** • Air leak in the intake manifold, or in the EGR or PCM system • Base engine mechanical fault that affects only one cylinder • Fuel delivery component fault that affects only one cylinder (i.e., a contaminated, dirty or sticking fuel injector) • Ignition system problem (coil or plug) that affects one cylinder
DTC: P0304 **2T CCM; MIL: Yes** **1999, 2000, 2001, 2002** **Models:** Grand Vitara, XL-7 **Engines:** 2.5L VIN 6, 2.7L VIN 9 **Transmissions:** A/T, M/T	**Cylinder 4 Misfire Detected** DTC P0335 and P0340 not set, BARO sensor more than 75 kPa, fuel level from 15-85%, IAT signal more than 18°F, ECT sensor from 18-230°F, engine running, and the PCM detected a misfire in one cylinder in the 200 (Catalyst) or 1000-rpm (High Emissions) range. **Note: If the misfire is severe, the MIL will flash on/off on the 1st trip!** **Possible Causes:** • Air leak in the intake manifold, or in the EGR or PCM system • Base engine mechanical fault that affects only one cylinder • Fuel delivery component fault that affects only one cylinder (i.e., a contaminated, dirty or sticking fuel injector) • Ignition system problem (coil or plug) that affects one cylinder
DTC: P0305 **2T CCM; MIL: Yes** **1999, 2000, 2001, 2002** **Models:** Grand Vitara, XL-7 **Engines:** 2.5L VIN 6, 2.7L VIN 9 **Transmissions:** A/T, M/T	**Cylinder 5 Misfire Detected** DTC P0335 and P0340 not set, BARO sensor more than 75 kPa, fuel level from 15-85%, IAT signal more than 18°F, ECT sensor from 18-230°F, engine running, and the PCM detected a misfire in one cylinder in the 200 (Catalyst) or 1000-rpm (High Emissions) range. **Note: If the misfire is severe, the MIL will flash on/off on the 1st trip!** **Possible Causes:** • Air leak in the intake manifold, or in the EGR or PCM system • Base engine mechanical fault that affects only one cylinder • Fuel delivery component fault that affects only one cylinder (i.e., a contaminated, dirty or sticking fuel injector) • Ignition system problem (coil or plug) that affects one cylinder
DTC: P0306 **2T CCM; MIL: Yes** **1999, 2000, 2001, 2002** **Models:** Grand Vitara, XL-7 **Engines:** 2.5L VIN 6, 2.7L VIN 9 **Transmissions:** A/T, M/T	**Cylinder 6 Misfire Detected** DTC P0335 and P0340 not set, BARO sensor more than 75 kPa, fuel level from 15-85%, IAT signal more than 18°F, ECT sensor from 18-230°F, engine running, and the PCM detected a misfire in one cylinder in the 200 (Catalyst) or 1000-rpm (High Emissions) range. **Note: If the misfire is severe, the MIL will flash on/off on the 1st trip!** **Possible Causes:** • Air leak in the intake manifold, or in the EGR or PCM system • Base engine mechanical fault that affects only one cylinder • Fuel delivery component fault that affects only one cylinder (i.e., a contaminated, dirty or sticking fuel injector) • Ignition system problem (coil or plug) that affects one cylinder

DTC	Trouble Code Title, Conditions & Possible Causes
DTC: P0327 **1T CCM; MIL: Yes** **1999, 2000, 2001, 2002** **Models:** Vitara, XL-7 **Engines:** 1.6L VIN 0, 2.0L VIN 5, 2.7L VIN 9 **Transmissions:** A/T, M/T	**Knock Sensor Circuit Low Input** Key on or engine running, and the PCM detected an unexpected "high" voltage condition (0.90v or less) on the Knock Sensor circuit. **Possible Causes:** • Knock sensor signal circuit is shorted to sensor ground • Knock sensor signal circuit is shorted to chassis ground • Knock sensor is damaged or has failed • PCM has failed
DTC: P0328 **1T CCM; MIL: Yes** **1999, 2000, 2001, 2002** **Models:** Vitara, XL-7 **Engines:** 1.6L VIN 0, 2.0L VIN 5, 2.7L VIN 9 **Transmissions:** A/T, M/T	**Knock Sensor Circuit High Input** Key on or engine running, and the PCM detected an unexpected "low" voltage condition (3.98v or less) on the Knock Sensor circuit. **Possible Causes:** • Knock sensor signal circuit open between the sensor and PCM • Knock sensor signal circuit is shorted to VREF or system power • Knock sensor ground circuit is open • Knock sensor is damaged or has failed • PCM has failed
DTC: P0335 **1T CCM; MIL: Yes** **1996, 1997, 1998, 1999, 2000, 2001, 2002** **Models:** Esteem, Swift, X-90 **Engines:** 1.3L VIN 2, 1.6L VIN 3, 1.6L VIN 0, 1.6L VIN 1, 1.8L VIN 4, 2.0L VIN 5 **Transmissions:** A/T, M/T	**Crankshaft Position Sensor Circuit Malfunction** Engine cranking, and the PCM did not detect any CKP signals for two seconds during the CCM test. **Possible Causes:** • CKP sensor positive (+) circuit is open or shorted to ground • CKP sensor negative (-) circuit is open or shorted to ground • CKP sensor is damaged or has failed • PCM has failed
DTC: P0335 **1T CCM; MIL: Yes** **1999, 2000, 2001, 2002** **Models:** Grand Vitara, XL-7 **Engines:** 2.5L VIN 6, 2.7L VIN 9 **Transmissions:** A/T, M/T	**Crankshaft Position Sensor Circuit Malfunction** Engine started, and the PCM did not detect any CKP signals for 3 seconds or after 100 CMP pulses were received in the CCM test. **Possible Causes:** • CKP sensor positive (+) circuit is open or shorted to ground • CKP sensor negative (-) circuit is open or shorted to ground • CKP sensor is damaged or has failed • PCM has failed
DTC: P0335 **1T CCM; MIL: Yes** **1996, 1997, 1998** **Models:** Sidekick, Vitara **Engines:** 1.6L VIN 0, 1.8L VIN 2 **Transmissions:** A/T, M/T	**Crankshaft Position Sensor Circuit Malfunction** Engine started, engine running with 20 or more CMP sensor signals received, and the PCM did not detect any CKP signals in the test. **Possible Causes:** • CKP sensor positive (+) circuit is open or shorted to ground • CKP sensor negative (-) circuit is open or shorted to ground • CKP sensor is damaged or has failed • PCM has failed
DTC: P0340 **1T CCM; MIL: Yes** **1996, 1997, 1998, 1999, 2000, 2001, 2002** **Models:** Esteem, Swift, Vitara **Engines:** 1.3L VIN 2, 1.6L VIN 3, 1.6L VIN 0, 1.8L VIN 4, 2.0L VIN 5 **Transmissions:** A/T, M/T	**Camshaft Position Sensor Circuit Malfunction** Engine started, engine running with CKP sensor signals received, and the PCM did not detect CMP sensor signals for 5 seconds during the CCM test. **Possible Causes:** • CMP sensor positive (+) circuit is open or shorted to ground • CMP sensor negative (-) circuit is open or shorted to ground • CMP sensor is damaged or has failed • PCM has failed
DTC: P0340 **1T CCM; MIL: Yes** **1999, 2000, 2001, 2002** **Models:** Grand Vitara, XL-7 **Engines:** 2.5L VIN 6, 2.7L VIN 9 **Transmissions:** A/T, M/T	**Camshaft Position Sensor Circuit Malfunction** Engine cranking, engine start signal received (from starter circuit), and the PCM did not detect any CMP sensor signals for 5 seconds. **Possible Causes:** • CMP sensor signal circuit is open between sensor and PCM • CMP sensor signal circuit is shorted to ground • CMP sensor power circuit is open (check power from the relay) • CMP sensor is damaged or has failed • PCM has failed

DTC	Trouble Code Title, Conditions & Possible Causes
DTC: P0340 **1T CCM; MIL: Yes** **1996, 1997, 1998** **Models:** Sidekick, X-90 **Engines:** 1.6L VIN 0, 1.6L VIN 1, 1.8L VIN 2 **Transmissions:** A/T, M/T	**Camshaft Position Sensor Circuit Malfunction** Engine cranking, engine start signal received (from starter circuit), and the PCM did not detect any CMP sensor signals for 3 seconds. **Possible Causes:** • CMP sensor signal circuit is open between sensor and PCM • CMP sensor signal circuit is shorted to ground • CMP sensor power circuit is open (check power from the relay) • CMP sensor is damaged or has failed • PCM has failed
DTC: P0400 **2T CCM; MIL: Yes** **1996, 1997, 1998, 1999, 2000, 2001, 2002** **Models:** Esteem, Sidekick, Swift, X-90, Vitara **Engines:** 1.3L VIN 2, 1.6L VIN 3, 1.8L VIN 4, 1.6L VIN 0, 1.6L VIN 1, 2.0L VIN 5 **Transmissions:** A/T	**EGR System Fault (Too Much or Too Little Flow)** Engine started, fuel level over 25%, ambient temperature over 14°F, IAT signal less than 122°F, vehicle driven at a speed of 50-55 mph for 2 minutes in 'D' range, ECT sensor from 158-230°F, then the engine speed was increased to 4000 rpm in 'D2' range, followed by a deceleration period without the brakes applied to an engine speed of 1500 rpm with Fuel Cut enabled, the PCM detected too large or too small an amount of change in the MAP sensor signal in the test. **Possible Causes:** • EGR stepper motor control circuit(s) open or shorted to ground • EGR stepper motor assembly is damaged or has failed • Exhaust connection to EGR valve is clogged or restricted • Catalytic converter or exhaust system is clogged or restricted • PCM has failed
DTC: P0400 **2T CCM; MIL: Yes** **1996, 1997, 1998, 1999, 2000, 2001, 2002** **Models:** Esteem, Sidekick, Swift, Vitara, X-90 **Engines:** 1.3L VIN 2, 1.6L VIN 0, 1.6L VIN 1, 1.6L VIN 3, 1.8L VIN 4, 2.0L VIN 5 **Transmissions:** M/T	**EGR System Fault (Too Much or Too Little Flow)** Engine started, fuel level over 25%, BARO sensor over 75 kPa, IAT sensor less than 122°F, vehicle driven at a speed of 50-55 mph for 2 minutes in 5th gear, ECT sensor from 158-230°F, then the engine speed was increased to 4000 rpm in 2nd gear, followed by a deceleration period without the brakes applied to an engine speed of 1500 rpm with Fuel Cut enabled, the PCM detected too large or too small an amount of change in the MAP sensor signal in the test. **Possible Causes:** • EGR stepper motor control circuit(s) open or shorted to ground • EGR stepper motor assembly is damaged or has failed • Exhaust connection to EGR valve is clogged or restricted • Catalytic converter or exhaust system is clogged or restricted • PCM has failed
DTC: P0400 **2T CCM; MIL: Yes** **1999, 2000, 2001, 2002** **Models:** Grand Vitara, XL-7 **Engines:** 2.5L VIN 6, 2.7L VIN 9 **Transmissions:** A/T, M/T	**EGR System Fault (Too Much or Too Little Flow)** Engine started, BARO sensor over 75 kPa, IAT signal more than 18°F, ECT sensor from 18-230°F, vehicle driven at a speed of 35-40 mph with the throttle steady for 3 minutes, followed by a deceleration period with Fuel Cut enabled, and the PCM detected the change in the MAP sensor signal was too small after the EGR valve was opened momentarily, and then closed during the EGR flow test. **Possible Causes:** • EGR stepper motor control circuit(s) open or shorted to ground • EGR stepper motor assembly is damaged or has failed • Exhaust connection to EGR valve is clogged or restricted • Catalytic converter or exhaust system is clogged or restricted • PCM has failed
DTC: P0403 **1T CCM; MIL: Yes** **1999, 2000, 2001, 2002** **Models:** Grand Vitara, Vitara, XL-7 **Engines:** 1.6L VIN 0, 2.0L VIN 5, 2.7L VIN 9 **Transmissions:** A/T, M/T	**EGR Solenoid Control Circuit Malfunction** Engine started, IAT sensor over 18°F, ECT sensor from 18-230°F, BARO sensor over 75 kPa, and the PCM detected an unexpected voltage condition on the EGR solenoid control circuit during the test. **Possible Causes:** • One or more EGR stepper motor circuits is open • One or more EGR stepper motor circuits is shorted to ground • EGR stepper motor power circuit is open (check the J/B fuse) • EGR stepper motor is damaged or has failed • PCM has failed
DTC: P0420 **2T CCM; MIL: Yes** **1996, 1997, 1998, 1999, 2000** **Models:** Esteem, Swift **Engines:** 1.3L VIN 2, 1.6L VIN 3 **Transmissions:** A/T, M/T	**Catalyst Efficiency Below Normal (Bank 1)** DTC P0106, P0107, P0108, P0111, P0112, P0113, P0116, P0117, P0118, P0131, P0132, P0133, P0134, P0135, P0136, P0141, P0335, P0461, P0463, P0500, P1450 and P1451 not set, IAT signal less than 122°F, ambient temperature over 14°F, vehicle driven at over 35 mph for 1 minute, then at 55-60 mph for 5 minutes, and the PCM detected the HO2S-12 and HO2S-11 signals were too similar. **Possible Causes:** • Air leaks at the exhaust manifold or in the exhaust pipes • Catalytic converter is damaged, contaminated or has failed • Front HO2S or rear HO2S is contaminated with fuel or moisture • Front HO2S and/or the rear HO2S is loose in the mounting hole • Front HO2S is older (aged) than the rear HO2S (HO2S is lazy)

DTC	Trouble Code Title, Conditions & Possible Causes
DTC: P0420 **2T CCM; MIL: Yes** **1999, 2000, 2001, 2002** **Models:** Esteem, Sidekick, Vitara, X-90 **Engines:** 1.8L VIN 4, 1.6L VIN 0, 1.6L VIN 1, 1.8L VIN 2, 2.0L VIN 5 **Transmissions:** A/T, M/T	**Catalyst Efficiency Below Normal (Bank 1)** DTC P0101, P0102, P0103, P0111, P0112, P0113, P0116, P0117, P0118, P0131, P0132, P0133, P0134, P0135, P0136, P0141, P0335, P0461, P0463, P0500, P1450 and P1451 not set, IAT signal from 6.8-158°F, ambient temperature over 14°F, vehicle driven at over 35 mph for 1 minute, then at 55-60 mph for 5 minutes, and the PCM detected the HO2S-12 and HO2S-11 signals were too similar. **Possible Causes:** • Air leaks at the exhaust manifold or in the exhaust pipes • Catalytic converter is damaged, contaminated or has failed • Front HO2S or rear HO2S is contaminated with fuel or moisture • Front HO2S and/or the rear HO2S is loose in the mounting hole • Front HO2S is older (aged) than the rear HO2S (HO2S is lazy)
DTC: P0420 **2T CCM; MIL: Yes** **1996, 1997, 1998, 1999, 2000, 2001** **Models:** Swift **Engines:** 1.3L VIN 2 **Transmissions:** A/T, M/T	**Catalyst Efficiency Below Normal (Bank 1)** DTC P0106, P0107, P0108, P0111, P0112, P0113, P0116, P0117, P0118, P0131, P0132, P0133, P0134, P0135, P0136, P0141, P0335, P0461, P0463, P0500, P1450 and P1451 not set, IAT signal from 6.8-158°F, ambient temperature over 14°F, vehicle driven at over 35 mph for 1 minute, then at 55-60 mph for 5 minutes, and the PCM detected the HO2S-12 and HO2S-11 signals were too similar. **Possible Causes:** • Air leaks at the exhaust manifold or in the exhaust pipes • Catalytic converter is damaged, contaminated or has failed • Front HO2S or rear HO2S is contaminated with fuel or moisture • Front HO2S and/or the rear HO2S is loose in the mounting hole • Front HO2S is older (aged) than the rear HO2S (HO2S is lazy)
DTC: P0420 **2T CCM; MIL: Yes** **1999, 2000, 2001, 2002** **Models:** Grand Vitara, XL-7 **Engines:** 2.5L VIN 6, 2.7L VIN 9 **Transmissions:** A/T, M/T	**Catalyst Efficiency Below Normal (Bank 1)** DTC P0101, P0102, P0103, P0111-P0113, P0116-P0118, P0131, P0132, P0133, P0134, P0135, P0136, P0141, P0151, P0152, P0153, P0154, P0155, P0156, P0161, P0335, P0461, P0463, P0500, P1450 and P1451 not set, IAT signal more than 18°F, ECT sensor from 1858-230°F, BARO sensor over 75 kPa, vehicle driven at over 35 mph, then at 55-60 mph for 5 minutes; and the PCM detected the HO2S-12 and HO2S-11 signals were too similar. **Possible Causes:** • Air leaks at the exhaust manifold or in the exhaust pipes • Catalytic converter is damaged, contaminated or has failed • Front HO2S or rear HO2S is contaminated with fuel or moisture • Front HO2S and/or the rear HO2S is loose in the mounting hole • Front HO2S is older (aged) than the rear HO2S (HO2S is lazy)
DTC: P0430 **2T CCM; MIL: Yes** **1999, 2000, 2001, 2002** **Models:** Grand Vitara, XL-7 **Engines:** 2.5L VIN 6, 2.7L VIN 9 **Transmissions:** A/T, M/T	**Catalyst Efficiency Below Normal (Bank 2)** DTC P0101, P0102, P0103, P0111-P0113, P0116-P0118, P0131, P0132, P0133, P0134, P0135, P0136, P0141, P0151, P0152, P0153, P0154, P0155, P0156, P0161, P0335, P0461, P0463, P0500, P1450 and P1451 not set, IAT signal more than 18°F, ECT sensor from 158°F-230°F, BARO sensor over 75 kPa, vehicle driven at over 35 mph, then at 55-60 mph for 5 minutes; and the PCM detected the HO2S-22 and HO2S-21 signals were too similar. **Possible Causes:** • Air leaks at the exhaust manifold or in the exhaust pipes • Catalytic converter is damaged, contaminated or has failed • Front HO2S or rear HO2S is contaminated with fuel or moisture • Front HO2S and/or the rear HO2S is loose in the mounting hole • Front HO2S is older (aged) than the rear HO2S (HO2S is lazy)
DTC: P0440 **2T CCM; MIL: Yes** **1996, 1997, 1998, 1999, 2000, 2001, 2002** **Models:** Esteem, Sidekick, Swift, Vitara, Grand Vitara, XL-7, X-90 **Engines:** 1.3L VIN 2, 1.6L VIN 3, 1.8L VIN 4, 1.6L VIN 0, 1.6L VIN 1, 1.8L VIN 2, 2.0L VIN 5, 2.7L VIN 9 **Transmissions:** A/T, M/T	**EVAP System Malfunction** Engine started, vehicle driven to a speed of 35-40 for 20 minutes under conditions that do not include high-load, high engine speed, rapid acceleration or deceleration events, then at a constant speed of 30-40 mph for 3 minutes, fuel level from 25-75%, BARO sensor over 75 kPa, ambient temperature over 14°F, IAT sensor less than 158°F, ECT sensor from 158°F-230°F, then after the pressure in the EVAP system reached a predetermined value, and with the Purge valve command closed, the PCM detected a leak in the system. **Possible Causes:** • Canister air valve (solenoid) is damaged or has failed • Charcoal canister is clogged, loaded with fuel or moisture • Fuel filler cap loose, cross-threaded, incorrect part or damaged • Fuel tank pressure sensor is damaged or has failed • Fuel tank or fuel tank sender assembly 'O' ring is leaking • Fuel tank vapor line(s) blocked, damaged or disconnected • Tank pressure control solenoid is contaminated or damaged • Purge solenoid control circuit is open or shorted to ground • Purge solenoid power circuit is open (test FI fuse in relay box) • Purge solenoid is contaminated, damaged or has failed • PCM has failed

DTC	Trouble Code Title, Conditions & Possible Causes
DTC: P0443 **2T CCM; MIL: Yes** **1996, 1997, 1998** **Models:** Esteem, Sidekick, Swift, X-90 **Engines:** 1.3L VIN 2, 1.6L VIN 0, 1.6L VIN 1, 1.6L VIN 3, 1.8L VIN 2 **Transmissions:** A/T, M/T	**EVAP Canister Purge Solenoid Circuit Malfunction** Engine started, IAT signal less than 122°F, ambient temperature over 14°F, engine idle speed stable (±100 rpm), fuel level from 25-75%, all accessory loads off, and the PCM detected an unexpected voltage condition on the EVAP Canister Purge solenoid circuit. **Possible Causes:** • Purge solenoid control circuit is open or shorted to ground • Purge solenoid power circuit is open (check IG fuse in the J/B) • Purge solenoid is damaged or has failed • PCM has failed
DTC: P0443 **2T CCM; MIL: Yes** **1999, 2000, 2001, 2002** **Models:** Vitara, XL-7 **Engines:** 1.6L VIN 0, 2.0L VIN 5, 2.7L VIN 9 **Transmissions:** A/T, M/T	**EVAP Canister Purge Solenoid Circuit Malfunction** Engine started, IAT signal less than 122°F, ambient temperature over 14°F, engine running at a stable idle speed, fuel level from 25-75%, all accessory loads off, and the PCM detected an unexpected voltage condition on the EVAP Canister Purge solenoid circuit. **Possible Causes:** • Purge solenoid control circuit is open or shorted to ground • Purge solenoid power circuit is open (check IG fuse in the J/B) • Purge solenoid is damaged or has failed • PCM has failed
DTC: P0450 **2T CCM; MIL: Yes** **1997, 1998, 1999, 2000, 2001, 2002** **Models:** Esteem, Sidekick, Swift, Vitara, Grand Vitara, XL-7, X-90 **Engines:** 1.3L VIN 2, 1.6L VIN 0, 1.6L VIN 1, 1.6L VIN 3, 1.8L VIN 4, 1.8L VIN 2, 2.0L VIN 5, 2.7L VIN 9 **Transmissions:** A/T, M/T	**EVAP Pressure Sensor Circuit Malfunction** Engine started, then driven in stop and go conditions for 5 minutes, fuel level from 25-75%, ambient temperature over 14°F, IAT signal less than 122°F, and the PCM detected an unexpected voltage condition on the Pressure sensor circuit during the CCM test. **Possible Causes:** • Pressure sensor signal circuit is open or shorted to ground • Pressure sensor signal is shorted to VREF or system power • Pressure sensor is damaged or has failed • PCM has failed
DTC: P0451 **2T CCM; MIL: Yes** **1997, 1998, 1999, 2000, 2001, 2002** **Models:** Esteem, Swift, X-90 **Engines:** 1.3L VIN 2, 1.6L VIN 1, 1.6L VIN 3, 1.8L VIN 4 **Transmissions:** A/T, M/T	**EVAP Pressure Sensor Performance** Engine started, vehicle driven to a speed over 45 mph for 6 minutes at an engine speed from 2500-3000 rpm, followed by a deceleration period of 3 seconds, fuel level from 15-85%, BARO sensor more than 75 kPa, IAT sensor more than 158°F, ambient temperature over 14°F, ECT sensor from 158-230°F, and the PCM detected too small of a Pressure sensor change under these test conditions. **Possible Causes:** • Pressure sensor signal circuit is open or shorted to ground • Pressure sensor signal is shorted to VREF or system power • Pressure sensor is damaged or has failed • PCM has failed
DTC: P0451 **2T CCM; MIL: Yes** **1999, 2000, 2001, 2002** **Models:** Vitara, Grand Vitara, XL-7 **Engines:** 1.6L VIN 0, 2.5L VIN 6, 2.7L VIN 9 **Transmissions:** A/T, M/T	**EVAP Pressure Sensor Performance** Engine started, vehicle driven to a speed over 45 mph for 6 minutes at an engine speed from 2500-3000 rpm, followed by a deceleration period of 3 seconds, fuel level from 15-85%, BARO sensor more than 75 kPa, IAT sensor more than 158°F, ambient temperature over 14°F, ECT sensor from 158-230°F, and the PCM detected too small of a Pressure sensor change under these test conditions. **Possible Causes:** • Pressure sensor signal circuit is open or shorted to ground • Pressure sensor signal is shorted to VREF or system power • Pressure sensor is damaged or has failed • PCM has failed
DTC: P0455 **2T CCM; MIL: Yes** **1996, 1997, 1998, 1999, 2000, 2001, 2002** **Models:** Esteem, Sidekick, Swift, Grand Vitara, Vitara, XL-7 **Engines:** 1.3L VIN 2, 1.6L VIN 3, 1.8L VIN 4, 1.8L VIN 2, 2.5L VIN 6, 1.6L VIN 0, 2.0L VIN 5, 2.7L VIN 9 **Transmissions:** A/T, M/T	**EVAP System Large Leak (0.080") Detected** Cold engine startup, vehicle driven at 35-55 mph for 2 minutes, then returned to idle speed, ambient temperature over 14°F, ECT signal from 18-230°F, and the PCM detected a large change in the FTP sensor indicating a large leak (0.080") present during the leak test. **Possible Causes:** • Canister vent (CV) solenoid is stuck open • EVAP canister tube, EVAP canister purge outlet tube or EVAP return tube disconnected or cracked, or canister is damaged • EVAP canister purge valve stuck closed, or canister damaged • Fuel filler cap missing, loose (not tightened) or the wrong part • Fuel vapor hoses/tubes blocked or restricted, or fuel vapor control valve tube or fuel vapor vent valve assembly blocked • Fuel tank pressure (FTP) sensor has failed (mechanical fault) • Fuel tank control valve is contaminated, damaged or has failed

DTC	Trouble Code Title, Conditions & Possible Causes
DTC: P0461 **2T CCM; MIL: Yes** **1996, 1997, 1998, 1999, 2000, 2001, 2002** **Models:** Esteem, Sidekick, Swift, Vitara, Grand Vitara, XL-7, X-90 **Engines:** 1.3L VIN 2, 1.6L VIN 3, 1.6L VIN 0, 1.6L VIN 1, 1.8L VIN 2, 1.8L VIN 4, 2.5L VIN 6, 2.0L VIN 5, 2.7L VIN 9 **Transmissions:** A/T, M/T	**Fuel Level Sensor Performance** DTC P0461 not set, engine started, then after the vehicle was driven a predetermined amount of miles, the PCM detected the fuel level signal value changed less than 0.14v, or the PCM detected a signal from the Fuel Level sensor that was too "low" or too "high" during the CCM Rationality test. **Possible Causes:** • Fuel tank empty or overfull (fuel sender is stuck mechanically) • Wrong fuel gauge is installed, or instrument panel is damaged • Fuel gauge sender unit is damaged or has failed • PCM has failed
DTC: P0463 **2T CCM; MIL: Yes** **1996, 1997, 1998, 1999, 2000, 2001, 2002** **Models:** Esteem, Sidekick, Swift, Vitara, Grand Vitara, XL-7, X-90 **Engines:** 1.3L VIN 2, 1.6L VIN 3, 1.6L VIN 0, 1.6L VIN 1, 1.8L VIN 2, 1.8L VIN 4, 2.5L VIN 6, 2.0L VIN 5, 2.7L VIN 9 **Transmissions:** A/T, M/T	**Fuel Level Sensor Circuit High Input** Key on or engine running, and the PCM detected the Fuel Level Sensor signal indicated more than 4.60v during the CCM test. **Possible Causes:** • Fuel level signal circuit is open between the splice and the I/P • Wrong fuel gauge is installed, or instrument panel is damaged • Fuel gauge sender unit is damaged or has failed • PCM has failed
DTC: P0480 **2 CCM; MIL: Yes** **1997, 1998, 1999, 2000, 2001, 2002** **Models:** Esteem, Swift **Engines:** 1.3L VIN 2, 1.6L VIN 3, 1.8L VIN 4 **Transmissions:** A/T, M/T	**Radiator Fan Control System Performance** Engine started, engine running with the ECT sensor indicated a temperature of less than 200°F, and the PCM detected the Radiator Fan Control circuit indicated a "low" voltage condition during the test. **Possible Causes:** • Radiator fan relay control circuit is open • Radiator fan relay control circuit is shorted to ground • Radiator fan relay power circuit is open (check the relay fuse) • Radiator fan relay is damaged or has failed • PCM has failed
DTC: P0481 **2T CCM; MIL: Yes** **1996, 1997, 1998, 1999, 2000, 2001, 2002** **Models:** Esteem **Engines:** 1.6L VIN 3, 1.8L VIN 4 **Transmissions:** A/T, M/T	**A/C Cooling Relay Control Circuit Malfunction** Engine started, engine running, A/C switch in "off" position, ECT sensor signal less than 230°F, and the PCM detected the Radiator Fan Control circuit indicated a "low" voltage condition during the test. **Possible Causes:** • Condenser relay control circuit is open • Condenser fan relay control circuit is shorted to ground • Condenser fan relay power circuit is open (check relay fuse) • Radiator fan relay is damaged or has failed • PCM has failed
DTC: P0500 **1T CCM; MIL: Yes** **1996, 1997, 1998, 1999, 2000, 2001, 2002** **Models:** Esteem, Sidekick, Swift, Vitara, Grand Vitara, XL-7, X-90 **Engines:** 1.3L VIN 2, 1.6L VIN 3, 1.6L VIN 0, 1.6L VIN 1, 1.8L VIN 2, 1.8L VIN 4, 2.5L VIN 6, 2.0L VIN 5, 2.7L VIN 9 **Transmissions:** A/T, M/T	**Vehicle Speed Sensor Circuit Malfunction** Engine started, vehicle driven in 'D' range for 1 minute, or under "fuel cut" conditions, and the PCM did not detect any VSS signals. **Possible Causes:** • VSS signal circuit is open or shorted to ground • VSS ground circuit is open • VSS signal circuit from Combo Meter open or shorted to ground • VSS is damaged or has failed (drive gear may be damaged) • PCM has failed • TSB TS4-23 (6/96) contains information related to this code
DTC: P0505 **2T CCM; MIL: Yes** **1996, 1997, 1998, 1999, 2000, 2001, 2002** **Models:** Esteem, Sidekick, Swift, Vitara, Grand Vitara, XL-7, X-90 **Engines:** 1.3L VIN 2, 1.6L VIN 3, 1.8L VIN 4, 1.6L VIN 0, 1.8L VIN 2, 1.6L VIN 1, 2.5L VIN 6, 2.0L VIN 5, 2.7L VIN 9 **Transmissions:** A/T, M/T	**Idle Speed Control System** Engine started, ECT sensor signal more than 140°F, engine running at idle speed with the throttle closed, and the PCM detected the Actual idle speed was more than or less than the Target idle speed by too great an amount during the test. **Possible Causes:** • IAC valve control circuit is open or shorted to ground • IAC valve power circuit is open (check power from the relay) • IAC valve ground circuit is open • IAC valve is damaged or has failed • PCM has failed

DTC	Trouble Code Title, Conditions & Possible Causes
DTC: P0506 **2T CCM; MIL: Yes** **1997, 1998, 1999, 2000** **Models:** Esteem, Swift **Engines:** 1.3L VIN 2, 1.6L VIN 3, 1.8L VIN 4 **Transmissions:** A/T, M/T	**Idle Speed Control System Lower Than Expected** Engine started, engine running at idle speed, throttle switch is "on", IAT signal less than 158°F, ambient temperature over 14°F, ECT sensor from 158-230°F, and the PCM detected the Actual idle speed was over 100 rpm less than the Desired idle speed in the test. **Possible Causes:** • IAC valve control circuit is open or shorted to ground • IAC valve power circuit is open (intermittent fault) • IAC valve ground circuit is open (intermittent fault) • Throttle plate is carbon fouled (it may need to be cleaned) • PCM has failed
DTC: P0507 **2T CCM; MIL: Yes** **1997, 1998, 1999, 2000** **Models:** Esteem, Swift **Engines:** 1.3L VIN 2, 1.6L VIN 3, 1.8L VIN 4 **Transmissions:** A/T, M/T	**Idle Speed Control System Higher Than Expected** Engine started, engine running at idle speed, throttle switch is "on", IAT signal less than 122°F, ambient temperature over 14°F, ECT sensor from 158-230°F, and the PCM detected the Actual idle speed was over 100 rpm more than the Desired idle speed in the test. **Possible Causes:** • IAC valve control circuit is open (intermittent fault) • IAC valve power circuit is open (intermittent fault) • IAC valve ground circuit is open (intermittent fault) • Throttle plate is carbon fouled (it may need to be cleaned) • PCM has failed
DTC: P0510 **1T CCM; MIL: Yes** **1996, 1997, 1998** **Models:** Esteem, Sidekick, Swift, X-90 **Engines:** 1.3L VIN 2, 1.6L VIN 1, 1.6L VIN 3, 1.6L VIN 0, 1.8L VIN 2 **Transmissions:** A/T, M/T	**Closed Throttle Position Switch Circuit Malfunction** Engine started, engine running at cruise speed, then back to idle speed, IAT signal less than 122°F, ambient temperature over 14°F, and the PCM did not detect any change in the Closed Throttle Position switch status under these operating condition in the test. **Possible Causes:** • Closed throttle position switch signal circuit is open or grounded • Closed throttle position switch signal circuit is shorted to power • Closed throttle position switch or TP sensor damaged or failed • PCM has failed
DTC: P0601 **1T CCM; MIL: Yes** **1996, 1997, 1998, 1999, 2000, 2001, 2002** **Models:** Esteem, Sidekick, Swift, Vitara, Grand Vitara, XL-7, X-90 **Engines:** 1.3L VIN 2, 1.6L VIN 3, 1.8L VIN 4, 1.6L VIN 0, 1.8L VIN 2, 1.6L VIN 1, 2.5L VIN 6, 2.0L VIN 5, 2.7L VIN 9 **Transmissions:** A/T, M/T	**PCM Memory Check Sum Error** Key on, and the PCM detected a Memory Check Sum Error in the initial check during the CCM test. **Note: The engine may not start when this trouble code is set.** **Possible Causes:** • Clear the trouble code, and recheck for the same code. • If this trouble code resets, the PCM must be replaced
DTC: P0603 **1T; MIL: Yes** **1996, 1997, 1998, 1999, 2000, 2001, 2002** **Models:** Esteem, Sidekick, Swift, X-90 **Engines:** 1.3L VIN 2, 1.6L VIN 0, 1.6L VIN 1, 1.6L VIN 3, 1.8L VIN 2, 1.8L VIN 4 **Transmissions:** A/T, M/T	**PCM EEPROM Memory Error** Key on, and the PCM detected an EEPROM Memory Error. **Possible Causes:** • The contents of the EEPROM have changed • Clear the trouble code, and recheck for the same code. • If this trouble code resets, the PCM must be replaced
DTC: P0705 **2T CCM; MIL: Yes** **1996, 1997, 1998, 1999, 2000, 2001, 2002** **Models:** Esteem, Sidekick, Swift, X-90, Vitara, Grand Vitara, XL-7 **Engines:** 1.6L VIN 2, 1.6L VIN 3, 1.8L VIN 4, 1.6L VIN 0, 1.8L VIN 2, 1.6L VIN 1, 2.5L VIN 6, 2.0L VIN 5, 2.7L VIN 9 **Transmissions:** A/T	**Transmission Range Sensor Circuit Malfunction** Engine started, then driven to a speed of over 8 mph, and the PCM did not detect 'R', 'N', 'D' or 2nd gear TR switch inputs for 5 seconds, or it detected multiple TR switch signals during the CCM test. **Note: This is a 1-trip code multiple TR switch inputs are detected.** **Possible Causes:** • TR range switch signal is open • TR range switch signal shorted to another switch position signal • TR range switch is damaged or has failed • PCM has failed

DTC	Trouble Code Title, Conditions & Possible Causes
DTC: P0715 **2T CCM; MIL: Yes** **1996, 1997, 1998, 1999, 2000, 2001, 2002** **Models:** Esteem, Vitara, Grand Vitara, XL-7 **Engines:** 1.6L VIN 3, 1.8L VIN 4, 2.5L VIN 6, 2.0L VIN 5, 2.7L VIN 9 **Transmissions:** A/T	**TCM Input Speed Sensor Circuit Malfunction** Engine started, vehicle driven at of speed of 30 mph in 'D', 2nd or 'L', and then at over 30 mph in Drive for 30 seconds, and the TCM did not detect any ISS (A/T VSS) signals during the test. **Note: The Output Speed sensor resistance is 387-475 ohms at 68°F.** **Possible Causes:** • ISS signal (+) circuit is open or shorted to ground • ISS signal (-) circuit is open or shorted to ground • ISS is damaged or has failed • TCM has failed
DTC: P0720 **2T CCM; MIL: Yes** **1996, 1997, 1998, 1999, 2000, 2001, 2002** **Models:** Esteem, Sidekick, Swift, Vitara, Grand Vitara, XL-7, X-90 **Engines:** 1.6L VIN 2, 1.6L VIN 0, 1.6L VIN 1, 1.6L VIN 3, 1.8L VIN 2, 1.8L VIN 4, 2.5L VIN 6, 2.0L VIN 5, 2.7L VIN 9 **Transmissions:** A/T	**TCM Output Speed Sensor Circuit Malfunction** Engine started, vehicle driven at of speed of 30 mph in 'D', 2nd or 'L', and then at over 30 mph in Drive for 30 seconds, and the TCM did not detect any OSS signals during the CCM the test. **Note: The Output Speed sensor resistance is 648-792 ohms at 68°F.** **Possible Causes:** • OSS signal (+) circuit is open or shorted to ground • OSS signal (-) circuit is open or shorted to ground • OSS is damaged or has failed • TCM has failed
DTC: P0725 **2T CCM; MIL: Yes** **1996, 1997, 1998, 1999, 2000, 2001, 2002** **Models:** Esteem, Sidekick, X-90 **Engines:** 1.6L VIN 3, 1.8L VIN 4, 1.6L VIN 0, 1.8L VIN 2, 1.6L VIN 1 **Transmissions:** A/T	**TCM Engine Speed Input Circuit Malfunction** Engine started, vehicle driven to a speed of over 30 mph, and the TCM did not receive any engine speed signals on the ignition coil circuit during the CCM test. **Possible Causes:** • Engine speed signal to the TCM is open • Engine speed signal to the TCM is shorted to ground • Engine speed signal to the TCM is shorted to power • TCM has failed
DTC: P0741 **2T CCM; MIL: Yes** **1996, 1997, 1998, 1999, 2000, 2001, 2002** **Models:** Sidekick, Esteem, Grand Vitara, Vitara, XL-7, X-90 **Engines:** 1.6L VIN 3, 1.8L VIN 4, 1.6L VIN 0, 1.6L VIN 1, 1.8L VIN 2, 2.0L VIN 5, 2.5L VIN 6, 2.7L VIN 9 **Transmissions:** A/T	**TCC Solenoid Performance (Stuck Off)** Engine started, vehicle driven in 'D' range (4th gear), and the PCM detected the difference in rpm between the engine and A/T input speeds was too large with the TCC commanded "on", or with the engine running in 'D' range (3rd gear), the difference between the engine and A/T input speeds was too small during the test. **Possible Causes:** • TCC solenoid has a mechanical failure • TCC solenoid has a hydraulic failure • PCM has failed
DTC: P0743 **1T CCM; MIL: Yes** **1996, 1997, 1998, 1999, 2000, 2001, 2002** **Models:** Esteem, Sidekick, Grand Vitara, Vitara, XL-7, X-90 **Engines:** 1.6L VIN 3, 1.8L VIN 4, 1.6L VIN 0, 1.6L VIN 1, 1.8L VIN 2, 2.0L VIN 5, 2.5L VIN 6, 2.7L VIN 9 **Transmissions:** A/T	**Torque Converter Clutch Solenoid Circuit Malfunction** Engine started, engine running in closed loop, O/D Cut Switch "on" (O/D Lamp ON), gearshift in Drive at least 20 seconds, vehicle driven in 4th gear in 'D' range at less than a 40% throttle opening at a vehicle speed of over 45 mph or more for 10 seconds. This step needs to be repeated at least three times for this code to set. **Note: The TCC solenoid resistance is 11-15 ohms at 68°F.** **Possible Causes:** • TCC solenoid control circuit is open or shorted to ground • TCC solenoid control circuit is shorted to system power (B+) • TCC solenoid power circuit is open (check main relay & fuse) • TCC solenoid is damaged or has failed • PCM has failed
DTC: P0745 **2T CCM; MIL: Yes** **1996, 1997, 1998, 1999, 2000, 2001, 2002** **Models:** Esteem **Engines:** 1.6L VIN 3, 1.8L VIN 4 **Transmissions:** A/T	**Pressure Control Solenoid Circuit Malfunction** Engine started, engine running, and the PCM detected the current level of the PCS circuit was too "high" or too "low" during the test. **Note: The Pressure control solenoid resistance is 3-7 ohms at 68°F.** **Possible Causes:** • PCS control circuit is open or shorted to ground • PCS ground circuit is open • PCS is damaged or has failed • PCM has failed

DTC	Trouble Code Title, Conditions & Possible Causes
DTC: P0751 **2T CCM; MIL: Yes** **1996, 1997, 1998, 1999, 2000, 2001, 2002** **Models:** Esteem, Sidekick, Grand Vitara, Vitara, XL-7, X-90 **Engines:** 1.6L VIN 3, 1.8L VIN 4, 1.6L VIN 0, 1.6L VIN 1, 1.8L VIN 2, 2.0L VIN 5, 2.5L VIN 6, 2.7L VIN 9 **Transmissions:** A/T	**A/T Shift Solenoid 'A' Performance (Stuck Off)** Engine started, vehicle driven to a speed of over 30 mph, and the PCM detected a transmission gear ratio equal to 3rd gear with 2nd gear commanded "on" or a transmission gear ratio equal to 2nd gear with 3rd gear commanded "on" during the CCM Rationality test. **Possible Causes:** • SSA is stuck in "off" position (mechanical problem) • SSA is damaged is damaged or has failed (mechanical fault) • SSA has a hydraulic problem • PCM has failed
DTC: P0753 **1T CCM; MIL: Yes** **1996, 1997, 1998, 1999, 2000, 2001, 2002** **Models:** Esteem, Sidekick, Swift, Vitara, Grand Vitara, XL-7, X-90 **Engines:** 1.6L VIN 2, 1.6L VIN 3, 1.6L VIN 0, 1.6L VIN 1, 1.8L VIN 2, 1.8L VIN 4, 2.0L VIN 5, 2.5L VIN 6, 2.7L VIN 9 **Transmissions:** A/T	**A/T Shift Solenoid 'A' Circuit Malfunction** Engine started, vehicle driven to over 40 mph fro 10 seconds, and the PCM detected a "high" voltage condition on the SSA circuit with the solenoid commanded "off", or the SSA CCM detected a "low" voltage condition with the SSA commanded "on" in the CCM test. **Note: The Shift Solenoid 'A' resistance is 11-15 ohms at 68°F.** **Possible Causes:** • SSA control circuit is open or shorted to ground • SSA ground circuit is open, or the SSA power circuit is open • SSA is damaged or has failed • PCM has failed
DTC: P0756 **2T CCM; MIL: Yes** **1996, 1997, 1998, 1999, 2000, 2001, 2002** **Models:** Esteem, Sidekick, Grand Vitara, Vitara, XL-7, X-90 **Engines:** 1.6L VIN 3, 1.8L VIN 4, 1.6L VIN 0, 1.8L VIN 2, 1.6L VIN 1, 2.5L VIN 6, 2.0L VIN 5, 2.7L VIN 9**Transmissions:** A/T	**A/T Shift Solenoid 'B' Performance (Stuck Off)** Engine started, gear selector in 'D' position for 20 seconds, then after the vehicle was driven through a 1st, 2nd, 3rd and 4th gear shift change, and the PCM detected a transmission gear ratio equal to 4th gear with 3rd gear commanded "on" (SSA is "off", SSB is "on"), or a transmission gear ratio equal to 3rd gear with 4th gear commanded "on" (SSA is "off", SSB is "off") during the CCM test. **Possible Causes:** • SSB is stuck in "off" position (mechanical problem) • SSB is damaged is damaged or has failed (mechanical fault) • SSB has a hydraulic problem • PCM has failed
DTC: P0756 **2T CCM; MIL: Yes** **2000, 2001, 2002** **Models:** Swift **Engines:** 1.3L VIN 2 **Transmissions:** A/T	**A/T Shift Solenoid 'B' Performance (Stuck Off)** Engine started, then accelerated to 25-35 mph at medium position for 5 seconds, then accelerate to 45-55 mph at half-throttle position for at least 5 seconds, and the TCM detected a problem with the operation of the Shift Solenoid 'B' (SSB) control during the CCM test. **Possible Causes:** • SSB is stuck in "off" position (mechanical problem) • SSB is damaged is damaged or has failed (mechanical fault) • SSB has a hydraulic problem • PCM has failed
DTC: P0758 **1T CCM; MIL: Yes** **1996, 1997, 1998, 1999, 2000, 2001, 2002** **Models:** Esteem, Sidekick, Grand Vitara, Vitara, XL-7, X-90 **Engines:** 1.6L VIN 3, 1.6L VIN 0, 1.6L VIN 1, 1.8L VIN 2, 1.8L VIN 4, 2.0L VIN 5, 2.5L VIN 6, 2.7L VIN 9 **Transmissions:** A/T	**A/T Shift Solenoid 'B' Circuit Malfunction** Engine started, vehicle driven to over 40 mph fro 10 seconds, and the PCM detected a "high" voltage condition on the SSB circuit with the solenoid commanded "off", or the CCM detected a "low" voltage condition with the SSB commanded "on" during the CCM test. **Note: The Shift Solenoid 'B' resistance is 11-15 ohms at 68°F.** **Possible Causes:** • SSB control circuit is open or shorted to ground • SSB ground circuit is open, or the SSA power circuit is open • SSB is damaged or has failed • PCM has failed
DTC: P0771 **2T CCM; MIL: Yes** **1996, 1997, 1998** **Models:** Esteem **Engines:** 1.6L VIN 3 **Transmissions:** A/T	**Torque Converter Clutch System Performance** Engine started, engine running in closed loop, O/D Cut Switch "on" (O/D Lamp ON), gearshift in Drive for 20 seconds, vehicle driven in 4th gear in 'O' range at more than a 10% throttle opening at a speed of 15-30 mph for 10 seconds; then turn the O/D Cut Switch "off" and drive in 4th gear in 'O' range at less than a 40% throttle opening at over 45 mph for 10 seconds, **Possible Causes:** • TCC solenoid valve is damaged, leaking or allowing slipping • Internal transmission problems (i.e., TCC control valve is damaged or has failed, or a fluid passage is clogged or leaking) • Torque converter clutch is damaged or has failed • PCM has failed

DTC	Trouble Code Title, Conditions & Possible Causes
DTC: P0773 **2T CCM; MIL: Yes** **1996, 1997, 1998** **Models:** Esteem **Engines:** 1.6L VIN 3 **Transmissions:** A/T	**Torque Converter Clutch Solenoid Circuit Malfunction** Engine started, engine running in closed loop, vehicle driven in 4th gear for 2-3 minutes, and the PCM detected a "high" voltage condition on the TCC solenoid circuit with the TCC commanded "on", or it detected a "low" voltage condition while the TCC was off. **Note: The TCC solenoid resistance is 11-15 ohms at 68°F.** **Possible Causes:** • TCC solenoid control circuit is open or shorted to ground • TCC solenoid control circuit is shorted to system power (B+) • TCC solenoid power circuit is open (check main relay & fuse) • TCC solenoid is damaged or has failed • PCM has failed
DTC: P0780 **2T CCM; MIL: Yes** **1996, 1997, 1998, 1999** **Models:** Swift **Engines:** 1.3L VIN 2 **Transmissions:** A/T	**Torque Converter Clutch Solenoid Circuit Malfunction** Engine started, engine running in closed loop, vehicle driven in 4th gear for 2-3 minutes, and the PCM detected a "high" voltage condition on the TCC solenoid circuit with the TCC commanded "on", or it detected a "low" voltage condition while the TCC was off. **Note: The TCC solenoid resistance is 11-15 ohms at 68°F.** **Possible Causes:** • TCC solenoid control circuit is open or shorted to ground • TCC solenoid control circuit is shorted to system power (B+) • TCC solenoid power circuit is open (check main relay & fuse) • TCC solenoid is damaged or has failed • PCM has failed

Gas Engine OBD II Trouble Code List (P1xxx Codes)

DTC: P1250 **2T CCM; MIL: Yes** **1996, 1997** **Models:** Swift **Engines:** 1.3L VIN 2 **Transmissions:** A/T, M/T	**EFE Heater Circuit Malfunction** Engine started (cold), BARO sensor more than 75 kPa, IAT sensor less than 122°F, ambient temperature over 14°F, engine running and the PCM detected a "high" voltage condition on the EFE Heater control circuit, or with the engine running at hot idle, it detected a "low" voltage condition on the EFE Heater control circuit in the test. **Possible Causes:** • EFE heater control circuit is open or shorted to ground • EFE heater ground circuit or power circuit is open (from relay) • EFE heater control circuit is shorted to system power (B+) • EFE heater is damaged or has failed • PCM has failed
DTC: P1408 **2T CCM; MIL: Yes** **1996, 1997, 1998** **Models:** Sidekick, X-90 **Engines:** 1.6L VIN 0, 1.6L VIN 1, 1.8L VIN 2 **Transmissions:** A/T, M/T	**Manifold Differential Pressure Sensor Performance** Engine started, engine running in closed loop, IAT sensor more than 41°F, and the PCM detected the MDP sensor signal was more than 4.60v with the engine load lower than a calibrated amount, or it detected the MDP sensor signal was less than 0.20v with the engine load more than a calibrated amount during the CCM Rationality test. **Possible Causes:** • MDP sensor signal circuit is open or shorted to ground • MDP sensor vacuum source in intake manifold is clogged • MDP sensor is damaged, out-of-calibration or has failed • PCM has failed
DTC: P1408 **2T CCM; MIL: Yes** **1999, 2000, 2001, 2002** **Models:** Vitara, Grand Vitara, XL-7 **Engines:** 1.8L VIN 0, 2.0L VIN 5, 2.5L VIN 6, 2.7L VIN 9 **Transmissions:** A/T, M/T	**Manifold Absolute Pressure Sensor Performance** Engine started, vehicle driven at idle and cruise speed for 1 minute, ECT sensor signal from 14-158°F, and the PCM detected the MAP sensor signal was more than 4.16v or less than 0.5v for 5 seconds. **Possible Causes:** • MDP sensor signal circuit is open or shorted to ground • MDP sensor vacuum source in intake manifold is clogged • MDP sensor is damaged, out-of-calibration or has failed • PCM has failed
DTC: P1410 **2T CCM; MIL: Yes** **1998, 1999, 2000, 2001, 2002** **Models:** Esteem, X-90 **Engines:** 1.6L VIN 3, 1.6L VIN 1, 1.8L VIN 4 **Transmissions:** A/T, M/T	**Fuel Tank Pressure Control Solenoid Circuit Malfunction** Key on or engine running, ECT signal more than 158°F, IAT signal less than 122°F, ambient air temperature over 14°F, and the PCM detected an unexpected voltage on the FTP Control solenoid circuit. **Note: The FTP Control solenoid resistance is 28-39 ohms at 68°F.** **Possible Causes:** • FTP control solenoid circuit is open or shorted to ground • FTP control solenoid power circuit is open (check the relay) • FTP control solenoid is damaged, or has failed • PCM has failed
DTC: P1410 **2T CCM; MIL: Yes** **1996, 1997, 1998** **Models:** Sidekick **Engines:** 1.6L VIN 0, 1.8L VIN 2 **Transmissions:** A/T, M/T	**Fuel Tank Pressure Control Solenoid Circuit Malfunction** Key on or engine running, ECT signal more than 158°F, IAT signal less than 122°F, ambient air temperature over 14°F, and the PCM detected an unexpected voltage on the FTP Control solenoid circuit. **Note: The FTP Control solenoid resistance is 28-39 ohms at 68°F.** **Possible Causes:** • FTP control solenoid circuit is open or shorted to ground • FTP control solenoid power circuit is open (check the relay) • FTP control solenoid is damaged, or has failed • PCM has failed
DTC: P1410 **2T CCM; MIL: Yes** **1996, 1997, 1998, 1999, 2000, 2001** **Models:** Swift **Engines:** 1.3L VIN 2 **Transmissions:** A/T, M/T	**Fuel Tank Pressure Control Solenoid Circuit Malfunction** Key on or engine running, ECT signal more than 158°F, IAT signal less than 122°F, ambient air temperature over 14°F, and the PCM detected an unexpected voltage on the FTP Control solenoid circuit. **Note: The FTP Control solenoid resistance is 28-39 ohms at 68°F.** **Possible Causes:** • FTP control solenoid circuit is open or shorted to ground • FTP control solenoid power circuit is open (check the relay) • FTP control solenoid is damaged, or has failed • PCM has failed

DTC	Trouble Code Title, Conditions & Possible Causes
DTC: P1410 **2T CCM; MIL: Yes** **1999, 2000, 2001, 2002** **Models:** Vitara, Grand Vitara, XL-7 **Engines:** 1.8L VIN 0, 2.0L VIN 5, 2.5L VIN 6, 2.7L VIN 9 **Transmissions:** A/T, M/T	**Fuel Tank Pressure Control Solenoid Circuit Malfunction** Engine running, IAT signal from 6.8-158°F, altitude less than 9150 feet, vehicle driven to a speed of over 25 mph, and the PCM detected an unexpected voltage on the FTP Control solenoid circuit. **Note: The FTP Control solenoid resistance is 28-39 ohms at 68°F.** **Possible Causes:** • FTP control solenoid circuit is open or shorted to ground • FTP control solenoid power circuit is open (check the relay) • FTP control solenoid is damaged, or has failed • PCM has failed
DTC: P1450 **1T CCM; MIL: Yes** **1996, 1997, 1998, 1999, 2000,** **2001, 2002** **Models:** Esteem, Swift **Engines:** 1.3L VIN 2, 1.6L VIN 3, 1.8L VIN 4 **Transmissions:** A/T, M/T	**Barometric Pressure Sensor Circuit Malfunction** Key on or engine running, and the PCM detected the BARO sensor indicated more than 136 kPa, or less than 33 kPa in the CCM test. **Possible Causes:** • BARO sensor signal circuit is open or shorted to ground • BARO sensor signal circuit shorted to VREF or system power • BARO sensor is contaminated, damaged or has failed • PCM has failed
DTC: P1450 **2T CCM; MIL: Yes** **1996, 1997, 1998** **Models:** Sidekick, X-90 **Engines:** 1.6L VIN 0, 1.6L VIN 1, 1.8L VIN 2 **Transmissions:** A/T, M/T	**Barometric Pressure Sensor Circuit Malfunction** Key on or engine running, and the PCM detected the BARO sensor indicated more than 136 kPa, or less than 33 kPa in the CCM test. **Possible Causes:** • BARO sensor signal circuit is open or shorted to ground • BARO sensor signal circuit shorted to VREF or system power • BARO sensor is contaminated, damaged or has failed • PCM has failed
DTC: P1450 **2T CCM; MIL: Yes** **1999, 2000, 2001, 2002** **Models:** Vitara, Grand Vitara, XL-7 **Engines:** 1.8L VIN 0, 2.0L VIN 5, 2.5L VIN 6, 2.7L VIN 9 **Transmissions:** A/T, M/T	**Barometric Pressure Sensor Circuit Malfunction** Key on or engine running, and the PCM detected the BARO sensor indicated more than 136 kPa, or less than 33 kPa in the CCM test. **Possible Causes:** • BARO sensor signal circuit is open or shorted to ground • BARO sensor signal circuit shorted to VREF or system power • BARO sensor is contaminated, damaged or has failed • PCM has failed
DTC: P1451 **2T CCM; MIL: Yes** **1996, 1997, 1998, 1999, 2000,** **2001, 2002** **Models:** Esteem, Swift **Engines:** 1.3L VIN 2, 1.6L VIN 3, 1.8L VIN 4 **Transmissions:** A/T, M/T	**Barometric Pressure Sensor Performance** Engine cranking for 2 seconds, or engine runtime over one minute, and the PCM detected more than a 66 kPa difference between the BARO pressure and intake manifold pressure during the CCM test. **Possible Causes:** • BARO sensor signal circuit is open or shorted to ground • BARO sensor signal circuit shorted to VREF or system power • BARO sensor is contaminated, damaged or has failed • PCM has failed
DTC: P1451 **2T CCM; MIL: Yes** **1996, 1997, 1998** **Models:** Sidekick, X-90 **Engines:** 1.6L VIN 0, 1.6L VIN 1, 1.8L VIN 2 **Transmissions:** A/T, M/T	**Barometric Pressure Sensor Performance** Engine cranking for 2 seconds, or engine runtime over one minute, and the PCM detected too large a difference (over 66 kPa) between the BARO pressure and intake manifold pressure in the test. **Possible Causes:** • BARO sensor signal circuit is open or shorted to ground • BARO sensor signal circuit shorted to VREF or system power • BARO sensor is contaminated, damaged or has failed • PCM has failed
DTC: P1451 **2T CCM; MIL: Yes** **1999, 2000, 2001, 2002** **Models:** Vitara, Grand Vitara, XL-7 **Engines:** 1.6L VIN 0, 2.0L VIN 5, 2.5L VIN 6, 2.7L VIN 9 **Transmissions:** A/T, M/T	**Barometric Pressure Sensor Performance** Engine cranking for 2 seconds, or engine runtime over one minute, and the PCM detected too large a difference (over 66 kPa) between the BARO pressure and intake manifold pressure in the test. **Possible Causes:** • BARO sensor signal circuit is open or shorted to ground • BARO sensor signal circuit shorted to VREF or system power • BARO sensor is contaminated, damaged or has failed • PCM has failed

DTC	Trouble Code Title, Conditions & Possible Causes
DTC: P1460 **2T CCM; MIL: Yes** **1996, 1997** **Models:** Esteem, Swift **Engines:** 1.3L VIN 2, 1.6L VIN 3 **Transmissions:** A/T, M/T	**Radiator Fan Control System Circuit Malfunction** Engine started, engine running with the ECT sensor indicating a temperature of less than 200°F, and the PCM detected a "low" voltage condition on the Radiator Fan Control circuit during the test. **Possible Causes:** • Radiator fan relay control circuit is open or shorted to ground • Radiator fan relay ground circuit or power circuit is open • Radiator fan relay is damaged or has failed • PCM has failed
DTC: P1500 **2T CCM; MIL: Yes** **1996, 1997, 1998, 1999, 2000,** **2001, 2002** **Models:** Esteem, Sidekick, Swift, Vitara, Grand Vitara, XL-7, X-90 **Engines:** 1.3L VIN 2, 1.6L VIN 3, 1.8L VIN 4, 1.6L VIN 0, 1.6L VIN 1, 1.8L VIN 2, 2.0L VIN 5, 2.5L VIN 6, 2.7L VIN 9 **Transmissions:** A/T, M/T	**Engine Starter Signal Circuit Malfunction** Engine cranking, and the PCM detected an unexpected "low" voltage condition on the Engine Starter signal circuit, or with the engine running, it detected a "high" voltage condition on the Engine Starter signal circuit during the CCM test. **Possible Causes:** • Engine starter signal circuit is open or shorted to ground • Engine starter signal circuit is shorted to system power (B+) • PCM has failed
DTC: P1510 **1T CCM; MIL: Yes** **1996, 1997, 1998, 1999, 2000,** **2001, 2002** **Models:** Esteem, Sidekick, Swift, Vitara, Grand Vitara, XL-7, X-90 **Engines:** 1.3L VIN 2, 1.6L VIN 3, 1.8L VIN 4, 1.6L VIN 0, 1.6L VIN 1, 1.8L VIN 2, 2.0L VIN 5, 2.5L VIN 6, 2.7L VIN 9 **Transmissions:** A/T, M/T	**PCM Backup Power Supply Circuit Malfunction** Engine started, engine runtime one minute, and the PCM detected an unexpected low voltage on the Backup Power Supply circuit. **Possible Causes:** • Backup power supply circuit is open (check the Dome fuse) • Battery terminal connections may be loose or corroded • PCM has failed
DTC: P1530 **1T CCM; MIL: Yes** **1996, 1997, 1998, 1999, 2000,** **2001, 2002** **Models:** Esteem, Sidekick, Swift, X-90 **Engines:** 1.3L VIN 2, 1.6L VIN 0, 1.6L VIN 1, 1.6L VIN 3, 1.8L VIN 4 **Transmissions:** A/T, M/T	**Ignition Timing Adjustment Switch Circuit Malfunction** Engine started, engine running, and PCM detected an unexpected "low" voltage condition on the ITA switch circuit during the CCM test. **Possible Causes:** • ITA switch circuit is shorted to sensor or chassis ground somewhere between the PCM and the test connector • PCM has failed
DTC: P1600 **1T PCM; MIL: Yes** **1996, 1997, 1998, 1999, 2000,** **2001, 2002** **Models:** Esteem, Sidekick, X-90 **Engines:** 1.6L VIN 3, 1.6L VIN 0, 1.6L VIN 1, 1.8L VIN 4, 1.8L VIN 2 **Transmissions:** A/T, M/T	**PCM/TCM Serial Communication Circuit Malfunction** Engine started, engine running, and the PCM detected an irregular voltage condition on the PCM to TCM serial data circuit, or it detected a Check Sum Error occurred during the test. **Possible Causes:** • TCM to PCM serial data circuit is open • TCM to PCM serial data circuit is shorted to ground • TCM or PCM has failed
DTC: P1700 **1T CCM; MIL: Yes** **1996, 1997, 1998, 1999, 2000** **Models:** Esteem, Sidekick, X-90 **Engines:** 1.6L VIN 3, 1.6L VIN 0, 1.8L VIN 2, 1.6L VIN 1 **Transmissions:** A/T String	**TCM Throttle Position Sensor Circuit Malfunction** Engine started, engine runtime 30 seconds and the PCM detected the TP sensor signal indicated 0.00v or a specified time limit. **Note: The TP sensor signal is shared with the PCM on this circuit.** **Possible Causes:** • TP sensor signal circuit is open between the PCM and the TCM • TP sensor signal circuit is shorted to ground • TP sensor signal circuit is shorted to VREF or system power • PCM or TCM has failed

DTC	Trouble Code Title, Conditions & Possible Causes
DTC: P1702 **1T CCM; MIL: Yes** **2001, 2002** **Models:** Esteem **Engines:** 1.8L VIN 4 **Transmissions:** A/T	**TCM Internal Module Memory Check Sum Error** Key on, and the TCM detected a Memory Check Sum Error in the initial check during the CCM test. **Note: The engine may not start when this trouble code is set.** **Possible Causes:** • Clear the trouble code, and recheck for the same code. • If this trouble code resets, the TCM must be replaced
DTC: P1705 **1T CCM; MIL: Yes** **1996, 1997, 1998, 1999, 2000** **Models:** Esteem, Sidekick, X-90 **Engines:** 1.6L VIN 3, 1.6L VIN 0, 1.6L VIN 1, 1.8L VIN 2 **Transmissions:** A/T	**TCM ECT Sensor Circuit Malfunction** Engine started, engine runtime over 15 minutes, and the PCM detected an unexpected "low" low voltage (0.00v) condition, or an unexpected "high" voltage condition (9-14v) on the ECT sensor data circuit from the PCM to the TCM during the CCM test. **Possible Causes:** • ECT sensor data circuit is open between the PCM and TCM • ECT sensor data circuit is shorted to ground • ECT sensor data circuit is shorted to VREF or system power • PCM or TCM has failed
DTC: P1710 **1T CCM; MIL: Yes** **1996, 1997, 1998** **Models:** Sidekick **Engines:** 1.6L VIN 0, 1.8L VIN 2 **Transmissions:** A/T	**Speed Sensor Backup Signal Circuit Malfunction** Engine started, vehicle driven at more than 1 mph, and the PCM did not detect a VSS signal on the Speed Sensor Backup Signal circuit. **Possible Causes:** • VSS (backup) signal circuit is open between sensor and TCM • VSS (backup) signal is shorted to ground • VSS signal is damaged or has failed • Speedometer cable, drive gear or driven gear is damaged • TCM has failed
DTC: P1715 **1T CCM; MIL: Yes** **1996, 1997, 1998** **Models:** Sidekick, X-90 **Engines:** 1.6L VIN 0, 1.6L VIN 1 **Transmissions:** A/T	**Park Neutral Position Switch Circuit Malfunction** Engine started, engine running, and the PCM detected an engine speed change of 300-500 rpm with the VSS input at zero (0) mph and the TR switch indicating "off"; or the PCM detected the TR switch indicated "on" for a specific amount of time with the engine running at higher than a specified speed and load conditions. **Possible Causes:** • PNP switch signal circuit is open or shorted to ground • PNP switch signal circuit is shorted to VREF or system power • PNP switch signal circuit is shorted to another input circuit • PNP switch is damaged or has failed • PCM has failed
DTC: P1717 **2T CCM; MIL: Yes** **1996, 1997, 1998, 1999, 2000,** **2001, 2002** **Models:** Esteem **Engines:** 1.6L VIN 3, 1.8L VIN 4 **Transmissions:** A/T	**A/T Transmission Range Signal Circuit Malfunction** Engine started, engine running under cruise speed and engine load conditions, and the PCM did not detect a 'D' position signal, but did detect a Park/Neutral position signal during the CCM test. **Possible Causes:** • TCM data circuit to the PCM is open or shorted to ground • TR switch 'D', 'R', '2' or 'L' signal circuit is open • TCM power or ground circuit is open • TR switch is damaged or has failed • TCM or PCM has failed
DTC: P1875 **2T CCM; MIL: Yes** **1999, 2000, 2001, 2002** **Models:** Esteem **Engines:** 1.6L VIN 3, 1.8L VIN 4 **Transmissions:** A/T	**PCM Torque Reduction Circuit Malfunction** Engine started, engine running, and the PCM detected an unexpected "low" voltage condition on the PCM Torque Reduction circuit (5-volt) during the CCM test. **Possible Causes:** • Engine Torque control signal circuit is open • Engine Torque control signal is shorted to ground • TCM (controller) has failed

DTC	Trouble Code Title, Conditions & Possible Causes
DTC: P1875 **2T CCM; MIL: Yes** **1996, 1997, 1998** **Models:** Sidekick **Engines:** 1.6L VIN 0, 1.8L VIN 2 **Transmissions:** A/T	**4WD Low Switch Circuit Malfunction** Engine started, engine running, and the PCM detected an Output Speed sensor signal speed ratio from the 4WD Low Switch with the switch turned "off" or an OSS output speed ratio in a range other than the amount detected when the 4WD switch was first turned "on". **Possible Causes:** • 4WD switch signal circuit is open or shorted to ground • 4WD switch signal circuit is shorted to VREF or system power • 4WD switch is damaged or has failed • TCM or PCM has failed
DTC: P1875 **2T CCM; MIL: Yes** **1999, 2000, 2001, 2002** **Models:** Vitara, Grand Vitara, XL-7 **Engines:** 1.6L VIN 0, 2.0L VIN 5, 2.5L VIN 6, 2.7L VIN 9 **Transmissions:** A/T	**4WD Low Switch Circuit Malfunction** Engine started, engine running, and the PCM detected an Output Speed sensor signal speed ratio from the 4WD Low Switch with the switch turned "off" or an OSS output speed ratio in a range other than the amount detected when the 4WD switch was first turned "on". **Possible Causes:** • 4WD switch signal circuit is open or shorted to ground • 4WD switch signal circuit is shorted to VREF or system power • 4WD switch is damaged or has failed • TCM or PCM has failed

TOYOTA
DIAGNOSTIC TROUBLE CODES

14

TABLE OF CONTENTS

OBD II VEHICLE APPLICATIONS

TOYOTA

4Runner
1996–2006
2.7L I4 MFI 3RZ-FE .... VIN M
3.4L V6 MFI 5VZ-FE .... VIN N
4.7L V8 MFI 2UZ-FE ....VIN T
4.0L V6 MFI 1GR-FE .... VIN U

Avalon
1996–2006
3.0L V6 MFI 1MZ-FE ....VIN F
3.5L V6 MFI 2GR-FE ....VIN K

Camry
1996–2006
2.2L I4 MFI 5S-FE .... VIN G
3.0L V6 MFI 1MZ-FE ....VIN F
2.2L I4 MFI 5S-FNE .... VIN N
2.4L I4 MFI 2AZ-FE .... VIN E
3.3L V6 MFI 3MZ-FE ....VIN A

Camry Solara
1999–2006
2.2L I4 MFI 5S-FE .... VIN G
3.0L V6 MFI 1MZ-FE ....VIN F
2.4L I4 MFI 2AZ-FE .... VIN E
3.3L V6 MFI 3MZ-FE ....VIN A

Celica
1996–2005
2.2L I4 MFI 5S-FE .... VIN G
1.8L I4 MFI 7A-FE ....VIN B
1.8L I4 MFI 1ZZ-FE ....VIN R
1.8L I4 MFI 2ZZ-GE ....VIN Y

Corolla
1996–2006
1.6L I4 MFI 4A-FE ....VIN A
1.8L I4 MFI 7A-FE ....VIN B
1.8L I4 MFI 1ZZ-FE ....VIN R

Echo
2000–2005
1.5L I4 MFI 1NZ-FE ....VIN T

Highlander
2001–2006
2.4L I4 MFI 2AZ-FE .... VIN D
3.0L V6 MFI 1MZ-FE ....VIN F
3.3L V6 MFI 3MZ-FE ....VIN A

Land Cruiser
1996–2006
4.5L I6 MFI 1FZ-FE ....VIN J
4.7L V8 MFI 2UZ-FE ....VIN T

Matrix
2003–2006
1.8L I4 MFI 1ZZ-FE ....VIN R
1.8L I4 MFI 2ZZ-GE ....VIN Y

MR2
2000–2005
1.8L I4 MFI 1ZZ-FE ....VIN R

Paseo
1996–1997
1.5L I4 MFI 5E-FE ....VIN C

Previa
1996–1997
2.4L I4 MFI 2TZ-FZE ....VIN K

Prius
2001–2006
1.5L I4 MFI 1NZ-FXE ....VIN K
1.5L I4 MFI 1NZ-FXE ....VIN B

RAV 4
1996–2006
2.0L I4 MFI 3S-FE ....VIN P
2.0L I4 MFI 1AZ-FE .... VIN H
2.4L I4 MFI 2AZ-FE .... VIN D

Sequoia
2001–2006
4.7L V8 MFI 2UZ-FE ....VIN T

Sienna
1998–2006
3.0L V6 MFI 1MZ-FE ....VIN F
3.3L V6 MFI 3MZ-FE ....VIN A

Supra
1996–1998
3.0L I6 MFI 2JZ-GE .... VIN D
3.0L I6 MFI 2JZ-GTE ....VIN E

T 100
1996–1998
2.7L I4 MFI 3RZ-FE .... VIN M
3.4L V6 MFI 5VZ-FE .... VIN N

Tacoma
1996–2006
2.4L I4 MFI 2RZ-FE ....VIN L
2.7L I4 MFI 3RZ-FE .... VIN M
3.4L V6 MFI 5VZ-FE .... VIN N

Tercel
1996–1998
1.5L I4 MFI 5E-FE .... VIN C

Tundra
2000–2006
3.4L V6 MFI 5VZ-FE .... VIN N
4.7L V8 MFI 2UZ-FE ....VIN T

Yaris
2006
1.5L I4 MFI 1NZ-FE ....VIN T

DIAGNOSTIC TROUBLE CODES

OBD II Trouble Code List (P0xxx Codes)

DTC	Trouble Code Title, Conditions & Possible Causes
DTC: P2A00 **2T CCM, MIL: Yes** **2003, 2004s, 2005, 2006** **Models:** Avalon, Camry, Highlander, Sienna, Tacoma, Tundra **Engines:** 2.4L VIN D, 2.4L VIN L, 2.7L VIN M, 3.0L VIN F, 3.4L VIN N **Transmissions:** All	**Air Fuel Sensor (Bank 1 Sensor 1) Signal Slow Response Conditions:** Vehicle driven at cruise speed at over 1400 rpm in closed loop at 60 mph, and the PCM detected an unexpected voltage condition on the Bank 1 Air Fuel Sensor 1 (AFS1) circuit. **Possible Causes:** • A/F sensor connector is damaged or loose • A/F sensor circuit is open or shorted, or the sensor has failed • A/F sensor heater is damaged or has failed • A/F sensor heater relay circuit is open or the relay has failed • Fuel delivery component has failed (fuel pressure regulator, one or more fuel injectors is leaking or severely restricted) • Induction system problems (air leaks or restricted air filter) • PCM has failed
DTC: P2A03 **2T CCM, MIL: Yes** **2003, 2004, 2005, 2006** **Models:** Avalon, Camry, Highlander, Sienna **Engines:** 3.0L VIN F **Transmissions:** All	**Air Fuel Sensor (Bank 2 Sensor 1) Signal Slow Response Conditions:** Vehicle driven at cruise speed at over 1400 rpm in closed loop at 60 mph, and the PCM detected an unexpected voltage condition on the Bank 2 Air Fuel Sensor 1 (AFS1) circuit. **Possible Causes:** • A/F sensor connector is damaged or loose • A/F sensor circuit is open or shorted, or the sensor has failed • A/F sensor heater is damaged or has failed • A/F sensor heater relay circuit is open or the relay has failed • Fuel delivery component has failed (fuel pressure regulator, one or more fuel injectors is leaking or severely restricted) • Induction system problems (air leaks or restricted air filter) • PCM has failed
DTC: P0010 **2T CCM, MIL: Yes** **2003, 2004, 2005, 2006** **Models:** Avalon, Camry, Echo, Yaris, Highlander, MR2, Sienna **Engines:** 1.5L VIN T, 1.8L VIN R, 3.0L VIN F **Transmissions:** All	**VVT Oil Control Circuit Malfunction (Bank 1) Conditions:** Key on or engine running; and the PCM detected an unexpected voltage condition on the VVT Oil Control Valve Bank 1 circuit. The VVT system controls the intake camshaft in order to provide optimal valve timing during all conditions based signals from the ECT, IAT and TP sensor. The VVT regulates the intake camshaft angle using oil pressure through the Oil Control Valve. This results in the relative position between the camshaft and crankshaft to become optimal. The result is higher torque, better fuel economy and low emissions. **Possible Causes:** • OCV assembly connector is damaged or loose • OCV assembly control circuit is open or shorted to ground • OCV assembly is damaged or has failed • PCM has failed
DTC: P0011 **2T CCM, MIL: Yes** **2003, 2004, 2005, 2006** **Models:** Avalon, Camry, Echo, Yaris, Highlander, MR2, Sienna **Engines:** 1.5L VIN T, 1.8L VIN R, 3.0L VIN F **Transmissions:** All	**Camshaft Position 'A' Over-Advanced Or System Performance (Bank 1) Conditions:** Engine started, ECT sensor more than 158°F, vehicle driven at an engine speed of 400-4000 rpm, and the PCM detected the valve timing did not change from the "current" valve timing, or the valve timing remain fixed during testing. The VVT system controls the intake camshaft in order to provide optimal valve timing during all conditions based signals from the ECT, IAT and TP sensor. The VVT regulates the intake camshaft angle using oil pressure through the Oil Control Valve. This results in the relative position between the camshaft and crankshaft to become optimal. The result is better engine torque, fuel economy and lower emissions. **Possible Causes:** • Engine valve timing malfunction • Camshaft timing oil control valve unit is damaged or has failed • PCM or VVT ECM has failed
DTC: P0012 **2T CCM, MIL: Yes** **2003, 2004, 2005, 2006** **Models:** Avalon, Camry, Echo, Yaris, Highlander, MR2, Sienna **Engines:** 1.5L VIN T, 1.8L VIN R, 3.0L VIN F **Transmissions:** All	**Camshaft Position 'A' Over-Retarded (Bank 1) Conditions:** Engine started, ECT sensor more than 158°F, vehicle driven at an engine speed of 400-4000 rpm, and the PCM detected the valve timing did not change from the "current" valve timing, or that the valve timing remain fixed during the test period. The VVT system controls the intake camshaft in order to provide optimal valve timing during all conditions. The VVT regulates the intake camshaft angle using oil pressure through the OCV. This causes the relative position between the camshaft and crankshaft to become optimal. The result is improved engine torque, better fuel economy and low emissions. **Possible Causes:** • Engine valve timing malfunction • Camshaft timing oil control valve unit is damaged or has failed • PCM or VVT ECM has failed

DTC	Trouble Code Title, Conditions & Possible Causes
DTC: P0016 **2T CCM, MIL: Yes** **2003, 2004, 2005, 2006** **Models:** Avalon, Camry, Echo, Yaris, Highlander, MR2, Sienna **Engines:** 1.5L VIN T, 1.8L VIN R, 3.0L VIN F **Transmissions:** All	**Camshaft Position-Crankshaft Position (Bank 1 Sensor A) Conditions:** Engine started, engine running, and the PCM detected a deviation between the crankshaft position sensor signal and the VVT Sensor 1 signal during the test period. The crankshaft position (NE) sensor consists of a magnet, iron core and pickup coil. The NE sensor signal plate, installed on the crankshaft-timing pulley, has 34 teeth. This sensor generates 34 signals for each engine revolution. The PCM detects the crankshaft angle and engine speed based on the NE signal. It detects the correct cylinder based on signals from the VVT 1 sensor along signals from the crankshaft position sensor. **Possible Causes:** • Engine valve timing problem • Engine timing belt mechanical problem (skipped teeth or belt) • PCM has failed
DTC: P0018 **2T CCM, MIL: Yes** **2003, 2004, 2005, 2006** **Models:** Avalon, Camry, Echo, Yaris, Highlander, MR2, Sienna **Engines:** 3.0L VIN F **Transmissions:** All	**Camshaft Position-Crankshaft Position (Bank 2 Sensor A) Conditions:** Engine started, engine running, and the PCM detected a deviation between the crankshaft position sensor signal and the VVT Sensor 2 signal during the test period. The crankshaft position (NE) sensor consists of a magnet, iron core and pickup coil. The NE sensor signal plate, installed on the crankshaft-timing pulley, has 34 teeth. This sensor generates 34 signals for each engine revolution. The PCM detects the crankshaft angle and engine speed based on the NE signal. It detects the correct cylinder based on signals from the VVT 2 sensor along signals from the crankshaft position sensor. **Possible Causes:** • Engine valve timing problem • Engine timing belt mechanical problem (skipped teeth or belt) • PCM has failed
DTC: P0020 **2T CCM, MIL: Yes** **2003, 2004, 2005, 2006** **Models:** Avalon, Camry, Echo, Yaris, Highlander, MR2, Sienna **Engines:** 3.0L VIN F **Transmissions:** All	**Camshaft Position Sensor Actuator 'A' Circuit (Bank 2) Conditions:** Key on or engine running; and the PCM detected an unexpected voltage condition on the Camshaft Position Sensor 'A' Bank 2 circuit. The VVT system controls the intake camshaft in order to provide optimal valve timing during all conditions based signals from the ECT, IAT and TP sensor. The VVT regulates the intake camshaft angle using oil pressure through the OCV. The result is that relative position between the camshaft and crankshaft becomes optimal. The engine has better torque, fuel economy and lower emissions. **Possible Causes:** • OCV assembly connector is damaged or loose • OCV assembly control circuit is open or shorted to ground • OCV assembly is damaged or has failed • PCM has failed
DTC: P0021 **2T CCM, MIL: Yes** **2003, 2004, 2005, 2006** **Models:** Avalon, Camry, Echo, Yaris, Highlander, MR2, Sienna **Engines:** 3.0L VIN F **Transmissions:** All	**Camshaft Position 'A' Timing Over-Advanced Or System Performance (Bank 2) Conditions:** Engine started, ECT sensor more than 158°F, vehicle driven at an engine speed of 400-4000 rpm, and the PCM detected the valve timing did not change from the "current" valve timing, or the valve timing remain fixed during testing. The VVT system controls the intake camshaft in order to provide optimal valve timing during all conditions. The VVT regulates the intake camshaft angle using oil pressure through the Oil Control Valve. The result is that relative position between the camshaft and crankshaft becomes optimal. The engine has better torque, fuel economy and lower emissions. **Possible Causes:** • Engine valve timing malfunction • Camshaft timing oil control valve unit is damaged or has failed • PCM or VVT ECM has failed
DTC: P0022 **2T CCM, MIL: Yes** **2003, 2004, 2005, 2006** **Models:** Avalon, Camry, Echo, Yaris, Highlander, MR2, Sienna **Engines:** 3.0L VIN F **Transmissions:** All	**Camshaft Position 'A' Timing Over-Retarded (Bank 2) Conditions:** Engine started, ECT sensor over 158°F, engine speed of 400-4000 rpm, and the PCM detected the valve timing did not change from the "current" valve timing, or the valve timing remain fixed. The VVT system controls the intake camshaft in order to provide optimal valve timing during all conditions. The VVT regulates the intake camshaft angle using oil pressure through the OCV. The result is that relative position between the camshaft and crankshaft becomes optimal. The engine has better torque, fuel economy and lower emissions. **Possible Causes:** • Engine valve timing malfunction • Camshaft timing oil control valve unit is damaged or has failed • PCM or VVT ECM has failed
DTC: P0031 **1T CCM, MIL: Yes** **2003, 2004, 2005, 2006** **Models:** 4Runner, Echo, Yaris, Land Cruiser, MR2, Sequoia, Tacoma, Tundra **Engines:** All **Transmissions:** All	**Heated Oxygen Sensor (Bank 1 Sensor 1) Heater Circuit Low Code Conditions:** Engine started, and the PCM detected the HO2S-11 heater control circuit indicated less than 0.20 amps during the CCM test period. **Possible Causes:** • HO2S-11 heater control circuit is open • HO2S-11 heater assembly is damaged or has failed • PCM has failed

DTC	Trouble Code Title, Conditions & Possible Causes
DTC: P0032 **1T CCM, MIL: Yes** **2003, 2004, 2005, 2006** **Models:** 4Runner, Echo, Yaris, Land Cruiser, MR2, Sequoia, Tacoma, Tundra **Engines:** All **Transmissions:** All	**Heated Oxygen Sensor (Bank 1 Sensor 1) Heater Circuit High Input Conditions:** Engine started, and the PCM detected the HO2S-11 heater control circuit indicated more than 2.0 amps during the CCM test period. **Possible Causes:** • HO2S-11 heater control circuit is shorted to ground • HO2S-11 heater assembly is damaged or has failed • PCM has failed
DTC: P0036 **1T CCM, MIL: Yes** **2003, 2004, 2005, 2006** **Models:** Highlander, Tacoma **Engines:** 2.4L VIN L, 2.7L VIN M **Transmissions:** All	**Air Fuel Sensor (Bank 1 Sensor 2) Heater Circuit Malfunction Conditions:** Engine started, and the PCM detected the HO2S-12 heater control circuit indicated more than 2.35 amps or less than 0.20 amps. **Possible Causes:** • HO2S-12 heater control circuit is open or shorted to ground • HO2S-12 heater assembly is damaged or has failed • HO2S-12 heater power circuit open (test power from EFI relay) • PCM has failed
DTC: P0036 **1T CCM, MIL: Yes** **2003, 2004, 2005, 2006** **Models:** Avalon, Camry, Highlander, Sienna **Engines:** 3.0L VIN F **Transmissions:** All	**Heated Oxygen Sensor (Bank 1 Sensor 2) Heater Circuit Malfunction Conditions:** Engine started, and the PCM detected the HO2S-12 heater control circuit indicated more than 2.35 amps or less than 0.20 amps. **Possible Causes:** • HO2S-12 heater control circuit is open or shorted to ground • HO2S-12 heater assembly is damaged or has failed • HO2S-12 heater power circuit open (test power from EFI relay) • PCM has failed
DTC: P0037 **2T CCM, MIL: Yes** **2003, 2004, 2005, 2006** **Models:** 4Runner, Echo, Yaris, Land Cruiser, MR2, Sequoia, Tacoma, Tundra **Engines:** All **Transmissions:** All	**Heated Oxygen Sensor (Bank 1 Sensor 2) Heater Circuit Low Input Conditions:** Engine started, and the PCM detected the HO2S-12 heater control circuit indicated less than 0.20 amps during the CCM test period. **Possible Causes:** • HO2S-12 heater control circuit is open • HO2S-12 heater assembly is damaged or has failed • PCM has failed
DTC: P0038 **2T CCM, MIL: Yes** **2003, 2004, 2005, 2006** **Models:** 4Runner, Echo, Yaris, Land Cruiser, MR2, Sequoia, Tacoma, Tundra **Engines:** All **Transmissions:** All	**Heated Oxygen Sensor (Bank 1 Sensor 2) Heater Circuit High Input Conditions:** Engine started, and the PCM detected the HO2S-12 heater control circuit indicated more than 2.0 amps during the CCM test period. **Possible Causes:** • HO2S-12 heater control circuit is shorted to ground • HO2S-12 heater assembly is damaged or has failed • PCM has failed
DTC: P0051 **2T CCM, MIL: Yes** **2003, 2004, 2005, 2006** **Models:** 4Runner, Land Cruiser, MR2, Sequoia, Tundra **Engines:** 1.8L VIN R, 4.0L VIN U, 4.7L VIN T **Transmissions:** All	**Heated Oxygen Sensor (Bank 2 Sensor 1) Heater Circuit Low Input Conditions:** Engine started, and the PCM detected the HO2S-21 heater control circuit indicated less than 0.20 amps during the CCM test period. **Possible Causes:** • HO2S-21 heater control circuit is open • HO2S-21 heater assembly is damaged or has failed • PCM has failed
DTC: P0052 **2T CCM, MIL: Yes** **2003, 2004, 2005, 2006** **Models:** 4Runner, Land Cruiser, MR2, Sequoia, Tundra **Engines:** 1.8L VIN R, 4.0L VIN U, 4.7L VIN T **Transmissions:** All	**Heated Oxygen Sensor (Bank 2 Sensor 1) Heater Circuit High Input Conditions:** Engine started, and the PCM detected the HO2S-21 heater control circuit indicated more than 2.0 amps during the CCM test period. **Possible Causes:** • HO2S-21 heater control circuit is shorted to ground • HO2S-21 heater assembly is damaged or has failed • PCM has failed
DTC: P0057 **2T CCM, MIL: Yes** **2003, 2004, 2005, 2006** **Models:** 4Runner, Land Cruiser, Sequoia, Tundra **Engines:** 4.0L VIN U, 4.7L VIN T **Transmissions:** All	**Heated Oxygen Sensor (Bank 2 Sensor 2) Heater Circuit Low Input Conditions:** Engine started, and the PCM detected the HO2S-22 heater control circuit indicated less than 0.20 amps during the CCM test period. **Possible Causes:** • HO2S-22 heater control circuit is open • HO2S-22 heater assembly is damaged or has failed • PCM has failed

DTC	Trouble Code Title, Conditions & Possible Causes
DTC: P0058 **2T CCM, MIL: Yes** **2003, 2004, 2005, 2006** **Models:** 4Runner, Land Cruiser, Sequoia, Tundra **Engines:** 4.0L VIN U, 4.7L VIN T **Transmissions:** All	**Heated Oxygen Sensor (Bank 2 Sensor 2) Heater Circuit High Input Conditions:** Engine started, and the PCM detected the HO2S-22 heater control circuit indicated more than 2.0 amps during the CCM test period. **Possible Causes:** • HO2S-22 heater control circuit is shorted to ground • HO2S-22 heater assembly is damaged or has failed • PCM has failed
DTC: P0100 **1T CCM, MIL: Yes** **1995** **Models:** Avalon, Camry, Land Cruiser, Previa, T100, Tacoma, Tercel **Engines:** 1.5L VIN E, 2.4L VIN A, 2.4L VIN R, 2.7L VIN U, 3.0L VIN G, 3.4L VIN V, 4.5L VIN D **Transmissions:** All	**Mass Airflow Sensor Circuit Malfunction Conditions:** Engine running at under 4000 rpm, and the PCM detected an unexpected voltage condition on the MAF sensor circuit. The MAF sensor on this engine includes a hot wire assembly with an air temperature sensor, platinum hot wire and control unit mounted in a plastic housing. This airflow meter works on the principle that the hot wire and temperature sensor located in the intake air bypass of the housing detect any changes in the (incoming) air temperature. **Possible Causes:** • MAF sensor signal circuit is open, shorted to ground or power • MAF sensor ground circuit is open between sensor and ground • MAF sensor power circuit is open (check the power to the relay) • MAF sensor has failed, or the PCM has failed
DTC: P0100 **1T CCM, MIL: Yes** **1996, 1997, 1998, 1999, 2000, 2001, 2002, 2003, 2004, 2005, 2006** **Models:** All **Engines:** All **Transmissions:** All	**Mass Airflow Sensor Circuit Malfunction Conditions:** Engine started, engine running at under 4000 rpm, and the PCM detected an unexpected low or high voltage condition on the Mass Airflow (MAF) sensor circuit for over 3 seconds during the CCM test. The MAF sensor on this engine includes a hot wire assembly with an air temperature sensor, platinum hot wire and control unit mounted in a plastic housing. This airflow meter works on the principle that the hot wire and temperature sensor located in the intake air bypass of the housing detect any changes in the (incoming) air temperature. **Possible Causes:** • MAF sensor signal circuit is open, shorted to ground or power • MAF sensor ground circuit is open between sensor and ground • MAF sensor power circuit is open (check the power to the relay) • MAF sensor has failed, or the PCM has failed
DTC: P0101 **2T CCM, MIL: Yes** **1995** **Models:** Avalon, Camry, Land Cruiser, Previa, T100, Tacoma, Tercel **Engines:** 1.5L VIN E, 2.4L VIN A, 2.4L VIN R, 2.7L VIN U, 3.0L VIN G, 3.4L VIN V, 4.5L VIN D **Transmissions:** All	**Mass Airflow Sensor Signal Range/Performance Conditions:** DTC P0100 not set; engine speed under 900 rpm, throttle valve closed, ECT sensor more than 158°F, and the PCM detected the MAF sensor was more than 2.20v; or with the engine speed over 1500 rpm, the throttle valve closed and the TP sensor above 0.63v, the MAF sensor was below 1.06v. This airflow meter works on the principle where a temperature sensor and hot wire in the intake air bypass area detect any changes in the incoming air. **Possible Causes:** • MAF sensor signal circuit is open, shorted to ground or power • MAF sensor is contaminated, damaged or it has failed • PCM has failed
DTC: P0101 **2T CCM, MIL: Yes** **1996, 1997, 1998, 1999, 2000, 2001, 2002, 2003, 2004, 2005, 2006** **Engines:** All **Transmissions:** All	**Mass Airflow Sensor Signal Range/Performance Conditions:** DTC P0100 not set, engine speed under 900 rpm, throttle valve closed, ECT sensor over 158°F, and the PCM detected the MAF was above 2.20v; or with the engine speed over 1500 rpm, the throttle valve closed and the TP sensor over 0.63v, the MAF sensor was less than 1.06v. This airflow meter works on the principle that the hot wire and temperature sensor in the intake air bypass of the housing detect any changes in the (incoming) air temperature **Possible Causes:** • MAF sensor signal circuit is open, shorted to ground or power • MAF sensor is contaminated, damaged or it has failed • PCM has failed
DTC: P0102 **2T CCM, MIL: Yes** **2003, 2004, 2005, 2006** **Models:** 4Runner, Echo, Yaris, Land Cruiser, Sequoia, Tacoma, Tundra **Engines:** 1.5L VIN T, 3.4L VIN N, 4.0L VIN U, 4.7L VIN T **Transmissions:** All	**Mass Airflow Sensor Circuit Low Input Conditions:** DTC P0100 not set, engine started, and the PCM detected an unexpected low voltage condition on the MAF sensor circuit during the CCM test period. This airflow meter works on the principle that the hot wire and temperature sensor located in the intake air bypass of the housing detect any changes in the (incoming) air temperature **Possible Causes:** • MAF sensor signal circuit is open or shorted to ground • MAF sensor is contaminated, damaged or it has failed • PCM has failed

DTC	Trouble Code Title, Conditions & Possible Causes
DTC: P0103 **2T CCM, MIL: Yes** **2003, 2004, 2005, 2006** **Models:** 4Runner, Echo, Yaris, Land Cruiser, Sequoia, Tacoma, Tundra **Engines:** 1.5L VIN T, 3.4L VIN N, 4.0L VIN U, 4.7L VIN T **Transmissions:** All	**Mass Airflow Sensor Circuit High Input Conditions:** DTC P0100 not set, engine started, and the PCM detected an unexpected high voltage condition on the MAF sensor circuit during the CCM test period. This airflow meter works on the principle that the hot wire and temperature sensor located in the intake air bypass of the housing detect any changes in the (incoming) air temperature **Possible Causes:** • MAF sensor signal circuit is shorted to power • MAF sensor is contaminated, damaged or it has failed • PCM has failed
DTC: P0105 **1T CCM, MIL: Yes** **1995** **Models:** Paseo **Engines:** 1.5L VIN E **Transmissions:** All	**Manifold Air Pressure Sensor Circuit Malfunction Conditions:** Key on or engine running; and the PCM detected the MAP sensor signal was less than 0 kPa or more than 130 kPa in the CCM test. **Possible Causes:** • MAP sensor signal circuit is open, shorted to ground or power • MAP sensor ground circuit is open between sensor and ground • MAP sensor is damaged or has failed • PCM has failed
DTC: P0105 **1T CCM, MIL: Yes** **1996, 1997, 1998, 1999, 2000, 2001, 2002** **Models:** Paseo, Tercel, Camry, Camry Solara **Engines:** 1.5L VIN E, 2.2L VIN G, 2.4L VIN D **Transmissions:** All	**Manifold Air Pressure Sensor Circuit Malfunction Conditions:** Key on or engine running; and the PCM detected the MAP sensor signal was less than 0 kPa or more than 130 kPa in the CCM test. **Possible Causes:** • MAP sensor signal circuit is open, shorted to ground or power • MAP sensor ground circuit is open between sensor and ground • MAP sensor is damaged or has failed • PCM has failed
DTC: P0106 **1T CCM, MIL: Yes** **1995** **Models:** Paseo **Engines:** 1.5L VIN E **Transmissions:** All	**Manifold Air Pressure Sensor Signal Range/Performance Conditions:** DTC P0105 not set, engine started, engine speed at 400-1000 rpm, closed throttle switch indicating "on", ECT sensor input over 158°F, and the PCM detected the MAP sensor input was more than 3.3v for 10 seconds, or it detected it was less than 1.0v at under 2500 rpm with a VTA sensor input of more than 1.82v for 5 seconds. **Possible Causes:** • MAP sensor source vacuum line is leaking or restricted • MAP sensor source vacuum line is plugged at intake manifold • MAP sensor is damaged, out-of-calibration or has failed • PCM has failed
DTC: P0106 **1T CCM, MIL: Yes** **1996, 1997, 1998, 1999, 2000, 2001, 2002** **Models:** Paseo, Tercel, Camry, Camry Solara **Engines:** 1.5L VIN E, 2.2L VIN G, 2.4L VIN D **Transmissions:** All	**Manifold Air Pressure Sensor Signal Range/Performance Conditions:** DTC P0105 not set, engine started, engine speed at 400-1000 rpm, closed throttle switch indicating "on", ECT sensor input over 158°F, and the PCM detected the MAP sensor was above 3.3v (10 sec's), or it less below 1.0v at under 2500 rpm (VTA over 1.82v for 5 sec's). **Possible Causes:** • MAP sensor source vacuum line is leaking or restricted • MAP sensor source vacuum line is plugged at intake manifold • MAP sensor is damaged, out-of-calibration or has failed • PCM has failed
DTC: P0110 **1T CCM, MIL: Yes** **1995** **Models:** Avalon, Camry, Land Cruiser, Previa, T100, Tacoma, Tercel **Engines:** 1.5L VIN E, 2.4L VIN A, 2.4L VIN R, 2.7L VIN U, 3.0L VIN G, 3.4L VIN V, 4.5L VIN D **Transmissions:** All	**Intake Air Temperature Sensor Circuit Malfunction Conditions:** Key on or engine running; and the PCM detected an unexpected "low" or "high" voltage on the IAT sensor circuit (Scan Tool reads less than -40°F or low voltage or more than 284°F for high voltage). **Note: The IAT sensor is built into the MAF sensor on some engines.** **Possible Causes:** • IAT sensor signal circuit is open between the sensor and PCM • IAT sensor signal circuit is shorted to ground or to VREF • IAT sensor ground circuit is open between sensor and ground • IAT sensor is contaminated, damaged or has failed • PCM has failed
DTC: P0110 **1T CCM, MIL: Yes** **1996, 1997, 1998, 1999, 2000, 2001, 2002, 2003, 2004, 2005, 2006** **Models:** All **Engines:** All **Transmissions:** All	**Intake Air Temperature Sensor Circuit Malfunction Conditions:** Engine started, and the PCM detected an unexpected low or high voltage on the IAT sensor circuit (Scan Tool reads -40°F or 284°F). This sensor is located inside the MAF sensor. **Possible Causes:** • IAT sensor signal circuit is open, shorted to ground or VREF • IAT sensor ground circuit is open between sensor and ground • IAT sensor is contaminated, damaged or has failed • PCM has failed

DTC	Trouble Code Title, Conditions & Possible Causes
DTC: P0112 **1T CCM, MIL: Yes** 2003, 2004, 2005, 2006 **Models:** 4Runner, Echo, Yaris, Land Cruiser, Sequoia, Tacoma, Tundra **Engines:** 1.5L VIN T, 3.4L VIN N, 4.0L VIN U, 4.7L VIN T **Transmissions:** All	**Intake Air Temperature Sensor Circuit Low Input Conditions:** Key on or engine running; and the PCM detected an unexpected low voltage condition on the IAT sensor circuit for over 500 ms. **Possible Causes:** • IAT sensor connector is damaged (it may be shorted internally) • IAT sensor ground circuit is shorted to ground • IAT sensor is damaged or has failed • PCM has failed
DTC: P0113 **1T CCM, MIL: Yes** 2003, 2004, 2005, 2006 **Models:** 4Runner, Echo, Yaris, Land Cruiser, Sequoia, Tacoma, Tundra **Engines:** 1.5L VIN T, 3.4L VIN N, 4.0L VIN U, 4.7L VIN T **Transmissions:** All	**Intake Air Temperature Sensor Circuit High Input Conditions:** Key on or engine running; and the PCM detected an unexpected high voltage condition on the IAT sensor circuit for over 500 ms. **Possible Causes:** • IAT sensor connector is damaged (it may be open internally) • IAT sensor ground circuit is open • IAT sensor is damaged or has failed • PCM has failed
DTC: P0115 **2T CCM, MIL: Yes** 1995 **Models:** Avalon, Camry, Land Cruiser, Previa, T100, Tacoma, Tercel **Engines:** 1.5L VIN E, 2.4L VIN A, 2.4L VIN R, 2.7L VIN U, 3.0L VIN G, 3.4L VIN V, 4.5L VIN D **Transmissions:** All	**Engine Coolant Temperature Sensor Circuit Malfunction Conditions:** Key on or engine running; and the PCM detected an unexpected "low" or "high" voltage on the ECT sensor circuit (Scan Tool reads less than -40°F or low voltage or more than 284°F for high voltage). **Possible Causes:** • ECT sensor signal circuit is open, shorted to ground or VREF • ECT sensor ground circuit is open between sensor and ground • ECT sensor is contaminated, damaged or has failed • PCM has failed
DTC: P0115 **2T CCM, MIL: Yes** 1996, 1997, 1998, 1999, 2000, 2001, 2002, 2003, 2004, 2005, 2006 **Models:** All **Engines:** All **Transmissions:** All	**Engine Coolant Temperature Sensor Circuit Malfunction Conditions:** Key on or engine running; and the PCM detected an unexpected "low" or "high" voltage on the ECT sensor circuit (Scan Tool reads less than -40°F or low voltage or more than 284°F for high voltage). **Possible Causes:** • IAT sensor signal circuit is open, shorted to ground or VREF • ECT sensor ground circuit is open between sensor and ground • ECT sensor is contaminated, damaged or has failed • PCM has failed
DTC: P0116 **2T CCM, MIL: Yes** 1996, 1997, 1998, 1999, 2000, 2001, 2002 **Models:** All **Engines:** All **Transmissions:** All	**Engine Coolant Temperature Sensor Range/Performance Conditions:** Engine runtime over 20 minutes; and the PCM detected the ECT sensor indicated less than 95°F during the test. **Note: Check the condition and operation of the Cooling system.** **Possible Causes:** • Check for a low coolant level or an incorrect coolant mixture • ECT sensor signal circuit or ground circuit has high resistance • ECT sensor is contaminated, damaged or out-of-calibration • PCM has failed
DTC: P0116 **2T CCM, MIL: Yes** 1995 **Models:** Avalon, Camry, Land Cruiser, Previa, T100, Tacoma, Tercel **Engines:** 1.5L VIN E, 2.4L VIN A, 2.4L VIN R, 2.7L VIN U, 3.0L VIN G, 3.4L VIN V, 4.5L VIN D **Transmissions:** All	**Engine Coolant Temperature Sensor Range/Performance Conditions:** Engine started, engine runtime over 20 minutes, and the PCM detected the ECT sensor indicated less than 95°F during the test. **Note: Check the condition and operation of the Cooling system.** **Possible Causes:** • Check for a low coolant level or an incorrect coolant mixture • ECT sensor signal circuit has high resistance • ECT sensor ground circuit has high resistance • ECT sensor is contaminated, damaged or out-of-calibration • PCM has failed
DTC: P0116 **2T CCM, MIL: Yes** 2003, 2004, 2005, 2006 **Models:** All **Engines:** All **Transmissions:** All	**Engine Coolant Temperature Sensor Range/Performance Conditions:** Engine started, ECT sensor from 95°F to 140°F, IAT sensor more than 19.9°F, vehicle driven with several changes in the VSS signals, and the PCM detected the ECT sensor signal did not increase more than 37.4°F after engine was started and the test period completed. **Possible Causes:** • Check for problems in the cooling system (i.e., coolant, the fan) • ECT sensor signal circuit or ground circuit has high resistance • ECT sensor is contaminated, damaged or has failed • PCM has failed

DTC	Trouble Code Title, Conditions & Possible Causes
DTC: P0117 **2T CCM, MIL: Yes** **2003, 2004, 2005, 2006** **Models:** 4Runner, Echo, Yaris, Land Cruiser, Sequoia, Tacoma, Tundra **Engines:** 1.5L VIN T, 3.4L VIN N, 4.0L VIN U, 4.7L VIN T **Transmissions:** All	**Engine Coolant Temperature Sensor Circuit Low Input Conditions:** Key on or engine running; and the PCM detected an unexpected low voltage condition on the ECT sensor (Scan Tool reads below -40°F). **Possible Causes:** • ECT sensor signal circuit is shorted to ground • ECT sensor is damaged or has failed • PCM has failed
DTC: P0117 **2T CCM, MIL: Yes** **2003, 2004, 2005, 2006** **Models:** 4Runner, Echo, Yaris, Land Cruiser, Sequoia, Tacoma, Tundra **Engines:** 1.5L VIN T, 3.4L VIN N, 4.0L VIN U, 4.7L VIN T **Transmissions:** All	**Engine Coolant Temperature Sensor Circuit High Input Conditions:** Key on or engine running; and the PCM detected an unexpected high voltage condition on the ECT sensor (Scan Tool reads over 284°F). **Possible Causes:** • ECT sensor signal circuit is open • ECT sensor ground circuit is open • ECT sensor is damaged or has failed • PCM has failed
DTC: P0120 **2T CCM, MIL: Yes** **1995** **Models:** Avalon, Camry, Land Cruiser, Previa, T100, Tacoma, Tercel **Engines:** 1.5L VIN E, 2.4L VIN A, 2.4L VIN R, 2.7L VIN U, 3.0L VIN G, 3.4L VIN V, 4.5L VIN D **Transmissions:** All	**TP Sensor or Switch 'A' Circuit Malfunction Conditions:** Key on or engine running; and the PCM detected the TP sensor indicated less than 0.1v with the closed throttle position switch off (in open position), or the TP sensor input indicated 4.9v at any time. **Possible Causes:** • TP sensor signal circuit open or shorted to ground • TP sensor ground circuit is open • TP sensor power circuit is open (check VREF circuit at PCM) • TP sensor is damaged or has failed • PCM has failed
DTC: P0120 **2T CCM, MIL: Yes** **1996, 1997, 1998, 1999, 2000, 2001, 2002** **Models:** All **Engines:** All **Transmissions:** All	**TP Sensor or Switch 'A' Circuit Malfunction Conditions:** Key on or engine running; and the PCM detected the TP sensor indicated less than 0.1v with the closed throttle position switch off (open position), or the TP sensor was 4.9v at any time. **Possible Causes:** • TP sensor signal circuit is open or shorted to ground • TP sensor ground circuit is open, or the power circuit is open • TP sensor is damaged or has failed
DTC: P0120 **2T CCM, MIL: Yes** **2003, 2004, 2005, 2006** **Models:** Avalon, Camry, Camry Solara, Celica, Corolla, Echo, Yaris, Highlander, Matrix, MR2, Prius, RAV4, Sienna **Engines:** 1.5L VIN B, 1.5L VIN T, 1.8L VIN R, 1.8L VIN Y, 2.0L VIN K, 2.4L VIN D, 3.0L VIN F, 4.7L VIN T **Transmissions:** All	**TP Sensor or Switch 'A' Circuit Malfunction Conditions:** Engine started, and the PCM detected the TP sensor signal (VTA) was under 0.10v or over 4.90v. The TP sensor, mounted on the throttle body, detects the Throttle Valve opening angle (0.30v-0.70v closed). The PCM uses this signal for A/F ratio correction and power increase changes during all modes of engine operation. **Possible Causes:** • VTA sensor signal circuit is open or shorted to ground • VTA sensor ground circuit is open • VTA power circuit (VREF) is open • TP sensor is damaged or has failed • PCM has failed
DTC: P0120 **2T CCM, MIL: Yes** **2003, 2004, 2005, 2006** **Models:** 4Runner, Avalon, Camry, Highlander, Land Cruiser, Sequoia, Sienna, Tacoma, Tundra **Engines:** 2.4L VIN L, 2.7L VIN M engines, 3.0L VIN F, 4.0L VIN U, 4.7L VIN T **Transmissions:** All	**Throttle Pedal Position Sensor/Switch 'A' Circuit Malfunction Conditions** Key on or engine running; and the PCM detected the TP sensor 'A' Signal indicated less than 0.10v with the throttle position closed (the switch is open), or the TP sensor signal indicated more than 4.9v at any time. The Electric TP Sensor is mounted on the throttle body. It has two sensors (the electrical throttle system does not use a cable). **Possible Causes:** • TP sensor signal circuit open is or shorted to ground • TP sensor ground circuit is open • TP sensor power circuit is open (test VREF circuit at the PCM) • TP sensor is damaged or has failed • PCM has failed

DTC	Trouble Code Title, Conditions & Possible Causes
DTC: P0121 **2T CCM, MIL: Yes** **1995** **Models:** Avalon, Camry, Land Cruiser, Previa, T100, Tacoma, Tercel **Engines:** 1.5L VIN E, 2.4L VIN A, 2.4L VIN R, 2.7L VIN U, 3.0L VIN G, 3.4L VIN V, 4.5L VIN D **Transmissions:** All	**TP Sensor or TP Switch 'A' Signal Range/Performance Conditions:** Engine started, VSS input exceeds 19 mph at least once, and the PCM detected that the TP sensor input was out of the applicable range with the VSS reading between 0 and 30 mph. **Possible Causes:** • TP sensor signal circuit open or shorted to ground (intermittent) • TP sensor is loose at it mounting or the throttle is binding • TP sensor is damaged or has failed (perform a sweep test) • PCM has failed
DTC: P0121 **2T CCM, MIL: Yes** **1996, 1997, 1998, 1999, 2000, 2001, 2002** **Models:** All **Engines:** All **Transmissions:** All	**TP Sensor or TP Switch 'A' Signal Range/Performance Conditions:** Vehicle speed more than 19 mph at least once and the PCM detected the TP sensor input was out of the applicable range with the VSS input reading between 30 mph and 0 mph. **Possible Causes:** • TP sensor signal circuit open or shorted to ground (intermittent) • TP sensor is loose at it mounting or the throttle is binding • TP sensor is damaged or has failed (perform a sweep test) • PCM has failed
DTC: P0121 **2T CCM, MIL: Yes** **2003, 2004, 2005, 2006** **Models:** Avalon, Camry, Camry Solara, Celica, Corolla, Echo, Yaris, Highlander, Matrix, MR2, Prius, Sienna **Engines:** All **Transmissions:** All	**TP Sensor or TP Switch 'A' Signal Range/Performance Conditions:** Vehicle speed more than 19 mph at least once; and the PCM detected the TP sensor input was out of the applicable range with the VSS input reading between 30 mph and 0 mph. **Possible Causes:** • TP sensor signal circuit open or shorted to ground (intermittent) • TP sensor is loose at it mounting or the throttle is binding • TP sensor is damaged or has failed (perform a sweep test) • PCM has failed
DTC: P0121 **2T CCM, MIL: Yes** **2003, 2004, 2005, 2006** **Models:** 4Runner, Land Cruiser, RAV4, Sequoia, Tacoma, Tundra **Engines:** 2.0L VIN K, 2.4L VIN L, 2.7L VIN M, 3.4L VIN N, 4.0L VIN U, 4.7L VIN T **Transmissions:** All	**Throttle Pedal Position Sensor Switch 'A' Signal Range/Performance Conditions:** Engine started; and the PCM detected the difference between the TP sensor VTA1 and VTA2 signal was out-of-range. The TP sensor, mounted on the throttle body, detects the Throttle Valve opening angle (about 0.70v with the throttle closed). The PCM uses the VTA signal for air/fuel ratio and power increase correction. The Electric TP Sensor is mounted on the throttle body. It has two sensors (the electrical throttle system does not use a cable). **Possible Causes:** • TP sensor connector is damaged or it is open • TP sensor is damaged or has failed • PCM has failed
DTC: P0122 **2T CCM, MIL: Yes** **2003, 2004, 2005, 2006** **Models:** 4Runner, Land Cruiser, Sequoia, Tacoma, Tundra **Engines:** 3.4L VIN N engine, 4.0L VIN U, 4.7L VIN T **Transmissions:** All	**Throttle Pedal Position Sensor/Switch 'A' Circuit Low Input Conditions:** Engine started, and the PCM detected an unexpected low voltage (below 0.20v) on the VTA1 signal circuit. The TP sensor detects the Throttle Valve opening angle (0.70v with the throttle closed). The Electric TP Assembly has 2 sensors (this system does not use a throttle cable). **Possible Causes:** • Throttle control motor and sensor unit is damaged or failed • TP sensor VC (VREF) circuit is open, or the VTA1 signal circuit is shorted to ground • PCM has failed
DTC: P0123 **2T CCM, MIL: Yes** **2003, 2004, 2005, 2006** **Models:** 4Runner, Land Cruiser, Sequoia, Tacoma, Tundra **Engines:** 3.4L VIN N engine, 4.0L VIN U, 4.7L VIN T **Transmissions:** All	**Throttle Pedal Position Sensor/Switch 'A' Circuit High Input Conditions:** Engine started, and the PCM detected an unexpected high voltage (over 4.80v) on the VTA1 signal circuit. The TP sensor detects the Throttle Valve opening angle (0.70v with the throttle closed). The Electric TP Assembly has 2 sensors (this system does not use a throttle cable). **Possible Causes:** • Throttle control motor and sensor unit is damaged or failed • TP sensor VC (VREF) circuit is shorted to the VTA1 circuit • VTA1 ground circuit is open or the VTA1 signal circuit is open • PCM has failed
DTC: P0125 **1T CCM, MIL: Yes** **1995** **Models:** Avalon, Camry, Land Cruiser, Previa, T100, Tacoma, Tercel **Engines:** 1.5L VIN E, 2.4L VIN A, 2.4L VIN R, 2.7L VIN U, 3.0L VIN G, 3.4L VIN V, 4.5L VIN D **Transmissions:** All	**Insufficient Coolant Temperature For Closed Loop: Conditions:** DTC P0115 and P0116 not set, engine started, engine running, ECT sensor more than 140°F, vehicle driven to a speed of 25-62 mph at an engine speed over 1500 rpm, throttle valve not fully closed, and the PCM detected the HO2S signal (internal value) did not change. **Possible Causes:** • Check the operation of the thermostat (it may be stuck open) • ECT sensor signal circuit has high resistance • ECT sensor is out-of-calibration, skewed, or it has failed • Inspect for low coolant level or an incorrect coolant mixture

DTC	Trouble Code Title, Conditions & Possible Causes
DTC: P0125 **1T CCM, MIL: Yes** **1996, 1997, 1998, 1999, 2000, 2001, 2002** **Models:** All **Engines:** All **Transmissions:** All	**Insufficient Coolant Temperature For Closed Loop: Conditions:** **California Models** DTC P0115 and P0116 not set, engine started, ECT sensor more than 140°F, vehicle driven to a speed of 25-62 mph at an engine speed over 1500 rpm, throttle valve not fully closed, and the PCM detected the A/FS or HO2S signal (internal value) did not change. **Federal Models** Engine started, ECT sensor more than 140°F, vehicle driven to a speed of 25-62 mph at an engine speed over 1400 rpm, throttle valve not fully closed, and the PCM detected the HO2S-11 input did not exceed 450 mv at least during a test period of 1.5 minutes. **Possible Causes:** • Check the operation of the thermostat (it may be stuck open) • ECT sensor signal circuit has high resistance • ECT sensor has failed • Inspect for low coolant level or an incorrect coolant mixture
DTC: P0125 **1T CCM, MIL: Yes** **2003, 2004, 2005, 2006** **Models:** All **Engines:** All **Transmissions:** All	**Insufficient Coolant Temperature For Closed Loop Conditions:** **California Models** ECT sensor less than 19.4°F, engine started, engine runtime over 20 minutes, and the PCM detected the ECT sensor indicated 68°F or less; or with the ECT sensor from 19.4°F to 50.0°F at startup, engine runtime over 5 minutes, the PCM detected the ECT sensor indicated 68°F or less; or with the ECT sensor more than 19.4°F at startup, and the engine runtime over 5 minutes, the PCM detected the ECT sensor signal was 68°F or less; or with the ECT sensor over 50°F at startup, engine runtime over 2 minutes, the PCM detected the ECT sensor signal did not reach 86°F during the CCM test period. **Possible Causes:** • Check the operation of the thermostat (it may be stuck open) • ECT sensor signal circuit has high resistance • ECT sensor has failed • Inspect for low coolant level for an incorrect coolant mixture
DTC: P0128 **2T CCM, MIL: Yes** **2000, 2001, 2002, 2003, 2004, 2005, 2006** **Models:** All **Engines:** All **Transmissions:** All	**Thermostat System Malfunction Conditions:** Engine started, ECT sensor signal below 140°F at startup, and the PCM detected the ECT sensor did not reach 167°F after the warmup period expired (engine runtime 5-10 minutes). **Possible Causes:** • Check the operation of the thermostat (it may be stuck open) • ECT sensor is out-of-calibration or skewed • Inspect for low coolant level or for an incorrect coolant mixture
DTC: P0130 **2T CCM, MIL: Yes** **1995** **Models:** Avalon, Camry, Land Cruiser, Previa, T100, Tacoma, Tercel **Engines:** 1.5L VIN E, 2.4L VIN A, 2.4L VIN R, 2.7L VIN U, 3.0L VIN G, 3.4L VIN V, 4.5L VIN D **Transmissions:** All	**HO2S-11 (Bank 1 Sensor 1) Circuit Malfunction Conditions:** Engine started, engine warmup completed, engine idling in closed loop, and the PCM detected the HO2S-11 signal was fixed at 400 mv or higher, or that it was fixed at less than 550 mv during the test. **Possible Causes:** • HO2S signal circuit is open between the sensor and the PCM • HO2S signal circuit is shorted to sensor or chassis ground • HO2S signal circuit is shorted to VREF or system power (B+) • HO2S is damaged, contaminated or it has failed
DTC: P0130 **2T CCM, MIL: Yes** **1996, 1997, 1998, 1999, 2000, 2001, 2002** **Models:** All **Engines:** All **Transmissions:** All	**HO2S-11 (Bank 1 Sensor 1) Circuit Malfunction Conditions:** Engine warmup completed, engine idling in closed loop, and the PCM detected the HO2S-11 signal was fixed at 400 mv or higher, or that it was fixed at less than 550 mv during the test. **Possible Causes:** • HO2S signal circuit is open between the sensor and the PCM • HO2S signal circuit is shorted to sensor or chassis ground • HO2S signal circuit is shorted to VREF or system power (B+) • HO2S is damaged, contaminated or it has failed • PCM has failed
DTC: P0130 **2T CCM, MIL: Yes** **2003, 2004, 2005, 2006** **Models:** 4Runner, Celica, Corolla, Echo, Yaris, Land Cruiser, Matrix, MR2, Prius, Sequoia, Tundra **Engines:** 1.5L VIN B, 1.5L VIN T, 1.8L VIN R, 1.8L VIN Y, 4.0L VIN U, 4.7L VIN T **Transmissions:** All	**HO2S-11 (Bank 1 Sensor 1) Circuit Malfunction Conditions:** Engine started, engine warmup completed, engine idling in closed loop, and the PCM detected the HO2S-11 signal was fixed at 400 mv or higher, or that it was fixed at less than 550 mv during the test. **Possible Causes:** • HO2S signal circuit is open between the sensor and the PCM • HO2S signal circuit is shorted to sensor ground or to VREF • HO2S is damaged, contaminated or it has failed

DTC	Trouble Code Title, Conditions & Possible Causes
DTC: P0133 **2T O2S, MIL: Yes** **1995** **Models:** Avalon, Camry, Land Cruiser, Previa, T100, Tacoma, Tercel **Engines:** 1.5L VIN E, 2.4L VIN A, 2.4L VIN R, 2.7L VIN U, 3.0L VIN G, 3.4L VIN V, 4.5L VIN D **Transmissions:** All	**HO2S-11 (Bank 1 Sensor 1) Slow Response Conditions:** Engine started, engine idling in closed loop, and the PCM detected the HO2S-11 response time from rich-to-lean or from lean-to-rich was 1.1 second or longer during the CCM test. **Note: This fault must be detected at least 3 times at idle speed.** **Possible Causes:** • HO2S signal circuit is open or shorted to ground • HO2S element is contaminated, or HO2S heater has failed • Intake air leaks, exhaust manifold leaks or PCV system leaks • MAF sensor out of calibration (it may be dirty or contaminated)
DTC: P0133 **2T O2S, MIL: Yes** **1996, 1997, 1998, 1999, 2000, 2001, 2002** **Models:** All **Engines:** All **Transmissions:** All	**HO2S-11 (Bank 1 Sensor 1) Slow Response Conditions:** Engine started, engine idling in closed loop, and the PCM detected the HO2S-11 response time from rich-to-lean or from lean-to-rich was 1.1 second or longer during the CCM test. **Note: This fault must be detected at least 3 times at idle speed.** **Possible Causes:** • HO2S signal circuit is open or shorted to ground • HO2S element is contaminated, or HO2S heater has failed • Intake air leaks, exhaust manifold leaks or PCV system leaks • MAF sensor out of calibration (it may be dirty or contaminated)
DTC: P0133 **2T O2S, MIL: Yes** **2003, 2004, 2005, 2006** **Models:** 4Runner, Celica, Corolla, Echo, Yaris, Land Cruiser, Matrix, MR2, Prius, Sequoia, Tundra **Engines:** 1.5L VIN B, 1.5L VIN T, 1.8L VIN R, 1.8L VIN Y, 4.0L VIN U, 4.7L VIN T **Transmissions:** All	**HO2S-11 (Bank 1 Sensor 1) Slow Response Conditions:** Engine started, engine idling in closed loop, and the PCM detected the HO2S-11 response time from rich-to-lean or from lean-to-rich was 1.1 second or longer during the CCM test. **Note: This fault must be detected at least 3 times at idle speed.** **Possible Causes:** • HO2S signal circuit is open or shorted to ground • HO2S element is contaminated, or HO2S heater has failed • Intake air leaks, exhaust manifold leaks or PCV system leaks • MAF sensor out of calibration (it may be dirty or contaminated)
DTC: P0134 **2T O2S, MIL: Yes** **2003, 2004, 2005, 2006** **Models:** Avalon, Camry, Highlander, Sienna, Tacoma **Engines:** 2.4L VIN D, 2.4L VIN L, 2.7L VIN M, 3.0L VIN F **Transmissions:** All	**Air Fuel Sensor 1 (Bank 1 Sensor 1) Circuit No Activity Detected Conditions:** Engine started, engine runtime over 140 seconds, vehicle driven at a steady speed of 25-81 mph at over 1500 rpm with the throttle valve open, and the PCM detected the A/FS-11 signal did not indicate rich (more than 450 mv) after 65 seconds under these conditions. **Possible Causes:** • Air leaks present in the PCV valve, hoses or hose connections • A/FS1 connector is damaged or loose • A/FS1 signal circuit is open or shorted to ground • A/FS1 is damaged, contaminated or it has failed • Fuel Control component faults (sticking injector, low pressure) • Gas leaks in the exhaust system in front of the Oxygen sensor • PCM has failed
DTC: P0134 **2T O2S, MIL: Yes** **2003, 2004, 2005, 2006** **Models:** 4Runner, Land Cruiser, MR2, Sequoia, Tacoma, Tundra **Engines:** 1.8L VIN R, 2.7L VIN M, 3.4L VIN N, 4.0L VIN U, 4.7L VIN T **Transmissions:** All	**HO2S-11 (Bank 1 Sensor 1) Circuit No Activity Detected Conditions:** Engine started, engine runtime over 140 seconds, vehicle driven at a steady speed of 25-81 mph at over 1500 rpm with the throttle valve open, and the PCM detected the HO2S-11 signal did not indicate rich (more than 450 mv) after 65 seconds under these conditions. **Possible Causes:** • Air leaks present in the PCV valve, hoses or hose connections • Fuel Control component faults (sticking injector, low pressure) • Gas leaks in the exhaust system in front of the Oxygen sensor • HO2S connector is damaged or loose • HO2S signal circuit is open or shorted to ground • HO2S is damaged, contaminated or it has failed • PCM has failed
DTC: P0135 **2T CCM, MIL: Yes** **1995** **Models:** Avalon, Camry, Land Cruiser, Previa, T100, Tacoma, Tercel **Engines:** 1.5L VIN E, 2.4L VIN A, 2.4L VIN R, 2.7L VIN U, 3.0L VIN G, 3.4L VIN V, 4.5L VIN D **Transmissions:** All	**HO2S-11 (Bank 1 Sensor 1) Heater Circuit Malfunction Conditions:** Engine started, engine running, and the PCM detected the HO2S-11 heater current exceeded 2 amps, or that it was 0.25 amps or less. **Possible Causes:** • HO2S heater control circuit is open or shorted to ground • HO2S heater control circuit is shorted to power • HO2S heater power circuit is open (check power from the relay) • HO2S heater is damaged or has failed • PCM has failed

DTC	Trouble Code Title, Conditions & Possible Causes
DTC: P0135 **2T CCM, MIL: Yes** **1996, 1997, 1998, 1999, 2000, 2001, 2002** **Models:** All **Engines:** All **Transmissions:** All	**HO2S-11 (Bank 1 Sensor 1) Heater Circuit Malfunction Conditions:** Engine started, engine running, and the PCM detected the HO2S-11 heater current exceeded 2 amps, or that it was 0.25 amps or less. **Possible Causes:** • HO2S heater control circuit is open or shorted to ground • HO2S heater control circuit is shorted to power • HO2S heater power circuit is open (check power from the relay) • HO2S heater is damaged or has failed • PCM has failed
DTC: P0135 **2T CCM, MIL: Yes** **2003, 2004, 2005, 2006** **Models:** Celica, Corolla, Matrix, MR2, Prius **Engines:** 1.5L VIN B, 1.8L VIN R, 1.8L VIN Y **Transmissions:** All	**HO2S-11 (Bank 1 Sensor 1) Heater Circuit Malfunction Conditions:** Engine started, engine running, and the PCM detected the HO2S-11 heater current exceeded 2 amps, or that it was 0.25 amps or less. **Possible Causes:** • HO2S heater control circuit is open or shorted to ground • HO2S heater control circuit is shorted to power • HO2S heater power circuit is open (check power from the relay) • HO2S heater is damaged or has failed • PCM has failed
DTC: P0135 **2T CCM, MIL: Yes** **2003, 2004, 2005, 2006** **Models:** Avalon, Camry, Highlander, Sienna, Tacoma **Engines:** 2.4L VIN D, 2.4L VIN L, 2.7L VIN M, 3.0L VIN F **Transmissions:** All	**Air Fuel Sensor 1 (Bank 1 Sensor 1) Heater Circuit Malfunction Conditions:** Engine started, engine running, and the PCM detected the A/FS-11 heater current indicated over 8 amps, or it was less than 0.25 amps. **Possible Causes:** • A/FS1 heater control circuit is open or shorted to ground • A/FS1 heater control circuit is shorted to power • A/FS1 heater power circuit is open (check power from relay) • A/FS1 heater is damaged or has failed • PCM has failed
DTC: P0136 **2T CCM, MIL: Yes** **1995** **Models:** Avalon, Camry, Land Cruiser, Previa, T100, Tacoma, Tercel **Engines:** 1.5L VIN E, 2.4L VIN A, 2.4L VIN R, 2.7L VIN U, 3.0L VIN G, 3.4L VIN V, 4.5L VIN D **Transmissions:** All	**HO2S-12 (Bank 1 Sensor 2) Circuit Malfunction Conditions:** Engine started, vehicle driven to a speed of over 25 mph while in closed loop, and the PCM detected the HO2S-12 signal remained fixed at more than 400 mv, or that it was fixed at less than 600 mv. **Possible Causes:** • HO2S signal circuit is open or shorted to ground • HO2S signal circuit shorted to VREF or system power (B+) • HO2S is damaged, contaminated or it has failed • PCM has failed
DTC: P0136 **2T CCM, MIL: Yes** **1996, 1997, 1998, 1999, 2000, 2001, 2002** **Engines:** All **Transmissions:** All	**HO2S-12 (Bank 1 Sensor 2) Circuit Malfunction Code Conditions:** Vehicle driven to a speed of over 25 mph while in closed loop, and the PCM detected the HO2S-12 signal remained fixed at more than 400 mv, or it was fixed at less than 600 mv. **Possible Causes:** • HO2S signal circuit is open or shorted to ground • HO2S signal circuit shorted to VREF or system power (B+) • HO2S is damaged, contaminated or it has failed
DTC: P0136 **2T CCM, MIL: Yes** **2003, 2004, 2005, 2006** **Models:** 4Runner, Celica, Corolla, Echo, Yaris, Land Cruiser, Matrix, MR2, Prius, Sequoia, Tundra **Engines:** 1.5L VIN B, 1.5L VIN T, 1.8L VIN R, 1.8L VIN Y, 4.7L VIN T **Transmissions:** All	**HO2S-12 (Bank 1 Sensor 2) Circuit Malfunction Conditions:** Vehicle driven to a speed of over 25 mph while in closed loop, and the PCM detected the HO2S-12 signal was fixed at more than 400 mv, or that it was fixed at less than 600 mv. **Possible Causes:** • HO2S signal circuit is open or shorted to ground • HO2S signal circuit shorted to VREF or system power (B+) • HO2S is damaged, contaminated or it has failed
DTC: P0136 **2T CCM, MIL: Yes** **2003, 2004, 2005, 2006** **Models:** Avalon, Camry, Highlander, RAV4, Sienna, Tacoma, Tundra **Engines:** 2.0L VIN K, 2.4L VIN D, 2.4L VIN L, 2.7L VIN M, 3.0L VIN F, 3.4L VIN N **Transmissions:** All	**HO2S-12 (Bank 1 Sensor 2) Circuit Malfunction Conditions:** Engine started, vehicle driven to a speed of over 31 mph while in closed loop, and the PCM detected the HO2S-12 signal indicated more than 400 mv, or it indicated less than 500 mv during the test. **Possible Causes:** • HO2S signal circuit is open or shorted to ground • HO2S signal circuit shorted to VREF or system power (B+) • HO2S is damaged, contaminated or it has failed

DTC	Trouble Code Title, Conditions & Possible Causes
DTC: P0141 **2T CCM, MIL: Yes** **1995** **Models:** Avalon, Camry, Land Cruiser, Previa, T100, Tacoma, Tercel **Engines:** 1.5L VIN E, 2.4L VIN A, 2.4L VIN R, 2.7L VIN U, 3.0L VIN G, 3.4L VIN V, 4.5L VIN D **Transmissions:** All	**HO2S-12 (Bank 1 Sensor 2) Heater Circuit Malfunction Conditions:** Engine started, engine running, and the PCM detected the HO2S-12 heater current exceeded 2 amps, or that it was 0.25 amps or less. **Possible Causes:** • HO2S heater control circuit is open or shorted to ground • HO2S heater control circuit is shorted to power • HO2S heater power circuit is open (check power from the relay) • HO2S heater is damaged or has failed • PCM has failed
DTC: P0141 **2T CCM, MIL: Yes** **1996, 1997, 1998, 1999, 2000, 2001, 2002** **Models:** All **Engines:** All **Transmissions:** All	**HO2S-12 (Bank 1 Sensor 2) Heater Circuit Malfunction Conditions:** Engine started, engine running, and the PCM detected the HO2S-12 heater current exceeded 2 amps, or that it was 0.25 amps or less. **Possible Causes:** • HO2S heater control circuit is open, shorted to ground or power • HO2S heater power circuit is open (check power from the relay) • HO2S heater is damaged or has failed • PCM has failed
DTC: P0141 **2T CCM, MIL: Yes** **2003, 2004, 2005, 2006** **Models:** Celica, Corolla, Echo, Yaris, Matrix, MR2, Prius, RAV4 **Engines:** 1.5L VIN B, 1.5L VIN T, 1.8L VIN R, 1.8L VIN Y, 2.0L VIN K **Transmissions:** All	**HO2S-12 (Bank 1 Sensor 2) Heater Circuit Malfunction Conditions:** Engine started, engine running, and the PCM detected the HO2S-12 heater current exceeded 2 amps, or that it was 0.25 amps or less. **Possible Causes:** • HO2S heater control circuit is open, shorted to ground or power • HO2S heater power circuit is open (check power from the relay) • HO2S heater is damaged or has failed • PCM has failed
DTC: P0150 **2T CCM, MIL: Yes** **1995** **Models:** Avalon, Camry **Engines:** 3.0L VIN G **Transmissions:** All	**HO2S-21 (Bank 2 Sensor 1) Circuit Malfunction Conditions:** Engine warmup completed, engine idling in closed loop, and the PCM detected the HO2S-21 signal was fixed at 400 mv or higher, or that it was fixed from 400-550 mv during the test. **Possible Causes:** • HO2S signal circuit is open between the sensor and the PCM • HO2S signal circuit is shorted to sensor or chassis ground • HO2S is damaged, contaminated or it has failed
DTC: P0150 **2T CCM, MIL: Yes** **2003, 2004, 2005, 2006** **Models:** 4Runner, Land Cruiser, MR2, Sequoia, Tundra **Engines:** 1.8L VIN R, 4.0L VIN U, 4.7L VIN T **Transmissions:** All	**HO2S-21 (Bank 2 Sensor 1) Circuit Malfunction Conditions:** Engine warmup completed, engine idling in closed loop, and the PCM detected the HO2S-21 signal was fixed at 400 mv or higher, or that it was fixed at less than 550 mv during the test. **Possible Causes:** • HO2S signal circuit is open between the sensor and the PCM • HO2S signal circuit is shorted to sensor ground or to VREF • HO2S is damaged, contaminated or it has failed
DTC: P0150 **2T CCM, MIL: Yes** **1996, 1997, 1998, 1999, 2000, 2001, 2002** **Models:** Camry, Camry Solara, Land Cruiser, Sequoia, Sienna, Supra, Tundra **Engines:** 3.0L VIN D, 3.0L VIN E, 3.0L VIN F, 4.7L VIN T **Transmissions:** All	**HO2S-21 (Bank 2 Sensor 1) Circuit Malfunction Conditions:** Engine started, engine warmup completed, engine idling in closed loop, and the PCM detected the HO2S-21 signal was fixed at 400 mv or higher, or that it was fixed from 400-550 mv during the test. **Possible Causes:** • HO2S signal circuit is open between the sensor and the PCM • HO2S signal circuit is shorted to sensor or chassis ground • HO2S is damaged, contaminated or it has failed • PCM has failed
DTC: P0153 **2T O2S, MIL: Yes** **1995** **Models:** Avalon, Camry **Engines:** 3.0L VIN G **Transmissions:** All	**HO2S-21 (B2 S1) Slow Response Conditions:** Engine started, engine idling in closed loop, and the PCM detected the HO2S-21 response time from rich-to-lean or from lean-to-rich was 1.1 second or longer during the CCM test. **Note: This fault must be detected at least 3 times at idle speed.** **Possible Causes:** • HO2S signal circuit is open or shorted to ground • HO2S element is contaminated, or HO2S heater has failed • Intake air leaks, exhaust manifold leaks or PCV system leaks • MAF sensor out of calibration (it may be dirty or contaminated)

DTC	Trouble Code Title, Conditions & Possible Causes
DTC: P0153 **2T O2S, MIL: Yes** **1996, 1997, 1998, 1999, 2000, 2001, 2002** **Models:** Camry, Camry Solara, Land Cruiser, Sequoia, Sienna, Supra, Tundra **Engines:** 3.0L VIN D, 3.0L VIN E, 3.0L VIN F, 4.7L VIN T	**HO2S-21 (B2 S1) Slow Response Conditions:** Engine started, engine idling in closed loop, and the PCM detected the HO2S-21 response time from rich-to-lean or from lean-to-rich was 1.1 second or longer during the CCM test. **Note: This fault must be detected at least 3 times at idle speed.** **Possible Causes:** • HO2S signal circuit is open or shorted to ground • HO2S element is contaminated, or HO2S heater has failed • Intake air leaks, exhaust manifold leaks or PCV system leaks • MAF sensor out of calibration (it may be dirty or contaminated)
DTC: P0154 **2T O2S, MIL: Yes** **2003, 2004, 2005, 2006** **Models:** Avalon, Camry, Highlander, Sienna **Engines:** 3.0L VIN F **Transmissions:** All	**Air Fuel Sensor 2 (Bank 2 Sensor 1) Circuit No Activity Detected Conditions:** Engine started, engine runtime over 140 seconds, vehicle driven at a steady speed of 25-81 mph at over 1500 rpm with the throttle valve open, and the PCM detected the Bank 2 A/FS-21 signal did not indicate rich (more than 450 mv) after 65 seconds during the test. **Possible Causes:** • Air leaks present in the PCV valve, hoses or hose connections • A/FS1 connector is damaged or loose • A/FS1 signal circuit is open or shorted to ground • A/FS1 is damaged, contaminated or it has failed • Fuel Control component faults (sticking injector, low pressure) • Gas leaks in the exhaust system in front of the Oxygen sensor
DTC: P0154 **2T O2S, MIL: Yes** **2003, 2004, 2005, 2006** **Models:** 4Runner, Land Cruiser, MR2, Sequoia, Tundra **Engines:** 1.8L VIN R, 4.0L VIN U, 4.7L VIN T **Transmissions:** All	**HO2S-21 (Bank 2 Sensor 1) Circuit No Activity Detected Conditions:** Engine runtime over 140 seconds, VSS from 25-81 mph at over 1500 rpm with throttle valve open, and the PCM detected the HO2S-21 signal did not go over 450 mv after 65 seconds. **Possible Causes:** • Air leaks present in the PCV valve, hoses or hose connections • Fuel Control component faults (sticking injector, low pressure) • Gas leaks in the exhaust system in front of the Oxygen sensor • HO2S-21 signal circuit is open or shorted to ground • HO2S-21 is damaged, contaminated or it has failed • PCM has failed
DTC: P0155 **2T CCM, MIL: Yes** **1995** **Models:** Avalon, Camry **Engines:** 3.0L VIN G **Transmissions:** All	**HO2S-21 (B2 S1) Heater Circuit Malfunction Conditions:** Engine started, engine running, and the PCM detected the HO2S-21 heater current exceeded 2 amps, or that it was 0.25 amps or less. **Possible Causes:** • HO2S heater control circuit is open, shorted to ground or power • HO2S heater power circuit is open (check power from the relay) • HO2S heater is damaged or has failed • PCM has failed
DTC: P0155 **2T CCM, MIL: Yes** **1996, 1997, 1998, 1999, 2000, 2001, 2002** **Models:** Camry, Camry Solara, Land Cruiser, Sequoia, Sienna, Supra, Tundra **Engines:** 3.0L VIN D, 3.0L VIN E, 3.0L VIN F, 4.7L VIN T	**HO2S-21 (B2 S1) Heater Circuit Malfunction Conditions:** Engine started, engine running, and the PCM detected the HO2S-21 heater current exceeded 2 amps, or that it was 0.25 amps or less. **Possible Causes:** • HO2S heater control circuit is open or shorted to ground • HO2S heater control circuit is shorted to power • HO2S heater power circuit is open (check power from the relay) • HO2S heater is damaged or has failed • PCM has failed
DTC: P0155 **2T CCM, MIL: Yes** **2003, 2004, 2005, 2006** **Models:** Avalon, Camry, Highlander, Sienna **Engines:** 3.0L VIN F **Transmissions:** All	**Air Fuel Sensor 2 (Bank 2 Sensor 1) Heater Circuit Malfunction Conditions:** Engine started, engine running, and the PCM detected the A/FS-21 heater current indicated over 8 amps, or it was less than 0.25 amps. **Possible Causes:** • A/FS1 heater control circuit is open, shorted to ground or power • A/FS1 heater power circuit is open (check power from relay) • A/FS1 heater is damaged or has failed • PCM has failed

DTC	Trouble Code Title, Conditions & Possible Causes
DTC: P0156 **2T CCM, MIL: Yes** **1996, 1997, 1998, 1999, 2000, 2001, 2002, 2003, 2004, 2005, 2006** **Models:** 4Runner, Land Cruiser, RAV4, Sequoia, Tundra **Engines:** 2.0L VIN K, 4.0L VIN U, 4.7L VIN T **Transmissions:** All	**HO2S-22 (Bank 2 Sensor 2) Circuit Malfunction Conditions:** Engine started, engine warmup completed, engine idling in closed loop, and the PCM detected the HO2S-22 signal was fixed at 400 mv or higher, or that it was fixed from 400-550 mv during the test. **Possible Causes:** • HO2S signal circuit is open between the sensor and the PCM • HO2S signal circuit is shorted to sensor or chassis ground • HO2S is damaged, contaminated or it has failed
DTC: P0161 **2T CCM, MIL: Yes** **1996, 1997, 1998, 1999, 2000, 2001, 2002** **Models:** Land Cruiser, Sequoia, Tundra **Engines:** 4.7L VIN T **Transmissions:** All	**HO2S-22 (B2 S2) Heater Circuit Malfunction Conditions:** Engine started, engine running, and the PCM detected the HO2S-22 heater current exceeded 2 amps, or that it was 0.25 amps or less. **Possible Causes:** • HO2S heater control circuit is open, shorted to ground or power • HO2S heater power circuit is open (check power from the relay) • HO2S heater is damaged or has failed • PCM has failed
DTC: P0161 **2T CCM, MIL: Yes** **2003, 2004, 2005, 2006** **Models:** 4Runner, RAV4, Land Cruiser, Sequoia, Tundra **Engines:** 2.0L VIN K, 4.0L VIN U, 4.7L VIN T **Transmissions:** All	**HO2S-22 (B2 S2) Heater Circuit Malfunction Conditions:** Engine started, engine running, and the PCM detected the HO2S-22 heater current exceeded 2 amps, or it was less than 0.25 amps. **Possible Causes:** • HO2S heater control circuit is open, shorted to ground or power • HO2S heater power circuit is open (check power from the relay) • HO2S heater is damaged or has failed • PCM has failed
DTC: P0170 **2T FUEL, MIL: Yes** **1995** **Models:** Avalon, Camry, Land Cruiser, Previa, T100, Tacoma, Tercel **Engines:** 1.5L VIN E, 2.4L VIN A, 2.4L VIN R, 2.7L VIN U, 3.0L VIN G, 3.4L VIN V, 4.5L VIN D **Transmissions:** All	**Fuel System Too Rich or Too Lean (Bank 1) Conditions:** DTC P0100, P0101, P0110, P0115, P0120, P0121, P0130, P0133, P0136, P0135, P0136, P0141, P0153, P0155, P0201-206, P0300, P0301-P0306, P0401, P0402 and P0441 not set, engine running in closed loop at a stable engine speed, and the PCM detected the lean or rich fuel trim correction value was more than or less than a calibrated limit in memory. **Possible Causes:** • Air leaks present in the exhaust manifold or exhaust pipes • Air being drawn in from leaks in engine gaskets or other seals • Base engine "mechanical" fault affecting one or more cylinders • Fuel control sensor is out of calibration (i.e., ECT, IAT or MAF) • Fuel delivery system supplying too much or too little fuel at idle or cruise (e.g., faulty fuel pump or dirty, restricted fuel filter) • One or more fuel injectors is dirty, leaking or stuck open/closed • HO2S element is contaminated, damaged or it has failed
DTC: P0171 **2T FUEL, MIL: Yes** **1996, 1997, 1998, 1999, 2000, 2001, 2002** **Models:** All **Engines:** All **Transmissions:** All	**Fuel System Too Lean (Bank 1) Conditions:** DTC P0100, P0101, P0105, P0110, P0115, P0120, P0121, P0130, P0133, P0136, P0135, P0136, P0141, P0151, P0156, P0161, P0300, P0301-P0306, P0440, P0500 and P0505 not set, ECT sensor more than 158°F, vehicle driven at a constant speed of less than 62 mph with the engine speed over 1500 rpm, and the PCM detected the lean fuel trim correction value was over the limit. **Possible Causes:** • A/FS or HO2S is contaminated, deteriorated or it has failed • Air leaks after the MAF sensor, or in the EGR or PCV system • Base engine "mechanical" fault affecting one or more cylinders • Exhaust leaks located in front of the A/FS or HO2S location • Fuel system supplying too little fuel during cruise or idle (faulty fuel pump or fuel filter) • Fuel injector (one or more) dirty or pressure regulator has failed • Vehicle driven low on fuel or until it ran out of fuel
DTC: P0171 **2T FUEL, MIL: Yes** **2003, 2004, 2005, 2006** **Models:** Avalon, Camry, Highlander, Sienna **Engines:** 2.4L VIN D, 3.0L VIN F **Transmissions:** All	**Fuel System Too Lean (Bank 1) Conditions:** DTC P0100, P0101, P0105, P0110, P0115, P0120, P0121, P0130, P0133, P0136, P0135, P0136, P0141, P0151, P0156, P0161, P0300, P0301-P0306, P0440, P0500 and P0505 not set, vehicle speed less than 62 mph with the engine speed over 1500 rpm, ECT sensor more than 158°F, and the PCM detected the lean fuel trim correction value was over the limit. **Possible Causes:** • A/FS or HO2S is contaminated, deteriorated or it has failed • Air leaks after the MAF sensor, or in the EGR or PCV system • Base engine "mechanical" fault affecting one or more cylinders • Exhaust leaks located in front of the A/FS or HO2S location • Fuel system supplying too little fuel during cruise or idle (faulty fuel pump or fuel filter) • Fuel injector (one or more) dirty or pressure regulator has failed • Vehicle driven low on fuel or until it ran out of fuel

DTC	Trouble Code Title, Conditions & Possible Causes
DTC: P0171 **2T FUEL, MIL: Yes** **2003, 2004, 2005, 2006** **Models:** 4Runner, Celica, Corolla, Echo, Yaris, Land Cruiser, Matrix, MR2, Prius, RAV4, Sequoia, Tacoma, Tundra **Engines:** 1.5L VIN B, 1.5L VIN T, 1.8L VIN R, 1.8L VIN Y, 2.0L VIN K, 2.4L VIN L, 2.7L VIN M, 3.4L VIN N, 4.0L VIN U, 4.7L VIN T **Transmissions:** All	**Fuel System Too Lean (Bank 1) Conditions:** DTC P0100, P0101, P0105, P0110, P0115, P0120, P0121, P0130, P0133, P0136, P0135, P0136, P0141, P0151, P0156, P0161, P0300, P0301-P0306, P0440, P0500 and P0505 not set, ECT sensor more than 158°F, vehicle driven at a constant speed of less than 62 mph with the engine speed over 1500 rpm, and the PCM detected the lean fuel trim correction value was over the limit. **Possible Causes:** • A/FS or HO2S is contaminated, deteriorated or it has failed • Air leaks after the MAF sensor, or in the EGR or PCV system • Base engine "mechanical" fault affecting one or more cylinders • Exhaust leaks located in front of the A/FS or HO2S location • Fuel control sensor is out of calibration (i.e., ECT, IAT or MAP) • Fuel delivery system supplying too little fuel during cruise or idle periods (e.g., faulty fuel pump or dirty, restricted fuel filter) • Fuel injector (one or more) dirty or pressure regulator has failed • Vehicle driven low on fuel or until it ran out of fuel
DTC: P0172 **2T FUEL, MIL: Yes** **1996, 1997, 1998, 1999, 2000, 2001, 2002** **Models:** All **Engines:** All **Transmissions:** All	**Fuel System Too Rich (Bank 1) Conditions:** DTC P0100, P0101, P0105, P0110, P0115, P0120, P0121, P0130, P0133, P0136, P0135, P0136, P0141, P0151, P0156, P0161, P0300, P0301-P0306, P0440, P0500 and P0505 not set, ECT sensor more than 158°F, vehicle driven at a constant speed of less than 62 mph with the engine speed over 1500 rpm, and the PCM detected the rich fuel trim correction value was over the limit. **Possible Causes:** • A/FS or HO2S is contaminated, deteriorated or it has failed • Base engine "mechanical" fault affecting one or more cylinders • EVAP system component has failed or canister fuel saturated • Exhaust leaks located in front of the A/FS or HO2S location • Fuel control sensor is out of calibration (i.e., ECT, IAT or MAF) • Fuel delivery system supplying too much fuel during cruise or idle periods (e.g., faulty fuel pump, or faulty pressure regulator) • Fuel injector(s) is leaking or stuck partially open (one or more)
DTC: P0172 **2T FUEL, MIL: Yes** **2003, 2004, 2005, 2006** **Models:** 4Runner, Celica, Corolla, Echo, Yaris, Land Cruiser, Matrix, MR2, Prius, RAV4, Sequoia, Tacoma, Tundra **Engines:** 1.5L VIN B, 1.5L VIN T, 1.8L VIN R, 1.8L VIN Y, 2.0L VIN K, 2.4L VIN L, 2.7L VIN M, 3.4L VIN N, 4.0L VIN U, 4.7L VIN T **Transmissions:** All	**Fuel System Too Rich (Bank 1) Conditions:** DTC P0100, P0101, P0105, P0110, P0115, P0120, P0121, P0130, P0133, P0136, P0135, P0136, P0141, P0151, P0156, P0161, P0300, P0301-P0306, P0440, P0500 and P0505 not set, ECT sensor more than 158°F, vehicle driven at a constant speed of less than 62 mph with the engine speed over 1500 rpm, and the PCM detected the rich fuel trim correction value was over the limit. **Possible Causes:** • A/FS or HO2S is contaminated, deteriorated or it has failed • Base engine "mechanical" fault affecting one or more cylinders • EVAP system component has failed or canister fuel saturated • Exhaust leaks located in front of the A/FS or HO2S location • Fuel control sensor is out of calibration (i.e., ECT, IAT or MAF) • Fuel delivery system supplying too much fuel during cruise or idle periods (e.g., faulty fuel pump, or faulty pressure regulator) • Fuel injector(s) is leaking or stuck partially open (one or more)
DTC: P0172 **2T FUEL, MIL: Yes** **2003, 2004, 2005, 2006** **Models:** Avalon, Camry, Camry Solara, Highlander, Sienna **Engines:** 2.4L VIN E, 3.0L VIN F **Transmissions:** All	**Fuel System Too Rich (Bank 1) Conditions:** DTC P0100, P0101, P0105, P0110, P0115, P0120, P0121, P0130, P0133, P0136, P0135, P0136, P0141, P0151, P0156, P0161, P0300, P0301-P0306, P0440, P0500 and P0505 not set, ECT sensor more than 158°F, vehicle driven at a constant speed of less than 62 mph with the engine speed over 1500 rpm, and the PCM detected the rich fuel trim correction value was over the limit. **Possible Causes:** • A/FS or HO2S is contaminated, deteriorated or it has failed • Base engine "mechanical" fault affecting one or more cylinders • EVAP system component has failed or canister fuel saturated • Exhaust leaks located in front of the A/FS or HO2S location • Fuel control sensor is out of calibration (i.e., ECT, IAT or MAF) • Fuel delivery system supplying too much fuel during cruise or idle periods (e.g., faulty fuel pump, or faulty pressure regulator) • Fuel injector(s) is leaking or stuck partially open (one or more)

DTC	Trouble Code Title, Conditions & Possible Causes
DTC: P0174 **2T FUEL, MIL: Yes** **2001, 2002, 2003, 2004, 2005, 2006** **Models:** 4Runner, Camry, Camry Solara, Highlander, RAV4, Sequoia, Sienna, Tundra **Engines:** 2.0L VIN K, 3.0L VIN F, 4.0L VIN U, 4.7L VIN T **Transmissions:** All	**Fuel System Too Lean (Bank 2) Conditions:** DTC P0100, P0101, P0105, P0110, P0115, P0120, P0121, P0136, P0141, P0151, P0156, P0161, P0300, P0301-P0306, P0440, P0500, P0505, P1130, P1133, P1135, 1150, 53 and P1155 not set, ECT sensor more than 158°F, vehicle driven at a constant speed of less than 62 mph with the engine speed over 1500 rpm, and the PCM detected the lean fuel trim correction value was over the limit. **Possible Causes:** • A/FS or HO2S is contaminated, deteriorated or it has failed • Air leaks after the MAF sensor, or in the EGR or PCV system • Base engine "mechanical" fault affecting one or more cylinders • Exhaust leaks located in front of the A/FS or HO2S location • Fuel control sensor is out of calibration (i.e., ECT, IAT or MAF) • Fuel delivery system supplying too little fuel during cruise or idle periods (e.g., faulty fuel pump or dirty, restricted fuel filter) • Fuel injector (one or more) dirty or pressure regulator has failed • MAF sensor is contaminated, out-of-calibration or damaged • Vehicle driven low on fuel or until it ran out of fuel
DTC: P0174 **2T FUEL, MIL: Yes** **2003, 2004, 2005, 2006** **Models:** Avalon, MR2 **Engines:** 1.8L VIN R, 3.0L VIN F **Transmissions:** All	**Fuel System Too Lean (Bank 2) Conditions:** DTC P0100, P0101, P0105, P0110, P0115, P0120, P0121, P0136, P0141, P0151, P0156, P0161, P0300, P0301-P0306, P0440, P0500, P0505, P1130, P1133, P1135, 1150, 53 and P1155 not set, ECT sensor more than 158°F, vehicle driven at a constant speed of less than 62 mph with the engine speed over 1500 rpm, and the PCM detected the lean fuel trim correction value was over the limit. **Possible Causes:** • A/FS or HO2S is contaminated, deteriorated or it has failed • Air leaks after the MAF sensor, or in the EGR or PCV system • Base engine "mechanical" fault affecting one or more cylinders • Exhaust leaks located in front of the A/FS or HO2S location • Fuel control sensor is out of calibration (i.e., ECT, IAT or MAF) • Fuel delivery system supplying too little fuel during cruise or idle periods (e.g., faulty fuel pump or dirty, restricted fuel filter) • Fuel injector (one or more) dirty or pressure regulator has failed • MAF sensor is contaminated, out-of-calibration or damaged • Vehicle driven low on fuel or until it ran out of fuel
DTC: P0175 **2T FUEL, MIL: Yes** **2001, 2002, 2003, 2004, 2005, 2006** **Models:** 4Runner, Camry, Camry Solara, Highlander, RAV4, Sequoia, Sienna, Tundra **Engines:** 3.0L VIN F, 2.0L VIN K, 4.0L VIN U, 4.7L VIN T **Transmissions:** All	**Fuel System Too Rich (Bank 2) Conditions:** DTC P0100, P0101, P0105, P0110, P0115, P0120, P0121, P0136, P0141, P0151, P0156, P0161, P0300, P0301-P0306, P0440, P0500, P0505, P1130, P1133, P1135, 1150, 53 and P1155 not set, ECT sensor more than 158°F, vehicle driven at a constant speed of less than 62 mph with the engine speed over 1500 rpm, and the PCM detected the rich fuel trim correction value was over the limit. **Possible Causes:** • A/FS or HO2S is contaminated, deteriorated or it has failed • Base engine "mechanical" fault affecting one or more cylinders • EVAP system component has failed or canister fuel saturated • Exhaust leaks located in front of the A/FS or HO2S location • Fuel control sensor is out of calibration (i.e., ECT, IAT or MAF) • Fuel delivery system supplying too much fuel during cruise or idle periods (e.g., faulty fuel pump, or faulty pressure regulator) • Fuel injector(s) is leaking or stuck partially open (one or more)
DTC: P0175 **2T FUEL, MIL: Yes** **2003, 2004, 2005, 2006** **Models:** Avalon, MR2 **Engines:** 1.8L VIN R, 3.0L VIN F **Transmissions:** All	**Fuel System Too Rich (Bank 2) Conditions:** DTC P0100, P0101, P0105, P0110, P0115, P0120, P0121, P0136, P0141, P0151, P0156, P0161, P0300, P0301-P0306, P0440, P0500, P0505, P1130, P1133, P1135, 1150, 53 and P1155 not set, ECT sensor more than 158°F, vehicle driven at a constant speed of less than 62 mph with the engine speed over 1500 rpm, and the PCM detected the rich fuel trim correction value was over the limit. **Possible Causes:** • A/FS or HO2S is contaminated, deteriorated or it has failed • Base engine "mechanical" fault affecting one or more cylinders • EVAP system component has failed or canister fuel saturated • Exhaust leaks located in front of the A/FS or HO2S location • Fuel control sensor is out of calibration (i.e., ECT, IAT or MAF) • Fuel delivery system supplying too much fuel during cruise or idle periods (e.g., faulty fuel pump, or faulty pressure regulator) • Fuel injector(s) is leaking or stuck partially open (one or more)

DTC	Trouble Code Title, Conditions & Possible Causes
DTC: P0201 **2T CCM, MIL: Yes** **1995** **Models:** Previa **Engines:** 2.4L VIN A **Transmissions:** All	**Fuel Injector 1 Misfire Detected Conditions:** Engine started, engine speed over 500 rpm, and the PCM detected an incorrect voltage condition on the Fuel Injector 1 control circuit. **Possible Causes:** • Base engine mechanical fault that affects one or more cylinders • Ignition system problem (coil or plug) that affects one cylinder • Injector control circuit is open or shorted to ground • Injector power circuit is open (check the power from the relay) • Fuel injector is damaged or has failed
DTC: P0202 **2T CCM, MIL: Yes** **1995** **Models:** Previa **Engines:** 2.4L VIN A **Transmissions:** All	**Fuel Injector 2 Misfire Detected Conditions:** Engine started, engine speed over 500 rpm, and the PCM detected an incorrect voltage condition on the Fuel Injector 2 control circuit. **Possible Causes:** • Base engine mechanical fault that affects one or more cylinders • Ignition system problem (coil or plug) that affects one cylinder • Injector control circuit is open or shorted to ground • Injector power circuit is open (check the power from the relay) • Fuel injector is damaged or has failed
DTC: P0203 **2T CCM, MIL: Yes** **1995** **Models:** Previa **Engines:** 2.4L VIN A **Transmissions:** All	**Fuel Injector 3 Misfire Detected Conditions:** Engine started, engine speed over 500 rpm, and the PCM detected an incorrect voltage condition on the Fuel Injector 3 control circuit. **Possible Causes:** • Base engine mechanical fault that affects one or more cylinders • Ignition system problem (coil or plug) that affects one cylinder • Injector control circuit is open or shorted to ground • Injector power circuit is open (check the power from the relay) • Fuel injector is damaged or has failed
DTC: P0204 **2T CCM, MIL: Yes** **1995** **Models:** Previa **Engines:** 2.4L VIN A **Transmissions:** All	**Fuel Injector 4 Misfire Detected Conditions:** Engine started, engine speed over 500 rpm, and the PCM detected an incorrect voltage condition on the Fuel Injector 4 control circuit. **Possible Causes:** • Base engine mechanical fault that affects one or more cylinders • Ignition system problem (coil or plug) that affects one cylinder • Injector control circuit is open or shorted to ground • Injector power circuit is open (check the power from the relay) • Fuel injector is damaged or has failed
DTC: P0220 **2T CCM, MIL: Yes** **2003, 2004, 2005, 2006** **Models:** 4Runner, Land Cruiser, Sequoia, Tacoma, Tundra **Engines:** 3.4L VIN N engine, 4.0L VIN U, 4.7L VIN T **Transmissions:** All	**Throttle Pedal Position Sensor/Switch 'B' Circuit Malfunction Conditions:** Key on or engine running; and the PCM detected the TP sensor 'B' Signal indicated less than 0.50v with the throttle position closed (the switch is open), or the TP sensor signal indicated more than 4.9v at any time. The Electric TP Sensor is mounted on the throttle body. It has two sensors (the electrical throttle system does not use a cable). **Possible Causes:** • TP sensor signal circuit open or shorted to ground • TP sensor ground circuit is open • TP sensor power circuit is open (check VREF circuit at PCM) • TP sensor is damaged or has failed • PCM has failed
DTC: P0220 **2T CCM, MIL: Yes** **2003, 2004, 2005, 2006** **Models:** 4Runner, Land Cruiser, Sequoia, Tacoma, Tundra **Engines:** 3.4L VIN N engine, 4.0L VIN U, 4.7L VIN T **Transmissions:** All	**Throttle Pedal Position Sensor Switch B Circuit Low Input Conditions:** Key on or engine running; and the PCM detected an unexpected low voltage condition (less than 0.50v) on the VTA2 circuit. The Electric TP Sensor is mounted on the throttle body. It has two sensors (the electrical throttle system does not use a cable). **Possible Causes:** • Electric TP sensor connector is damaged or shorted • Electric TP sensor circuit is shorted to ground • Electric TP sensor is damaged or has failed • PCM has failed
DTC: P0223 **2T CCM, MIL: Yes** **2003, 2004, 2005, 2006** **Models:** 4Runner, Land Cruiser, Sequoia, Tacoma, Tundra **Engines:** 3.4L VIN N engine, 4.0L VIN U, 4.7L VIN T **Transmissions:** All	**Throttle Pedal Position Sensor Switch B Circuit High Input Conditions:** Key on or engine running; and the PCM detected an unexpected high voltage condition (more than 4.97v) on the VTA2 circuit. The Electric TP Sensor is mounted on the throttle body. It has two sensors (the electrical throttle system does not use a cable). **Possible Causes:** • Electric TP sensor connector is damaged or open • Electric TP sensor circuit is open or shorted to VREF • Electric TP sensor is damaged or has failed • PCM has failed

DTC	Trouble Code Title, Conditions & Possible Causes
DTC: P0230 **2T CCM, MIL: Yes** **2003, 2004, 2005, 2006** **Models:** 4Runner, Land Cruiser, Sequoia, Tundra **Engines:** 4.0L VIN U, 4.7L VIN T **Transmissions:** All	**Fuel Pump Primary Circuit Malfunction Conditions:** Engine started; and the PCM detected an unexpected voltage on the Fuel Pump Primary control circuit (from ST terminal to Starter Relay coil and to the STA terminal of the PCM). **Possible Causes:** • Circuit opening relay is damaged or has failed • Fuel pump relay control circuit is open or shorted to ground • Fuel pump relay is damaged or has failed • Fuel pump is damaged or has failed • PCM has failed
DTC: P0300 **2T MISFIRE, MIL: Yes** **1995** **Models:** Avalon, Camry, Land Cruiser, Previa, T100, Tacoma, Tercel **Engines:** 1.5L VIN E, 2.4L VIN A, 2.4L VIN R, 2.7L VIN U, 3.0L VIN G, 3.4L VIN V, 4.5L VIN D **Transmissions:** All	**Multiple Cylinder Misfire Detected Conditions:** DTC P0100, P0101, P0105, P0110, P0115, P0120, P0121, P0125, P0335, P0340, P0500, P0505 and P0510 not set, engine started, vehicle driven to a speed of over 3 mph for 1 minute, and the PCM detected a misfire rate of 1-2% (High Emissions 2T), or a misfire rate of 6-30% (Catalyst Damaging 1T) in two or more cylinders. **Note: If the misfire is severe, the MIL will flash on/off on the 1st trip!** **Possible Causes:** • Air leak in the intake manifold, or in the EGR or PCV system • Base engine mechanical fault that affects one or more cylinders • Erratic or interrupted CKP or CMP sensor signals • Fuel delivery component fault that affects one or more cylinders (e.g., a contaminated, dirty or sticking fuel injector) • Ignition system problem (coil or plug) in one or more cylinders
DTC: P0301 **2T MISFIRE, MIL: Yes** **1995** **Models:** Avalon, Camry, Land Cruiser, Previa, T100, Tacoma, Tercel **Engines:** 1.5L VIN E, 2.4L VIN A, 2.4L VIN R, 2.7L VIN U, 3.0L VIN G, 3.4L VIN V, 4.5L VIN D **Transmissions:** All	**Cylinder 1 Misfire Detected Conditions:** DTC P0100, P0101, P0105, P0110, P0115, P0120, P0121, P0335, P0340 and :P0500 not set, engine running, vehicle speed over 3 mph, and the PCM detected a misfire condition in Cylinder 1 in the 200 (Catalyst) or 1000-rpm (High Emissions) revolution range. **Note: If the misfire is severe, the MIL will flash on/off on the 1st trip!** **Possible Causes:** • Air leak in the intake manifold, or in the EGR or PCV system • Base engine mechanical fault that affects only one cylinder • Air leak in the intake manifold, or in the EGR or PCV system • Base engine mechanical fault that affects only one cylinder • Fuel component fault that affects only one cylinder (e.g., a dirty or sticking fuel injector) • Ignition system problem (coil or plug) that affects one cylinder
DTC: P0302 **2T MISFIRE, MIL: Yes** **1995** **Models:** Avalon, Camry, Land Cruiser, Previa, T100, Tacoma, Tercel **Engines:** 1.5L VIN E, 2.4L VIN A, 2.4L VIN R, 2.7L VIN U, 3.0L VIN G, 3.4L VIN V, 4.5L VIN D **Transmissions:** All	**Cylinder 2 Misfire Detected Conditions:** DTC P0100, P0101, P0105, P0110, P0115, P0120, P0121, P0335, P0340 and P0500 not set, engine running, vehicle speed over 3 mph, and the PCM detected a misfire condition in Cylinder 2 in the 200 (Catalyst) or 1000-rpm (High Emissions) revolution range. **Note: If the misfire is severe, the MIL will flash on/off on the 1st trip!** **Possible Causes:** • Air leak in the intake manifold, or in the EGR or PCV system • Base engine mechanical fault that affects only one cylinder • Fuel component fault that affects only one cylinder (e.g., a dirty or sticking fuel injector) • Ignition system problem (coil or plug) that affects one cylinder
DTC: P0303 **2T MISFIRE, MIL: Yes** **1995** **Models:** Avalon, Camry, Land Cruiser, Previa, T100, Tacoma, Tercel **Engines:** 1.5L VIN E, 2.4L VIN A, 2.4L VIN R, 2.7L VIN U, 3.0L VIN G, 3.4L VIN V, 4.5L VIN D **Transmissions:** All	**Cylinder 3 Misfire Detected Conditions:** DTC P0100, P0101, P0105, P0110, P0115, P0120, P0121, P0335, P0340 and P0500 not set, engine running, vehicle speed over 3 mph, and the PCM detected a misfire condition in Cylinder 3 in the 200 (Catalyst) or 1000-rpm (High Emissions) revolution range. **Note: If the misfire is severe, the MIL will flash on/off on the 1st trip!** **Possible Causes:** • Air leak in the intake manifold, or in the EGR or PCV system • Base engine mechanical fault that affects only one cylinder • Fuel component fault that affects only one cylinder (e.g., a dirty or sticking fuel injector) • Ignition system problem (coil or plug) that affects one cylinder
DTC: P0304 **2T MISFIRE, MIL: Yes** **1995** **Models:** Avalon, Camry, Land Cruiser, Previa, T100, Tacoma, Tercel **Engines:** 1.5L VIN E, 2.4L VIN A, 2.4L VIN R, 2.7L VIN U, 3.0L VIN G, 3.4L VIN V, 4.5L VIN D **Transmissions:** All	**Cylinder 4 Misfire Detected Conditions:** DTC P0100, P0101, P0105, P0110, P0115, P0120, P0121, P0335, P0340 and P0500 not set, engine running, vehicle speed over 3 mph, and the PCM detected a misfire condition in Cylinder 4 in the 200 (Catalyst) or 1000-rpm (High Emissions) revolution range. **Note: If the misfire is severe, the MIL will flash on/off on the 1st trip!** **Possible Causes:** • Air leak in the intake manifold, or in the EGR or PCV system • Base engine mechanical fault that affects only one cylinder • Fuel component fault that affects only one cylinder (e.g., a dirty or sticking fuel injector) • Ignition system problem (coil or plug) that affects one cylinder

DTC	Trouble Code Title, Conditions & Possible Causes
DTC: P0305 **2T MISFIRE, MIL: Yes** **1995** **Models:** Avalon, Camry, Land Cruiser, T100, Tacoma **Engines:** 3.0L VIN G, 3.4L VIN V, 4.5L VIN D **Transmissions:** All	**Cylinder 5 Misfire Detected Conditions:** DTC P0100, P0101, P0105, P0110, P0115, P0120, P0121, P0335, P0340 and P0500 not set, engine running, vehicle speed over 3 mph, and the PCM detected a misfire condition in Cylinder 5 in the 200 (Catalyst) or 1000-rpm (High Emissions) revolution range. **Note:** If the misfire is severe, the MIL will flash on/off on the 1st trip! **Possible Causes:** • Air leak in the intake manifold, or in the EGR or PCV system • Base engine mechanical fault that affects only one cylinder • Fuel component fault that affects only one cylinder (e.g., a dirty or sticking fuel injector) • Ignition system problem (coil or plug) that affects one cylinder
DTC: P0306 **2T MISFIRE, MIL: Yes** **1995** **Models:** Avalon, Camry, Land Cruiser, T100, Tacoma **Engines:** 3.0L VIN G, 3.4L VIN V, 4.5L VIN D **Transmissions:** All	**Cylinder 6 Misfire Detected Conditions:** DTC P0100, P0101, P0105, P0110, P0115, P0120, P0121, P0335, P0340 and P0500 not set, engine running, vehicle speed over 3 mph, and the PCM detected a misfire condition in Cylinder 6 in the 200 (Catalyst) or 1000-rpm (High Emissions) revolution range. **Note:** If the misfire is severe, the MIL will flash on/off on the 1st trip! **Possible Causes:** • Air leak in the intake manifold, or in the EGR or PCV system • Base engine mechanical fault that affects only one cylinder • Fuel component fault that affects only one cylinder (e.g., a dirty or sticking fuel injector) • Ignition system problem (coil or plug) that affects one cylinder
DTC: P0300 **2T MISFIRE, MIL: Yes** **1996, 1997, 1998, 1999, 2000, 2001, 2002** **Models:** All **Engines:** All **Transmissions:** All	**Multiple Cylinder Misfire Detected Conditions:** No PCM codes set, engine running, VSS signal over 3 mph, and the PCM detected a misfire rate of 1-2% (High Emissions 2T), or a misfire rate of 6-30% (Catalyst Damaging 1T) in 2 or more cylinders. Note: If the misfire is severe, the MIL will flash on/off on the 1st trip! **Possible Causes:** • Air leak in the intake manifold, or in the EGR or PCV system • Base engine mechanical fault that affects one or more cylinders • Erratic or interrupted CKP or CMP sensor signals • Fuel delivery component fault that affects one or more cylinders (e.g., a contaminated, dirty or sticking fuel injector) • Ignition system problem (coil or plug) in one or more cylinders • TSB EG011-01 (8/01) contains information related to this code
DTC: P0301 **2T MISFIRE, MIL: Yes** **1996, 1997, 1998, 1999, 2000, 2001, 2002** **Models:** All **Engines:** All **Transmissions:** All	**Cylinder 1 Misfire Detected Conditions:** DTC P0100, P0101, P0105, P0110, P0115, P0120, P0121, P0335, P0340 and P0500 not set, engine running, vehicle speed over 3 mph, and the PCM detected a misfire condition in Cylinder 1 in the 200 (Catalyst) or 1000-rpm (High Emissions) revolution range. **Note:** If the misfire is severe, the MIL will flash on/off on the 1st trip! **Possible Causes:** • Air leak in the intake manifold, or in the EGR or PCV system • Base engine mechanical fault that affects only one cylinder • Fuel delivery component fault that affects only one cylinder (e.g., a contaminated, dirty or sticking fuel injector) • Ignition system problem (coil or plug) that affects one cylinder
DTC: P0302 **2T MISFIRE, MIL: Yes** **1996, 1997, 1998, 1999, 2000, 2001, 2002** **Models:** All **Engines:** All **Transmissions:** All	**Cylinder 2 Misfire Detected Conditions:** DTC P0100, P0101, P0105, P0110, P0115, P0120, P0121, P0335, P0340 and P0500 not set, engine running, vehicle speed over 3 mph, and the PCM detected a misfire condition in Cylinder 2 in the 200 (Catalyst) or 1000-rpm (High Emissions) revolution range. **Note:** If the misfire is severe, the MIL will flash on/off on the 1st trip! **Possible Causes:** • Air leak in the intake manifold, or in the EGR or PCV system • Base engine mechanical fault that affects only one cylinder • Fuel delivery component fault that affects only one cylinder (e.g., a contaminated, dirty or sticking fuel injector) • Ignition system problem (coil or plug) that affects one cylinder
DTC: P0303 **2T MISFIRE, MIL: Yes** **1996, 1997, 1998, 1999, 2000, 2001, 2002** **Models:** All **Engines:** All **Transmissions:** All	**Cylinder 3 Misfire Detected Conditions:** DTC P0100, P0101, P0105, P0110, P0115, P0120, P0121, P0335, P0340 and P0500 not set, engine running, vehicle speed over 3 mph, and the PCM detected a misfire condition in Cylinder 3 in the 200 (Catalyst) or 1000-rpm (High Emissions) revolution range. **Note:** If the misfire is severe, the MIL will flash on/off on the 1st trip! **Possible Causes:** • Air leak in the intake manifold, or in the EGR or PCV system • Base engine mechanical fault that affects only one cylinder • Fuel delivery component fault that affects only one cylinder (e.g., a contaminated, dirty or sticking fuel injector) • Ignition system problem (coil or plug) that affects one cylinder

DTC	Trouble Code Title, Conditions & Possible Causes
DTC: P0304 **2T MISFIRE, MIL: Yes** **1996, 1997, 1998, 1999, 2000,** **2001, 2002** **Models:** All **Engines:** All **Transmissions:** All	**Cylinder 4 Misfire Detected Conditions:** DTC P0100, P0101, P0105, P0110, P0115, P0120, P0121, P0335, P0340 and P0500 not set, engine running, vehicle speed over 3 mph, and the PCM detected a misfire condition in Cylinder 4 in the 200 (Catalyst) or 1000-rpm (High Emissions) revolution range. **Note: If the misfire is severe, the MIL will flash on/off on the 1st trip!** **Possible Causes:** • Air leak in the intake manifold, or in the EGR or PCV system • Base engine mechanical fault that affects only one cylinder • Fuel delivery component fault that affects only one cylinder (e.g., a contaminated, dirty or sticking fuel injector) • Ignition system problem (coil or plug) that affects one cylinder
DTC: P0305 **2T MISFIRE, MIL: Yes** **1996, 1997, 1998, 1999, 2000,** **2001, 2002** **Models:** 4Runner, Avalon, Highlander, Land Cruiser, Sequoia, Sienna, Tacoma, Tundra **Engines:** 3.0L VIN D, 3.0L VIN E, 3.0L VIN F, 3.4L VIN N, 4.5L VIN J, 4.7L VIN T **Transmissions:** All	**Cylinder 5 Misfire Detected Conditions:** DTC P0100, P0101, P0105, P0110, P0115, P0120, P0121, P0335, P0340 and P0500 not set, engine running, vehicle speed over 3 mph, and the PCM detected a misfire condition in Cylinder 5 in the 200 (Catalyst) or 1000-rpm (High Emissions) revolution range. **Note: If the misfire is severe, the MIL will flash on/off on the 1st trip!** **Possible Causes:** • Air leak in the intake manifold, or in the EGR or PCV system • Base engine mechanical fault that affects only one cylinder • Fuel delivery component fault that affects only one cylinder (e.g., a contaminated, dirty or sticking fuel injector) • Ignition system problem (coil or plug) that affects one cylinder
DTC: P0306 **2T MISFIRE, MIL: Yes** **1996, 1997, 1998, 1999, 2000,** **2001, 2002** **Models:** 4Runner, Avalon, Highlander, Land Cruiser, Sequoia, Sienna, Tacoma, Tundra **Engines:** 3.0L VIN D, 3.0L VIN E, 3.0L VIN F, 3.4L VIN N, 4.5L VIN J, 4.7L VIN T **Transmissions:** All	**Cylinder 6 Misfire Detected Conditions:** DTC P0100, P0101, P0105, P0110, P0115, P0120, P0121, P0335, P0340 and P0500 not set, engine running, vehicle speed over 3 mph, and the PCM detected a misfire condition in Cylinder 6 in the 200 (Catalyst) or 1000-rpm (High Emissions) revolution range. **Note: If the misfire is severe, the MIL will flash on/off on the 1st trip!** **Possible Causes:** • Air leak in the intake manifold, or in the EGR or PCV system • Base engine mechanical fault that affects only one cylinder • Fuel delivery component fault that affects only one cylinder (e.g., a contaminated, dirty or sticking fuel injector) • Ignition system problem (coil or plug) that affects one cylinder
DTC: P0307 **2T MISFIRE, MIL: Yes** **1998, 1999, 2000, 2001, 2002** **Models:** Land Cruiser, Sequoia, Tundra **Engines:** 4.7L VIN T **Transmissions:** All	**Cylinder 7 Misfire Detected Conditions:** DTC P0100, P0101, P0105, P0110, P0115, P0120, P0121, P0335, P0340 and P0500 not set, engine running, vehicle speed over 3 mph, and the PCM detected a misfire condition in Cylinder 7 in the 200 (Catalyst) or 1000-rpm (High Emissions) revolution range. **Note: If the misfire is severe, the MIL will flash on/off on the 1st trip!** **Possible Causes:** • Air leak in the intake manifold, or in the EGR or PCV system • Base engine mechanical fault that affects only one cylinder • Fuel delivery component fault that affects only one cylinder (e.g., a contaminated, dirty or sticking fuel injector) • Ignition system problem (coil or plug) that affects one cylinder
DTC: P0308 **2T MISFIRE, MIL: Yes** **1998, 1999, 2000, 2001, 2002** **Models:** Land Cruiser, Sequoia, Tundra **Engines:** 4.7L VIN T **Transmissions:** All; Model clarification string	**Cylinder 8 Misfire Detected Conditions:** DTC P0100, P0101, P0105, P0110, P0115, P0120, P0121, P0335, P0340 and P0500 not set, engine running, vehicle speed over 3 mph, and the PCM detected a misfire condition in Cylinder 8 in the 200 (Catalyst) or 1000-rpm (High Emissions) revolution range. **Note: If the misfire is severe, the MIL will flash on/off on the 1st trip!** **Possible Causes:** • Air leak in the intake manifold, or in the EGR or PCV system • Base engine mechanical fault that affects only one cylinder • Fuel delivery component fault that affects only one cylinder (e.g., a contaminated, dirty or sticking fuel injector) • Ignition system problem (coil or plug) that affects one cylinder
DTC: P0300 **2T MISFIRE, MIL: Yes** **2001, 2002, 2003, 2004, 2005,** **2006** **Models:** Prius **Engines:** All **Transmissions:** All	**Multiple Cylinder Misfire Detected Conditions:** **Trouble Code Conditions** DTC P0100, P0101, P0115, P0116, P0120, P0121, P0335, P0340 and P0500 not set, engine running, VSS signal over 3 mph, and the PCM detected a misfire rate of 1-2% (High Emissions 2T), or a misfire rate of 6-30% (Catalyst Damaging 1T) in two or more cylinder. **Note: If the misfire is severe, the MIL will flash on/off on the 1st trip!** **Possible Causes:** • Air leak in the intake manifold, or in the EGR or PCV system • Base engine mechanical fault that affects one or more cylinders • Erratic or interrupted CKP or CMP sensor signals • Fuel delivery component fault that affects one or more cylinders (e.g., a contaminated, dirty or sticking fuel injector) • Ignition system problem (coil or plug) in one or more cylinders • TSB EG006-02 (1/02) contains information related to this code

DTC	Trouble Code Title, Conditions & Possible Causes
DTC: P0301 **2T MISFIRE, MIL: Yes** **2001, 2002, 2003, 2004, 2005, 2006** **Models:** Prius **Engines:** All **Transmissions:** All	**Cylinder 1 Misfire Detected Conditions:** DTC P0100, P0101, P0115, P0116, P0120, P0121, P0335, P0340 and P0500 not set, engine running, vehicle speed over 3 mph, and the PCM detected a misfire condition in Cylinder 1 in the 200 (Catalyst) or 1000-rpm (High Emissions) revolution range. **Note: If the misfire is severe, the MIL will flash on/off on the 1st trip!** **Possible Causes:** • Air leak in the intake manifold, or in the EGR or PCV system • Base engine mechanical fault that affects only one cylinder • Fuel delivery component fault that affects only one cylinder (e.g., a contaminated, dirty or sticking fuel injector) • Ignition system problem (coil or plug) that affects one cylinder • TSB EG006-02 (1/02) contains information related to this code
DTC: P0302 **2T MISFIRE, MIL: Yes** **2001, 2002, 2003, 2004, 2005, 2006** **Models:** Prius **Engines:** All **Transmissions:** All	**Cylinder 2 Misfire Detected Conditions:** DTC P0100, P0101, P0115, P0116, P0120, P0121, P0335, P0340 and P0500 not set, engine running, vehicle speed over 3 mph, and the PCM detected a misfire condition in Cylinder 2 in the 200 (Catalyst) or 1000-rpm (High Emissions) revolution range. **Note: If the misfire is severe, the MIL will flash on/off on the 1st trip!** **Possible Causes:** • Air leak in the intake manifold, or in the EGR or PCV system • Base engine mechanical fault that affects only one cylinder • Fuel delivery component fault that affects only one cylinder (e.g., a contaminated, dirty or sticking fuel injector) • Ignition system problem (coil or plug) that affects one cylinder • TSB EG006-02 (1/02) contains information related to this code
DTC: P0303 **2T MISFIRE, MIL: Yes** **2001, 2002, 2003, 2004, 2005, 2006** **Models:** Prius **Engines:** All **Transmissions:** All	**Cylinder 3 Misfire Detected Conditions:** DTC P0100, P0101, P0115, P0116, P0120, P0121, P0335, P0340 and P0500 not set, engine running, vehicle speed over 3 mph, and the PCM detected a misfire condition in Cylinder 3 in the 200 (Catalyst) or 1000-rpm (High Emissions) revolution range. **Note: If the misfire is severe, the MIL will flash on/off on the 1st trip!** **Possible Causes:** • Air leak in the intake manifold, or in the EGR or PCV system • Base engine mechanical fault that affects only one cylinder • Fuel delivery component fault that affects only one cylinder (e.g., a contaminated, dirty or sticking fuel injector) • Ignition system problem (coil or plug) that affects one cylinder • TSB EG006-02 (1/02) contains information related to this code
DTC: P0304 **2T MISFIRE, MIL: Yes** **2001, 2002, 2003, 2004, 2005, 2006** **Models:** Prius **Engines:** All **Transmissions:** All	**Cylinder 4 Misfire Detected Conditions:** DTC P0100, P0101, P0115, P0116, P0120, P0121, P0335, P0340 and P0500 not set, engine running, vehicle speed over 3 mph, and the PCM detected a misfire condition in Cylinder 4 in the 200 (Catalyst) or 1000-rpm (High Emissions) revolution range. **Note: If the misfire is severe, the MIL will flash on/off on the 1st trip!** **Possible Causes:** • Air leak in the intake manifold, or in the EGR or PCV system • Base engine mechanical fault that affects only one cylinder • Fuel delivery component fault that affects only one cylinder (e.g., a contaminated, dirty or sticking fuel injector) • Ignition system problem (coil or plug) that affects one cylinder • TSB EG006-02 (1/02) contains information related to this code
DTC: P0300 **2T MISFIRE, MIL: Yes** **2003, 2004, 2005, 2006** **Models:** 4Runner, Avalon, Camry, Camry Solara, Celica, Corolla, Highlander, Land Cruiser, Matrix, MR2, RAV4, Sequoia, Sienna, Tacoma, Tundra **Engines:** All **Transmissions:** All	**Multiple Cylinder Misfire Detected Conditions:** **Trouble Code Conditions** DTC P0100, P0101, P0102, P0103, P0105, P0110, P0112, P0113, P0115, P0117, P0118, P0120, P0121, P0122, P0123, P0125, P0335, P0340, P0500, P0505 and P0510 not set, engine started, vehicle driven to a speed of over 3 mph for 1 minute, and the PCM detected a misfire rate of 1-2% (High Emissions 2T), or a misfire rate of 6-30% (Catalyst Damaging 1T) in two or more cylinders. **Note: If the misfire is severe, the MIL will flash on/off on the 1st trip! Look at the misfire ratio for all of the cylinders on the Scan Tool. The cylinder with the highest misfire ratio should be checked first!** **Possible Causes:** • Air leak in the intake manifold, or in the EGR or PCV system • Base engine mechanical fault that affects two or more cylinders • EGR valve is stuck open or the PCV system has a vacuum leak • Fuel delivery component fault that affects two or more cylinders (e.g., contaminated, dirty or sticking fuel injectors) • Ignition system fault (coil or plug) that affects several cylinders • Mass airflow meter is contaminated, or its signal is out of range

DTC	Trouble Code Title, Conditions & Possible Causes

DTC: P0301
2T MISFIRE, MIL: Yes
2003, 2004, 2005, 2006
Models: 4Runner, Avalon, Camry, Camry Solara, Celica, Corolla, Highlander, Land Cruiser, Matrix, MR2, RAV4, Sequoia, Sienna, Tacoma, Tundra
Engines: All
Transmissions: All
	Cylinder 1 Misfire Detected Conditions:
	Trouble Code Conditions
	DTC P0100, P0101, P0102, P0103, P0105, P0110, P0112, P0113, P0115, P0117, P0118, P0120, P0121, P0122, P0123, P0125, P0335, P0340, P0500, P0505 and P0510 not set, engine started, vehicle driven to a speed of over 3 mph for 1 minute, and the PCM detected a misfire rate of 1-2% (High Emissions 2T), or a misfire rate of 6-30% (Catalyst Damaging 1T) in Cylinder 1.
	Note: If the misfire is severe, the MIL will flash on/off on the 1st trip!
	Possible Causes:
	   • Base engine mechanical fault that affects only Cylinder 1
	   • EGR valve is stuck open or the PCV system has a vacuum leak
	   • Fuel component fault that affects only Cylinder 1 (a contaminated or sticking injector)
	   • Ignition system problem (coil or plug) that affects Cylinder 1

DTC: P0302
2T MISFIRE, MIL: Yes
2003, 2004, 2005, 2006
Models: 4Runner, Avalon, Camry, Camry Solara, Celica, Corolla, Highlander, Land Cruiser, Matrix, MR2, RAV4, Sequoia, Sienna, Tacoma, Tundra
Engines: All
Transmissions: All
	Cylinder 2 Misfire Detected Conditions:
	Trouble Code Conditions
	DTC P0100, P0101, P0102, P0103, P0105, P0110, P0112, P0113, P0115, P0117, P0118, P0120, P0121, P0122, P0123, P0125, P0335, P0340, P0500, P0505 and P0510 not set, engine started, vehicle driven to a speed of over 3 mph for 1 minute, and the PCM detected a misfire rate of 1-2% (High Emissions 2T), or a misfire rate of 6-30% (Catalyst Damaging 1T) in Cylinder 2.
	Note: If the misfire is severe, the MIL will flash on/off on the 1st trip!
	Possible Causes:
	   • Base engine mechanical fault that affects only Cylinder 2
	   • EGR valve is stuck open or the PCV system has a vacuum leak
	   • Fuel component fault that affects only Cylinder 2 (a contaminated or sticking injector)
	   • Ignition system problem (coil or plug) that affects Cylinder 2

DTC: P0303
2T MISFIRE, MIL: Yes
2003, 2004, 2005, 2006
Models: 4Runner, Avalon, Camry, Camry Solara, Celica, Corolla, Highlander, Land Cruiser, Matrix, MR2, RAV4, Sequoia, Sienna, Tacoma, Tundra
Engines: All
Transmissions: All
	Cylinder 3 Misfire Detected Conditions:
	Trouble Code Conditions
	DTC P0100, P0101, P0102, P0103, P0105, P0110, P0112, P0113, P0115, P0117, P0118, P0120, P0121, P0122, P0123, P0125, P0335, P0340, P0500, P0505 and P0510 not set, engine started, vehicle driven to a speed of over 3 mph for 1 minute, and the PCM detected a misfire rate of 1-2% (High Emissions 2T), or a misfire rate of 6-30% (Catalyst Damaging 1T) in Cylinder 3.
	Note: If the misfire is severe, the MIL will flash on/off on the 1st trip!
	Possible Causes:
	   • Base engine mechanical fault that affects only Cylinder 3
	   • EGR valve is stuck open or the PCV system has a vacuum leak
	   • Fuel component fault that affects only Cylinder 3 (a contaminated or sticking injector)
	   • Ignition system problem (coil or plug) that affects Cylinder 3

DTC: P0304
2T MISFIRE, MIL: Yes
2003, 2004, 2005, 2006
Models: 4Runner, Avalon, Camry, Camry Solara, Celica, Corolla, Highlander, Land Cruiser, Matrix, MR2, RAV4, Sequoia, Sienna, Tacoma, Tundra
Engines: All
Transmissions: All
	Cylinder 4 Misfire Detected Conditions:
	Trouble Code Conditions
	DTC P0100, P0101, P0102, P0103, P0105, P0110, P0112, P0113, P0115, P0117, P0118, P0120, P0121, P0122, P0123, P0125, P0335, P0340, P0500, P0505 and P0510 not set, engine started, vehicle driven to a speed of over 3 mph for 1 minute, and the PCM detected a misfire rate of 1-2% (High Emissions 2T), or a misfire rate of 6-30% (Catalyst Damaging 1T) in Cylinder 4.
	Note: If the misfire is severe, the MIL will flash on/off on the 1st trip!
	Possible Causes:
	   • Base engine mechanical fault that affects only Cylinder 4
	   • EGR valve is stuck open or the PCV system has a vacuum leak
	   • Fuel component fault that affects only Cylinder 4 (a contaminated or sticking injector)
	   • Ignition system problem (coil or plug) that affects Cylinder 4

DTC: P0305
2T MISFIRE, MIL: Yes
2003, 2004, 2005, 2006
Models: 4Runner, Avalon, Camry, Camry Solara, Highlander, Land Cruiser, Sequoia, Sienna, Tacoma, Tundra
Engines: 3.0L VIN F, 3.4L VIN N, 4.7L VIN T
Transmissions: All
	Cylinder 5 Misfire Detected Conditions:
	Trouble Code Conditions
	DTC P0100, P0101, P0102, P0103, P0105, P0110, P0112, P0113, P0115, P0117, P0118, P0120, P0121, P0122, P0123, P0125, P0335, P0340, P0500, P0505 and P0510 not set, engine started, vehicle driven to a speed of over 3 mph for 1 minute, and the PCM detected a misfire rate of 1-2% (High Emissions 2T), or a misfire rate of 6-30% (Catalyst Damaging 1T) in Cylinder 5.
	Note: If the misfire is severe, the MIL will flash on/off on the 1st trip!
	Possible Causes:
	   • Base engine mechanical fault that affects only Cylinder 5
	   • EGR valve is stuck open or the PCV system has a vacuum leak
	   • Fuel delivery component fault that affects only Cylinder 5 (e.g., a contaminated, dirty or sticking fuel injector)
	   • Ignition system problem (coil or plug) that affects Cylinder 5

DTC	Trouble Code Title, Conditions & Possible Causes
DTC: P0306 **2T MISFIRE, MIL: Yes** **2003, 2004, 2005, 2006** **Models:** 4Runner, Avalon, Camry, Camry Solara, Highlander, Land Cruiser, Sequoia, Sienna, Tacoma, Tundra **Engines:** 3.0L VIN F, 3.4L VIN N, 4.7L VIN T **Transmissions:** All	**Cylinder 6 Misfire Detected Conditions:** **Trouble Code Conditions** DTC P0100, P0101, P0102, P0103, P0105, P0110, P0112, P0113, P0115, P0117, P0118, P0120, P0121, P0122, P0123, P0125, P0335, P0340, P0500, P0505 and P0510 not set, engine started, vehicle driven to a speed of over 3 mph for 1 minute, and the PCM detected a misfire rate of 1-2% (High Emissions 2T), or a misfire rate of 6-30% (Catalyst Damaging 1T) in Cylinder 6. **Note: If the misfire is severe, the MIL will flash on/off on the 1st trip!** **Possible Causes:** • Base engine mechanical fault that affects only Cylinder 6 • EGR valve is stuck open or the PCV system has a vacuum leak • Fuel delivery component fault that affects only Cylinder 6 (e.g., a contaminated, dirty or sticking fuel injector) • Ignition system problem (coil or plug) that affects Cylinder 6 • Mass airflow meter is contaminated, or its signal is out of range
DTC: P0307 **2T MISFIRE, MIL: Yes** **2003, 2004, 2005, 2006** **Models:** 4Runner, Land Cruiser, Sequoia, Tundra **Engines:** 4.7L VIN T **Transmissions:** All	**Cylinder 7 Misfire Detected Conditions:** **Trouble Code Conditions** DTC P0100, P0101, P0102, P0103, P0105, P0110, P0112, P0113, P0115, P0117, P0118, P0120, P0121, P0122, P0123, P0125, P0335, P0340, P0500, P0505 and P0510 not set, engine started, vehicle driven to a speed of over 3 mph for 1 minute, and the PCM detected a misfire rate of 1-2% (High Emissions 2T), or a misfire rate of 6-30% (Catalyst Damaging 1T) in Cylinder 7. **Note: If the misfire is severe, the MIL will flash on/off on the 1st trip!** **Possible Causes:** • Base engine mechanical fault that affects only Cylinder 7 • EGR valve is stuck open or the PCV system has a vacuum leak • Fuel delivery component fault that affects only Cylinder 7 (e.g., a contaminated, dirty or sticking fuel injector) • Ignition system problem (coil or plug) that affects Cylinder 7 • Mass airflow meter is contaminated, or its signal is out of range
DTC: P0308 **2T MISFIRE, MIL: Yes** **2003, 2004, 2005, 2006** **Models:** 4Runner, Land Cruiser, Sequoia, Tundra **Engines:** 4.7L VIN T **Transmissions:** All	**Cylinder 8 Misfire Detected Conditions:** **Trouble Code Conditions** DTC P0100, P0101, P0102, P0103, P0105, P0110, P0112, P0113, P0115, P0117, P0118, P0120, P0121, P0122, P0123, P0125, P0335, P0340, P0500, P0505 and P0510 not set, engine started, vehicle driven to a speed of over 3 mph for 1 minute, and the PCM detected a misfire rate of 1-2% (High Emissions 2T), or a misfire rate of 6-30% (Catalyst Damaging 1T) in Cylinder 8. **Note: If the misfire is severe, the MIL will flash on/off on the 1st trip!** **Possible Causes:** • Base engine mechanical fault that affects only Cylinder 8 • EGR valve is stuck open or the PCV system has a vacuum leak • Fuel delivery component fault that affects only Cylinder 8 (e.g., a contaminated, dirty or sticking fuel injector) • Ignition system problem (coil or plug) that affects Cylinder 8 • Mass airflow meter is contaminated, or its signal is out of range
DTC: P0325 **1T CCM, MIL: Yes** **1995** **Models:** Previa, T100, Tacoma **Engines:** 2.4L VIN A, 2.4L VIN R, 2.4L VIN U, 2.7L VIN U **Transmissions:** All	**Knock Sensor 1 Circuit Malfunction Conditions:** Engine started, vehicle driven with the engine speed over 1200 rpm, and the PCM detected an unexpected voltage condition on the Knock Sensor 1 (KS1) circuit during the CCM test. **Possible Causes:** • Verify that the Knock Sensor (KS) is tightened to specification • Knock sensor signal circuit is open or shorted to ground • Knock sensor signal circuit is shorted to VREF or system power • Knock sensor is damaged or has failed • PCM has failed
DTC: P0325 **1T CCM, MIL: Yes** **1995** **Models:** Avalon, Camry, Land Cruiser, T100, Tacoma **Engines:** 3.0L VIN G, 3.4L VIN V, 4.5L VIN D **Transmissions:** All	**Knock Sensor 1 Circuit Malfunction Conditions:** Engine started, vehicle driven with the engine speed over 2000 rpm, and the PCM detected an unexpected voltage condition on the Knock Sensor 1 (KS1) circuit during the CCM test. **Possible Causes:** • Verify that the Knock Sensor (KS) is tightened to specification • Knock sensor signal circuit is open or shorted to ground • Knock sensor signal circuit is shorted to VREF or system power • Knock sensor is damaged or has failed • PCM has failed

DTC	Trouble Code Title, Conditions & Possible Causes
DTC: P0325 **1T CCM, MIL: Yes** **1996, 1997, 1998, 1999, 2000,** **2001, 2002** **Models:** All **Engines:** 1.5L VIN C, 1.6L VIN A, 1.6L VIN B, 1.8L VIN B, 1.8L VIN R, 2.0L VIN P, 2.2L VIN G, 2.4L VIN D, 2.4L VIN K, 2.4L VIN L, 2.7L VIN M **Transmissions:** All	**Knock Sensor 1 Circuit Malfunction Conditions:** Engine started, vehicle driven with the engine speed over 1200 rpm, and the PCM detected an unexpected voltage condition on the Knock Sensor 1 (KS1) circuit during the CCM test. **Possible Causes:** • Verify that the Knock Sensor (KS) is tightened to specification • Knock sensor signal circuit is open or shorted to ground • Knock sensor signal circuit is shorted to VREF or system power • Knock sensor is damaged or has failed • PCM has failed
DTC: P0325 **1T CCM, MIL: Yes** **1996, 1997, 1998, 1999, 2000,** **2001, 2002** **Models:** All **Engines:** 3.0L VIN D, 3.0L VIN E, 3.0L VIN F, 3.4L VIN N, 4.5L VIN J, 4.7L VIN T **Transmissions:** All	**Knock Sensor 1 Circuit Malfunction Conditions:** Engine started, vehicle driven with the engine speed over 2000 rpm, and the PCM detected an unexpected voltage condition on the Knock Sensor 1 (KS1) circuit during the CCM test. **Possible Causes:** • Verify that the Knock Sensor (KS) is tightened to specification • Knock sensor signal circuit is open or shorted to ground • Knock sensor signal circuit is shorted to VREF or system power • Knock sensor is damaged or has failed • PCM has failed
DTC: P0325 **1T CCM, MIL: Yes** **2003, 2004, 2005, 2006** **Models:** All **Engines:** All **Transmissions:** All	**Knock Sensor 1 Circuit Malfunction Conditions:** Engine started, vehicle driven with the engine speed over 2000 rpm, and the PCM did not detect a signal on the Knock Sensor 1 circuit. **Possible Causes:** • Knock sensor signal circuit is open or shorted to ground • Knock sensor signal circuit is shorted to VREF or system power • Knock sensor is damaged, not tightened properly or has failed • PCM has failed
DTC: P0326 **1T CCM, MIL: Yes** **1995** **Models:** Land Cruiser **Engines:** All **Transmissions:** All	**Knock Sensor 2 Circuit Malfunction Conditions:** Engine started, engine running at 1200 rpm or higher, and the PCM detected an unexpected voltage condition on the Knock Sensor 2 (KS2 circuit during the CCM test. **Possible Causes:** • Verify that the Knock Sensor (KS) is tightened to specification • Knock sensor signal circuit is open or shorted to ground • Knock sensor signal circuit is shorted to VREF or system power • Knock sensor is damaged or has failed • PCM has failed
DTC: P0327 **1T CCM, MIL: Yes** **2003, 2004, 2005, 2006** **Models:** Tacoma, Tundra **Engines:** 3.4L VIN N **Transmissions:** All	**Knock Sensor 1 Circuit Low Input (Bank 1) Conditions:** Engine started, engine speed from 1500-5500 rpm, and the PCM detected an unexpected low voltage on the Knock Sensor 1 circuit. **Possible Causes:** • Verify that the Knock Sensor (KS) is tightened to specification • Knock sensor signal circuit is shorted to ground • Knock sensor is damaged or has failed • PCM has failed
DTC: P0328 **1T CCM, MIL: Yes** **2003, 2004, 2005, 2006** **Models:** Tacoma, Tundra **Engines:** 3.4L VIN N **Transmissions:** All	**Knock Sensor 1 Circuit High Input (Bank 1) Conditions:** Engine started, engine speed from 1500-5500 rpm, and the PCM detected an unexpected high voltage on the Knock Sensor 1 circuit. **Possible Causes:** • Verify that the Knock Sensor (KS) is tightened to specification • Knock sensor signal circuit is open or shorted to power • Knock sensor is damaged or has failed • PCM has failed
DTC: P0330 **1T CCM, MIL: Yes** **1995** **Models:** Avalon, Camry, Land Cruiser, Previa, T100, Tacoma, Tercel **Engines:** 3.0L VIN G, 3.4L VIN V, 4.5L VIN D **Transmissions:** All	**Knock Sensor 2 Circuit Malfunction Conditions:** Engine started, vehicle driven with the engine speed over 1200 rpm, and the PCM detected an unexpected voltage condition on the Knock Sensor 1 (KS1) circuit during the CCM test. **Possible Causes:** • Verify that the Knock Sensor (KS) is tightened to specification • Knock sensor signal circuit is open or shorted to ground • Knock sensor signal circuit is shorted to VREF or system power • Knock sensor is damaged or has failed • PCM has failed

DTC	Trouble Code Title, Conditions & Possible Causes
DTC: P0330 **2T CCM, MIL: Yes** **1996, 1997, 1998, 1999, 2000, 2001, 2002** **Models:** All **Engines:** 3.0L VIN D, 3.0L VIN E, 3.0L VIN F, 3.4L VIN N, 4.5L VIN J, 4.7L VIN T **Transmissions:** All	**Knock Sensor 2 Circuit Malfunction Conditions:** Engine started, vehicle driven with the engine speed over 2000 rpm, and the PCM detected an unexpected voltage condition on the Knock Sensor 1 (KS1) circuit during the CCM test. **Possible Causes:** • Verify that the Knock Sensor (KS) is tightened to specification • Knock sensor signal circuit is open, shorted to ground or VREF • Knock sensor is damaged or has failed • PCM has failed
DTC: P0330 **1T CCM, MIL: Yes** **2003, 2004, 2005, 2006** **Models:** 4Runner, Avalon, Camry, Land Cruiser, Sequoia, Tundra **Engines:** 3.0L VIN F, 4.0L VIN U, 4.7L VIN T **Transmissions:** All	**Knock Sensor 2 Circuit Malfunction Conditions:** Engine started, vehicle driven with the engine speed over 2000 rpm, and the PCM did not detect a signal on the Knock Sensor 2 circuit. **Possible Causes:** • Knock sensor signal circuit is open, shorted to ground or VREF • Knock sensor is damaged, not tightened properly or has failed • PCM has failed
DTC: P0332 **1T CCM, MIL: Yes** **2003, 2004, 2005, 2006** **Models:** Tacoma, Tundra **Engines:** 3.4L VIN N **Transmissions:** All	**Knock Sensor 2 Circuit Low Input (Bank 2) Conditions:** Engine started, engine speed from 1500-5500 rpm, and the PCM detected an unexpected low voltage on the Knock Sensor 2 circuit. **Possible Causes:** • Verify that the Knock Sensor (KS) is tightened to specification • Knock sensor signal circuit is shorted to ground • Knock sensor had failed, or the PCM has failed
DTC: P0333 **1T CCM, MIL: Yes** **2003, 2004, 2005, 2006** **Models:** Tacoma, Tundra **Engines:** 3.4L VIN N **Transmissions:** All	**Knock Sensor 2 Circuit High Input (Bank 2) Conditions:** Engine speed from 1500-5500 rpm, and the PCM detected an unexpected high voltage on the Knock Sensor 2 circuit. **Possible Causes:** • Verify that the Knock Sensor (KS) is tightened to specification • Knock sensor signal circuit is open or shorted to power • Knock sensor has failed, or the PCM has failed
DTC: P0335 **2T CCM, MIL: Yes** **1995** **Models:** Avalon, Camry, Land Cruiser, Previa, T100, Tacoma, Tercel **Engines:** 1.5L VIN E, 2.4L VIN A, 2.4L VIN R, 2.7L VIN U, 3.0L VIN G, 3.4L VIN V, 4.5L VIN D **Transmissions:** All	**Crankshaft Position Sensor 'A' Circuit Malfunction Conditions:** Engine cranking; and the PCM did not detect any CKP Sensor 'A' signals, or with the engine speed over 600 rpm, it did not receive any CKP sensor signals, or the CKP signal was lost. **Possible Causes:** • CKP Sensor 'A' signal circuit is open or shorted to ground • CKP Sensor 'A' signal ground circuit is open • CKP Sensor 'A' signal is shorted to VREF or system power • CKP Sensor 'A' is damaged or has failed • PCM has failed
DTC: P0335 **1T CCM, MIL: Yes** **1996, 1997, 1998, 1999, 2000, 2001, 2002, 2003, 2004, 2005, 2006** **Models:** All **Engines:** All **Transmissions:** All	**Crankshaft Position Sensor 'A' Circuit Malfunction Conditions:** Engine cranking; and the PCM did not detect any CKP Sensor 'A' signals, or with the engine speed over 600 rpm, it did not receive any CKP sensor signals, or the CKP signal was lost. **Possible Causes:** • CKP Sensor 'A' signal circuit is open, shorted to ground or shorted to system power • CKP Sensor 'A' signal ground circuit is open • CKP Sensor 'A' is damaged or has failed
DTC: P0336 **2T CCM, MIL: Yes** **1996, 1997** **Models:** 4Runner, T100, Tacoma **Engines:** 2.7L VIN M **Transmissions:** All	**Crankshaft Position Sensor 'A' Range/Performance Conditions:** Engine running at idle or cruise speed for one minute, and the PCM detected a variation between the CKP Sensor and the CMP sensor signals. **Possible Causes:** • Base engine "mechanical" problem (e.g., valve timing is wrong) • Distributor installation is incorrect • PCM has failed

DTC	Trouble Code Title, Conditions & Possible Causes
DTC: P0339 **2T CCM, MIL: Yes** **2003, 2004, 2005, 2006** **Models:** 4Runner, Camry, Camry Solara, Land Cruiser, MR2, Sequoia, Tacoma, Tundra **Engines:** 1.8L VIN R, 2.4L VIN D, 2.4L VIN L, 3.4L VIN N, 4.0L VIN U, 4.7L VIN T **Transmissions:** All	**Crankshaft Position Sensor 'A' Circuit Intermittent Conditions:** Engine started, STA signal indicating "off", engine runtime over 3 seconds since STA switched from "on" to "off", engine speed over 1000 rpm, and the PCM did not detect any CKP Sensor 'A' signals for 500 ms. The crankshaft position (NE) sensor consists of a magnet, iron core and pickup coil. The NE sensor signal plate, which has 34 teeth, installed on the crankshaft-timing pulley. This sensor generates 34 signals for each engine revolution. The PCM detects the crankshaft angle and engine speed based on the NE signal. **Possible Causes:** • CKP sensor signal circuit is open, shorted to ground or power • CKP Sensor signal ground circuit is open • Crankshaft timing pulley is damaged or out of alignment • CKP Sensor has failed, or the PCM has failed
DTC: P0340 **2T CCM, MIL: Yes** **1995** **Models:** Avalon, Camry, Land Cruiser, Previa, T100, Tacoma, Tercel **Engines:** 1.5L VIN E, 2.4L VIN A, 2.4L VIN R, 2.7L VIN U, 3.0L VIN G, 3.4L VIN V, 4.5L VIN D **Transmissions:** All	**Camshaft Position Sensor Circuit Malfunction Conditions:** Engine cranking; and the PCM did not detect any CMP sensor signals, or with the engine speed over 600 rpm, it did not detect any CMP signals or the CMP signal was interrupted. **Possible Causes:** • CMP sensor signal circuit is open or shorted to ground • CMP sensor signal ground circuit is open • CMP sensor signal is shorted to VREF or system power • CMP sensor is damaged or has failed • PCM has failed
DTC: P0340 **2T CCM, MIL: Yes** **1996, 1997, 1998, 1999, 2000, 2001, 2002** **Models:** All **Engines:** All **Transmissions:** All	**Camshaft Position Sensor Circuit Malfunction Conditions:** Engine cranking; and the PCM did not detect any CMP sensor signals, or with the engine speed over 600 rpm, it did not detect any CMP signals, or the CMP signal was interrupted. **Possible Causes:** • CMP sensor signal circuit is open or shorted to ground • CMP sensor signal ground circuit is open • CMP sensor signal is shorted to VREF or system power • CMP sensor is damaged or has failed • PCM has failed • TSB EG010-02 (4/02) contains information related to this code
DTC: P0340 **2T CCM, MIL: Yes** **2003, 2004, 2005, 2006** **Models:** All **Engines:** All **Transmissions:** All	**Camshaft Position Sensor Circuit Malfunction Conditions:** Engine cranking; and the PCM did not detect any CMP sensor signals, or with the engine speed over 600 rpm, it did not detect any CMP signals, or the CMP signal was interrupted. **Possible Causes:** • CMP sensor signal circuit is open, shorted to ground or power • CMP sensor signal ground circuit is open • CMP sensor has failed, or the PCM has failed
DTC: P0341 **2T CCM, MIL: Yes** **2003, 2004, 2005, 2006** **Models:** All **Engines:** All **Transmissions:** All	**Camshaft Position Sensor 'A' Signal Range/Performance Conditions:** Engine cranking; and the PCM detected twelve (12) or more Camshaft Position Sensor 'A' (Bank 1) signals during the test. **Possible Causes:** • CMP sensor signal circuit is open, shorted to ground or power • CMP sensor pulley is damaged, or timing belt has jumped teeth • CMP sensor is damaged or has failed • PCM has failed
DTC: P0345 **2T CCM, MIL: Yes** **2003, 2004, 2005, 2006** **Models:** Avalon, Camry, Highlander, Sienna **Engines:** 3.0L VIN F **Transmissions:** All	**Camshaft Position Sensor 'A' Signal Range/Performance Conditions:** Engine cranking; and the PCM detected twelve (12) or more CMP Sensor 'A' (Bank 2) signals during the test. The Left Hand VVT Camshaft Position sensor consists of a magnet, and a circuit board in which a Magnetic Resistive (MR) device is mounted. The VVT signal plate includes three (3) protrusions on its outer surface. **Possible Causes:** • VVT sensor signal circuit is open, shorted to ground or power • VVT sensor pulley is damaged, or timing belt has jumped teeth • VVT sensor has failed, or the PCM has failed
DTC: P0346 **2T CCM, MIL: Yes** **2003, 2004, 2005, 2006** **Models:** Avalon, Camry, Highlander, Sienna **Engines:** 3.0L VIN F **Transmissions:** All	**Camshaft Position Sensor 'A' Signal Range/Performance Conditions:** Engine cranking; and the PCM detected twelve (12) or more CMP Sensor 'A' (Bank 2) signals during the test. The Left Hand VVT Camshaft Position sensor consists of a magnet, and a circuit board in which a Magnetic Resistive (MR) device is mounted. The VVT signal plate includes three (3) protrusions on its outer surface. **Possible Causes:** • VVT sensor signal circuit is open, shorted to ground or power • VVT sensor pulley is damaged, or timing belt has jumped teeth • VVT sensor had failed, or the PCM has failed

DTC	Trouble Code Title, Conditions & Possible Causes
DTC: P0351 **1T CCM, MIL: Yes** **2003, 2004, 2005, 2006** **Models:** 4Runner, Avalon, Camry, Camry Solara, Echo, Yaris, Highlander, Land Cruiser, Matrix, MR2, Sienna, Sequoia, Tacoma, Tundra **Engines:** 1.5L VIN T, 1.8L VIN R, 2.4L VIN D, 2.4L VIN L, 2.7L VIN M, 3.0L VIN F, 4.0L VIN U, 4.7L VIN T **Transmissions:** All	**Ignition Coil No. 1 Primary/Secondary Circuit Malfunction Conditions:** Engine started, and the PCM did not detect a change in the IGF signal on the Ignition Coil No. 1 IGF circuit. This engine uses a Direct Ignition (DI) system where one coil is used to fire one cylinder. The coil high-energy secondary wire is connected to one spark plug. If P0351 to P0356 are all set, check for an open/shorted IGF circuit. **Possible Causes:** • IGT1 circuit is open or shorted to ground • Ignition Coil No. 1 is damaged or it has failed • Problem present in the Ignition System • PCM has failed
DTC: P0352 **1T CCM, MIL: Yes** **2003, 2004, 2005, 2006** **Models:** 4Runner, Avalon, Camry, Camry Solara, Echo, Yaris, Highlander, Land Cruiser, Matrix, MR2, Sienna, Sequoia, Tacoma, Tundra **Engines:** 1.5L VIN T, 1.8L VIN R, 2.4L VIN D, 2.4L VIN L, 2.7L VIN M, 3.0L VIN F, 4.0L VIN U, 4.7L VIN T **Transmissions:** All	**Ignition Coil No. 2 Primary/Secondary Circuit Malfunction Conditions:** Engine started, and the PCM did not detect a change in the IGF signal on the Ignition Coil No. 2 IGF circuit. This engine uses a Direct Ignition (DI) system where one coil is used to fire one cylinder. The coil high-energy secondary wire is connected to one spark plug. If P0351 to P0356 are all set, check for an open/shorted IGF circuit. **Possible Causes:** • IGT2 circuit is open or shorted to ground • Ignition Coil No. 2 is damaged or it has failed • Problem present in the Ignition System • PCM has failed
DTC: P0353 **1T CCM, MIL: Yes** **2003, 2004, 2005, 2006** **Models:** 4Runner, Avalon, Camry, Camry Solara, Echo, Yaris, Highlander, Land Cruiser, Matrix, MR2, Sienna, Sequoia, Tacoma, Tundra **Engines:** 1.5L VIN T, 1.8L VIN R, 2.4L VIN D, 2.4L VIN L, 2.7L VIN M, 3.0L VIN F, 4.0L VIN U, 4.7L VIN T **Transmissions:** All	**Ignition Coil No. 3 Primary/Secondary Circuit Malfunction Conditions:** Engine started, and the PCM did not detect a change in the IGF signal on the Ignition Coil No. 3 IGF circuit. This engine uses a Direct Ignition (DI) system where one coil is used to fire one cylinder. The coil high-energy secondary wire is connected to one spark plug. If P0351 to P0356 are all set, check for an open/shorted IGF circuit. **Possible Causes:** • IGT3 circuit is open or shorted to ground • Ignition Coil No. 3 is damaged or it has failed • Problem present in the Ignition System • PCM has failed
DTC: P0354 **1T CCM, MIL: Yes** **2003, 2004, 2005, 2006** **Models:** 4Runner, Avalon, Camry, Camry Solara, Echo, Yaris, Highlander, Land Cruiser, Matrix, MR2, Sienna, Sequoia, Tacoma, Tundra **Engines:** 1.5L VIN T, 1.8L VIN R, 2.4L VIN D, 2.4L VIN L, 2.7L VIN M, 3.0L VIN F, 4.0L VIN U, 4.7L VIN T **Transmissions:** All	**Ignition Coil No. 4 Primary/Secondary Circuit Malfunction Conditions:** Engine started, and the PCM did not detect a change in the IGF signal on the Ignition Coil No. 4 IGF circuit. This engine uses a Direct Ignition (DI) system where one coil is used to fire one cylinder. The coil high-energy secondary wire is connected to one spark plug. If P0351 to P0356 are all set, check for an open/shorted IGF circuit. **Possible Causes:** • IGT4 circuit is open or shorted to ground • Ignition Coil No. 4 is damaged or it has failed • Problem present in the Ignition System • PCM has failed
DTC: P0355 **1T CCM, MIL: Yes** **2003, 2004, 2005, 2006** **Models:** 4Runner, Avalon, Camry, Camry Solara, Highlander, Land Cruiser, Sequoia, Sienna, Tundra **Engines:** 3.0L VIN F, 4.0L VIN U, 4.7L VIN T **Transmissions:** All	**Ignition Coil No. 5 Primary/Secondary Circuit Malfunction Conditions:** Engine started, and the PCM did not detect a change in the IGF signal on the Ignition Coil No. 5 IGF circuit. This engine uses a Direct Ignition (DI) system where one coil is used to fire one cylinder. The coil high-energy secondary wire is connected to one spark plug. If P0351 to P0356 are all set, check for an open/shorted IGF circuit. **Possible Causes:** • IGT5 circuit is open or shorted to ground • Ignition Coil No. 5 is damaged or it has failed • Problem present in the Ignition System • PCM has failed

DTC	Trouble Code Title, Conditions & Possible Causes
DTC: P0356 **1T CCM, MIL: Yes** **2003, 2004, 2005, 2006** **Models:** 4Runner, Avalon, Camry, Camry Solara, Highlander, Land Cruiser, Sequoia, Sienna, Tundra **Engines:** 3.0L VIN F, 4.0L VIN U, 4.7L VIN T **Transmissions:** All	**Ignition Coil No. 6 Primary/Secondary Circuit Malfunction Conditions:** Engine started, and the PCM did not detect a change in the IGF signal on the Ignition Coil No. 6 IGF circuit. This engine uses a Direct Ignition (DI) system where one coil is used to fire one cylinder. The coil high-energy secondary wire is connected to one spark plug. If P0351 to P0356 are all set, check for an open/shorted IGF circuit. **Possible Causes:** • IGT6 circuit is open or shorted to ground • Ignition Coil No. 6 is damaged or it has failed • Problem present in the Ignition System • PCM has failed
DTC: P0357 **1T CCM, MIL: Yes** **2003, 2004, 2005, 2006** **Models:** 4Runner, Land Cruiser, Sequoia, Tundra **Engines:** 4.7L VIN T **Transmissions:** All	**Ignition Coil No. 7 Primary/Secondary Circuit Malfunction Conditions:** Engine started, and the PCM did not detect a change in the IGF signal on the Ignition Coil No. 7 IGF circuit. This engine uses a Direct Ignition (DI) system where one coil is used to fire one cylinder. The coil high-energy secondary wire is connected to one spark plug. If P0351 to P0358 are all set, check for an open/shorted IGF circuit. **Possible Causes:** • IGT7 circuit is open or shorted to ground • Ignition Coil No. 7 is damaged or it has failed • Problem present in the Ignition System • PCM has failed
DTC: P0358 **1T CCM, MIL: Yes** **2003, 2004, 2005, 2006** **Models:** 4Runner, Land Cruiser, Sequoia, Tundra **Engines:** 4.7L VIN T **Transmissions:** All	**Ignition Coil No. 8 Primary/Secondary Circuit Malfunction Conditions:** Engine started, and the PCM did not detect a change in the IGF signal on the Ignition Coil No. 8 IGF circuit. This engine uses a Direct Ignition (DI) system where one coil is used to fire one cylinder. The coil high-energy secondary wire is connected to one spark plug. If P0351 to P0358 are all set, check for an open/shorted IGF circuit. **Possible Causes:** • IGT8 circuit is open or shorted to ground • Ignition Coil No. 8 is damaged or it has failed • Problem present in the Ignition System • PCM has failed
DTC: P0385 **2T CCM, MIL: Yes** 1996, 1997 **Models:** Supra **Engines:** 3.0L VIN D **Transmissions:** All	**Crankshaft Position Sensor 'B' Circuit Malfunction Conditions:** Engine started, engine speed more than 600 rpm, it did not detect any Crankshaft Position (CKP) Sensor 'B' signals during the test. **Possible Causes:** • CKP Sensor 'B' signal circuit is open, shorted to ground or shorted to system power • CKP Sensor 'B' signal ground circuit is open • CKP Sensor 'B' has failed, or the PCM has failed
DTC: P0401 **2T EGR, MIL: Yes** 1995 **Models:** Avalon, Camry, Land Cruiser, Previa, T100, Tacoma, Tercel **Engines:** 1.5L VIN E, 2.4L VIN A, 2.4L VIN R, 2.7L VIN U, 3.0L VIN G, 3.4L VIN V, 4.5L VIN D **Transmissions:** All	**EGR System Insufficient Flow Detected Conditions:** Engine started, vehicle driven at a speed of over 50 mph with the engine running in closed loop at a steady throttle for 3-5 minutes, and the PCM detected the EGR gas temperature signal was not 106-140°F higher than the ambient air temperature signal in the test. **Possible Causes:** • EGR gas temperature sensor circuit open or shorted to VREF • EGR gas temperature sensor is damaged or has failed • EGR valve is stuck in closed position • EGR valve vacuum hose is disconnected or leaking • PCM has failed
DTC: P0401 **2T EGR, MIL: Yes** **1996, 1997, 1998, 1999, 2000, 2001, 2002** **Models:** All **Engines:** All **Transmissions:** All	**EGR System Insufficient Flow Detected Conditions:** Engine started, vehicle driven at a speed of over 50 mph with the engine running in closed loop at a steady throttle for 3-5 minutes, and the PCM detected the EGR sensor signal was not 106-140°F higher than the ambient air temperature in the EGR Monitor test. **Possible Causes:** • EGR gas temperature sensor circuit open or shorted to VREF • EGR gas temperature sensor is damaged or has failed • EGR valve is stuck in closed position • EGR valve vacuum hose is disconnected or leaking
DTC: P0401 **2T EGR, MIL: Yes** **2003, 2004, 2005, 2006** **Models:** Tacoma **Engines:** 2.4L VIN L, 2.7L VIN M **Transmissions:** All	**EGR System Insufficient Flow Detected Conditions:** Engine started, vehicle driven at a speed of over 50 mph with the engine running in closed loop at a steady throttle for 3-5 minutes, and the PCM detected the EGR sensor signal was not 106-140°F higher than the ambient air temperature in the EGR Monitor test. **Possible Causes:** • EGR gas temperature sensor circuit open or shorted to VREF • EGR gas temperature sensor is damaged or has failed • EGR valve is stuck in closed position • EGR valve vacuum hose is disconnected or leaking

DTC	Trouble Code Title, Conditions & Possible Causes
DTC: P0402 **2T EGR, MIL: Yes** **1995** **Models:** Avalon, Camry, Land Cruiser, Previa, T100, Tacoma, Tercel **Engines:** 1.5L VIN E, 2.4L VIN A, 2.4L VIN R, 2.7L VIN U, 3.0L VIN G, 3.4L VIN V, 4.5L VIN D **Transmissions:** All	**Excessive EGR Flow Detected Conditions:** Engine started cold (ECT sensor signal less than 86°F), engine running without load at less than 4000 rpm, and the PCM detected the EGR sensor indicated a high value during the EGR Cutoff Test, or it detected the EGR valve was open during all driving conditions. **Possible Causes:** • EGR gas temperature sensor circuit open or shorted to ground • EGR gas temperature sensor is damaged or has failed • EGR Vacuum Switching Valve (VSV) is damaged or has failed • EGR valve stuck partially or EGR valve stuck fully open • EGR VSV control circuit is open or shorted to system power • PCM has failed
DTC: P0402 **2T EGR, MIL: Yes** **1996, 1997, 1998, 1999, 2000, 2001, 2002** **Models:** All **Engines:** All **Transmissions:** All	**Excessive EGR Flow Detected Conditions:** Engine started cold (ECT sensor signal less than 86°F), engine running without load at less than 4000 rpm, and the PCM detected the EGR sensor indicated a high value during the EGR Cutoff test; or it detected the EGR valve was open during all driving conditions. **Possible Causes:** • EGR gas temperature sensor circuit open or shorted to ground • EGR gas temperature sensor is damaged or has failed • EGR Vacuum Switching Valve (VSV) is damaged or has failed • EGR valve stuck partially or EGR valve stuck fully open • EGR VSV control circuit is open or shorted to system power • PCM has failed
DTC: P0402 **2T EGR, MIL: Yes** **2003, 2004, 2005, 2006** **Models:** Tacoma **Engines:** 2.4L VIN L, 2.7L VIN M **Transmissions:** All	**Excessive EGR Flow Detected Conditions:** Engine started cold (ECT sensor signal less than 86°F), engine running without load at less than 4000 rpm, and the PCM detected the EGR sensor indicated a high value during the EGR Cutoff test; or it detected the EGR valve was open during all driving conditions. **Possible Causes:** • EGR gas temperature sensor circuit open or shorted to ground • EGR gas temperature sensor is damaged or has failed • EGR Vacuum Switching Valve (VSV) is damaged or has failed • EGR valve stuck partially or EGR valve stuck fully open • EGR VSV control circuit is open or shorted to system power • PCM has failed
DTC: P0420 **2T CAT, MIL: Yes** **1995** **Models:** Avalon, Camry, Land Cruiser, Previa, T100, Tacoma, Tercel **Engines:** 1.5L VIN E, 2.4L VIN A, 2.4L VIN R, 2.7L VIN U, 3.0L VIN G, 3.4L VIN V, 4.5L VIN D **Transmissions:** All	**Catalyst Efficiency Below Normal (Bank 1) Conditions:** DTC P0100, P0101, P0105, P0110, P0115, P0120, P0121, P0335, P0340 and P0500 not set, vehicle driven to a speed of 45-60 mph at 2500-3000 rpm in closed loop for 3-5 minutes, and the PCM detected the voltage amplitudes of the rear and front HO2S were too similar. **Possible Causes:** • Air leaks at the exhaust manifold or in the exhaust pipes • Catalytic converter is damaged, contaminated or has failed • Front HO2S or rear HO2S is contaminated with fuel or moisture • Front HO2S or the rear HO2S is loose in its mounting hole • Front HO2S is older (aged) than the rear HO2S (HO2S is lazy)
DTC: P0420 **2T CAT, MIL: Yes** **1996, 1997, 1998, 1999, 2000, 2001, 2002** **Models:** All **Engines:** All **Transmissions:** Al	**Catalyst Efficiency Below Normal (Bank 1) Conditions:** DTC P0100, P0101, P0105, P0110, P0115, P0120, P0121, P0335, P0340 and P0500 not set, engine started, vehicle driven to a speed of 45-60 mph at 2500-3000 rpm in closed loop for 3-5 minutes, and the PCM detected the voltage amplitudes of the rear HO2S-12 and front HO2S-11 were similar during the Catalyst Monitor test. **Possible Causes:** • Air leaks at the exhaust manifold or in the exhaust pipes • Catalytic converter is damaged, contaminated or has failed • Front HO2S or rear HO2S is contaminated with fuel or moisture • Front HO2S or the rear HO2S is loose in its mounting hole • Front HO2S is older (aged) than the rear HO2S (HO2S is lazy)
DTC: P0420 **2T CAT, MIL: Yes** **2003, 2004, 2005, 2006** **Models:** All **Engines:** All **Transmissions:** All	**Catalyst Efficiency Below Normal (Bank 1) Conditions:** DTC P0100, P0101, P0102, P0103, P0110, P0112, P0113, P0115, P0116, P0117, P0118, P0120, P0121, P0122, P0123, P0335, P0340 and P0500 not set, engine started, vehicle driven to a speed of 45-60 mph at 2500-3000 rpm in closed loop for 3-5 minutes, and the PCM detected too much variation in the voltage amplitudes of the HO2S-12 signal (Bank 1). **Possible Causes:** • Catalytic converter is damaged, contaminated or has failed • Front A/FS or rear HO2S is contaminated with fuel or moisture • Front A/FS or the rear HO2S is loose in its mounting hole • Front A/FS is older (aged) than the rear HO2S (HO2S is lazy) • Gas leaks at the exhaust manifold or in the exhaust pipes

DTC	Trouble Code Title, Conditions & Possible Causes
DTC: P0430 **2T CAT, MIL: Yes** **1996, 1997, 1998, 1999, 2000, 2001, 2002** **Models:** Land Cruiser, Sequoia, Tundra **Engines:** 4.7L VIN T **Transmissions:** All	**Catalyst Efficiency Below Normal (Bank 2) Conditions:** DTC P0100, P0101, P0105, P0110, P0115, P0120, P0121, P0335, P0340 and P0500 not set, vehicle speed from 45-60 mph at 2500-3000 rpm in closed loop for 3-5 minutes, and the PCM detected the voltage amplitudes of the rear HO2S-22 and front HO2S-21 were similar. **Possible Causes:** • Air leaks at the exhaust manifold or in the exhaust pipes • Catalytic converter is damaged, contaminated or has failed • Front HO2S or rear HO2S is contaminated with fuel or moisture • Front HO2S or the rear HO2S is loose in its mounting hole • Front HO2S is older (aged) than the rear HO2S (HO2S is lazy)
DTC: P0430 **2T CAT, MIL: Yes** **2003, 2004, 2005, 2006** **Models:** 4Runner, Avalon, Camry, Land Cruiser, RAV4, Sequoia, Tundra **Engines:** 2.0L VIN K, 3.0L VIN F, 4.0L VIN U, 4.7L VIN T **Transmissions:** All	**Catalyst Efficiency Below Normal (Bank 2) Conditions:** DTC P0100, P0101, P0102, P0103, P0110, P0112, P0113, P0115, P0116, P0117, P0118, P0120, P0121, P0122, P0123, P0335, P0340 and P0500 not set, engine started, vehicle driven to a speed of 45-60 mph at 2500-3000 rpm in closed loop for 3-5 minutes, and the PCM detected too much variation in the voltage amplitudes of the Rear HO2S-12 for Bank 2 during the Catalyst Monitor test. **Possible Causes:** • Catalytic converter is damaged, contaminated or has failed • Front A/FS or rear HO2S is contaminated with fuel or moisture • Front A/FS or the rear HO2S is loose in its mounting hole • Front A/FS is older (aged) than the rear HO2S (HO2S is lazy) • Gas leaks at the exhaust manifold or in the exhaust pipes
DTC: P0440 **2T EVAP, MIL: Yes** **1996, 1997, 1998, 1999, 2000, 2001, 2002** **Models:** All **Engines:** All **Transmissions:** All	**EVAP Control System Large Leak (0.080") Detected Conditions:** Engine started cold (ECT sensor signal less 86°F), engine runtime over 20 minutes in closed loop, vehicle driven to a speed of 55-60 mph, and the PCM detected the fuel tank pressure indicated the same value as the atmospheric pressure in the EVAP Monitor test. **Possible Causes:** • Charcoal canister is clogged, loaded with fuel or with moisture • Fuel filler cap missing, loose (not tightened) or the wrong part • Fuel tank over-fill check valve cracked or damaged • Fuel tank seal leaking, fuel tank cracked or damaged/leaking • Fuel vapor hoses/tubes blocked or restricted, or fuel vapor control valve tube or fuel vapor vent valve assembly blocked • Vacuum hose or tubing cracked, damaged or disconnected • Vapor Pressure sensor incorrectly installed • PCM has failed
DTC: P0440 **2T EVAP, MIL: Yes** **2003, 2004, 2005, 2006** **Models:** Celica, Corolla, Matrix, RAV4 **Engines:** 1.8L VIN R, 1.8L VIN Y, 2.0L VIN K **Transmissions:** All	**EVAP Control System Large Leak (0.080") Detected Conditions:** Engine started cold (ECT sensor signal less 86°F), engine runtime over 20 minutes in closed loop, vehicle driven to a speed of 55-60 mph, and the PCM detected the fuel tank pressure indicated the same value as the atmospheric pressure in the EVAP Monitor test. **Possible Causes:** • Charcoal canister is clogged, loaded with fuel or with moisture • Fuel filler cap missing, loose (not tightened) or the wrong part • Fuel tank over-fill check valve cracked or damaged • Fuel tank seal leaking, fuel tank cracked or damaged/leaking • Fuel vapor hoses/tubes blocked or restricted, or fuel vapor control valve tube or fuel vapor vent valve assembly blocked • Vacuum hose or tubing cracked, damaged or disconnected • Vapor Pressure sensor incorrectly installed • PCM has failed
DTC: P0441 **2T EVAP, MIL: Yes** **1995** **Models:** Avalon, Camry, Land Cruiser, Previa, T100, Tacoma, Tercel **Engines:** 1.5L VIN E, 2.4L VIN A, 2.4L VIN R, 2.7L VIN U, 3.0L VIN G, 3.4L VIN V, 4.5L VIN D **Transmissions:** All	**EVAP System Incorrect Purge Flow Detected Conditions:** Cold engine startup requirement met (ECT sensor signal less 86°F), engine running at cruise speed for over 3 minutes, VSS input from 55-60 mph, and the PCM detected the canister pressure did not decrease during purge or it remained too low during purge cutoff conditions. **Possible Causes:** • Charcoal canister is clogged, loaded with fuel or with moisture • Fuel tank over-fill check valve cracked or damaged • Fuel tank seal leaking, fuel tank cracked or damaged/leaking • Fuel vapor hoses/tubes blocked or restricted, or fuel vapor control valve tube or fuel vapor vent valve assembly blocked • Vacuum hose or tubing cracked, damaged or disconnected • VSV for the canister purge solenoid is open or shorted to ground, or the purge solenoid is damaged or has failed (closed) • VSV for vapor pressure sensor circuit is open, shorted to ground, or sensor has failed • PCM has failed

DTC	Trouble Code Title, Conditions & Possible Causes
DTC: P0441 **2T EVAP, MIL: Yes** **1996, 1997, 1998, 1999, 2000, 2001, 2002, 2003, 2004, 2005, 2006** **Models:** All **Engines:** All **Transmissions:** All	**EVAP System Incorrect Purge Flow Detected Conditions:** ECT sensor less than 86°F at startup, vehicle driven at 55-60 mph for 2-3 minutes, and the PCM detected the EVAP canister pressure did not decrease during purge conditions, or it remained too low during purge cutoff conditions. The vapor pressure sensor, VSV for the canister closed valve (CCV) and the VSV for the vapor-switching valve are used to detect EVAP system faults. The PCM closes the CCV and opens the VSV for vapor switching valve to cause an increase in vacuum in the EVAP system. Once the vacuum reaches a certain point, the PCM closes the VSV to test system operation. **Possible Causes:** • Charcoal canister is clogged, loaded with fuel or with moisture • Fuel tank over-fill check valve cracked or damaged • Fuel tank seal leaking, fuel tank cracked or damaged/leaking • Fuel vapor hoses/tubes blocked or restricted, or fuel vapor control valve tube or fuel vapor vent valve assembly blocked • Vacuum hose or tubing cracked, damaged or disconnected • Vapor pressure sensor is damaged or has failed • VSV circuit for the canister purge, VSV for the CCV or the VSV for the pressure switching valve is open or shorted to ground • VSV for the vapor pressure sensor circuit is open or shorted to ground, or the vapor pressure sensor is damaged or has failed
DTC: P0442 **2T EVAP, MIL: Yes** **1998, 1999, 2000, 2001, 2002, 2003, 2004, 2005, 2006** **Models:** All **Engines:** All **Transmissions:** All	**EVAP System Small Leak (0.040") Detected Conditions:** Engine started; IAT sensor signal from 39-86°F, fuel tank level from 25-75% for 10 seconds, and the PCM detected the EVAP system was unable to hold a specified vacuum level for a set period of time. After the system is purged, the PCM shuts off the VSV for the purge valve to seal the vacuum in the system, and then monitors the increase in pressure in the system. The pressure should increases slowly. If it increases at too fast a rate, then this code is set. **Possible Causes:** • Canister Purge valve is damaged, leaking or has failed • Charcoal canister is loaded with fuel or moisture • Fuel filler cap loose, cross-threaded, incorrect part or damaged • Fuel tank is cracked (leaking), or a leak exists in the 'O' ring • Fuel tank pressure sensor is damaged or has failed • Fuel vapor line(s), fuel pipes or hoses damaged or leaking • PCM has failed
DTC: P0442 **2T EVAP, MIL: Yes** **1996, 1997, 1998, 1999, 2000, 2001, 2002, 2003, 2004, 2005, 2006** **Models:** All **Engines:** All **Transmissions:** All	**EVAP Vent Control Solenoid Circuit Malfunction Conditions:** Engine started, engine running at cruise speed under light load conditions, VSV for vapor pressure switching valve "on", and the PCM detected a lack of vacuum continuity between the vapor pressure sensor, the charcoal canister and the fuel tank; or with the VSV for the vapor pressure switching valve "off", it detected the pressure in the fuel tank remained at atmospheric pressure; or with the VSV for the CCV "on", it detected the pressure in the charcoal canister and the fuel tank remained near atmospheric pressure. The PCM closes the VSV for the vapor-switching valve, and this action blocks any air from entering the fuel tank side of the system. The pressure rise on the fuel tank under these conditions is minimal. If there was no change in pressure, the PCM determines the VSV for the vapor-switching valve did not close, and it will set this code. **Possible Causes:** • Charcoal canister is clogged, loaded with fuel or with moisture • Fuel tank over-fill check valve cracked or damaged • Fuel tank seal leaking, fuel tank cracked, damaged or leaking • Fuel vapor hoses/tubes blocked or restricted, or fuel vapor control valve tube or fuel vapor vent valve assembly blocked • Vacuum hose or tubing cracked, damaged or disconnected • Vapor pressure sensor is damaged or has failed • VSV circuit for the canister purge, VSV for the CCV or the VSV for the pressure switching valve is open or shorted to ground • VSV for the vapor pressure sensor circuit is open or shorted to ground, or the vapor pressure sensor is damaged or has failed
DTC: P0450 **2T CCM, MIL: Yes** **1996, 1997, 1998, 1999, 2000, 2001, 2002, 2003, 2004, 2005, 2006** **Models:** All **Engines:** All **Transmissions:** All	**EVAP Vapor Pressure Sensor Circuit Malfunction Conditions:** Engine started, engine runtime less than 10 seconds since startup, and the PCM detected the Vapor Pressure sensor value was less than -3.5 kPa (-1.0 in. Hg), or the Vapor Pressure sensor was more than or equal to 1.5 kPa (0.4 in. Hg) during testing. The PCM uses the Vapor Pressure Sensor, VSV for the Canister Closed valve and VSV for the Pressure Switching valve to find faults in this system. **Possible Causes:** • Vapor pressure sensor signal circuit open or shorted to ground • Vapor pressure sensor ground circuit is open • Vapor pressure sensor power circuit is open • Vapor pressure sensor is damaged or has failed • PCM has failed

DTC	Trouble Code Title, Conditions & Possible Causes
DTC: P0451 **2T CCM, MIL: Yes** **1996, 1997, 1998, 1999, 2000, 2001, 2002, 2003, 2004, 2005, 2006** **Models:** All **Engines:** All **Transmissions:** All	**EVAP Vapor Pressure Sensor Range/Performance Conditions:** Engine started, engine at idle speed with the VSS indicating 0 mph, VSV for Vapor Switching valve "off", and the PCM detected too much change in the pressure sensor value, or the pressure sensor value equaled the opening value of the charcoal canister. The PCM uses the Vapor Pressure Sensor, VSV for the Canister Closed valve and VSV for Pressure Switching valve to find faults in the system. **Possible Causes:** • Vapor pressure sensor vacuum hoses loose or damaged • Vapor pressure sensor is damaged or has failed • PCM has failed
DTC: P0452 **2T CCM, MIL: Yes** **2003, 2004, 2005, 2006** **Models:** 4Runner, Echo, Yaris, Land Cruiser, Sequoia, Tacoma, Tundra **Engines:** All **Transmissions:** All	**EVAP Vapor Pressure Sensor Circuit Low Input Conditions:** Engine started; VSV for vapor pressure switching valve "off"; and the PCM detected an unexpected low voltage condition on the vapor pressure sensor circuit during the CCM test. **Possible Causes:** • Vapor pressure sensor connector is damaged or open • Vapor pressure sensor circuit is open • Vapor pressure sensor is damaged or has failed • PCM has failed
DTC: P0453 **2T CCM, MIL: Yes** **2003, 2004, 2005, 2006** **Models:** 4Runner, Echo, Yaris, Land Cruiser, Sequoia, Tacoma, Tundra **Engines:** All **Transmissions:** All	**EVAP Vapor Pressure Sensor Circuit High Input Conditions:** Engine started; VSV for vapor pressure switching valve "off"; and the PCM detected an unexpected high voltage condition on the vapor pressure sensor circuit during the CCM test. **Possible Causes:** • Vapor pressure sensor connector is damaged or shorted • Vapor pressure sensor circuit is shorted to VREF • Vapor pressure sensor is damaged or has failed • PCM has failed
DTC: P0456 **2T EVAP, MIL: Yes** **2003, 2004, 2005, 2006** **Models:** 4Runner, Camry, Camry Solara, Echo, Yaris, Land Cruiser, Matrix, MR2, RAV4, Sequoia, Sienna, Tacoma, Tundra **Engines:** All **Transmissions:** All	**EVAP System Very Small Leak (0.020") Detected Conditions:** Engine started; IAT sensor signal from 39-86°F, fuel tank level from 25-75% for 10 seconds, and the PCM detected the EVAP system was unable to hold a specified vacuum level for a set period of time. After the system is purged, the PCM shuts off the VSV for the purge valve to seal the vacuum in the system, and then monitors the increase in pressure in the system. The pressure should increase slowly. If it increases at too fast a rate, this code is set. **Possible Causes:** • Canister Purge valve is damaged, leaking or has failed • Charcoal canister is loaded with fuel or moisture • Fuel filler cap loose, cross-threaded, incorrect part or damaged • Fuel tank is cracked (leaking), or a leak exists in the 'O' ring • Fuel tank pressure sensor is damaged or has failed • Fuel tank overfill check valve is cracked or is damaged • Fuel vapor line(s), fuel pipes or hoses damaged or leaking • PCM has failed
DTC: P0500 **2T EVAP, MIL: Yes** **1995** **Models:** Avalon, Camry, Land Cruiser, Previa, T100, Tacoma, Tercel **Engines:** 1.5L VIN E, 2.4L VIN A, 2.4L VIN R, 2.7L VIN U, 3.0L VIN G, 3.4L VIN V, 4.5L VIN D **Transmissions:** A/T	**Vehicle Speed Sensor Circuit Malfunction Conditions:** Engine started, ECT sensor more than 158°F, vehicle driven with the engine speed from 1500-5500 rpm, and the PCM did not receive any VSS signals from the Combination Meter during the CCM test. **Possible Causes:** • VSS signal circuit is open between the meter and the PCM • VSS signal circuit shorted to ground between meter and PCM • VSS signal circuit shorted to VREF or system power • Combination Meter is damaged or has failed • PCM has failed
DTC: P0500 **1T CCM, MIL: Yes** **1996** **Models:** All **Engines:** All **Transmissions:** A/T	**Vehicle Speed Sensor Circuit Malfunction Conditions:** Engine started, ECT sensor more than 158°F, P/N switch indicating "off", vehicle driven at an engine speed of 1500-3500 rpm, and the PCM did not receive any VSS signals from the Combination Meter. **Possible Causes:** • VSS signal circuit is open between the meter and the PCM • VSS signal circuit shorted to ground between meter and PCM • VSS signal circuit shorted to VREF or system power • Combination Meter is damaged or has failed • PCM has failed

DTC	Trouble Code Title, Conditions & Possible Causes
DTC: P0500 **1T CCM, MIL: Yes** **1997, 1998, 1999, 2000, 2001, 2002** **Models:** All **Engines:** All **Transmissions:** A/T	**Vehicle Speed Sensor Circuit Malfunction Conditions:** Engine started, engine speed more than 2350 rpm, P/N switch indicating "off" for 1 second, throttle angle equal to or less than 13 degrees (°), and the PCM did not detect any VSS signals during the test period (Scan Tool PID ECT = Electronic Controlled Transaxle). **Possible Causes:** • VSS signal circuit is open, shorted to ground or to power (B+) • Combination Meter is damaged or has failed • PCM has failed • TSB EG004-02 (2/02) contains information related to this code
DTC: P0500 **1T CCM, MIL: Yes** **1996, 1997, 1998, 1999, 2000, 2001, 2002** **Models:** All **Engines:** All **Transmissions:** M/T	**Vehicle Speed Sensor Circuit Malfunction Conditions:** Engine started, ECT sensor signal over 158°F, vehicle driven with the engine speed from 1500-5500 rpm, and the PCM did not receive any VSS signals from the Combination Meter. **Possible Causes:** • VSS signal circuit is open between the meter and the PCM • VSS signal circuit shorted to ground or to VREF or power (B+) • Combination Meter is damaged or has failed • PCM has failed
DTC: P0500 **2T CCM, MIL: Yes** **2003, 2004, 2005, 2006** **Models:** 4Runner, Camry, Camry Solara, Celica, Corolla, Echo, Yaris, Land Cruiser, Matrix, MR2, RAV4, Sequoia, Tacoma, Tundra **Engines:** 1.5L VIN T, 1.8L VIN R, 1.8L VIN Y, 2.0L VIN K, 2.4L VIN D, 2.4L VIN L, 2.7L VIN M, 3.4L VIN N, 4.0L VIN U, 4.7L VIN T **Transmissions:** All	**Vehicle Speed Sensor Circuit Malfunction Conditions:** Engine started, vehicle driven with the engine speed from 1500-5500 rpm and back to idle several times, and the PCM did not receive any VSS signals during the test. The VSS (No.1) assembly outputs a 4-pulse signal for every revolution of the rotor shaft, which is generated by the transmission output shaft via the driven gear. **Possible Causes:** • VSS signal circuit is open between the meter and the PCM • VSS signal circuit shorted to ground between meter and PCM • VSS No. 1 is damaged or has failed • Combination Meter is damaged or has failed • PCM has failed
DTC: P0500 **1T CCM, MIL: Yes** **2003, 2004, 2005, 2006** **Models:** Avalon, Camry, Highlander, Sienna **Engines:** 3.0L VIN F **Transmissions:** A/T	**Vehicle Speed Sensor Circuit Malfunction Conditions:** No TP sensor codes set, engine runtime 2 seconds with the ECT sensor more than 132°F and IAT sensor more than 50°F, P/N switch indicating 'P' or 'N', TP angle less than 13° with the engine speed less than 2350 rpm; or TP angle less than 21° with the engine speed less than 2680 rpm; or TP angle less than 30° with the engine speed less than 2835 rpm; or TP angle less than 30° with the engine speed less than 3250 rpm; and the PCM detected the engine speed was equal or more than the VSS signal speed for 500 ms during testing. **Possible Causes:** • ABS speed sensor connector is damaged or open • ABS speed sensor circuit is open or shorted to ground • ABS speed sensor is damaged or has failed • Combination Meter is damaged or has failed • PCM or the ABS controller has failed
DTC: P0503 **2T CCM, MIL: Yes** **2003, 2004, 2005, 2006** **Models:** 4Runner, Land Cruiser, MR2, Sequoia, Tundra **Engines:** All **Transmissions:** All	**Vehicle Speed Sensor 'A' Signal Erratic Or Intermittent Conditions:** Engine started, vehicle driven through several transmission shifts and braking events; and the PCM detected an interruption in the VSS signal from the Combination Meter, or the signal was too high. The speed sensor for the ABS detects the wheel speeds and sends the signals to the ABS ECU. The ECU converts these signals into a 4-pulse signal and outputs this signal to the Combination Meter. **Possible Causes:** • ABS ECU is damaged or has failed • Combination Meter is damaged or has failed • VSS signal circuit is open between the meter and the PCM • VSS signal circuit shorted to ground between meter and PCM
DTC: P0504 **2T CCM, MIL: Yes** **2003, 2004, 2005, 2006** **Models:** 4Runner, Land Cruiser, MR2, Sequoia, Tundra **Engines:** All **Transmissions:** All	**Brake Switch 'A' To 'B' Correlation Malfunction Conditions:** Key on or engine running; brake pedal released, and the PCM detected the STP signal indicated "off" while the ST1 signal also indicated "off". The stoplight switch signal is used to prevent the engine from stalling when the brakes are applied suddenly. The stoplight switch uses a duplex system (STP and ST1 signals). **Possible Causes:** • Stoplight switch signal circuit is shorted to power • Stoplight switch assembly is damaged or shorted • PCM has failed

DTC	Trouble Code Title, Conditions & Possible Causes
DTC: P0505 **2T CCM, MIL: Yes** **1995** **Models:** Avalon, Camry, Land Cruiser, Previa, T100, Tacoma, Tercel **Engines:** 1.5L VIN E, 2.4L VIN A, 2.4L VIN R, 2.7L VIN U, 3.0L VIN G, 3.4L VIN V, 4.5L VIN D **Transmissions:** All	**Idle Control System Malfunction Conditions:** Engine started, engine running at idle speed in closed loop, and the PCM detected the Actual Idle Speed was more than 100-200 rpm above or below the Target Idle Speed. **Note: RSO is the acronym for Rotary Solenoid (IAC) Open Coil and RSC is the acronym for the Rotary Solenoid (IAC) Close Coil.** **Possible Causes:** • RSC or RSO control circuit is open, shorted to ground or power • RSO or RSO power circuit is open (test power from EFI relay) • IAC valve is contaminated, damaged or has failed
DTC: P0505 **2T CCM, MIL: Yes** **1996, 1997, 1998, 1999, 2000, 2001, 2002** **Models:** All **Engines:** All **Transmissions:** All	**Idle Control System Malfunction Conditions:** Engine started, engine running at idle speed in closed loop, and the PCM detected the Actual Idle Speed was more than 100-200 rpm above or below the Target Idle Speed. **Note: RSO is the acronym for Rotary Solenoid (IAC) Open Coil and RSC is the acronym for the Rotary Solenoid (IAC) Close Coil.** **Possible Causes:** • RSC/RSO connector is damaged, open or shorted • RSC or RSO control circuit is open, shorted to ground or power • RSO or RSO power circuit is open (test power from EFI relay) • IAC valve is contaminated
DTC: P0505 **2T CCM, MIL: Yes** **2003, 2004, 2005, 2006** **Models:** 4Runner, Avalon, Camry, Camry, Solara, Celica, Corolla, Echo, Yaris, Land Cruiser, Matrix, MR2, RAV4, Sequoia, Tacoma, Tundra **Engines:** 1.5L VIN T, 1.8L VIN R, 1.8L VIN Y, 2.0L VIN K, 2.4L VIN D, 2.4L VIN L, 2.7L VIN M, 3.0L VIN F, 4.0L VIN U, 4.7L VIN T **Transmissions:** All	**Idle Control System Malfunction Conditions:** Engine started, engine running at idle speed n closed loop, and the PCM detected the Actual Idle Speed was more than 100-200 rpm above or below the Target Idle Speed. A Rotary solenoid type of ISC valve is located in front of the air intake chamber and intake air bypassing the throttle valve is directed to the Intake Air Control (IAC) valve via a passage. The PCM controls the idle speed by regulating the amount of intake air volume that bypasses the throttle valve. **Possible Causes:** • Air Induction system leaks (check for intake manifold leaks) • Air leaks in the PCV system (at the valve or its related hoses) • Throttle body assembly is damaged or has failed • PCM has failed
DTC: P0510 **2T CCM, MIL: Yes** **1995** **Models:** Avalon, Camry, Land Cruiser, Previa, T100, Tacoma, Tercel **Engines:** 1.5L VIN E, 2.4L VIN A, 2.4L VIN R, 2.7L VIN U, 3.0L VIN G, 3.4L VIN V, 4.5L VIN D **Transmissions:** All	**Closed Throttle Position Switch Circuit Malfunction Conditions:** Engine started, and the PCM detected throttle switch input did not change from Off to On during a normal driving period. Refer to Circuit Test or code repair chart to test code. **Possible Causes:** • Closed throttle position switch signal circuit is open or grounded • Closed throttle position switch signal circuit is shorted to power • Closed throttle position switch or TP sensor damaged or failed • PCM has failed
DTC: P0510 **2T CCM, MIL: Yes** **1996, 1997** **Models:** All **Engines:** All **Transmissions:** All	**Closed Throttle Position Switch Circuit Malfunction Conditions:** Engine started, vehicle driven with VSS signals received, and the PCM did not detect any change in the Closed Throttle Position switch status (from off to on, or from on to off). **Possible Causes:** • Closed throttle position switch signal circuit is open or grounded • Closed throttle position switch signal circuit is shorted to power • Closed throttle position switch or TP sensor damaged or failed • PCM has failed
DTC: P0511 **2T CCM, MIL: Yes** **2003, 2004, 2005, 2006** **Models:** Avalon, Camry, Echo, Yaris, Highlander, Matrix, MR2, Sienna, Tacoma **Engines:** 1.5L VIN T, 1.8L VIN R, 2.4L VIN D, 2.4L VIN L, 2.7L VIN M, 3.0L VIN F **Transmissions:** All	**Idle Control Valve Circuit Malfunction Conditions:** Engine running at idle speed n closed loop, and the PCM detected an unexpected voltage condition on the IAC valve circuit. A Rotary solenoid valve, in front of the air intake chamber, allows intake air to bypass the throttle valve and be directed to the Intake Air Control (IAC) valve through a passage. This configuration allows the PCM to control the engine idle speed by regulating the amount of intake air volume that bypasses the throttle valve. **Possible Causes:** • IAC valve connector is damaged or loose • IAC valve control circuit is open, shorted to ground or power • IAC valve is damaged or has failed • PCM has failed

DTC	Trouble Code Title, Conditions & Possible Causes
DTC: P0513 **2T PCM, MIL: Yes** **2003, 2004, 2005, 2006** **Models:** Avalon, Camry, Highlander, Sienna **Engines:** 3.0L VIN F **Transmissions:** All	**Unmatched Key Code Detected Conditions:** Key inserted, and the PCM detected that a key with an unregistered key code had been inserted into the ignition lock. **Possible Causes:** • An unmatched ignition key was inserted
DTC: P0550 **1T CCM, MIL: Yes** **2000, 2001, 2002** **Models:** Echo, Yaris **Engines:** 1.5L VIN T **Transmissions:** All	**Power Steering Pressure Sensor Circuit Malfunction Conditions:** Key on or engine running; and the PCM detected the Power Steering Pressure (PSP) sensor was less than 0.26v, or the PCM detected the PSP sensor was over 4.90v during the test. **Possible Causes:** • PSP sensor connector is damaged, open or shorted • PSP sensor signal circuit is open, shorted to ground or VREF • PSP sensor ground circuit is open • PSP sensor is damaged or has failed • PCM has failed
DTC: P0552 **1T CCM, MIL: Yes** **2003, 2004, 2005, 2006** **Models:** Echo, Yaris **Engines:** 1.5L VIN T **Transmissions:** All	**Power Steering Pressure Sensor Circuit Low Input Conditions:** Key on or engine running; and the PCM detected the Power Steering Pressure (PSP) sensor was less than 0.26v during the CCM test. **Possible Causes:** • PSP sensor signal circuit is open or shorted to ground • PSP sensor ground circuit is open • PSP sensor is damaged or has failed • PCM has failed
DTC: P0553 **1T CCM, MIL: Yes** **2003, 2004, 2005, 2006** **Models:** Echo, Yaris **Engines:** 1.5L VIN T **Transmissions:** All	**Power Steering Pressure Sensor Circuit High Input Conditions:** Key on or engine running; and the PCM detected the Power Steering Pressure (PSP) sensor was more than 4.90v during the CCM test. **Possible Causes:** • PSP sensor signal circuit is shorted to power • PSP sensor ground circuit is open • PSP sensor is damaged or has failed • PCM has failed
DTC: P0560 **2T CCM, MIL: Yes** **2003, 2004, 2005, 2006** **Models:** All **Engines:** All **Transmissions:** All	**System Voltage (Backup Power Circuit) Malfunction Conditions:** Key on or engine running; and the PCM detected an unexpected low voltage condition on the Backup Power Circuit during the test. **Possible Causes:** • Battery backup circuit is open between battery and the PCM • PCM has failed
DTC: P0571 **2T CCM, MIL: Yes** **2003, 2004, 2005, 2006** **Models:** Tacoma, Tundra **Engines:** 3.4L VIN N **Transmissions:** All	**Brake Switch 'A' Circuit Malfunction Conditions:** Engine started, vehicle driven to cruise speed and then back to idle speed several times, and the PCM did not detect any change in the Brake Switch 'A' circuit status. The signal from this switch is used to determine when the brakes have been applied, and to determine the Fuel Cutoff engine speed during some types of braking operations. **Possible Causes:** • Stoplight switch signal circuit is shorted to power • Stoplight switch assembly is damaged or shorted • PCM has failed
DTC: P0604 **1T PCM, MIL: Yes** **2003, 2004, 2005, 2006** **Models:** 4Runner, Land Cruiser, MR2, Sequoia, Tacoma, Tundra **Engines:** All **Transmissions:** All	**PCM Internal Control Module Random Access Memory Processing Error Conditions:** Key on, and the PCM detected a processing error in the Internal Control Module Random Access Memory (RAM) function. **Possible Causes:** • Clear the codes and retest for this code. If the same code resets, substitute a known good control module and retest. If the trouble code is gone, the original PCM has failed. • TSB TC002-03 (6/03) contains information related to this code
DTC: P0606 **1T PCM, MIL: Yes** **2003, 2004, 2005, 2006** **Models:** 4Runner, Echo, Yaris, Land Cruiser, MR2, Sequoia, Tacoma, Tundra **Engines:** All **Transmissions:** All	**ECM/PCM Processing Error Conditions:** Key on, and the PCM detected a processing error occurred. **Possible Causes:** • Clear the codes and retest for this code. If the same code resets, substitute a known good control module and retest. If the trouble code is gone, the original PCM has failed. • TSB TC002-03 (6/03) contains information related to this code

DTC	Trouble Code Title, Conditions & Possible Causes
DTC: P0607 **1T PCM, MIL: Yes** **2003, 2004, 2005, 2006** **Models:** 4Runner, Echo, Yaris, Land Cruiser, MR2, Sequoia, Tacoma, Tundra **Engines:** All **Transmissions:** All	**Control Module Performance Conditions:** Key on, and the PCM detected a performance problem occurred. **Possible Causes:** • Clear the codes and retest for this code. If the same code resets, substitute a known good control module and retest. If the trouble code is gone, the original PCM has failed. • TSB TC002-03 (6/03) contains information related to this code
DTC: P0617 **1T CCM, MIL: Yes** **2003, 2004, 2005, 2006** **Models:** 4Runner, Echo, Yaris, Land Cruiser, MR2, Sequoia, Tacoma, Tundra **Engines:** All **Transmissions:** All	**Starter Relay Circuit High Input Conditions:** Engine started, engine speed over 1000 rpm, system voltage over 10.5v, and the PCM detected the Starter Motor signal indicated high. **Possible Causes:** • Park/Neutral switch assembly is damaged or it has failed • Ignition switch is damaged or has failed• PCM has failed
DTC: P0657 **1T CCM, MIL: Yes** **2003, 2004, 2005, 2006** **Models:** 4Runner, Land Cruiser, MR2, Sequoia, Tacoma, Tundra **Engines:** 1.8L VIN R, 3.4L VIN N, 4.0L VIN U, 4.7L VIN T **Transmissions:** All	**PCM Actuator Supply Voltage Circuit Malfunction Conditions:** Key on or engine running; and the PCM detected an unexpected voltage condition on the Actuator Supply Voltage circuit. **Possible Causes:** • Actuator supply voltage circuit is open • Clear the codes and retest for this code. If the same code resets, substitute a known good control module and retest. If the trouble code is gone, the original PCM has failed. • TSB TC002-03 (6/03) contains information related to this code
DTC: P0705 **2T CCM, MIL: Yes** **2003, 2004, 2005, 2006** **Models:** Avalon, Camry, Highlander, Sienna, Tacoma, Tundra **Engines:** 2.4L VIN D, 2.4L VIN L, 2.7L VIN M, 3.0L VIN F, 3.4L VIN N **Transmissions:** A/T	**A/T Range Sensor Circuit (PRNDL) Malfunction Conditions:** Key on or engine running; and the PCM detected simultaneous "on" signals (N, 2, L or R) from the Transmission Range sensor circuit. The P/N switch indicates "on" whenever the shift lever is in the 'N' or 'P' position. When it is "on", the NSW circuit to the PCM is grounded to chassis ground through the starter motor relay, and reads 0.00v. When the shift lever is in 'R', 'D' or 'L' position, the switch is "off" and the NSW circuit reads 12.0v. When the shift lever is moved from the 'N' to the 'D' position, the PCM uses this signal to air/fuel ratio correction and idle speed control (estimated control) functions. **Possible Causes:** • Park/Neutral switch assembly is shorted • Park/Neutral switch assembly is damaged or has failed • PCM has failed.
DTC: P0710 **1T CCM, MIL: Yes** **1996, 1997, 1998, 1999, 2000, 2001, 2002** **Models:** All **Engines:** All **Transmissions:** A/T	**Transmission Fluid Temperature Sensor Circuit Malfunction Conditions:** Engine started, and the PCM detected the TFT sensor indicated less than -40°F, or after the engine runtime exceeded 15 minutes, it indicated more than 300°F during the CCM test. **Possible Causes:** • TFT sensor signal circuit is open, shorted to ground or shorted to system power • TFT sensor is damaged or has failed • PCM has failed
DTC: P0710 **1T CCM, MIL: Yes** **2003, 2004, 2005, 2006** **Models:** Echo, Yaris, RAV4, Highlander, Sienna **Engines:** 1.5L VIN T engine, 2.0L VIN K, 3.0L VIN F **Transmissions:** A/T	**Transmission Fluid Temperature Sensor Circuit Malfunction Conditions:** Engine started, and the PCM detected the TFT sensor indicated less than -40°F, or with the engine runtime over 15 minutes, it detected the TFT sensor indicated more than 300°F. **Possible Causes:** • TFT sensor signal circuit is open, shorted to ground or shorted to system power • TFT sensor is damaged or has failed • PCM has failed
DTC: P0711 **1T CCM, MIL: Yes** **1996, 1997, 1998, 1999, 2000, 2001, 2002** **Models:** All **Engines:** All **Transmissions:** A/T	**Transmission Fluid Temperature Sensor Performance Conditions:** Engine started, engine runtime over 15 seconds, and the PCM detected the ambient air temperature and transmission fluid temperature varied by more than 40°F, or after an engine runtime of 20 minutes and 6.2 miles were traveled, it indicated less than 50°F. **Possible Causes:** • TFT sensor signal circuit or ground circuit has high resistance • TFT sensor has failed, or the PCM has failed

DTC	Trouble Code Title, Conditions & Possible Causes
DTC: P0711 **1T CCM, MIL: Yes** **2003, 2004, 2005, 2006** **Models:** Echo, Yaris, RAV4 **Engines:** 1.5L VIN T engine, 2.0L VIN K **Transmissions:** A/T	**Transmission Fluid Temperature Sensor Performance Conditions:** ECT and IAT sensor signals more than 14°F, engine runtime over 12 seconds, then after the vehicle was driven for more than 6.2 miles and the engine runtime exceeded 20 minutes, the PCM detected the TFT sensor remained at a value less than 14°F during the CCM test. **Possible Causes:** • TFT sensor signal circuit has high resistance • TFT sensor is damaged or has failed (it may be contaminated) • PCM has failed
DTC: P0712 **1T CCM, MIL: Yes** **2003, 2004, 2005, 2006** **Models:** Echo, Yaris **Engines:** 1.5L VIN T **Transmissions:** A/T	**Transmission Fluid Temperature Sensor Circuit Low Input Conditions:** Engine started, engine running, and the PCM detected the Transmission Fluid Temperature (TFT) sensor indicated a value of more than 284°F for 500 ms in the test. **Possible Causes:** • TFT sensor signal circuit is shorted • TFT sensor is damaged or has failed • PCM has failed
DTC: P0713 **1T CCM, MIL: Yes** **2003, 2004, 2005, 2006** **Models:** Echo, Yaris **Engines:** 1.5L VIN T **Transmissions:** A/T	**Transmission Fluid Temperature Sensor Circuit High Input Conditions:** Engine started, engine running, and the PCM detected the TFT sensor indicated a value of less than -40°F for 500 ms in t he test. **Possible Causes:** • TFT sensor signal circuit is open • TFT sensor is damaged or has failed • PCM has failed
DTC: P0715 **1T CCM, MIL: Yes** **1998, 1999, 2000, 2001, 2002** **Models:** Land Cruiser, Sequoia, Tundra **Engines:** All **Transmissions:** A/T	**A/T Input Shaft/Direct Clutch Speed Sensor Circuit Malfunction Conditions:** Engine started, P/N switch indicating "off", vehicle driven in 1st, 2nd or 3rd gear or in Overdrive, the SS1, SS2 and SL (shift valves) and VSS all operating normally, and the PCM detected the ISS signal indicated 100 rpm or less during the CCM Rationality test. **Possible Causes:** • O/D direct clutch speed sensor circuit is open • O/D direct clutch speed sensor circuit is shorted to ground • O/D direct clutch speed sensor is damaged or has failed • PCM has failed
DTC: P0717 **1T CCM, MIL: Yes** **2003, 2004, 2005, 2006** **Models:** Avalon, Camry, Highlander, Sienna **Engines:** 3.0L VIN F **Transmissions:** A/T	**A/T Input Shaft Speed Sensor 'A' Circuit No Signal Conditions:** No SS1, SS2, SL shift solenoid or VSS codes set, engine started, P/N switch indicating "off", gear position indicating 1st, 2nd, 3rd gear, or in O/D, no gear change occurring, and the PCM detected the Input Shaft Speed sensor was below 300 rpm, or over 1000 rpm for 4 seconds. **Possible Causes:** • Direct clutch connector is damaged (it may be open or shorted) • Direct clutch sensor signal circuit is open or shorted • Direct clutch speed sensor is damaged or has failed • PCM has failed
DTC: P0717 **1T CCM, MIL: Yes** **2003, 2004, 2005, 2006** **Models:** Echo, Yaris **Engines:** 1.5L VIN T **Transmissions:** A/T	**A/T Turbine Shaft Speed Sensor Circuit No Signal Conditions:** No Shift Solenoid or P/N codes set, engine started, P/N switch indicating "off", gear position indicating 2nd, 3rd gear, or in O/D, no gear change occurring, and the PCM detected the Turbine Shaft Speed (ISS) sensor indicated less than 300 rpm, or more than 1000 rpm for 4 seconds. The PCM detects the rotation speed of the input turbine, and compares the signals from the input turbine speed (NT) sensor to the counter gear speed sensor (NC). The PCM uses this signal to detect the shift time so that it can control the engine torque and hydraulic pressure in response to various driving conditions. **Possible Causes:** • Input shaft speed sensor is damaged or loose • Input shaft speed sensor signal (NT) circuit is open or shorted • Input shaft speed sensor is damaged or has failed • PCM has failed
DTC: P0724 **2T CCM, MIL: Yes** **2003, 2004, 2005, 2006** **Models:** Avalon, Camry, Highlander, Sienna, Tacoma, Tundra **Engines:** 2.4L VIN D, 2.4L VIN L, 2.7L VIN M, 3.0L VIN F, 3.4L VIN N **Transmissions:** A/T	**Brake Switch 'B' Circuit High Input Conditions:** Engine started, vehicle driven to cruise speed and then back to idle speed at least 30 times, and the PCM did not detect any change in the Brake Switch 'A' circuit status. The STP 'B' switch signal is used to determine when the brakes have been applied, and to determine the Fuel Cutoff engine speed during periods with the brakes applied. **Possible Causes:** • Stoplight switch signal circuit is shorted to power • Stoplight switch assembly is damaged or shorted • PCM has failed

DTC	Trouble Code Title, Conditions & Possible Causes
DTC: P0724 **1T CCM, MIL: Yes** **2003, 2004, 2005, 2006** **Models:** Echo, Yaris, MR2 **Engines:** 1.5L VIN T, 1.8L VIN R **Transmissions:** A/T	**Brake Switch 'B' Circuit High Input Conditions:** Engine started, vehicle driven to cruise speed and then back to idle speed at least 30 times, and the PCM did not detect any change in the Brake Switch 'A' circuit status. The STP 'B' switch signal is used to determine when the brakes have been applied, and to determine the Fuel Cutoff engine speed during periods with the brakes applied. **Possible Causes:** • Stoplight switch signal circuit is shorted to power • Stoplight switch assembly is damaged or shorted • PCM has failed
DTC: P0724 **1T CCM, MIL: Yes** **2003, 2004, 2005, 2006** **Models:** Avalon, Camry, Echo, Yaris, Highlander, Sienna **Engines:** 1.5L VIN T, 3.0L VIN F **Transmissions:** A/T	**A/T Torque Converter Clutch Shift Solenoid Performance Conditions:** Engine started, vehicle driven to over 50 mph and then back to idle speed several times, and the PCM detected that TCC lockup did not occur, or that the TCC remained in lockup position in the "off" range. The PCM uses signals from the CKP, MAF and TP sensors to monitor engagement of the Lockup Clutch solenoid to find a fault. **Possible Causes:** • Lockup clutch solenoid is damaged or has failed Shift solenoid SL is damaged or has failed (mechanical fault) • Shift solenoid SL is stuck in "on" or "off" position • Valve body is blocked or stuck
DTC: P0743 **2T CCM, MIL: Yes** **2003, 2004, 2005, 2006** **Models:** Avalon, Camry, Echo, Yaris, Highlander, Sienna **Engines:** 1.5L VIN T, 3.0L VIN F **Transmissions:** A/T	**A/T Torque Converter Clutch Shift Solenoid Circuit Malfunction Conditions:** Vehicle driven to over 50 mph and then back to idle speed several times; and the PCM detected an unexpected voltage on the Shift Solenoid (SL) valve control circuit. The Shift Solenoid valve is turned On/Off" through commands from the PCM to control the hydraulic pressure acting on the lockup relay valve (and it controls the operation of the lockup clutch). **Possible Causes:** • Shift solenoid SL connector is damaged or loose • Shift solenoid SL control circuit is open or shorted to ground • Shift solenoid SL is damaged or has failed (electrical fault) • PCM has failed
DTC: P0750 **2T CCM, MIL: Yes** **1995** **Models:** Avalon, Camry, Land Cruiser, Previa, T100, Tacoma, Tercel **Engines:** 1.5L VIN E, 2.4L VIN A, 2.4L VIN R, 2.7L VIN U, 3.0L VIN G, 3.4L VIN V, 4.5L VIN D **Transmissions:** A/T	**A/T Shift Solenoid 1 or 'A' Malfunction (Mechanical) Conditions:** Engine started, vehicle driven under normal driving conditions, and the PCM detected the Actual gear ratio and the Required gear ratio did not match during the CCM Rationality test. **Possible Causes:** • A/T component problems (i.e., in clutch, brake or gears) • SS1 or SSA is damaged, stuck "open" or stuck "closed" • Transmission valve body is clogged, dirty or stuck
DTC: P0750 **2T CCM, MIL: Yes** **1996, 1997, 1998, 1999, 2000, 2001, 2002** **Models:** All **Engines:** All **Transmissions:** A/T	**A/T Shift Solenoid 1 or 'A' Malfunction (Mechanical) Conditions:** Engine started, vehicle driven under normal driving conditions, and the PCM detected the Actual gear ratio and the Required gear ratio did not match during the CCM Rationality test. **Possible Causes:** • A/T component problems (i.e., in clutch, brake or gears) • SS1 or SSA is damaged, stuck "open" or stuck "closed" • Transmission valve body is clogged, dirty or stuck
DTC: P0750 **2T CCM, MIL: Yes** **2003, 2004, 2005, 2006** **Models:** Celica, Corolla, Matrix, RAV4 **Engines:** 1.8L VIN R, 1.8L VIN Y, 2.0L VIN K **Transmissions:** A/T	**A/T Shift Solenoid 'A' Malfunction (Mechanical) Conditions:** Engine started, vehicle driven under normal driving conditions, and the PCM detected the Actual gear ratio and the Required gear ratio did not match during the CCM Rationality test. **Possible Causes:** • A/T component problems (i.e., in clutch, brake or gears) • SSA is damaged, stuck "open" or stuck "closed" • Transmission valve body is clogged, dirty or stuck
DTC: P0751 **8T CCM, MIL: Yes** **2003, 2004, 2005, 2006** **Models:** Avalon, Camry, Echo, Yaris, Highlander, Sienna **Engines:** 1.5L VIN T, 3.0L VIN F **Transmissions:** A/T	**A/T Shift Solenoid 'A' Signal Range/Performance Conditions:** Engine started, vehicle driven to a speed over 50 mph, and the PCM detected the Actual gear position did not match the Desired gear position during the CCM test period. The PCM uses inputs from the VSS and Direct Clutch speed sensor to determine the actual gear position (i.e., 1st, 2nd, 3rd or O/D gear). **Possible Causes:** • SSA control circuit is open or shorted to ground • SSA control circuit is shorted to system power (B+) • SSA is damaged or has failed (an electrical fault) • PCM has failed

DTC	Trouble Code Title, Conditions & Possible Causes
DTC: P0753 **8T CCM, MIL: Yes** **1995** **Models:** Avalon, Camry, Land Cruiser, Previa, T100, Tacoma, Tercel **Engines:** 1.5L VIN E, 2.4L VIN A, 2.4L VIN R, 2.7L VIN U, 3.0L VIN G, 3.4L VIN V, 4.5L VIN D **Transmissions:** A/T	**A/T Shift Solenoid 1 or 'A' Circuit Malfunction Conditions:** Engine started, engine running during normal driving conditions; and the PCM detected the Shift Solenoid 1 or 'A' (SS1/SSA) control circuit voltage was "high" with the solenoid "on", or the SS1 or SSA control circuit voltage was "low" with the SS1/SSA commanded "off". **Possible Causes:** • SS1 or SSA control circuit is open or shorted to ground • SS1or SSA control circuit is shorted to system power (B+) • SS1 or SSA is damaged or has failed (an electrical fault) • PCM has failed
DTC: P0753 **8T CCM, MIL: Yes** **1996, 1997, 1998, 1999, 2000, 2001, 2002** **Models:** All **Engines:** All **Transmissions:** A/T	**A/T Shift Solenoid 1 or 'A' Circuit Malfunction Conditions:** Engine started, engine running during normal driving conditions; and the PCM detected the Shift Solenoid 1 or 'A' (SS1/SSA) control circuit voltage was "high" with the solenoid "on", or the SS1 or SSA control circuit voltage was "low" with the SS1/SSA commanded "off". **Possible Causes:** • SS1 or SSA control circuit is open, shorted to ground or shorted to system power (B+) • SS1 or SSA has failed, or the PCM has failed
DTC: P0753 **8T CCM, MIL: Yes** **2003, 2004, 2005, 2006** **Models:** Avalon, Camry, Highlander, Sienna **Engines:** 3.0L VIN F **Transmissions:** A/T	**A/T Shift Solenoid 'A' Circuit Malfunction Conditions:** Engine started, engine running during normal driving conditions; and the PCM detected an unexpected voltage condition on the Shift Solenoid 'A' control circuit during the CCM test. **Possible Causes:** • SSA control circuit is open, shorted to ground or to power (B+) • SSA is damaged or has failed (an electrical fault) • PCM has failed
DTC: P0753 **8T CCM, MIL: Yes** **2003, 2004, 2005, 2006** **Models:** Celica, Corolla, Matrix, RAV4 **Engines:** 1.8L VIN R, 1.8L VIN Y, 2.0L VIN K **Transmissions:** A/T	**A/T Shift Solenoid 'A' Circuit Malfunction Conditions:** Engine running during normal driving conditions; and the PCM detected an unexpected voltage condition on the Shift Solenoid 'A' control circuit during the CCM test. **Possible Causes:** • SSA control circuit is open, shorted to ground or to power (B+) • SSA is damaged or has failed (an electrical fault) • PCM has failed
DTC: P0755 **2T CCM, MIL: Yes** **1995** **Models:** Avalon, Camry, Land Cruiser, Previa, T100, Tacoma, Tercel **Engines:** 1.5L VIN E, 2.4L VIN A, 2.4L VIN R, 2.7L VIN U, 3.0L VIN G, 3.4L VIN V, 4.5L VIN D **Transmissions:** A/T	**A/T Shift Solenoid 'B' Malfunction (Mechanical) Conditions:** Engine started, vehicle driven under normal driving conditions, and the PCM detected the Actual gear ratio and the Required gear ratio did not match during the CCM Rationality test. **Possible Causes:** • A/T component problems (i.e., in clutch, brake or gears) • SSB is damaged, stuck "open" or stuck "closed" • Transmission valve body is clogged, dirty or stuck
DTC: P0755 **2T CCM, MIL: Yes** **1996, 1997, 1998, 1999, 2000, 2001, 2002** **Models:** All **Engines:** All **Transmissions:** A/T	**A/T Shift Solenoid 'B' Malfunction (Mechanical) Conditions:** Engine started, vehicle driven under normal driving conditions, and the PCM detected the Actual gear ratio and the Required gear ratio did not match during the CCM Rationality test. **Possible Causes:** • A/T component problems (i.e., in clutch, brake or gears) • SSB is damaged, stuck "open" or stuck "closed" • Transmission valve body is clogged, dirty or stuck
DTC: P0755 **8T CCM, MIL: Yes** **2003, 2004, 2005, 2006** **Models:** Celica, Corolla, Matrix **Engines:** 1.8L VIN R, 1.8L VIN Y **Transmissions:** A/T	**A/T Shift Solenoid 'B' Malfunction (Mechanical) Conditions:** Engine started, vehicle driven under normal driving conditions, and the PCM detected the Actual gear ratio and the Required gear ratio did not match during the CCM Rationality test. **Possible Causes:** • A/T component problems (i.e., in clutch, brake or gears) • SSB is damaged, stuck "open" or stuck "closed" • Transmission valve body is clogged, dirty or stuck

DTC	Trouble Code Title, Conditions & Possible Causes
DTC: P0756 **8T CCM, MIL: Yes** **2003, 2004, 2005, 2006** **Models:** Avalon, Camry, Echo, Yaris, Highlander, Sienna **Engines:** 1.5L VIN T, 3.0L VIN F **Transmissions:** A/T	**A/T Shift Solenoid 'B' Signal Range/Performance Conditions:** Vehicle driven to a speed over 50 mph, and the PCM detected the Actual gear position did not match the Desired gear position. The PCM uses inputs from the VSS and Direct Clutch speed sensor to determine the actual gear position (i.e., 1st, 2nd, 3rd or O/D gear). **Possible Causes:** • SSB control circuit is open, shorted to ground or to power (B+) • SSB is damaged or has failed (an electrical fault) • PCM has failed
DTC: P0758 **8T CCM, MIL: Yes** **1995** **Models:** Avalon, Camry, Land Cruiser, Previa, T100, Tacoma, Tercel **Engines:** 1.5L VIN E, 2.4L VIN A, 2.4L VIN R, 2.7L VIN U, 3.0L VIN G, 3.4L VIN V, 4.5L VIN D **Transmissions:** A/T	**A/T Shift Solenoid 2 or 'B' Circuit Malfunction Conditions:** Engine started, engine running during normal driving conditions, and the PCM detected the Shift Solenoid 2 or 'B' (SS2/SSB) control circuit voltage was "high" with the solenoid "on", or the SS2 or SSB control circuit voltage was "low" with the SS2/SSB commanded "off". **Possible Causes:** • SS2 or SSB control circuit is open or shorted to ground • SS2 or SSB control circuit is shorted to system power (B+) • SS2 or SSB is damaged or has failed (an electrical fault) • PCM has failed
DTC: P0758 **1T CCM, MIL: Yes** **1996, 1997, 1998, 1999, 2000,** **2001, 2002** **Models:** All **Engines:** All **Transmissions:** A/T	**A/T Shift Solenoid 2 or 'B' Circuit Malfunction Conditions:** Engine started, engine running during normal driving conditions, and the PCM detected the Shift Solenoid 2 or 'B' (SS2/SSB) control circuit voltage was "high" with the solenoid "on", or the SS2 or SSB control circuit voltage was "low" with the SS2/SSB commanded "off". **Note: This problem must occur eight (8) times during one trip before this code is triggered.** **Possible Causes:** • SS2 or SSB control circuit is open or shorted to ground • SS2 or SSB control circuit is shorted to system power (B+) • SS2 or SSB is damaged or has failed (an electrical fault) • PCM has failed
DTC: P0758 **8T CCM, MIL: Yes** **2003, 2004, 2005, 2006** **Models:** Celica, Corolla, Matrix, RAV4 **Engines:** 1.8L VIN R, 1.8L VIN Y, 2.0L VIN K **Transmissions:** A/T	**A/T Shift Solenoid 'B' Circuit Malfunction Conditions:** Engine started, engine running during normal driving conditions, and the PCM detected an unexpected voltage condition on the Shift Solenoid 'B' control circuit during the CCM test. **Possible Causes:** • SSB control circuit is open, shorted to ground or to power (B+) • SSB is damaged or has failed (an electrical fault) • PCM has failed
DTC: P0758 **8T CCM, MIL: Yes** **2003, 2004, 2005, 2006** **Models:** Avalon, Camry, Highlander, Sienna **Engines:** 2.4L VIN D, 3.0L VIN F **Transmissions:** A/T	**A/T Shift Solenoid 'B' Circuit Malfunction Conditions:** Engine started, engine running during normal driving conditions, and the PCM detected an unexpected voltage condition on the Shift Solenoid 'B' control circuit during the CCM test. **Possible Causes:** • SSB control circuit is open, shorted to ground or to power (B+) • SSB is damaged or has failed (an electrical fault) • PCM has failed
DTC: P0765 **8T CCM, MIL: Yes** **2003, 2004, 2005, 2006** **Models:** RAV4 **Engines:** 2.0L VIN K **Transmissions:** A/T	**A/T Shift Solenoid 'D' Malfunction (Mechanical) Conditions:** Engine started, engine running during normal driving conditions, and the PCM detected an unexpected voltage condition on the Shift Solenoid 'D' control circuit during the CCM test. **Possible Causes:** • SL1 or SL2 is damaged, stuck "open" or stuck "closed" • S4 is damaged, stuck "open" or stuck "closed" • Transmission valve body is clogged, dirty or stuck
DTC: P0768 **8T CCM, MIL: Yes** **2003, 2004, 2005, 2006** **Models:** RAV4 **Engines:** 2.0L VIN K **Transmissions:** A/T	**A/T Shift Solenoid 'D' Circuit Malfunction Conditions:** Engine started, engine running during normal driving conditions, and the PCM detected an unexpected voltage condition on the Shift Solenoid 'D' control circuit during the CCM test. **Possible Causes:** • SL1, SL2 or S4 circuit is open, shorted to ground or to power • SL1, SL2 or S4 assembly is damaged or has failed • PCM has failed

DTC	Trouble Code Title, Conditions & Possible Causes
DTC: P0770 **2T CCM, MIL: Yes** **1995** **Models:** Avalon, Camry, Land Cruiser, Previa, T100, Tacoma, Tercel **Engines:** 1.5L VIN E, 2.4L VIN A, 2.4L VIN R, 2.7L VIN U, 3.0L VIN G, 3.4L VIN V, 4.5L VIN D **Transmissions:** A/T	**A/T Shift Solenoid 'E' (SL) Malfunction (Mechanical) Conditions:** Engine started, vehicle driven at a speed of over 50 mph for 1-3 minutes, and the PCM detected the transmission lockup function did not occur in the "lockup" range, or the "lockup" function was "on" during periods when it should have been "off" during the CCM test. **Note: The A/T clutch or A/T brake clutch will slip with this code set.** **Possible Causes:** • A/T component problems (i.e., in clutch, brake or gears) • SL (shift lockup) is damaged, stuck "open" or stuck "closed" • Transmission valve body is clogged, dirty or stuck
DTC: P0770 **1T CCM, MIL: Yes** **1996, 1997, 1998, 1999, 2000, 2001, 2002** **Models:** All **Engines:** All **Transmissions:** A/T	**A/T Shift Solenoid 'E' (SL) Malfunction (Mechanical) Conditions:** Engine started, vehicle driven at a speed of over 50 mph for 1-3 minutes, and the PCM detected the transmission lockup function did not occur in the "lockup" range, or the "lockup" function was "on" during periods when it should have been "off" during the CCM test. **Note: The A/T clutch or A/T brake clutch will slip with this code set.** **Possible Causes:** • A/T component problems (i.e., in clutch, brake or gears) • SL (shift lockup) is damaged, stuck "open" or stuck "closed" • Transmission valve body is clogged, dirty or stuck
DTC: P0770 **1T CCM, MIL: Yes** **2003, 2004, 2005, 2006** **Models:** Celica, Corolla, Matrix, RAV4 **Engines:** 1.8L VIN R, 1.8L VIN Y, 2.0L VIN K **Transmissions:** A/T	**A/T Shift Solenoid 'E' (SL) Malfunction (Mechanical) Conditions:** Engine started, vehicle driven at a speed of over 50 mph for 1-3 minutes, and the PCM detected the transmission lockup function did not occur in the "lockup" range, or the "lockup" function was "on" during periods when it should have been "off" during the CCM test. **Note: The A/T clutch or A/T brake clutch will slip with this code set.** **Possible Causes:** • A/T component problems (i.e., in clutch, brake or gears) • SL (shift lockup) is damaged, stuck "open" or stuck "closed" • Transmission valve body is clogged, dirty or stuck
DTC: P0773 **1T CCM, MIL: Yes** **1995** **Models:** Avalon, Camry, Land Cruiser, Previa, T100, Tacoma, Tercel **Engines:** 1.5L VIN E, 2.4L VIN A, 2.4L VIN R, 2.7L VIN U, 3.0L VIN G, 3.4L VIN V, 4.5L VIN D **Transmissions:** A/T	**A/T Shift Solenoid 'E' (SL) Circuit Malfunction Conditions:** Engine started, engine running during normal driving conditions, and the PCM detected the Shift Lockup (SL) control circuit voltage was "high" with the solenoid "on", or the SL control circuit voltage was "low" with the SL commanded "off" during the CCM Rationality test. **Possible Causes:** • SL control circuit is open or shorted to ground • SL control circuit is shorted to system power (B+) • SL is damaged or has failed (an electrical fault) • PCM has failed
DTC: P0773 **1T CCM, MIL: Yes** **1996, 1997, 1998, 1999, 2000, 2001** **Models:** All **Engines:** All **Transmissions:** A/T	**A/T Shift Solenoid 'E' (SL) Circuit Malfunction Conditions:** Engine started, engine running during normal driving conditions, and the PCM detected the Shift Lockup (SL) control circuit voltage was "high" with the solenoid "on", or the SL control circuit voltage was "low" with the SL commanded "off" during the CCM Rationality test. **Possible Causes:** • SL control circuit is open or shorted to ground • SL control circuit is shorted to system power (B+) • SL is damaged or has failed (an electrical fault) • PCM has failed
DTC: P0773 **1T CCM, MIL: Yes** **2001** **Models:** Avalon, Sienna **Engines:** All **Transmissions:** A/T	**A/T Shift Solenoid 'E' (SL) Circuit Malfunction Conditions:** Engine started, engine running during normal driving conditions, and the PCM detected the Shift Lockup (SL) control circuit voltage was "high" with the solenoid "on", or the SL control circuit voltage was "low" with the SL commanded "off" during the CCM Rationality test. **Possible Causes:** • SL control circuit is open, shorted to ground or to power (B+) • SL is damaged or has failed (an electrical fault) • PCM has failed • TSB EG16-01 (12/01) contains information related to this code

DTC	Trouble Code Title, Conditions & Possible Causes
DTC: P0773 **1T CCM, MIL: Yes** **2002** **Models:** All **Engines:** All **Transmissions:** A/T	**A/T Shift Solenoid 'E' (SL) Circuit Malfunction Conditions:** Engine started, engine running during normal driving conditions, and the PCM detected the Shift Lockup (SL) control circuit voltage was "high" with the solenoid "on", or the SL control circuit voltage was "low" with the SL commanded "off" during the CCM Rationality test. **Possible Causes:** • SL control circuit is open or shorted to ground • SL control circuit is shorted to system power (B+) • SL is damaged or has failed (an electrical fault) • PCM has failed
DTC: P0773 **1T CCM, MIL: Yes** **2003, 2004, 2005, 2006** **Models:** Celica, Corolla, Matrix, RAV4 **Engines:** 1.8L VIN R, 1.8L VIN Y, 2.0L VIN K **Transmissions:** A/T	**A/T Shift Solenoid 'E' (SL) Circuit Malfunction Conditions:** Engine started, engine running during normal driving conditions, and the PCM detected the Shift Lockup (SL) control circuit voltage was "high" with the solenoid "on", or the SL control circuit voltage was "low" with the SL commanded "off" during the CCM Rationality test. **Possible Causes:** • SL solenoid control circuit is open or shorted to ground • SL solenoid control circuit is shorted to system power (B+) • SL solenoid is damaged or has failed (an electrical fault) • PCM has failed
DTC: P0787 **1T CCM, MIL: Yes** **2003, 2004, 2005, 2006** **Models:** Echo, Yaris **Engines:** 1.5L VIN T **Transmissions:** A/T	**A/T Shift Timing Solenoid (ST) Circuit Low Input Conditions:** Engine started, engine running during normal driving conditions, and the PCM detected an unexpected low voltage condition on the Shift Timing Solenoid (SL) control circuit at least (4) times during testing. **Possible Causes:** • ST solenoid connector is damaged or shorted • SL solenoid control circuit is shorted to ground • SL solenoid is damaged or has failed • PCM has failed
DTC: P0788 **1T CCM, MIL: Yes** **2003, 2004, 2005, 2006** **Models:** Echo, Yaris **Engines:** 1.5L VIN T **Transmissions:** A/T	**A/T Shift Timing Solenoid (ST) Circuit High Input Conditions:** Engine started, engine running during normal driving conditions, and the PCM detected an unexpected high voltage condition on the Shift Timing Solenoid (SL) control circuit at least (4) times during testing. **Possible Causes:** • ST solenoid connector is damaged or loose • SL solenoid control circuit is open or shorted to system power • SL solenoid is damaged or has failed • PCM has failed
DTC: P0850 **1T CCM, MIL: Yes** **2003, 2004, 2005, 2006** **Models:** Avalon, Camry, Highlander, Sienna, Tacoma, Tundra **Engines:** 2.4L VIN D, 2.4L VIN L, 2.7L VIN M, 3.0L VIN F, 3.4L VIN N **Transmissions:** A/T	**A/T Park/Neutral Switch Circuit Malfunction Conditions:** Engine started, vehicle at 1500-2500 rpm at a speed over 43 mph for at least 30 seconds, and the PCM detected a continuous "on" "N" signal from the P/N switch. The P/N switch indicates "on" whenever the shift lever is in the 'N' or 'P' position. When it is "on", the NSW circuit to the PCM is grounded to chassis ground through the starter motor relay, and reads 0.00v. When the shift lever is in 'R', 'D' or 'L' position, the switch is "off" and the NSW circuit reads 12.0v. When the shift lever is moved from the 'N' to the 'D' position, the PCM uses this signal to air/fuel ratio correction and idle speed control functions. **Possible Causes:** • Park/Neutral switch assembly is shorted • Park/Neutral switch assembly is damaged or has failed • PCM has failed.
DTC: P0973 **1T CCM, MIL: Yes** **2003, 2004, 2005, 2006** **Models:** Echo, Yaris **Engines:** 1.5L VIN T **Transmissions:** A/T	**A/T Shift Solenoid 1 Circuit Low Input Conditions:** Engine started, engine running during normal driving conditions, and the PCM detected an unexpected low voltage condition on the Shift Solenoid 1 control circuit at least (4) times. **Possible Causes:** • SS1 (solenoid) connector is damaged or shorted • SS1 (solenoid) control circuit is shorted to ground • SS1 (solenoid) is damaged or has failed • PCM has failed
DTC: P0974 **1T CCM, MIL: Yes** **2003, 2004, 2005, 2006** **Models:** Echo, Yaris **Engines:** 1.5L VIN T **Transmissions:** A/T	**A/T Shift Solenoid 1 Circuit High Input Conditions:** Engine started, engine running during normal driving conditions, and the PCM detected an unexpected high voltage condition on the Shift Solenoid 1 control circuit at least (4) times. **Possible Causes:** • SS1 (solenoid) connector is damaged or open • SS1 (solenoid) control circuit is open or shorted to power (B+) • SS1 (solenoid) is damaged or has failed • PCM has failed

DTC	Trouble Code Title, Conditions & Possible Causes
DTC: P0976 **1T CCM, MIL: Yes** **2003, 2004, 2005, 2006** **Models:** Echo, Yaris **Engines:** 1.5L VIN T **Transmissions:** A/T	**A/T Shift Solenoid 2 Circuit Low Input Conditions:** Engine started, engine running during normal driving conditions, and the PCM detected an unexpected low voltage condition on the Shift Solenoid 2 control circuit at least (4) times during testing. **Possible Causes:** • SS2 (solenoid) connector is damaged or shorted • SS2 (solenoid) control circuit is shorted to ground • SS2 (solenoid) is damaged or has failed • PCM has failed
DTC: P0977 **1T CCM, MIL: Yes** **2003, 2004, 2005, 2006** **Models:** Echo, Yaris **Engines:** 1.5L VIN T **Transmissions:** A/T	**A/T Shift Solenoid 2 Circuit High Input Conditions:** Engine started, engine running during normal driving conditions, and the PCM detected an unexpected high voltage condition on the Shift Solenoid 2 control circuit at least (4) times during testing. **Possible Causes:** • SS2 (solenoid) connector is damaged or open • SS2 (solenoid) control circuit is open or shorted to power (B+) • SS2 (solenoid) is damaged or has failed • PCM has failed

OBD II Trouble Code List (P1xxx Codes)

DTC	Trouble Code Title, Conditions & Possible Causes
DTC: P1100 **1T CCM, MIL: Yes** **1995, 1996, 1997** **Models:** Supra **Engines:** 3.0L VIN E **Transmissions:** All	**Barometric Pressure Sensor Circuit Malfunction Conditions:** Key on or engine running; and the PCM detected an unexpected voltage condition on the Barometric Pressure (BARO) sensor circuit during the CCM test. **Possible Causes:** • BARO sensor signal circuit is open or shorted to ground • BARO sensor signal circuit shorted to VREF or system power • BARO sensor power circuit is open between sensor and PCM • BARO sensor ground circuit is open between sensor and PCM • BARO sensor is damaged or has failed • PCM has failed
DTC: P1120 **1T CCM, MIL: Yes** **1998, 1999, 2000, 2001, 2002** **Models:** Supra, Land Cruiser, Sequoia, Tundra **Engines:** 3.0L VIN D, 4.7L VIN T **Transmissions:** All	**Accelerator Pedal Position Sensor Circuit Malfunction Conditions:** Engine started; and the PCM detected the APP sensor VPA reading was equal to or less than 0.2v with a VPA2 reading of equal to or less than 0.5v; or the VPA reading was equal to or more than 4.7v; or the VPA reading indicated from 0.2-1.8v with the VPA2 reading equal to or more than 4.97v; or the VPA reading minus the VPA2 reading was less than 0.02v; or the VPA 2 reading minus the VPA reading was less than 0.02v for 5 seconds. **Possible Causes:** • APP sensor circuit is open or shorted to ground • APP sensor circuit is shorted to VREF • APP sensor power circuit is open between sensor and the PCM • APP sensor ground circuit is open between sensor and PCM • APP sensor is damaged or has failed • PCM has failed
DTC: P1121 **1T CCM, MIL: Yes** **1998, 1999, 2000, 2001, 2002** **Models:** Supra, Land Cruiser, Sequoia, Tundra **Engines:** 3.0L VIN D, 4.7L VIN T **Transmissions:** All	**ETCS Accelerator Pedal Position Sensor Performance Conditions:** Engine started, engine running, and the PCM detected the difference between the Electronic Throttle Control System (ETCS) APP sensor VPA and VPA2 readings was less than 0.7v or 1.7v for more than 2 seconds during the CCM test. **Possible Causes:** • APP sensor is damaged or has failed • Throttle assembly or throttle linkage is binding or has failed • PCM has failed
DTC: P1125 **1T CCM, MIL: Yes** **1998, 1999, 2000, 2001, 2002** **Models:** Supra, Land Cruiser, Sequoia, Tundra **Engines:** 3.0L VIN D, 4.7L VIN T **Transmissions:** All	**ETCS Throttle Control Motor Circuit Malfunction Conditions:** Engine started, engine running, and the PCM detected the motor duty cycle was equal to or more than 80% with a current level less than 0.5A; or the throttle motor current level was more than 16A, or the motor current level was equal to or more than 7A for 600 ms. **Possible Causes:** • ETCS throttle motor control (+) circuit is open, shorted to ground or to power (B+) • ETCS throttle motor control (-) circuit is open, shorted to ground or to power (B+) • ETCS throttle motor is damaged or has failed • PCM has failed
DTC: P1125 **1T CCM, MIL: Yes** **2001, 2002, 2003, 2004, 2005, 2006** **Models:** Prius **Engines:** 1.5L VIN B **Transmissions:** All	**ETCS Throttle Control Motor Circuit Malfunction Conditions:** Engine started, engine running, and the PCM detected the motor duty cycle was equal to or more than 80% with a current level less than 0.5A; or the throttle motor current level was more than 16A, or the motor current level was equal to or more than 7A for 600 ms. **Possible Causes:** • ETCS throttle motor control (+) circuit is open, shorted to ground or to power (B+) • ETCS throttle motor control (-) circuit is open, shorted to ground or to power (B+) • ETCS throttle motor is damaged or has failed • PCM has failed
DTC: P1126 **1T CCM, MIL: Yes** **1998, 1999, 2000, 2001, 2002** **Models:** Supra, Land Cruiser, Sequoia, Tundra **Engines:** 3.0L VIN, 4.7L VIN T **Transmissions:** All	**Magnetic Clutch Circuit Malfunction Conditions:** Engine started; and the PCM detected the magnetic clutch current was equal to or greater than 1.4 amps, or it was in a range of 0.8 amps to 1.4 amps for 1.5 seconds during the test. **Possible Causes:** • ETCS magnetic clutch (+) circuit is open, shorted to ground or to power (B+) • ETCS magnetic clutch (-) circuit is open, shorted to ground or to power (B+) • ETCS magnetic clutch is damaged or has failed • PCM has failed

DTC	Trouble Code Title, Conditions & Possible Causes
DTC: P1126 **1T CCM, MIL: Yes** **2003, 2004, 2005, 2006** **Models:** MR2 **Engines:** 1.8L VIN R **Transmissions:** All	**Magnetic Clutch Circuit Malfunction Conditions:** Engine started; and the PCM detected the magnetic clutch current was equal to or greater than 1.4 amps, or it was in a range of 0.8 amps to 1.4 amps for 1.5 seconds during the test. **Possible Causes:** • ETCS magnetic clutch (+) circuit is open, shorted to ground or to power (B+) • ETCS magnetic clutch (-) circuit is open, shorted to ground or to power (B+) • ETCS magnetic clutch is damaged or has failed • PCM has failed
DTC: P1127 **1T CCM, MIL: Yes** **1998, 1999, 2000, 2001, 2002** **Models:** Supra, Land Cruiser, Sequoia, Tundra **Engines:** 3.0L VIN D, 4.7L VIN T **Transmissions:** All	**ETCS Actuator Power Source Circuit Malfunction Conditions:** Key on and the PCM detected a fault in the Electric Throttle Control system power source. The PCM shuts off power to the throttle motor and magnetic clutch (throttle valve is fully closed by a return spring, so the accelerator pedal can be opened with the throttle valve). **Possible Causes:** • Battery connections are dirty or loose • ETCS power source circuit is open (check the ETCS fuse) • PCM has failed
DTC: P1127 **1T CCM, MIL: Yes** **2001, 2002, 2003, 2004, 2005, 2006** **Models:** Prius **Engines:** 1.5L VIN B **Transmissions:** All	**ETCS Actuator Power Source Circuit Malfunction Conditions:** Key on, and the PCM detected an unexpected voltage condition on the Electric Throttle Control System (ETCS) power source circuit. **Note: The PCM shuts off power to the throttle motor and magnetic clutch (the accelerator pedal can be opened with the throttle valve).** **Possible Causes:** • Battery connections are dirty or loose • ETCS power source circuit is open (check the ETCS fuse) • PCM has failed
DTC: P1128 **1T CCM, MIL: Yes** **1998, 1999, 2000, 2001, 2002** **Models:** Supra, Land Cruiser, Sequoia, Tundra **Engines:** 3.0L VIN D, 4.7L VIN T **Transmissions:** All	**ETCS Throttle Control Motor Lock Malfunction Conditions:** Engine started, engine running, and the PCM detected the throttle control motor position was "locked" during normal operation of the throttle control motor during the CCM test. **Possible Causes:** • ETCS throttle motor is damaged or has failed • Throttle body assembly is damaged or stuck closed • PCM has failed
DTC: P1128 **1T CCM, MIL: Yes** **2001, 2002, 2003, 2004, 2005, 2006** **Models:** Prius **Engines:** 1.5L VIN B **Transmissions:** All	**ETCS Throttle Control Motor Lock Malfunction Conditions:** Engine started, engine running, and the PCM detected the throttle control motor position was "locked" during normal operation of the throttle control motor during the CCM test. **Possible Causes:** • ETCS throttle motor is damaged or has failed • Throttle body assembly is damaged or stuck • PCM has failed
DTC: P1129 **1T CCM, MIL: Yes** **1998, 1999, 2000, 2001, 2002** **Models:** Supra, Land Cruiser, Sequoia, Tundra **Engines:** 3.0L VIN D, 4.7L VIN T **Transmissions:** All	**Electric Throttle Control System (ETCS) Malfunction Conditions:** Engine started, engine running, and the PCM detected the Actual throttle opening angle varied too much from the Target throttle opening angle during the CCM Rationality test. **Possible Causes:** • ETCS (system) is damaged or has failed • PCM has failed
DTC: P1129 **1T CCM, MIL: Yes** **2001, 2002, 2003, 2004, 2005, 2006** **Models:** Prius **Engines:** 1.5L VIN B **Transmissions:** All	**Electric Throttle Control System (ETCS) Malfunction Conditions:** Engine started, engine running, and the PCM detected the Actual throttle opening angle varied too much from the Target throttle opening angle during the CCM Rationality test. **Possible Causes:** • ETCS (system) is damaged or has failed • PCM has failed
DTC: P1130 **2T O2S, MIL: Yes** **1998, 1999, 2000, 2001, 2002** **Models:** All **Engines:** All **Transmissions:** All	**A/F Sensor-11 (Bank 1 Sensor 1) Circuit Malfunction Conditions:** DTC P1135 not set, engine started, engine warmup completed, and the PCM detected the A/F sensor signal was more than 3.8v or that it was less than 2.8v; or the A/F sensor signal was fixed at 3.30v; or the PCM detected an unexpected "high" or "low" voltage condition in the A/F sensor signal circuit during the CCM test. **Possible Causes:** • A/FS signal circuit or ground circuit is open (intermittent fault) • A/FS is contaminated, damaged or has failed • PCM has failed

DTC	Trouble Code Title, Conditions & Possible Causes
DTC: P1130 2T O2S, MIL: Yes 2003, 2004, 2005, 2006 Models: RAV4 Engines: All Transmissions: All	**A/F Sensor-11 (Bank 1 Sensor 1) Circuit Malfunction Conditions:** DTC P1135 not set, engine started, engine warmup completed, and the PCM detected the A/F sensor signal was more than 3.8v or that it was less than 2.8v; or the A/F sensor signal was fixed at 3.30v; or the PCM detected an unexpected "high" or "low" voltage condition in the A/F sensor signal circuit during the CCM test. **Possible Causes:** • A/FS signal circuit or ground circuit is open (intermittent fault) • A/FS is contaminated, damaged or has failed • PCM has failed
DTC: P1133 1T O2S, MIL: Yes 1998, 1999, 2000, 2001, 2002 Models: All Engines: All Transmissions: All	**A/F Sensor-11 (Bank 1 Sensor 1) Slow Response Conditions:** DTC P1135 not set, vehicle speed over 38 mph at over 1600 rpm for 1-3 minutes, and the PCM detected the A/F sensor response times deteriorated below an acceptable value. **Possible Causes:** • A/F sensor signal circuit or ground circuit is open • A/F sensor power circuit is open • A/F sensor is contaminated, damaged or has failed • Intake air leaks, exhaust manifold leaks or PCV system leaks • MAF sensor out of calibration (it may be dirty or contaminated) • PCM has failed
DTC: P1133 2T O2S, MIL: Yes 2003, 2004, 2005, 2006 Models: RAV4 Engines: All Transmissions: All	**A/F Sensor-11 (Bank 1 Sensor 1) Slow Response Conditions:** DTC P1135 not set, engine started, vehicle driven to a speed of over 38 mph at over 1600 rpm for 1-3 minutes, and the PCM detected the A/F sensor response times deteriorated below an acceptable value. **Possible Causes:** • A/F sensor signal circuit or ground circuit is open • A/F sensor power circuit is open • A/F sensor is contaminated, damaged or has failed • Intake air leaks, exhaust manifold leaks or PCV system leaks • MAF sensor out of calibration (it may be dirty or contaminated) • PCM has failed
DTC: P1135 2T CCM, MIL: Yes 1998, 1999, 2000, 2001, 2002 Models: All Engines: All Transmissions: All	**A/F Sensor-11 (Bank 1 Sensor 1) Heater Circuit Malfunction Conditions:** Engine started, engine running, and the PCM detected the A/F Sensor-11 (A/FS-11) heater current was less than 0.25A, or that it was more than 8.0A at anytime during the CCM test. **Possible Causes:** • A/F sensor heater circuit or ground circuit is open • A/F sensor heater power circuit is open • A/F sensor heater is damaged or has failed • PCM has failed
DTC: P1135 2T CCM, MIL: Yes 2003, 2004, 2005, 2006 Models: RAV4 Engines: All Transmissions: All	**A/F Sensor-11 (Bank 1 Sensor 1) Heater Circuit Malfunction Conditions:** Engine started; and the PCM detected the A/F Sensor-11 (A/FS-11) heater current was less than 0.25A, or that it was more than 8.0A at anytime during the CCM test. **Possible Causes:** • A/F sensor heater circuit or ground circuit is open • A/F sensor heater power circuit is open • A/F sensor heater is damaged or has failed • PCM has failed
DTC: P1150 2T CCM, MIL: Yes 1998, 1999, 2000, 2001, 2002 Models: All Engines: All Transmissions: All	**A/F Sensor-21 (Bank 2 Sensor 1) Circuit Malfunction Conditions:** DTC P1155 not set, engine running in closed loop, and the PCM detected the A/F sensor signal was more than 3.8v or less than 2.8v; or the A/F sensor signal was fixed at 3.30v; or it detected an unexpected "high" or "low" voltage condition on the A/F Sensor signal circuit. **Possible Causes:** • A/FS signal circuit or ground circuit is open (intermittent fault) • A/FS power circuit is open (an intermittent fault) • A/FS is contaminated, damaged or has failed • PCM has failed
DTC: P1150 2T CCM, MIL: Yes 2003, 2004, 2005, 2006 Models: RAV4 Engines: All Transmissions: All	**A/F Sensor-21 (Bank 2 Sensor 1) Circuit Malfunction Conditions:** DTC P1155 not set; engine running in closed loop, and the PCM detected the A/F sensor signal was more than 3.8v or less than 2.8v; or the A/F sensor signal was fixed at 3.30v; or it detected an unexpected "high" or "low" voltage condition on the A/F Sensor signal circuit. **Possible Causes:** • A/FS signal circuit or ground circuit is open (intermittent fault) • A/FS power circuit is open (an intermittent fault) • A/FS is contaminated, damaged or has failed • PCM has failed

DTC	Trouble Code Title, Conditions & Possible Causes
DTC: P1153 2T O2S, MIL: Yes 1998, 1999, 2000, 2001, 2002 **Models:** All **Engines:** All **Transmissions:** All	**A/F Sensor-21 (Bank 2 Sensor 1) Slow Response Conditions:** DTC P1155 not set; vehicle speed over 38 mph at over 1600 rpm for 1-3 minutes, and the PCM detected the A/FS-21 response time had deteriorated below an acceptable value. **Possible Causes:** • A/F sensor signal circuit or ground circuit is open • A/F sensor power circuit is open • A/F sensor is contaminated, damaged or has failed • Intake air leaks, exhaust manifold leaks or PCV system leaks • MAF sensor out of calibration (it may be dirty or contaminated)
DTC: P1153 2T O2S, MIL: Yes 2003, 2004, 2005, 2006 **Models:** RAV4 **Engines:** All **Transmissions:** All	**A/F Sensor-21 (Bank 2 Sensor 1) Slow Response Conditions:** DTC P1155 not set; vehicle speed over 38 mph at over 1600 rpm for 1-3 minutes, and the PCM detected the A/FS-21 response time had deteriorated below an acceptable value. **Possible Causes:** • A/F sensor signal circuit or ground circuit is open • A/F sensor power circuit is open • A/F sensor is contaminated, damaged or has failed • Intake air leaks, exhaust manifold leaks or PCV system leaks • MAF sensor out of calibration (it may be dirty or contaminated) • PCM has failed
DTC: P1155 2T CCM, MIL: Yes 1998, 1999, 2000, 2001, 2002 **Models:** All **Engines:** All **Transmissions:** All	**A/F Sensor-21 (Bank 2 Sensor 1) Heater Circuit Malfunction Conditions:** Engine started, engine running, and the PCM detected the A/F Sensor-21 (A/FS-21) heater current was less than 0.25A, or that it was more than 8.0A at anytime during the CCM test. **Possible Causes:** • A/F sensor heater circuit or ground circuit is open • A/F sensor heater power circuit is open • A/F sensor heater is damaged or has failed • PCM has failed
DTC: P1155 2T CCM, MIL: Yes 2003, 2004, 2005, 2006 **Models:** RAV4 **Engines:** All **Transmissions:** All	**A/F Sensor-21 (Bank 2 Sensor 1) Heater Circuit Malfunction Conditions:** Engine started, engine running, and the PCM detected the A/F Sensor-21 (A/FS-21) heater current was less than 0.25A, or that it was more than 8.0A at anytime during the CCM test. **Possible Causes:** • A/F sensor heater circuit or ground circuit is open • A/F sensor heater power circuit is open • A/F sensor heater is damaged or has failed • PCM has failed
DTC: P1200 2T CCM, MIL: Yes 1996, 1997 **Models:** Supra **Engines:** All **Transmissions:** All	**Fuel Pump Relay/ECU Circuit Malfunction Conditions:** Engine speed less than 1000 rpm; and the PCM detected an unexpected voltage condition on the fuel pump, fuel pump relay ECU input circuit, or on the fuel pump Diagnostic circuit. **Possible Causes:** • Fuel Pump ECU control circuit is open or shorted to ground • Fuel Pump ECU control circuit is shorted to system power (B+) • Fuel Pump ECU power circuit is open (check power from relay) • Fuel Pump ECU is damaged or has failed • Fuel Pump is damaged or has failed • PCM has failed
DTC: P1300 1T CCM, MIL: Yes 1995 **Models:** Avalon, Camry, Land Cruiser, Previa, T100, Tacoma, Tercel **Engines:** 1.5L VIN E, 2.4L VIN A, 2.4L VIN R, 2.7L VIN U, 3.0L VIN G, 3.4L VIN V, 4.5L VIN D **Transmissions:** All	**Igniter Circuit Malfunction Conditions:** Engine started; and the PCM did not detect any IGF signals after detecting two IGT signals. **Note: The fault must occur for 3 IGF signals after 6 IGT signals in order to set this code.** **Possible Causes:** • IGT or IGF signal circuit is open or shorted to ground • Igniter is damaged or has failed • PCM has failed
DTC: P1300 1T CCM, MIL: Yes 1996, 1997, 1998, 1999 **Models:** Celica, Camry **Engines:** 1.8L VIN B, 2.2L VIN G **Transmissions:** All	**Igniter Circuit Malfunction Conditions:** Engine started, engine running, and the PCM did not detect any IGF signals after detecting four IGT1 signals during the CCM test. **Possible Causes:** • IGT or IGF signal circuit is open between the igniter and PCM • IGT or IGF signal circuit is shorted to ground • Igniter is damaged or has failed • PCM has failed

DTC	Trouble Code Title, Conditions & Possible Causes
DTC: P1300 **1T CCM, MIL: Yes** 2000, 2001, 2002 **Models:** Camry, Camry Solara **Engines:** 2.2L VIN G, 2.2L VIN N **Transmissions:** All	**Igniter No. 1 Circuit Malfunction Conditions:** Engine started, engine running, and the PCM did not detect any IGF signals after detecting four IGT1 signals during the CCM test. **Possible Causes:** • IGT or IGF signal circuit is open between the igniter and PCM • IGT or IGF signal circuit is shorted to ground • Igniter is damaged or has failed • PCM has failed
DTC: P1300 **1T CCM, MIL: Yes** 1997, 1998, 1999, 2000 **Models:** 4Runner, RAV4, T100, Tacoma 2.0L VIN K, 2.0L VIN P, 2.2L VIN S, **Engines:** 2.4L VIN L, 2.7L VIN M **Transmissions:** All	**Igniter No. 1 Circuit Malfunction Conditions:** Engine started; and the PCM did not detect any IGF signals after detecting four IGT1 signals during the CCM test. **Possible Causes:** • IGT or IGF signal circuit is open between the igniter and PCM • IGT or IGF signal circuit is shorted to ground • Igniter is damaged or has failed • PCM has failed
DTC: P1300 **1T CCM, MIL: Yes** 1996, 1997, 1998, 1999 **Models:** Avalon, Land Cruiser, Previa, T100, Tacoma, Tercel **Engines:** All **Transmissions:** All	**Igniter Circuit Malfunction Conditions:** Engine started; and the PCM did not detect any IGF signals after detecting two or more IGT signals during the CCM test. **Note: This problem must occur for 3 IGF signals after 6 IGT signals in order to set this trouble code and activate the MIL.** **Possible Causes:** • IGT or IGF signal circuit is open or shorted to ground • Igniter is damaged or has failed • PCM has failed
DTC: P1300 **1T CCM, MIL: Yes** 1997, 1998, 1999, 2000, 2001, 2002 **Models:** Camry, Camry Solara, 4Runner **Engines:** 3.0L VIN F, 3.4L VIN N **Transmissions:** All	**Igniter Circuit Malfunction Conditions:** Engine started, engine running, and the PCM did not detect any IGF signals after detecting four IGT1 signals during the CCM test. **Possible Causes:** • IGT or IGF signal circuit is open between the igniter and PCM • IGT or IGF signal circuit is shorted to ground • Igniter is damaged or has failed • PCM has failed
DTC: P1300 **1T CCM, MIL: Yes** 2000, 2001, 2002 **Models:** Avalon, Echo, Yaris, MR2, Highlander, Prius, Sienna, Land Cruiser, Sequoia, Tundra **Engines:** All **Transmissions:** All	**Igniter No. 1 Circuit Malfunction Conditions:** Engine started, engine running, and the PCM did not detect any IGF1 signals after detecting two IGT1 signals during the CCM test. **Note: This vehicle uses a Coil-On-Plug design Ignition System.** **Possible Causes:** • IGT or IGF signal circuit is open or shorted to ground • Igniter is damaged or has failed • PCM has failed
DTC: P1300 **1T CCM, MIL: Yes** 2001, 2002, 2003, 2004, 2005, 2006 **Models:** RAV4 **Engines:** 2.0L VIN K **Transmissions:** All	**Igniter No. 1 Circuit Malfunction Conditions:** Engine started, engine running, and the PCM did not detect any IGF1 signals after detecting two IGT1 signals during the CCM test. **Possible Causes:** • IGF circuit is open or shorted to ground • IGT1 circuit is open or shorted to ground • Igniter is damaged or has failed • PCM has failed
DTC: P1300 **1T CCM, MIL: Yes** 1998, 1999, 2000 **Models:** Sienna **Engines:** 3.0L VIN F **Transmissions:** All	**Igniter No. 1 Circuit Malfunction Conditions:** Engine started; and the PCM did not detect any IGF1 signals after detecting 4 IGT1 signals. **Possible Causes:** • IGT or IGF signal circuit is open between the igniter and PCM • IGT or IGF signal circuit is shorted to ground • Igniter is damaged or has failed • PCM has failed
DTC: P1300 **1T CCM, MIL: Yes** 2000, 2001, 2002 **Models:** Tacoma **Engines:** 2.4L VIN L **Transmissions:** All	**Igniter No. 1 Circuit Malfunction Conditions:** Engine started; and the PCM did not detect any IGF1 signals after detecting 2 IGT1 signals. **Possible Causes:** • IGT or IGF signal circuit is open or shorted to ground • Igniter is damaged or has failed • PCM has failed

DTC	Trouble Code Title, Conditions & Possible Causes
DTC: P1300 **1T CCM, MIL: Yes** **2000, 2001, 2002, 2003, 2004, 2005, 2006** **Models:** Celica, Corolla, Matrix, Prius **Engines:** All **Transmissions:** All	**Igniter No. 1 Circuit Malfunction Conditions:** Engine started, engine running, and the PCM did not detect any IGF signals after detecting four IGT3 signals. This engine uses a Direct Ignition (DI) system where one coil is used to fire one cylinder. The coil high-energy secondary wire is connected to one spark plug. If P1300 to P1315 are all set, check for an open or shorted IGF circuit. **Possible Causes:** • IGT or IGF1 signal circuit is open between the igniter and PCM • IGT or IGF1 signal circuit is shorted to ground • Igniter is damaged or has failed • PCM has failed
DTC: P1305 **1T CCM, MIL: Yes** **2000, 2001, 2002** **Models:** Avalon, Echo, Yaris, MR2, Highlander, Sienna, Land Cruiser, Sequoia, Tundra **Engines:** All **Transmissions:** All	**Igniter No. 2 Circuit Malfunction Conditions:** Engine started; and the PCM did not detect any IGF2 signals after detecting two IGT2 signals. **Note: This vehicle uses a Coil-On-Plug design Ignition System.** **Possible Causes:** • IGT or IGF signal circuit is open or shorted to ground • Igniter is damaged or has failed • PCM has failed
DTC: P1305 **1T CCM, MIL: Yes** **2000, 2001, 2002, 2003, 2004, 2005, 2006** **Models:** Celica, Corolla, Matrix, Prius **Engines:** All **Transmissions:** All	**Igniter No. 2 Circuit Malfunction Conditions:** Engine started; and the PCM did not detect any IGF signals after detecting four IGT3 signals. This engine uses a Direct Ignition (DI) system where one coil is used to fire one cylinder. The coil high-energy secondary wire is connected to one spark plug. If P1300 to P1315 are both set, check for an open or shorted condition in the IGF circuit. **Possible Causes:** • IGT or IGF2 signal circuit is open between the igniter and PCM • IGT or IGF2 signal circuit is shorted to ground • Igniter is damaged or has failed • PCM has failed
DTC: P1305 **1T CCM, MIL: Yes** **2001, 2002, 2003, 2004, 2005, 2006** **Models:** RAV4 **Engines:** 2.0L VIN K **Transmissions:** All	**Igniter No. 2 Circuit Malfunction Conditions:** Engine started; and the PCM did not detect any IGF2 signals after detecting 2 IGT2 signals. **Possible Causes:** • IGF circuit is open or shorted to ground • IGT2 circuit is open or shorted to ground • Igniter is damaged or has failed • PCM has failed
DTC: P1305 **1T CCM, MIL: Yes** **2000, 2001, 2002** **Models:** Tacoma **Engines:** 2.4L VIN L **Transmissions:** All	**Igniter No. 2 Circuit Malfunction Conditions:** Engine started; and the PCM did not detect any IGF signals after detecting 2 IGT2 signals. **Possible Causes:** • IGT or IGF signal circuit is open or shorted to ground • Igniter is damaged or has failed • PCM has failed
DTC: P1310 **1T CCM, MIL: Yes** **1997, 1998, 1999, 2000** **Models:** 4Runner, RAV4, T100, Tacoma **Engines:** 2.0L VIN K, 2.0L VIN P, 2.2L VIN S, 2.7L VIN M **Transmissions:** All	**Igniter No. 2 Circuit Malfunction Conditions:** Engine started; and the PCM did not detect any IGF signals after detecting 4 IGT2 signals. **Possible Causes:** • IGT or IGF signal circuit is open between the igniter and PCM • IGT or IGF signal circuit is shorted to ground • Igniter is damaged or has failed • PCM has failed
DTC: P1310 **1T CCM, MIL: Yes** **2001, 2002, 2003, 2004, 2005, 2006** **Models:** RAV4 **Engines:** 2.0L VIN K **Transmissions:** All	**Igniter No. 3 Circuit Malfunction Conditions:** Engine started; and the PCM did not detect any IGF3 signals after detecting 2 IGT3 signals. **Possible Causes:** • IGF circuit is open or shorted to ground • IGT3 circuit is open or shorted to ground • Igniter is damaged or has failed • PCM has failed
DTC: P1310 **1T CCM, MIL: Yes** **1998, 1999, 2000** **Models:** Sienna **Engines:** 3.0L VIN F **Transmissions:** All	**Igniter No. 2 Circuit Malfunction Conditions:** Engine started; and the PCM did not detect any IGF2 signals after detecting 4 IGT2 signals. **Possible Causes:** • IGT or IGF signal circuit is open between the igniter and PCM • IGT or IGF signal circuit is shorted to ground • Igniter has failed, or the PCM has failed

DTC	Trouble Code Title, Conditions & Possible Causes
DTC: P1310 **1T CCM, MIL: Yes** **2000, 2001, 2002** **Models:** Tacoma **Engines:** 2.4L VIN L **Transmissions:** All	**Igniter No. 3 Circuit Malfunction Conditions:** Engine started; and the PCM did not detect any IGF3 signals after detecting 2 IGT3 signals. **Possible Causes:** • IGT or IGF signal circuit is open or shorted to ground • Igniter is damaged or has failed • PCM has failed
DTC: P1310 **1T CCM, MIL: Yes** **2000, 2001, 2002** **Models:** Avalon, Echo, Yaris, MR2, Highlander, Sienna, Land Cruiser, Sequoia, Tundra **Engines:** All **Transmissions:** All	**Igniter No. 3 Circuit Malfunction Conditions:** Engine started; and the PCM did not detect any IGF3 signals after detecting two IGT3 signals during the test. **Note: This vehicle uses a Coil-On-Plug design Ignition System.** **Possible Causes:** • IGT or IGF signal circuit is open or shorted to ground • Igniter is damaged or has failed •
DTC: P1310 **1T CCM, MIL: Yes** **2000, 2001, 2002, 2003, 2004, 2005, 2006** **Models:** Celica, Corolla, Matrix, Prius **Engines:** All **Transmissions:** All	**Igniter No. 3 Circuit Malfunction Conditions:** Engine started; and the PCM did not detect any IGF signals after detecting four IGT3 signals. This engine uses a Direct Ignition (DI) system where one coil is used to fire one cylinder. The coil high-energy secondary wire is connected to one spark plug. If P1300 to P1315 are all set, check for an open or shorted IGF circuit. **Possible Causes:** • IGT or IGF3 signal circuit is open between the igniter and PCM • IGT or IGF3 signal circuit is shorted to ground • Igniter is damaged or has failed • PCM has failed
DTC: P1315 **1T CCM, MIL: Yes** **2000, 2001, 2002** **Models:** Tacoma **Engines:** 2.4L VIN L **Transmissions:** All	**Igniter No. 4 Circuit Malfunction Conditions:** Engine started; and the PCM did not detect any IGF4 signals after detecting 2 IGT4 signals. **Possible Causes:** • IGT or IGF signal circuit is open or shorted to ground • Igniter is damaged or has failed • PCM has failed
DTC: P1315 **1T CCM, MIL: Yes** **2001, 2002, 2003, 2004, 2005, 2006** **Models:** RAV4 **Engines:** 2.0L VIN K **Transmissions:** All	**Igniter No. 4 Circuit Malfunction Conditions:** Engine started; and the PCM did not detect any IGF4 signals after detecting 2 IGT4 signals. **Possible Causes:** • IGF circuit is open or shorted to ground • IGT4 circuit is open or shorted to ground • Igniter is damaged or has failed • PCM has failed
DTC: P1315 **1T CCM, MIL: Yes** **2000, 2001, 2002** **Models:** Avalon, Echo, Yaris, MR2, Highlander, Sienna, Land Cruiser, Sequoia, Tundra **Engines:** All **Transmissions:** All	**Igniter No. 4 Circuit Malfunction Conditions:** Engine started; and the PCM did not detect any IGF4 signals after detecting 2 IGT4 signals during the CCM test. **Note: This vehicle uses a Coil-On-Plug design Ignition System.** **Possible Causes:** • IGT or IGF signal circuit is open or shorted to ground • Igniter is damaged or has failed • PCM has failed
DTC: P1315 **1T CCM, MIL: Yes** **2000, 2001, 2002, 2003, 2004, 2005, 2006** **Models:** Celica, Corolla, Matrix, Prius **Engines:** All **Transmissions:** All	**Igniter No. 4 Circuit Malfunction Conditions:** Engine started; and the PCM did not detect any IGF signals after detecting four IGT4 signals. This engine uses a Direct Ignition (DI) system where one coil is used to fire one cylinder. The coil high-energy secondary wire is connected to one spark plug. If P1300 to P1315 are all set, check for an open or shorted IGF circuit. **Possible Causes:** • IGT or IGF4 signal circuit is open between the igniter and PCM • IGT or IGF4 signal circuit is shorted to ground • Igniter is damaged or has failed • PCM has failed
DTC: P1320 **1T CCM, MIL: Yes** **2000, 2001, 2002** **Models:** Avalon, Highlander, Land Cruiser, Sequoia, Sienna, Tundra **Engines:** 3.0L VIN F, 4.7L VIN T **Transmissions:** All	**Igniter No. 5 Circuit Malfunction Conditions:** Engine started; and the PCM did not detect any IGF5 signals after detecting IGT5 signals during the CCM test. **Note: This vehicle uses a Coil-On-Plug design Ignition System.** **Possible Causes:** • IGT or IGF signal circuit is open between the igniter and PCM • IGT or IGF signal circuit is shorted to ground • Igniter is damaged or has failed • PCM has failed

DTC	Trouble Code Title, Conditions & Possible Causes
DTC: P1325 **1T CCM, MIL: Yes** **2000, 2001, 2002** **Models:** Avalon, Highlander, Land Cruiser, Sequoia, Sienna, Tundra **Engines:** 3.0L VIN F, 4.7L VIN T **Transmissions:** All	**Igniter No. 6 Circuit Malfunction Conditions:** Engine started; and the PCM did not detect any IGF6 signals after detecting IGT6 signals during the CCM test. **Note: This vehicle uses a Coil-On-Plug design Ignition System.** **Possible Causes:** • IGT or IGF signal circuit is open between the igniter and PCM • IGT or IGF signal circuit is shorted to ground • Igniter is damaged or has failed • PCM has failed
DTC: P1330 **1T CCM, MIL: Yes** **1998, 1999, 2000, 2001, 2002** **Models:** Land Cruiser, Sequoia, Tundra **Engines:** 4.7L VIN T **Transmissions:** All	**Igniter No. 7 Circuit Malfunction Conditions:** Engine started; and the PCM did not detect any IGF7 signals after detecting IGT7 signals during the CCM test. **Note: This vehicle uses a Coil-On-Plug design Ignition System.** **Possible Causes:** • IGT or IGF signal circuit is open between the igniter and PCM • IGT or IGF signal circuit is shorted to ground • Igniter is damaged or has failed • PCM has failed
DTC: P1335 **1T CCM, MIL: Yes** **1995** **Models:** Avalon, Camry, Land Cruiser, Previa, T100, Tacoma, Tercel **Engines:** 1.5L VIN E, 2.4L VIN A, 2.4L VIN R, 2.7L VIN U, 3.0L VIN G, 3.4L VIN V, 4.5L VIN D **Transmissions:** All	**Crankshaft Position Sensor Circuit Malfunction Conditions:** Engine speed over 1000 rpm, and the PCM did not detect any CKP (NE) sensor signals. **Possible Causes:** • CKP sensor (+) circuit is open or shorted to ground • CKP sensor (-) circuit is open or shorted to ground • CKP sensor (+), (-) circuit is shorted to VREF or system power • CKP sensor is damaged or has failed • PCM has failed
DTC: P1335 **1T CCM, MIL: Yes** **1996, 1997, 1998, 1999, 2000, 2001, 2002, 2003, 2004, 2005, 2006** **Models:** All **Engines:** All **Transmissions:** All	**Crankshaft Position Sensor Circuit Malfunction Conditions:** Engine speed over 1000 rpm, and the PCM did not detect any CKP (NE) sensor signals for 500 ms. This code may be set by an intermittent loss of the CKP sensor signal. **Possible Causes:** • CKP sensor (+) circuit is open or shorted to ground • CKP sensor (-) circuit is open or shorted to ground • CKP sensor (+), (-) circuit is shorted to VREF or system power • CKP sensor is damaged or has failed • PCM has failed
DTC: P1340 **1T CCM, MIL: Yes** **1998, 1999, 2000, 2001, 2002** **Models:** Land Cruiser, Sequoia, Tundra **Engines:** 4.7L VIN T **Transmissions:** All	**Igniter No. 8 Circuit Malfunction Conditions:** Engine started; and the PCM did not detect any IGF8 signals after detecting IGT8 signals during the CCM test. **Note: This vehicle uses a Coil-On-Plug design Ignition System.** **Possible Causes:** • IGT or IGF signal circuit is open between the igniter and PCM • IGT or IGF signal circuit is shorted to ground • Igniter is damaged or has failed • PCM has failed
DTC: P1345 **1T CCM, MIL: Yes** **2000, 2001, 2002** **Models:** Echo, Yaris **Engines:** All **Transmissions:** All	**Variable Valve Timing, CMP Sensor Circuit Malfunction Conditions:** Engine cranking for 4 seconds, and the PCM did not detect a signal from the VVT sensor; or with the engine speed over 600 rpm, the PCM did not detect VVT signal for 5 seconds. **Possible Causes:** • VVT sensor (+) signal circuit is open or shorted to ground • VVT sensor (-) signal circuit is open or shorted to ground • VVT sensor is damaged or has failed • PCM has failed
DTC: P1346 **1T CCM, MIL: Yes** **2000, 2001, 2002, 2003, 2004, 2005, 2006** **Models:** Echo, Yaris, Celica, Corolla, Highlander, Matrix, MR2, Prius, RAV4 **Engines:** All **Transmissions:** All	**Variable Valve Timing, CMP Sensor Range/Performance Conditions:** Engine started; and the PCM detected a deviation between the CKP sensor and the VVT sensor signals due to a mechanical fault in the timing belt or the VVT sensor during testing. **Possible Causes:** • Worn timing belt (i.e., the belt may be skipping teeth) • PCM has failed

DTC	Trouble Code Title, Conditions & Possible Causes
DTC: P1349 **1T CCM, MIL: Yes** **2000, 2001, 2002, 2003, 2004, 2005, 2006** **Models:** Echo, Yaris, Celica, Corolla, Matrix, MR2, Prius, RAV4 **Engines:** All **Transmissions:** All	**Variable Valve Timing System (Bank 1) Conditions:** Engine speed from 400-4000 rpm in closed loop, and the PCM detected the valve timing did not change from its initial position, or the current valve timing was fixed during the CCM test. **Possible Causes:** • Oil control valve is damaged or has failed • Valve timing is not correct • PCM has failed
DTC: P1400 **1T CCM, MIL: Yes** **1995, 1996, 1997, 1998** **Models:** Supra **Engines:** 3.0L VIN E **Transmissions:** All	**Sub-Throttle Position Sensor Circuit Malfunction Conditions:** Engine started; and the PCM detected the Sub-Throttle Position sensor (VTA2) signal was less than 0.25v with the Closed Throttle Position switch "off", or more than 4.90v at any time. **Possible Causes:** • VTA2 signal circuit is open or shorted to ground • Sub-Throttle position sensor VREF or ground circuit is open • Sub-Throttle position sensor is damaged or has failed • PCM has failed
DTC: P1401 **2T CCM, MIL: Yes** **1995, 1996, 1997, 1998** **Models:** Supra **Engines:** 3.0L VIN E **Transmissions:** All	**Sub-Throttle Position Sensor Range/Performance Conditions:** Vehicle speed over 3 mph, and the PCM detected the difference between the Throttle Position sensor angle and Sub-Throttle Position sensor angle was over 35 degrees. **Possible Causes:** • Sub-Throttle position sensor is damaged or has failed • Throttle linkage or throttle body is binding or sticking • PCM has failed
DTC: P1405 **2T CCM, MIL: Yes** **1995, 1996, 1997, 1998** **Models:** Supra **Engines:** 3.0L VIN E **Transmissions:** All	**Turbo Pressure Sensor Circuit Malfunction Conditions:** Engine started, engine running, and the PCM detected an unexpected voltage condition on the Turbo Boost Pressure sensor circuit during the CCM test. **Possible Causes:** • Boost Pressure sensor signal circuit open or shorted to ground • Boost Pressure sensor power or ground circuit is open • Boost Pressure sensor is damaged or has failed • PCM has failed
DTC: P1406 **2T CCM, MIL: Yes** **1995, 1996, 1997, 1998** **Models:** Supra **Engines:** 3.0L VIN E **Transmissions:** All	**Turbo Pressure Sensor Circuit Malfunction Conditions:** Engine started, engine running, MAF sensor more than 1.3 g/sec, and the PCM detected the Turbo Pressure sensor signal was less than 1.2v; or with the MAF sensor less than 0.45 g/sec, the Turbo Pressure sensor indicated more than 4.2v during the CCM test. **Possible Causes:** • Boost Pressure sensor signal circuit open or shorted to ground • Boost Pressure sensor power or ground circuit is open • Boost Pressure sensor is damaged or has failed • PCM has failed
DTC: P1410 **2T CCM, MIL: Yes** **1998, 1999, 2000, 2001, 2002** **Models:** Avalon, Camry, Camry Solara **Engines:** 3.0L VIN F **Transmissions:** All	**EGR Valve Position Sensor Circuit Malfunction Conditions:** Key on or engine running; and the PCM detected an unexpected voltage condition on the EGR EVP sensor circuit during the test. **Possible Causes:** • EVP sensor signal circuit is open between sensor and the PCM • EVP sensor signal circuit is shorted to ground • EVP sensor power (VREF) circuit is open • EVP sensor ground circuit is open • EVP sensor is damaged or has failed • PCM has failed
DTC: P1411 **2T CCM, MIL: Yes** **1998, 1999, 2000, 2001, 2002** **Models:** Avalon, Camry, Camry Solara **Engines:** 3.0L VIN F **Transmissions:** All	**EGR Valve Position Sensor Range/Performance Conditions:** Engine started; ECT sensor signal below 131°F, and the PCM detected the EGR sensor signal was under 0.35v or it indicated a value equal to or greater than 1.65v for 7 seconds. **Possible Causes:** • EVP sensor signal circuit is open between sensor and the PCM • EVP sensor signal circuit is shorted to ground • EVP sensor power (VREF) circuit is open • EVP sensor ground circuit is open • EVP sensor is damaged or has failed • PCM has failed

DTC	Trouble Code Title, Conditions & Possible Causes
DTC: P1430 1T CCM, MIL: Yes **2001, 2002, 2003, 2004, 2005, 2006** Models: Prius Engines: All Transmissions: A/T	**Vacuum Sensor for Absorber & Catalyst System Circuit Malfunction Conditions:** Engine started; and the PCM detected an unexpected voltage condition on the Vacuum Sensor Absorber circuit. **Note: The Scan Tool reads 0 kPa or over 130 kPa if this code sets.** **Possible Causes:** • Vacuum sensor signal circuit is open or shorted to ground • Vacuum sensor signal circuit is shorted to VREF • Vacuum sensor power (VREF) circuit is open • Vacuum sensor is damaged or has failed • PCM has failed
DTC: P1431 2T CCM, MIL: Yes **2001, 2002, 2003, 2004, 2005, 2006** Models: Prius Engines: All Transmissions: A/T	**Vacuum Sensor for Absorber & Catalyst Performance Conditions:** Key on, engine stopped, ECT sensor more than 32°F, VSV for HC Absorber "off", and the PCM detected the PIM input was less than 1.20v; or engine running at over 1000 rpm, VSV for the HC Absorber "on", ECT sensor more than 32°F, and the PCM detected the PIM input indicated more than 3.96v. **Note: If DTC P0110, P0115, P0120, P0121, P1430 and P1431 are set at the same time, the "common" ground circuit may be open.** **Possible Causes:** • Vacuum sensor line (vacuum line) is damaged or disconnected • Vacuum sensor is damaged or has failed • PCM has failed
DTC: P1436 1T CCM, MIL: Yes **2001, 2002, 2003, 2004, 2005, 2006** Models: Prius Engines: All Transmissions: A/T	**Variable Valve Malfunction Conditions:** Cold engine startup (ECT sensor from 14-104°F), engine running, ECT sensor more than 113°F, and the PCM detected the Bypass valve operation was not performed correctly during the CCM test. **Possible Causes:** • Front exhaust pipe is damaged or leaking • Variable Valve actuator is damaged or has failed • Vacuum line to the Variable Valve actuator is damaged or off
DTC: P1437 1T CCM, MIL: Yes **2001, 2002, 2003, 2004, 2005, 2006** Models: Prius Engines: All Transmissions: A/T	**Vacuum Line Malfunction Conditions:** ECT sensor signal from 14-104°F at startup (cold engine); ECT sensor over 113°F, engine load factor more than 30%, and the PCM detected an unusual negative pressure amount. **Possible Causes:** • Check valve is clogged, damaged or has failed • Vacuum sensor line (vacuum line) is damaged or disconnected • Vacuum sensor is damaged or has failed • VSV for the HC Absorber and Catalyst System has failed
DTC: P1455 1T CCM, MIL: Yes **2001, 2002, 2003, 2004, 2005, 2006** Models: Prius Engines: All Transmissions: A/T	**Vapor Reducing Fuel Tank Small Leak Detected Conditions:** Engine started, engine running, VSV for the Purge Flow switching valve "on", and the PCM detected the value of the vapor density of the air that flows from the EVAP VSV (vacuum switching valve) to the intake manifold was too high. **Possible Causes:** • Fuel system component problems • HO2S is contaminated, damaged or has failed • Hose and/or pipe is damaged or has failed • Ignition system component problems • Mass airflow meter is damaged or has failed • VVT System is damaged or has failed • PCM has failed
DTC: P1500 1T CCM, MIL: Yes **1995** Models: Avalon, Camry, Land Cruiser, Previa, T100, Tacoma, Tercel Engines: 1.5L VIN E, 2.4L VIN A, 2.4L VIN R, 2.7L VIN U, 3.0L VIN G, 3.4L VIN V, 4.5L VIN D Transmissions: All	**Starter Signal Circuit Malfunction Conditions:** Engine cranking; and the PCM detected that it did not receive a signal from the starter signal circuit during the test period. **Possible Causes:** • Engine starter signal circuit is open or shorted to ground • Engine starter signal circuit is shorted to system power (B+) • Starter has failed • PCM has failed
DTC: P1500 1T CCM, MIL: Yes **1996, 1997, 1998, 1999, 2000, 2001, 2002** Models: All Engines: All Transmissions: All	**Starter Signal Circuit Malfunction Conditions:** Engine cranking and the PCM detected it did not receive a signal on the starter signal circuit. **Possible Causes:** • Engine starter signal circuit is open or shorted to ground • Engine starter signal circuit is shorted to system power (B+) • Starter has failed • PCM has failed

DTC	Trouble Code Title, Conditions & Possible Causes
DTC: P1510 **1T CCM, MIL: Yes** **1996, 1997** **Models:** Previa **Engines:** All **Transmissions:** A/T	**Boost Pressure Control Circuit Malfunction (SC Enabled) Conditions:** Engine running with the Super magnetic clutch relay "on", and the PCM detected the Intake Air volume flow rate was too high or too low, or with the engine speed at over 2800 rpm, it detected the Intake Air volume flow rate was too low. **Possible Causes:** • Supercharger magnetic clutch relay circuit open or shorted • Supercharger magnetic clutch control circuit open or shorted • Supercharger magnetic clutch relay or magnetic clutch is damaged or has failed • Supercharger bypass valve had failed, or the PCM has failed
DTC: P1511 **2T CCM, MIL: Yes** **1995, 1996, 1997, 1998** **Models:** Supra **Engines:** 3.0L VIN E **Transmissions:** All	**Turbo Boost Pressure Low Malfunction Conditions:** Engine speed more than 2600 rpm in closed loop, then during a WOT event, the PCM detected +740 mmHg or more of intake pipe pressure during the CCM Rationality test. **Possible Causes:** • Actuator for the Waste Gate valve, Intake Air Control valve, Exhaust Gas Control valve is damaged, binding or has failed • Intake Air system is clogged or leaking • VSV control circuit for the Waste Gate valve, Intake Air Control valve, Exhaust Gas Control valve is open • PCM has failed
DTC: P1512 **2T CCM, MIL: Yes** **1995, 1996, 1997, 1998** **Models:** Supra **Engines:** 3.0L VIN E **Transmissions:** All	**Turbo Boost Pressure High Malfunction Conditions:** Engine speed less than 3400 rpm in closed loop, then during a WOT event, the PCM detected +150 mmHg or less of intake pipe pressure during the CCM Rationality test. **Possible Causes:** • Waste Gate valve, Intake Air Control valve, Exhaust Gas Control valve has failed • Intake Air system is clogged or leaking • VSV control circuit for the Waste Gate valve, Intake Air Control valve, Exhaust Gas Control valve is shorted to ground • PCM has failed
DTC: P1520 **1T CCM, MIL: Yes** **1998, 1999, 2000, 2001, 2002,** **2003, 2004, 2005, 2006** **Models:** All **Engines:** All **Transmissions:** All	**Stop Light Switch Circuit Malfunction Conditions:** Vehicle driven to a speed of over 19 mph several times; and the PCM detected the Stop Light signal status did not change at least once under these operating conditions. **Possible Causes:** • Stop light switch is shorted to ground • Stop light switch is damaged or has failed • PCM has failed
DTC: P1525 **1T CCM, MIL: Yes** **2001, 2002, 2003, 2004, 2005,** **2006** **Models:** Prius **Engines:** All **Transmissions:** A/T	**Resolver Circuit Malfunction Conditions:** Engine started, vehicle driven to a speed of over 12 mph for at least 16 seconds, and the PCM did not detect any vehicle speed signals from the HV ECU SPDO circuit during the CCM test. **Possible Causes:** • HV ECU SPDO circuit is open or shorted to ground • HV ECU is damaged or has failed • PCM has failed
DTC: P1525 **1T CCM, MIL: Yes** **1999** **Models:** Land Cruiser **Engines:** All **Transmissions:** All	**Cruise Control Switch Circuit Malfunction Conditions:** Engine started, vehicle driven to a speed of over 35 mph, and the PCM detected an unexpected voltage condition on the Cruise Control switch circuit during the CCM test. **Possible Causes:** • Cruise control switch signal circuit is shorted to ground • Cruise control switch is damaged or has failed • PCM has failed
DTC: P1566 **1T CCM, MIL: Yes** **1999** **Models:** Land Cruiser **Engines:** All **Transmissions:** All	**Cruise Control Main Switch Circuit Malfunction Conditions:** Engine started, vehicle driven to a speed of over 35 mph, and the PCM detected an unexpected voltage condition on the Cruise Control Main switch circuit during the CCM test. **Possible Causes:** • Cruise control Main switch signal circuit is open • Cruise control Main switch is damaged or has failed • PCM has failed
DTC: P1600 **1T CCM, MIL: Yes** **1995** **Models:** Avalon, Camry, Land Cruiser, Previa, T100, Tacoma, Tercel **Engines:** 1.5L VIN E, 2.4L VIN A, 2.4L VIN R, 2.7L VIN U, 3.0L VIN G, 3.4L VIN V, 4.5L VIN D **Transmissions:** All	**PCM Battery Backup Circuit Malfunction Conditions:** Key on, and the PCM detected an unexpected voltage in the Battery Backup circuit (KAM circuit) during the CCM test. **Note:** The PCM will not store any other codes with this code set. **Possible Causes:** • Battery backup circuit is open (check EFI fuse and fuse link) • Battery terminals are corroded or loose • PCM has failed

DTC	Trouble Code Title, Conditions & Possible Causes
DTC: P1600 **1T CCM, MIL: Yes** **1996, 1997, 1998, 1999, 2000, 2001, 2002, 2003, 2004, 2005, 2006** **Models:** All **Engines:** All **Transmissions:** All	**PCM Battery Backup Circuit Malfunction Conditions:** Key on, and the PCM detected an unexpected voltage in the Battery Backup circuit (KAM circuit) during the CCM test. **Note: The PCM will not store any other codes with this code set.** **Possible Causes:** • Battery backup circuit is open (check EFI fuse and fuse link) • Battery terminals are corroded or loose • PCM has failed
DTC: P1605 **1T CCM, MIL: Yes** **1995, 1996** **Models:** Avalon, Camry, Land Cruiser, Previa, T100, Tacoma, Tercel **Engines:** 1.5L VIN E, 2.4L VIN A, 2.4L VIN R, 2.7L VIN U, 3.0L VIN G, 3.4L VIN V, 4.5L VIN D **Transmissions:** All	**Knock Control CPU Malfunction Conditions:** Engine started, engine running, and the PCM detected a problem in the Knock Control portion of the controller during the test period. **Possible Causes:** • Clear the codes and determine if this code resets. If the same trouble code (P1605 resets), the PCM has failed. • TSB TC002-03 (6/03) contains information related to this code
DTC: P1630 **1T CCM, MIL: Yes** **1998** **Models:** Supra **Engines:** All **Transmissions:** All	**Traction Control System Malfunction Conditions:** Engine runtime over 5 seconds, and the PCM detected an unexpected voltage condition on the Traction Control system circuit, or it received a signal that a TRAC problem existed. **Possible Causes:** • ETC+ or ETC signal circuit is open, shorted to ground or power • EFI+ or EFI- signal circuit is open, shorted to ground or power • Throttle Control ECU or the PCM has failed
DTC: P1633 **1T CCM, MIL: Yes** **1998** **Models:** Supra **Engines:** All **Transmissions:** All	**Engine Throttle Control System Circuit Malfunction Conditions:** Engine started, engine running, and the PCM detected a problem in the ETCS portion of the circuit located in the engine control module. **Possible Causes:** • Clear the codes and determine if this code resets. If the same trouble code (P1633 resets), the PCM has failed. • TSB TC002-03 (6/03) contains information related to this code
DTC: P1633 **1T CCM, MIL: Yes** **2001, 2002, 2003, 2004, 2005, 2006** **Models:** Prius **Engines:** All **Transmissions:** A/T	**ECM ETCS Circuit Malfunction Conditions:** Key on or engine running; and the PCM detected that it lost communication with the ECU ETCS circuit for 1.5 seconds. **Possible Causes:** • ETCS signal circuit to the PCM is open • ETCS signal circuit to the PCM is shorted to ground or power • ECU ETCU has failed or the PCM has failed • TSB TC002-03 (6/03) contains information related to this code
DTC: P1636 **1T CCM, MIL: Yes** **2001, 2002, 2003, 2004, 2005, 2006** **Models:** Prius **Engines:** All **Transmissions:** All	**HV ECU Malfunction Conditions:** Key on or engine running; and the PCM detected that it lost communication with the HV ECU module for 1.5 seconds. **Possible Causes:** • HV ECU signal (+) or (-) circuit to PCM is open • HV ECU signal (+) or (-) circuit to PCM is shorted to ground • HV ECU has failed or the PCM has failed • TSB TC002-03 (6/03) contains information related to this code
DTC: P1637 **1T CCM, MIL: Yes** **2001, 2002, 2003, 2004, 2005, 2006** **Models:** Prius **Engines:** All **Transmissions:** A/T	**EGSTP Signal Malfunction Conditions:** Key on or engine running; and the PCM did not detect any EGSTP signals from the HV ECU for 2 seconds during the CCM test. **Possible Causes:** • EGSTP signal circuit to PCM is open • EGSTP signal circuit to PCM is shorted to ground • EGSTP signal circuit to PCM is shorted to power • HV ECU or the PCM has failed
DTC: P1645 **1T CCM, MIL: Yes** **2000, 2001, 2002, 2003, 2004, 2005, 2006** **Models:** Celica, Corolla, Matrix, MR2 **Engines:** 1.8L VIN R, 1.8L VIN Y **Transmissions:** All	**Body ECU Malfunction Conditions:** Key on or engine running; and the PCM did not detect any signals from the Body or A/C ECU for a period of 3 seconds during the test. **Possible Causes:** • A/C ECU had failed (no communication fault) • Body Control ECU has failed (no communication fault) • Communication data bus circuit is open or shorted to ground • Communication data bus circuit is shorted to system power

DTC	Trouble Code Title, Conditions & Possible Causes
DTC: P1646 **1T CCM, MIL: Yes** **2003, 2004, 2005, 2006** **Models:** MR2 **Engines:** 1.8L VIN R **Transmissions:** All	**Transmission Control ECU Malfunction Conditions:** Key on or engine running; and the PCM did not detect any signals from the Transmission Control ECU for 3 seconds during the test. **Possible Causes:** • Communication data bus circuit is open or shorted to ground • Communication data bus circuit is shorted to system power • Transmission Control ECU has failed (no communication fault)
DTC: P1656 **1T CCM, MIL: Yes** **2000, 2001, 2002, 2003, 2004,** **2005, 2006** **Models:** Celica, Corolla, Matrix, MR2, Prius, RAV4 **Engines:** All **Transmissions:** All	**Oil Control Valve Circuit Malfunction (Bank 1) Conditions:** Key on or engine running; and the PCM detected an unexpected voltage condition on the Oil Control Valve (OCV) circuit in the test. **Possible Causes:** • OCV signal (+) circuit is open or shorted to ground • OCV signal (-) circuit is open or shorted to ground • OCV (valve) is damaged or has failed • PCM has failed
DTC: P1663 **1T CCM, MIL: Yes** **2001, 2002** **Models:** Prius **Engines:** All **Transmissions:** A/T	**Oil Control Valve Circuit Malfunction (Bank 2) Conditions:** Key on or engine running; and the PCM detected an unexpected voltage condition on the Oil Control Valve (OCV) circuit in the test. **Possible Causes:** • OCV signal (+) circuit is open or shorted to ground • OCV signal (-) circuit is open or shorted to ground • OCV (valve) is damaged or has failed • PCM has failed
DTC: P1690 **1T CCM, MIL: Yes** **2000, 2001, 2002, 2003, 2004,** **2005, 2006** **Models:** Celica, Corolla, Matrix **Engines:** All **Transmissions:** All	**Oil Control Valve Circuit Malfunction Conditions:** Key on or engine running; and the PCM detected an unexpected voltage condition on the Oil Control Valve (OCV) circuit in the test. **Possible Causes:** • OCV (valve) signal (+) or (-) circuit is open or shorted to ground • OCV (valve) signal circuit is shorted to VREF or system power • OCV (valve) is damaged or has failed • PCM had failed
DTC: P1692 **1T CCM, MIL: Yes** **1998, 1999, 2000, 2001, 2002,** **2003, 2004, 2005, 2006** **Models:** Celica, Corolla, Matrix **Engines:** All **Transmissions:** All	**Oil Control Valve Circuit Malfunction (Open) Conditions:** Engine speed less than 6000 rpm, and after the PCM switched the locker arm from low speed to high speed, it detected the Oil Pressure switch signal indicated "on" for 5 seconds. **Possible Causes:** • OCV (valve) connector is damaged or open • OCV (valve) control circuit is open • Oil control valve for the VVTL is damaged or has failed • Oil pressure switch for the VVTL is damaged or has failed • PCM has failed
DTC: P1693 **1T CCM, MIL: Yes** **1998, 1999, 2000, 2001, 2002,** **2003, 2004, 2005, 2006** **Models:** Celica, Corolla, Matrix **Engines:** All **Transmissions:** All	**Oil Control Valve Circuit Malfunction (Closed) Conditions:** Engine started, ECT sensor more than 140°F engine speed less than 6000 rpm, then after the PCM switched the locker arm from low speed to high speed, the PCM detected the Oil Pressure switch signal indicated "off" for over 1 second or more during the CCM test. **Possible Causes:** • OCV (valve) control circuit is shorted to ground • Oil control valve for the VVTL is damaged or has failed • Oil pressure switch for the VVTL is damaged or has failed • PCM has failed
DTC: P1700 **2T CCM, MIL: Yes** **1995, 1996, 1997, 1998** **Models:** Avalon, Camry, Land Cruiser, Previa, T100, Tacoma, Tercel **Engines:** 1.5L VIN E, 2.4L VIN A, 2.4L VIN R, 2.7L VIN U, 3.0L VIN G, 3.4L VIN V, 4.5L VIN D **Transmissions:** A/T	**Vehicle Speed Sensor '2' Circuit Malfunction Conditions:** Vehicle driven to a speed of over 6 mph, P/N switch indicating "off" (not in Park or Neutral position) for over 4 seconds, TR switch indicating a position other than Neutral, and the PCM did not detect any VSS2 signals after detecting at least 4 VSS1 signals. **Note: This problem must occur at least 500 times (continuously) in order for this trouble code to set.** **Possible Causes:** • VSS signal (+) or (-) circuit is open, shorted to ground or power • VSS is damaged or has failed • PCM has failed

DTC	Trouble Code Title, Conditions & Possible Causes
DTC: P1705 **1T CCM, MIL: Yes** **1995** **Models:** Avalon, Camry, Land Cruiser, Previa, T100, Tacoma, Tercel **Engines:** 1.5L VIN E, 2.4L VIN A, 2.4L VIN R, 2.7L VIN U, 3.0L VIN G, 3.4L VIN V, 4.5L VIN D **Transmissions:** A/T	**A/T Direct Clutch Speed Sensor Circuit Malfunction Conditions:** Vehicle running, P/N switch at "off" and the PCM detected a Direct Clutch Speed sensor output of 300 rpm or less. **Possible Causes:** • Direct clutch speed sensor signal (+) or (-) circuit is open • Direct clutch speed sensor signal (+) or (-) shorted to ground • Direct clutch speed sensor (+) or (-) signal is shorted to power • Direct clutch is damaged or has failed • PCM has failed
DTC: P1725 **1T CCM, MIL: Yes** **2000, 2001, 2002, 2003, 2004, 2005, 2006** **Models:** Echo, Yaris, RAV4 **Engines:** 1.5L VIN T, 2.0L VIN K **Transmissions:** A/T	**A/T Turbine Shaft Speed Sensor Circuit Malfunction Conditions:** No Shift Solenoid or P/N codes set, engine started, P/N switch indicating "off", gear position indicating 2nd, 3rd gear, or in O/D, no gear change occurring, and the PCM detected the Turbine Shaft Speed (ISS) sensor indicated less than 300 rpm for 4 seconds. **Possible Causes:** • Input shaft speed sensor signal (NT) circuit is open or shorted • Input shaft speed sensor is damaged or has failed • PCM has failed
DTC: P1730 **1T CCM, MIL: Yes** **2003, 2004, 2005, 2006** **Models:** RAV4 **Engines:** 2.0L VIN K **Transmissions:** A/T	**A/T Revolution Speed Sensor Circuit Malfunction Conditions:** No Shift Solenoid or Neutral codes set, engine started, gear position indicating 2nd, 3rd gear or O/D, no gear change occurring, and the PCM detected the Revolution Speed (NC) sensor indicated less than 300 rpm for 5 seconds. The PCM detects the rotation speed of the counter gear, and compares the signals from the Direct Clutch speed (NT) sensor to the counter gear speed sensor (NC). The PCM uses this signal to detect the shift time so that it can control the engine torque and hydraulic pressure in response to various conditions. **Possible Causes:** • Counter gear speed sensor (NC) circuit is open or shorted • Counter gear speed sensor is damaged or has failed • PCM has failed
DTC: P1760 **1T CCM, MIL: Yes** 2000, 2001, 2002 **Models:** Echo, Yaris **Engines:** 1.5L VIN T **Transmissions:** A/T	**A/T Linear Shift Solenoid Circuit Malfunction Conditions:** Engine speed over 500 rpm, P/N switch indicating off, and the PCM detected an unexpected low voltage (under 0.20v) or unexpected high voltage (12.0v) on the SLT solenoid circuit. **Possible Causes:** • SLT shift solenoid control circuit is open or shorted to ground • SLT shift solenoid is damaged or has failed • PCM has failed
DTC: P1760 **1T CCM, MIL: Yes** **2003, 2004, 2005, 2006** **Models:** Celica, Corolla, Matrix, RAV4 **Engines:** 1.8L VIN R, 2.0L VIN K **Transmissions:** A/T	**A/T Linear Shift Solenoid Circuit Malfunction Conditions:** Engine started, engine speed over 500 rpm, P/N switch indicating off, and the PCM detected an unexpected low voltage (under 0.20v) or unexpected high voltage (12.0v) on the SLT solenoid circuit. **Possible Causes:** • SLT shift solenoid control circuit is open or shorted to ground • SLT shift solenoid is damaged or has failed • PCM has failed
DTC: P1765 **1T CCM, MIL: Yes** **1995** **Models:** Avalon, Camry, Land Cruiser, Previa, T100, Tacoma, Tercel **Engines:** 1.5L VIN E, 2.4L VIN A, 2.4L VIN R, 2.7L VIN U, 3.0L VIN G, 3.4L VIN V, 4.5L VIN D	**A/T Linear Shift Solenoid Circuit Malfunction Conditions:** Engine speed over 500 rpm, P/N switch indicating off, and the PCM detected that the Linear Shift Solenoid valve current flow was less than 0.2 amps during the CCM test. **Possible Causes:** • SLT shift solenoid control circuit is open or shorted to ground • Linear shift solenoid control circuit is shorted to power • Linear shift solenoid is damaged or has failed • PCM has failed
DTC: P1780 **1T CCM, MIL: Yes** **1995** **Models:** Avalon, Camry, Land Cruiser, Previa, T100, Tacoma, Tercel **Engines:** 1.5L VIN E, 2.4L VIN A, 2.4L VIN R, 2.7L VIN U, 3.0L VIN G, 3.4L VIN V, 4.5L VIN D **Transmissions:** All	**Park/Neutral Position Switch Circuit Malfunction Conditions:** Engine started, engine runtime over 30 seconds, and the PCM detected two or more P/N inputs (Drive, Neutral, 2nd, Low or Reverse) at the same time; or with the engine speed from 1500-2500 rpm with the VSS indicating over 50 mph, it detected the P/N switch indicated "on" for 30 seconds during the CCM test. **Possible Causes:** • P/N switch is shorted to ground • P/N switch is out-of-adjustment, damaged or has failed • PCM has failed

DTC	Trouble Code Title, Conditions & Possible Causes
DTC: P1780 **1T CCM, MIL: Yes** **1996, 1997, 1998, 1999, 2000, 2001, 2002** **Models:** All **Engines:** All **Transmissions:** All	**Park/Neutral Position Switch Circuit Malfunction Conditions:** Engine started, engine runtime over 30 seconds, and the PCM detected two or more P/N inputs (Drive, Neutral, 2nd, Low or Reverse) at the same time; or with the engine speed from 1500-2500 rpm with the VSS indicating over 50 mph, it detected the P/N switch indicated "on" for 30 seconds during the CCM test. **Possible Causes:** • P/N switch is shorted to ground or wiring harness is shorted • P/N switch is out-of-adjustment, damaged or has failed • PCM has failed
DTC: P1780 **1T CCM, MIL: Yes** **2003, 2004, 2005, 2006** **Models:** Celica, Corolla, Matrix, RAV4 **Engines:** 1.8L VIN R, 1.8L VIN Y, 2.0L VIN K **Transmissions:** All	**Park/Neutral Position Switch Circuit Malfunction Conditions:** Engine started, engine runtime over 30 seconds, and the PCM detected two or more P/N inputs (Neutral, 2nd, Low or Reverse) at the same time; or with the engine speed from 1500-5000 rpm with the VSS indicating over 50 mph, MAP sensor at 300 mmHg or more, it detected the P/N switch indicated "on" for 30 seconds in the test. **Possible Causes:** • P/N switch is shorted to ground or wiring harness is shorted • P/N switch is out-of-adjustment, damaged or has failed • PCM has failed
DTC: P1780 **1T CCM, MIL: Yes** 2000, 2001, 2002 **Models:** Echo, Yaris **Engines:** 1.5L VIN T **Transmissions:** All	**A/T Shift Solenoid Valve Circuit Malfunction Conditions:** Engine speed over 500 rpm, P/N switch indicating off, and the PCM detected an unexpected low voltage or unexpected high voltage (12v) on the Shift Solenoid Valve (ST) control circuit. **Possible Causes:** • ST shift solenoid connector is damaged or loose • ST shift solenoid control circuit is open or shorted to ground • ST shift solenoid is damaged or has failed • PCM has failed

OBD II Trouble Code List (P2xxx Codes)

DTC	Trouble Code Title, Conditions & Possible Causes
DTC: P2102 **1T CCM, MIL: Yes** **2003, 2004, 2005, 2006** **Models:** 4Runner, Land Cruiser, MR2, Sequoia, Tacoma, Tundra **Engines:** 1.8L VIN R, 3.4L VIN N, 4.0L VIN U, 4.7L VIN T **Transmissions:** All	**Throttle Actuator Control Motor Circuit Low Input Conditions:** Engine started, throttle control motor output duty cycle at 80% or higher, and the PCM detected the throttle motor current was less than 0.5 amps for 2 seconds. The PCM controls the motor position in order to open and close the throttle valve. The opening angle of the throttle valve is sensed by the TP sensor mounted on the throttle body. The PCM uses the TP sensor signal to control the Throttle Valve opening angle (throttle motor) to respond to driving conditions. **Possible Causes:** • Throttle motor connector is damaged or open • Throttle control motor circuit is open • Throttle control motor is damaged or has failed • PCM has failed
DTC: P2103 **1T CCM, MIL: Yes** **2003, 2004, 2005, 2006** **Models:** 4Runner, Land Cruiser, MR2, Sequoia, Tacoma, Tundra **Engines:** 1.8L VIN R, 3.4L VIN N, 4.0L VIN U, 4.7L VIN T **Transmissions:** All	**Throttle Actuator Control Motor Circuit High Input Conditions:** Engine started, throttle control motor output duty cycle at 80% or higher, and the PCM detected the throttle motor current was more than 10.0 amps for 600 ms. The PCM controls the motor position in order to open and close the throttle valve. The opening angle of the throttle valve is sensed by the TP sensor mounted on the throttle body. The PCM uses the TP sensor signal to control the Throttle Valve opening angle (throttle motor) to respond to driving conditions. **Possible Causes:** • Throttle motor connector is damaged or shorted • Throttle control motor circuit is shorted • Throttle control motor is damaged or has failed • PCM has failed
DTC: P2111 **1T CCM, MIL: Yes** **2003, 2004, 2005, 2006** **Models:** 4Runner, Land Cruiser, MR2, Sequoia, Tacoma, Tundra **Engines:** 1.8L VIN R, 3.4L VIN N, 4.0L VIN U, 4.7L VIN T **Transmissions:** All	**Throttle Actuator Control System Stuck Open Conditions:** Key on or engine started, and the PCM detected the throttle control motor position is stuck open. The PCM controls the motor position in order to open and close the throttle valve. The opening angle of the throttle valve is sensed by the TP sensor mounted on the throttle body. The PCM uses the TP sensor signal to control the Throttle Valve opening angle (throttle motor) to respond to driving conditions. **Possible Causes:** • Throttle control motor circuit is open • Throttle control motor is damaged or has failed • Throttle body or throttle valve is damaged or has failed
DTC: P2112 **1T CCM, MIL: Yes** **2003, 2004, 2005, 2006** **Models:** 4Runner, Land Cruiser, MR2, Sequoia, Tacoma, Tundra **Engines:** 1.8L VIN R, 3.4L VIN N, 4.0L VIN U, 4.7L VIN T **Transmissions:** All	**Throttle Actuator Control System Stuck Closed Conditions:** Key on or engine started, and the PCM detected the throttle control motor position is stuck closed. The PCM controls the motor position in order to open and close the throttle valve. The opening angle of the throttle valve is sensed by the TP sensor mounted on the throttle body. The PCM uses the TP sensor signal to control the Throttle Valve opening angle (throttle motor) to respond to driving conditions. **Possible Causes:** • Throttle control motor circuit is shorted • Throttle control motor is damaged or has failed • Throttle body or throttle valve is damaged or has failed
DTC: P2118 **1T CCM, MIL: Yes** **2003, 2004, 2005, 2006** **Models:** 4Runner, Land Cruiser, MR2, Sequoia, Tacoma, Tundra **Engines:** 1.8L VIN R, 3.4L VIN N, 4.0L VIN U, 4.7L VIN T **Transmissions:** All	**Throttle Actuator Control Motor Current Performance Conditions:** Key on or engine started, and the PCM detected an unexpected low voltage condition (open circuit) on the ETCS power source circuit. Battery positive voltage is applied to the +BM circuit of the PCM under both Key on and Key off conditions. **Possible Causes:** • ETCS power source circuit is open • PCM has failed
DTC: P2119 **1T CCM, MIL: Yes** **2003, 2004, 2005, 2006** **Models:** 4Runner, Land Cruiser, MR2, Sequoia, Tacoma, Tundra **Engines:** 1.8L VIN R, 3.4L VIN N, 4.0L VIN U, 4.7L VIN T **Transmissions:** All	**Throttle Actuator Control Throttle Body Performance Conditions:** Engine started, and the PCM detected Actual throttle opening angle continued to vary greatly from the Target opening angel. The idle speed on this vehicle is controlled by the Electronic Throttle Control system (ETCS). This system includes a throttle control motor to operate the throttle valve, a throttle position sensor to detect the accelerator pedal position, and the PCM to control the ETCS and one-valve design of throttle body. The PCM controls this motor in order to control the throttle valve opening to achieve its target speed. **Possible Causes:** • ETCS throttle control system • PCM has failed

DTC	Trouble Code Title, Conditions & Possible Causes
DTC: P2120 **1T CCM, MIL: Yes** **2003, 2004, 2005, 2006** **Models:** 4Runner, Land Cruiser, MR2, Sequoia, Tacoma, Tundra **Engines:** 1.8L VIN R, 3.4L VIN N, 4.0L VIN U, 4.7L VIN T **Transmissions:** All	**Throttle Pedal Position Sensor/Switch 'D' Circuit Malfunction Conditions:** Engine started, and the PCM detected VPA1 signal indicated less than 0.20v while the VPA2 signal indicated over 0.97 degrees, or the VPA1 signal indicated more than 4.80v for 500 ms. This system (ETCS) does not use a throttle cable. The Accelerator Pedal Position (APP) sensor is mounted on the accelerator pedal bracket. It includes two sensors to detect the accelerator position, and to detect any faults in the APP sensor or its related circuits. **Possible Causes:** • APP sensor signal circuit is open or shorted to ground • APP sensor is damaged or has failed • PCM has failed
DTC: P2121 **1T CCM, MIL: Yes** **2003, 2004, 2005, 2006** **Models:** 4Runner, Land Cruiser, MR2, Sequoia, Tacoma, Tundra **Engines:** 1.8L VIN R, 3.4L VIN N, 4.0L VIN U, 4.7L VIN T **Transmissions:** All	**Accelerator Pedal Position Sensor Signal Performance Conditions:** Engine started, IDL signal "off", and the PCM detected the difference between the VPA and VPA2 signal was out-of-range for 2 seconds. This system (ETCS) does not use a throttle cable. The Accelerator Pedal Position (APP) sensor is mounted on the accelerator pedal bracket. It includes two sensors to detect the accelerator position, and to detect any faults in the APP sensor or its related circuits. **Possible Causes:** • APP sensor is damaged or has failed • PCM has failed
DTC: P2122 **1T CCM, MIL: Yes** **2003, 2004, 2005, 2006** **Models:** 4Runner, Land Cruiser, Sequoia, Tacoma, Tundra **Engines:** 3.4L VIN N, 4.0L VIN U, 4.7L VIN **Transmissions:** All	**Throttle Pedal Position Sensor/Switch 'D' Circuit Low Input Conditions:** Engine started, and the PCM detected VPA1 signal was less than 0.20v while the VPA2 signal indicated over 0.97 degrees for 500 ms. This system (ETCS) does not use a throttle cable. The Accelerator Pedal Position (APP) sensor is mounted on the accelerator pedal bracket. It includes two sensors to detect the accelerator position, and to detect any faults in the APP sensor or its related circuits. **Possible Causes:** • APP sensor signal circuit is shorted to ground • APP sensor is damaged or has failed • PCM has failed
DTC: P2123 **1T CCM, MIL: Yes** **2003, 2004, 2005, 2006** **Models:** 4Runner, Land Cruiser, Sequoia, Tacoma, Tundra **Engines:** 3.4L VIN N, 4.0L VIN U, 4.7L VIN T **Transmissions:** All	**Throttle Pedal Position Sensor/Switch 'D' Circuit High Input Conditions:** Engine started, and the PCM detected VPA1 signal indicated over 4.80v for 2 seconds. This system (ETCS) does not use a throttle cable. The Accelerator Pedal Position (APP) sensor is mounted on the accelerator pedal bracket. It includes two sensors to detect the accelerator position, and to detect any faults in the APP sensor. **Possible Causes:** • APP sensor signal circuit is open • APP sensor is damaged or has failed • PCM has failed
DTC: P2125 **1T CCM, MIL: Yes** **2003, 2004, 2005, 2006** **Models:** 4Runner, Land Cruiser, Sequoia, Tacoma, Tundra **Engines:** 3.4L VIN N, 4.0L VIN U, 4.7L VIN T **Transmissions:** All	**Throttle Pedal Position Sensor/Switch 'E' Circuit Malfunction Conditions:** Engine started, and the PCM detected VPA2 signal indicated less than 0.50v while the VPA1 signal indicated over 0.97 degrees, or the VPA1 signal was more than 4.80v or less than 0.20v for 500 ms. This system (ETCS) does not use a throttle cable. The Accelerator Pedal Position (APP) sensor is mounted on the accelerator pedal bracket. It includes two sensors to detect the accelerator position, and to detect any faults in the APP sensor or its circuits. **Possible Causes:** • APP sensor signal circuit is open or shorted to ground • APP sensor is damaged or has failed • PCM has failed
DTC: P2127 **1T CCM, MIL: Yes** **2003, 2004, 2005, 2006** **Models:** 4Runner, Land Cruiser, Sequoia, Tacoma, Tundra **Engines:** 3.4L VIN N, 4.0L VIN U, 4.7L VIN T **Transmissions:** All	**Throttle Pedal Position Sensor/Switch 'E' Circuit Low Input Conditions:** Engine started, and the PCM detected VPA2 signal was less than 0.20v while the VPA1 signal indicated over 0.97 degrees for 500 ms. The ETCS does not use a throttle cable. The Accelerator Pedal Position sensor is mounted on the accelerator pedal bracket. It includes two sensors to detect the accelerator position or any faults in the APP sensor or its circuits. **Possible Causes:** • APP sensor signal circuit is shorted to ground • APP sensor has failed, or the PCM has failed
DTC: P2128 **1T CCM, MIL: Yes** **2003, 2004, 2005, 2006** **Models:** 4Runner, Land Cruiser, Sequoia, Tacoma, Tundra **Engines:** 3.4L VIN N, 4.0L VIN U, 4.7L VIN T **Transmissions:** All	**Throttle Pedal Position Sensor/Switch 'E' Circuit High Input Conditions:** Engine started, and the PCM detected VPA1 signal was over 4.80v or under 0.20v for 2 seconds. This system (ETCS) does not use a throttle cable. The Accelerator Pedal Position (APP) sensor is mounted on the accelerator pedal bracket. It includes two sensors to detect the accelerator position, and any faults in the APP sensor. **Possible Causes:** • APP sensor signal circuit is open • APP sensor is damaged or has failed • PCM has failed

DTC	Trouble Code Title, Conditions & Possible Causes
DTC: P2135 **1T CCM, MIL: Yes** **2003, 2004, 2005, 2006** **Models:** 4Runner, Land Cruiser, Sequoia, Tundra **Engines:** 4.0L VIN U, 4.7L VIN T **Transmissions:** All	**Throttle Pedal Position Sensor/Switch 'A'/'B' Voltage Correlation Conditions:** Engine started, and the PCM detected the value of the VPA1 signal less the VPA2 was less than 0.02v, or the VPA1 signal was less than 0.20v with the VPA2 signal less than 0.50v for 400-500 ms. This system (ETCS) does not use a throttle cable. The Accelerator Pedal Position (APP) sensor is mounted on the accelerator pedal bracket. It includes two sensors to detect the accelerator position, and to detect any faults present in the APP sensor. **Possible Causes:** • APP sensor signal circuit is open • APP sensor is damaged or has failed • PCM has failed
DTC: P2138 **1T CCM, MIL: Yes** **2003, 2004, 2005, 2006** **Models:** 4Runner, Land Cruiser, Sequoia, Tundra **Engines:** 4.0L VIN U, 4.7L VIN T **Transmissions:** All	**Throttle Pedal Position Sensor/Switch 'D'/'E' Voltage Correlation Conditions:** Engine started, and the PCM detected the value of the VPA1 signal less the VPA2 was less than 0.02v, or the VPA1 signal was less than 0.20v with the VPA2 signal less than 0.50v for 2 seconds. This system (ETCS) does not use a throttle cable. The Accelerator Pedal Position (APP) sensor is mounted on the accelerator pedal bracket. It includes two sensors to detect the accelerator position, and to detect any faults present in the APP sensor. **Possible Causes:** • APP sensor signal circuit is open • APP sensor is damaged or has failed • PCM has failed
DTC: P2716 **1T CCM, MIL: Yes** **2003, 2004, 2005, 2006** **Models:** Echo, Yaris **Engines:** 1.5L VIN T **Transmissions:** A/T	**A/T Pressure Control Solenoid 'D' Circuit Malfunction Conditions:** Engine speed over 500 rpm, P/N switch indicating off, and the PCM detected an unexpected low voltage (under 0.20v) on the Shift Solenoid Valve (SLT) control circuit, or it detected an unexpected high voltage (over 12.0v) on the SLT solenoid control circuit during the CCM test. **Possible Causes:** • SLT shift solenoid connector is damaged or loose • SLT shift solenoid control circuit is open or shorted to ground • SLT shift solenoid is damaged or has failed • PCM has failed
DTC: P2195 **1T CCM, MIL: Yes** **2003, 2004, 2005, 2006** **Models:** Avalon, Camry, Echo, Yaris, Highlander, Sienna, Tacoma **Engines:** 1.5L VIN T, 2.4L VIN D, 2.4L VIN L, 2.7L VIN M, 3.0L VIN F **Transmissions:** All	**Air Fuel Sensor 1 (Bank 1 Sensor 1) Signal Stuck "Lean" Conditions:** Vehicle speed from 25-87 mph at over 1500 rpm with the throttle valve open, and the PCM detected the Air Fuel sensor signal indicated more than 3.80v for 10 seconds. **Possible Causes:** • A/FS1 signal circuit is open or shorted to ground • A/FS1 is damaged, contaminated or it has failed • Air induction system is severely restricted • Fuel Control component problems (e.g., low fuel pressure, or one or more severely restricted fuel injectors) • PCM has failed
DTC: P2196 **1T CCM, MIL: Yes** **2003, 2004, 2005, 2006** **Models:** Avalon, Camry, Echo, Yaris, Highlander, Sienna **Engines:** 1.5L VIN T, 2.4L VIN D, 2.4L VIN L, 2.7L VIN M, 3.0L VIN F **Transmissions:** All	**Air Fuel Sensor 1 (Bank 1 Sensor 1) Signal Stuck "Rich" Conditions:** Vehicle speed from 25-87 mph at over 1500 rpm with the throttle valve open, and the PCM detected the Air Fuel sensor signal indicated more than 2.80v for 10 seconds. **Possible Causes:** • A/FS1 signal circuit is open or shorted to ground • A/FS1 is damaged, contaminated or it has failed • Air induction system is leaking (check for PCV system leaks) • Fuel component problem (high fuel pressure, leaking regulator or a leaking injector) • PCM has failed
DTC: P2196 **1T CCM, MIL: Yes** **2003, 2004, 2005, 2006** **Models:** Avalon, Camry, Echo, Yaris, Highlander, Sienna **Engines:** 1.5L VIN T, 2.4L VIN D, 3.0L VIN F **Transmissions:** All	**Air Fuel Sensor 1 (Bank 2 Sensor 1) Signal Stuck "Lean" Conditions:** Vehicle speed from 25-87 mph at over 1500 rpm with the throttle valve open, and the PCM detected the Air Fuel sensor signal indicated more than 3.80v for 10 seconds. **Possible Causes:** • A/FS1 signal circuit is open or shorted to ground • A/FS1 is damaged, contaminated or it has failed • Air induction system is severely restricted • Fuel component problem (e.g., low fuel pressure, or a severely restricted fuel injector) • PCM has failed

DTC	Trouble Code Title, Conditions & Possible Causes
DTC: P2198 **1T CCM, MIL: Yes** **2003, 2004, 2005, 2006** **Models:** Avalon, Camry, Echo, Yaris, Highlander, Sienna **Engines:** 1.5L VIN T, 2.4L VIN D, 3.0L VIN F **Transmissions:** All	**Air Fuel Sensor 1 (Bank 2 Sensor 1) Signal Stuck "Rich" Conditions:** Vehicle driven at a steady speed of 25-87 mph at over 1500 rpm with the throttle valve open, and the PCM detected the Air Fuel sensor signal indicated more than 2.80v for 10 seconds. **Possible Causes:** • A/FS1 signal circuit is open or shorted to ground • A/FS1 is damaged, contaminated or it has failed • Air induction system is leaking (check for PCV system leaks) • Fuel Control component problems (e.g., high fuel pressure, a leaking pressure regulator, one or more leaking fuel injectors) • PCM has failed
DTC: P2237 **1T CCM, MIL: Yes** **2003, 2004, 2005, 2006** **Models:** Avalon, Camry, Highlander, Sienna, Tacoma **Engines:** 2.4L VIN L, 2.7L VIN M, 3.0L VIN F **Transmissions:** All	**Air Fuel Sensor 1 (Bank 1 Sensor 1) Pumping Current Signal Open Conditions:** Vehicle speed from 25-87 mph at over 1500 rpm with the throttle valve open, and the PCM detected the A/F Sensor AF+ signal was less than 0.50v or more than 4.80v for 5 seconds. **Possible Causes:** • A/FS1 signal circuit is open or shorted to ground • A/FS1 is damaged, contaminated or it has failed • A/FS1 heater assembly is damaged or its circuit has failed • A/FS1 heater relay is damaged or has failed • PCM has failed
DTC: P2240 **1T CCM, MIL: Yes** **2003, 2004, 2005, 2006** **Models:** Avalon, Camry, Highlander, Sienna **Engines:** 3.0L VIN F **Transmissions:** All	**Air Fuel Sensor 1 (Bank 2 Sensor 1) Pumping Current Signal Open Conditions:** Engine started, vehicle driven at a steady speed of 25-87 mph at over 1500 rpm with the throttle valve open, and the PCM detected the Air Fuel Sensor AF+ signal indicated less than 0.50v or indicated more than 4.80v for 5 seconds. **Possible Causes:** • A/FS1 signal circuit is open or shorted to ground • A/FS1 is damaged, contaminated or it has failed • A/FS1 heater assembly is damaged or its circuit has failed • A/FS1 heater relay is damaged or has failed • PCM has failed
DTC: P2725 **1T CCM, MIL: Yes** **2003, 2004, 2005, 2006** **Models:** Avalon, Camry, Highlander, Sienna **Engines:** 3.0L VIN F **Transmissions:** All	**A/T Pressure Control Solenoid 'E' Circuit Malfunction Conditions:** Engine started, engine warmup period completed, gearshift selector in 'P' or 'N', engine speed 500 rpm or more, and the PCM detected that current flowed to the Shift Solenoid (SLN) control circuit for over 1 second. The Shift Solenoid SLN controls the hydraulic pressure acting on the accumulator control valve when gears are shifted in order to provide smooth gear shifting. **Possible Causes:** • Shift solenoid (SLN) connector is damaged or loose • Shift solenoid (SLN) control circuit is open or shorted to ground • Shift solenoid (SLN) is damaged or has failed (electrical fault) • PCM has failed
DTC: P3190 **1T CCM, MIL: Yes** **2001, 2002, 2003, 2004, 2005, 2006** **Models:** Prius **Engines:** All **Transmissions:** A/T	**Poor Engine Power Conditions:** Engine started, engine running at a stable speed 3-5 minutes, HV ECU communicating with the PCM, engine not in start mode, engine target torque at a fixed value or higher, and the PCM detected the ratio of estimated torque to the target torque was less than 20%. **Possible Causes:** • Airflow meter is damaged or has failed • Base engine mechanical problems are present • Camshaft position or crankshaft position sensor is damaged • Intake air leaks or restrictions are present • Fuel pressure is too low or too high • Throttle body is damaged or has failed • Water temperature sensor is damaged or has failed • PCM has failed
DTC: P3190 **1T CCM, MIL: Yes** **2001, 2002, 2003, 2004, 2005, 2006** **Models:** Prius **Engines:** All **Transmissions:** A/T	**Engine Does Not Start Conditions:** Engine cranking at a stable speed for a fixed length of time, HV ECU communication with the PCM valid, and the PCM detected the engine remained in engine start mode (it did not start). **Possible Causes:** • Airflow meter is damaged or has failed • Base engine mechanical problems are present • Camshaft position or crankshaft position sensor is damaged • Fuel pressure is too low or too high • Throttle body is damaged or has failed • Vehicle is out of fuel • PCM has failed

TOYOTA
COMPONENT TESTING

TABLE OF CONTENTS

Component Locations

4RUNNER

2.7L I4 MFI 3RZ-FE VIN M

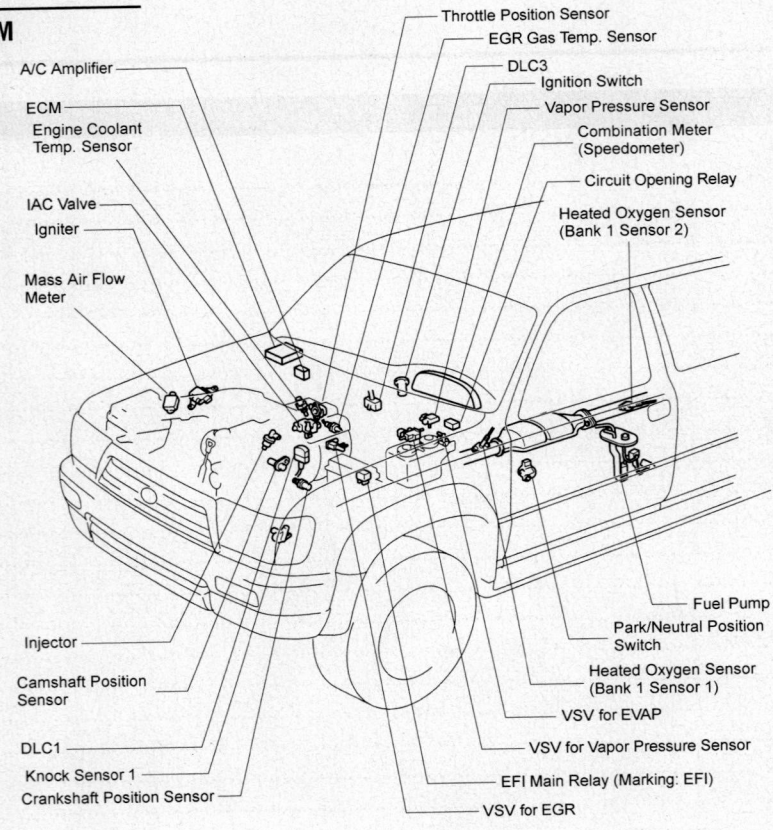

Throttle Position Sensor
EGR Gas Temp. Sensor
DLC3
Ignition Switch
Vapor Pressure Sensor
Combination Meter (Speedometer)
Circuit Opening Relay
Heated Oxygen Sensor (Bank 1 Sensor 2)
A/C Amplifier
ECM
Engine Coolant Temp. Sensor
IAC Valve
Igniter
Mass Air Flow Meter
Fuel Pump
Park/Neutral Position Switch
Heated Oxygen Sensor (Bank 1 Sensor 1)
Injector
VSV for EVAP
Camshaft Position Sensor
VSV for Vapor Pressure Sensor
DLC1
EFI Main Relay (Marking: EFI)
Knock Sensor 1
Crankshaft Position Sensor
VSV for EGR

29157_TOYO_G0001

3.4L V6 MFI 5VZ-FE VIN N

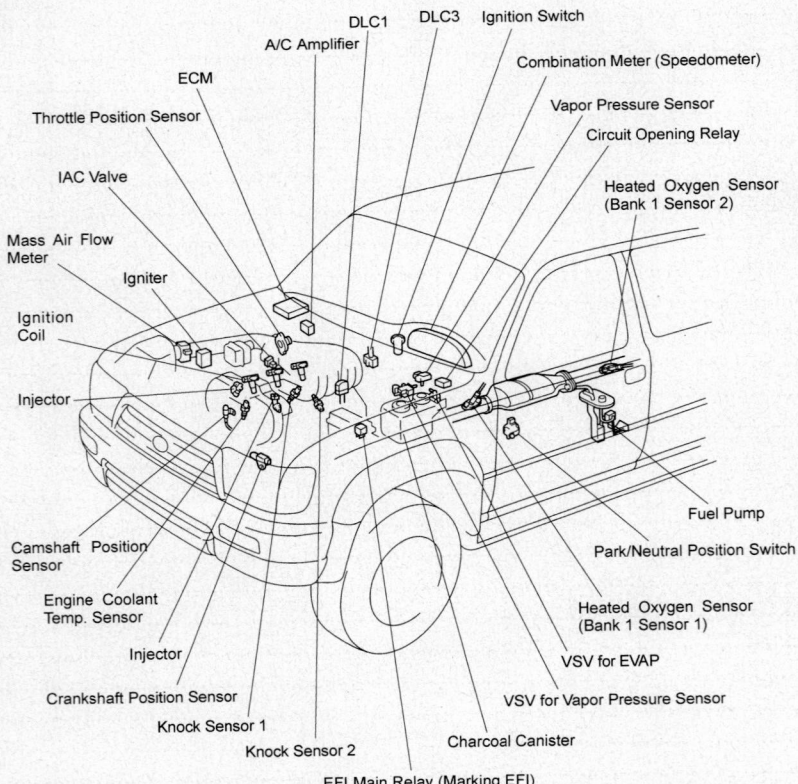

DLC1
DLC3
Ignition Switch
A/C Amplifier
Combination Meter (Speedometer)
ECM
Vapor Pressure Sensor
Throttle Position Sensor
Circuit Opening Relay
IAC Valve
Heated Oxygen Sensor (Bank 1 Sensor 2)
Mass Air Flow Meter
Igniter
Ignition Coil
Injector
Fuel Pump
Park/Neutral Position Switch
Camshaft Position Sensor
Engine Coolant Temp. Sensor
Heated Oxygen Sensor (Bank 1 Sensor 1)
Injector
VSV for EVAP
Crankshaft Position Sensor
VSV for Vapor Pressure Sensor
Knock Sensor 1
Charcoal Canister
Knock Sensor 2
EFI Main Relay (Marking EFI)

29157_TOYO_G0002

4.7L V8 MFI 2UZ-FE VIN T

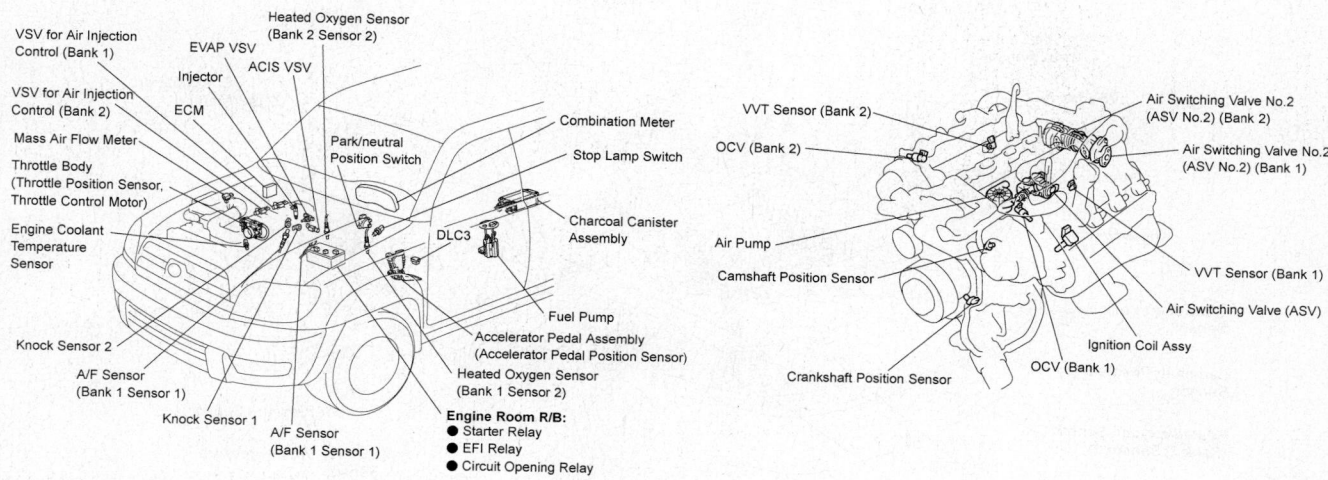

29157_TOYO_G0003

4.0L V6 MFI 1GR-FE VIN U

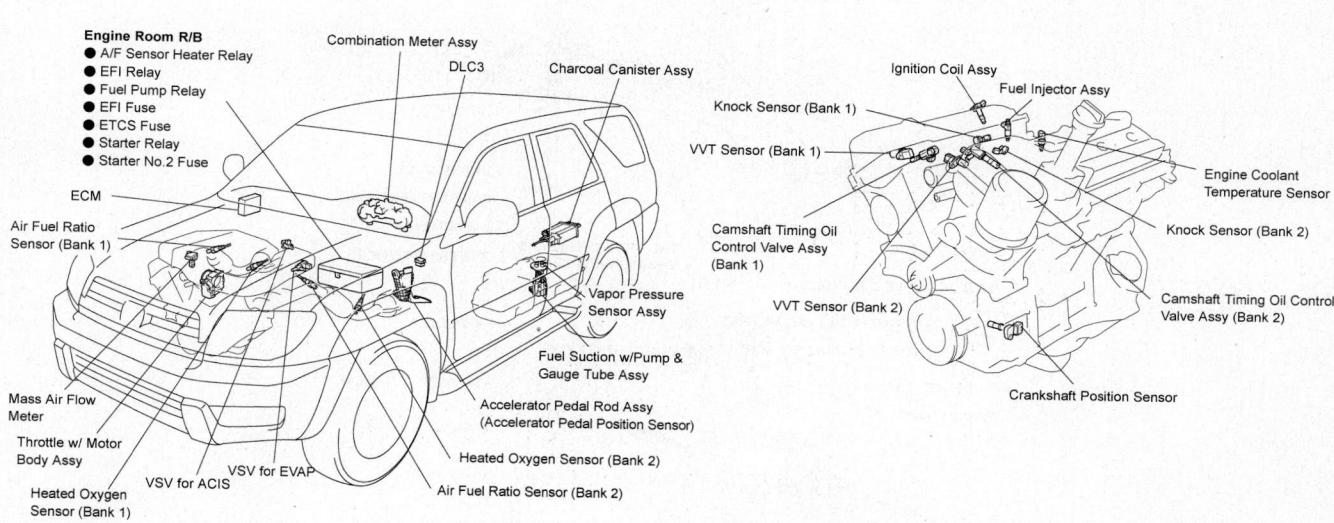

29157_TOYO_G0004

AVALON

3.0L V6 MFI 1MZ-FE VIN F

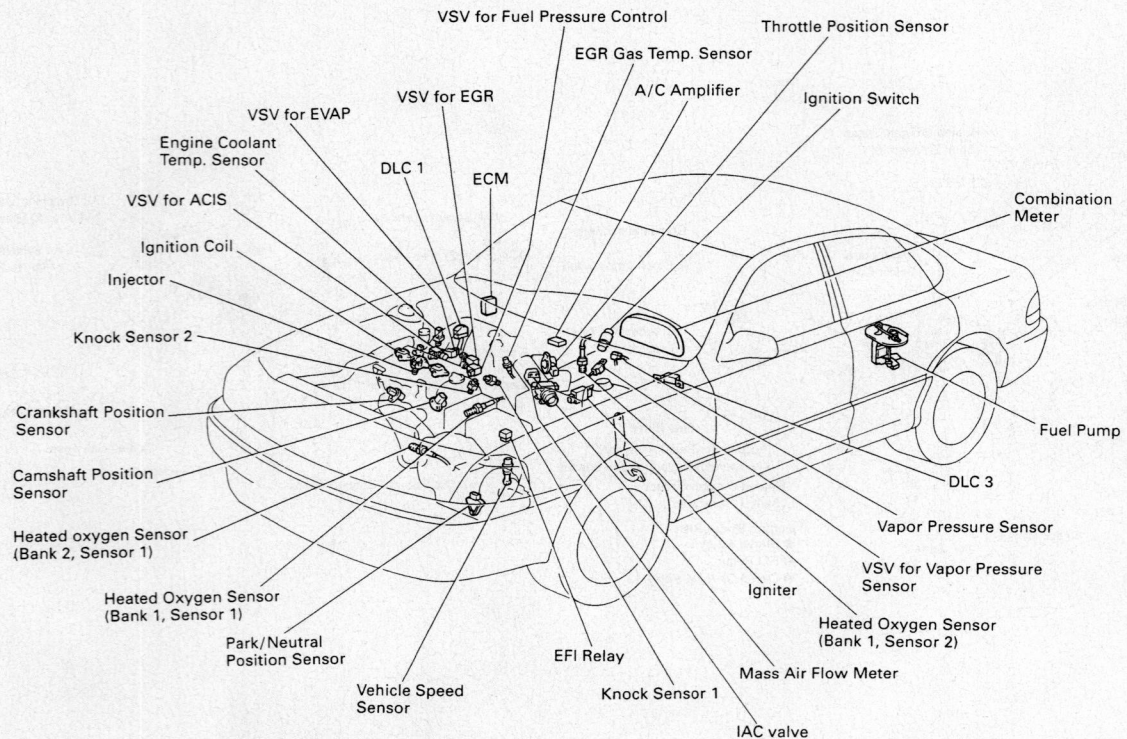

VSV for Fuel Pressure Control
Throttle Position Sensor
EGR Gas Temp. Sensor
VSV for EGR
A/C Amplifier
Ignition Switch
VSV for EVAP
Engine Coolant Temp. Sensor
DLC 1
ECM
VSV for ACIS
Ignition Coil
Injector
Knock Sensor 2
Crankshaft Position Sensor
Camshaft Position Sensor
Heated oxygen Sensor (Bank 2, Sensor 1)
Heated Oxygen Sensor (Bank 1, Sensor 1)
Park/Neutral Position Sensor
Vehicle Speed Sensor
EFI Relay
Knock Sensor 1
Mass Air Flow Meter
IAC valve
Igniter
Heated Oxygen Sensor (Bank 1, Sensor 2)
VSV for Vapor Pressure Sensor
Vapor Pressure Sensor
DLC 3
Fuel Pump
Combination Meter

29157_TOYO_G0005

3.5L V6 MFI 2GR-FE VIN K

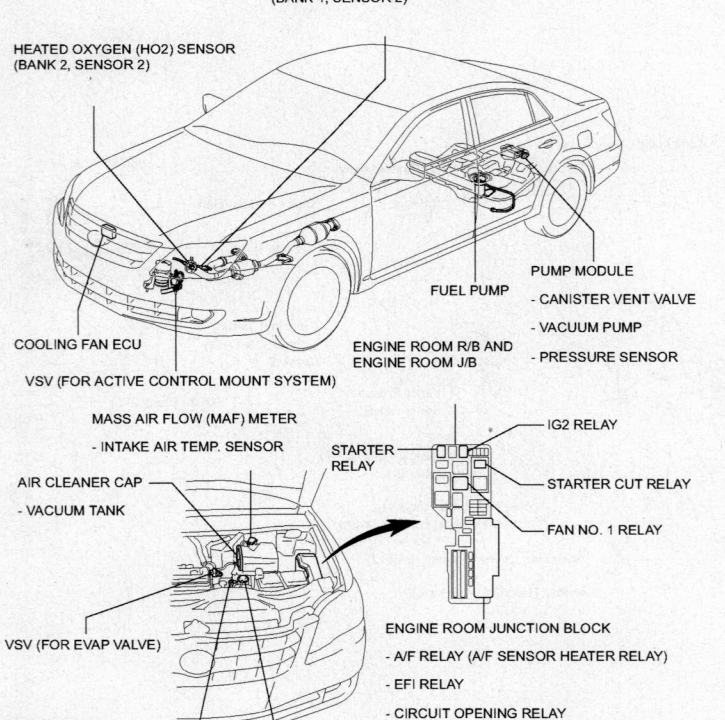

HEATED OXYGEN (HO2) SENSOR (BANK 1, SENSOR 2)
HEATED OXYGEN (HO2) SENSOR (BANK 2, SENSOR 2)
COOLING FAN ECU
VSV (FOR ACTIVE CONTROL MOUNT SYSTEM)
FUEL PUMP
ENGINE ROOM R/B AND ENGINE ROOM J/B
PUMP MODULE
- CANISTER VENT VALVE
- VACUUM PUMP
- PRESSURE SENSOR

MASS AIR FLOW (MAF) METER
- INTAKE AIR TEMP. SENSOR
AIR CLEANER CAP
- VACUUM TANK
STARTER RELAY
IG2 RELAY
STARTER CUT RELAY
FAN NO. 1 RELAY
VSV (FOR EVAP VALVE)
ENGINE ROOM JUNCTION BLOCK
- A/F RELAY (A/F SENSOR HEATER RELAY)
- EFI RELAY
- CIRCUIT OPENING RELAY
VSV (FOR AIR INTAKE CONTROL SYSTEM (AICS))
ACTUATOR (FOR AIR INTAKE CONTROL SYSTEM (AICS))

29157_TOYO_G0006

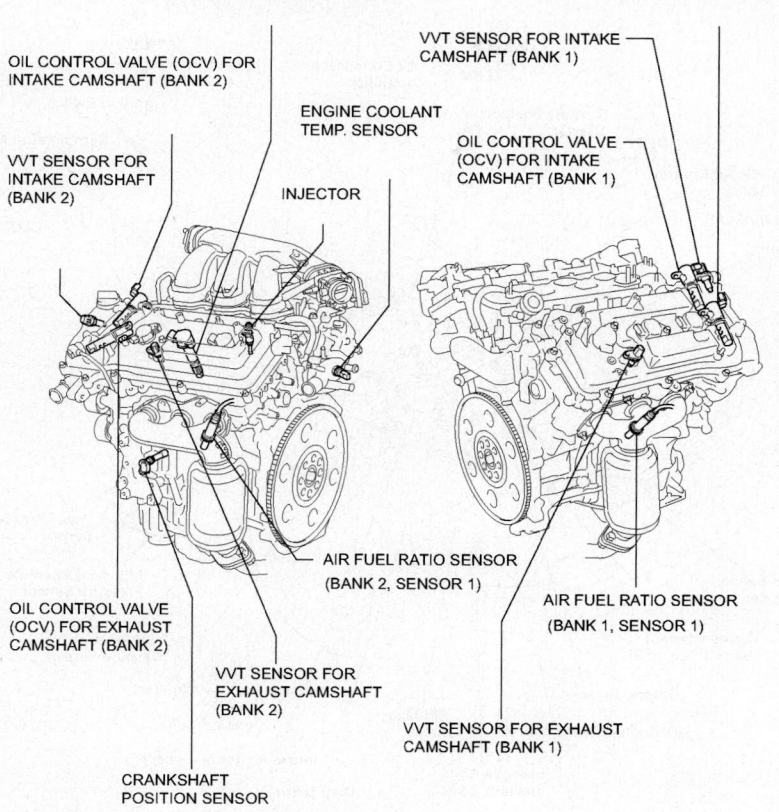

OIL CONTROL VALVE (OCV) FOR
EXHAUST CAMSHAFT (BANK 1)

IGNITION COIL WITH IGNITER

VVT SENSOR FOR INTAKE
CAMSHAFT (BANK 1)

OIL CONTROL VALVE (OCV) FOR
INTAKE CAMSHAFT (BANK 2)

ENGINE COOLANT
TEMP. SENSOR

OIL CONTROL VALVE
(OCV) FOR INTAKE
CAMSHAFT (BANK 1)

VVT SENSOR FOR
INTAKE CAMSHAFT
(BANK 2)

INJECTOR

AIR FUEL RATIO SENSOR
(BANK 2, SENSOR 1)

AIR FUEL RATIO SENSOR
(BANK 1, SENSOR 1)

OIL CONTROL VALVE
(OCV) FOR EXHAUST
CAMSHAFT (BANK 2)

VVT SENSOR FOR
EXHAUST CAMSHAFT
(BANK 2)

VVT SENSOR FOR EXHAUST
CAMSHAFT (BANK 1)

CRANKSHAFT
POSITION SENSOR

29157_TOYO_G0007

ACIS ACTUATOR (FOR ACOUSTIC
CONTROL INDUCTION SYSTEM (ACIS))

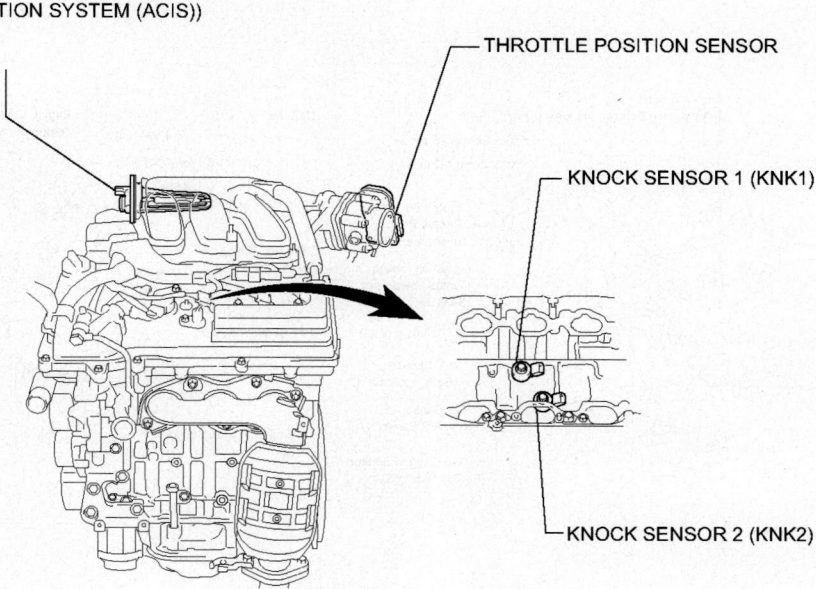

THROTTLE POSITION SENSOR

KNOCK SENSOR 1 (KNK1)

KNOCK SENSOR 2 (KNK2)

29157_TOYO_G0008

CAMRY

2.2L I4 MFI 5S-FE VIN G

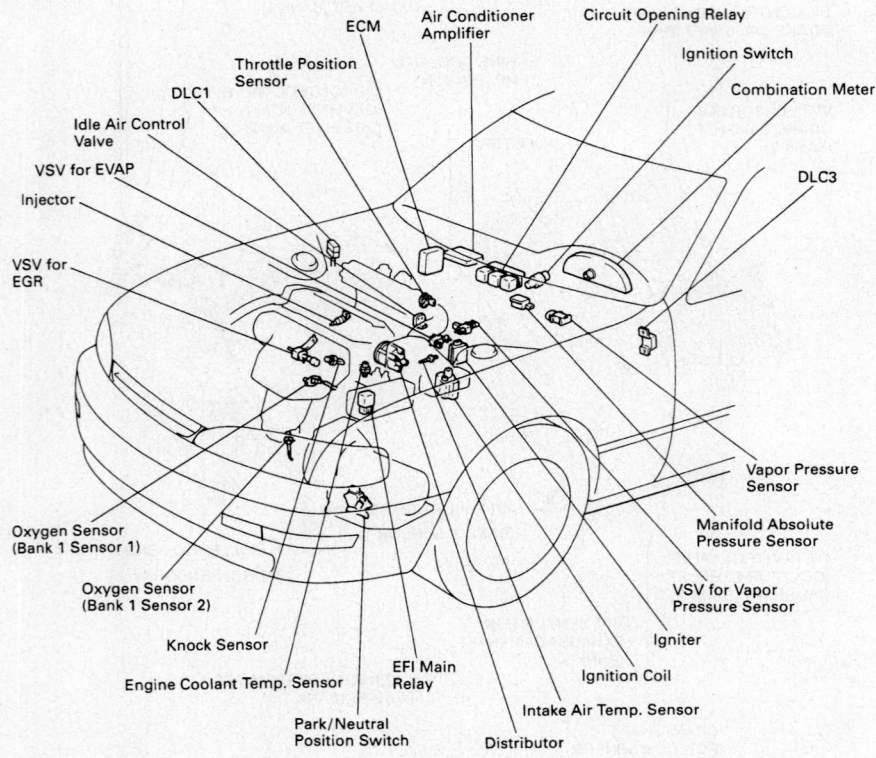

ECM
Air Conditioner Amplifier
Circuit Opening Relay
Ignition Switch
Combination Meter
Throttle Position Sensor
DLC1
Idle Air Control Valve
VSV for EVAP
Injector
VSV for EGR
DLC3
Oxygen Sensor (Bank 1 Sensor 1)
Oxygen Sensor (Bank 1 Sensor 2)
Knock Sensor
Engine Coolant Temp. Sensor
Park/Neutral Position Switch
EFI Main Relay
Distributor
Intake Air Temp. Sensor
Ignition Coil
Igniter
VSV for Vapor Pressure Sensor
Manifold Absolute Pressure Sensor
Vapor Pressure Sensor

29157_TOYO_G0009

3.0L V6 MFI 1MZ-FE VIN F
3.3L V6 MFI 3MZ-FE VIN A

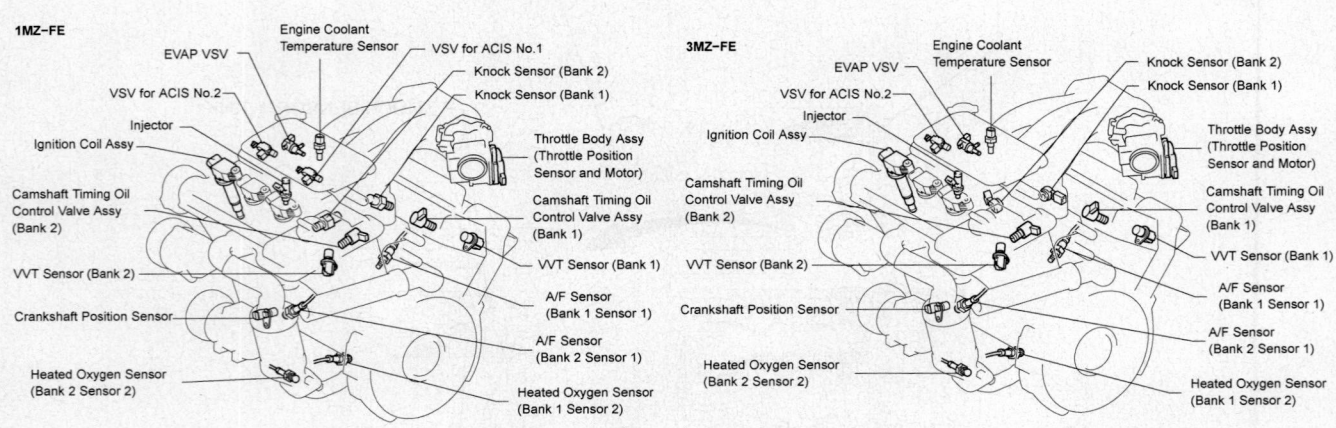

1MZ-FE
EVAP VSV
Engine Coolant Temperature Sensor
VSV for ACIS No.1
VSV for ACIS No.2
Knock Sensor (Bank 2)
Knock Sensor (Bank 1)
Injector
Ignition Coil Assy
Throttle Body Assy (Throttle Position Sensor and Motor)
Camshaft Timing Oil Control Valve Assy (Bank 2)
Camshaft Timing Oil Control Valve Assy (Bank 1)
VVT Sensor (Bank 2)
VVT Sensor (Bank 1)
Crankshaft Position Sensor
A/F Sensor (Bank 1 Sensor 1)
A/F Sensor (Bank 2 Sensor 1)
Heated Oxygen Sensor (Bank 2 Sensor 2)
Heated Oxygen Sensor (Bank 1 Sensor 2)

3MZ-FE
EVAP VSV
Engine Coolant Temperature Sensor
VSV for ACIS No.2
Knock Sensor (Bank 2)
Knock Sensor (Bank 1)
Injector
Ignition Coil Assy
Throttle Body Assy (Throttle Position Sensor and Motor)
Camshaft Timing Oil Control Valve Assy (Bank 2)
Camshaft Timing Oil Control Valve Assy (Bank 1)
VVT Sensor (Bank 2)
VVT Sensor (Bank 1)
Crankshaft Position Sensor
A/F Sensor (Bank 1 Sensor 1)
A/F Sensor (Bank 2 Sensor 1)
Heated Oxygen Sensor (Bank 2 Sensor 2)
Heated Oxygen Sensor (Bank 1 Sensor 2)

29157_TOYO_G00010

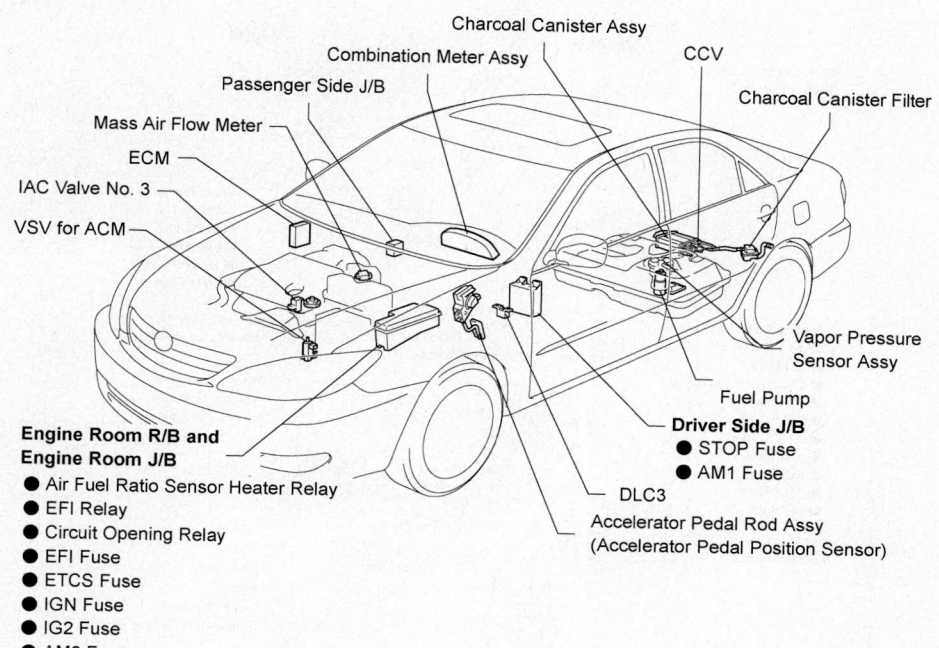

Charcoal Canister Assy

Combination Meter Assy

CCV

Passenger Side J/B

Charcoal Canister Filter

Mass Air Flow Meter

ECM

IAC Valve No. 3

VSV for ACM

Vapor Pressure
Sensor Assy

Fuel Pump

**Engine Room R/B and
Engine Room J/B**
- Air Fuel Ratio Sensor Heater Relay
- EFI Relay
- Circuit Opening Relay
- EFI Fuse
- ETCS Fuse
- IGN Fuse
- IG2 Fuse
- AM2 Fuse

Driver Side J/B
- STOP Fuse
- AM1 Fuse

DLC3

Accelerator Pedal Rod Assy
(Accelerator Pedal Position Sensor)

29157_TOYO_G0011

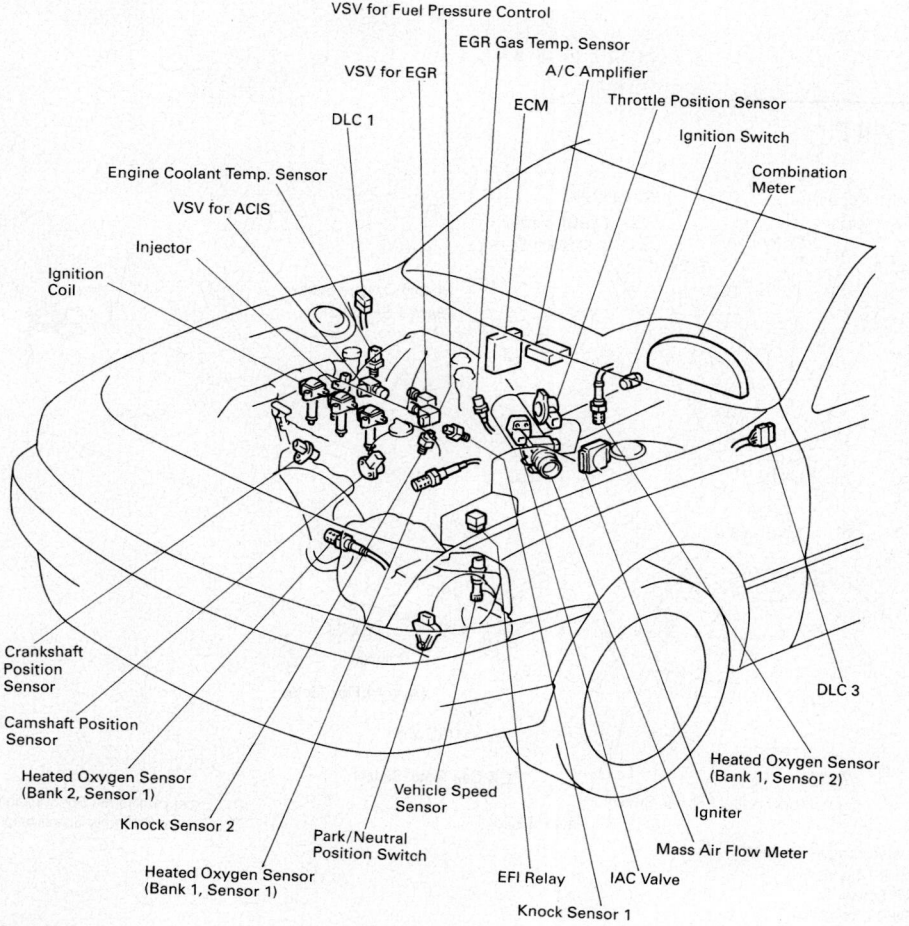

VSV for Fuel Pressure Control

EGR Gas Temp. Sensor

VSV for EGR

A/C Amplifier

ECM

Throttle Position Sensor

DLC 1

Ignition Switch

Engine Coolant Temp. Sensor

Combination
Meter

VSV for ACIS

Injector

Ignition
Coil

Crankshaft
Position
Sensor

DLC 3

Camshaft Position
Sensor

Heated Oxygen Sensor
(Bank 2, Sensor 1)

Heated Oxygen Sensor
(Bank 1, Sensor 2)

Knock Sensor 2

Vehicle Speed
Sensor

Igniter

Park/Neutral
Position Switch

Mass Air Flow Meter

Heated Oxygen Sensor
(Bank 1, Sensor 1)

EFI Relay

IAC Valve

Knock Sensor 1

29157_TOYO_G0012

2.4L I4 MFI 2AZ-FE VIN E

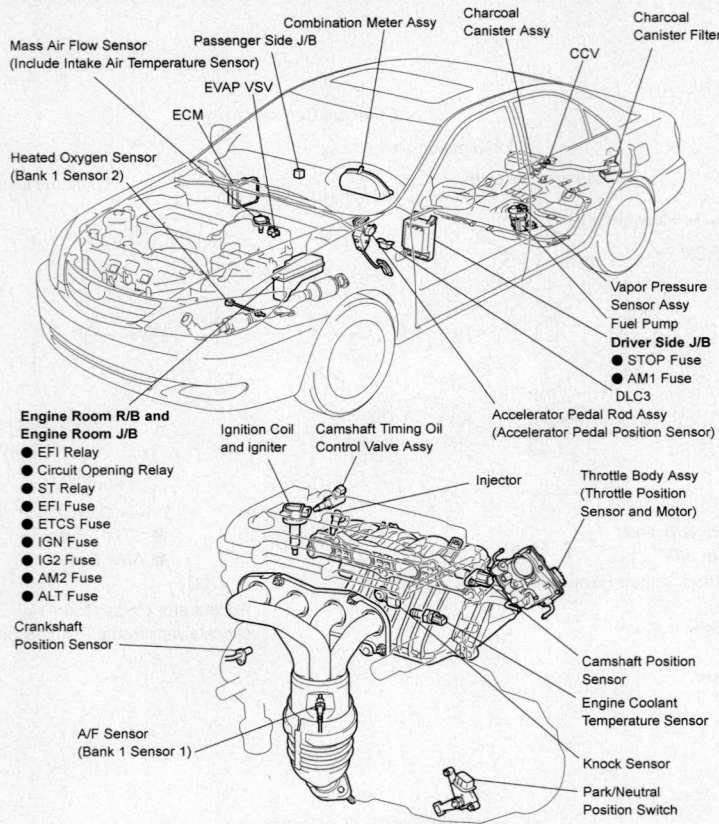

Mass Air Flow Sensor (Include Intake Air Temperature Sensor)

Combination Meter Assy

Passenger Side J/B

Charcoal Canister Assy

Charcoal Canister Filter

CCV

EVAP VSV

ECM

Heated Oxygen Sensor (Bank 1 Sensor 2)

Vapor Pressure Sensor Assy

Fuel Pump

Driver Side J/B
● STOP Fuse
● AM1 Fuse
DLC3

Accelerator Pedal Rod Assy (Accelerator Pedal Position Sensor)

Engine Room R/B and Engine Room J/B
● EFI Relay
● Circuit Opening Relay
● ST Relay
● EFI Fuse
● ETCS Fuse
● IGN Fuse
● IG2 Fuse
● AM2 Fuse
● ALT Fuse

Ignition Coil and igniter

Camshaft Timing Oil Control Valve Assy

Injector

Throttle Body Assy (Throttle Position Sensor and Motor)

Crankshaft Position Sensor

Camshaft Position Sensor

Engine Coolant Temperature Sensor

Knock Sensor

A/F Sensor (Bank 1 Sensor 1)

Park/Neutral Position Switch

29157_TOYO_G0013

CAMRY SOLARA

3.0L V6 MFI 1MZ-FE VIN F

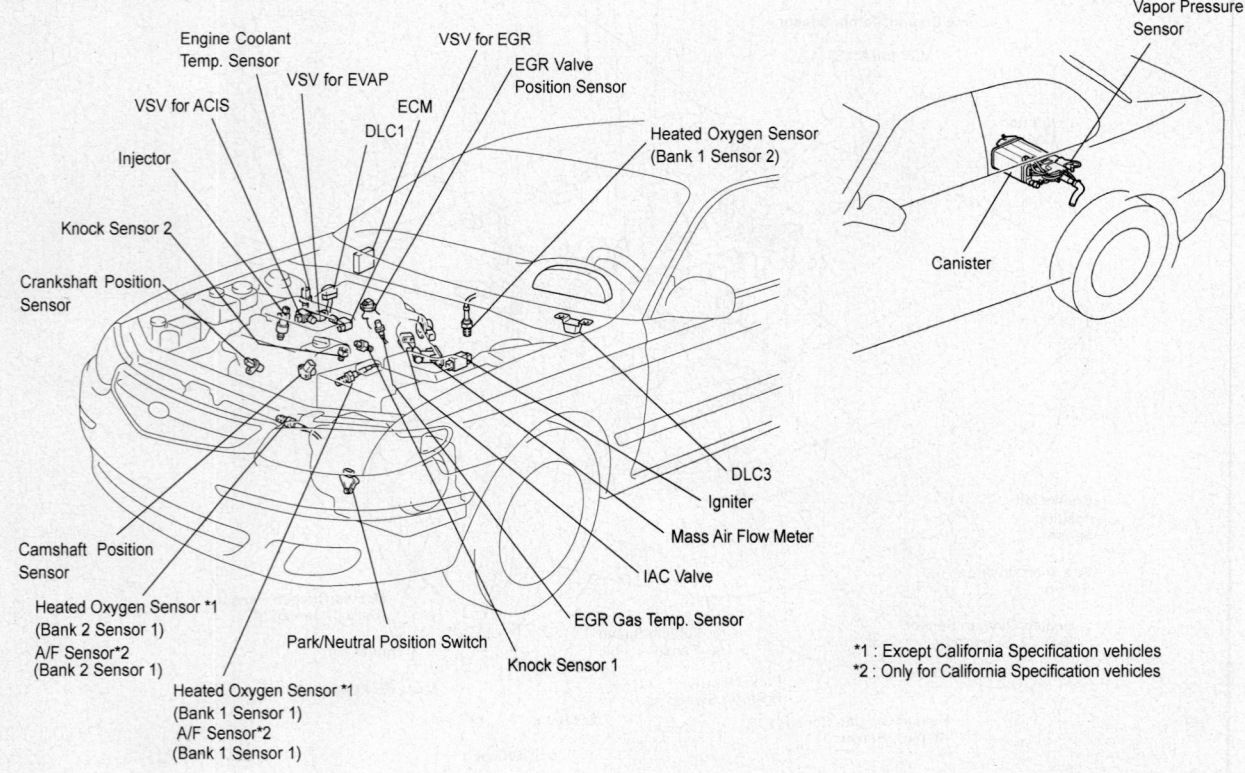

Engine Coolant Temp. Sensor

VSV for EVAP

VSV for EGR

EGR Valve Position Sensor

VSV for ACIS

ECM

DLC1

Injector

Heated Oxygen Sensor (Bank 1 Sensor 2)

Knock Sensor 2

Vapor Pressure Sensor

Crankshaft Position Sensor

Canister

Camshaft Position Sensor

DLC3

Igniter

Heated Oxygen Sensor *1 (Bank 2 Sensor 1)
A/F Sensor*2 (Bank 2 Sensor 1)

Park/Neutral Position Switch

Knock Sensor 1

Heated Oxygen Sensor *1 (Bank 1 Sensor 1)
A/F Sensor*2 (Bank 1 Sensor 1)

EGR Gas Temp. Sensor

IAC Valve

Mass Air Flow Meter

*1 : Except California Specification vehicles
*2 : Only for California Specification vehicles

29157_TOYO_G0014

2.2L I4 MFI 5S-FE VIN G

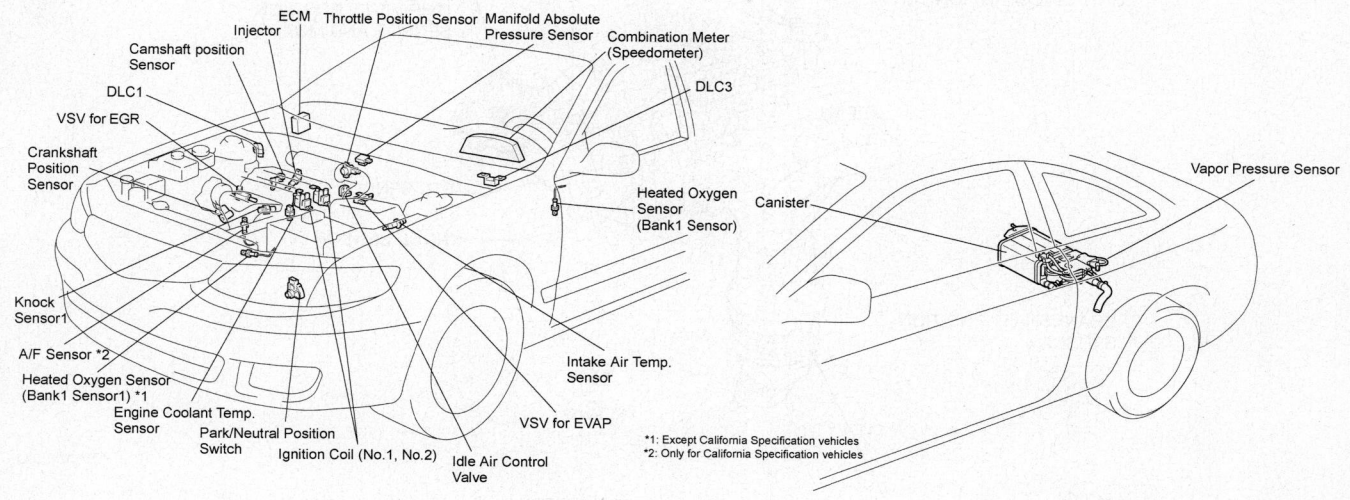

Camshaft position Sensor
ECM
Injector
Throttle Position Sensor
Manifold Absolute Pressure Sensor
Combination Meter (Speedometer)
DLC1
VSV for EGR
DLC3
Crankshaft Position Sensor
Heated Oxygen Sensor (Bank1 Sensor)
Canister
Vapor Pressure Sensor
Knock Sensor1
A/F Sensor *2
Heated Oxygen Sensor (Bank1 Sensor1) *1
Engine Coolant Temp. Sensor
Park/Neutral Position Switch
Ignition Coil (No.1, No.2)
Idle Air Control Valve
VSV for EVAP
Intake Air Temp. Sensor

*1: Except California Specification vehicles
*2: Only for California Specification vehicles

29157_TOYO_G0015

2.4L I4 MFI 2AZ-FE VIN E

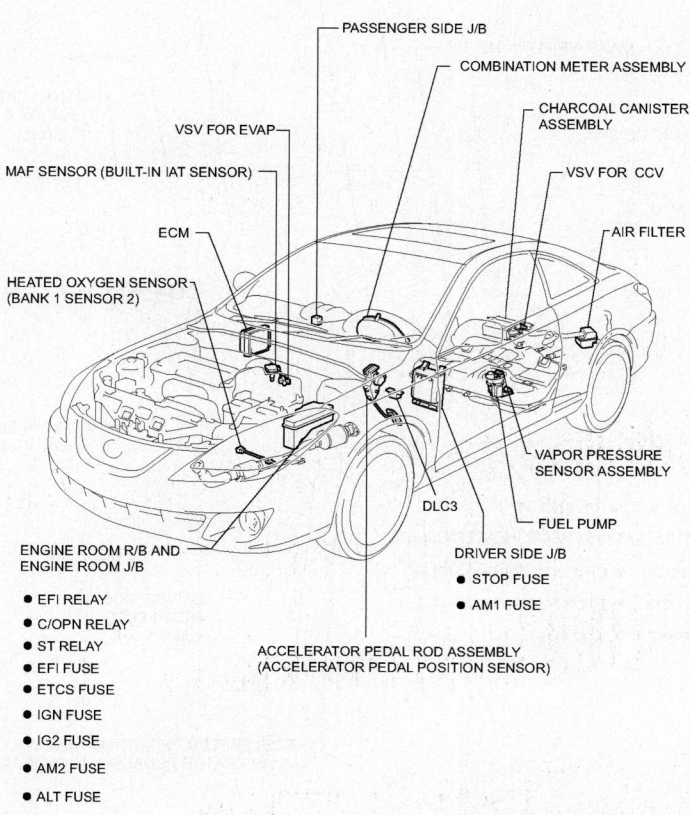

PASSENGER SIDE J/B
COMBINATION METER ASSEMBLY
VSV FOR EVAP
CHARCOAL CANISTER ASSEMBLY
MAF SENSOR (BUILT-IN IAT SENSOR)
VSV FOR CCV
ECM
AIR FILTER
HEATED OXYGEN SENSOR (BANK 1 SENSOR 2)
VAPOR PRESSURE SENSOR ASSEMBLY
DLC3
FUEL PUMP
ENGINE ROOM R/B AND ENGINE ROOM J/B
DRIVER SIDE J/B
● EFI RELAY
● STOP FUSE
● C/OPN RELAY
● AM1 FUSE
● ST RELAY
● EFI FUSE
● ETCS FUSE
ACCELERATOR PEDAL ROD ASSEMBLY (ACCELERATOR PEDAL POSITION SENSOR)
● IGN FUSE
● IG2 FUSE
● AM2 FUSE
● ALT FUSE

29157_TOYO_G0016

2.4L I4 MFI 2AZ-FE VIN E (cont'd)

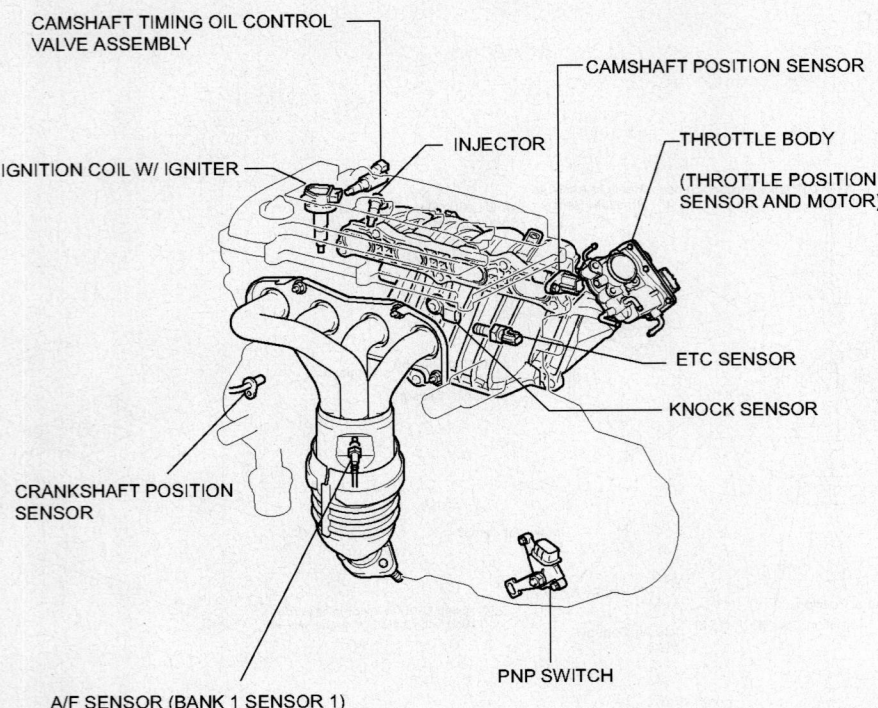

CAMSHAFT TIMING OIL CONTROL VALVE ASSEMBLY

CAMSHAFT POSITION SENSOR

INJECTOR

THROTTLE BODY
(THROTTLE POSITION SENSOR AND MOTOR)

IGNITION COIL W/ IGNITER

ETC SENSOR

KNOCK SENSOR

CRANKSHAFT POSITION SENSOR

PNP SWITCH

A/F SENSOR (BANK 1 SENSOR 1)

29157_TOYO_G0017

3.3L V6 MFI 3MZ-FE VIN A

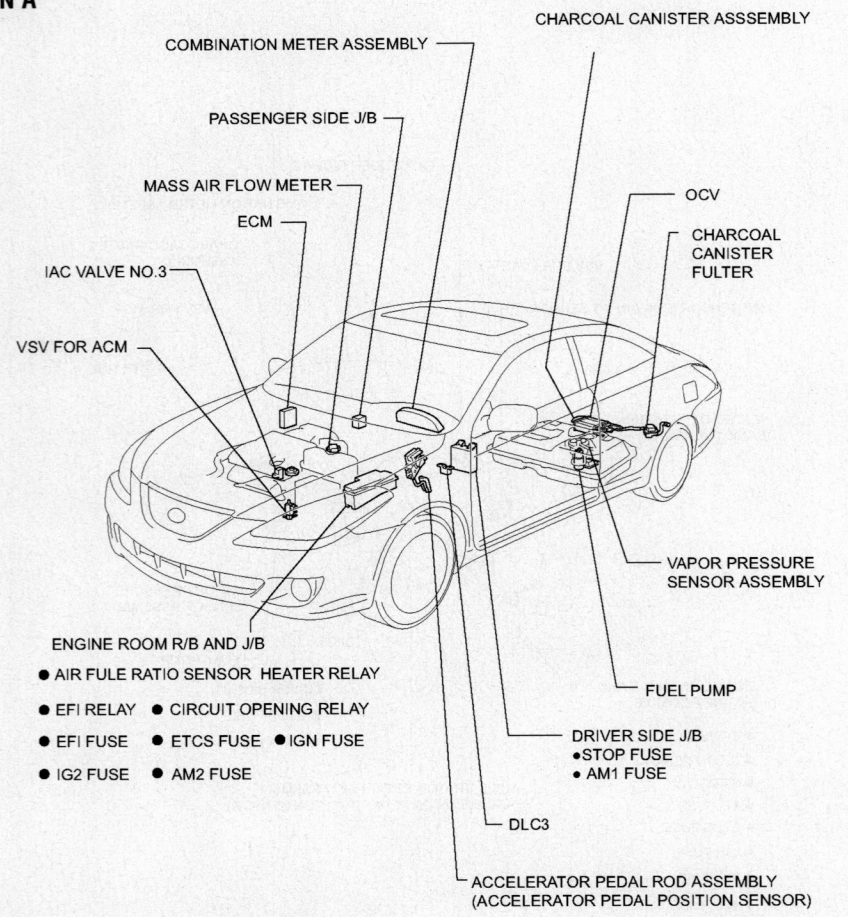

CHARCOAL CANISTER ASSSEMBLY

COMBINATION METER ASSEMBLY

PASSENGER SIDE J/B

MASS AIR FLOW METER

ECM

OCV

CHARCOAL CANISTER FULTER

IAC VALVE NO.3

VSV FOR ACM

VAPOR PRESSURE SENSOR ASSEMBLY

ENGINE ROOM R/B AND J/B
● AIR FULE RATIO SENSOR HEATER RELAY
● EFI RELAY   ● CIRCUIT OPENING RELAY
● EFI FUSE   ● ETCS FUSE   ● IGN FUSE
● IG2 FUSE   ● AM2 FUSE

FUEL PUMP

DRIVER SIDE J/B
●STOP FUSE
● AM1 FUSE

DLC3

ACCELERATOR PEDAL ROD ASSEMBLY
(ACCELERATOR PEDAL POSITION SENSOR)

29157_TOYO_G0018

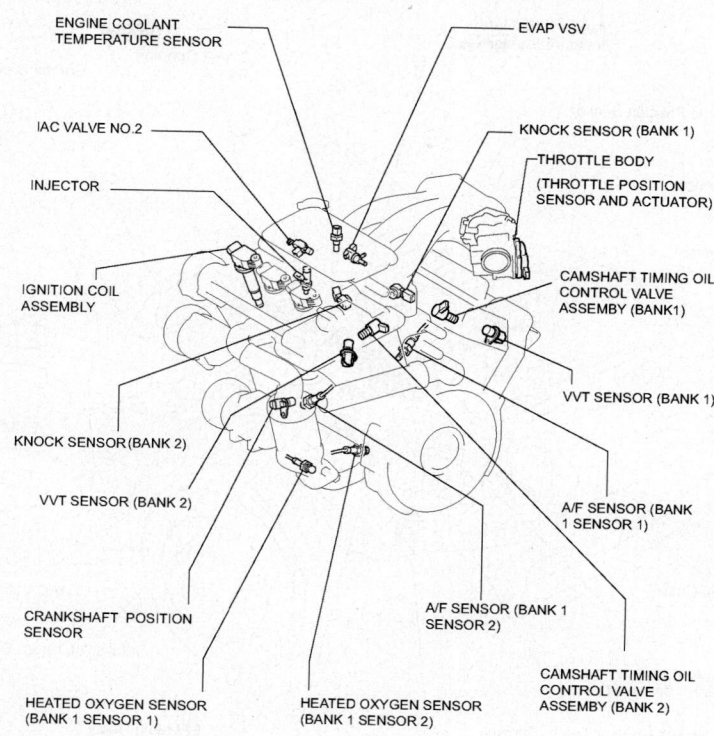

29157_TOYO_G0019

CELICA

2.2L I4 MFI 5S-FE VIN G

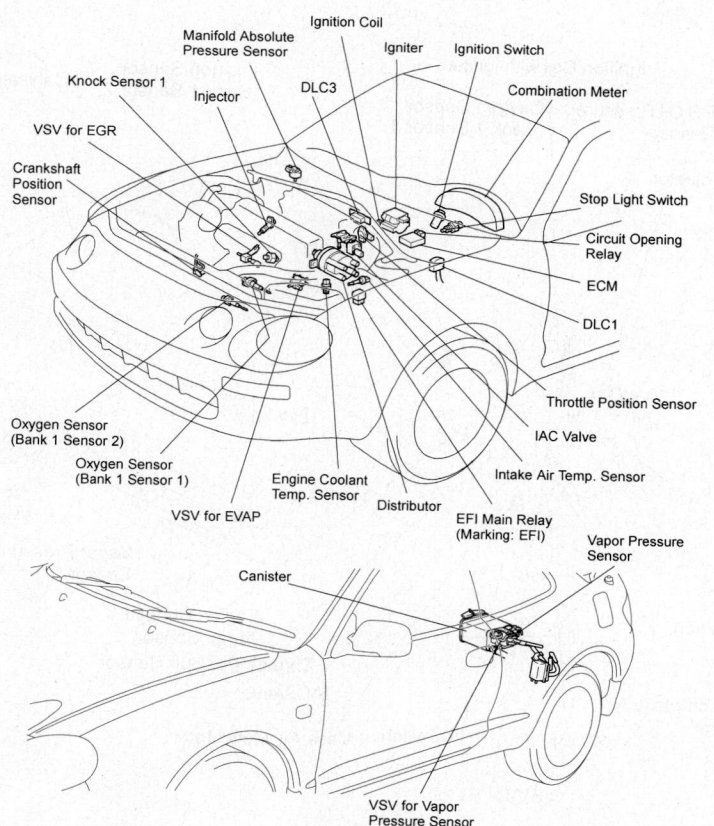

29157_TOYO_G0020

1.8L I4 MFI 7A-FE VIN B

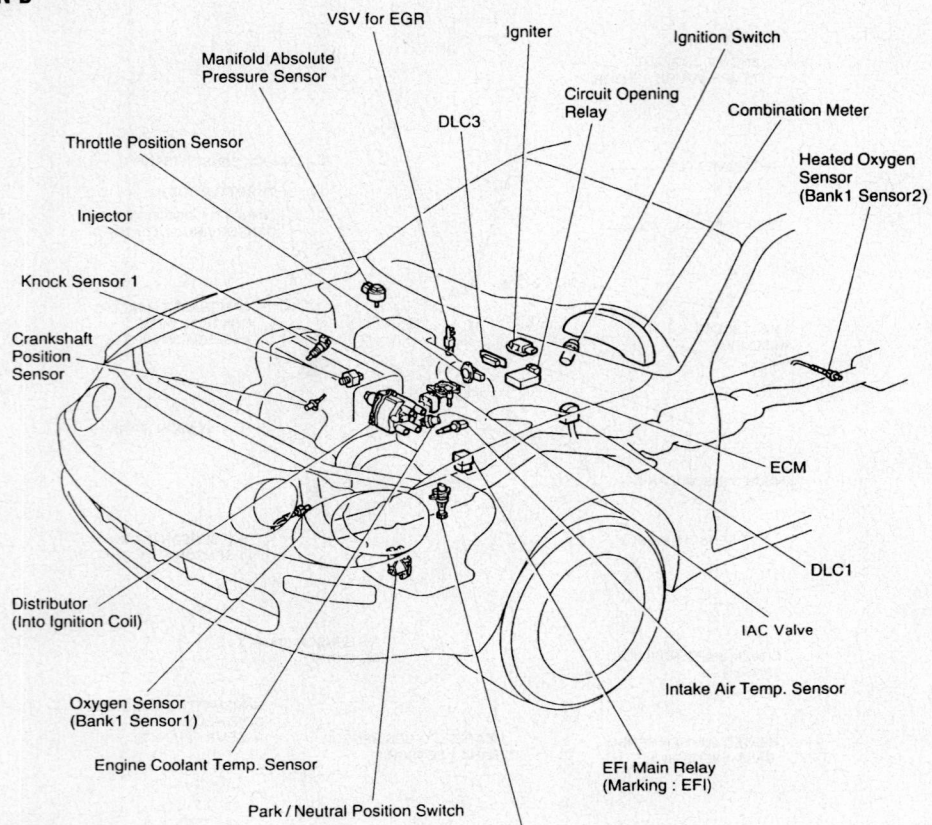

VSV for EGR

Igniter

Ignition Switch

Manifold Absolute
Pressure Sensor

Circuit Opening
Relay

Combination Meter

DLC3

Throttle Position Sensor

Heated Oxygen
Sensor
(Bank1 Sensor2)

Injector

Knock Sensor 1

Crankshaft
Position
Sensor

ECM

Distributor
(Into Ignition Coil)

DLC1

IAC Valve

Intake Air Temp. Sensor

Oxygen Sensor
(Bank1 Sensor1)

Engine Coolant Temp. Sensor

EFI Main Relay
(Marking : EFI)

Park / Neutral Position Switch

29157_TOYO_G0021

1.8L I4 MFI 1ZZ-FE VIN R

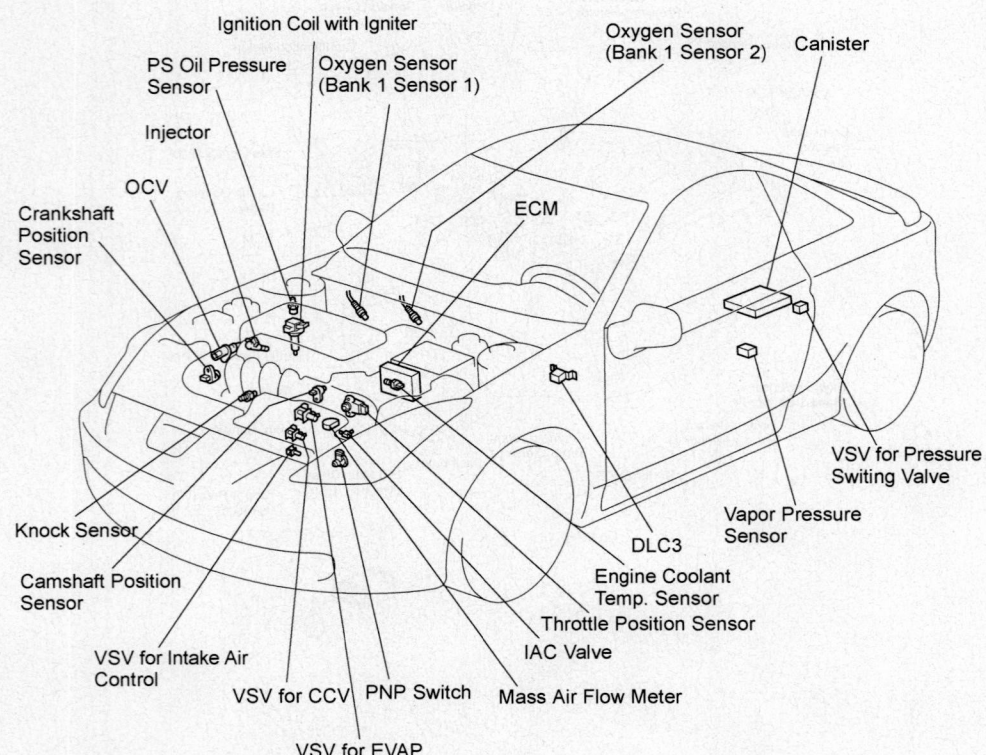

Ignition Coil with Igniter

Oxygen Sensor
(Bank 1 Sensor 2)

Canister

PS Oil Pressure
Sensor

Oxygen Sensor
(Bank 1 Sensor 1)

Injector

OCV

ECM

Crankshaft
Position
Sensor

VSV for Pressure
Switing Valve

Vapor Pressure
Sensor

Knock Sensor

DLC3

Camshaft Position
Sensor

Engine Coolant
Temp. Sensor

Throttle Position Sensor

VSV for Intake Air
Control

IAC Valve

VSV for CCV   PNP Switch

Mass Air Flow Meter

VSV for EVAP

29157_TOYO_G0022

1.8L I4 MFI 2ZZ-GE VIN Y

Engine Room J/B:
- EFI Relay
- Circuit Opening Relay
- ST Relay
- EFI Fuse
- EFI No. 1 Fuse
- EFI No. 2 Fuse
- ST Relay

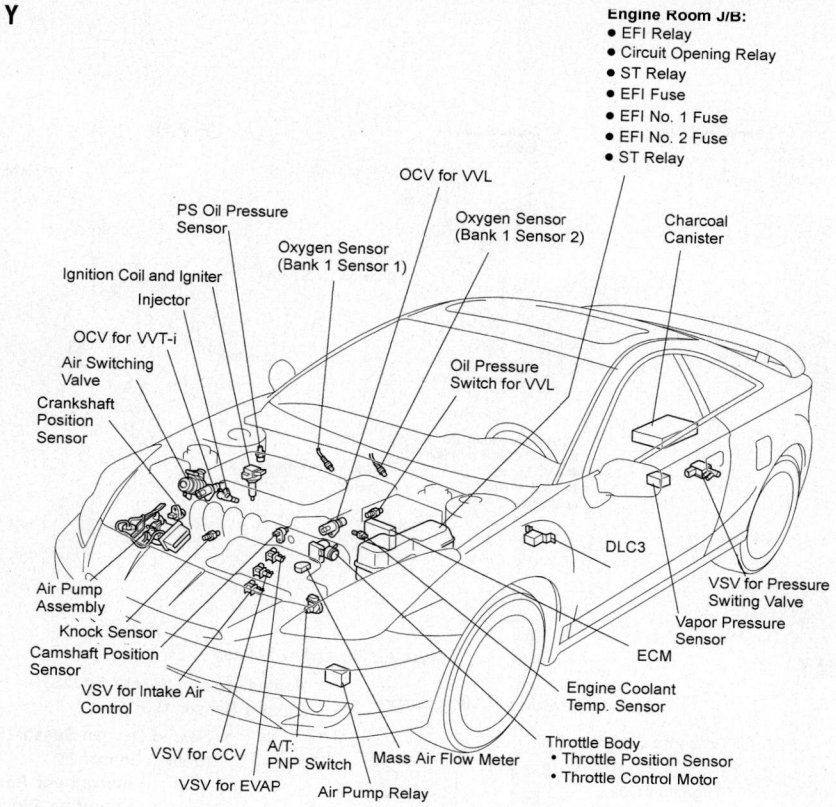

OCV for VVL
PS Oil Pressure Sensor
Oxygen Sensor (Bank 1 Sensor 1)
Oxygen Sensor (Bank 1 Sensor 2)
Charcoal Canister
Ignition Coil and Igniter
Injector
OCV for VVT-i
Air Switching Valve
Crankshaft Position Sensor
Oil Pressure Switch for VVL
Air Pump Assembly
Knock Sensor
Camshaft Position Sensor
VSV for Intake Air Control
VSV for CCV
A/T: PNP Switch
VSV for EVAP
Air Pump Relay
Mass Air Flow Meter
DLC3
VSV for Pressure Switing Valve
Vapor Pressure Sensor
ECM
Engine Coolant Temp. Sensor
Throttle Body
• Throttle Position Sensor
• Throttle Control Motor

29157_TOYO_G0023

COROLLA

1.6L I4 MFI 4A-FE VIN A
1.8L I4 MFI 7A-FE VIN B

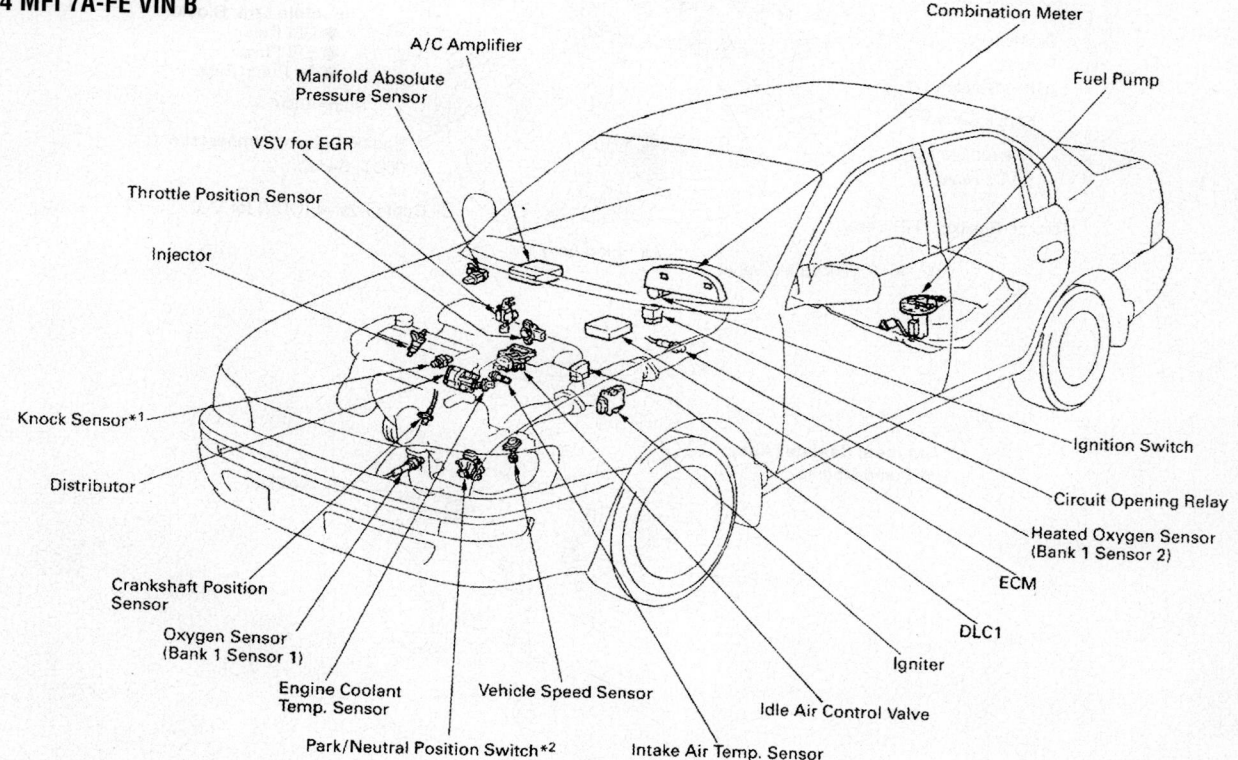

A/C Amplifier
Manifold Absolute Pressure Sensor
VSV for EGR
Throttle Position Sensor
Injector
Combination Meter
Fuel Pump
Knock Sensor*1
Distributor
Crankshaft Position Sensor
Oxygen Sensor (Bank 1 Sensor 1)
Engine Coolant Temp. Sensor
Vehicle Speed Sensor
Park/Neutral Position Switch*2
Intake Air Temp. Sensor
Idle Air Control Valve
Igniter
DLC1
ECM
Heated Oxygen Sensor (Bank 1 Sensor 2)
Circuit Opening Relay
Ignition Switch

29157_TOYO_G0024

1.8L I4 MFI 1ZZ-FE VIN R

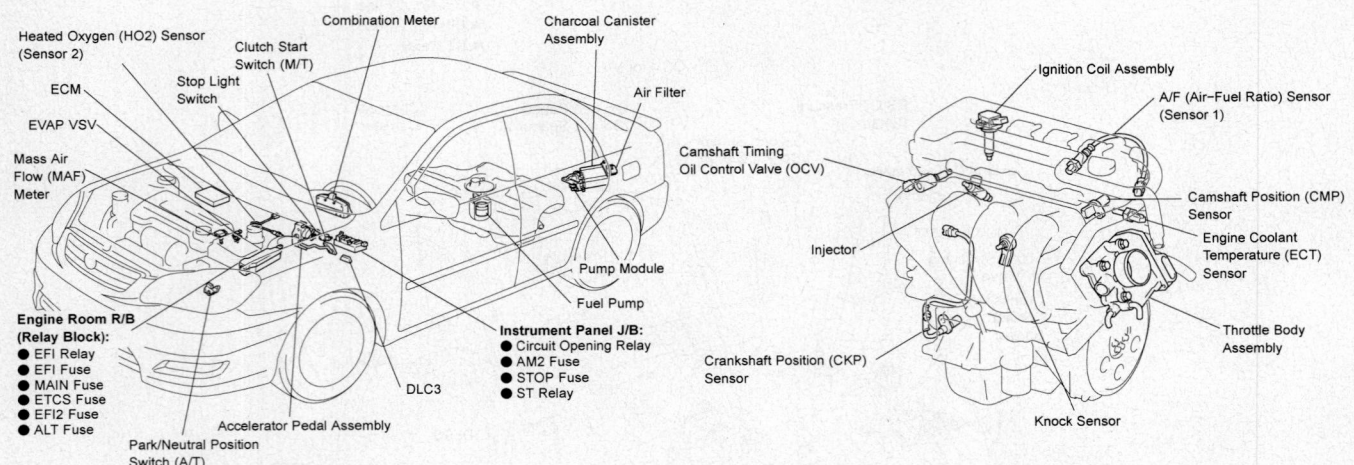

Heated Oxygen (HO2) Sensor (Sensor 2)
ECM
EVAP VSV
Mass Air Flow (MAF) Meter
Stop Light Switch
Clutch Start Switch (M/T)
Combination Meter
Charcoal Canister Assembly
Air Filter

Engine Room R/B (Relay Block):
● EFI Relay
● EFI Fuse
● MAIN Fuse
● ETCS Fuse
● EFI2 Fuse
● ALT Fuse

Park/Neutral Position Switch (A/T)
Accelerator Pedal Assembly
DLC3

Pump Module
Fuel Pump

Instrument Panel J/B:
● Circuit Opening Relay
● AM2 Fuse
● STOP Fuse
● ST Relay

Ignition Coil Assembly
A/F (Air-Fuel Ratio) Sensor (Sensor 1)
Camshaft Timing Oil Control Valve (OCV)
Injector
Crankshaft Position (CKP) Sensor
Knock Sensor
Camshaft Position (CMP) Sensor
Engine Coolant Temperature (ECT) Sensor
Throttle Body Assembly

29157_TOYO_G0025

1.8L I4 MFI 2ZZ-GE VIN Y

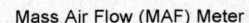
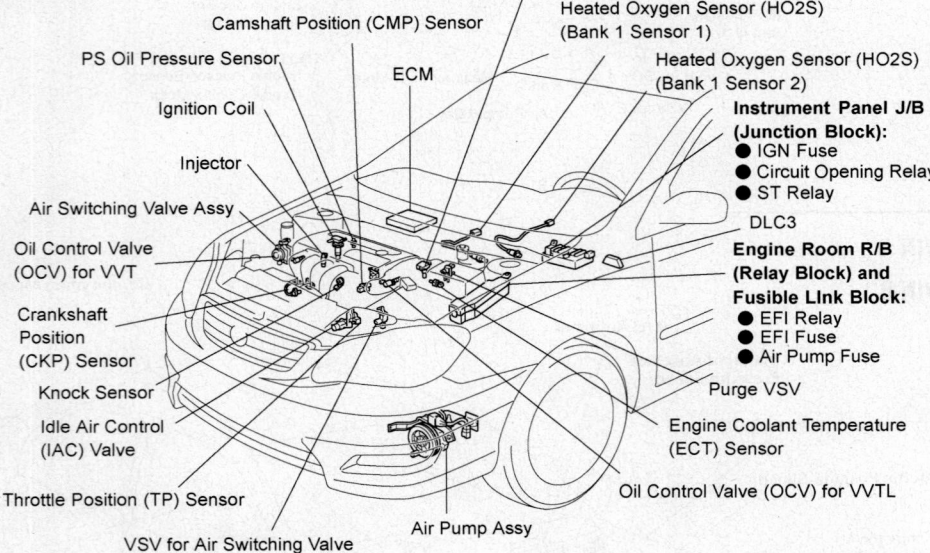

Mass Air Flow (MAF) Meter
Camshaft Position (CMP) Sensor
PS Oil Pressure Sensor
ECM
Heated Oxygen Sensor (HO2S) (Bank 1 Sensor 1)
Heated Oxygen Sensor (HO2S) (Bank 1 Sensor 2)
Ignition Coil
Injector
Air Switching Valve Assy
Oil Control Valve (OCV) for VVT
Crankshaft Position (CKP) Sensor
Knock Sensor
Idle Air Control (IAC) Valve
Throttle Position (TP) Sensor
VSV for Air Switching Valve
Air Pump Assy

Instrument Panel J/B (Junction Block):
● IGN Fuse
● Circuit Opening Relay
● ST Relay

DLC3

Engine Room R/B (Relay Block) and Fusible Link Block:
● EFI Relay
● EFI Fuse
● Air Pump Fuse

Purge VSV
Engine Coolant Temperature (ECT) Sensor
Oil Control Valve (OCV) for VVTL

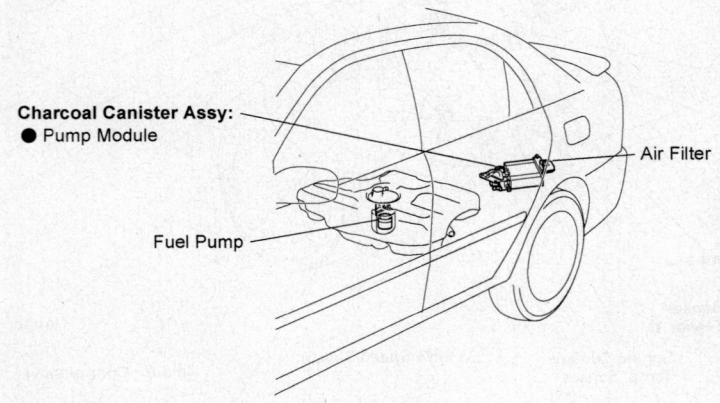

Charcoal Canister Assy:
● Pump Module

Fuel Pump
Air Filter

29157_TOYO_G0026

ECHO

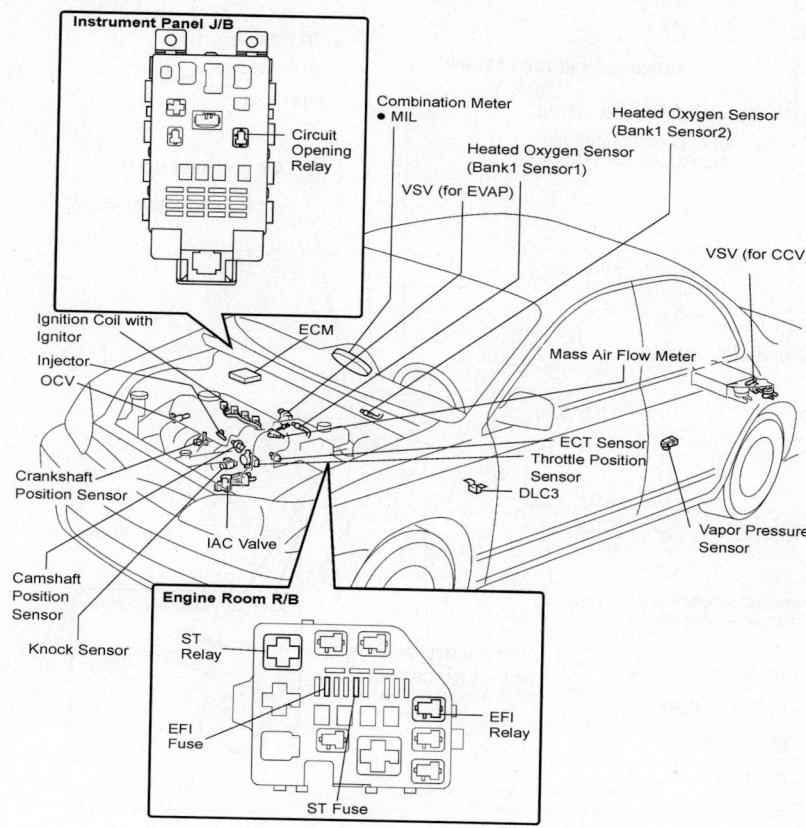

Instrument Panel J/B

Circuit Opening Relay

Combination Meter
● MIL

Heated Oxygen Sensor (Bank1 Sensor2)

Heated Oxygen Sensor (Bank1 Sensor1)

VSV (for EVAP)

VSV (for CCV)

Ignition Coil with Ignitor

ECM

Injector

OCV

Mass Air Flow Meter

Crankshaft Position Sensor

ECT Sensor
Throttle Position Sensor

DLC3

IAC Valve

Vapor Pressure Sensor

Camshaft Position Sensor

Engine Room R/B

Knock Sensor

ST Relay

EFI Fuse

EFI Relay

ST Fuse

29157_TOYO_G0027

HIGHLANDER

2.4L I4 MFI 2AZ-FE VIN D

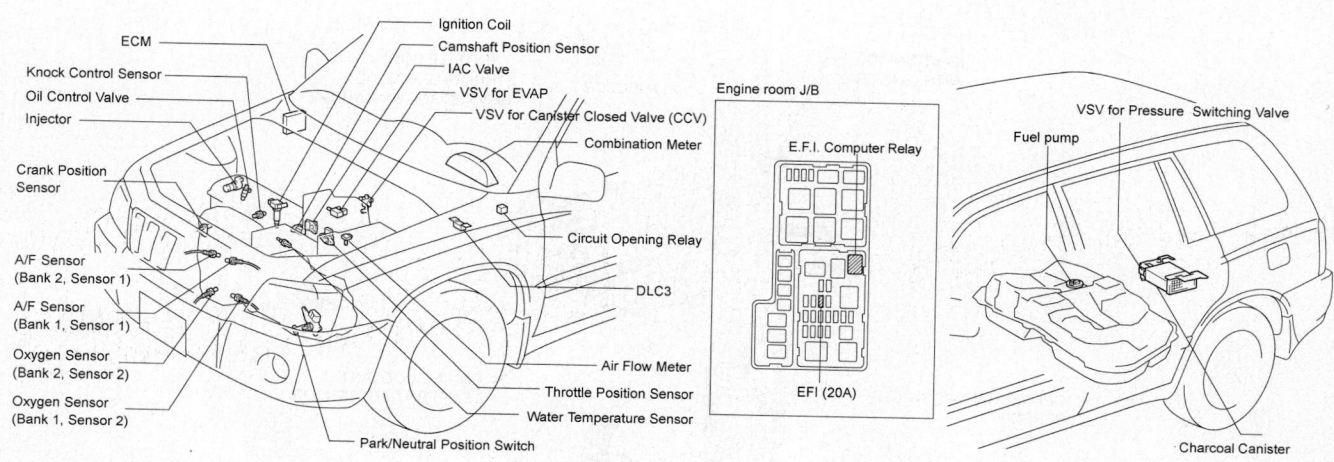

ECM

Ignition Coil

Knock Control Sensor

Camshaft Position Sensor

Oil Control Valve

IAC Valve

Injector

VSV for EVAP

VSV for Canister Closed Valve (CCV)

Crank Position Sensor

Combination Meter

Circuit Opening Relay

Engine room J/B

VSV for Pressure Switching Valve

Fuel pump

E.F.I. Computer Relay

A/F Sensor (Bank 2, Sensor 1)

A/F Sensor (Bank 1, Sensor 1)

Oxygen Sensor (Bank 2, Sensor 2)

DLC3

Air Flow Meter

Throttle Position Sensor

Oxygen Sensor (Bank 1, Sensor 2)

Water Temperature Sensor

EFI (20A)

Park/Neutral Position Switch

Charcoal Canister

29157_TOYO_G0028

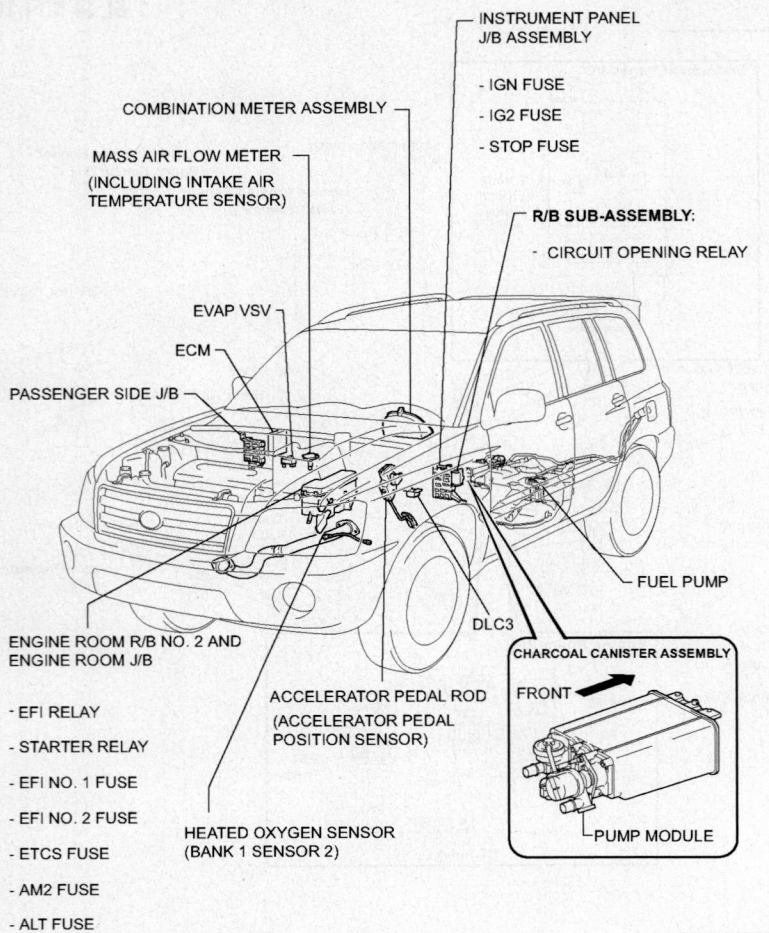

INSTRUMENT PANEL
J/B ASSEMBLY

- IGN FUSE
- IG2 FUSE
- STOP FUSE

COMBINATION METER ASSEMBLY

MASS AIR FLOW METER
(INCLUDING INTAKE AIR
TEMPERATURE SENSOR)

R/B SUB-ASSEMBLY:

- CIRCUIT OPENING RELAY

EVAP VSV

ECM

PASSENGER SIDE J/B

FUEL PUMP

DLC3

ENGINE ROOM R/B NO. 2 AND
ENGINE ROOM J/B

ACCELERATOR PEDAL ROD
(ACCELERATOR PEDAL
POSITION SENSOR)

- EFI RELAY
- STARTER RELAY
- EFI NO. 1 FUSE
- EFI NO. 2 FUSE
- ETCS FUSE
- AM2 FUSE
- ALT FUSE
- STARTER FUSE

HEATED OXYGEN SENSOR
(BANK 1 SENSOR 2)

CHARCOAL CANISTER ASSEMBLY

FRONT

PUMP MODULE

29157_TOYO_G0029

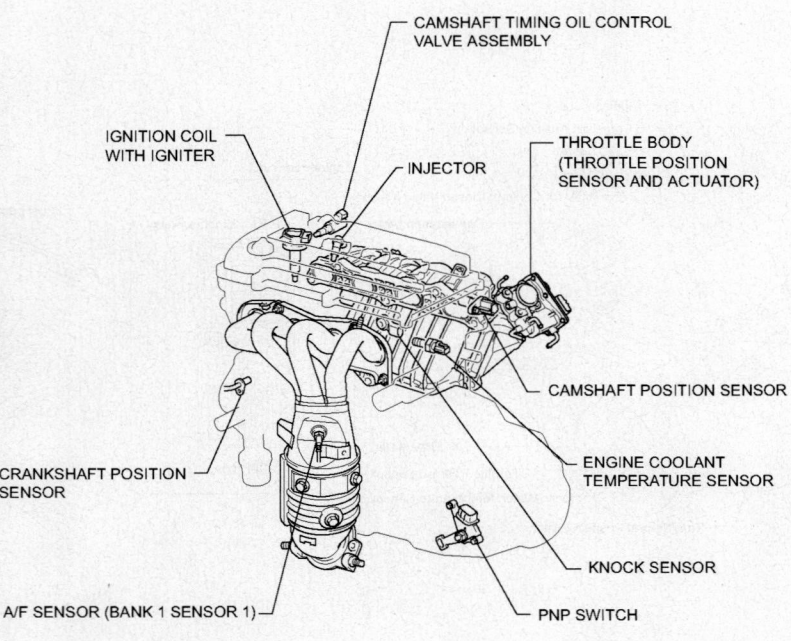

CAMSHAFT TIMING OIL CONTROL
VALVE ASSEMBLY

IGNITION COIL
WITH IGNITER

INJECTOR

THROTTLE BODY
(THROTTLE POSITION
SENSOR AND ACTUATOR)

CAMSHAFT POSITION SENSOR

CRANKSHAFT POSITION
SENSOR

ENGINE COOLANT
TEMPERATURE SENSOR

KNOCK SENSOR

A/F SENSOR (BANK 1 SENSOR 1)

PNP SWITCH

29157_TOYO_G0030

3.0L V6 MFI 1MZ-FE VIN F

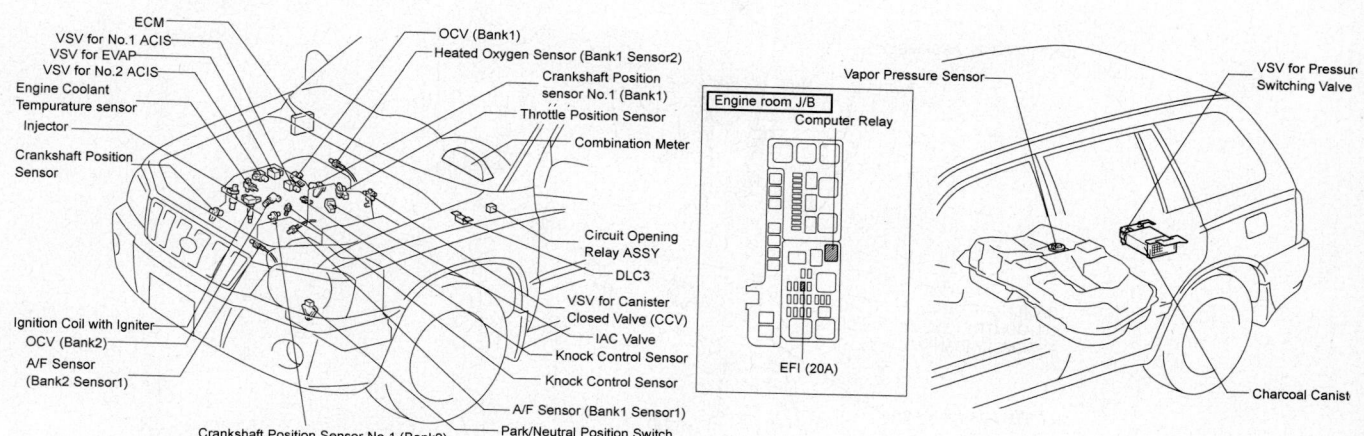

ECM
VSV for No.1 ACIS
VSV for EVAP
VSV for No.2 ACIS
Engine Coolant
Tempurature sensor
Injector
Crankshaft Position
Sensor

OCV (Bank1)
Heated Oxygen Sensor (Bank1 Sensor2)
Crankshaft Position
sensor No.1 (Bank1)
Throttle Position Sensor
Combination Meter

Circuit Opening
Relay ASSY
DLC3
VSV for Canister
Closed Valve (CCV)
IAC Valve
Knock Control Sensor
Knock Control Sensor

Ignition Coil with Igniter
OCV (Bank2)
A/F Sensor
(Bank2 Sensor1)

Crankshaft Position Sensor No.1 (Bank2)

A/F Sensor (Bank1 Sensor1)
Park/Neutral Position Switch

Engine room J/B
Computer Relay

Vapor Pressure Sensor

VSV for Pressure
Switching Valve

Charcoal Canist

EFI (20A)

29157_TOYO_G0031

3.3L V6 MFI 3MZ-FE VIN A

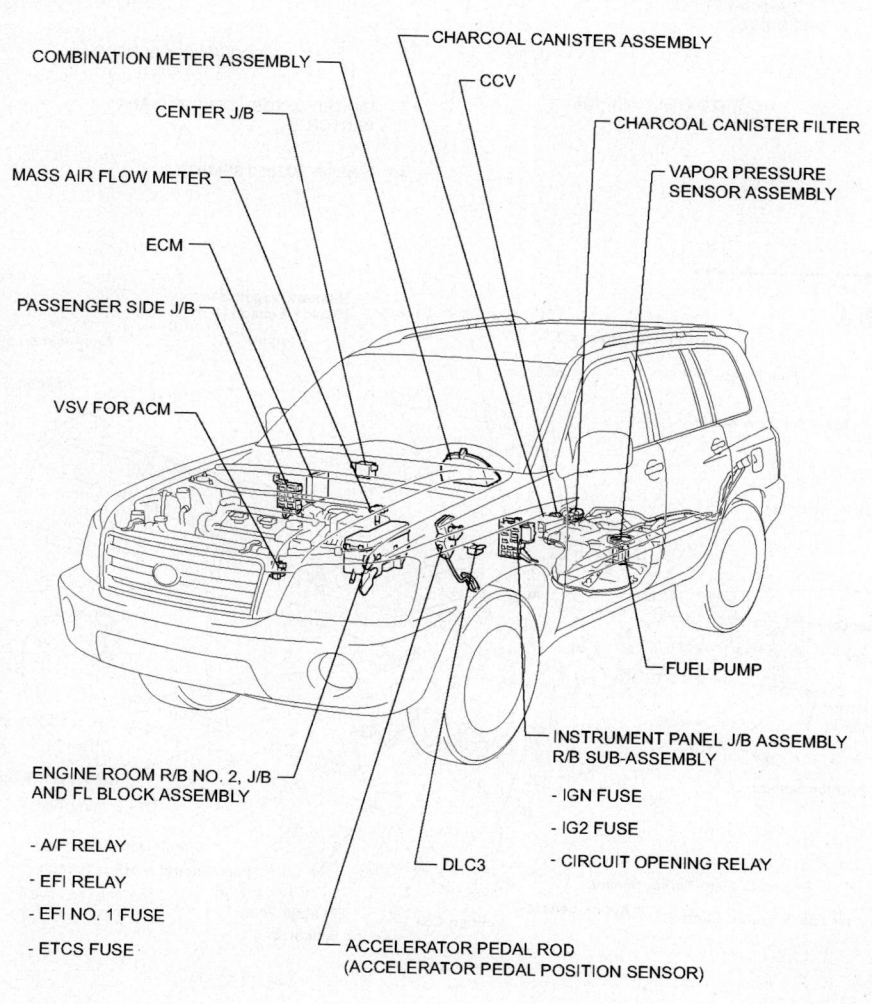

COMBINATION METER ASSEMBLY
CENTER J/B
MASS AIR FLOW METER
ECM
PASSENGER SIDE J/B
VSV FOR ACM

CHARCOAL CANISTER ASSEMBLY
CCV
CHARCOAL CANISTER FILTER
VAPOR PRESSURE
SENSOR ASSEMBLY

FUEL PUMP

INSTRUMENT PANEL J/B ASSEMBLY
R/B SUB-ASSEMBLY

- IGN FUSE
- IG2 FUSE
- CIRCUIT OPENING RELAY

ENGINE ROOM R/B NO. 2, J/B
AND FL BLOCK ASSEMBLY

- A/F RELAY
- EFI RELAY
- EFI NO. 1 FUSE
- ETCS FUSE

DLC3

ACCELERATOR PEDAL ROD
(ACCELERATOR PEDAL POSITION SENSOR)

29157_TOYO_G0032

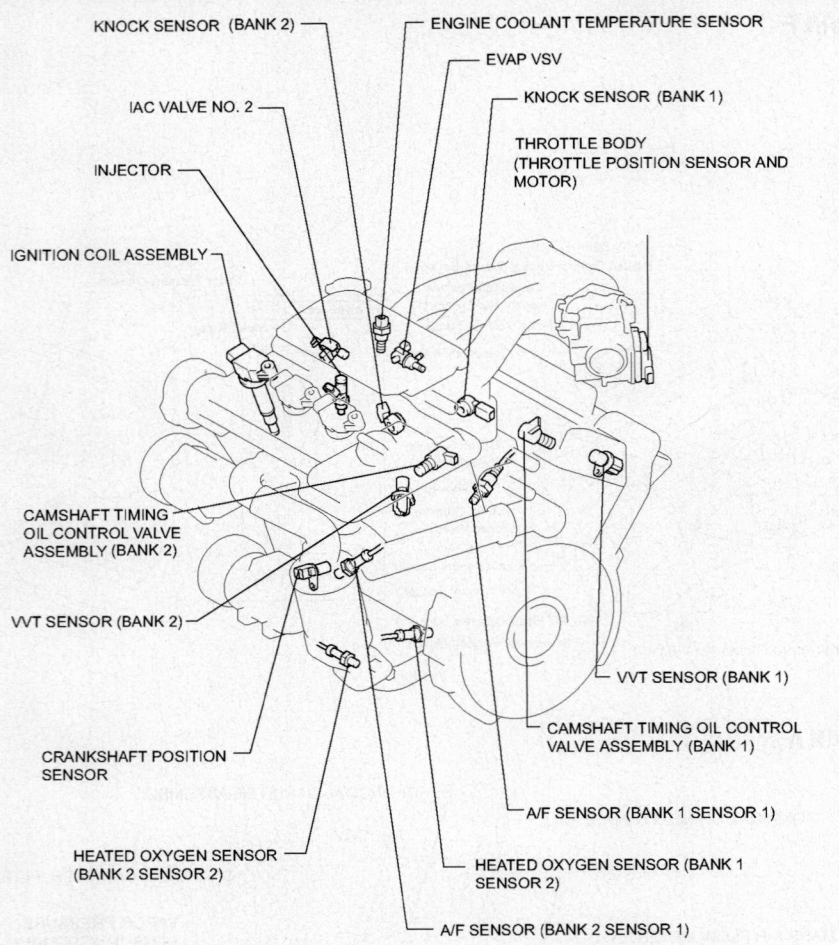

KNOCK SENSOR (BANK 2)

ENGINE COOLANT TEMPERATURE SENSOR

EVAP VSV

IAC VALVE NO. 2

KNOCK SENSOR (BANK 1)

INJECTOR

THROTTLE BODY
(THROTTLE POSITION SENSOR AND
MOTOR)

IGNITION COIL ASSEMBLY

CAMSHAFT TIMING
OIL CONTROL VALVE
ASSEMBLY (BANK 2)

VVT SENSOR (BANK 2)

VVT SENSOR (BANK 1)

CAMSHAFT TIMING OIL CONTROL
VALVE ASSEMBLY (BANK 1)

CRANKSHAFT POSITION
SENSOR

A/F SENSOR (BANK 1 SENSOR 1)

HEATED OXYGEN SENSOR
(BANK 2 SENSOR 2)

HEATED OXYGEN SENSOR (BANK 1
SENSOR 2)

A/F SENSOR (BANK 2 SENSOR 1)

29157_TOYO_G0033

LAND CRUISER

4.5L I6 MFI 1FZ-FE VIN J

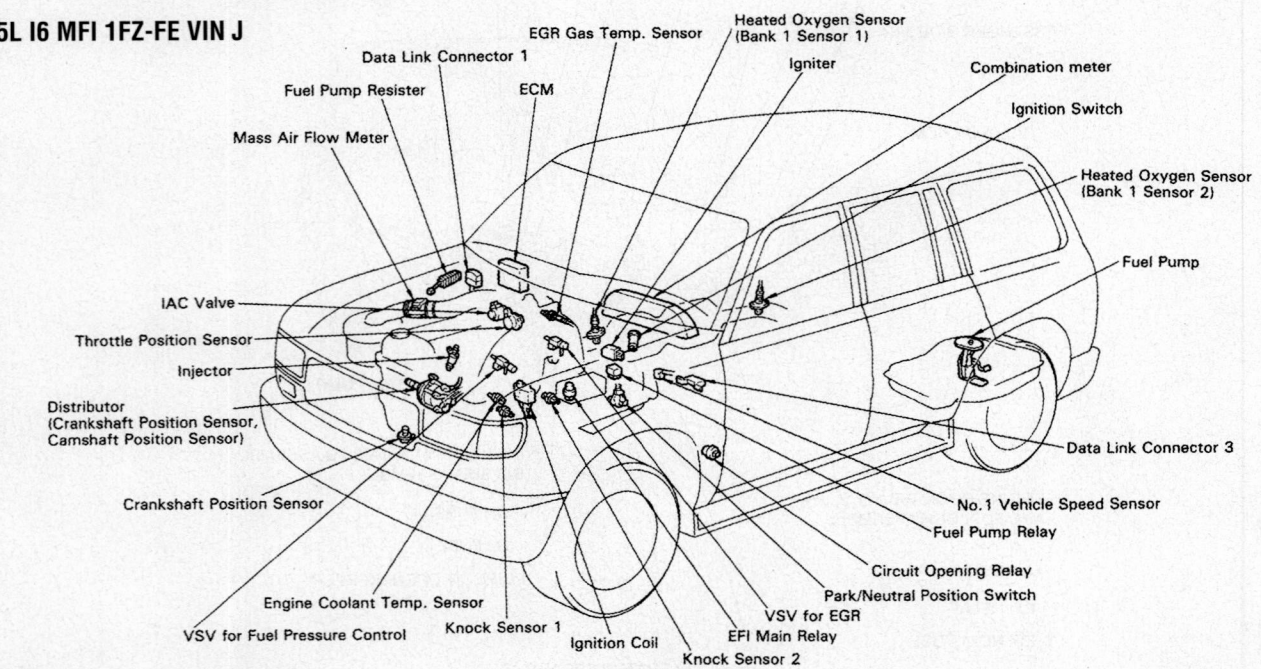

EGR Gas Temp. Sensor

Heated Oxygen Sensor
(Bank 1 Sensor 1)

Data Link Connector 1

Igniter

Combination meter

Fuel Pump Resister

ECM

Ignition Switch

Mass Air Flow Meter

Heated Oxygen Sensor
(Bank 1 Sensor 2)

Fuel Pump

IAC Valve

Throttle Position Sensor

Injector

Distributor
(Crankshaft Position Sensor,
Camshaft Position Sensor)

Data Link Connector 3

Crankshaft Position Sensor

No. 1 Vehicle Speed Sensor

Fuel Pump Relay

Circuit Opening Relay

Park/Neutral Position Switch

Engine Coolant Temp. Sensor

VSV for EGR

VSV for Fuel Pressure Control

Knock Sensor 1

Ignition Coil

EFI Main Relay

Knock Sensor 2

29157_TOYO_G0034

4.7L V8 MFI 2UZ-FE VIN T

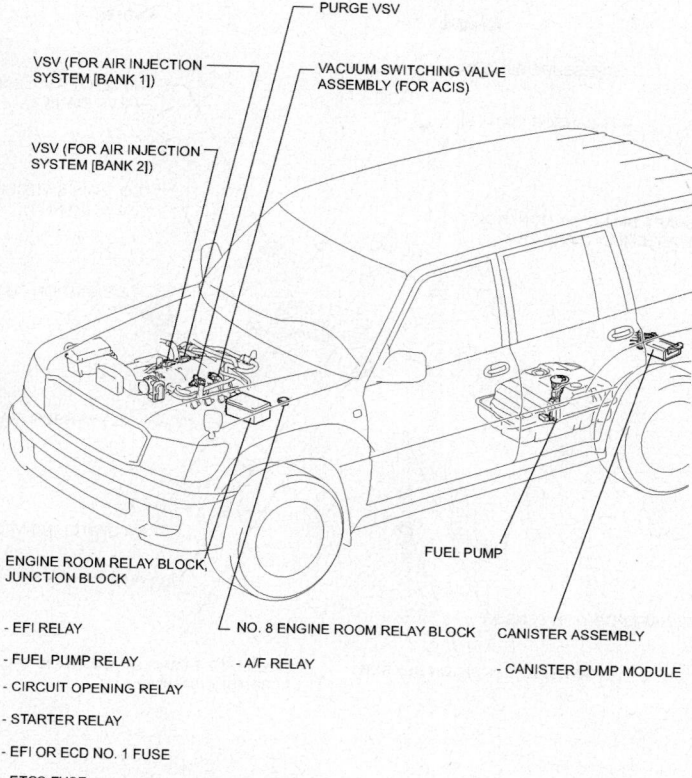

PURGE VSV

VSV (FOR AIR INJECTION
SYSTEM [BANK 1])

VACUUM SWITCHING VALVE
ASSEMBLY (FOR ACIS)

VSV (FOR AIR INJECTION
SYSTEM [BANK 2])

FUEL PUMP

ENGINE ROOM RELAY BLOCK,
JUNCTION BLOCK

- EFI RELAY

- FUEL PUMP RELAY

- CIRCUIT OPENING RELAY

- STARTER RELAY

- EFI OR ECD NO. 1 FUSE

- ETCS FUSE

NO. 8 ENGINE ROOM RELAY BLOCK

- A/F RELAY

CANISTER ASSEMBLY

- CANISTER PUMP MODULE

29157_TOYO_G0035

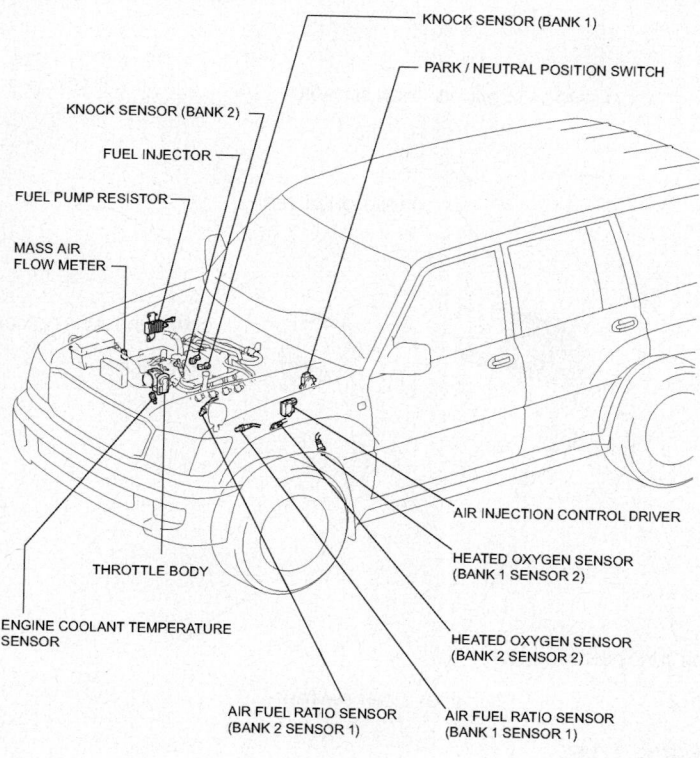

KNOCK SENSOR (BANK 1)

PARK / NEUTRAL POSITION SWITCH

KNOCK SENSOR (BANK 2)

FUEL INJECTOR

FUEL PUMP RESISTOR

MASS AIR
FLOW METER

AIR INJECTION CONTROL DRIVER

HEATED OXYGEN SENSOR
(BANK 1 SENSOR 2)

THROTTLE BODY

HEATED OXYGEN SENSOR
(BANK 2 SENSOR 2)

ENGINE COOLANT TEMPERATURE
SENSOR

AIR FUEL RATIO SENSOR
(BANK 2 SENSOR 1)

AIR FUEL RATIO SENSOR
(BANK 1 SENSOR 1)

29157_TOYO_G0036

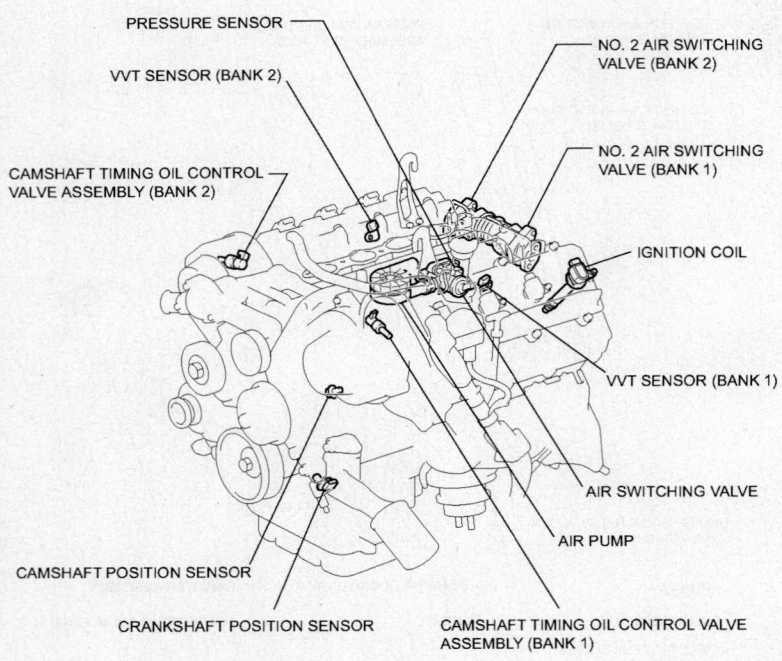

PRESSURE SENSOR

VVT SENSOR (BANK 2)

CAMSHAFT TIMING OIL CONTROL
VALVE ASSEMBLY (BANK 2)

NO. 2 AIR SWITCHING
VALVE (BANK 2)

NO. 2 AIR SWITCHING
VALVE (BANK 1)

IGNITION COIL

VVT SENSOR (BANK 1)

AIR SWITCHING VALVE

AIR PUMP

CAMSHAFT POSITION SENSOR

CRANKSHAFT POSITION SENSOR

CAMSHAFT TIMING OIL CONTROL VALVE
ASSEMBLY (BANK 1)

29157_TOYO_G0037

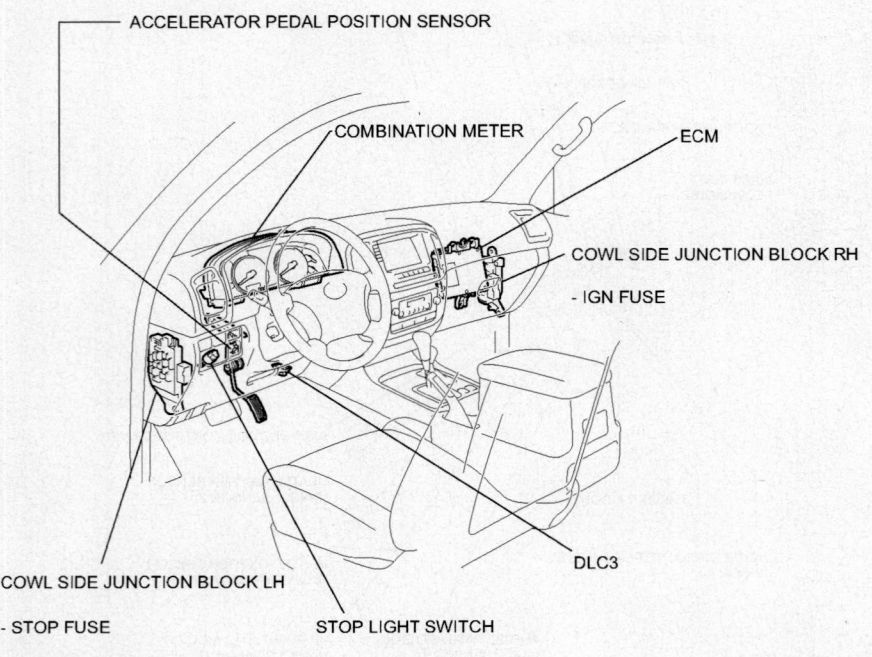

ACCELERATOR PEDAL POSITION SENSOR

COMBINATION METER

ECM

COWL SIDE JUNCTION BLOCK RH

- IGN FUSE

DLC3

COWL SIDE JUNCTION BLOCK LH

- STOP FUSE

STOP LIGHT SWITCH

- EFI OR ECD NO. 2 FUSE

1.8L I4 MFI 1ZZ-FE VIN R

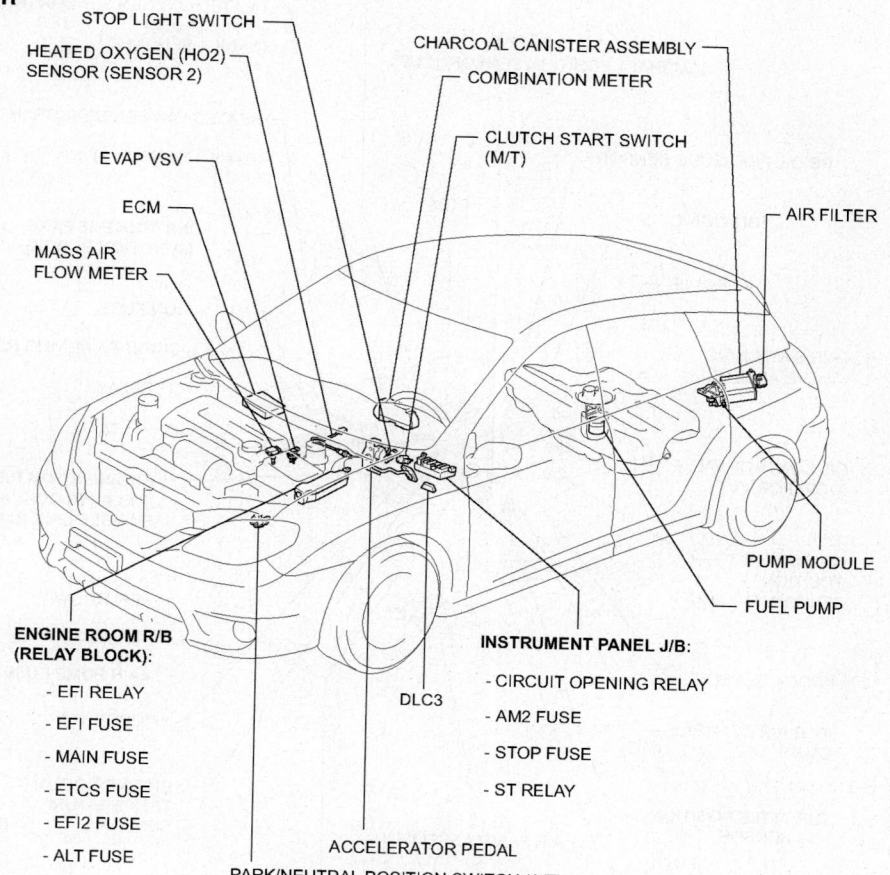

STOP LIGHT SWITCH

HEATED OXYGEN (HO2)
SENSOR (SENSOR 2)

EVAP VSV

ECM

MASS AIR
FLOW METER

CHARCOAL CANISTER ASSEMBLY

COMBINATION METER

CLUTCH START SWITCH
(M/T)

AIR FILTER

PUMP MODULE

FUEL PUMP

**ENGINE ROOM R/B
(RELAY BLOCK):**

- EFI RELAY

- EFI FUSE

- MAIN FUSE

- ETCS FUSE

- EFI2 FUSE

- ALT FUSE

DLC3

INSTRUMENT PANEL J/B:

- CIRCUIT OPENING RELAY

- AM2 FUSE

- STOP FUSE

- ST RELAY

ACCELERATOR PEDAL

PARK/NEUTRAL POSITION SWITCH (A/T)

29157_TOYO_G0039

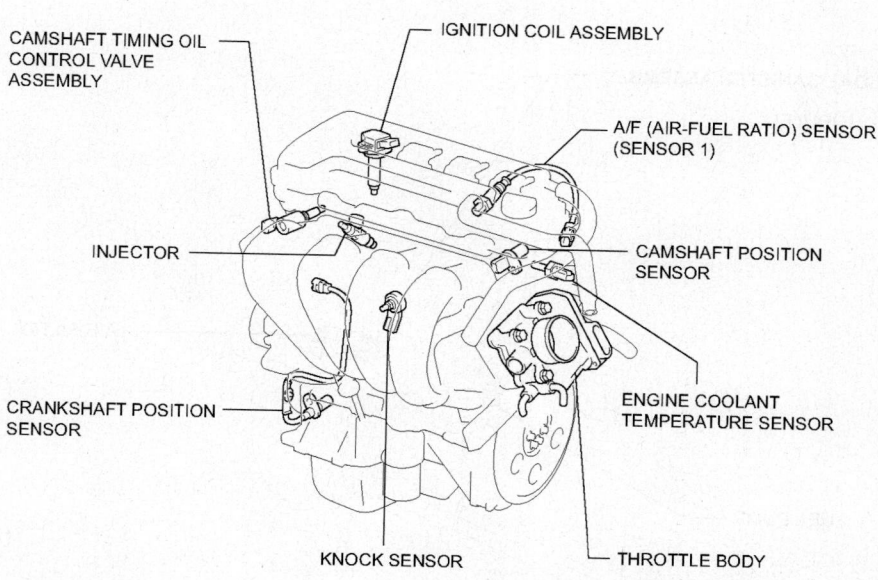

CAMSHAFT TIMING OIL
CONTROL VALVE
ASSEMBLY

IGNITION COIL ASSEMBLY

A/F (AIR-FUEL RATIO) SENSOR
(SENSOR 1)

INJECTOR

CAMSHAFT POSITION
SENSOR

CRANKSHAFT POSITION
SENSOR

ENGINE COOLANT
TEMPERATURE SENSOR

KNOCK SENSOR

THROTTLE BODY

29157_TOYO_G0040

1.8L I4 MFI 2ZZ-GE VIN Y

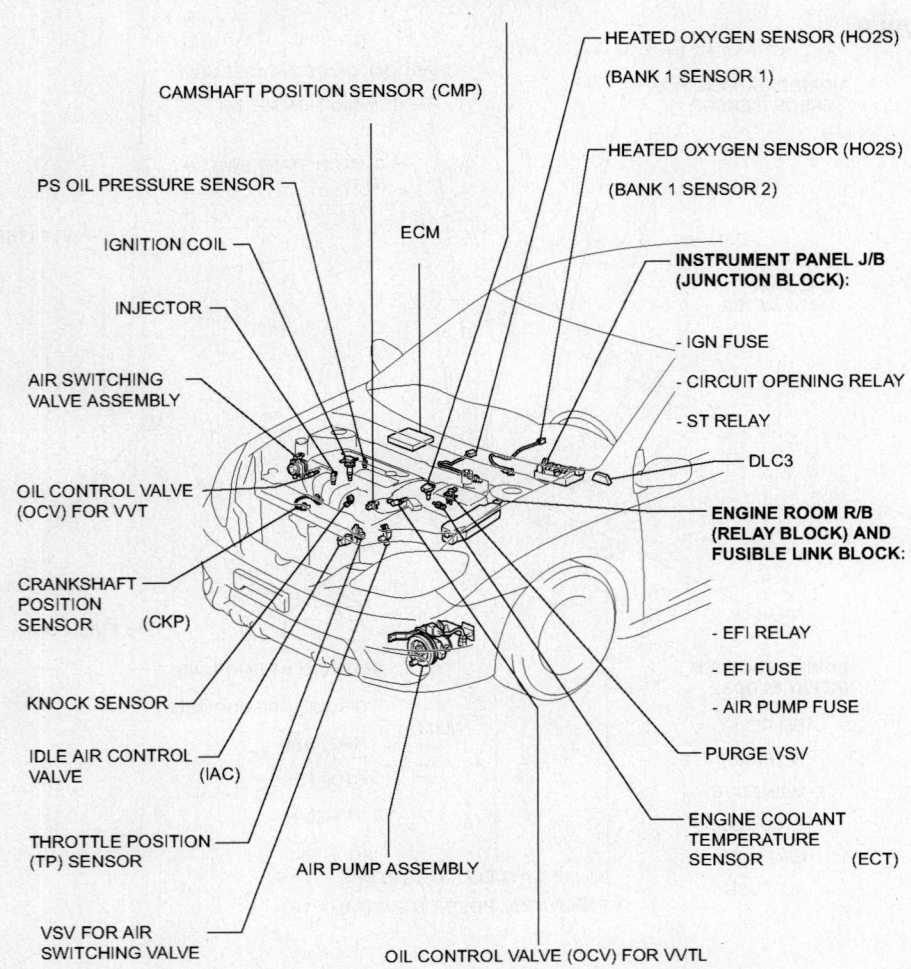

MASS AIR FLOW METER (MAF)

HEATED OXYGEN SENSOR (HO2S)
(BANK 1 SENSOR 1)

CAMSHAFT POSITION SENSOR (CMP)

HEATED OXYGEN SENSOR (HO2S)
(BANK 1 SENSOR 2)

PS OIL PRESSURE SENSOR

ECM

IGNITION COIL

**INSTRUMENT PANEL J/B
(JUNCTION BLOCK):**

INJECTOR

- IGN FUSE

AIR SWITCHING
VALVE ASSEMBLY

- CIRCUIT OPENING RELAY

- ST RELAY

OIL CONTROL VALVE
(OCV) FOR VVT

- DLC3

**ENGINE ROOM R/B
(RELAY BLOCK) AND
FUSIBLE LINK BLOCK:**

CRANKSHAFT
POSITION
SENSOR   (CKP)

- EFI RELAY

- EFI FUSE

KNOCK SENSOR

- AIR PUMP FUSE

IDLE AIR CONTROL
VALVE   (IAC)

PURGE VSV

THROTTLE POSITION
(TP) SENSOR

ENGINE COOLANT
TEMPERATURE
SENSOR   (ECT)

AIR PUMP ASSEMBLY

VSV FOR AIR
SWITCHING VALVE

OIL CONTROL VALVE (OCV) FOR VVTL

29157_TOYO_G0041

CHARCOAL CANISTER ASSEMBLY:

- PUMP MODULE

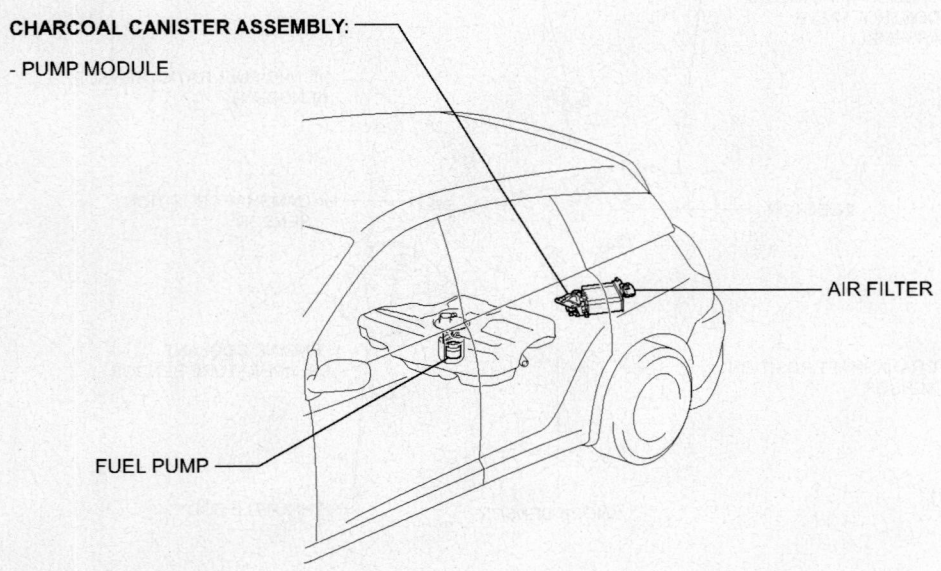

AIR FILTER

FUEL PUMP

29157_TOYO_G0042

MR2

1.8L I4 MFI 1ZZ-FE VIN R

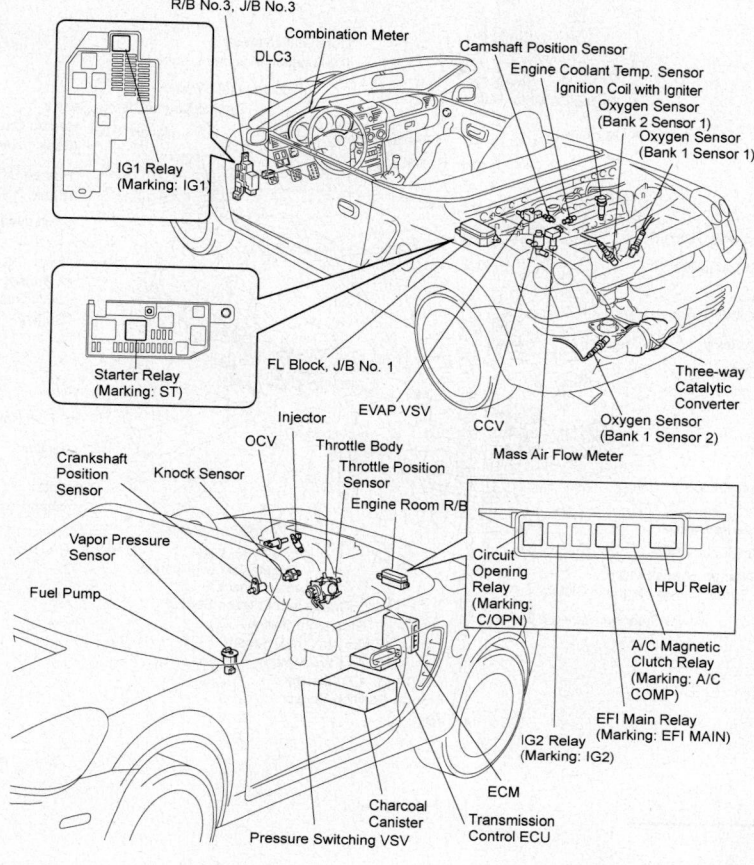

R/B No.3, J/B No.3
Combination Meter
DLC3
Camshaft Position Sensor
Engine Coolant Temp. Sensor
Ignition Coil with Igniter
Oxygen Sensor (Bank 2 Sensor 1)
Oxygen Sensor (Bank 1 Sensor 1)

IG1 Relay (Marking: IG1)

Starter Relay (Marking: ST)

FL Block, J/B No. 1
EVAP VSV
CCV
Three-way Catalytic Converter
Oxygen Sensor (Bank 1 Sensor 2)
Mass Air Flow Meter

Injector
OCV
Throttle Body
Throttle Position Sensor
Engine Room R/B

Crankshaft Position Sensor
Knock Sensor

Vapor Pressure Sensor

Fuel Pump

Circuit Opening Relay (Marking: C/OPN)
HPU Relay
A/C Magnetic Clutch Relay (Marking: A/C COMP)
EFI Main Relay (Marking: EFI MAIN)
IG2 Relay (Marking: IG2)

Charcoal Canister
ECM
Transmission Control ECU
Pressure Switching VSV

29157_TOYO_G0043

PASEO

1.5L I4 MFI 5E-FE VIN C

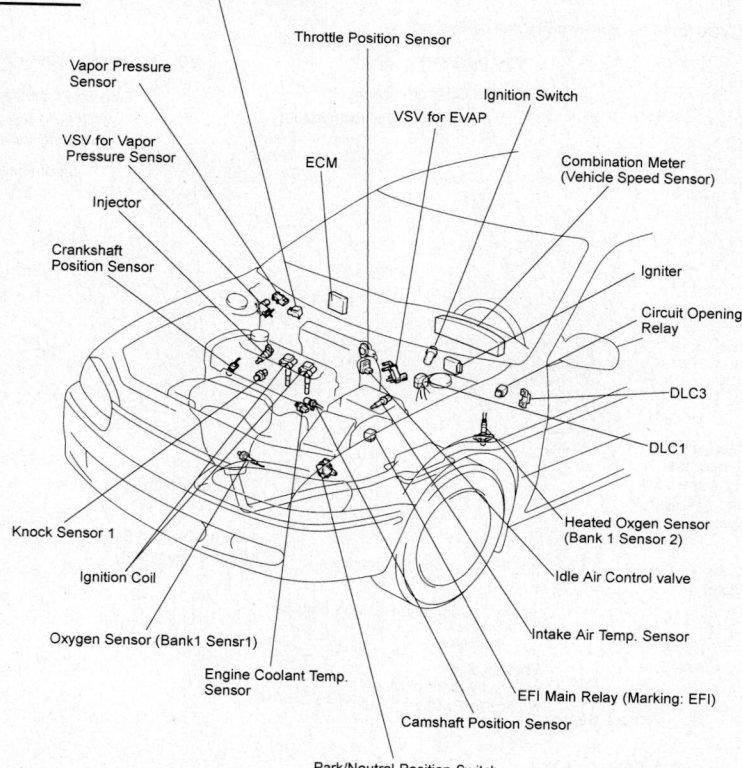

Manifold Absolute Pressure Sensor
Throttle Position Sensor
Vapor Pressure Sensor
Ignition Switch
VSV for EVAP
VSV for Vapor Pressure Sensor
ECM
Combination Meter (Vehicle Speed Sensor)
Injector
Crankshaft Position Sensor
Igniter
Circuit Opening Relay
DLC3
DLC1
Heated Oxgen Sensor (Bank 1 Sensor 2)
Knock Sensor 1
Idle Air Control valve
Ignition Coil
Intake Air Temp. Sensor
Oxygen Sensor (Bank1 Sensr1)
Engine Coolant Temp. Sensor
EFI Main Relay (Marking: EFI)
Camshaft Position Sensor
Park/Neutral Position Switch

29157_TOYO_G0044

PREVIA

2.4L I4 MFI 2TZ-FZE VIN K

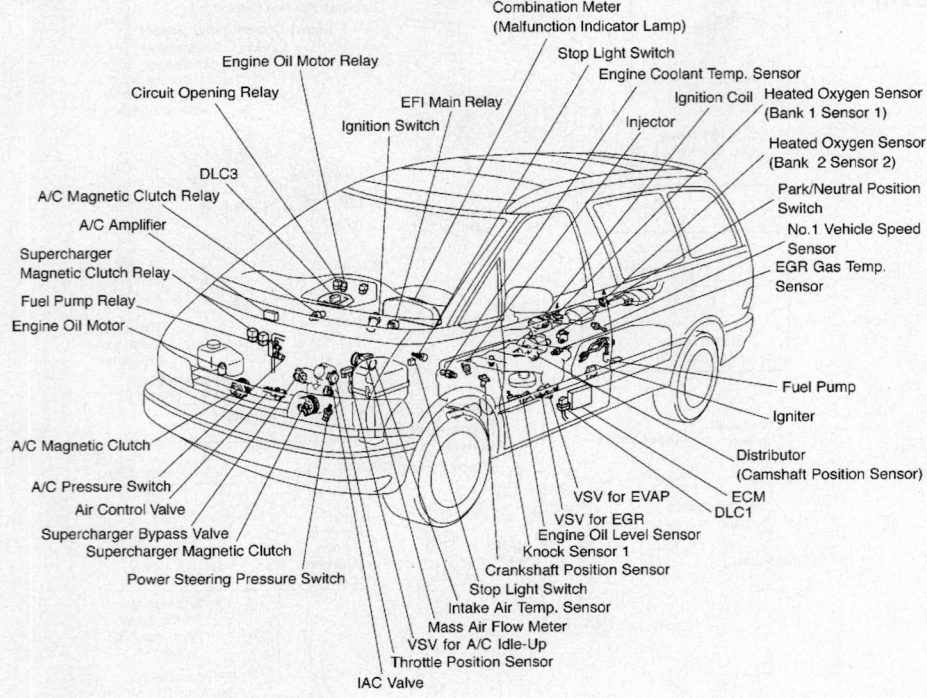

Combination Meter
(Malfunction Indicator Lamp)

Stop Light Switch

Engine Coolant Temp. Sensor

Engine Oil Motor Relay

Ignition Coil

Heated Oxygen Sensor
(Bank 1 Sensor 1)

Circuit Opening Relay

EFI Main Relay

Injector

Heated Oxygen Sensor
(Bank 2 Sensor 2)

Ignition Switch

Ignition Switch

DLC3

Park/Neutral Position
Switch

A/C Magnetic Clutch Relay

No.1 Vehicle Speed
Sensor

A/C Amplifier

EGR Gas Temp.
Sensor

Supercharger
Magnetic Clutch Relay

Fuel Pump Relay

Engine Oil Motor

Fuel Pump

Igniter

A/C Magnetic Clutch

Distributor
(Camshaft Position Sensor)

A/C Pressure Switch

ECM

Air Control Valve

DLC1

Supercharger Bypass Valve

VSV for EVAP

Supercharger Magnetic Clutch

VSV for EGR

Power Steering Pressure Switch

Engine Oil Level Sensor

Knock Sensor 1

Crankshaft Position Sensor

Stop Light Switch

Intake Air Temp. Sensor

Mass Air Flow Meter

VSV for A/C Idle-Up

Throttle Position Sensor

IAC Valve

29157_TOYO_G0045

PRIUS

1.5L I4 MFI 1NZ-FXE VIN K

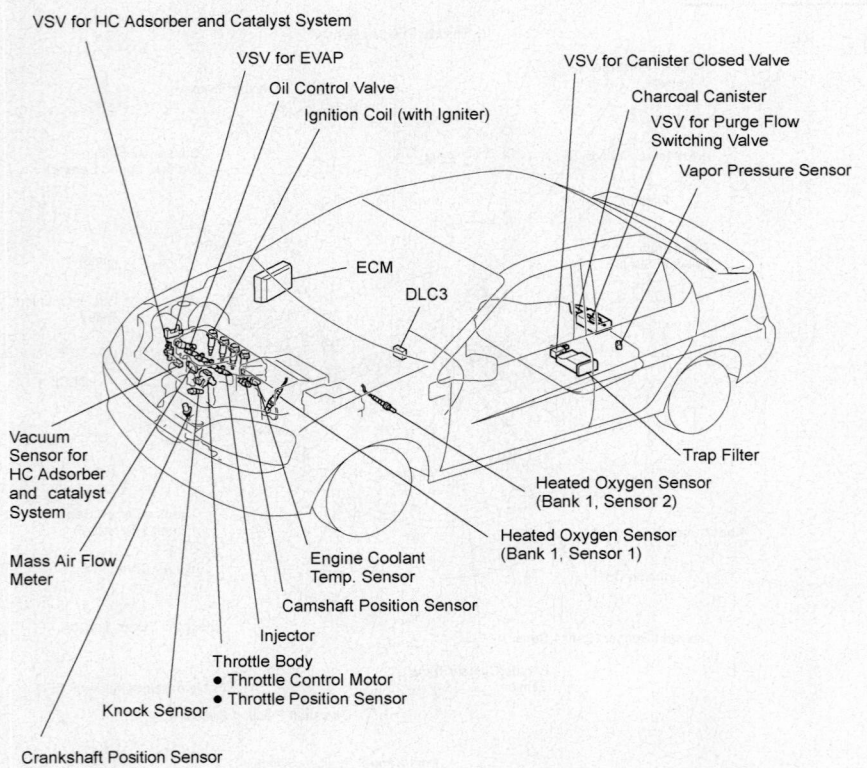

VSV for HC Adsorber and Catalyst System

VSV for EVAP

VSV for Canister Closed Valve

Oil Control Valve

Charcoal Canister

Ignition Coil (with Igniter)

VSV for Purge Flow
Switching Valve

Vapor Pressure Sensor

ECM

DLC3

Vacuum
Sensor for
HC Adsorber
and catalyst
System

Trap Filter

Heated Oxygen Sensor
(Bank 1, Sensor 2)

Mass Air Flow
Meter

Heated Oxygen Sensor
(Bank 1, Sensor 1)

Engine Coolant
Temp. Sensor

Camshaft Position Sensor

Injector

Throttle Body
● Throttle Control Motor
● Throttle Position Sensor

Knock Sensor

Crankshaft Position Sensor

29157_TOYO_G0046

1.5L I4 MFI 1NZ-FXE VIN B

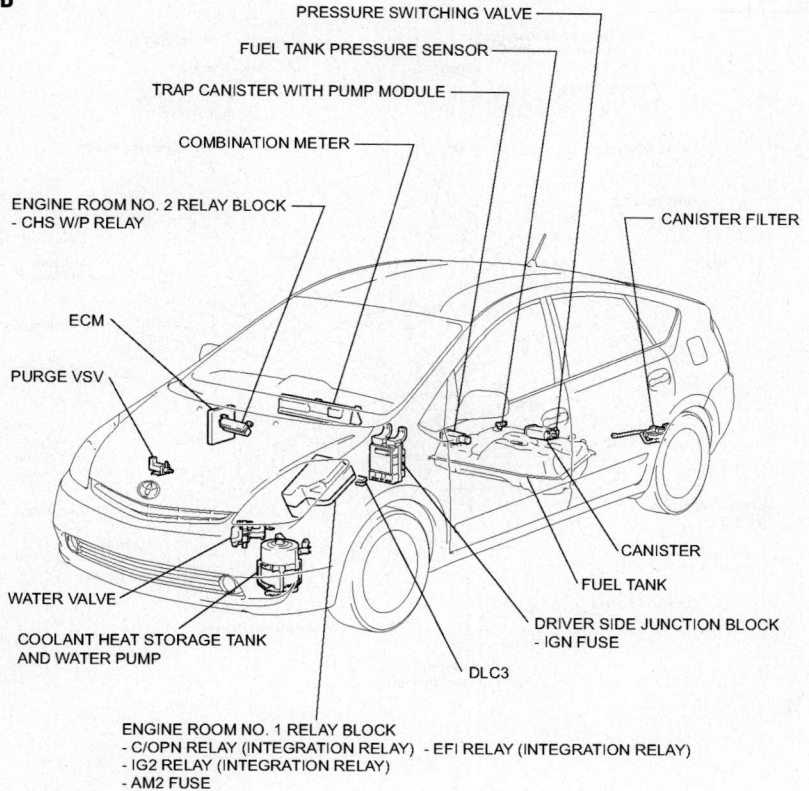

PRESSURE SWITCHING VALVE

FUEL TANK PRESSURE SENSOR

TRAP CANISTER WITH PUMP MODULE

COMBINATION METER

ENGINE ROOM NO. 2 RELAY BLOCK
- CHS W/P RELAY

CANISTER FILTER

ECM

PURGE VSV

WATER VALVE

COOLANT HEAT STORAGE TANK
AND WATER PUMP

DLC3

CANISTER

FUEL TANK

DRIVER SIDE JUNCTION BLOCK
- IGN FUSE

ENGINE ROOM NO. 1 RELAY BLOCK
- C/OPN RELAY (INTEGRATION RELAY) - EFI RELAY (INTEGRATION RELAY)
- IG2 RELAY (INTEGRATION RELAY)
- AM2 FUSE
- EFI FUSE

29157_TOYO_G0047

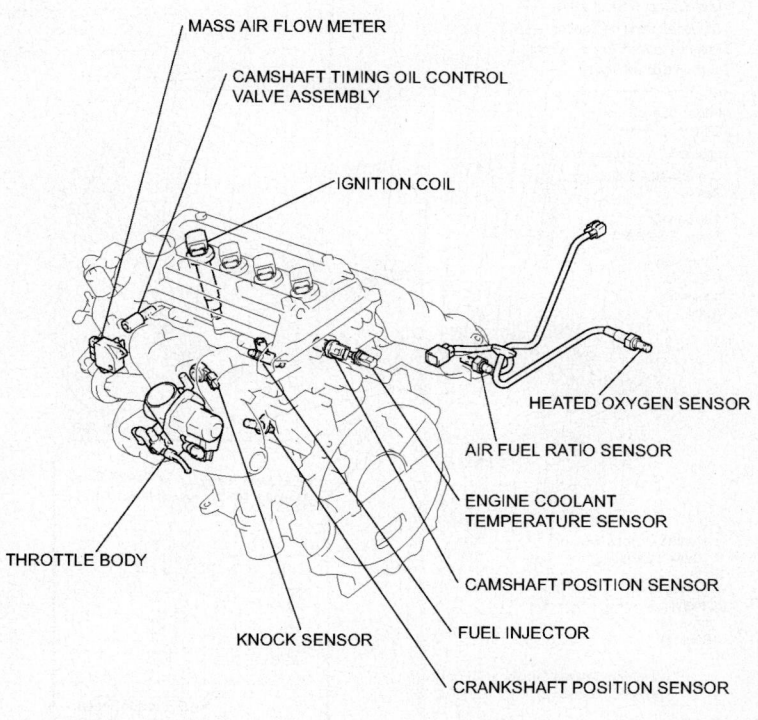

MASS AIR FLOW METER

CAMSHAFT TIMING OIL CONTROL
VALVE ASSEMBLY

IGNITION COIL

HEATED OXYGEN SENSOR

AIR FUEL RATIO SENSOR

ENGINE COOLANT
TEMPERATURE SENSOR

CAMSHAFT POSITION SENSOR

THROTTLE BODY

FUEL INJECTOR

KNOCK SENSOR

CRANKSHAFT POSITION SENSOR

29157_TOYO_G0048

2.0L I4 MFI 3S-FE VIN P

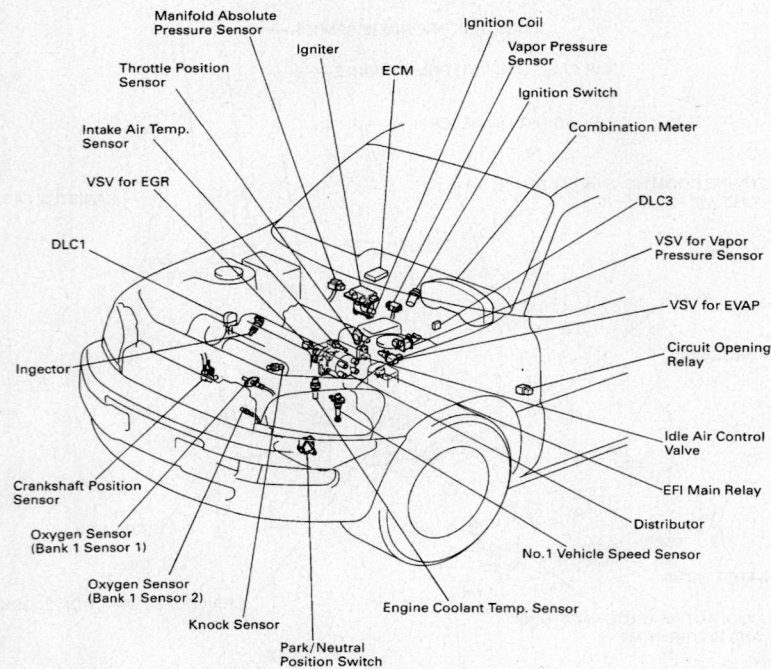

Manifold Absolute Pressure Sensor
Igniter
ECM
Ignition Coil
Vapor Pressure Sensor
Ignition Switch
Combination Meter
Throttle Position Sensor
Intake Air Temp. Sensor
VSV for EGR
DLC1
DLC3
VSV for Vapor Pressure Sensor
VSV for EVAP
Circuit Opening Relay
Ingector
Idle Air Control Valve
Crankshaft Position Sensor
EFI Main Relay
Distributor
Oxygen Sensor (Bank 1 Sensor 1)
No.1 Vehicle Speed Sensor
Oxygen Sensor (Bank 1 Sensor 2)
Knock Sensor
Engine Coolant Temp. Sensor
Park/Neutral Position Switch

29157_TOYO_G0049

2.0L I4 MFI 1AZ-FE VIN H

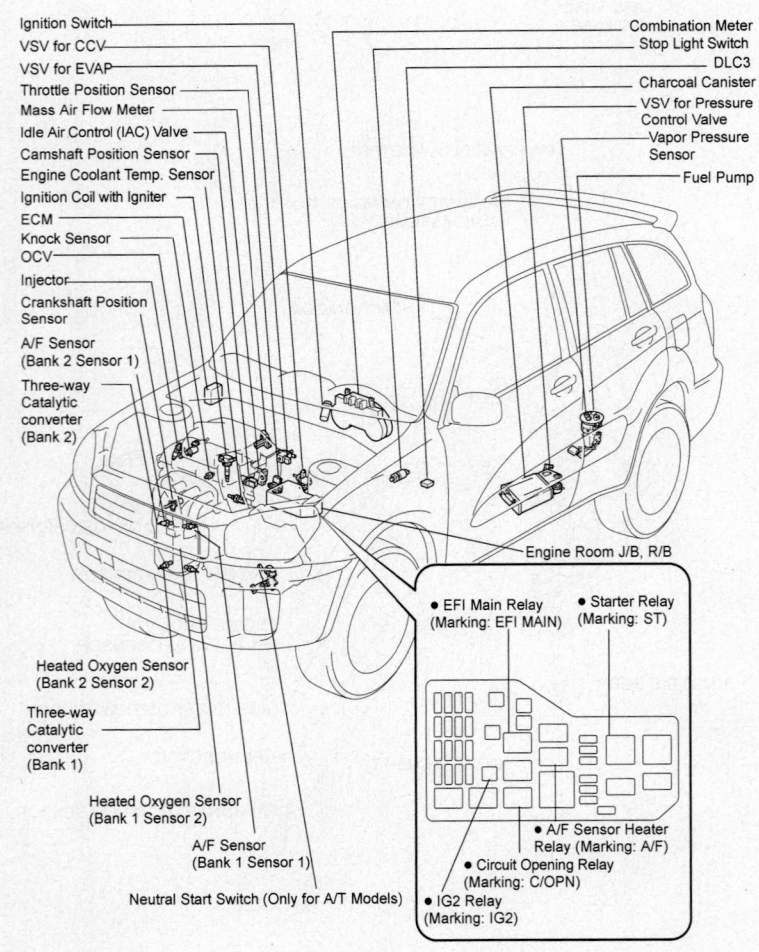

Ignition Switch
VSV for CCV
VSV for EVAP
Throttle Position Sensor
Mass Air Flow Meter
Idle Air Control (IAC) Valve
Camshaft Position Sensor
Engine Coolant Temp. Sensor
Ignition Coil with Igniter
ECM
Knock Sensor
OCV
Injector
Crankshaft Position Sensor
A/F Sensor (Bank 2 Sensor 1)
Three-way Catalytic converter (Bank 2)

Combination Meter
Stop Light Switch
DLC3
Charcoal Canister
VSV for Pressure Control Valve
Vapor Pressure Sensor
Fuel Pump

Heated Oxygen Sensor (Bank 2 Sensor 2)
Three-way Catalytic converter (Bank 1)
Heated Oxygen Sensor (Bank 1 Sensor 2)
A/F Sensor (Bank 1 Sensor 1)
Neutral Start Switch (Only for A/T Models)

Engine Room J/B, R/B

• EFI Main Relay (Marking: EFI MAIN)
• Starter Relay (Marking: ST)
• A/F Sensor Heater Relay (Marking: A/F)
• Circuit Opening Relay (Marking: C/OPN)
• IG2 Relay (Marking: IG2)

29157_TOYO_G0050

2.4L I4 MFI 2AZ-FE VIN D

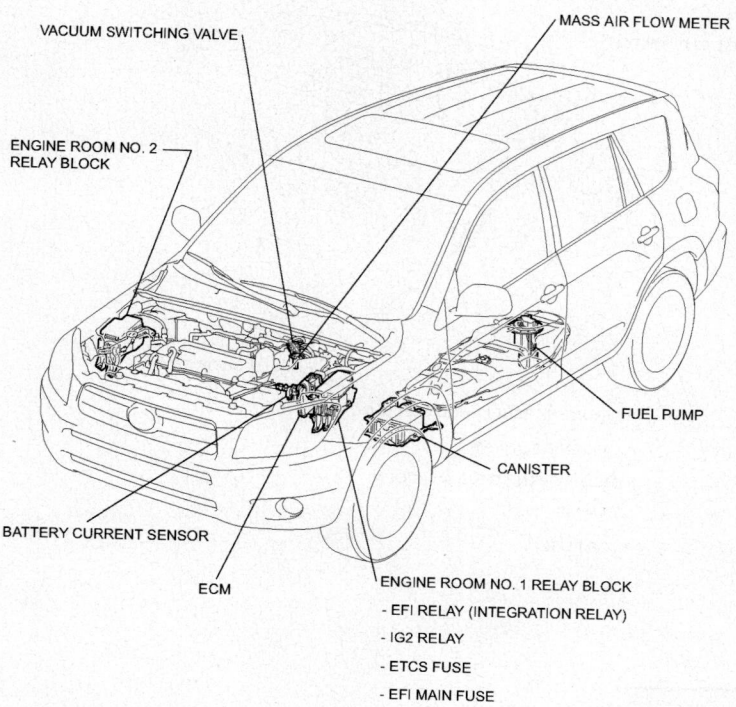

VACUUM SWITCHING VALVE

MASS AIR FLOW METER

ENGINE ROOM NO. 2
RELAY BLOCK

FUEL PUMP

CANISTER

BATTERY CURRENT SENSOR

ECM

ENGINE ROOM NO. 1 RELAY BLOCK

- EFI RELAY (INTEGRATION RELAY)
- IG2 RELAY
- ETCS FUSE
- EFI MAIN FUSE

29157_TOYO_G0051

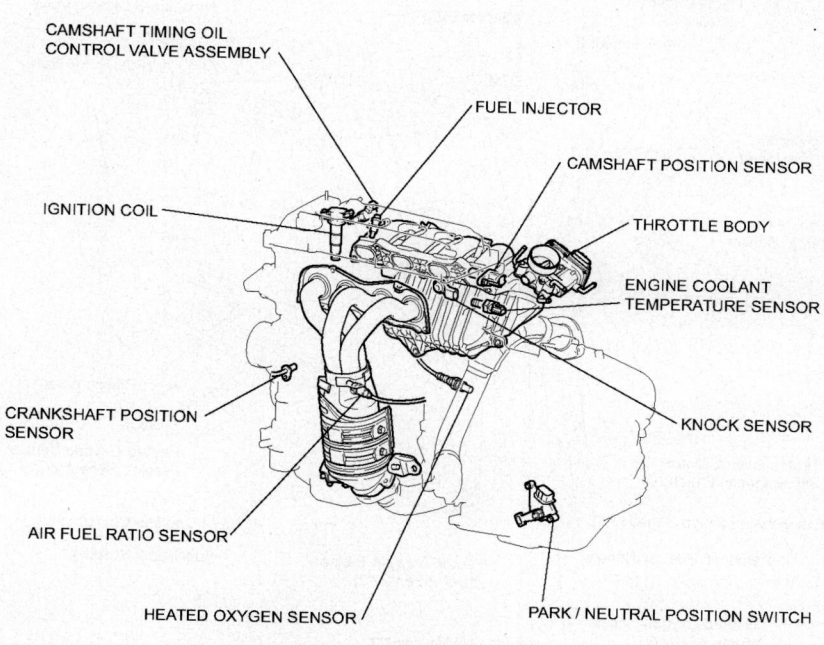

CAMSHAFT TIMING OIL
CONTROL VALVE ASSEMBLY

FUEL INJECTOR

CAMSHAFT POSITION SENSOR

IGNITION COIL

THROTTLE BODY

ENGINE COOLANT
TEMPERATURE SENSOR

CRANKSHAFT POSITION
SENSOR

KNOCK SENSOR

AIR FUEL RATIO SENSOR

HEATED OXYGEN SENSOR

PARK / NEUTRAL POSITION SWITCH

29157_TOYO_G0052

COMBINATION METER

STOP LIGHT SWITCH

DLC3

ACCELERATOR PEDAL ROD

INSTRUMENT PANEL JUNCTION BLOCK

- STARTER RELAY (ST)

- CIRCUIT OPENING RELAY (C/OPN)

- IGN FUSE

- STOP FUSE

29157_TOYO_G0053

SEQUOIA

4.7L V8 MFI 2UZ-FE VIN T

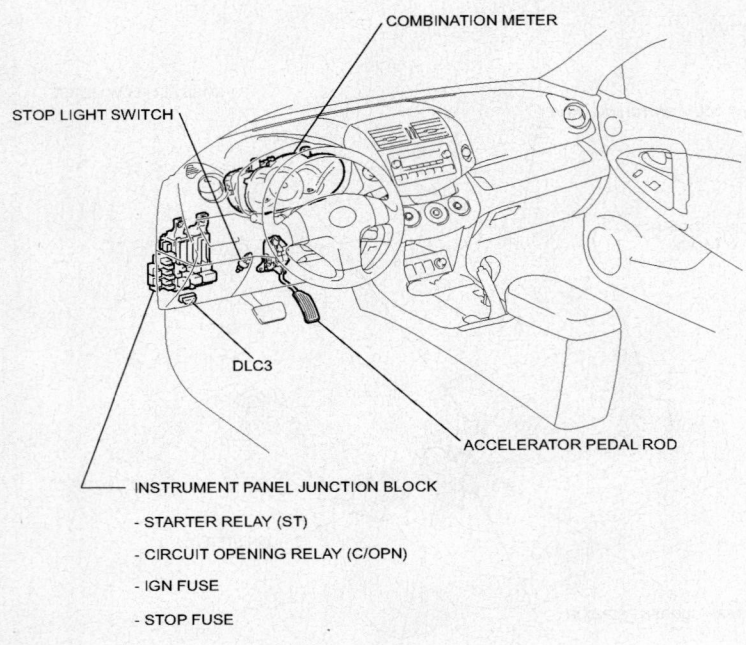

Park/Neutral Position Switch

Combination Meter

Accelerator Pedal
Position Sensor

Throttle Position
Sensor

Knock Sensor 2

VSV for EVAP

VSV for
Pressure Switching Valve

Ignition Coil
with Igniter

ECM

Fuel Pump Relay

Injector

Mass Air
Flow Meter

Engine Coolant
Temp. Sensor

Vapor Pressure Sensor

DLC3

Throttle Control Motor
(with Magnetic Clutch)

Heated Oxygen Sensor
(Bank 1 Sensor 2)

Crankshaft Position Sensor

VSV for
Canister Closed Valve

Camshaft Position Sensor

Fuel Pump Resister

Heated Oxygen Sensor
(Bank 2 Sensor 2)

Heated Oxygen Sensor
(Bank 2 Sensor 1)

Heated Oxygen Sensor
(Bank 1 Sensor 1)

Knock Sensor 1

29157_TOYO_G0054

SIENNA

3.0L V6 MFI 1MZ-FE VIN F

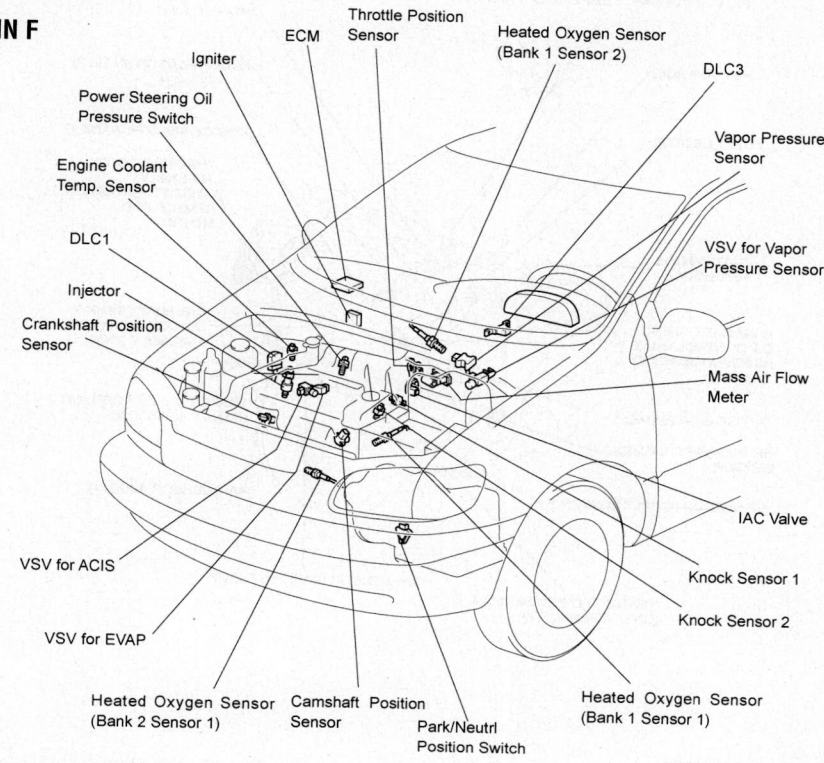

ECM

Throttle Position Sensor

Heated Oxygen Sensor (Bank 1 Sensor 2)

DLC3

Igniter

Power Steering Oil Pressure Switch

Engine Coolant Temp. Sensor

Vapor Pressure Sensor

DLC1

VSV for Vapor Pressure Sensor

Injector

Crankshaft Position Sensor

Mass Air Flow Meter

IAC Valve

VSV for ACIS

Knock Sensor 1

Knock Sensor 2

VSV for EVAP

Heated Oxygen Sensor (Bank 2 Sensor 1)

Camshaft Position Sensor

Park/Neutrl Position Switch

Heated Oxygen Sensor (Bank 1 Sensor 1)

29157_TOYO_G0055

3.3L V6 MFI 3MZ-FE VIN A

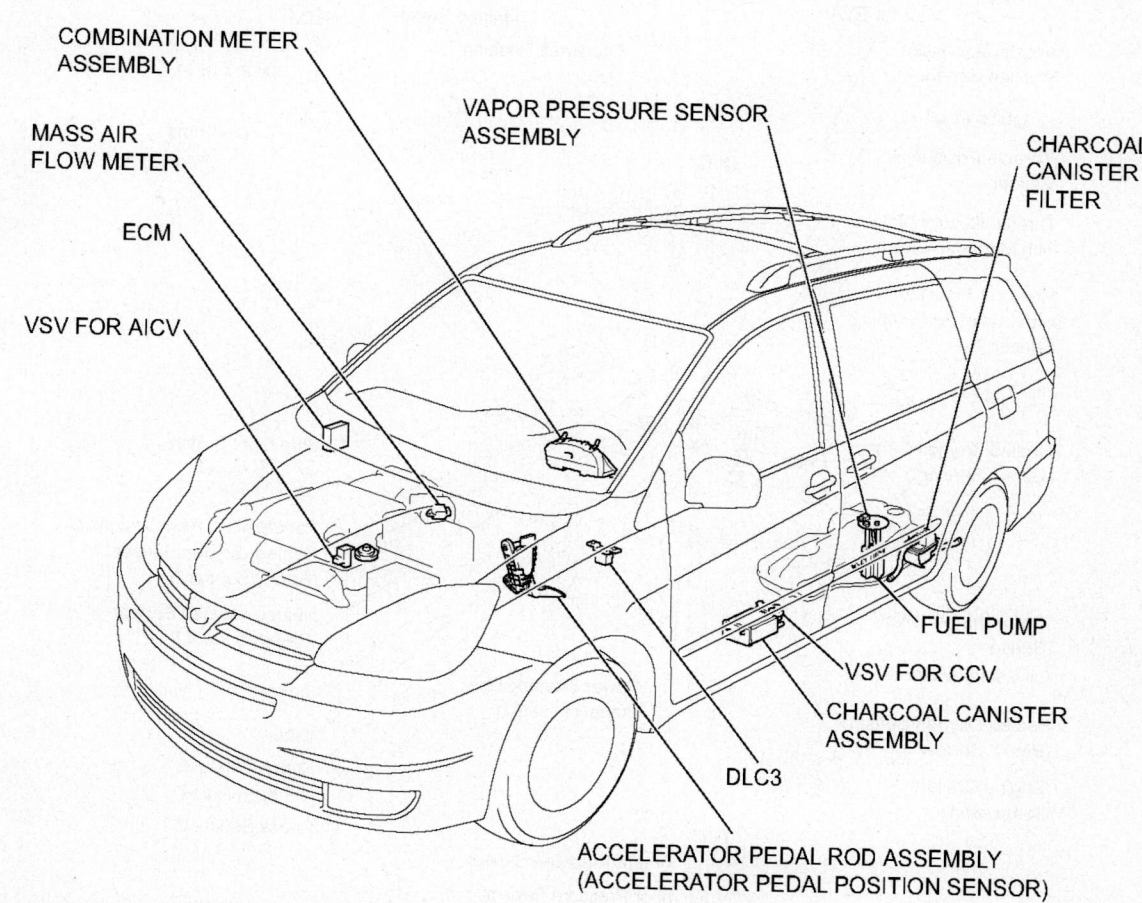

COMBINATION METER ASSEMBLY

VAPOR PRESSURE SENSOR ASSEMBLY

CHARCOAL CANISTER FILTER

MASS AIR FLOW METER

ECM

VSV FOR AICV

FUEL PUMP

VSV FOR CCV

CHARCOAL CANISTER ASSEMBLY

DLC3

ACCELERATOR PEDAL ROD ASSEMBLY (ACCELERATOR PEDAL POSITION SENSOR)

29157_TOYO_G0056

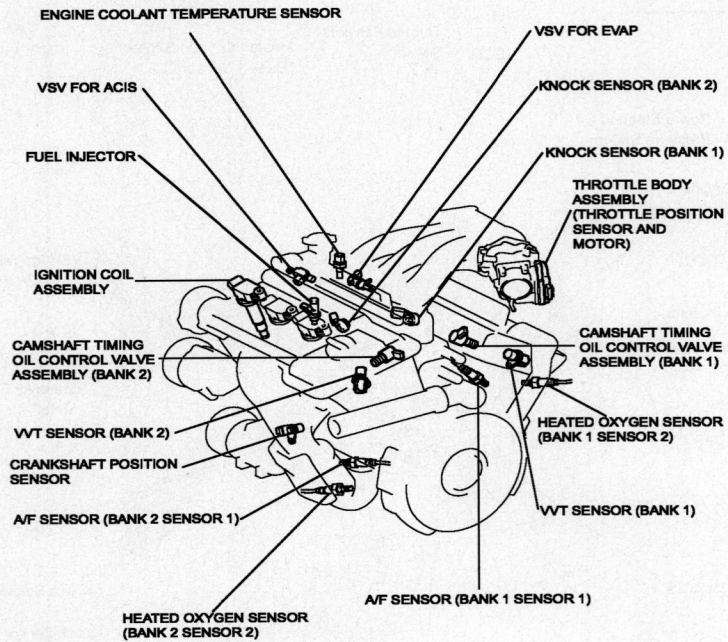

ENGINE COOLANT TEMPERATURE SENSOR

VSV FOR EVAP

VSV FOR ACIS

KNOCK SENSOR (BANK 2)

FUEL INJECTOR

KNOCK SENSOR (BANK 1)

THROTTLE BODY ASSEMBLY (THROTTLE POSITION SENSOR AND MOTOR)

IGNITION COIL ASSEMBLY

CAMSHAFT TIMING OIL CONTROL VALVE ASSEMBLY (BANK 2)

CAMSHAFT TIMING OIL CONTROL VALVE ASSEMBLY (BANK 1)

VVT SENSOR (BANK 2)

HEATED OXYGEN SENSOR (BANK 1 SENSOR 2)

CRANKSHAFT POSITION SENSOR

VVT SENSOR (BANK 1)

A/F SENSOR (BANK 2 SENSOR 1)

A/F SENSOR (BANK 1 SENSOR 1)

HEATED OXYGEN SENSOR (BANK 2 SENSOR 2)

29157_TOYO_G0057

SUPRA

3.0L I6 MFI 2JZ-GE VIN D

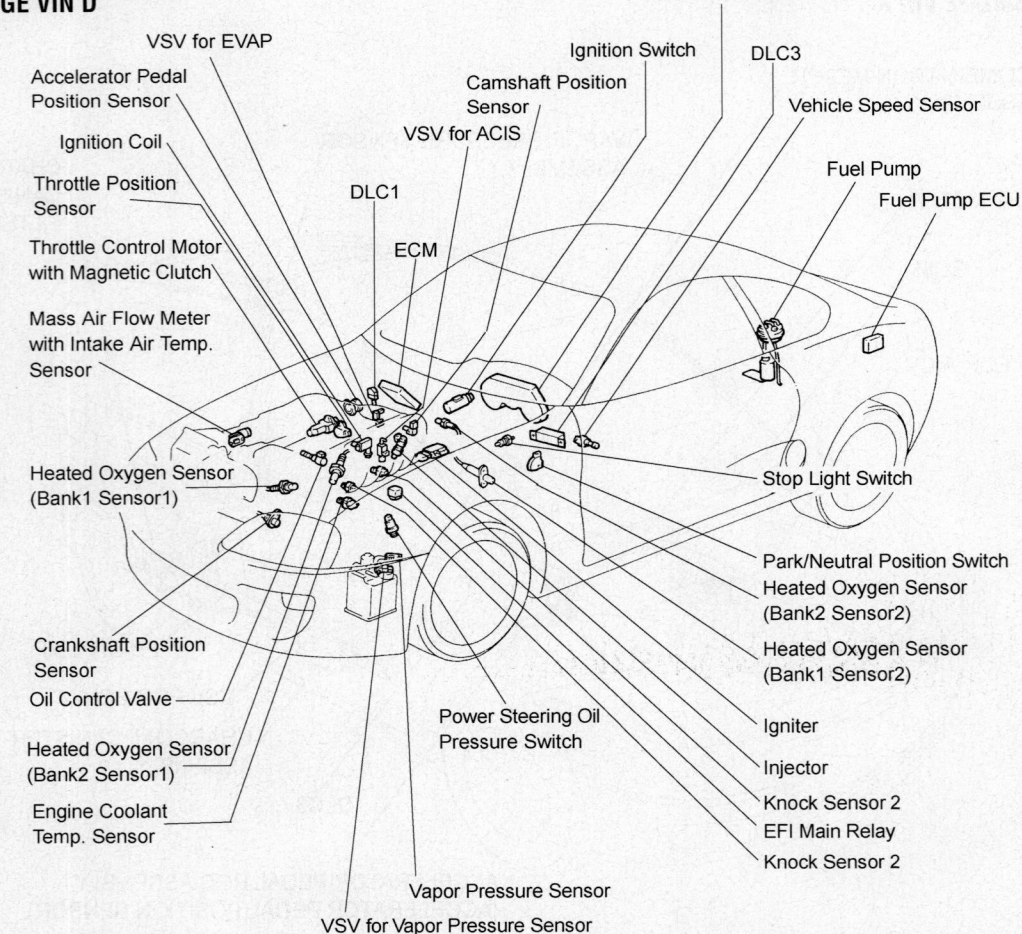

VSV for EVAP

Accelerator Pedal Position Sensor

Ignition Coil

Throttle Position Sensor

Throttle Control Motor with Magnetic Clutch

Mass Air Flow Meter with Intake Air Temp. Sensor

Heated Oxygen Sensor (Bank1 Sensor1)

Crankshaft Position Sensor

Oil Control Valve

Heated Oxygen Sensor (Bank2 Sensor1)

Engine Coolant Temp. Sensor

DLC1

ECM

VSV for ACIS

Camshaft Position Sensor

Ignition Switch

Combination Meter

DLC3

Vehicle Speed Sensor

Fuel Pump

Fuel Pump ECU

Stop Light Switch

Park/Neutral Position Switch

Heated Oxygen Sensor (Bank2 Sensor2)

Heated Oxygen Sensor (Bank1 Sensor2)

Igniter

Injector

Knock Sensor 2

EFI Main Relay

Knock Sensor 2

Power Steering Oil Pressure Switch

Vapor Pressure Sensor

VSV for Vapor Pressure Sensor

29157_TOYO_G0058

3.0L I6 MFI 2JZ-GTE VIN E

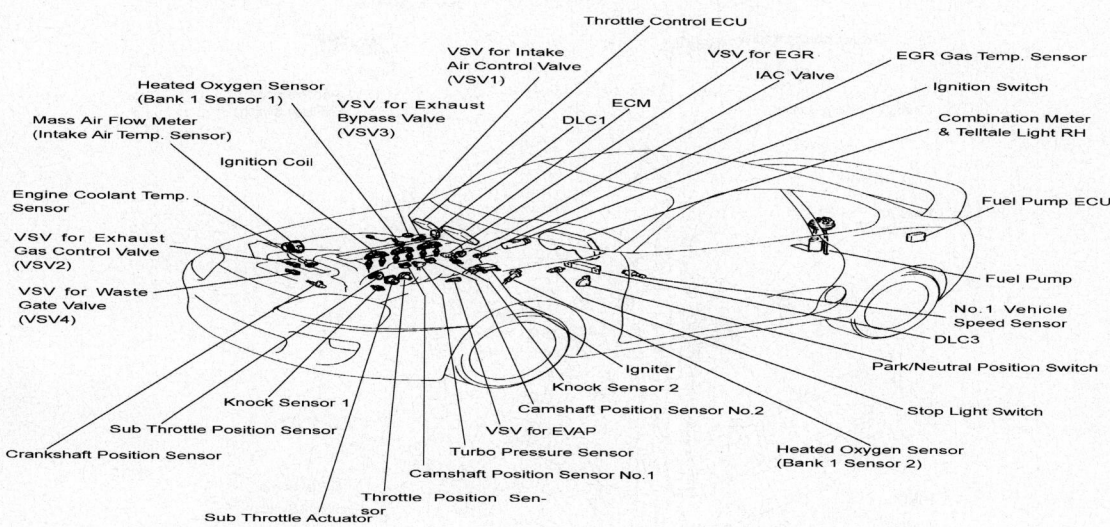

Throttle Control ECU
VSV for Intake Air Control Valve (VSV1)
VSV for EGR
IAC Valve
EGR Gas Temp. Sensor
Heated Oxygen Sensor (Bank 1 Sensor 1)
VSV for Exhaust Bypass Valve (VSV3)
ECM
DLC1
Ignition Switch
Mass Air Flow Meter (Intake Air Temp. Sensor)
Combination Meter & Telltale Light RH
Ignition Coil
Engine Coolant Temp. Sensor
Fuel Pump ECU
VSV for Exhaust Gas Control Valve (VSV2)
Fuel Pump
VSV for Waste Gate Valve (VSV4)
No.1 Vehicle Speed Sensor
DLC3
Park/Neutral Position Switch
Igniter
Knock Sensor 2
Stop Light Switch
Knock Sensor 1
Camshaft Position Sensor No.2
Sub Throttle Position Sensor
VSV for EVAP
Heated Oxygen Sensor (Bank 1 Sensor 2)
Crankshaft Position Sensor
Turbo Pressure Sensor
Camshaft Position Sensor No.1
Throttle Position Sensor
Sub Throttle Actuator

29157_TOYO_G0059

T 100

2.7L I4 MFI 3RZ-FE VIN M

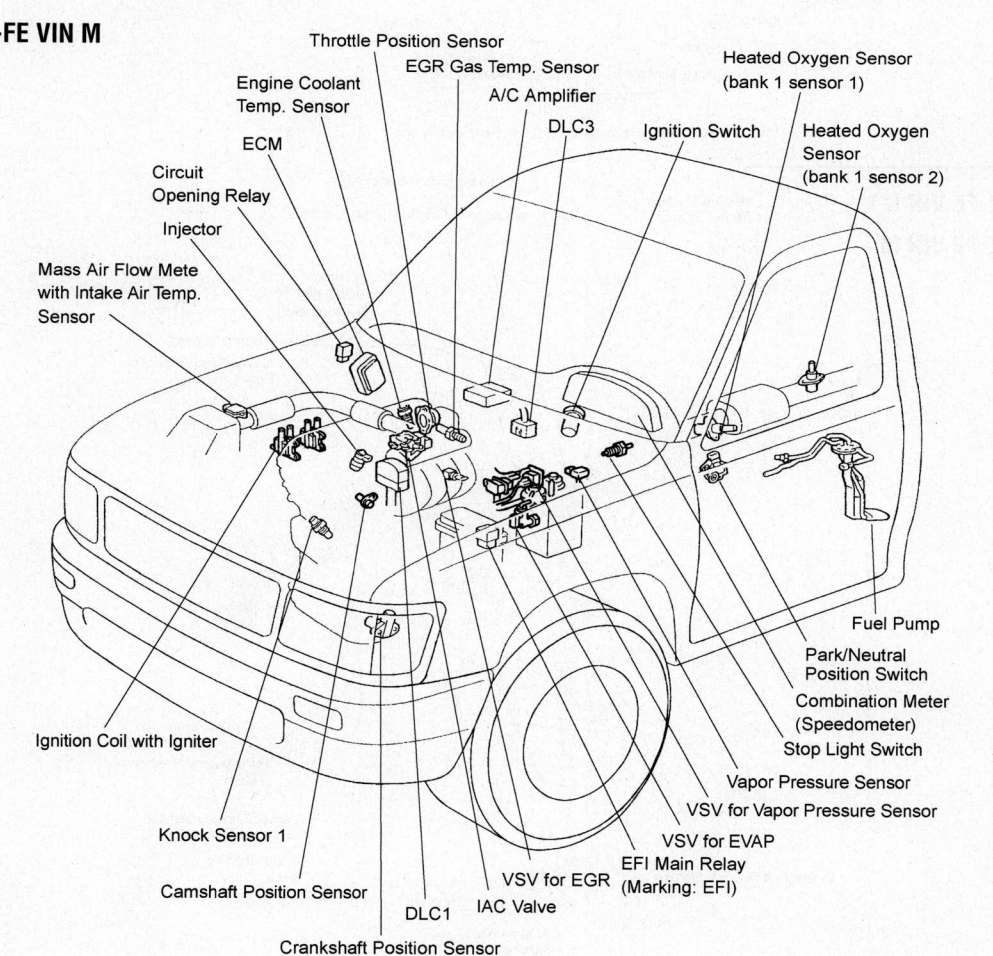

Throttle Position Sensor
EGR Gas Temp. Sensor
A/C Amplifier
DLC3
Ignition Switch
Heated Oxygen Sensor (bank 1 sensor 1)
Engine Coolant Temp. Sensor
Heated Oxygen Sensor (bank 1 sensor 2)
ECM
Circuit Opening Relay
Injector
Mass Air Flow Mete with Intake Air Temp. Sensor
Fuel Pump
Park/Neutral Position Switch
Combination Meter (Speedometer)
Stop Light Switch
Vapor Pressure Sensor
VSV for Vapor Pressure Sensor
Ignition Coil with Igniter
VSV for EVAP
EFI Main Relay (Marking: EFI)
Knock Sensor 1
VSV for EGR
IAC Valve
Camshaft Position Sensor
DLC1
Crankshaft Position Sensor

29157_TOYO_G0060

3.4L V6 MFI 5VZ-FE VIN N

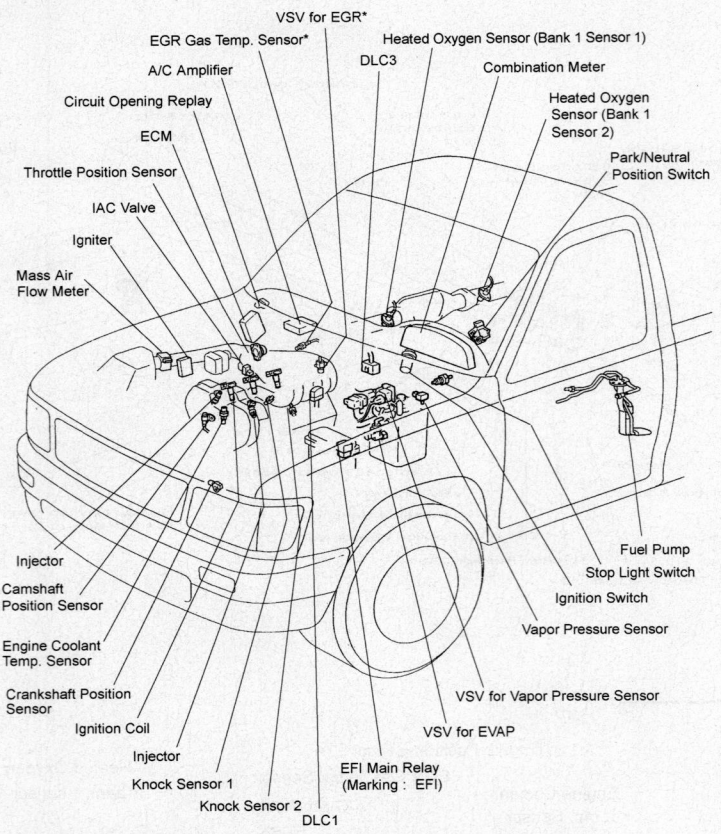

VSV for EGR*
EGR Gas Temp. Sensor*
A/C Amplifier
Heated Oxygen Sensor (Bank 1 Sensor 1)
DLC3
Combination Meter
Circuit Opening Replay
Heated Oxygen Sensor (Bank 1 Sensor 2)
ECM
Throttle Position Sensor
Park/Neutral Position Switch
IAC Valve
Igniter
Mass Air Flow Meter
Injector
Camshaft Position Sensor
Fuel Pump
Stop Light Switch
Ignition Switch
Engine Coolant Temp. Sensor
Vapor Pressure Sensor
Crankshaft Position Sensor
Ignition Coil
VSV for Vapor Pressure Sensor
Injector
VSV for EVAP
Knock Sensor 1
Knock Sensor 2
EFI Main Relay (Marking : EFI)
DLC1

*: Only for 2WD models with a load capacity of 0.5 ton and regular cab.

29157_TOYO_G0061

TACOMA

2.4L I4 MFI 2RZ-FE VIN L
2.7L I4 MFI 3RZ-FE VIN M

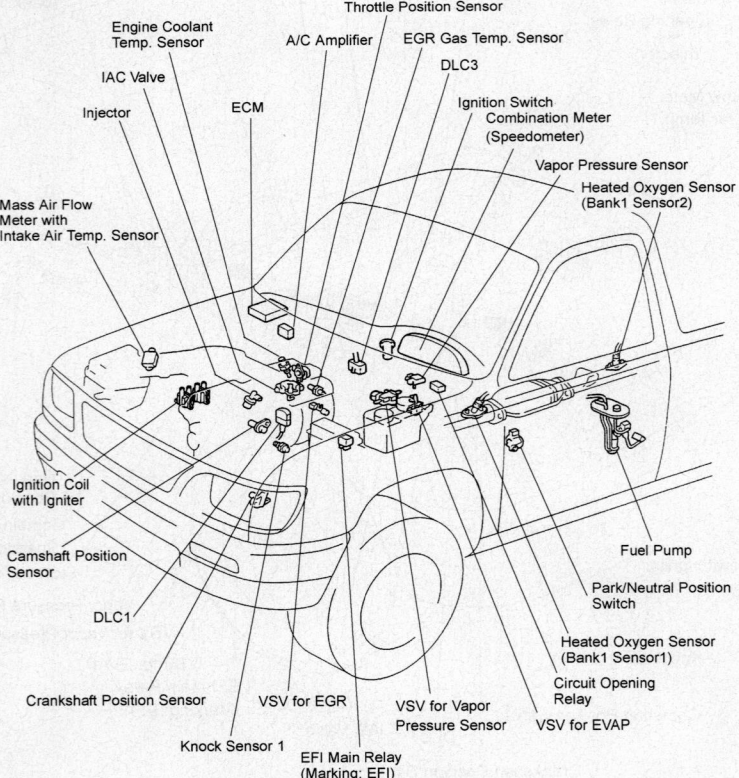

Throttle Position Sensor
Engine Coolant Temp. Sensor
A/C Amplifier
EGR Gas Temp. Sensor
IAC Valve
DLC3
Injector
ECM
Ignition Switch
Combination Meter (Speedometer)
Vapor Pressure Sensor
Mass Air Flow Meter with Intake Air Temp. Sensor
Heated Oxygen Sensor (Bank1 Sensor2)
Ignition Coil with Igniter
Camshaft Position Sensor
Fuel Pump
Park/Neutral Position Switch
DLC1
Heated Oxygen Sensor (Bank1 Sensor1)
Circuit Opening Relay
Crankshaft Position Sensor
VSV for EGR
VSV for Vapor Pressure Sensor
VSV for EVAP
Knock Sensor 1
EFI Main Relay (Marking: EFI)

29157_toyo_G0062

3.4L V6 MFI 5VZ-FE VIN N

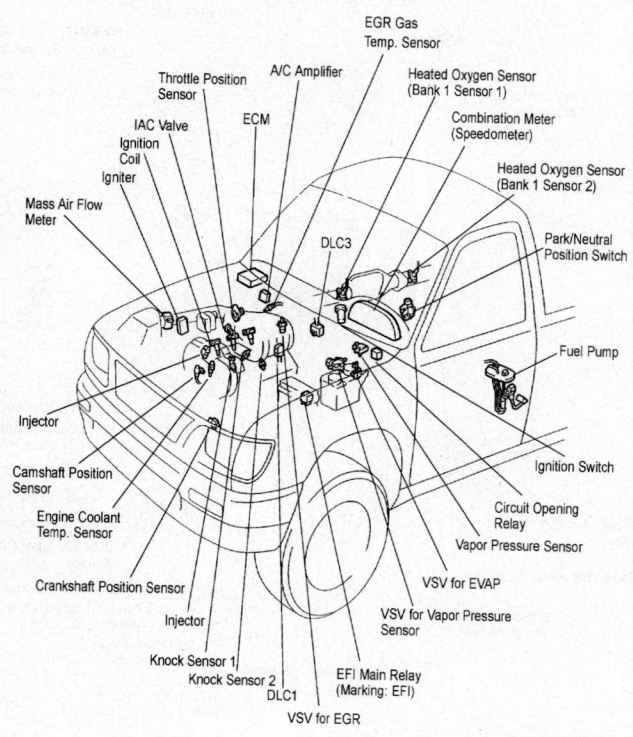

EGR Gas Temp. Sensor

A/C Amplifier

Heated Oxygen Sensor (Bank 1 Sensor 1)

Throttle Position Sensor

Combination Meter (Speedometer)

ECM

Heated Oxygen Sensor (Bank 1 Sensor 2)

IAC Valve

Ignition Coil

Igniter

DLC3

Park/Neutral Position Switch

Mass Air Flow Meter

Fuel Pump

Injector

Ignition Switch

Camshaft Position Sensor

Circuit Opening Relay

Engine Coolant Temp. Sensor

Vapor Pressure Sensor

Crankshaft Position Sensor

VSV for EVAP

Injector

VSV for Vapor Pressure Sensor

Knock Sensor 1

Knock Sensor 2

EFI Main Relay (Marking: EFI)

DLC1

VSV for EGR

29157_TOYO_G0063

TERCEL

1.5L I4 MFI 5E-FE VIN C

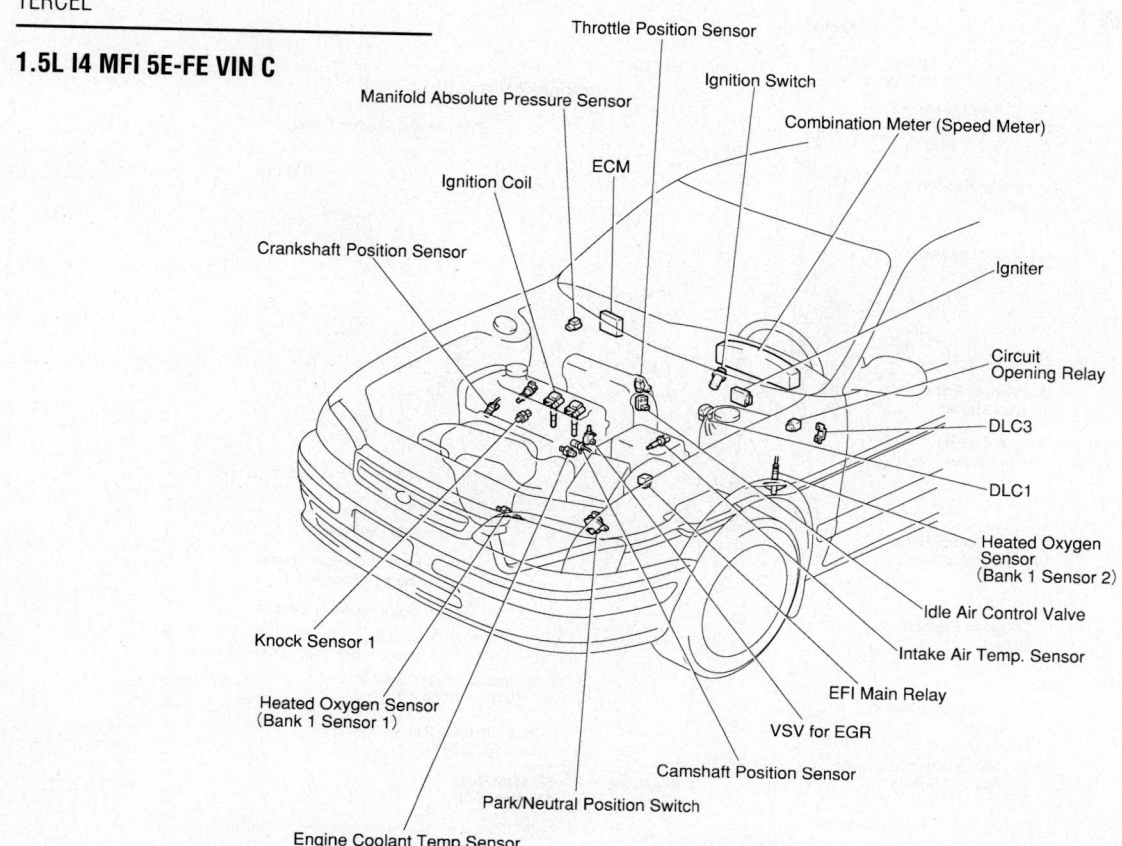

Throttle Position Sensor

Manifold Absolute Pressure Sensor

Ignition Switch

Combination Meter (Speed Meter)

Ignition Coil

ECM

Crankshaft Position Sensor

Igniter

Circuit Opening Relay

DLC3

DLC1

Heated Oxygen Sensor (Bank 1 Sensor 2)

Idle Air Control Valve

Intake Air Temp. Sensor

Knock Sensor 1

EFI Main Relay

Heated Oxygen Sensor (Bank 1 Sensor 1)

VSV for EGR

Camshaft Position Sensor

Park/Neutral Position Switch

Engine Coolant Temp.Sensor

29157_TOYO_G0064

TUNDRA

3.4L V6 MFI 5VZ-FE VIN N

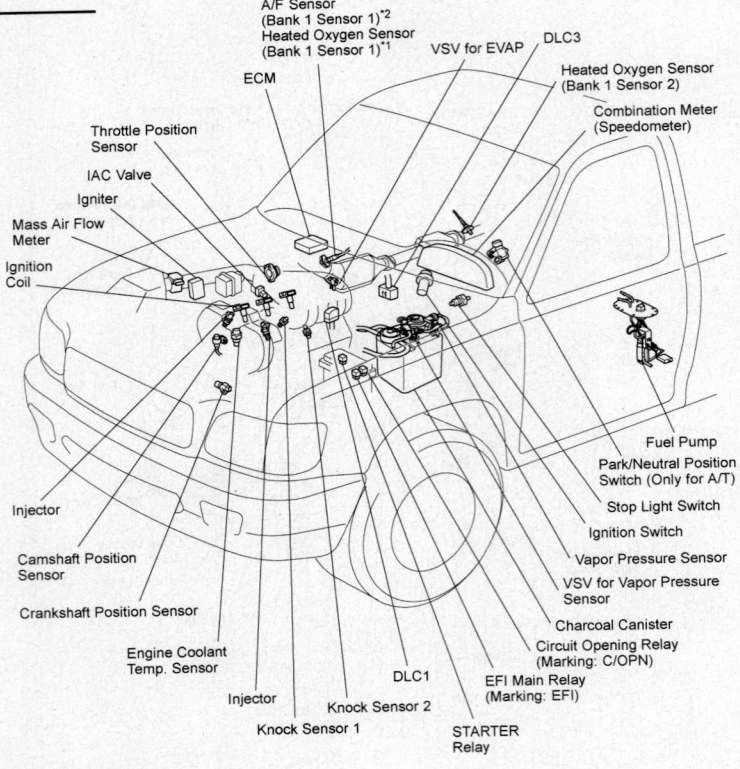

A/F Sensor
(Bank 1 Sensor 1)*2
Heated Oxygen Sensor
(Bank 1 Sensor 1)*1

VSV for EVAP   DLC3

Heated Oxygen Sensor
(Bank 1 Sensor 2)

Combination Meter
(Speedometer)

ECM

Throttle Position
Sensor

IAC Valve

Igniter

Mass Air Flow
Meter

Ignition
Coil

Injector

Camshaft Position
Sensor

Crankshaft Position Sensor

Engine Coolant
Temp. Sensor

Injector

DLC1

Knock Sensor 2

Knock Sensor 1

STARTER
Relay

Fuel Pump
Park/Neutral Position
Switch (Only for A/T)
Stop Light Switch
Ignition Switch
Vapor Pressure Sensor
VSV for Vapor Pressure
Sensor
Charcoal Canister
Circuit Opening Relay
(Marking: C/OPN)
EFI Main Relay
(Marking: EFI)

*1: Except Calif.
*2: Only for Calif.

29157_TOYO_G0065

4.7L V8 MFI 2UZ-FE VIN T

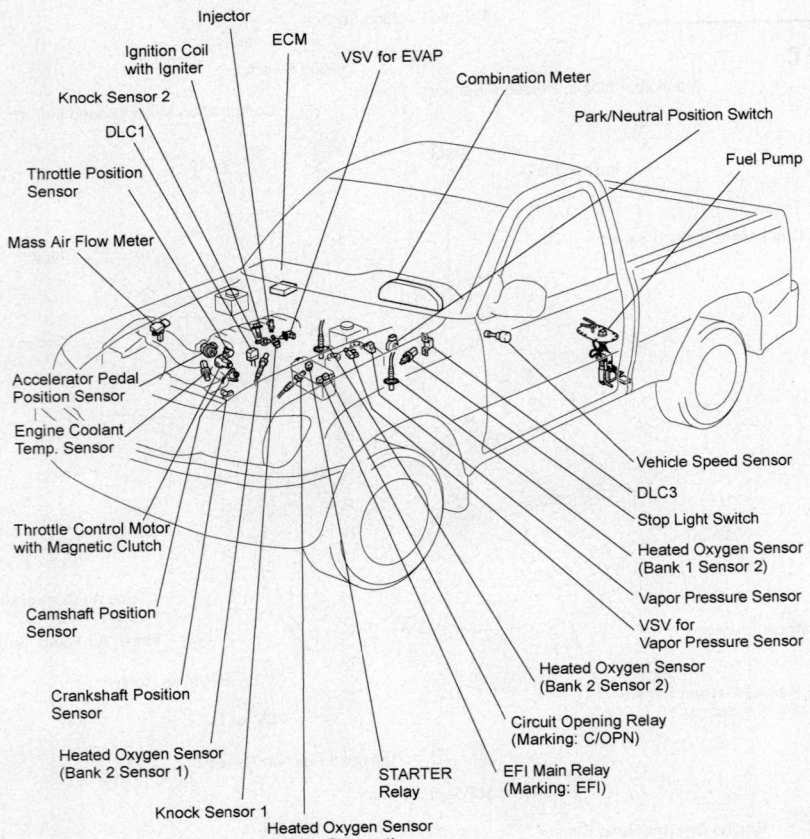

Injector

Ignition Coil
with Igniter

ECM

VSV for EVAP

Combination Meter

Park/Neutral Position Switch

Fuel Pump

Knock Sensor 2

DLC1

Throttle Position
Sensor

Mass Air Flow Meter

Accelerator Pedal
Position Sensor

Engine Coolant
Temp. Sensor

Throttle Control Motor
with Magnetic Clutch

Camshaft Position
Sensor

Crankshaft Position
Sensor

Heated Oxygen Sensor
(Bank 2 Sensor 1)

Knock Sensor 1

Heated Oxygen Sensor
(Bank 1 Sensor 1)

STARTER
Relay

EFI Main Relay
(Marking: EFI)

Circuit Opening Relay
(Marking: C/OPN)

Heated Oxygen Sensor
(Bank 2 Sensor 2)

VSV for
Vapor Pressure Sensor

Vapor Pressure Sensor

Heated Oxygen Sensor
(Bank 1 Sensor 2)

Stop Light Switch

DLC3

Vehicle Speed Sensor

29157_TOYO_G0066

YARIS

1.5L I4 MFI 1NZ-FE VIN T

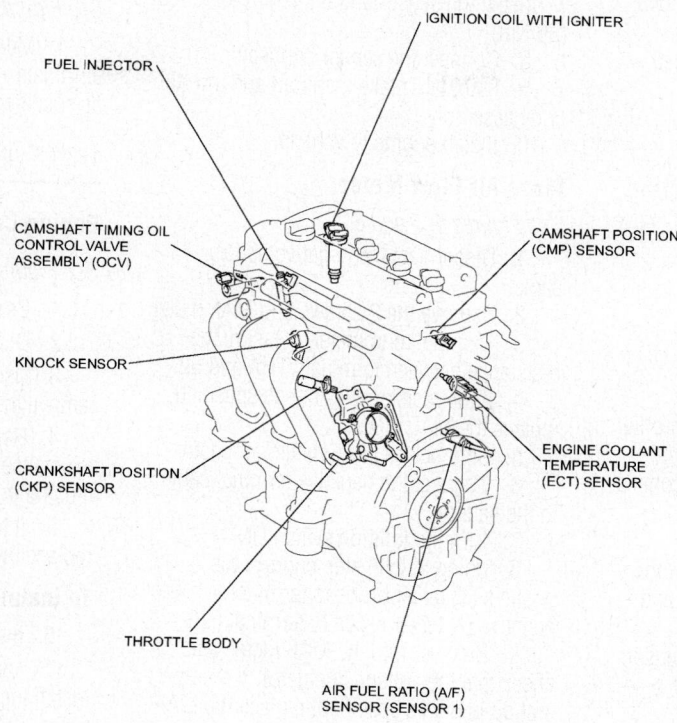

FUEL INJECTOR

IGNITION COIL WITH IGNITER

CAMSHAFT TIMING OIL CONTROL VALVE ASSEMBLY (OCV)

CAMSHAFT POSITION (CMP) SENSOR

KNOCK SENSOR

ENGINE COOLANT TEMPERATURE (ECT) SENSOR

CRANKSHAFT POSITION (CKP) SENSOR

THROTTLE BODY

AIR FUEL RATIO (A/F) SENSOR (SENSOR 1)

Component Testing

1AZ-FE VIN H

Engine Coolant Temperature Sensor

See Figures 1 and 2.

1. Drain engine coolant.
2. Disconnect the sensor connector.
3. Using a 19 mm deep socket wrench, remove the sensor and gasket.
4. Using an ohmmeter, measure the resistance between terminals. Refer to the illustration.
5. If the resistance is not as specified, replace the sensor.

To install:

6. Install a new gasket to the sensor.
7. Using a 19 mm deep socket wrench, install the sensor. Tighten the sensor to 15 ft. lbs. (20 Nm).
8. Connect the sensor connector.
9. Refill with engine coolant.

Idle Air Control

1. Remove throttle body.
2. Remove the 3 screws, IAC valve and gasket.
3. Check that the IAC valve is half-opened.
4. Connect the IAC valve connector.
5. Turn the ignition switch ON.

29157_TOYO_G0067

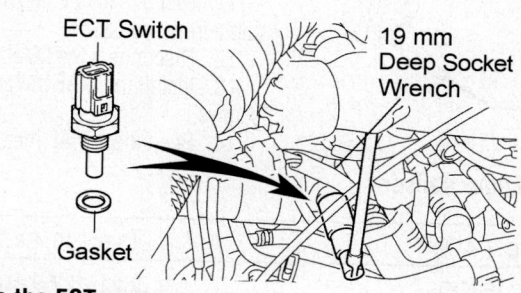

ECT Switch

19 mm Deep Socket Wrench

Gasket

Fig. 1 Remove the ECT sensor

29157_TOYO_G0068

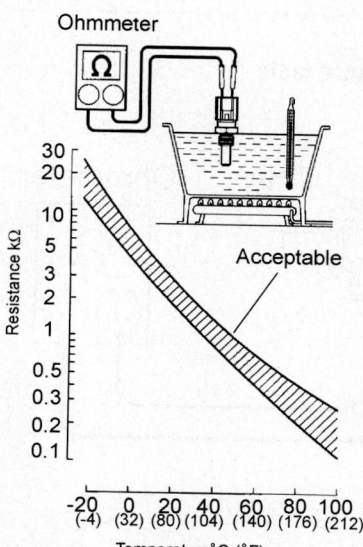

Ohmmeter

Acceptable

Resistance KΩ

30
20
10
5
3
2
1
0.5
0.3
0.2
0.1

-20 0 20 40 60 80 100
(-4) (32) (80) (104) (140) (176) (212)

Temperature°C (°F)

Fig. 2 ECT temperature/resistance graph

29157_TOYO_G0069

6. Check that the IAC valve operates in sequence, half open, fully close, fully open and then half open, within 0.5 seconds.

7. If the operation is not as specified, replace the IAC valve.

To install:

8. Turn the ignition switch OFF.

9. Disconnect the IAC valve connector.

10. Install a new gasket and the IAC valve with the 3 screws.

11. Reinstall throttle body.

Knock Sensor

See Figure 3.

1. Remove engine from vehicle.

2. Remove intake manifold and throttle body assembly.

3. Disconnect the knock sensor connector.

4. Remove the sensor.

5. Using an ohmmeter, check that there is no continuity between the terminal and body.

6. If there is continuity, replace the sensor.

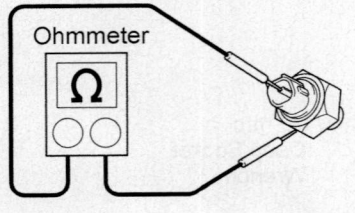

Fig. 3 Testing the knock sensor

-20 °C (-4°F)	13.6 - 18.4 kΩ
20°C (68°F)	2.21 - 2.69 kΩ
60°C (140°F)	0.49 - 0.67 kΩ

Fig. 4 Temperature/Resistance table

To install:

7. Install the sensor and tighten to 29 ft. lbs. (39 Nm).

8. Connect the sensor connector.

9. Reinstall intake manifold and throttle body assembly.

10. Install engine to vehicle.

Mass Air Flow Meter

See Figures 4, 5 and 6.

1. Disconnect the negative battery cable.

2. Remove the 2 screws and MAF meter.

3. Using an ohmmeter, measure the resistance between terminals THA and E2.

4. If the resistance is not as specified, replace the MAF meter.

5. Connect the MAF meter connector.

6. Connect the negative terminal cable to the battery.

7. Turn the ignition switch ON.

8. Using a voltmeter, connect the positive (+) tester probe to terminal VG, and negative (-) tester probe to terminal E2G.

9. Blow air into the MAF meter, and check that the voltage fluctuates. If the operation is not as specified, replace the MAF meter.

10. Turn the ignition switch OFF.

11. Disconnect the negative terminal cable from the battery.

12. Disconnect the MAF meter connector.

13. Install the MAF meter with the 2 screws.

14. Reconnect MAF meter connector.

Heated Oxygen Sensor

See Figures 7 and 8.

1. Measure the resistance between terminals +B and HT. If the resistance is not as specified, replace the oxygen sensor.

1FZ-FE VIN J

Engine Coolant Temperature Sensor

See Figures 9 and 10.

1. Drain engine coolant.

2. Disconnect the sensor connector.

3. Using a 19 mm deep socket wrench, remove the sensor and gasket.

4. Using an ohmmeter, measure the resistance between terminals. Refer to the illustration.

5. If the resistance is not as specified, replace the sensor.

To install:

6. Install a new gasket to the sensor.

7. Using a 19 mm deep socket wrench, install the sensor. Tighten the sensor to 17.5 ft. lbs. (24.5 Nm).

8. Connect the sensor connector.

9. Refill with engine coolant.

Idle Air Control Valve

See Figure 11.

1. Apply battery voltage to terminals B1 and B2, and while repeatedly grounding S1-S2-S3-S4-S1 in sequence, and check that the valve moves toward the closed position.

2. Apply battery voltage to terminals B1 and B2, and while repeatedly grounding S4-S3-S2-S1-S4 in sequence, and check that the valve moves toward the open position.

3. If operation is not as specified, replace the IAC valve.

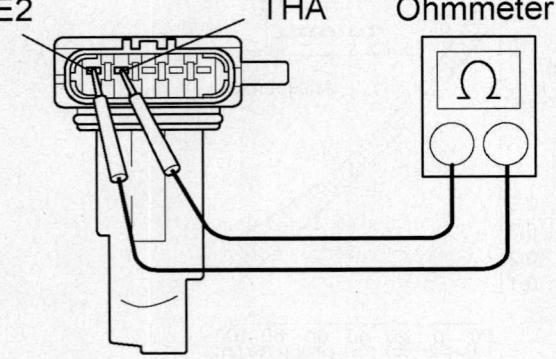

Fig. 5 Terminals THA and E2

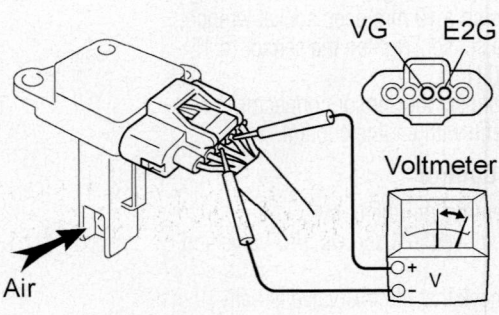

Fig. 6 Terminals VG and E2G

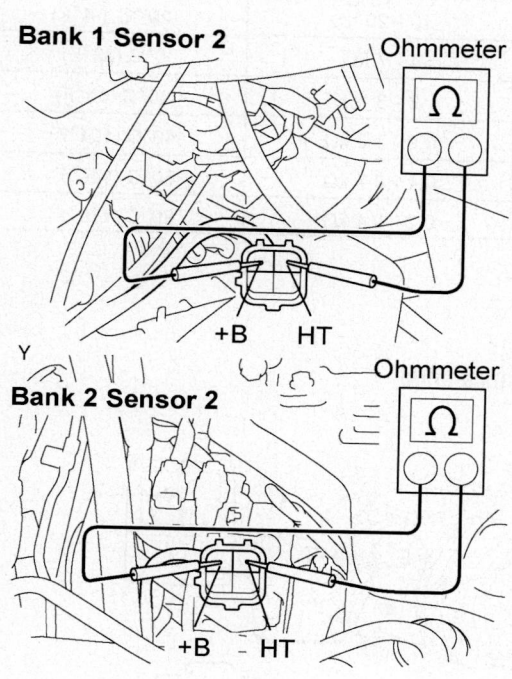

Fig. 7 Oxygen sensor testing

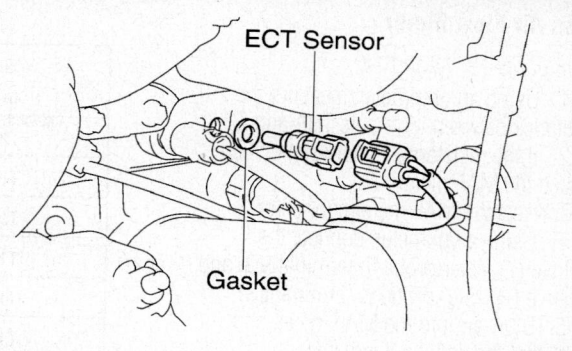

Fig. 9 Remove the ECT sensor

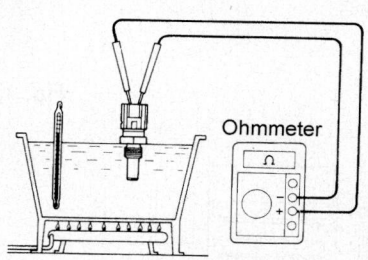

20°C (68°F)	11 - 16 Ω
800°C (1,472°F)	23 - 32 Ω

Fig. 8 Temperature/resistance table

Knock Sensor

1. Disconnect the knock sensor connector.

2. Remove the knock sensor.

3. Using an ohmmeter, check that there is no continuity between the terminal and body.

4. If there is continuity, replace the sensor.

To install:

5. Install the knock sensor and tighten to 33 ft. lbs. (44 Nm).

6. Connect the knock sensor connector.

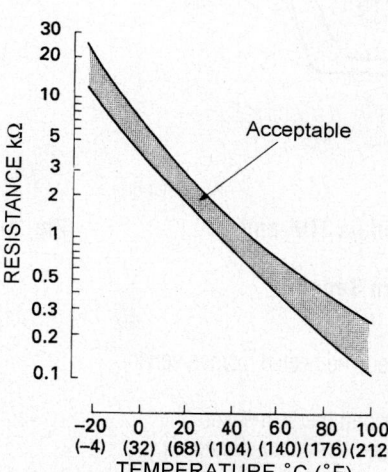

Fig. 10 ECT temperature/resistance graph

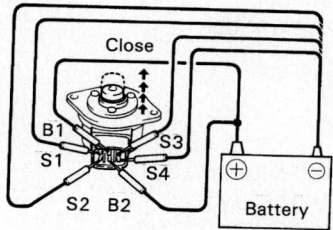

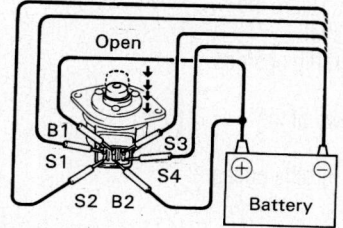

Fig. 11 Testing the IAC valve

Mass Air Flow Meter

See Figures 12, 13 and 14.

1. Using an ohmmeter, measure the resistance between terminals THA and E2.
2. If the resistance is not as specified, replace the MAF meter.
3. Connect the MAF meter connector.
4. Using a voltmeter, connect the positive (+) tester probe to terminal VG, and negative (-) tester probe to terminal E3.
5. Blow air into the MAF meter, and check that the voltage fluctuates.
6. If operation is not as specified, replace the MAF meter.

Between terminals	Resistance	Temperature
THA - E2	10 - 20 kΩ	-20 °C (-4°F)
THA - E2	4 - 7 kΩ	0°C (32°F)
THA - E2	2 - 3 kΩ	20°C (68°F)
THA - E2	0.9 - 1.3 kΩ	40°C (104°F)
THA - E2	0.4 - 0.7 kΩ	60°C (140°F)
THA - E2	0.2 - 0.4 kΩ	80°C (176°F)

29157_TOYO_G0080

Fig. 13 Temperature/resistance table

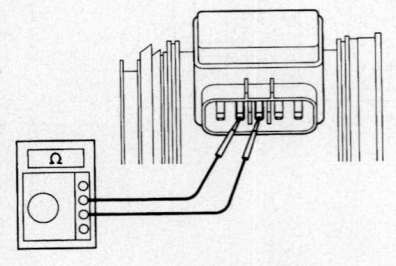

29157_TOYO_G0079

Fig. 12 Terminals THA and E2

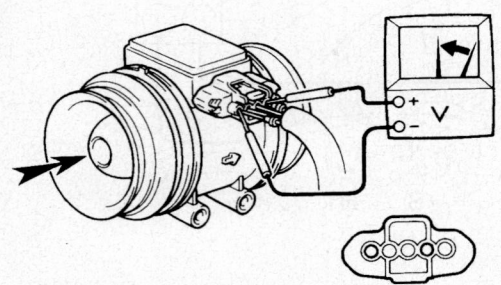

29157_TOYO_G0081

Fig. 14 Terminals VG and E3

Heated Oxygen Sensor

See Figure 15.

1. Disconnect the heated oxygen sensor connector.
2. Using an ohmmeter, measure the resistance between terminals +B and HT.
3. Resistance should be 11 - 16 ohms at 20°C (68°F) If resistance is not as specified, replace the heated oxygen sensor.
4. Reconnect the heated oxygen sensor connector.

Throttle Position Sensor

See Figures 16 through 19.

1. Apply vacuum to the throttle opener.
2. Insert a 0.50 mm (0.020 in.) or 0.75 mm (0.030 in.) feeler gauge between the throttle stop screw and stop lever.
3. Using an ohmmeter, measure the resistance between each terminal.
4. If necessary, adjust throttle position sensor as follows:
   a. Loosen the 2 set screws of the sensor.
   b. Apply vacuum to the throttle opener.

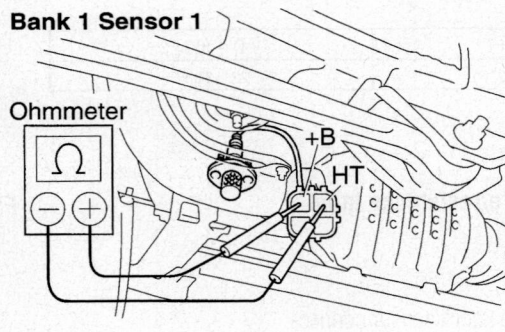

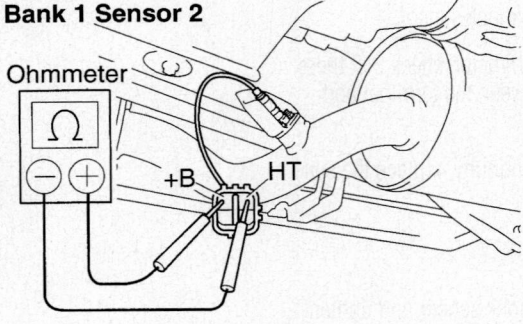

29157_TOYO_G0082

Fig. 15 Testing the Heated Oxygen Sensors

c. Insert a 0.62 mm (0.024 in.) feeler gauge between the throttle stop screw and stop lever.

d. Connect the test probe of an ohmmeter to the terminals

e. IDL and E2 of the sensor.

f. Gradually turn the sensor clockwise until the ohmmeter deflects, and secure it with the 2 set screws.

g. Recheck the continuity between terminals IDL and E2.

1GR-FE VIN U

Mass Air Flow Meter

See Figure 20.

1. Apply battery voltage across terminals 1 (+B) and 2 (E2G).

2. Connect the positive (+) tester prove to terminal 3 (VG), and negative (-) tester prove to terminal 2 (E2G).

3. Blow air into the MAF meter, and check if the voltage fluctuates.

4. If operation is not as specified, replace the MAF meter.

5. Using an ohmmeter, measure resistance between the terminals 4 (THA) and 5 (E2). Compare to values given in the illustration.

6. If operation is not as specified, replace the MAF meter.

Knock Sensor

1. Using an ohmmeter, measure the resistance between terminals. Resistance should be 120–280 KOhms.

2. If the resistance is not as specified, replace the sensor.

Engine Coolant Temperature Sensor

See Figure 21.

1. Using an ohmmeter, measure the resistance between terminals.

2. If the resistance is not as specified, replace the sensor.

Throttle Position Sensor

1. Connect the hand-held tester to the DLC3.

2. Turn the ignition switch ON.

3. When turning the accelerator pedal position sensor lever to the full-open position, check that the percentage of throttle valve opening angle. Standard percentage of the valve opening angle: 60 % or more

4. If operation is not as specified, check that the accelerator pedal position sensor, the wiring and the ECM.

1MZ-FE VIN F

Engine Coolant Temperature Sensor

See Figure 22.

1. Drain engine coolant.

2. Disconnect the ECT sensor connector.

3. Using a 19 mm deep socket wrench, remove the ECT sensor and gasket.

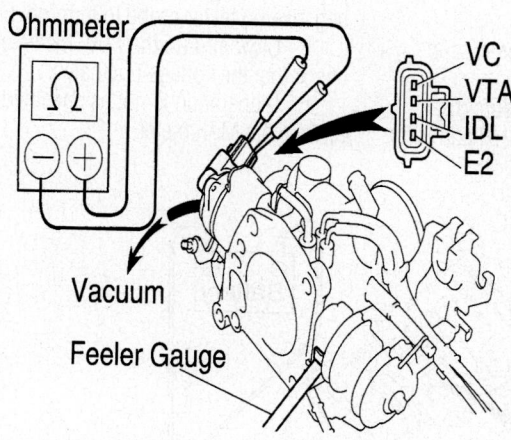

Fig. 16 Throttle Position Sensor testing

29157_TOYO_G0083

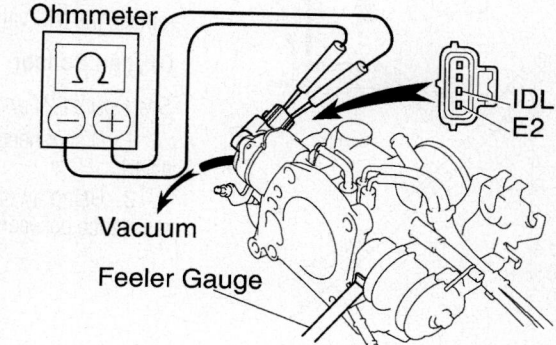

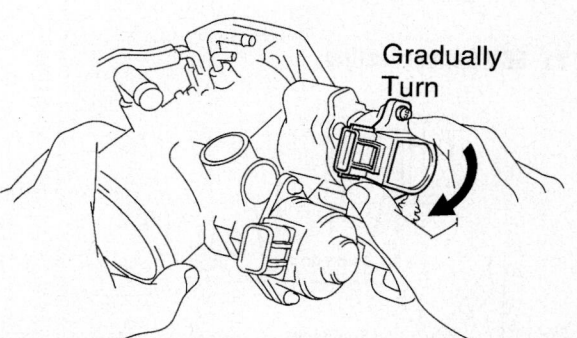

Fig. 18 Throttle Position Sensor adjustment

29157_TOYO_G0085

Clearance between lever and stop screw	Between terminals	Resistance
0 mm (0 in.)	VTA - E2	0.2 - 5.7 kΩ
0.50 mm (0.020 in.)	IDL - E2	2.3 kΩ or less
0.75 mm (0.030 in.)	IDL - E2	Infinity
Throttle valve fully open	VTA - E2	2.0 - 10.2 kΩ
-	VC - E2	2.5 - 5.9 kΩ

Clearance between lever and stop screw	Continuity (IDL - E2)
0.50 mm (0.020 in.)	Continuity
0.75 mm (0.030 in.)	No continuity

Fig. 17 Resistance table

29157_TOYO_G0084

Fig. 19 Recheck the TPS adjustment

29157_TOYO_G0086

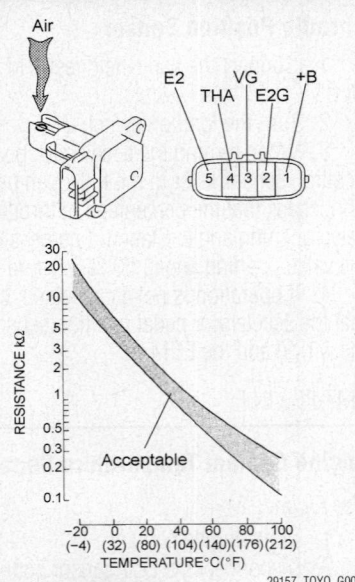

Fig. 20 MAF Testing

29157_TOYO_G0087

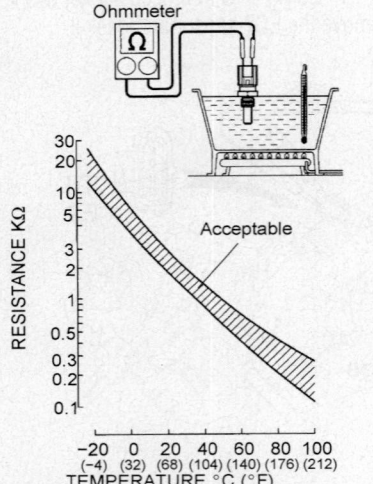

Fig. 21 ECT Sensor testing

29157_TOYO_G0088

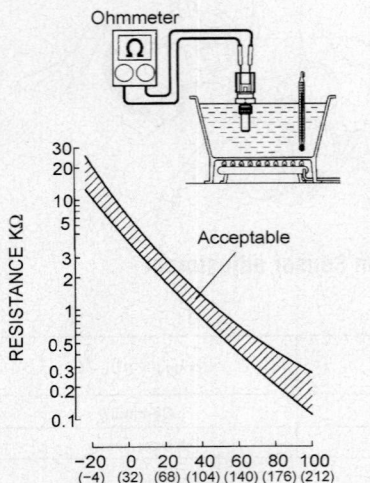

Fig. 22 ECT testing

29157_TOYO_G0088

4. Using an ohmmeter, measure the resistance between the terminals.

5. If the resistance is not as specified, replace the ECT sensor.

To install:

6. Install a new gasket to the ECT sensor.

7. Using a 19 mm deep socket, install the ECT sensor. Tighten to 14 ft. lbs. (20 Nm).

8. Connect the ECT sensor connector.

9. Refill with engine coolant.

Idle Air Control Valve

See Figure 23.

1. Connect the positive (+) lead from the battery to terminal +B and negative (-) lead to terminal RSC, and check that the valve is closed.

2. Connect the positive (+) lead from the battery to terminal +B and negative (-) lead to terminal RSO, and check that the valve is open.

3. If operation is not as specified, replace the IAC valve.

Oxygen Sensor

See Figures 24 and 25.

1. Disconnect the oxygen sensor connector.

2. Using an ohmmeter, measure the resistance between the terminals +B and HT.

Resistance specification: 11 - 16 ohms at 20°C (68°F).

3. If the resistance is not as specified, replace the sensor.

4. Reconnect the oxygen sensor connector.

Throttle Position Sensor

See Figures 26 and 27.

1. Apply vacuum to the throttle opener.

2. Disconnect the sensor connector.

3. Insert a feeler gauge between the throttle stop screw and stop lever.

4. Using an ohmmeter, measure the resistance between each terminal.

5. If the resistance is not as specified, replace the sensor.

6. Reconnect the sensor connector.

Mass Air Flow Meter

See Figures 28, 29 and 30.

1. Using an ohmmeter, measure the resistance between terminals THA and E2.

2. If the resistance is not as specified, replace the MAF meter.

3. Connect the MAF meter connector.

4. Using a voltmeter, connect the positive (+) tester probe to terminal VG, and negative (-) tester probe to terminal E3.

5. Blow air into the MAF meter, and check that the voltage fluctuates.

6. If operation is not as specified, replace the MAF meter.

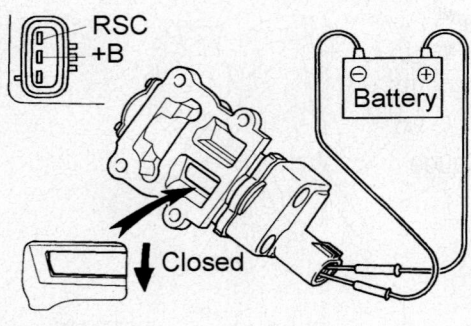

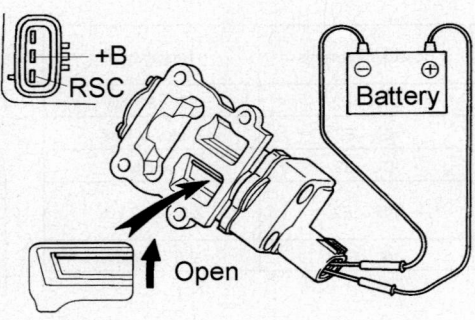

Fig. 23 IAC valve testing

29157_TOYO_G0090

Bank 1 Sensor 1

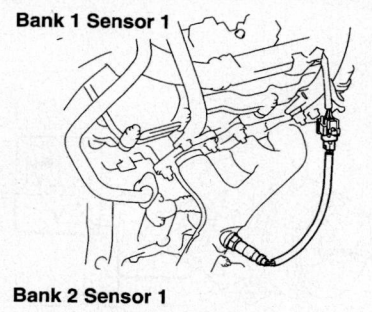

Bank 2 Sensor 1

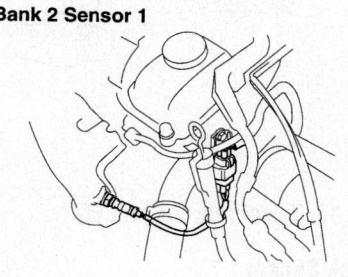

29157_TOYO_G0091

Fig. 24 Oxygen Sensor Connector locations

Bank 1 Sensor 1

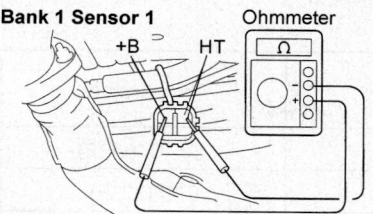

Bank 2 Sensor 1

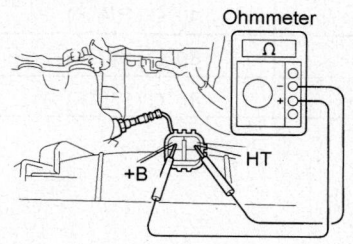

29157_TOYO_G0092

Fig. 25 Oxygen Sensor testing

1NZ-FE VIN T

See Figure 31.

Engine Coolant Temperature Sensor

1. Drain engine coolant.
2. Disconnect the ECT sensor connector.
3. Using a 19 mm deep socket wrench, remove the ECT sensor and gasket.
4. Using an ohmmeter, measure the resistance between terminals 1 and 2.
5. If the resistance is not as specified, replace the ECT sensor.

To install:

6. Install a new gasket to the ECT sensor.
7. Using a 19 mm deep socket, install the ECT sensor. Tighten to 14 ft. lbs. (20 Nm).
8. Connect the ECT sensor connector.
9. Refill with engine coolant.

Idle Air Control Valve

See Figure 32.

1. Check that the IAC valve is half open.
2. Connect the IAC valve connector to the IAC valve.
3. Disconnect the ECT sensor connector from the ECT sensor.
4. Turn the ignition switch ON.
5. Check that the IAC valve moves.
6. Repeat connecting and disconnecting of IAC valve connector several times and check the operation of the valve.
7. If operation is not as specified, replace the IAC valve.
8. Turn the ignition switch OFF.
9. Connect the ECT sensor connector to the ECT sensor.

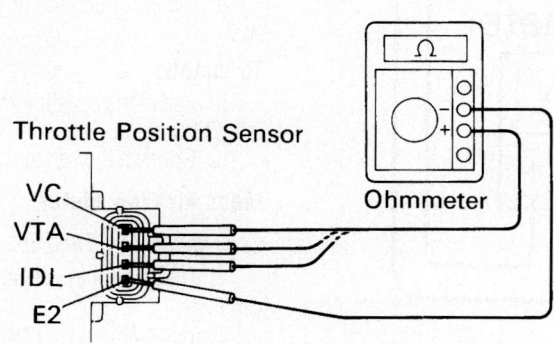

29157_TOYO_G0093

Fig. 26 TPS testing

Clearance between lever and stop screw	Between terminals	Resistance
0 mm (0 in.)	VTA - E2	0.28 - 6.4 kΩ
0.35 mm (0.014 in.)	IDL - E2	0.5 kΩ or less
0.70 mm (0.028 in.)	IDL - E2	Infinity
Throttle valve fully open	VTA - E2	2.0 - 11.6 kΩ
-	VC - E2	2.7 - 7.7 kΩ

29157_TOYO_G0094

Fig. 27 TPS resistance table

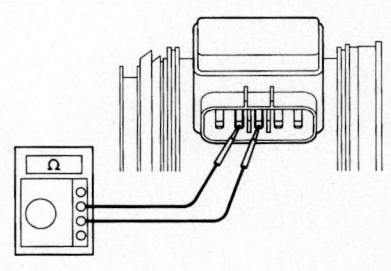

29157_TOYO_G0079

Fig. 28 Terminals THA and E2

Between terminals	Resistance	Temperature
THA - E2	10 - 20 kΩ	-20°C (-4°F)
THA - E2	4 - 7 kΩ	0°C (32°F)
THA - E2	2 - 3 kΩ	20°C (68°F)
THA - E2	0.9 - 1.3 kΩ	40°C (104°F)
THA - E2	0.4 - 0.7 kΩ	60°C (140°F)
THA - E2	0.2 - 0.4 kΩ	80°C (176°F)

29157_TOYO_G0080

Fig. 29 Temperature/resistance table

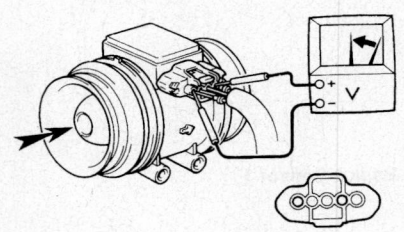

29157_TOYO_G0081

Fig. 30 Terminals VG and E3

Knock Sensor

See Figure 33.

1. Disconnect the knock sensor connector.
2. Remove the sensor.
3. Using an ohmmeter, check that there is no continuity between the terminal and body.
4. If there is continuity, replace the sensor.

To install:

5. Install the sensor and tighten to 29 ft. lbs. (39 Nm).
6. Connect the sensor connector.

Mass Air Flow Meter

See Figures 34, 35 and 36.

1. Disconnect the negative battery cable.
2. Remove the 2 screws and MAF meter.
3. Using an ohmmeter, measure the resistance between terminals THA and E2.
4. If the resistance is not as specified, replace the MAF meter.
5. Connect the MAF meter connector.
6. Connect the negative terminal cable to the battery.

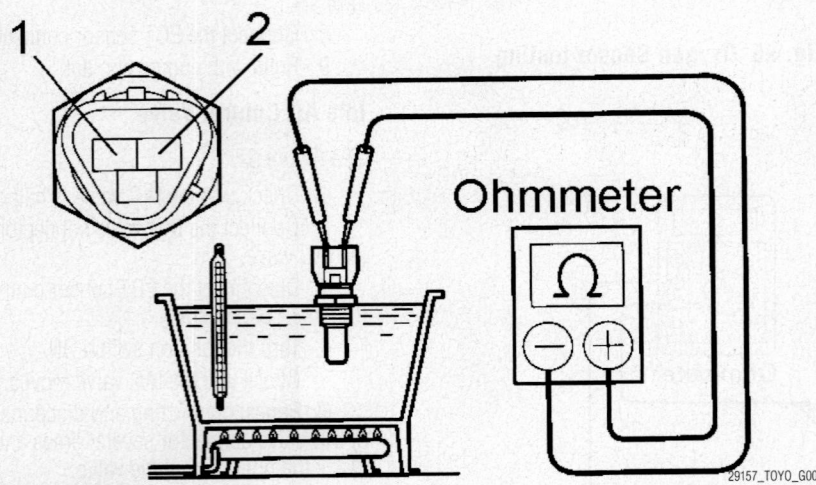

Fig. 31 ECT testing

29157_TOYO_G0095

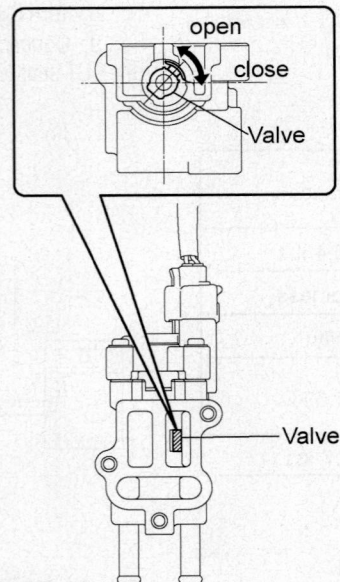

29157_TOYO_G0096

Fig. 32 Idle Air Control Valve

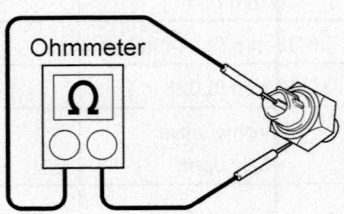

29157_TOYO_G0070

Fig. 33 Testing the knock sensor

-20 °C (-4°F)	13.6 - 18.4 kΩ
20°C (68°F)	2.21 - 2.69 kΩ
60°C (140°F)	0.49 - 0.67 kΩ

Fig. 34 Temperature/Resistance table

29157_TOYO_G0071

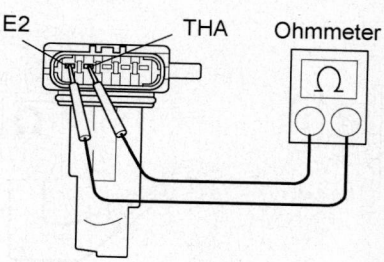

29157_TOYO_G0072

Fig. 35 Terminals THA and E2

1NZ-FXE VIN K,B

Accelerator Pedal Position Sensor

See Figures 41, 42 and 43.

1. Disconnect the sensor connector.

2. Using an ohmmeter, measure the resistance between each terminal.

Engine Coolant Temperature Sensor

See Figure 44.

1. Using an ohmmeter, measure the resistance between terminals.

2. If the resistance is not as specified, replace the sensor.

Knock Sensor

See Figure 45.

1. Disconnect the knock sensor connector.

2. Remove the sensor.

3. Using an ohmmeter, check that there is no continuity between the terminal and body.

4. If there is continuity, replace the sensor.

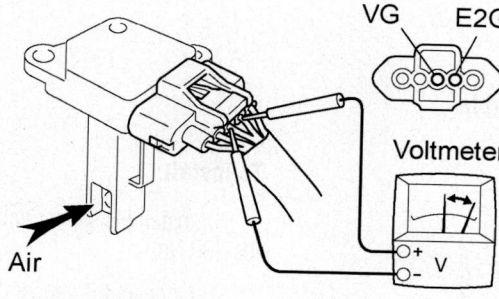

29157_TOYO_G0073

Fig. 36 Terminals VG and E2G

7. Turn the ignition switch ON.

8. Using a voltmeter, connect the positive (+) tester probe to terminal VG, and negative (-) tester probe to terminal E2G.

9. Blow air into the MAF meter, and check that the voltage fluctuates. If the operation is not as specified, replace the MAF meter.

10. Turn the ignition switch OFF.

11. Disconnect the negative terminal cable from the battery.

12. Disconnect the MAF meter connector.

13. Install the MAF meter with the 2 screws.

14. Reconnect MAF meter connector.

Heated Oxygen Sensor

See Figures 37 and 38.

1. Measure the resistance between terminals +B and HT. If the resistance is not as specified, replace the oxygen sensor.

Throttle Position Sensor

See Figures 39 and 40.

1. Disconnect the TPS connector.

2. Using an ohmmeter, measure the resistance between each terminal.

3. If the resistance is not as specified, replace the TPS.

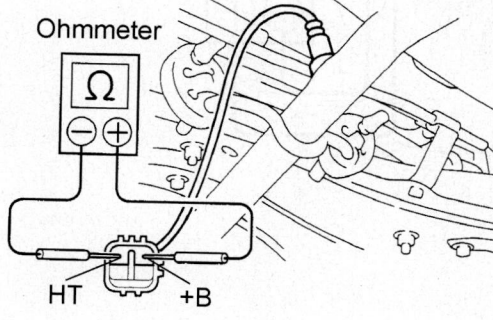

Fig. 37 Oxygen sensor testing

29157_TOYO_G0097

20°C (68°F)	11 - 16 Ω
800°C (1,472°F)	23 - 32 Ω

Fig. 38 Temperature/resistance table

29157_TOYO_G0075

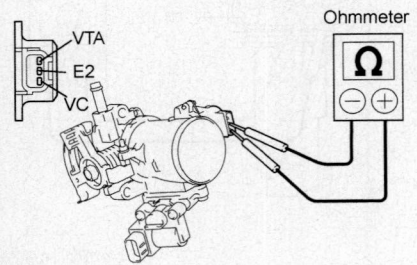

Clearance between lever and stop screw	Between terminals	Resistance
0 mm (0 in.)	VTA - E2	0.2 - 5.7 kΩ
Throttle valve fully open	VTA - E2	2.0 - 10.2 kΩ
Always	VC - E2	2.5 - 5.9 kΩ

29157_TOYO_G0098

Fig. 39 TPS terminal identification

29157_TOYO_G0099

Fig. 40 Resistance table

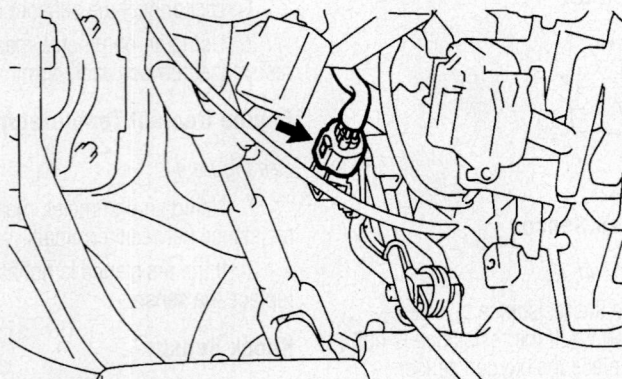

29157_TOYO_G0100

Fig. 41 Connector location

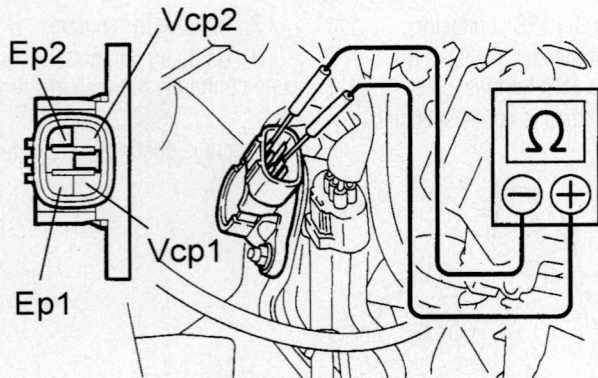

29157_TOYO_G0101

Fig. 42 Connector pin identification

Resistance: at 20 °C (68 °F)

Terminals	Resistance
V_{CP1} - E_{P1} V_{CP2} - E_{P2}	1.5 - 6.0 kΩ

29157_TOYO_G0102

Fig. 43 Resistance table

To install:

5. Install the sensor and tighten to 29 ft. lbs. (39 Nm).

6. Connect the sensor connector.

Mass Air Flow Meter

See Figures 46, 47 and 48.

1. Disconnect the negative battery cable.

2. Remove the 2 screws and MAF meter.

3. Using an ohmmeter, measure the resistance between terminals THA and E2.

4. If the resistance is not as specified, replace the MAF meter.

5. Connect the MAF meter connector.

6. Connect the negative terminal cable to the battery.

7. Turn the ignition switch ON.

8. Using a voltmeter, connect the positive (+) tester probe to terminal VG, and negative (-) tester probe to terminal E2G.

9. Blow air into the MAF meter, and check that the voltage fluctuates. If the operation is not as specified, replace the MAF meter.

10. Turn the ignition switch OFF.

11. Disconnect the negative terminal cable from the battery.

12. Disconnect the MAF meter connector.

13. Install the MAF meter with the 2 screws.

14. Reconnect MAF meter connector.

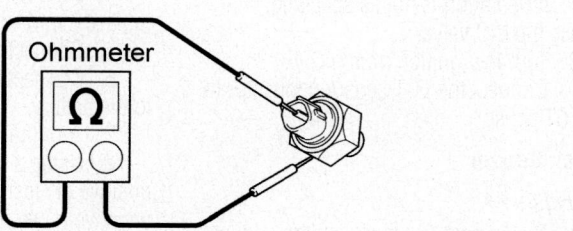

Fig. 44 ECT Sensor testing

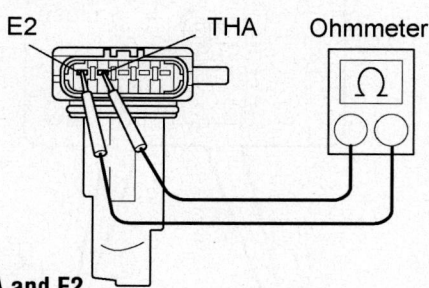

Fig. 45 Testing the knock sensor

-20 °C (-4°F)	13.6 - 18.4 kΩ
20°C (68°F)	2.21 - 2.69 kΩ
60°C (140°F)	0.49 - 0.67 kΩ

Fig. 46 Temperature/Resistance table

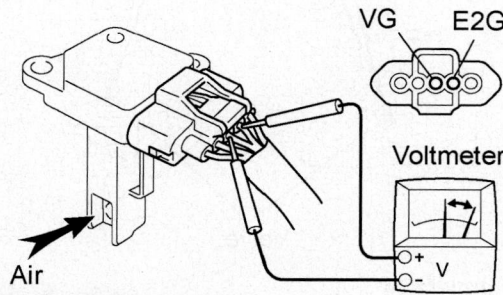

Fig. 47 Terminals THA and E2

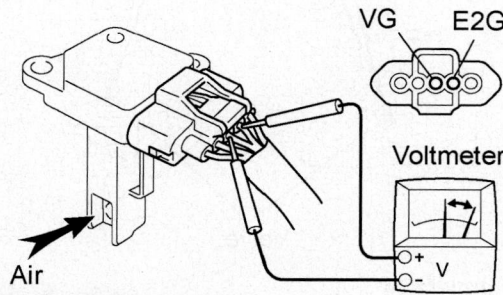

Fig. 48 Terminals VG and E2G

Heated Oxygen Sensor

See Figures 49 and 50.

1. Measure the resistance between terminals +B and HT. If the resistance is not as specified, replace the oxygen sensor.

Throttle Position Sensor

See Figure 51.

1. Disconnect the throttle position sensor connector.

2. Using an ohmmeter, measure the resistance between terminals VC and E2.

3. Resistance: 1.2 - 3.2 KOhms at 20°C (68°F) If the resistance is not as specified, replace the throttle position sensor.

4. Reconnect the throttle position sensor connector.

1ZZ-FE VIN R

Engine Coolant Temperature Sensor

See Figure 52.

1. Drain engine coolant.

2. Disconnect the ECT sensor connector.

3. Using a 19 mm deep socket wrench, remove the ECT sensor and gasket.

4. Using an ohmmeter, measure the resistance between terminals 1 and 2.

5. If the resistance is not as specified, replace the ECT sensor.

To install:

6. Install a new gasket to the ECT sensor.

7. Using a 19 mm deep socket, install the ECT sensor. Tighten to 14 ft. lbs. (20 Nm).

8. Connect the ECT sensor connector.

9. Refill with engine coolant.

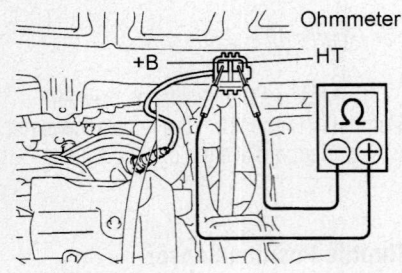

Fig. 49 Oxygen sensor testing

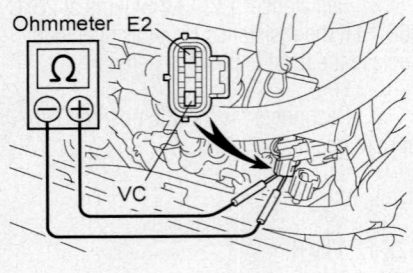

Fig. 51 Connector pin identification

| 20°C (68°F) | 11 - 16 Ω |
| 800°C (1,472°F) | 23 - 32 Ω |

Fig. 50 Temperature/resistance table

Idle Air Control Valve

See Figure 53.

1. Check that the IAC valve is half open.
2. Connect the IAC valve connector to the IAC valve.
3. Disconnect the ECT sensor connector from the ECT sensor.
4. Turn the ignition switch ON.
5. Check that the IAC valve moves.
6. Repeat connecting and disconnecting of IAC valve connector several times and check the operation of the valve.
7. If operation is not as specified, replace the IAC valve.
8. Turn the ignition switch OFF.
9. Connect the ECT sensor connector to the ECT sensor.

Knock Sensor

See Figure 54.

1. Disconnect the knock sensor connector.
2. Remove the sensor.
3. Using an ohmmeter, check that there is no continuity between the terminal and body.
4. If there is continuity, replace the sensor.

To install:

5. Install the sensor and tighten to 29 ft. lbs. (39 Nm).
6. Connect the sensor connector.

Mass Air Flow Meter

See Figures 55, 56 and 57.

1. Disconnect the negative battery cable.
2. Remove the 2 screws and MAF meter.
3. Using an ohmmeter, measure the resistance between terminals THA and E2.
4. If the resistance is not as specified, replace the MAF meter.
5. Connect the MAF meter connector.
6. Connect the negative terminal cable to the battery.
7. Turn the ignition switch ON.
8. Using a voltmeter, connect the positive (+) tester probe to terminal VG, and negative (-) tester probe to terminal E2G.
9. Blow air into the MAF meter, and check that the voltage fluctuates. If the operation is not as specified, replace the MAF meter.
10. Turn the ignition switch OFF.
11. Disconnect the negative terminal cable from the battery.

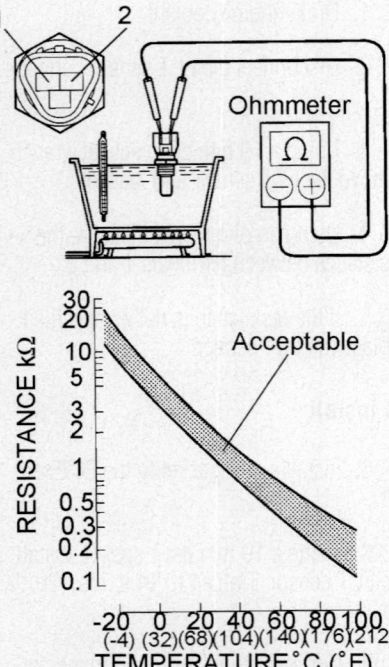

Fig. 52 ECT testing

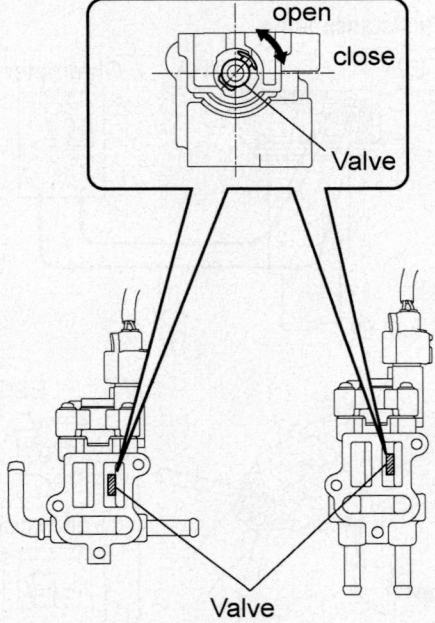

Fig. 53 Idle Air Control Valve

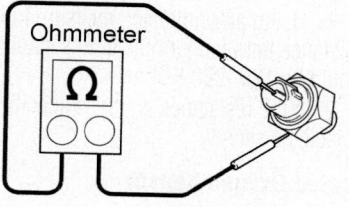

Fig. 54 Testing the knock sensor

-20 °C (-4°F)	13.6 - 18.4 kΩ
20°C (68°F)	2.21 - 2.69 kΩ
60°C (140°F)	0.49 - 0.67 kΩ

Fig. 55 Temperature/Resistance table

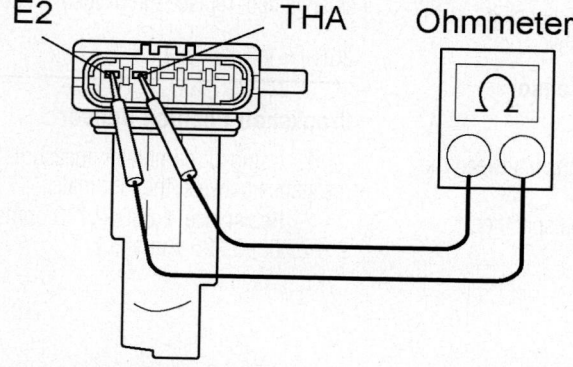

Fig. 56 Terminals THA and E2

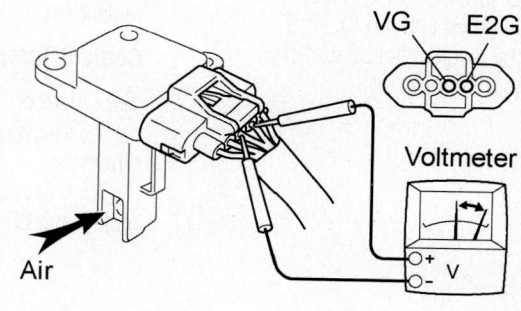

Fig. 57 Terminals VG and E2G

12. Disconnect the MAF meter connector.

13. Install the MAF meter with the 2 screws.

14. Reconnect MAF meter connector.

Heated Oxygen Sensor

See Figure 58.

1. Measure the resistance between terminals +B and HT. Specification is 5–10 ohms.

2. If the resistance is not as specified, replace the oxygen sensor.

Throttle Position Sensor

See Figures 59 and 60.

1. Disconnect the TPS connector.

2. Using an ohmmeter, measure the resistance between each terminal.

3. If the resistance is not as specified, replace the TPS.

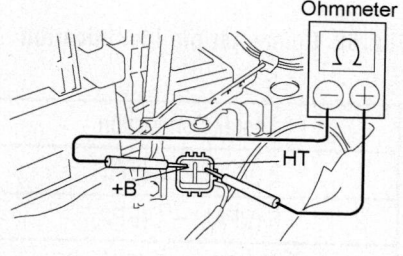

Fig. 58 Oxygen sensor testing

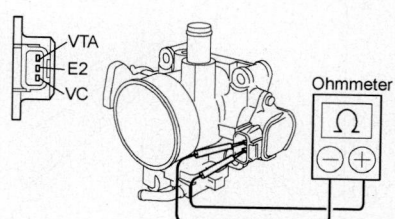

Fig. 59 Throttle Position Sensor

Clearance between lever and stop screw	Between terminals	Resistance
0 mm (0 in.)	VTA - E2	0.2 - 5.7 kΩ
Throttle valve fully open	VTA - E2	2.0 - 10.2 kΩ
Always	VC - E2	2.5 - 5.9 kΩ

Fig. 60 Resistance table

2AZ-FE VIN E

Mass Air Flow Meter

See Figure 61.

1. Apply battery voltage across terminals 1 (+B) and 2 (E2G).

2. Using a voltmeter, connect the positive (+) tester probe to terminal VG, and negative (-) tester probe to terminal E2G.

3. Blow air into the MAF meter, and check that the voltage fluctuates.

4. Measure the resistance between terminals 4 (THA) and 5 (E2).

5. If the result is not as specified, replace the MAF meter.

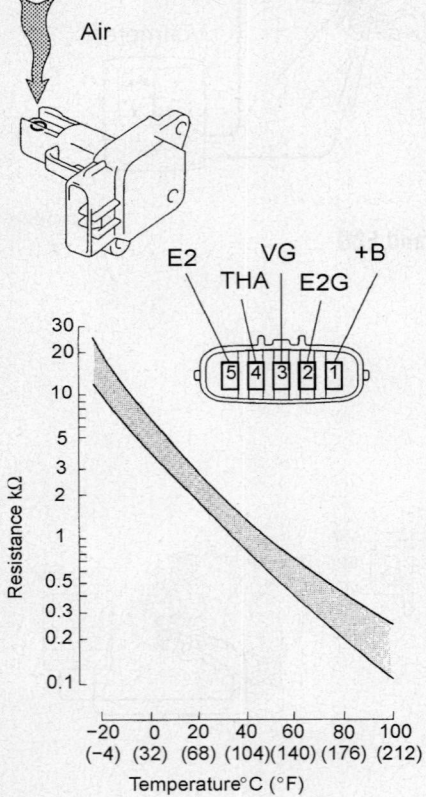

Fig. 61 Connector pin identification and Resistance table

Accelerator Pedal Position Sensor

See Figures 62 and 63.

1. Measure the resistance between the terminals.

2. If the result is not as specified, replace the pedal assy.

Throttle Assembly

See Figures 64 and 65.

1. Measure the resistance between the terminals.

2. If the result is not as specified, replace the throttle body assy.

Coolant Temperature Sensor

See Figure 66.

1. Measure the resistance between the terminals.

2. If the result is not as specified, replace the ECT.

Knock Sensor

1. Using an ohmmeter, measure the resistance between terminals. Resistance should be 120–280 KOhms.

2. If the resistance is not as specified, replace the sensor.

Heated Oxygen Sensor

See Figures 67 and 68.

1. Measure the resistance between terminals +B and HT. If the resistance is not as specified, replace the oxygen sensor.

2GR-FE VIN K

Crankshaft Position Sensor

1. Using an ohmmeter, measure the resistance between the terminals.

2. Resistance: 1,630–2,740 ohms cold and 2,065–3,225 ohms hot.

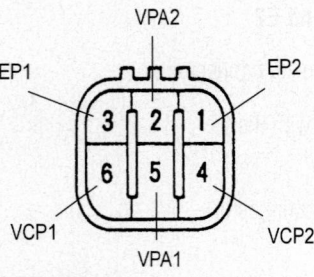

Fig. 62 Connector pin identification

Tester Connection	Specified Condition
2 (VPA2) – 3 (EP1)	5.0 kΩ or less
5 (VPA1) – 1 (EP2)	5.0 kΩ or less
6 (VCP1) – 3 (EP1)	2.25 to 4.75 kΩ
4 (VCP2) – 1 (EP2)	2.25 to 4.75 kΩ

Fig. 63 Resistance table

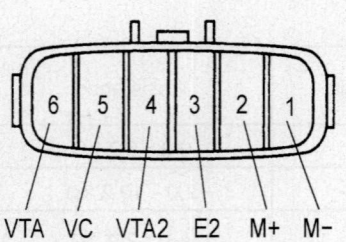

Fig. 64 Connector pin identification

Tester Connection	Condition	Specified Condition
2 (M+) – 1 (M−)	20°C (68°F)	0.3 to 100 Ω
5 (VC) – 3 (E2)	20°C (68°F)	1.2 to 3.2 kΩ

Fig. 65 Resistance table

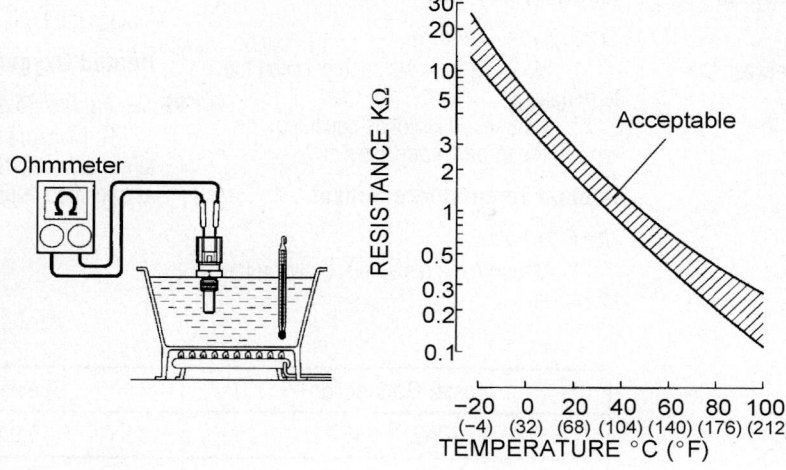

29157_TOYO_G0088

Fig. 66 ECT testing

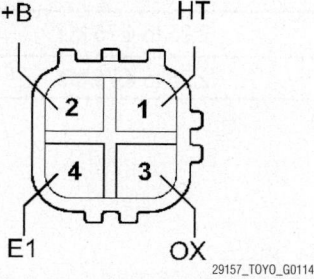

+B   HT

E1   OX

29157_TOYO_G0114

**Fig. 67 Connector terminal identifica-
tion**

Coolant Temperature Sensor

See Figure 69.

1. Measure the resistance between the terminals.
2. If the result is not as specified, replace the ECT.

Knock Sensor

1. Using an ohmmeter, measure the resistance between terminals. Resistance should be 120–280 KOhms.
2. If the resistance is not as specified, replace the sensor.

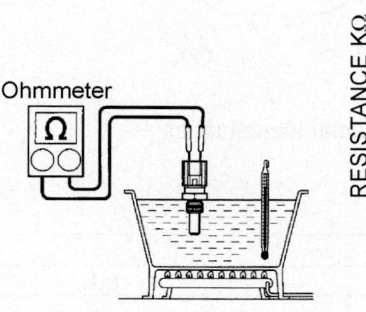

Fig. 69 ECT testing

| 20°C (68°F) | 11 - 16 Ω |
| 800°C (1,472°F) | 23 - 32 Ω |

29157_TOYO_G0075

Fig. 68 Temperature/resistance table

Mass Air Flow Meter

See Figure 70.

1. Apply battery voltage across terminals 1 (+B) and 2 (E2G).
2. Using a voltmeter, connect the positive (+) tester probe to terminal VG, and the negative (-) tester probe to terminal E2G.
3. Blow air into the MAF meter, and check that the voltage fluctuates.
4. Measure the resistance between terminals 4 (THA) and 5 (E2).
5. If the result is not as specified, replace the MAF meter.

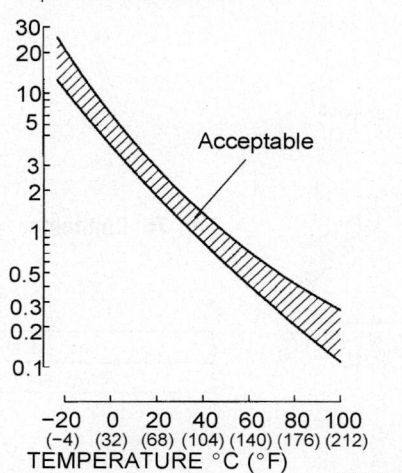

29157_TOYO_G0088

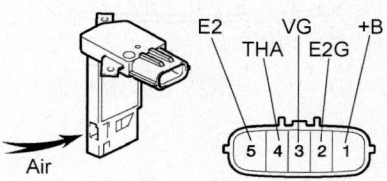

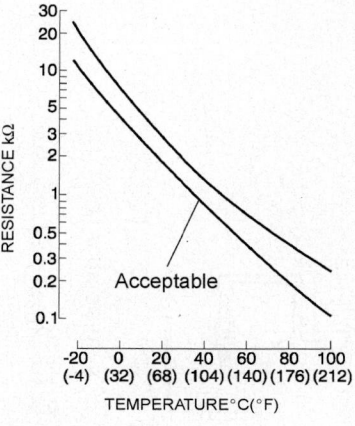

29157_TOYO_G0113

**Fig. 70 Connector pin identification
and Resistance table**

Accelerator Pedal Position Sensor

See Figures 71 and 72.

1. Measure the resistance between the terminals.

2. If the result is not as specified, replace the pedal assy.

Throttle Assembly

See Figures 73 and 74.

1. Measure the resistance between the terminals.

2. If the result is not as specified, replace the throttle body assy.

Coolant Temperature Sensor

See Figure 75.

1. Measure the resistance between the terminals.

2. If the result is not as specified, replace the ECT.

Heated Oxygen Sensor

See Figures 76 and 77.

1. Measure the resistance between terminals +B and HT. If the resistance is not as specified, replace the oxygen sensor.

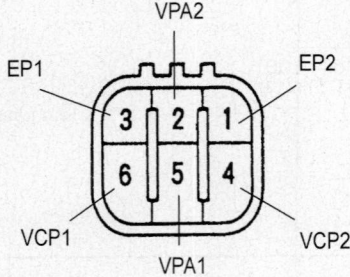

29157_TOYO_G0109

Fig. 71 Connector pin identification

Tester Connection	Specified Condition
2 (VPA2) – 3 (EP1)	5.0 kΩ or less
5 (VPA1) – 1 (EP2)	5.0 kΩ or less
6 (VCP1) – 3 (EP1)	2.25 to 4.75 kΩ
4 (VCP2) – 1 (EP2)	2.25 to 4.75 kΩ

29157_TOYO_G0110

Fig. 72 Resistance table

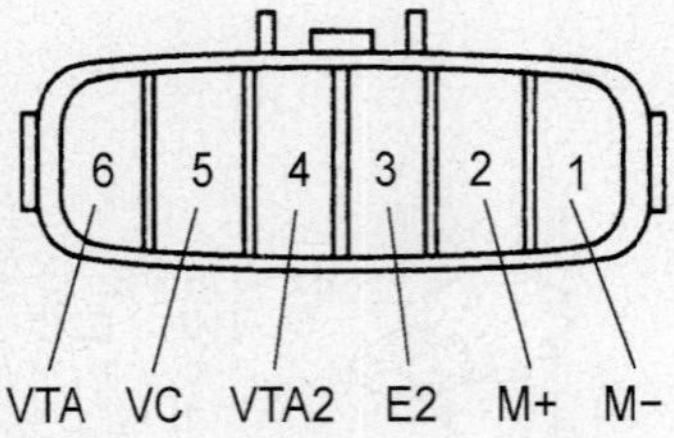

29157_TOYO_G111

Fig. 73 Connector pin identification

Tester Connection	Condition	Specified Condition
2 (M+) – 1 (M–)	20°C (68°F)	0.3 to 100 Ω
5 (VC) – 3 (E2)	20°C (68°F)	1.2 to 3.2 kΩ

29157_TOYO_G0112

Fig. 74 Resistance table

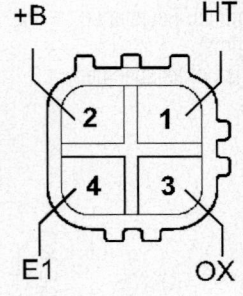

29157_TOYO_G0114

Fig. 76 Connector terminal identification

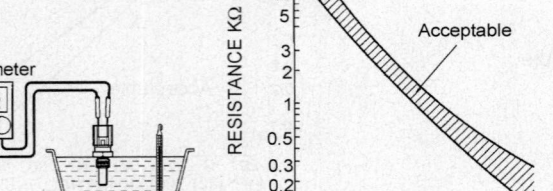

29157_TOYO_G0088

20°C (68°F)	11 - 16 Ω
800°C (1,472°F)	23 - 32 Ω

Fig. 75 ECT testing

29157_TOYO_G0066

Fig. 77 Temperature/resistance table

Coolant Temperature Sensor

See Figure 78.

1. Measure the resistance between the terminals.
2. If the result is not as specified, replace the ECT.

Knock Sensor

See Figure 79.

1. Disconnect the knock sensor connector.
2. Remove the sensor.
3. Using an ohmmeter, check that there is no continuity between the terminal and body.
4. If there is continuity, replace the sensor.

To install:

5. Install the sensor and tighten to 33 ft. lbs. (44 Nm).
6. Connect the sensor connector.

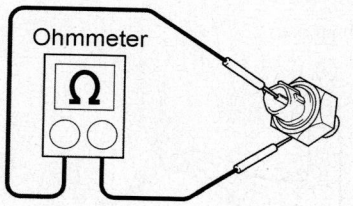

Fig. 79 Testing the knock sensor

29157_TOYO_G0070

Mass Air Flow Meter

See Figures 80 and 81.

1. Disconnect the negative battery cable.
2. Disconnect the MAF meter connector.
3. Remove the 2 bolts, MAF meter and gasket.
4. Using an ohmmeter, measure the resistance between terminals THA and E2.
5. If the resistance is not as specified, replace the MAF meter.
6. Connect the MAF meter connector.
7. Connect the negative (-) terminal cable to the battery.
8. Turn the ignition switch ON.
9. Using a voltmeter, connect the positive (+) tester probe to terminal VG, and negative (-) tester probe to terminal E2G.
10. Blow air into the MAF meter, and check that the voltage fluctuates. If operation

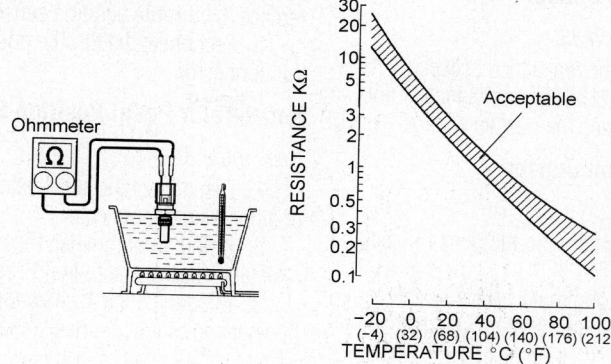

Fig. 78 ECT testing

29157_TOYO_G0088

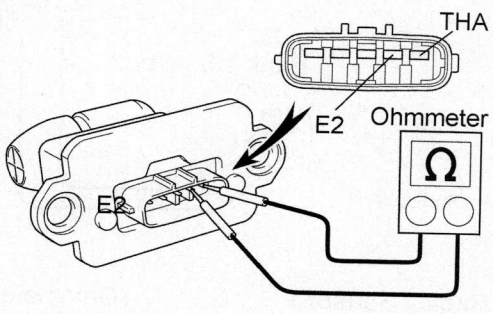

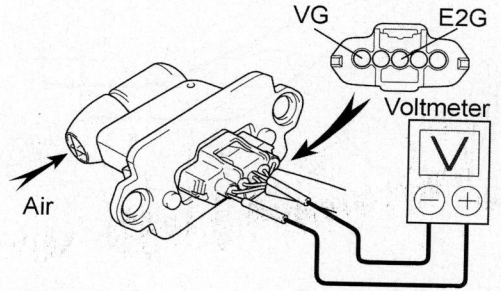

Fig. 80 Testing the MAF meter

29157_TOYO_G0115

Terminals	Resistance	Temperature
THA-E2	13.6 - 18.4 kΩ	-20 °C (-4°F)
THA-E2	2.21 - 2.69 kΩ	20°C (68°F)
THA-E2	0.493 - 0.667 kΩ	60°C (140°F)

Fig. 81 Resistance table

29157_TOYO_G0116

is not as specified, replace the MAF meter.
11. Turn the ignition switch OFF.
12. Disconnect the negative (-) terminal cable from the battery.
13. Disconnect the MAF meter connector.

14. Install the gasket to the MAF meter.
15. Install the MAF meter with the 2 bolts.
16. Connect the negative battery cable.

Heated Oxygen Sensor

See Figures 82 and 83.

1. Measure the resistance between terminals +B and HT. If the resistance is not as specified, replace the oxygen sensor.

Throttle Position Sensor

See Figure 84.

1. Disconnect the throttle position sensor connector.
2. Using an ohmmeter, measure the resistance between terminal VC and E2. Resistance: 1.2 - 3.2 KOhms at 20°C (68°F).

3. If the resistance is not as specified, replace the throttle position sensor.
4. Reconnect the throttle position sensor connector.

Accelerator Pedal Position Sensor

See Figure 85.

1. Disconnect the accelerator pedal position sensor connector.
2. Using an ohmmeter, measure the resistance between terminal VC and E2.
3. Reconnect the accelerator pedal position sensor connector. Resistance: 1.2 - 3.2 KOhms at 20°C (68°F).

4. If the resistance is not as specified, replace the accelerator pedal position sensor.

2JZ-GTE VIN E

Coolant Temperature Sensor

See Figure 86.

1. Measure the resistance between the terminals.
2. If the result is not as specified, replace the ECT.

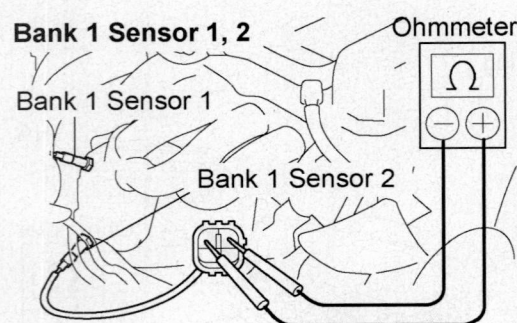

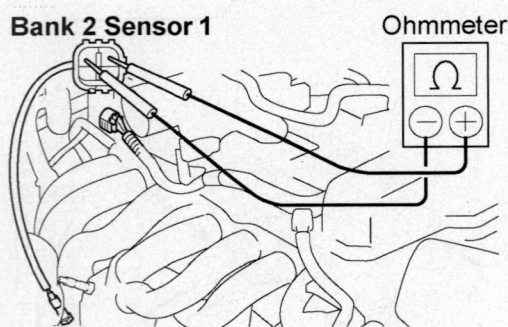

Fig. 82 Testing the Oxygen Sensors

20°C (68°F)	11 - 16 Ω
800°C (1,472°F)	23 - 32 Ω

29157_TOYO_G0075

Fig. 83 Temperature/resistance table

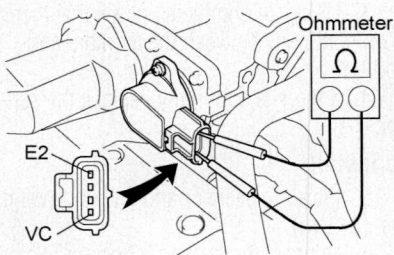

29157_TOYO_G0066

Fig. 84 Testing the TPS

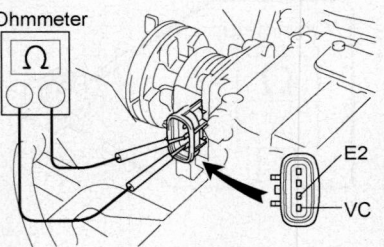

29157_TOYO_G0119

Fig. 85 Testing the Accelerator Pedal Position Sensor

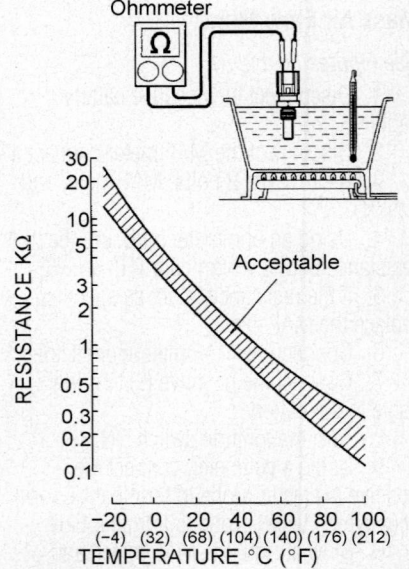

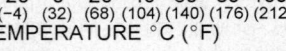

29157_TOYO_G0088

Fig. 86 ECT testing

EGR Gas Temperature Sensor

See Figures 87 and 88.

1. Remove the EGR gas temperature sensor.

2. Using an ohmmeter, measure the resistance between the terminals.

3. If the resistance is not as specified, replace the sensor. Tighten to 14 ft. lbs. (20 Nm).

Knock Sensor

1. Disconnect the knock sensor connector.

2. Remove the sensor.

3. Using an ohmmeter, check that there is no continuity between the terminal and body.

4. If there is continuity, replace the sensor.

To install:

5. Install the sensor and tighten to 33 ft. lbs. (44 Nm).

6. Connect the sensor connector.

Mass Air Flow Meter

See Figures 89, 90 and 91.

1. Using an ohmmeter, measure the resistance between terminals THA and E2.

2. If the resistance is not as specified, replace the MAF meter.

3. Connect the MAF meter connector.

4. Using a voltmeter, connect the positive (+) tester probe to terminal VG, and negative (-) tester probe to terminal E21.

5. Blow air into the MAF meter, and check that the voltage fluctuates.

6. If operation is not as specified, replace the MAF meter.

Heated Oxygen Sensor

BANK 1 SENSOR 1

See Figure 92.

1. Disconnect the oxygen sensor connector.

2. Using an ohmmeter, measure the resistance between the terminals +B and HT. Resistance: 11 - 16 ohms at 20°C (68°F)

3. If the resistance is not as specified, replace the sensor. Tighten to 14 ft. lbs. (20 Nm).

4. Reconnect the oxygen sensor connector.

BANK 1 SENSOR 2

1. Remove the driver's seat.

2. Take out the console box side of the floor carpet.

3. Disconnect the oxygen sensor connector.

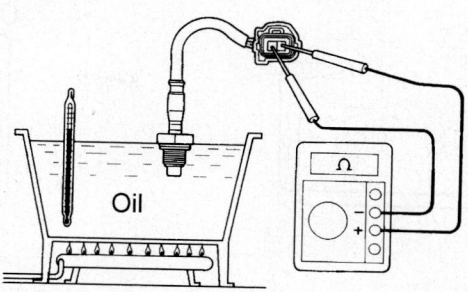

Fig. 87 Testing the EGR Gas Temperature Sensor

29157_TOYO_G0120

At 50°C (122°F)	64 - 97 kΩ
At 100°C (212°F)	11 - 16 kΩ
At 150°C (302°F)	2 - 4 kΩ

Fig. 88 Resistance/Temperature Table

29157_TOYO_G0121

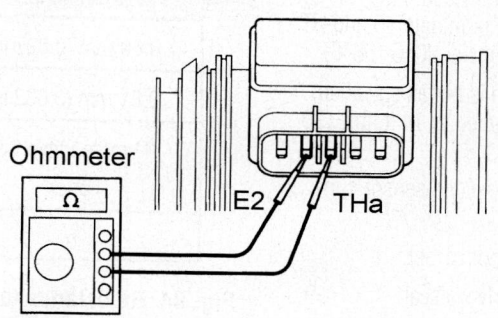

Fig. 89 Terminals THA and E2

29157_TOYO_G0122

Temperature	Resistance
-20°C (-4°F)	10 - 20 kΩ
0°C (32°F)	4 - 7 kΩ
20°C (68°F)	2 - 3 kΩ
40°C (104°F)	0.9 - 1.3 kΩ
60°C (140°F)	0.4 - 0.7 kΩ
80°C (176°F)	0.2 - 0.4 kΩ

Fig. 90 Temperature/resistance table

29157_TOYO_G0123

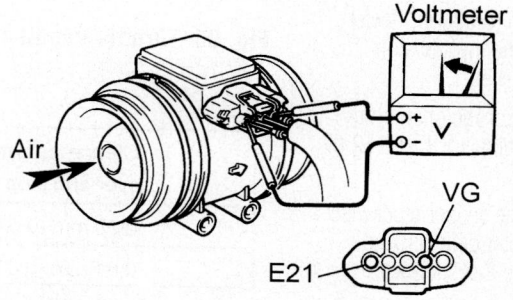

Fig. 91 Terminals VG and E21

29157_TOYO_G0124

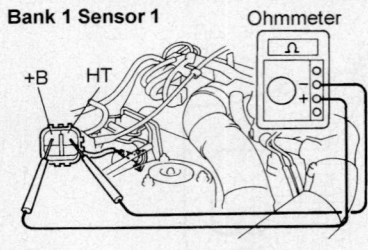

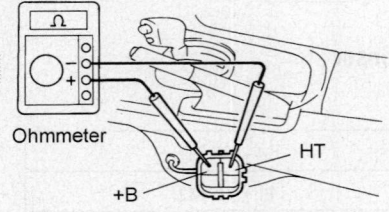

Fig. 92 Testing the Oxygen Sensors

4. Using an ohmmeter, measure the resistance between the terminals +B and HT. Resistance: 11 - 16 ohms at 20°C (68°F)

5. If the resistance is not as specified, replace the sensor. Tighten to 14 ft. lbs. (20 Nm).

6. Reconnect the oxygen sensor connector.

7. Reinstall the floor carpet.

8. Reinstall the driver's seat.

Throttle Position Sensor

See Figures 93 through 96.

1. Apply vacuum to the throttle opener.

2. Insert a 0.69 mm (0.027 in.) or 0.81 mm (0.032 in.) feeler gauge between the throttle stop screw and stop lever.

3. Using an ohmmeter, measure the resistance between each terminal.

4. If necessary, adjust throttle position sensor as follows:

   a. Loosen the 2 set screws of the sensor.

   b. Apply vacuum to the throttle opener.

   c. Insert a 0.77 mm (0.030 in.) feeler gauge between the throttle stop screw and stop lever.

   d. Connect the test probe of an ohmmeter to the terminals IDL and E2 of the sensor.

   e. Gradually turn the sensor clockwise until the ohmmeter deflects, and secure it with the 2 set screws.

   f. Recheck the continuity between terminals IDL and E2.

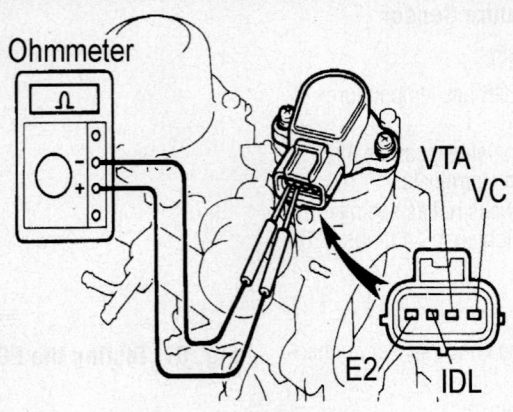

Fig. 93 Throttle Position Sensor testing

Clearance between lever and stop screw	Between terminals	Resistance
0 mm (0 in.)	VTA - E2	0.34 - 6.3 kΩ
0.69 mm (0.027 in.)	IDL - E2	0.5 kΩ or less
0.81 mm (0.032 in.)	IDL - E2	Infinity
Throttle valve fully open	VTA - E2	2.4 - 11.2 kΩ
-	VC - E2	3.1 - 7.2 kΩ

Fig. 94 Resistance table

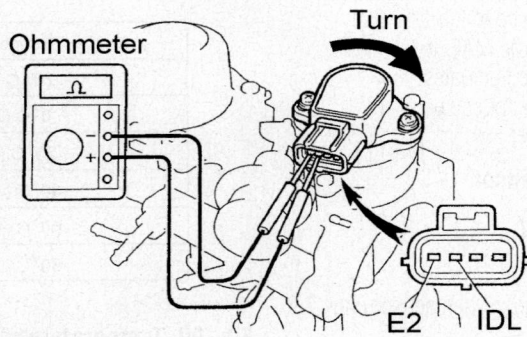

Fig. 95 Throttle Position Sensor adjustment

Clearance between lever and stop screw	Continuity (IDL - E2)
0.69 mm (0.027 in.)	Continuity
0.81 mm (0.032 in.)	No continuity

Fig. 96 Recheck the TPS adjustment

Coolant Temperature Sensor

See Figure 97.

1. Measure the resistance between the terminals.
2. If the result is not as specified, replace the ECT.

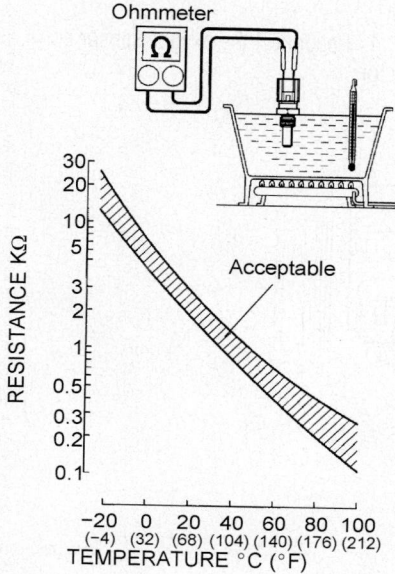

Fig. 97 ECT testing

EGR Gas Temperature Sensor

See Figures 98 and 99.

1. Remove the EGR gas temperature sensor.
2. Using an ohmmeter, measure the resistance between the terminals.
3. If the resistance is not as specified, replace the sensor. Tighten to 14 ft. lbs. (20 Nm).

Idle Air Control Valve

EXCEPT CALIFORNIA SPECIFICATION

See Figures 100, 101 and 102.

1. Disconnect the IAC valve connector.
2. Using an ohmmeter, measure resistance between terminal +B and the other (RSC, RSO) terminals.
3. If resistance is not as specified, replace the IAC valve.
4. Connect the positive (+) lead from the battery to terminal +B and negative (−) to terminal RSC and check that the valve is closed.
5. Connect the positive (+) lead from the battery to terminal +B and negative (−) to terminal RSO and check that the valve is open.

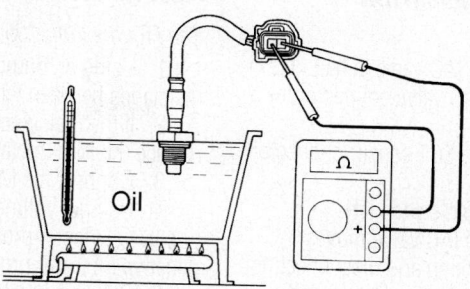

Fig. 98 Testing the EGR Gas Temperature Sensor

29157_TOYO_G0120

At 50°C (122°F)	64 - 97 kΩ
At 100°C (212°F)	11 - 16 kΩ
At 150°C (302°F)	2 - 4 kΩ

Fig. 99 Resistance/Temperature Table

29157_TOYO_G0121

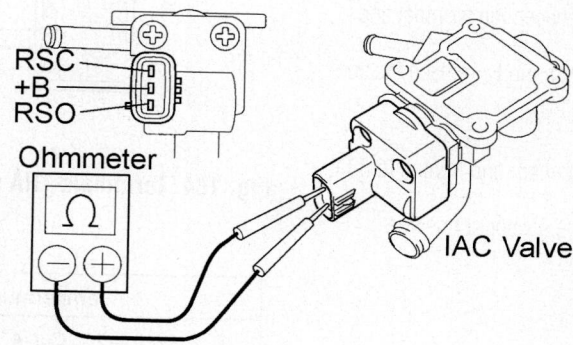

Fig. 100 IAC valve terminal identification

29157_TOYO_G0130

Cold	17.0 - 24.5 Ω
Hot	21.5 - 28.5 Ω

Fig. 101 Resistance table

29157_TOYO_G0131

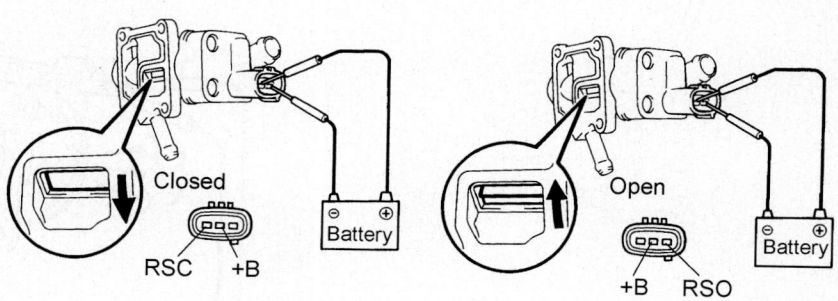

29157_TOYO_G0132

Fig. 102 Checking IAC valve operation

CALIFORNIA SPECIFICATION

See Figure 103.

1. Check that the IAC valve is half open.
2. Connect the IAC valve connector to the IAC valve.
3. Disconnect the ECT sensor connector from the ECT sensor.
4. Turn the ignition switch ON.
5. Check that the IAC valve moves.
6. Repeat connecting and disconnecting of IAC valve connector several times and check the operation of the valve.
7. If operation is not as specified, replace the IAC valve.
8. Turn the ignition switch OFF.
9. Connect the ECT sensor connector to the ECT sensor.

Knock Sensor

1. Disconnect the knock sensor connector.
2. Remove the sensor.
3. Using an ohmmeter, check that there is no continuity between the terminal and body.
4. If there is continuity, replace the sensor.

To install:

5. Install the sensor and tighten to 33 ft. lbs. (44 Nm).
6. Connect the sensor connector.

Mass Air Flow Meter

See Figures 104, 105 and 106.

1. Using an ohmmeter, measure the resistance between terminals THA and E2.
2. If the resistance is not as specified, replace the MAF meter.
3. Connect the MAF meter connector.
4. Using a voltmeter, connect the positive (+) tester probe to terminal VG, and negative (-) tester probe to terminal E3.
5. Blow air into the MAF meter, and check that the voltage fluctuates.
6. If operation is not as specified, replace the MAF meter.

Heated Oxygen Sensor

See Figure 107.

1. Disconnect the oxygen sensor connector.
2. Using an ohmmeter, measure the resistance between the terminals +B and HT. Resistance: 11 - 16 ohms at 20°C (68°F)
3. If the resistance is not as specified, replace the sensor. Tighten to 14 ft. lbs. (20 Nm).
4. Reconnect the oxygen sensor connector.

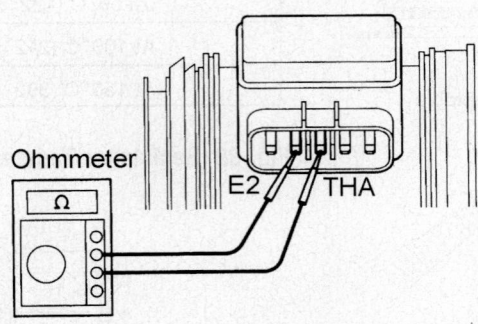

Fig. 104 Terminals THA and E2

29157_TOYO_G0134

Temperature	Resistance
-20 °C (-4°F)	10 - 20 kΩ
0°C (32°F)	4 - 7 kΩ
20°C (68°F)	2 - 3 kΩ
40°C (104°F)	0.9 - 1.3 kΩ
60°C (140°F)	0.4 - 0.7 kΩ
80°C (176°F)	0.2 - 0.4 kΩ

29157_TOYO_G0123

Fig. 105 Temperature/resistance table

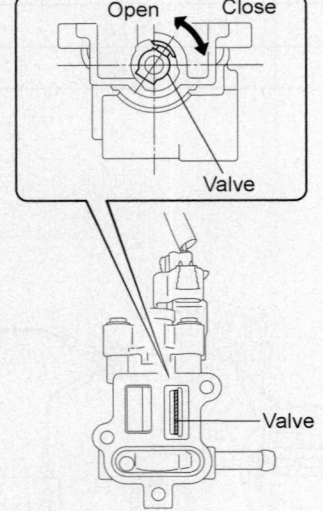

29157_TOYO_G0133

Fig. 103 Idle Air Control Valve

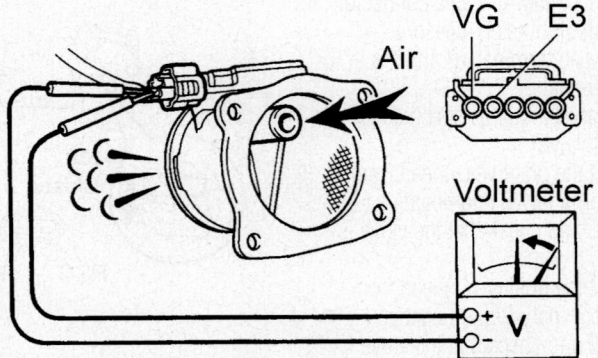

29157_TOYO_G0135

Fig. 106 Terminals VG and E3

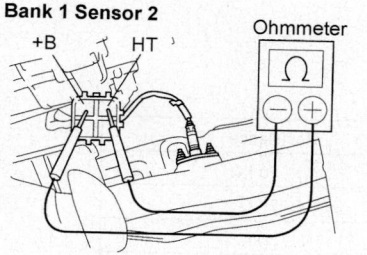

Fig. 107 Testing the Oxygen Sensors

Throttle Position Sensor

See Figures 108 and 109.

1. Apply vacuum to the throttle opener.
2. Insert a thickness gauge between the throttle stop screw and stop lever.
3. Using an ohmmeter, measure the resistance between each terminal.

2TZ-FZE VIN K

Coolant Temperature Sensor

See Figure 110.

1. Measure the resistance between the terminals.
2. If the result is not as specified, replace the ECT.

Idle Air Control Valve

See Figure 111.

1. Connect the positive (+) lead from the battery to terminal +B and negative (-) to terminal RSC and check that the valve is closed.
2. Connect the positive (+) lead from the battery to terminal +B and negative (-) to terminal RSO and check that the valve is open.

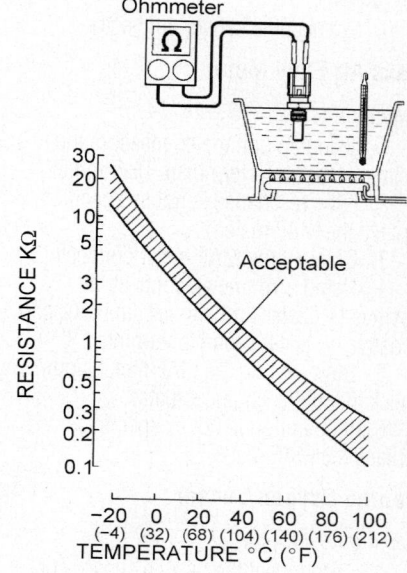

Fig. 110 ECT testing

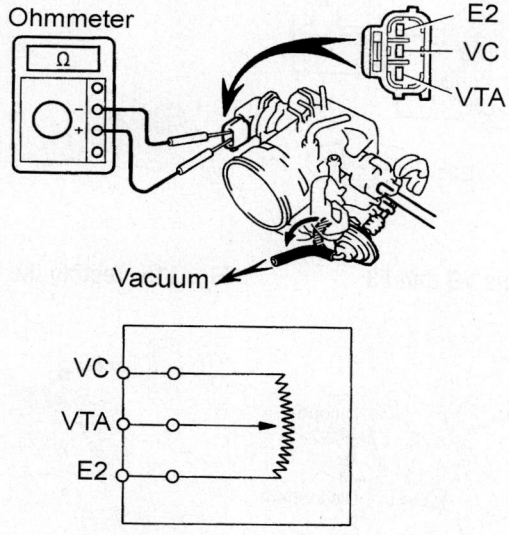

Fig. 108 TPS testing

Clearance between lever and stop screw	Between terminals	Resistance
0 mm (0 in.)	VTA - E2	0.2 - 5.7 kΩ
Throttle valve fully open	VTA - E2	2.0 - 10.2 kΩ
-	VC - E2	2.5 - 5.9 kΩ

Fig. 109 Resistance table

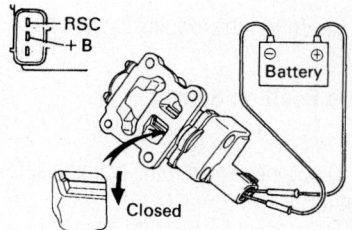

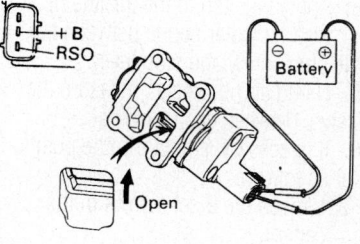

Fig. 111 Checking IAC valve operation

Knock Sensor

1. Disconnect the knock sensor connector.

2. Remove the sensor.

3. Using an ohmmeter, check that there is no continuity between the terminal and body.

4. If there is continuity, replace the sensor.

To install:

5. Install the sensor and tighten to 33 ft. lbs. (44 Nm).

6. Connect the sensor connector.

Mass Air Flow Meter

See Figures 112, 113 and 114.

1. Using an ohmmeter, measure the resistance between terminals THA and E2.

2. If the resistance is not as specified, replace the MAF meter.

3. Connect the MAF meter connector.

4. Using a voltmeter, connect the positive (+) tester probe to terminal VG, and negative (-) tester probe to terminal E3.

5. Blow air into the MAF meter, and check that the voltage fluctuates.

6. If operation is not as specified, replace the MAF meter.

Heated Oxygen Sensor

See Figure 115.

1. Disconnect the oxygen sensor connector.

2. Using an ohmmeter, measure the resistance between the terminals +B and HT. Resistance: 11 - 16 ohms at 20°C (68°F).

3. If the resistance is not as specified, replace the sensor. Tighten to 14 ft. lbs. (20 Nm).

4. Reconnect the oxygen sensor connector.

Throttle Position Sensor

See Figures 116 thru 119.

1. Disconnect the throttle position sensor connector.

2. Disconnect the vacuum hose from the throttle opener.

3. Apply vacuum to the throttle opener.

4. Insert a feeler gauge between the throttle stop screw and stop lever.

5. Using an ohmmeter, measure the resistance between each terminal.

6. If necessary, adjust throttle position sensor as follows:

    a. Loosen the 2 set screws of the sensor.

    b. Apply vacuum to the throttle opener.

    c. Insert a 0.73 mm (0.029 in.) feeler gauge between the throttle stop screw and stop lever.

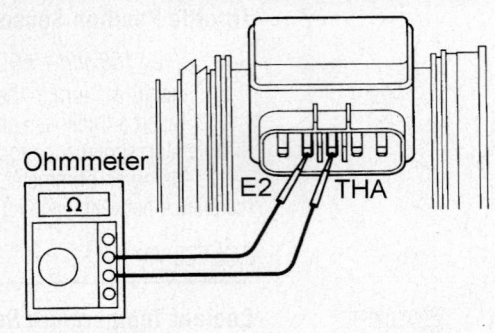

Fig. 112 Terminals THA and E2

Temperature	Resistance
-20°C (-4°F)	10 - 20 kΩ
0°C (32°F)	4 - 7 kΩ
20°C (68°F)	2 - 3 kΩ
40°C (104°F)	0.9 - 1.3 kΩ
60°C (140°F)	0.4 - 0.7 kΩ
80°C (176°F)	0.2 - 0.4 kΩ

Fig. 113 Temperature/resistance table

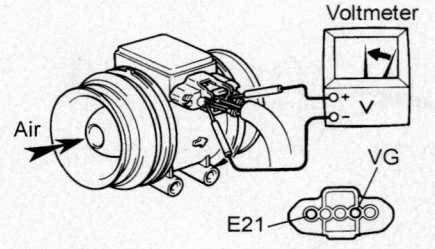

Fig. 114 Terminals VG and E3

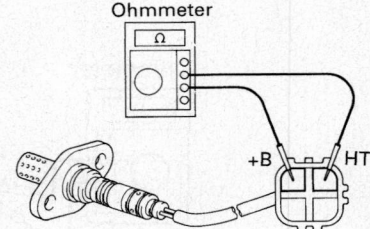

Fig. 115 Testing the Oxygen Sensor

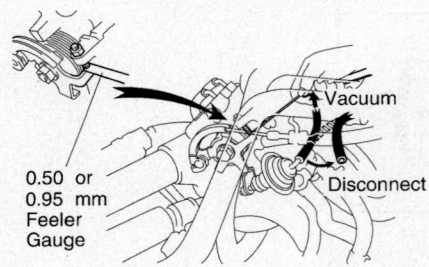

Fig. 116 TPS testing

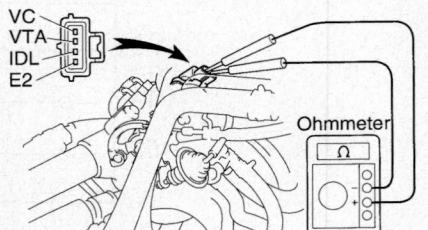

Clearance between lever and stop screw	Between terminals	Resistance
0 mm (0 in.)	VTA - E2	0.2 - 5.7 kΩ
0.50 mm (0.20 in.)	IDL - E2	2.3 kΩ or less
0.95 mm (0.037 in.)	IDL - E2	Infinity
Throttle valve fully open	VTA - E2	2.0 - 10.2 kΩ
-	VC - E2	2.5 - 5.9 kΩ

Fig. 117 Resistance table

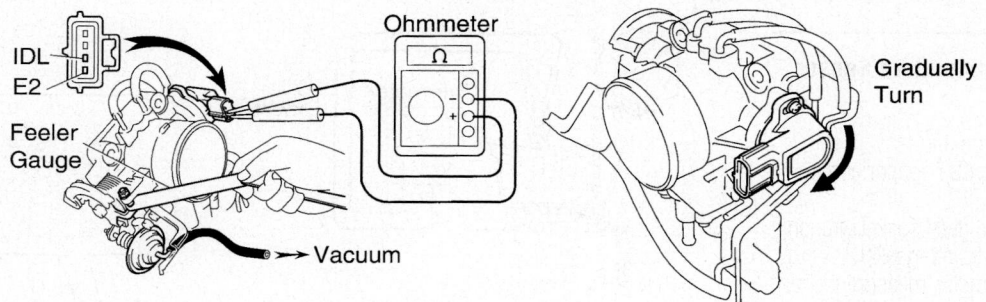

Fig. 118 TPS adjustment

29157_TOYO_G0143

Clearance between lever and stop screw	Continuity (IDL - E2)
0.50 mm (0.020 in.)	Continuity
0.95 mm (0.037 in.)	No continuity

29157_TOYO_G0144

Fig. 119 Resistance table

d. Connect the test probe of an ohmmeter to the terminals IDL and E2 of the sensor.

e. Gradually turn the sensor clockwise until the ohmmeter deflects, and secure it with the 2 set screws.

f. Recheck the continuity between terminals IDL and E2.

2UZ-FE VIN T

Mass Air Flow Meter

See Figure 120.

1. Apply battery voltage across terminals 3 (+B) and 4 (E2G).

2. Connect the positive (+) tester probe to terminal 5 (VG), and the negative (-) tester probe to terminal 4 (E2G).

3. Blow air into the MAF meter, and check if the voltage fluctuates.

4. Using an ohmmeter, measure the resistance between terminals 2 (THA) and 1 (E2).

Knock Sensor

1. Using an ohmmeter, measure the resistance between terminals. Resistance should be 120–280 KOhms.

2. If the resistance is not as specified, replace the sensor.

Coolant Temperature Sensor

See Figure 121.

1. Measure the resistance between the terminals.

2. If the result is not as specified, replace the ECT.

Heated Oxygen Sensor

1. Disconnect the oxygen sensor connector.

2. Using an ohmmeter, measure the resistance between the terminals +B and HT. Resistance: 11 - 16 ohms at 20°C (68°F)

3. If the resistance is not as specified, replace the sensor. Tighten to 14 ft. lbs. (20 Nm).

4. Reconnect the oxygen sensor connector.

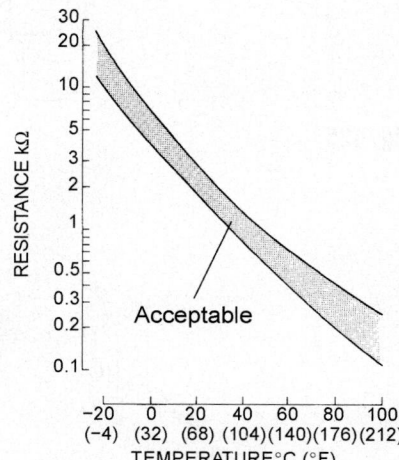
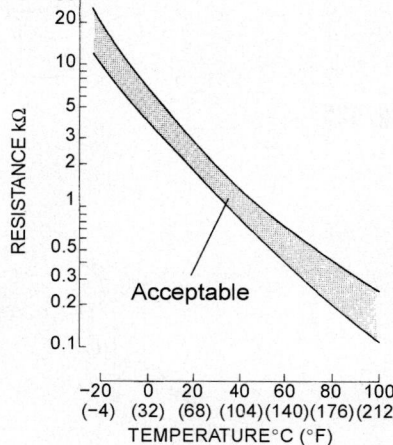

Fig. 120 MAF meter testing

29157_TOYO_G0145

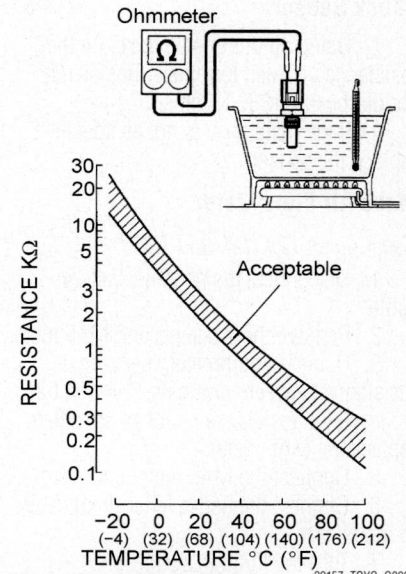

Fig. 121 ECT testing

29157_TOYO_G0088

2ZZ-FE VIN Y

Engine Coolant Temperature Sensor

See Figure 122.

1. Drain engine coolant.
2. Disconnect the ECT sensor connector.
3. Using a 19 mm deep socket wrench, remove the ECT sensor and gasket.
4. Using an ohmmeter, measure the resistance between terminals 1 and 2.
5. If the resistance is not as specified, replace the ECT sensor.

To install:

6. Install a new gasket to the ECT sensor.
7. Using a 19 mm deep socket, install the ECT sensor. Tighten to 14 ft. lbs. (20 Nm).
8. Connect the ECT sensor connector.
9. Refill with engine coolant.

Idle Air Control Valve

See Figure 123.

1. Check that the IAC valve is half open.
2. Connect the IAC valve connector to the IAC valve.
3. Disconnect the ECT sensor connector from the ECT sensor.
4. Turn the ignition switch ON.
5. Check that the IAC valve moves.
6. Repeat connecting and disconnecting of IAC valve connector several times and check the operation of the valve.
7. If operation is not as specified, replace the IAC valve.
8. Turn the ignition switch OFF.
9. Connect the ECT sensor connector to the ECT sensor.

Knock Sensor

1. Using an ohmmeter, measure the resistance between terminals. Resistance should be 120–280 KOhms.
2. If the resistance is not as specified, replace the sensor.

Mass Air Flow Meter

See Figures 124, 125 and 126.

1. Disconnect the negative battery cable.
2. Remove the 2 screws and MAF meter.
3. Using an ohmmeter, measure the resistance between terminals THA and E2.
4. If the resistance is not as specified, replace the MAF meter.
5. Connect the MAF meter connector.
6. Connect the negative terminal cable to the battery.
7. Turn the ignition switch ON.

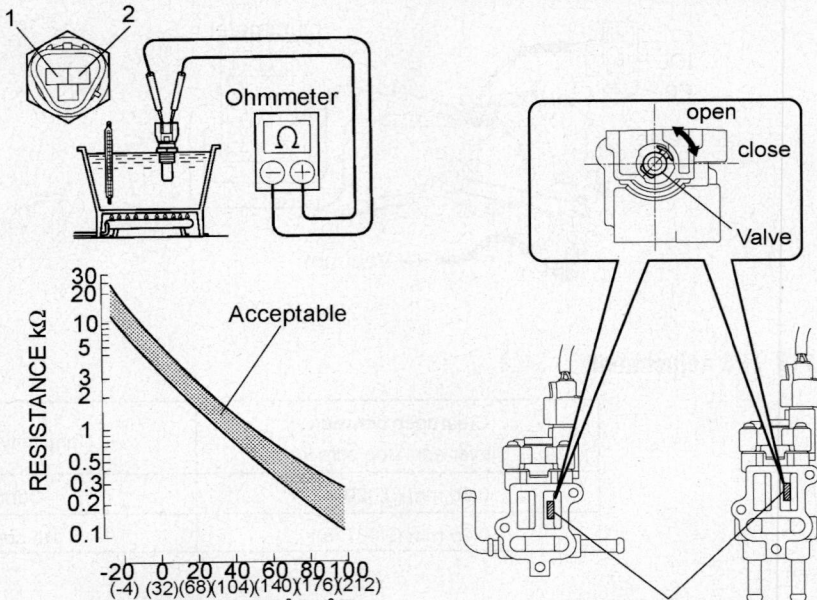

29157_TOYO_G0095

Fig. 122 ECT testing

29157_TOYO_G0105

Fig. 123 Idle Air Control Valve

-20 °C (-4°F)	13.6 - 18.4 kΩ
20°C (68°F)	2.21 - 2.69 kΩ
60°C (140°F)	0.49 - 0.67 kΩ

29157_TOYO_G0071

Fig. 124

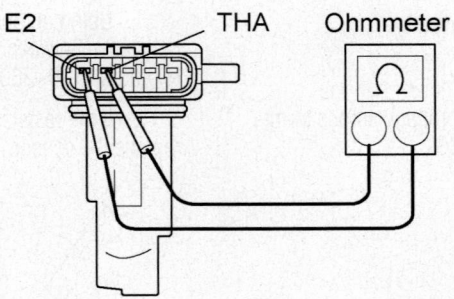

29157_TOYO_G0072

Fig. 125

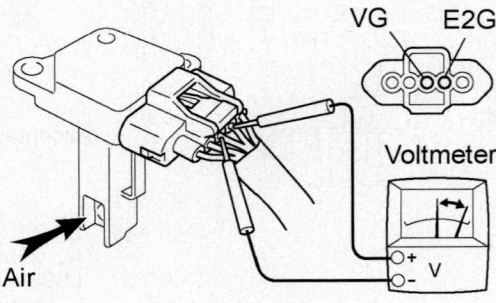

29157_TOYO_G0073

Fig. 126

8. Using a voltmeter, connect the positive (+) tester probe to terminal VG, and negative (-) tester probe to terminal E2G.

9. Blow air into the MAF meter, and check that the voltage fluctuates. If the operation is not as specified, replace the MAF meter.

10. Turn the ignition switch OFF.

11. Disconnect the negative terminal cable from the battery.

12. Disconnect the MAF meter connector.

13. Install the MAF meter with the 2 screws.

14. Reconnect MAF meter connector.

Heated Oxygen Sensor

See Figure 127.

1. Measure the resistance between terminals +B and HT. Specification is 5–10 ohms.

2. If the resistance is not as specified, replace the oxygen sensor.

Throttle Position Sensor

See Figures 128 and 129.

1. Disconnect the TPS connector.

2. Using an ohmmeter, measure the resistance between each terminal.

3. If the resistance is not as specified, replace the TPS.

3MZ-FE VIN A

Engine Coolant Temperature Sensor

See Figure 130.

1. Drain engine coolant.

2. Disconnect the ECT sensor connector.

3. Using a 19 mm deep socket wrench, remove the ECT sensor and gasket.

4. Using an ohmmeter, measure the resistance between the terminals.

5. If the resistance is not as specified, replace the ECT sensor.

To install:

6. Install a new gasket to the ECT sensor.

7. Using a 19 mm deep socket, install the ECT sensor. Tighten to 14 ft. lbs. (20 Nm).

8. Connect the ECT sensor connector.

9. Refill with engine coolant.

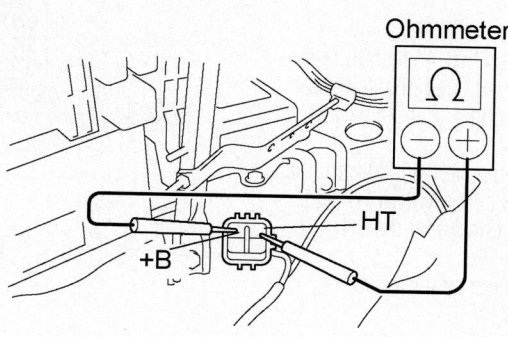

29157_TOYO_G0106

Fig. 127 Oxygen sensor testing

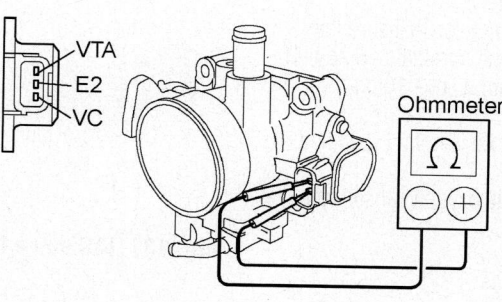

29157_TOYO_G0066

Fig. 128 TPS terminal identification

Clearance between lever and stop screw	Between terminals	Resistance
0 mm (0 in.)	VTA - E2	0.2 - 5.7 kΩ
Throttle valve fully open	VTA - E2	2.0 - 10.2 kΩ
Always	VC - E2	2.5 - 5.9 kΩ

Fig. 129 Resistance table

29157_TOYO_G0066

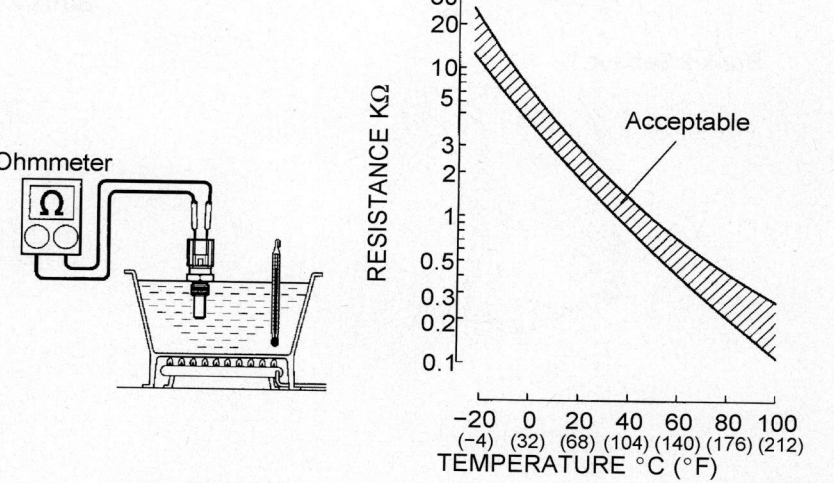

29157_TOYO_G0066

Fig. 130 ECT testing

Idle Air Control Valve

See Figure 131.

1. Connect the positive (+) lead from the battery to terminal +B and negative (-) lead to terminal RSC, and check that the valve is closed.

2. Connect the positive (+) lead from the battery to terminal +B and negative (-) lead to terminal RSO, and check that the valve is open.

3. If operation is not as specified, replace the IAC valve.

Oxygen Sensor

See Figures 132 and 133.

1. Disconnect the oxygen sensor connector.

2. Using an ohmmeter, measure the resistance between the terminals +B and HT. Resistance specification: 11 - 16 ohms at 20°C (68°F).

3. If the resistance is not as specified, replace the sensor.

4. Reconnect the oxygen sensor connector.

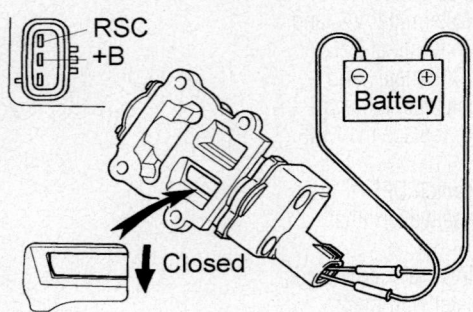
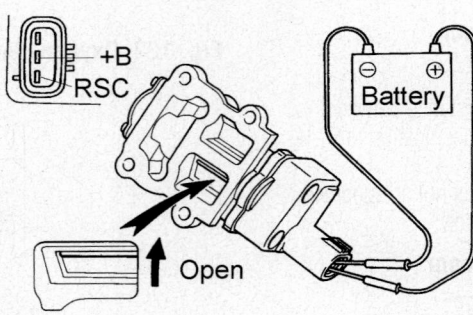

Fig. 131 IAC valve testing

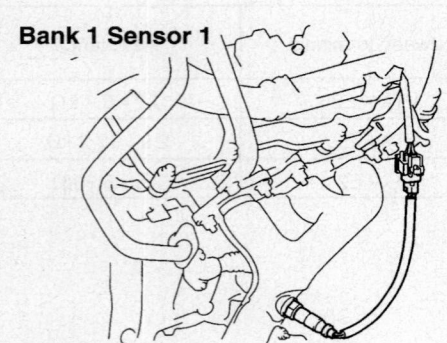
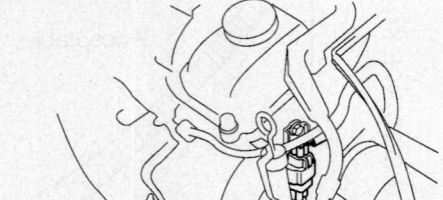

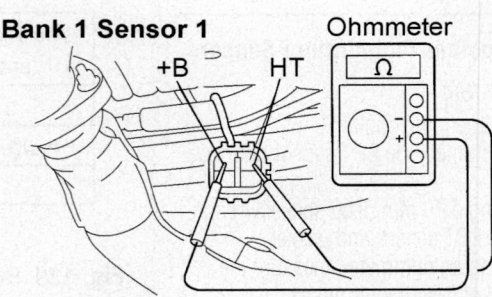
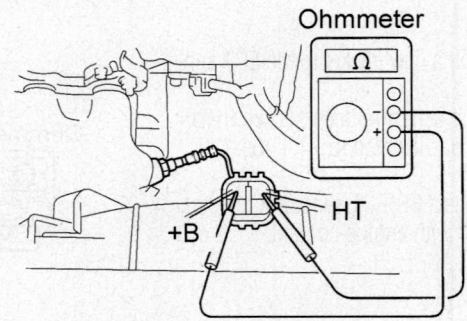

Fig. 132 Oxygen Sensor Connector locations   **Fig. 133 Oxygen Sensor testing**

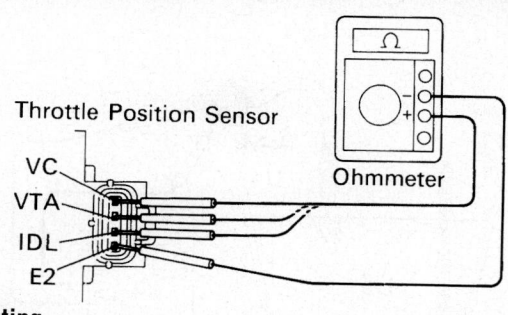

Fig. 134 TPS testing

29157_TOYO_G0093

Clearance between lever and stop screw	Between terminals	Resistance
0 mm (0 in.)	VTA - E2	0.28 - 6.4 kΩ
0.35 mm (0.014 in.)	IDL - E2	0.5 kΩ or less
0.70 mm (0.028 in.)	IDL - E2	Infinity
Throttle valve fully open	VTA - E2	2.0 - 11.6 kΩ
-	VC - E2	2.7 - 7.7 kΩ

Fig. 135 TPS resistance table

29157_TOYO_G0094

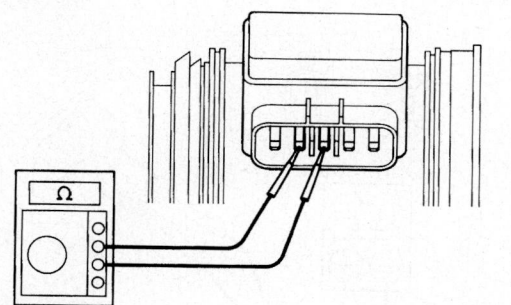

Fig. 136 Terminals THA and E2

29157_TOYO_G0079

Between terminals	Resistance	Temperature
THA - E2	10 - 20 kΩ	-20 °C (-4°F)
THA - E2	4 - 7 kΩ	0°C (32°F)
THA - E2	2 - 3 kΩ	20°C (68°F)
THA - E2	0.9 - 1.3 kΩ	40°C (104°F)
THA - E2	0.4 - 0.7 kΩ	60°C (140°F)
THA - E2	0.2 - 0.4 kΩ	80°C (176°F)

Fig. 137 Temperature/resistance table

29157_TOYO_G0080

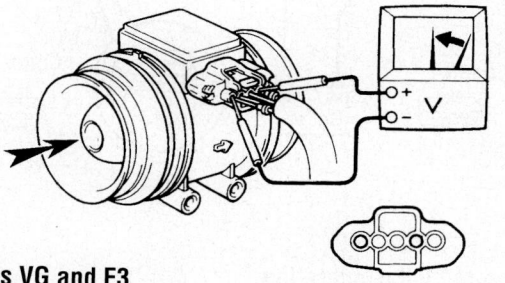

Fig. 138 Terminals VG and E3

29157_TOYO_G0081

Throttle Position Sensor

See Figures 134 and 135.

1. Apply vacuum to the throttle opener.
2. Disconnect the sensor connector.
3. Insert a feeler gauge between the throttle stop screw and stop lever.
4. Using an ohmmeter, measure the resistance between each terminal.
5. If the resistance is not as specified, replace the sensor.
6. Reconnect the sensor connector.

Mass Air Flow Meter

See Figures 136, 137 and 138.

1. Using an ohmmeter, measure the resistance between terminals THA and E2.
2. If the resistance is not as specified, replace the MAF meter.
3. Connect the MAF meter connector.
4. Using a voltmeter, connect the positive (+) tester probe to terminal VG, and negative (-) tester probe to terminal E3.
5. Blow air into the MAF meter, and check that the voltage fluctuates.
6. If operation is not as specified, replace the MAF meter.

3RZ-FE VIN M

Coolant Temperature Sensor

See Figure 139.

1. Measure the resistance between the terminals.
2. If the result is not as specified, replace the ECT.

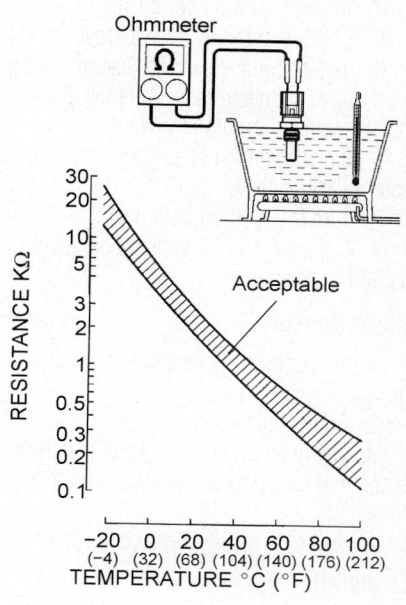

Fig. 139 ECT testing

29157_TOYO_G0088

EGR Gas Temperature Sensor

See Figures 140 and 141.

1. Remove the EGR gas temperature sensor.
2. Using an ohmmeter, measure the resistance between the terminals.
3. If the resistance is not as specified, replace the sensor. Tighten to 14 ft. lbs. (20 Nm).

Idle Air Control Valve

EXCEPT CALIFORNIA SPECIFICATION

See Figures 142, 143 and 144.

1. Disconnect the IAC valve connector.
2. Using an ohmmeter, measure resistance between terminal +B and the other (RSC, RSO) terminals.
3. If resistance is not as specified, replace the IAC valve.
4. Connect the positive (+) lead from the battery to terminal +B and negative (-) to terminal RSC and check that the valve is closed.
5. Connect the positive (+) lead from the battery to terminal +B and negative (-) to terminal RSO and check that the valve is open.

CALIFORNIA SPECIFICATION

See Figure 145.

1. Check that the IAC valve is half open.
2. Connect the IAC valve connector to the IAC valve.
3. Disconnect the ECT sensor connector from the ECT sensor. .
4. Turn the ignition switch ON.
5. Check that the IAC valve moves.
6. Repeat connecting and disconnecting of IAC valve connector several times and check the operation of the valve.
7. If operation is not as specified, replace the IAC valve.
8. Turn the ignition switch OFF.
9. Connect the ECT sensor connector to the ECT sensor.

Knock Sensor

1. Disconnect the knock sensor connector.
2. Remove the sensor.
3. Using an ohmmeter, check that there is no continuity between the terminal and body.
4. If there is continuity, replace the sensor.

To install:

5. Install the sensor and tighten to 33 ft. lbs. (44 Nm).
6. Connect the sensor connector.

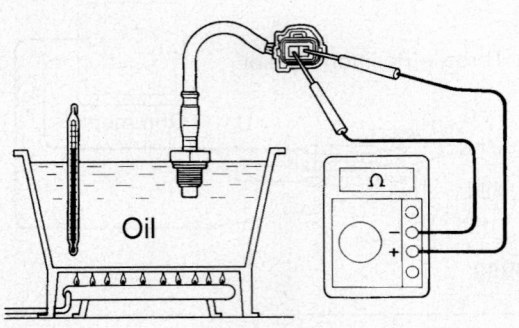

Fig. 140 Testing the EGR Gas Temperature Sensor

29157_TOYO_G0120

At 50°C (122°F)	64 - 97 kΩ
At 100°C (212°F)	11 - 16 kΩ
At 150°C (302°F)	2 - 4 kΩ

Fig. 141 Resistance/Temperature Table

29157_TOYO_G0121

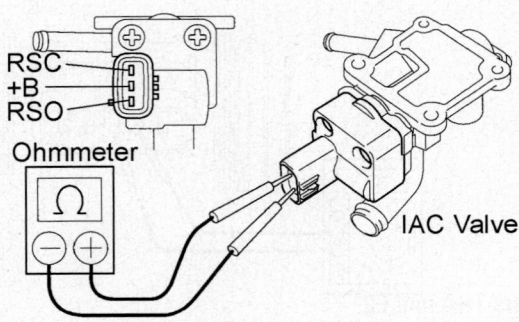

Fig. 142 IAC valve terminal identification

29157_TOYO_G0130

Cold	17.0 - 24.5 Ω
Hot	21.5 - 28.5 Ω

Fig. 143 Resistance table

29157_TOYO_G0131

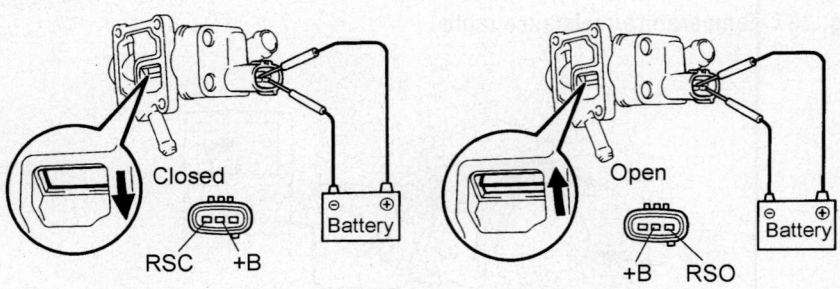

29157_TOYO_G0132

Fig. 144 Checking IAC valve operation

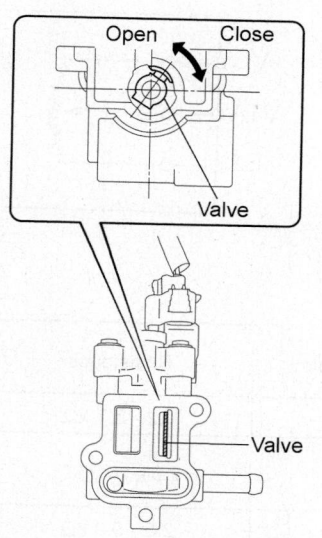

Fig. 145 **Idle Air Control Valve**

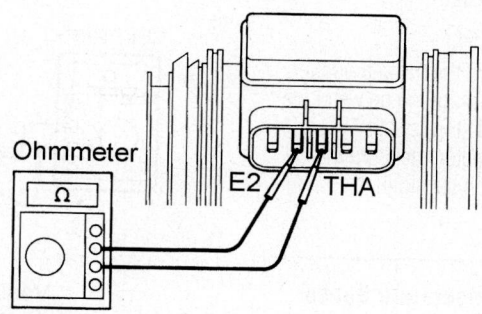

Fig. 146 **Terminals THA and E2**

Temperature	Resistance
-20°C (-4°F)	10 - 20 kΩ
0°C (32°F)	4 - 7 kΩ
20°C (68°F)	2 - 3 kΩ
40°C (104°F)	0.9 - 1.3 kΩ
60°C (140°F)	0.4 - 0.7 kΩ
80°C (176°F)	0.2 - 0.4 kΩ

Fig. 147 **Temperature/resistance table**

Mass Air Flow Meter

See Figures 146, 147 and 148.

1. Using an ohmmeter, measure the resistance between terminals THA and E2.

2. If the resistance is not as specified, replace the MAF meter.

3. Connect the MAF meter connector.

4. Using a voltmeter, connect the positive (+) tester probe to terminal VG, and negative (-) tester probe to terminal E3.

5. Blow air into the MAF meter, and check that the voltage fluctuates.

6. If operation is not as specified, replace the MAF meter.

Heated Oxygen Sensor

See Figure 149.

1. Disconnect the oxygen sensor connector.

2. Using an ohmmeter, measure the resistance between the terminals +B and HT. Resistance: 11 - 16 ohms at 20°C (68°F)

3. If the resistance is not as specified, replace the sensor. Tighten to 14 ft. lbs. (20 Nm).

4. Reconnect the oxygen sensor connector.

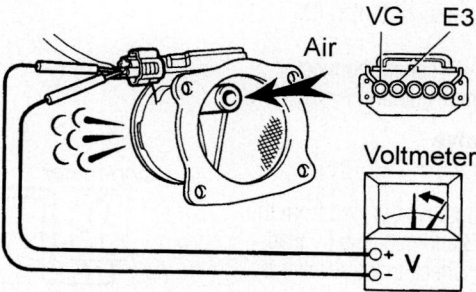

Fig. 148 **Terminals VG and E3**

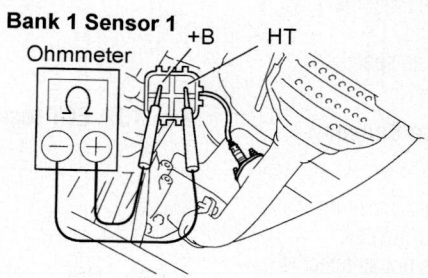

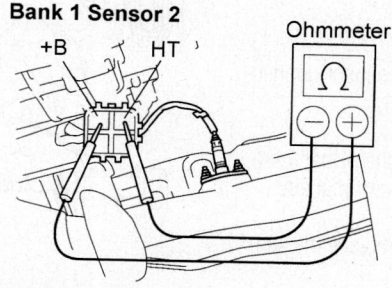

Fig. 149 **Testing the Oxygen Sensors**

Throttle Position Sensor

See Figures 150 and 151.

1. Apply vacuum to the throttle opener.
2. Insert a thickness gauge between the throttle stop screw and stop lever.
3. Using an ohmmeter, measure the resistance between each terminal.

3S-FE VIN P

Engine Coolant Temperature Sensor

See Figure 152.

1. Drain engine coolant.
2. Disconnect the ECT sensor connector.
3. Using a 19 mm deep socket wrench, remove the ECT sensor and gasket.
4. Using an ohmmeter, measure the resistance between terminals 1 and 2.
5. If the resistance is not as specified, replace the ECT sensor.

To install:

6. Install a new gasket to the ECT sensor.
7. Using a 19 mm deep socket, install the ECT sensor. Tighten to 18 ft. lbs. (25 Nm).
8. Connect the ECT sensor connector.
9. Refill with engine coolant.

Idle Air Control Valve

See Figure 153.

1. Connect the positive (+) lead from the battery to terminal +B and negative (-) lead to terminal RSC, and check that the valve is closed.
2. Connect the positive (+) lead from the battery to terminal +B and negative (-) lead to terminal RSO, and check that the valve is open.
3. If operation is not as specified, replace the IAC valve.

Intake Air Temperature Sensor

See Figure 154.

1. Using an ohmmeter, measure the resistance between the terminals.
2. If the resistance is not as specified, replace the sensor.

Knock Sensor

1. Disconnect the knock sensor connector.
2. Remove the sensor.
3. Using an ohmmeter, check that there is no continuity between the terminal and body.
4. If there is continuity, replace the sensor.
5. Connect the sensor connector.

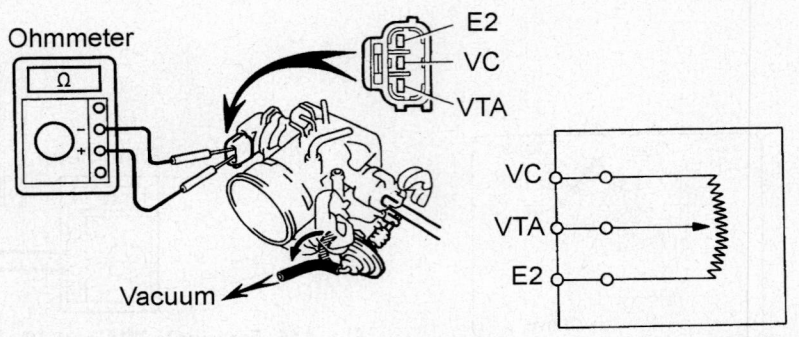

Fig. 150 TPS testing

29157_TOYO_G0137

Clearance between lever and stop screw	Between terminals	Resistance
0 mm (0 in.)	VTA - E2	0.2 - 5.7 kΩ
Throttle valve fully open	VTA - E2	2.0 - 10.2 kΩ
-	VC - E2	2.5 - 5.9 kΩ

Fig. 151 Resistance table

29157_TOYO_G0138

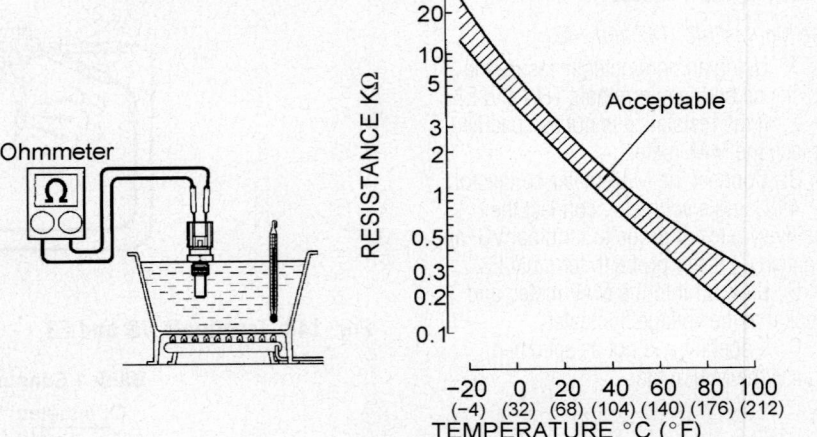

Fig. 152 ECT testing

29157_TOYO_G0088

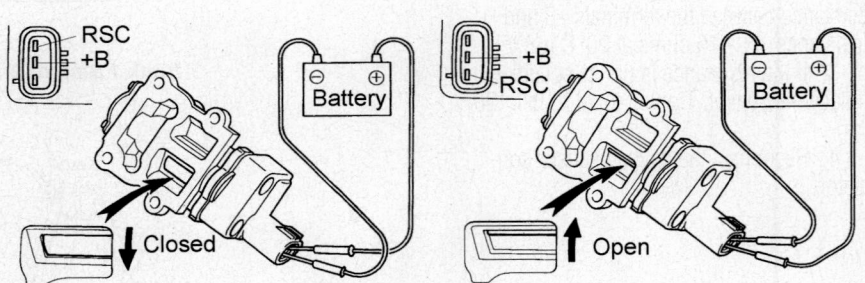

Fig. 153 IAC valve testing

29157_TOYO_G0090

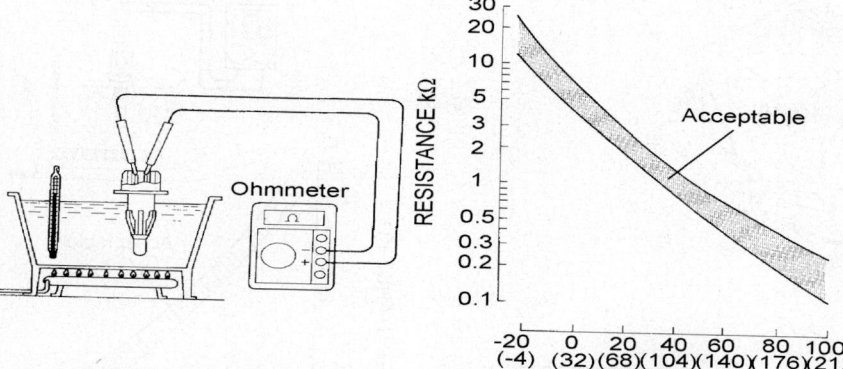

Fig. 154 Testing the IAT sensor

29157_TOYO_G0146

voltage under ambient atmospheric pressure.

9. Apply vacuum to the MAP sensor in 13.3 kPa (100 mmHg, 3.94 in.Hg) segments to 66.7 kPa (500 mmHg, 19.69 in.Hg).

10. Measure the voltage drop from step above for each segment.

11. If operation is not as specified, replace the MAP sensor.

12. Reconnect the vacuum hose to the air intake chamber.

Oxygen Sensor

See Figure 158.

1. Disconnect the oxygen sensor connector.

2. Using an ohmmeter, measure the resistance between the terminals +B and HT. Resistance specification: 11 - 16 ohms at 20°C (68°F).

3. If the resistance is not as specified, replace the sensor.

4. Reconnect the oxygen sensor connector.

Throttle Position Sensor

See Figures 159 and 160.

1. Disconnect the TPS connector.

2. Using an ohmmeter, measure the resistance between each terminal.

3. If the resistance is not as specified, replace the TPS.

Manifold Absolute Pressure Sensor

See Figures 155, 156 and 157.

1. Disconnect the MAP sensor connector.

2. Turn the ignition switch ON.

3. Using a voltmeter measure the voltage between connector terminals VC and E2 of the wiring harness side.

4. Turn the ignition switch to LOCK.

5. Reconnect the MAP sensor connector.

6. Turn the ignition switch ON.

7. Disconnect the vacuum hose on the air intake chamber

8. Connect a voltmeter to terminals PIM and E2 of the ECM, and measure the output

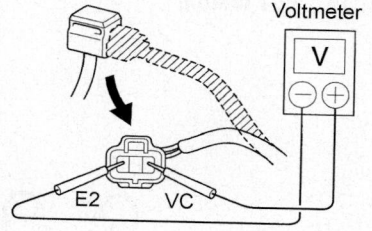

Fig. 155 MAP Sensor power input testing

29157_TOYO_G0147

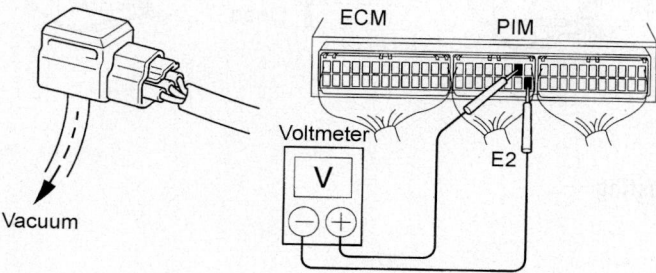

Fig. 156 MAP Sensor power output testing

29157_TOYO_G0148

Applied Vacuum kPa (mmHg / in.Hg)	13.3 (100 / 3.94)	26.7 (200 / 7.87)	40.0 (300 / 11.81)	53.5 (400 / 15.75)	66.7 (500 / 19.69)
Voltage drop V	0.3 - 0.5	0.7 - 0.9	1.1 - 1.3	1.5 - 1.7	1.9 - 2.1

Fig. 157 Vacuum/voltage table

29157_TOYO_G0149

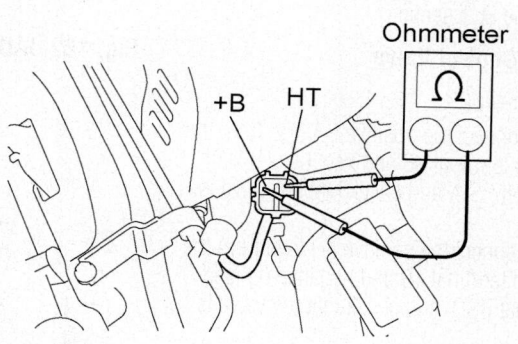

Fig. 158 Oxygen Sensor testing

29157_TOYO_G0150

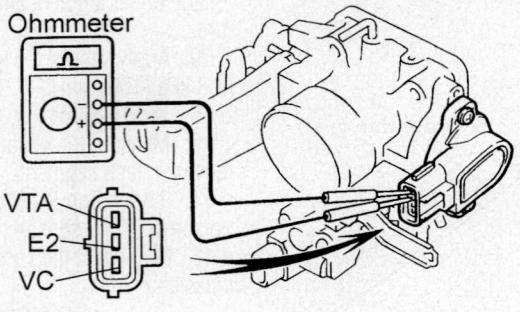

Fig. 159 TPS testing

Clearance between lever and stop screw	Between terminals	Resistance
0 mm (0 in.)	VTA - E2	0.2 - 5.7 kΩ
Throttle valve fully open	VTA - E2	2.0 - 10.2 kΩ
Always	VC - E2	2.5 - 5.9 kΩ

Fig. 160 Resistance table

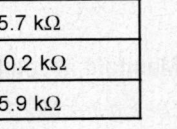

Fig. 161 ECT testing

4A-FE VIN A

Engine Coolant Temperature Sensor

See Figure 161.

1. Drain engine coolant.
2. Disconnect the ECT sensor connector.
3. Using a 19 mm deep socket wrench, remove the ECT sensor and gasket.
4. Using an ohmmeter, measure the resistance between terminals 1 and 2.
5. If the resistance is not as specified, replace the ECT sensor.

Idle Air Control Valve

See Figure 162.

1. Connect the positive (+) lead from the battery to terminal +B and negative (-) lead to terminal RSC, and check that the valve is closed.
2. Connect the positive (+) lead from the battery to terminal +B and negative (-) lead to terminal RSO, and check that the valve is open.
3. If operation is not as specified, replace the IAC valve.

Intake Air Temperature Sensor

See Figure 163.

1. Using an ohmmeter, measure the resistance between the terminals.
2. If the resistance is not as specified, replace the sensor.

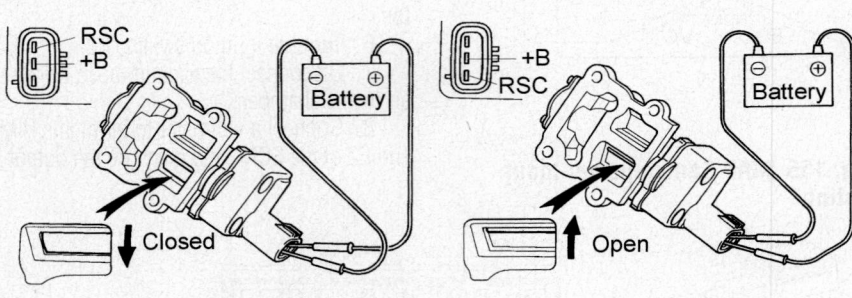

Fig. 162 IAC valve testing

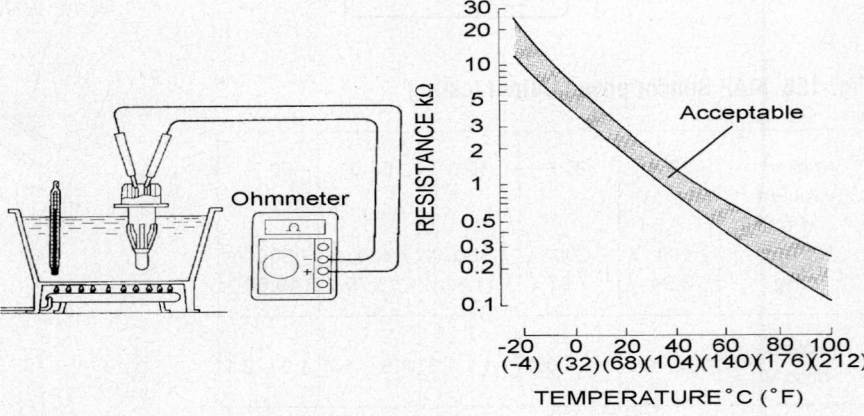

Fig. 163 Testing the IAT sensor

Knock Sensor

1. Disconnect the knock sensor connector.

2. Remove the sensor.

3. Using an ohmmeter, check that there is no continuity between the terminal and body.

4. If there is continuity, replace the sensor.

5. Connect the sensor connector.

Manifold Absolute Pressure Sensor

See Figures 164, 165 and 166.

1. Disconnect the MAP sensor connector.

2. Turn the ignition switch ON.

3. Using a voltmeter measure the voltage between connector terminals VC and E2 of the wiring harness side.

4. Turn the ignition switch to LOCK.

5. Reconnect the MAP sensor connector.

6. Turn the ignition switch ON.

7. Disconnect the vacuum hose on the air intake chamber

8. Connect a voltmeter to terminals PIM and E2 of the ECM, and measure the output voltage under ambient atmospheric pressure.

9. Apply vacuum to the MAP sensor in 13.3 kPa (100 mmHg, 3.94 in.Hg) segments to 66.7 kPa (500 mmHg, 19.69 in.Hg).

10. Measure the voltage drop from step above for each segment.

11. If operation is not as specified, replace the MAP sensor.

12. Reconnect the vacuum hose to the air intake chamber.

Oxygen Sensor

See Figure 167.

1. Remove the right front seat and floor carpeting.

2. Disconnect the oxygen sensor connector.

3. Using an ohmmeter, measure the resistance between the terminals +B and HT. Resistance specification: 11 - 16 ohms at 20°C (68°F).

4. If the resistance is not as specified, replace the sensor. Tighten to 32 ft. lbs. (44 Nm).

5. Reconnect the oxygen sensor connector.

6. Install the right front seat and floor carpeting.

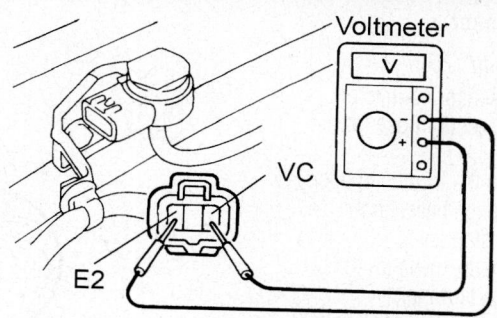

Fig. 164 MAP Sensor power input testing

29157_TOYO_G0151

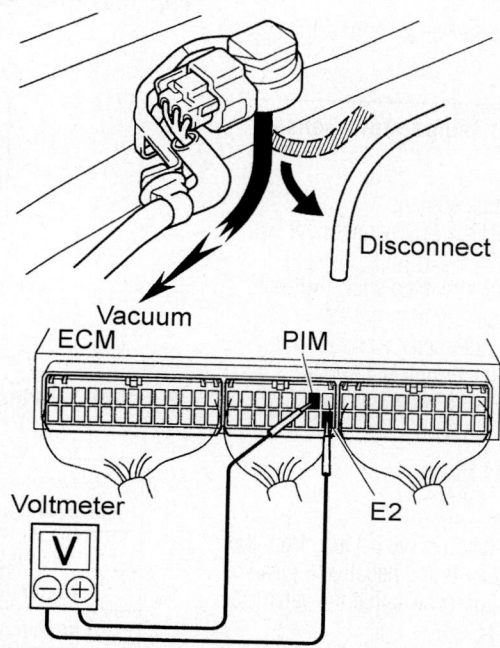

Fig. 165 MAP Sensor power output testing

29157_TOYO_G0152

Applied Vacuum kPa	13.3	26.7	40.0	53.5	66.7
$\left(\begin{array}{c}\text{mm Hg}\\\text{in. Hg}\end{array}\right)$	$\left(\begin{array}{c}100\\3.94\end{array}\right)$	$\left(\begin{array}{c}200\\7.87\end{array}\right)$	$\left(\begin{array}{c}300\\11.81\end{array}\right)$	$\left(\begin{array}{c}400\\15.75\end{array}\right)$	$\left(\begin{array}{c}500\\19.69\end{array}\right)$
Voltage drop V	0.3 - 0.5	0.7 - 0.9	1.1 - 1.3	1.5 - 1.7	1.9 - 2.1

Fig. 166 Vacuum/voltage table

29157_TOYO_G0153

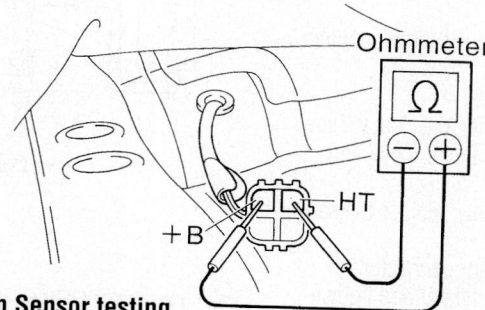

Fig. 167 Oxygen Sensor testing

29157_TOYO_G0154

Throttle Position Sensor

See Figures 168, 169 and 170.

1. Disconnect the sensor connector.
2. Disconnect the vacuum hose from the throttle opener.
3. Apply vacuum to the throttle opener.
4. Insert a feeler gauge between the throttle stop screw and stop lever.
5. Using an ohmmeter, measure the resistance between each terminal.
6. If the resistance is not as specified, replace the sensor.
7. Reconnect the vacuum hose to the throttle opener.
8. Reconnect the sensor connector.

5E-FE VIN C

Engine Coolant Temperature Sensor

See Figure 171.

1. Drain engine coolant.
2. Disconnect the ECT sensor connector.
3. Using a 19 mm deep socket wrench, remove the ECT sensor and gasket.
4. Using an ohmmeter, measure the resistance between terminals 1 and 2.
5. If the resistance is not as specified, replace the ECT sensor.

Idle Air Control Valve

See Figure 172.

1. Connect the positive (+) lead from the battery to terminal +B and negative (−) lead to terminal RSC, and check that the valve is closed.
2. Connect the positive (+) lead from the battery to terminal +B and negative (−) lead to terminal RSO, and check that the valve is open.
3. If operation is not as specified, replace the IAC valve.

Intake Air Temperature Sensor

See Figure 173.

1. Using an ohmmeter, measure the resistance between the terminals.
2. If the resistance is not as specified, replace the sensor.

Knock Sensor

1. Disconnect the knock sensor connector.
2. Remove the sensor.
3. Using an ohmmeter, check that there is no continuity between the terminal and body.
4. If there is continuity, replace the sensor.
5. Connect the sensor connector.
6. Install the knock sensor and tighten to 33 ft. lbs (45 Nm).

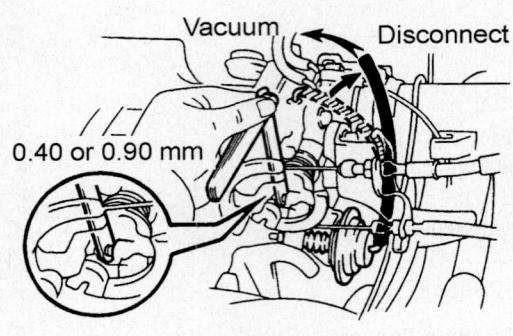

Fig. 168 Place feeler gauge for TPS testing

29157_TOYO_G0155

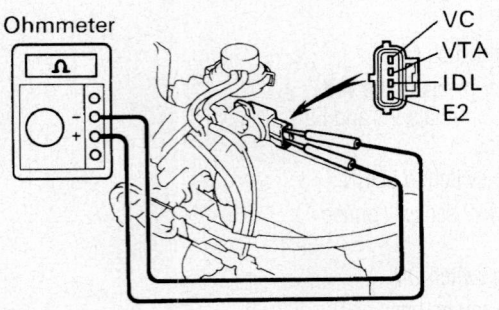

Fig. 169 TPS testing

29157_TOYO_G0156

Clearance between lever and stop screw	Between terminals	Resistance
0 mm (0 in.)	VTA - E2	0.2 - 5.7 kΩ
0.40 mm (0.016 in.)	IDL-E2	2.3 kΩ or less
0.90 mm (0.035 in.)	IDL-E2	Infinity
Throttle valve fully open	VTA - E2	2.0 - 10.2 kΩ
-	VC - E2	2.5 - 5.9 kΩ

29157_TOYO_G0157

Fig. 170 Resistance table

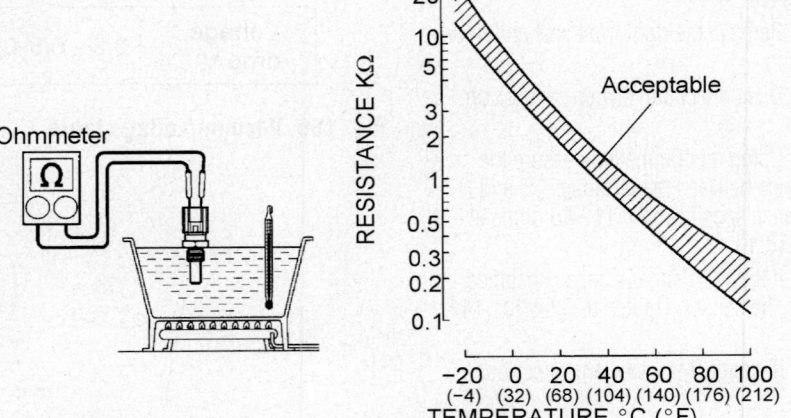

Fig. 171 ECT testing

29157_TOYO_G0088

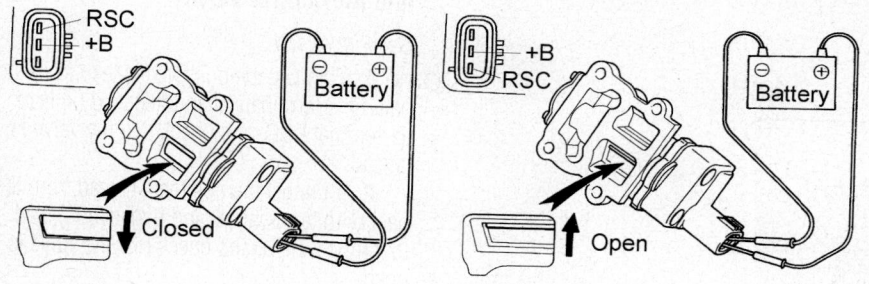

Fig. 172 IAC valve testing

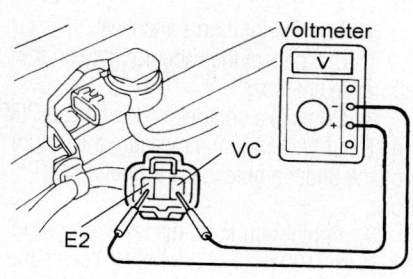

Fig. 173 Testing the IAT sensor

9. Apply vacuum to the MAP sensor in 13.3 kPa (100 mmHg, 3.94 in.Hg) segments to 66.7 kPa (500 mmHg, 19.69 in.Hg).

10. Measure the voltage drop from step above for each segment.

11. If operation is not as specified, replace the MAP sensor.

12. Reconnect the vacuum hose to the air intake chamber.

Oxygen Sensor

See Figure 177.

1. Remove the right front seat and floor carpeting.

2. Disconnect the oxygen sensor connector.

3. Using an ohmmeter, measure the resistance between the terminals +B and HT. Resistance specification: 11 - 16 ohms at 20°C (68°F).

4. If the resistance is not as specified, replace the sensor. Tighten to 32 ft. lbs. (44 Nm).

5. Reconnect the oxygen sensor connector.

6. Install the right front seat and floor carpeting.

Throttle Position Sensor

See Figures 178 and 179.

1. Disconnect the TPS connector.

2. Using an ohmmeter, measure the resistance between each terminal.

3. If the resistance is not as specified, replace the TPS.

Manifold Absolute Pressure Sensor

See Figures 174, 175 and 176.

1. Disconnect the MAP sensor connector.

2. Turn the ignition switch ON.

3. Using a voltmeter measure the voltage between connector terminals VC and E2 of the wiring harness side.

4. Turn the ignition switch to LOCK.

5. Reconnect the MAP sensor connector.

6. Turn the ignition switch ON.

7. Disconnect the vacuum hose on the air intake chamber.

8. Connect a voltmeter to terminals PIM and E2 of the ECM, and measure the output voltage under ambient atmospheric pressure.

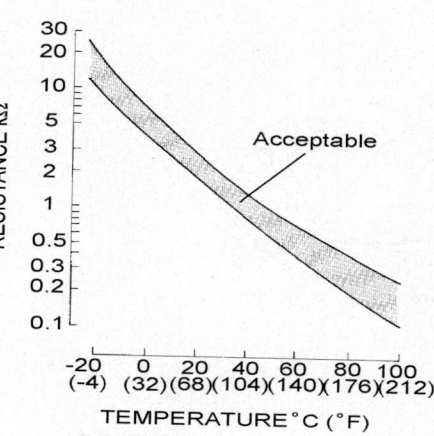

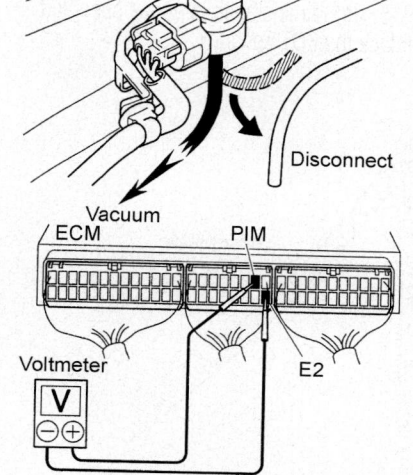

Fig. 174 MAP Sensor power input testing

Fig. 175 MAP Sensor power output testing

Applied Vacuum kPa (mm Hg) (in. Hg)	13.3 (100) (3.94)	26.7 (200) (7.87)	40.0 (300) (11.81)	53.5 (400) (15.75)	66.7 (500) (19.69)
Voltage drop V	0.3 - 0.5	0.7 - 0.9	1.1 - 1.3	1.5 - 1.7	1.9 - 2.1

Fig. 176 Vacuum/voltage table

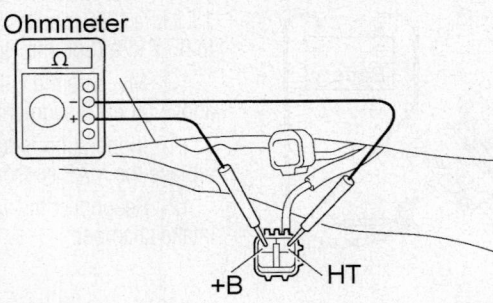

Fig. 177 Oxygen Sensor testing

29157_TOYO_G0158

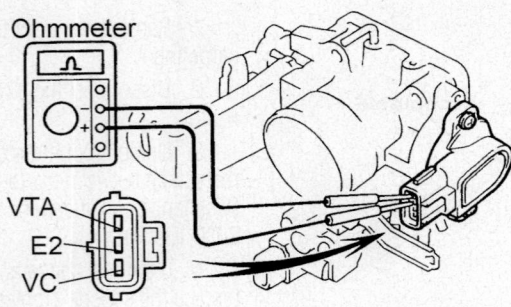

Fig. 178 TPS testing

29157_TOYO_G0159

Clearance between lever and stop screw	Between terminals	Resistance
0 mm (0 in.)	VTA - E2	0.2 - 5.7 kΩ
Throttle valve fully open	VTA - E2	2.0 - 10.2 kΩ
Always	VC - E2	2.5 - 5.9 kΩ

29157_TOYO_G0099

Fig. 179

5S-FE VIN G

Engine Coolant Temperature Sensor

See Figure 180.

1. Drain engine coolant.
2. Disconnect the ECT sensor connector.
3. Using a 19 mm deep socket wrench, remove the ECT sensor and gasket.
4. Using an ohmmeter, measure the resistance between terminals 1 and 2.
5. If the resistance is not as specified, replace the ECT sensor.

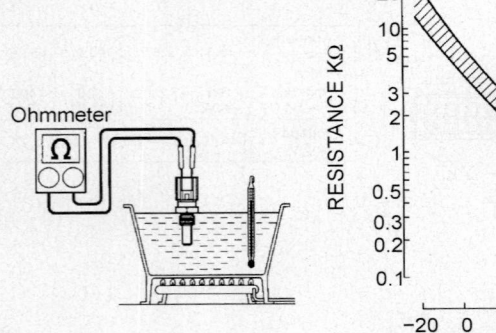

Fig. 180 ECT testing

29157_TOYO_G0088

Idle Air Control Valve

See Figure 181.

1. Connect the positive (+) lead from the battery to terminal +B and negative (-) lead to terminal RSC, and check that the valve is closed.
2. Connect the positive (+) lead from the battery to terminal +B and negative (-) lead to terminal RSO, and check that the valve is open.
3. If operation is not as specified, replace the IAC valve.

Intake Air Temperature Sensor

See Figure 182.

1. Using an ohmmeter, measure the resistance between the terminals.
2. If the resistance is not as specified, replace the sensor.

Knock Sensor

1. Disconnect the knock sensor connector.
2. Remove the sensor.
3. Using an ohmmeter, check that there is no continuity between the terminal and body.
4. If there is continuity, replace the sensor.
5. Connect the sensor connector.
6. Install the knock sensor and tighten to 32 ft. lbs (44 Nm).

Manifold Absolute Pressure Sensor

See Figures 183, 184 and 185.

1. Disconnect the MAP sensor connector.
2. Turn the ignition switch ON.
3. Using a voltmeter measure the voltage between connector terminals VC and E2 of the wiring harness side.
4. Turn the ignition switch to LOCK.
5. Reconnect the MAP sensor connector.
6. Turn the ignition switch ON.
7. Disconnect the vacuum hose on the air intake chamber
8. Connect a voltmeter to terminals PIM and E2 of the ECM, and measure the output voltage under ambient atmospheric pressure.
9. Apply vacuum to the MAP sensor in 13.3 kPa (100 mmHg, 3.94 in.Hg) segments to 66.7 kPa (500 mmHg, 19.69 in.Hg).
10. Measure the voltage drop from step above for each segment.
11. If operation is not as specified, replace the MAP sensor.
12. Reconnect the vacuum hose to the air intake chamber.

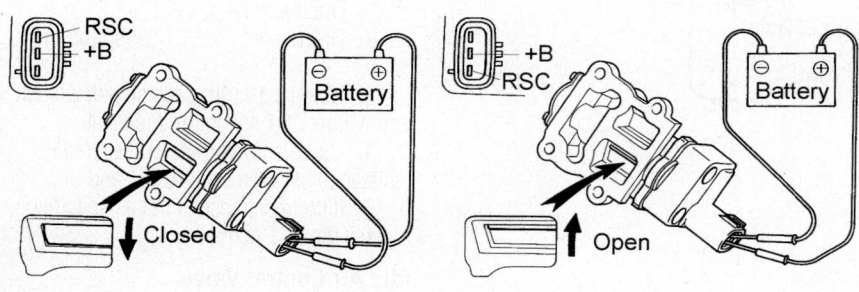

Fig. 181 IAC valve testing

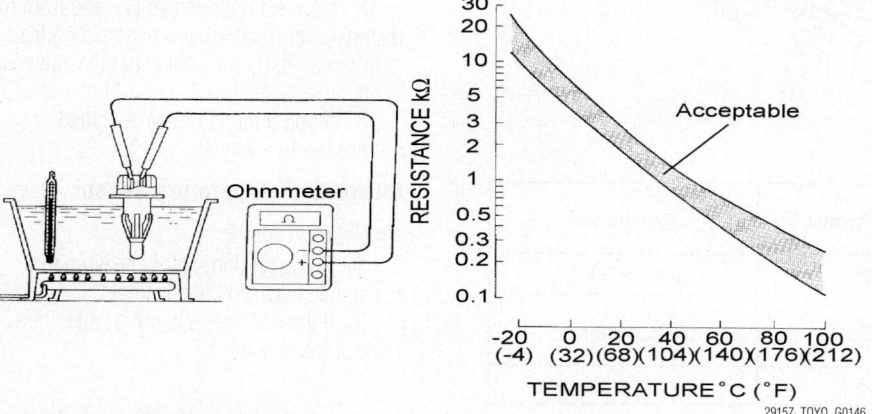

Fig. 182 Testing the IAT sensor

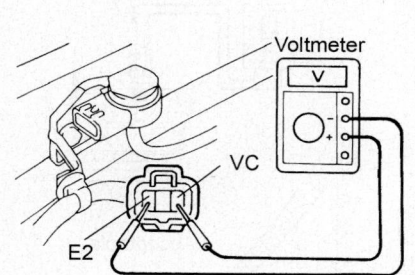

Fig. 183 MAP Sensor power input testing

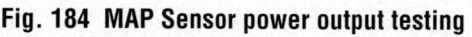
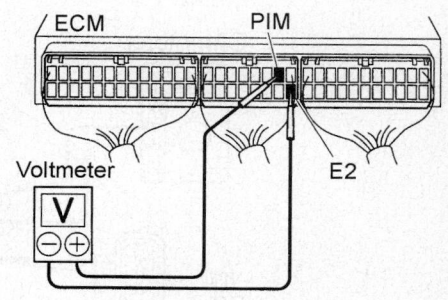

Fig. 184 MAP Sensor power output testing

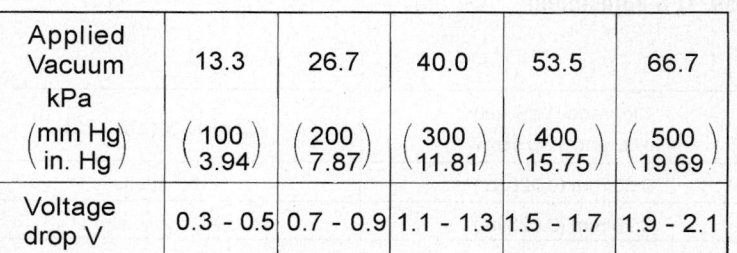

Applied Vacuum kPa	13.3	26.7	40.0	53.5	66.7
(mm Hg / in. Hg)	(100 / 3.94)	(200 / 7.87)	(300 / 11.81)	(400 / 15.75)	(500 / 19.69)
Voltage drop V	0.3 - 0.5	0.7 - 0.9	1.1 - 1.3	1.5 - 1.7	1.9 - 2.1

Fig. 185 Vacuum/voltage table

Oxygen Sensor

See Figure 186.

1. Disconnect the oxygen sensor connector.

2. Using an ohmmeter, measure the resistance between the terminals +B and HT. Resistance specification: 11 - 16 ohms at 20°C (68°F).

3. If the resistance is not as specified, replace the sensor.

4. Reconnect the oxygen sensor connector.

Throttle Position Sensor

See Figures 187 thru 190.

1. Apply vacuum to the throttle opener.

2. Using an ohmmeter, measure the resistance between each terminal.

3. If necessary, adjust the TPS as follows:

   a. Loosen the 2 set screws of the sensor.

   b. Insert a 0.60 mm (0.024 in.) feeler gauge between the throttle stop screw and stop lever.

   c. Connect the tester probe of an ohmmeter to the terminals IDL and E2 of the sensor.

   d. Gradually turn the sensor clockwise until the ohmmeter deflects, and secure it with the 2 set screws.

   e. Recheck the continuity between terminals IDL and E2.

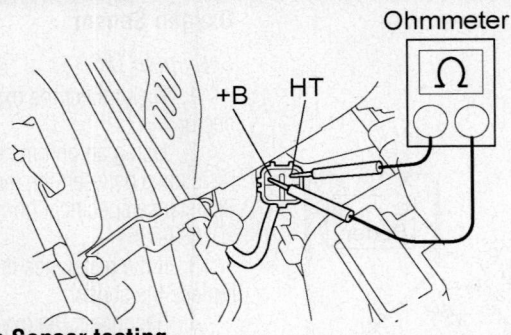

Fig. 186 Oxygen Sensor testing

29157_TOYO_G0150

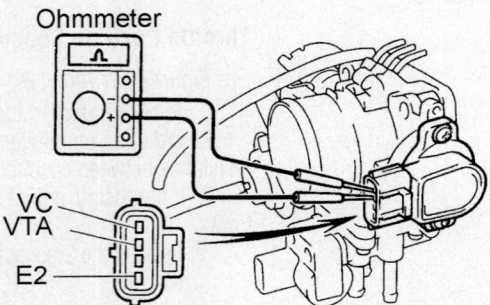

Fig. 187 TPS testing

29157_TOYO_G0160

Clearance between lever and stop screw	Between terminals	Resistance
0 mm (0 in.)	VTA - E2	0.2 - 5.7 kΩ
Throttle valve fully open	VTA - E2	2.0 - 10.2 kΩ
-	VC - E2	2.5 - 5.9 kΩ

Fig. 188 Resistance table

29157_TOYO_G0161

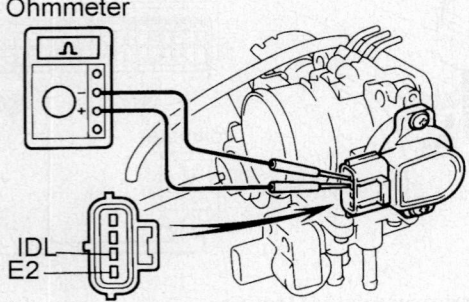

Fig. 189 TPS adjustment

29157_TOYO_G0162

Clearance between lever and stop screw	Continuity (IDL - E2)
0.50 mm (0.020 in.)	Continuity
0.70 mm (0.028 in.)	No continuity

Fig. 190 Resistance table

29157_TOYO_G0163

5S-FNE VIN N

Engine Coolant Temperature Sensor

See Figure 191.

1. Drain engine coolant.
2. Disconnect the ECT sensor connector.
3. Using a 19 mm deep socket wrench, remove the ECT sensor and gasket.
4. Using an ohmmeter, measure the resistance between terminals 1 and 2.
5. If the resistance is not as specified, replace the ECT sensor.

Idle Air Control Valve

See Figure 192.

1. Connect the positive (+) lead from the battery to terminal +B and negative (-) lead to terminal RSC, and check that the valve is closed.
2. Connect the positive (+) lead from the battery to terminal +B and negative (-) lead to terminal RSO, and check that the valve is open.
3. If operation is not as specified, replace the IAC valve.

Intake Air Temperature Sensor

See Figure 193.

1. Using an ohmmeter, measure the resistance between the terminals.
2. If the resistance is not as specified, replace the sensor.

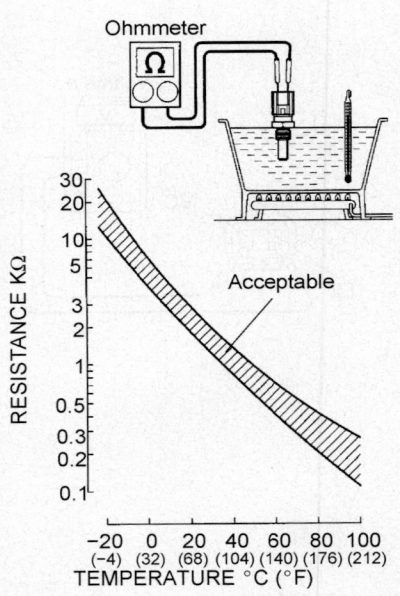

Fig. 191 ECT testing

29157_TOYO_G0066

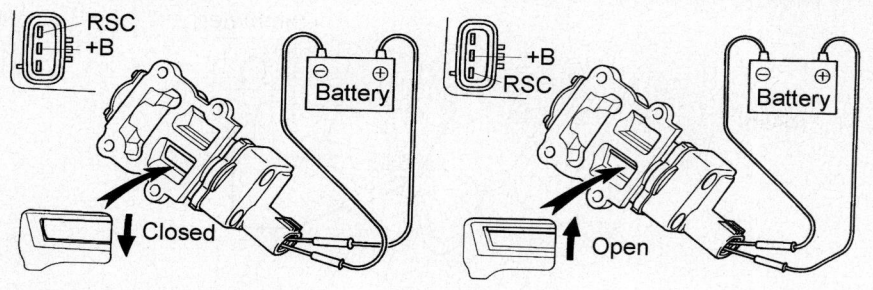

Fig. 192 IAC valve testing

29157_TOYO_G0066

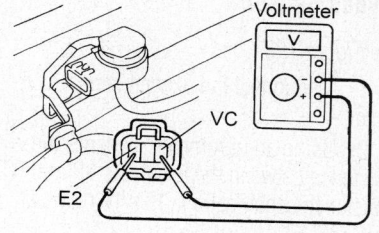

29157_TOYO_G0066

Fig. 194 MAP Sensor power input testing

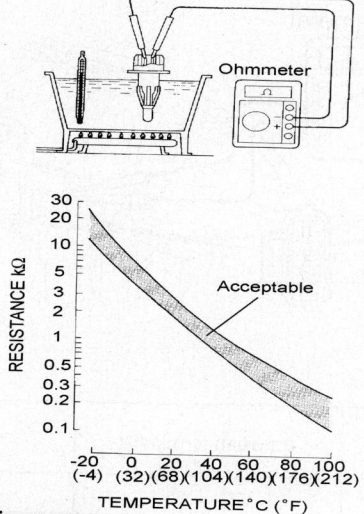

Fig. 193 Testing the IAT sensor

29157_TOYO_G0146

Knock Sensor

1. Disconnect the knock sensor connector.

2. Remove the sensor.

3. Using an ohmmeter, check that there is no continuity between the terminal and body.

4. If there is continuity, replace the sensor.

5. Connect the sensor connector.

6. Install the knock sensor and tighten to 32 ft. lbs (44 Nm).

Manifold Absolute Pressure Sensor

See Figures 194, 195 and 196.

1. Disconnect the MAP sensor connector.

2. Turn the ignition switch ON.

3. Using a voltmeter measure the voltage between connector terminals VC and E2 of the wiring harness side.

4. Turn the ignition switch to LOCK.

5. Reconnect the MAP sensor connector.

6. Turn the ignition switch ON.

7. Disconnect the vacuum hose on the air intake chamber

8. Connect a voltmeter to terminals PIM and E2 of the ECM, and measure the output voltage under ambient atmospheric pressure.

9. Apply vacuum to the MAP sensor in 13.3 kPa (100 mmHg, 3.94 in.Hg) segments to 66.7 kPa (500 mmHg, 19.69 in.Hg).

10. Measure the voltage drop from step above for each segment.

11. If operation is not as specified, replace the MAP sensor.

12. Reconnect the vacuum hose to the air intake chamber.

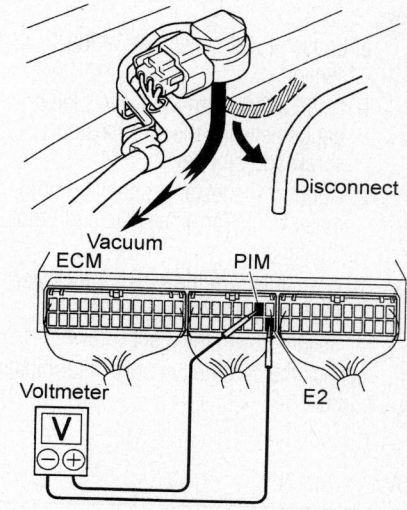

Fig. 195 MAP Sensor power output testing

29157_TOYO_G0152

Applied Vacuum kPa (mm Hg)(in. Hg)	13.3 (100)(3.94)	26.7 (200)(7.87)	40.0 (300)(11.81)	53.5 (400)(15.75)	66.7 (500)(19.69)
Voltage drop V	0.3 - 0.5	0.7 - 0.9	1.1 - 1.3	1.5 - 1.7	1.9 - 2.1

Fig. 196 Vacuum/voltage table

29157_TOYO_G0153

Oxygen Sensor

See Figure 197.

1. Disconnect the oxygen sensor connector.

2. Using an ohmmeter, measure the resistance between the terminals +B and HT. Resistance specification: 11 - 16 ohms at 20°C (68°F).

3. If the resistance is not as specified, replace the sensor.

4. Reconnect the oxygen sensor connector.

Throttle Position Sensor

See Figure 198 thru 201.

1. Apply vacuum to the throttle opener.

2. Using an ohmmeter, measure the resistance between each terminal.

3. If necessary, adjust the TPS as follows:

   a. Loosen the 2 set screws of the sensor.

   b. Insert a 0.60 mm (0.024 in.) feeler gauge between the throttle stop screw and stop lever.

   c. Connect the tester probe of an ohmmeter to the terminals IDL and E2 of the sensor.

   d. Gradually turn the sensor clockwise until the ohmmeter deflects, and secure it with the 2 set screws.

4. Recheck the continuity between terminals IDL and E2.

5VZ-FE VIN N

Engine Coolant Temperature Sensor

See Figure 202.

1. Drain engine coolant.

2. Disconnect the ECT sensor connector.

3. Using a 19 mm deep socket wrench, remove the ECT sensor and gasket.

4. Using an ohmmeter, measure the resistance between terminals 1 and 2.

5. If the resistance is not as specified, replace the ECT sensor.

To install:

6. Install a new gasket to the ECT sensor.

7. Using a 19 mm deep socket, install the ECT sensor. Tighten to 14 ft. lbs. (20 Nm).

8. Connect the ECT sensor connector.

9. Refill with engine coolant.

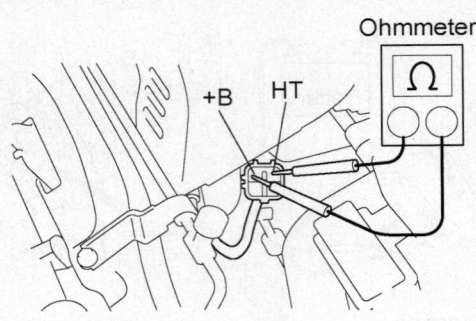

Fig. 197 Oxygen Sensor testing

29157_TOYO_G0150

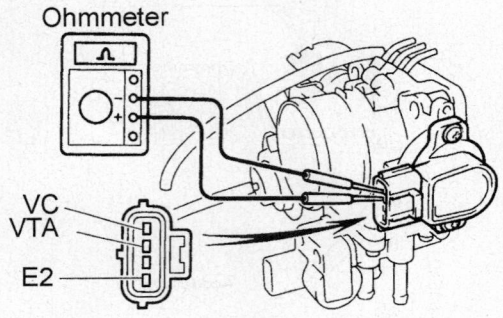

Fig. 198 TPS testing

29157_TOYO_G0160

Clearance between lever and stop screw	Between terminals	Resistance
0 mm (0 in.)	VTA - E2	0.2 - 5.7 kΩ
Throttle valve fully open	VTA - E2	2.0 - 10.2 kΩ
-	VC - E2	2.5 - 5.9 kΩ

Fig. 199 Resistance table

29157_TOYO_G0161

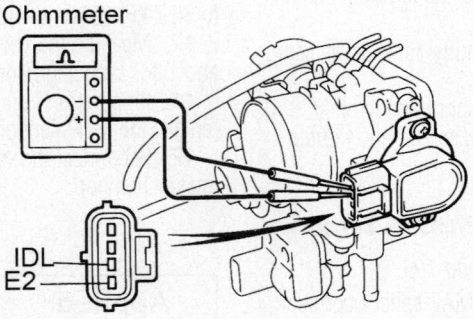

Fig. 200 TPS adjustment

29157_TOYO_G0162

Clearance between lever and stop screw	Continuity (IDL - E2)
0.50 mm (0.020 in.)	Continuity
0.70 mm (0.028 in.)	No continuity

Fig. 201 Resistance table

29157_TOYO_G0163

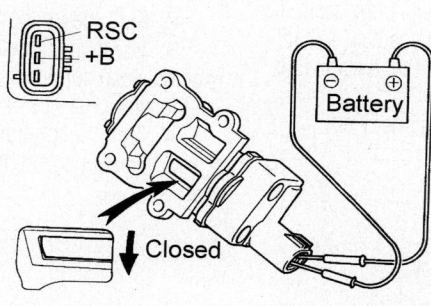

Fig. 202 ECT testing

29157_TOYO_G0088

Idle Air Control Valve

See Figure 203.

1. Connect the positive (+) lead from the battery to terminal +B and negative (-) lead to terminal RSC, and check that the valve is closed.

2. Connect the positive (+) lead from the battery to terminal +B and negative (-) lead to terminal RSO, and check that the valve is open.

3. If operation is not as specified, replace the IAC valve.

Knock Sensor

1. Disconnect the knock sensor connector.

2. Remove the sensor.

3. Using an ohmmeter, check that there is no continuity between the terminal and body.

4. If there is continuity, replace the sensor.

5. Connect the sensor connector.

6. Install the knock sensor and tighten to 29 ft. lbs (39 Nm).

Mass Air Flow Meter

See Figures 204, 205 and 206.

1. Disconnect the MAF meter connector.

2. Using an ohmmeter, measure the resistance between terminals THA and E2.

3. If the resistance is not as specified, replace the MAF meter.

4. Connect the MAF meter connector.

5. Using a voltmeter, connect the positive (+) tester probe to terminal VG, and negative (-) tester probe to terminal E2G.

6. Turn the ignition switch ON.

7. Blow air into the MAF meter, and check that the voltage fluctuates.

8. If operation is not as specified, replace the MAF meter.

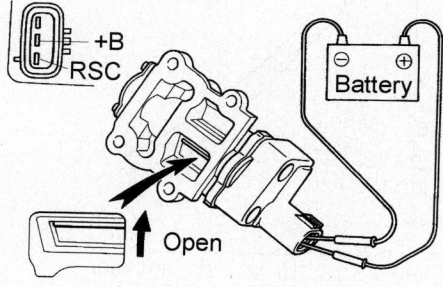

Fig. 203 IAC valve testing

29157_TOYO_G0090

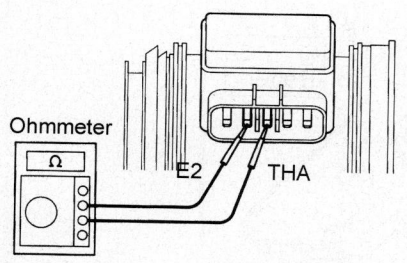

29157_TOYO_G0164

Fig. 204 MAF terminals THA and E2

Terminals	Resistance	Temperature
THA - E2	10 - 20 kΩ	-20°C (-4°F)
THA - E2	4 - 7 kΩ	0°C (32°F)
THA - E2	2 - 3 kΩ	20°C (68°F)
THA - E2	0.9 - 103 kΩ	40°C (104°F)
THA - E2	0.4 - 0.7 kΩ	60°C (140°F)
THA - E2	0.2 - 0.4 kΩ	80°C (176°F)

Fig. 205 Resistance/temperature table

29157_TOYO_G0165

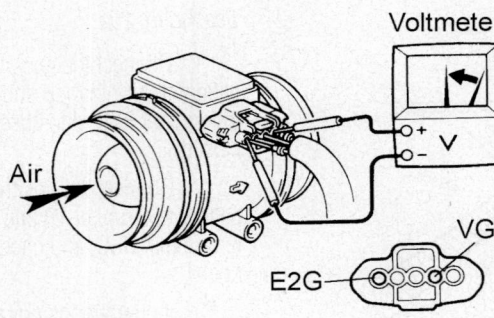

Fig. 206 MAF terminals VG and E2G

29157_TOYO_G0166

7A-FE VIN B

Engine Coolant Temperature Sensor

See Figure 207.

1. Drain engine coolant.
2. Disconnect the ECT sensor connector.
3. Using a 19 mm deep socket wrench, remove the ECT sensor and gasket.
4. Using an ohmmeter, measure the resistance between terminals 1 and 2.
5. If the resistance is not as specified, replace the ECT sensor.

Idle Air Control Valve

See Figure 208.

1. Connect the positive (+) lead from the battery to terminal +B and negative (-) lead to terminal RSC, and check that the valve is closed.
2. Connect the positive (+) lead from the battery to terminal +B and negative (-) lead to terminal RSO, and check that the valve is open.
3. If operation is not as specified, replace the IAC valve.

Intake Air Temperature Sensor

See Figure 209.

1. Using an ohmmeter, measure the resistance between the terminals.
2. If the resistance is not as specified, replace the sensor.

Knock Sensor

1. Disconnect the knock sensor connector.
2. Remove the sensor.
3. Using an ohmmeter, check that there is no continuity between the terminal and body.
4. If there is continuity, replace the sensor.
5. Connect the sensor connector.

Manifold Absolute Pressure Sensor

See Figures 210, 211 and 212.

1. Disconnect the MAP sensor connector.
2. Turn the ignition switch ON.
3. Using a voltmeter measure the voltage between connector terminals VC and E2 of the wiring harness side.
4. Turn the ignition switch to LOCK.
5. Reconnect the MAP sensor connector.

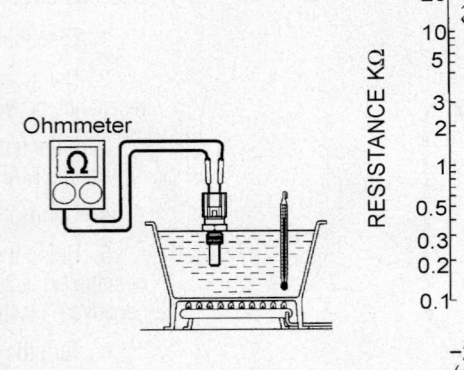

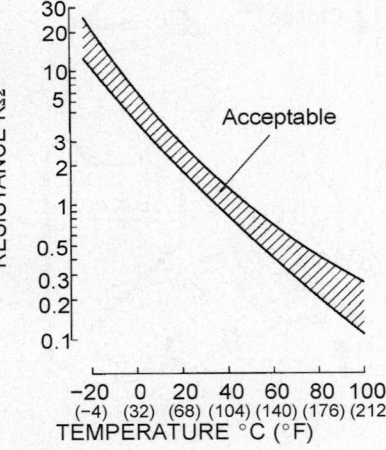

Fig. 207 ECT testing

29157_TOYO_G0088

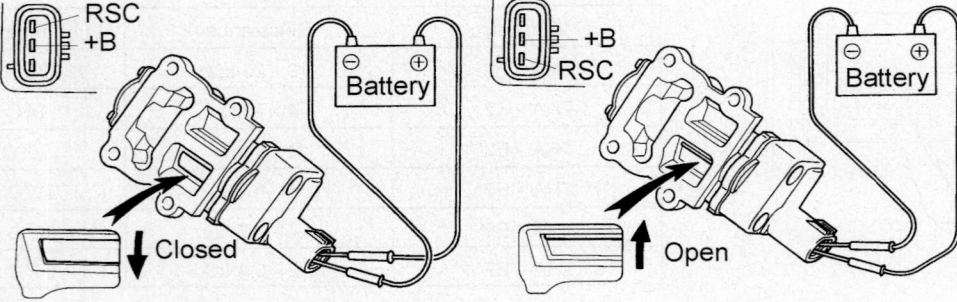

Fig. 208 IAC valve testing

29157_TOYO_G0090

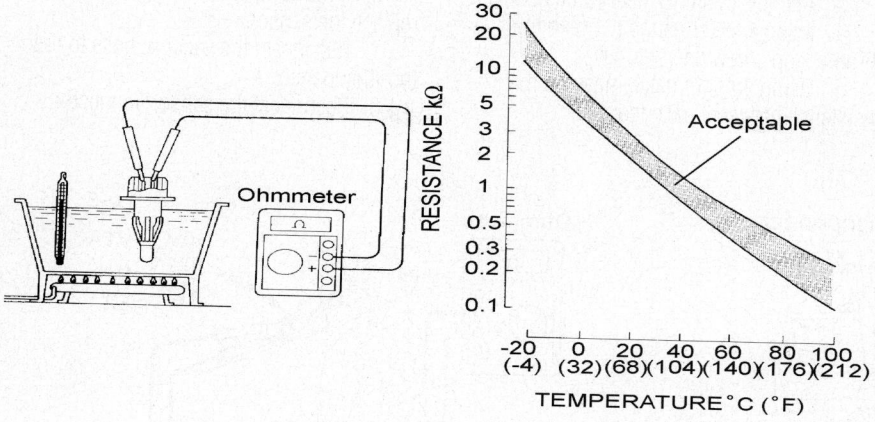

Fig. 209 IAT sensor testing

29157_TOYO_G0146

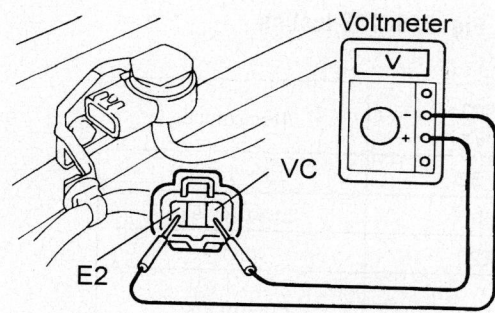

Fig. 210 MAP Sensor power input testing

29157_TOYO_G0151

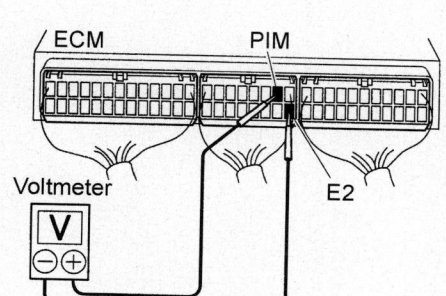

Fig. 211 MAP Sensor power output testing

29157_TOYO_G0152

Applied Vacuum kPa (mm Hg) (in. Hg)	13.3 (100) (3.94)	26.7 (200) (7.87)	40.0 (300) (11.81)	53.5 (400) (15.75)	66.7 (500) (19.69)
Voltage drop V	0.3 - 0.5	0.7 - 0.9	1.1 - 1.3	1.5 - 1.7	1.9 - 2.1

Fig. 212 Vacuum/voltage table

29157_TOYO_G0153

6. Turn the ignition switch ON.

7. Disconnect the vacuum hose on the air intake chamber

8. Connect a voltmeter to terminals PIM and E2 of the ECM, and measure the output voltage under ambient atmospheric pressure.

9. Apply vacuum to the MAP sensor in 13.3 kPa (100 mmHg, 3.94 in.Hg) segments to 66.7 kPa (500 mmHg, 19.69 in.Hg).

10. Measure the voltage drop from step above for each segment.

11. If operation is not as specified, replace the MAP sensor.

12. Reconnect the vacuum hose to the air intake chamber.

Oxygen Sensor

See Figure 213.

1. Remove the right front seat and floor carpeting.

2. Disconnect the oxygen sensor connector.

3. Using an ohmmeter, measure the resistance between the terminals +B and HT. Resistance specification: 11 - 16 ohms at 20°C (68°F).

4. If the resistance is not as specified, replace the sensor. Tighten to 32 ft. lbs. (44 Nm).

5. Reconnect the oxygen sensor connector.

6. Install the right front seat and floor carpeting.

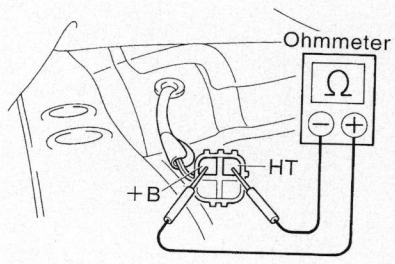

Fig. 213 Oxygen Sensor testing

29157_TOYO_G0154

Throttle Position Sensor

See Figures 214, 215 and 216.

1. Disconnect the sensor connector.
2. Disconnect the vacuum hose from the throttle opener.
3. Apply vacuum to the throttle opener.
4. Insert a feeler gauge between the throttle stop screw and stop lever.
5. Using an ohmmeter, measure the resistance between each terminal.
6. If the resistance is not as specified, replace the sensor.
7. Reconnect the vacuum hose to the throttle opener.
8. Reconnect the sensor connector.

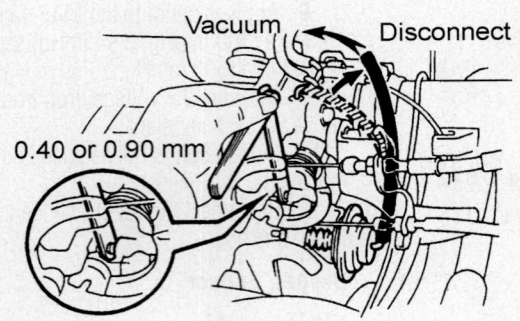

29157_TOYO_G0155

Fig. 214 Place feeler gauge for TPS testing

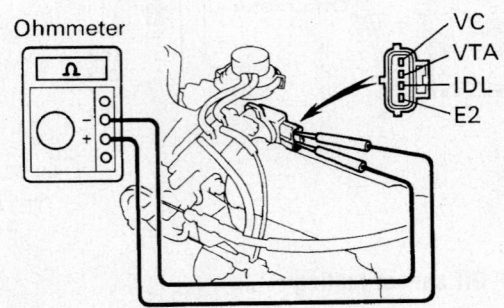

29157_TOYO_G0156

Fig. 215 TPS testing

Clearance between lever and stop screw	Between terminals	Resistance
0 mm (0 in.)	VTA - E2	0.2 - 5.7 kΩ
0.40 mm (0.016 in.)	IDL-E2	2.3 kΩ or less
0.90 mm (0.035 in.)	IDL-E2	Infinity
Throttle valve fully open	VTA - E2	2.0 - 10.2 kΩ
-	VC - E2	2.5 - 5.9 kΩ

29157_TOYO_G0157

Fig. 216 Resistance table

TOYOTA
PIN CHARTS

16

TABLE OF CONTENTS

PIN CHARTS

Introduction

A Pin Voltage Table is a term used to describe a table that identifies PCM pins, wire colors of the PCM circuits, circuit descriptions and "known good" values for devices that connect to the PCM. These tables include the following information:

- Signals from various sensors (ECT, IAT, MAP, TPS, etc.)
- Signals from various switches (PNP, PSP, WOT, etc.)
- Signals from oxygen sensors (O2S, HO2S)
- Signals from output devices (IAC, INJ, TCC, etc.)
- Power & ground signals

Pin Voltage Tables

Information contained within the Pin Voltage Tables can be used to:

- Test circuits for open, short to power or short to ground faults
- Check the operation of a component before or after a repair
- Check the operation of a component or system by viewing signals on PCM input/output circuits with a DVOM or Lab Scope

Using a Breakout Box

There are several Breakout Box (BOB) designs available for use to test the PCM and its input and output circuits. However, all of them require removal of the wire harness to the PCM so that the BOB can be installed between the PCM and wire harness connector. Several breakout boxes require the use of overlays in order to allow the tool to be used on more than one year or engine type. Always verify that the correct adapter and overlays are used to prevent connection to the wrong circuits and a misdiagnosis.

Power and Ground Circuit Checks

Measurements made at the BOB are accomplished via test leads and probes from the DVOM or a Lab Scope. If any of the terminals on the PCM or BOB are damaged or loose, test measurements made at the Breakout Box will be inaccurate. To verify the PCM battery power and ground circuits are normal (correct) at the BOB, test the condition of the circuit between the battery negative (-) post and these circuits prior to starting a test sequence.

Diagnosis with Pin Voltage Tables

See Figure 1.

Once an actual PCM pin voltage reading is recorded, it can be compared to an example from a vehicle with "known good" values. In the example shown the Value at Hot Idle for the EVP sensor signal (0.4v) is the "known good" value.

Wire Color Changes

Every effort has been made to obtain and list the correct circuit wire colors for all vehicles. However, running changes from the vehicle manufacturer can cause the wrong colors to be listed.

PCM Pin #	W/Color	Circuit Description (60-Pin)	Value at Hot Idle
27	BN/LG	EVP Sensor Signal	0.4v

Fig. 1 Example

AVALON PIN CHARTS

1995 3.0L V6 VIN G (A/T-ECT) 16 Pin Connector

PCM Pin #	Wire Color	Circuit Description (16 Pin)	Value at Hot Idle
1	WT/BL	A/C Idle-Up Solenoid	A/C Off: 12v, On: 1v
2 (Cal)	BK/RD	EVAP Solenoid Control	12v or 0v
3	GN/RD	MIL (lamp) Control	MIL Off: 12v, On: 1v
5	GY/BK	Check Connector	12-14v
6	RD/YL	Intake Air Solenoid	12v or 0v
7	RD/BK	MAF Sensor Ground	<0.050v
10	PK	HO2S-21 (B2 S1) Heater	1v (Heater On)
11	BL/BK	HO2S-11 (B1 S1) Heater	1v (Heater On)
12	BK/BL	EGR Solenoid Control (VSV)	12v or 0v
14	GN/YL	EGR Gas Temp. Sensor	3.5-4.0v
16	BR	Sensor Ground	<0.050v

1995 3.0L V6 VIN G (A/T-ECT) 22 Pin Connector

PCM Pin #	Wire Color	Circuit Description (22 Pin)	Value at Hot Idle
1	BL/RD	Sensor VREF (VC)	4.9-5.1v
4	WT/BL	OD Clutch Speed Sensor (-)	AC pulse signals
5	RD	CKP Sensor Signal (NE+)	AC pulse signals
6	GN	CKP Sensor Signal (NE-)	<0.050v
7	BK/YL	TP Sensor Signal	0.3-0.8v
8	RD	MAF Sensor Signal	1.1-1.5v
9	YL/BL	OD Clutch Speed Sensor (+)	AC pulse signals
13	RD/BL	HO2S-11 (B1 S1) Signal	0.1-1.1v
14	WT	Knock Sensor 1 Signal	<0.075v AC
15	WT	Knock Sensor 2 Signal	<0.075v AC
16	BK/WT	CMP Sensor Signal (G+)	AC pulse signals
17	BL	CMP Sensor Signal (G-)	<0.050v
18	GN/RD	Circuit Opening Relay (FC)	0-3v, off-idle: 12v
19	RD/BL	HO2S-21 (B2 S1) Signal	0.1-1.1v
20	GN/BK	ECT Sensor Signal	At 180°F: 0.51v
21	BL/BK	IAT Sensor Signal	At 100°F: 2.60v
22	BR	Sensor Ground	<0.050v

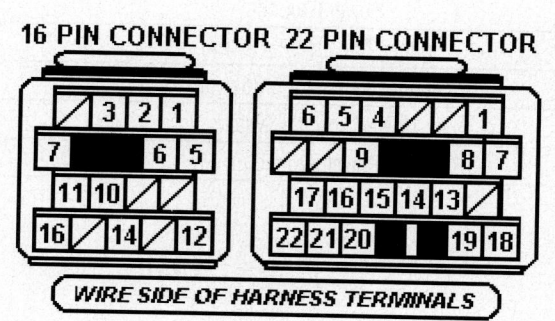

16 PIN CONNECTOR   22 PIN CONNECTOR

WIRE SIDE OF HARNESS TERMINALS

Pin Connector Graphic

05533_ADIA_G000

1995 3.0L V6 VIN G (A/T-ECT) 28 Pin Connector

PCM Pin #	Wire Color	Circuit Description (28 Pin)	Value at Hot Idle
5	GN/BK	A/C Amplifier Signal (ACT)	Clutch On: 12v, Off: 1.5v
9	RD	Cooling Fan Relay	Relay Off: 12v, On: 1v
12	PK/YL	Vehicle Speed Sensor	At 55 mph: 48 Hz
13	BK	Tachometer Signal (TACO)	Pulse Signals
14	WT/BL	Battery Direct	12-14v
20	BK/YL	A/C Amplifier Signal (AC1)	Clutch On: 1.5v, Off: 12v
21	BK/RD	Defogger/Light Idle Up Signal	Switch On: 12v, Off: 0v
23	BK/OR	EFI Main Relay Power	12-14v
24	GN/WT	Stop Light Switch Signal	Brake Off: 0v, On: 12v
25	PK/BK	HO2S-12 (B1 S2) Heater	1v (Heater On)
26	BK	HO2S-12 (B1 S2) Signal	0.1-1.1v
28	WT	Data Link Connector	12v

1995 3.0L V6 VIN G (A/T-ECT) 34 Pin Connector

PCM Pin #	Wire Color	Circuit Description (34 Pin)	Value at Hot Idle
3, 27	YL, BL	A/T-ECT Solenoid SL-, SL+	In Lockup: 12-14v
5	GN	Injector 6 Control	2.0-3.3 ms
6	RD	Injector 5 Control	2.0-3.3 ms
7	BL	Injector 4 Control	2.0-3.3 ms
8	GY	Injector 3 Control	2.0-3.3 ms
9	YL	Injector 2 Control	2.0-3.3 ms
10	WT	Injector 1 Control	2.0-3.3 ms
11	PK	A/T-ECT Solenoid (S1)	In 3rd or OD: 1v
12	WT/RD	Igniter Signal (IGF)	Digital Signal: 0-5-0v
13	BK/WT	Starter Switch Signal	9-11v (cranking)
14	BK/WT	Neutral Start Switch	In P/N: 9-11v (cranking)
15	GY/BK	Igniter Transistor 3 Control	7% duty cycle
16	YL/RD	Igniter Transistor 2 Control	7% duty cycle
17	PK/BK	A/T-ECT Solenoid (S2)	1st or OD: 1v
22, 23	YL, GN	IAC Signals (RSC, RSO)	Pulse Signals
24	WT/GN	Igniter Transistor 1 Control	7% duty cycle
25	BK/RD	Fuel Pressure Up Solenoid	1v (at hot restart)
26	BL/BK	Igniter Transistor 4 Control	7% duty cycle
28, 33-34	WT/BK	Power Ground	<0.1v
29	GN/RD	Igniter Transistor 6 Control	7% duty cycle
30	RD/BK	Igniter Transistor 5 Control	7% duty cycle
32	LB	Closed Throttle Switch	<0.1v
33	WT/BK	Power Ground	<0.1v

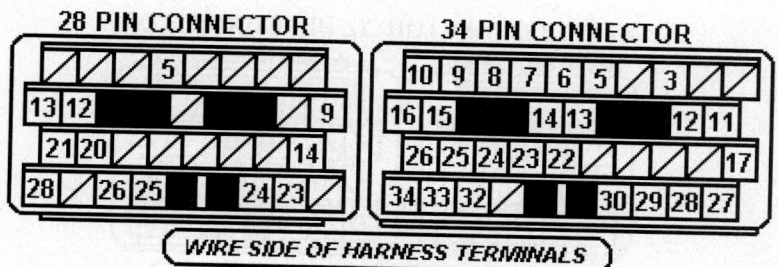

28 PIN CONNECTOR   **34 PIN CONNECTOR**

WIRE SIDE OF HARNESS TERMINALS

05533_ADIA_G001

Pin Connector Graphic

1996 3.0L V6 VIN F (A/T-ECT) 16 Pin Connector

PCM Pin #	Wire Color	Circuit Description (16 Pin)	Value at Hot Idle
2 (Cal)	BK/RD	EVAP Purge Solenoid (VSV)	12v or 0v
3	GN/RD	MIL (lamp) Control	MIL Off: 12v, On: 1v
5	GY/BK	Check Connector	12-14v
6	RD/YL	Intake Air Solenoid	12v or 0v
7	RD/BK	MAF Sensor Ground	<0.050v
8	WT/BL	EVAP Vapor Pressure (VSV)	12v or 0v
10	BL/RD	HO2S-21 (B2 S1) Heater	1v (Heater On)
11	BL/BK	HO2S-11 (B1 S1) Heater	1v (Heater On)
12	BK/WT	EGR Solenoid Control (VSV)	12v or 0v
13	PK	EVAP Vapor Pressure Sensor	2.9-3.1v (with hose off)
14	GN/RD	EGR Gas Temp. Sensor	3.5-4.0v
16	BR	Sensor Ground	<0.050v

1996 3.0L V6 VIN F (A/T-ECT) 22 Pin Connector

PCM Pin #	Wire Color	Circuit Description (22 Pin)	Value at Hot Idle
1	BL/RD	Sensor VREF (VC)	4.9-5.1v
4	WT/BL	OD Clutch Speed Sensor (-)	AC pulse signals
5	WT	CKP Sensor Signal (NE+)	AC pulse signals
6	OR/BL	CKP Sensor Signal (NE-)	<0.050v
7	BK/YL	TP Sensor Signal	0.3-0.8v
8	RD	MAF Sensor Signal	1.1-1.5v
9	YL/BL	OD Clutch Speed Sensor (+)	AC pulse signals
13	WT	HO2S-11 (B1 S1) Signal	0.1-1.1v
14	WT	Knock Sensor 1 Signal	<0.075v AC
15	WT	Knock Sensor 2 Signal	<0.075v AC
16	WT/BL	CMP Sensor Signal (G+)	AC pulse signals
18	GN	Circuit Opening Relay (FC)	0-3v, off-idle: 12v
19	RD/BL	HO2S-21 (B2 S1) Signal	0.1-1.1v
20	GN/BK	ECT Sensor Signal	At 180°F: 0.51v
21	BL/BK	IAT Sensor Signal	At 100°F: 2.60v
22	BR	Sensor Ground	<0.050v

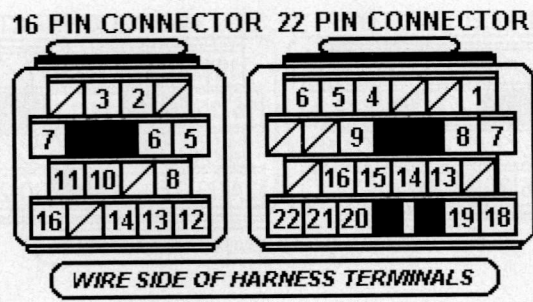

05533_ADIA_G002

Pin Connector Graphic
1996 3.0L V6 VIN F (A/T-ECT) 28 Pin Connector

PCM Pin #	Wire Color	Circuit Description (28 Pin)	Value at Hot Idle
5	PK	A/C Amplifier Signal (ACT)	Clutch On: 12v, Off: 1.5v
9	RD	Cooling Fan Relay	Relay Off: 12v, On: 1v
12	PK/YL	Vehicle Speed Sensor	At 55 mph: 48 Hz
13	BK	Tachometer Signal (TACO)	Pulse Signals
14	BK/YL	Battery Direct	12-14v
20	BK/YL	A/C Amplifier Signal (AC1)	Clutch On: 1.5v, Off: 12v
21	BK/RD	Defogger Idle Up Signal	Switch On: 12v, Off: 0v
23	BK/OR	EFI Main Relay Power	12-14v
24	GN/WT	Stop Light Switch Signal	Brake Off: 0v, On: 12v
25	PK/BK	HO2S-12 (B1 S2) Heater	1v (Heater On)
26	BK	HO2S-12 (B1 S2) Signal	0.1-1.1v
28	WT	Data Link Connector	12v

1996 3.0L V6 VIN F (A/T-ECT) 34 Pin Connector

PCM Pin #	Wire Color	Circuit Description (34 Pin)	Value at Hot Idle
3	YL/GN	A/T-ECT Solenoid (SL-)	In Lockup: 12-14v
5	GN	Injector 6 Control	2.0-3.3 ms
6	RD	Injector 5 Control	2.0-3.3 ms
7	BL	Injector 4 Control	2.0-3.3 ms
8	GY	Injector 3 Control	2.0-3.3 ms
9	YL	Injector 2 Control	2.0-3.3 ms
10	WT	Injector 1 Control	2.0-3.3 ms
11	PK	A/T-ECT Solenoid (S1)	In 3rd or OD: 1v
12	WT/RD	Igniter Signal (IGF)	Digital Signal: 0-5-0v
13	BK/RD	Starter Switch Signal	9-11v (cranking)
14	BK/WT	Neutral Start Switch	In P/N: 9-11v (cranking:
15	GY/BK	Igniter Transistor 3 Control	7% duty cycle
16	YL/RD	Igniter Transistor 2 Control	7% duty cycle
17	PK/BK	A/T-ECT Solenoid (S2)	1st or OD: 1v
22	YL/BK	IAC Signal (RSC)	Pulse Signals
23	GN/BK	IAC Signal (RSO)	Pulse Signals

1996 3.0L V6 VIN F (A/T-ECT) 34 Pin Connector, *continued*

24	BL/BK	Igniter Transistor 1 Control	7% duty cycle
25	BK/YL	Fuel Pressure Up Solenoid	Hot Restart: 12-14v
27	BL/YL	A/T-ECT Solenoid (SL+)	In Lockup: 12-14v
28, 33-34	WT/BK	Power Ground	<0.1v
32	BL/WT	Closed Throttle Switch	1v, off-idle: 12v

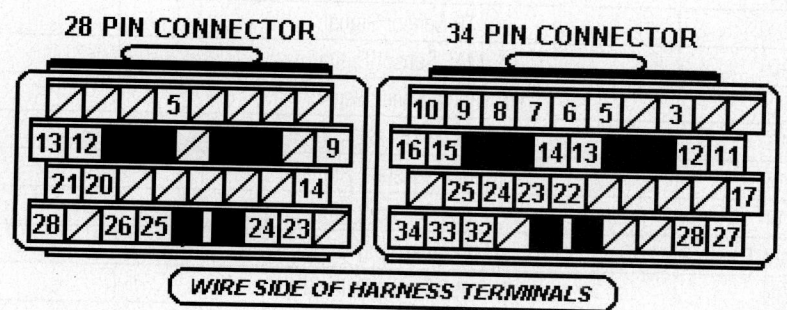

Pin Connector Graphic

05533_ADIA_G003

1997 3.0L V6 MFI VIN F (A/T-ECT) 16 Pin Connector

PCM Pin #	Wire Color	Circuit Description (16 Pin)	Value at Hot Idle
2	BK/RD	EVAP Purge Solenoid (VSV)	12v or 0v
3	GN/RD	MIL (lamp) Control	MIL Off: 12v, On: 1v
5	BL	Data Link Connector	12-14v
6	BL	Intake Air Solenoid	12v or 0v
7	RD/BK	MAF Sensor Ground	<0.050v
8	WT/BL	EVAP Vapor Pressure (VSV)	12v or 0v
9	BK/YL	Cooling Fan Relay	Relay Off: 12v, On: 1v
10	BL/RD	HO2S-21 (B2 S1) Heater	1v (Heater on)
11	BL/BK	HO2S-11 (B1 S1) Heater	1v (Heater on)
12	BK/WT	EGR Solenoid Control (VSV)	12v or 0v
13	PK	EVAP Vapor Pressure Sensor	2.9-3.1v (with hose off)
14	GN/RD	EGR Gas Temp. Sensor	3.5-4.0v
15	BK/OR	EGR Valve Position Sensor	0.4-1.6v
16	BR	Sensor Ground	<0.050v

1997 3.0L V6 MFI VIN F (A/T-ECT) 22 Pin Connector

PCM Pin #	Wire Color	Circuit Description (22 Pin)	Value at Hot Idle
1	BL/RD	Sensor VREF (VC)	4.9-5.1v
2	YL/GN	A/T-ECT Solenoid (SLN-)	During shifting: 12v
4	BL	OD Clutch Speed Sensor (-)	AC pulse signals
5	BK/RD	CKP Sensor Signal (NE+)	AC pulse signals
6	BL	CKP Sensor Signal (NE-)	<0.050v
7	BK/YL	TP Sensor Signal	0.3-0.8v
8	RD	MAF Sensor Signal	1.1-1.5v
9	YL	OD Clutch Speed Sensor (+)	AC pulse signals
13	WT	HO2S-11 (B1 S1) Signal	0.1-1.1v
14	WT	Knock Sensor 1 Signal	<0.075v AC
15	WT	Knock Sensor 2 Signal	<0.075v AC
17	BK/WT	CMP Sensor Signal (G+)	AC pulse signals
18	GN	Circuit Opening Relay (FC)	0-3v, off-idle: 12v
19	BL/BK	HO2S-21 (B2 S1) Signal	0.1-1.1v
20	GN/BK	ECT Sensor Signal	At 180°F: 0.51v
21	BL/BK	IAT Sensor Signal	At 100°F: 2.60v
22	BR	Sensor Ground	<0.050v

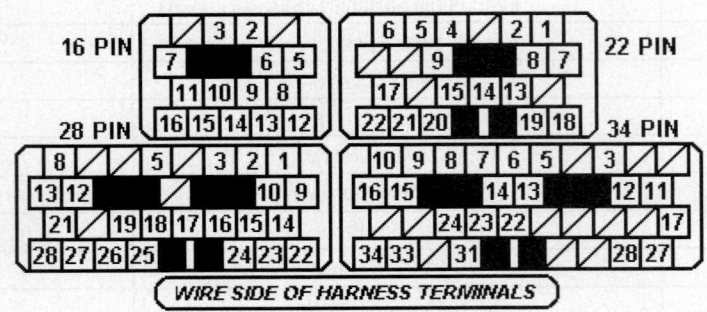

WIRE SIDE OF HARNESS TERMINALS

05533_ADIA_G004

Pin Connector Graphic

1997 3.0L V6 MFI VIN F (A/T-ECT) 28 Pin Connector

PCM Pin #	Wire Color	Circuit Description (28 Pin)	Value at Hot Idle
1	YL/BL	A/T Select Switch Low	In Low: 12v, Others: 0v
2	BK/RD	Mirror Heater Switch Signal	Switch On: 12v, Off: 0v
3	GN	Tail Light Switch Signal	Switch On: 12v, Off: 0v
5	PK	A/C Amplifier Signal (ACT)	Clutch On: 12v, Off: 1.5v
8	WT	Data Link Connector	12v
9	RD	Cooling Fan Relay	Relay Off: 12v, On: 1v
10	GN/YL	A/T Select Switch 2nd	In 2nd: 12v, Others: 0v
12	PK	Vehicle Speed Sensor	At 55 mph: 48 Hz
13	BK	Tachometer Signal (TACO)	Pulse Signals
14	BK/YL	Battery Direct	12-14v
15	RD/BK	A/T Select Switch Reverse	In 'R': 12v, Others: 0v
16	BK/YL	A/C Amplifier Signal (AC1)	Clutch On: 1.5v, Off: 12v
17	PK/BK	HO2S-12 (B1 S2) Heater	1v (Heater On)

1997 3.0L V6 MFI VIN F (A/T-ECT) 28 Pin Connector, *continued*

18	BK	HO2S-12 (B1 S2) Signal	0.1-1.1v
19	BR/WT	ABS/Traction ECU (NEO)	Pulse Signals
21	BK/RD	Defogger Idle Up Signal	Switch On: 12v, Off: 0v
22	BL/RD	A/T Pattern Selector Switch	Norm: 0v, PWR: 12v
23	LG	EFI Main Relay Power	12-14v
24	GN/WT	Stop Light Switch Signal	Brake Off: 0v, On: 12v
25	PK	ABS/Traction ECU (EFI-)	Pulse Signals
26	LG	ABS/Traction ECU (EFI+)	Pulse Signals
27	GN/RD	ABS/Traction ECU (TRC-)	Pulse Signals
28	GN	ABS/Traction ECU (TRC+)	Pulse Signals

1997 3.0L V6 MFI VIN F (A/T-ECT) 34 Pin Connector

PCM Pin #	Wire Color	Circuit Description (34 Pin)	Value at Hot Idle
3	GN	A/T-ECT Solenoid (SLN-)	During shifting: 12v
5	GN	Injector 6 Control	2.0-3.3 ms
6	RD	Injector 5 Control	2.0-3.3 ms
7	BL	Injector 4 Control	2.0-3.3 ms
8	BK	Injector 3 Control	2.0-3.3 ms
9	YL	Injector 2 Control	2.0-3.3 ms
10	WT	Injector 1 Control	2.0-3.3 ms
11	PK	A/T-ECT Solenoid (S1)	In 3rd or OD: 1v
12	WT/RD	Igniter Signal (IGF)	Digital Signal: 0-5-0v
13	BK/RD	Starter Switch Signal	9-11v (cranking)
14	BK/WT	Neutral Start Switch	In P/N: 9-11v (cranking)
15	GN/RD	Igniter Transistor 3 Control	7% duty cycle
16	YL/RD	Igniter Transistor 2 Control	7% duty cycle
17	GN/YL	A/T-ECT Solenoid (S2)	1st or OD: 1v
22	YL/BK	IAC Signal (RSC)	Pulse Signals
23	GN/BK	IAC Signal (RSO)	Pulse Signals
24	BL/BK	Igniter Transistor 1 Control	7% duty cycle
27	BL/YL	A/T-ECT Solenoid (SL)	In Lockup: 12-14v
28	WT/BK	Power Ground	<0.1v
31	PK/BK	PSP Switch Signal	Straight: 12v, Turning: 0v
33	WT/BK	Power Ground	<0.1v
34	WT/BK	Power Ground	<0.1v

1998 3.0L V6 W/O EIS-TC VIN F Federal 16 Pin Connector

PCM Pin #	Wire Color	Circuit Description (16 Pin)	Value at Hot Idle
2	BK/RD	EVAP Purge Solenoid (VSV)	12v or 0v
3	GN/RD	MIL (lamp) Control	MIL Off: 12v, On: 1v
5	BL	Data Link Connector	12v
6	RD/YL	Intake Air Control Solenoid	12v or 0v
7	RD/BK	MAF Sensor Ground	<0.050v
8	WT/BL	EVAP Vapor Pressure (VSV)	12v or 0v
9	BK/YL	Cooling Fan 1 Relay	Relay Off: 12v, On: 1v
10	BL/RD	HO2S-21 (B2 S1) Heater	1v (Heater on)
11	BL/BK	HO2S-11 (B1 S1) Heater	1v (Heater on)
12	BK/WT	EGR Solenoid Control (VSV)	12v or 0v
13	PK	EVAP Vapor Pressure Sensor	2.9-3.1v (with hose off)
14	GN/RD	EGR Gas Temp. Sensor	3.5-4.0v
15	RD/WT	EGR Valve Position Sensor	0.4-1.6v
16	BR	Sensor Ground	<0.050v

1998 3.0L V6 W/O EIS-TC VIN F Federal 22 Pin Connector

PCM Pin #	Wire Color	Circuit Description (22 Pin)	Value at Hot Idle
1	BL/RD	Sensor VREF (VC)	4.9-5.1v
2	YL/GN	A/T-ECT Solenoid (SL-)	In Lockup: 12-14v
4	BL	OD Clutch Speed Signal (-)	AC pulse signals
5	BK/RD	CKP Sensor Signal (NE+)	AC pulse signals
6	BL	CKP Sensor Signal (NE-)	<0.050v
7	BK/YL	TP Sensor Signal	0.3-0.8v
8	RD	MAF Sensor Signal	1.1-1.5v
9	YL	OD Clutch Speed Signal (+)	AC pulse signals
13	WT	HO2S-11 (B1 S1) Signal	0.1-1.1v
14	WT	Knock Sensor 1 Signal	<0.075v AC
15	WT	Knock Sensor 2 Signal	<0.075v AC
17	BK/WT	CMP Sensor Signal (G+)	AC pulse signals
18	GN	Circuit Opening Relay (FC)	0-3v, off-idle: 12v
19	RD/BL	HO2S-21 (B2 S1) Signal	0.1-1.1v
20	GN/BK	ECT Sensor Signal	At 180°F: 0.51v
21	BL/BK	IAT Sensor Signal	At 100°F: 2.60v
22	BR	Sensor Ground	<0.050v

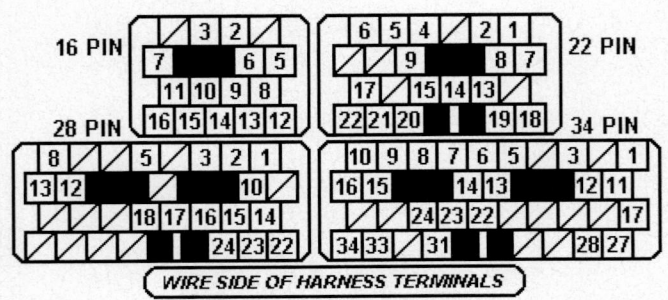

WIRE SIDE OF HARNESS TERMINALS

05533_ADIA_G005

1998 3.0L V6 W/O EIS-TC VIN F Federal 28 Pin Connector

PCM Pin #	Wire Color	Circuit Description (28 Pin)	Value at Hot Idle
1	YL	A/T Select Switch Low	In Low: 12v, Others: 0v
2	BK/RD	Mirror Heater Switch Signal	Switch On: 12v, Off: 0v
3	GN	Tail Light Switch Signal	Switch On: 12v, Off: 0v
5	PK	A/C Amplifier Signal (ACT)	Clutch On: 12v, Off: 1.5v
8	WT	Data Link Connector	Activated: Pulses
10	GN/YL	A/T Select Switch 2nd	In 2nd: 12v, Others: 0v
12	PK	Vehicle Speed Sensor	At 55 mph: 48 Hz
13	BK	Tachometer Signal (TACO)	Pulses
14	BK/YL	Battery Direct	12-14v
15	RD/BK	A/T Select Switch Reverse	In 'R': 12v, Others: 0v
16	BK/YL	A/C Amplifier Signal (AC1)	Clutch On: 1.5v, Off: 12v
17	PK/BK	HO2S-12 (B1 S2) Heater	1v (Heater on)
18	BK	HO2S-12 (B1 S2) Signal	0.1-1.1v
22	BL/RD	A/T Pattern Selector Switch	Norm: 0v, PWR: 12v
23	LG	EFI Main Relay Power	12-14v
24	GN/WT	Stop Light Switch Signal	Brake Off: 0v, On: 12v

1998 3.0L V6 W/O EIS-TC VIN F Federal 34 Pin Connector

PCM Pin #	Wire Color	Circuit Description (34 Pin)	Value at Hot Idle
1	BR	Sensor Ground	<0.050v
3	GN	A/T-ECT Solenoid (SL-)	In Lockup: 12-14v
5	GN	Injector 6 Control	2.0-3.3 ms
6	RD	Injector 5 Control	2.0-3.3 ms
7	BL	Injector 4 Control	2.0-3.3 ms
8	BK	Injector 3 Control	2.0-3.3 ms
9	YL	Injector 2 Control	2.0-3.3 ms
10	WT	Injector 1 Control	2.0-3.3 ms
11	PK	A/T-ECT Solenoid (S1)	In 3rd or OD: 1v
12	WT/RD	Igniter Signal (IGF)	Digital Signal: 0-5-0v
13	BK/RD	Starter Switch Signal	9-11v (cranking)
14	BK/WT	Neutral Start Switch	In P/N: 9-11v (cranking)
15	GN/RD	Igniter Transistor 3 Control	7% duty cycle
16	YL/RD	Igniter Transistor 2 Control	7% duty cycle
17	GN/YL	A/T-ECT Solenoid (S2)	1st or OD: 1v
22	YL/BK	IAC Signal (RSC)	Pulse Signals
23	GN/BK	IAC Signal (RSO)	Pulse Signals
24	BL/BK	Igniter Transistor 1 Control	7% duty cycle
27	BL/YL	A/T-ECT Solenoid (SL+)	In Lockup: 12-14v
28	WT/BK	Power Ground	<0.1v
31	PK/BK	PSP Switch Signal	Straight: 12v, Turning: 0v
33	WT/BK	Power Ground	<0.1v
34	WT/BK	Power Ground	<0.1v

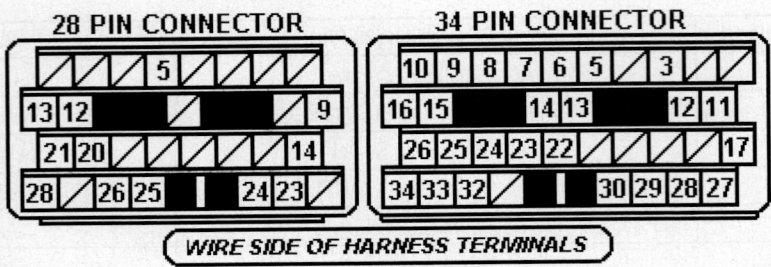

Pin Connector Graphic

1998-99 3.0L V6 W/EIS-TC CAL VIN F 17 Pin Connector

PCM Pin #	Wire Color	Circuit Description (17 Pin)	Value at Hot Idle
4	BK/RD	Transponder Amplifier Code	Inserting key: pulses
5	YL	Transponder Amplifier Signal	Inserting key: pulses
10	GN	Transponder Amplifier Signal	Inserting key: pulses
11	RD/YL	Unlock Warning Switch	No key: 4-5v
13	BR	Sensor Shield Ground	<0.050v
16	PK	Security Indicator Light	Digital Signal

1998-99 3.0L V6 W/EIS-TC CAL VIN F 22 Pin Connector

PCM Pin #	Wire Color	Circuit Description (22 Pin)	Value at Hot Idle
1	BK/YL	Battery Direct	12-14v
2	BK/OR	Ignition Switch Power	12-14v
3	GN	Circuit Opening Relay (FC)	0-3v, off-idle: 12v
6	GN/RD	MIL (lamp) Control	MIL Off: 12v, On: 1v
7	BK/RD	Starter Switch Signal	9-11v (cranking)
8	BK/OR	EFI Main Relay Power	12-14v
9	WT/BL	EVAP Vapor Pressure (VSV)	12v or 0v
11	WT	SIL Signal (Scan Tool)	Digital Signal
13	GN	ABS/Traction ECU (TRC+)	Pulses
14	LG	ABS/Traction ECU (EFI+)	Pulses
15	GN/WT	Stop Light Switch Signal	Brake Off: 0v, On: 12v
16	LG	EFI Main Relay Power	12-14v
17	PK	EVAP Vapor Pressure Sensor	2.9-3.1v (with hose off)
18	BK/RD	Mirror Heater Switch Signal	Switch On: 12v, Off: 0v
19	GN	Tail Light Switch Signal	Switch On: 12v, Off: 0v
20	GN/RD	ABS/Traction ECU (TRC-)	Pulses
21	PK	ABS/Traction ECU (EFI-)	Pulses

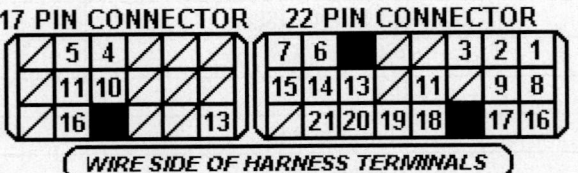

Pin Connector Graphic

1998-99 3.0L V6 W/EIS-TC CAL VIN F 24 Pin Connector

PCM Pin #	Wire Color	Circuit Description (24 Pin)	Value at Hot Idle
1	WT/BK	Power Ground	<0.1v
2	BL/RD	Sensor VREF	4.9-5.1v
3	BL/RD	AFS-11 (B1 S1) Heater	1v (Heater On)
4	BK/RD	AFS-21 (B2 S1) Heater	1v (Heater On)
5	WT	Injector 1 Control	2.0-3.3 ms
6	YL	Injector 2 Control	2.0-3.3 ms
7	BK/RD	EVAP Purge Solenoid (VSV)	12v or 0v
8	WT/BK	Sensor Ground	<0.050v
9	PK/BK	PSP Switch Signal	Straight: 12v, Turning: 0v
10	RD	MAF Sensor Signal	1.1-1.5v
11	BK/WT	AFS-11 (B1 S1) Signal (+)	3.0-3.6v
12	BR	AFS-21 (B2 S1) Signal (+)	3.0-3.6v
13	GN/RD	EGR Gas Temp. Sensor	3.5-4.0v
14	GN/BK	ECT Sensor Signal	At 180°F: 0.51v
16	BK/RD	CKP Sensor Signal (NE+)	AC pulse signals
17	BR	Shield Ground	<0.050v
18	BR	Sensor Ground	<0.050v
19	RD/BK	MAF Sensor Ground	<0.050v
20	BL	AFS-11 (B1 S1) Signal (AF-)	Fixed at 3v
21	BK/RD	AFS-21 (B2 S2) Signal (-)	Fixed at 3v
22	BL/BK	IAT Sensor Signal	At 100°F: 2.60v
23	BK/YL	TP Sensor Signal	0.3-0.8v
24	BL	CKP/CMP Sensor Ground (-)	<0.050v

1998-99 3.0L V6 W/EIS-TC CAL VIN F 28 Pin Connector

PCM Pin #	Wire Color	Circuit Description (28 Pin)	Value at Hot Idle
8	BK	HO2S-12 (B1 S2) Signal	0.1-1.1v
9	PK/BK	HO2S-12 (B1 S2) Heater	1v (Heater on)
13	PK	A/C Amplifier Signal (ACT)	Clutch On: 12v, Off: 1.5v
14	BL/RD	A/C Amplifier Signal (THWO)	Pulse Signals
16	BR/WT	ABS/Traction ECU (NEO)	Pulse Signals
20	BK/WT	Neutral Start Switch	In P/N: 9-11v (cranking)
22	PK	Vehicle Speed Sensor	At 55 mph: 48 Hz
25	BK/YL	A/C Amplifier Signal (AC1)	Clutch On: 1.5v, Off: 12v
27	BK	Tachometer Signal (TACO)	Pulse Signal

Standard Colors and Abbreviations

Abbreviation	Color	Abbreviation	Color	Abbreviation	Color
BK	Black	GY	Gray	RD	Red
BL	Blue	GN	Green	TN	Tan
BR	Brown	LG	Light Green	VT	Violet
DB	Dark Blue	OR	Orange	WT	White
DG	DK Green	PK	Pink	YL	Yellow

1998-99 3.0L V6 W/EIS-TC Calif. VIN F 31 Pin Connector

PCM Pin #	Wire Color	Circuit Description (31 Pin)	Value at Hot Idle
1	BK	Injector 3 Control	2.0-3.3 ms
2	BL	Injector 4 Control	2.0-3.3 ms
3	RD	Injector 5 Control	2.0-3.3 ms
4	GN	Injector 6 Control	2.0-3.3 ms
6	BL	Data Link Connector (TE)	12-14v
10	BK/WT	CMP Sensor Signal (G+)	AC pulse signals
11	BL/BK	Igniter Transistor 1 Control	7% duty cycle
12	YL/RD	Igniter Transistor 2 Control	7% duty cycle
13	GN/RD	Igniter Transistor 3 Control	7% duty cycle
15	YL/BK	IAC Signal (RSC)	Pulse Signals
16	GN/BK	IAC Signal (RSO)	Pulse Signals
17	RD/YL	ACIS Solenoid Control (VSV)	12v or 0v
18	BK/WT	EGR Solenoid Control (VSV)	12v or 0v
21	WT/BK	Power Ground	<0.1v
22	RD/WT	EGR Valve Position Sensor	0.4-1.6v
23 ('98)	BR	Sensor Ground	<0.050v
25	WT/RD	Igniter Signal (IGF)	Digital Signal: 0-5-0v
27	WT	Knock Sensor 1 Signal	<0.075v AC
28	WT	Knock Sensor 2 Signal	<0.075v AC
29	BK/YL	Cooling Fan 1 Relay	Relay Off: 12v, On: 1v
30	WT/BK	Power Ground	<0.1v
31	WT/BK	Power Ground	<0.1v

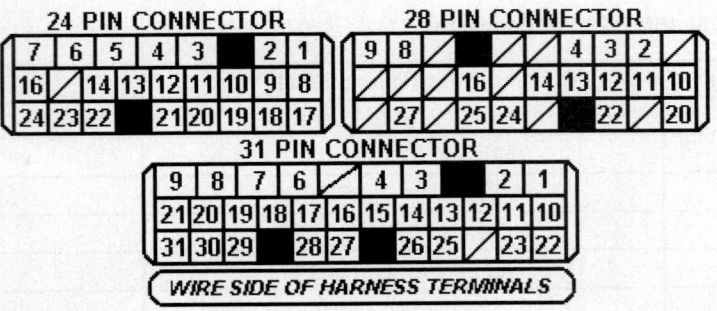

Pin Connector Graphic

1998-99 3.0L V6 W/EIS VIN F Federal 17 Pin Connector

PCM Pin #	Wire Color	Circuit Description (17 Pin)	Value at Hot Idle
4	BK/RD	Transponder Amplifier Code	Inserting key: pulses
5	YL	Transponder Amplifier Signal	Inserting key: pulses
10	GN	Transponder Amplifier Signal	Inserting key: pulses
11	RD/YL	Unlock Warning Switch	No Key: 4-5v
13	BR	Sensor Shield Ground	<0.050v
16	PK	Security Indicator Light	Pulses

1998-99 3.0L V6 W/EIS VIN F Federal 22 Pin Connector

PCM Pin #	Wire Color	Circuit Description (22 Pin)	Value at Hot Idle
1	BK/YL	Battery Direct	12-14v
2	BK/OR	Ignition Switch Power	12-14v
3	GN	Circuit Opening Relay (FC)	0-3v, off-idle: 12v
6	GN/RD	MIL (lamp) Control	MIL Off: 12v, On: 1v
7	BK/RD	Starter Switch Signal	9-11v (cranking)
8	BK/OR	EFI Main Relay Power	12-14v
9	WT/BL	EVAP Vapor Pressure (VSV)	12v or 0v
11	WT	SIL Signal (Scan Tool)	Digital Signal
13	GN	ABS/Traction ECU (TRC+)	Pulses
14	LG	ABS/Traction ECU (EFI+)	Pulses
15	GN/WT	Stop Light Switch Signal	Brake Off: 0v, On: 12v
16	LG	EFI Main Relay Power	12-14v
17	PK	EVAP Vapor Pressure Sensor	2.9-3.1v (with hose off)
18	BK/RD	Mirror Heater Switch Signal	Heater On: 12-14v
19	GN	Tail Light Switch Signal	Switch On: 12v, Off: 0v
20	GN/RD	ABS/Traction ECU (TRC-)	Pulses
21	PK	ABS/Traction ECU (EFI-)	Pulses

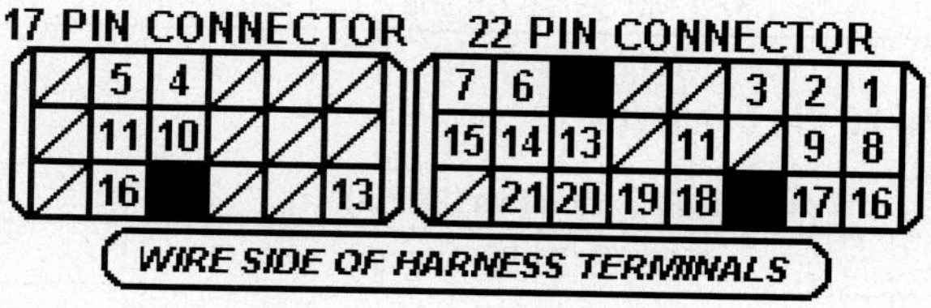

Pin Connector Graphic

05533_ADIA_G006

1998-99 3.0L V6 W/EIS VIN F Federal 24 Pin Connector

PCM Pin #	Wire Color	Circuit Description (24 Pin)	Value at Hot Idle
2	BL/RD	Sensor VREF	4.9-5.1v
3	BL/BK	HO2S-11 (B1 S1) Heater	1v (Heater on)
4	BL/RD	HO2S-21 (B2 S1) Heater	1v (Heater on)
5	WT	Injector 1 Control	2.0-3.3 ms
6	YL	Injector 2 Control	2.0-3.3 ms
7	BK/RD	EVAP Purge Solenoid (VSV)	12v or 0v
9	PK/BK	PSP Switch Signal	Straight: 12v, Turning: 0v
10	RD	MAF Sensor Signal	1.1-1.5v
11	WT	HO2S-11 (B1 S1) Signal	0.1-1.1v
12	RD/BL	HO2S-21 (B2 S1) Signal	0.1-1.1v
13	GN/RD	EGR Gas Temp. Sensor	3.5-4.0v
14	GN/BK	ECT Sensor Signal	At 180°F: 0.51v
16	BK/RD	CKP Sensor Signal (NE+)	AC pulse signals
17	BR	Shield Ground	<0.050v
18	BR	Sensor Ground	<0.050v
19	RD/BK	MAF Sensor Ground	<0.050v
22	BL/BK	IAT Sensor Signal	At 100°F: 2.60v
23	BK/YL	TP Sensor Signal	0.3-0.8v
24	BL	CKP/CMP Sensor Ground (-)	<0.050v

1998-99 3.0L V6 W/EIS VIN F Federal 28 Pin Connector

PCM Pin #	Wire Color	Circuit Description (28 Pin)	Value at Hot Idle
8	BK	HO2S-12 (B1 S2) Signal	0.1-1.1v
9	PK/BK	HO2S-12 (B1 S2) Heater	1v (Heater on)
13	PK	A/C Amplifier Signal (ACT)	Clutch On: 12v, Off: 1.5v
14	BL/RD	A/C Amplifier Signal (THWO)	Pulse Signals
16	BR/WT	ABS/Traction ECU (NEO)	Pulse Signals
20	BK/WT	Neutral Start Switch	In P/N: 9-11v (cranking)
22	PK	Vehicle Speed Sensor	At 55 mph: 48 Hz
25	BK/YL	A/C Amplifier Signal (AC1)	Clutch On: 1.5v, Off: 12v
27	BK	Tachometer Signal (TACO)	Pulse Signals

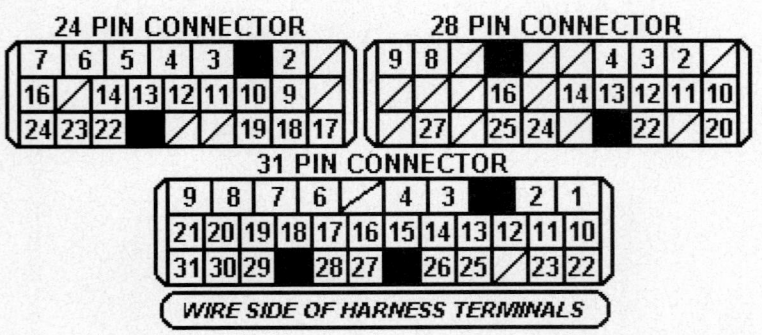

05533_ADIA_G008

Pin Connector Graphic

Standard Colors and Abbreviations

Abbreviation	Color	Abbreviation	Color	Abbreviation	Color
BK	Black	GY	Gray	RD	Red
BL	Blue	GN	Green	TN	Tan
BR	Brown	LG	Light Green	VT	Violet
DB	Dark Blue	OR	Orange	WT	White
DG	DK Green	PK	Pink	YL	Yellow

1998-99 3.0L V6 W/EIS VIN F Federal 31 Pin Connector

PCM Pin #	Wire Color	Circuit Description (31 Pin)	Value at Hot Idle
1	BK	Injector 3 Control	2.0-3.3 ms
2	BL	Injector 4 Control	2.0-3.3 ms
3	RD	Injector 5 Control	2.0-3.3 ms
4	GN	Injector 6 Control	2.0-3.3 ms
5	---	Not Used	---
6	BL	Data Link Connector (TE)	12v
7-9	---	Not Used	---
10	BK/WT	CMP Sensor Signal (G+)	AC pulse signals
11	BL/BK	Igniter Transistor 1 Control	7% duty cycle
12	YL/RD	Igniter Transistor 2 Control	7% duty cycle
13	GN/RD	Igniter Transistor 3 Control	7% duty cycle
14	---	Not Used	---
15	YL/BK	IAC Signal (RSC)	Pulse Signals
16	GN/BK	IAC Signal (RSO)	Pulse Signals
17	RD/YL	ACIS Solenoid Control (VSV)	12v or 0v
18	BK/WT	EGR Solenoid Control (VSV)	12v or 0v
19-20	---	Not Used	---
21	WT/BK	Power Ground	<0.1v
22	RD/WT	EGR Valve Position Sensor	0.4-1.6v
23 ('98)	BR	Sensor Ground	<0.050v
25	WT/RD	Igniter Signal (IGF)	Digital Signal: 0-5-0v
26	---	Not Used	---
27	WT	Knock Sensor 1 Signal	<0.075v AC
28	WT	Knock Sensor 2 Signal	<0.075v AC
29	BK/YL	Cooling Fan 1 Relay Control	Relay Off: 12v, On: 1v
30	WT/BK	Power Ground	<0.1v
31	WT/BK	Power Ground	<0.1v

24 PIN CONNECTOR

7	6	5	4	3	■	2	/
16	/	14	13	12	11	10	9	/
24	23	22	■	/	/	19	18	17

28 PIN CONNECTOR

9	8	■	/	/	4	3	2	/
/	/	16		14	13	12	11	10
/	27	/	25	24	/	■	22	/	20

31 PIN CONNECTOR

9	8	7	6	/	4	3	■	2	1
21	20	19	18	17	16	15	14	13	12	11	10
31	30	29	■	28	27		26	25	/	23	22

WIRE SIDE OF HARNESS TERMINALS

05533_ADIA_G008

Pin Connector Graphic

2000-01 3.0L V6 DOHC VIN F (A/T-ECT) E4 31P Connector

PCM Pin #	Wire Color	Circuit Description (31 Pin)	Value at Hot Idle
1	YL	Injector 3 Control	1.6-2-9 ms
2	WT	Injector 4 Control	1.6-2-9 ms
3	GN	Injector 5 Control	1.6-2-9 ms
4	GN	Injector 6 Control	1.6-2.9 ms
5	RD	VVT Solenoid Control (OC1-)	12v or 0v
6	RD/BK	VVT Solenoid Control (OC1+)	12v or 0v
7	PK	A/T-ECT Solenoid (S1)	In 3rd or OD: 1v
8-9	---	Not Used	---
10	BK/WT	CMP Sensor Signal (RH+)	AC pulse signals
11	GY	Igniter Transistor 1 Control	6°, at 55 mph: 8° dwell
12	BK/RD	Igniter Transistor 2 Control	6°, at 55 mph: 8° dwell
13	LG/BK	Igniter Transistor 3 Control	6°, at 55 mph: 8° dwell
14	BL/YL	Igniter Transistor 4 Control	6°, at 55 mph: 8° dwell
15	BL	Igniter Transistor 5 Control	6°, at 55 mph: 8° dwell
16	LG	Igniter Transistor 6 Control	6°, at 55 mph: 8° dwell
17	RD/YL	ACIS 1 Control (VSV)	12v or 0v
18	RD/WT	VVT Solenoid Control (OC2-)	12v or 0v
19	YL	A/T-ECT Solenoid (SLN-)	Pulse Signals
20	PK/BL	A/T-ECT Solenoid (SLN+)	Pulse Signals
21	WT/BK	Power Ground (E01)	<0.1v
22	BK/WT	CMP Sensor Signal (LH+)	AC pulse signals
23	GN	O/D Clutch Speed Sensor (-)	AC pulse signals
24	PK/BL	O/D Clutch Speed Sensor (+)	AC pulse signals
25	BK/YL	Igniter Signal (IGF)	Digital Signal: 0-5-0v
26	GN/BK	IAC Signal (RSO)	Pulse Signals
27	WT	Knock Sensor 1 Signal	0.075v AC
28	WT	Knock Sensor 2 Signal	0.075v AC
29	RD/BL	VVT Solenoid Control (OC2+)	12v or 0v
30	WT/BK	Power Ground (E03)	<0.1v
31	WT/BK	Power Ground (E04)	<0.1v

E4 31-PIN CONNECTOR

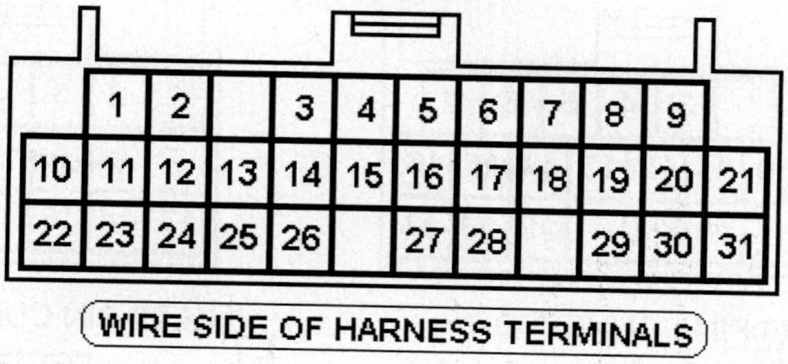

	1	2		3	4	5	6	7	8	9	
10	11	12	13	14	15	16	17	18	19	20	21
22	23	24	25	26		27	28		29	30	31

WIRE SIDE OF HARNESS TERMINALS

05533_ADIA_G009

Pin Connector Graphic

2000-01 3.0L V6 DOHC VIN F (A/T-ECT) E5 24P Connector

PCM Pin #	Wire Color	Circuit Description (24 Pin)	Value at Hot Idle
1	WT/BK	Power Ground (E04)	<0.1v
2	BL/RD	Sensor VREF (VC)	4.9-5.1v
3	BK/RD	AFS-11 (B1 S1) Heater	1v (Heater On)
4	BK/RD	AFS-21 (B2 S1) Heater	1v (Heater On)
5	BL	Injector 1 Control	1.6-2-9 ms
6	RD	Injector 2 Control	1.6-2-9 ms
7	BK/RD	EVAP Purge Solenoid (VSV)	12v or 0v
8	WT/BK	Power Ground (E05)	<0.1v
9	BK/RD	PSP Switch Signal	Straight: 12v, Turned: 0v
10	RD	MAF Sensor Signal	1-1.1v
11	BK/RD	AFS-11 (B1 S1) Signal.(+)	3.0-3.6v
12	BK/WT	AFS-21 (B2 S1) Signal (+)	3.0-3.6v
13	---	Not Used	---
14	GN/YL	ECT Sensor Signal	At 180°F: 0.51v
15	WT/RD	VSV ACIS 2 Control	12v or 0v
16	BK/WT	CKP Sensor Signal (NE+)	310-330 Hz
17	BR	Shield Ground (E1)	<0.050v
18	WT	Sensor Ground	<0.050v
19	RD/BK	MAF Sensor Ground (E2G)	<0.050v
20	BR	AFS-11 (B1 S1) Signal (AF-)	3.0-3.6v
21	LB	AFS-21 (B2 S1) Signal (-)	3.0-3.6v
22	BL/BK	IAT Sensor Signal	0.5-3.4v
23	LG	TP Sensor Signal	0.53-1.27v
24	BL	CKP/CMP Sensor Ground (-)	<0.050v

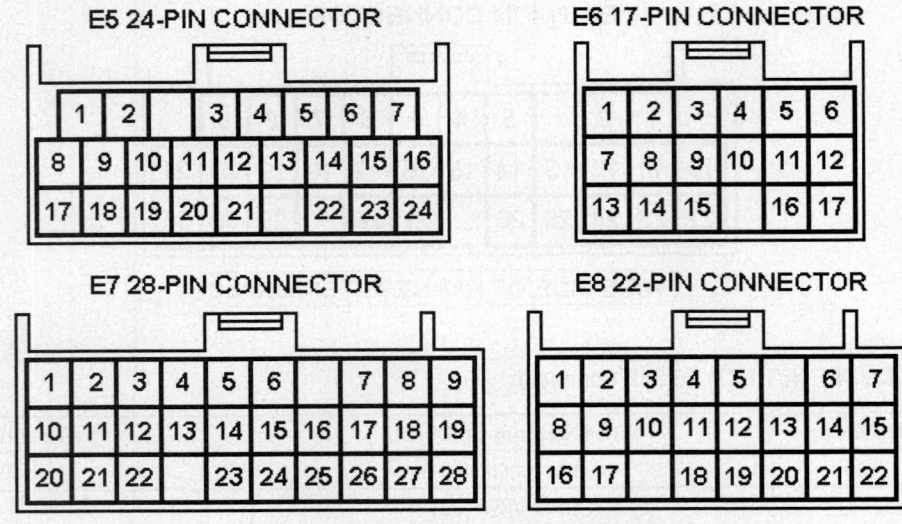

E5 24-PIN CONNECTOR

E6 17-PIN CONNECTOR

E7 28-PIN CONNECTOR

E8 22-PIN CONNECTOR

WIRE SIDE OF HARNESS TERMINALS

05533_ADIA_G010

Pin Connector Graphic

2000-01 3.0L V6 DOHC VIN F (A/T-ECT) E6 17P Connector

PCM Pin #	Wire Color	Circuit Description (17 Pin)	Value at Hot Idle
1	GN/YL	A/T-ECT Solenoid (S2)	1st or OD: 1v
2-7	---	Not Used	---
8	RD/BK	A/T Select Switch Reverse	In 'R': 12v, Others: 0v
9-11	---	Not Used	---
12	GN/OR	Overdrive Main Switch	Switch Off: 12v, On: 1v
13	YL	A/T Select Switch Low	In Low: 12v, Others: 0v
14	GN/YL	A/T Select Switch 2nd	In 2nd: 12v, Others: 0v
15	GN/WT	A/T-ECT Solenoid (SL+)	In Lockup: 12-14v

2000-01 3.0L V6 DOHC VIN F (A/T-ECT) E7 28P Connector

PCM Pin #	Wire Color	Circuit Description (28 Pin)	Value at Hot Idle
1-2	---	Not Used	---
3	PK/BK	EVAP Vapor Pressure (VSV)	12v or 0v
4	GN	Tail Light Switch Signal	Switch On: 12v, Off: 0v
5	BL	Data Link Connector (TE1)	12-14v
6	LG/RD	A/C Magnetic Clutch (ACMG)	Clutch Off: 0v, On: 12v
7	YL/RD	A/C Amplifier Signal (AC1)	Clutch On: 1.5v, Off: 12v
8	BK	HO2S-12 (B1 S2) Signal	0.1-1.1v
9	LG	HO2S-12 (B1 S2) Heater	1v (Heater On)
10-12	---	Not Used	---
13	BK	Mirror Heater Switch Signal	Switch Off: 1.5v, On: 12v
14	YL/GN	A/C Amplifier Signal (THWO)	Pulse Signals
15	---	Not Used	---
16	YL/BK	ABS/Traction NEO Signal	Pulse Signals
17	---	Not Used	---
18	BL/YL	TXCT Ignition Signal	Inserting key: pulses

2000-01 3.0L V6 DOHC VIN F (A/T-ECT) E7 28P Connector, *continued*

19	---	Not Used	---
20	BK/WT	Neutral Start Switch (NSW)	In P/N: 9-11v (cranking)
21	---	Not Used	---
22	VT/WT	Speedometer Indicator	At 55 mph: 48 Hz
23	RD/YL	Unlock Warning Switch	Key In: 1.5v, Out: 4.5v
24	GY	Cruise Control Signal (OD1)	At Cruise in OD: 12v
25	YL	Cruise Control Signal (IDLO)	1.5v, off-idle: 12v
26	GN	Cooling Fan 1 Relay Control	Relay Off: 12v, On: 1v
27	BK	Tachometer Signal (TACO)	Pulse Signals
28	PK/GN	Ignition Switch Code	Inserting key: pulses

2000-01 3.0L V6 DOHC VIN F (A/T-ECT) E8 22P Connector

PCM Pin #	Wire Color	Circuit Description (22 Pin)	Value at Hot Idle
1	BK/RD	Direct Battery	12-14v
2	BK/OR	Ignition Switch Power	12-14v
3	GN/BK	Circuit Opening Relay (FC)	0-0.3v, off-idle: 12-14v
4	WT	SIL Signal (Scan Tool)	Digital Signal
5	---	Not Used	---
6	BK/YL	MIL (lamp) Control	MIL Off: 12v, On: 1v
7	BK/RD	Starter Signal (STA)	Cranking: 9-11v
8	BL/OR	EFI Main Relay Power	12-14v
9	GN/OR	Overdrive Lamp Control	At Cruise in OD: 1v
10	GN	EVAP Canister Closed Valve	12v or 0v
11-12	---	Not Used	---
13	BR	ABS/Traction: TRC (+) Signal	DC pulse signals
14	PK	ABS/Traction: ENG (+) Signal	DC pulse signals
15	GN/WT	Brake Switch Signal	Brake Off: 0v, On: 12v
16	BK/WT	EFI Main Relay Power	12-14v
17	PK	EVAP Vapor Pressure Sensor	2.9-3.1v (with hose off)
18-19	---	Not Used	---
20	GN	ABS/Traction: TRC (-) Signal	DC pulse signals
21	GY	ABS/Traction: ENG (-) Signal	DC pulse signals
22	VT/WT	Security Indicator Light	Inserting key: pulses

2002-06 3.0L V6 DOHC VIN F (A/T-ECT) E4 31P Connector

PCM Pin #	Wire Color	Circuit Description (31 Pin)	Value at Hot Idle
1	YL	Injector 3 Control	1.6-2-9 ms
2	WT	Injector 4 Control	1.6-2-9 ms
3	BL/RD	Injector 5 Control	1.6-2-9 ms
4	GN	Injector 6 Control	1.6-2.9 ms
5	GR	VVT Solenoid Control (OC1-)	12v or 0v
6	RD/BK	VVT Solenoid Control (OC1+)	12v or 0v
7	PK	A/T-ECT Solenoid (S1)	In 3rd or OD: 1v
8-9	---	Not Used	---
10	BK/WT	CMP Sensor Signal (RH+)	AC pulse signals
11	GY/RD	Igniter Transistor 1 Control	6°, at 55 mph: 8° dwell
12	YL/GY	Igniter Transistor 2 Control	6°, at 55 mph: 8° dwell
13	GY/YL	Igniter Transistor 3 Control	6°, at 55 mph: 8° dwell
14	BL/YL	Igniter Transistor 4 Control	6°, at 55 mph: 8° dwell
15	BL	Igniter Transistor 5 Control	6°, at 55 mph: 8° dwell
16	BK	Igniter Transistor 6 Control	6°, at 55 mph: 8° dwell
17	RD/YL	ACIS 1 Control (VSV)	12v or 0v
18	RD/WT	VVT Solenoid Control (OC2-)	12v or 0v
19	YL	A/T-ECT Solenoid (SLN-)	Pulse Signals
20	PK/BL	A/T-ECT Solenoid (SLN+)	Pulse Signals
21	WT/BK	Power Ground (E01)	<0.1v
22	BK/WT	CMP Sensor Signal (LH+)	AC pulse signals
23	GN	O/D Clutch Speed Sensor (-)	AC pulse signals
24	PK/BL	O/D Clutch Speed Sensor (+)	AC pulse signals
25	BK/YL	Igniter Signal (IGF)	Digital Signal: 0-5-0v
26	GN/BK	IAC Signal (RSO)	Pulse Signals
27	WT	Knock Sensor 1 Signal	0.075v AC
28	WT	Knock Sensor 2 Signal	0.075v AC
29	RD/BL	VVT Solenoid Control (OC2+)	12v or 0v
30	WT/BK	Power Ground (E03)	<0.1v
31	WT/BK	Power Ground (E02)	<0.1v

E4 31-PIN CONNECTOR

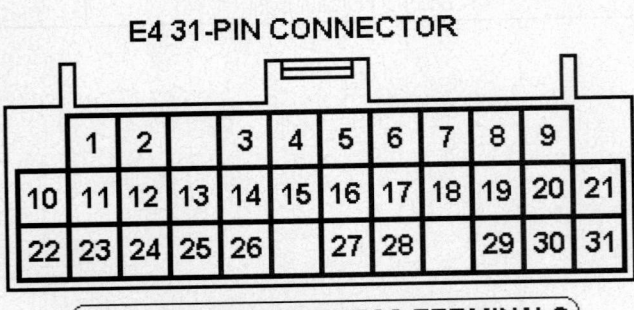

WIRE SIDE OF HARNESS TERMINALS

05533_ADIA_G009

Pin Connector Graphic

2002-06 3.0L V6 DOHC VIN F (A/T-ECT) E5 24P Connector

PCM Pin #	Wire Color	Circuit Description (24 Pin)	Value at Hot Idle
1	WT/BK	Power Ground (E04)	<0.1v
2	BL/RD	Sensor VREF (VC)	4.9-5.1v
3	GN	AFS-11 (B1 S1) Heater	1v (Heater On)
4	BK/RD	AFS-21 (B2 S1) Heater	1v (Heater On)
5	BL	Injector 1 Control	1.6-2-9 ms
6	RD	Injector 2 Control	1.6-2-9 ms
7	BK	EVAP Purge Solenoid (VSV)	12v or 0v
8	WT/BK	Power Ground (E05)	<0.1v
9	GN	PSP Switch Signal	Straight: 12v, Turned: 0v
10	RD	MAF Sensor Signal	1-1.1v
11	GN	AFS-11 (B1 S1) Signal (+)	3.0-3.6v
12	BK/WT	AFS-21 (B2 S1) Signal (+)	3.0-3.6v
13	---	Not Used	---
14	GN/YL	ECT Sensor Signal	At 180°F: 0.51v
15	WT/RD	VSV ACIS 2 Control	12v or 0v
16	BK/WT	CKP Sensor Signal (NE+)	310-330 Hz
17	BR	Shield Ground (E1)	<0.050v
18	WT	Sensor Ground	<0.050v
19	RD/BK	MAF Sensor Ground (E2)	<0.050v
20	RD	AFS-11 (B1 S1) Signal (AF-)	3.0-3.6v
21	BL	AFS-21 (B2 S1) Signal (-)	3.0-3.6v
22	BL/BK	IAT Sensor Signal	0.5-3.4v
23	LG	TP Sensor Signal	0.53-1.27v
24	BL	CKP/CMP Sensor Ground (-)	<0.050v

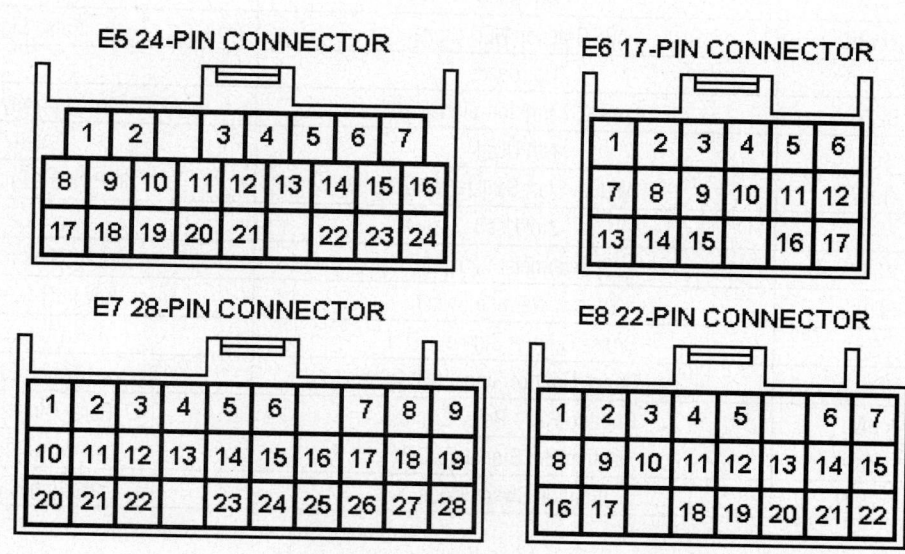

E5 24-PIN CONNECTOR

E6 17-PIN CONNECTOR

E7 28-PIN CONNECTOR

E8 22-PIN CONNECTOR

WIRE SIDE OF HARNESS TERMINALS

Pin Connector Graphic

05533_ADIA_G010

2002-06 3.0L V6 DOHC VIN F (A/T-ECT) E6 17P Connector

PCM Pin #	Wire Color	Circuit Description (17 Pin)	Value at Hot Idle
1	GN/YL	A/T-ECT Solenoid (S2)	1st or OD: 1v
2-7	---	Not Used	---
8	RD/BK	A/T Select Switch Reverse	In 'R': 12v, Others: 0v
9-10	---	Not Used	---
11	WT/BL	Active Control Engine Mount (VSV)	12v or 0v
12	GN/OR	Overdrive Main Switch	Switch Off: 12v, On: 1v
13	YL	A/T Select Switch Low	In Low: 12v, Others: 0v
14	GN/YL	A/T Select Switch 2nd	In 2nd: 12v, Others: 0v
15	GN/WT	A/T-ECT Solenoid (SL+)	In Lockup: 12-14v

2002-06 3.0L V6 DOHC VIN F (A/T-ECT) E7 28P Connector

PCM Pin #	Wire Color	Circuit Description (28 Pin)	Value at Hot Idle
1	BR	Power Ground (EOM)	<0.1v
2	---	Not Used	---
3	PK/BK	EVAP Vapor Pressure (VSV)	12v or 0v
4	GN	Tail Light Switch Signal	Switch On: 12v, Off: 0v
5	LG/RD	Data Link Connector (TE1)	12-14v
6	LG/BK	A/C Magnetic Clutch (ACMG)	Clutch Off: 0v, On: 12v
7	YL/RD	A/C Amplifier Signal (AC1)	Clutch On: 1.5v, Off: 12v
8	BK	HO2S-12 (B1 S2) Signal	0.1-1.1v
9	LG	HO2S-12 (B1 S2) Heater	1v (Heater On)
10-12	---	Not Used	---
13	BK	Mirror Heater Switch Signal	Switch Off: 1.5v, On: 12v
14	YL/GN	A/C Amplifier Signal (THWO)	Pulse Signals
15	---	Not Used	---
16	YL/BK	ABS/Traction NEO Signal	Pulse Signals
17	---	Not Used	---
18	BL/YL	TXCT Ignition Signal	Inserting key: pulses
19	---	Not Used	---
20	BK/WT	Neutral Start Switch (NSW)	In P/N: 9-11v (cranking)
21	---	Not Used	---
22	VT/WT	Speedometer Indicator	At 55 mph: 48 Hz
23	RD/YL	Unlock Warning Switch	Key In: 1.5v, Out: 4.5v
24	GY	Cruise Control Signal (OD1)	At Cruise in OD: 12v
25	YL	Cruise Control Signal (IDLO)	1.5v, off-idle: 12v
26	GN	Cooling Fan 1 Relay Control	Relay Off: 12v, On: 1v
27	BK	Tachometer Signal (TACO)	Pulse Signals
28	PK/GN	Ignition Switch Code	Inserting key: pulses

2002-06 3.0L V6 DOHC VIN F (A/T-ECT) E8 22P Connector

PCM Pin #	Wire Color	Circuit Description (22 Pin)	Value at Hot Idle
1	BK/RD	Direct Battery	12-14v
2	BK/OR	Ignition Switch Power	12-14v
3	GN/BK	Circuit Opening Relay (FC)	0-0.3v, off-idle: 12-14v
4	WT	SIL Signal (Scan Tool)	Digital Signal
6	BK/YL	MIL (lamp) Control	MIL Off: 12v, On: 1v
7	BK/RD	Starter Signal (STA)	Cranking: 9-11v
8	BL/OR	EFI Main Relay Power	12-14v
9	GN/OR	Overdrive Lamp Control	At Cruise in OD: 1v
10	GN	EVAP Canister Closed Valve	12v or 0v
13	BR	ABS/Traction: TRC (+) Signal	DC pulse signals
14	VT	ABS/Traction: ENG (+) Signal	DC pulse signals
15	GN/WT	Brake Switch Signal	Brake Off: 0v, On: 12v
16	BK/WT	EFI Main Relay Power	12-14v
17	PK	EVAP Vapor Pressure Sensor	2.9-3.1v (with hose off)
20	GN/WT	ABS/Traction: TRC (-) Signal	DC pulse signals
21	GY	ABS/Traction: ENG (-) Signal	DC pulse signals
22	VT/WT	Security Indicator Light	Inserting key: pulses

CAMRY PIN CHARTS

1990-91 2.0L (A/T) VIN S 10 Pin Connector

PCM Pin #	Wire Color	Circuit Description (10 Pin)	Value at Hot Idle
1	BK/WT	A/T Select Switch Neutral	In 'N': 12v, Others: 0v
2	RD/WT	Check Connector	12-14v
3	BK	A/T: Starter Switch Signal	9-11v (cranking)
3	BK/WT	M/T: Starter Switch Signal	9-11v (cranking)
4	WT	Injector Pair 1 & 3 Control	2.0-3.3 ms
5	WT/BR	Power Ground	<0.1v
7	BR	Sensor Ground	<0.050v
8	WT	Igniter Signal (IGT)	Digital Signal: 0-5-0v
9	YL	Injector Pair 2 & 4 Control	2.0-3.3 ms
10	WT/BK	Power Ground	<0.1v

1990-91 2.0L (A/T) VIN S 18 Pin Connector

PCM Pin #	Wire Color	Circuit Description (18 Pin)	Value at Hot Idle
1	WT	Distributor Signal (NE+)	AC pulse signals
3	RD	Distributor Signal (G+)	AC pulse signals
4	BK	Distributor Signal (G-)	<0.050v
5	WT/RD	Igniter Signal (IGF)	Digital Signal: 0-5-0v
6	BL	Closed Throttle Switch	<0.1v
7	YL/GN	Check Connector	12-14v
8	GN/RD	MIL (lamp) Control	MIL Off: 12v, On: 1v
9	YL	ISC Motor 1	Pulse Signals
10	GN	ECT Sensor Signal	At 180°F: 0.51v
11	BK/RD	TP Sensor Signal (VTA)	0.3-0.8v
12 (Cal)	BL/BK	EGR Gas Temp. Sensor	3.5-4.0v
13	WT	Main O2S Signal	0.1-1.1v
14	BR	Sensor Ground	<0.050v
15	BK/WT	A/C Magnetic Clutch (ACMG)	Clutch Off: 0v, On: 12v
16	BL/BK	A/C Amplifier Signal (ACT)	Clutch On: 12v, Off: 1.5v
17 (Cal)	RD/BL	Sub O2S Signal	0.1-1.1v
18	GN/BK	ISC Motor 2	Pulse Signals

1990-91 2.0L (A/T) VIN S 14-Pin Connector

PCM Pin #	Wire Color	Circuit Description (14-Pin)	Value at Hot Idle
1	WT/RD	EFI Main Relay Power	12-14v
2	WT/BL	Battery Direct	12-14v
3	YL/RD	IAT Sensor Signal	At 100°F: 2.60v
4	YL/BL	Airflow Meter Signal	2.5v
5	BL/RD	Sensor VREF	4.9-5.1v
6	BL/BK	A/T Signal (L1)	5v, at WOT: 0v
7	YL/RD	A/T Signal (L3)	0v, at WOT: 5v
8	WT/RD	EFI Main Relay Power	12-14v
9	GN	Defogger/Light Idle Up Signal	Switch On: 12v, Off: 0v

1990-91 2.0L (A/T) VIN S 14-Pin Connector, *continued*

10	PK/YL	Vehicle Speed Sensor	At 55 mph: 48 Hz
11	GN/WT	Stop Light Switch Signal	Brake Off: 0v, On: 12v
12	WT/BK	Sensor Ground	<0.050v
13	YL/GN	A/T Signal (L2)	5v, at WOT: 5v
14	YL/BK	A/T Signal (IDL)	0v, at WOT: 12v

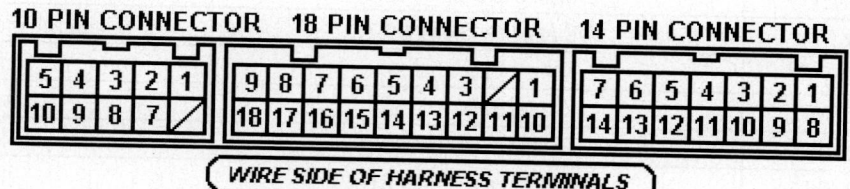

Pin Connector Graphic

05533_ADIA_G011

1992-93 2.2L VIN S (A/T-ECT) 16 Pin Connector

PCM Pin #	Wire Color	Circuit Description (16 Pin)	Value at Hot Idle
1	RD	Sensor VREF	4.9-5.1v
2	BK/YL	MAP Sensor Signal	1-1.5v
3	BL/BK	IAT Sensor Signal	At 100°F: 2.60v
4	LG	ECT Sensor Signal	At 180°F: 0.51v
5 (Cal)	RD/BL	Sub O2S Signal	0.1-1.1v
6	WT	Main O2S Signal	0.1-1.1v
7	BR/BK	Check Connector	12-14v
8	RD/WT	Check Connector	12-14v
9	BR	Sensor Ground	<0.050v
10 (Cal)	GY	EGR Gas Temp. Sensor	3.5-4.0v
11	BK	TP Sensor Signal	0.3-0.8v
12	BL	Closed Throttle Switch	1v, at off-idle: 12v
13	WT	Knock Sensor Signal	<0.075v AC
14	GN/WT	Check Connector	12-14v
15	GY	Check Connector	12-14v
16	BR	Sensor Ground	<0.050v

1992-93 2.2L VIN S (A/T-ECT) 22 Pin Connector

PCM Pin #	Wire Color	Circuit Description (22 Pin)	Value at Hot Idle
1	WT/BL	Battery Direct	12-14v
2	BK/RD	Defogger/Light Idle Up Signal	Switch On: 12v, Off: 0v
4	GN/WT	Stop Light Switch Signal	Brake Off: 0v, On: 12v
5	GN/RD	MIL (lamp) Control	MIL Off: 12v, On: 1v
7	GY/BL	Main Overdrive Switch	Switch Off: 12-14v
8	BK/YL	A/C Amplifier Signal (AC1)	Clutch On: 1.5v, Off: 12v
9	PK/YL	Vehicle Speed Sensor	At 55 mph: 48 Hz
11	BK/WT	Starter Switch Signal	9-11v (cranking)
12	BK/OR	EFI Main Relay Power	12-14v
13	BK/OR	EFI Main Relay Power B1+	12-14v
14	GN/RD	Circuit Opening Relay (FC)	0-3v
20	PK	Cruise Control ECU	At Cruise in OD: 12v
21	GN/BK	A/C Amplifier Signal (ACT)	Clutch On: 12v, Off: 1.5v
22	BK/WT	A/T Neutral Start Switch	In P/N: 9-11v (cranking)

Standard Colors and Abbreviations

Abbreviation	Color	Abbreviation	Color	Abbreviation	Color
BK	Black	GY	Gray	RD	Red
BL	Blue	GN	Green	TN	Tan
BR	Brown	LG	Light Green	VT	Violet
DB	Dark Blue	OR	Orange	WT	White
DG	DK Green	PK	Pink	YL	Yellow

1992-93 2.2L VIN S (A/T-ECT) 26 Pin Connector

PCM Pin #	Wire Color	Circuit Description (26 Pin)	Value at Hot Idle
1	BL/YL	A/T-ECT Solenoid (SL)	In Lockup: 12-14v
2	PK	A/T-ECT Solenoid (S1)	In 1st: 12-14v
3	WT/RD	Igniter Signal (IGF)	Digital Signal: 0-5-0v
4	RD	Distributor Signal (NE+)	AC pulse signals
5	BL	Distributor Signal (NE-)	<0.050v
6	OR	A/T Select Switch 2nd	In 2nd: 12v, Others: 0v
7	LG	A/C Idle Up Solenoid	A/C Off: 12v, On: 1v
9	GN/RD	ISC Signal (ISCC)	Pulse Signals
10	GN/YL	ISC Signal (ISCO)	Pulse Signals
11	YL	Injector Pair 2 & 4 Control	2.0-3.3 ms
12	WT	Injector Pair 1 & 3 Control	2.0-3.3 ms
13	WT/BK	Power Ground	<0.1v
14	BR	Sensor Ground	<0.050v
15	PK/BL	A/T-ECT Solenoid (S2)	In 2nd: 12v, Others: 0v
16	GN/BK	A/T-ECT Speed Sensor	Moving: varies 0-5v
17	BK	Distributor Signal (G-)	<0.050v
18	YL	Distributor Signal (G+)	AC pulse signals

1992-93 2.2L VIN S (A/T-ECT) 26 Pin Connector, *continued*

19	YL/BL	A/T Select Switch Low	In Low: 12v, Others: 0v
20	WT	Igniter Signal (IGT)	Digital Signal: 0-5-0v
22	BL/RD	A/T Pattern Selector Switch	Norm: 0v, PWR: 12v
23	GN	EGR Solenoid Control (VSV)	12v or 0v
26	WT/BK	Power Ground	<0.1v

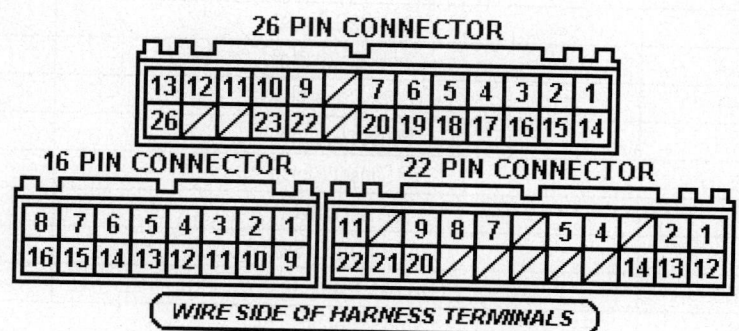

05533_ADIA_G012

Pin Connector Graphic

Standard Colors and Abbreviations

Abbreviation	Color	Abbreviation	Color	Abbreviation	Color
BK	Black	GY	Gray	RD	Red
BL	Blue	GN	Green	TN	Tan
BR	Brown	LG	Light Green	VT	Violet
DB	Dark Blue	OR	Orange	WT	White
DG	DK Green	PK	Pink	YL	Yellow

1992-93 2.2L I4 MFI VIN S (M/T) 12 Pin Connector

PCM Pin #	Wire Color	Circuit Description (12 Pin)	Value at Hot Idle
1	BK/OR	EFI Main Relay Power B1+	12-14v
2	WT/BL	Battery Direct	12-14v
3	BK/YL	A/C Amplifier Signal (AC1)	Clutch On: 1.5v, Off: 12v
4	GN/RD	Circuit Opening Relay (FC)	0-3v, off-idle: 12v
6	GN/BK	A/C Amplifier Signal (ACT)	Clutch On: 12v, Off: 1.5v
7	BK/OR	EFI Main Relay Power	12-14v
8	GN/RD	MIL (lamp) Control	MIL Off: 12v, On: 1v
11	PK/YL	Vehicle Speed Sensor	At 55 mph: 48 Hz
12	BK/RD	Defogger/Light Idle Up Signal	Switch On: 12v, Off: 0v

1992-93 2.2L I4 MFI VIN S (M/T) 16 Pin Connector

PCM Pin #	Wire Color	Circuit Description (16 Pin)	Value at Hot Idle
1 (Cal)	RD/BL	Sub O2S Signal	0.1-1.1v
2	BK/YL	MAP Sensor Signal	1-1.5v
3	BL/BK	IAT Sensor Signal	At 100°F: 2.60v
4	LG	ECT Sensor Signal	At 180°F: 0.51v
5	WT	Knock Sensor Signal	<0.075v AC
6	WT	Main O2S Signal	0.1-1.1v
7	GN/WT	Check Connector	12-14v
8	RD/WT	Check Connector	12-14v
9	BR	Sensor Ground	<0.050v
10	BK	TP Sensor Signal	0.3-0.8v
11	RD	Sensor VREF	4.9-5.1v
12	BL	Closed Throttle Switch	<0.1v
13 (Cal)	GN/RD	EGR Gas Temp. Sensor	3.5-4.0v
15	GN	Check Connector	12-14v
16	BR	Sensor Ground	<0.050v

1992-93 2.2L I4 MFI VIN S (M/T) 26 Pin Connector

PCM Pin #	Wire Color	Circuit Description (26 Pin)	Value at Hot Idle
1	BK/YL	A/C Idle Up Solenoid	A/C Off: 12v, On: 1v
2	BK/WT	Starter Switch Signal	9-11v (cranking)
3	WT/RD	Igniter Signal (IGF)	Varies: 1-3v
4	RD	Distributor Signal (NE+)	AC pulse signals
5	YL	Distributor Signal (G+)	AC pulse signals
9	GN/RD	ISC Signal (ISCC)	Pulse Signals
10	GN/YL	ISC Signal (ISCO)	Pulse Signals
12	WT	Injector Pair 1 & 3 Control	2.0-3.3 ms
13	WT/BK	Power Ground	<0.1v
17	BL	Distributor Signal (NE-)	<0.050v
18	BK	Distributor Signal (G-)	<0.050v
22	WT	Igniter Signal (IGT)	Digital Signal: 0-5-0v
23	GN	EGR Solenoid Control (VSV)	12v or 0v
24	BR	Sensor Ground	<0.050v
25	YL	Injector Pair 2 & 4 Control	2.0-3.3 ms
26	WT/BK	Power Ground	<0.1v

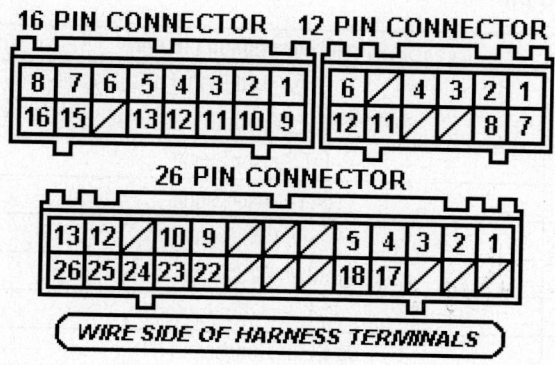

16 PIN CONNECTOR   **12 PIN CONNECTOR**

26 PIN CONNECTOR

WIRE SIDE OF HARNESS TERMINALS

05533_ADIA_G013

Pin Connector Graphic
1994-95 2.2L I4 Federal M/T VIN S 26 Pin

PCM Pin #	Wire Color	Circuit Description (26 Pin)	Value at Hot Idle
3	WT/RD	Igniter Signal (IGF)	Digital Signal: 0-5-0v
4 (Cal)	RD	Distributor Signal (NE+)	AC pulse signals
4 (Fed)	RD	Distributor Signal (NE+)	AC pulse signals
5 (Cal)	BL	Distributor Signal (G2+)	AC pulse signals
5 (Fed)	BL	Distributor Signal (NE-)	<0.050v
7	LG	A/C Idle Up Solenoid	12v or 0v
8	BK/RD	Fuel Pressure Up Solenoid	12v or 0v
9	GN/RD	IAC Signal (ISCC)	Pulse Signals
10	GN/YL	IAC Signal (ISCO)	Pulse Signals
11	YL	Injector 2 Control	2.0-3.3 ms
12	WT	Injector 1 Control	2.0 - 5.0ms
13	WT/BK	Power Ground	<0.1v
14	BR	Sensor Ground	<0.050v
17	BK	Distributor Signal (G-)	<0.050v
18 (Cal)	YL	Distributor Signal (G1+)	AC pulse signals
18 (Fed)	YL	Distributor Signal (G+)	AC pulse signals
20	WT	Igniter Signal (IGT)	Digital Signal: 0-5-0v
23	GN	EGR Solenoid Control (VSV)	12v or 0v
24	RD/BK	Injector 4 Control	2.0-3.3 ms
25	RD/BL	Injector 3 Control	2.0 - 5.0ms
26	WT/BK	Power Ground	<0.1v

1994-95 2.2L I4 Federal M/T VIN S 16 Pin

PCM Pin #	Wire Color	Circuit Description (16 Pin)	Value at Hot Idle
1	RD	Sensor VREF (VC)	4.9-5.1v
2	BK/YL	MAP Sensor Signal	1-1.5v
3	BL/BK	IAT Sensor Signal	At 100°F: 2.8v
4	LG	ECT Sensor Signal	At 180°F: 0.6v
5	RD/BL	Sub O2S Signal	0.1-1.1v
6	WT	Main O2S Signal	0.1-1.1v
7	BR/BK	Data Link Connector	12-14v
8	RD/WT	Data Link Connector (TN)	12-14v
9	BR	Sensor Ground	<0.050v
10	GY	EGR Gas Temp. Sensor	3.5-4.0v
11	BK	TP Sensor Signal	0.5v
12	BL	Closed Throttle Switch	<0.1v
13	WT	Knock Sensor	<0.075v AC
14	GN/WT	Data Link Connector (TN)	12-14v
15	GY	Data Link Connector (TN)	12-14v

Standard Colors and Abbreviations

Abbreviation	Color	Abbreviation	Color	Abbreviation	Color
BK	Black	GY	Gray	RD	Red
BL	Blue	GN	Green	TN	Tan
BR	Brown	LG	Light Green	VT	Violet
DB	Dark Blue	OR	Orange	WT	White
DG	DK Green	PK	Pink	YL	Yellow

1994-95 2.2L I4 Federal M/T VIN S 22 Pin

PCM Pin #	Wire Color	Circuit Description (22 Pin)	Value at Hot Idle
1	WT/BL	Battery Direct	12-14v
2	BK/RD	Defogger/Light Idle Up Signal	Switch On: 12v, Off: 0v
4	GN/WT	Stop Light Switch Signal	Brake Off: 0v, On: 12v
5	GN/RD	MIL (lamp) Control	MIL Off: 12v, On: 1v
9	PK/YL	Vehicle Speed Sensor	At 55 mph: 48 Hz
10	BK/YL	A/C Amplifier Signal (AC1)	Clutch On: 1.5v, Off: 12v
11	BK/WT	Starter Switch Signal	9-11v (cranking)
12	BK/OR	EFI Main Relay Power	12-14v
13	BK/OR	EFI Main Relay Power B1+	12-14v
14	GN/RD	Circuit Opening Relay (FC)	0-3v, off-idle: 12v
21	GN/BK	A/C Amplifier Signal (ACT)	Clutch On: 12v, Off: 1.5v
22	BK/WT	A/T Neutral Start Switch	In P/N: 9-11v (cranking)

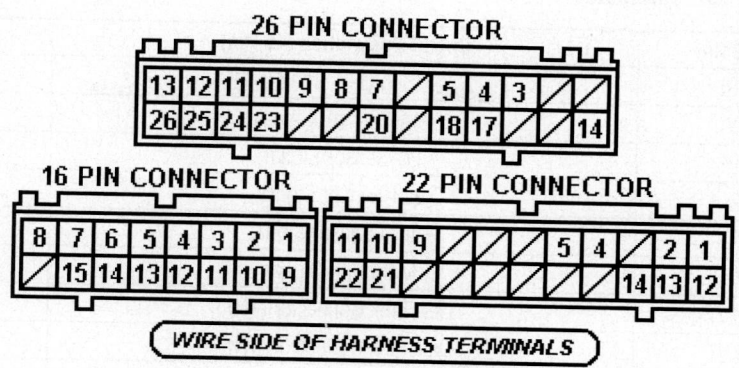

05533_ADIA_G014

Pin Connector Graphic
Standard Colors and Abbreviations

Abbreviation	Color	Abbreviation	Color	Abbreviation	Color
BK	Black	GY	Gray	RD	Red
BL	Blue	GN	Green	TN	Tan
BR	Brown	LG	Light Green	VT	Violet
DB	Dark Blue	OR	Orange	WT	White
DG	DK Green	PK	Pink	YL	Yellow

1994-95 2.2L Cal M/T VIN S 26P Connector

PCM Pin #	Wire Color	Circuit Description (26 Pin)	Value at Hot Idle
1	LG	A/C Idle Up Solenoid	A/C On: 1v
2	BK/WT	Starter Switch Signal	9-11v (cranking)
3	WT/RD	Igniter Signal (IGF)	Digital Signal: 0-5-0v
4	RD	Distributor Signal (NE+)	AC pulse signals
5	YL	Distributor Signal (G+)	AC pulse signals
9	GN/RD	IAC Signal (ISCC)	Pulse Signals
10	GN/YL	IAC Signal (ISCO)	Pulse Signals
12	WT	Injector Pair 1 & 3 Control	2.0 - 5.0ms
13	WT/BK	Power Ground	<0.1v
17	BL	Distributor Signal (NE-)	<0.050v
18	BK	Distributor Signal (G-)	<0.050v
22	WT	Igniter Signal (IGT)	Digital Signal: 0-5-0v
23	GN	EGR Solenoid Control (VSV)	12v or 0v
24	BR	Sensor Ground	<0.050v
25	YL	Injector Pair 2 & 4 Control	2.0-3.3 ms
26	WT/BK	Power Ground	<0.1v

1994-95 2.2L Cal M/T VIN S 16P Connector

PCM Pin #	Wire Color	Circuit Description (16 Pin)	Value at Hot Idle
1	RD/BL	Sub O2S Signal	0.1-1.1v
2	BK/YL	MAP Sensor Signal	0.9-1.1v
3	BL/BK	IAT Sensor Signal	At 100°F: 2.8v
4	LG	ECT Sensor Signal	At 180°F: 0.6v
5	WT	Knock Sensor Signal	<0.075v AC
6	WT	Main O2S Signal	0.1-1.1v
7	GN/WT	Data Link Connector (TN)	12-14v
8	RD/WT	Data Link Connector (TN)	12-14v
9	BR	Sensor Ground	<0.050v
10	BK	TP Sensor Signal	0.5v
11	RD	Sensor VREF (VC)	4.9-5.1v
12	BL	Closed Throttle Switch	<0.1v
13	GY	EGR Gas Temp. Sensor	3.5-4.0v
15	GY	Data Link Connector	12-14v
16	BR	Sensor Ground	<0.050v

1994-95 2.2L Cal M/T VIN S 12P Connector

PCM Pin #	Wire Color	Circuit Description (12 Pin)	Value at Hot Idle
2	WT/BL	Battery Direct	12-14v
3	BK/YL	A/C Amplifier Signal (AC1)	Clutch On: 1.5v, Off: 12v
4	GN/RD	Circuit Opening Relay (FC)	0-3v, off-idle: 12v
6	GN/BK	A/C Amplifier Signal (ACT)	Clutch On: 12v, Off: 1.5v
7	BK/OR	EFI Main Relay Power	12-14v
8 ('94)	BK/OR	EFI Main Relay Power	12-14v
9	GN/RD	MIL (lamp) Control	MIL Off: 12v, On: 1v
11	PK/YL	Vehicle Speed Sensor	At 55 mph: 48 Hz
12	BK/RD	Defogger/Light Idle Up Signal	Switch On: 12v, Off: 0v

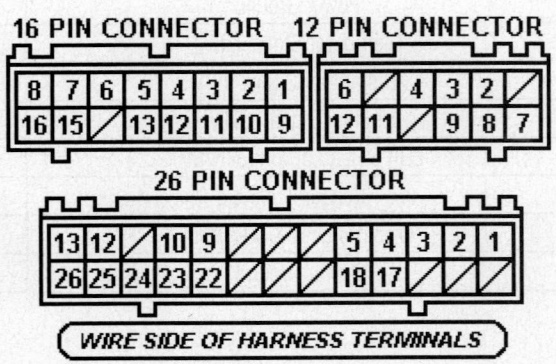

Pin Connector Graphic

05533_ADIA_G015

1996 2.2L I4 MFI VIN G (A/T) 26 Pin Connector

PCM Pin #	Wire Color	Circuit Description (26 Pin)	Value at Hot Idle
1	BL/YL	A/T-ECT Solenoid (SL)	In Lockup: 12-14v
2	PK	A/T-ECT Solenoid (S1)	In 3rd or OD: 1v
3	WT/RD	Igniter Signal (IGF)	Digital Signal: 0-5-0v
4 (TMC)	RD	CKP Sensor Signal (NE+)	AC pulse signals
4 (TMM)	WT/BL	CKP Sensor Signal (NE+)	AC pulse signals
5 (TMC)	BK/WT	Distributor Signal (G2+)	AC pulse signals
5 (TMM)	BL/GN	Distributor Signal (G2+)	AC pulse signals
9	GN/RD	IAC Signal (RSC)	Pulse Signals
10	GN/YL	IAC Signal (RSO)	Pulse Signals
11	YL	Injector 2 Control	2.0-3.3 ms
12	WT	Injector 1 Control	2.0-3.3 ms
13	WT/BK	Power Ground	<0.1v
14	BR	Sensor Ground	<0.050v
17 (TMC)	GN	CKP Sensor Signal (NE-)	<0.050v
17 (TMM)	OR/BL	CKP Sensor Signal (NE-)	<0.050v
20	WT	Igniter Signal (IGT)	Digital Signal: 0-5-0v
22	YL/BK	EVAP Purge Solenoid (VSV)	12v or 0v
23	GN	EGR Solenoid Control (VSV)	12v or 0v
24	RD/BK	Injector 4 Control	2.0-3.3 ms
25	RD/BL	Injector 3 Control	2.0-3.3 ms
26	WT/BK	Power Ground	<0.1v

1996 2.2L I4 MFI VIN G (A/T) 16 Pin Connector

PCM Pin #	Wire Color	Circuit Description (16 Pin)	Value at Hot Idle
1	RD	Sensor VREF (VC)	4.9-5.1v
2	BK/YL	MAP Sensor Signal	1-1.5v
3	BL/BK	IAT Sensor Signal	At 100°F: 2.8v
4	LG	ECT Sensor Signal	At 180°F: 0.6v
5	RD/BL	O2S-12 (B1 S2) Signal	0.1-1.1v
6	WT	O2S-11 (B1 S1) Signal	0.1-1.1v
7	BL/YL	EVAP Vapor Pressure Sensor	2.9-3.1v (with hose off)
8	BK/RD	EVAP Vapor Pressure (VSV)	12v or 0v
9	BR	Sensor Ground	<0.050v
11	BK	TP Sensor Signal	0.3-0.8v
13	WT	Knock Sensor Signal	<0.075v AC
15	GY	Data Link Connector	12-14v
16 (TMC)	BR	Power Ground	<0.1v
16 (TMM)	WT/BK	Power Ground	<0.1v

Standard Colors and Abbreviations

Abbreviation	Color	Abbreviation	Color	Abbreviation	Color
BK	Black	GY	Gray	RD	Red
BL	Blue	GN	Green	TN	Tan
BR	Brown	LG	Light Green	VT	Violet
DB	Dark Blue	OR	Orange	WT	White
DG	DK Green	PK	Pink	YL	Yellow

1996 2.2L I4 MFI VIN G (A/T) 22 Pin Connector

PCM Pin #	Wire Color	Circuit Description (22 Pin)	Value at Hot Idle
1	WT/BL	Battery Direct	12-14v
2	BK/RD	Defogger/Light Idle Up Signal	Switch On: 12v, Off: 0v
4	GN/WT	Stop Light Switch Signal	Brake Off: 0v, On: 12v
5	GN/RD	MIL (lamp) Control	MIL Off: 12v, On: 1v
6	BL/RD	A/T Pattern Selector Switch	Norm: 0v, PWR: 12v
7	GN/OR	Overdrive Main Switch	Switch Off: 12v
9	PK/YL	Vehicle Speed Sensor	At 55 mph: 48 Hz
10	BK/YL	A/C Amplifier Signal (AC1)	Clutch On: 1.5v, Off: 12v
11	BK/WT	Starter Switch Signal	9-11v (cranking)
12	BK/OR	EFI Main Relay Power	12-14v
14	GN/RD	Circuit Opening Relay (FC)	0-3v, off-idle: 12v
16	WT	Data Link Connector	12-14v
17	RD/BK	A/T Select Switch Reverse	In 'R': 12v, Others: 0v
18	OR	A/T Select Switch 2nd	In 2nd: 12v, Others: 0v
19	YL/BL	A/T Select Switch Low	In Low: 12v, Others: 0v
21	GN/BK	A/C Amplifier Signal (ACT)	Clutch On: 12v, Off: 1.5v
22	BK/WT	A/T Neutral Start Switch	In P/N: 9-11v (cranking)

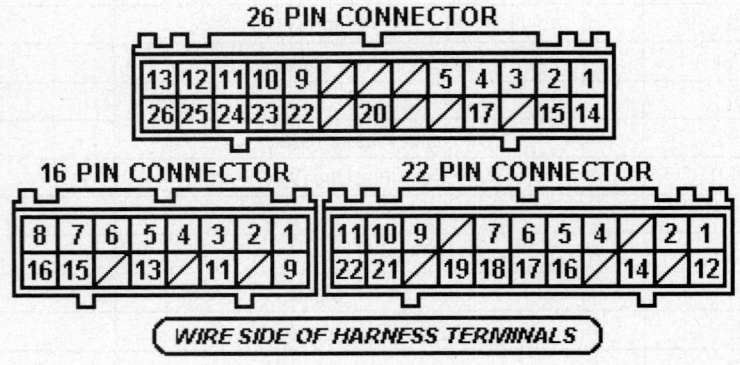

Pin Connector Graphic

05533_ADIA_G016

1997 2.2L I4 MFI VIN G (A/T) 26 Pin Connector

PCM Pin #	Wire Color	Circuit Description (26 Pin)	Value at Hot Idle
1	PK	A/T-ECT Solenoid (SL)	In Lockup: 12-14v
2 (Cal)	GN	HTAF Heater Control	1v (Heater On)
3	WT/RD	Igniter Signal (IGF)	Digital Signal: 0-5-0v
4	BK/RD	CKP Sensor Signal (NE+)	AC pulse signals
5	BK/WT	CMP Sensor Signal (G+)	AC pulse signals
6	BL/BK	A/T-ECT Solenoid (S2)	1st or OD: 1v
7	PK	A/T-ECT Solenoid (S1)	In 3rd or OD: 1v
8	BL/YL	HO2S-11 (B1 S1) Heater	1v (Heater On)
9	BK/OR	IAC Signal (ISCC)	Pulse Signals
10	WT	IAC Signal (ISCO)	Pulse Signals
11	RD	Injector 2 Control	2.0-3.3 ms
12	BL	Injector 1 Control	2.0-3.3 ms
13	BR	Power Ground	<0.1v
14	BR	Sensor Ground	<0.050v
15 (Cal)	BR	Power Ground	<0.1v
17	BL	CKP Sensor Signal (NE-)	<0.050v
18	WT/BL	A/C System (Lock In)	A/C On: 12v
19	YL/RD	Igniter Transistor 2 Control	Digital Signal: 0-5-0v
20	BK	Igniter Transistor 1 Control	Digital Signal: 0-5-0v
21	PK/BK	HO2S-12 (B1 S2) Heater	1v (Heater On)
22	PK/WT	EVAP Purge Solenoid (VSV)	12v or 0v
23	PK/BK	EGR Solenoid Control (VSV)	12v or 0v
24	WT	Injector 4 Control	2.0-3.3 ms
25	YL	Injector 3 Control	2.0-3.3 ms
26	BR	Power Ground	<0.1v

1997 2.2L I4 MFI VIN G (A/T) 16 Pin Connector

PCM Pin #	Wire Color	Circuit Description (16 Pin)	Value at Hot Idle
1	YL	Sensor VREF	4.9-5.1v
2	BK/YL	MAP Sensor Signal	1-1.5v
3	YL/BK	IAT Sensor Signal	At 100°F: 2.8v
4	GN/BK	ECT Sensor Signal	At 180°F: 0.6v
5	BK	HO2S-12 (B1 S2) Signal	0.1-1.1v
6	WT	HO2S-11 (B1 S1) Signal	0.1-1.1v
6 (Cal)	WT	AFS-11 (B1 S1) Signal (AF+)	3.0-3.6v
7	PK	EVAP Vapor Pressure Sensor	2.9-3.1v (with hose off)
8	PK	EVAP Vapor Pressure (VSV)	12v or 0v
9	BR	Sensor Ground	<0.050v
10	BL/WT	A/C Evaporator Temp. Signal	1.4-1.8v (temp. >59°F)
11	LG	TP Sensor Signal	0.3-0.8v
12	BK/BL	PSP Switch Signal	Straight: 12v, Turning: 0v
13	WT	Knock Sensor Signal	<0.075v AC
14 (Cal)	OR	AFS-11 (B1 S1) Signal (AF-)	Fixed at 3v
15	BL/WT	Data Link Connector	12-14v
16	BR	Power Ground	<0.1v

1997 2.2L I4 MFI VIN G (A/T) 22 Pin Connector

PCM Pin #	Wire Color	Circuit Description (22 Pin)	Value at Hot Idle
1	BK/YL	Battery Direct	12-14v
2	BK/RD	Defogger/Light Idle Up Signal	Switch On: 12v, Off: 0v
3	BL/RD	Cruise Control ECU	At Cruise in OD: 12v
4	GN/WT	Stop Light Switch Signal	Brake Off: 0v, On: 12v
5	GN/RD	MIL (lamp) Control	MIL Off: 12v, On: 1v
7	GN/OR	Overdrive Main Switch	Switch Off: 12v
8	BK/OR	Tachometer Signal (TACO)	Pulse Signals
9	PK/WT	Vehicle Speed Sensor	At 55 mph: 48 Hz
10	RD/BK	A/C Switch Signal	Clutch On: 1.5v, Off: 12v
11 (TMC)	GY	Starter Switch Signal	9-11v (cranking)
11 (TMM)	BK/WT	Starter Switch Signal	9-11v (cranking)
12	BK/YL	EFI Main Relay Power	12-14v
13	GN	A/C Dual Press Switch	Switch Off: 12v
14	GN/RD	Circuit Opening Relay (FC)	0-3v, off-idle: 12v
15	RD/WT	A/C Lock In Signal	A/C On: 12v
16	WT	Data Link Connector	No Scan Tool: 0v
17	RD/BK	A/T Select Switch Reverse	In 'R': 12v, Others: 0v
18 (TMC)	BL/WT	A/T Select Switch Second	In 2nd: 12v, Others: 0v
18 (TMM)	OR	A/T Select Switch Second	In 2nd: 12v, Others: 0v
19	YL	A/T Select Switch Low	In Low: 12v, Others: 0v
20	YL/BK	Cruise Control ECU	At Cruise in OD: 12v
21	BL/YL	A/C Magnetic Clutch (ACMG)	Clutch Off: 0v, On: 12v
22	BK/WT	A/T Neutral Start Switch	In P/N: 9-11v (cranking)

Note: TMC indicates a vehicle produced by Toyota Motor Corp. (Japan) and TMM indicates a vehicle produced by Toyota Motor Manufacturer (USA).

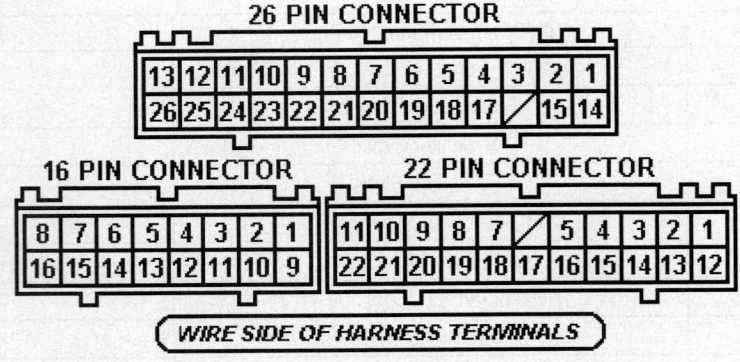

Pin Connector Graphic

05533_ADIA_G017

1998-99 2.2L I4 VIN G (A/T) 26 Pin Connector

PCM Pin #	Wire Color	Circuit Description (26 Pin)	Value at Hot Idle
1	PK	A/T-ECT Solenoid (SL)	In Lockup: 12-14v
2 (Cal)	GN	AFS-11 (B1 S1) Heater	1v (Heater On)
3	WT/RD	Igniter Signal (IGF)	Digital Signal: 0-5-0v
4	BK/RD	CKP Sensor Signal (NE+)	AC pulse signals
5	BK/WT	CMP Sensor Signal (G+)	AC pulse signals
6	BL/BK	A/T-ECT Solenoid (S2)	1st or OD: 1v
7	PK	A/T-ECT Solenoid (S1)	In 3rd or OD: 1v
8	BL/YL	HO2S-11 (B1 S1) Heater	1v (Heater On)
9	BK/OR	IAC Signal (ISCC)	Pulses
10	WT	IAC Signal (ISCO)	Pulses
11	RD	Injector 2 Control	2.0-3.3 ms
12	BL	Injector 1 Control	2.0-3.3 ms
13	BR	Power Ground	<0.1v
14	BR	Power Ground	<0.1v
15 (Cal)	BR	Power Ground	<0.1v
17	BL	CKP/CMP Sensor Ground (-)	<0.050v
18	WT/BL	A/C Lock In Signal	A/C On: 12v
19	YL/RD	Igniter Transistor 2 Control	Digital Signal: 0-5-0v
20	BK	Igniter Transistor 1 Control	Digital Signal: 0-5-0v
21	PK/BK	HO2S-12 (B1 S2) Heater	1v (Heater On)
22	PK/WT	EVAP Purge Solenoid (VSV)	12v or 0v
23	PK/BK	EGR Solenoid Control (VSV)	12v or 0v
24	WT	Injector 4 Control	2.0-3.3 ms
25	YL	Injector 3 Control	2.0-3.3 ms
26	BR	Power Ground	<0.1v

1998-99 2.2L I4 VIN G (A/T) 16 Pin Connector

PCM Pin #	Wire Color	Circuit Description (16 Pin)	Value at Hot Idle
1	YL	Sensor VREF (VC)	4.9-5.1v
2	BK/YL	MAP Sensor Signal	1-1.5v
3	YL/BK	IAT Sensor Signal	At 100°F: 2.8v
4	GN/BK	ECT Sensor Signal	At 180°F: 0.6v
5	BK	HO2S-12 (B1 S2) Signal	0.1-1.1v
6 (TMC)	BL	AFS-11 (B1 S1) Signal (AF+)	3.0-3.6v
6 (TMM)	WT	AFS-11 (B1 S1) Signal (AF+)	3.0-3.6v
6	WT	HO2S-11 (B1 S1) Signal	0.1-1.1v
7	PK	EVAP Vapor Pressure Sensor	2.9-3.1v (with hose off)
8	PK	EVAP Vapor Pressure (VSV)	12v or 0v
9	BR	Sensor Ground	<0.050v
10	BL/RD	A/C Evaporator Temp. Sensor	EVAP Temp. 59°F: 2.4v
11	LG	TP Sensor Signal	0.3-0.8v
12	BK/BL	PSP Switch Signal	Straight: 12v, Turning: 0v
13	WT	Knock Sensor Signal	<0.075v AC
14 (TMC)	BK/WT	AFS-11 (B1 S1) Signal (AF-)	Fixed at 3v
14 (TMM)	OR	AFS-11 (B1 S1) Signal (AF-)	Fixed at 3v
15	BL/WT	Data Link Connector (TE)	12-14v
16	BR	Power Ground	<0.1v

1998-99 2.2L I4 VIN G (A/T) 22 Pin Connector

PCM Pin #	Wire Color	Circuit Description (22 Pin)	Value at Hot Idle
1	BK/YL	Battery Direct	12-14v
2	BK/RD	Defogger/Light Idle Up Signal	Switch On: 12v, Off: 0v
3	BL/RD	Cruise Control Signal (IDLO)	1.5v, off-idle: 12v
4	GN/WT	Stop Light Switch Signal	Brake Off: 0v, On: 12v
5	GN/RD	MIL (lamp) Control	MIL Off: 12v, On: 1v
7	GN/OR	Overdrive "Off" Indicator	Light Off: 12v
8	BK/OR	Tachometer Signal (TACO)	Pulses
9	PK/WT	Vehicle Speed Sensor	At 55 mph: 48 Hz
10	RD/BK	A/C Switch Signal	Clutch On: 1.5v, Off: 12v
11 (TMC)	GY	Starter Switch Signal	9-11v (cranking)
11 (TMM)	BK/OR	Starter Switch Signal	9-11v (cranking)
12	BK/YL	EFI Main Relay Power	12-14v
13	GN	A/C Dual Press Switch	Switch Off: 12v
14	GN/RD	Circuit Opening Relay (FC)	0-3v
15	RD/WT	A/C Lock In Signal	A/C On: 12v
16	WT	SIL Signal (Scan Tool)	Digital Signal
17	RD/BK	A/T Select Switch Reverse	In 'R': 12v, Others: 0v
18 (TMC)	BL/WT	A/T Select Switch 2nd	In 2nd: 12v, Others: 0v
18 (TMM)	OR	A/T Select Switch 2nd	In 2nd: 12v, Others: 0v

1998-99 2.2L I4 VIN G (A/T) 22 Pin Connector, *continued*

19	YL	A/T Select Switch Low	In Low: 12v, Others: 0v
20	YL/BK	Cruise Control Signal (OD1)	At Cruise in OD: 12v
21	BL/YL	A/C Magnetic Clutch (ACMG)	Clutch Off: 0v, On: 12v
22	BK/WT	A/T Neutral Start Switch	In P/N: 9-11v (cranking)

Note: TMC indicates a vehicle produced by Toyota Motor Corp. (Japan) and TMM indicates a vehicle produced by Toyota Motor Manufacturer (USA).

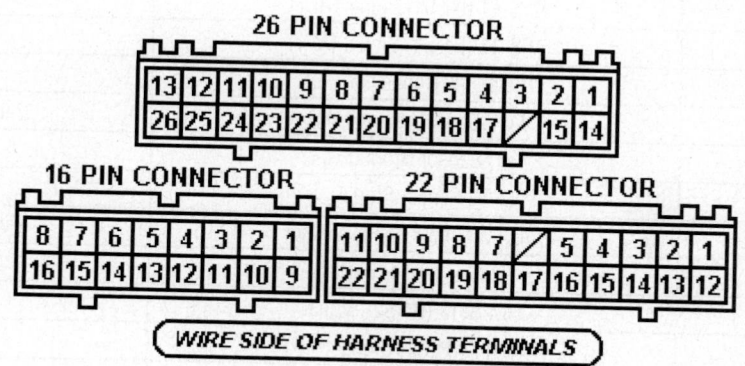

05533_ADIA_G018

Pin Connector Graphic

1998-99 2.2L I4 VIN G (A/T) 26 Pin Connector

PCM Pin #	Wire Color	Circuit Description (26 Pin)	Value at Hot Idle
1 (Fed)	BL/YL	HO2S-11 (B1 S1) Heater	1v (Heater On)
2 (Cal)	GN	AFS-11 (B1 S1) Heater	1v (Heater On)
3	PK/WT	EVAP Purge Solenoid (VSV)	12v or 0v
4	BK/BL	PSP Switch Signal	Straight: 12v, Turning: 0v
6	BK/OR	IAC Signal (ISCC)	Pulse Signals
7	WT	IAC Signal (ISCO)	Pulse Signals
8	PK	A/T-ECT Solenoid (S1)	In 3rd or OD: 1v
9	WT	Injector 4 Control	2.0-3.3 ms
10	YL	Injector 3 Control	2.0-3.3 ms
11	RD	Injector 2 Control	2.0-3.3 ms
12	BL	Injector 1 Control	2.0-3.3 ms
13	BR	Power Ground	<0.1v
14	PK/BK	HO2S-12 (B1 S2) Heater	1v (Heater On)
15 (Cal)	BR	Power Ground	<0.1v
17	WT/RD	Igniter Signal (IGF)	Digital Signal: 0-5-0v
19	WT/BL	A/C Lock In Signal	A/C On: 12v
20	PK	A/T-ECT Solenoid (SL)	In Lockup: 12-14v
21	BL/BK	A/T-ECT Solenoid (S2)	1st or OD: 1v
22	YL/RD	Igniter Transistor 2 Control	Digital Signal: 0-5-0v
23	BK	Igniter Transistor 1 Control	Digital Signal: 0-5-0v
24	BR	Sensor Ground	<0.050v
25	BR	Power Ground	<0.1v
26	BR	Power Ground	<0.1v

1998-99 2.2L I4 VIN G (A/T) 16 Pin Connector

PCM Pin #	Wire Color	Circuit Description (16 Pin)	Value at Hot Idle
1	YL	Sensor VREF	4.9-5.1v
2	BK/YL	MAP Sensor Signal	1-1.5v
3	YL/BK	IAT Sensor Signal	At 100°F: 2.8v
4	GN/BK	ECT Sensor Signal	At 180°F: 0.6v
5	WT	HO2S-11 (B1 S1) Signal	0.1-1.1v
6 (TMC)	BL	AFS-11 (B1 S1) Signal (AF+)	3.0-3.6v
6 (TMM)	WT	AFS-11 (B1 S1) Signal (AF+)	3.0-3.6v
7	BL/WT	Data Link Connector (TE)	12-14v
8	PK	EVAP Vapor Pressure Sensor	2.9-3.1v (with hose off)
9	BR	Sensor Ground	<0.050v
10	LG	TP Sensor Signal	0.3-0.8v
11	BL/RD	A/C Evaporator Temp. Sensor	EVAP Temp. 59°F: 2.4v
12	WT	Knock Sensor Signal	<0.075v AC
13	BK	HO2S-12 (B1 S2) Signal	0.1-1.1v
14 (TMC)	BK/WT	AFS-11 (B1 S1) Signal (AF-)	Fixed at 3v
14 (TMM)	OR	AFS-11 (B1 S1) Signal (AF-)	Fixed at 3v
15	PK/BK	EGR Solenoid Control (VSV)	1v (Heater On)
16	PK	EVAP Vapor Pressure (VSV)	12v or 0v

1998-99 2.2L I4 VIN G (A/T) 22 Pin Connector

PCM Pin #	Wire Color	Circuit Description (22 Pin)	Value at Hot Idle
1	BK/RD	Ignition Switch Power	12-14v
2	BK/YL	Battery Direct	12-14v
3	BL/RD	Cruise Control Signal (IDLO)	1.5v, off-idle: 12v
4	GN/RD	MIL (lamp) Control	MIL Off: 12v, On: 1v
5	GN/OR	Overdrive "Off" Indicator	LED Off: 12v, On: 1v
6	WT	SIL Signal (Scan Tool)	Digital Signal
7	BK/OR	Tachometer Signal (TACO)	Pulses
8	PK/WT	Vehicle Speed Sensor	At 55 mph: 48 Hz
9	GN/WT	Stop Light Switch Signal	Brake Off: 0v, On: 12v
10	RD/BK	A/C Switch Signal	Clutch On: 1.5v, Off: 12v
11 (TMC)	GY	Starter Switch Signal	9-11v (cranking)
11 (TMM)	BK/OR	Starter Switch Signal	9-11v (cranking)
12	BK/YL	EFI Main Relay Power	12-14v
13	BK/RD	Defogger/Light Idle Up Signal	Switch On: 12v, Off: 0v
14	GN/RD	Circuit Opening Relay (FC)	0-3v, off-idle: 12v
15	YL	A/T Select Switch Low	In Low: 12v, Others: 0v
16 (TMC)	BL/WT	A/T Select Switch 2nd	In 2nd: 12v, Others: 0v
16 (TMM)	OR	A/T Select Switch 2nd	In 2nd: 12v, Others: 0v
17	RD/BK	A/T Select Switch Reverse	In 'R': 12v, Others: 0v
18	YL/BK	Cruise Control Signal (OD1)	At Cruise in OD: 12v
19	GN	A/C Dual Pressure Switch	A/C Off: 12v, On: 1v
20	RD/WT	A/C Lock In Signal	A/C On: 12v
21	BL/YL	A/C Magnetic Clutch (ACMG)	Clutch Off: 0v, On: 12v
22	BK/WT	A/T Neutral Start Switch	In P/N: 9-11v (cranking)

Note: TMC indicates a vehicle produced by Toyota Motor Corp. (Japan) and TMM indicates a vehicle produced by Toyota Motor Manufacturer (USA).

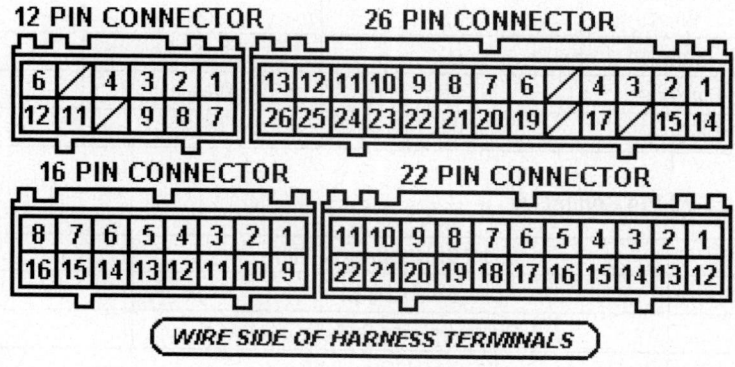

05533_ADIA_G019

Pin Connector Graphic

1998-99 2.2L I4 VIN G (A/T) 12 Pin Connector

PCM Pin #	Wire Color	Circuit Description (12 Pin)	Value at Hot Idle
1	RD/YL	Theft Deterrent ECU	Pulse Signals
2	BR	Sensor Ground	<0.050v
3	RD/BL	Transponder Amplifier Signal	Inserting key: pulses
4	BL/BK	Unlock Warning Switch	No Key: 4-5v
6	BL	CKP/CMP Sensor Ground (-)	<0.050v
7	BK/WT	EFI Main Relay Power	12-14v
8	GN/WT	Transponder Amplifier Code	Inserting key: pulses
9	BL/YL	Transponder Amplifier Signal	Inserting key: pulses
11	BK/WT	CMP Sensor Signal (G+)	AC pulse signals
12	BK/RD	CKP Sensor Signal (NE+)	AC pulse signals

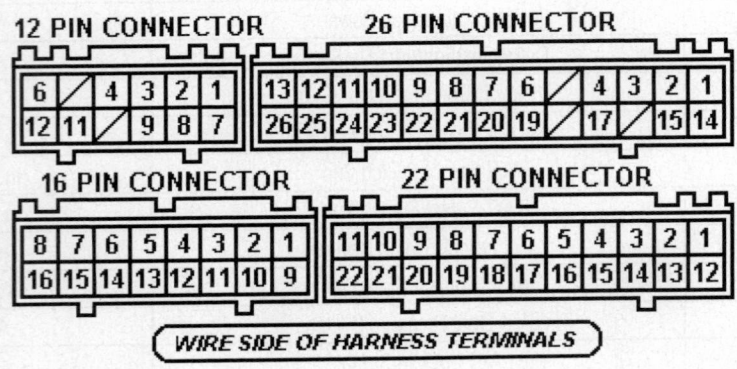

05533_ADIA_G019

Pin Connector Graphic

Standard Colors and Abbreviations

Abbreviation	Color	Abbreviation	Color	Abbreviation	Color
BK	Black	GY	Gray	RD	Red
BL	Blue	GN	Green	TN	Tan
BR	Brown	LG	Light Green	VT	Violet
DB	Dark Blue	OR	Orange	WT	White
DG	DK Green	PK	Pink	YL	Yellow

2000-01 2.2L I4 VIN G (A/T) E7 22 Pin Connector

PCM Pin #	Wire Color	Circuit Description (22 Pin)	Value at Hot Idle
1	BK/YL	Battery Direct	12-14v
3	GN/RD	Circuit Opening Relay (FC)	0-3v, off-idle: 12v
4	WT	SIL Signal (Scan Tool)	Pulse Signals
5	PK	Overdrive "Off" Indicator	Light Off: 12v
6	GN/RD	MIL (lamp) Control	MIL Off: 12v, On: 1v
7 (TMC)	GY	Starter Switch Signal	9-11v (cranking)
7 (TMM)	BK/OR	Starter Switch Signal	9-11v (cranking)
9	YL/BK	Cruise Control Signal (OD1)	At Cruise in OD: 12v
10	BL/RD	Cruise Control Signal (IDLO)	1.5v, off-idle: 12v

2000-01 2.2L I4 VIN G (A/T) E7 22 Pin Connector, *continued*

15	GN/WT	Stop Light Switch Signal	Brake Off: 0v, On: 12v
16	BK/YL	EFI Main Relay Power	12-14v
18	BK/RD	Defogger/Light Switch Signal	Switch On: 12v, Off: 0v
22	BR	Power Ground (EOM)	<0.1v

2000-01 2.2L I4 VIN G (A/T) E8 28 Pin Connector

PCM Pin #	Wire Color	Circuit Description (28 Pin)	Value at Hot Idle
2	RD/BK	A/T Select Switch Reverse	In 'R': 12v, Others: 0v
3 (TMC)	BL/WT	A/T Select Switch 2nd	In 2nd: 12v, Others: 0v
3 (TMM)	OR	A/T Select Switch 2nd	In 2nd: 12v, Others: 0v
5	BL/WT	Data Link Connector (TE1)	12-14v
6	RD/BK	A/C Switch Signal	Clutch On: 1.5v, Off: 12v
7	GN/OR	Overdrive Main Switch	Switch Off: 12v
8	RD/WT	A/C Lock In Signal	A/C On: 12v
10	PK	EVAP Vapor Pressure Sensor	2.9-3.1v (with hose off)
12	YL	A/T Select Switch Low	In Low: 12v, Others: 0v
17	GN	A/C Dual Pressure Switch	A/C Off: 12v, On: 1v
20	BK/WT	A/T Neutral Start Switch	In P/N: 9-11v (cranking)
22	VT/WT	Speedometer Indicator	At 55 mph: 48 Hz
27	BK/OR	Tachometer Signal (TACO)	Pulse Signals

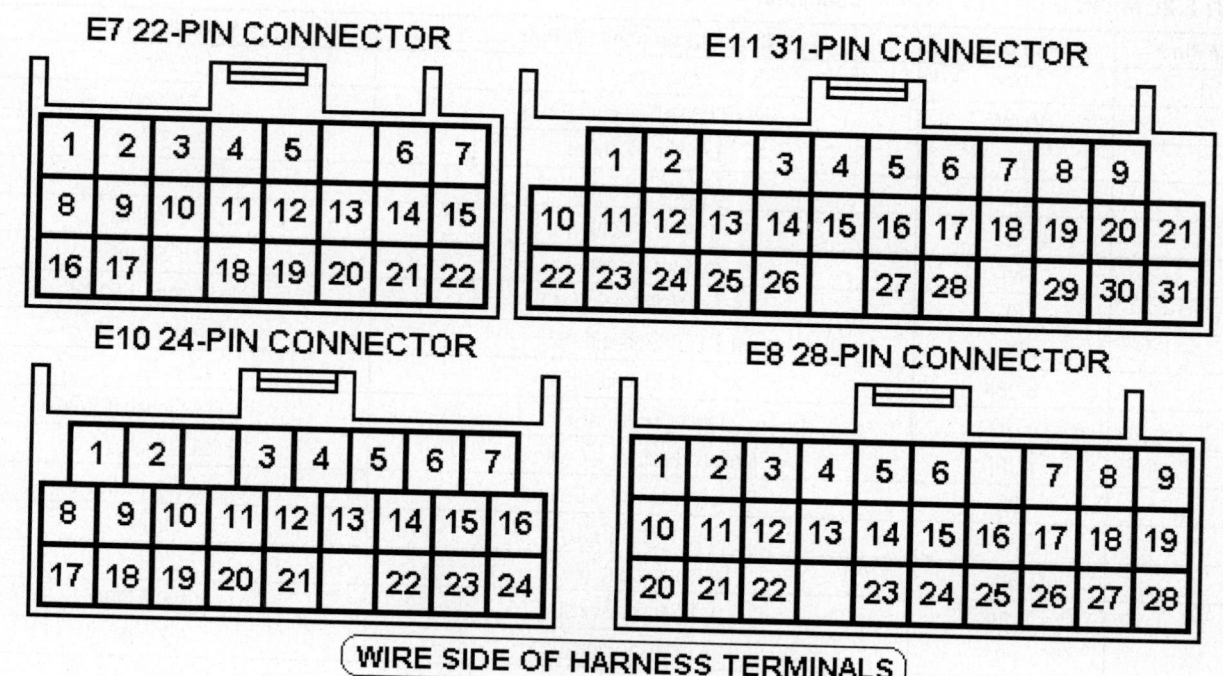

WIRE SIDE OF HARNESS TERMINALS

Pin Connector Graphic

05533_ADIA_G020

2000-01 2.2L I4 VIN G (A/T) E10 24 Pin Connector

PCM Pin #	Wire Color	Circuit Description (24 Pin)	Value at Hot Idle
2	YL	Sensor VREF (VC)	4.9-5.1v
3	PK/BK	HO2S-12 (B1 S2) Heater	1v (Heater On)
4 (Cal)	GN	HAF1A-11 (B1 S1) Heater	1v (Heater On)
4 (Fed)	BL/YL	HO2S-11 (B1 S1) Heater	1v (Heater On)
5	BL/YL	A/C Magnetic Clutch (ACMG)	Clutch Off: 0v, On: 12v
6	PK/BK	EGR Solenoid Control (VSV)	12v or 0v
7	BL/RD	EVAP Canister Closed Valve	12v or 0v
8 (Cal)	BR	Power Ground	<0.1v
9	BK/YL	MAP Sensor Signal	1-1.5v
12	BK/BL	PSP Switch Signal	Straight: 12v, Turning: 0v
13	BK	HO2S-12 (B1 S2) Signal	0.1-1.1v
14 (Cal)	BL	AFS-11 (B1 S1) Signal (AF+)	Fixed at 3.0v
14 (Fed)	WT	HO2S-11 (B1 S1) Signal	0.1-1.1v
16	BK/WT	CMP Sensor Signal (G+)	AC pulse signals
17	BR	Power Ground (E1)	<0.1v
18	BR	Sensor Ground (E2)	<0.050v
22 (Cal)	BK/WT	AFS-11 (B1 S1) Signal (AF-)	Fixed at 3v
23	BL	CKP/CMP Sensor Ground (-)	<0.050v
24	BK/RD	CKP Sensor Signal (NE+)	AC pulse signals

2000-01 2.2L I4 VIN G (A/T) E11 31 Pin Connector

PCM Pin #	Wire Color	Circuit Description (31 Pin)	Value at Hot Idle
1	VT	EVAP Vapor Pressure (VSV)	12v or 0v
2	VT/WT	EVAP Purge Solenoid (VSV)	12v or 0v
3	BL	Injector 1 Control	2.0-3.3 ms
4	RD	Injector 2 Control	2.0-3.3 ms
5	YL	Injector 3 Control	2.0-3.3 ms
6	WT	Injector 4 Control	2.0-3.3 ms
7	PK	A/T-ECT Solenoid (SL)	In Lockup: 12-14v
8	PK	A/T-ECT Solenoid (S2)	In 3rd or OD: 1v
9	BL/BK	A/T-ECT Solenoid (S1)	1st or OD: 1v
10	WT/RD	Igniter Signal (IGF)	Digital Signal: 0-5-0v
11	YL/RD	Igniter Transistor 2 Control	Digital Signal: 0-5-0v
12	BK	Igniter Transistor 1 Control	Digital Signal: 0-5-0v
20	WT	IAC Signal (RSD)	Pulses
21	BR	Sensor Ground (E01)	<0.1v
22	YL/BK	IAT Sensor Signal	At 100°F: 2.8v
23	LG	TP Sensor Signal	0.3-0.8v
24	GN/BK	ECT Sensor Signal	At 180°F: 0.6v
25	BL/RD	A/C Evaporator Temp. Sensor	EVAP Temp. 59°F: 2.4v
26	WT/BL	A/C M/C & Lock In Signal	A/C On: 12v, Off: 0v
28	WT	Knock Sensor Signal	<0.075v AC
30	BR	Power Ground (E03)	<0.1v
31	BR	Power Ground (E02)	<0.1v

Note: TMC indicates a vehicle produced by Toyota Motor Corp. (Japan) and TMM indicates a vehicle produced by Toyota Motor Manufacturer (USA).

2000-01 2.2L I4 VIN G (A/T) E7 22 Pin Connector

PCM Pin #	Wire Color	Circuit Description (22 Pin)	Value at Hot Idle
1	BK/YL	Battery Direct	12-14v
2	BK/RD	Ignition Switch Power	12-14v
3	GN/RD	Circuit Opening Relay (FC)	0-3v, off-idle: 12v
4	WT	SIL Signal (Scan Tool)	Digital Signal
5	PK	Overdrive "Off" Indicator	Light Off: 12v
6	GN/RD	MIL (lamp) Control	MIL Off: 12v, On: 1v
7 (TMC)	GY	Starter Switch Signal	9-11v (cranking)
7 (TMM)	BK/OR	Starter Switch Signal	9-11v (cranking)
8	BK/WT	EFI Main Relay Power	12-14v
9	YL/BK	Cruise Control Signal (OD1)	At Cruise in OD: 12v
10	BL/RD	Cruise Control Signal (IDLO)	1.5v, off-idle: 12v
15	GN/WT	Stop Light Switch Signal	Brake Off: 0v, On: 12v
16	BK/YL	EFI Main Relay Power	12-14v
18	BK/RD	Defogger/Light Switch Signal	Switch On: 12v, Off: 0v
22	BR	Power Ground (EOM)	<0.1v

2000-01 2.2L I4 VIN G (A/T) E8 28 Pin Connector

PCM Pin #	Wire Color	Circuit Description (28 Pin)	Value at Hot Idle
2	RD/BK	A/T Select Switch Reverse	In 'R': 12v, Others: 0v
3 (TMC)	BL/WT	A/T Select Switch 2nd	In 2nd: 12v, Others: 0v
3 (TMM)	OR	A/T Select Switch 2nd	In 2nd: 12v, Others: 0v
4	BK/YL	A/C Amplifier Signal (AC1)	Clutch On: 1.5v, Off: 12v
5	BL/WT	Data Link Connector (TE1)	12-14v
6	RD/BK	A/C Switch Signal	Clutch On: 1.5v, Off: 12v
7	GN/OR	Overdrive Main Switch	Switch Off: 12v
8	RD/WT	A/C Lock In Signal	A/C On: 12v
10	PK	EVAP Vapor Pressure Sensor	2.9-3.1v (with hose off)
12	YL	A/T Select Switch Low	In Low: 12v, Others: 0v
13	LG/BK	A/C Amplifier Signal (ACT)	Clutch On: 12v, Off: 1.5v
14	PK	A/C Amplifier Signal (THWO)	Pulse Signals
17	GN	A/C Dual Pressure Switch	A/C Off: 12v, On: 1v
18	BL/YL	Transponder Amplifier Signal	Inserting key: pulses
19	RD/BL	Transponder Amplifier Signal	Inserting key: pulses
20	BK/WT	A/T Neutral Start Switch	In P/N: 9-11v (cranking)
22	VT/WT	Speedometer Indicator	At 55 mph: 48 Hz
25	BL/BK	Unlock Warning Switch	Inserting key: pulses
26	RD/YL	Theft Deterrent ECU Signal	No key inserted: pulses
27	BK/OR	Tachometer Signal (TACO)	Pulse Signals
28	GN/WT	Transponder Amplifier Code	Inserting key: pulses

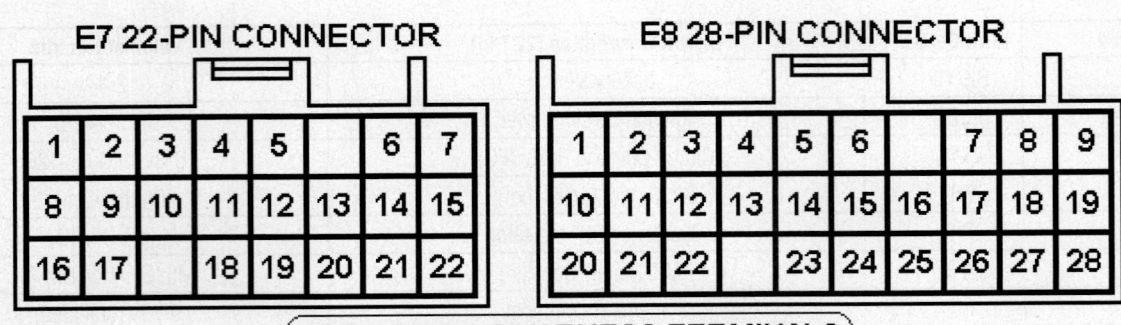

E7 22-PIN CONNECTOR

E8 28-PIN CONNECTOR

WIRE SIDE OF HARNESS TERMINALS

05533_ADIA_G021

Pin Connector Graphic

2000-01 2.2L I4 VIN G (A/T) E10 24 Pin Connector

PCM Pin #	Wire Color	Circuit Description (24 Pin)	Value at Hot Idle
1	PK/BK	EGR Solenoid Control (VSV)	12v or 0v
2	YL	Sensor VREF (VC)	4.9-5.1v
3	BL/RD	EVAP Canister Closed Valve	12v or 0v
4	PK	EVAP Vapor Pressure (VSV)	12v or 0v
5	PK/BK	HO2S-12 (B1 S2) Heater	1v (Heater On)
6 (Cal)	GN	AFS-11 (B1 S1) Heater	1v (Heater On)
6 (Fed)	BL/YL	HO2S-11 (B1 S1) Heater	1v (Heater On)
7	PK/WT	EVAP Purge Solenoid (VSV)	12v or 0v
8	BR	Power Ground (E04)	<0.1v
11	BK/BL	PSP Switch Signal	Straight: 12v, Turning: 0v
13	BK	HO2S-12 (B1 S2) Signal	0.1-1.1v
14 (Cal)	BL	AFS-11 (B1 S1) Signal (AF+)	Fixed at 3.3v
14 (Fed)	WT	HO2S-11 (B1 S1) Signal	0.1-1.1v
15	BK/WT	CMP Sensor Signal (G+)	AC pulse signals
16	BK/RD	CKP Sensor Signal (NE+)	AC pulse signals
17	BR	Power Ground (E1)	<0.1v
18	BR	Sensor Ground (E2)	<0.050v
19	BL/YL	A/C Magnetic Clutch (ACMG)	Clutch Off: 0v, On: 12v
21	BK/YL	MAP Sensor Signal	1-1.5v
23 (Cal)	BK/WT	AFS-11 (B1 S1) Signal (AF-)	Fixed at 3v
24	BL	CKP/CMP Sensor Ground (-)	<0.050v

2000-01 2.2L I4 VIN G (A/T) E11 31 Pin Connector

PCM Pin #	Wire Color	Circuit Description (31 Pin)	Value at Hot Idle
1	BL	Injector 1 Control	2.0-3.3 ms
2	RD	Injector 2 Control	2.0-3.3 ms
3	YL	Injector 3 Control	2.0-3.3 ms
4	WT	Injector 4 Control	2.0-3.3 ms
6	PK	A/T-ECT Solenoid (SL)	In Lockup: 12-14v
7	PK	A/T-ECT Solenoid (S1)	In 3rd or OD: 1v
8	BL/BK	A/T-ECT Solenoid (S2)	1st or OD: 1v
10	BK	Igniter Transistor 1 Control	Digital Signal: 0-5-0v

2000-01 2.2L I4 VIN G (A/T) E11 31 Pin Connector, *continued*

11	YL/RD	Igniter Transistor 2 Control	Digital Signal: 0-5-0v
12	WT/RD	Igniter Signal (IGF)	Digital Signal: 0-5-0v
13	BL/RD	A/C Evaporator Temp. Sensor	EVAP Temp. 59°F: 2.4v
18	WT	IAC Signal (RSD)	Pulses
21	BR	Sensor Ground	<0.050v
22	GN/BK	ECT Sensor Signal	At 180°F: 0.6v
23	YL/BK	IAT Sensor Signal	At 100°F: 2.8v
24	LG	TP Sensor Signal	0.3-0.8v
25	WT/BL	A/C M/C & Lock In Signal	A/C On: 12v, Off: 0v
27	WT	Knock Sensor Signal	<0.075v AC
30	BR	Power Ground (E03)	<0.1v
31	BR	Power Ground (E02)	<0.1v

Note: TMC indicates a vehicle produced by Toyota Motor Corp. (Japan) and TMM indicates a vehicle produced by Toyota Motor Manufacturer (USA).

2002-06 2.4L I4 VIN E W/EIS (A/T) E6 31 Pin Connector

PCM Pin #	Wire Color	Circuit Description (31 Pin)	Value at Hot Idle
1	BK/RD	EFI Main Relay Output (B+)	12-14v
2	BL/RD	Battery Direct (+BM)	12-14v
3	BK/YL	Battery Direct (BATT)	12-14v
4	VT	EVAP Vapor Pressure (VSV)	12v or 0v
5	BK/OR	Tachometer Signal (TACO)	DC pulse signals
6-7, 20, 24, 26	---	Not Used	---
8	BK/WT	EFI Main Relay Control	Relay On: 12v, Off: 0v
9	BK/OR	Ignition Switch Power (B+)	12-14v
10	GN/RD	Circuit Opening Relay (FC)	0-3v, off-idle: 12v
11	WT	SIL Signal (Scan Tool)	Digital Signal
12	GN	Tail Light Switch Signal	Switch Off: 1.5v, On: 12v
13	BK/YL	Mirror Heater Switch Signal	Switch Off: 1.5v, On: 12v
14	PK/BK	Data Link Connector (TC)	12-14v
15	BR	Power Ground (EOM)	<0.1v
18	GN/RD	MIL (lamp) Control	MIL Off: 12v, On: 1v
19	RD	Scan Tool (WFSE)	12v
21	PK	EVAP Vapor Pressure Sensor	2.9-3.1v (with hose off)
22	BL/YL	APP Sensor (VPA) Signal	0.8-1.2v
23	WT/RD	APP Sensor (VPA2) Signal	0.8-1.2v
25	RD	APP Sensor VREF	4.5-5.5v
27	RD	APP Sensor VREF	4.5-5.5
28	LG/BK	APP Sensor Ground (EPA)	<0.050v
29	LG	APP Sensor Ground (EPA2)	<0.050v

2002-06 2.4L I4 VIN E W/EIS (A/T) E7 35 Pin Connector

PCM Pin #	Wire Color	Circuit Description (35 Pin)	Value at Hot Idle
1-3, 5-6, 12-13	---	Not Used	---
4	BL	HO2S-12 (B1 S2) Heater	Heater On: <1v
7	OR	Overdrive Indicator Control	LED Off: 12v, On: 0v
8	YL	A/T Select Switch Low Signal	In Low: 12v, Others: 0v
9	BL/WT	A/T Select Switch 2nd Signal	In 2nd: 12v, Others: 0v
10	WT/BL	A/T Select Switch 'D' Signal	In Drive: 12v, Others: 0v
11	RD/BK	A/T Select Switch 'R' Signal	In 'R': 12v, Others: 0v
14	YL/GN	Hot Engine Lamp Control	Lamp Off: 12v, On: 1v
15	GN/WT	Transponder Amplifier Code	Inserting key: pulses
16	VT	Transponder IMLD Signal	Inserting key: pulses
17	VT/WT	Speedometer Indicator	Moving: 0-5-0v
18, 20-21	---	Not Used	---
19	GN/WT	Stop Light Switch Signal	Brake Off: 0v, On: 12v
22	BK	HO2S-12 (B1 S2) Signal	0.1-1.1v
23-25, 28, 32	---	Not Used	---
26	BL/YL	Theft Deterrent ECU Signal	No key inserted: pulses
27	RD/BL	Transponder Amplifier Code	Inserting key: pulses
29	GN/OR	Overdrive Main Switch	Switch Off: 12v, On: 0v
31	PK/BL	A/C Switch Signal	AC On: 9-14v
33	BK	A/C Switch Signal (Auto AC)	AC On: 9-14v
34	BL	Transponder Amplifier Signal	Key Inserted: <1.5v

2002-06 2.4L I4 VIN E W/EIS (A/T) E8 32 Pin Connector

PCM Pin #	Wire Color	Circuit Description (24 Pin)	Value at Hot Idle
1	BR	Power Ground (E1)	<0.1v
2	BL/WT	A/C Magnetic Clutch (ACMG)	A/C On: <3.0v
4	WT	Throttle Body Motor (M-)	DC pulse signals
5	BK	Throttle Body Motor (M+)	DC pulse signals
6	WT/BK	Sensor Ground (ME01)	<0.050v
7	WT/BK	Power Ground (E03)	<0.1v
8	WT/BK	Power Ground (E02)	<0.1v
9	GN/WT	Cooling Fan Relay Control	Relay Off: 12v, On: 1v
10	RD/WT	PSP Switch Signal	Straight: 12v, Turning: 0v
11	BK/RD	EVAP Solenoid Control (VSV)	12v or 0v
12	GN	EVAP Canister Closed Valve	12v or 0v
17	BR	Sensor Shield Ground	<0.050v
15	YL	Cam Timing Oil Valve (OCV-)	Pulse signals
16	BK/WT	Cam Timing Oil Valve (OCV+)	Pulse signals
30	YL/BK	Cooling Fan Relay Control	Relay Off: 12v, On: 1v

2002-06 2.4L I4 VIN E W/ESI (A/T) E9 35 Pin Connector

PCM Pin #	Wire Color	Circuit Description (35 Pin)	Value at Hot Idle
1	WT	Knock Sensor Signal	<0.075v AC
2-3	---	Not Used	---
4	BK/RD	AFS-11 (B1 S1) Heater	1v (Heater On)
5	---	Not Used	---
6	WT/BK	Power Ground (E05)	<0.1v
7	WT/BK	Power Ground (E04)	<0.1v
8	BK/YL	Neutral Start Switch	In P/N: 9-11v (cranking)
9	BK/WT	Starter Switch Signal (STA)	9-11v (cranking)
10	---	Not Used	---
11	YL	A/T-ECT Solenoid (DSL)	In Lockup: 12-14v
12-15	---	Not Used	---
16	BL/RD	A/T-ECT Solenoid (SL2-)	Moving in 3rd or OD: <1v
17	BL/YL	A/T-ECT Solenoid (SL2+)	Moving in 3rd or OD: <1v
18	PK	A/T-ECT Solenoid (SL1-)	Moving in 1st Gear: <1v
19	RD/BK	A/T-ECT Solenoid (SL1+)	Moving in 1st Gear: <1v
18	WT	IAC Signal (RSD)	Pulses
23	OR	AFS-11 (B1 S1) Signal (AF+)	3.0-3.6v
24	RD	MAF Sensor Signal (VG)	1.1-1.5v
26	RD	Counter Gear Speed (NC+)	AC pulse signals
27	BL	Turbine Speed Sensor (NT+)	AC pulse signals
28-30	---	Not Used	---
31	WT	AFS-11 (B1 S1) Signal (AF-)	Fixed at 3v
32	BL/WT	MAF Sensor Ground (E2G)	<0.050v
33	---	Not Used	---
34	GN	Counter Gear Speed (NC-)	AC pulse signals
35	LG	Turbine Speed Sensor (NT-)	AC pulse signals

2002-06 2.4L I4 VIN E W/EIS (A/T) E10 34 Pin Connector

PCM Pin #	Wire Color	Circuit Description (34 Pin)	Value at Hot Idle
1	BL	Injector 1 Control	2.0-3.3 ms
2	RD	Injector 2 Control	2.0-3.3 ms
3	YL	Injector 3 Control	2.0-3.3 ms
4	WT	Injector 4 Control	2.0-3.3 ms
5	---	Not Used	---
6	WT/BK	Power Ground (E02)	<0.1v
7	WT/BK	Power Ground (E01)	<0.1v
8	RD/WT	Igniter Transistor 1 Control	7% duty cycle
9	PK	Igniter Transistor 2 Control	7% duty cycle
10	LG/BK	Igniter Transistor 3 Control	7% duty cycle
11	BL/YL	Igniter Transistor 2 Control	7% duty cycle
12	---	Not Used	---
13	YL	Sensor VREF (VC)	4.9-5.1v
14-15	---	Not Used	---
16	YL/BK	A/T-ECT Solenoid (SLT-)	During shifting: 12v
17	YL/RD	A/T-ECT Solenoid (SLT+)	During shifting: 12v
18	YL	TP Sensor VREF (VC)	4.9-5.1v
19	GN/YL	ECT Sensor Signal	At 180°F: 0.51v
20	LB	IAT Sensor Signal	At 100°F: 2.60v
21	LG	TP Sensor Signal	0.4-1.0v
22	---	Not Used	---
23	WT/RD	Igniter Signal (IGF)	Digital Signal: 0-5-0v
24-25	---	Not Used	---
26	BK/WT	CMP Sensor Signal (G22+)	AC pulse signals
27	RD	CKP Sensor Signal (NE+)	AC pulse signals
28	BR	Power Ground (E2)	<0.1v
29	---	Not Used	---
30 (TMC)	GN	A/T ATF Sensor Signal	<1.5v at 239°F
30 (TMM)	GN/RD	A/T ATF Sensor Signal	<1.5v at 239°F
31	BK/RD	TP Sensor 2 Signal (VTA2)	2.0-2.9v
32-33	---	Not Used	---
34	GN	CKP/CMP Sensor Ground	<0.050v

1990-91 2.5L V6 VIN V (A/T-ECT) 26 Pin Connector

PCM Pin #	Wire Color	Circuit Description (26 Pin)	Value at Hot Idle
1	BL	Distributor Signal (NE+)	AC pulse signals
2	YL	Distributor Signal (G2+)	AC pulse signals
3	WT/RD	Igniter Signal (IGF)	Digital Signal: 0-5-0v
4	RD/YL	ISC Motor (ISC4)	Pulses
5	BL/GN	ISC Motor (ISC3)	Pulses
6	GY/YL	ISC Motor (ISC2)	Pulses
7	WT/BL	ISC Motor (ISC1)	Pulses
8	RD/GN	Main O2S Heater	1v (Heater On)
9	BK/RD	Fuel Pressure Up Solenoid	12v or 0v

1990-91 2.5L V6 VIN V (A/T-ECT) 26 Pin Connector, *continued*

10	GN/BK	Cold Start Injector Control	1v (at cold startup)
11	YL	Injector Pair 2 & 3 Control	2.0-3.3 ms
12	WT	Injector Pair 1 & 6 Control	2.0-3.3 ms
13	WT/BK	Power Ground	<0.1v
14	BK	Distributor Signal (G-)	<0.050v
15	RD	Distributor Signal (G+)	AC pulse signals
17	BL/YL	A/T-ECT Solenoid (SL)	In Lockup: 12-14v
18	PK/RD	A/T-ECT Solenoid (S2)	1st or OD: 1v
19	PK	A/T-ECT Solenoid (S1)	In 3rd or OD: 1v
20	WT/GN	Igniter Signal (IGT)	Digital Signal: 0-5-0v
22	BL/BK	A/C Acceleration Cut	A/C On: 4.5-5.5v
24	BR	Sensor Ground	<0.050v
25	GY	Injector Pair 4 & 5 Control	2.0-3.3 ms
26	WT/BK	Power Ground	<0.1v

1990-91 2.5L V6 VIN V (A/T-ECT) 16 Pin Connector

PCM Pin #	Wire Color	Circuit Description (16 Pin)	Value at Hot Idle
1	BL/RD	Airflow Meter Signal	1.5-2.5v
2	YL/RD	Sensor VREF	4.9-5.1v
3	YL	IAT Sensor Signal	At 100°F: 2.60v
4	GN	ECT Sensor Signal	At 180°F: 0.51v
5	RD/BL	Sub O2S Signal	0.1-1.1v
6	WT	Main O2S Signal	0.1-1.1v
7	BL/RD	A/T Pattern Selector Switch	Norm: 0v, PWR: 12v
8	GN/RD	Check Connector	12-14v
9	BR	Sensor Ground	<0.050v
10	GY	EGR Gas Temp. Sensor	3.5-4.0v
11	BK	TP Sensor Signal	0.3-0.8v
12	BL	Closed Throttle Switch	0-3v, off-idle: 12v
13	GN/WT	Stop Light Switch Signal	Brake Off: 0v, On: 12v
14	WT	Knock Sensor Signal	<0.075v AC
15	YL/GN	Check Connector	12-14v

1990-91 2.5L V6 VIN V (A/T-ECT) 22 Pin Connector

PCM Pin #	Wire Color	Circuit Description (22 Pin)	Value at Hot Idle
1	WT/BL	Battery Direct	12-14v
2	BK/OR	Ignition Switch Power	12-14v
3	RD/BK	A/T Select Switch Reverse	In 'R': 12v, Others: 0v
4	BK/YL	EFI Main Relay Power	12-14v
5	GY	MIL (lamp) Control	MIL Off: 12v, On: 1v
6 (Cal)	BL/WT	Sub O2S Signal	0.1-1.1v
7 ('91)	BR/BK	A/T Select Switch Drive	In 'D': 12v, Others: 0v
8 ('91)	GN/BK	A/T-ECT Speed Sensor	Moving: varies 0-5v
9	PK/YL	Vehicle Speed Sensor	At 55 mph: 48 Hz
10	BK/GN	A/C Magnetic Clutch (ACMG)	Clutch Off: 0v, On: 12v
11	BK	A/T: Starter Switch Signal	9-11v (cranking)
11	BK/WT	M/T: Starter Switch Signal	9-11v (cranking)
12	WT/RD	EFI Main Relay Power	12-14v
13	WT/RD	EFI Main Relay Power	12-14v
14	YL/BL	A/T Select Switch Low	In Low: 12v, Others: 0v
15	OR	A/T Select Switch 2nd	In 2nd: 12v, Others: 0v
16	RD	A/T: Select Switch Neutral	In Neutral: 8-14v
20	GN/OR	Main Overdrive Switch	Switch Off: 12-14v
21	YL/BK	Cruise Control ECU	At Cruise in OD: 12v
22	BK/WT	A/T: Neutral Drive Switch	In P/N: 9-11v (cranking)

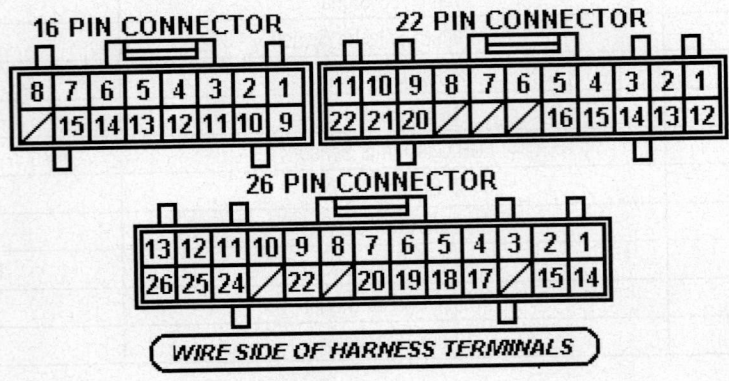

Pin Connector Graphic

1992-93 3.0L VIN G (A/T-ECT) 16 Pin Connector

PCM Pin #	Wire Color	Circuit Description (16 Pin)	Value at Hot Idle
1	BL/RD	Sensor VREF (VC)	4.9-5.1v
2	YL/BL	Air Flow Meter Signal	2.5-4.5v
3	BL/BK	IAT Sensor Signal	At 100°F: 2.60v
4	GN	ECT Sensor Signal	At 180°F: 0.51v
5	WT	O2S-11 (B1 S1) Signal	0.1-1.1v
6	WT	Knock Sensor 1 Signal	<0.075v AC
7	GY	Check Connector	12-14v
8	GY/BK	Check Connector	12-14v

1992-93 3.0L VIN G (A/T-ECT) 16 Pin Connector, *continued*

9	BR	Sensor Ground	<0.050v
10	BK	TP Sensor Signal	0.3-0.8v
11	BL	Closed Throttle Switch	1v, at off-idle: 12v
12 (Cal)	GY	EGR Gas Temp. Sensor	3.5-4.0v
13	RD/BL	O2S-12 (B1 S2) Signal	0.1-1.1v
14	BK	Knock Sensor 2 Signal	<0.075v AC
15	GN/WT	Check Connector	12-14v
16	GN/BK	A/T-ECT Speed Sensor	Moving: varies 0-5v

1992-93 3.0L VIN G (A/T-ECT) 22 Pin Connector

PCM Pin #	Wire Color	Circuit Description (22 Pin)	Value at Hot Idle
1	BK/OR	EFI Main Relay Power B1+	12-14v
2	WT/BL	Battery Direct	12-14v
3	RD/YL	EFI Main Relay Power	12-14v
4	GN/RD	MIL (lamp) Control	MIL Off: 12v, On: 1v
6	GN/BK	A/C Amplifier Signal (ACT)	Clutch On: 12v, Off: 1.5v
7	BK/YL	A/C Amplifier Signal (AC1)	Clutch On: 1.5v, Off: 12v
8	PK/YL	Vehicle Speed Sensor	At 55 mph: 48 Hz
9	GY/BL	Main Overdrive Switch	Switch Off: 12-14v
11	BK/WT	A/T: Starter Switch Signal	9-11v (cranking)
12	BK/OR	EFI Main Relay Power	12-14v
13	BK/OR	Ignition Switch Power	12-14v
14	GN/WT	Stop Light Switch Signal	Brake Off: 0v, On: 12v
18	YL/BK	Cruise Control ECU	At Cruise in OD: 12v
19	RD/BK	A/T Select Switch Reverse	In 'R': 12v, Others: 0v
20	BL/RD	A/T Pattern Selector Switch	Norm: 0v, PWR: 12v
22	BK/WT	A/T Neutral Start Switch	In P/N: 9-11v (cranking)

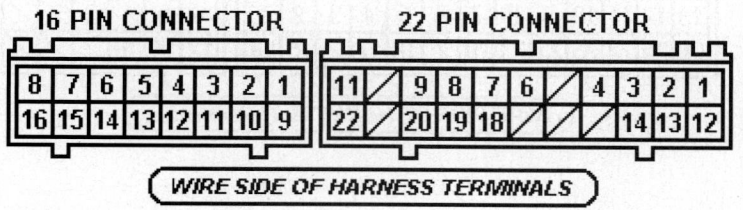

16 PIN CONNECTOR   **22 PIN CONNECTOR**

8	7	6	5	4	3	2	1
16	15	14	13	12	11	10	9

11	/	9	8	7	6	/	4	3	2	1
22	/	20	19	18	/	/	/	14	13	12

WIRE SIDE OF HARNESS TERMINALS

05533_ADIA_G023

Pin Connector Graphic

1992-93 3.0L VIN G (A/T-ECT) 12 Pin Connector

PCM Pin #	Wire Color	Circuit Description (12 Pin)	Value at Hot Idle
2	BK/RD	Fuel Pressure Up Solenoid	12v or 0v
3	LG	Intake Air Control Solenoid	12v or 0v
5 (Cal)	WT	Sub O2S Signal	0.1-1.1v
6	YL	Distributor Signal (G-)	<0.050v
10	BL	Distributor Signal (G2+)	AC pulse signals
11	RD	Distributor Signal (G1+)	AC pulse signals
12	BK	Distributor Signal (NE-)	<0.050v

1992-93 3.0L VIN G (A/T-ECT) 26 Pin Connector

PCM Pin #	Wire Color	Circuit Description (26 Pin)	Value at Hot Idle
1 (Cal)	PK/BK	HO2S Heater	1v (Heater On)
2	OR	A/T Select Switch 2nd	In 2nd: 12v, Others: 0v
3	YL/BL	A/T Select Switch Low	In Low: 12v, Others: 0v
4	RD/BK	ISC Signal 4	Pulse Signals
5	BL/RD	ISC Signal 3	Pulse Signals
6	GN/WT	ISC Signal 2	Pulse Signals
7	WT/BL	ISC Signal 1	Pulse Signals
8	BL/YL	A/T-ECT Solenoid (SL)	In Lockup: 12-14v
9	PK/BL	A/T-ECT Solenoid (S2)	In 2nd: 12v, Others: 0v
10	PK	A/T-ECT Solenoid (S1)	In 1st: 12-14v
11	YL	Injector 2 Control	2.0-3.3 ms
12	WT	Injector 1 Control	2.0-3.3 ms
13	WT/BK	Power Ground	<0.1v
14	BK	Check Connector	12-14v
15	RD/WT	Check Connector	12-14v
17	WT/RD	Igniter Signal (IGF)	Digital Signal: 0-5-0v
18	WT	Igniter Signal (IGT)	Digital Signal: 0-5-0v
20	GN	Injector 6 Control	2.0-3.3 ms
21	RD	Injector 5 Control	2.0-3.3 ms
22	BL	Injector 4 Control	2.0-3.3 ms
23	GY	Injector 3 Control	2.0-3.3 ms
24	BR	Sensor Ground	<0.050v
25	GN	Cold Start Injector Control	1v (at cold startup)
26	WT/BK	Power Ground	<0.1v

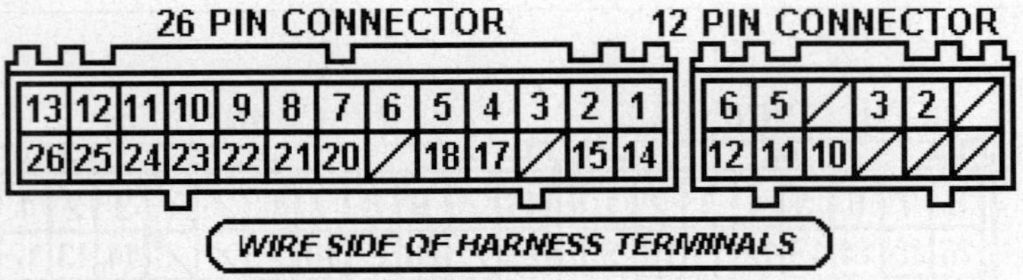

26 PIN CONNECTOR   **12 PIN CONNECTOR**

```
13 12 11 10 9 8 7 6 5 4 3 2 1      6 5 / 3 2
26 25 24 23 22 21 20 / 18 17 / 15 14    12 11 10 / /
```

WIRE SIDE OF HARNESS TERMINALS

05533_ADIA_G024

Pin Connector Graphic

1994-95 3.0L V6 MFI VIN G (A/T) 16 Pin Connector

PCM Pin #	Wire Color	Circuit Description (16 Pin)	Value at Hot Idle
1	WT/BL	A/C Idle Up Solenoid	12v or 0v
3	GN/RD	MIL (lamp) Control	MIL Off: 12v, On: 1v
5	GY/BK	Check Connector	12-14v
6	RD/YL	Intake Air Control Solenoid	12v or 0v
7	RD/BK	MAF Sensor Ground	<0.050v
10	BL/RD	HO2S-21 (B2 S1) Heater	1v (Heater On)
11	PK/BK	HO2S-11 (B1 S1) Heater	1v (Heater On)
12	BK/BL	EGR Solenoid Control (VSV)	12v or 0v
14 ('94)	GN/YL	EGR Gas Temp. Sensor	3.5-4.0v
14 ('95)	GN/BK	EGR Gas Temp. Sensor	3.5-4.0v
16	BR	Sensor Ground	<0.050v

1994-95 3.0L V6 MFI VIN G (A/T) 22 Pin Connector

PCM Pin #	Wire Color	Circuit Description (22 Pin)	Value at Hot Idle
1	BL/RD	Sensor VREF (VC)	4.9-5.1v
5	RD	CKP Sensor Signal (NE+)	AC pulse signals
6	GN	CKP Sensor Signal (NE-)	<0.050v
7	BK/YL	TP Sensor Signal	0.3-0.8v
8	RD	MAF Sensor Signal	1.1-1.5v
13	RD/BL	HO2S-11 (B1 S1) Signal	0.1-1.1v
14	WT	Knock Sensor 1 Signal	<0.075v AC
15	WT	Knock Sensor 2 Signal	<0.075v AC
16	BK/WT	CMP Sensor Signal (G+)	AC pulse signals
17	BL	CMP Sensor Signal (G-)	<0.050v
18	GN/RD	Circuit Opening Relay (FC)	0-3v, off-idle: 12v
19	RD/BL	HO2S-21 (B2 S1) Signal	0.1-1.1v
20	GN/BK	ECT Sensor Signal	At 180°F: 0.51v
21	BL/BK	IAT Sensor Signal	At 100°F: 2.60v
22	BR	Sensor Ground	<0.050v

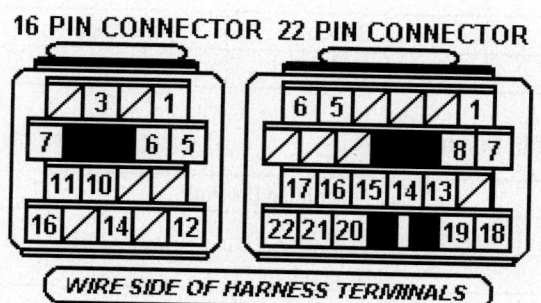

16 PIN CONNECTOR 22 PIN CONNECTOR

WIRE SIDE OF HARNESS TERMINALS

05533_ADIA_G025

Pin Connector Graphic

1994-95 3.0L V6 MFI VIN G (A/T) 28 Pin Connector

PCM Pin #	Wire Color	Circuit Description (28 Pin)	Value at Hot Idle
5	GN/BK	A/C Amplifier Signal (ACT)	Clutch On: 12v, Off: 1.5v
9 ('95)	RD	Cooling Fan Relay	Fan On: <1v
11	GN/WT	Data Link Connector (TN)	12-14v
12	PK/YL	Vehicle Speed Sensor	At 55 mph: 48 Hz
13 ('95)	BK	Tachometer Signal (TACO)	Pulses
14	WT/BL	Battery Direct	12-14v
20	BK/YL	A/C Amplifier Signal (AC1)	Clutch On: 1.5v, Off: 12v
21	BK/RD	Idle-Up Signal	Load On: 12v, Off: 0v
22	BK/OR	EFI Main Relay Power B1+	12-14v
23	BK/OR	EFI Main Relay Power	12-14v
24 ('95)	GN/WT	Stop Light Switch Signal	Brake Off: 0v, On: 12v
25	PK/BK	HO2S-12 (B1 S2) Heater	12v or 0v
26	BK	HO2S-12 (B1 S2) Signal	0.1-1.1v
28	WT	Data Link Connector (TN)	12-14v

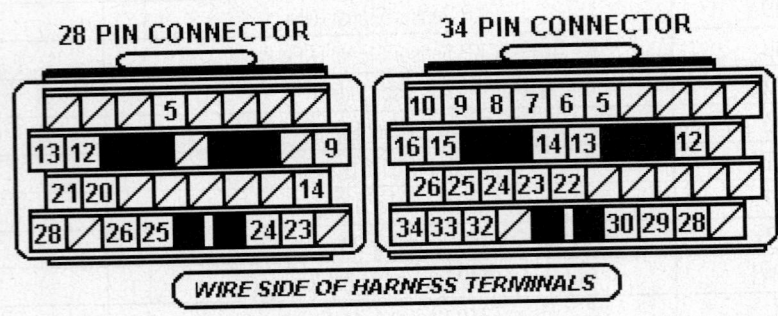

28 PIN CONNECTOR   **34 PIN CONNECTOR**

WIRE SIDE OF HARNESS TERMINALS

05533_ADIA_G026

Pin Connector Graphic

Standard Colors and Abbreviations

Abbreviation	Color	Abbreviation	Color	Abbreviation	Color
BK	Black	GY	Gray	RD	Red
BL	Blue	GN	Green	TN	Tan
BR	Brown	LG	Light Green	VT	Violet
DB	Dark Blue	OR	Orange	WT	White
DG	DK Green	PK	Pink	YL	Yellow

1994-95 3.0L V6 MFI VIN G (A/T) 34 Pin Connector

PCM Pin #	Wire Color	Circuit Description (34 Pin)	Value at Hot Idle
5	GN	Injector 6 Control	2.0-3.3 ms
6	RD	Injector 5 Control	2.0-3.3 ms
7	BL	Injector 4 Control	2.0-3.3 ms
8	GY	Injector 3 Control	2.0-3.3 ms
9	YL	Injector 2 Control	2.0-3.3 ms
10	WT	Injector 1 Control	2.0-3.3 ms
12	WT/RD	Igniter Signal (IGF)	Digital Signal: 0-5-0v
13	BK/WT	Starter Switch Signal	9-11v (cranking)

1994-95 3.0L V6 MFI VIN G (A/T) 34 Pin Connector, *continued*

14	BK/WT	Neutral Start Switch	In P/N: 9-11v (cranking)
15	GY/BK	Igniter Transistor 3 Control	7% duty cycle
16	GY/BK	Igniter Transistor 2 Control	7% duty cycle
22	YL/BK	IAC Signal (RSC)	Pulse Signals
23	GN/BK	IAC Signal (RSO)	Pulse Signals
24	WT/GN	Igniter Transistor 1 Control	7% duty cycle
25	BK/RD	Fuel Pressure Up Solenoid	1v (at hot restart)
26	BL/BK	Igniter Transistor 4 Control	7% duty cycle
28	WT/BK	Power Ground	<0.1v
29	GN/RD	Igniter Transistor 6 Control	7% duty cycle
30	RD/BK	Igniter Transistor 5 Control	7% duty cycle
32	BL	Closed Throttle Switch	1v, at off-idle: 12v
33	WT/BK	Power Ground	<0.1v
34	WT/BK	Power Ground	<0.1v

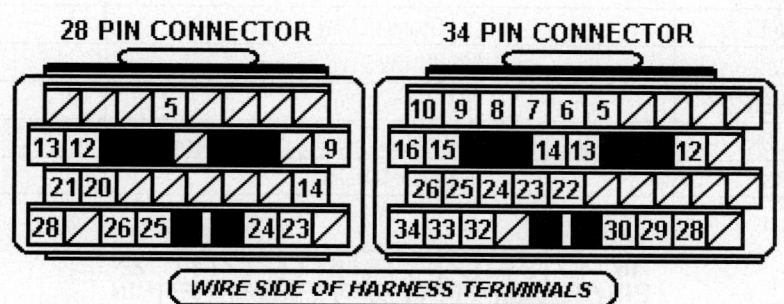

28 PIN CONNECTOR   **34 PIN CONNECTOR**

WIRE SIDE OF HARNESS TERMINALS

05533_ADIA_G026

Pin Connector Graphic

1996 3.0L V6 VIN F (A/T-ECT) 16 Pin Connector

PCM Pin #	Wire Color	Circuit Description (16 Pin)	Value at Hot Idle
3	GN/RD	MIL (lamp) Control	MIL Off: 12v, On: 1v
5	GY/BK	Check Connector	12-14v
6	RD/YL	Intake Air Control Solenoid	12v or 0v
7	RD/BK	MAF Sensor Ground	<0.050v
10	BL/RD	HO2S-21 (B2 S1) Heater	1v (Heater On)
11	BL/BK	HO2S-11 (B1 S1) Heater	1v (Heater On)
12	BK/WT	EGR Solenoid Control (VSV)	12v or 0v
14	GN/YL	EGR Gas Temp. Sensor	3.5-4.0v
16	BR	Sensor Ground	<0.050v

1996 3.0L V6 VIN F (A/T-ECT) 22 Pin Connector

PCM Pin #	Wire Color	Circuit Description (22 Pin)	Value at Hot Idle
1	BL/RD	Sensor VREF (VC)	4.9-5.1v
4	WT/BL	OD Clutch Speed Signal (-)	AC pulse signals
5 (TMC)	RD	CKP Sensor Signal (NE+)	AC pulse signals
5 (TMM)	WT	CKP Sensor Signal (NE+)	AC pulse signals
6 (TMC)	GN	CKP Sensor Signal (NE-)	<0.050v
6 (TMM)	OR	CKP Sensor Signal (NE-)	<0.050v
7	BK/YL	TP Sensor Signal	0.3-0.8v
8	RD	MAF Sensor Signal	1.1-1.5v
9	YL/BL	OD Clutch Speed Signal (+)	AC pulse signals
13	WT	HO2S-11 (B1 S1) Signal	0.1-1.1v
14	WT	Knock Sensor 1 Signal	<0.075v AC
15	WT	Knock Sensor 2 Signal	<0.075v AC
16 (TMC)	BK/WT	CMP Sensor Signal (G+)	AC pulse signals
16 (TMM)	WT/BL	CMP Sensor Signal (G+)	AC pulse signals
18	GN/RD	Circuit Opening Relay (FC)	0-3v, off-idle: 12v
19	RD/BL	HO2S-21 (B2 S1) Signal	0.1-1.1v
20	GN/BK	ECT Sensor Signal	At 180°F: 0.51v
21	BL/BK	IAT Sensor Signal	At 100°F: 2.60v
22	BR	Sensor Ground	<0.050v

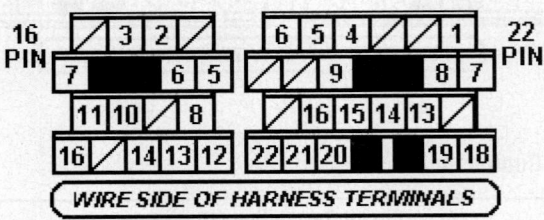

05533_ADIA_G027

Pin Connector Graphic

1996 3.0L V6 VIN F (A/T-ECT) 28 Pin Connector

PCM Pin #	Wire Color	Circuit Description (28 Pin)	Value at Hot Idle
5	GN/BK	A/C Amplifier Signal (ACT)	Clutch On: 12v, Off: 1.5v
7	YL/BK	Cruise Control ECU	At Cruise in OD: 12v
9	BK/RD	Cooling Fan Relay	Relay Off: 12v, On: 1v
12	PK/YL	Vehicle Speed Sensor	At 55 mph: 48 Hz
13	BK	Tachometer Signal (TACO)	Pulse Signals
14	WT/BL	Battery Direct	12-14v
20	BK/YL	A/C Amplifier Signal (AC1)	Clutch On: 1.5v, Off: 12v
21	BK/RD	Defogger Idle Up Signal	Switch On: 12v, Off: 0v
23	BK/OR	EFI Main Relay Power	12-14v
24	GN/WT	Stop Light Switch Signal	Brake Off: 0v, On: 12v
25	PK/BK	HO2S-12 (B1 S2) Heater	1v (Heater On)
26	BK	HO2S-12 (B1 S2) Signal	0.1-1.1v
28	WT	Data Link Connector	No Scan Tool: 0v

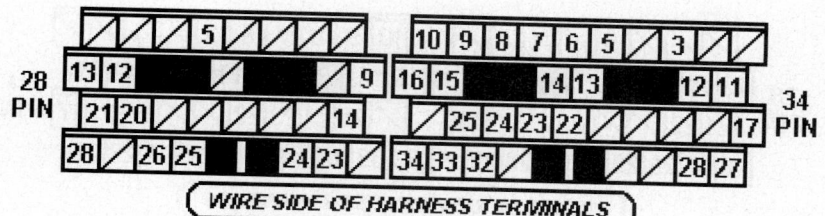

WIRE SIDE OF HARNESS TERMINALS

05533_ADIA_G028

Pin Connector Graphic
Standard Colors and Abbreviations

Abbreviation	Color	Abbreviation	Color	Abbreviation	Color
BK	Black	GY	Gray	RD	Red
BL	Blue	GN	Green	TN	Tan
BR	Brown	LG	Light Green	VT	Violet
DB	Dark Blue	OR	Orange	WT	White
DG	DK Green	PK	Pink	YL	Yellow

1996 3.0L V6 VIN F (A/T-ECT) 34 Pin Connector

PCM Pin #	Wire Color	Circuit Description (34 Pin)	Value at Hot Idle
3	YL/GN	A/T-ECT Solenoid (SLN-)	SLN shifts: 9-14v
5	GN	Injector 6 Control	2.0-3.3 ms
6	RD	Injector 5 Control	2.0-3.3 ms
7	BL	Injector 4 Control	2.0-3.3 ms
8	GY	Injector 3 Control	2.0-3.3 ms
9	YL	Injector 2 Control	2.0-3.3 ms
10	WT	Injector 1 Control	2.0-3.3 ms
11	PK	A/T-ECT Solenoid (S1)	SLN shifts: 9-14v
12	WT/RD	Igniter Signal (IGF)	Digital Signal: 0-5-0v
13	BK/WT	Starter Switch Signal	9-11v (cranking)
14	BK/WT	Neutral Start Switch	In P/N: 9-11v (cranking)
15 (TMC)	GY/BK	Igniter Transistor 3 Control	7% duty cycle
15 (TMM)	GN/RD	Igniter Transistor 3 Control	7% duty cycle
16	YL/RD	Igniter Transistor 2 Control	7% duty cycle
17	PK/BK	A/T-ECT Solenoid (S2)	In Lockup: 12v
22	YL/BK	IAC Signal (RSC)	Pulse Signals
23	GN/BK	IAC Signal (RSO)	Pulse Signals
24 (TMC)	WT/GN	Igniter Transistor 1 Control	7% duty cycle
24 (TMM)	BL/BK	Igniter Transistor 1 Control	7% duty cycle
25	BK/RD	Fuel Pressure Up Solenoid	1v (at hot restart)
27	BK/YL	A/T-ECT Solenoid (SL)	In Lockup: 12v
28	BR	Power Ground	<0.1v
32	BL/WT	Closed Throttle Switch	1v, at off-idle: 12v
33	WT/BK	Power Ground	<0.1v
34	WT/BK	Power Ground	<0.1v

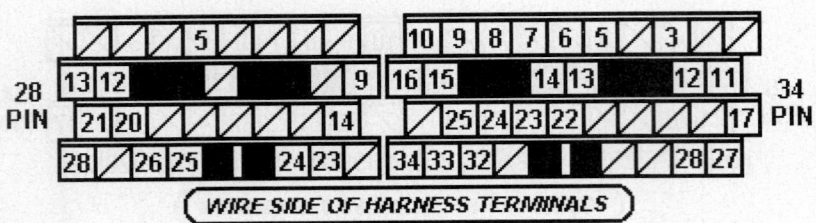

28 PIN   34 PIN

WIRE SIDE OF HARNESS TERMINALS

05533_ADIA_G028

Pin Connector Graphic

1997 3.0L V6 MFI VIN F W/EIS (A/T-ECT) 16 Pin Connector

PCM Pin #	Wire Color	Circuit Description (16 Pin)	Value at Hot Idle
2	LG	EVAP Purge Solenoid (VSV)	12v or 0v
3	GN/RD	MIL (lamp) Control	MIL Off: 12v, On: 1v
5	BL/WT	Data Link Connector	12-14v
6	RD/YL	Intake Air Control Solenoid	12v or 0v
7	RD/BK	MAF Sensor Ground	<0.050v
8	WT/RD	EVAP Vapor Pressure (VSV)	12v or 0v
9	GN/WT	Cooling Fan 1 Relay	Relay Off: 12v, On: 1v
10	YL/RD	HO2S-21 (B2 S1) Heater	1v (Heater On)
11	BL/BK	HO2S-11 (B1 S1) Heater	1v (Heater On)
12	YL/GN	EGR Solenoid Control (VSV)	12v or 0v
13	BL/RD	EVAP Vapor Pressure Sensor	2.9-3.1v (with hose off)
14	GN/YL	EGR Gas Temp. Sensor	3.5-4.0v
15	WT/GN	EGR Valve Position Sensor	1.1-1.9v
16	BR	Sensor Ground	<0.050v

1997 3.0L V6 MFI VIN F W/EIS (A/T-ECT) 22 Pin Connector

PCM Pin #	Wire Color	Circuit Description (22 Pin)	Value at Hot Idle
1	YL	Sensor VREF (VC)	4.9-5.1v
2	BK/YL	A/T-ECT Solenoid (SLN-)	During shifting: 12v
4	WT/BL	OD Clutch Speed Signal (-)	AC pulse signals
5	BK/RD	CKP Sensor Signal (NE+)	AC pulse signals
6	BL	CKP Sensor Signal (NE-)	<0.050v
7	BL	TP Sensor Signal	0.3-0.8v
8	PK	MAF Sensor Signal	1.1-1.5v
9	RD	OD Clutch Speed Signal (+)	AC pulse signals
13	WT	HO2S-11 (B1 S1) Signal	0.1-1.1v
14	WT	Knock Sensor 1 Signal	<0.075v AC
15	WT	Knock Sensor 2 Signal	<0.075v AC
17	BK/WT	CMP Sensor Signal (G+)	AC pulse signals
18	GN/RD	Circuit Opening Relay (FC)	0-3v, off-idle: 12v
19	RD/BL	HO2S-21 (B2 S1) Signal	0.1-1.1v
20	GN/BK	ECT Sensor Signal	At 180°F: 0.51v
21	BL/YL	IAT Sensor Signal	At 100°F: 2.60v
22	BR	Sensor Ground	<0.050v

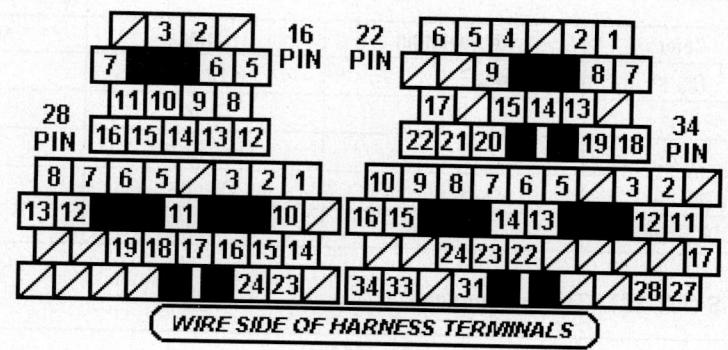

05533_ADIA_G029

Pin Connector Graphic

1997 3.0L V6 MFI VIN F W/EIS (A/T-ECT) 28 Pin Connector

PCM Pin #	Wire Color	Circuit Description (28 Pin)	Value at Hot Idle
1	YL	A/T Select Switch Low	In Low: 12v, Others: 0v
2	BK/YL	Mirror Heater Switch Signal	Switch On: 12v, Off: 0v
3	GN/OR	Tail Light Switch Signal	Switch On: 12v, Off: 0v
5	GN/BK	A/C Amplifier Signal (ACT)	Clutch On: 12v, Off: 1.5v
6	GN/OR	Overdrive Main Switch	Switch Off: 12v
7	YL/BK	Cruise Control ECU	At Cruise in OD: 12v
8	WT	Data Link Connector	12v
10	BL/WT	A/T Select Switch 2nd	In 2nd: 12v, Others: 0v
11	BL/RD	Cruise Control ECU	At Cruise in OD: 12v
12	PK/WT	Vehicle Speed Sensor	At 55 mph: 48 Hz
13	BK/OR	Tachometer Signal (TACO)	Pulses
14	BK/YL	Battery Direct	12-14v
15	RD/BK	A/T Select Switch Reverse	In 'R': 12v, Others: 0v
16	BK/YL	A/C Amplifier Signal (AC1)	Clutch On: 1.5v, Off: 12v
17	PK/BK	HO2S-12 (B1 S2) Heater	1v (Heater On)
18	BK	HO2S-12 (B1 S2) Signal	0.1-1.1v
19	BR/WT	ABS/Traction ECU (NEO)	Pulse Signals
23	BK/YL	EFI Main Relay Power	12-14v
24	GN/WT	Stop Light Switch Signal	Brake Off: 0v, On: 12v

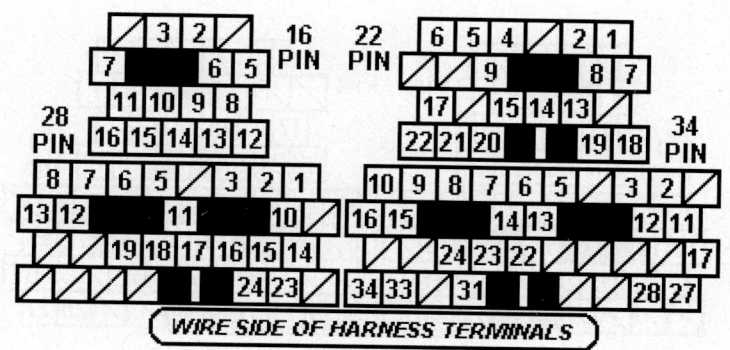

05533_ADIA_G029

Pin Connector Graphic

Standard Colors and Abbreviations

Abbreviation	Color	Abbreviation	Color	Abbreviation	Color
BK	Black	GY	Gray	RD	Red
BL	Blue	GN	Green	TN	Tan
BR	Brown	LG	Light Green	VT	Violet
DB	Dark Blue	OR	Orange	WT	White
DG	DK Green	PK	Pink	YL	Yellow

1997 3.0L V6 MFI VIN F W/EIS (A/T-ECT) 34 Pin Connector

PCM Pin #	Wire Color	Circuit Description (34 Pin)	Value at Hot Idle
2	BR	Shield Ground	<0.050v
3	WT/BL	A/T-ECT Solenoid (SLN)	During shifting: 12v
5	GN	Injector 6 Control	2.0-3.3 ms
6	RD/BL	Injector 5 Control	2.0-3.3 ms
7	WT	Injector 4 Control	2.0-3.3 ms
8	YL	Injector 3 Control	2.0-3.3 ms
9	RD	Injector 2 Control	2.0-3.3 ms
10	BL	Injector 1 Control	2.0-3.3 ms
11	PK	A/T-ECT Solenoid (S1)	In 3rd or OD: 1v
12	WT/RD	Igniter Signal (IGF)	Digital Signal: 0-5-0v
13	GY	Starter Switch Signal	9-11v (cranking)
14	BK/WT	A/T Neutral Start Switch	In P/N: 9-11v (cranking)
15	GN/BK	Igniter Transistor 3 Control	7% duty cycle
16	BR/YL	Igniter Transistor 2 Control	7% duty cycle
17	BL/BK	A/T-ECT Solenoid (S2)	1st or OD: 1v
22	YL/BK	IAC Signal (RSC)	Pulse Signals
23	RD/WT	IAC Signal (RSO)	Pulse Signals
24	GY	Igniter Transistor 1 Control	7% duty cycle
27	PK/BL	A/T-ECT Solenoid (SL)	In Lockup: 12-14v
28	BR	Power Ground	<0.1v
31	BK/BL	PSP Switch Signal	Straight: 12v, Turning: 0v
33	BR	Power Ground	<0.1v
34	BR	Power Ground	<0.1v

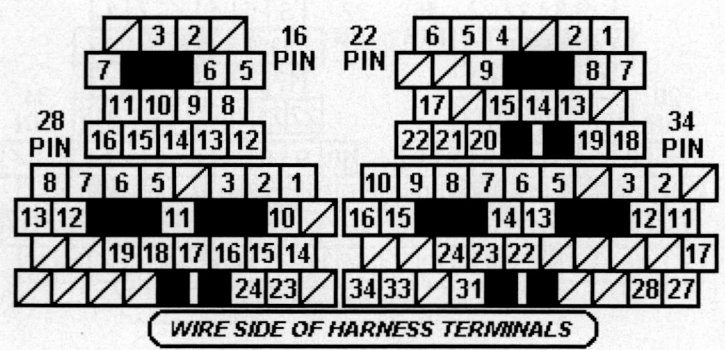

WIRE SIDE OF HARNESS TERMINALS

05533_ADIA_G029

Pin Connector Graphic

Standard Colors and Abbreviations

Abbreviation	Color	Abbreviation	Color	Abbreviation	Color
BK	Black	GY	Gray	RD	Red
BL	Blue	GN	Green	TN	Tan
BR	Brown	LG	Light Green	VT	Violet
DB	Dark Blue	OR	Orange	WT	White
DG	DK Green	PK	Pink	YL	Yellow

1998-99 3.0L V6 VIN F W/EIS-TC (A/T) 17 Pin Connector

PCM Pin #	Wire Color	Circuit Description (17 Pin)	Value at Hot Idle
4	GN/WT	Transponder Amplifier Code	Inserting key: pulses
5	RD/BL	Transponder Amplifier Signal	Inserting key: pulses
10	BL/YL	Transponder Amplifier Signal	Inserting key: pulses
11	BL/BK	Unlock Warning Switch	No Key: 4-5v
13	BR	Sensor Data Link Ground	<0.050v
16	RD/YL	Theft Deterrent ECU	Pulse Signals

1998-99 3.0L V6 VIN F W/EIS-TC (A/T) 22 Pin Connector

PCM Pin #	Wire Color	Circuit Description (22 Pin)	Value at Hot Idle
1	BK/YL	Battery Direct	12-14v
2	BK/RD	Ignition Switch Power	12-14v
3	GN/RD	Circuit Opening Relay (FC)	0-3v, off-idle: 12v
6	GN/RD	MIL (lamp) Control	MIL Off: 12v, On: 1v
7	GY	Starter Switch Signal	9-11v (cranking)
7	BK/OR	Starter Switch Signal	9-11v (cranking)
8	BK/WT	EFI Main Relay Power	12-14v
9	WT/RD	EVAP Vapor Pressure (VSV)	12v or 0v
9	PK	EVAP Vapor Pressure (VSV)	12v or 0v
11	WT	SIL Signal (Scan Tool)	Digital Signal
13	LG	ABS/Traction ECU (TRC+)	Pulse Signals
14	WT	ABS/Traction ECU (EFI+)	Pulse Signals
15	GN/WT	Stop Light Switch Signal	Brake Off: 0v, On: 12v
16	BK/YL	EFI Main Relay Power	12-14v
17 (TMC)	BL/RD	EVAP Vapor Pressure Sensor	2.9-3.1v (with hose off)
17 (TMM)	PK	EVAP Vapor Pressure Sensor	2.9-3.1v (with hose off)
18	BK/YL	Mirror Heater Switch Signal	Switch On: 12v, Off: 0v
19	GN/OR	Tail Light Switch Signal	Switch On: 12v, Off: 0v
20	BL	ABS/Traction ECU (TRC-)	Pulse Signals
21	WT	ABS/Traction ECU (EFI-)	Pulse Signals

1998-99 3.0L V6 VIN F W/EIS-TC (A/T) 28 Pin Connector

PCM Pin #	Wire Color	Circuit Description (28 Pin)	Value at Hot Idle
2	RD/BK	A/T Select Switch Reverse	In 'R': 12v, Others: 0v
3 (TMC)	BL/WT	A/T Select Switch 2nd	In 2nd: 12v, Others: 0v
3 (TMM)	OR	A/T Select Switch 2nd	In 2nd: 12v, Others: 0v
4	BL/RD	Cruise Control Signal (IDLO)	1.5v, off-idle: 12v
8	BK	HO2S-12 (B1 S2) Signal	0.1-1.1v
9	PK/BK	HO2S-12 (B1 S2) Heater	1v (Heater On)
10	GN/OR	Overdrive Main Switch	Switch Off: 12v
12	YL	A/T Select Switch Low	In Low: 12v, Others: 0v
13	GN/BK	A/C Amplifier Signal (ACT)	Clutch On: 12v, Off: 1.5v
14	PK	A/C Amplifier Signal (THWO)	Pulse Signals
16	BR/WT	ABS/Traction ECU (NEO)	Pulse Signals
20	BK/WT	Neutral Start Switch	In P/N: 9-11v (cranking)
22	PK/WT	Vehicle Speed Sensor	At 55 mph: 48 Hz
24	YL/BK	Cruise Control Signal (OD1)	At Cruise in OD: 12v
25	BK/YL	A/C Amplifier Signal (AC1)	Clutch On: 1.5v, Off: 12v
27	BK/OR	Tachometer Signal (TACO)	Pulses

1998-99 3.0L V6 VIN F W/EIS-TC (A/T) 24 Pin Connector

PCM Pin #	Wire Color	Circuit Description (24 Pin)	Value at Hot Idle
1	BR	Sensor Ground	<0.050v
2	YL	Sensor VREF (VC)	4.9-5.1v
3 (Cal)	BK/RD	AFS-11 (B1 S1) Heater	1v (Heater On)
3 (Fed)	BL/BK	HO2S-11 (B1 S1) Heater	1v (Heater On)
4 (Cal)	BK/WT	AFS-21 (B2 S1) Heater	1v (Heater On)
4 (Fed)	YL/RD	HO2S-21 (B2 S1) Heater	1v (Heater On)
5	BL	Injector 1 Control	2.0-3.3 ms
6	RD	Injector 2 Control	2.0-3.3 ms
7	LG	EVAP Purge Solenoid (VSV)	12v or 0v
8	BR	Power Ground	<0.1v
9	BK/BL	PSP Switch Signal	Straight: 12v, Turning: 0v
10	PK	MAF Sensor Signal	1.1-1.5v
11 (TCM)	GN	AFS-11 (B1 S1) Signal (AF+)	3.0-3.6v
11 (TMM)	BR	AFS-11 (B1 S1) Signal (AF+)	3.0-3.6v
11 (Fed)	WT	HO2S-11 (B1 S1) Signal	0.1-1.1v
12 (TMC)	BK/WT	AFS-21 (B2 S1) Signal (AF+)	3.0-3.6v
12 (TMM)	BL	AFS-21 (B2 S1) Signal (AF+)	3.0-3.6v
12 (Fed)	BK	HO2S-21 (B2 S1) Signal	0.1-1.1v
13	GN/YL	EGR Gas Temp. Sensor	3.5-4.0v
14	GN/BK	ECT Sensor Signal	At 180°F: 0.51v
16	BK/RD	CKP Sensor Signal (NE+)	AC pulse signals
17	BR	Shield Ground	<0.050v
18	BR	Sensor Ground	<0.050v
19	RD/BK	MAF Sensor Ground	<0.050v
20 (TMC)	BK/RD	AFS-11 (B1 S1) Signal (AF-)	Fixed at 3v

1998-99 3.0L V6 VIN F W/EIS-TC (A/T) 24 Pin Connector, *continued*

20 (TMM)	RD	AFS-11 (B1 S1) Signal (AF-)	Fixed at 3v
21 (TMC)	BK/WT	AFS-21 (B2 S1) Signal (AF-)	Fixed at 3v
21 (TMM)	BL	AFS-21 (B2 S1) Signal (AF-)	Fixed at 3v
22	BL/YL	IAT Sensor Signal	At 100°F: 2.60v
23	BL	TP Sensor Signal	0.3-1.1v
24	BL	CKP/CMP Sensor Ground (-)	<0.050v

Pin Connector Graphic

05533_ADIA_G006

1998-99 3.0L V6 VIN F W/EIS-TC (A/T) 31 Pin Connector

PCM Pin #	Wire Color	Circuit Description (31 Pin)	Value at Hot Idle
1	YL	Injector 3 Control	2.0-3.3 ms
2	WT	Injector 4 Control	2.0-3.3 ms
3	RD/BL	Injector 5 Control	2.0-3.3 ms
4	GN	Injector 6 Control	2.0-3.3 ms
6	BL/WT	Data Link Connector (TE)	12-14v
7	PK	A/T-ECT Solenoid (S1)	In 3rd or OD: 1v
8	BL/BK	A/T-ECT Solenoid (S2)	1st or OD: 1v
9	PK/BL	A/T-ECT Solenoid (SL)	In Lockup: 12-14v
10	BK/WT	CMP Sensor Signal (G+)	AC pulse signals
11	GY	Igniter Transistor 1 Control	7% duty cycle
12	BR/YL	Igniter Transistor 2 Control	7% duty cycle
13	GN/BK	Igniter Transistor 3 Control	7% duty cycle
14	RD	OD Clutch Speed Signal (+)	AC pulse signals
15	YL/BK	IAC Signal (RSC)	Pulse Signals
16	RD/WT	IAC Signal (RSO)	Pulse Signals
17	RD/YL	ACIS Solenoid Control (VSV)	12v or 0v
18	YL/GN	EGR Solenoid Control (VSV)	12v or 0v
19	BK/YL	A/T-ECT Solenoid (SLN-)	During shifting: 12v
20	WT/BL	A/T-ECT Solenoid (SLN+)	During shifting: 12v
21	BR	Power Ground	<0.1v
22	WT/GN	EGR Valve Position Sensor	0.4-1.6v
23	BR	A/T: Sensor Ground	<0.050v
24 (Fed)	BR	M/T: Sensor Ground	<0.050v
25	WT/RD	Igniter Signal (IGF)	Digital Signal: 0-5-0v
26	GN	OD Clutch Speed Signal (-)	AC pulse signals
27	WT	Knock Sensor 1 Signal	<0.075v AC
28	WT	Knock Sensor 2 Signal	<0.075v AC
29	GN/WT	Cooling Fan 1 Relay	Relay Off: 12v, On: 1v
30	WT/BK	Power Ground	<0.1v
31	WT/BK	Power Ground	<0.1v

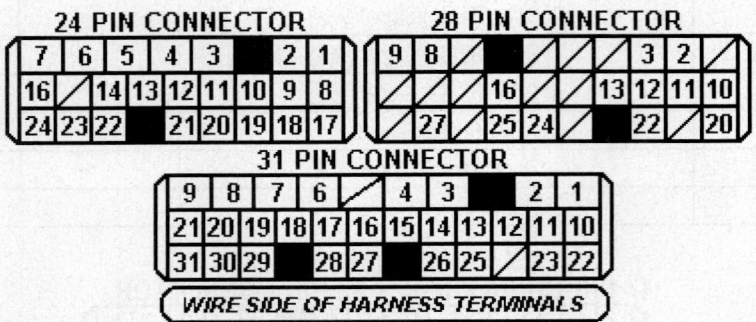

24 PIN CONNECTOR   **28 PIN CONNECTOR**

31 PIN CONNECTOR

WIRE SIDE OF HARNESS TERMINALS

05533_ADIA_G030

Pin Connector Graphic
1998 3.0L V6 VIN F W/O EIS-TC Federal 28 Pin Connector

PCM Pin #	Wire Color	Circuit Description (28 Pin)	Value at Hot Idle
1	YL	A/T Select Switch Low	In Low: 12v, Others: 0v
2	GN/OR	Tail Light Switch Signal	Switch On: 12v, Off: 0v
3	BK/YL	Mirror Heater Switch Signal	Switch On: 12v, Off: 0v
5	GN/BK	A/C Amplifier Signal (ACT)	Clutch On: 12v, Off: 1.5v
6	GN/OR	Overdrive Main Switch	Switch Off: 12v
7	YL/BK	Cruise Control Signal (OD1)	At Cruise in OD: 12v
8	WT	Data Link Connector	No Scan Tool: 0v
10 (TMC)	BL/WT	A/T Select Switch 2nd	In 2nd: 12v, Others: 0v
10 (TMM)	OR	A/T Select Switch 2nd	In 2nd: 12v, Others: 0v
11	BL/RD	Cruise Control Signal (IDLO)	1.5v, off-idle: 12v
12	PK/WT	Vehicle Speed Sensor	At 55 mph: 48 Hz
13	BK/OR	Tachometer Signal (TACO)	Pulses
14	BK/YL	Battery Direct	12-14v
15	RD/BK	A/T Select Switch Reverse	In 'R': 12v, Others: 0v
16	BK/YL	A/C Amplifier Signal (AC1)	Clutch On: 1.5v, Off: 12v
17	PK/BK	HO2S-12 (B1 S2) Heater	1v (Heater On)
18	BK	HO2S-12 (B1 S2) Signal	0.1-1.1v
23	BK/YL	EFI Main Relay Power	12-14v
24	GN/WT	Stop Light Switch Signal	Brake Off: 0v, On: 12v

1998 3.0L V6 VIN F W/O EIS-TC Federal 16 Pin Connector

PCM Pin #	Wire Color	Circuit Description (16 Pin)	Value at Hot Idle
2	LG	EVAP Purge Solenoid (VSV)	12v or 0v
3	GN/RD	MIL (lamp) Control	MIL Off: 12v, On: 1v
5	BL/WT	Data Link Connector	12-14v
6	RD/YL	Intake Air Control Solenoid	12v or 0v
7	RD/BK	MAF Sensor Ground	<0.050v
8	WT/RD	EVAP Vapor Pressure (VSV)	12v or 0v
9	GN/WT	Cooling Fan 1 Relay	Relay Off: 12v, On: 1v
10	YL/RD	HO2S-21 (B2 S1) Heater	1v (Heater On)
11	BL/BK	HO2S-11 (B1 S1) Heater	1v (Heater On)
12	YL/GN	EGR Solenoid Control (VSV)	12v or 0v

1998 3.0L V6 VIN F W/O EIS-TC Federal 16 Pin Connector, *continued*

13	BL/RD	EVAP Vapor Pressure Sensor	2.9-3.1v (with hose off)
14	GN/YL	EGR Gas Temp. Sensor	3.5-4.0v
15	WT/GN	EGR Valve Position Sensor	1.1-1.9v
16	BR	Sensor Ground	<0.050v

Note: TMC indicates a vehicle produced by Toyota Motor Corp. (Japan) and TMM indicates a vehicle produced by Toyota Motor Manufacturer (USA).

1998 3.0L V6 VIN F W/O EIS-TC Federal 22 Pin Connector

PCM Pin #	Wire Color	Circuit Description (22 Pin)	Value at Hot Idle
1	YL	Sensor VREF (VC)	4.9-5.1v
2	BK/YL	A/T-ECT Solenoid (SLN-)	During shifting: 12v
4	GN	OD Clutch Speed Signal (-)	AC pulse signals
5	BK/RD	CKP Sensor Signal (NE+)	AC pulse signals
6	BL	CKP Sensor Signal (NE-)	<0.050v
7	BL	TP Sensor Signal	0.3-0.8v
8	PK	MAF Sensor Signal	1.1-1.5v
9	RD	OD Clutch Speed Signal (+)	AC pulse signals
13	WT	HO2S-11 (B1 S1) Signal	0.1-1.1v
14	WT	Knock Sensor 2 Signal	<0.075v AC
15	WT	Knock Sensor 1 Signal	<0.075v AC
17	BK/WT	CMP Sensor Signal (G+)	AC pulse signals
18	GN/RD	Circuit Opening Relay (FC)	0-3v, off-idle: 12v
19	BK	HO2S-21 (B2 S1) Signal	0.1-1.1v
20	GN/BK	ECT Sensor Signal	At 180°F: 0.51v
21	BL/YL	IAT Sensor Signal	At 100°F: 2.60v
22	BR	Sensor Ground	<0.050v

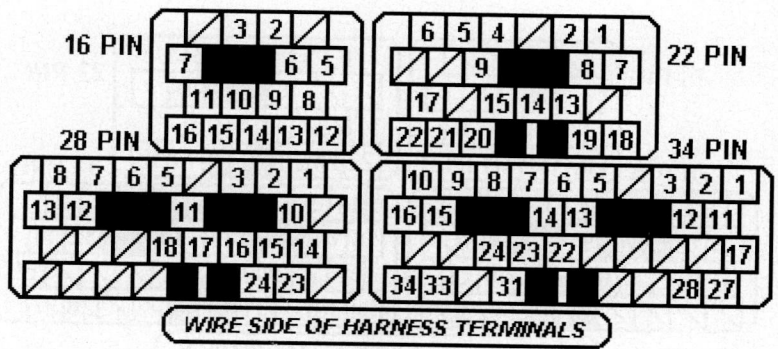

Pin Connector Graphic

05533_ADIA_G031

1998 3.0L V6 VIN F W/O EIS-TC Federal 28 Pin Connector

PCM Pin #	Wire Color	Circuit Description (28 Pin)	Value at Hot Idle
1	BR	A/T: Sensor Ground	<0.1v
2	BR	M/T: Sensor Ground	<0.1v
3	WT/BL	A/T-ECT Solenoid (SLN+)	During shifting: 12v
5	GN	Injector 6 Control	2.0-3.3 ms
6	RD/BL	Injector 5 Control	2.0-3.3 ms
7	WT	Injector 4 Control	2.0-3.3 ms
8	YL	Injector 3 Control	2.0-3.3 ms
9	RD	Injector 2 Control	2.0-3.3 ms
10	BL	Injector 1 Control	2.0-3.3 ms
11	PK	A/T-ECT Solenoid (S1)	In 3rd or OD: 1v
12	WT/RD	Igniter Signal (IGF)	Digital Signal: 0-5-0v
13	GY	Starter Switch Signal	9-11v (cranking)
13	BK/OR	Starter Switch Signal	9-11v (cranking)
14	BK/WT	A/T Neutral Start Switch	In P/N: 9-11v (cranking)
15	GN/BK	Igniter Transistor 3 Control	7% duty cycle
16	BR/YL	Igniter Transistor 2 Control	7% duty cycle
17	BL/BK	A/T-ECT Solenoid (S2)	1st or OD: 1v
22	YL/BK	IAC Signal (RSC)	Pulse Signals
23	RD/WT	IAC Signal (RSO)	Pulse Signals
24	GY	Igniter Transistor 1 Control	7% duty cycle
27	PK/BL	A/T-ECT Solenoid (SL)	In Lockup: 12-14v
28	BR	Power Ground	<0.1v
31	BK/BL	PSP Switch Signal	Straight: 12v, Turning: 0v
33	BR	Power Ground	<0.1v
34	BR	Power Ground	<0.1v

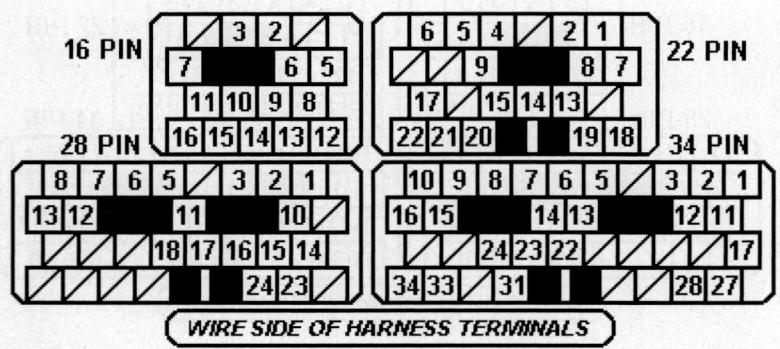

05533_ADIA_G031

Pin Connector Graphic

Standard Colors and Abbreviations

Abbreviation	Color	Abbreviation	Color	Abbreviation	Color
BK	Black	GY	Gray	RD	Red
BL	Blue	GN	Green	TN	Tan
BR	Brown	LG	Light Green	VT	Violet
DB	Dark Blue	OR	Orange	WT	White
DG	DK Green	PK	Pink	YL	Yellow

2000-01 3.0L V6 VIN F W/EIS-TC (A/T) E11 31P Connector

PCM Pin #	Wire Color	Circuit Description (31 Pin)	Value at Hot Idle
1	YL	Injector 3 Control	1.6-2.9 ms
2	WT	Injector 4 Control	1.6-2.9 ms
3	RD/BL	Injector 5 Control	1.6-2.9 ms
4	GN	Injector 6 Control	1.6-2.9 ms
6	BL/WT	Data Link Connector (TE1)	12-14v
7	VT	A/T-ECT Solenoid (S1)	In 3rd or OD: 1v
8	BL/BK	A/T-ECT Solenoid (S2)	1st or OD: 1v
9	PK/BL	A/T-ECT Solenoid (SL)	In Lockup: 12-14v
10	BK/WT	CMP Sensor Signal (G+)	AC pulse signals
11	GY	Igniter Transistor 1 Control	7% duty cycle
12	BR/YL	Igniter Transistor 2 Control	7% duty cycle
13	GN/BK	Igniter Transistor 3 Control	7% duty cycle
14	RD	OD Clutch Speed Signal (+)	AC pulse signals
15	YL/BK	IAC Signal (RSC)	Pulse Signals
16	RD/WT	IAC Signal (RSO)	Pulse Signals
17	RD/YL	ACIS Solenoid Control (VSV)	12v or 0v
18	YL/GN	EGR Solenoid Control (VSV)	12v or 0v
19	BK/YL	A/T-ECT Solenoid (SLN-)	During shifting: 12v
20	WT/BL	A/T-ECT Solenoid (SLN+)	During shifting: 12v
21	BR	Power Ground (E01)	<0.1v
22	WT/GN	EGR Valve Position Sensor	0.4-1.6v
23	GN/OR	Overdrive (OD) Main Switch	Switch Off: 12v, On: 1v
25	WT/RD	Igniter Signal (IGF)	Digital Signal: 0-5-0v
26	GN	OD Clutch Speed Signal (-)	AC pulse signals
27	WT	Knock Sensor 1 Signal	<0.075v AC
28	WT	Knock Sensor 2 Signal	<0.075v AC
29	GN/WT	Cooling Fan 1 Relay	Relay Off: 12v, On: 1v
30	WT/BK	Power Ground (E03)	<0.1v
31	WT/BK	Power Ground (E02)	<0.1v

E11 31-PIN CONNECTOR

WIRE SIDE OF HARNESS TERMINALS

05533_ADIA_G032

Pin Connector Graphic

2000-01 3.0L V6 VIN F W/EIS-TC (A/T) E7 22P Connector

PCM Pin #	Wire Color	Circuit Description (22 Pin)	Value at Hot Idle
1	BK/YL	Battery Direct	12-14v
2	BK/RD	Ignition Switch Power	12-14v
3	GN/RD	Circuit Opening Relay (FC)	0-3v, off-idle: 12v
6	GN/RD	MIL (lamp) Control	MIL Off: 12v, On: 1v
7 (TMC)	GY	Starter Switch Signal	9-11v (cranking)
7 (TMM)	BK/OR	Starter Switch Signal	9-11v (cranking)
8	BK/RD	EFI Main Relay Power	12-14v
9	PK	Overdrive "Off" Indicator	LED Off: 12v, On: 1v
11	WT	SIL Signal (Scan Tool)	Digital Signal
13	LG	ABS/Traction ECU (TRC+)	Pulse Signals
14	WT	ABS/Traction ECU (EFI+)	Pulse Signals
15	GN/WT	Stop Light Switch Signal	Brake Off: 0v, On: 12v
16	BK/YL	EFI Main Relay Power	12-14v
17	PK	EVAP Vapor Pressure Sensor	2.9-3.1v (with hose off)
18	BK/YL	Mirror Heater Switch Signal	Switch Off: 1.5v, On: 12v
19	GN/OR	Tail Light Switch Signal	Switch Off: 1.5v, On: 12v
20	BL	ABS/Traction ECU (TRC-)	Pulse Signals
21	BK	ABS/Traction ECU (EFI-)	Pulse Signals

2000-01 3.0L V6 VIN F W/EIS-TC (A/T) E8 28P Connector

PCM Pin #	Wire Color	Circuit Description (28 Pin)	Value at Hot Idle
1	WT/RD	EVAP Vapor Pressure (VSV)	12v or 0v
2	RD/BK	A/T Select Switch Reverse	In 'R': 12v, Others: 0v
3 (TMC)	BL/WT	A/T Select Switch 2nd	In 2nd: 12v, Others: 0v
3 (TMM)	OR	A/T Select Switch 2nd	In 2nd: 12v, Others: 0v
4	BL/RD	Cruise Control Signal (IDLO)	1.5v, off-idle: 12v
5	BL/WT	EVAP Canister Closed Valve	12v or 0v
8	BK	HO2S-12 (B1 S2) Signal	0.1-1.1v
9	PK/GN	HO2S-12 (B1 S2) Heater	1v (Heater On)
12	YL	A/T Select Switch Low	In Low: 12v, Others: 0v
13	GN/BK	A/C Amplifier Signal (ACT)	Clutch On: 12v, Off: 1.5v
14	VT	A/C Amplifier Signal (THWO)	Pulse Signals

2000-01 3.0L V6 VIN F W/EIS-TC (A/T) E8 28P Connector, *continued*

16	BR/WT	ABS/Traction ECU (NEO)	Pulse Signals
20	BK/WT	A/T Neutral Start Switch	In P/N: 9-11v (cranking)
22	PK/WT	Speedometer Indicator	At 55 mph: 48 Hz
24	YL/BK	Cruise Control Signal (OD1)	At Cruise in OD: 12v
25	BK/YL	A/C Amplifier Signal (AC1)	Clutch On: 1.5v, Off: 12v
27	BK/OR	Tachometer Signal (TACO)	Pulse Signals

2000-01 3.0L V6 VIN F W/EIS-TC (A/T) E9 17P Connector

PCM Pin #	Wire Color	Circuit Description (17 Pin)	Value at Hot Idle
1-3, 6-9	---	Not Used	---
4	GN/WT	Transponder Amplifier Code	Inserting key: pulses
5	RD/BL	Transponder Amplifier Signal	Inserting key: pulses
12, 15-16	---	Not Used	---
10	BL/YL	Transponder Amplifier Signal	Inserting key: pulses
11	BL/BK	Unlock Warning Switch	Inserting key: pulses
13	BR	Power Ground (EOM)	<0.1v
16	RD/YL	Theft Deterrent ECU Signal	No key inserted: pulses

2000-01 3.0L V6 VIN F W/EIS-TC (A/T) E10 24P Connector

PCM Pin #	Wire Color	Circuit Description (24 Pin)	Value at Hot Idle
1 (TMC)	BR	Sensor Ground (E04)	<0.050v
2	YL	Sensor VREF (VC)	4.9-5.1v
3 (Cal)	BK/RD	AFS-11 (B1 S1) Heater	1v (Heater On)
3 (Fed)	BL/BK	HO2S-11 (B1 S1) Heater	1v (Heater On)
4	YL/RD	HO2S-21 (B2 S1) Heater	1v (Heater On)
5	BL	Injector 1 Control	1.6-2.9 ms
6	RD	Injector 2 Control	1.6-2.9 ms
7	LG	EVAP Purge Solenoid (VSV)	12v or 0v
8	BR	Power Ground (E05)	<0.1v
9	BK/BL	PSP Switch Signal	Straight: 12v, Turning: 0v
10	PK	MAF Sensor Signal	1.1-1.5v
11 (Cal)	BR	AFS-11 (B1 S1) Signal (AF+)	3.0-3.6v
12 (Fed)	BK	HO2S-21 (B2 S1) Signal	0.1-1.1v
13	GN/YL	EGR Gas Temp. Sensor	3.5-4.0v
14	GN/BK	ECT Sensor Signal	At 180°F: 0.51v
16	BK/RD	CKP Sensor Signal (NE+)	AC pulse signals
17	BR	Power Ground (E1)	<0.050v
18	BR	Sensor Ground (E2)	<0.050v
19	RD/BK	MAF Sensor Ground	<0.050v
20 (Cal)	BK/RD	AFS-11 (B1 S1) Signal (AF-)	Fixed at 3v
22	BL/YL	IAT Sensor Signal	At 100°F: 2.60v
23	BL	TP Sensor Signal	0.3-1.1v
24	BL	CKP/CMP Sensor Ground (-)	<0.050v

Note: TMC indicates a vehicle produced by Toyota Motor Corp. (Japan) and TMM indicates a vehicle produced by Toyota Motor Manufacturer (USA).

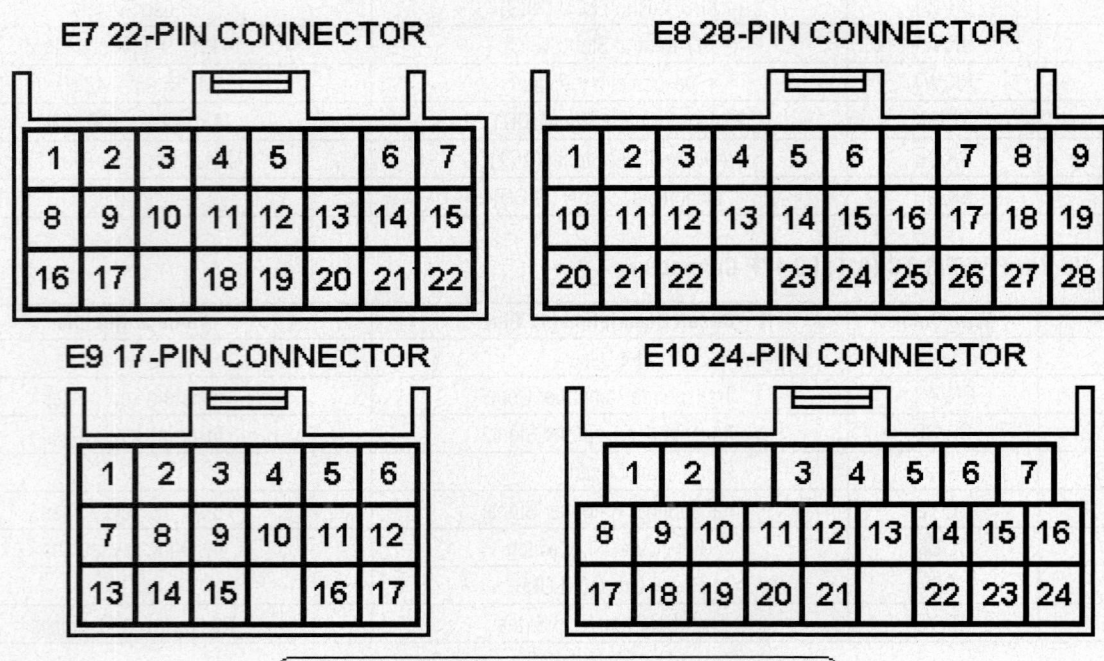

WIRE SIDE OF HARNESS TERMINALS

05533_ADIA_G033

Pin Connector Graphic

2002-06 3.0L V6 VIN F W/EIS-TC (A/T) E6 31P Connector

PCM Pin #	Wire Color	Circuit Description (31 Pin)	Value at Hot Idle
1	BK/RD	EFI Main Relay Power	12-14v
2-3	---	Not Used	---
4	VT	EVAP Vapor Pressure (VSV)	12v or 0v
5	BK/OR	Tachometer Signal (TACO)	DC pulse signals
6-7	---	Not Used	---
8	BK/WT	EFI Main Relay Power	12-14v
9	BK/OR	Ignition Switch Power (B+)	12-14v
10	GN/RD	Circuit Opening Relay (FC)	0-3v, off-idle: 12v
11	PK/BK	Data Link Connector (TC)	12-14v
12	GN/RD	MIL (lamp) Control	MIL Off: 12v, On: 1v
13	GN	Tail Light Switch Signal	Switch Off: 0v, On: 12v
14	BK/YL	Mirror Heater Switch Signal	Switch Off: 1.5v, On: 12v
15	BR	Power Ground (EOM)	<0.1v
16	GN/OR	Overdrive (OD) Main Switch	Switch Off: 12v, On: 1v
17	BR/WT	(NEO)	---
18	WT	SIL Signal (Scan Tool)	Digital Signal
19	RD	Scan Tool (WFSE)	12v
20	---	Not Used	---
21	PK	EVAP Vapor Pressure Sensor	2.9-3.1v (with hose off)
22	BL/YL	APP Sensor Signal (VPA)	1.1-1.5v
23	WT/RD	APP Sensor Signal (VPA2)	0.9-2.3v
24	WT	ABS/Traction ECU (ENG+)	Pulse Signals

2002-06 3.0L V6 VIN F W/EIS-TC (A/T) E6 31P Connector, *continued*

25	GN	ABS/Traction ECU (TRC+)	Pulse Signals
26	RD	APP Sensor VREF (VCPA)	4.5-5.5v
27	BR/RD	APP Sensor VREF (VCP2)	4.5-5.5v
28	LG/BK	APP Sensor Ground (EPA)	<0.050v
29	LG	APP Sensor Ground (EPA2)	<0.050v
30	BK	ABS/Traction ECU (ENG-)	Pulse Signals
31	BL	ABS/Traction ECU (TRC-)	Pulse Signals

Note: TMC indicates a vehicle produced by Toyota Motor Corp. (Japan) and TMM indicates a vehicle produced by Toyota Motor Manufacturer (USA).

E6 31-Pin Connector

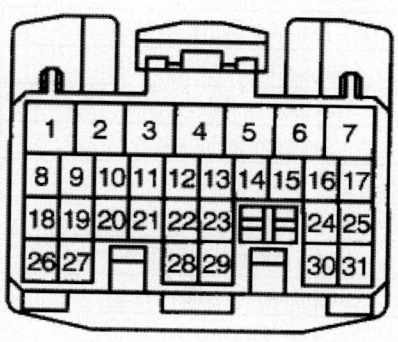

Pin Connector Graphic

05533_ADIA_G034

2002-06 3.0L V6 VIN F W/EIS-TC (A/T) E7 35P Connector

PCM Pin #	Wire Color	Circuit Description (35 Pin)	Value at Hot Idle
1, 4-5	---	Not Used	---
2	BK/YL	Battery Direct (BATT)	12-14v
3	BL/WT	A/C Magnetic Clutch (ACMG)	A/C On: <3.0v
6	BL/RD	Battery Direct (+BM)	12-14v
7	OR	Overdrive "Off" Indicator	LED Off: 12v, On: 1v
8	YL	A/T Select Switch Low Signal	In Low: 12v, Others: 0v
9	BL/WT	A/T Select Switch 2nd Signal	In 2nd: 12v, Others: 0v
10	WT/BL	A/T Select Switch 'D' Signal	In 'D': 12v, Others: 0v
11	RDBK	A/T Select Switch 'R' Signal	In 'R': 12v, Others: 0v
12-13	---	Not Used	---
14	YL/GN	A/C Amplifier Signal (THWO)	Pulse Signals
15	GN/WT	Transponder Amplifier Code	Inserting key: pulses
16	VT	Transponder IMLD Signal	Inserting key: pulses
17	VT/WT	Speedometer Indicator	Moving: 0-5-0v
18, 20-25	---	Not Used	---
19	GN/WT	Stop Light Switch Signal	Brake Off: 0v, On: 12v
26	BL/YL	Theft Deterrent ECU Signal	No key inserted: pulses
27	RD/BL	Transponder Amplifier Signal	Inserting key: pulses
28-30	---	Not Used	---
31	PK/BL	A/C Switch Signal (Auto AC)	A/C On: 9-14v
31	WT	A/C Switch Signal (Manual AC)	A/C On: 9-14v
32, 35	---	Not Used	---
33	BK	A/C Switch Signal (Auto AC)	A/C On: 9-14v
33	YL/BK	A/C Switch Signal (Manual AC)	A/C On: 9-14v
34	BL	Transponder Amplifier Signal	Key Inserted: <1.5v

2002-06 3.0L V6 VIN F W/EIS-TC (A/T) E8 32P Connector

PCM Pin #	Wire Color	Circuit Description (32 Pin)	Value at Hot Idle
1	BR	Power Ground (E1)	<0.050v
2	WT	Throttle Body Motor (M-)	DC pulse signals
3	BK	Throttle Body Motor (M+)	DC pulse signals
4	WT/BL	Power Ground (ME01)	<0.1v
5	YL	HO2S-12 (B1 S2) Heater	1v (Heater On)
6	BL	HO2S-22 (B2 S2) Heater	1v (Heater On)
7	WT/BK	Power Ground (E03)	<0.1v
8	GN/WT	Cooling Fan Relay Control	Relay Off: 12v, On: 1v
9, 13-16	---	Not Used	---
10	RD/WT	PSP Switch Signal	Straight: 12v, Turning: 0v
11	LG	EGR Solenoid Control (VSV)	12v or 0v
12	BL	EVAP Canister Closed Valve	12v or 0v
17	BR	Shield Ground (GE01)	<0.050v
18-26	---	Not Used	---
27	BK/WT	CMP Sensor Signal (G22+)	AC pulse signals
28-32	---	Not Used	---

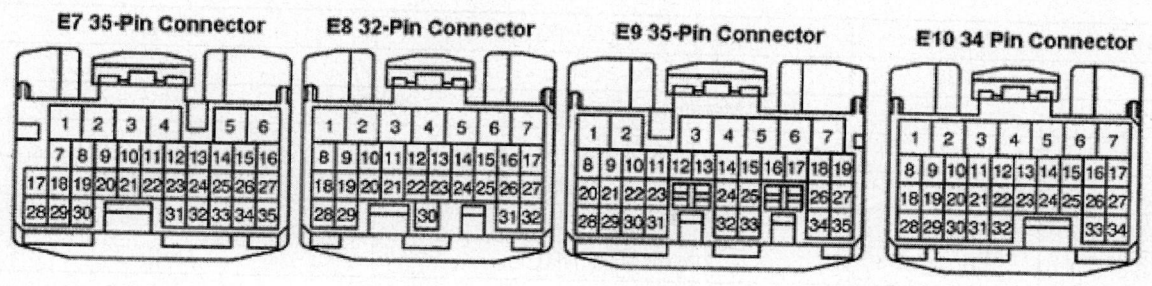

E7 35-Pin Connector   E8 32-Pin Connector   E9 35-Pin Connector   E10 34 Pin Connector

05533_ADIA_G035

Pin Connector Graphic
2002-06 3.0L V6 VIN F W/EIS-TC (A/T) E9 35P Connector

PCM Pin #	Wire Color	Circuit Description (35 Pin)	Value at Hot Idle
1	BK	Knock Sensor 1 Signal	<0.075v AC
2	WT	Knock Sensor 2 Signal	<0.075v AC
3	BK/RD	AFS-21 (B2 S1) Heater	1v (Heater On)
4	BK/WT	AFS-11 (B1 S1) Heater	1v (Heater On)
5	GN	Injector 6 Control	1.6-2.9 ms
6	WT/BK	Power Ground (E05)	<0.1v
7	WT/BK	Sensor Ground (E04)	<0.050v
8	BK/YL	Neutral Start Switch	In P/N: 9-11v (cranking)
9	BK/WT	Starter Switch Signal (STA)	9-11v (cranking)
10-12	---	Not Used	---
13	BL	A/T-ECT Overdrive Solenoid	Moving in OD: 12-14v
14-15, 25	---	Not Used	---
16	BL/RD	A/T-ECT Solenoid (SL2)	Moving in 3rd or OD: 1v
17	BL/YL	A/T-ECT Solenoid (SL2+)	Moving in 3rd or OD: 1v
18	PK	A/T-ECT Solenoid (SL1-)	Moving in 1st Gear: 1v
19	RD/BK	A/T-ECT Solenoid (SL1+)	Moving in 1st Gear: 1v
20	YL/GN	EGR Solenoid Control (VSV)	12v or 0v
21	WT	HO2S-11 (B1 S1) Signal	0.1-1.1v
22	BR	AFS-11 (B1 S1) Signal (AFR+)	3.0-3.6v
23	OR	AFS-21 (B2 S1) Signal (AFL)	3.0-3.6v
24	RD	MAF Sensor Signal (VG)	1.1-1.5v
26	RD	Clutch Speed Signal (NC+)	AC pulse signals
27	Rd	Clutch Speed Signal (NT+)	AC pulse signals
28	WT/GN	EGR Valve Position Sensor	0.4-1.6v
29	OR	HO2S-22 (B2 S2) Signal	0.1-1.1v
30	BK/RD	AFS-11 (B1 S1) Signal (AF-)	Fixed at 3v
31	WT	AFS-11 (B1 S1) Signal (AF-)	Fixed at 3v
32	BL/WT	MAF Sensor Ground (EG2)	<0.050v
34	GN	Clutch Speed Signal (NC-)	AC pulse signals
35	BL	Clutch Speed Signal (NT-)	AC pulse signals

2002-06 3.0L V6 VIN F W/EIS-TC (A/T) E10 34P Connector

PCM Pin #	Wire Color	Circuit Description (34 Pin)	Value at Hot Idle
1	BL	Injector 1 Control	2.0-3.3 ms
2	RD	Injector 2 Control	2.0-3.3 ms
3	YL	Injector 3 Control	2.0-3.3 ms
4	WT	Injector 4 Control	2.0-3.3 ms
5	RD/BL	Injector 5 Control	1.6-2.9 ms
6	WT/BK	Power Ground (E02)	<0.050v
7	WT/BK	Power Ground (E01)	<0.050v
8	RD/WT	Igniter Transistor 1 Control	7% duty cycle
9	PK	Igniter Transistor 2 Control	7% duty cycle
10	LG/BK	Igniter Transistor 3 Control	7% duty cycle
11	BL/YL	Igniter Transistor 4 Control	7% duty cycle
12	GN/RD	Igniter Transistor 5 Control	7% duty cycle
13	BL	Igniter Transistor 6 Control	7% duty cycle
14, 22, 26	---	Not Used	---
15	RD/YL	A/C Amplifier Signal (ACIS)	Clutch On: 12v, Off: 1.5v
16	YL/BK	A/T-ECT Solenoid (STN-)	During Shifting: 12v
17	YL/RD	A/T-ECT Solenoid (STN+)	During shifting: 12v
18	YL	Sensor VREF (VC)	4.9-5.1v
19	GN/BK	ECT Sensor Signal	At 180°F: 0.51v
20	LB	IAT Sensor Signal	At 100°F: 2.60v
21	LG	TP Sensor Signal (VTA1)	0.3-1.1v
23	WT/RD	Igniter Signal (IGF)	Digital Signal: 0-5-0v
25	WT	A/C Amplifier Signal (AICV)	Clutch On: 1.5v, Off: 12v
27	RD	CKP Sensor Signal (NE+)	AC pulse signals
28	BR	Power Ground (E2)	<0.1v
29	GN/YL	EGR Gas Temperature Sensor	---
30 (TMC)	GN	A/T ATF Sensor Signal	<1.5v at 239°F
30 (TMM)	GN/RD	A/T ATF Sensor Signal	<1.5v at 239°F
31	BK/RD	TP Sensor 2 Signal (VTA2)	2.0-2.9v
32-33	---	Not Used	---
34	GN	CKP/CMP Sensor Ground (-)	<0.050v

CELICA PIN CHARTS

1990-93 2.0L MFI VIN S (M/T) 16 Pin Connector

PCM Pin #	Wire Color	Circuit Description (16 Pin)	Value at Hot Idle
1	PK/BL	Sensor VREF	0-5v
2	GY/BK	Airflow Meter Signal	1.5-2.5v
3	GY	IAT Sensor Signal	At 100°F: 2.60v
4	RD	ECT Sensor Signal	At 180°F: 0.51v
5 ('90-'91)	WT	Knock Sensor Signal	<0.075v AC
5 ('92-'93)	RD/WT	Turbo Pressure Sensor	2.5-4.5v
6	WT	O2S Signal	0.1-1.1v
8	PK/YL	Check Connector	12-14v
9	BR	Sensor Ground	<0.050v
10 (Cal)	BL	EGR Gas Temp. Sensor	3.5-4.0v
11	PK/BK	TP Sensor Signal	0.3-0.8v
12	PK	Closed Throttle Switch	1v, at off-idle: 12v
13	RD/WT	Turbo Pressure Sensor	2.5-4.5v
13 ('92-'93)	WT	Knock Sensor Signal	<0.075v AC
14	WT	O2S Check	0.1-1.1v
15	OR	Check Connector	12-14v
16 ('90-'91)	BK	Distributor Signal (G-)	<0.050v

1990-93 2.0L MFI VIN S (M/T) 22 Pin Connector

PCM Pin #	Wire Color	Circuit Description (22 Pin)	Value at Hot Idle
1	PK	Battery Direct	12-14v
2	OR	Defogger/Light Idle Up Signal	Switch On: 12v, Off: 0v
4	GN/WT	Stop Light Switch Signal	Brake Off: 0v, On: 12v
5	PK	MIL (lamp) Control	MIL Off: 12v, On: 1v
6	GN/RD	Fuel Pump Relay	Relay Off: 12v, On: 1v
9	BL/WT	Vehicle Speed Sensor	At 55 mph: 48 Hz
10	BK/WT	A/C Magnetic Clutch (ACMG)	Clutch Off: 0v, On: 12v
11	BK	Starter Switch Signal	9-11v (cranking)
12	BK/YL	EFI Main Relay Power	12-14v
13	BK/YL	EFI Main Relay Power	12-14v
14 ('92-'93)	GN	Circuit Opening Relay (FC)	0-3v, off-idle: 12v
21	GN/BK	A/C Amplifier Signal (ACT)	Clutch On: 12v, Off: 1.5v

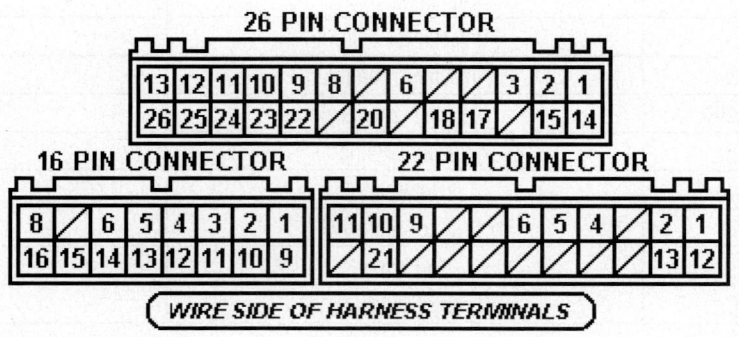

Pin Connector Graphic

05533_ADIA_G036

1990-91 2.0L MFI VIN S (M/T) 26 Pin Connector

PCM Pin #	Wire Color	Circuit Description (26 Pin)	Value at Hot Idle
1	YL	Distributor Signal (NE+)	AC pulse signals
2	BL	Distributor Signal (G2+)	AC pulse signals
3	BK/YL	Distributor Signal (IGF)	Digital Signal: 0-5-0v
4-5, 7	---	Not Used	---
6	BL/RD	Turbo Pressure Solenoid	12v or 0v
8	PK/BK	HO2S Heater	1v (Heater On)
9	GN/WT	ISC Motor (ISC1)	Pulse Signals
10	GN	Cold Start Injector Control	1v (at cold startup)
11	GN	Injector 2 Control	2.0-3.3 ms
12	RD	Injector 1 Control	2.0-3.3 ms
13	WT/BK	Power Ground	<0.1v
14	BR	Sensor Ground	<0.050v
15	GN	Distributor Signal (G1+)	AC pulse signals
16, 19, 21	---	Not Used	---
17	GN	Circuit Opening Relay (FC)	0-3v, off-idle: 12v
18	LG	TN-VIS Solenoid	12v or 0v
20	WT	Igniter Signal (IGT)	0.74-0.76v
22	GN/BK	ISC Motor (ISC2)	Pulse Signals
23	GN/RD	EGR Solenoid Control (VSV)	12v or 0v
24	BL	Injector 4 Control	2.0-3.3 ms
25	YL	Injector 3 Control	2.0-3.3 ms
26	WT/BK	Power Ground	<0.1v

1992-93 2.0L I4 VIN S (M/T) 26 Pin Connector

PCM Pin #	Wire Color	Circuit Description (26 Pin)	Value at Hot Idle
1	BL/GN	TN-VIS Solenoid	12v or 0v
2	BL/RD	Turbo Pressure Solenoid	12v or 0v
3	BK/YL	Distributor Signal (IGF)	0.72-0.74v
4	YL	Distributor Signal (NE+)	AC pulse signals
5	BL	Distributor Signal (G2+)	AC pulse signals
6	GN/RD	EGR Solenoid Control (VSV)	12v or 0v
7	GN	Cold Start Injector Control	1v (at cold startup)
8	PK/BK	HO2S Heater	1v (Heater On)
9	GN/WT	ISC Signal (RSC)	Pulse Signals
10	GN/BK	ISC Signal (RSO)	Pulse Signals
11	GN	Injector 2 Control	2.0-3.3 ms
12	RD	Injector 1 Control	2.0-3.3 ms
13	WT/BK	Power Ground	<0.1v
14	BR	Sensor Ground	<0.050v
15-16	---	Not Used	---
17	BK	Distributor Signal (G-)	<0.050v
19	---	Not Used	---
18	RD	Distributor Signal (G1+)	AC pulse signals

1992-93 2.0L I4 VIN S (M/T) 26 Pin Connector, *continued*

20	WT	Igniter Signal (IGT)	0.74-0.76v
21-23	---	Not Used	---
24	BL	Injector 4 Control	2.0-3.3 ms
25	YL	Injector 3 Control	2.0-3.3 ms
26	WT/BK	Power Ground	<0.1v

1990-93 1.6L I4 MFI VIN A (All) 16 Pin Connector

PCM Pin #	Wire Color	Circuit Description (16 Pin)	Value at Hot Idle
1	BK/YL	EFI Main Relay Power	12-14v
2	PK	Battery Direct	12-14v
4	GN	Circuit Opening Relay (FC)	0-3v, off-idle: 12v
7	GN/BK	A/C Amplifier Signal (ACT)	Clutch On: 12v, Off: 1.5v
8	OR	Data Link Connector	12-14v
9	BK/YL	EFI Main Relay Power	12-14v
10	PK	MIL (lamp) Control	MIL Off: 12v, On: 1v
12	BK/WT	A/C Magnetic Clutch (ACMG)	Clutch Off: 0v, On: 12v
13	BL/WT	Vehicle Speed Sensor	At 55 mph: 48 Hz
14 ('90-'91)	YL/BK	Overdrive Solenoid	Solenoid Off: 12v, On: 1v
15 ('92-'93)	YL/BK	Overdrive Solenoid	Solenoid Off: 12v, On: 1v
16	PK/YL	Data Link Connector	12-14v

1990-93 1.6L I4 MFI VIN A (All) 26 Pin Connector

PCM Pin #	Wire Color	Circuit Description (26 Pin)	Value at Hot Idle
1 (Cal)	BL/BK	EGR Solenoid Control (VSV)	12v or 0v
2	BK	A/T: Starter Switch Signal	9-11v (cranking)
3	RD	ECT Sensor Signal	At 180°F: 0.51v
4	GY/BK	MAP Sensor Signal	1-1.5v
5	GY	IAT Sensor Signal	At 100°F: 2.60v
6	BK	Igniter Signal (IGT)	0.74-0.76v
7	BK/YL	Igniter Signal (IGF)	0.72-0.74v
8	RD	Distributor Signal (G1+)	AC pulse signals
9	BK	Distributor Signal (G-)	<0.050v
10 (Cal)	WT	O2S-11 Signal	0.1-1.1v
10 (Fed)	WT	HO2S-11 Signal	0.1-1.1v
11	BK	Starter Switch Signal	9-11v (cranking)
12	WT	Injector 1 & 3 Control	2.0-3.3 ms
13	WT/BK	Power Ground	<0.1v
14	BL	ISC (+) Motor Solenoid	Pulse Signals
15 (Cal)	YL/BK	Overdrive Solenoid	12v or 0v
15 (Fed)	PK/BK	HO2S-11 Heater	1v (Heater On)
16, 22, 24	BR	Sensor Ground	<0.050v
17	PK/BK	Wide Open Throttle Switch	1v, at off-idle: 12v
18	PK/BL	Vacuum Sensor VREF	4.9-5.1v
19	PK	Closed Throttle Switch	1v, at off-idle: 12v
20 (Cal)	BL	EGR Gas Temp. Sensor	3.5-4.0v
21	WT	Distributor Signal (NE+)	AC pulse signals
25	YL	Injector 2 & 4 Control	2.0-3.3 ms
26	WT/BK	Power Ground	<0.1v

26 PIN CONNECTOR

```
13 12 11 10 9 8 7 6 5 4 3 2 1
26 25 24 /  22 21 20 19 18 17 16 15 14
```

16 PIN CONNECTOR

```
8 7 / / 4 / 2 1
16 15 14 13 12 / 10 9
```

WIRE SIDE OF HARNESS TERMINALS

05533_ADIA_G037

Pin Connector Graphic

Standard Colors and Abbreviations

Abbreviation	Color	Abbreviation	Color	Abbreviation	Color
BK	Black	GY	Gray	RD	Red
BL	Blue	GN	Green	TN	Tan
BR	Brown	LG	Light Green	VT	Violet
DB	Dark Blue	OR	Orange	WT	White
DG	DK Green	PK	Pink	YL	Yellow

1994-95 1.8L I4 MFI VIN A (A/T-ECT) 16 Pin Connector

PCM Pin #	Wire Color	Circuit Description (16 Pin)	Value at Hot Idle
1	RD	Sensor VREF	4.9-5.1v
2	GY/BK	MAP Sensor Signal	1-1.5v
3	BL/RD	IAT Sensor Signal	At 100°F: 2.60v
4	GN	ECT Sensor Signal	At 180°F: 0.51v
5	WT	Sub O2S Signal	0.1-1.1v
6	WT	Main O2S Signal	0.1-1.1v
7	BK	Data Link Connector	12-14v
8	RD/WT	Data Link Connector	12-14v
9	BR	Sensor Ground	<0.050v
10 (Cal)	BK/RD	EGR Gas Temp. Sensor	3.5-4.0v
11	BK/WT	TP Sensor Signal	0.3-0.8v
12	BL/WT	Closed Throttle Switch	1v, at off-idle: 12v
13	WT	Knock Sensor Signal	<0.075v AC
14	LG	Data Link Connector	12-14v
15	YL	Data Link Connector	12-14v

1994-95 1.8L I4 MFI VIN A (A/T-ECT) 22 Pin Connector

PCM Pin #	Wire Color	Circuit Description (22 Pin)	Value at Hot Idle
1	PK	Battery Direct	12-14v
2	OR	Defogger/Light Idle Up Signal	Switch On: 12v, Off: 0v
4	GN/WT	Stop Light Switch Signal	Brake Off: 0v, On: 12v
5	RD/GN	MIL (lamp) Control	MIL Off: 12v, On: 1v
9	OR	Vehicle Speed Sensor	At 55 mph: 48 Hz
10	BL/BK	A/C Amplifier Signal (ACT)	Clutch On: 12v, Off: 1.5v
11	BK	Starter Switch Signal	9-11v (cranking)
12	BK/YL	EFI Main Relay Power	12-14v
14	GN/RD	Circuit Opening Relay (FC)	0-3v, off-idle: 12v
21	GN/YL	A/C Amplifier Signal (AC1)	Clutch On: 1.5v, Off: 12v
22	BK/YL	A/T Neutral Start Switch	In P/N: 9-11v (cranking)

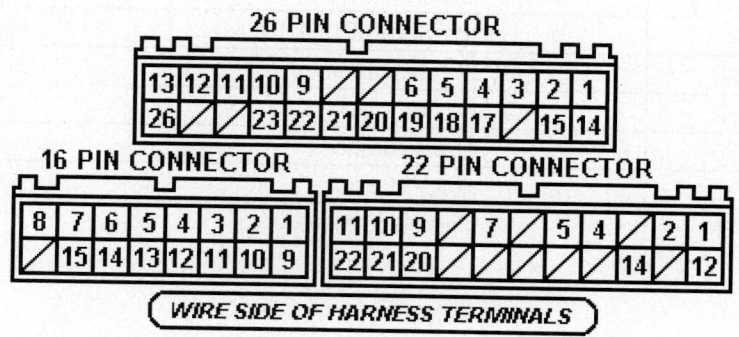

Pin Connector Graphic

05533_ADIA_G038

26 PIN CONNECTOR   **16 PIN CONNECTOR**

13	12	11	10	9	8	7	6	5	4	3	2	1
26	25	24	/	22	21	20	19	18	17	16	15	14

8	7	/	/	4	/	2	1
16	15	14	13	12	/	10	9

WIRE SIDE OF HARNESS TERMINALS

05533_ADIA_G037

Pin Connector Graphic
Standard Colors and Abbreviations

Abbreviation	Color	Abbreviation	Color	Abbreviation	Color
BK	Black	GY	Gray	RD	Red
BL	Blue	GN	Green	TN	Tan
BR	Brown	LG	Light Green	VT	Violet
DB	Dark Blue	OR	Orange	WT	White
DG	DK Green	PK	Pink	YL	Yellow

1994-95 1.8L I4 MFI VIN A (A/T-ECT) 26 Pin Connector

PCM Pin #	Wire Color	Circuit Description (26 Pin)	Value at Hot Idle
1	BL/YL	A/T-ECT Solenoid (SL)	In Lockup: 12-14v
2	BL/RD	A/T-ECT Solenoid (S1)	In 3rd or OD: 1v
3	BK/YL	Igniter Signal (IGF)	Digital Signal: 0-5-0v
4	YL	Distributor Signal (NE+)	AC pulse signals
5	BK	Distributor Signal (G-)	<0.050v
6	PK/GN	A/T Select Switch 2nd	In 2nd: 12v, Others: 0v
9	GN/WT	ISC Signal (RSC)	Pulse Signals
10	BK/WT	ISC Signal (RSO)	Pulse Signals
11	YL	Injector 2 Control	2.0-3.3 ms
12	WT	Injector 1 Control	2.0-3.3 ms
13	WT/BK	Power Ground	<0.1v
14	BR	Sensor Ground	<0.050v
15	BR/YL	A/T-ECT Solenoid (S2)	1st or OD: 1v
17	BK	Distributor Signal (NE-)	<0.050v
18	RD	Distributor Signal (G+)	AC pulse signals
19	PK/RD	A/T Select Switch Low	In Low: 12v, Others: 0v
20	BK	Igniter Signal (IGT)	Digital Signal: 0-5-0v
22	BL/GN	A/T Select Switch Reverse	In 'R': 12v, Others: 0v
23	BL/BK	EGR Solenoid Control (VSV)	12v or 0v
24	BK/RD	Injector 4 Control	2.0-3.3 ms
25	BK/YL	Injector 3 Control	2.0-3.3 ms
26	WT/BK	Power Ground	<0.1v

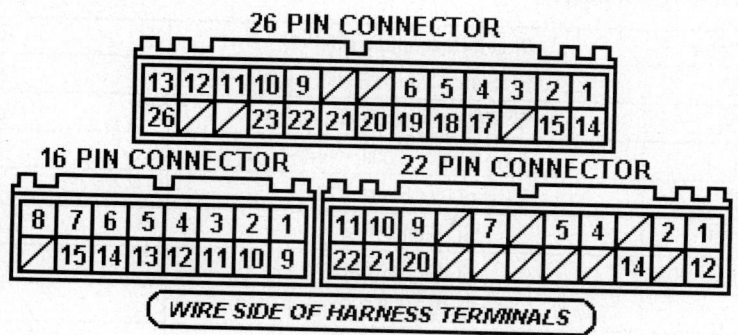

Pin Connector Graphic

Pin Connector Graphic
Standard Colors and Abbreviations

Abbreviation	Color	Abbreviation	Color	Abbreviation	Color
BK	Black	GY	Gray	RD	Red
BL	Blue	GN	Green	TN	Tan
BR	Brown	LG	Light Green	VT	Violet
DB	Dark Blue	OR	Orange	WT	White
DG	DK Green	PK	Pink	YL	Yellow

1994-95 1.8L I4 VIN A (M/T) 16 Pin Connector

PCM Pin #	Wire Color	Circuit Description (16 Pin)	Value at Hot Idle
2	GY/BK	MAP Sensor Signal	2.5-4.5v
3	BL/RD	IAT Sensor Signal	At 100°F: 2.60v
4	RD	ECT Sensor Signal	At 180°F: 0.51v
4	GN	ECT Sensor Signal	At 180°F: 0.51v
5	WT	Sub O2S Signal	0.1-1.1v
6	WT	Main O2S Signal	0.1-1.1v
7	LG	Data Link Connector	12-14v
8	RD/WT	Data Link Connector	12-14v
9	BR	Sensor Ground	<0.050v
10	BK/WT	TP Sensor Signal	0.3-0.8v
11	RD	Sensor VREF	4.9-5.1v
12	BL/WT	Closed Throttle Switch	1v, at off-idle: 12v
13	BK/RD	EGR Gas Temp. Sensor	3.5-4.0v
14	WT	Knock Sensor Signal	<0.075v AC
15	YL	Data Link Connector	12-14v

1994-95 1.8L I4 VIN A (M/T) 26 Pin Connector

PCM Pin #	Wire Color	Circuit Description (26 Pin)	Value at Hot Idle
2	BK	Starter Switch Signal	9-11v (cranking)
3	BK/YL	Igniter Signal (IGF)	Digital Signal: 0-5-0v
4	YL	Distributor Signal (NE+)	AC pulse signals
5	RD	Distributor Signal (G1+)	AC pulse signals
9	GN/WT	ISC Signal (RSC)	Pulses
10	WT/BK	ISC Signal (RSO)	Pulses
12	BK/YL	Injector Pair 1 & 3 Control	2.0-3.3 ms
13	WT/BK	Power Ground	<0.1v
16	BL/BK	EGR Solenoid Control (VSV)	12v or 0v
17	BL	Distributor Signal (NE-)	<0.050v
18	BK	Distributor Signal (G-)	<0.050v
22	WT	Igniter Signal (IGT)	Digital Signal: 0-5-0v
23	RD/YL	A/C Idle Up Solenoid	12v or 0v
24	BR	Sensor Ground	<0.050v
25	BK/RD	Injector Pair 2 & 4 Control	2.0-3.3 ms
26	WT/BK	Power Ground	<0.1v

1994-95 1.8L I4 VIN A (M/T) 12 Pin Connector

PCM Pin #	Wire Color	Circuit Description (12 Pin)	Value at Hot Idle
2	PK	Battery Direct	12-14v
4	GN/RD	Circuit Opening Relay (FC)	0-3v, off-idle: 12v
6	GN/YL	A/C Amplifier Signal (ACT)	Clutch On: 12v, Off: 1.5v
7	BK/YL	EFI Main Relay Power	12-14v
8	RD/GN	MIL (lamp) Control	MIL Off: 12v, On: 1v
10	BL/BK	A/C Amplifier Signal (AC1)	Clutch On: 1.5v, Off: 12v
11	OR	Vehicle Speed Sensor	At 55 mph: 48 Hz
12	OR	Mirror Heater Switch Signal	Switch On: 12v, Off: 0v

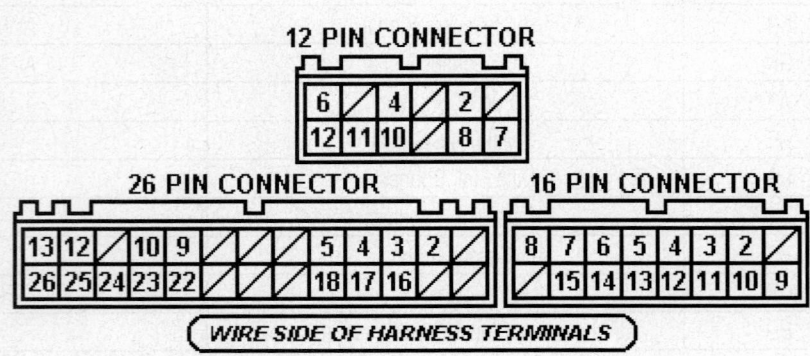

05533_ADIA_G039

Pin Connector Graphic

1996-97 1.8L I4 MFI VIN B (A/T-ECT) 16 Pin Connector

PCM Pin #	Wire Color	Circuit Description (16 Pin)	Value at Hot Idle
1	RD	Sensor VREF	4.9-5.1v
2	BL	MAP Sensor Signal	2.5-4.5v
3	BL/RD	IAT Sensor Signal	At 100°F: 2.60v
4	GN	ECT Sensor Signal	At 180°F: 0.51v
5	WT	HO2S-12 (B1 S2) Signal	0.1-1.1v
6	WT	HO2S-11 (B1 S1) Signal	0.1-1.1v
9	BR	Sensor Ground	<0.050v
11	BK/WT	TP Sensor Signal	0.3-0.8v
13	WT	Knock Sensor Signal	<0.075v AC
15	YL	Data Link Connector	12-14v
16	BR	Power Ground	<0.1v

1996-97 1.8L I4 MFI VIN B (A/T-ECT) 22 Pin Connector

PCM Pin #	Wire Color	Circuit Description (22 Pin)	Value at Hot Idle
1	PK	Battery Direct	12-14v
2	OR	Defogger/Light Idle Up Signal	Switch On: 12v, Off: 0v
4	GN/WT	Stop Light Switch Signal	Brake Off: 0v, On: 12v
5	RD/BK	MIL (lamp) Control	MIL Off: 12v, On: 1v
7	GY/BL	Overdrive Main Switch	Switch Off: 12v
9	OR	Vehicle Speed Sensor	At 55 mph: 48 Hz
10	BL/BK	A/C Amplifier Signal (ACT)	Clutch On: 12v, Off: 1.5v
11	BK	Starter Switch Signal	9-11v (cranking)
12	BK/RD	EFI Main Relay Power	12-14v
14	GN/RD	Circuit Opening Relay (FC)	12-14v
16	WT	Data Link Connector (SDL)	No Scan Tool: 0v
17	RD/WT	A/T Select Switch Reverse	In 'R': 12v, Others: 0v
18	YL/GN	A/T Select Switch 2nd	In 2nd: 12v, Others: 0v
19	YL/RD	A/T Select Switch Low	In Low: 12v, Others: 0v
20	PK	Cruise Control ECU	At Cruise in OD: 12v
21	GN/YL	A/C Amplifier Signal (AC1)	Clutch On: 1.5v, Off: 12v
22	BK/YL	A/T Neutral Start Switch	In P/N: 9-11v (cranking)

Standard Colors and Abbreviations

Abbreviation	Color	Abbreviation	Color	Abbreviation	Color
BK	Black	GY	Gray	RD	Red
BL	Blue	GN	Green	TN	Tan
BR	Brown	LG	Light Green	VT	Violet
DB	Dark Blue	OR	Orange	WT	White
DG	DK Green	PK	Pink	YL	Yellow

1996-97 1.8L I4 MFI VIN B (A/T-ECT) 26 Pin Connector

PCM Pin #	Wire Color	Circuit Description (26 Pin)	Value at Hot Idle
1	BL/YL	A/T-ECT Solenoid (SL)	In Lockup: 12-14v
2	BL/RD	A/T-ECT Solenoid (S1)	In 3rd or OD: 1v
3	GN/YL	Igniter Signal (IGF)	Digital Signal: 0-5-0v
4	OR	CKP Sensor Signal (NE+)	AC pulse signals
5	BK	Distributor Signal (G-)	<0.050v
9	GN/WT	ISC Signal (RSC)	Pulse Signals
10	BK/WT	ISC Signal (RSO)	Pulse Signals
11	BK/RD	Injector 2 Control	2.0-3.3 ms
12	BK/WT	Injector 1 Control	2.0-3.3 ms
13	BR	Power Ground	<0.1v
14	BR	Sensor Ground	<0.050v
15	BR/YL	A/T-ECT Solenoid (S2)	1st or OD: 1v
17	WT	CKP Sensor Signal (NE-)	<0.050v
19	PK	HO2S-12 (B1 S2) Heater	1v (Heater On)
20	WT	Igniter Signal (IGT)	Digital Signal: 0-5-0v
23	BK/BK	EGR Solenoid Control (VSV)	12v or 0v
24	BK/RD	Injector 4 Control	2.0-3.3 ms
25	BK/WT	Injector 3 Control	2.0-3.3 ms
26	BR	Power Ground	<0.1v

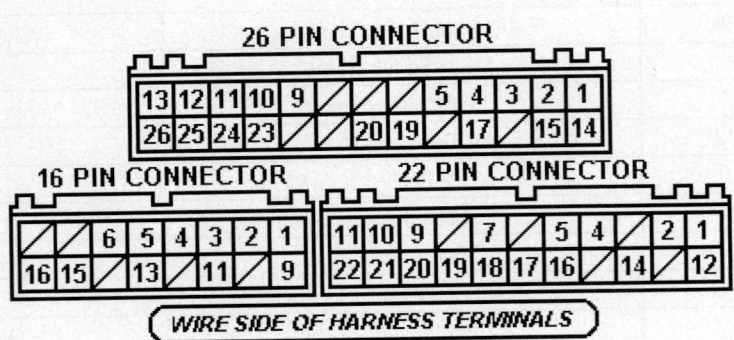

Pin Connector Graphic

2000-05 1.8L I4 1ZZ-FE DOHC VIN R (All) E2 22P Connector

PCM Pin #	Wire Color	Circuit Description (22 Pin)	Value at Hot Idle
1	WT	Battery Direct	12-14v
2	---	Not Used	---
3	GN/RD	Circuit Opening Relay (FC)	0-3v, off-idle: 12v
4	BL/BK	EVAP Vapor Pressure Sensor	Fuel Cap Off: 2.9-3.7v
5	BR/YL	Overdrive Main Switch	Switch Off: 12v, On: 0v
6	LG/RD	Multiplex Meter Input (MPX2)	Digital Signals
7	WT/BK	Power Ground (E03)	<0.1v
8	BK/OR	Ignition Switch Power	12-14v
9	---	Not Used	---
10	BL/WT	Cruise Control Signal (IDLO)	1.5v, off-idle: 12v
11	WT	SIL Signal (Scan Tool)	Digital Signal

2000-05 1.8L I4 1ZZ-FE DOHC VIN R (All) E2 22P Connector, *continued*

12	YL/BK	A/C Magnetic Clutch (ACMG)	Clutch Off: 0v, On: 12v
13-14	---	Not Used	---
15	RD/BK	MIL (lamp) Control	MIL Off: 12v, On: 1v
16	BK/RD	EFI Main Relay Power	12-14v
17-22	---	Not Used	---

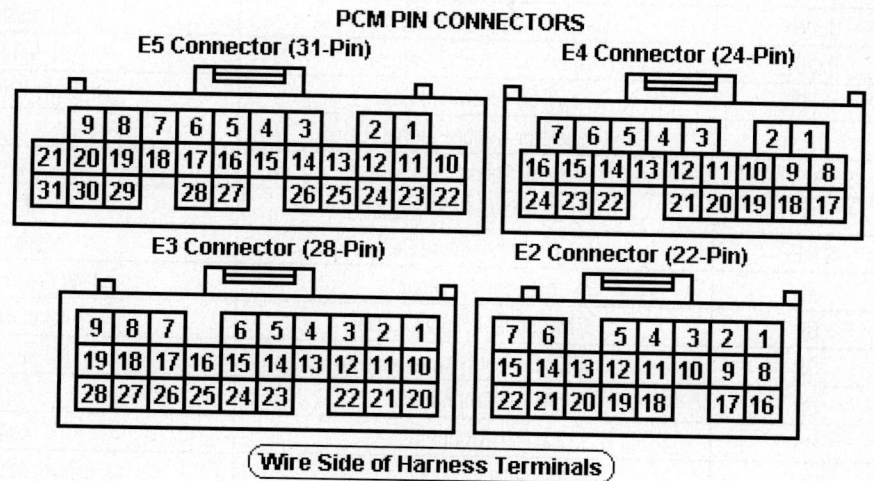

PCM PIN CONNECTORS

Wire Side of Harness Terminals

05533_ADIA_G041

Pin Connector Graphic

2000-05 1.8L I4 1ZZ-FE DOHC VIN R (All) E3 28P Connector

PCM Pin #	Wire Color	Circuit Description (28 Pin)	Value at Hot Idle
1	BK	Cruise Control Signal (OD1)	At Cruise in OD: 12v
2	RD/BK	A/T Select Switch Reverse	In 'R': 12v, Others: 0v
3	BL/WT	Cooling Fan 3 Relay	Relay Off: 12v, On: 1v
4	BL/YL	Cooling Fan 1 Relay	Relay Off: 12v, On: 1v
5	PK/BK	Data Link Connector (TC)	12-14v
6	GN/WT	Stop Light Switch Signal	Brake Off: 0v, On: 12v
7, 13-15	---	Not Used	---
8	PK	Center Airbag Assembly	Digital Signals
9	RD/BL	ACIS Solenoid Control (VSV)	12v or 0v
10	LG/RD	Multiplex Meter Input (MPX1)	Digital Signals
11	BL	Starter Switch Signal	9-11v (cranking)
12	WT/BK	Power Ground (EC)	<0.1v
16	YL/GN	HO2S-12 (B1 S2) Heater	1v (Heater On)
17, 19-20	---	Not Used	---
18	OR	A/C Dual Pressure Switch	A/C Off: 12v, On: 1v
21	BL/BK	EFI Main Relay Power	12-14v
22	WT/RD	Speedometer Indicator	At 55 mph: 48 Hz
23	GN/OR	EVAP Vapor Pressure (VSV)	12v or 0v
24	BL/WT	A/T Select Switch Drive	In 'D': 12v, Others: 0v
25	WT	HO2S-12 (B1 S2) Signal	0.1-1.1v
27	BR/WT	Tachometer Signal (TACO)	Pulse Signals
28	GN	A/C Magnetic Clutch & Lock Sensor	A/C Off: 12v, On: 1v

2000-05 1.8L I4 1ZZ-FE DOHC VIN R (All) E4 24P Connector

PCM Pin #	Wire Color	Circuit Description (24 Pin)	Value at Hot Idle
1	YL/GN	MAF Sensor Ground (EVG)	<0.050v
2	RD	Sensor VREF	4.9-5.1v
3	YL/RD	HO2S-11 (B1 S1) Heater	1v (Heater On)
4	GN/OR	EVAP Purge Solenoid (VSV)	12v or 0v
6	BL/BK	Cam Timing Oil Control VVT-	AC pulse signals
7	BL/WT	Cam Timing Oil Control VVT+	AC pulse signals
8	PK/BL	A/T: Select Switch Neutral	In P/N: 9-11v (cranking)
9	YL/BK	A/T Select Switch Low	In Low: 12v, Others: 0v
10	YL	Generator Control Signal (L)	12v
11	GN/WT	MAF Sensor Signal	1.1-1.5v
12	BK	HO2S-11 (B1 S1) Signal	0.1-1.1v
13	GY/BL	A/T Oil Temperature Sensor	At 230°F: 0.95v
14	GN	ECT Sensor Signal	1-4v (varies with temp.)
15	BL	CMP Sensor Signal (G2+)	AC pulse signals
16	OR	CKP Sensor Signal (NE+)	AC pulse signals
17	BR	Power Ground (E1)	<0.1v
18	BR	Sensor Ground (E2)	<0.050v
19	BL/YL	A/T Select Switch 2nd	In 2nd: 12v, Others: 0v
20	PK	A/T: Select Switch Park	In P/N: 9-11v (cranking)
21	GN	Oil Pressure Switch Signal	Open: 12v, Closed: 0v
22	BL/RD	IAT Sensor Signal	1-4v (varies with temp.)
23	BK/WT	TP Sensor Signal	0.3-1.0v
24	WT	CKP/CMP Sensor Ground (-)	<0.050v

2000-05 1.8L I4 1ZZ-FE DOHC VIN R (All) E5 31P Connector

PCM Pin #	Wire Color	Circuit Description (31 Pin)	Value at Hot Idle
1	RD	Injector 1 Control	2.0-3.3 ms
2	RD/BL	Injector 2 Control	2.0-3.3 ms
3	RD/WT	Injector 3 Control	2.0-3.3 ms
4	RD/BK	Injector 4 Control	2.0-3.3 ms
5	YL/GN	A/T-ECT Solenoid (SLT-)	Pulse signals
6	YL/RD	A/T-ECT Solenoid (SLT+)	Pulse signals
7	GN	A/T-ECT Solenoid (SL1+)	Pulse signals
8	BR/YL	A/T-ECT Solenoid (S1)	In 3rd or OD: 1v
9	VT	A/T-ECT Solenoid (SL1-)	Pulse Signals
10	RD/BK	Igniter Transistor 1 Control	Digital Signal: 0-5-0v
11	RD/WT	Igniter Transistor 2 Control	Digital Signal: 0-5-0v
12	GN/RD	Igniter Transistor 3 Control	Digital Signal: 0-5-0v
13	RD/YL	Igniter Transistor 4 Control	Digital Signal: 0-5-0v
15	BK	A/T: ECT VSS (-) Signal	Pulse signals
16	WT/BL	A/T: ECT VSS (+) Signal	Pulse signals
17	PK/WT	EVAP Canister Closed Valve	12v or 0v
18	BK/WT	IAC Valve Signal (RSO)	Pulse signals
19	GN/WT	A/T-ECT Solenoid (SL)	In Lockup: 12-14v

2000-05 1.8L I4 1ZZ-FE DOHC VIN R (All) E5 31P Connector, *continued*

20	YL	A/T-ECT Solenoid (S2)	1st or OD: 1v
21	WT/BK	Power Ground (E01)	<0.1v
22	YL/BK	OIL Pressure Switch	Switch On: 12-14v
23	WT	Camshaft Timing Oil Control Valve (-)	Pulse signals
24	GN/OR	Camshaft Timing Oil Control Valve (+)	Pulse signals
25	BK/YL	Igniter Signal (IGF)	Digital Signal: 0-5-0v
27	WT	Knock Sensor Signal	<0.075v AC
28	PK	PSP Switch Signal	Straight: 12v, Turning: 0v
29	BR/WT	A/T-ECT ST Signal	Pulse signals
31	WT/BK	Power Ground (E02)	<0.1v

PCM PIN CONNECTORS

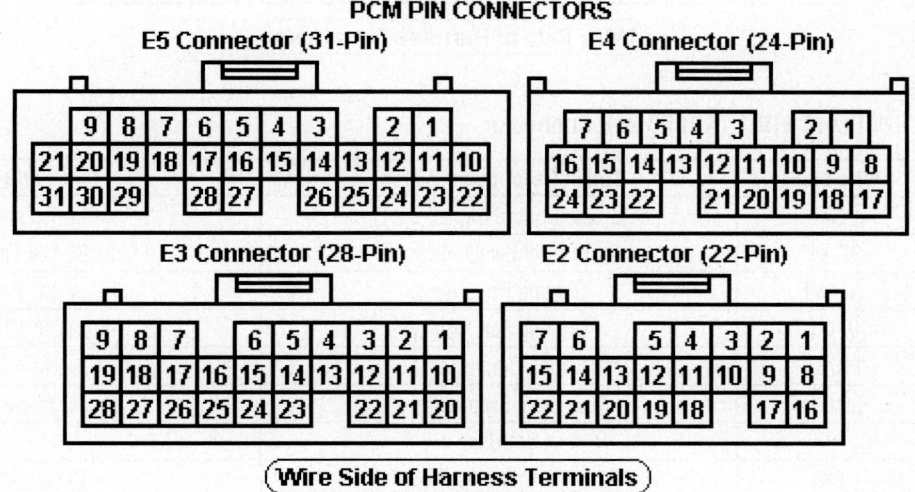

Pin Connector Graphic

2000-05 1.8L I4 2ZZ-GE DOHC VIN Y (All) E2 22P Connector

PCM Pin #	Wire Color	Circuit Description (22 Pin)	Value at Hot Idle
1	WT	Battery Direct	12-14v
2	---	Not Used	---
3	GN/RD	Circuit Opening Relay (FC)	0-3v, off-idle: 12v
4	BL/BK	EVAP Vapor Pressure Sensor	Fuel Cap Off: 2.9-3.7v
5	BR/YL	Overdrive Main Switch	Open: 12v, Closed: 0v
6	LG/RD	Multiplex Meter Input (MPX2)	Digital Signals
7	WT/BK	Power Ground (E03)	<0.1v
8	BK/OR	Ignition Switch Power	12-14v
9	BL	A/T: Trans. Shift Switch SFTU	Switch Off: 12v
10	BL/WT	Cruise Control Signal (IDLO)	1.5v, off-idle: 12v
11	WT	SIL Signal (Scan Tool)	Digital Signals
12	YL/BK	A/C Magnetic Clutch (ACMG)	Clutch Off: 0v, On: 12v
13-14	---	Not Used	---
15	RD/BK	MIL (lamp) Control	MIL Off: 12v, On: 1v
16	BK/RD	EFI Main Relay Power	12-14v
17	PK	A/T: Trans. Shift Switch SFTD	Switch Off: 12v
18-22	---	Not Used	---

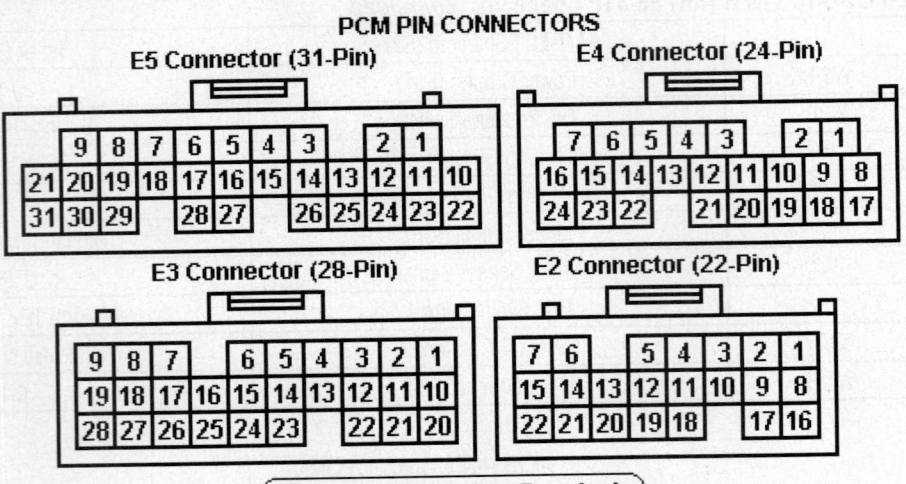

PCM PIN CONNECTORS

Wire Side of Harness Terminals

05533_ADIA_G041

Pin Connector Graphic

2000-05 1.8L I4 2ZZ-GE DOHC VIN Y (All) E3 28P Connector

PCM Pin #	Wire Color	Circuit Description (28 Pin)	Value at Hot Idle
1	BK	Cruise Control Signal (OD1)	At Cruise in OD: 12v
2	RD/BK	A/T Select Switch Reverse	In 'R': 12v, Others: 0v
3	BL/WT	Cooling Fan 3 Relay	Relay Off: 12v, On: 1v
4	BL/YL	Cooling Fan 1 Relay	Relay Off: 12v, On: 1v
5	PK/BK	Data Link Connector (TC)	12-14v
6	GN/WT	Stop Light Switch Signal	Brake Off: 0v, On: 12v
7	---	Not Used	---
8	PK	Center Airbag Assembly	Digital Signals
9	RD/BL	ACIS Solenoid Control (VSV)	12v or 0v
10	LG/RD	Multiplex Meter Input (MPX1)	Digital Signals
11	BL	Starter Switch Signal	9-11v (cranking)
12	WT/BK	Power Ground (EC)	<0.1v
13	---	Not Used	---
14	BK/WT	Cruise Control (D) ECU Input	Pulse Signals
15	---	Not Used	---
16	YL/GN	HO2S-12 (B1 S2) Heater	1v (Heater On)
17	---	Not Used	---
18	OR	A/C Dual Pressure Switch	A/C Off: 12v, On: 1v
19-20	---	Not Used	---
21	BL/BK	EFI Main Relay Power	12-14v
22	WT/RD	Speedometer Indicator	At 55 mph: 48 Hz
23	GN/OR	EVAP Vapor Pressure (VSV)	12v or 0v
24	BL/WT	A/T Select Switch Drive	In 'D': 12v, Others: 0v
25	WT	HO2S-12 (B1 S2) Signal	0.1-1.1v
26	---	Not Used	---
27	BR/WT	Tachometer Signal (TACO)	Pulse Signals
28	GN	A/C Magnetic Clutch & Lock Sensor	A/C Off: 12v, On: 1v

2000-05 1.8L I4 2ZZ-GE DOHC VIN Y (All) E4 24P Connector

PCM Pin #	Wire Color	Circuit Description (24 Pin)	Value at Hot Idle
1	YL/GN	MAF Sensor Ground	<0.050v
2	RD	Sensor VREF	4.9-5.1v
3	YL/RD	HO2S-11 (B1 S1) Heater	1v (Heater On)
4	GN/OR	EVAP Purge Solenoid (VSV)	12v or 0v
6	BL/BK	Cam/Time Oil (OVL-) Valve	DC pulse signals
7	BL/WT	Cam/Time Oil (OVL+) Valve	DC pulse signals
8	PK/BL	A/T: Select Switch Neutral	In P/N: 9-11v (cranking)
9	YL/BK	A/T Select Switch Low	In Low: 12v, Others: 0v
10	YL	Generator Control Signal (L)	12v
11	GN/WT	MAF Sensor Signal	1-3v
12	BK	HO2S-11 (B1 S1) Signal	0.1-1.1v
13	GY/BL	A/T Oil Temperature Sensor	At 230°F: 0.95v
14	GN	ECT Sensor Signal	1-4v (varies with temp.)
15	BL	CMP Sensor Signal (G2+)	AC pulse signals
16	OR	CKP Sensor Signal (NE+)	AC pulse signals
17	BR	Power Ground (E1)	<0.050v
18	BR	Sensor Ground (E2)	<0.050v
19	BL/YL	A/T Select Switch 2nd	In 2nd: 12v, Others: 0v
20	PK	A/T: Select Switch Park	In P/N: 9-11v (cranking)
21	GY	OIL Level Warning Switch	Open: 12v, Closed: 0v
22	BL/RD	IAT Sensor Signal	1-4v (varies with temp.)
23	BK/WT	TP Sensor Signal	1.1-1.9v
24	WT	CKP/CMP Sensor Ground (-)	<0.050v

2000-05 1.8L I4 2ZZ-GE DOHC VIN Y (All) E5 31P connector

PCM Pin #	Wire Color	Circuit Description (31 Pin)	Value at Hot Idle
1	RD	Injector 1 Control	2.0-3.3 ms
2	RD/BL	Injector 2 Control	2.0-3.3 ms
3	RD/WT	Injector 3 Control	2.0-3.3 ms
4	RD/BK	Injector 4 Control	2.0-3.3 ms
5	YL/GN	A/T-ECT Solenoid (SLT-)	Pulse signals
6	YL/RD	A/T ECT Solenoid (SLT+)	Pulse signals
7, 9	GN, VT	A/T ECT Solenoid (SL1+, -)	Pulse signals
8, 20	BR, OR	A/T ECT Solenoid (SL2+, -)	Pulse signals
10	RD/BK	Igniter Transistor 1 Control	Digital Signal: 0-5-0v
11	RD/WT	Igniter Transistor 2 Control	Digital Signal: 0-5-0v
12	GN/RD	Igniter Transistor 3 Control	Digital Signal: 0-5-0v
13	RD/YL	Igniter Transistor 4 Control	Digital Signal: 0-5-0v
14	PK/BK	OD Clutch Speed Signal (+)	AC pulse signals
15	BK	A/T ECT VSS (-) Signal	Pulse signals
16	WT/BL	A/T ECT VSS (+) Signal	Pulse signals
17	PK/WT	EVAP Canister Closed Valve	12v or 0v
18	BK/WT	IAC Valve Signal (RSO)	Pulse signals
19	PK/BL	A/T ECT Solenoid (DSL)	In Lockup: 12-14v
21	WT/BK	Power Ground (E01)	<0.1v
22	YL/BK	OIL Pressure Switch	Open: 12v, Closed: 0v
23	WT	Cam Timing Oil Valve (OCV-)	Pulse signals
24	GN/OR	Cam Timing Oil Valve (OCV+)	Pulse signals
25	BK/YL	Igniter Signal (IGF)	Pulse signals
26	GN/WT	OD Clutch Speed Signal (-)	AC pulse signals
27	WT	Knock Sensor Signal	<0.075v AC
28	PK	PSP Switch Signal	Straight: 12v, Turning: 0v
29	YL	A/T-ECT Solenoid (S4)	In Drive: 1v
31	WT/BK	Power Ground (E02)	<0.1v

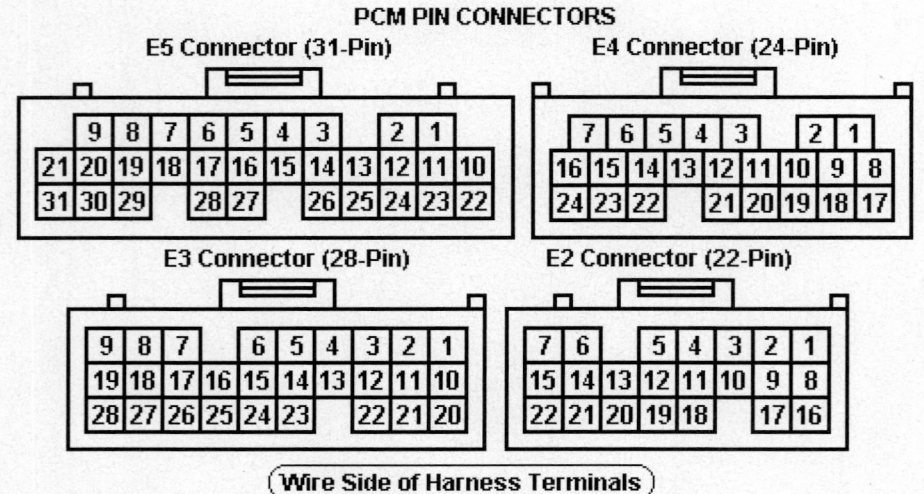

Pin Connector Graphic (View is into Wire Side of Harness)

05533_ADIA_G041

1990-91 2.2L VIN S (All) 16 Pin Connector

PCM Pin #	Wire Color	Circuit Description (16 Pin)	Value at Hot Idle
1	PK/BL	Vacuum Sensor VREF	4.9-5.1v
2	GN/BK	Vacuum Sensor Signal	2.5-4.5v
3	GN/RD	IAT Sensor Signal	At 100°F: 2.60v
4	RD	ECT Sensor Signal	At 180°F: 0.51v
6	WT	Main O2S Signal	0.1-1.1v
7	BK	Check Connector	12-14v
8	PK/YL	Check Connector	12-14v
9	BR	Sensor Ground	<0.050v
10 (Cal)	BL	EGR Gas Temp. Sensor	3.5-4.0v
11	PK/BK	TP Sensor Signal	0.3-0.8v
12	PK	Closed Throttle Switch	1v, at off-idle: 12v
14 (Cal)	WT	Sub O2S Signal	0.1-1.1v
15	OR	Check Connector	12-14v
16	BK	Distributor Signal (G-)	<0.050v

1990-91 2.2L VIN S (All) 22 Pin Connector

PCM Pin #	Wire Color	Circuit Description (22 Pin)	Value at Hot Idle
1	PK	Battery Direct	12-14v
2	OR	Defogger/Light Idle Up Signal	Switch On: 12v, Off: 0v
4	GN/WT	Stop Light Switch Signal	Brake Off: 0v, On: 12v
5	PK	MIL (lamp) Control	MIL Off: 12v, On: 1v
7	GY/BL	Main Overdrive Switch	Switch Off: 12-14v
8	BK/YL	A/C Amplifier Signal (AC1)	Clutch On: 1.5v, Off: 12v
9	BL/WT	Vehicle Speed Sensor	At 55 mph: 48 Hz
10	BK/WT	A/C Magnetic Clutch (ACMG)	Clutch Off: 0v, On: 12v
11	BK	A/T: Starter Switch Signal	9-11v (cranking)
11	BK/WT	M/T: Starter Switch Signal	9-11v (cranking)
12	BK/YL	EFI Main Relay Power B1+	12-14v
13	BK/YL	EFI Main Relay Power	12-14v
14	GN	Circuit Opening Relay (FC)	0-3v, off-idle: 12v
20	PK	Cruise Control ECU	At Cruise in OD: 12v
21	GN/BK	A/C Amplifier Signal (ACT)	Clutch On: 12v, Off: 1.5v
22	BK/WT	A/T Neutral Start Switch	In P/N: 9-11v (cranking)
22	BK	M/T: Neutral Start Switch	In 'N': 9-11v (cranking)

Standard Colors and Abbreviations

Abbreviation	Color	Abbreviation	Color	Abbreviation	Color
BK	Black	GY	Gray	RD	Red
BL	Blue	GN	Green	TN	Tan
BR	Brown	LG	Light Green	VT	Violet
DB	Dark Blue	OR	Orange	WT	White
DG	DK Green	PK	Pink	YL	Yellow

1990-91 2.2L VIN S 26 Pin (All) Connector

PCM Pin #	Wire Color	Circuit Description (26 Pin)	Value at Hot Idle
1	WT	Distributor Signal (NE+)	AC pulse signals
3	BK/YL	Igniter Signal (IGF)	Digital Signal: 0-5-0v
9	GN/WT	ISC Signal (ISCC)	Pulse Signals
10	BK/WT	ISC Signal (ISCO)	Pulse Signals
11	YL	Injector 2 & 4 Control	2.0-3.3 ms
12	WT	Injector 1 & 3 Control	2.0-3.3 ms
13	WT/BK	Power Ground	<0.1v
14	BR	Sensor Ground	<0.050v
15	RD	Distributor Signal (G+)	AC pulse signals
16	BR	Sensor Ground	<0.050v
18	PK/GN	A/T Select Switch 2nd	In 2nd: 12v, Others: 0v
19	PK/RD	A/T Select Switch Low	In Low: 12v, Others: 0v
20	BK	Igniter Signal (IGT)	0.72-0.74v
23	BL/BK	EGR Solenoid Control (VSV)	12v or 0v
26	WT/BK	Power Ground	<0.1v

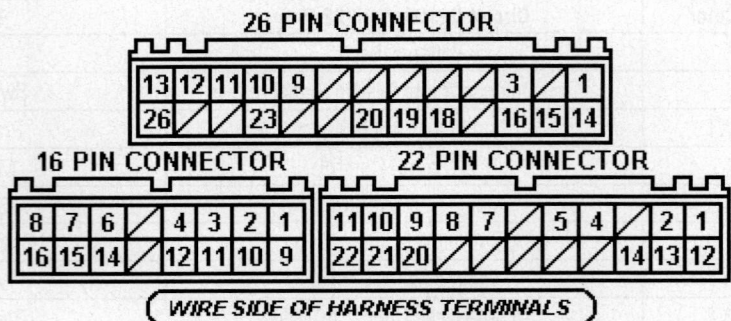

Pin Connector Graphic

1990-91 2.2L MFI VIN S 10 Pin Connector

PCM Pin #	Wire Color	Circuit Description (10 Pin)	Value at Hot Idle
1	BK/WT	A/T Neutral Start Switch	In P/N: 9-11v (cranking)
1	BK	M/T: Neutral Start Switch	In 'N': 9-11v (cranking)
3	BK	A/T: Starter Switch Signal	9-11v (cranking)
3	BK/WT	M/T: Starter Switch Signal	9-11v (cranking)
4	WT	Injector 1 & 3 Control	2.0-3.3 ms
5	WT/BK	Power Ground	<0.1v
6	BL/BK	EGR Solenoid Control (VSV)	12v or 0v
7	BR	Sensor Ground	<0.050v
8	WT	Igniter Signal (IGT)	0.72-0.74v
9	YL	Injector 2 & 4 Control	2.0-3.3 ms
10	WT/BK	Power Ground	<0.1v

1990-91 2.2L MFI VIN S 14-Pin Connector

PCM Pin #	Wire Color	Circuit Description (14-Pin)	Value at Hot Idle
1	BK/YL	EFI Main Relay Power B1+	12-14v
2	PK	Battery Direct	12-14v
3	PK/YL	Check Connector	12-14v
4	GN	Circuit Opening Relay (FC)	0-3v, off-idle: 12v
5	OR	Defogger/Light Idle Up Signal	Switch On: 12v, Off: 0v
6	GN/WT	Stop Light Switch Signal	Brake Off: 0v, On: 12v
7	BK/YL	A/C Amplifier Signal (AC1)	Clutch On: 1.5v, Off: 12v
8	BK/YL	EFI Main Relay Power	12-14v
9	GN	MIL (lamp) Control	MIL Off: 12v, On: 1v
11	BK/WT	A/C Magnetic Clutch (ACMG)	Clutch Off: 0v, On: 12v
12	BL/WT	Vehicle Speed Sensor	At 55 mph: 48 Hz
13	GN/BK	A/C Amplifier Signal (ACT)	Clutch On: 12v, Off: 1.5v
14	BL/GN	Short Pin	Open: 12v, Closed: 0v

1991 2.2L MFI VIN S (All) 18 Pin Connector

PCM Pin #	Wire Color	Circuit Description (18 Pin)	Value at Hot Idle
1	RD	ECT Sensor Signal	At 180°F: 0.51v
2	GY/BK	Vacuum Sensor Signal	1-1.5v
3	GN/RD	IAT Sensor Signal	At 100°F: 2.60v
4	OR	Check Connector	12-14v
5	BK/YL	Igniter Signal (IGF)	Digital Signal: 0-5-0v
6	RD	Distributor Signal (G+)	AC pulse signals
7	BK	Distributor Signal (G-)	<0.050v
8	WT	Main O2S Signal	0.1-1.1v
9	GN/WT	ISC Signal (ISCC)	Pulse Signals
10	BR	Sensor Ground	<0.050v
11	PK/BK	A/T: TP Sensor Signal	0.3-0.8v
11	PK/BK	M/T: PSW Signal	1v, at off-idle: 12v
12	PK/BL	Vacuum Sensor VREF	4.9-5.1v
13	PK	Closed Throttle Switch	1v, at off-idle: 12v
14 (Cal)	BL	EGR Gas Temp. Sensor	3.5-4.0v
15	WT	Distributor Signal (NE+)	AC pulse signals
16	BR	Sensor Ground	<0.050v
17 (Cal)	WT	Sub O2S Signal	0.1-1.1v
18	BK/WT	ISC Signal (ISCO)	Pulse Signals

Pin Connector Graphic

1990-91 2.2L MFI VIN S (A/T-ECT) 16 Pin Connector

PCM Pin #	Wire Color	Circuit Description (16 Pin)	Value at Hot Idle
1	PK/BL	Vacuum Sensor VREF	4.9-5.1v
2	GN/BK	Vacuum Sensor Signal	2.5-4.5v
3	GN/RD	IAT Sensor Signal	At 100°F: 2.60v
4	RD	ECT Sensor Signal	At 180°F: 0.51v
6	WT	Main O2S Signal	0.1-1.1v
7	BK	Check Connector	12-14v
8	PK/YL	Check Connector	12-14v
9	BR	Sensor Ground	<0.050v
10 (Cal)	BL	EGR Gas Temp. Sensor	3.5-4.0v
11	PK/BK	TP Sensor Signal	0.3-0.8v
12	PK	Closed Throttle Switch	1v, at off-idle: 12v
14 (Cal)	WT	Sub O2S Signal	0.1-1.1v
15	OR	Check Connector	12-14v
16	BK	Distributor Signal (G-)	<0.050v

1990-91 2.2L MFI VIN S (A/T-ECT) 22 Pin Connector

PCM Pin #	Wire Color	Circuit Description (22 Pin)	Value at Hot Idle
1	PK	Battery Direct	12-14v
2	OR	Defogger/Light Idle Up Signal	Switch On: 12v, Off: 0v
4	GN/WT	Stop Light Switch Signal	Brake Off: 0v, On: 12v
5	PK	MIL (lamp) Control	MIL Off: 12v, On: 1v
7	GY/BL	Main Overdrive Switch	Switch Off: 12-14v
8	BK/YL	A/C Amplifier Signal (AC1)	Clutch On: 1.5v, Off: 12v
9	BL/WT	Vehicle Speed Sensor	At 55 mph: 48 Hz
10	BK/WT	A/C Magnet Cutout Relay	A/C Off: 12v, On: 1v
11	BK	A/T: Starter Switch Signal	9-11v (cranking)
12	BK/YL	EFI Main Relay Power B1+	12-14v
13	BK/YL	EFI Main Relay Power	12-14v
14	GN	Circuit Opening Relay (FC)	0-3v, off-idle: 12v
20	PK	Cruise Control ECU	At Cruise in OD: 12v
21	GN/BK	A/C Amplifier Signal (ACT)	Clutch On: 12v, Off: 1.5v
22	BK/WT	A/T Neutral Start Switch	In P/N: 9-11v (cranking)

Standard Colors and Abbreviations

Abbreviation	Color	Abbreviation	Color	Abbreviation	Color
BK	Black	GY	Gray	RD	Red
BL	Blue	GN	Green	TN	Tan
BR	Brown	LG	Light Green	VT	Violet
DB	Dark Blue	OR	Orange	WT	White
DG	DK Green	PK	Pink	YL	Yellow

1990-91 2.2L MFI VIN S (A/T-ECT) 26 Pin Connector

PCM Pin #	Wire Color	Circuit Description (26 Pin)	Value at Hot Idle
1	WT	Distributor Signal (NE+)	AC pulse signals
2	BL/GN	A/T Pattern Selector Switch	Norm: 0v, PWR: 12v
3	BK/YL	Igniter Signal (IGF)	Digital Signal: 0-5-0v
4	BL/YL	A/T-ECT Solenoid (SL)	In Lockup: 12-14v
5	BR/YL	A/T-ECT Solenoid (S2)	1st or OD: 1v
6	BL/RD	A/T-ECT Solenoid (S1)	In 3rd or OD: 1v
9	GN/WT	ISC Signal (ISCC)	Pulse Signals
10	BK/WT	ISC Signal (ISCO)	Pulse Signals
11	YL	Injector 2 & 4 Control	2.0-3.3 ms
12	WT	Injector 1 & 3 Control	2.0-3.3 ms
13	WT/BK	Power Ground	<0.1v
14	BR	Sensor Ground	<0.050v
15	RD	Distributor Signal (G+)	AC pulse signals
16	BR	Sensor Ground	<0.050v
17	BL/RD	A/T-ECT Speed Sensor	Moving: 0-5-0v
18	PK/GN	A/T Select Switch 2nd	In 2nd: 12v, Others: 0v
19	PK/RD	A/T Select Switch Low	In Low: 12v, Others: 0v
20	BK	Igniter Signal (IGT)	Digital Signal: 0-5-0v
23	BL/BK	EGR Solenoid Control (VSV)	12v or 0v
26	WT/BK	Power Ground	<0.1v

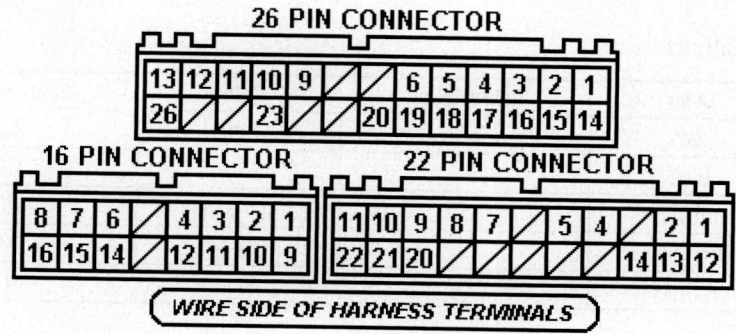

05533_ADIA_G044

Pin Connector Graphic

1992-93 2.2L I4 VIN S (A/T) 16 Pin Connector

PCM Pin #	Wire Color	Circuit Description (16 Pin)	Value at Hot Idle
1	PK/BL	Sensor VREF	4.9-5.1v
2	GY/BK	MAP Sensor Signal	2.5-4.5v
3	GY	IAT Sensor Signal	At 100°F: 2.60v
4	RD	ECT Sensor Signal	At 180°F: 0.51v
5	WT	Sub O2S Signal	0.1-1.1v
6	WT	Main O2S Signal	0.1-1.1v
7	BK	Data Link Connector	12-14v
8	PK/YL	Data Link Connector	12-14v
9	BR	Sensor Ground	<0.050v
10 (Cal)	RD	EGR Gas Temp. Sensor	3.5-4.0v
11	PK/BK	TP Sensor Signal	0.3-0.8v
12	PK	Closed Throttle Switch	1v, at off-idle: 12v
13	WT	Knock Sensor Signal	<0.075v AC
14	LG	Data Link Connector	12-14v
15	YL	Data Link Connector	12-14v

1992-93 2.2L I4 VIN S (A/T) 22 Pin Connector

PCM Pin #	Wire Color	Circuit Description (22 Pin)	Value at Hot Idle
1	PK	Battery Direct	12-14v
2	OR	Defogger/Light Idle Up Signal	Switch On: 12v, Off: 0v
4	GN/WT	Stop Light Switch Signal	Brake Off: 0v, On: 12v
5	GN/WT	MIL (lamp) Control	MIL Off: 12v, On: 1v
9	BL/RD	Vehicle Speed Sensor	At 55 mph: 48 Hz
10	BL/BK	A/C Amplifier Signal (ACT)	Clutch On: 12v, Off: 1.5v
11	BK	Starter Switch Signal	9-11v (cranking)
12	BK/YL	EFI Main Relay Power	12-14v
14	GN	Circuit Opening Relay (FC)	0-3v, off-idle: 12v
21	GN/YL	A/C Amplifier Signal (AC1)	Clutch On: 1.5v, Off: 12v
22	BK/WT	A/T Neutral Start Switch	In P/N: 9-11v (cranking)

Standard Colors and Abbreviations

Abbreviation	Color	Abbreviation	Color	Abbreviation	Color
BK	Black	GY	Gray	RD	Red
BL	Blue	GN	Green	TN	Tan
BR	Brown	LG	Light Green	VT	Violet
DB	Dark Blue	OR	Orange	WT	White
DG	DK Green	PK	Pink	YL	Yellow

1992-93 2.2L I4 VIN S (A/T) 26 Pin Connector

PCM Pin #	Wire Color	Circuit Description (26 Pin)	Value at Hot Idle
1	BL/YL	A/T-ECT Solenoid (SL)	In Lockup: 12-14v
2	PK	A/T-ECT Solenoid (S1)	In 3rd or OD: 1v
3	BK/YL	Igniter Signal (IGF)	Digital Signal: 0-5-0v

1992-93 2.2L I4 VIN S (A/T) 26 Pin Connector, *continued*

4	YL	Distributor Signal (NE+)	AC pulse signals
5	BK	Distributor Signal (G-)	<0.050v
6	PK/GN	A/T Select Switch 2nd	In 2nd: 12v, Others: 0v
9	GN/WT	ISC Signal (RSC)	Pulse Signals
10	BK/WT	ISC Signal (RSO)	Pulse Signals
11	YL	Injector 2 & 4 Control	2.0-3.3 ms
12	WT	Injector 1 & 3 Control	2.0-3.3 ms
13	WT/BK	Power Ground	<0.1v
14	BR	Sensor Ground	<0.050v
15	BR/YL	A/T-ECT Solenoid (S2)	1st or OD: 1v
17	BK	Distributor Signal (NE-)	<0.050v
18	RD	Distributor Signal (G+)	AC pulse signals
19	PK/RD	A/T Select Switch Low	In Low: 12v, Others: 0v
20	BK	Igniter Signal (IGT)	Digital Signal: 0-5-0v
22	BL/GN	A/T Select Switch Reverse	In 'R': 12v, Others: 0v
23	BK/YL	EGR Solenoid Control (VSV)	12v or 0v
26	WT/BK	Power Ground	<0.1v

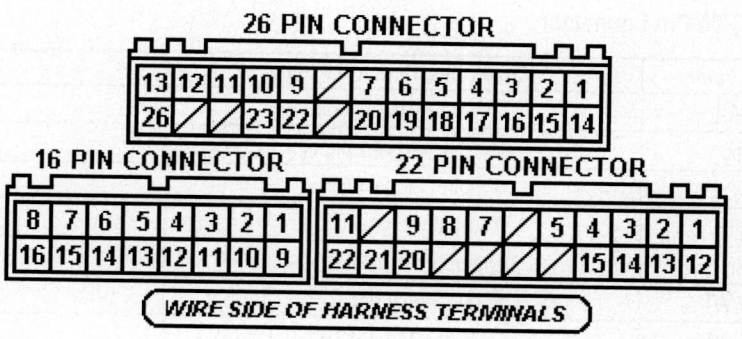

05533_ADIA_G045

Pin Connector Graphic

1992-93 2.2L I4 VIN S (M/T) 12 Pin Connector

PCM Pin #	Wire Color	Circuit Description (12 Pin)	Value at Hot Idle
1	BK/YL	EFI Main Relay Power B1+	12-14v
2	PK	Battery Direct	12-14v
3	BK/YL	A/C Amplifier Signal (AC1)	Clutch On: 1.5v, Off: 12v
4	GN	Circuit Opening Relay (FC)	0-3v, off-idle: 12v
6	GN/BL	A/C Amplifier Signal (ACT)	Clutch On: 12v, Off: 1.5v
7	BK/YL	EFI Main Relay Power	12-14v
8	PK	MIL (lamp) Control	MIL Off: 12v, On: 1v
10	BK/WT	A/C Magnetic Clutch (ACMG)	Clutch Off: 0v, On: 12v
11	BL/RD	Vehicle Speed Sensor	At 55 mph: 48 Hz
12	OR	Defogger/Light Idle Up Signal	Switch On: 12v, Off: 0v

1992-93 2.2L I4 VIN S (M/T) 16 Pin Connector

PCM Pin #	Wire Color	Circuit Description (16 Pin)	Value at Hot Idle
1	WT	Sub O2S Signal	0.1-1.1v
2	GY/BK	MAP Sensor Signal	1-1.5v
3	GY	IAT Sensor Signal	At 100°F: 2.60v
4	RD	ECT Sensor Signal	At 180°F: 0.51v
5	WT	Knock Sensor Signal	<0.075v AC
6	WT	Main O2S Signal	0.1-1.1v
7	BL/GN	Data Link Connector	12-14v
8	PK/YL	Data Link Connector	12-14v
9	BR	Sensor Ground	<0.050v
10	PK/BK	TP Sensor Signal	0.3-0.8v
11	PK/BL	MAP Sensor VREF	4.9-5.1v
12	PK	Closed Throttle Switch	1v, at off-idle: 12v
13 (Cal)	BL	EGR Gas Temp. Sensor	3.5-4.0v
14	YL/RD	A/C Evaporator Temp. Signal	1.4-1.8v (temp. >59°F)
15	BK	Data Link Connector	12-14v
16	BR	Sensor Ground	<0.050v

1992-93 2.2L I4 VIN S (M/T) 26 Pin Connector

PCM Pin #	Wire Color	Circuit Description (26 Pin)	Value at Hot Idle
1	YL/BL	A/C Idle Up Solenoid	A/C On: 1v
2	BK	M/T: Clutch Start Signal	9-11v (cranking)
3	BK/YL	Igniter Signal (IGF)	Digital Signal: 0-5-0v
4	YL	Distributor Signal (NE+)	AC pulse signals
5	RD	Distributor Signal (G+)	AC pulse signals
9	GN/WT	ISC Signal (ISCC)	Pulse Signals
10	BK/WT	ISC Signal (ISCO)	Pulse Signals
12	WT	Injector 1 & 3 Control	2.0-3.3 ms
13	WT/BK	Power Ground	<0.1v
17	BL	Distributor Signal (NE-)	<0.050v
18	BK	Distributor Signal (G-)	<0.050v
22	BK	Igniter Signal (IGT)	Digital Signal: 0-5-0v
23	BK/YL	EGR Solenoid Control (VSV)	12v or 0v
24	BR	Sensor Ground	<0.050v
25	YL	Injector 2 & 4 Control	2.0-3.3 ms
26	WT/BK	Power Ground	<0.1v

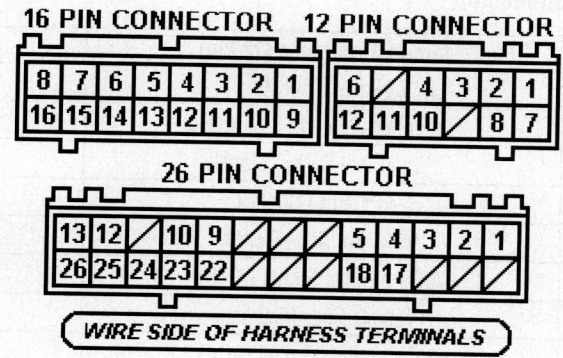

05533_ADIA_G046

Pin Connector Graphic

1994 2.2L I4 MFI VIN S (All) 16 Pin Connector

PCM Pin #	Wire Color	Circuit Description (16 Pin)	Value at Hot Idle
1	RD	Sensor VREF	4.9-5.1v
2	GY/BK	MAP Sensor Signal	1-1.5v
3	BL/RD	IAT Sensor Signal	At 100°F: 2.60v
4	GN	ECT Sensor Signal	At 180°F: 0.51v
5 (Cal)	WT	Sub O2S Signal	0.1-1.1v
6 (Fed)	WT	Main O2S Signal	0.1-1.1v
7	BK	Data Link Connector	12-14v
8	RD/WT	Data Link Connector	12-14v
9	BR	Sensor Ground	<0.050v
10 (Cal)	BK/RD	EGR Gas Temp. Sensor	3.5-4.0v
11	BK/WT	TP Sensor Signal	0.3-0.8v
12	BL/WT	Closed Throttle Switch	1v, at off-idle: 12v
13	WT	Knock Sensor Signal	<0.075v AC
14	LG	Data Link Connector	12-14v
15	YL	Data Link Connector	12-14v

1994 2.2L I4 MFI VIN S (All) 22 Pin Connector

PCM Pin #	Wire Color	Circuit Description (22 Pin)	Value at Hot Idle
1	PK	Battery Direct	12-14v
2	OR	Defogger/Light Idle Up Signal	Switch On: 12v, Off: 0v
4	GN/WT	Stop Light Switch Signal	Brake Off: 0v, On: 12v
5	RD/GN	MIL (lamp) Control	MIL Off: 12v, On: 1v
7	GY/BL	Main Overdrive Switch	Switch Off: 12v, On: 0v
8	BL/YL	A/C Amplifier Signal (A/TS)	A/C On: 1.5v
9	BL/RD	Vehicle Speed Sensor	At 55 mph: 48 Hz
10	BL/BK	A/C Amplifier Signal (ACA)	A/C On: 1.5v
11	BK	Starter Switch Signal	9-11v (cranking)
12	BK/YL	EFI Main Relay Power	12-14v
14	GN	Circuit Opening Relay (FC)	0-3v, off-idle: 12v
18	PK/RD	A/T Select Switch 2nd	In 2nd: 12v, Others: 0v
19	BR/YL	A/T Select Switch Low	In Low: 12v, Others: 0v
20	GN/RD	A/T-ECT ECU Signal	Pulse Signals
21	GN/YL	A/C Amplifier Signal (AC1)	Clutch On: 1.5v, Off: 12v
22	BK/WT	Neutral Start Switch	In P/N: 9-11v (cranking)

Standard Colors and Abbreviations

Abbreviation	Color	Abbreviation	Color	Abbreviation	Color
BK	Black	GY	Gray	RD	Red
BL	Blue	GN	Green	TN	Tan
BR	Brown	LG	Light Green	VT	Violet
DB	Dark Blue	OR	Orange	WT	White
DG	DK Green	PK	Pink	YL	Yellow

1994 2.2L I4 MFI VIN S (All) 26 Pin Connector

PCM Pin #	Wire Color	Circuit Description (26 Pin)	Value at Hot Idle
1	BL/YL	A/T-ECT Solenoid (SL)	In Lockup: 12-14v
2	PK	A/T-ECT Solenoid (S1)	In 3rd or OD: 1v
3	BK/YL	Igniter Signal (IGF)	Digital Signal: 0-5-0v
4 (TMC)	YL	Distributor Signal (NE+)	AC pulse signals
4 (TMM)	YL	Distributor Signal (NE+)	AC pulse signals
5 (TMC)	BK	Distributor Signal (G2+)	AC pulse signals
5 (TMM)	BK	Distributor Signal (NE-)	<0.050v
7	RD/YL	A/C Idle Up Solenoid	A/C Off: 12v, On: 1v
8	BL/RD	Fuel Pressure Up Solenoid	1v (at hot restart)
9	GN/WT	ISC Signal (ISCC)	Pulse Signals
10	BK/WT	ISC Signal (ISCO)	Pulse Signals
11 (TMC)	BK/WT	Injector 2 Control	2.0-3.3 ms
11 (TMM)	BK/RD	Injector 2 & 4 Control	2.0-3.3 ms
12 (TMC)	BK/YL	Injector 1 Control	2.0-3.3 ms
12 (TMM)	BK/BL	Injector 1 & 3 Control	2.0-3.3 ms

1994 2.2L I4 MFI VIN S (All) 26 Pin Connector, *continued*

13	WT/BK	Power Ground	<0.1v
14	BR	Sensor Ground	<0.050v
15	BR/YL	A/T-ECT Solenoid (S2)	1st or OD: 1v
17	BK	Distributor Signal (G-)	<0.050v
18 (TMC)	RD	Distributor Signal (G1+)	AC pulse signals
18 (TMM)	RD	Distributor Signal (G+)	AC pulse signals
20	WT	Igniter Signal (IGT)	Digital Signal: 0-5-0v
23	BL/BK	EGR Solenoid Control (VSV)	12v or 0v
24 (TMC)	BK/RD	Injector 4 Control	2.0-3.3 ms
25 (TMM)	BK/RD	Injector 3 Control	2.0-3.3 ms
26	WT/BK	Power Ground	<0.1v

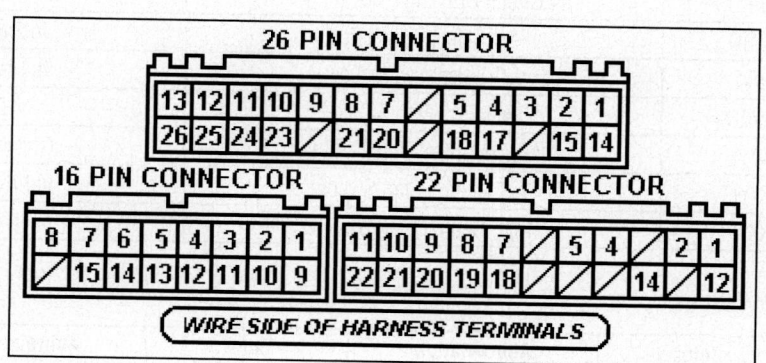

05533_ADIA_G047

Pin Connector Graphic

1995 2.2L I4 MFI VIN S (All) 16 Pin Connector

PCM Pin #	Wire Color	Circuit Description (16 Pin)	Value at Hot Idle
1	RD	Sensor VREF	4.9-5.1v
2	GY/BK	MAP Sensor Signal	2.5-4.5v
3	BL/RD	IAT Sensor Signal	At 100°F: 2.60v
4	GN	ECT Sensor Signal	At 180°F: 0.51v
5	WT	Sub O2S Signal	0.1-1.1v
6	WT	Main O2S Signal	0.1-1.1v
7	BK	Data Link Connector	12-14v
8	RD/WT	Data Link Connector	12-14v
9	BR	Sensor Ground	<0.050v
10	BK/RD	EGR Gas Temp. Sensor	3.5-4.0v
11	BK/WT	TP Sensor Signal	0.3-0.8v
12	BL/WT	Closed Throttle Switch	1v, at off-idle: 12v
13	WT	Knock Sensor Signal	<0.075v AC
14	LG	Data Link Connector	12-14v
15	YL	Data Link Connector	12-14v

1995 2.2L I4 MFI VIN S (All) 22 Pin Connector

PCM Pin #	Wire Color	Circuit Description (22 Pin)	Value at Hot Idle
1	PK	Battery Direct	12-14v
2	OR	Defogger/Light Idle Up Signal	Switch On: 12v, Off: 0v
4	GN/WT	Stop Light Switch Signal	Brake Off: 0v, On: 12v
5	RD/GN	MIL (lamp) Control	MIL Off: 12v, On: 1v
7	GY/BL	Main Overdrive Switch	Switch Off: 12-14v
8	BL/YL	A/C Amplifier A/TS Signal	A/C On: <1.5v
9	OR	Vehicle Speed Sensor	At 55 mph: 48 Hz
10	BL/BK	A/C Amplifier ACA Signal	A/C On: <1.5v
11	BK	Starter Switch Signal	9-11v (cranking)
12	BK/YL	EFI Main Relay Power	12-14v
14	GN/RD	Circuit Opening Relay (FC)	0-3v, off-idle: 12v
18	PK/RD	A/T Select Switch 2nd	In 2nd: 12v, Others: 0v
19	BR/YL	A/T Select Switch Low	In Low: 12v, Others: 0v
20	GN/RD	A/T-ECT ECU Signal	Pulse Signals
21	GN/YL	A/C Amplifier Signal (AC1)	Clutch On: 1.5v, Off: 12v
22	BK/YL	A/T Neutral Start Switch	In P/N: 9-11v (cranking)
22	BK/YL	M/T: Neutral Start Switch	In 'N': 9-11v (cranking)

Standard Colors and Abbreviations

Abbreviation	Color	Abbreviation	Color	Abbreviation	Color
BK	Black	GY	Gray	RD	Red
BL	Blue	GN	Green	TN	Tan
BR	Brown	LG	Light Green	VT	Violet
DB	Dark Blue	OR	Orange	WT	White
DG	DK Green	PK	Pink	YL	Yellow

1995 2.2L I4 MFI VIN S (All) 26 Pin Connector

PCM Pin #	Wire Color	Circuit Description (26 Pin)	Value at Hot Idle
1	BL/YL	A/T-ECT Solenoid (SL)	In Lockup: 12-14v
2	PK	A/T-ECT Solenoid (S1)	In 3rd or OD: 1v
3	BK/YL	Igniter Signal (IGF)	Digital Signal: 0-5-0v
4 (TMC)	YL	Distributor Signal (NE+)	AC pulse signals
4 (TMM)	YL	Distributor Signal (NE+)	AC pulse signals
5 (TMC)	BK	Distributor Signal (G2+)	AC pulse signals
5 (TMM)	BK	Distributor Signal (NE-)	<0.050v
7	RD/YL	A/C Idle Up Solenoid	A/C Off: 12v, On: 1v
8	BL/RD	Fuel Pressure Up Solenoid	1v (at hot restart)
9	GN/WT	ISC Signal (ISCC)	Pulse Signals
10	BK/WT	ISC Signal (ISCO)	Pulse Signals
11 (TMC)	BK/WT	Injector 2 Control	2.0-3.3 ms
11 (TMM)	BK/RD	Injector 2 & 4 Control	2.0-3.3 ms
12 (TMC)	BK/BL	Injector 1 Control	2.0-3.3 ms
12 (TMM)	BK/YL	Injector 1 & 3 Control	2.0-3.3 ms

1995 2.2L I4 MFI VIN S (All) 26 Pin Connector, *continued*

13	WT/BK	Power Ground	<0.1v
14	BR	Sensor Ground	<0.050v
15	BR/YL	A/T-ECT Solenoid (S2)	1st or OD: 1v
17 (TMC)	BK	Distributor G Signal	AC pulse signals
17 (TMM)	BK	Distributor Signal (G-)	<0.050v
18 (TMC)	RD	Distributor Signal (G1+)	AC pulse signals
18 (TMM)	RD	Distributor Signal (G+)	AC pulse signals
20	WT	Igniter Signal (IGT)	Digital Signal: 0-5-0v
21	GN/BK	Water Temperature Switch	Switch On: 0.1v
23	BL/BK	EGR Solenoid Control (VSV)	12v or 0v
24 (TMC)	BK/YL	Injector 4 Control	2.0-3.3 ms
25 (TMM)	BK/RD	Injector 3 Control	2.0-3.3 ms
26	WT/BK	Power Ground	<0.1v

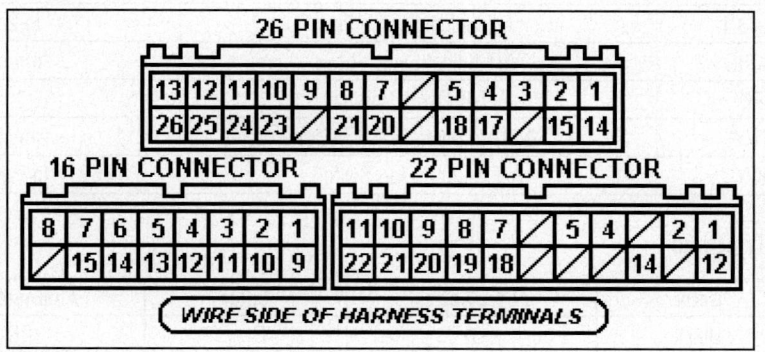

05533_ADIA_G048

Pin Connector Graphic

1996-97 2.2L I4 VIN G (All) 16 Pin Connector

PCM Pin #	Wire Color	Circuit Description (16 Pin)	Value at Hot Idle
1	RD	Sensor VREF	4.9-5.1v
2	BL	MAP Sensor Signal	1-1.5v
3	BL/RD	IAT Sensor Signal	At 100°F: 2.60v
4	GN	ECT Sensor Signal	At 180°F: 0.51v
5	WT	HO2S-12 (B1 S2) Signal	0.1-1.1v
6	WT	HO2S-11 (B1 S1) Signal	0.1-1.1v
9	BR	Sensor Ground	<0.050v
11	BK/WT	TP Sensor Signal	0.3-0.8v
13	WT	Knock Sensor Signal	<0.075v AC
15	YL	Data Link Connector	12-14v

1996-97 2.2L I4 VIN G (All) 22 Pin Connector

PCM Pin #	Wire Color	Circuit Description (22 Pin)	Value at Hot Idle
1	PK	Battery Direct	12-14v
2	OR	Defogger/Light Idle Up Signal	Switch On: 12v, Off: 0v
4	GN/WT	Stop Light Switch Signal	Brake Off: 0v, On: 12v
5	RD/BK	MIL (lamp) Control	MIL Off: 12v, On: 1v
7	GY/BL	Main Overdrive Switch	Switch Off: 12-14v
8	BL/YL	A/C Amplifier Signal (A/TS)	A/C On: 1.5v
9	OR	Vehicle Speed Sensor	At 55 mph: 48 Hz
10	BL/BK	A/C Amplifier Signal (ACA)	A/C On: 1.5v
11	BK	Starter Switch Signal	9-11v (cranking)
12	BK/RD	EFI Main Relay Power	12-14v
14	GN/RD	Circuit Opening Relay (FC)	0-3v, off-idle: 12v
16	WT	SIL Signal (Scan Tool)	Digital Signal
17	RD/WT	A/T Select Switch Reverse	In 'R': 12v, Others: 0v
18	YL/GN	A/T Select Switch 2nd	In 2nd: 12v, Others: 0v
19	YL/RD	A/T Select Switch Low	In Low: 12v, Others: 0v
20	PK	A/T-ECT ECU Signal	Pulse Signals
21	GN/YL	A/C Amplifier Signal (AC1)	Clutch On: 1.5v, Off: 12v
22	BK/YL	A/T Neutral Start Switch	In P/N: 9-11v (cranking)

Standard Colors and Abbreviations

Abbreviation	Color	Abbreviation	Color	Abbreviation	Color
BK	Black	GY	Gray	RD	Red
BL	Blue	GN	Green	TN	Tan
BR	Brown	LG	Light Green	VT	Violet
DB	Dark Blue	OR	Orange	WT	White
DG	DK Green	PK	Pink	YL	Yellow

1996-97 2.2L I4 VIN G (All) 26 Pin Connector

PCM Pin #	Wire Color	Circuit Description (26 Pin)	Value at Hot Idle
1	YL	A/T-ECT Solenoid (SL)	In Lockup: 12-14v
2	PK	A/T-ECT Solenoid (S1)	In 3rd or OD: 1v
3	BK/YL	Igniter Signal (IGF)	Digital Signal: 0-5-0v
4	OR	CKP Sensor Signal (NE+)	AC pulse signals
5	BK	Distributor Signal (G2+)	AC pulse signals
9	GN/WT	ISC Signal (ISCC)	Pulse Signals
10	BK/WT	ISC Signal (ISCO)	Pulse Signal
11	BK/WT	Injector 2 Control	2.0-3.3 ms
12	BL	Injector 1 Control	2.0-3.3 ms
13	BR	Power Ground	<0.1v
14	BR	Sensor Ground	<0.050v
15	BR/YL	A/T-ECT Solenoid (S2)	1st or OD: 1v
17	WT	CKP Sensor Signal (NE-)	<0.050v

1996-97 2.2L I4 VIN G (All) 26 Pin Connector, *continued*

20	WT	Igniter Signal (IGT)	Digital Signal: 0-5-0v
23	BL/BK	EGR Solenoid Control (VSV)	12v or 0v
24	BK/YL	Injector 4 Control	2.0-3.3 ms
25	BK/RD	Injector 3 Control	2.0-3.3 ms
26	BR	Power Ground	<0.1v

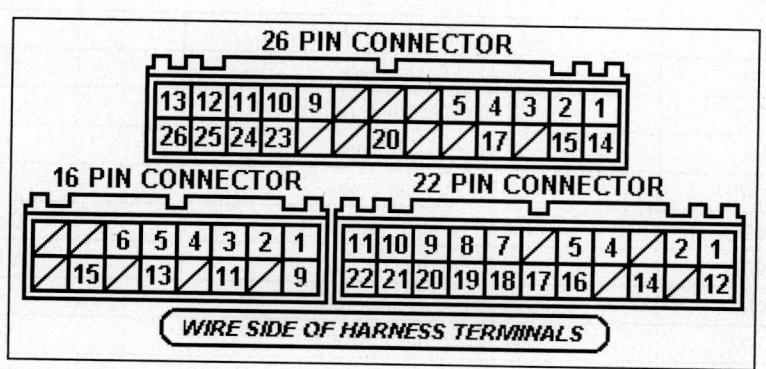

Pin Connector Graphic

1998-99 2.2L I4 VIN G (All) 16 Pin Connector

PCM Pin #	Wire Color	Circuit Description (16 Pin)	Value at Hot Idle
1	RD	Sensor VREF	4.9-5.1v
2	BL	MAP Sensor Signal	1-1.5v
3	BL/RD	IAT Sensor Signal	At 100°F: 2.60v
4	GN	ECT Sensor Signal	At 180°F: 0.51v
5	WT	O2S-12 (B1 S2) Signal	0.1-1.1v
6	WT	O2S-11 (B1 S1) Signal	0.1-1.1v
7	RD/WT	EVAP Vapor Pressure Sensor	2.9-3.1v (with hose off)
8	YL/BK	VSV Vapor Press Solenoid	12v or 0v
9	BR	Sensor Ground	<0.050v
11	BK/WT	TP Sensor Signal	0.3-0.8v
13	WT	Knock Sensor Signal	<0.075v AC
15	YL	Data Link Connector (TE)	12-14v
16	BR	Power Ground	<0.1v

05533_ADIA_G049

1998-99 2.2L I4 VIN G (All) 22 Pin Connector

PCM Pin #	Wire Color	Circuit Description (22 Pin)	Value at Hot Idle
1	PK	Battery Direct	12-14v
2	OR	Defogger/Light Idle Up Signal	Switch On: 12v, Off: 0v
4	GN/WT	Stop Light Switch Signal	Brake Off: 0v, On: 12v
5	RD/BK	MIL (lamp) Control	MIL Off: 12v, On: 1v
7	GY/BL	Main Overdrive Switch	Switch Off: 12-14v
8	BL/YL	A/C Amplifier Signal (A/TS)	A/C On: 1.5v
9	OR	Vehicle Speed Sensor	At 55 mph: 48 Hz
10	BL/BK	A/C Amplifier Signal (AC1)	Clutch On: 1.5v, Off: 12v
11	BK	Starter Switch Signal	9-11v (cranking)
12	BK/RD	EFI Main Relay Power	12-14v
14	GN/RD	Circuit Opening Relay (FC)	0-3v, off-idle: 12v
15	GN/BK	A/C Pressure Switch	Switch Closed: 0.1v
16	WT	Data Link Connector (SDL)	0v
17	RD/WT	A/T Select Switch Reverse	In 'R': 12v, Others: 0v
18	YL/GN	A/T Select Switch 2nd	In 2nd: 12v, Others: 0v
19	YL/RD	A/T Select Switch Low	In Low: 12v, Others: 0v
20	PK	A/T-ECT ECU (OD1) Signal	Pulses
21	GN/YL	A/C Amplifier Signal (ACT)	Clutch On: 12v, Off: 1.5v
22	BK/YL	A/T Neutral Start Switch	In P/N: 9-11v (cranking)

Standard Colors and Abbreviations

Abbreviation	Color	Abbreviation	Color	Abbreviation	Color
BK	Black	GY	Gray	RD	Red
BL	Blue	GN	Green	TN	Tan
BR	Brown	LG	Light Green	VT	Violet
DB	Dark Blue	OR	Orange	WT	White
DG	DK Green	PK	Pink	YL	Yellow

1998-99 2.2L I4 VIN G (All) 26 Pin Connector

PCM Pin #	Wire Color	Circuit Description (26 Pin)	Value at Hot Idle
1	YL/BK	A/T-ECT Solenoid (SL)	In Lockup: 12-14v
2	PK	A/T-ECT Solenoid (S1)	In 3rd or OD: 1v
3	BK/YL	Igniter Signal (IGF)	Digital Signal: 0-5-0v
4	OR	CKP Sensor Signal (NE+)	AC pulse signals
5	BK	Distributor Signal (G2+)	AC pulse signals
9	GN/WT	ISC Signal (ISCC)	Pulse Signals
10	BK/WT	ISC Signal (ISCO)	Pulse Signals
11	BK/WT	Injector 2 Control	2.0-3.3 ms
12	BL	Injector 1 Control	2.0-3.3 ms
13	BR	Power Ground	<0.1v
14	BR	Sensor Ground	<0.050v
15	BR/YL	A/T-ECT Solenoid (S2)	1st or OD: 1v
17	WT	CKP/CMP Sensor Ground (-)	<0.050v
20	WT	Igniter Signal (IGT)	Digital Signal: 0-5-0v
22	GN/BK	EVAP Purge Solenoid (VSV)	12v or 0v
23	BL/BK	EGR Solenoid Control (VSV)	12v or 0v
24	BK/YL	Injector 4 Control	2.0-3.3 ms
25	BK/RD	Injector 3 Control	2.0-3.3 ms
26	BR	Power Ground	<0.1v

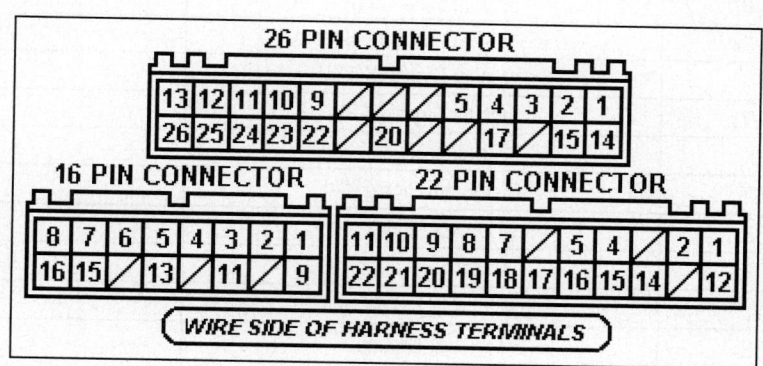

05533_ADIA_G050

Pin Connector Graphic

COROLLA PIN CHARTS

1990-91 1.6L I4 4A-FE MFI VIN A 16 Pin Connector

PCM Pin #	Wire Color	Circuit Description (16 Pin)	Value at Hot Idle
1	BK/RD	EFI Main Relay Power	12-14v
2	RD/WT	Battery Direct	12-14v
4	GN/RD	Circuit Opening Relay (FC)	0-3v, off-idle: 12v
5	BK	Defogger/Light Idle Up Signal	Switch On: 12v, Off: 0v
6 ('91)	GN	Cold Start Injector Control	1v (at cold startup)
7	RD/YL	A/C Amplifier Signal (ACT)	Clutch On: 12v, Off: 1.5v
8	BL/WT	Check Connector	12-14v
9	BK/RD	EFI Main Relay Power	12-14v
10	RD/YL	MIL (lamp) Control	MIL Off: 12v, On: 1v
12	BK/WT	A/C Magnetic Clutch (ACMG)	Clutch Off: 0v, On: 12v
13	PK/WT	Vehicle Speed Sensor	At 55 mph: 48 Hz
16	RD/WT	Check Connector	12-14v

1990-91 1.6L I4 4A-FE MFI VIN A 26 Pin Connector

PCM Pin #	Wire Color	Circuit Description (26 Pin)	Value at Hot Idle
1 (Cal)	BK/YL	EGR Solenoid Control (VSV)	12v or 0v
2	BK/WT	A/T Neutral Start Switch	In P/N: 9-11v (cranking)
3	WT	ECT Sensor Signal	At 180°F: 0.51v
4	GN/RD	Vacuum Sensor Signal	2.5-4.5v
5	YL/BK	IAT Sensor Signal	At 100°F: 2.60v
6	BK	Igniter Signal (IGT)	0.72-0.74v
7	BK/YL	Igniter Signal (IGF)	0.74-0.76v
8	BK	Distributor Signal (G+)	AC pulse signals
9	WT	Distributor Signal (G-)	<0.050v
10	BK	O2S Signal	0.1-1.1v
11	BK	Starter Switch Signal	9-11v (cranking)
12	YL	Injector Pair 1 & 3 Control	2.0-3.3 ms
13	BR	Power Ground	<0.1v
14	BK	ISC Solenoid Control	Pulse Signals
15	RD	Main Overdrive Switch	Switch Off: 12-14v
16	BR	Sensor Ground	<0.050v
17	LG	Wide Open Throttle Switch	1v, at off-idle: 12v
18	YL	Vacuum Sensor VREF	4.9-5.1v
19	BL	Closed Throttle Switch	1v, at off-idle: 12v
20 (Cal)	PK	EGR Gas Temp. Sensor	3.5-4.0v
21	RD	Distributor Signal (NE+)	AC pulse signals
22	BR	Sensor Ground	<0.050v
24	BR	Sensor Ground	<0.050v
25	YL	Injector Pair 2 & 4 Control	2.0-3.3 ms
26	BR	Power Ground	<0.1v

26 PIN CONNECTOR   16 PIN CONNECTOR

13	12	11	10	9	8	7	6	5	4	3	2	1
26	25	24	/	22	21	20	19	18	17	16	15	14

8	7	/	5	4	/	2	1
16	/	/	13	12	/	10	9

WIRE SIDE OF HARNESS TERMINALS

Pin Connector Graphic

05533_ADIA_G051

1990-91 1.6L 4A-GE VIN A (All) 12 Pin Connector

PCM Pin #	Wire Color	Circuit Description (12 Pin)	Value at Hot Idle
1	BK/RD	EFI Main Relay Power B1+	12-14v
2	RD/WT	Battery Direct	12-14v
5	GN/WT	Stop Light Switch Signal	Brake Off: 0v, On: 12v
6	RD/YL	A/C Amplifier Signal (ACT)	Clutch On: 12v, Off: 1.5v
7	BK/RD	EFI Main Relay Power	12-14v
8	RD	MIL (lamp) Control	MIL Off: 12v, On: 1v
10	BK/WT	A/C Magnetic Clutch (ACMG)	Clutch Off: 0v, On: 12v
11	PK/WT	Vehicle Speed Sensor	At 55 mph: 48 Hz

1990-91 1.6L 4A-GE VIN A (All) 16 Pin Connector

PCM Pin #	Wire Color	Circuit Description (16 Pin)	Value at Hot Idle
1	YL/BK	Airflow Meter VREF	5-9v
2	YL/BL	Airflow Meter Signal	1.5-2.5v
3	BK	IAT Sensor Signal	At 100°F: 2.60v
4	WT	ECT Sensor Signal	At 180°F: 0.51v
5 (Cal)	BK	Sub HO2S Signal	0.1-1.1v
6 (Fed)	BK	Main HO2S Signal	0.1-1.1v
7	RD/GN	Check Connector	12-14v
8	RD/WT	Check Connector	12-14v
9	BR	Sensor Ground	<0.050v
10	RD	TP Sensor Signal	0.3-0.8v
11	YL	Sensor VREF (VC)	4.9-5.1v
12	BL	Closed Throttle Switch	1v, at off-idle: 12v
13 (Cal)	RD/GN	EGR Gas Temp. Sensor	3.5-4.0v
14	BK	Knock Sensor Signal	<0.075v AC
16	BR	Sensor Ground	<0.050v

1990-91 1.6L 4A-GE VIN A (All) 26 Pin Connector

PCM Pin #	Wire Color	Circuit Description (26 Pin)	Value at Hot Idle
1	LG	Fuel Pressure Up Solenoid	1v (at hot restart)
2	BK/WT	Starter Switch Signal	9-11v (cranking)
3	BK/YL	Igniter Signal (IGF)	Digital Signal: 0-5-0v
4	BK/RD	Distributor Signal (NE+)	AC pulse signals
5	BK	Distributor Signal (G+)	AC pulse signals
9	BK/WT	A/C Idle Up Solenoid	A/C On: 1v
11 (Cal)	RD/BK	Sub HO2S Heater	1v (Heater On)
12	GN/RD	Injector Pair 3 & 4 Control	2.0-3.3 ms
13	BR	Power Ground	<0.1v
14 (Fed)	RD/BK	Main HO2S Heater	1v (Heater On)
15	BL/RD	Check Connector	12-14v
16 (Cal)	GN/RD	EGR Solenoid Control (VSV)	12v or 0v
17	WT	Distributor Signal (G-)	<0.050v
22	BK	Igniter Signal (IGT)	Digital Signal: 0-5-0v
24	BR	Sensor Ground	<0.050v
25	GN/YL	Injector Pair 1 & 2 Control	2.0-3.3 ms
26	BR	Power Ground	<0.1v

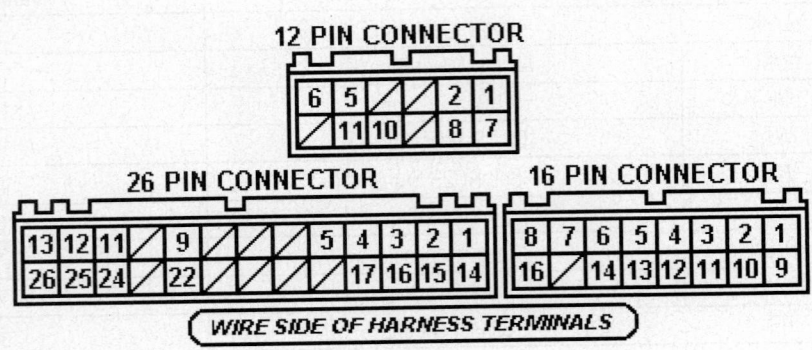

05533_ADIA_G052

Pin Connector Graphic

1992 1.6L I4 MFI VIN A (All) 16 Pin Connector

PCM Pin #	Wire Color	Circuit Description (16 Pin)	Value at Hot Idle
1	BK/RD	EFI Main Relay Power	12-14v
2	RD/WT	Battery Direct	12-14v
4	GN/RD	Circuit Opening Relay (FC)	0-3v, off-idle: 12v
5	BK	Defogger/Light Idle Up Signal	Switch On: 12v, Off: 0v
6	GN	Cold Start Injector Control	1v (at cold startup)
7	RD/YL	A/C Amplifier Signal (ACT)	Clutch On: 12v, Off: 1.5v
8	BL/WT	Check Connector	12-14v
9	BK/RD	EFI Main Relay Power	12-14v
10	RD	MIL (lamp) Control	MIL Off: 12v, On: 1v
12	BK/WT	A/C Magnetic Clutch (ACMG)	Clutch Off: 0v, On: 12v
13	PK/WT	Vehicle Speed Sensor	At 55 mph: 48 Hz
16	RD/WT	Check Connector	12-14v

1992 1.6L I4 MFI VIN A (All) 26 Pin Connector

PCM Pin #	Wire Color	Circuit Description (26 Pin)	Value at Hot Idle
1 (Cal)	BK/YL	EGR Solenoid Control (VSV)	12v or 0v
2	BK/WT	A/T Neutral Start Switch	In P/N: 9-11v (cranking)
3	WT	ECT Sensor Signal	At 180°F: 0.51v
4	GN/RD	Vacuum Sensor Signal	2.5-4.5v
5	YL/BK	IAT Sensor Signal	At 100°F: 2.60v
6	BK	Igniter Signal (IGT)	Digital Signal: 0-5-0v
7	BK/YL	Igniter Signal (IGF)	Digital Signal: 0-5-0v
8	BK	Distributor Signal (G+)	AC pulse signals
9	WT	Distributor Signal (G-)	<0.050v
10	BK	O2S-11 Signal	0.1-1.1v
11	BK	Starter Switch Signal	9-11v (cranking)
12	YL	Injector Pair 1 & 3 Control	2.0-3.3 ms
13	BR	Power Ground	<0.1v
14	BK	ISC Solenoid Control	Pulse Signals
15	RD	Main Overdrive Switch	Switch Off: 12v, On: 0v
16	BR	Sensor Ground	<0.050v
17	LG	Wide Open Throttle Switch	1v, at off-idle: 12v
18	YL	Sensor VREF (VC)	4.9-5.1v
19	BL	Closed Throttle Switch	1v, at off-idle: 12v
20 (Cal)	PK	EGR Gas Temp. Sensor	3.5-4.0v
21	RD	Distributor Signal (NE+)	AC pulse signals
22	BR	Sensor Ground	<0.050v
24	BR	Sensor Ground	<0.050v
25	YL	Injector Pair 2 & 4 Control	2.0-3.3 ms
26	BR	Power Ground	<0.1v

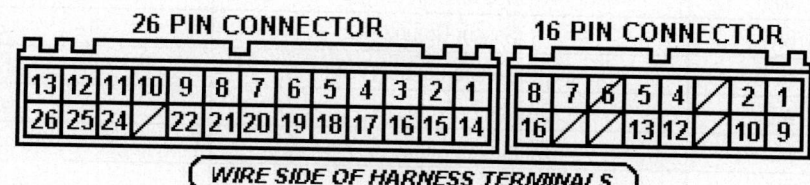

Pin Connector Graphic

05533_ADIA_G053

1993-94 1.6L I4 MFI VIN A (All) 12 Pin Connector

PCM Pin #	Wire Color	Circuit Description (12 Pin)	Value at Hot Idle
1	BK/RD	EFI Main Relay Power B1+	12-14v
2	RD/WT	Battery Direct	12-14v
4	GN/RD	Circuit Opening Relay (FC)	0-3v, off-idle: 12v
6	RD/YL	A/C Amplifier Signal (ACT)	Clutch On: 12v, Off: 1.5v
7	BK/RD	EFI Main Relay Power	12-14v
8	RD/YL	MIL (lamp) Control	MIL Off: 12v, On: 1v
10	YL/BL	A/C Amplifier Signal (AC1)	Clutch On: 1.5v, Off: 12v
11	PK/WT	Vehicle Speed Sensor	At 55 mph: 48 Hz
12	BK	Defogger/Light Idle Up Signal	Switch On: 12v, Off: 0v

1993-94 1.6L I4 MFI VIN A (All) 16 Pin Connector

PCM Pin #	Wire Color	Circuit Description (16 Pin)	Value at Hot Idle
2	GN/RD	MAP Sensor Signal	1-1.5v
3	YL/BK	IAT Sensor Signal	At 100°F: 2.60v
4	WT	ECT Sensor Signal	At 180°F: 0.51v
5 (Cal)	BK	Sub O2S Signal	0.1-1.1v
6 (Fed)	BK	Main O2S Signal	0.1-1.1v
7	GN/WT	Check Connector	12-14v
8	RD/WT	Check Connector	12-14v
9	BR	Sensor Ground	<0.050v
10	LG	TP Sensor Signal	0.3-0.8v
11	YL	Sensor VREF	4.9-5.1v
12	BL	Closed Throttle Switch	1v, at off-idle: 12v
13 (Cal)	WT/BL	EGR Gas Temp. Sensor	3.5-4.0v
15	BL/WT	Check Connector	12-14v
16	BR	Sensor Ground	<0.050v

1993-94 1.6L I4 MFI VIN A (All) 26 Pin Connector

PCM Pin #	Wire Color	Circuit Description (26 Pin)	Value at Hot Idle
2	BK	Starter Switch Signal	9-11v (cranking)
3	BK/YL	Igniter Signal (IGF)	Digital Signal: 0-5-0v
4	BK	Distributor Signal (NE+)	AC pulse signals
5	RD	Distributor Signal (G-)	<0.050v
9	BK/WT	ISC Signal (RSC)	Pulse Signals
10	BK/BL	ISC Signal (RSO)	Pulse Signals
12	BK	Injector Pair 1 & 3 Control	2.0-3.3 ms
13	BR	Power Ground	<0.1v
15	BK/WT	A/T Neutral Start Switch	In P/N: 9-11v (cranking)
15	BR	Sensor Ground	<0.050v
16 (Cal)	BK/RD	EGR Solenoid Control (VSV)	12v or 0v
17	WT	Distributor Signal (NE-)	<0.050v
18	GN	Distributor Signal (G+)	AC pulse signals
22	BK	Igniter Signal (IGT)	Digital Signal: 0-5-0v
23	WT/RD	A/C Idle Up Solenoid	A/C On: 1v
24	BR	Sensor Ground	<0.050v
25	BK/RD	Injector Pair 2 & 4 Control	2.0-3.3 ms
26	BR	Power Ground	<0.1v

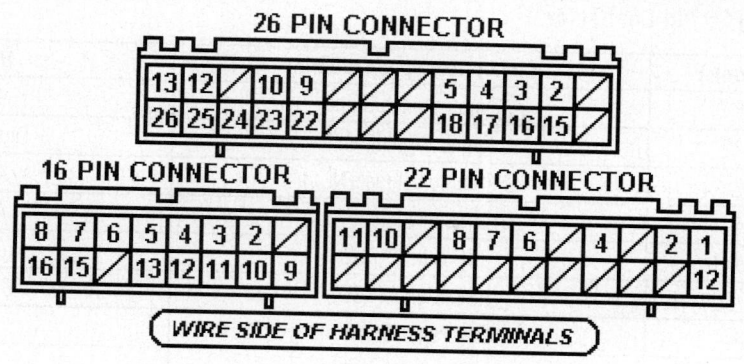

05533_ADIA_G054

Pin Connector Graphic
1995 1.6L I4 MFI VIN A (All) 16 Pin Connector

PCM Pin #	Wire Color	Circuit Description (16 Pin)	Value at Hot Idle
2	GN/RD	MAP Sensor Signal	1-1.5v
3	YL/BK	IAT Sensor Signal	At 100°F: 2.60v
4	WT	ECT Sensor Signal	At 180°F: 0.51v
5 (Cal)	BK	Sub O2S Signal	0.1-1.1v
6 (Fed)	BK	Main O2S Signal	0.1-1.1v
7	GN/WT	Check Connector	12-14v
8	RD/WT	Check Connector	12-14v
9	BR	Sensor Ground	<0.050v
10	LG	TP Sensor Signal	0.3-0.8v
11	YL	Sensor VREF	4.9-5.1v
12	BL	Closed Throttle Switch	1v, at off-idle: 12v
13 (Cal)	WT/BL	EGR Gas Temp. Sensor	3.5-4.0v
15	BL/WT	Check Connector	12-14v

1995 1.6L I4 MFI VIN A (All) 22 Pin Connector

PCM Pin #	Wire Color	Circuit Description (22 Pin)	Value at Hot Idle
2	RD/WT	Battery Direct	12-14v
4	GN/RD	Circuit Opening Relay (FC)	0-3v, off-idle: 12v
6	RD/YL	A/C Amplifier Signal (ACT)	Clutch On: 12v, Off: 1.5v
7	BK/RD	EFI Main Relay Power	12-14v
8	RD/YL	MIL (lamp) Control	MIL Off: 12v, On: 1v
10	YL/BL	A/C Amplifier Signal (AC1)	Clutch On: 1.5v, Off: 12v
11	PK/WT	Vehicle Speed Sensor	At 55 mph: 48 Hz
12	BK	Defogger/Light Idle Up Signal	Switch On: 12v, Off: 0v

1995 1.6L I4 MFI VIN A (All) 26 Pin Connector

PCM Pin #	Wire Color	Circuit Description (26 Pin)	Value at Hot Idle
2	BK	Starter Switch Signal	9-11v (cranking)
3	BK/YL	Igniter Signal (IGF)	Digital Signal: 0-5-0v
4	BK	Distributor Signal (NE+)	AC pulse signals
5	RD	Distributor Signal (G1+)	AC pulse signals
9	BK/WT	ISC Signal (RSC)	Pulses
10	BK/BL	ISC Signal (RSO)	Pulses
12	BK	Injector Pair 1 & 3 Control	2.0-3.3 ms
13	BR	Power Ground	<0.1v
15	BK/WT	A/T: Neutral Start Switch	In P/N: 9-11v (cranking)
15	BR	M/T: Neutral Start Switch	In 'N': 9-11v (cranking)
16 (Cal)	BK/RD	EGR Solenoid Control (VSV)	12v or 0v
17	WT	Distributor Signal (NE-)	<0.050v
18	GN	Distributor Signal (G-)	<0.050v
22	BK	Igniter Signal (IGT)	Digital Signal: 0-5-0v
23	WT/RD	Idle Up Solenoid Control	A/C On: 1v, Off: 12v
24	BR	Sensor Ground	<0.050v·
25	BK/RD	Injector Pair 2 & 4 Control	2.0-3.3 ms
26	BR	Power Ground	<0.1v

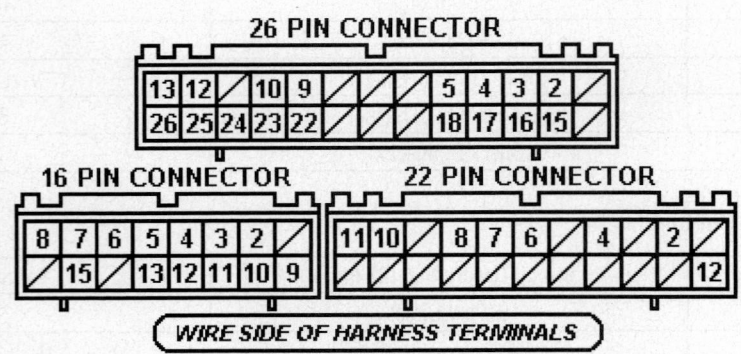

Pin Connector Graphic

05533_ADIA_G055

1996-97 1.6L I4 MFI VIN A (All) 16 Pin Connector

PCM Pin #	Wire Color	Circuit Description (16 Pin)	Value at Hot Idle
1	YL	Sensor VREF	4.9-5.1v
2	GN/RD	MAP Sensor Signal	1-1.5v
3	YL/BK	IAT Sensor Signal	At 100°F: 2.60v
4	WT	ECT Sensor Signal	At 180°F: 0.51v
5	BK	HO2S-12 (B1 S2) Signal	0.1-1.1v
6	BK	O2S-11 (B1 S1) Signal	0.1-1.1v
9	BR	Sensor Ground	<0.050v
11	LG	TP Sensor Signal	0.3-3.1v
13	BK	Knock Sensor Signal	<0.075v AC
15	BL/WT	Data Link Connector	12-14v
16	BR	Power Ground	<0.1v

1996-97 1.6L I4 MFI VIN A (All) 22 Pin Connector

PCM Pin #	Wire Color	Circuit Description (22 Pin)	Value at Hot Idle
1	RD/WT	Battery Direct	12-14v
2	GN	Tail Light Switch Signal	Switch On: 12v, Off: 0v
3	BK	Defogger Idle Up Signal	Switch On: 12v, Off: 0v
4	BK/WT	Stop Light Switch Signal	Brake Off: 0v, On: 12v
5	RD/YL	MIL (lamp) Control	MIL Off: 12v, On: 1v
7	LG	Overdrive Main Switch	Switch Off: 12v
9	PK/WT	Vehicle Speed Sensor	At 55 mph: 48 Hz
10 (TMMC)	YL/RD	A/C Amplifier Signal (AC1)	Clutch On: 1.5v, Off: 12v
10 (NUMMI)	YL/BL	A/C On Signal ('97)	Clutch On: 1.5v, Off: 12v
11	BK	Starter Switch Signal	9-11v (cranking)
12 (TMMC)	BK	Main Relay B+	12-14v
12 (NUMMI)	BK/RD	Main Relay B+ ('97)	12-14v
13	BL/BK	A/T Select Switch Drive	In 'D': 12v, Others: 0v
14	GN/RD	Circuit Opening Relay (FC)	0-3v, off-idle: 12v
16	BK	Data Link Connector	12-14v
17	RD/BK	A/T Select Switch Reverse	In 'R': 12v, Others: 0v
18	GN/RD	A/T Select Switch 2nd	In 2nd: 12v, Others: 0v
19	LG	A/T Select Switch Low	In Low: 12v, Others: 0v
20	RD	Cruise Control ECU	At Cruise in OD: 12v
21	RD/YL	A/C Amplifier Signal (ACT)	Clutch On: 12v, Off: 1.5v
22	BK/WT	A/T Neutral Start Switch	In P/N: 9-11v (cranking)

Standard Colors and Abbreviations

Abbreviation	Color	Abbreviation	Color	Abbreviation	Color
BK	Black	GY	Gray	RD	Red
BL	Blue	GN	Green	TN	Tan
BR	Brown	LG	Light Green	VT	Violet
DB	Dark Blue	OR	Orange	WT	White
DG	DK Green	PK	Pink	YL	Yellow

1996-97 1.6L I4 MFI VIN A (All) 26 Pin Connector

PCM Pin #	Wire Color	Circuit Description (26 Pin)	Value at Hot Idle
1	BL/YL	A/T-ECT Solenoid (SL)	In Lockup: 12-14v
2	PK	A/T-ECT Solenoid (S1)	In 3rd or OD: 1v
3	BK/YL	Igniter Signal (IGF)	Digital Signal: 0-5-0v
4	BK	CKP Sensor Signal (NE+)	AC pulse signals
5	BK	Distributor Signal (G2+)	AC pulse signals
9	BK/WT	ISC Signal (RSC)	Pulse Signals
10	BK/BL	ISC Signal (RSO)	Pulse Signals
11	BK/RD	Injector 2 Control	2.0-3.3 ms
12	YL	Injector 1 Control	2.0-3.3 ms
13	BR	Power Ground	<0.1v
14	BR	Sensor Ground	<0.050v
15	BR/YL	A/T-ECT Solenoid (S2)	1st or OD: 1v
17	WT	CKP Sensor Signal (NE-)	<0.050v
19	BL/BK	HO2S-11 (B1 S1) Heater	1v (Heater On)
20	YL/GN	Igniter Signal (IGT)	Digital Signal: 0-5-0v
23	BK/WT	EGR Solenoid Control (VSV)	12v or 0v
24	BK	Injector 4 Control	2.0-3.3 ms
25	BK/WT	Injector 3 Control	2.0-3.3 ms
26	BR	Power Ground	<0.1v

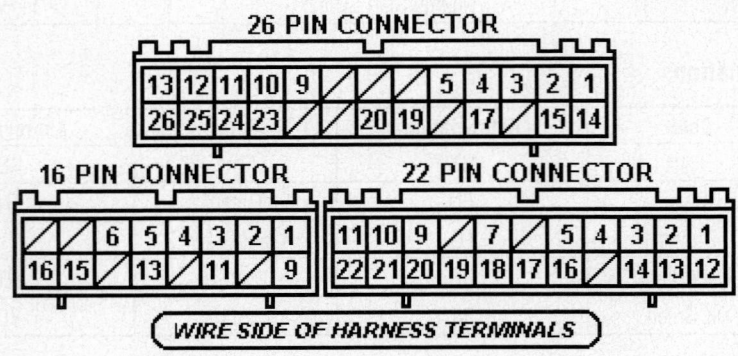

Pin Connector Graphic

1993-94 1.8L VIN A (A/T, ECT) 16 Pin Connector

PCM Pin #	Wire Color	Circuit Description (16 Pin)	Value at Hot Idle
1	YL	Sensor VREF	4.9-5.1v
2	GN/RD	MAP Sensor Signal	1-1.5v
3	YL/BK	IAT Sensor Signal	At 100°F: 2.60v
4	WT	ECT Sensor Signal	At 180°F: 0.51v
5 (Cal)	BK	Sub O2S Signal	0.1-1.1v
6 (Fed)	BK	Main O2S Signal	0.1-1.1v
7	BK	Check Connector	12-14v
8	RD/WT	Check Connector	12-14v
9	BR	Sensor Ground	<0.050v
10 (Cal)	WT/BL	EGR Gas Temp. Sensor	3.5-4.0v

1993-94 1.8L VIN A (A/T, ECT) 16 Pin Connector, *continued*

11	LG	TP Sensor Signal	0.3-0.8v
12	BL	Closed Throttle Switch	1v, at off-idle: 12v
13	BK	Knock Sensor Signal	<0.075v AC
14	GN/WT	Check Connector	12-14v
15	BL/WT	Check Connector	12-14v
16	BR	Sensor Ground	<0.050v

1993-94 1.8L VIN A (A/T, ECT) 22 Pin Connector

PCM Pin #	Wire Color	Circuit Description (22 Pin)	Value at Hot Idle
1	RD/WT	Battery Direct	12-14v
2	BK	Defogger/Light Idle Up Signal	Switch On: 12v, Off: 0v
4	GN/WT	Stop Light Switch Signal	Brake Off: 0v, On: 12v
5	RD/YL	MIL (lamp) Control	MIL Off: 12v, On: 1v
7	LG	Main Overdrive Switch	Switch Off: 12-14v
9	PK/WT	Vehicle Speed Sensor	At 55 mph: 48 Hz
10	YL/BL	A/C Amplifier Signal (AC1)	Clutch On: 1.5v, Off: 12v
11	BK	Starter Switch Signal	9-11v (cranking)
12	BK/RD	EFI Main Relay Power	12-14v
13	BK/RD	EFI Main Relay Power B1+	12-14v
14	GN/RD	Circuit Opening Relay (FC)	0-3v, off-idle: 12v
20	RD	Cruise Control ECU	At Cruise in OD: 12v
21	RD/YL	A/C Amplifier Signal (ACT)	Clutch On: 12v, Off: 1.5v
22	BK/WT	Neutral Start Switch	In P/N: 9-11v (cranking)

Standard Colors and Abbreviations

Abbreviation	Color	Abbreviation	Color	Abbreviation	Color
BK	Black	GY	Gray	RD	Red
BL	Blue	GN	Green	TN	Tan
BR	Brown	LG	Light Green	VT	Violet
DB	Dark Blue	OR	Orange	WT	White
DG	DK Green	PK	Pink	YL	Yellow

1993-94 1.8L VIN A (A/T, ECT) 26 Pin Connector

PCM Pin #	Wire Color	Circuit Description (26 Pin)	Value at Hot Idle
1	BL/YL	A/T-ECT Solenoid (SL)	In Lockup: 12v
2	PK	A/T-ECT Solenoid (S1)	In 3rd or OD: 1v
3	BK/YL	Igniter Signal (IGF)	Digital Signal: 0-5-0v
4	BK	Distributor Signal (NE+)	AC pulse signals
5	GN	Distributor Signal (G-)	<0.050v
6	GN/RD	A/T Select Switch 2nd	In 2nd: 12v, Others: 0v
9	BK/WT	IAC Signal (RSC)	Pulse Signals
10	BK/BL	IAC Signal (RSO)	Pulse Signals
11	BK/RD	Injector Pair 2 & 4 Control	2.0-3.3 ms
12	BK	Injector Pair 1 & 3 Control	2.0-3.3 ms
13	BR	Power Ground	<0.1v
14	BR	Sensor Ground	<0.050v
15	BR/YL	A/T-ECT Solenoid (S2)	1st or OD: 1v
17	WT	Distributor Signal (NE-)	<0.050v
18	RD	Distributor Signal (G1)	AC pulse signals
19	LG	A/T Select Switch Low	In Low: 12v, Others: 0v
20	BK	Igniter Signal (IGT)	Digital Signal: 0-5-0v
21	WT/RD	A/C Idle Up Solenoid	A/C On: 1v, Off: 12v
22	RD/BK	A/T Select Switch Reverse	In 'R': 12v, Others: 0v
23 (Cal)	BK/RD	EGR Solenoid Control (VSV)	12v or 0v
26	BR	Power Ground	<0.1v

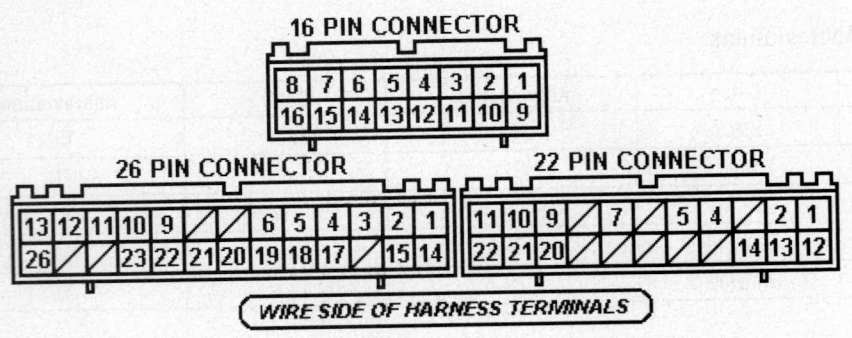

Pin Connector Graphic

1993-94 1.8L I4 MFI VIN A (M/T) 12 Pin Connector

PCM Pin #	Wire Color	Circuit Description (12 Pin)	Value at Hot Idle
1	BK/RD	EFI Main Relay Power B1+	12-14v
2	RD/WT	Battery Direct	12-14v
4	GN/RD	Circuit Opening Relay (FC)	0-3v, off-idle: 12v
6	RD/YL	A/C Amplifier Signal (ACT)	Clutch On: 12v, Off: 1.5v
7	BK/RD	EFI Main Relay Power	12-14v
8	RD/YL	MIL (lamp) Control	MIL Off: 12v, On: 1v
10	YL/BL	A/C Amplifier Signal (AC1)	Clutch On: 1.5v, Off: 12v
11	PK/WT	Vehicle Speed Sensor	At 55 mph: 48 Hz
12	BK	Defogger/Light Idle Up Signal	Switch On: 12v, Off: 0v

1993-94 1.8L I4 MFI VIN A (M/T) 16 Pin Connector

PCM Pin #	Wire Color	Circuit Description (16 Pin)	Value at Hot Idle
2	GN/RD	MAP Sensor Signal	1-1.5v
3	YL/BK	IAT Sensor Signal	At 100°F: 2.60v
4	WT	ECT Sensor Signal	At 180°F: 0.51v
5 (Cal)	BK	Sub O2S Signal	0.1-1.1v
6 (Fed)	BK	Main O2S Signal	0.1-1.1v
7	GN/WT	Check Connector	12-14v
8	RD/WT	Check Connector	12-14v
9	BR	Sensor Ground	<0.050v
10	LG	TP Sensor Signal	0.3-0.8v
11	YL	Sensor VREF	4.9-5.1v
12	BL	Closed Throttle Switch	1v, at off-idle: 12v
13 (Cal)	WT/BL	EGR Gas Temp. Sensor	3.5-4.0v
14	BK	Knock Sensor Signal	<0.075v AC
15	BL/WT	Check Connector	12-14v
16	BR	Sensor Ground	<0.050v

1993-94 1.8L I4 MFI VIN A (M/T) 26 Pin Connector

PCM Pin #	Wire Color	Circuit Description (26 Pin)	Value at Hot Idle
2	BK	Starter Switch Signal	9-11v (cranking)
3	BK/YL	Igniter Signal (IGF)	Digital Signal: 0-5-0v
4	BK	Distributor Signal (NE+)	AC pulse signals
5	GN	Distributor Signal (G1+)	AC pulse signals
9	BK/WT	IAC Signal (RSC)	Pulse Signals
10	BK/BL	IAC Signal (RSO)	Pulse Signals
12	BK/RD	Injector Pair 1 & 3 Control	2.0-3.3 ms
13	BR	Power Ground	<0.1v
16 (Cal)	BK/RD	EGR Solenoid Control (VSV)	12v or 0v
17	WT	Distributor Signal (NE-)	<0.050v
18	GN	Distributor Signal (G-)	<0.050v
22	BK	Igniter Signal (IGT)	Digital Signal: 0-5-0v
23	WT/RD	ISC Solenoid Control	Pulse Signals
24	BR	Sensor Ground	<0.050v
25	BK/RD	Injector Pair 2 & 4 Control	2.0-3.3 ms
26	BR	Power Ground	<0.1v

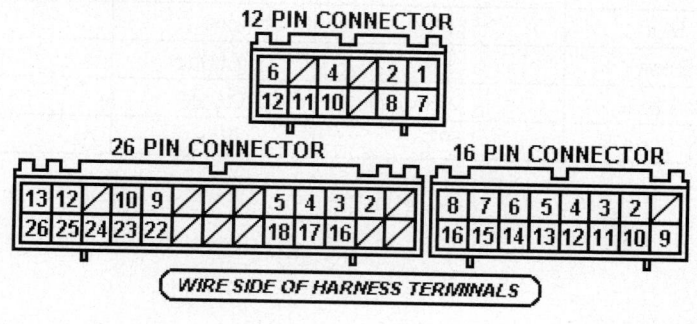

WIRE SIDE OF HARNESS TERMINALS

Pin Connector Graphic

05533_ADIA_G058

1995 1.8L I4 VIN A (A/T-ECT) 16 Pin Connector

PCM Pin #	Wire Color	Circuit Description (16 Pin)	Value at Hot Idle
1	YL	Sensor VREF	4.9-5.1v
2	GN/RD	MAP Sensor Signal	1-1.5v
3	YL/BK	IAT Sensor Signal	At 100°F: 2.60v
4	WT	ECT Sensor Signal	At 180°F: 0.51v
5 (Cal)	BK	Sub O2S Signal	0.1-1.1v
6 (Fed)	BK	Main O2S Signal	0.1-1.1v
7	BK	Check Connector	12-14v
8	RD/WT	Check Connector	12-14v
9	BR	Sensor Ground	<0.050v
10 (Cal)	WT/BL	EGR Gas Temp. Sensor	3.5-4.0v
11	LG	TP Sensor Signal	0.3-0.8v
12	BL	Closed Throttle Switch	1v, at off-idle: 12v
13	BK	Knock Sensor Signal	<0.075v AC
14	GN/WT	Check Connector	12-14v
15	BL/WT	Check Connector	12-14v

1995 1.8L I4 VIN A (A/T-ECT) 22 Pin Connector

PCM Pin #	Wire Color	Circuit Description (22 Pin)	Value at Hot Idle
1	RD/WT	Battery Direct	12-14v
2	BK	Tail Light Switch Signal	Switch Off: 12v, On: 1v
3	GN	Defogger Idle Up Signal	Switch On: 12v, Off: 0v
5	RD/YL	MIL (lamp) Control	MIL Off: 12v, On: 1v
7	LG	Main Overdrive Switch	Switch Off: 12v, On: 0v
9	PK/WT	Vehicle Speed Sensor	At 55 mph: 48 Hz
10	YL/BL	A/C Amplifier Signal (AC1)	Clutch On: 1.5v, Off: 12v
11	BK	Starter Switch Signal	9-11v (cranking)
12	BK/RD	EFI Main Relay Power	12-14v
14	GN/RD	Circuit Opening Relay (FC)	0-3v, off-idle: 12v
20	RD	Cruise Control ECU	At Cruise in OD: 12v
21	RD/YL	A/C Amplifier Signal (ACT)	Clutch On: 12v, Off: 1.5v
22	BK/WT	Neutral Start Switch	In PK or N: 12v

Standard Colors and Abbreviations

Abbreviation	Color	Abbreviation	Color	Abbreviation	Color
BK	Black	GY	Gray	RD	Red
BL	Blue	GN	Green	TN	Tan
BR	Brown	LG	Light Green	VT	Violet
DB	Dark Blue	OR	Orange	WT	White
DG	DK Green	PK	Pink	YL	Yellow

1995 1.8L I4 VIN A (A/T-ECT) 26 Pin Connector

PCM Pin #	Wire Color	Circuit Description (26 Pin)	Value at Hot Idle
1	BL/YL	A/T-ECT Solenoid (SL)	In Lockup: 12-14v
2	PK	A/T-ECT Solenoid (S1)	In 3rd or OD: 1v
3	BK/YL	Igniter Signal (IGF)	Digital Signal: 0-5-0v
4	BK	Distributor Signal (NE+)	AC pulse signals
5	BK	Distributor Signal (G2+)	AC pulse signals
6	GN/RD	A/T Select Switch 2nd	In 2nd: 12v, Others: 0v
7	BL/BK	Sub HO2S Heater	1v (Heater On)
9	BK/WT	ISC Signal (RSC)	Pulse Signals
10	BK/BL	ISC Signal (RSO)	Pulse Signals
11	BK/RD	Injector 2 Control	2.0-3.3 ms
12	BK/YL	Injector 1 Control	2.0-3.3 ms
13	BR	Power Ground	<0.1v
14	BR	Sensor Ground	<0.050v
15	GN/WT	A/T-ECT Solenoid (S2)	1st or OD: 1v
17	WT	Distributor Signal (NE-)	<0.050v
18	RD	Distributor Signal (G1)	AC pulse signals
19	LG	A/T Select Switch Low	In Low: 12v, Others: 0v
20	BK	Igniter Signal (IGT)	Digital Signal: 0-5-0v
21	WT/RD	A/C Idle Up Solenoid	A/C Off: 12v, On: 1v
22	WT/BK	A/T Select Switch Reverse	In 'R': 12v, Others: 0v
23 (Cal)	BK/RD	EGR Solenoid Control (VSV)	12v or 0v
24	BK	Injector 4 Control	2.0-3.3 ms
25	BK/WT	Injector 3 Control	2.0-3.3 ms
26	BR	Power Ground	<0.1v

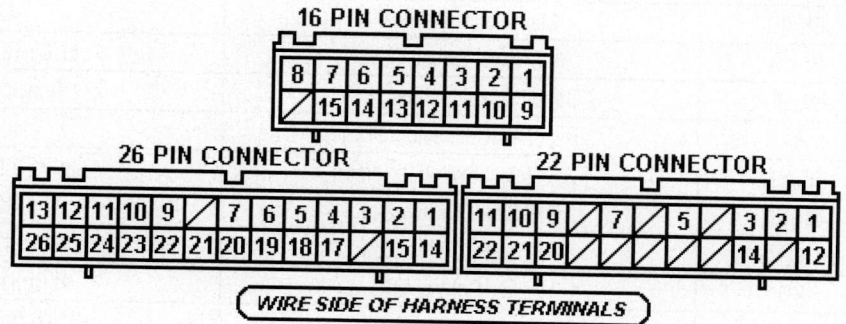

Pin Connector Graphic

05533_ADIA_G059

1996-97 1.8L I4 MFI VIN B (A/T-ECT) 16 Pin Connector

PCM Pin #	Wire Color	Circuit Description (16 Pin)	Value at Hot Idle
1	YL	Sensor VREF	4.9-5.1v
2	GN/RD	MAP Sensor Signal	1-1.5v
3	YL/BK	IAT Sensor Signal	At 100°F: 2.60v
4	WT	ECT Sensor Signal	At 180°F: 0.51v
5	BK	HO2S-12 (B1 S2) Signal	0.1-1.1v
6	BK	O2S-11 (B1 S1) Signal	0.1-1.1v
9	BR	Sensor Ground	<0.050v
11	LG	TP Sensor Signal	0.3-0.8v
13	BK	Knock Sensor Signal	<0.075v AC
15	BL/WT	Data Link Connector	12-14v
16	BR	Power Ground	<0.1v

1996-97 1.8L I4 MFI VIN B (A/T-ECT) 22 Pin Connector

PCM Pin #	Wire Color	Circuit Description (22 Pin)	Value at Hot Idle
1	RD/WT	Battery Direct	12-14v
2	GN	Tail Light Switch Signal	Switch On: 12v, Off: 0v
3	BK	Defogger Idle Up Signal	Switch On: 12v, Off: 0v
4	GN/WT	Stop Light Switch Signal	Brake Off: 0v, On: 12v
5	RD/YL	MIL (lamp) Control	MIL Off: 12v, On: 1v
7	LG	Overdrive Main Switch	Switch Off: 12v
9	PK/WT	Vehicle Speed Sensor	At 55 mph: 48 Hz
10 (TMMC)	YL/RD	A/C Cutout Signal	A/C On: 4-6v
10 (NUMMI)	YL/BL	A/C Cutout Signal (1997)	A/C On: 4-6v
11	BK	Starter Switch Signal	9-11v (cranking)
12 (TMMC)	BK	Main Relay B+	12-14v
12 (NUMMI)	BK/RD	Main Relay B+ (1997)	12-14v
13	BL/BK	A/T Select Switch Drive	In 'D': 12v, Others: 0v
14	GN/RD	Circuit Opening Relay (FC)	0-3v, off-idle: 12v
16	BK	Data Link Connector (SDL)	0v
17	RD/BK	A/T Select Switch Reverse	In 'R': 12v, Others: 0v
18	GN/RD	A/T Select Switch 2nd	In 2nd: 12v, Others: 0v
19	LG	A/T Select Switch Low	In Low: 12v, Others: 0v
20	RD	Cruise Control ECU	At Cruise in OD: 12v
21	RD/YL	A/C Amplifier Signal (ACT)	Clutch On: 12v, Off: 1.5v
22	BK/WT	A/T Neutral Start Switch	In PK or N: 12v

Standard Colors and Abbreviations

Abbreviation	Color	Abbreviation	Color	Abbreviation	Color
BK	Black	GY	Gray	RD	Red
BL	Blue	GN	Green	TN	Tan
BR	Brown	LG	Light Green	VT	Violet
DB	Dark Blue	OR	Orange	WT	White
DG	DK Green	PK	Pink	YL	Yellow

1996-97 1.8L I4 MFI VIN B (A/T-ECT) 26 Pin Connector

PCM Pin #	Wire Color	Circuit Description (26 Pin)	Value at Hot Idle
1	BL/YL	A/T-ECT Solenoid (SL)	In Lockup: 12-14v
2	PK	A/T-ECT Solenoid (S1)	In 3rd or OD: 1v
3	BL/YL	Igniter Signal (IGF)	Digital Signal: 0-5-0v
4	BK	CKP Sensor Signal (NE+)	AC pulse signals
5	BK	Distributor Signal (G2+)	AC pulse signals
9	BK/WT	ISC Signal (RSC)	Pulse Signals
10	BK/BL	ISC Signal (RSO)	Pulse Signals
11	BK/RD	Injector 2 Control	2.0-3.3 ms
12	YL	Injector 1 Control	2.0-3.3 ms
13	BR	Power Ground	<0.1v
14	BR	Sensor Ground	<0.050v
15	BR/YL	A/T-ECT Solenoid (S2)	1st or OD: 1v
17	WT	CKP Sensor Signal (NE-)	<0.050v
19	BL/BK	HO2S-12 (B1 S2) Heater	1v (Heater On)
20	YL/GN	Igniter Signal (IGT)	Digital Signal: 0-5-0v
23	BK/WT	EGR Solenoid Control (VSV)	12v or 0v
24	BK	Injector 4 Control	2.0-3.3 ms
25	BK/WT	Injector 3 Control	2.0-3.3 ms
26	BR	Power Ground	<0.1v

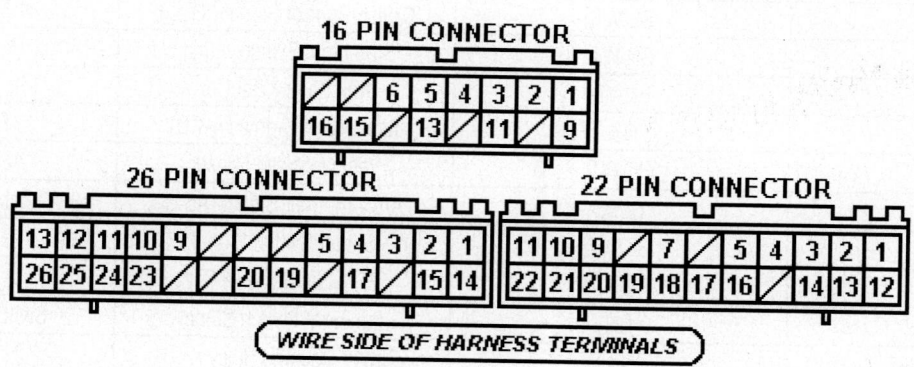

Pin Connector Graphic

05533_ADIA_G060

1998-99 1.8L I4 MFI VIN R (A/T-ECT) 16 Pin Connector

PCM Pin #	Wire Color	Circuit Description (16 Pin)	Value at Hot Idle
1	YL	Sensor VREF	4.9-5.1v
2	LG/RD	MAP Sensor Signal	1-1.5v
3	YL/BK	IAT Sensor Signal	At 100°F: 2.60v
4	WT	ECT Sensor Signal	At 180°F: 0.51v
5	RD	HO2S-12 (B1 S2) Signal	0.1-1.1v
6	WT	HO2S-11 (B1 S1) Signal	0.1-1.1v
7	BL	EVAP Vapor Pressure Sensor	2.9-3.1v (with hose off)
8	PK	HO2S-11 (B1 S1) Heater	1v (Heater On)
9	BR	Sensor Ground	<0.050v
11	LG	TP Sensor Signal	0.3-0.8v
12	RD/WT	Cruise Control Signal (OD1)	At Cruise in OD: 12v
13	BK	Knock Sensor Signal	<0.075v AC
14	PK	HO2S-12 (B1 S2) Heater	1v (Heater On)
15	BL/WT	Data Link Connector	12-14v
16	BR	Power Ground	<0.1v

1998-99 1.8L I4 MFI VIN R (A/T-ECT) 22 Pin Connector

PCM Pin #	Wire Color	Circuit Description (22 Pin)	Value at Hot Idle
1	RD/WT	Battery Direct	12-14v
2	BK	Defogger Idle Up Switch	Switch On: 12v, Off: 0v
3	BL/WT	Cruise Control Signal (IDLO)	1.5v, off-idle: 12v
4	GN/WT	Stop Light Switch Signal	Brake Off: 0v, On: 12v
5	RD/YL	MIL (lamp) Control	MIL Off: 12v, On: 1v
8	BK	Tachometer Signal (TACO)	Pulse Signals
9	PK/WT	Vehicle Speed Sensor	At 55 mph: 48 Hz
10	YL/RD	A/C Amplifier Signal (AC1)	Clutch On: 1.5v, Off: 12v
11	BK/WT	Starter Switch Signal	9-11v (cranking)
12	BK	EFI Main Relay Power	12-14v
13	GN	Tail Light Switch Signal	Switch On: 12v, Off: 0v
14	GN/RD	Circuit Opening Relay (FC)	0-3v, off-idle: 12v
16	WT	SIL Signal (Scan Tool)	Digital Signal
17	RD/BK	A/T Select Switch Reverse	In 'R': 12v, Others: 0v
18	GN/RD	A/T Select Switch 2nd	In 2nd: 12v, Others: 0v
19	GN/BK	A/T Select Switch Low	In Low: 12v, Others: 0v
21	RD/BL	A/C Amplifier Signal (ACT)	Clutch On: 12v, Off: 1.5v
22	LG	Overdrive Main Switch	Switch Off: 12v, On: 0v

Standard Colors and Abbreviations

Abbreviation	Color	Abbreviation	Color	Abbreviation	Color
BK	Black	GY	Gray	RD	Red
BL	Blue	GN	Green	TN	Tan
BR	Brown	LG	Light Green	VT	Violet
DB	Dark Blue	OR	Orange	WT	White
DG	DK Green	PK	Pink	YL	Yellow

1998-99 1.8L I4 MFI VIN R (A/T-ECT) 26 Pin Connector

PCM Pin #	Wire Color	Circuit Description (26 Pin)	Value at Hot Idle
1	BK/WT	A/T Neutral Start Switch	In P/N: 9-11v (cranking)
4	BK	CKP Sensor Signal (NE+)	AC pulse signals
5	BK	Distributor Signal (G2+)	AC pulse signals
6	BL/RD	PSP Switch Signal	Straight: 12v, Turning: 0v
7	RD	EVAP Vapor Pressure (VSV)	12v or 0v
8	BL/BK	EVAP Purge Solenoid (VSV)	12v or 0v
10	BK/BL	ISC Signal (RSO)	Pulse Signals
11	BK/RD	Injector 2 Control	2.0-3.3 ms
12	YL	Injector 1 Control	2.0-3.3 ms
13	BR	Sensor Ground	<0.050v
14	BR	Power Ground	<0.1v
16	BL/YL	Igniter Signal (IGF)	Digital Signal: 0-5-0v
17	WT	CKP/CMP Sensor Ground (-)	<0.050v
19	RD/BL	Igniter Transistor 2 Control	Digital Signal: 0-5-0v
20	YL/GN	Igniter Transistor 1 Control	Digital Signal: 0-5-0v
21	BL/YL	A/T-ECT Solenoid (SL)	In Lockup: 12-14v
22	BR/YL	A/T-ECT Solenoid (S2)	1st or OD: 1v
23	PK	A/T-ECT Solenoid (S1)	In 3rd or OD: 1v
24	BK	Injector 4 Control	2.0-3.3 ms
25	WT	Injector 3 Control	2.0-3.3 ms
26	BR	Sensor Ground	<0.050v

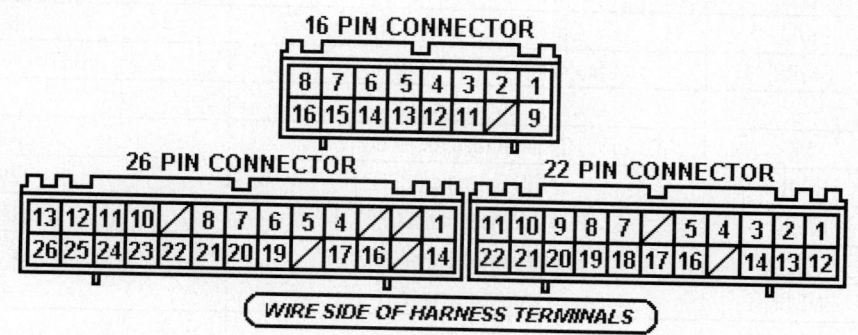

Pin Connector Graphic

05533_ADIA_G061

2000-06 1.8L I4 MFI VIN R (All) E2 22P Connector

PCM Pin #	Wire Color	Circuit Description (22 Pin)	Value at Hot Idle
1	WT	Battery Direct	12-14v
2	---	Not Used	---
3	GN/RD	Circuit Opening Relay (FC)	0-3v, off-idle: 12v
4	PK	EVAP Vapor Pressure Sensor	2.9-3.1v (with hose off)
5	BR/YL	Overdrive "Off" Indicator	Light Off: 12v, On: 1v
6	GN/OR	Multiplex Meter Input (MPX2)	Digital Signals
7	WT/BK	Power Ground (E03)	<0.1v
8	BK/OR	Ignition Switch Power (B+)	12-14v
9	---	Not Used	---
10	BL/WT	Cruise Control Signal (IDLO)	1.5v, off-idle: 12v
11	WT	SIL Signal (Scan Tool)	Digital Signals
12-14	---	Not Used	---
15	WT	MIL (lamp) Control	MIL Off: 12v, On: 1v
16	BK/RD	EFI No. 1 Fuse Power	12-14v
17-22	---	Not Used	---

2000-06 1.8L I4 MFI VIN R (All) E4 28P Connector

PCM Pin #	Wire Color	Circuit Description (28 Pin)	Value at Hot Idle
1	BK	Cruise Control Signal (OD1)	At Cruise in OD: 12v
2	RD/BK	Backup Light Switch Signal	In 'R': 12v, Others: 0v
3	BL/WT	Cooling Fan Relay 3 Control	Relay Off: 12v, On: 1v
4	BL/YL	Cooling Fan Relay 1 Control	Relay Off: 12v, On: 1v
5	PK/BL	Data Link Connector (TC)	12-14v
6	GN/WT	Stop Light Switch Signal (STP)	Brake Off: 0v, On: 12v
7	---	Not Used	---
8	PK	Center Airbag Assembly	Digital Signals
9	RD/BL	ACIS Solenoid (VSV)	12v or 0v
10	LG/RD	Multiplex Meter Input (MPX1)	Digital Signals
11	BL	Starter Switch Signal (STA)	9-11v (cranking)
12	WT/BK	Power Ground (EC)	<0.1v
13-15	---	Not Used	---
16	YL/GN	HO2S-12 (B1 S2) Heater	1v (Heater On)
17	---	Not Used	---
18	OR	A/C Dual Pressure Switch	A/C Off: 12v, On: 1v
19	BR/YL	Overdrive Lamp Control	At Cruise in OD: 1v
20	---	Not Used	---
21	BL/BK	EFI Main Relay Power	12-14v
22	WT/RD	Vehicle Speed Sensor (SPD)	At 55 mph: 48 Hz
23	GN/OR	EVAP Vapor Pressure (VSV)	12v or 0v
24	BK/OR	A/T Select Switch Drive	In 'D': 12v, Others: 0v
25	WT	HO2S-12 (B1 S2) Signal	0.1-1.1v
26	---	Not Used	---
27	BR/WT	Tachometer Signal (TACO)	Pulse Signals
28	GN	A/C Magnetic Clutch (LCK1)	Clutch Off: 0v, On: 12v

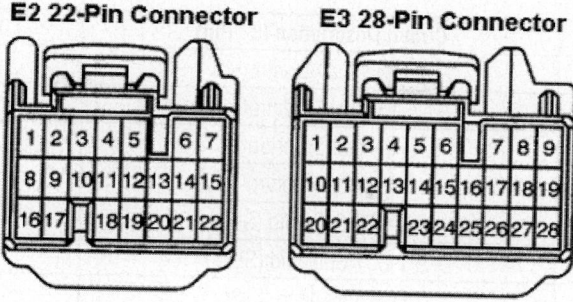

E2 22-Pin Connector   **E3 28-Pin Connector**

05533_ADIA_G062

Pin Connector Graphic
2000-06 1.8L I4 MFI VIN R (All) E4 24P Connector

PCM Pin #	Wire Color	Circuit Description (24 Pin)	Value at Hot Idle
1	YL/GN	MAF Sensor Ground (EVG)	<0.050v
2	RD	Sensor VREF (VC)	4.9-5.1v
3	YL/RD	HO2S-11 (B1 S1) Heater	1v (Heater On)
4	GN/OR	EVAP Purge Solenoid (VSV)	12v or 0v
8	PK/BL	A/T Neutral Start Switch	In 'N': 12v, Others: 0v
9	YL/BK	A/T Low Gear Position Switch	In 'LL': 12v, Others: 0v
11	GN/WT	MAF Sensor Signal (VG)	1.0-1.5v
12	BK	HO2S-11 (B1 S1) Signal	0.1-1.1v
13	GN/BL	TFT (fluid) Sensor (THO)	1-4v (varies with temp.)
14	GN	ECT Sensor Signal (THW)	1-4v (varies with temp.)
15	BL	CMP Sensor Signal (G2)	AC pulse signals
16, 24	BK, WT	CKP Sensor Signal (NE+), (NE-)	AC pulse signals
17	**BR**	**Power Ground (E1)**	<0.1v
18	**BR**	**Power Ground (E02)**	<0.1v
19	**BL/YL**	**A/T 2nd Gear Position Switch**	In '2L': 12v, Others: 0v
20	**PK**	**A/T Park Position Switch**	In 'P': 12v, Others: 0v
22	**BL/RD**	**IAT Sensor Signal (THA)**	1-4v (varies with temp.)
23	BK/WT	TP Sensor Signal (VTA)	1.1-1.9v

2000-06 1.8L I4 MFI VIN R (All) E5 31P Connector

PCM Pin #	Wire Color	Circuit Description (31 Pin)	Value at Hot Idle
1	RD	Injector 1 Control	2.0-3.3 ms
2	RD/BL	Injector 2 Control	2.0-3.3 ms
3	RD/WT	Injector 3 Control	2.0-3.3 ms
4	RD/BK	Injector 4 Control	2.0-3.3 ms
5	YL/GN	A/T-ECT Solenoid (SLT-)	Pulse Signals
6	YL/RD	A/T-ECT Solenoid (SLT+)	Pulse Signals
7	WT/BK	Sensor Ground (E03)	<0.050v
8	BR/YL	A/T-ECT Solenoid (S1)	In 3rd or OD: 1v
10	RD/BK	Igniter Transistor 1 Control	Digital Signal: 0-5-0v
11	RD/WT	Igniter Transistor 2 Control	Digital Signal: 0-5-0v
12	GN/RD	Igniter Transistor 3 Control	Digital Signal: 0-5-0v
13	RD/YL	Igniter Transistor 4 Control	Digital Signal: 0-5-0v
15, 16	BK, WT/BL	Turbine Speed Sensor (NT-), (NT+)	AC pulse signals
17	VT/WT	EVAP Canister Closed Valve (VSV)	12v or 0v
18	BK/WT	IAC Valve Signal (RSO)	
19	GN/WT	A/T-ECT Solenoid (SL)	In Lockup: 12-14v
20	YL	A/T-ECT Solenoid (S2)	1st or OD: 1v
21	WT/BK	Power Ground (E01)	<0.1v
22	YL/BK	OIL Pressure Switch (MOPS)	Closed: 1v, Open: 12v
23, 24	WT, GN/OR	Cam/Time Oil Control Valve (-), (+)	Pulse signals
25	BL/YL	Igniter Signal (IGF)	Digital Signal: 0-5-0v
27	WT	Knock Sensor Signal (KNK1)	<0.075v AC
28	PK	PSP Switch Signal	Straight: 12v, Turning: 0v
29	BR/WT	A/T-ECT Solenoid (ST)	Pulse Signals
31	WT/BK	Sensor Ground (E02)	<0.050v

E4 24-Pin Connector   **E5 31-Pin Connector**

05533_ADIA_G063

Pin Connector Graphic

CRESSIDA PIN CHARTS

1990-92 3.0L I6 MFI VIN M (A/T) 16 Pin Connector

PCM Pin #	Wire Color	Circuit Description (16 Pin)	Value at Hot Idle
1	RD/WT	Sensor VREF	KOER: 4.5-5.5v
2	GN/YL	Airflow Meter Signal	1.5-2.5v
3	YL	IAT Sensor Signal	At 100°F: 2.60v
4 ('90)	RD/BL	ECT Sensor Signal	At 180°F: 0.51v
4	GN	ECT Sensor Signal	At 180°F: 0.51v
5 (Cal)	BK	Sub O2S Signal	0.1-1.1v
6	BK	Main O2S Signal	0.1-1.1v
7	GN/RD	Check Connector	12-14v
8	GY	Check Connector	12-14v
9	BR	Sensor Ground	<0.050v
10 (Cal)	GN/WT	EGR Gas Temp. Sensor	3.5-4.0v
11	YL	TP Sensor Signal	0.3-0.8v
12	RD	Closed Throttle Switch	1v, at off-idle: 12v
13	GN/BK	Stop Light Switch Signal	Brake Off: 0v, On: 12v
14	BK	Knock Sensor Signal	<0.075v AC
15	GN/BK	Check Connector	12-14v
16	BK	Distributor Signal (G-)	<0.050v

1990-92 3.0L I6 MFI VIN M (A/T) 22 Pin Connector

PCM Pin #	Wire Color	Circuit Description (22 Pin)	Value at Hot Idle
1	WT/RD	Battery Direct	12-14v
2	BK/WT	Ignition Switch Power	12-14v
3	BL/WT	A/T Pattern Selector Switch	Norm: 0v, PWR: 12v
4 (91)	GY	EFI Main Relay Power	12-14v
4	GN	EFI Main Relay Power	12-14v
5 (90)	YL/RD	MIL (lamp) Control	MIL Off: 12v, On: 1v
5	RD	MIL (lamp) Control	MIL Off: 12v, On: 1v
6	GN/RD	Fuel Pump Relay	Relay Off: 12v, On: 1v
7	LG	Check Connector	12-14v
8	GN/WT	A/T-ECT Speed Sensor	Moving: 0-5-0v
9	PK/WT	Vehicle Speed Sensor	At 55 mph: 48 Hz
10	BK/WT	A/C Magnetic Clutch (ACMG)	Clutch Off: 0v, On: 12v
11	BK	Starter Switch Signal	9-11v (cranking)
12	BK/RD	EFI Main Relay Power B1+	12-14v
13	BK/RD	EFI Main Relay Power	12-14v
14	OR	A/T Select Switch 2nd	In 2nd: 12v, Others: 0v
15	RD	A/T Pattern Selector Switch	Norm: 0v, PWR: 12v
16	YL/BL	A/T Select Switch Low	In Low: 12v, Others: 0v
20	PK/GN	Main Overdrive Switch	Switch Off: 12v, On: 0v
21	PK/BL	Cruise Control ECU	At Cruise in OD: 12v
22	BK/WT	Neutral Start Switch	In P/N: 9-11v (cranking)

1990-92 3.0L I6 MFI VIN M (A/T) 26 Pin Connector

PCM Pin #	Wire Color	Circuit Description (26 Pin)	Value at Hot Idle
1	WT	Distributor Signal (NE+)	AC pulse signals
2	RD	Distributor Signal (G2+)	AC pulse signals
3	BK/YL	Igniter Signal (IGF)	Digital Signal: 0-5-0v
4	GN/OR	ISC Motor (ISC4)	Pulse Signals
5	GN/YL	ISC Motor (ISC3)	Pulse Signals
6	GN/RD	ISC Motor (ISC2)	Pulse Signals
7	GN/WT	ISC Motor (ISC1)	Pulse Signals
8 (Cal)	BK/WT	HO2S Heater	1v (Heater On)
10	GN/RD	EGR Solenoid Control (VSV)	12v or 0v
11	WT	Injector Pair 2 & 6 Control	2.0-3.3 ms
12	YL	Injector Pair 1 & 4 Control	2.0-3.3 ms
13	BR	Power Ground	<0.1v
14	BR	Sensor Ground	<0.050v
15	GN	Distributor Signal (G1+)	AC pulse signals
16	GN/RD	Defogger/Light Idle Up Signal	Switch On: 12v, Off: 0v
17	PK/GN	A/T-ECT Solenoid (SL)	In Lockup: 12-14v
18	BK	A/T-ECT Solenoid (S2)	1st or OD: 1v
19	GY	A/T-ECT Solenoid (S1)	In 3rd or OD: 1v
20	YL/GN	Igniter Signal (IGT)	0.72-0.74v
21	GN/OR	Cruise Control ECU	At Cruise in OD: 12v
24	YL	Cold Start Injector Control	1v (at cold startup)
25	BL	Injector Pair 3 & 5 Control	2.0-3.3 ms
26	BR	Power Ground	<0.1v

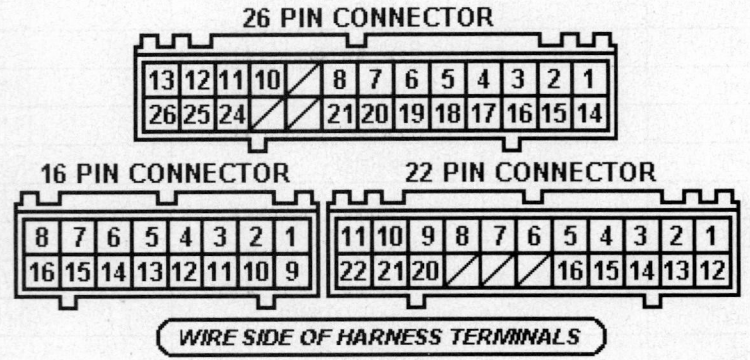

Pin Connector Graphic

ECHO PIN CHARTS

2000 1.5L I4 DOHC MFI VIN T (All) E4 22 Pin Connector

PCM Pin #	Wire Color	Circuit Description (22 Pin)	Value at Hot Idle
1	BK/YL	Battery Direct	12-14v
2	YL	Water Temp. Cool Indicator	Indicator On: 12, Off: 0v
3	BL	EVAP Canister Closed Valve	12v or 0v
5	YL/RD	MIL (lamp) Control	MIL Off: 12v, On: 1v
6	YL/RD	Water Temp. Hot Indicator	Indicator On: 12v, Off: 0v
8	BK	Tachometer Signal (TACO)	Pulse Signals
9	VT/WT	Speed Signal from C/Meter	At 55 mph: 48 Hz
10	BK/WT	A/C Amplifier Signal (AC1)	Clutch On: 1.5v, Off: 12v
11	BK	Starter Switch Signal	9-11v (cranking)
12	BK/RD	EFI Main Relay Power	12-14v
14	BK	Circuit Opening Relay (FC)	0-3v, off-idle: 12v
16	LG/BK	SIL Signal (Scan Tool)	Digital Signal
17	RD/YL	A/T Select Switch Reverse	In 'R': 12v, Others: 0v
18	GN	A/T Select Switch 2nd	In 2nd: 12v, Others: 0v
19	GN/WT	A/T Select Switch Low	In Low: 12v, Others: 0v
20	BL/WT	Mirror Heater Switch Signal	Switch On: 12-14v
21	BK	A/C Amplifier Signal (ACT)	Clutch On: 12v, Off: 1.5v
22	BK	A/T Neutral Start Switch	Cranking: 9-11v

2000 1.5L I4 DOHC MFI VIN T A/T E5 12 Pin Connector

PCM Pin #	Wire Color	Circuit Description (12 Pin)	Value at Hot Idle
1	WT/BL	A/T-ECT Solenoid (SLT+)	Pulse Signals
2	BK/WT	A/T-ECT Solenoid (ST)	Pulse Signals
3	WT/GN	A/T-ECT Solenoid (S1)	In 3rd or OD: 1v
4	GN/OR	A/T: Overdrive "Off" Indicator	Light Off: 12v
5	RD	Turbine Speed Sensor (NT+)	AC pulse signals
6	GN	A/T-ECT Solenoid (SL)	In Lockup: 12-14v
7	WT	A/T-ECT Solenoid (SLT-)	Pulse Signals
9	WT/RD	A/T: ECT Solenoid (S2)	1st or OD: 1v
10	YL	A/T Oil Temperature Sensor	At 230°F: 0.95v
11	BK	Turbine Speed Sensor (NT-)	AC pulse signals
12	GN/OR	A/T: Overdrive Main Switch	Switch Off: 12v

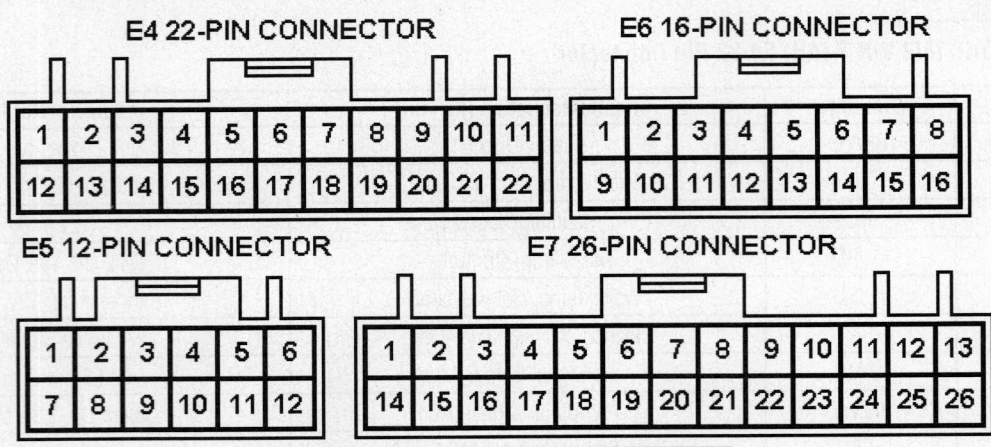

E4 22-PIN CONNECTOR

E6 16-PIN CONNECTOR

E5 12-PIN CONNECTOR

E7 26-PIN CONNECTOR

WIRE SIDE OF HARNESS TERMINALS

05533_ADIA_G065

Pin Connector Graphic
2000 1.5L I4 DOHC MFI VIN T (All) E6 16 Pin Connector

PCM Pin #	Wire Color	Circuit Description (16 Pin)	Value at Hot Idle
1	RD/WT	Sensor VREF (VC)	4.9-5.1v
2	PK	MAF Sensor Signal (VG)	1.1-1.5v
3	YL/BK	IAT Sensor Signal	1-4v (varies with temp.)
4	RD/BL	ECT Sensor Signal	1-4v (varies with temp.)
5	BK	HO2S-12 (B1 S2) Signal	0.1-1.1v
6	RD	HO2S-11 (B1 S1) Signal	0.1-1.1v
7	BL/BK	Maximum Hot Switch	Switch On: 12v, Off: 0v
8	BK/OR	HO2S-11 (B1 S1) Heater	1v (Heater On)
9	BR	Sensor Ground (E2)	<0.050v
10	VT	MAF Sensor Ground (EVG)	<0.050v
11	YL/RD	TP Sensor Signal (VTA)	1.1-1.9v
12	YL	PSP Switch Signal	Straight: 12v, Turning: 0v
13	WT	Knock Sensor Signal	<0.075v AC
14	BK	EVAP Vapor Pressure Sensor	Fuel Cap Off: 2.9-3.7v
15	PK/BL	Data Link Connector (TC)	12-14v
16	WT	HO2S-12 (B1 S2) Heater	1v (Heater On)

2000 1.5L I4 DOHC MFI VIN T (All) E7 26 Pin Connector

PCM Pin #	Wire Color	Circuit Description (26 Pin)	Value at Hot Idle
1	BR	Power Ground (E03)	<0.1v
2	BK/RD	IAC Signal (RSD)	Pulse Signals
3	YL	Igniter Signal (IGF)	Pulse Signal: 0-5-0v
4	BL	Generator Control Signal (L)	12v
5	RD/YL	A/T Select Switch Drive	In 'D': 12v, Others: 0v
6	GN/WT	Stop Light Switch Signal	Brake Off: 0v, On: 12v
7	---	Not Used	---
8	WT/BL	Water Temperature Switch	Switch Off: 12v
9	WT/GN	EVAP Purge Solenoid (VSV)	12v or 0v
10	RD/WT	Camshaft Timing Oil Control Valve (+)	AC pulse signals

2000 1.5L I4 DOHC MFI VIN T (All) E7 26 Pin Connector, *continued*

11	BK/YL	Injector 2 Control	2.0-3.3 ms
12	BK/OR	Injector 1 Control	2.0-3.3 ms
13	BR	Power Ground (E1)	<0.1v
14	BR	Power Ground (E01)	<0.1v
15	BR	Case Ground	<0.050v
16	WT	CKP/CMP Sensor Ground (-)	<0.050v
17	OR	CKP Sensor Signal (NE+)	AC pulse signals
18	BK	CMP Sensor Signal (G2+)	AC pulse signals
19	GN/YL	Igniter Transistor 4 Control	Digital Signal: 0-5-0v
20	GN/OR	Igniter Transistor 3 Control	Digital Signal: 0-5-0v
21	GN/BK	Igniter Transistor 2 Control	Digital Signal: 0-5-0v
22	GN/RD	Igniter Transistor 1 Control	Digital Signal: 0-5-0v
23	RD/BK	Camshaft Timing Oil Control Valve (-)	AC pulse signals
24	BK/BL	Injector 4 Control	2.0-3.3 ms
25	BK/WT	Injector 3 Control	2.0-3.3 ms
26	BR	Power Ground (E02)	<0.1v

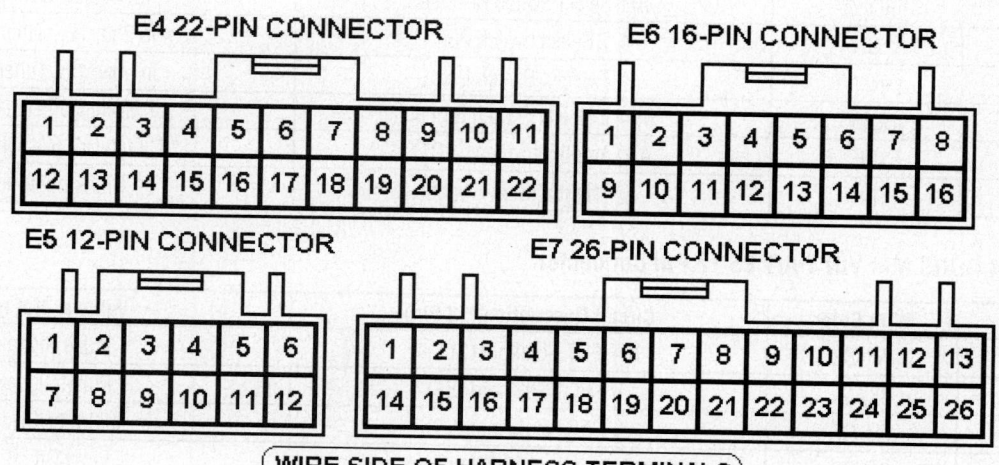

Pin Connector Graphic

05533_ADIA_G065

2001-02 1.5L I4 DOHC MFI VIN T (All) E4 22 Pin Connector

PCM Pin #	Wire Color	Circuit Description (22 Pin)	Value at Hot Idle
1	BK/YL	Battery Direct	12-14v
2	YL	Water Temp. Cool Indicator	Indicator On: 12, Off: 0v
3	BL	EVAP Canister Closed Valve	12v or 0v
4	---	Not Used	---
5	YL/RD	MIL (lamp) Control	MIL Off: 12v, On: 1v
6	YL/RD	Water Temp. Hot Indicator	Indicator On: 12v, Off: 0v
8	BK	Tachometer Signal (TACO)	Pulse Signals
9	VT/WT	Speed Signal from C/Meter	At 55 mph: 48 Hz
10	BK/WT	A/C Amplifier Signal (AC1)	Clutch On: 1.5v, Off: 12v
11	BK	Starter Switch Signal	9-11v (cranking)
12	BK/RD	EFI Main Relay Power	12-14v
13	---	Not Used	---
14	BK	Circuit Opening Relay (FC)	0-3v, off-idle: 12v
15	---	Not Used	---
16	LG/BK	SIL Signal (Scan Tool)	Digital Signal
17	RD/YL	A/T Select Switch Reverse	In 'R': 12v, Others: 0v
18	GN	A/T Select Switch 2nd	In 2nd: 12v, Others: 0v
19	GN/WT	A/T Select Switch Low	In Low: 12v, Others: 0v
20	BL/WT	Mirror Heater Switch Signal	Switch On: 12-14v
21	BK	A/C Amplifier Signal (ACT)	Clutch On: 12v, Off: 1.5v
22	BK	A/T Neutral Start Switch	Cranking: 9-11v

2001-02 1.5L I4 DOHC MFI VIN T A/T E5 12 Pin Connector

PCM Pin #	Wire Color	Circuit Description (12 Pin)	Value at Hot Idle
1	WT/BL	A/T-ECT Solenoid (SLT+)	Pulse Signals
2	BK/WT	A/T-ECT Solenoid (ST)	Pulse Signals
3	WT/GN	A/T-ECT Solenoid (S1)	In 3rd or OD: 1v
4	GN/OR	A/T: Overdrive "Off" Indicator	Light Off: 12v
5	RD	Turbine Speed Sensor (NT+)	AC pulse signals
6	GN	A/T-ECT Solenoid (SL)	In Lockup: 12-14v
7	WT	A/T-ECT Solenoid (SLT-)	Pulse Signals
9	WT/RD	A/T: ECT Solenoid (S2)	1st or OD: 1v
10	YL	A/T Oil Temperature Sensor	At 230°F: 0.95v
11	BK	Turbine Speed Sensor (NT-)	AC pulse signals
12	GN/OR	A/T: Overdrive Main Switch	Switch Off: 12v

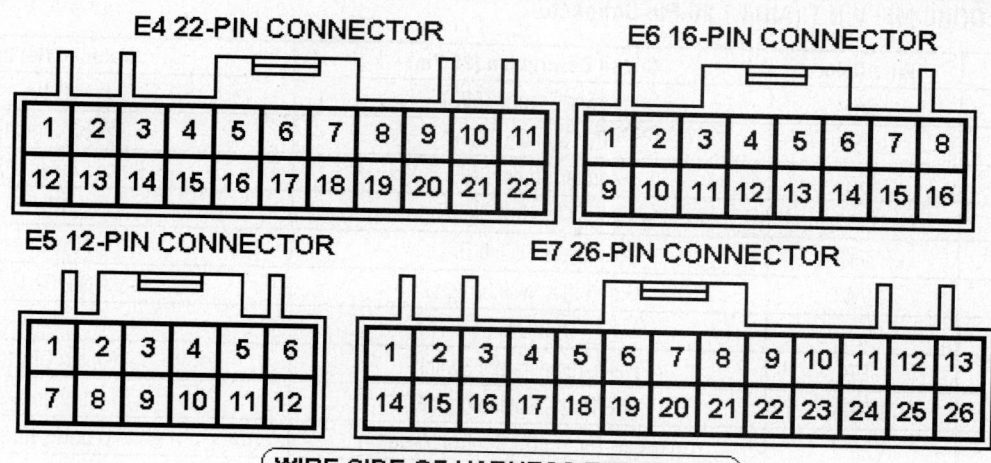

E4 22-PIN CONNECTOR

E6 16-PIN CONNECTOR

E5 12-PIN CONNECTOR

E7 26-PIN CONNECTOR

WIRE SIDE OF HARNESS TERMINALS

Pin Connector Graphic

05533_ADIA_G065

2001-02 1.5L I4 DOHC MFI VIN T (All) E6 16 Pin Connector

PCM Pin #	Wire Color	Circuit Description (16 Pin)	Value at Hot Idle
1	RD/WT	Sensor VREF (VC)	4.9-5.1v
2	PK	MAF Sensor Signal (VG)	1.1-1.5v
3	YL/BK	IAT Sensor Signal	1-4v (varies with temp.)
4	RD/BL	ECT Sensor Signal	1-4v (varies with temp.)
5	BK	HO2S-12 (B1 S2) Signal	0.1-1.1v
6	WT	HO2S-11 (B1 S1) Signal	0.1-1.1v
7	BL/BK	Maximum Hot Switch	Switch On: 12v, Off: 0v
8	OR	HO2S-11 (B1 S1) Heater	1v (Heater On)
9	BR	Sensor Ground (E2)	<0.050v
10	VT	MAF Sensor Ground (EVG)	<0.050v
11	YL/RD	TP Sensor Signal (VTA)	1.1-1.9v
12	YL	PSP Switch Signal	Straight: 12v, Turning: 0v
13	WT	Knock Sensor Signal	<0.075v AC
14	BK	EVAP Vapor Pressure Sensor	Fuel Cap Off: 2.9-3.7v
15	PK/BL	Data Link Connector (TC)	12-14v
16	WT	HO2S-12 (B1 S2) Heater	1v (Heater On)

2001-02 1.5L I4 DOHC MFI VIN T (All) E7 26 Pin Connector

PCM Pin #	Wire Color	Circuit Description (26 Pin)	Value at Hot Idle
1	BR	Power Ground (E03)	<0.1v
2	BK/RD	IAC Signal (RSD)	Pulse Signals
3	YL	Igniter Signal (IGF)	Pulse Signal: 0-5-0v
4	BL	Generator Control Signal (L)	12v
5	RD/YL	A/T Select Switch Drive	In 'D': 12v, Others: 0v
6	GN/WT	Stop Light Switch Signal	Brake Off: 0v, On: 12v
7	---	Not Used	---
8	WT/BL	Water Temperature Switch	Switch Off: 12v
9	WT/GN	EVAP Purge Solenoid (VSV)	12v or 0v
10	RD/WT	Camshaft Timing Oil Control Valve (+)	AC pulse signals
11	BK/YL	Injector 2 Control	2.0-3.3 ms
12	BK/OR	Injector 1 Control	2.0-3.3 ms
13	BR	Power Ground (E1)	<0.1v
14	BR	Power Ground (E01)	<0.1v
15	BR	Case Ground	<0.050v
16	WT	CKP/CMP Sensor Ground (-)	<0.050v
17	OR	CKP Sensor Signal (NE+)	AC pulse signals
18	BK	CMP Sensor Signal (G2+)	AC pulse signals
19	GN/YL	Igniter Transistor 4 Control	Digital Signal: 0-5-0v
20	GN/OR	Igniter Transistor 3 Control	Digital Signal: 0-5-0v
21	GN/BK	Igniter Transistor 2 Control	Digital Signal: 0-5-0v
22	GN/RD	Igniter Transistor 1 Control	Digital Signal: 0-5-0v
23	RD/BK	Camshaft Timing Oil Control Valve (-)	AC pulse signals
24	BK/BL	Injector 4 Control	2.0-3.3 ms
25	BK/WT	Injector 3 Control	2.0-3.3 ms
26	BR	Power Ground (E02)	<0.1v

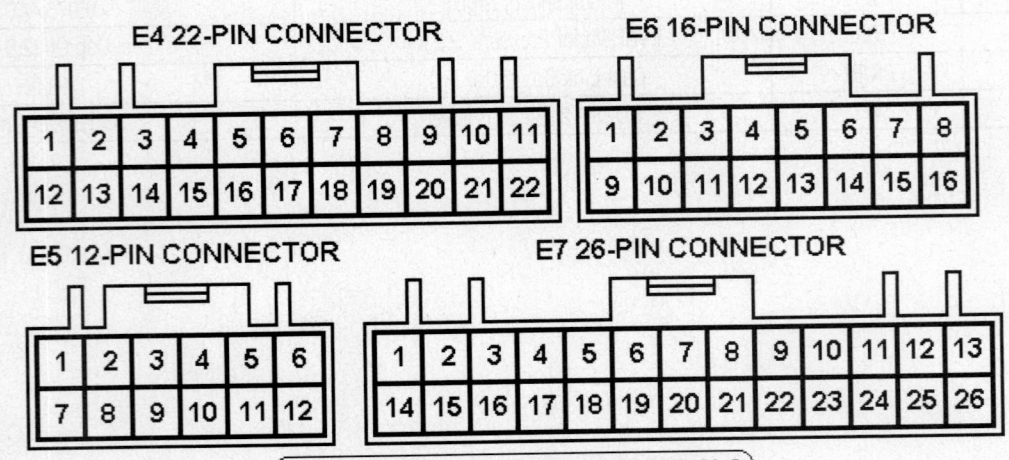

Pin Connector Graphic

05533_ADIA_G065

2003-2005 1.5L I4 DOHC MFI VIN T (All) E4 31 Pin Connector

PCM Pin #	Wire Color	Circuit Description (31 Pin)	Value at Hot Idle
1	BK/RD	EFI Main Relay Power	12-14v
2	BL/BK	Maximum Hot Switch (MHSW)	Switch Open: 0v, Closed: 12v
3	BK/YL	Battery Direct	12-14v
4	---	Not Used	---
5	BK	Tachometer Signal (TACO)	Pulse Signals
6	YL/RD	Cooling Fan 1 Relay (CF)	Relay Off: 12v, On: 1v
7	WT/BL	A/C Single Pressure Switch (FAN)	Switch Open: 12v, Closed: 1v
8-9	---	Not Used	---
10	BK	Circuit Opening Relay (FC)	0-3v, off-idle: 12v
11	YL/RD	MIL (lamp) Control	MIL Off: 12v, On: 1v
12	BL/WT	Heater Sub1 Relay Control (ELS)	Relay Off: 12v, On: 1v
13	---	Not Used	---
14	YL/BK	Center Airbag Assembly	Digital Signals
15-17	---	Not Used	---
18	LG/BK	Scan Tool Signal (SIL)	Digital Signal
19	RD	Scan Tool (WFSE)	12v
20	PK/BL	Data Link Connector (TC)	12-14v
21	BK	EVAP Vapor Pressure Sensor	Fuel Cap Off: 2.9-3.7v
22-31	---	Not Used	---

2003-2005 1.5L I4 DOHC MFI VIN T (All) E5 35 Pin Connector

PCM Pin #	Wire Color	Circuit Description (35 Pin)	Value at Hot Idle
1-2	---	Not Used	---
3	YL/RD	Water Temperature Cool Indicator	Indicator On: 12v, Off: 0v
4	BK/RD	HO2S-12 (B1 S2) Heater	1v (Heater On)
5-6	---	Not Used	---
7	GN/OR	A/T Overdrive "Off" Indicator	Light Off: 12v
8	GN/WT	A/T Select Switch Low	In Low: 12v, Others: 0v
9	GN	A/T Select Switch 2nd	In 2nd: 12v, Others: 0v
10	RD/YL	A/T Select Switch Drive	In 'D': 12v, Others: 0v
11	RD/WT	A/T Select Switch Reverse	In 'R': 12v, Others: 0v
12	---	Not Used	---
13	YL	PSP Switch Signal	Straight: 12v, Turning: 0v
14-16	---	Not Used	---
17	VT/WT	Speed Signal from Combination Meter	At 55 mph: 48 Hz
18	---	Not Used	---
19	GN/WT	Stop Light Switch Signal	Brake Off: 0v, On: 12v
20-27	---	Not Used	---
28	YL/RD	Water Temperature Hot Indicator	Indicator On: 12v, Off: 0v
29	GN/OR	A/T: Overdrive Main Switch	Switch Off: 12v
30	---	Not Used	---
31	BK/WT	A/C Amplifier Signal (AC)	Clutch On: 1.5v, Off: 12v
32	---	Not Used	---
33	BK	A/C Amplifier Signal (ACT)	Clutch On: 12v, Off: 1.5v
34-35	---	Not Used	---

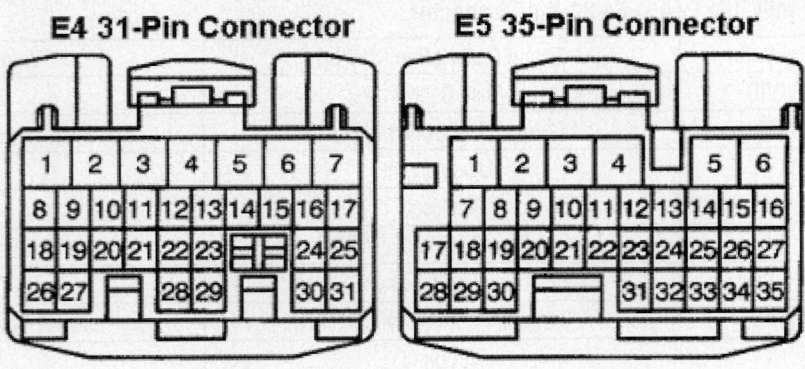

E4 31-Pin Connector

E5 35-Pin Connector

05533_ADIA_G066

Pin Connector Graphic
2003-2005 1.5L I4 DOHC MFI VIN T (All) E6 34 Pin Connector

PCM Pin #	Wire Color	Circuit Description (34 Pin)	Value at Hot Idle
1	BK/OR	Injector 1 Control	2.0-3.3 ms
2	BK/YL	Injector 2 Control	2.0-3.3 ms
3	BK/WT	Injector 3 Control	2.0-3.3 ms
4	BK/BL	Injector 4 Control	2.0-3.3 ms
5	BK/RD	IAC Signal (RSD)	Pulse Signals
6	BR	Power Ground (E02)	<0.1v
7	BR	Power Ground (E01)	<0.1v
8	GN/RD	Igniter Transistor 1 Control	Digital Signal: 0-5-0v
9	GN/BK	Igniter Transistor 2 Control	Digital Signal: 0-5-0v
10	GN/OR	Igniter Transistor 3 Control	Digital Signal: 0-5-0v
11	GN/YL	Igniter Transistor 4 Control	Digital Signal: 0-5-0v
12	WT/GN	EVAP Purge Solenoid (VSV)	12v or 0v
16	WT	A/T-ECT Solenoid (SLT-)	Pulse Signals
17	WT/BL	A/T-ECT Solenoid (SLT+)	Pulse Signals
18	RD/WT	Sensor VREF (VC)	4.9-5.1v
19	RD/BL	ECT Sensor Signal (THW)	1-4v (varies with temp.)
20	YL/BK	IAT Sensor Signal	1-4v (varies with temp.)
21	YL/RD	TP Sensor Signal (VTA)	1.1-1.9v
23	YL	Igniter Signal (IGF)	Pulse Signal: 0-5-0v
26	BK	CMP Sensor Signal (G2+)	AC pulse signals
27	OR	CKP Sensor Signal (NE+)	AC pulse signals
28	BR	Sensor Ground (E2)	<0.050v
30	YL/GN	A/T Oil Temperature Sensor (OIL)	At 230°F: 0.95v
34	WT	CKP Sensor Ground (NE-)	<0.050v

2003-2005 1.5L I4 DOHC MFI VIN T (All) E7 35 Pin Connector

PCM Pin #	Wire Color	Circuit Description (35 Pin)	Value at Hot Idle
1	WT	Knock Sensor Signal (KNK1)	<0.075v AC
4	BK/RD	HO2S-11 (B1 S1) Heater	1v (Heater On)
5	BR	Power Ground (E03)	<0.1v
7	BL	Generator Control Signal	12-0-12v
8	BK	A/T Neutral Start Switch (NSW)	Cranking: 9-11v
9	BK/YL	Starter Switch Signal (STA)	9-11v (cranking)
12	BK/WT	A/T-ECT Solenoid (ST)	Pulse Signals
13	GN	A/T-ECT Solenoid (SL)	In Lockup: 12-14v
14	WT/RD	A/T: ECT Solenoid (S2)	1st or OD: 1v
15	WT/GN	A/T-ECT Solenoid (S1)	In 3rd or OD: 1v
21	BK	HO2S-12 (B1 S2) Signal	0.1-1.1v
23	WT	HO2S-11 (B1 S1) Signal	0.1-1.1v
24	PK	MAF Sensor Signal (VG)	1.1-1.5v
27	RD	Turbine Speed Sensor (NT+)	AC pulse signals
28	BR	Power Ground (EC)	<0.1v
29	VT	PSP Switch Signal	Straight: 12v, Turning: 0v
32	VT	MAF Sensor Ground (EVG)	<0.050v
35	BK	Turbine Speed Sensor (NT-)	AC pulse signals

E6 34-Pin Connector   **E7 35-Pin Connector**

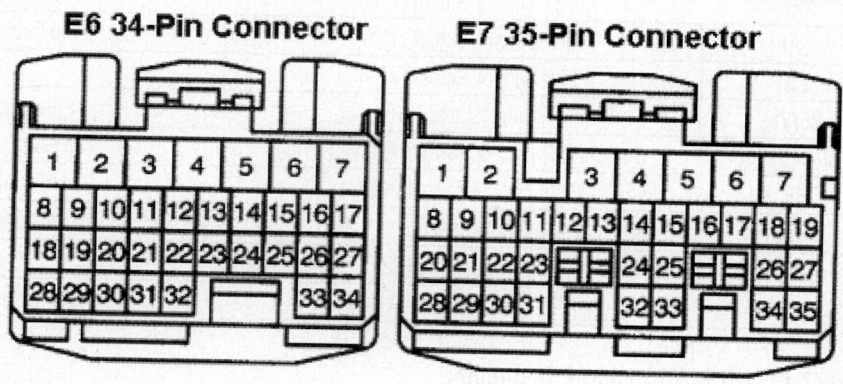

Pin Connector Graphic

05533_ADIA_G067

MATRIX PIN CHARTS

2003-2006 1.8L DOHC 1ZZ-FE VIN R (All) E3 34P Connector

PCM Pin #	Wire Color	Circuit Description (34 Pin)	Value at Hot Idle
1	YL	Injector 1 Control	2.0-3.3 ms
2	BK	Injector 2 Control	2.0-3.3 ms
3	WT	Injector 3 Control	2.0-3.3 ms
4	BL	Injector 4 Control	2.0-3.3 ms
5	BK/BL	IAC Air Control Valve Signal (RSD)	Pulse Signals
6	WT/BK	Power Ground (E02)	<0.1v
7	WT/BK	Power Ground (E01)	<0.1v
8	RD/BL	Igniter Transistor 1 Control	Digital Signal: 0-5-0v
9	YL/GN	Igniter Transistor 2 Control	Digital Signal: 0-5-0v
10	GY	Igniter Transistor 3 Control	Digital Signal: 0-5-0v
11	WT	Igniter Transistor 4 Control	Digital Signal: 0-5-0v
12	BL/BK	EVAP Vapor Pressure (VSV)	12v or 0v
13	YL	A/C Pressure Switch (HP)	A/C Off: 12v, On: 1v
14	BK/YL	Camshaft Timing Oil Control Valve (-) Left	AC pulse signals
15	YL	Camshaft Timing Oil Control Valve (+) Left	AC pulse signals
16, 17	PK, RD/WT	A/T-ECT Solenoid (SLT-), (SLT+)	Pulse Signals
18	YL	Sensor VREF (VC)	4.9-5.1v
19	WT	ECT Sensor Signal (THW)	1-4v (varies with temp.)
20	YL/BK	IAT Sensor Signal (THA)	1-4v (varies with temp.)
21	LG	TP Sensor Signal (VTA)	0.3-0.9v
23	BL/YL	Igniter Signal (IGF)	Pulse signal: 0-5-0v
24, 25	BL/YL, BK	Camshaft Timing Oil Control Valve (-), (+)	AC pulse signals
26	BK	CMP Sensor Signal (G22+)	AC pulse signals
27, 34	BK, WT	CKP Sensor Signal (NE+), (NE-)	AC pulse signals
28	BR	Power Ground (E2)	<0.1v
30	WT/BL	TFT (fluid) Sensor (THO)	1-4v (varies with temp.)

2003-2006 1.8L DOHC 1ZZ-FE VIN R (All) E4 35P Connector

PCM Pin #	Wire Color	Circuit Description (35 Pin)	Value at Hot Idle
1	BK, WT	Knock Sensor Signal (+), (-)	<0.075v AC
4	PK	HO2S-11 (B1 S1) Heater	1v (Heater On)
5	WT/BK	Power Ground (E03)	<0.1v
7	BR	Power Ground (E1)	<01v
8	RD	Neutral Start Switch (NSW)	Cranking: 0-3v
9	BK	Starter Signal (STA)	Cranking: 9-11v
11	BK	A/T-ECT Solenoid (DSL)	In Lockup: 12-14v
12	BL/OR	A/T-ECT Solenoid (ST)	In Lockup: 12-14v
13	BL/WT	A/T-ECT Solenoid (SL)	In Lockup: 12-14v
15	RD/YL	A/T-ECT Solenoid (S1)	In 3rd or OD: 1v
16, 17	BL, BL/OR	A/T-ECT Solenoid (SL2-), (SL2+)	Moving in 3rd or OD: <1v
18, 19	BL, RD/YL	A/T-ECT Solenoid (SL1-), SL1+)	Moving in 1st Gear: <1v

2003-2006 1.8L DOHC 1ZZ-FE VIN R (All) E4 35P Connector, *continued*

21	WT	HO2S-21 (B2 S1) Signal (OX1B)	0.1-1.1v
23	RD	HO2S-11 (B1 S1) Signal (OX1A)	0.1-1.1v
24	GN	MAF Sensor Signal (VG)	1-3v
28	WT/BK	Power Ground (EC)	<0.1v
29	BL/RD	PSP Switch Signal (PSW)	Straight: 12v, Turning: 0v
32	BL/WT	MAF Sensor Ground (EVG)	<0.050v
33	BK/WT	OIL Pressure Switch Signal (OSW)	Closed: 1v, Open: 12v

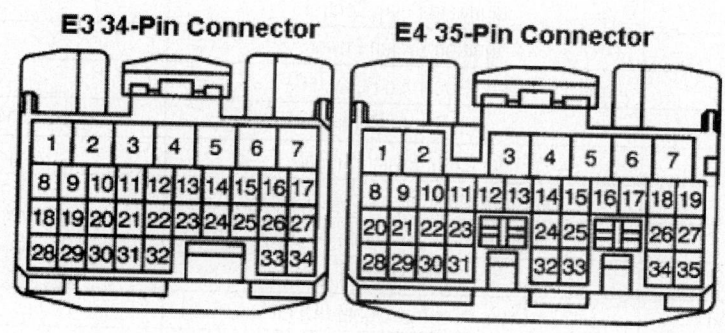

E3 34-Pin Connector   **E4 35-Pin Connector**

05533_ADIA_G068

Pin Connector Graphic

2003-2006 1.8L DOHC 1ZZ-FE VIN R (All) E5 35P Connector

PCM Pin #	Wire Color	Circuit Description (35 Pin)	Value at Hot Idle
1	BL	EVAP Canister Closed Valve	12v or 0v
2	GN/WT	A/C Magnetic Clutch Relay (ACMG)	Relay Off: 12v, On: 1v
3	---	Not Used	---
4	PK/BK	HO2S-21 (B2 S1) Heater (HT1B)	1v (Heater On)
5	BK	Tachometer Signal (TACO)	Pulse Signals
6	---	Not Used	---
7	LG	Overdrive Indicator Control (ODLP)	Indicator Off: 12v, On: 1v
8	LG/BK	A/T Select Switch Low Signal (L)	In Low: 12v, Others: 0v
9	LG	A/T Select Switch 2nd Signal (2L)	In 2nd: 12v, Others: 0v
10	BL	A/T Select Switch Drive Signal (DL)	In Drive: 12v, Others: 0v
11	RD/BK	A/T Select Switch Reverse (R)	In Reverse: 12v, Others: 0v
12-13	---	Not Used	---
14	YL/RD	Hot Light Indicator Control	Indicator Off: 12v, On: 1v
15-16	---	Not Used	---
17	VT/WT	Speedometer Indicator	At 55 mph: 48 Hz
18	RD/YL	Overdrive Signal to C/C Module (OD1)	0-12-0v
19	GN/WT	Stop Light Switch Signal (STP)	Brake Off: 0v, On: 12v
20-28	---	Not Used	---
29	LG/BK	Overdrive Main Switch Signal (ODMS)	Switch Off: 12v
30	---	Not Used	---
31	YL	A/C Pressure Switch Signal (ACIS)	Switch Open, 12v, Closed: 0v
32	BK/BL	A/C Thermistor Signal	1.4-1.8v (temp. >59°F)
33	GN/WT	A/C Switch Signal (ACLD)	A/C On: 12v, Off: 0v
34-35	---	Not Used	---

2003-2006 1.8L DOHC 1ZZ-FE VIN R (All) E6 31P Connector

PCM Pin #	Wire Color	Circuit Description (31 Pin)	Value at Hot Idle
1	BK	EFI Main Relay Power	12-14v
2	---	Not Used	---
3	RD/WT	Battery Direct	12-14v
4	RD	EVAP Pressure Switching (VSV)	12v or 0v
5	---	Not Used	---
6	LG	Cooling Fan 1 Relay (CF)	Relay Off: 12v, On: 1v
7	LG/BK	Cooling Fan 2 Relay (FAN)	Relay Off: 12v, On: 1v
8	BK/WT	EFI Main Relay Control	Relay On: 12v, Off: 1v
9	RD	Ignition Switch Power	12-14v
10	GN/RD	Circuit Opening Relay (FC)	0-3v, off-idle: 12v
11	YL/RD	MIL (lamp) Control	MIL Off: 12v, On: 1v
12	GN	Tail Light Switch Signal	Switch On: 12v, Off: 0v
13	WT	Mirror Heater/Defrost Switch Signal	Heater On: 12-14v
14	YL	Center Airbag Assembly (F/PS)	Digital Signals
15	---	Not Used	---
16	BL/WT	Cruise Control Signal (IDLO)	1.5v, off-idle: 12v
17	---	Not Used	---
18	BL/RD	SIL Signal (Scan Tool)	Digital Signal
19	PK	Scan Tool (WFSE)	12v
20	PK/BK	Data Link Connector (TC)	12-14v
21	BL	EVAP Vapor Pressure Sensor	2.9-3.1v (with hose off)
22-31	---	Not Used	---

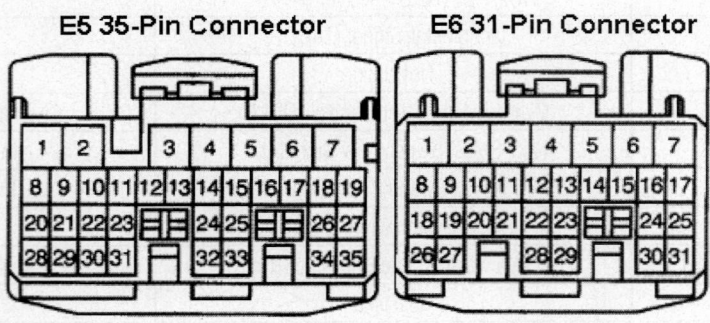

E5 35-Pin Connector   E6 31-Pin Connector

05533_ADIA_G069

Pin Connector Graphic

2003-2006 1.8L DOHC 2ZZ-GE VIN Y (All) E3 34P Connector

PCM Pin #	Wire Color	Circuit Description (34 Pin)	Value at Hot Idle
1	YL	Injector 1 Control	2.0-3.3 ms
2	BK	Injector 2 Control	2.0-3.3 ms
3	WT	Injector 3 Control	2.0-3.3 ms
4	BL	Injector 4 Control	2.0-3.3 ms
5	BK/BL	IAC Air Control Valve Signal (RSD)	Pulse Signals
6	WT/BK	Power Ground (E02)	<0.1v
7	WT/BK	Power Ground (E01)	<0.1v
8	RD/BL	Igniter Transistor 1 Control	Digital Signal: 0-5-0v

2003-2006 1.8L DOHC 2ZZ-GE VIN Y (All) E3 34P Connector, *continued*

9	YL/GN	Igniter Transistor 2 Control	Digital Signal: 0-5-0v
10	GY	Igniter Transistor 3 Control	Digital Signal: 0-5-0v
11	WT	Igniter Transistor 4 Control	Digital Signal: 0-5-0v
12	BL/BK	EVAP Vapor Pressure (VSV)	12v or 0v
13	YL	A/C Pressure Switch (HP)	A/C Off: 12v, On: 1v
14	BK/YL	Camshaft Timing Oil Control Valve (-) Left	AC pulse signals
15	YL	Camshaft Timing Oil Control Valve (+) Left	AC pulse signals
16, 17	PK, RD/WT	A/T-ECT Solenoid (SLT-), (SLT+)	Pulse Signals
18	YL	Sensor VREF (VC)	4.9-5.1v
19	WT	ECT Sensor Signal (THW)	1-4v (varies with temp.)
20	YL/BK	IAT Sensor Signal (THA)	1-4v (varies with temp.)
21	LG	TP Sensor Signal (VTA)	0.3-0.9v
23	BL/YL	Igniter Signal (IGF)	Pulse signal: 0-5-0v
24, 25	BL/YL, BK	Camshaft Timing Oil Control Valve (-), (+)	AC pulse signals
26	BK	CMP Sensor Signal (G22+)	AC pulse signals
27, 34	BK, WT	CKP Sensor Signal (NE+), (NE-)	AC pulse signals
28	BR	Power Ground (E2)	<0.1v
30	WT/BL	TFT (fluid) Sensor (THO)	1-4v (varies with temp.)

2003-2006 1.8L DOHC 2ZZ-GE VIN Y (All) E4 35P Connector

PCM Pin #	Wire Color	Circuit Description (35 Pin)	Value at Hot Idle
1	BK, WT	Knock Sensor Signal (+), (-)	<0.075v AC
4	PK	HO2S-11 (B1 S1) Heater	1v (Heater On)
5	WT/BK	Power Ground (E03)	<0.1v
7	BR	Power Ground (E1)	<01v
8	RD	Neutral Start Switch (NSW)	Cranking: 0-3v
9	BK	Starter Signal (STA)	Cranking: 9-11v
11	BK	A/T-ECT Solenoid (DSL)	In Lockup: 12-14v
12	BL/OR	A/T-ECT Solenoid (ST)	In Lockup: 12-14v
13	BL/WT	A/T-ECT Solenoid (SL)	In Lockup: 12-14v
15	RD/YL	A/T-ECT Solenoid (S1)	In 3rd or OD: 1v
16, 17	BL, BL/OR	A/T-ECT Solenoid (SL2-), (SL2+)	Moving in 3rd or OD: <1v
18, 19	BL, RD/YL	A/T-ECT Solenoid (SL1-), SL1+)	Moving in 1st Gear: <1v
21	WT	HO2S-21 (B2 S1) Signal (OX1B)	0.1-1.1v
23	RD	HO2S-11 (B1 S1) Signal (OX1A)	0.1-1.1v
24	GN	MAF Sensor Signal (VG)	1-3v
28	WT/BK	Power Ground (EC)	<0.1v
29	BL/RD	PSP Switch Signal (PSW)	Straight: 12v, Turning: 0v
32	BL/WT	MAF Sensor Ground (EVG)	<0.050v
33	BK/WT	OIL Pressure Switch Signal (OSW)	Closed: 1v, Open: 12v

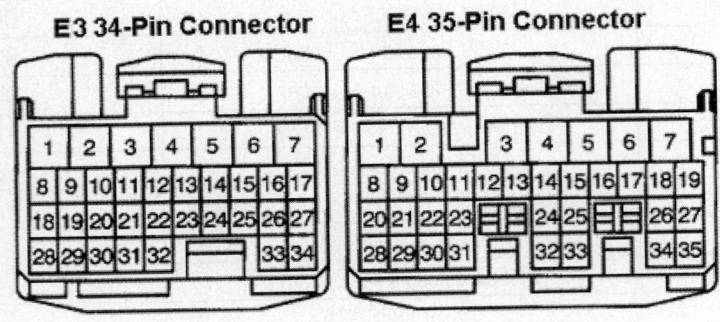

E3 34-Pin Connector   **E4 35-Pin Connector**

05533_ADIA_G068

Pin Connector Graphic
2003-2006 1.8L DOHC 2ZZ-GE VIN Y (All) E5 35P Connector

PCM Pin #	Wire Color	Circuit Description (35 Pin)	Value at Hot Idle
1	BL	EVAP Canister Closed Valve	12v or 0v
2	GN/WT	A/C Magnetic Clutch Relay (ACMG)	Relay Off: 12v, On: 1v
3	---	Not Used	---
4	PK/BK	HO2S-21 (B2 S1) Heater (HT1B)	1v (Heater On)
5	BK	Tachometer Signal (TACO)	Pulse Signals
6	---	Not Used	---
7	LG	Overdrive Indicator Control (ODLP)	Indicator Off: 12v, On: 1v
8	LG/BK	A/T Select Switch Low Signal (L)	In Low: 12v, Others: 0v
9	LG	A/T Select Switch 2nd Signal (2L)	In 2nd: 12v, Others: 0v
10	BL	A/T Select Switch Drive Signal (DL)	In Drive: 12v, Others: 0v
11	RD/BK	A/T Select Switch Reverse (R)	In Reverse: 12v, Others: 0v
12-13	---	Not Used	---
14	YL/RD	Hot Light Indicator Control	Indicator Off: 12v, On: 1v
15-16	---	Not Used	---
17	VT/WT	Speedometer Indicator	At 55 mph: 48 Hz
18	RD/YL	Overdrive Signal to C/C Module (OD1)	0-12-0v
19	GN/WT	Stop Light Switch Signal (STP)	Brake Off: 0v, On: 12v
20-28	---	Not Used	---
29	LG/BK	Overdrive Main Switch Signal (ODMS)	Switch Off: 12v
30	---	Not Used	---
31	YL	A/C Pressure Switch Signal (ACIS)	Switch Open, 12v, Closed: 0v
32	BK/BL	A/C Thermistor Signal	1.4-1.8v (temp. >59°F)
33	GN/WT	A/C Switch Signal (ACLD)	A/C On: 12v, Off: 0v
34-35	---	Not Used	---

2003-2006 1.8L DOHC 2ZZ-GE VIN Y (All) E6 31P Connector

PCM Pin #	Wire Color	Circuit Description (31 Pin)	Value at Hot Idle
1	BK	EFI Main Relay Power	12-14v
2	---	Not Used	---
3	RD/WT	Battery Direct	12-14v
4	RD	EVAP Pressure Switching (VSV)	12v or 0v
5	---	Not Used	---
6	LG	Cooling Fan 1 Relay (CF)	Relay Off: 12v, On: 1v
7	LG/BK	Cooling Fan 2 Relay (FAN)	Relay Off: 12v, On: 1v
8	BK/WT	EFI Main Relay Control	Relay On: 12v, Off: 1v
9	RD	Ignition Switch Power	12-14v
10	GN/RD	Circuit Opening Relay (FC)	0-3v, off-idle: 12v
11	YL/RD	MIL (lamp) Control	MIL Off: 12v, On: 1v
12	GN	Tail Light Switch Signal	Switch On: 12v, Off: 0v
13	WT	Mirror Heater/Defrost Switch Signal	Heater On: 12-14v
14	YL	Center Airbag Assembly (F/PS)	Digital Signals
15	---	Not Used	---
16	BL/WT	Cruise Control Signal (IDLO)	1.5v, off-idle: 12v
17	---	Not Used	---
18	BL/RD	SIL Signal (Scan Tool)	Digital Signal
19	PK	Scan Tool (WFSE)	12v
20	PK/BK	Data Link Connector (TC)	12-14v
21	BL	EVAP Vapor Pressure Sensor	2.9-3.1v (with hose off)
22-31	---	Not Used	---

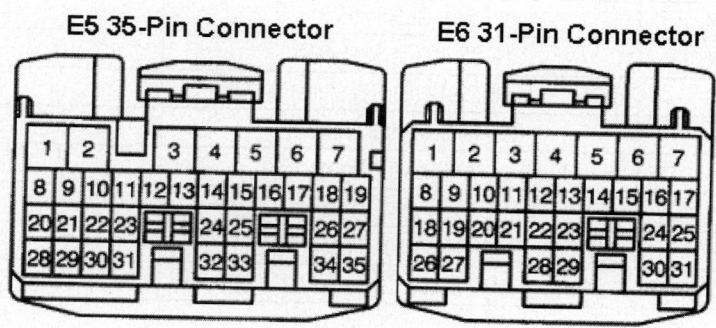

E5 35-Pin Connector   E6 31-Pin Connector

05533_ADIA_G069

Pin Connector Graphic

MR2 SPYDER PIN CHARTS

2000 1.8L DOHC 1ZZ-FE VIN R (M/T) E3 28P Connector

PCM Pin #	Wire Color	Circuit Description (28 Pin)	Value at Hot Idle
3	PK/GN	Cooling Fan 3 Relay	Relay Off: 12v, On: 1v
4	BL	Cooling Fan 2 Relay	Relay Off: 12v, On: 1v
5	PK/BK	Data Link Connector (TC)	12-14v
6	GN	Stop Light Switch Signal	Brake Off: 0v, On: 12v
8	BK/WT	Center Airbag Assembly	Digital Signals
10	BL	Multiplex Communication (1)	Digital Signals
11	BK	Starter Switch Signal	9-11v (cranking)
16	RD/YL	ABS ECU Signal	Pulse Signals
18	WT/BL	A/C Dual Pressure Switch	A/C Off: 12v, On: 1v
21	GY	EFI Main Relay Control	Relay Off: 12v, On: 1v
22	VT/WT	Speedometer Indicator	At 55 mph: 48 Hz
23	BL/WT	EVAP Vapor Pressure (VSV)	12v or 0v
27	WT	Tachometer Signal (TACO)	Pulse Signals
28	WT/RD	A/C Magnetic Clutch & Lock Sensor	A/C Off: 12v, On: 1v

2000 1.8L DOHC 1ZZ-FE VIN R (M/T) E4 24P Connector

PCM Pin #	Wire Color	Circuit Description (24 Pin)	Value at Hot Idle
1	BK/RD	MAF Sensor Ground (EVG)	<0.050v
2	BL/RD	Sensor VREF (VC)	4.9-5.1v
3	BK/YL	HO2S-11 (B1 S1) Heater	1v (Heater On)
4	WT	EVAP Purge Solenoid (VSV)	12v or 0v
5	BK/WT	HO2S-21 (B2 S1) Heater	1v (Heater On)
6	WT/BK	Power Ground (E03)	<0.1v
8	BL/BK	HO2S-12 (B1 S2) Heater	1v (Heater On)
9	BK	HO2S-12 (B1 S2) Signal	0.1-1.1v
10	YL	Generator Control Signal (L)	12v
11	VT	MAF Sensor Signal	1-3v
12	BK	HO2S-11 (B1 S1) Signal	0.1-1.1v
14	RD/BL	ECT Sensor Signal	1-4v (varies with temp.)
15	WT	CMP Sensor Signal (G2+)	AC pulse signals
16	WT	CKP Sensor Signal (NE+)	AC pulse signals
17	BR	Power Ground (E1)	<01v
18	BR	Sensor Ground (E2)	<0.050v
21	BK	HO2S-21 (B2 S1) Signal	0.1-1.1v
22	YL/BK	IAT Sensor Signal	1-4v (varies with temp.)
23	YL/GN	TP Sensor Signal	1.1-1.9v
24	BK	CKP/CMP Sensor Ground (-)	<0.050v

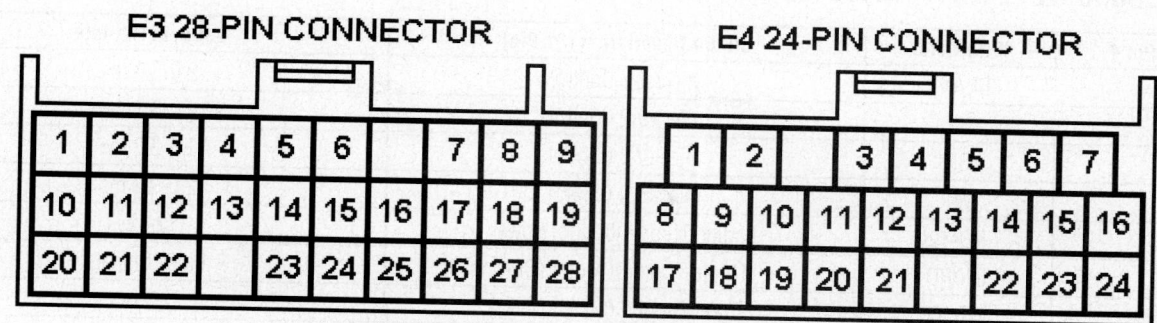

WIRE SIDE OF HARNESS TERMINALS

05533_ADIA_G070

Pin Connector Graphic
2000 1.8L DOHC 1ZZ-FE VIN R (M/T) E2 22P Connector

PCM Pin #	Wire Color	Circuit Description (22 Pin)	Value at Hot Idle
1	BK/YL	Battery Direct	12-14v
2	WT/GN	Immobilizer Security Indicator	LED Off: 0v, On: 12v
3	GN/RD	Circuit Opening Relay (FC)	0-3v, off-idle: 12v
4	GN/BK	EVAP Vapor Pressure Sensor	2.9-3.1v (with hose off)
6	BK	Multiplex Communication (2)	Digital Signals
8	BK/RD	Ignition Switch Power	12-14v
11	WT/GN	SIL Signal (Scan Tool)	Digital Signal
12	BL	A/C Magnetic Clutch (ACMG)	Clutch Off: 0v, On: 12v
15	YL/RD	MIL (lamp) Control	MIL Off: 12v, On: 1v
16	BK	EFI Main Relay Power	12-14v

2000 1.8L DOHC 1ZZ-FE VIN R (M/T) E5 31P Connector

PCM Pin #	Wire Color	Circuit Description (31 Pin)	Value at Hot Idle
1	BK/WT	Injector 1 Control	2.0-3.3 ms
2	BK	Injector 2 Control	2.0-3.3 ms
3	BL	Injector 3 Control	2.0-3.3 ms
4	WT	Injector 4 Control	2.0-3.3 ms
6	RD/YL	Engine Throttle Control Power	12-14v
7	GN/RD	Throttle Control Motor (CL-)	Pulse Signals
8, 31	WT/BK	Power Ground (ME01, ME2)	<0.1v
9	GY	Motor Control Shield Ground	<0.050v
10	BK	Igniter Transistor 1 Control	Digital Signal: 0-5-0v
11	BL	Igniter Transistor 2 Control	Digital Signal: 0-5-0v
12	BL/WT	Igniter Transistor 3 Control	Digital Signal: 0-5-0v
13	BL	Igniter Transistor 4 Control	Digital Signal: 0-5-0v
17	BL	EVAP Canister Closed Valve	12v or 0v
18	GN	IAC Valve Signal (RSO)	Pulse Signals
19	RD	Throttle Control Motor (M+)	Pulse Signals
20	YL/RD	Throttle Control Motor (CL+)	Pulse Signals
21	WT/BK	Power Ground (E01)	<0.1v
22	YL/BK	OIL Pressure Switch	Closed: 1v, Open: 12v
23	WT	Camshaft Timing Oil Control Valve (-)	AC pulse signals
24	RD	Camshaft Timing Oil Control Valve (+)	AC pulse signals
25	BK/YL	Igniter Signal (IGF)	Pulse signal: 0-5-0v
27	BK	Knock Sensor Signal	<0.075v AC
29	GN	Throttle Control Motor (M-)	Pulse Signals

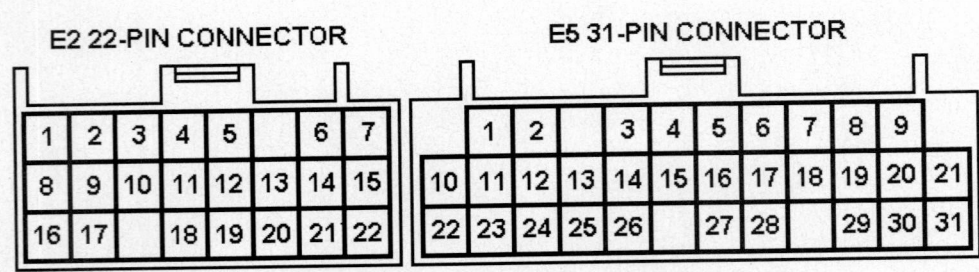

E2 22-PIN CONNECTOR   E5 31-PIN CONNECTOR

WIRE SIDE OF HARNESS TERMINALS

05533_ADIA_G071

Pin Connector Graphic

2001-05 1.8L DOHC 1ZZ-FE VIN R (M/T) E2 22P Connector

PCM Pin #	Wire Color	Circuit Description (22 Pin)	Value at Hot Idle
1	BK/YL	Battery Direct	12-14v
2	WT/GN	Immobilizer Security Indicator	LED Off: 0v, On: 12v
3	GN/RD	Circuit Opening Relay (FC)	0-3v, off-idle: 12v
4	GN/BK	EVAP Vapor Pressure Sensor	2.9-3.1v (with hose off)
5	BK	Multiplex Meter Input (MPX2)	Digital Signals
6	BK	Multiplex Communication (2)	Digital Signals
7	BL/BK	A/T Neutral Start Switch (NSW)	Cranking: 9-11v

2001-05 1.8L DOHC 1ZZ-FE VIN R (M/T) E2 22P Connector, *continued*

8	BK/RD	Ignition Switch Power	12-14v
9-10	---	Not Used	---
11	WT/GN	SIL Signal (Scan Tool)	Digital Signal
12	BL	A/C Magnetic Clutch (ACMG)	Clutch Off: 0v, On: 12v
13-14	---	Not Used	---
15	YL/RD	MIL (lamp) Control	MIL Off: 12v, On: 1v
16	BK	EFI Main Relay Power	12-14v

2001-05 1.8L DOHC 1ZZ-FE VIN R (M/T) E3 28P Connector

PCM Pin #	Wire Color	Circuit Description (28 Pin)	Value at Hot Idle
1	BK/YL	Start Relay Control (STD)	Relay Off: 12v, On: 1v
2	---	Not Used	---
3	YL/GN	Cooling Fan 3 Relay (CF)	Relay Off: 12v, On: 1v
4	BL	Cooling Fan 2 Relay (FAN)	Relay Off: 12v, On: 1v
5	PK/BK	Data Link Connector (TC)	12-14v
6	GN	Stop Light Switch Signal	Brake Off: 0v, On: 12v
7	---	Not Used	---
8	BK/WT	Center Airbag Assembly (F/PS)	Digital Signals
9	---	Not Used	---
10	BL	Multiplex Meter Input (MPX1)	Digital Signals
11	BK	Starter Switch Signal	9-11v (cranking)
12-15	---	Not Used	---
16	RD/YL	ABS ECU Signal	Digital Signals
17	---	Not Used	---
18	WT/BL	A/C Dual Pressure Switch (PRE)	A/C Off: 12v, On: 1v
19	WT/G	Power Steering ECU (PSCT)	Pulse Signals
20	---	Not Used	---
21	GY	EFI Main Relay Control	Relay Off: 12v, On: 1v
22	VT/WT	Speedometer Indicator	At 55 mph: 48 Hz
23	BL/WT	EVAP Vapor Pressure (VSV)	12v or 0v
24	---	Not Used	---
25	BL/OR	Power Steering ECU (PS)	Pulse Signals
26	---	Not Used	---
27	WT	Tachometer Signal (TACO)	Pulse Signals
28	WT/RD	A/C Magnetic Clutch & Lock Sensor	A/C Off: 12v, On: 1v

E2 22-Pin Connector   **E3 28-Pin Connector**

Pin Connector Graphic

05533_ADIA_G062

2001-05 1.8L DOHC 1ZZ-FE VIN R (M/T) E4 24P Connector

PCM Pin #	Wire Color	Circuit Description (24 Pin)	Value at Hot Idle
1	BK/RD	MAF Sensor Ground (EVG)	<0.050v
2	BL/RD	Sensor VREF (VC)	4.9-5.1v
3	BK/YL	HO2S-11 (B1 S1) Heater	1v (Heater On)
4	WT	EVAP Purge Solenoid (VSV)	12v or 0v
5	BK/WT	HO2S-21 (B2 S1) Heater (HT2A)	1v (Heater On)
6	WT/BK	Power Ground (E03)	<0.1v
7	---	Not Used	---
8	BL/BK	HO2S-12 (B1 S2) Heater	1v (Heater On)
9	BK	HO2S-12 (B1 S2) Signal	0.1-1.1v
10	YL	Generator Control Signal (RL)	12-0-12v
11	VT	MAF Sensor Signal (VG)	1-3v
12	BK	HO2S-11 (B1 S1) Signal	0.1-1.1v
13	YL/BK	TP Sensor Signal (VTA2)	0.3-0.9v
14	RD/BL	ECT Sensor Signal (THW)	1-4v (varies with temp.)
15	WT	CMP Sensor Signal (G2+)	AC pulse signals
16, 24	WT, BK	CKP Sensor Signal (NE+), (NE-)	AC pulse signals
17	BR	Power Ground (E1)	<01v
18	BR	Sensor Ground (E2)	<0.050v
19	BK/BL	Accelerator Pedal Position Sensor (VPA2)	1.1-1.9v
20	BL	Accelerator Pedal Position Sensor (VPA)	1.1-1.9v
21	BK	HO2S-21 (B2 S1) Signal (OX2A)	0.1-1.1v
22	YL/BK	IAT Sensor Signal (THA)	1-4v (varies with temp.)
23	YL/GN	TP Sensor Signal (VTA)	0.3-0.9v

2001-05 1.8L DOHC 1ZZ-FE VIN R (M/T) E5 31P Connector

PCM Pin #	Wire Color	Circuit Description (31 Pin)	Value at Hot Idle
1	BK/WT	Injector 1 Control	2.0-3.3 ms
2	BK	Injector 2 Control	2.0-3.3 ms
3	BL	Injector 3 Control	2.0-3.3 ms
4	WT	Injector 4 Control	2.0-3.3 ms
6	RD/YL	Engine Throttle Control Power (+BM)	12-14v
7, 20	GN, YL/RD	Throttle Control Motor (CL-), (CL+)	Pulse Signals
8	WT/BK	Power Ground (ME01, ME2)	<0.1v
9	GY	Motor Control Shield Ground (GE01)	<0.050v
10	BK	Igniter Transistor 1 Control	Digital Signal: 0-5-0v
11	BL	Igniter Transistor 2 Control	Digital Signal: 0-5-0v
12	BL/WT	Igniter Transistor 3 Control	Digital Signal: 0-5-0v
13	BL	Igniter Transistor 4 Control	Digital Signal: 0-5-0v
14, 28	BL, RD/BL	Controller Area Network (+), (-)	0-7-0v
15, 30	PK, BL/OR	Sequential M/T Signal (NEO), (CSMT)	Digital Signals
17	BL	EVAP Canister Closed Valve	12v or 0v
18	GN	IAC Valve Signal (RSO)	Pulse Signals
19	RD	Throttle Control Motor (M+)	Pulse Signals
21	WT/BK	Power Ground (E01)	<0.1v
22	YL/BK	OIL Pressure Switch Signal (MOPS)	Closed: 1v, Open: 12v
23, 24	WT, RD	Camshaft Timing Oil Control Valve (-), (+)	AC pulse signals
25	BK/YL	Igniter Signal (IGF)	Pulse signal: 0-5-0v
27	BK	Knock Sensor Signal	<0.075v AC
29	GN	Throttle Control Motor (M-)	Pulse Signals
31	WT/BK	Power Ground (E02)	<0.1v

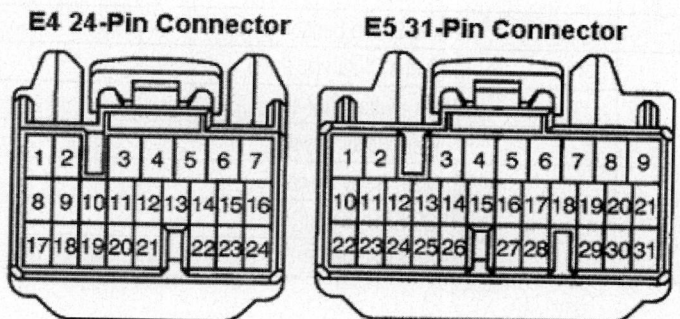

Pin Connector Graphic

05533_ADIA_G063

1991-92 2.0L I4 Turbo VIN S (M/T) 16 Pin Connector

PCM Pin #	Wire Color	Circuit Description (16 Pin)	Value at Hot Idle
1	RD/BL	Sensor VREF	4.9-5.1v
2	GY/BK	Airflow Meter Signal	1.5-2.5v
3	YL	IAT Sensor Signal	At 100°F: 2.60v
4	RD	ECT Sensor Signal	At 180°F: 0.51v
5	BK	Knock Sensor Signal	<0.075v AC
6	WT	HO2S Signal	0.1-1.1v
8	RD/WT	Check Connector	12-14v
9	BR	Sensor Ground	<0.050v
10 (Cal)	BK/YL	EGR Gas Temp. Sensor	3.5-4.0v
11	WT	TP Sensor Signal	0.3-0.8v
12	PK	Closed Throttle Switch	1v, at off-idle: 12v
13	GN/RD	Turbo Charger Press. Sensor	2.5-4.5v
14	BK	Check Connector	12-14v
15	OR	Check Connector	12-14v
16	RD	Distributor Signal (G-)	<0.050v

1991-92 2.0L I4 Turbo VIN S (M/T) 22 Pin Connector

PCM Pin #	Wire Color	Circuit Description (22 Pin)	Value at Hot Idle
1	WT/RD	Battery Direct	12-14v
2	BK	Defogger/Light Idle Up Signal	Switch On: 12v, Off: 0v
4	GN/WT	Stop Light Switch Signal	Brake Off: 0v, On: 12v
5	GN/WT	MIL (lamp) Control	MIL Off: 12v, On: 1v
6	BL/RD	Fuel Pump Relay	Relay Off: 12v, On: 1v
8	GN/YL	ABS ECU Signal	Pulses
9	PK/WT	Vehicle Speed Sensor	At 55 mph: 48 Hz
10	BK/WT	A/C Magnetic Clutch (ACMG)	Clutch Off: 0v, On: 12v
11	RD	Starter Switch Signal	9-11v (cranking)
12	BK/YL	EFI Main Relay Power	12-14v
13	BK/YL	EFI Main Relay Power	12-14v
14	BL/YL	PSP ECU Signal	Pulse Signals
15	BK	PSP ECU Signal	Pulse Signals

Standard Colors and Abbreviations

Abbreviation	Color	Abbreviation	Color	Abbreviation	Color
BK	Black	GY	Gray	RD	Red
BL	Blue	GN	Green	TN	Tan
BR	Brown	LG	Light Green	VT	Violet
DB	Dark Blue	OR	Orange	WT	White
DG	DK Green	PK	Pink	YL	Yellow

1991-92 2.0L I4 Turbo VIN S (M/T) 26 Pin Connector

PCM Pin #	Wire Color	Circuit Description (26 Pin)	Value at Hot Idle
1	BK	Distributor Signal (NE+)	AC pulse signals
2	WT	Distributor Signal (G2+)	AC pulse signals
3	WT/RD	Distributor Signal (IGF)	Digital Signal: 0-5-0v
6	BL/RD	Turbo Pressure Solenoid	12v or 0v
8	PK/BK	HO2S-11 (B1 S1) Heater	1v (Heater On)
9	GN/WT	ISC Motor (ISC1)	Pulse Signals
10	GN	Cold Start Injector Control	1v (at cold startup)
11	WT/BL	Injector 2 Control	2.0-3.3 ms
12	RD/WT	Injector 1 Control	2.0-3.3 ms
13	BR	Power Ground	<0.1v
14	BR	Sensor Ground	<0.050v
15	GN	Distributor Signal (G1+)	AC pulse signals
17	GN	Circuit Opening Relay (FC)	0-3v, off-idle: 12v
18	LG	Intake Solenoid Signal (TVIS)	12v or 0v
20	WT	Igniter Signal (IGT)	Digital Signal: 0-5-0v
22	GN/BK	ISC Motor (ISC2)	Pulse Signals
23	GN/RD	EGR Solenoid Control (VSV)	12v or 0v
24	GN/YL	Injector 4 Control	2.0-3.3 ms
25	BK/WT	Injector 3 Control	2.0-3.3 ms
26	BR	Power Ground	<0.1v

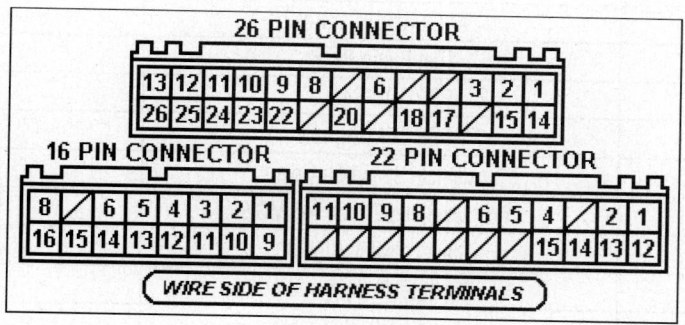

05533_ADIA_G072

Pin Connector Graphic

1993 2.0L I4 Turbo VIN S (M/T) 16 Pin Connector

PCM Pin #	Wire Color	Circuit Description (16 Pin)	Value at Hot Idle
1	RD/BL	Sensor VREF	4.9-5.1v
2	YL/BL	MAP Sensor Signal	3.5-3.1v
3	BK	IAT Sensor Signal	At 100°F: 2.60v
4	RD	ECT Sensor Signal	At 180°F: 0.51v
5	BL/BK	Turbo Charger Press. Sensor	2.5-4.5v
6	BK	HO2S Signal	0.1-1.1v
8	RD/WT	Check Connector	12-14v
9	BR	Sensor Ground	<0.050v
10 (Cal)	BK/YL	EGR Gas Temp. Sensor	3.5-4.0v
10 (Fed)	BK/YL	Short Pin	Open: 12v, Closed: 0v
11	WT	TP Sensor Signal	0.3-0.8v
12	PK	Closed Throttle Switch	1v, at off-idle: 12v
13	BK	Knock Sensor Signal	<0.075v AC
14	PK/BL	Check Connector	12-14v
15	BK	Check Connector	12-14v

1993 2.0L I4 Turbo VIN S (M/T) 22 Pin Connector

PCM Pin #	Wire Color	Circuit Description (22 Pin)	Value at Hot Idle
1	WT/RD	Battery Direct	12-14v
2	BK	Defogger/Light Idle Up Signal	Switch On: 12v, Off: 0v
4	GN/WT	Stop Light Switch Signal	Brake Off: 0v, On: 12v
5	GN/WT	MIL (lamp) Control	MIL Off: 12v, On: 1v
6	BL/RD	Fuel Pump Relay	Relay Off: 12v, On: 1v
7	BL/YL	PSP ECU Signal	Pulse Signals
8	BK	PSP ECU Signal	Pulse Signals
9	PK/WT	Vehicle Speed Sensor	At 55 mph: 48 Hz
10	BK/WT	A/C Magnetic Clutch (ACMG)	Clutch Off: 0v, On: 12v
11	RD	Starter Switch Signal	9-11v (cranking)
12	BK/YL	EFI Main Relay Power	12-14v
13	BK/YL	EFI Main Relay Power B1+	12-14v
14	GN/RD	Circuit Opening Relay (FC)	0-3v, off-idle: 12v
15	GN/YL	ABS ECU Signal	Pulse Signals

Standard Colors and Abbreviations

Abbreviation	Color	Abbreviation	Color	Abbreviation	Color
BK	Black	GY	Gray	RD	Red
BL	Blue	GN	Green	TN	Tan
BR	Brown	LG	Light Green	VT	Violet
DB	Dark Blue	OR	Orange	WT	White
DG	DK Green	PK	Pink	YL	Yellow

1993 2.0L I4 Turbo VIN S (M/T) 26 Pin Connector

PCM Pin #	Wire Color	Circuit Description (26 Pin)	Value at Hot Idle
1	GN/BK	Intake Solenoid Signal (TVIS)	12v or 0v
2	BL/OR	Turbo Pressure Solenoid	12v or 0v
3	WT/RD	Distributor Signal (IGF)	Digital Signal: 0-5-0v
4	BK	Distributor Signal (NE+)	AC pulse signals
5	WT	Distributor Signal (G2+)	AC pulse signals
6	BK/RD	EGR Solenoid Control (VSV)	12v or 0v
7	GN	Cold Start Injector Control	1v (at cold startup)
8	RD/WT	HO2S-11 (B1 S1) Heater	1v (Heater On)
9	GN	ISC Signal (RSC)	Pulses
10	RD	ISC Signal (RSO)	Pulses
11	WT/BL	Injector 2 Control	2.0-3.3 ms
12	RD/WT	Injector 1 Control	2.0-3.3 ms
13	BR	Power Ground	<0.1v
14	BR	Sensor Ground	<0.050v
17	RD	Distributor Signal (G-)	<0.050v
20	WT	Igniter Signal (IGT)	Digital Signal: 0-5-0v
24	GN/YL	Injector 4 Control	2.0-3.3 ms
25	BK/WT	Injector 3 Control	2.0-3.3 ms
26	BR	Power Ground	<0.1v

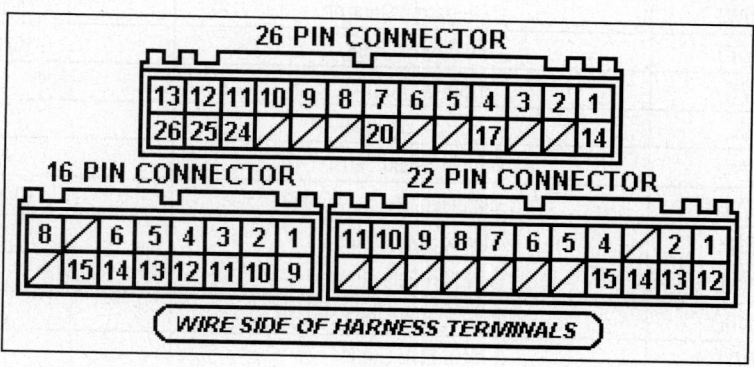

Pin Connector Graphic

05533_ADIA_G073

1994-95 2.0L I4 Turbo VIN S (M/T) 16 Pin Connector

PCM Pin #	Wire Color	Circuit Description (16 Pin)	Value at Hot Idle
1	RD/BL	Sensor VREF	4.9-5.1v
2	YL/RD	MAP Sensor Signal	1.5-2.5v
3	BK	IAT Sensor Signal	At 100ºF: 2.60v
4	RD	ECT Sensor Signal	At 180ºF: 0.51v
5	BL/BK	Turbo Pressure Sensor	2.5-4.5v
6	BK	HO2S Signal	0.1-1.1v
8	RD/WT	Data Link Connector	12-14v
9	BR	Sensor Ground	<0.050v
10	BK/YL	EGR Gas Temp. Sensor	3.5-4.0v
11	WT	TP Sensor Signal	0.3-0.8v
12	WT	Closed Throttle Switch	0-3v, off-idle: 12v
13	BK	Knock Sensor Signal	<0.075v AC
14	PK/BL	Data Link Connector	12-14v
15	BK	Data Link Connector	12-14v

1994-95 2.0L I4 Turbo VIN S (M/T) 22 Pin Connector

PCM Pin #	Wire Color	Circuit Description (22 Pin)	Value at Hot Idle
1	WT/RD	Battery Direct	12-14v
2	BK	Defogger/Light Idle Up Signal	Switch On: 12v, Off: 0v
4	GN/WT	Stop Light Switch Signal	Brake Off: 0v, On: 12v
5	GN/WT	MIL (lamp) Control	MIL Off: 12v, On: 1v
6	BL/RD	Fuel Pump Relay	Relay Off: 12v, On: 1v
7	BL/YL	PSP ECU Signal	Pulse Signals
8	BK	PSP ECU Signal	Pulse Signals
9	PK/WT	Vehicle Speed Sensor	At 55 mph: 48 Hz
10	BK/WT	A/C Magnetic Clutch (ACMG)	Clutch Off: 0v, On: 12v
11	RD	Starter Switch Signal	9-11v (cranking)
12	BK/YL	EFI Main Relay Power	12-14v
14	GN/RD	Circuit Opening Relay (FC)	0-3v, off-idle: 12v
15	GN/YL	ABS ECU Signal	Pulse Signals

Standard Colors and Abbreviations

Abbreviation	Color	Abbreviation	Color	Abbreviation	Color
BK	Black	GY	Gray	RD	Red
BL	Blue	GN	Green	TN	Tan
BR	Brown	LG	Light Green	VT	Violet
DB	Dark Blue	OR	Orange	WT	White
DG	DK Green	PK	Pink	YL	Yellow

1994-95 2.0L I4 Turbo VIN S (M/T) 26 Pin Connector

PCM Pin #	Wire Color	Circuit Description (26 Pin)	Value at Hot Idle
1	GN/BK	TN-VIS Solenoid Control	12v or 0v
2	BL/OR	Turbo Pressure Solenoid	12v or 0v
3	WT/RD	Distributor Signal (IGF)	Digital Signal: 0-5-0v
4	BK	Distributor Signal (NE+)	AC pulse signals
5	WT	Distributor Signal (G2+)	AC pulse signals
6	BK/RD	EGR Solenoid Control (VSV)	12v or 0v
7	GN	Cold Start Injector Control	1v (at cold startup)
8	RD/WT	HO2S-11 (B1 S1) Heater	1v (Heater On)
9	GN	ISC Signal (RSC)	Pulse Signals
10	RD	ISC Signal (RSO)	Pulse Signals
11	WT/BL	Injector 2 Control	2.0-3.3 ms
12	RD/WT	Injector 1 Control	2.0-3.3 ms
13	BR	Power Ground	<0.1v
14	BR	Sensor Ground	<0.050v
17	RD	Distributor Signal (G-)	<0.050v
18	GN	Distributor Signal (G1+)	AC pulse signals
20	WT	Igniter Signal (IGT)	Digital Signal: 0-5-0v
24	GN/YL	Injector 4 Control	2.0-3.3 ms
25	BK/WT	Injector 3 Control	2.0-3.3 ms
26	BR	Power Ground	<0.1v

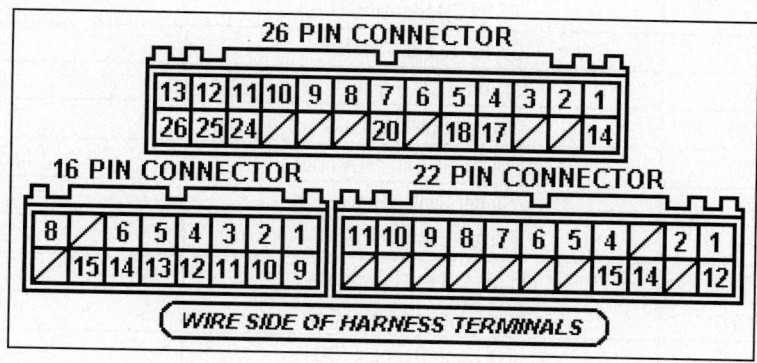

Pin Connector Graphic

05533_ADIA_G074

1991-92 2.2L I4 MFI VIN S (A/T-ECT) 16 Pin Connector

PCM Pin #	Wire Color	Circuit Description (16 Pin)	Value at Hot Idle
1	RD/BL	Sensor VREF	4.9-5.1v
2	GN/RD	MAP Sensor Signal	1-1.5v
3	YL	IAT Sensor Signal	At 100°F: 2.60v
4	RD	ECT Sensor Signal	At 180°F: 0.51v
6	BK	Main O2S Signal	0.1-1.1v
7	BR/BK	Check Connector	12-14v
8	RD/WT	Check Connector	12-14v
9	BR	Sensor Ground	<0.050v
10 (Cal)	BK/YL	EGR Gas Temp. Sensor	3.5-4.0v
11	WT	TP Sensor Signal	0.3-0.8v
12	PK	Closed Throttle Switch	1v, at off-idle: 12v
14 (Cal)	BK	Sub O2S Signal	0.1-1.1v
15	OR	Check Connector	12-14v
16	BK	Distributor Signal (G-)	<0.050v

1991-92 2.2L I4 MFI VIN S (A/T-ECT) 22 Pin Connector

PCM Pin #	Wire Color	Circuit Description (22 Pin)	Value at Hot Idle
1	WT/RD	Battery Direct	12-14v
2	BK	Defogger/Light Idle Up Signal	Switch On: 12v, Off: 0v
4	GN/WT	Stop Light Switch Signal	Brake Off: 0v, On: 12v
5	GN/WT	MIL (lamp) Control	MIL Off: 12v, On: 1v
6	BL/YL	PSP ECU Signal (CT)	Pulse Signals
7	LG	Overdrive Main Switch	Switch Off: 12v
8	BK	ABS ECU Signal	Pulse Signals
9	PK/WT	Vehicle Speed Sensor	At 55 mph: 48 Hz
10	BK/WT	A/C Magnetic Clutch (ACMG)	Clutch Off: 0v, On: 12v
11	BK	Starter Switch Signal	9-11v (cranking)
12	BK/YL	EFI Main Relay Power	12-14v
13	BK/YL	EFI Main Relay Power B1+	12-14v
14	GN/RD	Circuit Opening Relay (FC)	0-3v, off-idle: 12v
20	PK/BK	Cruise Control Signal (OD1)	At Cruise in OD: 12v
21	PK/GN	A/C Amplifier Signal (ACT)	Clutch On: 12v, Off: 1.5v
22	BK/WT	A/T Neutral Start Switch	In P/N: 9-11v (cranking)

Standard Colors and Abbreviations

Abbreviation	Color	Abbreviation	Color	Abbreviation	Color
BK	Black	GY	Gray	RD	Red
BL	Blue	GN	Green	TN	Tan
BR	Brown	LG	Light Green	VT	Violet
DB	Dark Blue	OR	Orange	WT	White
DG	DK Green	PK	Pink	YL	Yellow

1991-92 2.2L I4 MFI VIN S (A/T-ECT) 26 Pin Connector

PCM Pin #	Wire Color	Circuit Description (26 Pin)	Value at Hot Idle
1	WT	Distributor Signal (NE+)	AC pulse signals
2	BL/GN	A/T Pattern Selector Switch	Norm: 0v, PWR: 12v
3	WT/RD	Igniter Signal (IGF)	0.74-0.76v
4	GY/BK	A/T-ECT Solenoid (SL)	In Lockup: 12-14v
5	BR/WT	A/T-ECT Solenoid (S2)	1st or OD: 1v
6	PK	A/T-ECT Solenoid (S1)	In 3rd or OD: 1v
7	GN	Cold Start Injector Control	1v (at cold startup)
8	GN/BK	Fuel Pressure Up Solenoid	1v (at hot restart)
9	GN	ISC Signal (ISCC)	Pulse Signals
10	RD	ISC Signal (ISCO)	Pulse Signals
11	WT	Injector Pair 2 & 4 Control	2.0-3.3 ms
12	WT	Injector Pair 1 & 3 Control	2.0-3.3 ms
13	WT/BK	Power Ground	<0.1v
14	BR	Sensor Ground	<0.050v
15	RD	Distributor Signal (G+)	AC pulse signals
16	BR	Sensor Ground	<0.050v
17	GN/BK	A/T-ECT Speed Sensor	Moving: 0-5-0v
18	LG	A/T Select Switch 2nd	In 2nd: 12v, Others: 0v
19	GN/RD	A/T Select Switch Low	In Low: 12v, Others: 0v
20	WT	Igniter Signal (IGT)	0.72-0.74v
23	BL/OR	EGR Solenoid Control (VSV)	12v or 0v
26	BR	Power Ground	<0.1v

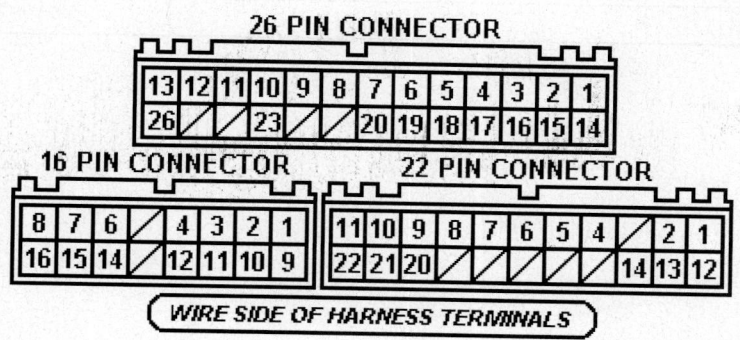

Pin Connector Graphic

05533_ADIA_G075

1991-92 2.2L I4 MFI VIN S (M/T) 10 Pin Connector

PCM Pin #	Wire Color	Circuit Description (10 Pin)	Value at Hot Idle
1	GN/BK	Fuel Pressure Up Solenoid	1v (at hot restart)
2	GN	Cold Start Injector Control	1v (at cold startup)
3	RD/WT	Starter Switch Signal	9-11v (cranking)
4	WT	Injector Pair 1 & 3 Control	2.0-3.3 ms
5	BR	Power Ground	<0.1v
6	BL/OR	EGR Solenoid Control (VSV)	12v or 0v
7	BR	Sensor Ground	<0.050v
8	WT	Igniter Signal (IGT)	0.72-0.74v
9	WT	Injector Pair 2 & 4 Control	2.0-3.3 ms
10	BR	Power Ground	<0.1v

1991-92 2.2L I4 MFI VIN S (M/T) 14-Pin Connector

PCM Pin #	Wire Color	Circuit Description (14-Pin)	Value at Hot Idle
1	BK/YL	EFI Main Relay Power	12-14v
2	WT/RD	Battery Direct	12-14v
3	PK/YL	Check Connector	12-14v
4	GN/RD	Circuit Opening Relay (FC)	0-3v, off-idle: 12v
5	OR	Defogger/Light Idle Up Signal	Switch On: 12v, Off: 0v
7	BL/YL	PSP ECU Signal (CT)	Pulse Signals
8	BK/YL	EFI Main Relay Power	12-14v
9	GN/WT	MIL (lamp) Control	MIL Off: 12v, On: 1v
11	BK/WT	A/C Magnetic Clutch (ACMG)	Clutch Off: 0v, On: 12v
12	PK/WT	Vehicle Speed Sensor	At 55 mph: 48 Hz
14	BK	PSP ECU Signal	Pulse Signals

1991-92 2.2L I4 MFI VIN S (M/T) 18 Pin Connector

PCM Pin #	Wire Color	Circuit Description (18 Pin)	Value at Hot Idle
1	RD	ECT Sensor Signal	At 180°F: 0.51v
2	GN/RD	MAP Sensor Signal	1-1.5v
3	YL	IAT Sensor Signal	At 100°F: 2.60v
4	OR	Check Connector	12-14v
5	WT/RD	Igniter Signal (IGF)	0.74-0.76v
6	RD	Distributor Signal (G1+)	AC pulse signals
7	BK	Distributor Signal (G-)	<0.050v
8	BK	Main O2S Signal	0.1-1.1v
9	GN	ISC Signal (ISCC)	Pulse Signals
10	BR	Sensor Ground	<0.050v
11	RD/GN	Wide Open Throttle Switch	1v, at off-idle: 12v
12	RD/BL	Sensor VREF	4.9-5.1v
13	PK	Closed Throttle Switch	1v, at off-idle: 12v
14 (Cal)	BK/YL	EGR Gas Temp. Sensor	3.5-4.0v
15	WT	Distributor Signal (NE+)	AC pulse signals
16	BR	Sensor Ground	<0.050v
17 (Cal)	BK	Sub O2S Signal	0.1-1.1v
18	RD	ISC Signal (ISCO)	Pulse Signals

10 PIN CONNECTOR   18 PIN CONNECTOR   14 PIN CONNECTOR

```
 5  4  3  2  1        9 8 7 6 5 4 3 2 1         7 /  5 4 3 2 1
10  9  8  7  6       18 17 16 15 14 13 12 11 10  14 / 12 11 / 9 8
```

WIRE SIDE OF HARNESS TERMINALS

Pin Connector Graphic

05533_ADIA_G076

1993-95 2.2L I4 MFI VIN S (A/T-ECT) 16 Pin Connector

PCM Pin #	Wire Color	Circuit Description (16 Pin)	Value at Hot Idle
1	RD/BL	Sensor VREF	4.9-5.1v
2	GN/RD	MAP Sensor Signal	1-1.5v
3	YL	IAT Sensor Signal	At 100°F: 2.60v
4	RD	ECT Sensor Signal	At 180°F: 0.51v
5 (Cal)	BK	Sub O2S Signal	0.1-1.1v
6	BK	Main O2S Signal	0.1-1.1v
7	BR/BK	Data Link Connector	12-14v
8	RD/WT	Data Link Connector	12-14v
9	BR	Sensor Ground	<0.050v
10 (Cal)	BK/YL	EGR Gas Temp. Sensor	3.5-4.0v
11	WT	TP Sensor Signal	0.3-0.8v
12	PK	Closed Throttle Switch	1v, at off-idle: 12v
13	BK	Knock Sensor Signal	<0.075v AC
14	PK/BL	Data Link Connector	12-14v
15	BK	Data Link Connector	12-14v
16 (93)	BR	Sensor Ground	<0.050v

1993-95 2.2L I4 MFI VIN S (A/T-ECT) 22 Pin Connector

PCM Pin #	Wire Color	Circuit Description (22 Pin)	Value at Hot Idle
1	WT/RD	Battery Direct	12-14v
2	BK	Defogger/Light Idle Up Signal	Switch On: 12v, Off: 0v
3	RD/WT	A/C Evaporator Temperature Signal	1.4-1.8v (temp. >59°F)
4	GN/WT	Stop Light Switch Signal	Brake Off: 0v, On: 12v
5	GN/WT	MIL (lamp) Control	MIL Off: 12v, On: 1v
7	LG	Overdrive Main Switch	Switch Off: 12v
8	BK/WT	A/C Amplifier Signal (AC1)	Clutch On: 1.5v, Off: 12v
9	PK/WT	Vehicle Speed Sensor	At 55 mph: 48 Hz
11	BK	Starter Switch Signal	9-11v (cranking)
12	BK/YL	EFI Main Relay Power	12-14v
13 (93)	BK/YL	EFI Main Relay Power B1+	12-14v
14	GN/RD	Circuit Opening Relay (FC)	0-3v, off-idle: 12v
15	BL/YL	PSP ECU Signal	Pulse Signals
16	BK	PSP ECU Signal	Pulse Signals
20	PK/BK	Cruise Control Signal (OD1)	At Cruise in OD: 12v
21	PK/GN	A/C Amplifier Signal (ACT)	Clutch On: 12v, Off: 1.5v
22	RD/WT	Neutral Start Switch	In P/N: 9-11v (cranking)

Standard Colors and Abbreviations

Abbreviation	Color	Abbreviation	Color	Abbreviation	Color
BK	Black	GY	Gray	RD	Red
BL	Blue	GN	Green	TN	Tan
BR	Brown	LG	Light Green	VT	Violet
DB	Dark Blue	OR	Orange	WT	White
DG	DK Green	PK	Pink	YL	Yellow

1993-95 2.2L I4 MFI VIN S (A/T-ECT) 26 Pin Connector

PCM Pin #	Wire Color	Circuit Description (26 Pin)	Value at Hot Idle
1	GY/BK	A/T-ECT Solenoid (SL)	In Lockup: 12-14v
2	YL/RD	A/T-ECT Solenoid (S1)	In 3rd or OD: 1v
2 (93)	PK	A/T-ECT Solenoid (S1)	In 3rd or OD: 1v
3	WT/RD	Igniter Signal (IGF)	Digital Signal: 0-5-0v
4	WT	Distributor Signal (NE+)	AC pulse signals
5	GN	Distributor Signal (NE-)	<0.050v
6	LG	A/T Select Switch 2nd	In 2nd: 12v, Others: 0v
7	GN	A/C Idle Up Solenoid	A/C Off: 12v, On: 1v
8	GN/BK	Fuel Pressure Up Solenoid	1v (at hot restart)
9	GN	ISC Signal (ISCC)	Pulse Signals
10	RD	ISC Signal (ISCO)	Pulse Signals
11	WT	Injector Pair 2 & 4 Control	2.0-3.3 ms
12	WT	Injector Pair 1 & 3 Control	2.0-3.3 ms
13	BR	Power Ground	<0.1v
14	BR	Sensor Ground	<0.050v
15	BR/WT	A/T-ECT Solenoid (S2)	1st or OD: 1v

1993-95 2.2L I4 MFI VIN S (A/T-ECT) 26 Pin Connector, *continued*

16	GN/BK	A/T-ECT Speed Sensor	Moving: 0-5-0v
17	BK	Distributor Signal (G-)	<0.050v
18	RD	Distributor Signal (G+)	AC pulse signals
19	GN/RD	A/T Select Switch Low	In Low: 12v, Others: 0v
20	WT	Igniter Signal (IGT)	0.72-0.74v
23	BL/OR	EGR Solenoid Control (VSV)	12v or 0v
26	BR	Power Ground	<0.1v

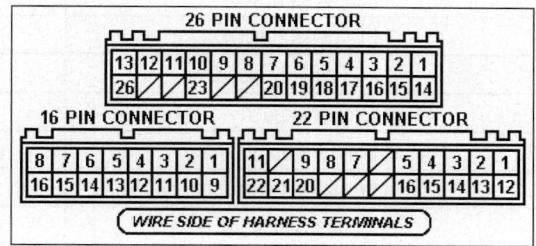

05533_ADIA_G077

Pin Connector Graphic

1993-95 2.2L I4 MFI VIN S (M/T) 12 Pin Connector

PCM Pin #	Wire Color	Circuit Description (12 Pin)	Value at Hot Idle
1 (93)	BK/YL	EFI Main Relay Power B1+	12-14v
2	WT/RD	Battery Direct	12-14v
3	BK/WT	A/C Amplifier Signal (AC1)	Clutch On: 1.5v, Off: 12v
4	GN/RD	Circuit Opening Relay (FC)	0-3v, off-idle: 12v
6	PK/GN	A/C Amplifier Signal (ACT)	Clutch On: 12v, Off: 1.5v
7	BK/YL	EFI Main Relay Power	12-14v
8	GN/WT	MIL (lamp) Control	MIL Off: 12v, On: 1v
11	PK/WT	Vehicle Speed Sensor	At 55 mph: 48 Hz
12	BK	Defogger/Light Idle Up Signal	Switch On: 12v, Off: 0v

1993-95 2.2L I4 MFI VIN S (M/T) 16 Pin Connector

PCM Pin #	Wire Color	Circuit Description (16 Pin)	Value at Hot Idle
1	BK	Sub O2S Signal	0.1-1.1v
2	GN/RD	MAP Sensor Signal	1-1.5v
3	YL	IAT Sensor Signal	At 100°F: 2.60v
4	RD	ECT Sensor Signal	At 180°F: 0.51v
5	BK	Knock Sensor Signal	<0.075v AC
6	BK	Main O2S Signal	0.1-1.1v
7	PK/BL	Data Link Connector	12-14v
8 (94)	RD/WT	Data Link Connector	12-14v
9	BR	Sensor Ground	<0.050v
10	WT	TP Sensor Signal	0.3-0.8v
11	RD/BL	MAP Sensor VREF	4.9-5.1v
12	PK	Closed Throttle Switch	1v, at off-idle: 12v
13 (Cal)	BK/YL	EGR Gas Temp. Sensor	3.5-4.0v
14	RD/WT	A/C Evaporator Temperature Signal	1.4-1.8v (temp. >59°F)
15	BK	Data Link Connector	12-14v
16	BR	Sensor Ground	<0.050v

1993-95 2.2L I4 MFI VIN S (M/T) 26 Pin Connector

PCM Pin #	Wire Color	Circuit Description (26 Pin)	Value at Hot Idle
1	GN	A/C Idle Up Solenoid	A/C On: 1v, Off: 12v
2	RD/WT	Starter Switch Signal	9-11v (cranking)
3	WT/RD	Igniter Signal (IGF)	Digital Signal: 0-5-0v
4	WT	Distributor Signal (NE+)	AC pulse signals
5	RD	Distributor Signal (G+)	AC pulse signals
7	BK	PSP ECU Signal	Pulse Signals
9	GN	ISC Signal (ISCC)	Pulse Signals
10	RD	ISC Signal (ISCO)	Pulse Signals
12	WT	Injector Pair 1 & 3 Control	2.0-3.3 ms
13	BR	Power Ground	<0.1v
14	BK/YL	Fuel Pressure Up Solenoid	1v (at hot restart)
14 (93)	GN/BK	Fuel Pressure Up Solenoid	1v (at hot restart)
17	GN	Distributor Signal (NE-)	<0.050v
18	BK	Distributor Signal (G-)	<0.050v
21	BL/YL	PSP ECU Signal	Straight: 12v, Turning: 0v
22	WT	Igniter Signal (IGT)	Digital Signal: 0-5-0v
23	BL/OR	EGR Solenoid Control (VSV)	12v or 0v
24	BR	Sensor Ground	<0.050v
25	WT	Injector Pair 2 & 4 Control	2.0-3.3 ms
26	BR	Power Ground	<0.1v

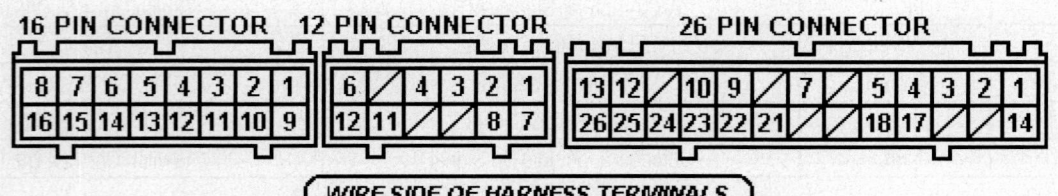

16 PIN CONNECTOR   12 PIN CONNECTOR   26 PIN CONNECTOR

WIRE SIDE OF HARNESS TERMINALS

05533_ADIA_G078

Pin Connector Graphic

PASEO PIN CHARTS

1992 1.5L VIN E (A/T-ECT) 16 Pin Connector

PCM Pin #	Wire Color	Circuit Description (16 Pin)	Value at Hot Idle
1	GN/RD	Sensor VREF	4.9-5.1v
2	PK	MAP Sensor Signal	1-1.5v
3	GN/BK	IAT Sensor Signal	At 100°F: 2.60v
4	GN/BK	ECT Sensor Signal	At 180°F: 0.51v
6	BK	O2S Signal	0.1-1.1v
7	BL	Data Link Connector	12-14v
8	GN/YL	Data Link Connector	12-14v
9	BR	Sensor Ground	<0.050v
10 (Cal)	BL/RD	EGR Gas Temp. Sensor	3.5-4.0v
11	YL/GN	TP Sensor Signal	0.3-0.8v
12	YL/BK	Closed Throttle Switch	1v, at off-idle: 12v
15	PK	Data Link Connector	12-14v
16	BR	Sensor Ground	<0.050v

1992 1.5L VIN E (A/T-ECT) 22 Pin Connector

PCM Pin #	Wire Color	Circuit Description (22 Pin)	Value at Hot Idle
1	WT/RD	Battery Direct	12-14v
2	GN/RD	A/C High Press. Switch	Switch Off: 12v
4	GN/WT	Stop Light Switch Signal	Brake Off: 0v, On: 12v
5	GN/WT	MIL (lamp) Control	MIL Off: 12v, On: 1v
7	RD/WT	Overdrive Main Switch	Switch Off: 12v, On: 0v
9	YL	Vehicle Speed Sensor	At 55 mph: 48 Hz
10	PK	A/C Magnetic Clutch (ACMG)	Clutch Off: 0v, On: 12v
11	BK	Starter Switch Signal	9-11v (cranking)
12	BK/OR	EFI Main Relay Power	12-14v
13	BK/OR	EFI Main Relay Power B1+	12-14v
14	GN	Circuit Opening Relay (FC)	0-3v, off-idle: 12v
20	RD/GN	Cruise Control ECU	At Cruise in OD: 12v
21	GN/BK	A/C Amplifier Signal (ACT)	Clutch On: 12v, Off: 1.5v
22	BK/BL	Neutral Start Switch	In P/N: 9-11v (cranking)

Standard Colors and Abbreviations

Abbreviation	Color	Abbreviation	Color	Abbreviation	Color
BK	Black	GY	Gray	RD	Red
BL	Blue	GN	Green	TN	Tan
BR	Brown	LG	Light Green	VT	Violet
DB	Dark Blue	OR	Orange	WT	White
DG	DK Green	PK	Pink	YL	Yellow

1992 1.5L VIN E (A/T-ECT) 26 Pin Connector

PCM Pin #	Wire Color	Circuit Description (26 Pin)	Value at Hot Idle
1	PK/BK	A/T-ECT Solenoid (SL)	In Lockup: 12-14v
2	PK	A/T-ECT Solenoid (S1)	In 3rd or OD: 1v
3	GN/YL	Igniter Signal (IGF)	Digital Signal: 0-5-0v
4	RD	Distributor Signal (NE+)	AC pulse signals
5	GN	Distributor Signal (G+)	AC pulse signals
6	OR	A/T Select Switch 2nd	In 2nd: 12v, Others: 0v
7	BL	Water Temperature Switch	Switch On: 0.1v
8	RD/GN	Throttle Opener Solenoid	12v or 0v
9	BL/WT	Fuel Pressure Up Solenoid	1v (at hot restart)
10	GN/RD	A/C Idle Up Solenoid	A/C On: 1v
11	YL	Injector Pair 2 & 4 Control	2.0-3.3 ms
12	WT	Injector Pair 1 & 3 Control	2.0-3.3 ms
13	BR	Power Ground	<0.1v
14	BR	Sensor Ground	<0.050v
15	PK/GN	A/T-ECT Solenoid (S2)	1st or OD: 1v
16	GN/BK	A/T-ECT Speed Sensor	Moving: 0-5-0v
17	WT	Distributor Signal (NE-)	<0.050v
18	BK	Distributor Signal (G-)	<0.050v
19	YL/BL	A/T Select Switch Low	In Low: 12v, Others: 0v
20	LG	Igniter Signal (IGT)	Digital Signal: 0-5-0v
23	PK/BL	EGR Solenoid Control (VSV)	12v or 0v
26	BR	Power Ground	<0.1v

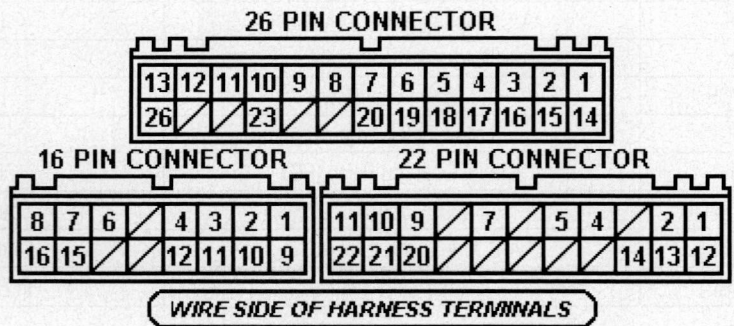

05533_ADIA_G079

Pin Connector Graphic

1992 1.5L I4 MFI VIN E (M/T) 16 Pin Connector

PCM Pin #	Wire Color	Circuit Description (16 Pin)	Value at Hot Idle
1	BK/OR	EFI Main Relay Power B1+	12-14v
2	WT/RD	Battery Direct	12-14v
4	GN	Circuit Opening Relay (FC)	0-3v, off-idle: 12v
5	GN/RD	Headlight Switch	Lights On: <0.1v
7 (Cal)	RD/GN	Throttle Opener Solenoid	12v or 0v
8	PK	Data Link Connector	12-14v
9	BK/OR	EFI Main Relay Power	12-14v
10	GN/WT	MIL (lamp) Control	MIL Off: 12v, On: 1v

1992 1.5L I4 MFI VIN E (M/T) 16 Pin Connector, *continued*

11	PK	A/C Magnetic Clutch (ACMG)	Clutch Off: 0v, On: 12v
13	YL	Vehicle Speed Sensor	At 55 mph: 48 Hz
14	GN/BK	A/C Amplifier Signal (ACT)	Clutch On: 12v, Off: 1.5v
16	GN/YL	Data Link Connector	12-14v

1992 1.5L I4 MFI VIN E (M/T) 26 Pin Connector

PCM Pin #	Wire Color	Circuit Description (26 Pin)	Value at Hot Idle
1	BL/WT	Fuel Pressure Up Solenoid	1v (at hot restart)
3	GN/BK	ECT Sensor Signal	At 180°F: 0.51v
4	PK	MAP Sensor Signal	1-1.5v
5	GN/BK	IAT Sensor Signal	At 100°F: 2.60v
6	LG	Igniter Signal (IGT)	Digital Signal: 0-5-0v
7	GN/YL	Igniter Signal (IGF)	Digital Signal: 0-5-0v
8	GN	Distributor Signal (G1+)	AC pulse signals
9	BK	Distributor Signal (G-)	<0.050v
10	BK	O2S Signal	0.1-1.1v
11	BK	Starter Switch Signal	9-11v (cranking)
12	WT	Injector Pair 1 & 3 Control	2.0-3.3 ms
13	BR	Power Ground	<0.1v
14	GN/RD	A/C Idle Up Solenoid	A/C On: 1v
15 (Cal)	PK/BL	EGR Solenoid Control (VSV)	12v or 0v
16	BR	Sensor Ground	<0.050v
15	YL/GN	TP Sensor Signal	0.3-0.8v
18	GN/RD	Sensor VREF	4.9-5.1v
19	YL/BK	Closed Throttle Switch	1v, at off-idle: 12v
20	WT	Distributor Signal (NE-)	<0.050v
21	RD	Distributor Signal (NE+)	AC pulse signals
23 (Cal)	BL/RD	EGR Gas Temp. Sensor	3.5-4.0v
24	BR	Sensor Ground	<0.050v
25	WT	Injector Pair 2 & 4 Control	2.0-3.3 ms
26	BR	Power Ground	<0.1v

26 PIN CONNECTOR **16 PIN CONNECTOR**

| 13 | 12 | 11 | 10 | 9 | 8 | 7 | 6 | 5 | 4 | 3 | / | 1 |
| 26 | 25 | 24 | 23 | / | 21 | 20 | 19 | 18 | 17 | 16 | 15 | 14 |

| 8 | 7 | / | 5 | 4 | / | 2 | 1 |
| 16 | / | 14 | 13 | / | 11 | 10 | 9 |

WIRE SIDE OF HARNESS TERMINALS

Pin Connector Graphic

05533_ADIA_G080

1993-94 1.5L I4 MFI VIN E (A/T-ECT) 16 Pin Connector

PCM Pin #	Wire Color	Circuit Description (16 Pin)	Value at Hot Idle
1	GN/RD	Sensor VREF	4.9-5.1v
2	PK	MAP Sensor Signal	1-1.5v
3	GN/BK	IAT Sensor Signal	At 100°F: 2.60v
4	GN/BK	ECT Sensor Signal	At 180°F: 0.51v
6	BK	O2S Signal	0.1-1.1v
7	BL	Data Link Connector	12-14v
8	GN/YL	Data Link Connector	12-14v
9	BR	Sensor Ground	<0.050v
10 (Cal)	BL	EGR Gas Temp. Sensor	3.5-4.0v
11	YL/GN	TP Sensor Signal	0.3-0.8v
12	BL/RD	Closed Throttle Switch	1v, at off-idle: 12v
15	PK	Data Link Connector	12-14v
16	BR	Sensor Ground	<0.050v

1993-94 1.5L I4 MFI VIN E (A/T-ECT) 22 Pin Connector

PCM Pin #	Wire Color	Circuit Description (22 Pin)	Value at Hot Idle
1	WT/RD	Battery Direct	12-14v
2	GN/RD	A/C High Pressure Switch	Switch Off: 12v
4	GN/WT	Stop Light Switch Signal	Brake Off: 0v, On: 12v
5	GN/WT	MIL (lamp) Control	MIL Off: 12v, On: 1v
7	RD/WT	Overdrive Main Switch	Switch Off: 12v
9	YL	Vehicle Speed Sensor	At 55 mph: 48 Hz
10	PK/BK	A/C Magnetic Clutch (ACMG)	Clutch Off: 0v, On: 12v
11	BK/RD	Starter Switch Signal	9-11v (cranking)
12	BK/OR	EFI Main Relay Power	12-14v
13	BK/OR	EFI Main Relay Power B1+	12-14v
14	GN	Circuit Opening Relay (FC)	0-3v, off-idle: 12v
20	RD/GN	Cruise Control ECU	At Cruise in OD: 12v
21	GN/BK	A/C Amplifier Signal (ACT)	Clutch On: 12v, Off: 1.5v
22	BK/WT	A/T Neutral Start Switch	In P/N: 9-11v (cranking)

Standard Colors and Abbreviations

Abbreviation	Color	Abbreviation	Color	Abbreviation	Color
BK	Black	GY	Gray	RD	Red
BL	Blue	GN	Green	TN	Tan
BR	Brown	LG	Light Green	VT	Violet
DB	Dark Blue	OR	Orange	WT	White
DG	DK Green	PK	Pink	YL	Yellow

1993-94 1.5L I4 MFI VIN E (A/T-ECT) 26 Pin Connector

PCM Pin #	Wire Color	Circuit Description (26 Pin)	Value at Hot Idle
1	PK/BK	A/T-ECT Solenoid (SL)	In Lockup: 12-14v
2	PK	A/T-ECT Solenoid (S1)	In 3rd or OD: 1v
3	GN/YL	Igniter Signal (IGF)	Digital Signal: 0-5-0v
4	RD	Distributor Signal (NE+)	AC pulse signals
5	GN	Distributor Signal (G+)	AC pulse signals
6	OR	A/T Select Switch 2nd	In 2nd: 12v, Others: 0v
7	BL	Water Temperature Switch	Switch On: 0.1v
8	PK/BL	Throttle Opener Solenoid	12v or 0v
9	BL/WT	Fuel Pressure Up Solenoid	1v (at hot restart)
10	GN/RD	A/C Idle Up Solenoid	A/C On: 1v
11	YL	Injector Pair 2 & 4 Control	2.0-3.3 ms
12	WT	Injector Pair 1 & 3 Control	2.0-3.3 ms
13	BR	Power Ground	<0.1v
14	BR	Sensor Ground	<0.050v
15	PK/BK	A/T-ECT Solenoid (S2)	1st or OD: 1v
16	GN/BK	A/T-ECT Speed Sensor	Moving: 0-5-0v
17	WT	Distributor Signal (NE-)	<0.050v
18	BK	Distributor Signal (G-)	<0.050v
19	YL/BL	A/T Select Switch Low	In Low: 12v, Others: 0v
20	LG	Igniter Signal (IGT)	Digital Signal: 0-5-0v
23	PK/BL	EGR Solenoid Control (VSV)	12v or 0v
26	BR	Power Ground	<0.1v

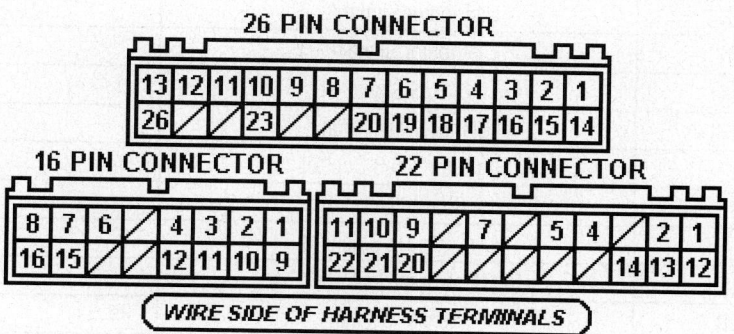

05533_ADIA_G079

Pin Connector Graphic

Standard Colors and Abbreviations

Abbreviation	Color	Abbreviation	Color	Abbreviation	Color
BK	Black	GY	Gray	RD	Red
BL	Blue	GN	Green	TN	Tan
BR	Brown	LG	Light Green	VT	Violet
DB	Dark Blue	OR	Orange	WT	White
DG	DK Green	PK	Pink	YL	Yellow

1993-94 1.5L I4 MFI VIN E (M/T) 16 Pin Connector

PCM Pin #	Wire Color	Circuit Description (16 Pin)	Value at Hot Idle
1 (93)	BK/OR	EFI Main Relay Power B1+	12-14v
2	WT/RD	Battery Direct	12-14v
4	GN	Circuit Opening Relay (FC)	0-3v, off-idle: 12v
5	GN/RD	Headlight Switch Signal	Lights On: 0.1v
5	RD/BL	Headlight Switch Signal	Lights On: 0.1v
7 (Cal)	RD/GN	Throttle Opener Solenoid	12v or 0v
8	PK	Data Link Connector	12-14v
9	BK/OR	EFI Main Relay Power	12-14v
10	GN/WT	MIL (lamp) Control	MIL Off: 12v, On: 1v
11	PK/BK	A/C Magnetic Clutch (ACMG)	Clutch Off: 0v, On: 12v
13	YL	Vehicle Speed Sensor	At 55 mph: 48 Hz
14	GN/BK	A/C Amplifier Signal (ACT)	Clutch On: 12v, Off: 1.5v
16	GN/YL	Data Link Connector	12v

1993-94 1.5L I4 MFI VIN E (M/T) 26 Pin Connector

PCM Pin #	Wire Color	Circuit Description (26 Pin)	Value at Hot Idle
1	BL/WT	Fuel Pressure Up Solenoid	1v (at hot restart)
2 (94)	BK/RD	M/T Neutral Start Switch	In P/N: 9-11v (cranking)
3	GN/BK	ECT Sensor Signal	At 180°F: 0.51v
4	PK	MAP Sensor Signal	1-1.5v
5	GN/BK	IAT Sensor Signal	At 100°F: 2.60v
6	LG	Igniter Signal (IGT)	Digital Signal: 0-5-0v
7	GN/YL	Igniter Signal (IGF)	Digital Signal: 0-5-0v
8	GN	Distributor Signal (G+)	AC pulse signals
9	BK	Distributor Signal (G-)	<0.050v
10	BK	O2S Signal	0.1-1.1v
11	BK/RD	Starter Switch Signal	9-11v (cranking)
12	WT	Injector 1 & 3 Control	2.0-3.3 ms
13	BR	Power Ground	<0.1v
14	GN/RD	A/C Idle Up Solenoid	A/C On: 1v
15 (Cal)	PK/BL	EGR Solenoid Control (VSV)	12v or 0v
16	BR	Sensor Ground	<0.050v
17	YL/GN	TP Sensor Signal	0.3-0.8v
18	GN/RD	Sensor VREF	4.9-5.1v
19	BL/RD	Closed Throttle Switch	1v, at off-idle: 12v
20	WT	Distributor Signal (NE-)	<0.050v
21	RD	Distributor Signal (NE+)	AC pulse signals
23 (Cal)	BL	EGR Gas Temp. Sensor	3.5-4.0v
24	BR	Sensor Ground	<0.050v
25	YL	Injector 2 & 4 Control	2.0-3.3 ms
26	BR	Power Ground	<0.1v

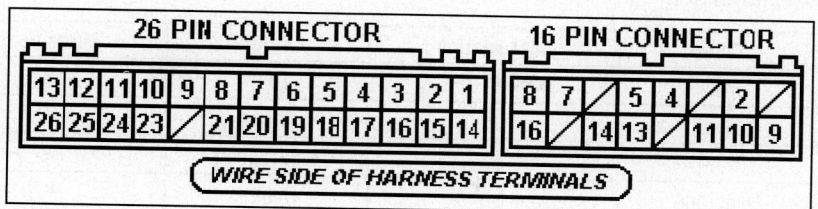

05533_ADIA_G081

Pin Connector Graphic
Standard Colors and Abbreviations

Abbreviation	Color	Abbreviation	Color	Abbreviation	Color
BK	Black	GY	Gray	RD	Red
BL	Blue	GN	Green	TN	Tan
BR	Brown	LG	Light Green	VT	Violet
DB	Dark Blue	OR	Orange	WT	White
DG	DK Green	PK	Pink	YL	Yellow

1995 1.5L I4 VIN E California (A/T-ECT) 12 Pin Connector

PCM Pin #	Wire Color	Circuit Description (12 Pin)	Value at Hot Idle
1	GN/WT	Stop Light Switch Signal	Brake Off: 0v, On: 12v
2	WT/RD	Battery Direct	12-14v
3	RD/GN	Cruise Control ECU	At Cruise in OD: 12v
4	GN	Circuit Opening Relay (FC)	0-3v, off-idle: 12v
6	GN/BK	A/C Amplifier Signal (ACT)	Clutch On: 12v, Off: 1.5v
7	BK/OR	EFI Main Relay Power	12-14v
8	GN/WT	MIL (lamp) Control	MIL Off: 12v, On: 1v
9	RD/WT	Overdrive Main Switch	Switch Off: 12v
10	PK/BK	A/C Amplifier Signal (AC1)	Clutch On: 1.5v, Off: 12v
11	YL	Vehicle Speed Sensor	At 55 mph: 48 Hz

1995 1.5L I4 VIN E California (A/T-ECT) 16 Pin Connector

PCM Pin #	Wire Color	Circuit Description (16 Pin)	Value at Hot Idle
1	BL	Data Link Connector	12-14v
2	PK	MAP Sensor Signal	1-1.5v
3	BL/BK	IAT Sensor Signal	At 100°F: 2.60v
4	GN/BK	ECT Sensor Signal	At 180°F: 0.51v
5	WT	Sub HO2S Signal	0.1-1.1v
6	BK	Main HO2S Signal	0.1-1.1v
7	PK/WT	Data Link Connector	12-14v
8	GN/YL	Data Link Connector	12-14v
9	BR	Sensor Ground	<0.050v
10	YL/GN	TP Sensor Signal	0.3-0.8v
11	GN/RD	Sensor VREF	4.9-5.1v
12	YL/BK	Closed Throttle Switch	1v, at off-idle: 12v
13	BL/RD	EGR Gas Temp. Sensor	3.5-4.0v
14	BK	Knock Sensor Signal	<0.075v AC
15	BL	Data Link Connector	12-14v

Standard Colors and Abbreviations

Abbreviation	Color	Abbreviation	Color	Abbreviation	Color
BK	Black	GY	Gray	RD	Red
BL	Blue	GN	Green	TN	Tan
BR	Brown	LG	Light Green	VT	Violet
DB	Dark Blue	OR	Orange	WT	White
DG	DK Green	PK	Pink	YL	Yellow

1995 1.5L I4 VIN E California (A/T-ECT) 26 Pin Connector

PCM Pin #	Wire Color	Circuit Description (26 Pin)	Value at Hot Idle
1	PK/BK	A/T-ECT Solenoid (SL)	In Lockup: 12-14v
2	BK/WT	Starter Switch Signal	9-11v (cranking)
3	GN/RD	Igniter Signal (IGF)	Digital Signal: 0-5-0v
4	BK	CKP Sensor Signal (NE+)	AC pulse signals
7	PK/BK	A/T-ECT Solenoid (S2)	1st or OD: 1v
9	BL/RD	ISC Signal (RSC)	Pulse Signals
10	BL	ISC Signal (RSO)	Pulse Signals
11	WT	Sub HO2S Heater	1v (Heater On)
12	WT	Injector Pair 1 & 3 Control	2.0-3.3 ms
13	BR	Power Ground	<0.1v
14	WT/GN	Main HO2S Heater	1v (Heater On)
15	BK	Neutral Start Switch	In P/N: 9-11v (cranking)
16	BK/WT	EGR Solenoid Control (VSV)	12v or 0v
17	GN	CMP Sensor Signal (G+)	AC pulse signals
18	RD	CMP Sensor Signal (G-)	<0.050v
19	OR	A/T Select Switch 2nd	In 2nd: 12v, Others: 0v
20	YL/BL	A/T Select Switch Low	In Low: 12v, Others: 0v
21	BL/YL	Igniter Transistor 2 Control	Digital Signal: 0-5-0v
22	LG	Igniter Transistor 1 Control	Digital Signal: 0-5-0v
23	GN/BK	A/T-ECT Speed Sensor	Moving: 0-5-0v
24	BR	Sensor Ground	<0.050v
25	YL	Injector Pair 2 & 4 Control	2.0-3.3 ms
26	PK	A/T-ECT Solenoid (S1)	In 3rd or OD: 1v

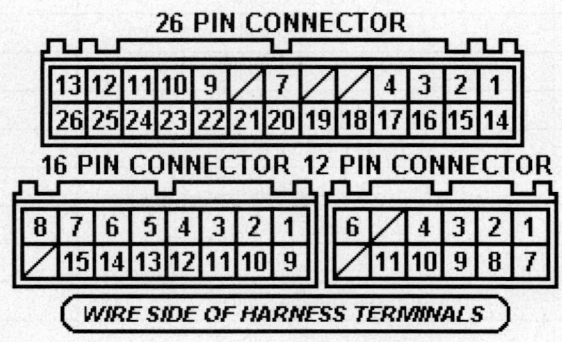

05533_ADIA_G082

Pin Connector Graphic

1995 1.5L I4 MFI VIN E Fed (A/T-ECT) 16 Pin Connector

PCM Pin #	Wire Color	Circuit Description (16 Pin)	Value at Hot Idle
1	GN/RD	Sensor VREF	4.9-5.1v
2	PK	MAP Sensor Signal	1-1.5v
3	GN/BK	IAT Sensor Signal	At 100°F: 2.60v
4	GN/BK	ECT Sensor Signal	At 180°F: 0.51v
6	BK	O2S Signal	0.1-1.1v
7	BL	Data Link Connector	12-14v
8	GN/YL	Data Link Connector	12-14v
9	BR	Sensor Ground	<0.050v
10	BL/RD	EGR Gas Temp. Sensor	3.5-4.0v
11	YL/GN	TP Sensor Signal	0.3-0.8v
12	BL/RD	Closed Throttle Switch	1v, at off-idle: 12v
15	PK	Data Link Connector	12-14v

1995 1.5L I4 MFI VIN E Fed (A/T-ECT) 22 Pin Connector

PCM Pin #	Wire Color	Circuit Description (22 Pin)	Value at Hot Idle
1	WT/RD	Battery Direct	12-14v
2	GN/RD	A/C Fan Relay	Relay Off: 12v, On: 1v
4	GN/WT	Stop Light Switch Signal	Brake Off: 0v, On: 12v
5	GN/WT	MIL (lamp) Control	MIL Off: 12v, On: 1v
7	RD/WT	Overdrive Main Switch (OD2)	Switch Off: 12v
9	YL	Vehicle Speed Sensor	At 55 mph: 48 Hz
10	PK/BK	A/C Amplifier Signal (AC1)	Clutch On: 1.5v, Off: 12v
11	BK/WT	Starter Switch Signal	9-11v (cranking)
12	BK/OR	EFI Main Relay Power	12-14v
14	GN	Circuit Opening Relay (FC)	0-3v, off-idle: 12v
20	RD/GN	Overdrive Main Switch (OD1)	Switch Off: 12v
21	GN/BK	A/C Amplifier Signal (ACT)	Clutch On: 12v, Off: 1.5v
22	BK	A/T Neutral Start Switch	In P/N: 9-11v (cranking)

Standard Colors and Abbreviations

Abbreviation	Color	Abbreviation	Color	Abbreviation	Color
BK	Black	GY	Gray	RD	Red
BL	Blue	GN	Green	TN	Tan
BR	Brown	LG	Light Green	VT	Violet
DB	Dark Blue	OR	Orange	WT	White
DG	DK Green	PK	Pink	YL	Yellow

1995 1.5L I4 MFI VIN E Fed (A/T-ECT) 26 Pin Connector

PCM Pin #	Wire Color	Circuit Description (26 Pin)	Value at Hot Idle
1	GN/RD	A/T-ECT Solenoid (SL)	In Lockup: 12-14v
2	PK/BL	A/T-ECT Solenoid (S1)	In 3rd or OD: 1v
3	GN/YL	Igniter Signal (IGF)	Digital Signal: 0-5-0v
4	RD	Distributor Signal (NE+)	AC pulse signals
5	GN	Distributor Signal (G+)	AC pulse signals
6	OR	A/T Select Switch 2nd	In 2nd: 12v, Others: 0v
7	BL	Water Temperature Switch	Switch On: 0.1v
8	RD/GN	Throttle Opener Solenoid	12v or 0v
9	BL/WT	Fuel Pressure Up Solenoid	1v (at hot restart)
10	GN/RD	IAC Solenoid (IAC)	Pulse Signals
11	YL	Injector Pair 2 & 4 Control	2.0-3.3 ms
12	WT	Injector Pair 1 & 3 Control	2.0-3.3 ms
13	BR	Power Ground	<0.1v
14	BR	Sensor Ground	<0.050v
15	PK/BK	A/T-ECT Solenoid (S2)	1st or OD: 1v
16	GN/BK	A/T-ECT Speed Sensor	Moving: 0-5-0v
17	WT	Distributor Signal (NE-)	<0.050v
18	BK	Distributor Signal (G-)	<0.050v
19	YL/BL	A/T Select Switch Low	In Low: 12v, Others: 0v
20	LG	Igniter Transistor 1 Control	Digital Signal: 0-5-0v
23	PK/BL	EGR Solenoid Control (VSV)	12v or 0v
26	BR	Power Ground	<0.1v

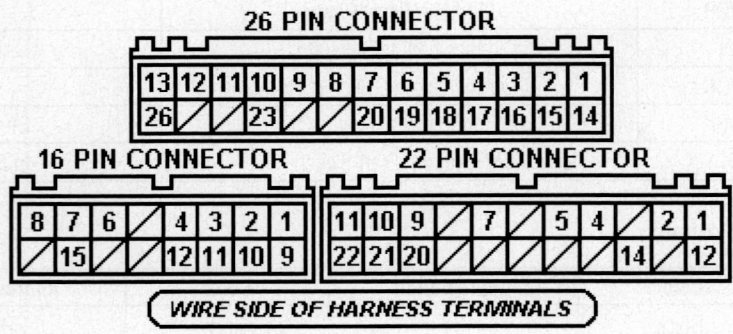

05533_ADIA_G083

Pin Connector Graphic

1995 1.5L I4 MFI VIN E California (M/T) 16 Pin Connector

PCM Pin #	Wire Color	Circuit Description (16 Pin)	Value at Hot Idle
2	WT/RD	Battery Direct	12-14v
4	GN	Circuit Opening Relay (FC)	0-3v, off-idle: 12v
5	WT/RD	Sub HO2S Heater	1v (Heater On)
6	WT/GN	Main HO2S Heater	1v (Heater On)
7	PK/WT	Data Link Connector	12-14v
8	GN	Data Link Connector	12-14v
9	BK/OR	EFI Main Relay Power	12-14v
10	GN/WT	MIL (lamp) Control	MIL Off: 12v, On: 1v

1995 1.5L I4 MFI VIN E California (M/T) 16 Pin Connector, *continued*

12	PK/BK	A/C Amplifier Signal (AC1)	Clutch On: 1.5v, Off: 12v
13	YL	Vehicle Speed Sensor	At 55 mph: 48 Hz
14	GN/BK	A/C Amplifier Signal (ACT)	Clutch On: 12v, Off: 1.5v
15	BK/WT	EGR Solenoid Control (VSV)	12v or 0v
16	GN/YL	Data Link Connector	12-14v

1995 1.5L I4 MFI VIN E California (M/T) 26 Pin Connector

PCM Pin #	Wire Color	Circuit Description (26 Pin)	Value at Hot Idle
1	BL/RD	ISC Signal (RSC)	Pulses
3	GN/BK	ECT Sensor Signal	At 180°F: 0.51v
4	PK	MAP Sensor Signal	1-1.5v
5	PK/YL	Igniter Transistor 2 Control	Digital Signal: 0-5-0v
6	GN	Igniter Transistor 1 Control	Digital Signal: 0-5-0v
7	GN/RD	Igniter Signal (IGF)	Digital Signal: 0-5-0v
8	BK	Knock Sensor Signal	<0.075v AC
9	GN	CKP Sensor Signal (NE-)	<0.050v
10	BK	Main HO2S Signal	0.1-1.1v
11	BK/WT	Starter Switch Signal	9-11v (cranking)
12	GN	Injectors 1 & 3 Control	2.0-3.3 ms
13	BR	Power Ground	<0.1v
14	BL/RD	ISC Signal (RSO)	Pulse Signals
15	BL/BK	IAT Sensor Signal	At 100°F: 2.60v
16	BR	Sensor Ground	<0.050v
17	YL/GN	TP Sensor Signal	0.3-0.8v
18	GN/RD	Sensor VREF	4.9-5.1v
19	YL/BK	Closed Throttle Switch	1v, at off-idle: 12v
20	RD	CMP Sensor Signal (G+)	AC pulse signals
21	BK	CKP Sensor Signal (NE+)	AC pulse signals
22	BL/RD	EGR Gas Temp. Sensor	3.5-4.0v
23	WT	Sub HO2S Signal	0.1-1.1v
24	BR	Power Ground	<0.1v
25	YL	Injectors 2 & 4 Control	2.0-3.3 ms

Pin Connector Graphic

05533_ADIA_G084

1995 1.5L I4 MFI VIN E Federal (M/T) 16 Pin Connector

PCM Pin #	Wire Color	Circuit Description (16 Pin)	Value at Hot Idle
2	WT/RD	Battery Direct	12-14v
4	GN	Circuit Opening Relay (FC)	0-3v, off-idle: 12v
5	RD/GN	Headlight Switch	Switch On: 12v, Off: 0v
7	RD/GN	Throttle Opener Solenoid	12v or 0v
8	PK	Data Link Connector	12-14v
9	BK/OR	EFI Main Relay Power	12-14v
10	GN/WT	MIL (lamp) Control	MIL Off: 12v, On: 1v
11	PK/BK	A/C Amplifier Signal (AC1)	Clutch On: 1.5v, Off: 12v
13	YL	Vehicle Speed Sensor	At 55 mph: 48 Hz
14	GN/BK	A/C Amplifier Signal (ACT)	Clutch On: 12v, Off: 1.5v
16	GN/YL	Data Link Connector	12-14v

1995 1.5L I4 MFI VIN E Federal (M/T) 26 Pin Connector

PCM Pin #	Wire Color	Circuit Description (26 Pin)	Value at Hot Idle
1	BL/WT	Fuel Pressure Up Solenoid	12v or 0v
2	BK/RD	Neutral Start Switch	In P/N: 9-11v (cranking)
3	GN/BK	ECT Sensor Signal	At 180°F: 0.51v
4	PK	MAP Sensor Signal	1-1.5v
5	GN/BK	IAT Sensor Signal	At 100°F: 2.60v
6	LG	Igniter Signal (IGT)	Digital Signal: 0-5-0v
7	GN/YL	Igniter Signal (IGF)	Digital Signal: 0-5-0v
8	GN	Distributor Signal (G+)	Pulses
9	BK	Distributor Signal (G-)	<0.050v
10	BK	O2S Signal	0.1-1.1v
11	BK/RD	Starter Switch Signal	9-11v (cranking)
12	WT	Injectors 1 & 3 Control	2.0-3.3 ms
13	BR	Power Ground	<0.1v
14	GN/RD	Throttle Opener Solenoid	12v or 0v
15	BK/WT	EGR Solenoid Control (VSV)	12v or 0v
24	BR	Sensor Ground	<0.050v
17	YL/GN	TP Sensor Signal	0.3-0.8v
18	GN/RD	Sensor VREF	4.9-5.1v
19	BL/RD	Closed Throttle Switch	1v, at off-idle: 12v
20	WT	Distributor Signal (NE-)	<0.050v
21	RD	Distributor Signal (NE+)	AC pulse signals
23	BR	EGR Gas Temp. Sensor	3.5-4.0v
24	BR	Sensor Ground	<0.050v
25	YL	Injectors 2 & 4 Control	2.0-3.3 ms
26	BR	Power Ground	<0.1v

26 PIN CONNECTOR   **16 PIN CONNECTOR**

| 13 | 12 | 11 | 10 | 9 | 8 | 7 | 6 | 5 | 4 | 3 | 2 | 1 |
| 26 | 25 | 24 | 23 | / | 21 | 20 | 19 | 18 | 17 | 16 | 15 | 14 |

| 8 | 7 | / | 5 | 4 | / | 2 | / |
| 16 | / | 14 | 13 | / | 11 | 10 | 9 |

WIRE SIDE OF HARNESS TERMINALS

05533_ADIA_G085

Pin Connector Graphic

1996 1.5L I4 MFI VIN C (A/T-ECT) 16 Pin Connector

PCM Pin #	Wire Color	Circuit Description (16 Pin)	Value at Hot Idle
1	GN/RD	Sensor VREF	4.9-5.1v
2	PK	MAP Sensor Signal	1-1.5v
3	BL/BK	IAT Sensor Signal	At 100°F: 2.60v
4	GN/BK	ECT Sensor Signal	At 180°F: 0.51v
5	WT	HO2S-12 (B1 S2) Signal	0.1-1.1v
6	BK	HO2S-11 (B1 S1) Signal	0.1-1.1v
9	BR	Sensor Ground	<0.050v
10	BK	Knock Sensor Signal	<0.075v AC
11	YL/GN	TP Sensor Signal	0.3-0.8v
12	YL/BK	Closed Throttle Switch	1v, at off-idle: 12v
15	GN	Data Link Connector	12-14v
16	BR	Power Ground	<0.1v

1996 1.5L I4 MFI VIN C (A/T-ECT) 22 Pin Connector

PCM Pin #	Wire Color	Circuit Description (22 Pin)	Value at Hot Idle
1	WT/RD	Battery Direct	12-14v
4	GN/WT	Stop Light Switch Signal	Brake Off: 0v, On: 12v
5	GY/BL	MIL (lamp) Control	MIL Off: 12v, On: 1v
7	RD/WT	Overdrive Main Switch	Switch Off: 12v, On: 0v
8	RD/BK	A/T Select Switch Reverse	In 'R': 12v, Others: 0v
9	YL	Vehicle Speed Sensor	At 55 mph: 48 Hz
10	GN/BK	A/C Amplifier Signal (AC1)	Clutch On: 1.5v, Off: 12v
11	BK/WT	Starter Switch Signal	9-11v (cranking)
12	BK/RD	EFI Main Relay Power	12-14v
14	GN/BK	Circuit Opening Relay (FC)	0-3v, off-idle: 12v
15	GN/BK	A/T-ECT Speed Sensor	Moving: 0-5-0v
16	WT	Data Link Connector (SDL)	No Scan Tool: 0v
18	GN/RD	A/T Select Switch 2nd	In 2nd: 12v, Others: 0v
19	LG	A/T Select Switch Low	In Low: 12v, Others: 0v
20	YL/RD	Cruise Control ECU	At Cruise in OD: 12v
21	BL	A/C Amplifier Signal (ACT)	Clutch On: 12v, Off: 1.5v
22	BK	A/T Neutral Start Switch	In P/N: 9-11v (cranking)

Standard Colors and Abbreviations

Abbreviation	Color	Abbreviation	Color	Abbreviation	Color
BK	Black	GY	Gray	RD	Red
BL	Blue	GN	Green	TN	Tan
BR	Brown	LG	Light Green	VT	Violet
DB	Dark Blue	OR	Orange	WT	White
DG	DK Green	PK	Pink	YL	Yellow

1996 1.5L I4 MFI VIN C (A/T-ECT) 26 Pin Connector

PCM Pin #	Wire Color	Circuit Description (26 Pin)	Value at Hot Idle
1	PK/BK	A/T-ECT Solenoid (SL)	In Lockup: 12-14v
2	PK	A/T-ECT Solenoid (S1)	In 3rd or OD: 1v
3	GN/BK	Igniter Signal (IGF)	Digital Signal: 0-5-0v
4	BK	CKP Sensor Signal (NE+)	AC pulse signals
5	RD	CMP Sensor Signal (G+)	AC pulse signals
6	WT/BL	HO2S-12 (B1 S2) Heater	1v (Heater On)
9	BL/RD	IAC Signal (RSC)	Pulse Signals
10	BL	IAC Signal (RSO)	Pulse Signals
11	YL	Injector Pair 2 & 4 Control	2.0-3.3 ms
12	GN	Injector Pair 1 & 3 Control	2.0-3.3 ms
13	BR	Power Ground	<0.1v
14	BR	Shield Ground	<0.050v
15	PK/GN	A/T-ECT Solenoid (S2)	1st or OD: 1v
17	GN	CKP/CMP Sensor Ground (-)	<0.050v
19	WT/RD	HO2S-11 (B1 S1) Heater	1v (Heater On)
20	LG	Igniter Transistor 1 Control	Digital Signal: 0-5-0v
21	BL/YL	Igniter Transistor 2 Control	Digital Signal: 0-5-0v
23	BK/WT	EGR Solenoid Control (VSV)	12v or 0v

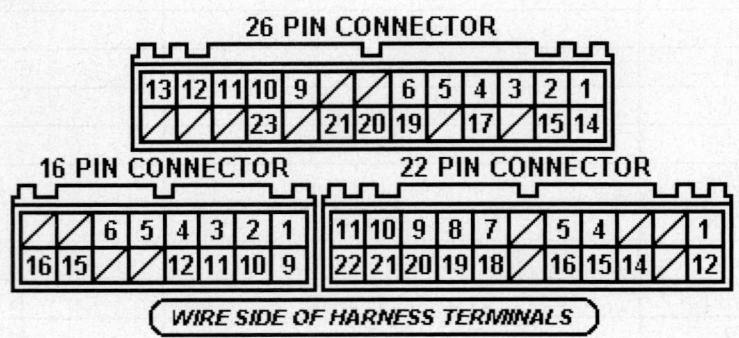

Pin Connector Graphic

1996 1.5L I4 MFI VIN C (M/T) 16 Pin Connector

PCM Pin #	Wire Color	Circuit Description (16 Pin)	Value at Hot Idle
2	WT/RD	Battery Direct	12-14v
4	GN/BK	Circuit Opening Relay (FC)	0-3v, off-idle: 12v
5	WT/RD	HO2S-12 (B1 S2) Heater	1v (Heater On)
6	WT/BL	HO2S-11 (B1 S1) Heater	1v (Heater On)

1996 1.5L I4 MFI VIN C (M/T) 16 Pin Connector, *continued*

8	GN	Data Link Connector	12-14v
9	BK/RD	EFI Main Relay Power	12-14v
10	GN/BL	MIL (lamp) Control	MIL Off: 12v, On: 1v
12	GN/BK	A/C Amplifier Signal (AC1)	Clutch On: 1.5v, Off: 12v
13	YL	Vehicle Speed Sensor	At 55 mph: 48 Hz
14	BL	A/C Amplifier Signal (ACT)	Clutch On: 12v, Off: 1.5v
15	BK/WT	EGR Solenoid Control (VSV)	12v or 0v
16	WT	Data Link Connector	12v

1996 1.5L I4 MFI VIN C (M/T) 26 Pin Connector

PCM Pin #	Wire Color	Circuit Description (26 Pin)	Value at Hot Idle
1	BL/RD	ISC Signal (RSC)	Pulses
3	GN/BK	ECT Sensor Signal	At 180°F: 0.51v
4	PK	MAP Sensor Signal	1-1.5v
5	BL/YL	Igniter Transistor 2 Control	Digital Signal: 0-5-0v
6	LG	Igniter Transistor 1 Control	Digital Signal: 0-5-0v
7	GN/BK	Igniter Signal (IGF)	Digital Signal: 0-5-0v
8	BK	Knock Sensor Signal	<0.075v AC
9	GN	CKP Sensor Signal (NE-)	<0.050v
10	BK	HO2S-11 (B1 S1) Signal	0.1-1.1v
11	BK/WT	Starter Switch Signal	9-11v (cranking)
12	GN	Injectors 1 & 3 Control	2.0-3.3 ms
13	BR	Power Ground	<0.1v
14	BL	ISC Signal (RSO)	Pulses
15	BL/BK	IAT Sensor Signal	At 100°F: 2.60v
16	BR	Sensor Ground	<0.050v
17	YL/GN	TP Sensor Signal	0.3-0.8v
18	GN/RD	Sensor VREF	4.9-5.1v
19	YL/BK	Closed Throttle Switch	0-3v, off-idle: 12v
20	RD	CMP Sensor Signal (G+)	AC pulse signals
21	BK	CKP Sensor Signal (NE+)	AC pulse signals
23	WT	HO2S-12 (B1 S2) Signal	0.1-1.1v
24	BR	Sensor Ground	<0.050v
25	YL	Injectors 2 & 4 Control	2.0-3.3 ms
26	BR	Power Ground	<0.1v

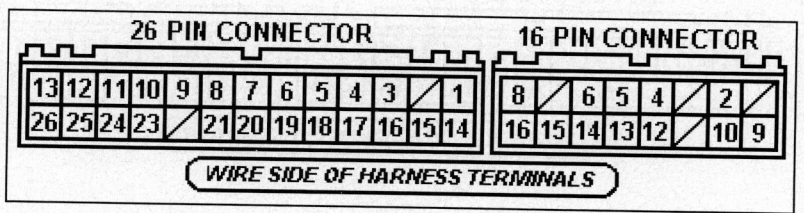

Pin Connector Graphic

1997-99 1.5L I4 VIN C (All) 16 Pin Connector

PCM Pin #	Wire Color	Circuit Description (16 Pin)	Value at Hot Idle
1	GN/WT	Stop Light Switch Signal	Brake Off: 0v, On: 12v
2	PK	MAP Sensor Signal	1-1.5v
3	BL/BK	IAT Sensor Signal	At 100°F: 2.60v
4	BK/RD	ECT Sensor Signal	At 180°F: 0.51v
5	WT	HO2S-12 (B1 S2) Signal	0.1-1.1v
6	BK	O2S-11 (B1 S1) Signal	0.1-1.1v
7	BL/WT	EVAP Purge Solenoid (VSV)	12v or 0v
8	GN/YL	EVAP Vapor Pressure (VSV)	12v or 0v
9	BR	Sensor Ground	<0.050v
10	YL/GN	TP Sensor Signal	0.3-0.8v
11	GN/RD	Sensor VREF	4.9-5.1v
12	YL/BK	EVAP Vapor Pressure Sensor	2.5-3.7v
13	WT	SIL Signal (Scan Tool)	Digital Signal
14	BK	Knock Sensor Signal	<0.075v AC
15	GN	Data Link Connector (TE)	12-14v
16	BR	Power Ground	<0.1v

1997-99 1.5L I4 VIN C (All) 12 Pin Connector

PCM Pin #	Wire Color	Circuit Description (12 Pin)	Value at Hot Idle
1	GY/BL	Cruise Control Signal (IDLO)	1.5v, off-idle: 12v
2	WT/RD	Battery Direct	12-14v
3	YL/RD	Cruise Control Signal (OD1)	At Cruise in OD: 12v
4	GN/BK	Circuit Opening Relay (FC)	0-3v, off-idle: 12v
6	BL	A/C Amplifier Signal (ACT)	Clutch On: 12v, Off: 1.5v
7	BK/RD	EFI Main Relay Power	12-14v
8	GY/BL	MIL (lamp) Control	MIL Off: 12v, On: 1v
9	RD/WT	Overdrive Main Switch	Switch Off: 12v
10	GN/BK	A/C Amplifier Signal (AC1)	Clutch On: 1.5v, Off: 12v
11	YL	Vehicle Speed Sensor	At 55 mph: 48 Hz

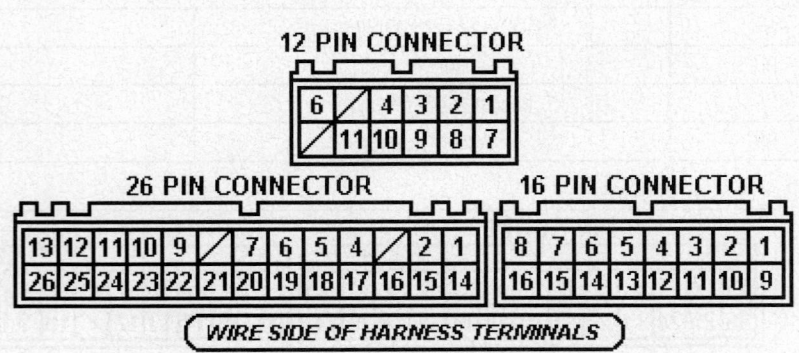

Pin Connector Graphic

05533_ADIA_G088

Standard Colors and Abbreviations

Abbreviation	Color	Abbreviation	Color	Abbreviation	Color
BK	Black	GY	Gray	RD	Red
BL	Blue	GN	Green	TN	Tan
BR	Brown	LG	Light Green	VT	Violet
DB	Dark Blue	OR	Orange	WT	White
DG	DK Green	PK	Pink	YL	Yellow

1997-99 1.5L I4 VIN C (All) 26 Pin Connector

PCM Pin #	Wire Color	Circuit Description (26 Pin)	Value at Hot Idle
1	GY	Igniter Transistor 1 Control	Digital Signal: 0-5-0v
2	BK/WT	Starter Switch Signal	9-11v (cranking)
4	BK	CKP Sensor Signal (NE+)	AC pulse signals
5	LG	A/T Select Switch Low	In Low: 12v, Others: 0v
6	GN/RD	A/T Select Switch 2nd	In 2nd: 12v, Others: 0v
7	RD/BL	Igniter Signal (IGF)	Digital Signal: 0-5-0v
9 ('97)	BL/RD	ISC Signal (RSC)	Pulse Signals
10 ('97)	BL	ISC Signal (RSO)	Pulse Signals
10 ('98-'99)	BL	ISC Signal (RSD)	Pulse Signals
11	RD	Injector 3 Control	2.0-3.3 ms
12	GN	Injector 1 Control	2.0-3.3 ms
13	BR	Sensor Ground	<0.050v
14	BL/YL	Igniter Transistor 2 Control	Digital Signal: 0-5-0v
15	BK	Neutral Start Switch	In P/N: 9-11v (cranking)
15	BK/WT	Neutral Start Switch	In P/N: 9-11v (cranking)
16	GN/BK	A/T-ECT Speed Sensor	Moving: 0-5-0v
17	GN	CKP/CMP Sensor Ground (-)	<0.050v
18	RD	CMP Sensor Signal (G+)	AC pulse signals
19	PK/GN	A/T-ECT Solenoid (S2)	1st or OD: 1v
20	PK	A/T-ECT Solenoid (S1)	In 3rd or OD: 1v
21	WT/RD	HO2S-12 (B1 S2) Heater	1v (Heater On)
22	RD/BK	A/T Select Switch Reverse	In 'R': 12v, Others: 0v
23	PK/BK	A/T-ECT Solenoid (SL)	In Lockup: 12-14v
24	BL	Injector 4 Control	2.0-3.3 ms
25	YL	Injector 2 Control	2.0-3.3 ms
26	BR	Shield Ground	<0.050v

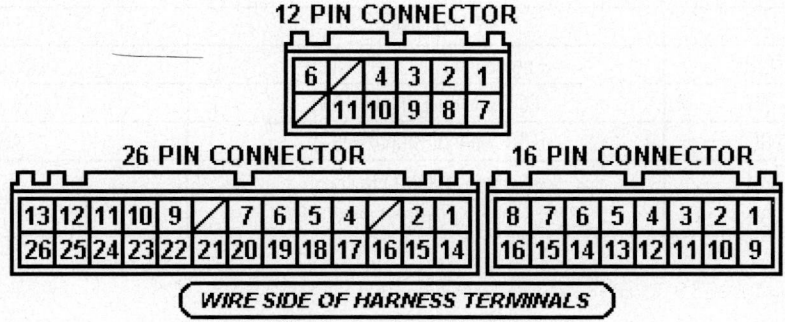

Pin Connector Graphic

PRIUS PIN CHARTS

2000-06 1.5L I4 Hybrid MFI VIN B (A/T) E7 31P Connector

PCM Pin #	Wire Color	Circuit Description (31 Pin)	Value at Hot Idle
1	BL/WT	Injector 3 Control	2.0-3.3 ms
2	RD/WT	Injector 4 Control	2.0-3.3 ms
3	WT/BK	Power Ground (E03)	<0.1v
4	LG	Fan Relay 1 & 2 Control	Relay Off: 12v, On: 1v
5, 15-20	---	Not Used	---
6	GY	ATC Motor Power Supply	12-14v
7	PK	Throttle Control Motor (M-)	Pulse Signals
8	BL	Throttle Control Motor (M+)	Pulse Signals
9	WT/BK	ETC Motor Ground (ME01)	<0.1v
10	RD	CMP Sensor Signal (G2)	AC pulse signals
11	YL/GN	Igniter Transistor 1 Control	Digital Signal: 0-5-0v
12	WT	Igniter Transistor 2 Control	Digital Signal: 0-5-0v
13	GN	Igniter Transistor 3 Control	Digital Signal: 0-5-0v
14	YL	Igniter Transistor 4 Control	Digital Signal: 0-5-0v
21	WT/BK	Power Ground (E01)	<0.1v
22, 26-27	---	Not Used	---
23	YL/RD	Camshaft Timing Oil Control Valve (+)	AC pulse signals
24	WT/GN	Camshaft Timing Oil Control Valve (-)	AC pulse signals
25	BK/RD	Igniter Signal (IGF)	Pulse Signal: 0-5-0v
28	BK	Knock Sensor Signal (KNK1)	<0.075v AC
29	RD/BL	EVAP Purge Solenoid (VSV)	12v or 0v
30	GY	Motor Shield Ground (GE01)	<0.050v
31	WT/BK	Power Ground (E02)	<0.1v

2000-06 1.5L I4 Hybrid MFI VIN B (A/T) E8 24P Connector

PCM Pin #	Wire Color	Circuit Description (24 Pin)	Value at Hot Idle
1, 7-8	---	Not Used	---
2	YL/RD	Sensor VREF (VC)	4.9-5.1v
3	RD/WT	Battery Direct	12-14v
4	BK	EFI Relay Power	12-14v
5	YL	Injector 1 Control	2.0-3.3 ms
6	BK/RD	Injector 2 Control	2.0-3.3 ms
9	GN/RD	Circuit Opening Relay (FC)	0-3v, off-idle: 12v
10	GN	MAF Sensor Signal (VG)	0.3-0.9v
11, 15, 20	---	Not Used	---
12	YL/BK	Engine Oil Pressure Switch	Open: 12v, Closed: 1v
14	WT	ECT Sensor Signal (THW)	1-4v (varies with temp.)
16	RD	CKP Sensor Signal (NE+)	AC pulse signals
17	BR	Power Ground (E1)	<0.1v
18	BR	Sensor Ground (E2)	<0.050v

2000-06 1.5L I4 Hybrid MFI VIN B (A/T) E8 24P Connector, *continued*

19	RD	MAF Sensor Ground (EVG)	<0.050v
21	BL	TP Sensor Signal 2 (VTA2)	1.1-1.9v
22	RD/BK	IAT Sensor Signal (THA)	1-4v (varies with temp.)
23	PK	TP Sensor Signal (VTA)	2.0-2.9v
24	GN	CKP/CMP Sensor Ground (-)	<0.050v

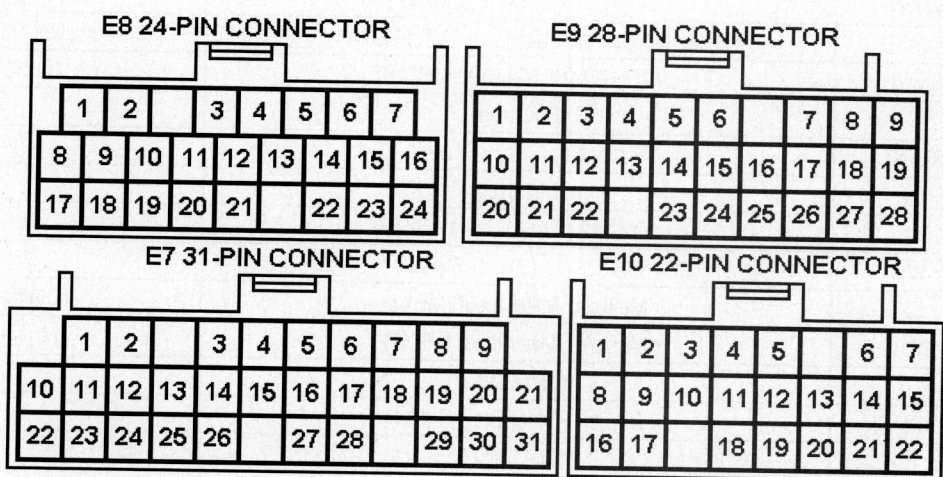

Pin Connector Graphic

05533_ADIA_G089

2000-06 1.5L I4 Hybrid MFI VIN T (A/T) E9 28 Pin Connector

PCM Pin #	Wire Color	Circuit Description (28 Pin)	Value at Hot Idle
1-5	---	Not Used	---
6	PK/BK	Data Link Connector (TC)	12-14v
7, 9	---	Not Used	---
8	WT/BL	SIL Signal (Scan Tool)	Digital Signal
10	OR	Hybrid ECU Speed Signal	Moving: pulse signals
11	BK/RD	EVAP Purge Switching Valve (VSV)	12v or 0v
12	---	Not Used	---
13	BL	EVAP Canister Closed Valve (VSV)	12v or 0v
14	VT/WT	Hybrid ECU HCLS Signal	1.3-1.9 with vacuum "on"
15, 17	---	Not Used	---
16	RD/YL	Hybrid ECU ESTP Signal	12-14v
18	BR	Hybrid ECU HTE- Signal	Pulse Signals
19	YL	Hybrid ECU HTE+ Signal	Pulse Signals
18	BR	Hybrid ECU THE+ Signal	Pulse Signals
19	YL	Hybrid ECU THE- Signal	Pulse Signals
20	BK/WT	HC Absorber Catalyst Signal	12-14v
21	GN	Hybrid ECU GO Signal	Pulse Signals
22	BL/RD	EVAP Vapor Pressure Sensor (HCC)	2.9-3.1v (with hose off)
23	WT/GN	Outside Air Temperature Sensor (TAM)	Varies: 1-4v
24	---	Not Used	---
25	GN/RD	EFI Main Relay Control	Relay Off: 12v, On: 1v
27	BL	Hybrid ECU ETH- Signal	Pulse Signals
28	PK	Hybrid ECU ETH+ Signal	Pulse Signals

2000-06 1.5L I4 Hybrid MFI VIN T (A/T) E10 22 Pin Connector

PCM Pin #	Wire Color	Circuit Description (22 Pin)	Value at Hot Idle
1	PK/BL	HO2S-11 (B1 S1) Heater	1v (Heater On)
2	---	Not Used	---
3	LG	A/C Amplifier Signal (NEO)	Clutch On: 1.5v, Off: 12v
4	---	Not Used	---
5	VT/WT	Speed Signal from C/Meter	At 55 mph: 48 Hz
6	GN//RD	Malfunction Indicator Lamp (MIL) Control	Indicator Off: 12v, On: 1v
7	GN/YL	HO2S-12 (B1 S2) Heater (HT1B)	1v (Heater On)
8	---	Not Used	---
9	BK/WT	Ignition Switch Power	12-14v
10	---	Not Used	---
11	YL	HO2S-12 (B1 S2) Signal (OX1B)	0.1-1.1v
12	WT	HO2S-11 (B1 S1) Signal	0.1-1.1v
13	GY	Multiplex Meter Input (MPX2)	Digital Signals
14	GY/BL	Multiplex Meter Input (MPX1)	Digital Signals
15	---	Not Used	---
16	BR	Oxygen Sensor Ground (E11)	<0.050v
17-21	---	Not Used	---
22	PK/GN	A/C Amplifier Signal (ACT)	Clutch On: 12v, Off: 1.5v

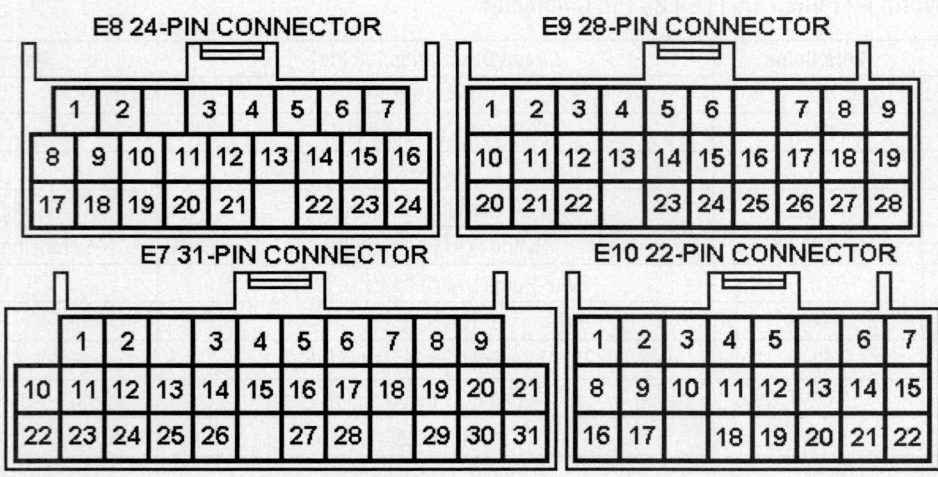

Pin Connector Graphic

05533_ADIA_G089

SUPRA PIN CHARTS

1990-92 3.0L I6 MFI VIN M (All) 10 Pin Connector

PCM Pin #	Wire Color	Circuit Description (10 Pin)	Value at Hot Idle
1	BK/WT	A/T Neutral Start Switch	In P/N: 9-11v (cranking)
1	BK/YL	M/T Clutch Start Switch	In 'N': 9-11v (cranking)
2	GN	Cold Start Injector Control	1v (at cold startup)
3	BK/BL	Starter Switch Signal	9-11v (cranking)
4	WT/BL	Injector Pair 1 & 4 Control	2.0-3.3 ms
5	WT/BK	Power Ground	<0.1v
6 (Cal)	RD/BL	Sub HO2S Heater	1v (Heater On)
7	BR	Sensor Ground	<0.050v
8	WT	Injector Pair 3 & 5 Control	2.0-3.3 ms
9	WT/RD	Injector Pair 2 & 6 Control	2.0-3.3 ms
10	WT/BK	Power Ground	<0.1v

1990-92 3.0L I6 MFI VIN M (All) 18 Pin Connector

PCM Pin #	Wire Color	Circuit Description (18 Pin)	Value at Hot Idle
1	GN	ECT Sensor Signal	At 180°F: 0.51v
2	GN/BK	Igniter Signal (IGF)	0.74-0.76v
3	LG	Igniter Signal (IGT)	0.72-0.74v
4	BL	Distributor Signal (NE+)	AC pulse signals
5	YL	Distributor Signal (G2+)	AC pulse signals
6	RD	Distributor Signal (G1+)	AC pulse signals
7	BK	Distributor Signal (G-)	<0.050v
8	GN/WT	ISC Motor (ISC2)	Pulse Signals
9	WT/YL	ISC Motor (ISC1)	Pulse Signals
10	WT	Knock Sensor Signal	<0.075v AC
11 (Cal)	RD/BL	Sub O2S Signal	0.1-1.1v
12 (Cal)	BK/YL	EGR Gas Temp. Sensor	3.5-4.0v
13	YL/BL	Closed Throttle Switch	0-3v, off-idle: 12v
14	WT/RD	TP Sensor Signal	0.3-0.8v
15	RD/GN	Check Connector	12-14v
16	GY	Check Connector	12-14v
17	RD/BK	ISC Motor (ISC4)	Pulse Signals
18	BL/RD	ISC Motor (ISC3)	Pulse Signals

Standard Colors and Abbreviations

Abbreviation	Color	Abbreviation	Color	Abbreviation	Color
BK	Black	GY	Gray	RD	Red
BL	Blue	GN	Green	TN	Tan
BR	Brown	LG	Light Green	VT	Violet
DB	Dark Blue	OR	Orange	WT	White
DG	DK Green	PK	Pink	YL	Yellow

1990-92 3.0L I6 MFI VIN M (All) 24 Pin Connector

PCM Pin #	Wire Color	Circuit Description (24 Pin)	Value at Hot Idle
1	BK/OR	Ignition Switch Power	12-14v
1	GY	Ignition Switch Power	12-14v
2	BK/YL	Battery Direct	12-14v
3	BL/RD	Sensor VREF	4.9-5.1v
4	GN/RD	Airflow Meter Signal	1.5-2.5v
5	LG	IAT Sensor Signal	At 100°F: 2.60v
6	YL	Fuel Pump Relay	Relay Off: 12v, On: 1v
7	PK	Vehicle Speed Sensor	At 55 mph: 48 Hz
8	RD/BK	EGR Solenoid Control (VSV)	12v or 0v
9	BK/OR	EFI Main Relay Power	12-14v
11	GN/BK	Intake Air Control Solenoid	12v or 0v
12 (Fed)	WT	Main O2S Signal	0.1-1.1v
13	BK/RD	EFI Main Relay Power B1+	12-14v
14	BK/RD	EFI Main Relay Power	12-14v
15 (Cal)	BR	Sensor Ground	<0.050v
16	RD/YL	Headlight Relay Control	Relay On: <1v
17	RD/YL	A/T-ECT ECU Signal	Pulse Signals
17	RD/BL	A/T-ECT ECU Signal	Pulse Signals
18	PK	Defogger/Light Idle Up Signal	Switch On: 12v, Off: 0v
19	GY/GN	MIL (lamp) Control	MIL Off: 12v, On: 1v
20	BL/RD	A/C Magnetic Clutch (ACMG)	Clutch Off: 0v, On: 12v
21	RD/WT	TCM Signal (L3)	0v, at WOT: 5v
22	RD	TCM Signal (L2)	5v, at WOT: 5v
23	BK	TCM Signal (L1)	5v, at WOT: 0v
24	BR	Sensor Ground	<0.050v

24 PIN CONNECTOR

12	11	/	9	8	7	6	5	4	3	2	1
24	23	22	21	20	19	18	17	16	15	14	13

10 PIN CONNECTOR

5	4	3	2	1
10	9	8	7	6

18 PIN CONNECTOR

9	8	7	6	5	4	3	2	1
18	17	16	15	14	13	12	11	10

WIRE SIDE OF HARNESS TERMINALS

05533_ADIA_G090

Pin Connector Graphic

1993-95 3.0L I6 MFI VIN J (All) 40 Pin Connector

PCM Pin #	Wire Color	Circuit Description (40 Pin)	Value at Hot Idle
1	BK/OR	Ignition Switch Power	12-14v
2	PK	Speed Sensor 1 Signal	Moving: 0-5-0v
3	YL	Kickdown Switch	Switch On: <0.1v
4	GN/WT	Stop Light Switch Signal	Brake Off: 0v, On: 12v
6	BL/BK	MIL (lamp) Control	MIL Off: 12v, On: 1v
9	GN/RD	A/T Select Switch 2nd	In 2nd: 12v, Others: 0v
10	GN/BK	A/T Select Switch Low	In Low: 12v, Others: 0v
12	BR/BK	Cruise Control ECU	At Cruise in OD: 12v
15	RD/YL	Defogger Idle Up Signal	Switch On: 12v, Off: 0v
17	GY/RD	Data Link Connector	12-14v
18	GN/YL	A/T Pattern Selector Switch	Norm: 0v, PWR: 12v
19	PK/GN	Data Link Connector	12-14v
20	YL/BL	Data Link Connector	12-14v
21	GN	Fuel Pump ECU	Pulse Signals
22	PK/WT	Fuel Pump ECU	Pulse Signals
23	WT/GN	A/C Magnetic Clutch (ACMG)	Clutch Off: 0v, On: 12v
24	GY	EFI Main Relay Power	12-14v
25	WT/BL	Manual Indicator Light	Light Off: 12-14v
28	PK/GN	Overdrive "Off" Indicator	LED Off: 12v, On: 1v
30 (Cal)	RD/BL	Sub O2S Signal	0.1-1.1v
31	BK/RD	EFI Main Relay Power	12-14v
32	BK/RD	EFI Main Relay Power	12-14v
33	BK/YL	Battery Direct	12-14v
34	BL/RD	A/C Amplifier Signal (AC1)	Clutch On: 1.5v, Off: 12v
35 (95)	BR/WT	PSP Switch Signal	Straight: 12v, Turning: 0v
36 (Cal)	BR/WT	Sub HO2S Heater	1v (Heater On)

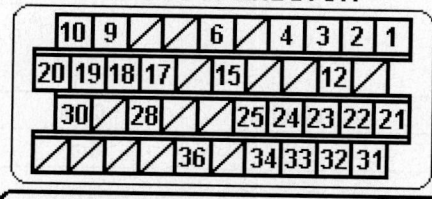

40 PIN CONNECTOR

WIRE SIDE OF HARNESS TERMINALS

Pin Connector Graphic

05533_ADIA_G091

1993-95 3.0L I6 MFI VIN J (All) 80 Pin Connector

PCM Pin #	Wire Color	Circuit Description (80 Pin)	Value at Hot Idle
3	GN	Vehicle Speed 2 Signal (-)	At 55 mph: 48 Hz
4	BR	Sensor Ground	<0.050v
7	GN	Distributor Signal (G-)	<0.050v
8	BK/RD	A/T-ECT Solenoid S3	In Lockup: 12-14v
9	RD/BL	A/T-ECT Solenoid (S2)	1st or OD: 1v
10	WT/RD	A/T-ECT Solenoid (S1)	In 3rd or OD: 1v
15	RD/BK	Injector 6 Control	2.0-3.3 ms
16	RD	Injector 5 Control	2.0-3.3 ms
17	RD/WT	Injector 4 Control	2.0-3.3 ms
18	RD/GN	Injector 3 Control	2.0-3.3 ms
19	RD/YL	Injector 2 Control	2.0-3.3 ms
20	RD/BL	Injector 1 Control	2.0-3.3 ms
23	RD	Vehicle Speed 2 Signal (+)	At 55 mph: 48 Hz
24	OR	A/T Oil Temperature Sensor	Varies: 0.5-4.5v
25	WT	Distributor Signal (G2+)	AC pulse signals
26	RD	Distributor Signal (G1+)	AC pulse signals
27	BK	Distributor Signal (NE+)	AC pulse signals
28 (94)	BR	Sensor Ground	<0.050v
28	BL/OR	Data Link Connector	12-14v
29	LG	Data Link Connector	12-14v
32	RD/GN	ISC Motor (ISC4)	Pulse Signals
33	GN/OR	ISC Motor (ISC3)	Pulse Signals
34	GN/WT	ISC Motor (ISC2)	Pulse Signals
35	PK/YL	ISC Motor (ISC1)	Pulse Signals
36	WT/BL	Fuel Pressure Up Solenoid	1v (at hot restart)
38 (94)	GN/RD	Exhaust Bypass Solenoid	12v or 0v
39	GN/YL	Exhaust Control Solenoid	12v or 0v
40 (94)	GN/BK	Intake Air Control Solenoid	12v or 0v

Standard Colors and Abbreviations

Abbreviation	Color	Abbreviation	Color	Abbreviation	Color
BK	Black	GY	Gray	RD	Red
BL	Blue	GN	Green	TN	Tan
BR	Brown	LG	Light Green	VT	Violet
DB	Dark Blue	OR	Orange	WT	White
DG	DK Green	PK	Pink	YL	Yellow

1993-95 3.0L I6 MFI VIN J (All) 80 Pin Connector

PCM Pin #	Wire Color	Circuit Description (80 Pin)	Value at Hot Idle
41	BL/RD	Sensor VREF	4.9-5.1v
43	YL	TP Sensor Signal	0.3-0.8v
44	BL/YL	ECT Sensor Signal	At 180°F: 0.51v
45	PK/BL	IAT Sensor Signal	At 100°F: 2.60v
46	BR/YL	EGR Gas Temp. Sensor	3.5-4.0v

1993-95 3.0L I6 MFI VIN J (All) 80 Pin Connector, *continued*

47	RD/BL	Rear O2S Signal	0.1-1.1v
48	WT	Front O2S Signal	0.1-1.1v
49	WT	Knock Sensor 2 Signal	<0.075v AC
50	WT	Knock Sensor 1 Signal	<0.075v AC
57	RD/WT	Igniter Signal (IGT)	Digital Signal: 0-5-0v
58	RD/YL	Igniter Signal (IGF)	Digital Signal: 0-5-0v
60 (94)	BL/WT	Waste Gate Solenoid	12v or 0v
64	RD	Closed Throttle Switch	1v, at off-idle: 12v
65	BR/BK	Sensor Ground	<0.050v
66	GN/BK	Airflow Meter Signal	1.5-2.5v
69	BR	Sensor Ground	<0.050v
72	BK/YL	Rear HO2S Heater	1v (Heater On)
73	BK/BL	Front HO2S Heater	1v (Heater On)
74	PK	EVAP Purge Solenoid (VSV)	12v or 0v
75	PK	EGR Solenoid Control (VSV)	12v or 0v
76	BK/WT	Neutral Start Switch	In P/N: 9-11v (cranking)
77	BK	Starter Switch Signal	9-11v (cranking)
79	BR	Power Ground	<0.1v
80	BR	Power Ground	<0.1v

80 PIN CONNECTOR

WIRE SIDE OF HARNESS TERMINALS

05533_ADIA_G092

Pin Connector Graphic

1996 3.0L I6 MFI VIN D (All) 40 Pin Connector

PCM Pin #	Wire Color	Circuit Description (40 Pin)	Value at Hot Idle
1	BK/OR	Ignition Switch Power	12-14v
2	PK	Speed Sensor 1 Signal	Moving: 0-5-0v
3	YL	Kickdown Switch	Open: 12v, Closed: 0v
4	GN/WT	Stop Light Switch Signal	Brake Off: 0v, On: 12v
6	BL/BK	MIL (lamp) Control	MIL Off: 12v, On: 1v
8	GN	Data Link Connector (SDL)	No Scan Tool: 0v
9	GN/RD	A/T Select Switch 2nd	In 2nd: 12v, Others: 0v
10	GN/BK	A/T Select Switch Low	In Low: 12v, Others: 0v
12	BR/BK	Cruise Control ECU	At Cruise in OD: 12v
15	RD/YL	Idle Up Diode Signal	Switch On: 12v, Off: 0v
18	GN/YL	A/T Pattern Selector Switch	Norm: 0v, PWR: 12v
20	YL/BL	Data Link Connector	12-14v
21	GN	Fuel Pump (Control) ECU	Pulse Signals
22	PK/WT	Fuel Pump (DI) ECU	Pulse Signals
23	WT/GN	A/C Magnetic Clutch (ACMG)	Clutch Off: 0v, On: 12v
24	GY	EFI Main Relay Power	12-14v
25	WT/BL	Manual Indicator Light	Light Off: 12-14v
28	PK/GN	Overdrive Main Switch	Switch Off: 12v, On: 0v
30	RD/BL	Sub HO2S Signal	0.1-1.1v
31	BK/RD	EFI Main Relay Power	12-14v
33	BK/YL	Battery Direct	12-14v
34	BL/RD	A/C Amplifier Signal (AC1)	Clutch On: 1.5v, Off: 12v
35	BR/WT	PSP Switch Signal	Straight: 12v, Turning: 0v
36	BR/WT	Sub HO2S Heater	1v (Heater On)

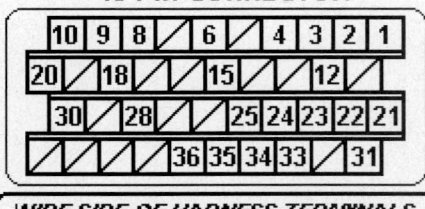

40 PIN CONNECTOR

WIRE SIDE OF HARNESS TERMINALS

05533_ADIA_G093

Pin Connector Graphic

1996 3.0L I6 MFI VIN D (All) 80 Pin Connector

PCM Pin #	Wire Color	Circuit Description (80 Pin)	Value at Hot Idle
3	GN	Vehicle Speed 2 Sensor (-)	Moving: 0-5-0v
5	WT	CKP Sensor Signal (NE2-)	AC pulse signals
6	BK	CKP Sensor Signal (NE2+)	AC pulse signals
7	GN	Distributor Signal (G-)	<0.050v
8	BK/RD	A/T-ECT Solenoid S3	In Lockup: 12-14v
9	RD/BL	A/T-ECT Solenoid (S2)	1st or OD: 1v
10	WT/RD	A/T-ECT Solenoid (S1)	In 3rd or OD: 1v
15	RD/BK	Injector 6 Control	2.0-3.3 ms

1996 3.0L I6 MFI VIN D (All) 80 Pin Connector, *continued*

16	RD	Injector 5 Control	2.0-3.3 ms
17	RD/WT	Injector 4 Control	2.0-3.3 ms
18	RD/GN	Injector 3 Control	2.0-3.3 ms
19	RD/YL	Injector 2 Control	2.0-3.3 ms
20	RD/BL	Injector 1 Control	2.0-3.3 ms
23	RD	Vehicle Speed 2 Sensor (+)	Moving: 0-5-0v
24	OR	A/T Oil Temperature Sensor	Varies: 0.5-4.5v
25	WT	Distributor Signal (G2+)	AC pulse signals
26	RD	Distributor Signal (G1+)	AC pulse signals
27	BK	Distributor Signal (NE+)	AC pulse signals
30	BR	MAF Sensor Ground	<0.050v
32	RD/GN	ISC Motor (ISC4)	Pulse Signals
33	GN/OR	ISC Motor (ISC3)	Pulse Signals
34	GN/WT	ISC Motor (ISC2)	Pulse Signals
35	PK/YL	ISC Motor (ISC1)	Pulse Signals
36	WT/BL	Fuel Pressure Up Solenoid	1v (at hot restart)
39	GN/BK	Intake Air Control Solenoid	12v or 0v

80 PIN CONNECTOR

WIRE SIDE OF HARNESS TERMINALS

Pin Connector Graphic

05533_ADIA_G094

1996 3.0L I6 MFI VIN D (All) 80 Pin Connector

PCM Pin #	Wire Color	Circuit Description (80 Pin)	Value at Hot Idle
41	BL/RD	Sensor VREF (VC)	4.9-5.1v
43	YL	TP Sensor Signal	0.3-0.8v
44	BL/YL	ECT Sensor Signal	At 180°F: 0.51v
45	PK/BL	IAT Sensor Signal	At 100°F: 2.60v
46	BR/YL	EGR Gas Temp. Sensor	3.5-4.0v
47	RD/BL	Rear HO2S Signal	0.1-1.1v
48	WT	Front HO2S Signal	0.1-1.1v
49	WT	Knock Sensor 2 Signal	<0.075v AC
50	WT	Knock Sensor 1 Signal	<0.075v AC
57	RD/WT	Igniter Signal (IGT)	Digital Signal: 0-5-0v
58	RD/YL	Igniter Signal (IGF)	Digital Signal: 0-5-0v
64	RD	Closed Throttle Switch	1v, at off-idle: 12v
65	BR/BK	Sensor Ground	<0.050v
66	YL/RD	Mass Airflow Sensor	1.0-1.8v
69	BR	Shield Ground	<0.050v
72	BK/YL	Rear HO2S Heater	1v (Heater On)
73	BK/BL	Front HO2S Heater	1v (Heater On)
74	PK	EVAP Purge Solenoid (VSV)	12v or 0v
75	PK	EGR Solenoid Control (VSV)	1v (Heater On)
76	BK/WT	A/T Neutral Start Switch	In P/N: 9-11v (cranking)
77	BK	Starter Switch Signal	9-11v (cranking)
78	BR	Power Ground	<0.1v
79	BR	Power Ground	<0.1v
80	BR	Power Ground	<0.1v

Standard Colors and Abbreviations

Abbreviation	Color	Abbreviation	Color	Abbreviation	Color
BK	Black	GY	Gray	RD	Red
BL	Blue	GN	Green	TN	Tan
BR	Brown	LG	Light Green	VT	Violet
DB	Dark Blue	OR	Orange	WT	White
DG	DK Green	PK	Pink	YL	Yellow

1997 3.0L I6 MFI VIN D (All) 40 Pin Connector

PCM Pin #	Wire Color	Circuit Description (40 Pin)	Value at Hot Idle
1	BK/OR	Ignition Switch Power	12-14v
2	PK	Speed Sensor 1 Signal	Moving: 0-5-0v
3	YL	Kickdown Switch	Open: 12v, Closed: 0v
4	GN/WT	Stop Light Switch Signal	Brake Off: 0v, On: 12v
6	BL/BK	MIL (lamp) Control	MIL Off: 12v, On: 1v
8	GN	Data Link Connector (SDL)	0v
9	GN/RD	A/T Select Switch 2nd	In 2nd: 12v, Others: 0v
10	GN/BK	A/T Select Switch Low	In Low: 12v, Others: 0v
12	GN/BK	Cruise Control ECU	At Cruise in OD: 12v

1997 3.0L I6 MFI VIN D (All) 40 Pin Connector, *continued*

15	RD/YL	Idle Up Diode Signal	Switch On: 12v, Off: 0v
18	GN/YL	A/T Pattern Selector Switch	Norm: 0v, PWR: 12v
20	YL/GN	Data Link Connector	12-14v
21	GN	Fuel Pump (Control) ECU	Pulse Signals
22	PK/WT	Fuel Pump (DI) ECU	Pulse Signals
23	WT/GN	A/C Magnetic Clutch (ACMG)	Clutch Off: 0v, On: 12v
24	BK/YL	EFI Main Relay Power	12-14v
25	WT/BL	Manual Indicator Light	Light Off: 12v, On: 1v
28	PK	Overdrive Main Switch	Switch Off: 12v
30	RD/BL	Sub HO2S Signal	0.1-1.1v
31	BK/RD	EFI Main Relay Power	12-14v
32	BR/WT	PSP Switch Signal	Straight: 12v, Turning: 0v
33	BK/WT	Battery Direct	12-14v
34	BL/RD	A/C Amplifier Signal (AC1)	Clutch On: 1.5v, Off: 12v
36	BR/WT	Sub HO2S Heater	1v (Heater On)

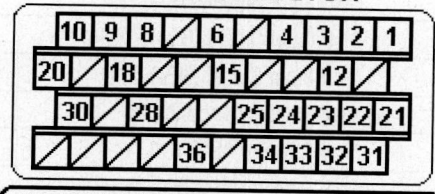

40 PIN CONNECTOR

WIRE SIDE OF HARNESS TERMINALS

Pin Connector Graphic

05533_ADIA_G095

1997 3.0L I6 MFI VIN D (All) 80 Pin Connector

PCM Pin #	Wire Color	Circuit Description (80 Pin)	Value at Hot Idle
3	GN	Vehicle Speed 2 Sensor (-)	Moving: 0-5-0v
4	BK	CKP Sensor Signal (NE2+)	AC pulse signals
5	WT	CKP Sensor Signal (NE2-)	<0.050v
7	GN	Distributor Signal (G-)	<0.050v
8	BK/RD	A/T-ECT Solenoid (SL)	In Lockup: 12-14v
9	RD/BL	A/T-ECT Solenoid (S2)	1st or OD: 1v
10	WT/RD	A/T-ECT Solenoid (S1)	In 3rd or OD: 1v
15	RD/BK	Injector 6 Control	2.0-3.3 ms
16	RD	Injector 5 Control	2.0-3.3 ms
17	RD/WT	Injector 4 Control	2.0-3.3 ms
18	BL	Injector 3 Control	2.0-3.3 ms
19	BL/RD	Injector 2 Control	2.0-3.3 ms
20	RD/BL	Injector 1 Control	2.0-3.3 ms
23	RD	Vehicle Speed 2 Sensor (+)	Moving: 0-5-0v
24	BL/BK	A/T Oil Temperature Sensor	0.5-4.5v
25	WT	Distributor Signal (G2+)	AC pulse signals
26	RD	Distributor Signal (G1+)	AC pulse signals
27	BK	Distributor Signal (NE+)	AC pulse signals
28	BR	MAF Sensor Ground	<0.050v
32	RD/BK	ISC Motor (ISC4)	Pulse Signals
33	BL/BK	ISC Motor (ISC3)	Pulse Signals
34	GN/WT	ISC Motor (ISC2)	Pulse Signals
35	YL/BK	ISC Motor (ISC1)	Pulse Signals
39	GN/YL	Intake Air Control Solenoid	12v or 0v

1997 3.0L I6 MFI VIN D (All) 80 Pin Connector

PCM Pin #	Wire Color	Circuit Description (80 Pin)	Value at Hot Idle
41	BL/RD	Sensor VREF (VC)	4.9-5.1v
43	YL	TP Sensor Signal	0.3-0.8v
44	BL/YL	ECT Sensor Signal	At 180°F: 0.51v
45	GN/BK	IAT Sensor Signal	At 100°F: 2.60v
46	BR/YL	EGR Gas Temp. Sensor	3.5-4.0v
47	RD/BL	HO2S-21 (B2 S1) Signal	0.1-1.1v
48	WT	HO2S-11 (B1 S1) Signal	0.1-1.1v
49	WT	Knock Sensor 2 Signal	<0.075v AC
50	WT	Knock Sensor 1 Signal	<0.075v AC
57	RD/WT	Igniter Signal (IGT)	Digital Signal: 0-5-0v
58	RD/YL	Igniter Signal (IGF)	Digital Signal: 0-5-0v
64	RD/BK	Closed Throttle Switch	1v, at off-idle: 12v
65	WT/BK	Sensor Ground	<0.050v
66	YL/RD	Mass Airflow Sensor	1.0-1.8v
69	BR	Shield Ground	<0.050v
71	GN	HO2S-11 (B1 S1) Heater	1v (Heater On)
72	BK/YL	HO2S-21 (B2 S1) Heater	1v (Heater On)

1997 3.0L I6 MFI VIN D (All) 80 Pin Connector, *continued*

73	WT/BL	Fuel Pressure Up Solenoid	1v (at hot restart)
74	PK	EVAP Purge Solenoid (VSV)	12v or 0v
75	PK	EGR Solenoid Control (VSV)	1v (Heater On)
76	BK/WT	A/T Neutral Start Switch	In P/N: 9-11v (cranking)
77	BK	Starter Switch Signal	9-11v (cranking)
78	BR	Power Ground	<0.1v
79	BR	Power Ground	<0.1v
80	BR	Power Ground	<0.1v

80 PIN CONNECTOR

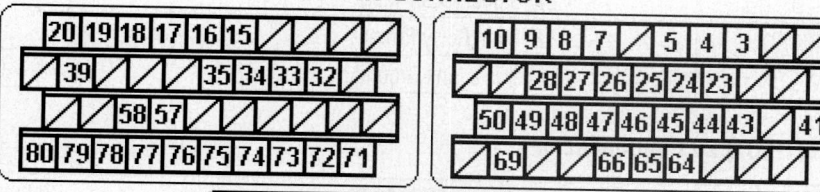

WIRE SIDE OF HARNESS TERMINALS

05533_ADIA_G096

Pin Connector Graphic

1998 3.0L I6 MFI VIN D (A/T-ECT) 31 Pin Connector

PCM Pin #	Wire Color	Circuit Description (31 Pin)	Value at Hot Idle
1	BL	Injector 3 Control	2.0-3.3 ms
2	RD/WT	Injector 4 Control	2.0-3.3 ms
3	RD	Injector 5 Control	2.0-3.3 ms
4	RD/BK	Injector 6 Control	2.0-3.3 ms
5	GN/YL	Intake Air Control Solenoid	12v or 0v
7	BK	Throttle Control Motor (M-)	Pulse Signals
8	WT	Throttle Control Motor (M+)	Pulse Signals
9	BR	Sensor Ground	<0.050v
10	BL	CMP Sensor Signal (G+)	AC pulse signals
11	RD/WT	Igniter Transistor 1 Control	Digital Signal: 0-5-0v
12	LG	Igniter Transistor 2 Control	Digital Signal: 0-5-0v
13	GN/RD	Igniter Transistor 3 Control	Digital Signal: 0-5-0v
16	GN	PSP Switch Signal	Straight: 12v, Turning: 0v
17	YL/BK	Cam Timing Oil signal (OCV-)	Pulse Signals
18	WT/RD	Cam Timing Oil signal (OCV+)	Pulse Signals
19	YL	Throttle Control Motor (CL-)	Pulse Signals
20	BL	Throttle Control Motor (CL+)	Pulse Signals
21	WT/BR	Power Ground	<0.1v
22	WT	CKP/CMP Sensor Ground (-)	<0.050v
23	BK	CKP Sensor Signal (NE+)	AC pulse signals
24	BK/WT	Neutral Start Switch	In P/N: 9-11v (cranking)
25	RD/YL	Igniter Signal (IGF)	Digital Signal: 0-5-0v
27	WT	Knock Sensor 2 Signal	<0.075v AC
28	WT	Knock Sensor 1 Signal	<0.075v AC
30	BR	Shield Ground	<0.050v
31	WT/BR	Power Ground	<0.1v

1998 3.0L I6 MFI VIN D (A/T-ECT) 22 Pin Connector

PCM Pin #	Wire Color	Circuit Description (22 Pin)	Value at Hot Idle
1	BK/WT	Battery Direct	12-14v
4	GN	Fuel Pump (Control) ECU	Pulse Signals
5	PK/WT	Fuel Pump (DI) ECU	Pulse Signals
6	BL/BK	MIL (lamp) Control	MIL Off: 12v, On: 1v
7	BL/RD	EFI Main Relay Power BM+	12-14v
8	BK/RD	EFI Main Relay Power B2+	12-14v
9	BK/OR	Ignition Switch Power	12-14v
10	BK/YL	EFI Main Relay Power	12-14v
11	BK	SIL Signal (Scan Tool)	Digital Signal
16	BK/RD	EFI Main Relay Power	12-14v
22	WT/BK	ECM/Data Link Ground	<0.050v

1998 3.0L I6 MFI VIN D (A/T-ECT) 28 Pin Connector

PCM Pin #	Wire Color	Circuit Description (28 Pin)	Value at Hot Idle
1	BL/RD	A/C Amplifier Signal (AC1)	Clutch On: 1.5v, Off: 12v
2	BK	Starter Switch Signal	9-11v (cranking)
4	GN/WT	Stop Light Switch Signal	Brake Off: 0v, On: 12v
5	RD	Data Link Connector	12-14v
8	WT	HO2S-12 (B1 S2) Signal	0.1-1.1v
13	WT/GN	A/C Magnetic Clutch (ACMG)	Clutch Off: 0v, On: 12v
16	RD/BK	A/T Select Switch Reverse	In 'R': 12v, Others: 0v
17	GN/RD	A/T Select Switch Drive	In 'D': 12v, Others: 0v
18	BR/YL	EVAP Vapor Pressure Sensor	2.5-3.1v (with hose off)
23	BL	Cruise Control ECU	At Cruise in OD: 12v
24	RD/YL	Cruise Control ECU	At Cruise in OD: 12v
25	WT	Vehicle Speed Sensor	At 55 mph: 48 Hz
26	RD/YL	Idle Up Diode Signal	Switch On: 12v, Off: 0v
28	PK	Overdrive Main Switch	Switch Off: 12v, On: 0v

1998 3.0L I6 MFI VIN D (A/T-ECT) 17 Pin Connector

PCM Pin #	Wire Color	Circuit Description (17 Pin)	Value at Hot Idle
1	WT/RD	A/T-ECT Solenoid (S1)	In 3rd or OD: 1v
2	RD/BL	A/T-ECT Solenoid (S2)	1st or OD: 1v
4	RD	OD Clutch Sensor Signal (+)	Pulse Signals
5	RD/YL	Speed Sensor 2 Signal (+)	Pulse Signals
7	GN/WT	A/T-ECT Solenoid (SLU+)	Pulse Signals
8	YL/GN	A/T-ECT Solenoid (SLN+)	Pulse Signals
9	GN/RD	A/T-ECT Solenoid (SLT+)	Pulse Signals
10	RD	OD Clutch Sensor Signal (-)	Pulse Signals
11	BL/YL	Speed Sensor Signal 2 (-)	Pulse Signals
13	BL/RD	A/T-ECT Solenoid (SLU-)	Pulse Signals

1998 3.0L I6 MFI VIN D (A/T-ECT) 17 Pin Connector, *continued*

14	PK	A/T-ECT Solenoid (SLN-)	Pulse Signals
15	RD/BK	A/T-ECT Solenoid (SLT-)	Pulse Signals
16	OR	C/C Indicator Light	Light Off: 12v, On: 1v
17	BL/BK	A/T Oil Temperature Sensor	0.5-4.5v

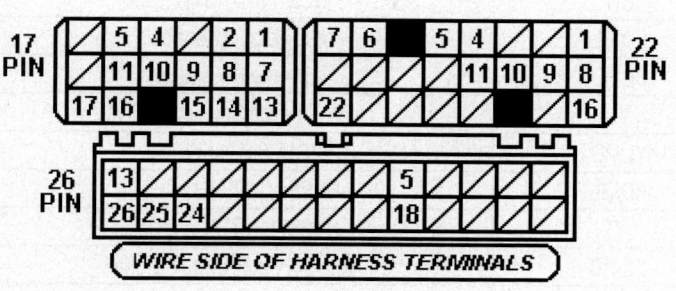

05533_ADIA_G097

Pin Connector Graphic

1998 3.0L I6 MFI VIN D (A/T-ECT) 26 Pin Connector

PCM Pin #	Wire Color	Circuit Description (26 Pin)	Value at Hot Idle
5	GN/YL	A/T Pattern Selector Switch	Norm: 0v, PWR: 12v
13	RD/BK	EVAP Vapor Pressure (VSV)	12v or 0v
18	WT/BL	Manual Indicator Light	Light Off: 12-14v
24	WT	HO2S-22 (B2 S2) Signal	0.1-1.1v
25	GN/YL	HO2S-22 (B2 S2) Heater	1v (Heater On)
26	BL/WT	HO2S-12 (B1 S2) Heater	1v (Heater On)

1998 3.0L I6 MFI VIN D (A/T-ECT) 24 Pin Connector

PCM Pin #	Wire Color	Circuit Description (24 Pin)	Value at Hot Idle
1	WT/BK	Power Ground	<0.1v
2	BL/RD	Sensor VREF	4.9-5.1v
3	BK/YL	HO2S-21 (B2 S1) Heater	1v (Heater On)
4	GN	HO2S-11 (B1 S1) Heater	1v (Heater On)
5	RD/BL	Injector 1 Control	2.0-3.3 ms
6	BL/RD	Injector 2 Control	2.0-3.3 ms
7	YL	EVAP Purge Solenoid (VSV)	12v or 0v
10	YL/RD	MAF Sensor Signal	1.1-1.5v
11	RD/BL	HO2S-21 (B2 S1) Signal	0.1-1.1v
12	WT	HO2S-11 (B1 S1) Signal	0.1-1.1v
13	BL/RD	Mirror Heater Switch Signal	Heater On: 12-14v
14	BL	ECT Sensor Signal	At 180°F: 0.51v
15	GN	Accel Position Sensor (VPA)	0.25-0.9v
16	WT	Accel Position Sensor (VPA2)	1.8-2.7v
17	BR	Sensor Ground	<0.050v
18	BR	Sensor Ground	<0.050v
19	BR	MAF Sensor Ground	<0.050v
20	BL/YL	A/T Select Switch 2nd	In 2nd: 12v, Others: 0v
21	GN/BK	A/T Select Switch Low	In Low: 12v, Others: 0v
22	GN/WT	IAT Sensor Signal	At 100°F: 2.60v
23	YL	TP Sensor (VTA1)	0.4-1.0v
24	RD/BK	TP Sensor (VTA2)	2.0-2.9v

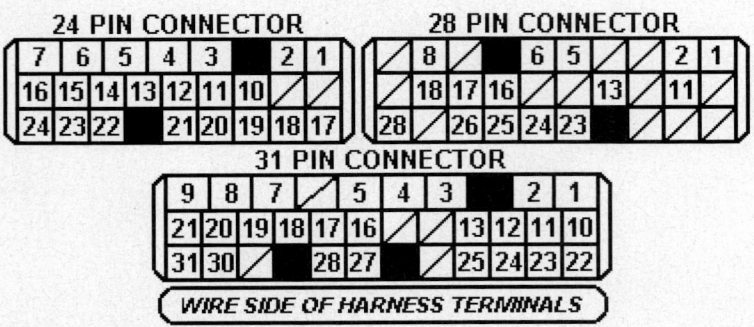

05533_ADIA_G098

Pin Connector Graphic

1990-92 3.0L Turbo I6 VIN M (A/T) 16 Pin Connector

PCM Pin #	Wire Color	Circuit Description (16 Pin)	Value at Hot Idle
1	BL/RD	Sensor VREF	4.9-5.1v
2	GN/RD	Airflow Meter Signal	1.5-2.5v
3	LG	IAT Sensor Signal	At 100°F: 2.60v
4	GN	ECT Sensor Signal	At 180°F: 0.51v
5	WT	Knock Sensor 1 Signal	<0.075v AC
6	WT	O2S-11 (B1 S1) Signal	0.1-1.1v
7	YL/BK	Oil Pressure Switch	Switch Closed: 0.1v
8	GY	Check Connector	12-14v

1990-92 3.0L Turbo I6 VIN M (A/T) 16 Pin Connector, *continued*

9	BR	Sensor Ground	<0.050v
11	WT/RD	TP Sensor Signal	0.3-0.8v
12	YL/BL	Closed Throttle Switch	1v, at off-idle: 12v
13	WT	Knock Sensor 2 Signal	<0.075v AC
15	RD/GN	Check Connector	12-14v
16	BK	Distributor Signal (G-)	<0.050v

1990-92 3.0L Turbo I6 VIN M (A/T) 22 Pin Connector

PCM Pin #	Wire Color	Circuit Description (22 Pin)	Value at Hot Idle
1	BK/YL	Battery Direct	12-14v
2	BK/OR	Ignition Switch Power	12-14v
2 ('90-'91)	GY	Ignition Switch Power	12-14v
4	BK/OR	EFI Main Relay Power	12-14v
5	GY/GN	MIL (lamp) Control	MIL Off: 12v, On: 1v
6	YL	Fuel Pump Relay	Relay Off: 12v, On: 1v
7	GN	Circuit Opening Relay (FC)	0-3v, off-idle: 12v
8	PK	Defogger Diode Signal	Switch On: 12v, Off: 0v
9	PK	Vehicle Speed Sensor	At 55 mph: 48 Hz
10	BL/RD	A/C Magnetic Clutch (ACMG)	Clutch Off: 0v, On: 12v
11	BK/BL	Starter Switch Signal	9-11v (cranking)
12	BK/RD	EFI Main Relay Power	12-14v
13	BK/RD	EFI Main Relay Power B1+	12-14v
17	RD/WT	L3 Signal to TCM	0v, at WOT: 5v
18	RD	L2 Signal to TCM	5v, at WOT: 5v
19	BK	L1 Signal to TCM	5v, at WOT: 0v
20	RD/BL	A/T-ECT ECU Signal	Pulse Signals
21	RD/YL	Headlight Diode Signal	Switch On: 12v, Off: 0v
22	BK/WT	A/T Neutral Start Switch	In P/N: 9-11v (cranking)
22	BK/YL	M/T Clutch Start Switch	In 'N': 9-11v (cranking)

Standard Colors and Abbreviations

Abbreviation	Color	Abbreviation	Color	Abbreviation	Color
BK	Black	GY	Gray	RD	Red
BL	Blue	GN	Green	TN	Tan
BR	Brown	LG	Light Green	VT	Violet
DB	Dark Blue	OR	Orange	WT	White
DG	DK Green	PK	Pink	YL	Yellow

1990-92 3.0L Turbo I6 VIN M (All) 26 Pin Connector

PCM Pin #	Wire Color	Circuit Description (26 Pin)	Value at Hot Idle
1	BL	Distributor Signal (NE+)	AC pulse signals
2	YL	Distributor Signal (G2+)	AC pulse signals
3	GN/BK	Igniter Signal (IGF)	Digital Signal: 0-5-0v
4	RD/BK	ISC Motor (ISC4)	Pulse Signals
5	BL/RD	ISC Motor (ISC3)	Pulse Signals
6	GN/WT	ISC Motor (ISC2)	Pulse Signals
7	WT/YL	ISC Motor (ISC1)	Pulse Signals
8	RD/BK	EGR Solenoid Control (VSV)	1v (Heater On)
9	RD/BL	Fuel Pressure Up Solenoid	1v (at hot restart)
10	RD/BL	HO2S-11 (B1 S1) Heater	1v (Heater On)
11	WT/RD	Injector Pair 2 & 6 Control	2.0-3.3 ms
12	WT/BL	Injector Pair 1 & 4 Control	2.0-3.3 ms
13	WT/BK	Power Ground	<0.1v
14	BR	Sensor Ground	<0.050v
15	RD	Distributor Signal (G1+)	AC pulse signals
16	BK/YL	EGR Gas Temp. Sensor	3.5-4.0v
18	BL/WT	Igniter Signal (IGB)	Digital Signal: 0-5-0v
19	BL/BK	Igniter Signal (IGA)	Digital Signal: 0-5-0v
20	LG	Igniter Signal (IGT)	Digital Signal: 0-5-0v
24	GN	Cold Start Injector Control	1v (at cold startup)
25	WT	Injector Pair 3 & 5 Control	2.0-3.3 ms
26	WT/BK	Power Ground	<0.1v

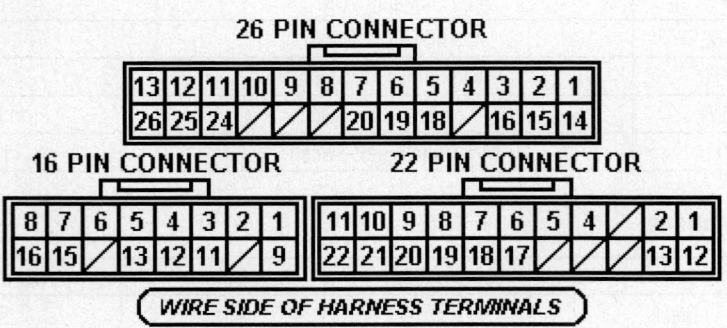

05533_ADIA_G099

Pin Connector Graphic

1993 3.0L Twin/Turbo I6 VIN J (All) 40 Pin Connector

PCM Pin #	Wire Color	Circuit Description (40 Pin)	Value at Hot Idle
1	BK/OR	Ignition Switch Power	12-14v
2	PK	Speed Sensor 1 Signal	Moving: 0-5-0v
3	YL	Kickdown Switch	Switch Open: 12v
4	GN/WT	Stop Light Switch Signal	Brake Off: 0v, On: 12v
6	BL/BK	MIL (lamp) Control	MIL Off: 12v, On: 1v
7	RD/BK	A/T Select Switch Reverse	In 'R': 12v, Others: 0v
9	GN/RD	A/T Select Switch 2nd	In 2nd: 12v, Others: 0v
10	GN/BK	A/T Select Switch Low	In Low: 12v, Others: 0v

1993 3.0L Twin/Turbo I6 VIN J (All) 40 Pin Connector, *continued*

11	YL/RD	ABS /Traction ECU	Pulse Signals
12	BR/BK	Cruise Control ECU	At Cruise in OD: 12v
13	WT/RD	Traction Control (TRC-)	Pulse Signals
14	OR	Traction Control (TRC+)	Pulse Signals
15	RD/YL	Defogger Diode Signal	Switch On: 12v, Off: 0v
16	BK/WT	Tachometer Signal (TACO)	Pulse Signals
17	GY/RD	Data Link Connector	12-14v
18	GN/YL	A/T Pattern Selector Switch	Norm: 0v, PWR: 12v
19	PK/GN	Data Link Connector	12-14v
20	YL/BL	Data Link Connector	12-14v
21	GN	Fuel Pump ECU	Pulse Signals
22	PK/WT	Fuel Pump ECU	Pulse Signals
23	WT/GN	A/C Magnetic Clutch (ACMG)	Clutch Off: 0v, On: 12v
24	GY	EFI Main Relay Power	12-14v
25	WT/BL	Manual Indicator Light	Light Off: 12-14v
28	PK/GN	Overdrive "Off" Indicator	Switch Off: 12v
31	BK/RD	EFI Main Relay Power	12-14v
32	BK/RD	EFI Main Relay Power	12-14v
33	BK/YL	Battery Direct	12-14v
34	BL/RD	A/C Amplifier Signal (AC1)	Clutch On: 1.5v, Off: 12v
38	PK/BL	Traction Control (NEO)	Pulses

40 PIN CONNECTOR

10	9	/	7	6	/	4	3	2	1
20	19	18	17	16	15	14	13	12	11
/	/	28	27	26	25	24	23	22	21
/	/	38	/	/	/	34	33	32	31

WIRE SIDE OF HARNESS TERMINALS

Pin Connector Graphic

05533_ADIA_G100

1993 3.0L Twin/Turbo I6 VIN J (All) 80 Pin Connector

PCM Pin #	Wire Color	Circuit Description (80 Pin)	Value at Hot Idle
1	YL	OD Clutch Speed Sensor (-)	Pulses
3	GN	Vehicle Speed Sensor 2 (-)	Pulses
4	BR	Sensor Ground	<0.050v
5	BL	Distributor Signal (G2+)	AC pulse signals
6	OR	Distributor Signal (G1+)	AC pulse signals
7	BK/RD	Distributor Signal (NE+)	AC pulse signals
9	RD/BL	A/T-ECT Solenoid (S2)	1st or OD: 1v
10	WT/RD	A/T-ECT Solenoid (S1)	In 3rd or OD: 1v
12	GN/RD	A/T-ECT Solenoid (SLT-)	Pulse Signals
13	YL/GN	A/T-ECT Solenoid (SLN-)	Pulse Signals
14	GN/BK	A/T-ECT Solenoid (SLU-)	Pulse Signals
15	RD/BK	Injector 6 Control	2.0-3.3 ms
16	RD	Injector 5 Control	2.0-3.3 ms
17	RD/WT	Injector 4 Control	2.0-3.3 ms
18	RD/GN	Injector 3 Control	2.0-3.3 ms
19	RD/YL	Injector 2 Control	2.0-3.3 ms
20	RD/BL	Injector 1 Control	2.0-3.3 ms
21	BL	OD Clutch Speed Sensor (+)	Pulse Signals
23	RD	Vehicle Speed Sensor 2 (+)	Pulse Signals
24	OR	A/T Oil Temperature Sensor	0.5-4.5v
25	BK/WT	Distributor Signal (G2-)	<0.050v
26	WT	Distributor Signal (G1-)	<0.050v
27	BR	Distributor Signal (NE-)	<0.050v
28	BR	MAF Sensor Ground	<0.050v
29	LG	Data Link Connector	12-14v
31	WT/GN	ECT Solenoid (SLT+)	Distributor (NE-) Signal
32	RD/GN	ISC Motor (ISC4)	Pulse Signals
33	GN/OR	ISC Motor (ISC3)	Pulse Signals
34	GN/WT	ISC Motor (ISC2)	Pulse Signals
35	PK/YL	ISC Motor (ISC1)	Pulse Signals
38	GN/RD	Exhaust Bypass Solenoid	12v or 0v
39	GN/YL	EGR Solenoid Control (VSV)	1v (Heater On)
40	GN/BK	Intake Air Control Solenoid	12v or 0v

Standard Colors and Abbreviations

Abbreviation	Color	Abbreviation	Color	Abbreviation	Color
BK	Black	GY	Gray	RD	Red
BL	Blue	GN	Green	TN	Tan
BR	Brown	LG	Light Green	VT	Violet
DB	Dark Blue	OR	Orange	WT	White
DG	DK Green	PK	Pink	YL	Yellow

1993 3.0L Twin/Turbo I6 VIN J (All) 80 Pin Connector

PCM Pin #	Wire Color	Circuit Description (80 Pin)	Value at Hot Idle
41	BL/RD	Sensor VREF (VC)	4.9-5.1v
42	YL/BL	TP Sensor (VTA2)	2.0-2.9v
43	YL	TP Sensor (VTA1)	0.4-1.0v
44	BL/YL	ECT Sensor Signal	At 180°F: 0.51v
45	PK/BL	IAT Sensor Signal	At 100°F: 2.60v
46	BR/YL	EGR Gas Temp. Sensor	3.5-4.0v
47	RD/BL	Rear HO2S Signal	0.1-1.1v
48	WT	Front HO2S Signal	0.1-1.1v
49	WT	Knock Sensor 2 Signal	<0.075v AC
50	WT	Knock Sensor 1 Signal	<0.075v AC
52	BL/OR	Igniter Transistor 6 Control	7% duty cycle
53	PK/BK	Igniter Transistor 5 Control	7% duty cycle
54	GY/GN	Igniter Transistor 4 Control	7% duty cycle
55	GY/BK	Igniter Transistor 3 Control	7% duty cycle
56	BK/OR	Igniter Transistor 2 Control	7% duty cycle
57	RD/WT	Igniter Transistor 1 Control	7% duty cycle
58	RD/YL	Igniter Signal (IGF)	Digital Signal: 0-5-0v
60	BL/WT	Waste Gate Solenoid	12v or 0v
62	BK/YL	Vacuum Sensor Signal	1-1.5v
63	GY/RD	Closed Throttle Switch (IDL2)	1v, at off-idle: 12v
64	RD	Closed Throttle Switch (IDL1)	1v, at off-idle: 12v
65	BR/BK	Sensor Ground	<0.050v
66	GN/BK	MAF Sensor Signal	1.1-1.5v
69	BR	Sensor Ground	<0.050v
71	BK/BL	Front HO2S Heater	1v (Heater On)
72	BR/WT	Rear HO2S Heater	1v (Heater On)
73	WT/BL	Fuel Pressure Up Solenoid	1v (at hot restart)
74	PK	EVAP Purge Solenoid (VSV)	12v or 0v
75	PK	EGR Solenoid Control (VSV)	1v (Heater On)
76	BK/WT	Neutral Start Switch	In P/N: 9-11v (cranking)
77	BK	Starter Switch Signal	9-11v (cranking)
79	BR	Power Ground	<0.1v
80	BR	Power Ground	<0.1v

80 PIN CONNECTOR

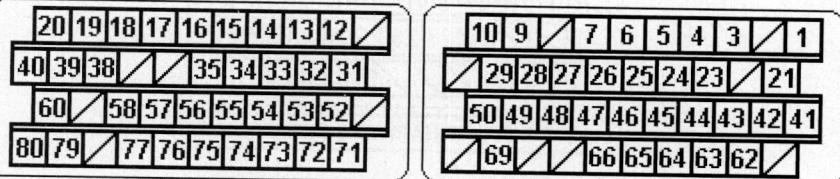

WIRE SIDE OF HARNESS TERMINALS

Pin Connector Graphic

05533_ADIA_G101

1994-95 3.0L Twin/Turbo I6 VIN J 40 Pin Connector

PCM Pin #	Wire Color	Circuit Description (40 Pin)	Value at Hot Idle
1	BK/OR	Ignition Switch Power	12-14v
2	PK	Speed Sensor 1 Signal	Moving: 0-5-0v
3	YL	Kickdown Switch	Switch Closed: 12v
4	GN/WT	Stop Light Switch Signal	Brake Off: 0v, On: 12v
6	BL/BK	MIL (lamp) Control	MIL Off: 12v, On: 1v
7	RD/BK	A/T Select Switch Reverse	In 'R': 12v, Others: 0v
9	GN/RD	A/T Select Switch 2nd	In 2nd: 12v, Others: 0v
10	GN/BK	A/T Select Switch Low	In Low: 12v, Others: 0v
11	YL/RD	ABS/Traction ECU	Pulses
12	BR/BK	Cruise Control ECU	At Cruise in OD: 12v
13	WT/RD	Traction Control (TRC-)	Pulse Signals
14	OR	Traction Control (TRC+)	Pulse Signals
15	RD/YL	Idle Up Diode Signal	Switch On: 12v, Off: 0v
16	BK/WT	Tachometer Signal (TACO)	Pulse Signals
17	GY/RD	Data Link Connector	12-14v
18	GN/YL	A/T Pattern Selector Switch	Norm: 0v, PWR: 12v
19	PK/GN	Data Link Connector	12-14v
20	YL/BL	Data Link Connector	12-14v
21	GN	Fuel Pump ECU	Pulse Signals
22	PK/WT	Fuel Pump ECU	Pulse Signals
23	WT/GN	A/C Magnetic Clutch (ACMG)	Clutch Off: 0v, On: 12v
24	GY	EFI Main Relay Power	12-14v
25	WT/BL	Manual Indicator Light	Light Off: 12-14v
26	WT	EFI ECU Signal (-)	Pulse Signals
27	BK	EFI ECU Signal (+)	Pulse Signals
28	PK/GN	Overdrive "Off" Indicator	LED Off: 12v, On: 1v
31	BK/RD	EFI Main Relay Power	12-14v
32 (94)	BK/RD	EFI Main Relay Power B1+	12-14v
33	BK/YL	Battery Direct	12-14v
34	BL/RD	A/C Amplifier Signal (AC1)	Clutch On: 1.5v, Off: 12v
38	PK/BL	Traction Control (NEO)	Pulses

40 PIN CONNECTOR

```
10  9  /  7  6  /  4  3  2  1
20 19 18 17 16 15 14 13 12 11
/  / 28 27 26 25 24 23 22 21
/  / 38  /  /  / 34 33 32 31
```

WIRE SIDE OF HARNESS TERMINALS

05533_ADIA_G102

Pin Connector Graphic

1994-95 3.0L Twin/Turbo I6 VIN J 80 Pin Connector

PCM Pin #	Wire Color	Circuit Description (80 Pin)	Value at Hot Idle
1	YL	OD Clutch Speed Sensor (-)	Pulse Signals
3	GN	Vehicle Speed Sensor 2 (-)	Pulse Signals
4 (94)	BR	Sensor Ground	<0.050v
5	BL	CMP Sensor Signal (G2-)	<0.050v
6	OR	CMP Sensor Signal (G1-)	<0.050v
7	BR	CKP Sensor Signal (NE-)	<0.050v
7 (94)	BK/RD	CKP Sensor Signal (NE-)	<0.050v
9	RD/BL	A/T-ECT Solenoid (S2)	1st or OD: 1v
10	WT/RD	A/T-ECT Solenoid (S1)	In 3rd or OD: 1v
12	GN/RD	A/T-ECT Solenoid (SLT-)	Pulse Signals
13	YL/GN	A/T-ECT Solenoid (SLN-)	Pulse Signals
14	GN/BK	A/T-ECT Solenoid (SLU-)	Pulse Signals
15	RD/BK	Injector 6 Control	2.0-3.3 ms
16	RD	Injector 5 Control	2.0-3.3 ms
17	RD/WT	Injector 4 Control	2.0-3.3 ms
18	RD/GN	Injector 3 Control	2.0-3.3 ms
19	RD/YL	Injector 2 Control	2.0-3.3 ms
20	RD/BL	Injector 1 Control	2.0-3.3 ms
21	BL	OD Clutch Speed Sensor (+)	Pulse Signals
23	RD	Vehicle Speed Sensor 2 (+)	Pulse Signals
24	OR	A/T Oil Temperature Sensor	0.5-4.5v
25	BK/WT	CMP Sensor Signal (G2+)	AC pulse signals
26	WT	CMP Sensor Signal (G1+)	AC pulse signals
27	BK/RD	CKP Sensor Signal (NE+)	AC pulse signals
27 (94)	BR	CKP Sensor Signal (NE+)	AC pulse signals
28	BR	MAF Sensor Ground	<0.050v
29	LG	Data Link Connector	12-14v
31	WT/GN	A/T-ECT Solenoid (SLT+)	Pulses
32	RD/GN	ISC Motor (ISC4)	Pulse Signals
33	GN/OR	ISC Motor (ISC3)	Pulse Signals
34	GN/WT	ISC Motor (ISC2)	Pulse Signals
35	PK/YL	ISC Motor (ISC1)	Pulse Signals
38	GN/RD	Exhaust Bypass Solenoid	12v or 0v
39	GN/YL	EGR Solenoid Control (VSV)	1v (Heater On)
40	GN/BK	Intake Air Control Solenoid	12v or 0v

Standard Colors and Abbreviations

Abbreviation	Color	Abbreviation	Color	Abbreviation	Color
BK	Black	GY	Gray	RD	Red
BL	Blue	GN	Green	TN	Tan
BR	Brown	LG	Light Green	VT	Violet
DB	Dark Blue	OR	Orange	WT	White
DG	DK Green	PK	Pink	YL	Yellow

16-210 TOYOTA PIN CHARTS

1994-95 3.0L Twin/Turbo I6 VIN J 80 Pin Connector

PCM Pin #	Wire Color	Circuit Description (80 Pin)	Value at Hot Idle
41	BL/RD	Main Sensor VREF (VC)	4.9-5.1v
42	YL/BL	Sub TP Sensor (VTA2)	2.0-2.9v
43	YL	Main TP Sensor (VTA1)	0.4-1.0v
44	BL/YL	ECT Sensor Signal	At 180°F: 0.51v
45	PK/BL	IAT Sensor Signal	At 100°F: 2.60v
46	BR/YL	EGR Gas Temp. Sensor	3.5-4.0v
47	RD/BL	Rear HO2S Signal	0.1-1.1v
48	WT	Front HO2S Signal	0.1-1.1v
49	WT	Knock Sensor 2 Signal	<0.075v AC
50	WT	Knock Sensor 1 Signal	<0.075v AC
52 (94)	RD/WT	Igniter Transistor 6 Control	Digital Signal: 0-5-0v
52	RD/YL	Igniter Transistor 6 Control	Digital Signal: 0-5-0v
53	BL/OR	Igniter Transistor 5 Control	Digital Signal: 0-5-0v
54	GY/GN	Igniter Transistor 4 Control	Digital Signal: 0-5-0v
55	GY/BK	Igniter Transistor 3 Control	Digital Signal: 0-5-0v
56	BK/OR	Igniter Transistor 2 Control	Digital Signal: 0-5-0v
57	RD/WT	Igniter Transistor 1 Control	Digital Signal: 0-5-0v
58	RD/YL	Igniter Signal (IGF)	Digital Signal: 0-5-0v
60	BL/WT	Waste Gate Solenoid	12v or 0v
62	BK/YL	Turbo Pressure Sensor	2.5-4.5v
63	GY/RD	Closed Throttle Switch (IDL2)	1v, at off-idle: 12v
64	RD	Closed Throttle Switch (IDL1)	1v, at off-idle: 12v
65	BR/BK	Sensor Ground	<0.050v
66	GN/BK	MAF Sensor Signal	1.1-1.5v
69 (94)	BR	Sensor Ground	<0.050v
71	BK/BL	Front HO2S Heater	1v (Heater On)
72	BR/WT	Rear HO2S Heater	1v (Heater On)
73	WT/BL	Fuel Pressure Up Solenoid	1v (at hot restart)
74	PK	EVAP Purge Solenoid (VSV)	12v or 0v
75	PK	EGR Solenoid Control (VSV)	1v (Heater On)
76	BK/WT	Neutral Start Switch	In P/N: 9-11v (cranking)
77	BK	Starter Switch Signal	9-11v (cranking)
79	BR	Power Ground	<0.1v
80	BR	Power Ground	<0.1v

80 PIN CONNECTOR

WIRE SIDE OF HARNESS TERMINALS

05533_ADIA_G103

Pin Connector Graphic

1996 3.0L Twin/Turbo I6 VIN E (All) 40 Pin Connector

PCM Pin #	Wire Color	Circuit Description (40 Pin)	Value at Hot Idle
1	BK/OR	Ignition Switch Power	12-14v
2	PK	Speed Sensor 1 Signal	Moving: 0-5-0v
3	YL	Kickdown Switch	Closed: 12v, Open: 0v
4	GN/WT	Stop Light Switch Signal	Brake Off: 0v, On: 12v
6	BL/BK	MIL (lamp) Control	MIL Off: 12v, On: 1v
7	RD/BK	A/T Select Switch Reverse	In 'R': 12v, Others: 0v
9	GN/RD	A/T Select Switch 2nd	In 2nd: 12v, Others: 0v
10	GN/BK	A/T Select Switch Low	In Low: 12v, Others: 0v
11	YL/RD	ABS /Traction ECU	Pulse Signals
12	BR/BK	Cruise Control ECU	At Cruise in OD: 12v
13	WT/RD	Traction Control (TRC-)	Pulse Signals
14	OR	Traction Control (TRC+)	Pulse Signals
15	RD/YL	Idle Up Diode Signal	Switch On: 12v, Off: 0v
16	BK/WT	Tachometer Signal (TACO)	Pulse Signals
17	GY/RD	Data Link Connector	12-14v
18	GN/YL	A/T Pattern Selector Switch	Norm: 0v, PWR: 12v
19	PK/GN	Data Link Connector	12-14v
20	YL/BL	Data Link Connector	12-14v
21	GN	Fuel Pump ECU	Pulse Signals
22	PK/WT	Fuel Pump ECU	Pulse Signals
23	WT/GN	A/C Magnetic Clutch (ACMG)	Clutch Off: 0v, On: 12v
24	GY	EFI Main Relay Power	12-14v
25	WT/BL	Manual Indicator Light	Light Off: 12-14v
26	WT	Traction Control ECU (EFI-)	Pulse Signals
27	BK	Traction Control ECU (EFI+)	Pulse Signals
28	PK/GN	Overdrive Main Switch	Switch Off: 12v
31	BK/RD	EFI Main Relay Power	12-14v
33	BK/YL	Battery Direct	12-14v
34	BL/RD	A/C Amplifier Signal (AC1)	Clutch On: 1.5v, Off: 12v
38	PK/BL	Traction Control (NEO)	Pulse Signals

40 PIN CONNECTOR

WIRE SIDE OF HARNESS TERMINALS

Pin Connector Graphic

05533_ADIA_G102

1996 3.0L Twin/Turbo I6 VIN E (All) 80 Pin Connector

PCM Pin #	Wire Color	Circuit Description (80 Pin)	Value at Hot Idle
1	YL	OD Clutch Speed Sensor (-)	Pulses
3	GN	Vehicle Speed Sensor 2 (-)	At 55 mph: 48 Hz
5	BL	CMP Sensor Signal (G2-)	<0.050v
6	OR	CMP Sensor Signal (G1-)	<0.050v
7	BR	CKP Sensor Signal (NE-)	<0.050v
9	RD/BL	A/T-ECT Solenoid (S2)	1st or OD: 1v
10	WT/RD	A/T-ECT Solenoid (S1)	In 3rd or OD: 1v
12	GN/RD	A/T-ECT Solenoid (SLT-)	Pulse Signals
13	YL/GN	A/T-ECT Solenoid (SLN-)	Pulse Signals
14	GN/BK	A/T-ECT Solenoid (SLU-)	Pulse Signals
15	RD/BK	Injector 6 Control	2.0-3.3 ms
16	RD	Injector 5 Control	2.0-3.3 ms
17	RD/WT	Injector 4 Control	2.0-3.3 ms
18	RD/GN	Injector 3 Control	2.0-3.3 ms
19	RD/YL	Injector 2 Control	2.0-3.3 ms
20	RD/BL	Injector 1 Control	2.0-3.3 ms
21	BL	OD Clutch Speed Sensor (+)	Pulse Signals
23	RD	Vehicle Speed Sensor 2 (+)	Pulse Signals
24	OR	A/T Oil Temperature Sensor	0.5-4.5v
25	BK/WT	CMP Sensor Signal (G2+)	AC pulse signals
26	WT	CMP Sensor Signal (G1+)	AC pulse signals
27	BK/RD	CKP Sensor Signal (NE+)	AC pulse signals
28	BR	MAF Sensor Ground	<0.050v
29	LG	Data Link Connector	12-14v
31	WT/GN	A/T-ECT Solenoid (SLT+)	Pulse Signals
32	RD/GN	ISC Motor (ISC4)	Pulse Signals
33	GN/OR	ISC Motor (ISC3)	Pulse Signals
34	GN/WT	ISC Motor (ISC2)	Pulse Signals
35	PK/YL	ISC Motor (ISC1)	Pulse Signals
38	GN/RD	Exhaust Bypass Solenoid	12v or 0v
39	GN/YL	EGR Solenoid Control (VSV)	1v (Heater On)
40	GN/BK	Intake Air Control Solenoid	12v or 0v

Standard Colors and Abbreviations

Abbreviation	Color	Abbreviation	Color	Abbreviation	Color
BK	Black	GY	Gray	RD	Red
BL	Blue	GN	Green	TN	Tan
BR	Brown	LG	Light Green	VT	Violet
DB	Dark Blue	OR	Orange	WT	White
DG	DK Green	PK	Pink	YL	Yellow

1996 3.0L Twin/Turbo I6 VIN E (All) 80 Pin Connector

PCM Pin #	Wire Color	Circuit Description (80 Pin)	Value at Hot Idle
41	BL/RD	Main Sensor VREF (VC)	4.9-5.1v
42	YL/BL	Sub TP Sensor (VTA2)	2.0-2.9v
43	YL	Main TP Sensor (VTA1)	0.4-1.0v
44	BL/YL	ECT Sensor Signal	At 180ºF: 0.51v
45	PK/BL	IAT Sensor Signal	At 100ºF: 2.60v
46	BR/YL	EGR Gas Temp. Sensor	3.5-4.0v
47	RD/BL	HO2S-12 (B1 S2) Signal	0.1-1.1v
48	WT	HO2S-11 (B1 S1) Signal	0.1-1.1v
49	WT	Knock Sensor 2 Signal	<0.075v AC
50	WT	Knock Sensor 1 Signal	<0.075v AC
52	RD/YL	Igniter Transistor 6 Control	7% duty cycle
53	BL/OR	Igniter Transistor 5 Control	7% duty cycle
54	GY/GN	Igniter Transistor 4 Control	7% duty cycle
55	GY/BK	Igniter Transistor 3 Control	7% duty cycle
56	BK/OR	Igniter Transistor 2 Control	7% duty cycle
57	RD/WT	Igniter Transistor 1 Control	7% duty cycle
58	RD/YL	Igniter Signal (IGF)	Digital Signal: 0-5-0v
60	BL/WT	Waste Gate Solenoid	12v or 0v
62	BK/YL	Turbo Pressure Sensor	2.5-4.5v
63	GY/RD	Closed Throttle Switch (IDL2)	1v, at off-idle: 12v
64	RD	Closed Throttle Switch (IDL1)	1v, at off-idle: 12v
65	BR/BK	Sensor Ground	<0.050v
66	GN/BK	Vacuum Sensor Signal	1-1.5v
69	BR/BK	Sensor Ground	<0.050v
71	BK/BL	HO2S-11 (B1 S1) Heater	1v (Heater On)
72	BR/WT	HO2S-12 (B1 S2) Heater	1v (Heater On)
73	WT/BL	Fuel Pressure Up Solenoid	1v (at hot restart)
74	PK	EVAP Purge Solenoid (VSV)	12v or 0v
75	PK	EGR Solenoid Control (VSV)	1v (Heater On)
76	BK/WT	Neutral Start Switch	In P/N: 9-11v (cranking)
77	BK	Starter Switch Signal	9-11v (cranking)
79	BR	Power Ground	<0.1v
80	BR	Power Ground	<0.1v

80 PIN CONNECTOR

WIRE SIDE OF HARNESS TERMINALS

Pin Connector Graphic

05533_ADIA_G103

1997-98 3.0L Twin/Turbo I6 MFI VIN E (All) 40 Pin Connector

PCM Pin #	Wire Color	Circuit Description (40 Pin)	Value at Hot Idle
1	BK/OR	Ignition Switch Power	12-14v
2	PK	Speed Sensor 1 Signal	Moving: 0-5-0v
3	YL	Kickdown Switch	Closed: 12v, Closed: 0v
4	GN/WT	Stop Light Switch Signal	Brake Off: 0v, On: 12v
6	BL/BK	MIL (lamp) Control	MIL Off: 12v, On: 1v
7	RD/BK	A/T Select Switch Reverse	In 'R': 12v, Others: 0v
8	GN	Data Link Connector	No Scan Tool: 0v
9	GN/RD	A/T Select Switch 2nd	In 2nd: 12v, Others: 0v
10	GN/BK	A/T Select Switch Low	In Low: 12v, Others: 0v
11	YL/RD	ABS ECU Signal	Pulse Signals
12	BR/BK	Cruise Control ECU	At Cruise in OD: 12v
13	BR	Throttle Control ECU (ETC-)	Pulse Signals
14	YL	Throttle Control ECU (ETC+)	Pulse Signals
15	RD/YL	Idle Up Diode Signal	Load On: 12v
16	BK/WT	Tachometer Signal (TACO)	Pulse Signals
18	GN/YL	A/T Pattern Selector Switch	Norm: 0v, PWR: 12v
20	BL	Data Link Connector	12-14v
21	GN	Fuel Pump ECU	Pulse Signals
22	PK/WT	Fuel Pump ECU	Pulse Signals
23	WT/GN	A/C Magnetic Clutch (ACMG)	Clutch Off: 0v, On: 12v
24	GY	EFI Main Relay Power	12-14v
25	WT/BL	Manual Indicator Light	Light Off: 12v, On: 1v
26	WT	Traction Control ECU (EFI-)	Pulse Signals
27	BK	Traction Control ECU (EFI+)	Pulse Signals
28	PK/GN	Overdrive Main Switch	Switch Off: 12v
31	BK/RD	EFI Main Relay Power	12-14v
33	BK/YL	Battery Direct	12-14v
34	BL/RD	A/C Amplifier Signal (AC1)	Clutch On: 1.5v, Off: 12v
38	PK/BL	Traction Control ECU (NEO)	Pulse Signals
39	GN/OR	Throttle Control ECU (VTO2)	Pulse Signals
40	PK/YL	Throttle Control ECU (VTO1)	Pulse Signals

40 PIN CONNECTOR

WIRE SIDE OF HARNESS TERMINALS

05533_ADIA_G104

Pin Connector Graphic

Standard Colors and Abbreviations

Abbreviation	Color	Abbreviation	Color	Abbreviation	Color
BK	Black	GY	Gray	RD	Red
BL	Blue	GN	Green	TN	Tan
BR	Brown	LG	Light Green	VT	Violet
DB	Dark Blue	OR	Orange	WT	White
DG	DK Green	PK	Pink	YL	Yellow

1997-98 3.0L Twin/Turbo I6 MFI VIN E (All) 80 Pin Connector

PCM Pin #	Wire Color	Circuit Description (80 Pin)	Value at Hot Idle
1	YL	OD Clutch Speed Sensor (-)	Pulse Signals
3	GN	Vehicle Speed 2 Sensor (-)	Pulse Signals
6	OR	CMP Sensor Signal (G1-)	<0.050v
7	BR	CKP/CMP Sensor Ground (-)	<0.050v
9	RD/BL	A/T-ECT Solenoid (S2)	1st or OD: 1v
10	WT/RD	A/T-ECT Solenoid (S1)	In 3rd or OD: 1v
12	GN/RD	A/T-ECT Solenoid (SLT-)	Pulse Signals
13	YL/GN	A/T-ECT Solenoid (SLN-)	Pulse Signals
14	GN/BK	A/T-ECT Solenoid (SLU-)	Pulse Signals
15	RD/BK	Injector 6 Control	2.0-3.3 ms
16	RD	Injector 5 Control	2.0-3.3 ms
17	RD/WT	Injector 4 Control	2.0-3.3 ms
18	RD/GN	Injector 3 Control	2.0-3.3 ms
19	RD/YL	Injector 2 Control	2.0-3.3 ms
20	RD/BL	Injector 1 Control	2.0-3.3 ms
21	BL	OD Clutch Speed Sensor (+)	Pulse Signals
23	RD	Vehicle Speed 2 Sensor (+)	Pulse Signals
24	OR	A/T Oil Temperature Sensor	0.5-4.5v
25	BK/WT	CMP Sensor Signal (G2+)	AC pulse signals
26	WT	CMP Sensor Signal (G1+)	AC pulse signals
27	BK/RD	CKP Sensor Signal (NE+)	AC pulse signals
28	BR	MAF Sensor Ground	<0.050v
31	WT/GN	ECT Solenoid (SLT+)	Pulse Signals
32	RD/GN	ISC Motor (ISC4)	Pulse Signals
33	GN/OR	ISC Motor (ISC3)	Pulse Signals
34	GN/WT	ISC Motor (ISC2)	Pulse Signals
35	PK/YL	ISC Motor (ISC1)	Pulse Signals
38	GN/RD	Exhaust Bypass Solenoid	12v or 0v
39	GN/YL	EGR Solenoid Control (VSV)	1v (Heater On)
40	GN/BK	Intake Air Control Solenoid	12v or 0v

80 PIN CONNECTOR

WIRE SIDE OF HARNESS TERMINALS

05533_ADIA_G105

Pin Connector Graphic

1997-98 3.0L Twin/Turbo I6 MFI VIN E (All) 80 Pin Connector

PCM Pin #	Wire Color	Circuit Description (80 Pin)	Value at Hot Idle
41	BL/RD	Sensor VREF (VC)	4.9-5.1v
42	YL/BL	Sub TP Sensor (VTA2)	2.0-2.9v
43	YL	Main TP Sensor (VTA1)	0.4-1.0v
44	BL/YL	ECT Sensor Signal	At 180°F: 0.51v
45	PK/BL	IAT Sensor Signal	At 100°F: 2.60v
46	BR/YL	EGR Gas Temp. Sensor	3.5-4.0v
47	RD/BL	HO2S-12 (B1 S2) Signal	0.1-1.1v
48	WT	HO2S-11 (B1 S1) Signal	0.1-1.1v
49	WT	Knock Sensor 2 Signal	<0.075v AC
50	WT	Knock Sensor 1 Signal	<0.075v AC
51	WT/BL	Throttle (FAIL) ECU	Pulses
52	RD	Igniter Transistor 6 Control	7% duty cycle
53	BL	Igniter Transistor 5 Control	7% duty cycle
54	BK/RD	Igniter Transistor 4 Control	7% duty cycle
55	LG	Igniter Transistor 3 Control	7% duty cycle
56	WT/RD	Igniter Transistor 2 Control	7% duty cycle
57	RD/WT	Igniter Transistor 1 Control	7% duty cycle
58	RD/YL	Igniter Signal (IGF)	Digital Signal: 0-5-0v
60	BL/WT	Waste Gate Solenoid	12v or 0v
62	BK/YL	Turbo Pressure Sensor	2.5-4.5v
63	GY/RD	Closed Throttle Switch (IDL2)	1v, at off-idle: 12v
64	RD	Closed Throttle Switch (IDL1)	1v, at off-idle: 12v
65	WT/BK	Sensor Ground	<0.050v
66	YL/RD	MAF Sensor Signal	1.1-1.5v
67	PK	Throttle Signal (EFIF)	Pulse Signals
69	BR	Shield Ground	<0.050v
71	BK/BL	HO2S-11 (B1 S1) Heater	1v (Heater On)
72	BR/WT	HO2S-12 (B1 S2) Heater	1v (Heater On)
73	WT/BL	Fuel Pressure Up Solenoid	1v (at hot restart)
74	PK	EVAP Purge Solenoid (VSV)	12v or 0v

1997-98 3.0L Twin/Turbo I6 MFI VIN E (All) 80 Pin Connector, *continued*

75	PK	EGR Solenoid Control (VSV)	1v (Heater On)
76	BK/WT	A/T Neutral Start Switch	In P/N: 9-11v (cranking)
77	BK	Starter Switch Signal	9-11v (cranking)
78	BR	Power Ground	<0.1v
79	BR	Power Ground	<0.1v
80	BR	Power Ground	<0.1v

Standard Colors and Abbreviations

Abbreviation	Color	Abbreviation	Color	Abbreviation	Color
BK	Black	GY	Gray	RD	Red
BL	Blue	GN	Green	TN	Tan
BR	Brown	LG	Light Green	VT	Violet
DB	Dark Blue	OR	Orange	WT	White
DG	DK Green	PK	Pink	YL	Yellow

TERCEL PIN CHARTS

1990 1.5L I4 MFI VIN E (All) 10 Pin Connector

PCM Pin #	Wire Color	Circuit Description (10 Pin)	Value at Hot Idle
1	BK/WT	A/T Neutral Start Switch	In P/N: 9-11v (cranking)
1	BK/RD	M/T Clutch Switch	In 'N': 9-11v (cranking)
3	BK	Starter Switch Signal	9-11v (cranking)
4	WT	Injectors 1 & 3 Control	2.0-3.3 ms
5	BR	Power Ground	<0.1v
6 (Cal)	PK/BL	EGR Solenoid Control (VSV)	1v (Heater On)
7	BR	Sensor Ground	<0.050v
8	LG	Igniter Signal (IGT)	0.72-074v
9	YL	Injectors 2 & 4 Control	2.0-3.3 ms
10	BR	Power Ground	<0.1v

1990 1.5L I4 MFI VIN E (All) 14-Pin Connector

PCM Pin #	Wire Color	Circuit Description (14-Pin)	Value at Hot Idle
1	BK/OR	EFI Main Relay Power	12-14v
2	GN/RD	Battery Direct	12-14v
4	GN	Circuit Opening Relay (FC)	0-3v, off-idle: 12v
8	BK/OR	EFI Main Relay Power	12-14v
9	GN/WT	MIL (lamp) Control	MIL Off: 12v, On: 1v
11	BK/RD	A/C Magnetic Clutch (ACMG)	Clutch Off: 0v, On: 12v
12	YL/BL	Vehicle Speed Sensor	At 55 mph: 48 Hz

1990 1.5L I4 MFI VIN E (All) 18 Pin Connector

PCM Pin #	Wire Color	Circuit Description (18 Pin)	Value at Hot Idle
1	GN/BK	ECT Sensor Signal	At 180°F: 0.51v
2	PK	Vacuum Sensor Signal	1-1.5v
3	GN/BK	IAT Sensor Signal	At 100°F: 2.60v
4	PK	Data Link Connector	12-14v
5	GN/YL	Igniter Signal (IGF)	0.74-0.76v
6	RD	Distributor Signal (G1+)	AC pulse signals
7	BK	Distributor Signal (G-)	<0.050v
8	BK	O2S-11 (B1 S1) Signal	0.1-1.1v
9	GN/YL	Idle Up Solenoid Control	Load On: 12v, Off: 0v
10	BR	Sensor Ground	<0.050v
11	YL/GN	Wide Open Throttle Switch	1v, at off-idle: 12v
12	GN/RD	Vacuum Sensor VREF	4.9-5.1v
13	YL/RD	Closed Throttle Switch	1v, at off-idle: 12v
14 (Cal)	BL/RD	EGR Gas Temp. Sensor	3.5-4.0v
15	WT	Distributor Signal (NE+)	AC pulse signals
16	BR	Sensor Ground	<0.050v
17	GN/YL	Data Link Connector	12-14v
18	BL/WT	Fuel Pressure Up Solenoid	12v or 0v

10 PIN CONNECTOR   18 PIN CONNECTOR   14 PIN CONNECTOR

5	4	3	/	1
10	9	8	7	6

9	8	7	6	5	4	3	2	1
18	17	16	15	14	13	12	11	10

/	/	/	/	4	/	2	1
		12	11	/		9	8

WIRE SIDE OF HARNESS TERMINALS

05533_ADIA_G106

Pin Connector Graphic
Standard Colors and Abbreviations

Abbreviation	Color	Abbreviation	Color	Abbreviation	Color
BK	Black	GY	Gray	RD	Red
BL	Blue	GN	Green	TN	Tan
BR	Brown	LG	Light Green	VT	Violet
DB	Dark Blue	OR	Orange	WT	White
DG	DK Green	PK	Pink	YL	Yellow

1991-94 1.5L I4 MFI VIN E (All) 16 Pin Connector

PCM Pin #	Wire Color	Circuit Description (16 Pin)	Value at Hot Idle
1	BK/OR	EFI Main Relay Power B1+	12-14v
2	WT/RD	Battery Direct	12-14v
4	GN	Circuit Opening Relay (FC)	0-3v, off-idle: 12v
7 (Cal)	RD/GN	Throttle Opener Solenoid	12v or 0v
8	PK	Data Link Connector	12-14v
9	BK/OR	EFI Main Relay Power	12-14v
10	GN/WT	MIL (lamp) Control	MIL Off: 12v, On: 1v
12	PK/BK	A/C Magnetic Clutch (ACMG)	Clutch Off: 0v, On: 12v
13	YL	Vehicle Speed Sensor	At 55 mph: 48 Hz
14	GN/BK	A/C Amplifier Signal (ACT)	Clutch On: 12v, Off: 1.5v
16	GN/YL	Data Link Connector	12v

1991-94 1.5L I4 MFI VIN E (All) 26 Pin Connector

PCM Pin #	Wire Color	Circuit Description (26 Pin)	Value at Hot Idle
1	BL/WT	Fuel Pressure Up Solenoid	1v (at hot restart)
2	BK/WT	A/T Neutral Start Switch	In P/N: 9-11v (cranking)
2	BK/RD	M/T: Neutral Start Switch	In 'N': 9-11v (cranking)
3	GN/BK	ECT Sensor Signal	At 180°F: 0.51v
4	PK	MAP Sensor Signal	1-1.5v
5	GN/BK	IAT Sensor Signal	At 100°F: 2.60v
6	LG	Igniter Signal (IGT)	Digital Signal: 0-5-0v
7	GN/YL	Igniter Signal (IGF)	Digital Signal: 0-5-0v
9	BK	Distributor Signal (NE-)	<0.050v
10	BK	O2S-11 (B1 S1) Signal	0.1-1.1v
11	BK/RD	Starter Switch Signal	9-11v (cranking)
12	WT	Injectors 1 & 3 Control	2.0-3.3 ms
13	BR	Power Ground	<0.1v
14	GN/RD	Idle-Up Solenoid Control	Load On: 1v
15 (Cal)	PK/BL	EGR Solenoid Control (VSV)	1v (Heater On)
16	BR	Sensor Ground	<0.050v
17	YL/GN	TP Sensor Signal	0.3-0.8v
18	GN/RD	Sensor VREF	4.9-5.1v
19	BL/RD	Closed Throttle Switch	1v, at off-idle: 12v
21	WT	Distributor Signal (NE+)	AC pulse signals
22	BR	Sensor Ground	<0.050v
23 (Cal)	BL/RD	EGR Gas Temp. Sensor	3.5-4.0v
24	BR	Sensor Ground	<0.050v
25	WT	Injectors 2 & 4 Control	2.0-3.3 ms
26	BR	Power Ground	<0.1v

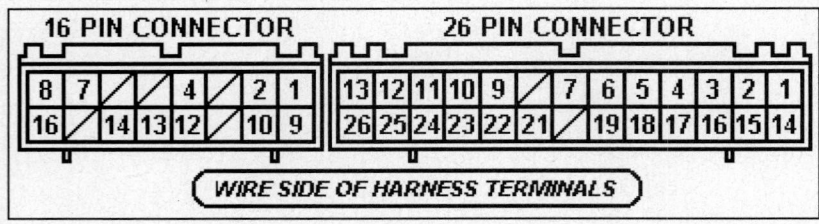

05533_ADIA_G107

Pin Connector Graphic

Standard Colors and Abbreviations

Abbreviation	Color	Abbreviation	Color	Abbreviation	Color
BK	Black	GY	Gray	RD	Red
BL	Blue	GN	Green	TN	Tan
BR	Brown	LG	Light Green	VT	Violet
DB	Dark Blue	OR	Orange	WT	White
DG	DK Green	PK	Pink	YL	Yellow

1995-96 1.5L I4 MFI VIN C, VIN E (All) 16 Pin Connector

PCM Pin #	Wire Color	Circuit Description (16 Pin)	Value at Hot Idle
2	RD/WT	Battery Direct	12-14v
3	BK	Overdrive Main Switch	Switch Off: 12v, On: 0v
4	GN/BK	Circuit Opening Relay (FC)	0-3v, off-idle: 12v
5	WT/RD	HO2S-12 (B1 S2) Heater	1v (Heater On)
6	WT/BK	HO2S-11 (B1 S1) Heater	1v (Heater On)
8	GN	Data Link Connector	12v
9	BK/RD	EFI Main Relay Power	12-14v
10	GN/BL	MIL (lamp) Control	MIL Off: 12v, On: 1v
11	RD	Overdrive Solenoid	Solenoid Off: 12v, On: 1v
12	BL	A/C Amplifier Signal (AC1)	Clutch On: 1.5v, Off: 12v
13	YL	Vehicle Speed Sensor	At 55 mph: 48 Hz
14	GN/BK	A/C Amplifier Signal (ACT)	Clutch On: 12v, Off: 1.5v
15	BK/WT	EGR Solenoid Control (VSV)	1v (Heater On)
16	WT	Data Link Connector (SDL)	0v

1995-96 1.5L I4 MFI VIN C, VIN E (All) 26 Pin Connector

PCM Pin #	Wire Color	Circuit Description (26 Pin)	Value at Hot Idle
1	BL/RD	IAC Signal (RSC)	Pulse Signals
2	BK	Neutral Start Switch	In P/N: 9-11v (cranking)
3	GN/BK	ECT Sensor Signal	At 180°F: 0.51v
4	PK	MAP Sensor Signal	1-1.5v
5	BL/YL	Igniter Transistor 2 Control	Digital Signal: 0-5-0v
6	LG	Igniter Transistor 1 Control	Digital Signal: 0-5-0v
7	GN/BK	Igniter Signal (IGF)	Digital Signal: 0-5-0v
8	BK	Knock Sensor Signal	<0.075v AC
9	WT	CMP Sensor Signal (G+)	AC pulse signals
10	BK	HO2S-11 (B1 S1) Signal	0.1-1.1v
11	BK/WT	Starter Switch Signal	9-11v (cranking)
12	GN	Injectors 1 & 3 Control	2.0-3.3 ms
13	BR	Power Ground	<0.1v
14	BL	IAC Signal (RSO)	Pulse Signals
15	BL/BK	IAT Sensor Signal	At 100°F: 2.60v
16	BR	Sensor Ground	<0.050v
17	YL/GN	TP Sensor Signal	0.3-0.8v
18	GN/RD	Sensor VREF	4.9-5.1v
19	YL/BK	Closed Throttle Switch	1v, at off-idle: 12v
20	RD	CMP Sensor Signal (G-)	AC pulse signals
21	BK	CKP Sensor Signal (NE+)	AC pulse signals
23	WT	HO2S-12 (B1 S2) Signal	0.1-1.1v
24	BR	Shield Ground	<0.050v
25	YL	Injectors 2 & 4 Control	2.0-3.3 ms
26	BR	Power Ground	<0.1v

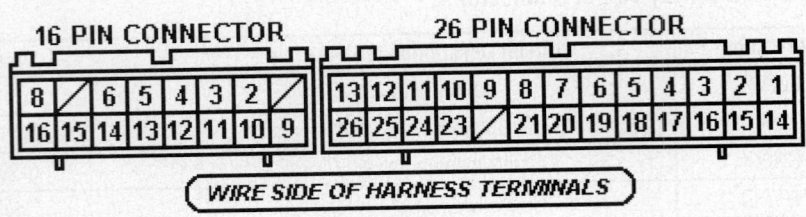

05533_ADIA_G108

Pin Connector Graphic
Standard Colors and Abbreviations

Abbreviation	Color	Abbreviation	Color	Abbreviation	Color
BK	Black	GY	Gray	RD	Red
BL	Blue	GN	Green	TN	Tan
BR	Brown	LG	Light Green	VT	Violet
DB	Dark Blue	OR	Orange	WT	White
DG	DK Green	PK	Pink	YL	Yellow

1997-99 1.5L I4 MFI VIN C (All) 12 Pin Connector

PCM Pin #	Wire Color	Circuit Description (12 Pin)	Value at Hot Idle
2	WT/RD	Battery Direct	12-14v
4	GN/BK	Circuit Opening Relay (FC)	1v, at off-idle: 12v
6	BL	A/C Amplifier Signal (ACT)	Clutch On: 12v, Off: 1.5v
7	BK/RD	EFI Main Relay Power	12-14v
8	GY/BL	MIL (lamp) Control	MIL Off: 12v, On: 1v
10	GN/BK	A/C Amplifier Signal (AC1)	Clutch On: 1.5v, Off: 12v
11	YL	Vehicle Speed Sensor	At 55 mph: 48 Hz

1997-99 1.5L I4 MFI VIN C (All) 26 Pin Connector

PCM Pin #	Wire Color	Circuit Description (26 Pin)	Value at Hot Idle
1	GY	Igniter Transistor 1 Control	Digital Signal: 0-5-0v
2	BK/WT	Starter Switch Signal	9-11v (cranking)
4	BK	CKP Sensor Signal (NE+)	AC pulse signals
7	RD/BL	Igniter Signal (IGF)	Digital Signal: 0-5-0v
9 (97)	BL/RD	IAC Signal (RSC)	Pulse Signals
10 (97)	BL	IAC Signal (RSO)	Pulse Signals
10 (98-'99)	BL	IAC Signal (RSD)	Pulse Signals
11	RD	Injector 3 Control	2.0-3.3 ms
12	GN	Injector 1 Control	2.0-3.3 ms
13	BR	Power Ground	<0.1v
14	BL/YL	Igniter Transistor 2 Control	Digital Signal: 0-5-0v
15	BK	A/T Neutral Start Switch	In P/N: 9-11v (cranking)
15	BK/WT	M/T: Neutral Start Switch	In 'N': 9-11v (cranking)
17	GN	CKP/CMP Sensor Ground (-)	<0.050v

1997-99 1.5L I4 MFI VIN C (All) 26 Pin Connector, *continued*

18	RD	CMP Sensor Signal (G+)	AC pulse signals
21	WT/RD	HO2S-12 (B1 S2) Heater	1v (Heater On)
24	BL	Injector 4 Control	2.0-3.3 ms
25	YL	Injector 2 Control	2.0-3.3 ms
26	BR	Shield Ground	<0.050v

1997-99 1.5L I4 MFI VIN C (All) 16 Pin Connector

PCM Pin #	Wire Color	Circuit Description (16 Pin)	Value at Hot Idle
2	PK	MAP Sensor Signal	1-1.5v
3	BL/BK	IAT Sensor Signal	At 100°F: 2.60v
4	BK/RD	ECT Sensor Signal	At 180°F: 0.51v
5	WT	HO2S-12 (B1 S2) Signal	0.1-1.1v
6	BK	O2S-11 (B1 S1) Signal	0.1-1.1v
7	BL/WT	EVAP Purge Solenoid (VSV)	12v or 0v
8	GN/YL	EVAP Vapor Pressure (VSV)	12v or 0v
9	BR	Sensor Ground	<0.050v
10	YL/GN	TP Sensor Signal	0.3-0.8v
11	GN/RD	Sensor VREF	4.9-5.1v
12	YL/BK	EVAP Vapor Pressure Sensor	12v or 0v
13	WT	SIL Signal (Scan Tool)	Digital Signal
14	BK	Knock Sensor Signal	<0.075v AC
15	GN	Data Link Connector (TE)	12-14v
16	BR	Power Ground	<0.1v

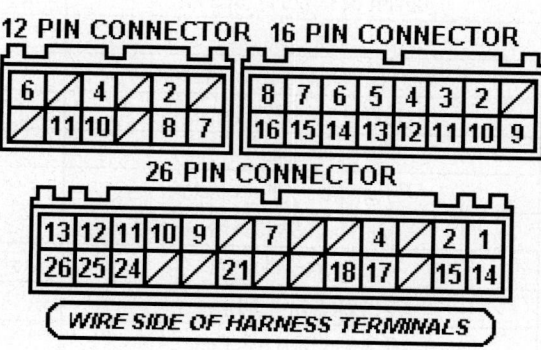

05533_ADIA_G109

Pin Connector Graphic

YARIS PIN CHARTS

2006 1.5L I4 DOHC MFI VIN T (All) E4 31 Pin Connector

PCM Pin #	Wire Color	Circuit Description (31 Pin)	Value at Hot Idle
1	BK/RD	EFI Main Relay Power	12-14v
2	BL/BK	Maximum Hot Switch (MHSW)	Switch Open: 0v, Closed: 12v
3	BK/YL	Battery Direct	12-14v
4	---	Not Used	---
5	BK	Tachometer Signal (TACO)	Pulse Signals
6	YL/RD	Cooling Fan 1 Relay (CF)	Relay Off: 12v, On: 1v
7	WT/BL	A/C Single Pressure Switch (FAN)	Switch Open: 12v, Closed: 1v
8-9	---	Not Used	---
10	BK	Circuit Opening Relay (FC)	0-3v, off-idle: 12v
11	YL/RD	MIL (lamp) Control	MIL Off: 12v, On: 1v
12	BL/WT	Heater Sub1 Relay Control (ELS)	Relay Off: 12v, On: 1v
13	---	Not Used	---
14	YL/BK	Center Airbag Assembly	Digital Signals
15-17	---	Not Used	---
18	LG/BK	Scan Tool Signal (SIL)	Digital Signal
19	RD	Scan Tool (WFSE)	12v
20	PK/BL	Data Link Connector (TC)	12-14v
21	BK	EVAP Vapor Pressure Sensor	Fuel Cap Off: 2.9-3.7v
22-31	---	Not Used	---

2006 1.5L I4 DOHC MFI VIN T (All) E5 35 Pin Connector

PCM Pin #	Wire Color	Circuit Description (35 Pin)	Value at Hot Idle
1-2	---	Not Used	---
3	YL/RD	Water Temperature Cool Indicator	Indicator On: 12v, Off: 0v
4	BK/RD	HO2S-12 (B1 S2) Heater	1v (Heater On)
5-6	---	Not Used	---
7	GN/OR	A/T Overdrive "Off" Indicator	Light Off: 12v
8	GN/WT	A/T Select Switch Low	In Low: 12v, Others: 0v
9	GN	A/T Select Switch 2nd	In 2nd: 12v, Others: 0v
10	RD/YL	A/T Select Switch Drive	In 'D': 12v, Others: 0v
11	RD/WT	A/T Select Switch Reverse	In 'R': 12v, Others: 0v
12	---	Not Used	---
13	YL	PSP Switch Signal	Straight: 12v, Turning: 0v
14-16	---	Not Used	---
17	VT/WT	Speed Signal from Combination Meter	At 55 mph: 48 Hz
18	---	Not Used	---
19	GN/WT	Stop Light Switch Signal	Brake Off: 0v, On: 12v
20-27	---	Not Used	---
28	YL/RD	Water Temperature Hot Indicator	Indicator On: 12v, Off: 0v

2006 1.5L I4 DOHC MFI VIN T (All) E5 35 Pin Connector, *continued*

29	GN/OR	A/T: Overdrive Main Switch	Switch Off: 12v
30	---	Not Used	---
31	BK/WT	A/C Amplifier Signal (AC)	Clutch On: 1.5v, Off: 12v
32	---	Not Used	---
33	BK	A/C Amplifier Signal (ACT)	Clutch On: 12v, Off: 1.5v
34-35	---	Not Used	---

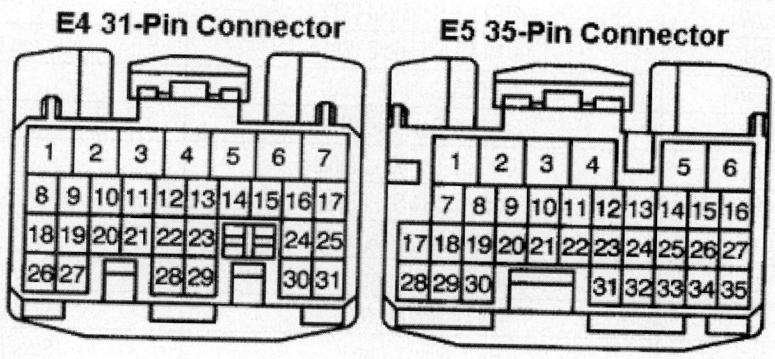

05533_ADIA_G066

Pin Connector Graphic

2006 1.5L I4 DOHC MFI VIN T (All) E6 34 Pin Connector

PCM Pin #	Wire Color	Circuit Description (34 Pin)	Value at Hot Idle
1	BK/OR	Injector 1 Control	2.0-3.3 ms
2	BK/YL	Injector 2 Control	2.0-3.3 ms
3	BK/WT	Injector 3 Control	2.0-3.3 ms
4	BK/BL	Injector 4 Control	2.0-3.3 ms
5	BK/RD	IAC Signal (RSD)	Pulse Signals
6	BR	Power Ground (E02)	<0.1v
7	BR	Power Ground (E01)	<0.1v
8	GN/RD	Igniter Transistor 1 Control	Digital Signal: 0-5-0v
9	GN/BK	Igniter Transistor 2 Control	Digital Signal: 0-5-0v
10	GN/OR	Igniter Transistor 3 Control	Digital Signal: 0-5-0v
11	GN/YL	Igniter Transistor 4 Control	Digital Signal: 0-5-0v
12	WT/GN	EVAP Purge Solenoid (VSV)	12v or 0v
16	WT	A/T-ECT Solenoid (SLT-)	Pulse Signals
17	WT/BL	A/T-ECT Solenoid (SLT+)	Pulse Signals
18	RD/WT	Sensor VREF (VC)	4.9-5.1v
19	RD/BL	ECT Sensor Signal (THW)	1-4v (varies with temp.)
20	YL/BK	IAT Sensor Signal	1-4v (varies with temp.)
21	YL/RD	TP Sensor Signal (VTA)	1.1-1.9v
23	YL	Igniter Signal (IGF)	Pulse Signal: 0-5-0v
26	BK	CMP Sensor Signal (G2+)	AC pulse signals
27	OR	CKP Sensor Signal (NE+)	AC pulse signals
28	BR	Sensor Ground (E2)	<0.050v
30	YL/GN	A/T Oil Temperature Sensor (OIL)	At 230°F: 0.95v
34	WT	CKP Sensor Ground (NE-)	<0.050v

2006 1.5L I4 DOHC MFI VIN T (All) E7 35 Pin Connector

PCM Pin #	Wire Color	Circuit Description (35 Pin)	Value at Hot Idle
1	WT	Knock Sensor Signal (KNK1)	<0.075v AC
4	BK/RD	HO2S-11 (B1 S1) Heater	1v (Heater On)
5	BR	Power Ground (E03)	<0.1v
7	BL	Generator Control Signal	12-0-12v
8	BK	A/T Neutral Start Switch (NSW)	Cranking: 9-11v
9	BK/YL	Starter Switch Signal (STA)	9-11v (cranking)
12	BK/WT	A/T-ECT Solenoid (ST)	Pulse Signals
13	GN	A/T-ECT Solenoid (SL)	In Lockup: 12-14v
14	WT/RD	A/T: ECT Solenoid (S2)	1st or OD: 1v
15	WT/GN	A/T-ECT Solenoid (S1)	In 3rd or OD: 1v
21	BK	HO2S-12 (B1 S2) Signal	0.1-1.1v
23	WT	HO2S-11 (B1 S1) Signal	0.1-1.1v
24	PK	MAF Sensor Signal (VG)	1.1-1.5v
27	RD	Turbine Speed Sensor (NT+)	AC pulse signals
28	BR	Power Ground (EC)	<0.1v
29	VT	PSP Switch Signal	Straight: 12v, Turning: 0v
32	VT	MAF Sensor Ground (EVG)	<0.050v
35	BK	Turbine Speed Sensor (NT-)	AC pulse signals

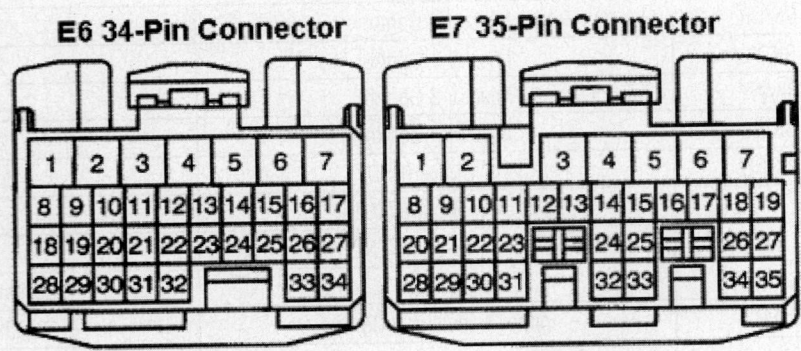

E6 34-Pin Connector   **E7 35-Pin Connector**

05533_ADIA_G067

Pin Connector Graphic

HIGHLANDER PIN CHARTS

2001-06 2.4L I4 VIN D 2AZ-FE (A/T) E5 22 Pin Connector

PCM Pin #	Wire Color	Circuit Description (22 Pin)	Value at Hot Idle
1	BK/WT	Direct Battery	12-14v
2	BK/OR	Ignition Switch Power (IGSW) (IGSW)	12-14v
3	GN/RD	Circuit Opening Relay (FC)	0-3v, at off-idle: 12v
4	WT	SIL Signal (Scan Tool)	Digital Signals
5	BK/RD	O/D OFF Indicator (ODLP)	LED Off: 12v, On: 1v
6	BL/RD	Malfunction Indicator Lamp Control	MIL Off: 12v, On: 1v
7	YL/GN	A/C Magnetic Clutch Relay (ACMG)	Relay Off: 0v, On: 12v
8	PK/BL	EFI Main Relay Control	Relay Off: 0v, On: 12v
9	VT	ABS/Traction ECU (NEO)	Pulse Signals
10-11	---	Not Used	---
12	YL	Body Control ECU Signal (MPX1)	Digital Signals
13	GN/BK	Traction Control Signal (TRC+)	Pulse Signals
14	WT/BL	Traction Control Signal (ENG+)	Pulse Signals
15	GN/WT	Stop Light Switch Signal	Brake Off: 0v, On: 12v
16	WT	EFI Main Relay B+	12-14v
17	WT/BL	Data Link Connector (WFSE)	12v
18	---	Not Used	---
19	BL/WT	Sliding Roof ECU Signal (MPX2)	12v or 0v
20	YL/BK	Traction Control Signal (TRC-)	Pulse Signals
21	WT/GN	Traction Control Signal (ENG-)	Pulse Signals
22	WT/BK	Power Ground (EOM)	<0.1v

2001-06 2.4L I4 VIN D 2AZ-FE (A/T) E6 28 Pin Connector

PCM Pin #	Wire Color	Circuit Description (28 Pin)	Value at Hot Idle
1	RD/BL	Fan Control Relay 3 Control	Relay Off: 12v, On: 1v
2	RD/BK	A/T Select Switch Reverse	In 'R': 12v, Others: 0v
3-4	---	Not Used	---
5	PK/BK	Data Link Connector 3 Signal (TC)	12v
6	GN/YL	EVAP Vapor Pressure Valve (VSV)	12v or 0v
7	BK/OR	OD Main Switch (ODMS)	Switch Off: 12v, On: 1v
8-9	---	Not Used	---
10	GN/WT	A/T Select Switch Park	In 'P': 12v, Others: 0v
11	RD/WT	A/T Select Switch Neutral	In 'N': 12v, Others: 0v
12	GN/BK	Fan Control Relay 2 Control (PR2)	Relay Off: 12v, On: 1v
13-14	---	Not Used	---
15	RD/YL	Fan Control Relay 1 Control (FAN)	Relay Off: 12v, On: 1v
16-17	---	Not Used	---
18	GY	Transponder Amplifier Signal (RXCK)	Inserting key: pulses
19	PK	Transponder Amplifier Signal (TXCK)	Inserting key: pulses
20	BL/WT	A/T Neutral Start Signal (NSW)	In P/N: 0-3.0v
21	BL/BK	EVAP Vapor Pressure Sensor (PTNK)	2.5-3.1v (hose off)
22	WT/RD	Speedometer Indicator (SPD)	At 55 mph: 48 Hz
23	BL	Unlock Warning Switch	Inserting key: pulses
24	WT	Cruise Control Signal (OD1)	At Cruise in OD: 12v
25	PK/BL	Cruise Control ECU (IDLO)	1.5v, off-idle: 12v
26	LTGN	Theft Deterrent Indicator Control (IMLD)	LED Off: 12v, On: 1v
27	WT/GN	Tachometer Signal (TACO)	DC pulse signals
28	GY/BL	Transponder Amplifier Signal (Code)	Inserting key: pulses

2001-06 2.4L I4 VIN D 2AZ-FE (A/T) E7 17 Pin Connector

PCM Pin #	Wire Color	Circuit Description (17 Pin)	Value at Hot Idle
1	GN/RD	A/T Select Switch Drive	In 'D': 12v, Others: 0v
2	PK	EVAP Canister Closed Valve (CCV)	12v or 0v
3	---	Not Used	---
4	YL	Generator Control (RL)	12v
5	YL/GN	Engine Oil Pressure Switch (MOPS)	Open: 12v, Closed: 0v
6	RD/BL	A/T-ECT Solenoid DLS	Driving in Lockup: 12v
7	BL/YL	A/T Select Switch Second	In 2nd: 12v, Others: 0v
8	GY/BL	Center Airbag Sensor Assembly (F/PS)	Digital Signals
9-10	---	Not Used	---
11	GN/YL	PSP Switch Signal (PSW)	Straight: 12v, Turning: 0v
12	RD/YL	A/T-ECT Solenoid (S4)	Driving in OD: 12v
13	BL/WT	A/T Select Switch Low	In Low: 12v, Others: 0v
14	BK	HO2S-12 (B1 S2) Heater (HT2B)	Heater Off: 12v, On: 1v
15	WT	HO2S-12 (B1 S2) Signal (OX2B)	0.1-1.1v
16	BL	Starter Switch Signal (STA)	In P/N: 0-3.0v
17	---	Not Used	---

2001-06 2.4L I4 VIN D 2AZ-FE (A/T) E8 24 Pin Connector

PCM Pin #	Wire Color	Circuit Description (24 Pin)	Value at Hot Idle
1	WT/BK	Power Ground (E04)	<0.1v
2	BL/RD	Sensor VREF (VC)	4.9-5.1v
3	RD/BL	HO2S-22 (B2 S2) Heater (HT1B)	1v (Heater on)
4	BK/RD	AFS-21 (B1 S1) Heater (HAF2A)	1v (Heater on)
5	BK	AFS-21 (B2 S1) Heater (HAF1A)	1v (Heater on)
6	BK/RD	EVAP Purge Solenoid (VSV)	12v or 0v
7	---	Not Used	---
8	WT/BK	Power Ground (E05)	<0.1v
9-10	---	Not Used	---
11	BK	HO2S-22 (B2 S2) Signal (OX1B)	0.1-1.1v
12	YL/GN	MAF Sensor Signal (VG)	1.1-1.8v
13	YL	AFS-21 (B2 S1) Signal (AF2+)	Fixed at 3.3v
14	GN	AFS-11 (B1 S1) Signal (+)	Fixed at 3.3v
15	YL	CMP Sensor Signal (G22+)	AC pulse signals
16	PK	CKP Sensor Signal (NE+)	AC pulse signals
17	BR	Power Ground (E1)	<0.050v
18	WT	Sensor Ground (E2)	<0.050v
19-20	---	Not Used	---
21	RD/BK	MAF Sensor Ground (E2G)	<0.050v
22	BR	AFS-21 (B2 S1) Signal (AF2-)	Fixed at 3.3v
23	RD	AFS-11 (B1 S1) Signal (AF1A-)	Fixed at 3.3v
24	BL	CKP/CMP Sensor Signal (-)	<0.050v

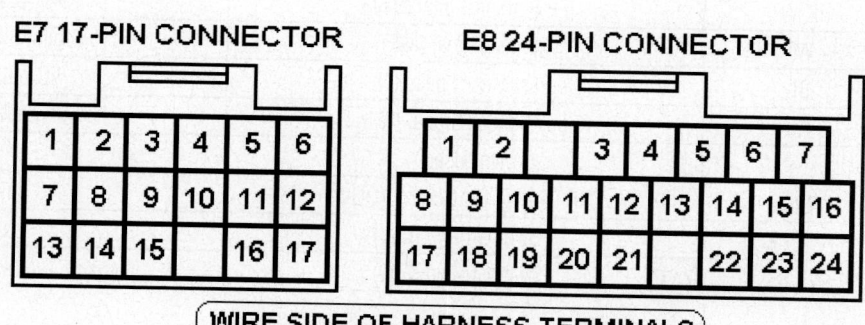

Pin Connector Graphic

05533_ADIA_G110

2001-06 2.4L I4 VIN D 2AZ-FE (A/T) E9 31 Pin Connector

PCM Pin #	Wire Color	Circuit Description (31 Pin)	Value at Hot Idle
1	RD/BL	Injector 1 Control	2.0-3.3 ms
2	YL	Injector 2 Control	2.0-3.3 ms
3	BK/WT	Injector 3 Control	2.0-3.3 ms
4	BL	Injector 4 Control	2.0-3.3 ms
5	BK/RD	A/T-ECT Solenoid (SLT-)	Pulse Signals
6	GN/YL	A/T-ECT Solenoid (SLT+)	Pulse Signals
7	BL/BK	A/T-ECT Solenoid (SL1+)	Pulse Signals
8	RD/BK	A/T-ECT Solenoid (SL2+)	Pulse Signals
9	BL/WT	A/T-ECT Solenoid (SL1-)	Pulse Signals
10	BK/YL	Igniter Transistor 1 Control	6°, 55 mph: 8° dwell
11	YL/GN	Igniter Transistor 2 Control	6°, 55 mph: 8° dwell
12	BL/YL	Igniter Transistor 3 Control	6°, 55 mph: 8° dwell
13	BL/RD	Igniter Transistor 4 Control	6°, 55 mph: 8° dwell
14	WT/RD	Counter Gear Speed (NC+)	AC pulse signals
15	LG	Turbine Speed Sensor (NT-)	AC pulse signals
16	PK	Turbine Speed Sensor (NT+)	AC pulse signals
17	GN/YL	A/T Oil Temperature Signal	At 230°F: <1.5v
18	GN/BK	Idle Air Control Valve (RSD)	Pulse Signals
19	GN/WT	Cam Timing Oil Control (OCV+)	AC pulse signals
20	RD/WT	A/T-ECT Solenoid (SL2-)	At 68°F: 4-5v
21	WT/BK	Power Ground (E01)	<0.1v
22	WT/BL	ECT Sensor Signal (THW)	At 180°F: 0.51v
23	RD/YL	IAT Sensor Signal (THA)	At 100°F: 2.60v
24	BK/WT	TP Sensor Signal (VTA)	0.3-0.8v
25	WT	Igniter Signal (IGF)	Digital Signal: 0-5-0v
26	BL	Counter Gear Speed (NC-)	AC pulse signals
27	WT	Knock Sensor Signal (KNK1)	<0.075v AC
28	---	Not Used	---
29	YL/RD	Cam Timing Oil Control (OCV-)	AC pulse signals
30	WT/BK	Power Ground (E03)	<0.1v
31	WT/BK	Power Ground (E02)	<0.1v

Pin Connector Graphic

05533_ADIA_G111

2001-06 3.0L V6 1MZ-FE VIN F (A/T) E5 22 Pin Connector

PCM Pin #	Wire Color	Circuit Description (22 Pin)	Value at Hot Idle
1	BK/WT	Battery Direct	12-14v
2	BK/OR	Ignition Switch Power (IGSW)	12-14v
3	GN/RD	Circuit Opening Relay (FC)	0-3v, off-idle: 12v
4	WT	SIL Signal (Scan Tool)	Digital Signal
5	GN/YL	EVAP Vapor Pressure Valve (VSV)	12v or 0v
6	BL/RD	Malfunction Indicator Lamp Control	MIL Off: 12v, On: 1v
7	BL	Starter Switch Signal	In P/N: 0-3.0v
8	PK/BL	EFI Main Relay Control	Relay Off: 0v, On: 12v
9	BK/RD	Overdrive "Off" LED Indicator	LED Off: 12v, On: 1v
10	PK	EVAP Canister Closed Valve (CCV)	12v or 0v
11	YL	Multiplex Meter Input (MPX1)	Digital Signals
12	GY/BL	Center Airbag Sensor Assembly (F/PS)	12v or 0v
13	GN/BK	ABS/Traction Signal (TRC+)	Pulse Signals
14	WT/BL	ABS/Traction Signal (ENG+)	Pulse Signals
15	GN/WT	Stop Light Switch Signal	Brake Off: 0v, On: 12v
16	WT	EFI Main Relay Power	12-14v
17	BL/BK	EVAP Vapor Pressure Sensor (PTNK)	2.9-3.7v (hose off)
18	BL/WT	Multiplex Meter Input (MPX2)	Digital Signals
19	---	Not Used	---
20	YL/BK	ABS/Traction Signal (TRC-)	Pulse Signals
21	WT/GN	ABS/Traction Signal (ENG-)	Pulse Signals
22	LG	Theft Deterrent Control (IMLD)	LED Off: 12v, On: 1v

2001-06 3.0L V6 1MZ-FE VIN F (A/T) E6 28 Pin Connector

PCM Pin #	Wire Color	Circuit Description (28 Pin)	Value at Hot Idle
1	WT/BK	Power Ground (EOM)	<0.1v
2	---	Not Used	---
3	GN/YL	EVAP Vapor Pressure Valve (VSV)	12v or 0v
4	---	Not Used	---
5	PK/BK	Data Link Connector Signal (D3)	12v
6	YL/GN	A/C Magnetic Clutch Relay (ACMG)	Relay Off: 0v, On: 12v
7	---	Not Used	---
8	BK	HO2S-12 (B1 S2) Signal (OXS)	0.1-1.1v
9	BL/RD	HO2S-12 (B1 S2) Heater (HTS)	1v (Heater on)
10-14	---	Not Used	---
15	YL	Generator Control Signal (RL)	12v
16	VT	ABS/Traction ECU (NEO)	Pulse Signals
17	---	Not Used	---
18	GY	Transponder Amplifier Signal (TXCK)	Inserting key: pulses
19	PK	Transponder Amplifier Signal (RXCK)	Inserting key: pulses
20	BL/WT	A/T: Neutral Start Switch (NSW)	In P/N: 0-3.0v
21	---	Not Used	---
22	WT/RD	Speedometer Indicator (SPD)	At 55 mph: 48 Hz
23	BL	Unlock Warning Switch (KSW)	Inserting key: pulses
24	WT	Cruise Control Signal (OD1)	At Cruise in OD: 12v
25	PK/BL	Cruise Control Signal (IDLO)	1.5v, off-idle: 12v
26	RD/BL	Cooling Fan Relay 1 or 2 (CF)	Relay Off: 12v, On: 1v
27	WT/GN	Tachometer Signal (TACO)	Pulse Signals
28	GY/BL	Transponder Amplifier Signal (Code)	Inserting key: pulses

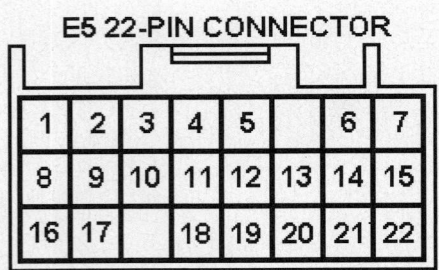

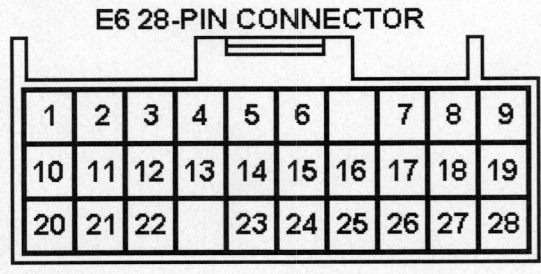

05533_ADIA_G112

Pin Connector Graphic

2001-06 3.0L V6 1MZ-FE VIN F (A/T) E7 17 Pin Connector

PCM Pin #	Wire Color	Circuit Description (17 Pin)	Value at Hot Idle
1	RD/YL	A/T-ECT Solenoid (S4)	Driving in OD: 12v
2	BR/RD	A/T ECT Solenoid (SLT-)	Pulse Signals
3	GN/YL	A/T ECT Solenoid (SLT+)	Pulse Signals
4	GN	Turbine Speed Shaft Signal (NT-)	AC pulse signals
5	PK	Turbine Speed Shaft Signal (NT+)	AC pulse signals
6	---	Not Used	---

2001-06 3.0L V6 1MZ-FE VIN F (A/T) E7 17 Pin Connector, *continued*

7	RD/WT	A/T Select Switch Neutral	In 'N': 12v, Others: 0v
8	RD/BK	A/T Select Switch Reverse	In 'R': 12v, Others: 0v
9	RD/YL	A/T Select Switch Park	In Park: 12v, Others: 0v
10	---	Not Used	---
11	BK	Air Conditioning Signal (VSV)	Pulse Signals
12	---	Not Used	---
13	BL/RD	A/T Select Switch Low	In Low: 12v, Others: 0v
14	BL/YL	A/T Select Switch Second	In 2nd: 12v, Others: 0v
15	---	Not Used	---
16	GN/BK	A/T Select Switch Drive	In 'D': 12v, Others: 0v
17	YL/GN	Engine Oil Pressure Switch (MOPS)	Open: 12v, Closed: 0v

2001-06 3.0L V6 1MZ-FE VIN F (A/T) E8 24 Pin Connector

PCM Pin #	Wire Color	Circuit Description (24 Pin)	Value at Hot Idle
1	WT/BK	Sensor Ground (E04)	<0.050v
2	BL/RD	Sensor VREF (VC)	4.9-5.1v
3	BK/RD	AFS-11 (B1 S1) Heater (HAFR)	1v (Heater on)
4	BK	AFS-21 (B2 S1) Heater (HAFL)	1v (Heater on)
5	RD/BL	Injector 1 Control	2.0-3.3 ms
6	YL	Injector 2 Control	2.0-3.3 ms
7	BK/RD	EVAP Purge Solenoid (VSV)	12v or 0v
8	WT/BK	Power Ground (E05)	<0.1v
9	GN/YL	PSP Switch Signal (PSW)	Straight: 12v, Turning: 0v
10	YL/GN	MAF Sensor Signal (VG)	1.1-1.5v
11	BL	AFS-11 (B1 S1) signal (AFR+)	3.0-3.6v
12	GN	AFS-21 (B2 S1) Signal (AFL+)	3.0-3.6v
13	GN/RD	ATF Oil Temperature Signal	At 68°F: 4-5v
14	WT/BL	ECT Sensor Signal (THW)	At 180°F: 0.51v
15	WT/RD	ACIS Solenoid 2 Control (VSV)	12v or 0v
16	PK	CKP Sensor Signal (NE+)	AC pulse signals
17	BR	Power Ground (E1)	<0.1v
18	WT	Sensor Ground (E2)	<0.050v
19	RD/BK	MAF Sensor Ground (E2G)	<0.050v
20	YL	AFS-11 (B1 S1) Signal (AFR-)	Fixed at 3.3v
21	RD	AFS-21 (B2 S1) Signal (AFL-)	Fixed at 3.3v
22	RD/YL	IAT Sensor Signal (THA)	At 100°F: 2.60v
23	BK/WT	TP Sensor Signal (VTA)	0.3-1.1v
24	BL	VVT Sensor (-) Ground	<0.050v

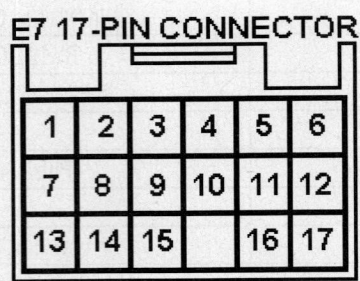

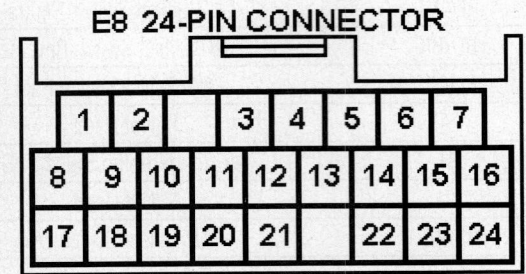

05533_ADIA_G113

Pin Connector Graphic

2001-06 3.0L V6 1MZ-FE VIN F (A/T) E9 31 Pin Connector

PCM Pin #	Wire Color	Circuit Description (31 Pin)	Value at Hot Idle
1	BK/WT	Injector 3 Control	2.0-3.3 ms
2	BL	Injector 4 Control	2.0-3.3 ms
3	BK/RD	Injector 5 Control	2.0-3.3 ms
4	GN	Injector 6 Control	2.0-3.3 ms
5	GN/YL	Cam Timing Oil Valve (OC1-)	Pulse signals
6	WT/GN	Cam Timing Oil Valve (OC1+)	Pulse signals
7	RD/BL	A/T-ECT Solenoid (DSL)	In Lockup: 12-14v
8	RD/WT	A/T ECT Solenoid (SL2-)	During shifting: 12v
9	RD/BK	A/T ECT Solenoid (SL2+)	During shifting: 12v
10	PK	VVT Sensor Left Hand (VV2-)	Pulse signals
11	BK/YL	Igniter Transistor 1 Control	7% duty cycle
12	YL/GN	Igniter Transistor 2 Control	7% duty cycle
13	BL/YL	Igniter Transistor 3 Control	7% duty cycle
14	BL/RD	Igniter Transistor 4 Control	7% duty cycle
15	YL	Igniter Transistor 5 Control	7% duty cycle
16	GN/RD	Igniter Transistor 3 Control	7% duty cycle
17	RD/YL	ACIS 1 Solenoid Control (VSV)	12v or 0v
18	YL/RD	Cam Timing Oil Valve (OC2-)	Pulse signals
19	BL/WT	A/T ECT Solenoid (SL1-)	Pulse Signals
20	BL/BK	A/T ECT Solenoid (SL1+)	Pulse Signals
21	WT/BK	Power Ground (E01)	<0.1v
22	RD	VVT Sensor Left Hand (VV2+)	Pulse signals
23	BL	Counter Gear Speed (NC-)	Pulse signals
24	WT/RD	Counter Gear Speed (NC+)	Pulse signals
25	WT	Igniter Signal (IGF)	Digital Signal: 0-5-0v
26	GN/BK	Idle Air Control Valve (RSO)	Pulse Signals
27	WT	Knock Sensor 1 Signal (KNKR)	<0.075v AC
28	BK	Knock Sensor 2 Signal (KNKL)	<0.075v AC
29	GN/WT	Cam Timing Oil Valve (OC2+)	Pulse signals
30	WT/BK	Power Ground (E03)	<0.1v
31	WT/BK	Power Ground (E02)	<0.1v

E9 31-PIN CONNECTOR

1	2		3	4	5	6	7	8	9
10	11	12	13	14	15	16	17	18	19	20	21
22	23	24	25	26		27	28		29	30	31

WIRE SIDE OF HARNESS TERMINALS

Pin Connector Graphic

05533_ADIA_G114

1990 4.0L I6 MFI VIN F (A/T) 10 Pin Connector

PCM Pin #	Wire Color	Circuit Description (10 Pin)	Value at Hot Idle
1	BL	Check Connector	12-14v
2	BL/RD	Cold Start Injector Control	1v (at cold startup)
3	GY	HO2S-11 (B1 S1) Heater	1v (Heater on)
4	WT/BL	Injectors 1 & 2 & 3 Control	2.0-3.3 ms
5	BR	Power Ground	<0.1v
6	PK	Water Temperature Switch	Open: 12v, Closed: 0v
7	BR	Sensor Ground	<0.050v
8	OR	HO2S-12 (B1 S2) Heater	1v (Heater on)
9	YL	Injectors 4 & 5 & 6 Control	2.0-3.3 ms
10	BR	Power Ground	<0.1v

1990 4.0L I6 MFI VIN F (A/T) 18 Pin Connector

PCM Pin #	Wire Color	Circuit Description (18 Pin)	Value at Hot Idle
1	RD/WT	ECT Sensor Signal (THW)	At 180°F: 0.51v
2	BL/WT	HO2S-12 (B1 S2) Signal	0.1-1.1v
3 (Cal)	GN/YL	EGR Gas Temperature Sensor	3.5-4.0v
4	GN/WT	Distributor Signal (NE+)	AC pulse signals
6	GN/RD	Distributor Signal (G+)	AC pulse signals
7	GN/BK	Distributor Signal (G-)	AC pulse signals
8	RD/BK	ISC Signal (ISC2)	Pulse Signals
9	BL/BK	ISC Signal (ISC1)	Pulse Signals
10	BR	Sensor Ground	<0.050v
11	WT	HO2S-11 (B1 S1) Signal	0.1-1.1v
12	BR/BK	Sensor Ground	<0.050v
13	GN/WT	Closed Throttle Switch	1v, off-idle: 12v
14	WT/RD	TP Sensor Signal (VTA)	0.3-0.8v
15	BK/WT	Check Connector	12-14v
16	GN	Check Connector	12-14v
17	BL/YL	ISC Signal (ISC4)	Pulse Signals
18	YL	ISC Signal (ISC3)	Pulse Signals

Standard Colors and Abbreviations

Abbreviation	Color	Abbreviation	Color	Abbreviation	Color
BK	Black	GY	Gray	RD	Red
BL	Blue	GN	Green	TN	Tan
BR	Brown	LG	LT Green	VT	Violet
DB	Dark Blue	OR	Orange	WT	White
DG	DK Green	PK	Pink	YL	Yellow

1990 4.0L I6 MFI VIN F (A/T) 24 Pin Connector

PCM Pin #	Wire Color	Circuit Description (24 Pin)	Value at Hot Idle
1	BK/BL	Ignition Switch Power (IGSW)	12-14v
2	RD/YL	Direct Battery	12-14v
3	BL/RD	Sensor VREF	4.9-5.1v
4	GN/BL	Airflow Meter Signal	2-4v
5	BL/YL	IAT Sensor Signal (THA)	At 100°F: 2.60v
6	BL/RD	Fuel Pressure Up Solenoid	1v (at hot restart)
7	GN/BK	Vehicle Speed Sensor	At 55 mph: 48 Hz
8	BL/WT	EGR Solenoid Control (VSV)	12v or 0v
9	YL/RD	EFI Main Relay	12-14v
11	BK/GN	Igniter Signal (IGF)	Digital Signal: 0-5-0v
12	BK/GN	Igniter Signal (IGF)	Digital Signal: 0-5-0v
13	YL/RD	EFI Main Relay B1+	12-14v
14	YL/RD	EFI Main Relay B+	12-14v
16	BL	Air Injection Solenoid	12v or 0v
17	BK/RD	Starter Switch Signal	9-11v (cranking)
18	BK/WT	Neutral Start Switch	In P/N: 0-3.0v
19	YL/GN	Malfunction Indicator Lamp Control	MIL Off: 12v, On: 1v
20	BK/WT	A/C Magnetic Clutch Relay (ACMG)	Relay Off: 0v, On: 12v
21	GN/WT	Stop Light Switch Signal	Brake Off: 0v, On: 12v
22	YL/BK	4WD Indicator Light	Off: 12-14v, On: 0.1v
24	BR/BK	Sensor Ground	<0.050v

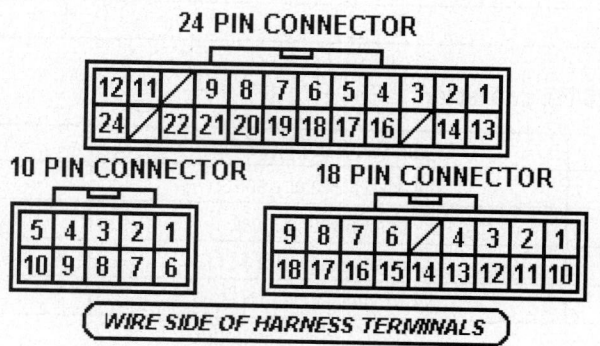

24 PIN CONNECTOR

10 PIN CONNECTOR   **18 PIN CONNECTOR**

WIRE SIDE OF HARNESS TERMINALS

05533_ADIA_G115

Pin Connector Graphic

1991-92 4.0L I6 MFI VIN F (A/T) 12 Pin Connector

PCM Pin #	Wire Color	Circuit Description (12 Pin)	Value at Hot Idle
1	YL/RD	EFI Main Relay	12-14v
2	RD/YL	Direct Battery	12-14v
5	GN/WT	Stop Light Switch Signal	Brake Off: 0v, On: 12v
6	BK/BL	Ignition Switch Power (IGSW)	12-14v
7	YL/RD	EFI Main Relay B+	12-14v
8	YL/WT	Malfunction Indicator Lamp Control	MIL Off: 12v, On: 1v
10	BK/WT	A/C Magnetic Clutch Relay (ACMG)	Relay Off: 0v, On: 12v
11	BK/RD	Vehicle Speed Sensor	At 55 mph: 48 Hz
12	YL/BK	4WD Indicator Light	Off: 12-14v, On: 0.1v

1991-92 4.0L I6 MFI VIN F (A/T) 16 Pin Connector

PCM Pin #	Wire Color	Circuit Description (16 Pin)	Value at Hot Idle
1	GN/RD	Air Flow Meter VREF	4.9-5.1v
2	GN/BL	Airflow Meter Signal	2-4v
3	RD	EFI Main Relay	12-14v
4	RD/WT	ECT Sensor Signal (THW)	At 180°F: 0.51v
5 (Cal)	WT	Sub HO2S Signal	0.1-1.1v
6	BL/WT	Main HO2S Signal	0.1-1.1v
7	BK/WT	Check Connector	12-14v
8	GN	Check Connector	12-14v
9	BR	Sensor Ground	<0.050v
10	GN/BK	TP Sensor Signal (VTA)	0.3-0.8v
11	BL/YL	IAT Sensor Signal (THA)	At 100°F: 2.60v
12	GN/WT	Closed Throttle Switch	1v, off-idle: 12v
13 (Cal)	GN/YL	EGR Gas Temperature Sensor	3.5-4.0v
15	BL	Check Connector	12-14v
16	BR/BK	Sensor Ground	<0.050v

Standard Colors and Abbreviations

Abbreviation	Color	Abbreviation	Color	Abbreviation	Color
BK	Black	GY	Gray	RD	Red
BL	Blue	GN	Green	TN	Tan
BR	Brown	LG	LT Green	VT	Violet
DB	Dark Blue	OR	Orange	WT	White
DG	DK Green	PK	Pink	YL	Yellow

1991-92 4.0L I6 MFI VIN F (A/T) 26 Pin Connector

PCM Pin #	Wire Color	Circuit Description (26 Pin)	Value at Hot Idle
1	OR	Sub HO2S Heater Control	1v (Heater on)
2	BK/RD	Starter Switch Signal	In P/N: 0-3.0v
3	BK/GN	Igniter Signal (IGF)	Digital Signal: 0-5-0v
4	GN/WT	Distributor Signal (NE+)	AC pulse signals
5	BL/RD	Fuel Pressure Up Solenoid	1v (at hot restart)
7	BL/YL	ISC Signal (ISC4)	Pulse Signals
8	YL	ISC Signal (ISC3)	Pulse Signals
9	RD/BK	ISC Signal (ISC2)	Pulse Signals
10	BL/BK	ISC Signal (ISC1)	Pulse Signals
11	BL/RD	Cold Start Injector Control	1v (at cold startup)
12	WT/BL	Injectors 1 & 2 & 3 Control	2.0-3.3 ms
13	BR	Power Ground	<0.1v
14	GY	Main HO2S Heater Control	1v (Heater on)
15	BK/WT	Neutral Start Switch	In P/N: 0-3.0v
16	BL/WT	EGR Solenoid Control (VSV)	12v or 0v
17	BL	Distributor Signal (G-)	<0.050v
18	GN/RD	Distributor Signal (G+)	AC pulse signals

1991-92 4.0L I6 MFI VIN F (A/T) 26 Pin Connector, *continued*

21	PK	Water Temperature Switch	Switch Closed: 0.1v
22	BK/GN	Igniter Signal (IGF)	Digital Signal: 0-5-0v
23	BL	Air Injection Solenoid	12-14v
24	BR/BK	Sensor Ground	<0.050v
25	YL	Injectors 4 & 5 & 6 Control	2.0-3.3 ms
26	BR	Power Ground	<0.1v

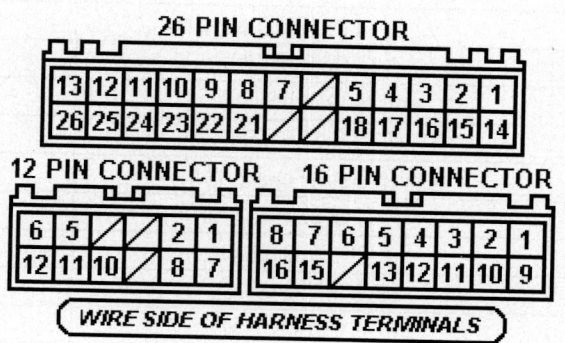

Pin Connector Graphic

1993-94 4.5L I6 MFI VIN D (A/T-ECT) 16 Pin Connector

PCM Pin #	Wire Color	Circuit Description (16 Pin)	Value at Hot Idle
1	GN/RD	Air Flow Meter VREF	4.9-5.1v
2	GN/BL	Airflow Meter Signal	2-4v
3	BL/YL	IAT Sensor Signal (THA)	At 100°F: 2.60v
4	RD/WT	ECT Sensor Signal (THW)	At 180°F: 0.51v
5	BL/WT	HO2S-11 (B1 S1) Signal	0.1-1.1v
6	WT	Knock Sensor 1 Signal	<0.075v AC
7	BL	Data Link Connector	12-14v
8 (Cal)	GN/YL	EGR Gas Temperature Sensor	3.5-4.0v
9	BR/BK	Sensor Ground	<0.050v
10	GN/BK	TP Sensor Signal (VTA)	0.3-0.8v
11	GN/WT	Closed Throttle Switch	1v, off-idle: 12v
12	BK/YL	Data Link Connector	12-14v
13	WT	HO2S-12 (B1 S2) Signal	0.1-1.1v
14	BL/WT	Knock Sensor 2 Signal	<0.075v AC
15	BL/YL	Data Link Connector	12-14v
16	BR/BK	Sensor Ground	<0.050v

1993-94 4.5L I6 MFI VIN D (A/T-ECT) 22 Pin Connector

PCM Pin #	Wire Color	Circuit Description (22 Pin)	Value at Hot Idle
1	YL/RD	EFI Main Relay B1+	12-14v
2	RD/YL	Direct Battery	12-14v
3	RD	EFI Main Relay	12-14v
4	YL/WT	Malfunction Indicator Lamp Control	MIL Off: 12v, On: 1v
7	BK/WT	A/C Magnetic Clutch Relay (ACMG)	Relay Off: 0v, On: 12v
8	BK/RD	Vehicle Speed Sensor	At 55 mph: 48 Hz
9	BL	A/T-ECT TCM (VA)	Pulses
10	BL/YL	A/T-ECT TCM (NE)	Pulses
11	BK/RD	Starter Switch Signal	In P/N: 0-3.0v
12	YL/RD	EFI Main Relay B+	12-14v
13	BK/BL	Ignition Switch Power (IGSW)	12-14v
14	GN/WT	Stop Light Switch Signal	Brake Off: 0v, On: 12v
15	BL/WT	A/T-ECT TCM (ESA3)	4.5-5.5v
16	BL/RD	A/T-ECT TCM (ESA2)	4.5-5.5v
17	BL/GN	A/T-ECT TCM (ESA1)	4.5-5.5v
18	YL/GN	A/T-ECT TCM (ECT)	Pulses
19	BL/BK	A/T-ECT TCM (ECT2)	At 180°F: 0.51v
20	BL/OR	A/T-ECT TCM (ECT1)	KOEO: 12-14v
22	BK/WT	Neutral Start Switch	In P/N: 0-3.0v

Standard Colors and Abbreviations

Abbreviation	Color	Abbreviation	Color	Abbreviation	Color
BK	Black	GY	Gray	RD	Red
BL	Blue	GN	Green	TN	Tan
BR	Brown	LG	LT Green	VT	Violet
DB	Dark Blue	OR	Orange	WT	White
DG	DK Green	PK	Pink	YL	Yellow

1993-94 4.5L I6 MFI VIN D (A/T-ECT) 12 Pin Connector

PCM Pin #	Wire Color	Circuit Description (12 Pin)	Value at Hot Idle
1	OR	HO2S-11 (B1 S1) Heater	1v (Heater on)
2	GY	HO2S-12 (B1 S2) Heater	1v (Heater on)
6	BL	Distributor Signal (G-)	<0.050v
10	GN	Distributor Signal (G2+)	AC pulse signals
11	RD	Distributor Signal (G1+)	AC pulse signals
12	WT	Distributor Signal (NE+)	AC pulse signals

1993-94 4.5L I6 MFI VIN D (A/T-ECT) 26 Pin Connector

PCM Pin #	Wire Color	Circuit Description (26 Pin)	Value at Hot Idle
1	YL/BL	Injector 6 Control	2.0-3.3 ms
2	YL/RD	Injector 5 Control	2.0-3.3 ms
3	WT/RD	Fuel Pump Relay Control	Relay Off: 12v, On: 1v
4	BL/YL	ISC Signal (ISC4)	Pulse Signals

1993-94 4.5L I6 MFI VIN D (A/T-ECT) 26 Pin Connector, *continued*

5	YL	ISC Signal (ISC3)	Pulse Signals
6	RD/BK	ISC Signal (ISC2)	Pulse Signals
7	BL/BK	ISC Signal (ISC1)	Pulse Signals
8	BL	Air Injection Solenoid	12-14v
19	YL/GN	EVAP Purge Solenoid	12-14v
10	YL	Injector 4 Control	2.0-3.3 ms
11	WT/RD	Injector 2 Control	2.0-3.3 ms
12	WT/BL	Injector 1 Control	2.0-3.3 ms
13	BR	Power Ground	<0.1v
15	OR	Data Link Connector	12-14v
16	BK/WT	Data Link Connector	12-14v
17	BK/GN	Igniter Signal (IGF)	Digital Signal: 0-5-0v
21	BL/RD	Fuel Pressure Up Solenoid	1v (at hot restart)
22	BL/WT	EGR Solenoid Control (VSV)	12v or 0v
23	BK/GN	Igniter Signal (IGF)	Digital Signal: 0-5-0v
24	BR/BK	Sensor Ground	<0.050v
25	WT/GN	Injector 3 Control	2.0-3.3 ms
26	BR	Power Ground	<0.1v

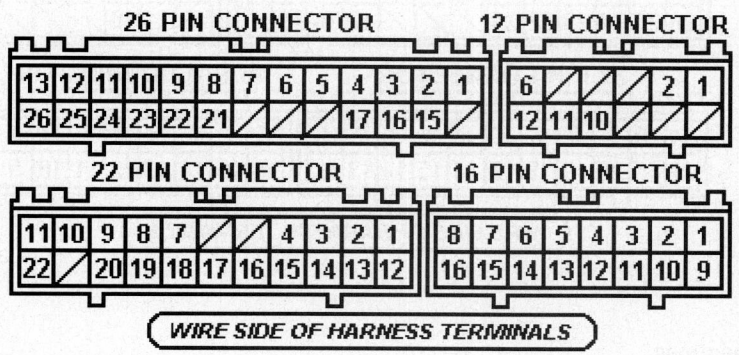

Pin Connector Graphic

1995-97 4.5L I6 VIN D, VIN J (A/T-ECT) 12 Pin Connector

PCM Pin #	Wire Color	Circuit Description (12 Pin)	Value at Hot Idle
1	GY/OR	HO2S-11 (B1 S1) Heater	1v (Heater on)
2	PK	A/T Vehicle Speed Sensor (+)	Pulse Signals
4	BL	CKP Sensor Signal (NE+)	AC pulse signals
5	BR	CKP Sensor Signal (NE-)	<0.050v
6	BL	CMP Sensor Signal (G-)	<0.050v
7	GY/OR	HO2S-12 (B1 S2) Heater	1v (Heater on)
8	PK/GN	A/T Vehicle Speed Sensor (-)	Pulse Signals
10	GN	CMP Sensor Signal (G2+)	AC pulse signals
11	RD	CMP Sensor Signal (G1+)	AC pulse signals
12	WT	CMP Sensor Signal (G+)	AC pulse signals

1995-97 4.5L I6 VIN D, VIN J (A/T-ECT) 16 Pin Connector

PCM Pin #	Wire Color	Circuit Description (16 Pin)	Value at Hot Idle
1	RD/GN	TP Sensor VREF	4.9-5.1v
2	GN	MAF Sensor Signal	1.1-1.5v
3	BL/YL	IAT Sensor Signal (THA)	At 100°F: 2.60v
4	RD/WT	ECT Sensor Signal (THW)	At 180°F: 0.51v
5	PK/BL	HO2S-11 (B1 S1) Signal	0.1-1.1v
6	WT	Knock Sensor 1 Signal	<0.075v AC
7	GN/OR	Data Link Connector	12-14v
9	BR/BK	Sensor Ground	<0.050v
10	GN/BK	TP Sensor Signal (VTA)	0.3-0.8v
11	GN/WT	Closed Throttle Switch	1v, off-idle: 12v
12	RD/WT	A/T Oil Temperature Sensor	At 68°F: 4-5v
13	WT	HO2S-12 (B1 S2) Signal	0.1-1.1v
14	BK	Knock Sensor 2 Signal	<0.075v AC
16	BR/BK	Sensor Ground	<0.050v

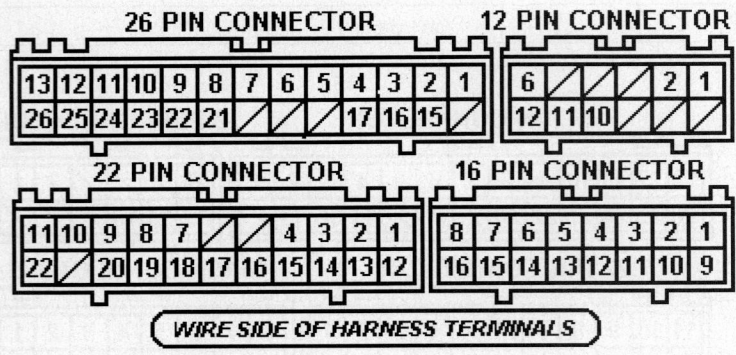

Pin Connector Graphic

Standard Colors and Abbreviations

Abbreviation	Color	Abbreviation	Color	Abbreviation	Color
BK	Black	GY	Gray	RD	Red
BL	Blue	GN	Green	TN	Tan
BR	Brown	LG	LT Green	VT	Violet
DB	Dark Blue	OR	Orange	WT	White
DG	DK Green	PK	Pink	YL	Yellow

1995-97 4.5L I6 VIN D, VIN J (A/T-ECT) 22 Pin Connector

PCM Pin #	Wire Color	Circuit Description (22 Pin)	Value at Hot Idle
1	BK/BL	Ignition Switch Power (IGSW)	12-14v
2	RD/YL	Direct Battery	12-14v
3	RD	EFI Main Relay	12-14v
4	YL/RD	Malfunction Indicator Lamp Control	MIL Off: 12v, On: 1v
5	YL/BL	A/T Oil Temperature Lamp	Lamp Off: 12v
6	WT	Data Link Connector (SDL)	0v

1995-97 4.5L I6 VIN D, VIN J (A/T-ECT) 22 Pin Connector, *continued*

7	BK/WT	A/C Magnetic Clutch Relay (ACMG)	Relay Off: 0v, On: 12v
8	BL/WT	Vehicle Speed Sensor	At 55 mph: 48 Hz
9	BK/BL	4WD Detection Transfer (L4)	Switch Closed: 12v
10	BK	Tachometer Signal (TACO)	Pulse Signals
11	BK/RD	Starter Switch Signal	In P/N: 0-3.0v
12	YL	EFI Main Relay B+	12-14v
13	WT/RD	Fuel Pump Relay	Relay Off: 12v, On: 1v
14	GN/WT	Stop Light Switch Signal	Brake Off: 0v, On: 12v
15	RD/BK	A/T Select Switch Reverse	In 'R': 12v, Others: 0v
16	OR	A/T Select Switch 2nd	In 2nd: 12v, Others: 0v
17	GN/WT	A/T Select Switch Low	In Low: 12v, Others: 0v
18	GN/OR	Cruise Control ECU	At Cruise in OD: 12v
19	PK/BL	Overdrive Main Switch	Switch Off: 12v, On: 1v
20	PK/BK	A/T Pattern Select Switch	Norm: 0v, PWR: 12v
21	YL/BL	4WD Detection Transfer (N)	Open: 12v, Closed: 0v
22	BK/WT	Neutral Start Switch	In P/N: 0-3.0v

1995-97 4.5L I6 VIN D, VIN J (A/T-ECT) 26 Pin Connector

PCM Pin #	Wire Color	Circuit Description (26 Pin)	Value at Hot Idle
1	YL/RD	Injector 5 Control	2.0-3.3 ms
2	YL	Injector 4 Control	2.0-3.3 ms
3	OR	A/T Pattern Select Switch	Norm: 0v, PWR: 12v
4	BL/YL	ISC Signal (ISC4)	Pulse Signals
5	GN/YL	ISC Signal (ISC3)	Pulse Signals
6	RD/BK	ISC Signal (ISC2)	Pulse Signals
7	RD/GN	ISC Signal (ISC1)	Pulse Signals
8	RD/BL	A/T-ECT Solenoid (SL)	In Lockup: 12-14v
9	RD/YL	A/T-ECT Solenoid (S2)	1st or OD: 1v
10	RD	A/T-ECT Solenoid (S1)	3rd or OD: 1v
11	WT/RD	Injector 2 Control	2.0-3.3 ms
12	WT/BL	Injector 1 Control	2.0-3.3 ms
13	BR	Power Ground	<0.1v
14	RD/WT	Circuit Opening Relay (FC)	0-3v, off-idle: 12v
15	YL/BL	Injector 6 Control	2.0-3.3 ms
16	BR	Power Ground	<0.1v
17	BK/YL	Igniter Signal (IGF)	Digital Signal: 0-5-0v
18	RD/WT	A/T Select Switch 2nd	2nd: 12-14v
19	GN/YL	EGR Gas Temperature Sensor	3.5-4.0v
21	BL/RD	Fuel Pressure Up Solenoid	1v (at hot restart)
22	BL/WT	EGR Solenoid Control (VSV)	12v or 0v
23	BK/GN	Igniter Signal (IGF)	Digital Signal: 0-5-0v
24	BR/BK	Sensor Ground	<0.050v
25	WT/GN	Injector 3 Control	2.0-3.3 ms
26	BR	Power Ground	<0.1v

Standard Colors and Abbreviations

Abbreviation	Color	Abbreviation	Color	Abbreviation	Color
BK	Black	GY	Gray	RD	Red
BL	Blue	GN	Green	TN	Tan
BR	Brown	LG	LT Green	VT	Violet
DB	Dark Blue	OR	Orange	WT	White
DG	DK Green	PK	Pink	YL	Yellow

1998 4.7L V8 VIN T (A/T-ECT) 17 Pin Connector

PCM Pin #	Wire Color	Circuit Description (17 Pin)	Value at Hot Idle
4	BL/RD	Direct Clutch Speed Input (+)	Pulse Signals
5	RD	A/T Vehicle Speed Sensor (+)	Pulse Signals
10	BL/WT	Direct Clutch Speed Input (-)	Pulse Signals
11	GN	A/T Vehicle Speed Sensor (-)	Pulse Signals
17	GN/YL	A/T Oil Temperature Sensor	At 68°F: 4-5v

1998 4.7L V8 VIN T (A/T-ECT) 28 Pin Connector

PCM Pin #	Wire Color	Circuit Description (28 Pin)	Value at Hot Idle
1	BK	Overdrive Main Switch	Switch Off: 12v, On: 1v
2	RD/BK	A/T Select Switch Reverse	In 'R': 12v, Others: 0v
3	GN	A/T Select Switch 2nd	In 2nd: 12v, Others: 0v
4	GN/BK	A/T Select Switch Low	In Low: 12v, Others: 0v
5	PK/RD	Data Link Connector	12-14v
6	GN/WT	Stop Light Switch Signal	Brake Off: 0v, On: 12v
7	RD/BK	HO2S-22 (B2 S2) Heater	1v (Heater on)
8	BL	HO2S-12 (B1 S2) Heater	1v (Heater on)
9	YL/BK	C/C Indicator Lamp	Lamp On: 0.1v
10	BL	EVAP Vapor Pressure Valve (VSV)	12v or 0v
11	BL/WT	A/T Pattern Select Switch	Norm: 0v, PWR: 12v
12	GN/WT	Idle-Up Signal	Load On: 1v, Off: 12v
13	BL/BK	A/C Amplifier Signal (ACT)	Clutch On: 12v, Off: 1.5v
14	YL/BK	A/C Amplifier Signal (THWO)	Pulse Signals
15	PK	Vehicle Speed Sensor	At 55 mph: 48 Hz
16	BK	Tachometer Signal (TACO)	Pulse Signals
17	BK/RD	Starter Switch Signal	In P/N: 0-3.0v
18	BK	HO2S-12 (B1 S2) Signal	0.1-1.1v
19	GN/YL	A/T Select Switch Drive	In Drive: 12-14v
20	BK/WT	Start Circuit Signal	Cranking: 9-11v
21	WT/BK	Sensor Ground	<0.050v
22	BL/BK	EVAP Vapor Pressure Sensor (PTNK)	2.5-3.1v (with cap off)
25	WT/GN	A/C Amplifier Signal (AC1)	Clutch On: 1.5v, Off: 12v
26	OR	A/T Oil Temperature Lamp	Lamp Off: 12v, On: 1v
27	WT	HO2S-22 (B2 S2) Signal	0.1-1.1v

17 PIN CONNECTOR   22 PIN CONNECTOR

WIRE SIDE OF HARNESS TERMINALS

Pin Connector Graphic
1998 4.7L V8 VIN T (A/T-ECT) 24 Pin Connector

05533_ADIA_G118

PCM Pin #	Wire Color	Circuit Description (24 Pin)	Value at Hot Idle
1	WT/BK	Power Ground	<0.1v
2	BL/RD	Sensor VREF	4.9-5.1v
3	YL	HO2S-21 (B2 S1) Heater	1v (Heater on)
4	RD	HO2S-11 (B1 S1) Heater	1v (Heater on)
5	YL	Injector 1 Control	2.0-3.3 ms
6	BK	Injector 2 Control	2.0-3.3 ms
7	BL/BK	EVAP Purge Solenoid (VSV)	12v or 0v
8	RD	ABS ECU Signal	Pulse Signals
9	RD/BK	Accelerator Position Sensor 2	1.8-2.7v
10	BL/YL	MAF Sensor Signal (VG)	1.1-1.5v
11	WT	HO2S-21 (B2 S1) Signal	0.1-1.1v
12	BK	HO2S-11 (B1 S1) Signal	0.1-1.1v
13	RD/YL	TP Sensor Signal (VTA1)	0.4-1.0v
14	GN/BK	ECT Sensor Signal (THW)	At 180°F: 0.51v
17	BR	Sensor Ground	<0.050v
18	BR/WT	Sensor Ground	<0.050v
19	GN/WT	MAF Sensor Ground (EVG)	<0.050v
20	YL/BK	TP Sensor Signal (VTA2)	2.0-2.9v
21	RD	Accelerator Position Sensor 1	0.25-0.90v
22	YL/BK	IAT Sensor Signal (THA)	At 100°F: 2.60v

1998 4.7L V8 VIN T (A/T-ECT) 22 Pin Connector

PCM Pin #	Wire Color	Circuit Description (22 Pin)	Value at Hot Idle
1	RD/YL	Direct Battery	12-14v
2	YL	A/T Pattern Select Switch	2nd: 12-14v
3	BL/RD	A/T Select Switch 2nd	In 2nd: 12v, Others: 0v
4	GN/RD	Fuel Pump Signal (DI)	Pulse Signals
5	GN/WT	Fuel Control Switch Signal	Open: 12v, Closed: 0v
6	WT	Malfunction Indicator Lamp Control	MIL Off: 12v, On: 1v
7	YL/BK	BM (+) Signal	12-14v
8	BK/YL	EFI Main Relay B1+	12-14v
9	BK/RD	Ignition Switch Power (IGSW)	12-14v
10	BK/WT	EFI Main Relay	12-14v
11	PK/WT	SIL Signal (Scan Tool)	12v
12	BL/BK	Transponder Amplifier Signal (Code)	Inserting key: pulses
13	PK/GN	Transponder Amplifier Signal (RXCK)	Inserting key: pulses
14	RD/YL	Transponder Amplifier Signal (RXCK)	Inserting key: pulses
15	YL/GN	A/T Select Switch Park	In P/N: 0-3.0v
16	BK/YL	EFI Main Relay B+	12-14v
18	YL/RD	A/C Amplifier On Signal	Clutch On: 1.5v, Off: 12v
20	RD/BK	Unlock Warning Switch	No Key: 4-5v
21	GN/RD	LED Signal	LED Off: 12v, On: 1v
22	BK/BL	4WD Detection Transfer (L4)	Open: 12v, Closed: 0v

1998 4.7L V8 VIN T (A/T-ECT) 31 Pin Connector

PCM Pin #	Wire Color	Circuit Description (31 Pin)	Value at Hot Idle
1	BL	Injector 3 Control	2.0-3.3 ms
2	RD	Injector 4 Control	2.0-3.3 ms
3	GN	Injector 5 Control	2.0-3.3 ms
4	RD/BL	Injector 6 Control	2.0-3.3 ms
5	WT	Injector 7 Control	2.0-3.3 ms
6	BK/WT	Injector 8 Control	2.0-3.3 ms
7	WT	Throttle Control Motor (M-)	Pulse Signals
8	RD	Throttle Control Motor (M+)	Pulse Signals
9	WT/BK	Power Ground	<0.1v
10	RD	CMP Sensor Signal (G+)	AC pulse signals
11	BK	Igniter Transistor 1 Control	7% duty cycle
12	RD	Igniter Transistor 2 Control	7% duty cycle
13	BL	Igniter Transistor 3 Control	7% duty cycle
14	GN	Igniter Transistor 4 Control	7% duty cycle
15	YL	Igniter Transistor 5 Control	7% duty cycle
16	BK/YL	Igniter Transistor 6 Control	7% duty cycle
17	WT	Knock Sensor 2 Signal	<0.075v AC
18	BK	Knock Sensor 1 Signal	<0.075v AC
21	WT/BK	Power Ground	<0.1v
22	GN	CMP/CKP Sensor Signal (-)	<0.050v
23	BL	CKP Sensor Signal (NE+)	AC pulse signals

1998 4.7L V8 VIN T (A/T-ECT) 31 Pin Connector, *continued*

24	BL	Throttle Control Motor (CL−)	Pulse Signals
25	BK/BL	Igniter Transistor 7 Control	7% duty cycle
26	BL/BK	Igniter Transistor 8 Control	7% duty cycle
27	BK/WT	Igniter Signal (IGF1)	Digital Signal: 0-5-0v
28	BK/RD	Igniter Signal (IGF2)	Digital Signal: 0-5-0v
29	GN	Throttle Control Motor (CL+)	Pulse Signals
30	BR	Shield Ground	<0.050v
31	WT/BK	Power Ground	<0.1v

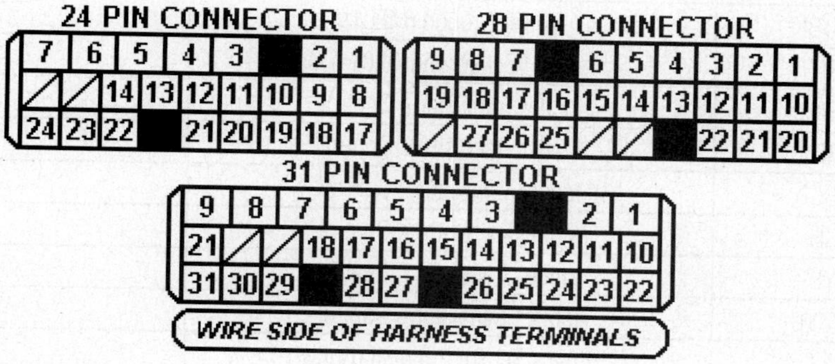

24 PIN CONNECTOR

28 PIN CONNECTOR

31 PIN CONNECTOR

WIRE SIDE OF HARNESS TERMINALS

05533_ADIA_G119

Pin Connector Graphic

1999-2002 4.7L 2UZ-FE V8 VIN T E5 31P Connector

PCM Pin #	Wire Color	Circuit Description (31 Pin)	Value at Hot Idle
1	BL	Injector 3 Control	2.0-3.3 ms
2	RD	Injector 4 Control	2.0-3.3 ms
3	GN	Injector 5 Control	2.0-3.3 ms
4	RD/BL	Injector 6 Control	2.0-3.3 ms
5	WT	Injector 7 Control	2.0-3.3 ms
6	BK/WT	Injector 8 Control	2.0-3.3 ms
7	WT	Throttle Control Motor (M-)	Pulse Signals
8	RD	Throttle Control Motor (M+)	Pulse Signals
9	WT/BK	Power Ground (ME01)	<0.1v
10	RD	CMP Sensor Signal (G+)	AC pulse signals
11	BK	Igniter Transistor 1 Control	7% duty cycle
12	RD	Igniter Transistor 2 Control	7% duty cycle
13	BL	Igniter Transistor 3 Control	7% duty cycle
14	GN	Igniter Transistor 4 Control	7% duty cycle
15	YL	Igniter Transistor 5 Control	7% duty cycle
16	BK/YL	Igniter Transistor 6 Control	7% duty cycle
17	WT	Knock Sensor 2 Signal (right)	<0.075v AC
18	BK	Knock Sensor 1 Signal (left)	<0.075v AC
19-20	---	Not Used	---
21	WT/BK	Power Ground (E01)	<0.1v
22	GN	CMP/CKP Sensor Signal (-)	<0.050v
23	BL	CKP Sensor Signal (NE+)	AC pulse signals
24	BL	Throttle Control Motor (CL-)	Pulse Signals
25	BK/BL	Igniter Transistor 7 Control	7% duty cycle
26	BL/BK	Igniter Transistor 8 Control	7% duty cycle
27	BK/WT	Igniter Signal (IGF1)	Digital Signal: 0-5-0v
28	BK/RD	Igniter Signal (IGF2)	Digital Signal: 0-5-0v
29	GN	Throttle Control Motor (CL+)	Pulse Signals
30	BR	Shield Ground (GE01)	<0.050v
31	WT/BK	Power Ground (E02)	<0.1v

E5 31-PIN CONNECTOR

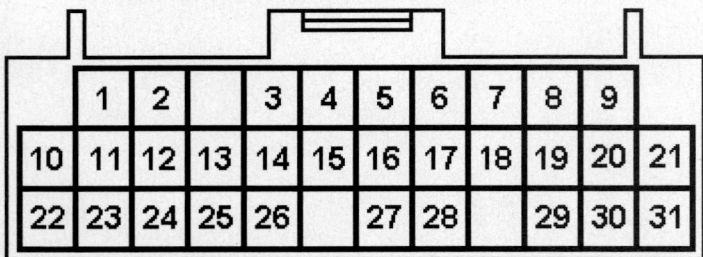

WIRE SIDE OF HARNESS TERMINALS

05533_ADIA_G120

Pin Connector Graphic

1999-2002 4.7L 2UZ-FE V8 VIN T E6 24P Connector

PCM Pin #	Wire Color	Circuit Description (24 Pin)	Value at Hot Idle
1	WT/BK	Power Ground (E03)	<0.1v
2	BL/RD	Sensor VREF (VC)	4.9-5.1v
3	YL	HO2S-21 (B2 S1) Heater	1v (Heater on)
4	RD	HO2S-11 (B1 S1) Heater	1v (Heater on)
5	YL	Injector 1 Control	2.0-3.3 ms
6	BK	Injector 2 Control	2.0-3.3 ms
7	BL/BK	EVAP Purge Solenoid (VSV)	12v or 0v
8	RD	ABS ECU Signal	Digital Signals
9	RD/BK	Accelerator Position Sensor 2	1.8-2.7v
10	BL/YL	Mass Airflow Sensor (VG)	1.1-1.5v
11	WT	HO2S-21 (B2 S1) Signal	0.1-1.1v
12	BK	HO2S-11 (B1 S1) Signal	0.1-1.1v
13	RD/YL	TP Sensor Signal (VTA)	0.4-1.0v
14	GN/BK	ECT Sensor Signal (THW)	At 180°F: 0.51v
15-16	---	Not Used	---
17	BR	Power Ground (E1)	<0.1v
18	BR/WT	Sensor Ground (E2)	<0.050v
19	GN/WT	MAF Sensor Ground (EVG)	<0.050v
20	YL/BK	TP Sensor Signal (VTA2)	2.0-2.9v
21	RD	Accelerator Position Sensor 1	0.3-0.90v
22	YL/BK	IAT Sensor Signal (THA)	At 100°F: 2.60v
23 (01'-02')	RD/BK	HO2S-22 (B2 S2) Heater	1v (Heater on)
24 (01'-02')	BL	HO2S-12 (B1 S2) Heater	1v (Heater on)

1999-2002 4.7L V8 2UZ-FE VIN T E7 17P Connector

PCM Pin #	Wire Color	Circuit Description (17 Pin)	Value at Hot Idle
1	RD	A/T ECT Solenoid (S1)	In 3rd or OD: 1v
2	WT	A/T ECT Solenoid (S2)	In 1st or OD: 1v
3	GN	A/T ECT Solenoid (SL)	In Lockup: 12-14v
4	BL/RD	Direct Clutch Speed Input (+)	AC pulse signals
5	RD	A/T Vehicle Speed Sensor (+)	AC pulse signals
6-8	---	Not Used	---
9	GN/WT	A/T-ECT Solenoid (SLT+)	Pulse Signals
10	BL/WT	Direct Clutch Speed Input (-)	AC pulse signals
11	GN	A/T Vehicle Speed Sensor (-)	Pulse Signals
12-14	---	Not Used	---
15	GNBK	A/T-ECT Solenoid (SLT-)	Pulse Signals
16	---	Not Used	---
17	GN/YL	A/T Oil Temperature Sensor	At 68°F: 4-5v

E6 24-PIN CONNECTOR

E7 17-PIN CONNECTOR

WIRE SIDE OF HARNESS TERMINALS

05533_ADIA_G121

Pin Connector Graphic

1999-2002 4.7L 2UZ-FE V8 VIN T E8 28P Connector

PCM Pin #	Wire Color	Circuit Description (28 Pin)	Value at Hot Idle
5	PK/RD	Data Link Connector (TE1)	12-14v
6	GN/WT	Stop Light Switch Signal	Brake Off: 0v, On: 12v
7	RD/BK	HO2S-22 (B2 S2) Heater	1v (Heater on)
8	BL	HO2S-12 (B1 S2) Heater	1v (Heater on)
9	---	Not Used	---
10	BL	EVAP Vapor Pressure Valve (VSV)	12v or 0v
11	---	Not Used	---
12	GN/WT	Defogger & Tail Light Switch	Switch Off: 0v, On: 12v
13	BL/BK	A/C Amplifier Signal (ACT)	Clutch On: 12v, Off: 1.5v
14	YL/BK	A/C Amplifier Signal (THWO)	A/C Off: 12v, On: 1v
15	VT	Vehicle Speed Sensor	At 55 mph: 48 Hz
16	BK	Tachometer Signal (TACO)	Pulse Signals
17	BK/RD	Starter Switch Signal	9-11v (cranking)
18	BK	HO2S-12 (B1 S2) Signal	0.1-1.1v
19	---	Not Used	---
20	BK/WT	Neutral Start Switch (NSW)	In P/N: 0-3.0v
21	WT/BK	Power Ground (EOM)	<0.1v
22	BL/BK	EVAP Vapor Pressure Sensor (PTNK)	2.5-3.1v (with cap off)
23-24	---	Not Used	---
25	WT/GN	A/C Amplifier Signal (AC1)	Clutch On: 1.5v, Off: 12v
26	---	Not Used	---
27	WT	HO2S-22 (B2 S2) Signal	0.1-1.1v
28	---	Not Used	---

1999-2002 4.7L 2UZ-FE V8 VIN T E9 22P Connector

PCM Pin #	Wire Color	Circuit Description (22 Pin)	Value at Hot Idle
1	RD/YL	Direct Battery	12-14v
2-3	---	Not Used	---
4	GN/RD	Fuel Pump Signal (DI)	12-14v
5	GN/WT	Fuel Control Switch Signal	Closed: 0v, Open: 12v
6	WT	Malfunction Indicator Lamp Control	MIL Off: 12v, On: 1v
7	YL/BK	BM (+) Power	12-14v
8	BK/YL	EFI Main Relay B1+	12-14v

1999-2002 4.7L 2UZ-FE V8 VIN T E9 22P Connector, *continued*

9	BK/RD	Ignition Switch Power (IGSW)	12-14v
10	BK/WT	EFI Main Relay Control	Relay Off: 0v, On: 12v
11	PK/WT	SIL Signal (Scan Tool)	12v
12	BL/BK	Transponder Amplifier Signal (Code)	Inserting key: pulses
13	VT/GN	Transponder Amplifier Signal (RXCK)	Inserting key: pulses
14	RD/YL	Transponder Amplifier Signal (RXCK)	Inserting key: pulses
15	---	Not Used	---
16	BK/YL	EFI Main Relay B+	12-14v
18-19	---	Not Used	---
20	RD/BK	Unlock Warning Switch	Key In: 1.5v, Out: 4-5v
21	GN/RD	Center ECU LED Signal	LED Off: 12v, On: 1v
22	---	Not Used	---

E8 28-PIN CONNECTOR

1	2	3	4	5	6		7	8	9
10	11	12	13	14	15	16	17	18	19
20	21	22		23	24	25	26	27	28

E9 22-PIN CONNECTOR

1	2	3	4	5		6	7
8	9	10	11	12	13	14	15
16	17		18	19	20	21	22

Pin Connector Graphic

05533_ADIA_G122

2003-2006 4.7L 2UZ-FE V8 VIN T E5 34 Pin Connector

PCM Pin #	Wire Color	Circuit Description (34 Pin)	Value at Hot Idle
1	YL	Injector 1 Control	2.0-3.3 ms
2	BK	Injector 2 Control	2.0-3.3 ms
3	BL	Injector 3 Control	2.0-3.3 ms
4	RD	Injector 4 Control	2.0-3.3 ms
5	GN	Injector 5 Control	2.0-3.3 ms
6	WT/BK	Power Ground (E02)	<0.1v
7	WT/BK	Power Ground (E01)	<0.1v
8	RD	Igniter Transistor 2 Control	7% duty cycle
9	BK	Igniter Transistor 1 Control	7% duty cycle
10	BL/BK	Igniter Transistor 8 Control	7% duty cycle
11	GN	Igniter Transistor 4 Control	7% duty cycle
12	YL	Igniter Transistor 5 Control	7% duty cycle
13	BK/BL	Igniter Transistor 7 Control	7% duty cycle
15	RD/GN	A/C Relay Control (ACCR)	Relay Off: 12v, On: 1v
16	BK/WT	Neutral Start Switch (NSW)	In P/N: 0-3.0v
17	BL/RD	Starter Switch Signal (STA)	In P/N: 0-3.0v
18	BL/RD	Sensor VREF (VC)	4.9-5.1v
19	GN/BK	ECT Sensor Signal (THW)	At 180°F: 0.51v
20	YL/BK	IAT Sensor Signal (THA)	At 100°F: 2.60v
21	RD/YL	TP Sensor Signal (VTA1)	0.4-1.0v
22, 32	---	Not Used	---
23	BK/RD	Igniter Signal (IGF2)	Digital Signal: 0-5-0v
24	BK/WT	Igniter Signal (IGF1)	Digital Signal: 0-5-0v
25	BL	Igniter Transistor 3 Control	7% duty cycle
26	BK/YL	Igniter Transistor 6 Control	7% duty cycle
27	BL/RD	EVAP Canister Closed Valve (CCV)	12v or 0v
28	BR/WT	Sensor Ground (E2)	<0.050v
29	GN/WT	MAF Sensor Ground (E2G)	<0.050v
30	BL/YL	Mass Airflow Sensor (VG)	1.1-1.5v
31	YL/BK	TP Sensor Signal (VTA2)	2.0-2.9v
33	GN/WT	Fuel Pump Relay Control (FPR)	Relay Off: 12v, On: 1v
34	BL/BK	EVAP Purge Solenoid (VSV)	12v or 0v

2003-2006 4.7L 2UZ-FE V8 VIN T E6 35 Pin Connector

PCM Pin #	Wire Color	Circuit Description (35 Pin)	Value at Hot Idle
1	BK	Knock Sensor 1 Signal (KNK1 - left)	<0.075v AC
2	WT	Knock Sensor 2 Signal (KNK2 - right)	<0.075v AC
3	RD/BL	Injector 6 Control	2.0-3.3 ms
4	RD	HO2S-11 (B1 S1) Heater (HT1A)	1v (Heater on)
5	BL	HO2S-12 (B1 S2) Heater (HT1B)	1v (Heater on)
7-8, 20	---	Not Used	---
9	BK/WT	Start Signal (STAR)	In P/N: 0-3.0v
10	WT	ECT Solenoid Control (S2)	12v or 0v
11	RD	ECT Solenoid Control (S1)	12v or 0v

2003-2006 4.7L 2UZ-FE V8 VIN T E6 35 Pin Connector, *continued*

12	GN/BK	ECT Solenoid Control (SLT-)	12v or 0v
13	GN/WT	ECT Solenoid Control (SLT+)	12v or 0v
16	PK/BK	A/T Solenoid Control (SL2-)	Pulse Signals
17	PK/BK	A/T Solenoid Control (SL2+)	Pulse Signals
18	RD/WT	A/T Solenoid Control (SL1-)	Pulse Signals
19	RD/BL	A/T Solenoid Control (SL1+)	Pulse Signals
21	WT	HO2S-22 (B2 S2) Signal (OX2B)	0.1-1.1v
22	WT	HO2S-21 (B2 S1) Signal (OX2A)	0.1-1.1v
23	BK	HO2S-11 (B1 S1) Signal (OX1A)	0.1-1.1v
24	BL	A/T Oil Temperature Sensor 2 (THO2)	At 68°F: 4-5v
25	RD/BK	HO2S-22 (B2 S2) Heater (HT2B)	1v (Heater on)
26	RD	ECT Vehicle Speed Sensor (SP2+)	Pulse Signals
27	BL	ECT Turbine Speed Sensor (NT-)	Pulse Signals
28	---	Not Used	---
29	BK	HO2S-12 (B1 S2) Signal (OX1B)	0.1-1.1v
30-31	---	Not Used	---
32	GN/YL	A/T Oil Temperature Sensor 1 (THO1)	At 68°F: 4-5v
33	YL	HO2S-21 (B2 S1) Heater (HT2A)	1v (Heater on)
34	GN	ECT Vehicle Speed Sensor (SP2-)	Pulse Signals
35	WT	ECT Turbine Speed Sensor (NT+)	Pulse Signals

2003-2006 4.7L 2UZ-FE V8 VIN T E7 32 Pin Connector

PCM Pin #	Wire Color	Circuit Description (32 Pin)	Value at Hot Idle
1	BR	Power Ground (E1)	<0.1v
2	WT	Throttle Control Motor (M-)	Pulse Signals
3	RD	Throttle Control Motor (M+)	Pulse Signals
4	WT/BK	Power Ground (ME01)	<0.1v
5	BK/WT	Injector 8 Control	2.0-3.3 ms
6	WT	Injector 7 Control	2.0-3.3 ms
7	WT/BK	Power Ground (E03)	<0.1v
11	YL/GN	Neutral Detection Switch (L4)	Switch Open: 0v, Closed: 12v
12	BK/WT	Start Switch Signal (STSW)	Cranking: 9-11v
13-14, 18-20	---	Not Used	---
15	BK	A/T Solenoid Control (SLU-)	Pulse Signals
16	PK/GN	A/T Solenoid Control (SLU+)	Pulse Signals
17	BR	Shield Ground (GE01)	<0.050v
21	BK/OR	Generator Control (RL)	12v
22, 26	---	Not Used	---
23	BL	A/C Lock Sensor (LCK)	12v or 0v
24	GN	CKP Sensor Signal (NE-)	<0.050v
25	BL	CKP Sensor Signal (NE+)	AC pulse signals
27	RD	CMP Sensor Signal (G2+)	AC pulse signals
28-31	---	Not Used	---
32	GN	CMP Sensor Signal (G2-)	AC pulse signals

2003-2006 4.7L V8 2UZ-FE VIN T E8 35 Pin Connector

PCM Pin #	Wire Color	Circuit Description (35 Pin)	Value at Hot Idle
1	WT/BK	Power Ground (HP)	<0.1v
2, 7-8	---	Not Used	---
3	GN	A/T Select Switch 2nd Signal (2L)	In 2nd: 12v, Others: 0v
4	BK/BL	Low Detection Switch (L4)	Switch Open: 0v, Closed: 12v
5	BL/WT	ECT Pattern Switch 2nd Position (SNW1)	2nd Position: 12v
6	YL/BK	ETCS Power (+BM)	12-14v
9	GN/BK	Shift Lock ECU Control (D)	12v or 0v
10	GN/YL	Shift Lock ECU Control (D)	12v or 0v
11	RD/BK	A/T Select Switch Reverse Signal	In 'R': 1v, Others: 12v
12, 15-16	---	Not Used	---
13	YL/RD	A/C Magnetic Clutch Relay (ACMG)	Relay Off: 0v, On: 12v
14	YL/GN	A/C Amplifier Signal (THWO)	A/C Off: 12v, On: 1v
17	VT	Vehicle Speed Sensor (SPD)	At 55 mph: 48 Hz
18	PK/BK	Body Control ECU Signal (MPX1)	Digital Signals
19	GN/WT	Stop Light Switch Signal	Brake Off: 0v, On: 12v
20-22, 25	---	Not Used	---
23	GN/RD	Shift Lock ECU Control (4)	12v or 0v
26	YL	Transponder Amplifier Signal (IMD)	Inserting key: pulses
27	WT	Transponder Amplifier Signal (IMI)	Inserting key: pulses
28	BL/WT	ECT Pattern Switch Power Signal	Power Position: 12v
29	BK	Body Control ECU Signal (MPX2)	Digital Signals
30	---	Not Used	---
31	BL/BK	A/C Amplifier Signal (ACT)	Relay Off: 12v, On: 1v
32	PK/BK	A/C Amplifier Signal (THE)	A/C Off: 12v, On: 1v
33	GN/BK	A/C Switch Signal (ACLD)	A/C On: 12v, Off: 0v
34-35	---	Not Used	---

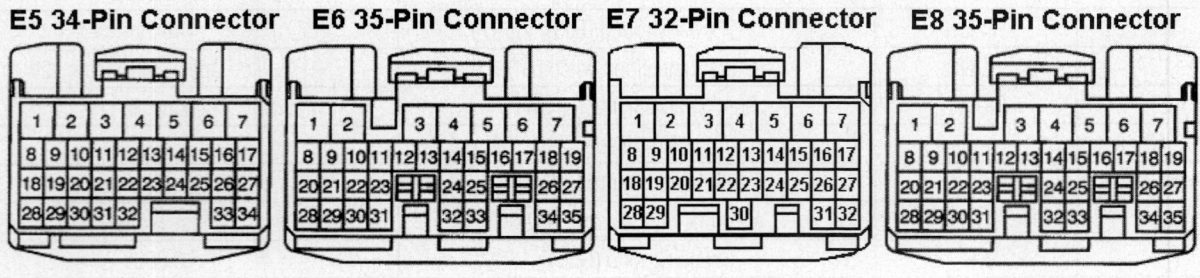

E5 34-Pin Connector   **E6 35-Pin Connector**   **E7 32-Pin Connector**   **E8 35-Pin Connector**

05533_ADIA_G123

Pin Connector Graphic

2003-2006 4.7L 2UZ-FE V8 VIN T E9 31 Pin Connector

PCM Pin #	Wire Color	Circuit Description (31 Pin)	Value at Hot Idle
1	BK/YL	EFI Main Relay Power (EFI Fuse)	12-14v
2	BK/YL	EFI Main Relay Power (EFI Fuse)	12-14v
3	BK/RD	Direct Battery	12-14v
4	BL	EVAP Pressure Switching Valve (VSV)	12v or 0v
5	BK	Tachometer Signal (TACO)	Pulse Signals
6	GN/WT	A/T Select Switch Park Signal (P)	In 'P': 12v, Others: 0v
7	GN/RD	A/T Select Switch Neutral Signal (N)	In 'N': 12v, Others: 0v
8	BK/WT	EFI Main Relay Control	Relay Off: 0v, On: 12v
9	BK/RD	Ignition Switch Power (IGSW)	12-14v
10	BK/WT	Circuit Opening Relay (FC)	0-3v, off-idle: 12v
11	WT	Malfunction Indicator Lamp Control	MIL Off: 12v, On: 1v
12	GN/WT	Body Control ECU	Digital Signals
13	YL/RD	Horn Relay Control	Relay Off: 12v, On: 1v
14	BK	Center Airbag Assembly (F/PS)	Digital Signals
15	WT/BK	Power Ground (EOM)	<0.1v
16	---	Not Used	---
17	WT	Transponder Amplifier Signal (NEO)	Inserting key: pulses
18	VT/WT	SIL Signal (Scan Tool)	Transmitting: pulses
19	WT/RD	Data Link Connector (WFSE)	N/A
20	PK/BK	Data Link Connector (TC)	12v
21	BL/BK	EVAP Vapor Pressure Sensor (PTNK)	2.5-3.1v (with cap off)
22	RD	Accelerator Position Sensor 1 (VPA)	0.3-0.90v
23	RD/BK	Accelerator Position Sensor 2 (VPA2)	1.8-2.7v
24	RD	Traction Control Engine Signal (ENG+)	Pulse Signals
25	YL	Traction Control Signal (TRC+)	Pulse Signals
26	BL/RD	Accelerator Pedal Position Sensor 1 VREF	4.9-5.1v
27	WT	Accelerator Pedal Position Sensor 2 VREF	4.9-5.1v
28	BR/WT	Accelerator Pedal Position Sensor Ground	<0.050v
29	WT/RD	Accelerator Position Sensor 2 (EPA2)	1.8-2.7v
30	GN	Traction Control Engine Signal (ENG-)	Pulse Signals
31	BL	Traction Control Signal (TRC-)	Pulse Signals

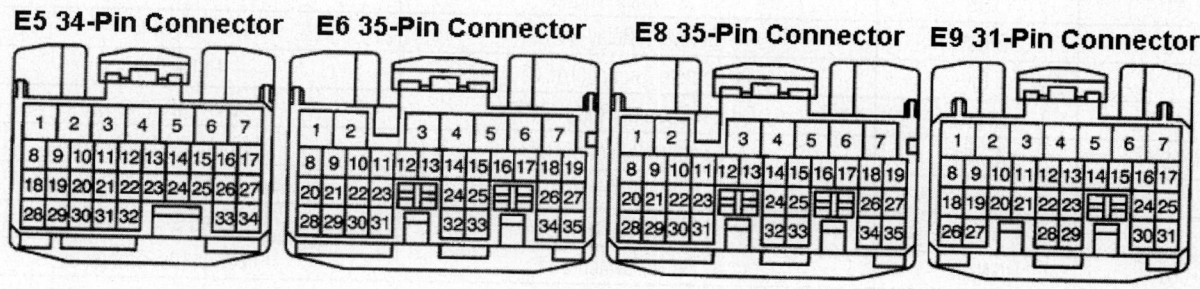

E5 34-Pin Connector   **E6 35-Pin Connector**   **E8 35-Pin Connector**   **E9 31-Pin Connector**

Pin Connector Graphic

05533_ADIA_G124

RAV4 PIN CHARTS

1996-98 2.0L I4 MFI VIN P (All) 16 Pin Connector

PCM Pin #	Wire Color	Circuit Description (16 Pin)	Value at Hot Idle
1	YL	Sensor VREF (VC)	4.9-5.1v
2	GN/BK	MAP Sensor Signal	Idling: 1-1.5v
3	YL/BK	IAT Sensor Signal (THA)	At 100°F: 2.60v
4	WT	ECT Sensor Signal (THW)	At 180°F: 0.51v
5	RD	HO2S-12 (B1 S2) Signal	0.1-1.1v
6	BK	HO2S-11 (B1 S1) Signal	0.1-1.1v
6 (98' - Cal)	WT	AFS-11 (B1 S1) Signal (+)	Fixed at 3.3v
6 (98' - Fed)	WT	HO2S-11 (B1 S1) Signal	0.1-1.1v
7	BL/YL	EVAP Vapor Pressure Sensor (PTNK)	2.5-3.1v (with cap off)
8	RD/WT	EVAP Vapor Pressure Valve (VSV)	12v or 0v
9	BR	Sensor Ground	<0.050v
10	BK	Knock Sensor Signal	<0.075v AC
11	BL/RD	TP Sensor Signal (VTA)	0.3-0.8v
12	RD/BL	4WD Speed Sensor (RR-)	Pulse Signals
13	BL/YL	4WD Speed Sensor (RR+)	Pulse Signals
14 (98' - Cal)	BK	AFS-11 (B1 S1) Signal (-)	Fixed at 3.0v
15	BR	Sensor Ground	<0.050v
16	YL/BK	A/T ECT Solenoid (SL)	In Lockup: 12-14v

1996-98 2.0L I4 MFI VIN P (All) 22 Pin Connector

PCM Pin #	Wire Color	Circuit Description (22 Pin)	Value at Hot Idle
1	RD/WT	Direct Battery	12-14v
2	BL/RD	4WD ECT Solenoid (SLD-)	Pulse Signals
3	BL/BK	4WD ECT Solenoid (SLD+)	Pulse Signals
4	GN/WT	Stop Light Switch Signal	Brake Off: 0v, On: 12v
5	GN/RD	Malfunction Indicator Lamp Control	MIL Off: 12v, On: 1v
6	BL/WT	Data Link Connector	12-14v
7	LG	Overdrive Main Switch	Switch Off: 12v, On: 1v
8	RD/BK	A/T Select Switch Reverse	In 'R': 12v, Others: 0v
9	PK/WT	Vehicle Speed Sensor	At 55 mph: 48 Hz
10	YL/GN	A/C Amplifier On Signal	Clutch On: 1.5v, Off: 12v
11	BK	Starter Switch Signal	In P/N: 0-3.0v
12	BK/WT	EFI Main Relay B+	12-14v
13	BK	Tachometer Signal (TACO)	Pulse Signals
14	GN/RD	Circuit Opening Relay (FC)	0-3v, off-idle: 12v
15	RD/GN	A/T Pattern Select Switch	Norm: 0v, PWR: 12v
16	WT	SIL Signal (Scan Tool)	0v
17	BK	Integration Relay	Load On: 12-14v
18	GN/BK	A/T Select Switch 2nd	In 2nd: 12v, Others: 0v
19	LG	A/T Select Switch Low	In Low: 12v, Others: 0v
20	YL/BK	Cruise Control ECU	At Cruise in OD: 12v
21	RD/YL	A/C Amplifier Signal (ACT)	Clutch On: 12v, Off: 1.5v
22	BK/WT	Neutral Start Switch (NSW)	Cranking: 9-11v

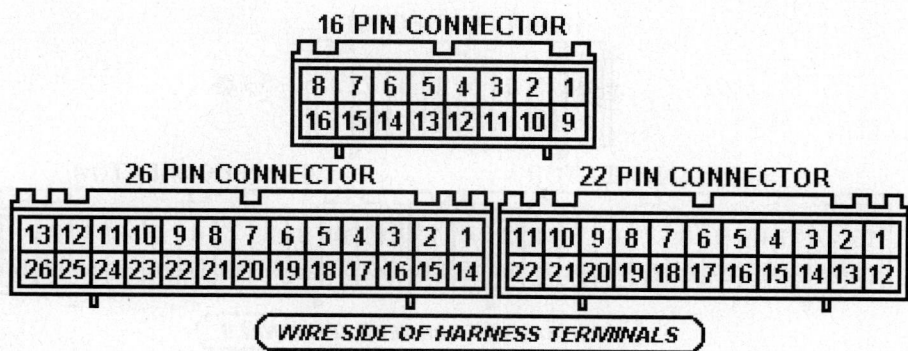

05533_ADIA_G125

Pin Connector Graphic
1996-98 2.0L I4 MFI VIN P (All) 26 Pin Connector

PCM Pin #	Wire Color	Circuit Description (26 Pin)	Value at Hot Idle
1	BR	Sensor Ground	<0.050v
2 (Cal)	RD	AFRS-11 (B1 S1) Heater	1v (Heater on)
2 (Fed)	RD	HO2S-11 (B1 S1) Heater	1v (Heater on)
3	BL/YL	Igniter Signal (IGF)	Digital Signal: 0-5-0v
4	RD	CKP Sensor Signal (NE+)	AC pulse signals
5	BK	CMP Sensor Signal (G+)	AC pulse signals
6	BL	Cruise Control ECU (IDLO)	1.5v, off-idle: 12v
7	WT/RD	4WD Speed Sensor (FR-)	Pulse Signals
8	YL/RD	4WD Speed Sensor (FR+)	Pulse Signals
9	BK/YL	ISC Signal (ISCC)	Pulse Signals
10	BK/BL	ISC Signal (ISCO)	Pulse Signals
11	BK	Injector 2 Control	2.0-3.3 ms
12	BK/RD	Injector 1 Control	2.0-3.3 ms
13	BR	Power Ground	<0.1v
14	BR	Sensor Ground	<0.050v
15	PK	A/T-ECT Solenoid (S2)	1st or OD: 1v
16	LG	A/T-ECT Solenoid (ST)	Pulse Signals
17	PK/BK	A/T-ECT Solenoid (S1)	In 3rd or OD: 1v
18	WT	CKP/CMP Sensor Signal (-)	<0.050v
19	RD/WT	HO2S-12 (B1 S2) Heater	1v (Heater on)
20	BK	Igniter Transistor 1 Control	6°, at 55 mph: 8° dwell
21	BK	Igniter Transistor 2 Control	6°, at 55 mph: 8° dwell
22	PK	EVAP Purge Solenoid (VSV)	12v or 0v
23	BK/WT	EGR Solenoid Control (VSV)	12v or 0v
24	BK/BL	Injector 4 Control	2.0-3.3 ms
25	BK/YL	Injector 3 Control	2.0-3.3 ms
26	BR	Power Ground	<0.1v

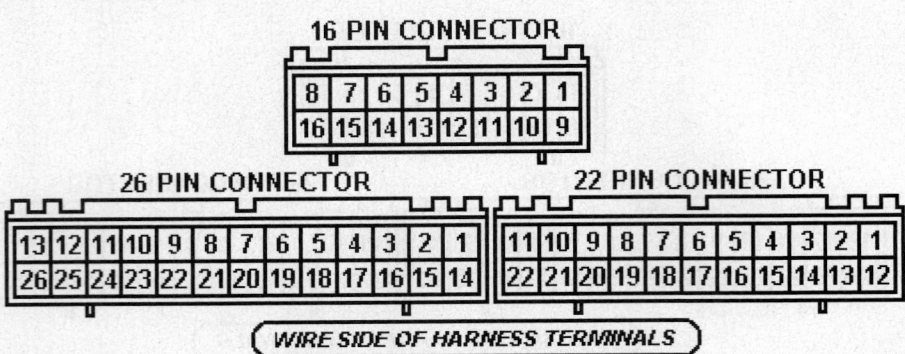

Pin Connector Graphic
1999-2000 2.0L I4 MFI VIN P (AII) E4 26P Connector

PCM Pin #	Wire Color	Circuit Description (26 Pin)	Value at Hot Idle
1	BR	Sensor Ground	<0.050v
2 (Cal)	RD	AFRS-11 (B1 S1) Heater	1v (Heater on)
2 (Fed)	RD	HO2S-11 (B1 S1) Heater	1v (Heater on)
3	BL/YL	Igniter Signal (IGF)	Digital Signal: 0-5-0v
4	RD	CKP Sensor Signal (NE+)	AC pulse signals
5	BK	CMP Sensor Signal (G+)	AC pulse signals
6	BL	Cruise Control ECU (IDLO)	1.5v, off-idle: 12v
7-8	---	Not Used	---
9	BK/YL	ISC Signal (ISCC)	Pulse Signals
10	BK/BL	ISC Signal (ISCO)	Pulse Signals
11	BK	Injector 2 Control	2.0-3.3 ms
12	BK/RD	Injector 1 Control	2.0-3.3 ms
13	BR	Power Ground	<0.1v
14	BR	Sensor Ground	<0.050v
15-17	---	Not Used	---
18	WT	CKP/CMP Sensor Signal (-)	<0.050v
19	RD/WT	HO2S-12 (B1 S2) Heater	1v (Heater on)
20	BK	Igniter Transistor 1 Control	6°, at 55 mph: 8° dwell
21	BK	Igniter Transistor 2 Control	6°, at 55 mph: 8° dwell
22	PK	EVAP Purge Solenoid (VSV)	12v or 0v
23	BK/WT	EGR Solenoid Control (VSV)	12v or 0v
24	BK/BL	Injector 4 Control	2.0-3.3 ms
25	BK/YL	Injector 3 Control	2.0-3.3 ms
26	BR	Power Ground	<0.1v

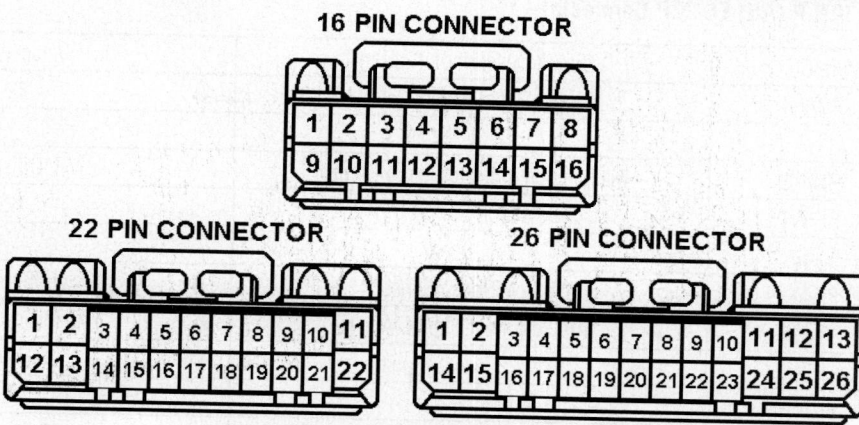

16 PIN CONNECTOR

22 PIN CONNECTOR

26 PIN CONNECTOR

WIRE SIDE OF HARNESS TERMINALS

Pin Connector Graphic

05533_ADIA_G126

1999-2000 2.0L I4 MFI VIN P (All) E6 16P Connector

PCM Pin #	Wire Color	Circuit Description (16 Pin)	Value at Hot Idle
1	YL	Sensor VREF	4.9-5.1v
2	GN/BK	MAP Sensor Signal	Idling: 1-1.5v
3	YL/BK	IAT Sensor Signal (THA)	At 100°F: 2.60v
4	WT	ECT Sensor Signal (THW)	At 180°F: 0.51v
5	RD	HO2S-12 (B1 S2) Signal	0.1-1.1v
6 (Cal)	WT	AFS-11 (B1 S1) Signal (+)	Fixed at 3.3v
6 (Fed)	WT	HO2S-11 (B1 S1) Signal	0.1-1.1v
7	BL/YL	EVAP Vapor Pressure Sensor (PTNK)	2.5-3.1v (with cap off)
8	RD/WT	EVAP Vapor Pressure Valve (VSV)	12v or 0v
9	BR	Sensor Ground	<0.050v
10	BK	Knock Sensor Signal	<0.075v AC
11	BL/RD	TP Sensor Signal (VTA)	0.3-0.8v
12-13	---	Not Used	---
14 (Cal)	BK	AFS-11 (B1 S1) Signal (-)	Fixed at 3.0v
15	BR	Sensor Ground	<0.050v
16	---	Not Used	---

1999-2000 2.0L I4 MFI VIN P (All) E6 22P Connector

PCM Pin #	Wire Color	Circuit Description (22 Pin)	Value at Hot Idle
1	RD/WT	Direct Battery	12-14v
2-4	---	Not Used	---
5	GN/RD	Malfunction Indicator Lamp Control	MIL Off: 12v, On: 1v
6	BL/WT	Data Link Connector	12-14v
7-8	---	Not Used	---
9	PK/WT	Vehicle Speed Sensor	At 55 mph: 48 Hz
10	YL/GN	A/C Amplifier Signal (AC1)	Clutch On: 1.5v, Off: 12v
11	BK	Starter Switch Signal	In P/N: 0-3.0v
12	BK/WT	EFI Main Relay B+	12-14v
13	BK	Tachometer Signal (TACO)	Pulse Signals
14	GN/RD	Circuit Opening Relay (FC)	0-3v, off-idle: 12v
15, 18-19	---	Not Used	---
16	WT	SIL Signal (Scan Tool)	12v
17	BK	Integration Relay Signal	Load On: 12-14v
20	YL/BK	Cruise Control Signal (OD1)	At Cruise in OD: 12v
21	RD/YL	A/C Amplifier Signal (ACT)	Clutch On: 12v, Off: 1.5v
22	BK/WT	Start Circuit Signal	Cranking: 9-11v

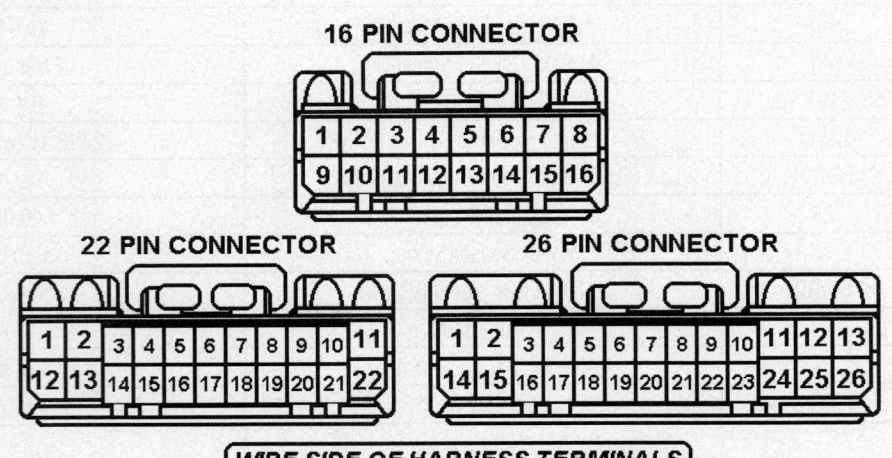

05533_ADIA_G126

Pin Connector Graphic

Standard Colors and Abbreviations

Abbreviation	Color	Abbreviation	Color	Abbreviation	Color
BK	Black	GY	Gray	RD	Red
BL	Blue	GN	Green	TN	Tan
BR	Brown	LG	LT Green	VT	Violet
DB	Dark Blue	OR	Orange	WT	White
DG	DK Green	PK	Pink	YL	Yellow

2001-06 2.0L I4 (All) VIN K E4 31 Pin Connector

PCM Pin #	Wire Color	Circuit Description (31 Pin)	Value at Hot Idle
1	BK/RD	Injector 1 Control	2.0-3.3 ms
2	BK	Injector 2 Control	2.0-3.3 ms
3	WT	Injector 3 Control	2.0-3.3 ms
4	RD	Injector 4 Control	2.0-3.3 ms
5	YL/GN	A/T Solenoid (SLT-)	Pulse Signals
6	YL/RD	A/T Solenoid (SLT+)	Pulse Signals
7	GN/WT	A/T Solenoid (SL1+)	Pulse Signals
8	PK	A/T: Solenoid (SL2+)	Pulse Signals
9	VT	A/T Solenoid (SL1-)	Pulse Signals
10	GN/BK	Igniter Transistor 1 Control	6% duty cycle
11	BL/BK	Igniter Transistor 2 Control	6% duty cycle
12	GN	Igniter Transistor 3 Control	6% duty cycle
13	BK	Igniter Transistor 4 Control	6% duty cycle
14	RD	Counter Gear Speed Input (+)	AC pulse signals
15	B	Turbine Speed Sensor Input -	AC pulse signals
16	WT	Turbine Speed Sensor Input +	AC pulse signals
17	GY	A/T: Trans. Oil Temp. Sensor	At 230°F: 0.95v
18	BK/BL	Idle Air Control Valve (RSD)	Pulse Signals
19	WT/RD	Cam Timing Oil Valve (OCV+)	Pulse signals
20	YL/BK	A/T Solenoid (SL2-)	Pulse Signals
21	WT/BK	Power Ground (E01)	<0.1v
22	WT	ECT Sensor Signal (THW)	At 180°F: 0.51v
23	RD/WT	IAT Sensor Signal (THA)	At 100°F: 2.60v
24	BL/RD	TP Sensor Signal (VTA)	0.3-1.0v
25	RD	Igniter Signal (IGF)	Digital Signal: 0-5-0v
26	GN	Counter Gear Speed Input (-)	AC pulse signals
27	WT	Knock Sensor Signal (KNK1)	<0.075v AC
28	---	Not Used	---
29	WT	Cam Timing Oil Valve (OCV-)	Pulse signals
30	WT/BK	Power Ground (E03)	<0.1v
31	WT/BK	Power Ground (E02)	<0.1v

E4 31-PIN CONNECTOR

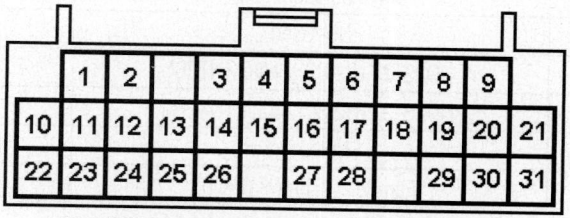

(WIRE SIDE OF HARNESS TERMINALS)

Pin Connector Graphic

05533_ADIA_G127

2001-06 2.0L I4 VIN H (All) E5 24 Pin Connector

PCM Pin #	Wire Color	Circuit Description (24 Pin)	Value at Hot Idle
1	WT/BK	Power Ground (E04)	<0.1v
2	YL	Sensor VREF (VC)	4.9-5.1v
3	RD/BK	HO2S-11 (B1 S2) Heater (HT1B)	Heater Off: 0v, On: 12v
4	WT/RD	AFS-21 (B2 S1) Heater (HAF2A)	Heater Off: 0v, On: 12v
5	WT/BL	AFS-11 (B1 S1) Heater (HAF1A)	Heater Off: 0v, On: 12v
6	PK	EVAP Purge Solenoid (VSV)	12v or 0v
7	---	Not Used	---
8	WT/BK	Power Ground (E05)	<0.1v
9-10	---	Not Used	---
11	BK	HO2S-11 (B1 S2) Signal (OX1B)	0.1-1.1v
12	BL/WT	MAF Sensor Signal (VG)	1.0-1.5v
13	WT	AFS-21 (B2 S1) Sensor (AF2A+)	Fixed at 3.3v
14	GN	AFS-11 (B1 S1) Sensor (AF1A+)	Fixed at 3.3v
15	YL	CMP Sensor Signal (G22+)	AC pulse signals
16	RD	CKP Sensor Signal (NE+)	AC pulse signals
17	BR	Power Ground (E1)	<0.1v
18	BR	Sensor Ground (E2)	<0.050v
19-20	---	Not Used	---
21	BL	MAF Sensor Ground (E2G)	<0.050v
22	BK	AFS-21 (B2 S1) Sensor (AF2A-)	Fixed at 3.0v
23	RD	AFS-11 (B1 S1) Sensor (AF1A-)	Fixed at 3.0v
24	PK	CKP/CMP Sensor Ground (-)	<0.050v

2001-06 2.0L I4 VIN H (All) E6 17 Pin Connector

PCM Pin #	Wire Color	Circuit Description (17 Pin)	Value at Hot Idle
1	YL	Sensor VREF	4.9-5.1v
2	YL/GN	EVAP Canister Closed Valve (VSV)	12v or 0v
3	GN/OR	EVAP Vapor Pressure Valve (VSV)	12v or 0v
4-5	---	Not Used	---
6	RD/WT	A/T-ECT Solenoid (DSL)	In Lockup: 12-14v
7-10	---	Not Used	---
11	PK	PSP Switch Signal (PSW)	Straight: 12v, Turning: 0v
12	YL	A/T Solenoid (S4)	Pulse Signals
13	---	Not Used	---
14	RD/YL	HO2S-22 (B2 S2) Heater (HT2B)	Heater Off: 12v, On: 1v
15	WT	HO2S-22 (B2 S2) Signal (OX2B)	0.1-1.1v
16	---	Not Used	---
17	LG/BK	Fan Relay Control (FAN)	Relay Off: 12v, On: 1v

E5 24-PIN CONNECTOR

1	2		3	4	5	6	7
8	9	10	11	12	13	14	15	16
17	18	19	20	21		22	23	24

E6 17-PIN CONNECTOR

1	2	3	4	5	6
7	8	9	10	11	12
13	14	15		16	17

WIRE SIDE OF HARNESS TERMINALS

Pin Connector Graphic

2001-06 2.0L I4 VIN H (All) E7 28 Pin Connector

05533_ADIA_G128

PCM Pin #	Wire Color	Circuit Description (28 Pin)	Value at Hot Idle
1	BL/BK	A/T Select Switch Drive	In Drive: 12-14v
2	RD/BK	A/T Select Switch Reverse	In 'R': 12v, Others: 0v
3	LG/BK	A/T Select Switch 2nd	In 2nd: 12v, Others: 0v
4	YL/GN	A/C Amplifier (AC1) Signal	Relay Off: 1.5v, On: 12v
5	PK/BK	Data Link Connector (TC)	12v
6	---	Not Used	---
7	YL/BK	Cruise Control Signal (OD1)	At Cruise in OD: 12v
8-11	---	Not Used	---
12	LG	A/T Select Switch Low	In Low: 12v, Others: 0v
13	---	Not Used	---
14	GN	Engine Hot Indicator Control	Indicator Off: 12v, On: 1v
15	RD/YL	A/C Amplifier Signal (ACMG)	Relay Off: 12v, On: 1v
16	PK	Oil Pressure Lamp Control	Lamp Off: 12v, On: 1v
17-19	---	Not Used	---
20	BK/WT	Neutral Start Switch Signal (NSW)	In P/N: 0v, Others: 12v
21	BL/BK	EVAP Vapor Pressure Sensor (PTNK)	2.5-3.1v (with cap off)
22	VT/WT	Vehicle Speed Sensor	At 55 mph: 48 Hz
23-24	---	Not Used	---
25	BL	Cruise Control Signal (IDLO)	1.5v, off-idle: 12v
26	---	Not Used	---
27	BK	Tachometer Signal (TACO)	DC pulse signals
28	---	Not Used	---

2001-06 2.0L I4 VIN H (All) E8 22 Pin Connector

PCM Pin #	Wire Color	Circuit Description (22 Pin)	Value at Hot Idle
1	RD/WT	Direct Battery	12-14v
2	BK/OR	Ignition Switch Power (IGSW)	12-14v
3	GN/RD	Circuit Opening Relay (FC)	Relay Off: 12v, On: 1v
4	WT	SIL Signal (Scan Tool)	12v
5	LG/BK	Overdrive On Lamp Control	Lamp Off: 12v, On: 1v
6	GN/RD	Malfunction Indicator Lamp Control	MIL Off: 12v, On: 1v
7	BK/RD	Starter Switch Signal (STA)	Cranking: 9-11v
8	GN/WT	EFI Main Relay Control	Relay Off: 0v, On: 12v
9	---	Not Used	---
10	LG	Overdrive Main Switch	Switch Off: 12v, On: 1v
11	BL	Center Airbag Assembly (F/PS)	Digital Signals
12-13	---	Not Used	---
14	GY	Combination Meter Hot Light	Light Off: 12v, On: 1v
15	GN/WT	Stop Light Switch Signal	Brake Off: 0v, On: 12v
16	BK/RD	Circuit Opening Relay (FC)	12-14v
17	PK	Data Link Connector (WFSE)	12v
18	PK	Defroster Switch & Taillight Switch Signal	Switch Off: 0v, On: 12v
19-21	---	Not Used	---
22	BR	Power Ground (EOM)	<0.1v

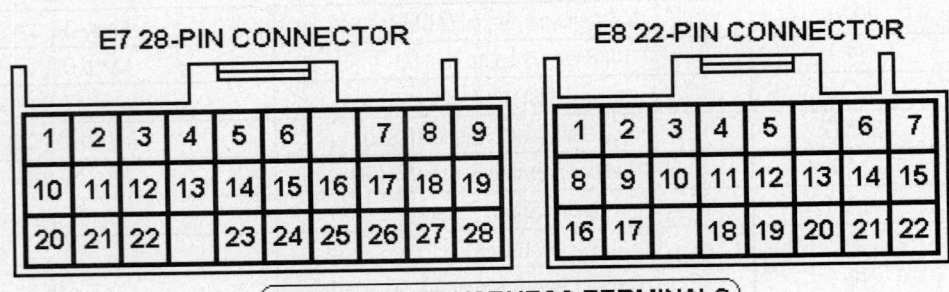

05533_ADIA_G129

Pin Connector Graphic

2001-02 4.7L V8 VIN T (All) E4 22 Pin Connector

PCM Pin #	Wire Color	Circuit Description (22 Pin)	Value at Hot Idle
1	BR/RD	Direct Battery	12-14v
2	PK	Neutral Start Switch (NSW)	In P/N: 0-3.0v
3	PK/BK	TC Signal to D6 DLC3	12v
4	VT	Fuel Pump Relay Control	Relay Off: 12v, On: 1v
5	BK/WT	EFI Main Relay Control	Relay Off: 0v, On: 12v
6	GN/OR	Fuel Control Switch Signal	Closed: 0v, Open: 12v
7	WT/GN	BM (+) Power	12-14v
8	BK/YL	EFI Main Relay B1+	12-14v
9	BK/RD	Ignition Switch Power (IGSW)	12-14v
10	OR	Power Ground (EOM)	<0.1v
12	YL	Transponder Amplifier Signal (RXCK)	Inserting key: pulses
13	LG	Transponder Amplifier Signal (RXCK)s	Inserting key: pulses
14	RD/YL	Transponder Amplifier Signal (Code)	Inserting key: pulses
15	GN/YL	Stop Light Switch Signal	Brake Off: 0v, On: 12v
16	BK/YL	EFI Main Relay B1+	12-14v
17	GN/RD	SIL Signal (Scan Tool)	Transmitting: pulses
18	BL	Center Airbag Sensor Assembly (F/PS)	Digital Signals
19	WT/RD	Data Link Connector (WFSE)	N/A
20	YL/BK	A/C Amplifier Signal (THWO)	A/C Off: 12v, On: 1v
21	PK/BL	A/C Amplifier Signal (AC1)	Clutch On: 1.5v, Off: 12v
22	LG/BK	A/C Amplifier Signal (ACT)	Clutch On: 12v, Off: 1.5v

2001-02 4.7L V8 VIN T (A/T) E5 28 Pin Connector

PCM Pin #	Wire Color	Circuit Description (28 Pin)	Value at Hot Idle
1	LG	A/T Select Switch Low	In 'L': 1v, Others: 12v
2	BL	A/T Select Switch 2nd	In 2nd: 12v, Others: 0v
3	BK/YL	A/T Select Switch Reverse	In 'R': 12v, Others: 0v
4	YL/GN	Tachometer Signal (TACO)	Pulse Signals
5	GN/OR	VSS Signal (Combo Meter)	At 55 mph: 48 Hz
6	GN	Detection Switch Transfer	Open: 12v, Closed: 1v
7	BK	Starter Switch Signal	In P/N: 0-3.0v
8	BL/YL	4WD ECU C/C Signal	12v
9	BL/RD	4WD Detection Transfer (L4)	Open: 12v, Closed: 0v
12	GN/WT	Defogger & Tail Light Switch	Switch Off: 0v, On: 12v
15	BL/WT	OD Main Switch ODMS Input	Open: 12v, Closed: 1v
17	YL	ABS/BA/TRAC/VSC ECU	Digital Signals (NEO)
18	RD	TRAC Engine (ENG-) Signal	Pulse Signals
19	GN	TRAC TRC (TRC-) Signal	Pulse Signals
20	WT/BL	EVAP Canister Closed Valve (CCV)	12v or 0v
21	PK/BL	EVAP Vapor Pressure Valve (VSV)	12v or 0v
22	RD/GN	EVAP Vapor Pressure Sensor (PTNK)	2.5-3.1v (with cap off)
23	VT/WT	Malfunction Indicator Lamp Control	MIL Off: 12v, On: 1v
24	YL/RD	A/T: Oil Temp. Lamp Control	Lamp Off: 12v, On: 1v
25	LG/BK	Security Indictor Light (LED)	LED Off: 12v, On: 1v
26	BL/OR	Overdrive Lamp Control	At Cruise in OD: 1v
27	PK	TRAC Engine (ENG+) Signal	Pulse Signals
28	RD/WT	TRAC TRC (TRC+) Signal	Pulse Signals

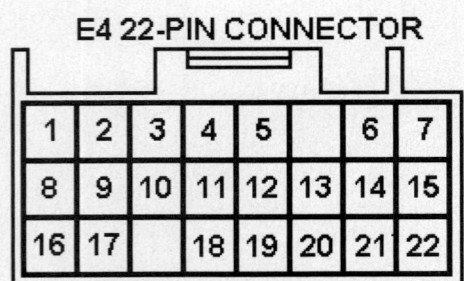

E4 22-PIN CONNECTOR

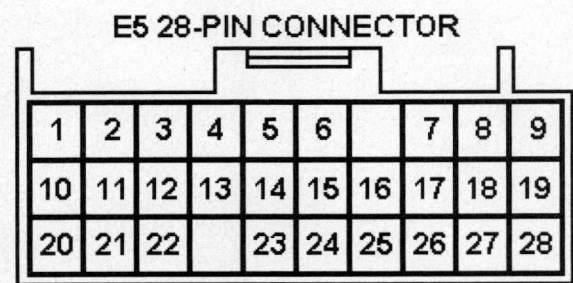

E5 28-PIN CONNECTOR

05533_ADIA_G130

Pin Connector Graphic

2001-02 4.7L V8 VIN T (A/T) E6 17 Pin Connector

PCM Pin #	Wire Color	Circuit Description (17 Pin)	Value at Hot Idle
1	RD	A/T ECT Solenoid (S1)	In 3rd or OD: 1v
2	WT/BL	A/T ECT Solenoid (S2)	In 1st or OD: 1v
3	GN/RD	A/T ECT Solenoid (SL)	In Lockup: 12-14v
4	WT	Direct Clutch Speed Input (+)	AC pulse signals
5	YL/RD	A/T: VSS Signal 2 (SP2+)	AC pulse signals
9	BK/RD	A/T-ECT Solenoid (SLT+)	Pulse Signals

2001-02 4.7L V8 VIN T (A/T) E6 17 Pin Connector, *continued*

10	BK	Direct Clutch Speed Input (-)	AC pulse signals
11	WT/RD	A/T: VSS Signal 2 (SP2-)	AC pulse signals
15	GN/YL	A/T-ECT Solenoid (SLT-)	AC pulse signals
17	RD/YL	A/T Oil Temperature Sensor	At 68°F: 4-5v

2001-02 4.7L V8 VIN T (A/T) E7 24 Pin Connector

PCM Pin #	Wire Color	Circuit Description (24 Pin)	Value at Hot Idle
1	WT/BK	Power Ground (E03)	<0.1v
2	GN/BK	Sensor VREF (VC)	4.9-5.1v
3	BL/RD	HO2S-21 (B2 S1) Heater	1v (Heater on)
4	BL	HO2S-11 (B1 S1) Heater	1v (Heater on)
5	RD	Injector 1 Control	2.0-3.3 ms
6	WT	Injector 2 Control	2.0-3.3 ms
7	WT/GN	EVAP Purge Solenoid (VSV)	12v or 0v
8	---	Not Used	---
9	BL/YL	Accelerator Position Sensor 2	1.8-2.7v
10	RD/WT	Mass Airflow Sensor (VG)	1.1-1.5v
11	WT	HO2S-21 (B2 S1) Signal	0.1-1.1v
12	BK	HO2S-11 (B1 S1) Signal	0.1-1.1v
13	BK/YL	TP Sensor Signal (VTA)	0.4-1.0v
14	GN/YL	ECT Sensor Signal (THW)	At 180°F: 0.51v
15	RD	HO2S-22 (B2 S2) Signal	0.1-1.1v
16	GN	HO2S-12 (B1 S2) Signal	0.1-1.1v
17	BR	Power Ground (E1)	<0.1v
18	GN/WT	Sensor Ground (E2)	<0.050v
19	BK/WT	MAF Sensor Ground (EVG)	<0.050v
20	PK/BL	TP Sensor Signal (VTA2)	2.0-2.9v
21	GN/RD	Accelerator Position Sensor 1	0.3-0.90v
22	YL/GN	IAT Sensor Signal (THA)	At 100°F: 2.60v
23	YL	HO2S-22 (B2 S2) Heater	1v (Heater on)
24	WT/RD	HO2S-12 (B1 S2) Heater	1v (Heater on)

E6 17-PIN CONNECTOR   **E7 24-PIN CONNECTOR**

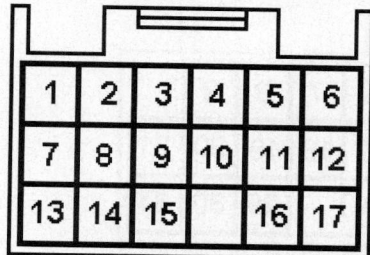
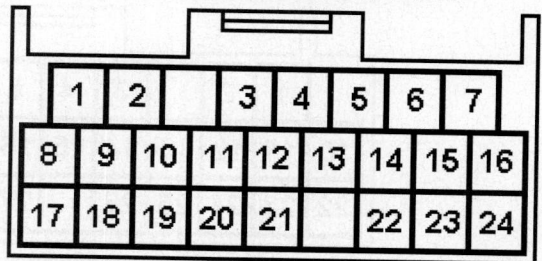

WIRE SIDE OF HARNESS TERMINALS

05533_ADIA_G131

Pin Connector Graphic

2001-02 4.7L V8 VIN T (A/T) E8 31 Pin Connector

PCM Pin #	Wire Color	Circuit Description (31 Pin)	Value at Hot Idle
1	GN	Injector 3 Control	2.0-3.3 ms
2	RD/BL	Injector 4 Control	2.0-3.3 ms
3	BL	Injector 5 Control	2.0-3.3 ms
4	YL	Injector 6 Control	2.0-3.3 ms
5	BL/RD	Injector 7 Control	2.0-3.3 ms
6	RD/WT	Injector 8 Control	2.0-3.3 ms
7	PK	Throttle Control Motor (M-)	Pulse Signals
8	VT	Throttle Control Motor (M+)	Pulse Signals
9	WT/BK	Power Ground (ME01)	<0.1v
10	YL	CMP Sensor Signal (G+)	AC pulse signals
11	BK/BL	Igniter Transistor 1 Control	7% duty cycle
12	LG/BK	Igniter Transistor 2 Control	7% duty cycle
13	GN/BK	Igniter Transistor 3 Control	7% duty cycle
14	RD/WT	Igniter Transistor 4 Control	7% duty cycle
15	GN/WT	Igniter Transistor 5 Control	7% duty cycle
16	PK/BL	Igniter Transistor 6 Control	7% duty cycle
17	BK	Knock Sensor 2 Signal (right)	<0.075v AC
18	GY	Knock Sensor 1 Signal (left)	<0.075v AC
19	BL/WT	Throttle Control Motor (CL-)	Pulse Signals
21	WT/BK	Power Ground (E01)	<0.1v
22	RD	CKP Sensor Signal (NE-)	<0.050v
23	GN	CKP Sensor Signal (NE+)	AC pulse signals
24	BL	CMP Sensor Signal (G-)	AC pulse signals
25	GN/RD	Igniter Transistor 7 Control	7% duty cycle
26	LG	Igniter Transistor 8 Control	7% duty cycle
27	BK/RD	Igniter Signal (IGF1)	Digital Signal: 0-5-0v
28	BK/WT	Igniter Signal (IGF2)	Digital Signal: 0-5-0v
29	GN	Throttle Control Motor (CL+)	Pulse Signals
30	BL/RD	Shield Ground (GE01)	<0.050v
31	WT/BK	Power Ground (E02)	<0.1v

E8 31-PIN CONNECTOR

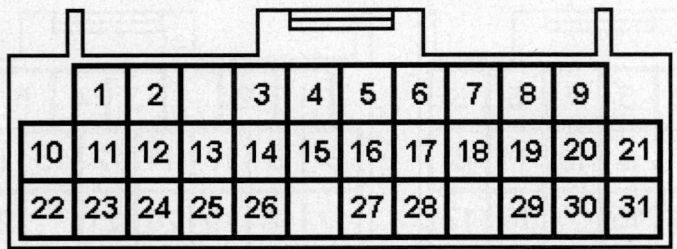

WIRE SIDE OF HARNESS TERMINALS

05533_ADIA_G132

Pin Connector Graphic

2003-2006 4.7L V8 VIN T E4 31 Pin Connector

PCM Pin #	Wire Color	Circuit Description (31 Pin)	Value at Hot Idle
1	BK/YL	EFI Main Relay Power (+B)	12-14v
2	BK/YL	EFI Main Relay Power (+B2)	12-14v
3	BK/YL	Direct Battery	12-14v
4	PK/BL	EVAP Pressure Switching Valve (VSV)	12v or 0v
5	YL/GN	Tachometer Signal (TACO)	Pulse Signals
6	GN/WT	A/T Select Switch Park Signal (P)	In 'P': 12v, Others: 0v
7	GN/RD	A/T Select Switch Neutral Signal (N)	In 'N': 12v, Others: 0v
8	BK/WT	EFI Main Relay Control	Relay Off: 0v, On: 12v
9	BK/OR	Ignition Switch Power (IGSW)	12-14v
10	GN/OR	Circuit Opening Relay (FC)	0-3v, off-idle: 12v
11	VT/WT	Malfunction Indicator Lamp Control	MIL Off: 12v, On: 1v
12	GN/YL	Taillight Switch Signal (ELS)	Taillights Off: 0v, On: 12v
13	YL/RD	Horn Relay Control	Relay Off: 12v, On: 1v
14	BL	Center Airbag Assembly (F/PS)	Digital Signals
15	OR	Power Ground (EOM)	<0.1v
16	---	Not Used	---
17	YL	Transponder Amplifier Signal (NEO)	Inserting key: pulses
18	GN/RD	SIL Signal (Scan Tool)	Transmitting: pulses
19	RD	Data Link Connector (WFSE)	N/A
20	PK/BK	Data Link Connector (TC)	12v
21	RD/GN	EVAP Vapor Pressure Sensor (PTNK)	2.5-3.1v (with cap off)
22	GN/RD	Accelerator Position Sensor 1 (VPA)	0.3-0.90v
23	BL/YL	Accelerator Position Sensor 2 (VPA2)	1.8-2.7v
24	PK	Traction Control Engine Signal (ENG+)	Pulse Signals
25	BL	Traction Control Signal (TRC+)	Pulse Signals
26	GN/BK	Accelerator Pedal Position Sensor 1 VREF	4.9-5.1v
27	BL/RD	Accelerator Pedal Position Sensor 2 VREF	4.9-5.1v
28	GN/WT	Accelerator Pedal Position Sensor (EPA)	<0.050v
29	BL/BK	Accelerator Position Sensor 2 (EPA2)	1.8-2.7v
30	RD	Traction Control Engine Signal (ENG-)	Pulse Signals
31	LG	Traction Control Signal (TRC-)	Pulse Signals

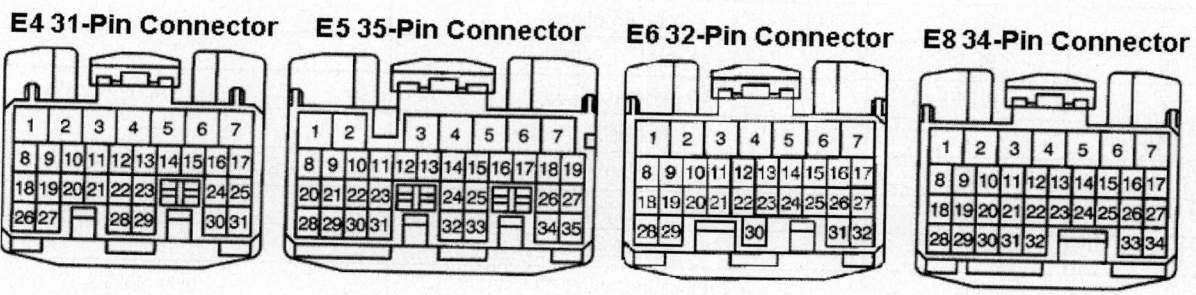

E4 31-Pin Connector   **E5 35-Pin Connector**   **E6 32-Pin Connector**   **E8 34-Pin Connector**

05533_ADIA_G133

Pin Connector Graphic

2003-2006 4.7L V8 VIN T E5 35 Pin Connector

PCM Pin #	Wire Color	Circuit Description (35 Pin)	Value at Hot Idle
1	WT/BK	Power Ground (HP)	<0.1v
2	BL	A/C Magnetic Clutch Relay (ACMG)	Relay Off: 0v, On: 12v
3	GN	A/T Select Switch 2nd Signal (2L)	In 2nd: 12v, Others: 0v
4	BL/RD	Low Detection Switch (L4)	Switch Open: 0v, Closed: 12v
5	BL/WT	ECT Pattern Switch 2nd Position (SNW1)	2nd Position: 12v
6	WT/GN	ETCS Power (+BM)	12-14v
7-8	---	Not Used	---
9	GN/BK	Shift Lock ECU Control (2)	12v or 0v
10	GN/YL	Shift Lock ECU Control (D)	12v or 0v
11	RD/BK	A/T Select Switch Reverse Signal	In 'R': 1v, Others: 12v
12, 15-16	---	Not Used	---
14	WT/RD	A/C Amplifier Signal (THWO)	A/C Off: 12v, On: 1v
17	VT	Vehicle Speed Sensor (SPD)	At 55 mph: 48 Hz
18	PK/BK	Body Control ECU Signal (MPX1)	Digital Signals
19	GN/YL	Stop Light Switch Signal	Brake Off: 0v, On: 12v
20-22, 25	---	Not Used	---
23	GN/RD	Shift Lock ECU Control (4)	12v or 0v
26	BL/YL	Transponder Amplifier Signal (IMD)	Inserting key: pulses
27	PK	Transponder Amplifier Signal (IMI)	Inserting key: pulses
28	BL/WT	ECT Pattern Switch Power Signal	Power Position: 12v
29	BK	Body Control ECU Signal (MPX2)	Digital Signals
30, 34-35	---	Not Used	---
31	GR/BL	A/C Amplifier Signal (A/CS)	Relay Off: 12v, On: 1v
32	PK/BK	A/C Amplifier Signal (THE)	A/C Off: 12v, On: 1v
33	LG/BK	A/C Switch Signal (ACLD)	A/C On: 12v, Off: 0v

2003-2006 4.7L V8 VIN T E6 32 Pin Connector

PCM Pin #	Wire Color	Circuit Description (32 Pin)	Value at Hot Idle
1	BR	Power Ground (E1)	<0.1v
2	PK	Throttle Control Motor (M-)	Pulse Signals
3	VT	Throttle Control Motor (M+)	Pulse Signals
4	WT/BK	Power Ground (ME01)	<0.1v
5	RD/WT	Injector 8 Control	2.0-3.3 ms
6	BL/RD	Injector 7 Control	2.0-3.3 ms
7	WT/BK	Power Ground (E03)	<0.1v
8-10	---	Not Used	---
11	YL/GN	Neutral Detection Switch (L4)	Switch Open: 0v, Closed: 12v
12	BK	Start Switch Signal (STSW)	Cranking: 9-11v
13-14	---	Not Used	---
15	BK	A/T Solenoid Control (SLU-)	Pulse Signals
16	PK/GN	A/T Solenoid Control (SLU+)	Pulse Signals
17	BL/RD	Shield Ground (GE01)	<0.050v
18-20	---	Not Used	---
21	BK/OR	Generator Control (RL)	12v

2003-2006 4.7L V8 VIN T E6 32 Pin Connector, *continued*

22, 26	---	Not Used	---
23	BL	A/C Lock Sensor (LCK)	12v or 0v
24	RD	CKP Sensor Signal (NE-)	<0.050v
25	GN	CKP Sensor Signal (NE+)	AC pulse signals
27	YL	CMP Sensor Signal (G2+)	AC pulse signals
28-31	---	Not Used	---
32	BL	CMP Sensor Signal (G2-)	AC pulse signals

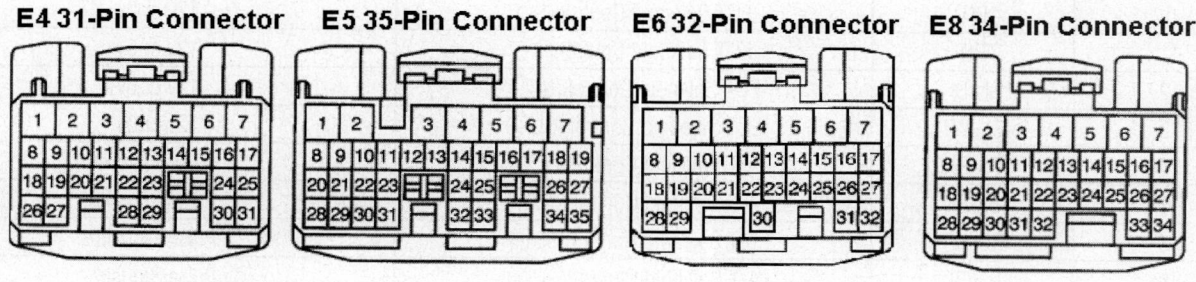

Pin Connector Graphic

05533_ADIA_G133

2003-2006 4.7L V8 VIN T E7 35 Pin Connector

PCM Pin #	Wire Color	Circuit Description (35 Pin)	Value at Hot Idle
1	GY	Knock Sensor 1 Signal (KNK1)	<0.075v AC
2	BK	Knock Sensor 2 Signal (KNK2)	<0.075v AC
3	YL	Injector 6 Control	2.0-3.3 ms
4	BL/RD	HO2S-11 (B1 S1) Heater (HT1A)	1v (Heater on)
5	WT/RD	HO2S-12 (B1 S2) Heater (HT1B)	1v (Heater on)
7-8, 20	---	Not Used	---
9	PK	Start Signal (STAR)	In P/N: 0-3.0v
10	WT	ECT Solenoid Control (S2)	12v or 0v
11	RD	ECT Solenoid Control (S1)	12v or 0v
12	GN/BK	ECT Solenoid Control (SLT-)	12v or 0v
13	GN/WT	ECT Solenoid Control (SLT+)	12v or 0v
16	PK/BK	A/T Solenoid Control (SL2-)	Pulse Signals
17	PK/BK	A/T Solenoid Control (SL2+)	Pulse Signals
18	RD/WT	A/T Solenoid Control (SL1-)	Pulse Signals
19	RD/BL	A/T Solenoid Control (SL1+)	Pulse Signals
21	RD	HO2S-22 (B2 S2) Signal (OX2B)	0.1-1.1v
22	WT	HO2S-21 (B2 S1) Signal (OX2A)	0.1-1.1v
23	BK	HO2S-11 (B1 S1) Signal (OX1A)	0.1-1.1v
24	BL	A/T Oil Temperature Sensor 2 (THO2)	At 68°F: 4-5v
25	YL	HO2S-22 (B2 S2) Heater (HT2B)	1v (Heater on)
26	RD	ECT Vehicle Speed Sensor (SP2+)	Pulse Signals
27	BL	ECT Turbine Speed Sensor (NT-)	Pulse Signals
28	---	Not Used	---
29	GN	HO2S-12 (B1 S2) Signal (OX1B)	0.1-1.1v
30-31	---	Not Used	---
32	GN/YL	A/T Oil Temperature Sensor 1 (THO1)	At 68°F: 4-5v
33	BL	HO2S-21 (B2 S1) Heater (HT2A)	1v (Heater on)
34	GN	ECT Vehicle Speed Sensor (SP2-)	Pulse Signals
35	WT	ECT Turbine Speed Sensor (NT+)	Pulse Signals

2003-2006 4.7L V8 VIN T E8 34 Pin Connector

PCM Pin #	Wire Color	Circuit Description (34 Pin)	Value at Hot Idle
1	RD	Injector 1 Control	2.0-3.3 ms
2	WT	Injector 2 Control	2.0-3.3 ms
3	GN	Injector 3 Control	2.0-3.3 ms
4	RD/BK	Injector 4 Control	2.0-3.3 ms
5	BL	Injector 5 Control	2.0-3.3 ms
6	WT/BK	Power Ground (E02)	<0.1v
7	WT/BK	Power Ground (E01)	<0.1v
8	LG/BK	Igniter Transistor 2 Control	7% duty cycle
9	BK/BL	Igniter Transistor 1 Control	7% duty cycle
10	LG	Igniter Transistor 8 Control	7% duty cycle
11	RD/WT	Igniter Transistor 4 Control	7% duty cycle
12	GN/WT	Igniter Transistor 5 Control	7% duty cycle

2003-2006 4.7L V8 VIN T E8 34 Pin Connector, *continued*

13	GN/RD	Igniter Transistor 7 Control	7% duty cycle
15	BK/OR	A/C Relay Control (ACCR)	Relay Off: 12v, On: 1v
16	PK	Neutral Start Switch (NSW)	In P/N: 0-3.0v
17	BL/RD	Starter Switch Signal (STA)	In P/N: 0-3.0v
18	GN/BK	Sensor VREF (VC)	4.9-5.1v
19	GN/YL	ECT Sensor Signal (THW)	At 180°F: 0.51v
20	YL/GN	IAT Sensor Signal (THA)	At 100°F: 2.60v
21	BK/YL	TP Sensor Signal (VTA1)	0.4-1.0v
22, 32	---	Not Used	---
23	BK/WT	Igniter Signal (IGF2)	Digital Signal: 0-5-0v
24	BK/RD	Igniter Signal (IGF1)	Digital Signal: 0-5-0v
25	GN/BK	Igniter Transistor 3 Control	7% duty cycle
26	PK/BL	Igniter Transistor 6 Control	7% duty cycle
27	WT/BL	EVAP Canister Closed Valve (CCV)	12v or 0v
28	GN/WT	Sensor Ground (E2)	<0.050v
29	BK/WT	MAF Sensor Ground (E2G)	<0.050v
30	RD/WT	Mass Airflow Sensor (VG)	1.1-1.5v
31	PK/BL	TP Sensor Signal (VTA2)	2.0-2.9v
33	VT	Fuel Pump Relay Control (FPR)	Relay Off: 12v, On: 1v
34	BL/BK	EVAP Purge Solenoid (VSV)	12v or 0v

4RUNNER PIN CHARTS

1990-92 2.4L I4 MFI VIN R (A/T-ECT) 16 Pin Connector

PCM Pin #	Wire Color	Circuit Description (16 Pin)	Value at Hot Idle
1	GN/YL	TP Sensor VREF	4.9-5.1v
2	GN/BK	Airflow Meter VREF	4.9-5.1v
3	YL/GN	IAT Sensor Signal (THA)	At 100°F: 2.60v
4	GN/BL	ECT Sensor Signal (THW)	At 180°F: 0.51v
5	BK	Knock Sensor Signal	<0.075v AC
6	BK	HO2S-11 (B1 S1) Signal	0.1-1.1v
7	GN/YL	4WD Oil Temperature Sensor	At 230°F: <1.5v
8	YL	Check Connector	12-14v
9	BR/BK	Sensor Ground	<0.050v
10	YL/BL	Airflow Meter Signal	0.5-2.5v
11	YL	TP Sensor Signal (VTA)	0.3-0.8v
12	YL/BL	Closed Throttle Switch	1v, off-idle: 12v
13 (Cal)	GN/WT	EGR Gas Temperature Sensor	3.5-4.0v
15	PK/WT	Check Connector	12-14v

1990-92 2.4L I4 MFI VIN R (A/T-ECT) 22 Pin Connector

PCM Pin #	Wire Color	Circuit Description (22 Pin)	Value at Hot Idle
1	BK/GN	Direct Battery	12-14v
4	BL/WT	A/T Oil Temperature Indicator	Lamp Off: 12v, On: 1v
5	PK	Malfunction Indicator Lamp Control	MIL Off: 12v, On: 1v
6	GN/WT	Stop Light Switch Signal	Brake Off: 0v, On: 12v
7	RD/BL	A/T Pattern Select Switch	Norm: 0v, PWR: 12v
8	PK/GN	4WD Indicator Switch	Switch Off: 12v, On: 1v
9	GN/BL	Vehicle Speed Sensor	At 55 mph: 48 Hz
11	BK/WT	Starter Switch Signal	9-11v (cranking)
12	WT/RD	EFI Main Relay Power	12-14v
13	WT/RD	EFI Main Relay Power	12-14v
15	BR/BK	Sensor Ground	<0.050v
16	YL/GN	Overdrive Main Switch	Switch Off: 12v, On: 1v
20	PK	Check Connector	12v
21	YL/RD	Cruise Control ECU	At Cruise in OD: 12v

Standard Colors and Abbreviations

Abbreviation	Color	Abbreviation	Color	Abbreviation	Color
BK	Black	GY	Gray	RD	Red
BL	Blue	GN	Green	TN	Tan
BR	Brown	LG	LT Green	VT	Violet
DB	Dark Blue	OR	Orange	WT	White
DG	DK Green	PK	Pink	YL	Yellow

1990-92 2.4L I4 MFI VIN R (A/T-ECT) 26 Pin Connector

PCM Pin #	Wire Color	Circuit Description (26 Pin)	Value at Hot Idle
1	BK/OR	Distributor Signal (NE+)	AC pulse signals
2	GN	Cold Start Injector Control	1v (at cold startup)
3	BK/YL	Igniter Signal (IGF)	Digital Signal: 0-5-0v
4	RD/YL	A/T-ECT Solenoid (S4)	Pulse Signals
5	RD/GN	A/T-ECT Solenoid (S3)	Pulse Signals
6	RD/WT	A/T-ECT Solenoid (S2)	1st or OD: 1v
7	BL/RD	A/T-ECT Solenoid (S1)	3rd or OD: 1v
10	PK/GN	HO2S-11 (B1 S1) Heater	1v (Heater on)
11	GN	Fuel Pressure Up Solenoid	1v (at hot restart)
12	WT/RD	Injector Pair 1 & 3 Control	2.0-3.3 ms
13	BR	Power Ground	<0.1v
15	BR	Shield Ground	<0.050v
16	BL	A/T Speed Sensor Signal	Moving: 0-5-0-5v
17	PK/WT	A/T Select Switch Low	In Low: 12v, Others: 0v
18	PK/GN	A/T Select Switch 2nd	In 2nd: 12v, Others: 0v
19	PK/RD	A/T Select Switch Neutral	In 'N': 12v, Others: 0v
20	RD	4WD Detection Transfer (L4)	Open: 12v, Closed: 0v
21	BK/BL	Igniter Signal (IGT)	Digital Signal: 0-5-0v
23	GN/RD	Intake Air Solenoid Control	12v or 0v
25	WT	Injector Pair 2 & 4 Control	2.0-3.3 ms
26	BR	Power Ground	<0.1v

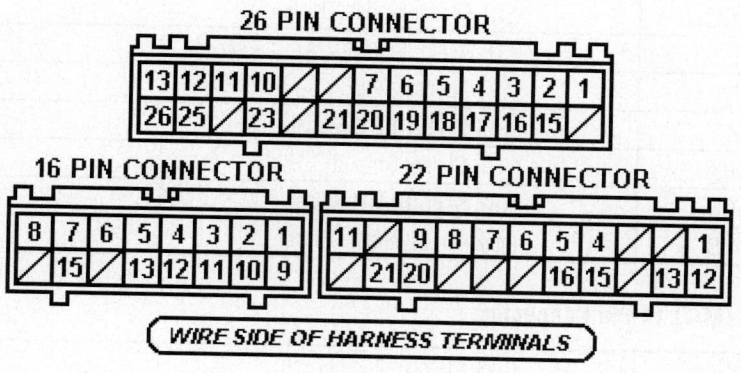

Pin Connector Graphic

Standard Colors and Abbreviations

Abbreviation	Color	Abbreviation	Color	Abbreviation	Color
BK	Black	GY	Gray	RD	Red
BL	Blue	GN	Green	TN	Tan
BR	Brown	LG	LT Green	VT	Violet
DB	Dark Blue	OR	Orange	WT	White
DG	DK Green	PK	Pink	YL	Yellow

05533_ADIA_G134

1990-92 2.4L I4 MFI VIN R (M/T) 10 Pin Connector

PCM Pin #	Wire Color	Circuit Description (10 Pin)	Value at Hot Idle
1	BK/WT	A/T Neutral Start Switch	In P/N: 0-3.0v
1	BK	M/T Clutch Start Switch	9-11v (cranking)
2	GN	Cold Start Injector Control	1v (at cold startup)
3	BK/WT	Starter Switch Signal	9-11v (cranking)
4	WT/RD	Injector Pair 1 & 3 Control	2.0-3.3 ms
5	BR	Power Ground	<0.1v
6	YL	Check Connector	12-14v
7	BR	Sensor Ground	<0.050v
8	BK/BL	Igniter Signal (IGT)	Digital Signal: 0-5-0v
9	WT	Injector Pair 2 & 4 Control	2.0-3.3 ms
10	BR	Power Ground	<0.1v

1990-92 2.4L I4 MFI VIN R (M/T) 18 Pin Connector

PCM Pin #	Wire Color	Circuit Description (18 Pin)	Value at Hot Idle
1	BK/OR	Distributor Signal (NE+)	AC pulse signals
2	BK	Knock Sensor Signal	<0.075v AC
3 (Cal)	GN/WT	EGR Gas Temperature Sensor	3.5-4.0v
5	BK/YL	Igniter Signal (IGF)	Digital Signal: 0-5-0v
6	YL/BL	Closed Throttle Switch	1v, off-idle: 12v
7	PK/WT	Check Connector	12-14v
8	PK	Malfunction Indicator Lamp Control	MIL Off: 12v, On: 1v
9	GN	Fuel Pressure Up Solenoid	1v (at hot restart)
10	GN/BL	ECT Sensor Signal (THW)	At 180°F: 0.51v
11	YL	TP Sensor Signal (VTA)	0.3-0.8v
12	GN/YL	Sensor VREF	4.9-5.1v
13	BK	HO2S-11 (B1 S1) Signal	0.1-1.1v
14	BR/BK	Sensor Ground	<0.050v
15	PK/GN	HO2S-11 (B1 S1) Heater	1v (Heater on)
17	GN/RD	Intake Air Solenoid Control	12v or 0v

1990-92 2.4L I4 MFI VIN R (M/T) 14 Pin Connector

PCM Pin #	Wire Color	Circuit Description (14 Pin)	Value at Hot Idle
1	WT/RD	EFI Main Relay Power	12-14v
2	GN/BK	Direct Battery	12-14v
3	YL/GN	IAT Sensor Signal (THA)	At 100°F: 2.60v
4	YL/BL	Airflow Meter Signal	0.5-2.5v
5	GN/BK	Airflow Meter VREF	4.9-5.1v
8	WT/RD	EFI Main Relay Power	12-14v
9	GN/WT	Stop Light Switch Signal	Brake Off: 0v, On: 12v
10	GN/BL	Vehicle Speed Sensor	At 55 mph: 48 Hz
11	GN/YL	4WD Indicator Switch	Switch Off: 12v, On: 1v
12	BR	Sensor Ground	<0.050v
14	YL/RD	Overdrive Relay Control	Relay Off: 12v, On: 1v

Pin Connector Graphic

1993 2.4L I4 MFI VIN R (A/T-ECT) 26 Pin Connector

PCM Pin #	Wire Color	Circuit Description (26 Pin)	Value at Hot Idle
1	GN	Cold Start Injector Control	1v (at cold startup)
2	PK/GN	Main O2S Heater Control	1v (Heater on)
3	BK/YL	Igniter Signal (IGF)	Digital Signal: 0-5-0v
4	BK/OR	Distributor Signal (NE+)	AC pulse signals
5	RD/GN	A/T-ECT Solenoid (S3)	Pulses
6	RD/WT	A/T-ECT Solenoid (S2)	1st or OD: 1v
7	BL/RD	A/T-ECT Solenoid (S1)	3rd or OD: 1v
8	PK	EGR Solenoid Control (VSV)	12v or 0v
9	GN/RD	Intake Air Solenoid Control	12v or 0v
10	GN	Fuel Pressure Up Solenoid	1v (at hot restart)
11	WT	Injector Pair 2 & 4 Control	2.0-3.3 ms
12	WT/RD	Injector Pair 1 & 3 Control	2.0-3.3 ms
13	BR	Power Ground	<0.1v
14	BR	Sensor Ground	<0.050v
15	GN/RD	Sub O2S Heater Control	1v (Heater on)
16	BL	A/T Speed Sensor Signal	Moving: 0-5-0-5v
17	PK/WT	A/T Select Switch Low	In Low: 12v, Others: 0v
18	PK/GN	A/T Select Switch 2nd	In 2nd: 12v, Others: 0v
19	PK/RD	A/T Select Switch Neutral	In 'N': 12v, Others: 0v
20	RD	4WD Detection Transfer (L4)	Open: 12v, Closed: 0v
22	BK/BL	Igniter Signal (IGT)	Digital Signal: 0-5-0v
26	BR	Power Ground	<0.1v

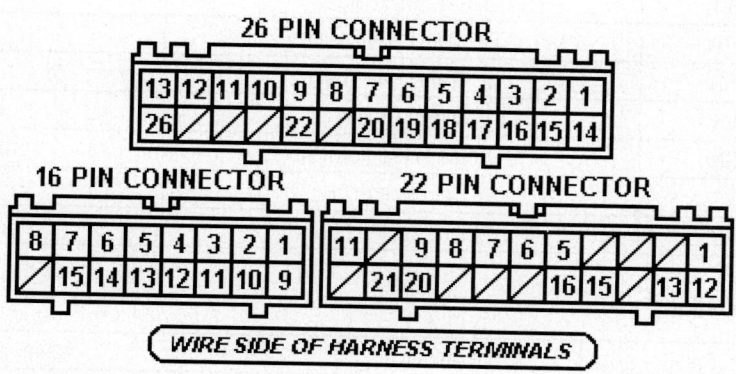

Pin Connector Graphic

Standard Colors and Abbreviations

Abbreviation	Color	Abbreviation	Color	Abbreviation	Color
BK	Black	GY	Gray	RD	Red
BL	Blue	GN	Green	TN	Tan
BR	Brown	LG	LT Green	VT	Violet
DB	Dark Blue	OR	Orange	WT	White
DG	DK Green	PK	Pink	YL	Yellow

1993 2.4L I4 MFI VIN R (A/T-ECT) 16 Pin Connector

PCM Pin #	Wire Color	Circuit Description (16 Pin)	Value at Hot Idle
1	GN/YL	TP Sensor VREF	4.9-5.1v
2	YL/BL	Airflow Meter Signal	0.5-2.5v
3	YL/GN	IAT Sensor Signal (THA)	At 100°F: 2.60v
4	GN/BL	ECT Sensor Signal (THW)	At 180°F: 0.51v
5	WT	Sub O2S Signal	0.1-1.1v
6	BK	Main O2S Signal	0.1-1.1v
7	BK	Knock Sensor Signal	<0.075v AC
8	YL	Check Connector	12-14v
9	BR/BK	Sensor Ground	<0.050v
10	YL/BL	Airflow Meter VREF	4.9-5.1v
11	YL	TP Sensor Signal (VTA)	0.3-0.8v
12	YL/BL	Closed Throttle Switch	1v, off-idle: 12v
13 (Cal)	GN/WT	EGR Gas Temperature Sensor	3.5-4.0v
14	PK/GN	Check Connector	12-14v
15	PK/WT	Check Connector	12-14v

1993 2.4L I4 MFI VIN R (A/T-ECT) 22 Pin Connector

PCM Pin #	Wire Color	Circuit Description (22 Pin)	Value at Hot Idle
1	BK/GN	Direct Battery	12-14v
5	PK	Malfunction Indicator Lamp Control	MIL Off: 12v, On: 1v
6	GN/WT	Stop Light Switch Signal	Brake Off: 0v, On: 12v
7	RD/BL	A/T Pattern Select Switch	Norm: 0v, PWR: 12v
8	PK/GN	4WD Indicator Switch	Switch Off: 12v, On: 1v
9	GN/BL	Vehicle Speed Sensor	At 55 mph: 48 Hz
11	BK/WT	Starter Switch Signal	In P/N: 0-3.0v
12	WT/RD	EFI Main Relay Power	12-14v
13	WT/RD	EFI Main Relay Power	12-14v
15	BR/BK	Sensor Ground	<0.050v
16	YL/GN	Overdrive Main Switch	Switch Off: 12v, On: 1v
20	PK	Check Connector	12-14v
21	YL/RD	Cruise Control ECU	At Cruise in OD: 12v

Standard Colors and Abbreviations

Abbreviation	Color	Abbreviation	Color	Abbreviation	Color
BK	Black	GY	Gray	RD	Red
BL	Blue	GN	Green	TN	Tan
BR	Brown	LG	LT Green	VT	Violet
DB	Dark Blue	OR	Orange	WT	White
DG	DK Green	PK	Pink	YL	Yellow

1993 2.4L I4 MFI VIN R (M/T) 16 Pin Connector

PCM Pin #	Wire Color	Circuit Description (16 Pin)	Value at Hot Idle
1	PK/GN	Main O2S Heater Control	1v (Heater on)
2	BK/WT	Starter Switch Signal	9-11v (cranking)
3	BK/YL	Igniter Signal (IGF)	Digital Signal: 0-5-0v
4	BK/OR	Distributor Signal (NE+)	AC pulse signals
9	GN/RD	Intake Air Solenoid Control	12v or 0v
10	GN	Fuel Pressure Up Solenoid	1v (at hot restart)
11	GN	Cold Start Injector Control	1v (at cold startup)
12	WT	Injector Pair 2 & 4 Control	2.0-3.3 ms
13	BR	Power Ground	<0.1v
14	RD/GN	Sub O2S Heater Control	1v (Heater on)
15	BK/WT	A/T: Neutral Start Switch	In P/N: 0-3.0v
15	BK	M/T: Clutch Start Switch	9-11v (cranking)

1993 2.4L I4 MFI VIN R (M/T) 26 Pin Connector

PCM Pin #	Wire Color	Circuit Description (26 Pin)	Value at Hot Idle
1	YL/GN	IAT Sensor Signal (THA)	At 100°F: 2.60v
2	YL/BL	Airflow Meter Signal	0.5-2.5v
3	GN/BK	Airflow Meter VREF	4.9-5.1v
4	GN/BL	ECT Sensor Signal (THW)	At 180°F: 0.51v
5	WT	Sub O2S Signal	0.1-1.1v
6	BK	Main O2S Signal	0.1-1.1v
7	PK/GN	Data Link Connector	12-14v
8	YL	Data Link Connector	12-14v
9	BR/BK	Sensor Ground	<0.050v
10	YL	TP Sensor Signal (VTA)	0.3-0.8v
11	GN/YL	TP Sensor VREF	4.9-5.1v
12	YL/BL	Closed Throttle Switch	1v, off-idle: 12v
13 (Cal)	GN/WT	EGR Gas Temperature Sensor	3.5-4.0v
14	BK	Knock Sensor Signal	<0.075v AC
15	PK/GN	Data Link Connector	12-14v
16	BR/BK	Sensor Ground	<0.050v
22	BK/BL	Igniter Signal (IGT)	Digital Signal: 0-5-0v
23	PK	EGR Solenoid Control (VSV)	12v or 0v
24	BR	Sensor Ground	<0.050v
25	WT	Injector Pair 1 & 3 Control	2.0-3.3 ms
26	BR	Power Ground	<0.1v

1993 2.4L I4 MFI VIN R (M/T) 12 Pin Connector

PCM Pin #	Wire Color	Circuit Description (12 Pin)	Value at Hot Idle
1	WT/RD	EFI Main Relay Power	12-14v
2	BK/GN	Direct Battery	12-14v
6	BR/YL	4WD Indicator Switch	Switch Off: 12v, On: 1v
7	WT/RD	EFI Main Relay Power	12-14v
8	PK	Malfunction Indicator Lamp Control	MIL Off: 12v, On: 1v
11	GN/BL	Vehicle Speed Sensor	At 55 mph: 48 Hz
12	GN/WT	Stop Light Switch Signal	Brake Off: 0v, On: 12v

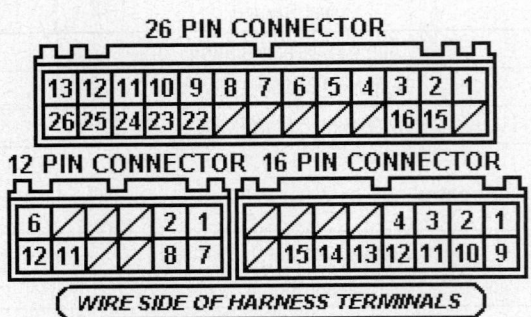

05533_ADIA_G137

Pin Connector Graphic

1994-95 2.4L I4 MFI VIN R (A/T-ECT) 26 Pin Connector

PCM Pin #	Wire Color	Circuit Description (26 Pin)	Value at Hot Idle
1	GN	Cold Start Injector Control	1v (at cold startup)
2	PK/GN	Main O2S Heater Control	1v (Heater on)
3	BK/YL	Igniter Signal (IGF)	Digital Signal: 0-5-0v
4	BK/OR	Distributor Signal (NE+)	AC pulse signals
5	RD/GN	A/T-ECT Solenoid (S3)	Pulse Signals
6	RD/WT	A/T-ECT Solenoid (S2)	1st or OD: 1v
7	BL/RD	A/T-ECT Solenoid (S1)	3rd or OD: 1v
9	GN/RD	Intake Air Solenoid Control	12v or 0v
10	GN	Fuel Pressure Up Solenoid	1v (at hot restart)
11	WT	Injector Pair 2 & 4 Control	2.0-3.3 ms
12	WT/RD	Injector Pair 1 & 3 Control	2.0-3.3 ms
13	BR	Power Ground	<0.1v
14	BR	Sensor Ground	<0.050v
19	BL	A/T Speed Sensor Signal	Moving: 0-5-0-5v
20	BK/BL	Igniter Signal (IGT)	Digital Signal: 0-5-0v
21	PK/WT	A/T Select Switch Low	In Low: 12v, Others: 0v
22	PK/GN	A/T Select Switch 2nd	In 2nd: 12v, Others: 0v
23	PK/RD	A/T Select Switch Neutral	In 'N': 12v, Others: 0v
26	BR	Power Ground	<0.1v

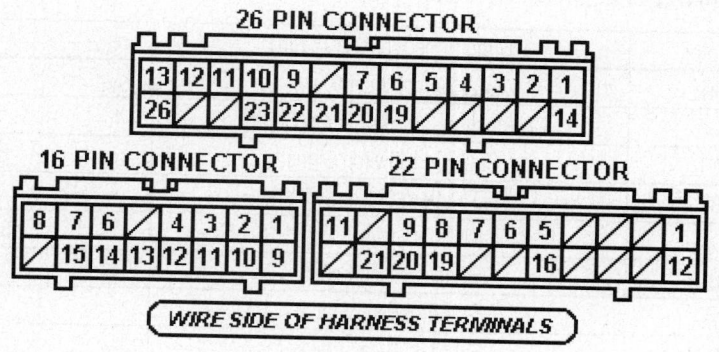

WIRE SIDE OF HARNESS TERMINALS

05533_ADIA_G138

Pin Connector Graphic
Standard Colors and Abbreviations

Abbreviation	Color	Abbreviation	Color	Abbreviation	Color
BK	Black	GY	Gray	RD	Red
BL	Blue	GN	Green	TN	Tan
BR	Brown	LG	LT Green	VT	Violet
DB	Dark Blue	OR	Orange	WT	White
DG	DK Green	PK	Pink	YL	Yellow

1994-95 2.4L I4 MFI VIN R (A/T-ECT) 16 Pin Connector

PCM Pin #	Wire Color	Circuit Description (16 Pin)	Value at Hot Idle
1	GN/YL	Sensor VREF	4.9-5.1v
2	YL/BL	Airflow Meter Signal	0.5-2.5v
3	YL/GN	IAT Sensor Signal (THA)	At 100°F: 2.60v
4	GN/BL	ECT Sensor Signal (THW)	At 180°F: 0.51v
6	BK	HO2S-11 (B1 S1) Heater	0.1-1.1v
7	BK	Knock Sensor Signal	<0.075v AC
8	YL	Check Connector	12-14v
9	BR/BK	Sensor Ground	<0.050v
10	GN/BK	Airflow Meter VREF	4.9-5.1v
11	YL	TP Sensor Signal (VTA)	0.3-0.8v
12	YL/BL	Closed Throttle Switch	1v, off-idle: 12v
13 (Cal)	GN/WT	EGR Gas Temperature Sensor	3.5-4.0v
14	PK/GN	Check Connector	12-14v
15	PK/WT	Check Connector	12-14v

1994-95 2.4L I4 MFI VIN R (A/T-ECT) 22 Pin Connector

PCM Pin #	Wire Color	Circuit Description (22 Pin)	Value at Hot Idle
1	BK/GN	Direct Battery	12-14v
5	PK	Malfunction Indicator Lamp Control	MIL Off: 12v, On: 1v
6	GN/WT	Stop Light Switch Signal	Brake Off: 0v, On: 12v
7	RD/BL	A/T Pattern Select Switch	Norm: 0v, PWR: 12v
8	PK/GN	4WD Indicator Switch	Switch Off: 12v, On: 1v
9	GN/BL	Vehicle Speed Sensor	At 55 mph: 48 Hz
11	BK/YL	Starter Switch Signal	In P/N: 0-3.0v
12	WT/RD	EFI Main Relay Power	12-14v
16	YL/GN	Overdrive Main Switch	Switch Off: 12v, On: 1v
19	RD	4WD Detection Transfer (L4)	Open: 12v, Closed: 0v
20	PK	Check Connector	12-14v
21	YL/RD	Cruise Control ECU	At Cruise in OD: 12v

Standard Colors and Abbreviations

Abbreviation	Color	Abbreviation	Color	Abbreviation	Color
BK	Black	GY	Gray	RD	Red
BL	Blue	GN	Green	TN	Tan
BR	Brown	LG	LT Green	VT	Violet
DB	Dark Blue	OR	Orange	WT	White
DG	DK Green	PK	Pink	YL	Yellow

1994-95 2.4L I4 MFI VIN R (M/T) 16 Pin Connector

PCM Pin #	Wire Color	Circuit Description (16 Pin)	Value at Hot Idle
1	YL/GN	IAT Sensor Signal (THA)	At 100°F: 2.60v
2	YL/BL	Airflow Meter Signal	0.5-2.5v
3	GN/BK	Airflow Meter VREF	4.9-5.1v
4	GN/BL	ECT Sensor Signal (THW)	At 180°F: 0.51v
5 (Cal)	WT	Sub O2S Signal	0.1-1.1v
6	BK	Main O2S Signal	0.1-1.1v
7	PK/GN	Data Link Connector	12-14v
8	YL	Data Link Connector	12-14v
9	BR/BK	Sensor Ground	<0.050v
10	YL	TP Sensor Signal (VTA)	0.3-0.8v
11	GN/YL	TP Sensor VREF	4.9-5.1v
12	YL/BL	Closed Throttle Switch	1v, off-idle: 12v
13 (Cal)	GN/WT	EGR Gas Temperature Sensor	3.5-4.0v
14	BK	Knock Sensor Signal	<0.075v AC
15	PK/WT	Data Link Connector	12-14v

1994-95 2.4L I4 MFI VIN R (M/T) 26 Pin Connector

PCM Pin #	Wire Color	Circuit Description (26 Pin)	Value at Hot Idle
1	PK/GN	Main O2S Heater Control	1v (Heater on)
2	BK/WT	Starter Switch Signal	In P/N: 0-3.0v
3	BK/YL	Igniter Signal (IGF)	Digital Signal: 0-5-0v
4	BK/OR	Distributor Signal (NE+)	AC pulse signals
9	GN/RD	Intake Air Solenoid Control	12v or 0v
10	GN	Fuel Pressure Up Solenoid	1v (at hot restart)
11	GN	Cold Start Injector Control	1v (at cold startup)
12	WT/RD	Injector Pair 1 & 3 Control	2.0-3.3 ms
13	BR	Power Ground	<0.1v
14 (Cal)	RD/GN	Sub O2S Heater Control	1v (Heater on)
15	BK	Clutch Start Switch	9-11v (cranking)
22	BK/BL	Igniter Signal (IGT)	Digital Signal: 0-5-0v
23	PK	EGR Solenoid Control (VSV)	12v or 0v
24	BR	Sensor Ground	<0.050v
25	WT	Injector Pair 2 & 4 Control	2.0-3.3 ms
26	BR	Power Ground	<0.1v

1994-95 2.4L I4 MFI VIN R (M/T) 12 Pin Connector

PCM Pin #	Wire Color	Circuit Description (12 Pin)	Value at Hot Idle
2	BK/GN	Direct Battery	12-14v
6	GN/WT	4WD Indicator Switch	Switch Off: 12v, On: 1v
7	WT/RD	EFI Main Relay Power	12-14v
8	PK	Malfunction Indicator Lamp Control	MIL Off: 12v, On: 1v
11	GN/BL	Vehicle Speed Sensor	At 55 mph: 48 Hz
12	GN/WT	Stop Light Switch Signal	Brake Off: 0v, On: 12v

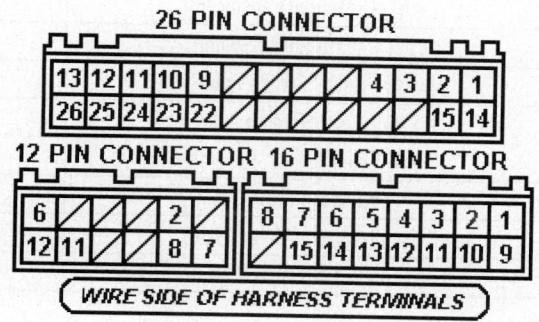

Pin Connector Graphic

05533_ADIA_G139

1996 2.7L I4 MFI VIN M (A/T-ECT) 16 Pin Connector

PCM Pin #	Wire Color	Circuit Description (16 Pin)	Value at Hot Idle
1	GN/YL	TP Sensor VREF	4.9-5.1v
2	BK/RD	MAF Sensor Signal	1.1-1.8v
3	RD/WT	MAF Sensor Ground (E2G)	<0.050v
4	GN/RD	ECT Sensor Signal (THW)	At 180ºF: 0.51v
5	WT	HO2S-11 (B1 S1) Signal	0.1-1.1v
6	GY	Knock Sensor Signal	<0.075v AC
7	PK/WT	Data Link Connector	12-14v
9	BR/BK	Sensor Ground	<0.050v
10	YL	TP Sensor Signal (VTA)	0.3-0.8v
11	YL/BL	Closed Throttle Switch	1v, off-idle: 12v
12	YL/GN	IAT Sensor Signal (THA)	At 100ºF: 2.60v
13	RD	HO2S-12 (B1 S2) Signal	0.1-1.1v
14	PK	EGR Gas Temperature Sensor	3.5-4.0v

1996 2.7L I4 MFI VIN M (A/T-ECT) 22 Pin Connector

PCM Pin #	Wire Color	Circuit Description (22 Pin)	Value at Hot Idle
2	BK/WT	Direct Battery	12-14v
3	YL	A/T Oil Temperature Indicator	Lamp Off: 12v, On: 1v
4	PK	Malfunction Indicator Lamp Control	MIL Off: 12v, On: 1v
5	BL/OR	Overdrive Main Switch	Switch Off: 12v, On: 1v
6	BL/BK	A/C Amplifier Signal (ACT)	Clutch On: 12v, Off: 1.5v
7	BL/YL	A/C Amplifier On Signal	Clutch On: 1.5v, Off: 12v
8	GN/OR	Vehicle Speed Sensor	At 55 mph: 48 Hz
10	GN/BK	4WD Indicator Switch	Switch Off: 12v, On: 1v
11	BK/WT	Starter Switch Signal	In P/N: 0-3.0v
12	WT/BL	EFI Main Relay Power	12-14v
13	RD/GN	4WD Detection Transfer (N)	Open: 12v, Closed: 0v
14	GN/RD	A/T Pattern Select Switch	Norm: 0v, PWR: 12v
15	LG	A/T Select Switch Low	In Low: 12v, Others: 0v
16	PK/RD	A/T Select Switch 2nd	In 2nd: 12v, Others: 0v
17	RD/YL	A/T Select Switch Reverse	In 'R': 12v, Others: 0v
18	BR/YL	Cruise Control ECU	At Cruise in OD: 12v
19	WT	Data Link Connector (SDL)	0v
21	GN/WT	Stop Light Switch Signal	Brake Off: 0v, On: 12v

Standard Colors and Abbreviations

Abbreviation	Color	Abbreviation	Color	Abbreviation	Color
BK	Black	GY	Gray	RD	Red
BL	Blue	GN	Green	TN	Tan
BR	Brown	LG	LT Green	VT	Violet
DB	Dark Blue	OR	Orange	WT	White
DG	DK Green	PK	Pink	YL	Yellow

1996 2.7L I4 MFI VIN M (A/T-ECT) 12 Pin Connector

PCM Pin #	Wire Color	Circuit Description (12 Pin)	Value at Hot Idle
1	BK/WT	EVAP Purge Solenoid (VSV)	12v or 0v
2	RD/BK	EVAP Vapor Pressure Valve (VSV)	12v or 0v
4	WT/RD	A/T Vehicle Speed Sensor (-)	Pulse Signals
5	BL	Distributor Signal (G-)	<0.050v
6	GN	CKP Sensor Signal (NE-)	<0.050v
10	YL/RD	A/T Vehicle Speed Sensor (+)	Pulse Signals
11	PK	Distributor Signal (G1+)	AC pulse signals
12	RD	CKP Sensor Signal (NE+)	AC pulse signals

1996 2.7L I4 MFI VIN M (A/T-ECT) 26-PK Connector

PCM Pin #	Wire Color	Circuit Description (26 Pin)	Value at Hot Idle
3	PK	HO2S-11 (B1 S1) Heater	1v (Heater on)
6	BK	IAC Signal (RSC)	Pulse Signals
7	BK/RD	Idle Air Control Valve (RSO)	Pulse Signals
8	RD/GN	A/T-ECT Solenoid (SL)	In Lockup: 12-14v
9	LG	A/T-ECT Solenoid (S2)	1st or OD: 1v
10	GN/RD	A/T-ECT Solenoid (S1)	3rd or OD: 1v
11	WT	Injectors 2 & 4 Control	2.0-3.3 ms
12	RD	Injectors 1 & 3 Control	2.0-3.3 ms
13	BR	Power Ground	<0.1v
14	GN/YL	Circuit Opening Relay (FC)	0-3v, off-idle: 12v
16	RD/WT	HO2S-12 (B1 S2) Heater	1v (Heater on)
17	BK/YL	Igniter Signal (IGF)	Digital Signal: 0-5-0v
18	GN/BK	EVAP Vapor Pressure Sensor (PTNK)	2.5-3.1v (with cap off)
21	PK/YL	A/T Oil Temperature Sensor	At 230°F: <1.5v
22	PK/BK	EGR Solenoid Control (VSV)	12v or 0v
23	BK/BL	Igniter Signal (IGT)	Digital Signal: 0-5-0v
24	BR	Sensor Ground	<0.050v
25	BR	Power Ground	<0.1v
26	BR	Power Ground	<0.1v

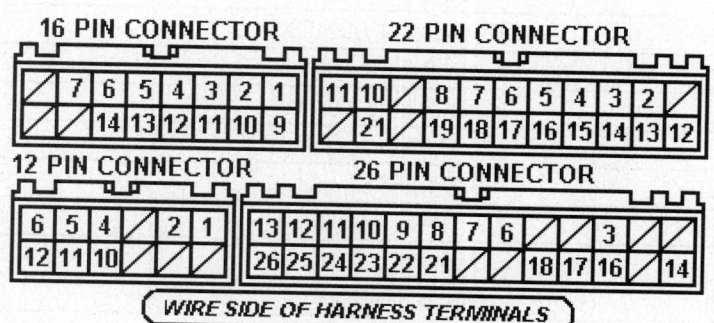

Pin Connector Graphic

05533_ADIA_G140

Standard Colors and Abbreviations

Abbreviation	Color	Abbreviation	Color	Abbreviation	Color
BK	Black	GY	Gray	RD	Red
BL	Blue	GN	Green	TN	Tan
BR	Brown	LG	LT Green	VT	Violet
DB	Dark Blue	OR	Orange	WT	White
DG	DK Green	PK	Pink	YL	Yellow

1996 2.7L I4 MFI VIN M (M/T) 16 Pin Connector

PCM Pin #	Wire Color	Circuit Description (16 Pin)	Value at Hot Idle
1	GN/YL	TP Sensor VREF	4.9-5.1v
2	BK/RD	MAF Sensor Signal	Idling: 1-1.5v
3	PK	EGR Gas Temperature Sensor	3.5-4.0v
4	GN/RD	ECT Sensor Signal (THW)	At 180°F: 0.51v
5	RD	HO2S-12 (B1 S2) Signal	0.1-1.1v
6	WT	HO2S-11 (B1 S1) Signal	0.1-1.1v
7	YL/GN	IAT Sensor Signal (THA)	At 100°F: 2.60v
9	BR/BK	Sensor Ground	<0.050v
10	GN/BK	EVAP Vapor Pressure Sensor (PTNK)	2.5-3.1v (with cap off)
11	YL	TP Sensor Signal (VTA)	0.3-0.8v
12	YL/BL	Closed Throttle Switch	1v, off-idle: 12v
13	GY	Knock Sensor Signal	<0.075v AC
15	PK/WT	Data Link Connector	12v
16	RD/WT	MAF Meter Return	<0.050v

1996 2.7L I4 MFI VIN M (M/T) 26 Pin Connector

PCM Pin #	Wire Color	Circuit Description (26 Pin)	Value at Hot Idle
2	PK	HO2S-11 (B1 S1) Heater	1v (Heater on)
3	BK/YL	Igniter Signal (IGF)	Digital Signal: 0-5-0v
4	RD	CKP Sensor Signal (NE+)	AC pulse signals
5	PK	Distributor Signal (G1+)	AC pulse signals
6	PK/BK	EGR Solenoid Control (VSV)	12v or 0v
9	BK	IAC Signal (RSC)	Pulse Signals
10	BK/RD	Idle Air Control Valve (RSO)	Pulse Signals
11	WT	Injector Pair 2 & 4 Control	2.0-3.3 ms
12	RD	Injector Pair 1 & 3 Control	2.0-3.3 ms
13, 14	BR	Power Ground	<0.1v
15	RD/WT	HO2S-12 (B1 S2) Heater	1v (Heater on)
17	GN	CKP Sensor Signal (NE-)	AC pulse signals
18	BL	Distributor Signal (G-)	<0.050v
20	BK/BL	Igniter Signal (IGT)	Digital Signal: 0-5-0v
21	RD/BK	EVAP Vapor Pressure Valve (VSV)	12v or 0v
23	BK/WT	EVAP Purge Solenoid (VSV)	12v or 0v
25, 26	BR	Power Ground	<0.1v

1996 2.7L I4 MFI VIN M (M/T) 22 Pin Connector

PCM Pin #	Wire Color	Circuit Description (22 Pin)	Value at Hot Idle
1	BK/WT	Direct Battery	12-14v
5	PK	Malfunction Indicator Lamp Control	MIL Off: 12v, On: 1v
7	WT	Data Link Connector (SDL)	0v
8	BL/BK	A/C Amplifier Signal (ACT)	Clutch On: 12v, Off: 1.5v
9	GN/OR	Vehicle Speed Sensor	At 55 mph: 48 Hz
10	BL/YL	A/C Amplifier Signal (AC1)	Clutch On: 1.5v, Off: 12v
11	BK/WT	Starter Switch Signal	In P/N: 0-3.0v
12	WT/BL	EFI Main Relay Power	12-14v
14	GN/YL	Circuit Opening Relay (FC)	0-3v, off-idle: 12v
20	GN/WT	Stop Light Switch Signal	Brake Off: 0v, On: 12v
21	GN/BK	4WD Indicator Switch	Switch Off: 12v, On: 1v

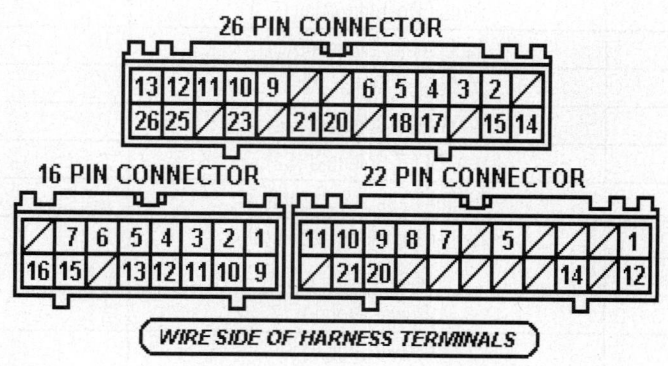

05533_ADIA_G141

Pin Connector Graphic

1997 2.7L I4 MFI VIN M (All) 16 Pin Connector

PCM Pin #	Wire Color	Circuit Description (16 Pin)	Value at Hot Idle
1	GN/YL	Sensor VREF	4.9-5.1
2	BK/RD	MAF Sensor Signal	1.1-1.8v
3	YL/GN	IAT Sensor Signal (THA)	At 100°F: 2.60v
4	GN/RD	ECT Sensor Signal (THW)	At 180°F: 0.51v
5	WT	HO2S-11 (B1 S1) Signal	0.1-1.1v
7	GN/BK	Data Link Connector	12-14v
8	RD/BK	EVAP Vapor Pressure Valve (VSV)	12v or 0v
9	BR/BK	Sensor Ground	<0.050v
10	YL	TP Sensor Signal (VTA)	0.3-0.8v
11	PK	EGR Gas Temperature Sensor	3.5-4.0v
12	GY	Knock Sensor Signal	<0.075v AC
13	RD	HO2S-12 (B1 S2) Signal	0.1-1.1v
15	PK/BK	EGR Solenoid Control (VSV)	12v or 0v

1997 2.7L I4 MFI VIN M (All) 26 Pin Connector

PCM Pin #	Wire Color	Circuit Description (26 Pin)	Value at Hot Idle
1	RD/WT	HO2S-12 (B1 S2) Heater	1v (Heater on)
2	PK	HO2S-11 (B1 S1) Heater	1v (Heater on)
3	BK/WT	EVAP Purge Solenoid	12-14v
4	YL/BL	Closed Throttle Switch	1v, off-idle: 12v
6	BK	IAC Signal (RSC)	Pulse Signals
7	BK/RD	Idle Air Control Valve (RSO)	Pulse Signals
8	GN/RD	A/T-ECT Solenoid (S1)	3rd or OD: 1v
9	RD/BL	Injector 4 Control	2.0-3.3 ms
10	GN	Injector 3 Control	2.0-3.3 ms
11	WT	Injector 2 Control	2.0-3.3 ms
12	RD	Injector 1 Control	2.0-3.3 ms
13	BR	Power Ground	<0.1v
14	GN/YL	Circuit Opening Relay (FC)	0-3v, off-idle: 12v
15	BK	Tachometer Signal (TACO)	Pulse Signals
17	BK/YL	Igniter Signal (IGF)	Digital Signal: 0-5-0v
20	RD/GN	A/T-ECT Solenoid (SL)	In Lockup: 12-14v
21	LG	A/T-ECT Solenoid (S2)	1st or OD: 1v
22	YL/BK	Igniter Transistor 2 Control	6°, at 55 mph: 8° dwell
23	BK/BL	Igniter Transistor 1 Control	6°, at 55 mph: 8° dwell
24	BR	Sensor Ground	<0.050v
25	BR	Power Ground	<0.1v
26	BR	Power Ground	<0.1v

1997 2.7L I4 MFI VIN M (All) 12 Pin Connector

PCM Pin #	Wire Color	Circuit Description (12 Pin)	Value at Hot Idle
1	BK/WT	EVAP Purge Solenoid	12-14v
2	OR	A/T: Defogger Idle-Up Signal	Switch On: 12v
3	WT/RD	A/T Vehicle Speed Sensor (-)	AC pulse signals
5	PK	CMP Sensor Signal (G-)	<0.050v
6	OR	CKP Sensor Signal (NE-)	<0.050v
7	RD/WT	MAF Meter Ground	<0.050v
9	YL/RD	A/T Vehicle Speed Sensor (+)	AC pulse signals
10	GN/BK	EVAP Vapor Pressure Sensor (PTNK)	2.5-3.1v
11	PK	CMP Sensor Signal (G+)	AC pulse signals
12	RD	CKP Sensor Signal (NE+)	AC pulse signals

1997 2.7L I4 MFI VIN M (All) 22 Pin Connector

PCM Pin #	Wire Color	Circuit Description (22 Pin)	Value at Hot Idle
2	BK/WT	Direct Battery	12-14v
3	YL	A/T Oil Temperature Indicator	Lamp Off: 12v, On: 1v
4	PK	Malfunction Indicator Lamp Control	MIL Off: 12v, On: 1v
5	BL/OR	Overdrive Main Switch	Switch Off: 12v, On: 1v
6	BL/BK	A/C Amplifier Signal (ACT)	Clutch On: 12v, Off: 1.5v

1997 2.7L I4 MFI VIN M (All) 22 Pin Connector, *continued*

7	BL/YL	A/C Amplifier Signal (AC1)	Clutch On: 1.5v, Off: 12v
8	GN/OR	Vehicle Speed Sensor	At 55 mph: 48 Hz
10	GN/BK	4WD Indicator Switch	Switch Off: 12v, On: 1v
11	BK/WT	Starter Switch Signal	In P/N: 0-3.0v
12	WT/BL	EFI Main Relay B+	12-14v
13	RD/GN	4WD Detection Transfer (N)	Open: 12v, Closed: 0v
14	GN/RD	AT ECT Pattern Select Switch	Norm: 0v, PWR: 12v
15	LG	A/T Select Switch Neutral	In 'N': 12v, Others: 0v
16	PK/RD	A/T Select Switch Low	In Low: 12v, Others: 0v
17	RD/YL	A/T Select Switch Reverse	In 'R': 12v, Others: 0v
18	BR/YL	Cruise Control ECU	At Cruise in OD: 12v
19	WT	SIL Signal (Scan Tool)	12v
21	GN/WT	Stop Light Switch Signal	Brake Off: 0v, On: 12v

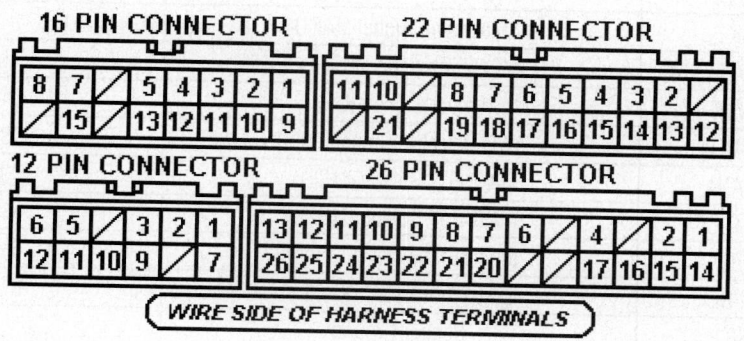

05533_ADIA_G142

Pin Connector Graphic

1998-99 2.7L I4 MFI VIN M (All) 16 Pin Connector

PCM Pin #	Wire Color	Circuit Description (16 Pin)	Value at Hot Idle
1	GN/YL	TP Sensor VREF	4.9-5.1v
2	BK/RD	MAF Sensor Signal	1.1-1.8v
3	YL/GN	IAT Sensor Signal (THA)	At 100°F: 2.60v
4	GN/RD	ECT Sensor Signal (THW)	At 180°F: 0.51v
5	WT	HO2S-11 (B1 S1) Signal	0.1-1.1v
7	GN/BK	Data Link Connector	12-14v
8	RD/BK	EVAP Vapor Pressure Valve (VSV)	12v or 0v
9	BR/BK	Sensor Ground	<0.050v
10	YL	TP Sensor Signal (VTA)	0.3-0.8v
11	PK	EGR Gas Temperature Sensor	3.5-4.0v
12	GY	Knock Sensor Signal	<0.075v AC
13	RD	HO2S-12 (B1 S2) Signal	0.1-1.1v
15	PK/BL	EGR Solenoid Control (VSV)	12v or 0v

1998-99 2.7L I4 MFI VIN M (All) 26 Pin Connector

PCM Pin #	Wire Color	Circuit Description (26 Pin)	Value at Hot Idle
1	RD/WT	HO2S-12 (B1 S2) Heater	1v (Heater on)
2	PK	HO2S-11 (B1 S1) Heater	1v (Heater on)
3	BK/WT	EVAP Purge Solenoid (VSV)	12v or 0v
4	YL/GN	Closed Throttle Switch	1v, off-idle: 12v
6	PK/BK	IAC Signal (RSC)	Pulse Signals
7	BK/RD	Idle Air Control Valve (RSO)	Pulse Signals
8	GN/RD	AT ECT Solenoid (S1)	3rd or OD: 1v
9	RD/BL	Injector 4 Control	2.0-3.3 ms
10	GN	Injector 3 Control	2.0-3.3 ms
11	WT	Injector 2 Control	2.0-3.3 ms
12	RD	Injector 1 Control	2.0-3.3 ms
13	BR	Power Ground	<0.1v
14	GN/YL	Circuit Opening Relay (FC)	0-3v, off-idle: 12v
15	BK	Tachometer Signal (TACO)	Pulse Signals
16	PK/YL	A/T Oil Temperature Sensor	At 68°F: 4-5v
17	BK/YL	Igniter Signal (IGF)	Digital Signal: 0-5-0v
18	BL/RD	4WD Detection Transfer (L4)	Switch Closed: 0.1v
20	RD/GN	A/T ECT Solenoid (SL)	In Lockup: 12-14v
21	LG	A/T ECT Solenoid (S2)	1st or OD: 1v
22	YL/BK	Igniter Transistor 2 Control	6°, at 55 mph: 8° dwell
23	BK/BL	Igniter Transistor 1 Control	6°, at 55 mph: 8° dwell
24	BR	Sensor Ground	<0.050v
25	BR	Power Ground	<0.1v
26	BR	Power Ground	<0.1v

1998-99 2.7L I4 MFI VIN M (All) 12 Pin Connector

PCM Pin #	Wire Color	Circuit Description (12 Pin)	Value at Hot Idle
2	OR	A/T: Defogger Idle-up Signal	Load On: 12v
3	WT/RD	A/T ECT Speed Sensor (-)	Pulses
5	PK	CMP Sensor Signal (G-)	AC pulse signals
6	OR	CKP Sensor Signal (NE-)	AC pulse signals
7	RD/WT	MAF Meter Ground	0.050v
9	YL/RD	A/T ECT Speed Sensor (+)	Pulses
10	GN/BK	EVAP Vapor Pressure Sensor (PTNK)	2.5-3.1v
11	PK	CMP Sensor Signal (G+)	AC pulse signals
12	RD	CKP Sensor Signal (NE+)	AC pulse signals

1998-99 2.7L I4 MFI VIN M (All) 22 Pin Connector

PCM Pin #	Wire Color	Circuit Description (22 Pin)	Value at Hot Idle
2	BL/RD	Direct Battery	12-14v
3	YL	A/T Oil Temperature Indicator	Light Off: 12v, On: 1v
4	PK	Malfunction Indicator Lamp Control	MIL Off: 12v, On: 1v
5	BL/OR	Overdrive Main Switch	Switch Off: 12v, On: 1v

1998-99 2.7L I4 MFI VIN M (All) 22 Pin Connector, *continued*

6	BL/BK	A/C Amplifier Signal (ACT)	Clutch On: 12v, Off: 1.5v
7	BL/YL	A/C Amplifier Signal (AC1)	Clutch On: 1.5v, Off: 12v
8	GN/OR	Vehicle Speed Sensor	At 55 mph: 48 Hz
10	GN/BK	4WD Detection Switch	Switch Off: 12v, On: 1v
10 ('99)	BL	4WD Detection Switch	Switch Off: 12v, On: 1v
11	BK/WT	Starter Switch Signal	In P/N: 0-3.0v
12	WT/BL	EFI Main Relay B+	12-14v
13	RD/GN	4WD Detection Transfer (N)	Open: 12v, Closed: 0v
14	GN/RD	AT ECT Pattern Select Switch	Norm: 0v, PWR: 12v
15	LG	A/T Select Switch Low	In Low: 12v, Others: 0v
16	PK/RD	A/T Select Switch 2nd	In 2nd: 12v, Others: 0v
17	RD/YL	A/T Select Switch Reverse	In 'R': 12v, Others: 0v
18	BR/YL	Cruise Control ECU	At Cruise in OD: 12v
19	WT	SIL Signal (Scan Tool)	12v
21	GN/WT	Stop Light Switch Signal	Brake Off: 0v, On: 12v
22	BK	Start Circuit Signal	Cranking: 9-11v

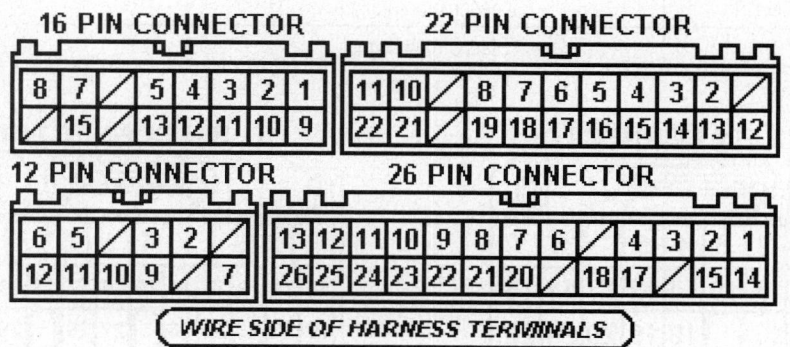

05533_ADIA_G143

Pin Connector Graphic

2000 2.7L I4 MFI VIN M California (All) 31 Pin Connector

PCM Pin #	Wire Color	Circuit Description (31 Pin)	Value at Hot Idle
1	RD	Injector 1 Control	2.0-3.3 ms
2	WT	Injector 2 Control	2.0-3.3 ms
3	GN	Injector 3 Control	2.0-3.3 ms
4	RD/BL	Injector 4 Control	2.0-3.3 ms
5, 6, 21, 31	WT/BK	Power Ground	<0.1v
7	LG/RD	AT ECT Solenoid (S1)	3rd or OD: 1v
8	BL/WT	A/T ECT Solenoid (S2)	1st or OD: 1v
9	GN/RD	A/T ECT Solenoid (SL)	In Lockup: 12-14v
10	BK/YL	Igniter Signal (IGF)	Digital Signal: 0-5-0v
11	BK/BL	Igniter Transistor 1 Control	6°, at 55 mph: 8° dwell
12	YL/BK	Igniter Transistor 2 Control	6°, at 55 mph: 8° dwell
13	BL	Igniter Transistor 3 Control	6°, at 55 mph: 8° dwell
14	BL/YL	Igniter Transistor 4 Control	6°, at 55 mph: 8° dwell
15	BK/RD	Idle Air Control Valve (RSO)	Pulse Signals
26	PK/YL	A/T Oil Temperature Sensor	At 68°F: 4-5v
28	GY	Knock Sensor Signal	<0.075v AC

2000 2.7L I4 MFI VIN M California (All) 22 Pin Connector

PCM Pin #	Wire Color	Circuit Description (22 Pin)	Value at Hot Idle
1	BL/RD	Direct Battery	12-14v
5	YL	A/T: Oil Temp Lamp Indicator	Lamp Off: 12v, On: 1v
6	PK	Malfunction Indicator Lamp Control	MIL Off: 12v, On: 1v
7	BK/WT	Starter Switch Signal	In P/N: 0-3.0v
9	RD/GN	4WD Detection Transfer (N)	Open: 12v, Closed: 0v
10	BL/OR	O/D OFF (lamp) Indicator	Light Off: 12v, On: 1v
11	BR/YL	A/T Select Switch Drive	In Drive: 12-14v
12	WT	SIL Signal (Scan Tool)	0v
13	BL/YL	A/C Amplifier Signal (AC1)	Clutch On: 1.5v, Off: 12v
14	BL/BK	A/C Amplifier Signal (ACT)	Clutch On: 12v, Off: 1.5v
15	GN/WT	Stop Light Switch Signal	Brake Off: 0v, On: 12v
16	WT/BL	EFI Main Relay B+	12-14v
17	GN/RD	AT ECT Pattern Select Switch	Norm: 0v, PWR: 12v
19	YL/BK	4WD Detection Transfer (L4)	Switch Closed: 0.1v
20	OR	A/T: Defogger Idle-up Signal	Load On: 12v
21	GN/OR	Speedometer Indicator	At 55 mph: 48 Hz
22	BK	Start Circuit Signal	Cranking: 9-11v

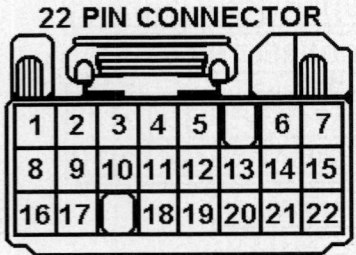

WIRE SIDE OF HARNESS TERMINALS

05533_ADIA_G144

Pin Connector Graphic

2000 2.7L I4 MFI VIN M California (All) 28 Pin Connector

PCM Pin #	Wire Color	Circuit Description (28 Pin)	Value at Hot Idle
2	RD/YL	A/T Select Switch Reverse	In 'R': 12v, Others: 0v
3	LG	A/T Select Switch Low	In Low: 12v, Others: 0v
4	PK/RD	A/T Select Switch 2nd	In 2nd: 12v, Others: 0v
5	YL/GN	Closed Throttle Switch	1v, off-idle: 12v
6	GN/YL	Circuit Opening Relay (FC)	0-3v, off-idle: 12v
7	LG/BK	Data Link Connector	12-14v
8	YL	EVAP Vapor Pressure Sensor (PTNK)	2.5-3.1v
10	BL	4WD Detection Switch	Switch Off: 12v, On: 1v
13	BK	Tachometer Signal (TACO)	Pulses
14	YL/RD	A/T ECT Speed Sensor (+)	Pulses
23	WT/RD	A/T ECT Speed Sensor (-)	Pulses
25	BL/WT	Overdrive Main Switch	Switch Off: 12v, On: 1v
28	BL/RD	PSP Switch Signal (PSW)	Straight: 12v, Turned: 0v

2000 2.7L I4 MFI VIN M California (All) 24 Pin Connector

PCM Pin #	Wire Color	Circuit Description (24 Pin)	Value at Hot Idle
2	GN/YL	TP Sensor VREF	4.9-5.1v
3	RD/WT	HO2S-12 (B1 S2) Heater	1v (Heater on)
4	RD	AFRS-11 (B1 S1) Heater	1v (Heater on)
5	PK/BL	EGR Solenoid Control (VSV)	12v or 0v
7	RD/BK	EVAP Vapor Pressure Valve (VSV)	12v or 0v
8	BK/WT	EVAP Purge Solenoid (VSV)	12v or 0v
9	YL	TP Sensor Signal (VTA)	0.3-0.8v
10	RD	HO2S-12 (B1 S2) Signal	0.1-1.1v
11	BK	AFS-11 (B1 S1) Signal (+)	Fixed at 3.3v
12	GN/RD	ECT Sensor Signal (THW)	At 180°F: 0.51v
14	BK/RD	MAF Sensor Signal	1.1-1.8v
15	RD	CMP Sensor Signal	AC pulse signals
16	RD	CKP Sensor Signal	AC pulse signals
17	BR	Sensor Ground	<0.050v
18	BL/BK	Sensor Ground	<0.050v
19	PK	EGR Gas Temperature Sensor	3.5-4.0v
20	WT	AFS-11 (B1 S1) Signal (-)	Fixed at 3.3v
21	YL/GN	IAT Sensor Signal (THA)	At 100°F: 2.60v
22	B/YL	MAF Meter Ground	0.050v
24	GN	CKP/CMP Sensor Signal (G-)	0.050v

WIRE SIDE OF HARNESS TERMINALS

Pin Connector Graphic

05533_ADIA_G145

2000 2.7L I4 MFI VIN M Federal (All) 22 Pin Connector

PCM Pin #	Wire Color	Circuit Description (22 Pin)	Value at Hot Idle
2	BL/RD	Direct Battery	12-14v
3	YL	A/T: Oil Temp. Lamp Indicator	Lamp Off: 12v, On: 1v
4	PK	Malfunction Indicator Lamp Control	MIL Off: 12v, On: 1v
5	BL/OR	O/D OFF (lamp) Indicator	Light Off: 12v, On: 1v
6	BL/BK	A/C Amplifier Signal (ACT)	Clutch On: 12v, Off: 1.5v
7	BL/YL	A/C Amplifier Signal (AC1)	Clutch On: 1.5v, Off: 12v
8	GN/OR	Speedometer Indicator	At 55 mph: 48 Hz
10	BL	4WD Detection Switch	Switch Off: 12v, On: 1v
11	BK/WT	Starter Switch Signal	In P/N: 0-3.0v
12	WT/BL	EFI Main Relay B+	12-14v
13	RD/GN	4WD Detection Transfer (N)	Open: 12v, Closed: 0v
14	GN/RD	AT ECT Pattern Select Switch	Norm: 0v, PWR: 12v
15	LG	A/T Select Switch Low	In Low: 12v, Others: 0v
16	PK/RD	A/T Select Switch 2nd	In 2nd: 12v, Others: 0v
17	RD/YL	A/T Select Switch Reverse	In 'R': 12v, Others: 0v
18	BR/YL	Cruise Control ECU	At Cruise in OD: 12v
19	WT	SIL Signal (Scan Tool)	12v
21	GN/WT	Stop Light Switch Signal	Brake Off: 0v, On: 12v
22	BK	Start Circuit Signal	Cranking: 9-11v

2000 2.7L I4 MFI VIN M Federal (All) 26 Pin Connector

PCM Pin #	Wire Color	Circuit Description (26 Pin)	Value at Hot Idle
1	RD/WT	HO2S-12 (B1 S2) Heater	1v (Heater on)
2	PK	HO2S-11 (B1 S1) Heater	1v (Heater on)
3	BK/WT	EVAP Purge Solenoid (VSV)	12v or 0v
4	YL/GN	Closed Throttle Switch	1v, off-idle: 12v
6	PK/BL	IAC Signal (RSC)	Pulse Signals
7	BK/RD	Idle Air Control Valve (RSO)	Pulse Signals
8	GN/RD	AT ECT Solenoid (S1)	3rd or OD: 1v
9	RD/BL	Injector 4 Control	2.0-3.3 ms
10	GN	Injector 3 Control	2.0-3.3 ms
11	WT	Injector 2 Control	2.0-3.3 ms
12	RD	Injector 1 Control	2.0-3.3 ms
13	WT/BK	Power Ground	<0.1v
14	GN/YL	Circuit Opening Relay (FC)	0-3v, off-idle: 12v
15	BK	Tachometer Signal (TACO)	Pulses
16	PK/YL	A/T Oil Temperature Sensor	At 68°F: 4-5v
17	BK/YL	Igniter Signal (IGF)	Digital Signal: 0-5-0v
18	BL/RD	4WD Detection Transfer (L4)	Switch Closed: 0.1v
20	RD/GN	A/T ECT Solenoid (SL)	In Lockup: 12-14v
21	LG	A/T ECT Solenoid (S2)	1st or OD: 1v
22	YL/BK	Igniter Transistor 2 Control	6°, at 55 mph: 8° dwell

2000 2.7L I4 MFI VIN M Federal (All) 26 Pin Connector, *continued*

23	BK/BL	Igniter Transistor 1 Control	6°, at 55 mph: 8° dwell
24	BR	Sensor Ground	<0.050v
25	WT/BK	Power Ground	<0.1v
26	WT/BK	Power Ground	<0.1v

2000 2.7L I4 MFI VIN M Federal (All) 12 Pin Connector

PCM Pin #	Wire Color	Circuit Description (12 Pin)	Value at Hot Idle
1	BL/WT	Overdrive Main Switch	Switch Off: 12v, On: 1v
2	OR	A/T: Defogger Idle-up Signal	Load On: 12v
3	WT/RD	ECT Speed (-) Sensor	Pulses
6	GN	CKP/CMP Sensor Signal (G-)	0.050v
7	RD/WT	MAF Meter Ground	0.050v
9	YL/RD	A/T ECT Speed Sensor (+)	Pulses
10	LG/BK	EVAP Vapor Pressure Sensor (PTNK)	2.5-3.1v
11	RD	CMP Sensor Signal (G+)	AC pulse signals
12	RD	CKP Sensor Signal (NE+)	AC pulse signals

2000 2.7L I4 MFI VIN M Federal (All) 16 Pin Connector

PCM Pin #	Wire Color	Circuit Description (16 Pin)	Value at Hot Idle
1	GN/YL	TP Sensor VREF	4.9-5.1v
2	BK/RD	MAF Sensor Signal	1.1-1.8v
3	YL/GN	IAT Sensor Signal (THA)	At 100°F: 2.60v
4	GN/RD	ECT Sensor Signal (THW)	At 180°F: 0.51v
5	WT	HO2S-11 (B1 S1) Signal	0.1-1.1v
7	LG/BK	Data Link Connector	12-14v
8	RD/BK	EVAP Vapor Pressure Valve (VSV)	12v or 0v
9	BL/BK	Sensor Ground	<0.050v
10	YL	TP Sensor Signal (VTA)	0.3-0.8v
11	PK	EGR Gas Temperature Sensor	3.5-4.0v
12	GY	Knock Sensor Signal	<0.075v AC
13	RD	HO2S-12 (B1 S2) Signal	0.1-1.1v
15	PK/BL	EGR Solenoid Control (VSV)	12v or 0v

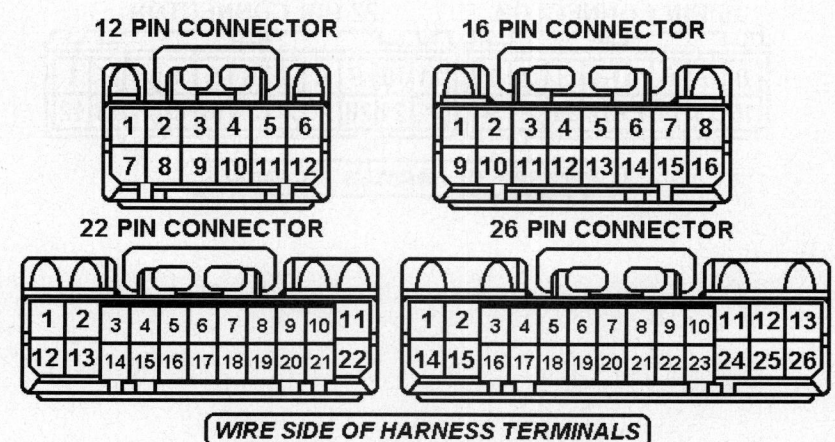

Pin Connector Graphic

05533_ADIA_G146

1990 3.0L V6 MFI VIN V (A/T-ECT) 26 Pin Connector

PCM Pin #	Wire Color	Circuit Description (26 Pin)	Value at Hot Idle
1	WT	Distributor Signal (NE+)	AC pulse signals
2	RD	Distributor Signal (G1+)	AC pulse signals
3	BK/YL	Igniter Signal (IGF)	Digital Signal: 0-5-0v
4	GN/RD	AT ECT Solenoid S4	Pulses
5	YL/BK	AT ECT Solenoid S3	Pulses
6	BK	A/T ECT Solenoid (S2)	1st or OD: 1v
7	WT	AT ECT Solenoid (S1)	3rd or OD: 1v
8	GN	Fuel Pressure Up Solenoid	1v (at hot restart)
9	GN	Cold Start Injector Control	1v (at cold startup)
10	PK/GN	HO2S Heater Control	1v (Heater on)
11	BR	Sensor Ground	<0.050v
12	WT/RD	Injectors 1 & 3 & 5 Control	1.6-2.9 ms
13	BR	Power Ground	<0.1v
14	GN	Distributor Signal (G-)	AC pulse signals
15	BK	Distributor Signal (G2+)	AC pulse signals
16	BR/RD	AT ECT Speed Sensor	Moving: 0-5-0-5v
17	PK/WT	A/T Select Switch Low	In Low: 12v, Others: 0v
18	PK/GN	A/T Select Switch 2nd	In 2nd: 12v, Others: 0v
19	PK/RD	A/T Select Switch Neutral	In 'N': 12v, Others: 0v
20	YL/RD	4WD Detection Transfer (L4)	Switch Closed: 0.1v
21	BK/BL	Igniter Signal (IGT)	Digital Signal: 0-5-0v
22	PK	EGR Solenoid Control (VSV)	12v or 0v
23	GN/RD	Intake Air Solenoid Control	12v or 0v
24	BK/RD	A/C Idle-Up Solenoid	A/C Off: 12v, On: 1v
25	WT	Injectors 2 & 4 & 6 Control	1.6-2.9 ms
26	BR	Power Ground	<0.1v

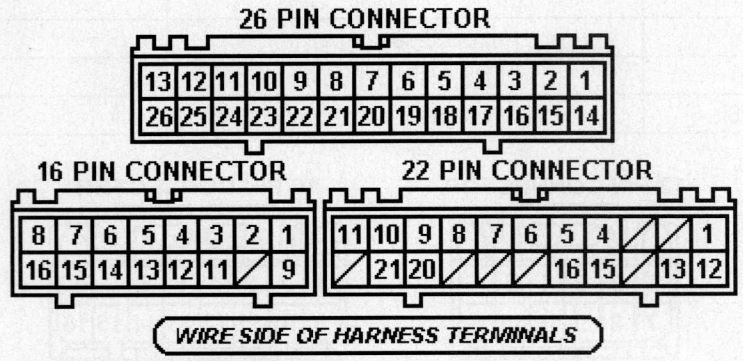

Pin Connector Graphic

05533_ADIA_G147

Standard Colors and Abbreviations

Abbreviation	Color	Abbreviation	Color	Abbreviation	Color
BK	Black	GY	Gray	RD	Red
BL	Blue	GN	Green	TN	Tan
BR	Brown	LG	LT Green	VT	Violet
DB	Dark Blue	OR	Orange	WT	White
DG	DK Green	PK	Pink	YL	Yellow

1990 3.0L V6 MFI VIN V (A/T-ECT) 16 Pin Connector

PCM Pin #	Wire Color	Circuit Description (16 Pin)	Value at Hot Idle
1	GN/BK	Sensor VREF	4.9-5.1v
2	YL/BL	Airflow Meter Signal	0.5-2.5v
3	YL/GN	IAT Sensor Signal (THA)	At 100°F: 2.60v
4	GN/BL	ECT Sensor Signal (THW)	At 180°F: 0.51v
5	BK	Knock Sensor Signal	<0.075v AC
6	BK	HO2S-11 (B1 S1) Signal	0.1-1.1v
7	GN/BK	4WD Oil Temperature Sensor	At 68°F: 4-5v
8	YL/GN	Check Connector	12-14v
9	BR/BK	Sensor Ground	<0.050v
11	YL	TP Sensor Signal (VTA)	0.3-0.8v
12	YL/BL	Closed Throttle Switch	1v, off-idle: 12v
13 (Cal)	GN/WT	EGR Gas Temperature Sensor	3.5-4.0v
14	LG	Transfer Fluid Temp. Sensor	At 230°F: <1.5v
15	PK/WT	Check Connector	12-14v
16	PK	Coolant Temperature Switch	Switch Closed: 0.1v

1990 3.0L V6 MFI VIN V (A/T-ECT) 22 Pin Connector

PCM Pin #	Wire Color	Circuit Description (22 Pin)	Value at Hot Idle
1	BK/GN	Direct Battery	12-14v
4	BL/GN	A/T Oil Temperature Indicator	Light Off: 12v, On: 1v
5	PK	Malfunction Indicator Lamp Control	MIL Off: 12v, On: 1v
6	GN/WT	Stop Light Switch Signal	Brake Off: 0v, On: 12v
7	GN/OR	AT ECT Pattern Select Switch	Norm: 0v, PWR: 12v
8	PK/GN	4WD Indicator Switch	Switch Off: 12v, On: 1v
9	GN/BL	Vehicle Speed Sensor	At 55 mph: 48 Hz
10	BK/WT	A/C Amplifier On Signal	Clutch On: 1.5v, Off: 12v
11	BK/WT	Starter Switch Signal	In P/N: 0-3.0v
12	WT/RD	EFI Main Relay B+	12-14v
13	WT/RD	EFI Main Relay B1+	12-14v
15	BR/BK	Sensor Ground	<0.050v
16	YL/GN	Overdrive Main Switch	Switch Off: 12v, On: 1v
20	PK	Check Connector	12-14v
21	YL/RD	Cruise Control ECU	At Cruise in OD: 12v

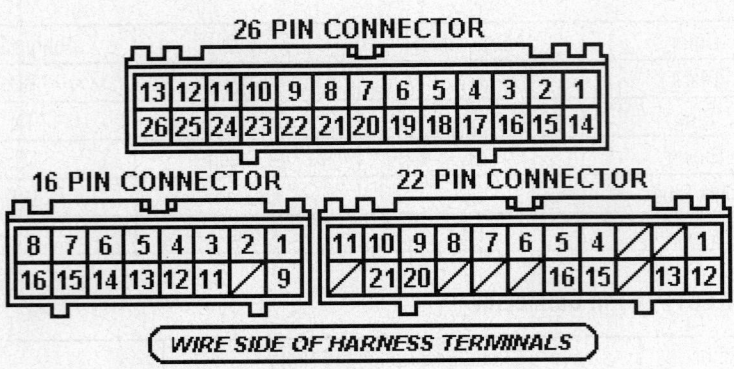

05533_ADIA_G147

Pin Connector Graphic
Standard Colors and Abbreviations

Abbreviation	Color	Abbreviation	Color	Abbreviation	Color
BK	Black	GY	Gray	RD	Red
BL	Blue	GN	Green	TN	Tan
BR	Brown	LG	LT Green	VT	Violet
DB	Dark Blue	OR	Orange	WT	White
DG	DK Green	PK	Pink	YL	Yellow

1990 3.0L V6 MFI VIN V (All) 10 Pin Connector

PCM Pin #	Wire Color	Circuit Description (10 Pin)	Value at Hot Idle
2	GN	Cold Start Injector Control	1v (at cold startup)
3	BK/WT	Starter Switch Signal	In P/N: 0-3.0v
4	WT/RD	Injectors 1 & 3 & 5 Control	1.6-2.9 ms
5	BR	Power Ground	<0.1v
7	BR	Sensor Ground	<0.050v
8	BL/BK	Igniter Signal (IGT)	0.72-0.74v
9	BK/WT	Injectors 2 & 4 & 6 Control	1.6-2.9 ms
10	BR	Power Ground	<0.1v

1990 3.0L V6 MFI VIN V (All) 18 Pin Connector

PCM Pin #	Wire Color	Circuit Description (18 Pin)	Value at Hot Idle
1	GN/BL	ECT Sensor Signal (THW)	At 180°F: 0.51v
2	BK/YL	Distributor IGF Signal	0.74-0.76v
3	PK/GN	HO2S Heater Control	1v (Heater on)
4	WT	Distributor Signal (NE+)	AC pulse signals
5	BK	Distributor Signal (G2+)	AC pulse signals
6	RD	Distributor Signal (G1+)	AC pulse signals
7	GN	Distributor Signal (G-)	AC pulse signals
8	GN	Fuel Pressure Up Solenoid	1v (at hot restart)
10	BR/BK	Sensor Ground	<0.050v
12	PK	EGR Solenoid Control (VSV)	12v or 0v
13	YL/BL	Closed Throttle Switch	1v, off-idle: 12v
14	YL	TP Sensor Signal (VTA)	0.3-0.8v

1990 3.0L V6 MFI VIN V (All) 18 Pin Connector, *continued*

15	PK/WT	Check Connector	12-14v
16	YL/GN	Check Connector	12-14v
17	BK/RD	A/C Idle-Up Solenoid	12-14v
18 (Cal)	GN/WT	EGR Gas Temperature Sensor	3.5-4.0v
18 (Fed)	GN/WT	Short Pin	N/A

1990 3.0L V6 MFI VIN V (All) 24 Pin Connector

PCM Pin #	Wire Color	Circuit Description (24 Pin)	Value at Hot Idle
2	BK/GN	Direct Battery	12-14v
3	GN/BK	Airflow Meter VREF	4.9-5.1v
4	YL/BL	Airflow Meter Signal	0.5-2.5v
5	YL/GN	IAT Sensor Signal (THA)	At 100°F: 2.60v
6	GN/WT	Stop Light Switch Signal	Brake Off: 0v, On: 12v
7	GN/BL	Vehicle Speed Sensor	At 55 mph: 48 Hz
12	BK	HO2S-11 (B1 S1) Signal	0.1-1.1v
13	WT/RD	EFI Main Relay B1+	12-14v
14	WT/RD	EFI Main Relay B+	12-14v
15	PK	Water Temperature Switch	Switch Closed: 0.1v
16	GN/WT	4WD Indicator Switch	Switch Off: 12v, On: 1v
17	BK	Knock Sensor Signal	<0.075v AC
18	GN/RD	Intake Air Solenoid Control	12v or 0v
19	PK	Malfunction Indicator Lamp Control	MIL Off: 12v, On: 1v
20	BK/WT	A/C Magnetic Clutch Relay (ACMG)	Relay Off: 0v, On: 12v
24	BR/BK	Sensor Ground	<0.050v

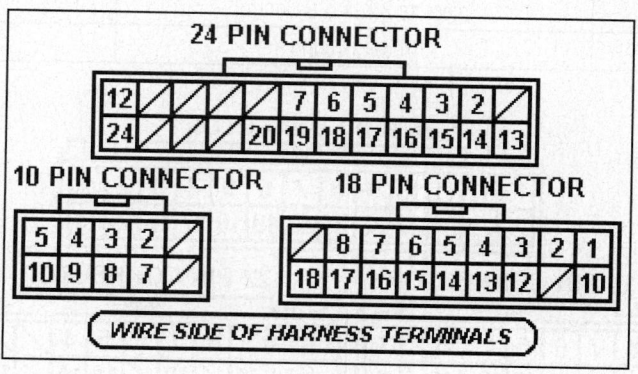

Pin Connector Graphic

05533_ADIA_G148

1991-92 3.0L V6 MFI VIN V (A/T-ECT) 26 Pin Connector

PCM Pin #	Wire Color	Circuit Description (26 Pin)	Value at Hot Idle
1	WT	Distributor Signal (NE+)	AC pulse signals
2	RD	Distributor Signal (G1+)	AC pulse signals
3	BK/YL	Igniter Signal (IGF)	Digital Signal: 0-5-0v
4	GN/RD	AT ECT Solenoid S4	Pulses
5	YL/BK	AT ECT Solenoid S3	Pulses
6	BK	A/T ECT Solenoid (S2)	1st or OD: 1v
7	WT	AT ECT Solenoid (S1)	3rd or OD: 1v
8	GN	Fuel Pressure Up Solenoid	1v (at hot restart)
9	GN	Cold Start Injector Control	1v (at cold startup)
10	PK/GN	HO2S Heater Control	1v (Heater on)
11	BR	Sensor Ground	<0.050v
12	WT/RD	Injectors 1 & 3 & 5 Control	1.6-2.9 ms
13	BR	Power Ground	<0.1v
14	GN	Distributor Signal (G-)	AC pulse signals
15	BK	Distributor Signal (G2+)	AC pulse signals
16	BR/RD	AT ECT Speed Sensor	Moving: 0-5-0-5v
17	PK/WT	A/T Select Switch Low	In Low: 12v, Others: 0v
18	PK/GN	A/T Select Switch 2nd	In 2nd: 12v, Others: 0v
19	PK/RD	A/T Select Switch Neutral	In 'N': 12v, Others: 0v
20	YL/RD	4WD Detection Transfer (L4)	Switch Closed: 0.1v
21	BK/BL	Igniter Signal (IGT)	Digital Signal: 0-5-0v
22	PK	EGR Solenoid Control (VSV)	12v or 0v
23	GN/RD	Intake Air Solenoid Control	12v or 0v
24	BK/RD	A/C Idle-Up Solenoid	A/C Off: 12v, On: 1v
25	WT	Injectors 2 & 4 & 6 Control	1.6-2.9 ms
26	BR	Power Ground	<0.1v

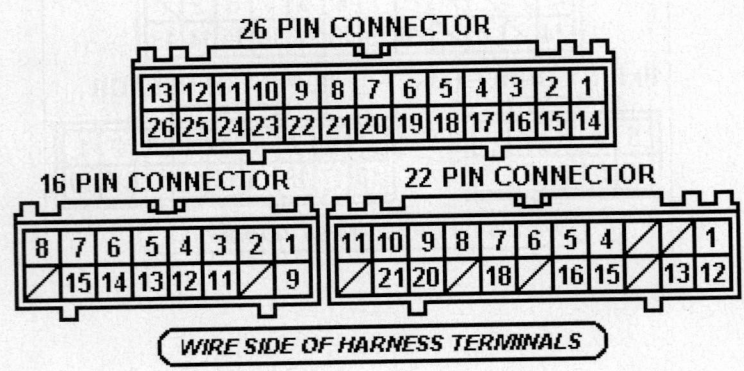

Pin Connector Graphic

1991-92 3.0L V6 MFI VIN V (A/T-ECT) 16 Pin Connector

PCM Pin #	Wire Color	Circuit Description (16 Pin)	Value at Hot Idle
1	GN/BK	Airflow Meter VREF	4.9-5.1v
2	RD	Airflow Meter Signal	0.5-2.5v
3	YL/BK	IAT Sensor Signal (THA)	At 100°F: 2.60v
4	GN/BL	ECT Sensor Signal (THW)	At 180°F: 0.51v

1991-92 3.0L V6 MFI VIN V (A/T-ECT) 16 Pin Connector, *continued*

5	BK	Knock Sensor Signal	<0.075v AC
6	BK	HO2S-11 (B1 S1) Signal	0.1-1.1v
7	GN/BK	4WD Oil Temperature Sensor	At 68°F: 4-5v
8	YL/GN	Check Connector	12-14v
9	BR/BK	Sensor Ground	<0.050v
11	YL	TP Sensor Signal (VTA)	0.3-0.8v
12	YL/BL	Closed Throttle Switch	1v, off-idle: 12v
13 (Cal)	GN/WT	EGR Gas Temperature Sensor	3.5-4.0v
14	LG	Transfer Fluid Temp. Sensor	At 230°F: <1.5v
15	PK/WT	Check Connector	12-14v

1991-92 3.0L V6 MFI VIN V (A/T-ECT) 22 Pin Connector

PCM Pin #	Wire Color	Circuit Description (22 Pin)	Value at Hot Idle
1	BK/GN	Direct Battery	12-14v
4	BL/GN	A/T Oil Temperature Indicator	Light Off: 12v, On: 1v
5	PK	Malfunction Indicator Lamp Control	MIL Off: 12v, On: 1v
6	GN/WT	Stop Light Switch Signal	Brake Off: 0v, On: 12v
7	GN/OR	AT ECT Pattern Select Switch	Norm: 0v, PWR: 12v
8	PK/GN	4WD Indicator Switch	Switch Off: 12v, On: 1v
9	GN/BL	Vehicle Speed Sensor	At 55 mph: 48 Hz
10	BK/WT	A/C Amplifier Signal	A/C Off: 12v, On: 1v
11	BK/WT	Starter Switch Signal	In P/N: 0-3.0v
12	WT/RD	EFI Main Relay B+	12-14v
13	WT/RD	EFI Main Relay B1+	12-14v
15	BR/BK	Sensor Ground	<0.050v
16	YL/GN	Overdrive Main Switch	Switch Off: 12v, On: 1v
18	WT/BK	2WD Select (SEL1)	<0.1v
20	PK	Check Connector	12-14v
21	YL/RD	Cruise Control ECU	At Cruise in OD: 12v

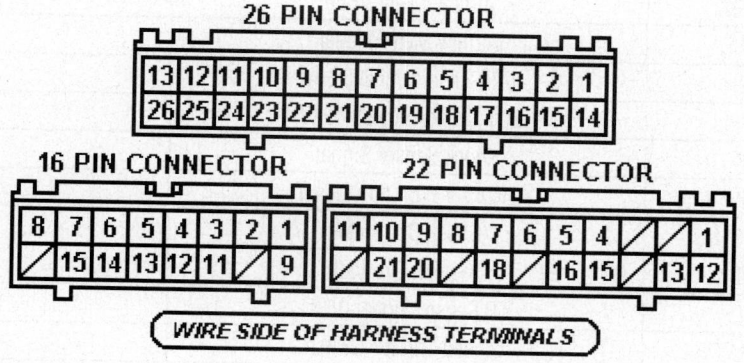

Pin Connector Graphic

05533_ADIA_G149

Standard Colors and Abbreviations

Abbreviation	Color	Abbreviation	Color	Abbreviation	Color
BK	Black	GY	Gray	RD	Red
BL	Blue	GN	Green	TN	Tan
BR	Brown	LG	LT Green	VT	Violet
DB	Dark Blue	OR	Orange	WT	White
DG	DK Green	PK	Pink	YL	Yellow

1991-92 3.0L V6 MFI VIN V (All) 26 Pin Connector

PCM Pin #	Wire Color	Circuit Description (26 Pin)	Value at Hot Idle
1	WT	Distributor Signal (NE+)	AC pulse signals
2	RD	Distributor Signal (G1+)	AC pulse signals
3	BK/YL	Igniter Signal (IGF)	Digital Signal: 0-5-0v
8	GN	Fuel Pressure Up Solenoid	1v (at hot restart)
9	GN	Cold Start Injector Control	1v (at cold startup)
10	PK/GN	HO2S Heater Control	1v (Heater on)
11	BR	Sensor Ground	<0.050v
12	WT/RD	Injectors 1 & 3 & 5 Control	1.6-2.9 ms
13	BR	Power Ground	<0.1v
14	GN	Distributor Signal (G-)	AC pulse signals
15	BK	Distributor Signal (G2+)	AC pulse signals
21	BK/BL	Igniter Signal (IGT)	Digital Signal: 0-5-0v
22	PK	EGR Solenoid Control (VSV)	12v or 0v
23	GN/RD	Intake Air Solenoid Control	12v or 0v
24	BK/RD	A/C Idle-Up Solenoid	A/C Off: 12v, On: 1v
25	WT	Injectors 2 & 4 & 6 Control	1.6-2.9 ms
26	BR	Power Ground	<0.1v

1991-92 3.0L V6 MFI VIN V (All) 16 Pin Connector

PCM Pin #	Wire Color	Circuit Description (16 Pin)	Value at Hot Idle
1	GN/BK	Airflow Meter VREF	4.9-5.1v
2	YL/RD	Airflow Meter Signal	0.5-2.5v
3	YL/BK	IAT Sensor Signal (THA)	At 100°F: 2.60v
4	GN/BL	ECT Sensor Signal (THW)	At 180°F: 0.51v
5	BK	Knock Sensor Signal	<0.075v AC
6	BK	HO2S-11 (B1 S1) Signal	0.1-1.1v
8	YL/GN	Check Connector	12-14v
9	BR/BK	Sensor Ground	<0.050v
11	YL	TP Sensor Signal (VTA)	0.3-0.8v
12	YL/WT	Closed Throttle Switch	1v, off-idle: 12v
13 (Cal)	GN/WT	EGR Gas Temperature Sensor	3.5-4.0v
15	PK/WT	Check Connector	12-14v

1991-92 3.0L V6 MFI VIN V (All) 22 Pin Connector

PCM Pin #	Wire Color	Circuit Description (22 Pin)	Value at Hot Idle
1	BK/GN	Direct Battery	12-14v
5	PK	Malfunction Indicator Lamp Control	MIL Off: 12v, On: 1v
6	GN/WT	Stop Light Switch Signal	Brake Off: 0v, On: 12v
8	PK/GN	4WD Indicator Switch	Switch Off: 12v, On: 1v
9	GN/BL	Vehicle Speed Sensor	At 55 mph: 48 Hz
10	BK/WT	A/C Magnetic Clutch Relay (ACMG)	Relay Off: 0v, On: 12v
11	BK/WT	Starter Switch Signal	In P/N: 0-3.0v
12	WT/RD	EFI Main Relay B+	12-14v
13	WT/RD	EFI Main Relay B1+	12-14v
15	BR/BK	Sensor Ground	<0.050v
17	WT/BK	4WD Select (SEL2)	<0.1v
18	WT/BK	2WD Select (SEL1)	<0.1v

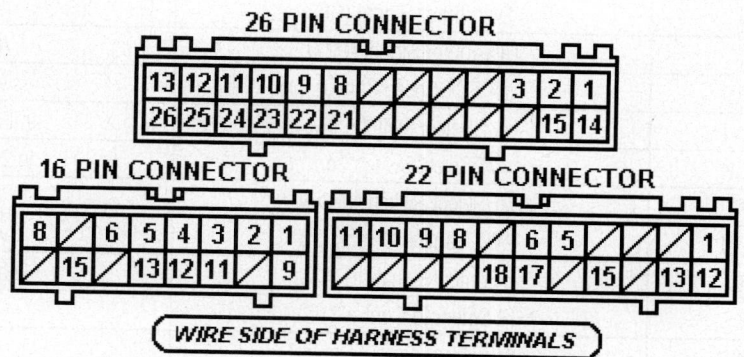

Pin Connector Graphic

05533_ADIA_G150

1993-95 3.0L V6 MFI VIN V (All) 26 Pin Connector

PCM Pin #	Wire Color	Circuit Description (26 Pin)	Value at Hot Idle
1	WT	Distributor Signal (NE+)	AC pulse signals
2	RD	Distributor Signal (G1+)	AC pulse signals
3	BK/YL	Igniter Signal (IGF)	Digital Signal: 0-5-0v
4	GN/RD	AT ECT Solenoid S4	Pulses
5	YL/BK	AT ECT Solenoid S3	Pulses
6	BK	A/T ECT Solenoid (S2)	1st or OD: 1v
7	WT	AT ECT Solenoid (S1)	3rd or OD: 1v
8	GN	Fuel Pressure Up Solenoid	1v (at hot restart)
9	GN	Cold Start Injector Control	1v (at cold startup)
10	PK/GN	Main HO2S Heater Control	1v (Heater on)
11	BR	Sensor Ground	<0.050v
12	WT/RD	Injectors 1 & 3 & 5 Control	1.6-2.9 ms
13	BR	Power Ground	<0.1v
14	GN	Distributor Signal (G-)	AC pulse signals
15	BK	Distributor Signal (G2+)	AC pulse signals
16	BR/RD	AT ECT Speed Sensor	Varies: 0-5-0-5v
17	PK/WT	A/T Select Switch Low	In Low: 12v, Others: 0v
18	PK/GN	A/T Select Switch 2nd	In 2nd: 12v, Others: 0v
19	PK/RD	A/T Select Switch Neutral	In 'N': 12v, Others: 0v
20	YL/RD	4WD Transfer Detect (L4)	Switch Closed: 0.1v
21	BK/BL	Igniter Signal (IGT)	Digital Signal: 0-5-0v
22	PK	EGR Solenoid Control (VSV)	12v or 0v
23	GN/RD	Intake Air Solenoid Control	12v or 0v
24	BK/RD	A/C Idle-Up Solenoid	A/C Off: 12v, On: 1v
25	WT	Injectors 2 & 4 & 6 Control	1.6-2.9 ms
26	BR	Power Ground	<0.1v

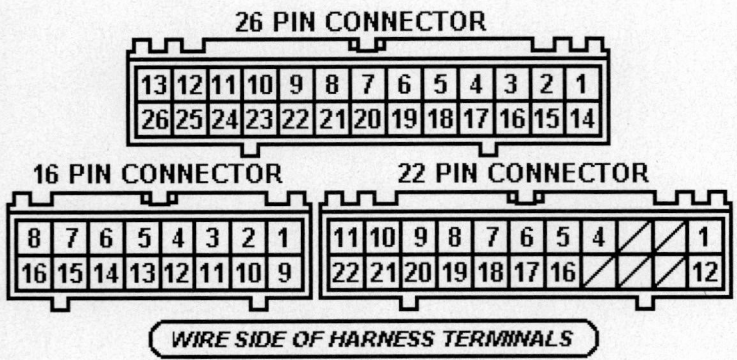

Pin Connector Graphic

1993-95 3.0L V6 MFI VIN V (All) 16 Pin Connector

PCM Pin #	Wire Color	Circuit Description (16 Pin)	Value at Hot Idle
1	GN/BK	Airflow Meter VREF	4.9-5.1v
2	YL/RD	Airflow Meter Signal	0.5-2.5v
3	YL/BK	IAT Sensor Signal (THA)	At 100°F: 2.60v
4	GN/BL	ECT Sensor Signal (THW)	At 180°F: 0.51v

1993-95 3.0L V6 MFI VIN V (All) 16 Pin Connector, *continued*

5	BK	Knock Sensor Signal	<0.075v AC
6	BK	Main HO2S Signal	0.1-1.1v
7	GN/BK	A/T Oil Temperature Sensor	At 68°F: 4-5v
8	YL/GN	Check Connector	12-14v
9	BR/BK	Sensor Ground	<0.050v
10 (Cal)	WT	Sub HO2S Signal	0.1-1.1v
11	YL	TP Sensor Signal (VTA)	0.3-0.8v
12	YL/WT	Closed Throttle Switch	1v, off-idle: 12v
13 (Cal)	GN/WT	EGR Gas Temperature Sensor	3.5-4.0v
14	LG	Transfer Fluid Temp. Sensor	At 230°F: <1.5v
15	PK/WT	Check Connector	12-14v

1993-95 3.0L V6 MFI VIN V (All) 22 Pin Connector

PCM Pin #	Wire Color	Circuit Description (22 Pin)	Value at Hot Idle
1	BK/GN	Direct Battery	12-14v
4	BL/GN	A/T Oil Temperature Indicator	Light Off: 12v, On: 1v
5	PK	Malfunction Indicator Lamp Control	MIL Off: 12v, On: 1v
6	GN/WT	Stop Light Switch Signal	Brake Off: 0v, On: 12v
7	GN/OR	AT ECT Pattern Select Switch	Norm: 0v, PWR: 12v
8	GN/WT	4WD Indicator Switch	Switch Off: 12v, On: 1v
9	GN/BL	Vehicle Speed Sensor	At 55 mph: 48 Hz
10	BK/WT	A/C Magnetic Clutch Relay (ACMG)	Relay Off: 0v, On: 12v
11	BK/WT	MT: Starter Switch Signal	In P/N: 0-3.0v
11	BK/YL	A/T: Starter Switch Signal	In P/N: 0-3.0v
12	WT/RD	EFI Main Relay B+	12-14v
16	YL/GN	Overdrive Main Switch	Switch Off: 12v, On: 1v
17	BR	4WD MT: Select (SEL2)	<0.1v
18	BR	2WD A/T: Select (SEL1)	<0.1v
19	BL/BK	A/C Amplifier Signal (ACT)	Clutch On: 12v, Off: 1.5v
20	PK	Check Connector	12-14v
21	YL/RD	Cruise Control ECU	At Cruise in OD: 12v
22 (Cal)	RD/GN	Sub HO2S Heater	1v (Heater on)

Standard Colors and Abbreviations

Abbreviation	Color	Abbreviation	Color	Abbreviation	Color
BK	Black	GY	Gray	RD	Red
BL	Blue	GN	Green	TN	Tan
BR	Brown	LG	LT Green	VT	Violet
DB	Dark Blue	OR	Orange	WT	White
DG	DK Green	PK	Pink	YL	Yellow

1996-97 3.4L V6 MFI VIN N (A/T-ECT) 28 Pin Connector

PCM Pin #	Wire Color	Circuit Description (28 Pin)	Value at Hot Idle
1	RD/YL	A/T Select Switch Reverse	In 'R': 12v, Others: 0v
2	PK/RD	A/T Select Switch 2nd	In 2nd: 12v, Others: 0v
3	GN	A/T Select Switch Low	In Low: 12v, Others: 0v
4	YL	A/T Oil Temperature Sensor	At 68°F: 4-5v
5	BL/BK	A/C Amplifier Signal (ACT)	Clutch On: 12v, Off: 1.5v
6	BL/OR	Overdrive Main Switch	Switch Off: 12v, On: 1v
7	BR/YL	Cruise Control ECU	At Cruise in OD: 12v
10	GN/RD	AT ECT Pattern Select Switch	Norm: 0v, PWR: 12v
12	GN/OR	Vehicle Speed Sensor	At 55 mph: 48 Hz
14	BK/WT	Direct Battery	12-14v
17	RD/GN	Transfer Detect Switch (N)	Switch Closed: 0.1v
18	WT	Data Link Connector (SDL)	0v
20	BL/YL	A/C Amplifier Signal (AC1)	Clutch On: 1.5v, Off: 12v
22	WT/BL	EFI Main Relay B+	12-14v
25	GN/WT	Stop Light Switch Signal	Brake Off: 0v, On: 12v
26	GN/BK	4WD Indicator Switch	Switch On: 12v

1996-97 3.4L V6 MFI VIN N (A/T-ECT) 22 Pin Connector

PCM Pin #	Wire Color	Circuit Description (22 Pin)	Value at Hot Idle
1	GN/BK	Sensor VREF	4.9-5.1v
4	WT/RD	AT ECT Speed (+) Signal	At 55 mph: 48 Hz
5	PK	CKP Sensor Signal (NE+)	AC pulse signals
6	PK	CKP Sensor Signal (NE-)	AC pulse signals
7	BK/YL	TP Sensor Signal (VTA)	0.3-0.8v
8	RD/WT	Mass Airflow Sensor	1.1-1.8v
9	WT/RD	AT ECT Speed (-) Signal	At 55 mph: 48 Hz
10	YL	CMP Sensor Signal (G+)	AC pulse signals
11	BL	CMP Sensor Signal (G-)	AC pulse signals
12	PK/YL	A/T Oil Temperature Sensor	AT 68°F: 4-5v
13	WT	HO2S-11 (B1 S1) Signal	0.1-1.1v
14	YL/GN	IAT Sensor Signal (THA)	At 100°F: 2.60v
15	YL/BK	EVAP Vapor Pressure Sensor (PTNK)	2.5-3.1v
16	GY	Knock Sensor 2 Signal	<0.075v AC
17	BK	Knock Sensor 1 Signal	<0.075v AC
18	BK/WT	Power Ground	<0.1v
19	RD	HO2S-12 (B1 S2) Signal	0.1-1.1v
20	GN/RD	ECT Sensor Signal (THW)	At 180°F: 0.51v
22	BR/BK	Sensor Ground	<0.050v

1996-97 3.4L V6 MFI VIN N (A/T-ECT) 16 Pin Connector

PCM Pin #	Wire Color	Circuit Description (16 Pin)	Value at Hot Idle
1	YL/GN	Closed Throttle Switch	1v, off-idle: 12v
3	PK	Malfunction Indicator Lamp Control	MIL Off: 12v, On: 1v
4	GN/YL	Circuit Opening Relay (FC)	0-3v, off-idle: 12v
5	PK/WT	Data Link Connector	12-14v
13	PK/BK	EVAP Vapor Pressure Valve (VSV)	12v or 0v
15	WT/GN	EVAP Purge Solenoid (VSV)	12v or 0v
16	BR	Sensor Ground	<0.050v

1996-97 3.4L V6 MFI VIN N (A/T-ECT) 34-Pin Connector

PCM Pin #	Wire Color	Circuit Description (34-Pin)	Value at Hot Idle
1	BR	Power Ground	<0.1v
5	YL	Injector 6 Control	1.6-2.9 ms
6	BL	Injector 5 Control	1.6-2.9 ms
7	RD/BK	Injector 4 Control	1.6-2.9 ms
8	GN	Injector 3 Control	1.6-2.9 ms
9	WT	Injector 2 Control	1.6-2.9 ms
10	RD	Injector 1 Control	1.6-2.9 ms
11	GN/RD	AT ECT Solenoid (S1)	3rd or OD: 1v
12	BK/YL	Igniter Signal (IGF)	Digital Signal: 0-5-0v
13	BK/WT	Starter Switch Signal	In P/N: 0-3.0v
15	RD/WT	HO2S-12 (B1 S2) Heater	1v (Heater on)
16	PK/GN	HO2S-11 (B1 S1) Heater	1v (Heater on)
17	LG	A/T ECT Solenoid (S2)	1st or OD: 1v
22	BK/RD	IAC Signal (RSC)	Pulse Signals
23	BR/RD	Idle Air Control Valve (RSO)	Pulse Signals
24	BK/BL	Igniter Transistor 1 Control	7% duty cycle
25	BR/YL	Igniter Transistor 2 Control	7% duty cycle
26	BK/WT	Igniter Transistor 3 Control	7% duty cycle
27	RD/GN	A/T ECT Solenoid (SL)	In Lockup: 12-14v
29	GN	4WD Detection Transfer (L4)	Switch Closed: 0.1v
33	BR	Power Ground	<0.1v
34	BR	Power Ground	<0.1v

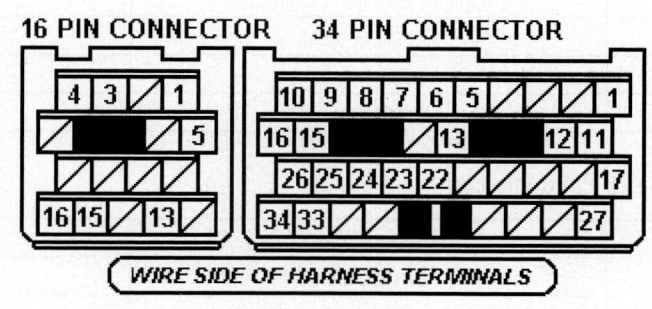

Pin Connector Graphic

1996-97 3.4L V6 MFI VIN N (M/T) 16 Pin Connector

PCM Pin #	Wire Color	Circuit Description (16 Pin)	Value at Hot Idle
1 ('96)	GN/BK	Sensor VREF	4.9-5.1v
1 ('97)	GN/YL	Sensor VREF	4.9-5.1v
2 ('96)	RD/WT	MAF Sensor Signal	1.1-1.8v
2 ('97)	BK/RD	MAF Sensor Signal	1.1-1.8v
3	GY	Knock Sensor 2 Signal	<0.075v AC
4	GN/RD	ECT Sensor Signal (THW)	At 180°F: 0.51v
5	WT	HO2S-11 (B1 S1) Signal	0.1-1.1v
6	BK	Knock Sensor 1 Signal	<0.075v AC
7	WT	Data Link Connector	12-14v
8	BK/WT	MAF Sensor Ground (E2G)	<0.050v
9	BR/BK	Sensor Ground	<0.050v
10	BK/YL	TP Sensor Signal (VTA)	0.3-0.8v
12	YL/GN	IAT Sensor Signal (THA)	At 100°F: 2.60v
13	RD	HO2S-12 (B1 S2) Signal	0.1-1.1v

1996-97 3.4L V6 MFI VIN N (M/T) 26 Pin Connector

PCM Pin #	Wire Color	Circuit Description (26 Pin)	Value at Hot Idle
3	PK/BK	EVAP Vapor Pressure Valve (VSV)	12v or 0v
4	YL/GN	Closed Throttle Switch	1v, off-idle: 12v
5 ('96)	BK/WT	EVAP Purge Solenoid (VSV)	12-14v
5 ('97)	WT/GN	EVAP Purge Solenoid (VSV)	12-14v
6	BK/RD	IAC Signal (RSC)	Pulse Signals
7	BR/RD	Idle Air Control Valve (RSO)	Pulse Signals
8	YL	Injector 6 Control	1.6-2.9 ms
9	BL	Injector 5 Control	1.6-2.9 ms
10	RD/BK	Injector 4 Control	1.6-2.9 ms
11	WT	Injector 2 Control	1.6-2.9 ms
12	RD	Injector 1 Control	1.6-2.9 ms
13	BR	Power Ground	<0.1v
14	GN/YL	Circuit Opening Relay (FC)	0-3v, off-idle: 12v
17	BK/YL	Igniter Signal (IGF)	7% duty cycle
21	BK/WT	Igniter Transistor 3 Control	7% duty cycle
22	BR/YL	Igniter Transistor 2 Control	7% duty cycle
23	BK/BL	Igniter Transistor 1 Control	7% duty cycle
24	BR	Sensor Ground	<0.050v
25	GN	Injector 3 Control	1.6-2.9 ms
26	BR	Power Ground	<0.1v

1996-97 3.4L V6 MFI VIN N (M/T) 22 Pin Connector

PCM Pin #	Wire Color	Circuit Description (22 Pin)	Value at Hot Idle
2	BK/WT	Direct Battery	12-14v
4	PK	Malfunction Indicator Lamp Control	MIL Off: 12v, On: 1v
6	BL/BK	A/C Amplifier Signal (ACT)	Clutch On: 12v, Off: 1.5v

1996-97 3.4L V6 MFI VIN N (M/T) 22 Pin Connector, *continued*

7	BL/YL	A/C Amplifier Signal (AC1)	Clutch On: 1.5v, Off: 12v
8	GN/OR	Vehicle Speed Sensor	At 55 mph: 48 Hz
9	GN/BK	4WD Detection Switch	Switch On: 12v, Off: 0v
11	BK/WT	Starter Switch Signal	In P/N: 0-3.0v
12	WT/BL	EFI Main Relay Power	12-14v
19	WT	Data Link Connector (SDL)	0v
20	GN/WT	Stop Light Switch Signal	Brake Off: 0v, On: 12v

1996-97 3.4L V6 MFI VIN N (M/T) 12 Pin Connector

PCM Pin #	Wire Color	Circuit Description (12 Pin)	Value at Hot Idle
3	PK	HO2S-11 (B1 S1) Heater	1v (Heater on)
4	YL/BK	EVAP Vapor Pressure Sensor (PTNK)	2.5-3.1v (with cap off)
5	BL	CMP Sensor Signal (G-)	<0.050v
6	PK	CKP Sensor Signal (NE-)	<0.050v
7	BR	Power Ground	<0.1v
9	RD/WT	HO2S-12 (B1 S2) Heater	1v (Heater on)
11	YL	CMP Sensor Signal (G+)	AC pulse signals
12	PK	CKP Sensor Signal (NE+)	AC pulse signals

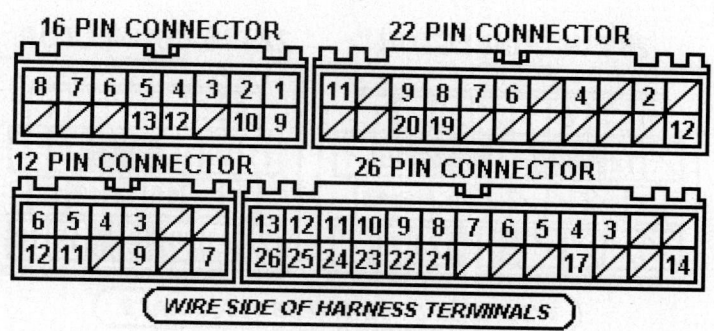

05533_ADIA_G153

Pin Connector Graphic

1998 3.4L V6 MFI VIN N (All) 28 Pin Connector

PCM Pin #	Wire Color	Circuit Description (28 Pin)	Value at Hot Idle
1	LG	A/T Select Switch Low	In Low: 12v, Others: 0v
5	BL/BK	A/C Amplifier Signal (ACT)	Clutch On: 12v, Off: 1.5v
6	BL/OR	Overdrive Main Switch	Switch Off: 12v, On: 1v
7	BR/YL	Cruise Control ECU	At Cruise in OD: 12v
8	WT	SIL Signal (Scan Tool)	12v
10	PK/RD	A/T Select Switch 2nd	In 2nd: 12v, Others: 0v
11	YL/GN	Closed Throttle Switch	1v, off-idle: 12v
12	GN/OR	Vehicle Speed Sensor	At 55 mph: 48 Hz
14	BL/RD	Direct Battery	12-14v
15	RD/YL	A/T Select Switch Reverse	In 'R': 12v, Others: 0v
16	BL/YL	A/C Amplifier Signal (AC1)	Clutch On: 1.5v, Off: 12v
23	WT/BL	EFI Main Relay Power	12-14v
24	GN/WT	Stop Light Switch Signal	Brake Off: 0v, On: 12v

1998 3.4L V6 MFI VIN N (All) 22 Pin Connector

PCM Pin #	Wire Color	Circuit Description (22 Pin)	Value at Hot Idle
1	GN/BK	Sensor VREF	4.9-5.1v
4	WT/RD	A/T ECT Speed Sensor (-)	Pulse Signals
5	PK	CKP Sensor Signal (NE+)	Pulse Signals
6	PK	CKP Sensor Signal (NE-)	Pulse Signals
7	BK/YL	TP Sensor Signal (VTA)	0.3-0.8v
8	RD/WT	MAF Sensor Signal	1.1-1.8v
9	YL/RD	A/T ECT Speed Sensor (+)	Pulse Signals
12	GN/RD	AT ECT Pattern Select Switch	Norm: 0v, PWR: 12v
13	WT	HO2S-11 (B1 S1) Signal	0.1-1.1v
14	GY	Knock Sensor 2 Signal	<0.075v AC
15	BK	Knock Sensor 1 Signal	<0.075v AC
17	BL	CMP Sensor Signal (G+)	AC pulse signals
18	GN/YL	Circuit Opening Relay (FC)	0-3v, off-idle: 12v
19	RD	HO2S-12 (B1 S2) Signal	0.1-1.1v
20	GN	ECT Sensor Signal (THW)	At 180°F: 0.51v
21	YL/GN	IAT Sensor Signal (THA)	At 100°F: 2.60v
22	BR/BK	Sensor Ground	<0.050v

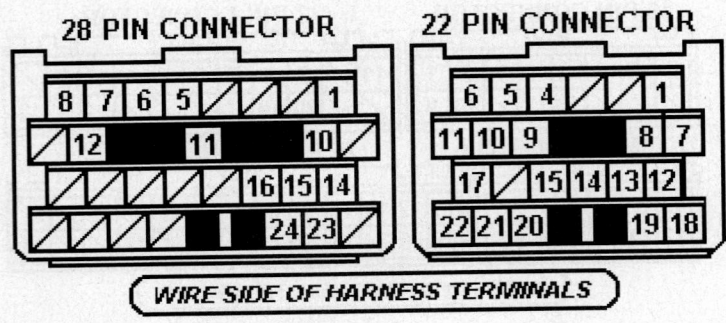

28 PIN CONNECTOR **22 PIN CONNECTOR**

WIRE SIDE OF HARNESS TERMINALS

05533_ADIA_G154

Pin Connector Graphic

1998 3.4L V6 MFI VIN N (All) 16 Pin Connector

PCM Pin #	Wire Color	Circuit Description (16 Pin)	Value at Hot Idle
2	WT/GN	EVAP Purge Solenoid (VSV)	12v or 0v
3	PK	Malfunction Indicator Lamp Control	MIL Off: 12v, On: 1v
5	PK/WT	Data Link Connector	12-14v
6	YL	A/T Oil Temperature Sensor	At 68°F: 4-5v
7	BK/WT	MAF Sensor Ground (E2G)	<0.050v
8	PK/BK	EVAP Vapor Pressure Valve (VSV)	12v or 0v
9	RD/GN	4WD Detection Transfer (N)	Open: 12v, Closed: 0v
10	RD/WT	HO2S-12 (B1 S2) Heater	1v (Heater on)
11	PK/BL	HO2S-11 (B1 S1) Heater	1v (Heater on)
13	YL/BK	EVAP Vapor Pressure Sensor (PTNK)	2.5-3.1v (with cap off)
16	BR	Sensor Ground	<0.050v

1998 3.4L V6 MFI VIN N (All) 34-Pin Connector

PCM Pin #	Wire Color	Circuit Description (34-Pin)	Value at Hot Idle
1	GN/BK	4WD Detection Switch	Switch On: 12v
2	BL/RD	4WD Detection Transfer (L4)	Switch Closed: 0.1v
5	YL	Injector 6 Control	1.6-2.9 ms
6	BL	Injector 5 Control	1.6-2.9 ms
7	RD/BK	Injector 4 Control	1.6-2.9 ms
8	GN	Injector 3 Control	1.6-2.9 ms
9	WT	Injector 2 Control	1.6-2.9 ms
10	RD	Injector 1 Control	1.6-2.9 ms
11	GN/RD	AT ECT Solenoid (S1)	S1: 3rd or OD: 1v
12	BK/YL	Igniter Signal (IGF)	Digital Signal: 0-5-0v
13	BK/WT	Starter Switch Signal	9-11v (cranking)
14	BK	Neutral Start Switch	In P/N: 0-3.0v
15	BK/WT	Igniter Transistor 3 Control	7% duty cycle
16	BK/YL	Igniter Transistor 2 Control	7% duty cycle
17	GN	A/T ECT Solenoid (S2)	S1: 3rd or OD: 1v
22	BK/RD	IAC Signal (RSC)	Pulse Signals
23	BR	Idle Air Control Valve (RSO)	Pulse Signals
24	BK/BL	Igniter Transistor 1 Control	7% duty cycle
27	GN/RD	A/T ECT Solenoid (SL)	In Lockup: 12-14v
28	BR	Power Ground	<0.1v
30	PK/YL	A/T Oil Temperature Sensor	At 68°F: 4-5v
31	PK	PSP Switch Signal (PSW)	Straight: 12v, Turned: 0v
33	BR	Power Ground	<0.1v
34	BR	Power Ground	<0.1v

1999-2000 3.4L V6 (All) VIN N E9 31 Pin Connector

PCM Pin #	Wire Color	Circuit Description (31 Pin)	Value at Hot Idle
1	GN	Injector 3 Control	1.6-2-9 ms
2	RD/BK	Injector 4 Control	1.6-2-9 ms
3	BL	Injector 5 Control	1.6-2-9 ms
4	YL	Injector 6 Control	1.6-2.9 ms
5	---	Not Used	---
6	PK/WT	Data Link Connector	12-14v
7	LG/RD	AT ECT Solenoid (S1)	3rd or OD: 1v
8	BL/WT	A/T ECT Solenoid (S2)	1st or OD: 1v
9	GN/RD	A/T ECT Solenoid (SL)	In Lockup: 12-14v
10	RD	CMP Sensor Signal (G+)	AC pulse signals
11	BK/BL	Igniter Transistor 1 Control	6°, at 55 mph: 8° dwell
12	BL/YL	Igniter Transistor 2 Control	6°, at 55 mph: 8° dwell
13	GN/WT	Igniter Transistor 3 Control	6°, at 55 mph: 8° dwell
14	YL/RD	A/T ECT Speed Sensor (+)	Pulse Signals
15	BK/RD	IAC Signal (RSC)	Pulse Signals
16	BR/RD	Idle Air Control Valve (RSO)	Pulse Signals
17-20	---	Not Used	---
21	WT/BK	Power Ground (E01)	<0.1v
22-24	---	Not Used	---
25	BK/YL	Igniter Signal (IGF)	Digital Signal: 0-5-0v
26	WT/RD	A/T ECT Speed Sensor (-)	Pulse Signals
27	BK	Knock Sensor 1 Signal	<0.075 VAC
28	GY	Knock Sensor 2 Signal	<0.075 VAC
29	---	Not Used	---
30	WT/BK	Power Ground (E03)	<0.1v
31	WT/BK	Power Ground (E02)	<0.1v

E9 31-PIN CONNECTOR

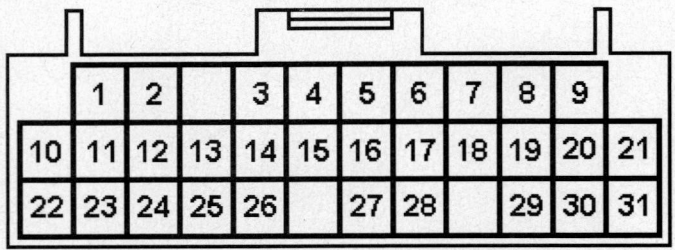

WIRE SIDE OF HARNESS TERMINALS

05533_ADIA_G155

Pin Connector Graphic

Standard Colors and Abbreviations

Abbreviation	Color	Abbreviation	Color	Abbreviation	Color
BK	Black	GY	Gray	RD	Red
BL	Blue	GN	Green	TN	Tan
BR	Brown	LG	LT Green	VT	Violet
DB	Dark Blue	OR	Orange	WT	White
DG	DK Green	PK	Pink	YL	Yellow

1999-2000 3.4L V6 VIN N (All) E10 24 Pin Connector

PCM Pin #	Wire Color	Circuit Description (24 Pin)	Value at Hot Idle
1, 3	---	Not Used	---
2	GN/BK	Sensor VREF	4.9-5.1v
4 (Cal)	BL	HAFR-11 (B1 S1) Heater	1v (Heater on)
4 (Fed)	PK/BL	HO2S-11 (B1 S1) Signal	0.1-1.1v
5	RD	Injector 1 Control	1.6-2-9 ms
6	WT	Injector 2 Control	1.6-2-9 ms
7	WT/GN	EVAP Purge Solenoid (VSV)	12v or 0v
8 (Cal)	WT/BK	Power Ground	<0.1v
9	PK	PSP Switch Signal (PSW)	Straight: 12v, Turned: 0v
10	RD/WT	MAF Sensor Signal	1-1.1v
12 (Cal)	WT	AFR-11 (B1 S1) Signal (+)	Fixed at 3.3v
12 (Fed)	WT	HO2S-11 (B1 S1) Heater	1v (Heater on)
13	---	Not Used	---
14	GN	ECT Sensor Signal (THW)	0.5-0.6v
15	PK/YL	A/T Oil Temperature Sensor	At 68°F: 4-5v
16	RD	CKP Sensor Signal (NE+)	AC pulse signals
17	BR	Shield Ground	<0.050v
18	BL/BK	Sensor Ground	<0.050v
19	BK/WT	MAF Sensor Ground (E2G)	<0.050v
20	---	Not Used	---
21 (Cal)	BK	AFR-11 (B1 S1) Signal (-)	Fixed at 3.3v
22	YL/GN	IAT Sensor Signal (THA)	0.5-3.4v
23	BK/YL	TP Sensor Signal (VTA)	0.53-1.27v
24	GN	CMP & CKP Sensor Return	<0.050v

1999-2000 3.4L V6 VIN N (All) E11 17 Pin Connector

PCM Pin #	Wire Color	Circuit Description (17 Pin)	Value at Hot Idle
1-3	---	Not Used	---
4	GY/RD	Transponder Amplifier Signal (Code)	Inserting key: pulses
5	RD/BK	Transponder Amplifier Signal (RXCK)	Inserting key: pulses
6-9	---	Not Used	---
10	PK/BK	Transponder Amplifier Signal (RXCK)	Inserting key: pulses
11	YL/RD	Unlock Warning Switch	No Key: 4-5v
12-15	---	Not Used	---
16	BL	Theft Security Indicator Light	Digital Signal

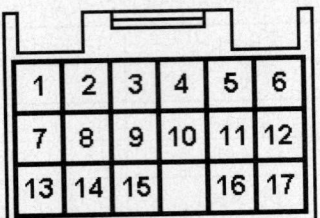

E10 24-PIN CONNECTOR

1	2		3	4	5	6	7
8	9	10	11	12	13	14	15	16
17	18	19	20	21		22	23	24

E11 17-PIN CONNECTOR

1	2	3	4	5	6
7	8	9	10	11	12
13	14	15		16	17

WIRE SIDE OF HARNESS TERMINALS

05533_ADIA_G156

Pin Connector Graphic
1999-2000 3.4L V6 (All) VIN N E12 28 Pin Connector

PCM Pin #	Wire Color	Circuit Description (28 Pin)	Value at Hot Idle
2	RD/YL	A/T Select Switch Reverse	In 'R': 12v, Others: 0v
3	PK/RD	A/T Select Switch 2nd	In 2nd: 12v, Others: 0v
4	YL/GN	Cruise Control ECU	At Cruise in OD: 12v
7	YL	A/T Oil Temperature Indicator	Lamp Off: 12v, On: 1v
8	RD	HO2S-12 (B1 S2) Signal	0.1-1.1v
9	RD/WT	HO2S-12 (B1 S2) Heater	1v (Heater on)
10	BL/OR	Overdrive "Off" Indicator	Lamp Off: 12v, On: 1v
11	GN/RD	AT ECT Pattern Select Switch	Norm: 0v, PWR: 12v
12	LG	A/T Select Switch Low	In Low: 12v, Others: 0v
13	BL/BK	A/C Amplifier Signal (ACT)	Clutch On: 12v, Off: 1.5v
17	RD/GN	4WD Detection Transfer (N)	Open: 12v, Closed: 0v
18	BK/BL	4WD ADD Indicator Switch	Open: 12v, Closed: 0v
19	BL/RD	4WD Detection Transfer (L4)	Switch Closed: 0.1v
20	BK	A/T Neutral Start Signal	In P/N: 0-3.0v
22	GN/OR	Speedometer Indicator	At 55 mph: 48 Hz
24	BR/YL	Overdrive Main Switch	Switch Off: 12v, On: 1v
25	BL/YL	A/C Amplifier Signal (AC1)	Clutch On: 1.5v, Off: 12v

1999-2000 3.4L V6 VIN N (All) E14 22 Pin Connector

PCM Pin #	Wire Color	Circuit Description (22 Pin)	Value at Hot Idle
1	BL/RD	Direct Battery	12-14v
2	BK/BL	Ignition Switch	12-14v
3	GN/YL	Circuit Opening Relay (FC)	Relay On: 1v, Off: 12v
4-5	---	Not Used	---
6	PK	Malfunction Indicator Lamp Control	MIL Off: 12v, On: 1v
7	BK/WT	Starter Switch Signal	9-11v (cranking)
8	GY/BK	EFI Main Relay Control	Relay Off: 0v, On: 12v
9	WT/RD	EVAP Vapor Pressure Valve (VSV)	12v or 0v
10	---	Not Used	---
11	WT	SIL Signal (Scan Tool)	0v
12	WT/BK	Power Ground	<0.1v

1999-2000 3.4L V6 VIN N (All) E14 22 Pin Connector, *continued*

13-14	---	Not Used	---
15	GN/WT	Brake Switch Signal	Brake Off: 0v, On: 12v
16	WT/BL	EFI Main Relay Power	12-14v
17	YL	EVAP Vapor Pressure Sensor (PTNK)	2.9-3.7v (hose off)
18-22	---	Not Used	---

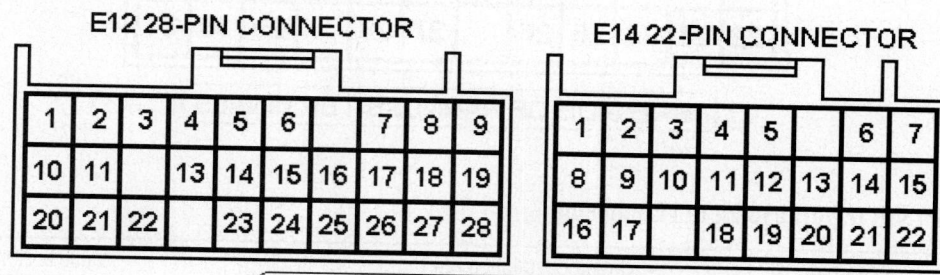

E12 28-PIN CONNECTOR

E14 22-PIN CONNECTOR

WIRE SIDE OF HARNESS TERMINALS

05533_ADIA_G157

Pin Connector Graphic

2001-02 3.4L V6 MFI VIN N (All) E9 31 Pin Connector

PCM Pin #	Wire Color	Circuit Description (31 Pin)	Value at Hot Idle
1	LG	A/T ECT Solenoid (SL)	In Lockup: 12-14v
2	BL/WT	A/T ECT Solenoid (S2)	1st or OD: 1v
3	PK/BL	A/T ECT Solenoid (S3)	3rd or OD: 1v
4	WT/BK	Power Ground (E01)	<0.1v
5	WT/BK	Power Ground (E02)	<0.1v
6	WT/BK	Power Ground (ME01)	<0.1v
7	WT/BK	Power Ground (E03)	<0.1v
8	WT/BK	Power Ground (E04)	<0.1v
9	RD	Throttle Control Motor (M+)	Pulse Signals
10	BK/WT	MAF Sensor Ground (E2G)	<0.050v
11	BL/BK	Sensor Ground (E2)	<0.050v
12	RD/WT	MAF Sensor Signal (VG)	1-1.1v
13	YL/GN	IAT Sensor Signal (THA)	0.5-3.4v
14	WT	AFR-11 (B1 S1) Signal (+)	Fixed at 3.3v
15	BK/YL	TP Sensor Signal (VTA1)	0.53-1.27v
16	---	Not Used	---
17	BR	Shield Ground (GE01)	<0.050v
18	GN	ECT Sensor Signal (THW)	0.5-0.6v
19	BL/RD	ATF Fluid Temp. Input (THO)	At 68°F: 4-5v
20, 28, 30	---	Not Used	---
21	BL	HAFR-11 (B1 S1) Heater	1v (Heater on)
22	GY	Knock Sensor 2 Signal	<0.075 VAC
23	BK	Knock Sensor 1 Signal	<0.075 VAC
24	YL	EVAP Vapor Pressure Sensor (PTNK)	2.9-3.7v (hose off)
25	GN/BK	Sensor VREF	4.9-5.1v
26	BK	AFR-11 (B1 S1) Signal (-)	Fixed at 3.3v
27	BK	HO2S-12 (B1 S2) Signal	0.1-1.1v
29	GN/YL	HO2S-12 (B1 S2) Heater	1v (Heater on)
31	RD	Throttle Control Motor (M-)	Pulse Signals

E9 31-PIN CONNECTOR

WIRE SIDE OF HARNESS TERMINALS

05533_ADIA_G155

Pin Connector Graphic
2001-02 3.4L V6 MFI VIN N (All) E10 24 Pin Connector

PCM Pin #	Wire Color	Circuit Description (24 Pin)	Value at Hot Idle
1	BL	Injector 5 Control	1.6-2-9 ms
2	BK/YL	Igniter Signal (IGF)	Digital Signal: 0-5-0v
3	RD/BK	Injector 4 Control	1.6-2-9 ms
4	GN	Injector 3 Control	1.6-2-9 ms
5	WT	Injector 2 Control	1.6-2-9 ms
6	RD	Injector 1 Control	1.6-2-9 ms
7	GY	Data Link Connector (TC)	12-14v
8	YL	Injector 6 Control	1.6-2.9 ms
9	GN/WT	Igniter Transistor 3 Control	6°, at 55 mph: 8° dwell
10	BL/YL	Igniter Transistor 2 Control	6°, at 55 mph: 8° dwell
11	BK/BL	Igniter Transistor 1 Control	6°, at 55 mph: 8° dwell
12	BL	CKP Sensor Signal (NE+)	AC pulse signals
13	RD	CMP Sensor Signal (G+)	AC pulse signals
14	WT/GN	EVAP Purge Solenoid (VSV)	12v or 0v
15-16	---	Not Used	---
17	BR	Power Ground (E1)	<0.1v
18	PK	PSP Switch Signal (PSW)	Straight: 12v, Turned: 0v
19	BL/BK	EVAP Canister Closed Valve (CCV)	12v or 0v
20	RD/YL	EVAP Vapor Pressure Valve (VSV)	12v or 0v
21	GN	CMP & CKP Sensor Ground	<0.050v
22	WT/RD	A/T ECT Speed Sensor (-)	Pulse Signals
23	YL/RD	A/T ECT Speed Sensor (+)	Pulse Signals
24	---	Not Used	---

2001-02 3.4L V6 MFI VIN N (All) E11 17 Pin Connector

PCM Pin #	Wire Color	Circuit Description (17 Pin)	Value at Hot Idle
1, 3	---	Not Used	---
2	YL	Throttle Control Motor (CL+)	Pulse Signals
4	GN/YL	TP Sensor 2 (VTA2)	2.0-2.9v
5	RD/YL	A/T-ECT Solenoid (SLT+)	Pulse Signals
6-9	---	Not Used	---
8	BL	Throttle Control Motor (CL-)	Pulse Signals

2001-02 3.4L V6 MFI VIN N (All) E11 17 Pin Connector, *continued*

10	GY	Accel Position Sensor (VPA)	0.25-0.9v
11	YL/BK	A/T-ECT Solenoid (SLT-)	Pulse Signals
12-14	---	Not Used	---
15	BL	Accel Position Sensor (VPA2)	1.8-2.7v
16-17	---	Not Used	---

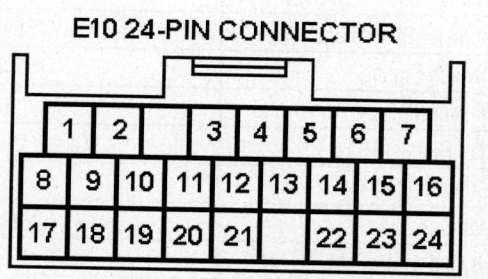

05533_ADIA_G156

Pin Connector Graphic

2001-02 3.4L V6 MFI VIN N (All) E12 28 Pin Connector

PCM Pin #	Wire Color	Circuit Description (28 Pin)	Value at Hot Idle
1	GN/RD	A/T: ECT Power Select Signal	Norm: 0v, PWR: 12v
2	BL/RD	4WD Detection Transfer (L4)	Switch Closed: 0.1v
3	BK	Neutral Start Switch Signal	In P/N: 9-12v (cranking)
4	LG	A/T Select Switch Low	In Low: 12v, Others: 0v
5	VT	A/T Select Switch Drive	In Drive: 12-14v
6	WT/BL	A/T Select Switch Reverse	In 'R': 12v, Others: 0v
8	VT/WT	ABS/BA/TRAC/VSC ECU	Digital Signals (NEO)
11	BK/RD	4WD Detection Transfer (N)	Open: 12v, Closed: 0v
13	BK/BL	4WD ADD Indicator Switch	Switch Closed: 0.1v
14	BL/WT	Overdrive Main Switch	Switch Off: 12v, On: 1v
18, 19	PK, BL	TRAC Engine (-), (+) Signals	Pulse Signals
20	GN/YL	O/D OFF (lamp) Indicator	Light Off: 12v, On: 1v
21	YL	A/T: Oil Temp. Lamp Indicator	Lamp Off: 12v, On: 1v
22	GN	Cruise Control Switch Signal	Main Switch On: 1v
25	WT/RD	2WD Detect Transfer Switch	Open: 0v, Closed: 12v
25	GY/BK	4WD Detect Transfer Switch	Open: 0v, Closed: 12v
27, 28	YL, BR	TRAC TRC (-), (+) Signals	Pulse Signals

2001-02 3.4L V6 MFI VIN N (All) E14 22 Pin Connector

PCM Pin #	Wire Color	Circuit Description (22 Pin)	Value at Hot Idle
1	WT/BL	EFI Main Relay Power	12-14v
2	VT	Malfunction Indicator Lamp Control	MIL Off: 12v, On: 1v
3	BL/BK	A/C Amplifier Signal (ACT)	Clutch On: 12v, Off: 1.5v
4	GY/BK	EFI Main Relay Control	Relay Off: 0v, On: 12v
5	GY/RD	Transponder Amplifier Signal (Code)	Inserting key: pulses
6	GN/OR	Combination Meter SP1 Input	At 55 mph: 48 Hz
7	BK/WT	Starter Switch Signal	In P/N: 0-3.0v
8	GN	ETC Power (B+)	12-14v
9	BL/YL	A/C Amplifier Signal (AC1)	Clutch On: 1.5v, Off: 12v
10	WT/BK	Power Ground (EOM)	<0.1v
11	YL/RD	Unlock Warning Switch	No Key: 4-5v
12	BL	Theft Security Indicator Light	Theft LED On: 12v
13	RD/WT	Data Link Connector (WFSE)	N/A
14	WT	SIL Signal (Scan Tool)	Digital Signal
15	BK/BL	Ignition Switch	12-14v
16	BL/RD	Direct Battery	12-14v
17	GY/RD	Cruise Control Indicator	Lamp On: 1v, Off: 12v
18, 19	RD, PK	Transponder Amplifier Signal (RXCK)s	Inserting key: pulses
20	GN/WT	Stop Light Switch Signal	Brake Off: 0v, On: 12v
22	GN/YL	Circuit Opening Relay (FC)	Relay On: 1v, Off: 12v

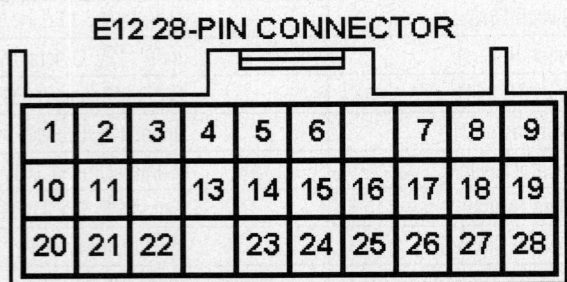

E12 28-PIN CONNECTOR

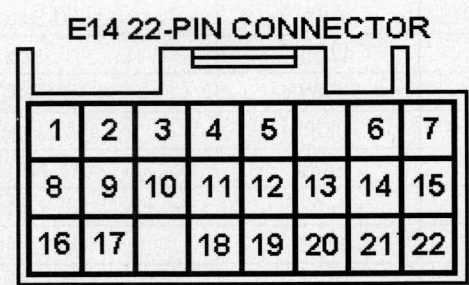

E14 22-PIN CONNECTOR

05533_ADIA_G158

Pin Connector Graphic

2003-2006 4.0L V6 VIN U E4 34 Pin Connector

PCM Pin #	Wire Color	Circuit Description (34 Pin)	Value at Hot Idle
1	RD/BL	Injector 1 Control	2.0-3.3 ms
2	BK	Injector 2 Control	2.0-3.3 ms
3	LG	Injector 3 Control	2.0-3.3 ms
4	GN	Injector 4 Control	2.0-3.3 ms
5	YL	Injector 5 Control	2.0-3.3 ms
6, 7	WT/BK	Power Ground (E02), (E01)	<0.1v
8	YL/RD	Igniter Transistor 1 Control	7% duty cycle
9	PK/BL	Igniter Transistor 2 Control	7% duty cycle
10	LG	Igniter Transistor 3 Control	7$ duty cycle
11	LG/BK	Igniter Transistor 4 Control	7% duty cycle
12	GY	Igniter Transistor 5 Control	7% duty cycle
13	BL	Igniter Transistor 6 Control	7% duty cycle
14	WT/GN	A/C Relay Control (ACCR)	Relay Off: 12v, On: 1v
15	WT/BL	ACIS Control (VSV)	12v or 0v
16	PK	Neutral Start Switch (NSW)	In P/N: 0-3.0v
17	BK/YL	Starter Switch Signal (STA)	In P/N: 0-3.0v
18	BL/RD	Sensor VREF (VC)	4.9-5.1v
19	BK/BL	ECT Sensor Signal (THW)	At 180°F: 0.51v
20	RD/BK	IAT Sensor Signal (THA)	At 100°F: 2.60v
21	GN/BK	TP Sensor Signal (VTA1)	0.4-1.0v
22-23, 25-26	---	Not Used	---
24	WT/RD	Igniter Signal (IGF1)	Digital Signal: 0-5-0v
27	RD/GN	EVAP Canister Closed Valve (CCV)	12v or 0v
28	BR	Sensor Ground (E2)	<0.050v
29	RD/WT	MAF Sensor Ground (E2G)	<0.050v
30	RD/YL	Mass Airflow Sensor (VG)	1.1-1.5v
31	GN/WT	TP Sensor Signal (VTA2)	2.0-2.9v
33	YL/BK	Fuel Pump Relay Control (FPR)	Relay Off: 12v, On: 1v
34	GN/YL	EVAP Purge Solenoid (VSV)	12v or 0v

2003-2006 4.0L V6 VIN U E5 35 Pin Connector

PCM Pin #	Wire Color	Circuit Description (35 Pin)	Value at Hot Idle
1	BK	Knock Sensor 1 Signal (KNK1)	<0.075v AC
2	GN	Knock Sensor 2 Signal (KNK2)	<0.075v AC
3	BL	Injector 6 Control	2.0-3.3 ms
4	BK/WT	AFS-11 (B2 S1) Heater (HAFL)	1v (Heater on)
5	RD/BL	AFS-11 (B1 S1) Heater (HAFR)	1v (Heater on)
6, 7	WT/BK	Power Ground (E05), (E04)	<0.050v
8	GN	4WD Switch Signal (4WD)	In 4WD: 12v
9	PK	Start Signal (STAR)	In P/N: 0-3.0v
10	BL/WT	ECT Solenoid Control (S2)	12v or 0v
11	BL/RD	ECT Solenoid Control (S1)	12v or 0v
12	BL/BK	ECT Solenoid Control (SLT-)	12v or 0v
13	GN/YL	ECT Solenoid Control (SLT+)	12v or 0v
15	GN	ECT Solenoid Control (SL)	12v or 0v
16	PK/BL	A/T Solenoid Control (SL2-)	Pulse Signals
17	PK/BK	A/T Solenoid Control (SL2+)	Pulse Signals
18	RD/WT	A/T Solenoid Control (SL1-)	Pulse Signals
19	RD/BL	A/T Solenoid Control (SL1+)	Pulse Signals
20	RD	Knock Sensor 2 Ground	<0.050v
21	WT	HO2S-12 (B1 S2) Signal (OX1B)	0.1-1.1v
22	PK	AFS-11 (B1 S1) Signal (AFR+)	3.0-3.6v
23	YL	AFS-21 (B2 S1) Signal (AFL+)	3.0-3.6v
24	BL	A/T Oil Temperature Sensor 2 (THO2)	At 68°F: 4-5v
25	GN	HO2S-12 (B1 S2) Heater (HT1B)	1v (Heater on)
26	GN	ECT Vehicle Speed Sensor (SP2+)	Pulse Signals
27	WT/RD	Turbine Speed Sensor (NCO+)	Pulse Signals
28	WT	Knock Sensor 1 Ground	<0.050v
29	BK	HO2S-22 (B2 S2) Signal (OX2B)	0.1-1.1v
30	BL	AFS-11 (B1 S1) Signal (AFR-)	Fixed at 3.3v
31	BR	AFS-21 (B2 S1) Signal (AFL-)	Fixed at 3.3v
32	GN/YL	A/T Oil Temperature Sensor 1 (THO1)	At 68°F: 4-5v
33	BL	HO2S-22 (B2 S2) Heater (HT2B)	1v (Heater on)
34	RD	ECT Vehicle Speed Sensor (SP2-)	Pulse Signals
35	YL/RD	Turbine Speed Sensor (NCO-)	Pulse Signals

2003-2006 4.0L V6 VIN U E6 32 Pin Connector

PCM Pin #	Wire Color	Circuit Description (32 Pin)	Value at Hot Idle
1	BR	Power Ground (E1)	<0.1v
2	BL	Throttle Control Motor (M-)	Pulse Signals
3	PK	Throttle Control Motor (M+)	Pulse Signals
4	WT/BK	Power Ground (ME01)	<0.1v
5-6	---	Not Used	---
7	WT/BK	Power Ground (E03)	<0.1v
8-9	---	Not Used	---
10	GN/WT	PSP Switch Signal (PSW)	Straight: 12v, Turning: 0v

2003-2006 4.0L V6 VIN U E6 32 Pin Connector, *continued*

11	YL/GN	Neutral Detection Switch (L4)	Switch Open: 0v, Closed: 12v
12	BL/YL	Start Switch Signal (STSW)	Cranking: 9-11v
13	BL/RD	Camshaft Timing Control Valve LH (OC2-)	Pulse Signals
14	BL/WT	Camshaft Timing Control Valve LH (OC2+)	Pulse Signals
15	BL/BK	Camshaft Timing Control Valve RH (OC1-)	Pulse Signals
16	GN/YL	Camshaft Timing Control Valve RH (OC1+)	Pulse Signals
17	BR	Shield Ground (GE01)	<0.050v
18-20	---	Not Used	---
21	BK/OR	Generator Control (RL)	12v
22	---	Not Used	---
23	RD/YL	A/C Lock Sensor (LCK)	12v or 0v
24	WT	CKP Sensor Signal (NE-)	<0.050v
25	BK	CKP Sensor Signal (NE+)	AC pulse signals
26	YL	Variable Valve Timing Sensor LH (VV2+)	AC pulse signals
27	RD	Variable Valve Timing Sensor RH (VV1+)	AC pulse signals
28-31	---	Not Used	---
32	WT	CMP Sensor Signal (G2-)	AC pulse signals

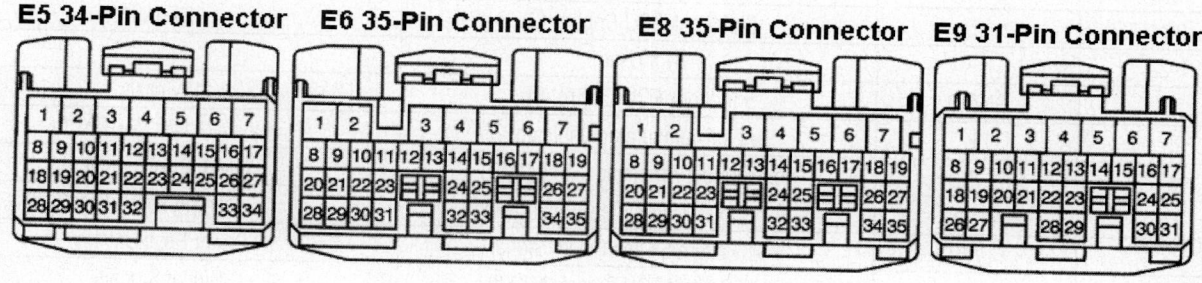

E5 34-Pin Connector   **E6 35-Pin Connector**   **E8 35-Pin Connector**   **E9 31-Pin Connector**

05533_ADIA_G124

Pin Connector Graphic

E4 34-Pin Connector   **E5 35-Pin Connector**   **E6 32-Pin Connector**   **E8 31-Pin Connector**

05533_ADIA_G159

Pin Connector Graphic

2003-2006 4.0L V6 VIN U E7 35 Pin Connector

PCM Pin #	Wire Color	Circuit Description (35 Pin)	Value at Hot Idle
1	WT/BK	Power Ground (HP)	<0.1v
2	BK/YL	A/C Magnetic Clutch Relay (ACMG)	Relay Off: 0v, On: 12v
3	GN	Transmission Control Switch Signal (2L)	In 2nd: 12v, Others: 0v
4	WT/BL	Low Detection Switch (L4)	Switch Open: 0v, Closed: 12v
5	BL/WT	ECT Pattern Switch 2nd Position (SNW1)	2nd Position: 12v
6	GN	ETCS Power (+BM)	12-14v
7	---	Not Used	---
8	BL/YL	Shift Lock ECU Control (L)	12v or 0v
9	PK/BL	4WD Low Indicator Control	Indicator Off: 12v, On: 0v
10	GN/YL	Park Neutral Position Switch (D)	In 'D': 12v, Others: 0v
11	RD/YL	A/T Select Switch Reverse (R)	In 'R': 12v, Others: 0v
12	---	Not Used	---
14	BL/BK	A/C Amplifier Signal (THWO)	A/C Off: 12v, On: 1v
15-16	---	Not Used	---
17	VT/RD	Vehicle Speed Sensor (SPD)	At 55 mph: 48 Hz
18	PK/BK	Body Control ECU Signal (MPX1)	Digital Signals
19	GN/YL	Stop Light Switch Signal	Brake Off: 0v, On: 12v
20	BL	Shift Lock ECU Control (3)	12v or 0v
21-22	---	Not Used	---
23	GN/RD	Shift Lock ECU Control (4)	12v or 0v
25	PK	A/T Oil Temperature Indicator	Indicator Off: 12v, On: 1v
26	BL/RD	Transponder Amplifier Signal (IMD)	Inserting key: pulses
27	WT/RD	Transponder Amplifier Signal (IMI)	Inserting key: pulses
28	BL/WT	ECT Pattern Switch Power Signal	Power Position: 12v
29	BK	Body Control ECU Signal (MPX2)	Digital Signals
30	---	Not Used	---
31	GR/BK	A/C Amplifier Signal (A/CS)	Relay Off: 12v, On: 1v
32	GY/GN	A/C Amplifier Signal (THE)	A/C Off: 12v, On: 1v
33	BK/RD	A/C Switch Signal (ACLD)	A/C On: 12v, Off: 0v
34-35	---	Not Used	---

2003-2006 4.0L V6 VIN U E8 31 Pin Connector

PCM Pin #	Wire Color	Circuit Description (31 Pin)	Value at Hot Idle
1	BK	EFI Main Relay Power (+B)	12-14v
2	BK	EFI Main Relay Power (+B2)	12-14v
3	BL	Direct Battery	12-14v
4	PK/BL	EVAP Pressure Switching Valve (VSV)	12v or 0v
5	BK/WT	Tachometer Signal (TACO)	Pulse Signals
6	GN/WT	A/T Select Switch Park Signal (P)	In 'P': 12v, Others: 0v
7	GN/RD	A/T Select Switch Neutral Signal (N)	In 'N': 12v, Others: 0v
8	WT/GN	EFI Main Relay Control	Relay Off: 0v, On: 12v
9	BK/OR	Ignition Switch Power (IGSW)	12-14v
10	GR/BK	Circuit Opening Relay (FC)	0-3v, off-idle: 12v
11	RD/BK	Malfunction Indicator Lamp Control	MIL Off: 12v, On: 1v
12	YL/GN	Defogger Switch Signal (ELS)	Defogger Off: 0v, On: 12v
13	GN	Taillight Switch Signal (ELS2)	Taillights Off: 12v, On: 1v
14	BL	Center Airbag Assembly (F/PS)	Digital Signals
15	WT/BK	Power Ground (EOM)	<0.1v
16	---	Not Used	---
17	PK	Transponder Amplifier Signal (NEO)	Inserting key: pulses
18	RD/YL	SIL Signal (Scan Tool)	Transmitting: pulses
19	RD/WT	Data Link Connector (WFSE)	12v
20	PK/BL	Data Link Connector (TC)	12v
21	OR	EVAP Vapor Pressure Sensor (PTNK)	2.5-3.1v (with cap off)
22	WT/RD	Accelerator Position Sensor 1 (VPA)	0.3-0.90v
23	RD/BK	Accelerator Position Sensor 2 (VPA2)	1.8-2.7v
24	RD	Traction Control Engine Signal (ENG+)	Pulse Signals
25	BK	Traction Control Signal (TRC+)	Pulse Signals
26	BK/YL	Accelerator Pedal Position Sensor 1 VREF	4.9-5.1v
27	WT/BL	Accelerator Pedal Position Sensor 2 VREF	4.9-5.1v
28	LG/BK	Accelerator Pedal Position Sensor (EPA)	<0.050v
29	VT/WT	Accelerator Position Sensor 2 (EPA2)	1.8-2.7v
30	WT	Traction Control Engine Signal (ENG-)	Pulse Signals
31	YL	Traction Control Signal (TRC-)	Pulse Signals

2003-2006 4.7L V8 VIN T E4 34 Pin Connector

PCM Pin #	Wire Color	Circuit Description (34 Pin)	Value at Hot Idle
1	RD/BL	Injector 1 Control	2.0-3.3 ms
2	BL	Injector 2 Control	2.0-3.3 ms
3	WT	Injector 3 Control	2.0-3.3 ms
4	VT	Injector 4 Control	2.0-3.3 ms
5	GN	Injector 5 Control	2.0-3.3 ms
6	WT/BK	Power Ground (E02)	<0.1v
7	WT/BK	Power Ground (E01)	<0.1v
8	LG/BK	Igniter Transistor 2 Control	7% duty cycle
9	LG	Igniter Transistor 1 Control	7% duty cycle
10	GN/BK	Igniter Transistor 8 Control	7% duty cycle
11	BL/YL	Igniter Transistor 4 Control	7% duty cycle
12	BK/YL	Igniter Transistor 5 Control	7% duty cycle
13	GN/WT	Igniter Transistor 7 Control	7% duty cycle
14, 22, 32	---	Not Used	---
15	RD/GN	A/C Relay Control (ACCR)	Relay Off: 12v, On: 1v
16	PK	Neutral Start Switch (NSW)	In P/N: 0-3.0v
17	BK/YL	Starter Switch Signal (STA)	In P/N: 0-3.0v
18	BL/RD	Sensor VREF (VC)	4.9-5.1v
19	RD/BL	ECT Sensor Signal (THW)	At 180ºF: 0.51v
20	YL/BK	IAT Sensor Signal (THA)	At 100ºF: 2.60v
21	YL	TP Sensor Signal (VTA1)	0.4-1.0v
23	RD/WT	Igniter Signal (IGF2)	Digital Signal: 0-5-0v
24	RD/YL	Igniter Signal (IGF1)	Digital Signal: 0-5-0v
25	BL/WT	Igniter Transistor 3 Control	7% duty cycle
26	GN	Igniter Transistor 6 Control	7% duty cycle
27	RD/GN	EVAP Canister Closed Valve (CCV)	12v or 0v
28	BR	Sensor Ground (E2)	<0.050v
29	BK/WT	MAF Sensor Ground (E2G)	<0.050v
30	WT/RD	Mass Airflow Sensor (VG)	1.1-1.5v
31	RD/BK	TP Sensor Signal (VTA2)	2.0-2.9v
33	GN/RD	Fuel Pump Relay Control (FPR)	Relay Off: 12v, On: 1v
34	YL/RD	EVAP Purge Solenoid (VSV)	12v or 0v

2003-2006 4.7L V8 VIN T E5 35 Pin Connector

PCM Pin #	Wire Color	Circuit Description (35 Pin)	Value at Hot Idle
1	BK	Knock Sensor 1 Signal (KNK1)	<0.075v AC
2	WT	Knock Sensor 2 Signal (KNK2)	<0.075v AC
3	RD	Injector 6 Control	2.0-3.3 ms
4	LG	HO2S-11 (B1 S1) Heater (HT1A)	1v (Heater on)
5	GN/YL	HO2S-12 (B1 S2) Heater (HT1B)	1v (Heater on)
6	BR	4WD Switch Ground	<0.050v
6-8	---	Not Used	---
9	PK	Start Signal (STAR)	In P/N: 0-3.0v
10	WT	ECT Solenoid Control (S2)	12v or 0v

2003-2006 4.7L V8 VIN T E5 35 Pin Connector, *continued*

11	RD	ECT Solenoid Control (S1)	12v or 0v
12	GN/BK	ECT Solenoid Control (SLT-)	12v or 0v
13	GN/WT	ECT Solenoid Control (SLT+)	12v or 0v
15	GN	ECT Solenoid Control (SL)	12v or 0v
16	PK/BL	A/T Solenoid Control (SL2-)	Pulse Signals
17	PK/BK	A/T Solenoid Control (SL2+)	Pulse Signals
18	RD/WT	A/T Solenoid Control (SL1-)	Pulse Signals
19	RD/BL	A/T Solenoid Control (SL1+)	Pulse Signals
20, 28, 30-31	---	Not Used	---
21	WT	HO2S-22 (B2 S2) Signal (OX2B)	0.1-1.1v
22	GN	HO2S-21 (B2 S1) Signal (OX2A)	0.1-1.1v
23	RD	HO2S-11 (B1 S1) Signal (OX1A)	0.1-1.1v
24	BL	A/T Oil Temperature Sensor 2 (THO2)	At 68°F: 4-5v
25	PK	HO2S-22 (B2 S2) Heater (HT2B)	1v (Heater on)
26	RD	ECT Vehicle Speed Sensor (SP2+)	Pulse Signals
27	BL	ECT Turbine Speed Sensor (NT+)	Pulse Signals
29	BK	HO2S-12 (B1 S2) Signal (OX1B)	0.1-1.1v
32	GN/YL	A/T Oil Temperature Sensor 1 (THO1)	At 68°F: 4-5v
33	BK/BL	HO2S-21 (B2 S1) Heater (HT2A)	1v (Heater on)
34	GN	ECT Vehicle Speed Sensor (SP2-)	Pulse Signals
35	WT	ECT Turbine Speed Sensor (NT-)	Pulse Signals

2003-2006 4.7L V8 VIN T E6 32 Pin Connector

PCM Pin #	Wire Color	Circuit Description (32 Pin)	Value at Hot Idle
1	BR	Power Ground (E1)	<0.1v
2	GN	Throttle Control Motor (M-)	Pulse Signals
3	RD	Throttle Control Motor (M+)	Pulse Signals
4	WT/BK	Power Ground (ME01)	<0.1v
5	BR	Injector 8 Control	2.0-3.3 ms
6	WT	Injector 7 Control	2.0-3.3 ms
7	WT/BK	Power Ground (E03)	<0.1v
8-10	---	Not Used	---
11	YL/GN	Neutral Detection Switch (L4)	Switch Open: 0v, Closed: 12v
12	BL/YL	Start Switch Signal (STSW)	Cranking: 9-11v
13-14, 18-20	---	Not Used	---
15	BK	A/T Solenoid Control (SLU-)	Pulse Signals
16	PK/GN	A/T Solenoid Control (SLU+)	Pulse Signals
17	BR	Shield Ground (GE01)	<0.050v
21	BK/OR	Generator Control (RL)	12v
22, 26, 28-31	---	Not Used	---
23	RD/YL	A/C Lock Sensor (LCK)	12v or 0v
24	YL	CKP Sensor Signal (NE-)	<0.050v
25	BL	CKP Sensor Signal (NE+)	AC pulse signals
27	BK	CMP Sensor Signal (G2+)	AC pulse signals
32	WT	CMP Sensor Signal (G2-)	AC pulse signals

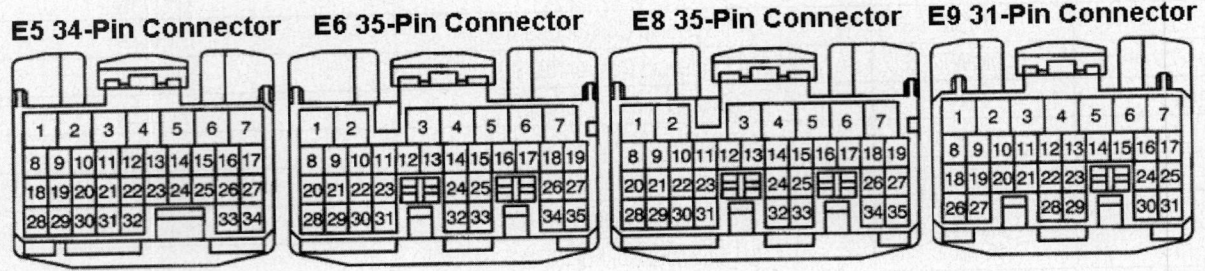

Pin Connector Graphic

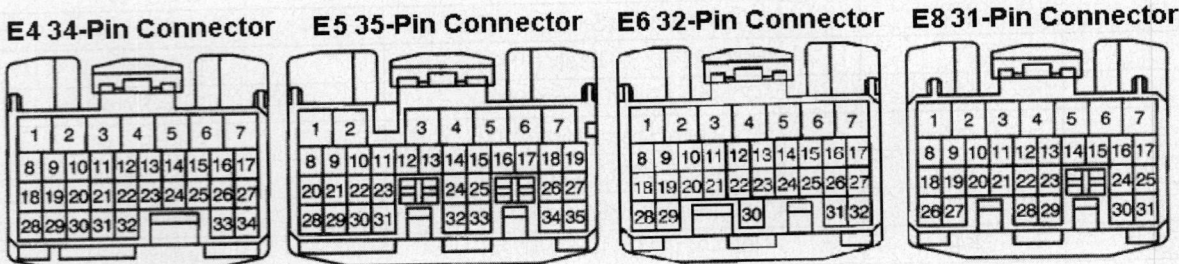

Pin Connector Graphic

2003-2006 4.7L V8 VIN T E7 35 Pin Connector

PCM Pin #	Wire Color	Circuit Description (35 Pin)	Value at Hot Idle
1	WT/BK	Power Ground (HP)	<0.1v
2	BL	A/C Magnetic Clutch Relay (ACMG)	Relay Off: 0v, On: 12v
3	GN	A/T Select Switch 2nd Signal (2L)	In 2nd: 12v, Others: 0v
4	WT/BL	Low Detection Switch (L4)	Switch Open: 0v, Closed: 12v
5	BL/WT	ECT Pattern Switch 2nd Position (SNW1)	2nd Position: 12v
6	GN	ETCS Power (+BM)	12-14v
7	---	Not Used	---
8	RD/BL	Shift Lock ECU Control (L)	12v or 0v
9	GN/BK	Shift Lock ECU Control (2)	12v or 0v
10	GN/YL	Shift Lock ECU Control (D)	12v or 0v
11	RD/BK	A/T Select Switch Reverse (R)	In 'R': 1v, Others: 12v
12	---	Not Used	---
14	BL/BK	A/C Amplifier Signal (THWO)	A/C Off: 12v, On: 1v
15-16	---	Not Used	---
17	VT/RD	Vehicle Speed Sensor (SPD)	At 55 mph: 48 Hz
18	PK/BK	Body Control ECU Signal (MPX1)	Digital Signals
19	GN/YL	Stop Light Switch Signal	Brake Off: 0v, On: 12v
20	GN	Shift Lock ECU Control (3)	12v or 0v
21-22	---	Not Used	---
23	GN/RD	Shift Lock ECU Control (4)	12v or 0v
25	---	Not Used	---
26	BL/RD	Transponder Amplifier Signal (IMD)	Inserting key: pulses

2003-2006 4.7L V8 VIN T E7 35 Pin Connector, *continued*

27	WT/RD	Transponder Amplifier Signal (IMI)	Inserting key: pulses
28	BL/WT	ECT Pattern Switch Power Signal	Power Position: 12v
29	BK	Body Control ECU Signal (MPX2)	Digital Signals
30	---	Not Used	---
31	BR/BK	A/C Amplifier Signal (A/CS)	Relay Off: 12v, On: 1v
32	GY/GN	A/C Amplifier Signal (THE)	A/C Off: 12v, On: 1v
33	BK/RD	A/C Switch Signal (ACLD)	A/C On: 12v, Off: 0v
34-35	---	Not Used	---

2003-2006 4.7L V8 VIN T E8 31 Pin Connector

PCM Pin #	Wire Color	Circuit Description (31 Pin)	Value at Hot Idle
1	BK	EFI Main Relay Power (+B)	12-14v
2	BK	EFI Main Relay Power (+B2)	12-14v
3	BL	Direct Battery	12-14v
4	PK/BL	EVAP Pressure Switching Valve (VSV)	12v or 0v
5	BK/WT	Tachometer Signal (TACO)	Pulse Signals
6	GN/WT	A/T Select Switch Park Signal (P)	In 'P': 12v, Others: 0v
7	GN/RD	A/T Select Switch Neutral Signal (N)	In 'N': 12v, Others: 0v
8	WT/GN	EFI Main Relay Control	Relay Off: 0v, On: 12v
9	BK/OR	Ignition Switch Power (IGSW)	12-14v
10	GN/BK	Circuit Opening Relay (FC)	0-3v, off-idle: 12v
11	RD/BK	Malfunction Indicator Lamp Control	MIL Off: 12v, On: 1v
12	BL/RD	Taillight Switch Signal (ELS)	Taillights Off: 0v, On: 12v
13	YL/RD	Horn Relay Control	Relay Off: 12v, On: 1v
14	BL	Center Airbag Assembly (F/PS)	Digital Signals
15	OR	Power Ground (EOM)	<0.1v
16	---	Not Used	---
17	PK	Transponder Amplifier Signal (NEO)	Inserting key: pulses
18	RD/YL	SIL Signal (Scan Tool)	Transmitting: pulses
19	RD/WT	Data Link Connector (WFSE)	12v
20	PK/BL	Data Link Connector (TC)	12v
21	OR	EVAP Vapor Pressure Sensor (PTNK)	2.5-3.1v (with cap off)
22	WT/RD	Accelerator Position Sensor 1 (VPA)	0.3-0.90v
23	RD/BK	Accelerator Position Sensor 2 (VPA2)	1.8-2.7v
24	RD	Traction Control Engine Signal (ENG+)	Pulse Signals
25	BK	Traction Control Signal (TRC+)	Pulse Signals
26	BK/YL	Accelerator Pedal Position Sensor 1 VREF	4.9-5.1v
27	WT/BL	Accelerator Pedal Position Sensor 2 VREF	4.9-5.1v
28	LG/BK	Accelerator Pedal Position Sensor (EPA)	<0.050v
29	VT/WT	Accelerator Position Sensor 2 (EPA2)	1.8-2.7v
30	WT	Traction Control Engine Signal (ENG-)	Pulse Signals
31	YL	Traction Control Signal (TRC-)	Pulse Signals

PICKUP PIN CHARTS

1990-92 2.4L I4 MFI VIN R (A/T-ECT) 26 Pin Connector

PCM Pin #	Wire Color	Circuit Description (26 Pin)	Value at Hot Idle
1	BK/OR	Distributor Signal (NE+)	AC pulse signals
2	GN	Cold Start Injector	1v (at cold startup)
3	BK/YL	Igniter Signal (IGF)	0.74-0.76v
5	RD/GN	A/T-ECT Solenoid (S3)	In Lockup: 12-14v
6	RD/WT	A/T-ECT Solenoid (S2)	1st or OD: 1v
7	BL/RD	A/T-ECT Solenoid (S1)	3rd or OD: 1v
9 (Cal)	RD/GN	Sub O2S Heater Control	Heater Off: 12v, On: 1v
10	PK/GN	Main O2S Heater Control	Heater Off: 12v, On: 1v
11	GN	Fuel Pressure Up Solenoid	1v (at hot startup)
12	WT/RD	Injector Pair 1 & 3 Control	1.6-2-9 ms
13	BR	Power Ground	<0.1v
15	BR	Sensor Ground	<0.050v
16	BL	A/T Vehicle Speed Sensor	Moving: 0-5-0v
17	PK/WT	A/T Select Switch Low	In Low: 12v, Others: 0v
18	PK/GN	A/T Select Switch 2nd	In 2nd: 12v, Others: 0v
19	VT/RD	A/T Select Switch Neutral	In 'N': 12v, Others: 0v
20	RD	4WD Detection Transfer (L4)	Open: 12v, Closed: 0v
21	BK/BL	Igniter Signal (IGF)	Digital Signal: 0-5-0v
22	PK	EGR Solenoid Control (VSV)	12v or 0v
23	GN/RD	Intake Air Solenoid Control	12v or 0v
24	BK/RD	A/C Idle-Up Solenoid	A/C Off: 12v, On: 1v
25	WT	Injector Pair 2 & 4 Control	1.6-2-9 ms
26	BR	Power Ground	<0.1v

1990-92 2.4L I4 MFI VIN R (A/T-ECT) 22 Pin Connector

PCM Pin #	Wire Color	Circuit Description (22 Pin)	Value at Hot Idle
1	BK/GN	Battery Direct	12-14v
4	BL/WT	A/T Oil Temperature Lamp	Lamp Off: 12v, On: 1v
5	PK	MIL (lamp) Control	MIL Off: 12v, On: 1v
6	GN/WT	Stop Light Switch Signal	Brake Off: 0v, On: 12v
7	RD/BL	A/T Pattern Select Switch	Norm: 0v, PWR: 12v
8	PK/GN	4WD Selector Switch	Switch Off: 12v, On: 1v
9	GN/BL	Vehicle Speed Sensor	At 55 mph: 48 Hz
11	BK/WT	Starter Switch Signal	9-11v (cranking)
12	WT/RD	EFI Main Relay Power	12-14v
13	WT/RD	EFI Main Relay Power	12-14v
15	BR/BK	Sensor Ground	<0.050v
16	YL/GN	Overdrive Main Switch	Switch Off: 12v, On: 1v
20	PK	Check Connector	12-14v
21	YL/RD	Cruise Control ECU (OD1)	At Cruise in OD: 12v

1990-92 2.4L I4 MFI VIN R (A/T-ECT) 16 Pin Connector

PCM Pin #	Wire Color	Circuit Description (16 Pin)	Value at Hot Idle
1	GN/YL	Sensor VREF (VC)	4.9-5.1v
2	GN/BK	Airflow Meter VREF	4.9-5.1v
3	YL/GN	IAT Sensor Signal (THA)	At 100°F: 2.60v
4	GN/BL	ECT Sensor Signal (THW)	At 180°F: 0.51v
5	BK	Knock Sensor Signal	<0.075v AC
6	BK	Main O2S Signal	0.1-1.1v
7	GN/YL	4WD Oil Temperature Sensor	At 68°F: 4-5v
8	YL	Check Connector	12-14v
9	BR/BK	Sensor Ground	<0.050v
10	YL/BL	Airflow Meter Signal	0.5-2.5v
11	YL	TP Sensor Signal (VTA)	0.3-0.8v
12	YL/BL	Closed Throttle Switch	1v, off-idle: 12v
13	GN/WT	EGR Gas Temperature Sensor (THG)	3.5-4.0v
14 (Cal)	WT	Sub O2S Signal	0.1-1.1v
15	PK/WT	Check Connector	12-14v

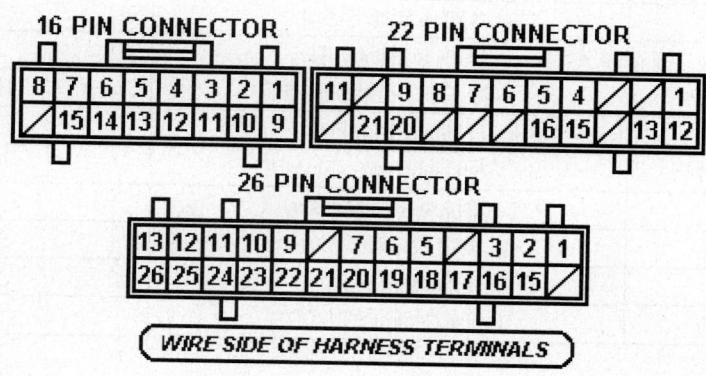

05533_ADIA_G160

Pin Connector Graphic

Standard Colors and Abbreviations

Abbreviation	Color	Abbreviation	Color	Abbreviation	Color
BK	Black	GY	Gray	RD	Red
BL	Blue	GN	Green	TN	Tan
BR	Brown	LG	LT Green	VT	Violet
DB	Dark Blue	OR	Orange	WT	White
DG	DK Green	PK	Pink	YL	Yellow

1990 2.4L I4 MFI 22R-E VIN R (M/T) 10 Pin Connector

PCM Pin #	Wire Color	Circuit Description (10 Pin)	Value at Hot Idle
1	BK	M/T Clutch Start Switch	9-11v (cranking)
2	GN	Cold Start Injector	1v (at cold startup)
3	BK/WT	Starter Switch Signal	9-11v (cranking)
4	WT/RD	Injector Pair 1 & 3 Control	1.6-2-9 ms
5	BR	Power Ground	<0.1v
6	YL	Check Connector	12-14v
7	BR	Sensor Ground	<0.050v
8	BK/BL	Igniter Signal (IGF)	Digital Signal: 0-5-0v
9	WT	Injector Pair 2 & 4 Control	1.6-2-9 ms
10	BR	Power Ground	<0.1v

1990 2.4L I4 MFI 22R-E VIN R (M/T) 14 Pin Connector

PCM Pin #	Wire Color	Circuit Description (14-Pin)	Value at Hot Idle
1	WT/RD	EFI Main Relay Power	12-14v
2	GN/BK	Battery Direct	12-14v
3	YL/GN	IAT Sensor Signal (THA)	At 100°F: 2.60v
4	YL/BL	Airflow Meter Signal	0.5-2.5v
5	GN/BK	Airflow Meter VREF	4.9-5.1v
7 (Cal)	RD/GN	Sub O2S Heater Control	Heater Off: 12v, On: 1v
8	WT/RD	EFI Main Relay Power	12-14v
9	GN/WT	Stop Light Switch Signal	Brake Off: 0v, On: 12v
10	GN/BL	Vehicle Speed Sensor	At 55 mph: 48 Hz
11	GN/YL	4WD Selector Switch	Switch Off: 12v, On: 1v
12	BR/BK	Sensor Ground	<0.050v
14	YL/RD	Cruise Control ECU (OD1)	At Cruise in OD: 12v

1990 2.4L I4 MFI 22R-E VIN R (M/T) 18 Pin Connector

PCM Pin #	Wire Color	Circuit Description (18 Pin)	Value at Hot Idle
1	BK/OR	Distributor Signal (NE+)	AC pulse signals
2	BK	Knock Sensor Signal	<0.075v AC
3	GN/WT	EGR Gas Temperature Sensor (THG)	3.5-4.0v
4 (Cal)	WT	Sub O2 Sensor Signal	0.1-1.1v
5	BK/YL	Igniter Signal (IGF)	Digital Signal: 0-5-0v
6	YL/BL	Closed Throttle Switch	1v, off-idle: 12v
7	PK/WT	Check Connector	12-14v
8	PK	MIL (lamp) Control	MIL Off: 12v, On: 1v
9	GN	Fuel Pressure Up Solenoid	1v (at hot startup)
10	GN/BL	ECT Sensor Signal (THW)	At 180°F: 0.51v
11	YL	TP Sensor Signal (VTA)	0.3-0.8v
12	GN/YL	Sensor VREF (VC)	4.9-5.1v
13	BK	Main O2S Signal	0.1-1.1v

1990 2.4L I4 MFI 22R-E VIN R (M/T) 18 Pin Connector, *continued*

14	BR/BK	Sensor Ground	<0.050v
15	PK/GN	Main O2S Heater Control	Heater Off: 12v, On: 1v
16	PK	EGR Solenoid Control (VSV)	12v or 0v
17	GN/RD	Intake Air Solenoid Control	12v or 0v
18	BK/RD	A/C Idle-Up Solenoid	A/C Off: 12v, On: 1v

Pin Connector Graphic

05533_ADIA_G161

1991-92 2.4L I4 MFI 22R-E VIN R (M/T) 10 Pin Connector

PCM Pin #	Wire Color	Circuit Description (10 Pin)	Value at Hot Idle
1	BK	M/T Clutch Start Switch	9-11v (cranking)
2	GN	Cold Start Injector	1v (at cold startup)
3	BK/WT	Starter Switch Signal	9-11v (cranking)
4	WT/RD	Injector Pair 1 & 3 Control	1.6-2-9 ms
5	BR	Power Ground	<0.1v
6	YL	Check Connector	12-14v
7	BR	Sensor Ground	<0.050v
8	BK/BL	Igniter Signal (IGF)	Digital Signal: 0-5-0v
9	WT	Injector Pair 2 & 4 Control	1.6-2-9 ms
10	BR	Power Ground	<0.1v

1991-92 2.4L I4 MFI 22R-E VIN R (M/T) 14 Pin Connector

PCM Pin #	Wire Color	Circuit Description (14-Pin)	Value at Hot Idle
1	WT/RD	EFI Main Relay B1+	12-14v
2	GN/BK	Battery Direct	12-14v
3	YL/GN	IAT Sensor Signal (THA)	At 100°F: 2.60v
4	YL/BL	Airflow Meter Signal	0.5-2.5v
5	GN/BK	Airflow Meter VREF	4.9-5.1v
7 (Cal)	RD/GN	Sub O2S Heater Control	Heater Off: 12v, On: 1v
8	WT/RD	EFI Main Relay (B+)	12-14v
9	GN/WT	Stop Light Switch Signal	Brake Off: 0v, On: 12v
10	GN/BL	Vehicle Speed Sensor	At 55 mph: 48 Hz
12	BR/BK	Sensor Ground	<0.050v
14	YL/RD	Cruise Control ECU	At Cruise in OD: 12v

1991-92 2.4L I4 MFI 22R-E VIN R (M/T) 18 Pin Connector

PCM Pin #	Wire Color	Circuit Description (18 Pin)	Value at Hot Idle
1	BK/OR	Distributor Signal (NE+)	AC pulse signals
2	BK	Knock Sensor Signal	<0.075v AC
3	GN/WT	EGR Gas Temperature Sensor (THG)	3.5-4.0v
4 (Cal)	WT	Sub O2 Sensor Signal	0.1-1.1v
5	BK/YL	Igniter Signal (IGF)	Digital Signal: 0-5-0v
6	YL/BL	Closed Throttle Switch	1v, off-idle: 12v
7	PK/WT	Check Connector	12-14v
8	PK	MIL (lamp) Control	MIL Off: 12v, On: 1v
9	GN	Fuel Pressure Up Solenoid	1v (at hot startup)
10	GN/BL	ECT Sensor Signal (THW)	At 180°F: 0.51v
11	YL	TP Sensor Signal (VTA)	0.3-0.8v
12	GN/YL	Sensor VREF (VC)	4.9-5.1v
13	BK	Main O2S Signal	0.1-1.1v
14	BR/BK	Sensor Ground	<0.050v
15	PK/GN	Main O2S Heater Control	Heater Off: 12v, On: 1v
16	PK	EGR Solenoid Control (VSV)	12v or 0v
17	GN/RD	Intake Air Solenoid Control	12v or 0v

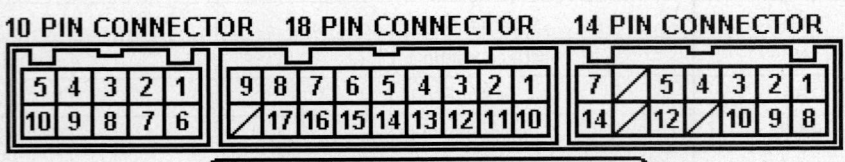

05533_ADIA_G162

Pin Connector Graphic

1991-92 2.4L I4 MFI VIN R (M/T) 16 Pin Connector

PCM Pin #	Wire Color	Circuit Description (16 Pin)	Value at Hot Idle
1	WT/RD	EFI Main Relay Power	12-14v
2	BK/GN	Battery Direct	12-14v
6	PK	EGR Solenoid Control (VSV)	12v or 0v
7	GN	Fuel Pressure Up Solenoid	1v (at hot startup)
8	PK/WT	Check Connector	12-14v
9	WT/RD	EFI Main Relay Power	12-14v
10	PK	MIL (lamp) Control	MIL Off: 12v, On: 1v
12	GN/WT	Stop Light Switch Signal	Brake Off: 0v, On: 12v
13	GN/BL	Vehicle Speed Sensor	At 55 mph: 48 Hz
14	YL/GN	Overdrive Main Switch	Switch Off: 12v, On: 1v
15	GN/RD	Intake Air Solenoid Control	12v or 0v
16	YL	Check Connector	12v

1991-92 2.4L I4 MFI VIN R (M/T) 26 Pin Connector

PCM Pin #	Wire Color	Circuit Description (26 Pin)	Value at Hot Idle
1	PK/GN	Main O2S Heater Control	Heater Off: 12v, On: 1v
2	BK/WT	Starter Switch Signal	9-11v (cranking)
3	GN/BL	ECT Sensor Signal (THW)	At 180°F: 0.51v
4	YL/BL	Airflow Meter Signal	0.5-2.5v
5	YL/GN	IAT Sensor Signal (THA)	At 100°F: 2.60v
6	BK/BL	Igniter Signal (IGF)	Digital Signal: 0-5-0v
7	BK/YL	Igniter Signal (IGF)	Digital Signal: 0-5-0v
8	GN/WT	EGR Gas Temperature Sensor (THG)	3.5-4.0v
9	GN/BK	Airflow Meter VREF	4.9-5.1v
10	BK	Main O2S Signal	0.1-1.1v
11	GN	Cold Start Injector	1v (at cold startup)
12	WT/RD	Injector 1 & 3 Control	1.6-2-9 ms
13	BR	Power Ground	<0.1v
14 (Cal)	RD/GN	Sub O2S Heater Control	Heater Off: 12v, On: 1v
15	BK	Clutch Start Switch	Clutch Out: 12v, In: 0v
16, 22	BR/BK	Sensor Ground	<0.050v
17	YL	TP Sensor Signal (VTA)	0.3-0.8v
18	GN/YL	Sensor VREF (VC)	4.9-5.1v
19	YL/BL	Closed Throttle Switch	1v, off-idle: 12v
21	BK/OR	Distributor Signal (NE+)	AC pulse signals
23 (Cal)	WT	Sub O2S Signal	0.1-1.1v
24	BR	Sensor Ground	<0.050v
25	WT	Injector 2 & 4 Control	1.6-2-9 ms
26	BR	Power Ground	<0.1v

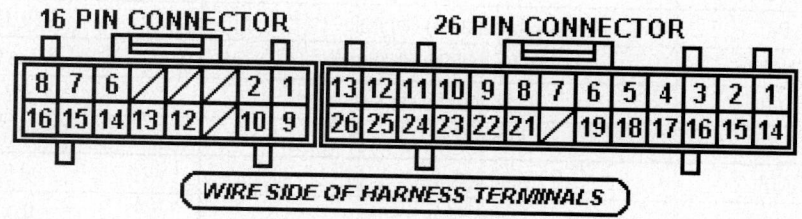

Pin Connector Graphic

05533_ADIA_G163

1993-95 2.4L I4 MFI VIN R (A/T-ECT) 26 Pin Connector

PCM Pin #	Wire Color	Circuit Description (26 Pin)	Value at Hot Idle
1	GN	Cold Start Injector	1v (at cold startup)
2	PK/GN	Main O2S Heater Control	Heater Off: 12v, On: 1v
3	BK/YL	Igniter Signal (IGF)	Digital Signal: 0-5-0v
4	BK/OR	Distributor Signal (NE+)	AC pulse signals
5	RD/GN	A/T-ECT Solenoid (S3)	In Lockup: 12-14v
6	RD/WT	A/T-ECT Solenoid (S2)	1st or OD: 1v
7	BL/RD	A/T-ECT Solenoid (S1)	3rd or OD: 1v
8	PK	EGR Solenoid Control (VSV)	12v or 0v
9	GN/RD	Intake Air Solenoid Control	12v or 0v
10	GN	Fuel Pressure Up Solenoid	1v (at hot startup)
11	WT	Injector Pair 2 & 4 Control	1.6-2-9 ms
12	WT/RD	Injector Pair 1 & 3 Control	1.6-2-9 ms
13	BR	Power Ground	<0.1v
14	BR	Sensor Ground	<0.050v
15 (Cal)	RD/GN	Sub-HO2S Heater Control	Heater Off: 12v, On: 1v
19	BL	A/T Vehicle Speed Sensor	Moving: 0-5-0v
20	BK/BL	Igniter Signal (IGF)	Digital Signal: 0-5-0v
21	PK/WT	A/T Select Switch Low	In Low: 12v, Others: 0v
22	PK/GN	A/T Select Switch 2nd	In 2nd: 12v, Others: 0v
23	PK/RD	A/T Select Switch Neutral	In 'N': 12v, Others: 0v
26	BR	Power Ground	<0.1v

1993-95 2.4L I4 MFI VIN R (A/T-ECT) 22 Pin Connector

PCM Pin #	Wire Color	Circuit Description (22 Pin)	Value at Hot Idle
1	BK/GN	Battery Direct	12-14v
5	PK	MIL (lamp) Control	MIL Off: 12v, On: 1v
6	GN/WT	Stop Light Switch Signal	Brake Off: 0v, On: 12v
7	RD/BL	A/T Pattern Select Switch	Norm: 0v, PWR: 12v
8	PK/GN	4WD Selector Switch	Switch Off: 12v, On: 1v
9	GN/BL	Vehicle Speed Sensor	At 55 mph: 48 Hz
11	BK/WT	Starter Switch Signal	9-11v (cranking)
12	WT/RD	EFI Main Relay (B+)	12-14v
13	WT/RD	EFI Main Relay B1+	12-14v
15	BR/BK	Sensor Ground	<0.050v
16	YL/GN	Overdrive Main Switch	Switch Off: 12v, On: 1v
19	RD	4WD Detection Transfer (L4)	Open: 12v, Closed: 0v
20	PK	Data Link Connector	12-14v
21	YL/RD	Cruise Control ECU	At Cruise in OD: 12v

1993-95 2.4L I4 MFI VIN R (A/T-ECT) 16 Pin Connector

PCM Pin #	Wire Color	Circuit Description (16 Pin)	Value at Hot Idle
1	GN/YL	Sensor VREF (VC)	4.9-5.1v
2	YL/BL	Airflow Meter Signal	0.5-2.5v
3	YL/GN	IAT Sensor Signal (THA)	At 100°F: 2.60v
4	GN/BL	ECT Sensor Signal (THW)	At 180°F: 0.51v
5 (Cal)	WT	Sub O2S Signal	0.1-1.1v
6	BK	Main O2S Signal	0.1-1.1v
7	BK	Knock Sensor Signal	<0.075v AC
8	YL	Data Link Connector	12-14v
9	BR/BK	Sensor Ground	<0.050v
10	GN/BK	Airflow Meter VREF	4.9-5.1v
11	YL	TP Sensor Signal (VTA)	0.3-0.8v
12	YL/BL	Closed Throttle Switch	1v, off-idle: 12v
13	GN/WT	EGR Gas Temperature Sensor (THG)	3.5-4.0v
14	PK/GN	Data Link Connector	12-14v
15	PK/WT	Data Link Connector	12-14v

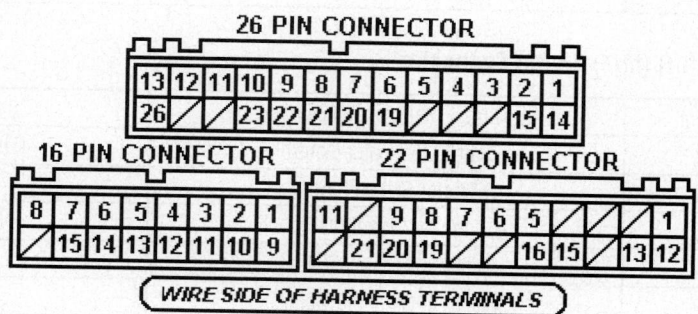

Pin Connector Graphic

05533_ADIA_G164

Standard Colors and Abbreviations

Abbreviation	Color	Abbreviation	Color	Abbreviation	Color
BK	Black	GY	Gray	RD	Red
BL	Blue	GN	Green	TN	Tan
BR	Brown	LG	LT Green	VT	Violet
DB	Dark Blue	OR	Orange	WT	White
DG	DK Green	PK	Pink	YL	Yellow

1993-95 2.4L I4 MFI 22R-E VIN R (M/T) 12 Pin Connector

PCM Pin #	Wire Color	Circuit Description (12 Pin)	Value at Hot Idle
1	WT/RD	EFI Main Relay B1+	12-14v
2	GN/BK	Battery Direct	12-14v
6	GN/YL	4WD Selector Switch	Switch Off: 12v, On: 1v
7	WT/RD	EFI Main Relay (B+)	12-14v
8	PK	MIL (lamp) Control	MIL Off: 12v, On: 1v
11	GN/BL	Vehicle Speed Sensor	At 55 mph: 48 Hz
12	GN/WT	Stop Light Switch Signal	Brake Off: 0v, On: 12v

1993-95 2.4L I4 MFI 22R-E VIN R (M/T) 16 Pin Connector

PCM Pin #	Wire Color	Circuit Description (16 Pin)	Value at Hot Idle
1	YL/GN	IAT Sensor Signal (THA)	At 100°F: 2.60v
2	YL/BL	Airflow Meter Signal	0.5-2.5v
3	GN/BK	Airflow Meter VREF	4.9-5.1v
4	GN/BL	ECT Sensor Signal (THW)	At 180°F: 0.51v
5 (Cal)	WT	Sub O2S Signal	0.1-1.1v
6	BK	Main O2S Signal	0.1-1.1v
7	PK/GN	Data Link Connector	12-14v
8	YL	Data Link Connector	12-14v
9	BR/BK	Sensor Ground	<0.050v
10	YL	TP Sensor Signal (VTA)	0.3-0.8v
11	GN/YL	Sensor VREF (VC)	4.9-5.1v
12	YL/BL	Closed Throttle Switch	1v, off-idle: 12v
13	GN/WT	EGR Gas Temperature Sensor (THG)	3.5-4.0v
14	BK	Knock Sensor Signal	<0.075v AC
15	PK/WT	Data Link Connector	12-14v
16	BR/BK	Data Link Connector	<0.050v

1993-95 2.4L I4 MFI 22R-E VIN R (M/T) 26 Pin Connector

PCM Pin #	Wire Color	Circuit Description (26 Pin)	Value at Hot Idle
1	PK/GN	Main O2S Heater Control	Heater Off: 12v, On: 1v
2	BK/WT	Starter Switch Signal	9-11v (cranking)
3	BK/YL	Igniter Signal (IGF)	Digital Signal: 0-5-0v
4	BK/OR	Distributor Signal (NE+)	AC pulse signals
9	GN/RD	Intake Air Solenoid Control	12v or 0v
10	GN	Fuel Pressure Up Solenoid	1v (at hot startup)
11	GN	Cold Start Injector	1v (at cold startup)
12	WT/RD	Injector Pair 1 & 3 Control	1.6-2-9 ms
13	BR	Power Ground	<0.1v
14 (Cal)	RD/GN	Sub O2S Heater Control	Heater Off: 12v, On: 1v
15	BK	M/T Clutch Start Switch	Clutch Out: 12v, In: 0v
22	BK/BL	Igniter Signal (IGF)	Digital Signal: 0-5-0v
23	PK	EGR Solenoid Control (VSV)	12v or 0v
24	BR/BK	Sensor Ground	<0.050v
25	WT	Injector Pair 2 & 4 Control	1.6-2-9 ms
26	BR	Power Ground	<0.1v

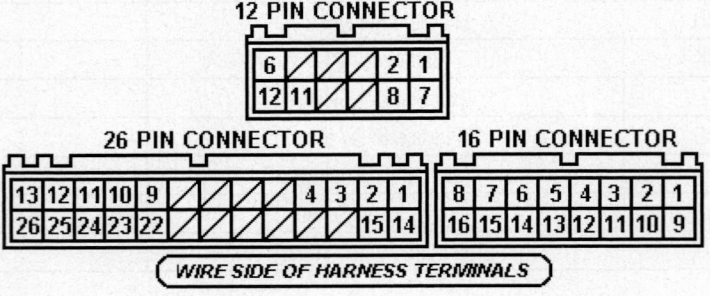

Pin Connector Graphic

05533_ADIA_G165

1993-95 2.4L I4 MFI VIN R (M/T) 16 Pin Connector

PCM Pin #	Wire Color	Circuit Description (16 Pin)	Value at Hot Idle
1	WT/RD	EFI Main Relay B1+	12-14v
2	BK/GN	Battery Direct	12-14v
5	PK	EGR Solenoid Control (VSV)	12v or 0v
6	GN	Fuel Pressure Up Solenoid	1v (at hot startup)
7	PK/GN	Data Link Connector	12-14v
8	PK/WT	Data Link Connector	12-14v
9	WT/RD	EFI Main Relay (B+)	12-14v
10	PK	MIL (lamp) Control	MIL Off: 12v, On: 1v
11	BR/BK	Sensor Ground	<0.050v
12	GN/WT	Stop Light Switch Signal	Brake Off: 0v, On: 12v
13	GN/BL	Vehicle Speed Sensor	At 55 mph: 48 Hz
14	YL/GN	Overdrive Main Switch	Switch Off: 12v, On: 1v
15	GN/RD	Intake Air Solenoid Control	12v or 0v
16	YL	Data Link Connector	12-14v

1993-95 2.4L I4 MFI VIN R (M/T) 26 Pin Connector

PCM Pin #	Wire Color	Circuit Description (26 Pin)	Value at Hot Idle
1	PK/GN	Main O2S Heater Control	Heater Off: 12v, On: 1v
2	BK/WT	Neutral Start Switch	In P/N: 9-11v (cranking)
3	GN/BL	ECT Sensor Signal (THW)	At 180°F: 0.51v
4	YL/BL	Airflow Meter Signal	0.5-2.5v
5	GN/BK	Airflow Meter VREF	4.9-5.1v
6	BK/BL	Igniter Signal (IGF)	Digital Signal: 0-5-0v
7	BK/YL	Igniter Signal (IGF)	Digital Signal: 0-5-0v
10	BK	Main O2S Signal	0.1-1.1v
11	BK/WT	M/T: Starter Switch Signal	9-11v (cranking)
11	BK/YL	A/T: Starter Switch Signal	9-11v (cranking)
12	WT/RD	Injector 1 & 3 Control	1.6-2-9 ms
13	BR	Power Ground	<0.1v
14 (Cal)	RD/GN	Sub O2S Heater Control	Heater Off: 12v, On: 1v
15	GN	Cold Start Injector	1v (at cold startup)
16	BR/BK	Sensor Ground	<0.050v
17	YL	TP Sensor Signal (VTA)	0.3-0.8v
18	GN/YL	Sensor VREF (VC)	4.9-5.1v
19	YL/BL	Closed Throttle Switch	1v, off-idle: 12v
20	YL/GN	IAT Sensor Signal (THA)	At 100°F: 2.60v
21	BK/OR	Distributor Signal (NE+)	AC pulse signals
22	GN/WT	EGR Gas Temperature Sensor (THG)	3.5-4.0v
23 (Cal)	WT	Sub O2S Signal	0.1-1.1v
24	BR	Sensor Ground	<0.050v
25	WT	Injector 2 & 4 Control	1.6-2-9 ms
26	BR	Power Ground	<0.1v

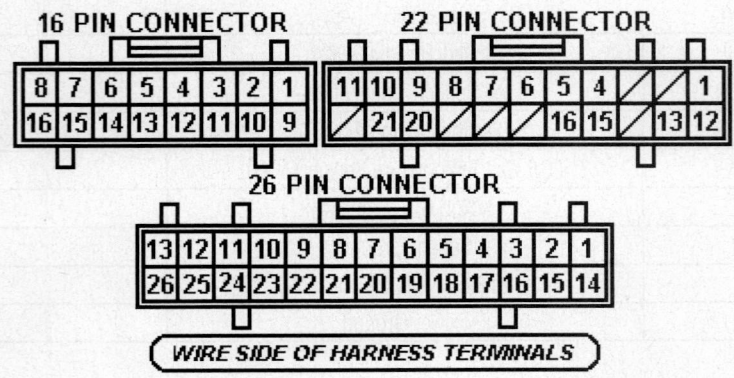

05533_ADIA_G166

Pin Connector Graphic

1990 3.0L V6 MFI VIN V (A/T-ECT) 26 Pin Connector

PCM Pin #	Wire Color	Circuit Description (26 Pin)	Value at Hot Idle
1	WT	Distributor Signal (NE+)	AC pulse signals
2	RD	Distributor Signal (G1+)	AC pulse signals
3	BK/YL	Igniter Signal (IGF)	0.74-0.76v
4	GN/RD	A/T-ECT Solenoid (S4)	In Lockup: 12-14v
5	YL/BK	A/T-ECT Solenoid (S3)	In Lockup: 12-14v
6	BK	A/T-ECT Solenoid (S2)	1st or OD: 1v
7	PK	A/T-ECT Solenoid (S1)	3rd or OD: 1v
8	GN	Fuel Pressure Up Solenoid	1v (at hot startup)
9	GN	Cold Start Injector	1v (at cold startup)
10	PK/GN	HO2S-11 (B1 S1) Heater	Heater Off: 12v, On: 1v
11	BR	Sensor Ground	<0.050v
12	WT/RD	Injectors 1 & 3 & 5 Control	1.6-2.9 ms
13	BR	Power Ground	<0.1v
14	GN	Distributor Signal (GN-)	<0.050v
15	BK	Distributor Signal (G2+)	AC pulse signals
16	BR/RD	A/T Vehicle Speed Sensor	Moving: 0-5-0v
17	PK/WT	A/T Select Switch Low	In Low: 12v, Others: 0v
18	PK/GN	A/T Select Switch 2nd	In 2nd: 12v, Others: 0v
19	PK/RD	A/T Select Switch Neutral	In 'N': 12v, Others: 0v
20	YL/RD	4WD Detection Transfer (L4)	Open: 12v, Closed: 0v
21	BK/BL	Igniter Signal (IGF)	Digital Signal: 0-5-0v
22	PK	EGR Solenoid Control (VSV)	12v or 0v
23	GN/RD	Intake Air Solenoid Control	12v or 0v
24	BK/RD	A/C Idle-Up Solenoid	A/C Off: 12v, On: 1v
25	WT	Injectors 2 & 4 & 6 Control	1.6-2.9 ms
26	BR	Power Ground	<0.1v

05533_ADIA_G167

Pin Connector Graphic

Standard Colors and Abbreviations

Abbreviation	Color	Abbreviation	Color	Abbreviation	Color
BK	Black	GY	Gray	RD	Red
BL	Blue	GN	Green	TN	Tan
BR	Brown	LG	LT Green	VT	Violet
DB	Dark Blue	OR	Orange	WT	White
DG	DK Green	PK	Pink	YL	Yellow

1990 3.0L V6 MFI VIN V (A/T-ECT) 16 Pin Connector

PCM Pin #	Wire Color	Circuit Description (16 Pin)	Value at Hot Idle
1	GN/BK	Airflow Meter VREF	4.9-5.1v
2	YL/BL	Airflow Meter Signal	0.5-2.5v
3	YL/GN	IAT Sensor Signal (THA)	At 100°F: 2.60v
4	GN/BL	ECT Sensor Signal (THW)	At 180°F: 0.51v
5	BK	Knock Sensor Signal	<0.075v AC
6	BK	Main O2S Signal	0.1-1.1v
7	GN/BK	4WD Oil Temperature Sensor	At 68°F: 4-5v
8	YL/GN	Check Connector	12-14v
9	BR/BK	Sensor Ground	<0.050v
10 (Cal)	RD/GN	Sub O2S Signal	0.1-1.1v
11	YL	TP Sensor Signal (VTA)	0.3-0.8v
12	YL/BL	Closed Throttle Switch	1v, off-idle: 12v
13	GN/WT	EGR Gas Temperature Sensor (THG)	3.5-4.0v
14	LG	Transfer Oil Temp. Sensor	At 68°F: 4-5v
15	PK/WT	Check Connector	12-14v
16	PK	Coolant Temperature Switch	Open: 12v, Closed: 0v

1990 3.0L V6 MFI VIN V (A/T-ECT) 22 Pin Connector

PCM Pin #	Wire Color	Circuit Description (22 Pin)	Value at Hot Idle
1	BK/GN	Battery Direct	12-14v
4	PK/GN	A/T Oil Temperature Lamp	Lamp Off: 12v, On: 1v
5	PK	MIL (lamp) Control	MIL Off: 12v, On: 1v
6	GN/WT	Stop Light Switch Signal	Brake Off: 0v, On: 12v
7	GN/OR	A/T Pattern Select Switch	Norm: 0v, PWR: 12v
8	PK/GN	4WD Selector Switch	Switch Off: 12v, On: 1v
9	GN/BL	Vehicle Speed Sensor	At 55 mph: 48 Hz
10	BK/RD	A/C Magnetic Clutch (ACMG)	Clutch Off: 0v, On: 12v
11	BK/WT	Starter Switch Signal	9-11v (cranking)
12	WT/RD	EFI Main Relay (B+)	12-14v
13	WT/RD	EFI Main Relay B1+	12-14v
15	BR/BK	Sensor Ground	<0.050v
16	YL/GN	Overdrive Main Switch	Switch Off: 12v, On: 1v
20	PK	Check Connector	12-14v
21	YL/RD	Cruise Control ECU	At Cruise in OD: 12v

Standard Colors and Abbreviations

Abbreviation	Color	Abbreviation	Color	Abbreviation	Color
BK	Black	GY	Gray	RD	Red
BL	Blue	GN	Green	TN	Tan
BR	Brown	LG	LT Green	VT	Violet
DB	Dark Blue	OR	Orange	WT	White
DG	DK Green	PK	Pink	YL	Yellow

1990 3.0L V6 MFI 3VZ-E VIN V (M/T) 10 Pin Connector

PCM Pin #	Wire Color	Circuit Description (10 Pin)	Value at Hot Idle
1	BK/WT	A/T Neutral Start Switch	In P/N: 9-11v (cranking)
1	BK	M/T Clutch Start Switch	9-11v (cranking)
2	GN	Cold Start Injector	1v (at cold startup)
3	BK/WT	Starter Switch Signal	9-11v (cranking)
4	WT/RD	Injector Pair 1 & 3 Control	1.6-2.9 ms
5	BR	Power Ground	<0.1v
6	YL	Check Connector	12-14v
7	BR	Sensor Ground	<0.050v
8	BK/BL	Igniter Signal (IGF)	Digital Signal: 0-5-0v
9	WT	Injector Pair 2 & 4 Control	1.6-2.9 ms
10	BR	Power Ground	<0.1v

1990 3.0L V6 MFI 3VZ-E VIN V (M/T) 14-Pin Connector

PCM Pin #	Wire Color	Circuit Description (14-Pin)	Value at Hot Idle
1	WT/RD	EFI Main Relay B1+	12-14v
2	GN/BK	Battery Direct	12-14v
3	YL/GN	IAT Sensor Signal (THA)	At 100°F: 2.60v
4	YL/BL	Airflow Meter Signal	0.5-2.5v
5	GN/BK	Airflow Meter VREF	4.9-5.1v
7 (Cal)	RD/GN	Sub O2S Heater Control	Heater Off: 12v, On: 1v
8	WT/RD	EFI Main Relay (B+)	12-14v
9	GN/WT	Stop Light Switch Signal	Brake Off: 0v, On: 12v
10	GN/BL	Vehicle Speed Sensor	At 55 mph: 48 Hz
11	GN/YL	4WD Selector Switch	Switch Off: 12v, On: 1v
12	BR/BK	Sensor Ground	<0.050v
14	YL/RD	Cruise Control ECU	At Cruise in OD: 12v

1990 3.0L V6 MFI 3VZ-E VIN V (M/T) 18 Pin Connector

PCM Pin #	Wire Color	Circuit Description (18 Pin)	Value at Hot Idle
1	BK/OR	Distributor Signal (NE+)	AC pulse signals
2	BK	Knock Sensor Signal	<0.075v AC
3	GN/WT	EGR Gas Temperature Sensor (THG)	3.5-4.0v
4 (Cal)	WT	Sub O2S Signal	0.1-1.1v
5	BK/YL	Igniter Signal (IGF)	Digital Signal: 0-5-0v
6	YL/BL	Closed Throttle Switch	1v, off-idle: 12v

1990 3.0L V6 MFI 3VZ-E VIN V (M/T) 18 Pin Connector, *continued*

7	PK/WT	Check Connector	12-14v
8	PK	MIL (lamp) Control	MIL Off: 12v, On: 1v
9	GN	Fuel Pressure Up Solenoid	1v (at hot startup)
10	GN/BL	ECT Sensor Signal (THW)	At 180°F: 0.51v
11	YL	TP Sensor Signal (VTA)	0.3-0.8v
12	GN/YL	Sensor VREF (VC)	4.9-5.1v
13	BK	Main O2S Signal	0.1-1.1v
14	BR/BK	Sensor Ground	<0.1v
15	PK/GN	Main O2S Heater Control	Heater Off: 12v, On: 1v
16	PK	EGR Solenoid Control (VSV)	12v or 0v
17	GN/RD	Intake Air Solenoid Control	12v or 0v
18	BK/RD	A/C Idle-Up Solenoid	A/C Off: 12v, On: 1v

10 PIN CONNECTOR   18 PIN CONNECTOR   14 PIN CONNECTOR

WIRE SIDE OF HARNESS TERMINALS

05533_ADIA_G161

Pin Connector Graphic

1991-92 3.0L V6 MFI VIN V (All) 16 Pin Connector

PCM Pin #	Wire Color	Circuit Description (16 Pin)	Value at Hot Idle
1	GN/BK	Airflow Meter VREF	4.9-5.1v
2	YL/BL	Airflow Meter Signal	0.5-2.5v
3	YL/GN	IAT Sensor Signal (THA)	At 100°F: 2.60v
3 ('92)	YL/BL	IAT Sensor Signal (THA)	At 100°F: 2.60v
4	GN/BL	ECT Sensor Signal (THW)	At 180°F: 0.51v
5	BK	Knock Sensor Signal	<0.075v AC
6	BK	Main O2S Signal	0.1-1.1v
7	GN/BK	A/T Oil Temperature Sensor	At 68°F: 4-5v
8	YL/GN	Check Connector	12-14v
9	BR/BK	Sensor Ground	<0.050v
10 (Cal)	RD/GN	Sub O2S Signal	0.1-1.1v
11	YL	TP Sensor Signal (VTA)	0.3-0.8v
12	YL/BL	Closed Throttle Switch	1v, off-idle: 12v
13	GN/WT	EGR Gas Temperature Sensor (THG)	3.5-4.0v
13 (Fed)	GN/WT	Short Pin	Open: 12v, Closed: 0v
14	LG	Transfer Oil Temp. Sensor	At 68°F: 4-5v
15	PK/WT	Check Connector	12-14v

1991-92 3.0L V6 MFI VIN V (All) 22 Pin Connector

PCM Pin #	Wire Color	Circuit Description (22 Pin)	Value at Hot Idle
1	BK/GN	Battery Direct	12-14v
4 ('91)	PK/GN	A/T Oil Temperature Lamp	Lamp Off: 12v, On: 1v
4 ('92)	BL/WT	A/T Oil Temperature Lamp	Lamp Off: 12v, On: 1v
5	PK	MIL (lamp) Control	MIL Off: 12v, On: 1v
6	GN/WT	Stop Light Switch Signal	Brake Off: 0v, On: 12v
7	GN/OR	A/T Pattern Select Switch	Norm: 0v, PWR: 12v
8	PK/GN	A/T 4WD Indicator Switch	Switch Off: 12v, On: 1v
8	GN/WT	M/T 4WD Indicator Switch	Switch Off: 12v, On: 1v
9	GN/BL	Vehicle Speed Sensor	At 55 mph: 48 Hz
10	BK/WT	A/C Amplifier Signal (AC1)	Clutch On: 1.5v, Off: 12v
11	BK/WT	M/T: Starter Switch Signal	9-11v (cranking)
11	BK/RD	A/T: Starter Switch Signal	9-11v (cranking)
12	WT/RD	EFI Main Relay (B+)	12-14v
13	WT/RD	EFI Main Relay B1+	12-14v
15	BR/BK	Sensor Ground	<0.050v
16	YL/GN	Overdrive Main Switch	Switch Off: 12v, On: 1v
17	GY	4WD M/T Select (SEL2)	Open: 12v, Closed: 0v
18	WT/BK	2WD A/T Select (SEL1)	Open: 12v, Closed: 0v
20	YL/WT	Check Connector	12-14v
21	YL/RD	Cruise Control ECU	At Cruise in OD: 12v

1991-92 3.0L V6 MFI VIN V (All) 26 Pin Connector

PCM Pin #	Wire Color	Circuit Description (26 Pin)	Value at Hot Idle
1	WT	Distributor Signal (NE+)	AC pulse signals
2	RD	Distributor Signal (G1+)	AC pulse signals
3	BK/YL	Igniter Signal (IGF)	Digital Signal: 0-5-0v
4	GN/RD	A/T-ECT Solenoid (S4)	In Lockup: 12-14v
5	YL/BK	A/T-ECT Solenoid (S3)	In Lockup: 12-14v
6	BK	A/T-ECT Solenoid (S2)	1st or OD: 1v
7	WT	A/T-ECT Solenoid (S1)	3rd or OD: 1v
8	GN	Fuel Pressure Up Solenoid	1v (at hot startup)
9	GN	Cold Start Injector	1v (at cold startup)
10	PK/GN	HO2S Heater Control	Heater Off: 12v, On: 1v
11	BR	Sensor Ground	<0.050v
12	WT/RD	Injectors 1 & 3 & 5 Control	1.6-2.9 ms
13	BR	Power Ground	<0.1v
14	GN	Distributor Signal (GN-)	<0.050v
15	BK	Distributor Signal (G2+)	AC pulse signals
16	BR/RD	A/T Vehicle Speed Sensor	Moving: 0-5-0v
17	PK/WT	A/T Select Switch Low	In Low: 12v, Others: 0v
18	PK/GN	A/T Select Switch 2nd	In 2nd: 12v, Others: 0v
19	PK/RD	A/T Select Switch Neutral	In 'N': 12v, Others: 0v
20	YL/RD	4WD Detection Transfer (L4)	Open: 12v, Closed: 0v
21	BL/BK	Igniter Signal (IGT)	Digital Signal: 0-5-0v

1991-92 3.0L V6 MFI VIN V (All) 26 Pin Connector, *continued*

22	PK	EGR Solenoid Control (VSV)	12v or 0v
23	GN/RD	Intake Air Solenoid Control	12v or 0v
24	BK/RD	A/C Amplifier Signal (ACT)	Clutch On: 12v, Off: 1.5v
25	WT	Injectors 2 & 4 & 6 Control	1.6-2.9 ms
26	BR	Power Ground	<0.1v

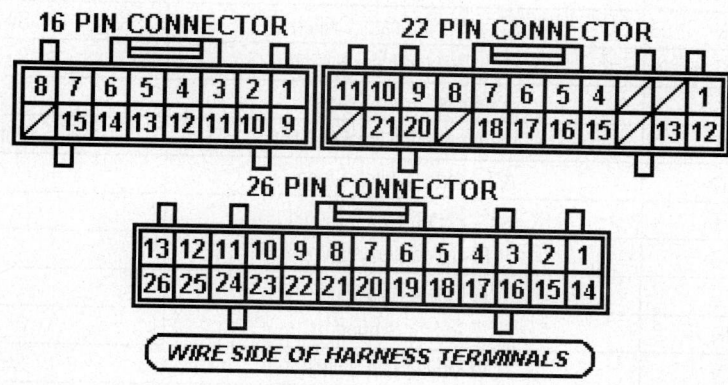

Pin Connector Graphic

05533_ADIA_G168

1993 3.0L V6 MFI VIN V (All) 16 Pin Connector

PCM Pin #	Wire Color	Circuit Description (16 Pin)	Value at Hot Idle
1	GN/BK	Airflow Meter VREF	4.9-5.1v
2	YL/RD	Airflow Meter Signal	0.5-2.5v
3	YL/BK	IAT Sensor Signal (THA)	At 100°F: 2.60v
4	GN/BL	ECT Sensor Signal (THW)	At 180°F: 0.51v
5	BK	Knock Sensor Signal	<0.075v AC
6	BK	Main O2S Signal	0.1-1.1v
7	GN/BK	A/T Oil Temperature Sensor	At 68°F: 4-6v
8	YL/GN	Data Link Connector	12-14v
9	BR/BK	Sensor Ground	<0.050v
10 (Cal)	RD/GN	Sub O2S Signal	0.1-1.1v
11	YL	TP Sensor Signal (VTA)	0.3-0.8v
12	YL/WT	Closed Throttle Switch	1v, off-idle: 12v
13	GN/WT	EGR Gas Temperature Sensor (THG)	3.5-4.0v
13 (Fed)	GN/WT	Short Pin	Open: 12v, Closed: 0v
14	LG	4WD Oil Temperature Sensor	At 68°F: 4-6v
15	PK/WT	Data Link Connector	12-14v
16	PK/GN	Data Link Connector	12-14v

1993 3.0L V6 MFI VIN V (All) 22 Pin Connector

PCM Pin #	Wire Color	Circuit Description (22 Pin)	Value at Hot Idle
1	BK/GN	Battery Direct	12-14v
4	BL/WT	A/T Oil Temperature Lamp	Lamp Off: 12v, On: 1v
5	PK	MIL (lamp) Control	MIL Off: 12v, On: 1v
6	GN/WT	Stop Light Switch Signal	Brake Off: 0v, On: 12v
7	GN/OR	A/T Pattern Select Switch	Norm: 0v, PWR: 12v
8	PK/GN	A/T 4WD Indicator Switch	Switch Off: 12v, On: 1v
8	GN/WT	M/T 4WD Indicator Switch	Switch Off: 12v, On: 1v
9	GN/BL	Vehicle Speed Sensor	At 55 mph: 48 Hz
10	BK/WT	A/C Amplifier Signal (AC1)	Clutch On: 1.5v, Off: 12v
11	BK/RD	A/T: Starter Switch Signal	9-11v (cranking)
11	BK/WT	M/T: Starter Switch Signal	9-11v (cranking)
12	WT/RD	EFI Main Relay (B+)	12-14v
13	WT/RD	EFI Main Relay B1+	12-14v
15	BR/BK	Sensor Ground	<0.050v
16	YL/GN	Overdrive Main Switch	Switch Off: 12v, On: 1v
17	BR	4WD M/T: Select (SEL2)	Open: 12v, Closed: 0v
18	BR	2WD A/T: Select (SEL1)	Open: 12v, Closed: 0v
19	BL/BK	A/C Amplifier Signal (ACT)	Clutch On: 12v, Off: 1.5v
20	YL/WT	Data Link Connector	12-14v
21	YL/RD	Cruise Control ECU	At Cruise in OD: 12v

1993 3.0L V6 MFI VIN V (All) 26 Pin Connector

PCM Pin #	Wire Color	Circuit Description (26 Pin)	Value at Hot Idle
1	WT	Distributor Signal (NE+)	AC pulse signals
2	RD	Distributor Signal (G1+)	AC pulse signals
3	BK/YL	Igniter Signal (IGF)	Digital Signal: 0-5-0v
4	GN/RD	A/T-ECT Solenoid (S4)	In Lockup: 12-14v
5	YL/BK	A/T-ECT Solenoid (S3)	In Lockup: 12-14v
6	BK	A/T-ECT Solenoid (S2)	1st or OD: 1v
7	WT	A/T-ECT Solenoid (S1)	3rd or OD: 1v
8	GN	Fuel Pressure Up Solenoid	1v (at hot startup)
9	GN	Cold Start Injector	1v (at cold startup)
10	PK/GN	HO2S-11 (B1 S1) Heater	Heater Off: 12v, On: 1v
11	BR	Sensor Ground	<0.050v
12	WT/RD	Injectors 1 & 3 & 5 Control	1.6-2-9 ms
13	BR	Power Ground	<0.1v
14	GN	Distributor Signal (GN-)	AC pulse signals
15	BK	Distributor Signal (G2+)	AC pulse signals
16	BR/RD	A/T Vehicle Speed Sensor	Moving: 0-5-0v
17	PK/WT	A/T Select Switch Low	In Low: 12v, Others: 0v
18	PK/GN	A/T Select Switch 2nd	In 2nd: 12v, Others: 0v
19	PK/RD	A/T Select Switch Neutral	In 'N': 12v, Others: 0v
20	YL/RD	4WD Detection Transfer (L4)	Open: 12v, Closed: 0v
21	BL/BK	Igniter Signal (IGT)	Digital Signal: 0-5-0v

1993 3.0L V6 MFI VIN V (All) 26 Pin Connector, *continued*

22	PK	EGR Solenoid Control (VSV)	12v or 0v
23	GN/RD	Intake Air Solenoid Control	12v or 0v
24	BK/RD	A/C Idle-Up Signal	A/C Off: 12v, On: 1v
25	WT	Injectors 2 & 4 & 6 Control	1.6-2-9 ms
26	BR	Power Ground	<0.1v

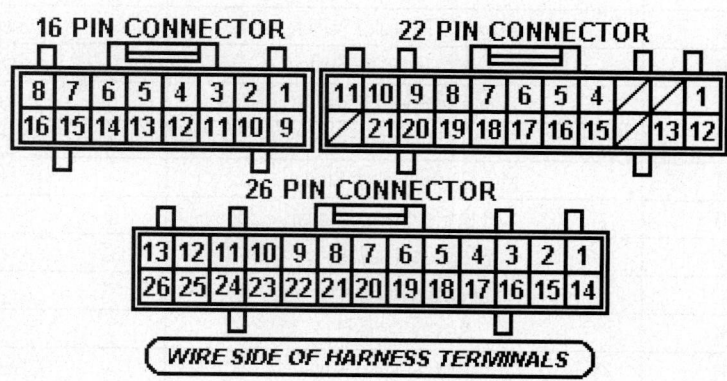

Pin Connector Graphic

1994-95 3.0L V6 MFI VIN V (All) 16 Pin Connector

PCM Pin #	Wire Color	Circuit Description (16 Pin)	Value at Hot Idle
1	GN/BK	Airflow Meter VREF	4.9-5.1v
2	YL/RD	Airflow Meter Signal	0.5-2.5v
3	YL/BK	IAT Sensor Signal (THA)	At 100°F: 2.60v
4	GN/BL	ECT Sensor Signal (THW)	At 180°F: 0.51v
5	BK	Knock Sensor Signal	<0.075v AC
6	BK	HO2S-11 (B1 S1) Signal	0.1-1.1v
7	GN/BK	A/T Oil Temperature Sensor	At 68°F: 4-5v
8	YL/GN	Data Link Connector	12-14v
9	BR/BK	Sensor Ground	<0.050v
10	WT	HO2S-12 (B1 S2) Signal	0.1-1.1v
11	YL	TP Sensor Signal (VTA)	0.3-0.8v
12	YL/WT	Closed Throttle Switch	1v, off-idle: 12v
13	GN/WT	EGR Gas Temperature Sensor (THG)	3.5-4.0v
13 (Fed)	GN/WT	Short Pin	Open: 12v, Closed: 0v
14	LG	Transfer Oil Temp. Sensor	At 68°F: 4-5v
15	PK/WT	Data Link Connector	12-14v
16	PK/GN	Data Link Connector	12-14v

1994-95 3.0L V6 MFI VIN V (All) 22 Pin Connector

PCM Pin #	Wire Color	Circuit Description (22 Pin)	Value at Hot Idle
1	BK/GN	Battery Direct	12-14v
4	BL/WT	A/T Oil Temperature Lamp	Lamp Off: 12v, On: 1v
5	PK	MIL (lamp) Control	MIL Off: 12v, On: 1v
6	GN/WT	Stop Light Switch Signal	Brake Off: 0v, On: 12v
7	GN/OR	A/T Pattern Select Switch	Norm: 0v, PWR: 12v
8	PK/GN	A/T 4WD Indicator Switch	Switch Off: 12v, On: 1v
8	GN/WT	M/T 4WD Indicator Switch	Switch Off: 12v, On: 1v
9	GN/BL	Vehicle Speed Sensor	At 55 mph: 48 Hz
10	BK/WT	A/C Amplifier Signal (AC1)	Clutch On: 1.5v, Off: 12v
11	BK/WT	M/T: Starter Switch Signal	9-11v (cranking)
11	BK/RD	A/T: Starter Switch Signal	9-11v (cranking)
12	WT/RD	EFI Main Relay (B+)	12-14v
13	WT/RD	EFI Main Relay B1+	12-14v
15	BR/BK	Sensor Ground	<0.050v
16	YL/GN	Overdrive Main Switch	Switch Off: 12v, On: 1v
17	BR	4WD M/T: Select (SEL2)	Open: 12v, Closed: 0v
18	BR	2WD A/T: Select (SEL1)	Open: 12v, Closed: 0v
19	BL/BK	A/C Amplifier Signal (ACT)	Clutch On: 12v, Off: 1.5v
20	YL/WT	Data Link Connector	12-14v
21	YL/RD	Cruise Control ECU	At Cruise in OD: 12v
22	RD/GN	HO2S-12 (B1 S2) Heater	Heater Off: 12v, On: 1v

1994-95 3.0L V6 MFI VIN V (All) 26 Pin Connector

PCM Pin #	Wire Color	Circuit Description (26 Pin)	Value at Hot Idle
1	WT	Distributor Signal (NE+)	AC pulse signals
2	RD	Distributor Signal (G1+)	AC pulse signals
3	BK/YL	Igniter Signal (IGF)	Digital Signal: 0-5-0v
4	GN/RD	A/T-ECT Solenoid (S4)	In Lockup: 12-14v
5	YL/BK	A/T-ECT Solenoid (S3)	In Lockup: 12-14v
6	BK	A/T-ECT Solenoid (S2)	1st or OD: 1v
7	WT	A/T-ECT Solenoid (S1)	3rd or OD: 1v
8	GN	Fuel Pressure Up Solenoid	1v (at hot startup)
9	GN	Cold Start Injector	1v (at cold startup)
10	PK/GN	HO2S-11 (B1 S1) Heater	Heater Off: 12v, On: 1v
11	BR	Sensor Ground	<0.050v
12	WT/RD	Injectors 1 & 3 & 5 Control	1.6-2-9 ms
13	BR	Power Ground	<0.1v
14	GN	Distributor Signal (GN-)	AC pulse signals
15	BK	Distributor Signal (G2+)	AC pulse signals
16	BR/RD	A/T Vehicle Speed Sensor	Moving: 0-5-0v
17	PK/WT	A/T Select Switch Low	In Low: 12v, Others: 0v
18	PK/GN	A/T Select Switch 2nd	In 2nd: 12v, Others: 0v
19	PK/RD	A/T Select Switch Neutral	In 'N': 12v, Others: 0v
20	YL/RD	4WD Detection Transfer (L4)	Open: 12v, Closed: 0v

1994-95 3.0L V6 MFI VIN V (All) 26 Pin Connector, *continued*

21	BL/BK	Igniter Signal (IGT)	Digital Signal: 0-5-0v
22	PK	EGR Solenoid Control (VSV)	12v or 0v
23	GN/RD	Intake Air Solenoid Control	12v or 0v
24	BK/RD	A/C Idle-Up Solenoid	A/C Off: 12v, On: 1v
25	WT	Injectors 2 & 4 & 6 Control	1.6-2-9 ms
26	BR	Power Ground	<0.1v

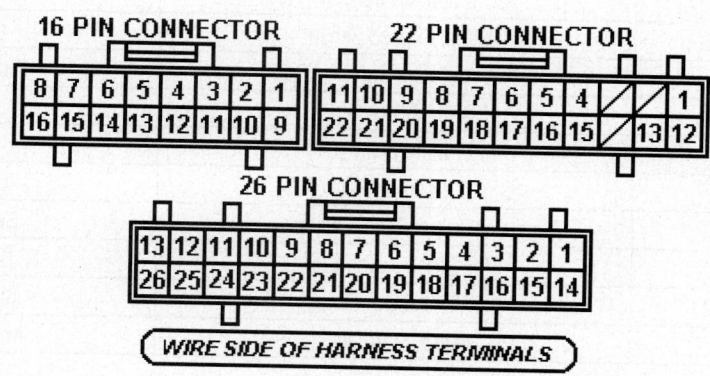

05533_ADIA_G170

Pin Connector Graphic

Standard Colors and Abbreviations

Abbreviation	Color	Abbreviation	Color	Abbreviation	Color
BK	Black	GY	Gray	RD	Red
BL	Blue	GN	Green	TN	Tan
BR	Brown	LG	LT Green	VT	Violet
DB	Dark Blue	OR	Orange	WT	White
DG	DK Green	PK	Pink	YL	Yellow

TACOMA PIN CHARTS

1995 2.4L I4 MFI VIN U (All) 22 Pin Connector

PCM Pin #	Wire Color	Circuit Description (22 Pin)	Value at Hot Idle
1	BK/YL	Battery Direct	12-14v
4	GN/OR	A/T Pattern Select Switch	Norm: 0v, PWR: 12v
5	PK	MIL (lamp) Control	MIL Off: 12v, On: 1v
7	WT	Data Link Connector (SDL)	0v
8	BL/BK	A/C Amplifier Signal (ACT)	Clutch On: 12v, Off: 1.5v
9	GN/OR	Vehicle Speed Sensor	At 55 mph: 48 Hz
10	BK/YL	A/C Amplifier Signal (AC1)	Clutch On: 1.5v, Off: 12v
11	BK/WT	Starter Switch Signal	9-11v (cranking)
12	WT/RD	EFI Main Relay (B+)	12-14v
14	GN/YL	Circuit Opening Relay (FC)	0-3v, at off-idle: 12v
17	BL	Overdrive Main Switch	Switch Off: 12v, On: 1v
20	GN/WT	Stop Light Switch Signal	Brake Off: 0v, On: 12v
21	GN/WT	4WD Selector Switch	Switch Off: 12v, On: 1v
22	BK	Neutral Start Switch	In P/N: 9-11v (cranking)

1995 2.4L I4 MFI VIN U (All) 16 Pin Connector

PCM Pin #	Wire Color	Circuit Description (16 Pin)	Value at Hot Idle
1	GN/YL	Sensor VREF	4.9-5.1v
2	GY/RD	MAF Sensor Signal (VG)	1.1-1.8v
3	PK	EGR Gas Temperature Sensor (THG)	3.5-4.0v
4	GN/RD	ECT Sensor Signal (THW)	At 180°F: 0.51v
5	BK	HO2S-12 (B1 S2) Signal	0.1-1.1v
6	WT	HO2S-11 (B1 S1) Signal	0.1-1.1v
7	YL/GN	IAT Sensor Signal (THA)	At 100°F: 2.60v
9, 16	BR, BR/WT	Sensor Ground	<0.050v
11	YL	TP Sensor Signal (VTA)	0.3-0.8v
12	YL/BL	Closed Throttle Switch	1v, off-idle: 12v
13	BK	Knock Sensor Signal	<0.075v AC
15	PK/WT	Data Link Connector	12-14v

1995 2.4L I4 MFI VIN U (All) 26 Pin Connector

PCM Pin #	Wire Color	Circuit Description (26 Pin)	Value at Hot Idle
1	BK/WT	A/C Idle-Up Solenoid	A/C Off: 12v, On: 1v
2	PK/GN	HO2S-11 (B1 S1) Heater	Heater Off: 12v, On: 1v
3	BK/YL	Igniter Signal (IGF)	Digital Signal: 0-5-0v
4, 17	RD, GN	CKP Sensor Signal (NE+), (NE-)	AC pulse signals
5	YL	Distributor Signal (G2+)	AC pulse signals
6	PK/BK	EGR Solenoid Control (VSV)	12v or 0v
8	GN	Fuel Pressure Up Solenoid	1v (at hot startup)
9	BK	IAC Signal (RSC)	Pulse Signals
10	BK/RD	IAC Signal (RSO)	Pulse Signals
11	WT	Injector Pair 2 & 4 Control	1.6-2-9 ms

1995 2.4L I4 MFI VIN U (All) 26 Pin Connector, *continued*

12	WT/RD	Injector Pair 1 & 3 Control	1.6-2-9 ms
13, 14	BR	Power Ground	<0.1v
15	RD/WT	HO2S-12 (B1 S2) Heater	Heater Off: 12v, On: 1v
18	BL	Distributor Signal (GN-)	<0.050v
20	BK/BL	Igniter Signal (IGF)	Digital Signal: 0-5-0v
23	WT/GN	EVAP Purge Solenoid (VSV)	12v or 0v
25, 26	BR	Power Ground	<0.1v

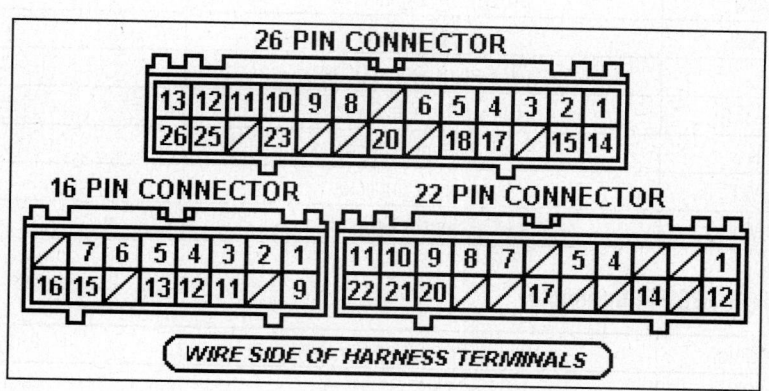

Pin Connector Graphic

1996-97 2.4L I4 MFI VIN L (All) 22 Pin Connector

PCM Pin #	Wire Color	Circuit Description (22 Pin)	Value at Hot Idle
1	BK/YL	Battery Direct	12-14v
5	PK	MIL (lamp) Control	MIL Off: 12v, On: 1v
6	BL	Cruise Control ECU	At Cruise in OD: 12v
7	WT	Data Link Connector (SDL)	0v
8	BL/BK	A/C Amplifier Signal (ACT)	Clutch On: 12v, Off: 1.5v
9	GN/OR	Vehicle Speed Sensor	At 55 mph: 48 Hz
10	BL/YL	A/C Amplifier Signal (AC1)	Clutch On: 1.5v, Off: 12v
11	BK/WT	Starter Switch Signal	9-11v (cranking)
12	WT/RD	EFI Main Relay (B+)	12-14v
14	GN/YL	Circuit Opening Relay (FC)	0-3v, at off-idle: 12v
15	PK/WT	A/T Select Switch Low	In Low: 12v, Others: 0v
16	PK	A/T Select Switch 2nd	In 2nd: 12v, Others: 0v
17	BL	Overdrive Main Switch	Switch Off: 12v, On: 1v
19	GN/RD	Taillight Switch Signal (ELS)	Switch Off: 0v, On: 12v
20	GN/WT	Stop Light Switch Signal	Brake Off: 0v, On: 12v
22	BK	Neutral Start Switch	In P/N: 9-11v (cranking)

1996-97 2.4L I4 MFI VIN L (All) 16 Pin Connector

PCM Pin #	Wire Color	Circuit Description (16 Pin)	Value at Hot Idle
1	GN/YL	Sensor VREF	4.9-5.1v
2	GY/RD	MAF Sensor Signal (VG)	1.1-1.8v
3	PK	EGR Gas Temperature Sensor (THG)	3.5-4.0v
4	GN/RD	ECT Sensor Signal (THW)	At 180°F: 0.51v
5	BK	HO2S-12 (B1 S2) Signal	0.1-1.1v
6	WT	HO2S-11 (B1 S1) Signal	0.1-1.1v
7	YL/GN	IAT Sensor Signal (THA)	At 100°F: 2.60v
9	BR/BK	Sensor Ground	<0.050v
11	YL	TP Sensor Signal (VTA)	0.3-0.8v
12	YL/BL or BL	Closed Throttle Switch	1v, off-idle: 12v
13	BK	Knock Sensor Signal	<0.075v AC
15	PK/WT	Data Link Connector	12-14v
16	BR/WT	MAF Sensor Ground	<0.050v

1996-97 2.4L I4 MFI VIN L (All) 26 Pin Connector

PCM Pin #	Wire Color	Circuit Description (26 Pin)	Value at Hot Idle
2	PK/GN	HO2S-11 (B1 S1) Heater	Heater Off: 12v, On: 1v
3	BK/YL	Igniter Signal (IGF)	Digital Signal: 0-5-0v
4	RD, GN	CKP Sensor Signal (NE+), (NE-)	AC pulse signals
5, 18	YL, BL	Distributor Signal (G2+), (GN-)	AC pulse signals
6	PK/BK	EGR Solenoid Control (VSV)	12v or 0v
9	BK	IAC Signal (RSC)	Pulse Signals
10	BK/RD	IAC Signal (RSO)	Pulse Signals
11	WT	Injector Pair 2 & 4 Control	1.6-2-9 ms
12	WT/RD	Injector Pair 1 & 3 Control	1.6-2-9 ms
13, 14	BR	Power Ground	<0.1v
15	RD/WT	HO2S-12 (B1 S2) Heater	Heater Off: 12v, On: 1v
20	BK/BL	Igniter Signal (IGF)	Digital Signal: 0-5-0v
23	WT/GN	EVAP Purge Solenoid (VSV)	12v or 0v
25, 26	BR	Power Ground	<0.1v

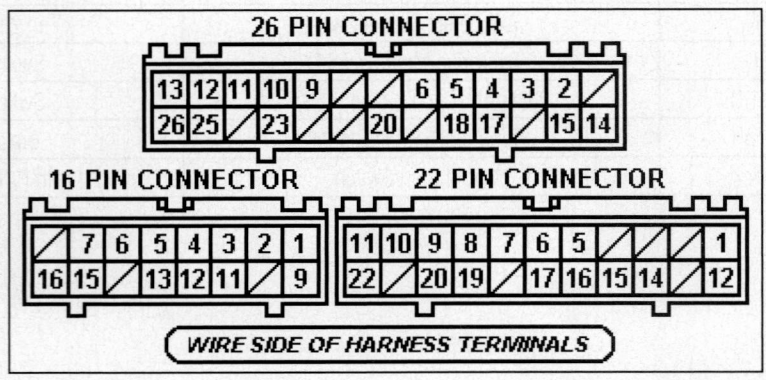

Pin Connector Graphic

05533_ADIA_G172

1998-99 2.4L I4 MFI VIN L (All) 22 Pin Connector

PCM Pin #	Wire Color	Circuit Description (22 Pin)	Value at Hot Idle
2	BK/YL	Battery Direct	12-14v
4	PK/RD	MIL (lamp) Control	MIL Off: 12v, On: 1v
6	BL/BK	A/C Amplifier Signal (ACT)	Clutch On: 12v, Off: 1.5v
7	BL/YL	A/C Amplifier Signal (AC1)	Clutch On: 1.5v, Off: 12v
8	GN/OR	Speedometer Indicator	At 55 mph: 48 Hz
11	BK/WT	Starter Switch Signal	9-11v (cranking)
12	WT/RD	EFI Main Relay (B+)	12-14v
15	PK/WT	A/T Select Switch Low	In Low: 12v, Others: 0v
16	PK	A/T Select Switch 2nd	In 2nd: 12v, Others: 0v
17	BL	Overdrive Main Switch	Switch Off: 12v, On: 1v
19	WT	SIL (Scan Tool) Signal	12v
21	GN/WT	Stop Light Switch Signal	Brake Off: 0v, On: 12v
22	BK/RD	A/T Neutral Start Switch	In P/N: 9-11v (cranking)

1998-99 2.4L I4 MFI VIN L (All) 16 Pin Connector

PCM Pin #	Wire Color	Circuit Description (16 Pin)	Value at Hot Idle
1	GN/YL	Sensor VREF (VC)	4.9-5.1v
2	GY/RD	MAF Sensor Signal (VG)	1.1-1.8v
3	YL/GN	IAT Sensor Signal (THA)	At 100°F: 2.60v
4	GN/RD	ECT Sensor Signal (THW)	At 180°F: 0.51v
5	WT	HO2S-11 (B1 S1) Signal	0.1-1.1v
7	PK/WT	Data Link Connector	12-14v
8	GY/GN	EVAP Vapor Pressure (VSV)	12v or 0v
9	BR/BK	Sensor Ground	<0.050v
10	YL	TP Sensor Signal (VTA)	0.3-0.8v
11	PK	EGR Gas Temperature Sensor (THG)	3.5-4.0v
12	BK	Knock Sensor Signal	<0.075v AC
13	BK	HO2S-12 (B1 S2) Signal	0.1-1.1v
15	PK/BK	EGR Solenoid Control (VSV)	12v or 0v

1998-99 2.4L I4 MFI VIN L (All) 12 Pin Connector

PCM Pin #	Wire Color	Circuit Description (12 Pin)	Value at Hot Idle
2	GN/RD	Taillight Switch Signal	Switch Off: 0v, On: 12v
5	BL	CMP Sensor Signal (GN-)	<0.050v
6	GN	CKP Sensor Signal (NE-)	<0.050v
7	BR/WT	MAF Sensor Ground (E2G)	<0.050v
10	RD/YL	EVAP Vapor Pressure Sensor	2.5-3.1v (with fuel cap off)
11	YL	CMP Sensor Signal (G2+)	AC pulse signals
12	RD	CKP Sensor Signal (NE+)	AC pulse signals

Standard Colors and Abbreviations

Abbreviation	Color	Abbreviation	Color	Abbreviation	Color
BK	Black	GY	Gray	RD	Red
BL	Blue	GN	Green	TN	Tan
BR	Brown	LG	LT Green	VT	Violet
DB	Dark Blue	OR	Orange	WT	White
DG	DK Green	PK	Pink	YL	Yellow

1998-99 2.4L I4 MFI VIN L (All) 26 Pin Connector

PCM Pin #	Wire Color	Circuit Description (26 Pin)	Value at Hot Idle
1	RD/WT	HO2S-12 (B1 S2) Heater	Heater Off: 12v, On: 1v
2	PK/GN	HO2S-11 (B1 S1) Heater	Heater Off: 12v, On: 1v
2 ('99)	RD/BK	HO2S-11 (B1 S1) Heater	Heater Off: 12v, On: 1v
3	WT/GN	EVAP Purge Solenoid (VSV)	12v or 0v
4	GN/RD	Cruise Control ECU	At Cruise in OD: 12v
6	BK	IAC Signal (RSC)	Pulse Signals
7	BK/RD	IAC Signal (RSO)	Pulse Signals
9	RD/BL	Injector 4 Control	1.6-2-9 ms
10	RD	Injector 3 Control	1.6-2-9 ms
11	WT	Injector 2 Control	1.6-2-9 ms
12	WT/RD	Injector 1 Control	1.6-2-9 ms
13	BR	Power Ground	<0.1v
14	GN/YL	Circuit Opening Relay (FC)	0-3v, at off-idle: 12v
15	BK	Tachometer Signal (TACO)	Pulse Signals
17	BK/YL	Igniter Signal (IGF)	Digital Signal: 0-5-0v
22	BR/YL	Igniter Transistor 2 Control	Digital Signal: 0-5-0v
23	BK/BL	Igniter Transistor 1 Control	Digital Signal: 0-5-0v
24	BR	Shield Ground	<0.050v
25	BR	Power Ground	<0.1v
26	BR	Power Ground	<0.1v

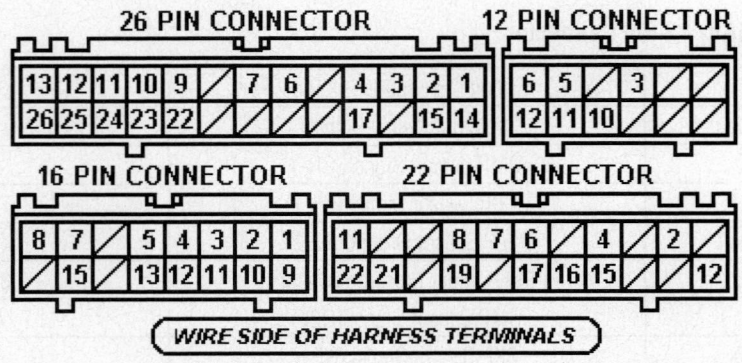

05533_ADIA_G173

Pin Connector Graphic

2000 2.4L I4 VIN L California (All) E5 22 Pin Connector

PCM Pin #	Wire Color	Circuit Description (22 Pin)	Value at Hot Idle
1	BK/YL	Direct Battery	12-14v
2	RD/BK	A/T Select Switch Reverse	In 'R': 12v, Others: 0v
3	PK/WT	A/T Select Switch Low	In Low: 12v, Others: 0v
4	PK	A/T Select Switch 2nd	In 2nd: 12v, Others: 0v
5	PK	A/T Oil Temperature Indicator	Lamp Off: 12v, On: 1v
6	VT/RD	MIL (lamp) Control	MIL Off: 12v, On: 1v
7	BK/WT	Starter Switch Signal	9-11v (cranking)
8	---	Not Used	---
9	BL	A/T Park (lamp) Indicator	Lamp Off: 12v, On: 1v
10	BL/OR	O/D OFF Indicator (ODLP)	Lamp Off: 12v, On: 1v
11	BR/YL	A/T Select Switch Drive	In 'D': 12v, Others: 0v
12	WT	SIL (Scan Tool) Signal	12v
13	BL/YL	A/C Amplifier Signal (AC1)	Clutch On: 1.5v, Off: 12v
14	BL/BK	A/C Amplifier Signal (ACT)	Clutch On: 12v, Off: 1.5v
15	GN/WT	Stop Light Switch Signal	Brake Off: 0v, On: 12v
16	WT/RD	EFI Main Relay (B+)	12-14v
17	GN	A/T Pattern Select Switch	Norm: 0v, PWR: 12v
18-19	---	Not Used	---
20	GN/RD	Defogger Idle-up Signal	Load On: 12v, Off: 0v
21	GN/OR	Speedometer Indicator (SP1)	At 55 mph: 48 Hz
22	BK/RD	A/T: Neutral Start Signal	In P/N: 9-11v (cranking)

2000 2.4L I4 VIN L California (All) E6 28 Pin Connector

PCM Pin #	Wire Color	Circuit Description (28 Pin)	Value at Hot Idle
1-4	---	Not Used	---
5	LG/RD	Cruise Control ECU (IDLO)	1.5v, off-idle: 12v
6	GN/YL	Circuit Opening Relay (FC)	0-3v, at off-idle: 12v
7	VT/WT	DLC 7 Signal (TC)	12v
8	RD/YL	EVAP Vapor Pressure Sensor	2.5-3.1v (with fuel cap off)
9-12	---	Not Used	---
13	BK	Tachometer Signal (TACO)	Pulse Signals
14-23	---	Not Used	---
24	---	Not Used	---
25	GN	OD Main Switch (ODMS)	Switch Off: 12v, On: 1v
26-27	---	Not Used	---
28	YL	PSP Switch Signal	Straight: 12v, Turning: 0v

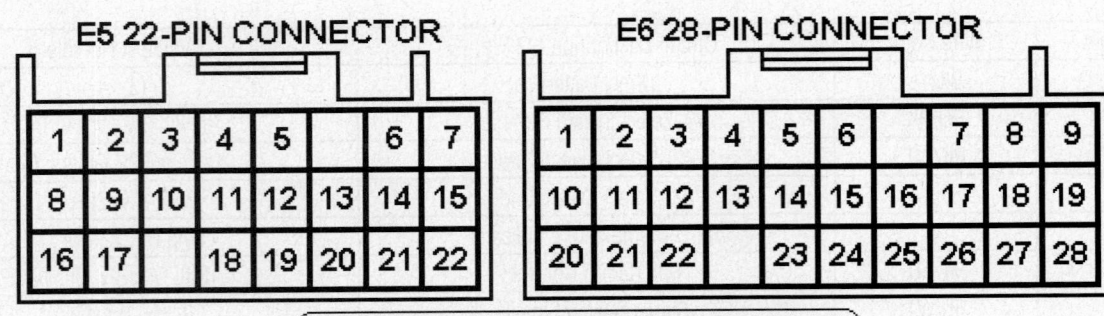

E5 22-PIN CONNECTOR   E6 28-PIN CONNECTOR

WIRE SIDE OF HARNESS TERMINALS

05533_ADIA_G174

Pin Connector Graphic
2000 2.4L I4 VIN L California (All) E7 24 Pin Connector

PCM Pin #	Wire Color	Circuit Description (24 Pin)	Value at Hot Idle
2	GN/YL	Sensor VREF (VC)	4.9-5.1v
3	RD/WT	HO2S-12 (B1 S2) Heater	Heater Off: 12v, On: 1v
4	WT	AFS-11 (B1 S1) Heater	Heater Off: 12v, On: 1v
6	WT/GN	EVAP Purge Solenoid (VSV)	12v or 0v
7	GY/GN	EVAP Vapor Pressure (VSV)	12v or 0v
9	BK/BL	TP Sensor Signal (VTA)	0.3-0.8v
10	RD	HO2S-12 (B1 S2) Signal	0.1-1.1v
11	WT	AFS-11 (B1 S1) Signal (+)	Fixed at 3.3v
12	GN/RD	ECT Sensor Signal (THW)	At 180°F: 0.51v
14	GY	MAF Sensor Signal (VG)	1.1-1.8v
15	YL	CMP Sensor Signal	AC pulse signals
16	RD	CKP Sensor Signal	AC pulse signals
17	BR	Shield Ground (E1)	<0.050v
18	BR/BK	Sensor Ground (E2)	<0.050v
20	RD	AFS-11 (B1 S1) Signal (-)	Fixed at 3.3v
21	YL/GN	IAT Sensor Signal (THA)	At 100°F: 2.60v
22	BK/WT	MAF Sensor Ground (E2G)	<0.050v
24	GN	CKP/CMP Sensor Signal (-)	<0.050v

2000 2.4L I4 VIN L California (All) E8 31 Pin Connector

PCM Pin #	Wire Color	Circuit Description (31 Pin)	Value at Hot Idle
1	WT/RD	Injector 1 Control	2.0-3.3 ms
2	WT	Injector 2 Control	2.0-3.3 ms
3	RD	Injector 3 Control	2.0-3.3 ms
4	RD/BL	Injector 4 Control	2.0-3.3 ms
5, 6	WT/BK	Power Ground (E03), (E04)	<0.1v
7	PK/YL	A/T-ECT Solenoid (S1)	3rd or OD: 1v
8	LG	A/T-ECT Solenoid (S2)	1st or OD: 1v
9	RD/GN	A/T-ECT Solenoid (SL)	In Lockup: 12-14v
10	BK/YL	Igniter Signal (IGF)	Digital Signal: 0-5-0v
11	BK/BL	Igniter Transistor 1 Control	6°, 55 mph: 8° dwell
12	BK/WT	Igniter Transistor 2 Control	6°, 55 mph: 8° dwell
13	BL/RD	Igniter Transistor 3 Control	6°, 55 mph: 8° dwell

2000 2.4L I4 VIN L California (All) E8 31 Pin Connector, *continued*

14	BL/YL	Igniter Transistor 4 Control	6°, 55 mph: 8° dwell
15	BK/RD	IAC Control (RSD)	Pulse Signals
21	WT/BK	Power Ground (E01)	<0.1v
26	YL/RD	A/T Oil Temperature Sensor	At 68°F: 4-5v
28	GY	Knock Sensor Signal	<0.075v AC
31	WT/BK	Power Ground (E02)	<0.1v

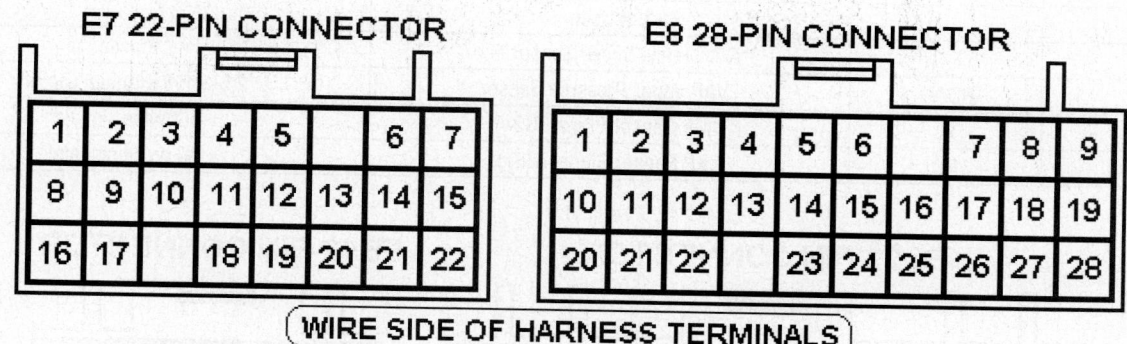

05533_ADIA_G175

Pin Connector Graphic

2000 2.4L I4 MFI VIN L Federal (All) E5 22 Pin Connector

PCM Pin #	Wire Color	Circuit Description (22 Pin)	Value at Hot Idle
1	---	Not Used	---
2	BK/YL	Direct Battery	12-14v
3	PK	A/T Oil Temperature Indicator	Lamp Off: 12v, On: 1v
4	PK/RD	MIL (lamp) Control	MIL Off: 12v, On: 1v
5	BL/OR	O/D OFF Indicator (ODLP)	Lamp Off: 12v, On: 1v
6	BL/BK	A/C Amplifier Signal (ACT)	Clutch On: 12v, Off: 1.5v
7	BL/YL	A/C Amplifier Signal (AC1)	Clutch On: 1.5v, Off: 12v
8	GN/OR	Speedometer Indicator	At 55 mph: 48 Hz
9-10	---	Not Used	---
11	BK/WT	Starter Switch Signal	9-11v (cranking)
12	WT/RD	EFI Main Relay (B+)	12-14v
13	---	Not Used	---
14	GN	A/T Pattern Select Switch	Norm: 0v, PWR: 12v
15	PK/WT	A/T Select Switch Low	In Low: 12v, Others: 0v
16	PK	A/T Select Switch 2nd	In 2nd: 12v, Others: 0v
17	RD/BK	A/T Select Switch Reverse	In 'R': 12v, Others: 0v
18	RD/YL	Cruise Control ECU (OD1)	At Cruise in OD: 12v
19	WT	SIL (Scan Tool) Signal	Digital Signals
20	---	Not Used	---
21	GN/WT	Stop Light Switch Signal	Brake Off: 0v, On: 12v
22	BK/RD	Start Circuit Signal	Cranking: 9-11v

2000 2.4L I4 MFI VIN L Federal (All) E6 12 Pin Connector

PCM Pin #	Wire Color	Circuit Description (12 Pin)	Value at Hot Idle
1	GN	Overdrive Main Switch	Switch Off: 12v, On: 1v
2	GN/RD	Defogger Idle-up Signal	Load On: 12v, Off: 0v
3	WT/RD	A/T Vehicle Speed Sensor (-)	Pulse Signals
5	BL	CMP Sensor Signal (GN-)	<0.050v
6	GN	CKP Sensor Signal (NE-)	<0.050v
7	BR/WT	MAF Meter Ground	<0.050v
8	---	Not Used	---
9	YL/RD	A/T Vehicle Speed Sensor (+)	Pulses
10	RD/YL	EVAP Vapor Pressure Sensor	2.5-3.1v (with fuel cap off)
11	YL	CMP Sensor Signal (G2+)	AC pulse signals
12	RD	CKP Sensor Signal (NE+)	AC pulse signals

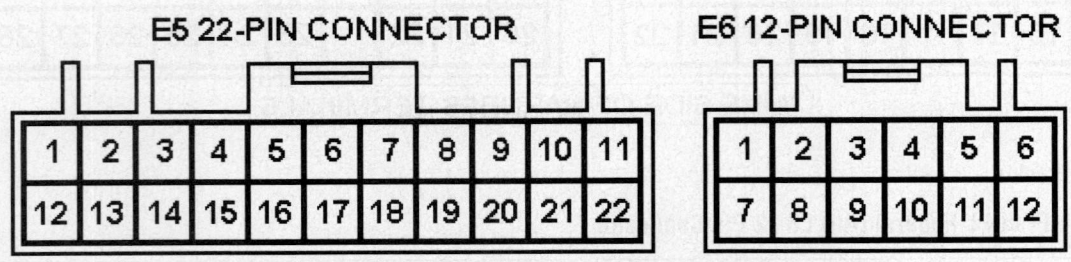

05533_ADIA_G176

Pin Connector Graphic

2000 2.4L I4 MFI VIN L Federal (All) E7 16 Pin Connector

PCM Pin #	Wire Color	Circuit Description (16 Pin)	Value at Hot Idle
1	GN/YL	Sensor VREF (VC)	4.9-5.1v
2	GY/RD	MAF Sensor Signal (VG)	1.1-1.8v
3	YL/GN	IAT Sensor Signal (THA)	At 100°F: 2.60v
4	GN/RD	ECT Sensor Signal (THW)	At 180°F: 0.51v
5	WT	HO2S-11 (B1 S1) Signal	0.1-1.1v
7	PK/WT	Data Link Connector	12-14v
8	GY/GN	EVAP Vapor Pressure (VSV)	12v or 0v
9	BR/BK	Sensor Ground	<0.050v
10	YL	TP Sensor Signal (VTA)	0.3-0.8v
11	PK	EGR Gas Temperature Sensor (THG)	3.5-4.0v
12	GY	Knock Sensor Signal	<0.075v AC
13	BK	HO2S-12 (B1 S2) Signal	0.1-1.1v
15	PK/BK	EGR Solenoid Control (VSV)	12v or 0v

2000 2.4L I4 MFI VIN L Federal (All) E8 26 Pin Connector

PCM Pin #	Wire Color	Circuit Description (26 Pin)	Value at Hot Idle
1	RD/WT	HO2S-12 (B1 S2) Heater	Heater Off: 12v, On: 1v
2	RD/BK	HO2S-11 (B1 S1) Heater	Heater Off: 12v, On: 1v
3	WT/GN	EVAP Purge Solenoid (VSV)	12v or 0v
4	LG/RD	Cruise Control ECU (IDLO)	1.5v, off-idle: 12v
6	BK	IAC Signal (RSC)	Pulse Signals

2000 2.4L I4 MFI VIN L Federal (All) E8 26 Pin Connector, *continued*

7	BK/RD	IAC Signal (RSO)	Pulse Signals
8	PK/YL	A/T-ECT Solenoid (S1)	3rd or OD: 1v
9	RD/BL	Injector 4 Control	2.0-3.3 ms
10	RD	Injector 3 Control	2.0-3.3 ms
11	WT	Injector 2 Control	2.0-3.3 ms
12	WT/RD	Injector 1 Control	2.0-3.3 ms
13	BR	Power Ground	<0.1v
14	GN/YL	Circuit Opening Relay (FC)	0-3v, at off-idle: 12v
15	BK	Tachometer Signal (TACO)	Pulse Signals
16	YL/RD	A/T Oil Temperature Sensor	At 68°F: 4-5v
17	BK/YL	Igniter Signal (IGF)	Digital Signal: 0-5-0v
20	RD/GN	A/T-ECT Solenoid (SL)	In Lockup: 12-14v
21	LG	A/T-ECT Solenoid (S2)	1st or OD: 1v
22	BR/YL	Igniter Transistor 2 Control	6°, 55 mph: 8° dwell
23	BK/BL	Igniter Transistor 1 Control	6°, 55 mph: 8° dwell
24	BR	Sensor Ground	<0.050v
25	BR	Power Ground	<0.1v
26	BR	Power Ground	<0.1v

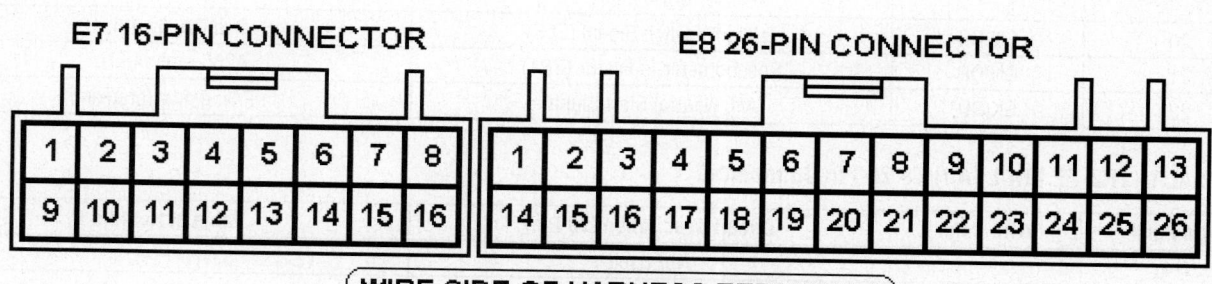

E7 16-PIN CONNECTOR   **E8 26-PIN CONNECTOR**

WIRE SIDE OF HARNESS TERMINALS

05533_ADIA_G177

Pin Connector Graphic

2001 2.4L I4 2RZ-FE VIN L (All) E5 22 Pin Connector

PCM Pin #	Wire Color	Circuit Description (22 Pin)	Value at Hot Idle
1	BK/YL	Direct Battery	12-14v
2	RD/BK	A/T Select Switch Reverse	In 'R': 12v, Others: 0v
3	PK/WT	A/T Select Switch Low	In Low: 12v, Others: 0v
4	PK	A/T Select Switch 2nd	In 2nd: 12v, Others: 0v
5	PK	A/T Oil Temperature Indicator	Lamp Off: 12v, On: 1v
6	VT/RD	MIL (lamp) Control	MIL Off: 12v, On: 1v
7	BK/WT	Starter Switch Signal	In P/N: 9-11v (cranking)
8	---	Not Used	---
9	BL	A/T Park (lamp) Indicator	Lamp Off: 12v, On: 1v
10	BL/OR	O/D OFF Indicator (ODLP)	Lamp Off: 12v, On: 1v
11	BR/YL	A/T Select Switch Drive	In 'D': 12v, Others: 0v
12	WT	SIL (Scan Tool) Signal	12v
13	BL/YL	A/C Amplifier Signal (AC1)	Clutch On: 1.5v, Off: 12v
14	BL/BK	A/C Amplifier Signal (ACT)	Clutch On: 12v, Off: 1.5v
15	GN/WT	Stop Light Switch Signal	Brake Off: 0v, On: 12v
16	WT/RD	EFI Main Relay (B+)	12-14v
17	GN	A/T Pattern Select Switch	Norm: 0v, PWR: 12v
18-19	---	Not Used	---
20	GN/RD	Taillight Switch Signal (ELS)	Switch Off: 0v, On: 12v
21	GN/OR	Speedometer Indicator (SP1)	At 55 mph: 48 Hz
22	BK/RD	A/T: Neutral Start Signal	In P/N: 9-11v (cranking)

2001 2.4L I4 2RZ-FE VIN L (All) E6 28 Pin Connector

PCM Pin #	Wire Color	Circuit Description (28 Pin)	Value at Hot Idle
1-4	---	Not Used	---
5	LG/RD	Cruise Control ECU (IDLO)	1.5v, off-idle: 12v
6	WT/BL	Circuit Opening Relay (FC)	0-3v, at off-idle: 12v
7	VT/WT	DLC 7 Signal (TC)	12v
8	RD/YL	EVAP Vapor Pressure Sensor	2.5-3.1v (with fuel cap off)
9-12	---	Not Used	---
13	BK	Tachometer Signal (TACO)	Pulse Signals
14-24	---	Not Used	---
25	GN	OD Main Switch (ODMS)	Switch Off: 12v, On: 1v
26-27	---	Not Used	---
28	YL	PSP Switch Signal	Straight: 12v, Turning: 0v

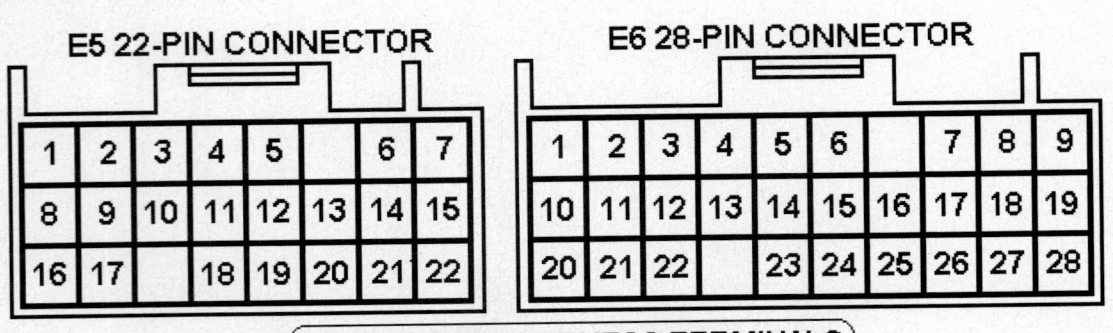

WIRE SIDE OF HARNESS TERMINALS

05533_ADIA_G174

Pin Connector Graphic

2001 2.4L I4 2RZ-FE VIN L (All) E7 24 Pin Connector

PCM Pin #	Wire Color	Circuit Description (24 Pin)	Value at Hot Idle
1	RD/BL	EVAP Closed Canister Vent	12v or 0v
2	GN/YL	Sensor VREF (VC)	4.9-5.1v
3	RD/WT	HO2S-12 (B1 S2) Heater	Heater Off: 12v, On: 1v
4	WT	AFS-11 (B1 S1) Heater	Heater Off: 12v, On: 1v
6	WT/GN	EVAP Purge Solenoid (VSV)	12v or 0v
7	GY/GN	EVAP Vapor Pressure (VSV)	12v or 0v
8	---	Not Used	---
9	YL	TP Sensor Signal (VTA)	0.3-0.8v
10	RD	HO2S-12 (B1 S2) Signal	0.1-1.1v
11	VT	AFS-11 (B1 S1) Signal (+)	Fixed at 3.3v
12	GN/RD	ECT Sensor Signal (THW)	At 180°F: 0.51v
13	---	Not Used	---
14	GY	MAF Sensor Signal (VG)	1.1-1.8v
15	RD	CMP Sensor Signal	AC pulse signals
16	BL	CKP Sensor Signal	AC pulse signals
17	BR	Shield Ground (E1)	<0.050v
18	LG	Sensor Ground (E2)	<0.050v
19	---	Not Used	---
20	PK	AFS-11 (B1 S1) Signal (-)	Fixed at 3.3v
21	YL/GN	IAT Sensor Signal (THA)	At 100°F: 2.60v
22	BK/WT	MAF Sensor Ground (E2G)	<0.050v
23	---	Not Used	---
24	GN	CKP/CMP Sensor Signal (-)	<0.050v

2001 2.4L I4 2RZ-FE VIN L (All) E8 31 Pin Connector

PCM Pin #	Wire Color	Circuit Description (31 Pin)	Value at Hot Idle
1	WT/RD	Injector 1 Control	2.0-3.3 ms
2	WT	Injector 2 Control	2.0-3.3 ms
3	RD	Injector 3 Control	2.0-3.3 ms
4	RD/BL	Injector 4 Control	2.0-3.3 ms
5	WT/BK	Power Ground (E03)	<0.1v
6	WT/BK	Power Ground (E04)	<0.1v
7	PK/YL	A/T-ECT Solenoid (S1)	3rd or OD: 1v
8	LG	A/T-ECT Solenoid (S2)	1st or OD: 1v
9	RD/GN	A/T-ECT Solenoid (SL)	In Lockup: 12-14v
10	BK/YL	Igniter Signal (IGF)	Digital Signal: 0-5-0v
11	BK/BL	Igniter Transistor 1 Control	6°, 55 mph: 8° dwell
12	BL	Igniter Transistor 2 Control	6°, 55 mph: 8° dwell
13	BL/RD	Igniter Transistor 3 Control	6°, 55 mph: 8° dwell
14	BL/YL	Igniter Transistor 4 Control	6°, 55 mph: 8° dwell
15	BK/RD	IAC Control (RSD)	Pulse Signals
16-20	---	Not Used	---
21	WT/BK	Power Ground (E01)	<0.1v
26	YL/RD	A/T Oil Temperature Sensor	At 68°F: 4-5v
27	---	Not Used	---
28	BK/YL	Knock Sensor Signal	<0.075v AC
29-30	---	Not Used	---
31	WT/BK	Power Ground (E02)	<0.1v

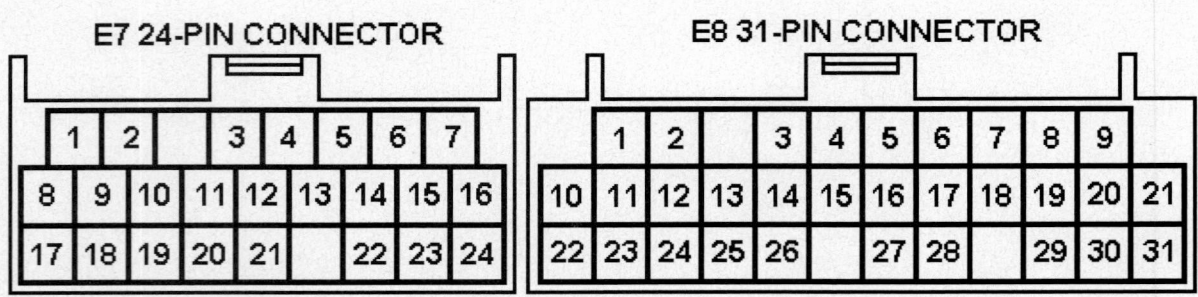

E7 24-PIN CONNECTOR   E8 31-PIN CONNECTOR

05533_ADIA_G174

Pin Connector Graphic

2002-06 2.4L I4 2RZ-FE VIN L (All) E5 22 Pin Connector

PCM Pin #	Wire Color	Circuit Description (22 Pin)	Value at Hot Idle
1	BK/YL	Direct Battery	12-14v
2	RD/BK	A/T Select Switch Reverse	In 'R': 12v, Others: 0v
3	PK/WT	A/T Select Switch Low	In Low: 12v, Others: 0v
4	PK	A/T Select Switch 2nd	In 2nd: 12v, Others: 0v
5	PK	A/T Oil Temperature Indicator	Lamp Off: 12v, On: 1v
6	VT/RD	Malfunction Indicator Lamp (MIL) Control	MIL Off: 12v, On: 1v
7	GN	Starter Switch Signal (STA)	In P/N: 9-11v (cranking)
8	---	Not Used	---

2002-06 2.4L I4 2RZ-FE VIN L (All) E5 22 Pin Connector, *continued*

9	BL	A/T Park (lamp) Indicator	Lamp Off: 12v, On: 1v
10	BL/OR	O/D OFF Indicator (ODLP)	Lamp Off: 12v, On: 1v
11	BR/YL	A/T Select Switch Drive	In 'D': 12v, Others: 0v
12	WT	SIL (Scan Tool) Signal	12v
13	BL/YL	A/C Amplifier Signal (AC1)	Clutch On: 1.5v, Off: 12v
14	BL/BK	A/C Amplifier Signal (ACT)	Clutch On: 12v, Off: 1.5v
15	GN/WT	Stop Light Switch Signal	Brake Off: 0v, On: 12v
16	WT/RD	EFI Main Relay (B+)	12-14v
17	GN	A/T Pattern Select Switch	Norm: 0v, PWR: 12v
18-19	---	Not Used	---
20	GN/RD	Taillight Switch Signal (ELS)	Switch Off: 0v, On: 12v
21	GN/OR	Speedometer Indicator (SP1)	At 55 mph: 48 Hz
22	BK/RD	A/T: Neutral Start Signal (NSW)	In P/N: 9-11v (cranking)

2002-06 2.4L I4 2RZ-FE VIN L (All) E6 28 Pin Connector

PCM Pin #	Wire Color	Circuit Description (28 Pin)	Value at Hot Idle
1-4	---	Not Used	---
5	LG/RD	Cruise Control ECU (IDLO)	1.5v, off-idle: 12v
6	WT/BL	Circuit Opening Relay (FC)	0-3v, at off-idle: 12v
7	YL/BK	DLC 7 Signal (TC)	12v
8	RD/YL	EVAP Vapor Pressure Sensor (PTNK)	2.5-3.1v (with fuel cap off)
9-12	---	Not Used	---
13	LG/BK	Tachometer Signal (TACO)	Pulse Signals
14-24	---	Not Used	---
25	GN	OD Main Switch (ODMS)	Switch Off: 12v, On: 1v
26-27	---	Not Used	---
28	YL	PSP Switch Signal (PSSW)	Straight: 12v, Turning: 0v

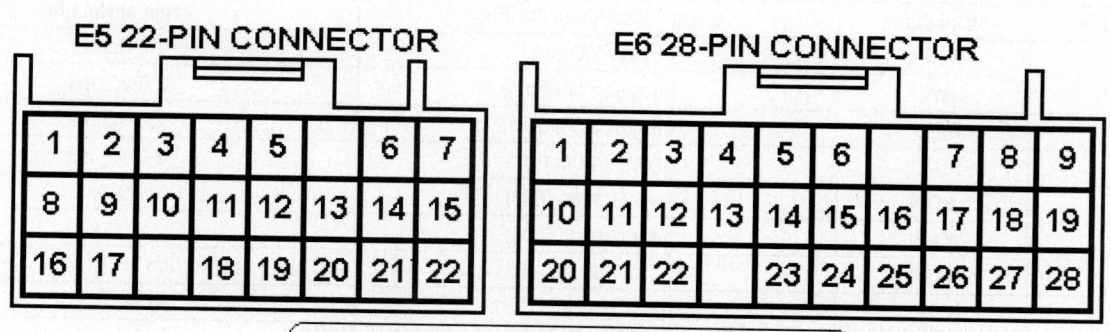

Pin Connector Graphic

05533_ADIA_G174

2002-06 2.4L I4 2RZ-FE VIN L (All) E7 24 Pin Connector

PCM Pin #	Wire Color	Circuit Description (24 Pin)	Value at Hot Idle
1	RD/BL	EVAP Closed Canister Vent (VSV)	12v or 0v
2	GN/YL	Sensor VREF (VC)	4.9-5.1v
3	RD/WT	HO2S-12 (B1 S2) Heater (HTS)	Heater Off: 0v, On: 12v
4	WT	AFS-11 (B1 S1) Heater (AFHT)	Heater Off: 0v, On: 12v
5	---	Not Used	---
6	WT/GN	EVAP Purge Solenoid (VSV)	12v or 0v
7	GN/BK	EVAP Vapor Pressure (VSV)	12v or 0v
8	---	Not Used	---
9	YL	TP Sensor Signal (VTA)	0.3-0.8v
10	BK	HO2S-12 (B1 S2) Signal (OXS)	0.1-1.1v
11	VT	AFS-11 (B1 S1) Signal (AF+)	Fixed at 3.3v
12	GN/RD	ECT Sensor Signal (THW)	At 180°F: 0.51v
13	---	Not Used	---
14	GY	MAF Sensor Signal (VG)	1.1-1.8v
15	RD	CMP Sensor Signal (G2+)	AC pulse signals
16	BL	CKP Sensor Signal (NE+)	AC pulse signals
17	BR	Shield Ground (E1)	<0.050v
18	BL/BK	Sensor Ground (E2)	<0.050v
19	---	Not Used	---
20	PK	AFS-11 (B1 S1) Signal (AF-)	Fixed at 3.3v
21	YL/GN	IAT Sensor Signal (THA)	At 100°F: 2.60v
22	BK/WT	MAF Sensor Ground (E2G)	<0.050v
23	---	Not Used	---
24	GN	CKP/CMP Sensor Signal (NE-)	<0.050v

2002-06 2.4L I4 2RZ-FE VIN L (All) E8 31 Pin Connector

PCM Pin #	Wire Color	Circuit Description (31 Pin)	Value at Hot Idle
1	WT/RD	Injector 1 Control	2.0-3.3 ms
2	WT	Injector 2 Control	2.0-3.3 ms
3	RD	Injector 3 Control	2.0-3.3 ms
4	RD/BL	Injector 4 Control	2.0-3.3 ms
5	WT/BK	Power Ground (E03)	<0.1v
6	WT/BK	Power Ground (E04)	<0.1v
7	PK/YL	A/T-ECT Solenoid (S1)	3rd or OD: 1v
8	LG	A/T-ECT Solenoid (S2)	1st or OD: 1v
9	RD/GN	A/T-ECT Solenoid (SL)	In Lockup: 12-14v
10	BK/YL	Igniter Signal (IGF)	Digital Signal: 0-5-0v
11	BK/BL	Igniter Transistor 1 Control	6°, 55 mph: 8° dwell
12	BL	Igniter Transistor 2 Control	6°, 55 mph: 8° dwell
13	BL/RD	Igniter Transistor 3 Control	6°, 55 mph: 8° dwell
14	BL/YL	Igniter Transistor 4 Control	6°, 55 mph: 8° dwell
15	BK/RD	Idle Air Control Valve (RSO)	Pulse Signals
16-20	---	Not Used	---
21	WT/BK	Power Ground (E01)	<0.1v
22-25	---	Not Used	---
26	YL/RD	A/T Oil Temperature Sensor	At 68°F: 4-5v

2002-06 2.4L I4 2RZ-FE VIN L (All) E8 31 Pin Connector, *continued*

27	---	Not Used	---
28	BK	Knock Sensor Signal (KNK)	<0.075v AC
29-30	---	Not Used	---
31	WT/BK	Power Ground (E02)	<0.1v

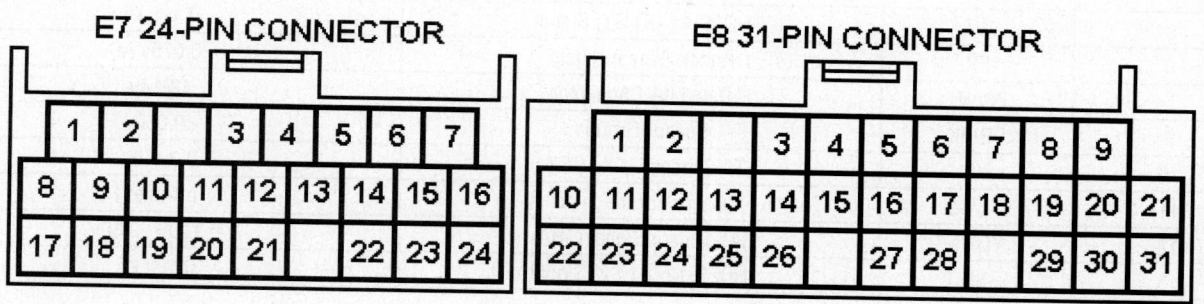

E7 24-PIN CONNECTOR   **E8 31-PIN CONNECTOR**

05533_ADIA_G178

Pin Connector Graphic

1995 2.7L I4 MFI VIN U (A/T-ECT) 22 Pin Connector

PCM Pin #	Wire Color	Circuit Description (22 Pin)	Value at Hot Idle
1	---	Not Used	---
2	BK/YL	Battery Direct	12-14v
3	PK	A/T Oil Temperature Indicator	Lamp Off: 12v, On: 1v
4	PK	MIL (lamp) Control	MIL Off: 12v, On: 1v
5	BL/OR	Overdrive Main Switch	Switch Off: 12v, On: 1v
6	BL/BK	A/C Amplifier Signal (ACT)	Clutch On: 12v, Off: 1.5v
7	BK/YL	A/C Amplifier Signal (AC1)	Clutch On: 1.5v, Off: 12v
8	GN/OR	Vehicle Speed Sensor	At 55 mph: 48 Hz
9	---	Not Used	---
10	GN/WT	4WD Selector Switch	Switch Off: 12v, On: 1v
11	BK/WT	Starter Switch Signal	9-11v (cranking)
12	WT/RD	EFI Main Relay (B+)	12-14v
13	RD/GN	A/T Select Switch Park	In Park: 12v, Others: 0v
14	GN/OR	A/T Pattern Select Switch	Norm: 0v, PWR: 12v
15	PK/WT	A/T Select Switch Low	In Low: 12v, Others: 0v
16	PK	A/T Select Switch 2nd	In 2nd: 12v, Others: 0v
17	---	Not Used	---
18	BL	Cruise Control ECU (OD1)	At Cruise in OD: 12v
19	WT	Data Link Connector (SDL)	0v
20	---	Not Used	---
21	GN/WT	Stop Light Switch Signal	Brake Off: 0v, On: 12v
22	BK	Neutral Start Switch	In P/N: 9-11v (cranking)

1995 2.7L I4 MFI VIN U (A/T-ECT) 16 Pin Connector

PCM Pin #	Wire Color	Circuit Description (16 Pin)	Value at Hot Idle
1	GN/YL	Sensor VREF	4.9-5.1v
2	GY/RD	MAF Sensor Signal (VG)	1.1-1.8v
3	BR/WT	Sensor Ground	<0.050v
4	GN/RD	ECT Sensor Signal (THW)	At 180°F: 0.51v
5	WT	HO2S-11 (B1 S1) Signal	0.1-1.1v
6	BK	Knock Sensor Signal	<0.075v AC
7	PK/WT	Data Link Connector	12-14v
9	BR/BK	Sensor Ground	<0.050v
10	YL	TP Sensor Signal (VTA)	0.3-0.8v
11	YL/BL	Closed Throttle Switch	1v, off-idle: 12v
12	YL/GN	IAT Sensor Signal (THA)	At 100°F: 2.60v
13	BK	HO2S-12 (B1 S2) Signal	0.1-1.1v
14	PK	EGR Gas Temperature Sensor (THG)	3.5-4.0v

Standard Colors and Abbreviations

Abbreviation	Color	Abbreviation	Color	Abbreviation	Color
BK	Black	GY	Gray	RD	Red
BL	Blue	GN	Green	TN	Tan
BR	Brown	LG	LT Green	VT	Violet
DB	Dark Blue	OR	Orange	WT	White
DG	DK Green	PK	Pink	YL	Yellow

1995 2.7L I4 MFI VIN U (A/T-ECT) 26 Pin Connector

PCM Pin #	Wire Color	Circuit Description (26 Pin)	Value at Hot Idle
1	---	Not Used	---
2	BK/WT	A/C Idle-Up Solenoid	A/C Off: 12v, On: 1v
3	PK/GN	HO2S-11 (B1 S1) Heater	Heater Off: 12v, On: 1v
4	GN	Fuel Pressure Up Solenoid	1v (at hot startup)
5	---	Not Used	---
6	BK	IAC Signal (RSC)	Pulse Signals
7	BK/RD	IAC Signal (RSO)	Pulse Signals
8	RD/GN	A/T-ECT Solenoid (SL)	In Lockup: 12-14v
9	LG	A/T-ECT Solenoid (S2)	1st or OD: 1v
10	PK/YL	A/T-ECT Solenoid (S1)	3rd or OD: 1v
11	WT	Injector Pair 2 & 4 Control	1.6-2-9 ms
12	WT/RD	Injector Pair 1 & 3 Control	1.6-2-9 ms
13	BR	Power Ground	<0.1v
14	GN/YL	Circuit Opening Relay (FC)	0-3v, at off-idle: 12v
15	---	Not Used	---
16	RD/WT	HO2S-12 (B1 S2) Heater	Heater Off: 12v, On: 1v
17	BK/YL	Igniter Signal (IGF)	Digital Signal: 0-5-0v
18-19	---	Not Used	---
20	GN	4WD Detection Transfer (L4)	Open: 12v, Closed: 0v
21	YL/RD	A/T Oil Temperature Sensor	At 68°F: 4-5v
22	PK/BK	EGR Solenoid Control (VSV)	12v or 0v

1995 2.7L I4 MFI VIN U (A/T-ECT) 26 Pin Connector, *continued*

23	BK/BL	Igniter Signal (IGF)	Digital Signal: 0-5-0v
24	BR	Sensor Ground	<0.050v
25	BR	Power Ground	<0.1v
26	BR	Power Ground	<0.1v

1995 2.7L I4 MFI VIN U (A/T-ECT) 12 Pin Connector

PCM Pin #	Wire Color	Circuit Description (12 Pin)	Value at Hot Idle
1	WT/GN	EVAP Purge Solenoid (VSV)	12v or 0v
2-3	---	Not Used	---
4	WT/RD	A/T Vehicle Speed Sensor (-)	Pulse Signals
5	BL	Distributor Signal (GN-)	<0.050v
6	GN	CKP Sensor Signal (NE-)	<0.050v
7-9	---	Not Used	---
10	YL/RD	A/T Vehicle Speed Sensor (+)	Pulses
11	YL	Distributor Signal (G2+)	AC pulse signals
12	RD	CKP Sensor Signal (NE+)	AC pulse signals

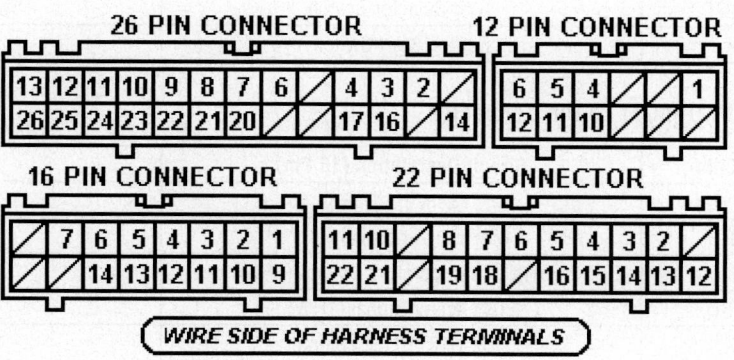

05533_ADIA_G179

Pin Connector Graphic

1995 2.7L I4 MFI 3RZ-FE VIN U (M/T) 22 Pin Connector

PCM Pin #	Wire Color	Circuit Description (22 Pin)	Value at Hot Idle
1	BK/YL	Battery Direct	12-14v
2-4	---	Not Used	---
5	PK	MIL (lamp) Control	MIL Off: 12v, On: 1v
6	---	Not Used	---
7	WT	Data Link Connector (SDL)	0v
8	BL/BK	A/C Amplifier Signal (ACT)	Clutch On: 12v, Off: 1.5v
9	GN/OR	Vehicle Speed Sensor	At 55 mph: 48 Hz
10	BK/YL	A/C Amplifier Signal (AC1)	Clutch On: 1.5v, Off: 12v
11	BK/WT	Starter Switch Signal	9-11v (cranking)
12	WT/RD	EFI Main Relay (B+)	12-14v
13	---	Not Used	---
14	GN/YL	Circuit Opening Relay (FC)	0-3v, at off-idle: 12v
15-16	---	Not Used	---
17	BL	Overdrive Main Switch	Switch Off: 12v, On: 1v
18-19	---	Not Used	---
20	GN/WT	Stop Light Switch Signal	Brake Off: 0v, On: 12v
21	GN/WT	4WD Selector Switch	Switch Off: 12v, On: 1v
22	BK	Neutral Start Switch	9-11v (cranking)

1995 2.7L I4 MFI 3RZ-FE VIN U (M/T) 16 Pin Connector

PCM Pin #	Wire Color	Circuit Description (16 Pin)	Value at Hot Idle
1	GN/YL	Sensor VREF	4.9-5.1v
2	GY/RD	MAF Sensor Signal (VG)	1.1-1.8v
3	PK	EGR Gas Temperature Sensor (THG)	3.5-4.0v
4	GN/RD	ECT Sensor Signal (THW)	At 180°F: 0.51v
5	BK	HO2S-12 (B1 S2) Signal	0.1-1.1v
6	WT	HO2S-11 (B1 S1) Signal	0.1-1.1v
7	YL/GN	IAT Sensor Signal (THA)	At 100°F: 2.60v
8	---	Not Used	---
9	BR/BK	Sensor Ground	<0.050v
10	---	Not Used	---
11	YL	TP Sensor Signal (VTA)	0.3-0.8v
12	YL/BL	Closed Throttle Switch	1v, off-idle: 12v
13	BK	Knock Sensor Signal	<0.075v AC
14	---	Not Used	---
15	PK/WT	Data Link Connector	12-14v
16	BR/WT	Sensor Ground	<0.050v

Standard Colors and Abbreviations

Abbreviation	Color	Abbreviation	Color	Abbreviation	Color
BK	Black	GY	Gray	RD	Red
BL	Blue	GN	Green	TN	Tan
BR	Brown	LG	LT Green	VT	Violet
DB	Dark Blue	OR	Orange	WT	White
DG	DK Green	PK	Pink	YL	Yellow

1995 2.7L I4 MFI 3RZ-FE VIN U (M/T) 26 Pin Connector

PCM Pin #	Wire Color	Circuit Description (26 Pin)	Value at Hot Idle
1	BK/WT	A/C Idle-Up Solenoid	A/C Off: 12v, On: 1v
2	PK/GN	HO2S-11 (B1 S1) Heater	Heater Off: 12v, On: 1v
3	BK/YL	Igniter Signal (IGF)	Digital Signal: 0-5-0v
4	RD	CKP Sensor Signal	AC pulse signals
5	YL	Distributor Signal (G2+)	AC pulse signals
6	PK/BK	EGR Solenoid Control (VSV)	12v or 0v
8	GN	Fuel Pressure Up Solenoid	1v (at hot startup)
9	BK	IAC Signal (RSC)	Pulse Signals
10	BK/RD	IAC Signal (RSO)	Pulse Signals
11	WT	Injector Pair 2 & 4 Control	1.6-2-9 ms
12	WT/RD	Injector Pair 1 & 3 Control	1.6-2-9 ms
13	BR	Power Ground	<0.1v
14	BR	Sensor Ground	<0.050v
15	RD/WT	HO2S-12 (B1 S2) Heater	Heater Off: 12v, On: 1v
17	GN	CKP Sensor Signal (NE-)	<0.050v
18	BL	Distributor Signal (GN-)	<0.050v
20	BK/BL	Igniter Signal (IGF)	Digital Signal: 0-5-0v
23	WT/GN	EVAP Purge Solenoid (VSV)	12v or 0v
25	BR	Power Ground	<0.1v
26	BR	Power Ground	<0.1v

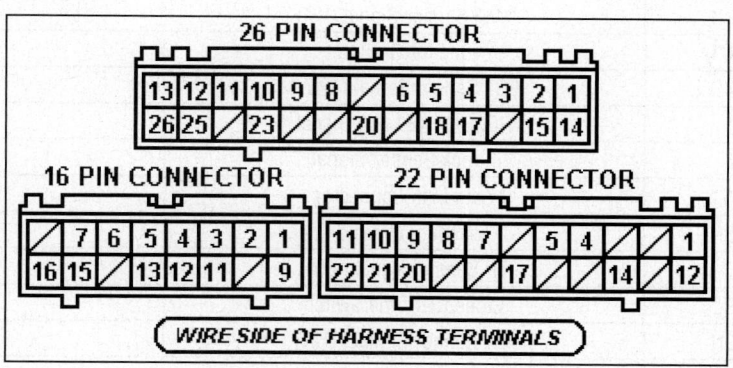

05533_ADIA_G171

Pin Connector Graphic

1996 2.7L I4 MFI VIN M (A/T-ECT) 22 Pin Connector

PCM Pin #	Wire Color	Circuit Description (22 Pin)	Value at Hot Idle
2	BK/YL	Battery Direct	12-14v
3	PK	A/T Oil Temperature Indicator	Lamp Off: 12v, On: 1v
4	PK	MIL (lamp) Control	MIL Off: 12v, On: 1v
5	BL/OR	Overdrive Main Switch	Switch Off: 12v, On: 1v
6	BL/BK	A/C Amplifier Signal (ACT)	Clutch On: 12v, Off: 1.5v
7	BL/YL	A/C Amplifier Signal (AC1)	Clutch On: 1.5v, Off: 12v
8	GN/OR	Vehicle Speed Sensor	At 55 mph: 48 Hz
10	GN/WT	4WD Selector Switch	Switch Off: 12v, On: 1v
11	BK/WT	Starter Switch Signal	9-11v (cranking)
12	WT/RD	EFI Main Relay (B+)	12-14v
14	GN/OR	A/T Pattern Select Switch	Norm: 0v, PWR: 12v
15	PK/WT	A/T Select Switch Low	In Low: 12v, Others: 0v
16	PK	A/T Select Switch 2nd	In 2nd: 12v, Others: 0v
18	BL	Cruise Control ECU (OD1)	At Cruise in OD: 12v
19	WT	Data Link Connector (SDL)	0v
21	GN/WT	Stop Light Switch Signal	Brake Off: 0v, On: 12v
22	BK	Neutral Start Switch	In P/N: 9-11v (cranking)

1996 2.7L I4 MFI VIN M (A/T-ECT) 16 Pin Connector

PCM Pin #	Wire Color	Circuit Description (16 Pin)	Value at Hot Idle
1	GN/YL	Sensor VREF	4.9-5.1v
2	GY/RD	MAF Sensor Signal (VG)	1.1-1.8v
3	BR/WT	Sensor Ground	<0.050v
4	GN/RD	ECT Sensor Signal (THW)	At 180°F: 0.51v
5	WT	HO2S-11 (B1 S1) Signal	0.1-1.1v
6	BK	Knock Sensor Signal	<0.075v AC
7	PK/WT	Data Link Connector	12-14v
9	BR/BK	Sensor Ground	<0.050v
10	YL	TP Sensor Signal (VTA)	0.1-1.0v
11	YL/BL	Closed Throttle Switch	1v, off-idle: 12v
12	YL/GN	IAT Sensor Signal (THA)	At 100°F: 2.60v
13	BK	HO2S-12 (B1 S2) Signal	0.1-1.1v
14	PK	EGR Gas Temperature Sensor (THG)	3.5-4.0v

Standard Colors and Abbreviations

Abbreviation	Color	Abbreviation	Color	Abbreviation	Color
BK	Black	GY	Gray	RD	Red
BL	Blue	GN	Green	TN	Tan
BR	Brown	LG	LT Green	VT	Violet
DB	Dark Blue	OR	Orange	WT	White
DG	DK Green	PK	Pink	YL	Yellow

1996 2.7L I4 MFI VIN M (A/T-ECT) 26 Pin Connector

PCM Pin #	Wire Color	Circuit Description (26 Pin)	Value at Hot Idle
3	PK/GN	HO2S-11 (B1 S1) Heater	Heater Off: 12v, On: 1v
6	BK	IAC Signal (RSC)	Pulse Signals
7	BK/RD	IAC Signal (RSO)	Pulse Signals
8	RD/GN	A/T-ECT Solenoid (SL)	In Lockup: 12-14v
9	LG	A/T-ECT Solenoid (S2)	1st or OD: 1v
10	PK/YL	A/T-ECT Solenoid (S1)	3rd or OD: 1v
11	WT	Injector Pair 2 & 4 Control	1.6-2-9 ms
12	WT/RD	Injector Pair 1 & 3 Control	1.6-2-9 ms
13	BR	Power Ground	<0.1v
14	GN/YL	Circuit Opening Relay (FC)	0-3v, at off-idle: 12v
16	RD/WT	HO2S-12 (B1 S2) Heater	Heater Off: 12v, On: 1v
17	BK/YL	Igniter Signal (IGF)	Digital Signal: 0-5-0v
20	GN	4WD Detection Transfer (L4)	Open: 12v, Closed: 0v
21	YL/RD	A/T Oil Temperature Sensor	At 68°F: 4-5v
22	PK/BK	EGR Solenoid Control (VSV)	12v or 0v
23	BK/BL	Igniter Signal (IGF)	Digital Signal: 0-5-0v
24	BR	Sensor Ground	<0.050v
25	BR	Power Ground	<0.1v
26	BR	Power Ground	<0.1v

1996 2.7L I4 MFI VIN M (A/T-ECT) 12 Pin Connector

PCM Pin #	Wire Color	Circuit Description (12 Pin)	Value at Hot Idle
1	WT/GN	EVAP Purge Solenoid (VSV)	12v or 0v
4	WT/RD	A/T Vehicle Speed Sensor (-)	Pulses
5	BL	Distributor Signal (GN-)	<0.050v
6	GN	CKP Sensor Signal (NE-)	<0.050v
10	YL/RD	A/T Vehicle Speed Sensor (+)	Pulse Signal
11	YL	Distributor Signal (G2+)	AC pulse signals
12	RD	CKP Sensor Signal (NE+)	AC pulse signals

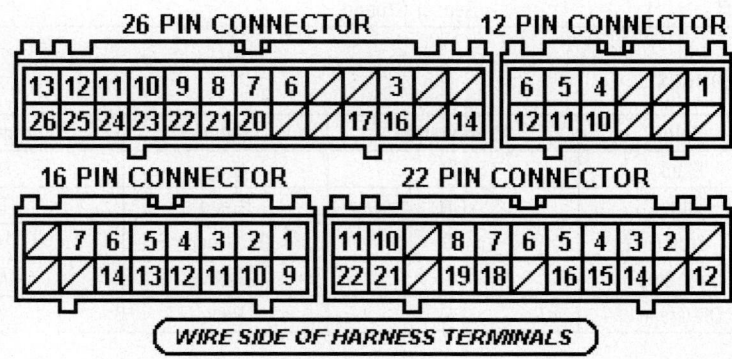

Pin Connector Graphic

05533_ADIA_G180

1996 2.7L I4 MFI 3RZ-FE VIN M (M/T) 22 Pin Connector

PCM Pin #	Wire Color	Circuit Description (22 Pin)	Value at Hot Idle
1	BK/YL	Battery Direct	12-14v
4	GN/OR	A/T Pattern Select Switch	Norm: 0v, PWR: 12v
5	PK	MIL (lamp) Control	MIL Off: 12v, On: 1v
6	BL	Cruise Control ECU (OD1)	At Cruise in OD: 12v
7	WT	Data Link Connector (SDL)	0v
8	BL/BK	A/C Amplifier Signal (ACT)	Clutch On: 12v, Off: 1.5v
9	GN/OR	Vehicle Speed Sensor	At 55 mph: 48 Hz
10	BK/YL	A/C Amplifier Signal (AC1)	Clutch On: 1.5v, Off: 12v
11	BK/WT	Starter Switch Signal	9-11v (cranking)
12	WT/RD	EFI Main Relay (B+)	12-14v
14	GN/YL	Circuit Opening Relay (FC)	0-3v, at off-idle: 12v
17	BL	Overdrive Main Switch	Switch Off: 12v, On: 1v
20	GN/WT	Stop Light Switch Signal	Brake Off: 0v, On: 12v
21	GN/WT	4WD Selector Switch	Switch Off: 12v, On: 1v
22	BK	Neutral Start Switch	9-11v (cranking)

1996 2.7L I4 MFI 3RZ-FE VIN M (M/T) 16 Pin Connector

PCM Pin #	Wire Color	Circuit Description (16 Pin)	Value at Hot Idle
1	GN/YL	Sensor VREF	4.9-5.1v
2	GY/RD	MAF Sensor Signal (VG)	1.1-1.8v
3	PK	EGR Gas Temperature Sensor (THG)	3.5-4.0v
4	GN/RD	ECT Sensor Signal (THW)	At 180°F: 0.51v
5	BK	HO2S-12 (B1 S2) Signal	0.1-1.1v
6	WT	HO2S-11 (B1 S1) Signal	0.1-1.1v
7	YL/GN	IAT Sensor Signal (THA)	At 100°F: 2.60v
9	BR/BK	Sensor Ground	<0.050v
11	YL	TP Sensor Signal (VTA)	0.3-0.8v
12	YL/BL	Closed Throttle Switch	1v, off-idle: 12v
13	BK	Knock Sensor Signal	<0.075v AC
15	PK/WT	Data Link Connector	12-14v
16	BR/WT	Sensor Ground	<0.050v

Standard Colors and Abbreviations

Abbreviation	Color	Abbreviation	Color	Abbreviation	Color
BK	Black	GY	Gray	RD	Red
BL	Blue	GN	Green	TN	Tan
BR	Brown	LG	LT Green	VT	Violet
DB	Dark Blue	OR	Orange	WT	White
DG	DK Green	PK	Pink	YL	Yellow

1996 2.7L I4 MFI 3RZ-FE VIN M (M/T) 26 Pin Connector

PCM Pin #	Wire Color	Circuit Description (26 Pin)	Value at Hot Idle
2	PK/GN	HO2S-11 (B1 S1) Heater	Heater Off: 12v, On: 1v
3	BK/YL	Igniter Signal (IGF)	Digital Signal: 0-5-0v
4	RD	CKP Sensor Signal (NE+)	AC pulse signals
5	YL	Distributor Signal (G2+)	AC pulse signals
6	PK/BK	EGR Solenoid Control (VSV)	12v or 0v
9	BK	IAC Signal (RSC)	Pulse Signals
10	BK/RD	IAC Signal (RSO)	Pulse Signals
11	WT	Injector Pair 2 & 4 Control	1.6-2-9 ms
12	WT/RD	Injector Pair 1 & 3 Control	1.6-2-9 ms
13	BR	Power Ground	<0.1v
14	BR	Shield Ground	<0.050v
15	RD/WT	HO2S-12 (B1 S2) Heater	Heater Off: 12v, On: 1v
17	GN	CKP Sensor Signal (NE-)	<0.050v
18	BL	Distributor Signal (GN-)	<0.050v
20	BK/BL	Igniter Signal (IGF)	Digital Signal: 0-5-0v
23	WT/GN	EVAP Purge Solenoid (VSV)	12v or 0v
25	BR	Power Ground	<0.1v
26	BR	Power Ground	<0.1v

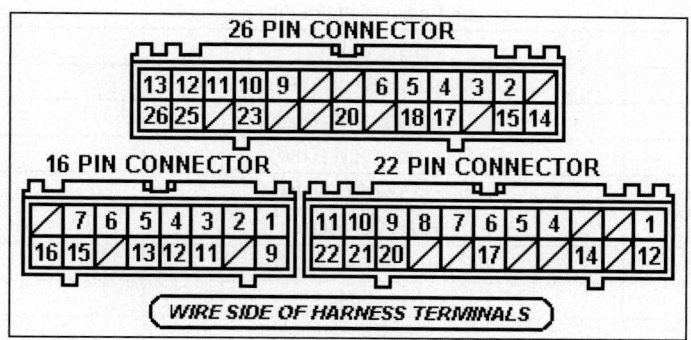

05533_ADIA_G181

Pin Connector Graphic

1997 2.7L I4 MFI VIN M (All) 22 Pin Connector

PCM Pin #	Wire Color	Circuit Description (22 Pin)	Value at Hot Idle
2	BK/YL	Battery Direct	12-14v
4	PK	MIL (lamp) Control	MIL Off: 12v, On: 1v
5	BL/OR	Overdrive Main Switch	Switch Off: 12v, On: 1v
6	BL/BK	A/C Amplifier Signal (ACT)	Clutch On: 12v, Off: 1.5v
7	BL/YL	A/C Amplifier Signal (AC1)	Clutch On: 1.5v, Off: 12v
8	GN/OR	Vehicle Speed Sensor	At 55 mph: 48 Hz
10	GN/WT	4WD Selector Switch	Switch Off: 12v, On: 1v
11	BK/WT	Starter Switch Signal	9-11v (cranking)
12	WT/RD	EFI Main Relay (B+)	12-14v
13	RD/GN	4WD Detection Transfer (N)	Open: 12v, Closed: 0v
15	PK/WT	A/T Select Switch Low	In Low: 12v, Others: 0v
16	PK	A/T Select Switch 2nd	In 2nd: 12v, Others: 0v
17	RD/BK	A/T Select Switch Reverse	In 'R': 12v, Others: 0v
18	RD/YL	Cruise Control ECU (OD1)	At Cruise in OD: 12v
19	WT	Data Link Connector (SDL)	0v
21	GN/WT	Stop Light Switch Signal	Brake Off: 0v, On: 12v
22	BK	Neutral Start Switch	9-11v (cranking)

1997 2.7L I4 MFI VIN M (All) 16 Pin Connector

PCM Pin #	Wire Color	Circuit Description (16 Pin)	Value at Hot Idle
1	GN/YL	Sensor VREF	4.9-5.1v
2	GY/RD	MAF Sensor Signal (VG)	1.1-1.8v
3	YL/GN	IAT Sensor Signal (THA)	At 100°F: 2.60v
4	GN/RD	ECT Sensor Signal (THW)	At 180°F: 0.51v
5	WT	HO2S-11 (B1 S1) Signal	0.1-1.1v
7	PK/WT	Data Link Connector	12-14v
8	GY/GN	EVAP Vapor Pressure (VSV)	12v or 0v
9	BR/BK	Sensor Ground	<0.050v
10	YL	TP Sensor Signal (VTA)	0.3-0.8v
11	PK	EGR Gas Temperature Sensor (THG)	3.5-4.0v
12	BK	Knock Sensor Signal	<0.075v AC
13	BK	HO2S-12 (B1 S2) Signal	0.1-1.1v
15	PK/BK	EGR Solenoid Control (VSV)	12v or 0v

1997 2.7L I4 MFI VIN M (All) 12 Pin Connector

PCM Pin #	Wire Color	Circuit Description (12 Pin)	Value at Hot Idle
2	GN/RD	Taillight Switch Signal	Switch Off: 0v, On: 12v
3	WT/RD	AT Vehicle Speed Signal (-)	Pulse Signals
5	BL	CMP Sensor Signal (GN-)	<0.050v
6	GN	CKP Sensor Signal (NE-)	<0.050v
7	BR/WT	MAF Sensor Ground	<0.050v
9	YL/RD	AT Vehicle Speed Signal (+)	Pulse Signals
10	RD/YL	EVAP Vapor Pressure Sensor	2.5-3.1v (with fuel cap off)
11	YL	CMP Sensor Signal (G2+)	AC pulse signals
12	RD	CKP Sensor Signal (NE+)	AC pulse signals

1997 2.7L I4 MFI VIN M (All) 26 Pin Connector

PCM Pin #	Wire Color	Circuit Description (26 Pin)	Value at Hot Idle
1	RD/WT	HO2S-12 (B1 S2) Heater	Heater Off: 12v, On: 1v
2	PK/GN	HO2S-11 (B1 S1) Heater	Heater Off: 12v, On: 1v
3	WT/GN	EVAP Purge Solenoid (VSV)	12v or 0v
4	BL	Cruise Control ECU (IDLO)	1.5v, off-idle: 12v
6	BK	IAC Signal (RSC)	Pulse Signals
7	BK/RD	IAC Signal (RSO)	Pulse Signals
8	PK/YL	A/T-ECT Solenoid (S1)	3rd or OD: 1v
9	YL/RD	Injector 4 Control	1.6-2-9 ms
10	WT/GN	Injector 3 Control	1.6-2-9 ms
11	WT	Injector 2 Control	1.6-2-9 ms
12	WT/RD	Injector 1 Control	1.6-2-9 ms
13	BR	Power Ground	<0.1v
14	GN/YL	Circuit Opening Relay (FC)	0-3v, at off-idle: 12v
15	BK	Tachometer Signal (TACO)	Pulse Signals
16	YL/RD	A/T Oil Temperature Sensor	At 68°F: 4-5v
17	BK/YL	Igniter Signal (IGF)	Digital Signal: 0-5-0v
18	GN	4WD Detection Transfer (L4)	Open: 12v, Closed: 0v
22	BR/YL	Igniter Transistor 2 Control	Digital Signal: 0-5-0v
23	BK/BL	Igniter Transistor 1 Control	Digital Signal: 0-5-0v
24	BR	Shield Ground	<0.050v
25	BR	Power Ground	<0.1v
26	BR	Power Ground	<0.1v

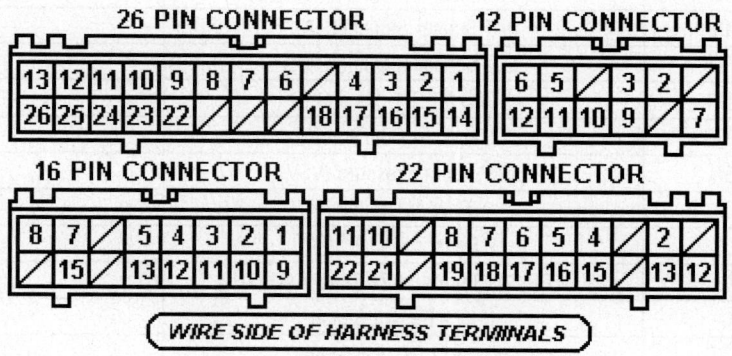

05533_ADIA_G182

Pin Connector Graphic

1998-99 2.7L I4 MFI VIN M (All) 22 Pin Connector

PCM Pin #	Wire Color	Circuit Description (22 Pin)	Value at Hot Idle
2	BK/YL	Battery Direct	12-14v
3	PK	A/T Oil Temperature Indicator	Light On: 0.1v, Off: 12v
4	PK/RD	MIL (lamp) Control	MIL Off: 12v, On: 1v
5	BL/OR	Overdrive Main Switch	Switch Off: 12v, On: 1v
6	BL/BK	A/C Amplifier Signal (ACT)	Clutch On: 12v, Off: 1.5v
7	BL/YL	A/C Amplifier Signal (AC1)	Clutch On: 1.5v, Off: 12v
8	GN/OR	Speedometer Indicator	At 55 mph: 48 Hz
10	RD/WT	4WD Selector Switch	Switch Off: 12v, On: 1v
11	BK/WT	Starter Switch Signal	9-11v (cranking)
12	WT/RD	EFI Main Relay (B+)	12-14v
13	RD/GN	4WD Detection Transfer (N)	Open: 12v, Closed
14	GN	A/T Pattern Select Switch	Norm: 0v, PWR: 12v
15	PK/WT	A/T Select Switch Low	In Low: 12v, Others: 0v
16	PK	A/T Select Switch 2nd	In 2nd: 12v, Others: 0v
17	RD/BK	A/T Select Switch Reverse	In 'R': 12v, Others: 0v
18	RD/YL	Cruise Control ECU (OD1)	At Cruise in OD: 12v
19	WT	SIL (Scan Tool) Signal	12v
21	GN/WT	Stop Light Switch Signal	Brake Off: 0v, On: 12v
22	BK/RD	Start Circuit Signal	Cranking: 9-11v

1998-99 2.7L I4 MFI VIN M (All) 16 Pin Connector

PCM Pin #	Wire Color	Circuit Description (16 Pin)	Value at Hot Idle
1	GN/YL	Sensor VREF	4.9-5.1v
2	GY/RD	MAF Sensor Signal (VG)	1.1-1.8v
3	YL/GN	IAT Sensor Signal (THA)	At 100°F: 2.60v
4	GN/RD	ECT Sensor Signal (THW)	At 180°F: 0.51v
5	WT	HO2S-11 (B1 S1) Signal	0.1-1.1v
7	PK/WT	Data Link Connector	12-14v
8	GY/GN	EVAP Vapor Pressure (VSV)	12v or 0v
9	BR/BK	Sensor Ground	<0.050v
10	YL	TP Sensor Signal (VTA)	0.3-0.8v
11	PK	EGR Gas Temperature Sensor (THG)	3.5-4.0v
12	BK	Knock Sensor Signal	<0.075v AC
13	BK	HO2S-12 (B1 S2) Signal	0.1-1.1v
15	PK/BK	EGR Solenoid Control (VSV)	12v or 0v

1998-99 2.7L I4 MFI VIN M (All) 12 Pin Connector

PCM Pin #	Wire Color	Circuit Description (12 Pin)	Value at Hot Idle
2 ('98)	GN/RD	Taillight Switch Signal	Lights On: 12v
3	WT/RD	AT Vehicle Speed Signal (-)	Pulse Signals
5	BL	CMP Sensor Signal (GN-)	<0.050v
6	GN	CKP Sensor Signal (NE-)	<0.050v
7	BR/WT	MAF Sensor Ground (E2G)	<0.050v
9	YL/RD	AT Vehicle Speed Signal (+)	Pulse Signals
10	RD/YL	EVAP Vapor Pressure Sensor	2.5-3.1v (with fuel cap off)
11	YL	CMP Sensor Signal (G2+)	AC pulse signals
12	RD	CKP Sensor Signal (NE+)	AC pulse signals

1998-99 2.7L I4 MFI VIN M (All) 26 Pin Connector

PCM Pin #	Wire Color	Circuit Description (26 Pin)	Value at Hot Idle
1	RD/WT	HO2S-12 (B1 S2) Heater	Heater Off: 12v, On: 1v
2	PK/GN	HO2S-11 (B1 S1) Heater	Heater Off: 12v, On: 1v
2 ('99)	RD/BK	HO2S-11 (B1 S1) Heater	Heater Off: 12v, On: 1v
3	WT/GN	EVAP Purge Solenoid (VSV)	12v or 0v
4	GN/RD	Cruise Control ECU (IDLO)	1.5v, off-idle: 12v
6	BK	IAC Signal (RSC)	Pulse Signals
7	BK/RD	IAC Signal (RSO)	Pulse Signals
8	PK/YL	A/T-ECT Solenoid (S1)	3rd or OD: 1v
9	RD/BL	Injector 4 Control	1.6-2-9 ms
10	RD	Injector 3 Control	1.6-2-9 ms
11	WT	Injector 2 Control	1.6-2-9 ms
12	WT/RD	Injector 1 Control	1.6-2-9 ms
13	BR	Power Ground	<0.1v
14	GN/YL	Circuit Opening Relay (FC)	0-3v, at off-idle: 12v
15	BK	Tachometer Signal (TACO)	Pulse Signals
16	YL/RD	A/T Oil Temperature Sensor	At 68°F: 4-5v
17	BK/YL	Igniter Signal (IGF)	Digital Signal: 0-5-0v
18	GY	4WD Detection Transfer (L4)	Open: 12v, Closed: 0v
20	RD/GN	A/T-ECT Solenoid (SL)	In Lockup: 12-14v
21	LG	A/T-ECT Solenoid (S1)	3rd or OD: 1v
22	BR/YL	Igniter Transistor 2 Control	Digital Signal: 0-5-0v
23	BK/BL	Igniter Transistor 1 Control	Digital Signal: 0-5-0v
24	BR	Shield Ground	<0.050v
25	BR	Power Ground	<0.1v
26	BR	Power Ground	<0.1v

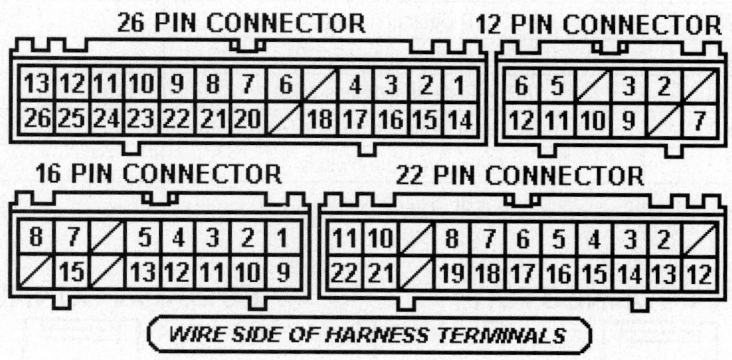

05533_ADIA_G183

Pin Connector Graphic

2000 2.7L I4 VIN M California (All) E5 22P Connector

PCM Pin #	Wire Color	Circuit Description (22 Pin)	Value at Hot Idle
1	BK/YL	Direct Battery	12-14v
2	RD/BK	A/T Select Switch Reverse	In 'R': 12v, Others: 0v
3	PK/WT	A/T Select Switch Low	In Low: 12v, Others: 0v
4	PK	A/T Select Switch 2nd	In 2nd: 12v, Others: 0v
5	PK	A/T Oil Temperature Indicator	Lamp Off: 12v, On: 1v
6	PK/RD	MIL (lamp) Control	MIL Off: 12v, On: 1v
7	BK/WT	Starter Switch Signal	9-11v (cranking)
9	RD/GN	4WD Detection Transfer (N)	Open: 12v, Closed: 0v
10	BL/OR	O/D OFF Indicator (ODLP)	Lamp Off: 12v, On: 1v
11	RD/YL	Cruise Control ECU (OD1)	At Cruise in OD: 12v
12	WT	SIL (Scan Tool) Signal	12v
13	BL/YL	A/C Amplifier Signal (AC1)	Clutch On: 1.5v, Off: 12v
14	BL/BK	A/C Amplifier Signal (ACT)	Clutch On: 12v, Off: 1.5v
15	GN/WT	Stop Light Switch Signal	Brake Off: 0v, On: 12v
16	WT/RD	EFI Main Relay (B+)	12-14v
17	GN	A/T Pattern Select Switch	Norm: 0v, PWR: 12v
19	GY	4WD Detection Transfer (L4)	Open: 12v, Closed: 0v
21	GN/OR	Speedometer Indicator	At 55 mph: 48 Hz
22	BK/RD	Start Circuit Signal	Cranking: 9-11v

2000 2.7L I4 VIN M California (All) E6 28P Connector

PCM Pin #	Wire Color	Circuit Description (28 Pin)	Value at Hot Idle
5	LG/RD	Cruise Control ECU (IDLO)	1.5v, off-idle: 12v
6	GN/YL	Circuit Opening Relay (FC)	0-3v, at off-idle: 12v
7	PK/WT	Data Link Connector	12-14v
8	RD/YL	EVAP Vapor Pressure Sensor	2.5-3.1v (with fuel cap off)
10	RD/WT	4WD Detection Switch	Switch Off: 12v, On: 1v
13	BK	Tachometer Signal (TACO)	Pulse Signals
14	YL/RD	A/T Vehicle Speed Sensor (+)	AC pulse signals
23	WT/RD	A/T Vehicle Speed Sensor (-)	AC pulse signals
25	GN	Overdrive Main Switch	Switch Off: 12v, On: 1v
28	YL	PSP Switch Signal	Straight: 12v, Turning: 0v

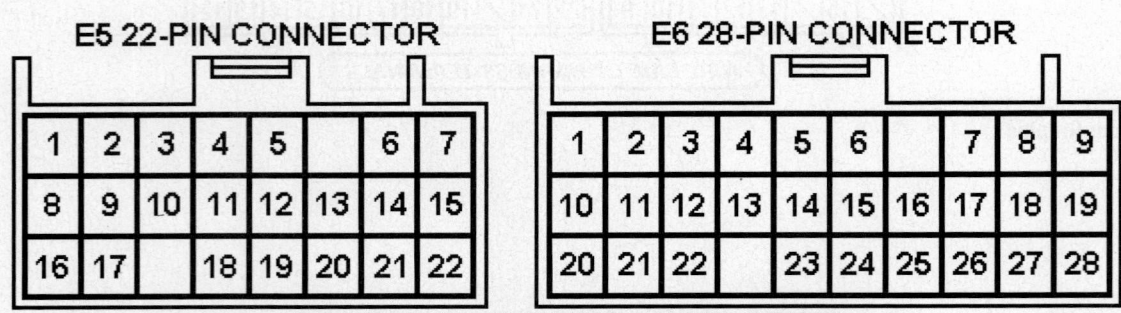

E5 22-PIN CONNECTOR

1	2	3	4	5		6	7
8	9	10	11	12	13	14	15
16	17		18	19	20	21	22

E6 28-PIN CONNECTOR

1	2	3	4	5	6		7	8	9
10	11	12	13	14	15	16	17	18	19
20	21	22		23	24	25	26	27	28

WIRE SIDE OF HARNESS TERMINALS

05533_ADIA_G184

Pin Connector Graphic

2000 2.7L I4 VIN M California (All) E7 24P Connector

PCM Pin #	Wire Color	Circuit Description (24 Pin)	Value at Hot Idle
2	GN/YL	Sensor VREF (VC)	4.9-5.1v
3	RD/WT	HO2S-12 (B1 S2) Heater	Heater Off: 12v, On: 1v
4	WT	AFRS-11 (B1 S1) Heater	Heater Off: 12v, On: 1v
5	PK/BK	EGR Solenoid Control (VSV)	12v or 0v
6	WT/GN	EVAP Purge Solenoid (VSV)	12v or 0v
7	GY/GN	EVAP Vapor Pressure (VSV)	12v or 0v
9	BK/BL	TP Sensor Signal (VTA)	0.3-0.8v
10	BK	HO2S-12 (B1 S2) Signal	0.1-1.1v
11	WT	AFS-11 (B1 S1) Signal (AF+)	Fixed at 3.3v
12	GN/RD	ECT Sensor Signal (THW)	At 180°F: 0.51v
14	GY/RD	MAF Sensor Signal (VG)	1.1-1.8v
15	YL	CMP Sensor Signal	AC pulse signals
16	RD	CKP Sensor Signal	AC pulse signals
17, 18	BR	Sensor Ground	<0.050v
19	PK	EGR Gas Temperature Sensor (THG)	3.5-4.0v
20	RD	AFS-11 (B1 S1) Signal (AF-)	Fixed at 3.3v
21	YL/GN	IAT Sensor Signal (THA)	At 100°F: 2.60v
22	BR/WT	MAF Sensor Ground (E2G)	<0.050v
24	GN	CKP/CMP Sensor Signal (-)	<0.050v

2000 2.7L I4 VIN M California (All) E8 31P Connector

PCM Pin #	Wire Color	Circuit Description (31 Pin)	Value at Hot Idle
1	WT/RD	Injector 1 Control	2.0-3.3 ms
2	WT	Injector 2 Control	2.0-3.3 ms
3	RD	Injector 3 Control	2.0-3.3 ms
4	RD/BL	Injector 4 Control	2.0-3.3 ms
5, 21, 31	BR	Power Ground	<0.1v
6	WT/BK	Sensor Ground	<0.050v
7	PK/YL	A/T-ECT Solenoid (S1)	3rd or OD: 1v
8	LG	A/T-ECT Solenoid (S2)	1st or OD: 1v
9	RD/GN	A/T-ECT Solenoid (SL)	In Lockup: 12-14v
10	BK/YL	Igniter Signal (IGF)	Digital Signal: 0-5-0v
11	BK/BL	Igniter Transistor 1 Control	6°, 55 mph: 8° dwell
12	BK/WT	Igniter Transistor 2 Control	6°, 55 mph: 8° dwell
13	BL/RD	Igniter Transistor 3 Control	6°, 55 mph: 8° dwell
14	BL/YL	Igniter Transistor 4 Control	6°, 55 mph: 8° dwell
15	BK/RD	IAC Control (RSD)	Pulse Signals
26	YL/RD	A/T Oil Temperature Sensor	At 68°F: 4-5v
28	GY	Knock Sensor Signal	<0.075v AC

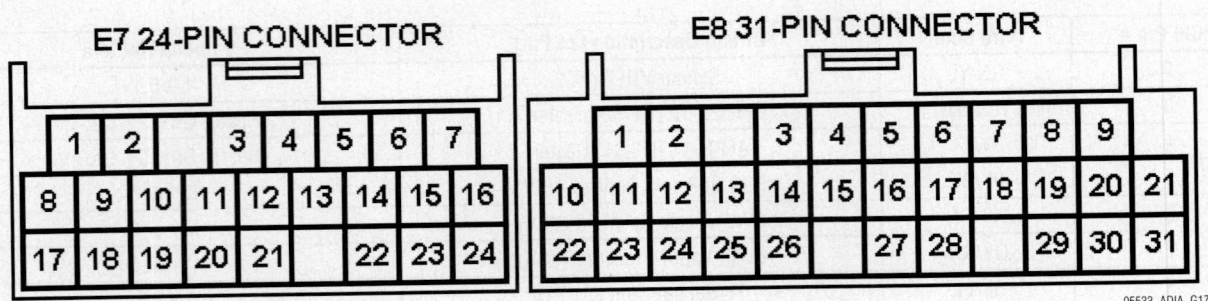

E7 24-PIN CONNECTOR   E8 31-PIN CONNECTOR

05533_ADIA_G178

Pin Connector Graphic
2000 2.7L I4 VIN M Federal (All) E5 22 Pin Connector

PCM Pin #	Wire Color	Circuit Description (22 Pin)	Value at Hot Idle
2	BK/YL	Direct Battery	12-14v
3	PK	A/T Oil Temperature Indicator	Lamp Off: 12v, On: 1v
4	PK/RD	MIL (lamp) Control	MIL Off: 12v, On: 1v
5	BL/OR	O/D OFF Indicator (ODLP)	Lamp Off: 12v, On: 1v
6	BL/BK	A/C Amplifier Signal (ACT)	Clutch On: 12v, Off: 1.5v
7	BL/YL	A/C Amplifier Signal (AC1)	Clutch On: 1.5v, Off: 12v
8	GN/OR	Speedometer Indicator	At 55 mph: 48 Hz
10	RD/WT	4WD Detection Switch	Switch Off: 12v, On: 1v
11	BK/WT	Starter Switch Signal	9-11v (cranking)
12	WT/RD	EFI Main Relay Power	12-14v
13	RD/GN	4WD Detection Transfer (N)	Switch Closed: 0v
14	GN	A/T Pattern Select Switch	Norm: 0v, PWR: 12v
15	PK/WT	A/T Select Switch Low	In Low: 12v, Others: 0v
16	PK	A/T Select Switch 2nd	In 2nd: 12v, Others: 0v
17	RD/BK	A/T Select Switch Reverse	In 'R': 12v, Others: 0v
18	BR/YL	Cruise Control ECU (OD1)	At Cruise in OD: 12v
19	WT	SIL (Scan Tool) Signal	12v
21	GN/WT	Stop Light Switch Signal	Brake Off: 0v, On: 12v
22	BK/RD	Start Circuit Signal	Cranking: 9-11v

2000 2.7L I4 VIN M Federal (All) E6 12 Pin Connector

PCM Pin #	Wire Color	Circuit Description (12 Pin)	Value at Hot Idle
1	GN	Overdrive Main Switch	Switch Off: 12v, On: 1v
3	WT/RD	A/T Vehicle Speed Sensor (-)	AC pulse signals
5	BL	CMP Sensor Signal (GN-)	<0.050v
6	GN	CKP Sensor Signal (NE-)	<0.050v
7	BR/WT	MAF Sensor Ground (E2G)	<0.050v
9	YL/RD	A/T Vehicle Speed Sensor (+)	AC pulse signals
10	RD/YL	EVAP Vapor Pressure Sensor	2.5-3.1v (with fuel cap off)
11	YL	CMP Sensor Signal (G2+)	AC pulse signals
12	RD	CKP Sensor Signal (NE+)	AC pulse signals

E5 22-PIN CONNECTOR

E6 12-PIN CONNECTOR

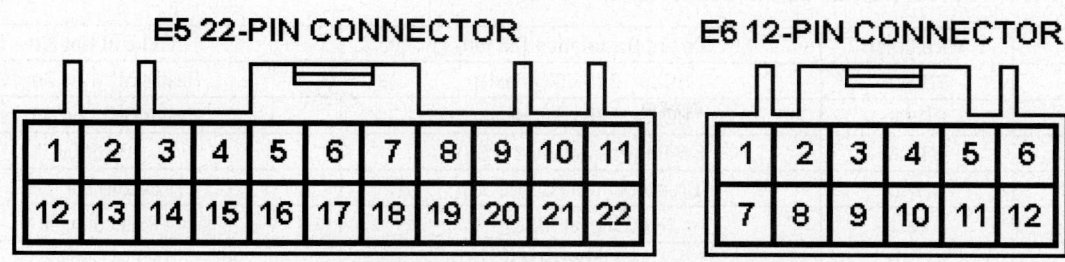

WIRE SIDE OF HARNESS TERMINALS

05533_ADIA_G185

Pin Connector Graphic

2000 2.7L I4 VIN M Federal (All) E7 16 Pin Connector

PCM Pin #	Wire Color	Circuit Description (16 Pin)	Value at Hot Idle
1	GN/YL	Sensor VREF (VC)	4.9-5.1v
2	GY/RD	MAF Sensor Signal (VG)	1.1-1.8v
3	YL/GN	IAT Sensor Signal (THA)	At 100°F: 2.60v
4	GN/RD	ECT Sensor Signal (THW)	At 180°F: 0.51v
5	WT	HO2S-11 (B1 S1) Signal	0.1-1.1v
7	PK/WT	Data Link Connector	12v
8	GY/GN	EVAP Vapor Pressure (VSV)	12v or 0v
9	BR/BK	Sensor Ground	<0.050v
10	YL	TP Sensor Signal (VTA)	0.3-0.8v
11	PK	EGR Gas Temperature Sensor (THG)	3.5-4.0v
12	GY	Knock Sensor Signal	<0.075v AC
13	BK	HO2S-12 (B1 S2) Signal	0.1-1.1v
15	PK/BK	EGR Solenoid Control (VSV)	12v or 0v

2000 2.7L I4 VIN M Federal (All) E8 26 Pin Connector

PCM Pin #	Wire Color	Circuit Description (26 Pin)	Value at Hot Idle
1	RD/WT	HO2S-12 (B1 S2) Heater	Heater Off: 12v, On: 1v
2	RD/BK	HO2S-11 (B1 S1) Heater	Heater Off: 12v, On: 1v
3	WT/GN	EVAP Purge Solenoid (VSV)	12v or 0v
4	LG/RD	Cruise Control ECU (IDLO)	1.5v, off-idle: 12v
6	BK	IAC Signal (RSC)	Pulse Signals
7	BK/RD	IAC Signal (RSO)	Pulse Signals
8	PK/YL	A/T-ECT Solenoid (S1)	3rd or OD: 1v
9	RD/BL	Injector 4 Control	2.0-3.3 ms
10	RD	Injector 3 Control	2.0-3.3 ms
11	WT	Injector 2 Control	2.0-3.3 ms
12	WT/RD	Injector 1 Control	2.0-3.3 ms
13	BR	Power Ground	<0.1v
14	GN/YL	Circuit Opening Relay (FC)	0-3v, at off-idle: 12v
15	BK	Tachometer Signal (TACO)	Pulse Signals
16	YL/RD	A/T Oil Temperature Sensor	At 68°F: 4-5v
17	BK/YL	Igniter Signal (IGF)	Digital Signal: 0-5-0v
18	GY	4WD Detection Transfer (L4)	Open: 12v, Closed: 0v
20	RD/GN	A/T-ECT Solenoid (SL)	In Lockup: 12-14v
21	LG	A/T-ECT Solenoid (S2)	1st or OD: 1v
22	BR/YL	Igniter Transistor 2 Control	6°, 55 mph: 8° dwell
23	BK/BL	Igniter Transistor 1 Control	6°, 55 mph: 8° dwell
24	BR	Sensor Ground	<0.050v
25, 26	BR	Power Ground	<0.1v

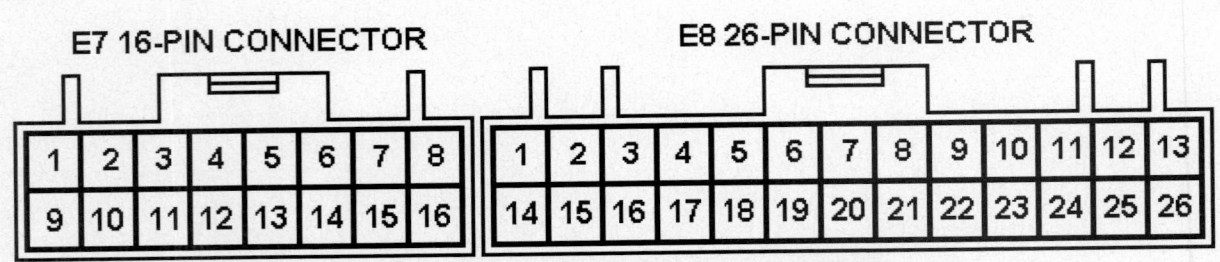

E7 16-PIN CONNECTOR   **E8 26-PIN CONNECTOR**

WIRE SIDE OF HARNESS TERMINALS

05533_ADIA_G177

Pin Connector Graphic

2001-06 2.7L I4 3RZ-FE VIN M (All) E5 22 Pin Connector

PCM Pin #	Wire Color	Circuit Description (22 Pin)	Value at Hot Idle
1	BK/YL	Direct Battery	12-14v
2	RD/BK	A/T Select Switch Reverse	In 'R': 12v, Others: 0v
3	RD	A/T Select Switch Low	In Low: 12v, Others: 0v
4	PK	A/T Select Switch 2nd	In 2nd: 12v, Others: 0v
5	OR	A/T Oil Temperature Indicator	Lamp Off: 12v, On: 1v
6	VT/RD	Malfunction Indicator Lamp (MIL) Control	MIL Off: 12v, On: 1v
7	GN	Starter Switch Signal (STA)	9-11v (cranking)
8	---	Not Used	---
9	BL	A/T P/N Position Lamp (TFN)	Lamp Off: 12v, On: 1v

2001-06 2.7L I4 3RZ-FE VIN M (All) E5 22 Pin Connector, *continued*

10	BL/OR	O/D OFF Indicator (ODLP)	Lamp Off: 12v, On: 1v
11	BR/YL	Cruise Control Signal (OD1)	At Cruise in OD: 12v
12	WT	SIL (Scan Tool) Signal	12v
13	BL/YL	A/C Amplifier Signal (AC1)	Clutch On: 1.5v, Off: 12v
14	BL/BK	A/C Amplifier Signal (ACT)	Clutch On: 12v, Off: 1.5v
15	GN/WT	Stop Light Switch Signal	Brake Off: 0v, On: 12v
16	WT/RD	EFI Main Relay (B+)	12-14v
17	GN	A/T Pattern Select Switch	Norm: 0v, PWR: 12v
18	---	Not Used	---
19	GY	4WD Detection Transfer (L4)	Open: 12v, Closed: 0v
20	GN/RD	Taillight Switch Signal (ELS)	Switch Off: 0v, On: 12v
21	GN/OR	Speedometer Indicator (SP1)	At 55 mph: 48 Hz
22	BK/YL	A/T Neutral Start Switch Signal (NSW)	9-11v (cranking)

2001-06 2.7L I4 3RZ-FE VIN M (All) E6 28 Pin Connector

PCM Pin #	Wire Color	Circuit Description (28 Pin)	Value at Hot Idle
1-4	---	Not Used	---
5	LG/RD	Cruise Control ECU (IDLO)	1.5v, off-idle: 12v
6	WT/BL	Circuit Opening Relay (FC)	0-3v, at off-idle: 12v
7	RD	DLC3 D7 Signal (TC)	12v
8	RD/YL	EVAP Vapor Pressure Sensor (PTNK)	2.5-3.1v (with fuel cap off)
9	---	Not Used	---
10	GNBK	4WD Switch Signal (4WD)	4WD Switch On: 12v
11-12	---	Not Used	---
13	LG/BK	Tachometer Signal (TACO)	Pulse Signals
14	YL/RD	A/T Vehicle Speed Sensor (+)	AC pulse signals
15-17	---	Not Used	---
18	GN/BK	4WD ADD Detection Switch	Open: 12v, Closed: 0v
19	---	Not Used	---
20-22	---	Not Used	---
23	WT/RD	A/T Vehicle Speed Sensor (-)	AC pulse signals
24	---	Not Used	---
25	GN	OD Main Switch (ODMS)	Switch Off: 12v, On: 1v
26-27	---	Not Used	---
28	YL	PSP Switch Signal (PSW)	Straight: 12v, Turning: 0v

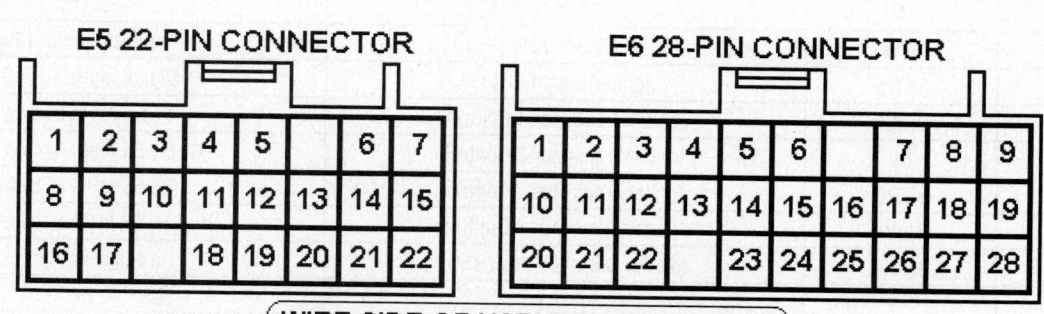

E5 22-PIN CONNECTOR   E6 28-PIN CONNECTOR

WIRE SIDE OF HARNESS TERMINALS

Pin Connector Graphic

05533_ADIA_G184

2001-06 2.7L I4 3RZ-FE VIN M (All) E7 24 Pin Connector

PCM Pin #	Wire Color	Circuit Description (24 Pin)	Value at Hot Idle
1	RD/BL	EVAP Closed Canister Vent (VSV)	12v or 0v
2	GN/YL	Sensor VREF (VC)	4.9-5.1v
3	RD/WT	HO2S-12 (B1 S2) Heater (HTS)	Heater Off: 12v, On: 1v
4	WT	AFS-11 (B1 S1) Heater (AFHT)	Heater Off: 12v, On: 1v
5	RD/BK	EGR Solenoid Control (VSV)	12v or 0v
6	WT/GN	EVAP Purge Solenoid (VSV)	12v or 0v
7	GN/BK	EVAP Vapor Pressure (VSV)	12v or 0v
8	---	Not Used	---
9	YL	TP Sensor Signal (VTA)	0.3-0.8v
10	BK	HO2S-12 (B1 S2) Signal (OXS)	0.1-1.1v
11	VT	AFS-11 (B1 S1) Signal (AF+)	Fixed at 3.3v
12	GN/RD	ECT Sensor Signal (THW)	At 180°F: 0.51v
13	---	Not Used	---
14	GY	MAF Sensor Signal (VG)	1.1-1.8v
15	RD	CMP Sensor Signal (G2+)	AC pulse signals
16	BL	CKP Sensor Signal (NE+)	AC pulse signals
17	BR	Power Ground (E1)	<0.050v
18	BL/BK	Sensor Ground (E2)	<0.050v
19	PK/BL	EGR Temperature Sensor (THG)	3.5-4.0v
20	PK	AFS-11 (B1 S1) Signal (AF-)	Fixed at 3.3v
21	YL/GN	IAT Sensor Signal (THA)	At 100°F: 2.60v
22	BK/WT	MAF Sensor Ground (EVG)	<0.050v
23	---	Not Used	---
24	GN	CKP/CMP Sensor Signal (NE-)	<0.050v

2001-06 2.7L I4 3RZ-FE VIN M (All) E8 31 Pin Connector

PCM Pin #	Wire Color	Circuit Description (31 Pin)	Value at Hot Idle
1	WT/RD	Injector 1 Control	2.0-3.3 ms
2	WT	Injector 2 Control	2.0-3.3 ms
3	RD	Injector 3 Control	2.0-3.3 ms
4	RD/BL	Injector 4 Control	2.0-3.3 ms
5	WT/BK	Power Ground (E03)	<0.1v
6	WT/BK	Power Ground (E04)	<0.1v
7	VT	A/T-ECT Solenoid (S1)	3rd or OD: 1v
8	LG	A/T-ECT Solenoid (S2)	1st or OD: 1v
9	RD/WT	A/T-ECT Solenoid (SL)	In Lockup: 12-14v
10	BK/YL	Igniter Signal (IGF)	Digital Signal: 0-5-0v
11	BK/BL	Igniter Transistor 1 Control	6°, 55 mph: 8° dwell
12	BL	Igniter Transistor 2 Control	6°, 55 mph: 8° dwell
13	BL/RD	Igniter Transistor 3 Control	6°, 55 mph: 8° dwell
14	BL/YL	Igniter Transistor 4 Control	6°, 55 mph: 8° dwell
15	BK/RD	Idle Air Control Valve (RSD)	Pulse Signals
16-20	---	Not Used	---
21	WT/BK	Power Ground (E01)	<0.1v
22-25	---	Not Used	---
26	YL/RD	A/T Oil Temperature Sensor	At 68°F: 4-5v

2001-06 2.7L I4 3RZ-FE VIN M (All) E8 31 Pin Connector, *continued*

27	---	Not Used	---
28	GY	Knock Sensor Signal (KNK)	<0.075v AC
29-30	---	Not Used	---
31	WT/BK	Power Ground (E02)	<0.1v

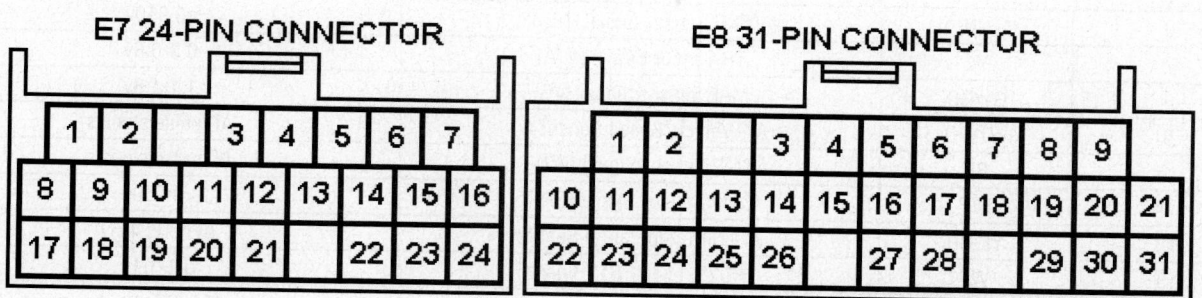

05533_ADIA_G178

Pin Connector Graphic

1995 3.4L V6 MFI VIN V (A/T-ECT) 28 Pin Connector

PCM Pin #	Wire Color	Circuit Description (28 Pin)	Value at Hot Idle
1	---	Not Used	---
2	PK	A/T Select Switch 2nd	In 2nd: 12v, Others: 0v
3	PK/WT	A/T Select Switch Low	In Low: 12v, Others: 0v
4	PK	A/T Oil Temperature Lamp	Lamp Off: 12v, On: 1v
5	BL/BK	A/C Amplifier Signal (ACT)	Clutch On: 12v, Off: 1.5v
6	BL/OR	Overdrive Main Switch	Switch Off: 12v, On: 1v
7	BL/YL	Cruise Control ECU	At Cruise in OD: 12v
10	GN/OR	A/T Pattern Select Switch	Norm: 0v, PWR: 12v
12	GN/OR	Vehicle Speed Sensor	At 55 mph: 48 Hz
14	BK/YL	Battery Direct	12-14v
17	RD/GN	4WD Detection Transfer (N)	Open: 12v, Closed: 0v
18	WT	Data Link Connector	12-14v
20	BL/BK	A/C Amplifier Signal (AC1)	Clutch On: 1.5v, Off: 12v
22	WT/RD	EFI Main Relay (B+)	12-14v
25	GN/WT	Stop Light Switch Signal	Brake Off: 0v, On: 12v
26	GN/WT	4WD Selector Switch	Open: 12v, Closed: 0v

1995 3.4L V6 MFI VIN V (A/T-ECT) 22 Pin Connector

PCM Pin #	Wire Color	Circuit Description (22 Pin)	Value at Hot Idle
1	GN/BK	Sensor VREF	4.9-5.1v
2-3	---	Not Used	---
4	WT/RD	A/T Vehicle Speed Sensor (+)	AC pulse signals
5	RD	CKP Sensor Signal (NE+)	AC pulse signals
6	GN	CKP Sensor Signal (NE-)	<0.050v
7	YL	TP Sensor Signal (VTA)	0.3-0.8v
8	GY/RD	MAF Sensor Signal (VG)	1.1-1.8v
9	WT/RD	A/T Vehicle Speed Sensor (-)	AC pulse signals
10	BK	CMP Sensor Signal (G2+)	AC pulse signals
11	WT	CMP Sensor Signal (GN-)	<0.050v
12	YL/RD	A/T Oil Temperature Sensor	At 68°F: 4-5v
13	WT	HO2S-11 (B1 S1) Signal	0.1-1.1v
14	YL/GN	IAT Sensor Signal (THA)	At 100°F: 2.60v
16	GY	Knock Sensor 2 Signal	<0.075v AC
17	BK	Knock Sensor 1 Signal	<0.075v AC
18, 22	BR/WT	Sensor Ground	<0.050v
19	BK	HO2S-12 (B1 S2) Signal	0.1-1.1v
20	GN/RD	ECT Sensor Signal (THW)	At 180°F: 0.51v
21	PK/GN	EGR Gas Temperature Sensor (THG)	3.5-4.0v

Standard Colors and Abbreviations

Abbreviation	Color	Abbreviation	Color	Abbreviation	Color
BK	Black	GY	Gray	RD	Red
BL	Blue	GN	Green	TN	Tan
BR	Brown	LG	LT Green	VT	Violet
DB	Dark Blue	OR	Orange	WT	White
DG	DK Green	PK	Pink	YL	Yellow

1995 3.4L V6 MFI VIN V (A/T-ECT) 16 Pin Connector

PCM Pin #	Wire Color	Circuit Description (16 Pin)	Value at Hot Idle
1-2	---	Not Used	---
3	PK	MIL (lamp) Control	MIL Off: 12v, On: 1v
4	GN/YL	Circuit Opening Relay (FC)	0-3v, at off-idle: 12v
5	PK/WT	Data Link Connector	12-14v
6-7	---	Not Used	---
8	RD/WT	EGR Solenoid Control (VSV)	12v or 0v
9	RD/BK	Fuel Pressure Up Solenoid	1v (at hot startup)
10	BK/RD	A/C Idle-Up Solenoid	A/C Off: 12v, On: 1v
11-14	---	Not Used	---
15	WT/GN	EVAP Purge Solenoid (VSV)	12v or 0v
16	BR	Sensor Ground	<0.050v

1995 3.4L V6 MFI VIN V (A/T-ECT) 34-Pin Connector

PCM Pin #	Wire Color	Circuit Description (34-Pin)	Value at Hot Idle
1	BR	Sensor Ground	<0.050v
2-4	---	Not Used	---
5	YL/BK	Injector 6 Control	1.6-2.9 ms
6	WT/BL	Injector 5 Control	1.6-2.9 ms
7	YL/RD	Injector 4 Control	1.6-2.9 ms
8	WT/GN	Injector 3 Control	1.6-2.9 ms
9	WT	Injector 2 Control	1.6-2.9 ms
10	WT/RD	Injector 1 Control	1.6-2.9 ms
11	PK/YL	A/T-ECT Solenoid (S1)	3rd or OD: 1v
12	BK/YL	Igniter Signal (IGF)	Digital Signal: 0-5-0v
13	BK/WT	Starter Switch Signal	9-11v (cranking)
14	BK	Neutral Start Switch	In P/N: 9-11v (cranking)
15	RD/WT	HO2S-12 (B1 S2) Heater	Heater Off: 12v, On: 1v
16	PK/GN	HO2S-11 (B1 S1) Heater	Heater Off: 12v, On: 1v
17	LG	A/T-ECT Solenoid (S2)	1st or OD: 1v
18-21	---	Not Used	---
22	BK/RD	IAC Signal (RSC)	Pulse Signals
23	BR/RD	IAC Signal (RSO)	Pulse Signals
24	BK/BL	Igniter Transistor 1 Control	Digital Signal: 0-5-0v
25	BR/BK	Igniter Transistor 2 Control	Digital Signal: 0-5-0v
26	BK/WT	Igniter Transistor 3 Control	Digital Signal: 0-5-0v
27	RD/GN	A/T-ECT Solenoid (SL)	In Lockup: 12-14v
29	GN	4WD Detection Transfer (L4)	Open: 12v, Closed: 0v
30-31	---	Not Used	---
32	YL/BL	Closed Throttle Switch	1v, off-idle: 12v
33	BR	Power Ground	<0.1v
34	BR	Power Ground	<0.1v

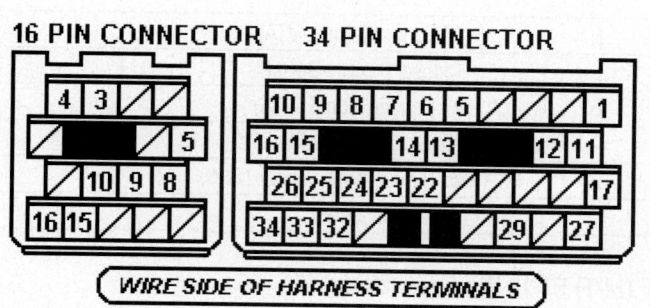

Pin Connector Graphic

05533_ADIA_G186

1995 3.4L V6 MFI VIN V 5VZ-FE (M/T) E5 22 Pin Connector

PCM Pin #	Wire Color	Circuit Description (22 Pin)	Value at Hot Idle
1	---	Not Used	---
2	BK/YL	Battery Direct	12-14v
3	---	Not Used	---
4	PK	MIL (lamp) Control	MIL Off: 12v, On: 1v
5	---	Not Used	---
6	BL/BK	A/C Amplifier Signal (ACT)	Clutch On: 12v, Off: 1.5v
7	BK/YL	A/C Amplifier Signal (AC1)	Clutch On: 1.5v, Off: 12v
8	GN/OR	Vehicle Speed Sensor	At 55 mph: 48 Hz
9	GN/WT	4WD Selector Switch	Open: 12v, Closed: 0v
10	---	Not Used	---
11	BK/WT	Starter Switch Signal	9-11v (cranking)
12	WT/RD	EFI Main Relay (B+)	12-14v
13-18	---	Not Used	---
19	WT	Data Link Connector (SDL)	0v
20	GN/WT	Stop Light Switch Signal	Brake Off: 0v, On: 12v
21-22	---	Not Used	---

1995 3.4L V6 MFI VIN V 5VZ-FE (M/T) E7 16 Pin Connector

PCM Pin #	Wire Color	Circuit Description (16 Pin)	Value at Hot Idle
1	GN/BK	Sensor VREF (VC)	4.9-5.1v
2	GY/RD	MAF Sensor Signal (VG)	1.1-1.8v
3	GY	Knock Sensor 2 Signal	<0.075v AC
4	GN/RD	ECT Sensor Signal (THW)	At 180°F: 0.51v
5	WT	HO2S-11 (B1 S1) Signal	0.1-1.1v
6	BK	Knock Sensor 1 Signal	<0.075v AC
7	RD	Data Link Connector	12-14v
8	BR/WT	Sensor Ground	<0.050v
9	BR/YL	Sensor Ground	<0.050v
10	YL	TP Sensor Signal (VTA)	0.3-0.8v
11	YL/BL	Closed Throttle Switch	1v, off-idle: 12v
12	YL/GN	IAT Sensor Signal (THA)	At 100°F: 2.60v
13	BK	HO2S-12 (B1 S2) Signal	0.1-1.1v
14	PK/GN	EGR Gas Temperature Sensor (THG)	3.5-4.0v
15-16	---	Not Used	---

1995 3.4L V6 MFI VIN V 5VZ-FE (M/T) E7 12 Pin Connector

PCM Pin #	Wire Color	Circuit Description (12 Pin)	Value at Hot Idle
1-2	---	Not Used	---
3	PK/GN	HO2S-11 (B1 S1) Heater	Heater Off: 12v, On: 1v
4	---	Not Used	---
5	BL	CMP Sensor Signal (GN-)	<0.050v
6	GN	CKP Sensor Signal (NE-)	<0.050v
7	BR	Sensor Ground	<0.050v
8	---	Not Used	---
9	RD/GN	HO2S-12 (B1 S2) Heater	Heater Off: 12v, On: 1v

1995 3.4L V6 MFI VIN V 5VZ-FE (M/T) E7 12 Pin Connector, *continued*

10	---	Not Used	---
11	YL	CMP Sensor Signal (G2+)	AC pulse signals
12	RD	CKP Sensor Signal (NE+)	AC pulse signals

1995 3.4L V6 MFI VIN V 5VZ-FE (M/T) E8 26 Pin Connector

PCM Pin #	Wire Color	Circuit Description (26 Pin)	Value at Hot Idle
2	BK/RD	A/C Idle-Up Solenoid	A/C Off: 12v, On: 1v
5	WT/GN	EVAP Purge Solenoid (VSV)	12v or 0v
6	BK/RD	IAC Signal (RSC)	Pulse Signals
7	BR/RD	IAC Signal (RSO)	Pulse Signals
8	YL/BK	Injector 6 Control	1.6-2.9 ms
9	WT/BL	Injector 5 Control	1.6-2.9 ms
10	YL/RD	Injector 4 Control	1.6-2.9 ms
11	WT	Injector 2 Control	1.6-2.9 ms
12	WT/RD	Injector 1 Control	1.6-2.9 ms
13	BR	Power Ground	<0.1v
14	GN/YL	Circuit Opening Relay (FC)	0-3v, at off-idle: 12v
17	BK/YL	Igniter Signal (IGF)	Digital Signal: 0-5-0v
18	RD/WT	EGR Solenoid Control (VSV)	12v or 0v
19	RD/BK	Fuel Pressure Up Solenoid	1v (at hot startup)
21	BK/WT	Igniter Transistor 3 Control	Digital Signal: 0-5-0v
22	BR/YL	Igniter Transistor 2 Control	Digital Signal: 0-5-0v
23	BK/BL	Igniter Transistor 1 Control	Digital Signal: 0-5-0v
24	BR	Power Ground	<0.1v
25	WT/GN	Injector 3 Control	1.6-2.9 ms
26	BR	Power Ground	<0.1v

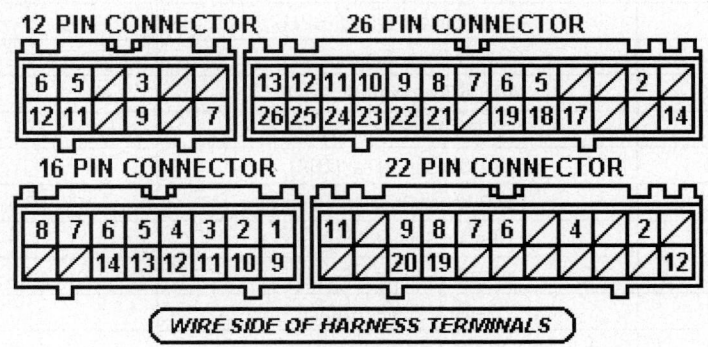

Pin Connector Graphic

05533_ADIA_G187

1996 3.4L V6 MFI VIN N (A/T-ECT) E5 28 Pin Connector

PCM Pin #	Wire Color	Circuit Description (28 Pin)	Value at Hot Idle
1	RD/BK	A/T Select Switch Reverse	In 'R': 12v, Others: 0v
2	PK	A/T Select Switch 2nd	In 2nd: 12v, Others: 0v
3	PK/WT	A/T Select Switch Low	In Low: 12v, Others: 0v
4	PK	A/T Oil Temperature Lamp	Lamp Off: 12v, On: 1v
5	BL/BK	A/C Amplifier Signal (ACT)	Clutch On: 12v, Off: 1.5v
6	BL/OR	Overdrive Main Switch	Open: 12v, Closed: 0v
7	BL	Cruise Control ECU	At Cruise in OD: 12v
10	GN/OR	A/T Pattern Select Switch	Norm: 0v, PWR: 12v
12	GN/OR	Vehicle Speed Sensor	At 55 mph: 48 Hz
14	BK/YL	Battery Direct	12-14v
17	RD/GN	4WD Detection Transfer (N)	Open: 12v, Closed: 0v
18	WT	Data Link Connector	12-14v
20	BL/YL	A/C Amplifier Signal (AC1)	Clutch On: 1.5v, Off: 12v
22	WT/RD	EFI Main Relay (B+)	12-14v
25	GN/WT	Stop Light Switch Signal	Brake Off: 0v, On: 12v
26	GN/WT	4WD Selector Switch	Clutch On: 1.5v, Off: 12v

1996 3.4L V6 MFI VIN N (A/T-ECT) E7 22 Pin Connector

PCM Pin #	Wire Color	Circuit Description (22 Pin)	Value at Hot Idle
1	GN/BK	Sensor VREF	4.9-5.1v
4	WT/RD	AT Vehicle Speed Signal (-)	AC pulse signals
5	RD	CKP Sensor Signal (NE+)	AC pulse signals
6	GN	CKP Sensor Signal (NE-)	<0.050v
7	YL	TP Sensor Signal (VTA)	0.3-0.8v
8	GY/RD	MAF Sensor Signal (VG)	1.1-1.8v
9	YL/RD	AT Vehicle Speed Signal (+)	AC pulse signals
10	BK	CMP Sensor Signal (G2+)	AC pulse signals
11	WT	CMP Sensor Signal (GN-)	<0.050v
12	YL/RD	A/T Oil Temperature Sensor	At 68°F: 4-5v
13	WT	HO2S-11 (B1 S1) Signal	0.1-1.1v
14	YL/GN	IAT Sensor Signal (THA)	At 100°F: 2.60v
15	RD/YL	EVAP Vapor Pressure Sensor	2.5-3.1v (with fuel cap off)
16	GY	Knock Sensor 2 Signal	<0.075v AC
17	BK	Knock Sensor 1 Signal	<0.075v AC
18, 22	BR/WT	Sensor, Power Ground	<0.050v
19	BK	HO2S-12 (B1 S2) Signal	0.1-1.1v
20	GN/RD	ECT Sensor Signal (THW)	At 180°F: 0.51v
21	PK/GN	EGR Gas Temperature Sensor (THG)	3.5-4.0v

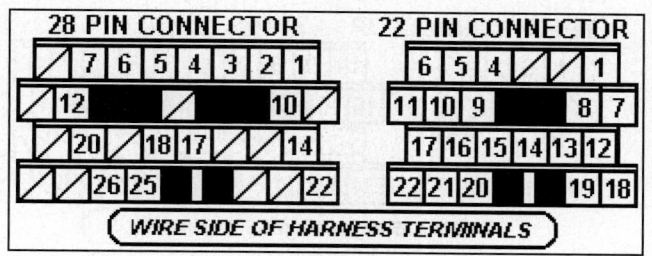

28 PIN CONNECTOR   22 PIN CONNECTOR

WIRE SIDE OF HARNESS TERMINALS

05533_ADIA_G188

Pin Connector Graphic

1996 3.4L V6 MFI VIN N (A/T-ECT) E6 16 Pin Connector

PCM Pin #	Wire Color	Circuit Description (16 Pin)	Value at Hot Idle
1-2, 6-7	---	Not Used	---
3	PK	MIL (lamp) Control	MIL Off: 12v, On: 1v
4	GN/YL	Circuit Opening Relay (FC)	0-3v, at off-idle: 12v
5	PK/WT	Data Link Connector	12v
8	RD/WT	EGR Solenoid Control (VSV)	12v or 0v
9	RD/BK	Fuel Pressure Up Solenoid	1v (at hot startup)
10	BK/RD	A/C Idle-Up Solenoid	A/C Off: 12v, On: 1v
11-12, 14	---	Not Used	---
13	WT/GN	EVAP Purge Solenoid (VSV)	12v or 0v
15	WT/GN	EVAP Vapor Pressure Sensor	2.5-3.1v (with fuel cap off)
16	BR	Shield Ground	<0.050v

1996 3.4L V6 MFI VIN N (A/T-ECT) E8 34-Pin Connector

PCM Pin #	Wire Color	Circuit Description (34-Pin)	Value at Hot Idle
1	BR	Power Ground	<0.1v
5	YL/BK	Injector 6 Control	1.6-2.9 ms
6	WT/BL	Injector 5 Control	1.6-2.9 ms
7	YL/RD	Injector 4 Control	1.6-2.9 ms
8	WT/GN	Injector 3 Control	1.6-2.9 ms
9	WT	Injector 2 Control	1.6-2.9 ms
10	WT/RD	Injector 1 Control	1.6-2.9 ms
11	PK/YL	A/T-ECT Solenoid (S1)	3rd or OD: 1v
12	BK/YL	Igniter Signal (IGF)	Digital Signal: 0-5-0v
13	BK/WT	Starter Switch Signal	9-11v (cranking)
14	BK	Neutral Start Switch	In P/N: 9-11v (cranking)
15	RD/WT	HO2S-12 (B1 S2) Heater	Heater Off: 12v, On: 1v
16	PK/GN	HO2S-11 (B1 S1) Heater	Heater Off: 12v, On: 1v
17	GN	A/T-ECT Solenoid (S2)	1st or OD: 1v
22	BK/RD	IAC Signal (RSC)	Pulse Signals
23	BR/RD	IAC Signal (RSO)	Pulse Signals
24	BK/BL	Igniter Transistor 1 Control	Digital Signal: 0-5-0v
25	BR/YL	Igniter Transistor 2 Control	Digital Signal: 0-5-0v
26	BK/WT	Igniter Transistor 3 Control	Digital Signal: 0-5-0v
27	RD/GN	A/T-ECT Solenoid (SL)	In Lockup: 12-14v
29	GN	4WD Detection Transfer (L4)	Open: 12v, Closed: 0v
32	YL/BL	Closed Throttle Switch	1v, off-idle: 12v
33	BR	Power Ground	<0.1v
34	BR	Power Ground	<0.1v

16 PIN CONNECTOR   34 PIN CONNECTOR

```
16 PIN CONNECTOR                34 PIN CONNECTOR
 4  3 //                   10 9 8 7 6 5 // // //      1
// ///////// 5             16 15 ▓▓▓ 14 13 ▓▓ 12 11
 // 10  9  8               26 25 24 23 22 // // //    17
16 15 // 13                34 33 32 // ▌ // 29 // 27
```

WIRE SIDE OF HARNESS TERMINALS

05533_ADIA_G189

Pin Connector Graphic
1996 3.4L V6 MFI 5VZ-FE VIN N (M/T) E5 22 Pin Connector

PCM Pin #	Wire Color	Circuit Description (22 Pin)	Value at Hot Idle
1, 3, 5	---	Not Used	---
2	BK/YL	Battery Direct	12-14v
4	PK	MIL (lamp) Control	MIL Off: 12v, On: 1v
6	BL/BK	A/C Amplifier Signal (ACT)	Clutch On: 12v, Off: 1.5v
7	BL/YL	A/C Amplifier Signal (AC1)	Clutch On: 1.5v, Off: 12v
8	GN/OR	Vehicle Speed Sensor	At 55 mph: 48 Hz
9	GN/WT	4WD Selector Switch	Open: 12v, Closed: 0v
10, 13-18	---	Not Used	---
11	BK/WT	Starter Switch Signal	9-11v (cranking)
12	WT/RD	EFI Main Relay (B+)	12-14v
19	WT	Data Link Connector (SDL)	0v
20	GN/WT	Stop Light Switch Signal	Brake Off: 0v, On: 12v
21-22	---	Not Used	---

1996 3.4L V6 MFI 5VZ-FE VIN N (M/T) E7 16 Pin Connector

PCM Pin #	Wire Color	Circuit Description (16 Pin)	Value at Hot Idle
1	GN/BK	Sensor VREF	4.9-5.1v
2	GY/RD	MAF Sensor Signal (VG)	1.1-1.8v
3	GY	Knock Sensor 2 Signal	<0.075v AC
4	GN/RD	ECT Sensor Signal (THW)	At 180°F: 0.51v
5	WT	HO2S-11 (B1 S1) Signal	0.1-1.1v
6	BK	Knock Sensor 1 Signal	<0.075v AC
7	RD	Data Link Connector	12v
8	BR/WT	MAF Sensor Ground (EVG)	<0.050v
9	BR/BK	Sensor Ground	<0.050v
10	YL	TP Sensor Signal (VTA)	0.3-0.8v
11	YL/BL	Closed Throttle Switch	1v, off-idle: 12v
12	YL/GN	IAT Sensor Signal (THA)	At 100°F: 2.60v
13	BK	HO2S-12 (B1 S2) Signal	0.1-1.1v
14	PK/GN	EGR Gas Temperature Sensor (THG)	3.5-4.0v
15-16	---	Not Used	---

1996 3.4L V6 MFI 5VZ-FE VIN N (M/T) E6 12 Pin Connector

PCM Pin #	Wire Color	Circuit Description (12 Pin)	Value at Hot Idle
1-2	---	Not Used	---
3	PK/GN	HO2S-11 (B1 S1) Heater	Heater Off: 12v, On: 1v
4	RD/YL	EVAP Vapor Pressure Sensor	2.5-3.1v (with fuel cap off)

1996 3.4L V6 MFI 5VZ-FE VIN N (M/T) E6 12 Pin Connector, *continued*

5	BL	CMP Sensor Signal (GN-)	<0.050v
6	GN	CKP Sensor Signal (NE-)	<0.050v
7	BR	Power Ground	<0.1v
8	---	Not Used	---
9	RD/GN	HO2S-12 (B1 S2) Heater	Heater Off: 12v, On: 1v
11	YL	CMP Sensor Signal (G2+)	AC pulse signals
12	RD	CKP Sensor Signal (NE+)	AC pulse signals

1996 3.4L V6 MFI 5VZ-FE VIN N (M/T) E7 26 Pin Connector

PCM Pin #	Wire Color	Circuit Description (26 Pin)	Value at Hot Idle
2	BK/RD	A/C Idle-Up Solenoid	A/C Off: 12v, On: 1v
3	WT/GN	EVAP Purge Solenoid (VSV)	12v or 0v
5	GY/GN	EVAP Vapor Pressure (VSV)	12v or 0v
6	BK/RD	IAC Signal (RSC)	Pulse Signals
7	BR/RD	IAC Signal (RSO)	Pulse Signals
8	YL/BK	Injector 6 Control	1.6-2.9 ms
9	WT/BL	Injector 5 Control	1.6-2.9 ms
10	YL/RD	Injector 4 Control	1.6-2.9 ms
11	WT	Injector 2 Control	1.6-2.9 ms
12	WT/RD	Injector 1 Control	1.6-2.9 ms
13	BR	Power Ground	<0.1v
14	GN/YL	Circuit Opening Relay (FC)	0-3v, at off-idle: 12v
17	BK/YL	Igniter Signal (IGF)	Digital Signal: 0-5-0v
18	RD/WT	EGR Solenoid Control (VSV)	12v or 0v
19	RD/BK	Fuel Pressure Up Solenoid	1v (at hot startup)
21	BK/WT	Igniter Transistor 3 Control	Digital Signal: 0-5-0v
22	BR/YL	Igniter Transistor 2 Control	Digital Signal: 0-5-0v
23	BK/BL	Igniter Transistor 1 Control	Digital Signal: 0-5-0v
24	BR	Shield Ground	<0.050v
25	WT/GN	Injector 3 Control	1.6-2.9 ms
26	BR	Power Ground	<0.1v

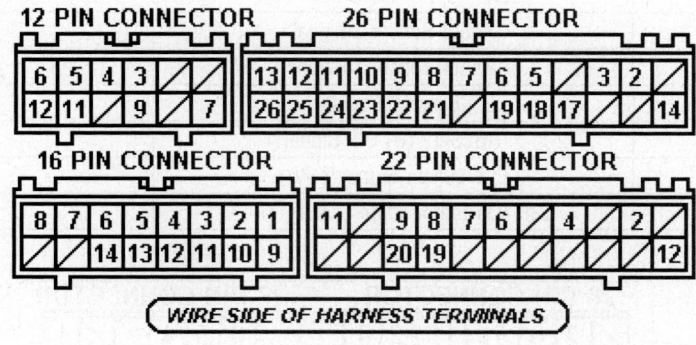

Pin Connector Graphic

05533_ADIA_G190

1997 3.4L V6 MFI VIN N (A/T-ECT) 28 Pin Connector

PCM Pin #	Wire Color	Circuit Description (28 Pin)	Value at Hot Idle
1	RD/BK	A/T Select Switch Reverse	In 'R': 12v, Others: 0v
2	PK	A/T Select Switch 2nd	In 2nd: 12v, Others: 0v
3	PK/WT	A/T Select Switch Low	In Low: 12v, Others: 0v
4	PK	A/T Oil Temperature Lamp	Lamp Off: 12v, On: 1v
5	BL/BK	A/C Amplifier Signal (ACT)	Clutch On: 12v, Off: 1.5v
6	BL/OR	Overdrive Main Switch	Switch Off: 12v, On: 1v
7	BL	Cruise Control ECU	At Cruise in OD: 12v
10	GN/OR	A/T Pattern Select Switch	Norm: 0v, PWR: 12v
12	GN/OR	Vehicle Speed Sensor	At 55 mph: 48 Hz
14	BK/YL	Battery Direct	12-14v
17	RD/GN	4WD Detection Transfer (N)	Open: 12v, Closed
18	WT	SIL (Scan Tool) Signal	12v
20	BL/YL	A/C Amplifier Signal (AC1)	Clutch On: 1.5v, Off: 12v
22	WT/RD	EFI Main Relay (B+)	12-14v
25	GN/WT	Stop Light Switch Signal	Brake Off: 0v, On: 12v
26	GN/WT	4WD Selector Switch	Switch Off: 12v, On: 1v

1997 3.4L V6 MFI VIN N (A/T-ECT) 22 Pin Connector

PCM Pin #	Wire Color	Circuit Description (22 Pin)	Value at Hot Idle
1	GN/BK	Sensor VREF	4.9-5.1v
4	WT/RD	A/T Vehicle Speed Sensor (-)	AC pulse signals
5	RD	CKP Sensor Signal (NE+)	AC pulse signals
6	GN	CKP Sensor Signal (NE-)	<0.050v
7	YL	TP Sensor Signal (VTA)	0.3-0.8v
8	GY/RD	MAF Sensor Signal (VG)	1.1-1.8v
9	YL/RD	A/T Vehicle Speed Sensor (+)	AC pulse signals
10	YL	CMP Sensor Signal (G2+)	AC pulse signals
11	BL	CMP Sensor Signal (GN-)	<0.050v
12	YL/RD	A/T Oil Temperature Sensor	At 68°F: 4-5v
13	WT	HO2S-11 (B1 S1) Signal	0.1-1.1v
14	YL/GN	IAT Sensor Signal (THA)	At 100°F: 2.60v
15	RD/YL	EVAP Vapor Pressure Sensor	2.5-3.1v (with fuel cap off)
16	GY	Knock Sensor 2 Signal	<0.075v AC
17	BK	Knock Sensor 1 Signal	<0.075v AC
18, 22	BR/WT	Sensor Ground	<0.050v
19	BK	HO2S-12 (B1 S2) Signal	0.1-1.1v
20	GN/RD	ECT Sensor Signal (THW)	At 180°F: 0.51v
21	PK/GN	EGR Gas Temperature Sensor (THG)	3.5-4.0v

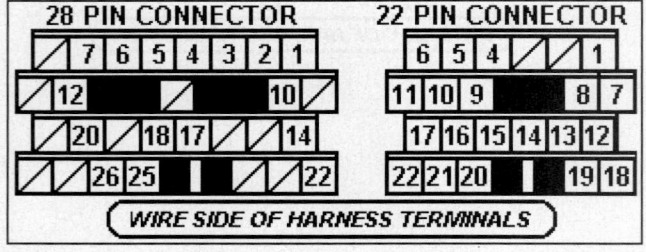

Pin Connector Graphic

1997 3.4L V6 MFI VIN N (A/T-ECT) 16 Pin Connector

PCM Pin #	Wire Color	Circuit Description (16 Pin)	Value at Hot Idle
1	BL	Cruise Control ECU	At Cruise in OD: 12v
3	PK	MIL (lamp) Control	MIL Off: 12v, On: 1v
4	GN/YL	Circuit Opening Relay (FC)	0-3v, at off-idle: 12v
5	PK/WT	Data Link Connector	12-14v
8	RD/WT	EGR Solenoid Control (VSV)	12v or 0v
13	GY/GN	EVAP Vapor Pressure (VSV)	12v or 0v
15	WT/GN	EVAP Purge Solenoid (VSV)	12v or 0v
16	BR	Shield Ground	<0.050v

1997 3.4L V6 MFI VIN N (A/T-ECT) 34-Pin Connector

PCM Pin #	Wire Color	Circuit Description (34-Pin)	Value at Hot Idle
1	BR	Power Ground	<0.1v
5	YL/BK	Injector 6 Control	1.6-2.9 ms
6	WT/BL	Injector 5 Control	1.6-2.9 ms
7	YL/RD	Injector 4 Control	1.6-2.9 ms
8	WT/GN	Injector 3 Control	1.6-2.9 ms
9	WT	Injector 2 Control	1.6-2.9 ms
10	WT/RD	Injector 1 Control	1.6-2.9 ms
11	PK/YL	A/T-ECT Solenoid (S1)	3rd or OD: 12-14v
12	BK/YL	Igniter Signal (IGF)	Digital Signal: 0-5-0v
13	BK/WT	Starter Switch Signal	9-11v (cranking)
14	BK	Neutral Start Switch	In P/N: 9-11v (cranking)
15	RD/WT	HO2S-12 (B1 S2) Heater	Heater Off: 12v, On: 1v
16	PK/GN	HO2S-11 (B1 S1) Heater	Heater Off: 12v, On: 1v
17	GN	A/T-ECT Solenoid (S2)	1st or OD: 12-14v
22	BK/RD	IAC Signal (RSC)	Pulse Signals
23	BR/RD	IAC Signal (RSO)	Pulse Signals
24	BK/BL	Igniter Transistor 1 Control	Digital Signal: 0-5-0v
25	BR/YL	Igniter Transistor 2 Control	Digital Signal: 0-5-0v
26	BK/WT	Igniter Transistor 3 Control	Digital Signal: 0-5-0v
27	RD/GN	A/T-ECT Solenoid (SL)	In Lockup: 12-14v
29	GN	4WD Detection Transfer (L4)	Open: 12v, Closed: 0v
33	BR	Power Ground	<0.1v
34	BR	Power Ground	<0.1v

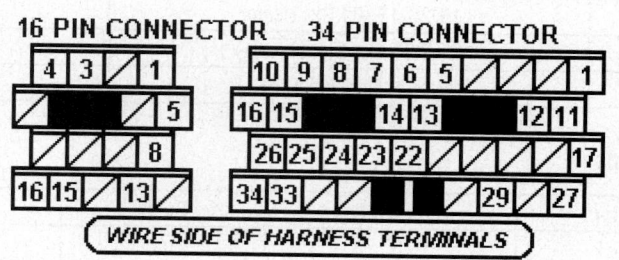

Pin Connector Graphic

05533_ADIA_G191

1997 3.4L V6 MFI 5VZ-FE VIN N (M/T) 22 Pin Connector

PCM Pin #	Wire Color	Circuit Description (22 Pin)	Value at Hot Idle
2	BK/YL	Battery Direct	12-14v
4	PK	MIL (lamp) Control	MIL Off: 12v, On: 1v
6	BL/BK	A/C Amplifier Signal (ACT)	Clutch On: 12v, Off: 1.5v
7	BL/YL	A/C Amplifier Signal (AC1)	Clutch On: 1.5v, Off: 12v
8	GN/OR	Vehicle Speed Sensor	At 55 mph: 48 Hz
9	GN/WT	4WD Selector Switch	Switch Off: 12v, On: 1v
11	BK/WT	Starter Switch Signal	9-11v (cranking)
12	WT/RD	EFI Main Relay (B+)	12-14v
19	WT	SIL (Scan Tool) Signal	12v
20	GN/WT	Stop Light Switch Signal	Brake Off: 0v, On: 12v

1997 3.4L V6 MFI 5VZ-FE VIN N (M/T) 16 Pin Connector

PCM Pin #	Wire Color	Circuit Description (16 Pin)	Value at Hot Idle
1	GN/BK	Sensor VREF	4.9-5.1v
2	GY/RD	MAF Sensor Signal (VG)	1.1-1.8v
3	GY	Knock Sensor 2 Signal	<0.075v AC
4	GN/RD	ECT Sensor Signal (THW)	At 180°F: 0.51v
5	WT	HO2S-11 (B1 S1) Signal	0.1-1.1v
6	BK	Knock Sensor 1 Signal	<0.075v AC
7	RD	Data Link Connector	12-14v
8	BR/WT	MAF Sensor Ground	<0.050v
9	BR/BK	Sensor Ground	<0.050v
10	YL	TP Sensor Signal (VTA)	0.3-0.8v
12	YL/GN	IAT Sensor Signal (THA)	At 100°F: 2.60v
13	BK	HO2S-12 (B1 S2) Signal	0.1-1.1v
14	PK/GN	EGR Gas Temperature Sensor (THG)	3.5-4.0v

1997 3.4L V6 MFI 5VZ-FE VIN N (M/T) 12 Pin Connector

PCM Pin #	Wire Color	Circuit Description (12 Pin)	Value at Hot Idle
3	PK/GN	HO2S-11 (B1 S1) Heater	Heater Off: 12v, On: 1v
4	RD/YL	EVAP Vapor Pressure Sensor	2.5-3.1v (with fuel cap off)
5	BL	CMP Sensor Signal (GN-)	<0.050v
6	GN	CKP Sensor Signal (NE-)	<0.050v
7	BR	Power Ground	<0.1v
9	RD/GN	HO2S-12 (B1 S2) Heater	Heater Off: 12v, On: 1v
11	YL	CMP Sensor Signal (G2+)	AC pulse signals
12	RD	CKP Sensor Signal (NE+)	AC pulse signals

Standard Colors and Abbreviations

Abbreviation	Color	Abbreviation	Color	Abbreviation	Color
BK	Black	GY	Gray	RD	Red
BL	Blue	GN	Green	TN	Tan
BR	Brown	LG	LT Green	VT	Violet
DB	Dark Blue	OR	Orange	WT	White
DG	DK Green	PK	Pink	YL	Yellow

1997 3.4L V6 MFI 5VZ-FE VIN N (M/T) 26 Pin Connector

PCM Pin #	Wire Color	Circuit Description (26 Pin)	Value at Hot Idle
3	WT/GN	EVAP Vapor Pressure (VSV)	12v or 0v
4	BL	Cruise Control ECU	At Cruise in OD: 12v
5	WT/GN	EVAP Purge Solenoid (VSV)	12v or 0v
6	BK/RD	IAC Signal (RSC)	Pulse Signals
7	BR/RD	IAC Signal (RSO)	Pulse Signals
8	YL/BK	Injector 6 Control	1.6-2.9 ms
9	WT/BL	Injector 5 Control	1.6-2.9 ms
10	YL/RD	Injector 4 Control	1.6-2.9 ms
11	WT	Injector 2 Control	1.6-2.9 ms
12	WT/RD	Injector 1 Control	1.6-2.9 ms
13	BR	Power Ground	<0.1v
14	GN/YL	Circuit Opening Relay (FC)	0-3v, at off-idle: 12v
17	BK/YL	Igniter Signal (IGF)	Digital Signal: 0-5-0v
18	RD/WT	EGR Solenoid Control (VSV)	12v or 0v
21	BK/WT	Igniter Transistor 3 Control	Digital Signal: 0-5-0v
22	BR/YL	Igniter Transistor 2 Control	Digital Signal: 0-5-0v
23	BK/BL	Igniter Transistor 1 Control	Digital Signal: 0-5-0v
24	BR	Shield Ground	<0.050v
25	WT/GN	Injector 3 Control	1.6-2.9 ms
26	BR	Power Ground	<0.1v

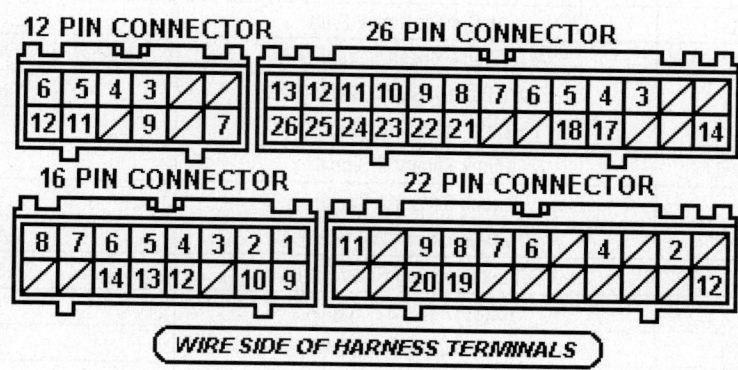

Pin Connector Graphic

05533_ADIA_G192

1998-99 3.4L V6 MFI VIN N (A/T-ECT) 28 Pin Connector

PCM Pin #	Wire Color	Circuit Description (28 Pin)	Value at Hot Idle
1	PK/WT	A/T Select Switch Low	In Low: 12v, Others: 0v
5	BL/BK	A/C Amplifier Signal (ACT)	Clutch On: 12v, Off: 1.5v
6	BL/OR	Overdrive Main Switch	Switch Off: 12v, On: 1v
7	RD/YL	Cruise Control ECU (OD1)	At Cruise in OD: 12v
8	WT	Data Link Connector	12-14v
10	PK	A/T Select Switch 2nd	In 2nd: 12v, Others: 0v
11	GN/RD	Cruise Control ECU (IDLO)	1.5v, off-idle: 12v
12	GN/OR	Speedometer Indicator	At 55 mph: 48 Hz
14	BK/YL	Battery Direct	12-14v
15	RD/BK	A/T Select Switch Reverse	In 'R': 12v, Others: 0v
16	BL/YL	A/C Amplifier Signal (AC1)	Clutch On: 1.5v, Off: 12v
23	WT/RD	EFI Main Relay (B+)	12-14v
24	GN/WT	Stop Light Switch Signal	Brake Off: 0v, On: 12v

1998-99 3.4L V6 MFI VIN N (A/T-ECT) 22 Pin Connector

PCM Pin #	Wire Color	Circuit Description (22 Pin)	Value at Hot Idle
1	GN/BK	Sensor VREF	4.9-5.1v
4	WT/RD	A/T Vehicle Speed Sensor (-)	AC pulse signals
5	RD	CKP Sensor Signal (NE+)	AC pulse signals
6	GN	CKP/CMP Sensor Signal (GN-)	<0.050v
7	YL	TP Sensor Signal (VTA)	0.3-0.8v
8	GY/RD	MAF Sensor Signal (VG)	1.1-1.8v
9	YL/RD	A/T Vehicle Speed Sensor (+)	AC pulse signals
12	GN	A/T Pattern Select Switch	Norm: 0v, PWR: 12v
13	WT	HO2S-11 (B1 S1) Signal	0.1-1.1v
14	GY	Knock Sensor 2 Signal	<0.075v AC
15	BK	Knock Sensor 1 Signal	<0.075v AC
17	YL	CMP Sensor Signal (G2+)	AC pulse signals
18	GN/YL	Circuit Opening Relay (FC)	0-3v, at off-idle: 12v
19	BK	HO2S-12 (B1 S2) Signal	0.1-1.1v
20	GN/RD	ECT Sensor Signal (THW)	At 180°F: 0.51v
21	YL/GN	IAT Sensor Signal (THA)	At 100°F: 2.60v
22	BR/BK	Sensor Ground	<0.050v

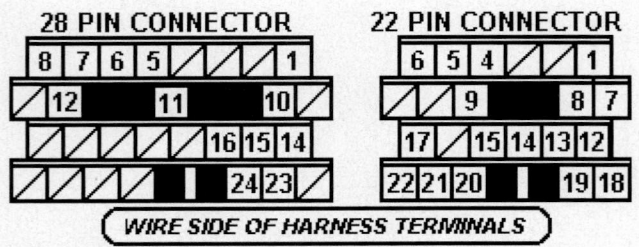

Pin Connector Graphic

05533_ADIA_G193

1998-99 3.4L V6 MFI VIN N (A/T-ECT) 16 Pin Connector

PCM Pin #	Wire Color	Circuit Description (16 Pin)	Value at Hot Idle
2	WT/GN	EVAP Purge Solenoid (VSV)	12v or 0v
3	PK/RD	MIL (lamp) Control	MIL Off: 12v, On: 1v
5	RD	Data Link Connector	12-14v
6	PK	A/T Oil Temperature Lamp	Lamp Off: 12v, On: 1v
7	BR/WT	MAF Sensor Ground (E2G)	<0.050v
8	GY/GN	EVAP Vapor Pressure (VSV)	12v or 0v
9	RD/GN	4WD Detection Transfer (N)	Open: 12v, Closed
10	RD/WT	HO2S-12 (B1 S2) Heater	Heater Off: 12v, On: 1v
11	PK/GN	HO2S-11 (B1 S1) Heater	Heater Off: 12v, On: 1v
12	RD/WT	EGR Solenoid Control (VSV)	12v or 0v
13	RD/YL	EVAP Vapor Pressure Sensor	2.5-3.1v (with fuel cap off)
14	PK/GN	EGR Gas Temperature Sensor (THG)	3.5-4.0v
16	BR	Shield Ground	<0.050v

1998-99 3.4L V6 MFI VIN N (A/T-ECT) 34-Pin Connector

PCM Pin #	Wire Color	Circuit Description (34-Pin)	Value at Hot Idle
1	RD/WT	4WD Selector Switch	Switch Off: 12v, On: 1v
2	GN	4WD Detection Transfer (L4)	Open: 12v, Closed: 0v
2	GY	4WD Detection Transfer (L4)	Open: 12v, Closed: 0v
5	BL	Injector 6 Control	1.6-2.9 ms
6	WT/BL	Injector 5 Control	1.6-2.9 ms
7	BL/RD	Injector 4 Control	1.6-2.9 ms
8	WT/GN	Injector 3 Control	1.6-2.9 ms
9	WT	Injector 2 Control	1.6-2.9 ms
10	WT/RD	Injector 1 Control	1.6-2.9 ms
11	PK/YL	A/T-ECT Solenoid (S1)	3rd or OD: 1v
12	BK/YL	Igniter Signal (IGF)	Digital Signal: 0-5-0v
13	BK/WT	Starter Switch Signal	9-11v (cranking)
14	BK/RD	Neutral Start Switch	In P/N: 9-11v (cranking)
15	BK/WT	Igniter Transistor 3 Control	Digital Signal: 0-5-0v
16	BR/YL	Igniter Transistor 2 Control	Digital Signal: 0-5-0v
17	GN	A/T-ECT Solenoid (S2)	1st or OD: 1v
22	BK/RD	IAC Signal (RSC)	Pulse Signals
23	BR/RD	IAC Signal (RSO)	Pulse Signals
24	BK/BL	Igniter Transistor 1 Control	Digital Signal: 0-5-0v
27	RD/GN	A/T-ECT Solenoid (SL)	In Lockup: 12-14v
28	BR	Power Ground	<0.1v
30	YL/RD	A/T Oil Temperature Sensor	At 68°F: 4-5v
31	YL	PSP Switch Signal	Straight: 12v, Turning: 0v
33	BR	Power Ground	<0.1v
34	BR	Power Ground	<0.1v

2000 3.4L V6 DOHC VIN N California E3 22 Pin Connector

PCM Pin #	Wire Color	Circuit Description (22 Pin)	Value at Hot Idle
1	BK/RD	Direct Battery	12-14v
2	BK/OR	Ignition Switch Power	12-14v
3	PK	Circuit Opening Relay (FC)	0-3v, at off-idle: 12v
6	PK/GN	MIL (lamp) Control	MIL Off: 12v, On: 1v
7	BK/WT	Starter Switch Signal	9-11v (cranking)
8	BK/YL	EFI Main Relay	12-14v
9	GY/RD	EVAP Vapor Pressure (VSV)	12v or 0v
11	WT	SIL (Scan Tool) Signal	12v
15	GN/WT	Stop Light Switch Signal	Brake Off: 0v, On: 12v
16	WT/BL	EFI Main Relay (B+)	12-14v
17	RD/GN	EVAP Vapor Pressure Sensor	2.9-3.1v (with fuel cap off)

2000 3.4L V6 DOHC VIN N California E6 31 Pin Connector

PCM Pin #	Wire Color	Circuit Description (31 Pin)	Value at Hot Idle
1	GN	Injector 3 Control	1.6-2-9 ms
2	RD/BK	Injector 4 Control	1.6-2-9 ms
3	BL	Injector 5 Control	1.6-2-9 ms
4	YL	Injector 6 Control	1.6-2.9 ms
6	RD	Data Link Connector	12-14v
7	RD	A/T-ECT Solenoid (S1)	3rd or OD: 1v
8	WT/BL	A/T-ECT Solenoid (S2)	1st or OD: 1v
9	GN/RD	A/T-ECT Solenoid (SL)	In Lockup: 12-14v
10	BL	CMP Sensor Signal (G2)	AC pulse signals
11	BK/BL	Igniter Transistor 1 Control	6°, 55 mph: 8° dwell
12	RD/BL	Igniter Transistor 2 Control	6°, 55 mph: 8° dwell
13	LG	Igniter Transistor 3 Control	6°, 55 mph: 8° dwell
14	YL/RD	A/T Vehicle Speed Sensor (+)	AC pulse signals
15	BK/RD	IAC Signal (RSC)	Pulse Signals
16	RD/WT	IAC Signal (RSO)	Pulse Signals
21	BR	Power Ground	<0.1v
25	BK/YL	Igniter Signal (IGF)	Digital Signal: 0-5-0v
26	WT/RD	A/T Vehicle Speed Sensor (-)	AC pulse signals
27	BK	Knock Sensor 1 Signal	<0.075v AC
28	GY	Knock Sensor 2 Signal	<0.075v AC
30	WT/BK	Power Ground (E03)	<0.1v
31	WT/BK	Power Ground (E02)	<0.1v

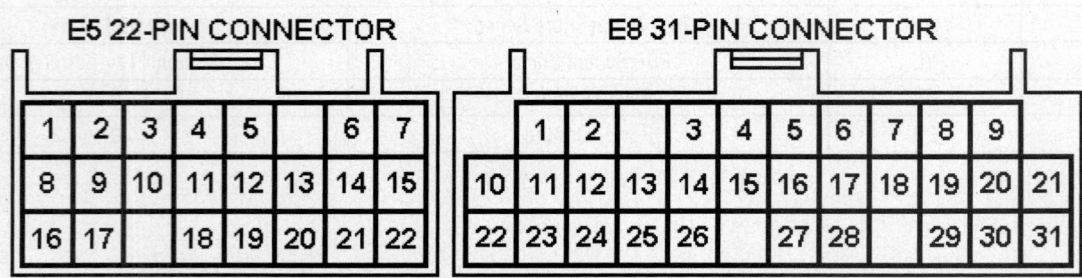

Pin Connector Graphic

05533_ADIA_G194

2000 3.4L V6 DOHC VIN N California E4 28 Pin Connector

PCM Pin #	Wire Color	Circuit Description (28 Pin)	Value at Hot Idle
2	RD/BK	A/T Select Switch Reverse	In 'R': 12v, Others: 0v
3	BL	A/T Select Switch 2nd	In 2nd: 12v, Others: 0v
4	YL/GN	Cruise Control ECU (IDLO)	1.5v, off-idle: 12v
7	YL/RD	A/T Oil Temperature Indicator	Lamp Off: 12v, On: 1v
8	RD	HO2S-12 (B1 S2) Signal	0.1-1.1v
9	RD/WT	HO2S-12 (B1 S2) Heater	Heater Off: 12v, On: 1v
10	BL/OR	O/D "Off" Indicator (ODLP)	Lamp Off: 12v, On: 1v
12	LG	A/T Select Switch Low	In Low: 12v, Others: 0v
13	BL/BK	A/C Amplifier Signal (ACT)	Clutch On: 12v, Off: 1.5v
17	RD/YL	4WD Detection Transfer (N)	Open: 12v, Closed: 0v
18	GY	4WD ADD Indicator Switch	Open: 12v, Closed: 0v
19	BL/RD	4WD Detection Transfer (L4)	Open: 12v, Closed: 0v
20	PK	Starter Switch Signal	9-11v (cranking)
22	GN/OR	Speedometer Indicator	At 55 mph: 48 Hz
24	BR/YL	Cruise Control ECU (OD1)	At Cruise in OD: 12v
25	BL/YL	A/C Amplifier Signal (AC1)	Clutch On: 1.5v, Off: 12v

2000 3.4L V6 DOHC VIN N California E5 24 Pin Connector

PCM Pin #	Wire Color	Circuit Description (24 Pin)	Value at Hot Idle
2	GN/BK	Sensor VREF	4.9-5.1v
4	YL	HAFR-11 (B1 S1) Heater	Heater Off: 12v, On: 1v
5	RD	Injector 1 Control	1.6-2-9 ms
6	WT	Injector 2 Control	1.6-2-9 ms
7	WT/GN	EVAP Purge Solenoid (VSV)	12v or 0v
8	BR	Power Ground	<0.1v
9	YL/RD	PSP Switch Signal	Straight: 12v, Turning: 0v
10	RD/WT	MAF Sensor Signal (VG)	1-1.1v
12	GN	AFS-11 (B1 S1) Signal (AF+)	Fixed at 3.3v
14	GN	ECT Sensor Signal (THW)	0.5-0.6v
15	RD/YL	A/T Oil Temperature Sensor	At 68°F: 4-5v
16	PK	CKP Sensor Signal (NE+)	AC pulse signals
17	BR	Shield Ground (E1)	<0.050v
18	LG	Sensor Ground (E2)	<0.050v
19	BK/WT	MAF Sensor Ground (E2G)	<0.050v
21	RD	AFS-11 (B1 S1) Signal (AF-)	Fixed at 3.3v
22	YL/GN	IAT Sensor Signal (THA)	0.5-3.4v
23	BK/YL	TP Sensor Signal (VTA)	0.53-1.27v
24	PK	CMP & CKP Sensor Return	<0.050v

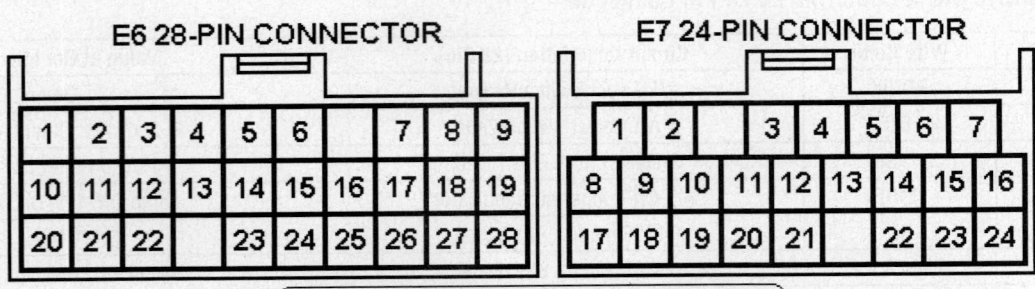

WIRE SIDE OF HARNESS TERMINALS

05533_ADIA_G195

Pin Connector Graphic

2000 3.4L V6 MFI VIN N Federal E3 28 Pin Connector

PCM Pin #	Wire Color	Circuit Description (28 Pin)	Value at Hot Idle
1	LG	A/T Select Switch Low	In Low: 12v, Others: 0v
2	BL/OR	Overdrive Main Switch	Switch Off: 12v, On: 1v
5	BL/BK	A/C Amplifier Signal (ACT)	Clutch On: 12v, Off: 1.5v
7	BR/YL	Cruise Control ECU (OD1)	At Cruise in OD: 12v
8	WT	SIL (Scan Tool) Signal	12v
10	BL	A/T Select Switch 2nd	In 2nd: 12v, Others: 0v
11	YL/GN	Cruise Control ECU (IDLO)	1.5v, off-idle: 12v
12	GN/OR	Speedometer Indicator	At 55 mph: 48 Hz
14	BK/RD	Direct Battery	12-14v
15	RD/BK	A/T Select Switch Reverse	In 'R': 12v, Others: 0v
16	BL/YL	A/C Amplifier Signal (AC1)	Clutch On: 1.5v, Off: 12v
23	WT/BL	EFI Main Relay (B+)	12-14v
24	GN/WT	Stop Light Switch Signal	Brake Off: 0v, On: 12v

2000 3.4L V6 MFI VIN N Federal E4 16 Pin Connector

PCM Pin #	Wire Color	Circuit Description (16 Pin)	Value at Hot Idle
2	WT/GN	EVAP Purge Solenoid (VSV)	12v or 0v
3	PK/GN	MIL (lamp) Control	MIL Off: 12v, On: 1v
5	RD	Data Link Connector	12-14v
6	YL/RD	A/T Oil Temperature Sensor	At 68°F: 4-5v
7	BK/WT	MAF Sensor Ground (E2G)	<0.050v
8	GY/RD	EVAP Vapor Pressure (VSV)	12v or 0v
9	RD/YL	4WD Detection Transfer (N)	Switch Closed: 0v
10	RD/WT	HO2S-12 (B1 S2) Heater	Heater Off: 12v, On: 1v
11	PK/BL	HO2S-11 (B1 S1) Heater	Heater Off: 12v, On: 1v
13	RD/GN	EVAP Vapor Pressure Sensor	2.5-3.1v (with fuel cap off)
16	BR	Sensor Ground	<0.050v

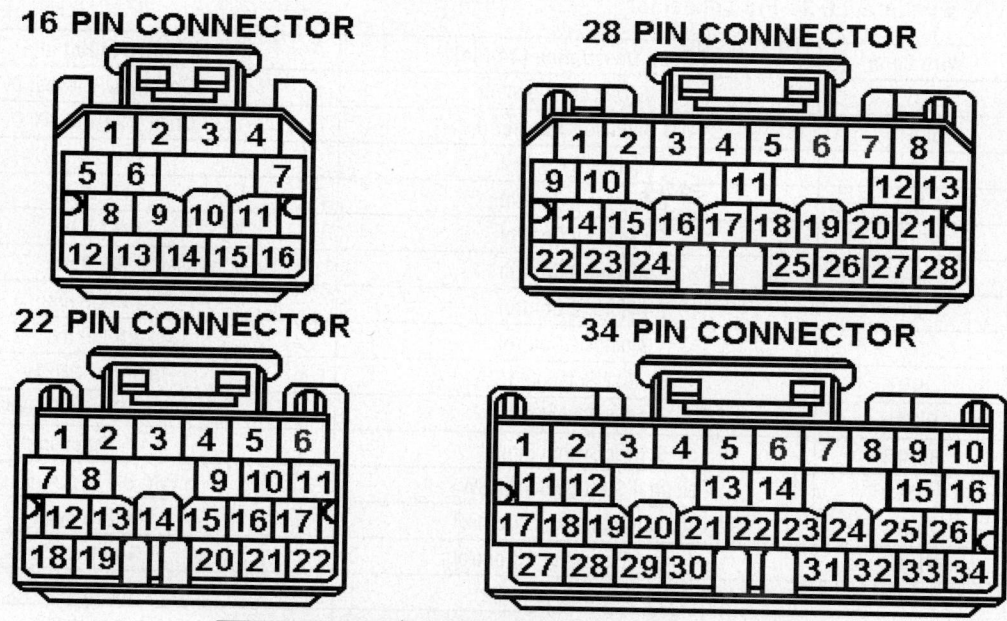

16 PIN CONNECTOR

28 PIN CONNECTOR

22 PIN CONNECTOR

34 PIN CONNECTOR

WIRE SIDE OF HARNESS TERMINALS

05533_ADIA_G196

Pin Connector Graphic

2000 3.4L V6 MFI VIN N Federal E5 22 Pin Connector

PCM Pin #	Wire Color	Circuit Description (22 Pin)	Value at Hot Idle
1	GN/BK	Sensor VREF	4.9-5.1v
4	WT/RD	A/T Vehicle Speed Sensor (-)	AC pulse signals
5	PK	CKP Sensor Signal (NE+)	AC pulse signals
6	PK	CKP Sensor Signal (NE-)	<0.050v
7	BK/YL	TP Sensor Signal (VTA)	0.3-0.8v
8	RD/WT	MAF Sensor Signal (VG)	1.1-1.8v
9	YL/RD	A/T Vehicle Speed Sensor (+)	AC pulse signals
13	WT	HO2S-11 (B1 S1) Signal	0.1-1.1v
14	GY	Knock Sensor 2 Signal	<0.075v AC
15	BK	Knock Sensor 1 Signal	<0.075v AC
17	BL	CMP Sensor Signal (G2+)	AC pulse signals
18	YL	Circuit Opening Relay (FC)	0-3v, at off-idle: 12v
19	RD	HO2S-12 (B1 S2) Signal	0.1-1.1v
20	GN	ECT Sensor Signal (THW)	At 180°F: 0.51v
21	YL/GN	IAT Sensor Signal (THA)	At 100°F: 2.60v
22	GN/WT	Sensor Ground (E2)	<0.050v

2000 3.4L V6 MFI VIN N Federal E6 34-Pin Connector

PCM Pin #	Wire Color	Circuit Description (34-Pin)	Value at Hot Idle
1	GY	4WD Detection Switch	Open: 12v, Closed: 0v
2	BL/RD	4WD Detection Transfer (L4)	Open: 12v, Closed: 0v
5	YL	Injector 6 Control	1.6-2.9 ms
6	BL	Injector 5 Control	1.6-2.9 ms
7	RD/BK	Injector 4 Control	1.6-2.9 ms
8	GN	Injector 3 Control	1.6-2.9 ms
9	WT	Injector 2 Control	1.6-2.9 ms
10	RD	Injector 1 Control	1.6-2.9 ms
11	RD	A/T-ECT Solenoid (S1)	In 3rd or OD: 1v
12	BK/YL	Igniter Signal (IGF)	Digital Signal: 0-5-0v
13	BK/WT	Starter Switch Signal	9-11v (cranking)
14	PK	Neutral Start Switch (NSW)	In P/N: 9-11v (cranking)
15	LG	Igniter Transistor 3 Control	7% duty cycle
16	RD/BL	Igniter Transistor 2 Control	7% duty cycle
17	WT/BL	A/T-ECT Solenoid (S2)	In 3rd or OD: 1v
22	BK/RD	IAC Signal (RSC)	Pulse Signals
23	RD/WT	IAC Signal (RSO)	Pulse Signals
24	BK/BL	Igniter Transistor 1 Control	7% duty cycle
27	GN/RD	A/T-ECT Solenoid (SL)	In Lockup: 12-14v
28	BR	Power Ground	<0.1v
30	PK/YL	A/T Oil Temperature Sensor	At 68°F: 4-5v
31	YL/RD	PSP Switch Signal	Straight: 12v, Turning: 0v
33, 34	BR	Power Ground	<0.1v

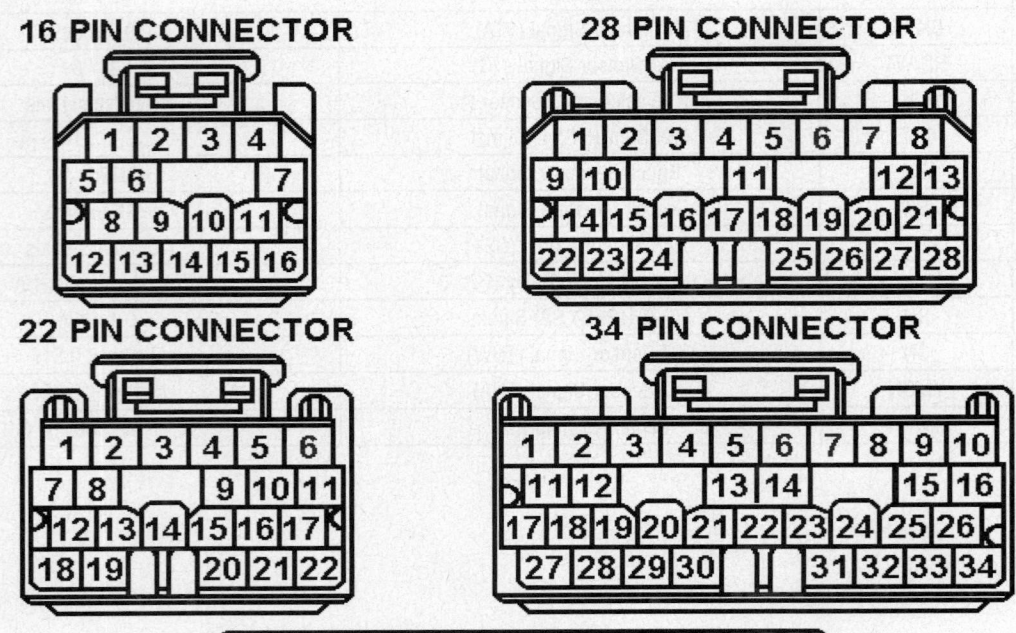

Pin Connector Graphic

2001 3.4L V6 DOHC MFI VIN N (All) E8 31 Pin Connector

PCM Pin #	Wire Color	Circuit Description (31 Pin)	Value at Hot Idle
1	GN	Injector 3 Control	1.6-2-9 ms
2	RD/BK	Injector 4 Control	1.6-2-9 ms
3	BL	Injector 5 Control	1.6-2-9 ms
4	YL	Injector 6 Control	1.6-2.9 ms
6	RD	DLC1 D1 Signal (TE1)	12v
7	VT	A/T-ECT Solenoid (S1)	3rd or OD: 1v
8	LG	A/T-ECT Solenoid (S2)	1st or OD: 1v
9	RD/WT	A/T-ECT Solenoid (SL)	In Lockup: 12-14v
10	RD	CMP Sensor Signal (G2)	AC pulse signals
11	BK/BL	Igniter Transistor 1 Control	6°, 55 mph: 8° dwell
12	GN/BK	Igniter Transistor 2 Control	6°, 55 mph: 8° dwell
13	BK/WT	Igniter Transistor 3 Control	6°, 55 mph: 8° dwell
14	YL/RD	A/T Vehicle Speed Sensor (+)	AC pulse signals
15	BK/RD	IAC Signal (RSC)	Pulse Signals
16	BL/BK	IAC Signal (RSO)	Pulse Signals
17-20	---	Not Used	---
21	WT/BK	Power Ground (E01)	<0.1v
22-24	---	Not Used	---
25	BK/YL	Igniter Signal (IGF)	Digital Signal: 0-5-0v
26	WT/RD	A/T Vehicle Speed Sensor (-)	AC pulse signals
27	BK	Knock Sensor 1 Signal	<0.075v AC
28	GY	Knock Sensor 2 Signal	<0.075v AC
29	---	Not Used	---
30	WT/BK	Power Ground (E03)	<0.1v
31	WT/BK	Power Ground (E02)	<0.1v

2001 3.4L V6 DOHC MFI VIN N (All) E5 22 Pin Connector

PCM Pin #	Wire Color	Circuit Description (22 Pin)	Value at Hot Idle
1	BK/YL	Direct Battery	12-14v
2	BK/WT	Ignition Switch Power	12-14v
3	WT/BL	Circuit Opening Relay (FC)	0-3v, at off-idle: 12v
6	PK/GN	MIL (lamp) Control	MIL Off: 12v, On: 1v
7	VT/RD	Starter Switch Signal	9-11v (cranking)
8	BK/OR	EFI Main Relay Control	Relay Off: 0v, On: 12v
9	GN/BK	EVAP Vapor Pressure (VSV)	12v or 0v
10	---	Not Used	---
11	WT	SIL (Scan Tool) Signal	12v
12-14	---	Not Used	---
15	GN/WT	Stop Light Switch Signal	Brake Off: 0v, On: 12v
16	WT/RD	EFI Main Relay (B+)	12-14v
17	RD/YL	EVAP Vapor Pressure Sensor	2.9-3.1v (with fuel cap off)
18-22	---	Not Used	---

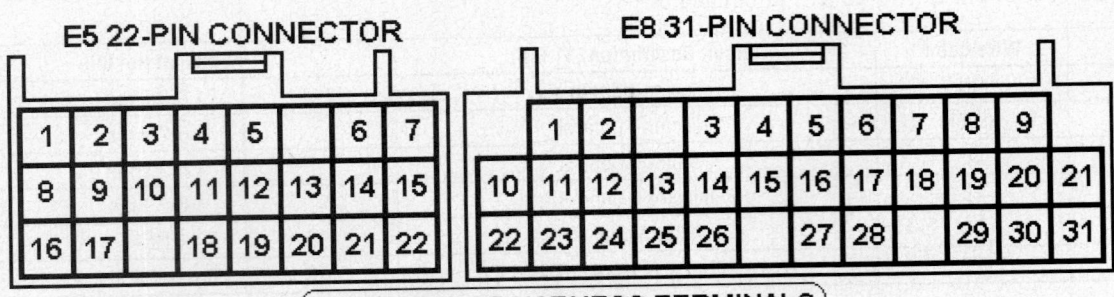

E5 22-PIN CONNECTOR

E8 31-PIN CONNECTOR

WIRE SIDE OF HARNESS TERMINALS

05533_ADIA_G197

Pin Connector Graphic

2001 3.4L V6 DOHC MFI VIN N (All) E7 24 Pin Connector

PCM Pin #	Wire Color	Circuit Description (24 Pin)	Value at Hot Idle
2	GN/BK	Sensor VREF (VC)	4.9-5.1v
4	WT	AFS-11 (B1 S1) Heater	Heater Off: 12v, On: 1v
5	WT/RD	Injector 1 Control	1.6-2-9 ms
6	BK	Injector 2 Control	1.6-2-9 ms
7	WT/GN	EVAP Purge Solenoid (VSV)	12v or 0v
8	WT/BK	Power Ground	<0.1v
9	BK	PSP Switch Signal	Straight: 12v, Turning: 0v
10	GY	MAF Sensor Signal (VG)	1-1.1v
12	VT	AFS-11 (B1 S1) Signal (AF+)	Fixed at 3.3v
14	GN/RD	ECT Sensor Signal (THW)	0.5-0.6v
15	YL/RD	A/T Oil Temperature Sensor	At 68°F: 4-5v
16	BL	CKP Sensor Signal (NE+)	AC pulse signals
17	BR	Shield Ground (E1)	<0.050v
18	LG	Sensor Ground (E2)	<0.050v
19	BK/WT	MAF Sensor Ground (E2G)	<0.050v
21	PK	AFS-11 (B1 S1) Signal (AF-)	Fixed at 3.3v
22	YL/GN	IAT Sensor Signal (THA)	0.5-3.4v
23	YL	TP Sensor Signal (VTA)	0.53-1.27v
24	GN	CMP/CKP Sensor Ground (-)	<0.050v

2001 3.4L V6 DOHC MFI VIN N (All) E6 28 Pin Connector

PCM Pin #	Wire Color	Circuit Description (28 Pin)	Value at Hot Idle
2	RD/BK	A/T Select Switch Reverse	In 'R': 12v, Others: 0v
3	PK	A/T Select Switch 2nd	In 2nd: 12v, Others: 0v
4	LG/RD	Cruise Control ECU (IDLO)	1.5v, off-idle: 12v
7	OR	A/T Oil Temperature Indicator	Lamp Off: 12v, On: 1v
8	BK	HO2S-12 (B1 S2) Signal	0.1-1.1v
9	RD/WT	HO2S-12 (B1 S2) Heater	Heater Off: 12v, On: 1v
10	BL/OR	O/D OFF Indicator (ODLP)	Lamp Off: 12v, On: 1v
12	RD	A/T Select Switch Low	In Low: 12v, Others: 0v
13	BL/BK	A/C Amplifier Signal (ACT)	Clutch On: 12v, Off: 1.5v
17	BL	A/T Park Indicator (TFN)	Lamp Off: 12v, On: 1v
18	GN/BK	4WD ADD Indicator Switch	Open: 12v, Closed: 0v
19	GY	4WD Detection Transfer (L4)	Open: 12v, Closed: 0v

2001 3.4L V6 DOHC MFI VIN N (All) E6 28 Pin Connector, *continued*

20	BK/YL	Starter Switch Signal (NSW)	9-11v (cranking)
22	GN/OR	Speedometer Indicator (SPI)	At 55 mph: 48 Hz
24	RD/YL	Cruise Control ECU (OD1)	At Cruise in OD: 12v
25	BL/YL	A/C Amplifier Signal (AC1)	Clutch On: 1.5v, Off: 12v

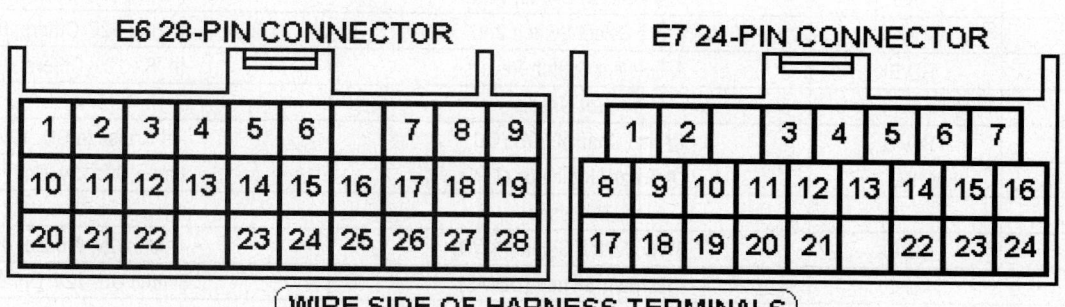

05533_ADIA_G198

Pin Connector Graphic

2002-06 3.4L V6 DOHC MFI VIN N (All) E5 22 Pin Connector

PCM Pin #	Wire Color	Circuit Description (22 Pin)	Value at Hot Idle
1	BK/RD	EFI Main Relay Control (+B)	Relay Off: 0v, On: 12v
2	GN/OR	MIL (lamp) Control	MIL Off: 12v, On: 1v
3	BL/BK	A/C Amplifier Signal (ACT)	Clutch On: 12v, Off: 1.5v
5	---	Not Used	---
6	GN/OR	Speedometer Indicator (SPI)	At 55 mph: 48 Hz
7	GN	Starter Switch Signal (STA)	In P/N: 0-3v
8	---	Not Used	---
9	BL/YL	A/C Amplifier Signal (AC1)	Clutch On: 1.5v, Off: 12v
10-11	---	Not Used	---
13	YL/RD	Scan Tool (WFSE)	12v
14	WT	SIL (Scan Tool) Signal	12v
15	BK/WT	Ignition Switch (IGSW)	12-14v
16	BK/YL	Direct Battery (BATT)	12-14v
17	LG/RD	Cruise Control ECU (IDLO)	1.5v, off-idle: 12v
18-19	---	Not Used	---
20	GN/WT	Stop Light Switch Signal (STP)	Brake Off: 0v, On: 12v
21	---	Not Used	---
22	WT/BL	Circuit Opening Relay (FC)	0-3v, at off-idle: 12v

2002-06 3.4L V6 DOHC MFI VIN N (All) E6 28 Pin Connector

PCM Pin #	Wire Color	Circuit Description (28 Pin)	Value at Hot Idle
1	GN	AT Pattern Select Switch	Norm: 0v, Power: 12v
2	GY	4WD Detection Transfer (L4)	Open: 12v, Closed: 0v
3	YL/GY	A/T Neutral Start Switch (NSW)	In P/N: 9-11v (while cranking)
4	RD	A/T Select Switch Low	In Low: 12v, Others: 0v
5	PK	A/T Select Switch 2nd	In 2nd: 12v, Others: 0v
6	RD/BK	A/T Select Switch Reverse	In 'R': 12v, Others: 0v
7-9	---	Not Used	---
10	RD/YL	Cruise Control ECU (OD1)	At Cruise in OD: 12v
11	BL	A/T Park Indicator (TFN)	Lamp Off: 12v, On: 1v
12	---	Not Used	---
13	GN/BK	4WD Switch Signal (4WD)	Open: 12v, Closed: 0v
14	GN	OD Main Switch (ODMS)	Switch Off: 12v, On: 0v
15-19	---	Not Used	---
20	OR	A/T Oil Temperature Indicator	Lamp Off: 12v, On: 1v
21-29	---	Not Used	---
30	WT/BK	Power Ground (E03)	<0.1v
31	WT/BK	Power Ground (E03)	<0.1v

2002-06 3.4L V6 DOHC MFI VIN N (All) E10 17 Pin Connector

PCM Pin #	Wire Color	Circuit Description (17 Pin)	Value at Hot Idle
1	---	Not Used	---
2	YL	Throttle Control Motor (CL+)	Pulses
3	---	Not Used	---
4	VT	TP Sensor Signal (VTA2)	2.0-2.9v
5-7	---	Not Used	---
8	BL	Throttle Control Motor (CL-)	Pulses
9	---	Not Used	---
10	GR	Accelerator Position Sensor (VPA)	0.25-0.90v
15	BL	Accelerator Position Sensor (VPA2)	1.8-2.7v
16-17	---	Not Used	---

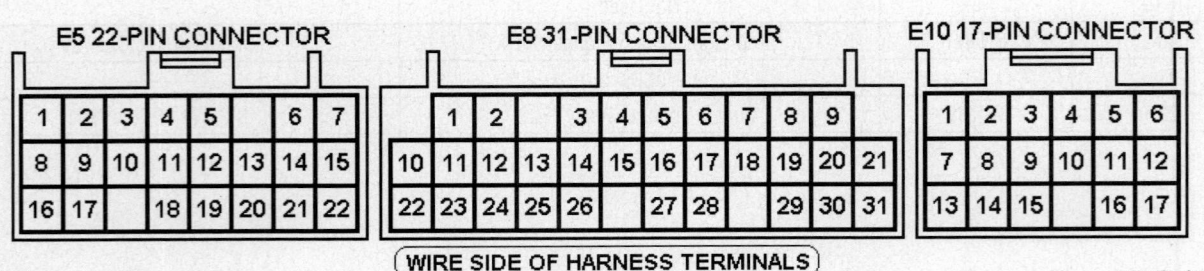

WIRE SIDE OF HARNESS TERMINALS

05533_ADIA_G199

Pin Connector Graphic

2002-06 3.4L V6 DOHC MFI VIN N (All) E7 24 Pin Connector

PCM Pin #	Wire Color	Circuit Description (24 Pin)	Value at Hot Idle
1	WT/BL	Injector 5 Control	1.6-2-9 ms
2	BK/YL	Igniter Signal (IGF)	Digital Signal: 0-5-0v
3	BL/RD	Injector 4 Control	1.6-2-9 ms
4	RD	Injector 3 Control	1.6-2-9 ms
5	BK	Injector 2 Control	1.6-2-9 ms
6	WT/RD	Injector 1 Control	1.6-2-9 ms
7	YL/BK	Data Link Connector (TC)	12v
8	BL	Injector 6 Control	1.6-2.9 ms
9	BK/WT	Igniter Transistor 3 Control	6°, 55 mph: 8° dwell
10	LG/RD	Igniter Transistor 2 Control	6°, 55 mph: 8° dwell
11	BK/BL	Igniter Transistor 1 Control	6°, 55 mph: 8° dwell
12	BL	CKP Sensor Signal (NE+)	AC pulse signals
13	RD	CMP Sensor Signal (G2)	AC pulse signals
14	WT/BK	EVAP Purge Solenoid (VSV)	12v or 0v
16	BL/BK	IAC Signal (RSO)	DC pulse signals
17	BR	Power Ground (E1)	<0.1v
18	BK	PSP Switch Signal (PSW)	Straight: 12v, Turning: 0v
19	PK/BL	EVAP Purge Solenoid (VSV)	12v or 0v
20	GN/BK	EVAP Vapor Pressure Solenoid (VSV)	12v or 0v
21	GN	CMP/CKP Sensor Ground (-)	<0.050v
22	WT/RD	AT Vehicle Speed Sensor (-)	Moving: AC pulse signals
23	YL/RD	AT Vehicle Speed Sensor (+)	Moving: AC pulse signals
24	BK/RD	Idle Speed Control Valve (RSD)	DC pulse signals

2002-06 3.4L V6 DOHC MFI VIN N (All) E8 31 Pin Connector

PCM Pin #	Wire Color	Circuit Description (31 Pin)	Value at Hot Idle
1	RD/WT	A/T-ECT Solenoid (SL)	In Lockup: 12-14v
2	LG	A/T-ECT Solenoid (S2)	1st or OD: 1v
3	VT	A/T-ECT Solenoid (S1)	3rd or OD: 1v
4, 5	WT/BK	Power Ground (E01), (E02)	<0.1v
6	WT/BK	Power Ground (ME01)	<0.1v
7, 8	WT/BK	Power Ground (E03), (E04)	<0.1v
9	RD	Throttle Control Motor (M+)	Pulse Signals
10	BK/WT	MAF Sensor Ground (E2G)	<0.050v
11	BL/BK	Sensor Ground (E2)	<0.050v
12	GN	MAF Sensor Signal (VG)	1-1.1v
13	YL/GN	IAT Sensor Signal (THA)	0.5-3.4v
14	VT	AFS-11 (B1 S1) Signal (AF+)	Fixed at 3.3v
15	YL	TP Sensor Signal (VTA)	0.53-1.27v
17	BR	Shield Ground (GE01)	<0.050v
18	GN/RD	ECT Sensor Signal (THW)	0.5-0.6v
19-20	---	Not Used	---
21	WT	AFS-11 (B1 S1) Heater (HAF1)	Heater Off: 12v, On: 1v
22	GY	Knock Sensor 2 Signal (KNK2)	<0.075v AC
23	BK	Knock Sensor 1 Signal (KNK1)	<0.075v AC
24	RD/YL	EVAP Vapor Pressure Sensor (PTNK)	2.9-3.1v (with fuel cap off)
25	GN/YL	Sensor VREF (VC)	4.9-5.1v
26	PK	AFS-11 (B1 S1) Signal (AF-)	Fixed at 3.3v
27	RD	HO2S-12 (B1 S2) Signal (OX2B)	0.1-1.1v
28, 30	---	Not Used	---
29	RD/WT	HO2S-12 (B1 S2) Heater (HT2B)	Heater Off: 12v, On: 1v
31	GN	Throttle Control Motor (M-)	Pulse Signals

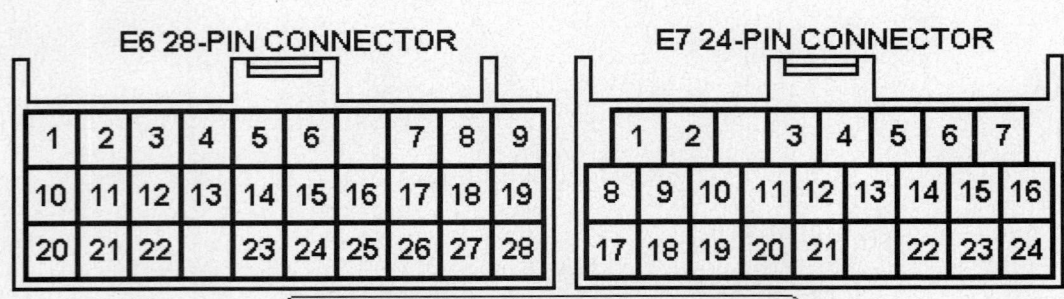

05533_ADIA_G198

Pin Connector Graphic

TUNDRA PIN CHARTS

2000 3.4L V6 MFI California VIN N E3 22 Pin Connector

PCM Pin #	Wire Color	Circuit Description (22 Pin)	Value at Hot Idle
1	BK/RD	Direct Battery	12-14v
2	BK/OR	Ignition Switch Power	12-14v
3	YL	Circuit Opening Relay (FC)	0-3v, at off-idle: 12v
6	PK/GN	MIL (lamp) Control	MIL Off: 12v, On: 1v
7	BK/WT	Starter Switch Signal	9-11v (cranking)
8	BK/YL	EFI Main Relay Control	Relay Off: 0v, On: 12v
11	WT	SIL (Scan Tool) Signal	12v
15	GN/WT	Stop Light Switch Signal	Brake Off: 0v, On: 12v
16	WT/BL	EFI Main Relay Power	12-14v
17	RD/GN	EVAP Vapor Pressure Sensor	2.9-3.1v (with fuel cap off)

2000 3.4L V6 MFI California VIN N E4 28 Pin Connector

PCM Pin #	Wire Color	Circuit Description (28 Pin)	Value at Hot Idle
2	RD/BK	A/T Select Switch Reverse	In 'R': 12v, Others: 0v
3	BL	A/T Select Switch 2nd	In 2nd: 12v, Others: 0v
4	YL/GN	Cruise Control ECU (IDLO)	1.5v, off-idle: 12v
7	YL/RD	A/T Oil Temperature Indicator	Lamp Off: 12v, On: 1v
8	RD	HO2S-12 (B1 S2) Signal	0.1-1.1v
9	RD/WT	HO2S-12 (B1 S2) Heater	Heater Off: 12v, On: 1v
10	BL/OR	O/D "Off" Indicator (ODLP)	Lamp Off: 12v, On: 1v
12	LG	A/T Select Switch Low	In Low: 12v, Others: 0v
13	BL/BK	A/C Amplifier Signal (ACT)	Clutch On: 12v, Off: 1.5v
17	RD/YL	4WD Detection Transfer (N)	Open: 12v, Closed: 0v
18	GY	4WD ADD Indicator Switch	Open: 12v, Closed: 0v
19	BL/RD	4WD Detection Transfer (L4)	Open: 12v, Closed: 0v
20	PK	Neutral Start Switch (NSW)	In P/N: 9-11v (cranking)
22	GN/OR	Speedometer Indicator	At 55 mph: 48 Hz
24	BR/YL	Cruise Control ECU (OD1)	At Cruise in OD: 12v
25	BL/YL	A/C Amplifier Signal (AC1)	Clutch On: 1.5v, Off: 12v

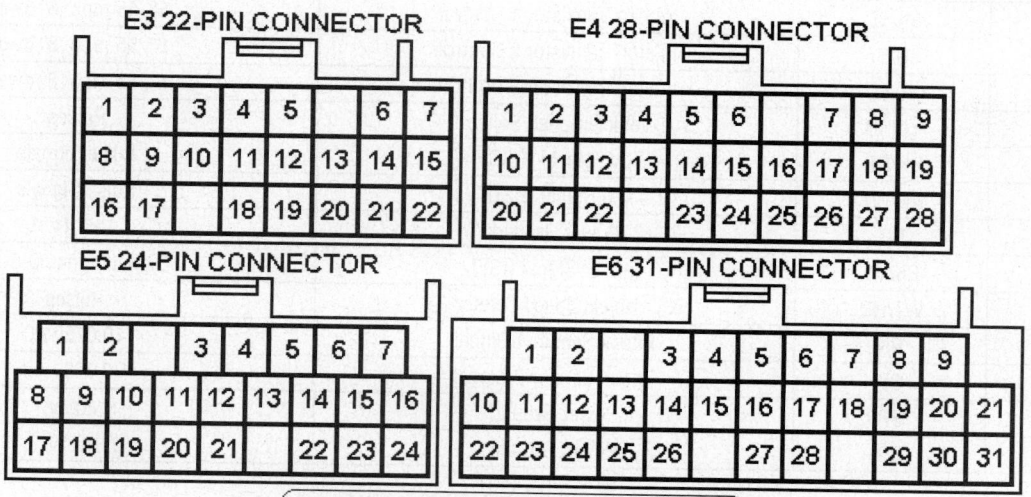

Pin Connector Graphic

2000 3.4L V6 MFI California VIN N E5 24 Pin Connector

PCM Pin #	Wire Color	Circuit Description (24 Pin)	Value at Hot Idle
2	GN/BK	Sensor VREF	4.9-5.1v
4	YL	AFS-11 (B1 S1) Heater	Heater Off: 12v, On: 1v
5	RD	Injector 1 Control	1.6-2-9 ms
6	WT	Injector 2 Control	1.6-2-9 ms
7	WT/GN	EVAP Purge Solenoid (VSV)	12v or 0v
8	BR	Power Ground	<0.1v
9	YL/RD	PSP Switch Signal	Straight: 12v, Turning: 0v
10	RD/WT	MAF Sensor Signal (VG)	1-1.1v
12	GN	AFS-11 (B1 S1) Signal (AF+)	Fixed at 3.3v
14	GN	ECT Sensor Signal (THW)	0.5-0.6v
15	RD/YL	A/T Oil Temperature Sensor	At 68°F: 4-5v
16	PK	CKP Sensor Signal (NE+)	AC pulse signals
17	BR	Power Ground	<0.050v
18	GN/WT	Sensor Ground	<0.050v
19	BK/WT	MAF Sensor Ground (E2G)	<0.050v
21	RD	AFS-11 (B1 S1) Signal (AF-)	Fixed at 3.3v
22	YL/GN	IAT Sensor Signal (THA)	0.5-3.4v
23	BK/YL	TP Sensor Signal (VTA)	0.53-1.27v
24	PK	CMP & CKP Sensor Return	<0.050v

2000 3.4L V6 MFI California VIN N E6 31 Pin Connector

PCM Pin #	Wire Color	Circuit Description (31 Pin)	Value at Hot Idle
1	GN	Injector 3 Control	1.6-2-9 ms
2	RD/BK	Injector 4 Control	1.6-2-9 ms
3	BL	Injector 5 Control	1.6-2-9 ms
4	YL	Injector 6 Control	1.6-2.9 ms
6	RD	Data Link Connector	12-14v
7	RD	A/T-ECT Solenoid (S1)	3rd or OD: 1v
8	WT/BL	A/T-ECT Solenoid (S2)	1st or OD: 1v
9	GN/RD	A/T-ECT Solenoid (SL)	In Lockup: 12-14v
10	BL	CMP Sensor Signal (G2)	AC pulse signals
11	BK/BL	Igniter Transistor 1 Control	6°, 55 mph: 8° dwell
12	RD/BL	Igniter Transistor 2 Control	6°, 55 mph: 8° dwell
13	LG	Igniter Transistor 3 Control	6°, 55 mph: 8° dwell
14	YL/RD	A/T Vehicle Speed Sensor (+)	Pulses
15	BK/RD	IAC Signal (RSC)	Pulse Signals
16	RD/WT	IAC Signal (RSO)	Pulse Signals
21	BR	Power Ground	<0.1v
25	BK/YL	Igniter Signal (IGF)	Digital Signal: 0-5-0v
26	WT/RD	A/T Vehicle Speed Sensor (-)	Pulses
27	BK	Knock Sensor 1 Signal	<0.075v AC
28	GY	Knock Sensor 2 Signal	<0.075v AC
30	BR	Power Ground	<0.1v
31	BR	Power Ground	<0.1v

2000 3.4L V6 MFI Federal VIN N E3 28 Pin Connector

PCM Pin #	Wire Color	Circuit Description (28 Pin)	Value at Hot Idle
1	LG	A/T Select Switch Low	In Low: 12v, Others: 0v
2	BL/OR	O/D "Off" Indicator (ODLP)	Lamp Off: 12v, On: 1v
5	BL/BK	A/C Amplifier Signal (ACT)	Clutch On: 12v, Off: 1.5v
7	BR/YL	Cruise Control ECU (OD1)	At Cruise in OD: 12v
8	WT	SIL (Scan Tool) Signal	Digital Signals
10	BL	A/T Select Switch 2nd	In 2nd: 12v, Others: 0v
11	YL/GN	Cruise Control ECU (IDLO)	1.5v, off-idle: 12v
12	GN/OR	Speedometer Indicator	At 55 mph: 48 Hz
14	BK/RD	Direct Battery	12-14v
15	RD/BK	A/T Select Switch Reverse	In 'R': 12v, Others: 0v
16	BL/YL	A/C Amplifier Signal (AC1)	Clutch On: 1.5v, Off: 12v
23	WT/BL	EFI Main Relay Power	12-14v
24	GN/WT	Stop Light Switch Signal	Brake Off: 0v, On: 12v

2000 3.4L V6 MFI Federal VIN N E4 16 Pin Connector

PCM Pin #	Wire Color	Circuit Description (16 Pin)	Value at Hot Idle
2	WT/GN	EVAP Purge Solenoid (VSV)	12v or 0v
3	PK/GN	MIL (lamp) Control	MIL Off: 12v, On: 1v
5	RD	Data Link Connector	12-14v
6	YL/RD	A/T Oil Temperature Sensor	At 68°F: 4-5v
7	BK/WT	MAF Sensor Ground (E2G)	<0.050v
8	GY/RD	EVAP Vapor Pressure (VSV)	12v or 0v
9	RD/YL	4WD Detection Transfer (N)	Switch Closed: 0v
10	RD/WT	HO2S-12 (B1 S2) Heater	Heater Off: 12v, On: 1v
11	PK/BL	HO2S-11 (B1 S1) Heater	Heater Off: 12v, On: 1v
13	RD/GN	EVAP Vapor Pressure Sensor	2.5-3.1v (with fuel cap off)
16	BR	Sensor Ground	<0.050v

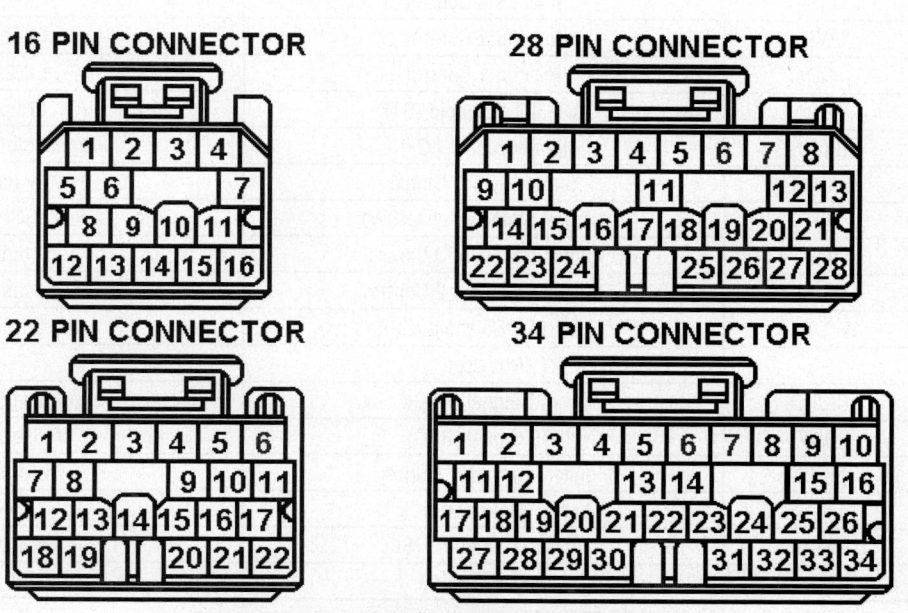

WIRE SIDE OF HARNESS TERMINALS

Pin Connector Graphic

05533_ADIA_G196

2000 3.4L V6 MFI Federal VIN N E5 22 Pin Connector

PCM Pin #	Wire Color	Circuit Description (22 Pin)	Value at Hot Idle
1	GN/BK	Sensor VREF	4.9-5.1v
2-3	---	Not Used	---
4	WT/RD	A/T Vehicle Speed Sensor (-)	AC pulse signals
5	PK	CKP Sensor Signal (NE+)	AC pulse signals
6	PK	CKP Sensor Signal (NE-)	<0.050v
7	BK/YL	TP Sensor Signal (VTA)	0.3-0.8v
8	RD/WT	MAF Sensor Signal (VG)	1.1-1.8v
9	YL/RD	A/T Vehicle Speed Sensor (+)	Pulses
10-12, 16	---	Not Used	---
13	WT	HO2S-11 (B1 S1) Signal	0.1-1.1v
14	GY	Knock Sensor 2 Signal	<0.075v AC
15	BK	Knock Sensor 1 Signal	<0.075v AC
17	BL	CMP Sensor Signal (G2+)	AC pulse signals
18	YL	Circuit Opening Relay (FC)	0-3v, at off-idle: 12v
19	RD	HO2S-12 (B1 S2) Signal	0.1-1.1v
20	GN	ECT Sensor Signal (THW)	At 180°F: 0.51v
21	YL/GN	IAT Sensor Signal (THA)	At 100°F: 2.60v
22	GN/WT	Sensor Ground	<0.050v

2000 3.4L V6 MFI Federal VIN N E6 34-Pin Connector

PCM Pin #	Wire Color	Circuit Description (34-Pin)	Value at Hot Idle
1	GY	4WD Detection Switch	Open: 12v, Closed: 0v
2	BL/RD	4WD Detection Transfer (L4)	Open: 12v, Closed: 0v
3-4	---	Not Used	---
5	YL	Injector 6 Control	1.6-2.9 ms
6	BL	Injector 5 Control	1.6-2.9 ms
7	RD/BK	Injector 4 Control	1.6-2.9 ms
8	GN	Injector 3 Control	1.6-2.9 ms
9	WT	Injector 2 Control	1.6-2.9 ms
10	RD	Injector 1 Control	1.6-2.9 ms
11	RD	A/T-ECT Solenoid (S1)	S1: 3rd or OD: 1v
12	BK/YL	Igniter Signal (IGF)	Digital Signal: 0-5-0v
13	BK/WT	Starter Switch Signal	9-11v (cranking)
14	PK	Neutral Start Switch (NSW)	In P/N: 9-11v (cranking)
15	LG	Igniter Transistor 3 Control	7% duty cycle
16	RD/BL	Igniter Transistor 2 Control	7% duty cycle
17	WT/BL	A/T-ECT Solenoid (S2)	S2: 3rd or OD: 1v
18-21	---	Not Used	---
22	BK/RD	IAC Signal (RSC)	Pulse Signals
23	RD/WT	IAC Signal (RSO)	Pulse Signals
24	BK/BL	Igniter Transistor 1 Control	7% duty cycle
25-26, 32	---	Not Used	---
27	GN/RD	A/T-ECT Solenoid (SL)	In Lockup: 12-14v
28	BR	Power Ground	<0.1v
30	PK/YL	A/T Oil Temperature Sensor	At 68°F: 4-5v

2000 3.4L V6 MFI Federal VIN N E6 34-Pin Connector, *continued*

31	YL/RD	PSP Switch Signal	Straight: 12v, Turning: 0v
33	BR	Power Ground	<0.1v
34	BR	Power Ground	<0.1v

2001-02 3.4L V6 DOHC VIN N (All) E3 22 Pin Connector

PCM Pin #	Wire Color	Circuit Description (22 Pin)	Value at Hot Idle
1	BK/RD	Direct Battery	12-14v
2	BK/OR	Ignition Switch Power	12-14v
3	YL	Circuit Opening Relay (FC)	0-3v, at off-idle: 12v
6	VT/GN	MIL (lamp) Control	MIL Off: 12v, On: 1v
7	BK/WT	Start Switch Signal (STA)	9-11v (cranking)
8	BK/YL	EFI Main Relay Control	Relay Off: 0v, On: 12v
9	GY/RD	EVAP Vapor Pressure (VSV)	12v or 0v
11	WT	SIL (Scan Tool) Signal	12v
15	GN/WT	Stop Light Switch Signal	Brake Off: 0v, On: 12v
16	WT/BL	EFI Main Relay (B+)	12-14v
17	RD/GN	EVAP Vapor Pressure Sensor	2.9-3.1v (with fuel cap off)

2001-02 3.4L V6 DOHC MFI VIN N E4 28 Pin Connector

PCM Pin #	Wire Color	Circuit Description (28 Pin)	Value at Hot Idle
2	RD/BK	A/T Select Switch Reverse	In 'R': 12v, Others: 0v
3	BL	A/T Select Switch 2nd	In 2nd: 12v, Others: 0v
4	YL/GN	Cruise Control ECU (IDLO)	1.5v, off-idle: 12v
6	RD	OD Main Switch (ODMS)	Switch Off: 12v, On: 1v
7	YL/RD	A/T Oil Temperature Indicator	Lamp Off: 12v, On: 1v
8	RD	HO2S-12 (B1 S2) Signal	0.1-1.1v
9	RD/WT	HO2S-12 (B1 S2) Heater	Heater Off: 12v, On: 1v
10	BL/OR	O/D OFF Indicator (ODLP)	Lamp Off: 12v, On: 1v
12	LG	A/T Select Switch Low	In Low: 12v, Others: 0v
13	BL/BK	A/C Amplifier Signal (ACT)	Clutch On: 12v, Off: 1.5v
17	RD/YL	4WD Detection Transfer (N)	Open: 12v, Closed: 0v
18	GY	4WD ADD Indicator Switch	Open: 12v, Closed: 0v
19	BL/RD	4WD Detection Transfer (L4)	Open: 12v, Closed: 0v
20	PK	Neutral Start Switch	In P/N: 9-11v (cranking)
22	GN/OR	Speedometer Indicator (SPI)	At 55 mph: 48 Hz
24	BR/YL	Cruise Control ECU (OD1)	At Cruise in OD: 12v
25	BL/YL	A/C Amplifier Signal (AC1)	Clutch On: 1.5v, Off: 12v

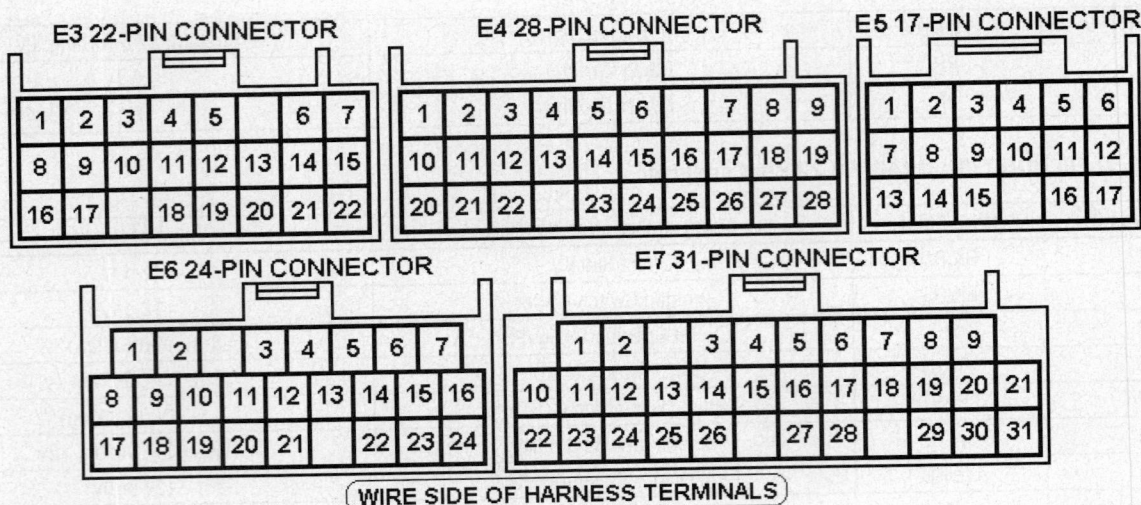

WIRE SIDE OF HARNESS TERMINALS

05533_ADIA_G201

Pin Connector Graphic
2001-02 3.4L V6 DOHC MFI VIN N (All) E5 17 Pin Connector

PCM Pin #	Wire Color	Circuit Description (17 Pin)	Value at Hot Idle
1, 3, 5-7	---	Not Used	---
2	YL	Throttle Control Motor (CL+)	Pulses
4	RD/BK	TP Sensor Signal (VTA2)	2.0-2.9v
8	BL	Throttle Control Motor (CL-)	Pulses
9	---	Not Used	---
10	RD/WT	Accelerator Position Sensor (VPA)	0.25-0.90v
15	BL	Accelerator Position Sensor (VPA2)	1.8-2.7v
16-17	---	Not Used	---

2001-02 3.4L V6 DOHC MFI VIN N E6 24 Pin Connector

PCM Pin #	Wire Color	Circuit Description (24 Pin)	Value at Hot Idle
2	GN/BK	Sensor VREF (VC)	4.9-5.1v
4	YL	AFS-11 (B1 S1) Heater	Heater Off: 12v, On: 1v
5	RD	Injector 1 Control	1.6-2-9 ms
6	WT	Injector 2 Control	1.6-2-9 ms
7	WT/GN	EVAP Purge Solenoid (VSV)	12v or 0v
8	WT/BK	Power Ground (E05)	<0.1v
9	YL/RD	PSP Switch Signal	Straight: 12v, Turning: 0v
10	RD/WT	MAF Sensor Signal (VG)	1-1.1v
12	GN	AFS-11 (B1 S1) Signal (AF+)	Fixed at 3.3v
14	GN	ECT Sensor Signal (THW)	0.5-0.6v
15	RD/YL	A/T Oil Temperature Sensor	At 68°F: 4-5v
16	VT	CKP Sensor Signal (NE+)	AC pulse signals
17	BR	Power Ground	<0.050v
18	GN/WT	Sensor Ground (E2)	<0.050v
19	BK/WT	MAF Sensor Ground (E2G)	<0.050v
21	RD	AFS-11 (B1 S1) Signal (AF-)	Fixed at 3.3v
22	YL/GN	IAT Sensor Signal (THA)	0.5-3.4v
23	BK/YL	TP Sensor Signal (VTA)	0.53-1.27v
24	PK	CMP/CKP Sensor Ground (-)	<0.050v

2001-02 3.4L V6 DOHC MFI VIN N E7 31 Pin Connector

PCM Pin #	Wire Color	Circuit Description (31 Pin)	Value at Hot Idle
1	GN	Injector 3 Control	1.6-2-9 ms
2	RD/BK	Injector 4 Control	1.6-2-9 ms
3	BL	Injector 5 Control	1.6-2-9 ms
4	YL	Injector 6 Control	1.6-2.9 ms
6	RD	Data Link Connector	12-14v
7	RD	A/T-ECT Solenoid (S1)	3rd or OD: 1v
8	WT/BL	A/T-ECT Solenoid (S2)	1st or OD: 1v
9	GN/RD	A/T-ECT Solenoid (SL)	In Lockup: 12-14v
10	BL	CMP Sensor Signal (G2)	AC pulse signals
11	BK/BL	Igniter Transistor 1 Control	6°, 55 mph: 8° dwell
12	RD/BL	Igniter Transistor 2 Control	6°, 55 mph: 8° dwell
13	LG	Igniter Transistor 3 Control	6°, 55 mph: 8° dwell
14	YL/RD	A/T Vehicle Speed Sensor (+)	AC pulse signals
15	BK/RD	IAC Signal (RSC)	Pulse Signals
16	RD/WT	IAC Signal (RSO)	Pulse Signals
21	BR	Power Ground (E01)	<0.1v
25	BK/YL	Igniter Signal (IGF)	Digital Signal: 0-5-0v
26	WT/RD	A/T Vehicle Speed Sensor (-)	AC pulse signals
27	BK	Knock Sensor 1 Signal	<0.075v AC
28	GY	Knock Sensor 2 Signal	<0.075v AC
30	BR	Power Ground (E03)	<0.1v
31	WT/BK	Power Ground (E02)	<0.1v

2003-2006 3.4L V6 DOHC VIN N (All) E3 22 Pin Connector

PCM Pin #	Wire Color	Circuit Description (22 Pin)	Value at Hot Idle
1	WT/BL	EFI Main Relay (B+)	12-14v
2	VT/GN	MIL (lamp) Control	MIL Off: 12v, On: 1v
3	BL/BK	A/C Amplifier Signal (ACT)	Clutch On: 12v, Off: 1.5v
4	BK/YL	EFI Main Relay Control	Relay Off: 0v, On: 12v
5	---	Not Used	---
6	GN/OR	Speedometer Indicator (SP1)	At 55 mph: 48 Hz
7	BK	Start Switch Signal (STA)	9-11v (cranking)
8	WT/GN	ETCS Power (+BM)	12-14v
9	BL/YL	A/C Amplifier Signal (AC1)	Clutch On: 1.5v, Off: 12v
10-12	---	Not Used	---
13	BL/WT	Data Link Connector (WFSE)	12v
14	WT	SIL (Scan Tool) Signal	12v
15	BK/OR	Ignition Switch Power (IGSW)	12-14v
16	BK/RD	Direct Battery	12-14v
17	YL	Cruise Control Indicator	Indicator Off: 12v, On: 1v
18-19	---	Not Used	---
20	GN/WT	Stop Light Switch Signal	Brake Off: 0v, On: 12v
21	RD/BK	Center Airbag Sensor Assembly (F/PS)	Digital Signals
22	YL	Circuit Opening Relay (FC)	0-3v, at off-idle: 12v

2003-2006 3.4L V6 DOHC MFI VIN N E4 28 Pin Connector

PCM Pin #	Wire Color	Circuit Description (28 Pin)	Value at Hot Idle
1, 7	---	Not Used	---
2	BL/RD	4WD Detection Transfer (L4)	Open: 12v, Closed: 0v
3	PK	Neutral Start Switch (NSW)	In 'P' or 'N': 0v
4	LG	A/T Select Switch Signal (L)	In 'L': 12v, Others: 0v
5	BL	A/T Select Switch Signal (2)	In 2nd: 12v, Others: 0v
6	RD/BL	A/T Select Switch Signal (R)	In 'R': 12v, Others: 0v
8	PK/BL	Traction Control ECU Signal (NEO)	Pulse Signals
9	BL/RD	Stop Light Switch (STI-)	Brake Off: 0v, On: 12v
10-12, 16-17	---	Not Used	---
13	GR	4WD Switch Signal (4WD)	4WD Switch On: 12v
14	RD	OD Main Switch (ODMS)	Switch Off: 12v, On: 1v
18	LG	Traction Control ECU Signal (ENG-)	Pulse Signals
19	LG/BK	Traction Control ECU Signal (ENG+)	Pulse Signals
20	BL/OR	O/D OFF Indicator (ODLP)	Lamp Off: 12v, On: 1v
21	YL/RD	A/T Oil Temperature Indicator (OILW)	Lamp Off: 12v, On: 1v
22	GN	Cruise Control Switch, Coast Control Switch	At Cruise in OD: 12v
23-24, 26	---	Not Used	---
25	WT/RD	A/T Select Switch Signal (D)	In 'D': 12v, Others: 0v
27	PK	Traction Control ECU Signal (TRC-)	Pulse Signals
28	VT/GN	Traction Control ECU Signal (TRC+)	Pulse Signals

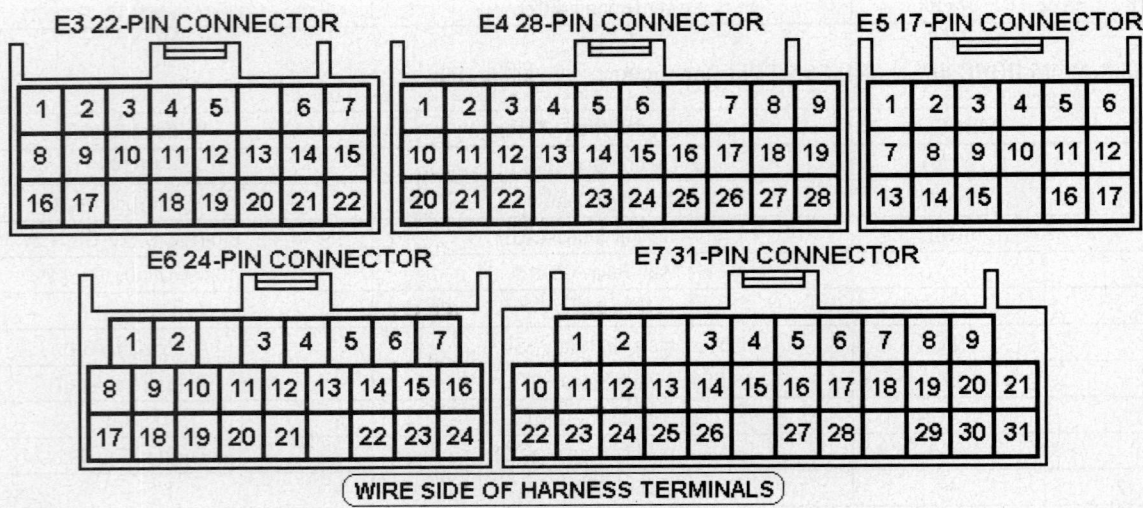

Pin Connector Graphic

2003-2006 3.4L V6 DOHC MFI VIN N (All) E5 17 Pin Connector

PCM Pin #	Wire Color	Circuit Description (17 Pin)	Value at Hot Idle
1, 3, 6-7, 9	---	Not Used	---
2	YL	Throttle Control Motor (CL+)	Pulses
4	RD/BK	TP Sensor Signal (VTA2)	0.3-1.0v
5	BK/RD	A/T-ECT Solenoid (SLT+)	Pulse Signals
8	BL	Throttle Control Motor (CL-)	Pulses
10	RD/WT	Accelerator Position Sensor (VPA)	0.3-1.0v

2003-2006 3.4L V6 DOHC MFI VIN N (All) E5 17 Pin Connector, *continued*

11	GNWT	A/T-ECT Solenoid (SLT-)	Pulse Signals
12-14, 16-17	---	Not Used	---
15	RD	Accelerator Position Sensor (VPA2)	0.3-1.0v

2003-2006 3.4L V6 DOHC MFI VIN N E6 24 Pin Connector

PCM Pin #	Wire Color	Circuit Description (24 Pin)	Value at Hot Idle
1	BL	Injector 5 Control	1.6-2-9 ms
2	BK/YL	Igniter Signal (IGF)	Digital Signal: 0-5-0v
3	RD/BK	Injector 4 Control	1.6-2-9 ms
4	GN	Injector 3 Control	1.6-2-9 ms
5	WT	Injector 2 Control	1.6-2-9 ms
6	RD	Injector 1 Control	1.6-2-9 ms
7	WT/GN	Data Link Connector (TC)	12v
8	YL	Injector 6 Control	1.6-2.9 ms
9	LG	Igniter Transistor 3 Control	6°, 55 mph: 8° dwell
10	RD/BL	Igniter Transistor 2 Control	6°, 55 mph: 8° dwell
11	BK/BL	Igniter Transistor 1 Control	6°, 55 mph: 8° dwell
12	RD	CKP Sensor Signal (NE+)	AC pulse signals
13	BL	CMP Sensor Signal (G2)	AC pulse signals
14	GN/WT	EVAP Purge Solenoid (VSV)	12v or 0v
15-16, 20	---	Not Used	---
17	BR	Power Ground (E01)	<0.1v
18	YL/RD	PSP Switch Signal (PSW)	Straight: 12v, Turning: 0v
19	BL/WT	EVAP Closed Canister Valve (VSV)	12v or 0v
21	GN	CMP/CKP Sensor Ground (NE-)	<0.050v
22	WT/RD	A/T Vehicle Speed Sensor (SP2-)	AC pulse signals
23	YL/RD	A/T Vehicle Speed Sensor (SP2+)	AC pulse signals

2003-2006 3.4L V6 DOHC MFI VIN N E7 31 Pin Connector

PCM Pin #	Wire Color	Circuit Description (31 Pin)	Value at Hot Idle
1	GN/RD	A/T-ECT Solenoid (SL)	In Lockup: 12-14v
2	WT/BL	A/T-ECT Solenoid (S2)	1st or OD: 1v
3	RD	A/T-ECT Solenoid (S1)	3rd or OD: 1v
4	WT/BK	Power Ground (E01)	<0.1v
5	WT/BK	Power Ground (E02)	<0.1v
6	BR	Power Ground (ME01)	<0.1v
7	WT/BK	Power Ground (E03)	<0.1v
8	WT/BK	Power Ground (E04)	<0.1v
9	GN	Throttle Control Motor (M+)	Pulse Signals
10	GN/WT	MAF Sensor Ground (E2G)	<0.050v
11	BK/WT	Sensor Ground (E2)	<0.050v
12	RD/YL	MAF Sensor Signal (VG)	1-1.1v
13	YL/GN	IAT Sensor Signal (THA)	0.2-1.0v
14	VT	AFS-11 (B1 S1) Signal (AF1+)	Fixed at 3.3v
15	BK/YL	TP Sensor Signal (VTA)	0.3-1.0v
16, 20	---	Not Used	---
17	BR	Shield Ground (GE01)	<0.050v
18	GN/YL	ECT Sensor Signal (THW)	0.5-0.6v
19	RD/YL	A/T Oil Temperature Sensor	At 68°F: 4-5v
21	YL	AFS-11 (B1 S1) Heater (HTAF1)	Heater Off: 12v, On: 1v
22	GR	Knock Sensor 2 Signal (KNK2)	<0.075v AC
23	BK	Knock Sensor 1 Signal (KNK1)	<0.075v AC
24	RD/BL	EVAP Vapor Pressure Sensor (PTNK)	2.9-3.1v (with fuel cap off)
25	GN/BK	Sensor VREF (VC)	4.9-5.1v
26	PK	AFS-11 (B1 S1) Signal (AF1-)	Fixed at 3.3v
27	RD	HO2S-12 (B1 S2) Signal (OX2B)	0.1-1.1v
28, 30	---	Not Used	---
29	RD/WT	HO2S-12 (B1 S2) Heater (HT2B)	Heater On: 12v, Off: 0v
31	RD	Throttle Control Motor (M-)	Pulse Signals

2000 4.7L V8 VIN T 2UZ-FE (All) E3 22 Pin Connector

PCM Pin #	Wire Color	Circuit Description (22 Pin)	Value at Hot Idle
1	BK/RD	Direct Battery	12-14v
4	BK/BL	Fuel Pump Relay	Relay Off: 12v, On: 1v
5	YL	Fuel Pump Switch	Switch On: 1v, Off: 12v
6	PK/GN	MIL (lamp) Control	MIL Off: 12v, On: 1v
7-	WT/GN	ETCS Power (+BM)	12-14v
8	WT/BL	EFI Main Relay Power	12-14v
9	BK/OR	Ignition Switch Power	12-14v
10	BK/YL	EFI Main Relay Control	Relay Off: 0v, On: 12v
11	WT	SIL (Scan Tool) Signal	12v
16	WT/BL	EFI Main Relay Power	12-14v

2000 4.7L V8 VIN T 2UZ-FE (All) E4 28 Pin Connector

PCM Pin #	Wire Color	Circuit Description (28 Pin)	Value at Hot Idle
1	BL/OR	Overdrive Main Switch	Switch Off: 12v, On: 1v
2	RD/BK	A/T Select Switch Reverse	In 'R': 12v, Others: 0v
3	BL	A/T Select Switch 2nd	In 2nd: 12v, Others: 0v
4	LG	A/T Select Switch Low	In Low: 12v, Others: 0v
5	GN	Data Link Connector	12-14v
6	GN/WT	Stop Light Switch Signal	Brake Off: 0v, On: 12v
7	YL	HO2S-22 (B2 S2) Heater	Heater Off: 12v, On: 1v
8	RD/YL	HO2S-12 (B1 S2) Heater	Heater Off: 12v, On: 1v
9	BL/RD	4WD Detection Transfer (L4)	Open: 12v, Closed: 0v
10	GY/RD	EVAP Vapor Pressure (VSV)	12v or 0v
12	GN/YL	Taillight Switch Signal	Switch Off: 0v, On: 12v
13	BL/BK	A/C Amplifier Signal (ACT)	Clutch On: 12v, Off: 1.5v
15	GN/OR	Speedometer Indicator	At 55 mph: 48 Hz
16	BL/WT	Tachometer Signal (TACO)	Pulse Signals
17	BK/WT	Starter Switch Signal	9-11v (cranking)
18	BK	HO2S-12 (B1 S2) Signal	0.1-1.1v
19	WT/RD	A/T Select Switch Drive	In 'D': 12v, Others: 0v
20	PK	A/T Neutral Start Switch	In P/N: 9-11v (cranking)
22	RD/GN	EVAP Vapor Pressure Sensor	2.5-3.1v (with fuel cap off)
25	BL/YL	A/C Amplifier Signal (AC1)	Clutch On: 1.5v, Off: 12v
26	YL/RD	A/T Oil Temperature Lamp	Lamp Off: 12v
27	WT	HO2S-22 (B2 S2) Signal	0.1-1.1v

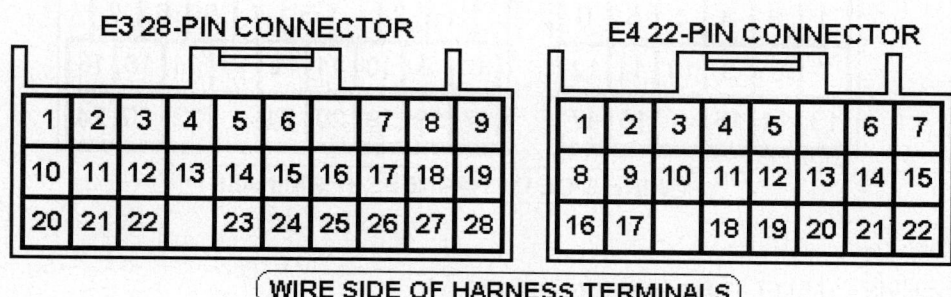

Pin Connector Graphic

2000 4.7L V8 2UZ-FE VIN T (All) E5 17-Pin Connector

PCM Pin #	Wire Color	Circuit Description (17-Pin)	Value at Hot Idle
1	RD	A/T-ECT Solenoid (S1)	S1: 3rd or OD: 1v
2	WT/BL	A/T-ECT Solenoid (S2)	S1: 3rd or OD: 1v
3	GN/RD	A/T-ECT Solenoid (SL)	In Lockup: 12-14v
4	WT	Direct Clutch Speed Input (+)	Pulse Signals
5	YL/RD	A/T Vehicle Speed Sensor (+)	AC pulse signals
9	BK/RD	A/T-ECT Solenoid (SLT+)	Pulse Signals
10	BK	Direct Clutch Speed Input (-)	Pulse Signals
11	WT/RD	A/T Vehicle Speed Sensor (-)	AC pulse signals
15	GN/YL	A/T-ECT Solenoid (SLT-)	Pulse Signals
17	RD/YL	A/T Oil Temperature Sensor	At 68°F: 4-5v

2000 4.7L V8 2UZ-FE VIN T (All) E6 24 Pin Connector

PCM Pin #	Wire Color	Circuit Description (24 Pin)	Value at Hot Idle
1	BR	Power Ground	<0.1v
2	GN/BK	Sensor VREF	4.9-5.1v
3	YL	HO2S-21 (B2 S1) Heater	Heater Off: 12v, On: 1v
4	RD	HO2S-11 (B1 S1) Heater	Heater Off: 12v, On: 1v
5	RD	Injector 1 Control	2.0-3.3 ms
6	WT	Injector 2 Control	2.0-3.3 ms
7	WT/GN	EVAP Purge Solenoid (VSV)	12v or 0v
9	RD/BK	Accelerator Position Sensor 2	1.8-2.7v
10	RD/WT	Mass Airflow Sensor	1.1-1.5v
11	WT	HO2S-21 (B2 S1) Signal	0.1-1.1v
12	BK	HO2S-11 (B1 S1) Signal	0.1-1.1v
13	BK/YL	TP Sensor Signal (VTA1)	0.4-1.0v
14	GN	ECT Sensor Signal (THW)	At 180°F: 0.51v
17	BR	Shield Ground	<0.050v
18	GN/WT	Sensor Ground (E2)	<0.050v
19	BK/WT	MAF Sensor Ground (EVG)	<0.050v
20	PK/BL	TP Sensor Signal (VTA2)	2.0-2.9v
21	GN/RD	Accelerator Position Sensor 1	0.25-0.90v
22	YL/GN	IAT Sensor Signal (THA)	At 100°F: 2.60v

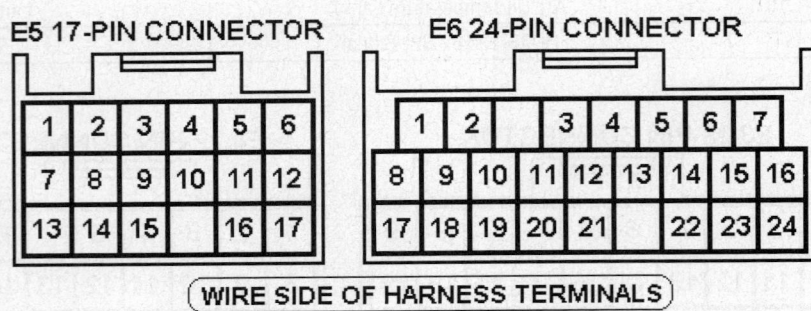

05533_ADIA_G203

Pin Connector Graphic

2000 4.7L V8 2UZ-FE VIN T (All) E7 31 Pin Connector

PCM Pin #	Wire Color	Circuit Description (31 Pin)	Value at Hot Idle
1	GN	Injector 3 Control	2.0-3.3 ms
2	RD/BK	Injector 4 Control	2.0-3.3 ms
3	BL	Injector 5 Control	2.0-3.3 ms
4	YL	Injector 6 Control	2.0-3.3 ms
5	BL/RD	Injector 7 Control	2.0-3.3 ms
6	GN/YL	Injector 8 Control	2.0-3.3 ms
7	WT	Throttle Control Motor (M-)	Pulse Signals
8	RD	Throttle Control Motor (M+)	Pulse Signals
9	BR	Power Ground	<0.1v
10	YL	CMP Sensor Signal (G2+)	AC pulse signals
11	BK/BL	COP Igniter 1 Control	7% duty cycle
12	LG/BK	COP Igniter 2 Control	7% duty cycle
13	BK/YL	COP Igniter 3 Control	7% duty cycle

2000 4.7L V8 2UZ-FE VIN T (All) E7 31 Pin Connector, *continued*

14	RD/WT	COP Igniter 4 Control	7% duty cycle
15	GN/WT	COP IGT 5 Control	7% duty cycle
16	PK/BL	COP IGT 6 Control	7% duty cycle
17	BK	Knock Sensor 2 Signal	<0.075v AC
18	GY	Knock Sensor 1 Signal	<0.075v AC
21	BR	Power Ground	<0.1v
22	RD	CMP/CKP Sensor Signal (-)	<0.050v
23	GN	CKP Sensor Signal (NE+)	AC pulse signals
24	BL/WT	Throttle Control Motor (CL-)	Pulses
25	PK	COP IGT 7 Control	7% duty cycle
26	LG	COP IGT 8 Control	7% duty cycle
27	BK/RD	Igniter Signal (IGF1)	Digital Signal: 0-5-0v
28	BK/WT	Igniter Signal (IGF2)	Digital Signal: 0-5-0v
29	BL/BK	Throttle Control Motor (CL+)	Pulse Signals
30	BL/RD	Shield Ground	<0.050v
31	BR	Power Ground	<0.1v

E7 31-PIN CONNECTOR

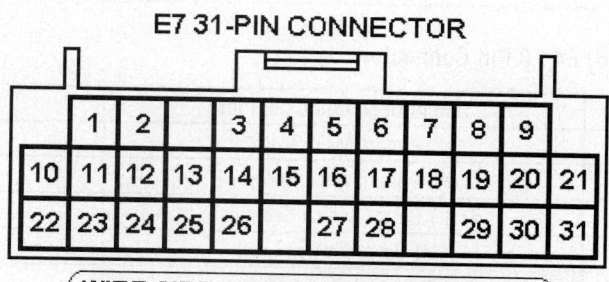

WIRE SIDE OF HARNESS TERMINALS

05533_ADIA_G204

Pin Connector Graphic

Standard Colors and Abbreviations

Abbreviation	Color	Abbreviation	Color	Abbreviation	Color
BK	Black	GY	Gray	RD	Red
BL	Blue	GN	Green	TN	Tan
BR	Brown	LG	LT Green	VT	Violet
DB	Dark Blue	OR	Orange	WT	White
DG	DK Green	PK	Pink	YL	Yellow

2001-02 4.7L V8 VIN T 2UZ-FE (All) E3 22 Pin Connector

PCM Pin #	Wire Color	Circuit Description (22 Pin)	Value at Hot Idle
1	BK/RD	Direct Battery (BATT)	12-14v
2	---	Not Used	---
3	WT/GN	Data Link Connector (TC)	12v
4	BK/BL	Fuel Pump Relay (FPR)	Relay Off: 12v, On: 1v
5	YL	Circuit Opening Relay (FC)	0-3v, at off-idle: 12v
6	VT/GN	MIL (lamp) Control	MIL Off: 12v, On: 1v
7	WT/GN	ETCS Power (+BM)	12-14v
8	WT/BL	EFI Main Relay Power (+B1)	12-14v
9	BK/OR	Ignition Switch Power (IGSW)	12-14v
10	BK/YL	EFI Main Relay Control	Relay Off: 0v, On: 12v
11	WT	SIL (Scan Tool) Signal	Digital Signals
12-15	---	Not Used	---
16	WT/BL	EFI Main Relay Power (+B1)	12-14v
17-18	---	Not Used	---
19	BL/WT	Data Link Connector (WFSE)	12v
20-22	---	Not Used	---

2001-02 4.7L V8 VIN T 2UZ-FE (All) E4 28 Pin Connector

PCM Pin #	Wire Color	Circuit Description (28 Pin)	Value at Hot Idle
1	BL/OR	Overdrive Main Switch	Switch Off: 12v, On: 1v
2	RD/BK	A/T Select Switch Reverse	In 'R': 12v, Others: 0v
3	BL	A/T Select Switch Second	In 2nd: 12v, Others: 0v
4	LG	A/T Select Switch Low	In Low: 12v, Others: 0v
5	GN	Data Link Connector	12v
6	GN/WT	Stop Light Switch Signal	Brake Off: 0v, On: 12v
7	YL	HO2S-22 (B2 S2) Heater	Heater Off: 12v, On: 1v
8	RD/YL	HO2S-12 (B1 S2) Heater	Heater Off: 12v, On: 1v
9	BL/RD	4WD Detection Transfer (L4)	Open: 12v, Closed: 0v
10	GY/RD	EVAP Vapor Pressure (VSV)	12v or 0v
11	---	Not Used	---
12	GN/YL	Taillight Switch (ELS)	Switch Off: 0v, On: 12v
13	BL/YL	A/C Amplifier Signal (ACT)	Clutch On: 1.5v, Off: 12v
14	---	Not Used	---
15	GN/OR	Speedometer Indicator (SPD)	At 55 mph: 48 Hz
16	BL/WT	Tachometer Signal (TACO)	Pulse Signals
17	BK/WT	Starter Switch Signal (STA)	9-11v (cranking)
18	---	Not Used	---
19	WT/RD	A/T Select Switch Drive	In 'D': 12v, Others: 0v
20	PK	Neutral Start Switch (NSW)	In 'P' or 'N': 0v
21	---	Not Used	---
22	RD/GN	EVAP Vapor Pressure Sensor	2.5-3.1v (with fuel cap off)
23	---	Not Used	---
24	YL/RD	AT Oil Temperature Lamp	Lamp Off: 12v, On: 1v
25	BL/YL	A/C Amplifier Signal (A/C)	Clutch On: 12v, Off: 1.5v
26	BL/OR	Overdrive Indicator Lamp	Lamp Off: 12v, On: 1v
27-28	---	Not Used	---

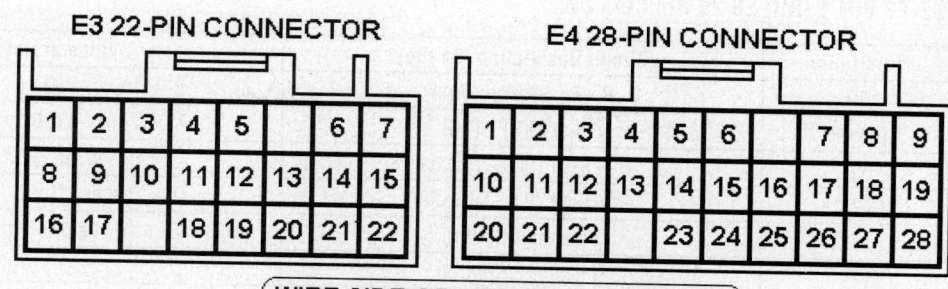

E3 22-PIN CONNECTOR   E4 28-PIN CONNECTOR

WIRE SIDE OF HARNESS TERMINALS

05533_ADIA_G205

Pin Connector Graphic
2001-02 4.7L V8 2UZ-FE VIN T (All) E5 17-Pin Connector

PCM Pin #	Wire Color	Circuit Description (17-Pin)	Value at Hot Idle
1	RD	A/T-ECT Solenoid (S1)	S1: 3rd or OD: 1v
2	WT/BL	A/T-ECT Solenoid (S2)	S1: 3rd or OD: 1v
3	GN/RD	A/T-ECT Solenoid (SL)	In Lockup: 12-14v
4	WT	Direct Clutch Speed Input (+)	Pulse Signals
5	YL/RD	A/T Vehicle Speed Sensor (+)	AC pulse signals
6-8	---	Not Used	---
9	BK/RD	A/T-ECT Solenoid (SLT+)	Pulse Signals
10	BK	Direct Clutch Speed Input (-)	Pulse Signals
11	WT/RD	A/T Vehicle Speed Sensor (-)	AC pulse signals
12-14	---	Not Used	---
15	GN/YL	A/T-ECT Solenoid (SLT-)	AC pulse signals
16	---	Not Used	---
17	RD/YL	A/T Oil Temperature Sensor	At 68°F: 4-5v

2001-02 4.7L V8 2UZ-FE VIN T (All) E6 24 Pin Connector

PCM Pin #	Wire Color	Circuit Description (24 Pin)	Value at Hot Idle
1	BR	Power Ground (E03)	<0.1v
2	GN/BK	Sensor VREF (VC)	4.9-5.1v
3	YL	HO2S-21 (B2 S1) Heater	Heater Off: 12v, On: 1v
4	GN/YL	HO2S-11 (B1 S1) Heater	Heater Off: 12v, On: 1v
5	RD	Injector 1 Control	2.0-3.3 ms
6	WT	Injector 2 Control	2.0-3.3 ms
7	WT/GN	EVAP Purge Solenoid (VSV)	12v or 0v
8	---	Not Used	---
9	BL	Accelerator Position Sensor (VPA2)	1.8-2.7v
10	RD/WT	Mass Airflow Sensor (VG)	1.1-1.5v
11	WT	HO2S-21 (B2 S1) Signal	0.1-1.1v
12	BK	HO2S-11 (B1 S1) Signal	0.1-1.1v
13	BK/YL	TP Sensor Signal (VTA)	0.4-1.0v
14	GN	ECT Sensor Signal (THW)	At 180°F: 0.51v
15	WT	HO2S-22 (B2 S2) Signal	0.1-1.1v
16	BK	HO2S-12 (B1 S2) Signal	0.1-1.1v
17	BR	Shield Ground (E1)	<0.050v
18	GN/WT	Sensor Ground (E2)	<0.050v
19	BK/WT	MAF Sensor Ground (EVG)	<0.050v
20	PK/BL	TP Sensor Signal (VTA2)	2.0-2.9v
21	GN/RD	Accelerator Position Sensor (VPA)	0.25-0.90v
22	YL/GN	IAT Sensor Signal (THA)	At 100°F: 2.60v
23	YL	HO2S-22 (B2 S2) Heater	Heater Off: 12v, On: 1v
24	RD/YL	HO2S-12 (B1 S2) Heater	Heater Off: 12v, On: 1v

2001-02 4.7L V8 2UZ-FE VIN T (All) E7 31 Pin Connector

PCM Pin #	Wire Color	Circuit Description (31 Pin)	Value at Hot Idle
1	GN	Injector 3 Control	2.0-3.3 ms
2	RD/BK	Injector 4 Control	2.0-3.3 ms
3	BL	Injector 5 Control	2.0-3.3 ms
4	YL	Injector 6 Control	2.0-3.3 ms
5	BL/RD	Injector 7 Control	2.0-3.3 ms
6	RD/WT	Injector 8 Control	2.0-3.3 ms
7	WT	Throttle Control Motor (M-)	Pulse Signals
8	RD	Throttle Control Motor (M+)	Pulse Signals
9	BR	Power Ground (ME01)	<0.1v
10	YL	CMP Sensor Signal (G2)	AC pulse signals
11	BK/BL	COP Igniter 1 Control	7% duty cycle
12	LG/BK	COP Igniter 2 Control	7% duty cycle
13	BK/YL	COP Igniter 3 Control	7% duty cycle
14	RD/WT	COP Igniter 4 Control	7% duty cycle
15	GN/WT	COP Igniter 5 Control	7% duty cycle
16	PK/BL	COP Igniter 6 Control	7% duty cycle
17	BK	Knock Sensor 2 Signal	<0.075v AC
18	GY	Knock Sensor 1 Signal	<0.075v AC
19-20	---	Not Used	---

2001-02 4.7L V8 2UZ-FE VIN T (All) E7 31 Pin Connector, *continued*

21	BR	Power Ground (E01)	<0.1v
22	RD	CKP Sensor Signal (NE-)	<0.050v
23	WT/GN	CKP Sensor Signal (NE+)	AC pulse signals
24	BL/WT	Throttle Control Motor (CL-)	Pulse Signals
25	PK	COP Igniter 7 Control	7% duty cycle
26	LG	COP Igniter 8 Control	7% duty cycle
27	BK/RD	Igniter Signal (IGF1)	Digital Signal: 0-5-0v
28	BK/WT	Igniter Signal (IGF2)	Digital Signal: 0-5-0v
29	BL/BK	Throttle Control Motor (CL+)	Pulses
30	BL/RD	Shield Ground (GE01)	<0.050v
31	BR	Power Ground (E02)	<0.1v

E7 31-PIN CONNECTOR

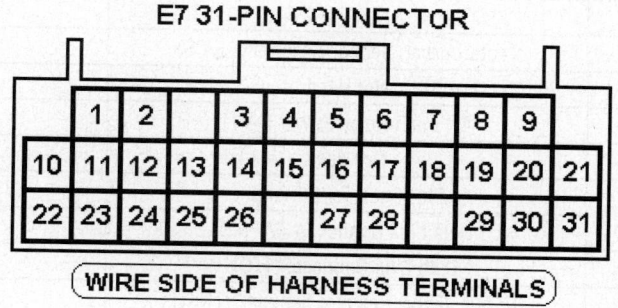

WIRE SIDE OF HARNESS TERMINALS

05533_ADIA_G204

Pin Connector Graphic

2003-2006 4.7L V8 2UZ-FE VIN T (All) E3 31 Pin Connector

PCM Pin #	Wire Color	Circuit Description (31 Pin)	Value at Hot Idle
1	WT/BL	EFI Main Relay Output (+B2)	12-14v
2	WT/BL	EFI Main Relay Output (+B)	12-14v
3	BK/RD	Direct Battery (BATT)	12-14v
4	BK/BL	Fuel Pump Relay (FPR)	Relay Off: 12v, On: 1v
5	BL/WT	Tachometer Signal (TACO)	Pulse Signals
7	WT/GN	ETCS Power (+BM)	12-14v
8	BK/YL	EFI Main Relay Power (+B1)	12-14v
9	BK/OR	Ignition Switch Power (IGSW)	12-14v
10	YL	Circuit Opening Relay (FC)	0-3v, at off-idle: 12v
11	VT/GN	Malfunction Indicator Lamp Control	MIL Off: 12v, On: 1v
12	GN/YL	Taillight Switch (ELS)	Switch Off: 0v, On: 12v
13	---	Not Used	---
14	RD/BK	Center Airbag Sensor Assembly (F/PS)	Digital Signals
15	---	Not Used	---
16	WT/BL	EFI Main Relay Power (+B1)	12-14v
17	---	Not Used	---
18	WT	SIL (Scan Tool) Signal	Digital Signals
19	BL/WT	Data Link Connector (WFSE)	12v
20	WT/GN	Data Link Connector (TC)	12v
21	RD/BL	EVAP Vapor Pressure Sensor (PTNK)	2.5-3.1v (with fuel cap off)
22	RD/GN	Accelerator Position Sensor (VPA)	0.25-0.90v
23	BL	Accelerator Position Sensor (VPA2)	1.8-2.7v
26	PK/GN	Accelerator Pedal Position Sensor (VCPA)	<0.050v
27	GN/RD	Accelerator Pedal Position Sensor (VCP2)	<0.050v
28	GN/WT	Accelerator Pedal Position Sensor (EPA)	<0.050v
29	LG/RD	Accelerator Position Sensor 2 (EPA2)	1.8-2.7v

2003-2006 4.7L V8 2UZ-FE VIN T (All) E4 34 Pin Connector

PCM Pin #	Wire Color	Circuit Description (34 Pin)	Value at Hot Idle
1	BL/WT	Power Ground (HP)	<0.1v
2	BL/YL	A/C Magnetic Clutch Relay (ACMG)	Relay Off: 0v, On: 12v
3	GN	Transmission Control Switch Signal (2L)	In 2nd: 12v, Others: 0v
4	BL/RD	Low Detection Switch (L4)	Switch Open: 0v, Closed: 12v
5	BL/WT	ECT Pattern Switch 2nd Position (SNW1)	2nd Position: 12v
6	WT/GN	ETCS Power (+BM)	12-14v
7	---	Not Used	---
8	BL/YL	Shift Lock ECU Control (L)	12v or 0v
9	YL/BK	Park Neutral Switch Signal (STAR)	In P/N: 9-12v
10	GN/YL	Park Neutral Position Switch (D)	In 'D': 12v, Others: 0v
11	RD/YL	A/T Select Switch Reverse (R)	In 'R': 12v, Others: 0v
12	---	Not Used	---
14	BL/BK	A/C Amplifier Signal (THWO)	A/C Off: 12v, On: 1v
15-16	---	Not Used	---
17	VT/RD	Vehicle Speed Sensor (SPD)	At 55 mph: 48 Hz
18	PK/BK	Body Control ECU Signal (MPX1)	Digital Signals
19	GN/WT	Stop Light Switch Signal (STP)	Brake Off: 0v, On: 12v

2003-2006 4.7L V8 2UZ-FE VIN T (All) E4 34 Pin Connector, *continued*

20	BL	Shift Lock ECU Control (3)	12v or 0v
21-22	---	Not Used	---
23	GN/RD	Shift Lock ECU Control (4)	12v or 0v
25	PK	A/T Oil Temperature Indicator	Indicator Off: 12v, On: 1v
26	BL/RD	Transponder Amplifier Signal (IMD)	Inserting key: pulses
27	WT/RD	Transponder Amplifier Signal (IMI)	Inserting key: pulses
28	BL/WT	ECT Pattern Switch Power Signal	Power Position: 12v
29	BK	Body Control ECU Signal (MPX2)	Digital Signals
30	---	Not Used	---
31	BL	A/C Amplifier Signal (A/CS)	Relay Off: 12v, On: 1v
32	BK/BL	A/C Amplifier Signal (THE)	A/C Off: 12v, On: 1v
33	BL/BK	A/C Switch Signal (ACLD)	A/C On: 12v, Off: 0v
34-35	---	Not Used	---

2003-2006 4.7L 2UZ-FE V8 VIN T E5 32 Pin Connector

PCM Pin #	Wire Color	Circuit Description (32 Pin)	Value at Hot Idle
1	BR	Power Ground (E1)	<0.1v
2	RD	Throttle Control Motor (M-)	Pulse Signals
3	WT	Throttle Control Motor (M+)	Pulse Signals
4	WT/BK	Power Ground (ME01)	<0.1v
5	RD/WT	Injector 8 Control	2.0-3.3 ms
6	BL/RD	Injector 7 Control	2.0-3.3 ms
7	WT/BK	Power Ground (E03)	<0.1v
11	YL/GN	Neutral Detection Switch (L4)	Switch Open: 0v, Closed: 12v
12	BK	Start Switch Signal (STSW)	Cranking: 9-11v
13-14, 18-20	---	Not Used	---
15	BK	A/T Solenoid Control (SLU-)	Pulse Signals
16	PK/GN	A/T Solenoid Control (SLU+)	Pulse Signals
17	BR	Shield Ground (GE01)	<0.050v
21	BK/OR	Generator Control (RL)	12v
22, 26, 28-31	---	Not Used	---
23	GN/WT	A/C Lock Sensor (LCK)	12v or 0v
24	RD	CKP Sensor Signal (NE-)	<0.050v
25	GN	CKP Sensor Signal (NE+)	AC pulse signals
27	YL	CMP Sensor Signal (G2+)	AC pulse signals
32	BL	CMP Sensor Signal (G2-)	AC pulse signals

2003-2006 4.7L V8 2UZ-FE VIN T (All) E6 35 Pin Connector

PCM Pin #	Wire Color	Circuit Description (35 Pin)	Value at Hot Idle
1	BK	Knock Sensor 1 Signal (KNK1 - left)	<0.075v AC
2	WT	Knock Sensor 2 Signal (KNK2 - right)	<0.075v AC
3	YL	Injector 6 Control	2.0-3.3 ms
4	GN/YL	HO2S-11 (B1 S1) Heater (HT1A)	Heater On: 12v, Off: 0v
5	RD/YL	HO2S-12 (B1 S2) Heater (HT1B)	Heater On: 12v, Off: 0v
6	YL/BK	ETCS Power (+BM)	12-14v
7, 12	---	Not Used	---
8	GR	4WD Switch Signal	In 4WD: 12v
9	GN/BK	Shift Lock ECU Control (D)	12v or 0v
10	GN/YL	Shift Lock ECU Control (D)	12v or 0v
11	RD/BK	A/T Select Switch Reverse Signal	In 'R': 1v, Others: 12v
13	YL/RD	A/C Magnetic Clutch Relay (ACMG)	Relay Off: 0v, On: 12v
14	YL/GN	A/C Amplifier Signal (THWO)	A/C Off: 12v, On: 1v
15-16, 20, 30	---	Not Used	---
17	VT	Vehicle Speed Sensor (SPD)	At 55 mph: 48 Hz
18	PK/BK	Body Control ECU Signal (MPX1)	Digital Signals
19	GN/WT	Stop Light Switch Signal	Brake Off: 0v, On: 12v
21	WT	HO2S-22 (B2 S2) Signal (OX2B)	0.1-1.1v
22	WT	HO2S-21 (B2 S1) Signal (OX2A)	0.1-1.1v
23	BK	HO2S-11 (B1 S1) Signal (OX1A)	0.1-1.1v
25	YL/BK	HO2S-22 (B2 S2) Heater (HT2B)	Heater On: 12v, Off: 0v
26	YL	Transponder Amplifier Signal (IMD)	Inserting key: pulses
27	WT	Transponder Amplifier Signal (IMI)	Inserting key: pulses
28	BL/WT	ECT Pattern Switch Power Signal	Power Position: 12v
29	BK	HO2S-12 (B1 S2) Signal (OX1B)	0.1-1.1v
31	BL/BK	A/C Amplifier Signal (ACT)	Relay Off: 12v, On: 1v
32	PK/BK	A/C Amplifier Signal (THE)	A/C Off: 12v, On: 1v
33	YL/GN	HO2S-21 (B2 S1) Heater (HT2A)	Heater On: 12v, Off: 0v
34-35	---	Not Used	---

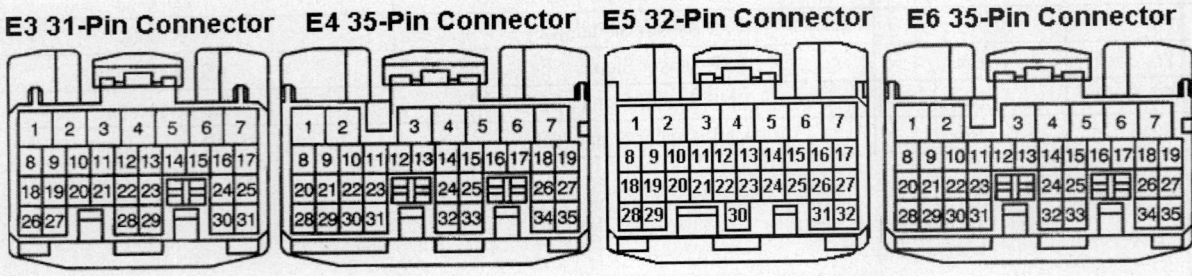

E3 31-Pin Connector   **E4 35-Pin Connector**   **E5 32-Pin Connector**   **E6 35-Pin Connector**

05533_ADIA_G206

Pin Connector Graphic

2003-2006 4.7L V8 VIN T E7 34 Pin Connector

PCM Pin #	Wire Color	Circuit Description (34 Pin)	Value at Hot Idle
1	RD	Injector 1 Control	2.0-3.3 ms
2	WT	Injector 2 Control	2.0-3.3 ms
3	GN	Injector 3 Control	2.0-3.3 ms

2003-2006 4.7L V8 VIN T E7 34 Pin Connector, *continued*

4	RD/BL	Injector 4 Control	2.0-3.3 ms
5	BL	Injector 5 Control	2.0-3.3 ms
6	WT/BK	Power Ground (E02)	<0.1v
7	WT/BK	Power Ground (E01)	<0.1v
8	LG/BK	Igniter Transistor 2 Control	7% duty cycle
9	GN/RD	Igniter Transistor 1 Control	7% duty cycle
10	LG	Igniter Transistor 8 Control	7% duty cycle
11	BK/BL	Igniter Transistor 4 Control	7% duty cycle
12	GN/WT	Igniter Transistor 5 Control	7% duty cycle
13	RD/WT	Igniter Transistor 7 Control	7% duty cycle
14	---	Not Used	---
15	GN/YL	A/C Relay Control (ACCR)	Relay Off: 12v, On: 1v
16	WT/BK	Neutral Start Switch (NSW)	In P/N: 0-3.0v
17	PK	Starter Switch Signal (STA)	In P/N: 0-3.0v
18	GN/BK	Sensor VREF (VC)	4.9-5.1v
19	GN	ECT Sensor Signal (THW)	At 180°F: 0.51v
20	YL/GN	IAT Sensor Signal (THA)	At 100°F: 2.60v
21	BK/YL	TP Sensor Signal (VTA1)	0.4-1.0v
22	---	Not Used	---
23	BL/WT	Igniter Signal (IGF2)	Digital Signal: 0-5-0v
24	BL/BK	Igniter Signal (IGF1)	Digital Signal: 0-5-0v
25	GN/BK	Igniter Transistor 3 Control	7% duty cycle
26	PK/BL	Igniter Transistor 6 Control	7% duty cycle
27	BL/WT	EVAP Canister Closed Valve (VSV)	12v or 0v
28	BK/WT	Sensor Ground (E2)	<0.050v
29	BK/YL	MAF Sensor Ground (E2G)	<0.050v
30	RD/WT	Mass Airflow Sensor (VG)	1.1-1.5v
31	PK/BL	TP Sensor Signal (VTA2)	2.0-2.9v
32	---	Not Used	---
33	GN/RD	Fuel Pump Relay Control (FPR)	Relay Off: 12v, On: 1v
34	YL/RD	EVAP Purge Solenoid (VSV)	12v or 0v

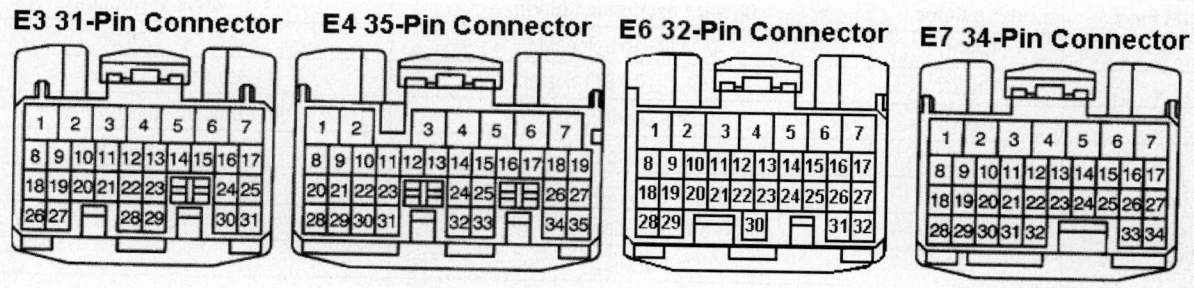

E3 31-Pin Connector   **E4 35-Pin Connector**   **E6 32-Pin Connector**   **E7 34-Pin Connector**

Pin Connector Graphic

05533_ADIA_G207

T-100 PIN CHARTS

1994-95 2.7L I4 3RZ-FE VIN U (M/T) 16 Pin Connector

PCM Pin #	Wire Color	Circuit Description (16 Pin)	Value at Hot Idle
1	GN/YL	Sensor VREF (VC)	4.9-5.1v
2	YL/RD	MAF Sensor Signal (VG)	1.1-1.8v
3	GN/WT	EGR Gas Temperature Sensor (THG)	3.5-4.0v
4	GN/BL	ECT Sensor Signal (THW)	At 180°F: 0.51v
5	WT	HO2S-12 (B1 S2) Signal	0.1-1.1v
6	BK	HO2S-11 (B1 S1) Signal	0.1-1.1v
7	YL/GN	IAT Sensor Signal (THA)	At 100°F: 2.60v
9	BR/BK	Sensor Ground	<0.050v
11	YL	TP Sensor Signal (VTA)	0.3-0.8v
12	YL/BL	Closed Throttle Switch	1v, off-idle: 12v
13	BK	Knock Sensor Signal	<0.075v AC
14	PK/GN	Data Link Connector	12-14v
15	PK/WT	Data Link Connector	12-14v
16	BR	Sensor Ground	<0.050v

1994-95 2.7L I4 3RZ-FE VIN U (M/T) 22 Pin Connector

PCM Pin #	Wire Color	Circuit Description (22 Pin)	Value at Hot Idle
1	BK/GN	Battery Direct	12-14v
5	PK	MIL (lamp) Control	MIL Off: 12v, On: 1v
7	WT	Data Link Connector	12v
8	BL/BK	A/C Amplifier Signal (ACT)	Clutch On: 12v, Off: 1.5v
9	GN	Vehicle Speed Sensor	At 55 mph: 48 Hz
10	BK/RD	A/C Amplifier Signal (AC1)	Clutch On: 1.5v, Off: 12v
11	BK/WT	Starter Switch Signal	9-11v (cranking)
12	WT/RD	EFI Main Relay Power	12-14v
14	GN/YL	Circuit Opening Relay (FC)	0-3v, at off-idle: 12v
20	GN/WT	Stop Light Switch Signal	Brake Off: 0v, On: 12v

1994 2.7L I4 3RZ-FE VIN U (M/T) 26 Pin Connector

PCM Pin #	Wire Color	Circuit Description (26 Pin)	Value at Hot Idle
1	BK/RD	A/C Idle-Up Solenoid	A/C Off: 12v, On: 1v
2	BK/GN	HO2S-11 (B1 S1) Heater	Heater Off: 12v, On: 1v
3	BK/YL	Igniter Signal (IGF)	Digital Signal: 0-5-0v
4	WT	Distributor Signal (NE+)	AC pulse signals
5	BK	Distributor Signal (G2+)	AC pulse signals
6 (Cal)	PK	EGR Solenoid Control (VSV)	12v or 0v
7, 16, 19	---	Not Used	---
8	GN	Fuel Pressure Up Solenoid	1v (at hot startup)
9	PK/YL	IAC Signal (RSC)	Pulse Signals
10	PK/RD	IAC Signal (RSO)	Pulse Signals
11	WT	Injector Pair 2 & 4 Control	1.6-2-9 ms
12	WT/RD	Injector Pair 1 & 3 Control	1.6-2-9 ms
13	BR	Power Ground	<0.1v
14	BR	Sensor Ground	<0.050v

1998-99 2.3L I4 VTEC VIN CG3 [All] 32P 'A' Connector, *continued*

15	RD/GN	HO2S-12 (B1 S2) Heater	Heater Off: 12v, On: 1v
17	BK	Distributor Signal (NE-)	<0.050v
18 ('94)	WT	Distributor Signal (GN-)	<0.050v
18 ('95)	GN	Distributor Signal (GN-)	<0.050v
20	BK/BL	Igniter Signal (IGF)	Digital Signal: 0-5-0v
21-22, 24	---	Not Used	---
23	WT/GN	EVAP Purge Solenoid (VSV)	12v or 0v
25	BR	Power Ground	<0.1v
26	BR	Power Ground	<0.1v

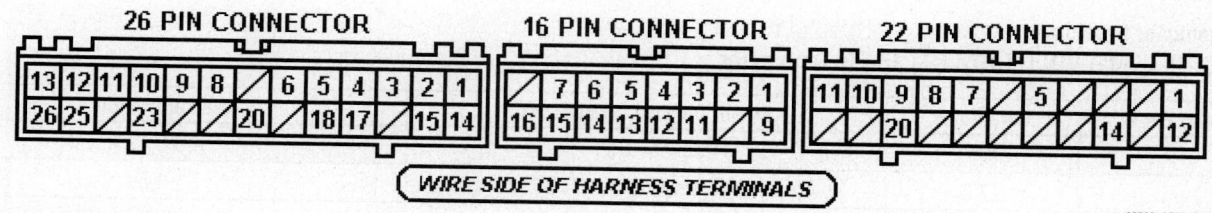

Pin Connector Graphic

05533_ADIA_G208

1995 2.7L I4 MFI VIN U (A/T-ECT) 12 Pin Connector

PCM Pin #	Wire Color	Circuit Description (12 Pin)	Value at Hot Idle
1	WT/GN	EVAP Purge Solenoid (VSV)	12v or 0v
4	BK/RD	A/T Vehicle Speed Sensor (-)	AC pulse signals
5	GN	Distributor Signal (GN-)	<0.050v
6	BK	CKP Sensor Signal (NE-)	<0.050v
10	BR/RD	A/T Vehicle Speed Sensor (+)	AC pulse signals
11	RD	Distributor Signal (G2+)	AC pulse signals
12	WT	CKP Sensor Signal (NE+)	AC pulse signals

1995 2.7L I4 MFI VIN U (A/T-ECT) 22 Pin Connector

PCM Pin #	Wire Color	Circuit Description (22 Pin)	Value at Hot Idle
2	BK/GN	Battery Direct	12-14v
4	PK	MIL (lamp) Control	MIL Off: 12v, On: 1v
5	YL/GN	Overdrive Main Switch	Switch Off: 12v, On: 1v
6	BL/BK	A/C Amplifier Signal (ACT)	Clutch On: 12v, Off: 1.5v
7	BK/RD	A/C Amplifier Signal (AC1)	Clutch On: 1.5v, Off: 12v
8	GN	Vehicle Speed Sensor	At 55 mph: 48 Hz
11	BK/WT	Starter Switch Signal	9-11v (cranking)
12	WT/RD	EFI Main Relay (B+)	12-14v
15	PK/WT	A/T Select Switch Low	In Low: 12v, Others: 0v
16	PK/GN	A/T Select Switch 2nd	In 2nd: 12v, Others: 0v
18	YL/RD	Cruise Control ECU (OD1)	At Cruise in OD: 12v
19	WT	Data Link Connector (SDL)	0v
21	GN/WT	Stop Light Switch Signal	Brake Off: 0v, On: 12v
22	BK/YL	Neutral Start Switch	In P/N: 9-11v (cranking)

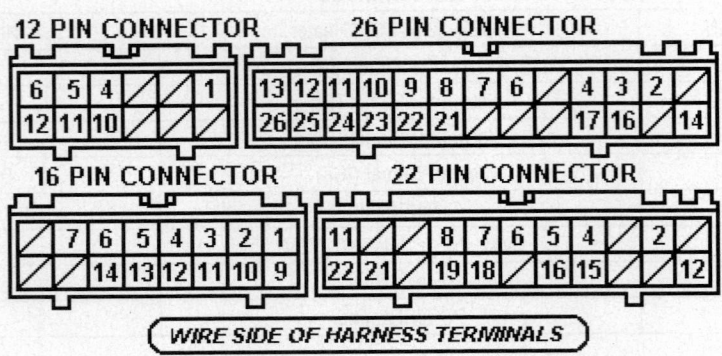

12 PIN CONNECTOR 26 PIN CONNECTOR

16 PIN CONNECTOR 22 PIN CONNECTOR

WIRE SIDE OF HARNESS TERMINALS

05533_ADIA_G209

Pin Connector Graphic
1995 2.7L I4 MFI VIN U (A/T-ECT) 16 Pin Connector

PCM Pin #	Wire Color	Circuit Description (16 Pin)	Value at Hot Idle
1	GN/YL	Sensor VREF	4.9-5.1v
2	YL/RD	MAF Sensor Signal (VG)	1.1-1.8v
3	BR	Sensor Ground	<0.050v
4	GN/YL	ECT Sensor Signal (THW)	At 180°F: 0.51v
5	BK	HO2S-11 (B1 S1) Signal	0.1-1.1v
6	BK	Knock Sensor Signal	<0.075v AC
7	PK/WT	Data Link Connector	12-14v
9	BR/BK	Sensor Ground	<0.050v
10	YL	TP Sensor Signal (VTA)	0.3-0.8v
11	YL/BL	Closed Throttle Switch	1v, off-idle: 12v
12	YL/GN	IAT Sensor Signal (THA)	At 100°F: 2.60v
13	WT	HO2S-12 (B1 S2) Signal	0.1-1.1v
14	GN/WT	EGR Gas Temperature Sensor (THG)	3.5-4.0v

1995 2.7L I4 MFI VIN U (A/T-ECT) 26 Pin Connector

PCM Pin #	Wire Color	Circuit Description (26 Pin)	Value at Hot Idle
2	BK/RD	A/C Idle-Up Solenoid	A/C Off: 12v, On: 1v
3	PK/GN	HO2S-11 (B1 S1) Heater	Heater Off: 12v, On: 1v
4	GN	Fuel Pressure Up Solenoid	1v (at hot startup)
6	PK/YL	IAC Signal (RSC)	Pulse Signals
7	PK/RD	IAC Signal (RSO)	Pulse Signals
8	YL/BK	A/T-ECT Solenoid (SL)	In Lockup: 12-14v
9	BK/WT	A/T-ECT Solenoid (S2)	1st or OD: 1v
10	WT	A/T-ECT Solenoid (S1)	3rd or OD: 1v
11	WT	Injector Pair 2 & 4 Control	1.6-2-9 ms
12	WT/RD	Injector Pair 1 & 3 Control	1.6-2-9 ms
13	BR	Power Ground	<0.1v
14	GN/YL	Circuit Opening Relay (FC)	0-3v, at off-idle: 12v
16	RD/GN	HO2S-12 (B1 S2) Heater	Heater Off: 12v, On: 1v
17	BK/YL	Igniter Signal (IGF)	Digital Signal: 0-5-0v
21	GN/BK	A/T Oil Temperature Sensor	At 68°F: 4-5v
22 (Cal)	PK	EGR Solenoid Control (VSV)	12v or 0v

1995 2.7L I4 MFI VIN U (A/T-ECT) 26 Pin Connector, *continued*

23	BK/BL	Igniter Signal (IGF)	Digital Signal: 0-5-0v
24	BR	Sensor Ground	<0.050v
25	BR	Power Ground	<0.1v
26	BR	Power Ground	<0.1v

Standard Colors and Abbreviations

Abbreviation	Color	Abbreviation	Color	Abbreviation	Color
BK	Black	GY	Gray	RD	Red
BL	Blue	GN	Green	TN	Tan
BR	Brown	LG	LT Green	VT	Violet
DB	Dark Blue	OR	Orange	WT	White
DG	DK Green	PK	Pink	YL	Yellow

1996-97 2.7L I4 MFI VIN M (A/T-ECT) 12 Pin Connector

PCM Pin #	Wire Color	Circuit Description (12 Pin)	Value at Hot Idle
1	WT/GN	EVAP Purge Solenoid (VSV)	12v or 0v
4	BK/RD	A/T Vehicle Speed Sensor (-)	AC pulse signals
5	GN	Distributor Signal (GN-)	<0.050v
6	BK	CKP Sensor Signal (NE-)	<0.050v
10	BR/RD	A/T Vehicle Speed Sensor (+)	Pulses
11	RD	Distributor Signal (G2+)	AC pulse signals
12	WT	CKP Sensor Signal (NE+)	AC pulse signals

1996-97 2.7L I4 MFI VIN M (A/T-ECT) 22 Pin Connector

PCM Pin #	Wire Color	Circuit Description (22 Pin)	Value at Hot Idle
2	BK/GN	Battery Direct	12-14v
4	PK	MIL (lamp) Control	MIL Off: 12v, On: 1v
5	YL/GN	Overdrive Main Switch	Switch Off: 12v, On: 1v
6	BL/BK	A/C Amplifier Signal (ACT)	Clutch On: 12v, Off: 1.5v
7	BK/RD	A/C Amplifier Signal (AC1)	Clutch On: 1.5v, Off: 12v
8	GN	Vehicle Speed Sensor	At 55 mph: 48 Hz
11	BK/WT	Starter Switch Signal	9-11v (cranking)
12	WT/RD	EFI Main Relay Power	12-14v
15	PK/WT	A/T Select Switch Low	In Low: 12v, Others: 0v
16	PK/GN	A/T Select Switch 2nd	In 2nd: 12v, Others: 0v
17	RD/BK	A/T Select Switch Reverse	In 'R': 12v, Others: 0v
19	WT	SIL (Scan Tool) Signal	12v
21	GN/WT	Stop Light Switch Signal	Brake Off: 0v, On: 12v
22	BK/YL	Neutral Start Switch	In P/N: 9-11v (cranking)

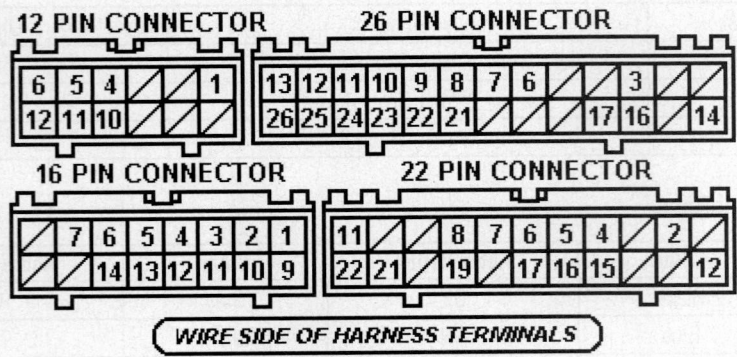

05533_ADIA_G210

Pin Connector Graphic

1996-97 2.7L I4 MFI VIN M (A/T-ECT) 16 Pin Connector

PCM Pin #	Wire Color	Circuit Description (16 Pin)	Value at Hot Idle
1	GN/YL	Sensor VREF	4.9-5.1v
2	YL/RD	MAF Sensor Signal (VG)	1.1-1.8v
3	BR	Sensor Ground	<0.050v
4	GN/YL	ECT Sensor Signal (THW)	At 180°F: 0.51v
5	BK	HO2S-11 (B1 S1) Signal	0.1-1.1v
6	BK	Knock Sensor Signal	<0.075v AC
7	PK/WT	Data Link Connector	12-14v
9	BR/BK	Sensor Ground	<0.050v
10	YL	TP Sensor Signal (VTA)	0.3-0.8v
11	YL/BL	Closed Throttle Switch	1v, off-idle: 12v
12	YL/GN	IAT Sensor Signal (THA)	At 100°F: 2.60v
13	WT	HO2S-12 (B1 S2) Signal	0.1-1.1v
14	GN/WT	EGR Gas Temperature Sensor (THG)	3.5-4.0v

1996-97 2.7L I4 MFI VIN M (A/T-ECT) 26 Pin Connector

PCM Pin #	Wire Color	Circuit Description (26 Pin)	Value at Hot Idle
3	PK/GN	HO2S-11 (B1 S1) Heater	Heater Off: 12v, On: 1v
6	PK/YL	IAC Signal (RSC)	Pulse Signals
7	PK/RD	IAC Signal (RSO)	Pulse Signals
8	YL/BK	A/T-ECT Solenoid (SL)	In Lockup: 12-14v
9	BK/WT	A/T-ECT Solenoid (S2)	1st or OD: 1v
10	WT	A/T-ECT Solenoid (S1)	3rd or OD: 1v
11	WT	Injector Pair 2 & 4 Control	1.6-2-9 ms
12	WT/RD	Injector Pair 1 & 3 Control	1.6-2-9 ms
13	BR	Power Ground	<0.1v
14	GN/YL	Circuit Opening Relay (FC)	0-3v, at off-idle: 12v
16	RD/GN	HO2S-12 (B1 S2) Heater	Heater Off: 12v, On: 1v
17	BK/YL	Igniter Signal (IGF)	Digital Signal: 0-5-0v
21	GN/BK	A/T Oil Temperature Sensor	At 68°F: 4-5v
22	PK	EGR Solenoid Control (VSV)	12v or 0v
23	BK/BL	Igniter Signal (IGF)	Digital Signal: 0-5-0v

1996-97 2.7L I4 MFI VIN M (A/T-ECT) 26 Pin Connector, *continued*

24	BR	Shield Ground	<0.050v
25	BR	Power Ground	<0.1v
26	BR	Power Ground	<0.1v

Standard Colors and Abbreviations

Abbreviation	Color	Abbreviation	Color	Abbreviation	Color
BK	Black	GY	Gray	RD	Red
BL	Blue	GN	Green	TN	Tan
BR	Brown	LG	LT Green	VT	Violet
DB	Dark Blue	OR	Orange	WT	White
DG	DK Green	PK	Pink	YL	Yellow

1996-97 2.7L I4 MFI 3RZ-FE VIN M (M/T) 22 Pin Connector

PCM Pin #	Wire Color	Circuit Description (22 Pin)	Value at Hot Idle
1	BK/GN	Battery Direct	12-14v
5	PK	MIL (lamp) Control	MIL Off: 12v, On: 1v
7	WT	SIL (Scan Tool) Signal	12v
8	BL/BK	A/C Amplifier Signal (ACT)	Clutch On: 12v, Off: 1.5v
9	GN	Vehicle Speed Sensor	At 55 mph: 48 Hz
10	BK/RD	A/C Amplifier Signal (AC1)	Clutch On: 1.5v, Off: 12v
11	BK/WT	Starter Switch Signal	9-11v (cranking)
12	WT/RD	EFI Main Relay (B+)	12-14v
14	GN/YL	Circuit Opening Relay (FC)	0-3v, at off-idle: 12v
20	GN/WT	Stop Light Switch Signal	Brake Off: 0v, On: 12v

1996-97 2.7L I4 MFI 3RZ-FE VIN M (M/T) 16 Pin Connector

PCM Pin #	Wire Color	Circuit Description (16 Pin)	Value at Hot Idle
1	GN/YL	Sensor VREF (VC)	4.9-5.1v
2	YL/RD	MAF Sensor Signal (VG)	1.1-1.8v
3	GN/WT	EGR Gas Temperature Sensor (THG)	3.5-4.0v
4	GN/YL	ECT Sensor Signal (THW)	At 180°F: 0.51v
5	WT	HO2S-12 (B1 S2) Signal	0.1-1.1v
6	BK	HO2S-11 (B1 S1) Signal	0.1-1.1v
7	YL/GN	IAT Sensor Signal (THA)	At 100°F: 2.60v
9	BR/BK	Sensor Ground	<0.050v
11	YL	TP Sensor Signal (VTA)	0.3-0.8v
12	YL/BL	Closed Throttle Switch	1v, off-idle: 12v
13	BK	Knock Sensor Signal	<0.075v AC
15	PK/WT	Data Link Connector	12-14v
16	BR	Sensor Ground	<0.050v

Standard Colors and Abbreviations

Abbreviation	Color	Abbreviation	Color	Abbreviation	Color
BK	Black	GY	Gray	RD	Red
BL	Blue	GN	Green	TN	Tan
BR	Brown	LG	LT Green	VT	Violet
DB	Dark Blue	OR	Orange	WT	White
DG	DK Green	PK	Pink	YL	Yellow

1996-97 2.7L I4 MFI 3RZ-FE VIN M (M/T) 26 Pin Connector

PCM Pin #	Wire Color	Circuit Description (26 Pin)	Value at Hot Idle
2	PK/GN	HO2S-11 (B1 S1) Heater	Heater Off: 12v, On: 1v
3	BK/YL	Igniter Signal (IGF)	Digital Signal: 0-5-0v
4	WT	CKP Sensor Signal (NE+)	AC pulse signals
5	RD	Distributor Signal (G2+)	AC pulse signals
6	PK	EGR Solenoid Control (VSV)	12v or 0v
9	PK/YL	IAC Signal (RSC)	Pulse Signals
10	PK/RD	IAC Signal (RSO)	Pulse Signals
11	WT	Injector Pair 2 & 4 Control	1.6-2-9 ms
12	WT/RD	Injector Pair 1 & 3 Control	1.6-2-9 ms
13	BR	Power Ground	<0.1v
14	BR	Shield Ground	<0.050v
15	RD/GN	HO2S-12 (B1 S2) Heater	Heater Off: 12v, On: 1v
17	BK	CKP Sensor Signal (NE-)	<0.050v
18	GN	Distributor Signal (GN-)	<0.050v
20	BK/BL	Igniter Signal (IGF)	Digital Signal: 0-5-0v
23	WT/GN	EVAP Purge Solenoid (VSV)	12v or 0v
25	BR	Power Ground	<0.1v
26	BR	Power Ground	<0.1v

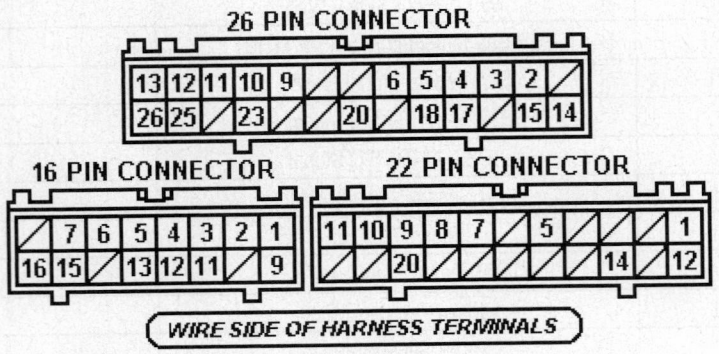

05533_ADIA_G211

Pin Connector Graphic
1998 2.7L I4 MFI VIN M (All) 12-Pin Connector

PCM Pin #	Wire Color	Circuit Description (12 Pin)	Value at Hot Idle
2	GN/RD	Taillight Switch Signal (ELS)	Lights On: 12v
3	BK/RD	A/T Vehicle Speed Sensor (-)	AC pulse signals
5	GN	CMP Sensor Signal (GN-)	<0.050v

1998 2.7L I4 MFI VIN M (All) 12-Pin Connector, *continued*

6	BK	CKP Sensor Signal (NE-)	<0.050v
7	BR	MAF Sensor Ground	<0.050v
9	BR/RD	A/T Vehicle Speed Sensor (+)	AC pulse signals
10	RD/BL	EVAP Vapor Pressure Sensor	2.5-3.1v (with fuel cap off)
11	RD	CMP Sensor Signal (G2+)	AC pulse signals
12	WT	CKP Sensor Signal (NE+)	AC pulse signals

1998 2.7L I4 MFI VIN M (All) 22 Pin Connector

PCM Pin #	Wire Color	Circuit Description (22 Pin)	Value at Hot Idle
2	BK/GN	Battery Direct	12-14v
4	PK	MIL (lamp) Control	MIL Off: 12v, On: 1v
5	YL/GN	Overdrive Main Switch	Switch Off: 12v, On: 1v
6	BL/BK	A/C Amplifier Signal (ACT)	Clutch On: 12v, Off: 1.5v
7	BK/RD	A/C Amplifier Signal (AC1)	Clutch On: 1.5v, Off: 12v
8	GN	Vehicle Speed Sensor	At 55 mph: 48 Hz
11	BK/WT	Starter Switch Signal	9-11v (cranking)
12	WT/RD	EFI Main Relay Power	12-14v
15	PK/WT	A/T Select Switch Low	In Low: 12v, Others: 0v
16	PK/GN	A/T Select Switch 2nd	In 2nd: 12v, Others: 0v
17	RD/BK	A/T Select Switch Reverse	In 'R': 12v, Others: 0v
19	WT	SIL (Scan Tool) Signal	12v
21	GN/WT	Stop Light Switch Signal	Brake Off: 0v, On: 12v
22	BK/YL	Neutral Start Switch	In P/N: 9-11v (cranking)

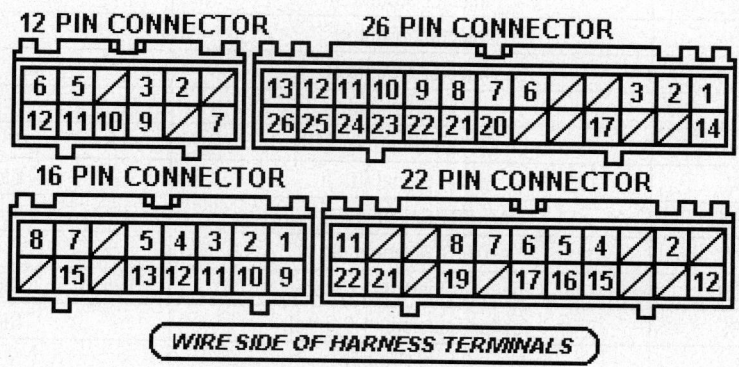

05533_ADIA_G212

Pin Connector Graphic

1998 2.7L I4 MFI VIN M (All) 16 Pin Connector

PCM Pin #	Wire Color	Circuit Description (16 Pin)	Value at Hot Idle
1	GN/YL	Sensor VREF (VC)	4.9-5.1v
2	YL/RD	MAF Sensor Signal (VG)	1.1-1.8v
3	YL/GN	IAT Sensor Signal (THA)	At 100°F: 2.60v
4	GN/YL	ECT Sensor Signal (THW)	At 180°F: 0.51v
5	BK	HO2S-11 (B1 S1) Signal	0.1-1.1v
7	PK/WT	Data Link Connector	12-14v
8	GN/RD	EVAP Vapor Pressure (VSV)	12v or 0v
9	BR/BK	Sensor Ground	<0.050v
10	YL	TP Sensor Signal (VTA)	0.3-0.8v
11	GN/WT	EGR Gas Temperature Sensor (THG)	3.5-4.0v
12	BK	Knock Sensor Signal	<0.075v AC
13	WT	HO2S-12 (B1 S2) Signal	0.1-1.1v
15	PK	EGR Solenoid Control (VSV)	12v or 0v

1998 2.7L I4 MFI VIN M (All) 26 Pin Connector

PCM Pin #	Wire Color	Circuit Description (26 Pin)	Value at Hot Idle
1	RD/GN	HO2S-12 (B1 S2) Heater	Heater Off: 12v, On: 1v
2	PK/GN	HO2S-11 (B1 S1) Heater	Heater Off: 12v, On: 1v
3	WT/GN	EVAP Purge Solenoid (VSV)	12v or 0v
6	PK/YL	IAC Signal (RSC)	Pulse Signals
7	PK/RD	IAC Signal (RSO)	Pulse Signals
8	WT	A/T-ECT Solenoid (S1)	3rd or OD: 1v
9	PK/BK	Injector 4 Control	1.6-2-9 ms
10	BK	Injector 3 Control	1.6-2-9 ms
11	WT	Injector 2 Control	1.6-2-9 ms
12	WT/RD	Injector 1 Control	1.6-2-9 ms
13	BR	Power Ground	<0.1v
14	GN/YL	Circuit Opening Relay (FC)	0-3v, at off-idle: 12v
17	BK/YL	Igniter Signal (IGF)	Digital Signal: 0-5-0v
20	YL/BK	A/T-ECT Solenoid (SL)	In Lockup: 12-14v
21	BK/WT	A/T-ECT Solenoid (S2)	1st or OD: 1v
22	BK/OR	Igniter Transistor 2 Control	Digital Signal: 0-5-0v
23	BK/BL	Igniter Transistor 1 Control	Digital Signal: 0-5-0v
24	BR	Shield Ground	<0.050v
25	BR	Power Ground	<0.1v
26	BR	Power Ground	<0.1v

Standard Colors and Abbreviations

Abbreviation	Color	Abbreviation	Color	Abbreviation	Color
BK	Black	GY	Gray	RD	Red
BL	Blue	GN	Green	TN	Tan
BR	Brown	LG	LT Green	VT	Violet
DB	Dark Blue	OR	Orange	WT	White
DG	DK Green	PK	Pink	YL	Yellow

1993-94 3.0L V6 MFI VIN V (All) 16 Pin Connector

PCM Pin #	Wire Color	Circuit Description (16 Pin)	Value at Hot Idle
1	GN/BK	Sensor VREF (VC)	4.9-5.1v
2	BL/YL	MAF Sensor Signal (VG)	1.1-1.8v
3	BL/RD	IAT Sensor Signal (THA)	At 100ºF: 2.60v
4	GN/BL	ECT Sensor Signal (THW)	At 180ºF: 0.51v
5	BK	Knock Sensor Signal	<0.075v AC
6	BK	HO2S-11 (B1 S1) Signal	0.1-1.1v
7	GN/BL	4WD Oil Temperature Sensor	At 68ºF: 4-5v
8	YL/GN	Check Connector	12-14v
9	BR	Sensor Ground	<0.050v
10	WT	HO2S-12 (B1 S2) Signal	0.1-1.1v
11	YL	TP Sensor Signal (VTA)	0.3-0.8v
12	YL/BL	Closed Throttle Switch	1v, off-idle: 12v
13 (Cal)	GN/WT	EGR Gas Temperature Sensor (THG)	3.5-4.0v
15	PK/WT	Data Link Connector	12-14v
16	PK/GN	Data Link Connector	12-14v

1993-94 3.0L V6 MFI VIN V (All) 22 Pin Connector

PCM Pin #	Wire Color	Circuit Description (22 Pin)	Value at Hot Idle
1	BK/GN	Battery Direct	12-14v
4	BL/WT	A/T Oil Temperature Lamp	Lamp Off: 12v, On: 1v
5	PK	MIL (lamp) Control	MIL Off: 12v, On: 1v
6	GN/WT	Stop Light Switch Signal	Brake Off: 0v, On: 12v
7	BL/RD	A/T Pattern Select Switch	Norm: 0v, PWR: 12v
8	PK/GN	4WD Selector Switch	Open: 12v, Closed: 0v
9	GN/BL	Vehicle Speed Sensor	At 55 mph: 48 Hz
10	BL/BK	A/C Amplifier Signal (ACT)	Clutch On: 12v, Off: 1.5v
11	BK/WT	Starter Switch Signal	9-11v (cranking)
12	WT/RD	EFI Main Relay Power	12-14v
13	WT/RD	EFI Main Relay Power	12-14v
15	BR/BK	Sensor Ground (E2)	<0.050v
16	YL/GN	Main Overdrive Switch	Switch Off: 12v, On: 1v
17	BR/WT	M/T Ground (SEL2)	<0.050v
17	BR/RD	A/T Short Pin Ground (SEL2)	<0.050v
18	BR/YL	A/T 2WD Ground (SEL1)	<0.050v
19	BK/WT	A/C Amplifier Signal (AC1)	Clutch On: 1.5v, Off: 12v
20	YL/WT	Data Link Connector	12v
21	YL/RD	Cruise Control ECU (OD1)	At Cruise in OD: 12v
22 (Cal)	RD/GN	HO2S-12 (B1 S2) Heater	Heater Off: 12v, On: 1v

1993-94 3.0L V6 MFI VIN V (All) 26 Pin Connector

PCM Pin #	Wire Color	Circuit Description (26 Pin)	Value at Hot Idle
1	WT	Distributor Signal (NE+)	AC pulse signals
2	RD	Distributor Signal (G1+)	AC pulse signals
3	BK/YL	Igniter Signal (IGF)	Digital Signal: 0-5-0v
4	WT/GN	EVAP Purge Solenoid (VSV)	12v or 0v
5	YL/BK	A/T-ECT Solenoid (S3)	In Lockup: 12-14v
6	BK	A/T-ECT Solenoid (S2)	1st or OD: 1v
7	WT	A/T-ECT Solenoid (S1)	3rd or OD: 1v
8	GN/YL	Fuel Pressure Up Solenoid	1v (at hot startup)
9	GN	Cold Start Injector	1v (at cold startup)
10	GY/BK	HO2S-11 (B1 S1) Heater	Heater Off: 12v, On: 1v
11	BR	Sensor Ground	<0.050v
12	WT/RD	Injectors 1 & 3 & 5 Control	1.6-2.9 ms
13	BR	Power Ground	<0.1v
14	GN	Distributor Signal (GN-)	<0.050v
15	BK	Distributor Signal (G2+)	AC pulse signals
16	BR/RD	A/T Vehicle Speed Sensor	Moving: 0-5-0v
17	PK/WT	A/T Select Switch Low	In Low: 12v, Others: 0v
18	PK/GN	A/T Select Switch 2nd	In 2nd: 12v, Others: 0v
19	PK/RD	A/T Select Switch Neutral	In 'N': 12v, Others: 0v
20	RD	4WD Detection Transfer (L4)	Open: 12v, Closed: 0v
21	BK/BL	Igniter Signal (IGF)	Digital Signal: 0-5-0v
22 (Cal)	PK	EGR Solenoid Control (VSV)	12v or 0v
23	GN/RD	Intake Air Solenoid Control	12v or 0v
24	BK/RD	A/C Idle-Up Solenoid	A/C Off: 12v, On: 1v
25	WT	Injectors 2 & 4 & 6 Control	1.6-2.9 ms
26	BR	Power Ground	<0.1v

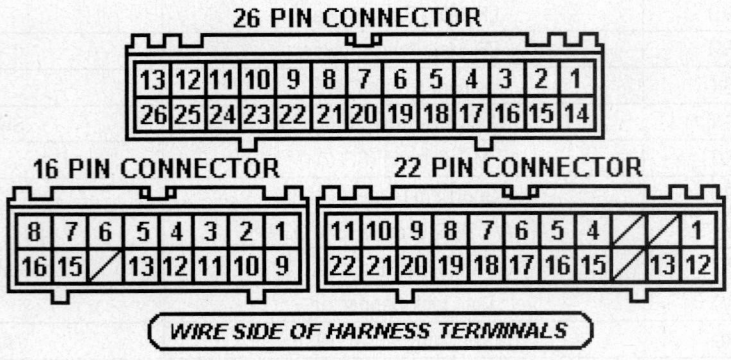

Pin Connector Graphic

1995 3.4L V6 MFI VIN V (A/T-ECT) 28 Pin Connector

PCM Pin #	Wire Color	Circuit Description (28 Pin)	Value at Hot Idle
2	PK/GN	A/T Select Switch 2nd	In 2nd: 12v, Others: 0v
3	PK/WT	A/T Select Switch Low	In Low: 12v, Others: 0v
4	LG	A/T Oil Temperature Lamp	Lamp Off: 12v, On: 1v
5	BL/BK	A/C Amplifier Signal (ACT)	Clutch On: 12v, Off: 1.5v

1995 3.4L V6 MFI VIN V (A/T-ECT) 28 Pin Connector, *continued*

6	YL/GN	Overdrive Indicator Lamp	Lamp Off: 12-14v
7	YL/RD	Cruise Control ECU (OD1)	At Cruise in OD: 12v
12	GN	Vehicle Speed Sensor	At 55 mph: 48 Hz
14	BK/GN	Battery Direct	12-14v
17	YL	A/T: Parking Indicator Lamp	Lamp Off: 12v, On: 1v
18	WT	SIL (Scan Tool) Signal	12v
20	BK/RD	A/C Amplifier Signal (AC1)	Clutch On: 1.5v, Off: 12v
22	WT/RD	EFI Main Relay Power	12-14v
25	GN/WT	Stop Light Switch Signal	Brake Off: 0v, On: 12v
26	PK/GN	4WD Selector Switch	Switch Off: 12v, On: 1v

1995 3.4L V6 MFI VIN V (A/T-ECT) 22 Pin Connector

PCM Pin #	Wire Color	Circuit Description (22 Pin)	Value at Hot Idle
1	GN/BK	Sensor VREF	4.9-5.1v
5	GN	CKP Sensor Signal (NE+)	AC pulse signals
6	BL	CKP Sensor Signal (NE-)	<0.050v
7	YL/BK	TP Sensor Signal (VTA)	0.3-0.8v
8	GY/RD	MAF Sensor Signal (VG)	1.1-1.8v
9	BR/RD	A/T Vehicle Speed Sensor	Moving: 0-5-0v
10	BK	CMP Sensor Signal (G2+)	AC pulse signals
11	WT	CMP Sensor Signal (GN-)	<0.050v
12	GN/BK	A/T Oil Temperature Sensor	At 230°F: <1.5v
13	WT	HO2S-11 (B1 S1) Signal	0.1-1.1v
14	YL/GN	IAT Sensor Signal (THA)	At 100°F: 2.60v
16	GY	Knock Sensor 2 Signal	<0.075v AC
17	BK	Knock Sensor 1 Signal	<0.075v AC
18	BR/WT	Sensor Ground	<0.050v
19	RD	HO2S-12 (B1 S2) Signal	0.1-1.1v
20	GN/YL	ECT Sensor Signal (THW)	At 180°F: 0.51v
21	PK	EGR Gas Temperature Sensor (THG)	3.5-4.0v
22	BR/BK	Sensor Ground	<0.050v

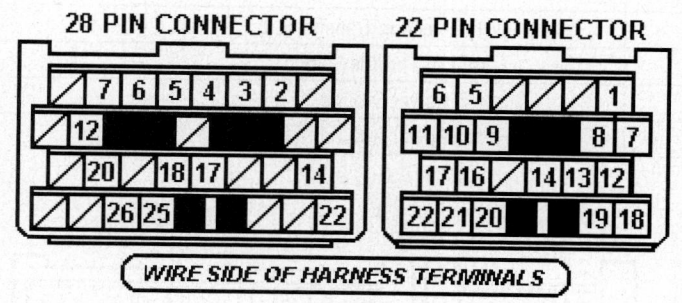

Pin Connector Graphic

05533_ADIA_G214

1995 3.4L V6 MFI VIN V (A/T-ECT) 16 Pin Connector

PCM Pin #	Wire Color	Circuit Description (16 Pin)	Value at Hot Idle
3	PK	MIL (lamp) Control	MIL Off: 12v, On: 1v
4	GN/YL	Circuit Opening Relay (FC)	0-3v, at off-idle: 12v
5	PK/WT	Data Link Connector	12-14v
8	RD/WT	EGR Solenoid Control (VSV)	12v or 0v
9	RD/BK	Fuel Pressure Up Solenoid	1v (at hot startup)
10	BK/RD	A/C Idle-Up Solenoid	A/C Off: 12v, On: 1v
15	WT/GN	EVAP Purge Solenoid (VSV)	12v or 0v
16	BR	Sensor Ground	<0.050v

1995 3.4L V6 MFI VIN V (A/T-ECT) 34-Pin Connector

PCM Pin #	Wire Color	Circuit Description (34-Pin)	Value at Hot Idle
1	BR	Power Ground	<0.1v
5	YL/BK	Injector 6 Control	1.6-2.9 ms
6	WT/BL	Injector 5 Control	1.6-2.9 ms
7	YL/RD	Injector 4 Control	1.6-2.9 ms
8	WT/GN	Injector 3 Control	1.6-2.9 ms
9	WT	Injector 2 Control	1.6-2.9 ms
10	WT/RD	Injector 1 Control	1.6-2.9 ms
11	WT	A/T-ECT Solenoid (S1)	3rd or OD: 1v
12	BK/YL	Igniter Signal (IGF)	Digital Signal: 0-5-0v
13	BK/WT	Starter Switch Signal	9-11v (cranking)
14	BK/OR	Neutral Start Switch	In P/N: 9-11v (cranking)
15	RD/GN	HO2S-12 (B1 S2) Heater	Heater Off: 12v, On: 1v
16	PK/GN	HO2S-11 (B1 S1) Heater	Heater Off: 12v, On: 1v
17	BK/WT	A/T-ECT Solenoid (S2)	1st or OD: 1v
22	BK/RD	IAC Signal (RSC)	Pulse Signals
23	BR/RD	IAC Signal (RSO)	Pulse Signals
24	BK/BL	Igniter Transistor 1 Control	Digital Signal: 0-5-0v
25	BR/BK	Igniter Transistor 2 Control	Digital Signal: 0-5-0v
26	BK/WT	Igniter Transistor 3 Control	Digital Signal: 0-5-0v
27	YL/BK	A/T-ECT Solenoid (SL)	In Lockup: 12-14v
29	RD	4WD Detection Transfer (L4)	Open: 12v, Closed: 0v
32	YL/BL	Closed Throttle Switch	1v, off-idle: 12v
33	BR	Power Ground	<0.1v
34	BR	Power Ground	<0.1v

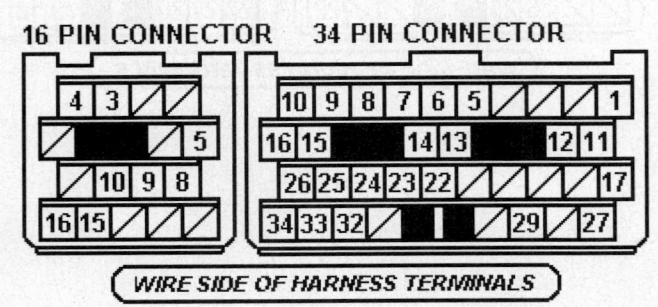

WIRE SIDE OF HARNESS TERMINALS

05533_ADIA_G186

Pin Connector Graphic

1995 3.4L V6 MFI VIN V (M/T) 12 Pin Connector

PCM Pin #	Wire Color	Circuit Description (12 Pin)	Value at Hot Idle
3	PK/GN	HO2S-11 (B1 S1) Heater	Heater Off: 12v, On: 1v
5	WT	CMP Sensor Signal (GN-)	<0.050v
6	BL	CKP Sensor Signal (NE-)	<0.050v
7	BR	Power Ground	<0.1v
9	RD/GN	HO2S-12 (B1 S2) Heater	Heater Off: 12v, On: 1v
11	BK	CMP Sensor Signal (G2+)	AC pulse signals
12	GN	CKP Sensor Signal (NE+)	AC pulse signals

1995 3.4L V6 MFI VIN V (M/T) 22 Pin Connector

PCM Pin #	Wire Color	Circuit Description (22 Pin)	Value at Hot Idle
2	BK/GN	Battery Direct	12-14v
4	PK	MIL (lamp) Control	MIL Off: 12v, On: 1v
6	BL/BK	A/C Amplifier Signal (ACT)	Clutch On: 12v, Off: 1.5v
7	BK/RD	A/C Amplifier Signal (AC1)	Clutch On: 1.5v, Off: 12v
8	GN	Vehicle Speed Sensor	At 55 mph: 48 Hz
9	PK/GN	4WD Selector Switch	Switch Off: 12v, On: 1v
11	BK/WT	Starter Switch Signal	9-11v (cranking)
12	WT/RD	EFI Main Relay Power	12-14v
19	WT	SIL (Scan Tool) Signal	12v
20	GN/WT	Stop Light Switch Signal	Brake Off: 0v, On: 12v

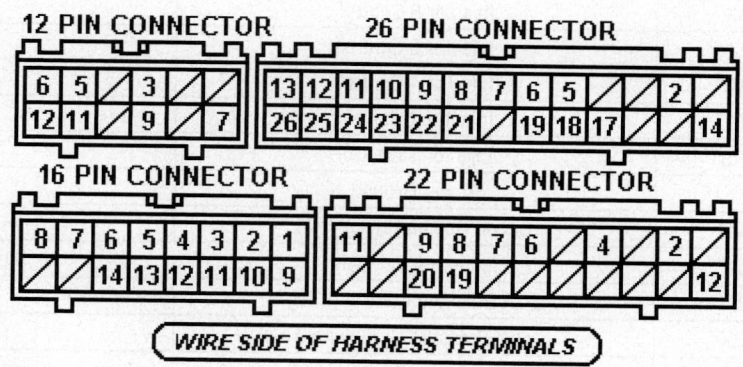

Pin Connector Graphic

05533_ADIA_G187

1995 3.4L V6 MFI VIN V (M/T) 16 Pin Connector

PCM Pin #	Wire Color	Circuit Description (16 Pin)	Value at Hot Idle
1	GN/YL	Sensor VREF (VC)	4.9-5.1v
2	GY/RD	MAF Sensor Signal (VG)	1.1-1.8v
3	GY	Knock Sensor 2 Signal	<0.075v AC
4	GN/YL	ECT Sensor Signal (THW)	At 180°F: 0.51v
5	WT	HO2S-11 (B1 S1) Signal	0.1-1.1v
6	BK	Knock Sensor 1 Signal	<0.075v AC
7	PK/WT	Data Link Connector	12-14v
8	BR/WT	Sensor Ground	<0.050v
9	BR	Sensor Ground	<0.050v
10	YL/BK	TP Sensor Signal (VTA)	0.3-0.8v
11	YL/BL	Closed Throttle Switch	1v, off-idle: 12v
12	YL/GN	IAT Sensor Signal (THA)	At 100°F: 2.60v
13	RD	HO2S-12 (B1 S2) Signal	0.1-1.1v
14	PK	EGR Gas Temperature Sensor (THG)	3.5-4.0v

1995 3.4L V6 MFI VIN V (M/T) 26 Pin Connector

PCM Pin #	Wire Color	Circuit Description (26 Pin)	Value at Hot Idle
2	BK/RD	A/C Idle-Up Solenoid	A/C Off: 12v, On: 1v
5	WT/GN	EVAP Purge Solenoid (VSV)	12v or 0v
6	BK/RD	IAC Signal (RSC)	Pulse Signals
7	BR/RD	IAC Signal (RSO)	Pulse Signals
8	YL/BK	Injector 6 Control	1.6-2.9 ms
9	WT/BL	Injector 5 Control	1.6-2.9 ms
10	YL/RD	Injector 4 Control	1.6-2.9 ms
11	WT	Injector 2 Control	1.6-2.9 ms
12	WT/RD	Injector 1 Control	1.6-2.9 ms
13	BR	Power Ground	<0.1v
14	GN/YL	Circuit Opening Relay (FC)	0-3v, at off-idle: 12v
17	BK/YL	Igniter Signal (IGF)	Digital Signal: 0-5-0v
18	RD/WT	EGR Solenoid Control (VSV)	12v or 0v
19	RD/BK	Fuel Pressure Up Solenoid	1v (at hot startup)
21	BK/WT	Igniter Transistor 3 Control	Digital Signal: 0-5-0v
22	BR/BK	Igniter Transistor 2 Control	Digital Signal: 0-5-0v
23	BK/BL	Igniter Transistor 1 Control	Digital Signal: 0-5-0v
24	BR	Sensor Ground	<0.050v
25	WT/GN	Injector 3 Control	1.6-2.9 ms
26	BR	Power Ground	<0.1v

1996 3.4L V6 MFI VIN N (A/T-ECT) 28 Pin Connector

PCM Pin #	Wire Color	Circuit Description (28 Pin)	Value at Hot Idle
1	RD/BK	A/T Select Switch Reverse	In 'R': 12v, Others: 0v
2	PK/GN	A/T Select Switch 2nd	In 2nd: 12v, Others: 0v
3	PK/WT	A/T Select Switch Low	In Low: 12v, Others: 0v
4	LG	A/T Oil Temperature Lamp	Lamp Off: 12v, On: 1v
5	BL/BK	A/C Amplifier Signal (ACT)	Clutch On: 12v, Off: 1.5v
6	YL/GN	Overdrive Main Switch	Switch Off: 12v, On: 1v

1996 3.4L V6 MFI VIN N (A/T-ECT) 28 Pin Connector, *continued*

7	YL/RD	Cruise Control ECU (OD1)	At Cruise in OD: 12v
12	GN	Vehicle Speed Sensor	At 55 mph: 48 Hz
14	BK/GN	Battery Direct	12-14v
17	YL	4WD Detection Transfer (N)	Open: 12v, Closed: 0v
18	WT	SIL (Scan Tool) Signal	12v
20	BK/RD	A/C Amplifier Signal (AC1)	Clutch On: 1.5v, Off: 12v
22	WT/RD	EFI Main Relay Power	12-14v
25	GN/WT	Stop Light Switch Signal	Brake Off: 0v, On: 12v
26	PK/GN	4WD Selector Switch	Switch Off: 12v, On: 1v

1996 3.4L V6 MFI VIN N (A/T-ECT) 22 Pin Connector

PCM Pin #	Wire Color	Circuit Description (22 Pin)	Value at Hot Idle
1	GN/BK	Sensor VREF (VC)	4.9-5.1v
5	GN	CKP Sensor Signal (NE+)	AC pulse signals
6	BL	CKP Sensor Signal (NE-)	<0.050v
7	YL/BK	TP Sensor Signal (VTA)	0.3-0.8v
8	GY/RD	MAF Sensor Signal (VG)	1.1-1.8v
9	BR/RD	A/T Vehicle Speed Sensor	Moving: 0-5-0v
10	BK	CMP Sensor Signal (G2+)	AC pulse signals
11	WT	CMP Sensor Signal (GN-)	<0.050v
12	GN/BK	A/T Oil Temperature Sensor	At 230°F: <1.5v
13	WT	HO2S-11 (B1 S1) Signal	0.1-1.1v
14	YL/GN	IAT Sensor Signal (THA)	At 100°F: 2.60v
16, 17	GY, BK	Knock Sensor 2 Signal	<0.075v AC
18	BR/WT	Sensor Ground	<0.050v
19 (2WD)	RD	HO2S-21 (B2 S1) Signal	0.1-1.1v
19 (4WD)	RD	HO2S-12 (B1 S2) Signal	0.1-1.1v
20	GN/YL	ECT Sensor Signal (THW)	At 180°F: 0.51v
21	PK	EGR Gas Temperature Sensor (THG)	3.5-4.0v
22	BR/BK	Sensor Ground	<0.050v

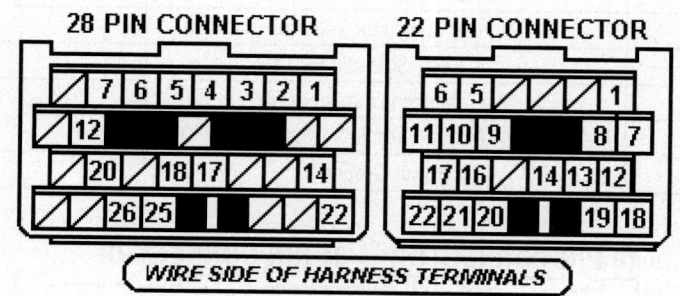

Pin Connector Graphic

05533_ADIA_G215

1996 3.4L V6 MFI VIN N (A/T-ECT) 16 Pin Connector

PCM Pin #	Wire Color	Circuit Description (16 Pin)	Value at Hot Idle
3	PK	MIL (lamp) Control	MIL Off: 12v, On: 1v
4	GN/YL	Circuit Opening Relay (FC)	0-3v, at off-idle: 12v
5	PK/WT	Data Link Connector	12-14v
8	RD/WT	EGR Solenoid Control (VSV)	12v or 0v
9	RD/BK	Fuel Pressure Up Solenoid	1v (at hot startup)
10	BK/RD	A/C Idle-Up Solenoid	A/C Off: 12v, On: 1v
15	WT/GN	EVAP Purge Solenoid (VSV)	12v or 0v
16	BR	Shield Ground	<0.050v

1996 3.4L V6 MFI VIN N (A/T-ECT) 34-Pin Connector

PCM Pin #	Wire Color	Circuit Description (34-Pin)	Value at Hot Idle
1	BR	Power Ground	<0.1v
5	YL/BK	Injector 6 Control	1.6-2.9 ms
6	WT/BL	Injector 5 Control	1.6-2.9 ms
7	YL/RD	Injector 4 Control	1.6-2.9 ms
8	WT/GN	Injector 3 Control	1.6-2.9 ms
9	WT	Injector 2 Control	1.6-2.9 ms
10	WT/RD	Injector 1 Control	1.6-2.9 ms
11	WT	A/T-ECT Solenoid (S1)	3rd or OD: 1v
12	BK/YL	Igniter Signal (IGF)	Digital Signal: 0-5-0v
13	BK/WT	Starter Switch Signal	9-11v (cranking)
14	BK/OR	Neutral Start Switch	In P/N: 9-11v (cranking)
15 (2WD)	RD/GN	HO2S-21 (B2 S1) Heater	Heater Off: 12v, On: 1v
15 (4WD)	RD/GN	HO2S-12 (B1 S2) Heater	Heater Off: 12v, On: 1v
16	PK/GN	HO2S-11 (B1 S1) Heater	Heater Off: 12v, On: 1v
17	BK/WT	A/T-ECT Solenoid (S2)	1st or OD: 1v
22	BK/RD	IAC Signal (RSC)	Pulse Signals
23	BR/RD	IAC Signal (RSO)	Pulse Signals
24	BK/BL	Igniter Transistor 1 Control	Digital Signal: 0-5-0v
25	BR/BK	Igniter Transistor 2 Control	Digital Signal: 0-5-0v
26	BK/WT	Igniter Transistor 3 Control	Digital Signal: 0-5-0v
27	YL/BK	A/T-ECT Solenoid (SL)	In Lockup: 12-14v
29	RD	4WD Detection Transfer (L4)	Open: 12v, Closed: 0v
32	YL/BL	Closed Throttle Switch	1v, off-idle: 12v
33 or 34	BR	Power Ground	<0.1v

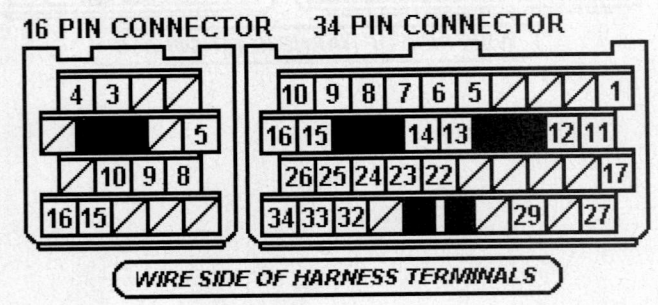

WIRE SIDE OF HARNESS TERMINALS

Pin Connector Graphic

1996 3.4L V6 MFI VIN N (M/T) 12 Pin Connector

PCM Pin #	Wire Color	Circuit Description (12 Pin)	Value at Hot Idle
3	PK/GN	HO2S-11 (B1 S1) Heater	Heater Off: 12v, On: 1v
5	WT	CMP Sensor Signal (GN-)	<0.050v
6	BL	CKP Sensor Signal (NE-)	<0.050v
7	BR	Power Ground	<0.1v
9	RD/GN	2WD HO2S-21 (B2 S1) HTR	Heater Off: 12v, On: 1v
9	RD/GN	4WD HO2S-12 (B1 S2) HTR	Heater Off: 12v, On: 1v
11	BK	CMP Sensor Signal (G2+)	AC pulse signals
12	GN	CKP Sensor Signal (NE+)	AC pulse signals

1996 3.4L V6 MFI VIN N (M/T) 22 Pin Connector

PCM Pin #	Wire Color	Circuit Description (22 Pin)	Value at Hot Idle
2	BK/GN	Battery Direct	12-14v
4	PK	MIL (lamp) Control	MIL Off: 12v, On: 1v
q	BL/BK	A/C Amplifier Signal (ACT)	Clutch On: 12v, Off: 1.5v
7	BK/RD	A/C Amplifier Signal (AC1)	Clutch On: 1.5v, Off: 12v
8	GN	Vehicle Speed Sensor	At 55 mph: 48 Hz
9	PK/GN	4WD Selector Switch	Switch Off: 12v, On: 1v
11	BK/WT	Starter Switch Signal	9-11v (cranking)
12	WT/RD	EFI Main Relay Power	12-14v
19	WT	SIL (Scan Tool) Signal	12v
20	GN/WT	Stop Light Switch Signal	Brake Off: 0v, On: 12v

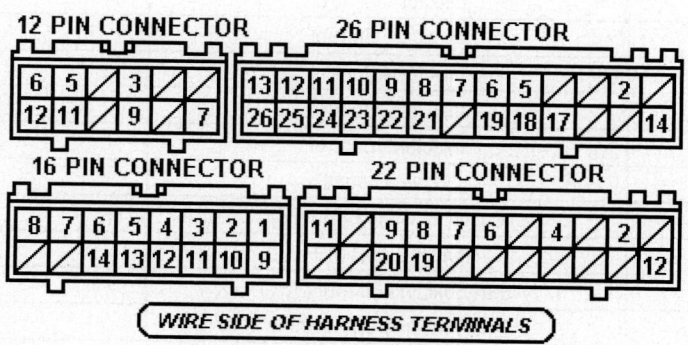

05533_ADIA_G187

Pin Connector Graphic

1996 3.4L V6 MFI VIN N (M/T) 16 Pin Connector

PCM Pin #	Wire Color	Circuit Description (16 Pin)	Value at Hot Idle
1	GN/BK	Sensor VREF	4.9-5.1v
2	GY/RD	MAF Sensor Signal (VG)	1.1-1.8v
3	GY	Knock Sensor 2 Signal	<0.075v AC
4	GN/YL	ECT Sensor Signal (THW)	At 180°F: 0.51v
5	WT	HO2S-11 (B1 S1) Signal	0.1-1.1v
6	BK	Knock Sensor 1 Signal	<0.075v AC
7	PK/WT	Data Link Connector	12-14v
8	BR/WT	MAF Sensor Ground	<0.050v
9	BR/BK	Sensor Ground	<0.050v
11	YL/BL	Closed Throttle Switch	1v, off-idle: 12v
10	YL/BK	TP Sensor Signal (VTA)	0.3-0.8v
12	YL/GN	IAT Sensor Signal (THA)	At 100°F: 2.60v
13 (2WD)	RD	HO2S-21 (B2 S1) Signal	0.1-1.1v
13 (4WD)	RD	HO2S-12 (B1 S2) Signal	0.1-1.1v
14	PK	EGR Gas Temperature Sensor (THG)	3.5-4.0v

1996 3.4L V6 MFI VIN N (M/T) 26 Pin Connector

PCM Pin #	Wire Color	Circuit Description (26 Pin)	Value at Hot Idle
2	BK/RD	A/C Idle-Up Solenoid	A/C Off: 12v, On: 1v
5	WT/GN	EVAP Purge Solenoid (VSV)	12v or 0v
6	BK/RD	IAC Signal (RSC)	Pulse Signals
7	BR/RD	IAC Signal (RSO)	Pulse Signals
8	YL/BK	Injector 6 Control	1.6-2.9 ms
9	WT/BL	Injector 5 Control	1.6-2.9 ms
10	YL/RD	Injector 4 Control	1.6-2.9 ms
11	WT	Injector 2 Control	1.6-2.9 ms
12	WT/RD	Injector 1 Control	1.6-2.9 ms
13	BR	Power Ground	<0.1v
14	GN/YL	Circuit Opening Relay (FC)	0-3v, at off-idle: 12v
17	BK/YL	Igniter Signal (IGF)	Digital Signal: 0-5-0v
18	RD/WT	EGR Solenoid Control (VSV)	12v or 0v
19	RD/BK	Fuel Pressure Up Solenoid	1v (at hot startup)
21	BK/WT	Igniter Transistor 3 Control	Digital Signal: 0-5-0v
22	BR/BK	Igniter Transistor 2 Control	Digital Signal: 0-5-0v
23	BK/BL	Igniter Transistor 1 Control	Digital Signal: 0-5-0v
24	BR	Shield Ground	<0.050v
25	WT/GN	Injector 3 Control	1.6-2.9 ms
26	BR	Power Ground	<0.1v

1997 3.4L V6 MFI VIN N (A/T-ECT) 28 Pin Connector

PCM Pin #	Wire Color	Circuit Description (28 Pin)	Value at Hot Idle
1	RD/BK	A/T Select Switch Reverse	In 'R': 12v, Others: 0v
2	PK/GN	A/T Select Switch 2nd	In 2nd: 12v, Others: 0v
3	PK/WT	A/T Select Switch Low	In Low: 12v, Others: 0v
4	LG	A/T Oil Temperature Lamp	Lamp Off: 12v, On: 1v
5	BL/BK	A/C Amplifier Signal (ACT)	Clutch On: 12v, Off: 1.5v

1997 3.4L V6 MFI VIN N (A/T-ECT) 28 Pin Connector, *continued*

6	YL/GN	Overdrive Main Switch	Switch Off: 12v, On: 1v
7	YL/RD	Cruise Control ECU (OD1)	At Cruise in OD: 12v
12	GN	Vehicle Speed Sensor	At 55 mph: 48 Hz
14	BK/GN	Battery Direct	12-14v
17	YL	4WD Detection Transfer (N)	Open: 12v, Closed
18	WT	Data Link Connector (SDL)	0v
20	BK/RD	A/C Amplifier Signal (AC1)	Clutch On: 1.5v, Off: 12v
22	WT/RD	EFI Main Relay Power	12-14v
25	GN/WT	Stop Light Switch Signal	Brake Off: 0v, On: 12v
26	PK/GN	4WD Selector Switch	Switch Off: 12v, On: 1v

1997 3.4L V6 MFI VIN N (A/T-ECT) 22 Pin Connector

PCM Pin #	Wire Color	Circuit Description (22 Pin)	Value at Hot Idle
1	GN/BK	Sensor VREF	4.9-5.1v
5	GN	CKP Sensor Signal (NE+)	AC pulse signals
6	BL	CKP Sensor Signal (NE-)	<0.050v
7	YL/BK	TP Sensor Signal (VTA)	0.3-0.8v
8	GY/RD	MAF Sensor Signal (VG)	1.1-1.8v
10	BK	CMP Sensor Signal (G2+)	AC pulse signals
11	WT	CMP Sensor Signal (GN-)	<0.050v
12	GN/BK	A/T Oil Temperature Sensor	At 230°F: <1.5v
13	WT	HO2S-11 (B1 S1) Signal	0.1-1.1v
14	YL/GN	IAT Sensor Signal (THA)	At 100°F: 2.60v
15	RD/BL	EVAP Vapor Pressure Sensor	2.5-3.1v (with fuel cap off)
16, 17	GY, BK	Knock Sensor 2, 1 Signal	<0.075v AC
18	BR/WT	Sensor Ground	<0.050v
19 (2WD)	RD	HO2S-21 (B2 S1) Signal	0.1-1.1v
19 (4WD)	RD	HO2S-12 (B1 S2) Signal	0.1-1.1v
20	GN/YL	ECT Sensor Signal (THW)	At 180°F: 0.51v
21	PK	EGR Gas Temperature Sensor (THG)	3.5-4.0v
22	BR/BK	Sensor Ground	<0.050v

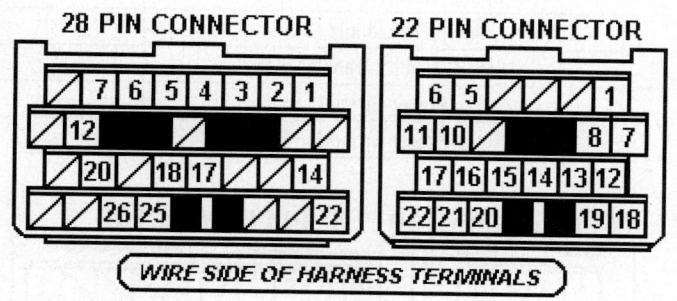

28 PIN CONNECTOR   22 PIN CONNECTOR

WIRE SIDE OF HARNESS TERMINALS

Pin Connector Graphic

05533_ADIA_G216

1997 3.4L V6 MFI VIN N (A/T-ECT) 16 Pin Connector

PCM Pin #	Wire Color	Circuit Description (16 Pin)	Value at Hot Idle
1	YL/BL	Cruise Control ECU (IDLO)	1.5v, off-idle: 12v
3	PK	MIL (lamp) Control	MIL Off: 12v, On: 1v
4	GN/YL	Circuit Opening Relay (FC)	0-3v, at off-idle: 12v
5	PK/WT	Data Link Connector	12-14v
8	RD/WT	EGR Solenoid Control (VSV)	12v or 0v
13	GN/RD	EVAP Vapor Pressure (VSV)	12v or 0v
15	WT/GN	EVAP Purge Solenoid (VSV)	12v or 0v
16	BR	Shield Ground	<0.050v

1997 3.4L V6 MFI VIN N (A/T-ECT) 34-Pin Connector

PCM Pin #	Wire Color	Circuit Description (34-Pin)	Value at Hot Idle
1	BR	Power Ground	<0.1v
5	YL/BK	Injector 6 Control	1.6-2.9 ms
6	WT/BL	Injector 5 Control	1.6-2.9 ms
7	YL/RD	Injector 4 Control	1.6-2.9 ms
8	WT/GN	Injector 3 Control	1.6-2.9 ms
9	WT	Injector 2 Control	1.6-2.9 ms
10	BK	Injector 1 Control	1.6-2.9 ms
11	WT	A/T-ECT Solenoid (S1)	3rd or OD: 1v
12	BK/YL	Igniter Signal (IGF)	Digital Signal: 0-5-0v
13	BK/WT	Starter Switch Signal	9-11v (cranking)
14	BK/OR	Neutral Start Switch	In P/N: 9-11v (cranking)
15	RD/GN	2WD HO2S-21 (B2 S1) HTR	Heater Off: 12v, On: 1v
15	RD/GN	4WD HO2S-12 (B1 S2) HTR	Heater Off: 12v, On: 1v
16	PK/GN	HO2S-11 (B1 S1) Heater	Heater Off: 12v, On: 1v
17	BK/WT	A/T-ECT Solenoid (S2)	1st or OD: 1v
22	BK/RD	IAC Signal (RSC)	Pulse Signals
23	BR/RD	IAC Signal (RSO)	Pulse Signals
24	BK/BL	Igniter Transistor 1 Control	Digital Signal: 0-5-0v
25	BR/BK	Igniter Transistor 2 Control	Digital Signal: 0-5-0v
26	BK/WT	Igniter Transistor 3 Control	Digital Signal: 0-5-0v
27	YL/BK	A/T-ECT Solenoid (SL)	In Lockup: 12-14v
29	RD/BK	4WD Detection Transfer (L4)	Open: 12v, Closed: 0v
33	BR	Power Ground	<0.1v
34	BR	Power Ground	<0.1v

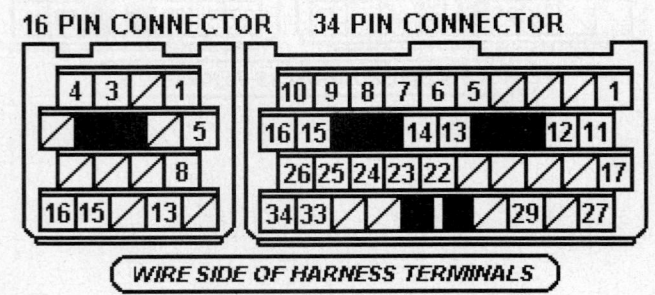

WIRE SIDE OF HARNESS TERMINALS

Pin Connector Graphic

05533_ADIA_G217

1997 3.4L V6 MFI VIN N (M/T) 12 Pin Connector

PCM Pin #	Wire Color	Circuit Description (12 Pin)	Value at Hot Idle
3	PK/GN	HO2S-11 (B1 S1) Heater	Heater Off: 12v, On: 1v
4	RD/BL	EVAP Vapor Pressure Sensor	2.5-3.1v (with fuel cap off)
5	WT	CMP Sensor Signal (GN-)	<0.050v
6	BL	CKP Sensor Signal (NE-)	<0.050v
7	BR	Power Ground	<0.1v
9	RD/GN	HO2S-21 (B2 S1) Heater	Heater Off: 12v, On: 1v
11	BK	CMP Sensor Signal (G2+)	AC pulse signals
12	GN	CKP Sensor Signal (NE+)	AC pulse signals

1997 3.4L V6 MFI VIN N (M/T) 22 Pin Connector

PCM Pin #	Wire Color	Circuit Description (22 Pin)	Value at Hot Idle
2	BK/GN	Battery Direct	12-14v
4	PK	MIL (lamp) Control	MIL Off: 12v, On: 1v
6	BL/BK	A/C Amplifier Signal (ACT)	Clutch On: 12v, Off: 1.5v
7	BK/RD	A/C Amplifier Signal (AC1)	Clutch On: 1.5v, Off: 12v
8	GN	Vehicle Speed Sensor	At 55 mph: 48 Hz
9	PK/GN	4WD Selector Switch	Switch Off: 12v, On: 1v
11	BK/WT	Starter Switch Signal	9-11v (cranking)
12	WT/RD	EFI Main Relay Power	12-14v
19	WT	SIL (Scan Tool) Signal	12v
20	GN/WT	Stop Light Switch Signal	Brake Off: 0v, On: 12v

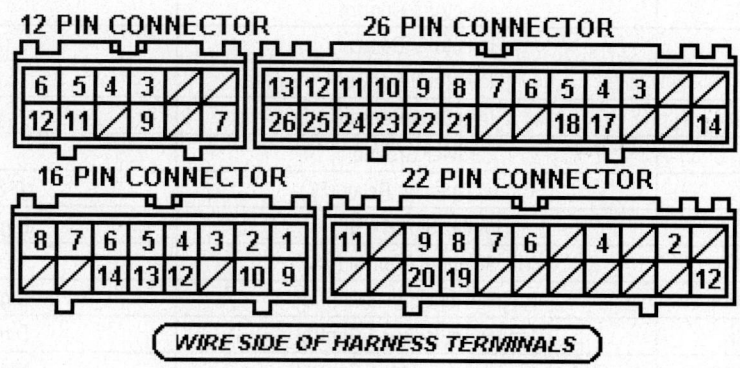

Pin Connector Graphic

05533_ADIA_G218

1997 3.4L V6 MFI VIN N (M/T) 16 Pin Connector

PCM Pin #	Wire Color	Circuit Description (16 Pin)	Value at Hot Idle
1	GN/BK	Sensor VREF	4.9-5.1v
2	GY/RD	MAF Sensor Signal (VG)	1.1-1.8v
3	GY	Knock Sensor 2 Signal	<0.075v AC
4	GN/YL	ECT Sensor Signal (THW)	At 180°F: 0.51v
5	WT	HO2S-11 (B1 S1) Signal	0.1-1.1v
6	BK	Knock Sensor 1 Signal	<0.075v AC
7	PK/WT	Data Link Connector	12-14v
8	BR/WT	MAF Sensor Ground	<0.050v
9	BR/BK	Sensor Ground	<0.050v
10	YL/BK	TP Sensor Signal (VTA)	0.3-0.8v
12	YL/GN	IAT Sensor Signal (THA)	At 100°F: 2.60v
13	RD	HO2S-21 (B2 S1) Signal	0.1-1.1v
14	PK	EGR Gas Temperature Sensor (THG)	3.5-4.0v

1997 3.4L V6 MFI VIN N (M/T) 26 Pin Connector

PCM Pin #	Wire Color	Circuit Description (26 Pin)	Value at Hot Idle
3	GN/RD	EVAP Vapor Pressure (VSV)	12v or 0v
4	YL/BL	Cruise Control ECU (IDLO)	1.5v, off-idle: 12v
5	WT/GN	EVAP Purge Solenoid (VSV)	12v or 0v
6	BK/RD	IAC Signal (RSC)	Pulse Signals
7	BR/RD	IAC Signal (RSO)	Pulse Signals
8	YL/BK	Injector 6 Control	1.6-2.9 ms
9	WT/BL	Injector 5 Control	1.6-2.9 ms
10	YL/RD	Injector 4 Control	1.6-2.9 ms
11	WT	Injector 2 Control	1.6-2.9 ms
12	WT/RD	Injector 1 Control	1.6-2.9 ms
13	BR	Power Ground	<0.1v
14	GN/YL	Circuit Opening Relay (FC)	0-3v, at off-idle: 12v
17	BK/YL	Igniter Signal (IGF)	Digital Signal: 0-5-0v
18	RD/WT	EGR Solenoid Control (VSV)	12v or 0v
21	BK/WT	Igniter Transistor 3 Control	Digital Signal: 0-5-0v
22	BR/BK	Igniter Transistor 2 Control	Digital Signal: 0-5-0v
23	BK/BL	Igniter Transistor 1 Control	Digital Signal: 0-5-0v
24	BR	Shield Ground	<0.050v
25	WT/GN	Injector 3 Control	1.6-2.9 ms
26	BR	Power Ground	<0.1v

1998 3.4L V6 MFI VIN N (A/T-ECT) 28 Pin Connector

PCM Pin #	Wire Color	Circuit Description (28 Pin)	Value at Hot Idle
1	PK/WT	A/T Select Switch Low	In Low: 12v, Others: 0v
5	BL/BK	A/C Amplifier Signal (ACT)	Clutch On: 12v, Off: 1.5v
6	YL/GN	Overdrive Main Switch	Switch Off: 12v, On: 1v
7	YL/RD	Cruise Control ECU (OD1)	At Cruise in OD: 12v
8	WT	SIL (Scan Tool) Signal	12v
10	PK/GN	A/T Select Switch 2nd	In 2nd: 12v, Others: 0v
11	YL/BL	Cruise Control ECU (IDLO)	1.5v, off-idle: 12v

1998 3.4L V6 MFI VIN N (A/T-ECT) 28 Pin Connector, *continued*

12	GN	Vehicle Speed Sensor	At 55 mph: 48 Hz
14	BK/GN	Battery Direct	12-14v
15	RD/BK	A/T Select Switch Reverse	In 'R': 12v, Others: 0v
16	BK/RD	A/C Amplifier Signal (AC1)	Clutch On: 1.5v, Off: 12v
23	WT/RD	EFI Main Relay (B+)	12-14v
24	GN/WT	Stop Light Switch Signal	Brake Off: 0v, On: 12v

1998 3.4L V6 MFI VIN N (A/T-ECT) 22 Pin Connector

PCM Pin #	Wire Color	Circuit Description (22 Pin)	Value at Hot Idle
1	GN/BK	Sensor VREF	4.9-5.1v
5	GN	CKP Sensor Signal (NE+)	AC pulse signals
6	BL	CKP/CMP Sensor Signal (-)	<0.050v
7	YL/BK	TP Sensor Signal (VTA)	0.3-0.8v
8	GY/RD	MAF Sensor Signal (VG)	1.1-1.8v
9	BR/RD	A/T Vehicle Speed Sensor	Moving: 0-5-0v
13	WT	HO2S-11 (B1 S1) Signal	0.1-1.1v
14	GY	Knock Sensor 2 Signal	<0.075v AC
15	BK	Knock Sensor 1 Signal	<0.075v AC
17	PK	CMP Sensor Signal (G2+)	AC pulse signals
18	GN/YL	Circuit Opening Relay (FC)	0-3v, at off-idle: 12v
19	RD	HO2S-12 (B1 S2) Signal	0.1-1.1v
20	GN/RD	ECT Sensor Signal (THW)	At 180°F: 0.51v
21	YL/GN	IAT Sensor Signal (THA)	At 100°F: 2.60v
22	BR/BK	Sensor Ground	<0.050v

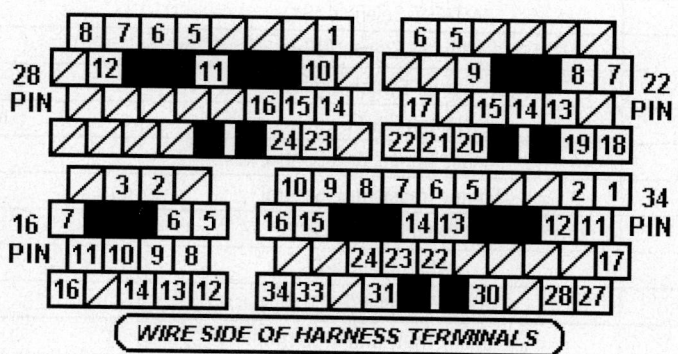

Pin Connector Graphic

05533_ADIA_G219

1998 3.4L V6 MFI VIN N (A/T-ECT) 16 Pin Connector

PCM Pin #	Wire Color	Circuit Description (16 Pin)	Value at Hot Idle
2	WT/GN	EVAP Purge Solenoid (VSV)	12v or 0v
3	PK	MIL (lamp) Control	MIL Off: 12v, On: 1v
5	PK/WT	Data Link Connector	12c
6	LG	A/T Oil Temperature Indicator	Light Off: 12v, On: 1v
7	BR/WT	MAF Meter Ground	<0.050v
8	GN/RD	EVAP Vapor Pressure (VSV)	12v or 0v
9	YL	4WD Detection Transfer (N)	Open: 12v, Closed
10	RD/GN	HO2S-12 (B1 S2) Heater	Heater Off: 12v, On: 1v
11	PK/GN	HO2S-11 (B1 S1) Heater	Heater Off: 12v, On: 1v
12	RD/WT	EGR Solenoid Control (VSV)	12v or 0v
13	RD/BL	EVAP Vapor Pressure Sensor	2.5-3.1v (with fuel cap off)
14	PK	EGR Gas Temperature Sensor (THG)	3.5-4.0v
16	BR	Shield Ground	<0.050v

1998 3.4L V6 MFI VIN N (A/T-ECT) 34-Pin Connector

PCM Pin #	Wire Color	Circuit Description (34-Pin)	Value at Hot Idle
1	PK/GN	4WD Selector Switch	Switch Off: 12v, On: 1v
2	RD/BK	4WD Detection Transfer (L4)	Open: 12v, Closed: 0v
5	YL/GN	Injector 6 Control	1.6-2.9 ms
6	WT/BL	Injector 5 Control	1.6-2.9 ms
7	YL/RD	Injector 4 Control	1.6-2.9 ms
8	WT/GN	Injector 3 Control	1.6-2.9 ms
9	WT	Injector 2 Control	1.6-2.9 ms
10	WT/RD	Injector 1 Control	1.6-2.9 ms
11	WT	A/T-ECT Solenoid (S1)	3rd or OD: 1v
12	BK/YL	Igniter Signal (IGF)	Digital Signal: 0-5-0v
13	BK/WT	Starter Switch Signal	9-11v (cranking)
14	BK/OR	Neutral Start Switch	In P/N: 9-11v (cranking)
15	BK/WT	Igniter Transistor 3 Control	Digital Signal: 0-5-0v
16	BR/BK	Igniter Transistor 2 Control	Digital Signal: 0-5-0v
17	RD/YL	A/T-ECT Solenoid (S2)	1st or OD: 1v
22	BK/RD	IAC Signal (RSC)	Pulse Signals
23	BR/RD	IAC Signal (RSO)	Pulse Signals
24	BK/BL	Igniter Transistor 1 Control	Digital Signal: 0-5-0v
27	YL/BK	A/T-ECT Solenoid (SL)	In Lockup: 12-14v
28	BR	Power Ground	<0.1v
30	GN/BK	A/T Oil Temperature Sensor	At 230°F: <1.5v
31	PK/BL	PSP Switch Signal	Straight: 12v, Turning: 0v
33	BR	Power Ground	<0.1v
34	BR	Power Ground	<0.1v

PREVIA PIN CHARTS

1991-93 2.4L I4 VIN A (A/T, M/T) 16 Pin Connector

PCM Pin #	Wire Color	Circuit Description (16 Pin)	Value at Hot Idle
1	BL/BK	Air Flow Meter VREF	4.9-5.1v
2	BL/YL	Air Flow Meter Signal	0.5-2.5v
3	BL/WT	IAT Sensor Signal	At 100°F: 2.60v
4	BL	ECT Sensor Signal	At 180°F: 0.51v
5	BK	O2S-12 (B1 S2) Signal	0.1.1-1.5v
6	BK	HO2S-11 (B1 S1) Signal	0.1.1-1.5v
8	GN	Data Link Connector	12-14v
9	BR/BK	Sensor Ground	<0.050v
10 (Cal)	BR/WT	EGR Gas Temperature Sensor	3.5-4.0v
11	YL/RD	TP Sensor Signal	0.3-0.8v
12	BL/YL	Closed Throttle Switch	0-3v, at off-idle: 12v
14	BK	Knock Sensor Signal	<0.075v AC
16	GN/WT	Data Link Connector	12-14v

1991-93 2.4L I4 VIN A (A/T, M/T) 22 Pin Connector

PCM Pin #	Wire Color	Circuit Description (22 Pin)	Value at Hot Idle
1	WT/RD	Direct Battery	12-14v
2	BK/RD	Ignition Switch Power	12-14v
4	RD/WT	Engine Oil Level Light	Light On: 1v, Off: 12v
5	GN/RD	MIL (lamp) Control	MIL Off: 12v, On: 1v
6	BK/BL	A/C Amplifier Signal (ACT)	Clutch On: 12v, Off: 1.5v
7	GN	Data Link Connector	12-14v
8	BR/BK	A/T ECT Speed Sensor	Moving: 0-5-0V
9	PK/WT	Vehicle Speed Sensor	At 55 mph: 48 Hz
10	BK/WT	A/C Magnetic Clutch (ACMG)	Clutch Off: 0v, On: 12v
11	BK	Starter Switch Signal	KOEC: 9-11v
12	BK/OR	EFI Main Relay B+	12-14v
13	BK/OR	EFI Main Relay B1+	12-14v
14	BK/BL	A/T Select Switch Low	In Low: 12v, Others: 0v
15	BK/RD	A/T Select Switch 2nd	In 2nd: 12v, Others: 0v
16	BK/YL	A/T Select Switch Neutral	In 'N': 12v, Others: 0v
17	YL/RD	Oil Level 2 Sensor	At 68°F: 4-5v
18	YL	Oil Level 1 Sensor	At 68°F: 4-5v
19	GN/WT	Stop Light Switch	Brake Off: 0v, On: 12v
20	BL/WT	Overdrive Main Switch	Switch Off: 12v, On: 1v
21	PK/YL	Cruise Control ECU	At Cruise in OD: 12v
22	BK/WT	Starter Switch Signal	Cranking: 0-3v
22	BK	Starter Switch Signal	Cranking: 0-3v

Standard Colors and Abbreviations

Abbreviation	Color	Abbreviation	Color	Abbreviation	Color
BK	Black	GY	Gray	RD	Red
BL	Blue	GN	Green	TN	Tan
BR	Brown	LG	LT Green	VT	Violet
DB	Dark Blue	OR	Orange	WT	White
DG	DK Green	PK	Pink	YL	Yellow

1991-93 2.4L I4 VIN A (A/T, M/T) 26 Pin Connector

PCM Pin #	Wire Color	Circuit Description (26 Pin)	Value at Hot Idle
1	GN/YL	A/T ECT Solenoid (SL)	In Lockup: 12-14v
3	BL/RD	Igniter Signal (IGF)	Digital Signal: 0-5-0v
4	WT	Distributor Signal (NE+)	AC pulse signals
5	GN	Distributor Signal (G2+)	AC pulse signals
6	BK	ISC Signal (ISC1)	Pulse Signals
7	RD	Engine Oil Feeder Motor	Pulse Signals
8	BK/BL	Engine Oil Feeder Relay	Relay Off: 12v, On: 1v
9	BL/BK	Igniter Signal (IGT)	Digital Signal: 0-5-0v
10	GN/BK	HO2S-11 (B1 S1) Heater	1v (Heater on)
11	RD/BK	Cold Start Injector	1v (at cold startup)
12	BK	Injectors 1 & 3 Control	1.6-2.9 ms
13	BR	Power Ground	<0.1v
14	BR/YL	A/T ECT Solenoid (S2)	1st or OD: 1v
15	BR/WT	A/T ECT Solenoid (S1)	3rd or OD: 1v
17	RD	Distributor Signal (G-)	<0.050v
18	BK	Distributor Signal (G1+)	AC pulse signals
19	WT	ISC Signal (ISC2)	Pulse Signals
23	YL	Fuel Pressure Up Solenoid	1v (at hot restart)
24	BR	Sensor Ground	<0.050v
25	BK/BL	Injectors 2 & 4 Control	1.6-2.9 ms
26	BR	Power Ground	<0.1v

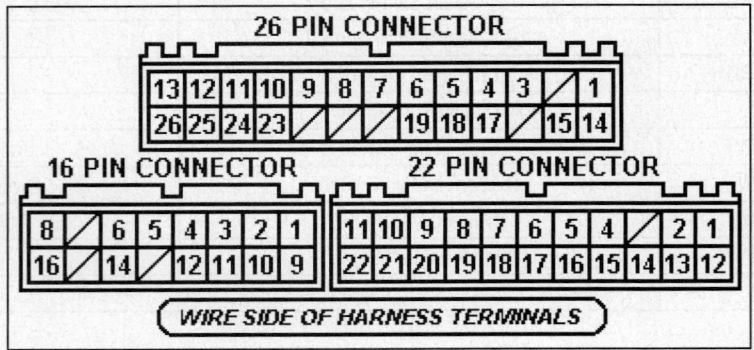

Pin Connector Graphic

05533_ADIA_G220

1994 2.4L I4 VIN A (A/T-ECT) 16 Pin Connector

PCM Pin #	Wire Color	Circuit Description (16 Pin)	Value at Hot Idle
1	BL/BK	Air Flow Meter VREF	4.9-5.1v
2	BL/YL	Air Flow Meter Signal	0.5-2.5v
3	BL/WT	IAT Sensor Signal	At 100°F: 2.60v
4	BL	ECT Sensor Signal	At 180°F: 0.51v
5	BK	O2S-12 (B1 S2) Signal	0.1.1-1.5v
6	BK	HO2S-11 (B1 S1) Signal	0.1.1-1.5v
8	GN	Data Link Connector	12-14v
9	BR/BK	Sensor Ground	<0.050v
10 (Cal)	BR/WT	EGR Gas Temperature Sensor	3.5-4.0v
11	YL/RD	TP Sensor Signal	0.3-0.8v
12	BL/YL	Closed Throttle Switch	0-3v, at off-idle: 12v
14	BK	Knock Sensor Signal	<0.075v AC
16	GN/WT	Data Link Connector	12v

1994 2.4L I4 VIN A (A/T-ECT) 22 Pin Connector

PCM Pin #	Wire Color	Circuit Description (22 Pin)	Value at Hot Idle
1	WT/RD	Direct Battery	12-14v
2	BK/RD	Ignition Switch Power	12-14v
4	RD/WT	Engine Oil Level Light	Light On: 1v, Off: 12v
5	GN/RD	MIL (lamp) Control	MIL Off: 12v, On: 1v
6	BK/BL	A/C Amplifier Signal (ACT)	Clutch On: 12v, Off: 1.5v
7	GN	Data Link Connector	12-14v
8	BR/BK	A/T ECT Speed Sensor	Moving: 0-5-0V
9	PK/WT	Vehicle Speed Sensor	At 55 mph: 48 Hz
10	BK/WT	A/C Magnetic Clutch (ACMG)	Clutch Off: 0v, On: 12v
11	BK	Starter Switch Signal	KOEC: 9-11v
12	BK/OR	EFI Main Relay B+	12-14v
13	BK/OR	EFI Main Relay B1+	12-14v
14	BK/BL	A/T Select Switch Low	In Low: 12v, Others: 0v
15	BK/RD	A/T Select Switch 2nd	In 2nd: 12v, Others: 0v
16	BK/YL	A/T Select Switch Neutral	In 'N': 12v, Others: 0v
17	YL/RD	Oil Level 2 Sensor	At 230°F: <1.5v
18	YL	Oil Level 1 Sensor	At 230°F: <1.5v
19	GN/WT	Stop Light Switch	Brake Off: 0v, On: 12v
20	BL/WT	Overdrive Main Switch	Switch Off: 12v, On: 1v
21	PK/YL	Cruise Control ECU	At Cruise in OD: 12v
22	BK/WT	Neutral Start Switch	In P/N: Cranking: 0-3v

Standard Colors and Abbreviations

Abbreviation	Color	Abbreviation	Color	Abbreviation	Color
BK	Black	GY	Gray	RD	Red
BL	Blue	GN	Green	TN	Tan
BR	Brown	LG	LT Green	VT	Violet
DB	Dark Blue	OR	Orange	WT	White
DG	DK Green	PK	Pink	YL	Yellow

1994 2.4L I4 VIN A (A/T-ECT) 26 Pin Connector

PCM Pin #	Wire Color	Circuit Description (26 Pin)	Value at Hot Idle
1	GN/YL	A/T ECT Solenoid (SL)	In Lockup: 12-14v
3	BL/RD	Igniter Signal (IGF)	Digital Signal: 0-5-0v
4	WT	Distributor Signal (NE+)	AC pulse signals
5	GN	Distributor Signal (G2+)	AC pulse signals
6	BK	ISC Signal (ISC1)	Pulse Signals
7	RD	Engine Oil Feeder Motor	Pulse Signals
8	BK/BL	Engine Oil Feeder Relay	Relay Off: 12v, On: 1v
9	BL/BK	Igniter Signal (IGT)	Digital Signal: 0-5-0v
10	GN/BK	HO2S-11 (B1 S1) Heater	1v (Heater on)
11	RD/BK	Cold Start Injector	1v (at cold startup)
12	BK	Injectors 1 & 3 Control	1.6-2.9 ms
13	BR	Power Ground	<0.1v
14	BR/YL	A/T ECT Solenoid (S2)	1st or OD: 1v
15	BR/WT	A/T ECT Solenoid (S1)	3rd or OD: 1v
17	RD	Distributor Signal (G-)	<0.050v
18	BK	Distributor Signal (G1+)	AC pulse signals
19	WT	ISC Signal (ISC2)	Pulse Signals
23	YL	Fuel Pressure Up Solenoid	1v (at hot restart)
24	BR	Sensor Ground	<0.050v
25	BK/BL	Injectors 2 & 4 Control	1.6-2.9 ms
26	BR	Power Ground	<0.1v

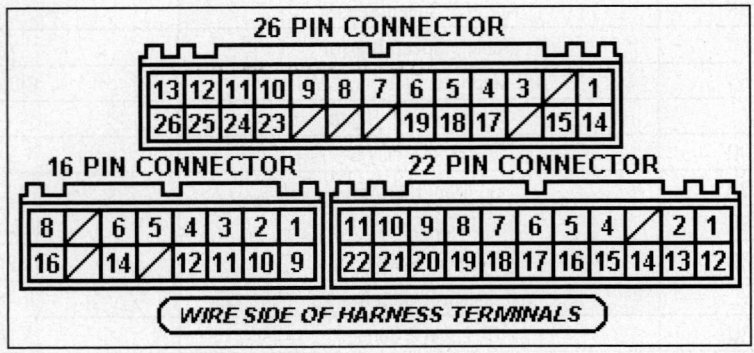

05533_ADIA_G220

Pin Connector Graphic

1994 2.4L VIN A (A/T-ECT) 16 Pin Connector

PCM Pin #	Wire Color	Circuit Description (16 Pin)	Value at Hot Idle
1	BL/BK	Air Flow Meter VREF	4.9-5.1v
2	GN/RD	Air Flow Meter Signal	0.5-2.5v
3	BL/WT	IAT Sensor Signal	At 100°F: 2.60v
4	BL	ECT Sensor Signal	At 180°F: 0.51v
5	BK	HO2S-11 (B1 S1) Signal	0.1-1.5v
6	BK	Knock Sensor Signal	<0.075v AC
7	GN/WT	Data Link Connector	12-14v
9	BR/BK	Sensor Ground	<0.050v
10	YL/RD	TP Sensor Signal	0.3-0.8v

1994 2.4L VIN A (A/T-ECT) 16 Pin Connector, *continued*

11	BL/YL	Closed Throttle Switch	0-3v, at off-idle: 12v
12 (Cal)	BR/WT	EGR Gas Temperature Sensor	3.5-4.0v
13	BK	HO2S-12 (B1 S2) Signal	0.1.1-1.1v
14	GN/WT	Air Flow Meter Ground	<0.050v
15	GN/BK	Data Link Connector	12-14v
16	BR/WT	A/T ECT Speed Sensor	Moving: 0-5-0V

1994 2.4L VIN A (A/T-ECT) 22 Pin Connector

PCM Pin #	Wire Color	Circuit Description (22 Pin)	Value at Hot Idle
2	WT/RD	Direct Battery	12-14v
3	BL/RD	A/C Amplifier Signal (ACT)	Clutch On: 12v, Off: 1.5v
4	GN/RD	MIL (lamp) Control	MIL Off: 12v, On: 1v
5	RD/GN	S/C Magnetic Clutch Relay	Relay Off: 12v, On: 1v
6	BK/BL	Engine Oil Feeder Relay	Relay Off: 12v, On: 1v
7	BK/WT	A/C Magnetic Clutch (ACMG)	Clutch Off: 0v, On: 12v
8	PK/WT	Vehicle Speed Sensor	At 55 mph: 48 Hz
9	BL/WT	Overdrive Main Switch	Switch Off: 12v, On: 1v
10	BK/BL	A/T Select Switch Low	In Low: 12v, Others: 0v
11	BK	Starter Switch Signal	KOEC: 9-11v
12	BK/OR	EFI Main Relay B+	12-14v
13	GN/OR	PSP Switch Signal	Straight: 12v, Turned: 0v
14	GN/WT	Stop Light Switch	Brake Off: 0v, On: 12v
15	RD/WT	Engine Oil Level Light	Light On: 1v, Off: 12v
16	RD/BK	EGR Solenoid Control	1v (Heater on)
17	RD	Engine Oil Feeder Motor	Pulse Signals
18	PK/YL	Cruise Control ECU	At Cruise in OD: 12v
21	BK/RD	A/T Select Switch 2nd	In 2nd: 12v, Others: 0v
22	BK/WT	Neutral Start Switch	In P/N: Cranking: 0-3v

Standard Colors and Abbreviations

Abbreviation	Color	Abbreviation	Color	Abbreviation	Color
BK	Black	GY	Gray	RD	Red
BL	Blue	GN	Green	TN	Tan
BR	Brown	LG	LT Green	VT	Violet
DB	Dark Blue	OR	Orange	WT	White
DG	DK Green	PK	Pink	YL	Yellow

1994 2.4L VIN A (A/T-ECT) 12 Pin Connector

PCM Pin #	Wire Color	Circuit Description (12 Pin)	Value at Hot Idle
2	BL/BK	Fuel Pump Relay Control	Relay Off: 12v, On: 1v
3	BK	Data Link Connector	12-14v
4	YL/RD	Oil Level Sensor 2 Signal	At 230°F: <1.5v
6	WT	Distributor Signal (NE-)	<0.050v
10	YL	Oil Level Sensor 1 Signal	At 230°F: <1.5v
11	GN	Distributor Signal (G+)	AC pulse signals
12	BK	Distributor Signal (NE+)	AC pulse signals

1994 2.4L VIN A (A/T-ECT) 26 Pin Connector

PCM Pin #	Wire Color	Circuit Description (26 Pin)	Value at Hot Idle
1	GN/BK	HO2S-11 (B1 S1) Heater	1v (Heater on)
2	GN/WT	HO2S-12 (B1 S2) Heater	1v (Heater on)
3	YL	S/C Bypass Control (AB3)	12v or 0v
4	YL/BK	S/C Bypass Control (AB2)	12v or 0v
5	RD/YL	S/C Bypass Control (AB1)	12v or 0v
6	BL	IAC Signal (ISCC)	Pulse Signals
7	BL/WT	IAC Signal (ISCO)	Pulse Signals
8	GN/YL	A/T ECT Solenoid (SL)	In Lockup: 12-14v
9	BR/YL	A/T ECT Solenoid (S2)	1st or OD: 1v
10	BR/WT	A/T ECT Solenoid (S1)	3rd or OD: 1v
11	BK/BL	Injectors 2 & 4 Control	1.6-2.9 ms
12	BK	Injectors 1 & 3 Control	1.6-2.9 ms
13	BR	Sensor Ground	<0.050v
14	BL/RD	Circuit Opening Relay	0-3v, at off-idle: 12v
16	GN/BK	S/C Bypass Control (AB4)	12v or 0v
17	BL/RD	Igniter Signal (IGF)	Digital Signal: 0-5-0v
18	BL/RD	A/C Amplifier Signal (ACT)	Clutch On: 12v, Off: 1.5v
19	RD/YL	EVAP Purge Solenoid (VSV)	12v or 0v
20	RD/WT	SCB Solenoid Control	12v or 0v
21	RD/BL	A/C Idle-Up Solenoid	A/C Off: 12v, On: 1v
22	BR	Power Ground	<0.1v
23	BL/BK	Igniter Signal (IGT)	Digital Signal: 0-5-0v
24	BR	Sensor Ground	<0.050v
25	BK/YL	A/C Amplifier Signal (AC1)	Clutch On: 1.5v, Off: 12v
26	BR	Power Ground	<0.1v

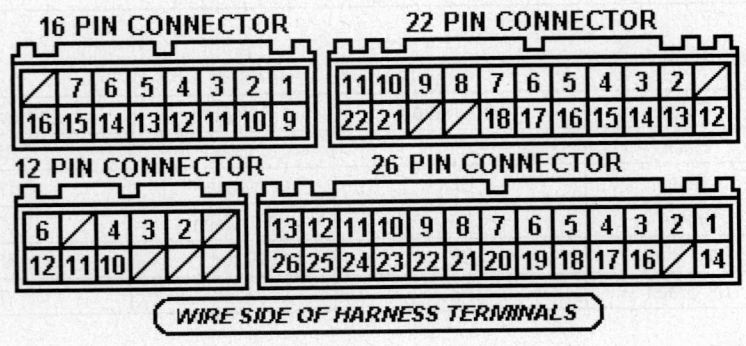

Pin Connector Graphic

1995 2.4L I4 VIN A (A/T-ECT) 16 Pin Connector

PCM Pin #	Wire Color	Circuit Description (16 Pin)	Value at Hot Idle
1	BL/BK	Sensor VREF	4.9-5.1v
2	BL/YL	MAF Sensor Signal	1.1-1.8v
3	BL/WT	IAT Sensor Signal	At 100°F: 2.60v
4	BL	ECT Sensor Signal	At 180°F: 0.51v
5	BK	O2S-12 (B1 S2) Signal	0.1.1-1.5v

1995 2.4L I4 VIN A (A/T-ECT) 16 Pin Connector, *continued*

6	BK	HO2S-11 (B1 S1) Signal	0.1.1-1.5v
8	GN	Data Link Connector	12-14v
9	BR/BK	Sensor Ground	<0.050v
10 (Cal)	BR/WT	EGR Gas Temperature Sensor	3.5-4.0v
11	YL/RD	TP Sensor Signal	0.3-0.8v
12	BL/YL	Closed Throttle Switch	0-3v, at off-idle: 12v
14	BK	Knock Sensor Signal	<0.075v AC
16	GN/WT	Data Link Connector	12-14v

1995 2.4L I4 VIN A (A/T-ECT) 22 Pin Connector

PCM Pin #	Wire Color	Circuit Description (22 Pin)	Value at Hot Idle
1	WT/RD	Direct Battery	12-14v
4	RD/WT	Engine Oil Level Light	Light On: 1v, Off: 12v
5	GN/RD	MIL (lamp) Control	MIL Off: 12v, On: 1v
6	BK/BL	A/C Amplifier Signal (ACT)	Clutch On: 12v, Off: 1.5v
7	GN	Data Link Connector	12-14v
8	BR/WT	A/T ECT Speed Sensor	Moving: 0-5-0v
9	PK/WT	Vehicle Speed Sensor	At 55 mph: 48 Hz
10	BK/WT	A/C Magnetic Clutch (ACMG)	Clutch Off: 0v, On: 12v
11	BK	Starter Switch Signal	Cranking: 0-3v
12	BK/OR	EFI Main Relay Power	12-14v
14	BK/BL	A/T Select Switch Low	In Low: 12v, Others: 0v
15	BK/RD	A/T Select Switch 2nd	In 2nd: 12v, Others: 0v
16	BK/YL	A/T Select Switch Neutral	In 'N': 12v, Others: 0v
17	YL/RD	Oil Level 2 Sensor	At 230°F: <1.5v
18	YL	Oil Level 1 Sensor	At 230°F: <1.5v
19	WT/RD	Stop Light Switch	Brake Off: 0v, On: 12v
20	BL/WT	Overdrive Main Switch	Switch Off: 12v, On: 1v
21	PK/YL	Cruise Control ECU	At Cruise in OD: 12v
22	BK/WT	Neutral Start Switch	In P/N: Cranking: 0-3v

Standard Colors and Abbreviations

Abbreviation	Color	Abbreviation	Color	Abbreviation	Color
BK	Black	GY	Gray	RD	Red
BL	Blue	GN	Green	TN	Tan
BR	Brown	LG	LT Green	VT	Violet
DB	Dark Blue	OR	Orange	WT	White
DG	DK Green	PK	Pink	YL	Yellow

1995 2.4L I4 VIN A (A/T-ECT) 26 Pin Connector

PCM Pin #	Wire Color	Circuit Description (26 Pin)	Value at Hot Idle
1	GN/YL	A/T ECT Solenoid (SL)	In Lockup: 12-14v
3	BL/RD	Igniter Signal (IGF)	Digital Signal: 0-5-0v
4	WT	Distributor Signal (NE+)	AC pulse signals
5	GN	Distributor Signal (G2+)	AC pulse signals
6	BK	ISC Signal (ISCC)	Pulse Signals
7	RD	Engine Oil Feeder Motor	Pulse Signals
8	BK/BL	Engine Oil Feeder Relay	Relay Off: 12v, On: 1v
9	BL/BK	Igniter Signal (IGT)	Digital Signal: 0-5-0v
10	GN/BK	HO2S-11 (B1 S1) Heater	1v (Heater on)
11	RD/BK	Cold Start Injector	1v (at cold startup)
12	BK	Injectors 1 & 3 Control	1.6-2.9 ms
13	BR	Power Ground	<0.1v
14	BR/YL	A/T ECT Solenoid (S2)	1st or OD: 1v
15	BR/WT	A/T ECT Solenoid (S1)	3rd or OD: 1v
17	RD	Distributor Signal (G-)	<0.050v
18	BK	Distributor Signal (G1+)	AC pulse signals
19	WT	ISC Signal (ISCO)	Pulse Signals
23	YL	Fuel Pressure Up Solenoid	1v (at hot restart)
24	BR	Sensor Ground	<0.050v
25	BK/BL	Injectors 2 & 4 Control	1.6-2.9 ms
26	BR	Power Ground	<0.1v

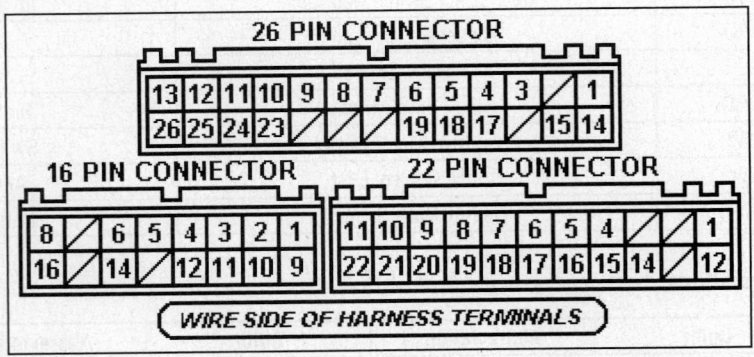

Pin Connector Graphic

1995 2.4L VIN A (A/T-ECT) 16 Pin Connector

PCM Pin #	Wire Color	Circuit Description (16 Pin)	Value at Hot Idle
1	BL/BK	Sensor VREF	4.9-5.1v
2	GN/RD	MAF Sensor Signal	1.1-1.8v
3	BL/WT	IAT Sensor Signal	At 100°F: 2.60v
4	BL	ECT Sensor Signal	At 180°F: 0.51v
5	BK	HO2S-11 (B1 S1) Signal	0.1.1-1.5v
6	BK	Knock Sensor Signal	<0.075v AC
7	GN/WT	Data Link Connector	12-14v
9	BR/BK	Sensor Ground	<0.050v
10	YL/RD	TP Sensor Signal	0.3-0.8v

1995 2.4L VIN A (A/T-ECT) 16 Pin Connector, *continued*

11	BL/YL	Closed Throttle Switch	0-3v, at off-idle: 12v
12 (Cal)	BR/WT	EGR Gas Temperature Sensor	3.5-4.0v
13	BK	HO2S-12 (B1 S2) Signal	0.1.1-1.5v
14	GN/WT	MAF Sensor Ground	<0.050v
15	GN/BK	Data Link Connector	12-14v
16	BR/WT	A/T ECT Speed Sensor	Moving: 0-5-0V

1995 2.4L VIN A (A/T-ECT) 22 Pin Connector

PCM Pin #	Wire Color	Circuit Description (22 Pin)	Value at Hot Idle
2	WT/RD	Direct Battery	12-14v
3	BL/RD	A/C Amplifier Signal (ACT)	Clutch On: 12v, Off: 1.5v
4	GN/RD	MIL (lamp) Control	MIL Off: 12v, On: 1v
5	RD/GN	S/C Magnetic Clutch Relay	Relay Off: 12v, On: 1v
6	BK/BL	Engine Oil Feeder Relay	Relay Off: 12v, On: 1v
7	BK/WT	A/C Magnetic Clutch (ACMG)	Clutch Off: 0v, On: 12v
8	PK/WT	Vehicle Speed Sensor	Moving: 0-5-0v
9	BL/WT	Overdrive Main Switch	Switch Off: 12v, On: 1v
10	BK/BL	A/T Select Switch Low	In Low: 12v, Others: 0v
11	BK	Starter Switch Signal	KOEC: 9-11v
12	BK/OR	EFI Main Relay Power	12-14v
13	GN/OR	PSP Switch Signal	Straight: 12v, Turned: 0v
14	GN/WT	Stop Light Switch	Brake Off: 0v, On: 12v
15	RD/WT	Engine Oil Level Light	Lamp On: 1v, Off: 12v
16	RD/BK	EGR Solenoid Control (VSV)	12v or 0v
17	RD	Engine Oil Feeder Motor	Pulse Signals
18	PK/YL	Cruise Control ECU	At Cruise in OD: 12v
21	BK/RD	A/T Select Switch 2nd	In 2nd: 12v, Others: 0v
22	BK/WT	Neutral Start Switch	In P/N: Cranking: 0-3v

Standard Colors and Abbreviations

Abbreviation	Color	Abbreviation	Color	Abbreviation	Color
BK	Black	GY	Gray	RD	Red
BL	Blue	GN	Green	TN	Tan
BR	Brown	LG	LT Green	VT	Violet
DB	Dark Blue	OR	Orange	WT	White
DG	DK Green	PK	Pink	YL	Yellow

1995 2.4L VIN A (A/T-ECT) 12 Pin Connector

PCM Pin #	Wire Color	Circuit Description (12 Pin)	Value at Hot Idle
1	RD	Defogger/Light Idle-Up Signal	Load On: 12v
2	BL/BK	Fuel Pump Relay Control	Relay Off: 12v, On: 1v
3	BK	Data Link Connector	12-14v
4	YL/RD	Oil Level Sensor Signal	At 230°F: <1.5v
6	WT	CKP Sensor Signal (NE-)	<0.050v
10	YL	Oil Level Sensor Signal	At 230°F: <1.5v
11	GN	Distributor Signal (G+)	AC pulse signals
12	BK	CKP Sensor Signal (NE+)	AC pulse signals

1995 2.4L VIN A (A/T-ECT) 26 Pin Connector

PCM Pin #	Wire Color	Circuit Description (26 Pin)	Value at Hot Idle
1	GN/BK	HO2S-11 (B1 S1) Heater	1v (Heater on)
2	GN/WT	HO2S-12 (B1 S2) Heater	1v (Heater on)
3	YL	S/C Bypass Control (AB3)	12v or 0v
4	YL/BK	S/C Bypass Control (AB2)	12v or 0v
5	RD/YL	S/C Bypass Control (AB1)	12v or 0v
6	BL	IAC Signal (RSC)	Pulse Signals
7	BL/WT	IAC Signal (RSO)	Pulse Signals
8	GN/YL	A/T ECT Solenoid (SL)	In Lockup: 12-14v
9	BR/YL	A/T ECT Solenoid (S2)	1st or OD: 1v
10	BR/WT	A/T ECT Solenoid (S1)	3rd or OD: 1v
11	BK/BL	Injectors 2 & 4 Control	1.6-2.9 ms
12	BK	Injectors 1 & 3 Control	1.6-2.9 ms
13, 24	BR	Sensor Ground	<0.050v
14	BL/RD	Circuit Opening Relay	0-3v, at off-idle: 12v
16	GN/BK	S/C Bypass Control (AB4)	12v or 0v
17	BL/RD	Igniter Signal (IGF)	Digital Signal: 0-5-0v
18	BL/RD	A/C Amplifier Signal (ACT)	Clutch On: 12v, Off: 1.5v
19	RD/YL	EVAP Purge Solenoid (VSV)	12v or 0v
20	RD/WT	SCB Solenoid Control	12v or 0v
21	RD/BL	A/C Idle-Up Solenoid	A/C Off: 12v, On: 1v
22	BR	Power Ground	<0.1v
23	BL/BK	Igniter Signal (IGT)	Digital Signal: 0-5-0v
25	BK/YL	A/C Amplifier Signal (AC1)	Clutch On: 1.5v, Off: 12v
26	BR	Power Ground	<0.1v

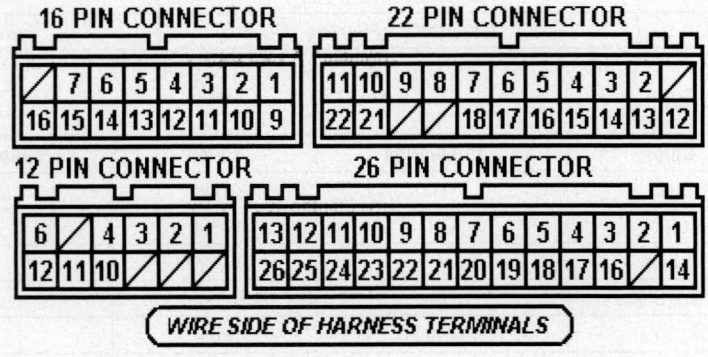

Pin Connector Graphic

05533_ADIA_G223

1996-97 2.4L I4 VIN K (A/T-ECT) 16 Pin Connector

PCM Pin #	Wire Color	Circuit Description (16 Pin)	Value at Hot Idle
1	BL/BK	Sensor VREF	4.9-5.1v
2	GN/RD	MAF Sensor Signal	1.1-1.8v
3	BL/WT	IAT Sensor Signal	At 100°F: 2.60v
4	BL	ECT Sensor Signal	At 180°F: 0.51v
5	BK	HO2S-11 (B1 S1) Signal	0.1.1-1.5v
6	BK	Knock Sensor Signal	<0.075v AC

1996-97 2.4L I4 VIN K (A/T-ECT) 16 Pin Connector, *continued*

7	GN/WT	Data Link Connector	12v
9	BR/BK	Sensor Ground	<0.050v
10	YL/RD	TP Sensor Signal	0.3-0.8v
11	BL/YL	Closed Throttle Switch	0-3v, at off-idle: 12v
12	BR/WT	EGR Gas Temperature Sensor	3.5-4.0v
13	BK	HO2S-12 (B1 S2) Signal	0.1.1-1.5v
14	GN/WT	MAF Sensor Ground	<0.050v
15	GN/BK	Data Link Connector	12-14v
16	BR/WT	A/T ECT Speed Sensor	Moving: 0-5-0V

1996-97 2.4L I4 VIN K (A/T-ECT) 22 Pin Connector

PCM Pin #	Wire Color	Circuit Description (22 Pin)	Value at Hot Idle
2	WT/RD	Direct Battery	12-14v
3	BL/RD	A/C Amplifier Signal (ACT)	Clutch On: 12v, Off: 1.5v
4	GN/RD	MIL (lamp) Control	MIL Off: 12v, On: 1v
5	RD/GN	S/C Magnetic Clutch Relay	Relay On: 12-14v
6	BK/BL	Engine Oil Feeder Relay	Relay Off: 12v, On: 1v
7	BK/WT	A/C Amplifier On Signal	Clutch On: 1.5v, Off: 12v
8	PK/WT	Vehicle Speed Sensor	At 55 mph: 48 Hz
9	BL/WT	Overdrive Main Switch	Switch Off: 12v, On: 1v
10	BK/BL	A/T Select Switch Low	In Low: 12v, Others: 0v
11	BK	Starter Switch Signal	KOEC: 9-11v
12	BK/OR	EFI Main Relay B+	12-14v
13	GN/OR	PSP Switch Signal	Straight: 12v, Turned: 0v
14	GN/WT	Stop Light Switch	Brake Off: 0v, On: 12v
15	RD/WT	Engine Oil Level Light	Light On: 1v, Off: 12v
16	RD/BK	EGR Solenoid Control	1v (Heater on)
17	RD	Engine Oil Feeder Motor	Pulse Signals
18	PK/YL	Cruise Control ECU	At Cruise in OD: 12v
19	BK	A/T Oil Temperature Indicator	Lamp On: 1v, Off: 12v
20	RD/YL	A/T Select Switch Reverse	In 'R': 12v, Others: 0v
21	BK/RD	A/T Select Switch 2nd	In 2nd: 12v, Others: 0v
22	BK/WT	Neutral Start Switch	In P/N: Cranking: 0-3v

Standard Colors and Abbreviations

Abbreviation	Color	Abbreviation	Color	Abbreviation	Color
BK	Black	GY	Gray	RD	Red
BL	Blue	GN	Green	TN	Tan
BR	Brown	LG	LT Green	VT	Violet
DB	Dark Blue	OR	Orange	WT	White
DG	DK Green	PK	Pink	YL	Yellow

1996-97 2.4L I4 VIN K (A/T-ECT) 12 Pin Connector

PCM Pin #	Wire Color	Circuit Description (12 Pin)	Value at Hot Idle
1	RD	Defogger/Light Idle-Up Signal	Load On: 12-14v
2	BL/BK	Fuel Pump Relay Control	Relay Off: 12v, On: 1v
3	BK	Data Link Connector (SDL)	0v
4	YL/RD	Engine Oil Level Sensor	At 230°F: <1.5v
6	WT	CKP Sensor Signal (NE-)	<0.050v
10	YL	Engine Oil Level Sensor	At 230°F: <1.5v
9	BK/BL	A/T Oil Temperature Sensor	At 230°F: <1.5v
11	GN, BK	Distributor Signal (G+)	AC pulse signals
12	BK	CKP Sensor Signal (NE+)	AC pulse signals

1996-97 2.4L I4 VIN K (A/T-ECT) 26 Pin Connector

PCM Pin #	Wire Color	Circuit Description (26 Pin)	Value at Hot Idle
1	GN/BK	HO2S-11 (B1 S1) Heater	1v (Heater on)
2	GN/WT	HO2S-12 (B1 S2) Heater	1v (Heater on)
3	YL	S/C Bypass Control (AB3)	12v or 0v
4	YL/BK	S/C Bypass Control (AB2)	12v or 0v
5	RD/YL	S/C Bypass Control (AB1)	12v or 0v
6	BL	IAC Signal (RSC)	Pulse Signals
7	BL/WT	IAC Signal (RSO)	Pulse Signals
8	GN/YL	A/T ECT Solenoid (SL)	In Lockup: 12-14v
9	BR/YL	A/T ECT Solenoid (S2)	S1: 3rd or OD: 1v
10	BR/WT	A/T ECT Solenoid (S1)	S1: 3rd or OD: 1v
11	BK/BL	Injectors 2 & 4 Control	1.6-2.9 ms
12	BK	Injectors 1 & 3 Control	1.6-2.9 ms
13, 24	BR	Sensor Ground	<0.050v
14	BL/RD	Circuit Opening Relay	0-3v, at off-idle: 12v
16	GN/BK	S/C Bypass Control (AB4)	12v or 0v
17	BL/RD	Igniter Signal (IGF)	Digital Signal: 0-5-0v
18	BL/RD	A/C Pressure Switch	Switch Closed: 0.1v
19	RD/YL	EVAP Purge Solenoid (VSV)	12v or 0v
20	RD/WT	SCB Solenoid Control	12v or 0v
21	RD/BL	A/C Idle-Up Solenoid	A/C Off: 12v, On: 1v
22, 26	BR	Power Ground	<0.1v
23	BL/BK	Igniter Signal (IGT)	Digital Signal: 0-5-0v
25	BK/YL	A/C Amplifier Signal (AC1)	Clutch On: 1.5v, Off: 12v

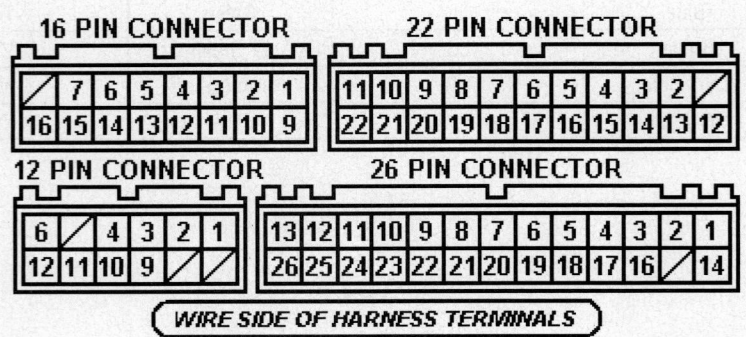

Pin Connector Graphic

SIENNA PIN CHARTS

1998 3.0L V6 MFI VIN F (A/T-ECT) 28 Pin Connector

PCM Pin #	Wire Color	Circuit Description (28 Pin)	Value at Hot Idle
1	YL	A/T Select Switch Low	In Low: 12v, Others: 0v
2	PK	Mirror Switch Signal	Switch Off: 0v, On: 12v
3	GN	Tail Light Signal	Switch Off: 0v, On: 12v
5	GN/BK	A/C Amplifier Signal (ACT)	Clutch On: 12v, Off: 1.5v
6	GN/OR	Overdrive Main Switch	Switch Off: 12v, On: 1v
7	YL/BK	Cruise Control ECU	At Cruise in OD: 12v
8	RD	SIL (Scan Tool Signal)	Digital Signal
10	OR	A/T Select Switch 2nd	In 2nd: 12v, Others: 0v
11	BL	Cruise Control (IDLO) ECU	1.5v, off-idle: 12v
12	PK/YL	Vehicle Speed Sensor	At 55 mph: 48 Hz
13	OR	Tachometer Signal (TACO)	Pulse Signals
14	BK/YL	Direct Battery	12-14v
15	RD/BK	A/T Select Switch Reverse	In 'R': 12v, Others: 0v
16	BK/YL	A/C Amplifier Signal (AC1)	Clutch On: 1.5v, Off: 12v
17	PK/BK	HO2S-12 (B1 S2) Heater	1v (Heater on)
18	WT	HO2S-12 (B1 S2) Signal	0.1.1-1.5v
23	BK/RD	EFI Main Relay B+	12-14v
24	GN/WT	Stop Light Switch	Brake Off: 0v, On: 12v

1998 3.0L V6 MFI VIN F (A/T-ECT) 16 Pin Connector

PCM Pin #	Wire Color	Circuit Description (16 Pin)	Value at Hot Idle
2	LG	EVAP Purge Solenoid (VSV)	12v or 0v
3	GN/RD	MIL (lamp) Control	MIL Off: 12v, On: 1v
5	BL/WT	Data Link Connector	12-14v
6	RD/YL	Intake Air Solenoid	12v or 0v
7	RD/BK	MAF Sensor Ground	<0.050v
8	WT/RD	EVAP Vapor Pressure (VSV)	12v or 0v
9	GN/WT	Cooling Fan Relay 1 & 2	Relay Off: 12v, On: 1v
10	YL/RD	HO2S-21 (B2 S1) Heater	1v (Heater on)
11	BL/BK	HO2S-11 (B1 S1) Heater	1v (Heater on)
13	BL/RD	EVAP Vapor Pressure Sensor	2.5-3.7v (with hose off)
16	BR	Shield Ground	<0.050v

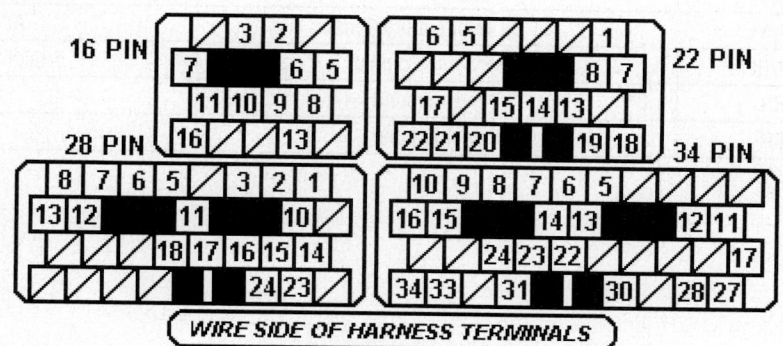

Pin Connector Graphic

1998 3.0L V6 MFI VIN F (A/T-ECT) 22 Pin Connector

PCM Pin #	Wire Color	Circuit Description (22 Pin)	Value at Hot Idle
1	YL	Sensor VREF	4.9-5.1v
5	BK/RD	CKP Sensor Signal (NE+)	AC pulse signals
6	BL	CKP/CMP Sensor Signal (-)	<0.050v
7	BL	TP Sensor Signal	0.3-0.8v
8	PK	MAF Sensor Signal	1.1-1.8v
13	WT	HO2S-11 (B1 S1) Signal	0.1.1-1.5v
14	WT	Knock Sensor Signal 1	<0.075v AC
15	WT	Knock Sensor Signal 2	<0.075v AC
17	BK/WT	CMP Sensor Signal (G+)	AC pulse signals
18	GN/RD	Circuit Opening Relay	0-3v, at off-idle: 12v
19	BK	HO2S-21 (B2 S1) Signal	0.1.1-1.5v
20	GN/BK	ECT Sensor Signal	At 180°F: 0.51v
21	BL/YL	IAT Sensor Signal	At 100°F: 2.60v
22	BR	Sensor Ground	<0.050v

1998 3.0L V6 MFI VIN F (A/T-ECT) 34-Pin Connector

PCM Pin #	Wire Color	Circuit Description (34-Pin)	Value at Hot Idle
5	GN	Injector 6 Control	1.6-2.9 ms
6	RD/BL	Injector 5 Control	1.6-2.9 ms
7	WT	Injector 4 Control	1.6-2.9 ms
8	YL	Injector 3 Control	1.6-2.9 ms
9	RD	Injector 2 Control	1.6-2.9 ms
10	BL	Injector 1 Control	1.6-2.9 ms
11	PK	A/T ECT Solenoid (S1)	3rd or OD: 1v
12	WT/RD	Igniter Signal (IGF)	Digital Signal: 0-5-0v
13	GY	Starter Switch Signal	1v (at cold startup)
14	BK/WT	Neutral Start Switch	In P/N: Cranking: 0-3v
15	GN/BK	Igniter Transistor 3 Control	7% duty cycle
16	BR/YL	Igniter Transistor 2 Control	7% duty cycle
17	BL/BK	A/T ECT Solenoid (S2)	1st or OD: 1v
22	YL/BK	IAC Signal (RSC)	Pulse Signals
23	RD/WT	IAC Signal (RSO)	Pulse Signals
24	GY	Igniter Transistor 1 Control	7% duty cycle
27	PK/BL	A/T ECT Solenoid (SL)	In Lockup: 12-14v
28	BR	Power Ground	<0.1v
30	GN/YL	A/T Oil Temperature Sensor	At 230°F: <1.5v
31	BK/BL	PSP Switch Signal	Straight: 12v, Turned: 0v
33	BR	Power Ground	<0.1v
34	BR	Power Ground	<0.1v

Standard Colors and Abbreviations

Abbreviation	Color	Abbreviation	Color	Abbreviation	Color
BK	Black	GY	Gray	RD	Red
BL	Blue	GN	Green	TN	Tan
BR	Brown	LG	LT Green	VT	Violet
DB	Dark Blue	OR	Orange	WT	White
DG	DK Green	PK	Pink	YL	Yellow

1999-2000 3.0L 1MZ-FE V6 VIN F E8 22 Pin Connector

PCM Pin #	Wire Color	Circuit Description (22 Pin)	Value at Hot Idle
1	RD/BK	Direct Battery	12-14v
2	BK/OR	Ignition Switch Power	12-14v
3	LG/RD	Circuit Opening Relay (FC)	0-3v, at off-idle: 12v
4-5	---	Not Used	---
6	GN/RD	MIL (lamp) Control	MIL Off: 12v, On: 1v
7 ('00)	GY	Starter Switch Signal	Cranking: 0-3v
8	YL/GN	EFI Main Relay Control	Relay Off: 0v, On: 12v
9	WT/RD	EVAP Vapor Pressure (VSV)	12v or 0v
10	---	Not Used	---
11	RD	SIL (Scan Tool Signal)	Digital Signal
12-14	---	Not Used	---
15	GN/WT	Stop Light Switch Signal	Brake Off: 0v, On: 12v
16	BK/RD	EFI Main Relay Power	12-14v
17	BL/RD	EVAP Vapor Pressure Sensor	2.9-3.1v (with hose off)
18	PK	Heated Mirror Switch Signal	Switch Off: 0v, On: 12v
19	GN	Rear Tail Light Switch Signal	Switch Off: 0v, On: 12v
20-22	---	Not Used	---

1999-2000 3.0L 1MZ-FE V6 VIN F E9 28 Pin Connector

PCM Pin #	Wire Color	Circuit Description (28 Pin)	Value at Hot Idle
1	---	Not Used	---
2	RD/BK	A/T Select Switch Reverse	In 'R': 12v, Others: 0v
3	OR	A/T Select Switch 2nd	In 2nd: 12v, Others: 0v
4	BL	Cruise Control ECU (IDLO)	1.5v, off-idle: 12v
5-7	---	Not Used	---
8	WT	HO2S-12 (B1 S2) Signal	0.1.1-1.5v
9	PK/BK	HO2S-12 (B1 S2) Heater	1v (Heater on)
10	---	Not Used	---
11-19	LG/BK	A/C Amplifier Signal (ACT)	Clutch On: 12v, Off: 1.5v
12	YL	A/T Select Switch Low	In Low: 12v, Others: 0v
20	BK/WT	Neutral Start Circuit Signal	In P/N: Cranking: 0-3v
21, 23	---	Not Used	---
22	VT/YL	Vehicle Speed Sensor	At 55 mph: 48 Hz
24	YL/BK	Cruise Control Signal (OD1)	At Cruise in OD: 12v
25	BK/YL	A/C Amplifier Signal (AC1)	Clutch On: 1.5v, Off: 12v
26, 28	---	Not Used	---
27	VT/RD	Tachometer Signal (TACO)	Pulse Signals
28	---	Not Used	---

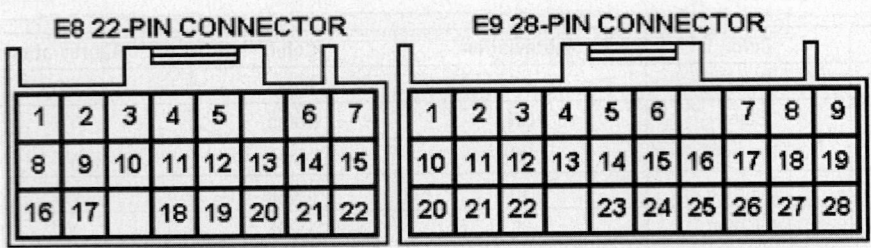

E8 22-PIN CONNECTOR   E9 28-PIN CONNECTOR

05533_ADIA_G226

Pin Connector Graphic

1999-2000 3.0L 1MZ-FE V6 VIN F E10 17-Pin Connector

PCM Pin #	Wire Color	Circuit Description (17-Pin)	Value at Hot Idle
1-3, 7-10	---	Not Used	---
4	BR/WT	Transponder Amplifier Code	Inserting key: pulses
5	BR/RD	Transponder Amplifier Signal	Inserting key: pulses
6	BR/YL	Transponder Amplifier Signal	Inserting key: pulses
11	BL/BK	Unlock Warning Switch	No Key: 4-5v
13 ('99)	BK/RD	Shield Ground	<0.050v
16	RD/WT	Theft Security Indicator Light	LED Off: 12v, On: 1v

1999-2000 3.0L 1MZ-FE V6 VIN F E11 24-Pin Connector

PCM Pin #	Wire Color	Circuit Description (24-Pin)	Value at Hot Idle
1	WT/BK	Power Ground (E04)	<0.1v
2	YL	Sensor VREF (VC)	4.9-5.1v
3 (Fed)	BL/BK	HO2S-11 (B1 S1) Signal	0.1.1-1.5v
3 (Cal)	BL	AFS-11 (B1 S1) Heater	1v (Heater on)
4 (Fed)	YL/RD	HO2S-21 (B2 S1) Signal	0.1.1-1.5v
4 (Cal)	BL	AFS-21 (B2 S1) Heater	1v (Heater on)
5	BL/RD	Injector 1 Control	1.6-2-9 ms
6	RD	Injector 2 Control	1.6-2-9 ms
7	LG	EVAP Purge Solenoid (VSV)	12v or 0v
8	WT/BK	Power Ground (E05)	<0.1v
9	BK/BL	PSP Switch Signal	Straight: 12v, Turned: 0v
10	PK	MAF Sensor Signal (VG)	1.1-1.5v
11 (Fed)	WT	HO2S-11 (B1 S1) Heater	1v (Heater on)
11 (Cal)	OR	AFS-11 (B1 S1) Signal (AF+)	Fixed at 3.3v
12 (Fed)	BK	HO2S-21 (B2 S1) Heater	1v (Heater on)
12 (Cal)	OR	AFS-21 (B2 S1) Signal (AF+)	Fixed at 3.3v
14	GN/BK	ECT Sensor Signal (THW)	0.5-0.6v
15	GN/YL	A/T: Fluid Temp. Input (THO)	At 68°F: 4-5v
16	BK/RD	CKP Sensor Signal (NE+)	AC pulse signals
17	BR	Shield Ground (E1)	<0.050v
18	BR	Sensor Ground (E2)	<0.050v
19	RD/BK	MAF Sensor Ground (E2G)	<0.050v
20 (Cal)	WT	AFS-11 (B1 S1) Signal (AF-)	Fixed at 3.3v
21 (Cal)	WT	AFS-21 (B2 S1) Signal (AF-)	Fixed at 3.3v
22	BL/YL	IAT Sensor Signal (THA)	0.5-3.4v
23	BL/WT	TP Sensor Signal (VTA1)	0.53-1.27v
24	BL	CMP/CKP Sensor Ground (-)	<0.050v

E10 17-PIN CONNECTOR   E11 24-PIN CONNECTOR

1	2	3	4	5	6
7	8	9	10	11	12
13	14	15		16	17

1	2		3	4	5	6	7
8	9	10	11	12	13	14	15	16
17	18	19	20	21		22	23	24

WIRE SIDE OF HARNESS TERMINALS

05533_ADIA_G227

Pin Connector Graphic
1999-2000 3.0L 1MZ-FE V6 VIN F E12 31-Pin Connector

PCM Pin #	Wire Color	Circuit Description (31-Pin)	Value at Hot Idle
1	YL	Injector 3 Control	1.6-2-9 ms
2	WT	Injector 4 Control	1.6-2-9 ms
3	RD/BL	Injector 5 Control	1.6-2-9 ms
4	GN	Injector 6 Control	1.6-2-9 ms
5	---	Not Used	---
6	BL/WT	DLC 1 Signal (TC)	12v
7	VT	A/T-ECT Shift Solenoid (S1)	In 3rd or OD: 1v
8	BL/BK	A/T-ECT Shift Solenoid (S2)	1st or OD: 1v
9	PK/BL	A/T-ECT Shift Solenoid (SL)	In Lockup: 12-14v
10	BK/WT	CMP Sensor Signal (G22+)	AC pulse signals
11	GY	IGT 1 Control	6°, 55 mph: 8° dwell
12	BR/YL	IGT 2 Control	6°, 55 mph: 8° dwell
13	LG/BK	IGT 3 Control	6°, 55 mph: 8° dwell
14	---	Not Used	---
15	YL/BK	IAC Signal (RSC)	Pulse Signals
16	RD/WT	IAC Signal (RSO)	Pulse Signals
17	RD/YL	ACIS Control (VSV)	12v or 0v
18	GN/RD	Overdrive Lamp Control	At Cruise in OD: 1v
19-20	---	Not Used	---
21	WT/BK	Power Ground (E01)	<0.1v
22	---	Not Used	---
23	G/OR	OD Main Sw. Input (ODMS)	Open: 12v, Closed: 1v
25	WT/RD	Igniter Signal (IGF)	Digital Signal: 0-5-0v
26	---	Not Used	---
27	WT	Knock Sensor 1 Signal (right)	<0.075v AC
28	WT	Knock Sensor 2 Signal (left)	<0.075v AC
29	GN/WT	Cooling Fan 1 Relay	Relay Off: 12v, On: 1v
30	WT/BK	Power Ground (E03)	<0.1v
31	WT/BK	Power Ground (E01)	<0.1v

E12 31-PIN CONNECTOR

```
        1   2      3   4   5   6   7   8   9
    10  11  12  13  14  15  16  17  18  19  20  21
    22  23  24  25  26      27  28      29  30  31
```

WIRE SIDE OF HARNESS TERMINALS

05533_ADIA_G228

Pin Connector Graphic
Standard Colors and Abbreviations

Abbreviation	Color	Abbreviation	Color	Abbreviation	Color
BK	Black	GY	Gray	RD	Red
BL	Blue	GN	Green	TN	Tan
BR	Brown	LG	LT Green	VT	Violet
DB	Dark Blue	OR	Orange	WT	White
DG	DK Green	PK	Pink	YL	Yellow

2001-06 3.0L 1MZ-FE V6 VIN F (A/T) E8 22 Pin Connector

PCM Pin #	Wire Color	Circuit Description (22 Pin)	Value at Hot Idle
1	RD/BK	Direct Battery	12-14v
2	BK/OR	Ignition Switch Power	12-14v
3	LG/RD	Circuit Opening Relay (FC)	0-3v, at off-idle: 12v
4	RD	SIL (Scan Tool Signal)	12v
5, 11	---	Not Used	---
6	BK/BL	MIL (lamp) Control	MIL Off: 12v, On: 1v
7	GY	Starter Switch Signal	Cranking: 0-3v
8	YL/GN	EFI Main Relay Control	Relay Off: 0v, On: 12v
9	GN/RD	OD Lamp Indicator Control	Lamp Off: 12v, On: 1v
10	BR/RD	EVAP Canister Closed Valve (VSV)	12v or 0v
12	BK	Center Airbag Sensor (F/PS)	Digital Signals
13	OR	TRAC Signal (TRC+)	Pulse Signals
14	GN	TRAC Engine Signal (ENG+)	Pulse Signals
15	GN/WT	Stop Light Switch Signal	Brake Off: 0v, On: 12v
16	BK/RD	EFI Main Relay (B+)	12-14v
17	BL/RD	EVAP Vapor Pressure Sensor (PTNK)	2.9-3.1v (with hose off)
18	PK	Heated Mirror Circuit	Heater On: 12-14v
19	GN	Electric Load Sensor Circuit	Lights On: 12-14v
20	BL/WT	TRAC Signal (TRC-)	Pulse Signals
21	WT	TRAC Engine Signal (ENG-)	Pulse Signals
22	RD/WT	Theft Deterrent Indicator (IMLD)	LED Off: 0v, On:

2001-06 3.0L 1MZ-FE V6 VIN F (A/T) E9 28 Pin Connector

PCM Pin #	Wire Color	Circuit Description (28 Pin)	Value at Hot Idle
1	BK/RD	Power Ground (EOM)	<0.1v
2, 10-12	---	Not Used	---
3	WT/RD	EVAP Pressure Switching Valve (VSV)	12v or 0v
4	GN	Tail Light Switch Signal	Switch Off: 0v, On: 12v
5	PK/BK	Data Link Connector (TC)	12v
6	RD/WT	A/C Amplifier Signal (ACMG)	Relay Off: 12v, On: 1v
7	BL/YL	A/C Amplifier Signal (AC)	Clutch On: 12v, Off: 1.5v
8	WT	HO2S-12 (B1 S2) Signal (OXS)	0.1.1-1.5v
9	BL/WT	HO2S-12 (B1 S2) Heater (HTS)	1v (Heater on)
13	VT	Mirror Heater Switch Signal	Switch Off: 0v, On: 12v
14	GY/BK	A/C Amplifier THWO Signal	A/C On: 12-14v
15-17, 21	---	Not Used	---
16	YL/BK	Transponder ECU Signal (NEO)	Inserting key: pulses
18	BR/YL	Transponder Amplifier Signal (TXCT)	Inserting key: pulses
19	BR/RD	Transponder Amplifier Signal (RXCK)	Inserting key: pulses
20	BK/WT	Neutral Start Switch Signal (NSW)	Cranking: 9-11v
22	VT/YL	Vehicle Speed Sensor	At 55 mph: 48 Hz
23	BL/BK	Unlock Warning Switch (KSW)	Switch Open:12v, Closed: 0v
24	BK/BL	Cruise Control Signal (OD1)	At Cruise in OD: 12v
25	BL	Cruise Control ECU (IDLO)	1.5v, off-idle: 12v
26	GN/WT	Cooling Fan Relay 1 (CF)	Relay Off: 12v, On: 1v
27	LG/BK	Tachometer Signal (TACO)	Pulse Signals
28	BR/WT	Transponder Amplifier Signal (Code)	Inserting key: pulses

E8 22-Pin Connector   E9 28-Pin Connector

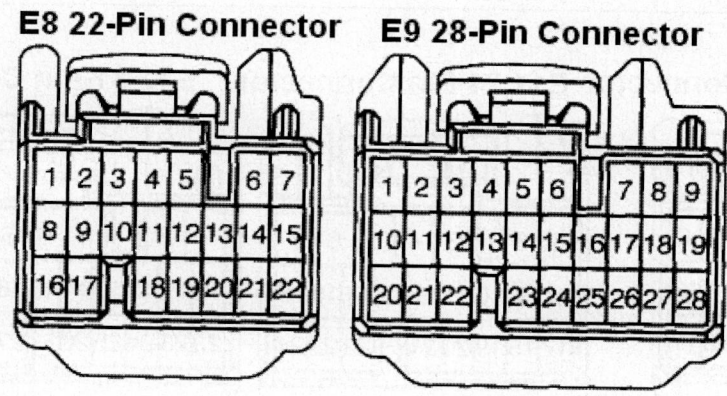

Pin Connector Graphic

05533_ADIA_G229

2001-06 3.0L 1MZ-FE V6 VIN F (A/T) E10 17-Pin Connector

PCM Pin #	Wire Color	Circuit Description (17-Pin)	Value at Hot Idle
1	BL/BK	A/T-ECT Shift Solenoid (S2)	1st or OD: 1v
2-7	---	Not Used	---
8	RD/BK	A/T Select Switch Signal (Reverse)	In 'R': 12v, Others: 0v
9-11	---	Not Used	---
12	GN/OR	Overdrive Main Switch Signal (ODMS)	Open: 12v, Closed: 1v
13	YL	A/T Select Switch Signal (Low)	In Low: 12v, Others: 0v
14	OR	A/T Select Switch Signal (2nd)	In 2nd: 12v, Others: 0v
15	PK/BL	A/T-ECT Shift Solenoid (SL)	In Lockup: 12-14v

2001-06 3.0L 1MZ-FE V6 VIN F (A/T) E11 24-Pin Connector

PCM Pin #	Wire Color	Circuit Description (24-Pin)	Value at Hot Idle
1	WT/BK	Power Ground (E04)	<0.1v
2	YL	Sensor VREF (VC)	4.9-5.1v
3	RD	AFS-11 (B1 S1) Heater (HAFR)	1v (Heater on)
4	BL	AFS-21 (B2 S1) Heater (HAFL)	1v (Heater on)
5	WT	Injector 1 Control	1.6-2-9 ms
6	BK	Injector 2 Control	1.6-2-9 ms
7	LG	EVAP Purge Solenoid (VSV)	12v or 0v
8	WT/BK	Power Ground (E05)	<0.1v
9	BK/BL	PSP Switch Signal (PS)	Straight: 12v, Turned: 0v
10	PK	MAF Sensor Signal (VG)	1.1-1.5v
11	BR	AFS-11 (B1 S1) Signal (AFR+)	Fixed at 3.3v
12	BL	AFS-21 (B2 S1) Signal (AFL+)	Fixed at 3.3v
13	GN/YL	Transmission Fluid Temperature Sensor	At 68°F: 4-5v
14	GN/BK	ECT Sensor Signal (THW)	0.5-0.6v
15	BK	Intake Air Control 2 (VSV)	12v or 0v
16	BK/WT	CKP Sensor Signal (NE+)	AC pulse signals
17	BR	DLC Ground (E1)	<0.050v
18	WT	Sensor Ground (E2)	<0.050v
19	RD/BK	MAF Sensor Ground (E2G)	<0.050v
20	BK/RD	AFR-11 (B1 S1) Signal (AFR-)	Fixed at 3.3v
21	BK/WT	AFR-21 (B2 S1) Signal (AFL-)	Fixed at 3.3v
22	BL/YL	IAT Sensor Signal (THA)	0.5-3.4v
23	BL/WT	TP Sensor Signal (VTA1)	0.53-1.27v
24	WT	CMP/CKP Sensor Ground (-)	<0.050v

E10 17-Pin Connector   E11 24-Pin Connector   E12 31-Pin Connector

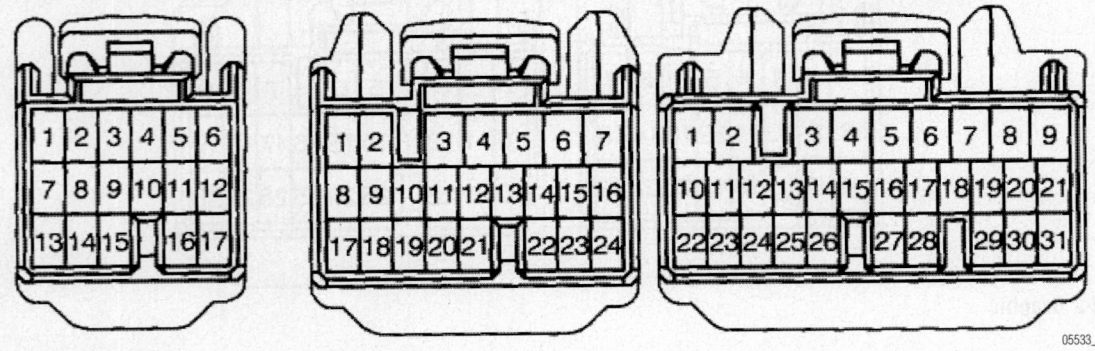

05533_ADIA_G230

Pin Connector Graphic

2001-06 3.0L 1MZ-FE V6 VIN F (A/T) E12 31-Pin Connector

PCM Pin #	Wire Color	Circuit Description (31-Pin)	Value at Hot Idle
1	BK	Injector 3 Control	1.6-2-9 ms
2	BL	Injector 4 Control	1.6-2-9 ms
3	RD	Injector 5 Control	1.6-2-9 ms
4	GN	Injector 6 Control	1.6-2.9 ms
5	BL/BK	VVT Solenoid RH Control (OC1-)	AC Pulse signals
6	RD/BK	VVT Solenoid RH Control (OC1+)	AC Pulse signals

2001-06 3.0L 1MZ-FE V6 VIN F (A/T) E12 31-Pin Connector, *continued*

7	WT/GN	A/T-ECT Solenoid S1	In 3rd or OD: 1v
8-9	---	Not Used	---
10	BK	Cam Timing Oil Control Valve RH (VV1+)	Pulse signals (17-18 Hz)
11	BL/RD	IGT 1 Control	6°, 55 mph: 8° dwell
12	GN/RD	IGT 2 Control	6°, 55 mph: 8° dwell
13	WT/RD	IGT 3 Control	6°, 55 mph: 8° dwell
14	BK/RD	IGT 4 Control	6°, 55 mph: 8° dwell
15	BL/WT	IGT 5 Control	6°, 55 mph: 8° dwell
16	GN/BK	IGT 6 Control	6°, 55 mph: 8° dwell
17	RD/YL	Intake Air Control Valve (ACIS)	12v or 0v
18	BK/WT	VVT Solenoid LH Control (OC2-)	AC Pulse signals
19	GN/WT	A/T-ECT Solenoid (SLN-)	Pulse Signals
20	BK/YL	A/T-ECT Solenoid (SLN+)	Pulse Signals
21	WT/BK	Power Ground (E01)	<0.1v
22	OR	Cam Timing Oil Control Valve LH (VV2+)	Pulse signals (17-18 Hz)
23	YL	Direct Clutch Speed Input (-)	AC pulse signals
24	BL	Direct Clutch Speed Input (+)	AC pulse signals
25	BK	Igniter Signal (IGF)	Digital Signal: 0-5-0v
26	RD/WT	Idle Air Control Valve (RSO)	Pulse Signals
27	WT	Knock Sensor 1 Signal (KNKR)	<0.075v AC
28	WT	Knock Sensor 2 Signal (KNKL)	<0.075v AC
29	RD	VVT Solenoid LH Control (OC2+)	AC Pulse signals
30	WT/BK	Power Ground (E03)	<0.1v
31	WT/BK	Power Ground (E02)	<0.1v

E10 17-Pin Connector E11 24-Pin Connector E12 31-Pin Connector

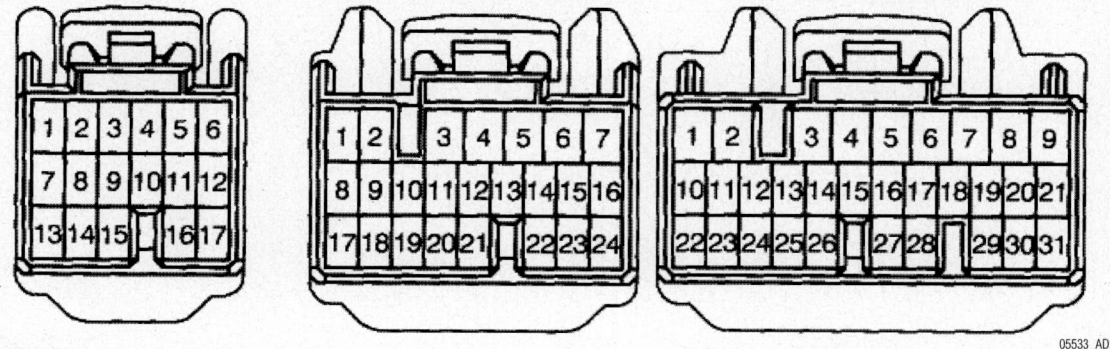

Pin Connector Graphic

05533_ADIA_G230